Springer Handbook
of Automation

Springer Handbooks provide a concise compilation of approved key information on methods of research, general principles, and functional relationships in physical sciences and engineering. The world's leading experts in the fields of physics and engineering will be assigned by one or several renowned editors to write the chapters comprising each volume. The content is selected by these experts from Springer sources (books, journals, online content) and other systematic and approved recent publications of physical and technical information.

The volumes are designed to be useful as readable desk reference books to give a fast and comprehensive overview and easy retrieval of essential reliable key information, including tables, graphs, and bibliographies. References to extensive sources are provided.

Part E Automation Management. The main aspects of automation management are covered: Cost effectiveness and economic reasons for the design, feasibility analysis, implementation, rationalization, use, and maintenance of particular automation; performance and functionality measures and criteria. Related also are the issues of how to manage automatically and control software maintenance, replacement, and upgrading; simplifying complex automation, and ethical concerns of automation.

Chapters 40 to 47

Part F Industrial Automation. This part begins with explanation of machine tool automation, including various types of numerical control (NC), flexible, and precision machinery for production, manufacturing, and assembly, digital and virtual industrial production, and detailed design, guidelines and application of automation in the principal industries, from aerospace and automotive to semi-conductor, mining, food, paper and wood industries. Chapters are also devoted to the design, control and operation of functions common to all industrial automation.

Chapters 48 to 60

Part G Infrastructure and Service Automation. Chapters in this part explain how automation is designed, selected, integrated, deployed, justified and applied; its challenges and emerging trends in how automation transforms, enables, and revolutionizes infrastructures and services, and in the construction of structures, roads and bridges; of smart buildings, smart roads and intelligent vehicles; cleaning of surfaces, tunnels and sewers; land, air, and space transportation; information, knowledge, learning, training, and library services; and in sports and entertainment.

Chapters 61 to 74

Part H Automation in Medical and Healthcare Systems. The main contributions of automation and the exponential penetration to the health and medical well being of individuals and societies are explained. First, the scientific and theoretical foundations of control and automation in biological and biomedical systems and mechanisms are introduced, then specific areas are described and analyzed. Available, proven, and emerging automation techniques in healthcare delivery and elimination of hospital and other medical errors are also addressed.

Chapters 75 to 82

Part I Home, Office, and Enterprise Automation. Functional automation areas at home, in the office, and in general enterprises are covered, including multi-enterprise networks. Chapters also cover the automation theories, techniques and practice, design, operation, challenges and emerging trends in education and learning, banking, commerce. An important dimension of the material compiled for this part is that it is useful for all other functional areas of automation.

Chapters 83 to 93

Part J Appendix. The concluding part of this Springer Handbook contains figures and tables with statistical information and summaries about automation applications and impacts in the main areas of industrial automation, service automation, and financial and e-commerce automation. A rich list of associations and of periodical publications around the world that focus on automation in its variety of related fields is also included for the benefit of readers worldwide.

Chapter 94

Springer Handbook

of Automation

Nof (Ed.)

With DVD-ROM, 1005 Figures, 222 in four color and 149 Tables

 Springer

Editor
Shimon Y. Nof
Purdue University
PRISM Center, and School of Industrial Engineering
315 N. Grant Street
West Lafayette IN 47907, USA
nof@purdue.edu

ISBN: 978-3-540-78830-0   e-ISBN: 978-3-540-78831-7
DOI 10.1007/978-3-540-78831-7
Springer Dordrecht Heidelberg London New York

Library of Congress Control Number: 2008934574

Production and typesetting: le-tex publishing services GmbH, Leipzig
Senior Manager Springer Handbook: Dr. W. Skolaut, Heidelberg
Typography and layout: schreiberVIS, Seeheim
Illustrations: Hippmann GbR, Schwarzenbruck
Cover design: eStudio Calamar S.L., Spain/Germany
Cover production: WMXDesign GmbH, Heidelberg
Printing and binding: Stürtz GmbH, Würzburg

Printed on acid free paper

Springer is part of Springer Science+Business Media (www.springer.com)

89/3180/YL   5 4 3 2 1 0

Dedication

This Springer Handbook is dedicated to all of us who collaborate with automation to advance humanity.

Foreword

Automation Is for Humans and for Our Environment

Preparing to write the Foreword for this outstanding *Springer Handbook of Automation*, I have followed Shimon Y. Nof's statement in his Preface vision: "The purpose of this Handbook is to understand automation knowledge and expertise for the solution of human society's significant challenges; automation provided answers in the past, and it will be harnessed to do so in the future." The *significant challenges* are becoming ever more complex, and learning how to address them with the help of automation is significant too. The publication of this *Handbook* with the excellent information and advice by a group of top international experts is, therefore, most timely and relevant.

The core of any automatic system is the idea of feedback, a simple principle governing any regulation process occurring in nature. The process of feedback governs the growth of living organisms and regulates an innumerable quantity of variables on which life is based, such as body temperature, blood pressure, cells concentration, and on which the interaction of living organisms with the environment is based, such as equilibrium, motion, visual coordination, response to stress and challenge, and so on.

Humans have always copied nature in the design of their inventions: feedback is no exception. The introduction of feedback in the design of man-made automation processes occurred as early as in the golden century of Hellenistic civilization, the third century BC. The scholar Ktesibios, who lived in Alexandria circa 240–280 BC and whose work has been handed to us only by the later roman architect Vitruvius, is credited for the invention of the first feedback device. He used feedback in the design of a water clock. The idea was to obtain a measure of time from the inspection of the position of a floater in a tank of water filled at constant velocity. To make this simple principle work, Ktesibios's challenge was to obtain a constant flow of water in the tank. He achieved this by designing a feedback device in which a conic floating valve serves the dual purpose of sensing the level of water in a compartment and of moderating the outflow of water.

The idea of using feedback to moderate the velocity of rotating devices eventually led to the design of the centrifugal governor in the 18th century. In 1787, T. Mead patented such a device for the regula-

Alberto Isidori
President IFAC

tion of the rotary motion of a wind mill, letting the sail area be decreased or increased as the weights in the centrifugal governor swing outward or, respectively, inward. The same principle was applied two years later, by M. Boulton and J. Watt, to control the steam inlet valve of a steam engine. The basic simple idea of proportional feedback was further refined in the middle of the 19th century, with the introduction of integral control to compensate for constant disturbances. W. von Siemens, in the 1880s, designed a governor in which integral action, achieved by means of a wheel-and-cylinder mechanical integrator, was deliberately introduced. The same principle of proportional and integral feedback gave rise, by the turning of the century, to the first devices for the automatic steering of ships, and became one of the enabling technologies that made the birth of aviation possible. The development of sensors, essential ingredients in any automatic control system, resulted in the creation of new companies.

The perception that feedback control and, in a wider domain, automation were taking the shape of an autonomous discipline, occurred at the time of the second world war, where the application to radar and artillery had a dramatic impact, and immediately after. By the early 1950s, the principles of this newborn discipline quickly became a core ingredient of most industrial engineering curricula, professional and academic societies were established, textbooks and handbooks became available. At the beginning of the 1960s, two new driving forces provoked an enormous leap ahead: the rush to space, and the advent of digital computers in the implementation of control system. The principles of optimal control, pioneered by R. Bellman and L. Pontryagin, became indispensable ingredients for the solution of the problem of soft landing on the moon and in manned space missions. Integrated computer control, introduced in 1959 by Texaco for set point adjustment and coordination of several local feedback loops in a refinery, quickly became the standard technique for controlling industrial processes.

Those years saw also the birth of an International Federation of Automatic Control (IFAC), as a multinational federation of scientific and/or engineering societies each of which represents, in its own nation, values and interests of scientists and professionals active in the field of automation and in related scientific disciplines. The purpose of such Federation, established in Heidelberg in 1956, is to facilitate growth and dissemination of knowledge useful to the development of automation and to its application to engineering and science. Created at a time of acute international tensions, IFAC was a precursor of the spirit of the so-called Helsinki agreements of scientific and technical cooperation between east and west signed in 1973. It represented, in fact, a sincere manifestation of interest, from scientists and professionals of the two confronting spheres of influence in which the world was split at that time, toward a true cooperation and common goals. This was the first opportunity, after the Second World War that scientists and engineers had of sharing complementary scientific and technological backgrounds, notably the early successes in the space race in the Soviet Union and the advent of electronic computers in the United States. The first President of IFAC was an engineer from the Unites States, while the first World Congress of the Federation was held in Moscow in 1960. The Federation currently includes 48 national member organizations, runs more than 60 scientific Conferences with a three-year periodicity, including a World Congress of Automatic Control, and publishes some of the leading Journals in the field.

Since then, three decades of steady progresses followed. Automation is now an essential ingredient in manufacturing, in petrochemical, pharmaceutical, and paper industry, in mining and metal industry, in conversion and distribution of energy, and in many services. Feedback control is indispensable and ubiquitous in automobiles, ships and aircrafts. Feedback control is also a key element of numerous scientific instruments as well as of consumer products, such as compact disc players. Despite of this pervasive role of automation in every aspect of the technology, its specific value is not always perceived as such and automation is often confused with other disciplines of engineering. The advent of robotics, in the late 1970s, is, in some sense, an exception to this, because the impact of robotics in modern manufacturing industry is under the eyes of everybody. However, also in this case there is a tendency to consider robotics and the associated impact on industry as an implementation of ideas and principles of computer engineering rather than principles of automation and feedback control.

In the recent years, though, automation and control have experienced a third, tumultuous expansion. Progresses in the automobile industry in the last decade have only been possible because of automation. Feedback control loops pervade our cars: steering, breaking, attitude stabilization, motion stabilization, combustion, emissions are all feedback controlled. This is a dramatic change that has revolutionized the way in which cars are conceived and maintained. Industrial robots have reached a stage of full maturity, but new generations of service robots are on their way. Four-legged and even two-legged autonomous walking machines are able to walk through rough terrains, service robot are able to autonomously interact with uncertain environment and adapt their mission to changing tasks, to explore hostile or hazardous environments and to perform jobs that would be otherwise dangerous for humans. Service robots assist elderly or disabled people and are about to perform routine services at home. Surgical robotics is a reality: minimally invasive micro robots are able to move within the body and to reach areas not directly accessible by standard techniques. Robots with haptic interfaces, able to return a force feedback to a remote human operator, make tele-surgery possible. New frontiers of automation encompass applications in agriculture, in recycling, in hazardous waste disposal, in environment protection, and in safe and reliable transportation.

At the dawn of the 20th century, the deterministic view of classical mechanics and some consequent positivistic philosophic beliefs that dominated the 19th century had been shaken by the advent of relativistic physics. Today, after a century dominated by the expansion of technology and, to some extent, by the belief that no technological goal was impossible to achieve, similar woes are feared. The clear perception that resources are limited, the uncertainty of the financial markets, the diverse rates of development among nations, all contribute to the awareness that the model of development followed in so far in the industrialized world will change. Today's wisdom and beliefs may not be the same tomorrow. All these expected changes might provide yet another great opportunity for automation. Automation will no longer be seen only as *automatic production*, but as a complex of technologies that guarantee reliability, flexibility, safety, for humans as well as for the environment. In a world of limited resources, automation can provide the answer to the challenges of a sustainable development. Automation has the opportunity of making a greater and even more significant impact on society. In the first half of the 20th century, the precepts of engineering and management helped solving economic recession and ease social anxiety. Similar opportunities and challenges are occurring today.

This leading-edge *Springer Handbook of Automation* will serve as a highly useful and powerful tool and companion to all modern-day engineers and managers in their respective profession. It comes at an appropriate time, and provides a fundamental core of basic principles, knowledge and experience by means of which engineers and managers will be able to quickly respond to changing automation needs and to find creative solutions to the challenges of today's and tomorrow's problems.

It has been a privilege for many members of IFAC to participate with Springer Publishers, Dr. Shimon Y. Nof, and the over 250 experts, authors and reviewers, in creating this excellent resource of automation knowledge and ideas. It provides also a full and comprehensive spectrum of current and prospective automation applications, in industry, agriculture, infrastructures, services, health care, enterprise and commerce. A number of recently developed concepts and powerful emerging techniques are presented here for the first time in an organized manner, and clearly illustrated by specialists in those fields. Readers of this original *Springer Handbook of Automation* are offered the opportunity to learn proven knowledge from underlying basic theory to cutting-edge applications in a variety of emerging fields.

Alberto Isidori
Rome, March 2009

Foreword

Automation Is at the Center of Human Progress

As I write this Foreword for the new *Springer Handbook of Automation*, the 2008 United States presidential elections are still in full swing. Not a day seems to go by without a candidate or newscaster opining on the impact of cheaper, offshore labor on the US economy. Similar debates are taking place in other developed countries around the globe.

Some argue that off-shoring jobs leads to higher unemployment and should be prohibited. Indeed some regions have passed legislation prohibiting their local agencies from moving work to lower cost locations.

Proponents argue off-shoring leads to lower unemployment. In their view freeing up of the labor force from lower skilled jobs allows more people to enter higher value jobs which are typically higher paying. This boosts incomes and in turn overall domestic consumption.

Then, what about automation? Is the displacement or augmentation of human labor with an automated machine bad for our economies, too? If so, let's ban it!

So, let's imagine a world in which automation didn't exist. . . .

To begin I wouldn't be writing this Foreword on my laptop computer since the highly sophisticated automation necessary to manufacture semiconductors wouldn't exist. That's okay I'll just use my old typewriter. Oops, the numerical controlled machines required to manufacture the typewriter's precision parts wouldn't exist. What about pencil and paper? Perhaps, but could I afford them given that there would be no sensors and controls needed to manufacture them in high volume?

IBM has been a leader and pioneer in many automation fields, both as a user and a provider of automation solutions. Beyond productivity and cost-effectiveness, automation also enables us to effectively monitor process quality, reveal to us opportunities for improvement and innovation, and assure product and service dependability and service-availability. Such techniques and numerous examples to advance with automation, as users and providers, are included in this *Springer Handbook of Automation*.

The expanding complexity and magnitude of high-priority society's problems, global needs and competition forcefully challenge organizations and companies. To succeed, they need to understand detailed knowledge

of many of the topics included in this *Springer Handbook of Automation*. Beyond an extensive reference resource providing the expert answers and solutions, readers and learners will be enriched from inspiration to innovate and create powerful applications for specific needs and challenges.

J. Bruce Harreld
Senior Vice President IBM

The best example I know is one I have witnessed first hand at IBM. Designing, developing, and manufacturing state-of-the art microprocessors have been a fundamental driver of our success in large computer and storage systems. Thirty years ago the manufacturing process for these microprocessors was fairly manual and not very capital intense. Today we manufacture microprocessors in a new state-of-the-art US$ 3 billion facility in East Fishkill, New York. This fabrication site contains the world's most advanced logistics and material handling system including real-time process control and fully automated workflow. The result is a completely *touchless* process that in turn allows us to produce the high quality, error free, and extremely fast microprocessors required for today's high end computing systems.

In addition to chapters devoted to a variety of industry and service automation topics, this *Springer Handbook of Automation* includes useful, well-organized information and examples on theory, tools, and their integration for successful, measurable results.

Automation is often viewed as impacting only the tangible world of physical products and facilities. Fortunately, this is completely wrong! Automation has also dramatically improved the way we develop software at IBM. Many years ago writing software was much like writing a report with each individual approaching the task quite differently and manually writing each line of code. Today, IBM's process for developing software is extremely automated with libraries of previously written code accessible to all of our programmers. Thus, once one person develops a program that performs a particular function, it is quickly shared and reused around the globe. This also allows us to pass a project to and from one team to the next so we can speed up cycle times for our clients by working on their projects 24 hours a day. The physical process of *writing* the lines of code has been replaced with pointing and clicking at objects on a computer screen. The result has been a dramatic reduction in mistakes with a concomitant increase in productivity. But we expect and anticipate even more from automation in support of our future, and the knowledge and guidelines on how to do it are described in this *Springer Handbook of Automation*.

The examples illustrated above highlight an important point. While we seldom touch automation, it touches us everyday in almost everything we do. Human progress is driven by day-to-day improvements in how we live. For more than one hundred years automation has been at the center of this exciting and meaningful journey. Since ancient history, humans have known how to benefit civilization with automation.

For engineers, scientists, managers and inventors, automation provides an exciting and important opportunity to implement ingenious human intelligence in automatic solutions for many needs, from simple applications, to difficult and complex requirements. Increasingly, multi-disciplinary cooperation in the study of automation helps in this creative effort, as detailed well in this *Springer Handbook of Automation*, including automatic control and mechatronics, nano-automation and collaborative, software-based automation concepts and techniques, from current and proven capabilities to emerging and forthcoming knowledge.

It is quite appropriate, therefore, that this original *Springer Handbook of Automation* has been published now. Its scope is vast and its detail deep. It covers the history as well as the social implications of automation. Then it dives into automation theory and techniques, design and modeling, and organization and management. Throughout the 94 chapters written by leading world experts, there are specific guidelines and examples of the application of automation in almost every facet of today's society and industry. Given this rich content I am confident that this *Handbook* will be useful not only to students and faculty but practitioners, researchers and managers across a wide range of fields and professions.

J. Bruce Harreld
Armonk, January 2009

Foreword

Dawn of Industrial Intelligent Robots

This *Handbook* is a significant educational, professional and research resource for anyone concerned about automation and robotics. It can serve well for global enterprises and for education globally. The impacts of automation in many fields have been and are essential for increasing the intelligence of services and of interaction with computers and with machines. Plenty of illustrations and statistics about the economics and sophistication impacts of automation are included in this *Handbook*.

Automation, in general, includes many computer and communication based applications, computer-integrated design, planning, management, decision support, informational, educational, and organizational resources, analytics and scientific applications, and more. There are also many automation systems involving robots. Robots have emerged from science fiction into industrial reality in the middle of the 20th Century, and are now available worldwide as reliable, industrially made, automated and programmable machines.

The field of robotics application is now expanding rapidly. As widely known, about 35% of industrial robots in the world are operating in Japan. In the 1970s, Japan started to introduce industrial robots, especially automotive spot welding robots, thereby establishing the industrial robot market. As the industries flourished and faced labor shortage, Japan introduced industrial robots vigorously. Industrial robots have since earned recognition as being able to perform repetitive jobs continuously, and produce quality products with reliability, convincing the manufacturing industry that it is keenly important to use them skillfully so as to achieve its global impact and competitiveness.

In recent years, the manufacturing industry faces severe cost competition, shorter lead-time, and skilled worker shortage in the aging society with lower birth rates. It is also required to manufacture many varieties of products in varied quantity. Against this backdrop, there is a growing interest in industrial intelligent robots as a new automation solution to these requirements. Intelligence here is not defined as human intelligence or a capacity to think, but as a capacity comparable to that of a skilled worker, with which a machine can be equipped.

Disadvantages of relatively simple, playback type robots without intelligent abilities result in relatively higher equipment costs for the elaborate peripheral equipment required, such as parts feeders and part positioning fixtures. Additionally for simpler robots, human workers must daily pre-position work-pieces in designated locations to operate the robots. In contrast, intelligent robots can address these requirements with their vision sensor, serving as the eye, and with their force sensor, serving as the hand providing sense of touch. These intelligent robots are much more effective and more useful. For instance, combined with machine tools as Robot Cells they can efficiently load/unload work-pieces to/from machine tools, thereby reducing machining costs substantially by enabling machine tools to operate long hours without disruptions. These successful solutions with industrial intelligent robots have established them as a key automation component to improve global competitiveness of the manufacturing industry. It signifies the dawn of the industrial intelligent robot.

Seiuemon Inaba
Chairman Fanuc Ltd.

Intelligent automation, including intelligent robots, can now help, as described very well in this *Springer Handbook of Automation*, not only with manufacturing, supply and production companies, but increasingly with security and emergency services; with healthcare delivery and scientific exploration; with energy exploration, production and delivery; and with a variety of home and special needs human services. I am most thankful for the efforts of all those who participated in the development of this useful *Springer Handbook of Automation* and contributed their expertise so that our future with automation and robotics will continue to bring prosperity.

Seiuemon Inaba
Oshino-mura, January 2009

Foreword

Automation Is the Science of Integration

In our understanding of the word *automation*, we used to think of manufacturing processes being run by machines without the need for human control or intervention. From the outset, the purpose of investing in automation has been to increase productivity at minimum cost and to assure uniform quality. Permanently assessing and exploiting the potential for automation in the manufacturing industries has, in fact, proved to be a sustainable strategy for responding to competition in the marketplace, thereby securing attractive jobs.

Automation equipment and related services constitute a large and rapidly growing market. Supply networks of component manufacturers and system integrators, allied with engineering skills for planning, implementing and operating advanced production facilities, are regarded as cornerstones of competitive manufacturing. Therefore, the emphasis of national and international initiatives aimed at strengthening the manufacturing base of economies is on holistic strategies for research and technical development, education, socio-economics and entrepreneurship.

Today, automation has expanded into almost every area of daily life: from smart products for everyday use, networked buildings, intelligent vehicles and logistics systems, service robots, to advanced healthcare and medical systems. In simplified terms, automation today can be considered as the combination of processes, devices and supporting technologies, coupled with advanced information and communication technology (ICT), where ICT is now evolving into the most important basic technology.

As a world-leading organization in the field of applied research, the Fraunhofer Society (Fraunhofer-Gesellschaft) has been a pioneer in relation to numerous technological innovations and novel system solutions in the broad field of automation. Its institutes have led the way in research, development and implementation of industrial robots and computer-integrated manufacturing systems, service robots for professional and domestic applications, advanced ICT systems for office automation and e-Commerce as well as automated residential and commercial buildings. Moreover, our research and development activities in advanced manufacturing and logistics as well as office and home automation have been accompanied by large-scale experiments and demonstration centers, the goal being to integrate, assess and showcase innovations in automation in real-world settings and application scenarios.

On the basis of this experience, we can state that, apart from research in key technologies such as sensors, actuators, process control and user interfaces, automation is first and foremost the science of integration, mastering the process from the specification, design and implementation through to the operation of

Hans-Jörg Bullinger
President
Fraunhofer Society

complex systems that have to meet the highest standards of functionality, safety, cost-effectiveness and usability. Therefore, scientists and engineers need to be experts in their respective disciplines while at the same time having the necessary knowledge and skills to create and operate large-scale systems.

The *Springer Handbook of Automation* is an excellent means of both educating students and also providing professionals with a comprehensive yet compact reference work for the field of automation. The Handbook covers the broad scope of relevant technologies, methods and tools and presents their use and integration in a wide selection of application contexts: from agricultural automation to surgical systems, transportation systems and business process automation.

I wish to congratulate the editor, Prof. Shimon Y. Nof, on succeeding in the difficult task of covering the multi-faceted field of automation and of organizing the material into a coherent and logically structured whole. The *Handbook* admirably reflects the connection between theory and practice and represents a highly worthwhile review of the vast accomplishments in the field. My compliments go to the many experts who have shared their insights, experience and advice in the individual chapters. Certainly, the Handbook will serve as a valuable tool and guide for those seeking to improve the capabilities of automation systems – for the benefit of humankind.

Hans-Jörg Bullinger
Munich, January 2009

Preface

We love automation when it does what we need and expect from it, like our most loyal partner: wash our laundry, count and deliver money bills, supply electricity where and when it is needed, search and display movies, maps, and weather forecasts, assemble and paint our cars, and more personally, image to diagnose our health problems, or tooth pains, cook our food, and photograph our journeys. Who would not love automation?

We hate automation and may even kick it when it fails us, like a betraying confidant: turn the key or push a button and nothing happens the way we anticipate – a car does not start, a TV does not display, our cellphone is misbehaving, the vending machine delivers the wrong item, or refuses to return change; planes are late *due to mechanical problem* and business transactions are lost or ignored *due to computer glitches*. Eventually those problems are fixed and we turn back to the first paragraph, loving it again.

We are amazed by automation and all those people behind it. Automation thrills us when we learn about its new abilities, better functions, more power, faster computing, smaller size, and greater reliability and precision. And we are fascinated by automation's marvels: in entertainment, communication, scientific discoveries; how it is applied to explore space and conquer difficult maladies of society, from medical and pharmaceutical automation solutions, to energy supply, distant education, smart transportation, and we are especially enthralled when we are not really sure how it works, but it works.

It all starts when we, as young children, observe and notice, perhaps we are bewildered, that a door automatically opens when we approach it, or when we are first driven by a train or bus, or when we notice the automatic sprinklers, or lighting, or home appliances: *How can it work on its own?* Yes, there is so much magic about automation.

This magic of automation is what inspired a large group of us, colleagues and friends from around the world, all connected by automation, to compile, develop, organize and present this unique *Springer Handbook of Automation*: Explain to our readers what automation is, how it works, how it is designed and built, where it is applied, and where and how it is going to be improved and be created even better; what are the scientific principles behind it, and what are emerging trends and challenges being confronted. All of it concisely yet comprehensively covered in the 94 chapters which are included in front of you, the readers.

Flying over beautiful Fall colored forest surrounding Binghamton, New York in the 1970s on my way to IBM's symposium on the future of computing, I was fascinated by the miracle of nature beneath the airplane: Such immense diversity of leaves' changing colors; such magically smooth, wavy movement of the leaves dancing with the wind, as if they are programmed with automatic control to responsively transmit certain messages needed by some unseen listeners. And the brilliance of the sunrays reflected in these beautiful dancing leaves (there must be some purpose to this automatically programmed beauty, I thought). More than once, during reading the chapters that follow, was I reminded of this unforgettable image of multi-layered, interconnected, interoperable, collaborative, responsive waves of leaves (and services): The take-home lesson from that symposium was that mainframe computers hit, about that time, a barrier – it was stated that faster computing was impossible since mainframes would not be able to cool off the heat they generated (unless a miracle happened). As we all know, with superb human ingenuity computing has overcome that barrier and other barriers, and progress, including fast personal computers, better software, and wireless computer communication, resulted in major performance and cost-effective improvements, such as client-server workstations, wireless access and local area networks (LAN), duo- and multi-core architectures, web-based Internetworking, grids, and more has been automated, and there is so much more yet to come. Thus, more intelligent automatic control and more responsive human–automation interfaces could be invented and deployed for the benefit of all.

Progress in distributed, networked, and collaborative control theory, computing, communication, and automation has enabled the emergence of e-Work, e-Business, e-Medicine, e-Service, e-Commerce, and many other significant e-Activities based on automation. It is not that our ancestors did not recognize the tremendous power and value of delegating effort to tools and machines, and furthermore, of synergy, team-work, collaborative interactions and decision-making, outsourcing and resource sharing, and in general, net-working. But only when efficient, reliable and scalable automation reached a certain level of maturity could it be designed into systems and infrastructures servicing effective supply and delivery networks, social networks, and multi-enterprise practices.

In their vision, enterprises expect to simplify their automation utilities and minimize their burden and cost, while increasing the value and usability of all deployed functions and acquirable information by their timely conversion into relevant knowledge, goods, and prac-tices. Streamlined knowledge, services and products would then be delivered through less effort, just when necessary and only to those clients or decision makers who truly need them. Whether we are in business and commerce or in service for society, the real purpose of automation is not merely *better computing* or *better au-tomation*, but *let us increase our competitive agility and service quality!*

This *Springer Handbook* achieves this purpose well. Throughout the 94 chapters, divided into ten main parts, with 125 tables, numerous equations, 1005 figures, and a vast number of references, with numerous guidelines, algorithms, and protocols, models, theories, techniques and practical principles and procedures, the 166 co-authors present proven knowledge, original analysis, best practices and authoritative expertise.

Plenty of case studies, creative examples and unique illustrations, covering topics of automation from the ba-sics and fundamentals to advanced techniques, cases and theories will serve the readers and benefit the students and researchers, engineers and managers, in-ventors, investors and developers.

Special Thanks

I wish to express my gratitude and thanks to our dis-tinguished Advisory Board members, who are leading international authorities, scholars, experts, and pioneers of automation, and who have guided the develop-ment of this *Springer Handbook* and shared with me their wisdom and advice along the challenging edi-torial process; to our distinguished authors and our esteemed reviewers, who are also leading experts, re-searchers, practitioners and pioneers of automation. Sadly, my personal friends and colleagues Professor Kazuo Tanie, Professor Heinz Erbe, and Professor Wal-ter Shaufelberger, who took active part in helping create this *Springer Handbook*, passed away before they could see it published. They left huge voids in our community and in my heart, but their legacy will continue.

All the chapters were reviewed thoroughly and anonymously by over 90 reviewers, and went through several critical reviews and revision cycles (each chap-ter was reviewed by at least five expert reviewers), to assure the accuracy, relevance, and high quality of the materials, which are presented in the *Springer Hand-book*. The reviewers included:

Kemal Altinkemer, Purdue University
Panos J. Antsaklis, University of Notre Dame
Hillel Bar-Gera, Ben-Gurion University, Israel
Ruth Bars, Budapest University of Technology and Economics, Hungary
Sigal Berman, Ben-Gurion University, Israel
Mark Bishop, Goldsmiths University of London, UK
Barrett Caldwell, Purdue University
Daniel Castro-Lacouture, Georgia Institute of Tech-nology
Enrique Castro-Leon, Intel Corporation
Xin W. Chen, Purdue University
Gary J. Cheng, Purdue University
George Chiu, Purdue University
Meerant Chokshi, Purdue University
Jae Woo Chung, Purdue University
Jason Clark, Purdue University
Rosalee Clawson, Purdue University
Monica Cox, Purdue University
Jose Cruz, Ohio State University
Juan Manuel De Bedout, GE Power Conversion Sys-tems
Menahem Domb, Amdocs, Israel
Vincent Duffy, Purdue University
Yael Edan, Ben-Gurion University, Israel
Aydan Erkmen, Middle East Technical University, Turky
Florin Filip, Academia Romana and National Insti-tute for R&D in Informatics, Romania
Gary Gear, Embry-Riddle University
Jackson He, Intel Corporation
William Helling, Indiana University
Steve Holland, GM R&D Manufacturing Systems Research
Chin-Yin Huang, Tunghai University, Taiwan
Samir Iqbal, University of Texas at Arlington
Alberto Isidori, Universita Roma, Italy
Nick Ivanescu, University Politehnica of Bucharest, Romania
Wootae Jeong, Korea Railroad Research Institute
Shawn Jordan, Purdue University

Stephen Kahne, Embry-Riddle University
Dimitris Kiritsis, EFPL, Switzerland
Hoo Sang Ko, Purdue University
Renata Konrad, Purdue University
Troy Kostek, Purdue University
Nicholas Kottenstette, University of Notre Dame
Diego Krapf, Colorado State University
Steve Landry, Purdue University
Marco Lara Gracia, University of Southern Indiana
Jean-Claude Latombe, Stanford University
Seokcheon Lee, Purdue University
Mark Lehto, Purdue University
Heejong Lim, LG Display, Korea
Bakhtiar B. Litkouhi, GM R&D Center
Yan Liu, Wright State University
Joachim Meyer, Ben Gurion University, Israel
Gaines E. Miles, Purdue University
Daiki Min, Purdue University
Jasmin Nof, University of Maryland
Myounggyu D. Noh, Chungnam National University, Korea
Nusa Nusawardhana, Cummins, Inc.
Tal Oron-Gilad, Ben Gurion University, Israel
Namkyu Park, Ohio University
Jimena Pascual, Universidad Catolica de Valparaiso, Chile
Anatol Pashkevich, Inst. Recherce en Communication et Cybernetique, Nantes, France
Gordon Pennock, Purdue University
Carlos Eduardo Pereira, Federal University of Rio Grande do Sul, Brazil
Guillermo Pinochet, Kimberly Clark Co., Chile
Arik Ragowsky, Wayne State University
Jackie Rees, Purdue University
Timothy I. Salsbury, Johnson Controls, Inc.
Gavriel Salvendy, Tsinghua University, China
Ivan G. Sears, GM Technical Center
Ramesh Sharda, Oklahoma State University
Mirosław J. Skibniewski, University of Maryland
Eugene Spafford, Purdue University
Jose M. Tanchoco, Purdue University
Mileta Tomovic, Purdue University
Jocelyn Troccaz, IMAG Institut d'Ingénierie de l'Information de Santé, France
Jay Tu, North Carolina State University
Juan Diego Velásquez, Purdue University
Sandor M. Veres, University of Southampton, UK
Matthew Verleger, Purdue University
Francois Vernadat, Cour des Comptes Europeenne, Luxembourg
Birgit Vogel-Heuser, University of Kassel, Germany

Edward Watson, Louisiana State University
James W. Wells, GM R&D Manufacturing Systems Research
Ching-Yi Wu, Purdue University
Moshe Yerushalmy, MBE Simulations, Israel
Yuehwern Yih, Purdue University
Sang Won Yoon, Purdue University
Yih-Choung Yu, Lafayette College
Firas Zahr, Cleveland Clinic Foundation

I wish to express my gratitude and appreciation also to my resourceful coauthors, colleagues and partners from IFAC, IFPR, IFIP, IIE, NSF, TRB, RIA, INFORMS, ACM, IEEE-ICRA, ASME, and PRISM Center at Purdue and PGRN, the PRISM Global Research Network, for all their support and cooperation leading to the successful creation of this *Springer Handbook*.

Special thanks to my late parents, Dr. Jacob and Yafa Berglas Nowomiast, whose brilliance, deep appreciation to scholarship, and inspiration keep enriching me; to my wife Nava for her invaluable support and wise advice; to Moriah, Jasmin, Jonathan, Haim, Daniel, Andrew, Chin-Yin, Jose, Moshe, Ed, Ruth, Pornthep, Juan Ernesto, Richard, Wootae, Agostino, Daniela, Tibor, Esther, Pat, David, Yan, Gad, Guillermo, Cristian, Carlos, Fay, Marco, Venkat, Masayuki, Hans, Laszlo, Georgi, Arturo, Yael, Dov, Florin, Herve, Gerard, Gavriel, Lily, Ted, Isaac, Dan, Veronica, Rolf, Yukio, Steve, Mark, Colin, Namkyu, Wil, Aditya, Ken, Hannah, Anne, Fang, Jim, Tom, Frederique, Alexandre, Coral, Tetsuo, and Oren, and to Izzy Vardinon, for sharing with me their thoughts, smiles, ideas and their automation expertise. Deep thanks also to Juan Diego Velásquez, to Springer-Verlag's Tom Ditzinger, Werner Skolaut and Heather King, and the le-tex team for their tremendous help and vision in completing this ambitious endeavor.

The significant achievements of humans with automation, in improving our life quality, innovating and solving serious problems, and enriching our knowledge; inspiring people to enjoy automation and provoking us to learn how to invent even better and greater automation solutions; the wonders and magic, opportunities and challenges with emerging and future automation – are all enormous. Indeed, automation is an essential and wonderful part of human civilization.

Shimon Yeshayahu Nof Nowomiast
West Lafayette, Indiana
May 2009

Advisory Board

Stephen Kahne

Embry-Riddle University
Prescott, AZ, USA
s.kahne@ieee.org

Stephen Kahne is Professor of Electrical Engineering at Embry-Riddle Aeronautical University in Prescott, Arizona where he was formerly Chancellor. Prior to coming to Embry-Riddle in 1995, he had been Chief Scientist at the MITRE Corporation. Dr. Kahne earned his BS degree from Cornell University and the MS and PhD degrees from the University of Illinois. Following a decade at the University of Minnesota, he was Professor at Case Western Reserve University, Professor and Dean of Engineering at Polytechnic Institute of New York, and Professor and President of the Oregon Graduate Center, Portland, Oregon. Dr. Kahne was a Division Director at the National Science Foundation in the early 1980s. He is a Fellow of the IEEE, AAAS, and IFAC. He was President of the IEEE Control Systems Society, a member of the IEEE Board of Directors of the IEEE in the 1980s, and President of IFAC in the 1990s.

Aditya P. Mathur

Purdue University
Department of Computer Science
West Lafayette, IN, USA
apm@cs.purdue.edu

Aditya Mathur received his PhD in 1977 from BITS, Pilani, India in Electrical Engineering. Until 1985 he was on the faculty at BITS where he spearheaded the formation of the first degree granting Computer Science department in India. In 1985 he moved briefly to Georgia Tech before joining Purdue University in 1987. Aditya is currently a Professor and Head in the Department of Computer Science where his research is primarily in the area of software engineering. He has made significant contributions in software testing and software process control and has authored three textbooks in the areas of programming, microprocessor architecture, and software testing.

Hak-Kyung Sung

Samsung Electronics
Mechatronics & Manufacturing
Technology Center
Suwon, Korea
hakksung@samsung.com

Hak-Kyung Sung received the Master degree in Mechanical Engineering from Yonsei University in Korea and the PhD degree in Control Engineering from Tokyo Institute of Technology in Japan, in 1985 and 1992, respectively. He is currently the Vice President in the Mechtronics & Manufacturing Technology Center, Samsung Electronics. His interests are in production engineering technology, such as robotics, control, and automation.

Gavriel Salvendy

Department of Industrial Engineering
Beijing, P.R. China

Gavriel Salvendy is Chair Professor and Head of the Department of Industrial Engineering at Tshinghua University, Beijing, Peoples Republic of China and Professor emeritus of Industrial Engineering at Purdue University. His research deals with the human aspects of design and operation of advanced computing systems requiring interaction with humans. In this area he has over 450 scientific publications and numerous books, including the Handbook of Industrial Engineering and Handbook of Human Factors and Ergonomics. He is a member of the USA National Academy of Engineering and the recipient of the John Fritz Medal.

George Stephanopoulos

Massachusetts Institute of Technology
Cambridge, MA, USA
geosteph@mit.edu

George Stephanopoulos is the A.D. Little Professor of Chemical Engineering and Director of LISPE (Laboratory for Intelligent Systems in Process Engineering) at MIT. He has also taught at the University of Minnesota (1974–1983) and National Technical University of Athens, Greece (1980–1984). His research interests are in process operations monitoring, analysis, diagnosis, control, and optimization. Recently he has extended his research to multi-scale modeling and design of materials and nanoscale structures with desired geometries. He is a member of the National Academy of Engineering, USA.

Kazuo Tanie (△)

Tokyo Metropolitan University
Human Mechatronics System Course,
Faculty of System Design
Tokyo, Japan

Professor Kazuo Tanie (1946–2007), received BE, MS, Dr. eng. in Mechanical Engineering from Waseda University. In 1971, he joined the Mechanical Engineering Laboratory (AIST-MITI), was Director of the Robotics Department and of the Intelligent Systems Institute of the National Institute of Advanced Industrial Science and Technology, Ministry of Economy, Trade, and Industry, where he led a large humanoid robotics program.

In addition, he held several academic positions in Japan, USA, and Italy. His research interests included tactile sensors, dexterous manipulation, force and compliance control for robotic arms and hands, virtual reality and telerobotics, human-robot coexisting systems, power assist systems and humanoids. Professor Tanie was active in IEEE Robotics and Automation Society, served as its president (2004–2005), and led several international conferences. One of the prominent pioneers of robotics in Japan, his leadership and skills led to major automation initiatives, including various walking robots, dexterous hands, seeing-eye robot (MEL Dog), rehabilitative and humanoid robotics, and network-based humanoid telerobotics.

Tibor Vámos

Hungarian Academy of Sciences
Computer and Automation Institute
Budapest, Hungary
vamos@sztaki.hu

Tibor Vámos graduated from the Budapest Technical University in 1949. Since 1986 he is Chairman of the Board, Computer and Automation Research Institute of the Hungarian Academy of Sciences, Budapest. He was President of IFAC 1981–1984 and is a Fellow of the IEEE, ECCAI, IFAC. Professor Vamos is Honorary President of the John v. Neumann Society and won the State Prize of Hungary in 1983, the Chorafas Prize in 1994, the Széchenyi Prize of Hungary in 2008 and was elected "The educational scientist of the year" in 2005. His main fields of interest cover large-scale systems in process control, robot vision, pattern recognition, knowledge-based systems, and epistemic problems. He is author and co-author of several books and about 160 papers.

François B. Vernadat

Université Paul Verlaine Metz
Laboratoire de Génie Industriel et
Productique de Metz (LGIPM)
Metz, France
Francois.Vernadat@eca.europa.eu

François Vernadat received the PhD in Electrical Engineering and Automatic Control from University of Clermont, France, in 1981. He has been a research officer at the National Research Council of Canada in the 1980s and at the Institut National de Recherche en Informatique et Automatique in France in the 1990s. He joined the University of Metz in 1995 as a full professor and founded the LGIPM research laboratory. His research interests include enterprise modeling, enterprise architectures, enterprise integration and interoperability. He is a member of IEEE and ACM and has been vice-chairman of several technical committees of IFAC. He has over 250 scientific papers in international journals and conferences.

Birgit Vogel-Heuser

University of Kassel
Faculty of Electrical Engineering/
Computer Science, Department Chair of
Embedded Systems
Kassel, Germany
vogel-heuser@uni-kassel.de

Birgit Vogel-Heuser graduated in Electrical Engineering and obtained her PhD in Mechanical Engineering from the RWTH Aachen in 1991. She worked nearly ten years in industrial automation for machine and plant manufacturing industry. After holding the Chair of Automation at the University of Hagen and the Chair of Automation/Process Control Engineering she is now head of the Chair of Embedded Systems at the University of Kassel. Her research work is focussed on improvement of efficiency in automation engineering for hybrid process and heterogeneous distributed embedded systems.

Andrew B. Whinston

The University of Texas at Austin
McCombs School of Business, Center for
Research in Electronic Commerce
Austin, TX, USA
abw@uts.cc.utexas.edu

Andrew Whinston is Hugh Cullen Chair Professor in the IROM department at the McCombs School of Business at the University of Texas at Austin. He is also the Director at the Center for Research in Electronic Commerce. His recent papers have appeared in Information Systems Research, Marketing Science, Management Science and the Journal of Economic Theory. In total he has published over 300 papers in the major economic and management journals and has authored 27 books. In 2005 he received the Leo Award from the Association for Information Systems for his long term research contribution to the information system field.

List of Authors

Nicoletta Adamo-Villani
Purdue University
Computer Graphics Technology
401 N. Grant Street
West Lafayette, IN 47907, USA
e-mail: *nadamovi@purdue.edu*

Panos J. Antsaklis
University of Notre Dame
Department of Electrical Engineering
205A Cushing
Notre Dame, IN 46556, USA
e-mail: *antsaklis.1@nd.edu*

Cecilia R. Aragon
Lawrence Berkeley National Laboratory
Computational Research Division
One Cyclotron Road, MS 50B-2239
Berkeley, CA 94720, USA
e-mail: *CRAragon@lbl.gov*

Neda Bagheri
Massachusetts Institute of Technology (MIT)
Department of Biological Engineering
77 Massachusetts Ave. 16-463
Cambridge, MA 02139, USA
e-mail: *nbagheri@mit.edu*

Greg Baiden
Laurentian University
School of Engineering
Sudbury, ON P3E 2C6, Canada
e-mail: *gbaiden@laurentian.ca*

Parasuram Balasubramanian
Theme Work Analytics Pvt. Ltd.
Gurukrupa, 508, 47th Cross, Jayanagar
Bangalore, 560041, India
e-mail: *balasubp@gmail.com*

P. Pat Banerjee
University of Illinois
Department of Mechanical
and Industrial Engineering
3029 Eng. Research Facility, 842 W. Taylor
Chicago, IL 60607-7022, USA
e-mail: *banerjee@uic.edu*

Ruth Bars
Budapest University of Technology
and Economics
Department of Automation
and Applied Informatics
Goldmann Gy. tér 3
1111 Budapest, Hungary
e-mail: *bars@aut.bme.hu*

Luis Basañez
Technical University of Catalonia (UPC)
Institute of Industrial and Control Engineering (IOC)
Av. Diagonal 647 planta 11
Barcelona 08028, Spain
e-mail: *luis.basanez@upc.edu*

Rashid Bashir
University of Illinois at Urbana-Champaign
Department of Electrical and Computer
Engineering and Bioengineering
208 North Wright Street
Urbana, IL 61801, USA
e-mail: *rbashir@uiuc.edu*

Wilhelm Bauer
Fraunhofer-Institute for Industrial Engineering IAO
Corporate Development and Work Design
Nobelstr. 12
70566 Stuttgart, Germany
e-mail: *wilhelm.bauer@iao.fraunhofer.de*

Gary R. Bertoline
Purdue University
Computer Graphics Technology
401 N. Grant St.
West Lafayette, IN 47907, USA
e-mail: *Bertoline@purdue.edu*

Christopher Bissell
The Open University
Department of Communication and Systems
Walton Hall
Milton Keynes, MK7 6AA, UK
e-mail: *c.c.bissell@open.ac.uk*

Richard Bossi
The Boeing Company
PO Box 363
Renton, WA 98057, USA
e-mail: *richard.h.bossi@boeing.com*

Martin Braun
Fraunhofer-Institute for Industrial Engineering IAO
Human Factors Engineering
Nobelstraße 12
70566 Stuttgart, Germany
e-mail: *martin.braun@iao.fraunhofer.de*

Sylvain Bruni
Aptima, Inc.
12 Gill St, Suite #1400
Woburn, MA 01801, USA
e-mail: *sbruni@aptima.com*

James Buttrick
The Boeing Company
BCA – Materials & Process Technology
PO Box #3707
Seattle, WA 98124, USA
e-mail: *james.n.buttrick@boeing.com*

Darwin G. Caldwell
Istituto Italiano Di Tecnologia
Department of Advanced Robotics
Via Morego 30
16163 Genova, Italy
e-mail: *Darwin.Caldwell@iit.it*

Brian Carlisle
Precise Automation
5665 Oak Knoll Lane
Auburn, CA 95602, USA
e-mail: *brian.carlisle@preciseautomation.com*

Dan L. Carnahan
Rockwell Automation
Department of Advanced Technology
1 Allen Bradley Drive
Mayfield Heights, OH 44124, USA
e-mail: *dlcarnahan@ra.rockwell.com*

Ángel R. Castaño
Universidad de Sevilla
Departamento de Ingeniería de Sistemas y Automática
Camino de los Descubrimientos
Sevilla 41092, Spain
e-mail: *castano@us.es*

Daniel Castro-Lacouture
Georgia Institute of Technology
Department of Building Construction
280 Ferst Drive
Atlanta, GA 30332-0680, USA
e-mail: *dcastro6@gatech.edu*

Enrique Castro-Leon
JF5-103, Intel Corporation
2111 NE 25th Avenue
Hillsboro, OR 97024, USA
e-mail: *Enrique.G.Castro-Leon@intel.com*

José A. Ceroni
Pontifica Universidad Católica de Valparaíso
School of Industrial Engineering
2241 Brazil Avenue
Valparaiso, Chile
e-mail: *jceroni@ucv.cl*

Deming Chen
University of Illinois, Urbana-Champaign
Electrical and Computer Engineering (ECE)
1308 W Main St.
Urbana, IL 61801, USA
e-mail: *dchen@illinois.edu*

Heping Chen
ABB Inc.
US Corporate Research Center
2000 Day Hill Road
Windsor, CT 06095, USA
e-mail: *heping.chen@us.abb.com*

Xin W. Chen
Purdue University
PRISM Center and School of Industrial Engineering
315 N. Grant Street
West Lafayette, IN 47907, USA
e-mail: *chen144@purdue.edu*

Benny C.F. Cheung
The Hong Kong Polytechnic University
Department of Industrial and Systems Engineering
Hung Hom
Kowloon, Hong Kong
e-mail: *mfbenny@inet.polyu.edu.hk*

Jaewoo Chung
Kyungpook National University
School of Business Administration
1370 Sankyuk-dong Buk-gu
Daegu, 702-701, South Korea
e-mail: *jaewooch@gmail.com*

Rodrigo J. Cruz Di Palma
Kimberly Clark, Latin American Operations
San Juan, 00919-1859, Puerto Rico
e-mail: *Rodrigo.J.Cruz@kcc.com*

Mary L. Cummings
Massachusetts Institute of Technology
Department of Aeronautics and Astronautics
77 Massachusetts Ave.
Cambridge, MA 02139, USA
e-mail: *missyc@mit.edu*

Christian Dannegger
Kandelweg 14
78628 Rottweil, Germany
e-mail: *cd@3a-solutions.com*

Steve Davis
Istituto Italiano Di Tecnologia
Department of Advanced Robotics
Via Morego 30
16163 Genova, Italy
e-mail: *steven.davis@iit.it*

Xavier Delorme
Ecole Nationale Supérieure des Mines
de Saint-Etienne
Centre Genie Industriel et Informatique (G2I)
158 cours Fauriel
42023 Saint-Etienne, France
e-mail: *delorme@emse.fr*

Alexandre Dolgui
Ecole Nationale Supérieure des Mines
de Saint-Etienne
Department of Industrial Engineering
and Computer Science
158, cours Fauriel
42023 Saint-Etienne, France
e-mail: *dolgui@emse.fr*

Alkan Donmez
National Institute of Standards and Technology
Manufacturing Engineering Laboratory
100 Bureau Drive
Gaithersburg, MD 20899, USA
e-mail: *alkan.donmez@nist.gov*

Francis J. Doyle III
University of California
Department of Chemical Engineering
Santa Barbara, CA 93106-5080, USA
e-mail: *frank.doyle@icb.ucsb.edu*

Yael Edan
Ben-Gurion University of the Negev
Department of Industrial Engineering
and Management
Beer Sheva 84105, Israel
e-mail: *yael@bgu.ac.il*

Thomas F. Edgar
University of Texas
Department of Chemical Engineering
1 University Station
Austin, TX 78712, USA
e-mail: *edgar@che.utexas.edu*

Norbert Elkmann
Fraunhofer IFF
Department of Robotic Systems
Sandtorstr. 22
39106 Magdeburg, Germany
e-mail: *norbert.elkmann@iff.fraunhofer.de*

Heinz-Hermann Erbe (△)
Technische Universität Berlin
Center for Human-Machine Systems
Franklinstrasse 28/29
10587 Berlin, Germany

Mohamed Essafi
Ecole des Mines de Saint-Etienne
Department Centre for Industrial Engineering
and Computer Science
Cours Fauriel
Saint-Etienne, France
e-mail: *essafi@emse.fr*

Florin-Gheorghe Filip
The Romanian Academy
Calea Victoriei 125
Bucharest, 010071, Romania
e-mail: *ffilip@acad.ro*

Markus Fritzsche
Fraunhofer IFF
Department of Robotic Systems
Sandtorstr. 22
39106 Magdeburg, Germany
e-mail: *markus.fritzsche@iff.fraunhofer.de*

Susumu Fujii
Sophia University
Graduate School of Science and Technology
7-1, Kioicho, Chiyoda
102-8554 Tokyo, Japan
e-mail: *susumu-f@sophia.ac.jp*

Christopher Ganz
ABB Corporate Research
Segelhofstr. 1
5405 Baden, Switzerland
e-mail: *christopher.ganz@ch.abb.com*

Mitsuo Gen
Waseda University
Graduate School of Information,
Production and Systems
2-7 Hibikino, Wakamatsu-ku
808-0135 Kitakyushu, Japan
e-mail: *gen@waseda.jp*

Birgit Graf
Fraunhofer IPA
Department of Robot Systems
Nobelstr. 12
70569 Stuttgart, Germany
e-mail: *birgit.graf@ipa.fraunhofer.de*

John O. Gray
Istituto Italiano Di Tecnologia
Department of Advanced Robotics
Via Morego 30
16163 Genova, Italy
e-mail: *John.Gray-2@manchester.ac.uk*

Rudiyanto Gunawan
National University of Singapore
Department of Chemical
and Biomolecular Engineering
4 Engineering Drive 4 Blk E5 #02-16
Singapore, 117576
e-mail: *chegr@nus.edu.sg*

Juergen Hahn
Texas A&M University
Artie McFerrin Dept. of Chemical Engineering
College Station, TX 77843-3122, USA
e-mail: *hahn@tamu.edu*

Kenwood H. Hall
Rockwell Automation
Department of Advanced Technology
1 Allen Bradley Drive
Mayfield Heights, OH 44124, USA
e-mail: *khhall@ra.rockwell.com*

Shufeng Han
John Deere
Intelligent Vehicle Systems
4140 114th Street
Urbandale, IA 50322, USA
e-mail: *hanshufeng@johndeere.com*

Nathan Hartman
Purdue University
Computer Graphics Technology
401 North Grant St.
West Lafayette, IN 47906, USA
e-mail: *nhartman@purdue.edu*

Yukio Hasegawa
Waseda University
System Science Institute
Tokyo, Japan
e-mail: *yukioh@green.ocn.ne.jp*

Jackson He
Intel Corporation
Digital Enterprise Group
2111 NE 25th Ave
Hillsboro, OR 97124, USA
e-mail: *jackson.he@intel.com*

Jenő Hetthéssy
Budapest University of Technology and Economics
Department of Automation
and Applied Informatics
Goldmann Gy. Tér 3
1111 Budapest, Hungary
e-mail: *jhetthessy@aut.bme.hu*

Karyn Holmes
Chevron Corp.
100 Northpark Blvd.
Covington, LA 70433, USA
e-mail: *kholmes@chevron.com*

Clyde W. Holsapple
University of Kentucky
School of Management,
Gatton College of Business and Economics
Lexington, KY 40506-0034, USA
e-mail: *cwhols@email.uky.edu*

Petr Horacek
Czech Technical University in Prague
Faculty of Electrical Engineering
Technicka 2
Prague, 16627, Czech Republic
e-mail: *horacek@fel.cvut.cz*

William J. Horrey
Liberty Mutual Research Institute for Safety
Center for Behavioral Sciences
71 Frankland Road
Hopkinton, MA 01748, USA
e-mail: *william.horrey@libertymutual.com*

Justus Hortig
Fraunhofer IFF
Department of Robotic Systems
Sandtorstr. 22
39106 Magdeburg, Germany
e-mail: *justus.hortig@iff.fraunhofer.de*

Chin-Yin Huang
Tunghai University
Industrial Engineering and Enterprise Information
Taichung, 407, Taiwan
e-mail: *huangcy@thu.edu.tw*

Yoshiharu Inaba
Fanuc Ltd.
Oshino-mura
401-0597 Yamanashi, Japan
e-mail: *inaba.yoshiharu@fanuc.co.jp*

Samir M. Iqbal
University of Texas at Arlington
Department of Electrical Engineering
500 S. Cooper St.
Arlington, TX 76019, USA
e-mail: *smiqbal@uta.edu*

Rolf Isermann
Technische Universität Darmstadt
Institut für Automatisierungstechnik,
Forschungsgruppe Regelungstechnik
und Prozessautomatisierung
Landgraf-Georg-Str. 4
64283 Darmstadt, Germany
e-mail: *risermann@iat.tu-darmstadt.de*

Kazuyoshi Ishii
Kanazawa Institute of Technology
Social and Industrial Management Systems
Yatsukaho 3-1
Hakusan City, Japan
e-mail: *ishiik@neptune.kanazawa-it.ac.jp*

Alberto Isidori
University of Rome "La Sapienza"
Department of Informatics and Sytematics
Via Ariosto 25
00185 Rome, Italy
e-mail: *albisidori@dis.uniroma1.it*

Nick A. Ivanescu
University Politechnica of Bucharest
Control and Computers
Spl. Independentei 313
Bucharest, 060032, Romania
e-mail: *nik@cimr.pub.ro*

Sirkka-Liisa Jämsä-Jounela
Helsinki University of Technology
Department of Biotechnology
and Chemical Technology
Espoo 02150, Finland
e-mail: *Sirkka-l@tkk.fi*

Bijay K. Jayaswal
Agilenty Consulting Group
3541 43rd Ave. S
Minneapolis, MN 55406, USA
e-mail: *bijay.jayaswal@agilenty.com*

Wootae Jeong
Korea Railroad Research Institute
360-1 Woram-dong
Uiwang 437-757, Korea
e-mail: *wjeong@krri.re.kr*

Timothy L. Johnson
General Electric
Global Research
786 Avon Crest Blvd.
Niskayuna, NY 12309, USA
e-mail: *johnsontl@nycap.rr.com*

Hemant Joshi
Research, Acxiom Corp.
CWY2002-7 301 E. Dave Ward Drive
Conway, AR 72032-7114, USA
e-mail: *hemant.joshi@acxiom.com*

Michael Kaplan
Ex Libris Ltd.
313 Washington Street
Newton, MA 02458, USA
e-mail: *michael.kaplan@exlibrisgroup.com*

Dimitris Kiritsis
Department STI-IGM-LICP
EPFL, ME A1 396, Station 9
1015 Lausanne, Switzerland
e-mail: *dimitris.kiritsis@epfl.ch*

Hoo Sang Ko
Purdue University
PRISM Center and School of Industrial Engineering
315 N Grant St.
West Lafayette, IN 47907, USA
e-mail: *ko0@purdue.edu*

Naoshi Kondo
Kyoto University
Division of Environmental Science and Technology,
Graduate School of Agriculture
Kitashirakawa-Oiwakecho
606-8502 Kyoto, Japan
e-mail: *kondonao@kais.kyoto-u.ac.jp*

Peter Kopacek
Vienna University of Technology
Intelligent Handling and Robotics – IHRT
Favoritenstrasse 9-11/E 325
1040 Vienna, Austria
e-mail: *kopacek@ihrt.tuwien.ac.at*

Nicholas Kottenstette
Vanderbilt University
Institute for Software Integrated Systems
PO Box 1829
Nashville, TN 37203, USA
e-mail: *nkottens@isis.vanderbilt.edu*

Eric Kwei
University of California, Santa Barbara
Department of Chemical Engineering
Santa Barbara, CA 93106, USA
e-mail: *kwei@engineering.ucsb.edu*

Siu K. Kwok
The Hong Kong Polytechnic University
Industrial and Systems Engineering
Yuk Choi Road
Kowloon, Hong Kong
e-mail: *mfskkwok@inet.polyu.edu.hk*

King Wai Chiu Lai
Michigan State University
Electrical and Computer Engineering
2120 Engineering Building
East Lansing, MI 48824, USA
e-mail: *kinglai@egr.msu.edu*

Dean F. Lamiano
The MITRE Corporation
Department of Communications
and Information Systems
7515 Colshire Drive
McLean, VA 22102, USA
e-mail: *dlamiano@mitre.org*

Steven J. Landry
Purdue University
School of Industrial Engineering
315 N. Grant St.
West Lafayette, IN 47906, USA
e-mail: slandry@purdue.edu

John D. Lee
University of Iowa
Department Mechanical and Industrial
Engineering, Human Factors Research National
Advanced Driving Simulator
Iowa City, IA 52242, USA
e-mail: jdlee@engineering.uiowa.edu

Tae-Eog Lee
KAIST
Department of Industrial and Systems Engineering
373-1 Guseong-dong, Yuseong-gu
Daejeon 305-701, Korea
e-mail: telee@kaist.ac.kr

Wing B. Lee
The Hong Kong Polytechnic University
Industrial and Systems Engineering
Yuk Choi Road
Kowloon, Hong Kong
e-mail: wb.lee@polyu.edu.hk

Mark R. Lehto
Purdue University
School of Industrial Engineering
315 North Grant Street
West Lafayette, IN 47907-2023, USA
e-mail: lehto@purdue.edu

Kauko Leiviskä
University of Oulu
Control Engineering Laboratory
Oulun Yliopisto 90014, Finland
e-mail: kauko.leiviska@oulu.fi

Mary F. Lesch
Liberty Mutual Research Institute for Safety
Center for Behavioral Sciences
71 Frankland Road
Hopkinton, MA 01748, USA
e-mail: mary.lesch@libertymutual.com

Jianming Lian
Purdue University
School of Electrical and Computer Engineering
465 Northwestern Avenue
West Lafayette, IN 47907-2035, USA
e-mail: jlian@purdue.edu

Lin Lin
Waseda University
Information, Production & Systems Research
Center
2-7 Hibikino, Wakamatsu-ku
808-0135 Kitakyushu, Japan
e-mail: linlin@aoni.waseda.jp

Laurent Linxe
Peugeot SA
Hagondang, France
e-mail: laurent.linxe@mpsa.com

T. Joseph Lui
Whirlpool Corporation
Global Product Organization
750 Monte Road
Benton Harbor, MI 49022, USA
e-mail: t_joseph_lui@whirlpool.com

Wolfgang Mann
Profactor Research and Solutions GmbH
Process Design and Automation,
Forschungszentrum
2444 Seibersdorf, Austria
e-mail: wolfgang.mann@profactor.at

Sebastian V. Massimini
The MITRE Corporation
7515 Colshire Drive
McLean, VA 22101, USA
e-mail: svm@mitre.org

Francisco P. Maturana
University/Company Rockwell Automation
Department Advanced Technology
1 Allen Bradley Drive
Mayfield Heights, OH 44124, USA
e-mail: fpmaturana@ra.rockwell.com

Henry Mirsky
University of California, Santa Barbara
Department of Chemical Engineering
Santa Barbara, CA 93106, USA
e-mail: *Mirsky@lifesci.ucsb.edu*

Sudip Misra
Indian Institute of Technology
School of Information Technology
Kharagpur, 721302, India
e-mail: *sudip_misra@yahoo.com*

Satish C. Mohleji
Center for Advanced Aviation System
Development (CAASD)
The MITRE Corporation
7515 Colshire Drive McLean, VA 22102-7508, USA
McLean, VA 22102-7508, USA
e-mail: *smohleji@mitre.org*

Gérard Morel
Centre de Recherche en Automatique Nancy (CRAN)
54506 Vandoeuvre, France
e-mail: *gerard.morel@cran.uhp-nancy.fr*

René J. Moreno Masey
University of Sheffield
Automatic Control and Systems Engineering
Mappin Street
Sheffield, S1 3JD, UK
e-mail: *cop07rjm@sheffield.ac.uk*

Clayton Munk
Boeing Commercial Airplanes
Material & Process Technology
Seattle, WA 98124-3307, USA
e-mail: *clayton.l.munk@boeing.com*

Yuko J. Nakanishi
Nakanishi Research and Consulting, LLC
93-40 Queens Blvd. 6A, Rego Park
New York, NY 11374, USA
e-mail: *nakanishi@transresearch.net*

Dana S. Nau
University of Maryland
Department of Computer Science
A.V. Williams Bldg.
College Park, MD 20742, USA
e-mail: *nau@cs.umd.edu*

Peter Neumann
Institut für Automation und Kommunikation
Werner-Heisenberg-Straße 1
39106 Magdeburg, Germany
e-mail: *peter.neumann@ifak.eu*

Shimon Y. Nof
Purdue University
PRISM Center and School of Industrial Engineering
West Lafayette, IN 47907, USA
e-mail: *nof@purdue.edu*

Anibal Ollero
Universidad de Sevilla
Departamento de Ingeniería de Sistemas y
Automática
Camino de los Descubrimientos
Sevilla 41092, Spain
e-mail: *aollero@cartuja.us.es*

John Oommen
Carleton University
School of Computer Science
1125 Colonel Bye Drive
Ottawa, K1S5B6, Canada
e-mail: *oommen@scs.carleton.ca*

Robert S. Parker
University of Pittsburgh
Department of Chemical
and Petroleum Engineering
1249 Benedum Hall
Pittsburgh, PA 15261, USA
e-mail: *rparker@pitt.edu*

Alessandro Pasetti
P&P Software GmbH
High Tech Center 1
8274 Tägerwilen, Switzerland
e-mail: *pasetti@pnp-software.com*

Anatol Pashkevich
Ecole des Mines de Nantes
Department of Automatic Control
and Production Systems
4 rue Alfred-Kastler
44307 Nantes, France
e-mail: *anatol.pashkevich@emn.fr*

Bozenna Pasik-Duncan
University of Kansas
Department of Mathematics
1460 Jayhawk Boulevard
Lawrence, KS 66045, USA
e-mail: *bozenna@math.ku.edu*

Peter C. Patton
Oklahoma Christian University
School of Engineering
PO Box 11000
Oklahoma City, OK 73136, USA
e-mail: *peter.patton@oc.edu*

Richard D. Patton
Lawson Software
380 Saint Peter St.
St. Paul, MN 55102-1313, USA
e-mail: *richard.patton@lawson.com*

Carlos E. Pereira
Federal University of Rio Grande do Sul (UFRGS)
Department Electrical Engineering
Av Osvaldo Aranha 103
Porto Alegre RS, 90035 190, Brazil
e-mail: *cpereira@ece.ufrgs.br*

Jean-François Pétin
Centre de Recherche en Automatique
de Nancy (CRAN)
54506 Vandoeuvre, France
e-mail: *jean-francois.petin@cran.uhp-nancy.fr*

Chandler A. Phillips
Wright State University
Department of Biomedical,
Industrial and Human Factors Engineering
3640 Colonel Glen Highway
Dayton, OH 45435-0001, USA
e-mail: *Chandler.Phillips@wright.edu*

Friedrich Pinnekamp
ABB Asea Brown Boveri Ltd.
Corporate Strategy
Affolternstrasse 44
8050 Zurich, Switzerland
e-mail: *friedrich.pinnekamp@ch.abb.com*

Daniel J. Power
University of Northern Iowa
College of Business Administration
Cedar Falls, IA 50614-0125, USA
e-mail: *daniel.power@uni.edu*

Damien Poyard
PCI/SCEMM
42030 Saint-Etienne, France
e-mail: *damien.poyard@pci.fr*

Srinivasan Ramaswamy
University of Arkansas at Little Rock
Department of Computer Science
2801 South University Ave
Little Rock, AR 72204, USA
e-mail: *srini@ieee.org*

Piercarlo Ravazzi
Politecnico di Torino
Department Manufacturing Systems
and Economics
C.so Duca degli Abruzzi 24
10129 Torino, Italy
e-mail: *piercarlo.ravazzi@polito.it*

Daniel W. Repperger
Wright Patterson Air Force Base
Air Force Research Laboratory
711 Human Performance Wing
Dayton, OH 45433-7022, USA
e-mail: *Daniel.repperger@wpafb.af.mil*

William Richmond
Western Carolina University
Accounting, Finance, Information Systems
and Economics
Cullowhee, NC 28723, USA
e-mail: *brichmond@email.wcu.eud*

Dieter Rombach
University of Kaiserslautern
Department of Computer Science, Fraunhofer
Institute for Experimental Software Engineering
67663 Kaiserslautern, Germany
e-mail: *rombach@iese.fraunhofer.de*

Shinsuke Sakakibara
Fanuc Ltd.
Oshino-mura
401-0597 Yamanashi, Japan
e-mail: *sakaki@i-dreamer.co.jp*

Timothy I. Salsbury
Johnson Controls, Inc.
Building Efficiency Research Group
507 E Michigan Street
Milwaukee, WI 53202, USA
e-mail: *tim.salsbury@gmail.com*

Branko Sarh
The Boeing Company – Phantom Works
5301 Bolsa Avenue
Huntington Beach, CA 92647, USA
e-mail: *branko.sarh@boeing.com*

Sharath Sasidharan
Marshall University
Department of Management and Marketing
One John Marshall Drive
Huntington, WV 25755, USA
e-mail: *sasidharan@marshall.edu*

Brandon Savage
GE Healthcare IT
Pollards Wood, Nightingales Lane
Chalfont St Giles, HP8 4SP, UK
e-mail: *Brandon.Savage@med.ge.com*

Manuel Scavarda Basaldúa
Kimberly Clark
Avda. del Libertador St. 498 Capital Federal
(C1001ABR)
Buenos Aires, Argentina
e-mail: *manuel.scavarda@kcc.com*

Walter Schaufelberger (△)
ETH Zurich
Institute of Automatic Control
Physikstrasse 3
8092 Zurich, Switzerland

Bobbie D. Seppelt
The University of Iowa
Mechanical and Industrial Engineering
3131 Seamans Centre
Iowa City, IA 52242, USA
e-mail: *bseppelt@engineering.uiowa.edu*

Ramesh Sharda
Oklahoma State University
Spears School of Business
Stillwater, OK 74078, USA
e-mail: *Ramesh.sharda@okstate.edu*

Keiichi Shirase
Kobe University
Department of Mechanical Engineering
1-1, Rokko-dai, Nada
657-8501 Kobe, Japan
e-mail: *shirase@mech.kobe-u.ac.jp*

Jason E. Shoemaker
University of California
Department of Chemical Engineering
Santa Barbara, CA 93106-5080, USA
e-mail: *jshoe@engr.ucsb.edu*

Moshe Shoham
Technion – Israel Institute of Technology
Department of Mechanical Engineering
Technion City
Haifa 32000, Israel
e-mail: *shoham@technion.ac.il*

Marwan A. Simaan
University of Central Florida
School of Electrical Engineering
and Computer Science
Orlando, FL 32816, USA
e-mail: *simaan@mail.ucf.edu*

Johannes A. Soons
National Institute of Standards and Technology
Manufacturing Engineering Laboratory
100 Bureau Drive
Gaithersburg, MD 20899-8223, USA
e-mail: *soons@nist.gov*

Dieter Spath
Fraunhofer-Institute for Industrial
Engineering IAO
Nobelstraße 12
70566 Stuttgart, Germany
e-mail: *dieter.spath@iao.fraunhofer.de*

Harald Staab
ABB AG, Corporate Research Center Germany
Robotics and Manufacturing
Wallstadter Str. 59
68526 Ladenburg, Germany
e-mail: *harald.staab@de.abb.com*

Petra Steffens
Fraunhofer Institute for Experimental
Software Engineering
Department Business Area e-Government
Fraunhofer-Platz 1
67663 Kaiserslautern, Germany
e-mail: *petra.steffens@iese.fraunhofer.de*

Jörg Stelling
ETH Zurich
Department of Biosystems Science
and Engineering
Mattenstr. 26
4058 Basel, Switzerland
e-mail: *joerg.stelling@bsse.ethz.ch*

Raúl Suárez
Technical University of Catalonia (UPC)
Institute of Industrial and Control Engineering (IOC)
Av. Diagonal 647 planta 11
Barcelona 08028, Spain
e-mail: *raul.suarez@upc.edu*

Kinnya Tamaki
Aoyama Gakuin University
School of Business Administration
Shibuya 4-4-25, Shibuya-ku
153-8366 Tokyo, Japan
e-mail: *ytamaki@a2en.aoyama.ac.jp*

Jose M.A. Tanchoco
Purdue University
School of Industrial Engineering
315 North Grant Street
West Lafayette, IN 47907-2023, USA
e-mail: *tanchoco@purdue.edu*

Stephanie R. Taylor
Department of Computer Science
Colby College, 5855 Mayflower Hill Dr.
Waterville, ME 04901, USA
e-mail: *srtaylor@colby.edu*

Peter Terwiesch
ABB Ltd.
8050 Zurich, Switzerland
e-mail: *peter.terwiesch@ch.abb.com*

Jocelyne Troccaz
CNRS – Grenoble University
Computer Aided Medical Intervention – TIMC
laboratory, IN3S – School of Medicine – Domaine
de la Merci
38700 La Tronche, France
e-mail: *jocelyne.troccaz@imag.fr*

Edward Tunstel
Johns Hopkins University
Applied Physics Laboratory,
Space Department
11100 Johns Hopkins Road
Laurel, MD 20723, USA
e-mail: *Edward.Tunstel@jhuapl.edu*

Tibor Vámos
Hungarian Academy of Sciences
Computer and Automation Institute
11 Lagymanyosi
1111 Budapest, Hungary
e-mail: *vamos@sztaki.hu*

István Vajk
Budapest University of Technology and Economics
Department of Automation
and Applied Informatics
1521 Budapest, Hungary
e-mail: *vajk@aut.bme.hu*

Gyula Vastag
Corvinus University of Budapest
Institute of Information Technology,
Department of Computer Science
13-15 Fõvám tér (Sóház)
1093 Budapest, Hungary
e-mail: *gyula.vastag@uni-corvinus.hu*

Juan D. Velásquez
Purdue University
PRISM Center and School of Industrial Engineering
315 N. Grant Street
West Lafayette, IN 47907, USA
e-mail: *jvelasqu@purdue.edu*

Matthew Verleger
Purdue University
Engineering Education
701 West Stadium Avenue
West Lafayette, IN 47907-2045, USA
e-mail: *matthew@mverleger.com*

François B. Vernadat
Université Paul Verlaine Metz
Laboratoire de Génie Industriel
et Productique de Metz (LGIPM)
Metz, France
e-mail: *Francois.Vernadat@eca.europa.eu*

Agostino Villa
Politecnico di Torino
Department Manufacturing Systems
and Economics
C. so Duca degli Abruzzi, 24
10129 Torino, Italy
e-mail: *agostino.villa@polito.it*

Birgit Vogel-Heuser
University of Kassel
Faculty of Electrical Engineering/Computer Science,
Department Chair of Embedded Systems
Wilhelmshöher Allee 73
34121 Kassel, Germany
e-mail: *vogel-heuser@uni-kassel.de*

Edward F. Watson
Louisiana State University
Information Systems and Decision Sciences
3183 Patrick F. Taylor Hall
Baton Rouge, LA 70803, USA
e-mail: *ewatson@lsu.edu*

Theodore J. Williams
Purdue University
College of Engineering
West Lafayette, IN 47907, USA
e-mail: *twilliam@purdue.edu*

Alon Wolf
Technion Israel Institute of Technology
Faculty of Mechanical Engineering
Haifa 32000, Israel
e-mail: *alonw@technion.ac.il*

Ning Xi
Michigan State University
Electrical and Computer Engineering
2120 Engineering Building
East Lansing, MI 48824, USA
e-mail: *xin@egr.msu.edu*

Moshe Yerushalmy
MBE Simulations Ltd.
10, Hamefalsim St.
Petach Tikva 49002, Israel
e-mail: *ymoshe@mbe-simulations.com*

Sang Won Yoon
Purdue University
PRISM Center and School of Industrial Engineering
315 N. Grant Street
West Lafayette, IN 47907-2023, USA
e-mail: *yoon6@purdue.edu*

Stanislaw H. Żak
Purdue University
School of Electrical and Computer Engineering
465 Northwestern Avenue
West Lafayette, IN 47907-2035, USA
e-mail: *zak@purdue.edu*

Contents

Part B Automation Theory and Scientific Foundations

Part C Automation Design: Theory, Elements, and Methods

Part D Automation Design: Theory and Methods for Integration

Part F Industrial Automation

Part H Automation in Medical and Healthcare Systems

Part I Home, Office, and Enterprise Automation

List of Abbreviations

α-HL	α-hemolysin
βCD	β-cyclodextrin
μC	micro controller
*FTTP	fault-tolerance time-out protocol
2-D	two-dimensional
3-D-CG	three-dimensional computer graphic
3-D	three-dimensional
3G	third-generation
3PL	third-party logistics
3SLS	three-stage least-square
4-WD	four-wheel-drive

A

A-PDU	application layer protocol data unit
A/D	analog-to-digital
AAAI	Association for the Advancement of Artificial Intelligence
AACC	American Automatic Control Council
AACS	automated airspace computer system
AAN	appliance area network
ABAS	aircraft-based augmentation system
ABB	Asea Brown Boveri
ABCS	automated building construction system
ABMS	agent-based management system
ABS	antilock brake system
AC/DC	alternating current/direct current
ACARS	aircraft communications addressing and reporting system
ACAS	aircraft collision avoidance system
ACAS	automotive collision avoidance system
ACCO	active control connection object
ACC	adaptive cruise control
ACC	automatic computer control
ACE	area control error
ACGIH	AmericanConference of Governmental Industrial Hygienists
ACH	automated clearing house
ACMP	autonomous coordinate measurement planning
ACM	Association for Computing Machinery
ACM	airport capacity model
ACN	automatic collision notification
ACT-R	adaptive control of thought-rational
AC	alternating-current
ADAS	advanced driver assistance system
ADA	Americans with Disabilities Act
ADC	analog-to-digital converter
ADS-B	automatic dependent surveillance-broadcast
ADSL	asymmetric digital subscriber line

ADT	admission/transfer/discharge
aecXML	architecture, engineering and construction extensive markup language
AFCS	automatic flight control system
AFM	atomic force microscopy
AFP	automated fiber placement
AF	application framework
AGC	automatic generation control
AGL	above ground level
AGV	autonomous guided vehicle
AHAM	Association of Home Appliance Manufacturers
AHP	analytical hierarchy process
AHS	assisted highway system
AIBO	artificial intelligence robot
AIDS	acquired immunodeficiency syndrome
AIM-C	accelerated insertion of materials-composite
AIMIS	agent interaction management system
AIMac	autonomous and intelligent machine tool
AI	artificial intelligence
ALB	assembly line balancing
ALD	atomic-layer deposition
ALU	arithmetic logic unit
AMHS	automated material-handling system
AMPA	autonomous machining process analyzer
ANFIS	adaptive neural-fuzzy inference system
ANN	artificial neural network
ANSI	American National Standards Institute
ANTS	Workshop on Ant Colony optimization and Swarm Intelligence
AOCS	attitude and orbit control system
AOC	airline operation center
AOI	automated optical inspection
AOP	aspect-oriented programming
AO	application object
APC	advanced process control
APFDS	autopilot/flight director system
API	applications programming interface
APL	application layer
APM	alternating pulse modulation
APO	advance planner and optimizer
APS	advanced planning and scheduling
APTMS	3-aminopropyltrimethoxysilane
APT	automatically programmed tool
APU	auxiliary power unit
APV	approach procedures with vertical guidance
AQ	as-quenched
ARCS	attention, relevance, confidence, satisfaction

ARIS	architecture for information systems
ARL	Applied Research Laboratory
ARPANET	advanced research projects agency net
ARPM	the application relationship protocol machine
ARSR	air route surveillance radar
ARTCC	air route traffic control center
ARTS	automated radar terminal system
aRT	acyclic real-time
AS/RC	automated storage/enterprise resource
AS/RS	automatic storage and retrieval system
ASAS	airborne separation assurance system
ASCII	American standard code for information interchange
ASDE	airport surface detection equipment
ASDI	aircraft situation display to industry
ASE	application service element
ASIC	application-specific IC
ASIMO	advanced step in innovation mobility
ASIP	application-specific instruction set processor
ASI	actuator sensor interface
ASME	American Society of Mechanical Engineers
ASP	application service provider
ASRS	automated storage and retrieval system
ASR	airport surveillance radar
ASSP	application-specific standard part
ASTD	American Society for Training and Development
ASW	American Welding Society
ASi	actuator sensor interface
AS	ancillary service
ATCBI-6	ARSR are ATC beacon interrogator
ATCSCC	air traffic control system command center
ATCT	air traffic control tower
ATC	available transfer capability
ATIS	automated terminal information service
ATL	automated tape layup
ATM	air traffic management
ATM	asynchronous transfer mode
ATM	automatic teller machine
ATPG	automatic test pattern generation
AT	adenine–thymine
AUTOSAR	automotive open system architecture
AUV	autonomous underwater vehicle
AVI	audio video interleaved
AWSN	ad hoc wireless sensor network
Aleph	automated library expandable program
awGA	adaptive-weight genetic algorithm
A&I	abstracting and indexing

B

B-rep	boundary representation
B2B	business-to-business

B2C	business-to-consumer
BAC	before automatic control
BALLOTS	bibliographic automation of large library operations using time sharing
BAP	Berth allocation planning
BAS	building automation systems
BA	balancing authority
BBS	bulletin-board system
BCC	before computer control
BCD	binary code to decimal
BDD	binary decision diagram
BDI	belief–desire–intention
BIM	building information model
BI	business intelligence
BLR	brick laying robot
BMP	best-matching protocol
BOL	beginning of life
BOM	bill of material
BPCS	basic process control system
BPEL	business process execution language
BPMN	business process modeling notation
BPM	business process management
BPO	business process outsourcing
BPR	business process reengineering
BP	broadcasting protocol
bp	base pair
BSS	basic service set
BST	biochemical systems theory
BS	base station

C

C2C	consumer-to-consumer
CAASD	center for advanced aviation system development
CAA	National Civil Aviation Authority
CAD/CAM	computer-aided design/manufacture
CADCS	computer aided design of control system
CAEX	computer aided engineering exchange
CAE	computer-aided engineering
CAI	computer-assisted (aided) instruction
CAMI	computer-assisted medical intervention
CAMP	collision avoidance metrics partnership
CAM	computer-aided manufacturing
CANbus	controller area network bus
CAN	control area network
CAOS	computer-assisted ordering system
CAPP	computer aided process planning
CAPS	computer-aided processing system
CASE	computer-aided software engineering
CAS	collision avoidance system
CAS	complex adaptive system
CAW	carbon arc welding
CA	conflict alert
CBM	condition-based maintenance
CBT	computer based training

CCC	Chinese Control Conference
CCD	charge-coupled device
CCGT	combined-cycle gas turbine
CCMP	create–collect–manage–protect
CCM	CORBA component model
CCP	critical control point
CCTV	closed circuit television
CCT	collaborative control theory
CDMS	conflict detection and management system
CDTI	cockpit display of traffic information
CDU	control display unit
CD	compact disc
CEC	Congress on Evolutionary Computation
CEDA	conflict and error detection agent
CEDM	conflict and error detection management
CEDM	conflict and error detection model
CEDP	conflict and error detection protocol
CEDP	conflict and error diagnostics and prognostics
CED	concurrent error detection
CEO	chief executive officers
CEPD	conflict and error prediction and detection
CERIAS	Center of Education and Research in Information Assurance and Security
CERN	European Organization for Nuclear Research
CERT	Computer Emergency Response Team
CE	Council Europe
CFD	computational fluid dynamics
CFG	context-free grammar
CFIT	controlled flight into terrain
cGMP	current good manufacturing practice
CG	computer graphics
CHAID	chi-square automatic interaction detector
CHART	Maryland coordinated highways action response team
CH	cluster-head
CIA	CAN in automation
CICP	coordination and interruption–continuation protocol
CIM	computer integrated manufacturing
CIO	chief information officer
CIP	common industrial protocol
CIRPAV	computer-integrated road paving
CLAWAR	climbing and walking autonomous robot
CLSI	Computer Library Services Inc.
CL	cutter location
CME	chemical master equation
CMI	computer-managed instruction
CML	case method learning
CMM	capability maturity model
CMS	corporate memory system
CMTM	control, maintenance, and technical management
CM	Clausius–Mossotti

CNC	computer numerical control
CNO	collaborative networked organization
CNS	collision notification system
CNS	communication, navigation, and surveillance
CNT	carbon nanotube
COA	cost-oriented automation
COBOL	common business-oriented language
COCOMO	constructive cost model
CODESNET	collaborative demand and supply network
COMET	collaborative medical tutor
COMSOAL	computer method of sequencing operations for assembly lines
COM	component object model
COP	coefficient of performance
COQ	cost of quality
CORBA	common object request broker architecture
CO	connection-oriented
CPA	closest point of approach
CPLD	complex programmable logic device
CPM	critical path method
CPOE	computerized provider order entry
CPU	central processing unit
CP	constraint programming
CP	coordination protocol
CQI	continuous quality improvement
CRF	Research Center of Fiat
CRM	customer relationship management
CRP	cooperation requirement planning
CRT	cathode-ray tube
cRT	cyclic real-time
CSCL	computer-supported collaborative learning
CSCW	computer-supported collaborative work
CSG	constructive solid geometry
CSR	corporate social responsibility
CSS	Control Systems Society
CSU	customer support unit
CSW	curve speed warning system
CTC	cluster tool controller
CTMC	cluster tool module communication
CT	computed tomography
CURV	cable-controlled undersea recovery vehicle
CVT	continuously variable transmission
CV	controlled variables
Co-X	collaborative tool for function X

D

D/A	digital-to-analog
D2D	discovery-to-delivery
DAC	digital-to-analog converter
DAFNet	data activity flow network

DAISY	differential algebra for identifiability of systems	DOT	US Department of Transportation
DAM	digital asset management	DO	device object
DARC	Duke Annual Robo-Climb Competition	DPA	discrete pursuit algorithm
DAROFC	direct adaptive robust output feedback controller	DPC	distributed process control
DARPA	Defense Advanced Research Projects Agency	DPIEM	distributed parallel integration evaluation method
DARSFC	direct adaptive robust state feedback controller	DP	decentralized periphery
DAS	driver assistance system	DRG	diagnostic related group
DA	data acquisition	DRR	digitally reconstructed radiograph
DB	database	DR	digital radiography
DCOM	distributed component object model	DSA	digital subtraction angiography
DCSS	dynamic case study scenario	DSDL	domain-specific design language
DCS	distributed control system	DSDT	distributed signal detection theoretic
DCS	disturbance control standard	DSL	digital subscriber line
DC	direct-current	DSL	domain-specific language
DDA	demand deposit account	DSN	distributed sensor network
DDC	direct digital control	DSP	digital signal processor
DEA	discrete estimator algorithm	DSRC	dedicated short-range communication
DEM	discrete element method	DSSS	direct sequence spread spectrum
DEP	dielectrophoretic	DSS	decision support system
DES	discrete-event system	DTC	direct torque control
DFBD	derived function block diagram	DTL	dedicated transfer line
DFI	data activity flow integration	DTP	desktop printing
DFM	design for manufacturing	DTSE	discrete TSE algorithm
DFT	discrete Fourier transform	DUC	distributable union catalog
DGC	DARPA Grand Challenge	DVD	digital versatile disk
DGPA	discretized generalized pursuit algorithm	DVI	digital visual interface
DGPS	differential GPS	DV	disturbance variables
DHCP	dynamic host configuration protocol	DXF	drawing interchange format
DHS	Department of Homeland Security	DoD	Department of Defense
DICOM	digital imaging and communication in medicine	DoS	denial of service
DIN	German Institute for Normalization		
DIO	digital input/output		

E

E-CAE	electrical engineering computer aided engineering
E-PERT	extended project estimation and review technique
E/H	electrohydraulic
EAI	enterprise architecture interface
EAP	electroactive polymer
EA	evolutionary algorithm
EBL	electron-beam lithography
EBM	evidence-based medicine
EBP	evidence-based practice
EBW	electron beam welding
ebXML	electronic business XML
EB	electron beam
ECG	electrocardiogram
ECU	electronic control unit
EC	European Community
EDA	electronic design automation
EDCT	expected departure clearance time
EDD	earliest due date
EDGE	enhanced data rates for GSM evolution

DISC	death inducing signalling complex
DLC	direct load control
DLF	Digital Library Foundation
DMC	dynamic matrix control
DME	distance measuring equipment
DMOD	distance modification
DMPM	data link mapping protocol machine
DMP	decision-making processes
DMP	dot matrix printer
DMSA/DMSN	distributed microsensor array and network
DMS	dynamic message sign
DM	decision-making
DNA	deoxyribonucleic acid
DNC	direct numerical control
DNS	domain name system
DOC	Department of Commerce
DOF	degrees of freedom
DOP	degree of parallelism

EDIFACT	Electronic Data Interchange for Administration, Commerce and Transport
EDI	electronic data interchange
EDPA	error detection and prediction algorithms
EDPVR	end-diastolic pressure–volume relationship
EDS	electronic die sorting
EDV	end-diastolic volume
EEC	European Economic Community
EEPROM	electrically erasable programmable read-only memory
EES	equipment engineering system
EFIS	electronic flight instrument system
EFSM	extended finite state machine
EFT	electronic funds transfer
EGNOS	European geostationary navigation overlay service
EHEDG	European Hygienic Engineering and Design Group
EICAS	engine indicating and crew alerting system
EIF	European Interoperability Framework
EII	enterprise information integration
EIS	executive information system
EIU	Economist Intelligence Unit
EI	Enterprise integration
eLPCO	e-Learning professional competency
ELV	end-of-life of vehicle
EL	electroluminescence
EMCS	energy management control systems
EMF	electromotive force
EMO	evolutionary multiobjective optimization
EMR	electronic medical record
EMS	energy management system
EOL	end-of-life
EPA	Environmental Protection Agency
EPC	engineering, procurement, and contsruction
EPGWS	enhanced GPWS
EPROM	erasable programmable read-only memory
EPSG	Ethernet PowerLink Standardization Group
EP	evolutionary programming
ERMA	electronic recording machine accounting
ERM	electronic resources management
ERP	enterprise resource planning
ESA	European Space Agency
ESB	enterprise service bus
ESD	electronic software delivery
ESD	emergency shutdown
ESL	electronic system-level
ESPVR	end-systolic pressure–volume relationship
ESP	electronic stability program
ESR	enterprise services repository

ESSENCE	Equation of State: Supernovae Trace Cosmic Expansion
ESS	extended service set
ES	enterprise system
ES	evolution strategy
ETA	estimated time of arrival
ETC	electronic toll collection
ETG	EtherCAT Technology Group
ETH	Swiss Federal Technical university
ETMS	enhanced traffic management system
ET	evolutionary technique
EURONORM	European Economic Community
EU	European Union
EVA	extravehicular activity
EVD	eigenvalue–eigenvector decomposition
EVM	electronic voting machine
EVS	enhanced vision system
EWMA	exponentially-weighted moving average
EWSS	e-Work support system
EXPIDE	extended products in dynamic enterprise
EwIS	enterprise-wide information system

F

FAA	US Federal Aviation Administration
fab	fabrication plant
FACT	fair and accurate credit transaction
FAF	final approach fix
FAL	fieldbus application layer
FAQ	frequently asked questions
FASB	Financial Accounting Standards Board
FAST	final approach spacing tool
FA	factory automation
FA	false alarm
FBA	flux balance analysis
FBD	function block diagram
FCAW	flux cored arc welding
FCC	flight control computer
FCW	forward collision warning
FCW	forward crash warning
FDA	US Food and Drug Administration
FDD	fault detection and diagnosis
FDL-CR	facility description language–conflict resolution
FDL	facility design language
FESEM	field-emission scanning electron microscope
FFT	fast Fourier transform
FHSS	frequency hopping spread spectrum
FIFO	first-in first-out
FIM	Fisher information matrix
FIPA	Foundation for Intelligent Physical Agents
FIRA	Federation of International Robot-Soccer Associations

FISCUS	Föderales Integriertes Standardisiertes Computer-Unterstütztes Steuersystem – federal integrated standardized computer-supported tax system
FIS	fuzzy inference system
fJSP	flexible jobshop problem
FK	forward kinematics
FLC	fuzzy logic control
FL	fuzzy-logic
FMCS	flight management computer system
FMC	flexible manufacturing cell
FMC	flight management computer
FMEA	failure modes and effects analysis
FMECA	failure mode, effects and criticality analysis
FMS	field message specification
FMS	flexible manufacturing system
FMS	flexible manufacturing system
FMS	flight management system
FM	Fiduccia–Mattheyses
FM	frequency-modulation
FOC	federation object coordinator
FOGA	Foundations of Genetic Algorithms
FOUP	front open unified pod
FOV	field of view
FPGA	field-programmable gate arrays
FPID	feedforward PID
FP	flooding protocol
FSK	frequency shift keying
FSM	finite-state machine
FSPM	FAL service protocol machine
FSSA	fixed structure stochastic automaton
FSS	flight service station
FSW	friction stir welding
FTA	fault tree analysis
FTC	fault tolerant control
FTE	flight technical error
FTE	full-time equivalent
FTL	flexible transfer line
FTP	file transfer protocol
FTSIA	fault-tolerance sensor integration algorithm
FTTP	fault tolerant time-out protocal
FW	framework

G

G2B	government-to-business
G2C	government-to-citizen
G2G	government-to-government
GAGAN	GEO augmented navigation
GAIA	geometrical analytic for interactive aid
GAMP	good automated manufacturing practice
GATT	General Agreement on Tariffs and Trade
GA	genetic algorithms
GBAS	ground-based augmentation system

GBIP	general purpose interface bus
GBS	goal-based scenario
GDP	gross domestic product
GDP	ground delay program
GDSII	graphic data system II
GDSS	group decision support system
GECCO	Genetic and Evolutionary Computation Conference
GEM	generic equipment model
GERAM	generalized enterprise reference architecture and methodology
GIS	geographic information system
GLS	GNSS landing system
GLUT4	activated Akt and PKCζ trigger glucose transporter
GMAW	gas metal arc welding
GMCR	graph model for conflict resolution
GNSS	global navigation satellite system
GPA	generalized pursuit algorithm
GPC	generalized predictive control
GPRS	general packet radio service
GPS	global positioning system
GPWS	ground-proximity warning system
GP	genetic programming
GRAI	graphes de résultats et activités interreliés
GRAS	ground regional augmentation system
GRBF	Gaussian RBF
GSM	global system for mobile communication
GTAW	gas tungsten arc welding
GUI	graphic user interface

H

HACCP	hazard analysis and critical control points
HACT	human–automation collaboration taxonomy
HAD	heterogeneous, autonomous, and distributed
HART	highway addressable remote transducer
HCI	human–computer interaction
HCS	host computer system
HDD	hard-disk drive
HEA	human error analysis
HEFL	hybrid electrode fluorescent lamp
HEP	human error probability
HERO	highway emergency response operator
HES	handling equipment scheduling
HFDS	Human Factors Design Standard
HF	high-frequency
HID	high-intensity discharge
HIS	hospital information system
HITSP	Healthcare Information Technology Standards Panel
HIT	healthcare information technology
HIV	human immunodeficiency virus
HJB	Hamilton–Jacobi–Bellman

HL7	Health Level 7
HMD	helmet-mounted display
HMI	human machine interface
HMM	hidden Markov model
HMS	hierarchical multilevel system
HOMO	highest occupied molecular orbital
HPC	high-performance computing
HPLC	high-performance liquid chromatography
HPSS	High-Performance Storage System
HPWREN	High-Performance Wireless Research and Education Network
HP	horsepower
HRA	human reliability analysis
HR	human resources
HSE	high speed Ethernet
HSI	human system interface
HSMS	high-speed message standard
HTN	hierarchical task network
HTTP	hypertext transfer protocol
HUD	heads up display
HUL	Harvard University Library
HVAC	heating, ventilation, air-conditioning
Hazop	hazardous operation
HiL	hardware-in-the-loop

I

i-awGA	interactive adaptive-weight genetic algorithm
I(P)AD	intelligent (power) assisting device
I/O	input/output
IAMHS	integrated automated material handling system
IAT	Institut Avtomatiki i Telemekhaniki
IAT	interarrival time
IB	internet banking
ICAO	International Civil Aviation Organization
ICORR	International Conference on Rehabilitation Robotics
ICRA	International Conference on Robotics and Automation
ICT	information and communication technology
IC	integrated circuit
IDEF	integrated definition method
IDL	Interactive Data Language
IDM	iterative design model
ID	identification
ID	instructional design
IEC	International Electrotechnical Commission
IFAC	International Federation of Automatic Control
IFC	industry foundation class
IFF	identify friend or foe
IFR	instrument flight rules

IGRT	image-guided radiation therapy
IGS	intended goal structure
IGVC	Intelligent Ground Vehicle Competition
IHE	integrating the healthcare enterprise
IIT	information interface technology
IK	inverse kinematics
ILS	instrument landing system
ILS	integrated library system
IL	instruction list
IMC	instrument meteorological condition
IMC	internal model controller
IML	inside mold line
IMM	interactive multiple model
IMRT	intensity modulated radiotherapy
IMS	infrastructure management service
IMT	infotronics and mechatronics technology
IMU	inertial measurement unit
INS	inertial navigation system
IO	inputoutput
IPA	intelligent parking assist
IPS	integrated pond system
IPv6	internet protocol version 6
IP	inaction–penalty
IP	industrial protocol
IP	integer programming
IP	intellectual property
IP	internet protocol
IRAF	Image Reduction and Analysis Facility
IRB	institutional review board
IRD	interactive robotic device
IROS	Intelligent Robots and Systems
IRR	internal rate of return
IRS1	insulin receptor substrate-1
IR	infrared
ISA	instruction set architecture
ISCIS	intra-supply-chain information system
iSCSI	Internet small computer system interface
ISDN	integrated services digital network
ISELLA	intrinsically safe lightweight low-cost arm
ISIC/MED	Intelligent Control/Mediterranean Conference on Control and Automation
ISM	industrial, scientific, and medical
ISO-OSI	International Standards Organization Open System Interconnection
ISO	International Organization for Standardization
ISO	independent system operator
ISP	internet service provider
ISS	input-to-state stability
IS	information system
ITC	information and communications technology
ITS	intelligent transportation system
IT	information technology
IVBSS	integrated vehicle-based safety system

IVI	Intelligent Vehicle Initiative
IV	intravenous

J

J2EE	Java to Enterprise Edition
JAUGS	joint architecture for unmanned ground system
JCL	job control language
JDBC	Java database connectivity
JDEM	Joint Dark Energy Mission
JDL	job description language
JIT	just-in-time
JLR	join/leave/remain
JPA	job performance aid
JPDO	joint planning and development office
JPL	Jet Propulsion Laboratory
JSR-001	Java specification request
Java RTS	Java real-time system
Java SE	Java standard runtime environment
JeLC	Japan e-Learning Consortium

K

KADS	knowledge analysis and documentation system
KCL	Kirchhoff's current law
KCM	knowledge chain management
KIF	knowledge interchange format
KISS	keep it simple system
KM	knowledge management
KPI	key performance indicators
KQML	knowledge query and manipulation language
KS	knowledge subsystem
KTA	Kommissiya Telemekhaniki i Avtomatiki
KVL	Kirchhoff's voltage law
KWMS	Kerry warehouse management system

L

LAAS	local-area augmentation system
LADARS	precision laser radar
LAN	local-area network
LA	learning automata
LBNL	Lawrence Berkeley National Laboratory
LBW	Laser beam welding
LC/MS	liquid-chromatography mass spectroscopy
LCD	liquid-crystal display
LCG	LHC computing grid
LCMS	learning contents management system
LCM	lane change/merge warning
LC	lean construction
LDW	lane departure warning
LDW	lateral drift warning system

LD	ladder diagram
LEACH	low-energy adaptive clustering hierarchy
LED	light-emitting diode
LEEPS	low-energy electron point source
LEO	Lyons Electronic Office
LES	logistic execution system
LFAD	light-vehicle module for LCM, FCW, arbitration, and DVI
LF	low-frequency
LHC	Large Hadron Collider
LHD	load–haul–dump
LIFO	last-in first-out
LIP	learning information package
LISI	levels of information systems interoperability
LISP	list processing
LLWAS	low-level wind-shear alert system
LMFD	left matrix fraction description
LMI	linear matrix inequality
LMPM	link layer mapping protocol machine
LMS	labor management system
LNAV	lateral navigation
LOA	levels of automation
LOCC	lines of collaboration and command
LOC	level of collaboration
LOINC	logical observation identifiers names and codes
LOM	learning object metadata/learning object reference model
LORANC	long-range navigational system
LPV	localizer performance with vertical guidance
LP	linear programming
LQG	linear-quadratic-Gaussian
LQR	linear quadratic regulator
LQ	linear quadratic
LS/AMC	living systems autonomic machine control
LS/ATN	living systems adaptive transportation network
LS/TS	Living Systems Technology Suite
LSL	low-level switch
LSST	Large Synoptic Survey Telescope
LSS	large-scale complex system
LS	language subsystem
LTI	linear time-invariant
LUMO	lowest unoccupied molecular orbital
LUT	look-up table
LVAD	left ventricular assist device
LVDT	linear variable differential transformer

M

m-SWCNT	metallic SWCNT
M/C	machining center
M2M	machine-to-machine

MAC	medium access control
MADSN	mobile-agent-based DSN
MAG	metal active gas
MAN	metropolitan area network
MAP	manufacturing assembly pilot
MAP	mean arterial pressure
MAP	missed approach point
MARC	machine-readable cataloging
MARR	minimum acceptable rate of return
MAS	multiagent system
MAU	medium attachment unit
MAV	micro air vehicle
MBP	Manchester bus powered
MCC	motor control center
MCDU	multiple control display unit
MCP	mode control panel
MCP	multichip package
MCS	material control system
MDF	medium-density fiber
MDI	manual data input
MDP	Markov decision process
MDS	management decision system
MD	missing a detection
MEMS	micro-electromechanical system
MEN	multienterprise network
MERP/C	ERP e-learning by MBE simulations with collaboration
MERP	Management Enterprise Resource Planning
MES	manufacturing execution system
METU	Middle East Technical University
MFD	multifunction display
MHA	material handling automation
MHEM	material handling equipment machine
MHIA	Material Handling Industry of America
MH	material handling
MIG	metal inert gas
MIMO	multi-input multi-output
MIP	mixed integer programming
MIS	management information system
MIS	minimally invasive surgery
MIT	Massachusetts Institute of Technology
MIT	miles in-trail
MKHC	manufacturing know-how and creativity
MLE	maximum-likelihood estimation
MMS	man–machine system
MMS	material management system
MOC	mine operation center
moGA	multiobjective genetic algorithm
MOL	middle of life
MOM	message-oriented middleware
MPAS	manufacturing process automation system
MPA	metabolic pathway analysis
MPC	model-based predictive control

mPDPTW	multiple pick up and delivery problem with time windows
MPEG	Motion Pictures Expert Group
MPLS	multi protocol label switching
MPS	master production schedule
MQIC	Medical Quality Improvement Consortium
MRI	magnetic resonance imaging
MRO	maintenance, repair, and operations
MRPII	material resource planning (2nd generation)
MRPI	material resource planning (1st generation)
MRP	manufacturing resources planning
MRR	material removal rate
MSAS	MTSAT satellite-based augmentation system
MSAW	minimum safe warning altitude
MSA	microsensor array
MSDS	material safety data sheet
MSI	multisensor integration
MSL	mean sea level
MTBF	mean time between failure
MTD	maximum tolerated dose
MTE	minimum transmission energy
MTSAT	multifunction transport satellite
MTTR	mean time to repair
MUX	multiplexor
MVFH	minimum vector field histogram
MV	manipulated variables
MWCNT	multi-walled carbon nanotube
MWKR	most work remaining
McTMA	multicenter traffic management advisor
Mcr	multi-approach to conflict resolution
MeDICIS	methodology for designing interenterprise cooperative information system
MidFSN	middleware for facility sensor network
Mips	million instructions per second
M&S	metering and spacing

N

NAE	National Academy of Engineering
NAICS	North American Industry Classification System
NASA	National Aeronautics and Space Administration
NASC	Naval Air Systems Command
NAS	National Airspace System
NATO	North Atlantic Treaty Organization
NBTI	negative-bias temperature instability
NCS	networked control system
NC	numerical control
NDB	nondirectional beacon
NDHA	National Digital Heritage Archive

NDI	nondestructive inspection	OCLC	Ohio College Library Center
NDRC	National Defence Research Committee	ODBC	object database connectivity
NEAT	Near-Earth Asteroid Tracking Program	ODE	ordinary differential equation
NEFUSER	neural-fuzzy system for error recovery	ODFI	originating depository financial institution
NEFUSER	neuro-fuzzy systems for error recovery	OECD	Organization for Economic Cooperation and Development
NEMA	National Electrical Manufacturers Association	OEE	overall equipment effectiveness
NEMS	nanoelectromechanical system	OEM	original equipment manufacturer
NERC	North American Electric Reliability Corporation	OGSA	open grid services architecture
NERSC	National Energy Research Scientific Computing Center	OHT	overhead hoist transporter
		OHT	overhead transport
NES	networked embedded system	OLAP	online analytical process
NFC	near field communication	OLE	object linking and embedding
NHTSA	National Highway Traffic Safety Administration	OML	outside mold line
		OMNI	office wheelchair with high manoeuvrability and navigational intelligence
NICU	neonatal intensive care unit		
NIC	network interface card		
NIR	near-infrared	OMS	order managements system
NISO	National Information Standards Organization	ONIX	online information exchange
		OOAPD	object-oriented analysis, design and programming
NIST	National Institute of Standards		
NLP	natural-language processing	OODB	object-oriented database
NNI	national nanotechnology initiative	OOM	object-oriented methodology
non RT	nonreal-time	OOOI	on, out, off, in
NP	nondeterministic polynomial-time	OOP	object-oriented programming
NPC	nanopore channel	OO	object-oriented
NPV	net present value	OPAC	online public access catalog
NP	nominal performance	OPC AE	OPC alarms and events
NRE	nonrecurring engineering	OPC XML-DA	OPC extensible markup language (XML) data access
nsGA	nondominated sorting genetic algorithm		
nsGA II	nondominated sorting genetic algorithm II	OPC	online process control
		OPM	object–process methodology
NSS	Federal Reserve National Settlement System	OQIS	online quality information system
		ORF	operating room of the future
NS	nominal stability	ORTS	open real-time operating system
NURBS	nonuniform rational B-splines	OR	operating room
NYSE	New York Stock Exchange	OR	operation research
NaroSot	Nano Robot World Cup Soccer Tournament	OSHA	Occupation Safety and Health Administration
NoC	network on chip	OSRD	Office of Scientific Research and Development
		OSTP	Office of Science and Technology Policy

O

		OS	operating system
O.R.	operations research	OTS	operator training systems
O/C	open-circuit	OWL	web ontology language
OAC	open architecture control		
OAGIS	open applications group		
OAI-PMH	open archieves initiative protocol for metadate harvesting		

P

OASIS	Organization for the Advancement of Structured Information Standards	P/D	pickup/delivery
		P/T	place/transition
OBB	oriented bounding box	PACS	picture archiving and communications system
OBEM	object-based equipment model		
OBS	on-board software	PAM	physical asset management
OBU	onboard unit	PAM	pulse-amplitude modulation
		PAN	personal area network

PARR	problem analysis resolution and ranking
PAT	process analytical technology
PAW	plasma arc welding
PBL	problem-based learning
PBPK	physiologically based pharmacokinetic
PCA	principal component analysis
PCBA	printed circuit board assembly
PCB	printed circuit board
PCFG	probabilistic context-free grammar
PCI	Peripheral Component Interconnect
PCR	polymerase chain reaction
PC	personal computer
PDA	personal digital assistant
PDC	predeparture clearance
PDDL	planning domain definition language
PDF	probability distribution function
pdf	probability distribution function
PDITC	1,4-phenylene diisothiocyanate
PDKM	product data and knowledge management
PDM	product data management
PDSF	Parallel Distributed Systems Facility
PDT	photodynamic therapy
PD	pharmacodynamics
PECVD	plasma enhanced chemical vapor deposition
PEID	product embedded information device
PERA	Purdue enterprise reference architecture
PERT/CPM	program evaluation and review technique/critical path method
PERT	project evaluation and review technique
PET	positron emission tomography
PE	pulse echo
PFS	precision freehand sculptor
PF	preference function
PGP	pretty good privacy
PHA	preliminary hazard analysis
PHERIS	public-health emergency response information system
PHR	personal healthcare record
PI3K	phosphatidylinositol-3-kinase
PID	proportional, integral, and derivative
PISA	Program for International Student Assessment
PI	proportional–integral
PKI	public-key infrastructure
PKM	parallel kinematic machine
PK	pharmacokinetics
PLA	programmable logic array
PLC	programmable logic controller
PLD	programmable logic device
PLM	product lifecycle management
PMC	process module controller
PMF	positioning mobile with respect to fixed
PM	process module
POMDP	partially observable Markov decision process

POS	point-of-sale
PPFD	photosynthetic photon flux density
PPS	problem processing subsystem
PRC	phase response curve
PROFIBUS-DP	process field bus–decentralized peripheral
PROMETHEE	preference ranking organization method for enrichment evaluation
PR	primary frequency
PSAP	public safety answering point
PSC	product services center
PSF	performance shaping factor
PSH	high-pressure switch
PSK	phase-shift keying
PSM	phase-shift mask
PS	price setting
PTB	German Physikalisch-Technische Bundesanstalt
PTO	power takeoff
PTP	point-to-point protocol
PTS	predetermined time standard
PWM	pulse-width-modulation
PXI	PCI extensions for instrumentation
ProVAR	professional vocational assistive robot
Prolog	programming in logics
P&ID	piping & instrumentation diagram

Q

QAM	quadrature amplitude modulation
QTI	question and test interoperability
QoS	quality of service

R

R.U.R.	Rossum's universal robots
R/T mPDPSTW	multiple pick up and delivery problem with soft time windows in real time
RAID	redundant array of independent disk
RAID	robot to assist the integration of the disabled
RAIM	receiver autonomous integrity monitoring
rALB	robot-based assembly line balancing
RAM	random-access memory
RAP	resource allocation protocol
RAS	recirculating aquaculture system
RA	resolution advisory
RBC	red blood cell
RBF	radial basis function
rcPSP	resource-constrained project scheduling problem
RCP	rapid control prototyping
RCRBF	raised-cosine RBF
RC	remote control
RC	repair center
RDB	relational database
RDCS	robust design computation system

RDCW FOT	Road Departure Crash Warning System Field Operational Test		RTOS	real-time operating system
RDCW	road departure crash warning		RTO	real-time optimization
RDF	resource description framework		RTO	regional transmission organization
RET	resolution enhancement technique		RTSJ	real-time specification for Java
RE	random environment		RT	radiotherapy
RFID	radiofrequency identification		RT	register transfer
RF	radiofrequency		rwGA	random-weight genetic algorithm
RGB	red–green–blue		RW	read/write
RGV	rail-guided vehicle		RZPR	power reserve
RHC	receding horizon control		RZQS	quick-start reserve
RHIO	regional health information organization		Recon	retrospective conversion
RIA	Robotics Industries Association		R&D	research and development
RISC	reduced instruction set computer			
RIS	real information system			
RI	reward–inaction			

S

RLG	Research Libraries Group		s-SWCNT	semiconducting MWCNT
RLG	ring-laser-gyro		S/C	short-circuit
RMFD	right matrix fraction description		SACG	Stochastic Adaptive Control Group
RMS	reconfigurable manufacturing systems		SADT	structured analysis and design technique
RMS	reliability, maintainability, and safety		SAGA	Standards und Architekturen für e-Government-Anwendungen – standards and architectures for e-Government applications
RMS	root-mean-square			
RM	real manufacturing			
RNAV	area navigation		sALB	simple assembly line balancing
RNA	ribonucleic acid		SAM	self-assembled monolayer
RNG	random-number generator		SAM	software asset management
RNP	required navigation performance		SAN	storage area network
ROBCAD	robotics computer aided design		SAO	Smithsonian Astrophysical Observatory
ROI	return on investment		SAW	submerged arc welding
ROM	range-of-motion		SA	situation awareness
ROT	runway occupancy time		SBAS	satellite-based augmentation system
ROV	remotely operated underwater vehicle		SBIR	small business innovation research
RO	read only		SBML	system biology markup language
RPC	remote procedure call		SCADA	supervisory control and data acquisition
RPM	revolutions per minute		SCARA	selective compliant robot arm
RPN	risk priority number		SCC	somatic cell count
RPS	real and physical system		SCM	supply chain management
RPTS	robot predetermined time standard		SCNM	slot communication network management
RPU	radar processing unit		SCN	suprachiasmatic nucleus
RPV	remotely piloted vehicle		SCORM	sharable content object reference model
RPW	ranked positioned weight		SCST	source-channel separation theorem
RP	reward–penalty		SDH	synchronous digital hierarchy
RRT	rapidly exploring random tree		SDSL	symmetrical digital subscriber line
RSEW	resistance seam welding		SDSS	Sloan Digital Sky Survey II
RSW	resistance spot welding		SDSS	spatial decision support system
RS	robust stability		SDS	sequential dynamic system
RT DMP	real-time decision-making processes		SDT	signal detection theory
RT-CORBA	real-time CORBA		SECS	semiconductor equipment communication standard
RTA	required time of arrival		SEC	Securities and Exchange Comission
RTDP	real-time dynamic programming		SEER	surveillance, epidemiology, and end result
RTD	resistance temperature detector			
RTE	real-time Ethernet		SEI	Software Engineering Institute
RTK GPS	real-time kinematic GPS		SELA	stochastic estimator learning algorithm
RTL	register transfer level		SEMI	Semiconductor Equipment and Material International
RTM	resin transfer molding			
RTM	robot time & motion method			

SEM	scanning electron microscopy
SEM	strategic enterprise management
SESAR	Single European Sky ATM research
SESS	steady and earliest starting schedule
SFC	sequential function chart
SFC	space-filling curve
SHMPC	shrinking horizon model predictive control
SIFT	scale-invariant feature transform
SIL	safety integrity level
SIM	single input module
SISO	single-input single-output
SIS	safety interlock system
SKU	stock keeping unit
SLAM	simultaneous localization and mapping technique
SLA	service-level agreement
SLIM-MAUD	success likelihood index method-multiattribute utility decomposition
SLP	storage locations planning
SL	sensitivity level
SMART	Shimizu manufacturing system by advanced robotics technology
SMAW	shielded metal arc welding
SMA	shape-memory alloys
SMC	sequential Monte Carlo
SME	small and medium-sized enterprises
SMIF	standard mechanical interface
SMS	short message service
SMTP	simple mail transfer protocol
SMT	surface-mounting technology
SNA	structural network analysis
SNIFS	Supernova Integral Field Spectrograph
SNLS	Supernova Legacy Survey
SNOMED	systematized nomenclature of medicine
SNfactory	Nearby Supernova Factory
SN	supernova
SOAP	simple object access protocol
SOA	service-oriented architecture
SOC	system operating characteristic
SOI	silicon-on-insulator
SONAR	sound navigation and ranging
SO	system operator
SPC	statistical process control
SPF/DB	superplastic forming/diffusion bonding
SPF	super plastic forming
SPIN	sensor protocol for information via negotiation
SPI	share price index
SQL	structured query language
SRAM	static random access memory
SRI	Stanford Research Institute
SRL	science research laboratory
SRM	supplier relationship management

SSADM	structured systems analysis and design method
SSA	stochastic simulation algorithm
ssDNA	single-strand DNA
SSH	secure shell
SSL	secure sockets layer
SSO	single sign-on
SSR	secondary surveillance radar
SSSI	single-sensor, single-instrument
SSV	standard service volume
SS	speed-sprayer
STARS	standard terminal automation replacement system
STAR	standard terminal arrival route
STA	static timing analysis
STCU	SmallTown Credit Union
STEM	science, technology, engineering, and mathematics
STM	scanning tunneling microscope
STTPS	single-truss tomato production system
ST	structured text
SUV	sports utility vehicle
SVM	support vector machine
SVS	synthetic vision system
SV	stroke volume
SWCNT	single-walled carbon nanotube
SWP	single-wafer processing
SW	stroke work
SaaS	software as a service
ServSim	maintenance service simulator
SiL	software-in-the-loop
Smac	second mitochondrial-activator caspase
SoC	system-on-chip
SoD	services-on-demand
SoS	systems of systems
SoTL	scholarship of teaching and learning
Sunfall	Supernova Factory Assembly Line
spEA	strength Pareto evolutionary algorithm
SysML	systems modeling language

T

TACAN	tactical air navigation
TALplanner	temporal action logic planner
TAP	task administration protocol
TAR	task allocation ratio
TA	traffic advisory
TB	terabytes
TCAD	technology computer-aided design
TCAS	traffic collision avoidance system
TCP/IP	transmission control protocol/internet protocol
TCP	transmission control protocol
TCS	telescope control system
TDMA	time-division multiple access

TEAMS	testability engineering and maintenance system		UGC	user generated content
TEG	timed event graph		UHF	ultrahigh-frequency
TEM	transmission electron microscope		UI	user interface
TER	tele-ultrasonic examination		UMDL	University of Michigan digital library
TFM	traffic flow management		UML	universal modeling language
THERP	technique for human error rate prediction		UMTS	universal mobile telecommunications system
THR	total hip replacement		UMTS	universal mobile telecommunications system
THW	time headway			
TIE/A	teamwork integration evaluator/agent		UN/CEFACT	United Nations Centre for Trade Facilitation and Electronic Business
TIE/MEMS	teamwork integration evaluator/MEMS			
TIE/P	teamwork integration evaluator/protocol		UN	United Nations
TIF	data information forwarding		UPC	universal product code
TIG	tungsten inert gas		UPMC	University of Pittsburgh Medical Center
TIMC	techniques for biomedical engineering and complexity management		UPS	uninterruptible power supply
			URET	user request evaluation tool
TLBP	transfer line balancing problem		URL	uniform resource locator
TLPlan	temporal logic planner		URM	unified resource management
TLX	task load index		UR	universal relay
TMA	traffic management advisor		USB	universal serial bus
TMC	traffic management center		USC	University of Southern California
TMC	transport module controller		UTLAS	University of Toronto Library Automation System
TMS	transportation management system			
TMU	traffic management unit		UT	ultrasonic testing
TOP	time-out protocol		UV	ultraviolet
TO	teleoperator		UWB	ultra wire band
TPN	trading process network			
TPS	throttle position sensor			
TPS	transaction processing system			

V

TRACON	terminal radar approach control		VAN-AP	VAN access point
TRIPS	trade related aspects of intellectual property rights		VAN	value-added network
			VAN	virtual automation network
TRV	total removal volume		VAV	variable-air-volume
TSCM	thin-seam continuous mining		VCR	video cassette recorder
TSE	total system error		VCT	virtual cluster tool
TSMP	time synchronized mesh protocol		VDL	VHF digital link
TSTP	transportation security training portal		VDU	visual display unit
TTC	time-to-collision		veGA	vector evaluated genetic algorithm
TTF	time to failure		VE	virtual environment
TTR	time to repair		VFD	variable-frequency drive
TTU	through transmission ultrasound		VFEI	virtual factory equipment interface
TU	transcriptional unit		VFR	visual flight rule
TV	television		VHDL	very high speed integrated circuit hardware description language
TestLAN	testers local area network			
			VHF	very high-frequency

U

			VICS	vehicle information and communication system
UAT	universal access transceiver		VII	vehicle infrastructure integration
UAV	unmanned aerial vehicle		VIS	virtual information system
UCMM	unconnected message manager		VLSI	very-large-scale integration
UCTE	Union for the Co-ordination of Transmission of Electricity		VMEbus	versa module eurobus
			VMIS	virtual machining and inspection system
UDDI	universal description, discovery, and integration		VMI	vendor-managed inventory
			VMM	virtual machine monitor
UDP	user datagram protocol		VMT	vehicle miles of travel
UEML	unified enterprise modeling language		VM	virtual machine

VM	virtual manufacturing
VNAV	vertical navigation
VNC	virtual network computing
VOD	virtual-object-destination
VORTAC	VOR tactical air navigation
VOR	VHF omnidirectional range
VPS	virtual physical system
VP	virtual prototyping
VRP	vehicle routing problem
VR	virtual reality
VSG	virtual service-oriented environment
VSP	vehicle scheduling problem
VSSA	variable structure stochastic automata
VTLS	Virginia Tech library system
VTW	virtual training workshop
VTx	virtualization technology
VoD	video on demand

W

W/WL	wired/wireless
WAAS	wide-area augmentation system
WAN	wide area network
WASCOR	WASeda construction robot
WBI	wafer burn-in
WBS	work breakdown structure
WBT	web-based training
WCDMA	wideband code division multiple access
WFMS	workflow management system
WI-Max	worldwide interoperability for microwave access
WIM	World-In-Miniatur
WIP	work-in-progress
WISA	wireless interface for sensors and actuator
WLAN	wireless local area network

WLN	Washington Library Network
WL	wireless LAN
WMS	warehouse management system
WMX	weight mapping crossover
WORM	write once and read many
WPAN	wireless personal area network
WSA	work safety analysis
WSDL	web services description language
WSN	wireless sensor network
WS	wage setting
WTO	World Trade Organization
WWII	world war 2
WWW	World Wide Web
WfMS	workflow management system
Wi-Fi	wireless fidelity

X

XIAP	X-linked inhibitor of apoptosis protein
XML	extensible mark-up language
XSLT	extensible stylesheet language transformation
XöV	XML for public administration

Y

Y2K	year-2000
YAG	Nd:yttrium–aluminum–garnet
ZDO	Zigbee device object

Z

ZVEI	Zentralverband Elektrotechnik- und Elektronikindustrie e.V.

Part A

Part A Development and Impacts of Automation

Development and Impacts of Automation. Part A The first part lays the conceptual foundations for the whole *Handbook* by explaining basic definitions of automation, its scope, its impacts and its meaning, from the views of prominent automation pioneers to a survey of concepts and applications around the world. The scope, evolution and development of automation are reviewed with illustrations, from prehistory throughout its development before and after the emergence of automatic control, during the Industrial Revolution, along the advancements in computing and communication, with and without robotics, and projections about the future of automation. Chapters in this part explain the significant influence of automation on our life: on individuals, organizations, and society; in economic terms and context; and impacts of precision, accuracy and reliability with automatic and automated equipment and operations.

1. Advances in Robotics and Automation: Historical Perspectives

Yukio Hasegawa

Historical perspectives are given about the impressive progress in automation. Automation, including robotics, has evolved by becoming useful and affordable. Methods have been developed to analyze and design better automation, and those methods have also been automated. The

most important issue in automation to make every effort to paying attention to all the details.

The bodies of human beings are smaller than those of wild animals. Our muscles, bones, and nails are smaller and weaker. However, human beings, fortunately, have larger brains and wisdom. Humans initially learned how to use tools and then started using machines to perform necessary daily operations. Without the help of these tools or machines we, as human beings, can no longer support our daily life normally.

Technology is making progress at an extremely high speed; for instance, about half a century ago I bought a camera for my own use; at that time, the price of a conventional German-made camera was very high, as much as 6 months income. However, the price of a similar quality camera now is the equivalent of only 2 weeks of the salary of a young person in Japan.

Seiko Corporation started production and sales of the world's first quartz watch in Japan about 40 years ago. At that time, the price of the watch was about 400 000 Yen. People used to tell me that such high-priced watches could only be purchased by a limited group of people with high incomes, such as airline pilots, company owners, etc. Today similar watches are sold in supermarkets for only 1000 Yen.

Furthermore, nowadays, we are moving towards the automation of information handling by using computers; for instance, at many railway stations, it is now common to see unmanned ticket consoles. Telephone exchanges have become completely automated and the cost to use telephone systems is now very low.

In recent years, robots have become commonplace for aiding in many different environments. Robots are machines which carry out motions and information handling automatically. In the 1970s I was asked to start conducting research on robots. One day, I was asked by the management of a Japanese company that wanted to start the sales of robots to determine whether such robots could be used in Japan. After analyzing robot motions by using a high-speed film analysis system, I reached the conclusion that the robot could be used both in Japan as well as in the USA.

After that work I developed a new motion analysis method named the robot predetermined time standard (RPTS). The RPTS method can be widely applied to robot operation system design and contributed to many robot operation system design projects.

In the USA, since the beginning of the last century, a lot of pioneers in human operation rationalization have made significant contributions. In 1911, Frederik Tailor proposed the scientific management method, which was later reviewed by the American Congress. Prof. Gilbreth of Purdue University developed the new motion analysis method, and contributed to the rationalization of human operations. Mr. Dancan of WOFAC Corporation proposed a human predetermined time standard (PTS) method, which was applied to human operation rationalizations worldwide.

In the robotic field, those contributions are only part of the solution, and people have understood that mechanical and control engineering are additionally important aspects. Therefore, analysis of human operations in robotic fields are combined with more analysis, design, and rationalization [1.1]. However, human operators play a very challenging role in operations. Therefore, study of the work involved is more important than the robot itself and I believe that industrial engineering is going to become increasingly important in the future [1.2]. Prof. Nof developed RTM, the robot time & motion computational method, which was applied in robot selection and program improvements, including mobile robots. Such techniques were then incorporated in ROBCAD, a computer aided design system to automate the design and implementation of robot installations and applications.

A number of years ago I had the opportunity to visit the USA to attend an international robot symposium. At that time the principle of "no hands in dies" was a big topic in America due to a serious problem with guaranteeing the safety of metal-stamping operations. People involved in the safety of metal-stamping operations could not decrease the accident rate in spite of their increasing efforts. The government decided that a new policy to fully automate stamping operations or to use additional devices to hold and place workpieces without inserting operators' hands between dies was needed. The decision was lauded by many stamping robot manufacturers. Many expected that about 50 000 pieces of stamping robots would be sold in the American market in a few years. At that time 700 000 stamping presses were used in the USA. In Japan, the forecast figure was modified to 20 000 pieces. The figure was not small and therefore we immediately organized a stamping robot development project team with government financial support. The project team was composed of ten people: three robot engineers, two stamping engineers, a stamping technology consultant, and four students. I also invited an expert who had previously been in charge of stamping robot development projects in Japan. A few years later, sales of stamping robots started, with very good sales (over 400 robots were sold in a few years).

However, the robots could not be used and were rather stored as inactive machines. I asked the person in charge for the reason for this failure and was told that designers had concentrated too much on the robot hardware development and overlooked analysis of the conditions of the stamping operations. Afterwards, our project team analyzed the working conditions of the stamping operations very carefully and classified them into 128 types. Finally the project team developed an operation analysis method for metal-stamping operations. In a few years, fortunately, by applying the method we were able to decrease the rate of metal-stamping operation accidents from 12 000 per year to fewer than 4000 per year. Besides metal-stamping operations, we worked on research projects for forgings and castings to promote labor welfare. Through those research endeavors we reached the conclusion that careful analysis of the operation is the most important issue for obtaining good results in the case of any type of operations [1.3].

I believe, from my experience, that the most important issue – not only in robot engineering but in all automation – is to make every effort to paying attention to all the details.

References

1.1   Y. Hasegawa: *Analysis of complicated operations for robotization*, SME Paper No. MS79–287 (1979)

1.2   Y. Hasegawa: Evaluation and economic justification. In: *Handbook of Industrial Robotics*, ed. by S.Y. Nof (Wiley, New York 1985) pp. 665–687

1.3   Y. Hasegawa: Analysis and classification of industrial robot characteristics, Ind. Robot Int. J. **1**(3), 106–111 (1974)

2. Advances in Industrial Automation: Historical Perspectives

Theodore J. Williams

Automation is a way for humans to extend the capability of their tools and machines. Self-operation by tools and machines requires four functions: Performance detection; process correction; adjustments due to disturbances; enabling the previous three functions without human intervention. Development of these functions evolved in history, and automation is the capability of causing machines to carry out a specific operation on command from external source. In chemical manufacturing and petroleum industries prior to 1940, most processing was in batch environment. The increasing demand for chemical and petroleum products by World War II and thereafter required different manufacturing setup, leading to continuous processing and efficiencies were achieved by automatic control and automation of process, flow and transfer. The increasing complexity of the control system for large plants necessitated applications of computers, which were introduced to the chemical industry in the 1960s. Automation has substituted computer-based control systems for most, if not all, control systems previously based on human-aided mechanical or pneumatic systems to the point that chemical and petroleum plant systems are now fully automatic to a very high degree. In addition, automation has replaced human effort, eliminates significant labor costs, and prevents accidents and injuries that might occur. The Purdue enterprise reference architecture (PERA) for hierarchical control structure, the hierarchy of personnel tasks, and plant operational management structure, as developed for large industrial plants, and a frameworks for automation studies are also illustrated.

Humans have always sought to increase the capability of their tools and their extensions, i. e., machines. A natural extension of this dream was making tools capable of self-operation in order to:

1. Detect when performance was not achieving the initial expected result
2. Initiate a correction in operation to return the process to its expected result in case of deviation from expected performance
3. Adjust ongoing operations to increase the machine's productivity in terms of (a) volume, (b) dimensional accuracy, (c) overall product quality or (d) ability to respond to a new previously unknown disturbance
4. Carry out the previously described functions without human intervention.

Item 1 was readily achieved through the development of sensors that could continuously or periodically measure the important variables of the process and signal the occurrence of variations in them. Item 2 was made possible next by the invention of controllers that convert knowledge of such variations into commands required to change operational variables and thereby return to the required operational results. The successful operation of any commercially viable process requires the solution of items 1 and 2.

The development of item 3 required an additional level of intelligence beyond items 1 and 2, i. e., the capability to make a comparison between the results achieved and the operating conditions used for a series of tests. Humans can, of course, readily perform this task. Accomplishing this task using a machine, however, requires the computational capability to compare successive sets of data, gather and interpret corrective results, and be able to apply the results obtained. For a few variables with known variations, this can be incorporated into the controller's design. However, for a large number of variables or when possible unknown ranges of responses may be present, a computer must be available.

Automation is the capability of causing a machine to carry out a specific operation on command from an external source. The nature of these operations may also be part of the external command received. The devise involved may likewise have the capability to respond to other external environmental conditions or signals when such responses are incorporated within its capabilities. Automation, in the sense used almost universally today in the chemical and petroleum industries, is taken to mean the complete or near-complete operation of chemical plants and petroleum refineries by digital computer

systems. This operation entails not only the monitoring and control of multiple flows of materials involved but also the coordination and optimization of these controls to achieve optimal production rate and/or the economic return desired by management. These systems are programmed to compensate, as far as the plant equipment itself will allow, for changes in raw material characteristics and availability and requested product flow rates and qualities.

In the early days of the chemical manufacturing and petroleum industries (prior to 1940), most processing was carried out in a *batch* environment. The needed ingredients were added together in a *kettle* and processed until the reaction or other desired action was completed. The desired product(s) were then sep-

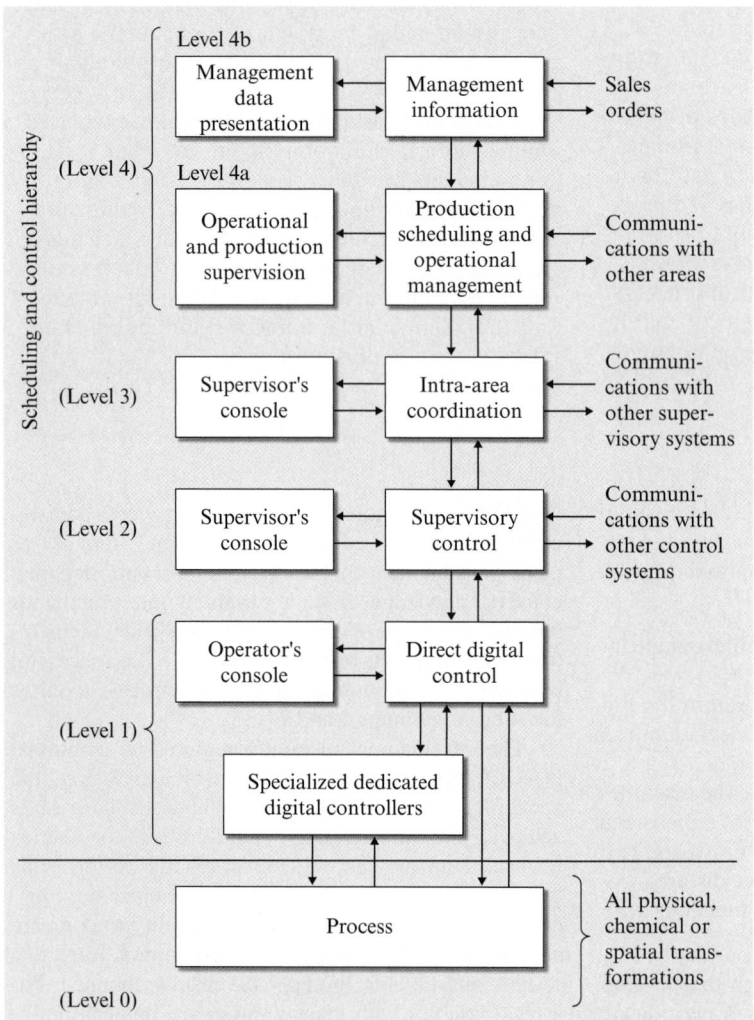

Fig. 2.1 The Purdue enterprise reference architecture (PERA). Hierarchical computer control structure for an industrial plant [2.1]

arated from the byproducts and unreacted materials by decanting, distilling, filtering or other applicable physical means. These latter operations are thus in contrast to the generally chemical processes of product formation. At that early time, the equipment and their accompanying methodologies were highly manpower dependent, particularly for those requiring coordination of the joint operation of related equipment, especially when succeeding steps involved transferring materials to different sets or types of equipment.

The strong demand for chemical and petroleum products generated by World War II and the following years of prosperity and rapid commercial growth required an entirely different manufacturing equipment setup. This led to the emergence of continuous processes where subsequent processes were continued in successive connected pieces of equipment, each devoted to a separate setup in the process. Thus a progression in distance to the succeeding equipment (rather than in time, in the same equipment) was now necessary. Since any specific piece of equipment or location in

the process train was then always used for the same operational stage, the formerly repeated filling, reacting, emptying, and cleaning operations in every piece of equipment was now eliminated. This was obviously much more efficient in terms of equipment usage. This type of operation, now called *continuous processing*, is in contrast to the earlier *batch processing* mode. However, the coordination of the now simultaneous operations connected together required much more accurate control of both operations to avoid the transmission of processing errors or *upsets* to downstream equipment.

Fortunately, our basic knowledge of the inherent chemical and physical properties of these processes had also advanced along with the development of the needed equipment and now allows us to adopt methodologies for assessing the quality and state of these processes during their operation, i.e., degree of completion, etc. Likewise, also fortunately, our basic knowledge of the technology of automatic control and its implementing equipment advanced along with knowledge of the

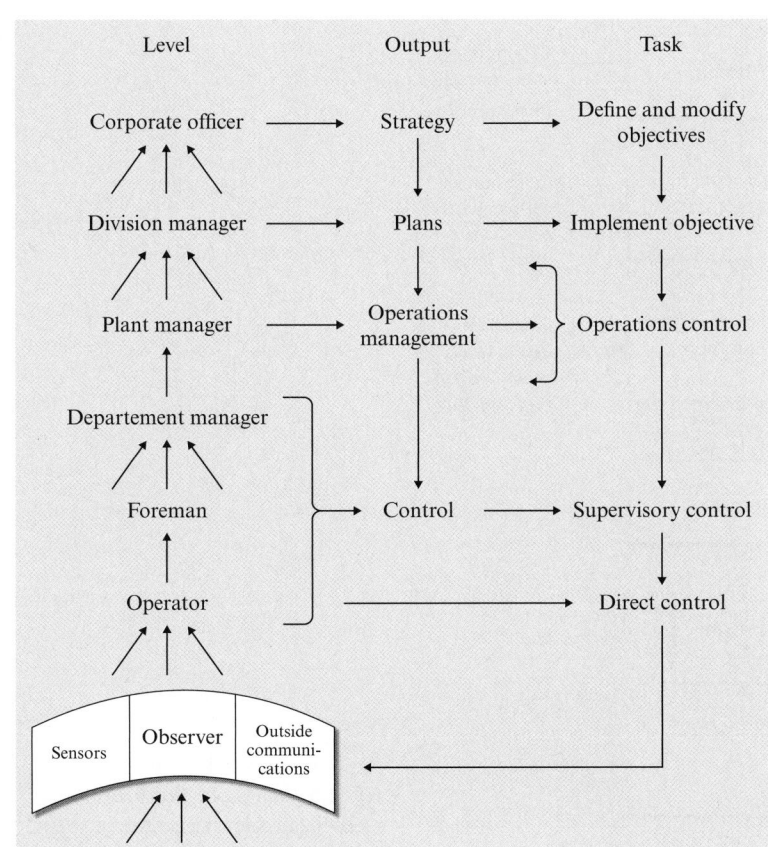

Fig. 2.2 Personnel task hierarchy in a large manufacturing plant

pneumatic and electronic techniques used to implement them. Pneumatic technology for the necessary control equipment was used almost exclusively from the original development of the technique to the 1920s until its replacement by the rapidly developing electronic techniques in the 1930s. This advanced type of equipment became almost totally electronic after the development of solid-state electronic technologies as the next advances. Pneumatic techniques were then used only where severe fire or explosive conditions prevented the use of electronics.

The overall complexity of the control systems for large plants made them objects for the consideration of the use of computers almost as soon as the early digital computers became practical and affordable. The

first computers for chemical plant and refinery control were installed in 1960, and they became quite prevalent by 1965. By now, computers are widely used in all large plant operations and in most small ones as well. If automation can be defined as the substitution of computer-based control systems for most, if not all, control systems previously based on human-aided mechanical or pneumatic systems, then for chemical and petroleum plant systems, we can now truly say that they are *fully automated*, to a very high degree.

As indicated above, a most desired byproduct of the automation of chemical and petroleum refining processes must be the replacement of human effort: first in directly handling the frequently dangerous chemical ingredients in the initiation of the process; second, that

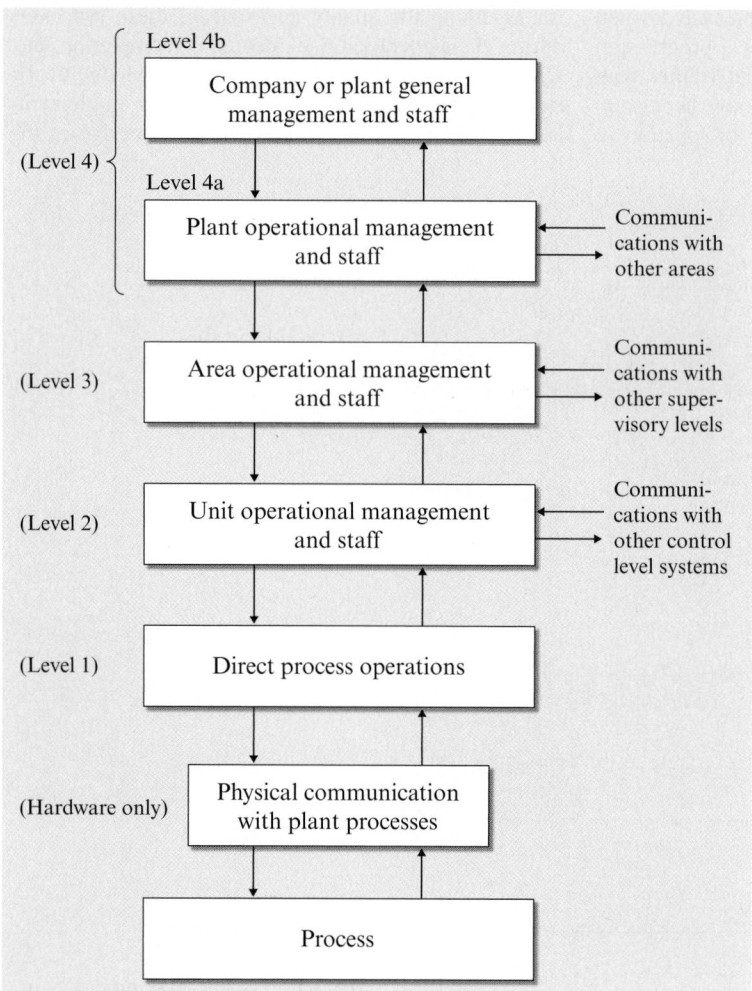

Fig. 2.3 Plant operational management hierarchical structure

of personally monitoring and controlling the carrying out and completion of these processes; and finally that of handling the resulting products. This omits the expenses involved in the employment of personnel for carrying out these tasks, and also prevents unnecessary accidents and injuries that might occur there. The staff at chemical plants and petroleum refineries has thus been dramatically decreased in recent years. In many locations this involves only a *watchman* role and an emergency maintenance function. This capability has further resulted in even further improvements in overall plant design to take full advantage of this new capability – a *synergy effect*. This *synergy effect* was next felt in the automation of the raw material acceptance

practices and the product distribution methodologies of these plants. Many are now connected directly to the raw material sources and their customers by pipelines, thus totally eliminating special raw material and product handling and packaging. Again, computers are widely used in the scheduling, monitoring, and controlling of all operations involved here.

Finally, it has been noted that there is a hierarchical relationship between the control of the industrial process plant unit automatic control systems and the duties of the successive levels of management in a large industrial plant from company management down to the final plant control actions [2.2–13]. It has also been shown that all actions normally taken by intermedi-

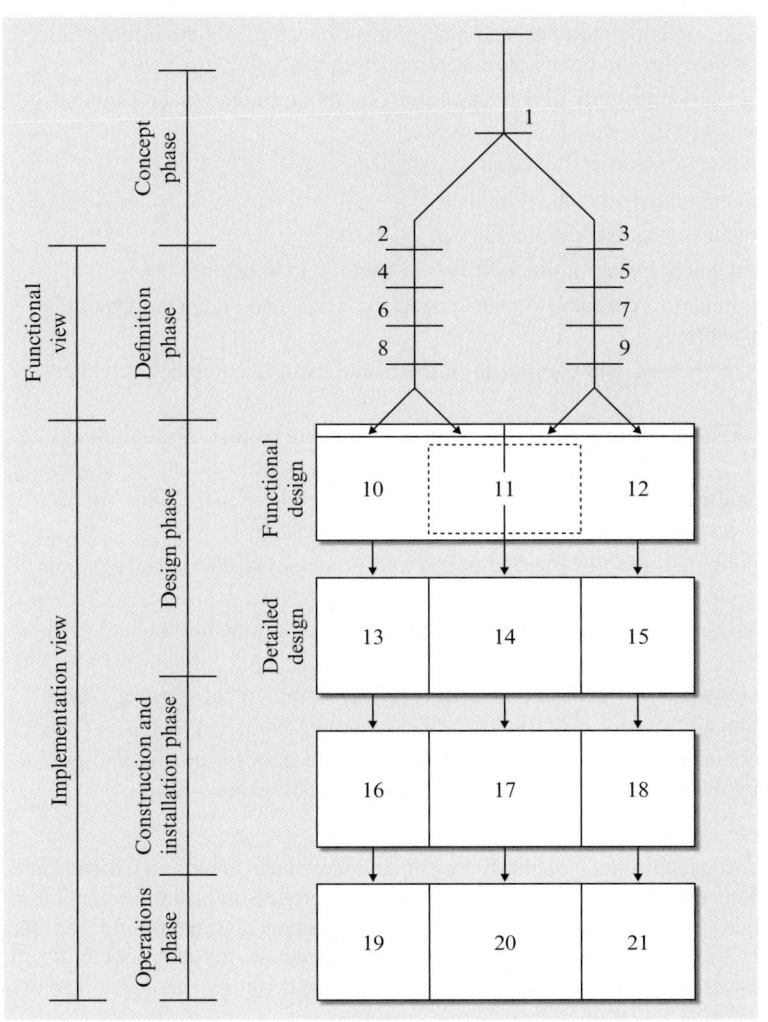

Fig. 2.4 Abbreviated sketch to represent the structure of the Purdue enterprise reference architecture

Table 2.1 Areas of interest for the architecture framework addressing development and implementation aids for automation studies (Fig. 2.4)

Area	Subjects of concern
1	Mission, vision and values of the company, operational philosophies, mandates, etc.
2	Operational policies related to the information architecture and its implementation
3	Operational strategies and goals related to the manufacturing architecture and its implementation
4	Requirements for the implementation of the information architecture to carry out the operational policies of the company
5	Requirements for physical production of the products or services to be generated by the company
6	Sets of tasks, function modules, and macrofunction modules required to carry out the requirements of the information architecture
7	Sets of production tasks, function modules, and macrofunctions required to carry out the manufacturing or service production mission of the company
8	Connectivity diagrams of the tasks, function modules, and macrofunction modules of the information network, probably in the form of data flow diagrams or related modeling methods
9	Process flow diagrams showing the connectivity of the tasks, function modules, and macrofunctions of the manufacturing processes involved
10	Functional design of the information systems architecture
11	Functional design of the human and organizational architecture
12	Functional design of the manufacturing equipment architecture
13	Detailed design of the equipment and software of the information systems architecture
14	Detailed design of the task assignments, skills development training courses, and organizations of the human and organizational architecture
15	Detailed design of components, processes, and equipment of the manufacturing equipment architecture
16	Construction, check-out, and commissioning of the equipment and software of the information systems architecture
17	Implementation of organizational development, training courses, and online skill practice for the human and organizational architecture
18	Construction, check-out, and commissioning of the equipment and processes of the manufacturing equipment architecture
19	Operation of the information and control system of the information systems architecture including its continued improvement
20	Continued organizational development and skill and human relations development training of the human and organizational architecture
21	Continued improvement of process and equipment operating conditions to increase quality and productivity, and to reduce costs involved for the manufacturing equipment architecture

ary plant staff in this hierarchy can be formulated into a computer-readable form for all operations that do not involve *innovation* or other problem-solving actions by plant staff. Figures 2.1–2.4 (with Table 2.1) illustrate this hierarchical structure and its components. See more on the history of automation and control in Chaps. 3 and 4; see further details on process industry automation in Chap. 31; on complex systems automation in Chap. 36; and on automation architecture for interoperability in Chap. 86.

References

2.1 T.J. Williams: *The Purdue Enterprise Reference Architecture* (Instrument Society of America, Pittsburgh 1992)

2.2 H. Li, T.J. Williams: Interface design for the Purdue Enterprise Reference Architecture (PERA) and methodology in e-Work, Prod. Plan. Control **14**(8), 704–719 (2003)

2.3 G.A. Rathwell, T.J. Williams: Use of Purdue Reference Architecture and Methodology in Industry (the Fluor Daniel Example). In: *Modeling and Methodologies for Enterprise Integration*, ed. by P. Bernus, L. Nemes (Chapman Hall, London 1996)

2.4 T.J. Williams, P. Bernus, J. Brosvic, D. Chen, G. Doumeingts, L. Nemes, J.L. Nevins, B. Vallespir, J. Vliestra, D. Zoetekouw: Architectures for integrating manufacturing activities and enterprises, Control Eng. Pract. **2**(6), 939–960 (1994)

2.5 T.J. Williams: One view of the future of industrial control, Eng. Pract. **1**(3), 423–433 (1993)

2.6 T.J. Williams: A reference model for computer integrated manufacturing (CIM). In: *Int. Purdue Workshop Industrial Computer Systems* (Instrument Society of America, Pittsburgh 1989)

2.7 T.J. Williams: *The Use of Digital Computers in Process Control* (Instrument Society of America, Pittsburgh 1984) p. 384

2.8 T.J. Williams: 20 years of computer control, Can. Control. Instrum. **16**(12), 25 (1977)

2.9 T.J. Williams: Two decades of change: a review of the 20-year history of computer control, Can. Control. Instrum. **16**(9), 35–37 (1977)

2.10 T.J. Williams: Trends in the development of process control computer systems, J. Qual. Technol. **8**(2), 63–73 (1976)

2.11 T.J. Williams: Applied digital control – some comments on history, present status and foreseen trends for the future, Adv. Instrum., Proc. 25th Annual ISA Conf. (1970) p. 1

2.12 T.J. Williams: Computers and process control, Ind. Eng. Chem. **62**(2), 28–40 (1970)

2.13 T.J. Williams: The coming years... The era of computing control, Instrum. Technol. **17**(1), 57–63 (1970)

Part A | 2

3. Automation:
What It Means to Us Around the World

Shimon Y. Nof

The meaning of the term *automation* is reviewed through its definition and related definitions, historical evolution, technological progress, benefits and risks, and domains and levels of applications. A survey of 331 people around the world adds insights to the current meaning of automation to people, with regard to: *What is your definition of automation? Where did you encounter automation first in your life?* and *What is the most important contribution of automation to society?* The survey respondents include 12 main aspects of the definition in their responses; 62 main types of first automation encounter; and 37 types of impacts, mostly benefits but also two benefit–risks combinations: replacing humans, and humans' inability to complete tasks by themselves. The most exciting contribution of automation found in the survey was to *encourage/inspire creative work; inspire newer solutions.* Minor variations were found in different regions of the world. Responses about the first automation encounter are somewhat related to the age of the respondent, e.g., pneumatic versus digital control, and to urban versus farming childhood environment. The chapter concludes with several emerging trends in bioinspired automation, collaborative control and automation, and risks to anticipate and eliminate.

3.1 The Meaning of Automation

What is the meaning of automation? When discussing this term and concept with many colleagues, leading experts in various aspects of automation, control theory, robotics engineering, and computer science during the development of this *Handbook of Automation*, many of them had different definitions; they even argued vehemently that *in their language, or their region of the world, or their professional domain*, automation has a unique meaning *and we are not sure it is the same meaning for other experts.* But there has been no doubt, no confusion, and no hesitation that automation is powerful; it has tremendous and amazing impact on civilization, on humanity, and it may carry risks.

So what is automation? This chapter introduces the meaning and definition of automation, at an introductory, overview level. Specific details and more theoretical definitions are further explained and illustrated throughout the following parts and chapters of this handbook. A survey of 331 participants from around the world was conducted and is presented in Sect. 3.5.

3.1.1 Definitions and Formalism

Automation, in general, implies *operating or acting, or self-regulating, independently, without human inter-*

vention. The term evolves from *automatos*, in Greek, meaning acting by itself, or by its own will, or spontaneously. Automation involves machines, tools, devices, installations, and systems that are all platforms developed by humans to perform a given set of activities without human involvement during those activities. But there are many variations of this definition. For instance, before modern automation (specifically defined in the modern context since about 1950s), *mechanization* was a common version of automation. When automatic control was added to mechanization as an intelligence feature, the distinction and advantages of automation became clear. In this chapter, we review these related definitions and their evolvement, and survey how people around the world perceive automation. Examples of automation are described, including ancient to early examples in Table 3.1, examples from the Industrial Revolution in Table 3.3, and modern and emerging examples in Table 3.4. From the general definition of automation, the automation formalism is presented in Fig. 3.1 with four main elements: platform, autonomy, process, and power source. Automation platforms are illustrated in Table 3.2.

This automation formalism can help us review some early examples that may also fall under the definition of automation (before the term *automation* was even coined), and differentiate from related terms, such as *mechanization, cybernetics, artificial intelligence,* and *robotics.*

Automaton
An *automaton* (plural: automata, or automatons) is an autonomous machine that contains its own power source and can perform without human intervention a complicated series of decisions and actions, in response to programs and external stimuli. Since the term automaton is used for a specific autonomous machine, tool or device, it usually does not include automation platforms such as automation infrastructure, automatic installations, or automation systems such as automation

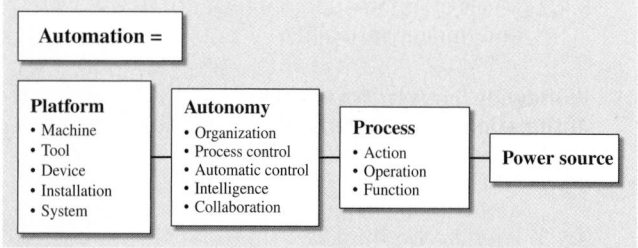

Fig. 3.1 Automation formalism. Automation comprises four basic elements. See representative illustrations of platforms, autonomy, process, and power source in Tables 3.1–3.2, 3.6, and the automation cases below, in Sect. 3.3

Table 3.1 Automation examples: ancient to early history

	Machine/system	Autonomous action/ function	Autonomy: control/ intelligence	Power source	Replacing	Process without human intervention
1.	Irrigation channels	Direct, regulate water flow	From-to, on-off gates, predetermined	Gravity	Manual watering	Water flow and directions
2.	Water supply by aqueducts over large distances	Direct, regulate water supply	From-to, on-off gates, predetermined	Gravity	Practically impossible	Water flow and directions
3.	Sundial clocks	Display current time	Predetermined timing	Sunlight	Impossible otherwise	Shadow indicating time
4.	Archytas' flying pigeon (4th century BC); Chinese mechanical orchestra (3rd century BC) Heron's mechanical chirping birds and moving dolls (1st century AD)	Flying; playing; chirping; moving	Predetermined sound and movements with some feedback	Heated air and steam (early hydraulics and pneumatics)	Real birds; human play	Mechanical bird or toy motions and sounds
5.	Ancient Greek temple automatic door opening	Open and close door	Preset states and positions with some feedback	Heated air, steam, water, gravity	Manual open and close	Door movements
6.	Windmills	Grinding grains	Predefined grinding	Winds	Animal and human power	Grinding process

Table 3.2 Automation platforms

Platform	Machine	Tool	Device	Installation	System	System of systems
Example	Mars lander	Sprinkler	Pacemaker	AS/RC (automated storage/ retrieval carousel)	ERP (enterprise resource planning)	Internet

software (even though some use the term *software automaton* to imply computing procedures). The scholar Al-Jazari from al-Jazira, Mesopotamia designed pioneering programmable automatons in 1206, as a set of dolls, or humanoid automata. Today, the most typical automatons are what we define as robots.

Table 3.3 Automation examples: Industrial Revolution to 1920

	Machine/system	Autonomous action/function	Autonomy: control/ intelligence	Power source	Replacing	Process without human intervention
1.	Windmills (17th century)	Flour milling	Feedback keeping blades always facing the wind	Winds	Nonfeedback windmills	Milling process
2.	Automatic pressure valve (Denis Papin, 1680)	Steam pressure in piston or engine	Feedback control of steam pressure	Steam	Practically impossible otherwise	Pressure regulation
3.	Automatic grist mill (Oliver Evans, 1784)	Continuous-flow flour production line	Conveyor speed control; milling process control	Water flow; steam	Human labor	Grains conveyance and milling process
4.	Flyball governor (James Watt, 1788)	Control of steam engine speed	Automatic feedback of centrifugal force for speed control	Steam	Human control	Speed regulation
5.	Steamboats, trains (18–19th century)	Transportation over very large distances	Basic speed and navigation controls	Steam	Practically impossible otherwise	Travel, freight hauling, conveyance
6.	Automatic loom (e.g., Joseph Jacquard, 1801)	Fabric weaving, including intricate patterns	Basic process control programs by interchangeable punched card	Steam	Human labor and supervision	Cloth weaving according to human design of fabric program
7.	Telegraph (Samuel Morse, 1837)	Fast delivery of text message over large distances	On-off, direction, and feedback	Electricity	Before tele-communication, practically impossible otherwise	Movement of text over wires

Table 3.3 (cont.)

	Machine/system	Autonomous action/function	Autonomy: control/intelligence	Power source	Replacing	Process without human intervention
8.	Semiautomatic assembly machines (Bodine Co., 1920)	Assembly functions including positioning, drilling, tapping, screw insertion, pressing	Connect/disconnect; process control	Electricity; compressed air through belts and pulleys	Human labor	Complex sequences of assembly operations
9.	Automatic automobile-chassis plant (A.O. Smith Co., 1920)	Chassis production	Parts, components, and products flow control, machining, and assembly process control	Electricity	Human labor and supervision	Manufacturing processes and part handling with better accuracy

Robot

A *robot* is a mechanical device that can be programmed to perform a variety of tasks of manipulation and locomotion under automatic control. Thus, a robot could also be an automaton. But unlike an automaton, a robot is usually designed for highly variable and flexible, purposeful motions and activities, and for specific operation domains, e.g., surgical robot, service robot, welding robot, toy robot, etc. General Motors implemented the first industrial robot, called UNIMATE, in 1961 for die-casting at an automobile factory in New Jersey. By now, millions of robots are routinely employed and integrated throughout the world.

Robotics

The science and technology of designing, building, and applying robots, computer-controlled mechanical devices, such as automated tools and machines. Science

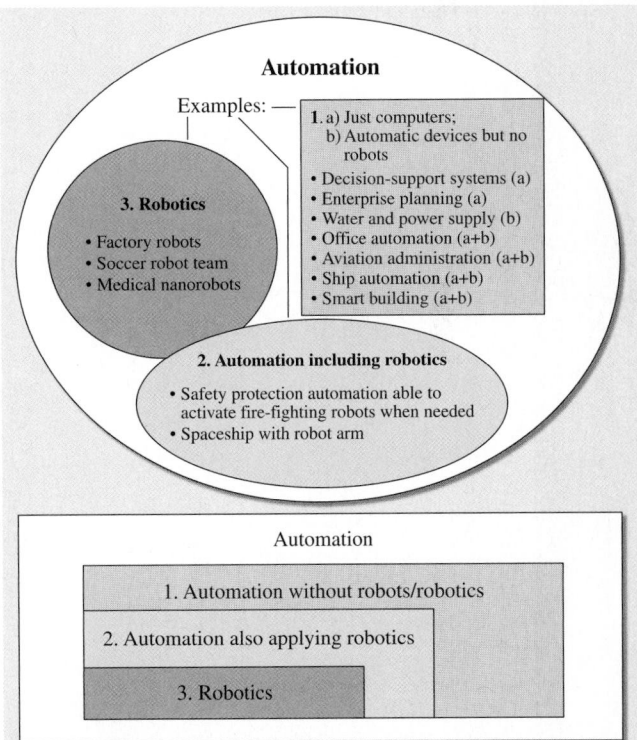

Fig. 3.2 The relation between robotics and automation: The scope of automation includes applications: (1a) with just computers, (1b) with various automation platforms and applications, but without robots; (2) automation including also some robotics; (3) automation with robotics

Part A | 3.1

Table 3.4 Automation examples: modern and emerging

	Machine/ system	Autonomous action/function	Autonomy: control/ intelligence	Power source	Replacing	Process without human intervention
1.	Automatic door opener	Opening and closing of doors triggered by sensors	Automatic control	Compressed air or electric motor	Human effort	Doors of buses, trains, buildings open and close by themselves
2.	Elevators, cranes	Lifting, carrying	On-off; feedback; preprogrammed or interactive	Hydraulic pumps; electric motors	Human climbing, carrying	Speed and movements require minimal supervision
3.	Digital computers	Data processing and computing functions	Variety of automatic and interactive control and operating systems; intelligent control	Electricity	Calculations at speeds, complexity, and with amounts of data that are humanly impossible	Cognitive and decision-making functions
4.	Automatic pilot	Steering aircraft or boat	Same as (3)	Electrical motors	Human pilot	Navigation, operations, e.g., landing
5.	Automatic transmission	Switch gears of power transmission	Automatic control	Electricity; hydraulic pumps	Manual transmission control	Engaging/ disengaging rotating gears
6.	Office automation	Document processing, imaging, storage, printing	Same as (3)	Electricity	Some manual work; some is practically impossible	Specific office procedures
7.	Multirobot factories	Robot arms and automatic devices perform variety of manufacturing and production processes	Optimal, adaptive, distributed, robust, self-organizing, collaborative, and other intelligent control	Hydraulic pumps, pneumatics, and electric motors	Human labor and supervision	Complex operations and procedures, including quality assurance

fiction author and scientist *Isaac Asimov* coined the term *robotics* in 1941 to describe the technology of robots and predicted the rise of a significant robot industry, e.g., in his foreword to [3.1]:

> *Since physics and most of its subdivisions routinely have the "-ics" suffix, I assumed that robotics was the proper scientific term for the systematic study of robots, of their construction, maintenance, and behavior, and that it was used as such.*

3.1.2 Robotics and Automation

Robotics is an important subset of automation (Fig. 3.2). For instance, of the 25 automation examples in Tables 3.1, 3.3 and 3.4, examples 4 in Table 3.1 and 7 in Table 3.4 are about robots. Beyond robotics, automation includes:

- Infrastructure, e.g., water supply, irrigation, power supply, telecommunication
- Nonrobot devices, e.g., timers, locks, valves, and sensors
- Automatic and automated machines, e.g., flour mills, looms, lathes, drills, presses, vehicles, and printers
- Automatic inspection machines, measurement workstations, and testers
- Installations, e.g., elevators, conveyors, railways, satellites, and space stations
- Systems, e.g., computers, office automation, Internet, cellular phones, and software packages.

Common to both robotics and automation are use of automatic control, and evolution with computing and communication progress. As in automation, robotics also relies on four major components, including a platform, autonomy, process, and power source, but in robotics, a robot is often considered a machine, thus the platform is mostly a machine, a tool or device, or a system of tools and devices. While robotics is, in a major way, about automation of motion and mobility, automation beyond robotics includes major areas based on software, decision-making, planning and optimization, collaboration, process automation, office automation, enterprise resource planning automation, and e-Services. Nevertheless, there is clearly an overlap between automation and robotics; while to most people a robot means a machine with certain automation intelligence, to many an intelligent elevator, or highly

Table 3.4 (cont.)

Machine/ system	Autonomous action/function	Autonomy: control/ intelligence	Power source	Replacing	Process without human intervention
8. Medical diagnostics (e.g., computerized tomography (CT), magnetic resonance imaging (MRI))	Visualization of medical test results in real time	Automatic control; automatic virtual reality	Electricity	Impossible otherwise	Noncontact testing and results presentation
9. Wireless services	Remote prognostics and automatic repair	Predictive control; interacting with radio frequency identification (RFID) and sensor networks	Electricity	Practically impossible otherwise	Monitoring remote equipment functions, executing self-repairs
10. Internet search engine	Finding requested information	Optimal control; multiagent control	Electricity	Human search, practically impossible	Search for specific information over a vast amount of data over worldwide systems

automated machine tool, or even a computer may also imply a robot.

Cybernetics

Cybernetics is the scientific study of control and communication in organisms, organic processes, and mechanical and electronic systems. It evolves from *kibernetes*, in Greek, meaning pilot, or captain, or governor, and focuses on applying technology to replicate or imitate biological control systems, often called today bioinspired, or system biology. *Cybernetics*, a book by Norbert Wiener, who is attributed with coining this word, appeared in 1948 and influenced artificial intelligence research. Cy-

bernetics overlaps with control theory and systems theory.

Cyber

Cyber- is a prefix, as in cybernetic, cybernation, or cyborg. Recently, cyber has assumed a meaning as a noun, meaning computers and information systems, virtual reality, and the Internet. This meaning has emerged because of the increasing importance of these automation systems to society and daily life.

Artificial Intelligence (AI)

Artificial intelligence (AI) is the ability of a machine system to perceive anticipated or unanticipated new

Table 3.5 Definitions of automation (after [3.2])

Source	Definition of automation
1. John Diebold, President, John Diebold & Associates, Inc.	*It is a means of organizing or controlling production processes to achieve optimum use of all production resources – mechanical, material, and human. Automation means optimization of our business and industrial activities.*
2. Marshall G. Nuance, VP, York Corp.	*Automation is a new word, and to many people it has become a scare word. Yet it is not essentially different from the process of improving methods of production which has been going on throughout human history.*
3. James B. Carey, President, International Union of Electrical Workers	*When I speak of automation, I am referring to the use of mechanical and electronic devices, rather than human workers, to regulate and control the operation of machines. In that sense, automation represents something radically different from the mere extension of mechanization. Automation is a new technology. Arising from electronics and electrical engineering.*
4. Joseph A. Beirne, President, Communications Workers of America	*We in the telephone industry have lived with mechanization and its successor automation for many years.*
5. Robert C. Tait, Senior VP, General Dynamics Corp.	*Automation is simply a phrase coined, I believe, by Del Harder of Ford Motor Co. in describing their recent supermechanization which represents an extension of technological progress beyond what has formerly been known as mechanization.*
6. Robert W. Burgess, Director, Census, Department of Commerce	*Automation is a new word for a now familiar process of expanding the types of work in which machinery is used to do tasks faster, or better, or in greater quantity.*
7. D.J. Davis, VP Manufacturing, Ford Motor Co.	*The automatic handling of parts between progressive production processes. It is the result of better planning, improved tooling, and the application of more efficient manufacturing methods, which take full advantage of the progress made by the machine-tool and equipment industries.*
8. Don G. Mitchell, President, Sylvania Electric Products, Inc.	*Automation is a more recent term for mechanization, which has been going on since the industrial revolution began. Automation comes in bits and pieces. First the automation of a simple process, and then gradually a tying together of several processes to get a group of subassembly complete.*

Table 3.6 Automation domains

Domain	Accounting	Agriculture	Banking	Chemical process	Communication	Construction	Design
Example	Billing software	Harvester	ATM (automatic teller machine)	Refinery	Print-press	Truck	CAD (computer-aided design)

Domain	Education	Engineering	Factory	Government	Healthcare	Home	Hospital
Example	Television	Simulation	AGV (automated guided vehicle)	Government web portals	Body scanner	Toaster	Drug delivery

Domain	Hospitality	Library	Logistics	Management	Manufacturing	Maritime	Military
Example	CRM (customer relations management)	Database	RFID (radio-frequency identification)	Financial analysis software	Assembly robot	Navigation	Intelligence satellite

Domain	Office	Post	Retail	Safety	Security	Service	Sports	Transportation
Example	Copying machine	Mail sorter	e-Commerce	Fire alarm	Motion detector	Vending machine	Tread mill	Traffic light

conditions, decide what actions must be performed under these conditions, and plan the actions accordingly. The main areas of AI study and application are knowledge-based systems, computer sensory systems, language processing systems, and machine learning. AI is an important part of automation, especially to characterize what is sometimes called intelligent automation (Sect. 3.4.2).

It is important to note that AI is actually human intelligence that has been implemented on machines, mainly through computers and communication. Its significant advantages are that it can function automatically, i.e., without human intervention during its operation/function; it can combine intelligence from many humans, and improve its abilities by automatic learning and adaptation; it can be automatically distributed, duplicated, shared, inherited, and if necessary, restrained and even deleted. With the advent of these abilities, remarkable progress has been achieved. There is also, however, an increasing risk of *running out of control* (Sect. 3.6), which must be considered carefully as with harnessing any other technology.

3.1.3 Early Automation

The creative human desire to develop automation, from ancient times, has been to recreate natural activities, either for enjoyment or for productivity with less human effort and hazard. It should be clear, however, that the following six imperatives have been proven about automation:

1. Automation has always been developed by people
2. Automation has been developed for the sake of people
3. The benefits of automation are tremendous
4. Often automation performs tasks that are impossible or impractical for humans
5. As with other technologies, care should be taken to prevent abuse of automation, and to eliminate the possibilities of unsafe automation
6. Automation is usually inspiring further creativity of the human mind.

The main evolvement of automation has followed the development of mechanics and fluidics, civil infrastructure and machine design, and since the 20th century, of computers and communication. Examples of ancient automation that follow the formal definition (Table 3.1) include flying and chirping birds, sundial clocks, irrigation systems, and windmills. They all include the four basic automation elements, and

have a clear autonomous process without human intervention, although they are mostly predetermined or predefined in terms of their control program and organization. But not all these ancient examples replace previously used human effort: Some of them would be impractical or even impossible for humans, e.g., displaying time, or moving large quantities of water by aqueducts over large distances. This observation is important, since, as evident from the definition surveys (Sect. 3.5):

1. In defining automation, over one-quarter of those surveyed associate automation with *replacing humans*, hinting somber connotation that humans are *losing* certain advantages. Many resources erroneously define automation as *replacement of human workers by technology*. But the definition is not about *replacing* humans, as many automation examples involve activities people cannot practically perform, e.g., complex and fast computing, wireless telecommunication, microelectronics manufacturing, and satellite-based positioning. The definition is about the autonomy of a system or process from human involvement and intervention during the process (independent of whether humans could or could not perform it themselves). Furthermore, automation is rarely disengaged from people, who must maintain and improve it (or at least replace its batteries).
2. Humans are always involved with automation, to a certain degree, from its development, to, at certain points, supervising, maintaining, repairing, and issuing necessary commands, e.g., *at which floor should this elevator stop for me?*

Describing automation, *Buckingham* [3.3] quotes Aristotle (384–322 BC): "When looms weave by themselves human's slavery will end." Indeed, the reliance on a process that can proceed successfully to completion autonomously, without human participation and intervention, is an essential characteristic of automation. But it took over 2000 years since Aristotle's prediction till the automatic loom was developed during the Industrial Revolution.

3.1.4 Industrial Revolution

Some scientists (e.g., *Truxal* [3.4]) define automation as applying machines or systems to execute tasks that involve more elaborate decision-making. Certain decisions were already involved in ancient automation, e.g., where to direct the irrigation water. More control sophistication was indeed developed later, beginning during the Industrial Revolution (see examples in Table 3.3).

During the Industrial Revolution, as shown in the examples, steam and later electricity became the main power sources of automation systems and machines, and autonomy of process and decision-making increasingly involved feedback control models.

3.1.5 Modern Automation

The term *automation* in its modern meaning was actually attributed in the early 1950s to D.S. Harder, a vice-president of the Ford Motor Company, who described it as a philosophy of manufacturing. Towards the 1950s, it became clear that automation could be viewed as substitution by mechanical, hydraulic, pneumatic, electric, and electronic devices for a combination of human efforts and decisions. Critics with humor referred to automation as *substitution of human error by mechanical error*. Automation can also be viewed as the combination of four fundamental principles: mechanization; process continuity; automatic control; and economic, social, and technological rationalization.

Mechanization

Mechanization is defined as the application of machines to perform work. Machines can perform various tasks, at different levels of complexity. When mechanization is designed with cognitive and decision-making functions, such as process control and automatic control, the modern term *automation* becomes appropriate. Some machines can be rationalized by benefits of safety and convenience. Some machines, based on their power, compactness, and speed, can accomplish tasks that could never be performed by human labor, no matter how much labor or how effectively the operation could be organized and managed. With increased availability and sophistication of power sources and of automatic control, the level of autonomy of machines and mechanical system created a distinction between mechanization and the more autonomous form of mechanization, which is automation (Sect. 3.4).

Process Continuity

Process continuity is already evident in some of the ancient automation examples, and more so in the Industrial Revolution examples (Tables 3.1 and 3.3). For instance, windmills could provide relatively uninterrupted cycles of grain milling. The idea of continuity

is to increase productivity, the useful output per labor-hour. Early in the 20th century, with the advent of mass production, it became possible to better organize workflow. Organization of production flow and assembly lines, and automatic or semiautomatic transfer lines increased productivity beyond mere mechanization. The emerging automobile industry in Europe and the USA in the early 1900s utilized the concept of moving work continuously, automatically or semiautomatically, to specialized machines and workstations. Interesting problems that emerged with flow automation included balancing the work allocation and regulating the flow.

Automatic Control

A key mechanism of automatic control is feedback, which is the regulation of a process according to its own output, so that the output meets the conditions of a predetermined, set objective. An example is the windmill that can adjust the orientation of it blades by feedback informing it of the changing direction of the current wind. Another example is the heating system that can stop and restart its heating or cooling process according to feedback from its thermostat. Watt's flyball governor applied feedback from the position of the rotating balls as a function of their rotating speed to automatically regulate the speed of the steam engine. Charles Babbage analytical engine for calculations applied the feedback principle in 1840 (see more on the development of automatic control in Chap. 4).

Automation Rationalization

Rationalization means a logical and systematic analysis, understanding, and evaluation of the objectives and constraints of the automation solution. Automation is rationalized by considering the technological and engineering aspects in the context of economic, social, and managerial considerations, including also: human factors and usability, organizational issues, environmental constraints, conservation of resources and energy, and elimination of waste (Chaps. 40 and 41).

Soon after automation enabled mass production in factories of the early 20th century and workers feared for the future of their jobs, the US Congress held hearings in which experts explained what automation means to them (Table 3.5). From our vantage point several generations later, it is interesting to read these definitions, while we already know about automation discoveries yet unknown at that time, e.g., laptop computers, robots, cellular telephones and personal digital assistants, the Internet, and more.

A relevant question is: Why automate? Several prominent motivations are the following, as has been indicated by the survey participants (Sect. 3.5):

1. *Feasibility*: Humans cannot handle certain operations and processes, either because of their scale, e.g., micro- and nanoparticles are too small, the amount of data is too vast, or the process happens too fast, for instance, missile guidance; microelectronics design, manufacturing and repair; and database search.
2. *Productivity*: Beyond feasibility, computers, automatic transfer machines, and other equipment can operate at such high speed and capacity that it would be practically impossible without automation, for instance, controlling consecutive, rapid chemical processes in food production; performing medicine tests by manipulating atoms or molecules; optimizing a digital image; and placing millions of colored dots on a color television (TV) screen.
3. *Safety*: Automation sensors and devices can operate well in environments that are unsafe for humans, for example, under extreme temperatures, nuclear radiation, or in poisonous gas.
4. *Quality and economy*: Automation can save significant costs on jobs performed without it, including consistency, accuracy, and quality of manufactured products and of services, and saving labor, safety, and maintenance costs.
5. *Importance to individuals, to organizations, and to society*: Beyond the above motivations, service- and knowledge-based automation reduces the need for middle managers and middle agents, thus reducing or eliminating the agency costs and removing layers of bureaucracy, for instance, Internet-based travel services and financial services, and direct communication between manufacturing managers and line operators, or cell robots. Remote supervision and telecollaboration change the nature, sophistication, skills and training requirements, and responsibility of workers and their managers. As automation gains intelligence and competencies, it takes over some employment skills and opens up new types of work, skills, and service requirements.
6. *Accessibility*: Automation enables better accessibility for all people, including disadvantaged and disabled people. Furthermore, automation opens up new types of employment for people with limitations, e.g., by integration of speech and vision recognition interfaces.
7. *Additional motivations*: Additional motivations are the competitive ability to integrate complex mechanization, advantages of modernization, convenience, and improvement in quality of life.

To be automated, a system must follow the motivations listed above. The modern and emerging automation examples in Table 3.4 and the automation cases in Sect. 3.3 illustrate these motivations, and the mechanization, process continuity, and automatic control features.

Certain limits and risks of automation need also be considered. Modern, computer-controlled automation must be programmable and conform to definable procedures, protocols, routines, and boundaries. The limits also follow the boundaries imposed by the four principles of automation. Can it be mechanized? Is there continuity in the process? Can automatic control be designed for it? Can it be rationalized? Theoretically, all continuous processes can be automatically controlled, but practically such automation must be rationalized first; for instance, jet engines may be continuously advanced on conveyors to assembly cells, but if the demand for these engines is low, there is no justification to automate their flow. Furthermore, all automation must be designed to operate within safe boundaries, so it does not pose hazards to humans and to the environment.

3.1.6 Domains of Automation

Some unique meanings of automation are associated with the domain of automation. Several examples of well-known domains are listed here:

- *Detroit automation* – Automation of transfer lines and assembly lines adopted by the automotive industry [3.5].
- *Flexible automation* – Manufacturing and service automation consisting of a group of processing stations and robots operating as an integrated system under computer control, able to process a variety of different tasks simultaneously, under automatic, adaptive control or learning control [3.5]. Also known as flexible manufacturing system (FMS), flexible assembly system, or robot cell, which are suitable for medium demand volume and medium variety of flexible tasks. Its purpose is to advance from mass production of products to more customer-oriented and customized supply. For higher flexibility with low demand volume, stand-

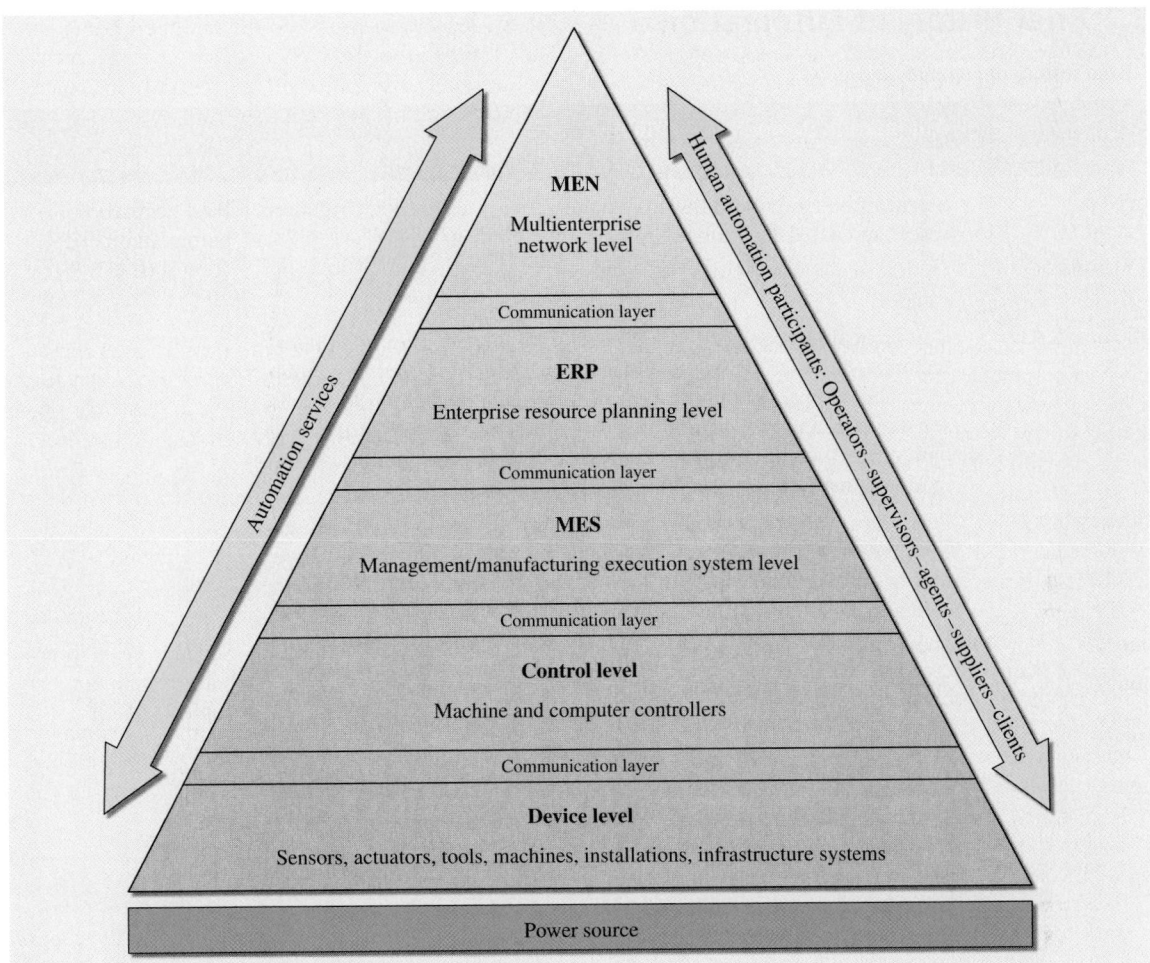

Fig. 3.3 The automation pyramid: organizational layers

alone numerically controlled (NC) machines and robots are preferred. For high demand volume with low task variability, automatic transfer lines are designed. The opposite of flexible automation is *fixed automation*, such as process-specific machine tools and transfer lines, lacking task flexibility. For mass customization (mass production with some flexibility to respond to variable customer demands), transfer lines with flexibility can be designed (see more on automation flexibility in Sect. 3.4).

- *Office automation* – Computer and communication machinery and software used to improve office procedures by digitally creating, collecting, storing, manipulating, displaying, and transmitting office information needed for accomplishing office tasks

and functions [3.6, 7]. Office automation became popular in the 1970s and 1980s when the desktop computer and the personal computer emerged.

Other examples of well-known domains of automation have been factory automation (e.g., [3.8]), healthcare automation (e.g., [3.9]), workflow automation (e.g., [3.10]), and service automation (e.g., [3.11]). More domain examples are illustrated in Table 3.6.

Throughout these different domains, automation has been applied for various organization functions. Five hierarchical layers of automation are shown in the automation pyramid (Fig. 3.3), which is a common depiction of how to organize automation implementation.

3.2 Brief History of Automation

Table 3.7 Brief history of automation events

Period	Automation inventions (examples)	Automation generation
Prehistory	Sterilization of food and water, cooking, ships and boats, irrigation, wheel and axle, flush toilet, alphabet, metal processing	First generation: before automatic control (BAC)
Ancient history	Optics, maps, water clock, water wheel, water mill, kite, clockwork, catapult	
First millennium AD	Central heating, compass, woodblock printing, pen, glass and pottery factories, distillation, water purification, wind-powered gristmills, feedback control, automatic control, automatic musical instruments, self-feeding and self-trimming oil lamps, chemotherapy, diversion dam, water turbine, mechanical moving dolls and singing birds, navigational instruments, sundial	
11th–15th century	Pendulum, camera, flywheel, printing press, rocket, clock automation, flow-control regulator, reciprocating piston engine, humanoid robot, programmable robot, automatic gate, water supply system, calibration, metal casting	
16th century	Pocket watch, Pascal calculator, machine gun, corn grinding machine	Second generation: before computer control (BCC)
17th century	Automatic calculator, pendulum clock, steam car, pressure cooker	
18th century	Typewriter, steam piston engine, Industrial Revolution early automation, steamboat, hot-air balloon, automatic flour mill	
19th century	Automatic loom, electric motor, passenger elevator, escalator, photography, electric telegraph, telephone, incandescent light, radio, x-ray machine, combine harvester, lead–acid battery, fire sprinkler system, player piano, electric street car, electric fan, automobile, motorcycle, dishwasher, ballpoint pen, automatic telephone exchange, sprinkler system, traffic lights, electric bread toaster	
Early 20th century	Airplane, automatic manufacturing transfer line, conveyor belt-based assembly line, analog computer, air conditioning, television, movie, radar, copying machine, cruise missile, jet engine aircraft, helicopter, washing machine, parachute, flip–flop circuit	

Automation has evolved, as described in Table 3.7, along three automation generations.

3.2.1 First Generation: Before Automatic Control (BAC)

Early automation is characterized by elements of process autonomy and basic decision-making autonomy, but without feedback, or with minimal feedback. The period is generally from prehistory till the 15th century. Some examples of basic automatic control can be found earlier than the 15th century, at least in conceptual design or mathematical definition. Automation examples of the first generation can also be found later, whenever automation solutions without automatic control could be rationalized.

3.2.2 Second Generation: Before Computer Control (BCC)

Automation with advantages of automatic control, but before the introduction and implementation of the

Table 3.7 (cont.)

Period	Automation inventions (examples)	Automation generation
1940s	Digital computer, Assembler programming language, transistor, nuclear reactor, microwave oven, atomic clock, barcode	Third generation: automatic computer control (ACC)
1950s	Mass-produced digital computer, computer operating system, FORTRAN programming language, automatic sliding door, floppy disk, hard drive, power steering, optical fiber, communication satellite, computerized banking, integrated circuit, artificial satellite, medical ultrasonics, implantable pacemaker	
1960s	Laser, optical disk, microprocessor, industrial robot, automatic teller machine (ATM), computer mouse, computer-aided design, computer-aided manufacturing, random-access memory, video game console, barcode scanner, radiofrequency identification tags (RFID), permanent press fabric, wide-area packet switching network	
1970s	Food processor, word processor, Ethernet, laser printer, database management, computer-integrated manufacturing, mobile phone, personal computer, space station, digital camera, magnetic resonance imaging, computerized tomography (CT), e-Mail, spreadsheet, cellular phone	
1980s	Compact disk, scanning tunneling microscope, artificial heart, deoxyribonucleic acid DNA fingerprinting, Internet transmission control protocol/Internet protocol TCP/IP, camcorder	
1990s	World Wide Web, global positioning system, digital answering machine, smart pills, service robots, Java computer language, web search, Mars Pathfinder, web TV	
2000s	Artificial liver, Segway personal transporter, robotic vacuum cleaner, self-cleaning windows, iPod, softness-adjusting shoe, drug delivery by ultrasound, Mars Lander, disk-on-key, social robots	

computer, especially the digital computer, belongs to this generation. Automatic control emerging during this generation offered better stability and reliability, more complex decision-making, and in general better control and automation quality. The period is between the 15th century and the 1940s. It would be generally difficult to rationalize in the future any automation with automatic control and without computers; therefore, future examples of this generation will be rare.

3.2.3 Third Generation: Automatic Computer Control (ACC)

The progress of computers and communication has significantly impacted the sophistication of automatic control and its effectiveness. This generation began in the 1940s and continues today. Further refinement of this generation can be found in Sect. 3.4, discussing the levels of automation.

See also Table 3.8 for examples discovered or implemented during the three automation generations.

Table 3.8 Automation generations

Generation		Example
BAC before automatic control prehistoric, ancient		Waterwheel
ABC automatic control before computers 16th century–1940		Automobile
CAC computer automatic control 1940–present	*Hydraulic automation* *Pneumatic automation* *Electrical automation* *Electronic automation* *Micro automation* *Nano automation* *Mobile automation* *Remote automation*	Hydraulic elevator Door open/shut Telegraph Microprocessor Digital camera Nanomemory Cellular phone Global positioning system (GPS)

3.3 Automation Cases

Ten automation cases are illustrated in this section to demonstrate the meaning and scope of automation in different domains.

3.3.1 Case A: Steam Turbine Governor (Fig. 3.4)

Source. Courtesy of Dresser-Rand Co., Houston (http://www.dresser-rand.com/).

Process. Operates a steam turbine used to drive a compressor or generator.

Platform. Device, integrated as a system with programmable logic controller (PLC).

Autonomy. Semiautomatic and automatic activation/deactivation and control of the turbine speed; critical speed-range avoidance; remote, auxiliary, and cascade speed control; loss of generator and loss of utility detection; hot standby ability; single and dual actuator control; programmable governor parameters via operator's screen and interface, and mobile computer. A manual mode is also available: The operator places

the system in run mode which opens the governor valve to the full position, then manually opens the T&T valve to idle speed, to warm up the unit. After warm-up, the operator manually opens the T&T valve to full position, and as the turbine's speed approaches the rated (desirable) speed, the governor takes control with the governor valve. In semiautomatic and automatic modes, once the operator places the system in run mode, the governor takes over control.

3.3.2 Case B: Bioreactor (Fig. 3.5)

Source. Courtesy of Applikon Biotechnology Co., Schiedam (http://www.pharmaceutical-technology.com/contractors/process_automation/applikon-technology/).

Process. Microbial or cell culture applications that can be validated, conforming with standards for equipment used in life science and food industries, such as good automated manufacturing practice (GAMP).

Platform. System or installation, including microreactors, single-use reactors, autoclavable glass bioreactors, and stainless-steel bioreactors.

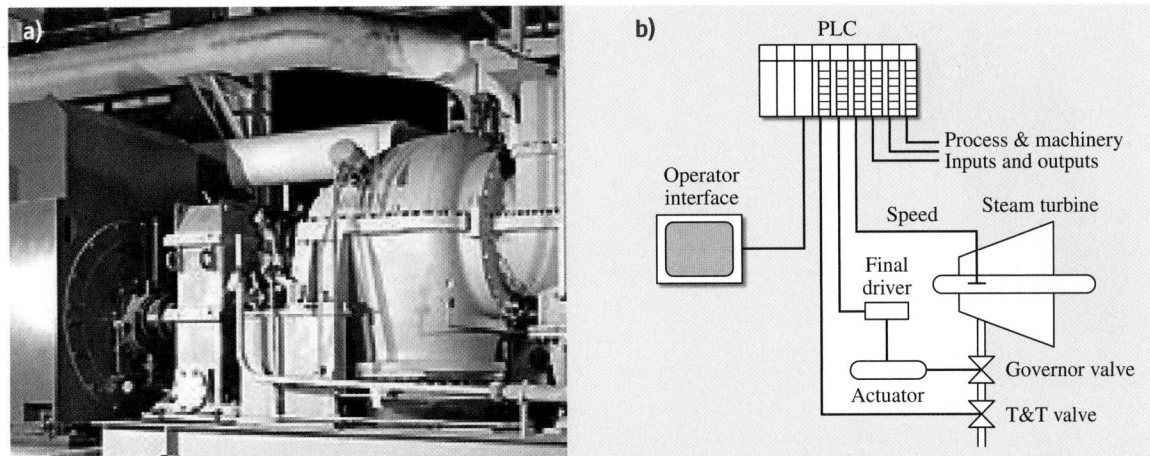

Fig. 3.4 (a) Steam turbine generator. **(b)** Governor block diagram (PLC: programmable logic controller; T&T: trip and throttle). A turbine generator designed for on-site power and distributed energy ranging from 0.5 to 100 MW. Turbine generator sets produce power for pulp and paper mills, sugar, hydrocarbon, petrochemical and process industries; palm oil, ethanol, waste-to-energy, other biomass burning facilities, and other installations (with permission from Dresser-Rand)

Autonomy. Bioreactor functions with complete measurement and control strategies and supervisory control and data acquisition (SCADA), including sensors and a cell retention device.

Fig. 3.5 Bioreactor system configured for microbial or cell culture applications. Optimization studies and screening and testing of strains and cell lines are of high importance in industry and research and development (R&D) institutes. Large numbers of tests are required and they must be performed in as short a time as possible. Tests should be performed so that results can be validated and used for further process development and production (with permission from Applikon Biotechnology)

3.3.3 Case C: Digital Photo Processing (Fig. 3.6)

Source. Adobe Systems Incorporated San Jose, California (http://adobe.com).

Process. Editing, enhancing, adding graphic features, removing stains, improving resolution, cropping and sizing, and other functions to process photo images.

Platform. Software system.

Autonomy. The software functions are fully automatic once activated by a user. The software can execute them semiautomatically under user control, or action series can be automated too.

3.3.4 Case D: Robotic Painting (Fig. 3.7)

Source. Courtesy of ABB Co., Zürich (http://www.ABB.com).

Process. Automatic painting under automatic control of car body movement, door opening and closing, paint-pump functions, fast robot motions to optimize finish quality, and minimize paint waste.

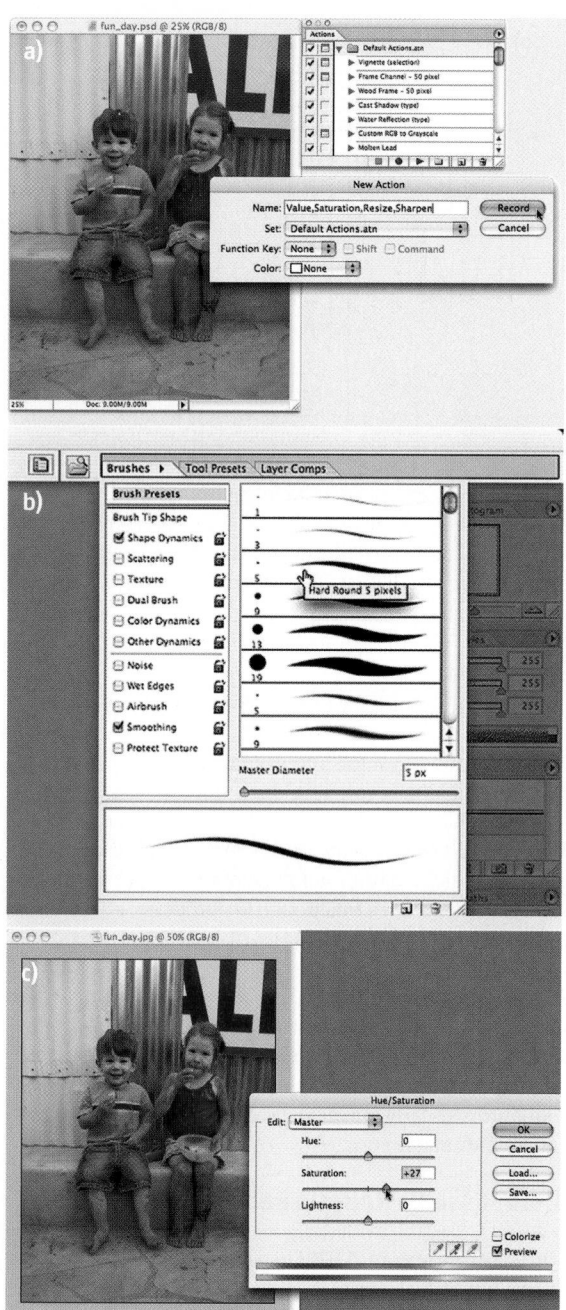

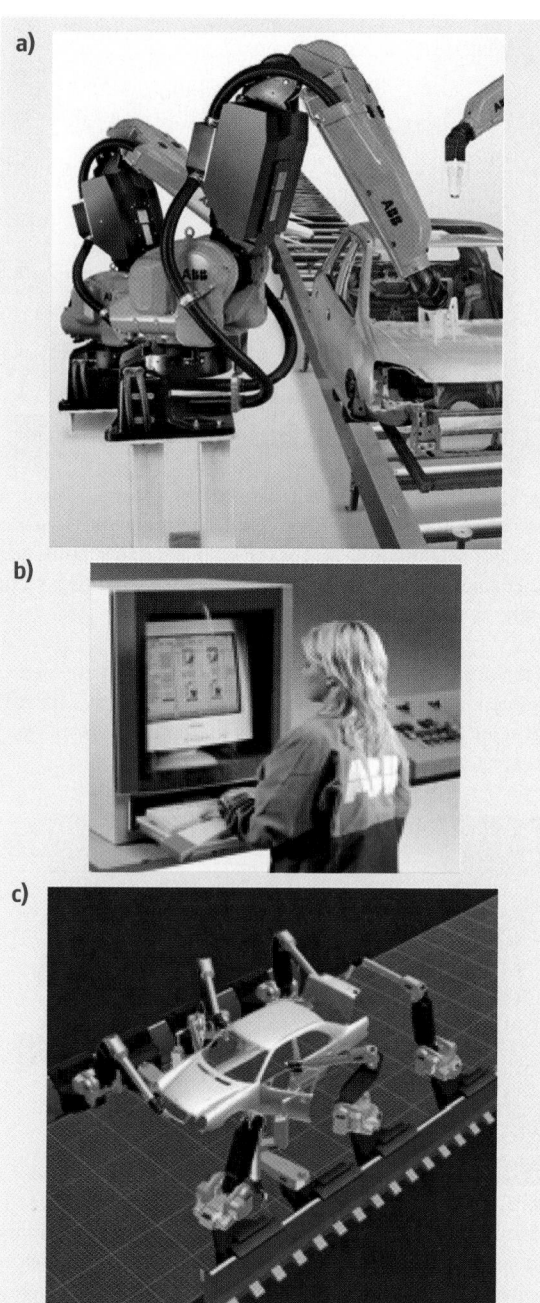

Fig. 3.6a–c Adobe Photoshop functions for digital image editing and processing: (**a**) automating functional actions, such as shadow, frame, reflection, and other visual effects; (**b**) selecting brush and palette for graphic effects; (**c**) setting color and saturation values (with permission from Adobe Systems Inc., 2008)

Fig. 3.7a–c A robotic painting line: (**a**) the facility; (**b**) programmer using interface to plan offline or online, experiment, optimize, and verify control programs for the line; (**c**) robotic painting facility design simulator (with permission from ABB)

Platform. Automatic tools, machines, and robots, including sensors, conveyors, spray-painting equipment, and integration with planning and programming software systems.

Autonomy. Flexibility of motions; collision avoidance; coordination of conveyor moves, robots' motions, and paint pump operations; programmability of process and line operations.

3.3.5 Case E: Assembly Automation (Fig. 3.8)

Source. Courtesy of Adapt Automation Inc., Santa Ana, California (www.adaptautomation.com/Medical.html).

Process. Hopper and two bowl feeders feed specimen sticks through a track to where they are picked and placed, two up, into a double nest on a 12-position indexed dial plate. Specimen pads are fed and placed on the sticks. Pads are fully seated and inspected. Rejected and good parts are separated into their respective chutes.

Platform. System of automatic tools, machines, and robots.

Autonomy. Automatic control through a solid-state programmable controller which operates the sequence of device operations with a control panel. Control programs include main power, emergency stop, manual, automatic, and individual operations controls.

3.3.6 Case F: Computer–Integrated Elevator Production (Fig. 3.9)

Process. Fabrication and production execution and management.

Platform. Automatic tools, machines, and robots, integrated with system-of-systems, comprising a production/manufacturing installation, with automated material handling equipment, fabrication and finishing machines and processes, and software and communication systems for production planning and control, robotic manipulators, and cranes. Human operators and supervisors are also included.

Autonomy. Automatic control, including knowledge-based control of laser, press, buffing, and sanding machines/cells; automated control of material handling

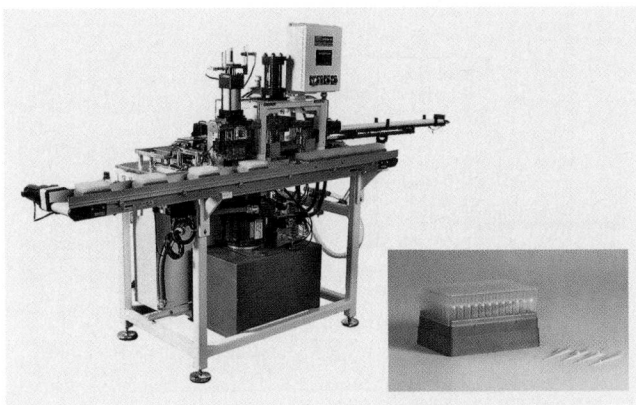

Fig. 3.8 Pharmaceutical pad-stick automatic assembly cell

and material flow, and of management information flow.

3.3.7 Case G: Water Treatment (Fig. 3.10)

Source. Rockwell Automation Co., Cleveland (www.rockwellautomation.com).

Process. Water treatment by reverse osmosis filtering system (Fig. 3.10a) and water treatment and disposal ((Fig. 3.10b). When preparing to filter impurities from the city water, the controllers activate the pumps, which in turn flush wells to clean water sufficiently before it flows through the filtering equipment (Fig. 3.10a) or activating complete system for removal of grit, sediments, and disposal of sludge to clean water supply (Fig. 3.10b).

Platform. Installation including water treatment plant with a network of pumping stations, integrated with programmable and supervisory control and data acquisition (SCADA) control, remote communication software system, and human supervisory interfaces.

Autonomy. Monitoring and tracking the entire water treatment and purification system.

3.3.8 Case H: Digital Document Workflow (Fig. 3.11)

Source. Xerox Co., Norwalk (http://www.xerox.com).

Process. On-demand customized processing, production, and delivery of print, email, and customized web sites.

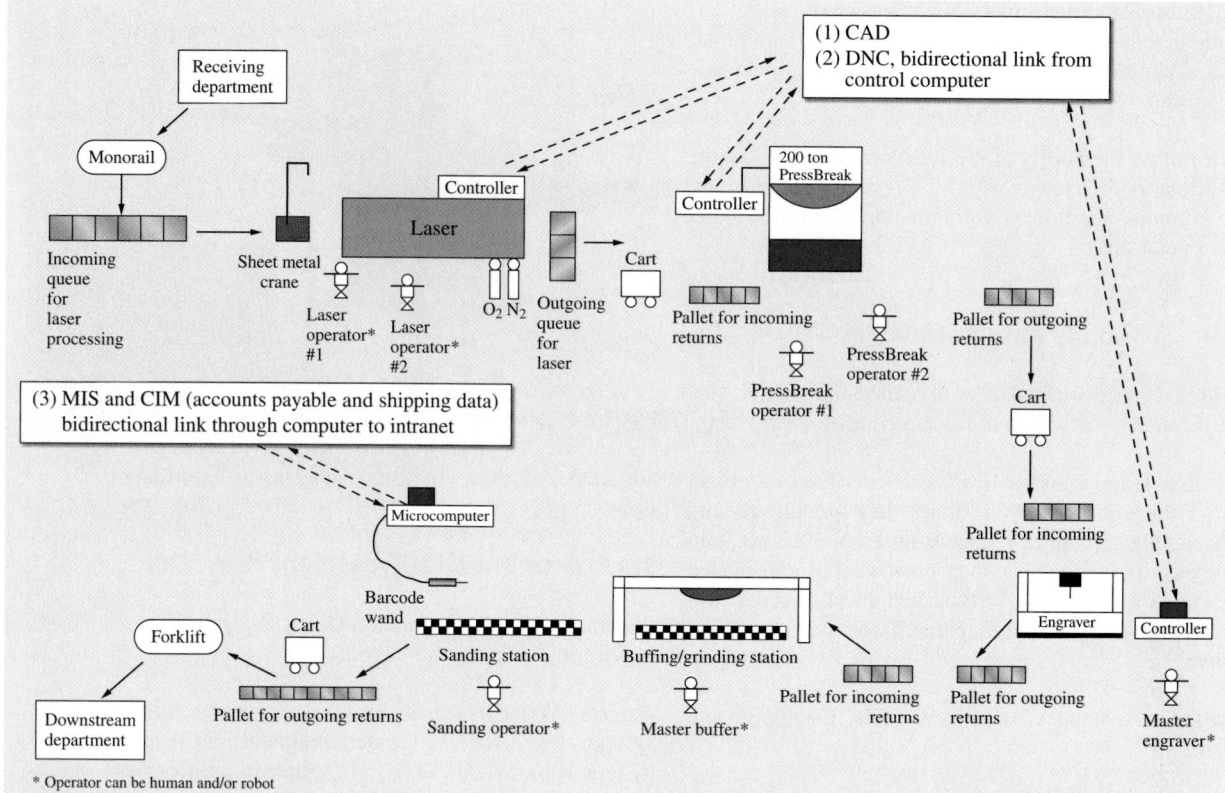

Fig. 3.9 Elevator swing return computer-integrated production system: Three levels of automation systems are integrated, including (1) link to computer-aided design (CAD) for individual customer elevator specifications and customized finish, (2) link to direct numerical control (DNC) of workcell machine and manufacturing activities, (3) link to management information system (MIS) and computer integrated manufacturing system (CIM) for accounting and shipping management (source: [3.12])

Platform. Network of devices, machines, robots, and software systems integrated within a system-of-systems with media technologies.

Autonomy. Integration of automatic workflow of document image capture, processing, enhancing, preparing, producing, and distributing.

3.3.9 Case I: Ship Building Automation (Fig. 3.12)

Source. [3.13]; Korea Shipbuilder's Association, Seoul (http://www.koshipa.or.kr); Hyundai Heavy Industries, Co., Ltd., Ulsan (http://www.hhi.co.kr).

Process. Shipbuilding manufacturing process control and automation; shipbuilding production, logistics, and service management; ship operations management.

Platform. Devices, tools, machines and multiple robots; system-of-software systems; system of systems.

Autonomy. Automatic control of manufacturing processes and quality assurance; automatic monitoring, planning and decision support software systems; integrated control and collaborative control for ship operations and bridge control of critical automatic functions of engine, power supply systems, and alarm systems.

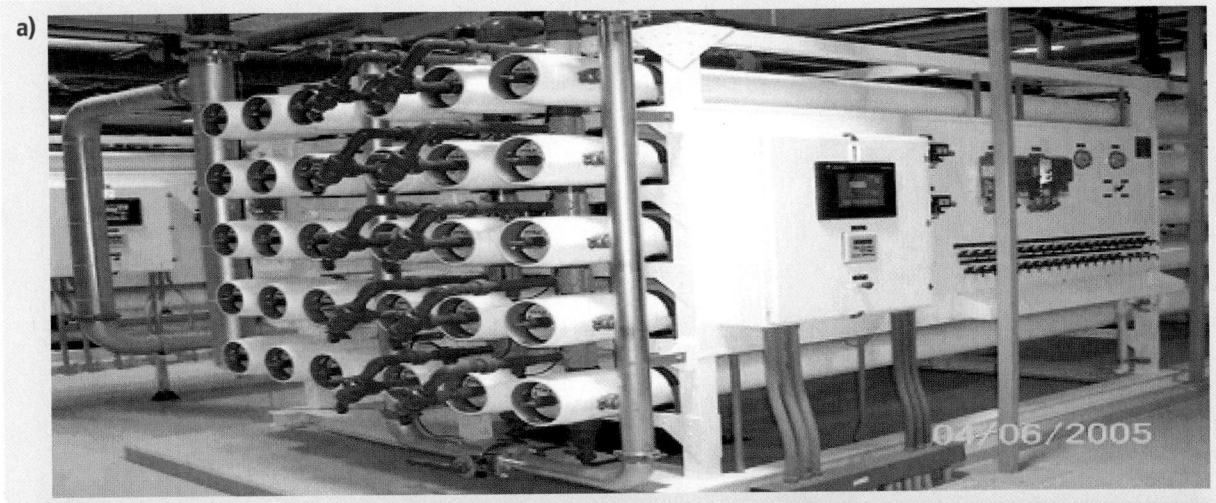

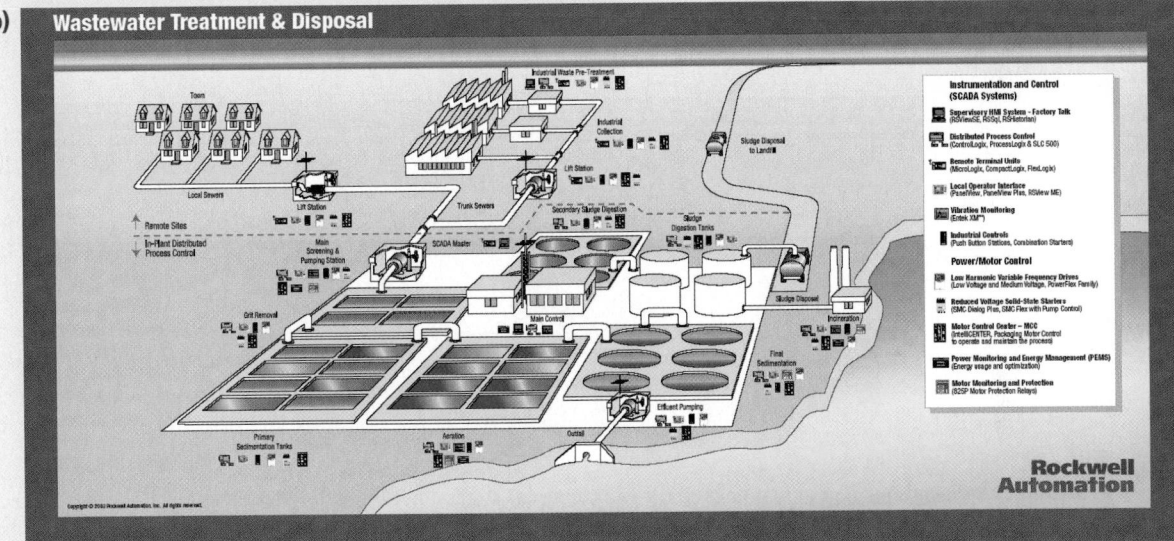

Fig. 3.10 (**a**) Municipal water treatment system in compliance with the Safe Drinking Water Federal Act (courtesy of City of Kewanee, IL; Engineered Fluid, Inc.; and Rockwell Automation Co.). (**b**) Wastewater treatment and disposal (courtesy of Rockwell Automation Co.)

3.3.10 Case J: Energy Power Substation Automation (Fig. 3.13)

Source. GE Energy Co., Atlanta (http://www.gepower.com/prod_serv/products/substation_automation/en/downloads/po.pdf).

Process. Automatically monitoring and activating backup power supply in case of breakdown in the power generation and distribution. Each substation automation platform has processing capacity to monitor and control thousands of input–output points and *intelligent electronic devices* over the network.

Platform. Devices integrated with a network of system-of-systems, including substation automation platforms, each communicating with and controlling thousands of power network devices.

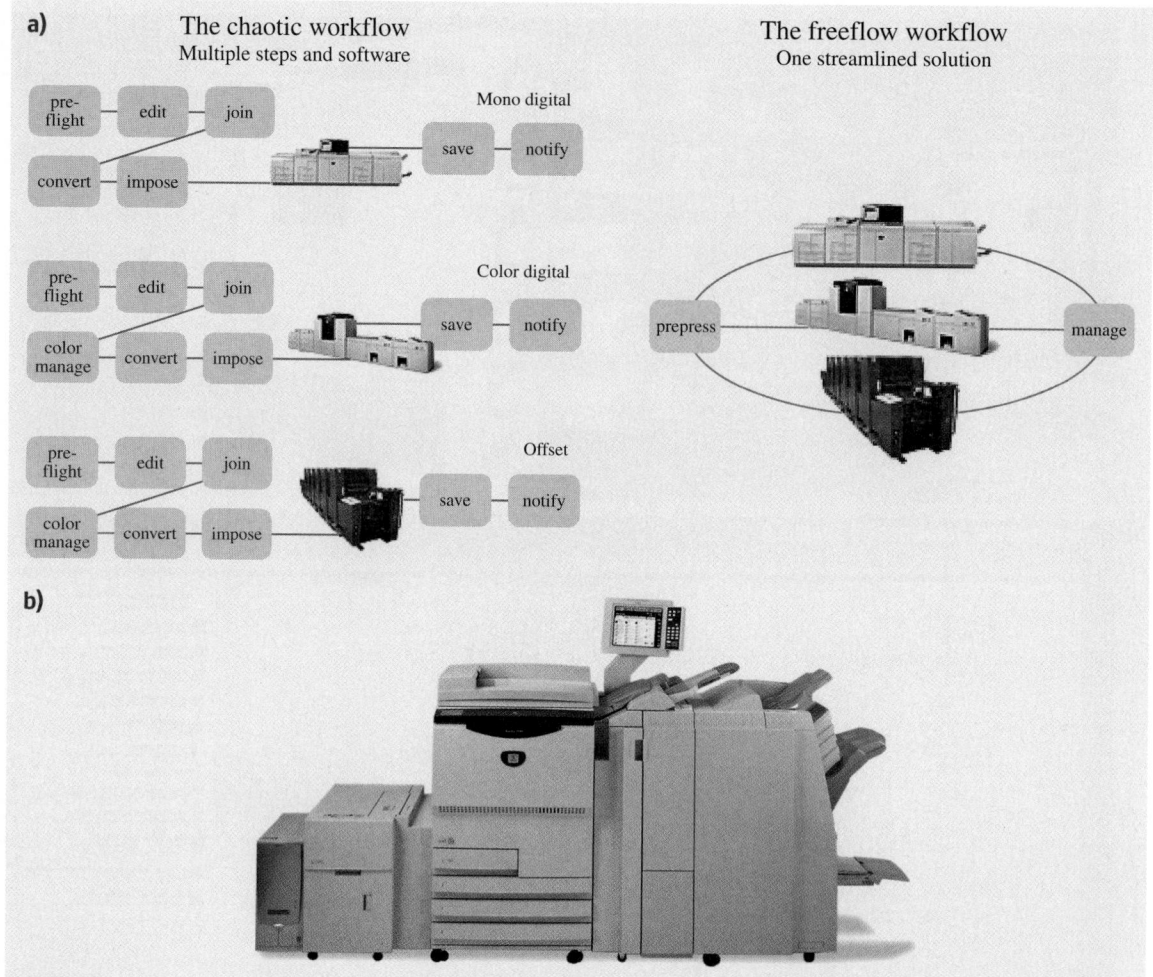

Fig. 3.11a,b Document imaging and color printing workflow: (**a**) streamlined workflow by FreeFlow. (**b**) *Detail* shows document scanner workstation with automatic document feeder and image processing (courtesy of Xerox Co., Norwalk)

Autonomy. Power generation, transmission, and distribution automation, including automatic steady voltage control, based on user-defined targets and settings; local/remote control of distributed devices; adjustment of control set-points based on control requests or control input values; automatic reclosure of tripped circuit breakers following momentary faults; automatic transfer of load and restoration of power to nonfaulty sections if possible; automatically locating and isolating faults to reduce customers' outage times; monitoring a network of substations and moving load off overloaded transformers to other stations as required.

These ten case studies cover a variety of automation domains. They also demonstrate different level of intelligence programmed into the automation application, different degrees of automation, and various types of automation flexibility. The meaning of these automation characteristics is explained in the next section.

a)

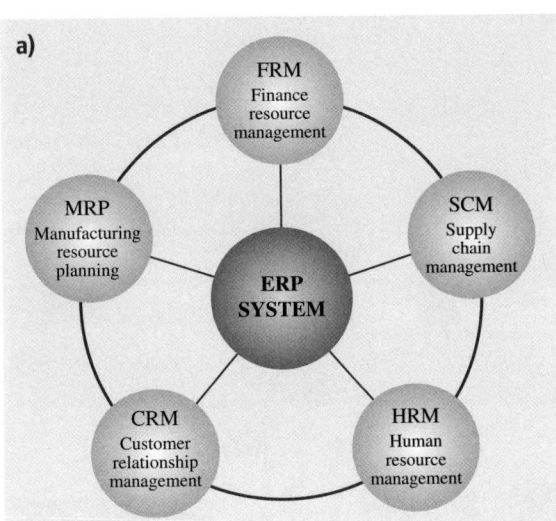

Fig. 3.12a–j Automation and control systems in ship-building: (**a**) production management through enterprise resource planning (ERP) systems. Manufacturing automation in shipbuilding examples: (**b**) overview; (**c**) automatic panel welding robots system; (**d**) sensor application in membrane tank fabrication; (**e**) propeller grinding process by robotic automation. Automatic ship operation systems examples: (**f**) overview; (**g**) alarm and monitoring system; (**h**) integrated bridge system; (**i**) power management system; (**j**) engine monitoring system (source: [3.13]) (with permission from Hyundai Heavy Industries)

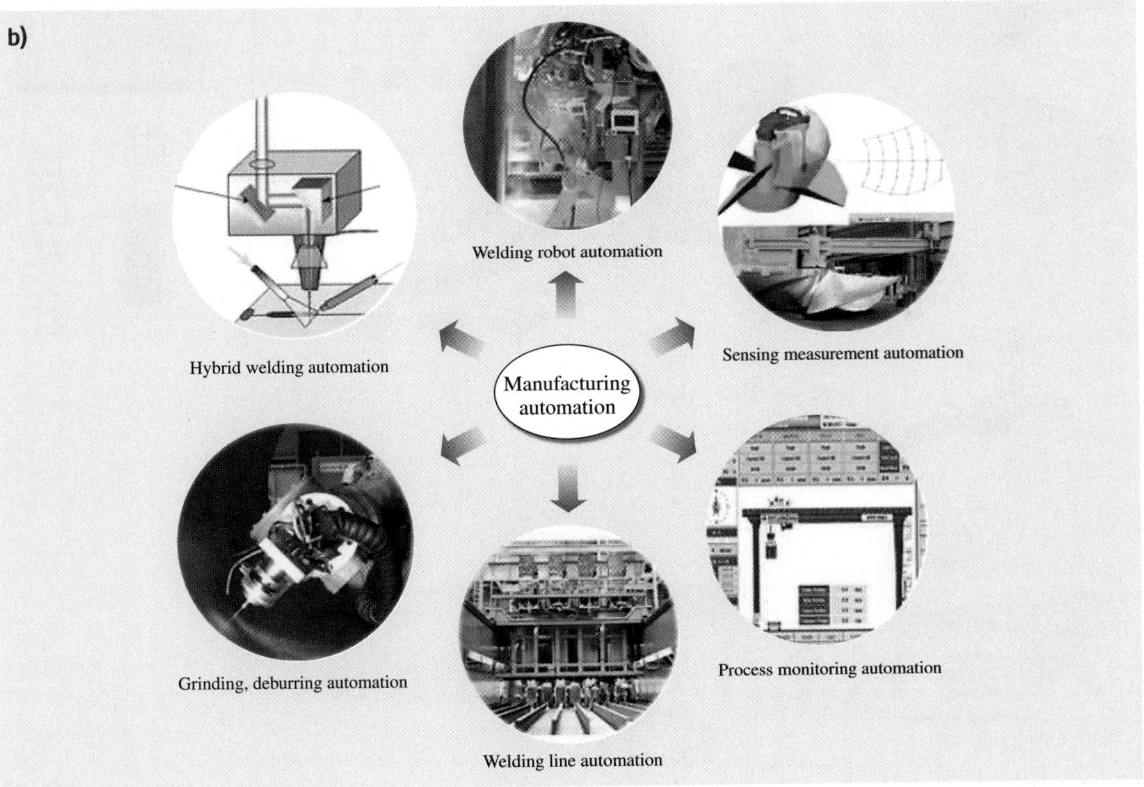

b)

Welding robot automation

Hybrid welding automation

Sensing measurement automation

Manufacturing automation

Grinding, deburring automation

Process monitoring automation

Welding line automation

Fig. 3.12a–j (cont.)

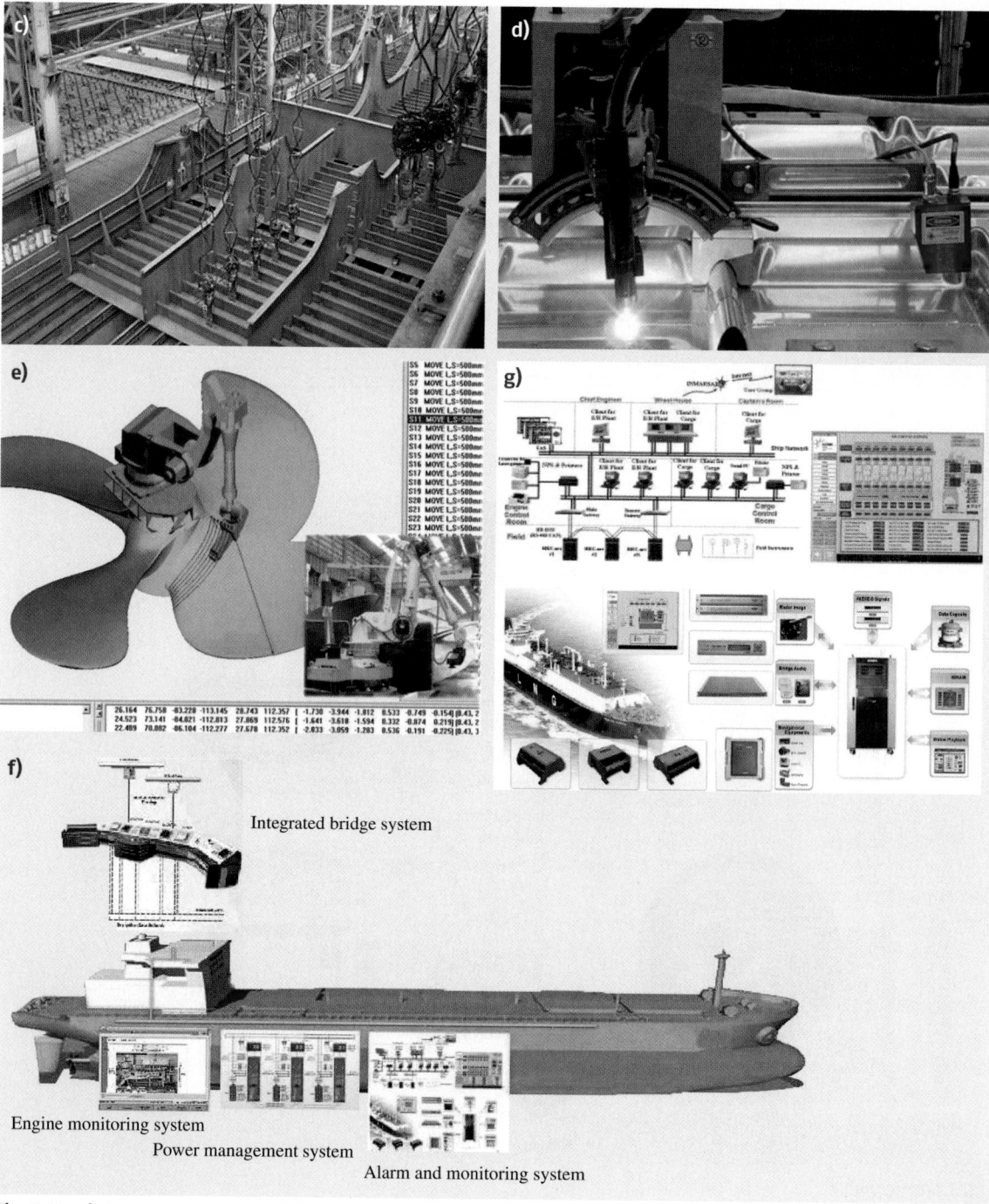

Fig. 3.12a–j (cont.)

h)

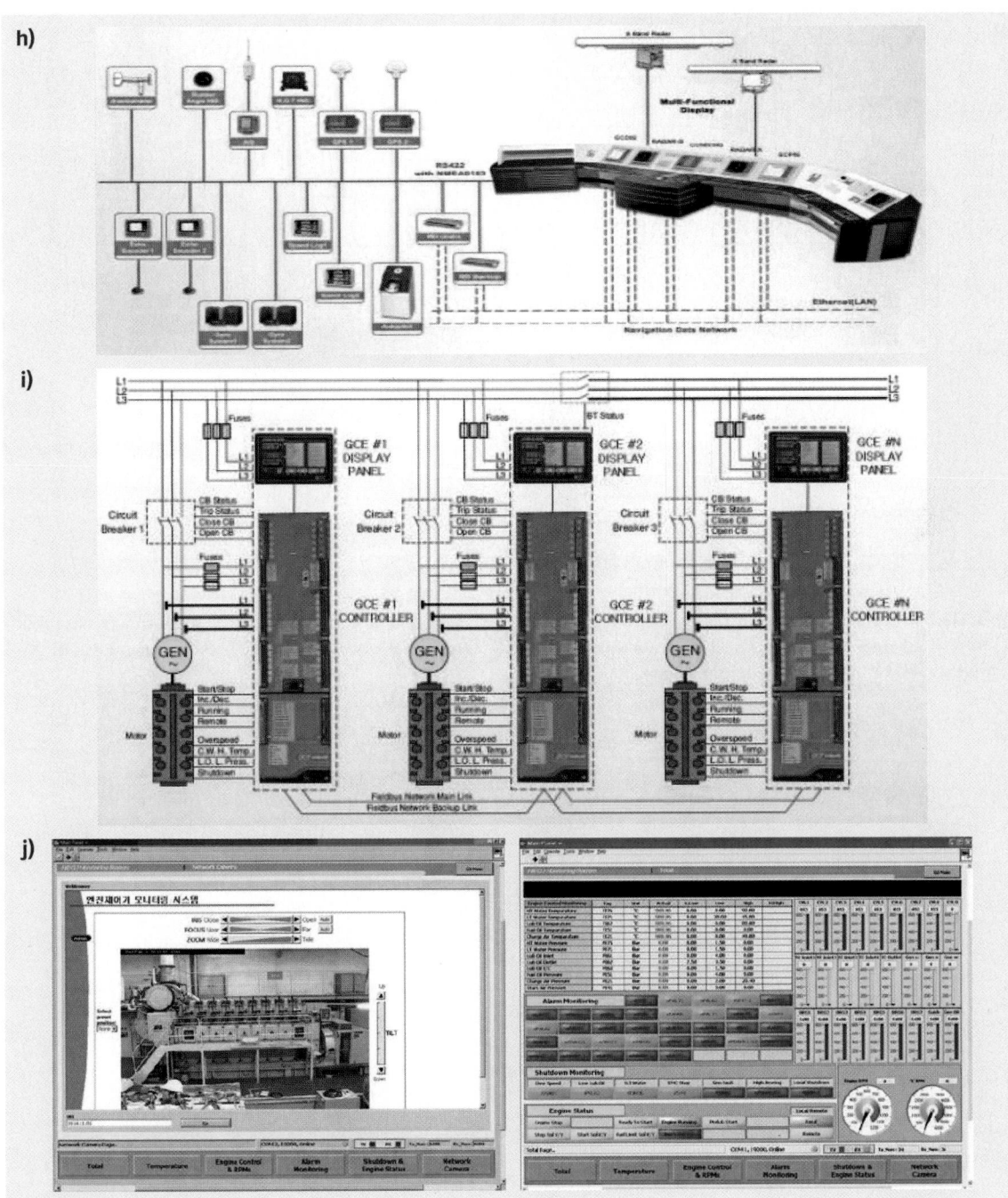

i)

j)

Fig. 3.12a–j (cont.)

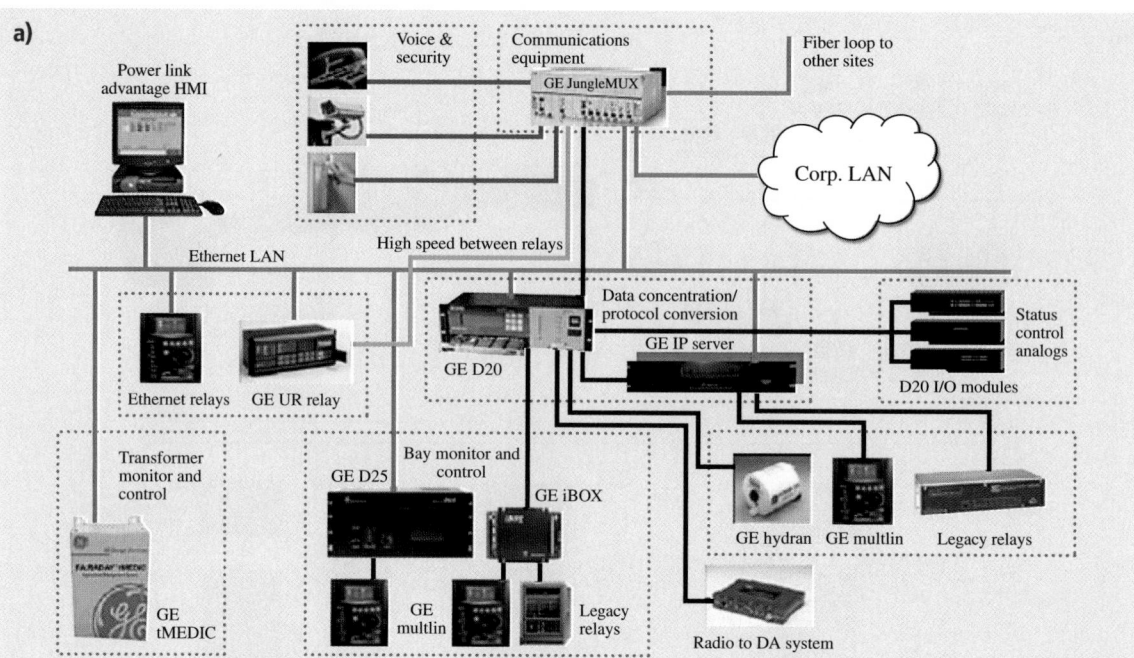

Fig. 3.13a,b Integrated power substation control system: (**a**) overview; (**b**) substation automation platform chassis (courtesy of GE Energy Co., Atlanta) (LAN – local area network, DA – data acquisition, UR – universal relay, MUX – multiplexor, HMI – human-machine interface)

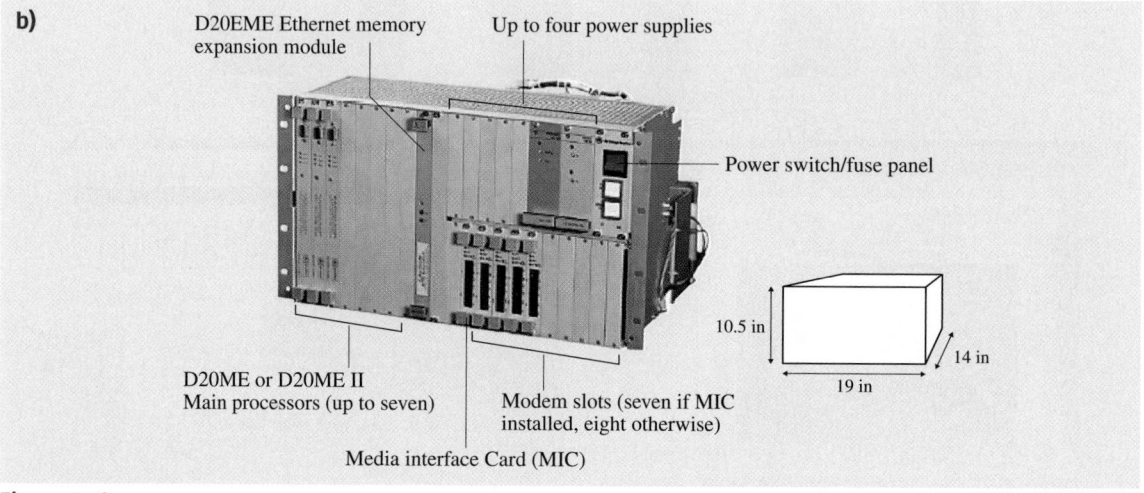

Fig. 3.13a,b (cont.)

3.4 Flexibility, Degrees, and Levels of Automation

Increasingly, solutions of society's problems cannot be satisfied by, and therefore cannot mean the automation of just a single, repeated process. By automating the networking and integration of devices and systems, they are able to perform different and variable tasks. Increasingly, this ability also requires cooperation (sharing of information and resources) and collaboration (sharing in the execution and responses) with other devices and systems. Thus, devices and systems have to be designed with inherent flexibility, which is motivated by the clients' requirements. With growing demand for service and product variety, there is also an increase in the expectations by users and customers for greater reliability, responsiveness, and smooth interoperability. Thus, the meaning of automation also involves the aspects of its flexibility, degree, and levels.

To enable design for flexibility, certain standards and measures have been and will continue to be established. Automation flexibility, often overlapping with the level of automation intelligence, depends on two main considerations:

1. The number of different states that can be assumed automatically
2. The length of time and amount of effort (setup process) necessary to respond and execute a change of state.

The number of different possible states and the cost of changes required are linked with two interrelated measures of flexibility: application flexibility and adaptation flexibility (Fig. 3.14). Both measures are concerned with the possible situations of the system and its environment. Automation solutions may address only switching between undisturbed, standard operations and nominal, variable situations, or can also aspire to respond when operations encounter disruptions and transitions, such as errors and conflicts, or significant design changes.

Application flexibility measures the number of different work states, scenarios, and conditions a system can handle. It can be defined as the probability that an arbitrary task, out of a given class of such tasks, can be carried out automatically. A relative comparison between the application flexibility of alternative designs is relevant mostly for the same domain of automation solutions. For instance, in Fig. 3.14 it is the domain of machining.

Adaptation flexibility is a measure of the time duration and the cost incurred for an automation device or system to transition from one given work state to another. Adaptation flexibility can also be measured only relatively, by comparing one automation device or system with another, and only for one defined change of state at a time. The change of state involves being in one possible state prior to the transition, and one possible state after it.

A relative estimate of the two flexibility measures (dimensions) for several implementations of machine tools automation is illustrated in Fig. 3.14. For generality, both measures are calibrated between 0 and 1.

3.4.1 Degree of Automation

Another dimension of automation, besides measures of its inherent flexibility, is the degree of automation. Automation can mean fully automatic or semiautomatic devices and systems, as exemplified in case A (Sect. 3.3.1), with the steam turbine speed governor, and in case E (Sect. 3.3.5), with a mix of robots and operators in elevator production. When a device or system is not fully automatic, meaning that some, or more frequent human intervention is required, they are con-

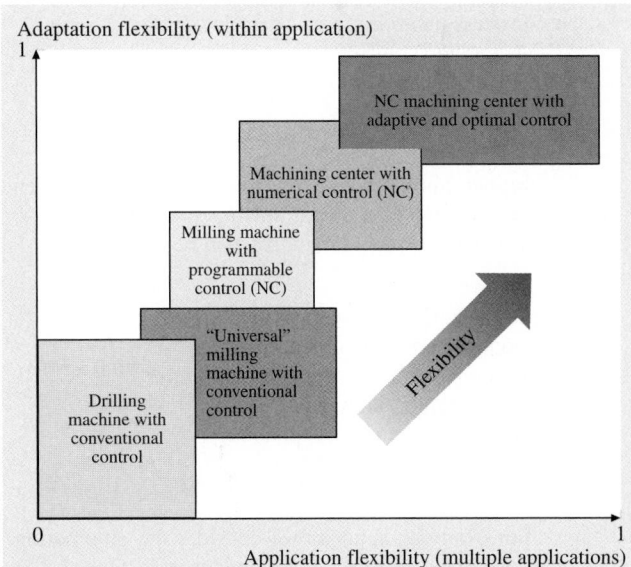

Fig. 3.14 Application flexibility and adaptation flexibility in machining automation (after [3.14])

sidered *automated*, or *semiautomatic*, equivalent terms implying *partial automation*.

A measure of the degree of automation, between fully manual to fully automatic, has been used to guide the design rationalization and compare between alternative solutions. Progression in the degree of automation in machining is illustrated in Fig. 3.14. The increase of automated partial functions is evident when comparing the drilling machine with the more flexible machines that can also drill and, in addition, are able to perform other processes such as milling.

The degree of automation can be defined as the fraction of automated functions out of the overall functions of an installation or system. It is calculated as the ratio between the number of automated operations, and the total number of operations that need to be performed, resulting in a value between 0 and 1. Thus, for a device or system with partial automation, where not all operations or functions are automatic, the degree of automation is less than 1. Practically, there are several methods to determine the degree of automation. The derived value requires a description of the method assumptions and steps. Typically, the degree of automation is associated with characteristics of:

1. Platform (device, system, etc.)
2. Group of platforms
3. Location, site
4. Plant, facility
5. Process and its scope
6. Process measures, e.g., operation cycle
7. Automatic control
8. Power source
9. Economic aspects
10. Environmental effects.

In determining the degree of automation of a given application, whether the following functions are also supposed to be considered must also be specified:

1. Setup
2. Organization, reorganization
3. Control and communication
4. Handling (of parts, components, etc.)
5. Maintenance and repair
6. Operation and process planning
7. Construction
8. Administration.

For example, suppose we consider the automation of a document processing system (case H, Sect. 3.3.8) which is limited to only the scanning process, thus omitting other workflow functions such as document feeding, joining, virtual inspection, and failure recovery. Then if the scanning is automatic, the degree of automation would be 1. However, if the other functions are also considered and they are not automatic, then the value would be less than 1.

Methods to determine the degree of automation divide into two categories:

● *Relative determination applying a graded scale*, containing all the functions of a defined domain process, relative to a defined system and the corresponding degrees of automation. For any given device or system in this domain, the degree of automation is found through comparison with the graded scale. This procedure is similar to other graded scales, e.g., Mohs' hardness scale, and Beaufort wind-speed scale. This method is illustrated in Fig. 3.15, which shows an example of the graded scale of mechanization and automation, following the scale developed by *Bright* [3.15].

● *Relative determination by a ratio between the autonomous and nonautonomous measures of reference*. The most common measure of reference is the number of decisions made during the process under consideration (Table 3.9). Other useful measures of reference for this determination are the comparative ratios of:

– Rate of service quality
– Human labor
– Time measures of effort
– Cycle time
– Number of mobility and motion functions
– Program steps.

To illustrate the method in reference to decisions made during the process, consider an example for case B (Sect. 3.3.2). Suppose in the bioreactor system process there is a total of seven decisions made automatically by the devices and five made by a human laboratory supervisor. Because these decisions are not similar, the degree of automation cannot be calculated simply as the ratio $7/(7+5) \approx 0.58$. The decisions must be weighted by their complexity, which is usually assessed by the number of control program commands or steps (Table 3.9). Hence, the degree of automation can be calculated as:

$$\text{degree of automation} = \frac{\left(\begin{array}{c}\text{sum of decision steps made} \\ \text{automatically by devices}\end{array}\right)}{\text{(total sum of decision steps made)}}$$
$$= 82/(82+178) \approx 0.32 \,.$$

From the worker		Through control-mechanism, testing determined work sequences		Through variable influences in the environment				Origin of the check	
						Reacts to the execution			
Variable		Fixed in the machine		React to signals		Selects from determined processes	Changes actions itself inside influences	Type of the machine reaction	
Manual		Mechanical (not done by hand)							Energy source
1	2	3	4	5	6	7	8	9	10	11	12	13	14	15	16	17	18	Step-No.

Column labels (steps 1–18):

1. Manual
2. Manual hand tools
3. Powered hand tools
4. Machine tools, manual controlled
5. Powered tool, fixed cycle, single function
6. Powered tool, programmed control with a sequence of functions
7. Machine system with remote control
8. Machine actuated by introduction of work-piece or material
9. Measures characteristics of the execution
10. Signals pre-selected values of measurement, includig error correction
11. Registers execution
12. Changes speed, position change and direction according to the measured signal
13. Segregates or rejects according to measurements
14. Identifies and selects operations
15. Corrects execution after the processing
16. Corrects execution while processing
17. Foresees the necessary working tasks, adjusts the execution
18. Prevents error and self-optimizes current execution

Steps of the automation

Fig. 3.15 Automation scale: scale for comparing grades of automation (after [3.15])

Now designers can compare this automation design against relatively more and less elaborate options. Rationalization will have to assess the costs, benefits, risks, and acceptability of the degree of automation for each alternative design.

Whenever the degree of automation is determined by a method, the following conditions must be followed:

1. The method is able to reproduce the same procedure consistently.

2. Comparison is only done between objective measures.

3. Degree values should be calibrated between 0 and 1 to simplify calculations and relative comparisons.

3.4.2 Levels of Automation, Intelligence, and Human Variability

There is obviously an inherent relation between the level of automation flexibility, the degree of automation, and the level of intelligence of a given automation ap-

Table 3.9 Degree of automation: calculation by the ratio of decision types

	Automatic decisions								Human decisions						Total
Decision number	1	2	3	4	5	6	7	Sum	8	9	10	11	12	Sum	
Complexity (number of program steps)	10	12	14	9	14	11	12	*82*	21	3	82	45	27	*178*	*260*

Part A | 3.4

Table 3.10 Levels of automation (updated and expanded after [3.16])

Level	Automation	Automated human attribute	Examples
A_0	Hand-tool; manual machine	None	Knife; scissors; wheelbarrow
A_1	Powered machine tools (non-NC)	Energy, muscles	Electric hand drill; electric food processor; paint sprayer
A_2	Single-cycle automatics and hand-feeding machines	Dexterity	Pipe threading machine; machine tools (non-NC)
A_3	Automatics; repeated cycles	Diligence	Engine production line; automatic copying lathe; automatic packaging; NC machine; pick-and-place robot
A_4	Self-measuring and adjusting; feedback	Judgment	*Feedback about product*: dynamic balancing; weight control. *Feedback about position*: pattern-tracing flame cutter; servo-assisted follower control; self-correcting NC machines; spray-painting robot
A_5	Computer control; automatic cognition	Evaluation	Rate of feed cutting; maintaining pH; error compensation; turbine fuel control; interpolator
A_6	Limited self-programming	Learning	Sophisticated elevator dispatching; telephone call switching systems; artificial neural network models
A_7	Relating cause from effects	Reasoning	Sales prediction; weather forecasting; lamp failure anticipation; actuarial analysis; maintenance prognostics; computer chess playing
A_8	Unmanned mobile machines	Guided mobility	Autonomous vehicles and planes; nano-flying exploration monitors
A_9	Collaborative networks	Collaboration	Collaborative supply networks; Internet; collaborative sensor networks
A_{10}	Originality	Creativity	Computer systems to compose music; design fabric patterns; formulate new drugs; play with automation, e.g., virtual-reality games
A_{11}	Human-needs and animal-needs support	Compassion	Bioinspired robotic seals (aquatic mammal) to help emotionally challenged individuals; social robotic pets
A_{12}	Interactive companions	Humor	Humorous gadgets, e.g., sneezing tissue dispenser; automatic systems to create/share jokes; interactive comedian robot

NC: numerically controlled; non-NC: manually controlled

plication. While there are no absolute measures of any of them, their meaning is useful and intriguing to inventors, designers, users, and clients of automation. Levels of automation are shown in Table 3.10 based on the intelligent human ability they represent.

It is interesting to note that the progression in our ability to develop and implement higher lev-els of automation follows the progress in our understanding of relatively more complex platforms; more elaborate control, communication, and solutions of computational complexity; process and operation programmability; and our ability to generate renewable, sustainable, and mobile power sources.

3.5 Worldwide Surveys: What Does Automation Mean to People?

With known human variability, we are all concerned about automation, enjoy its benefits, and wonder about its risks. But all individuals do not share our attitude towards automation equally, and it does not mean the same to everyone. We often hear people say:

- *I'll have to ask my grandson to program the video recorder.*
- *I hate my cell-phone.*
- *I can't imagine my life without a cell-phone.*
- *Those dumb computers.*
- *Sorry, I do not use elevators, I'll climb the six floors and meet you there!*

etc.

In an effort to explore the meaning of automation to people around the world, a random, nonscientific survey was conducted during 2007–2008 by the author, with the help of the following colleagues: Carlos Pereira (Brazil), Jose Ceroni (Chile), Alexandre Dolgui (France), Sigal Berman, Yael Edan, and Amit Gil (Israel), Kazayuhi Ishii, Masayuki Matsui, Jing Son, and Tetsuo Yamada (Japan), Jeong Wootae (Korea), Luis Basañez and Raúl Suárez Feijóo (Spain), Chin-Yin Huang (Taiwan), and Xin W. Chen (USA). Since the majority of the survey participants are students, undergraduate and graduate, and since they migrate globally, the respondents actually originate from all continents. In other words, while it is not a scientific survey, it carries a worldwide meaning.

Table 3.11 How do you define automation (do not use a dictionary)?

Definition	Asia–Pacific (%)	Europe + Israel (%)	North America (%)	South America (%)	World-wide (%)
1. Partially or fully replace human work [a]	6	43	18	5	27
2. Use machines/computers/robots to execute or help execute physical operations, computational commands or tasks	25	17	35	32	24
3. Work without or with little human participation	33	20	17	47	24
4. Improve work/system in terms of labor, time, money, quality, productivity, etc.	9	9	22	5	11
5. Functions and actions that assist humans	4	3	2	0	3
6. Integrated system of sensors, actuators, and controllers	6	2	2	0	3
7. Help do things humans cannot do	3	2	2	0	2
8. Promote human value	6	1	0	0	2
9. The mechanism of automatic machines	5	1	0	5	2
10. Information-technology-based organizational change	0	0	3	0	1
11. Machines with intelligence that works and knows what to do	1	1	0	5	1
12. Enable humans to perform multiple actions	0	1	0	0	0

[a] Note: This definition is inaccurate; most automation accomplishes work that humans cannot do, or cannot do effectively
* Respondents: 318 (244 undergraduates, 64 graduates, 10 others)

Table 3.12 When and where did you encounter and recognize automation first in your life (probably as a child)?

First encounter		Asia–Pacific (%)	Europe + Israel (%)	North America (%)	South America (%)	World-wide (%)
1.	Automated manufacturing machine, factory	12	9	30	32	16
2.	Vending machine: snacks, candy, drink, tickets	14	10	11	5	11
3.	Car, truck, motorcycle, and components	8	14	2	0	9
4.	Automatic door (pneumatic; electric)	14	2	2	5	5
5.	Toy	4	7	2	5	5
6.	Computer (software), e.g., Microsoft Office, e-Mail, programming language	1	3	13	5	4
7.	Elevator/escalator	8	3	2	0	3
8.	Movie, TV	7	2	0	5	3
9.	Robot	1	4	3	11	3
10.	Washing machine	1	6	0	0	3
11.	Automatic teller machine (ATM)	1	1	3	5	2
12.	Dishwasher	0	4	0	0	2
13.	Game machine	4	1	2	0	2
14.	Microwave	0	3	2	0	2
15.	Air conditioner	0	1	2	5	1
16.	Amusement park	1	1	0	0	1
17.	Automatic check-in at airports	0	0	3	0	1
18.	Automatic light	0	1	2	5	1
19.	Barcode scanner	0	0	5	0	1
20.	Calculator	0	1	2	0	1
21.	Clock/watch	0	1	2	0	1
22.	Agricultural combine	0	1	0	0	1
23.	Fruit classification machine	1	1	0	0	1
24.	Garbage truck	0	1	0	0	1
25.	Home automation	0	0	3	5	1
26.	Kitchen mixer	0	1	0	0	1
27.	Lego (with automation)	0	1	0	0	1
28.	Medical equipment in the birth-delivery room	0	1	0	0	1
29.	Milking machine	0	2	0	0	1
30.	Oven/toaster	0	2	0	0	1
31.	Pneumatic door of the school bus	0	1	2	0	1
32.	Tape recorder, player	1	1	0	5	1
33.	Telephone/answering machine	0	1	3	5	1
34.	Train (unmanned)	3	0	0	0	1
35.	X-ray machine	0	2	0	0	1
36.	Automated grinding machine for sharpening knives	1	0	0	0	0
37.	Automatic car wash	0	1	0	0	0
38.	Automatic bottle filling (with soda or wine)	0	0	2	0	0
39.	Automatic toll collection	0	0	2	0	0
40.	Bread machine	0	1	0	0	0
41.	Centrifuge	0	0	2	0	0
42.	Coffee machine	0	1	0	0	0

Table 3.12 (cont.)

First encounter		Asia Pacific (%)	Europe + Israel (%)	North America (%)	South America (%)	World- wide (%)
43.	Conveyor (deliver food to chickens)	0	1	0	0	0
44.	Electric shaver	0	1	0	0	0
45.	Food processor	0	1	0	0	0
46.	Fuse/breaker	0	1	0	0	0
47.	Kettle	0	1	0	0	0
48.	Library	1	0	0	0	0
49.	Light by electricity	0	1	0	0	0
50.	Luggage/baggage sorting machine	0	1	0	0	0
51.	Oxygen device	0	1	0	0	0
52.	Pulse recorder	0	1	0	0	0
53.	Radio	0	1	0	0	0
54.	Self-checkout machine	0	0	2	0	0
55.	Sprinkler	0	1	0	0	0
56.	Switch (power; light)	0	1	0	0	0
57.	Thermometer	0	1	0	0	0
58.	Traffic light	0	1	0	0	0
59.	Treadmill	0	0	2	0	0
60.	Ultrasound machine	0	1	0	0	0
61.	Video cassette recorder (VCR)	0	1	0	0	0
62.	Water delivery	0	1	0	0	0

* Respondents: 316 (249 undergraduates, 57 graduates, 10 others)

The survey population includes 331 respondents, from three general categories:

1. Undergraduate students of engineering, science, management, and medical sciences (251)
2. Graduate students from the same disciplines (70)
3. Nonstudents, experts, and novices in automation (10).

Three questions were posed in this survey:

● How do you define automation (do not use a dictionary)?
● When and where did you encounter and recognize automation first in your life (probably as a child)?
● What do you think is the major impact/contribution of automation to humankind (only one)?

The answers are summarized in Tables 3.11–3.13.

3.5.1 How Do We Define Automation?

The key answer to this was question was (Table 3.11, no. 3): operate without or with little human participation (24%). This answer reflects a meaning that corresponds well with the definition in the beginning of this chapter. It was the most popular response in Asia–Pacific and South America.

Overall, the 12 types of definition meanings follow three main themes: How automation works (answer nos. 2, 6, 9, 11, total 30%); automation replaces humans, works without them, or augments their functions (answer nos. 1, 3, 7, 10, total 54%); automation improves (answer nos. 4, 5, 8, 12, total 16%).

Interestingly, the overall most *popular* answer (answer no. 1; 27%) is a partially wrong answer. (It was actually the significantly most popular response only in Europe and Israel.) *Replacing human work* may represent legacy fear of automation. This answer lacks the recognition that most automation applications are performing tasks humans cannot accomplish. The latter is also a partial, yet positive, meaning of automation and is addressed by answer no. 7 (2%).

Answer nos. 2, 6, 9, and 11 (total 29%) represent a factual meaning of how automation is implemented. Answer 10, found only in North America responses, addresses a meaning of automation that narrows it down to

Table 3.13 What do you think is the major impact/contribution of automation to humankind (only one)?

Major impact or contribution	Asia–Pacific (%)	Europe + Israel (%)	North America (%)	South America (%)	World-wide (%)
1. Save time, increase productivity/efficiency; 24/7 operations	26	19	17	42	22
2. Advance everyday life/improve quality of life/convenience/ease of life/work	10	11	0	0	8
3. Save labor	17	5	3	0	8
4. Encourage/inspire creative work; inspire newer solutions	5	5	7	11	6
5. Mass production and service	0	5	17	5	6
6. Increase consistency/improve quality	5	5	2	16	5
7. Prevent people from dangerous activities	7	2	7	5	5
8. Detect errors in healthcare, flights, factories/reduce (human) errors	8	2	2	5	4
9. Medicine/medical equipment/medical system/biotechnology/healthcare	0	8	2	0	4
10. Save cost	7	2	5	0	4
11. Computer	0	5	3	0	3
12. Improve security and safety	0	7	2	0	3
13. Assist/work for people	3	3	0	5	2
14. Car	3	1	2	0	2
15. Do things that humans cannot do	1	2	3	5	2
16. Replace people; people lose jobs	2	2	2	0	2
17. Save lives	0	4	0	0	2
18. Transportation (e.g., train, traffic lights)	1	2	2	0	2
19. Change/improvement in (global) economy	0	1	3	0	1
20. Communication (devices)	0	1	5	0	1
21. Deliver products/service to more people	1	1	0	0	1
22. Extend life expectancy	0	1	0	0	1
23. Foundation of industry/growth of industry	0	2	0	0	1
24. Globalization and spread of culture and knowledge	0	0	3	0	1
25. Help aged/handicapped people	3	1	0	0	1
26. Manufacturing (machines)	1	1	2	0	1
27. Robot; industrial robot	0	0	5	0	1
28. Agriculture improvement	0	1	0	0	0
29. Banking system	0	0	2	0	0
30. Construction	0	0	2	0	0
31. Flexibility in manufacturing	0	1	0	0	0
32. Identify bottlenecks in production	0	0	2	0	0
33. Industrial revolution	0	1	0	0	0
34. Loom	0	0	2	0	0
35. People lose abilities to complete tasks	0	1	0	0	0
36. Save resources	0	1	0	0	0
37. Weather prediction	0	0	2	0	0

* Respondents: 330 (251 undergraduate, 70 graduate, 9 others)

information technology. Answer nos. 4, 5, and 12 (total 14%) imply that automation means improvements and assistance. Finally, answer no. 8, promote human value (2%, but found only in Asia–Pacific, and Europe and Israel) may reflect cultural meaning more than a definition.

In addition to the regional response variations mentioned above, it turns out that the first four answers in Table 3.11 comprise the majority of meaning in each region: 73% for Asia–Pacific; 89% for Europe and Israel and for South America; and 92% for North America (86% worldwide). In Asia–Pacific, four other answer types each comprised 4–6% of the 12 answer types.

3.5.2 When and Where Did We Encounter Automation First in Our Life?

The key answer to this question was: automated manufacturing machine or factory (16%), which was the top answer in North and South America, but only the third response in Asia–Pacific and in Europe and Israel. The top answer in Asia–Pacific is shared by vending machine and automatic door (14% each), and in Europe and Israel the top answer was car, truck, motorcycle, and components (14%). Overall, the 62 types of responses to this question represent a wide range of what automation means to us.

There are variations in responses between regions, with only two answers – no. 1, automated manufacturing machine, factory, and no. 2, vending machine: snacks, candy, drink, tickets – shared by at least three of the four regions surveyed. Automatic door is in the top three only in Asia–Pacific; car, truck, motorcycle, and components is in the top three only in Europe and Israel; computer is in the top three only in North

America; and robot is in the top three only in South America.

Of the 62 response types, almost 40 appear in only one region (or are negligible in other regions), and it is interesting to relate the regional context of the first encounter with automation. For instance:

- Answers no. 34, train (unmanned) and no. 36, automated grinding machine for sharpening knives, appear as responses mostly in Asia–Pacific (3%)
- Answers no. 12, dishwasher (4%); no. 35, x-ray machine (2%); and no. 58, traffic light (1%) appear as responses mostly in Europe and Israel
- Answers no. 19, barcode scanner (4%); no. 38, automatic bottle filling (2%); no. 39, automatic toll collection (2%); and no. 59, treadmill, (2%) appear as responses mostly in North America.

3.5.3 What Do We Think Is the Major Impact/Contribution of Automation to Humankind?

Thirty-seven types of impacts or benefits were found in the survey responses overall. The most inspiring impact and benefit of automation found is answer no. 4, *encourage/inspire creative work; inspire newer solutions* (6%).

The most popular response was: save time, increase productivity/efficiency; 24/7 operations (22%). The key risk identified, answer no. 16, was: replace people; people lose jobs (2%, but interestingly, this was not found in South America). Another risky impact identified is answer no. 35, people lose abilities to complete tasks, (1%, only in Europe and Israel). Nevertheless, the majority (98%) identified 35 types of positive impacts and benefits.

3.6 Emerging Trends

Many of us perceive the meaning of the automatic and automated factories and gadgets of the 20th and 21st century as outstanding examples of the human spirit and human ingenuity, no less than art; their disciplined organization and synchronized complex of carefully programmed functions and services mean to us harmonious expression, similar to good music (when they work).

Clearly, there is a mixture of emotions towards automation: Some of us are dismayed that humans cannot

usually be as accurate, or as fast, or as responsive, attentive, and indefatigable as automation systems and installations. On the other hand, we sometimes hear the word *automaton* or *robot* describing a person or an organization that lacks consideration and compassion, let alone passion. Let us recall that automation is made *by people and for the people*. But can it run away by its own autonomy and become undesirable? Future automation will advance in micro- and nanosystems and systems-of-systems. Bioinspired automation and bioinspired collaborative control theory will significantly improve artificial intelligence, and the quality of robotics and automation, as well as the engineering of their safety and security. In this context, it is interesting to examine the role of automation in the 20th and 21st centuries.

3.6.1 Automation Trends of the 20th and 21st Centuries

The US National Academy of Engineering, which includes US and worldwide experts, compiled the list shown in Table 3.14 as the top 20 achievements that *have shaped a century and changed the world* [3.17]. The table adds columns indicating the role that automation has played in each achievement, and clearly, automation has been relevant in all of them and essential to most of them.

The US National Academy of Engineers has also compiled a list of the grand challenges for the 21st cen-

tury. These challenges are listed in Table 3.15 with the anticipated and emerging role of automation in each. Again, automation is relevant to all of them and essential to most of them.

Some of the main trends in automation are described next.

3.6.2 Bioautomation

Bioinspired automation, also known as bioautomation or evolutionary automation, is emerging based on the trend of bioinspired computing, control, and AI. They influence traditional automation and artificial intelligence in the methods they offer for evolutionary machine learning, as opposed to what can be described as *generative methods* (sometimes called *creationist methods*) used in traditional programming and learning. In traditional methods, intelligence is typically programmed from top down: Automation engineers and programmers create and implement the automa-

Table 3.14 Top engineering achievements in the 20th century [3.17] and the role of automation

Achievement	Role of automation		
	Relevant		**Irrelevant**
	Essential	**Supportive**	
1. Electrification	×		
2. Automobile	×		
3. Airplane	×		
4. Water supply and distribution	×		
5. Electronics	×		
6. Radio and television	×		
7. Agricultural mechanization	×		
8. Computers	×		
9. Telephone	×		
10. Air conditioning and refrigeration	×		
11. Highways		×	
12. Spacecraft	×		
13. Internet	×		
14. Imaging	×		
15. Household appliances	×		
16. Health technologies	×		
17. Petroleum and petrochemical technologies	×		
18. Laser and fiber optics	×		
19. Nuclear technologies	×		
20. High-performance materials		×	

tion logic, and define the scope, functions, and limits of its intelligence. Bioinspired automation, on the other hand, is also created and implemented by automation engineers and programmers, but follows a bottom-up decentralized and distributed approach. Bioinspired techniques often involve a method of specifying a set of simple rules, followed and iteratively applied by a set of simple and autonomous manmade organisms. After several generations of rule repetition, initially manmade mechanisms of self-learning, self-repair, and self-organization enable self-evolution towards more complex behaviors. Complexity can result in unexpected behaviors, which may be robust and more reliable; can be counterintuitive compared with the original design; but can potentially become undesirable, out-of-control, and unsafe behaviors. This subject has been under intense research and examination in recent years.

Natural evolution and system biology (biology-inspired automation mechanisms for systems engineering) are the driving analogies of this trend: concurrent steps and rules of responsive selection, interdependent recombination, reproduction, mutation, reformation, adaptation, and death-and-birth can be defined, similar to how complex organisms function and evolve in nature. Similar automation techniques are used in genetic algorithms, artificial neural networks, swarm algorithms, and other emerging evolutionary automation

systems. Mechanisms of self-organization, parallelism, fault tolerance, recovery, backup, and redundancy are being developed and researched for future automation, in areas such as neuro-fuzzy techniques, biorobotics, digital organisms, artificial cognitive models and architectures, artificial life, bionics, and bioinformatics. See related topics in many following handbook chapters, particularly, 29, 75 and 76.

3.6.3 Collaborative Control Theory and e-Collaboration

Collaboration of humans, and its advantages and challenges are well known from prehistory and throughout history, but have received increased attention with the advent of communication technology. Significantly better enabled and potentially streamlined and even optimized through e-Collaboration (based on communication via electronic means), it is emerging as one of the most powerful trends in automation, with telecommunication, computer communication, and wireless communication influencing education and research, engineering and business, healthcare and service industries, and global society in general. Those developments, in turn, motivate and propel further applications and theoretical investigations into this highly intelligent level of automation (Table 3.10, level A_9 and higher).

Table 3.15 Grand engineering challenges for the 21st century [3.17] and the role of automation

Achievement	Anticipated and emerging role of automation		
	Essential	Relevant Supportive	Irrelevant
1. Make solar energy economical	×		
2. Provide energy from fusion	×		
3. Develop carbon sequestration methods	×		
4. Manage the nitrogen cycle	×		
5. Provide access to clean water	×		
6. Restore and improve urban infrastructure		×	
7. Advance health informatics	×		
8. Engineer better medicines	×		
9. Reverse-engineer the brain	×		
10. Prevent nuclear terror	×		
11. Secure cyberspace	×		
12. Enhance virtual reality	×		
13. Advance personalized learning	×		
14. Engineer the tools of scientific discovery	×		

Interesting examples of the e-Collaboration trend include *wikis*, which since the early 2000s have been increasingly adopted by enterprises as collaborative software, enriching static intranets and the Internet. Examples of e-Collaborative applications that emerged in the 1990s include project communication for coplanning, sharing the creation and editing of design documents as codesign and codocumentation, and mutual inspiration for collaborative innovation and invention through cobrainstorming.

Beyond human–human automation-supported collaboration through better and more powerful communication technology, there is a well-known but not yet fully understood trend for collaborative e-Work. Associated with this field is collaborative control theory (CCT), which is under development. Collaborative e-Work is motivated by significantly improved performance of humans leveraging their collaborative automatic agents. The latter, from software automata (e.g., constructive *bots* as opposed to spam and other destructive bots) to automation devices, multisensors, multiagents, and multirobots can operate in a parallel, autonomous cyberspace, thus multiplying our productivity and increasing our ability to design sustainable systems and operations. A related important trend is the emergence of *active middleware* for collaboration support of device networks and of human team networks and enterprises.

More about this subject area can be found in several chapters of this handbook, particularly in Chaps. 12, 14, 26, and 88.

3.6.4 Risks of Automation

As civilization increasingly depends on automation and looks for automation to support solutions of its serious problems, the risks associated with automation must be understood and eliminated. Failures of automation on a very large scale are most risky. Just a few examples of disasters caused by automation failures are nuclear accidents; power supply disruptions and blackout; Federal Aviation Administration control systems failures causing air transportation delays and shutdowns; cellular communication network failures; and water supply failures. The impacts of severe natural and manmade disasters on automated infrastructure are therefore the target of intense research and development. In addition, automation experts are challenged to apply automation to enable sustainability and better mitigate and eliminate natural and manmade disasters, such as security, safety, and health calamities.

3.6.5 Need for Dependability, Survivability, Security, and Continuity of Operation

Emerging efforts are addressing better automation dependability and security by structured backup and recovery of information and communication systems. For instance, with service orientation that is able to survive, automation can enable gradual and degraded services by sustaining critical continuity of operations until the repair, recovery, and resumption of full services. Automation continues to be designed with the goal of preventing and eliminating any conceivable errors, failures, and conflicts, within economic constraints. In addition, the trend of collaborative flexibility being designed into automation frameworks encourages reconfiguration tools that redirect available, safe resources to support the most critical functions, rather that designing absolutely failure-proof system.

With the trend towards collaborative, networked automation systems, dependability, survivability, security, and continuity of operations are increasingly being enabled by autonomous self-activities, such as:

- Self-awareness and situation awareness
- Self-configuration
- Self-explaining and self-rationalizing
- Self-healing and self-repair
- Self-optimization
- Self-organization
- Self-protection for security.

Other dimensions of emerging automation risks involve privacy invasion, electronic surveillance, accuracy and integrity concerns, intellectual and physical property protection and security, accessibility issues, confidentiality, etc. Increasingly, people ask about the meaning of automation, how can we benefit from it, yet find a way to contain its risks and powers. At the extreme of this concern is the *automation singularity* [3.18].

Automation singularity follows the evident acceleration of technological developments and discoveries. At some point, people ask, is it possible that superhuman machines can take over the human race? If we build them too autonomous, with collaborative ability to self-improve and self-sustain, would they not eventually be able to exceed human intelligence? In other words, superintelligent machines may autonomously, automatically, produce discoveries that are too complex for humans to comprehend; they may even act in ways that we consider out of control, chaotic, and even aimed at damaging and overpowering people. This emerging trend of thought will no doubt energize future research on how to prevent automation from running on its own without limits. In a way, this is the 21st century human challenge of *never play with fire*.

3.7 Conclusion

This chapter explores the meaning of automation to people around the world. After review of the evolution of automation and its influence on civilization, its main contributions and attributes, a survey is used to summarize highlights of the meaning of automation according to people around the world. Finally, emerging trends in automation and concerns about automation are also described. They can be summarized as addressing three general questions:

1. How can automation be improved and become more useful and dependable?
2. How can we limit automation from being too risky when it fails?
3. How can we develop better automation that is more autonomous and performs better, yet does not take over our humanity?

These topics are discussed in detail in the chapters of this handbook.

3.8 Further Reading

- U. Alon: *An Introduction to Systems Biology: Design Principles of Biological Circuits* (Chapman Hall, New York 2006)
- R.C. Asfahl: *Robots and Manufacturing Automation*, 2nd edn. (Wiley, New York 1992)
- M.V. Butz, O. Sigaud, G. Pezzulo, G. Baldassarre (Eds.): *Anticipatory Behavior in Adaptive Learning Systems: From Brains to Individual and Social Behavior* (Springer, Berlin Heidelberg 2007)
- Center for Chemical Process Safety (CCPS): *Guidelines for Safe and Reliable Instrumented Protective Systems* (Wiley, New York 2007)
- K. Collins: *PLC Programming For Industrial Automation* (Exposure, Goodyear 2007)
- C.H. Dagli, O. Ersoy, A.L. Buczak, D.L. Enke, M. Embrechts: *Intelligent Engineering Systems Through Artificial Neural Networks: Smart Systems Engineering*, Comput. Intell. Architect. Complex Eng. Syst., Vol. 17 (ASME, New York 2007)
- R.C. Dorf, R.H. Bishop: *Modern Control Systems* (Pearson, Upper Saddle River 2007)
- R.C. Dorf, S.Y. Nof (Eds.): *International Encyclopedia of Robotics and Automation* (Wiley, New York 1988)
- K. Elleithy (ed.): *Innovations and Advanced Techniques in Systems, Computing Sciences and Software Engineering* (Springer, Berlin Heidelberg 2008)
- K. Evans: *Programming of CNC Machines*, 3rd edn. (Industrial Press, New York 2007)
- M. Fewster, D. Graham: *Software Test Automation: Effective Use of Test Execution Tools* (Addison-Wesley, Reading 2000)
- P.G. Friedmann: *Automation and Control Systems Economics*, 2nd edn. (ISA, Research Triangle Park 2006)
- C.Y. Huang, S.Y. Nof: Automation technology. In: *Handbook of Industrial Engineering*, ed. by G. Salvendy (Wiley, New York 2001), 3rd edn., Chap. 5
- D.C. Jacobs, J.S. Yudken: *The Internet, Organizational Change and Labor: The Challenge of Virtualization* (Routledge, London 2003)
- S. Jaderstrom, J. Miller (Eds.): *Complete Office Handbook: The Definitive Reference for Today's Electronic Office*, 3rd edn. (Random House, New York 2002)
- T.R. Kochtanek, J.R. Matthews: *Library Information Systems: From Library Automation to Distributed Information Access Solutions* (Libraries Unlimited, Santa Barbara 2002)
- E.M. Marszal, E.W. Scharpf: *Safety Integrity Level Selection: Systematic Methods Including Layer of Protection Analysis* (ISA, Research Triangle Park 2002)
- A.P. Mathur: *Foundations of Software Testing* (Addison-Wesley, Reading 2008)
- D.F. Noble: *Forces of Production: A Social History of Industrial Automation* (Oxford Univ. Press, Cambridge 1986)
- R. Parasuraman, T.B. Sheridan, C.D. Wickens: A model for types and levels of human interac-

tion with automation, IEEE Trans. Syst. Man Cyber. **30**(3), 286–197 (2000)

- G.D. Putnik, M.M. Cunha (Eds.): *Encyclopedia of Networked and Virtual Organizations* (Information Science Reference, 2008)
- R.K. Rainer, E. Turban: *Introduction to Information Systems*, 2nd edn. (Wiley, New York 2009)
- R.L. Shell, E.L. Hall (Eds.): *Handbook of Industrial Automation* (CRC, Boca Raton 2000)
- T.B. Sheridan: *Humans and Automation: System Design and Research Issues* (Wiley, New York 2002)

- V. Trevathan: *A Guide to the Automation Body of Knowledge*, 2nd edn. (ISA, Research Triangle Park 2006)
- L. Wang, K.C. Tan: *Modern Industrial Automation Software Design* (Wiley-IEEE Press, New York 2006)
- N. Wiener: *Cybernetics, or the Control and Communication in the Animal and the Machine*, 2nd edn. (MIT Press, Cambridge 1965)
- T.J. Williams, S.Y. Nof: Control models. In: *Handbook of Indus trial Engineering*, ed. by G. Salvendy (Wiley, New York 1992), 2nd edn., Chap. 9

References

3.1   I. Asimov: Foreword. In: *Handbook of Industrial Robotics*, ed. by S.Y. Nof (Wiley, New York 1999), 2nd edn.

3.2   Hearings on automation and technological change, Subcommittee on Economic Stabilization of the Joint Committee on the Economic Report, US Congress, October 14–28 (1955)

3.3   W. Buckingham: *Automation: Its Impact on Business and People* (The New American Library, New York 1961)

3.4   J.G. Truxal: *Control Engineers' Handbook: Servomechanisms, Regulators, and Automatic Feedback Control Systems* (McGraw Hill, New York 1958)

3.5   M.P. Groover: *Automation, Production Systems, and Computer-Integrated Manufacturing*, 3rd edn. (Prentice Hall, Englewood Cliffs 2007)

3.6   D. Tapping, T. Shuker: *Value Stream Management for the Lean Office* (Productivity, Florence 2003)

3.7   S. Burton, N. Shelton: *Procedures for the Automated Office*, 6th edn. (Prentice Hall, Englewood Cliffs 2004)

3.8   A. Dolgui, G. Morel, C.E. Pereira (Eds.): INCOM'06, information control problems in manufacturing, Proc. 12th IFAC Symp., St. Etienne (2006)

3.9   R. Felder, M. Alwan, M. Zhang: *Systems Engineering Approach to Medical Automation* (Artech House, London 2008)

3.10   A. Cichocki, A.S. Helal, M. Rusinkiewicz, D. Woelk: *Workflow and Process Automation: Concepts and Technology* (Kluwer, Boston 1998)

3.11   B. Karakostas, Y. Zorgios: *Engineering Service Oriented Systems: A Model Driven Approach* (IGI Global, Hershey 2008)

3.12   G.M. Lenart, S.Y. Nof: Object-oriented integration of design and manufacturing in a laser processing cell, Int. J. Comput. Integr. Manuf. **10**(1–4), 29–50 (1997), special issue on design and implementation of CIM systems

3.13   K.-S. Min: Automation and control systems technology in korean shipbuilding industry: the state of the art and the future perspectives, Proc. 17th World Congr. IFAC, Seoul (2008)

3.14   S.Y. Nof, W.E. Wilhelm, H.J. Warnecke: *Industrial Assembly* (Chapman Hall, New York 1997)

3.15   J.R. Bright: *Automation and Management* (Harvard Univ. Press, Boston 1958)

3.16   G.H. Amber, P.S. Amber: *Anatomy of Automation* (Prentice Hall, Englewood Cliffs 1964)

3.17   NAE: US National Academy of Engineering, Washington (2008). http://www.engineeringchallenges.org/

3.18   Special Report: The singularity, IEEE Spectrum **45**(6) (2008)

4. A History of Automatic Control

Christopher Bissell

Automatic control, particularly the application of feedback, has been fundamental to the development of automation. Its origins lie in the level control, water clocks, and pneumatics/hydraulics of the ancient world. From the 17th century onwards, systems were designed for temperature control, the mechanical control of mills, and the regulation of steam engines. During the 19th century it became increasingly clear that feedback systems were prone to instability. A stability criterion was derived independently towards the end of the century by Routh in England and Hurwitz in Switzerland. The 19th century, too, saw the development of servomechanisms, first for ship steering and later for stabilization and autopilots. The invention of aircraft added (literally) a new dimension to the problem. Minorsky's theoretical analysis of ship control in the 1920s clarified the nature of three-term control, also being used for process applications by the 1930s. Based on servo and communications engineering developments of the 1930s, and driven by the need for high-performance gun control systems, the coherent body of theory known as *classical control* emerged during and just after WWII in the US, UK and elsewhere, as did cybernetics ideas. Meanwhile, an alternative approach to dynamic modeling had been developed in the USSR based on the approaches of Poincaré and Lyapunov.

Information was gradually disseminated, and *state-space* or *modern control* techniques, fuelled by Cold War demands for missile control systems, rapidly developed in both East and West. The immediate post-war period was marked by great claims for automation, but also great fears, while the digital computer opened new possibilities for automatic control.

4.1 Antiquity and the Early Modern Period

Feedback control can be said to have originated with the float valve regulators of the Hellenic and Arab worlds [4.1]. They were used by the Greeks and Arabs to control such devices as water clocks, oil lamps and wine dispensers, as well as the level of water in tanks. The precise construction of such systems is still not entirely clear, since the descriptions in the original Greek or Arabic are often vague, and lack illustrations. The best known Greek names are Ktsebios and Philon (third century BC) and Heron (first century AD) who were active in the eastern Mediterranean (Alexandria, Byzantium). The water clock tradition was continued in the Arab world as described in books by writers such as Al-Jazari (1203) and Ibn al-Sa-ati (1206), greatly influenced by the anonymous Arab author known as Pseudo-Archimedes of the ninth–tenth century AD, who makes specific reference to the Greek work of Heron and Philon. Float regulators in the tradition of Heron were also constructed by the three brothers Banu Musa in Baghdad in the ninth century AD.

The float valve level regulator does not appear to have spread to medieval Europe, even though translations existed of some of the classical texts by the above writers. It seems rather to have been reinvented during the industrial revolution, appearing in England, for

example, in the 18th century. The first independent European feedback system was the temperature regulator of Cornelius Drebbel (1572–1633). Drebbel spent most of his professional career at the courts of James I and Charles I of England and Rudolf II in Prague. Drebbel himself left no written records, but a number of contemporary descriptions survive of his invention. Essentially an alcohol (or other) thermometer was used to operate a valve controlling a furnace flue, and hence the temperature of an enclosure [4.2]. The device included screws to alter what we would now call the set point.

If level and temperature regulation were two of the major precursors of modern control systems, then

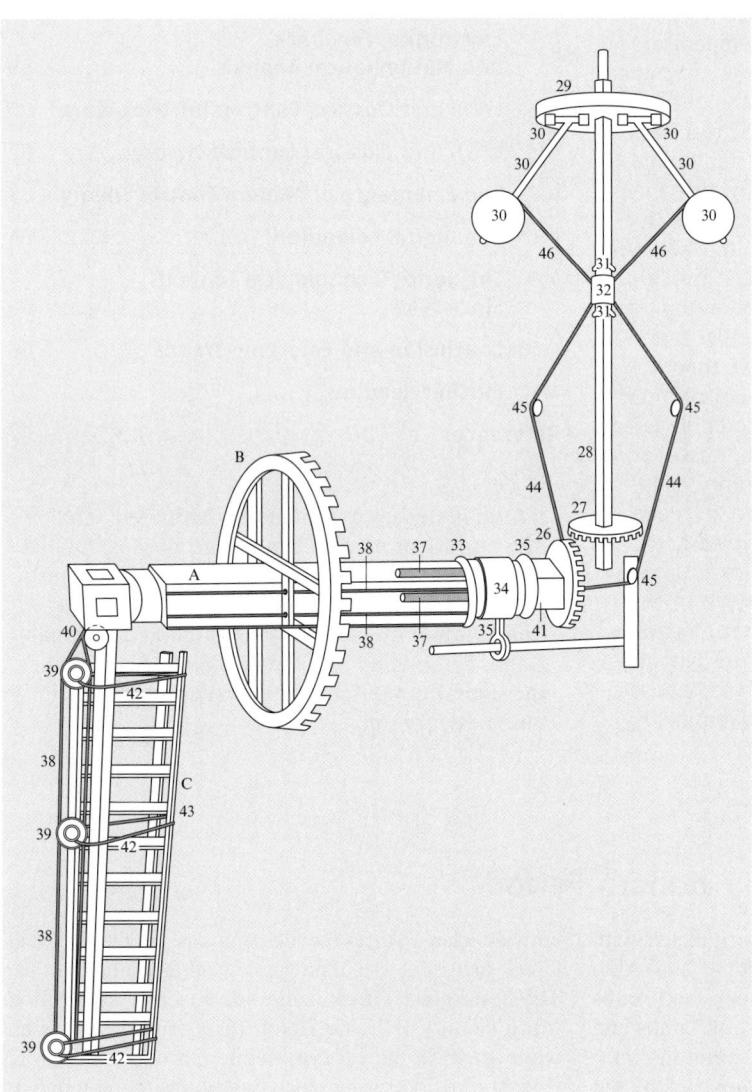

Fig. 4.1 Mead's speed regulator (after [4.1])

a number of devices designed for use with windmills pointed the way towards more sophisticated devices. During the 18th century the mill fantail was developed both to keep the mill sails directed into the wind and to automatically vary the angle of attack, so as to avoid excessive speeds in high winds. Another important device was the lift-tenter. Millstones have a tendency to separate as the speed of rotation increases, thus impairing the quality of flour. A number of techniques were developed to sense the speed and hence produce a restoring force to press the millstones closer together. Of these,

perhaps the most important were *Thomas Mead*'s devices [4.3], which used a centrifugal pendulum to sense the speed and – in some applications – also to provide feedback, hence pointing the way to the centrifugal governor (Fig. 4.1).

The first steam engines were the reciprocating engines developed for driving water pumps; James Watt's rotary engines were sold only from the early 1780s. But it took until the end of the decade for the centrifugal governor to be applied to the machine, following a visit by Watt's collaborator, Matthew Boulton, to

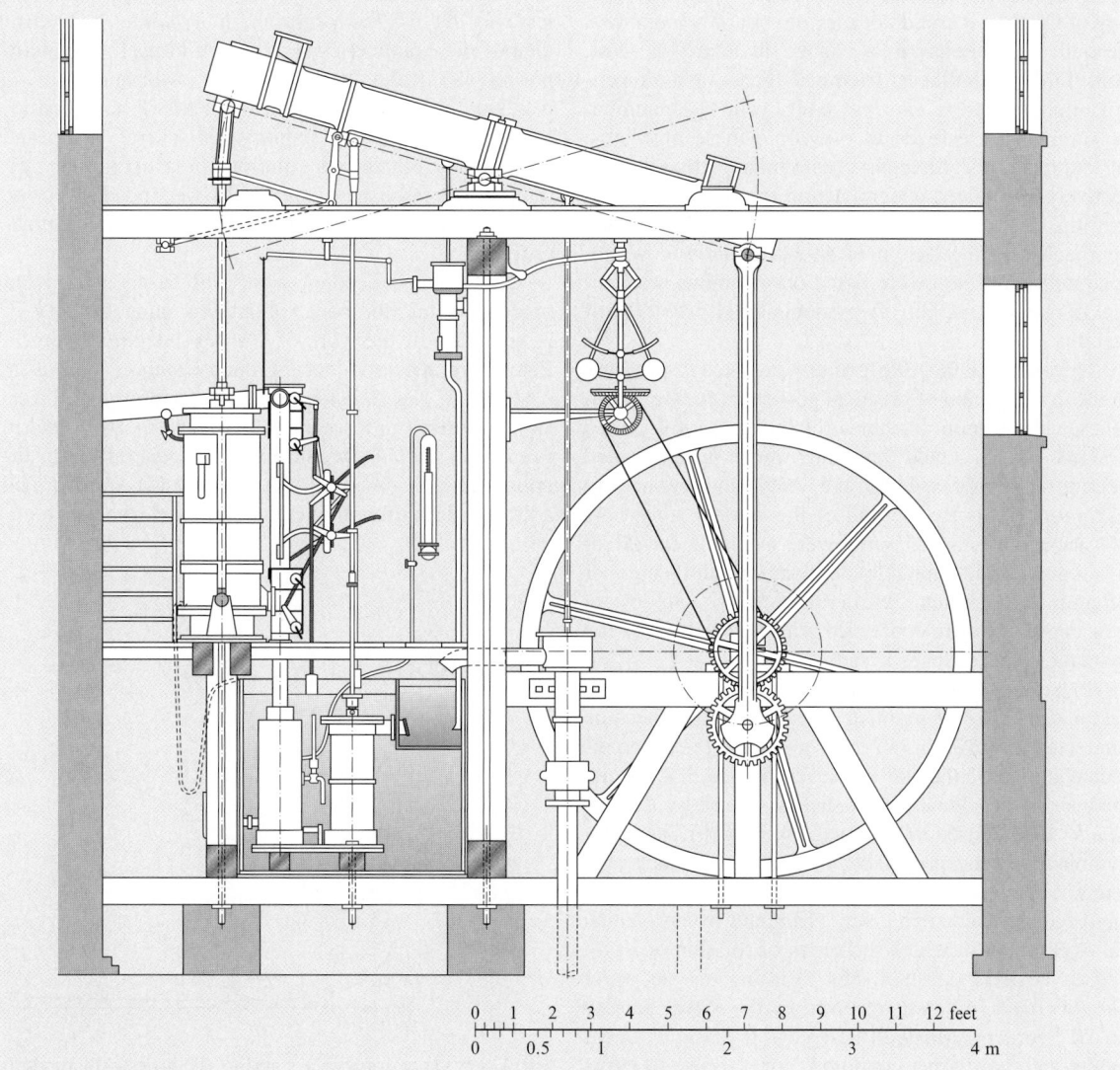

0   1   2   3   4   5   6   7   8   9   10   11   12 feet
0   0.5   1   2   3   4 m

Fig. 4.2 Boulton & Watt steam engine with centrifugal governor (after [4.1])

the Albion Mill in London where he saw a lift-tenter in action under the control of a centrifugal pendulum (Fig. 4.2). Boulton and Watt did not attempt to patent the device (which, as noted above, had essentially already been patented by Mead) but they did try unsuccessfully to keep it secret. It was first copied in 1793 and spread throughout England over the next ten years [4.4].

4.2 Stability Analysis in the 19th Century

With the spread of the centrifugal governor in the early 19th century a number of major problems became apparent. First, because of the absence of integral action, the governor could not remove offset: in the terminology of the time it could not *regulate* but only *moderate*. Second, its response to a change in load was slow. And thirdly, (nonlinear) frictional forces in the mechanism could lead to *hunting* (limit cycling). A number of attempts were made to overcome these problems: for example, the Siemens chronometric governor effectively introduced integral action through differential gearing, as well as mechanical amplification. Other approaches to the design of an *isochronous* governor (one with no offset) were based on ingenious mechanical constructions, but often encountered problems of stability.

Nevertheless the 19th century saw steady progress in the development of practical governors for steam engines and hydraulic turbines, including spring-loaded designs (which could be made much smaller, and operate at higher speeds) and relay (indirect-acting) governors [4.6]. By the end of the century governors of various sizes and designs were available for effective regulation in a range of applications, and a number of graphical techniques existed for steady-state design. Few engineers were concerned with the analysis of the dynamics of a feedback system.

In parallel with the developments in the engineering sector a number of eminent British scientists became interested in governors in order to keep a telescope directed at a particular star as the Earth rotated. A formal analysis of the dynamics of such a system by *George Bidell Airy*, Astronomer Royal, in 1840 [4.7] clearly demonstrated the propensity of such a feedback system to become unstable. In 1868 James Clerk Maxwell analyzed governor dynamics, prompted by an electrical experiment in which the speed of rotation of a coil had to be held constant. His resulting classic paper *On governors* [4.8] was received by the Royal Society on 20 February. Maxwell derived a third-order linear model and the correct conditions for stability in terms of the coefficients of the characteristic equation. Un-

able to derive a solution for higher-order models, he expressed the hope that the question would gain the attention of mathematicians. In 1875 the subject for the Cambridge University Adams Prize in mathematics was set as *The criterion of dynamical stability*. One of the examiners was Maxwell himself (prizewinner in 1857) and the 1875 prize (awarded in 1877) was won by Edward James Routh. *Routh* had been interested in dynamical stability for several years, and had already obtained a solution for a fifth-order system. In the published paper [4.9] we find derived the Routh version of the renowned Routh–Hurwitz stability criterion.

Related, independent work was being carried out in continental Europe at about the same time [4.5]. A summary of the work of I.A. Vyshnegradskii in St. Petersburg appeared in the French *Comptes Rendus de l'Academie des Sciences* in 1876, with the full version appearing in Russian and German in 1877, and in French in 1878/79. Vyshnegradskii (generally transliterated at the time as Wischnegradski) transformed a third-order differential equation model of a steam en-

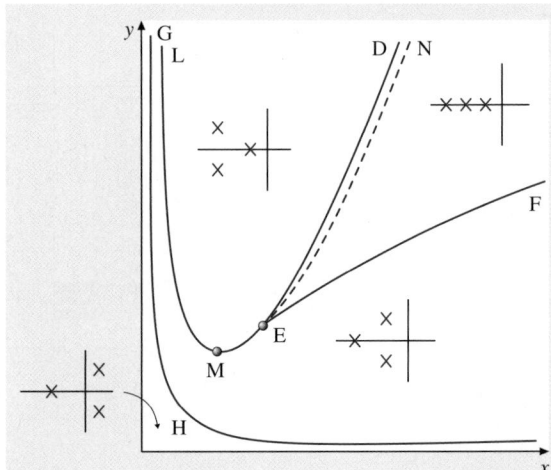

Fig. 4.3 Vyshnegradskii's stability diagram with modern pole positions (after [4.5])

gine with governor into a standard form

$$\varphi^3 + x\varphi^2 + y\varphi + 1 = 0 \,,$$

where x and y became known as the Vyshnegradskii parameters. He then showed that a point in the x–y plane defined the nature of the system transient response. Figure 4.3 shows the diagram drawn by Vyshnegradskii, to which typical pole constellations for various regions in the plane have been added.

In 1893 Aurel Boreslav Stodola at the Federal Polytechnic, Zurich, studied the dynamics of a high-pressure hydraulic turbine, and used Vyshnegradskii's method to assess the stability of a third-order model. A more re-alistic model, however, was seventh-order, and Stodola posed the general problem to a mathematician colleague *Adolf Hurwitz*, who very soon came up with his version of the Routh–Hurwitz criterion [4.10]. The two versions were shown to be identical by *Enrico Bompiani* in 1911 [4.11].

At the beginning of the 20th century the first general textbooks on the regulation of prime movers appeared in a number of European languages [4.12, 13]. One of the most influential was *Tolle*'s *Regelung der Kraftmaschine*, which went through three editions between 1905 and 1922 [4.14]. The later editions included the Hurwitz stability criterion.

4.3 Ship, Aircraft and Industrial Control Before WWII

The first ship steering engines incorporating feedback appeared in the middle of the 19th century. In 1873 Jean Joseph Léon Farcot published a book on servomotors in which he not only described the various designs developed in the family firm, but also gave an account of the general principles of position control. Another important maritime application of feedback control was in gun turret operation, and hydraulics were also extensively developed for transmission systems. Torpedoes, too, used increasingly sophisticated feedback systems for depth control – including, by the end of the century, gyroscopic action (Fig. 4.4).

During the first decades of the 20th century gyroscopes were increasingly used for ship stabilization and autopilots. Elmer Sperry pioneered the *active stabilizer*, the gyrocompass, and the gyroscope autopilot, filing various patents over the period 1907–1914. Sperry's autopilot was a sophisticated device: an inner loop controlled an electric motor which operated the steering engine, while an outer loop used a gyrocompass to sense the heading. Sperry also designed an *anticipator* to replicate the way in which an experienced helmsman would *meet* the helm (to prevent oversteering); the anticipator was, in fact, a type of adaptive control [4.16].

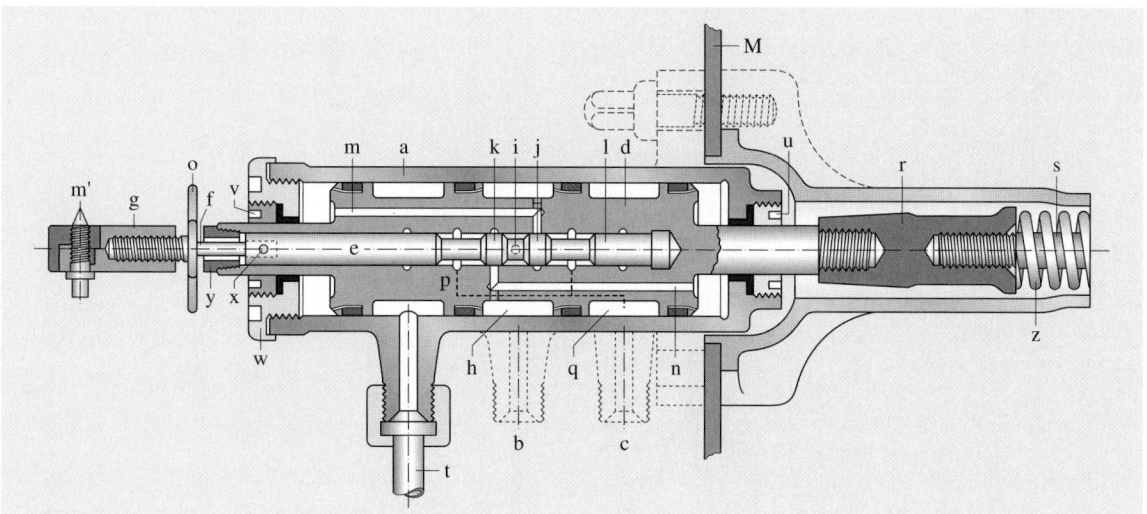

Fig. 4.4 Torpedo servomotor as fitted to Whitehead torpedoes around 1900 (after [4.15])

Sperry and his son Lawrence also designed aircraft autostabilizers over the same period, with the added complexity of three-dimensional control. *Bennett* describes the system used in an acclaimed demonstration in Paris in 1914 [4.17]:

For this system the Sperrys used four gyroscopes mounted to form a stabilized reference platform; a train of electrical, mechanical and pneumatic components detected the position of the aircraft relative to the platform and applied correction signals to the aircraft control surfaces. The stabilizer operated for both pitch and roll [. . .] The system

was normally adjusted to give an approximately deadbeat response to a step disturbance. The incorporation of derivative action [. . .] was based on Sperry's intuitive understanding of the behaviour of the system, not on any theoretical foundations. The system was also adaptive [. . .] adjusting the gain to match the speed of the aircraft.

Significant technological advances in both ship and aircraft stabilization took place over the next two decades, and by the mid 1930s a number of airlines were using Sperry autopilots for long-distance flights. However, apart from the stability analyses discussed

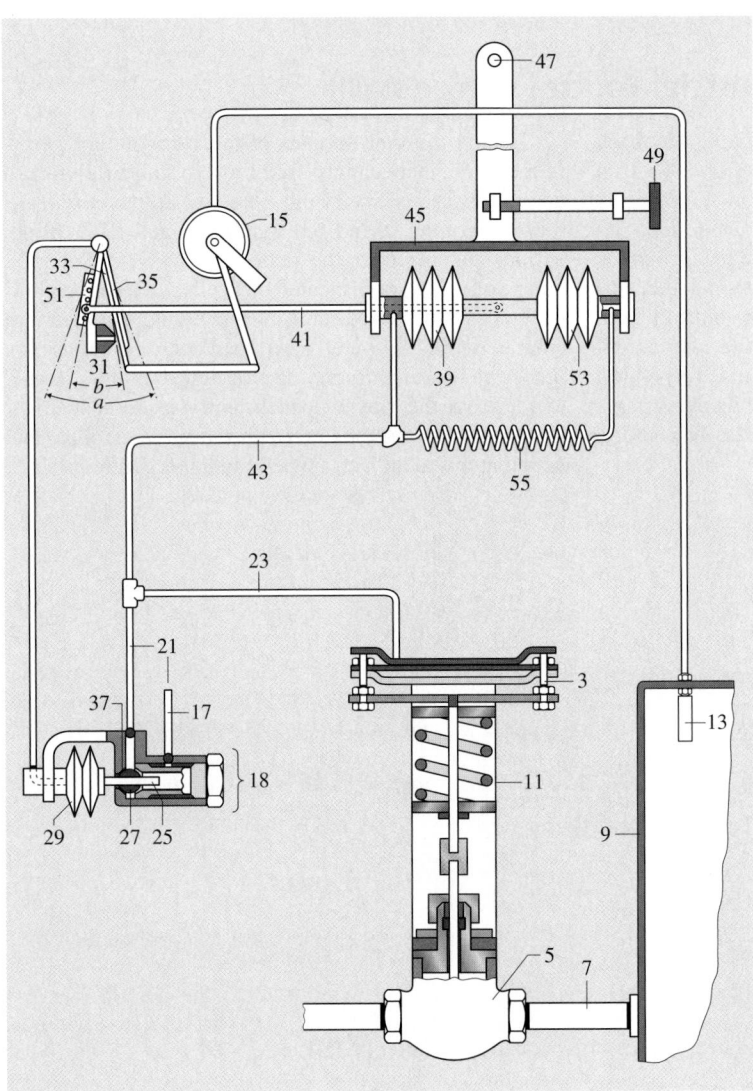

Fig. 4.5 The *Stabilog*, a pneumatic controller providing proportional and integral action [4.18]

in Sect. 4.2 above, which were not widely known at this time, there was little theoretical investigation of such feedback control systems. One of the earliest significant studies was carried out by *Nicholas Minorsky*, published in 1922 [4.19]. Minorsky was born in Russia in 1885 (his knowledge of Russian proved to be important to the West much later). During service with the Russian Navy he studied the ship steering problem and, following his emigration to the USA in 1918, he made the first theoretical analysis of automatic ship steering. This study clearly identified the way that control action should be employed: although Minorsky did not use the terms in the modern sense, he recommended an appropriate combination of proportional, derivative and integral action. Minorsky's work was not widely disseminated, however. Although he gave a good theoretical basis for closed loop control, he was writing in an age of heroic invention, when intuition and practical experience were much more important for engineering practice than theoretical analysis.

Important technological developments were also being made in other sectors during the first few decades of the 20th century, although again there was little theoretical underpinning. The electric power industry brought demands for voltage and frequency regulation; many processes using driven rollers required accurate speed control; and considerable work was carried out in a number of countries on systems for the accurate pointing of guns for naval and anti-aircraft gunnery. In the process industries, measuring instruments and pneumatic controllers of increasing sophistication were developed. Mason's *Stabilog* (Fig. 4.5), patented in 1933, included integral as well as proportional action, and by the end of the decade three-term controllers were available that also included *preact* or derivative control. Theoretical progress was slow, however, until the advances made in electronics and telecommunications in the 1920s and 30s were translated into the control field during WWII.

4.4 Electronics, Feedback and Mathematical Analysis

The rapid spread of telegraphy and then telephony from the mid 19th century onwards prompted a great deal of theoretical investigation into the behaviour of electric circuits. *Oliver Heaviside* published papers on his operational calculus over a number of years from 1888 onwards [4.20], but although his techniques produced valid results for the transient response of electrical networks, he was fiercely criticized by contemporary mathematicians for his lack of rigour, and ultimately he was blackballed by the establishment. It was not until the second decade of the 20th century that Bromwich, Carson and others made the link between Heaviside's operational calculus and Fourier methods, and thus proved the validity of Heaviside's techniques [4.21].

The first three decades of the 20th century saw important analyses of circuit and filter design, particularly in the USA and Germany. *Harry Nyquist* and Karl Küpfmüller were two of the first to consider the problem of the maximum transmission rate of telegraph signals, as well as the notion of *information* in telecommunications, and both went on to analyze the general stability problem of a feedback circuit [4.22]. In 1928 Küpfmüller analyzed the dynamics of an automatic gain control electronic circuit using feedback. He appreciated the dynamics of the feedback system, but his integral equation approach resulted only in a approximations and design diagrams, rather than a rigorous stability criterion. At about the same time in the USA, Harold Black was designing feedback amplifiers for transcontinental telephony (Fig. 4.6). In a famous epiphany on the Hudson River ferry in August 1927 he realized that negative feedback could reduce distortion at the cost of reducing overall gain. Black passed on the problem of the stability of such a feedback loop to his Bell Labs colleague *Harry Nyquist*, who published his celebrated frequency-domain encirclement criterion in 1932 [4.23]. Nyquist demonstrated, using results derived by Cauchy, that the key to stability is whether or not the open loop frequency response locus in the complex plane encircles (in Nyquist's original convention) the point $1 + i0$. One of the great advantages of this approach is that no analytical form of the open loop frequency response is required: a set of measured data points can be plotted without the need for a mathematical model. Another advantage is that, unlike the Routh–Hurwitz criterion, an assessment of the transient response can be made directly from the Nyquist plot in terms of gain and phase margins (how close the locus approaches the critical point).

Black's 1934 paper reporting his contribution to the development of the negative feedback amplifier included what was to become the standard closed-loop analysis in the frequency domain [4.24].

Part A | 4.4

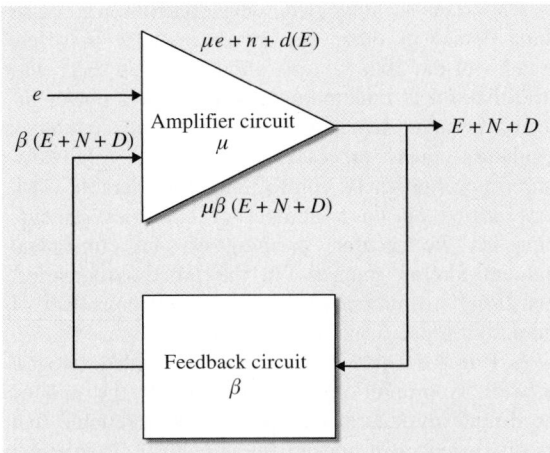

Fig. 4.6 Black's feedback amplifier (after [4.24])

The third key contributor to the analysis of feedback in electronic systems at Bell Labs was *Hendrik Bode* who worked on equalizers from the mid 1930s, and who demonstrated that attenuation and phase shift were related in any realizable circuit [4.25]. The dream of telephone engineers to build circuits with fast cutoff and low phase shift was indeed only a dream. It was Bode who introduced the notions of gain and phase margins, and redrew the Nyquist plot in its now conventional form with the critical point at $-1 + i0$. He also introduced the famous straight-line approximations to frequency response curves of linear systems plotted on log–log axes. *Bode* presented his methods in a classic text published immediately after the war [4.26].

If the work of the communications engineers was one major precursor of classical control, then the other

was the development of high-performance servos in the 1930s. The need for such servos was generated by the increasing use of analogue simulators, such as network analysers for the electrical power industry and differential analysers for a wide range of problems. By the early 1930s six-integrator differential analysers were in operation at various locations in the USA and the UK. A major center of innovation was MIT, where Vannevar Bush, Norbert Wiener and Harold Hazen had all contributed to design. In 1934 *Hazen* summarized the developments of the previous years in *The theory of servomechanisms* [4.27]. He adopted normalized curves, and parameters such as time constant and damping factor, to characterize servo-response, but he did not given any stability analysis: although he appears to have been aware of Nyquists's work, he (like almost all his contemporaries) does not appear to have appreciated the close relationship between a feedback servomechanism and a feedback amplifier.

The 1930s American work gradually became known elsewhere. There is ample evidence from prewar USSR, Germany and France that, for example, Nyquist's results were known – if not widely disseminated. In 1940, for example, *Leonhard* published a book on automatic control in which he introduced the inverse Nyquist plot [4.28], and in the same year a conference was held in Moscow during which a number of Western results in automatic control were presented and discussed [4.29]. Also in Russia, a great deal of work was being carried out on nonlinear dynamics, using an approach developed from the methods of Poincaré and Lyapunov at the turn of the century [4.30]. Such approaches, however, were not widely known outside Russia until after the war.

4.5 WWII and *Classical* Control: Infrastructure

Notwithstanding the major strides identified in the previous subsections, it was during WWII that a discipline of feedback control began to emerge, using a range of design and analysis techniques to implement high-performance systems, especially those for the control of anti-aircraft weapons. In particular, WWII saw the coming together of engineers from a range of disciplines – electrical and electronic engineering, mechanical engineering, mathematics – and the subsequent realisation that a common framework could be applied to all the various elements of a complex control system in order to achieve the desired result [4.18, 31].

The so-called fire control problem was one of the major issues in military research and development at the end of the 1930s. While not a new problem, the increasing importance of aerial warfare meant that the control of anti-aircraft weapons took on a new significance. Under manual control, aircraft were detected by radar, range was measured, prediction of the aircraft position at the arrival of the shell was computed, guns were aimed and fired. A typical system could involve up to 14 operators. Clearly, automation of the process was highly desirable, and achieving this was to require detailed research into such matters as the dynamics of the servomechanisms driving the gun aiming, the design of controllers, and the statistics of tracking aircraft possibly taking evasive action.

Government, industry and academia collaborated closely in the US, and three research laboratories were of prime importance. The Servomechanisms Laboratory at MIT brought together Brown, Hall, Forrester and others in projects that developed frequency-domain methods for control loop design for high-performance servos. Particularly close links were maintained with Sperry, a company with a strong track record in guidance systems, as indicated above. Meanwhile, at MIT's Radiation Laboratory – best known, perhaps, for its work on radar and long-distance navigation – researchers such as James, Nichols and Phillips worked on the further development of design techniques for auto-track radar for AA gun control. And the third institution of seminal importance for fire-control development was Bell Labs, where great names such as Bode, Shannon and Weaver – in collaboration with Wiener and Bigelow at MIT – attacked a number of outstanding problems, including the theory of smoothing and prediction for gun aiming. By the end of the war, most of the techniques of what came to be called classical control had been elaborated in these laboratories, and a whole series of papers and textbooks appeared in the late 1940s presenting this new discipline to the wider engineering community [4.32].

Support for control systems development in the United States has been well documented [4.18, 31]. The National Defence Research Committee (NDRC) was established in 1940 and incorporated into the Office of Scientific Research and Development (OSRD) the following year. Under the directorship of Vannevar Bush the new bodies tackled anti-aircraft measures, and thus the servo problem, as a major priority. Section D of the NDRC, devoted to Detection, Controls and Instruments was the most important for the development of feedback control. Following the establishment of the OSRD the NDRC was reorganised into divisions, and Division 7, Fire Control, under the overall direction of Harold Hazen, covered the subdivisions: ground-based anti-aircraft fire control; airborne fire control systems; servomechanisms and data transmission; optical rangefinders; fire control analysis; and navy fire control with radar.

Turning to the United Kingdom, by the outbreak of WWII various military research stations were highly active in such areas as radar and gun laying, and there were also close links between government bodies and industrial companies such as Metropolitan–Vickers,

British Thomson–Houston, and others. Nevertheless, it is true to say that overall coordination was not as effective as in the USA. A body that contributed significantly to the dissemination of theoretical developments and other research into feedback control systems in the UK was the so called Servo-Panel. Originally established informally in 1942 as the result of an initiative of Solomon (head of a special radar group at Malvern), it acted rather as a *learned society* with approximately monthly meetings from May 1942 to August 1945. Towards the end of the war meetings included contributions from the US.

Germany developed successful control systems for civil and military applications both before and during the war (torpedo and flight control, for example). The period 1938–1941 was particularly important for the development of missile guidance systems. The test and development center at Peenemünde on the Baltic coast had been set up in early 1936, and work on guidance and control saw the involvement of industry, the government and universities. However, there does not appear to have been any significant national coordination of R&D in the control field in Germany, and little development of high-performance servos as there was in the US and the UK. When we turn to the German situation outside the military context, however, we find a rather remarkable awareness of control and even cybernetics. In 1939 the *Verein Deutscher Ingenieure*, one of the two major German engineers' associations, set up a specialist committee on control engineering. As early as October 1940 the chair of this body *Herman Schmidt* gave a talk covering control engineering and its relationship with economics, social sciences and cultural aspects [4.33]. Rather remarkably, this committee continued to meet during the war years, and issued a report in 1944 concerning primarily control concepts and terminology, but also considering many of the fundamental issues of the emerging discipline.

The Soviet Union saw a great deal of prewar interest in control, mainly for industrial applications in the context of five-year plans for the Soviet command economy. Developments in the USSR have received little attention in English-language accounts of the history of the discipline apart from a few isolated papers. It is noteworthy that the *Kommissiya Telemekhaniki i Avtomatiki* (KTA) was founded in 1934, and the *Institut Avtomatiki i Telemekhaniki* (IAT) in 1939 (both under the auspices of the Soviet Academy of Sciences, which controlled scientific research through its network of institutes). The KTA corresponded with numerous western manufacturers of control equipment in the mid

1930s and translated a number articles from western journals. The early days of the IAT were marred, however, by the *Shchipanov affair*, a classic Soviet attack on a researcher for *pseudo-science*, which detracted from technical work for a considerable period of time [4.34]. The other major Russian center of research related to control theory in the 1930s and 1940s (if not for practical applications) was the University of Gorkii (now Nizhnii Novgorod), where Aleksandr Andronov and colleagues had established a center for the study of nonlinear dynamics during the 1930s [4.35]. Andronov was

in regular contact with Moscow during the 1940s, and presented the emerging control theory there – both the nonlinear research at Gorkii and developments in the UK and USA. Nevertheless, there appears to have been no coordinated wartime work on control engineering in the USSR, and the IAT in Moscow was evacuated when the capital came under threat. However, there does seem to have been an emerging control community in Moscow, Nizhnii Novgorod and Leningrad, and Russian workers were extremely well-informed about the open literature in the West.

4.6 WWII and *Classical* Control: Theory

Design techniques for servomechanisms began to be developed in the USA from the late 1930s onwards. In 1940 Gordon S. Brown and colleagues at MIT analyzed the transient response of a closed loop system in detail, introducing the *system operator* $1/(1 + \text{open loop})$ as functions of the Heaviside differential operator p. By the end of 1940 contracts were being drawn up between

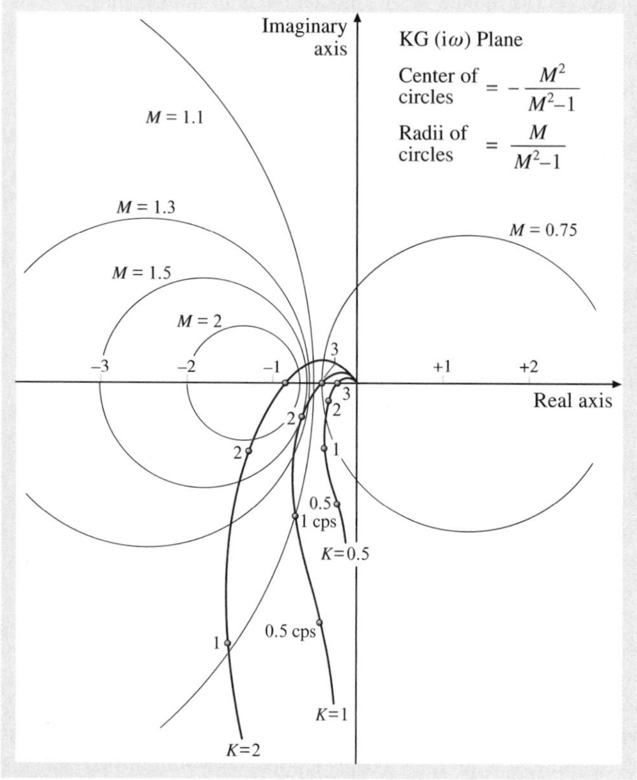

Fig. 4.7 Hall's M-circles (after [4.36])

the NDRC and MIT for a range of servo projects. One of the most significant contributors was *Albert Hall*, who developed classic frequency-response methods as part of his doctoral thesis, presented in 1943 and published initially as a confidential document [4.37] and then in the open literature after the war [4.36]. Hall derived the frequency response of a unity feedback servo as $KG(i\omega)/[1 + KG(i\omega)]$, applied the Nyquist criterion, and introduced a new way of plotting system response that he called M-circles (Fig. 4.7), which were later to inspire the Nichols Chart. As *Bennett* describes it [4.38]:

Hall was trying to design servosystems which were stable, had a high natural frequency, and high damping. [. . .] He needed a method of determining, from the transfer locus, the value of K that would give the desired amplitude ratio. As an aid to finding the value of K he superimposed on the polar plot curves of constant magnitude of the amplitude ratio. These curves turned out to be circles. . . By plotting the response locus on transparent paper, or by using an overlay of M-circles printed on transparent paper, the need to draw M-circles was obviated. . .

A second MIT group, known as the Radiation Laboratory (or RadLab) was working on auto-track radar systems. Work in this group was described after the war in [4.39]; one of the major innovations was the introduction of the Nichols chart (Fig. 4.8), similar to Hall's M-circles, but using the more convenient decibel measure of amplitude ratio that turned the circles into a rather different geometrical form.

The third US group consisted of those looking at smoothing and prediction for anti-aircraft weapons – most notably Wiener and Bigelow at MIT together with others, including Bode and Shannon, at Bell Labs. This work involved the application of correlation techniques to the statistics of aircraft motion. Although the prototype Wiener predictor was unsuccessful in attempts at practical application in the early 1940s, the general approach proved to be seminal for later developments.

Formal techniques in the United Kingdom were not so advanced. Arnold Tustin at Metropolitan–Vickers (Metro–Vick) worked on gun control from the late 1930s, but engineers had little appreciation of dynamics. Although they used harmonic response plots they appeared to have been unaware of the Nyquist criterion until well into the 1940s [4.40]. Other key researchers in the UK included *Whitely*, who proposed using the inverse Nyquist diagram as early as 1942, and introduced his *standard forms* for the design of various categories of servosystem [4.41]. In Germany, Winfried Oppelt, Hans Sartorius and Rudolf Oldenbourg were also coming to related conclusions about closed-loop design independently of allied research [4.42, 43].

The basics of sampled-data control were also developed independently during the war in several countries. The *z*-transform in all but name was described in a chapter by *Hurewizc* in [4.39]. Tustin in the UK developed the bilinear transformation for time series models, while Oldenbourg and Sartorius also used difference equations to model such systems.

From 1944 onwards the design techniques developed during the hostilities were made widely available in an explosion of research papers and text books – not only from the USA and the UK, but also from Germany and the USSR. Towards the end of the decade perhaps the final element in the classical control toolbox was added – *Evans*' root locus technique, which

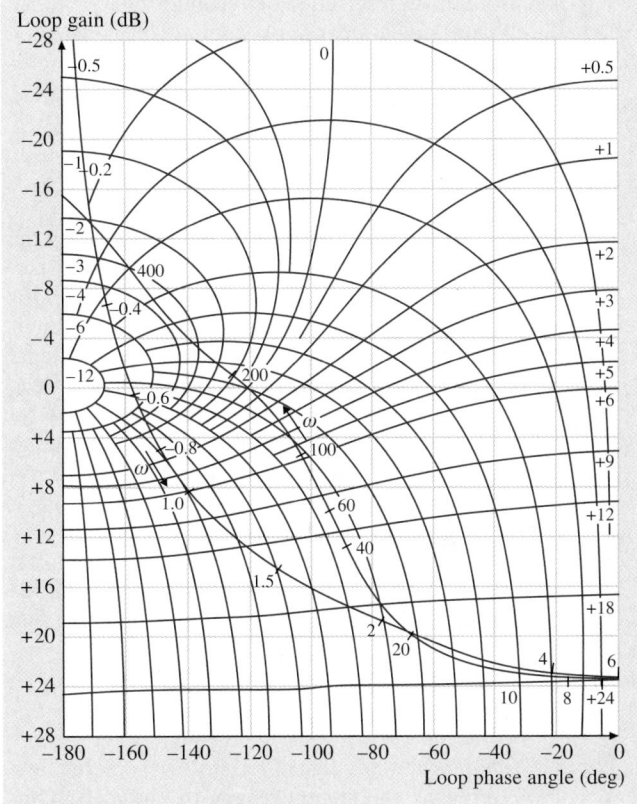

Fig. 4.8 Nichols Chart (after [4.38])

enabled plots of changing pole position as a function of loop gain to be easily sketched [4.44]. But a radically different approach was already waiting in the wings.

4.7 The Emergence of *Modern* Control Theory

The *modern* or *state space* approach to control was ultimately derived from original work by Poincaré and Lyapunov at the end of the 19th century. As noted above, Russians had continued developments along these lines, particularly during the 1920s and 1930s in centers of excellence in Moscow and Gorkii (now Nizhnii Novgorod). Russian work of the 1930s filtered slowly through to the West [4.45], but it was only in the post war period, and particularly with the introduction of cover-to-cover translations of the major Soviet journals, that researchers in the USA and elsewhere became familiar with Soviet work. But phase plane approaches had already been adopted by Western control engineers.

One of the first was *Leroy MacColl* in his early textbook [4.46].

The cold war requirements of control engineering centered on the control of ballistic objects for aerospace applications. Detailed and accurate mathematical models, both linear and nonlinear, could be obtained, and the classical techniques of frequency response and root locus – essentially approximations – were increasingly replaced by methods designed to optimize some measure of performance such as minimizing trajectory time or fuel consumption. Higher-order models were expressed as a set of first order equations in terms of the state variables. The state variables allowed for a more sophisticated representation of dynamic behaviour than the classical single-input single-output system modelled by a differential equation, and were suitable for multivariable problems. In general, we have in matrix form

$$x = \mathbf{A}x + \mathbf{B}u \,,$$
$$y = \mathbf{C}x \,,$$

where x are the state variables, u the inputs and y the outputs.

Automatic control developments in the late 1940s and 1950s were greatly assisted by changes in the engineering professional bodies and a series of international conferences [4.47]. In the USA both the American Society of Mechanical Engineers and the American Institute of Electrical Engineers made various changes to their structure to reflect the growing importance of servomechanisms and feedback control. In the UK similar changes took place in the British professional bodies, most notably the Institution of Electrical Engineers, but also the Institute of Measurement and Control and the mechanical and chemical engineering bodies. The first

conferences on the subject appeared in the late 1940s in London and New York, but the first truly international conference was held in Cranfield, UK in 1951. This was followed by a number of others, the most influential of which was the Heidelberg event of September 1956, organized by the joint control committee of the two major German engineering bodies, the VDE and VDI. The establishment of the International Federation of Automatic Control followed in 1957 with its first conference in Moscow in 1960 [4.48]. The Moscow conference was perhaps most remarkable for Kalman's paper *On the general theory of control systems* which identified the duality between multivariable feedback control and multivariable feedback filtering and which was seminal for the development of optimal control.

The late 1950s and early 1960s saw the publication of a number of other important works on dynamic programming and optimal control, of which can be singled out those by *Bellman* [4.49], *Kalman* [4.50–52] and *Pontryagin* and colleagues [4.53]. A more thorough discussion of control theory is provided in Chaps. 9, 11 and 10.

4.8 The Digital Computer

The introduction of digital technologies in the late 1950s brought enormous changes to automatic control. Control engineering had long been associated with computing devices – as noted above, a driving force for the development of servos was for applications in analogue computing. But the great change with the introduction of digital computers was that ultimately the approximate methods of frequency response or root locus design, developed explicitly to avoid computation, could be replaced by techniques in which accurate computation played a vital role.

There is some debate about the first application of digital computers to process control, but certainly the introduction of computer control at the Texaco Port Arthur (Texas) refinery in 1959 and the Monsanto ammonia plant at Luling (Louisiana) the following year are two of the earliest [4.54]. The earliest systems were supervisory systems, in which individual loops were controlled by conventional electrical, pneumatic or hydraulic controllers, but monitored and optimized by computer. Specialized process control computers followed in the second half of the 1960s, offering direct digital control (DDC) as well as supervisory control. In DDC the computer itself implements a discrete form of a control algorithm such as three-term control or other

procedure. Such systems were expensive, however, and also suffered many problems with programming, and were soon superseded by the much cheaper minicomputers of the early 1970s, most notably the Digital Equipment Corporation PDP series. But, as in so many other areas, it was the microprocessor that had the greatest effect. Microprocessor-based digital controllers were soon developed that were compact, reliable, included a wide selection of control algorithms, had good communications with supervisory computers, and comparatively easy to use programming and diagnostic tools via an effective operator interface. Microprocessors could also easily be built into specific pieces of equipment, such as robot arms, to provide dedicated position control, for example.

A development often neglected in the history of automatic control is the programmable logic controller (PLC). PLCs were developed to replace individual relays used for sequential (and combinational) logic control in various industrial sectors. Early plugboard devices appeared in the mid 1960s, but the first PLC proper was probably the Modicon, developed for General Motors to replace electromechanical relays in automotive component production. Modern PLCs offer a wide range of control options, including conventional

Fig. 4.9 The Modicon 084 PLC (after [4.55])

adaptive control the algorithm is modified according to circumstances. Adaptive control has a long history: so called gain scheduling, for example, when the gain of a controller is varied according to some measured parameter, was used well before the digital computer. (The classic example is in flight control, where the altitude affects aircraft dynamics, and needs therefore to be taken into account when setting gain.) Digital adaptive control, however, offers much greater possibilities for:

1. Identification of relevant system parameters
2. Making decisions about the required modifications to the control algorithm
3. Implementing the changes.

Optimal and robust techniques too, were developed, the most celebrated perhaps being the linear-quadratic-Gaussian (LQG) and H_∞ approaches from the 1960s onwards. Without digital computers these techniques, that attempt to optimize system rejection of disturbances (according to some measure of behaviour) while at the same time being resistant to errors in the model, would simply be mathematical curiosities [4.57].

A very different approach to control rendered possible by modern computers is to move away from purely mathematic models of system behaviour and controller algorithms. In fuzzy control, for example, control action is based on a set of rules expressed in terms of *fuzzy* variables. For example

> *IF the speed is "high"*
> *AND the distance to final stop is "short"*
> *THEN apply brakes "firmly".*

The fuzzy variables *high*, *short* and *firmly* can be translated by means of an appropriate computer program into effective control for, in this case, a train. Related techniques include *learning control* and *knowledge-based control*. In the former, the control system can *learn* about its environment using artificial intelligence techniques (AI) and modify its behaviour accordingly. In the latter, a range of AI techniques are applied to reasoning about the situation so as to provide appropriate control action.

closed loop control algorithms such as PID as well as the logic functions. In spite of the rise of the ruggedized PCs in many industrial applications, PLCs are still widely used owing to their reliability and familiarity (Fig. 4.9).

Digital computers also made it possible to implement the more advanced control techniques that were being developed in the 1960s and 1970s [4.56]. In

4.9 The Socio-Technological Context Since 1945

This short survey of the history of automatic control has concentrated on technological and, to some extent, institutional developments. A full social history of automatic control has yet to be written, although there are detailed studies of certain aspects. Here I shall merely indicate some major trends since WWII.

The wartime developments, both in engineering and in areas such as operations research, pointed the way towards the design and management af large-scale, complex, projects. Some of those involved in the wartime research were already thinking on a much larger scale. As early as 1949, in some rather prescient remarks at an ASME meeting in the fall of that year, *Brown* and *Campbell* said [4.58–60]:

> *We have in mind more a philosophic evaluation of systems which might lead to the improvement of product quality, to better coordination of plant operation, to a clarification of the economics related to new plant design, and to the safe operation of plants in our composite social-industrial community. [...] The conservation of raw materials used in a process often prompts reconsideration of control. The expenditure of power or energy in product manufacture is another important factor related to control. The protection of health of the population adjacent to large industrial areas against atmospheric poisoning and water-stream pollution is a sufficiently serious problem to keep us constantly alert for advances in the study and technique of automatic control, not only because of the human aspect, but because of the economy aspect.*

Many saw the new technologies, and the prospects of automation, as bringing great benefits to society; others were more negative. *Wiener*, for example, wrote [4.61]:

> *The modern industrial revolution is [...] bound to devalue the human brain at least in its simpler and more routine decisions. Of course, just as the skilled carpenter, the skilled mechanic, the skilled dressmaker have in some degree survived the first*

industrial revolution, so the skilled scientist and the skilled administrator may survive the second. However, taking the second revolution as accomplished, the average human of mediocre attainments or less has nothing to sell that it is worth anyone's money to buy.

It is remarkable how many of the wartime engineers involved in control systems development went on to look at social, economic or biological systems. In addition to Wiener's work on cybernetics, Arnold Tustin wrote a book on the application to economics of control ideas, and both Winfried Oppelt and Karl Küpfmüller investigated biological systems in the postwar period.

One of the more controversial applications of control and automation was the introduction of the computer numerical control (CNC) of machine tools from the late 1950s onwards. Arguments about increased productivity were contested by those who feared widespread unemployment. We still debate such issues today, and will continue to do so. *Noble*, in his critique of automation, particularly CNC, remarks [4.62]:

> *[...] when technological development is seen as politics, as it should be, then the very notion of progress becomes ambiguous: What kind of progress? Progress for whom? Progress for what? And the awareness of this ambiguity, this indeterminacy, reduces the powerful hold that technology has had upon our consciousness and imagination [...] Such awareness awakens us not only to the full range of technological possibilities and political potential but also to a broader and older notion of progress, in which a struggle for human fulfillment and social equality replaces a simple faith in technological deliverance....*

4.10 Conclusion and Emerging Trends

Technology is part of human activity, and cannot be divorced from politics, economics and society. There is no doubt that automatic control, at the core of automation, has brought enormous benefits, enabling modern production techniques, power and water supply, environmental control, information and communication technologies, and so on. At the same time automatic control has called into question the way we organize our societies, and how we run modern technological enterprises. Automated processes require much less human intervention, and there have been periods in the recent past when automation has been problematic in those parts of industrialized society that have traditionally relied on a large workforce for carrying out tasks that were subsequently automated. It seems unlikely that these socio-technological questions will be settled as we move towards the next generation of automatic control systems, such as the transformation of work through the use of information and communication technology ICT and the application of control ideas to this emerging field [4.63].

Future developments in automatic control are likely to exploit ever more sophisticated mathematical models for those applications amenable to exact technological modeling, plus a greater emphasis on human–machine systems, and further development of human behaviour modeling, including decision support and cognitive engineering systems [4.64]. As safety aspects of large-scale automated systems become ever more important, large scale integration, and novel ways of communicating between humans and machines, are likely to take on even greater significance.

4.11 Further Reading

- R. Bellman (Ed.): *Selected Papers on Mathematical Trends in Control Engineering* (Dover, New York 1964)
- C.C. Bissell: http://ict.open.ac.uk/classics (electronic resource)
- M.S. Fagen (Ed.): *A History of Engineering and Science in the Bell System: The Early Years (1875–1925)* (Bell Telephone Laboratories, Murray Hill 1975)
- M.S. Fagen (Ed.): *A History of Engineering and Science in the Bell System: National Service in War and Peace (1925–1975)* (Bell Telephone Laboratories, Murray Hill 1979)
- A.T. Fuller: *Stability of Motion*, ed. by E.J. Routh, reprinted with additional material (Taylor Francis, London 1975)
- A.T. Fuller: The early development of control theory, Trans. ASME J. Dyn. Syst. Meas. Control **98**, 109–118 (1976)
- A.T. Fuller: Lyapunov centenary issue, Int. J. Control **55**, 521–527 (1992)
- L.E. Harris: *The Two Netherlanders, Humphrey Bradley and Cornelis Drebbel* (Cambridge Univ. Press, Cambridge 1961)
- B. Marsden: *Watt's Perfect Engine* (Columbia Univ. Press, New York 2002)

- O. Mayr: *Authority, Liberty and Automatic Machinery in Early Modern Europe* (Johns Hopkins Univ. Press, Baltimore 1986)
- W. Oppelt: A historical review of autopilot development, research and theory in Germany, Trans ASME J. Dyn. Syst. Meas. Control **98**, 213–223 (1976)
- W. Oppelt: On the early growth of conceptual thinking in control theory – the German role up to 1945, IEEE Control Syst. Mag. **4**, 16–22 (1984)
- B. Porter: *Stability Criteria for Linear Dynamical Systems* (Oliver Boyd, Edinburgh, London 1967)
- P. Remaud: *Histoire de l'automatique en France 1850–1950* (Hermes Lavoisier, Paris 2007), in French
- K. Rörentrop: *Entwicklung der modernen Regelungstechnik* (Oldenbourg, Munich 1971), in German
- Scientific American: *Automatic Control* (Simon Shuster, New York 1955)
- J.S. Small: *The Analogue Alternative* (Routledge, London, New York 2001)
- G.J. Thaler (Ed.): *Automatic Control: Classical Linear Theory* (Dowden, Stroudsburg 1974)

References

4.1 O. Mayr: *The Origins of Feedback Control* (MIT, Cambridge 1970)

4.2 F.W. Gibbs: The furnaces and thermometers of Cornelius Drebbel, Ann. Sci. **6**, 32–43 (1948)

4.3 T. Mead: Regulators for wind and other mills, British Patent (Old Series) 1628 (1787)

4.4 H.W. Dickinson, R. Jenkins: *James Watt and the Steam Engine* (Clarendon Press, Oxford 1927)

4.5 C.C. Bissell: Stodola, Hurwitz and the genesis of the stability criterion, Int. J. Control **50**(6), 2313–2332 (1989)

4.6 S. Bennett: *A History of Control Engineering 1800–1930* (Peregrinus, Stevenage 1979)

4.7 G.B. Airy: On the regulator of the clock-work for effecting uniform movement of equatorials, Mem. R. Astron. Soc. **11**, 249–267 (1840)

4.8 J.C. Maxwell: On governors, Proc. R. Soc. **16**, 270–283 (1867)

4.9 E.J. Routh: *A Treatise on the Stability of a Given State of Motion* (Macmillan, London, 1877)

4.10 A. Hurwitz: Über die Bedingungen, unter welchen eine Gleichung nur Wurzeln mit negativen reellen

Teilen besitzt, Math. Ann. **46**, 273–280 (1895), in German

4.11   E. Bompiani: Sulle condizione sotto le quali un equazione a coefficienti reale ammette solo radici con parte reale negative, G. Mat. **49**, 33–39 (1911), in Italian

4.12   C.C. Bissell: The classics revisited – Part I, Meas. Control **32**, 139–144 (1999)

4.13   C.C. Bissell: The classics revisited – Part II, Meas. Control **32**, 169–173 (1999)

4.14   M. Tolle: *Die Regelung der Kraftmaschinen*, 3rd edn. (Springer, Berlin 1922), in German

4.15   O. Mayr: *Feedback Mechanisms* (Smithsonian Institution Press, Washington 1971)

4.16   T.P. Hughes: *Elmer Sperry: Inventor and Engineer* (Johns Hopkins Univ. Press, Baltimore 1971)

4.17   S. Bennett: *A History of Control Engineering 1800–1930* (Peregrinus, Stevenage 1979) p. 137

4.18   S. Bennett: *A History of Control Engineering 1930–1955* (Peregrinus, Stevenage 1993)

4.19   N. Minorsky: Directional stability of automatically steered bodies, Trans. Inst. Nav. Archit. **87**, 123–159 (1922)

4.20   O. Heaviside: *Electrical Papers* (Chelsea, New York 1970), reprint of the 2nd edn.

4.21   S. Bennett: *A History of Control Engineering 1800–1930* (Peregrinus, Stevenage 1979), Chap. 6

4.22   C.C. Bissell: Karl Küpfmüller: a German contributor to the early development of linear systems theory, Int. J. Control **44**, 977–89 (1986)

4.23   H. Nyquist: Regeneration theory, Bell Syst. Tech. J. **11**, 126–47 (1932)

4.24   H.S. Black: Stabilized feedback amplifiers, Bell Syst. Tech. J. **13**, 1–18 (1934)

4.25   H.W. Bode: Relations between amplitude and phase in feedback amplifier design, Bell Syst. Tech. J. **19**, 421–54 (1940)

4.26   H.W. Bode: *Network Analysis and Feedback Amplifier Design* (Van Nostrand, Princeton 1945)

4.27   H.L. Hazen: Theory of servomechanisms, J. Frankl. Inst. **218**, 283–331 (1934)

4.28   A. Leonhard: *Die Selbsttätige Regelung in der Elektrotechnik* (Springer, Berlin 1940), in German

4.29   C.C. Bissell: The *First All-Union Conference on Automatic Control*, Moscow, 1940, IEEE Control Syst. Mag. **22**, 15–21 (2002)

4.30   C.C. Bissell: A.A. Andronov and the development of Soviet control engineering, IEEE Control Syst. Mag. **18**, 56–62 (1998)

4.31   D. Mindell: *Between Human and Machine* (Johns Hopkins Univ. Press, Baltimore 2002)

4.32   C.C. Bissell: Textbooks and subtexts, IEEE Control Syst. Mag. **16**, 71–78 (1996)

4.33   H. Schmidt: Regelungstechnik – die technische Aufgabe und ihre wissenschaftliche, sozialpolitische und kulturpolitische Auswirkung, Z. VDI **4**, 81–88 (1941), in German

4.34   C.C. Bissell: Control Engineering in the former USSR: some ideological aspects of the early years, IEEE Control Syst. Mag. **19**, 111–117 (1999)

4.35   A.D. Dalmedico: Early developments of nonlinear science in Soviet Russia: the Andronov school at Gorky, Sci. Context **1/2**, 235–265 (2004)

4.36   A.C. Hall: Application of circuit theory to the design of servomechanisms, J. Frankl. Inst. **242**, 279–307 (1946)

4.37   A.C. Hall: *The Analysis and Synthesis of Linear Servomechanisms (Restricted Circulation)* (The Technology Press, Cambridge 1943)

4.38   S. Bennett: *A History of Control Engineering 1930–1955* (Peregrinus, Stevenage 1993) p. 142

4.39   H.J. James, N.B. Nichols, R.S. Phillips: *Theory of Servomechanisms*, Radiation Laboratory, Vol. 25 (McGraw-Hill, New York 1947)

4.40   C.C. Bissell: Pioneers of control: an interview with Arnold Tustin, IEE Rev. **38**, 223–226 (1992)

4.41   A.L. Whiteley: Theory of servo systems with particular reference to stabilization, J. Inst. Electr. Eng. **93**, 353–372 (1946)

4.42   C.C. Bissell: Six decades in control: an interview with Winfried Oppelt, IEE Rev. **38**, 17–21 (1992)

4.43   C.C. Bissell: An interview with Hans Sartorius, IEEE Control Syst. Mag. **27**, 110–112 (2007)

4.44   W.R. Evans: Control system synthesis by root locus method, Trans. AIEE **69**, 1–4 (1950)

4.45   A.A. Andronov, S.E. Khaikin: *Theory of Oscillators* (Princeton Univ. Press, Princeton 1949), translated and adapted by S. Lefschetz from Russian 1937 publication

4.46   L.A. MacColl: *Fundamental Theory of Servomechanisms* (Van Nostrand, Princeton 1945)

4.47   S. Bennett: The emergence of a discipline: automatic control 1940–1960, Automatica **12**, 113–121 (1976)

4.48   E.A. Feigenbaum: Soviet cybernetics and computer sciences, 1960, Commun. ACM **4**(12), 566–579 (1961)

4.49   R. Bellman: *Dynamic Programming* (Princeton Univ. Press, Princeton 1957)

4.50   R.E. Kalman: Contributions to the theory of optimal control, Bol. Soc. Mat. Mex. **5**, 102–119 (1960)

4.51   R.E. Kalman: A new approach to linear filtering and prediction problems, Trans. ASME J. Basic Eng. **82**, 34–45 (1960)

4.52   R.E. Kalman, R.S. Bucy: New results in linear filtering and prediction theory, Trans. ASME J. Basic Eng. **83**, 95–108 (1961)

4.53   L.S. Pontryagin, V.G. Boltyansky, R.V. Gamkrelidze, E.F. Mishchenko: *The Mathematical Theory of Optimal Processes* (Wiley, New York 1962)

4.54   T.J. Williams: Computer control technology – past, present, and probable future, Trans. Inst. Meas. Control **5**, 7–19 (1983)

4.55 C.A. Davis: *Industrial Electronics: Design and Application* (Merrill, Columbus 1973) p. 458

4.56 T. Williams, S.Y. Nof: Control models. In: *Handbook of Industrial Engineering*, 2nd edn., ed. by G. Salvendy (Wiley, New York 1992) pp. 211–238

4.57 J.C. Willems: In control, almost from the beginning until the day after tomorrow, Eur. J. Control **13**, 71–81 (2007)

4.58 G.S. Brown, D.P. Campbell: Instrument engineering: its growth and promise in process-control problems, Mech. Eng. **72**, 124–127 (1950)

4.59 G.S. Brown, D.P. Campbell: Instrument engineering: its growth and promise in process-control problems, Mech. Eng. **72**, 136 (1950)

4.60 G.S. Brown, D.P. Campbell: Instrument engineering: its growth and promise in process-control problems, Mech. Eng. **72**, 587–589 (1950), discussion

4.61 N. Wiener: *Cybernetics: Or Control and Communication in the Animal and the Machine* (Wiley, New York 1948)

4.62 D.F. Noble: *Forces of Production. A Social History of Industrial Automation* (Knopf, New York 1984)

4.63 S.Y. Nof: Collaborative control theory for e-Work, e-Production and e-Service, Annu. Rev. Control **31**, 281–292 (2007)

4.64 G. Johannesen: From control to cognition: historical views on human engineering, Stud. Inf. Control **16**(4), 379–392 (2007)

5. Social, Organizational, and Individual Impacts of Automation

Tibor Vámos

Society and information from the evolutionary early beginnings. The revolutionary novelties of our age: the possibility for the end of a human being in the role of draught animal and the symbolic representation of the individual and of his/her property by electronic means, free of distance and time constraints. As a consequence, changing human roles in production, services, organizations and innovation; changing society stratifications, human values, requirements in skills, individual conscience. New relations: centralization and decentralization, less hierarchies, discipline and autonomy, new employment relations, less job security, more free lance, working home, structural unemployment, losers and winners, according to age, gender, skills, social background. Education and training, levels, life long learning, changing methods of education. Role of memory and associative abilities. Changes reflected in linguistic relations, multilingual global society, developments and decays of regional and social vernaculars. The social-political arena, human rights, social philosophies, problems and perspectives of democracy. The global agora and global media rule. More equal or more divided society. Some typical society patterns: US, Europe, Far East, India, Latin America, Africa.

Part A | 5

Automation and closely related information systems are, naturally, innate and integrated ingredients of all kinds of objects, systems, and social relations of the present reality. This is the reason why this chapter treats the phenomena and problems of this automated/information society in a historical and structural framework much broader than any chapter on technology details. This process transforms traditional human work, offers freedom from burdensome physical and mental constraints and freedom for individuals and previously subjugated social layers, and creates new gaps and tensions. After detailed documentation of these phenomena, problems of education and culture are treated as main vehicles to adaptation. Legal aspects related to privacy, security, copyright, and patents are referred to, and then social philosophy, impacts of globalization, and new characteristics of information technology–society relations provide a conclusion of future prospects.

5.1 Scope of Discussion: Long and Short Range of Man–Machine Systems

Regarding the social effects of automation, a review of concepts is needed in the context of the purpose of this chapter. Automation, in general, and especially in our times, means all kinds of activity perfected by machines and not by the intervention of direct human control. This definition involves the use of some energy resources operating without human or livestock physical effort and with some kind of information system communicating the purpose of automated activity desired by humans and the automatic execution of the activity, i. e., its control.

This definition entails a widely extended view of automation, its relation to information, knowledge control systems, as well as the knowledge and practice of the related human factor. The human factor involves, practically, all areas of science and practice with respect to human beings: education, health, physical and mental abilities, instruments and virtues of cooperation (i. e., language and sociability), environmental conditions, short- and long-range ways of thinking, ethics, legal systems, various aspects of private life, and entertainment.

One of the major theses of this definitional and relational philosophy is the man–machine paradox: the human role in all kinds of automation is a continuously emerging constituent of man–machine symbiosis, with feedback to the same.

From this perspective, the inclusion of a discussion of early historical developments is not surprising. This evolution is of twin importance. First, due to the contradictory speeds of human and machine evolution, despite the fascinating results of machine technology, in most

About 375 BC	First automaton (Archytas dove, Syracuse)	1662	Elements of thermodynamics, Boyle
About 275	Archimedes	1663–68	Reflecting telescope, Gregory, Newton
About 60	Steampower, Heron	1665	Infinitesimal calculus, Newton, Leibniz
	Hodometer, Vitruvius	1666	Gravitation, Newton
About 1280 AD	Mechanical clocks	1673	Calculator, Leibniz
		1676	Universal joint, Hooke
1285–90	Windmills	1679	Pressure cooker, Papin
1328	First sawmill	1690	Light-wave theory, Huygens
1421	Hoisting gear	1689	Knitting machine, Lee
1475	Printing press	1698	Steam pump, Savery
1485–1519	Leonardo's technical designs	1712	Steam engine, Newcomen
1486	Copyright (Venice)	1718	Mercury thermometer, Fahrenheit
1500	Flush toilet	1722	Fire extinguisher, Hopffer
1510	Pocket watch, Henlein	1745	Leyden jar, capacitor, Kleist
1590	Compound microscope, Janssen	1758	Chromatic lens, Dolland
1593	Water thermometer (Galileo)	1764	Spinning jenny, Hargreaves
1608	Refracting telescope	1769	Steam engine, controlled by a centrifugal governor, Watt
1609	Kinematics of Galileo; Planetary motion, Kepler	1770	Talking machine and robot mechanisms, Kempelen
1614	Logarithms, Napier	1774	Electrical telegraph, Lesage
1606–28	Blood circulation, Harvey	1775	Flush toilet, Cummings
1620	Human powered submarine	1780	Bi-focal eyeglass, Franklin
1624	Slide rule, Oughtred	1784	Threshing machine, Meikle
1625	Blood transfusion, Denys	1785	Power loom, Cartwright; Torsion balance, Coulomb
1629	Steam turbine, Branca		
1636	Micrometer, Gascoigne	1790	Contact electricity, Galvani; Harmonic analysis, Fourier
1637	Analytic geometry, Descartes		
1642	Adding machine, Pascal	1792	Gas lighting, Murdoch
1643	Mercury barometer, Torricelli	1794	Ball bearings, Vaughan
1654	Probability, Fermat, Pascal, Huygens, J. Bernoulli	1799	Battery, Volta; Ohm's law, Volta, Cavendish, Ohm
1650	Air pump, Guericke		
1657	Pendulum clock, Huygens	1800	Loom, Jacquard

Fig. 5.1 Timeline of science and technology in Western civilization

applications the very slow progress of the user is critical. Incredibly sophisticated instruments are, and will be, used by creatures not too different from their ancestors of 1000 years ago. This contradiction appears not only in the significant abuse of automated equipment against other people but also in human user reasoning, which is sometimes hysterical and confused in its thinking.

The second significance of this paradox is in our perception of various changes: its reflection in problem solving, and in understanding the relevance of continuity and change, which is sometimes exaggerated, while other times it remains unrecognized in terms of its further effects.

These are the reasons why this chapter treats automation in a context that is wider and somehow deeper than usual, which appears to be necessary in certain application problems. The references relate to the twin peaks of the Industrial Revolution: the first, the classic period starting in the late 18th century with machine power; and the second, after World War II (WWII), in the information revolution.

Practically all measures of social change due to automation are hidden in the general indices of technology effects and, within those, in the effects of automation- and communication-related technology. Human response is, generally, a steady process, apart from dramatic events such as wars, social revolutions, and crises in the economy. Some special acceleration phenomena can be observed in periods of inflation and price changes, especially in terms of decreases in product prices for high technology, and changes in the composition of general price indices and in the spread of high-technology commodities; television (TV), color TV, mobile telephony, and Internet access are typical examples. These spearhead technologies rad-

1807	Steam ship, Fulton; Electric arc lamp, Davy	1868	The first paper on control theory, Maxwell; Air brakes, Westinghouse; Traffic light, Knight
1814	Spectroskopy, Frauenhofer; Photography, Niépce	1869	Periodic system, Mendeleev
1815	Miners lamp, Davy	1873	Theory of electromagnetism, Maxwell
1819	Stethoscope, Laënnec	1876	4-Cycle gas engine, Otto
1820	Electromagnetism, Oersted	1877	Phonograph, Edison
1825	Electromagnet, Sturgeon	1878	Lightbulb, Swan
1827	Microphone, Wheatstone	1879	Electrical locomotive, Siemens; Concept notation, Frege
1829	Locomotive, Stephenson; Typewriter, Burt	1880	Toilet paper; Seismograph, Milne
1830	Sewing machine, Thimmonier	1881	Metal detector, Bell; Roll film for cameras, Houston
1831	Electrical induction, Faraday	1884	Paper strip photo film, Eastman; Rayon, Chardonnay; Fountain pen, Waterman; Steam turbine, Parsons; Cash register, Ritty
1836	Analytical engine, Babbage		
1837	Telegraph, Morse		
1839	Rubber vulcanization, Goodyear; Photography, Daguerre; Bicycle, Niépce, MacMillan; Hydrogen fuel cell, Grove		
1841	Stapler, Slocum	1885	Automobile, Benz
1842	Programming Lady Lovelace, Ada Byron; Grain elevator, Dart	1886	Motorcycle, Daimler
		1887	Radar, Hertz; Gramaphone, Berliner; Contact lens, Flick
1843	Facsimile, Bain		
1845–51	Sewing machine, Howe, Singer; Vulcanized pneu, Thomson	1888	AC motor, Transformer, Tesla; Pneu, Dunlop
1850–86	Dishwasher, Houghton, Cochran	1891	Escalator, Reno
1852	Gyroscope, Foucault	1892	Diesel motor, Diesel
1854	Fiber optics, Tyndall	1893	Zipper, Judson
1856	Pasteurization, Pasteur	1894	Motion picture, Lumiere
1857	Sleeping car, Pullman	1895	X-ray, Röntgen
1861	Telephone, Bell; Safe elevator, Otis	1897–1916	Wireless, Marconi
1867	Practical typewriter, Scholes	1898	Radium, Curie

Fig. 5.1 (cont.)

1899–1901	Vacuum cleaner, Thurman, Booth	1928	Foundations of game theory, Neumann;
1901	Safety razor, Gilette;		Penicillin, Fleming;
	Radio receiver		Electric shaver, Schick
1902	Air conditioner, Carrier;	1930	Analog computer, Bush;
	Neon light, Claude		Jet engine, Whittle, von Ohain
1903	Airplane, Wright;	1931	Electron microscope, Knott, Ruska;
	Radioactivity, Rutherford		Undecidability theory, Gödel
1904	Teabag, Sullivan;	1932	Neutrons, positrons, Chadwick;
	Vacuum tube, Fleming		Polaroid photo, Land;
1905	Special relativity, Einstein		Zoom lens;
1906	Amplifier, audion, De Forest		Light meter;
1907	Bakelite, Baekeland;		Radio telescope, Jansky
	Color photography, Lumiere	1933	Frequency modulation, Armstrong;
1908	Model T, Ford;		Stereo recording
	Geiger counter, Geiger, Müller;	1934	Magnetic recording, Begun
	Artificial nitrates, Haber;	1935	Nylon, DuPont Labs;
	Gyrocompass, Sperry		Radar, Watson-Watt
1911	Engine electrical ignition, Kettering;	1936/37	Theoretical foundations of computer
	Helicopter, Cornu		science, Turing
1913	Atom model, Bohr	1938	Nuclear fission, Hahn, Straßmann;
1915	General relativity, Einstein		Foundations of information Theory, Shannon;
1916	Radio tuner		Ballpoint pen, Biro;
1918	Superheterodyne radio, Armstrong		Teflon, Plunkett;
1919	Short-wave radio;		First working turboprop;
	Flip-flop circuit;		Xerography, Carlson;
	Arc welder		Nescafe
1920	Robot concept, Capek	1939	First operational helicopter, Sikorsky;
1922	Insulin, Banting;		Electron microscope
	3-D movie	1940–49	Wiener filter, cybernetics, Wiener
1923–29	TV, Zworikin	1941	Computer, Zuse
1924	Dynamic loudspeaker, Rice, Kellog	1942	Computer, Atanasoff and Berry;
1925	Quantum mechanics, Heisenberg		Turboprop;
1927	Quartz clock;		Nuclear reactor, Fermi
	Technicolor	1944	Kidney dialysis, Kolff

Fig. 5.1 (cont.)

ically change lifestyles and social values but impact less slowly on the development of human motivations and, as a consequence, on several characteristics of in-dividual and social behavior. The differing speeds of advancement of technology and society will be reflected on later.

5.2 Short History

The history of automation is a lesson in the bilateral conditions of technology and society. In our times wider and deeper attention is focused on the impact of au-tomation on social relations. However, the progress of automation is arguably rather the result of social condi-tions.

It is generally known that automation was also present in antiquity; ingenious mechanisms operated impressive idols of deities: their gestures, winks, and opening the doors of their sanctuaries. Water-driven clocks applied the feedback principle for the correction of water-level effects. Sophisticated gearing, pumping, and elevating mechanisms helped the development of both human- and water-driven devices for construction, irrigation, traf-fic, and warfare. Water power was ubiquitous, wind power less so, and the invention of steam power more than 2000 years ago was not used for the obvious purpose of replacing human power and brute strength.

Historians contemplate the reasons why these given elements were not put together to create a more modern world based on the replacement of human and animal power. The hypothesis of the French Annales School of historians (named after their periodical, opened in 1929, and characterized by a new emphasis on geographical, economic, and social motifs of history, and less on events related to personal and empirical data) looks for social conditions: manpower acquired by slavery, especially following military operations, was economically the optimal energy resource. Even brute strength was, for a long time, more expensive, and for this reason was used for luxury and warfare more than for any other end, including agriculture. The application of the more efficient yoke for animal traction came into being only in the Middle Ages. Much later, arguments spoke for the better effect of human digging compared with an animal-driven plough [5.1–3].

Fire for heating and for other purposes was fed with wood, the universal material from which most objects were made, and that was used in industry for metal-producing furnaces. Coal was known but not generally used until the development of transport facilitated the joining of easily accessible coal mines with both industry centers and geographic points of high consumption. This high consumption and the accumulated wealth through commerce and population concentration were born in cities based on trade and manufacturing, creating a need for mass production of textiles. Hence, the first industrial application field of automation flourished with the invention of weaving machines and their punch-card control.

In the meantime, Middle Age and especially Renaissance mechanisms reached a level of sophistication

surpassed, basically, only in the past century. This social and secondary technological environment created the overall conditions for the Industrial Revolution in power resources (Fig. 5.1) [5.3–8]. This timeline is composed from several sources of data available on the Internet and in textbooks on the history of science and technology. It deliberately contains many disparate items to show the historical density foci, connections with everyday life comfort, and basic mathematical and physical sciences. Issues related to automation per se are sparse, due to the high level of embeddedness of the subject in the general context of progress. Some data are inconsistent. This is due to uncertainties in historical documents; data on first publications, patents, and first applications; and first acceptable and practically feasible demonstrations. However, the figure intends to give an overall picture of the scene and these uncertainties do not confuse the lessons it provides.

The timeline reflects the course of Western civilization. The great achievements of other, especially Chinese, Indian, and Persian, civilizations had to be omitted, since these require another deep analysis in terms of their fundamental impact on the origins of Western science and the reasons for their interruption. Current automation and information technology is the direct offspring of the Western timeline, which may serve as an apology for these omissions.

The whole process, until present times, has been closely connected with the increasing costs of manpower, competence, and education. Human requirements, welfare, technology, automation, general human values, and social conditions form an unbroken circle of multiloop feedback.

5.3 Channels of Human Impact

Automation and its related control technology have emerged as a partly hidden, natural ingredient of everyday life. This is the reason why it is very difficult to separate the progress of the technology concerned from general trends and usage. In the household of an average family, several hundred built-in processors are active but remain unobserved by the user. They are not easily distinguishable and countable, due to the rapid spread of multicore chips, multiprocessor controls, and communication equipment. The relevance of all of these developments is really expressed by their vegetative-like operation, similar to

the breathing function or blood circulation in the body.

An estimate of the effects in question can be given based on the automotive and aerospace industry. Recent medium-category cars contain about 50 electronic control units, high-class cars more than 70. Modern aircrafts are nearly fully automated; about 70% of all their functions are related to automatic operations and in several aerospace equipment even more. The limit is related to humans rather than to technology. Traffic control systems accounts for 30–35% of investment but provide a proportionally much larger return in terms of safety. These data change rapidly because proliferation decreases prices dramatically, as experienced in the cases of watches, mobile phones, and many other gadgets. On the one hand, the sophistication of the systems and by increasing the prices due to more comfort and luxury, on the other.

The silent intrusion of control and undetectable information technology into science and related transforming devices, systems, and methods of life can be observed in the past few decades in the great discoveries in biology and material science. The examples of three-dimensional (3-D) transparency technologies, ultrafast microanalysis, and nanotechnology observation into the nanometer, atomic world and picosecond temporal processes are partly listed on the timeline.

These achievements of the past half-century have changed all aspects of human-related sciences, e.g., psychology, linguistics, and social studies, but above all life expectancy, life values, and social conditions.

5.4 Change in Human Values

The most important, and all-determinant, effect of mechanization–automatization processes is the change of human roles [5.10]. This change influences social stratification and human qualities. The key problem is realizing freedom from hard, wearisome work, first as exhaustive physical effort and later as boring, dull activity. The first historical division of work created a class of clerical and administrative people in antiquity, a comparatively small and only relatively free group of people who were given spare energy for thinking.

The real revolutions in terms of mental freedom run parallel with the periods of the Industrial Revolution, and subsequently, the information–automation society. The latter is far from being complete, even in the most advanced parts of the world. This is the reason why no authentic predictions can be found regarding the possible consequences in terms of human history.

Slavery started to be banned by the time of the first Industrial Revolution, in England in 1772 [5.11, 12], in France in 1794, in the British Empire in 1834, and in the USA in 1865, serfdom in Russia in 1861, and worldwide abolition by consecutive resolutions in 1948, 1956, and 1965, mostly in the order of the development of mechanization in each country.

The same trend can be observed in prohibiting childhood work and ensuring equal rights for women. The minimum age for children to be allowed to work in various working conditions was first agreed on by a 1921 ILO (International Labour Organization) convention and was gradually refined until 1999, with increasingly restrictive, humanistic definitions. Childhood work under the age of 14 or 15 years and less rigorously under 16–18 years, was, practically, abolished in Europe, except for some regions in the Balkans.

In the USA, Massachusetts was the first state to regulate child labor; federal law came into place only in 1938 with the Federal Labor Standards Act, which has been modified many times since. The eradication of child labor slavery is a consequence of a radical change in human values and in the easy replacement of slave work by more efficient and reliable automation. The general need for higher education changed the status of children both in the family and society. This reason together with those mentioned above decreased the number of children dying in advanced countries; human life becomes much more precious after the defeat of infant mortality and the high costs of the required education period. The elevation of human values is a strong argument against all kinds of nostalgia back to the times before our automation–machinery world.

Also, legal regulations protecting women in work started in the 19th century with maternity- and health-

Table 5.1 Women in public life (due to elections and other changes of position the data are informative only) after [5.9]

Country	Members of national parliament 2005 (%)	Government ministers 2005/2006
Finland	38	8
France	12	6
Germany	30	6
Greece	14	3
Italy	12	2
Poland	20	1
Slovakia	17	0
Spain	36	8
Sweden	45	8

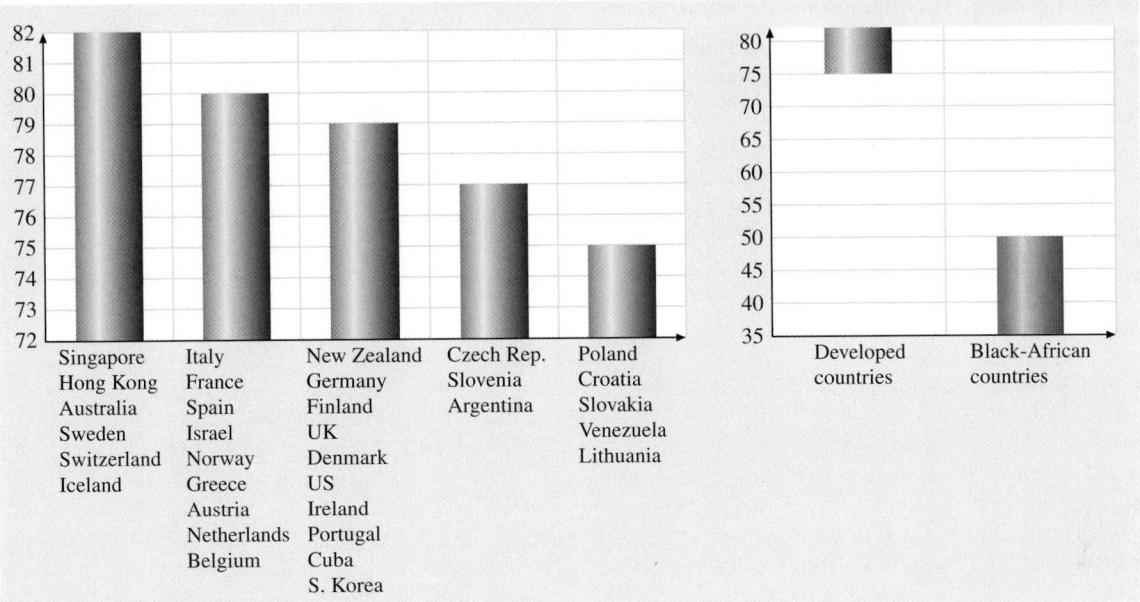

Fig. 5.2 Life expectancy and life conditions (after [5.14])

related laws and conventions. The progress of equal rights followed WWI and WWII, due to the need for female workforce during the wars and the advancement of technology replacing hard physical work. The correlation between gender equality and economic and society–cultural relations is well proven by the statistics of women in political power (Table 5.1) [5.9, 13].

The most important effect is a direct consequence of the statement that the human being is not a draught animal anymore, and this is represented in the role of physical power. Even in societies where women were sometimes forced to work harder than men, this situation was traditionally enforced by male physical superiority. Child care is much more a common bi-gender duty now, and all kinds of related burdens are supported by mass production and general services, based on automation. The doubled and active lifespan permits historically unparalleled multiplicity in life foci.

Another proof of the higher status of human values is the issue of safety at work [5.11, 12]. The ILO and the US Department of Labor issue deep analyses of injuries related to work, temporal change, and social work-related details. The figures show great improvement in high-technology workplaces and better-educated work-forces and the typical problems of low educated people, partly unemployed, partly employed under uncertain, dubious conditions. The drive for these values was the bilateral result of automatic equipment for production with automatic safety installations and stronger requirements for the human workforce.

All these and further measures followed the progress of technology and the consequent increase in the wealth of nations and regions. Life expectancy, clean water supplies, more free time, and opportunities for leisure, culture, and sport are clearly reflected in the figures of technology levels, automation, and wealth [5.15] (Figs. 5.2 and 5.3) [5.14, 16].

Life expectancy before the Industrial Revolution had been around 30 years for centuries. The social gap

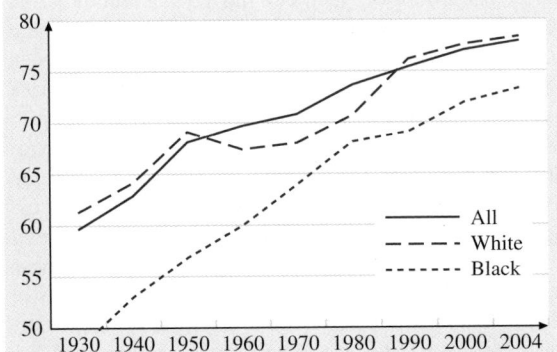

Fig. 5.3 Life expectancy at birth by race in the US (after [5.14])

Table 5.2 Relations of health and literacy (after [5.15])

Country	Approx. year	Life expectancy at birth (years)	Infant mortality to age 5 years per 1000 live births	Adult literacy (%)	Access to safe water (%)
Argentina	1960	65.2	72	91.0	51
	1980	69.6	38	94.4	58
	2001	74.1	19	96.9	94
Brazil	1960	54.9	177	61.0	32
	1980	62.7	80	74.5	56
	2001	68.3	36	87.3	87
Mexico	1960	57.3	134	65.0	38
	1980	66.8	74	82.2	50
	2001	73.4	29	91.4	88
Latin America	1960	56.5	154	74.0	35
	1980	64.7	79	79.9	53
	2001	70.6	34	89.2	86
East Asia	1960	39.2	198	n.a.	n.a.
	1980	60.0	82	68.8	n.a.
	2001	69.2	44	86.8	76

in life expectancy within one country's lowest and highest deciles, according to recent data from Hungary, is 19 years. The marked joint effects of progress are presented in Table 5.2 [5.17].

5.5 Social Stratification, Increased Gaps

Each change was followed, on the one hand, by mass tragedies for individuals, those who were not able to adapt, and by new gaps and tensions in societies, and on the other hand, by great opportunities in terms of social welfare and cultural progress, with new qualities of human values related to greater solidarity and personal freedom.

In each dynamic period of history, social gaps increase both within and among nations. Table 5.3 and Fig. 5.4 indicate this trend – in the figure markedly both with and without China – representing a promising future for all mankind, especially for long lagging developing countries, not only for a nation with a population of about one-sixth of the world [5.18]. This picture demonstrates the role of the Industrial Revolution and technological innovation in different parts of the world and also the very reasons why the only recipe for lagging regions is accelerated adaptation to the economic–social characteristics of successful historical choices.

The essential change is due to the two Industrial Revolutions, in relation to agriculture, industry, and services, and consequently to the change in professional and social distributions [5.16, 19]. The dramatic picture of the former is best described in the novels of Dickens, Balzac, and Stendhal, the transition to the second in Steinbeck and others. Recent social uncertainty dominates American literature of the past two decades.

This great change can be felt by the decease of distance. The troops of Napoleon moved at about the same speed as those of Julius Caesar [5.20], but mainland communication was accelerated in the USA between 1800 and 1850 by a factor of eight, and the usual 2 week passage time between the USA and Europe of the mid 19th century has decreased now by 50-fold. Similar figures can be quoted for numbers of traveling people and for prices related to automated mass production, especially for those items of high-technology consumer goods which are produced in their entirety by these technologies. On the other hand, regarding the prices of all items and services related to the human workforce, the opposite is true. Compensation in professions demanding higher education is through relative increase of salaries.

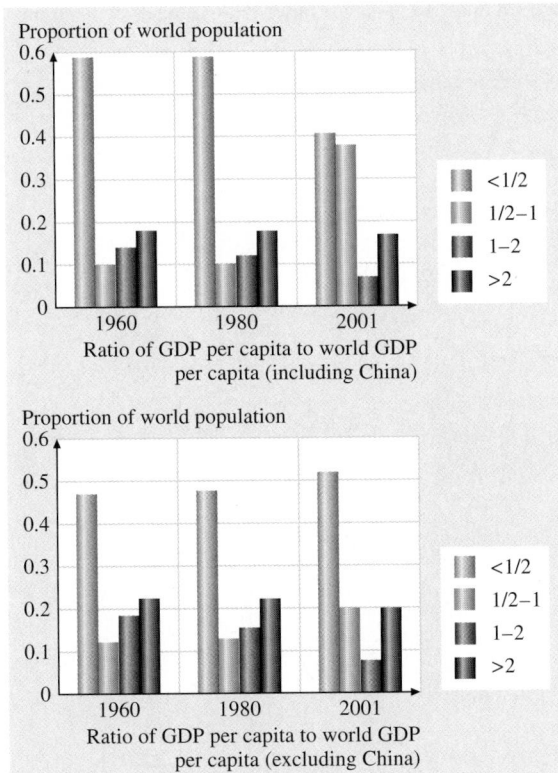

Fig. 5.4 World income inequality changes in relations of population and per capita income in proportions of the world distribution (after [5.21] and UN/DESA)

Due to these changed relations in distance and communication, cooperative and administrative relations have undergone a general transformation in the same sense, with more emphasis on the individual, more freedom from earlier limitations, and therefore, more personal contacts, less according to need and mandatory organization rather than current interests. The most

important phenomena are the emerging multinational production and service organizations. The increasing relevance of supranational and international political, scientific, and service organizations, international standards, guidelines, and fashions as driving forces of consumption attitudes, is a direct consequence of the changing technological background.

Figure 5.5 [5.22–24] shows current wages versus social strata and educational requirement distribution of the USA. Under the striking figures of large company CEOs (chief executive officers) and successful capitalists, who amount to about 1–1.5% of the population, the distribution of income correlates rather well with required education level, related responsibility, and adaptation to the needs of a continuously technologically advancing society. The US statistics are based on tax-refund data and reflect a rather growing disparity in incomes. Other countries with advanced economies show less unequal societies but the trend in terms of social gap for the time being seems to be similar. The disparity in jobs requiring higher education reflects a disparity in social opportunity on the one hand, but also probably a realistic picture of requirements on the other.

Figure 5.6 shows a more detailed and informative picture of the present American middle-class cross section.

A rough estimate of social breakdown before the automation–information revolution is composed of several different sources, as shown in Table 5.4 [5.25].

These dynamics are driven by finance and management, and this is the realistic reason for overvaluations in these professions. The entrepreneur now plays the role of the condottiere, pirate captain, discoverer, and adventurer of the Renaissance and later. These roles, in a longer retrospective, appear to be necessary in periods of great change and expansion, and will be consolidated in the new, emerging social order. The worse phenomena of these turbulent periods are the

Population in %	Income in 1000 US$/y	Class distribution		Education
1–2	200 and more	Capitalists, CEO, politicians, celebrities, etc.		?
15	200–60	Upper middle class	Professors managers	Graduate
30	60–30	Middle class	Professional sales, and support	Bachelor deg. significant skill
30	30–10	Lower middle class	Clerical, service, blue collar	Some college
22–23	10 and less	Poor underclass	Part time, unemployed	High school or less

Fig. 5.5 A rough picture of the US society (after [5.22–24])

Part A | 5.5

Table 5.3 The big divergence: developing countries versus developed ones, 1820–2001, (after [5.26] and United Nations Development of Economic and Social Affairs (UN/DESA))

	GDP per capita (1990 international Geary–Khamis dollars)					
	1820	1913	1950	1973	1980	2001
Developed world	1204	3989	6298	13 376	15 257	22 825
Eastern Europe	683	1695	2111	4988	5786	6027
Former USSR	688	1488	2841	6059	6426	4626
Latin America	692	1481	2506	4504	5412	5811
Asia	584	883	918	2049	2486	3998
China	600	552	439	839	1067	3583
India	533	673	619	853	938	1957
Japan	669	1387	1921	11 434	13 428	20 683
Africa	420	637	894	1410	1536	1489
	Ratio of GDP per capita to that of the developed world					
	1820	1913	1950	1973	1980	2001
Developed world	–	–	–	–	–	–
Eastern Europe	0.57	0.42	0.34	0.37	0.38	0.26
Former USSR	0.57	0.37	0.45	0.45	0.422	0.20
Latin America	0.58	0.37	0.40	0.34	0.35	0.25
Asia	0.48	0.22	0.15	0.15	0.16	0.18
China	0.50	0.14	0.07	0.06	0.07	0.16
India	0.44	0.17	0.10	0.06	0.06	0.09
Japan	0.56	0.35	0.30	0.85	0.88	0.91
Africa	0.35	0.16	0.14	0.11	0.10	0.07

Table 5.4 Social breakdown between the two world wars (rough, rounded estimations)

Country	Agriculture	Industry	Commerce	Civil servant and freelance	Domestic servant	Others
Finland	64	15	4	4	2	11
France	36	34	13	7	4	6
UK	6	46	19	10	10	9
Sweden	36	32	11	5	7	9
US	22	32	18	9	7	2

political adventurers, the dictators. The consequences of these new imbalances are important warnings in the directions of increased value of human and social relations.

The most important features of the illustrated changes are due to the transition from an agriculture-based society with remnants of feudalist burdens to an industrial one with a bourgeois–worker class, and now to an information society with a significantly different and mobile strata structure. The structural change in our age is clearly illustrated in Fig. 5.7 and in the investment policy of a typical country rapidly joining the welfare world, North Korea, in Fig. 5.8 [5.18].

Most organizations are applying less hierarchy. This is one effect of the general trend towards overall control modernization and local adaptation as leading principles of optimal control in complex systems. The concentration of overall control is a result of advanced, real-time information and measurement technology and related control theories, harmonizing with the local traditions and social relations. The principles developed in the control of industrial processes could find general validity in all kinds of complex systems, societies included.

The change of social strata and technology strongly affects organizational structures. The most characteris-

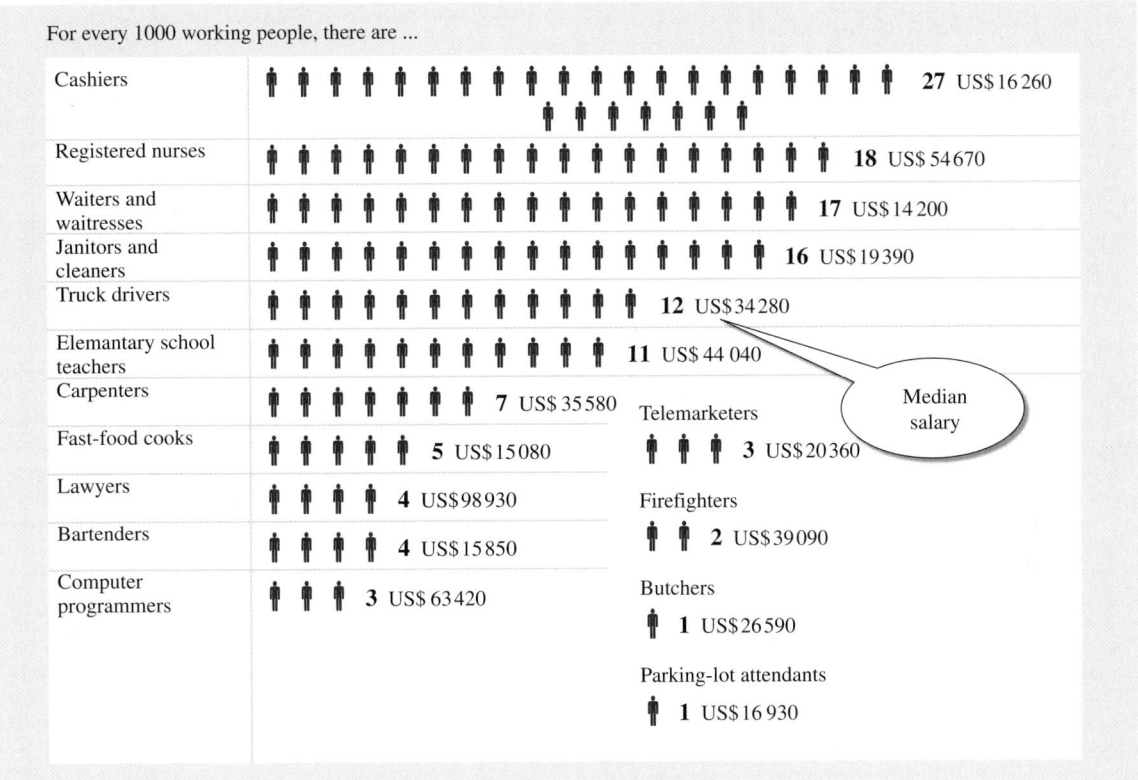

For every 1000 working people, there are ...

Cashiers	27 US$16 260
Registered nurses	18 US$54 670
Waiters and waitresses	17 US$14 200
Janitors and cleaners	16 US$19 390
Truck drivers	12 US$34 280
Elemantary school teachers	11 US$44 040
Carpenters	7 US$35 580
Fast-food cooks	5 US$15 080
Lawyers	4 US$98 930
Bartenders	4 US$15 850
Computer programmers	3 US$63 420
Telemarketers	3 US$20 360
Firefighters	2 US$39 090
Butchers	1 US$26 590
Parking-lot attendants	1 US$16 930

Median salary

Fig. 5.6 A characteristic picture of the modern society (after [5.22])

tic phenomenon is the individualization and localization of previous social entities on the one hand, and centralization and globalization on the other. Globalization has an usual meaning related to the entire globe, a general trend in very aspect of unlimited expansion, extension and proliferation in every other, more general dimensions.

The great change works in private relations as well. The great multigenerational family home model is over. The rapid change of lifestyles, entertainment technology, and semi-automated services, higher income standards, and longer and healthier lives provoke and allow the changing habits of family homes.

The development of home service robots will soon significantly enhance the individual life of handicapped persons and old-age care. This change may also place greater emphasis on human relations, with more freedom from burdensome and humiliating duties.

5.6 Production, Economy Structures, and Adaptation

Two important remarks should be made at this point. Firstly, the main effect of automation and information technology is not in the direct realization of these special goods but in a more relevant general elevation of any products and services in terms of improved qualities and advanced production processes. The computer and information services exports of India and Israel account for about 4% of their gross domestic product (GDP).

These very different countries have the highest figures of direct exports in these items [5.18].

The other remark concerns the effect on employment. Long-range statistics prove that this is more influenced by the general trends of the economy and by the adaptation abilities of societies. Old professions are replaced by new working opportunities, as demonstrated in Figs. 5.9 and 5.10 [5.22, 27, 28].

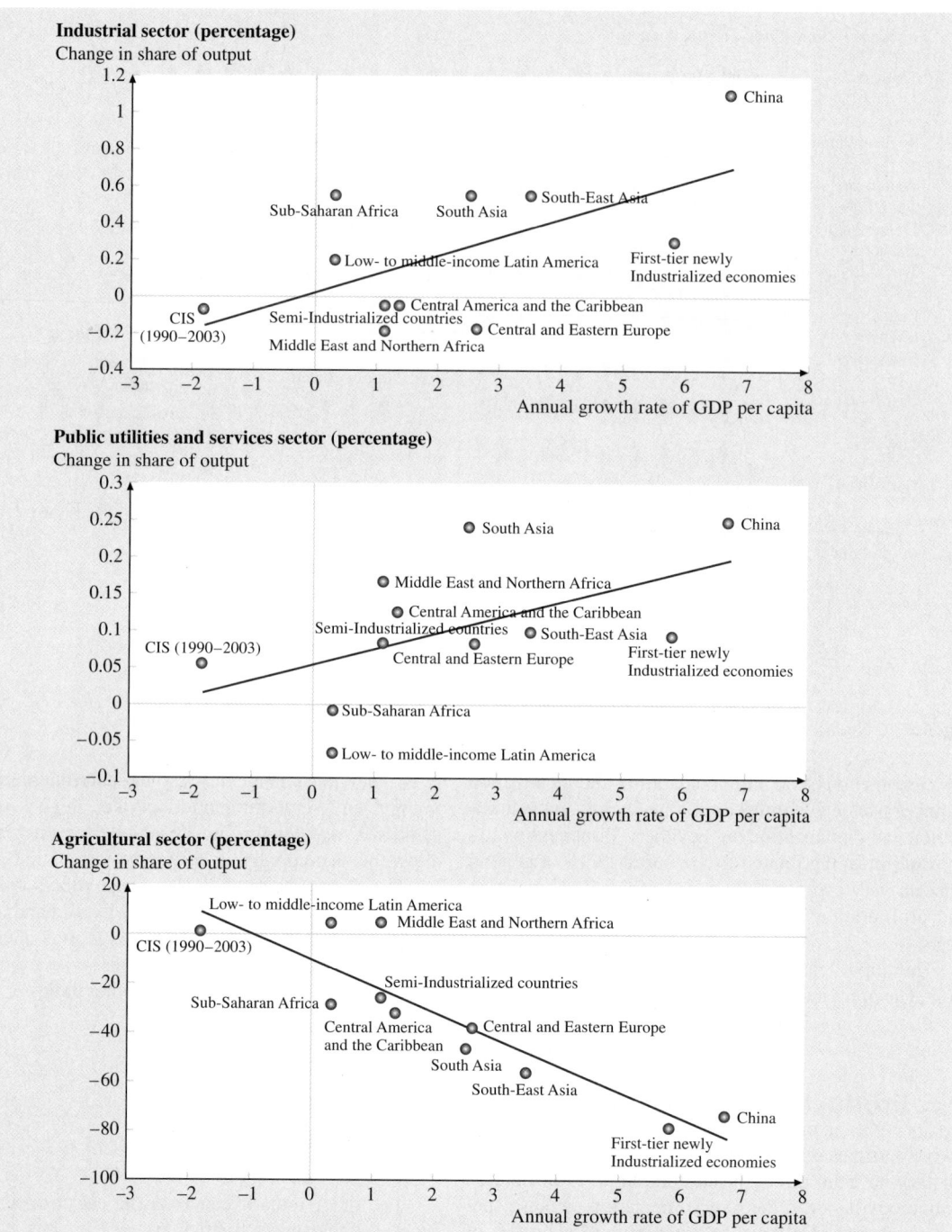

Fig. 5.7 Structural change and economic growth (after [5.18] and UN/DESA, based on United Nations Statistics Division, National Accounts Main Aggregates database. Structural change and economic growth)

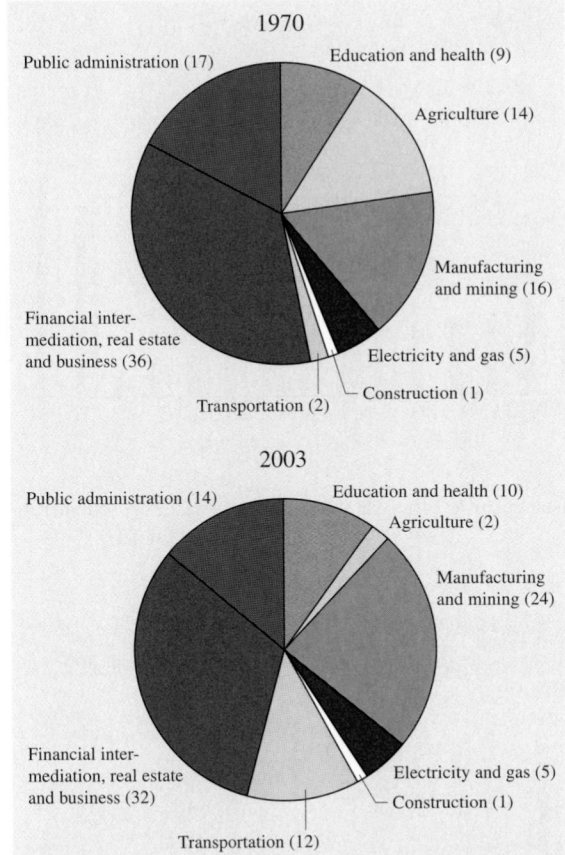

Fig. 5.8 Sector investment change in North Korea (after UN/DESA based on data from National Statistical Office, Republic of Korea Structural change and economic growth)

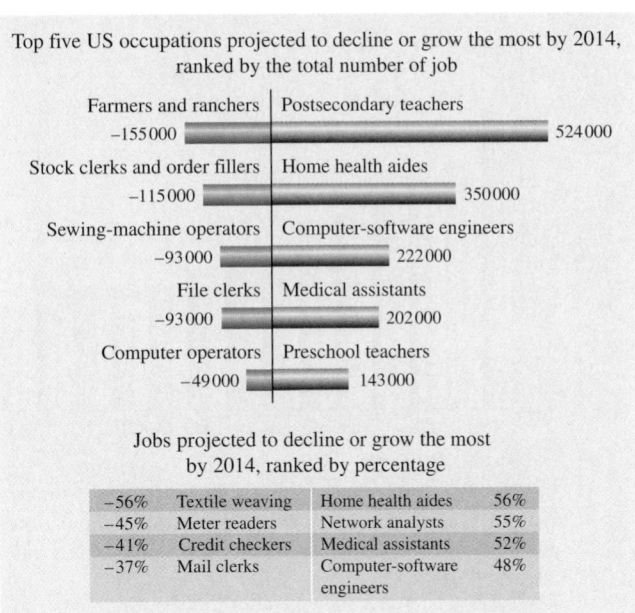

Fig. 5.9 Growing and shrinking job sectors (after [5.22])

majority of the population elevate its intellectual level to the new requirements from those of earlier animal and quasi-animal work? What will be the directions of adaptation to the new freedoms in terms of time, consumption, and choice of use, and misuse of possibilities given by the proliferation of science and technology?

These questions generate further questions: Should the process of adaptation, problem solving, be controlled or not? And, if so, by what means or organizations? And, not least, in what directions? What should be the control values? And who should decide about those, and how? Although questions like these have arisen in all historical societies, in the future, giving the

One aspect of changing working conditions is the evolution of teleworking and outsourcing, especially in digitally transferable services. Figure 5.11 shows the results of a statistical project closed in 2003 [5.29].

Due to the automation of production and servicing techniques, investment costs have been completely transformed from production to research, development, design, experimentation, marketing, and maintenance support activities. Also production sites have started to become mobile, due to the fast turnaround of production technologies. The main fixed property is knowhow and related talent [5.30, 31]. See also Chap. 6 on the economic costs of automation.

The open question regarding this irresistible process is the adaptation potential of mankind, which is closely related to the directions of adaptation. How can the

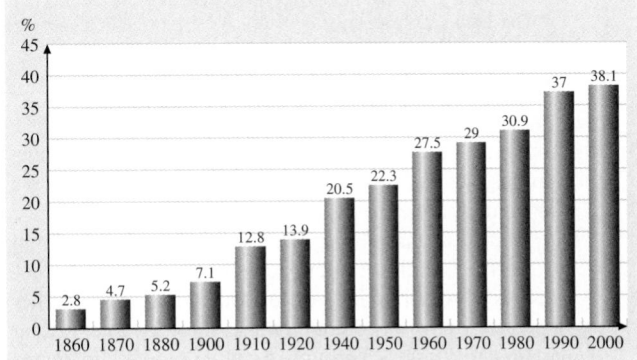

Fig. 5.10 White-collar workers in the USA (after [5.28])

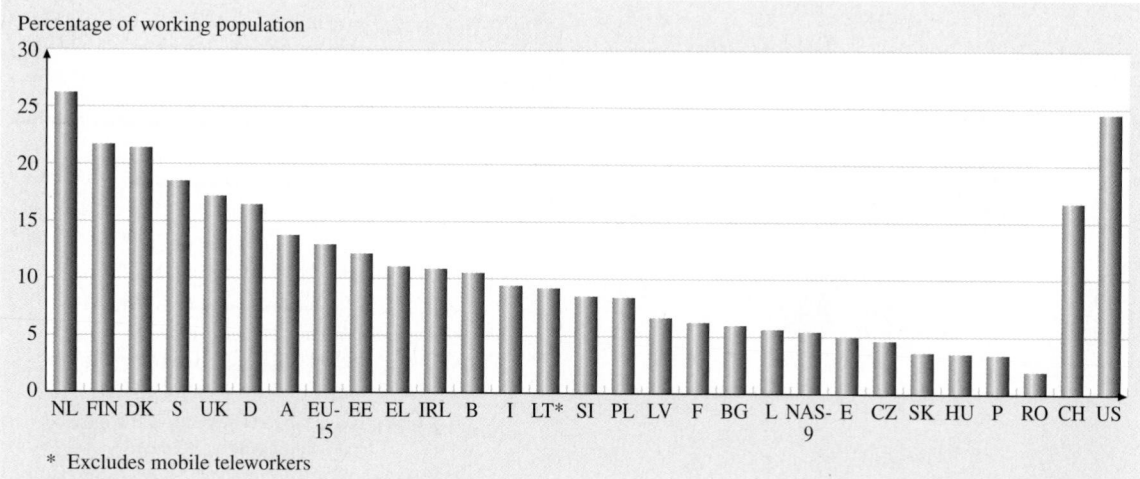

Fig. 5.11 Teleworking, percentage of working population at distance from office or workshop (after [5.29]) (countries, abbreviations according car identification, EU – European Union, NAS – newly associated countries of the EU)

Table 5.5 Coherence indices (after [5.29])

	Gross national income (GNI)/capita[a] (current thousand US$)	Corruption[b] (score, max. 10 [squeaky clean])	e-Readiness[c] (score, max. 10)
Canada	32.6	8.5	8.0
China	1.8	–	4.0
Denmark	47.4	9.5	8.0
Finland	37.5	9.6	8.0
France	34.8	7.4	7.7
Germany	34.6	8.0	7.9
Poland	7.1	3.7	5.5
Rumania	3.8	3.1[d]	4.0[d]
Russia	4.5	3.5	3.8
Spain	25.4	6.8	7.5
Sweden	41.0	9.2	8.0
Switzerland	54.9	9.1	7.9
UK	37.6	8.6	8.0

[a] according to the World Development Indicators of the World Bank, 2006,
[b] The 2006 Transparency, International Corruption Perceptions Index according to the Transparency International Survey,
[c] Economic Intelligence Unit [5.32],
[d] Other estimates

immensely greater freedom in terms of time and opportunities, the answers to these questions will be decisive for human existence.

Societies may be ranked nowadays by national product per capita, by levels of digital literacy, by estimates of corruption, by e-Readiness, and several other indicators. Not surprisingly, these show a rather strong coherence. Table 5.5 provide a small comparison, based on several credible estimates [5.29, 30, 33].

A recent compound comparison by the Economist Intelligence Unit (EIU) (Table 5.6) reflects the EIU e-Readiness rankings for 2007, ranking 69 countries in terms of six criteria. In order of importance, these are: consumer and business adoption; connectivity and technology infrastructure; business environment; social and cultural environment; government policy and vision; and legal and policy environment.

Table 5.6 The 2007 e-Readiness ranking

Economist Intelligence Unit e-Readiness rankings, 2007									
2007 e-Readiness rank (of 69)	2006 rank	Country	2007 e-Readiness score (of 10)	2006 score	2007 e-Readiness rank (of 69)	2006 rank	Country	2007 e-Readiness score (of 10)	2006 score
1	1	Denmark	8.88	9.00	36	37	Malaysia	5.97	5.60
2 (tie)	2	US	8.85	8.88	37	39	Latvia	5.88	5.30
2 (tie)	4	Sweden	8.85	8.74	38	39	Mexico	5.86	5.30
4	10	Hong Kong	8.72	8.36	39	36	Slovakia	5.84	5.65
5	3	Switzerland	8.61	8.81	40	34	Poland	5.80	5.76
6	13	Singapore	8.60	8.24	41	38	Lithuania	5.78	5.45
7	5	UK	8.59	8.64	42	45	Turkey	5.61	4.77
8	6	Netherlands	8.50	8.60	43	41	Brazil	5.45	5.29
9	8	Australia	8.46	8.50	44	42	Argentina	5.40	5.27
10	7	Finland	8.43	8.55	45	49	Romania	5.32	4.44
11	14	Austria	8.39	8.19	46 (tie)	43	Jamaica	5.05	4.67
12	11	Norway	8.35	8.35	46 (tie)	46	Saudi Arabia	5.05	5.03
13	9	Canada	8.30	8.37	48	44	Bulgaria	5.01	4.86
14	14	New Zealand	8.19	8.19	49	47	Thailand	4.91	4.63
15	20	Bermuda	8.15	7.81	50	48	Venezuela	4.89	4.47
16	18	South Korea	8.08	7.90	51	49	Peru	4.83	4.44
17	23	Taiwan	8.05	7.51	52	54	Jordan	4.77	4.22
18	21	Japan	8.01	7.77	53	51	Colombia	4.69	4.25
19	12	Germany	8.00	8.34	54 (tie)	53	India	4.66	4.04
20	17	Belgium	7.90	7.99	54 (tie)	56	Philippines	4.66	4.41
21	16	Ireland	7.86	8.09	56	57	China	4.43	4.02
22	19	France	7.77	7.86	57	52	Russia	4.27	4.14
23	22	Israel	7.58	7.59	58	55	Egypt	4.26	4.30
24	–	Malta[a]	7.56	–	59	58	Equador	4.12	3.88
25	25	Italy	7.45	7.14	60	61	Ukraine	4.02	3.62
26	24	Spain	7.29	7.34	61	59	Sri Lanka	3.93	3.75
27	26	Portugal	7.14	7.07	62	60	Nigeria	3.92	3.69
28	27	Estonia	6.84	6.71	63	67	Pakistan	3.79	3.03
29	28	Slovenia	6.66	6.43	64	64	Kazakhstan	3.78	3.22
30	31	Chile	6.47	6.19	65	66	Vietnam	3.73	3.12
31	32	Czech Rep.	6.32	6.14	66	63	Algeria	3.63	3.32
32	29	Greece	6.31	6.42	67	62	Indonesia	3.39	3.39
33	30	UAE	6.22	6.32	68	68	Azerbaijan	3.26	2.92
34	32	Hungary	6.16	6.14	69	65	Iran	3.08	3.15
35	35	South Africa	6.10	5.74					

[a] New to the annual rankings in 2007 (after EIU)

Part A | 5.6

5.7 Education

Radical change of education is enforced by the dramatic changes of requirements. The main directions are as follows:

- Population and generations to be educated
- Knowledge and skills to be learnt
- Methods and philosophy of education.

General education of the entire population was introduced with the Industrial Revolution and the rise of nation states, i.e., from the 18th to the end of the 19th centuries, starting with royal decrees and laws expressing a will and trend and concluding in enforced, pedagogically standardized, secular systems [5.35].

The related social structures and workplaces required a basic knowledge of reading and writing, first of names, simple sentences for professional and civil communication, and elements of arithmetic. The present requirement is much higher, defined (PISA, Program for International Student Assessment of the OECD) by understanding regular texts from the news, regulations, working and user instructions, elements of measurement, dimensions, and statistics.

Progress in education can be followed also as a consequence of technology sophistication, starting with four to six mandatory years of classes and continued by mandatory education from 6 to 18 years. The same is reflected in the figures of higher education beyond the mandatory education period [5.16, 30, 34].

For each 100 adults of tertiary-education age, 69 are enrolled in tertiary education programs in North America and Europe, compared with only 5 in sub-Saharan Africa and 10 in South and West Asia. Six countries host 67% of the world's foreign or mobile students: with 23% studying in the USA, followed by the UK (12%), Germany (11%), France (10%), Australia (7%), and Japan (5%).

An essential novelty lies in the rapid change of required knowledge content, due to the lifecycles of technology, see timeline of Fig. 5.1. The landmarks of technology hint at basic differences in the chemistry, physics, and mathematics of the components and, based

on the relevant new necessities in terms of basic and higher-level knowledge, are reflected in application demands. The same demand is mirrored in work-related training provided by employers.

The necessary knowledge of all citizens is also defined by the systems of democracy, and modern democracy is tied to market economy systems. This defines an elementary understanding of the constitutional–legal system, and the basic principles and practice of legal institutions. The concept of constitutional awareness is not bound to the existence of a canonized national constitution; it can be a consciousness, accord on fundamental social principles.

As stated, general education is a double requirement for historical development: professional knowledge for producers and users of technology, and services and civil culture as necessary conditions for democracy. These two should be unified to some extent in each person and generation. This provides another hint at the change from education of children and adolescents towards a well-designed, pedagogically renewed, socially regulated lifelong education schedule with mandatory basic requirements. There should also be mandatory requirements for each profession with greater responsibility and a wide spectrum of free opportunities, including in terms of retirement age (Table 5.7).

In advanced democracies this change strongly affects the principle of equal opportunities, and creates a probably unsolvable contradiction between increasing knowledge requirements, the available amount of different kinds of knowledge, maintenance and strengthening of the cultural–social coherence of societies, and the unavoidable leverage of education. Educating the social–cultural–professional elite and the masses of a democratic society, with given backgrounds in terms of talent, family, and social grouping, is the main concern of all responsible government policies. The US Ivy League universities, British Oxbridge, and the French Grand Écoles represent elite schools, mostly for a limited circle of young people coming from more highly educated, upper society layers.

Table 5.7 Lifelong learning. Percentage of the adult population aged 25–64 years participating in education and training (mostly estimated or reported values), after [5.34]

	1995	1996	1997	1998	1999	2000	2001	2002	2003	2004	2005
EU (25 countries)	–	–	–	–	–	7.5	7.5	7.6	9.0	9.9	10.2
EU (15 countries)	–	–	–	–	8.2	8.0	8.0	8.1	9.8	10.7	11.2
Euro area	4.5	5.1	5.1	–	5.6	5.4	5.2	5.3	6.5	7.3	8.1

The structures of education are also defined by the capabilities of private and public institutions: their regulation according to mandatory knowledge, subsidies depending on certain conditions, the ban on discrimination based on race or religion, and freedom of access for talented but poor people.

The structural variants of education depend on necessary and lengthening periods of professional education, and on the distribution of professional education between school and workplace. Open problems are the selection principles of educational quotas (if any), and the question of whether these should depend on government policy and/or be the responsibility of the individual, the family or educational institutions.

In the Modern Age, education and pedagogy have advanced from being a kind of affective, classical psychology-like quality to a science in the strong sense, not losing but strengthening the related human virtues. This new, science-type progression is strongly related to brain research, extended and advanced statistics and worldwide professional comparisons of different experiments. Brain research together with psychology provides a more reliable picture of development during different age periods. More is known on how conceptual, analogous, and logical thinking, memory, and processing of knowledge operate; what the coherences of special and general abilities are; what is genetically determined and definable; and what the possibilities of special training are. The problem is greater if there are deterministic features related to gender or other inherited conditions. Though these problems are particularly delicate issues, research is not excluded, nor should it be, although the results need to be scrutinized under very severe conditions of scientific validity.

The essential issue is the question of the necessary knowledge for citizens of today and tomorrow. A radical change has occurred in the valuation of traditional cultural values: memorizing texts and poems of acknowledged key authors from the past; the proportion of science- versus human-related subjects; and the role of physical culture and sport in education. The means of social discipline change within the class as a preparation for ethical and collegial cooperation, just like the abiding laws of the community.

The use of modern and developing instruments of education are overall useful innovations but no solution for the basic problems of individual and societal development. These new educational instruments include moving pictures, multimedia, all kinds of visual and auditive aids, animation, 3-D representation, question-answering automatic methods of teaching, freedom of learning schedules, mass use of interactive whiteboards, personal computers as a requisite for each student and school desk, and Internet-based support of remote learning. See Chap. 44 on *Education and Qualification* and Chap. 85 on *Automation in Education/Learning Systems* for additional information.

A special requirement is an international standard for automatic control, system science, and information technology. The International Federation of Automatic Control (IFAC), through its special interest committee and regular symposia, took the first steps in this direction from its start, 50 years ago [5.36]. Basic requirements could be set regarding different professional levels in studies of mathematics, algorithmics, control dynamics, networks, fundamentals of computing architecture and software, components (especially semiconductors), physics, telecommunication transmission and code theory, main directions of applications in system design, decision support and mechanical and sensory elements of complex automation, their fusion, and consideration of social impact.

All these disciplines change in their context and relevance during the lifetime of a professional generation, surviving at least three generations of their subject. This means greater emphasis on disciplinary basics and on the particular skill of adopting these for practical innovative applications, and furthermore on the disciplinary and quality ethics of work.

A major lesson of the current decade is not only the hidden spread of these techniques in every product, production process, and system but also the same spread of specialists in all kinds of professional activities.

All these phenomena and experiments display the double face of education in terms of an automated, communication-linked society. One of the surprising facts from the past few decades is the unexpected increase in higher-education enrolment for humanities, psychology, sociology, and similar curricula, and the decline in engineering- and science-related subjects. This orientation is somehow balanced by the burst of management education, though the latter has a trend to deepen knowledge in human aspects outweighing the previous, overwhelming organizational, structural knowledge.

Part A | 5.7

5.8 Cultural Aspects

The above contradictory, but socially relatively controlled, trend is a proof of the initial thesis: in the period of increasing automation, the human role is emerging more than ever before. This has a relevant linguistic meaning, too. Not only is knowledge of foreign languages (especially that of the modern lingua franca, English) gaining in importance, but so is the need for linguistic and metalinguistic instruments as well, i.e., a syncretistic approach to the development of sensitive communication facilities [5.37, 38].

The resulted plethora of these commodities is represented in the variations of goods, their usage, and in the spectra of quality. The abundance of supply is in accordance not only with material goods but also the mental market.

The end of the book was proclaimed about a decade ago. In the meantime the publication of books has mostly grown by a modest, few percentage points each year in most countries, in spite of the immense reading material available on the Internet. Recent global statistics indicate a growth of about 3–4% per year in the past period in juvenile books, the most sensitive category for future generations.

The rapid change of electronic entertainment media from cassettes to CD-DVD and MP3 semiconductor memories and the uncertainties around copyright problems made the market uncertain and confused. In the past 10 years the prices of video-cassettes have fallen by about 50%; the same happened to DVDs in the past 2 years.

All these issues initiate cultural programs for each age, each technological and cultural environment, and each kind of individual and social need, both maintaining some continuity and inducing continuous change.

On the other hand, the market has absorbed the share in the entertainment business with a rapidly changing focus on fashion-driven music, forgotten classics, professional tutoring, and great performances. The lesson is a naturally increasing demand together with more free time, greater income, and a rapidly changing world and human orientation. Adaptation is serviced by a great variety of different possibilities.

High and durable cultural quality is valued mostly later in time. The ratio of transitory low-brow cultural goods to high-brow permanent values has always been higher by orders of magnitude. Automatic, high-quality reproduction technology, unseen and unimaginable purchasing power, combined with cultural democracy is a product of automated, information-driven engineering development. The human response is a further question, and this is one reason why a nontechnical chapter has a place in this technology handbook.

5.9 Legal Aspects, Ethics, Standards, and Patents

5.9.1 Privacy

The close relations between continuity and change are most reflected in the legal environment: the embedding of new phenomena and legal requirements into the traditional framework of the law. This continuity is important because of the natural inertia of consolidated social systems and human attitudes. Due to this effect, both Western legal systems (Anglo-Saxon Common Law as a case-based system and continental rule-based legal practice) still have their common roots in Roman Law. In the progress of other civilizations towards an industrial and postindustrial society, these principles have been gradually accepted. The global process in question is now enforced, not by power, but by the same rationality of the present technology-created society [5.4, 39].

The most important legal issue is the combined task of warranting privacy and security. The privacy issue, despite having some antecedents in the Magna Carta and other documents of the Middle Ages, is a modern idea. It was originated for an equal-rights society and the concept of all kinds of private information properties of people. The modern view started with the paper entitled *The Right to Privacy* by *Warren* and *Brandeis*, at the advent of the 20th century [5.40]. The paper also defines the immaterial nature of the specific value related to privacy and the legal status of the material instrument of (re)production and of immaterial private property.

Three components of the subject are remarkable and all are related to the automation/communication issue: mass media, starting with high-speed wide-circulation printing, photography and its reproduction technologies, and a society based on the equal-rights principle.

In present times the motivations are in some sense contradictory: an absolute defense against any kind of intrusion into the privacy of the individual by alien power. The anxiety was generated by the real experience of the 20th century dictatorships, though executive terror and mass murder raged just before modern information instruments. On the other hand, user-friendly, efficient administration and security of the individual and of the society require well-organized data management and supervision. The global menace of terrorism, especially after the terrorist attacks of 11 September, 2001, has drawn attention to the development and introduction of all kinds of observational techniques.

The harmonization of these contradictory demands is given by the generally adopted principles of human rights, now with technology-supported principles:

- All kind of personal data, regarding race, religion, conscience, health, property, and private life are available only to the person concerned, accessible only by the individual or by legal procedure.
- All information regarding the interests of the citizen should be open; exemption can be made only by constitutional or equivalent, specially defined cases.
- The citizen should be informed about all kinds of access to his/her data with unalterable time and authorization stamps.

Table 5.8 Regulations concerning copyright and patent

Regulations: Article I, Section 8, Clause 8 of the US Constitution, also known as the Copyright Clause, gives Congress the power to enact statutes *To promote the Progress of Science and useful Arts, by securing for limited Times to Authors and Inventors the exclusive Right to their respective Writings and Discoveries.*

Congress first exercised this power with the enactment of the Copyright Act of 1790, and has changed and updated copyright statutes several times since. The Copyright Act of 1976, though it has been modified since its enactment, is currently the basis of copyright law in the USA.

The **Berne Convention for the Protection of Literary and Artistic Works**, usually known as the **Berne Convention**, is an international agreement about copyright, which was first adopted in Berne, Switzerland in 1886.

Paris Convention for the Protection of Industrial Property, signed in Paris, France, on March 20, 1883.

The **Agreement on Trade Related Aspects of Intellectual Property Rights** (**TRIPS**) is a treaty administered by the World Trade Organization (WTO) which sets down minimum standards for many forms of intellectual property (IP) regulation. It was negotiated at the end of the Uruguay Round of the General Agreement on Tariffs and Trade (GATT) treaty in 1994.

Specifically, TRIPS contains requirements that nations' laws must meet for: copyright rights, including the rights of performers, producers of sound recordings, and broadcasting organizations; geographical indications, including appellations of origin; industrial designs; integrated-circuit layout designs; patents; monopolies for the developers of new plant varieties; trademarks; trade address; and undisclosed or confidential information. TRIPS also specified enforcement procedures, remedies, and dispute-resolution procedures.

Patents in the modern sense originated in Italy in 1474. At that time the Republic of Venice issued a decree by which new and inventive devices, once they had been put into practice, had to be communicated to the Republic in order to obtain the right to prevent others from using them. England followed with the Statute of Monopolies in 1623 under King James I, which declared that patents could only be granted for "projects of new invention." During the reign of Queen Anne (1702–1714), the lawyers of the English Court developed the requirement that a written description of the invention must be submitted. These developments, which were in place during the colonial period, formed the basis for modern English and US patent law.

In the USA, during the colonial period and Articles of Confederation years (1778–1789), several states adopted patent systems of their own. The first congress adopted a Patent Act, in 1790, and the first patent was issued under this Act on July 31, 1790.

European patent law covers a wide range of legislations including national patent laws, the Strasbourg Convention of 1963, the European Patent Convention of 1973, and a number of European Union directives and regulations.

The perfection of the above principles is warranted by fast, broadband data links, and cryptographic and pattern recognition (identification) technology. The introduction of these tools and materialization of these principles strengthen the realization of the general, basic statement about the elevation effect of human consciousness. The indirect representation of the Self, and its rights and properties is a continuous, frequently used mirror of all this. And, through the same, the metaphoric high level of the mirror is a self-conscious intelligence test for beings. The system contributes to the legal consciousness of the advanced democracy.

The semi-automatic, data-driven system of administration separates the actions that can, and should be, executed in an automatic procedure by the control and evaluation of data, and follows the privacy and security principles. A well-operating system can save months of civilian inquiries, and hours and days of traveling, which constitutes a remarkable percentage of the administrative cost. A key aspect is the increased concentration on issues that require human judgment. In real human problems, human judgment is the final principle.

Citizens' indirect relationship with the authorities using automatic and telecommunication means evokes the necessity for natural language to be understood by machines, in written and verbal forms, and the creation of user-friendly, natural impression dialogues. The need for bidirectional translation between the language of law and natural communication, and translation into other languages, is emerging in wide, democratic usage. These research efforts are quickly progressing in several communities.

5.9.2 Free Access, Licence, Patent, Copyright, Royalty, and Piracy

Free access to information has different meanings. First of all, it is an achievement and new value of democracy: the right to access directly all information regarding an individual citizen's interests. Second, it entails a new relation to property that is immaterial, i. e., not physically decreased by alienation. Easy access changes the view regarding the right of the owner, the author. Though the classic legal concepts of patent and copyright are still valid and applied, the nonexistence of borders and the differences in local regulations and practice have opened up new discussions on the subject. Several companies and interest groups have been arguing for more liberal regulations. These arguments comprise the advertising interests of even more dynamic companies, the costs and further difficulties of safeguarding ownership rights, and the support of developing countries.

Table 5.8 presents an overview of the progress of these regulations.

5.10 Different Media and Applications of Information Automation

A contradictory trend in information technology is the development of individual user services, with and without centralized control and control of the individual user. The family of these services is characterized by decentralized input of information, centralized and decentralized storage and management, as well as unconventional automation combined with similarly strange freedom of access. Typical services are blog-type entertainment, individual announcements, publications, chat groups, other collaborative and companion search, private video communication, and advertisements. These all need well-organized data management, automatic, and desire-guided browsing search support and various identification and filtering services. All these are, in some sense, new avenues of information service automation and society organization in close interaction [5.26, 41, 42].

Two other services somehow belonging to this group are the information and economic power of very large search organizations, e.g., Google and Yahoo, and some minor, global, and national initiatives. Their automatic internal control has different search and grading mechanisms, fundamentally based on user statistics and subject classifications, but also on statistical categorizations and other pattern-recognition and machine-supported text-understanding instruments.

Wikipedia, YouTube, and similar initiatives have greater or lesser control: principally everybody can contribute to a vast and specific edifice of knowledge, controlled by the voluntary participants of the voluntary-access system. This social control appears to be, in some cases, competing in quality with traditional, professional encyclopedic-knowledge institutions. The contradictory trends survive further: social knowledge bases have started to establish some proved, controlled quality groups for, and of, professionals.

The entertainment/advertisement industry covers all these initiatives by its unprecedented financial interest and power, and has emerged as the leading economic power after the period of automotive and traffic-related industry that followed the iron and steel, textile, and agricultural supremacy. Views on this development are extremely different; in a future period can be judged if these resulted in a better-educated, more able society, or deteriorated essential cultural values.

Behind all these emerging and ruling trends operates the joint technology of automatic control, both in the form of instrumentation and in effects obeying the principles of feedback, multivariate, stochastic and nonlinear, and continuous and discrete system control. These principles are increasingly applied in modeling the social effects of these human–machine interactions, trying not only to understand but also to navigate the ocean of this new–old supernature.

5.11 Social Philosophy and Globalization

Automation is a global process, rapidly progressing in the most remote and least advanced corners of the world. Mass production and World Wide Web services enforce this, and no society can withstand this global trend; recent modernization revolution and its success in China and several other countries provide indisputable evidence. Change and the rapid speed in advanced societies and their most mobile layers create considerable tension among and within countries. Nevertheless, clever policies, if they are implemented, and the social movements evoked by this tension, result in a general progress in living qualities, best expressed by extending lifespans, decreasing famine regions, and the increasing responsibility displayed by those who are more influential. However, this overall historical process cannot protect against the sometimes long transitory sufferings, social clashes, unemployment, and other human social disasters [5.41, 43, 44].

Transitory but catastrophic phenomena are the consequence of minority feelings expressed in wide, national, religious, ideology-related movements with aggressive nature. The state of hopeless poverty is less irascible than the period of intolerance. The only general recommendation is given by *Neumann* [5.45]:

The only solid fact is that these difficulties are due to an evolution that, while useful and constructive, is also dangerous. Can we produce the required adjustments with the necessary speed? The most hopeful answer is that the human species has been subjected to similar tests before and seems to have a congenital ability to come through, after varying amounts of trouble. To ask in advance for a complete recipe would be unreasonable. We can specify only the human qualities required: patience, flexibility, and intelligence.

5.12 Further Reading

Journals and websites listed as references provide continuously further updated information. Recommended periodicals as basic theoretical and data sources are:

- *Philosophy and Public Affairs*, Blackwell, Princeton – quarterly
- *American Sociological Review*, American Sociological Association, Ohio State University, Columbia – bimonthly
- *Comparative Studies in Sociology and History*, Cambridge University Press, Cambridge/MA – quarterly
- *The American Statistician*, American Statistical Association – quarterly

- *American Journal of International Law*, American Society of International Law – quarterly
- *Economic Geography, Clark University*, Worcester/MA – quarterly
- *Economic History Review*, Blackwell, Princeton – three-yearly
- *Journal of Economic History*, Cambridge University Press, Cambridge/MA – quarterly
- *Journal of Labor Economics*, Society of Law Economists, University Chicago Press – quarterly
- *The Rand Journal of Economics*, The Rand Corporation, Santa Monica – quarterly

References

5.1   F. Braudel: *La Méditerranée et le monde méditer-ranéen à l'époque de Philippe II* (Armand Colin, Paris 1949, Deuxième édition, 1966)

5.2   F. Braudel: *Civilisation matérielle et capitalisme (XVe–XVIIIe siècle)*, Vol. 1 (Armand Colin, Paris 1967)

5.3   http://www.hyperhistory.com/online_n2/History_n2/a.html

5.4   http://www-groups.dcs.st-and.ac.uk/~history/Chronology

5.5   http://www.thocp.net/reference/robotics/robotics.html

5.6   Wikipedia, the Free Encyclopedia, http://en.wikipedia.org/wiki/

5.7   Encyclopaedia Britannica, 2006, DVD

5.8   J.D. Ryder, D.G. Fink: *Engineers and Electrons: A Century of Electrical Progress* (IEEE Press, New York 1984)

5.9   Public life and decision making – http://www.unece.org/stats/data.htm

5.10 K. Marx: *Grundrisse: Foundations of the Critique of Political Economy* (Penguin Classics, London 1973), translated by: M. Nicolaus

5.11 US Department of Labor, Bureau of Labor Statistics – http://stats.bls.gov/

5.12 ILO: International Labour Organization – http://www.ilo.org/public/english

5.13 Inter-Parliamentary Union – http://www.ipu.org/wmn-e/world.htm

5.14 Infoplease, Pearson Education – http://www.infoplease.com/

5.15 Economist, Apr. 26, 2003, p. 45

5.16 B.R. Mitchell: *European Historical Statistics 1750–1993* (Palgrave MacMillan, London 2000)

5.17 Hungarian Central Statisical Office: *Statistical Yearbook of Hungary 2005* (Statisztikai Kiadó, Budapest 2006)

5.18 World Economic and Social Survey, 2006, http://www.un.org/esa/policy/wess/

5.19 Federal Statistics of the US – http://www.fedstats.gov/

5.20 F. Braudel: *Civilisation matérielle, économie et capitalism, XVe–XVIIIe siècle* quotes a remark of Paul Valery

5.21 World Bank: World Development Indicators 2005 database

5.22 Time, America by the Numbers, Oct. 22, 2006

5.23 W. Thompson, J. Hickey: *Society in Focus* (Pearson, Boston 2004)

5.24 US Census Bureau – http://www.census.gov/

5.25 Hungarian Central Statisical Office: *Hungarian Statistical Pocketbook* (Magyar Statisztikai Zsebkönyv, Budapest 1937)

5.26 A. Maddison: *The World Economy: A Millennial Perspective* (Development Center Studies, Paris 2001)

5.27 E. Bowring: Post-Fordism and the end of work, Futures **34/2**, 159–172 (2002)

5.28 G. Michaels: Technology, complexity and information: The evolution on demand for office workers, Working Paper, http://econ-www.mit.edu

5.29 SIBIS (Statistical Indicators Benchmarking the Information Society) – http://www.sibis-eu.org/

5.30 World Development Indicators, World Bank, 2006

5.31 UNCTAD Handbook of Statistics, UN publ.

5.32 Economist Intelligence Unit, *Scattering the seeds of invention: the globalization of research and development* (Economist, London 2004), pp. 106–108

5.33 Transparency International, the global coalition against corruption – http://www.transparency.org/publications/gcr

5.34 http://epp.eurostat.ec.europa.eu

5.35 K.F. Ringer: *Education and Society in Modern Europe* (Indiana University Press, Bloomington 1934)

5.36 IFAC Publications of the Education, Social Effects Committee and of the Manufacturing and Logistic Systems Group – http://www.ifac-control.org/

5.37 http://www1.worldbank.org/education/edstats/

5.38 B. Moulton: The expanding role of hedonic methods in the official statistics of the United States, Working Paper (Bureau of Economic Analysis, Washington 2001)

5.39 E. Hayek: *Law, Legislation and Liberty* (University Chicago Press, Chicago 1973)

5.40 S. Warren, L. D. Brandeis: The Right to Privacy, Harv. Law Rev. **IV**(5), 193–220 (1890)

5.41 M. Castells: *The Information Age: Economy, Society and Culture* (Basil Blackwell, Oxford 2000)

5.42 Bureau of Economic Analysis, US Dept. Commerce, Washington, D.C. (2001)

5.43 M. Pohjola: The new economy: Facts, impacts and policies, Inf. Econ. Policy **14/2**, 133–144 (2002)

5.44 R. Fogel: Catching up with the American economy, Am. Econ. Rev. **89**(1), 1–21 (1999)

5.45 J. V. Neumann: Can we survive technology? Fortune **51**, 151–152 (1955)

6. Economic Aspects of Automation

Piercarlo Ravazzi, Agostino Villa

The increasing diffusion of automation in all sectors of the industrial world gives rise to a deep modification of labor organization and requires a new approach to evaluate industrial systems efficiency, effectiveness, and economic convenience. Until now, the evaluation tools and methods at disposal of industrial managers are rare and even complex. Easy-to-use criteria, possibly based on robust but simple models and concepts, appear to be necessary. This chapter gives an overview of concepts, based on the economic theory but revised in the light of industrial practice, which can be applied for evaluating the impact and effects of automation diffusion in enterprises.

Process automation spread through the industrial world in both production and services during the 20th century, and more intensively in recent decades. The conditions that assured its wide diffusion were first the development of electronics, then informatics, and today information and communication technologies (ICT), as demonstrated in Fig. 6.1a–c. Since the late 1970s, periods of large investment in automation, followed by periods of reflection with critical revision of previous implementations and their impact on revenue, have taken place. This periodic attraction and subsequent revision of automation applications is still occurring, mainly in small to mid-sized enterprises (SME) as well as in several large firms.

Paradigmatic could be the case of Fiat, which reached the highest level of automation in their assembly lines late in the 1980s, whilst during the subsequent decade it suffered a deep crisis in which investments in automation seemed to be unprofitable. However, the next period – the present one – is characterized by significant growth for which the high level of automation already at its disposal has been a driver.

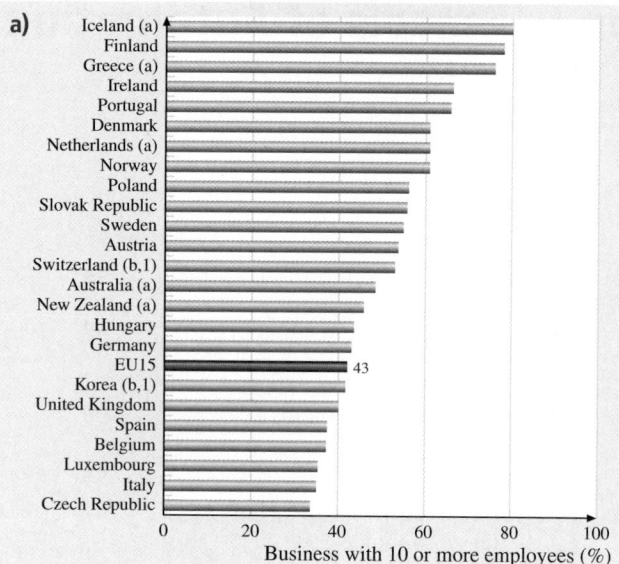

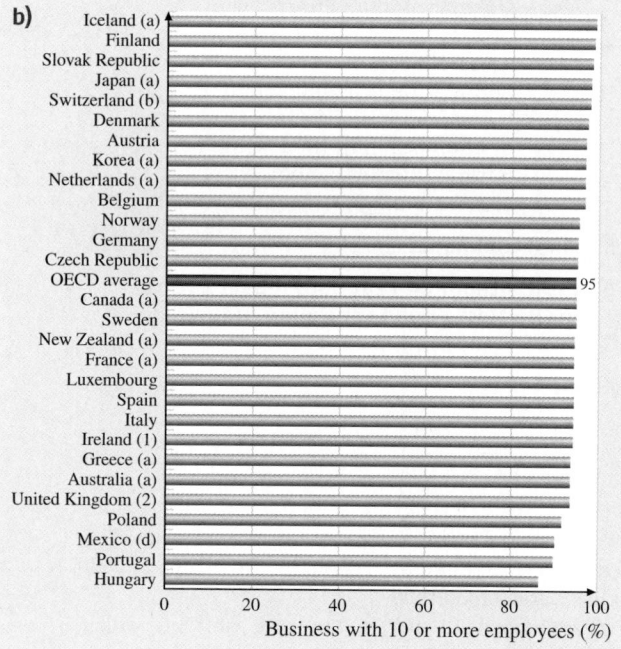

Fig. 6.1 (a) Use of information and communication technologies by businesses for returning filled forms to public authorities. The use of the Internet emphasizes the important role of transaction automation and the implementation of automatic control systems and services (source: OECD, ICT database, and Eurostat, community survey on ICT usage in households and by individuals, January 2008). **(b)** Business use of the Internet, 2007, as a percentage of businesses with ten or more employees. **(c)** ► Internet selling and purchasing by industry (2007). Percentage of businesses with ten or more employees in each industry group (source: OECD, ICT database, and Eurostat, community survey on ICT usage in enterprises, September 2008)

is rather specific since it was born together with – and as the instrument of – industrial automation. All other industrial sectors present a typical attitude of *strong but cautious interest* in automation. The principal motivation for this caution is the difficulty that managers face when evaluating the economic impact of automation on their own industrial organization. This difficulty results from the lack of simple methods to estimate economic impact and obtain an easily usable measure of automation revenue.

The aim of this chapter is to present an evaluation approach based on a compact and simple economic model to be used as a tool dedicated to SME managers, to analyze main effects of automation on production, labor, and costs.

The chapter is organized as follows. First, some basic concepts on which the evaluation of the automation effects are based are presented in Sect. 6.2. Then, a simple economic model, specifically developed for easy interpretation of the impact of automation on the enterprise, is discussed and its use for the analysis of some industrial situations is illustrated in Sect. 6.3. The most important effects of automation within the enterprise are considered in Sect. 6.4, in terms of its impact on production, incentivization and control of workers, and costs flexibility. In the final part of the chapter, mid-term effects of automation in the socioeconomic context are also analyzed. Considerations of such effects in some sectors of the Italian industrial system are discussed in Sect. 6.6, which can be considered as a typical example of the impact of automation on a developed industrial system, easily generalizable to other countries.

Automation implementation and perception of its convenience in the electronics sector is different from in the automotive sector. The electronics sector, however,

c)

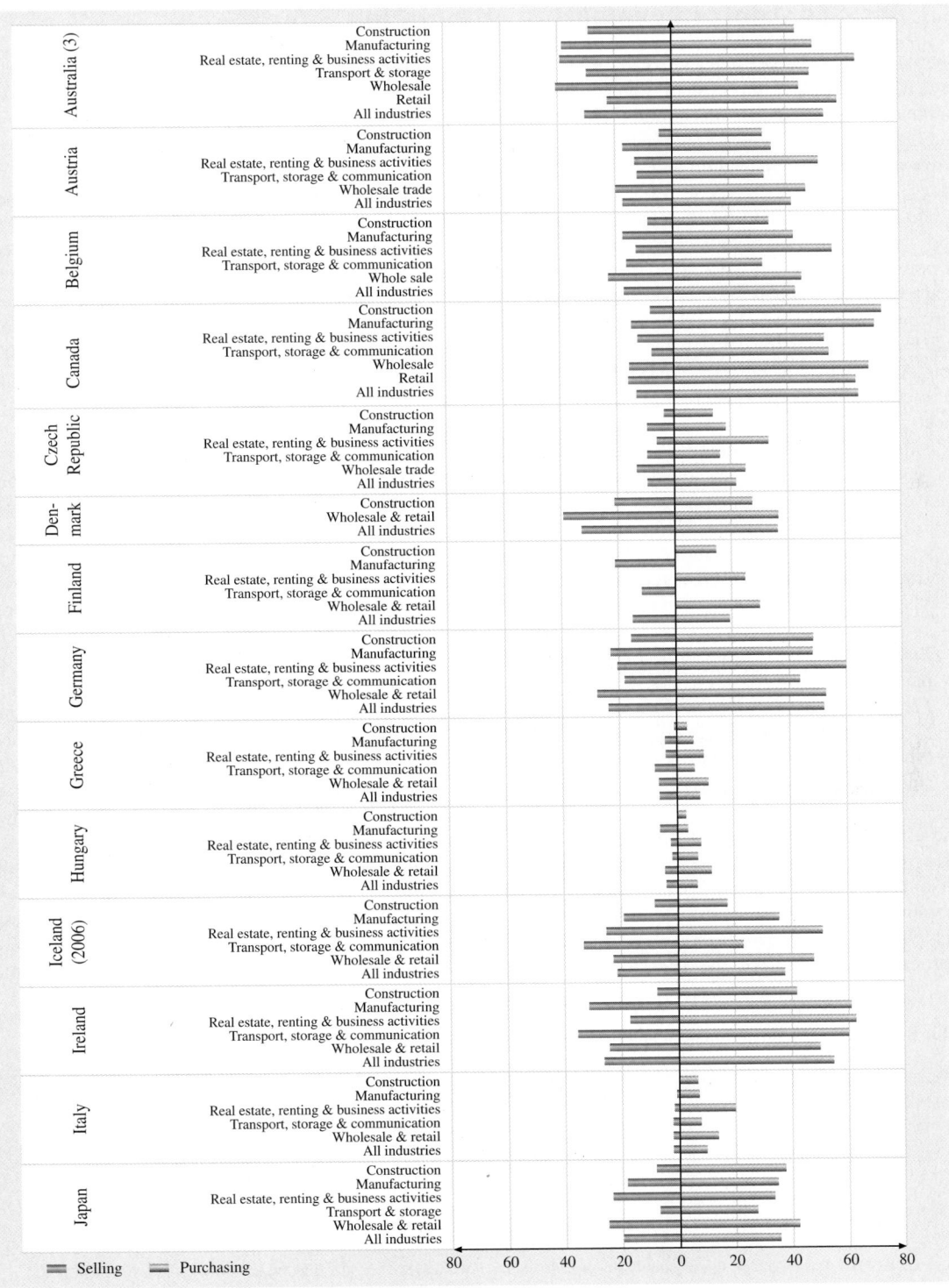

6.1 Basic Concepts in Evaluating Automation Effects

The desire of any SME manager is to be able to evaluate how to balance the cost of implementing some automated devices (either machining units or handling and moving mechanisms, or automated devices to improve production organization) and the related increase of revenue.

To propose a method for such an economic evaluation, it is first necessary to declare a simple catalogue of potential automation typologies, then to supply evidence of links between these typologies and the main variables of a SME which could be affected by process and labor modifications due to the applied automation. All variables to be analyzed and evaluated must be the usual ones presented in a standard balance sheet.

Analysis of a large number of SME clusters in ten European countries, developed during the collaborative demand and supply networks (CODESNET) project [6.1] funded by the European Commission, shows that the most important typologies of automation implementations in relevant industrial sectors can be classified as follows:

1. *Robotizing*, i.e., automation of manufacturing operations
2. *Flexibilitization*, i.e., *flexibility through automation*, by automating setup and supply
3. *Monitorizing*, i.e., *monitoring automation* through automating measures and operations control.

These three types of industrial automation can be related to effects on the process itself as well as on personnel. *Robotizing* allows the application of greater operation speed and calls for a reduced amount of direct work hours. *Flexibilitization* is crucial in mass customization, to reduce the lead time in the face of customer demands, by increasing the product mix, and by facilitating producer–client interaction. *Monitorizing* can indeed assure product quality for a wide range of final items through diffused control of work operations. Both automated flexibility and automated monitoring, however, require higher skills of personnel (Table 6.1).

However, a representation of the links between automation and either process attributes or personnel working time and skill, as outlined in Table 6.1, does not correspond to a method for evaluating the automation-induced profit in a SME, or in a larger enterprise. It only shows effects, whereas their impact on the SME balance sheet is what the manager wants to know.

To obtain this evaluation it is necessary:

1. To have clear that an investment in automation is generally relevant for any enterprise, and often critical for a SME, and typically can have an impact on mid/long-term revenue.
2. To realize that the success of an investment in terms of automation depends both on the amount of investment and on the reorganization of the workforce in the enterprise, which is a *microeconomic effect* (meaning to be estimated *within the enterprise*).
3. To understand that the impact of a significant investment in automation, made in an industrial sector, will surely have long-term and wide-ranging effects on employment at the *macroeconomic level* (i.e., at the level of the socioeconomic system or country).

All these effects must be interpreted by a single evaluation model which should be used:

- For a *microeconomic evaluation*, made by the enterprise manager, to understand how the two

Table 6.1 Links between automation and process/personnel in the firm

Automation typology induces ...	... effects on the process ...	... and effects on personnel
(a) Robotizing	Operation speed	Work reduction
(b) Flexibilitization	Response time to demand	Higher skills
(c) Monitorizing	Process accuracy and product quality	
Then automation calls for ...		**... and search for ...**
Investments		New labor positions
Investments and new labour positions should give rise to an expected target of production, conditioned on investments in high technologies and highly skill workforce utilization.		

above-mentioned *principal factors*, namely investment and workforce utilization, could affect the expected target of production, in case of a given automation implementation

- For a *macroeconomic evaluation*, to be done at the level of the industrial sector, to understand how relevant modification of personnel utilization, caused by the spread of automation, could be reflected in the socioeconomic system.

These are the two viewpoints according to which the automation economic impact will be analyzed upon the introduction of the above-mentioned interpretation model in Sect. 6.2.

6.2 The Evaluation Model

6.2.1 Introductory Elements of Production Economy

Preliminary, some definitions and notations from economics theory are appropriate (see the basic references [6.2–6]):

- A *production technique* is a combination of factors acquired by the market and applied in a product/service unit.
- *Production factors* will be simply limited to the *capital K* (i.e., industrial installation, manufacturing units, etc.), to *labor L*, and to *intermediate goods X* (i.e., goods and services acquired externally to contribute to production).
- A *production function* is given by the relation $Q = Q(K, L, X)$, which describes the output Q, *production of goods/services*, depending on the applied inputs.
- *Technological progress*, of which automation is the most relevant expression, must be incorporated into the capital K in terms of process and labor innovations through investments.
- *Technical efficiency* implies that a rational manager, when deciding on new investments, should make a choice among the available innovations that allow him to obtain the same increase of production without waste of inputs (e.g., if an innovation calls for K_0 units of capital and L_0 units of labor, and another one requires $L_1 > L_0$ units of labor, for the same capital and production, the former is to be preferred).
- *Economic efficiency* imposes that, if the combination of factors were different (e.g., for the same production level, the latter innovation has to use $K_1 < K_0$ capital units), then the rational manager's choice depends on the cost to be paid to implement the innovation, thus accounting also for production costs, not only quantity.

Besides these statements it has to be remarked that, according to economic theory, production techniques can be classified as either *fixed-coefficients technologies* or *flexible-coefficients technologies*. The former are characterized by nonreplaceable and strictly complementary factors, assuming that a given quantity of production can only be obtained by combining production factors at fixed rates, with the minimum quantities required by technical efficiency. The latter are characterized by the possibility of imperfect replacement of factors, assuming that the same production could be obtained through a variable, nonlinear combination of factors.

6.2.2 Measure of Production Factors

Concerning the measure of production factors, the following will be applied.

With regard to labor, the working time h (hours) done by personnel in the production system, is assumed to be a *homogeneous factor*, meaning that different skills can be taken into account through suitable weights.

In case of N persons in a shift and T shifts in unit time (e.g., day, week, month, etc.), the labor quantity L is given by

$$L = hNT . \tag{6.1}$$

In the following, the capital K refers to machines and installations in the enterprise. However, it could also be easily measured in the case of fixed-coefficient technologies: the capital stock, indeed, could be measured in terms of *standard-speed machine equivalent hours*. The capital K can also be further characterized by noting that, *over a short period*, it must be considered a fixed factor with respect to production quantity: excess capacity cannot be eliminated without suffering heavy losses.

Labor and intermediate goods should rather be variable factors with respect to produced quantities: excess stock of intermediate goods can be absorbed by reducing the next purchase, and excess workforce could be reduced gradually through turnover, or suddenly through dismissals or utilization of social measures in favor of unemployees.

In the long term, all production factors should be considered variables.

6.2.3 The Production Function Suggested by Economics Theory

The above-introduced concepts and measures allow to state the *flexible-coefficients production function*, by assuming (for the sake of simplicity and realism) that only the rate of intermediate goods over production is constant ($X/Q = b$)

$$Q = Q(K, L, X) = f(K, L) = X/b . \tag{6.2}$$

The production function is specified by the following properties:

- *Positive marginal productivity* (positive variation of production depending on the variation of a single factor, with the others being fixed, i. e., $\partial Q/\partial K > 0$, $\partial Q/\partial L > 0$)
- *Decreasing returns*, so that marginal productivity is a decreasing function with respect to any production factor, i. e., $\partial Q^2/\partial K^2 < 0$, $\partial Q^2/\partial L^2 < 0$.

In economics theory, from Clark and Marshall until now, the basis of production function analysis was the hypothesis of imperfect replacement of factors, assigning to each the law of decreasing returns.

The first-generation approach is due to *Cobb* and *Douglas* [6.7], who proposed a homogeneous production function whose factors can be additive in logarithmic form: $Q = AL^a K^b$, where the constant A summarizes all other factors except labor and capital.

This formulation has been used in empirical investigations, but with severe limitations.

The hypothesis of elasticity from the Cobb–Douglas function with respect to any production factor [such that $(\partial Q/Q)/(\partial L/L) = (\partial Q/Q)/(\partial K/K) = 1$], has been removed by *Arrow* et al. [6.8] and *Brown* and *De Cani* [6.9]. Subsequent criticism by *McFadden* [6.10] and *Uzawa* [6.11] gave rise to the more general form of variable elasticity function [6.12], up to the logarithmic expression due to *Christensen* et al. [6.13, 14], as clearly illustrated in the systematic survey due to *Nadiri* [6.15]. The strongest criticism on the flexible coefficient production function has been provided by *Shaikh* [6.16], but seems to have been ignored. The final step was that of abandoning direct estimation of the production function, and applying indirect estimation of the cost function [6.2, 17–20], up to the most recent theories of *Dievert* [6.21] and *Jorgenson* [6.22].

A significant modification to the analysis approach is possible based on the availability of large statistical databases of profit and loss accounts for enterprises, compared with the difficulty of obtaining data concerning production factor quantities. This approach does not adopt any explicit interpretation scheme, thus upsetting the approach of economics theory (deductive) and engineering (pragmatic), depending only on empirical verification. A correct technological–economic approach should reverse this sequence: with reference to production function analysis, it should be the joint task of the engineer and economist to propose a functional model including the typical parameters of a given production process. The econometric task should be applied to verify the proposed model based on estimation of the proposed parameters. In the following analysis, this latter approach will be adopted.

6.3 Effects of Automation in the Enterprise

6.3.1 Effects of Automation on the Production Function

The approach on which the following considerations are based was originally developed by *Luciano* and *Ravazzi* [6.23], assuming the extreme case of *production only using labor*, i. e., without employing capital (e.g., by using elemental production means). In this case, a typical human characteristic is that a worker can produce only at a rate that decreases with time during

his shift. So, the marginal work productivity is decreasing.

Then, taking account of the work time h of a worker in one shift, the decreasing efficiency of workers with time suggests the introduction of another measure, namely the *efficiency unit E*, given by

$$E = h^\alpha , \tag{6.3}$$

where $0 < \alpha < 1$ is the *efficiency elasticity* with respect to the hours worked by the worker.

Condition (6.3) includes the assumption of decreasing production rate versus time, because the derivative of E with respect to h is positive but the second derivative is negative.

Note that the efficiency elasticity can be viewed as a measure of the worker's strength.

By denoting λ_E as the *production rate of a work unit*, the production function (6.2) can be rewritten as

$$Q = \lambda_E ENT \ . \tag{6.4}$$

Then, substitution of (6.1) and (6.3) into (6.4), gives rise to a representation of the *average production rate*, which shows the decreasing value with worked hours

$$\lambda_L = Q/L = \lambda_E h^{\alpha-1} \ , \tag{6.5}$$

with $d\lambda_L/dh = (\alpha-1)\lambda_E h^{\alpha-2} < 0$.

Let us now introduce the capital K as the auxiliary instrument of work (a computer for intellectual work, an electric drill for manual work, etc.), but *without any process automation*.

Three questions arise:

1. How can capital be measured?
2. How can the effects produced by the association of capital and work be evaluated?
3. How can capital be included in the production function?

With regard to the first question, the usual industrial approach is to refer to the utilization time of the production instruments during the working shift. Then, let the capital K be expressed in terms of *hours of potential utilization* (generally corresponding to the working shift).

Utilization of more sophisticated tools (e.g., through the application of more automation) induces an increase in the production rate per hour. Denoting by $\gamma > 1$ a coefficient to be applied to the production rate λ_L, in order to measure its increase due to the effect of capital utilization, the average production rate per hour (6.5) can be rewritten as

$$\lambda_L = \gamma \lambda_E h^{\alpha-1} \ . \tag{6.6}$$

The above-mentioned effect of capital is the only one considered in economics theory (as suggested by the Cobb–Douglas function). However, another significant effect must also be accounted for: the capital's impact on the workers' strength in terms of labor (as mentioned in the second question above).

Automated systems can not only increase production rate, but can also strengthen labor efficiency elasticity, since they reduce physical and intellectual fatigue. To take account of this second effect, condition (6.6) can be reformulated by including a positive parameter $\delta > 0$ that measures the increase of labor efficiency and whose value is bounded by the condition $0 < (\alpha + \delta) < 1$ so as to maintain the hypothesis of decreasing production rate with time

$$\lambda_L = \gamma \lambda_E h^{\alpha+\delta-1} \ . \tag{6.7}$$

According to this model, a *labor-intensive technique* is defined as one in which capital and labor cooperate together, but in which the latter still dominates the former, meaning that a reduction in the labor marginal production rate still characterizes the production process $(\alpha + \delta < 1)$, even if it is reduced by the capital contribution $(\delta > 0)$.

The answer to the third question, namely how to include capital in the production function, strictly depends on the characteristics of the relevant machinery. Whilst the workers' nature can be modeled based on the assumption of decreasing production rates with time, production machinery does not operate in this way (one could only make reference to wear, although maintenance, which can prevent modification of the production rate, is reflected in capital cost).

On the contrary, it is the human operator who imposes his biological rhythm (e.g., the case of the speed of a belt conveyor that decreases in time during the working shift). This means that capital is linked to production through fixed coefficients: then the marginal production rate is not decreasing with the capital.

Indeed, a *decreasing utilization rate of capital* has to be accounted for as a consequence of the decreasing rate of labor. So, the hours of potential utilization of capital K have to be converted into productive hours through a *coefficient of capital utilization* θ and transformed into production through a *constant capital-to-production rate* parameter v

$$Q = \theta K/v = \theta \lambda_K K \ , \tag{6.8}$$

where $\lambda_K = 1/v$ is a measure of the capital constant productivity, while $0 < \theta < 1$ denotes the ratio between the effective utilization time of the process and the time during which it is available (i. e., the working shift).

Dividing (6.8) by L and substituting into (6.7), it follows that

$$\theta = v\gamma \lambda_E h^{\alpha+\delta-1} \ , \tag{6.9}$$

thus showing how the utilization rate of capital could fit the decreasing labor yield so as to link the mechanical rhythm of capital to the biological rhythm of labor.

Condition (6.9) leads to the first conclusion: *in labor-intensive systems (in which labor prevails over capital), decreasing yields occur, but depending only on the physical characteristics of workers and not on the constant production rates of production machinery.*

We should also remark on another significant consideration: *that new technologies also have the function of relieving labor fatigue by reducing undesirable effects due to marginal productivity decrease.*

This second conclusion gives a clear suggestion of the effects of automation concerning reduction of physical and intellectual fatigue. Indeed, automation implies dominance of capital over labor, thus constraining labor to a mechanical rhythm and removing the conditioning effects of biological rhythms.

This situation occurs when $\alpha + \delta = 1$, thus modifying condition (6.7) to

$$\lambda_L = Q/L = \gamma \lambda_E \,, \tag{6.10}$$

which, in condition (6.9), corresponds to $\theta = 1$, i.e., no pause in the labor rhythm.

In this case automation transforms the decreasing yield model into a constant yield model, i.e., the labor production rate is constant, as is the capital production rate, if capital is fully utilized during the work shift. Then, *capital-intensive processes* are defined as those that incorporate high-level automation, i.e., $\alpha + \delta \to 1$.

A number of examples of capital-intensive processes can be found in several industrial sectors, often concerning simple operations that have to be executed a very large number of times. A typical case, even if not often considered, are the new *intensive picking* systems in large-scale automated warehouses, with increasing diffusion in large enterprises as well as in industrial districts.

Section 6.6 provides an overview in several sectors of the two ratios (*capital/labor* and *production/labor*) that, according to the considerations above, can provide a measure of the effect of automation on production rate. Data are referred to the Italian economic/industrial system, but similar considerations could be drawn for other industrial systems in developed countries. Based on the authors' experience during the CODESNET project development, several European countries present aspects similar to those outlined in Sect. 6.6.

6.3.2 Effects of Automation on Incentivization and Control of Workers

Economic theory recognizes three main motivations that suggest that the enterprise can achieve greater wage efficiency than the one fixed by the market [6.24]:

1. The need to minimize costs for hiring and training workers by reducing voluntary resignations [6.25, 26]
2. The presence of information asymmetry between the workers and the enterprise (as only workers know their ability and diligence), so that the enterprise tries to engage the best elements from the market through ex ante incentives, and then to force qualified employees to contribute to the production process (the moral hazard problem) without resulting in too high supervision costs [6.27–29]
3. The specific features of production technologies that may force managers to allow greater autonomy to some worker teams, while paying an incentive in order to promote better participation in teamwork [6.30–32].

The first motivation will be discussed in Sect. 6.4, concerning the flexibility of labor costs, while the last one does not seem to be relevant. The second motivation appears to be crucial for labor-intensive systems, since system productivity cannot only be dependent on technologies and workers' physical characteristics, but also depends greatly on workers propensity to contribute.

So, system productivity is a function of wages, and maximum profit can no longer be obtained by applying the economic rule of marginal productivity equal to wages fixed by market. It is the obligation of the enterprise to determine wages so as to assure maximum profit.

Let the *production rate of a work unit* λ_E increase at a given rate in a first time interval in which the *wage* w_E *per work unit* plays a strongly incentive role, whilst it could increase subsequently at a lower rate owing to the reduction of the wages marginal utility, as modeled in the following expression, according to the economic hypothesis of the effort function

$$\lambda_E = \lambda_E(w_E), \quad \text{where } \lambda_E(0) = 0; \ d\lambda_E/dw_E > 0 \,;$$
$$\text{and } d^2\lambda_E/dw_E^2 \geq 0 \text{ if } w_E \geq \hat{w} \,;$$
$$d^2\lambda_E/dw_E^2 \leq 0 \text{ if } w_E \leq \hat{w} \,; \tag{6.11}$$

where $\hat{w}$ is the critical wages which forces a change of yield from increasing to decreasing rate.

In labor-intensive systems, the average production rate given by (6.7) can be reformulated as

$$\lambda_L = \gamma \lambda_E E/h \,. \tag{6.12}$$

Now, let $M = V_a - wL$ be the *contribution margin*, i.e., the difference between the production added value V_a and the labor cost: then the *unitary contribution margin per labor unit m* is defined by the rate of *M* over the labor *L*

$$m = M/L = \hat{p}\lambda_{\mathrm{L}} - w \,, \tag{6.13}$$

where $\hat{p} = (p - p_{\mathrm{X}})\beta$ is the difference between the sale price *p* and the cost p_{X} of a product's parts and materials, transformed into the final product according to the utilization coefficient $\beta = X/Q$.

It follows that $\hat{p}\lambda_{\mathrm{L}}$ is a measure of the added value, and that the wages w_{E} per work unit must be transformed into real wages through the rate of work units *E* over the work hours *h* of an employee during a working shift, according to

$$w = w_{\mathrm{E}}E/h \,. \tag{6.14}$$

The goal of the enterprise is to maximize *m*, in order to gain maximum profit, i.e.,

$$\max(m) = \left[\hat{p}\gamma\lambda_{\mathrm{E}}(w_{\mathrm{E}}) - w_{\mathrm{E}}\right]E/h \,. \tag{6.15}$$

The first-order optimal condition gives

$$\partial m/\partial h = \left[\hat{p}\gamma\lambda_{\mathrm{E}}(w_{\mathrm{E}}) - w_{\mathrm{E}}\right]\partial(E/h)/\partial h = 0$$
$$\Rightarrow \gamma\lambda_{\mathrm{E}}(w_{\mathrm{E}}) = w_{\mathrm{E}}/\hat{p} \,, \tag{6.16}$$
$$\partial m/\partial w_{\mathrm{E}} = (E/h)\left[\hat{p}\gamma(\partial\lambda_{\mathrm{E}}/\partial w_{\mathrm{E}}) - 1\right] = 0$$
$$\Rightarrow \hat{p}\gamma(\partial\lambda_{\mathrm{E}}/\partial w_{\mathrm{E}}) = 1 \,. \tag{6.17}$$

By substituting (6.17) into (6.16), the maximum-profit condition shows that the elasticity of productivity with respect to wages ε_{λ} will assume a value of unity

$$\varepsilon_{\lambda} = (\partial\lambda_{\mathrm{E}}/\partial w_{\mathrm{E}})(w_{\mathrm{E}}/\lambda_{\mathrm{E}}) = 1 \,. \tag{6.18}$$

So, the enterprise could maximize its profit by forcing the percentage variation of the efficiency wages to be equal to the percentage variation of the productivity $\partial\lambda_{\mathrm{E}}/\lambda_{\mathrm{E}} = \partial w_{\mathrm{E}}/w_{\mathrm{E}}$. If so, it could obtain the optimal values of wages, productivity, and working time.

As a consequence, the duration of the working shift is an endogenous variable, which shows why, in *labor-intensive systems*, the working hours for a worker can differ from the contractual values.

On the contrary, in *capital-intensive systems* with wide automation, it has been noted before that $E/h = 1$ and $\lambda_{\mathrm{E}} = \bar{\lambda}_{\mathrm{E}}$, because the mechanical rhythm prevails over the biological rhythm of work. In this case, efficiency wages do not exist, and the solution of maximum profit simply requires that wages be fixed at the minimum contractual level

$$\max(m) = \hat{p}\lambda_{\mathrm{L}} - w = \hat{p}\gamma\bar{\lambda}_{\mathrm{E}} - w \Rightarrow \min(w) \,. \tag{6.19}$$

As a conclusion, in labor-intensive systems, if λ_{E} could be either observed or derived from λ_{L}, incentive wages could be used to maximize profit by asking workers for optimal efforts for the enterprise. In capital-intensive systems, where automation has canceled out deceasing yield and mechanical rhythm prevails in the production process, worker incentives can no longer be justified. The only possibility is to reduce absenteeism, so that a share of salary should be reduced in case of negligence, not during the working process (which is fully controlled by automation), but outside.

In labor-intensive systems, as in personal service production, it could be difficult to measure workers' productivity: if so, process control by a supervisor becomes necessary.

In capital-intensive systems, automation eliminates this problem because the process is equipped with devices that are able to detect any anomaly in process operations, thus preventing inefficiency induced by negligent workers. In practice automation, by forcing fixed coefficients and full utilization of capital (process), performs itself the role of a working conditions supervisor.

6.3.3 Effects of Automation on Costs Flexibility

The transformation of a labor-intensive process into a capital-intensive one implies the modification of the cost structure of the enterprise by increasing the capital cost (that must be paid in the short term) while reducing labor costs.

Let the total cost C_{T} be defined by the costs of the three factors already considered, namely, intermediate goods, labor, and capital, respectively,

$$C_{\mathrm{T}} = p_{\mathrm{X}}X + wL + c_{\mathrm{K}}K \,, \tag{6.20}$$

where c_{K} denotes the unitary cost of capital.

Referring total cost to the production *Q*, the cost per production unit *c* can be stated by substituting the conditions (6.2) and (6.10) into (6.20), and assuming constant capital value in the short term

$$c = C_{\mathrm{T}}/Q = p_{\mathrm{X}}\beta + w/\lambda_{\mathrm{L}} + c_{\mathrm{K}}K/Q \,. \tag{6.21}$$

In labor-intensive systems, condition (6.21) can also be rewritten by using the efficiency wages w_{E}^{*} allocated in order to obtain optimal productivity $\lambda_{\mathrm{L}}^{*} = \gamma\lambda_{\mathrm{E}}(w_{\mathrm{E}}^{*})h^{*\alpha+\delta-1}$, as shown in Sect. 6.3.1,

$$c = p_{\mathrm{X}}\beta + w_{\mathrm{E}}^{*}/\lambda_{\mathrm{L}}^{*} + c_{\mathrm{K}}K/Q \,. \tag{6.22}$$

On the contrary, in capital-intensive systems, the presence of large amounts of automation induces the following effects:

1. Labor productivity λ_A surely greater than that which could be obtained in labor-intensive systems ($\lambda_A > \lambda_L^*$)
2. A salary w_A that does not require incentives to obtain optimum efforts from workers, but which implies an additional cost with respect to the minimum salary fixed by the market ($w_A \lessgtr w_E^*$), in order to select and train personnel
3. A positive correlation between labor productivity and production quantity, owing to the presence of qualified personnel who the enterprise do not like to substitute, even in the presence of temporary reductions of demand from the final product market

$$\lambda_A = \lambda_A(Q), \partial\lambda_A/\partial Q > 0, \partial^2\lambda_A/\partial Q^2 = 0$$

$$(6.23)$$

4. A significantly greater cost of capital, due to the higher cost of automated machinery, than that of a labor-intensive process ($c_{KA} > c_K$), even for the same useful life and same rate of interest of the loan.

According to these statements, the unitary cost in capital-intensive systems can be stated as

$$c_A = p_X\beta + w_A/\lambda_A(Q) + c_{KA}K/Q .   (6.24)$$

Denoting by *profit per product unit* π the difference between sale price and cost

$$\pi = p - c ,   (6.25)$$

the *relative advantage of automation D*, can be evaluated by the following condition, obtained by substituting (6.24) and then (6.22) into (6.25)

$$\begin{aligned} D &= \pi_A - \pi \\ &= w_E^*/\lambda_L^* - w_A/\lambda_A(Q) - (c_{KA} - c_K)K/Q . \end{aligned}$$

$$(6.26)$$

Except in extreme situations of large underutilization of production capacity (Q should be greater than the critical value Q_C), the inequality $w_E^*/\lambda_L^* > w_A/\lambda_A(Q)$ denotes the necessary condition that assures that automated production techniques can be economically efficient. In this case, the greater cost of capital can be counterbalanced by greater benefits in terms of labor cost per product unit.

In graphical terms, condition (6.26) could be illustrated as a function $D(Q)$ increasing with the production quantity Q, with positive values for production greater than the critical value Q_C. This means that large amounts of automation can be adopted for high-quantity (mass) production, because only in this case can the enterprise realize an increase in marginal productivity sufficient to recover the initial cost of capital. This result, however, shows that automation could be more risky than labor-intensive methods, since production variations induced by demand fluctuations could reverse the benefits. This risk, today, is partly reduced by mass-customized production, where automation and process programming can assure process flexibility able to track market evolutions [6.33].

6.4 Mid-Term Effects of Automation

6.4.1 Macroeconomics Effects of Automation: Nominal Prices and Wages

The analysis has been centered so far on microeconomics effects of automation on the firm's costs, assuming product prices are given, since they would be set in the market depending on demand–offer balance in a system with perfect competition.

Real markets however lack a Walrasian auctioneer and are affected by the incapacity of firms to know in advance (rational expectations) the market demand curve under *perfect competition*, and their own demand curve under *imperfect competition*. In the latter case, enterprises cannot maximize their profit on the basis of demand elasticity, as proposed by the economic the-

ory of imperfect competition [6.34–36]. Therefore, they cannot define their price and the related markup on their costs.

Below we suggest an alternative approach, which could be described as *technological–managerial*.

The balance price is not known a priori and price setting necessarily concerns enterprises: they have to submit to the market a price that they consider to be profitable, but not excessive because of the fear that competitors could block selling of all the scheduled production. Firms calculate the sale price p on the basis of a full unit cost c, including a minimum profit, considered as a *normal* capital remuneration.

The full cost is calculated corresponding to a scheduled production quantity Q^e that can be allegedly sold on the market, leaving a small share of productive capacity $\bar{Q}$ potentially unused

$$Q^e \leq \bar{Q} = \theta_K \lambda_K \bar{K} = \theta_K \bar{K}/v \,, \qquad (6.27)$$

where $\theta_K = 1$ in the presence of automation, while $v = 1/\lambda_K$ is – as stated above – the capital–product connection defining technology adopted by firms for full productive capacity utilization.

The difference $\bar{Q} - Q^e$ therefore represents the unused capacity that the firm plans to leave available for production demand above the forecast.

To summarize, the sales price is fixed by the enterprise, resorting to connection (6.21), relating the *break-even* point to Q^e

$$p = c(Q^e) = p_X \beta + w/\lambda_L + c_K v^e \,, \qquad (6.28)$$

where $v^e = \bar{K}/Q^e \geq v$ (for $Q^e \leq \bar{Q}$) is the programmed capital–product relationship.

In order to transfer this relation to the macroeconomics level, we have to express it in terms of added value (Q is the gross saleable production), as the *gross domestic product* (GDP) results from aggregation of the added values of firms. Therefore we define the added value as

$$PY = pQ - p_X X = (p - p_X \beta)Q \,,$$

where P represents the *prices general level* (the average price of goods that form part of the GDP) and Y the *aggregate supply* (that is, the GDP). From this relation, P is given by

$$P = (p - p_X \beta)/\theta_Y \,, \qquad (6.29)$$

where $\theta_Y = Y/Q$ measures the degree of vertical integration of the economic system, which in the medium/short term we can consider to be steady ($\theta_Y = \bar{\theta}_Y$).

By substituting relation (6.28) into (6.29), the price equation can be rewritten as

$$P = w/\hat{\lambda} + c_K \hat{v} \,, \qquad (6.30)$$

where work productivity and the capital/product ratio are expressed in terms of added value ($\hat{\lambda} = \theta_Y \lambda_L = Y/L$ and $\hat{v} = v^e/\theta_Y = \bar{K}/Y^e$).

In order to evaluate the effects of automation at the macroeconomic level, it is necessary to break up the capital unit cost c_K into its components:

- The initial purchase unit price of capital P_K^0
- The sample gross profit ρ^* sought by firms (as a percentage to be applied to the purchase price), which embodies both amortization rate d and the performance r^* requested by financiers to remunerate

debt (subscribed by bondholders) and risk capital (granted by owners).

So the following definition can be stated

$$c_K = \rho^* P_K^0 = \left[l^* P_K^0 + (1 - l^*) P_K\right]/a_{\bar{n}\,|r^*} \,. \qquad (6.31)$$

where

- $0 < l^* = D^*/(P_K^0 \bar{K}) < 1$ is the *leverage* (debt amount subscribed in capital stock purchase)
- $1 - l^*$ is the amount paid by owners
- $P_K > P_K^0$ is the *substitution price* of physical capital at the deadline and
- $a_{\bar{n}\,|r^*} = \frac{1-(1+r^*)^{-n}}{r^*}$ is the discounting back factor.

Relation (6.31) implies that the aim of the firm is to maintain unchanged the capital share amount initially brought by owners, while the debt amount is recovered to its face (book) value, as generally obligations are refunded at original monetary value and at a fixed interest rate stated in the contract.

It is also noteworthy that the indebtedness ratio l is fixed at its optimal level l^*, and the earning rate depends on it, $r^* = r(l^*)$, because r decreases as the debt-financed share of capital increases due to advantages obtained from the income tax deductibility of stakes [6.37, 38], and increases as a result of failure costs [6.39–41] and agency costs [6.42], which in turn grow as l grows. The optimal level l^* is obtained based on the balance between costs and marginal advantages.

The relation (6.31) can be rewritten by using a Taylor series truncated at the first term

$$c_K = (d + r^*)\left[l^* P_K^0 + (1 - l^*) P_K\right] \,, \qquad (6.32)$$

which, in the two extreme cases $P_K^0 = P_K$ and $P_K = P_K^0$, can be simplified to

$$c_K = (d + r^*) P_K \,, \qquad (6.33a)$$
$$c_K = (d + r^*) P_K^0 \,. \qquad (6.33b)$$

This solution implies the existence of monetary illusion, according to which capital monetary revaluation (following inflation) is completely abandoned, thus impeding owners from keeping their capital intact.

This irrational decision is widely adopted in practice by firms when inflation is low, as it rests upon the accounting procedure codified by European laws that calculated amortization and productivity on the basis of *book value*. Solution (6.33a) embodies two alternatives:

- Enterprises' decision to maintain physical capital undivided [6.43], or to recover the whole capital market value at the deadline, instead of being restricted to the capital value of the stakeholders, in order to guarantee its substitution without reducing existing production capacity; in this case the indebtedness with repayment to nominal value (at fixed rate) involves an extra profit Π for owners (resulting from debt devaluation), which for unity capital corresponds to the difference between relation (6.33a) and (6.32)

$$\Pi = (d + r^*)l^*\left(P_K - P_K^0\right), \tag{6.34}$$

as clarified by *Cohn* and *Modigliani* [6.44].

- Subscription of debts at variable interest rate, able to be adjusted outright to inflation rate to compensate completely for debt devaluation, according to Fisher's theory of interest; these possibilities should be rules out, as normally firms are insured against debt cost variation since they sign fixed-rate contracts.

Finally the only reason supporting the connection (6.33a) remains the first (accounting for inflation), but generally firms' behavior is intended to calculate c_K according to relation (6.33b) (accounting to *historical costs*). However, rational behavior should compel the use of (6.32), thereby avoiding the over- or underestimation of capital cost ex ante.

In summary, the prices equation can be written at a macroeconomics level by substituting (6.32) into (6.30)

$$P = w/\hat{\lambda} + (d + r^*)\hat{v}\left[l^* P_K^0 + (1 - l^*)P_K\right]. \tag{6.35a}$$

This relation is simplified according to the different aims of the enterprises:

1. *Keeping physical capital intact*, as suggested by accountancy for inflation, presuming that capital price moves in perfect accordance with product price ($P_K^0 = P_K = p_k P$ over $p_k = P_K/P > 1$ is the capital-related *price* compared with the product one)

$$P_a = \frac{w/\hat{\lambda}}{1 - (d + r^*)\hat{v}p_k} = \left(1 + \mu_a^*\right)w/\hat{\lambda}. \tag{6.35b}$$

2. *Integrity of capital conferred by owners*, as would be suggested by rational behavior ($P_K = p_k P > P_K^0$)

$$P_b = \frac{w/\hat{\lambda} + l^*(d + r^*)\hat{v}P_K^0}{1 - (1 - l^*)(d + r^*)\hat{v}p_k}. \tag{6.35c}$$

3. *Recovery of nominal value of capital*, so as only to take account of *historical costs* ($P_K = P_K^0$)

$$P_c = w/\hat{\lambda} + (d + r^*)\hat{v}P_K^0. \tag{6.35d}$$

Only in the particular case of (6.35b) can the price level be obtained by applying a steady profit-margin factor $(1 + \mu_a^*)$ to labor costs per product unit $(w/\hat{\lambda})$.

The *markup* μ^* desired by enterprises (the percentage calculated on variable work costs in order to recover fixed capital cost) results in the following expressions for the three cases above:

1. *Keeping physical capital intact*; in this case, the mark-up results independent of the nominal wage level w

$$\mu_a^* = P_a\hat{\lambda}/w - 1 = \frac{(d + r^*)\hat{v}p_k}{1 - (d + r^*)\hat{v}p_k}. \tag{6.36a}$$

2. *Integrity of capital conferred by owners,*

$$\mu_b^* = P_b\hat{\lambda}/w - 1$$
$$= \frac{(d + r^*)\hat{v}\left[l^* P_K^0\hat{\lambda}/w + (1 - l^*)p_k\right]}{1 - (1 - l^*)(d + r^*)\hat{v}p_k}. \tag{6.36b}$$

3. *Recovery of nominal value of capital,*

$$\mu_c^* = P_c\hat{\lambda}/w - 1 = (d + r^*)\hat{v}P_K^0\hat{\lambda}/w. \tag{6.36c}$$

Note that in case 1, enforcing automation generally implies adoption of manufacturing techniques whose relative cost p_k increases at a rate greater than proportionally with respect to the capital–product rate reduction $\hat{v}$. The desired markup must then be augmented in order to ensure coverage of capital. In relation (6.35b) this effect is compensated because of productivity $\hat{\lambda}$ growth due to greater automation, so that on the whole the effect of automation on the general price level is beneficial: for given nominal salary, automation reduces price level.

In cases 2 and 3 of rational behavior, referring to (6.35c) in which the enterprise is aware that its debt is to be refunded at its nominal value, and even more so in the particular case (6.35d) in which the firm is enduring monetary illusion, the desired *markup* is variable, a decreasing function of monetary wage.

Therefore the markup theory is a simplification limited to the case of maintaining physical capital intact, and neglecting effects of capital composition (debt refundable at nominal value).

Based on the previous prices equations it follows that an increase of nominal wages or profit rate sought by enterprises involves an increase in prices general level. In comparison with w, the elasticity is only unity in (6.35b) and diminishes increasingly when passing to (6.35c) and (6.35d).

A percentage increase of nominal salaries is therefore transferred on the level of prices in the same proportion – as stated by *markup* theory – only in the particular case of (6.35b). In the other cases the translation results less than proportional, as the sought *markup* decreases as w increases.

Nominal Prices and Wages: Some Conclusions

Elasticity of product price decreases if the term $\hat{v}P_K^0\hat{\lambda}/w$ increases, representing the ratio between the nominal value of capital ($P_K^0\bar{K}$) and labor cost (wL). Since automation necessarily implies a remarkable increase of this ratio, it results in minor prices sensibility to wages variation. In practice, automation implies beneficial effects on inflation: for increasing nominal wages, a high-capital-intensity economy is less subject to inflation *shocks* caused by wage rises.

6.4.2 Macroeconomics Effects of Automation in the Mid-Term: Actual Wages and Natural Unemployment

In order to analyze effects of automation on macroeconomic balance in the mid term, it is necessary to convey the former equations of prices in terms of real wages, dividing each member by P, in order to take account of $\omega = w/P$, which represents the maximum wage that firms are prepared to pay to employees without giving up their desired profit rate.

1. Keeping physical capital intact ($P_K^0 = P_K = p_k P$)

$$\omega_a = \hat{\lambda}[1 - (d+r^*)\hat{v}p_k] = \hat{\lambda}/(1+\mu_a^*) \quad (6.37a)$$

2. Wholeness of capital granted by owners ($P_K = p_k P > P_K^0$)

$$\omega_b = \hat{\lambda}\{1 - (d+r^*)\hat{v}[p_k - l^*(p_k - P_K^0/P)]\} \quad (6.37b)$$

3. Recovery of capital's nominal value ($P_K = P_K^0$)

$$\omega_c = \hat{\lambda}[1 - (d+r^*)\hat{v}P_K^0/P] \quad (6.37c)$$

In the borderline case of keeping physical capital intact (6.37a) only one level of real salary ω_a exists consistent with a specified level of capital productivity r^*, given work productivity $\hat{\lambda}$, amortization rate d, capital/product ratio $\hat{v}$ (corresponding to normal use of plants), and relative price p_k of capital with respect to product.

The value of ω_a can also be expressed as a link between work productivity and profit margin factor $(1+\mu_a^*)$ with variable costs, assuming that the desired *markup* is unchanging in comparison with prices.

In the rational case of corporate stock integrity and in the generally adopted case of recovery of capital nominal value, an increasing relation exists between real salary and general price level, with the desired markup being variable, as shown in connections (6.36b) and (6.36c).

The elasticity of ω with respect to P turns out to increase from (6.37a) to (6.37c)

$$\eta_a = (\partial\omega/\partial P)(P/\omega) = 0$$
$$< \eta_b = (\hat{\lambda}/\omega_b)l^*(d+r^*)\hat{v}P_K^0/P$$
$$< \eta_c = (\hat{\lambda}/\omega_c)(d+r^*)\hat{v}P_K^0/P$$
$$= \hat{\lambda}/\omega_c - 1 \underset{>}{\overset{<}{=}} 1 \quad \text{with} \quad \omega_c \underset{<}{\overset{>}{=}} \hat{\lambda}/2 . \quad (6.37d)$$

In cases 2 and 3 growth in the prices general level raises the added value share intended for normal capital remuneration, leaving firms a markup to be used in bargaining in order to grant employees a pay rise in real salary, even while keeping normal profit rate (the calculated level with programmed production) steady.

This markup of real wage bargaining depends on the choice of the capital recovery system: it is higher in the case of fund illusion and lower in the case in which capital stock is intended to be maintained undivided.

Remark: In all three cases above, the level of real salary depends on the degree of production process automation: in economic systems where this is high, productivity of work ($\hat{\lambda}$) and capital ($1/\hat{v}$) are necessarily higher than the related price of capital (p_k and P_K^0/P), so the real wage that the firms are willing to concede in bargaining is higher than the one affecting work-intensive economic systems. In substance, automation weakens firms' resistance in wage bargaining, making them more willing to concede higher real wages.

In order to qualify the positive role of automation we complete the mid-term period macroeconomic model with the wage equation from the workers' side.

Macroeconomics theory believes that, in the mid term, nominal wages w asked for by workers in syndicated or individual bargaining (or imposed by firms in order to select and extract optimal effort in production process) depend on three variables: the unemployment rate of man power u, other factors absorbed in a synthetic variable z, and the expected level of prices P^e.

So it is possible to write

$$w = \omega(u, z)P^e . \quad (6.38)$$

Part A | 6.4

Higher unemployment is supposed to weaken workers' contractual strength, compelling them to accept lower wages, and vice versa that a decrease of unemployment rate leads to requests for an increase of real wages ($\partial\omega/\partial u < 0$).

The variable z can express the effects of unemployment increase, which would compel workers to ask for a pay raise, because any termination would appear less risky in terms of *social salary* (the threshold above which an individual is compelled to work and under which he is not prepared to accept, at worst choosing unemployment). Similar effects would be induced by legal imposition of minimum pay and other forms of worker protection, which – making discharge more difficult – would strengthen workers' position in wage bargaining.

Regarding the expected level of prices, it is supposed that

$$\partial w/\partial P^e = w/P^e > 0 \Rightarrow (\partial w/\partial P^e)(P^e/w) = 1$$

to point out that rational subjects are interested in real wages (their buying power) and not nominal ones, so an increase in general level of prices would bring about an increase in nominal wages in the same proportion. Wages are however negotiated by firms on nominal terms, on the basis of a price foresight (P^e) for the whole length of the contract (generally more than 1 year), during which monetary wages are not corrected if $P \neq P^e$.

To accomplish this analysis, in the mid-term period, wage bargaining is supposed to have real wage as a subject (excluding systematic errors in expected inflation foresight), so that $P = P^e$ and (6.38) can be simplified to

$$\omega = \omega(u, z) \,. \tag{6.39}$$

Figure 6.2 illustrates this relation with a decreasing curve WS (*wage setting*), on Cartesian coordinates for a given level of z.

Two straight lines, PS (*price setting*), representing price equations (6.37a), considering mainly the borderline case of maintaining physical capital intact, are reported:

- The upper line PS_A refers to an economic system affected by a high degree of automation, or by technologies where $\theta_k = 1$ and work and capital productivity are higher, being able to contrast the capital's relative higher price; in this case firms are prone to grant higher real wages.

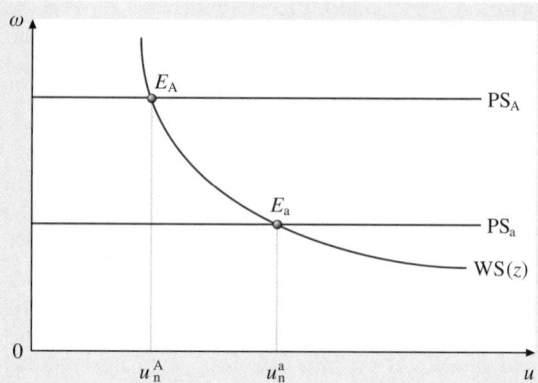

Fig. 6.2 Mid-term equilibrium with high and low automation level

- The lower line PS_a, characterizing an economy with a lower degree of automation, resulting in less willingness to grant high real wages.

The intersection between the curve WS and the line PS indicates one equilibrium point E for each economic system (where workers' plans are consistent with those of firms), corresponding to the case in which only one *natural* unemployment rate exists, obtained by balancing (6.37a) and (6.39)

$$\hat{\lambda}[1 - (d+r^*)\hat{v}p_k] = \omega(u, z) \Rightarrow u_n^A < u_n^a \,, \quad (6.40a)$$

assuming that $[1 - (d+r^*)\hat{v}p_k]d\hat{\lambda} - \hat{\lambda}(d+r^*)p_k d\hat{v} > \hat{\lambda}(d+r^*)\hat{v}dp_k$.

Remark: In the case of a highly automated system, wage bargaining in the mid-term period enables a natural unemployment rate smaller than that affecting a less automated economy: higher productivity makes enterprises more willing to grant higher real wages as a result of bargaining and the search for efficiency in production organizations, so that the economic system can converge towards a lower equilibrium rate of unemployment ($u_n^A < u_n^a$).

The uniqueness of this solution is valid only if enterprises aim to maintain physical capital undivided.

In the case of rational behavior intended to achieve the integrity of capital stock alone (6.37b), or business procedures oriented to recoup the accounting value of capital alone (6.37c), the natural unemployment rate theory is not sustainable. In fact, if we equalize these two relations to (6.39), respectively, we obtain in both cases a relation describing balance couples between unemployment rate and prices general level, indicating that the equilibrium rate is undetermined and therefore that the adjective *natural* is inappropriate

$$\hat{\lambda}\{1-(d+r^*)\hat{v}[p_{\mathrm{k}}-l^*(p_{\mathrm{k}}-P_{\mathrm{K}}^0/P)]\} = \omega(u,z)$$
$$\Rightarrow u_{\mathrm{n}} = u_{\mathrm{n}}^{\mathrm{b}}(P)\,, \tag{6.40b}$$
$$\hat{\lambda}[1-(d+r^*)\hat{v}P_{\mathrm{K}}^0/P] = \omega(u,z)$$
$$\Rightarrow u_{\mathrm{n}} = u_{\mathrm{n}}^{\mathrm{c}}(P)\,, \tag{6.40c}$$

where $\partial u_{\mathrm{n}}/\partial P < 0$, since an increase in P increases the margin necessary to recover capital cost and so enterprises can pay out workers a higher real salary $(\partial\omega/\partial P > 0)$, which in balance is compatible with a lower rate of unemployment.

Summing up, alternative enterprise behavior to that intended to maintain physical capital intact does not allow fixing univocally the natural rate of unemployment. In fact, highly automated economic systems cause a translation of the function $u_{\mathrm{n}} = u_{\mathrm{n}}(P)$ to a lower level: for an identical general level of prices the unemployment balance rate is lower since automation allows increased productivity and real wage that enterprises are prepared to pay.

6.4.3 Macroeconomic Effects of Automation in the Mid Term: Natural Unemployment and Technological Unemployment

The above optimistic conclusion of mid-term equilibrium being more profitable for highly automation economies must be validated in the light of constraints imposed by capital-intensive technologies, where the production rate is higher and yields are constant.

Assume that in the economic system an aggregated demand is guaranteed, such that all production volumes could be sold, either owing to fiscal and monetary politics oriented towards full employment or hoping that, in the mid term, the economy will spontaneously converge through monetary adjustments induced by variation of the price general level.

In a diffused automation context, a problem could result from a potential inconsistency between the unemployment natural rate u_{n} (obtained by imposing equality of expected and actual prices, related to labor market equilibrium) and the unemployment rate imposed by technology $\bar{u}$

$$\bar{u} = 1 - L^{\mathrm{d}}/L^{\mathrm{s}} = 1 - (\bar{Y}/\hat{\lambda})/L^{\mathrm{s}} \gtrless u_{\mathrm{n}}\,, \tag{6.41}$$

where $L^{\mathrm{d}} = \bar{Y}/\hat{\lambda}$ is the demand for labor, L^{s} is the labor supply, and $\bar{Y}$ is the potential production rate, compatible with full utilization of production capacity.

A justification of this conclusion can be seen by noting that automation calls for qualified technical personnel, who are still engaged by enterprises even during crisis periods, and of whom overtime work is required in case of expansion. Then, employment variation is not proportional to production variation, as clearly results from the law of *Okun* [6.45], and from papers by *Perry* [6.46] and *Tatom* [6.47].

Remark: From the definition of $\bar{u}$ it follows that economic systems with high automation level, and therefore characterized by high productivity of labor, present higher technological unemployment rates

$$(\partial\bar{u}/\partial\hat{\lambda})(\hat{\lambda}/\bar{u}) = (1-\bar{u})/\bar{u} > 0\,.$$

On one hand, automation reduces the natural unemployment rate u_{n}, but on the other it forces the technological unemployment rate $\bar{u}$ to increase. If it holds that $\bar{u} \leq u_{\mathrm{n}}$, the market is dominant and the economic system aims to converge in time towards an equilibrium characterized by higher real salary and lower (natural) unemployment rate, as soon as the beneficial effects of automation spread.

In Fig. 6.3, the previous Fig. 6.2 is modified under the hypothesis of a capital-intensive economy, by including two vertical lines corresponding to two potential unemployment rates imposed by technology ($\bar{u}$ and $\bar{u}_{\mathrm{A}}$). Note that $\bar{u}$ has been placed on the left of u_{n} whilst $\bar{u}_{\mathrm{A}}$ has been placed on the right of u_{n}. It can be seen that the labor market equilibrium (point E) dominates when technology implies a degree of automation compatible with a nonconstraining unemployment rate $\bar{u}$. Equilibrium could be obtained, on the contrary, when technology (for a given production capacity of the eco-

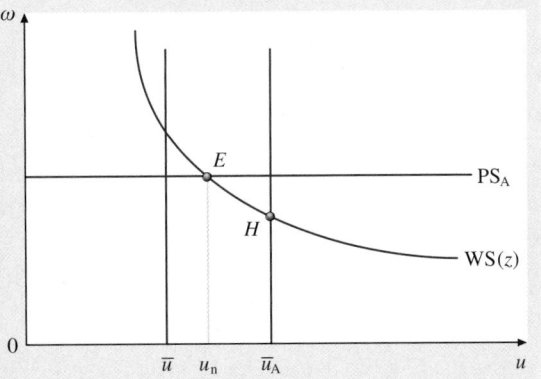

Fig. 6.3 Mid-term equilibrium with high automation level and two different technological unemployment rates

nomic system) cannot assure a low unemployment rate: $\bar{u}_A > u_n$.

The technological unemployment constraint implies a large weakness when wages are negotiated by workers, and it induces a constrained equilibrium (point H) in which the real salary perceived by workers is lower than that which enterprises are willing to pay. The latter can therefore achieve unplanned extra profits, so that automation benefits only generate capital owners' income which, in the share market, gives rise to share value increase.

Only a relevant production increase and equivalent aggregated demand could guarantee a reduction of $\bar{u}_A$, thus transferring automation benefits also to workers.

These conclusions can have a significant impact on the Fisher–Phillips curve (*Fisher* [6.48] and *Phillips* [6.49], first; theoretically supported by *Lipsey* [6.50] and then extended by *Samuelson* and *Solow* [6.51]) describing the trade-off between inflation and unemployment.

Assume that the quality constraint between expected prices and effective prices can, in the mid term, be neglected, in order to analyze the inflation dynamics. Then, substitute (6.38) into (6.35b), by assuming the hypothesis that maintaining capital intact implies constant markup

$$P = (1+\mu^*)\omega(u, z)P^e/\hat{\lambda} . \tag{6.42}$$

This relation shows that labor market equilibrium for imperfect information ($P^e \neq P$), implies a positive link between real and expected prices.

By dividing (6.42) by P_{-1}, the following relation results

$$1+\pi = (1+\pi^e)(1+\mu^*)\omega(u, z)/\hat{\lambda} , \tag{6.43a}$$

where:

- $\pi = P/P_{-1} - 1$ is the current inflation rate
- $\pi^e = P^e/P_{-1} - 1$ is the expected inflation rate.

Assume moderate inflation rates, such that

$$(1+\pi)/(1+\pi^e) \approx 1+\pi-\pi^e$$

and consider a linear relation between productivity and wages, such that it amounts to 100% if

$$z = u = 0 : \omega(u, z)/\hat{\lambda} = 1+\alpha_0 z - \alpha_1 u(\alpha_0, \alpha_1 > 0) .$$

Relation (6.43a) can be simplified to

$$\pi = \pi^e + (1+\mu^*)\alpha_0 z - (1+\mu^*)\alpha_1 u , \tag{6.43b}$$

which is a linear version of the Phillips curve; according to this form, the current inflation rate depends positively

on inflation expectation, markup level, and the variable z, and negatively on the unemployment rate.

For $\pi^e = 0$, the original curve which Phillips and Samuelson–Solow estimated for the UK and USA is obtained.

For $\pi^e = \pi_{-1}$ (i.e., *extrapolative expectations*) a link between the variation of inflation rate and the unemployment rate (accelerated Phillips curve), showing better interpolation of data observed since 1980s, is derived

$$\pi - \pi_{-1} = (1+\mu^*)\alpha_0 z - (1+\mu^*)\alpha_1 u . \tag{6.44}$$

Assuming $\pi = \pi^e = \pi_{-1}$ in (6.43b) and (6.44), an estimate of the natural unemployment rate (without systematic errors in mid-term forecasting) can be derived

$$u_n = \alpha_0 z/\alpha_1 . \tag{6.45}$$

Now, multiplying and dividing by the α_1 term $(1+\mu^*)\alpha_0 z$ in (6.44), and inserting (6.45) into (6.44), it results that, in the case of *extrapolative expectations*, the inflation rate reduces if the effective unemployment rate is greater than the natural one ($d\pi < 0$ if $u > u_n$); it increases in the opposite case ($d\pi > 0$ if $u < u_n$); and it is zero (constant inflation rate) if the effective unemployment rate is equal to the natural one ($d\pi = 0$ if $u = u_n$)

$$\pi - \pi_{-1} = -(1+\mu^*)\alpha_1(u - u_n) . \tag{6.46}$$

Remark: Relation (6.44) does not take into account effects induced by automation diffusion, which imposes on the economic system a technological unemployment $\bar{u}_A$ that increases with increasing automation.

It follows that any econometric estimation based on (6.44) no longer evaluates the natural unemployment rate (6.45) for $\pi - \pi_{-1} = 0$, because the latter varies in time, flattening the interpolating line (increasingly high unemployment rates related to increasingly lower real wages, as shown above). Then, as automation process spreads, the natural unemployment rate (which, without systematic errors, assures compatibility between the real salary paid by enterprises and the real salary either demanded by workers or supplied by enterprises for efficiency motivation) is no longer significant.

In Fig. 6.4a the Phillips curve for the Italian economic system, modified by considering inflation rate variations from 1953 to 2005 and the unemployment rate, is reported; the interpolation line is decreasing but very flat owing to $\bar{u}_A$ movement towards the right.

A natural unemployment rate between 7 and 8% seems to appear, but the intersection of the interpolation line with the abscissa is moved, as shown in the

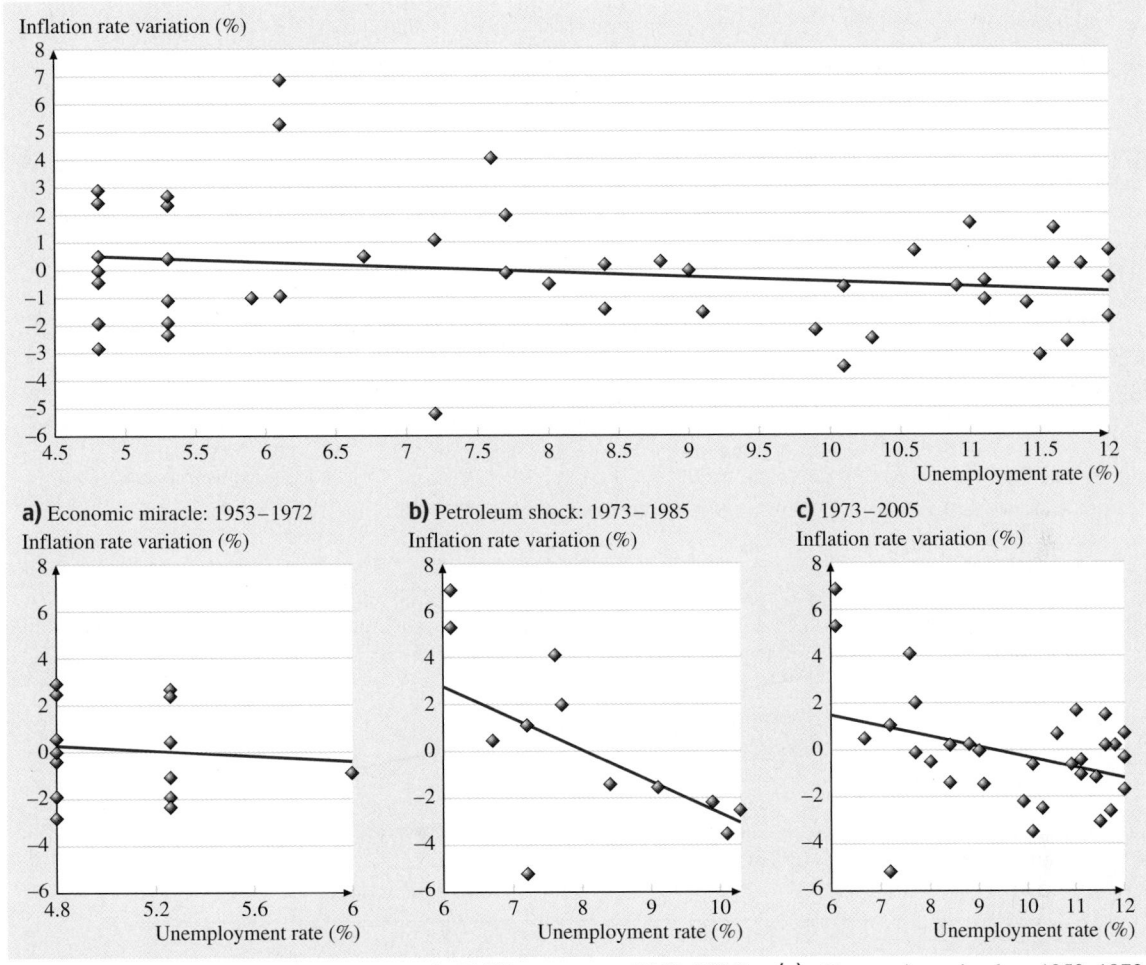

Fig. 6.4a–c Italy: Expectations-augmented Phillips curve (1953–2005). **(a)** Economic miracle: 1953–1972, **(b)** petroleum shock: 1973–1985, **(c)** 1973–2005

three representations where homogeneous data sets of the Italian economic evolution are considered.

Automation has a first significant impact during the oil crisis, moving the intersection from 5.2% for the first post-war 20 years up to 8% in the next period. Extending the time period up to 2005, intersection moves to 9%; the value could be greater, but it has been recently bounded by the introduction of temporary work contracts which opened new labor opportunities but lowered salaries.

Note that Figs. 6.4a–c are partial reproductions of the Fig. 6.4, considering three different intervals of unemployment rate values. This reorganization of data corresponds to the three different periods of the Italian economic growth: (a) the economic miracle (1953–

1972), (b) the petroleum shock (1973–1985), and (c) the period from the economic miracle until 2005 (1973–2005).

A more limited but comprehensive analysis, owing to lack of homogeneous historical data over the whole period, has been done also for two other important European countries, namely France (Fig. 6.5a) and Germany (Fig. 6.5b): the period considered ranges from 1977 to 2007.

The unemployment rate values have been recomputed so as to make all the historical series homogeneous. Owing to the limited availability of data for both countries, only two periods have been analyzed: the period 1977–1985, with the effects of the petroleum shock, and the whole period up to 2007, in order to

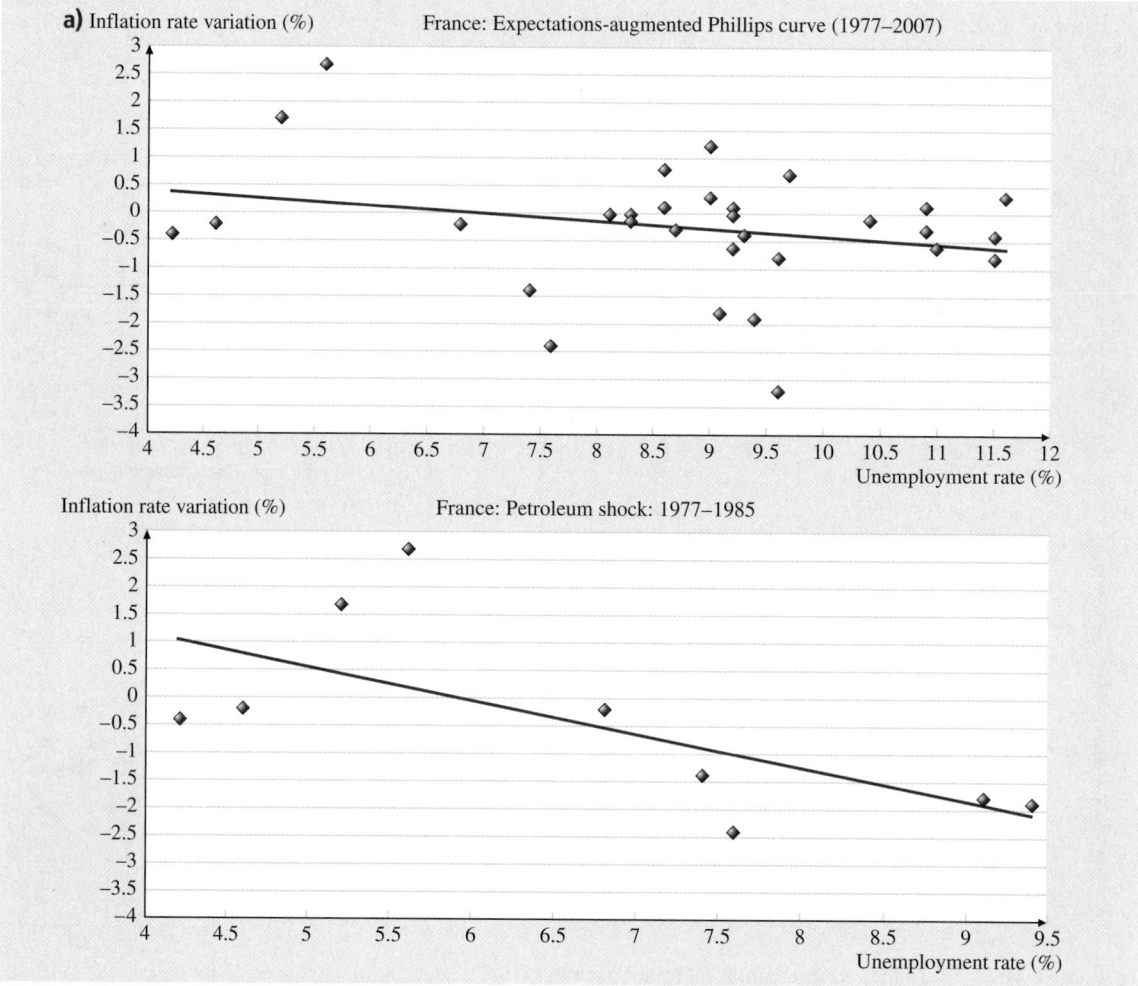

Fig. 6.5 (a) France: expectations-augmented Phillips curve (1977–2007) **(b)** ► Germany: expectations-augmented Phillips curve (1977–2007)

evaluate the increase of the natural unemployment rate from the intersection of the interpolation line with the abscissa.

As far as Fig. 6.5a is concerned, the natural unemployment rate is also increased in France – as in Italy – by more than one percentage point (from less than 6% to more than 7%). In the authors' opinion, this increase should be due to technology automation diffusion and consequent innovation of organization structures.

Referring to Fig. 6.5b, it can be noted that even in Germany the natural unemployment rate increased – more than in France and in Italy – by about three percentage points (from 2.5 to 5.5%). This effect is partly due to application of automated technologies (which

should explain about one percentage point, as in the other considered countries), and partly to the reunification of the two parts of German.

A Final Remark: Based on the considerations and data analysis above it follows that the natural unemployment rate, in an industrial system where capital intensive enterprises are largely diffused, is no longer significant, because enterprises do not apply methods to maintain capital intact, and because technological inflation tends to constraint the natural one.

Only structural opposing factors could slow this process, such as labor market reform, to give rise to new work opportunities, and mainly economic politics, which could increase the economy development

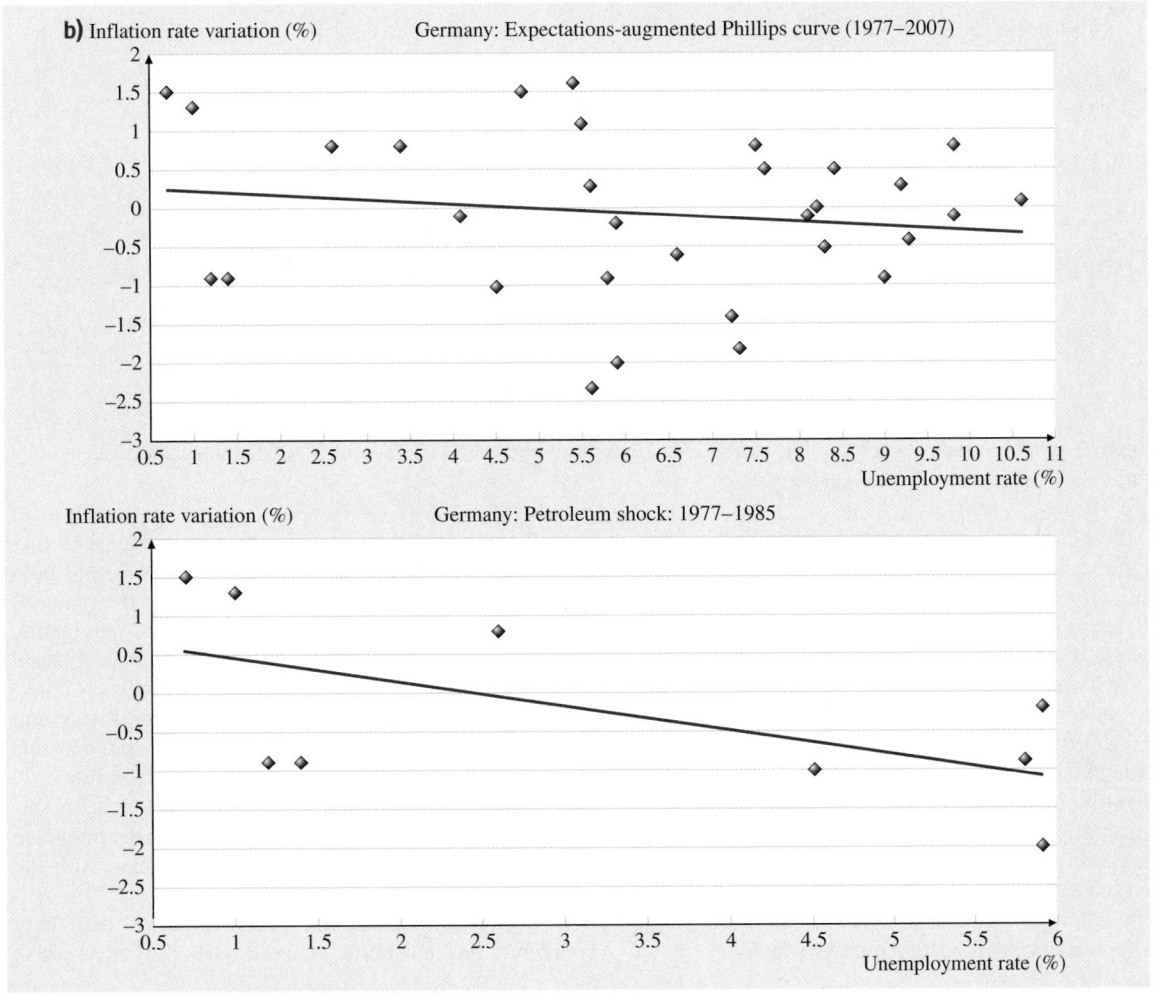

b) Inflation rate variation (%)   Germany: Expectations-augmented Phillips curve (1977–2007)

Inflation rate variation (%)   Germany: Petroleum shock: 1977–1985

trend more than the growth of productivity and labor supply. However, these aspects should be approached in a long-term analysis, and exceed the scope of this chapter.

6.5 Final Comments

Industrial automation is characterized by dominance of capital dynamics over the biological dynamics of human labor, thus increasing the production rate. Therefore automation plays a positive role in reducing costs related to both labor control, sometimes making monetary incentives useless, and supervisors employed to assure the highest possible utilization of personnel. It is automation itself that imposes the production rate and forces workers to correspond to this rate.

In spite of these positive effects on costs, the increased capital intensity implies greater rigidity of the cost structure: a higher capital cost depending on higher investment value, with consequent transformation of some variable costs into fixed costs. This induces a greater variance of profit in relation to production volumes and a higher risk of automation in respect to labor-intensive systems. However, the trade-off between automation yield and risk suggests that enterprises

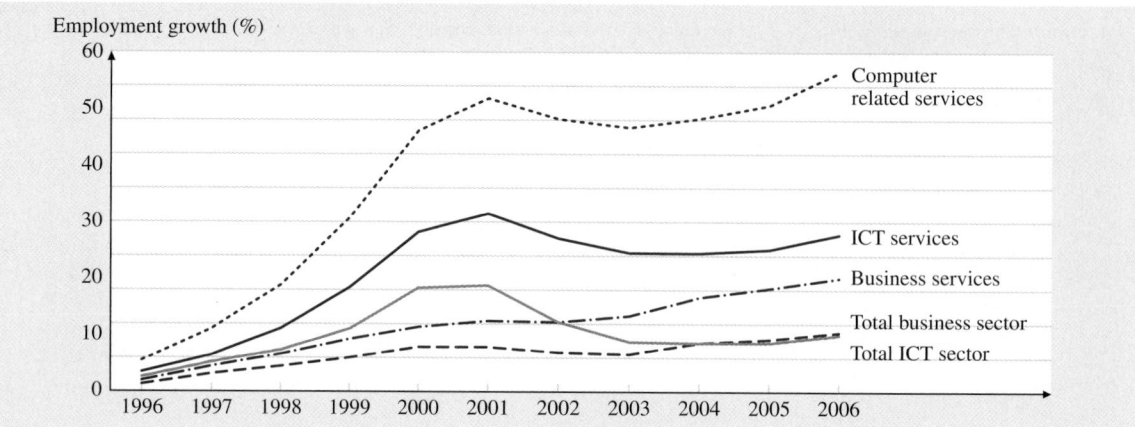

Fig. 6.6 Employment growth by sector with special focus on computer and ICT (source: OECD Information Technology Outlook 2008, based on STAN database)

should increase automation in order to obtain high utilization of production capacity. One positive effort in this direction has been the design and implementation of flexible automation during recent decades (e.g., see Chap. 50 on *Flexible and Precision Assembly*).

Moving from microeconomic to macroeconomic considerations, automation should reduce the effects on short-term inflation caused by nominal salary increases since it forces productivity to augment more than the markup necessary to cover higher fixed costs. In addition, the impact of automation depends on the book-keeping method adopted by the enterprise in order to recover the capital value: this impact is lower if a part of capital can be financed by debts and if the enterprise behavior is motivated by monetary illusion.

Over a mid-term period, automation has been recognized to have beneficial effects (Figs. 6.6–6.8) both on the real salary paid to workers and on the (natural) unemployment trend, if characteristics of automation technologies do not form an obstacle to the market trend towards equilibrium, i.e., negotiation between enterprises and trade unions. If, on the contrary, the system production capacity, for a compatible demand, prevents market convergence, a noncontractual equilibrium only dependent on technology capability could be established, to the advantage of enterprises and the prejudice of workers, thus reducing real wages and increasing the unemployment rate.

Empirical validation seems to show that Italy entered this technology-caused unemployment phase

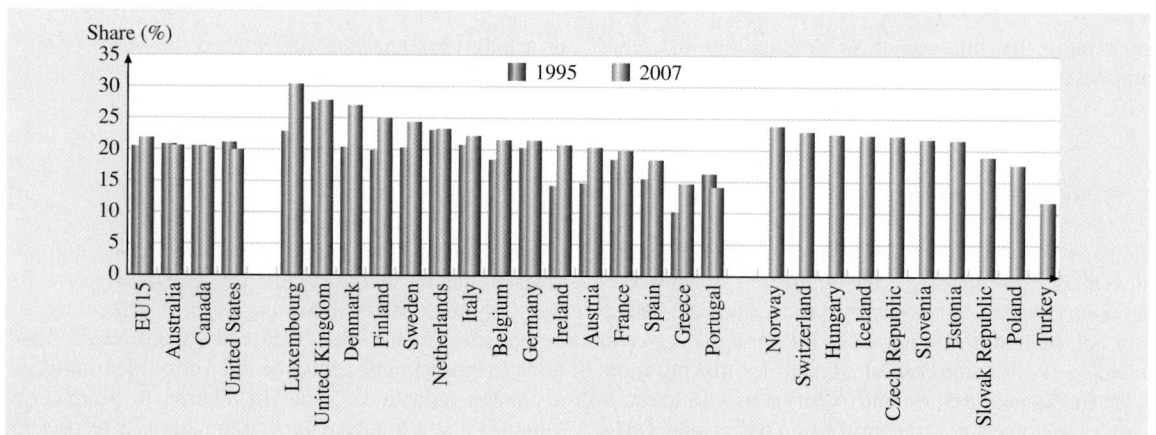

Fig. 6.7 Share of ICT-related occupations in the total economy from 1995 and 2007 (source: OECD IT Outlook 2008, forthcoming)

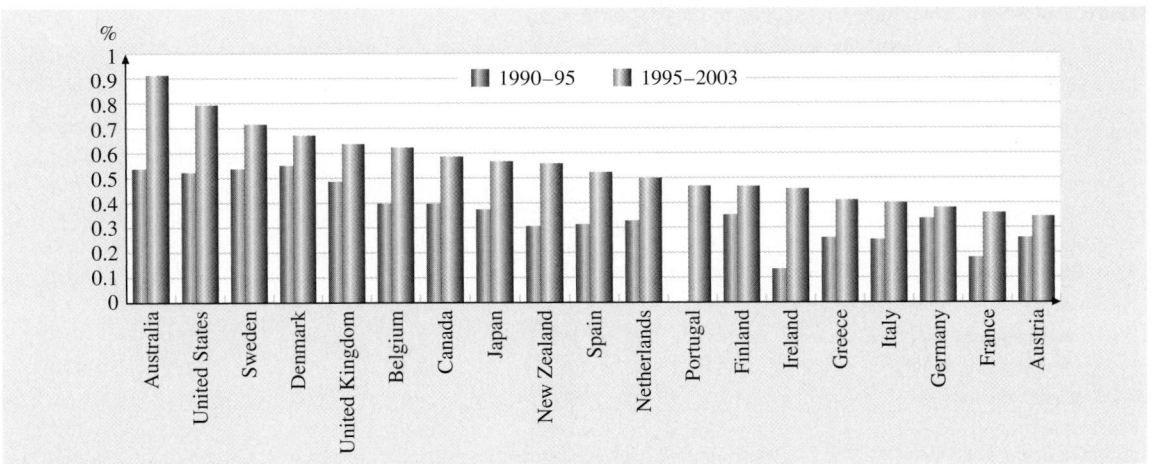

Fig. 6.8 Contributions of ICT investment to GDP growth, 1990–1995 and 1995–2003 in percentage points (source: OECD Productivity Database, September 2005)

during the last 20 years: this corresponds to a flatter Phillips curve because absence of inflation acceleration is related to natural unemployment rates that increase with automation diffusion, even if some structural modification of the labor market restrained this trend.

6.6 Capital/Labor and Capital/Product Ratios in the Most Important Italian Industrial Sectors

A list of the most important industrial sectors in Italy is reported in Table 6.2 (data collected by Mediobanca [6.52] for the year 2005). The following estimations of the model variables and rates have been adopted (as enforced by the available data): capital K is estimated through fixed assets; with reference to production, the added value V_a is considered; labor L is estimated in terms of number of workers.

Some interesting comments can be made based on Table 6.2, by using the production function models presented in the previous sections.

According to the authors' experience (based on their knowledge of the Italian industrial sectors), the following *anomalous sectors* can be considered:

● Sectors concerning *energy production and distribution* and *public services* for the following reasons: they consist of activities applying very high levels of automation; personnel applied in production are extremely scarce, whilst it is involved in organization and control; therefore anomalous values of the capital/labor ratio and of productivity result.

● The *transport* sector, because the very high capital/product ratio depends on the low level of the added value of enterprises that are operating with politically fixed (low) prices.

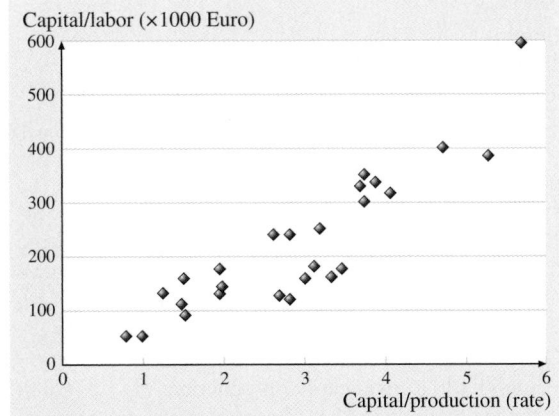

Fig. 6.9 Correlation between capital/labor ($\times 1000\,€$) and capital/production (ratios)

Table 6.2 Most important industrial sectors in Italy (after [6.52])

Industrial sector (year 2005)	Fixed assets (million euro) (a)	Added value (million euro) (b)	Number of workers (× 1000) (c)	Capital/ product (a/b)	Capital/ labor (× 1000 euro) (a/c)	Productivity (× 1000 euro) (b/c)
Industrial enterprises	309 689.6	84 373.4	933.8	3.7	331.6	90.4
Service enterprises	239 941.9	42 278.3	402.7	5.7	595.8	105.0
Clothing industry	2824.5	1882.9	30.4	1.5	93.0	62.0
Food industry: drink production	5041.7	1356.2	14.3	3.7	352.1	94.7
Food industry: milk & related products	1794.6	933.2	10.1	1.9	177.8	92.5
Food industry: alimentary preservation	2842.4	897.8	11.2	3.2	252.9	79.9
Food industry: confectionary	4316.1	1659.0	17.8	2.6	242.7	93.3
Food industry: others	5772.7	2070.2	23.9	2.8	241.4	86.6
Paper production	7730.9	1470.6	19.9	5.3	388.2	73.9
Chemical sector	16 117.7	4175.2	47.7	3.9	337.7	87.5
Transport means production	21 771.4	6353.3	121.8	3.4	178.8	52.2
Retail distribution	10 141.1	3639.8	84.2	2.8	120.4	43.2
Electrical household appliances	4534.7	1697.4	35.4	2.7	128.1	47.9
Electronic sector	9498.9	4845.4	65.4	2.0	145.1	74.0
Energy production/distribution	147 200.4	22 916.2	82.4	6.4	1786.8	278.2
Pharmaceuticals & cosmetics	9058.9	6085.2	56.3	1.5	160.8	108.0
Chemical fibers	2081.4	233.2	5.0	8.9	418.4	46.9
Rubber & cables	3868.3	1249.7	21.2	3.1	182.1	58.8
Graphic & editorial	2399.1	1960.4	18.0	1.2	133.3	108.9
Plant installation	1295.9	1683.4	23.2	0.8	56.0	72.7
Building enterprises	1337.7	1380.0	24.7	1.0	54.2	55.9
Wood and furniture	2056.1	689.2	12.8	3.0	160.2	53.7
Mechanical sector	18 114.1	9356.6	137.3	1.9	131.9	68.1
Hide and leather articles	1013.9	696.6	9.0	1.5	113.1	77.7
Products for building industry	11 585.2	2474.3	28.7	4.7	404.3	86.4
Public services	103 746.7	28 413.0	130.5	3.7	795.2	217.8
Metallurgy	18 885.6	5078.2	62.4	3.7	302.9	81.4
Textile	3539.5	1072.1	21.7	3.3	163.2	49.4
Transport	122 989.3	7770.5	142.1	15.8	865.2	54.7
Glass	2929.3	726.2	9.2	4.0	317.8	78.8

Notation: Labor-intensive sectors   Intermediate   Anomalous

- The *chemical fibres* sector, because at the time considered it was suffering a deep crisis with a large amount of unused production capacity, which gave rise to an anomalous (too high) capital/product ratio and anomalous (too small) value of productivity.

Sectors with a rate capital/production of 1.5 (such as the clothing industry, pharmaceutics and cosmetics, and hide and leather articles) has been considered as *intermediate sectors*, because automation is highly important in some working phases, whereas other working phases still largely utilize workers. These sectors (and the value 1.5 of the rate capital/production) are used as *separators* between capital-intensive systems (with high degrees of automation) and labor-intensive ones.

Sectors with a capital/production ratio of less than 1.5 have been considered as labor-intensive systems. Among these, note that the graphic sector is capital intensive, but available data for this sector are combined with the editorial sector, which in turn is largely labor intensive.

All other sectors can be viewed as capital-intensive sectors, even though it cannot be excluded that some working phases within their enterprises are still labor intensive.

Two potential correlations, if any, are illustrated in the next two figures: Fig. 6.9 shows the relations between the capital/labor and capital/production ratios, and Fig. 6.10 shows the relations between productivity and capital/labor ratio.

As shown in Fig. 6.9, the capital/labor ratio exhibits a clear positive correlation with the capital/production ratio; therefore they could be considered as alternative measures of capital intensity.

On the contrary, Fig. 6.10 shows that productivity does not present a clear correlation with capital. This could be motivated by the effects of other factors, including the utilization rate of production capacity and the nonuniform flexibility of the workforce, which could have effects on productivity.

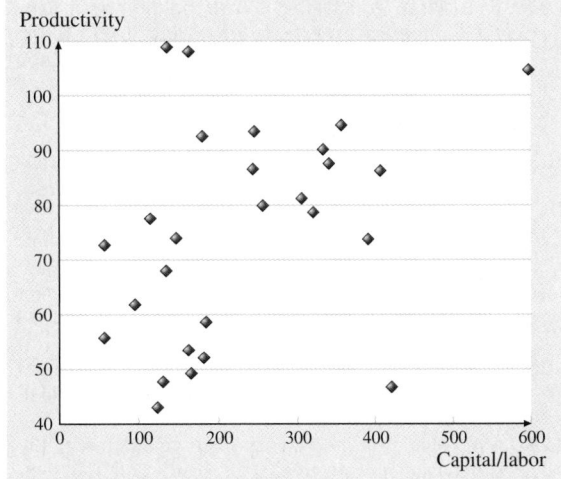

Fig. 6.10 Relation between productivity ($\times 100$ €) and capital/labor (rate)

References

6.1 CODESNET: Coordination Action No. IST-2002-506673 / Joint Call IST-NMP-1, A. Villa, coordinator. Web site address: www.codesnet.polito.it. (2004–2008)

6.2 R.W. Shephard: *Cost and Production Functions* (Princeton Univ. Press, Princeton 1953)

6.3 R.W. Shephard: *Theory of Cost and Production Functions* (Princeton Univ. Press, Princeton 1970)

6.4 R. Frisch: *Lois Techniques et Economiques de la Production* (Dunod, Paris 1963), (in French)

6.5 H. Uzawa: Duality principles in the theory of cost and production, Int. Econ. Rev. **5**, 216–220 (1964)

6.6 M. Fuss, D. Mc Fadden: *Production Economics: A Dual Approach to Theory and Application* (North-Holland, Amsterdam 1978)

6.7 C.W. Cobb, P.H. Douglas: A theory of production, Am. Econ. Rev. **18**, 139–165 (1928)

6.8 J.K. Arrow, H.B. Chenery, B.S. Minhas, R. Solow: Capital-labor substitution and economic efficiency, Rev. Econ. Stat. **63**, 225–47 (1961)

6.9 M. Brown, J.S. De Cani: Technological change and the distribution of income, Int. Econ. Rev. **4**, 289–95 (1963)

6.10 D. Mc Fadden: Further results on CES production functions, Rev. Econ. Stud. **30**, 73–83 (1963)

6.11 H. Uzawa: Production functions with constant elasticity of substitution, Rev. Econ. Stud. **29**, 291–99 (1962)

6.12 G.H. Hildebrand, T.C. Liu: *Manufacturing Production Functions in the United States* (State School of Industrial Labor Relations, New York 1965)

6.13 L.R. Christensen, D.W. Jorgenson, L.J. Lau: Conjugate duality and the transcendental logarithmic production function, Econometrica **39**, 255–56 (1971)

6.14 L.R. Christensen, D.W. Jorgenson, L.J. Lau: Transcendental logarithmic production frontier, Rev. Econ. Stat. **55**, 28–45 (1973)

6.15 I.M. Nadiri: Producers theory. In: *Handbook of Mathematical Economics*, Vol. II, ed. by K.J. Arrow, M.D. Intriligator (North-Holland, Amsterdam 1982)

6.16 A. Shaik: Laws of production and laws of algebra: the humbug production function, Rev. Econ. Stat. **56**, 115–20 (1974)

6.17 H. Hotelling: Edgeworth's taxation paradox and the nature of demand and supply functions, J. Polit. Econ. **40**, 577–616 (1932)

6.18 H. Hotelling: Demand functions with limited budgets, Econometrica **3**, 66–78 (1935)

6.19 P.A. Samuelson: *Foundations of Economic Analysis* (Harvard Univ. Press, Cambridge 1947)

6.20 P.A. Samuelson: Price of factors and goods in general equilibrium, Rev. Econ. Stud. **21**, 1–20 (1954)

6.21 W.E. Diewert: Duality approaches to microeconomic theory. In: *Handbook of Mathematical Economics*, Vol. II, ed. by K.J. Arrow, M.D. Intriligator (North-Holland, Amsterdam 1982)

6.22 D.W. Jorgenson: Econometric methods for modelling producer behaviour. In: *Handbook of Econometrics*, Vol. III, ed. by Z. Griliches, M.D. Intriligator (North-Holland, Amsterdam 1986)

Part A | 6

6.23   E. Luciano, P. Ravazzi: *I Costi nell'Impresa. Teoria Economica e Gestione Aziendale* (UTET, Torino 1997), (Costs in the Enterprise. Economic Theory and Industrial Management – in Italian)

6.24   A. Weiss: *Efficiency Wages* (Princeton Univ. Press, Princeton 1990)

6.25   J.E. Stiglitz: Wage determination and unemployment in LDC's: the labor turnover model, Q. J. Econ. **88**(2), 194–227 (1974)

6.26   S. Salop: A model of the natural rate of unemployment, Am. Econ. Rev. **69**(2), 117–25 (1979)

6.27   A. Weiss: Job queues and layoffs in labor markets with flexible wages, J. Polit. Econ. **88**, 526–38 (1980)

6.28   C. Shapiro, J.E. Stiglitz: Equilibrium unemployment as a worker discipline device, Am. Econ. Rev. **74**(3), 433–44 (1984)

6.29   G.A. Calvo: The inefficiency of unemployment: the supervision perspective, Q. J. Econ. **100**(2), 373–87 (1985)

6.30   G.A. Akerlof: Labor contracts as partial gift exchange, Q. J. Econ. **97**(4), 543–569 (1982)

6.31   G.A. Akerlof: Gift exchange and efficiency-wage theory: four views, Am. Econ. Rev. **74**(2), 79–83 (1984)

6.32   H. Miyazaki: Work, norms and involuntary unemployment, Q. J. Econ. **99**(2), 297–311 (1984)

6.33   D. Antonelli, N. Pasquino, A. Villa: Mass-customized production in a SME network, IFIP Int. Working Conf. APMS 2007 (Linkoping, 2007)

6.34   J. Robinson: *The Economics of Imperfect Competition* (Macmillan, London 1933)

6.35   E.H. Chamberlin: *The Theory of Monopolistic Competition* (Harvard Univ. Press, Harvard 1933)

6.36   P.W.S. Andrews: *On Competition in Economic Theory* (Macmillan, London 1964)

6.37   F. Modigliani, M. Miller: The cost of capital, corporation finance and the theory of investment, Am. Econ. Rev. **48**(3), 261–297 (1958)

6.38   F. Modigliani, M. Miller: Corporate income taxes and the cost of capital: a correction, Am. Econ. Rev. **53**, 433–443 (1963)

6.39   D.N. Baxter: Leverage, risk of ruin and the cost of capital, J. Finance **22**(3), 395–403 (1967)

6.40   K.H. Chen, E.H. Kim: Theories of corporate debt policy: a synthesis, J. Finance **34**(2), 371–384 (1979)

6.41   M.F. Hellwig: Bankruptcy, limited liability and the Modigliani–Miller theorem, Am. Econ. Rev. **71**(1), 155–170 (1981)

6.42   M. Jensen, W. Meckling: Theory of the firm: managerial behaviour, agency costs and ownership structure, J. Financial Econ. **3**(4), 305–360 (1976)

6.43   A.C. Pigou: Maintaining capital intact, Economica **45**, 235–248 (1935)

6.44   R.A. Cohn, F. Modigliani: Inflation, rational valuation and the market, Financial Anal. J. **35**, 24–44 (1979)

6.45   A.M. Okun: Potential GNP: its measurement and significance. In: *The Political Economy of Prosperity*, ed. by A.M. Okun (Brookings Institution, Washington 1970) pp. 132–145

6.46   G. Perry: Potential output and productivity, Brook. Pap. Econ. Activ. **8**, 11–60 (1977)

6.47   J.A. Tatom: *Economic Growth and Unemployment: A Reappraisal of the Conventional View* (Federal Reserve Bank of St. Louis Review, St. Louis 1978) pp. 16–22

6.48   I. Fisher: A statistical relation between unemployment and prices changes, Int. Labour Rev. **13**(6) 785–792 (1926), J. Polit. Econ. **81**(2), 596–602 (1973)

6.49   W.H. Phillips: The relation between unemployment and the rate of change of money wages rated in the United Kingdom: 1861–1957, Economica **25**(100), 283–299 (1958)

6.50   R.G. Lipsey: The relation between unemployment and the rate of change of money wages in UK: 1862-1957, Economica **27**, 1–32 (1960)

6.51   P.A. Samuelson, R.M. Solow: The problem of achieving and maintaining a stable price level: analytical aspects of anti-inflation policy, Am. Econ. Rev. **50**, 177–194 (1960)

6.52   Mediobanca: *Dati Cumulativi di 2010 Società Italiane* (Mediobanca, Milano 2006), (Cumulative Data of 2010 Italian Enterprises – in Italian)

7. Impacts of Automation on Precision

Alkan Donmez, Johannes A. Soons

Automation has significant impacts on the economy and the development and use of technology. In this chapter, the impacts of automation on precision, which also directly influences science, technology, and the economy, are discussed. As automation enables improved precision, precision also improves automation.

Following the definition of precision and the factors affecting it, the relationship between precision and automation is described. This chapter concludes with specific examples of how automation has improved the precision of manufacturing processes and manufactured products over the last decades.

7.1 What Is Precision?

Precision is the closeness of agreement between a series of individual measurements, values or results. For a manufacturing process, precision describes how well the process is capable of producing products with identical properties. The properties of interest can be the dimensions of the product, its shape, surface finish, color, weight, etc. For a device or instrument, precision describes the invariance of its output when operated with the same set of inputs. Measurement precision is defined by the *International Vocabulary of Metrology* as the [7.1]:

> ... *closeness of agreement between indications obtained by replicate measurements on the same or similar objects under specified conditions.*

In this definition, the *specified conditions* describe whether precision is associated with the repeatability or the reproducibility of the measurement process. Repeatability is the closeness of agreement between results of successive measurements of the same quantity carried out under the same conditions. These repeatability conditions include the measurement procedure, observer, instrument, environment, etc. Reproducibility is the closeness of the agreement between results of measurements carried out under changed measurement conditions. In computer science and mathematics, precision is often defined as a measure of the level of detail of a numerical quantity. This is usually expressed as the number of bits or decimal digits used to describe the quantity. In other areas, this aspect of precision is referred to as resolution: the degree to which nearly equal values of a quantity can be discriminated, the smallest measurable change in a quantity or the smallest controlled change in an output.

Precision is a necessary but not sufficient condition for accuracy. Accuracy is defined as the closeness of the agreement between a result and its *true* or intended value. For a manufacturing process, accuracy describes the closeness of agreement between the properties of the manufactured products and the properties defined in the product design. For a measurement, accuracy is the closeness of the agreement between the result of the measurement and a true value of the measurand – the quantity to be measured [7.1]. Accuracy is affected by both precision and bias. An instrument with an incorrect calibration table can be precise, but it would not be accurate. A challenge with the definition of accuracy is that the *true value* is a theoretical concept. In practice, there is a level of uncertainty associated with the true value due to the infinite amount of information required to describe the measurand completely. To the extent that it leaves room for interpretation, the incomplete definition of the measurand introduces uncertainty in the result of a measurement, which may or may not be significant relative to the accuracy required of the measurement; for example, suppose the measurand is the thickness of a sheet of metal. If this thickness is measured using a micrometer caliper, the result of the measurement may be called the best estimate of the *true value* (*true* in the sense that it satisfies the definition of the measurand.) However, had the micrometer caliper been applied to a different part of the sheet of material, the realized quantity would be different, with a different *true value* [7.2]. Thus the lack of information about where the thickness is defined introduces an uncertainty in the *true value*. At some level, every measurand or product design has such an *intrinsic* uncertainty.

7.2 Precision as an Enabler of Automation

Historically, precision is closely linked to automation through the concept of parts interchangeability. In more recent times, it can be seen as a key enabler of lean manufacturing practices. Interchangeable parts are parts that conform to a set of specifications that ensure that they can substitute each other. The concept of interchangeable parts radically changed the manufacturing system used in the first phase of the Industrial Revolution, the English system of manufacturing. The English system of manufacturing was based on the traditional artisan approach to making a product. Typically, a skilled craftsman would manufacture an individual product from start to finish before moving onto the next product. For products consisting of multiple parts, the parts were modeled, hand-fitted, and reworked to fit their counterparts. The craftsmen had to be highly skilled, there was no automation, and production was slow. Moreover, parts were not interchangeable. If a product failed, the entire product had to be sent to an expert craftsman to make custom repairs, including fabrication of replacement parts that would fit their counterparts.

Pioneering work on interchangeable parts occurred in the printing industry (movable precision type), clock and watch industry (toothed gear wheels), and armories (pulley blocks and muskets) [7.3]. In the mid to late 18th century, French General Jean Baptiste Va-

quette de Gribeauval promoted the use of standardized parts for key military equipment such as gun carriages and muskets. He realized that interchangeable parts would enable faster and more efficient manufacturing, while facilitating repairs in the field. The development was enabled by the introduction of two-dimensional mechanical drawings, providing a more accurate expression of design intent, and increasingly accurate gauges and templates (jigs), reducing the craftsman's room for deviations while allowing for lower skilled labor. In 1778, master gunsmith Honoré Blanc produced the first set of musket locks completely made from interchangeable parts. He demonstrated that the locks could be assembled from parts selected at random. Blanc understood the need for a hierarchy in measurement standards through the use of working templates for the various pieces of the lock and master copies to enable the reconstruction of the working templates in the case of loss or wear [7.3]. The use of semiskilled labor led to strong resistance from both craftsmen and the government, fearful of the growing independence of manufacturers. In 1806, the French government reverted back to the old system, using the argument that workers who do not function as a whole cannot produce harmonious products.

Thomas Jefferson, a friend of Blanc, promoted the new approach in the USA. Here the ideas led to the American system of manufacturing. The American system of manufacturing is characterized by the sequential application of specialized machinery and templates (jigs) to make large quantities of identical parts manufactured to a tolerance (see, e.g., [7.4]). Interchangeable parts allow the separation of parts production from assembly, enabling the development of the assembly line. The use of standardized parts furthermore facilitated the replacement of skilled labor and hand tools with specialized machinery, resulting in the economical and fast production of accurate parts.

The American system of manufacturing cannot exist without precision and standards. Firstly, the system requires a unified, standardized method of defining nominal part geometry and tolerances. The tolerances describe the maximum allowed deviations in actual part geometry and other properties that ensure proper functioning of the part, including interchangeability. Secondly, the system requires a quality control system, including sampling and acceptance rules, and gauges calibrated to a common standard to ensure that the parts produced are within tolerance. Thirdly, the system requires manufacturing processes capable of realizing parts that conform to tolerance. It is not surprising that the concept of interchangeable parts first came into widespread use in the watchmakers' industry, an area used to a high level of accuracy [7.5].

Precision remains a key requirement for automation. Precision eliminates fitting and rework, enabling automated assembly of parts produced across the globe. Precision improves agility by increasing the range of tasks that unattended manufacturing equipment can accomplish, while reducing the cost and time spent on production trials and incremental process improvements. Modern manufacturing principles such as lean manufacturing, agile manufacturing, just-in-time manufacturing, and zero-defect manufacturing cannot exist without manufacturing processes that are precise and well characterized.

Automated agile manufacturing, for example, is dependent upon the solution of several precision-related technical challenges. Firstly, as production machines become more agile, they also become more complex, yet precision must be maintained or improved for each of the increasing number of tasks that a machine can perform. The design, maintenance, and testing of these machines becomes more difficult as the level of agility increases. Secondly, the practice of trial runs and iterative accuracy improvements is not cost-effective when batch sizes decrease and new products are introduced at increasing speeds. Instead, the first and every part have to be produced on time and within tolerance. Accordingly, the characterization and improvement of the precision of each manufacturing process becomes a key requirement for competitive automated production.

7.3 Automation as an Enabler of Precision

As stated by *Portas, random results are the consequence of random procedures* [7.6]. In general, *random results* appear to be random due to a lack of understanding of cause-and-effect relationships and a lack of resources for controlling sources of variability; for example, an instrument may generate a measurement result that fluctuates over time. Closer inspection may reveal that the fluctuations result from environmental temperature variations that cause critical parts of the instrument to expand and deform. The apparent random variations can thus be reduced by tighter environmental temperature control, use of design principles and materials that make the device less sensitive to temperature

variations or application of temperature sensors and algorithms to compensate thermal errors in the instrument reading.

Automation has proven to be very effective in eliminating or minimizing variability. Automation reduces variability associated with human operation. Automation furthermore enables control of instruments, processes, and machines with a bandwidth, complexity, and resolution unattainable by human operators. While humans plan and supervise the operation of machines and instruments, the craftsmanship of the operator is no longer a dominant factor in the actual manufacturing or inspection process.

7.4 Cost and Benefits of Precision

Higher precision requires increased efforts to reduce sources of variability or their effect. Parts with tighter

tolerances are therefore more difficult to manufacture and more expensive to produce. In general, there is a belief that there exists a nearly exponential relationship between cost and precision, even when new equipment is not needed. However, greater precision does not necessarily imply higher cost when the total manufacturing enterprise, including the final product, is examined [7.7, 8].

The benefits of higher precision can be separated into benefits for product quality and benefits for manufacturing. Higher precision enables new products and new product capabilities. Other benefits are better product performance (e.g., longer life, higher loads, higher efficiency, less noise and wear, and better appearance and customer appeal), greater reliability, easier re-

pair (e.g., improved interchangeability of parts), and opportunities for fewer and smaller parts; for example, the improvements in the reliability and fuel efficiency of automobiles have to a large extent been enabled by increases in the precision of manufacturing processes and equipment. The benefits of higher precision for manufacturing include lower assembly cost (less selective assembly, elimination of *fitting* and rework, automated assembly), better interchangeability of parts sourced from multiple suppliers, lower inventory requirements, less time and cost spend on trial production, fewer rejects, and improved process consistency.

7.5 Measures of Precision

To achieve precision in a process means that the outcome of the process is highly uniform and predictable over a period of time. Since precision is an attribute of a series of entities or process outcomes, statistical methods and tools are used to describe precision. Traditional statistical measures such as mean and standard deviation are used to describe the average and dispersion of the characteristic parameters. International standards and technical reports provide guidance about how such statistical measures are applied for understanding of the short-term and long-term process behavior and for management and continuous improvement of processes [7.9–12].

Statistical process control is based on a comparison of current data with historical data. Historical data is used to build a model for the expected process behavior, including control limits for measurements of the output of the process. Data is then collected from the process and compared with the control limits to determine if the process is still behaving as expected. Process capability compares the output of an in-control process to the specification limits of the requested task. The *process capability index*, C_p, describes the process capability in

relation to specified tolerance

$$C_p = (U - L)/6\sigma , \tag{7.1}$$

where U is the upper specification limit, L is the lower specification limit, σ is the standard deviation of the dispersion (note that in the above equation 6σ corresponds to the reference interval of the dispersion for normal distribution; for other types of distribution the reference interval is determined based on the well-established statistical methods).

The *critical process capability index* C_{pk} also known as the *minimum process capability index*, describes the relationship between the proximity of the mean process parameter of interest to the specified tolerance

$$C_{pk} = \min(C_{pkL}, C_{pkU}) , \tag{7.2}$$

where

$$C_{pkU} = (U - \mu)/3\sigma \tag{7.3}$$

and

$$C_{pkL} = (\mu - L)/3\sigma , \tag{7.4}$$

and μ is the mean of the process parameter of interest.

7.6 Factors That Affect Precision

In case of manufacturing processes, there are many factors that affect the precision of the outcome. They are associated with expected and unexpected variations in environment, manufacturing equipment, and process as well as the operator of the equipment; for example,

ambient temperature changes over time or temperature gradients in space cause changes in performance of manufacturing equipment, which in turn causes variation in the outcome [7.13, 14]. Similarly, variations in workpiece material such as local hardness variations, residual stresses, deformations due to clamping or process-induced forces contribute to the variations in critical parameters of finished product. Process-induced variations include wear or catastrophic failures of cutting tools used in the process, thermal variations due to the interaction of coolant, workpiece, and the cutting tool, as well as variations in the set locations of tools used in the process (e.g., cutting tool offsets). In the case of manufacturing equipment, performance varia-

tions due to thermal deformations, static and dynamic compliances, influences of foundations, and ineffective maintenance are the contributors to the variations in product critical parameters. Finally, variations caused by the operator of the equipment due to insufficient training, motivation, care or information needed constitute the largest source of unexpected variations and therefore impact on the precision of the manufacturing process.

7.7 Specific Examples and Applications in Discrete Part Manufacturing

The effect of automation on improving of precision of discrete part manufacturing can be observed in many applications such as improvements in fabrication, assembly, and inspection of various components for high-value products. In this Section, one specific perspective is presented using the example of machine tools as the primary means of precision part fabrication.

7.7.1 Evolution of Numerical Control and Its Effects on Machine Tools and Precision

The development of numerically controlled machines represents a major revolution of automation in manufacturing industry. Metal-cutting machine tools are used to produce parts by removing material from a part *blank*, a block of raw material, according to the final desired shape of that part. In general, machine tools consist of components that hold the workpiece and the cutting tool. By providing relative motion between these two, a machine tool generates a cutting tool path which in turn generates the desired shape of the workpiece out of a part blank. In early-generation machine tools, the cutting tool motion is controlled manually (by crank wheels rotating the leadscrews), therefore the quality of the workpiece was mostly the result of the competence of the operator of the machine tool. Before the development of numerically controlled machine tools, complex contoured parts were made by drilling closely spaced holes along the desired contour and then manually finishing the resulting surface to obtain a specified surface finish. This process was very time consuming and prone to errors in locating the holes, which utilized cranks and leadscrews to control the orthogonal movements of the work table manually; for example, the best reported accuracy of airfoil shapes using such techniques was ± 0.175 mm [7.15]. Later

generation of machine tools introduced capabilities to move the cutting tool along a path by tracing a template using mechanical or hydraulic mechanisms, thus reducing reliance on operator competence [7.16]. On the other hand, creating accurate templates was still a main obstacle to achieving cost-effective precision manufacturing.

Around the late 1940s the US Air Force needed more precise parts for its high-performance (faster, highly maneuverable, and heavier) aircraft program (in the late 1940s the target was around ± 0.075 mm). There was no simple way to make wing panels to meet the new accuracy specifications. Manufacturing research community and industry had come up with a solution by introducing numerical control automation to general-purpose machine tools. In 1952, the first numerically controlled three-axis milling machine utilizing a paper tape for programmed instructions, vacuum-tube electronics, and relay-based memory was demonstrated by the Servomechanism Laboratory of the MIT [7.17]. This machine was able to move three axes in coordinated fashion with a speed of about 400 mm/min and a control resolution of 1.25 μm. The automation of machine tools was so effective in improving the accuracy and precision of complex-shaped aircraft components that by 1964 nearly 35 000 numerically controlled machine tools were in use in the USA.

Automation of machine tools by numerical control led to reduction of the need for complex fixtures, tooling, masters, and templates and replaced simple clamps, resulting in significant savings by industry. This was most important for complex parts where human error was likely to occur. With numerical control, once the control program was developed and checked for accuracy, the machine would work indefinitely making the same parts without any error.

7.7.2 Enablers to Improve Precision of Motion

Numerically controlled machine tools rely on sensors that detect positions of each machine component and convert them into digital information. Digital position information is used in control units to control actuators to position the cutting tool properly with respect to the workpiece being cut. The precision of such motion is determined by the resolution of the position sensor (feedback device), the digital control algorithm, and the mechanical and thermal behavior of the machine structural elements. Note that, contrary to manual machine tools, operator skill, experience, and dexterity are not part of the determining factors for the precision of motion. With proper design and environmental controls, it has been demonstrated that machine tools with numerical control can achieve levels of precision on the order of 1 μm or less [7.18, 19].

7.7.3 Modeling and Predicting Machine Behavior and Machining

In most material-removal-based manufacturing processes, the workpiece surfaces are generated as a time record of the position of the cutting tool with respect to the workpiece. The instantaneous position of the tool with respect to the workpiece is generated by the multiple axes of the manufacturing equipment moving in a coordinated fashion. Although the introduction of numerical control (NC) and later computer numerical control (CNC) removed the main source of variation in part quality – manual setups and operations – the complex structural nature of machines providing multi-degree-of-freedom motion and the influence

of changing thermal conditions within the structures as well as in the production environment still result in undesired variations, leading to reduced precision of products.

Specifically, machine tools are composed of multiple slides, rotary tables, and rotary joints, which are usually assembled on top of each other, each designed to move along a single axis of motion, providing either a translational or a rotational degree of freedom. In reality, each moving element of a machine tool has error motions in six degrees of freedom, three translations, and three rotations (Fig. 7.1). Depending on the number of axes of motion, a machine tool can therefore have as many as 30 individual error components. Furthermore, the construction of moving slides and their assemblies with respect to each other introduce additional error components such as squareness and parallelism between axes of motion.

Recognizing the significant benefits of automation provided by numerical control in eliminating random procedures and thus random behavior, in the last five decades many researchers have focused on understanding the fundamental deterministic behavior of error motions of machine tools caused by geometric and thermal influences such that they can be compensated by numerical control functions [7.20–22]. With the advances of robotics research in the 1980s, kinematic modeling of moving structures using homogeneous transformation matrices became a powerful tool for programming and controlling robotic devices [7.23]. Following these developments and assuming rigid-body motions, a general methodology for modeling geometric machine tool errors was introduced using homogeneous transformation matrices to define the relationships between individual error motions and the resulting position and orientation of the cutting tool with respect to the workpiece [7.24]. Kinematic models were further improved to describe the influences of the thermally induced error components of machine tool motions [7.25, 26].

7.7.4 Correcting Machine Errors

Automation of machine tool operation by computer numerical control and the modeling of machine tool systematic errors led to the creation of new hardware and software error compensation technologies enabling improvement of machine tool performance. Machine error compensation in the form of leadscrew pitch errors has been available since the early implementations of CNC. Such leadscrew error compensation is carried out

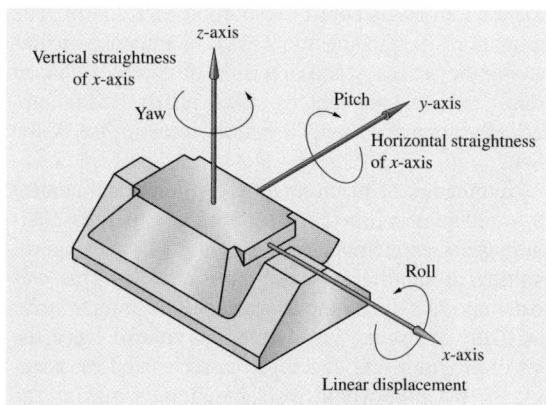

Fig. 7.1 Six error components of a machine slide

using error tables in the machine controller. When executing motion commands, the controller accesses these tables to adjust target positions used in motion servo algorithms (feedforward control). The leadscrew error compensation tables therefore provide one-dimensional error compensation. Modern machine controllers have more sophisticated compensation tables enabling two- or three-dimensional error compensation based on preprocess measurement of error motions. For more general error compensation capabilities, researchers have developed other means of interfacing with the controllers. One approach for such an interface was through hardware modification of the communication between the controller and the position feedback devices [7.27]. In this case, the position feedback signals are diverted to an external microcomputer, where they are counted to determine the instantaneous positions of the slides, and corresponding corrections were introduced by modifying the feedback signals before they are read by the machine controller. Similarly, the software approaches to error compensation were also implemented by interfacing with the CNC through the controller executive software and regular input/output (I/O) devices (such as parallel I/O) [7.28]. Generic functional diagrams depicting the two approaches are shown in Fig. 7.2a and b.

Real-time error compensation of geometric and thermally induced errors utilizing automated features of machine controllers has been reported in the literature to improve the precision of machine tools by up to an order of magnitude. Today's commercially available CNCs employ some of these technologies and cost-effectively transfer these benefits to the manufacturing end-users.

7.7.5 Closed-Loop Machining (Automation-Enabled Precision)

Beyond just machine tool control through CNC, automation has made significant inroads into manufacturing operations over the last several decades. From automated inspection using dedicated measuring systems (such as go/no-go gauges situated next to the production equipment) to more flexible and general-purpose inspection systems (such as coordinate measuring machines) automation has improved the quality control of manufacturing processes, thereby enabling more precise production.

Automation has even changed the paradigm of traditional quality control functions. Traditionally, the function of quality control in manufacturing has been the prevention of defective products being shipped to

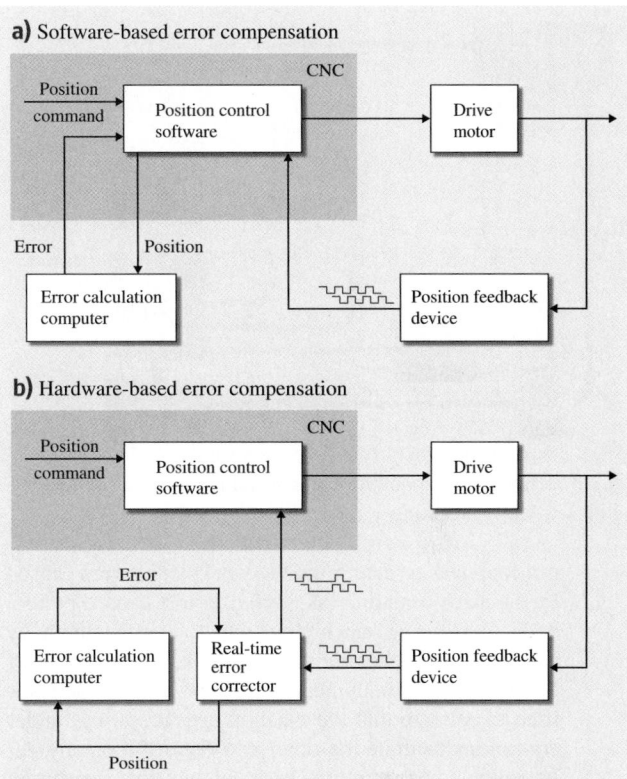

Fig. 7.2a,b Hardware and software error compensation approaches: (**a**) software-based error compensation and (**b**) hardware-based error compensation

the customers. Automation of machining, machine error correction, and part inspection processes have led to new quality control strategies in which real-time control of processes is possible based on real-time information about the machining process and equipment and the resulting part geometries.

In the mid 1990s, the Manufacturing Engineering Laboratory of the National Institute of Standards and Technology demonstrated such an approach in a research project called *Quality In Automation* [7.29]. A quality-control architecture was developed that consisted of three control loops around the machining process: real-time, process-intermittent, and postprocess control loops (Fig. 7.3).

The function of the real-time control loop was to monitor the machine tool and the machining process and to modify the cutting tool path, feed rate, and spindle speed in real time (based on models developed ahead of time) to achieve higher workpiece precision. The function of the process-intermittent con-

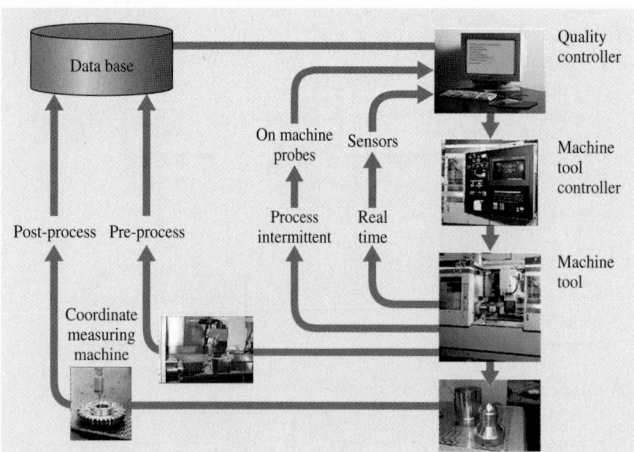

Fig. 7.3 A multilayered quality-control architecture for implementing closed-loop machining

trol loop was to determine the workpiece errors caused by the machining process, such as errors caused by tool deflection during machining, and to correct them by automatically generating a modified NC program for finishing cuts. Finally, the postprocess control loop was used to validate that the machining process was under control and to tune the other two control loops by detecting and correcting the residual systematic errors in the machining system.

7.7.6 Smart Machining

Enabled by automation, the latest developments in machining are leading the technology towards the realization of autonomous, smart machining systems. As described in the paragraphs above, continuous improvements in machining systems through NC and CNC as well as the implementations of various sensing and control technologies have responded to the continuous needs for higher-precision products at lower costs.

However, machining systems still require relatively long periods of trial-and-error processing to produce a given new product optimally. Machine tools still operate with NC programs, which provide the design intent of a product to be machined only partially at best. They have no information about the characteristics of the material to be machined. They require costly periodic maintenance to avoid unexpected breakdowns. These deficiencies increase cost and time to market, and reduce productivity.

Smart machining systems are envisioned to be capable of self-recognition, monitoring, and communication of their capabilities; self-optimization of their operations; self-assessment of the quality of their own work; and self-learning for performance improvement over time [7.30]. The underlying technologies are currently being developed by various research and development organizations; for example, a robust optimizer developed at the National Institute of Standards and Technology demonstrated a way to integrate machine tool performance information and process models with their associated uncertainties to determine the optimum operating conditions to achieve a particular set of objectives related to product precision, cycle time, and cost [7.31–33]. New sets of standards are being developed to define the data formats to communicate machine performance information and other machine characteristics [7.34, 35]. New methods to determine material properties under machining conditions (high strain rates and high temperatures) were developed to improve the machining models that are used in machining optimization [7.36]). New signal-processing algorithms are being developed to monitor the condition of machine spindles and predict failures before catastrophic breakdowns. It is expected that in the next 5–10 years smart machining systems will be available in the marketplace, providing manufacturers with cost-effective means of achieving high-precision products reliably.

7.8 Conclusions and Future Trends

Automation is a key enabler to achieve cost-effective, high-quality products and services to drive society's economical engine. The special duality relationship between automation and precision (each driving the other) escalate the effectiveness of automation in many fields. In this chapter this relationship was described from a relatively narrow perspective of discrete part fabrication. Tighter tolerances in product components that

lead to high-quality products are only made possible by a high degree of automation of the manufacturing processes. This is one of the reasons for the drive towards more manufacturing automation even in countries with low labor costs. The examples provided in this chapter can easily be extended to other economic and technological fields, demonstrating the significant effects of automation.

Recent trends and competitive pressures indicate that more knowledge has been generated about processes, which leads to the reduction of apparent nonsystematic variations. With increased knowledge and technical capabilities, producers are developing more complex, high-value products with smaller numbers of components and subassemblies. This trend leads to even more automation with less cost.

References

7.1 ISO/IEC Guide 99: *International Vocabulary of Metrology – Basic and General Concepts and Associated Terms (VIM)* (International Organization for Standardization, Genewa 2007)

7.2 ISO/IEC Guide 98: *Guide to the Expression of Uncertainty in Measurement (GUM)* (International Organization for Standardization, Genewa 1995)

7.3 C. Evans: *Precision Engineering: An Evolutionary View* (Cranfield Univ. Press, Cranfield 1989)

7.4 D.A. Hounshell: *From the American System to Mass Production, 1800–1932: the Development of Manufacturing Technology in the United States* (Johns Hopkins Univ. Press, Baltimore 1984)

7.5 D. Muir: *Reflections in Bullough's Pond; Economy and Ecosystem in New England* (Univ. Press of New England, Lebanon 2000)

7.6 J.B. Bryan: The benefits of brute strength, Prec. Eng. **2**(4), 173 (1980)

7.7 J.B. Bryan: Closer tolerances – economic sense, Ann. CIRP **19**(2), 115–120 (1971)

7.8 P.A. McKeown: Higher precision manufacturing and the British economy, Proc. Inst. Mech. Eng. **200**(B3), 147–165 (1986)

7.9 ISO/DIS 26303-1: *Capability Evaluation of Machining Processes on Metal-Cutting Machine Tools* (International Organization of Standardization, Genewa 2008)

7.10 ISO 21747: *Statistical Methods – Process Performance and Capability Statistics for Measured Quality Characteristics* (International Organization for Standardization, Genewa 2006)

7.11 ISO 22514-3: *Statistical Methods in Process Management – Capability and Performance – Part 3: Machine Performance Studies for Measured Data on Discrete Parts* (International Organization for Standardization, Genewa 2008)

7.12 ISO/TR 22514-4: *Statistical Methods in Process Management – Capability and Performance – Part 4: Process Capability Estimates and Performance Measures* (International Organization for Standardization, Genewa 2007)

7.13 ASME B89.6.2: *Temperature and Humidity Environment for Dimensional Measurement* (American Society of Mechanical Engineers, New York 1973)

7.14 J.B. Bryan: International status of thermal error research, Ann. CIRP **39**(2), 645–656 (1990)

7.15 G.S. Vasilah: The advent of numerical control 1951–1959, Manufact. Eng. **88**(1), 143–172 (1982)

7.16 D.B. Dallas: *Tool and Manufacturing Engineers Handbook*, 3rd edn. (Society of Manufacturing Engineers, McGraw-Hill, New York 1976)

7.17 J.F. Reintjes: *Numerical Control: Making a New Technology* (Oxford Univ. Press, New York 1991)

7.18 R. Donaldson, S.R. Patterson: Design and construction of a large vertical-axis diamond turning machine, Proc. SPIE 27th Annu. Int. Tech. Symp. Instrum. Disp., Vol. 23 (Lawrence Livermore National Laboratory, 1983), Report UCRL-89738

7.19 N. Taniguchi: The state of the art of nanotechnology for processing of ultraprecision and ultrafine products, Prec. Eng. **16**(1), 5–24 (1994)

7.20 R. Schultschick: The components of volumetric accuracy, Ann. CIRP **25**(1), 223–226 (1972)

7.21 R. Hocken, J.A. Simpson, B. Borchardt, J. Lazar, C. Reeve, P. Stein: Three dimensional metrology, Ann. CIRP **26**, 403–408 (1977)

7.22 V.T. Portman: Error summation in the analytical calculation of the lathe accuracy, Mach. Tool. **51**(1), 7–10 (1980)

7.23 R.P. Paul: *Robot Manipulators: Mathematics, Programming, and Control* (MIT Press, Cambridge 1981)

7.24 M.A. Donmez, C.R. Liu, M.M. Barash: A generalized mathematical model for machine tool errors, modeling, sensing, and control of manufacturing processes, Proc. ASME Winter Annu. Meet., PED, Vol. 23 (1986)

7.25 R. Venugopal, M.M. Barash: Thermal effects on the accuracy of numerically controlled machine tools, Ann. CIRP **35**(1), 255–258 (1986)

7.26 J.S. Chen, J. Yuan, J. Ni, S.M. Wu: Thermal error modeling for volumetric error compensation, Proc. ASME Winter Annu. Meet., PED, Vol. 55 (1992)

7.27 K.W. Yee, R.J. Gavin: *Implementing Fast Probing and Error Compensation on Machine Tools, NISTIR 4447* (The National Institute of Standards and Technology, Gaithersburg 1990)

7.28 M.A. Donmez, K. Lee, R. Liu, M. Barash: A real-time error compensation system for a computerized numerical control turning center, Proc. IEEE Int. Conf. Robot. Autom. (1986)

7.29 M.A. Donmez: Development of a new quality control strategy for automated manufacturing, Proc. Manufact. Int. (ASM, New York 1992)

7.30 L. Deshayes, L. Welsch, A. Donmez, R. Ivester, D. Gilsinn, R. Rhorer, E. Whitenton, F. Potra: Smart machining systems: issues and research trends, Proc. 12th CIRP Life Cycle Eng. Semin. (Grenoble 2005)

7.31   L. Deshayes, L.A. Welsch, R.W. Ivester, M.A. Donmez: Robust optimization for smart machining system: an enabler for agile manufacturing, Proc. ASME IMECE (2005)

7.32   R.W. Ivester, J.C. Heigel: Smart machining systems: robust optimization and adaptive control optimization for turning operations, Trans. North Am. Res. Inst. (NAMRI)/SME, Vol. 35 (2007)

7.33   J. Vigouroux, S. Foufou., L. Deshayes, J.J. Filliben, L.A. Welsch, M.A. Donmez: On tuning the design of an evolutionary algorithm for machining optimization problem, Proc. 4th Int. Conf. Inf. Control (Angers 2007)

7.34   ASME B5.59-1 (Draft): *Information Technology for Machine Tools – Part 1: Data Specification for Machine Tool Performance Tests* (American Society of Mechanical Engineers, New York 2007)

7.35   ASME B5.59-2 (Draft): *Information Technology for Machine Tools – Part 2: Data Specification for Properties of Machine Tools for Milling and Turning* (American Society of Mechanical Engineers, New York 2007)

7.36   T. Burns, S.P. Mates, R.L. Rhorer, E.P. Whitenton, D. Basak: Recent results from the NIST pulse-heated Kolsky bar, Proc. 2007 Annu. Conf. Expo. Exp. Appl. Mech. (Springfield 2007)

8. Trends in Automation

Peter Terwiesch, Christopher Ganz

The present chapter addresses automation as a major means for gaining and sustaining productivity advantages. Typical market environment factors for plant and mill operators are identified, and the analysis of current technology trends allows us to derive drivers for the automation industry.

A section on current trends takes a closer look at various aspects of integration and optimization. Integrating process and automation, safety equipment, but also information and engineering processes is analyzed for its benefit for owners during the lifecycle of an installation. Optimizing the operation through advanced control and plant asset monitoring to improve the plant performance is then presented as another trend that is currently being observed. The section covers system integration technologies such as IEC61850, wireless communication, fieldbuses, or plant data management. Apart from runtime system interoperability, the section also covers challenges in engineering integrated systems.

The section on the outlook into future trends addresses the issue of managing increased complexity in automation systems, takes a closer look at future control schemes, and takes an overall view on automation lifecycle planning.

Any work on prediction of the future is based on an extrapolation of current trends, and estimations of their future development. In this chapter we will therefore have a look at the trends that drive the automation industry and identify those developments that are in line with these drivers.

Like in all other areas of the industry, the future of automation is driven by market requirements on one hand and technology capabilities on the other hand. Both have undergone significant changes in recent years, and continue to do so.

In the business environment, globalization has led to increased worldwide competition. It is not only Western companies that use offshore production to lower their cost; it is more and

more also companies from upcoming regions such as China and India that go global and increase competition. The constant strive for increased productivity is inherent to all successful players in the market.

In this environment, automation technology benefits from the rapid developments in the information technology (IT) industry. Whereas some 15 years ago automation technology was mostly proprietary, today it builds on technology that is being applied in other fields. Boundaries that have clearly been defined due to the incompatibility of technologies are now fully transparent and allow the integration of various requirements throughout the value chain. Field-level data is distributed throughout the various networks that control a plant, both physically and economically, and can be used for analysis and optimization.

To achieve the desired return, companies need to exploit all possibilities to further improve their production or services. This affects all automation levels from field to enterprise optimization, all lifecycle stages from plant erection to dismantling, and all value chain steps from procurement to service.

In all steps, on all levels, automation may play a prominent role to optimize processes.

Part A | 8

8.1 Environment

8.1.1 Market Requirements

Today, even more than in the past, all players in the economy are constantly improving their competitiveness. Inventing and designing differentiating offerings is one key element to achieve this. Once conceived, these offerings need to be brought to market in the most efficient way.

To define the efficiency of a plant or service, we therefore define a measure to rate the various approaches to optimization: The overall equipment effectiveness (OEE). It defines how efficiently the equipment employed is performing its purpose.

Operational Excellence

Looking at the graph in Fig. 8.1, we can clearly see what factors influence a plant owner's return based on the operation of his plant (the graph does not include factors such as market conditions, product differentiation, etc.). The influencing factors are on the cost side, mainly the maintenance cost. Together with plant operation, maintenance quality then determines plant availability, performance, and production quality. From an automation perspective, other factors such as system architecture (redundancy) and system flexibility also have an influence on availability and performance. Operation costs, such as cost of energy/fuel, then have an influence on the product cost.

Future automation system developments must influence these factors positively in order to find wide acceptance in the market.

New Plant Construction

Optimizing plant operations by advanced automation applications is definitely an area where an owner gets most of his operational benefits. An example of the level of automation on plant operations can be seen in Figs. 8.2 and 8.3. When it comes to issues high on the priority list of automation suppliers, delivery costs are as high if not even higher. Although the main benefit of an advanced automation system is with the plant owner (or operator), the automation system is very often not directly sold to that organization, but to an engineering, procurement, and contsruction (EPC) contractor instead. And for these customers, price is one of the top decision criteria.

As automation systems are hardly ever sold off the shelf, but are designed for a specific plant, engineering costs are a major portion of the price of an automation system.

An owner who buys a new automation system looks seriously at the engineering capabilities of the supplier. The effect of efficient engineering on lowering the offer price is one key item that is taken into account. In today's fast developments in the industry, very often the ability to deliver in time is as important as bottom-line price. An owner is in many cases willing to pay a pre-

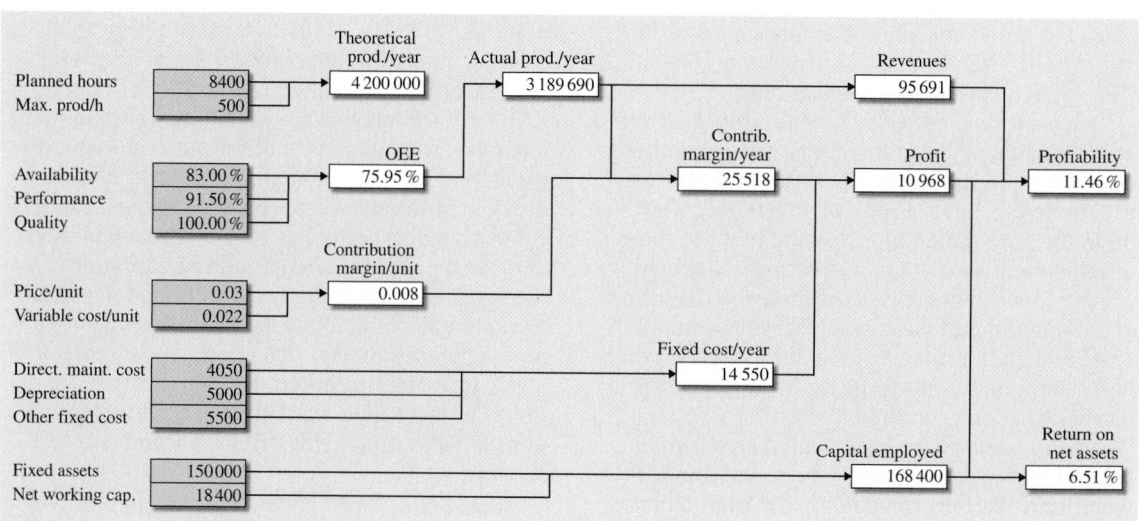

Fig. 8.1 Overall equipment effectiveness

Fig. 8.2 Control of plant operations in the past

mium for a short delivery time, but also for a reduced risk in project execution. Providing expertise from previous projects in an industry is required to keep the execution risk manageable. It also allows the automation company to continuously improve the application design, reuse previous solutions, and therefore increase the quality and reduce the cost of the offering. When talking about the future of automation, engineering will therefore be a major issue to cover.

Plant Upgrades and Extensions
Apart from newly installed greenfield projects, plant upgrades and extensions are becoming increasingly important in the automation business. Depending on the extent of the extension, the business case is similar to the greenfield approach, where an EPC is taking care of all the installations. In many cases however, the owner takes the responsibility of coordinating a plant upgrade. In this case, the focus is mostly on total cost of ownership.

Fig. 8.3 Trend towards fully automated control of plant operations

Furthermore, questions such as compatibility with already installed automation components, upgrade strategies, and integration of old and new components become important to obtain the optimal automation solution for the extended plant.

8.1.2 Technology

Increasingly, automation platforms are driven by information technology. While automation platforms in the past were fully proprietary systems, today they use common IT technology in most areas of the system [8.1]. On the one hand, this development greatly reduces development costs for such systems and eases procurement of off-the-shelf components. On the other hand, the lifecycle of a plant (and its major component the automation system), and IT technology greatly differs. Whereas plants follow investment cycles of 20–30 years, IT technology today at first sight has reached a product lifecycle of less than 1 year, although some underlying technologies may be as old as 10 years or more (e.g., component object model (COM) technology, an interface standard introduced by Microsoft in 1993).

Due to spare parts availability and the lifecycle of software, it is clear that an automation life span of 20 years is not achievable without intermediate updates of the system. A future automation system therefore needs to bridge this wide span of lifecycle expectations and provide means to follow technology in a manner that is safe and efficient for the plant.

Large investments such as field instrumentation, cabling and wiring, engineered control applications, and operational data need to be preserved throughout the lifecycle of the system. Maintaining an automation system as one of the critical assets of a plant needs to be taken into consideration when addressing plant lifecycle issues. In these considerations, striking the right balance between the benefits of standardized products, bringing quality, usability, cost, and training advantages, and customized solutions as the best solution for a given task may become critical.

8.1.3 Economical Trends

In today's economy, globalization is named as the driver for almost everything. However, there are some aspects apart from global price competitiveness that do have an influence on the future of automation.

Communication technology has enabled companies to spread more freely over the globe. While in the past a local company had to be more or less self-supporting (i.e., provide most functions locally), functions can today be distributed worldwide. Front- and back-office organizations no longer need to be under the same roof; development and production can be continents apart. Even within the same project, global organizations can contribute from various locations.

These organizations are interlinked by high-bandwidth communication. These communication links not only connect departments within a company, they also connect companies throughout the value chain. While in earlier days data between suppliers and customers were exchanged on paper in mail (with corresponding time lags), today's interactions between suppliers and customers are almost instant.

In today's business environment, distances as well as time are shorter, resulting in an increase in business interactions.

8.2 Current Trends

8.2.1 Integration

Looking at the trends and requirements listed in the previous Sections, there is one theme which supports these developments, and which is a major driver of the automation industry: integration. This term appears in various aspects of the discussions around requirements, in terms of horizontal, vertical, and temporal integration, among others. In this Section we will look at various trends in integration and analyze their effect on business.

Process Integration

Past approaches to first develop the process and then design the appropriate control strategy for it do not exploit the full advantages of today's advanced control capabilities. If we look at this under the overall umbrella of integration, there is a trend towards integrated design of process and control.

In many cases, more advanced automation solutions are required as constraints become tighter. Figures 8.4–8.6 show examples which highlight the greater degree and complexity of models (i.e., growing number of constraints, reduction in process buffers, and nonlinear dynamic models). There is an ongoing trend towards tighter hard constraints, imposed from regulating authorities. Health, safety, and especially environmental constraints are continuously becoming tighter.

Controllers today not only need to stabilize one control variable, or keep it within a range of the set-point. They also need to make sure that a control action does not violate environmental constraints by producing too much of a by-product (e.g., NO_x). Since many of these boundary conditions are penalized today, these control actions may easily result in significant financial losses or other severe consequences.

In addition to these hard constraints, more and more users want to optimize soft constraints, such as energy consumption (Fig. 8.7), or plant lifecycle. If one ramps up production in the fastest way possible (and maybe meet some market window or production deadline), energy consumed or plant lifecycle consumption due to increased stress on components may compensate these quick gains. An overall optimization problem that takes these soft constraints into account can therefore result in returns that are not obvious even to the experienced operator.

A controller that can keep a process in tighter constraints furthermore allows an owner to optimize process equipment. If the control algorithm can guarantee a tight operational band, process design can reduce buffers and spare capacity. Running the process closer

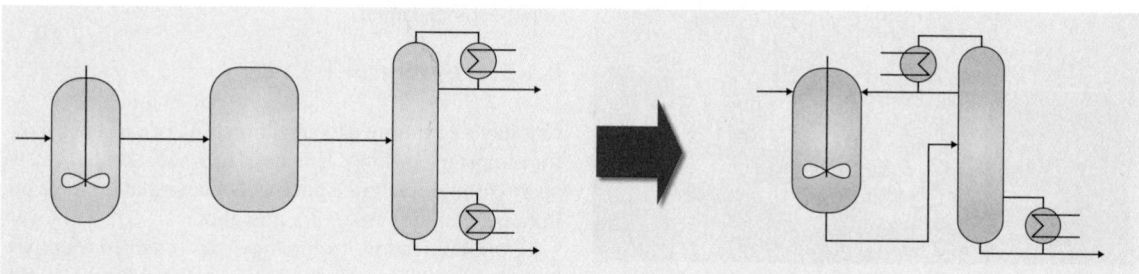

Fig. 8.4 Trend towards reduction of process buffers (e.g., supply chain, company, site, unit)

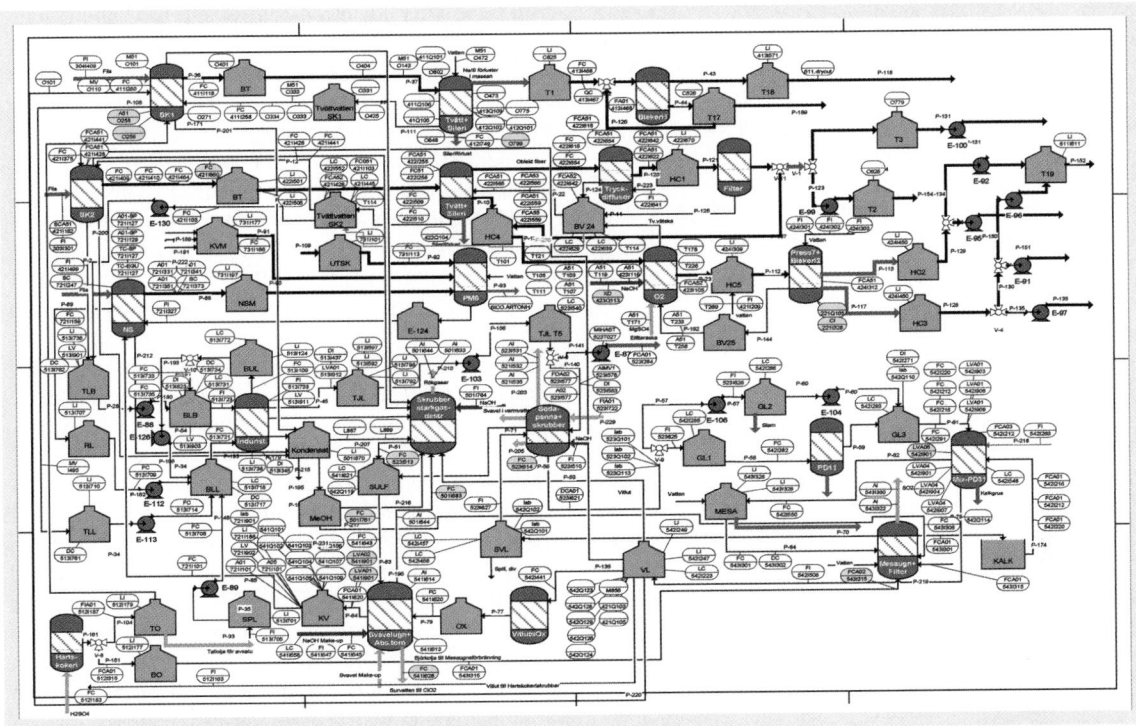

Fig. 8.5 Trend towards broader scope, more complex, and integrated online control, for example, in pulp operations

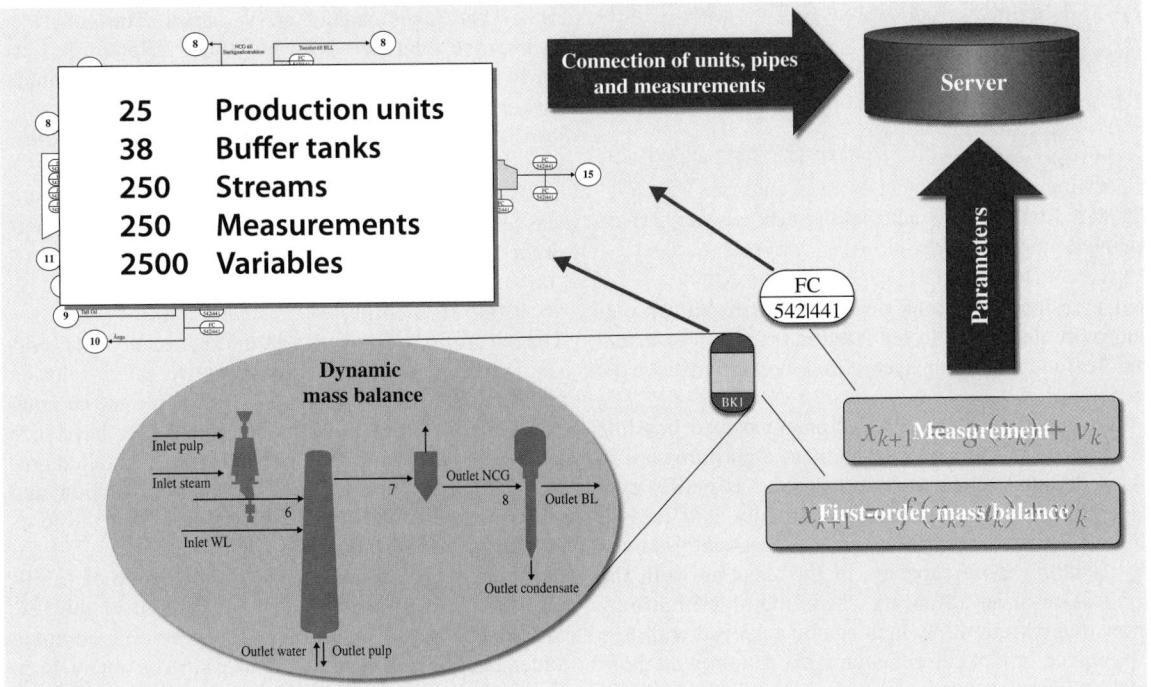

25	**Production units**
38	**Buffer tanks**
250	**Streams**
250	**Measurements**
2500	**Variables**

Dynamic mass balance

Inlet pulp
Inlet steam
Outlet NCG
Outlet BL
Inlet WL
Outlet condensate
Outlet water ↕ Outlet pulp

Connection of units, pipes and measurements

Server

Parameters

FC 542|441

BK1

x_{k+1} Measurement $g(x_k) - v_k$

x_{k+1} First-order mass balance k

Fig. 8.6 Trend towards a nonlinear dynamic model

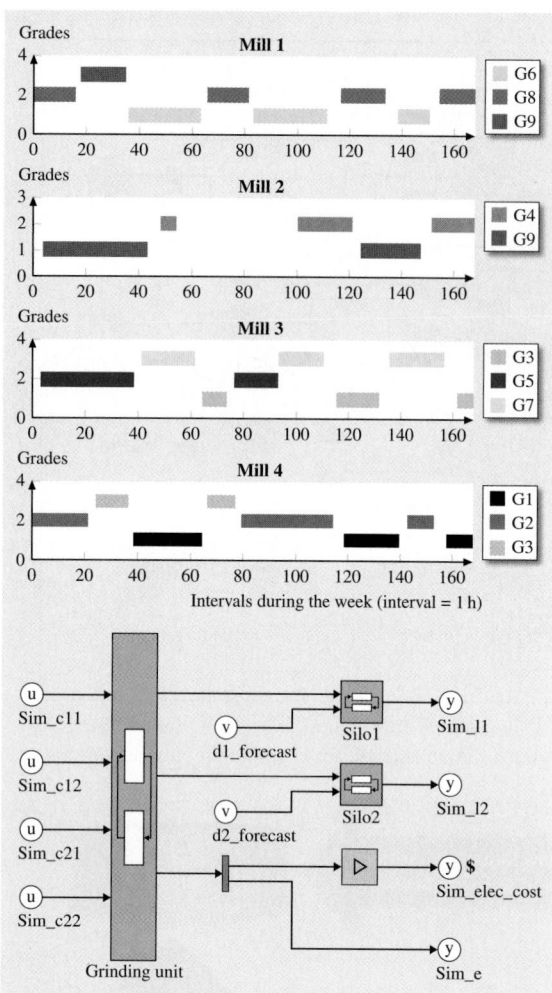

Fig. 8.7 Trend towards automated electrical energy management

By applying advanced control and scheduling algorithms, not only can an owner increase the productivity of the installed equipment, but he may also be able to reduce installed buffers. Intermediate storage tanks or queues can be omitted if an optimized control scheme considers a larger part of the production facility. In addition to reducing the investment costs by reducing equipment, the reduction of buffers also results in a reduction of work in progress, and in the end allows the owner to run the operation with less working capital.

Looking at the wide impact of more precise control algorithms (which in many cases implies more advanced algorithms) on OEE, we can easily conclude that once these capabilities in control system become more easily available, users will adopt them to their benefit.

Example: Thickness Control in Cold-Rolling Mills Using Adaptive MIMO Controller. In a cold-rolling mill, where a band of metal is rolled off an input coil, run through the mill to change its thickness, and then rolled onto an output coil, the torques of the coilers and the roll position are controlled to achieve a desired output thickness, uncoiler tension, and coiler tension. The past solution was to apply single-loop controllers for each variable together with feedforward strategies.

The approach taken in this case was to design an adaptive multiple-input/multiple-output (MIMO) controller that takes care of all variables. The goal was to improve tolerance over the whole strip length, improve quality during ramp-up/ramp-down, and enable higher speed based on better disturbance rejection. Figure 8.8 shows the results from the plant with a clear improvement by the new control scheme [8.2].

By applying the new control scheme, the operator was able to increase throughput and quality, two inputs of the OEE model shown in Fig. 8.1.

Integrated Safety

The integration of process and automation becomes critical in safety applications. Increasingly, safety-relevant actions are moved from process design into the automation system. With similar motivations as we have seen before, a plant owner may want to reduce installed process equipment in favor of an automation solution, and replace equipment that primarily serves safety purposes by an emergency shutdown system.

Today's automation systems are capable of fulfilling these requirements, and the evolution of the IEC 61508 standard [8.3] has helped to develop a common understanding throughout the world. The many local standards have mostly been replaced by IEC 61508's

to its design limitations results in either higher output, more flexibility, faster reaction or allows to install smaller (and mostly cheaper) components to achieve the same result.

A reduction of process equipment is also possible if advanced production management algorithms are in place. Manual scheduling of a process is hardly ever capable to load all equipment optimally, and the solution to a production bottleneck is frequently solved by installing more capacity. In this case as well, the application of an advanced scheduling algorithm may show that current throughput can be achieved with less equipment, or that current equipment can provide more capacity than expected.

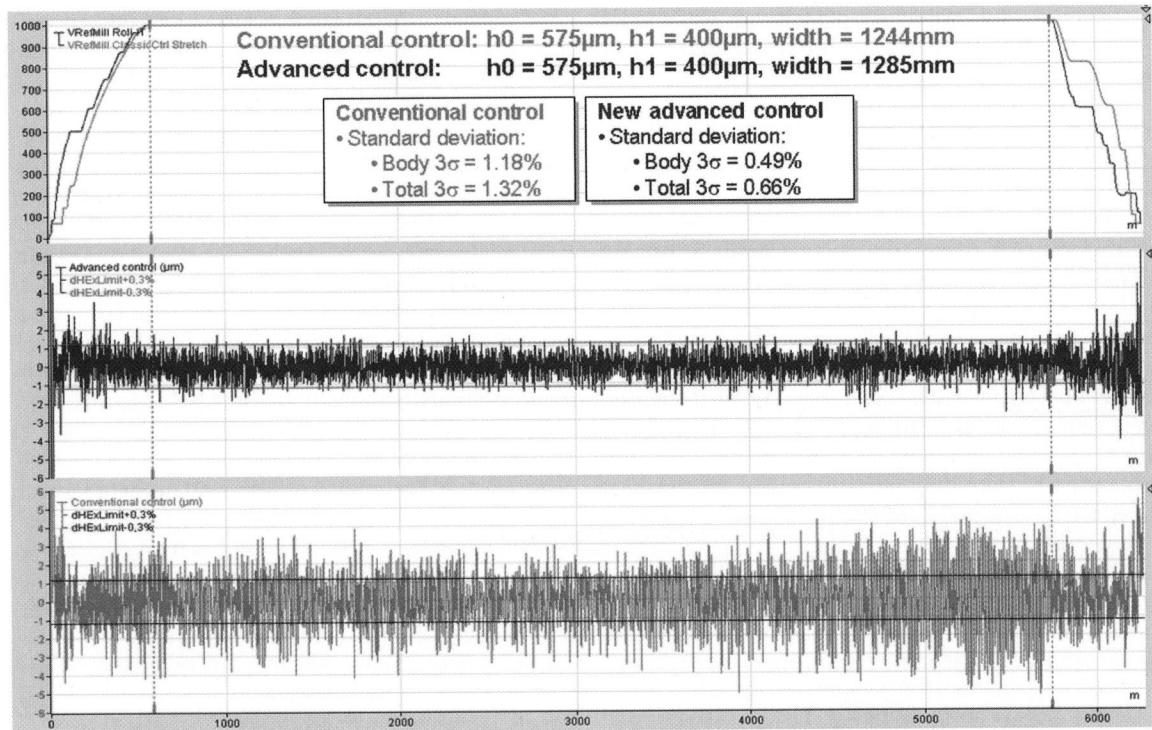

Fig. 8.8 Cold-rolling mill controller comparison

safety integrity level (SIL) requirements. Exceeding the scope of previous standards, ICE61508 not only defines device features that enable them to be used in safety critical applications, it also defines the engineering processes that need to be applied when designing electrical safety systems.

Many automation suppliers today provide a safety-certified variant of their controllers, allowing safety solutions to be tightly integrated into the automation system. Since in many cases these are specially designed and tested variants of the general-purpose controllers, they are perceived having a guaranteed higher quality with longer mean time between failures (MTBF) and/or shorter mean time to repair (MTTR). In some installations where high availability or high quality is required without the explicit need for a certified safety system, plant owners nevertheless choose the safety-certified variant of a system to achieve the desired quality attributes in their system.

A fully integrated safety system furthermore increases planning flexibility. In a fully integrated system, functionality can be moved between the safety controllers and regular controllers, allowing for a fully scalable system that provides the desired safety level.

For more information on safety in automation please refer to Chap. 39.

Information Integration
Device and System Integration
Intelligent Field Devices and their Integration. When talking about information integration, some words need to be spent on information sources in an automation system, i. e., on field devices.

Field devices today not only provide a process variable. Field devices today benefit from the huge advancements in miniaturization, which allows manufacturers to move measurement and even analysis functions from the distributed control system (DCS) into the field device. The data transmitted over fieldbuses not only consists of one single value, but of a whole set of information on the measurement. Quality as well as configuration information can be read directly from the device and can be used for advanced asset monitoring. The amount of information available from the field thus greatly increases and calls for extended processing capabilities in the control system.

Miniaturization and increased computing power also allow the integration of ultrafast control loops on the field level, i. e., within the field device, that are not feasible if the information has to traverse controllers and buses.

All these functions call for higher integration capabilities throughout the system. More data needs to be transferred not only from the field to the controller, but since much of the information is not required on the process control level but on the operations or even at the plant management level, information needs to be distributed further. Information management requirements are also increased and call for more and faster data processing.

In addition to the information exchanged online, intelligent field devices also call for an extended reach of engineering tools. Since these devices may include control functionality, planning an automation concept that spreads DCS controllers and field devices requires engineering tools that are capable of drawing the picture across systems and devices.

The increased capabilities of the field devices immediately create the need for standardization. The landscape that was common 20 years ago, where each vendor had his own proprietary integration standard, is gone. In addition to the fieldbus standards already widely in use today, we will look at IEC 61850 [8.4] as one of the industrial Ethernet-based standards that has recently evolved and gained wide market acceptance in short time.

Fieldbus. For some years now, the standard to communicate towards field devices is fieldbus technology [8.5], defined in IEC 61158 [8.6]. All major fieldbus technologies are covered in this standard, and depending on the geographical area and the industry, in most cases one or two of these implementations have evolved to be the most widely used in an area. In process automation, Foundation Fieldbus and Profibus are among the most prominent players.

Fieldbus provides the means for intelligent field devices to communicate their information to each other, or to the controller. It allows remote configuration as well as advanced diagnostics information.

In addition to the IEC 61158 protocols that are based on communication on a serial bus, the HART protocol (highway addressable remote transducer protocol, a master-slave field communication protocol) has evolved to be successful in existing, conventionally wired plants. HART adds serial communication on top of the standard 4–20 mA signal, allowing digital information to be transmitted over conventional wiring.

Fieldbus is the essential technology to further integrate more information from field devices into complex automation systems.

IEC 61850. IEC 61850 is a global standard for *communication networks and systems in substations*. It is a joint International Electrotechnical Commission (IEC) and American National Standards Institute (ANSI) standard, embraced by all major electrical vendors. In addition to just focusing on a communication protocol, IEC 61850 also defines a data model that comprises the context of the transmitted information. It is therefore one of the more successful approaches to achieve true interoperability between devices as well as tools from different vendors [8.7].

IEC 61850 defines all the information that can be provided by a control function through the definition of logical nodes. Substation automation devices can then implement one or several of these functions, and define their capabilities in standardized, extensible markup language (XML)-based data files, which in turn can be read by all IEC 61850-compliant tools [8.8].

System integration therefore becomes much faster than in the past. Engineering is done on an object-oriented level by linking functions together (Fig. 8.9). The configuration of the communication is then derived from these object structures without further manual engineering effort.

Due to the common approach by ANSI and IEC, by users and vendors, this standard was adopted very quickly and is today the common framework in substation automation around the world.

Once an owner has an electrical system that provides IEC 61850 integration, the integration into the plant DCS is an obvious request. To be able to not only communicate to an IEC 61850-based electrical system directly from the DCS, but also to make use of all object-oriented engineering information is a requirement that is becoming increasingly important for major DCS suppliers.

Wireless. When integrating field devices into an automation system, the availability of standard protocols is of great help, as we have seen in *Device and System Integration*. In the past this approach was very often either limited to new plant installations where cable trays were easily accessible, or resulted in very high installation costs. The success of the HART protocol is mainly due to the fact that it can reuse existing wiring [8.9].

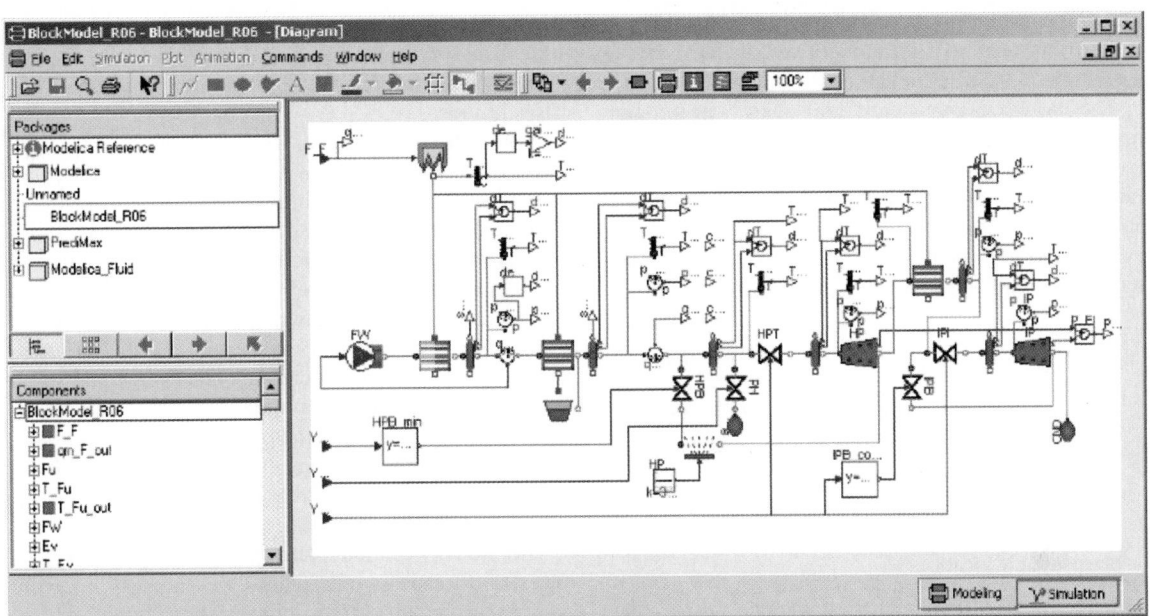

Fig. 8.9 Trend towards object-oriented modeling, e.g., visual flowsheet modeling; combined commodity models with proprietary knowledge; automatic generation of stand-alone executable code

The huge success of wireless technology in other areas of daily life raises the question of whether this technology can also be applied to automation problems. As recent developments have shown, this trend is continuously gaining momentum [8.10]. Different approaches are distinguished as a function of how power is supplied, e.g., by electrical cable, by battery or by power harvesting from their process environment, and by the way the communication is implemented.

As with *wired* communications, also wireless summarizes a variety of technologies which will be discussed in the following sections.

Close Range. In the very close range, serial communication today very often makes use of Bluetooth technology. Originating from mobile phone accessory integration, similar concepts can also be applied in situations in which devices need to communicate over a short distance, e.g., to upgrade firmware, or to read out diagnostics. It merely eliminates the need for a serial cable, and in many cases the requirement to have close-range interaction with a device has been removed completely, since it is connected to the system through other serial buses (such as a fieldbus) that basically allow the user to achieve the same results.

Another upcoming technology under the wireless umbrella is radio frequency identification (RFID).

These small chips are powered by the electromagnetic field set up to communicate by the sensing device. RFID chips are used to mark objects (e.g., items in a store), but can also be used to mark plant inventory and keep track of installed components. Chapter 49 discusses RFID technology in more detail.

RFID can not only be used to read out information about a device (such as a serial number or technical data), but to store data dynamically. The last maintenance activity can thus be stored on the device rather than in a plant database. The device keeps this information attached while being stored as spare part, even if it is disconnected from the plant network. When looking for spares, the one with the fewest past operating hours can therefore be chosen. RFID technology today allows for storage of increasing amounts of data, and in some cases is even capable of returning simple measurements from an integrated sensor.

Information stored on RFID chips is normally not accessed in real time through the automation system, but read out by the maintenance engineer walking through the plant or spare parts storage with the corresponding reading device. To display the full information on the device (online health information, data sheet, etc.) a laptop or tablet personal computer (PC) can then retrieve the information online through its wireless communication capability.

Mid-range. Apart from distributing online data to mobile operator terminals throughout the plant, WiFi has made its entrance also on the plant floor.

The aforementioned problem, where sensors cannot easily be wired to the main automation system, is increasingly being solved by the use of wireless communication, reducing the need and cost for additional cabling.

Applications where the installation of wired instruments is difficult are growing, including where:

- The device is in a remote location.
- The device is in an environment that does not allow for electrical signal cables, e.g., measurements on medium- or high-voltage equipment.
- The device is on the move, either rotating, or moving around as part of a production line.
- The device is only installed temporarily, either for commissioning, or for advanced diagnostics and precise fault location.

The wide range of applications has made wireless device communication one of the key topics in automation today.

Long Range. Once we leave the plant environment, wireless communication capabilities through GSM (global system for mobile communication) or more advanced third-generation (3G) communication technologies allow seamless integration of distributed automation systems.

Applications in this range are mostly found in distribution networks (gas, water, electricity), where small stations with low functionality are linked together in a large network with thousands of access points.

However, also operator station functionality can be distributed over long distances, by making thin-client capability available to handheld devices or mobile phones. To receive a plant alarm through SMS (short message service, a part of the GSM standard) and to be able to acknowledge it remotely is common practice in unmanned plants.

Nonautomation Data. In addition to real-time plant information that is conveyed thorough the plant automation system, plant operation requires much more information to run a plant efficiently. In normal operation as well as in abnormal conditions, a plant operator or a maintenance engineer needs to switch quickly between different views on the process. The process display shows the most typical view, and trend displays and event lists are commonly use to obtain the full picture

on the state of the process. To navigate quickly between these displays is essential. To call up the process display for a disturbed object directly from the alarm list saves critical time.

Once the device needs further analysis, this normally requires the availability of the plant documentation. Instead of flipping through hundreds of pages in documentation binders, it is much more convenient to directly open the electronic manual on the page where the failed pump is described together with possibilities to initiate the required maintenance actions.

Availability of the information in electronic format is today not an issue. Today, all plant documentation is provided in standard formats. However, the information is normally not linked. It is hardly possible to directly switch between related documents without manual search operations that look for the device's tag or name.

An object-oriented plant model that keeps references to all aspects of a plant object greatly helps in solving this problem. If in one location in the system, all views on the very same object are stored – process display, faceplate, event list, trend display, manufacturer instructions, but also maintenance records and inventory information – a more complete view of the plant state can be achieved.

The reaction to process problems can be much quicker, and personnel in the field can be guided to the source of the problem faster, thus resolving issues more efficiently and keeping the plant availability up. We will see in *Lifecycle Optimization* how maintenance efficiency can even be increased by advanced asset management methods.

Security. A general view on the future of automation systems would not be complete without covering the most prominent threat to the concepts presented so far: system security.

When systems were less integrated, decoupled from other information systems, or interconnected by 4–20 mA or binary input/output (I/O) signals, system security was limited to physical security, i.e., to prevent unauthorized people access the system by physical means (fences, building access, etc.).

Integrating more and more information systems into the automation system, and enabling them to distribute data to wherever it is needed (i.e., also to company headquarters through the Internet), security threats soon become a major concern to all plant owners.

The damage that can be caused to a business by negligence or deliberate intrusion is annoying when web sites are blocked by denial-of-service attacks. It is significant if it affects the financial system by spyware or phishing attacks, but it is devastating when a country's infrastructure is attacked. *Simply* bringing down the electricity system already has quite a high impact, but if a hacker gains access to an automation system, the plant can actually be damaged and be out of service for a significant amount of time. The damage to a modern society would be extremely high.

Security therefore has to be at the top of any list of priorities for any plant operator. Security measures in automation systems of the future need to be continuously increased without giving up some of the advantages of wider information integration.

In addition to technical measures to keep a plant secure, security management needs to be an integral part of any plant staff member's training, as is health and safety management today.

While security concerns for automation systems are valid and need to be addressed by the plant management, technical means and guidance on security-related processes are available today to secure control systems effectively [8.11]. Security concerns should therefore not be the reason for not using the benefits of information integration in plants and enterprises.

Engineering Integration

The increased integration of devices and systems from plant floor to enterprise management poses another challenge for automation engineers: information integration does not happen by itself, it requires significant engineering effort. This increased effort is a contradiction to the requirement for faster and lower-cost project execution. This dilemma can only be resolved by improved engineering environments. Chapter 86, enterprise integration and interoperability, delves deeper into this topic.

Today, all areas of plant engineering, starting at process design and civil engineering, are supported by specialized engineering tools. While their coupling was loose in the past, and results of an engineering phase were handed over on paper and very often typed into other tools again, the trend towards exchanging data in electronic format is obvious. Whoever has tried to exchange data between different types of engineering tools immediately faces the questions:

- What data?
- What format?

The question about what data can only be answered by the two parties exchanging data. The receiver only knows what data he needs, and the provider only knows what data she can provide. If the two parties are within different departments of the same company, an internal standard on data models can be agreed on, but when the exchange is between different business partners, this very often results in a per-project agreement.

In electrical systems, this issue has been addressed by IEC 61850. In addition to being a communication standard, it also covers a data model. Data objects (logical nodes) are defined by the standard, and engineering tools following the standards can easily integrate devices of various vendors without project specific agreements. The standard was even extended beyond electrical systems to cover hydropower plants (and also for wind generators in IEC 61400-25). So far, further extensions into other plant types or industries seem hardly feasible due to the variance and company internal-grown standards.

The discussion on the format today quickly turns into a spreadsheet-based solution. This approach is very common, and most tools provide export and/or import functionality in tabular form. However, this requires separate sheets for each object type, since the data fields may vary between objects. A format that supports a more object-oriented approach is required.

Recently, the most common approach is to go towards XML-based formats. IEC 61850 data is based on XML, and there are standardization tendencies that follow the same path. CAEX (computer aided engineering exchange, an engineering data format) according to IEC 62424 is just one example; PLCOpen XML or AutomationML are others.

The ability to agree on data standards between engineering tools greatly eases interaction between the various disciplines not only in automation engineering, but in plant engineering in general.

Once the data format is defined, there still remains the question of wording or language. Even when using the same language, two different engineering groups may call the same piece of information differently. A semantic approach to information processing may address this issue.

With some of these problems addressed in a nearer future, further optimization is possible by a more parallel approach to engineering. Since information is revised several times during plant design, working with early versions of the information is common. Updates of the information is normally required, and then updating the whole engineering chain is a challenge. To work on a common database is a trend that is evolving

Part A | 8.2

in plant design; but also in the design of the automation system, a common database to hold various aspects of automation engineering is an obvious idea. Once these larger engineering environments are in place, data exchange quickly becomes bidirectional. Modifications done in plant design affect the automation system, but also information from the automation database such as cabling or instrumentation details should be fed back into the plant database. This is only possible if the data exchange can be done without loss of information, otherwise data relations cannot be kept consistent.

Even if bidirectional data exchange is solved, more partners in complex projects easily result in multidirectional data exchange. Versioning becomes even more essential than in a single tool. Whether the successful solution of data exchange between two domains can be kept after each of the tools is released in a new version remains to be seen. The challenges in this area are still to be faced.

Customer Value

The overall value of integration for owners is apparent on various levels, as we have shown in the previous Sections.

The pressure to shorten projects and to bring down costs will increase the push for engineering data integration. This will also improve the owner's capability to maintain the plant later by continuously keeping the information up to date, therefore reducing the lifecycle cost of the plant.

The desire to operate the plant efficiently and keep downtimes low and production quality up will drive the urge to have real-time data integrated by connecting interoperable devices and systems on all levels of the automation system [8.12]. The ability to have common event and alarm lists, to operate various type of equipment from one operator workplace, and to obtain consistent asset information combined in one system are key enablers for operational excellence.

Security concerns require a holistic approach on the level of the whole plant, integrating all components into a common security framework, both technically and with regard to processes.

8.2.2 Optimization

The developments described up to now enable one further step in productivity increase that has only been partially exploited in the past. Having more information available at any point in an enterprise allows

for a typical control action: to close the loop and to optimize.

Control

Closest to the controlled process, closing the loop is the traditional field of automation. PID controllers (proportional-integral-derivative) mostly govern today's world of automation. Executed by programmable logic controllers (PLC) or DCS controllers, they do a fairly good job at keeping the majority of industrial processes stable.

However, even if much more advanced control schemes are available today, not even the ancient PID loops perform where they could if they were properly tuned. Controller tuning during commissioning is more of an art done by experienced experts than engineering science.

As we have already concluded in *Process Integration*, several advantages favor the application of advanced control algorithms. Their ability to keep processes stable in a narrower band allows either to choose smaller equipment to reach a given limit, or to increase the performance of existing equipment by running the process closer to boundaries.

However, controllers today are mostly designed based on knowledge of a predominantly fixed process, i.e., plant topology and behavior is assumed to be as-designed. This process knowhow is often depicted in a process model which is either used as part of the controller (e.g., model predictive control) or has been used to design the controller.

Once the process deviates from the predefined topology, controllers are soon at their limits. This can easily happen when sensors or communication links fail. This situation is today mostly solved by redundant design, but controllers that consider some amount of missing information may be an approach to increase the reliability of the control system even further. Controllers reacting more flexibly to changing boundary conditions will extend the plant's range of operation, but will also reduce predictability.

Another typical case of a plant deviating from the designed state is ageing or equipment degradation. Controllers that can handle this (e.g., adaptive control) can keep the process in an optimal state even if its components are not. Furthermore, a controller reacting on performance variations of the plant can not only adapt to it, but also convey this information to the maintenance personnel to allow for efficient plant management and optimization.

Plant Optimization

At a plant operation level, all the data generated by intelligent field devices and integrated systems comes together. To have more information available is positive, but to the plant operator it is also confusing. More devices generating more diverse alarms quickly flood a human operator's perception. More information does not per se improve the operation of a plant. Information needs to be turned into knowledge.

This knowledge is buried in large amounts of data, in the form of recorded analog trend signals as well as alarm and event information. Each signal in itself only tells a very small part of the story, but if a larger number of signals are analyzed by using advanced signal processing or model identification algorithms, they reveal information about the device, or the system observed.

This field is today known as asset monitoring. The term denotes anything from very simple use counters up to complex algorithms that derive system lifecycle information from measured data. In some cases, the internal state of a high-value asset can be assessed through the interpretation of signals that are available in the automation system already used in control schemes. If the decision is between applying some analysis software or to shut down the equipment, open it, and visually inspect, the software version can in many cases more directly direct the maintenance personnel towards the true fault of the equipment.

The availability of advanced asset monitoring algorithms allows for optimized operation of the plant. If component ageing can be calculated from measurements, optimizing control algorithms can put the load on less stressed components, or can trade asset lifecycle consumption against the quick return of a fast plant start-up.

The requirement to increase availability and production quality calls for advanced algorithms for asset monitoring and results in an asset optimization scheme that directly influences the plant operator's bottom line. The people operating the plant, be it in the operations department or in maintenance, are supported in their analysis of the situation and in their decisions by more advanced systems than are normally in operation today.

When it comes to discrete manufacturing plants, the optimization potential is as high as in continuous production. Advanced scheduling algorithms are capable of optimizing plant utilization and improving yield. If these algorithms are flexible to allow a rescheduling and

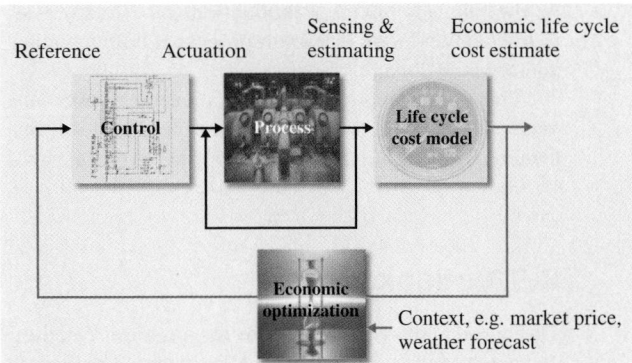

Fig. 8.10 Trend towards lifecycle optimization

production replanning in operation to accommodate urgent orders at runtime, plant efficiency can be optimized dynamically and have an even more positive effect on the bottom line.

Lifecycle Optimization

The optimization concepts presented so far enable a plant owner to optimize OEE on several levels (Fig. 8.10). We have covered online production optimization as well as predictive maintenance through advanced asset optimization tools. If the scope of the optimization can be extended to a whole fleet of plants and over a longer period of time, continuous plant improvement by collecting best practices and statistical information on all equipment and systems becomes feasible.

What does the automation system contribute to this level of optimization? Again, most data originates from the plant automation system's databases. In *Plant Optimization* we have even seen that asset monitoring provides information that goes beyond the raw signals measured by the sensors and can be used to draw conclusions on plant maintenance activities. From a fixed-schedule maintenance scheme where plant equipment is shut down based on statistical experience, asset monitoring can help moving towards a condition-based maintenance scheme. Either equipment operation can be extended towards more realistic schedules, or emergency plant shutdown can be avoided by early detecting equipment degradation and going into planned shutdown.

Interaction with maintenance management systems or enterprise resource planning systems is today evolving, supported by standards such as ISA95. Enterprise-wide information integration is substantial to

Part A | 8.2

be continuously on top of production and effectiveness, to track down inefficiencies in processes both technical and organizational.

These concepts have been presented over the years [8.13], but to really close the loop on that level requires significant investments by most industrial companies. Good examples of information integration on that level are airline operators and maintenance companies, where additional minutes used to service deficiencies in equipment become expensive. Failure to address these deficiencies become catastrophic and mission critical.

8.3 Outlook

The current trends presented in the previous Sections do show benefits, in some cases significant. The push to continue the developments along these lines will therefore most probably be sustained.

8.3.1 Complexity Increase

One general countertrend is also clearly visible in many areas: increased complexity to the extent that it becomes a limiting factor. System complexity does not only result in an increase in initial cost (more installed infrastructure, more engineering), but also in increased maintenance (IT support on plant floor). Both factors influence OEE (Fig. 8.1) negatively. To counter the perceived complexity in automation systems it is therefore important to facilitate wider distribution of the advanced concepts presented so far.

Modeling

Many of the solutions available today in either asset monitoring or advanced control rely on plant models. The availability of plant models for applications such as design or training simulation is also essential. However, plant models are highly dependent on the process installation, and need to be designed or at least tuned to every installation. Furthermore, the increased complexity of advanced industrial plants also calls for wider and more complex models. Model building and tuning is today still very expensive and requires highly skilled experts. There is a need common to different areas and industries to keep modeling affordable.

Reuse of models could address this issue in two dimensions:

● To reuse a model designed for one application in another, i. e., to build a controller design model based

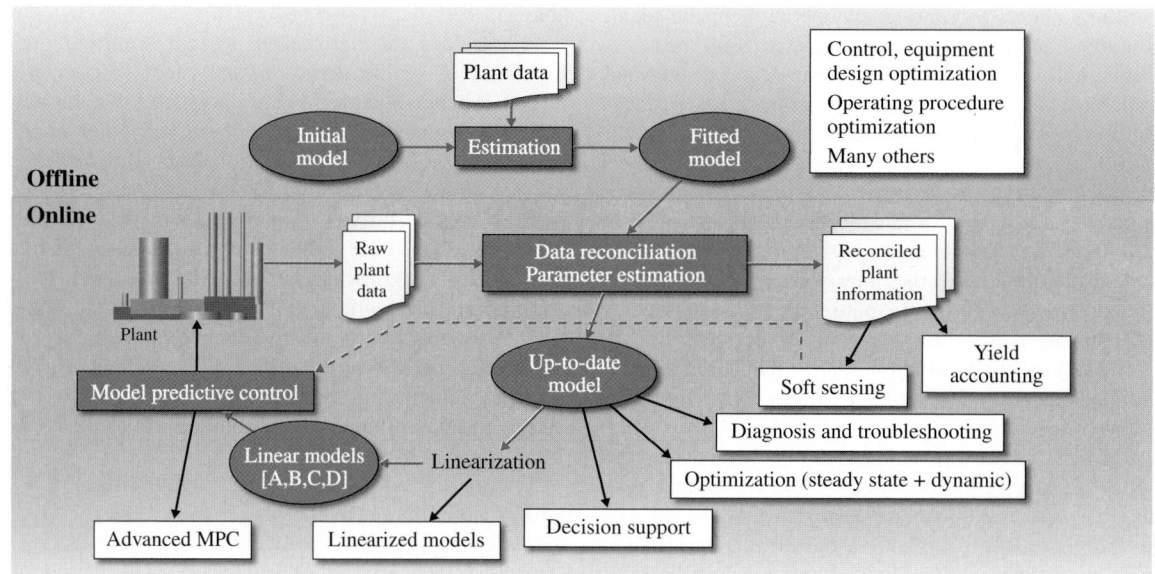

Fig. 8.11 Trend towards automation and control modeling and model reuse

on a model that was used for plant design, or derive the model used in a controller from one that is available for performance monitoring. The plant topology that connects the models can remain the same in all cases.

- To reuse models from project to project, an approach that can also be pursued with engineering solutions to bring down engineering costs (Fig. 8.11).

Operator Interaction

Although modern operator stations are capable of integrating much more information to the operator or maintenance personnel, the operator interaction is not always the most intuitive. In plant diagnostics and maintenance this may be acceptable, but for the operator it is often difficult to quickly perceive a plant situation, whether good or bad, and to act accordingly. This is mostly due to the fact that the operator interface was designed by the engineer who had as inputs the plans of the plant and the automation function, and not the plant environment where the operator needs to navigate.

An operator interface closer to the plant operator's natural environment (and therefore to his intuition) could improve the perception of the plant's current status.

One way of doing this is to display the status in a more intuitive manner. In an aircraft, the artificial horizon combines a large number of measurements in one very simple display, which can be interpreted intuitively by the pilot by just glancing at it. Its movement gives excellent feedback on the plane's dynamics. If we compare this simple display with a current plant operator station with process diagrams, alarm lists, and trend displays, it is obvious that plant dynamics cannot be perceived as easily. Some valuable time early in critical situations is therefore lost by analyzing numbers on a screen. To depict the plant status in more intuitive graphics could exploit humans' capability to interpret moving graphics in a qualitative way more efficiently than from numerical displays.

Automated Engineering

As we have pointed out in *Modeling*, designing and tuning models is a complex task. To design and tune advanced controllers is as complex. Even the effort to tune simple controllers is in reality very often skipped, and controller parameters are left at standard settings of 1.0 for any parameter.

In many cases these settings can be derived from plant parameters without the requirement to tune online on site. Drum sizes or process set-points are documented during plant engineering, and as we have seen in *Engineering Integration*, this data is normally available to the automation engineer. If a control loop's settings are automatically derived from the data found in the plant information, the settings will be much better than the standard values. To the commissioning engineer, this procedure hides some of the complexity of controller fine-tuning.

Whether it is possible to derive control loops automatically from the plant information received from the plant engineering tools remains to be seen. Very simple loops can be chosen based on standard configurations of pumps and valves, but a thorough check of the solution by an experienced automation engineer is still required.

On the other hand, the information contained in engineering data can be used to check the consistency of the manually designed code. If a plant topology model is available that was read out of the process & instrumentation diagram, also piping & instrumentation diagram (P&ID) tool information, automatic measures can be taken to check whether there is an influence of some sort (control logic, interlock logic) between a tank level and the feeding pump.

8.3.2 Controller Scope Extension

Today's control laws are designed on the assumption that the plant behaves as designed. Failed components are not taken into consideration, and deteriorating plant conditions (fouling, drift, etc.) are only to some extent compensated by controller action.

The coverage of nonstandard plant configurations in the design of controllers is rarely seen today. This is the case for advanced control schemes, but also for more advanced scheduling or batch solutions, consideration of these suboptimal plant states in the design of the automation system could improve plant availability. Although this may result in a reduction of quality, the production environment (i.e., the immediate market prices) can still make a lower-quality production useful. To detect whether this is the case, an integration between the business environment, with current cost of material, energy, and maybe even emissions, and the production information in the plant allows to solve optimization problems that optimize the bottom line directly.

8.3.3 Automation Lifecycle Planning

In the past, the automation system was an initial investment like many other installations in a plant.

It was maintained by replacing broken devices by spares, and kept its functionality throughout the years. This option is still available today. In addition to I/O cards, an owner needs to buy spare PCs like other spare parts that may difficult to buy on the market.

The other option an owner has is to continuously follow the technology trend and keep the automation system up to date. This results in much higher life-cycle cost, but against these costs is the benefit of always having the newest technology installed. This in turn requires automation system vendors to continuously provide functionality that improves the plant's performance, justifying the investment.

It is the owner's decision which way to go. It is not an easy decision and shows the importance of keeping total cost of ownership in mind also when purchasing the automation system.

8.4 Summary

Today's business environment as well as technology trends (i. e., robots) are continuously evolving at a fast pace (Fig. 8.12). To improve a plant's competitiveness, a modern automation system must make use of the advancements in technology to react to trends in the business world.

The reaction of the enterprise must be faster, at the lowest level to increase production and reduce downtime, and at higher levels to process customer orders efficiently and react to mid-term trends quickly. The data required for these decisions is mostly buried in the automation system; for dynamic operation it needs to be turned into information, which in turn needs to be processed quickly. To improve the situation with the automation system, different systems on all levels need to be integrated to allow for sophisticated information processing.

The availability of the full picture allows the optimization of single loops, plant operation, and economic performance of the enterprise.

Fig. 8.12 Trend towards more sophisticated robotics

The technologies that allow the automation system to be the core information processing system in a production plant are available today, are evolving quickly, and provide the means to bring the overall equipment effectiveness to new levels.

References

8.1   S. Behrendt, et al.: Integrierte Technologie-Roadmap Automation 2015+, ZVEI Automation (2006), in German

8.2   T. Hoernfeldt, A. Vollmer, A. Kroll: Industrial IT for Cold Rolling Mills: The next generation of Automation Systems and Solutions, IFAC Workshop New Technol. Autom. Metall. Ind. (Shanghai 2003)

8.3   IEC 61508, Functional safety of electrical/electronic/programmable electronic safety-related systems

8.4   IEC 61850, Communication networks and systems in substations

8.5   R. Zurawski: *The Industrial Information Technology Handbook* (CRC, Boca Raton 2005)

8.6   IEC 61158, Industrial communication networks – Fieldbus specifications

8.7   C. Brunner, K. Schwarz: Beyond substations – Use of IEC 61850 beyond substations, Praxis Profiline – IEC 61850 (April 2007)

8.8   K. Schwarz: Impact of IEC 61850 on system engineering, tools peopleware, and the role of the system integrator (2007) http://www.nettedautomation.com/download/IEC61850-Peopleware_2006-11-07.pdf

8.9   ARC Analysts: The top automation trends and technologies for 2008, ARC Strategies (2007)

8.10 G. Hale: People Power, InTech 01/08 (2208)

8.11 M. Naedele: Addressing IT Security for Critical Control Systems, 40th Hawaii Int. Conf. Syst. Sci. (HICSS–40) (Hawaii 2007)

8.12 E.F. Policastro: A Big Pill to Swallow, InTech 04/07 (2007) p. 16

8.13 Center for intelligent maintenance systems, www.nsf.gov/pubs/2002/nsf01168/nsf01168xx.htm

Part B Automation

Part B Automation Theory and Scientific Foundations

Automation Theory and Scientific Foundations. Part B Automation is based on control theory and intelligent control, although interestingly, automation existed before control theory was developed. The chapters in this part explain the theoretical aspects of automation and its scientific foundations, from the basics to advanced models and techniques; from simple feedback and feedforward automation functioning under certainty to fuzzy logic control, learning control automation, cybernetics, and artificial intelligence. Automation is also based on communication, and in this part this subject is explained from the fundamental communication between sensors and actuators, producers and consumers of signals and information, to automation of and with virtual reality; automation mobility and wireless communication of computers, devices, vehicles, flying objects and other location-based and geography-based automation. The theoretical and scientific knowledge about the human role in automation is covered from the human-oriented and human–centered aspects of automation to be applied and operated by humans, to the human role as supervisor and intelligent controller of automation systems and platforms. This part concludes with analysis and discussion on the limits of automation to the best of our current understanding.

9. Control Theory for Automation: Fundamentals

Alberto Isidori

In this chapter autonomous dynamical systems, stability, asymptotic behavior, dynamical systems with inputs, feedback stabilization of linear systems, feedback stabilization of nonlinear systems, and tracking and regulation are discussed to provide the foundation for control theory for automation.

Modern engineering systems are very complex and comprise a high number of interconnected subcomponents which, thanks to the remarkable development of communications and electronics, can be spread over broad areas and linked through data networks. Each component of this wide interconnected system is a complex system on its own and the good functioning of the overall system relies upon the possibility to efficiently control, estimate or monitor each one of these components. Each component is usually high dimensional, highly nonlinear, and hybrid in nature, and comprises electrical, mechanical or chemical components which interact with computers, decision logics, etc. The behavior of each subsystem is affected by the behavior of part or all of the other components of the system. The control of those complex systems can only be achieved in a decentralized mode, by appropriately designing local controllers for each individual component or small group of components. In this setup, the interactions between components are mostly treated as *commands*, dictated from one particular unit to another one, or as *disturbances*, generated by the operation of other interconnected units. The tasks of the various local controllers are then coordinated by some supervisory unit. Control and computational capabilities being distributed over the system, a steady exchange of data among the components is required, in order for the system to behave properly.

In this setup, each individual component (or small set of components) is viewed as a system whose behavior, in time, is determined or influenced by the behavior of other subsystems. Typically, the physical variables by

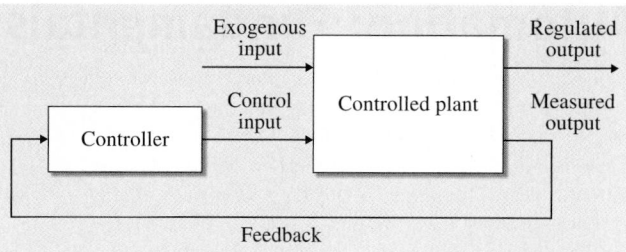

Fig. 9.1 Basic feedback loop

means of which this influence is exerted can be classified into two disjoint sets: one set consisting of all commands and/or disturbances generated by other components (which in this context are usually referred to as *exogenous inputs*) and another set consisting of all variables by means of which the accomplishment of the required tasks is actually imposed (which in this context are usually referred to as *control inputs*). The tasks in question typically comprise the case in which certain variables, called *regulated outputs*, are required to track the behavior of a set of exogenous commands. This leads to the definition, for the variables in question, of a *tracking error*, which should be kept as small as possible, in spite of the possible variation – in time – of the commands and in spite of all exogenous disturbances. The control input, in turn, is provided by a separate subsystem, the *controller*, which processes the information provided by a set of appropriate measurements (the *measured outputs*). The whole control configuration assumes – in this case – the form of a *feedback loop*, as shown in Fig. 9.1.

In any realistic scenario, the control goal has to be achieved in spite of a good number of phenomena which would cause the system to behave differently than expected. As a matter of fact, in addition to the exogenous phenomena already included in the scheme of Fig. 9.1, i.e., the exogenous commands and disturbances, a system may fail to behave as expected also because of endogenous causes, which include the case in which the controlled system responds differently as a consequence of poor knowledge about its behavior due to modeling errors, damages, wear, etc. The ability to handle large uncertainties successfully is one of the main, if not the single most important, reason for choosing the feedback configuration of Fig. 9.1.

To evaluate the overall performances of the system, a number of conventional criteria are chosen. First of all, it must be ensured that the behavior of the variables of the entire system is bounded. In fact, the feedback strategy, which is introduced for the purpose of offsetting exogenous inputs and to attenuate the effect of modeling error, may cause unbounded behaviors, which have to be avoided. Boundedness, and convergence to the desired behavior, are usually analyzed in conventional terms via the concepts of asymptotic *stability* and *steady-state behavior*, discussed in Sects. 9.2–9.3. Since the systems under considerations are systems with inputs (control inputs and exogenous inputs), the influence of such inputs on the behavior of a system also has to be assessed, as discussed in Sect. 9.4. The analytical tools developed in this way are then taken as a basis for the design of a controller, in which – usually – the control structure and free parameters are chosen in such a way as to guarantee that the overall configuration exhibits the desired properties in response to exogenous commands and disturbances and is sufficiently tolerant of any major source of uncertainty. This is discussed in Sects. 9.5–9.8.

9.1 Autonomous Dynamical Systems

In loose terms, a dynamical system is a way to describe how certain physical entities of interest, associated with a natural or artificial process, evolve in time and how their behavior is, or can be, influenced by the evolution of other variables. The most usual point of departure in the analysis of the behavior of a natural or artificial process is the construction of a mathematical model consisting of a set of equations expressing basic physical laws and/or constraints. In the most frequent case, when the study of evolution in time is the issue, the equations in question take the form of an ordinary differential equation, defined on a finite-dimensional Euclidean space. In this chapter, we shall review some fundamental facts underlying the analysis of the solutions of certain ordinary differential equations arising in the study of physical processes.

In this analysis, a convenient point of departure is the case of a mathematical model expressed by means of a first-order differential equation

$$\dot{x} = f(x) \, , \tag{9.1}$$

in which $x \in \mathbb{R}^n$ is a vector of variables associated with the physical entities of interest, usually referred to as the *state* of the system. A solution of the differential equation (9.1) is a differentiable function $\bar{x} : J \to \mathbb{R}^n$ defined on some interval $J \subset \mathbb{R}$ such that, for all $t \in J$,

$$\frac{d\bar{x}(t)}{dt} = f(\bar{x}(t)) \,.$$

If the map $f : \mathbb{R}^n \to \mathbb{R}^n$ is *locally Lipschitz*, i.e., if for every $x \in \mathbb{R}^n$ there exists a neighborhood U of x and a number $L > 0$ such that, for all x_1, x_2 in U,

$$|f(x_1) - f(x_2)| \le L|x_1 - x_2| \,,$$

then, for each $x_0 \in \mathbb{R}^n$ there exists two times $t^- < 0$ and $t^+ > 0$ and a solution $\bar{x}$ of (9.1), defined on the interval $(t^-, t^+) \subset \mathbb{R}$, that satisfies $\bar{x}(0) = x_0$. Moreover, if $\tilde{x} : (t^-, t^+) \to \mathbb{R}^n$ is any other solution of (9.1) satisfying $\tilde{x}(0) = x_0$, then necessarily $\tilde{x}(t) = \bar{x}(t)$ for all $t \in (t^-, t^+)$, that is, the solution $\bar{x}$ is unique. In general, the times $t^- < 0$ and $t^+ > 0$ may depend on the point x_0. For each x_0, there is a maximal open interval $(t_m^-(x_0), t_m^+(x_0))$ containing 0 on which is defined a solution $\bar{x}$ with $\bar{x}(0) = x_0$: this is the union of all open intervals on which there is a solution with $\bar{x}(0) = x_0$ (possibly, but not always, $t_m^-(x_0) = -\infty$ and/or $t_m^+(x_0) = +\infty$).

Given a differential equation of the form (9.1), associated with a locally Lipschitz map f, define a subset W of $\mathbb{R} \times \mathbb{R}^n$ as follows

$$W = \left\{ (t, x) : t \in \left(t_m^-(x), t_m^+(x) \right), x \in \mathbb{R}^n \right\} \,.$$

Then define on W a map $\phi : W \to \mathbb{R}^n$ as follows: $\phi(0, x) = x$ and, for each $x \in \mathbb{R}^n$, the function

$$\varphi_x : \left(t_m^-(x), t_m^+(x) \right) \to \mathbb{R}^n \,,$$
$$t \mapsto \phi(t, x)$$

is a solution of (9.1). This map is called the *flow* of (9.1). In other words, for each fixed x, the restriction of $\phi(t, x)$ to the subset of W consisting of all pairs (t, x) for which $t \in (t_m^-(x), t_m^+(x))$ is the unique (and maximally extended in time) solution of (9.1) passing through x at time $t = 0$.

A dynamical system is said to be *complete* if the set W coincides with the whole of $\mathbb{R} \times \mathbb{R}^n$.

Sometimes, a slightly different notation is used for the flow. This is motivated by the need tom express, within the same context, the flow of a system like (9.1) and the flow of another system, say $\dot{y} = g(y)$. In this case, the symbol ϕ, which represents the *map*, must be replaced by two different symbols, one denoting the

flow of (9.1) and the other denoting the flow of the other system. The easiest way to achieve this is to use the symbol x to represent the map that characterizes the flow of (9.1) and to use the symbol y to represent the map that characterizes the flow of the other system. In this way, the map characterizing the flow of (9.1) is written $x(t, x)$. This notation at first may seem confusing, because the same symbol x is used to represent the map and to represent the second argument of the map itself (the argument representing the initial condition of (9.1)), but this is somewhat inevitable. Once the notation has been understood, though, no further confusion should arise.

In the special case of a *linear* differential equation

$$\dot{x} = \mathbf{A}x \tag{9.2}$$

in which $\mathbf{A}$ is an $n \times n$ matrix of real numbers, the flow is given by

$$\phi(t, x) = e^{\mathbf{A}t}x \,,$$

where the matrix exponential $e^{\mathbf{A}t}$ is defined as the sum of the series

$$e^{\mathbf{A}t} = \sum_{i=0}^{\infty} \frac{t^i}{i!} \mathbf{A}^i \,.$$

Let S be a subset of $\mathbb{R}^n$. The set S is said to be *invariant* for (9.1) if, for all $x \in S$, $\phi(t, x)$ is defined for all $t \in (-\infty, +\infty)$ and

$$\phi(t, x) \in S \,, \quad \text{for all } t \in \mathbb{R} \,.$$

A set S is *positively* (resp. negatively) *invariant* if for all $x \in S$, $\phi(t, x)$ is defined for all $t \ge 0$ (resp. for all $t \le 0$) and $\phi(t, x) \in S$ for all such t.

Equation (9.1) defines a *dynamical system*. To reflect the fact that the map f does not depend on other independent entities (such as the time t or physical entities originated from *external* processes) the system in question is referred to an *autonomous* system. Complex autonomous systems arising in analysis and design of physical processes are usually obtained as a composition of simpler subsystems, each one modeled by equations of the form

$$\begin{aligned} \dot{x}_i &= f_i(x_i, u_i) \,, \\ y_i &= h_i(x_i, u_i) \,, \end{aligned} \quad i = 1, \dots, N \,,$$

in which $x_i \in \mathbb{R}^{n_i}$. Here $u_i \in \mathbb{R}^{m_i}$ and, respectively, $y_i \in \mathbb{R}^{p_i}$ are vectors of variables associated with physical entities by means of which the interconnection of various component parts is achieved.

9.2 Stability and Related Concepts

9.2.1 Stability of Equilibria

Consider an autonomous system as (9.1) and suppose that f is locally Lipschitz. A point $x_e \in \mathbb{R}^n$ is called an *equilibrium* point if $f(x_e) = 0$. Clearly, the constant function $x(t) = x_e$ is a solution of (9.1). Since solutions are unique, no other solution of (9.1) exists passing through x_e. The study of equilibria plays a fundamental role in analysis and design of dynamical systems. The most important concept in this respect is that of *stability*, in the sense of *Lyapunov*, specified in the following definition. For $x \in \mathbb{R}^n$, let $|x|$ denote the usual Euclidean norm, that is,

$$|x| = \left(\sum_{i=1}^{n} x_i^2 \right)^{1/2} .$$

Definition 9.1
An equilibrium x_e of (9.1) is *stable* if, for every $\varepsilon > 0$, there exists $\delta > 0$ such that

$$|x(0) - x_e| \leq \delta \Rightarrow |x(t) - x_e| \leq \varepsilon ,$$

$$\text{for all } t \geq 0 .$$

An equilibrium x_e of (9.1) is *asymptotically stable* if it is stable and, moreover, there exists a number $d > 0$ such that

$$|x(0) - x_e| \leq d \Rightarrow \lim_{t \to \infty} |x(t) - x_e| = 0 .$$

An equilibrium x_e of (9.1) is *globally asymptotically stable* if it is asymptotically stable and, moreover,

$$\lim_{t \to \infty} |x(t) - x_e| = 0 , \quad \text{for every } x(0) \in \mathbb{R}^n .$$

The most elementary, but rather useful in practice, result in stability analysis is described as follows. Assume that $f(x)$ is continuously differentiable and suppose, without loss of generality, that $x_e = 0$ (if not, change x into $\bar{x} := x - x_e$ and observe that $\bar{x}$ satisfies the differential equation $\dot{\bar{x}} = f(\bar{x} + x_e)$ in which now $\bar{x} = 0$ is an equilibrium). Expand $f(x)$ as follows

$$f(x) = \mathbf{A}x + \tilde{f}(x) , \tag{9.3}$$

in which

$$\mathbf{A} = \frac{\partial f}{\partial x}(0)$$

is the Jacobian matrix of $f(x)$, evaluated at $x = 0$, and by construction

$$\lim_{x \to 0} \frac{|\tilde{f}(x)|}{|x|} = 0 .$$

The linear system $\dot{x} = \mathbf{A}x$, with the matrix $\mathbf{A}$ defined as indicated, is called the *linear approximation* of the original nonlinear system (9.1) at the equilibrium $x = 0$.

Theorem 9.1
Let $x = 0$ be an equilibrium of (9.1). Suppose every eigenvalue of $\mathbf{A}$ has real part less than $-c$, with $c > 0$. Then, there are numbers $d > 0$ and $M > 0$ such that

$$|x(0)| \leq d \Rightarrow |x(t)| \leq M e^{-ct} |x(0)| ,$$

$$\text{for all } t \geq 0 . \tag{9.4}$$

In particular, $x = 0$ is asymptotically stable. If at least one eigenvalue of $\mathbf{A}$ has positive real part, the equilibrium $x = 0$ is not stable.

This property is usually referred to as the *principle of stability in the first approximation*. The equilibrium $x = 0$ is said to be *hyperbolic* if the matrix $\mathbf{A}$ has no eigenvalue with zero real part. Thus, it is seen from the previous Theorem that a hyperbolic equilibrium is either unstable or asymptotically stable.

The inequality on the right-hand side of (9.4) provides a useful bound on the norm of $x(t)$, expressed as a function of the norm of $x(0)$ and of the time t. This bound, though, is very special and restricted to the case of a hyperbolic equilibrium. In general, bounds of this kind can be obtained by means of the so-called *comparison functions*, which are defined as follows.

Definition 9.2
A continuous function $\alpha : [0, a) \to [0, \infty)$ is said to belong to class $\mathcal{K}$ if it is strictly increasing and $\alpha(0) = 0$. If $a = \infty$ and $\lim_{r \to \infty} \alpha(r) = \infty$, the function is said to belong to class $\mathcal{K}_\infty$. A continuous function $\beta : [0, a) \times [0, \infty) \to [0, \infty)$ is said to belong to class $\mathcal{KL}$ if, for each fixed s, the function

$$\alpha : [0, a) \to [0, \infty) ,$$

$$r \mapsto \beta(r, s)$$

belongs to class $\mathcal{K}$ and, for each fixed r, the function

$$\varphi : [0, \infty) \to [0, \infty) ,$$

$$s \mapsto \beta(r, s)$$

is decreasing and $\lim_{s \to \infty} \varphi(s) = 0$.

The composition of two class $\mathcal{K}$ (respectively, class $\mathcal{K}_\infty$) functions $\alpha_1(\cdot)$ and $\alpha_2(\cdot)$, denoted $\alpha_1(\alpha_2(\cdot))$ or $\alpha_1 \circ \alpha_2(\cdot)$, is a class $\mathcal{K}$ (respectively, class $\mathcal{K}_\infty$) function. If $\alpha(\cdot)$ is a class $\mathcal{K}$ function, defined on $[0, a)$ and $b = \lim_{r \to a} \alpha(r)$, there exists a unique *inverse* function, $\alpha^{-1} : [0, b) \to [0, a)$, namely a function satisfying

$$\alpha^{-1}(\alpha(r)) = r, \quad \text{for all } r \in [0, a)$$

and

$$\alpha(\alpha^{-1}(r)) = r, \quad \text{for all } r \in [0, b) .$$

Moreover, $\alpha^{-1}(\cdot)$ is a class $\mathcal{K}$ function. If $\alpha(\cdot)$ is a class $\mathcal{K}_\infty$ function, so is also $\alpha^{-1}(\cdot)$.

The properties of stability, asymptotic stability, and global asymptotic stability can be easily expressed in terms of inequalities involving comparison functions. In fact, it turns out that the equilibrium $x = 0$ is *stable* if and only if there exist a class $\mathcal{K}$ function $\alpha(\cdot)$ and a number $d > 0$ such that

$$|x(t)| \leq \alpha(|x(0)|) ,$$

for all $x(0)$ such that $|x(0)| \leq d$ and all $t \geq 0$,

the equilibrium $x = 0$ is *asymptotically stable* if and only if there exist a class $\mathcal{KL}$ function $\beta(\cdot, \cdot)$ and a number $d > 0$ such that

$$|x(t)| \leq \beta(|x(0)|, t) ,$$

for all $x(0)$ such that $|x(0)| \leq d$ and all $t \geq 0$,

and the equilibrium $x = 0$ is *globally asymptotically stable* if and only if there exist a class $\mathcal{KL}$ function $\beta(\cdot, \cdot)$ such that

$$|x(t)| \leq \beta(|x(0)|, t) , \quad \text{for all } x(0) \text{ and all } t \geq 0 .$$

9.2.2 Lyapunov Functions

The most important criterion for the analysis of the stability properties of an equilibrium is the criterion of Lyapunov. We introduce first the special form that this criterion takes in the case of a linear system.

Consider the autonomous linear system

$$\dot{x} = Ax$$

in which $x \in \mathbb{R}^n$. Any symmetric $n \times n$ matrix P defines a *quadratic form*

$$V(x) = x^\top P x .$$

The matrix P is said to be *positive definite* (respectively, positive semidefinite) if so is the associated quadratic form $V(x)$, i.e., if, for all $x \neq 0$,

$$V(x) > 0 , \quad \text{respectively } V(x) \geq 0 .$$

The matrix is said to be *negative definite* (respectively, negative semidefinite) if $-P$ is positive definite (respectively, positive semidefinite). It is easy to show that a matrix P is positive definite if (and only if) there exist positive numbers $\underline{a}$ and $\overline{a}$ satisfying

$$\underline{a}|x|^2 \leq x^\top P x \leq \overline{a}|x|^2 , \tag{9.5}$$

for all $x \in \mathbb{R}^n$. The property of a matrix P to be positive definite is usually expressed with the shortened notation $P > 0$ (which actually means $x^\top P x > 0$ for all $x \neq 0$).

In the case of linear systems, the criterion of Lyapunov is expressed as follows.

Theorem 9.2

The linear system $\dot{x} = Ax$ is asymptotically stable (or, what is the same, the eigenvalues of A have negative real part) if there exists a positive-definite matrix P such that the matrix

$$Q := PA + A^\top P$$

is negative definite. Conversely, if the eigenvalues of A have negative real part, then, for any choice of a negative-definite matrix Q, the linear equation

$$PA + A^\top P = Q$$

has a unique solution P, which is positive definite.

Note that, if $V(x) = x^\top P x$,

$$\frac{\partial V}{\partial x} = 2x^\top P$$

and hence

$$\frac{\partial V}{\partial x} Ax = x^\top (PA + A^\top P)x .$$

Thus, to say that the matrix $PA + A^\top P$ is negative definite is equivalent to say that the form

$$\frac{\partial V}{\partial x} Ax$$

is negative definite.

The general, nonlinear, version of the criterion of Lyapunov appeals to the existence of a positive definite, but not necessarily quadratic, function of x. The quadratic lower and upper bounds of (9.5) are therefore replaced by bounds of the form

$$\underline{\alpha}(|x|) \leq V(x) \leq \overline{\alpha}(|x|) , \tag{9.6}$$

in which $\underline{\alpha}(\cdot)$, $\overline{\alpha}(\cdot)$ are simply class $\mathcal{K}$ functions. The criterion in question is summarized as follows.

Theorem 9.3

Let $V : \mathbb{R}^n \to \mathbb{R}$ be a continuously differentiable function satisfying (9.6) for some pair of class $\mathcal{K}$ functions $\underline{\alpha}(\cdot)$, $\overline{\alpha}(\cdot)$. If, for some $d > 0$,

$$\frac{\partial V}{\partial x} f(x) \le 0 , \qquad \text{for all } |x| < d , \tag{9.7}$$

the equilibrium $x = 0$ of (9.1) is stable. If, for some class $\mathcal{K}$ function $\alpha(\cdot)$ and some $d > 0$,

$$\frac{\partial V}{\partial x} f(x) \le -\alpha(|x|) , \quad \text{for all } |x| < d , \tag{9.8}$$

the equilibrium $x = 0$ of (9.1) is locally asymptotically stable. If $\underline{\alpha}(\cdot)$, $\overline{\alpha}(\cdot)$ are class $\mathcal{K}_\infty$ functions and the inequality in (9.8) holds for all x, the equilibrium $x = 0$ of (9.1) is globally asymptotically stable.

A function $V(x)$ satisfying (9.6) and either of the subsequent inequalities is called a *Lyapunov function*. The inequality on the left-hand side of (9.6) is instrumental, together with (9.7), in establishing existence and boundedness of $x(t)$. A simple explanation of the arguments behind the criterion of Lyapunov can be obtained in this way. Suppose (9.7) holds. Then, if $x(0)$ is small, the differentiable function of time $V(x(t))$ is defined for all $t \ge 0$ and nonincreasing along the trajectory $x(t)$. Using the inequalities in (9.6) one obtains

$$\underline{\alpha}(|x(t)|) \le V(x(t)) \le V(x(0)) \le \overline{\alpha}(|x(0)|)$$

and hence $|x(t)| \le \underline{\alpha}^{-1} \circ \overline{\alpha}(|x(0)|)$, which establishes the stability of the equilibrium $x = 0$.

Similar arguments are very useful in order to establish the *invariance*, in positive time, of certain bounded subsets of $\mathbb{R}^n$. Specifically, suppose the various inequalities considered in Theorem 9.3 hold for $d = \infty$ and let Ω_c denote the set of all $x \in \mathbb{R}^n$ for which $V(x) \le c$, namely

$$\Omega_c = \{x \in \mathbb{R}^n : V(x) \le c\} .$$

A set of this kind is called a *sublevel set* of the function $V(x)$. Note that, if $\underline{\alpha}(\cdot)$ is a class $\mathcal{K}_\infty$ function, then Ω_c is a compact set for all $c > 0$. Now, if

$$\frac{\partial V(x)}{\partial x} f(x) < 0$$

at each point x of the boundary of Ω_c, it can be concluded that, for any initial condition in the interior of Ω_c, the solution $x(t)$ of (9.1) is defined for all $t \ge 0$ and

is such that $x(t) \in \Omega_c$ for all $t \ge 0$, that is, the set Ω_c is invariant in positive time. Indeed, existence and uniqueness are guaranteed by the local Lipschitz property so long as $x(t) \in \Omega_c$, because Ω_c is a compact set. The fact that $x(t)$ remains in Ω_c for all $t \ge 0$ is proved by contradiction. For, suppose that, for some trajectory $x(t)$, there is a time t_1 such that $x(t)$ is in the interior of Ω_c at all $t < t_1$ and $x(t_1)$ is on the boundary of Ω_c. Then,

$$V(x(t)) < c , \quad \text{for all } t < t_1 \quad \text{and} \quad V(x(t_1)) = c ,$$

and this contradicts the previous inequality, which shows that the derivative of $V(x(t))$ is strictly negative at $t = t_1$.

The criterion for asymptotic stability provided by the previous Theorem has a *converse*, namely, the existence of a function $V(x)$ having the properties indicated in Theorem 9.3 is *implied* by the property of asymptotic stability of the equilibrium $x = 0$ of (9.1). In particular, the following result holds.

Theorem 9.4

Suppose the equilibrium $x = 0$ of (9.1) is locally asymptotically stable. Then, there exist $d > 0$, a continuously differentiable function $V : \mathbb{R}^n \to \mathbb{R}$, and class $\mathcal{K}$ functions $\underline{\alpha}(\cdot)$, $\overline{\alpha}(\cdot)$, $\alpha(\cdot)$, such that (9.6) and (9.8) hold. If the equilibrium $x = 0$ of (9.1) is globally asymptotically stable, there exist a continuously differentiable function $V : \mathbb{R}^n \to \mathbb{R}$, and class $\mathcal{K}_\infty$ functions $\underline{\alpha}(\cdot)$, $\overline{\alpha}(\cdot)$, $\alpha(\cdot)$, such that (9.6) and (9.8) hold with $d = \infty$.

To conclude, observe that, if $x = 0$ is a hyperbolic equilibrium and all eigenvalues of $\mathbf{A}$ have negative real part, $|x(t)|$ is bounded, for small $|x(0)|$, by a class $\mathcal{KL}$ function $\beta(\cdot, \cdot)$ of the form

$$\beta(r, t) = M e^{-\lambda t} r .$$

If the equilibrium $x = 0$ of system (9.1) is globally asymptotically stable and, moreover, there exist numbers $d > 0$, $M > 0$, and $\lambda > 0$ such that

$$|x(t)| \le M e^{-\lambda t} |x(0)| , \qquad \text{for all } |x(0)| \le d$$
$$\text{and all } t \ge 0 ,$$

it is said that this equilibrium is *globally asymptotically and locally exponentially* stable. It can be shown that the equilibrium $x = 0$ of the nonlinear system (9.1) is globally asymptotically and locally exponentially stable if and only if there exists a continuously differentiable function $V(x) : \mathbb{R}^n \to \mathbb{R}$, and class $\mathcal{K}_\infty$ functions $\underline{\alpha}(\cdot)$, $\overline{\alpha}(\cdot)$, $\alpha(\cdot)$, and real numbers $\delta > 0$, $\underline{a} > 0$, $\overline{a} > 0$, $a > 0$, such that

$$\underline{\alpha}(|\boldsymbol{x}|) \le V(\boldsymbol{x}) \le \overline{\alpha}(|\boldsymbol{x}|) \,,$$

$$\frac{\partial V}{\partial \boldsymbol{x}} f(\boldsymbol{x}) \le -\alpha(|\boldsymbol{x}|) \,,$$

for all $\boldsymbol{x} \in \mathbb{R}^n$

and

$$\underline{\alpha}(s) = \underline{a}\, s^2 \,, \quad \alpha(s) = \overline{a}\, s^2 \,, \quad \alpha(s) = a\, s^2 \,,$$
$$\text{for all } s \in [0, \delta] \,.$$

9.3 Asymptotic Behavior

9.3.1 Limit Sets

In the analysis of dynamical systems, it is often important to determine whether or not, as time increases, the variables characterizing the motion asymptotically converge to special motions exhibiting some form of recurrence. This is the case, for instance, when a system possesses an asymptotically stable equilibrium: all motions issued from initial conditions in a neighborhood of this point converge to a special motion in which all variables remain constant. A constant motion, or more generally a periodic motion, is characterized by a property of recurrence that is usually referred to as *steady-state* motion or behavior.

The steady-state behavior of a dynamical system can be viewed as a kind of *limit* behavior, approached either as the *actual* time t tends to $+\infty$ or, alternatively, as the *initial* time t_0 tends to $-\infty$. Relevant in this regard are certain concepts introduced by *Birkhoff* in [9.1]. In particular, a fundamental role is played by the concept of ω-limit set of a given point, defined as follows. Consider an *autonomous* dynamical system such as (9.1) and let $\boldsymbol{x}(t, \boldsymbol{x}_0)$ denote its flow. Assume, in particular, that $\boldsymbol{x}(t, \boldsymbol{x}_0)$ is defined for all $t \ge 0$. A point $\boldsymbol{x}$ is said to be an ω-limit *point* of the motion

$\boldsymbol{x}(t, \boldsymbol{x}_0)$ if there exists a sequence of times $\{t_k\}$, with $\lim_{k \to \infty} t_k = \infty$, such that

$$\lim_{k \to \infty} \boldsymbol{x}(t_k, \boldsymbol{x}_0) = \boldsymbol{x} \,.$$

The ω-limit *set* of a point $\boldsymbol{x}_0$, denoted $\omega(\boldsymbol{x}_0)$, is *the union* of all ω-limit points of the motion $\boldsymbol{x}(t, \boldsymbol{x}_0)$ (Fig. 9.2).

If $\boldsymbol{x}_e$ is an asymptotically stable equilibrium, then $\boldsymbol{x}_e = \omega(\boldsymbol{x}_0)$ for all $\boldsymbol{x}_0$ in a neighborhood of $\boldsymbol{x}_e$. However, in general, an ω-limit point *is not* necessarily a limit of $\boldsymbol{x}(t, \boldsymbol{x}_0)$ as $t \to \infty$, because the function in question may not admit any limit as $t \to \infty$. It happens though, that if the motion $\boldsymbol{x}(t, \boldsymbol{x}_0)$ is *bounded*, then $\boldsymbol{x}(t, \boldsymbol{x}_0)$ asymptotically approaches *the set* $\omega(\boldsymbol{x}_0)$.

Lemma 9.1

Suppose there is a number M such that $|\boldsymbol{x}(t, \boldsymbol{x}_0)| \le M$ for all $t \ge 0$. Then, $\omega(\boldsymbol{x}_0)$ is a nonempty compact connected set, invariant under (9.1). Moreover, the distance of $\boldsymbol{x}(t, \boldsymbol{x}_0)$ from $\omega(\boldsymbol{x}_0)$ tends to 0 as $t \to \infty$.

It is seen from this that the set $\omega(\boldsymbol{x}_0)$ is filled by motions of (9.1) which are *defined, and bounded, for all backward and forward times*. The other remarkable feature is that $\boldsymbol{x}(t, \boldsymbol{x}_0)$ *approaches* $\omega(\boldsymbol{x}_0)$ as $t \to \infty$, in the sense that the distance *of the point* $\boldsymbol{x}(t, \boldsymbol{x}_0)$ (the value at time t of the solution of (9.1) starting in $\boldsymbol{x}_0$ at time $t = 0$) *to the set* $\omega(\boldsymbol{x}_0)$ tends to 0 as $t \to \infty$. A consequence of this property is that, in a system of the form (9.1), if *all* motions issued from a set B are bounded, all such motions asymptotically approach the set

$$\Omega = \bigcup_{\boldsymbol{x}_0 \in B} \omega(\boldsymbol{x}_0) \,.$$

However, the convergence of $\boldsymbol{x}(t, \boldsymbol{x}_0)$ to Ω is not guaranteed to be *uniform* in $\boldsymbol{x}_0$, even if the set B is compact. There is a larger set, though, which does have this property of uniform convergence. This larger set, known as the ω-limit set *of the set* B, is precisely defined as follows.

Consider again system (9.1), let B be a subset of $\mathbb{R}^n$, and suppose $\boldsymbol{x}(t, \boldsymbol{x}_0)$ is defined for all $t \ge 0$ and all

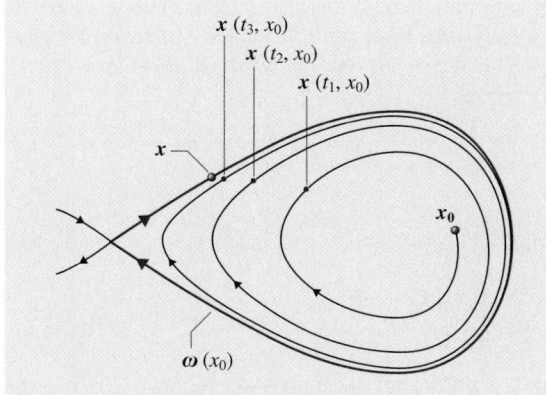

Fig. 9.2 The ω-limit set of a point x_0

$x_0 \in B$. The ω-limit set of B, denoted $\omega(B)$, is the set of all points x for which there exists a sequence of pairs $\{x_k, t_k\}$, with $x_k \in B$ and $\lim_{k \to \infty} t_k = \infty$ such that

$$\lim_{k \to \infty} x(t_k, x_k) = x .$$

It follows from the definition that, if B consists of only one single point x_0, all x_k in the definition above are necessarily equal to x_0 and the definition in question reduces to the definition of ω-limit set of a point, given earlier. It also follows that, if for some $x_0 \in B$ the set $\omega(x_0)$ is nonempty, all points of $\omega(x_0)$ are points of $\omega(B)$. Thus, in particular, if all motions with $x_0 \in B$ are bounded in positive time,

$$\bigcup_{x_0 \in B} \omega(x_0) \subset \omega(B) .$$

However, the converse inclusion is not true in general.

The relevant properties of the ω-limit set of a set, which extend those presented earlier in Lemma 9.1, can be summarized as follows [9.2].

Lemma 9.2

Let B be a nonempty bounded subset of $\mathbb{R}^n$ and suppose there is a number M such that $|x(t, x_0)| \leq M$ for all $t \geq 0$ and all $x_0 \in B$. Then $\omega(B)$ is a nonempty compact set, invariant under (9.1). Moreover, the distance of $x(t, x_0)$ from $\omega(B)$ tends to 0 as $t \to \infty$, uniformly in $x_0 \in B$. If B is connected, so is $\omega(B)$.

Thus, as is the case for the ω-limit set of a point, the ω-limit set of a bounded set B, being compact and invariant, is filled with motions which exist for all $t \in (-\infty, +\infty)$ and are bounded backward and forward in time. But, above all, the set in question is *uniformly* approached by motions with initial state $x_0 \in B$. An important corollary of the property of uniform convergence is that, if $\omega(B)$ is contained in the interior of B, then $\omega(B)$ is also asymptotically stable.

9.4 Dynamical Systems with Inputs

9.4.1 Input-to-State Stability (ISS)

In this section we show how to determine the stability properties of an interconnected system, on the basis of the properties of each individual component. The easiest interconnection to be analyzed is a cascade connection of two subsystems, namely a system of the

---

Lemma 9.3

Let B be a nonempty bounded subset of $\mathbb{R}^n$ and suppose there is a number M such that $|x(t, x_0)| \leq M$ for all $t \geq 0$ and all $x_0 \in B$. Then $\omega(B)$ is a nonempty compact set, invariant under (9.1). Suppose also that $\omega(B)$ is contained in the interior of B. Then, $\omega(B)$ is asymptotically stable, with a domain of attraction that contains B.

9.3.2 Steady-State Behavior

Consider now again system (9.1), with initial conditions in a closed subset $X \subset \mathbb{R}^n$. Suppose the set X is *positively invariant*, which means that, for any initial condition $x_0 \in X$, the solution $x(t, x_0)$ exists for all $t \geq 0$ and $x(t, x_0) \in X$ for all $t \geq 0$. The motions of this system are said to be *ultimately bounded* if there is a bounded subset B with the property that, for every compact subset X_0 of X, there is a time $T > 0$ such that $x(t, x_0) \in B$ for all $t \geq T$ and all $x_0 \in X_0$. In other words, if the motions of the system are ultimately bounded, every motion eventually enters and remains in the bounded set B.

Suppose the motions of (9.1) are ultimately bounded and let $B' \neq B$ be any other bounded subset with the property that, for every compact subset X_0 of X, there is a time $T > 0$ such that $x(t, x_0) \in B'$ for all $t \geq T$ and all $x_0 \in X_0$. Then, it is easy to check that $\omega(B') = \omega(B)$. Thus, in view of the properties described in Lemma 9.2 above, the following definition can be adopted [9.3].

Definition 9.3

Suppose the motions of system (9.1), with initial conditions in a closed and positively invariant set X, are ultimately bounded. A *steady-state* motion is any motion with initial condition $x(0) \in \omega(B)$. The set $\omega(B)$ is the *steady-state locus* of (9.1) and the *restriction* of (9.1) to $\omega(B)$ is the *steady-state behavior* of (9.1).

---

form

$$\dot{x} = f(x, z) ,$$
$$\dot{z} = g(z) , \tag{9.9}$$

with $x \in \mathbb{R}^n$, $z \in \mathbb{R}^m$ in which we assume $f(0, 0) = 0$, $g(0) = 0$.

If the equilibrium $x = 0$ of $\dot{x} = f(x, 0)$ is locally asymptotically stable and the equilibrium $z = 0$ of the lower subsystem is locally asymptotically stable then the equilibrium $(x, z) = (0, 0)$ of the cascade is locally asymptotically stable. However, in general, *global* asymptotic stability of the equilibrium $x = 0$ of $\dot{x} = f(x, 0)$ and *global* asymptotic stability of the equilibrium $z = 0$ of the lower subsystem *do not* imply *global* asymptotic stability of the equilibrium $(x, z) = (0, 0)$ of the cascade. To infer global asymptotic stability of the cascade, a stronger condition is needed, which expresses a property describing how – in the upper subsystem – the response $x(\cdot)$ is influenced by its input $z(\cdot)$.

The property in question requires that, when $z(t)$ is bounded over the semi-infinite time interval $[0, +\infty)$, then also $x(t)$ be bounded, and in particular that, if $z(t)$ asymptotically decays to 0, then also $x(t)$ decays to 0. These requirements altogether lead to the notion of *input-to-state stability*, introduced and studied in [9.4, 5]. The notion in question is defined as follows (see also [9.6, Chap. 10] for additional details). Consider a nonlinear system

$$\dot{x} = f(x, u) , \tag{9.10}$$

with state $x \in \mathbb{R}^n$ and input $u \in \mathbb{R}^m$, in which $f(0, 0) = 0$ and $f(x, u)$ is locally Lipschitz on $\mathbb{R}^n \times \mathbb{R}^m$. The input function $u : [0, \infty) \to \mathbb{R}^m$ of (9.10) can be any piecewise-continuous bounded function. The set of all such functions, endowed with the supremum norm

$$\|u(\cdot)\|_\infty = \sup_{t \geq 0} |u(t)|$$

is denoted by L_∞^m.

Definition 9.4
System (9.10) is said to be input-to-state stable if there exist a class $\mathcal{KL}$ function $\beta(\cdot, \cdot)$ and a class $\mathcal{K}$ function $\gamma(\cdot)$, called a gain function, such that, for any input $u(\cdot) \in L_\infty^m$ and any $x_0 \in \mathbb{R}^n$, the response $x(t)$ of (9.10) in the initial state $x(0) = x_0$ satisfies

$$|x(t)| \leq \beta(|x_0|, t) + \gamma(\|u(\cdot)\|_\infty) , \quad \text{for all } t \geq 0 . \tag{9.11}$$

It is common practice to replace the wording *input-to-state stable* with the acronym ISS. In this way, a system possessing the property expressed by (9.11) is said to be an ISS system. Since, for any pair $\beta > 0$,

$\gamma > 0$, $\max\{\beta, \gamma\} \leq \beta + \gamma \leq \max\{2\beta, 2\gamma\}$, an alternative way to say that a system is input-to-state stable is to say that there exists a class $\mathcal{KL}$ function $\beta(\cdot, \cdot)$ and a class $\mathcal{K}$ function $\gamma(\cdot)$ such that, for any input $u(\cdot) \in L_\infty^m$ and any $x_0 \in \mathbb{R}^n$, the response $x(t)$ of (9.10) in the initial state $x(0) = x_0$ satisfies

$$|x(t)| \leq \max\{\beta(|x_0|, t), \gamma(\|u(\cdot)\|_\infty)\} , \quad \text{for all } t \geq 0 . \tag{9.12}$$

The property, for a given system, of being input-to-state stable, can be given a characterization which extends the criterion of Lyapunov for asymptotic stability. The key tool for this analysis is the notion of *ISS-Lyapunov function*, defined as follows.

Definition 9.5
A C^1 function $V : \mathbb{R}^n \to \mathbb{R}$ is an ISS-Lyapunov function for system (9.10) if there exist class $\mathcal{K}_\infty$ functions $\underline{\alpha}(\cdot)$, $\overline{\alpha}(\cdot)$, $\alpha(\cdot)$, and a class $\mathcal{K}$ function $\chi(\cdot)$ such that

$$\underline{\alpha}(|x|) \leq V(x) \leq \overline{\alpha}(|x|) , \quad \text{for all } x \in \mathbb{R}^n \tag{9.13}$$

and

$$|x| \geq \chi(|u|) \Rightarrow \frac{\partial V}{\partial x} f(x, u) \leq -\alpha(|x|) , \quad \text{for all } x \in \mathbb{R}^n \text{and } u \in \mathbb{R}^m . \tag{9.14}$$

An alternative, equivalent, definition is the following one.

Definition 9.6
A C^1 function $V : \mathbb{R}^n \to \mathbb{R}$ is an ISS-Lyapunov function for system (9.10) if there exist class $\mathcal{K}_\infty$ functions $\underline{\alpha}(\cdot)$, $\overline{\alpha}(\cdot)$, $\alpha(\cdot)$, and a class $\mathcal{K}$ function $\sigma(\cdot)$ such that (9.13) holds and

$$\frac{\partial V}{\partial x} f(x, u) \leq -\alpha(|x|) + \sigma(|u|) , \quad \text{for all } x \in \mathbb{R}^n \text{and all } u \in \mathbb{R}^m . \tag{9.15}$$

The importance of the notion of ISS-Lyapunov function resides in the following criterion, which extends the criterion of Lyapunov for global asymptotic stability to systems with inputs.

Theorem 9.5
System (9.10) is input-to-state stable if and only if there exists an ISS-Lyapunov function.

The comparison functions appearing in the estimates (9.13) and (9.14) are useful to obtain an estimate of the gain function $\gamma(\cdot)$ which characterizes the bound (9.12). In fact, it can be shown that, if system (9.10) possesses an ISS-Lyapunov function $V(x)$, the sublevel set

$$\Omega_{\|u(\cdot)\|_\infty} = \{x \in \mathbb{R}^n : V(x) \le \overline{a}(\chi(\|u(\cdot)\|_\infty))\}$$

is invariant in positive time for (9.10). Thus, in view of the estimates (9.13), if the initial state of the system is initially inside this sublevel set, the following estimate holds

$$|x(t)| \le \underline{a}^{-1}\left(\overline{a}(\chi(\|u(\cdot)\|_\infty))\right), \quad \text{for all } t \ge 0,$$

and one can obtain an estimate of $\gamma(\cdot)$ as

$$\gamma(r) = \underline{a}^{-1} \circ \overline{a} \circ \chi(r).$$

In other words, establishing the existence of an ISS-Lyapunov function $V(x)$ is useful not only to check whether or not the system in question is input-to-state stable, but also to determine an estimate of the gain function $\gamma(\cdot)$. Knowing such estimate is important, as will be shown later, in using the concept of input-to-state stability to determine the stability of interconnected systems.

The following simple examples may help understanding the concept of input-to-state stability and the associated Lyapunov-like theorem.

Example 9.1: Consider a linear system

$$\dot{x} = Ax + Bu,$$

with $x \in \mathbb{R}^n$ and $u \in \mathbb{R}^m$ and suppose that all the eigenvalues of the matrix A have negative real part. Let $P > 0$ denote the unique solution of the Lyapunov equation $PA + A^\top P = -I$. Observe that the function $V(x) = x^\top Px$ satisfies

$$\underline{a}|x|^2 \le V(x) \le \overline{a}|x|^2,$$

for suitable $\underline{a} > 0$ and $\overline{a} > 0$, and that

$$\frac{\partial V}{\partial x}(Ax + Bu) \le -|x|^2 + 2|x||P||B||u|.$$

Pick any $0 < \varepsilon < 1$ and set

$$c = \frac{2}{1-\varepsilon}|P||B|, \quad \chi(r) = cr.$$

Then

$$|x| \ge \chi(|u|) \Rightarrow \frac{\partial V}{\partial x}(Ax + Bu) \le -\varepsilon|x|^2.$$

Thus, the system is input-to-state stable, with a gain function

$$\gamma(r) = (c\,\overline{a}/\underline{a})\,r$$

which is a *linear* function.

Consider now the simple nonlinear one-dimensional system

$$\dot{x} = -ax^k + x^p u,$$

in which $k \in \mathbb{N}$ is odd, $p \in \mathbb{N}$ satisfies $p < k$, and $a > 0$. Choose a candidate ISS-Lyapunov function as $V(x) = \frac{1}{2}x^2$, which yields

$$\frac{\partial V}{\partial x}f(x, u) =$$
$$- ax^{k+1} + x^{p+1}u \le -a|x|^{k+1} + |x|^{p+1}|u|.$$

Set $\nu = k - p$ to obtain

$$\frac{\partial V}{\partial x}f(x, u) \le |x|^{p+1}\left(-a|x|^\nu + |u|\right).$$

Thus, using the class $\mathcal{K}_\infty$ function $\alpha(r) = \varepsilon r^{k+1}$, with $\varepsilon > 0$, it is deduced that

$$\frac{\partial V}{\partial x}f(x, u) \le -\alpha(|x|)$$

provided that

$$(a - \varepsilon)|x|^\nu \ge |u|.$$

Taking, without loss of generality, $\varepsilon < a$, it is concluded that condition (9.14) holds for the class $\mathcal{K}$ function

$$\chi(r) = \left(\frac{r}{a - \varepsilon}\right)^{\frac{1}{\nu}}.$$

Thus, the system is input-to-state stable.

An important feature of the previous example, which made it possible to prove the system is input-to-state stable, is the inequality $p < k$. In fact, if this inequality does not hold, the system may fail to be input-to-state stable. This can be seen, for instance, in the simple example

$$\dot{x} = -x + xu.$$

To this end, suppose $u(t) = 2$ for all $t \ge 0$. The state response of the system, to this input, from the initial state $x(0) = x_0$ coincides with that of the autonomous system $\dot{x} = x$, i.e., $x(t) = e^t x_0$, which shows that the bound (9.11) cannot hold.

We conclude with an alternative characterization of the property of input-to-state stability, which is useful in many instances [9.7].

System (9.10) is input-to-state stable if and only if there exist class $\mathcal{K}$ functions $\gamma_0(\cdot)$ and $\gamma(\cdot)$ such that, for any input $u(\cdot) \in L_\infty^m$ and any $x_0 \in \mathbb{R}^n$, the response $x(t)$ in the initial state $x(0) = x_0$ satisfies

$$\|x(\cdot)\|_\infty \leq \max\{\gamma_0(|x_0|), \gamma(\|u(\cdot)\|_\infty)\},$$

$$\limsup_{t \to \infty} |x(t)| \leq \gamma(\limsup_{t \to \infty} |u(t)|).$$

9.4.2 Cascade Connections

The property of input-to-state stability is of paramount importance in the analysis of interconnected systems. The first application consists of the analysis of the *cascade connection*. In fact, the cascade connection of two input-to-state stable systems turns out to be input-to-state stable. More precisely, consider a system of the form (Fig. 9.3)

$$\dot{x} = f(x, z),$$
$$\dot{z} = g(z, u),\tag{9.16}$$

in which $x \in \mathbb{R}^n$, $z \in \mathbb{R}^m$, $f(0, 0) = 0$, $g(0, 0) = 0$, and $f(x, z)$, $g(z, u)$ are locally Lipschitz.

Theorem 9.7

Suppose that system

$$\dot{x} = f(x, z),\tag{9.17}$$

viewed as a system with input z and state x, is input-to-state stable and that system

$$\dot{z} = g(z, u),\tag{9.18}$$

viewed as a system with input u and state z, is input-to-state stable as well. Then, system (9.16) is input-to-state stable.

As an immediate corollary of this theorem, it is possible to answer the question of when the cascade connection (9.9) is globally asymptotically stable. In fact, if system

$$\dot{x} = f(x, z),$$

viewed as a system with input z and state x, is input-to-state stable and the equilibrium $z = 0$ of the lower subsystem is globally asymptotically stable, the equilibrium $(x, z) = (0, 0)$ of system (9.9) is globally

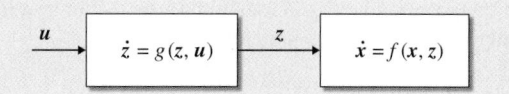

Fig. 9.3 Cascade connection

asymptotically stable. This is in particular the case if system (9.9) has the special form

$$\dot{x} = \mathbf{A}x + p(z),$$
$$\dot{z} = g(z),\tag{9.19}$$

with $p(0) = 0$ and the matrix $\mathbf{A}$ has all eigenvalues with negative real part. The upper subsystem of the cascade is input-to-state stable and hence, if the equilibrium $z = 0$ of the lower subsystem is globally asymptotically stable, so is the equilibrium $(x, z) = (0, 0)$ of the entire system.

9.4.3 Feedback Connections

In this section we investigate the stability property of nonlinear systems, and we will see that the property of input-to-state stability lends itself to a simple characterization of an important *sufficient condition* under which the feedback interconnection of two globally asymptotically stable systems remains globally asymptotically stable.

Consider the following interconnected system (Fig. 9.3)

$$\dot{x}_1 = f_1(x_1, x_2),$$
$$\dot{x}_2 = f_2(x_1, x_2, u),\tag{9.20}$$

in which $x_1 \in \mathbb{R}^{n_1}$, $x_2 \in \mathbb{R}^{n_2}$, $u \in \mathbb{R}^m$, and $f_1(0, 0) = 0$, $f_2(0, 0, 0) = 0$. Suppose that the first subsystem, viewed as a system with internal state x_1 and input x_2, is input-to-state stable. Likewise, suppose that the second subsystem, viewed as a system with internal state x_2 and inputs x_1 and u, is input-to-state stable. In view of the results presented earlier, the hypothesis of input-to-state stability of the first subsystem is equivalent to the existence of functions $\beta_1(\cdot, \cdot)$, $\gamma_1(\cdot)$, the first of class $\mathcal{KL}$

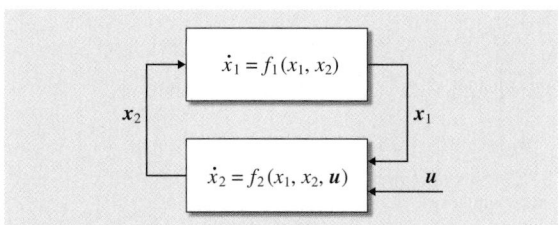

Fig. 9.4 Feedback connection

and the second of class $\mathcal{K}$, such that the response $\boldsymbol{x}_1(\cdot)$ to any input $\boldsymbol{x}_2(\cdot) \in L_\infty^{n_2}$ satisfies

$$|\boldsymbol{x}_1(t)| \leq \max\{\beta_1(\|\boldsymbol{x}_1(0)\|, t), \gamma_1(\|\boldsymbol{x}_2(\cdot)\|_\infty)\},$$
$$\text{for all } t \geq 0. \tag{9.21}$$

Likewise the hypothesis of input-to-state stability of the second subsystem is equivalent to the existence of three class functions $\beta_2(\cdot), \gamma_2(\cdot), \gamma_u(\cdot)$ such that the response $\boldsymbol{x}_2(\cdot)$ to any input $\boldsymbol{x}_1(\cdot) \in L_\infty^{n_1}, \boldsymbol{u}(\cdot) \in L_\infty^m$ satisfies

$$|\boldsymbol{x}_2(t)| \leq$$
$$\max\{\beta_2(\|\boldsymbol{x}_2(0)\|, t), \gamma_2(\|\boldsymbol{x}_1(\cdot)\|_\infty), \gamma_u(\|\boldsymbol{u}(\cdot)\|_\infty)\},$$
$$\text{for all } t \geq 0. \tag{9.22}$$

The important result for the analysis of the stability of the interconnected system (9.20) is that, if the composite function $\gamma_1 \circ \gamma_2(\cdot)$ is a *simple contraction*, i.e., if

$$\gamma_1(\gamma_2(r)) < r, \quad \text{for all } r > 0, \tag{9.23}$$

the system in question is input-to-state stable. This result is usually referred to as the *small-gain* theorem.

Theorem 9.8
If the condition (9.23) holds, system (9.20), viewed as a system with state $\boldsymbol{x} = (\boldsymbol{x}_1, \boldsymbol{x}_2)$ and input $\boldsymbol{u}$, is input-to-state stable.

The condition (9.23), i.e., the condition that the composed function $\gamma_1 \circ \gamma_2(\cdot)$ is a contraction, is usually referred to as the *small-gain condition*. It can be written in different alternative ways depending on how the functions $\gamma_1(\cdot)$ and $\gamma_2(\cdot)$ are estimated. For instance, if it is known that $V_1(\boldsymbol{x}_1)$ is an ISS-Lyapunov function for the upper subsystem of (9.20), i.e., a function such

$$\underline{\alpha}_1(|\boldsymbol{x}_1|) \leq V_1(\boldsymbol{x}_1) \leq \overline{\alpha}_1(|\boldsymbol{x}_1|),$$
$$|\boldsymbol{x}_1| \geq \chi_1(|\boldsymbol{x}_2|) \Rightarrow \frac{\partial V_1}{\partial \boldsymbol{x}_1} f_1(\boldsymbol{x}_1, \boldsymbol{x}_2) \leq -\alpha(|\boldsymbol{x}_1|),$$

then $\gamma_1(\cdot)$ can be estimated by

$$\gamma_1(r) = \underline{\alpha}_1^{-1} \circ \overline{\alpha}_1 \circ \chi_1(r).$$

Likewise, if $V_2(\boldsymbol{x}_2)$ is a function such that

$$\underline{\alpha}_2(|\boldsymbol{x}_2|) \leq V_2(\boldsymbol{x}_2) \leq \overline{\alpha}_2(|\boldsymbol{x}_2|),$$
$$|\boldsymbol{x}_2| \geq \max\{\chi_2(|\boldsymbol{x}_1|), \chi_u(|\boldsymbol{u}|)\} \Rightarrow$$
$$\frac{\partial V_2}{\partial \boldsymbol{x}_2} f_2(\boldsymbol{x}_1, \boldsymbol{x}_2, \boldsymbol{u}) \leq -\alpha(|\boldsymbol{x}_2|),$$

then $\gamma_2(\cdot)$ can be estimated by

$$\gamma_2(r) = \underline{\alpha}_2^{-1} \circ \overline{\alpha}_2 \circ \chi_2(r).$$

If this is the case, the small-gain condition of the theorem can be written in the form

$$\underline{\alpha}_1^{-1} \circ \overline{\alpha}_1 \circ \chi_1 \circ \underline{\alpha}_2^{-1} \circ \overline{\alpha}_2 \circ \chi_2(r) < r.$$

9.4.4 The Steady-State Response

In this subsection we show how the concept of steady state, introduced earlier, and the property of input-to-state stability are useful in the analysis of the *steady-state response* of a system to inputs generated by a separate autonomous dynamical system [9.8].

Example 9.2: Consider an n-dimensional, single-input, *asymptotically stable* linear system

$$\dot{z} = \mathbf{F}z + \mathbf{G}u \tag{9.24}$$

forced by the harmonic input $\boldsymbol{u}(t) = u_0 \sin(\omega t + \phi_0)$. A simple method to analyze the asymptotic behavior of (9.24) consists of viewing the forcing input $\boldsymbol{u}(t)$ as provided by an autonomous *signal generator* of the form

$$\dot{w} = \mathbf{S}w,$$
$$u = \mathbf{Q}w,$$

in which

$$\mathbf{S} = \begin{pmatrix} 0 & \omega \\ -\omega & 0 \end{pmatrix}, \quad \mathbf{Q} = \begin{pmatrix} 1 & 0 \end{pmatrix},$$

and in analyzing the state-state behavior of the associated *augmented* system

$$\dot{w} = \mathbf{S}w,$$
$$\dot{z} = \mathbf{F}z + \mathbf{G}\mathbf{Q}w. \tag{9.25}$$

As a matter of fact, let $\boldsymbol{\Pi}$ be the unique solution of the Sylvester equation $\boldsymbol{\Pi}\mathbf{S} = \mathbf{F}\boldsymbol{\Pi} + \mathbf{G}\mathbf{Q}$ and observe that the graph of the linear map $z = \boldsymbol{\Pi}w$ is an invariant subspace for the system (9.25). Since all trajectories of (9.25) approach this subspace as $t \to \infty$, the limit behavior of (9.25) is determined by the restriction of its motion to this invariant subspace.

Revisiting this analysis from the viewpoint of the more general notion of steady-state introduced earlier, let $W \subset \mathbb{R}^2$ be a set of the form

$$W = \{w \in \mathbb{R}^2 : \|w\| \leq c\}, \tag{9.26}$$

in which c is a fixed number, and suppose the set of initial conditions for (9.25) is $W \times \mathbb{R}^n$. This is in fact the case when the problem of evaluating the periodic response of (9.24) to harmonic inputs whose amplitude does not exceed a fixed number c is addressed. The set W is compact and invariant for the upper subsystem of (9.25) and, as is easy to check, the ω-limit set of W under the motion of the upper subsystem of (9.25) is the subset W itself.

The set $W \times \mathbb{R}^n$ is closed and positively invariant for the full system (9.25) and, moreover, since the lower subsystem of (9.25) is input-to-state stable, the motions of system of (9.25), for initial conditions taken in $W \times \mathbb{R}^n$, are ultimately bounded. It is easy to check that

$$\omega(\mathbf{B}) = \{(\boldsymbol{w}, z) \in \mathbb{R}^2 \times \mathbb{R}^n : \boldsymbol{w} \in W, z = \Pi \boldsymbol{w}\} \,,$$

i. e., that $\omega(\mathbf{B})$ is the graph of the restriction of the map $z = \Pi \boldsymbol{w}$ to the set W. The restriction of (9.25) to the invariant set $\omega(\mathbf{B})$ characterizes the steady-state behavior of (9.24) under the family of all harmonic inputs of fixed angular frequency ω and amplitude not exceeding c.

Example 9.3: A similar result, namely the fact that the *steady-state locus* is the *graph* of a map, can be reached if the *signal generator* is any nonlinear system, with initial conditions chosen in a compact invariant set W. More precisely, consider an augmented system of the form

$$\begin{aligned}\dot{\boldsymbol{w}} &= s(\boldsymbol{w}) \,, \\ \dot{z} &= \mathbf{F}z + \mathbf{G}q(\boldsymbol{w}) \,,\end{aligned} \tag{9.27}$$

in which $\boldsymbol{w} \in W \subset \mathbb{R}^r$, $\boldsymbol{x} \in \mathbb{R}^n$, and assume that: (i) all eigenvalues of F have negative real part, and (ii) the set W is a compact set, invariant for the the upper subsystem of (9.27).

As in the previous example, the ω-limit set of W under the motion of the upper subsystem of (9.27) is the subset W itself. Moreover, since the lower subsystem of (9.27) is input-to-state stable, the motions of system (9.27), for initial conditions taken in $W \times \mathbb{R}^n$, are ultimately bounded. It is easy to check that the steady-state locus of (9.27) is the graph of the map

$$\pi : W \to \mathbb{R}^n \,,$$
$$\boldsymbol{w} \mapsto \pi(\boldsymbol{w}) \,,$$

defined by

$$\pi(\boldsymbol{w}) = \lim_{T \to \infty} \int_{-T}^{0} \mathrm{e}^{-F\tau} \mathbf{G}q(\boldsymbol{w}(\tau, \boldsymbol{w})) \mathrm{d}\tau \,. \tag{9.28}$$

There are various ways in which the result discussed in the previous example can be generalized; for instance, it can be extended to describe the steady-state response of a nonlinear system

$$\dot{z} = f(z, \boldsymbol{u}) \tag{9.29}$$

in the neighborhood of a locally exponentially stable equilibrium point. To this end, suppose that $f(0, 0) = 0$ and that the matrix

$$\mathbf{F} = \left[\frac{\partial f}{\partial z}\right](0, 0)$$

has all eigenvalues with negative real part. Then, it is well known (see, e.g., [9.9, p. 275]) that it is always possible to find a compact subset $Z \subset \mathbb{R}^n$, which contains $z = 0$ in its interior and a number $\sigma > 0$ such that, if $|z_0| \in Z$ and $\|\boldsymbol{u}(t)\| \leq \sigma$ for all $t \geq 0$, the solution of (9.29) with initial condition $z(0) = z_0$ satisfies $|z(t)| \in Z$ for all $t \geq 0$. Suppose that the input $\boldsymbol{u}$ to (9.29) is produced, as before, by a signal generator of the form

$$\begin{aligned}\dot{\boldsymbol{w}} &= s(\boldsymbol{w}) \,, \\ \boldsymbol{u} &= q(\boldsymbol{w}) \,,\end{aligned} \tag{9.30}$$

with initial conditions chosen in a compact invariant set W and, moreover, suppose that, $\|q(\boldsymbol{w})\| \leq \sigma$ for all $\boldsymbol{w} \in W$. If this is the case, the set $W \times Z$ is positively invariant for

$$\begin{aligned}\dot{\boldsymbol{w}} &= s(\boldsymbol{w}) \,, \\ \dot{z} &= f(z, q(\boldsymbol{w})) \,,\end{aligned} \tag{9.31}$$

and the motions of the latter are ultimately bounded, with $\mathbf{B} = W \times Z$. The set $\omega(\mathbf{B})$ may have a complicated structure but it is possible to show, by means of arguments similar to those which are used in the proof of the center manifold theorem, that if Z and $\mathbf{B}$ are small enough, the set in question can still be expressed as the graph of a map $z = \pi(\boldsymbol{w})$. In particular, the graph in question is precisely the center manifold of (9.31) at $(0, 0)$ if $s(0) = 0$, and the matrix

$$S = \left[\frac{\partial s}{\partial \boldsymbol{w}}\right](0)$$

has all eigenvalues on the imaginary axis.

A common feature of the examples discussed above is the fact that the steady-state locus of a system of the form (9.31) can be expressed as the graph of a map $z = \pi(\boldsymbol{w})$. This means that, so long as this is the case, a system of this form has a *unique* well-defined *steady-state response* to the input $\boldsymbol{u}(t) = q(\boldsymbol{w}(t))$. As a matter of fact, the response in question is precisely $z(t) = \pi(\boldsymbol{w}(t))$. Of course, this may not always be the case and *multiple* steady-state responses to a given input may occur. In general, the following property holds.

Lemma 9.4

Let W be a compact set, invariant under the flow of (9.30). Let Z be a closed set and suppose that the motions of (9.31) with initial conditions in $W \times Z$ are ultimately bounded. Then, the steady-state locus of (9.31) is the graph of a set-valued map defined on the whole of W.

9.5 Feedback Stabilization of Linear Systems

9.5.1 Stabilization by Pure State Feedback

Consider a linear system, modeled by equations of the form

$$\dot{\boldsymbol{x}} = \mathbf{A}\boldsymbol{x} + \mathbf{B}\boldsymbol{u} ,$$
$$\boldsymbol{y} = \mathbf{C}\boldsymbol{x} , \tag{9.32}$$

in which $\boldsymbol{x} \in \mathbb{R}^n$, $\boldsymbol{u} \in \mathbb{R}^m$, and $\boldsymbol{y} \in \mathbb{R}^p$, and in which $\mathbf{A}$, $\mathbf{B}$, $\mathbf{C}$ are matrices with real entries.

We begin by analyzing the influence, on the response of the system, of control law of the form

$$\boldsymbol{u} = \mathbf{F}\boldsymbol{x} , \tag{9.33}$$

in which $\mathbf{F}$ is an $n \times m$ matrix with real entries. This type of control is usually referred to as *pure state feedback* or *memoryless state feedback*. The imposition of this control law on the first equation of (9.32) yields the autonomous linear system

$$\dot{\boldsymbol{x}} = (\mathbf{A} + \mathbf{BF})\boldsymbol{x} .$$

The purpose of the design is to choose $\mathbf{F}$ so as to obtain, if possible, a prescribed asymptotic behavior. In general, two options are sought: (i) the n eigenvalues of $(\mathbf{A} + \mathbf{BF})$ have negative real part, (ii) the n eigenvalues of $(\mathbf{A} + \mathbf{BF})$ coincide with the n roots of an arbitrarily fixed polynomial

$$p(\lambda) = \lambda^n + a_{n-1}\lambda^{n-1} + \cdots a_1\lambda + a_0$$

of degree n, with real coefficients. The first option is usually referred to as the *stabilization* problem, while the second is usually referred to as the *eigenvalue assignment* problem.

The conditions for the existence of solutions of these problems can be described as follows. Consider the $n \times (n + m)$ polynomial matrix

$$\mathbf{M}(\lambda) = \big((\mathbf{A} - \lambda\mathbf{I})\,\mathbf{B}\big) . \tag{9.34}$$

Definition 9.7

System (9.32) is said to be *stabilizable* if, for all λ which is an eigenvalue of $\mathbf{A}$ and has nonnegative real part, the matrix $\mathbf{M}(\lambda)$ has rank n. This system is said to be *controllable* if, for all λ which is an eigenvalue of $\mathbf{A}$, the matrix $\mathbf{M}(\lambda)$ has rank n.

The two properties thus identified determine the existence of solutions of the problem of stabilization and, respectively, of the problem of eigenvalue assignment. In fact, the following two results hold.

Theorem 9.9

There exists a matrix $\mathbf{F}$ such that $\mathbf{A} + \mathbf{BF}$ has all eigenvalues with negative real part if and only if system (9.32) is stabilizable.

Theorem 9.10

For any choice of a polynomial $p(\lambda)$ of degree n with real coefficients there exists a matrix $\mathbf{F}$ such that the n eigenvalues of $\mathbf{A} + \mathbf{BF}$ coincide with the n roots of $p(\lambda)$ if and only if system (9.32) is controllable.

The actual construction of the matrix $\mathbf{F}$ usually requires a preliminary transformation of the equations describing the system. As an example, we illustrate how this is achieved in the case of a single-input system, for the problem of eigenvalue assignment. If the input of a system is one dimensional, the system is controllable if and only if the $n \times n$ matrix

$$\mathbf{P} = \big(\mathbf{B} \ \ \mathbf{AB} \ \ \cdots \ \ \mathbf{A}^{n-1}\mathbf{B}\big) \tag{9.35}$$

is nonsingular. Assuming that this is the case, let γ denote the last row of $\mathbf{P}^{-1}$, that is, the unique solution of the set of equations

$$\gamma \mathbf{B} = \gamma \mathbf{AB} = \cdots = \gamma \mathbf{A}^{n-2} \mathbf{B} = 0,$$
$$\gamma \mathbf{A}^{n-1} \mathbf{B} = 1.$$

Then, simple manipulations show that the change of coordinates

$$\tilde{x} = \begin{pmatrix} \gamma \\ \gamma \mathbf{A} \\ \cdots \\ \gamma \mathbf{A}^{n-1} \end{pmatrix} x$$

transforms system (9.32) into a system of the form

$$\dot{\tilde{x}} = \tilde{\mathbf{A}} \tilde{x} + \tilde{\mathbf{B}} u,$$
$$y = \tilde{\mathbf{C}} \tilde{x} \qquad (9.36)$$

in which

$$\tilde{\mathbf{A}} = \begin{pmatrix} 0 & 1 & 0 & \cdots & 0 & 0 \\ 0 & 0 & 1 & \cdots & 0 & 0 \\ . & . & . & \cdots & . & . \\ 0 & 0 & 0 & \cdots & 0 & 1 \\ d_0 & d_1 & d_2 & \cdots & d_{n-2} & d_{n-1} \end{pmatrix}, \quad \tilde{\mathbf{B}} = \begin{pmatrix} 0 \\ 0 \\ \cdots \\ 0 \\ 1 \end{pmatrix}.$$

This form is known as *controllability canonical form* of the equations describing the system. If a system is written in this form, the solution of the problem of eigenvalue assignment is straightforward. If suffices, in fact, to pick a control law of the form

$$u = -(d_0 + a_0)\tilde{x}_1 - (d_1 + a_1)\tilde{x}_2 - \cdots$$
$$- (d_{n-1} + a_{n-1})\tilde{x}_n := \tilde{\mathbf{F}} \tilde{x} \qquad (9.37)$$

to obtain a system

$$\dot{\tilde{x}} = (\tilde{\mathbf{A}} + \tilde{\mathbf{B}} \tilde{\mathbf{F}}) \tilde{x}$$

in which

$$\tilde{\mathbf{A}} + \tilde{\mathbf{B}} \tilde{\mathbf{F}} = \begin{pmatrix} 0 & 1 & 0 & \cdots & 0 & 0 \\ 0 & 0 & 1 & \cdots & 0 & 0 \\ . & . & . & \cdots & . & . \\ 0 & 0 & 0 & \cdots & 0 & 1 \\ -a_0 & -a_1 & -a_2 & \cdots & -a_{n-2} & -a_{n-1} \end{pmatrix}.$$

The characteristic polynomial of this matrix coincides with the prescribed polynomial $p(\lambda)$ and hence the problem is solved. Rewriting the law (9.37) in the original coordinates, one obtains a formula that directly expresses the matrix $\mathbf{F}$ in terms of the parameters of the system (the $n \times n$ matrix $\mathbf{A}$ and the $1 \times n$ row

vector γ) and of the coefficients of the prescribed polynomial $p(\lambda)$

$$u = -\gamma \Big[(d_0 + a_0)\mathbf{I} + (d_1 + a_1)\mathbf{A} + \cdots$$
$$+ (d_{n-1} + a_{n-1})\mathbf{A}^{n-1} \Big] x$$
$$= -\gamma \Big[a_0 \mathbf{I} + a_1 \mathbf{A} + \cdots$$
$$+ a_{n-1} \mathbf{A}^{n-1} + \mathbf{A}^n \Big] x := \mathbf{F} x.$$

The latter is known as Ackermann's formula.

9.5.2 Observers and State Estimation

The imposition of a control law of the form (9.33) requires the availability of all n components of the state x of system (9.32) for measurement, which is seldom the case. Thus, the issue arises of when and how the components in question could be, at least asymptotically, estimated by means of an appropriate auxiliary dynamical system driven by the only variables that are actually accessible for measurement, namely the input u and the output y.

To this end, consider a n-dimensional system thus defined

$$\dot{\hat{x}} = \mathbf{A} \hat{x} + \mathbf{B} u + \mathbf{G}(y - \mathbf{C} \hat{x}), \qquad (9.38)$$

viewed as a system with state $\hat{x} \in \mathbb{R}^n$, driven by the *inputs* u and y. This system can be interpreted as a *copy* of the original dynamics of (9.32), namely

$$\dot{\hat{x}} = \mathbf{A} \hat{x} + \mathbf{B} u$$

corrected by a term proportional, through the $n \times p$ weighting matrix $\mathbf{G}$, to the *effect* that a possible difference between x and $\hat{x}$ has on the only available measurement. The idea is to determine $\mathbf{G}$ in such a way that x and $\hat{x}$ asymptotically converge. Define the difference

$$E = x - \hat{x},$$

which is called *observation error*. Simple algebra shows that

$$\dot{e} = (\mathbf{A} - \mathbf{GC})e.$$

Thus, the observation error obeys an autonomous linear differential equation, and its asymptotic behavior is completely determined by the eigenvalues of $(\mathbf{A} - \mathbf{GC})$. In general, two options are sought: (i) the n eigenvalues of $(\mathbf{A} - \mathbf{GC})$ have negative real part, (ii) the n eigenvalues of $(\mathbf{A} - \mathbf{GC})$ coincide with the n roots of an arbitrarily fixed polynomial of degree n having real coefficients. The first option is usually referred to as the *asymptotic state estimation* problem, while the second does not carry a special name.

Note that, if the eigenvalues of $(\mathbf{A} - \mathbf{GC})$ have negative real part, the state $\hat{x}$ of the auxiliary system (9.38) satisfies

$$\lim_{t \to \infty} [x(t) - \hat{x}(t)] = 0 ,$$

i. e., it asymptotically tracks the state $x(t)$ of (9.32) regardless of what the initial states $x(0)$, $\hat{x}(0)$ and the input $u(t)$ are. System (9.38) is called an asymptotic state estimator or a Luenberger *observer*.

The conditions for the existence of solutions of these problems can be described as follows. Consider the $(n + p) \times n$ polynomial matrix

$$\mathbf{N}(\lambda) = \begin{pmatrix} (\mathbf{A} - \lambda \mathbf{I}) \\ \mathbf{C} \end{pmatrix} . \tag{9.39}$$

Definition 9.8
System (9.32) is said to be *detectable* if, for all λ which is an eigenvalue of $\mathbf{A}$ and has nonnegative real part, the matrix $\mathbf{N}(\lambda)$ has rank n. This system is said to be *observable* if, for all λ which is an eigenvalue of $\mathbf{A}$, the matrix $\mathbf{N}(\lambda)$ has rank n.

Theorem 9.11
There exists a matrix $\mathbf{G}$ such that $\mathbf{A} - \mathbf{GC}$ has all eigenvalues with negative real part if and only if system (9.32) is detectable.

Theorem 9.12
For any choice of a polynomial $p(\lambda)$ of degree n with real coefficients there exists a matrix G such that the n eigenvalues of $\mathbf{A} - \mathbf{GC}$ coincide with the n roots of $p(\lambda)$ if and only if system (9.32) is observable.

In this case, also, the actual construction of the matrix $\mathbf{G}$ is made simple by transforming the equations describing the system. If the output of a system is one dimensional, the system is observable if and only if the $n \times n$ matrix

$$\mathbf{Q} = \begin{pmatrix} \mathbf{C} \\ \mathbf{CA} \\ \cdots \\ \mathbf{CA}^{n-1} \end{pmatrix} \tag{9.40}$$

is nonsingular. Let this be the case and let β denote the last column of $\mathbf{Q}^{-1}$, that is, the unique solution of the set of equations

$$\mathbf{C}\beta = \mathbf{CA}\beta = \cdots = \mathbf{CA}^{n-2}\beta = 0, \quad \mathbf{CA}^{n-1}\beta = 1 .$$

Then, simple manipulations show that the change of coordinates

$$\tilde{x} = \left(\mathbf{A}^{n-1}\beta \; \cdots \; \mathbf{A}\beta \; \beta \right)^{-1} x$$

transforms system (9.32) into a system of the form

$$\dot{\tilde{x}} = \tilde{\mathbf{A}}\tilde{x} + \tilde{\mathbf{B}}u ,$$
$$y = \tilde{\mathbf{C}}\tilde{x} \tag{9.41}$$

in which

$$\tilde{\mathbf{A}} = \begin{pmatrix} d_{n-1} & 1 & 0 & \cdots & 0 & 0 \\ d_{n-2} & 0 & 1 & \cdots & 0 & 0 \\ \cdot & & \cdot & \cdots & \cdot & \cdot \\ d_1 & 0 & 0 & \cdots & 0 & 1 \\ d_0 & 0 & 0 & \cdots & 0 & 0 \end{pmatrix} ,$$

$$\tilde{\mathbf{C}} = \begin{pmatrix} 1 & 0 & \cdots & 0 & 0 \end{pmatrix} .$$

This form is known as *observability canonical form* of the equations describing the system. If a system is written in this form, it is straightforward to write a matrix $\tilde{\mathbf{G}}$ assigning the eigenvalues to $(\tilde{\mathbf{A}} - \tilde{\mathbf{G}}\tilde{\mathbf{C}})$. If suffices, in fact, to pick a

$$\tilde{\mathbf{G}} = \begin{pmatrix} d_{n-1} + a_{n-1} \\ d_{n-2} + a_{n-2} \\ \cdots \\ d_0 + a_0 \end{pmatrix} \tag{9.42}$$

to obtain a matrix

$$\tilde{\mathbf{A}} - \tilde{\mathbf{G}}\tilde{\mathbf{C}} = \begin{pmatrix} -a_{n-1} & 1 & 0 & \cdots & 0 & 0 \\ -a_{n-2} & 0 & 1 & \cdots & 0 & 0 \\ \cdot & & \cdot & \cdots & \cdot & 0 \\ -a_1 & 0 & 0 & \cdots & 0 & 1 \\ -a_0 & 0 & 0 & \cdots & 0 & 0 \end{pmatrix} ,$$

whose characteristic polynomial coincides with the prescribed polynomial $p(\lambda)$.

9.5.3 Stabilization via Dynamic Output Feedback

Replacing, in the control law (9.33), the true state x by the estimate $\hat{x}$ provided by the asymptotic observer

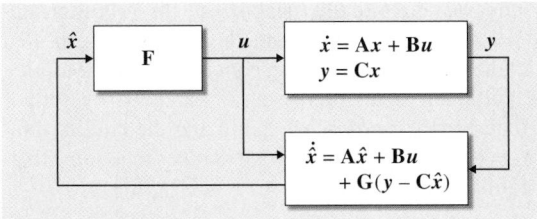

Fig. 9.5 Observer-based control

(9.38) yields a dynamic, output-feedback, control law of the form

$$u = \mathbf{F}\hat{x},$$
$$\dot{\hat{x}} = (\mathbf{A} + \mathbf{BF} - \mathbf{GC})\hat{x} + \mathbf{G}y. \qquad (9.43)$$

Controlling system (9.32) by means of (9.43) yields the closed-loop system (Fig. 9.5)

$$\begin{pmatrix} \dot{x} \\ \dot{\hat{x}} \end{pmatrix} = \begin{pmatrix} \mathbf{A} & \mathbf{BF} \\ \mathbf{GC} & \mathbf{A} + \mathbf{BF} - \mathbf{GC} \end{pmatrix} \begin{pmatrix} x \\ \hat{x} \end{pmatrix}. \qquad (9.44)$$

It is straightforward to check that the eigenvalues of the system thus obtained coincide with those of the two matrices $(\mathbf{A} + \mathbf{BF})$ and $(\mathbf{A} - \mathbf{GC})$. To this end, in fact, it suffices to replace $\hat{x}$ by $e = x - \hat{x}$, which changes system (9.44) into an equivalent system

$$\begin{pmatrix} \dot{x} \\ \dot{e} \end{pmatrix} = \begin{pmatrix} \mathbf{A} + \mathbf{BF} & -\mathbf{BF} \\ 0 & \mathbf{A} - \mathbf{GC} \end{pmatrix} \begin{pmatrix} x \\ e \end{pmatrix} \qquad (9.45)$$

in block-triangular form.

From this argument, it can be concluded that the dynamic feedback law (9.43) suffices to yield a closed-loop system whose $2n$ eigenvalues either have negative

real part (if system (9.32) is stabilizable and detectable) or even coincide with the roots of a pair of prescribed polynomials of degree n (if (9.32) is controllable and observable). In particular, the result in question can be achieved by means of a *separate* design of $\mathbf{F}$ and $\mathbf{G}$, the former to control the eigenvalues of $(\mathbf{A} + \mathbf{BF})$ and the latter to control the eigenvalues of $(\mathbf{A} - \mathbf{GC})$. This possibility is usually referred to as the *separation principle* for stabilization via (dynamic) output feedback.

It can be concluded from this argument that, if a system is stabilizable and detectable, there exists a dynamic, output feedback, law yielding a closed-loop system with all eigenvalues with negative real part. It is important to observe that also the converse of this property is true, namely the existence of a dynamic, output feedback, law yielding a closed-loop system with all eigenvalues with negative real part requires the controlled system to be *stabilizable and detectable*. The proof of this converse result is achieved by taking any arbitrary dynamic output-feedback law

$$\dot{\xi} = \bar{\mathbf{F}}\xi + \bar{\mathbf{G}}y,$$
$$u = \bar{\mathbf{H}}\xi + \bar{\mathbf{K}}y,$$

yielding a closed-loop system

$$\begin{pmatrix} \dot{x} \\ \dot{\xi} \end{pmatrix} = \begin{pmatrix} \mathbf{A} + \mathbf{B}\bar{\mathbf{K}}\mathbf{C} & \mathbf{B}\bar{\mathbf{H}} \\ \bar{\mathbf{G}}\mathbf{C} & \bar{\mathbf{F}} \end{pmatrix} \begin{pmatrix} x \\ \xi \end{pmatrix}$$

and proving, via the converse Lyapunov theorem for linear systems, that, if the eigenvalues of the latter have negative real part, *necessarily* there exist two matrices $\mathbf{F}$ and $\mathbf{G}$ such that the eigenvalues of $(\mathbf{A} + \mathbf{BF})$ and, respectively, $(\mathbf{A} - \mathbf{GC})$ have negative real part.

9.6 Feedback Stabilization of Nonlinear Systems

9.6.1 Recursive Methods for Global Stability

Stabilization of nonlinear systems is a very difficult task and general methods are not available. Only if the equations of the system exhibit a special structure do there exist systematic methods for the design of pure state feedback (or, if necessary, dynamic, output feedback) laws yielding global asymptotic stability of an equilibrium. In this section we review some of these special design procedures.

We begin by a simple *modular* property which can be recursively used to stabilize systems in *triangular form* (see [9.10, Chap. 9] for further details).

Lemma 9.5

Consider a system described by equations of the form

$$\dot{z} = f(z, \xi),$$
$$\dot{\xi} = q(z, \xi) + b(z, \xi)u, \qquad (9.46)$$

in which $(z, \xi) \in \mathbb{R}^n \times \mathbb{R}$, and the functions $f(z, \xi)$, $q(z, \xi)$, $b(z, \xi)$ are continuously differentiable functions. Suppose that $b(z, \xi) \neq 0$ for all (z, ξ) and that $f(0, 0) = 0$ and $q(0, 0) = 0$. If $z = 0$ is a globally asymptotically stable equilibrium of $\dot{z} = f(z, 0)$, there exists a differentiable function $u = u(z, \xi)$ with

$u(0,0) = 0$ such that the equilibrium at $(z, \boldsymbol{\xi}) = (0, 0)$

$$\dot{z} = f(z, \boldsymbol{\xi}),$$
$$\dot{\boldsymbol{\xi}} = q(z, \boldsymbol{\xi}) + b(z, \boldsymbol{\xi})u(z, \boldsymbol{\xi}),$$

is globally asymptotically stable.

The construction of the stabilizing feedback $\boldsymbol{u}(z, \boldsymbol{\xi})$ is achieved as follows. First of all observe that, using the assumption $b(z, \boldsymbol{\xi}) \neq 0$, the imposition of the preliminary feedback law

$$\boldsymbol{u}(z, \boldsymbol{\xi}) = \frac{1}{b(z, \boldsymbol{\xi})}(-q(z, \boldsymbol{\xi}) + v)$$

yields the simpler system

$$\dot{z} = f(z, \boldsymbol{\xi}),$$
$$\dot{\boldsymbol{\xi}} = v.$$

Then, express $f(z, \boldsymbol{\xi})$ in the form

$$f(z, \boldsymbol{\xi}) = f(z, 0) + p(z, \boldsymbol{\xi})\boldsymbol{\xi},$$

in which $p(z, \boldsymbol{\xi}) = [f(z, \boldsymbol{\xi}) - f(z, 0)]/\boldsymbol{\xi}$ is at least continuous.

Since by assumption $z = 0$ is a globally asymptotically stable equilibrium of $\dot{z} = f(z, 0)$, by the converse Lyapunov theorem there exists a smooth real-valued function $V(z)$, which is positive definite and proper, satisfying

$$\frac{\partial V}{\partial z} f(z, 0) < 0,$$

for all nonzero z. Now, consider the positive-definite and proper function

$$W(z, \boldsymbol{\xi}) = V(z) + \frac{1}{2}\boldsymbol{\xi}^2,$$

and observe that

$$\frac{\partial W}{\partial z}\dot{z} + \frac{\partial W}{\partial \boldsymbol{\xi}}\dot{\boldsymbol{\xi}} = \frac{\partial V}{\partial z} f(z, 0) + \frac{\partial V}{\partial z} p(z, \boldsymbol{\xi})\boldsymbol{\xi} + \boldsymbol{\xi}v.$$

Choosing

$$v = -\boldsymbol{\xi} - \frac{\partial V}{\partial z} p(z, \boldsymbol{\xi}) \tag{9.47}$$

yields

$$\frac{\partial W}{\partial z}\dot{z} + \frac{\partial W}{\partial \boldsymbol{\xi}}\dot{\boldsymbol{\xi}} = \frac{\partial V}{\partial z} f(z, 0) - \boldsymbol{\xi}^2 < 0,$$

for all nonzero $(z, \boldsymbol{\xi})$ and this, by the direct Lyapunov criterion, shows that the feedback law

$$\boldsymbol{u}(z, \boldsymbol{\xi}) = \frac{1}{b(z, \boldsymbol{\xi})}\left[-q(z, \boldsymbol{\xi}) - \boldsymbol{\xi} - \frac{\partial V}{\partial z} p(z, \boldsymbol{\xi})\right]$$

globally asymptotically stabilizes the equilibrium $(z, \boldsymbol{\xi}) = (0, 0)$ of the associated closed-loop system.

In the next Lemma (which contains the previous one as a particular case) this result is extended by showing that, for the purpose of stabilizing the equilibrium $(z, \boldsymbol{\xi}) = (0, 0)$ of system (9.46), it suffices to assume that the equilibrium $z = 0$ of

$$\dot{z} = f(z, \boldsymbol{\xi})$$

is *stabilizable* by means of a *virtual* control law $\boldsymbol{\xi} = v^\star(z)$.

Lemma 9.6
Consider again the system described by equations of the form (9.46). Suppose there exists a continuously differentiable function

$$\boldsymbol{\xi} = v^\star(z),$$

with $v^\star(0) = 0$, which globally asymptotically stabilizes the equilibrium $z = 0$ of $\dot{z} = f(z, v^\star(z))$. Then there exists a differentiable function $\boldsymbol{u} = \boldsymbol{u}(z, \boldsymbol{\xi})$ with $\boldsymbol{u}(0, 0) = 0$ such that the equilibrium at $(z, \boldsymbol{\xi}) = (0, 0)$

$$\dot{z} = f(z, \boldsymbol{\xi}),$$
$$\dot{\boldsymbol{\xi}} = q(z, \boldsymbol{\xi}) + b(z, \boldsymbol{\xi})u(z, \boldsymbol{\xi})$$

is globally asymptotically stable.

To prove the result, and to construct the stabilizing feedback, it suffices to consider the (globally defined) change of variables

$$y = \boldsymbol{\xi} - v^\star(z),$$

which transforms (9.46) into a system

$$\dot{z} = f(z, v^\star(z) + y),$$
$$\dot{y} = -\frac{\partial v^\star}{\partial z} f(z, v^\star(z) + y) + q(v^\star(z) + y, \boldsymbol{\xi})$$
$$+ b(v^\star(z) + y, \boldsymbol{\xi})u, \tag{9.48}$$

which meets the assumptions of Lemma 9.5, and then follow the construction of a stabilizing feedback as described. Using repeatedly the property indicated in Lemma 9.6 it is straightforward to derive the expression of a globally stabilizing feedback for a system in *triangular* form

$$\dot{z} = f(z, \xi_1),$$
$$\dot{\xi}_1 = q_1(z, \xi_1) + b_1(z, \xi_1)\xi_2,$$
$$\dot{\xi}_2 = q_2(z, \xi_1, \xi_2) + b_2(z, \xi_1, \xi_2)\xi_3,$$
$$\cdots$$
$$\dot{\xi}_r = q_r(z, \xi_1, \xi_2, \ldots, \xi_r) + b_r(z, \xi_1, \xi_2, \ldots, \xi_r)\boldsymbol{u}. \tag{9.49}$$

To this end, in fact, it suffices to assume that the equilibrium $z = 0$ of $\dot{z} = f(z, \xi)$ is stabilizable by means of a virtual law $\xi = v^\star(z)$, and that $b_1(z, \xi_1), b_2(z, \xi_1, \xi_2), \ldots, b_r(z, \xi_1, \xi_2, \ldots, \xi_r)$ are nowhere zero.

9.6.2 Semiglobal Stabilization via Pure State Feedback

The global stabilization results presented in the previous section are indeed conceptually appealing but the actual implementation of the feedback law requires the explicit knowledge of a Lyapunov function $V(z)$ for the system $\dot{z} = f(z, 0)$ (or for the system $\dot{z} = f(z, v^*(z))$ in the case of Lemma 9.6). This function, in fact, explicitly determines the structure of the feedback law which globally asymptotically stabilizes the system. Moreover, in the case of systems of the form (9.49) with $r > 1$, the computation of the feedback law is somewhat cumbersome, in that it requires to iterate a certain number of times the manipulations described in the proof of Lemmas 9.5 and 9.6. In this section we show how these drawbacks can be overcome, in a certain sense, if a less ambitious design goal is pursued, namely if instead of seeking global stabilization one is interested in a feedback law capable of asymptotically steering to the equilibrium point all trajectories which have origin in a *a priori fixed* (and hence possibly large) *bounded set*.

Consider again a system satisfying the assumptions of Lemma 9.5. Observe that $b(z, \xi)$, being continuous and nowhere zero, has a well-defined sign. Choose a simple control law of the form

$$u = -k \, \text{sign}(b) \, \xi \tag{9.50}$$

to obtain the system

$$\dot{z} = f(z, \xi),$$
$$\dot{\xi} = q(z, \xi) - k|b(z, \xi)|\xi. \tag{9.51}$$

Assume that the equilibrium $z = 0$ of $\dot{z} = f(z, 0)$ is globally asymptotically but also *locally exponentially* stable. If this is the case, then the linear approximation of the first equation of (9.51) at the point $(z, \xi) = (0, 0)$ is a system of the form

$$\dot{z} = \mathbf{F}z + \mathbf{G}\xi,$$

in which $\mathbf{F}$ is a Hurwitz matrix. Moreover, the linear approximation of the second equation of (9.51) at the point $(z, \xi) = (0, 0)$ is a system of the form

$$\dot{\xi} = \mathbf{Q}z + \mathbf{R}\xi - kb_0\xi,$$

in which $b_0 = |b(0, 0)|$. It follows that the linear approximation of system (9.51) at the equilibrium $(z, \xi) = (0, 0)$ is a linear system $\dot{x} = \mathbf{A}x$ in which

$$\mathbf{A} = \begin{pmatrix} \mathbf{F} & \mathbf{G} \\ \mathbf{Q} & (\mathbf{R} - kb_0) \end{pmatrix}.$$

Standard arguments show that, if the number k is large enough, the matrix in question has all eigenvalues with negative real part (in particular, as k increases, n eigenvalues approach the n eigenvalues of $\mathbf{F}$ and the remaining one is a real eigenvalue that tends to $-\infty$). It is therefore concluded, from the principle of stability in the first approximation, that if k is sufficiently large the equilibrium $(z, \xi) = (0, 0)$ of the closed-loop system (9.51) is locally asymptotically (actually locally exponentially) stable.

However, a stronger result holds. It can be proven that, for any arbitrary compact subset K of $\mathbb{R}^n \times \mathbb{R}$, there exists a number k^*, such that, for all $k \geq k^*$, the equilibrium $(z, \xi) = (0, 0)$ of the closed-loop system (9.51) is locally asymptotically stable and all initial conditions in K produce a trajectory that asymptotically converges to this equilibrium. In other words, the basin of attraction of the equilibrium $(z, \xi) = (0, 0)$ of the closed-loop system contains the set K. Note that the number k^* depends on the choice of the set K and, in principle, it increases as the size of K increases. The property in question can be summarized as follows (see [9.10, Chap. 9] for further details). A system

$$\dot{x} = f(x, u)$$

is said to be *semiglobally stabilizable* (an equivalent, but longer, terminology is *asymptotically stabilizable with guaranteed basin of attraction*) at a given point $\bar{x}$ if, for each *compact* subset $K \subset \mathbb{R}^n$, there exists a feedback law $u = u(x)$, which in general depends on K, such that in the corresponding closed-loop system

$$\dot{x} = f(x, u(x))$$

the point $x = \bar{x}$ is a locally asymptotically stable equilibrium, and

$$x(0) \in K \Rightarrow \lim_{t \to \infty} x(t) = \bar{x}$$

(i.e., the compact subset K is contained in the basin of attraction of the equilibrium $x = \bar{x}$). The result described above shows that system (9.46), under the said assumptions, is semiglobally stabilizable at $(z, \xi) = (0, 0)$, by means of a feedback law of the form (9.50).

The arguments just shown can be iterated to deal with a system of the form (9.49). In fact, it is easy to realize that, if the equilibrium $z = 0$ of $\dot{z} = f(z, 0)$ is globally asymptotically and also locally exponentially stable, if $q_i(z, \xi_1, \xi_2, \ldots, \xi_i)$ vanishes at $(z, \xi_1, \xi_2, \ldots, \xi_i) = (0, 0, 0, \ldots, 0)$ and $b_i(z, \xi_1, \xi_2, \ldots, \xi_i)$ is nowhere zero, for all $i = 1, \ldots, r$, system (9.49) is semiglobally stabilizable at the point $(z, \xi_1, \xi_2, \ldots, \xi_r) = (0, 0, 0, \ldots, 0)$, actually by means of a control law that has the following structure

$$u = \alpha_1 \xi_1 + \alpha_2 \xi_2 + \cdots + \alpha_r \xi_r .$$

The coefficients $\alpha_1, \ldots, \alpha_r$ that characterize this control law can be determined by means of recursive iteration of the arguments described above.

9.6.3 Semiglobal Stabilization via Dynamic Output Feedback

System (9.49) can be semiglobally stabilized, at the equilibrium $(z, \xi_1, \ldots, \xi_r) = (0, 0, \ldots, 0)$, by means of a simple feedback law, which is a linear function of the *partial state* $(\xi_1, \ldots, \xi_r)$. If these variables are not directly available for feedback, one may wish to use instead an estimate – as is possible in the case of linear systems – provided by a dynamical system driven by the measured output. This is actually doable if the output y of (9.49) coincides with the state variable ξ_1. For the purpose of stabilizing system (9.49) by means of dynamic output feedback, it is convenient to reexpress the equations describing this system in a simpler form, known as *normal form*. Set $\eta_1 = \xi_1$ and define

$$\eta_2 = q_1(z, \xi_1) + b_1(z, \xi_1)\xi_2 ,$$

by means of which the second equation of (9.49) is changed into $\dot{\eta}_1 = \eta_2$. Set now

$$\eta_3 = \frac{\partial(q_1 + b_1 \xi_2)}{\partial z} f(z, \xi_1)$$
$$+ \frac{\partial(q_1 + b_1 \xi_2)}{\partial \xi_1}[q_1 + b_1 \xi_2] + b_1[q_2 + b_2 \xi_3] ,$$

by means of which the third equation of (9.49) is changed into $\dot{\eta}_2 = \eta_3$. Proceeding in this way, it is easy to conclude that the system (9.49) can be changed into a system modeled by

$$\dot{z} = f(z, \eta_1) ,$$
$$\dot{\eta}_1 = \eta_2 ,$$
$$\dot{\eta}_2 = \eta_3 ,$$
$$\ldots$$
$$\dot{\eta}_r = q(z, \eta_1, \eta_2, \ldots, \eta_r) + b(z, \eta_1, \eta_2, \ldots, \eta_r)u ,$$
$$y = \eta_1 , \tag{9.52}$$

in which $q(0, 0, 0, \ldots, 0) = (0, 0, 0, \ldots, 0)$ and $b(z, \eta_1, \eta_2, \ldots, \eta_r)$ is nowhere zero.

It has been shown earlier that, if the equilibrium $z = 0$ of $\dot{z} = f(z, 0)$ is globally asymptotically and also locally exponentially stable, this system is semiglobally stabilizable, by means of a feedback law

$$u = h_1 \eta_1 + h_2 \eta_2 + \ldots + h_r \eta_r , \tag{9.53}$$

which is a linear function of the states $\eta_1, \eta_2, \ldots, \eta_r$. The feedback in question, if the coefficients are appropriately chosen, is able to steer at the equilibrium $(z, \eta_1, \ldots, \eta_r) = (0, 0, \ldots, 0)$ all trajectories with initial conditions in a given compact set K (whose size influences, as stressed earlier, the actual choice of the parameters $h_1, \ldots, h_r$). Note that, since all such trajectories will never exit, in positive time, a (possibly larger) compact set, there exists a number L such that

$$|h_1 \eta_1(t) + h_2 \eta_2(t) + \ldots + h_r \eta_r(t)| \leq L ,$$

for all $t \geq 0$ whenever the initial condition of the closed loop is in K. Thus, to the extent of achieving asymptotic stability with a basin of attraction including K, the feedback law (9.53) could be replaced with a (nonlinear) law of the form

$$u = \sigma_L(h_1 \eta_1 + h_2 \eta_2 + \ldots + h_r \eta_r) , \tag{9.54}$$

in which $\sigma(r)$ is any *bounded* function that coincides with r when $|r| \leq L$. The advantage of having a feedback law whose amplitude is guaranteed not to exceed a fixed bound is that, when the partial states η_i will be replaced by approximate estimates, possibly large errors in the estimates will not cause dangerously large control efforts.

Inspection of the equations (9.52) reveals that the state variables used in the control law (9.54) coincide with the measured output $\mathbf{y}$ and its derivatives with respect to time, namely

$$\eta_i = \mathbf{y}^{(i-1)} , \quad i = 1, 2, \ldots, r .$$

It is therefore reasonable to expect that these variables could be asymptotically estimated in some simple way by means of a dynamical system driven by the measured output itself. The system in question is actually of the form

$$\dot{\tilde{\eta}}_1 = \tilde{\eta}_2 - \kappa c_{r-1}(\mathbf{y} - \tilde{\eta}_1) ,$$
$$\dot{\tilde{\eta}}_2 = \tilde{\eta}_3 - \kappa^2 c_{r-2}(\mathbf{y} - \tilde{\eta}_1) ,$$
$$\ldots$$
$$\dot{\tilde{\eta}}_r = -\kappa^r c_0(\mathbf{y} - \tilde{\eta}_1) . \tag{9.55}$$

It is easy to realize that, if $\tilde{\eta}_1(t) = \mathbf{y}(t)$, then all $\tilde{\eta}_i(t)$, for $i = 2, \ldots, r$, coincide with $\eta_i(t)$. However, there is no a priori guarantee that this can be achieved and hence system (9.55) cannot be regarded as a true observer of the partial state $\eta_1, \ldots, \eta_r$ of (9.52). It happens, though, that if the reason why this partial state needs to be estimated is only the implementation of the feedback law (9.54), then an *approximate* observer such as (9.55) can be successfully used.

The fact is that, if the coefficients $c_0, \ldots, c_{r-1}$ are coefficients of a Hurwitz polynomial

$$p(\lambda) = \lambda^r + c_{r-1}\lambda^{r-1} + \ldots + c_1\lambda + c_0 ,$$

and if the parameter κ is sufficiently large, the *rough estimates* $\tilde{\eta}_i$ of η_i provided by (9.55) can be used to replace the true states η_i in the control law (9.54). This results in a controller, which is a dynamical system modeled by equations of the form (Fig. 9.6)

$$\dot{\tilde{\eta}} = \tilde{\mathbf{F}}\tilde{\eta} + \tilde{\mathbf{G}}y ,$$
$$u = \sigma_L(\mathbf{H}\eta) , \tag{9.56}$$

able to solve a problem of semiglobal stabilization for (9.52), if its parameters are appropriately chosen (see [9.6, Chap. 12] and [9.11, 12] for further details).

9.6.4 Observers and Full State Estimation

The design of observers for nonlinear systems modeled by equations of the form

$$\dot{x} = f(x, u) ,$$
$$y = h(x, u) , \tag{9.57}$$

with state $x \in \mathbb{R}^n$, input $u \in \mathbb{R}^m$, and output $y \in \mathbb{R}$ usually requires the preliminary transformation of the equations describing the system, in a form that suitably corresponds to the observability canonical form describe earlier for linear systems. In fact, a key requirement for the existence of observers is the existence of a global changes of coordinates $\tilde{x} = \Phi(x)$ carrying system (9.57) into a system of the form

$$\dot{\tilde{x}}_1 = \tilde{f}_1(\tilde{x}_1, \tilde{x}_2, u) ,$$
$$\dot{\tilde{x}}_2 = \tilde{f}_2(\tilde{x}_1, \tilde{x}_2, \tilde{x}_3, u) ,$$
$$\cdots$$
$$\dot{\tilde{x}}_{n-1} = \tilde{f}_{n-1}(\tilde{x}_1, \tilde{x}_2, \ldots, \tilde{x}_n, u) ,$$
$$\dot{\tilde{x}}_n = \tilde{f}_n(\tilde{x}_1, \tilde{x}_2, \ldots, \tilde{x}_n, u) ,$$
$$y = \tilde{h}(\tilde{x}_1, u) , \tag{9.58}$$

in which the $\tilde{h}(\tilde{x}_1, u)$ and $\tilde{f}_i(\tilde{x}_1, \tilde{x}_2, \ldots, \tilde{x}_{i+1}, u)$ satisfy

$$\frac{\partial \tilde{h}}{\partial \tilde{x}_1} \neq 0 , \quad \text{and} \quad \frac{\partial \tilde{f}_i}{\partial \tilde{x}_{i+1}} \neq 0 ,$$
$$\text{for all} \quad i = 1, \ldots, n-1 \tag{9.59}$$

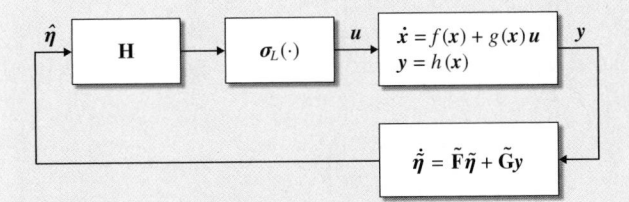

Fig. 9.6 Control via partial-state estimator

for all $\tilde{x} \in \mathbb{R}^n$, and all $u \in \mathbb{R}^m$. This form is usually referred to as the *uniform observability canonical form*.

The existence of canonical forms of this kind can be obtained as follows [9.13, Chap. 2]. Define – recursively – a sequence of real-valued functions $\varphi_i(x, u)$ as follows

$$\varphi_1(x, u) := h(x, u) ,$$
$$\vdots$$
$$\varphi_i(x, u) := \frac{\partial \varphi_{i-1}}{\partial x} f(x, u) ,$$

for $i = 1, \ldots, n$. Using these functions, define a sequence of i-vector-valued functions $\Phi_i(x, u)$ as follows

$$\Phi_i(x, u) = \begin{pmatrix} \varphi_1(x, u) \\ \vdots \\ \varphi_i(x, u) \end{pmatrix} ,$$

for $i = 1, \ldots, n$. Finally, for each of the $\Phi_i(x, u)$, compute the subspace

$$K_i(x, u) = \ker\left[\frac{\partial \Phi_i}{\partial x}\right]_{(x,u)} ,$$

in which $\ker[M]$ denotes the subspace consisting of all vectors v such that $Mv = 0$, that is the so-called null space of the matrix M. Note that, since the entries of the matrix

$$\frac{\partial \Phi_i}{\partial x}$$

are in general dependent on (x, u), so is its null space $K_i(x, u)$.

The role played by the objects thus defined in the construction of the change of coordinates yielding an observability canonical form is explained in this result.

Lemma 9.7

Consider system (9.57) and the map $\tilde{x} = \Phi(x)$ defined by

$$\Phi(x) = \begin{pmatrix} \varphi_1(x, 0) \\ \varphi_2(x, 0) \\ \vdots \\ \varphi_n(x, 0) \end{pmatrix}.$$

Suppose that $\Phi(x)$ has a globally defined and continuously differentiable inverse. Suppose also that, for all $i = 1, \ldots, n$,

$$\dim[K_i(x, u)] = n - i,$$

for all $u \in \mathbb{R}^m$

and for all $x \in \mathbb{R}^n$

$$K_i(x, u) = \text{independent of } u.$$

Then, system (9.57) is globally transformed, via $\Phi(x)$, into a system in uniform observability canonical form.

Once a system has been changed into its observability canonical form, an asymptotic observer can be built as follows. Take a *copy* of the dynamics of (9.58), corrected by an *innovation* term proportional to the difference between the output of (9.58) and the output of the copy. More precisely, consider a system of the form

$$\dot{\hat{x}}_1 = \tilde{f}_1(\hat{x}_1, \hat{x}_2, u) + \kappa c_{n-1}(y - h(\hat{x}_1, u)),$$
$$\dot{\hat{x}}_2 = \tilde{f}_2(\hat{x}_1, \hat{x}_2, \hat{x}_3, u) + \kappa^2 c_{n-2}(y - h(\hat{x}_1, u)),$$

$$\cdots$$

$$\dot{\hat{x}}_{n-1} = \tilde{f}_{n-1}(\hat{x}, u) + \kappa^{n-1} c_1(y - h(\hat{x}_1, u)),$$
$$\dot{\hat{x}}_n = \tilde{f}_n(\hat{x}, u) + \kappa^n c_0(y - h(\hat{x}_1, u)), \qquad (9.60)$$

in which κ and $c_{n-1}, c_{n-2}, \ldots, c_0$ are design parameters.

The state of the system thus defined is able to asymptotically track, no matter what the initial conditions $x(0)$, $\tilde{x}(0)$ and the input $u(t)$ are, the state of

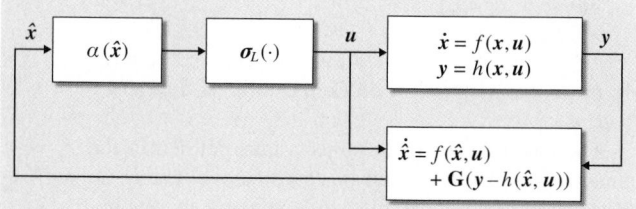

Fig. 9.7 Observer-based control for a nonlinear system

system (9.58) provided that the two following technical hypotheses hold:

(i) Each of the maps $\tilde{f}_i(\tilde{x}_1, \ldots, \tilde{x}_i, \tilde{x}_{i+1}, u)$, for $i = 1, \ldots, n$, is globally Lipschitz with respect to $(\tilde{x}_1, \ldots, \tilde{x}_i)$, uniformly in $\tilde{x}_{i+1}$ and u,

(ii) There exist two real numbers α, β, with $0 < \alpha < \beta$, such that

$$\alpha \leq \left| \frac{\partial \tilde{h}}{\partial \tilde{x}_1} \right| \leq \beta, \quad \text{and} \quad \alpha \leq \left| \frac{\partial \tilde{f}_i}{\partial \tilde{x}_{i+1}} \right| \leq \beta,$$

for all $i = 1, \ldots, n - 1$,

for all $\tilde{x} \in \mathbb{R}^n$, and all $u \in \mathbb{R}^m$.

Let the *observation error* be defined as

$$e_i = \hat{x}_i - \tilde{x}_i, \quad i = 1, 2, \ldots, n.$$

The fact is that, if the two assumptions above hold, there is a choice of the coefficients $c_0, c_1, \ldots, c_{n-1}$ and there is a number κ^* such that, if $\kappa \geq \kappa^*$, the observation error asymptotically decays to zero as time tends to infinity, regardless of what the initial states $\tilde{x}(0)$, $\hat{x}(0)$ and the input $u(t)$ are. For this reason the observer in question is called a *high-gain* observer (see [9.13, Chap. 6] for further details).

The availability of such an observer makes it possible to design a dynamic, output feedback, stabilizing control law, thus extending to the case of nonlinear systems the separation principle for stabilization of linear systems. In fact, consider a system in canonical form (9.58), rewritten as

$$\dot{\tilde{x}} = \tilde{f}(\tilde{x}, u),$$
$$y = \tilde{h}(\tilde{x}, u).$$

Suppose a feedback law is known $u = \alpha(\tilde{x})$ that globally asymptotic stabilizes the equilibrium point $\tilde{x} = 0$ of the closed-loop system

$$\dot{\tilde{x}} = \tilde{f}(\tilde{x}, \alpha(\tilde{x})).$$

Then, an output feedback controller of the form (Fig. 9.7)

$$\dot{\hat{x}} = \tilde{f}(\hat{x}, u) + \mathbf{G}[y - \tilde{h}(\hat{x}, u)],$$
$$u = \sigma_L(\alpha(\hat{x})),$$

whose dynamics are those of system (9.60) and $\sigma_L : \mathbb{R} \to \mathbb{R}$ is a bounded function satisfying $\sigma_L(r) = r$ for all $|r| \leq L$, is able to stabilize the equilibrium $(\tilde{x}, \hat{x}) = (0, 0)$ of the closed-loop system, with a basin of attraction that includes any a priori fixed compact set $K \times K$, if its parameters (the coefficients $c_0, c_1, \ldots, c_{n-1}$ and the parameter κ of (9.60) and the parameter L of $\sigma_L(\cdot)$) are appropriately chosen (see [9.13, Chap. 7] for details).

9.7 Tracking and Regulation

9.7.1 The Servomechanism Problem

A central problem in control theory is the design of feedback controllers so as to have certain outputs of a given plant *to track* prescribed reference trajectories. In any realistic scenario, this control goal has to be achieved in spite of a good number of phenomena which would cause the system to behave differently than expected. These phenomena could be endogenous, for instance, parameter variations, or exogenous, such as additional undesired inputs affecting the behavior of the plant. In numerous design problems, the trajectory to be tracked (or the disturbance to be rejected) is not available for measurement, nor is it known ahead of time. Rather, it is only known that this trajectory is simply an (undefined) member in a set of functions, for instance, the set of all possible solutions of an ordinary differential equation. Theses cases include the classical problem of the set-point control, the problem of active suppression of harmonic disturbances of unknown amplitude, phase and even frequency, the synchronization of nonlinear oscillations, and similar others.

In general, a tracking problem of this kind can be cast in the following terms. Consider a finite-dimensional, time-invariant, nonlinear system modeled by equations of the form

$$\dot{x} = f(w, x, u),$$
$$e = h(w, x),$$
$$y = k(w, x),\tag{9.61}$$

in which $x \in \mathbb{R}^n$ is a vector of state variables, $u \in \mathbb{R}^m$ is a vector of inputs used for *control* purposes, $w \in \mathbb{R}^s$ is a vector of inputs which cannot be controlled and include *exogenous* commands, exogenous disturbances, and model uncertainties, $e \in \mathbb{R}^p$ is a vector of *regulated* outputs which include tracking errors and any other variable that needs to be steered to 0, and $y \in \mathbb{R}^q$ is a vector of outputs that are available for *measurement* and hence used to feed the device that supplies the control action. The problem is to design a controller, which receives $y(t)$ as input and produces $u(t)$ as output, able to guarantee that, in the resulting closed-loop system, $x(t)$ remains bounded and

$$\lim_{t \to \infty} e(t) = 0,\tag{9.62}$$

regardless of what the exogenous input $w(t)$ actually is.

As observed at the beginning, $w(t)$ is not available for measurement, nor it is known ahead of time, but it is known *to belong to a fixed family* of functions of time, the family of all solutions obtained from a fixed ordinary differential equation of the form

$$\dot{w} = s(w)\tag{9.63}$$

as the corresponding initial condition $w(0)$ is allowed to vary on a prescribed set. This autonomous system is known as *the exosystem.*

The control law is to be provided by a system modeled by equations of the form

$$\dot{\xi} = \varphi(\xi, y),$$
$$u = \gamma(\xi, y),\tag{9.64}$$

with state $\xi \in \mathbb{R}^\nu$. The initial conditions $x(0)$ of the *plant* (9.61), $w(0)$ of the *exosystem* (9.63), and $\xi(0)$ of the *controller* (9.64) are allowed to range over fixed *compact* sets $X \subset \mathbb{R}^n$, $W \subset \mathbb{R}^s$, and $\varXi \subset \mathbb{R}^\nu$, respectively. All maps characterizing the model of the controlled plant, of the exosystem, and of the controller are assumed to be sufficiently differentiable.

The *generalized servomechanism problem* (or *problem of output regulation*) is to design a feedback controller of the form (9.64) so as to obtain a closed-loop system in which all trajectories are bounded and the regulated output $e(t)$ asymptotically decays to 0 as $t \to \infty$. More precisely, it is required that the composition of (9.61), (9.63), and (9.64), that is, the *autonomous* system

$$\dot{w} = s(w),$$
$$\dot{x} = f(w, x, \gamma(\xi, k(w, x))),$$
$$\dot{\xi} = \varphi(\xi, k(w, x)),\tag{9.65}$$

with output $e = h(w, x)$ be such that:

- The positive orbit of $W \times X \times \varXi$ is bounded, i.e., there exists a bounded subset S of $\mathbb{R}^s \times \mathbb{R}^n \times \mathbb{R}^\nu$ such that, for any $(w_0, x_0, \xi_0) \in W \times X \times \varXi$, the integral curve $(w(t), x(t), \xi(t))$ of (9.65) passing through (w_0, x_0, ξ_0) at time $t = 0$ remains in S for all $t \geq 0$.
- $\lim_{t \to \infty} e(t) = 0$, uniformly in the initial condition; i.e., for every $\varepsilon > 0$ there exists a time $\bar{t}$, depending only on ε and *not on* (w_0, x_0, ξ_0) such that the integral curve $(w(t), x(t), \xi(t))$ of (9.65) passing through (w_0, x_0, ξ_0) at time $t = 0$ satisfies $\|e(t)\| \leq \varepsilon$ for all $t \geq \bar{t}$.

9.7.2 Tracking and Regulation for Linear Systems

We show in this section how the servomechanism problem is treated in the case of linear systems. Let system (9.61) and exosystem (9.63) be linear systems, modeled by equations of the form

$$\dot{w} = \mathbf{S}w ,$$
$$\dot{x} = \mathbf{P}w + \mathbf{A}x + \mathbf{B}u ,$$
$$e = \mathbf{Q}w + \mathbf{C}x , \qquad (9.66)$$

and suppose that $y = e$, i.e., that regulated and measured variables coincide. We also consider, for simplicity, the case in which $m = 1$ and $p = 1$. Without loss of generality, it is assumed that all eigenvalues of $\mathbf{S}$ are simple and are on the imaginary axis.

A convenient point of departure for the analysis is the identification of conditions for the existence of a solution of the design problem. To this end, consider a dynamic, output-feedback controller

$$\dot{\xi} = \mathbf{F}\xi + \mathbf{G}e ,$$
$$u = \mathbf{H}\xi \qquad (9.67)$$

and the associated closed-loop system

$$\dot{w} = \mathbf{S}w ,$$
$$\begin{pmatrix} \dot{x} \\ \dot{\xi} \end{pmatrix} = \begin{pmatrix} \mathbf{P} \\ \mathbf{GQ} \end{pmatrix} w + \begin{pmatrix} \mathbf{A} & \mathbf{BH} \\ \mathbf{GC} & \mathbf{F} \end{pmatrix} \begin{pmatrix} x \\ \xi \end{pmatrix} . \qquad (9.68)$$

If the controller solves the problem at issue, all trajectories are bounded and $e(t)$ asymptotically decays to zero. Boundedness of all trajectories implies that all eigenvalues of

$$\begin{pmatrix} \mathbf{A} & \mathbf{BH} \\ \mathbf{GC} & \mathbf{F} \end{pmatrix} \qquad (9.69)$$

have nonpositive real part. However, if some of the eigenvalues of this matrix were on the imaginary axis, the property of boundedness of trajectories could be lost as a result of infinitesimal variations in the parameters of (9.66) and/or (9.67). Thus, only the case in which the eigenvalues of (9.69) have negative real part is of interest. If the controller is such that this happens, then *necessarily* the pair of matrices $(\mathbf{A}, \mathbf{B})$ is stabilizable and the pair of matrices $(\mathbf{A}, \mathbf{C})$ is detectable. Observe now that, if the matrix (9.69) has all eigenvalues with negative real part, system (9.68) has a well-defined steady state, which takes place on an invariant subspace (the steady-state locus). The latter, as shown earlier, is

necessarily the graph of a linear map, which expresses the x and ξ components of the state vector as functions of the w component. In other terms, the steady-state locus is the set of all triplets (w, x, ξ) in which w is arbitrary, while x and ξ are expressed as

$$x = \Pi w ,$$
$$\xi = \Sigma w ,$$

for some Π and Σ. These matrices, in turn, are solutions of the Sylvester equation

$$\begin{pmatrix} \Pi \\ \Sigma \end{pmatrix} \mathbf{S} = \begin{pmatrix} \mathbf{P} \\ \mathbf{GQ} \end{pmatrix} + \begin{pmatrix} \mathbf{A} & \mathbf{BH} \\ \mathbf{GC} & \mathbf{F} \end{pmatrix} \begin{pmatrix} \Pi \\ \Sigma \end{pmatrix} . \qquad (9.70)$$

All trajectories asymptotically converge to the steady state. Thus, in view of the expression thus found for the steady-state locus, it follows that

$$\lim_{t \to \infty} [x(t) - \Pi w(t)] = 0 ,$$
$$\lim_{t \to \infty} [\xi(t) - \Sigma w(t)] = 0 .$$

In particular, it is seen from this that

$$\lim_{t \to \infty} e(t) = \lim_{t \to \infty} [\mathbf{C}\Pi + \mathbf{Q}]w(t) .$$

Since $w(t)$ is a persistent function (none of the eigenvalues of $\mathbf{S}$ has a negative real part), it is concluded that the regulated variable $e(t)$ converges to 0 as $t \to \infty$ only if the map $e = \mathbf{C}x + \mathbf{Q}w$ is zero on the steady-state locus, i.e., if

$$0 = \mathbf{C}\Pi + \mathbf{Q} . \qquad (9.71)$$

Note that the Sylvester equation (9.70) can be split into two equations, the former of which

$$\Pi\mathbf{S} = \mathbf{P} + \mathbf{A}\Pi + \mathbf{BH}\Sigma ,$$

having set $\Gamma := \mathbf{H}\Sigma$, can be rewritten as

$$\Pi\mathbf{S} = \mathbf{A}\Pi + \mathbf{B}\Gamma + \mathbf{P} ,$$

while the second one, bearing in mind the constraint (9.71), reduces to

$$\Sigma\mathbf{S} = \mathbf{F}\Sigma .$$

These arguments have proven – *in particular* – that, if there exists a controller that controller solves the problem, necessarily there exists a pair of matrices Π, Γ such that

$$\Pi\mathbf{S} = \mathbf{A}\Pi + \mathbf{B}\Gamma + \mathbf{P}$$
$$0 = \mathbf{C}\Pi + \mathbf{Q} . \qquad (9.72)$$

The (linear) equations thus found are known as the *regulator equations* [9.14]. If, as observed above, the controller is required to solve the problem is spite of arbitrary (small) variations of the parameters of (9.66), the existence of solutions (9.72) is required to hold independently of the specific values of $\mathbf{P}$ and $\mathbf{Q}$. This occurs if and only if none of the eigenvalues of $\mathbf{S}$ is a root of

$$\det \begin{pmatrix} \mathbf{A} - \lambda\mathbf{I} & \mathbf{B} \\ \mathbf{C} & \mathbf{0} \end{pmatrix} = 0 . \tag{9.73}$$

This condition is usually referred to as the *nonresonance* condition.

In summary, it has been shown that, if there exists a controller that solves the servomechanism problem, necessarily the controlled plant (with $\boldsymbol{w} = 0$) is *stabilizable and detectable* and *none of the eigenvalues of* $\mathbf{S}$ *is a root of* (9.73). These *necessary* conditions turn out to be also *sufficient* for the existence of a controller that solves the servomechanism problem.

A procedure for the design of a controller is described below. Let

$$\psi(\lambda) = \lambda^s + d_{s-1}\lambda^{s-1} + \cdots + d_1\lambda + d_0$$

denote the minimal polynomial of $\mathbf{S}$. Set

$$\Phi = \begin{pmatrix} 0 & 1 & 0 & \cdots & 0 & 0 \\ 0 & 0 & 1 & \cdots & 0 & 0 \\ \cdot & \cdot & \cdot & \cdots & \cdot & \cdot \\ 0 & 0 & 0 & \cdots & 0 & 1 \\ -d_0 & -d_1 & -d_2 & \cdots & -d_{s-2} & -d_{s-1} \end{pmatrix} ,$$

$$G = \begin{pmatrix} 0 \\ 0 \\ \cdots \\ 0 \\ 1 \end{pmatrix} ,$$

$$H = \begin{pmatrix} 1 & 0 & 0 & \cdots & 0 & 0 \end{pmatrix} .$$

Let Π, Γ be a solution pair of (9.72) and note that the matrix

$$\Upsilon = \begin{pmatrix} \Gamma \\ \Gamma\mathbf{S} \\ \cdots \\ \Gamma\mathbf{S}^{s-1} \end{pmatrix}$$

satisfies

$$\Upsilon\mathbf{S} = \Phi\Upsilon , \quad \Gamma = H\Upsilon . \tag{9.74}$$

Define a controller as follows:

$$\dot{\xi} = \Phi\xi + \mathbf{G}e ,$$
$$\dot{\eta} = \mathbf{K}\eta + \mathbf{L}e ,$$
$$u = \mathbf{H}\xi + \mathbf{M}\eta , \tag{9.75}$$

in which the matrices $\Phi, \mathbf{G}, \mathbf{H}$ are those defined before and $\mathbf{K}, \mathbf{L}, \mathbf{M}$ are matrices to be determined. Consider now the associated closed-loop system, which can be written in the form

$$\dot{w} = \mathbf{S}w ,$$

$$\begin{pmatrix} \dot{x} \\ \dot{\xi} \\ \dot{\eta} \end{pmatrix} = \begin{pmatrix} \mathbf{P} \\ \mathbf{GQ} \\ \mathbf{LQ} \end{pmatrix} w + \begin{pmatrix} \mathbf{A} & \mathbf{BH} & \mathbf{BM} \\ \mathbf{GC} & \Phi & 0 \\ \mathbf{LC} & 0 & \mathbf{K} \end{pmatrix} \begin{pmatrix} x \\ \xi \\ \eta \end{pmatrix} . \tag{9.76}$$

By assumption, the pair of matrices $(\mathbf{A}, \mathbf{B})$ is stabilizable, the pair of matrices $(\mathbf{A}, \mathbf{C})$ is detectable, and none of the eigenvalues of $\mathbf{S}$ is a root of (9.73). As a consequence, in view of the special structure of $\Phi, \mathbf{G}, \mathbf{H}$, also the pair

$$\begin{pmatrix} \mathbf{A} & \mathbf{BH} \\ \mathbf{GC} & \Phi \end{pmatrix}, \begin{pmatrix} \mathbf{B} \\ 0 \end{pmatrix}$$

is stabilizable and the pair

$$\begin{pmatrix} \mathbf{A} & \mathbf{BH} \\ \mathbf{GC} & \Phi \end{pmatrix}, \begin{pmatrix} \mathbf{C} & 0 \end{pmatrix}$$

is detectable. This being the case, it is possible to pick $\mathbf{K}, \mathbf{L}, \mathbf{M}$ in such a way that all eigenvalues of

$$\begin{pmatrix} \mathbf{A} & \mathbf{BH} & \mathbf{BM} \\ \mathbf{GC} & \Phi & 0 \\ \mathbf{LC} & 0 & \mathbf{K} \end{pmatrix}$$

have negative real part.

As a result, all trajectories of (9.76) are bounded. Using (9.72) and (9.74) it is easy to check that the graph of the mapping

$$\pi : \boldsymbol{w} \to \begin{pmatrix} \Pi \\ \Upsilon \\ 0 \end{pmatrix} \boldsymbol{w}$$

is invariant for (9.76). This subspace is actually the steady-state locus of (9.76) and $\boldsymbol{e} = \mathbf{C}x + \mathbf{Q}w$ is zero on this subspace. Hence all trajectories of (9.76) are such that $\boldsymbol{e}(t)$ converges to 0 as $t \to \infty$.

The construction described above is insensitive to small arbitrary variations of the parameters, except for the case of parameter variations in the exosystem. The case of parameter variations in the exosystem requires a different design, as explained e.g., in [9.15]. A state-of-the-art discussion of the servomechanism problem for suitable classes of nonlinear systems can be found in [9.16].

9.8 Conclusion

This chapter has reviewed the fundamental methods and models of control theory as applied to automation. The following two chapters address further advancements in this area of automation theory.

References

9.1   G.D. Birkhoff: *Dynamical Systems* (Am. Math. Soc., Providence 1927)

9.2   J.K. Hale, L.T. Magalhães, W.M. Oliva: *Dynamics in Infinite Dimensions* (Springer, New York 2002)

9.3   A. Isidori, C.I. Byrnes: Steady-state behaviors in nonlinear systems with an application to robust disturbance rejection, Annu. Rev. Control **32**, 1–16 (2008)

9.4   E.D. Sontag: On the input-to-state stability property, Eur. J. Control **1**, 24–36 (1995)

9.5   E.D. Sontag, Y. Wang: On characterizations of the input-to-state stability property, Syst. Control Lett. **24**, 351–359 (1995)

9.6   A. Isidori: *Nonlinear Control Systems II* (Springer, London 1999)

9.7   A.R. Teel: A nonlinear small gain theorem for the analysis of control systems with saturations, IEEE Trans. Autom. Control **AC-41**, 1256–1270 (1996)

9.8   Z.P. Jiang, A.R. Teel, L. Praly: Small-gain theorem for ISS systems and applications, Math. Control Signal Syst. **7**, 95–120 (1994)

9.9   W. Hahn: *Stability of Motions* (Springer, Berlin, Heidelberg 1967)

9.10   A. Isidori: *Nonlinear Control Systems*, 3rd edn. (Springer, London 1995)

9.11   H.K. Khalil, F. Esfandiari: Semiglobal stabilization of a class of nonlinear systems using output feedback, IEEE Trans. Autom. Control **AC-38**, 1412–1415 (1993)

9.12   A.R. Teel, L. Praly: Tools for semiglobal stabilization by partial state and output feedback, SIAM J. Control Optim. **33**, 1443–1485 (1995)

9.13   J.P. Gauthier, I. Kupka: *Deterministic Observation Theory and Applications* (Cambridge Univ. Press, Cambridge 2001)

9.14   B.A. Francis, W.M. Wonham: The internal model principle of control theory, Automatica **12**, 457–465 (1976)

9.15   A. Serrani, A. Isidori, L. Marconi: Semiglobal nonlinear output regulation with adaptive internal model, IEEE Trans. Autom. Control **AC-46**, 1178–1194 (2001)

9.16   L. Marconi, L. Praly, A. Isidori: Output stabilization via nonlinear luenberger observers, SIAM J. Control Optim. **45**, 2277–2298 (2006)

10. Control Theory for Automation – Advanced Techniques

István Vajk, Jenő Hetthéssy, Ruth Bars

Analysis and design of control systems is a complex field. In order to develop appropriate concepts and methods to cover this field, mathematical models of the processes to be controlled are needed to apply. In this chapter mainly continuous-time linear systems with multiple input and multiple output (MIMO systems) are considered. Specifically, stability, performance, and robustness issues, as well as optimal control strategies are discussed in detail for MIMO linear systems. As far as system representations are concerned, transfer function matrices, matrix fraction descriptions, and state-space models are applied in the discussions. Several interpretations of all stabilizing controllers are shown for stable and unstable processes. Performance evaluation is supported by applying H_2 and H_∞ norms. As an important class for practical applications, predictive controllers are also discussed. In this case, according to the underlying implementation technique, discrete-time process models are considered. Transformation methods using state variable feedback are discussed, making the operation of nonlinear dynamic systems linear in the complete range of their operation. Finally, the sliding control concept is outlined.

10.1 MIMO Feedback Systems

This chapter on advanced automatic control for automation follows the previous introductory chapter. In this section continuous-time linear systems with multiple input and multiple output (MIMO systems) will be considered. As far as the mathematical models are concerned, transfer functions, matrix fraction descriptions, and state-space models will be used [10.1]. Regarding the notations concerned, transfer functions will always be denoted by explicitly showing the dependence of the complex frequency operator s, while variables in

Fig. 10.1 Distillation column in an oil-refinery plant

bold face represent vectors or matrices. Thus, $A(s)$ is a scalar transfer function, $\boldsymbol{A}(s)$ is a transfer function matrix, while $\mathbf{A}$ is a matrix.

Considering the structure of the systems to be discussed, feedback control systems will be studied. Feedback is the most inherent step to create practical

Fig. 10.2 Automated production line

Fig. 10.3 Rolling mill

control systems, as it allows one to change the dynamical and steady-state behavior of various processes to be controlled to match technical expectations [10.2–7]. In this chapter mainly continuous-time systems such as those in Figs. 10.1–10.3 will be discussed [10.8]. Note that special care should be taken to derive their appropriate discrete-time counterparts [10.9–12]. The well-known advantages of feedback structures, also called closed-loop systems, range from the *servo property* (i. e., to force the process output to follow a prescribed command signal) to effective *disturbance rejection* through *robustness* (the ability to achieve the control goals in spite of incomplete knowledge available on the process) and measurement *noise attenuation*. When designing a control system, however, *stability* should always remain the most important task. Figure 10.4 shows the block diagram of a conventional closed-loop system with negative feedback, where r is the set point, y is the controlled variable (output), u is the process input, d_i and d_o are the input and output disturbances acting on the process, respectively, while d_n represents the additive measurement noise [10.2].

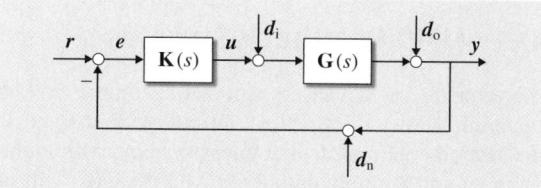

Fig. 10.4 Multivariable feedback system

In the figure $\mathbf{G}(s)$ denotes the transfer function matrix of the process and $\mathbf{K}(s)$ stands for the controller. For the designer of the closed-loop system, $\mathbf{G}(s)$ is given, while $\mathbf{K}(s)$ is the result of the design procedure. Note that $\mathbf{G}(s)$ is only a model of the process and serves here as the basis to design an appropriate $\mathbf{K}(s)$. In practice the signals driving the control system are delivered by a given process or technology and the control input is in fact applied to the given process.

The main design objectives are [10.2, 6, 13]:

- Closed-loop and internal stability (just as it will be addressed in this section)
- Good command following (servo property)
- Good disturbance rejection
- Good measurement noise attenuation.

In addition, to keep operational costs low, small process input values are preferred over large excursions in the control signal. Also, as the controller design is based on a model of the process, which always implies uncertainties, design procedures aiming at stability and desirable performance based on the nominal plant model should be extended to tolerate modeling uncertainties as well. Thus the list of the design objectives is to be completed by:

- Achieve reduced input signals
- Achieve robust stability
- Achieve robust performance.

Some of the above design objectives could be conflicting; however, the performance-related issues typically emerge in separable frequency ranges.

In this section linear multivariable feedback systems will be discussed with the following representations.

10.1.1 Transfer Function Models

Consider a linear process with n_u control inputs arranged into a $\mathbf{u} \in \mathbb{R}^{n_u}$ input vector and n_y outputs arranged into a $\mathbf{y} \in \mathbb{R}^{n_y}$ output vector. Then the transfer function matrix contains all possible transfer functions between any of the inputs and any of the outputs

$$\mathbf{y}(s) = \begin{pmatrix} y_1(s) \\ \vdots \\ y_{n_y-1}(s) \\ y_{n_y}(s) \end{pmatrix} = \mathbf{G}(s)\mathbf{u}(s)$$

$$= \begin{pmatrix} G_{1,1}(s) & G_{1,2}(s) & \cdots & G_{1,n_u}(s) \\ \vdots & \vdots & \ddots & \vdots \\ G_{n_y-1,1}(s) & G_{n_y-1,2}(s) & \cdots & G_{n_y-1,n_u}(s) \\ G_{n_y,1}(s) & G_{n_y,2}(s) & \cdots & G_{n_y,n_u}(s) \end{pmatrix}$$
$$\begin{pmatrix} u_1(s) \\ \vdots \\ u_{n_u-1}(s) \\ u_{n_u}(s) \end{pmatrix},$$

where s is the Laplace operator and $G_{k,l}(s)$ denotes the transfer function from the l-th component of the input $\mathbf{u}$ to the k-th component of the output $\mathbf{y}$. The transfer function approach has always been an emphasized modeling tool for control practice. One of the reasons is that the $G_{k,l}(s)$ transfer functions deliver the magnitude and phase frequency functions via a formal substitution of $G_{k,l}(s)\big|_{s=i\omega} = A_{k,l}(\omega)\mathrm{e}^{i\phi_{k,l}(\omega)}$. Note that for real physical processes $\lim_{\omega\to\infty} A_{k,l}(\omega) = 0$. The transfer function matrix $\mathbf{G}(s)$ is stable if each of its elements is a stable transfer function. Also, the transfer function matrix $\mathbf{G}(s)$ will be called proper if each of its elements is a proper transfer function.

10.1.2 State–Space Models

Introducing n_x state variables arranged into an $\mathbf{x} \in \mathbb{R}^{n_x}$ state vector, the state-space model of a MIMO system is given by the following equations

$$\dot{\mathbf{x}}(t) = \mathbf{A}\mathbf{x}(t) + \mathbf{B}\mathbf{u}(t),$$
$$\mathbf{y}(t) = \mathbf{C}\mathbf{x}(t) + \mathbf{D}\mathbf{u}(t),$$

where $\mathbf{A} \in \mathbb{R}^{n_x \times n_x}$, $\mathbf{B} \in \mathbb{R}^{n_x \times n_u}$, $\mathbf{C} \in \mathbb{R}^{n_y \times n_x}$, and $\mathbf{D} \in \mathbb{R}^{n_y \times n_u}$ are the system parameters [10.14, 15].

Important notions (state variable feedback, controllability, stabilizability, observability and detectability) have been introduced to support the deep analysis of state-space models [10.1, 2]. Roughly speaking a state-space representation is *controllable* if an arbitrary initial state can be moved to any desired state by suitable choice of control signals. In terms of state-space realizations, feedback means *state variable feedback* realized by a control law of $\mathbf{u} = -\mathbf{K}\mathbf{x}$, $\mathbf{K} \in \mathbb{R}^{n_u \times n_x}$. Regarding controllable systems, state variable feedback can relocate all the poles of the closed-loop system to arbitrary locations. If a system is not controllable, but the modes (eigenvalues) attached to the uncontrollable states are stable, the complete system is still *stabilizable*. A state-space realization is said to be *observable* if the initial state $x(0)$ can be determined from the output function $y(t)$, $0 \le t \le t_{final}$. A system is said to be *detectable* if the modes (eigenvalues) attached to the unobservable states are stable.

Using the Laplace transforms in the state-space model equations the relation between the state-space model and the transfer function matrix can easily be derived as

$$G(s) = C(sI - A)^{-1}B + D \, .$$

As far as the above relation is concerned the condition $\lim_{\omega \to \infty} A_{k,l}(\omega) = 0$ raised for real physical processes leads to $D = 0$. Note that the $G(s)$ transfer function contains only the controllable *and* observable subsystem represented by the state-space model $\{A, B, C, D\}$.

10.1.3 Matrix Fraction Description

Transfer functions can be factorized in several ways. As matrices, in general, do not commute, the matrix fraction description (MFD) form exists as a result of right and left factorization, respectively [10.2, 6, 13]

$$G(s) = B_R(s)A_R^{-1}(s) = A_L^{-1}(s)B_L(s) \, ,$$

where $A_R(s)$, $B_R(s)$, $A_L(s)$, and $B_L(s)$ are all stable transfer function matrices. In [10.2] it is shown that the right and left MFDs can be related to stabilizable and detectable state-space models, respectively. To outline the procedure consider first the right matrix fraction description (RMFD) $G(s) = B_R(s)A_R^{-1}(s)$. For the sake of simplicity the practical case of $D = 0$ will be considered. Assuming that $\{A, B\}$ is stabilizable, apply a state feedback to stabilize the closed-loop system using a gain matrix $K \in \mathbb{R}^{n_u \times n_x}$

$$u(t) = -Kx(t) \, ,$$

then the RMFD components can be derived in a straightforward way as

$$B_R(s) = C(sI - A + BK)^{-1}B \, ,$$
$$A_R(s) = I - K(sI - A + BK)^{-1}B \, .$$

It can be shown that $G(s) = B_R(s)A_R^{-1}(s)$ will not be a function of the stabilizing gain matrix K, however, the proof is rather involved [10.2]. Also, following the above procedure, both $B_R(s)$ and $A_R(s)$ will be stable transfer function matrices.

In a similar way, assuming that $\{A, C\}$ is detectable, apply a state observer to detect the closed-loop system using a gain matrix L. Then the left matrix fraction description (LMFD) components can be obtained as

$$B_L(s) = C(sI - A + LC)^{-1}B \, ,$$
$$A_L(s) = I - C(sI - A + LC)^{-1}L \, ,$$

being stable transfer function matrices. Again, $G(s) = A_L^{-1}(s)B_L(s)$ will be independent of L.

Concerning the coprime factorization, an important relation, the *Bezout identity*, will be used, which holds for the components of the RMFD and LMFD coprime factorization

$$\begin{pmatrix} X_L(s) & Y_L(s) \\ -B_L(s) & A_L(s) \end{pmatrix} \begin{pmatrix} A_R(s) & -Y_R(s) \\ B_R(s) & X_R(s) \end{pmatrix}$$
$$= \begin{pmatrix} A_R(s) & -Y_R(s) \\ B_R(s) & X_R(s) \end{pmatrix} \begin{pmatrix} X_L(s) & Y_L(s) \\ -B_L(s) & A_L(s) \end{pmatrix} = I \, ,$$

where

$$Y_R(s) = K(sI - A + BK)^{-1}L \, ,$$
$$X_R(s) = I + C(sI - A + BK)^{-1}L \, ,$$
$$Y_L(s) = K(sI - A + LC)^{-1}L \, ,$$
$$X_L(s) = I + K(sI - A + LC)^{-1}B \, .$$

Note that the Bezout identity plays an important role in control design. A good review on this can be found in [10.1]. Also note that a MFD factorization can be accomplished by using the Smith–McMillan form of $G(s)$ [10.1]. As a result of this procedure, however, $A_R(s)$, $B_R(s)$, $A_L(s)$, and $B_L(s)$ will be polynomial matrices. Moreover, both $A_R(s)$ and $A_L(s)$ will be diagonal matrices.

10.2 All Stabilizing Controllers

In general, a feedback control system follows the structure shown in Fig. 10.5, where the control configuration consists of two subsystems. In this general setup any of the subsystems $S_1(s)$ or $S_2(s)$ may play the role of the process or the controller [10.3]. Here $\{u_1, u_2\}$ and $\{y_1, y_2\}$ are multivariable external input and output sig-

nals in general sense, respectively. Moreover, $S_1(s)$ and $S_2(s)$ represent transfer function matrices according to

$$y_1(s) = S_1(s)[u_1(s) + y_2(s)] \, ,$$
$$y_2(s) = S_2(s)[u_2(s) + y_1(s)] \, .$$

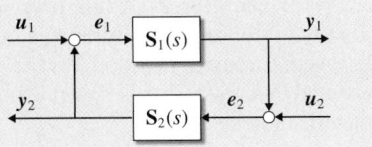

Fig. 10.5 A general feedback configuration

Being restricted to linear systems the closed-loop system is *internally stable* if and only if *all* the four entries of the transfer function matrix

$$\begin{pmatrix} \mathbf{H}_{11}(s) & \mathbf{H}_{12}(s) \\ \mathbf{H}_{21}(s) & \mathbf{H}_{22}(s) \end{pmatrix}$$

are asymptotically stable, where

$$\begin{aligned}
\begin{pmatrix} e_1(s) \\ e_2(s) \end{pmatrix} &= \begin{pmatrix} u_1(s) + y_2(s) \\ u_2(s) + y_1(s) \end{pmatrix} \\
&= \begin{pmatrix} \mathbf{H}_{11}(s) & \mathbf{H}_{12}(s) \\ \mathbf{H}_{21}(s) & \mathbf{H}_{22}(s) \end{pmatrix} \begin{pmatrix} u_1(s) \\ u_2(s) \end{pmatrix} \\
&= \begin{pmatrix} [\mathbf{I} - \mathbf{S}_2(s)\mathbf{S}_1(s)]^{-1} & [\mathbf{I} - \mathbf{S}_2(s)\mathbf{S}_1(s)]^{-1}\mathbf{S}_2(s) \\ [\mathbf{I} - \mathbf{S}_1(s)\mathbf{S}_2(s)]^{-1}\mathbf{S}_1(s) & [\mathbf{I} - \mathbf{S}_1(s)\mathbf{S}_2(s)]^{-1} \end{pmatrix} \\
&\quad \times \begin{pmatrix} u_1(s) \\ u_2(s) \end{pmatrix} .
\end{aligned}$$

Also, from

$$e_1(s) = u_1(s) + y_2(s) = u_1(s) + \mathbf{S}_2(s)e_2(s) ,$$
$$e_2(s) = u_2(s) + y_1(s) = u_2(s) + \mathbf{S}_1(s)e_1(s) ,$$

we have

$$\begin{aligned}
\begin{pmatrix} e_1 \\ e_2 \end{pmatrix} &= \begin{pmatrix} \mathbf{H}_{11}(s) & \mathbf{H}_{12}(s) \\ \mathbf{H}_{21}(s) & \mathbf{H}_{22}(s) \end{pmatrix} \begin{pmatrix} u_1(s) \\ u_2(s) \end{pmatrix} \\
&= \begin{pmatrix} \mathbf{I} & -\mathbf{S}_2(s) \\ -\mathbf{S}_1(s) & \mathbf{I} \end{pmatrix}^{-1} \begin{pmatrix} u_1 \\ u_2 \end{pmatrix} ,
\end{aligned}$$

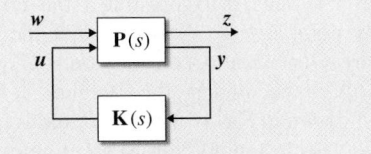

Fig. 10.6 General control system configuration

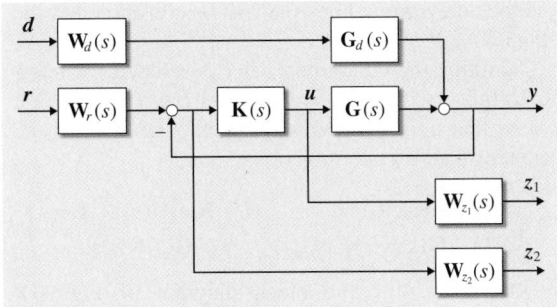

Fig. 10.7 A sample closed-loop control system

so for internal stability we need the transfer function matrix

$$\begin{aligned}
&\begin{pmatrix} [\mathbf{I} - \mathbf{S}_2(s)\mathbf{S}_1(s)]^{-1} & [\mathbf{I} - \mathbf{S}_2(s)\mathbf{S}_1(s)]^{-1}\mathbf{S}_2(s) \\ [\mathbf{I} - \mathbf{S}_1(s)\mathbf{S}_2(s)]^{-1}\mathbf{S}_1(s) & [\mathbf{I} - \mathbf{S}_1(s)\mathbf{S}_2(s)]^{-1} \end{pmatrix} \\
&= \begin{pmatrix} \mathbf{I} & -\mathbf{S}_2(s) \\ -\mathbf{S}_1(s) & \mathbf{I} \end{pmatrix}^{-1}
\end{aligned}$$

to be asymptotically stable [10.13].

In the control system literature a more practical, but still general closed-loop control scheme is considered, as shown in Fig. 10.6 with a generalized plant $\mathbf{P}(s)$ and controller $\mathbf{K}(s)$ [10.6, 13, 14]. In this configuration u and y represent the process input and output, respectively, w denotes external inputs (command signal, disturbance or noise), z is a set of signals representing the closed-loop performance in general sense. The controller $\mathbf{K}(s)$ is to be adjusted to ensure a stable closed-loop system with appropriate performance.

As an example Fig. 10.7 shows a possible control loop for tracking and disturbance rejection. Once the disturbance signal d and the command signal (or set point) signal r are combined to a vector-valued signal w, the block diagram can easily be redrawn to match the general scheme in Fig. 10.6. Note the $\mathbf{W}_d(s)$, $\mathbf{W}_r(s)$, and $\mathbf{W}_z(s)$ filters introduced just to shape the system performance. Since any closed-loop system can be redrawn to the general configuration shown in Fig. 10.5,

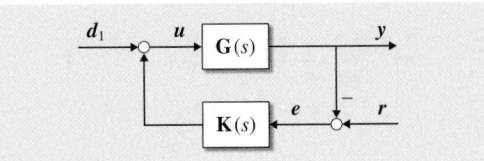

Fig. 10.8 Control system configuration including set point and input disturbance

Part B | 10.2

the block diagram in Fig. 10.8 will be considered in the sequel.

Adopting the condition earlier developed for internal stability with $\mathbf{S}_1(s) = \mathbf{G}(s)$ and $\mathbf{S}_2(s) = -\mathbf{K}(s)$ it is seen that now we need asymptotic stability for the following four transfer functions

$$\begin{pmatrix} [\mathbf{I}+\mathbf{K}(s)\mathbf{G}(s)]^{-1} & [\mathbf{I}+\mathbf{K}(s)\mathbf{G}(s)]^{-1}\mathbf{K}(s) \\ -[\mathbf{I}+\mathbf{G}(s)\mathbf{K}(s)]^{-1}\mathbf{G}(s) & [\mathbf{I}+\mathbf{G}(s)\mathbf{K}(s)]^{-1} \end{pmatrix} .$$

At the same time the block diagram of Fig. 10.8 suggests

$$e(s) = r(s) - y(s) = r(s) - \mathbf{G}(s)\mathbf{K}(s)e(s)$$
$$\Rightarrow e(s) = [\mathbf{I}+\mathbf{G}(s)\mathbf{K}(s)]^{-1}r(s) ,$$

which leads to

$$u(s) = \mathbf{K}(s)e(s) = \mathbf{K}(s)[\mathbf{I}+\mathbf{G}(s)\mathbf{K}(s)]^{-1}r(s)$$
$$= \mathbf{Q}(s)r(s) ,$$

where

$$\mathbf{Q}(s) = \mathbf{K}(s)[\mathbf{I}+\mathbf{G}(s)\mathbf{K}(s)]^{-1} .$$

It can easily be shown that, *in the case of a stable* $\mathbf{G}(s)$ *plant*, any stable $\mathbf{Q}(s)$ transfer function, in other words *Q parameter*, results in internal stability. Rearranging the above equation $\mathbf{K}(s)$ parameterized by $\mathbf{Q}(s)$ exhibits *all stabilizing controllers*

$$\mathbf{K}(s) = [\mathbf{I}-\mathbf{Q}(s)\mathbf{G}(s)]^{-1}\mathbf{Q}(s) .$$

This result is known as the *Youla parameterization* [10.13, 16]. Recalling $u(s) = \mathbf{Q}(s)r(s)$ and $y(s) = \mathbf{G}(s)u(s) = \mathbf{G}(s)\mathbf{Q}(s)r(s)$ allows one to draw the block diagram of the closed-loop system explicitly using $\mathbf{Q}(s)$ (Fig. 10.9). The control scheme shown in Fig. 10.9 satisfies $u(s) = \mathbf{Q}(s)r(s)$ and $y(s) = \mathbf{G}(s)\mathbf{Q}(s)r(s)$, moreover the process modeling uncertainties ($\mathbf{G}(s)$ of the physical process and $\mathbf{G}(s)$ of the model, as part of the controller are different) are also taken into account. This is the well-known internal model controller (IMC) scheme [10.17, 18].

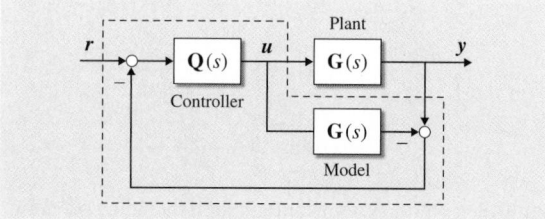

Fig. 10.9 Internal model controller

A quick evaluation for the Youla parameterization should point out a fundamental difference between designing an overall transfer function $\mathbf{T}(s)$ from the $r(s)$ reference signal to the $y(s)$ output signal using a nonlinear parameterization by $\mathbf{K}(s)$

$$y(s) = \mathbf{T}(s)r(s) = \mathbf{G}(s)\mathbf{K}(s)[\mathbf{I}+\mathbf{G}(s)\mathbf{K}(s)]^{-1}r(s)$$

versus a design by

$$y(s) = \mathbf{T}(s)r(s) = \mathbf{G}(s)\mathbf{Q}(s)r(s)$$

linear in $\mathbf{Q}(s)$.

Further analysis of the relation by $y(s) = \mathbf{T}(s)r(s) = \mathbf{G}(s)\mathbf{Q}(s)r(s)$ indicates that $\mathbf{Q}(s) = \mathbf{G}^{-1}(s)$ is a reasonable choice to achieve an ideal servo controller to ensure $y(s) = r(s)$. However, to intend to set $\mathbf{Q}(s) = \mathbf{G}^{-1}(s)$ is not a practical goal for several reasons [10.2]:

- Non-minimum-phase processes would exhibit unstable $\mathbf{Q}(s)$ controllers and closed-loop systems
- Problems concerning the realization of $\mathbf{G}^{-1}(s)$ are immediately seen regarding processes with positive relative degree or time delay
- The ideal servo property would destroy the disturbance rejection capability of the closed-loop system
- $\mathbf{Q}(s) = \mathbf{G}^{-1}(s)$ would lead to large control effort
- Effects of errors in modeling the real process by $\mathbf{G}(s)$ need further analysis.

Replacing the exact inverse $\mathbf{G}^{-1}(s)$ by an approximated inverse is in harmony with practical demands.

To discuss the concept of handling processes with time delay consider single-input single-output (SISO) systems and assume that

$$G(s) = G_p(s)e^{-sT_d} ,$$

where $G_p(s) = B(s)/A(s)$ is a proper transfer function with no time delay and $T_d > 0$ is the time delay. Recognizing that the time delay characteristics is not invertible

$$y(s) = T_p(s)e^{-sT_d}r(s) = G_p(s)Q(s)e^{-sT_d}r(s)$$

can be assigned as the overall transfer function to be achieved. Updating Fig. 10.9 for $G(s) = B(s)/A(s)e^{-sT_d}$, Fig. 10.10 illustrates the control scheme. A key point is here, however, that the parameterization by $Q(s)$ should consider only $G_p(s)$ to achieve $G_p(s)Q(s)$ specified by the designer. Note that, in the model shown in Fig. 10.10, uncertainties in $G_p(s)$ and in the time delay should both be taken into account when studying the closed-loop system.

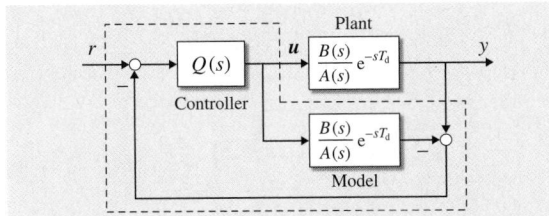

Fig. 10.10 IMC control of a plant with time delay

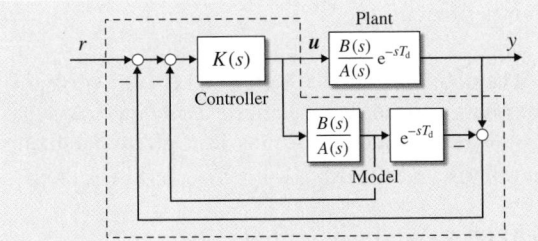

Fig. 10.11 Controller using Smith predictor

The control scheme in Fig. 10.10 has been immediately derived by applying the Youla parameterization concept for processes with time delay. The idea, however, of letting the time delay appear in the overall transfer function and restricting the design procedure to a process with no time delay is more than 50 years old and comes from *Smith* [10.19].

The fundamental concept of the design procedure called the *Smith predictor* is to set up a closed-loop system to control the output signal predicted ahead by the time delay. Then, to meet the causality requirement, the predicted output is delayed to derive the real system output. All these conceptional steps can be summarized in a control scheme; just redraw Fig. 10.10 to Fig. 10.11

with

$$Q(s) = K(s)[I + G_p(s)K(s)]^{-1} .$$

The fact that the output of the internal loop can be considered as the predicted value of the process output explains the name of the controller. Note that the Smith predictor is applicable for unstable processes as well.

In the case of *unstable plants*, stabilization of the closed-loop system needs a more involved discussion. In order to separate the unstable (in a more general sense, the undesired) poles both the plant and the controller transfer function matrices will be factorized to (right or left) coprime transfer functions

$$\mathbf{G}(s) = \mathbf{B}_R(s)\mathbf{A}_R^{-1}(s) = \mathbf{A}_L^{-1}(s)\mathbf{B}_L(s) ,$$

$$\mathbf{K}(s) = \mathbf{Y}_R(s)\mathbf{X}_R^{-1}(s) = \mathbf{X}_L^{-1}(s)\mathbf{Y}_L(s) ,$$

where $\mathbf{B}_R(s)$, $\mathbf{A}_R(s)$, $\mathbf{B}_L(s)$, $\mathbf{A}_L(s)$, $\mathbf{Y}_R(s)$, $\mathbf{X}_R(s)$, $\mathbf{Y}_L(s)$, and $\mathbf{X}_L(s)$ are all stable coprime transfer functions. Stability implies that $\mathbf{B}_R(s)$ should contain all the right half plane (RHP)-zeros of $\mathbf{G}(s)$, and $\mathbf{A}_R(s)$ should contain as RHP-zeros all the RHP-poles of $\mathbf{G}(s)$. Similar statements are valid for the left coprime pairs. As far as the internal stability analysis is concerned, assuming that $\mathbf{G}(s)$ is strictly proper and $\mathbf{K}(s)$ is proper, the coprime factorization offers the stability analysis via checking the stability of

$$\begin{pmatrix} \mathbf{A}_R(s) & -\mathbf{Y}_R(s) \\ \mathbf{B}_R(s) & \mathbf{X}_R(s) \end{pmatrix}^{-1} \quad \text{and} \quad \begin{pmatrix} \mathbf{X}_L(s) & \mathbf{Y}_L(s) \\ -\mathbf{B}_L(s) & \mathbf{A}_L(s) \end{pmatrix}^{-1},$$

respectively. According to the Bezout identity [10.6, 13, 20] there exist $\mathbf{X}_L(s)$ and $\mathbf{Y}_L(s)$ as stable transfer function matrices satisfying

$$\mathbf{X}_L(s)\mathbf{A}_R(s) + \mathbf{Y}_L(s)\mathbf{B}_R(s) = \mathbf{I} .$$

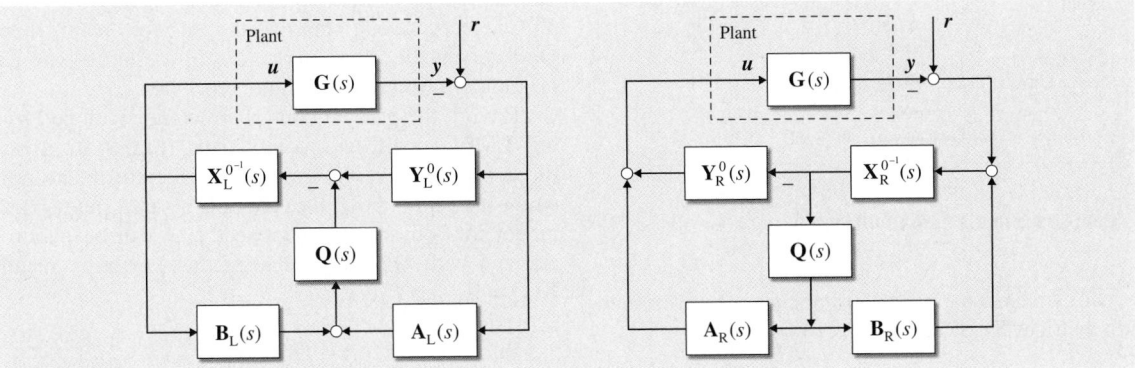

Fig. 10.12 Two different realizations of all stabilizing controllers for unstable processes

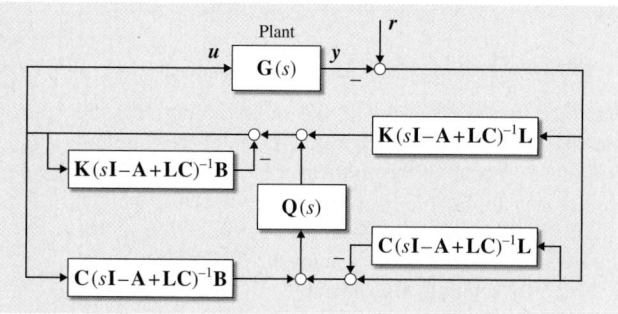

Fig. 10.13 State-space realization of all stabilizing controllers derived from LMFD components

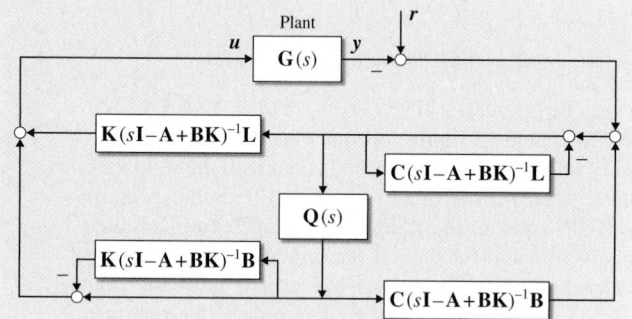

Fig. 10.14 State-space realization of all stabilizing controllers derived from RMFD components

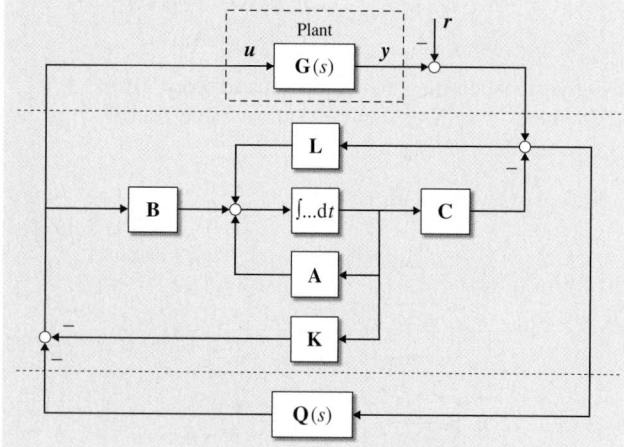

Fig. 10.15 State-space realization of all stabilizing controllers

In a similar way, a left coprime pair of transfer function matrices $\mathbf{X}_R(s)$ and $\mathbf{Y}_R(s)$ can be found by

$$\mathbf{B}_L(s)\mathbf{Y}_R(s) + \mathbf{A}_L(s)\mathbf{X}_R(s) = \mathbf{I} .$$

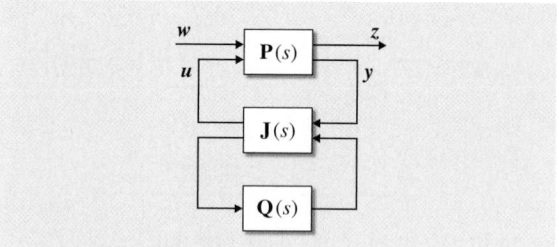

Fig. 10.16 General control system using Youla parameterization

The stabilizing $\mathbf{K}(s) = \mathbf{Y}_R(s)\mathbf{X}_R^{-1}(s) = \mathbf{X}_L^{-1}(s)\mathbf{Y}_L(s)$ controllers can be parameterized as follows. Assume that the Bezout identity results in a given stabilizing controller $\mathbf{K} = \mathbf{Y}_R^0(s)\mathbf{X}_R^{0\,-1}(s) = \mathbf{X}_L^{0\,-1}(s)\mathbf{Y}_L^0(s)$, then

$$\mathbf{X}_R(s) = \mathbf{X}_R^0(s) - \mathbf{B}_R(s)\mathbf{Q}(s) ,$$
$$\mathbf{Y}_R(s) = \mathbf{Y}_R^0(s) + \mathbf{A}_R(s)\mathbf{Q}(s) ,$$
$$\mathbf{X}_L(s) = \mathbf{X}_L^0(s) - \mathbf{Q}(s)\mathbf{B}_L(s) ,$$
$$\mathbf{Y}_L(s) = \mathbf{Y}_L^0(s) + \mathbf{Q}(s)\mathbf{A}_L(s) ,$$

delivers all stabilizing controllers parameterized by any stable proper $\mathbf{Q}(s)$ transfer function matrix with appropriate size.

Though the algebra of the controller design procedure may seem rather involved, in terms of block diagrams it can be interpreted in several ways. In Fig. 10.12 two possible realizations are shown to support the reader in comparing the results obtained for unstable processes with those shown earlier in Fig. 10.9 to control stable processes.

Another obvious interpretation of the general design procedure can also be read out from the realizations of Fig. 10.12. Namely, the immediate loops around $\mathbf{G}(s)$ along $\mathbf{Y}_L^0(s)$ and $\mathbf{X}_L^{0\,-1}(s)$ or along $\mathbf{X}_R^{0\,-1}(s)$ and $\mathbf{Y}_R^0(s)$, respectively, stabilize the unstable plant, then $\mathbf{Q}(s)$ serves the parameterization in a similar way as originally introduced for stable processes.

Having the general control structure developed using LMFD or RMFD components (Fig. 10.12 gives the complete review), respectively, we are in the position to show how the control of the state-space model introduced earlier in Sect. 10.1 can be parameterized with $\mathbf{Q}(s)$. To visualize this capability recall $\mathbf{K}(s) = \mathbf{X}_L^{-1}(s)\mathbf{Y}_L^0(s)$ and

$$\mathbf{B}_L(s) = \mathbf{C}(s\mathbf{I} - \mathbf{A} + \mathbf{LC})^{-1}\mathbf{B} ,$$
$$\mathbf{A}_L(s) = \mathbf{I} - \mathbf{C}(s\mathbf{I} - \mathbf{A} + \mathbf{LC})^{-1}\mathbf{L} ,$$

and apply these relations in the control scheme of Fig. 10.12 using LMFD components.

Similarly, recall $\mathbf{K}(s) = \mathbf{Y}_R^0(s)\mathbf{X}_R^{0\,-1}(s)$ and

$$\mathbf{B}_R(s) = \mathbf{C}(s\mathbf{I} - \mathbf{A} + \mathbf{BK})^{-1}\mathbf{B},$$

$$\mathbf{A}_R(s) = \mathbf{I} - \mathbf{K}(s\mathbf{I} - \mathbf{A} + \mathbf{BK})^{-1}\mathbf{B},$$

and apply these relations in the control scheme of Fig. 10.12 using RMFD components.

To complete the discussion on the various interpretations of the all stabilizing controllers, observe that the control schemes in Figs. 10.13 and 10.14 both use $4 \times n$ state variables to realize the controller dynamics beyond $\mathbf{Q}(s)$. As Fig. 10.15 illustrates, equivalent reduction of the block diagrams of Figs. 10.13 and 10.14, respectively, both lead to the realization of the all stabilizing controllers. Observe that the application of the

LMFD and RMFD components lead to identical control scheme.

In addition, any of the realizations shown in Figs. 10.12–10.15 can directly be redrawn to form the general control system scheme most frequently used in the literature to summarize the structure of the Youla parameterization. This general control scheme is shown in Fig. 10.16. In fact, the state-space realization by Fig. 10.15 follows the general control scheme shown in Fig. 10.16, assuming $z = 0$ and $w = r$. The transfer function $\mathbf{J}(s)$ itself is realized by the state estimator and state feedback using the gain matrices $\mathbf{L}$ and $\mathbf{K}$, as shown in Fig. 10.15. Note that Fig. 10.16 can also be derived from Fig. 10.6 by interpreting $\mathbf{J}(s)$ as a controller stabilizing $\mathbf{P}(s)$, thus allowing one to apply an additional all stabilizing $\mathbf{Q}(s)$ controller.

10.3 Control Performances

So far we have derived various closed-loop structures and parameterizations attached to them only to ensure internal stability. Stability, however, is not the only issue for the control system designer. To achieve goals in terms of the closed-loop performance needs further considerations [10.2, 6, 13, 17]. Just to see an example: in control design it is a widely posed requirement to ensure zero steady-state error while compensating step-like changes in the command or disturbance signals. The practical solution suggests one to insert an integrator into the loop. The same goal can be achieved while using the Youla parameterization, as well. To illustrate this action SISO systems will be considered. Apply stable $Q_1(s)$ and $Q_2(s)$ transfer functions to form

$$Q(s) = sQ_1(s) + Q_2(s).$$

Then Fig. 10.9 suggests the transfer function between r and $r - y$ to be

$$1 - Q(s)G(s).$$

To ensure

$$1 - [Q(s)G(s)]_{s=0} = 0$$

we need

$$Q_2(s)(0) = [G(0)]^{-1}.$$

Alternatively, using state models the selection according to

$$Q(0) = 1/[\mathbf{C}(-\mathbf{A} + \mathbf{BK} + \mathbf{LC})^{-1}\mathbf{B}]$$

will insert an integrator to the loop.

Several criteria exist to describe the required performances for the closed-loop performance. To be able to design closed-loop systems with various performance specifications, appropriate norms for the signals and systems involved should be introduced [10.13, 17].

10.3.1 Signal Norms

One possibility to characterize the closed-loop performance is to integrate various functions derived from the error signal. Assume that a generalized error signal $z(t)$ has been constructed. Then

$$\|z(t)\|_v = \left(\int_0^\infty |z|^v \, dt\right)^{1/v}$$

defines the L_v norm of $z(t)$ with v as a positive integer.

The relatively easy calculations required for the evaluations made the L_2 norm the most widely used criterion in control. A further advantage of the quadratic function is that energy represented by a given signal can also be taken into account in this way in many cases. Moreover, applying the Parseval's theorem the L_2

norm can be evaluated using the signal described in the frequency domain. Namely having $z(s)$ as the Laplace transform of $z(t)$

$$z(s) = \int_0^\infty z(t) \mathrm{e}^{-st} \, \mathrm{d}t$$

the Parseval's theorem offers the following closed form to calculate the L_2 norm as

$$\|z(t)\|_2 = \|z(s)\big|_{s=\mathrm{i}\omega}\|_2 = \|z(\mathrm{i}\omega)\|_2$$
$$\hat{=} \left(\frac{1}{2\pi} \int_{-\infty}^\infty |z(\mathrm{i}\omega)|^2 \, \mathrm{d}\omega \right)^{1/2} .$$

Another important selection for v takes $v \to \infty$, which results in

$$\|z(t)\|_\infty = \sup_t |z(t)|$$

and is interpreted as the largest or worst-case error.

10.3.2 System Norms

Frequency functions are extremely useful tools to analyze and design SISO closed-loop control systems. MIMO systems, however, exhibit an input-dependent, variable gain at a given frequency. Consider a MIMO system given by a transfer function matrix $\mathbf{G}(s)$ and driven by an input signal $\mathbf{w}$ and delivering an output signal z. The norm $\|z(\mathrm{i}\omega)\| = \|\mathbf{G}(\mathrm{i}\omega)\mathbf{w}(\mathrm{i}\omega)\|$ of the system output z depends on both the magnitude and direction of the input vector $\mathbf{w}(\mathrm{i}\omega)$, where $\|\ldots\|$ denotes Euclidean norm. The associated norms are therefore called *induced norms*. Bounds for $\|z(\mathrm{i}\omega)\|$ are given by

$$\underline{\sigma}(\mathbf{G}(\mathrm{i}\omega)) \le \frac{\|\mathbf{G}(\mathrm{i}\omega)\mathbf{w}(\mathrm{i}\omega)\|}{\|\mathbf{w}(\mathrm{i}\omega)\|} \le \overline{\sigma}(\mathbf{G}(\mathrm{i}\omega)) ,$$

where $\underline{\sigma}(\mathbf{G}(\mathrm{i}\omega))$ and $\overline{\sigma}(\mathbf{G}(\mathrm{i}\omega))$ denote the minimum and maximum values of the singular values of $\mathbf{G}(\mathrm{i}\omega)$, respectively. The most frequently used system norms are the H_2 and H_∞ norms defined as follows

$$\|\mathbf{G}\|_2 = \left(\frac{1}{2\pi} \int_{-\infty}^\infty \mathrm{trace}[\mathbf{G}^\top(-\mathrm{i}\omega)\mathbf{G}(\mathrm{i}\omega)] \, \mathrm{d}\omega \right)^{1/2}$$

and

$$\|\mathbf{G}\|_\infty = \sup_\omega \overline{\sigma}(\mathbf{G}(\mathrm{i}\omega)) .$$

It is clear that the system norms – as induced norms – can be expressed by using signal norms. Introducing

$g(t)$ as the unit impulse response of $\mathbf{G}(s)$ the Parseval's theorem suggests expressing the H_2 system norm by the L_2 signal norm

$$\|\mathbf{G}\|_2^2 = \int_0^\infty \mathrm{trace}[\mathbf{g}^\top(t)\mathbf{g}(t)] \, \mathrm{d}t .$$

Further on, the H_∞ norm can also be expressed as

$$\|\mathbf{G}\|_\infty = \sup_\omega \max_w \left(\frac{\|\mathbf{G}(\mathrm{i}\omega)\mathbf{w}\|}{\|\mathbf{w}\|} \right. \text{ where } \mathbf{w} \ne 0$$
$$\left. \text{and } \mathbf{w} \in \mathbb{C}^{n_w} \right) ,$$

where $\mathbf{w}$ denotes a complex-valued vector. For a dynamic system the above expression leads to

$$\|\mathbf{G}\|_\infty = \sup_w \left(\frac{\|z(t)\|_2}{\|\mathbf{w}(t)\|_2} \text{ where } \|\mathbf{w}(t)\|_2 \ne 0 \right) ,$$

if $\mathbf{G}(s)$ is stable and proper. The above expression means that the H_∞ norm can be expressed by L_2 signal norm.

Assume a linear system given by a state model $\{\mathbf{A}, \mathbf{B}, \mathbf{C}\}$ and calculate its H_2 and H_∞ norms. Transforming the state model to a transfer function matrix

$$\mathbf{G}(s) = \mathbf{C}(s\mathbf{I} - \mathbf{A})^{-1}\mathbf{B}$$

the H_2 norm is obtained by

$$\|\mathbf{G}\|_2^2 = \mathrm{trace}(\mathbf{C}\mathbf{P}_0\mathbf{C}^\top) = \mathrm{trace}(\mathbf{B}^\top \mathbf{P}_c \mathbf{B}) ,$$

where $\mathbf{P}_c$ and $\mathbf{P}_0$ are delivered by the solution of the Lyapunov equations

$$\mathbf{A}\mathbf{P}_0 + \mathbf{P}_0\mathbf{A}^\top + \mathbf{B}\mathbf{B}^\top = \mathbf{0} ,$$
$$\mathbf{P}_c\mathbf{A} + \mathbf{A}^\top \mathbf{P}_c + \mathbf{C}^\top \mathbf{C} = \mathbf{0} .$$

The calculation of the H_∞ norm can be performed via an iterative procedure, where in each step an H_2 norm is to be minimized. Assuming a stable system, construct the Hamiltonian matrix

$$\mathbf{H} = \begin{pmatrix} \mathbf{A} & \frac{1}{\gamma^2}\mathbf{B}\mathbf{B}^\top \\ -\mathbf{C}^\top\mathbf{C} & -\mathbf{A}^\top \end{pmatrix} .$$

For large γ the matrix $\mathbf{H}$ has n_x eigenvalues with negative real part and n_x eigenvalues with positive real part. As γ decreases these eigenvalues eventually hit the imaginary axis. Thus

$$\|\mathbf{G}\|_\infty = \inf_{\gamma > 0} (\gamma \in \mathbb{R} : \mathbf{H} \text{ has no eigenvalues}$$
$$\text{with zero real part}) .$$

Note that each step within the γ-iteration procedure is after all equivalent to solve an underlying Riccati equation. The solution of the Riccati equation will be detailed later on in Sect. 10.4.

So far stability issues have been discussed and signal and system norms, as performance measures, have been introduced to evaluate the overall operation of the closed-loop system. In the sequel the focus will be turned on design procedures resulting in both stable operation and expected performance. Controller design techniques to achieve appropriate performance measures via optimization procedures related to the H_2 and H_∞ norms will be discussed, respectively [10.6, 21].

10.4 H_2 Optimal Control

To start the discussion consider the general control system configuration shown in Fig. 10.6 describe the plant by the transfer function

$$\begin{pmatrix} z \\ y \end{pmatrix} = \begin{pmatrix} \mathbf{G}_{11} & \mathbf{G}_{12} \\ \mathbf{G}_{21} & \mathbf{G}_{22} \end{pmatrix} \begin{pmatrix} w \\ u \end{pmatrix}$$

or equivalently, by a state model

$$\begin{pmatrix} \dot{x} \\ z \\ y \end{pmatrix} = \begin{pmatrix} \mathbf{A} & \mathbf{B}_1 & \mathbf{B}_2 \\ \mathbf{C}_1 & \mathbf{0} & \mathbf{D}_{12} \\ \mathbf{C}_2 & \mathbf{D}_{21} & \mathbf{0} \end{pmatrix} \begin{pmatrix} x \\ w \\ u \end{pmatrix} .$$

Assume that $(\mathbf{A}, \mathbf{B}_1)$ is controllable, $(\mathbf{A}, \mathbf{B}_2)$ is stabilizable, $(\mathbf{C}_1, \mathbf{A})$ is observable, and $(\mathbf{C}_2, \mathbf{A})$ is detectable. For the sake of simplicity nonsingular $\mathbf{D}_{12}^\top \mathbf{D}_{12}$ and $\mathbf{D}_{21} \mathbf{D}_{21}^\top$ matrices, as well as $\mathbf{D}_{12}^\top \mathbf{C}_1 = \mathbf{0}$ and $\mathbf{D}_{21} \mathbf{B}_1^\top = \mathbf{0}$ will be considered.

Using a feedback via $\mathbf{K}(s)$

$$u = -\mathbf{K}(s) y ,$$

the closed-loop system becomes

$$z = \mathbf{F}[\mathbf{G}(s), \mathbf{K}(s)] w ,$$

where

$$\mathbf{F}[\mathbf{G}(s), \mathbf{K}(s)] = \mathbf{G}_{11}(s) - \mathbf{G}_{12}(s)[\mathbf{I} + \mathbf{K}(s)\mathbf{G}_{22}(s)]^{-1}$$
$$\mathbf{K}(s)\mathbf{G}_{21}(s) .$$

Aiming at designing optimal control in H_2 sense the $J_2 = \|\mathbf{F}(\mathbf{G}(i\omega), \mathbf{K}(i\omega))\|_2^2$ norm is to be minimized by a realizable $\mathbf{K}(s)$. Note that this control policy can be interpreted as a special case of the linear quadratic (LQ) control problem formulation. To show this relation assume a weighting matrix $\mathbf{Q}_x$ assigned for the state variables and a weighting matrix $\mathbf{R}_u$ assigned for the input variables. Choosing

$$\mathbf{C}_1 = \begin{pmatrix} \mathbf{Q}_x^{1/2} \\ \mathbf{0} \end{pmatrix} \quad \text{and} \quad \mathbf{D}_{12} = \begin{pmatrix} \mathbf{0} \\ \mathbf{R}_u^{1/2} \end{pmatrix}$$

and

$$z = \mathbf{C}_1 x + \mathbf{D}_{12} u$$

as an auxiliary variable the well-known LQ loss function can be reproduced with

$$\mathbf{Q}_x = \left(\mathbf{Q}_x^{1/2}\right)^\top \mathbf{Q}_x^{1/2} \quad \text{and} \quad \mathbf{R}_u = \left(\mathbf{R}_u^{1/2}\right)^\top \mathbf{R}_u^{1/2} .$$

Up to this point the feedback loop has been set up and the design problem has been formulated to find $\mathbf{K}(s)$ minimizing $J_2 = \|\mathbf{F}(\mathbf{G}(i\omega), \mathbf{K}(i\omega))\|_2^2$. Note that the optimal controller will be derived as a solution of the state-feedback problem. The optimization procedure is discussed below.

10.4.1 State-Feedback Problem

If all the state variables are available then the state variable feedback

$$u(t) = -\mathbf{K}_2 x(t)$$

is used with the gain

$$\mathbf{K}_2 = (\mathbf{D}_{12}^\top \mathbf{D}_{12})^{-1} \mathbf{B}_2^\top \mathbf{P}_c ,$$

where $\mathbf{P}_c$ represents the positive-definite or positive-semidefinite solution of the

$$\mathbf{A}^\top \mathbf{P}_c + \mathbf{P}_c \mathbf{A} - \mathbf{P}_c \mathbf{B}_2 \left(\mathbf{D}_{12}^\top \mathbf{D}_{12}\right)^{-1} \mathbf{B}_2^\top \mathbf{P}_c + \mathbf{C}_1^\top \mathbf{C}_1 = \mathbf{0}$$

Riccati equation. According to this control law the

$$\mathbf{A} - \mathbf{B}_2 \mathbf{K}_2$$

matrix will determine the closed-loop stability.

As far as the solution of the Riccati equation is concerned, an augmented problem setup can turn this task to an equivalent eigenvalue–eigenvector decomposition (EVD). In details, the EVD decomposition of the Hamiltonian matrix

$$\mathbf{H} = \begin{pmatrix} \mathbf{A} & -\mathbf{B}_2 \left(\mathbf{D}_{12}^\top \mathbf{D}_{12}\right)^{-1} \mathbf{B}_2^\top \\ -\mathbf{C}_1^\top \mathbf{C}_1 & -\mathbf{A}^\top \end{pmatrix}$$

will separate the eigenvectors belonging to stable and unstable eigenvalues, then the positive-definite $\mathbf{P}_c$ matrix can be calculated from the eigenvectors belonging to the stable eigenvalues. Denote Λ the diagonal matrix containing the stable eigenvalues and collect the associated eigenvectors to a block matrix

$$\begin{pmatrix} \mathbf{F} \\ \mathbf{G} \end{pmatrix} ,$$

i. e.,

$$\mathbf{H} \begin{pmatrix} \mathbf{F} \\ \mathbf{G} \end{pmatrix} = \begin{pmatrix} \mathbf{F} \\ \mathbf{G} \end{pmatrix} \Lambda .$$

Then it can be shown that the solution of the Riccati equation is obtained by

$$\mathbf{P}_c = \mathbf{G}\mathbf{F}^{-1} .$$

At the same time it should be noted that there exist further, numerically advanced procedures to find $\mathbf{P}_c$.

10.4.2 State–Estimation Problem

The optimal state estimation (state reconstruction) is the dual of the optimal control task [10.1, 4]. The estimated states are derived as the solution of the following differential equation:

$$\dot{\hat{x}}(t) = \mathbf{A}\hat{x}(t) + \mathbf{B}_2 u(t) + \mathbf{L}_2[y(t) - \mathbf{C}_2\hat{x}(t)] ,$$

where

$$\mathbf{L}_2 = \mathbf{P}_0 \mathbf{C}_2^\top (\mathbf{D}_{21}\mathbf{D}_{21}^\top)^{-1}$$

and the $\mathbf{P}_0$ matrix is the positive-definite or positive-semidefinite solution of the Riccati equation

$$\mathbf{P}_0\mathbf{A}^\top + \mathbf{A}\mathbf{P}_0 - \mathbf{P}_0\mathbf{C}_2^\top (\mathbf{D}_{21}\mathbf{D}_{21}^\top)^{-1}\mathbf{C}_2\mathbf{P}_0 + \mathbf{B}_1\mathbf{B}_1^\top = \mathbf{0} .$$

Note that the

$$\mathbf{A} - \mathbf{P}_0\mathbf{C}_2^\top (\mathbf{D}_{21}\mathbf{D}_{21}^\top)^{-1}\mathbf{C}_2$$

matrix characterizing the closed-loop system is stable, i. e., all its eigenvalues are on the left-hand half plane.

Remark 1: Putting the problem just discussed so far into a stochastic environment the above state estimation is also called a *Kalman filter*.

Remark 2: The gains $\mathbf{L}_2 \in \mathbb{R}^{n_x \times n_y}$ and $\mathbf{K}_2 \in \mathbb{R}^{n_u \times n_x}$ have been introduced and applied in earlier stages in this chapter to create the LMFD and RMFD descriptions, respectively. Here their optimal values have been derived in H_2 sense.

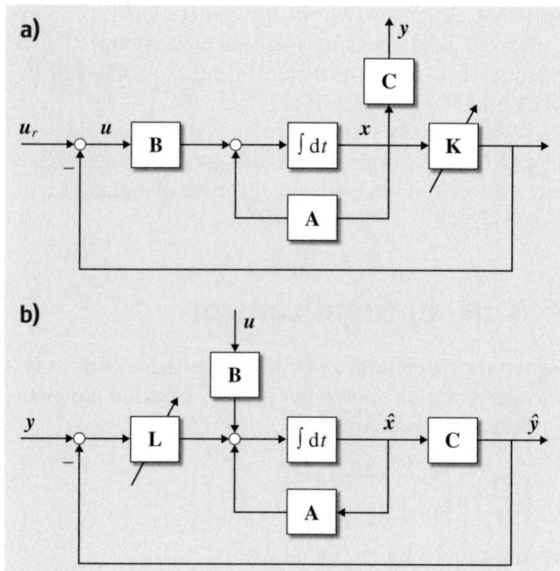

Fig. 10.17 Duality of state control and state estimation

Remark 3: State control and state estimation exhibit dual properties and share some common structural features. Comparing Fig. 10.17a and b it is seen that the structure of the state feedback control and that of the full order observer resemble each other to a large extent. The output signal, as well as the $\mathbf{L}$ and $\mathbf{C}$ matrices in the observer, play identical role as the control signal, as do the $\mathbf{B}$ and $\mathbf{K}$ matrices in state feedback control. Parameters in the matrices $\mathbf{L}$ and $\mathbf{K}$ are to be freely adjusted for the observer and for the state feedback control, respectively. In a sense, calculating the controller and observer feedback gain matrices represent dual problems. In this case duality means that any of the structures shown in Fig. 10.17a,b can be turned to its dual form by reversing the direction of the signal propagation, interchanging the input and output signals ($u \leftrightarrow y$), and transforming the summation points to signal nodes and vice versa.

10.4.3 Output–Feedback Problem

If the state variables are not available for feedback the optimal control law utilizes the reconstructed states. In case of designing optimal control in H_2 sense the control law $u(t) = -\mathbf{K}_2 x(t)$ is replaced by $u(t) = -\mathbf{K}_2 \hat{x}(t)$.

It is important to prove that the joint state estimation and control lead to stable closed-loop control. The proof is based on observing that the complete system satisfies the following state equations

$$
\begin{pmatrix} \dot{x} \\ \dot{x} - \dot{\hat{x}} \end{pmatrix} = \begin{pmatrix} \mathbf{A} - \mathbf{B}_2\mathbf{K}_2 & \mathbf{B}_2\mathbf{K}_2 \\ \mathbf{0} & \mathbf{A} - \mathbf{L}_2\mathbf{C}_2 \end{pmatrix} \begin{pmatrix} x \\ x - \hat{x} \end{pmatrix}
$$
$$
+ \begin{pmatrix} \mathbf{B}_1 \\ \mathbf{B}_1 - \mathbf{L}_2\mathbf{D}_{21} \end{pmatrix} w .
$$

This form clearly shows that the poles introduced by the controller and those introduced by the observer are separated from each other. The concept is therefore called the *separation principle*. The importance of this observation lies in the fact that, in the course of the design procedure the controller poles and the observer poles can be assigned independently from each other. Note that the control law $u(t) = -\mathbf{K}_2\hat{x}(t)$ still exhibits an optimal controller in the sense that $\|\mathbf{F}(\mathbf{G}(i\omega), \mathbf{K}(i\omega))\|_2$ is minimized.

10.5 H_∞ Optimal Control

Herewith below the optimal control in H_∞ sense will be discussed. The H_∞ optimal control minimizes the H_∞ norm of the overall transfer function of the closed-loop system

$$
J_\infty = \|\mathbf{F}(\mathbf{G}(i\omega), \mathbf{K}(i\omega))\|_\infty
$$

using a state variable feedback with a constant gain, where $\mathbf{F}[\mathbf{G}(s), \mathbf{K}(s)]$ denotes the overall transfer function matrix of the closed-loop system [10.2, 6, 6, 13]. To minimize J_∞ requires rather involved procedures. As one option, γ-iteration has already been discussed earlier. In short, as earlier discussions on the norms pointed out, the H_∞ norm can be calculated using L_2 norms by

$$
J_\infty \hat{=} \sup\left(\frac{\|z\|_2}{\|w\|_2} : \|w\|_2 \neq 0 \right) .
$$

10.5.1 State–Feedback Problem

If all the state variables are available then the state variable feedback

$$
u(t) = -\mathbf{K}_\infty x(t)
$$

is used with the gain $\mathbf{K}_\infty$ minimizing J_∞. Similarly to the H_2 optimal control discussed earlier, in each step of the γ-iteration $\mathbf{K}_\infty$ can be obtained via $\mathbf{P}_c$ as the symmetrical positive-definite or positive-semidefinite solution of the

$$
\mathbf{A}^\top \mathbf{P}_c + \mathbf{P}_c \mathbf{A} - \mathbf{P}_c \mathbf{B}_2 (\mathbf{D}_{12}^\top \mathbf{D}_{12})^{-1} \mathbf{B}_2^\top \mathbf{P}_c
$$
$$
+ \gamma^{-2} \mathbf{P}_c \mathbf{B}_1 \mathbf{B}_1^\top \mathbf{P}_c + \mathbf{C}_1^\top \mathbf{C}_1 = \mathbf{0}
$$

Riccati equation, provided that the matrix

$$
\mathbf{A} - \mathbf{B}_2 (\mathbf{D}_{12}^\top \mathbf{D}_{12})^{-1} \mathbf{B}_2^\top \mathbf{P}_c + \gamma^{-2} \mathbf{B}_1 \mathbf{B}_1^\top \mathbf{P}_c
$$

represents a stable system (i. e., all the eigenvalues are on the left-hand half plane). Once $\mathbf{P}_c$ belonging to the minimal γ value has been found the state variable feedback is realized by using the feedback gain matrix of

$$
\mathbf{K}_\infty = (\mathbf{D}_{12}^\top \mathbf{D}_{12})^{-1} \mathbf{B}_2^\top \mathbf{P}_c .
$$

10.5.2 State–Estimation Problem

The optimal state estimation in H_∞ sense requires to minimize

$$
J_\infty^0 \hat{=} \sup\left(\frac{\|z - \hat{z}\|_2}{\|w\|_2} : \|w\|_2 \neq 0 \right)
$$

as a function of $\mathbf{L}$. Again, minimization can be performed by γ-iteration. Specifically, $\gamma_{\min}$ is looked for to satisfy $J_\infty^0 < \gamma$ with $\gamma > 0$ for all w. To find the optimal $\mathbf{L}_\infty$ gain the symmetrical positive-definite or positive-semidefinite solution of the following Riccati equation is required

$$
\mathbf{P}_0 \mathbf{A}^\top + \mathbf{A} \mathbf{P}_0 - \mathbf{P}_0 \mathbf{C}_2^\top (\mathbf{D}_{12}\mathbf{D}_{12}^\top)^{-1} \mathbf{C}_2 \mathbf{P}_0
$$
$$
+ \gamma^{-2} \mathbf{P}_0 \mathbf{C}_1^\top \mathbf{C}_1 \mathbf{P}_0 + \mathbf{B}_1 \mathbf{B}_1^\top = \mathbf{0}
$$

provided that

$$
\mathbf{A} - \mathbf{P}_0 \mathbf{C}_2^\top (\mathbf{D}_{12}\mathbf{D}_{12}^\top)^{-1} \mathbf{C}_2 + \gamma^{-2} \mathbf{P}_0 \mathbf{C}_1^\top \mathbf{C}_1
$$

represents a stable system, i. e., it has all its eigenvalues on the left-hand half plane. Finding the solution $\mathbf{P}_0$ belonging to the minimal γ value, the optimal feedback gain matrix is obtained by

$$
\mathbf{L}_\infty = \mathbf{P}_0 \mathbf{C}_2^\top (\mathbf{D}_{21}\mathbf{D}_{21}^\top)^{-1} .
$$

Then

$$
\dot{\hat{x}}(t) = \mathbf{A}\hat{x}(t) + \mathbf{B}_2 u(t) + \mathbf{L}_\infty [y(t) - \mathbf{C}_2\hat{x}(t)]
$$

is in complete harmony with the filtering procedure obtained earlier for the state reconstruction in H_2 sense.

10.5.3 Output–Feedback Problem

If the state variables are not available for feedback then a $\mathbf{K}(s)$ controller satisfying $J_\infty < \gamma$ is looked for. This controller, similarly to the procedure followed by the H_2 optimal control design, can be determined in two phases: first the unavailable states are to be estimated, then state feedback driven by the estimated states is to be realized. As far as the state feedback is concerned, similarly to the H_2 optimal control law, the H_∞ optimal control is accomplished by

$$u(t) = -\mathbf{K}_\infty \hat{x}(t) \, .$$

However, the H_∞ optimal state estimation is more involved than the H_2 optimal state estimation. Namely the H_∞ optimal state estimation includes the worst-case estimation of the exogenous $\boldsymbol{w}$ input, and the feedback matrix $\mathbf{L}_\infty$ needs to be modified, too. The $\mathbf{L}_\infty^*$ modified feedback matrix takes the following form

$$\mathbf{L}_\infty^* = \left(\mathbf{I} - \gamma^{-2}\mathbf{P}_0\mathbf{P}_c\right)^{-1}\mathbf{L}_\infty$$
$$= \left(\mathbf{I} - \gamma^{-2}\mathbf{P}_0\mathbf{P}_c\right)^{-1}\mathbf{P}_0\mathbf{C}_2^\top(\mathbf{D}_{21}\mathbf{D}_{21}^\top)^{-1} \, ,$$

then the state estimation applies the above gain according to

$$\dot{\hat{x}}(t) = (\mathbf{A} + \mathbf{B}_1\gamma^{-2}\mathbf{B}_1^\top\mathbf{P}_c)\hat{x}(t) + \mathbf{B}_2\boldsymbol{u}(t)$$
$$+ \mathbf{L}_\infty^*[\boldsymbol{y}(t) - \mathbf{C}_2\hat{x}(t)] \, .$$

Reformulating the above results into a transfer function form gives

$$\mathbf{K}(s) = \mathbf{K}_\infty\left(s\mathbf{I} - \mathbf{A} - \mathbf{B}_1\gamma^{-2}\mathbf{B}_1^\top\mathbf{P}_c + \mathbf{B}_2\mathbf{K}_\infty\right.$$
$$\left. + \mathbf{L}_\infty^*\mathbf{C}_2\right)^{-1}\mathbf{L}_\infty^* \, .$$

The above $\mathbf{K}(s)$ controller satisfies the norm inequality $\|\mathbf{F}(\mathbf{G}(\mathrm{i}\omega), \mathbf{K}(\mathrm{i}\omega))\|_\infty < \gamma$ and it results in a stable control strategy if the following three conditions are satisfied [10.6]:

- $\mathbf{P}_c$ is a symmetrical positive-semidefinite solution of the algebraic Riccati equation

$$\mathbf{A}^\top\mathbf{P}_c + \mathbf{P}_c\mathbf{A} - \mathbf{P}_c\mathbf{B}_2\left(\mathbf{D}_{12}^\top\mathbf{D}_{12}\right)^{-1}\mathbf{B}_2^\top\mathbf{P}_c$$
$$+ \gamma^{-2}\mathbf{P}_c\mathbf{B}_1\mathbf{B}_1^\top\mathbf{P}_c + \mathbf{C}_1^\top\mathbf{C}_1 = \mathbf{0} \, ,$$

provided that

$$\mathbf{A} - \mathbf{B}_2(\mathbf{D}_{12}^\top\mathbf{D}_{12})^{-1}\mathbf{B}_2^\top\mathbf{P}_c + \gamma^{-2}\mathbf{B}_1\mathbf{B}_1^\top\mathbf{P}_c$$

is stable.

- $\mathbf{P}_0$ is a symmetrical positive-semidefinite solution of the algebraic Riccati equation

$$\mathbf{P}_0\mathbf{A}^\top + \mathbf{A}\mathbf{P}_0 - \mathbf{P}_0\mathbf{C}_2^\top\left(\mathbf{D}_{21}\mathbf{D}_{21}^\top\right)^{-1}\mathbf{C}_2\mathbf{P}_0$$
$$+ \gamma^{-2}\mathbf{P}_0\mathbf{C}_1^\top\mathbf{C}_1\mathbf{P}_0 + \mathbf{B}_1\mathbf{B}_1^\top = \mathbf{0} \, ,$$

provided that

$$\mathbf{A} - \mathbf{P}_0\mathbf{C}_2^\top\left(\mathbf{D}_{21}\mathbf{D}_{21}^\top\right)^{-1}\mathbf{C}_2 + \gamma^{-2}\mathbf{P}_0\mathbf{C}_1^\top\mathbf{C}_1$$

is stable.

- The largest eigenvalue of $\mathbf{P}_c\mathbf{P}_0$ is smaller than γ^2

$$\rho(\mathbf{P}_c\mathbf{P}_0) < \gamma^2 \, .$$

The H_∞ optimal output feedback control design procedure minimizes the $\|\mathbf{F}(\mathbf{G}(\mathrm{i}\omega), \mathbf{K}(\mathrm{i}\omega))\|_\infty$ norm via γ-iteration and while $\gamma_{\min}$ is looked for all the three conditions above should be satisfied. The optimal control in H_∞ sense is accomplished by $\mathbf{K}(s)$ belonging to $\gamma_{\min}$.

Remark: Now that we are ready to design optimal controllers in H_2 or H_∞ sense, respectively, it is worth devoting a minute to analyze what can be expected from these procedures. To compare the nature of the H_2 versus H_∞ norms a relation for $\|\mathbf{G}\|_2$ should be found, where the $\|\mathbf{G}\|_2$ norm is expressed by the singular values. It can be shown that

$$\|\mathbf{G}\|_2 = \left(\frac{1}{2\pi}\int_{-\infty}^{\infty}\sum_i\sigma_i^2(\mathbf{G}(\mathrm{i}\omega))\,\mathrm{d}\omega\right)^{1/2} \, .$$

Comparing now the above expression with

$$\|\mathbf{G}\|_\infty = \sup_\omega\overline{\sigma}(\mathbf{G}(\mathrm{i}\omega))$$

it is seen that $\|\mathbf{G}\|_\infty$ represents the largest possible singular value, while $\|\mathbf{G}\|_2$ represents the sum of all the singular values over all frequencies [10.6].

10.6 Robust Stability and Performance

When designing control systems, the design procedure needs a model of the process to be controlled. So far it has been assumed that the design procedure is based on a perfect model of the process. Stability analysis based on the nominal process model can be qualified as *nominal stability* (NS) analysis. Similarly, closed-loop performance analysis based on the nominal process model can be qualified as *nominal performance* (NP) analysis. It is evident, however, that some uncertainty is always present in the model. Moreover, an important purpose of using feedback is even to reduce the effects of uncertainty involved in the model. The classical approach introduced the notions of the phase margin and gain margin as measures to handle uncertainty. However, these measures are rather crude and contradictory [10.13, 22]. Though they work fine in a number of practical applications, they are not capable of supporting the design for processes exhibiting unusual frequency behavior (e.g., slightly damped poles). The postmodern era of control theory places special emphasis on modeling of uncertainties. Specifically, wide classes of structured, as well as additively or multiplicatively unstructured uncertainties have been introduced and taken into account in the design procedure. Modeling, analysis, and synthesis methods have been developed under the name *robust control* [10.17, 23, 24]. Note that the linear quadratic regulator (LQR) design method inherits some measures of robustness, however, in general the pure structure of the LQR regulators does not guarantee stability margins [10.1, 6].

As far as the unstructured uncertainties are concerned, let $\mathbf{G}_0(s)$ denote the nominal transfer function matrix of the process. Then the true plant behavior can be expressed by

$$\mathbf{G}(s) = \mathbf{G}_0(s) + \boldsymbol{\Delta}_a(s) ,$$
$$\mathbf{G}(s) = \mathbf{G}_0(s)[\mathbf{I} + \boldsymbol{\Delta}_i(s)] ,$$
$$\mathbf{G}(s) = [\mathbf{I} + \boldsymbol{\Delta}_o(s)]\mathbf{G}_0(s) ,$$

where $\boldsymbol{\Delta}_a(s)$ represents an additive perturbation, $\boldsymbol{\Delta}_i(s)$ an input multiplicative perturbation, and $\boldsymbol{\Delta}_o(s)$ an output multiplicative perturbation. These perturbations are assumed to be frequency independent with bounded $\|\boldsymbol{\Delta}_\bullet(s)\|_\infty$ norms concerning their size. Frequency dependence can easily be added to the perturbations by using appropriate pre- and post-filters.

All the above three perturbation models can be transformed to a common form

$$\mathbf{G}(s) = \mathbf{G}_0(s) + \mathbf{W}_1(s)\boldsymbol{\Delta}(s)\mathbf{W}_2(s) ,$$

where $\|\boldsymbol{\Delta}(s)\|_\infty \leq 1$. Uncertainties extend the general control system configuration outlined earlier in Fig. 10.6. The nominal plant now is extended by a block representing the uncertainties and the feedback is still applied in parallel as Fig. 10.18 shows. This standard model *removes* $\boldsymbol{\Delta}(s)$, as well as the $\mathbf{K}(s)$ controller from the closed-loop system and lets $\mathbf{P}(s)$ represent the

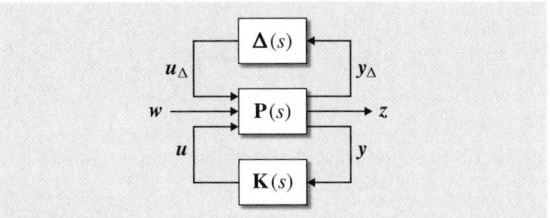

Fig. 10.18 Standard model of control system extended by uncertainties

rest of the components. It may involve additional output signals (z) and a set of external signals (w) including the set point.

Using a priori knowledge on the plant the concept of the uncertainty modeling can further be improved. Separating identical and independent technological components into groups the perturbations can be expressed as *structured uncertainties*

$$\boldsymbol{\Delta}(s) = \text{diag}\,[\boldsymbol{\Delta}_1(s), \boldsymbol{\Delta}_2(s), \ldots, \boldsymbol{\Delta}_r(s)] , \quad \text{where}$$
$$\|\boldsymbol{\Delta}_i(s)\|_\infty \leq 1 \quad i = 1 \ldots r .$$

Structured uncertainties clearly lead to less conservative design as the unstructured uncertainties may want to take care of perturbations never occurring in practice.

Consider the following control system (Fig. 10.19) as one possible realization of the standard model shown in Fig. 10.18. As a matter of fact here the common form of the perturbations is used. Derivation is also straightforward from the standard form of Fig. 10.18 with $z = 0$ and $w = r$. Handling the nominal plant and the feedback as one single unit described by $\mathbf{R}(s) = [\mathbf{I} + \mathbf{K}(s)\mathbf{G}_0(s)]^{-1}\mathbf{K}(s)$, condition for the robust stability can easily be derived by applying the *small gain theorem* (Fig. 10.20). The small gain theorem is the most fundamental result in robust stabilization under unstructured perturbations. According to the small gain theorem any closed-loop system consisting of two stable subsystems $\mathbf{G}_1(s)$ and $\mathbf{G}_2(s)$ results in stable closed-loop system provided that

$$\|\mathbf{G}_1(i\omega)\|_\infty \|\mathbf{G}_2(i\omega)\|_\infty < 1 .$$

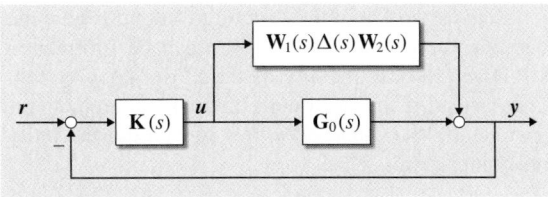

Fig. 10.19 Control system with uncertainties

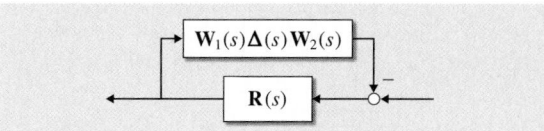

Fig. 10.20 Reduced control system with uncertainties

Applying the small gain theorem to the stability analysis of the system shown in Fig. 10.20, the condition

$$\|\mathbf{W}_2(i\omega)\mathbf{R}(i\omega)\mathbf{W}_1(i\omega)\mathbf{\Delta}(i\omega)\|_\infty < 1$$

guarantees closed-loop stability. As

$$\|\mathbf{W}_2(i\omega)\mathbf{R}(i\omega)\mathbf{W}_1(i\omega)\mathbf{\Delta}(i\omega)\|_\infty \leq$$
$$\|\mathbf{W}_2(i\omega)\mathbf{R}(i\omega)\mathbf{W}_1(i\omega)\|_\infty \|\mathbf{\Delta}\|_\infty \text{ and}$$
$$\|\mathbf{\Delta}(i\omega)\|_\infty \leq 1 ,$$

thus the stability condition reduces to

$$\|\mathbf{W}_2(i\omega)\mathbf{R}(i\omega)\mathbf{W}_1(i\omega)\|_\infty < 1 .$$

To support the closed-loop design procedure for robust stability introduce the γ-norm $\|\mathbf{W}_2(i\omega)\mathbf{R}(i\omega)$ $\mathbf{W}_1(i\omega) \mathbf{\Delta}(i\omega)\|_\infty = \gamma < 1$. Finding $\mathbf{K}(s)$ such that the γ-norm is kept at its minimum the *maximally stable robust controller* can be constructed.

The performance of a closed-loop system can be rather conservative in case of having structural information on the uncertainties. To avoid this drawback in the design procedure the so-called *structural singular value* is used instead of the H_∞ norm (being equal to the maximum of the singular value). The structured singular value of a matrix $\mathbf{M}$ is defined as

$$\mu_\Delta(\mathbf{M}) \widehat{=} \big[\min(k| \det(\mathbf{I} - k\mathbf{M}\Delta) = 0$$
$$\text{for structured } \Delta, \ \overline{\sigma}(\Delta) \leq 1) \big]^{-1} ,$$

where Δ has a block-diagonal form of $\Delta = \mathrm{diag}(\dots$ $\Delta_i \dots)$ and $\overline{\sigma}(\Delta) \leq 1$. This definition suggests the following interpretation: a large value of $\mu_\Delta(\mathbf{M})$ indicates that even a small perturbation can make the $\mathbf{I} - \mathbf{M}\Delta$ matrix singular. On the other hand a small value of $\mu_\Delta(\mathbf{M})$ represents favorable conditions in this sense. The structured singular value can be considered as the generalization of the maximal singular value [10.6, 13].

Using the notion of the structured singular value the condition for *robust stability* (RS) can be formulated as follows. Robust stability of the closed-loop system is guaranteed if the maximum of the structured singular value of $\mathbf{W}_2(i\omega)\mathbf{R}(i\omega)\mathbf{W}_1(i\omega)$ lies within the unity uncertainty radius

$$\sup_\omega \mu_\Delta(\mathbf{W}_2(i\omega)\mathbf{R}(i\omega)\mathbf{W}_1(i\omega)) < 1 .$$

Designing robust control systems is a far more involved task than testing robust stability. The design procedure minimizing the supremum of the structured singular value is called *structured singular value synthesis* or *μ-synthesis*. At this moment there is no direct method to synthesize a μ-optimal controller. Related algorithms to perform the minimization are discussed in the literature under the term DK-iteration [10.6]. In [10.6] not only a detailed discussion is presented, but also a MATLAB program is shown to provide better understanding of the iterations to improve the robust performance conditions.

So far the robust stability issue has been discussed in this section. It has been shown that the closed-loop system remains stable, i. e., it is robustly stable, if stability is guaranteed for all possible uncertainties. In a similar way, the notion of robust performance (RP) is to be worked out. The closed-loop system exhibits robust performance if the performance measures are kept within a prescribed limit even for all possible uncertainties, including the worst case, as well. Design considerations for the robust performance have been illustrated in Fig. 10.18.

As the system performance is represented by the signal z, robust performance analysis is based on investigating the relation between the external input signal $\mathbf{w}$ and the performance output z

$$z = \mathbf{F}[\mathbf{G}(s), \mathbf{K}(s), \Delta(s)]\mathbf{w} .$$

Defining a performance measure on the transfer function matrix $\mathbf{F}[\mathbf{G}(s), \mathbf{K}(s), \Delta(s)]$ by

$$J[\mathbf{F}(\mathbf{G}, \mathbf{K}, \Delta)]$$

the performance of the transfer function from the exogenous inputs $\mathbf{w}$ and to outputs z can be calculated. The maximum of the performance – even in the worst case possibly delivered by the uncertainties – can be evaluated by

$$\sup_\Delta \{J[\mathbf{F}(\mathbf{G}, \mathbf{K}, \Delta)]: \|\Delta\|_\infty < 1\} .$$

Based on this value the robust performance of the system can be judged. If the robust performance analysis is to be performed in H_∞ sense, the measure to be applied is

$$J_\infty = \|\mathbf{F}[\mathbf{G}(i\omega), \mathbf{K}(i\omega), \Delta(i\omega)]\|_\infty .$$

In this case the prespecified performance can be normalized and the limit can be selected as 1. So equivalently, the robust performance requirement can be formulated

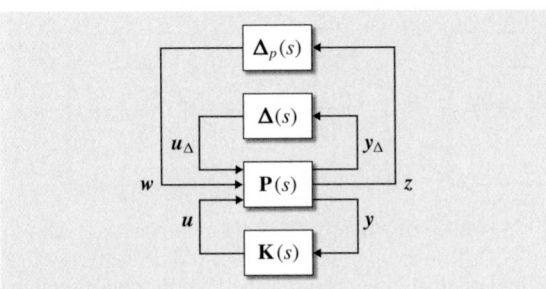

Fig. 10.21 Design for robust performance traced back to robust stability

as

$$\|\mathbf{F}[\mathbf{G}(i\omega),\ \mathbf{K}(i\omega),\ \Delta(i\omega)]\|_\infty < 1\ ,$$

$$\forall \|\Delta(i\omega)\|_\infty \leq 1\ .$$

Robust performance analysis can formally be traced back to robust stability analysis. In this case a fictitious Δ_p uncertainty block representing the nominal performance requirements should be inserted across $\boldsymbol{w}$ and $\boldsymbol{z}$ (Fig. 10.21). Then introducing the

$$\begin{pmatrix} \Delta_p & \mathbf{0} \\ \mathbf{0} & \Delta \end{pmatrix}$$

matrix of the uncertainties gives a pleasant way to trace back the robust performance problem to the robust stability problem.

If robust performance synthesis is used the performance measure must be minimized. In this case μ-optimal design problem can be solved as an extended robust stability design problem.

10.7 General Optimal Control Theory

In the previous sections design techniques have been presented to control linear or linearized plants. Minimization of L_2, H_2 or H_∞ loss functions all resulted in linear control strategies. In practice, however, both the processes and the control actions are mostly nonlinear, e.g., control inputs are typically constrained or saturated in various technologies, and time-optimal control needs to alter the control input instantaneously.

To cover a wider class of control problems the control tasks minimizing loss functions can be formulated in a more general framework [10.25–31]. Restricted to deterministic problems consider the following process to be controlled

$$\dot{\boldsymbol{x}}(t) = \boldsymbol{f}(\boldsymbol{x}(t), \boldsymbol{u}(t), t)\ , \quad 0 \leq t \leq T\ ,$$

$$\boldsymbol{x}(0)\colon \text{ given}\ ,$$

where $\boldsymbol{x}(t)$ denotes the state variables available for state feedback, and $\boldsymbol{u}(t)$ is the control input. The control performance is expressed via the loss function constructed by penalizing terms V_T and V

$$J = V_T(\boldsymbol{x}(T), T) + \int_0^T V[\boldsymbol{x}(t), \boldsymbol{u}(t), t]\, \mathrm{d}t\ .$$

Designing an optimal controller is equivalent to minimize the above loss function.

Denote by $J^*(\boldsymbol{x}(t), t)$ the optimal value of the loss function while the system is governed from an initial state $\boldsymbol{x}(0)$ to a final state $\boldsymbol{x}(T)$. The principle of *dynamic programming* [10.32] determines the optimal control law by

$$\min_{u \in U} \left\{ V[\boldsymbol{x}(t), \boldsymbol{u}(t), t] + \frac{\partial J^*[\boldsymbol{x}(t), t]}{\partial \boldsymbol{x}^\top} f[\boldsymbol{x}(t), \boldsymbol{u}(t), t] \right\}$$
$$= -\frac{\partial J^*[\boldsymbol{x}(t), t]}{\partial t}\ ,$$

where the optimal loss function satisfies

$$J^*[\boldsymbol{x}(T), T] = V_T[\boldsymbol{x}(T), T]\ .$$

The equation of the optimal control law is called the Hamilton–Jacobi–Bellman (HJB) equation in the control literature.

Note that the L_2 and H_2 optimal control policies discussed earlier can be regarded as the special case of the dynamic programming, where the process is linear and the loss function is quadratic, and moreover the control horizon is infinitely large and the control input is not restricted. Thus the linear system

$$\dot{\boldsymbol{x}}(t) = \mathbf{A}\boldsymbol{x}(t) + \mathbf{B}\boldsymbol{u}(t)$$

with the loss function

$$J = \frac{1}{2} \int_0^\infty \left(\boldsymbol{x}^\top \mathbf{Q}_x \boldsymbol{x} + \boldsymbol{u}^\top \mathbf{R}_u \boldsymbol{u} \right) \mathrm{d}t$$

requires the optimal control via state-variable feedback

$$\boldsymbol{u}(t) = -\mathbf{R}_u^{-1}\mathbf{B}^\top \mathbf{P}\boldsymbol{x}(t)\ ,$$

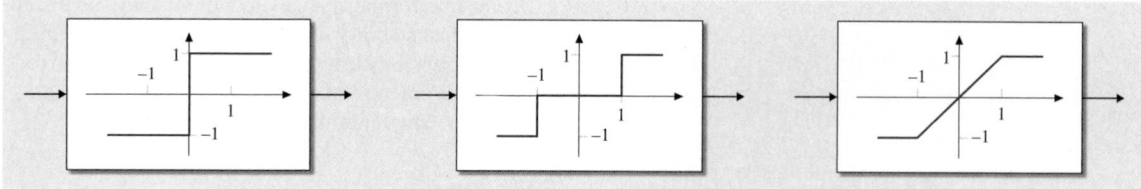

Fig. 10.22 Relay, relay with dead zone, and saturation function to be applied for each vector entry

where $\mathbf{Q}_x \geq \mathbf{0}$ and $\mathbf{R}_u > \mathbf{0}$, and finally the $\mathbf{P}$ matrix is derived as the solution of the following algebraic Riccati equation

$$\mathbf{A}^\top \mathbf{P} + \mathbf{P}\mathbf{A} - \mathbf{P}\mathbf{B}\mathbf{R}_u^{-1}\mathbf{B}^\top \mathbf{P} + \mathbf{Q}_x = \mathbf{0} .$$

At this point it is important to restate the importance of stabilizability and detectability conditions. Without satisfying these conditions, an optimizing controller merely optimizes the cost function and may not stabilize the closed-loop system. In particular, for the LQR problem just discussed, it is important to state that $\{\mathbf{A}, \mathbf{B}\}$ is stabilizable and $\{\mathbf{A}, \mathbf{Q}_x^{1/2}\}$ is detectable.

The HJB equation can be reformulated. To do so, introduce the auxiliary variable $\boldsymbol{\lambda}(t)$ along the optimal trajectory by

$$\boldsymbol{\lambda}(t) \mathrel{\hat{=}} \frac{\partial J[\boldsymbol{x}(t), t]}{\partial \boldsymbol{x}} .$$

Apply $\boldsymbol{\lambda}(t)$ to define the following Hamiltonian function

$$H(\boldsymbol{x}, \boldsymbol{u}, t) \mathrel{\hat{=}} V(\boldsymbol{x}, \boldsymbol{u}, t) + \boldsymbol{\lambda}^\top f(\boldsymbol{x}, \boldsymbol{u}, t) .$$

Then according to the Pontryagin *minimum principle*

$$\frac{\partial H}{\partial \boldsymbol{\lambda}} = \dot{\boldsymbol{x}}(t) ,$$

$$\frac{\partial H}{\partial \boldsymbol{x}} = -\dot{\boldsymbol{\lambda}}(t) ,$$

as well as

$$\boldsymbol{u}^*(t) = \arg \min_{u \in U} H$$

hold. (Note that if the control input is not restricted then the last equation can be written as $\partial H / \partial \boldsymbol{u} = 0$.)

Applying the minimum principle for time-invariant linear dynamic systems with constrained input ($|u_i| \leq 1$, $i = 1 \dots n_u$), various loss functions will lead to special optimal control laws. Having the Hamiltonian function in a general form of

$$H(\boldsymbol{x}, \boldsymbol{u}) = V(\boldsymbol{x}, \boldsymbol{u}) + \boldsymbol{\lambda}^\top (\mathbf{A}\boldsymbol{x} + \mathbf{B}\boldsymbol{u}) ,$$

various optimal control strategies can be formulated by assigning suitable $V(\boldsymbol{x}(t), t)$ loss functions:

- If the goal is to achieve minimal transfer time, then assign $V(\boldsymbol{x}, \boldsymbol{u}) = 1$.
- If the goal is to minimize fuel consumption, then assign $V(\boldsymbol{x}, \boldsymbol{u}) = \boldsymbol{u}^\top \operatorname{sign}(\boldsymbol{u})$.
- If the goal is to minimize energy consumption, then assign $V(\boldsymbol{x}, \boldsymbol{u}) = \frac{1}{2}\boldsymbol{u}^\top \boldsymbol{u}$.

Then the application of the minimum principle provides closed forms for the optimal control, namely:

- Minimal transfer time requires $\boldsymbol{u}^0(t) = -\operatorname{sign}(\mathbf{B}^\top \boldsymbol{\lambda})$ (relay control)
- Minimal fuel consumption requires $\boldsymbol{u}^0(t) = -\operatorname{sgzm}(\mathbf{B}^\top \boldsymbol{\lambda})$ (relay with dead zone)
- Minimal energy consumption requires $\boldsymbol{u}^0(t) = -\operatorname{sat}(\mathbf{B}^\top \boldsymbol{\lambda})$ (saturation).

These examples clarify that even for linear systems the optimal control policies can be (and typically are) nonlinear. The nonlinear relations participating in the above control laws are shown in Fig. 10.22.

Dynamic programming is a general concept allowing the exact mathematical handling of various control strategies. Apart from the simplest cases, however, the optimal control law needs demanding computer-aided calculations. In the next section a special class of optimal controllers will be considered. These controllers are called predictive controllers and they require restricted calculation complexity, however, result in good performance.

10.8 Model-Based Predictive Control

As we have seen so far most control system design methods assume a mathematical model of the process to be controlled. Given a process model and knowledge on the external system inputs the process output can be predicted to some extent. To characterize the behavior of the closed-loop system a combined loss function can be constructed from the values of the predicted process outputs and that of the associated control inputs. Control strategies minimizing this loss function are called *model-based predictive control* (MPC). Related algorithms using special loss functions can be interpreted as an LQ (linear quadratic) problem with finite horizon. The performance of well-tuned predictive control algorithms is outstanding for processes with dead time. Specific model-based predictive control algorithms are also known as dynamic matrix control (DMC), generalized predictive control (GPC), and receding horizon control (RHC) [10.33–41]. Due to the nature of the model-based control algorithms the discrete-time (sampled-data) version of the control algorithms will be discussed in the sequel. Also, most of the detailed discussion is reduced for SISO systems in this section.

The fundamental idea of predictive control can be demonstrated through the DMC algorithm [10.40], where the process output sample $y(k+1)$ is predicted by using all the available process input samples up to the discrete time instant k ($k = 0, 1, 2\ldots$) via a linear function *func*

$$\hat{y}(k+1) = func[u(k), u(k-1), u(k-2), \\ u(k-3), \ldots] \, .$$

Repeating the above one-step-ahead prediction for further time instants as

$$\hat{y}(k+2) = func[u(k+1), u(k), u(k-1), \\ u(k-2), \ldots] \, ,$$
$$\hat{y}(k+3) = func[u(k+2), u(k+1), u(k), \\ u(k-1), \ldots] \, ,$$
$$\vdots$$

requires the knowledge of future control actions $u(k+1), u(k+2), u(k+3), \ldots$, as well. Introduce the *free response* involving the future values of the process input obtained provided no change occurs in the control input at time k

$$y^*(k+1) = func[u(k) = u(k-1), u(k-1), \\ u(k-2), u(k-3), \ldots] \, ,$$
$$y^*(k+2) = func[u(k+1) = u(k-1), u(k) = \\ u(k-1), u(k-1), u(k-2), \ldots] \, ,$$
$$y^*(k+3) = func[u(k+2) = u(k-1), u(k+1) = \\ u(k-1), u(k) = u(k-1), u(k-1), \ldots] \, ,$$
$$\vdots$$

Using the free response just introduced above the predicted process outputs can be expressed by

$$\hat{y}(k+1) = s_1 \Delta u(k) + y^*(k+1)$$
$$\hat{y}(k+2) = s_1 \Delta u(k+1) + s_2 \Delta u(k) + y^*(k+2)$$
$$\hat{y}(k+3) = s_1 \Delta u(k+2) + s_2 \Delta u(k+1) + s_3 \Delta u(k) \\ + y^*(k+3) \, ,$$
$$\vdots$$

where

$$\Delta u(k+i) = u(k+i) - u(k+i-1) \, ,$$

and s_i denotes the i-th sample of the discrete-time-step response of the process. Now a more compact form of

Predicted value = Forced response + Free response

is looked for in vector/matrix form

$$\begin{pmatrix} \hat{y}(k+1) \\ \hat{y}(k+2) \\ \hat{y}(k+3) \\ \vdots \end{pmatrix} = \begin{pmatrix} s_1 & 0 & 0 & \ldots \\ s_2 & s_1 & 0 & \ldots \\ s_3 & s_2 & s_1 & \ldots \\ \vdots & \vdots & \vdots & \ddots \end{pmatrix} \begin{pmatrix} \Delta u(k) \\ \Delta u(k+1) \\ \Delta u(k+2) \\ \vdots \end{pmatrix}$$
$$+ \begin{pmatrix} y^*(k+1) \\ y^*(k+2) \\ y^*(k+3) \\ \vdots \end{pmatrix} \, .$$

Apply here the following notations

$$\mathbf{S} = \begin{pmatrix} s_1 & 0 & 0 & \dots \\ s_2 & s_1 & 0 & \dots \\ s_3 & s_2 & s_1 & \dots \\ \vdots & \vdots & \vdots & \ddots \end{pmatrix},$$

$$\hat{\mathbf{Y}} = [\hat{y}(k+1), \hat{y}(k+2), \hat{y}(k+3), \dots]^{\top},$$

$$\Delta \mathbf{U} = [\Delta u(k), \Delta u(k+1), \Delta u(k+2), \dots]^{\top},$$

$$\mathbf{Y}^* = [y^*(k+1), y^*(k+2), y^*(k+3), \dots]^{\top}.$$

Utilizing these notations

$$\hat{\mathbf{Y}} = \mathbf{S}\Delta\mathbf{U} + \mathbf{Y}^*$$

holds. Assuming that the reference signal (set point) $y_{\text{ref}}(k)$ is available for the future time instants, define the loss function to be minimized by

$$\sum_{i=1}^{N_y} \left[y_{\text{ref}}(k+i) - \hat{y}(k+i) \right]^2.$$

If no restriction for the control signal is taken into account the minimization leads to

$$\Delta \mathbf{U}_{\text{opt}} = \mathbf{S}^{-1}\left(\mathbf{Y}_{\text{ref}} - \mathbf{Y}^*\right),$$

where $\mathbf{Y}_{\text{ref}}$ has been constructed from the samples of the future set points.

The receding horizon control concept utilizes only the very first element of the $\Delta\mathbf{U}_{\text{opt}}$ vector according to

$$\Delta u(k) = \mathbf{1}\Delta\mathbf{U}_{\text{opt}},$$

where $\mathbf{1} = (1, 0, 0, \dots)$. Observe that RHC needs to re-calculate $\mathbf{Y}^*$ and to update $\Delta\mathbf{U}_{\text{opt}}$ in each step. The application of the RHC algorithm results in zero steady-state error; however, it requires considerable control effort while minimizing the related loss function.

Smoothing in the control input can be achieved:

- By extending the loss function with another component penalizing the control signal or its change
- By reducing the control horizon as follows

$$\Delta \mathbf{U} = [\Delta u(k), \Delta u(k+1), \Delta u(k+2), \dots,$$
$$\Delta u(k+N_u-1), 0, 0, \dots, 0]^{\top}.$$

Accordingly, define the loss function by

$$\sum_{i=1}^{N_y} \left[y_{\text{ref}}(k+i) - \hat{y}(k+i) \right]^2$$
$$+ \lambda \sum_{i=1}^{N_u} [u(k+i) - u(k+i-1)]^2,$$

then the control signal becomes

$$\Delta u(k) = \mathbf{1}(\mathbf{S}^{\top}\mathbf{S} + \lambda\mathbf{I})^{-1}\mathbf{S}^{\top}\left(\mathbf{Y}_{\text{ref}} - \mathbf{Y}^*\right),$$

where

$$\mathbf{S} = \begin{pmatrix} s_1 & 0 & \dots & 0 & 0 \\ s_2 & s_1 & \dots & 0 & 0 \\ s_3 & s_2 & \dots & 0 & 0 \\ \vdots & \vdots & \ddots & \vdots & \vdots \\ s_{N_y-1} & s_{N_y-2} & \dots & s_{N_y-N_u+1} & s_{N_y-N_u} \\ s_{N_y} & s_{N_y-1} & \dots & s_{N_y-N_u+2} & s_{N_y-N_u+1} \end{pmatrix}.$$

All the above relations can easily be modified to cover the control of processes with known time delay. Simply replace $y(k+1)$ by $y(k+d+1)$ to consider $y(k+d+1)$ as the earliest sample of the process output effected by the control action taken at time k, where $d > 0$ represents the discrete time delay.

As the idea of the model-based predictive control is quite flexible, a number of variants of the above discussed algorithms exist. The tuning parameters of the algorithm are the prediction horizon N_y, the control horizon N_u, and the λ penalizing factor. Predictions related to disturbances can also be included. Just as an example, a loss function

$$\left(\hat{\mathbf{Y}} - \mathbf{Y}_{\text{ref}}\right)^{\top}\mathbf{W}_y\left(\hat{\mathbf{Y}} - \mathbf{Y}_{\text{ref}}\right) + \mathbf{U}_{\text{c}}^{\top}\mathbf{W}_u\mathbf{U}_{\text{c}}$$

can be assigned to incorporate weighting matrices $\mathbf{W}_u$ and $\mathbf{W}_y$, respectively, and a reduced version of the control signals can be applied according to

$$\mathbf{U} = \mathbf{T}_{\text{c}}\mathbf{U}_{\text{c}},$$

where $\mathbf{T}_{\text{c}}$ is an a priori defined matrix typically containing zeros and ones. Then minimization of the loss function results in

$$\Delta u(k) = \mathbf{1}\mathbf{T}_{\text{c}}\left(\mathbf{T}_{\text{c}}^{\top}\mathbf{S}^{\top}\mathbf{W}_y\mathbf{S}\mathbf{T}_{\text{c}} + \mathbf{W}_u\right)^{-1}\mathbf{T}_{\text{c}}^{\top}\mathbf{S}^{\top}\mathbf{W}_y$$
$$\left(\mathbf{Y}_{\text{ref}} - \mathbf{Y}^*\right).$$

Constraints existing for the control input open a new class for the control algorithms. In this case a quadratic programming problem (conditional minimization of a quadratic loss function to satisfy control constraints represented by inequalities) should be solved in each step. In detail, the loss function

$$\left(\hat{\mathbf{Y}} - \mathbf{Y}_{\text{ref}}\right)^{\top}\mathbf{W}_y\left(\hat{\mathbf{Y}} - \mathbf{Y}_{\text{ref}}\right) + \mathbf{U}_{\text{c}}^{\top}\mathbf{W}_u\mathbf{U}_{\text{c}}$$

is to be minimized again by $\mathbf{U}_{\text{c}}$, where

$$\hat{\mathbf{Y}} = \mathbf{S}\mathbf{T}_{\text{c}}\mathbf{U}_{\text{c}} + \mathbf{Y}^*$$

under the constraint

$$\Delta u_{\min} \leq \Delta u(j) \leq \Delta u_{\max} \quad \text{or}$$

$$u_{\min} \leq u(j) \leq u_{\max} \, .$$

The classical DMC approach is based on the samples of the step response of the process. Obviously, the process model can also be represented by a unit impulse response, a state-space model or a transfer function. Consequently, beyond the control input, the process output prediction can utilize the process output, the state variables or the estimated state variables, as well. Note that the original DMC is an open-loop design method in nature, which should be extended by a closed-loop aspect or be combined with an IMC-compatible concept to utilize the advantages offered by the feedback concept.

A further remark relates to stochastic process models. As an example, the generalized predictive control concept [10.38, 39] applies the model

$$A(q^{-1})y(k) = B(q^{-1})u(k-d) + \frac{C(q^{-1})}{\Delta}\zeta_k \, .$$

where $A(q^{-1})$, $B(q^{-1})$, and $C(q^{-1})$ are polynomials of the backward shift operator q^{-1}, and $\Delta = 1 - q^{-1}$. Moreover, ζ_k is a discrete-time white-noise sequence. Then the conditional expected value of the loss function

$$E\left[(\hat{\boldsymbol{Y}} - \boldsymbol{Y}_{\mathrm{ref}})^{\top} \boldsymbol{W}_y (\hat{\boldsymbol{Y}} - \boldsymbol{Y}_{\mathrm{ref}}) + \boldsymbol{U}_{\mathrm{c}}^{\top} \boldsymbol{W}_u \boldsymbol{U}_{\mathrm{c}} | k \right]$$

is to be minimized by $\boldsymbol{U}_{\mathrm{c}}$. Note that model-based predictive control algorithms can be extended for MIMO and nonlinear systems.

While LQ design supposing infinite horizon provides stable performance, predictive control with finite horizon using receding horizon strategy lacks stability guarantees. Introduction of terminal penalty in the cost function including the quadratic deviations of the states from their final values is one way to ensure stable performance. Other methods leading to stable performance with detailed stability analysis, as well as proper handling of constraints, are discussed in [10.35, 36, 42], where mainly sufficient conditions have been derived for stability.

For real-time applications fast solutions are required. Effective numerical methods to solve optimization problems with reduced computational demand and suboptimal solutions have been developed [10.33].

MPC with linear constraints and uncertainties can be formulated as a multiparametric programming problem, which is a technique to obtain the solution of an optimization problem as a function of the uncertain parameters (generally the states). For the different ranges of the states the calculation can be executed offline [10.33, 43]. Different predictive control approaches for robust constrained predictive control of nonlinear systems are also in the forefront of interest [10.33, 44].

10.9 Control of Nonlinear Systems

In this section results available for linear control systems will be extended for a special class of nonlinear systems. For the sake of simplicity only SISO systems will be considered.

In practice all control systems exhibit nonlinear behavior to some extent [10.45, 46]. To avoid facing problems caused by nonlinear effects linear models around a nominal operating point are considered. In fact, most systems work in a linear region for small changes. However, at various operating points the linearized models are different from each other according to the nonlinear nature. In this section transformation methods will be discussed making the operation of nonlinear dynamic systems linear over the complete range of their operation. Clearly, this treatment, even though the original process to be controlled remains nonlinear, will allow us to apply all the design techniques developed for linear systems. As a common feature the

transformation methods developed for a special class of nonlinear systems all apply *state-variable feedback*. In the past decades a special tool, called Lie algebra, was developed by mathematicians to extend notions such as controllability or observability for nonlinear systems [10.45]. The formalism offered by the Lie algebra will not be discussed here; however, considerations behind the application of this methodology will be presented.

10.9.1 Feedback Linearization

Define the state vector $\boldsymbol{x} \in \mathbb{R}^n$ and the mappings $\{\boldsymbol{f}(\boldsymbol{x}), \boldsymbol{g}(\boldsymbol{x}) : \mathbb{R}^n \to \mathbb{R}^n\}$ as functions of the state vector. Then the Lie derivative of $\boldsymbol{g}(\boldsymbol{x})$ is defined by

$$L_f \boldsymbol{g}(\boldsymbol{x}) \hat{=} \frac{\partial \boldsymbol{g}(\boldsymbol{x})}{\partial \boldsymbol{x}^{\top}} \boldsymbol{f}(\boldsymbol{x})$$

and the Lie product of $g(x)$ and $f(x)$ is defined by

$$\mathrm{ad}_f g(x) \triangleq \frac{\partial g(x)}{\partial x^\top} f(x) - \frac{\partial f(x)}{\partial x^\top} g(x)$$
$$= L_f g(x) - \frac{\partial g(x)}{\partial x^\top} f(x) \,.$$

Consider now a SISO nonlinear dynamic system given by

$$\dot{x} = f(x) + g(x)u \,,$$
$$y = h(x) \,,$$

where $x = (x_1, x_2, \ldots, x_n)^\top$ is the state vector, u is the input, and y is the output, while f, g, and h are unknown smooth nonlinear functions with $\{f(x), g(x)\colon \mathbb{R}^n \to \mathbb{R}^n\}$, $\{h(x)\colon \mathbb{R}^n \to \mathbb{R}\}$, and $f(0) = 0$. The above SISO system has relative degree r at a point x_0 if:

- $L_g L_f^k h(x) = 0$ for all x in a neighborhood of x_0 and for all $k < r - 1$
- $L_g L_f^{r-1} h(x_0) \neq 0$.

It can be shown that the above system equation can be transformed to a form having identical structure for the first r entries

$$\dot{z}_1 = z_2 \,,$$
$$\dot{z}_2 = z_3 \,,$$
$$\vdots$$
$$\dot{z}_{r-1} = z_r \,,$$
$$\dot{z}_r = a(z) + b(z)u \,,$$
$$\dot{z}_{r+1} = q_{r+1}(z) \,,$$
$$\vdots$$
$$\dot{z}_n = q_n(z) \,,$$

and

$$y = z_1 \,,$$

where the last $n - r$ equations are called the zero dynamics. The above equation will be referred to later on as canonical form, where the new state vector is $z = (z_1, z_2, \ldots, z_n)^\top$. Using the Lie derivatives, $a(z)$ and $b(z)$ can be found by

$$a(z) = L_g L_f^{r-1} h(x) \,,$$
$$b(z) = L_f^r h(x) \,.$$

The normal form related to the original system equations can be defined by the diffeomorphism T as follows

$$z_1 = T_1(x) = y = h(x) \,,$$
$$z_2 = T_2(x) = \dot{y} = \frac{\partial h}{\partial x^\top} \dot{x} = L_f h(x) \,,$$
$$\vdots$$
$$z_r = T_r(x) = y^{(r)} = \frac{\partial L_f^{r-2} h(x)}{\partial x^\top} \dot{x} = L_f^{r-1} h(x) \,.$$

Assuming that

$$L_g h(x) = 0 \,,$$
$$L_g L_f h(x) = 0 \,,$$
$$\vdots$$
$$L_g L_f^{r-2} h(x) = 0 \,,$$

all the remaining

$$T_{r+1}(x), \ldots, T_n(x)$$

elements of the transformation matrix can be determined in a similar way. The geometric conditions for the existence of such a global normal form have been studied in [10.45].

Now, concerning the feedback linearization, the following result serves as a starting point for further analysis: a nonlinear system with the above assumptions can be locally transformed into a controllable linear system by state feedback. The transformation of coordinates can be achieved if and only if

$$\mathrm{rank}\{g(x_0), \mathrm{ad}_f g(x_0), \ldots, \mathrm{ad}_f^{n-1} g(x_0)\} = n$$

at a given x_0 point and

$$\{g, \mathrm{ad}_f g, \ldots, \mathrm{ad}_f^{n-2} g\}$$

is involutive near x_0. Introducing

$$v = \dot{z}_r$$

and using the canonical form it is seen that the feedback according to $u = [v - a(z)]/b(z)$ results in a linear relationship between v and y in such a way that v is simply the r-th derivative of y. In other words the linearizing feedback establishes a relation from v to y equivalent to r cascaded integrators. Note that the outputs of the integrators determine r states, while the remaining $(n - r)$ states correspond to the zero dynamics.

The importance of the feedback linearization lies in the fact that the linearized structure allows us to apply the wide selection in the control design methods available for linear systems. Before going into the details of such applications an engineering interpretation of the design aspects of the feedback linearization will be given. Specifically, the feedback linearization technique developed for nonlinear systems will be applied for linear plants.

10.9.2 Feedback Linearization Versus Linear Controller Design

Assume a single-input single-output *linear system* given by the following state-space representation

$$\dot{x} = \mathbf{A}x + \mathbf{B}u \,,$$

$$y = \mathbf{C}x \,.$$

The derivatives of the output can sequentially be generated as

$$y = \mathbf{C}x \,,$$
$$\dot{y} = \mathbf{C}\dot{x} = \mathbf{C}\mathbf{A}x \,,$$
$$\ddot{y} = \mathbf{C}\mathbf{A}\dot{x} = \mathbf{C}\mathbf{A}^2 x \,,$$
$$\vdots$$
$$y^{(r)} = \mathbf{C}\mathbf{A}^{r-1}\dot{x} = \mathbf{C}\mathbf{A}^r x + \mathbf{C}\mathbf{A}^{r-1}\mathbf{B}u \,,$$

where $\mathbf{C}\mathbf{A}^i\mathbf{B} = 0$ for $i < r-1$ and $\mathbf{C}\mathbf{A}^{r-1}\mathbf{B} \neq 0$ with r being the relative degree. Note that r is invariant to similarity transformations. Observe that this derivation is in close harmony with the canonical form derived earlier. The conditions of the exact state control ($r = n$) are that:

- The system is controllable: rank$(\mathbf{B}, \mathbf{A}\mathbf{B}, \mathbf{A}^2\mathbf{B}, \ldots) = n$
- The relative degree is equal to the system order ($r = n$), which could be interpreted as the involutive condition for the linear case.

A more direct relation between the feedback linearization and the linear control design is seen if the linear system is assumed to be given by an input–output (transfer function) representation

$$\frac{y}{u} = \frac{B_{n-r}(s)}{A_n(s)} \,,$$

where the transfer function of the process has appropriate $B_{n-r}(s)$ and $A_n(s)$ polynomials of the complex frequency operator s. The subindices refer to the degree of the polynomials. If the system has a relative degree

of r, the feedback generates

$$y^{(r)} = v \,,$$

which is formally equivalent to a series compensator of

$$C(s) = \frac{A_n(s)}{B_{n-r}(s)} \frac{1}{s^r} \,.$$

The above compensator is realizable in the sense that the numerator and the denominator polynomials are of identical order n. To satisfy practical needs when realizing a controller in a stable manner it is required that the zeros of the process must be in the left half plane. This condition is equivalent to the requirement that the zero dynamics remains exponentially stable as the feedback linearization has been performed.

10.9.3 Sliding–Mode Control

Coming back to the nonlinear case, now a solution is looked for to transform nonlinear dynamic systems via state-variable feedback to dynamic systems with relative degree of 1. Achieving this goal the nonlinear dynamic system with n state variables could be handled as one single integrator, which is evidently easy to control [10.47]. The key point of the solution is to create a fictitious system with relative degree of 1. Additionally, it will be explained that the internal dynamics of the closed-loop system can be defined by the designer.

For the sake of simplicity assume that the nonlinear system is given in controllable form by

$$y^{(n)} = f(x) + g(x)u \,,$$

where the state variables are

$$x = (y, \dot{y}, \ddot{y}, \ldots, y^{(n-1)}) \,.$$

Create the fictitious output signal as the linear combination of these state variables

$$z = h^\top x = h_0 y + h_1 \dot{y} + h_2 \ddot{y} + \ldots + y^{(n-1)} \,.$$

Now use the method elaborated earlier for the feedback linearization. Consider the derivative of the fictitious output signal and, taking $y^{(n)} = f(x) + g(x)u$ into account,

$$\dot{z} = h^\top \dot{x} = h_0 \dot{y} + h_1 \ddot{y} + h_2 \dddot{y} + \ldots + f(x) + g(x)u$$

is obtained. Observe that $\dot{z}$ appears to be a function of the input u, meaning that the relative degree of the fictitious system is 1. Now expressing u, and just as before, introducing the new input signal v,

$$u = \frac{1}{g(x)}\Big[v - h_0 \dot{y} - h_1 \ddot{y} - h_2 \dddot{y} - \ldots - h_{n-2} y^{(n-1)} - f(x) \Big]$$

is obtained. Consequently, the complete system can become realized as one single integrator by $\dot{z} = v$.

The internal dynamics of the closed-loop system is governed by the h_i coefficients. Using Laplace transforms and defining

$$H(s) = h_0 + h_1 s + \ldots + h_{n-2} s^{n-2} + s^{n-1},$$

$z(s)$ can be expressed by

$$z(s) = H(s)y(s).$$

Introduce the z_{ref} reference signal for z

$$\tilde{z}(s) = z(s) - z_{\mathrm{ref}}(s) = H(s)[y(s) - y_{\mathrm{ref}}(s)]$$
$$= H(s)\tilde{y}(s),$$

where

$$z_{\mathrm{ref}}(s) = H(s)y_{\mathrm{ref}}(s).$$

The question is, what happens to $\tilde{y}(t) = L^{-1}[\tilde{y}(s)]$ if $\tilde{z}(t) = L^{-1}[\tilde{z}(s)]$ tends to zero in steady state. Clearly, having all the roots of $H(s)$ on the left half plane, $\tilde{y}(t)$ will tend to zero together with $\tilde{z}(t)$ according to

$$\tilde{y}(s) = \frac{1}{H(s)}\tilde{z}(s).$$

Both $y(t) \rightarrow y_{\mathrm{ref}}(t)$ and $\tilde{z}(t) \rightarrow 0$ is highly expected from a well-designed control loop. In addition, the dynamic behavior of the $y(t) \rightarrow y_{\mathrm{ref}}(t)$ transient response depends on the location of the roots of $H(s)$. One possible setting for $H(s)$ is $H(s) = (s + \lambda)^{n-1}$.

Note that in practice, taking model uncertainties into account, the relation between v and z becomes $\dot{z} \approx v$. To see how to control an approximated integrator effectively recall the Lyapunov stability theory with the Lyapunov function

$$V(\tilde{z}(t)) = \frac{1}{2}\tilde{z}(t)^2.$$

Then $\tilde{z}(t) \rightarrow 0$ if $\dot{V}(\tilde{z}) = \dot{\tilde{z}}\tilde{z} < 0$; moreover, $\tilde{z}(t) = 0$ can be reached in finite time, if the above Lyapunov function satisfies

$$\dot{V}(\tilde{z}) < -\eta |\tilde{z}(t)|,$$

where η is some positive constant. By differentiating $V(\tilde{z})$ the above stability condition reduces to

$$\dot{\tilde{z}}(t) \leq -\eta \operatorname{sign}[\tilde{z}(t)].$$

The above control technique is called the *reaching law*. Here η is to be tuned by the control engineer to compensate the knowledge about the uncertainties involved.

According to the above considerations the control loop discussed earlier should be extended with a discontinuous component given by

$$v(t) = -K_s \operatorname{sign}[\tilde{z}(t)],$$

where $K_s > \eta$. To reduce the switching gain and to increase the speed of convergence, the discontinuous part of the controller can be extended by a proportional feedback

$$v(t) = -K_s \operatorname{sign}[\tilde{z}(t)] - K_p \tilde{z}(t).$$

Note that the application of the above control law may lead to chattering in the control signal. To avoid this undesired phenomenon the saturation function is used instead of the signum function in practice.

To conclude the discussion on sliding control it has been shown that this concept consists of two phases. In the first phase the sliding surface is to be reached (reaching mode), while in the second the system is controlled to move along the sliding surface (sliding mode). In fact, these two phases can be designed independently from each other. Reaching the sliding surface can be realized by appropriate switching elements. The way to reach the sliding surface can be modified via the parameters in the discontinuous part of the controller. Forcing the system to move along the sliding surface can be effected by assigning various parameters in the $H(s)$. The algorithm shown here can be regarded as a special version of the *variable-structure controller design*.

10.10 Summary

In this chapter advanced control system design methods have been discussed. Advanced methods reflect the current state of the art of the related applied research activity in the field and a number of advanced methods are available today for demanding control applications. One of the driving forces behind browsing among various advanced techniques has been the applicability of the control algorithms for practical applications. Starting with the stability issues, then covering performance and robustness issues, MIMO techniques, optimal control strategies, and predictive control algorithms have been discussed. The concept of feedback linearization for certain nonlinear systems has also been shown. Sliding control as a representative of the class of variable-structure controllers has been outlined. To check the detailed operation of the presented control techniques

in numerical examples the reader is kindly asked to use the various toolboxes [10.48].

References

10.1 T. Kailath: *Linear Systems* (Prentice Hall, Upper Saddle River 1980)

10.2 G.C. Goodwin, S.F. Graebe, M.E. Salgado: *Control System Design* (Prentice Hall, Upper Saddle River 2000)

10.3 W.S. Levine (Ed.): *The Control Handbook* (CRC, Boca Raton 1996)

10.4 B.G. Lipták (Ed.): *Instrument Engineers' Handbook, Process Control and Optimization*, 4th edn. (CRC, Boca Raton 2006)

10.5 P.K. Sinha: *Multivariable Control* (Marcel Dekker, New York 1984)

10.6 S. Skogestad, I. Postlethwaite: *Multivariable Feedback Control* (Wiley, New York 2005)

10.7 A.F. D'Souza: *Design of Control Systems* (Prentice Hall, Upper Saddle River 1988)

10.8 R. Bars, P. Colaneri, L. Dugard, F. Allgöwer, A. Kleimenow, C. Scherer: Trends in theory of control system design, 17th IFAC World Congr., Seoul, ed. by M.J. Chung, P. Misra (IFAC Coordinating Committee on Design Methods, San Francisco 2008) pp. 93–104

10.9 K.J. Åström, B. Wittenmark: *Computer-Controlled Systems: Theory and Design* (Prentice Hall, Upper Saddle River 1997)

10.10 G.C. Goodwin, R.H. Middleton: *Digital Control and Estimation: a Unified Approach* (Prentice Hall, Upper Saddle River 1990)

10.11 R. Isermann: *Digital Control Systems*, Vol. I. Fundamentals, Deterministic Control (Springer, Berlin Heidelberg 1989)

10.12 R. Isermann: *Digital Control Systems*, Vol. II. Stochastic Control, Multivariable Control, Adaptive Control, Applications (Springer, Berlin Heidelberg 1991)

10.13 J.M. Maciejowski: *Multivariable Feedback Design* (Addison-Wesley, Indianapolis 1989)

10.14 J.D. Aplevich: *The Essentials of Linear State-Space Systems* (Wiley, New York 2000)

10.15 R.A. Decarlo: *Linear Systems. A State Variable Approach with Numerical Implementation* (Prentice Hall, Upper Saddle River 1989)

10.16 D.C. Youla, H.A. Jabs, J.J. Bongiorno: Modern Wiener–Hopf design of optimal controller, IEEE Trans. Autom. Control **21**, 319–338 (1976)

10.17 M. Morari, E. Zafiriou: *Robust Process Control* (Prentice Hall, Upper Saddle River 1989)

10.18 E.C. Garcia, M. Morari: Internal model control: 1. A unifying review and some new results, Ind. Eng. Chem. Process Des. Dev. **21**, 308–323 (1982)

10.19 O.J.M. Smith: Close control of loops with dead time, Chem. Eng. Prog. **53**, 217–219 (1957)

10.20 V. Kučera: Diophantine equations in control – a survey, Automatica **29**, 1361–1375 (1993)

10.21 J.B. Burl: *Linear Optimal Control: H_2 and H_∞ Methods* (Addison-Wesley, Indianapolis 1999)

10.22 J.C. Doyle, B.A. Francis, A.R. Tannenbaum: *Feedback Control Theory* (Macmillan, London 1992)

10.23 K. Zhou, J.C. Doyle, K. Glover: *Robust and Optimal Control* (Prentice Hall, Upper Saddle River 1996)

10.24 M. Vidyasagar, H. Kimura: Robust controllers for uncertain linear multivariable systems, Automatica **22**, 85–94 (1986)

10.25 B.D.O. Anderson, J.B. Moore: *Optimal Control* (Prentice Hall, Upper Saddle River 1990)

10.26 A.E. Bryson, Y. Ho: *Applied Optimal Control* (Hemisphere/Wiley, New York 1975)

10.27 A.E. Bryson: *Dynamic Optimization* (Addison-Wesley, Indianapolis 1999)

10.28 H. Kwakernaak, R. Sivan: *Linear Optimal Control Systems* (Wiley-Interscience, New York 1972)

10.29 F.L. Lewis, V.L. Syrmos: *Optimal Control* (Wiley, New York 1995)

10.30 S.I. Lyashko: *Generalized Optimal Control of Linear Systems with Distributed Parameters* (Kluwer, Dordrecht 2002)

10.31 D.S. Naidu: *Optimal Control Systems* (CRC, Boca Raton 2003)

10.32 D.P. Bertsekas: *Dynamic Programming and Optimal Control*, Vol. I,II (Athena Scientific, Nashua 2001)

10.33 E.F. Camacho, C. Bordons: *Model Predictive Control* (Springer, Berlin Heidelberg 2004)

10.34 D.W. Clarke (Ed.): *Advances in Model-Based Predictive Control* (Oxford Univ. Press, Oxford 1994)

10.35 J.M. Maciejowski: *Predictive Control with Constraints* (Prentice Hall, Upper Saddle River 2002)

10.36 J.A. Rossiter: *Model-Based Predictive Control – a Practical Approach* (CRC, Boca Raton 2003)

10.37 R. Soeterboek: *Predictive Control – a Unified Approach* (Prentice Hall, Upper Saddle River 1992)

10.38 D.W. Clarke, C. Mohtadi, P.S. Tuffs: Generalised predictive control – Part 1. The basic algorithm, Automatica **23**, 137 (1987)

10.39 D.W. Clarke, C. Mohtadi, P.S. Tuffs: Generalised predictive control – Part 2. Extensions and interpretations, Automatica **23**, 149 (1987)

10.40 C.R. Cutler, B.L. Ramaker: Dynamic matrix control – a computer control algorithm, Proc. JACC (San Francisco 1980)

Part B | 10

10.41   C.E. Garcia, D.M. Prett, M. Morari: Model predictive control: theory and practice – a survey, Automatica **25**, 335–348 (1989)

10.42   D.Q. Mayne, J.B. Rawlings, C.V. Rao, P.O.M. Scokaert: Constrained model predictive control: stability and optimality, Automatica **36**, 789–814 (2000)

10.43   F. Borrelli: *Constrained Optimal Control of Linear and Hybrid Systems* (Springer, Berlin Heidelberg 2003)

10.44   T.A. Badgewell, S.J. Qin: *Nonlinear Predictive Control Chapter: Review of Nonlinear Model Predictive Control Application*, IEE Control Eng. Ser., Vol. 61, ed. by M.B. Cannon (Kouvaritabis, London 2001)

10.45   A. Isidori: *Nonlinear Control Systems* (Springer, Berlin Heidelberg 1995)

10.46   J.J.E. Slotine, W. Li: *Applied Nonlinear Control* (Prentice Hall, Upper Saddle River 1991)

10.47   V.I. Utkin: *Sliding Modes in Control and Optimization* (Springer, Berlin Heidelberg 1992)

10.48   *Control system toolbox for use with MATLAB. User's guide* (The Math Works Inc. 1998)

11. Control of Uncertain Systems

Jianming Lian, Stanislaw H. Żak

Novel direct adaptive robust state and output feedback controllers are presented for the output tracking control of a class of nonlinear systems with unknown system dynamics and disturbances. Both controllers employ a variable-structure radial basis function (RBF) network that can determine its structure dynamically to approximate unknown system dynamics. Radial basis functions are added or removed online in order to achieve the desired tracking accuracy and prevent to network redundancy. The raised-cosine RBF is employed to enable fast and efficient training and output evaluation of the RBF network. The direct adaptive robust output feedback controller is constructed by utilizing a high-gain observer to estimate the tracking error for the controller implementation. The closed-loop systems driven by the variable neural direct adaptive robust controllers are actually switched systems.

Part B | 11

Automation is commonly understood as the replacement of manual operations by computer-based methods. Automation is also defined as the condition of being automatically controlled or operated. Thus control is an essential ingredient of automation. The goal of control is to specify the controlled system inputs that force the system outputs to behave in a prespecified manner. This specification of the appropriate system inputs is realized by a controller being developed by a control engineer. In any controller design problem, the first step is to construct the so-called *truth model* of the dynamics of the process to be controlled, where the process is often referred to as the plant. Because the truth model contains all the relevant characteristics of the plant, it is too complicated to be used for the controller design and is mainly used as a simulation model to test the performance of the developed controller [11.1]. Thus, a simplified model that contains the essential features of the plant has to be derived to be used as design model for the controller design. However, it is often infeasible in real applications to obtain a quality mathematical model because the underlying dynamics of the plant may not be understood well enough. Thus, the derived mathematical model may contain uncertainties, which may come from lack of parameter values, either constant or time varying, or result from imperfect knowledge of system inputs. In addition, the inaccurate modeling can introduce uncertainties to the mathematical model as well. Examples of uncertain systems are robotic manipulators or chemical reactors [11.2]. In robotic manipulators, inertias as seen by the drive motors vary with the end-effector position and the load mass so that the robot's dynamical model varies with the robot's attitude. For chemical reactors, their transfer functions vary according to the mix of reagents and catalysts in the vessel and change as the reaction progresses [11.2]. Hence, effective approaches to the control of uncertain systems with high performance are in demand.

11.1 Background and Overview

One approach to the control of uncertain systems is so-called deterministic robust control. Deterministic robust controllers use fixed nonlinear feedback control to ensure the stability of the closed-loop system over a specified range of a class of parametric variations [11.3]. Deterministic control includes variable-structure control and Lyapunov min–max control. Variable structure control was first introduced by Emel'yanov et al. in the early 1960s. It is a nonlinear switching feedback control, which has discontinuity on one or more manifolds in the state space [11.4]. A particular type of variable-structure control is sliding mode control [11.5,6]. Under sliding mode control, the system states are driven to and are then constrained within a neighborhood of the intersection of all the switching manifolds. The Lyapunov min–max control was proposed in [11.7], where nonlinear controllers are developed based on Lyapunov functions and uncertainty bounds. Deterministic robust controllers can guarantee transient performance and final tracking accuracy by compensating for parametric uncertainties and input disturbances with robustifying components. However, they usually involve high-gain feedback or switching, which, in turn, results in high-frequency chattering in the responses of the controlled systems. On the other hand, high-frequency chattering may also excite the unmodeled high-frequency dynamics. Smoothing techniques to eliminate high-frequency chattering were proposed in [11.8]. The resulting controllers are continuous within a boundary layer in the neighborhood of the switching manifold so that the high-frequency chattering can be prevented. However, this is achieved at the price of degraded control performance.

Another effective approach to the control of uncertain systems is adaptive control [11.9–11]. Adaptive controllers are different from deterministic robust controllers because they have a learning mechanism that adjusts the controller's parameters automatically by adaptation laws in order to reduce the effect of the uncertainties. There are two kinds of adaptive controllers: indirect and direct adaptive controllers. In indirect adaptive control strategies, the plant's parameters are estimated online and the controller's parameters are adjusted based on these estimates, see [11.12, p. 14] and [11.13, p. 14]. In contrast, in direct adaptive control strategies, the controller's parameters are directly adjusted to improve a given performance index without the effort to identify the plant's parameters [11.12, p. 14]. Several adaptive controller design methodologies for uncertain systems have been introduced such as adaptive feedback linearization [11.14, 15], adaptive backstepping [11.11, 16–18], nonlinear damping and swapping [11.19] and switching adaptive control [11.20–22]. Adaptive controllers are capable of achieving asymptotic stabilization or tracking for systems subject to only parametric uncertainties without high-gain feedback. However, the adaptation laws of adaptive controllers may lead to instability even when small disturbances appears [11.23]. When adaptive controllers utilize function approximators to approximate unknown system dynamics, the robustness of the adaptation laws to approximation errors needs to be considered as well. The robustness issues make the applicability of adaptive controllers questionable, because there are always disturbances, internal or external, in real systems. To address this problem, robust adaptive controllers were developed to ensure the stability of adaptive controllers [11.23–26]. However, there is a disadvantage shared by adaptive controllers and robust adaptive controllers: their transient performance cannot be guaranteed and the final tracking accuracy usually depends on the approximation errors and disturbances. Thus, adaptive robust controllers, which effectively combine the design techniques of adaptive control and deterministic robust control, were proposed [11.27–33].

In particular, various adaptive (robust) control strategies for feedback linearizable uncertain systems have been proposed. A feedback linearizable system model can be transformed into equivalent linear models by a change of coordinates and a static-state feedback so that linear control design methods can be applied to achieve the desired performance. This approach has been successfully applied to the control of both single-input single-output (SISO) systems [11.10, 27–29, 32–40] and multi-input multi-output (MIMO) systems [11.41–46]. The above adaptive (robust) control strategies have been developed under the assumption that all system states are available in the controller implementation. However, in practical applications, only the system outputs are usually available. To overcome

Table 11.1 Limitations of different tracking control strategies

L1	Prior knowledge of f and/or g
L2	No disturbance
L3	Needs fuzzy rules describing the system operation
L4	Requires offline determination of the appropriate network structure
L5	Availability of the plant states
L6	Restrictive assumptions on the controller architecture
L7	Tracking performance depends on function approximation error

Table 11.2 Advantages of different tracking control strategies

A1	Uses system outputs only
A2	Guaranteed transient performance
A3	Guaranteed final tracking accuracy
A4	Avoids defining basis functions
A5	No need for offline neural network structure determination
A6	Removes the controller singularity problem completely

the problem of inaccessibility of the system states, output feedback controllers that employ state observers in feedback implementation were developed. In particular, a high-gain observer has been employed in the design of output feedback-based control strategies for nonlinear systems [11.38, 46–51]. The advantage of using a high-gain observer is that the control problem can be formulated in a standard singular perturbation format and then the singular perturbation theory can be applied to analyze the closed-loop system stability. The performance of the output feedback controller utilizing a high-gain observer would asymptotically approach the performance of the state feedback controller [11.49].

To deal with dynamical uncertainties, adaptive (robust) control strategies often involve certain types of function approximators to approximate unknown system dynamics. The use of fuzzy-logic systems for function approximation has been introduced [11.28, 29, 32, 33, 36, 39, 40, 42–44, 46]. However, the fuzzy rules required by the fuzzy-logic systems may not be available. On the other hand, one-layer neural-network-based adaptive (robust) control approaches have been reported [11.26, 27, 34, 38] that use radial basis function (RBF) networks to approximate unknown system dynamics. However, fixed-structure RBF networks require offline determination of the appropriate network structure, which is not suitable for the online operation. In [11.10, 35, 41], multilayer neural-network-based adaptive robust control strategies were proposed to avoid some limitations associated with one-layer neural network such as defining a basis function set or choosing some centers and variations of radial basis type of activation functions [11.35]. Although it is not required to define a basis function set for multilayer neural network, it is still necessary to predetermine the number of hidden neurons. Moreover, compared with multilayer neural networks, RBF networks are charac-

terized by simpler structures, faster computation time, and superior adaptive performance. Variable-structure neural-network-based adaptive (robust) controllers have recently been proposed for SISO feedback linearizable uncertain systems. In [11.52], a constructive wavelet-network-based adaptive state feedback controller was developed. In [11.53–56], variable-structure RBF networks are employed in the adaptive (robust) controller design. Variable-structure RBF networks preserve the advantages of the RBF network and, at the same time, overcome the limitations of fuzzy-logic systems and the fixed-structure RBF network. In [11.53], a growing RBF network was utilized for function approximation, and in [11.54–57] self-organizing RBF networks that can both grow and shrink were used. However, all these variable-structure RBF networks are subject to the problem of infinitely fast switching between different structures because there is no time constraint on two consecutive switchings. To overcome this problem, a dwelling-time requirement is introduced into the structure variation of the RBF network in [11.58].

In this chapter, the problem of output tracking control is considered for a class of SISO feedback linearizable uncertain systems modeled by

$$\begin{cases} \dot{x}_i = x_{i+1}, & i = 1, \ldots, n-1 \\ \dot{x}_n = f(\boldsymbol{x}) + g(\boldsymbol{x})u + d \\ y = x_1, \end{cases} \qquad (11.1)$$

where $\boldsymbol{x} = (x_1, \ldots, x_n)^\top \in \mathbb{R}^n$ is the state vector, $u \in \mathbb{R}$ is the input, $y \in \mathbb{R}$ is the output, d models the disturbance, and $f(\boldsymbol{x})$ and $g(\boldsymbol{x})$ are unknown functions with $g(\boldsymbol{x})$ bounded away from zero. A number of adaptive (robust) tracking control strategies have been reported in the literature. In this chapter, novel direct adaptive robust state and output feedback controllers are presented. Both controllers employ the variable-structure RBF network presented in [11.58], which is an improved version of the network considered in [11.56, 57], for

Table 11.3 Types of tracking control strategies

T1	Direct state feedback adaptive controller
T2	Direct output feedback adaptive controller
T3	Includes robustifying component
T4	Fuzzy logic system based function approximation
T5	Fixed-structure neural-network-based function approximation
T6	Multilayer neural-network-based function approximation
T7	Variable-structure neural-network-based function approximation

function approximation. This variable-structure RBF network avoids selecting basis functions offline by determining its structure online dynamically. It can add or remove RBFs according to the tracking performance in order to ensure tracking accuracy and prevent network redundancy simultaneously. Moreover, a dwelling-time requirement is imposed on the structure variation to avoid the problem of infinitely fast switching between different structures as in [11.52, 59]. The raised-cosine RBF presented in [11.60] is employed instead of the commonly used Gaussian RBF because the raised-cosine RBF has compact support, which can significantly reduce computations for the RBF network's training and output evaluation [11.61]. The direct adaptive robust output feedback controller is constructed by incorporating a high-gain observer to estimate the tracking error for the controller implementation. The closed-loop systems driven by the direct adaptive robust controllers are characterized by guaranteed transient performance and final tracking accuracy. The lists of limitations and advantages of different tracking control strategies found in the recent literature are given in Tables 11.1 and 11.2, respectively. In Table 11.3, different types of tracking control strategies are listed. In Table 11.4, these tracking control strategies are compared with each other. The control strategy in this chapter shares the same disadvantages and advantages as that in [11.56, 57].

Table 11.4 Comparison of different tracking control strategies

Controller types	Reference	Limitations							Advantages					
		L1	**L2**	**L3**	**L4**	**L5**	**L6**	**L7**	**A1**	**A2**	**A3**	**A4**	**A5**	**A6**
T2 T3	[11.50]	✓	✓				✓		✓	✓	✓			
	[11.62]	✓	✓						✓	✓	✓			
	[11.49]	✓	✓				✓		✓	✓	✓			
T1 T3 T4	[11.40]		✓	✓		✓				✓				
	[11.32]		✓	✓		✓				✓				
	[11.29]	✓	✓	✓		✓				✓				
T1 T4	[11.36]			✓		✓		✓		✓				
	[11.28]	✓	✓	✓		✓		✓		✓				
T1 T5	[11.37]	✓			✓	✓		✓	✓	✓				
	[11.63]		✓		✓	✓		✓		✓				✓
T1 T3 T5	[11.27]	✓	✓		✓	✓				✓				
	[11.34]	✓	✓		✓	✓				✓				
T2 T3 T5	[11.38]		✓		✓		✓		✓	✓				
T1 T6	[11.64]	✓				✓		✓	✓	✓	✓			
T1 T3 T6	[11.10]		✓			✓	✓			✓	✓			
	[11.35]					✓				✓	✓			✓
T1 T7	[11.54]		✓			✓		✓		✓		✓		
	[11.52]	✓	✓			✓		✓		✓		✓		
T1 T3 T7	[11.53]	✓	✓			✓				✓		✓		
	[11.55]		✓			✓				✓		✓		
T2 T3 T7	[11.56, 57]								✓	✓	✓	✓		

11.2 Plant Model and Notation

The system dynamics (11.1) can be represented in a canonical controllable form as

$$\begin{cases} \dot{x} = \mathbf{A}x + b\big(f(x) + g(x)u + d\big) , \\ y = cx , \end{cases} \tag{11.2}$$

where

$$\mathbf{A} = \begin{pmatrix} \mathbf{0}_{n-1} & \mathbf{I}_{n-1} \\ 0 & \mathbf{0}_{n-1}^{\top} \end{pmatrix} , \quad b = \begin{pmatrix} \mathbf{0}_{n-1} \\ 1 \end{pmatrix} ,$$

$$c^{\top} = \begin{pmatrix} 1 \\ \mathbf{0}_{n-1} \end{pmatrix} ,$$

and $\mathbf{0}_{n-1}$ denotes the $(n-1)$-dimensional zero vector. For the above system model, it is assumed in this chapter that $f(x)$ and $g(x)$ are unknown Lipschitz continuous functions. Without loss of generality, $g(x)$ is assumed to be strictly positive such that $0 < \underline{g} \leq g(x) \leq \overline{g}$, where $\underline{g}$ and $\overline{g}$ are lower and upper bounds of $g(x)$. The disturbance d could be in the form of $d(t)$, $d(x)$ or $d(x, t)$. It is assumed that d is Lipschitz-continuous in x and piecewise-continuous in t. It is also assumed that $|d| \leq d_0$, where d_0 is a known constant. The control objective is to develop a tracking control strategy such that the system output y tracks a reference signal y_d as accurately as possible. It is assumed that the desired trajectory y_d has bounded derivatives up to the n-th order, that is, $y_d^{(n)} \in \boldsymbol{\Omega}_{y_d}$, where $\boldsymbol{\Omega}_{y_d}$ is a compact subset of $\mathbb{R}$. The desired system state vector x_d is then defined as

$$x_d = \left(y_d, \dot{y}_d, \dots, y_d^{(n-1)} \right)^{\top} .$$

We have $x_d \in \boldsymbol{\Omega}_{x_d}$, where $\boldsymbol{\Omega}_{x_d}$ is a compact subset of $\mathbb{R}^n$. Let

$$e = y - y_d \tag{11.3}$$

denote the output tracking error and let

$$e = x - x_d = \left(e, \dot{e}, \dots, e^{(n-1)} \right)^{\top} \tag{11.4}$$

denote the system tracking error. Then the tracking error dynamics can be described as

$$\dot{e} = \mathbf{A}e + b\left[y^{(n)} - y_d^{(n)} \right] \tag{11.5}$$

$$= \mathbf{A}e + b\left[f(x) + g(x)u - y_d^{(n)} + d \right] . \tag{11.6}$$

Consider the following controller,

$$u_a = \frac{1}{\hat{g}(x)}\left[-\hat{f}(x) + y_d^{(n)} - ke \right] , \tag{11.7}$$

where $\hat{f}(x)$ and $\hat{g}(x)$ are approximations of $f(x)$ and $g(x)$, respectively, and k is selected such that $\mathbf{A}_m = \mathbf{A} - bk$ is Hurwitz. The controller u_a in (11.7) consists of a feedforward term $-\hat{f}(x) + y_d^{(n)}$ for model compensation and a linear feedback term $-ke$ for stabilization. Substituting (11.7) into (11.6), the tracking error dynamics become

$$\dot{e} = \mathbf{A}_m e + b\tilde{d} , \tag{11.8}$$

where

$$\tilde{d} = \left[f(x) - \hat{f}(x) \right] + \left[g(x) - \hat{g}(x) \right] u_a + d . \tag{11.9}$$

It follows from (11.8) that, if only u_a is applied to the plant, the tracking error does not converge to zero if $\tilde{d}$ is present. Therefore, an additional robustifying component is required to ensure the tracking performance in the presence of approximation errors and disturbances.

11.3 Variable-Structure Neural Component

In this section, the variable-structure RBF network employed to approximate $f(x)$ and $g(x)$ over a compact set $\boldsymbol{\Omega}_x \subset \mathbb{R}^n$ is first introduced. This variable-structure RBF network is an improved version of the self-organizing RBF network considered in [11.56], which in turn was adapted from [11.61]. The RBF adding and removing operations are improved and a dwelling time T_d is introduced into the structure variation of

the network to prevent fast switching between different structures.

The employed self-organizing RBF network has N different admissible structures, where N is determined by the design parameters discussed later. For each admissible structure illustrated in Fig. 11.1, the self-organizing RBF network consists of n input neurons, M_v hidden neurons, where $v \in \{1, \dots, N\}$

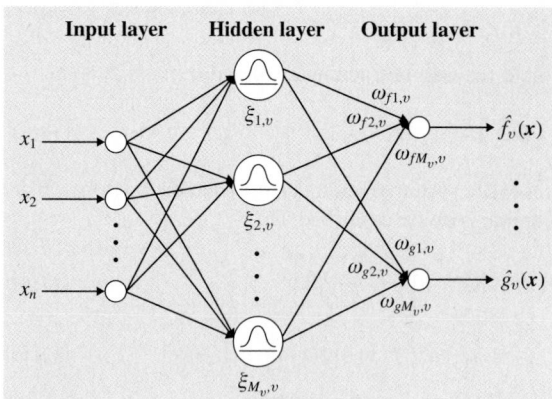

Fig. 11.1 Self-organizing radial basis function network

denotes the scalar index, and two output neurons corresponding to $\hat{f}_v(\boldsymbol{x})$ and $\hat{g}_v(\boldsymbol{x})$. For a given input $\boldsymbol{x} = (x_1, x_2, \ldots, x_n)^\top$, the output $\hat{f}_v(\boldsymbol{x})$ is represented as

$$\hat{f}_v(\boldsymbol{x}) = \sum_{j=1}^{M_v} \omega_{fj,v} \xi_{j,v}(\boldsymbol{x}) \tag{11.10}$$

$$= \sum_{j=1}^{M_v} \omega_{fj,v} \prod_{i=1}^{n} \psi\left(\frac{|x_i - c_{ij,v}|}{\delta_{ij,v}}\right), \tag{11.11}$$

where $\omega_{fj,v}$ is the adjustable weight from the j-th hidden neuron to the output neuron and $\xi_{j,v}(\boldsymbol{x})$ is the radial basis function for the j-th hidden neuron. The parameter $c_{ij,v}$ is the i-th coordinate of the center of $\xi_{j,v}(\boldsymbol{x})$, $\delta_{ij,v}$ is the radius of $\xi_{j,v}(\boldsymbol{x})$ in the i-th coordinate, and $\psi : [0, \infty) \to \mathbb{R}^+$ is the activation function. In the above, the symbol $\mathbb{R}^+$ denotes the set of nonnegative real numbers. Usually, the activation function ψ is constructed so that it is radially symmetric with respect to its center. The largest value of ψ is obtained when $x_i = c_{ij,v}$, and the value of ψ vanishes or becomes very small for large $|x_i - c_{ij,v}|$. Let

$$\boldsymbol{\omega}_{f,v} = \left(\omega_{f1,v}, \omega_{f2,v}, \ldots, \omega_{fM_v,v}\right)^\top \tag{11.12}$$

be the weight vector and let

$$\boldsymbol{\xi}_v(\boldsymbol{x}) = \left(\xi_{1,v}(\boldsymbol{x}), \xi_{2,v}(\boldsymbol{x}), \ldots, \xi_{M_v,v}(\boldsymbol{x})\right)^\top. \tag{11.13}$$

Then (11.11) can be rewritten as $\hat{f}_v(\boldsymbol{x}) = \boldsymbol{\omega}_{f,v}^\top \boldsymbol{\xi}_v(\boldsymbol{x})$, and the output $\hat{g}_v(\boldsymbol{x})$ can be similarly represented as $\hat{g}_v(\boldsymbol{x}) = \boldsymbol{\omega}_{g,v}^\top \boldsymbol{\xi}_v(\boldsymbol{x})$.

One of the most popular types of radial basis functions is the Gaussian RBF (GRBF) that has the form

$$\xi(x) = \exp\left(-\frac{(x-c)^2}{2\delta^2}\right). \tag{11.14}$$

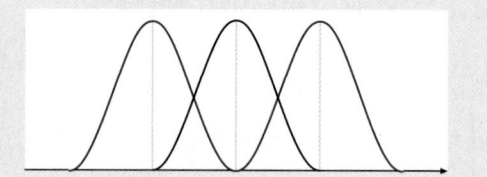

Fig. 11.2 Plot of one-dimensional (1-D) raised-cosine radial basis functions

The support of the GRBF is unbounded. The compact support of the RBF plays an important role in achieving fast and efficient training and output evaluation of the RBF network, especially as the size of the network and the dimensionality of the input space increase. Therefore, the raised-cosine RBF (RCRBF) presented in [11.60] which has compact support is employed herein. The one-dimensional raised-cosine RBF shown in Fig. 11.2 is described as

$$\xi(x) = \begin{cases} \frac{1}{2}\left[1 + \cos\left(\frac{\pi(x-c)}{\delta}\right)\right] & \text{if } |x - c| \le \delta \\ 0 & \text{if } |x - c| > \delta, \end{cases} \tag{11.15}$$

whose support is the compact set $[c - \delta, c + \delta]$. In the n-dimensional space, the raised-cosine RBF centered at $\boldsymbol{c} = [c_1, c_2, \ldots, c_n]$ with $\boldsymbol{\delta} = [\delta_1, \delta_2, \ldots, \delta_n]$ can be represented as the product of n one-dimensional raised-cosine RBFs

$$\xi(\boldsymbol{x}) = \prod_{i=1}^{n} \xi(x_i) \tag{11.16}$$

$$= \begin{cases} \prod_{i=1}^{n} \frac{1}{2}\left[1 + \cos\left(\frac{\pi(x_i - c_i)}{\delta_i}\right)\right] \\ \qquad \text{if } |x_i - c_i| \le \delta_i \text{ for all } i, \\ 0 \qquad \text{if } |x_i - c_i| > \delta_i \text{ for some } i. \end{cases} \tag{11.17}$$

A plot of a two-dimensional raised-cosine RBF is shown in Fig. 11.3.

Unlike fixed-structure RBF networks that require offline determination of the network structure, the employed self-organizing RBF network is capable of determining the parameters M_v, $c_{ij,v}$, and $\delta_{ij,v}$ dynamically according to the tracking performance. Detailed descriptions are given in the following subsections.

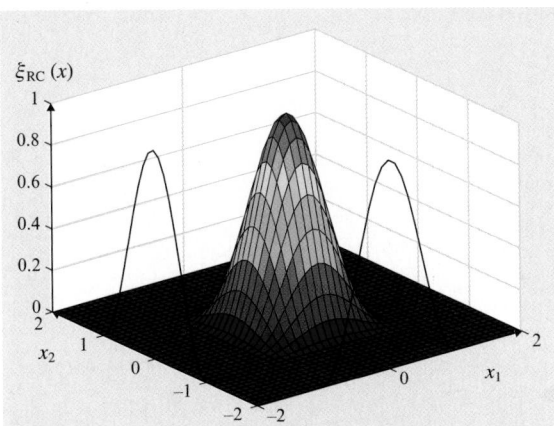

Fig. 11.3 Plot of a two-dimensional (2-D) raised-cosine radial basis function

11.3.1 Center Grid

Recall that the unknown functions are approximated over a compact set $\Omega_x \subset \mathbb{R}^n$. It is assumed that Ω_x can be represented as

$$\Omega_x = \left\{ x \in \mathbb{R}^n : x_l \le x \le x_u \right\} \tag{11.18}$$
$$= \left\{ x \in \mathbb{R}^n : x_{li} \le x_i \le x_{ui}, \ 1 \le i \le n \right\}, \tag{11.19}$$

where the n-dimensional vectors x_l and x_u denote lower and upper bounds of x, respectively. To locate the centers of RBFs inside the approximation region Ω_x, an n-dimensional center grid with layer hierarchy is utilized, where each grid node corresponds to the center of one RBF. The center grid is initialized with its nodes located at $(x_{l1}, x_{u1}) \times (x_{l2}, x_{u2}) \times \cdots \times (x_{ln}, x_{un})$, where $\times$ denotes the Cartesian product. The 2^n grid nodes of the initial center grid are referred to as boundary grid

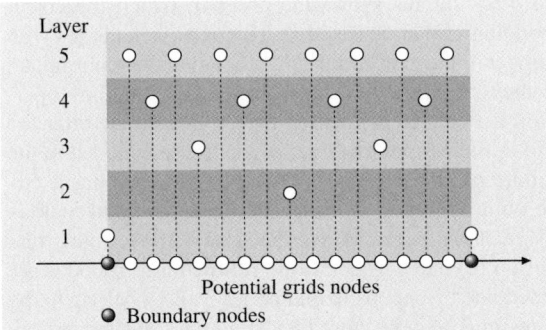

Fig. 11.4 Example of determining potential grid nodes in one coordinate

nodes and cannot be removed. Additional grid nodes will be added and then can be removed within this initial grid as the controlled system evolves in time. The centers of new RBFs can only be placed at the potential locations. The potential grid nodes are determined coordinate-wise. In each coordinate, the potential grid nodes of the first layer are the two fixed boundary grid nodes. The second layer has only one potential grid node in the middle of the boundary grid nodes. Then the potential grid nodes of the subsequent layers are in the middle of the adjacent potential grid nodes of all the previous layers. The determination of potential grid nodes in one coordinate is illustrated in Fig. 11.4.

11.3.2 Adding RBFs

As the controlled system evolves in time, the output tracking error e is measured. If the magnitude of e exceeds a predetermined threshold e_{max}, and if the dwelling time of the current network structure has been greater than the prescribed T_d, the network tries to add new RBFs at potential grid nodes, that is, add new grid nodes. First, the nearest-neighboring grid node, denoted $c_{(nearest)}$, to the current input x is located among existing grid nodes. Then the nearer-neighboring grid node denoted $c_{(nearer)}$ is located, where $c_{i(nearer)}$ is determined such that x_i is between $c_{i(nearest)}$ and $c_{i(nearer)}$. Next, the adding operation is performed for each coordinate independently.

In the i-th coordinate, if the distance between x_i and $c_{i(nearest)}$ is smaller than a prescribed threshold $d_{i(threshold)}$ or smaller than a quarter of the distance between $c_{i(nearest)}$ and $c_{i(nearer)}$, no new grid node is added in the i-th coordinate. Otherwise, a new grid node located at half of the sum of $c_{i(nearest)}$ and $c_{i(nearer)}$ is added in the i-th coordinate. The design parameter $d_{i(threshold)}$ specifies the minimum grid distance in the i-th coordinate. The above procedures for adding RBFs are illustrated with two-dimensional examples shown in Fig. 11.5. In case 1, no RBFs are added. In case 2, new grid nodes are added out of the first coordinate. In case 3, new grid nodes are added out of both coordinates. In summary, a new grid node is added in the i-th coordinate if the following conditions are satisfied:

1. $|e| > e_{max}$.
2. The elapsed time since last operation, adding or removing, is greater than T_d.
3. $\left| x_i - c_{i(nearest)} \right|$
   $> \max \left\{ \left| c_{i(nearest)} - c_{i(nearer)} \right| / 4, d_{i(threshold)} \right\}$.

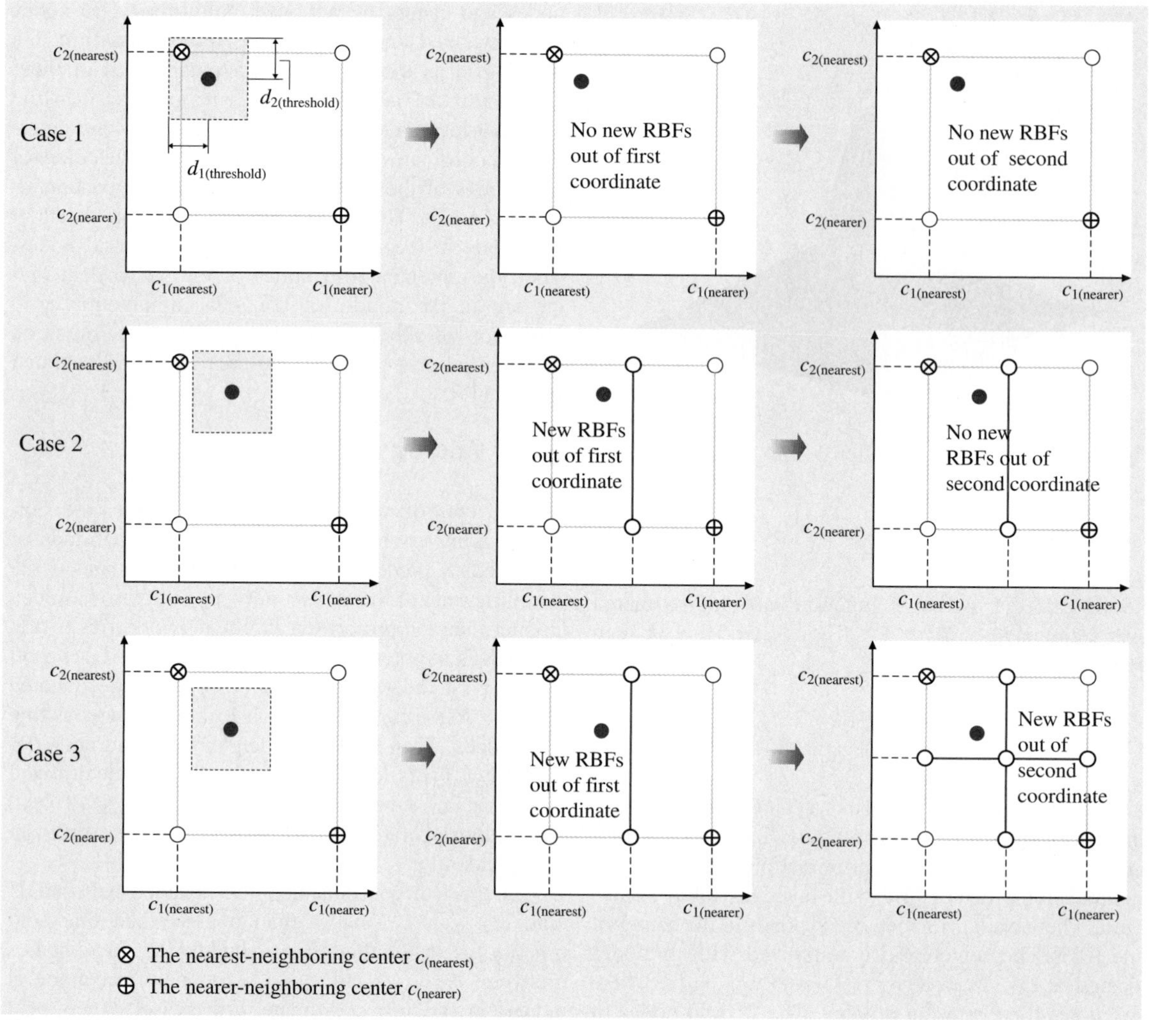

Fig. 11.5 Two-dimensional examples of adding RBFs

The layer of the i-th coordinate assigned to the newly added grid node is one level higher than the highest layer of the two adjacent existing grid nodes in the same coordinate. A possible scenario of formation of the layer hierarchy in one coordinate is shown in Fig. 11.6. The white circles denote potential grid nodes, and the black circles stand for existing grid nodes. The number in

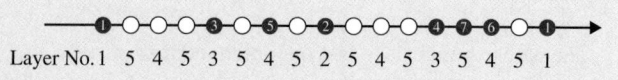

Layer No. 1   5 4 5 3 5 4 5 2 5 4 5 3 5 4 5 1

Fig. 11.6 Example of formation of the layer hierarchy in one coordinate

the black circles shows the order in which the corresponding grid node is added. The two black circles with number 1 are the initial grid nodes in this coordinate, so they are in the first layer. Suppose the adding operation is being implemented in this coordinate after the grid initialization. Then a new grid node is added in the middle of two boundary nodes 1 – see the black circle with number 2 in Fig. 11.6. This new grid node is assigned to the second layer because of the resulting resolution it yields. Then all the following grid nodes are added one by one. Note that nodes 3 and 4 belong to the same third layer because they yield the same resolution. On the other hand, node 5 belongs to the fourth layer because it yields higher resolution than nodes 2 and 3.

Nodes 6 and 7 are assigned to their layers in a similar fashion.

11.3.3 Removing RBFs

When the magnitude of the output tracking error e falls within the predetermined threshold e_{max} and the dwelling-time requirement has been satisfied, the network attempts to remove some of the existing RBFs, that is, some of the existing grid nodes, in order to avoid network redundancy. The RBF removing operation is also implemented for each coordinate independently. If $c_{i(nearest)}$ is equal to x_{li} or x_{ui}, then no grid node is removed from the i-th coordinate. Otherwise, the grid node located at $c_{i(nearest)}$ is removed from the i-th coordinate if this grid node is in the higher than or in the same layer as the highest layer of the two neighboring grid nodes in the same coordinate, and the distance between x_i and $c_{i(nearest)}$ is smaller than a fraction τ of the distance between $c_{i(nearest)}$ and $c_{i(nearer)}$, where

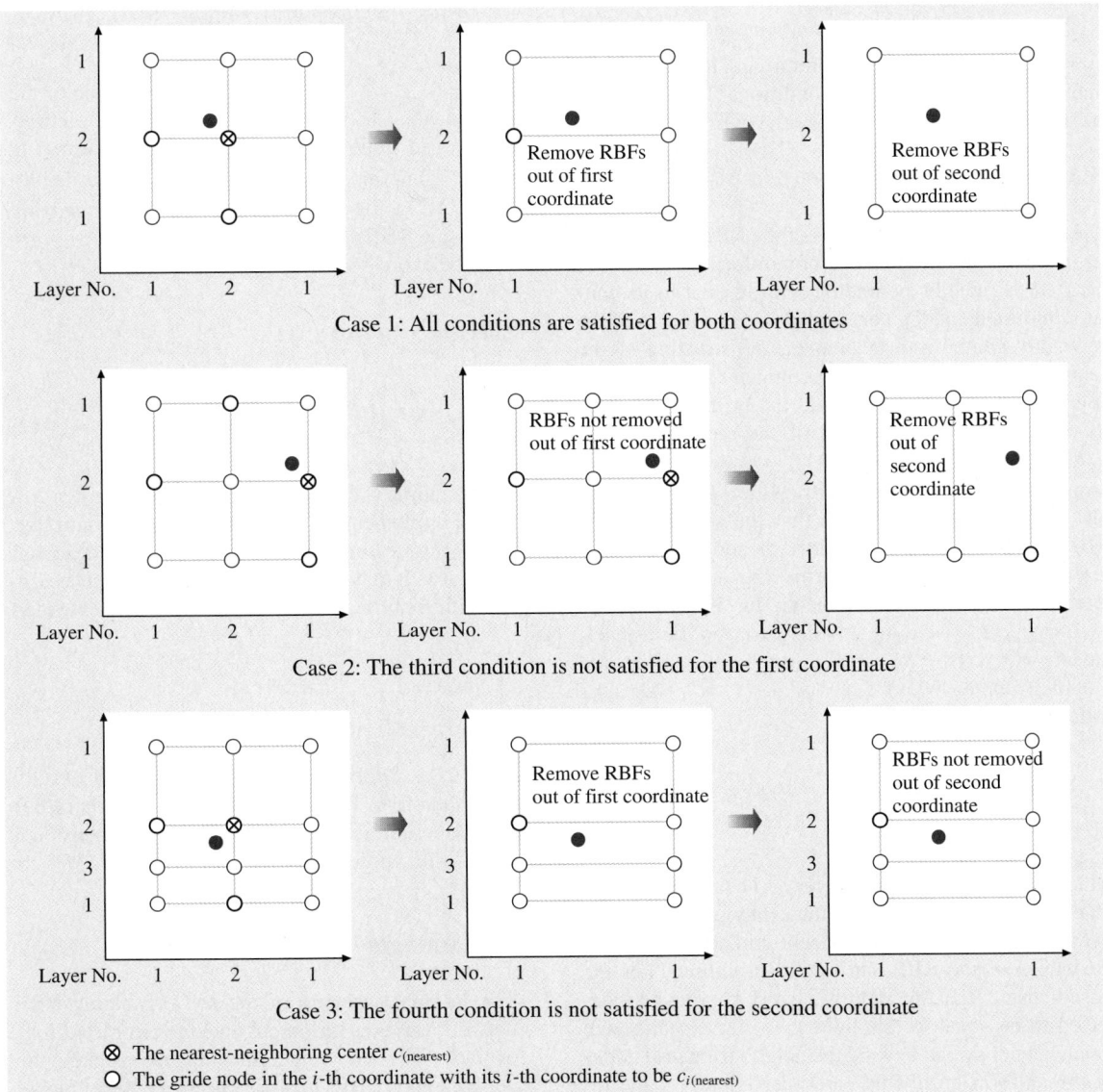

Fig. 11.7 Two-dimensional examples of removing RBFs

Part B | 11.3

the fraction τ is a design parameter between 0 and 0.5. The above conditions for the removing operation to take place in the i-th coordinate can be summarized as:

1. $|e| \leq e_{\max}$.
2. The elapsed time since last operation, adding or removing, is greater than T_d.
3. $c_{i(\text{nearest})} \notin (x_{li}, x_{ui})$.
4. The grid node in the i-th coordinate with its coordinate equal to $c_{i(\text{nearest})}$ is in a higher than or in the same layer as the highest layer of the two neighboring grid nodes in the same coordinate.
5. $\left| x_i - c_{i(\text{nearest})} \right| < \tau \left| c_{i(\text{nearest})} - c_{i(\text{nearer})} \right|$, $\tau \in (0, 0.5)$.

Two-dimensional examples of removing RBFs are illustrated in Fig. 11.7, where the conditions (1), (2), and (5) are assumed to be satisfied for both coordinates.

11.3.4 Uniform Grid Transformation

The determination of the radius of the RBF is much easier in a uniform grid than in a nonuniform grid because the RBF is radially symmetric with respect to its center. Unfortunately, the center grid used to locate RBFs is usually nonuniform. Moreover, the structure of the center grid changes after each adding or removing operation, which further complicates the problem. In order to simplify the determination of the radius, the one-to-one mapping $z(x) = [z_1(x_1), z_2(x_2), \ldots, z_n(x_n)]^\top$, proposed in [11.60], is used to transform the center grid into a uniform grid. Suppose that the self-organizing RBF network is now with the v-th admissible structure after the adding or removing operation and there are $M_{i,v}$ distinct elements in S_i, ordered as $c_{i(1)} < c_{i(2)} < \cdots < c_{i(M_{i,v})}$, where $c_{i(k)}$ is the k-th element with $c_{i(1)} = x_{li}$ and $c_{i(M_{i,v})} = x_{ui}$. Then the mapping function $z_i(x_i) : [x_{li}, x_{ui}] \to [1, M_{i,v}]$ takes the following form:

$$z_i(x_i) = k + \frac{x_i - c_{i(k)}}{c_{i(k+1)} - c_{i(k)}}, \quad c_{i(k)} \leq x_i < c_{i(k+1)}, \tag{11.20}$$

which maps $c_{i(k)}$ into the integer k. Thus, the transformation $z(x) : \Omega_x \to \mathbb{R}^n$ maps the center grid into a grid with unit spacing between adjacent grid nodes such that the radius of the RBF can be easily chosen. For the raised-cosine RBF, the radius in every coordinate is selected to be equal to one unit, that is, the radius will touch but not extend beyond the neighboring grid nodes in the uniform grid. This particular choice of the radius guarantees that for a given input x, the number of

nonzero raised-cosine RBFs in the uniform grid is at most 2^n.

To simplify the implementation, it is helpful to reorder the M_v grid nodes into a one-dimensional array of points using a scalar index j. Let the vector $q_v \in \mathbb{R}^n$ be the index vector of the grid nodes, where $q_v = (q_{1,v}, \ldots, q_{n,v})^\top$ with $1 \leq q_{i,v} \leq M_{i,v}$. Then the scalar index j can be uniquely determined by the index vector q_v, where

$$\begin{aligned} j = {} & (q_{n,v} - 1)M_{n-1,v} \cdots M_{2,v}M_{1,v} + \cdots \\ & + (q_{3,v} - 1)M_{2,v}M_{1,v} + (q_{2,v} - 1)M_{1,v} + q_{1,v}. \end{aligned} \tag{11.21}$$

Let $c_{j,v} = (c_{1j,v}, \ldots, c_{nj,v})^\top$ denote the location of the q_v-th grid node in the original grid. Then the corresponding grid node in the uniform grid is located at $z_{j,v} = z(c_{j,v}) = (q_{1,v}, \ldots, q_{n,v})^\top$. Using the scalar index j in (11.21), the output $\hat{f}_{i,v}(x)$ of the self-organizing raised-cosine RBF network implemented in the uniform grid can be expressed as

$$\begin{aligned} \hat{f}_v(x) &= \sum_{j=1}^{M_v} \omega_{fj,v} \xi_{j,v}(x) \\ &= \sum_{j=1}^{M_v} \omega_{fj,v} \prod_{i=1}^{n} \psi \left(\left| z_i(x_i) - q_{i,v} \right| \right), \tag{11.22} \end{aligned}$$

where the radius is one unit in each coordinate.

When implementing the output feedback controller, the state vector estimate $\hat{x}$ is used rather than the actual state vector x. It may happen that $\hat{x} \notin \Omega_x$. In such a case, the definition of the transformation (11.20) is extended as

$$\begin{cases} z_i(\hat{x}_i) = 1 & \text{if } \hat{x}_i < c_{i(1)} \\ z_i(\hat{x}_i) = M_{i,v} & \text{if } \hat{x}_i > c_{i(M_{i,v})}, \end{cases} \tag{11.23}$$

for $i = 1, 2, \ldots, n$. If $\hat{x} \in \Omega_x$, the transformation (11.20) is used. Therefore, it follows from (11.20) and (11.23) that the function $z(x)$ maps the whole n-dimensional space $\mathbb{R}^n$ into the compact set $[1, M_{1,v}] \times [1, M_{2,v}] \times \cdots \times [1, M_{n,v}]$.

11.3.5 Remarks

1. The internal structure of the self-organizing RBF network varies as the output tracking error trajectory evolves. When the output tracking error is large, the network adds RBFs in order to achieve better model compensation so that the large output tracking error

can be reduced. When, on the other hand, the output tracking error is small, the network removes RBFs in order to avoid a redundant structure. If the design parameter e_{max} is too large, the network may stop adding RBFs prematurely or even never adjust its structure at all. Thus, e_{max} should be at least smaller than $|e(t_0)|$. However, if e_{max} is too small, the network may keep adding and removing RBFs all the time and cannot approach a steady structure even though the output tracking error is already within the acceptable bound. In the worst case, the network will try to add RBFs forever. This, of course, leads to an unnecessary large network size and, at the same time, undesirable high computational cost. An appropriate e_{max} may be chosen by trial and error through numerical simulations.

2. The advantage of the raised-cosine RBF over the Gaussian RBF is the property of the compact support associated with the raised-cosine RBF. The number of terms in (11.22) grows rapidly with the increase of both the number of grid nodes M_i in each coordinate and the dimensionality n of the input space. For the GRBF network, all the terms will be nonzero due to the unbounded support, even though most of them are quite small. Thus, a lot of computations are required for the network's output evaluation, which is impractical for real-time applications, especially for higher-order systems. However, for the RCRBF network, most of the terms in (11.22) are zero and therefore do not have to be evaluated. Specifically, for a given input x, the number of nonzero raised-cosine RBFs in each coordinate is either one or two. Consequently, the number of nonzero terms in (11.22) is at most 2^n. This feature allows one to speed up the output evaluation of the network in comparison with a direct computation of (11.22) for the GRBF network. To illustrate the above discussion, suppose $M_i = 10$ and $n = 4$. Then the GRBF network will require 10^4 function evaluations, whereas the RCRBF network will only require 2^4 function evaluations, which is almost three orders of magnitude less than that required by the GRBF network. For a larger value of n and a finer grid, the saving of computations is even more dramatic. The same saving is also achieved for the network's training. When the weights of the RCRBF network are updated, there are also only 2^n weights to be updated for each output neuron, whereas $n \times M$ weights has to be updated for the GRBF network. Similar observations were also reported in [11.60, p. 6].

11.4 State Feedback Controller Development

The direct adaptive robust state feedback controller presented in this chapter has the form

$$
\begin{aligned}
u &= u_{a,v} + u_{s,v} \\
&= \frac{1}{\hat{g}_v(x)}\left(-\hat{f}_v(x) + y_d^{(n)} - ke\right) + u_{s,v} ,
\end{aligned} \tag{11.24}
$$

where $\hat{f}_v(x) = \boldsymbol{\omega}_{f,v}^\top \boldsymbol{\xi}_v(x)$, $\hat{g}_v(x) = \boldsymbol{\omega}_{g,v}^\top \boldsymbol{\xi}_v(x)$ and $u_{s,v}$ is the robustifying component to be described later. To proceed, let $\boldsymbol{\Omega}_{e_0}$ denote the compact set including all the possible initial tracking errors and let

$$
c_{e_0} = \max_{e \in \boldsymbol{\Omega}_{e_0}} \frac{1}{2} e^\top \mathbf{P}_m e , \tag{11.25}
$$

where $\mathbf{P}_m$ is the positive-definite solution to the continuous Lyapunov matrix equation $\mathbf{A}_m^\top \mathbf{P}_m + \mathbf{P}_m \mathbf{A}_m = -2\mathbf{Q}_m$ for $\mathbf{Q}_m = \mathbf{Q}_m^\top > 0$. Choose $c_e > c_{e_0}$ and let

$$
\boldsymbol{\Omega}_e = \left\{ e : \frac{1}{2} e^\top \mathbf{P}_m e \leq c_e \right\} . \tag{11.26}
$$

Then the compact set $\boldsymbol{\Omega}_x$ is defined as

$$
\boldsymbol{\Omega}_x = \left\{ x : x = e + x_d, e \in \boldsymbol{\Omega}_e, x_d \in \boldsymbol{\Omega}_{x_d} \right\} ,
$$

over which the unknown functions $f(x)$ and $g(x)$ are approximated.

For practical implementation, $\boldsymbol{\omega}_{f,v}$ and $\boldsymbol{\omega}_{g,v}$ are constrained, respectively, to reside inside compact sets $\boldsymbol{\Omega}_{f,v}$ and $\boldsymbol{\Omega}_{g,v}$ defined as

$$
\boldsymbol{\Omega}_{f,v} = \left\{ \boldsymbol{\omega}_{f,v} : \underline{\omega}_f \leq \omega_{fj,v} \leq \overline{\omega}_f, 1 \leq j \leq M_v \right\} \tag{11.27}
$$

and

$$
\boldsymbol{\Omega}_{g,v} = \left\{ \boldsymbol{\omega}_{g,v} : 0 < \underline{\omega}_g \leq \omega_{gj,v} \leq \overline{\omega}_g , 1 \leq j \leq M_v \right\} , \tag{11.28}
$$

where $\underline{\omega}_f$, $\overline{\omega}_f$, $\underline{\omega}_g$ and $\overline{\omega}_g$ are design parameters. Let $\boldsymbol{\omega}_{f,v}^*$ and $\boldsymbol{\omega}_{g,v}^*$ denote the *optimal* constant weight vectors corresponding to each admissible network structure, which are used only in the analytical analysis and defined, respectively, as

$$\boldsymbol{\omega}_{f,v}^* = \underset{\boldsymbol{\omega}_{f,v} \in \boldsymbol{\Omega}_{f,v}}{\arg\min} \ \underset{\boldsymbol{x} \in \boldsymbol{\Omega}_x}{\max} \left| f(\boldsymbol{x}) - \boldsymbol{\omega}_{f,v}^\top \boldsymbol{\xi}_v(\boldsymbol{x}) \right| \tag{11.29}$$

and

$$\boldsymbol{\omega}_{g,v}^* = \underset{\boldsymbol{\omega}_{g,v} \in \boldsymbol{\Omega}_{g,v}}{\arg\min} \ \underset{\boldsymbol{x} \in \boldsymbol{\Omega}_x}{\max} \left| g(\boldsymbol{x}) - \boldsymbol{\omega}_{g,v}^\top \boldsymbol{\xi}_v(\boldsymbol{x}) \right| . \tag{11.30}$$

For the controller implementation, let

$$d_f = \max_v \left(\underset{\boldsymbol{x} \in \boldsymbol{\Omega}_x}{\max} \left| f(\boldsymbol{x}) - \boldsymbol{\omega}_{f,v}^{*\top} \boldsymbol{\xi}_v(\boldsymbol{x}) \right| \right) \tag{11.31}$$

and

$$d_g = \max_v \left(\underset{\boldsymbol{x} \in \boldsymbol{\Omega}_x}{\max} \left| g(\boldsymbol{x}) - \boldsymbol{\omega}_{g,v}^{*\top} \boldsymbol{\xi}_v(\boldsymbol{x}) \right| \right) , \tag{11.32}$$

where $\max_v(\cdot)$ denotes the maximization taken over all admissible structures of the self-organizing RBF networks. Let $\boldsymbol{\phi}_{f,v} = \boldsymbol{\omega}_{f,v} - \boldsymbol{\omega}_{f,v}^*$ and $\boldsymbol{\phi}_{g,v} = \boldsymbol{\omega}_{g,v} - \boldsymbol{\omega}_{g,v}^*$, and let

$$c_f = \max_v \left(\underset{\boldsymbol{\omega}_{f,v}, \boldsymbol{\omega}_{f,v}^* \in \boldsymbol{\Omega}_{f,v}}{\max} \frac{1}{2\eta_f} \boldsymbol{\phi}_{f,v}^\top \boldsymbol{\phi}_{f,v} \right) \tag{11.33}$$

and

$$c_g = \max_v \left(\underset{\boldsymbol{\omega}_{g,v}, \boldsymbol{\omega}_{g,v}^* \in \boldsymbol{\Omega}_{g,v}}{\max} \frac{1}{2\eta_g} \boldsymbol{\phi}_{g,v}^\top \boldsymbol{\phi}_{g,v} \right) , \tag{11.34}$$

where η_f and η_g are positive design parameters often referred to as learning rates. It is obvious that c_f (or c_g) will decrease as η_f (or η_g) increases. Let $\sigma = \boldsymbol{b}^\top \mathbf{P}_m \boldsymbol{e}$. The following weight vector adaptation laws are employed, respectively, for the weight vectors $\boldsymbol{\omega}_f$ and $\boldsymbol{\omega}_g$,

$$\dot{\boldsymbol{\omega}}_{f,v} = \mathrm{Proj}\left[\boldsymbol{\omega}_{f,v}, \eta_f \sigma \boldsymbol{\xi}_v(\boldsymbol{x}) \right] \tag{11.35}$$

and

$$\dot{\boldsymbol{\omega}}_{g,v} = \mathrm{Proj}\left[\boldsymbol{\omega}_{g,v}, \eta_g \sigma \boldsymbol{\xi}_v(\boldsymbol{x}) u_{a,v} \right] , \tag{11.36}$$

where $\mathrm{Proj}(\boldsymbol{\omega}_v, \boldsymbol{\theta}_v)$ denotes $\mathrm{Proj}(\omega_{j,v}, \theta_{j,v})$ for $j = 1, \dots, M_v$ and

$$\mathrm{Proj}(\omega_{j,v}, \theta_{j,v}) = \begin{cases} 0 & \text{if } \omega_{j,v} = \underline{\omega} \text{ and } \theta_{j,v} < 0 , \\ 0 & \text{if } \omega_{j,v} = \overline{\omega} \text{ and } \theta_{j,v} > 0 , \\ \theta_{j,v} & \text{otherwise} , \end{cases} \tag{11.37}$$

is a discontinuous projection operator proposed in [11.65]. The robustifying component $u_{s,v}$ is designed as

$$u_{s,v} = -\frac{1}{g} k_{s,v} \, \mathrm{sat}\left(\frac{\sigma}{v} \right) , \tag{11.38}$$

where $k_{s,v} = d_f + d_g |u_{a,v}| + d_0$ and $\mathrm{sat}(\cdot)$ is the saturation function with small $v > 0$. Let

$$k_s = d_f + d_g \max_v \left(\max |u_{a,v}| \right) + d_0 , \tag{11.39}$$

where the inner maximization is taken over $\boldsymbol{e} \in \boldsymbol{\Omega}_e$, $\boldsymbol{x}_\mathrm{d} \in \boldsymbol{\Omega}_{x_\mathrm{d}}$, $y_\mathrm{d}^{(n)} \in \boldsymbol{\Omega}_{y_\mathrm{d}}$, $\boldsymbol{\omega}_{f,v} \in \boldsymbol{\Omega}_{f,v}$, and $\boldsymbol{\omega}_{g,v} \in \boldsymbol{\Omega}_{g,v}$.

It can be shown [11.58] that, for the plant (11.2) driven by the proposed direct adaptive robust state feedback controller (DARSFC) (11.24) with the robustifying component (11.38) and the adaptation laws (11.35) and (11.36), if one of the following conditions is satisfied:

- The dwelling time T_d of the self-organizing RBF network is selected such that

$$T_\mathrm{d} \geq \frac{1}{\mu} \ln\left(\frac{3}{2} \right) , \tag{11.40}$$

- The constants c_f, c_g, and v satisfy the inequality

$$0 < c_f + c_g < \frac{\exp(\mu T_\mathrm{d}) - 1}{3 - 2\exp(\mu T_\mathrm{d})} \frac{k_s v}{4\mu} , \tag{11.41}$$

where μ is the ratio of the minimal eigenvalve of $\mathbf{Q}_\mathrm{m}$ to the maximal eigenvalue of $\mathbf{P}_\mathrm{m}$. If η_f, η_g, and v are selected such that

$$c_e \geq \max \left\{ c_{e0} + c_f + c_g, \, 2\left(c_f + c_g + \frac{k_s v}{8\mu} \right) + c_f + c_g \right\} , \tag{11.42}$$

then $\boldsymbol{e}(t) \in \boldsymbol{\Omega}_e$ and $\boldsymbol{x}(t) \in \boldsymbol{\Omega}_x$ for $t \geq t_0$. Moreover, there exists a finite time $T \geq t_0$ such that

$$\frac{1}{2} \boldsymbol{e}^\top (t) \mathbf{P}_m \boldsymbol{e}(t) \leq 2\left(c_f + c_g + \frac{k_s v}{8\mu} \right) + c_f + c_g \tag{11.43}$$

for $t \geq T$. If, in addition, there exists a finite time $T_\mathrm{s} \geq t_0$ such that $v = v_\mathrm{s}$ for $t \geq T_\mathrm{s}$, then there exists a finite time $T \geq T_\mathrm{s}$ such that

$$\frac{1}{2} \boldsymbol{e}^\top (t) \mathbf{P}_m \boldsymbol{e}(t) \leq 2\left(c_f + c_g + \frac{k_s v}{8\mu} \right) \tag{11.44}$$

for $t \geq T$. It can be seen from (11.43) and (11.44) that the tracking performance is inversely proportional to η_f

and η_g, and proportional to v. Therefore, larger learning rates and smaller saturation boundary imply better tracking performance.

11.4.1 Remarks

1. For the above direct adaptive robust controller, the weight vector adaptation laws are synthesized together with the controller design. This is done for the purpose of reducing the output tracking error only. However, the adaptation laws are limited to be of gradient type with ceratin tracking errors as driving signals, which may not have as good convergence properties as other types of adaptation laws such as the ones based on the least-squares method [11.66]. Although this design methodology can achieve excellent output tracking performance, it may not achieve the convergence of the weight vectors. When good convergence of the weight vectors is a secondary goal to be achieved, an indirect adaptive robust controller [11.66] or an integrated direct/indirect adaptive robust control [11.65] have been proposed to overcome the problem of poor convergence associated with the direct adaptive robust controllers.
2. It seems to be desirable to select large learning rates and small saturation boundary based on (11.43) and (11.44). However, it is not desirable in practice to choose excessively large η_f and η_g. If η_f

and η_g are too large, fast adaptation could excite the unmodeled high-frequency dynamics that are neglected in the modeling. On the other hand, the selection of v cannot be too small either. Otherwise, the robustifying component exhibits high-frequency chattering, which may also excite the unmodeled dynamics. Moreover, smaller v requires higher bandwidth to implement the controller for small tracking error. To see this more clearly, consider the following first-order dynamics,

$$\dot{e} = ae + \left(f + gu - \dot{y}_d + d \right), \tag{11.45}$$

which is a special case of (11.6). Applying the following controller,

$$u = \frac{1}{\hat{g}} \left(-\hat{f} + \dot{y}_d - ke \right) - \frac{1}{g} k_s \, \text{sat} \left(\frac{\sigma}{v} \right), \tag{11.46}$$

where $\sigma = e$, one obtains

$$\dot{e} = -a_m e + \tilde{d} - \frac{g}{\underline{g}} k_s \, \text{sat} \left(\frac{e}{v} \right), \tag{11.47}$$

where $-a_m = a - k < 0$. When $|e| \leq v$, (11.47) becomes

$$\dot{e} = -\left(a_m + \frac{g}{\underline{g}} \frac{k_s}{v} \right) e + \tilde{d}, \tag{11.48}$$

which implies that smaller v results in higher controller bandwidth.

11.5 Output Feedback Controller Construction

The direct adaptive robust state feedback controller presented in the previous section requires the availability of the plant states. However, often in practice only the plant outputs are available. Thus, it is desirable to develop a direct adaptive robust output feedback controller (DAROFC) architecture. To overcome the problem of inaccessibility of the system states, the following high-gain observer [11.38, 49],

$$\dot{\hat{e}} = A\hat{e} + l\left(e - c\hat{e} \right), \tag{11.49}$$

is applied to estimate the tracking error e. The observer gain l is chosen as

$$l = \left(\frac{\alpha_1}{\epsilon}, \frac{\alpha_2}{\epsilon^2}, \ldots, \frac{\alpha_n}{\epsilon^n} \right)^\top, \tag{11.50}$$

where $\epsilon \in (0, 1)$ is a design parameter and α_i, $i = 1, 2, \ldots, n$, are selected so that the roots of the polynomial equation, $s^n + \alpha_1 s^{n-1} + \cdots + \alpha_{n-1} s + \alpha_n = 0$,

have negative real parts. The structure of the above high-gain tracking error observer is shown in Fig. 11.8. Substituting e with $\hat{e}$ in the controller u defined in (11.24) with (11.38) gives

$$\hat{u} = \hat{u}_{a,v} + \hat{u}_{s,v}, \tag{11.51}$$

where

$$\hat{u}_{a,v} = \frac{1}{\hat{g}_v(\hat{x})} \left[-\hat{f}_v(\hat{x}) + y_d^{(n)} - k\hat{e} \right] \tag{11.52}$$

and

$$\hat{u}_{s,v} = -\frac{1}{\underline{g}} \hat{k}_{s,v} \, \text{sat} \left(\frac{\hat{\sigma}}{v} \right), \tag{11.53}$$

with $\hat{x} = x_d + \hat{e}$, $\hat{k}_{s,v} = d_f + d_g |\hat{u}_{a,v}| + d_0$ and $\hat{\sigma} = b^\top P_m \hat{e}$. Let

$$\hat{k}_s = d_f + d_g \max_v \left(\max |\hat{u}_{a,v}| \right) + d_0,$$

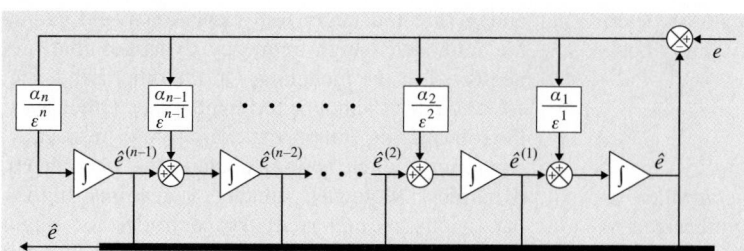

Fig. 11.8 Diagram of the high-gain observer

and the inner maximization is taken over $\hat{e} \in \Omega_{\hat{e}}$, $x_d \in \Omega_{x_d}$, $y_d^{(n)} \in \Omega_{y_d}$, $\omega_{f,v} \in \Omega_{f,v}$, and $\omega_{g,v} \in \Omega_{g,v}$. For the high-gain observer described by (11.49), there exist peaking phenomena [11.67]. Hence, the controller $\hat{u}$ defined in (11.51) cannot be applied to the plant directly. To eliminate the peaking phenomena, the saturation is introduced into the control input $\hat{u}$ in (11.51). Let

$$\Omega_{\hat{e}} = \left\{ e : \frac{1}{2} e^\top P_m e \le c_{\hat{e}} \right\},$$ (11.54)

where $c_{\hat{e}} > c_e$. Let

$$S \ge \max_v \left[\max \left| u \left(e, x_d, y_d^{(n)}, \omega_{f,v}, \omega_{g,v} \right) \right| \right],$$ (11.55)

where u is defined in (11.24) and the inner maximization is taken over $e \in \Omega_{\hat{e}}$, $x_d \in \Omega_{x_d}$, $y_d^{(n)} \in \Omega_{y_d}$, $\omega_{f,v} \in \Omega_{f,v}$, and $\omega_{g,v} \in \Omega_{g,v}$. Then the proposed direct adaptive robust output feedback controller takes the form

$$u^s = S \operatorname{sat} \left(\frac{\hat{u}_{a,v} + \hat{u}_{s,v}}{S} \right).$$ (11.56)

The adaptation laws for the weight vectors $\omega_{f,v}$ and $\omega_{g,v}$ change correspondingly and take the following new form, respectively

$$\dot{\omega}_{f,v} = \operatorname{Proj} \left[\omega_{f,v}, \eta_f \hat{\sigma} \xi_v(\hat{x}) \right]$$ (11.57)

and

$$\dot{\omega}_{g,v} = \operatorname{Proj} \left[\omega_{g,v}, \eta_g \hat{\sigma} \xi_v(\hat{x}) \hat{u}_{a,v} \right].$$ (11.58)

A block diagram of the above direct adaptive robust output feedback controller is shown in Fig. 11.9, while a block diagram of the closed-loop system is given in Fig. 11.10.

For the high-gain tracking error observer (11.49), it is shown in [11.68] that there exists a constant $\epsilon_1^* \in (0, 1)$ such that, if $\epsilon \in (0, \epsilon_1^*)$, then $\|e(t) - \hat{e}(t)\| \le \beta \epsilon$ with $\beta > 0$ for $t \in [t_0 + T_1(\epsilon), t_0 + T_3)$, where $T_1(\epsilon)$ is a finite time and $t_0 + T_3$ is the moment when the tracking error $e(t)$ leaves the compact set Ω_e for the first time. Moreover, we have $\lim_{\epsilon \to 0^+} T_1(\epsilon) = 0$ and $c_{e_1} = \frac{1}{2} e(t_0 + T_1(\epsilon))^\top P_m e(t_0 + T_1(\epsilon)) < c_e$. For the plant (11.2) driven by the direct adaptive robust output feedback controller given by (11.56) with the adaptation laws (11.57) and (11.58), if one of the following conditions is satisfied:

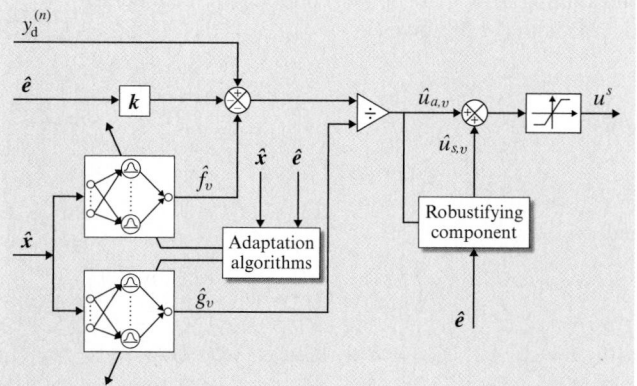

Fig. 11.9 Diagram of the direct adaptive robust output feedback controller (DAROFC)

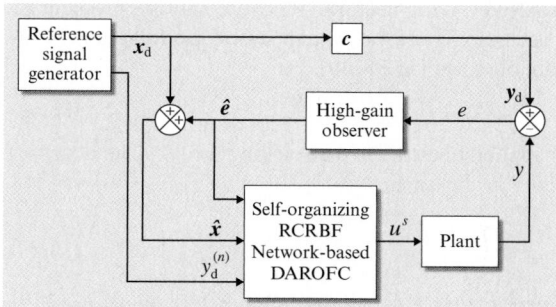

Fig. 11.10 Diagram of the closed-loop system driven by the output feedback controller

- The dwelling time T_d of the self-organizing RBF network is selected such that

$$T_d \geq \frac{1}{\mu} \ln\left(\frac{3}{2}\right),$$ (11.59)

- The constants c and v satisfy the inequality

$$0 < c_f + c_g < \frac{\exp(\mu T_d) - 1}{3 - 2\exp(\mu T_d)}\left(\frac{\hat{k}_s v}{4\mu} + r\epsilon\right),$$ (11.60)

and if η_f, η_g, and v are selected such that

$$c_e \geq c_{e_1} + c_f + c_g$$ (11.61)

and

$$c_e > 2\left(c_f + c_g + \frac{\hat{k}_s v}{8\mu}\right) + c_f + c_g,$$ (11.62)

there exists a constant $\epsilon^* \in (0, 1)$ such that, if $\epsilon \in (0, \epsilon^*)$, then $e(t) \in \Omega_e$ and $x(t) \in \Omega_x$ for $t \geq t_0$. Moreover, there

exists a finite time $T \geq t_0 + T_1(\epsilon)$ such that

$$\frac{1}{2}e(t)^\top \mathbf{P}_m e(t) \leq 2\left(c_f + c_g + \frac{\hat{k}_s v}{8\mu}\right) + r\epsilon + c_f + c_g$$ (11.63)

with some $r > 0$ for $t \geq T$. In addition, suppose that there exists a finite time $T_s \geq t_0 + T_1(\epsilon)$ such that $v = v_s$ for $t \geq T_s$. Then there exists a finite time $T \geq T_s$ such that

$$\frac{1}{2}e^\top(t)\mathbf{P}_m e(t) \leq 2\left(c_f + c_g + \frac{\hat{k}_s v}{8\mu}\right) + r\epsilon$$ (11.64)

for $t \geq T$. A proof of the above statement can be found in [11.58]. It can be seen that the performance of the output feedback controller approaches that of the state feedback controller as ϵ approaches zero.

11.6 Examples

In this section, two example systems are used to illustrate the features of the proposed direct adaptive robust controllers. In Example 11.1, a benchmark problem from the literature is used to illustrate the controller performance under different situations. Especially, the reference signal changes during the operation in order to demonstrate the advantage of the self-organizing RBF network. In Example 11.2, the Duffing forced oscillation system is employed to test the controller performance for time-varying systems.

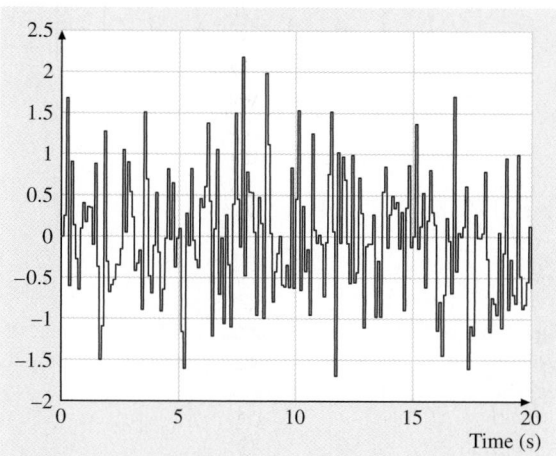

Fig. 11.11 Disturbance d in example 11.1

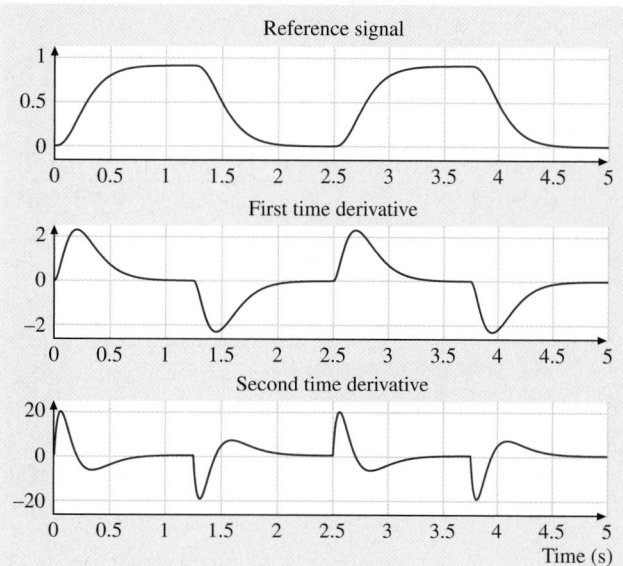

Fig. 11.12 Reference signal and its time derivatives

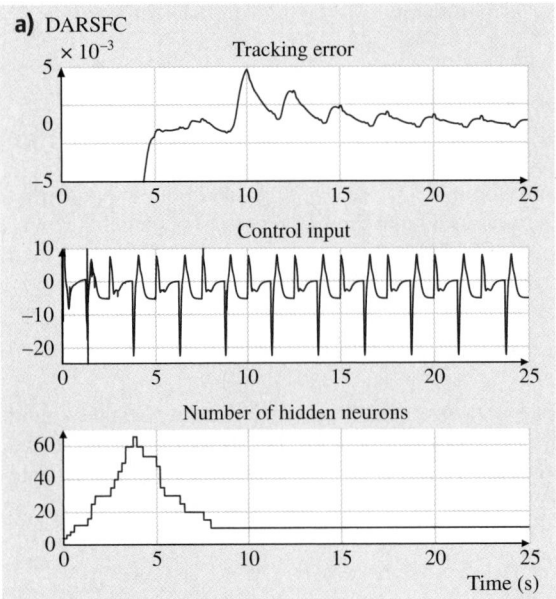

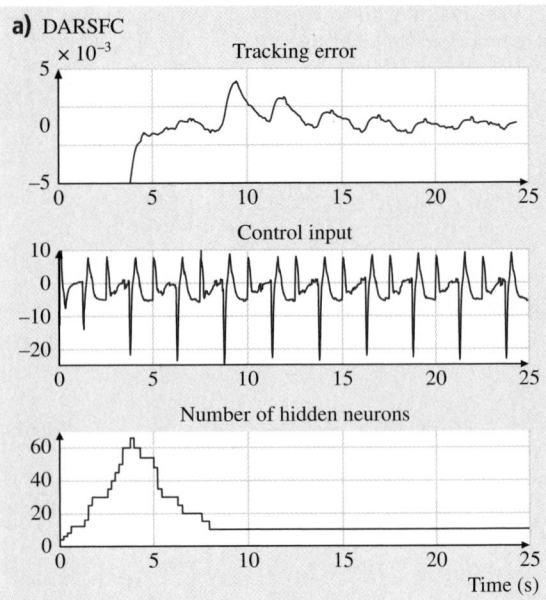

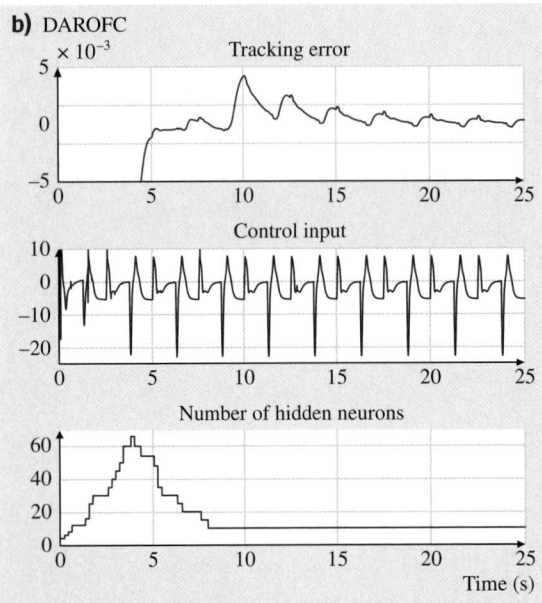

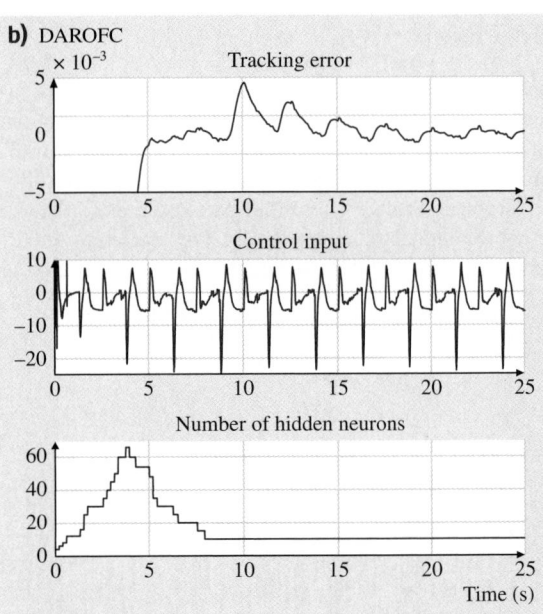

Fig. 11.13a,b Controller performance without disturbance in example 11.1. (**a**) State feedback controller, (**b**) output feedback controller

Fig. 11.14a,b Controller performance with disturbance in Example 11.1. (**a**) State feedback controller, (**b**) output feedback controller

Example 11.1: The nonlinear plant model used in this example is given by

$$\ddot{y} = f(y, \dot{y}) + g(y)u + d$$

$$= 16 \frac{\sin(4\pi y)}{4\pi y} \left(\frac{\sin(\pi \dot{y})}{\pi \dot{y}} \right)^2$$

$$+ \{2 + \sin[3\pi(y - 0.5)]\}u + d,$$

which, if $d = 0$, is the same plant model as in [11.27, 34, 38], used as a testbed for proposed controllers. It is easy to check that the above uncertain system dynamics are in the form of (11.2) with $x = [y, \dot{y}]^\top$. For the simulation, the disturbance d is selected to be band-limited white noise generated using SIMULINK (version 6.6) with noise power 0.05, sample time 0.1 s, and seed value 23 341, which is shown in Fig. 11.11.

The reference signal is the same as in [11.38], which is the output of a low-pass filter with the transfer function $(1 + 0.1s)^{-3}$, driven by a unity amplitude square-wave input with frequency 0.4 Hz and a time average of 0.5 s. The reference signal y_d and its derivatives $\dot{y}_d$ and $\ddot{y}_d$ are shown in Fig. 11.12. The grid boundaries for y and $\dot{y}$, respectively, are selected to be $(-1.5, 1.5)$ and $(-3.5, 3.5)$, that is, $x_l = (-1.5, -3.5)^\top$ and $x_u = (1.5, 3.5)^\top$. The rest of the network's parameters are $d_{\text{threshold}} = (0.2, 0.3)$, $e_{\max} = 0.005$, $T_d = 0.2$ s, $\overline{\omega}_f = 25$, $\underline{\omega}_f = -25$, $\overline{\omega}_g = 5$, $\underline{\omega}_g = 0.1$, and $\eta_f = \eta_g = 1000$. The controller's parameters are $k = (1, 2)$, $Q_m = 0.5I_2$, $d_f = 5$, $d_g = 2$, $d_0 = 3$, $v = 0.01$, and $S = 50$. The observer's parameters are $\epsilon = 0.001$, $\alpha_1 = 10$, and $\alpha_2 = 25$. The initial conditions are $y(0) = -0.5$ and $\dot{y}(0) = 2.0$. The controller performance without disturbance is shown in Fig. 11.13, whereas the controller performance in the presence of disturbance is illustrated in Fig. 11.14.

In order to demonstrate the advantages of the self-organizing RBF network in the proposed controller architectures, a different reference signal, $y_d(t) = \sin(2t)$, is applied at $t = 25$ s. It can be seen from Fig. 11.15 that the self-organizing RCRBF network-based direct adaptive robust output feedback controller performs very well for both reference signals. There is no need to adjust the network's or the controller's parameters offline when the new reference signal is applied. The self-organizing RBF network determines its structure dynamically by itself as the reference signal changes.

Example 11.2: In this example, the direct adaptive robust output feedback controller is tested on a time-varying system. The plant is the Duffing forced oscillation sys-

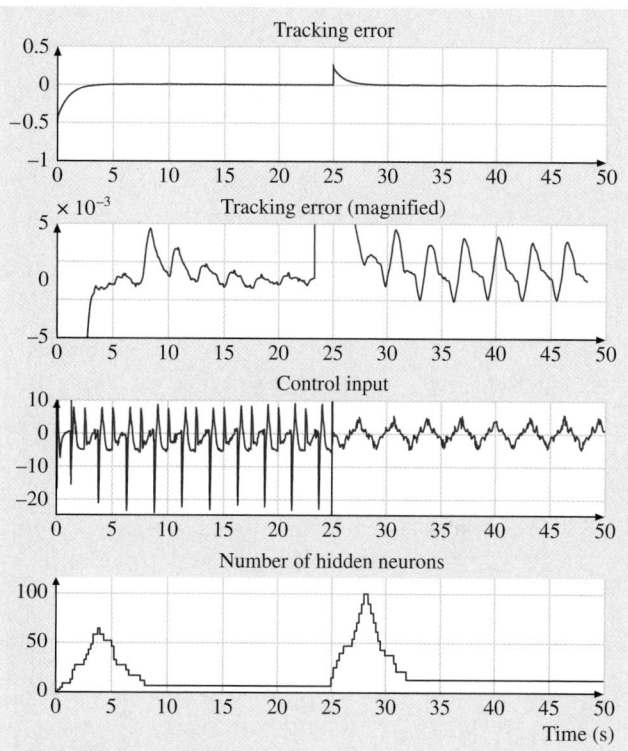

Fig. 11.15 Output feedback controller performance with varying reference signals in Example 11.1

tem [11.28] modeled by

$$\dot{x}_1 = x_2$$

$$\dot{x}_2 = -0.1x_2 - x_1^3 + 12\cos(t) + u.$$

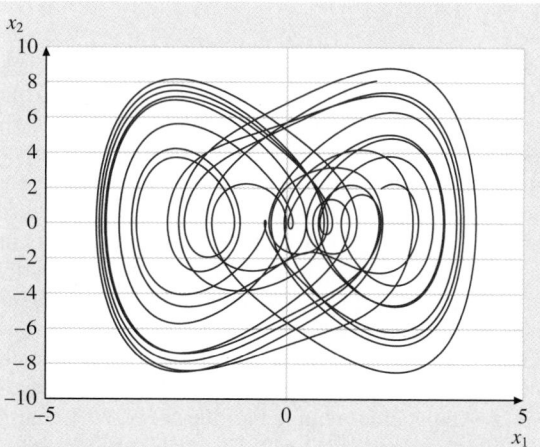

Fig. 11.16 Phase portrait of the uncontrolled system in Example 11.2

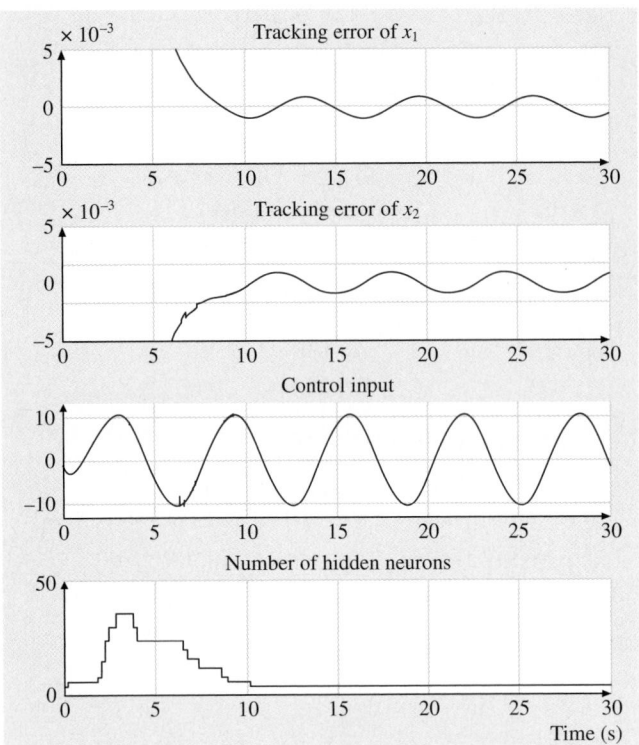

Fig. 11.18 Output feedback controller performance in Example 11.2

The phase portrait of the uncontrolled system is shown in Fig. 11.16 for $x_1(0) = x_2(0) = 2$, $t_0 = 0$,

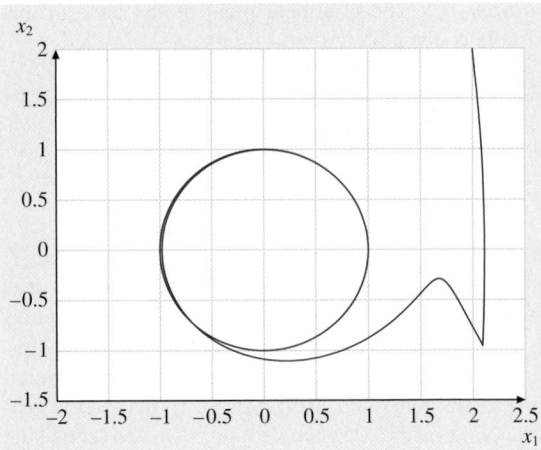

Fig. 11.17 Phase portrait of the closed-loop system driven by the output feedback controller in Example 11.2

and $t_f = 50$. The disturbance d is set to be zero.

The reference signal, $y_d(t) = \sin(t)$, is used, which is the unit circle in the phase plane. The grid boundaries for y and $\dot{y}$, respectively, are $[-2.5, 2.5]$ and $[-2.5, 2.5]$. The design parameters are chosen to be the same as in example 11.1 except that $e_{max} = 0.05$, $d_f = 15$, and $\nu = 0.001$. The phase portrait of the closed-loop system is shown in Fig. 11.17. It follows from Fig. 11.18 that the controller performs very well for this time-varying system.

11.7 Summary

Novel direct adaptive robust state and output feedback controllers have been presented for the output tracking control of a class of nonlinear systems with unknown system dynamics. The presented techniques incorporate a variable-structure RBF network to approximate the unknown system dynamics. The network structure varies as the output tracking error trajectory evolves in order to ensure tracking accuracy and, at the same time, avoid redundant network structure. The Gaussian RBF and the raised-cosine RBF are compared in the simulations. The property of compact support associated with the raised-cosine RBF results in significant reduction of computations required for the network's training and output evaluation [11.61]. This feature becomes especially important when the center grid becomes finer and the dimension of the network input becomes higher.

The effectiveness of the presented direct adaptive robust controllers are illustrated with two examples.

In order to evaluate and compare different proposed control strategies for the uncertain system given in (11.1), it is necessary to use performance measures. In the following, a list of possible performance indices [11.69] is given.

- Transient performance

$$e_M = \max_{t_0 \leq t \leq t_f} \left\{ |e(t)| \right\}$$

- Final tracking accuracy

$$e_F = \max_{t \in [t_f-2, t_f]} \left\{ |e(t)| \right\}$$

- Average tracking performance

$$L_2(e) = \sqrt{\frac{1}{t_f} \int_{t_0}^{t_f} |e(\tau)|^2 \, d\tau}$$

- Average control input

$$L_2(u) = \sqrt{\frac{1}{t_f} \int_{t_0}^{t_f} |u(\tau)|^2 \, d\tau}$$

- Degree of control chattering

$$c_u = \frac{L_2(\Delta u)}{L_2(u)},$$

where

$$L_2(\Delta u) = \sqrt{\frac{1}{N} \sum_{j=1}^{N} |u(j\Delta T) - u[(j-1)\Delta T]|^2}$$

The approach presented in this chapter has been used as a starting point towards the development of direct adaptive robust controllers for a class of MIMO uncertain systems in [11.58]. The MIMO uncertain system considered in [11.58] can be modeled by the following set of equations

$$\begin{cases} y_1^{(n_1)} = f_1(\boldsymbol{x}) + \sum_{j=1}^{p} g_{1j}(\boldsymbol{x})u_j + d_1 \\ y_2^{(n_2)} = f_2(\boldsymbol{x}) + \sum_{j=1}^{p} g_{2j}(\boldsymbol{x})u_j + d_2 \\ \vdots \\ y_p^{(n_p)} = f_p(\boldsymbol{x}) + \sum_{j=1}^{p} g_{pj}(\boldsymbol{x})u_j + d_p, \end{cases} \quad (11.65)$$

where $\boldsymbol{u} = (u_1, u_2, \ldots, u_p)^\top$ is the system input vector, $\boldsymbol{y} = (y_1, y_2, \ldots, y_p)^\top$ is the system output vector, $\boldsymbol{d} = (d_1, d_2, \ldots, d_p)^\top$ models the bounded disturbance, $\boldsymbol{x} = (\boldsymbol{x}_1^\top, \boldsymbol{x}_2^\top, \ldots, \boldsymbol{x}_p^\top)^\top \in \mathbb{R}^n$ is the system state vector with $\boldsymbol{x}_i = (y_i, \dot{y}_i, \ldots, y_i^{(n_i-1)})^\top$ and $n = \sum_{i=1}^{p} n_i$, and $f_i(\boldsymbol{x})$ and $g_{ij}(\boldsymbol{x})$ are unknown Lipschitz-continuous functions.

Part B | 11

References

11.1   S.H. Żak: *Systems and Control* (Oxford Univ. Press, New York 2003)

11.2   D.W. Clarke: Self-tuning control. In: *The Control Handbook*, ed. by W.S. Levine (CRC, Boca Raton 1996)

11.3   A.S.I. Zinober: *Deterministic Control of Uncertain Systems* (Peregrinus, London 1990)

11.4   R.A. DeCarlo, S.H. Żak, G.P. Matthews: Variabble structure control of nonlinear multivariable systems: A tutorial, Proc. IEEE **76**(3), 212–232 (1988)

11.5   V.I. Utkin: *Sliding Modes in Control and Optimization* (Springer, Berlin 1992)

11.6   C. Edwards, S.K. Spurgeon: *Sliding Mode Control: Theory and Applications* (Taylor Francis, London 1998)

11.7   S. Gutman: Uncertain dynamical systems – a Lyapunov min-max approach, IEEE Trans. Autom. Control **24**(3), 437–443 (1979)

11.8   M.J. Corless, G. Leitmann: Continuous state feedback guaranteeing uniform ultimate boundedness for uncertain dynamic systems, IEEE Trans. Autom. Control **26**(5), 1139–1144 (1981)

11.9   J.B. Pomet, L. Praly: Adaptive nonlinear regulation: Estimation from the Lyapunov equation, IEEE Trans. Autom. Control **37**(6), 729–740 (1992)

11.10   F.-C. Chen, C.-C. Liu: Adaptively controlling nonlinear continuous-time systems using multilayer neural networks, IEEE Trans. Autom. Control **39**(6), 1306–1310 (1994)

11.11   M. Krstic, I. Kanellakopoulos, P.V. Kokotovic: *Nonlinear and Adaptive Control Design* (Wiley, New York 1995)

11.12   K.S. Narendra, A.M. Annaswamy: *Stable Adaptive Systems* (Prentice Hall, Englewood Cliffs 1989)

11.13   K.J. Åström, B. Wittenmark: *Adaptive Control* (Addison-Wesley, Reading 1989)

11.14   S.S. Sastry, A. Isidori: Adaptive control of linearizable systems, IEEE Trans. Autom. Control **34**(11), 1123–1131 (1989)

11.15   I. Kanellakopoulos, P.V. Kokotovic, R. Marino: An extended direct scheme for robust adaptive nonlinear control, Automatica **27**(2), 247–255 (1991)

11.16   I. Kanellakopoulos, P.V. Kokotovic, A.S. Morse: Systematic design of adaptive controllers for feedback linearizable systems, IEEE Trans. Autom. Control **36**(11), 1241–1253 (1991)

11.17   D. Seto, A.M. Annaswamy, J. Baillieul: Adaptive control of nonlinear systems with a triangular structure, IEEE Trans. Autom. Control **39**(7), 1411–1428 (1994)

11.18   A. Kojic, A.M. Annaswamy: Adaptive control of nonlinearly parameterized systems with a triangular structure, Automatica **38**(1), 115–123 (2002)

11.19   M. Krstic, P.V. Kokotovic: Adaptive nonlineaer design with controller-identifier separation and swapping, IEEE Trans. Autom. Control **40**(3), 426–460 (1995)

11.20   A.S. Morse, D.Q. Mayne, G.C. Goodwin: Applications of hysteresis switching in parameter adaptive control, IEEE Trans. Autom. Control **34**(9), 1343–1354 (1992)

11.21   E.B. Kosmatopoulos, P.A. Ioannou: A switching adaptive controller for feedback linearizable systems, IEEE Trans. Autom. Control **44**(4), 742–750 (1999)

11.22   E.B. Kosmatopoulos, P.A. Ioannou: Robust switching adaptive control of multi-input nonlinear systems, IEEE Trans. Autom. Control **47**(4), 610–624 (2002)

11.23   J.S. Reed, P.A. Ioannou: Instability analysis and robust adaptive control of robotic manipulators, IEEE Trans. Robot. Autom. **5**(3), 381–386 (1989)

11.24   M.M. Polycarpou, P.A. Ioannou: A robust adaptive nonlinear control design, Automatica **32**(3), 423–427 (1996)

11.25   R.A. Freeman, M. Krstic, P.V. Kokotovic: Robustness of adaptive nonlinear control to bounded uncertainties, Automatica **34**(10), 1227–1230 (1998)

11.26   H. Xu, P.A. Ioannou: Robust adaptive control for a class of MIMO nonlinear systems with guaranteed error bounds, IEEE Trans. Autom. Control **48**(5), 728–742 (2003)

11.27   R.M. Sanner, J.J.E. Slotine: Gaussian networks for direct adaptive control, IEEE Trans. Neural Netw. **3**(6), 837–863 (1992)

11.28   L.-X. Wang: Stable adaptive fuzzy control of nonlinear systems, IEEE Trans. Fuzzy Syst. **1**(2), 146–155 (1993)

11.29   C.-Y. Su, Y. Stepanenko: Adaptive control of a class of nonlinear systems with fuzzy logic, IEEE Trans. Fuzzy Syst. **2**(4), 285–294 (1994)

11.30   M.M. Polycarpou: Stable adaptive neutral control scheme for nonlinear systems, IEEE Trans. Autom. Control **41**(3), 447–451 (1996)

11.31   L.-X. Wang: Stable adaptive fuzzy controllers with application to inverted pendulum tracking, IEEE Trans. Syst. Man Cybern. B **26**(5), 677–691 (1996)

11.32   J.T. Spooner, K.M. Passino: Stable adaptive control using fuzzy systems and neural networks, IEEE Trans. Fuzzy Syst. **4**(3), 339–359 (1996)

11.33   C.-H. Wang, H.-L. Liu, T.-C. Lin: Direct adaptive fuzzy-neutral control with state observer and supervisory controller for unknown nonlinear dynamical systems, IEEE Trans. Fuzzy Syst. **10**(1), 39–49 (2002)

11.34   E. Tzirkel-Hancock, F. Fallside: Stable control of nonlinear systems using neural networks, Int. J. Robust Nonlin. Control **2**(1), 63–86 (1992)

11.35   A. Yesildirek, F.L. Lewis: Feedback linearization using neural networks, Automatica **31**(11), 1659–1664 (1995)

11.36   B.S. Chen, C.H. Lee, Y.C. Chang: H^∞ tracking design of uncertain nonlinear SISO systems: Adaptive fuzzy approach, IEEE Trans. Fuzzy Syst. **4**(1), 32–43 (1996)

11.37   S.S. Ge, C.C. Hang, T. Zhang: A direct method for robust adaptive nonlinear control with guaranteed transient performance, Syst. Control Lett. **37**(5), 275–284 (1999)

11.38   S. Seshagiri, H.K. Khalil: Output feedback control of nonlinear systems using RBF neural networks, IEEE Trans. Neural Netw. **11**(1), 69–79 (2000)

11.39   M.I. El-Hawwary, A.L. Elshafei, H.M. Emara, H.A. Abdel Fattah: Output feedback control of a class of nonlinear systems using direct adaptive fuzzy controller, IEE Proc. Control Theory Appl. **151**(5), 615–625 (2004)

11.40   Y. Lee, S.H. Żak: Uniformly ultimately bounded fuzzy adaptive tracking controllers for uncertain systems, IEEE Trans. Fuzzy Syst. **12**(6), 797–811 (2004)

11.41   C.-C. Liu, F.-C. Chen: Adaptive control of nonlinear continuous-time systems using neural networks – general relative degree and MIMO cases, Int. J. Control **58**(2), 317–335 (1993)

11.42   R. Ordóñez, K.M. Passino: Stable multi-input multi-output adaptive fuzzy/neural control, IEEE Trans. Fuzzy Syst. **7**(3), 345–353 (2002)

11.43   S. Tong, J.T. Tang, T. Wang: Fuzzy adaptive control of multivariable nonlinear systems, Fuzzy Sets Syst. **111**(2), 153–167 (2000)

11.44   Y.-C. Chang: Robust tracking control of nonlinear MIMO systems via fuzzy approaches, Automatica **36**, 1535–1545 (2000)

11.45   Y.-C. Chang: An adaptive H^∞ tracking control for a class of nonlinear multiple-input-multiple-output (MIMO) systems, IEEE Trans. Autom. Control **46**(9), 1432–1437 (2001)

11.46   S. Tong, H.-X. Li: Fuzzy adaptive sliding-mode control for MIMO nonlinear systems, IEEE Trans. Fuzzy Syst. **11**(3), 354–360 (2003)

11.47   H.K. Khalil, F. Esfandiari: Semiglobal stabilization of a class of nonlinear systems using output feedback, IEEE Trans. Autom. Control **38**(9), 1412–1115 (1993)

11.48   H.K. Khalil: Robust servomechanism output feedback controllers for a class of feedback linearizable systems, Automatica **30**(10), 1587–1599 (1994)

11.49   H.K. Khalil: Adaptive output feedback control of nonlinear systems represented by input–output models, IEEE Trans. Autom. Control **41**(2), 177–188 (1996)

11.50   B. Aloliwi, H.K. Khalil: Robust adaptive output feedback control of nonlinear systems without persistence of excitation, Automatica **33**(11), 2025–2032 (1997)

11.51   S. Tong, T. Wang, J.T. Tang: Fuzzy adaptive output tracking control of nonlinear systems, Fuzzy Sets Syst. **111**(2), 169–182 (2000)

11.52   J.-X. Xu, Y. Tan: Nonlinear adaptive wavelet control using constructive wavelet networks, IEEE Trans. Neural Netw. **18**(1), 115–127 (2007)

11.53   S. Fabri, V. Kadirkamanathan: Dynamic structure neural networks for stable adaptive control of nonlinear systems, IEEE Trans. Neural Netw. **7**(5), 1151–1166 (1996)

11.54 G.P. Liu, V. Kadirkamanathan, S.A. Billings: Variable neural networks for adaptive control of nonlinear systems, IEEE Trans. Syst. Man Cybern. B **39**(1), 34–43 (1999)

11.55 Y. Lee, S. Hui, E. Zivi, S.H. Żak: Variable neural adaptive robust controllers for uncertain systems, Int. J. Adapt. Control Signal Process. **22**(8), 721–738 (2008)

11.56 J. Lian, Y. Lee, S.H. Żak: Variable neural adaptive robust control of uncertain systems, IEEE Trans. Autom. Control **53**(11), 2658–2664 (2009)

11.57 J. Lian, Y. Lee, S.D. Sudhoff, S.H. Żak: Variable structure neural network based direct adaptive robust control of uncertain systems, Proc. Am. Control Conf. (Seattle 2008) pp. 3402–3407

11.58 J. Lian, J. Hu, S.H. Żak: Adaptive robust control: A switched system approach, IEEE Trans. Autom. Control, to appear (2010)

11.59 J.P. Hespanha, A.S. Morse: Stability of switched systems with average dwell-time, Proc. 38th Conf. Decis. Control (Phoenix 1999) pp. 2655–2660

11.60 R.J. Schilling, J.J. Carroll, A.F. Al-Ajlouni: Approximation of nonlinear systems with radial basis function neural network, IEEE Trans. Neural Netw. **12**(1), 1–15 (2001)

11.61 J. Lian, Y. Lee, S.D. Sudhoff, S.H. Żak: Self-organizing radial basis function network for real-time approximation of continuous-time dy-

namical systems, IEEE Trans. Neural Netw. **19**(3), 460–474 (2008)

11.62 M. Jankovic: Adaptive output feedback control of nonlinear feedback linearizable systems, Int. J. Adapt. Control Signal Process. **10**, 1–18 (1996)

11.63 T. Zhang, S.S. Ge, C.C. Hang: Stable adaptive control for a class of nonlinear systems using a modified Lyapunov function, IEEE Trans. Autom. Control **45**(1), 129–132 (2000)

11.64 T. Zhang, S.S. Ge, C.C. Hang: Design and performance analysis of a direct adaptive controller for nonlinear systems, Automatica **35**(11), 1809–1817 (1999)

11.65 B. Yao: Integrated direct/indirect adaptive robust control of SISO nonlinear systems in semi-strict feedback form, Proc. Am. Control Conf., Vol. 4 (Denver 2003) pp. 3020–3025

11.66 B. Yao: Indirect adaptive robust control of SISO nonlinear systems in semi-strict feedback forms, Proc. 15th IFAC World Congr. (Barcelona 2002) pp. 1–6

11.67 F. Esfandiari, H.K. Khalil: Output feedback stabilization of fully linerizable systems, Int. J. Control **56**(5), 1007–1037 (1992)

11.68 N.A. Mahmoud, H.K. Khalil: Asymptotic regulation of minimum phase nonlinear systems using output feedback, IEEE Trans. Autom. Control **41**(10), 1402–1412 (1996)

11.69 B. Yao: *Lecture Notes from Course on Nonlinear Feedback Controller Design* (School of Mechanical Engineering, Purdue University, West Lafayette 2007)

Part B | 11

12. Cybernetics and Learning Automata

John Oommen, Sudip Misra

Stochastic learning automata are probabilistic finite state machines which have been used to model how biological systems can learn. The structure of such a machine can be fixed or can be changing with time. A learning automaton can also be implemented using action (choosing) probability updating rules which may or may not depend on estimates from the environment being investigated. This chapter presents an overview of the field of learning automata, perceived as a completely new paradigm for learning, and explains how it is related to the area of *cybernetics*.

Part B | 12

12.1 Basics

What is a *learning automaton*? What is *learning* all about? what are the different types of learning automata (LA) available? How are LA related to the general field of *cybernetics*? These are some of the fundamental issues that this chapter attempts to describe, so that we can understand the potential of the mechanisms, and their capabilities as primary tools which can be used to solve a host of very complex problems.

The Webster's dictionary defines *cybernetics* as:

> ... the science of communication and control theory that is concerned especially with the comparative study of automatic control systems (as the nervous system, the brain and mechanical–electrical communication systems).

The word *cybernetics* itself has its etymological origins in the Greek root *kybernan*, meaning *to steer* or *to govern*. Typically, as explained in the *Encyclopaedia Britannica*:

> *Cybernetics is associated with models in which a monitor compares what is happening to a system at various sampling times with some standard of what should be happening, and a controller adjusts the system's behaviour accordingly.*

Of course, the goal of the exercise is to design the *controller* so as to appropriately adjust the system's behavior. Modern cybernetics is an interdisciplinary field, which philosophically encompasses an ensemble of areas including neuroscience, computer science, cognition, control systems, and electrical networks.

The linguistic meaning of *automaton* is a self-operating machine or a mechanism that responds to a sequence of instructions in a certain way, so as to achieve a certain goal. The automaton either responds to a predetermined set of rules, or adapts to the environmental dynamics in which it operates. The latter types of automata are pertinent to this chapter, and are termed as *adaptive automata*. The term *learning* in psychology means the act of acquiring knowledge and modifying one's behavior based on the experience gained. Thus, in our case, the adaptive automaton we study in this chapter adapts to the responses from the environment through a series of interactions with it. It then attempts to learn the best action from a set of possible actions that are offered to it by the random stationary or nonstationary environment in which it operates. The automaton thus acts as a decision maker to arrive at the best action.

Well then, what do *learning automata* have to do with *cybernetics*? The answer to this probably lies in the results of the Russian pioneer *Tsetlin* [12.1, 2]. Indeed, when Tsetlin first proposed his theory of learning, his aim was to use the principles of automata theory to model how biological systems could learn. Little did he guess that his seminal results would lead to a completely new paradigm for learning, and a subfield of cybernetics.

The operations of the LA can be best described through the words of the pioneers *Narendra* and *Thathachar* [12.3, p. 3]:

> ... *a decision maker operates in the random environment and updates its strategy for choosing actions on the basis of the elicited response. The decision maker, in such a feedback configuration of decision maker (or automaton) and environment, is referred to as the learning automaton. The automaton has a finite set of actions, and corresponding to each action, the response of the environment can be either favorable or unfavorable with a certain probability.*

LA, thus, find applications in optimization problems in which an optimal action needs to be determined from a set of actions. It should be noted that, in this context, learning might be of best help only when there are high levels of *uncertainty* in the system in which the automaton operates. In systems with low levels of uncertainty, LA-based learning may not be a suitable tool of choice [12.3].

The first studies with LA models date back to the studies by mathematical psychologists such as *Bush* and *Mosteller* [12.4], and *Atkinson* et al. [12.5]. In 1961, the Russian mathematician, *Tsetlin* [12.1, 2] studied deterministic LA in detail. *Varshavskii* and *Vorontsova* [12.6]

introduced the stochastic variable structure versions of the LA. *Tsetlin*'s deterministic automata [12.1, 2] and *Varshavskii* and *Vorontsova*'s stochastic automata [12.6] were the major initial motivators of further studies in this area. Following them, several theoretical and experimental studies have been conducted by several researchers: Narendra, Thathachar, Lakshmivarahan, Obaidat, Najim, Poznyak, Baba, Mason, Papadimitriou, and Oommen, to mention a few. A comprehensive overview of research in the field of LA can be found in the classic text by *Narendra* and *Thathachar* [12.3], and in the recent special issue of *IEEE Transactions* [12.7].

It should be noted that none of the work described in this chapter is original. Most of the discussions, terminologies, and all the algorithms that are explained in this chapter are taken from the corresponding existing pieces of literature. Thus, the notation and terminology can be considered to be *off the shelf*, and fairly standard.

With regard to applications, the entire field of LA and stochastic learning, has had a myriad of applications [12.3, 8–11], which (apart from the many applications listed in these books) include solutions for problems in network and communications [12.12–15], network call admission, traffic control, quality-of-service routing, [12.16–18], distributed scheduling [12.19], training hidden Markov models [12.20], neural network adaptation [12.21], intelligent vehicle control [12.22], and even fairly theoretical problems such as graph partitioning [12.23].

We conclude this introductory section by emphasizing that this brief chapter should not be considered a comprehensive survey of the field of LA. In particular, we have not addressed the concept of LA which possess an infinite number of actions [12.24], systems which deal with *teachers* and *liars* [12.25], nor with any of the myriad issues that arise when we deal with networks of LA [12.11]. Also, the reader should not expect a mathematically deep exegesis of the field. Due to space limitations, the results available are merely cited. Additionally, while the results that are reported in the acclaimed books are merely alluded to, we give special attention to the more recent results – namely those which pertain to the discretized, pursuit, and estimator algorithms. Finally, we mention that the bibliography cited here is by no means comprehensive. It is brief and is intended to serve as a pointer to the representative papers in the theory and applications of LA.

12.2 A Learning Automaton

In the field of automata theory, an automaton can be defined as a quintuple consisting of a set of states, a set of outputs or actions, an input, a function that maps the current state and input to the next state, and a function that maps a current state (and input) into the current output [12.3, 8–11].

Definition 12.1
A LA is defined by a quintuple $\langle A, B, Q, F(\cdot, \cdot), G(\cdot) \rangle$, where:

(i) $A = \{\alpha_1, \alpha_2, \ldots, \alpha_r\}$ is the set of outputs or actions, and $\alpha(t)$ is the action chosen by the automaton at any instant t.
(ii) $B = \{\beta_1, \beta_2, \ldots, \beta_m\}$ is the set of inputs to the automaton. $\beta(t)$ is the input at any instant t. The set B can be finite or infinite. In this chapter, we consider the case when $m = 2$, i. e., when $B = \{0, 1\}$, where $\beta = 0$ represents the event that the LA has been rewarded, and $\beta = 1$ represents the event that the LA has been penalized.

(iii) $Q = \{q_1, q_2, \ldots, q_s\}$ is the set of finite states, where $q(t)$ denotes the state of the automaton at any instant t.
(iv) $F(\cdot, \cdot) : Q \times B \to Q$ is a mapping in terms of the state and input at the instant t, such that, $q(t+1) = F(q(t), \beta(t))$. It is called a *transition function*, i. e., a function that determines the state of the automaton at any subsequent time instant $t + 1$. This mapping can either be deterministic or stochastic.
(v) $G(\cdot)$ is a mapping $G : Q \to A$, and is called the *output function*. Depending on the state at a particular instant, this function determines the output of the automaton at the same instant as $\alpha(t) = G(q(t))$. This mapping can, again, be either deterministic or stochastic. Without loss of generality, G is deterministic.

If the sets Q, B, and A are all finite, the automaton is said be *finite*.

12.3 Environment

The environment E typically refers to the medium in which the automaton functions. The environment possesses all the external factors that affect the actions of the automaton. Mathematically, an environment can be abstracted by a triple $\langle A, C, B \rangle$. A, C, and B are defined as:

(i) $A = \{\alpha_1, \alpha_2, \ldots, \alpha_r\}$ is the set of actions.
(ii) $B = \{\beta_1, \beta_2, \ldots, \beta_m\}$ is the output set of the environment. Again, we consider the case when $m = 2$, i. e., with $\beta = 0$ representing a *reward*, and $\beta = 1$ representing a *penalty*.
(iii) $C = \{c_1, c_2, \ldots, c_r\}$ is a set of penalty probabilities, where element $c_i \in C$ corresponds to an input action α_i.

The process of learning is based on a learning loop involving the two entities: the random environment (RE), and the LA, as illustrated in Fig. 12.1. In the process of learning, the LA continuously interacts with the environment to process responses to its various actions (i. e., its choices). Finally, through sufficient interactions, the LA attempts to learn the optimal action offered by the RE. The actual process of learning is represented as a set of interactions between the RE and the LA.

The RE offers the automaton with a set of possible actions $\{\alpha_1, \alpha_2, \ldots, \alpha_r\}$ to choose from. The automaton chooses one of those actions, say α_i, which serves as an input to the RE. Since the RE is *aware* of the underlying penalty probability distribution of the system, depending on the *penalty probability* c_i corresponding to α_i, it *prompts* the LA with a reward (typically denoted by the value 0), or a *penalty* (typically denoted by the value 1). The reward/penalty information (corresponding to the action) provided to the LA helps it to choose the subsequent action. By repeating the above

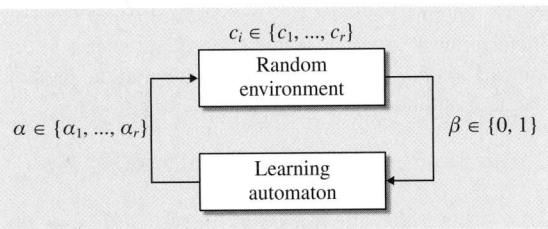

Fig. 12.1 The automaton–environment feedback loop

process, through a series of environment–automaton interactions, the LA finally attempts to learn the *optimal* action from the environment.

We now provide a few important definitions used in the field. $P(t)$ is referred to as the action probability vector, where, $P(t) = [p_1(t), p_2(t), \ldots, p_r(t)]^\top$, in which each element of the vector

$$p_i(t) = \Pr[\alpha(t) = \alpha_i], \quad i = 1, \ldots, r, \quad (12.1)$$

such that

$$\sum_{i=1}^{r} p_i(t) = 1 \quad \forall t.$$

Given an action probability vector, $P(t)$ at time t, the *average penalty* is

$$M(t) = E[\beta(t)|P(t)] = \Pr[\beta(t) = 1|P(t)]$$
$$= \sum_{i=1}^{r} \Pr[\beta(t) = 1|\alpha(t) = \alpha_i]\Pr[\alpha(t) = \alpha_i]$$
$$= \sum_{i=1}^{r} c_i p_i(t). \quad (12.2)$$

The average penalty for the *pure-chance* automaton is given by

$$M_0 = \frac{1}{r} \sum_{i=1}^{r} c_i. \quad (12.3)$$

As $t \to \infty$, if the average penalty $M(t) < M_0$, at least asymptotically, the automaton is generally considered to be better than the pure-chance automaton. $E[M(t)]$ is given by

$$E[M(t)] = E\{E[\beta(t)|P(t)]\} = E[\beta(t)]. \quad (12.4)$$

A LA that performs better than by pure chance is said to be *expedient*.

Definition 12.2
A LA is considered *expedient* if

$$\lim_{t \to \infty} E[M(t)] < M_0.$$

Definition 12.3
A LA is said to be *absolutely expedient* if $E[M(t+1)|P(t)] < M(t)$, implying that $E[M(t+1)] < E[M(t)]$.

Definition 12.4
A LA is considered *optimal* if $\lim_{t \to \infty} E[M(t)] = c_l$, where $c_l = \min_i\{c_i\}$.

Definition 12.5
A LA is considered ϵ-*optimal* if

$$\lim_{n \to \infty} E[M(t)] < c_l + \epsilon, \quad (12.5)$$

where $\epsilon > 0$, and can be arbitrarily small, by a suitable choice of some parameter of the LA.

It should be noted that no optimal LA exist. Marginally suboptimal performance, also termed above ϵ-optimal performance, is what LA researchers attempt to attain.

12.4 Classification of Learning Automata

12.4.1 Deterministic Learning Automata

An automaton is termed a *deterministic automaton*, if both the transition function $F(\cdot, \cdot)$ and the output function $G(\cdot)$ defined in Sect. 12.2 are deterministic. Thus, in a deterministic automaton, the subsequent state and action can be uniquely specified, provided that the present state and input are given.

12.4.2 Stochastic Learning Automata

If, however, either the transition function $F(\cdot, \cdot)$ or the output function $G(\cdot)$ are stochastic, the automaton is

termed a *stochastic automaton*. In such an automaton, if the current state and input are specified, the subsequent states and actions cannot be specified uniquely. In such a case, $F(\cdot, \cdot)$ only provides the probabilities of reaching the various states from a given state. Let $\mathbf{F}^{\beta_1}, \mathbf{F}^{\beta_2}, \ldots, \mathbf{F}^{\beta_m}$ denote the conditional probability matrices, where each of these conditional matrices $\mathbf{F}^\beta$ (for $\beta \in B$) is a $s \times s$ matrix, whose arbitrary element f_{ij}^β is

$$f_{ij}^\beta = \Pr[q(t+1) = q_j|q(t) = q_i, \beta(t) = \beta],$$
$$i, j = 1, 2, \ldots, s. \quad (12.6)$$

In (12.6), each element f_{ij}^{β} of the matrix $\mathbf{F}^{\beta}$ represents the probability of the automaton moving from state q_i to the state q_j on receiving an input signal β from the RE. $\mathbf{F}^{\beta}$ is a Markov matrix, and hence

$$\sum_{j=1}^{s} f_{ij}^{\beta} = 1 , \quad \text{where } \beta \in B; i = 1, 2, \ldots, s .$$

$$(12.7)$$

Similarly, in a stochastic automaton, if $G(\cdot)$ is stochastic, we have

$$g_{ij} = \Pr\{\alpha(t) = \alpha_j | q(t) = q_i\} , \quad i, j = 1, 2, \ldots, s ,$$

$$(12.8)$$

where g_{ij} represents the elements of the conditional probability matrix of dimension $s \times r$. Intuitively, g_{ij} denotes the probability that, when the automaton is in state q_i, it chooses the action α_j. As in (12.7), we have

$$\sum_{j=1}^{r} g_{ij} = 1 , \quad \text{for each row } i = 1, 2, \ldots, s . \quad (12.9)$$

Fixed Structure Learning Automata

In a stochastic LA, if the conditional probabilities f_{ij}^{β} and g_{ij} are constant, i. e., they do not vary with the time step t and the input sequence, the automaton is termed a *fixed structure stochastic automaton* (FSSA). The popular examples of these types of automata were proposed by *Tsetlin* [12.1, 2], *Krylov* [12.26], and *Krinsky* [12.27] – all of which are ϵ-optimal. Their details can be found in [12.3].

Variable Structure Learning Automata

Unlike the FSSA, *variable structure stochastic automata* (VSSA) are those in which the state transition probabilities are not fixed. In such automata, the state transitions or the action probabilities themselves are updated at every time instant using a suitable scheme. The transition probabilities f_{ij}^{β} and the output function g_{ij} vary with time, and the action probabilities are updated on the basis of the input. These automata are discussed here in the context of linear schemes, but the concepts discussed below can be extended to nonlinear updating schemes as well. The types of automata that update transition probabilities with time were introduced in 1963 by *Varshavskii* and *Vorontsova* [12.6]. A VSSA depends on a random-number generator for its implementation. The action

chosen is dependent on the action probability distribution vector, which is, in turn, updated based on the reward/penalty input that the automaton receives from the RE.

Definition 12.6
A VSSA is a quintuple $\langle Q, A, B, T \rangle$, where Q represents the different states of the automaton, A is the set of actions, B is the set of responses from the environment to the LA, G is the output function, and T is the action probability updating scheme $T : [0, 1]^r \times A \times B \mapsto [0, 1]^r$, such that

$$\boldsymbol{P}(t + 1) = T[\boldsymbol{P}(t), \alpha(t), \beta(t)] , \qquad (12.10)$$

where $\boldsymbol{P}(t)$ is the action probability vector.

Normally, VSSA involve the updating of both the state and action probabilities. For the sake of simplicity, in practice, it is assumed that in such automata, each state corresponds to a distinct action, in which case the action transition mapping G becomes the identity mapping, and the number of states, s, is equal to the number of actions, r ($s = r < \infty$).

VSSA can be analyzed using a discrete-time Markov process, defined on a suitable set of states. If a probability updating scheme T is time invariant, $\{\boldsymbol{P}(t)\}_{t \geq 0}$ is a discrete-homogenous Markov process, and the probability vector at the current time instant $\boldsymbol{P}(t)$ (along with $\alpha(t)$ and $\beta(t)$) completely determines $\boldsymbol{P}(t + 1)$. Hence, each distinct updating scheme, T, identifies a different type of learning algorithm, as follows:

- *Absorbing algorithms* are those in which the updating scheme, T, is chosen in such a manner that the Markov process has absorbing states;
- *Nonabsorbing algorithms* are those in which the Markov process has no absorbing states;
- *Linear algorithms* are those in which $\boldsymbol{P}(t + 1)$ is a linear function of $\boldsymbol{P}(t)$;
- *Nonlinear algorithms* are those in which $\boldsymbol{P}(t + 1)$ is a nonlinear function of $\boldsymbol{P}(t)$.

In a VSSA, if a chosen action α_i is rewarded, the probability for the current action is increased, and the probabilities for all other actions are decreased. On the other hand, if the chosen action α_i is penalized, the probability of the current action is decreased, whereas the probabilities for the rest of the actions could, typically, be increased. This leads to the following different types of learning schemes for VSSA:

- **Reward–penalty (RP):** In both cases, i.e., when the automaton is rewarded as well as penalized, the action probabilities are updated;
- **Inaction–penalty (IP):** When the automaton is penalized the action probability vector is updated, whereas when the automaton is rewarded the action probabilities are neither increased nor decreased;
- **Reward–inaction (RI):** The action probability vector is updated whenever the automaton is rewarded, and is unchanged whenever the automaton is penalized.

A LA is considered to be a *continuous automaton* if the probability updating scheme T is continuous, i.e., the probability of choosing an action can be any real number in the closed interval $[0, 1]$.

In a VSSA, if there are r actions operating in a stationary environment with $\beta = \{0, 1\}$, a general action probability updating scheme for a continuous automaton is described below. We assume that the action α_i is chosen, and thus, $\alpha(t) = \alpha_i$. The updated action probabilities can be specified as

$$\text{for } \beta(t) = 0 \, , \, \forall j \neq i \, , \quad p_j(t+1) = p_j(t) - g_j[\boldsymbol{P}(t)] \, ,$$
$$\text{for } \beta(t) = 1 \, , \, \forall j \neq i \, , \quad p_j(t+1) = p_j(t) + h_j[\boldsymbol{P}(t)] \, .$$
$$(12.11)$$

Since $\boldsymbol{P}(t)$ is a probability vector, $\sum_{j=1}^{r} p_j(t) = 1$. Therefore,

when $\beta(t) = 0$,

$$p_i(t+1) = p_i(t) + \sum_{j=1, j \neq i}^{r} g_j[\boldsymbol{P}(t)] \, ,$$

and when $\beta(t) = 1$,

$$p_i(t+1) = p_i(t) - \sum_{j=1, j \neq i}^{r} h_j[\boldsymbol{P}(t)] \, . \quad (12.12)$$

The functions h_j and g_j are *nonnegative* and *continuous* in $[0, 1]$, and obey

$$\forall i = 1, 2, \ldots, r \, , \, \forall P \in (0, 1)^R \, ,$$
$$0 < g_j(P) < p_j \, , \quad \text{and}$$
$$0 < \sum_{j=1, j \neq i}^{r} [p_j + h_j(P)] < 1 \, . \quad (12.13)$$

For *continuous linear* VSSA, the following four learning schemes are extensively studied in the literature. They are explained for the two-action case; their extension to the r-action case, where $r > 2$, is straightforward and can be found in [12.3]. The four learning schemes are:

- The linear reward–inaction scheme (L_{RI})
- The linear inaction–penalty scheme (L_{IP})
- The symmetric linear reward–penalty scheme (L_{RP})
- The linear reward–ϵ-penalty scheme ($L_{R-\epsilon P}$).

For a two-action LA, let

$$g_i[\boldsymbol{P}(t)] = a p_j(t) \quad \text{and}$$
$$h_j[\boldsymbol{P}(t)] = b(1 - p_j(t)) \, . \quad (12.14)$$

In (12.14), a and b are called the reward and penalty parameters, and they obey the following inequalities: $0 < a < 1$, $0 \leq b < 1$. Equation (12.14) will be used further to develop the action probability updating equations. The above-mentioned linear schemes are quite popular in LA because of their analytical tractability. They exhibit significantly different characteristics, as can be seen in Table 12.1.

The L_{RI} scheme was first introduced by *Norman* [12.28], and then studied by *Shapiro* and *Narendra* [12.29]. It is based on the principle that, whenever the automaton receives a favorable response (i.e., reward) from the environment, the action probabilities are

Table 12.1 Properties of the continuous learning schemes

Learning scheme	Learning parameters	Usefulness (good/bad)	Optimality	Ergodic/absorbing (when useful)
L_{RI}	$a > 0$, $b = 0$	Good	ϵ-optimal as $a \to 0$	Absorbing (stationary E)
L_{IP}	$a = 0$, $b > 0$	Very bad	Not even expedient	Ergodic (nonstationary E)
L_{RP} (symmetric)	$a = b$, $a, b > 0$	Bad	Never ϵ-optimal	Ergodic (nonstationary E)
$L_{R-\epsilon P}$	$a > 0$, $b \ll a$	Good	ϵ-optimal as $a \to 0$	Ergodic (nonstationary E)

updated in a *linear* manner, whereas if the automaton receives an unfavorable response (i. e., penalty) from the environment, they are unaltered.

The probability updating equations for this scheme can be simplified to

$$p_1(t+1) = p_1(t) + a[1 - p_1(t)],$$
$$\text{if } \alpha(t) = \alpha_1 \text{, and } \beta(t) = 0 \text{,}$$
$$p_1(t+1) = (1-a)p_1(t),$$
$$\text{if } \alpha(t) = \alpha_2 \text{, and } \beta(t) = 0 \text{,}$$
$$p_1(t+1) = p_1(t),$$
$$\text{if } \alpha(t) = \alpha_1 \text{ or } \alpha_2 \text{, and } \beta(t) = 1. \quad (12.15)$$

We see that, if action α_i is chosen, and a reward is received, the probability $p_i(t)$ is increased, and the other probability $p_j(t)$ (i. e., $j \neq i$) is decreased. If either α_1 or α_2 is chosen, and a penalty is received, $\boldsymbol{P}(t)$ is unaltered.

Equation (12.15) shows that the L_{RI} scheme has the vectors $[1, 0]^\top$ and $[0, 1]^\top$ as two absorbing states. Indeed, with probability 1, it gets absorbed into one of these absorbing states. Therefore, the convergence of the L_{RI} scheme is dependent on the nature of the initial conditions and probabilities. The scheme is not suitable for nonstationary environments. On the other hand, for stationary random environments, the L_{RI} scheme is both absolutely expedient, and ϵ-optimal [12.3].

The L_{IP} and L_{RP} schemes are devised similarly, and are omitted from further discussions. They, and their respective analysis, can be found in [12.3].

The so-called symmetry conditions for the functions $g(\cdot)$ and $h(\cdot)$ to lead to absolutely expedient LA are also derived in [12.3, 8].

Discretized Learning Automata

The VSSA algorithms presented in Sect. 12.4.2 are *continuous*, i. e., the action probabilities can assume any real value in the interval [0, 1]. In LA, the choice of an action is determined by a random-number generator (RNG). In order to increase the speed of convergence of these automata, *Thathachar* and *Oommen* [12.30] introduced the *discretized* algorithms for VSSA, in which they suggested the discretization of the probability space. The different properties (absorbing and ergodic) of these learning automata, and the updating schemes of action probabilities for these discretized automata (like their continuous counterparts), were later studied in detail by *Oommen* et al. [12.31–34].

Discretized automata can be perceived to be somewhat like a hybrid combination of FSSA and VSSA. Discretization is conceptualized by restricting the probability of choosing the actions to only a fixed number of values in the closed interval [0, 1]. Thus, the updating of the action probabilities is achieved in steps rather than in a continuous manner as in the case of continuous VSSA. Evidently, like FSSA, they possess finite sets, but because they have action probability vectors which are random vectors, they behave like VSSA.

Discretized LA can also be of two types:

(i) *Linear* – in which case the action probability values are uniformly spaced in the closed interval [0, 1]
(ii) *Nonlinear* – in which case the probability values are unequally spaced in the interval [0, 1] [12.30, 32–34].

Perhaps the greatest motivation behind discretization is overcoming the persistent limitation of con-

Part B | 12.4

Table 12.2 Properties of the discretized learning schemes

Learning scheme	Learning parameters	Usefulness (good/bad)	Optimality (as $N \to \infty$)	Ergodic/absorbing (when useful)
DL_{RI}	$N > 0$	Good	ϵ-optimal	Absorbing (stationary E)
DL_{IP}	$N > 0$	Very bad	Expedient	Ergodic (nonstationary E)
ADL_{IP}	$N > 0$	Good, sluggish	ϵ-optimal	Artificially absorbing (stationary environments)
DL_{RP}	$N > 0$	Reasonable	ϵ-optimal if $c_{min} < 0.5$	Ergodic (nonstationary E)
ADL_{RP}	$N > 0$	Good	ϵ-optimal	Artificially absorbing (stationary E)
MDL_{RP}	$N > 0$	Good	ϵ-optimal	Ergodic (nonstationary E)

tinuous learning automata, i.e., the slow rate of convergence. This is achieved by narrowing the underlying assumptions of the automata. Originally, the assumption was that the RNGs could generate real values with arbitrary precision. In the case of discretized LA, if an action probability is reasonably close to unity, the probability of choosing that action increases to unity (when the conditions are appropriate) directly, rather than asymptotically [12.30–34].

The second important advantage of discretization is that it is more practical in the sense that the RNGs used by continuous VSSA can only *theoretically* be assumed to adopt *any* value in the interval [0, 1], whereas almost all machine implementations of RNGs use pseudo-RNGs. In other words, the set of possible random values is not infinite in [0, 1], but finite.

Last, but not the least, discretization is also important in terms of implementation and representation. Discretized implementations of automata use integers for tracking the number of multiples of $1/N$ of the action probabilities, where N is the so-called *resolution parameter*. This not only increases the rate of convergence of the algorithm, but also reduces the time, in terms of the clock cycles it takes for the processor to do each iteration of the task, and the memory needed. Discretized algorithms have been proven to be both more time and space efficient than the continuous algorithms.

Similar to the continuous LA paradigm, the discretized versions, the DL_{RI}, DL_{IP}, and DL_{RP} automata, have also been reported. Their design, analysis, and properties are given in [12.30, 32–34], and are summarized in Table 12.2.

12.5 Estimator Algorithms

12.5.1 Rationale and Motivation

As we have seen so far, the rate of convergence of learning algorithms is one of the most important considerations, which was the primary reason for designing the family of *discretized* algorithms. With the same goal *Thathachar* and *Sastry* designed a new-class of algorithms, called the *estimator algorithms* [12.35–38], which have faster rate of convergence than all the previous families. These algorithms, like the previous ones, maintain and update an action probability vector. However, unlike the previous ones, these algorithms also keep running estimates for each action that is rewarded, using a *reward–estimate vector*, and then use those estimates in the probability updating equations. The reward estimates vector is, typically, denoted in the literature by $\hat{D}(t) = [\hat{d}_1(t), \dots, \hat{d}_r(t)]^\top$. The corresponding state vector is denoted by $Q(t) = \langle P(t), \hat{D}(t) \rangle$.

In a random environment, these algorithms help in choosing an action by increasing the confidence in the reward capabilities of the different actions; for example, these algorithms initially process each action a number of times, and then (in one version) could increase the probability of the action with the highest reward estimate [12.39]. This leads to a scheme with better accuracy in choosing the correct action. The previous nonestimator VSSA algorithms update the probability vector directly on the basis of the response of the environment to the automaton, where, depending on the type of vector updating scheme being used,

the probability of choosing a rewarded action in the subsequent time instant is increased, and the probabilities of choosing the other actions could be decreased. However, estimator algorithms update the probability vector based on both the estimate vector and the current feedback provided by the environment to the automaton. The environment influences the probability vector both directly and indirectly, the latter being as a result of the estimation of the reward estimates of the different actions. This may, thus, lead to increases in action probabilities different from the currently rewarded action.

Even though there is an added computational cost involved in maintaining the reward estimates, these estimator algorithms have an order of magnitude superior performance than the nonestimator algorithms previously introduced. *Lanctôt* and *Oommen* [12.31] further introduced the discretized versions of these estimator algorithms, which were proven to have an even faster rate of convergence.

12.5.2 Continuous Estimator Algorithms

Thathachar and *Sastry* introduced the class of continuous estimator algorithms [12.35–38] in which the probability updating scheme T is continuous, i.e., the probability of choosing an action can be any real number in the closed interval [0, 1]. As mentioned subsequently, the discretized versions of these algorithms were introduced by *Oommen* and his co-authors, *Lanc-*

tôt and *Agache* [12.31,40]. These algorithms are briefly explained in Sect. 12.5.3.

Pursuit Algorithm

The family of pursuit algorithms is a class of estimator algorithms that *pursue* an action that the automaton *currently* perceives to be the optimal one. The first pursuit algorithm, called the CP_{RP} algorithm, introduced by *Thathachar* and *Sastry* [12.36, 41], pursues the optimal action by changing the probability of the current optimal action whether it receives a reward or a penalty by the environment. In this case, the *currently perceived best action* is rewarded, and *its* action probability value is increased with a value directly proportional to its distance to unity, namely $1 - p_m(t)$, whereas the *less optimal actions* are penalized, and their probabilities decreased proportionally.

To start with, based on the probability distribution $P(t)$, the algorithm chooses an action $\alpha(t)$. Whether the response was a reward or a penalty, it increases that component of $P(t)$ which has the maximal current reward estimate, and it decreases the probability corresponding to the rest of the actions. Finally, the algorithm updates the running estimates of the reward probability of the action chosen, this being the principal idea behind keeping, and using, the running estimates. The estimate vector $\hat{D}(t)$ can be computed using the following formula which yields the maximum-likelihood estimate

$$\hat{d}_i(t) = \frac{W_i(t)}{Z_i(t)}, \quad \forall i = 1, 2, \ldots, r, \qquad (12.16)$$

where $W_i(t)$ is the number of times the action α_i has been rewarded until the current time t, and $Z_i(t)$ is the number of times α_i has been chosen until the current time t. Based on the above concepts, the CP_{RP} algorithm is formally given in [12.31, 39, 40].

The algorithm is similar in principle to the L_{RP} algorithm, because both the CP_{RP} and the L_{RP} algorithms increase/decrease the action probabilities of the vector, independent of whether the environment responds to the automaton with a reward or a penalty. The major difference lies in the way the reward estimates are maintained, used, and are updated on both reward/penalty. It should be emphasized that, whereas the nonpursuit algorithm moves the probability vector in the direction of the most recently rewarded action, the pursuit algorithm moves the probability vector in the direction of the action with the highest reward estimate. *Thathachar* and *Sastry* [12.41] have theoretically proven their ϵ-optimality, and experimentally proven that these pursuit algorithms are more accurate, and several orders of magnitude faster than the nonpursuit algorithms.

The reward–inaction version of this pursuit algorithm is also similar in design, and is described in [12.31, 40]. Other pursuit-like estimator schemes have also been devised and can be found in [12.40].

TSE Algorithm

A more advanced estimator algorithm, which we refer to as the TSE algorithm to maintain consistency with the existing literature [12.31, 39, 40], was designed by *Thathachar* and *Sastry* [12.37, 38].

Like the other estimator algorithms, the TSE algorithm maintains the running reward estimates vector $\hat{D}(t)$ and uses it to calculate the action probability vector $P(t)$. When an action $\alpha_i(t)$ is rewarded, according to the TSE algorithm, the probability components with a reward estimate greater than $\hat{d}_i(t)$ are treated differently from those components with a value lower than $\hat{d}_i(t)$. The algorithm does so by increasing the probabilities for all the actions that have a higher estimate than the estimate of the chosen action, and decreasing the probabilities of all the actions with a lower estimate. This is done with the help of an indicator function $S_{ij}(t)$ which assumes the value 1 if $\hat{d}_i(t) > \hat{d}_j(t)$ and the value 0 if $\hat{d}_i(t) \leq \hat{d}_j(t)$. Thus, the TSE algorithm uses both the probability vector $P(t)$ and the reward estimates vector $\hat{D}(t)$ to update the action probabilities. The algorithm is formally described in [12.39]. On careful inspection of the algorithm, it can be observed that $P(t+1)$ depends indirectly on the response of the environment to the automaton. The feedback from the environment changes the values of the components of $\hat{D}(t)$, which, in turn, affects the values of the functions $f(\cdot)$ and $S_{ij}(t)$ [12.31, 37–39].

Analyzing the algorithm carefully, we obtain three cases. If the i-th action is rewarded, the probability values of the actions with reward estimates higher than the reward estimate of the currently selected action are updated using [12.37]

$$p_j(t+1)$$
$$= p_j(t) - \lambda \left\{ f\left[\hat{d}_i(t) - \hat{d}_j(t)\right] \frac{[p_i(t) - p_j(t)p_i(t)]}{r - 1} \right\};$$
$$(12.17)$$

when $\hat{d}_i(t) < \hat{d}_j(t)$, since the function $f\left[\hat{d}_i(t) - \hat{d}_j(t)\right]$ is monotonic and increasing, $f\left[\hat{d}_i(t) - \hat{d}_j(t)\right]$ is seen to be negative. This leads to a higher value of $p_j(t+1)$ than that of $p_j(t)$, which indicates that the probability of choosing actions that have estimates greater than that of the estimates of the currently chosen action will increase.

For all the actions with reward estimates smaller than the estimate of the currently selected action, the probabilities are updated based on

$$p_j(t+1) = p_j(t) - \lambda f\left[\hat{d}_i(t) - \hat{d}_j(t)\right] p_j(t) \, . \quad (12.18)$$

The sign of the function $f\left[\hat{d}_i(t) - \hat{d}_j(t)\right]$ is negative, which indicates that the probability of choosing actions that have estimates less than that of the estimate of the currently chosen action will decrease.

Thathachar and *Sastry* have proven that the TSE algorithm is ϵ-optimal [12.37]. They have also experimentally shown that the TSE algorithm often converges several orders of magnitude faster than the L_{RI} scheme.

Generalized Pursuit Algorithm

Agache and *Oommen* [12.40] proposed a generalized version of the pursuit algorithm (CP_{RP}) proposed by *Thathachar* and *Sastry* [12.36, 41]. Their algorithm, called the *generalized pursuit algorithm* (GPA), generalizes Thathachar and Sastry's pursuit algorithm by pursuing all those actions that possess higher reward estimates than the chosen action. In this way the probability of choosing a wrong action is minimized. *Agache* and *Oommen* experimentally compared their pursuit algorithm with the existing algorithms, and found that their algorithm is the best in terms of the rate of convergence [12.40].

In the CP_{RP} algorithm, the probability of the best estimated action is maximized by first decreasing the probability of all the actions in the following manner [12.40]

$$p_j(t+1) = (1-\lambda)p_j(t) \, , \quad j = 1, 2, \ldots, r \, . \quad (12.19)$$

The sum of the action probabilities is made unity by the help of the probability mass Δ, which is given by [12.40]

$$\Delta = 1 - \sum_{j=1}^{r} p_j(t+1) = 1 - \sum_{j=1}^{r}(1-\lambda)p_j(t)$$

$$= 1 - \sum_{j=1}^{r} p_j(t) + \lambda \sum_{j=1}^{r} p_j(t) = \lambda \, . \quad (12.20)$$

Thereafter, the probability mass Δ is added to the probability of the best estimated action. The GPA algorithm, thus, equidistributes the probability mass Δ to the action estimated to be superior to the chosen action. This

gives us [12.40]

$$p_m(t+1) = (1-\lambda)p_m(t) + \Delta = (1-\lambda)p_m(t) + \lambda \, , \quad (12.21)$$

where $\hat{d}_m = \max_{j=1,2,\ldots,r}\left[\hat{d}_j(t)\right]$. Thus, the updating scheme is given by [12.40]

$$p_j(t+1) = (1-\lambda)p_j(t) + \frac{\lambda}{K(t)} \, ,$$
$$\text{if } \hat{d}_j(t) > \hat{d}_i(t) \, , \quad j \neq i \, ,$$
$$p_j(t+1) = (1-\lambda)p_j(t) \, ,$$
$$\text{if } \hat{d}_j(t) \leq \hat{d}_i(t) \, , \quad j \neq i \, ,$$
$$p_i(t+1) = 1 - \sum_{j \neq i} p_j(t+1) \, , \quad (12.22)$$

where $K(t)$ denotes the number of actions that have estimates greater than the estimate of the reward probability of the action currently chosen. The formal algorithm is omitted, but can be found in [12.40].

12.5.3 Discrete Estimator Algorithms

As we have seen so far, discretized LA are superior to their continuous counterparts, and the estimator algorithms are superior to the nonestimator algorithms in terms of the rate of convergence of the learning algorithms. Utilizing the previously proven capabilities of discretization in improving the speed of convergence of the learning algorithms, *Lanctôt* and *Oommen* [12.31] enhanced the pursuit and the TSE algorithms. This led to the designing of classes of learning algorithms, referred to in the literature as the *discrete estimator algorithms* (DEA) [12.31]. To this end, as done in the previous discrete algorithms, the components of the action probability vector are allowed to assume a finite set of discrete values in the closed interval [0, 1], which is, in turn, divided into a number of subintervals proportional to the resolution parameter N. Along with this, a reward estimate vector is maintained to keep an estimate of the reward probability of each action [12.31].

Lanctôt and *Oommen* showed that, for each member algorithm belonging to the class of DEAs to be ϵ-optimal, it must possess a pair of properties known as the *property of moderation* and the *monotone property*. Together these properties help prove the ϵ-optimality of any DEA algorithm [12.31].

Moderation Property

A DEA with r actions and a resolution parameter N is said to possess the *property of moderation* if the maximum magnitude by which an action probability can decrease per iteration is bounded by $1/(rN)$.

Monotone Property
Suppose there exists an index m and a time instant $t_0 < \infty$, such that $\hat{d}_m(t) > \hat{d}_j(t)$, $\forall j$ s.t. $j \neq m$ and $\forall t$ s.t. $t \geq t_0$, where $\hat{d}_m(t)$ is the maximal component of $\hat{D}(t)$. A DEA is said to possess the *monotone property* if there exists an integer N_0 such that, for all resolution parameters $N > N_0$, $p_m(t) \to 1$ with probability 1 as $t \to \infty$, where $p_m(t)$ is the maximal component of $P(t)$.

The discretized versions of the pursuit algorithm, and the TSE algorithm possessing the moderation and monotone properties, are presented in the next section.

Discrete Pursuit Algorithm
The discrete pursuit algorithm (formally described in [12.31]), is referred to as the DPA in the literature, and is similar to a great extent to its continuous pursuit counterpart, i.e., the CP_{RI} algorithm, except that the updates to the action probabilities for the DPA algorithm are made in discrete steps. Therefore, the equations in the CP_{RP} algorithm that involve multiplication by the learning parameter λ are substituted by the addition or subtraction of quantities proportional to the smallest step size.

As in the CP_{RI} algorithm, the DPA algorithm operates in three steps. If $\Delta = 1/(rN)$ (where N denotes the resolution, and r the number of actions) denotes the smallest step size, the integral multiples of Δ denote the step sizes in which the action probabilities are updated. Like the continuous reward–inaction algorithm, when the chosen action $\alpha(t) = \alpha_i$ is penalized, the action probabilities remain unchanged. However, when the chosen action $\alpha(t) = \alpha_i$ is rewarded, and the algorithm has not converged, the algorithm decreases, by the integral multiples of Δ, the action probabilities which do not correspond to the highest reward estimate.

Lanctôt and *Oommen* have shown that the DPA algorithm possesses the properties of moderation and monotonicity, and that it is thus ϵ-optimal [12.31]. They have also experimentally proved that, in different ranges of environments from simple to complex, the DPA algorithm is at least 60% faster than the CP_{RP} algorithm [12.31].

Discrete TSE Algorithm
Lanctôt and *Oommen* also discretized the TSE algorithm, and have referred to it as the *discrete TSE algorithm* (DTSE) [12.31]. Since the algorithm is based on the continuous version of the TSE algorithm, it obviously has the same level of intricacy, if not more. Lanctôt and Oommen theoretically proved that, like the DPA estimator algorithm, this algorithm also possesses the *moderation* and the *monotone* properties, while maintaining many of the qualities of the continuous TSE algorithm. They also provided the proof of convergence of this algorithm.

There are two notable parameters in the DTSE algorithm:

1. $\Delta = 1/(rN\theta)$, where N is the resolution parameter as before
2. θ, an integer representing the largest value by which any of the action probabilities can change in a single iteration.

A formal description of the DTSE algorithm is omitted here, but can be found in [12.31].

Discretized Generalized Pursuit Algorithm
Agache and *Oommen* [12.40] provided a discretized version of their GPA algorithm presented earlier. Their algorithm, called the *discretized generalized pursuit algorithm* (DGPA), also essentially generalizes *Thathachar* and *Sastry*'s pursuit algorithm [12.36, 41]. However, unlike the TSE, it pursues all those actions that possess higher reward estimates than the chosen action.

In essence, in any single iteration, the algorithm computes the number of actions that have higher reward estimates than the current chosen action, denoted by $K(t)$, whence the probability of all the actions that have estimates higher than the chosen action is increased by an amount $\Delta/K(t)$, and the probabilities for all the other actions are decreased by an amount $\Delta/(r - K(t))$, where $\Delta = 1/(rN)$ denotes the resolution step, and N the resolution parameter. The DGPA algorithm has been proven to possess the *moderation* and *monotone* properties, and is thus ϵ-optimal [12.40]. The detailed steps of the DGPA algorithm are omitted here.

12.5.4 Stochastic Estimator Learning Algorithm (SELA)

The SELA algorithm belongs to the class of discretized LA, and was proposed by *Vasilakos* and *Papadimitriou* [12.42]. It has, since then, been used for solving problems in the domain of computer networks [12.18, 43]. It is an ergodic scheme, which has the ability to converge to the optimal action irrespective of the distribution of the initial state [12.18, 42].

As before, let $A = \{\alpha_1, \alpha_2, \ldots, \alpha_r\}$ denote the set of actions and $B = \{0, 1\}$ denote the set of responses that can be provided by the environment, where $\beta(t)$ represents the feedback provided by the environment corresponding to a chosen action $\alpha(t)$ at time t. Let the probability of choosing the k-th action at the t-th time instant be $p_k(t)$. SELA updates the estimated environmental characteristics as the vector $E(t)$, which can be defined as $E(t) = \langle D(t), M(t), U(t) \rangle$, explained below.

$D(t) = \{d_1(t), d_2(t), \ldots, d_r(t)\}$ represents the vector of the reward estimates, where

$$d_k(t) = \sum_{i=1}^{W} \beta_k(t) . \qquad (12.23)$$

In (12.23), the numerator on the right-hand side represents the total rewards received by the LA in the window size representing the last W times a particular action α_k was selected by the algorithm. W is called the *learning window*.

The second parameter in $E(t)$ is called the *oldness vector*, and is represented as $M(t) = \{m_1(t), m_2(t), \ldots, m_r(t)\}$, where $m_k(t)$ represents the time passed (counted as the number of iterations) since the last time the action $\alpha_k(t)$ was selected.

The last parameter $U(t)$ is called the *stochastic estimator vector* and is represented as $U(t) = \{u_1(t), u_2(t), \ldots, u_r(t)\}$, where the stochastic estimate $u_i(t)$ of action α_i is calculated using

$$u_i(t) = d_i(t) + N[0, \sigma_i^2(t)] , \qquad (12.24)$$

where $N[0, \sigma_i^2(t)]$ represents a random number selected from a normal distribution that has a mean of 0 and a standard deviation of $\sigma_i(t) = \min\{\sigma_{max}, a \, m_i(t)\}$, and a is a parameter signifying the rate at which the stochastic estimates become independent, and σ_{max} represents the maximum possible standard deviation that the stochastic estimates can have. In symmetrically distributed noisy stochastic environments, SELA is shown to be ϵ-optimal, and has found applications for routing in ATM networks [12.18, 43].

12.6 Experiments and Application Examples

All the continuous and the discretized versions of the estimator algorithms presented above were experimentally evaluated [12.31, 40]. *Lanctôt* and *Oommen* [12.31] compared the rates of convergence between the discretized and the continuous versions of the pursuit and the TSE estimator algorithms. In their experiments, they required that their algorithm achieve a level of accuracy of not making any errors in convergence in 100 experiments. To initialize the reward estimates vector, 20 iterations were performed for each action. The experimental results for the TSE algorithms are summarized in Table 12.3. The corre-

sponding results of the pursuit algorithms are provided in Table 12.4 (the numbers indicate the number of iterations required to attain to convergence). The results show that the discretized TSE algorithm is faster (between 50–76%) than the continuous TSE algorithm. Similar observations were obtained for the pursuit algorithm. The discretized versions of the pursuit algorithms were found to be at least 60% faster than their continuous counterparts; for example, with $d_1 = 0.8$ and $d_2 = 0.6$, the continuous TSE algorithm required an average of 115 iterations to converge, whereas the discretized TSE took only 76. Another set of experimental

Table 12.3 The number of iterations until convergence in two-action environments for the TSE algorithms (after [12.31])

Probability of reward		Mean iterations	
Action 1	Action 2	Continuous	Discrete
0.800	0.200	28.8	24.0
0.800	0.400	37.0	29.0
0.800	0.600	115.0	76.0
0.800	0.700	400.0	380.0
0.800	0.750	2200.0	1200.0
0.800	0.775	8500.0	5600.0

Table 12.4 The number of iterations until convergence in two-action environments for the pursuit algorithms (after [12.31])

Probability of reward		Mean iterations	
Action 1	Action 2	Continuous	Discrete
0.800	0.200	22	22
0.800	0.400	22	39
0.800	0.600	148	125
0.800	0.700	636	357
0.800	0.750	2980	1290
0.800	0.775	6190	3300

Table 12.5 Comparison of the discrete and continuous estimator algorithms in a benchmark with ten-action environments (after [12.31])

Environment	Algorithm	Continuous	Discrete
E_A	Pursuit	1140	799
E_A	TSE	310	207
E_B	Pursuit	2570	1770
E_B	TSE	583	563

Table 12.6 Experimental comparison of the performance of the GPA and the DGPA algorithms in benchmark ten-action environments (after [12.40])

Environ-ment	GPA		DGPA	
	λ	Number of iterations	N	Number of iterations
E_A	0.0127	948.03	24	633.64
E_B	0.0041	2759.02	52	1307.76

Note: The reward probabilities for the actions are:
E_A : 0.7 0.5 0.3   0.2   0.4 0.5 0.4 0.3 0.5 0.2
E_B : 0.1 0.45 0.84 0.76 0.2 0.4 0.6 0.7 0.5 0.3

comparisons was performed between all the estimator algorithms presented so far in several ten-action environments [12.31]. Their results [12.31] are summarized in Table 12.5, and show that the TSE algorithm is much faster than the pursuit algorithm. Whereas the continuous pursuit algorithm required 1140 iterations to converge, the TSE algorithm took only 310. The same observation applies to their discrete versions. Similarly, it was observed that the discrete estimator algorithms were much faster than the continuous estimator algorithms; for example, for environment E_A, while the

continuous algorithm took 1140 iterations to converge, the discretized algorithm needed only 799 iterations. The GPA and the DGPA algorithms were compared for the benchmark ten-action environments. The results are summarized in Table 12.6. The DGPA algorithm was found to converge much faster than the GPA algorithm. This, once again, proves the superiority of the discretized algorithms over the continuous ones.

12.7 Emerging Trends and Open Challenges

Although the field of LA is relatively young, the *analytic* results that have been obtained are quite phenomenal. Simultaneously, however, it is also fair to assert that the tools available in the field have been far too underutilized in real-life problems.

We believe that the main areas of research that will emerge in the next few years will involve applying LA to a host of application domains. Here, as the saying goes, *the sky is the limit*, because LA can probably be used in *any* application where the parameters characterizing the underlying system are unknown and random. Some possible potential applications are listed below:

1. LA could be used in medicine to help with the diagnosis process.
2. LA have potential applications in intelligent tutorial (or tutorial-*like*) systems to assist in imparting imperfect knowledge to classrooms of students, where the teacher is also assumed to be imperfect. Some initial work is already available in this regard [12.44].

3. The use of LA in legal arguments and the associated decision-making processes is open.
4. Although LA have been used in some robotic applications, as far as we know, almost no work has been done for obstacle avoidance and intelligent path planning of real-life robots.
5. We are not aware of any results that use LA in the biomedical application domain. In particular, we believe that they can be fruitfully utilized for learning targets, and in the drug design phase.
6. One of the earliest applications of LA was in the routing of telephone calls over land lines, but the real-life application of LA in *wireless* and *multihop* networks is still relatively open.

We close this section by briefly mentioning that the main challenge in using LA for each of these application domains would be that of modeling what the *environment* and *automaton* are. Besides this, the practitioner would have to consider how the response of a particular solution can be interpreted as the reward/penalty for the automaton, or for the network of automata.

Part B | 12.7

12.8 Conclusions

In this chapter we have discussed most of the important learning mechanisms reported in the literature pertaining to learning automata (LA). After briefly stating the concepts of fixed structure stochastic LA, the families of continuous and discretized variable structure stochastic automata were discussed. The chapter, in particular, concentrated on the more recent results involving continuous and discretized pursuit and estimator algorithms. In each case we have briefly summarized the theoretical and experimental results of the different learning schemes.

References

12.1   M.L. Tsetlin: On the behaviour of finite automata in random media, Autom. Remote Control **22**, 1210–1219 (1962), Originally in Avtom. Telemekh. **22**, 1345–1354 (1961), in Russian

12.2   M.L. Tsetlin: *Automaton Theory and Modeling of Biological Systems* (Academic, New York 1973)

12.3   K.S. Narendra, M.A.L. Thathachar: *Learning Automata* (Prentice-Hall, Upper Saddle River 1989)

12.4   R.R. Bush, F. Mosteller: *Stochastic Models for Learning* (Wiley, New York 1958)

12.5   C.R. Atkinson, G.H. Bower, E.J. Crowthers: *An Introduction to Mathematical Learning Theory* (Wiley, New York 1965)

12.6   V.I. Varshavskii, I.P. Vorontsova: On the behavior of stochastic automata with a variable structure, Autom. Remote Control **24**, 327–333 (1963)

12.7   M.S. Obaidat, G.I. Papadimitriou, A.S. Pomportsis: Learning automata: theory, paradigms, and applications, IEEE Trans. Syst. Man Cybern. B **32**, 706–709 (2002)

12.8   S. Lakshmivarahan: *Learning Algorithms Theory and Applications* (Springer, New York 1981)

12.9   K. Najim, A.S. Poznyak: *Learning Automata: Theory and Applications* (Pergamon, Oxford 1994)

12.10   A.S. Poznyak, K. Najim: *Learning Automata and Stochastic Optimization* (Springer, Berlin 1997)

12.11   M.A.L.T. Thathachar, P.S. Sastry: *Networks of Learning Automata: Techniques for Online Stochastic Optimization* (Kluwer, Boston 2003)

12.12   S. Misra, B.J. Oommen: GPSPA: a new adaptive algorithm for maintaining shortest path routing trees in stochastic networks, Int. J. Commun. Syst. **17**, 963–984 (2004)

12.13   M.S. Obaidat, G.I. Papadimitriou, A.S. Pomportsis, H.S. Laskaridis: Learning automata-based bus arbitration for shared-medium ATM switches, IEEE Trans. Syst. Man Cybern. B **32**, 815–820 (2002)

12.14   B.J. Oommen, T.D. Roberts: Continuous learning automata solutions to the capacity assignment problem, IEEE Trans. Comput. C **49**, 608–620 (2000)

12.15   G.I. Papadimitriou, A.S. Pomportsis: Learning-automata-based TDMA protocols for broadcast communication systems with bursty traffic, IEEE Commun. Lett. **3**(3), 107–109 (2000)

12.16   A.F. Atlassis, N.H. Loukas, A.V. Vasilakos: The use of learning algorithms in atm networks call admission control problem: a methodology, Comput. Netw. **34**, 341–353 (2000)

12.17   A.F. Atlassis, A.V. Vasilakos: The use of reinforcement learning algorithms in traffic control of high speed networks. In: *Advances in Computational Intelligence and Learning* (Kluwer, Dordrecht 2002) pp. 353–369

12.18   A. Vasilakos, M.P. Saltouros, A.F. Atlassis, W. Pedrycz: Optimizing QoS routing in hierarchical ATM networks using computational intelligence techniques, IEEE Trans. Syst. Sci. Cybern. C **33**, 297–312 (2003)

12.19   F. Seredynski: Distributed scheduling using simple learning machines, Eur. J. Oper. Res. **107**, 401–413 (1998)

12.20   J. Kabudian, M.R. Meybodi, M.M. Homayounpour: Applying continuous action reinforcement learning automata (CARLA) to global training of hidden Markov models, Proc. ITCC'04 (Las Vegas 2004) pp. 638–642

12.21   M.R. Meybodi, H. Beigy: New learning automata based algorithms for adaptation of backpropagation algorithm parameters, Int. J. Neural Syst. **12**, 45–67 (2002)

12.22   C. Unsal, P. Kachroo, J.S. Bay: Simulation study of multiple intelligent vehicle control using stochastic learning automata, Trans. Soc. Comput. Simul. Int. **14**, 193–210 (1997)

12.23   B.J. Oommen, E.V. de St. Croix: Graph partitioning using learning automata, IEEE Trans. Comput. C **45**, 195–208 (1995)

12.24   G. Santharam, P.S. Sastry, M.A.L. Thathachar: Continuous action set learning automata for stochastic optimization, J. Franklin Inst. **331**(5), 607–628 (1994)

12.25   B.J. Oommen, G. Raghunath, B. Kuipers: Parameter learning from stochastic teachers and stochastic compulsive liars, IEEE Trans. Syst. Man Cybern. B **36**, 820–836 (2006)

12.26   V. Krylov: On the stochastic automaton which is asymptotically optimal in random medium, Autom. Remote Control **24**, 1114–1116 (1964)

12.27 V.I. Krinsky: An asymptotically optimal automaton with exponential convergence, Biofizika **9**, 484–487 (1964)

12.28 M.F. Norman: On linear models with two absorbing barriers, J. Math. Psychol. **5**, 225–241 (1968)

12.29 I.J. Shapiro, K.S. Narendra: Use of stochastic automata for parameter self-optimization with multi-modal performance criteria, IEEE Trans. Syst. Sci. Cybern. **SSC-5**, 352–360 (1969)

12.30 M.A.L. Thathachar, B.J. Oommen: Discretized reward–inaction learning automata, J. Cybern. Inf. Sci. **2**(1), 24–29 (1979)

12.31 J.K. Lanctôt, B.J. Oommen: Discretized estimator learning automata, IEEE Trans. Syst. Man Cybern. **22**, 1473–1483 (1992)

12.32 B.J. Oommen, J.P.R. Christensen: ϵ-optimal discretized linear reward–penalty learning automata, IEEE Trans. Syst. Man Cybern. B **18**, 451–457 (1998)

12.33 B.J. Oommen, E.R. Hansen: The asymptotic optimality of discretized linear reward–inaction learning automata, IEEE Trans. Syst. Man Cybern. **14**, 542–545 (1984)

12.34 B.J. Oommen: Absorbing and ergodic discretized two action learning automata, IEEE Trans. Syst. Man Cybern. **16**, 282–293 (1986)

12.35 P.S. Sastry: Systems of Learning Automata: Estimator Algorithms Applications. Ph.D. Thesis (Department of Electrical Engineering, Indian Institute of Science, Bangalore 1985)

12.36 M.A.L. Thathachar, P.S. Sastry: A new approach to designing reinforcement schemes for learning automata, Proc. IEEE Int. Conf. Cybern. Soc. (Bombay 1984)

12.37 M.A.L. Thathachar, P.S. Sastry: A class of rapidly converging algorithms for learning automata, IEEE Trans. Syst. Man Cybern. **15**, 168–175 (1985)

12.38 M.A.L. Thathachar, P.S. Sastry: Estimator algorithms for learning automata, Proc. Platin. Jubil. Conf. Syst. Signal Process. (Department of Electrical Engineering, Indian Institute of Science, Bangalore 1986)

12.39 M. Agache: Estimator Based Learning Algorithms. MSC Thesis (School of Computer Science, Carleton University, Ottawa 2000)

12.40 M. Agache, B.J. Oommen: Generalized pursuit learning schemes: new families of continuous and discretized learning automata, IEEE Trans. Syst. Man Cybern. B **32**(2), 738–749 (2002)

12.41 M.A.L. Thathachar, P.S. Sastry: Pursuit algorithm for learning automata. Unpublished paper that can be available from the authors

12.42 A.V. Vasilakos, G. Papadimitriou: Ergodic discretize destimator learning automata with high accuracy and high adaptation rate for nonstationary environments, Neurocomputing **4**, 181–196 (1992)

12.43 A.F. Atlasis, M.P. Saltouros, A.V. Vasilakos: On the use of a stochastic estimator learning algorithm to the ATM routing problem: a methodology, Proc. IEEE GLOBECOM (1998)

12.44 M.K. Hashem: Learning Automata-Based Intelligent Tutorial-Like Systems. Ph.D. Thesis (School of Computer Science, Carleton University, Ottawa 2007)

Part B | 12

13. Communication in Automation, Including Networking and Wireless

Nicholas Kottenstette, Panos J. Antsaklis

An introduction to the fundamental issues and limitations of communication and networking in automation is given. Digital communication fundamentals are reviewed and networked control systems together with teleoperation are discussed. Issues in both wired and wireless networks are presented.

Part B | 13

13.1 Basic Considerations

13.1.1 Why Communication Is Necessary in Automated Systems

Automated systems use local control systems that utilize sensor information in feedback loops, process this information, and send it as control commands to actuators to be implemented. Such closed-loop feedback control is necessary because of the uncertainties in the knowledge of the process and in the environmental conditions. Feedback control systems rely heavily on the ability to receive sensor information and send commands using wired or wireless communications.

In automated systems there is control supervision, and also health and safety monitoring via supervisory control and data acquisition (SCADA) systems. Values of important quantities (which may be temperatures, pressures, voltages, etc.) are sensed and transmitted to monitoring stations in control rooms. After processing

the information, decisions are made and supervisory commands are sent to change conditions such as set points or to engage emergency procedures. The data from sensors and set commands to actuators are sent via wired or wireless communication channels.

So, communication mechanisms are an integral part of any complex automated system.

13.1.2 Communication Modalities

In any system there are internal communication mechanisms that allow components to interact and exhibit a collective behavior, the system behavior; for example, in an electronic circuit, transistors, capacitors, resistances are connected so current can flow among them and the circuit can exhibit the behavior it was designed for. Such internal communication is an integral part of any system. At a higher level, subsystems that can each be quite complex interact via external communication links that may be wired or wireless. This is the case, for example, in antilock brake systems, vehicle stability systems, and engine and exhaust control systems in a car, or among unmanned aerial vehicles that communicate among themselves to coordinate their flight paths. Such external to subsystems communication is of prime interest in automated systems.

There are of course other types of communication, for example, machine to machine via mechanical links and human to machine, but here we will focus on electronic transmission of information and communication networks in automated systems.

Such systems are present in refineries, process plants, manufacturing, and automobiles, to mention but a few. Advances in computer and communication technologies coupled with lower costs are the main driving forces of communication methods in automated systems today. Digital communications, shared wired communication links, and wireless communications make up the communication networks in automated systems today.

In the following, after an introduction to digital communication fundamentals, the focus is on networked control systems that use shared communication links, which is common practice in automated systems.

13.2 Digital Communication Fundamentals

A digital communication system can generally be thought of as a system which allows either a continuous $x(t)$ or discrete random source of information to be transmitted through a channel to a given (set of) sink(s) (Fig. 13.1). The information that arrives at a given destination can be subject to delays, signal distortion, and noise. The digital communication channel typically is treated as a physical medium through which the information travels as an appropriately modulated analog signal, $s_m(t)$, subject to linear distortion and additive (typically Gaussian) noise $n(t)$. As is done in [13.1] we choose to use the simplified single-channel network shown in Fig. 13.1 in which the source encoder/decoder and channel encoder/decoder are separate entities. The design of the source encoder/decoder can usually be performed independently of the design of the channel encoder/decoder. This is possible due to the *source-channel separation theorem (SCST)* stated by *Shannon* [13.2], which states that, as long as the *average* information rate of bit/s from the source encoder R_s is strictly below the channel capacity C, information can be reliably transmitted with an appropriately designed

channel encoder. Conversely, if R_s is greater than or equal to C then it is impossible to send any information reliably. The interested reader should also see [13.3] for a more recent discussion as how the SCST relates to the single-channel case; [13.4] discusses the SCST as it applies to single-source broadcasting to many users, and [13.5] discusses how the SCST relates to many sources transmitting to one sink.

In Sect. 13.2.1 we will restate some of Shannon's key theorems as they relate to digital communication systems. With a clear understanding of the limitations and principles associated with digital communication systems we will address source encoder and decoder design in Sect. 13.2.2 and channel encoder and decoder design in the Appendix.

13.2.1 Entropy, Data Rates, and Channel Capacity

Entropy is a measure of uncertainty of a data source and is typically denoted by the symbol H. It can be seen as a measure of how many *bits* are required to describe

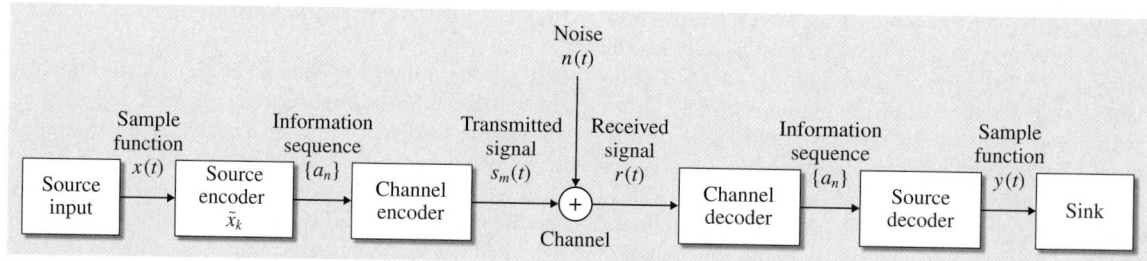

Fig. 13.1 Digital communication network with separate source and channel coding

a specific output *symbol* of the data source. Therefore, the natural unit of measure for entropy is bit/symbol and can also be used in terms of bit/s, depending on the context. Assuming that the source could have n outcomes in which each outcome has a probability p_i of occurrence the entropy has the form [13.2, Theorem 2]

$$H = -\sum_{i=1}^{n} p_i \log_2 p_i . \tag{13.1}$$

The entropy is greatest from a source where all symbols are equally likely; for example, given a 2 bit source in which each output symbol is {00, 01, 10, 11} with respective output probabilities $p_i = \left\{ \frac{p_0}{3}, \frac{p_0}{3}, \frac{p_0}{3}, 1 - p_0 \right\}$, it will have the following entropy, which is maximized when all outcomes are equally likely

$$H = -\frac{p_0}{3} \sum_{i=1}^{3} \log_2 \frac{p_0}{3} - (1 - p_0) \log_2(1 - p_0)$$

$$= -\frac{1}{4} \log_2 \left(\frac{1}{4} \right) = \log_2(4) = 2 , \tag{13.2}$$

$$p_0 = \frac{3}{4} .$$

Figure 13.2 shows a plot of entropy as a function of p_0; note that $H = 0$ bits when $p_0 = 0$; since the source would only generate the symbol 11 there is no need to actually transmit it to the receiver. Note that our 2 bit representations of our symbols is an inefficient choice; for example if $p_0 = 0.2$ we could represent this source with only 1 bit. This can be accomplished by encoding groups of symbols as opposed to considering individual symbols. By determining the redundancy of the source,

efficient compression algorithms can be derived, as discussed further in Sect. 13.2.2.

In digital communication theory we are typically concerned with describing the entropy of joint events $H(x, y)$ in which events x and y have, respectively, m and n possible outcomes with a joint probability of occurrence $p(x, y)$. The joint probability can be computed using

$$H(x, y) = -\sum_{i,j} p(i, j) \log_2 p(i, j) ,$$

in which it has been shown [13.2] that the following inequalities hold

$$H(x, y) \leq H(x) + H(y) \tag{13.3}$$
$$= H(x) + H_x(y) , \tag{13.4}$$
$$H(y) \geq H_x(y) . \tag{13.5}$$

Equality for (13.3) holds if and only if both events are independent. The uncertainty of y ($H(y)$) is never increased by knowledge of x ($H_x(y)$), as indicated by the conditional entropy inequality in (13.4). These measures provide a natural way of describing channel capacity when digital information is transmitted as an analog waveform through a channel which is subject to random noise. The effective rate of transmission, R, is the difference of the source entropy $H(x)$ from the average rate of conditional entropy, $H_y(x)$. Therefore, the channel capacity C is the maximum rate R achievable

$$R = H(x) - H_y(x) , \tag{13.6}$$
$$C = \max(H(x) - H_y(x)) . \tag{13.7}$$

This naturally leads to the discrete channel capacity theorem given by *Shannon* [13.2, Theorem 11]. The theorem states that, if a discrete source has entropy H that is less than the channel capacity C, their exists an encoding scheme such that data can be transmitted with an arbitrarily small frequency of errors (small equivocation), otherwise the equivocation will approach $H - C + \epsilon$, where $\epsilon > 0$ is arbitrarily small.

13.2.2 Source Encoder/Decoder Design

Source Data Compression
Shannon's fundamental theorem for a noiseless channel is the basis for understanding data compression algorithms. In [13.2, Theorem 9] he states that, for a given source with entropy H (bit/symbol) and channel capacity C (bit/s), a compression scheme exists such that one can transmit data at an average rate $R = \frac{C}{H} - \epsilon$

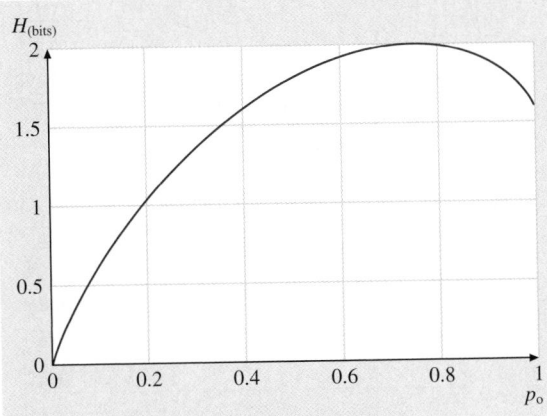

Fig. 13.2 Entropy of four-symbol source $p_i = \left\{ \frac{p_0}{3}, \frac{p_0}{3}, \frac{p_0}{3}, 1 - p_0 \right\}$

(symbol/second), where $\epsilon > 0$ is arbitrarily small; for example, if one had a 10 bit temperature measurement of a chamber which 99% of the time is at 25 °C and all other measurements are uniformly distributed for the remaining 1% of the time then you would only send a single bit to represent 25 °C instead of all 10 bits. Assuming that the capacity of the channel is 100 bit/s, then instead of sending data at an *average* rate of $10 = \frac{100}{10}$ measurements per second you will actually send data at an *average* rate of $99.1 = \left(0.99 \frac{100}{1} + 0.01 \frac{100}{10}\right)$ measurements per second.

Note that, as this applies to source coding theory, we can also treat the channel capacity C as the ideal H for the source, and so H is the actual bit rate achieved R for a given source. Then $R = \sum_{i=1}^{n} p_i n_i$, where p_i is the probability of occurrence for each code word of length n_i bit. When evaluating a source coding algorithm we can look at the *efficiency* of the algorithm, which is $100H/R\%$.

As seen in Fig. 13.2, if $p_0 = 0.19$ then $H = 1.0$ bit/symbol. If we used our initial encoding for the symbols, we would transmit on average 2 bits/symbol with an *efficiency* of 50%. We will discover that by

using a variable-length code and by making the following source encoder map $x_k = \{11, 00, 01, 10\} \rightarrow a_k = \{0, 01, 011, 111\}$ we can lower our average data rate to $R = 1.32$ bit/symbol, which improves the *efficiency* to 76%. Note that both mappings satisfy the *prefix condition* which requires that, for a given code word $\mathbf{C}_k$ of length k with bit elements $(b_1, b_2, \ldots, b_k)$, there is no other code word of length $l < k$ with elements $(b_1, b_2, \ldots, b_l)$ for $1 \le l < k$ [13.6]. Therefore, both codes satisfy the Kraft inequality [13.6, p. 93].

In order to get closer to the ideal $H = 1.0$ bit/symbol we will use the Huffman coding algorithm [13.6, pp. 95–99] and encode pairs of letters before transmission (which will naturally increase H to 2.0 bit/symbol − pair).

Figure 13.3 shows the resulting code words for transmitting pairs of symbols. We see that the encoding results in an *efficiency* of 95% in which $H = 2.0$ and the average achievable transmission rate is $R = 2.1$. The table is generated by sorting in descending order each code word pair and its corresponding probability of occurrence. Next, a tree is made in which pairs are generated by matching the two least probable events and

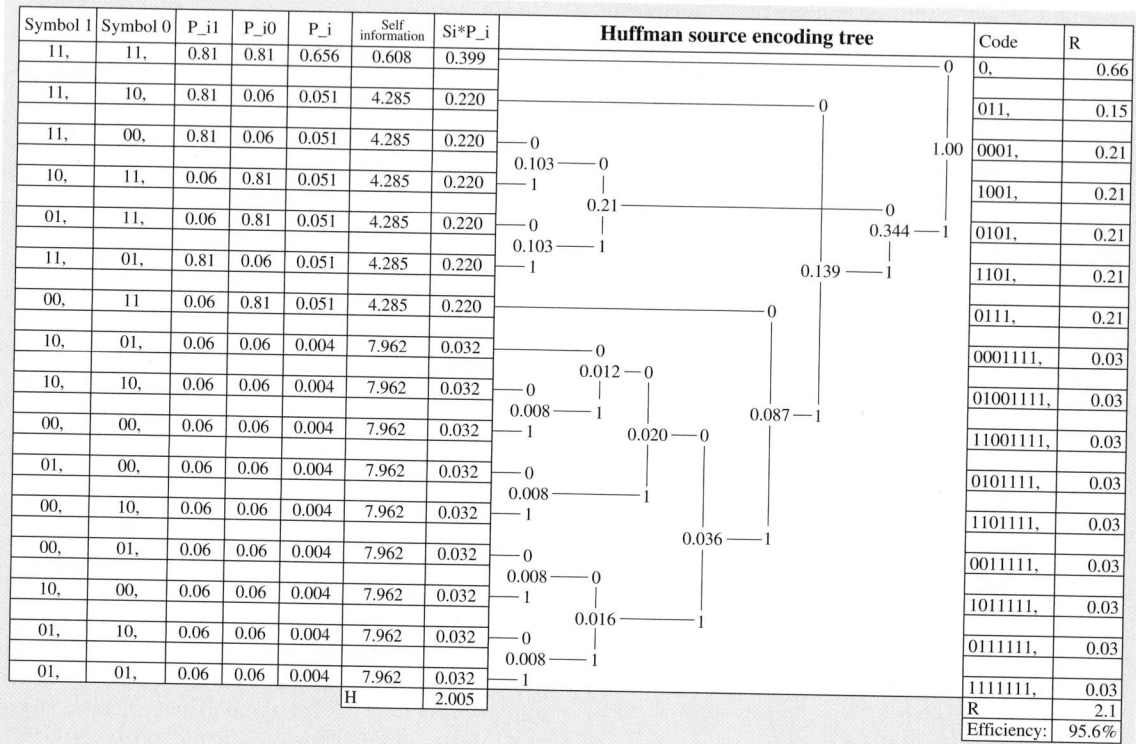

Symbol 1	Symbol 0	P_i1	P_i0	P_i	Self information	Si*P_i	Huffman source encoding tree	Code	R
11,	11,	0.81	0.81	0.656	0.608	0.399		0,	0.66
11,	10,	0.81	0.06	0.051	4.285	0.220		011,	0.15
11,	00,	0.81	0.06	0.051	4.285	0.220		0001,	0.21
10,	11,	0.06	0.81	0.051	4.285	0.220		1001,	0.21
01,	11,	0.06	0.81	0.051	4.285	0.220		0101,	0.21
11,	01,	0.81	0.06	0.051	4.285	0.220		1101,	0.21
00,	11	0.06	0.81	0.051	4.285	0.220		0111,	0.21
10,	01,	0.06	0.06	0.004	7.962	0.032		0001111,	0.03
10,	10,	0.06	0.06	0.004	7.962	0.032		01001111,	0.03
00,	00,	0.06	0.06	0.004	7.962	0.032		11001111,	0.03
01,	00,	0.06	0.06	0.004	7.962	0.032		0101111,	0.03
00,	10,	0.06	0.06	0.004	7.962	0.032		1101111,	0.03
00,	01,	0.06	0.06	0.004	7.962	0.032		0011111,	0.03
10,	00,	0.06	0.06	0.004	7.962	0.032		1011111,	0.03
01,	10,	0.06	0.06	0.004	7.962	0.032		0111111,	0.03
01,	01,	0.06	0.06	0.004	7.962	0.032		1111111,	0.03
					H	2.005		R	2.1
								Efficiency:	95.6%

Fig. 13.3 Illustration of Huffman encoding algorithm

are encoded with a corresponding 0 or 1. The probability of either event occurring is the sum of the two least probable events, as indicated. The tree continues to grow until all events have been accounted for. The code is simply determined by reading the corresponding 0 and 1 sequence from left to right.

Source Quantization

Due to the finite capacity (due to noise and limited bandwidth) of a digital communication channel, it is impossible to transmit an exact representation of a continuous signal from a source $x(t)$ since this would require an infinite number of bits. The question to be addressed is: how can the source be encoded in order to guarantee some minimal *distortion* of the signal when constrained by a given channel capacity C? For simplicity we will investigate the case when $x(t)$ is measured periodically at time T; the continuous sampled value is denoted as $x(k)$ and the quantized values is denoted as $\hat{x}(k)$. The *squared-error distortion* is a commonly used measure of distortion and is computed as

$$d\left(x_k, \hat{x}_k\right) = \left(x_k - \hat{x}_k\right)^2 . \tag{13.8}$$

Using X_n to denote n consecutive samples in a vector and $\hat{X}_n$ to denote the corresponding quantized samples, the corresponding distortion for the n samples is

$$d\left(X_n, \hat{X}_n\right) = \frac{1}{n} \sum_{k=1}^{n} d\left(x_k - \hat{x}_k\right) . \tag{13.9}$$

Assuming that the source is stationary, the expected value of the distortion of n samples is

$$D = E\left[d\left(X_n, \hat{X}_n\right)\right] = E\left[d\left(x_k - \hat{x}_k\right)\right] .$$

Given a memoryless and continuous random source X with a probability distribution function (pdf) $p(x)$ and a corresponding quantized amplitude alphabet $\hat{X}$ in which $x \in X$ and $\hat{x} \in \hat{X}$, we define the *rate distortion function* $R(D)$ as

$$R(D) \min_{p(\hat{x}|x): E[d(X, \hat{X})] \leq D} I\left(X; \hat{X}\right) , \tag{13.10}$$

in which $I(X; \hat{X})$ is denoted as the *mutual information* between X and $\hat{X}$ [13.7].

It has been shown that the *rate distortion function* for any memoryless source with zero mean and finite variance σ_x^2 can be bounded as follows

$$H(X) - \frac{1}{2} \log_2 2\pi e D \leq R(D) \leq \frac{1}{2} \log_2 \left(\frac{\sigma_x^2}{D}\right) ,$$

$$0 \leq D \leq \sigma_x^2 . \tag{13.11}$$

$H(X) = \int_{-\infty}^{\infty} p(x) \log p(x) \, dx$ is called the *differential entropy*. Note that the upper bound is the *rate distortion function* for a Gaussian source $H_g(X)$. Similarly, the bounds on the corresponding distortion rate function are

$$\frac{1}{2\pi e} 2^{-2(R - H(X))} \leq D(R) \leq 2^{-2R} \sigma_x^2 . \tag{13.12}$$

The *rate distortion function* for a band-limited Gaussian channel of width W normalized by σ_X^2 can be expressed in decibels as [13.6, pp. 104–108]

$$10 \log \frac{D_g(R)}{\sigma_x^2} = -\frac{3R}{W} . \tag{13.13}$$

Thus, decreasing the bandwidth of the source of information results in an exponential decrease in the *rate distortion function* for a given data rate R.

Similar to the grouped Huffman encoding algorithm, significant gains can be made by designing a quantizer $\hat{X} = Q(\cdot)$ for a vector X of individual scalar components $\{x_k, 1 \leq k \leq n\}$ which are described by the joint pdf $p(x_1, x_2, \ldots, x_n)$. The optimum quantizer is the one which can achieve the minimum distortion $D_n(R)$

$$D_n(R) \min_{Q(X)} E\left[d\left(X, \hat{X}\right)\right] . \tag{13.14}$$

As the dimension $n \to \infty$ it can be shown that $D(R) = D_n(R)$ in the limit [13.6, p. 116–117]. One method to implement such a vector quantization is the K-means algorithm [13.6, p.117].

13.3 Networked Systems Communication Limitations

As we have seen in our review of communication theory, there is no mathematical framework that guarantees a bounded deterministic fixed delay in transmitting information through a wireless or a wired medium. All digital representations of an analog waveform are transmitted with an average delay and variance, which is typically captured by its distortion measure. Clearly wired media tend to have a relative low degree of distortion when delivering information from a certain source to destination; for example, receiving digitally encoded data from a wired analog-to-digital converter, sent to a single digital controller at a fixed rate of 8 kbit/s, occurs with little data loss and distortion (i. e., only the least-significant bits tend to have errors). When sending digital information over a shared network, the problem becomes much more complex, in which the communication channel, medium access control (MAC) mechanism, and the data rate of each source on the network come into play [13.8]. Even to determine the average delay of a relatively simple MAC mechanism such as time-division multiple access (TDMA) is a fairly complex task [13.9]. In practice there are wired networking protocols which attempt to achieve a relatively constant delay profile by using a token to control access to the network, such as ControlNet and PROFIBUS-DP. Note that the control area network (CAN) offers a fixed priority scheme in which the highest-priority device will always gain access to the network, therefore allowing it to transmit data with the lowest average delay, whereas the lower-priority devices will have a corresponding increase in average delay [13.10, Fig. 4]. Protocols such as ControlNet and PROFIBUS-DP, how-ever, allow each member on the network an equal opportunity to transmit data within a given slot and can guarantee the same average delay for each node on a network for a given data rate. Usually the main source of variance in these delays is governed by the processing delays associated with the processors used on the network, and the additional higher-layer protocols which are built on top of these lower-layer protocols.

Wireless networks can perform as well as a wired network if the environmental conditions are ideal, for example, when devices have a clear line of sight for transmission, and are not subject to interference (high-gain microwave transmission stations). Unfortunately, devices which are used on a factory floor are more closely spaced and typically have isotropic antennas, which will lead to greater interference and variance of delays as compared with a wired network. Wireless token-passing protocols such as that described in [13.11] are a good choice to implement for control systems, since they limit interference in the network, which limits variance in delays, while providing a reasonable data throughput.

13.4 Networked Control Systems

One of the main advantages of using communication networks instead of point-to-point wired connections is the significantly reduced wiring together with the reduced failure rates of much lower connector numbers, which have significant cost implications in automated systems. Additional advantages include easier troubleshooting, maintenance, interoperability of devices, and integration of new devices added to the network [13.10]. Automated systems utilize digital shared communication networks. A number of communication protocols are used including Ethernet transmission control protocol/Internet protocol (TCP/IP), DeviceNet, ControlNet, WiFi, and Bluetooth. Each has different characteristics such as data speed and delays. Data are typically transmitted in packets of bits, for example an Ethernet IEEE 802.3 frame has a 112 or 176 bit header and a data field that must be at least 368 bits long.

Any automated system that uses shared digital wired or wireless communication networks must address certain concerns, including:

1. Bandwidth limitations, since any communication network can only carry a finite amount of information per unit of time

2. Delay jitter, since uncertainties in network access delay, or delay jitter, is commonly present
3. Packet dropouts, since transmission errors, buffer overflows due to congestion, or long transmission delays may cause packets to be dropped by the communication system.

All these issues are currently being addressed in ongoing research on networked control systems (NCS) [13.12].

13.4.1 Networked Control Systems

Figure 13.4 depicts a typical automation network in which two dedicated communication buses are used in order to control an overall process G_p with a dedicated controller G_c. The heavy solid line represents the control data network which provides timely sensor information y to G_c and distributes the appropriate control command u to the distributed controllers G_{c_i}. The heavy dashed solid line represents the monitor and configure data network, which allows the various controllers and sensors to be configured and monitored while G_p is being controlled. The control network usually has a lower data capacity but provides a fairly constant data delay with little variance in which field buses such as CAN, ControlNet, and PROFIBUS-DP are appropriate candidates. The monitoring and configuring network should have a higher data capacity but can tolerate more variance in its delays such that standard Ethernet or wireless networks using TCP/IP would be suitable. Sometimes the entire control network is monitored by a programmable logic controller (PLC) which acts as a gateway to the monitoring network as depicted in [13.10, Fig. 12]. However, there are advanced distributed controllers G_{c_i} which can both receive and deliver timely data over a control field bus such as CAN, yet still provide an Ethernet interface for configuration and monitoring. One such example is the πMFC, which is an advanced pressure-insensitive mass flow controller that provides both communication interfaces in which a low-cost low-power dual-processor architecture provides dedicated real-time control with advanced monitoring and diagnostic capabilities offloaded to the communications processor [13.13]. Although, not illustrated in this figure there is current research into establishing digital safety networks, as discussed in [13.10]. In particular the safety networks discussed are implemented over a serial–parallel line interface and implement the SafetyBUS p protocol.

Automated control systems with spatially distributed components have existed for several decades. Examples include chemical processes, refineries, power plants, and airplanes. In the past, in such systems the components were connected via hardwired connections and the systems were designed to bring all the information from the sensors to a central location, where the conditions were monitored and decisions were taken on how to act. The control policies were then implemented via the actuators, which could be valves, motors etc. Today's technology can put low-cost processing power at remote locations via microprocessors, and information can be transmitted reliably via shared digital networks or even wireless connections. These technology-driven changes are fueled by the high costs of wiring and the difficulty in introducing additional components into systems as needs change.

In 1983, Bosch GmbH began a feasibility study of using networked devices to control different functions in passenger cars. This appears to be one of the earliest efforts along the lines of modern networked control. The study bore fruit, and in February 1986 the innovative communications protocol of the control area network (CAN) was announced. By mid 1987, CAN

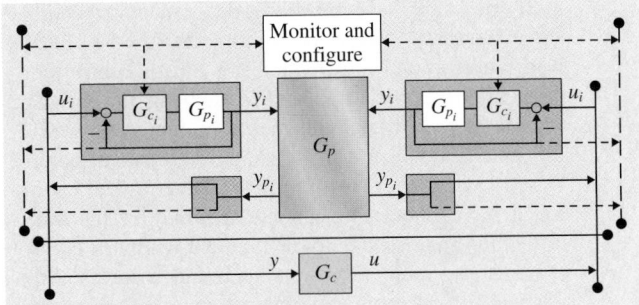

Fig. 13.4 Typical automation network

hardware in the form of Intel's 82526 chip had been introduced, and today virtually all cars manufactured in Europe include embedded systems integrated through CAN. Networked control systems are found in abundance in many technologies, and all levels of industrial systems are now being integrated through various types of data networks. Although networked control system technologies are now fairly mature in a variety of industrial applications, the recent trend toward integrating devices through wireless rather than wired communication channels has highlighted important potential application advantages as well as several challenging problems for current research.

These challenges involve the optimization of performance in the face of constraints on communication bandwidth, congestion, and contention for communication resources, delay, jitter, noise, fading, and the management of signal transmission power. While the greatest commercial impact of networked control systems to date has undoubtedly been in industrial implementations, recent research suggests great potential together with significant technical challenges in new applications to distributed sensing, reconnaissance and other military operations, and a variety of coordinated activities of groups of mobile robot agents. Taking a broad view of networked control systems we find that, in addition to the challenges of meeting real-time demands in controlling data flow through various feedback paths in the network, there are complexities associated with mobility and the constantly changing relative positions of agents in the network.

Networked control systems research lies primarily at the intersection of three research areas: control systems, communication networks and information theory, and computer science. Networked control systems research can greatly benefit from theoretical developments in information theory and computer science. The main difficulties in merging results from these different

fields of study have been the differences in emphasis in research so far. In information theory, delays
in the transmitted information are not of central concern, as it is more important to transmit the message
accurately even though this may sometimes involve significant delays in transmission. In contrast, in control
systems, delays are of primary concern. Delays are
much more important than the accuracy of the transmitted information due to the fact that feedback control
systems are quite robust to such inaccuracies. Similarly, in traditional computer science research, time
has not been a central issue since typical computer
systems were interacting with other computer systems
or a human operator and not directly with the physical world. Only recently have areas such as real-time
systems started addressing the issues of hard time constraints where the computer system must react within
specific time bounds, which is essential for embedded
processing systems that deal directly with the physical
world.

So far, researchers have focused primarily on a single loop and stability. Some fundamental results have
been derived that involve the minimum average bit
rate necessary to stabilize a linear time-invariant (LTI)
system.

An important result relates the minimum bit rate
R of feedback information needed for stability (for
a single-input linear system) to the fastest unstable
mode of the system via

$$R > \log_2 \exp \left(\sum \mathcal{R}(a_i) \right) . \tag{13.15}$$

Although progress has been made, much work remains to be done. In the case of a digital network over
which information is typically sent in packets, the minimum average rate is not the only guide to control
design. A transmitted packet typically contains a payload of tens of bytes, and so blocks of control data are
typically grouped together. This enters into the broader
set of research questions on the comparative value of
sending 1 bit/s or 1000 bits every 1000 s – for the same
average data rate. In view of typical actuator constraints,
an unstable system may not be able to recover after 1000 s.

An alternative measure is to see how infrequently
feedback information is needed to guarantee that
the system remains stable; see, for example, [13.14]
and [13.15], where this scheme has been combined
with model-based ideas for significant increases in the
periods during which the system is operating in an open-
loop fashion. Intermittent feedback is another way to
avoid taxing the networks that transmit sensor information. In this case, every so often the loop is closed
for a certain fixed or varying period of time [13.16].
This may correspond to opportunistic, bursty situations
in which the sensor sends bursts of information when
the network is available. The original idea of intermittent feedback was motivated by human motor control
considerations. There are strong connections with cooperative control, in which researchers have used spatial
invariance ideas to describe results on stability and performance [13.17]. If spatial invariance is not present,
then one may use the mathematical machinery of graph
theory to describe the interaction of systems/units and
to develop detailed models of groups of agents flying
in formation, foraging, cooperation in search of targets
or food, etc. An additional dimension in the wireless
case is to consider channels that vary with time, fade,
or disappear and reappear. The problem, of course, in
this case becomes significantly more challenging. Consensus approaches have also been used, which typically
assume rather simple dynamics for the agents and focus
on the topology considering fixed or time-varying links
in synchronous or asynchronous settings. Implementation issues in both hardware and software are at the
center of successful deployment of networked control
systems. Data integrity and security are also very important and may lead to special considerations in control
system design even at early stages.

Overall, single loop and stability have been emphasized and studied under quantization of sensor
measurements and actuator levels. Note that limits to
performance in networked control systems appear to be
caused primarily by delays and dropped packets. Other
issues being addressed by current research are actuator constraints, reliability, fault detection and isolation,
graceful degradation under failure, reconfigurable control, and ways to build increased degrees of autonomy
into networked control systems.

13.4.2 Teleoperation

An important area of networked control is teleoperation.
Teleoperation is the process of a human performing a remote task over a network with a *teleoperator* (TO).
Ideally, the TO's velocity ($f_{\text{top}}(t)$) should follow the human velocity commands ($f_{\text{hsi}}(t) = f_{\text{top}}(t - T)$) through
a *human system interface* (HSI) [13.18]. Force feed-
back from the TO ($e_{\text{top}}(t)$) is sent back to the HSI
($e_{\text{hsi}}(t) = e_{\text{top}}(t - T)$) in order for the operator to feel
immersed in the remote environment. The controller
(G_{top}) depicted in Fig. 13.5 is typically a proportional

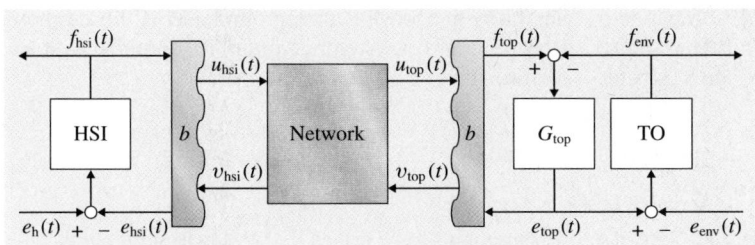

Fig. 13.5 Typical teleoperation network

derivative controller which maintains $f_{top}(t) = f_{env}(t)$ over a reasonably large bandwidth. The use of force feedback can lead to instabilities in the system due to small delays T in data transfer over the network. In order to recover stability the HSI velocity f_{hsi} and TO force e_{top} are encoded into *wave variables* [13.19], based on the wave port impedance b such that

$$u_{hsi}(t) = \frac{1}{\sqrt{2b}}(b f_{hsi}(t) + e_{hsi}(t)),$$ (13.16)

$$v_{top}(t) = \frac{1}{\sqrt{2b}}(b f_{top}(t) - e_{top}(t))$$ (13.17)

are transmitted over the network from the corresponding HSI and TO. As the delayed *wave variables*

are received ($u_{top}(t) = u_{hsi}(t - T)$, $v_{hsi}(t) = v_{top}(t - T)$), they are transformed back into the corresponding velocity and force variables ($f_{top}(t)$, $e_{hsi}(t)$) as follows

$$f_{top}(t) = \sqrt{\frac{2}{b}}u_{top}(t) - \frac{1}{b}e_{top}(t),$$ (13.18)

$$e_{hsi}(t) = b f_{hsi}(t) - \sqrt{2b}v_{hsi}(t).$$ (13.19)

Such a transformation allows the communication channel to remain *passive* for fixed time delays T and allows the teleoperation network to remain stable. The study of teleoperation continues to evolve for both the continuous- and discrete-time cases, as surveyed in [13.20].

13.5 Discussion and Future Research Directions

In summary, we have presented an overview of fundamental digital communication principles. In particular, we have shown that communication systems are effectively designed using a separation principle in which the source encoder and channel encoder can be designed separately. In particular, a source encoder can be designed to match the *uncertainty* (*entropy*) of a data source (H). All of the encoded data can then be effectively transmitted over a communication channel in which an appropriately designed channel encoder achieves the channel capacity C, which is typically determined by the modulation and noise introduced into the communication channel. As long as the channel capacity obeys $C > H$, then an average H symbols will be successfully received at the receiver. In *source data compression* we noted how to achieve a much higher average data rate by only using 1 bit to represent the temperature measurement of $25\,°C$ which occurs 99% of the time. In fact the average delay is roughly reduced from $10/100 = 0.1\,\text{s}$ to $(0.01 \cdot 10 + 0.99 \cdot 1)/100 = 0.0109\,\text{s}$. The key to designing an efficient automation communication network effectively is to understand the effective entropy H of the sys-

tem. Monitoring data, in which stability is not an issue, is a fairly straightforward task. When controlling a system the answer is not as clear; however, for deterministic channels (13.15) can serve as a guide for the classic control scheme. As the random behavior of the communication network becomes a dominating factor in the system an accurate analysis of how the delay and data dropouts occur is necessary. We have pointed the reader to texts which account for finite buffer size, and networking MAC to characterize communication delay and data dropouts [13.8, 9]. It remains to be shown how to incorporate such models effectively into the classic control framework in terms of showing stability, in particular when actuator limitations are present. It may be impossible to stabilize an unstable LTI system in any traditional stochastic framework when actuator saturation is considered. Teleoperation systems can cope with unknown fixed time delays in the case of *passive* networked control systems, by transmitting information using *wave variables*. We have extended the teleoperation framework to support lower-data-rate sampling and tolerate unknown time-varying delays and data dropouts without requiring any explicit knowledge of the communication channel model [13.21]. Confident that stability of these systems is preserved allows *much* greater

flexibility in choosing an appropriate MAC for our networked control system in order to optimize system performance.

13.6 Conclusions

Networked control systems over wired and wireless channels are becoming increasingly important in a wide range of applications. The area combines concepts and ideas from control and automation theory, communications, and computing. Although progress has been made in understanding important fundamental issues much

work remains to be done [13.12]. Understanding the effect of time-varying delays and designing systems to tolerate them is high priority. Research is needed to understand multiple interconnected systems over realistic channels that work together in a distributed fashion towards common goals with performance guarantee.

13.7 Appendix

13.7.1 Channel Encoder/Decoder Design

Denoting T (s) as the signal period, and W (Hz) as the bandwidth of a communication channel, we will use the ideal Nyquist rate assumption that $2TW$ symbols of $\{a_n\}$ can be transmitted with the analog wave forms $s_m(t)$ over the channel depicted in Fig. 13.1. We further assume that *independent* noise $n(t)$ is added to create the received signal $r(t)$. Then we can state the following

1. The actual rate of transmission is [13.2, Theorem 16]

$$R = H(s) - H(n) , \tag{13.20}$$

in which the channel capacity is the best signaling scheme which satisfies

$$C \max_{P(s_m)} H(s) - H(n) . \tag{13.21}$$

2. If we further assume the noise is white with power N and the signals are transmitted at power P then the channel capacity C (bit/s) is [13.2, Theorem 17]

$$C = W \log_2 \frac{P+N}{N} . \tag{13.22}$$

Various channel coding techniques have been devised in order to transmit digital information to achieve rates R which approach this channel capacity C with a correspondingly low bit error rate. Among these *bit error correcting* codes are block and convolutional codes in which the *Hamming code* [13.6, pp. 423–425] and the *Viterbi algorithm* [13.6, pp. 482–492] are classic examples for the respective implementations.

13.7.2 Digital Modulation

A linear filter can be described by its frequency response $H(f)$ and *real* impulse response $h(t)$ ($H^*(-f) = H(f)$). It can be represented in an equivalent low-pass form $H_l(f)$ in which

$$H_l(f - f_c) = \begin{cases} H(f), & f > 0 \\ 0, & f < 0 , \end{cases} \tag{13.23}$$

$$H_l^*(-f - f_c) = \begin{cases} 0, & f > 0 \\ H^*(-f), & f < 0 . \end{cases} \tag{13.24}$$

Therefore, with $H(f) = H_l(f - f_c) + H_l^*(f - f_c)$ the impulse response $h(t)$ can be written in terms of the *complex*-valued inverse transform of $H_l(f)$ ($h_l(t)$) [13.6, p. 153]

$$h(t) = 2\mathrm{Re}\left[h_l(t)e^{i2\pi f_c t}\right] . \tag{13.25}$$

Similarly the signal response $r(t)$ of a filtered input signal $s(t)$ through a linear filter $H(f)$ can be represented in terms of their low-pass equivalents

$$R_l(f) = S_l(f)H_l(f) . \tag{13.26}$$

Therefore it is mathematically convenient to discuss the transmission of equivalent low-pass signals through equivalent low-pass channels [13.6, p. 154].

Digital signals $s_m(t)$ consist of a set of analog wave-forms which can be described by an *orthonormal* set of waveforms $f_n(t)$. An *orthonormal* waveform satisfies

$$\langle f_i(t)f_j(t)\rangle_T = \begin{cases} 0, & i \neq j \\ 1, & i = j , \end{cases} \tag{13.27}$$

Table 13.1 Summary of PAM, PSK, and QAM

Modulation	$s_m(t)$	$f_1(t)$	$f_2(t)$
PAM	$s_m f_1(t)$	$\sqrt{\frac{2}{\mathcal{E}_g}} g(t) \cos 2\pi f_c t$	
PSK	$s_{m1} f_1(t) + s_{m2} f_2(t)$	$\sqrt{\frac{2}{\mathcal{E}_g}} g(t) \cos 2\pi f_c t$	$-\sqrt{\frac{2}{\mathcal{E}_g}} g(t) \sin 2\pi f_c t$
QAM	$s_{m1} f_1(t) + s_{m2} f_2(t)$	$\sqrt{\frac{2}{\mathcal{E}_g}} g(t) \cos 2\pi f_c t$	$-\sqrt{\frac{2}{\mathcal{E}_g}} g(t) \sin 2\pi f_c t$

Modulation	s_m	$d_{min}^{(e)}$
PAM	$(2m - 1 - M)d\sqrt{\frac{\mathcal{E}_g}{2}}$	$d\sqrt{2\mathcal{E}_g}$
PSK	$\sqrt{\frac{\mathcal{E}_g}{2}}\left[\cos\frac{2\pi}{M}(m-1), \sin\frac{2\pi}{M}(m-1)\right]$	$\sqrt{\mathcal{E}_g\left(1 - \cos\frac{2\pi}{M}\right)}$
QAM	$\sqrt{\frac{\mathcal{E}_g}{2}}[(2m_c - 1 - M)d, (2m_s - 1 - M)d]$	$d\sqrt{2\mathcal{E}_g}$

in which $\langle f(t)g(t)\rangle_T = \int_0^T f(t)g(t)\,dt$. The *Gram–Schmidt procedure* is a straightforward method to generate a set of *orthonormal* wave forms from a basis set of signals [13.6, p. 163].

Table 13.1 provides the corresponding orthonormal wave forms and minimum signal distances ($d_{min}^{(e)}$) for *pulse-amplitude modulation* (PAM), *phase-shift keying* (PSK), and *quadrature amplitude modulation* (QAM). Note that QAM is a combination of PAM and PSK in which $d_{min}^{(e)}$ is a special case of amplitude selection where $2d$ is the distance between adjacent signal amplitudes. Signaling amplitudes are in terms of the low-pass signal pulse shape $g(t)$ energy $\mathcal{E}_g = \langle g(t)g(t)\rangle_T$. The pulse shape is determined by the transmitting filter which typically has a *raised cosine* spectrum in order to minimize intersymbol interference at the cost of increased bandwidth [13.6, p. 559]. Each modulation scheme allows for M symbols in which $k = \log_2 M$ and N_0 is the average noise power per symbol transmission. Denoting P_M as the

probability of a symbol error and assuming that we use a Gray code, we can approximate the average bit error by $P_b \approx \frac{P_M}{k}$. The corresponding symbol errors are:

1. For M-ary PAM [13.6, p. 265]

$$P_M = \frac{2(M-1)}{M} Q\left(\sqrt{\frac{d^2 \mathcal{E}_g}{N_0}}\right) \tag{13.28}$$

2. For M-ary PSK [13.6, p. 270]

$$P_M \approx 2Q\left(\sqrt{\frac{\mathcal{E}_g}{N_0}} \sin\frac{\pi}{M}\right) \tag{13.29}$$

3. For QAM [13.6, p. 279]

$$P_M < (M-1)Q\left(\sqrt{\frac{\left[d_{min}^{(e)}\right]^2}{2N_0}}\right). \tag{13.30}$$

References

13.1 R. Gallager: *6.45 Principles of Digital Communication – I* (MIT, Cambridge 2002)

13.2 C.E. Shannon: A mathematical theory of communication, Bell Syst. Tech. J. **27**, 379–423 (1948)

13.3 S. Vembu, S. Verdu, Y. Steinberg: The source-channel separation theorem revisited, IEEE Trans. Inf. Theory **41**(1), 44–54 (1995)

13.4 M. Gasfpar, B. Rimoldi, M. Vetterli: To code, or not to code: lossy source–channel communication revisited, IEEE Trans. Inf. Theory **49**(5), 1147–1158 (2003)

13.5 H. El Gamal: On the scaling laws of dense wireless sensor networks: the data gathering channel, IEEE Trans. Inf. Theory **51**(3), 1229–1234 (2005)

13.6 J. Proakis: *Digital Communications*, 4th edn. (McGraw–Hill, New York 2000)

13.7 T.M. Cover, J.A. Thomas: *Elements of Information Theory* (Wiley, New York 1991)

13.8 M. Xie, M. Haenggi: *Delay–Reliability Tradeoffs in Wireless Networked Control Systems*, Lecture Notes in Control and Information Sciences (Springer, New York 2005)

13.9 K.K. Lee, S.T. Chanson: Packet loss probability for bursty wireless real-time traffic through delay model, IEEE Trans. Veh. Technol. **53**(3), 929–938 (2004)

13.10 J.R. Moyne, D.M. Tilbury: The emergence of industrial control networks for manufacturing control, diagnostics, and safety data, Proc. IEEE **95**(1), 29–47 (2007)

13.11   M. Ergen, D. Lee, R. Sengupta, P. Varaiya: WTRP – wireless token ring protocol, IEEE Trans. Veh. Technol. **53**(6), 1863–1881 (2004)

13.12   P.J. Antsaklis, J. Baillieul: Special issue: technology of networked control systems, Proc. IEEE **95**(1), 5–8 (2007)

13.13   A. Shajii, N. Kottenstette, J. Ambrosina: Apparatus and method for mass flow controller with network access to diagnostics, US Patent 6810308 (2004)

13.14   L.A. Montestruque, P.J. Antsaklis: On the model-based control of networked systems, Automatica **39**(10), 1837–1843 (2003)

13.15   L.A. Montestruque, P. Antsaklis: Stability of model-based networked control systems with time-varying transmission times, IEEE Trans. Autom. Control **49**(9), 1562–1572 (2004)

13.16   T. Estrada, H. Lin, P.J. Antsaklis: Model-based control with intermittent feedback, Proc. 14th

Mediterr. Conf. Control Autom. (Ancona 2006) pp.1–6

13.17   B. Recht, R. D'Andrea: Distributed control of systems over discrete groups, IEEE Trans. Autom. Control **49**(9), 1446–1452 (2004)

13.18   M. Kuschel, P. Kremer, S. Hirche, M. Buss: Lossy data reduction methods for haptic telepresence systems, Proc. Conf. Int. Robot. Autom., IEEE Cat. No. 06CH37729D (IEEE, Orlando 2006) pp.2933–2938

13.19   G. Niemeyer, J.-J.E. Slotine: Telemanipulation with time delays, Int. J. Robot. Res. **23**(9), 873–890 (2004)

13.20   P.F. Hokayem, M.W. Spong: Bilateral teleoperation: an historical survey, Automatica **42**(12), 2035–2057 (2006)

13.21   N. Kottenstette, P.J. Antsaklis: Stable digital control networks for continuous passive plants subject to delays and data dropouts, 46th IEEE Conf. Decis. Control (CDC) (IEEE, 2007)

14. Artificial Intelligence and Automation

Dana S. Nau

Artificial intelligence (AI) focuses on getting machines to do things that we would call intelligent behavior. Intelligence – whether artificial or otherwise – does not have a precise definition, but there are many activities and behaviors that are considered intelligent when exhibited by humans and animals. Examples include seeing, learning, using tools, understanding human speech, reasoning, making good guesses, playing games, and formulating plans and objectives. AI focuses on how to get machines or computers to perform these same kinds of activities, though not necessarily in the same way that humans or animals might do them.

To most readers, *artificial intelligence* probably brings to mind science-fiction images of robots or computers that can perform a large number of human-like activities: seeing, learning, using tools, understanding human speech, reasoning, making good guesses, playing games, and formulating plans and objectives. And indeed, AI research focuses on how to get machines or computers to carry out activities such as these. On the other hand, it is important to note that the goal of AI is not to simulate biological intelligence. Instead, the objective is to get machines to behave or think intelligently, regardless of whether or not the internal computational processes are the same as in people or animals.

Most AI research has focused on ways to achieve intelligence by manipulating symbolic representations of problems. The notion that symbol manipulation is sufficient for artificial intelligence was summarized by *Newell* and *Simon* in their famous physical-symbol system hypothesis: *A physical-symbol system has the*

necessary and sufficient means for general intelligent action and their heuristic search hypothesis [14.1]:

> *The solutions to problems are presented as symbol structures. A physical-symbol system exercises its intelligence in problem solving by search – that is – by generating and progressively modifying symbol structures until it produces a solution structure.*

On the other hand, there are several important topics of AI research – particularly machine-learning techniques such as neural networks and swarm intelligence – that are *subsymbolic* in nature, in the sense that they deal with vectors of real-valued numbers without attaching any explicit meaning to those numbers.

AI has achieved many notable successes [14.2]. Here are a few examples:

- Telephone-answering systems that understand human speech are now in routine use in many companies.

Part B | 14

- Simple room-cleaning robots are now sold as consumer products.
- Automated vision systems that read handwritten zip codes are used by the US Postal Service to route mail.
- Machine-learning techniques are used by banks and stock markets to look for fraudulent transactions and alert staff to suspicious activity.
- Several web-search engines use machine-learning techniques to extract information and classify data scoured from the web.
- Automated planning and control systems are used in unmanned aerial vehicles, for missions that are too *dull, dirty or dangerous* for manned aircraft.
- Automated planning and scheduling techniques were used by the National Aeronautics and Space Administration (NASA) in their famous Mars rovers.

AI is divided into a number of subfields that correspond roughly to the various kinds of activities mentioned in the first paragraph. Three of the most important subfields are discussed in other chapters: machine learning in Chaps. 12 and 29, computer vision in Chap. 20, and robotics in Chaps. 1, 78, 82, and 84. This chapter discusses other topics in AI, including search procedures (Sect. 14.1.1), logical reasoning (Sect. 14.1.2), reasoning about uncertain information (Sect. 14.1.3), planning (Sect. 14.1.4), games (Sect. 14.1.5), natural-language processing (Sect. 14.1.6), expert systems (Sect. 14.1.7), and AI programming (Sect. 14.1.8).

14.1 Methods and Application Examples

14.1.1 Search Procedures

Many AI problems require a trial-and-error search through a search space that consists of *states of the world* (or *states*, for short), to find a path to a state *s* that satisfies some *goal condition g*. Usually the set of states is finite but very large: far too large to give a list of all the states (as a control theorist might do, for example, when writing a state-transition matrix). Instead, an initial state s_0 is given, along with a set O of *operators* for producing new states from existing ones.

As a simple example, consider Klondike, the most popular version of solitaire [14.3]. As illustrated in Fig. 14.1a, the initial state of the game is determined by dealing 28 cards from a 52-card deck into an arrangement called the tableau; the other 28 cards then go into a pile called the stock. New states are formed from old ones by moving cards around according to the rules of the game; for example, in Fig. 14.1a there are two possible moves: either move the ace of hearts to one of the foundations and turn up the card beneath the ace as shown in Fig. 14.1b, or move three cards from the stock to the waste. The goal is to produce a state in which all of the cards are in the foundation piles, with each suit in a different pile, in numerical order from the ace at the bottom to the king at the top. A *solution* is any path (a sequence of moves, or equivalently, the sequence of states that these moves take us to) from the initial state to a goal state.

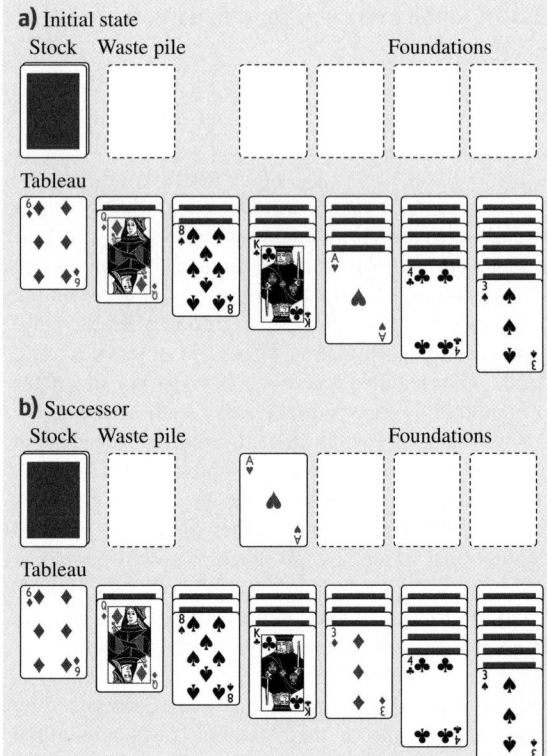

Fig. 14.1 (a) An initial state and (b) one of its two possible successors

Klondike has several characteristics that are typical of AI search problems:

- Each state is a combination of a finite set of features (in this case the cards and their locations), and the task is to find a path that leads from the initial state to a goal state.
- The rules for getting from one state to another can be represented using symbolic logic and discrete mathematics, but continuous mathematics is not as useful here, since there is no reasonable way to model the state space with continuous numeric functions.
- It is not clear a priori which paths, if any, will lead from the initial state to the goal states. The only obvious way to solve the problem is to do a trial-and-error search, trying various sequences of moves to see which ones might work.
- Combinatorial explosion is a big problem. The number of possible states in Klondike is well over 52!, which is many orders of magnitude larger than the number of atoms in the Earth. Hence a trial-and-error search will not terminate in a reasonable amount of time unless we can somehow restrict the search to a very small part of the search space – hopefully a part of the search space that actually contains a solution.
- In setting up the state space, we took for granted that the problem representation should correspond directly to the states of the physical system, but sometimes it is possible to make a problem much easier to solve by adapting a different representation; for example, [14.4] shows how to make Klondike much easier to solve by searching a different state space.

In many trial-and-error search problems, each solution path π will have a numeric measure $F(\pi)$ telling how desirable π is; for example, in Klondike, if we consider shorter solution paths to be more desirable than long ones, we can define $F(\pi)$ to be π's length. In such cases, we may be interested in finding either an *optimal* solution, i. e., a solution π such that $F(\pi)$ is as small as possible, or a *near-optimal* solution in which $F(\pi)$ is close to the optimal value.

Heuristic Search

The pseudocode in Fig. 14.2 provides an abstract model of state-space search. The input parameters include an initial state s_0 and a set of operators O. The procedure either fails, or returns a *solution path* π (i. e., a path from s_0 to a goal state).

```
1.  State-space-search(s_0; O)
2.      Active ← {⟨s_0⟩}
3.      while Active ≠ ∅ do
4.          choose a path  π = ⟨s_0,...,s_k⟩ ∈ Active and remove it from Active
5.          if s_k is a goal state then return π
6.          Successors ← {⟨s_0,...,s_k, o(s_k)⟩ : o ∈ O is applicable to s_k}
7.          optional pruning step: remove unpromising paths from Successors
8.          Active ← Active ∪ Successors
9.      repeat
10.     return failure
```

Fig. 14.2 An abstract model of state-space search. In line 6, $o(s_k)$ is the state produced by applying the operator o to the state s_k

As discussed earlier, we would like the search algorithm to focus on those parts of the state space that will lead to optimal (or at least near-optimal) solution paths. For this purpose, we will use a *heuristic function* $f(\pi)$ that returns a numeric value giving an approximate idea of how good a solution can be found by extending π, i. e.,

$$f(\pi) \approx \min\{F(\pi') :$$
$$\pi' \text{ is a solution path that is an extension of } \pi\} \ .$$

It is hard to give foolproof guidelines for writing heuristic functions. Often they can be very ad hoc: in the worst case, $f(\pi)$ may just be an arbitrary function that the user *hopes* will give reasonable estimates. However, often it works well to define an easy-to-solve *relaxation* of the original problem, i. e., a modified problem in which some of the constraints are weakened or removed. If π is a partial solution for the original problem, then we can compute $f(\pi)$ by extending π into a solution π' for the relaxed problem, and returning $F(\pi')$; for example, in the famous traveling-salesperson problem, $f(\pi)$ can be computed by solving a simpler problem called the assignment problem [14.5]. Here are several procedures that can make use of such a heuristic function:

- *Best-first search* means that at line 4 of the algorithm in Fig. 14.2, we always choose a path $\pi = \langle s_0, \ldots, s_k \rangle$ that has the smallest value $f(\pi)$ of any path we have seen so far. Suppose that at least one solution exists, that there are no infinite paths of finite cost, and that the heuristic function f has the following *lower-bound* property

$$f(\pi) \leq \min\{F(\pi') :$$
$$\pi' \text{ is a solution path that is an extension of } \pi\} \ .$$
$$(14.1)$$

Then best-first search will always return a solution π^* that minimizes $F(\pi^*)$. The well-known A* search procedure [14.6] is a special case of best-first search, with some modifications to handle situations where there are multiple paths to the same state.

Best-first search has the advantage that, if it chooses an obviously bad state s to explore next, it will not spend much time exploring the subtree below s. As soon as it reaches successors of s whose f-values exceed those of other states on the *Active* list, best-first search will go back to those other states. The biggest drawback is that best-first search must remember every state it has ever visited, hence its memory requirement can be huge. Thus, best-first search is more likely to be a good choice in cases where the state space is relatively small, and the difficulty of solving the problem arises for some other reason (e.g., a costly-to-compute heuristic function, as in [14.7]).

- In *depth-first branch and bound*, at line 4 the algorithm always chooses the longest path in *Active*; if there are several such paths then the algorithm chooses the one that has the smallest value for $f(\pi)$. The algorithm maintains a variable π^* that holds the best solution seen so far, and the pruning step in line 7 removes a path π iff $f(\pi) \geq F(\pi^*)$. If the state space is finite and acyclic, at least one solution exists, and (14.1) holds, then depth-first branch and bound is guaranteed to return a solution π^* that minimizes $F(\pi^*)$.

  The primary advantage of depth-first search is its low memory requirement: the number of nodes in *Active* will never exceed bd, where d is the length of the current path. The primary drawback is that, if it chooses the wrong state to look at next, it will explore the entire subtree below that state before returning and looking at the state's siblings. Depth-first search does better in cases where the likelihood of choosing the wrong state is small or the time needed to search the incorrect subtrees is not too great.

- *Greedy search* is a state-space search without any backtracking. It is accomplished by replacing line 8 with *Active* $\leftarrow \{\pi_1\}$, where π_1 is the path in *Successors* that minimizes $\{f(\pi') \mid \pi' \in$ *Successors*$\}$. *Beam search* is similar except that, instead of putting just one successor π_1 of π into *Active*, we put k successors $\pi_1, \ldots, \pi_k$ into *Active*, for some fixed k.

  Both greedy search and beam search will return very quickly once they find a solution, since neither of them will spend any time looking for better solutions. Hence they are good choices if the state space is large, most paths lead to solutions, and we are more interested in finding a solution quickly than in finding an optimal solution. However, if most paths do not lead to solutions, both algorithms may fail to find a solution at all (although beam search is more robust in this regard, since it explores several paths rather than just one path). In this case, it may work well to do a modified greedy search that backtracks and tries a different path every time it reaches a dead end.

Hill-Climbing

A *hill-climbing* problem is a special kind of search problem in which *every* state is a goal state. A hill-climbing procedure is like a greedy search, except that *Active* contains a single state rather than a single path; this is maintained in line 6 by inserting a single successor of the current state s_k into *Active*, rather than all of s_k's successors. In line 5, the algorithm terminates when none of s_k's successors looks better than s_k itself, i.e., when s_k has no successor s_{k+1} with $f(s_{k+1}) > f(s_k)$. There are several variants of the basic hill-climbing approach:

- *Stochastic hill-climbing and simulated annealing.* One difficulty with hill-climbing is that it will terminate in cases where s_k is a local minimum but not a global minimum. To prevent this from happening, a *stochastic hill-climbing* procedure does not always return when the test in line 5 succeeds. Probably the best known example is *simulated annealing*, a technique inspired by annealing in metallurgy, in which a material is heated and then slowly cooled. In simulated annealing, this is accomplished as follows. At line 5, if none of s_k's successors look better than s_k then the procedure will not necessarily terminate as in ordinary hill-climbing; instead it will terminate with some probability p_i, where i is the number of loop iterations and p_i grows monotonically with i.

- *Genetic algorithms.* A genetic algorithm is a modified version of hill-climbing in which successor states are generated not using the normal successor function, but instead using operators reminiscent of genetic recombination and mutation. In particular, *Active* contains k states rather than just one, each state is a string of symbols, and the operators O are computational analogues of genetic recombination and mutation. The termination criterion in line 5 is generally ad hoc; for example, the algorithm may terminate after a specified number of iterations, and return the best one of the states currently in *Active*.

Hill-climbing algorithms are good to use in problems where we want to find a solution very quickly, then continue to look for a better solution if additional time is available. More specifically, genetic algorithms are useful in situations where each solution can be represented as a string whose substrings can be combined with substrings of other solutions.

Constraint Satisfaction and Constraint Optimization

A constraint-satisfaction problem is a special kind of search problem in which each state is a set of assignments of values to variables $\{X_i\}_{i=1}^n$ that have finite domains $\{D_i\}_{i=1}^n$, and the objective is to assign values to the variables in such a way that some set of constraints is satisfied.

In the search space for a constraint-satisfaction problem, each state at depth i corresponds to an assignment of values to i of the n variables, and each branch corresponds to assigning a specific value to an unassigned variable. The search space is finite: the maximum length of any path from the root node is n since there are only n variables to assign values to. Hence a depth-first search works quite well for constraint-satisfaction problems. In this context, some powerful techniques have been formulated for choosing which variable to assign next, detecting situations where previous variable assignments will make it impossible to satisfy the remaining constraints, and even restructuring the problem into one that is easier to solve [14.8, Chap. 5].

A *constraint-optimization* problem combines a constraint-satisfaction problem with an objective function that one wants to optimize. Such problems can be solved by combining constraint-satisfaction techniques with the optimization techniques mentioned in *Heuristic Search*.

Applications of Search Procedures

Software using AI search techniques has been developed for a large number of commercial applications. A few examples include the following:

- Several universities routinely use constraint-satisfaction software for course scheduling.
- *Airline ticketing.* Finding the best price for an airline ticket is a constraint-optimization problem in which the constraints are provided by the airlines' various rules on what tickets are available at what prices under what conditions [14.9]. An example of software that works in this fashion is the ITA

software (itasoftware.com) system that is used by several airline-ticketing web sites, e.g., Orbitz (orbitz.com) and Kayak (kayak.com).
- *Scheduling and routing.* Companies such as ILOG (ilog.com) have developed software that uses search and optimization techniques for scheduling [14.10], routing [14.11], workflow composition [14.12], and a variety of other applications.
- *Information retrieval from the web.* AI search techniques are important in the web-searching software used at sites such as Google News [14.2].

Additional reading. For additional reading on search algorithms, see *Pearl* [14.13]. For additional details about constraint processing, see *Dechter* [14.14].

14.1.2 Logical Reasoning

A *logic* is a formal language for representing information in such a way that one can reason about what things are true and what things are false. The logic's *syntax* defines what the *sentences* are; and its *semantics* defines what those sentences mean in some *world*. The two best-known logical formalisms, propositional logic and first-order logic, are described briefly below.

Propositional Logic and Satisfiability

Propositional logic, also known as *Boolean algebra*, includes sentences such as $A \land B \Rightarrow C$, where A, B, and C are variables whose domain is $\{true, false\}$. Let w_1 be a world in which A and C are *true* and B is *false*, and let w_2 be a world in which all three of the Boolean variables are *true*. Then the sentence $A \land B \Rightarrow C$ is false in w_1 and true in w_2. Formally, we say that w_2 is a *model* of $A \land B \Rightarrow C$, or that it *entails* S_1. This is written symbolically as

$$w_2 \models A \land B \Rightarrow C \, .$$

The *satisfiability problem* is the following: given a sentence S of propositional logic, does there exist a world (i.e., an assignment of truth values to the variables in S) in which S is true? This problem is central to the theory of computation, because it was the very first computational problem shown to be *NP-complete*. Without going into a formal definition of *NP*-completeness, *NP* is, roughly, the set of all computational problems such that, if we are given a purported solution, we can check *quickly* (i.e., in a polynomial amount of computing time) whether the solution is correct. An *NP-complete* problem is a problem that is one of the hardest problems in *NP*, in the sense that

Part B | 14.1

solving any NP-complete problems would provide a solution to every problem in NP. It is conjectured that no NP-complete problem can be solved in a polynomial amount of computing time. There is a great deal of evidence for believing the conjecture, but nobody has ever been able to prove it. This is the most famous unsolved problem in computer science.

First-Order Logic

A much more powerful formalism is *first-order logic* [14.15], which uses the same logical connectives as in propositional logic but adds the following syntactic elements (and semantics, respectively): constant symbols (which denote the objects), variable symbols (which range over objects), function symbols (which represent functions), predicate symbols (which represent relations among objects), and the quantifiers $\forall x$ and $\exists x$, where x is any variable symbol (to specify whether a sentence is true for every value x or for at least one value of x).

First-order logic includes a standard set of *logical axioms*. These are statements that must be true in every possible world; one example is the transitive property of equality, which can be formalized as

$$\forall x \, \forall y \, \forall z \, (x = y \wedge y = z) \Rightarrow x = z \,.$$

In addition to the logical axioms, one can add a set of *nonlogical axioms* to describe what is true in a particular kind of world; for example, if we want to specify that there are exactly two objects in the world, we could do this by the following axioms, where a and b are constant symbols, and x, y, z are variable symbols

$$a \neq b \,, \tag{14.2a}$$

$$\forall x \, \forall y \, \forall z \, x = y \vee y = z \vee x = z \,. \tag{14.2b}$$

The first axiom asserts that there are at least two objects (namely a and b), and the second axiom asserts that there are no more than two objects.

First-order logic also includes a standard set of *inference rules*, which can be used to infer additional true statements. One example is *modus ponens*, which allows one to infer a statement Q from the pair of statements $P \Rightarrow Q$ and P.

The logical and nonlogical axioms and the rules of inference, taken together, constitute a *first-order theory*. If $\mathcal{T}$ is a first-order theory, then a *model* of $\mathcal{T}$ is any world in which $\mathcal{T}$'s axioms are true. (In science and engineering, a *mathematical model* generally means a formalism for some real-world phenomenon; but in mathematical logic, *model* means something very

different: the formalism is called a theory, and the the real-world phenomenon *itself* is a model of the theory.) For example, if $\mathcal{T}$ includes the nonlogical axioms given above, then a model of $\mathcal{T}$ is any world in which there are exactly two objects.

A *theorem* of $\mathcal{T}$ is defined recursively as follows: every axiom is a theorem, and any statement that can be produced by applying inference rules to theorems is also a theorem; for example, if $\mathcal{T}$ is any theory that includes the nonlogical axioms (14.2a) and (14.2b), then the following statement is a theorem of $\mathcal{T}$

$$\forall x \, x = a \vee x = b \,.$$

A fundamental property of first-order logic is *completeness*: for every first-order theory $\mathcal{T}$ and every statement S in $\mathcal{T}$, S is a theorem of $\mathcal{T}$ if and only if S is true in all models of $\mathcal{T}$. This says, basically, that first-order logical reasoning does exactly what it is supposed to do.

Nondeductive Reasoning

Deductive reasoning – the kind of reasoning used to derive theorems in first-order logic – consists of deriving a statement y as a consequence of a statement x. Such an inference is *deductively valid* if there is no possible situation in which x is true and y is false. However, several other kinds of reasoning have been studied by AI researchers. Some of the best known include *abductive reasoning* and *nonmonotonic reasoning*, which are discussed briefly below, and *fuzzy logic*, which is discussed later.

Nonmonotonic Reasoning. In most formal logics, deductive inference is *monotone*; i. e., adding a formula to a logical theory never causes something *not* to be a theorem that was a theorem of the original theory. Nonmonotonic logics allow deductions to be made from *beliefs* that may not always be true, such as the *default assumption* that birds can fly. In nonmonotonic logic, if b is a bird and we know nothing about b then we may conclude that b can fly; but if we later learn that b is an ostrich or b has a broken wing, then we will retract this conclusion.

Abductive Reasoning. This is the process of inferring x from y when x entails y. Although this can produce results that are incorrect within a formal deductive system, it can be quite useful in practice, especially when something is known about the probability of different causes of y; for example, the Bayesian reasoning described later can be viewed as a combination of deductive reasoning, abductive reasoning, and probabilities.

Applications of Logical Reasoning

The satisfiability problem has important applications in hardware design and verification; for example, electronic design automation (EDA) tools include satisfiability checking algorithms to check whether a given digital system design satisfies various criteria. Some EDA tools use first-order logic rather than propositional logic, in order to check criteria that are hard to express in propositional logic.

First-order logic provides a basis for automated reasoning systems in a number of application areas. Here are a few examples:

- *Logic programming*, in which mathematical logic is used as a programming language, uses a particular kind of first-order logic sentence called a Horn clause. Horn clauses are implications of the form $P_1 \wedge P_2 \wedge \ldots \wedge P_n \Rightarrow P_{n+1}$, where each P_i is an *atomic formula* (a predicate symbol and its argument list). Such an implication can be interpreted logically, as a statement that P_{n+1} is true if $P_1, \ldots, P_n$ are true, or procedurally, as a statement that a way to show or solve P_{n+1} is to show or solve $P_1, \ldots, P_n$. The best known implementation of logic programming is the programming language Prolog, described further below.
- *Constraint programming*, which combines logic programming and constraint satisfaction, is the basis for ILOG's CP Optimizer (http://www.ilog.com/products/cpoptimizer).
- The web ontology language (OWL) and DAML + OIL languages for semantic web markup are based on description logics, which are a particular kind of first-order logic.
- Fuzzy logic has been used in a wide variety of commercial products including washing machines, refrigerators, automotive transmissions and braking systems, camera tracking systems, etc.

14.1.3 Reasoning About Uncertain Information

Earlier in this chapter it was pointed out that AI systems often need to reason about discrete sets of states, and the relationships among these states are often nonnumeric. There are several ways in which uncertainty can enter into this picture; for example, various events may occur spontaneously and there may be uncertainty about whether they will occur, or there may be uncertainty about what things are currently true, or the degree to which they are true. The two best-known techniques for reasoning about such uncertainty are Bayesian probabilities and fuzzy logic.

Bayesian Reasoning

In some cases we may be able to model such situations probabilistically, but this means reasoning about discrete random variables, which unfortunately incurs a combinatorial explosion. If there are n random variables and each of them has d possible values, then the joint probability distribution function (PDF) will have d^n entries. Some obvious problems are (1) the worst-case time complexity of reasoning about the variables is $\Theta(d^n)$, (2) the worst-case space complexity is also $\Theta(d^n)$, and (3) it seems impractical to suppose that we can acquire accurate values for all d^n entries.

The above difficulties can be alleviated if some of the variables are known to be independent of each other; for example, suppose that the n random variables mentioned above can be partitioned into $\lceil n/k \rceil$ subsets, each containing at most k variables. Then the joint PDF for the entire set is the product of the PDFs of the subsets. Each of those has d^k entries, so there are only $\lceil n/k \rceil n^k$ entries to acquire and reason about.

Absolute independence is rare; but another property is more common and can yield a similar decrease in time and space complexity: *conditional* independence [14.16]. Formally, *a* is *conditionally independent of b given c* if $P(ab|c) = P(a|c)P(b|c)$.

Bayesian networks are graphical representations of conditional independence in which the network topology reflects knowledge about which events *cause* other events. There is a large body of work on these networks, stemming from seminal work by Judea Pearl. Here is a simple example due to *Pearl* [14.17]. Figure 14.3 represents the following hypothetical situation:

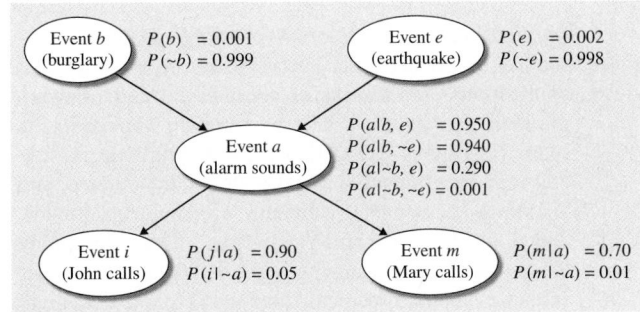

Fig. 14.3 A simple Bayesian network

My house has a burglar alarm that will usually go off (event a) if there's a burglary (event b), an earthquake (event e), or both, with the probabilities shown in Fig. 14.3. If the alarm goes off, my neighbor John will usually call me (event j) to tell me; and he may sometimes call me by mistake even if the alarm has not gone off, and similarly for my other neighbor Mary (event m); again the probabilities are shown in the figure.

The joint probability for each combination of events is the product of the conditional probabilities given in Fig. 14.3

$$
\begin{aligned}
P(b, e, a, j, m) = {}& P(b)P(e)P(a|b, e) \\
& \times P(j|a)P(m|a) , \\
P(b, e, a, j, \neg m) = {}& P(b)P(e)P(a|b, e)P(j|a) \\
& \times P(\neg m|a) , \\
P(b, \neg e, \neg a, j, \neg m) = {}& P(b)P(\neg e)P(\neg a|b, \neg e) \\
& \times P(j|\neg a)P(\neg m|\neg a) ,
\end{aligned}
$$

$$\cdots$$

Hence, instead of reasoning about a joint distribution with $2^5 = 32$ entries, we only need to reason about products of the five conditional distributions shown in the figure.

In general, probability computations can be done on Bayesian networks much more quickly than would be possible if all we knew was the joint PDF, by taking advantage of the fact that each random variable is conditionally independent of most of the other variables in the network. One important special case occurs when the network is acyclic (e.g., the example in Fig. 14.3), in which case the probability computations can be done in low-order polynomial time. This special case includes decision trees [14.8], in which the network is both acyclic and rooted. For additional details about Bayesian networks, see *Pearl* and *Russell* [14.16].

Applications of Bayesian Reasoning. Bayesian reasoning has been used successfully in a variety of applications, and dozens of commercial and freeware implementations exist. The best-known application is spam filtering [14.18, 19], which is available in several mail programs (e.g., Apple Mail, Thunderbird, and Windows Messenger), webmail services (e.g., gmail), and a plethora of third-party spam filters (probably the best-known is spamassassin [14.20]). A few other examples include medical imaging [14.21], document classification [14.22], and web search [14.23].

Fuzzy Logic

Fuzzy logic [14.24, 25] is based on the notion that, instead of saying that a statement P is true or false, we can give P a *degree of truth*. This is a number in the interval $[0, 1]$, where 0 means false, 1 means true, and numbers between 0 and 1 denote partial degrees of truth.

As an example, consider the action of moving a car into a parking space, and the statement *the car is in the parking space*. At the start, the car is not in the parking space, hence the statement's degree of truth is 0. At the end, the car is completely in the parking space, hence the statement's degree of is 1. Between the start and end of the action, the statement's degree of truth gradually increases from 0 to 1.

Fuzzy logic is closely related to fuzzy set theory, which assigns degrees of truth to set membership. This concept is easiest to illustrate with sets that are intervals over the real line; for example, Fig. 14.4 shows a set S having the following set membership function

$$
\text{truth}(x \in S) =
\begin{cases}
1 , & \text{if } 2 \le x \le 4 , \\
0 , & \text{if } x \le 1 \text{ or } x \ge 5 , \\
x - 1 , & \text{if } 1 < x < 2 , \\
5 - x , & \text{if } 4 < x < 5 .
\end{cases}
$$

The logical notions of conjunction, disjunction, and negation can be generalized to fuzzy logic as follows

$$
\begin{aligned}
\text{truth}(x \wedge y) &= \min[\text{truth}(x), \text{truth}(y)] ; \\
\text{truth}(x \vee y) &= \max[\text{truth}(x), \text{truth}(y)] ; \\
\text{truth}(\neg x) &= 1 - \text{truth}(x) .
\end{aligned}
$$

Fuzzy logic also allows other operators, more linguistic in nature, to be applied. Going back to the example of a full gas tank, if the degree of truth of *g is full* is d, then one might want to say that the degree of truth of *g is very full* is d^2. (Obviously, the choice of d^2 for *very* is subjective. For different users or different applications, one might want to use a different formula.) Degrees of truth are semantically distinct from probabilities, although the two concepts are often confused;

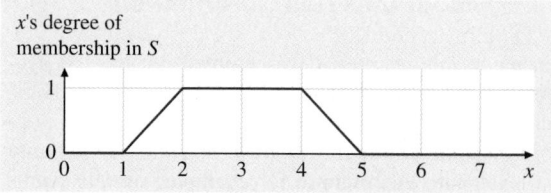

x's degree of membership in S

Fig. 14.4 A degree-of-membership function for a fuzzy set

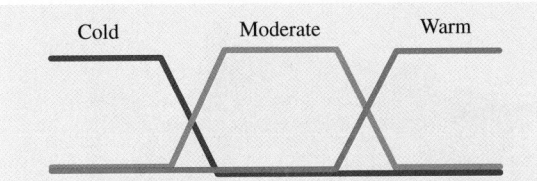

Fig. 14.5 Degree-of-membership functions for three over-lapping temperature ranges

for example, we could talk about the probability that someone would say the car is in the parking space, but this probability is likely to be a different number than the degree of truth for the statement that the car is in the parking space.

Fuzzy logic is controversial in some circles; e.g., many statisticians would maintain that probability is the only rigorous mathematical description of uncertainty. On the other hand, it has been quite successful from a practical point of view, and is now used in a wide variety of commercial products.

Applications of Fuzzy Logic. Fuzzy logic has been used in a wide variety of commercial products. Examples include washing machines, refrigerators, dishwashers, and other home appliances; vehicle subsystems such as automotive transmissions and braking systems; digital image-processing systems such as edge detectors; and some microcontrollers and microprocessors.

In such applications, a typical approach is to specify fuzzy sets that correspond to different subranges of a continuous variable; for instance, a temperature measurement for a refrigerator might have degrees of membership in several different temperature ranges, as shown in Fig. 14.5. Any particular temperature value will correspond to three degrees of membership, one for each of the three temperature ranges; and these degrees of membership could provide input to a control system to help it decide whether the refrigerator is too cold, too warm, or in the right temperature range.

14.1.4 Planning

In ordinary English, there are many different kinds of plans: project plans, floor plans, pension plans, urban plans, floor plans, etc. AI planning research focuses specifically on *plans of action*, i. e., [14.26]:

> ... *representations of future behavior ... usually a set of actions, with temporal and other constraints on them, for execution by some agent or agents.*

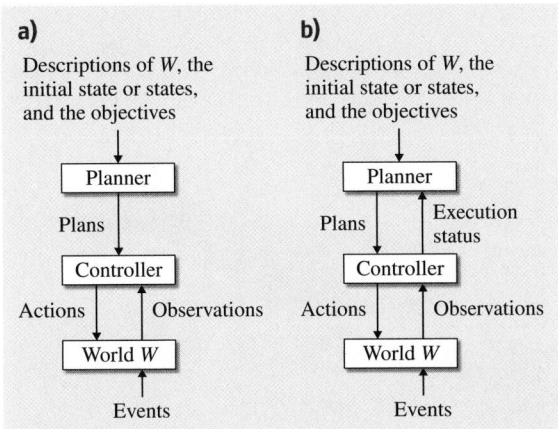

Fig. 14.6a,b Simple conceptual models for (**a**) offline and (**b**) online planning

Figure 14.6 gives an abstract view of the relationship between a planner and its environment. The planner's input includes a description of the world W in which the plan is to be executed, the initial state (or set of possible initial states) of the world, and the objectives that the plan is supposed to achieve. The planner produces a plan that is a set of instructions to a *controller*, which is the system that will execute the plan. In *offline planning*, the planner generates the entire plan, gives it to the controller, and exits. In *online planning*, plan generation and plan execution occur concurrently, and the planner gets feedback from the controller to aid it in generating the rest of the plan. Although not shown in the figure, in some cases the plan may go to a *scheduler* before going to the controller. The purpose of the scheduler is to make decisions about when to execute various parts of the plan and what resources to use during plan execution.

Examples. The following paragraphs include several examples of offline planners, including the sheet-metal bending planner in *Domain-Specific Planners*, and all of the planners in *Classical Planning* and *Domain-Configurable Planners*. One example of an online planner is the planning software for the Mars rovers in *Domain-Specific Planners*. The planner for the Mars rovers also incorporates a scheduler.

Domain-Specific Planners

A *domain-specific* planning system is one that is tailor-made for a given planning domain. Usually the design of the planning system is dictated primarily by the detailed requirements of the specific domain, and the

Fig. 14.7 One of the Mars rovers

Fig. 14.8 A sheet-metal bending machine

system is unlikely to work in any domain other other than the one for which it was designed.

Many successful planners for real-world applications are domain specific. Two examples are the autonomous planning system that controlled the Mars rovers [14.27] (Fig. 14.7), and the software for planning sheet-metal bending operations [14.28] that is bundled with Amada Corporation's sheet-metal bending machines (Fig. 14.8).

Classical Planning

Most AI planning research has been guided by a desire to develop principles that are *domain independent*, rather than techniques specific to a single planning domain. However, in order to make any significant headway in the development of such principles, it has proved necessary to make restrictions on what kinds of planning domains they apply to.

In particular, most AI planning research has focused on *classical* planning problems. In this class of planning problems, the world W is finite, fully observable, deterministic, and static (i.e., the world never changes except as a result of our actions); and the objective is to produce a finite sequence of actions that takes the world from some specific *initial* state to any of some set of *goal* states. There is a standard language, planning domain definition language (PDDL) [14.29], that can represent planning problems of this type, and there are dozens (possibly hundreds) of classical planning algorithms.

One of the best-known classical planning algorithms is GraphPlan [14.30], an iterative-deepening algorithm that performs the following steps in each iteration i:

1. Generate a *planning graph* of depth i. Without going into detail, the planning graph is basically the search space for a greatly simplified version of the planning problem that can be solved very quickly.
2. Search for a solution to the original unsimplified planning problem, but restrict this search to occur solely within the planning graph produced in step 1. In general, this takes much less time than an unrestricted search would take.

GraphPlan has been the basis for dozens of other classical planning algorithms.

Domain-Configurable Planners

Another important class of planning algorithms are the *domain-configurable planners*. These are planning systems in which the planning engine is domain independent but the input to the planner includes domain-specific information about how to do planning in the problem domain at hand. This information serves to constrain the planner's search so that the planner searches only a small part of the search space. There are two main types of domain-configurable planners:

- *Hierarchical task network* (HTN) planners such as O-Plan [14.31], SIPE-2 (system for interactive planning and execution) [14.32], and SHOP2 (simple hierarchical ordered planner 2) [14.33]. In these planners, the objective is described not as a set of goal states, but instead as a collection of *tasks* to perform. Planning proceeds by decomposing tasks into subtasks, subtasks into sub-subtasks, and so forth in a recursive manner until the planner reaches *primitive* tasks that can be performed using actions similar to those used in a classical planning system. To guide the decomposition process, the planner uses a collection of *methods* that give ways of decomposing tasks into subtasks.

- *Control-rule* planners such as temporal logic planner (TLPlan) [14.34] and temporal action logic planner (TALplanner) [14.35]. Here, the domain-specific knowledge is a set of rules that give conditions under which nodes can be pruned from the search space; for example, if the objective is to load a collection of boxes into a truck, one might write a rule telling the planner "do not pick up a box unless (1) it is not on the truck and (2) it is supposed to be on the truck." The planner does a forward search from the initial state, but follows only those paths that satisfy the control rules.

Planning with Uncertain Outcomes

One limitation of classical planners is that they cannot handle uncertainty in the outcomes of the actions. The best-known model of uncertainty in planning is the Markov decision process (MDP) model. MDPs are well known in engineering, but are generally defined over continuous sets of states and actions, and are solved using the tools of continuous mathematics. In contrast, the MDPs considered in AI research are usually discrete, with the relationships among the states and actions being symbolic rather than numeric (the latest version of PDDL [14.29] incorporates the ability to represent planning problems in this fashion):

- There is a set of states S and a set of actions A. Each state s has a reward $R(s)$, which is a numeric measure of the desirability of s. If an action a is applicable to s, then $C(a, s)$ is the cost of executing a in s.
- If we execute an action a in a state s, the outcome may be any state in S. There is a probability distribution over the outcomes: $P(s'|a, s)$ is the probability that the outcome will be s', with $\sum_{s' \in S} P(s'|a, s) = 1$.
- Starting from some *initial state* s_0, suppose we execute a sequence of actions that take the MDP from s_0 to some state s_1, then from s_1 to s_2, then from s_2 to s_3, and so forth. The sequence of states $h = \langle s_0, s_1, s_2, \ldots \rangle$ is called a *history*. In a *finite-horizon* problem, all of the MDP's possible histories are finite (i.e., the MDP ceases to operate after a finite number of state transitions). In an *infinite-horizon* problem, the histories are infinitely long (i.e., the MDP never stops operating).

- Each history h has a utility $U(h)$ that can be computed by summing the rewards of the states minus the costs of the actions

$$U(h) = \begin{cases} \sum_{i=0}^{n-1} R(s_i) - C(s_i, \pi(s_i)) + R(s_n) , \\ \quad \text{for finite-horizon problems} , \\ \sum_{i=0}^{\infty} \gamma^i R(s_i) - C(s_i, \pi(s_i)) \\ \quad \text{for infinite-horizon problems} . \end{cases}$$

In the equation for infinite-horizon problems, γ is a number between 0 and 1 called the *discount factor*. Various rationales have been offered for using discount factors, but the primary purpose is to ensure that the infinite sum will converge to a finite value.

- A *policy* is any function $\pi : S \to A$ that returns an action to perform in each state. (More precisely, π is a *partial* function from S to A. We do not need to define π at a state $s \in S$ unless π can actually generate a history that includes s.) Since the outcomes of the actions are probabilistic, each policy π induces a probability distribution over MDP's possible histories

$$P(h|\pi) = P(s_0)P(s_1|\pi(s_0), s_0)P(s_1|\pi(s_1), s_1) \\ \times P(s_2|\pi(s_2), s_2) \ldots$$

The *expected utility* of π is the sum, over all histories, of h's probability times its utility: $EU(\pi) = \sum_h P(h|\pi)U(h)$. Our objective is to generate a policy π having the highest expected utility.

Traditional MDP algorithms such as value iteration or policy iteration are difficult to use in AI planning problems, since these algorithms iterate over the entire set of states, which can be huge. Instead, the focus has been on developing algorithms that examine only a small part of the search space. Several such algorithms are described in [14.36]. One of the best-known is real-time dynamic programming (RTDP) [14.37], which works by repeatedly doing a forward search from the initial state (or the set of possible initial states), extending the frontier of the search a little further each time until it has found an acceptable solution.

Applications of Planning

The paragraph on *Domain-Specific Planners* gave several examples of successful applications of domain-specific planners. Domain-configurable HTN planners such as O-Plan, SIPE-2, and SHOP2 have been deployed in hundreds of applications; for example a system for controlling unmanned aerial vehicles (UAVs) [14.38] uses SHOP2 to decompose high-level objectives into low-level commands to the UAV's controller.

Because of the strict set of restrictions required for classical planning, it is not directly usable in most application domains. (One notable exception is a cybersecurity application [14.39].) On the other hand, several domain-specific or domain-configurable planners are based on generalizations of classical planning techniques. One example is the domain-specific Mars rover planning software mentioned in *Domain-Specific Planners*, which involved a generalization of a classical planning technique called *plan-space planning* [14.40, Chap. 5]. Some of the generalizations included ways to handle action durations, temporal constraints, and other problem characteristics. For additional reading on planning, see *Ghallab* et al. [14.40] and *LaValle* [14.41].

14.1.5 Games

One of the oldest and best-known research areas for AI has been classical games of strategy, such as chess, checkers, and the like. These are examples of a class of games called *two-player perfect-information zero-sum turn-taking games*. Highly successful decision-making algorithms have been developed for such games: Computer chess programs are as good as the best grandmasters, and many games – including most recently checkers [14.42] – are now completely solved.

A *strategy* is the game-theoretic version of a policy: a function from states into actions that tells us what move to make in any situation that we might en-

counter. Mathematical game theory often assumes that a player chooses an entire strategy in advance. However, in a complicated game such as chess it is not feasible to construct an entire strategy in advance of the game. Instead, the usual approach is to choose each move at the time that one needs to make this move.

In order to choose each move intelligently, it is necessary to get a good idea of the possible future consequences of that move. This is done by searching a *game tree* such as the simple one shown in Fig. 14.9. In this figure, there are two players whom we will call Max and Min. The square nodes represent states where it is Max's move, the round nodes represent states where it is Min's move, and the edges represent moves. The terminal nodes represent states in which the game has ended, and the numbers below the terminal nodes are the payoffs. The figure shows the payoffs for both Max and Min; note that they always sum to 0 (hence the name *zero-sum* games).

From von Neuman and Morgenstern's famous Minimax theorem, it follows that Max's *dominant* (i. e., best) strategy is, on each turn, to move to whichever state s has the highest *minimax value $m(s)$*, which is defined as follows

$$m(s) = \begin{cases} \text{Max's payoff at } s \,, \\ \quad \text{if } s \text{ is a terminal node} \,, \\ \max\{m(t) : t \text{ is a child of } s\} \,, \\ \quad \text{if it is Max's move at } s \,, \\ \min\{m(t) : t \text{ is a child of } s\} \,, \\ \quad \text{if it is Min's move at } s \,, \end{cases} \quad (14.3)$$

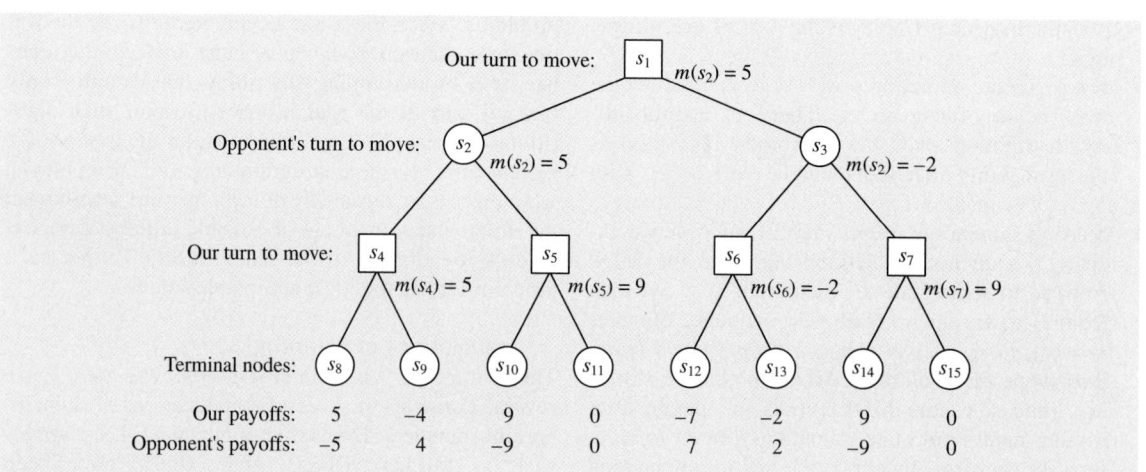

Fig. 14.9 A simple example of a game tree

where *child* means any immediate successor of s; for example, in Fig. 14.9,

$$m(s_2) = \min(\max(5, -4), \max(9, 0))$$
$$= \min(5, 9) = 5 ; \qquad (14.4)$$
$$m(s_3) = \min(\max(s_{12}), \max(s_{13}), \max(s_{14}),$$
$$\max(s_{15})) = \min(7, 0) = 0 . \qquad (14.5)$$

Hence Max's best move at s_1 is to move to s_2.

A brute-force computation of (14.3) requires searching every state in the game tree, but most nontrivial games have so many states that it is infeasible to explore more than a small fraction of them. Hence a number of techniques have been developed to speed up the computation. The best known ones include:

- *Alpha–beta pruning*, which is a technique for deducing that the minimax values of certain states cannot have any effect on the minimax value of s, hence those states and their successors do not need to be searched in order to compute s's minimax value. Pseudocode for the algorithm can be found in [14.8, 43], and many other places.

  In brief, the algorithm does a modified depth-first search, maintaining a variable α that contains the minimax value of the best move it has found so far for Max, and a variable β that contains the minimax value of the best move it has found so far for Min. Whenever it finds a move for Min that leads to a subtree whose minimax value is less than α, it does not search this subtree because Max can achieve at least α by making the best move that the algorithm found for Max earlier. Similarly, whenever the algorithm finds a move for Max that leads to a subtree whose minimax value exceeds β, it does not search this subtree because Min can achieve at least β by making the best move that the algorithm found for Min earlier.

  The amount of speedup provided by alpha–beta pruning depends on the order in which the algorithm visits each node's successors. In the worst case, the algorithm will do no pruning at all and hence will run no faster than a brute-force minimax computation, but in the best case, it provide an exponential speedup [14.43].

- *Limited-depth search*, which searches to an arbitrary *cutoff depth*, uses a *static evaluation function* $e(s)$ to estimate the utility values of the states at that depth, and then uses these estimates in (14.3) as if those states were terminal states and their estimated utility values were the exact utility values for those states [14.8].

Games with Chance, Imperfect Information, and Nonzero–Sum Payoffs

The game-tree search techniques outlined above do extremely well in perfect-information zero-sum games, and can be adapted to perform well in perfect-information games that include chance elements, such as backgammon [14.44]. However, game-tree search does less well in imperfect-information zero-sum games such as bridge [14.45] and poker [14.46]. In these games, the lack of imperfect information increases the effective branching factor of the game tree because the tree will need to include branches for all of the moves that the opponent *might* be able to make. This increases the size of the tree exponentially.

Second, the minimax formula implicitly assumes that the opponent will always be able to determine which move is best for them – an assumption that is less accurate in games of imperfect information than in games of perfect information, because the opponent is less likely to have enough information to be able to determine which move is best [14.47].

Some imperfect-information games are *iterated* games, i.e., tournaments in which two players will play the same game with each other again and again. By observing the opponent's moves in the previous *iterations* (i.e., the previous times one has played the game with this opponent), it is often possible to detect patterns in the opponent's behavior and use these patterns to make probabilistic predictions of how the opponent will behave in the next iteration. One example is Roshambo (rock–paper–scissors). From a game-theoretic point of view, the game is trivial: the best strategy is to play purely at random, and the expected payoff is 0. However, in practice, it is possible to do much better than this by observing the opponent's moves in order to detect and exploit patterns in their behavior [14.48]. Another example is poker, in which programs have been developed that play nearly as well as human champions [14.46]. The techniques used to accomplish this are a combination of probabilistic computations, game-tree search, and detecting patterns in the opponent's behavior [14.49].

Applications of Games

Computer programs have been developed to take the place of human opponents in so many different games of strategy that it would be impractical to list all of them here. In addition, game-theoretic techniques have application in several of the behavioral and social sciences, primarily in economics [14.50].

Highly successful computer programs have been written for chess [14.51], checkers [14.42, 52], bridge [14.45], and many other games of strategy [14.53]. AI game-searching techniques are being applied successfully to tasks such as business sourcing [14.54] and to games that are models of social behavior, such as the iterated prisoner's dilemma [14.55].

14.1.6 Natural-Language Processing

Natural-language processing (NLP) focuses on the use of computers to analyze and understand human (as opposed to computer) languages. Typically this involves three steps: part-of-speech tagging, syntactic parsing, and semantic processing. Each of these is summarized below.

Part-of-Speech Tagging

Part-of-speech tagging is the task of identifying individual words as nouns, adjectives, verbs, etc. This is an important first step in parsing written sentences, and it also is useful for *speech recognition* (i.e., recognizing spoken words) [14.56].

A popular technique for part-of-speech tagging is to use hidden Markov models (HMMs) [14.57]. A hidden Markov model is a finite-state machine that has states and probabilistic state transitions (i.e., at each state there are several different possible *next* states, with a different probability of going to each of them). The states themselves are not directly observable, but in each state the HMM emits a symbol that we can observe.

To use HMMs for part-of-speech tagging, we need an HMM in which each state is a pair (w, t), where w is a word in some finite lexicon (e.g., the set of all English words), and t is a part-of-speech tag such as *noun*, *adjective*, or *verb*. Note that, for each word w, there may be more than one possible part-of-speech tag, hence more than one state that corresponds to w; for example, the word *flies* could either be a plural noun (the insect), or a verb (the act of flying).

In each state (w, t), the HMM emits the word w, then transitions to one of its possible next states. As an example (adapted from [14.58]), consider the sentence, *Flies like a flower*. First, if we consider each of the words separately, every one of them has more than one possible part-of-speech tag:

Flies   could be a plural noun or a verb;
like   could be a preposition, adverb, conjunction, noun or verb;
a   could be an article or a noun, or a preposition;
flower could be a noun or a verb;

Here are two sequences of state transitions that could have produced the sentence:

- Start, (Flies, noun), (like, verb), (a, article), (flower, noun), End

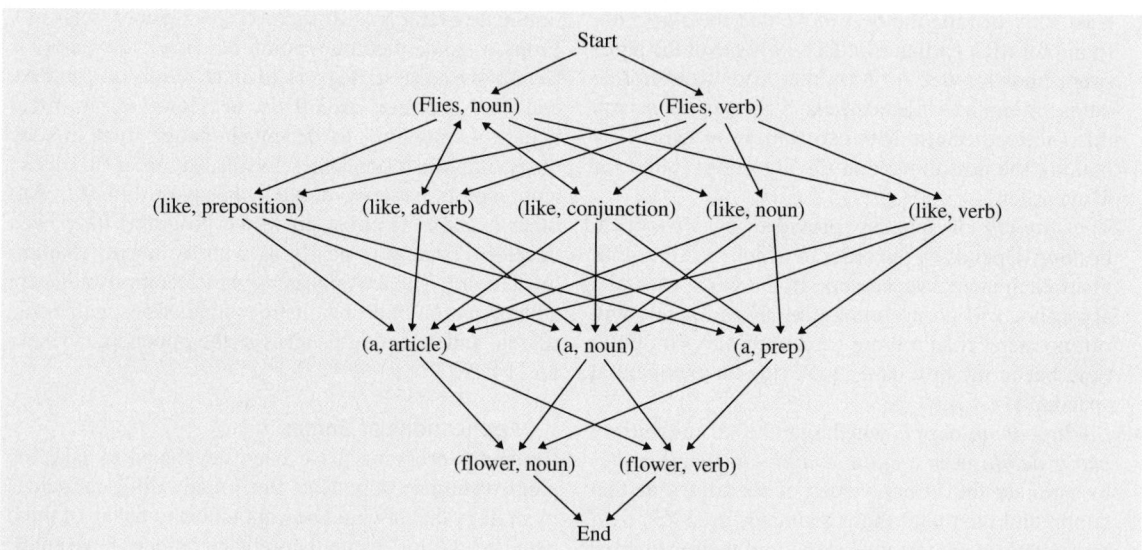

Fig. 14.10 A graphical representation of the set of all state transitions that might have produced the sentence *Flies like a flower*.

- Start, (Flies, verb), (like, preposition), (a, article), (flower, noun), End.

But there are many other state transitions that could also produce it; Fig. 14.10 shows all of them. If we know the probability of each state transition, then we can compute the probability of each possible sequence – which gives us the probability of each possible sequence of part-of-speech tags.

To establish the transition probabilities for the HMM, one needs a source of data. For NLP, these data sources are language *corpora* such as the Penn Treebank (http://www.cis.upenn.edu/ treebank/).

Context–Free Grammars

While HMMs are useful for part-of-speech tagging, it is generally accepted that they are not adequate for parsing entire sentences. The primary limitation is that HMMs, being finite-state machines, can only recognize *regular* languages, a language class that is too restricted to model several important syntactical features of human languages. A somewhat more adequate model can be provided by using *context-free grammars* [14.59].

In general, a *grammar* is a set of rewrite rules such as the following:

Sentence → NounPhrase VerbPhrase

NounPhrase → Article NounPhrase1

Article → the | a | an

. . .

The grammar includes both *nonterminal symbols* such as *NounPhrase*, which represents an entire noun phrase, and *terminal symbols* such as *the* and *an*, which represent actual words. A *context-free grammar* is a grammar in which the left-hand side of each rule is always a single nonterminal symbol (such as Sentence in the first rewrite rule shown above).

Context-free grammars can be used to *parse* sentences into parse trees such as the one shown in Fig. 14.11, and can also be used to generate sentences. A parsing algorithm (parser) is a procedure for searching through the possible ways of combining grammatical rules to find one or more parses (i.e., one or more trees similar to the one in Fig. 14.11) that match a given sentence.

Features. While context-free grammars are better at modeling the syntax of human languages than regular grammars, there are still important features of human

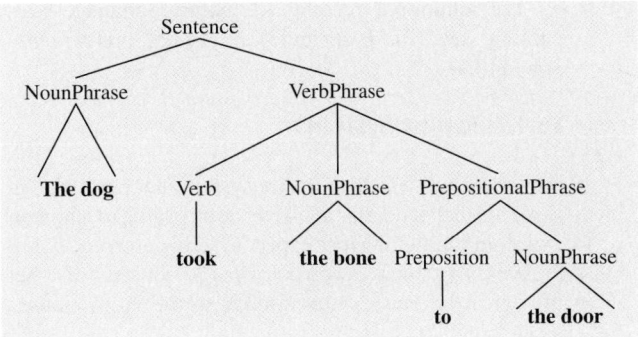

Fig. 14.11 A parse tree for the sentence *The dog took the bone to the door.*

languages that context-free grammars cannot handle well; for example, a pronoun should not be plural unless it refers to a plural noun. One way to handle these is to augment the grammar with a set of *features* that restrict the circumstances under which different rules can be used (e.g., to restrict a pronoun to be plural if its referent is also plural).

PCFGs. If a sentence has more than one parse, one of the parses might be more likely than the others: for example, *time flies* is more likely to be a statement about time than about insects. A *probabilistic context-free grammar* (PCFG) is a context-free grammar that is augmented by attaching a probability to each grammar rule to indicate how likely different possible parses may be.

PCFGs can be learned from a parsed language corpora in a manner somewhat similar (although more complicated) than learning HMMs [14.60]. The first step is to acquire CFG rules by reading them directly from the parsed sentences in the corpus. The second step is to try to assign probabilities to the rules, test the rules on a new corpus, and remove rules if appropriate (e.g., if they are redundant or if they do not work correctly).

Applications

NLP has a large number of applications. Some examples include automated language-translation services such as Babelfish, Google Translate, Freetranslation, Teletranslator and Lycos Translation [14.61], automated speech-recognition systems used in telephone call centers, systems for categorizing, summarizing, and retrieving text (e.g., [14.62, 63]), and automated evaluation of student essays [14.64].

For additional reading on natural-language processing, see *Wu, Hsu*, and *Tan* [14.65] and *Thompson* [14.66].

14.1.7 Expert Systems

An *expert system* is a software system that performs, in some specialized field, at a level comparable to a human expert in the field. Most expert systems are *rule-based systems*, i.e., their *expert knowledge* consists of a set of logical inference rules similar to the Horn clauses discussed in Sect. 14.1.2.

Often these rules also have probabilities attached to them; for example, instead of writing

if A_1 and A_2 **then** conclude A_3

one might write

if A_1 and A_2 **then** conclude A_3 with probability p_0 .

Now, suppose A_1 and A_2 are known to have probabilities p_1 and p_2, respectively, and to be stochastically independent so that $P(A_1 \wedge A_2) = p_1 p_2$. Then the rule would conclude $P(C) = p_0 p_1 p_2$.

If A_1 and A_2 are not known to be stochastically independent, or if there are several rules that conclude A_3, then the computations can get much more complicated. If there are n variables $A_1, \ldots, A_n$, then the worst case could require a computation over the entire joint distribution $P(A_1, \ldots, A_n)$, which would take exponential time and would require much more information than is likely to be available to the expert system.

In some of the early expert systems, the above complication was circumvented by assuming that various events were stochastically independent even when they were not. This made the computations tractable, but could lead to inaccuracies in the results. In more modern systems, conditional independence (Sect. 14.1.3) is used to obtain more accurate results in a computationally tractable manner.

Expert systems were quite popular in the early and mid-1980s, and were used successfully in a wide variety of applications. Ironically, this very success (and the hype resulting from it) gave many potential industrial users unrealistically high expectations of what expert systems might be able to accomplish for them, leading to disappointment when not all of these expectations were met. This led to a backlash against AI, the so-called *AI winter* [14.67], that lasted for some years. but in the meantime, it became clear that simple expert systems were more elaborate versions of the decision logic already used in computer programming; hence some of

the techniques of expert systems have become a standard part of modern programming practice.

Applications. Some of the better-known examples of expert-system applications include medical diagnosis [14.68], analysis of data gathered during oil exploration [14.69], analysis of DNA structure [14.70], configuration of computer systems [14.71], as well as a number of expert system *shell* (i. e., tools for building expert systems).

14.1.8 AI Programming Languages

AI programs have been written in nearly every programming language, but the most common languages for AI programming are Lisp, Prolog, C/C++, and Java.

Lisp

Lisp [14.72, 73] has many features that are useful for rapid prototyping and AI programming. These features include garbage collection, dynamic typing, functions as data, a uniform syntax, an interactive programming and debugging environment, ease of extensibility, and a plethora of high-level functions for both numeric and symbolic computations. As an example, Lisp has a built-in function, append, for concatenating two lists – but even if it did not, such a function could easily be written as follows:

```
(defun concatenate (x y)
  (if (null x)
      y
      (cons (first x)
        (concatenate (rest x) y))))
```

The above program is *tail-recursive*, i. e., the recursive call occurs at the very end of the program, and hence can easily be translated into a loop – a translation that most Lisp compilers perform automatically.

An argument often advanced in favor of conventional languages such as C++ and Java as opposed to Lisp is that they run faster, but this argument is largely erroneous. As of 2003, experimental comparisons showed compiled Lisp code to run nearly as fast as C++, and substantially faster than Java. (The speed comparison to Java might not be correct any longer, since a huge amount of work has been done since 2003 to improve Java compilers.) Probably the misconception about Lisp's speed arose from the fact that early Lisp systems ran Lisp code interpretively. Modern Lisp systems give users the option of running their code interpretively (which is useful for experimenting and debugging) or compiling their code (which provides much higher speed).

See [14.74] for a discussion of other advantages of Lisp. One notable disadvantage of Lisp is that, if one has a computer program written in a conventional language such as C, C++ or Java, it is difficult for such a program to call a Lisp program as a subroutine: one must run the Lisp program as a separate process in order to provide the Lisp execution environment. (On the other hand, Lisp programs can quite easily invoke subroutines written in conventional programming languages.)

Applications. Lisp was quite popular during the expert-systems boom of the mid-1980s, and several *Lisp machine* computer architectures were developed and marketed in which the entire operating system was written in Lisp. Ultimately these machines did not meet with long-term commercial success, as they were eventually surpassed by less-expensive, less-specialized hardware such as Sun workstations and Intel x86 machines.

On the other hand, development of software systems in Lisp has continued, and there are many current examples of Lisp applications. A few of them include the visual lisp extension language for the AutoCAD computer-aided design system (autodesk.com), the Elisp extension language for the Emacs editor (http://en.wikipedia.org/wiki/Emacs_Lisp) the Script-Fu plugins for the GNU Image Manipulation Program (GIMP), the Remote Agent software deployed on NASA's *Deep Space 1* spacecraft [14.75], the airline fare shopping engine used by Orbitz [14.9], the SHOP2 planning system [14.38], and the Yahoo Store e-commerce software. (As of 2003, about 20 000 Yahoo stores used this software. The author does not have access to more recent statistics.)

Prolog

Prolog [14.76] is based on the notion that a general theorem-prover can be used as a programming environment in which the program consists of a set of logical statements. As an example, here is a Prolog program for concatenating lists, analogous to the Lisp program given earlier

```
concatenate([],Y,Y).
concatenate([First|Rest],Y,[First|Z]) :-
concatenate(Rest,Y,Z).
```

To concatenate two lists [a,b] and [c], one asks the theorem prover if there exists a list Z that is their concatenation; and the theorem prover returns Z if it exists

```
?- concatenate([a,b],[c],Z)
Z=[a,b,c].
```

Alternatively, if one asks whether there are lists X and Y whose concatenation is a given list Z, then there are several possible values for X and Y, and the theorem prover will return *all* of them

```
?- concatenate(X,Y,[a,b])
X = []; Y = [a,b]
X = [a]; Y = [b]
X = [a,b]; Y = []
```

One of Prolog's biggest drawbacks is that several aspects of its programming style – for example, the lack of an assignment statement, and the automated backtracking – can require workarounds that feel unintuitive to most programmers. However, Prolog can be good for problems in which logic is intimately involved, or whose solutions have a succinct logical characterization.

Applications. Prolog became popular during the the expert-systems boom of the 1980s, and was used as the basis for the Japanese Fifth Generation project [14.77], but never achieved wide commercial acceptance. On the other hand, an extension of Prolog called constraint logic programming is important in several industrial applications (see *Constraint Satisfaction and Constraint Optimization*).

C, C++, and Java

C and C++ provide much less in the way of high-level programming constructs than Lisp, hence developing code in these languages can require much more effort. On the other hand, they are widely available and provide fast execution, hence they are useful for programs that are simple and need to be both portable and fast; for example, neural networks need very fast execution in order to achieve a reasonable learning rate, and a back-propagation procedure can be written in just a few pages of C or C++ code.

Java is a lower-level language than Lisp, but is higher-level than C or C++. It uses several ideas from Lisp, most notably garbage collection. As of 2003 it ran much more slowly than Lisp, but its speed has improved in the interim and it has the advantages of being highly portable and more widely known than Lisp.

14.2 Emerging Trends and Open Challenges

AI has gone through several periods of optimism and pessimism. The most recent period of pessimism was the *AI winter* mentioned in Sect. 14.1.7. AI has emerged from this period in recent years, primarily because of the following trends. First is the exponential growth improvement in computing power: computations that used to take days or weeks can now be done in minutes or seconds. Consequently, computers have become much better able to support the intensive computations that AI often requires. Second, the pervasive role of computers in everyday life is helping to erase the apprehension that has often been associated with AI in popular culture. Third, there have been huge advances in AI research itself. AI concepts such as search, planning, natural-language processing, and machine learning have developed mature theoretical underpinnings and extensive practical histories.

AI technology is widely expected to become increasingly pervasive in applications such as data mining, information retrieval (especially from the web), and prediction of human events (including anything from sports forecasting to economics to international conflicts).

During the next decade, it appears quite likely that AI will be able to make contributions to the behavioral and social sciences analogous to the contributions that computer science has made to the biological sciences during the past decade. To make this happen, one of the biggest challenges is the huge diversity among the various research fields that will be involved. These include behavioral and social sciences such as economics, political science, psychology, anthropology, and sociology, and technical disciplines such as AI, robotics, computational linguistics, game theory, and operations research. Researchers from these fields will need to forge a common understanding of principles, techniques, and objectives. Research laboratories are being set up to foster this goal (one example is the University of Maryland's Laboratory for Computational Cultural Dynamics (http://www.umiacs.umd.edu/research/LCCD/), which is co-directed by the author of this chapter), and several international conferences and workshops on the topic have been recently established [14.78, 79].

One of the biggest challenges that currently faces AI research is its fragmentation into a bewilderingly diverse collection of subdisciplines. Unfortunately, these subdisciplines are becoming rather insular, with their own research conferences and their own (sometimes idiosyncratic) notions of what constitutes a worthwhile research question or a significant result. The achievement of human-level AI will require integrating the best efforts among many different subfields of AI, and this in turn will require better communication amongst the researchers from these subfields. I believe that the field is capable of overcoming this challenge, and that human-level AI will be possible by the middle of this century.

References

14.1   A. Newell, H.A. Simon: Computer science as empirical inquiry: Symbols and search, Assoc. Comput. Mach. Commun. **19**(3), 113–126 (1976)

14.2   S. Hedberg: Proc. Int. Conf. Artif. Intell. (IJCAI) 03 conference highlights, AI Mag. **24**(4), 9–12 (2003)

14.3   C. Kuykendall: Analyzing solitaire, Science **283**(5403), 791 (1999)

14.4   R. Bjarnason, P. Tadepalli, A. Fern: Searching solitaire in real time, Int. Comput. Games Assoc. J. **30**(3), 131–142 (2007)

14.5   E. Horowitz, S. Sahni: *Fundamentals of Computer Algorithms* (Computer Science, Potomac 1978)

14.6   N. Nilsson: *Principles of Artificial Intelligence* (Morgan Kaufmann, San Francisco 1980)

14.7   D. Navinchandra: The recovery problem in product design, J. Eng. Des. **5**(1), 67–87 (1994)

14.8   S. Russell, P. Norvig: *Artificial Intelligence, A Modern Approach* (Prentice Hall, Englewood Cliffs 1995)

14.9   S. Robinson: Computer scientists find unexpected depths in airfare search problem, SIAM News **35**(1), 1–6 (2002)

14.10   C. Le Pape: Implementation of resource constraints in ilog schedule: a library for the development of constraint-based scheduling systems, Intell. Syst. Eng. **3**, 55–66 (1994)

14.11   P. Shaw: Using constraint programming and local search methods to solve vehicle routing problems, Proc. 4th Int. Conf. Princ. Pract. Constraint Program. (1998) pp. 417–431

14.12   P. Albertand, L. Henocque, M. Kleiner: Configuration based workflow composition, IEEE Int. Conf. Web Serv., Vol. 1 (2005) pp. 285–292

14.13   J. Pearl: *Heuristics* (Addison-Wesley, Reading 1984)

14.14   R. Dechter: *Constraint Processing* (Morgan Kaufmann, San Francisco 2003)

14.15   J. Shoenfield: *Mathematical Logic* (Addison-Wesley, Reading 1967)

14.16 J. Pearl, S. Russell: Bayesian networks. In: *Handbook of Brain Theory and Neural Networks*, ed. by M.A. Arbib (MIT Press, Cambridge 2003) pp.157–160

14.17 J. Pearl: *Probabilistic Reasoning in Intelligent Systems: Networks of Plausible Inference* (Morgan Kaufmann, San Fransisco 1988)

14.18 M. Sahami, S. Dumais, D. Heckerman, E. Horvitz: A bayesian approach to filtering junk e-Mail. In: *Learning for Text Categorization: Papers from the 1998 Workshop* AAAI Technical Report WS-98-05 (Madison 1998)

14.19 J.A. Zdziarski: *Ending Spam: Bayesian Content Filtering and the Art of Statistical Language Classification* (No Starch, San Francisco 2005)

14.20 D. Quinlan: BayesInSpamAssassin, http://wiki. apache.org/spamassassin/BayesInSpamAssassin (2005)

14.21 K.M. Hanson: Introduction to bayesian image analysis. In: *Medical Imaging: Image Processing*, Vol.1898, ed. by M.H. Loew, (Proc. SPIE, 1993) pp.716–732

14.22 L. Denoyer, P. Gallinari: Bayesian network model for semistructured document classification, Inf. Process. Manage. **40**, 807–827 (2004)

14.23 S. Fox, K. Karnawat, M. Mydland, S. Dumais, T. White: Evaluating implicit measures to improve web search, ACM Trans. Inf. Syst. **23**(2), 147–168 (2005)

14.24 L. Zadeh, G.J. Klir, Bo Yuan: *Fuzzy Sets, Fuzzy Logic, and Fuzzy Systems: Selected Papers by Lotfi Zadeh* (World Scientific, River Edge 1996)

14.25 G.J. Klir, U.S. Clair, B. Yuan: *Fuzzy Set Theory: Foundations and Applications* (Prentice Hall, Englewood Cliffs 1997)

14.26 A. Tate: Planning. In: *MIT Encyclopedia of the Cognitive Sciences*, (1999) pp. 652–653

14.27 T. Estlin, R. Castano, B. Anderson, D. Gaines, F. Fisher, M. Judd: Learning and planning for Mars rover science, Proc. Int. Joint Conf. Artif. Intell. (IJCAI) (2003)

14.28 S.K. Gupta, D.A. Bourne, K. Kim, S.S. Krishanan: Automated process planning for sheet metal bending operations, J. Manuf. Syst. **17**(5), 338–360 (1998)

14.29 A. Gerevini, D. Long: Plan constraints and preferences in pddl3: The language of the fifth international planning competition. Technical Report (University of Brescia, 2005), available at http://cs-www.cs.yale.edu/homes/dvm/papers/pddl-ipc5.pdf

14.30 A.L. Blum, M.L. Furst: Fast planning through planning graph analysis, Proc. Int. Joint Conf. Artif. Intell. (IJCAI) (1995) pp.1636–1642

14.31 A. Tate, B. Drabble, R. Kirby: *O-Plan2: An Architecture for Command, Planning and Control* (Morgan Kaufmann, San Francisco 1994)

14.32 D.E. Wilkins: *Practical Planning: Extending the Classical AI Planning Paradigm* (Morgan Kaufmann, San Mateo 1988)

14.33 D. Nau, T.-C. Au, O. Ilghami, U. Kuter, J.W. Murdock, D. Wu, F. Yaman: SHOP2: An HTN planning system, J. Artif. Intell. Res. **20**, 379–404 (2003)

14.34 F. Bacchus, F. Kabanza: Using temporal logics to express search control knowledge for planning, Artif. Intell. **116**(1/2), 123–191 (2000)

14.35 J. Kvarnström, P. Doherty: TALplanner: A temporal logic based forward chaining planner, Ann. Math. Artif. Intell. **30**, 119–169 (2001)

14.36 M. Fox, D. E. Smith: Special track on the 4th international planning competition. J. Artif. Intell. Res. (2006), available at http://www.jair.org/specialtrack.html

14.37 B. Bonet, H. Geffner: Labeled RTDP: Improving the convergence of real-time dynamic programming, Proc. 13th Int. Conf. Autom. Plan. Sched. (ICAPS) (AAAI, 2003), pp 12–21

14.38 D. Nau, T.-C. Au, O. Ilghami, U. Kuter, H. Muñoz-Avila, J.W. Murdock, D. Wu, F. Yaman: Applications of SHOP and SHOP2, IEEE Intell. Syst. **20**(2), 34–41 (2005)

14.39 M. Boddy, J. Gohde, J.T. Haigh, S. Harp: Course of action generation for cyber security using classical planning, Proc. 15th Int. Conf. Autom. Plan. Sched. (ICAPS) (2005)

14.40 M. Ghallab, D. Nau, P. Traverso: *Automated Planning: Theory and Practice* (Morgan Kaufmann, San Francisco 2004)

14.41 S.M. Lavalle: *Planning Algorithms* (Cambridge University Press, Cambridge 2006)

14.42 J. Schaeffer, N. Burch, Y. Björnsson, A. Kishimoto, M. Müller, R. Lake, P. Lu, S. Sutphen: Checkers is solved, Science **317**(5844), 1518–1522 (2007)

14.43 D.E. Knuth, R.W. Moore: An analysis of alpha-beta pruning, Artif. Intell. **6**, 293–326 (1975)

14.44 G. Tesauro: *Programming Backgammon Using Self-Teaching Neural Nets* (Elsevier, Essex 2002)

14.45 S.J.J. Smith, D.S. Nau, T. Throop: Computer bridge: A big win for AI planning, AI Mag. **19**(2), 93–105 (1998)

14.46 M. Harris: Laak-Eslami team defeats Polaris in man-machine poker championship, Poker News (2007), Available at www.pokernews.com/news/2007/7/laak-eslami-team-defeats-polaris-man-machine-poker-championship.htm

14.47 A. Parker, D. Nau, V.S. Subrahmanian: Overconfidence or paranoia? Search in imperfect-information games, Proc. Natl. Conf. Artif. Intell. (AAAI) (2006)

14.48 D. Billings: Thoughts on RoShamBo, Int. Comput. Games Assoc. J. **23**(1), 3–8 (2000)

14.49 B. Johanson: Robust strategies and counter-strategies: Building a champion level computer poker player. Master Thesis (University of Alberta, 2007)

14.50 S. Hart, R.J. Aumann (Eds.): *Handbook of Game Theory with Economic Applications 2* (North Holland, Amsterdam 1994)

Part B | 14

14.51 F.-H. Hsu: Chess hardware in deep blue, Comput. Sci. Eng. **8**(1), 50–60 (2006)

14.52 J. Schaeffer. *One Jump Ahead: Challenging Human Supremacy in Checkers* (Springer, Berlin, Heidelberg 1997)

14.53 J. Schaeffer: A gamut of games, AI Mag. **22**(3), 29–46 (2001)

14.54 T. Sandholm: Expressive commerce and its application to sourcing, Proc. Innov. Appl. Artif. Intell. Conf. (IAAI) (AAAI Press, Menlo Park 2006)

14.55 T.-C. Au, D. Nau: Accident or intention: That is the question (in the iterated prisoner's dilemma), Int. Joint Conf. Auton. Agents and Multiagent Syst. (AAMAS) (2006)

14.56 C. Chelba, F. Jelinek: Structured language modeling for speech recognition, Conf. Appl. Nat. Lang. Inf. Syst. (NLDB) (1999)

14.57 B.H. Juang, L.R. Rabiner: Hidden Markov models for speech recognition, Technometrics **33**(3), 251–272 (1991)

14.58 S. Lee, J. Tsujii, H. Rim: Lexicalized hidden Markov models for part-of-speech tagging, Proc. 18th Int. Conf. Comput. Linguist. (2000)

14.59 N. Chomsky: *Syntactic Structures* (Mouton, The Hague 1957)

14.60 E. Charniak: A maximum-entropy-inspired parser, Technical Report CS-99-12 (Brown University 1999)

14.61 F. Gaspari: *Online MT Services and Real Users Needs: An Empirical Usability Evaluation* (Springer, Berlin, Heidelberg 2004), pp 74–85

14.62 P. Jackson, I. Moulinier: *Natural Language Processing for Online Applications: Text Retrieval, Extraction, and Categorization* (John Benjamins, Amsterdam 2002)

14.63 M. Sahami, T.D. Heilman: A web-based kernel function for measuring the similarity of short text snippets, WWW '06: Proc. 15th Int. Conf. World Wide Web (New York 2006) pp. 377–386

14.64 J. Burstein, M. Chodorow, C. Leacock: Automated essay evaluation: the criterion online writing service, AI Mag. **25**(3), 27–36 (2004)

14.65 Zhi Biao Wu, Loke Soo Hsu, Chew Lim Tan: A survey of statistical approaches to natural language processing, Technical Report TRA4/92 (National University of Singapore 1992)

14.66 C. A. Thompson: A brief introduction to natural language processing for nonlinguists. In: *Learning Language in Logic*, Lecture Notes in Computer Science (LNCS) Ser., Vol. 1925, ed. by J. Cussens, S. Dzeroski (Springer, New York 2000) pp. 36–48

14.67 H. Havenstein: Spring comes to AI winter, Computerworld (2005), available at http://www.computerworld.com/action/article.do?command=viewArticleBasic&articleId=99691

14.68 B.G. Buchanan, E.H. Shortliffe (Eds.): *Rule-Based Expert Systems: The MYCIN Experiments of the Stanford Heuristic Programming Project* (Addison-Wesley, Reading 1984)

14.69 R.G. Smith, J.D. Baker: The dipmeter advisor system – a case study in commercial expert system development, Proc. Int. Joint Conf. Artif. Intell. (IJCAI) (1983) pp. 122–129

14.70 M. Stefik: Inferring DNA structures from segmentation data, Artif. Intell. **11**(1/2), 85–114 (1978)

14.71 J.P. McDermott: R1 ("XCON") at age 12: Lessons from an elementary school achiever, Artif. Intell. **59**(1/2), 241–247 (1993)

14.72 G. Steele: *Common Lisp: The Language*, 2nd edn. (Digital, Woburn 1990)

14.73 P. Graham: *ANSI Common Lisp* (Prentice Hall, Englewood Cliffs 1995)

14.74 P. Graham: *Hackers and Painters: Big Ideas from the Computer Age* (O'Reilly Media, Sebastopol 2004) pp. 165–180, also available at http://www.paulgraham.com/avg.html

14.75 N. Muscettola, P. Pandurang Nayak, B. Pell, B.C. Williams: Remote agent: To boldly go where no AI system has gone before, Artif. Intell. **103**(1/2), 5–47 (1998)

14.76 W. Clocksin, C. Mellish: *Programming in Prolog* (Springer, Berlin, Heidelberg 1981)

14.77 E. Feigenbaum, P. McCorduck: *The Fifth Generation: Artificial Intelligence and Japan's Computer Challenge to the World* (Addison-Wesley Longman, Boston 1983)

14.78 D. Nau, J. Wilkenfeld (Ed): Proc. 1st Int. Comput. Cult. Dyn. (ICCCD-2007) (AAAI Press, Menlo Park 2007)

14.79 H. Liu, J. Salerno, M. Young (Ed): Proc. 1st Int. Workshop Soc. Comput. Behav. Model. Predict. (Springer, Berlin, Heidelberg 2008)

15. Virtual Reality and Automation

P. Pat Banerjee

Virtual reality of human activities (e.g., in design, manufacturing, medical care, exploration or military operations) often concentrates on an automated interface between virtual reality (VR) technology and the theory and practice of these activities. In this chapter we focus mainly on the role of VR technology in developing this interface. Although the scope and range of applications is large, two illustrative areas (production/service applications and medical applications) are explained in some detail to offer some insight into the magnitude of the benefits and existing challenges.

15.1 Overview of Virtual Reality and Automation Technologies

Virtual manufacturing and automation first came into prominence in the early 1990s, in part as a result of the US Department of Defense Virtual Manufacturing Initiative. The topic broadly refers to the modeling of manufacturing systems and components with effective use of audio-visual and/or other sensory features to simulate or design alternatives for an actual manufacturing environment mainly through effective use of high-performance computers.

To understand the area, first we introduce a broader area of virtual reality (VR) and visual simulation systems. VR brings in a few exciting new developments. Firstly it provides a major redefinition of perspective projection, by introducing the concept of user-centered perspective. Traditional perspective projection is fixed, whereas in virtual reality one has the option to vary the perspective in real time, thus closely mimicking our natural experience of a three-dimensional (3-D) world. In the past few years, VR and visual simulation systems have made a huge impact in terms of industrial adoption; for example, Cyberedge Information Services conducted a survey [15.1] that provided the following details:

- Industry growth remains strong: 9.8% in 2003
- Industry value: US$ 42.6 billion worldwide
- Five-year forecast: industry value reaches US$ 78 billion in 2008
- Average system cost: US$ 356 310
- Number of systems sold in 2003: 338 000
- Companies involved in visual simulation worldwide: 17 334
- Most common visual display system: monoscopic desktop monitor
- Least common visual display system: autostereoscopic display
- Most common operating system: Microsoft Windows XP.

The survey [15.1] further comments on the top 17 applications of visual simulation systems from 1999–2003 as follows:

Table 15.1 Automation advantages of virtual reality (VR); * Unique to VR. General advantages of simulation, including less costs, less risks, less delay, and ability to study as yet unbuilt structures/objects are also valid

What is automated?	Application example	Advantage*
Application domain	• Testing a simulated mold of a new engine	• Better visualization of tested product
	• Developing battlefield tactics	• Greater ability to collaborate in planning and design of complex virtual situations
Human supervisory interface	• Training in dental procedures	• Detailed hands-on experience in a large simulated library of dental problems
	• Teleoperating robots	
Human feedback interface	• Improving the safety of a process	• Greater precision and quality of feedback through haptics
	• Learning the properties of surfaces	• Better understanding of simulated behaviors
Human application interaction	• Telerepair or teleassembly of remote objects	• Greater ability to obtain knowledge about and manipulate in remote global and space locations
		• Deeper sense of involvement and engagement by the interacting humans

1. Software development/testing
2. Computer aided design (CAD)/computer-aided manufacture (CAM) visualization or presentation
3. Postgraduate education (college)
4. Virtual prototype
5. Museum/exhibition
6. Design evaluation, general
7. Medical diagnostics
8. Undergraduate education (college)

9. Aerospace
10. Automobile, truck, heavy equipment
11. Games development
12. Collaborative work
13. Architecture
14. Medical training
15. Dangerous environment operations
16. Military operation training
17. Trade show exhibit.

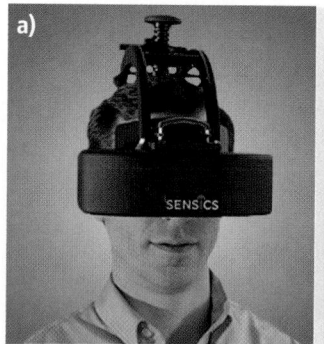 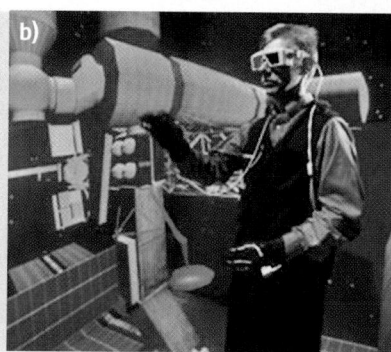 

Fig. 15.1 (a) Sensics piSight head-mounted display for virtual prototyping, training, data mining, and other applications (http://www.sensics.com/, December 2007). **(b)** Pinch is an interacting system for interacting in a virtual environment using sensors in each fingertip (http://www.fakespace.com/, December 2007). **(c)** Preview SR80 full-color SXGA 3-D stereoscopic head-mounted display (http://www.vrealities.com/, December 2007)

In each of these application areas, VR is advantageous in automating and improving three critical areas, as shown in Table 15.1. A smaller subsection of VR and visual simulation is virtual manufacturing and automation, which often concentrates on an interface between VR technology and manufacturing and automation theory and practice. In this chapter we concentrate mainly on the role of VR technology in developing this interface. Some areas that can benefit from development of virtual manufacturing include product design [15.2, 3], hazardous operations modeling [15.4, 5], production process modeling [15.6, 7], training [15.8, 9], education [15.10, 11], information visualization [15.12, 13], telecommunications [15.14, 15], and teletravel. Lately a number of surgical simulation applications have emerged.

VR is closely associated with an environment commonly known as a virtual environment (VE). VE systems differ from other previously developed computer-centered systems in the extent to which real-time interaction is facilitated, the perceived visual space is three- rather than two-dimensional, the human–machine interface is multimodal, and the operator is immersed in the computer-generated environment. The interactive, virtual image displays are enhanced by special processing and by nonvisual display modalities, such as auditory and haptic, to convince users that they are immersed in a synthetic space.

The means of simulating a VE today is through immersion in computer graphics coupled with acoustic interface and domain-independent interacting devices, such as wands, and domain-specific devices such as the steering and brakes for cars, or earthmovers or instrument clusters for airplanes (Fig. 15.1a–c). Immersion gives the feeling of depth, which is essential for a three-dimensional effect. Head-mounted displays, stereoscopic projectors, and retinal display are some of the technologies used for such environments.

15.2 Production/Service Applications

15.2.1 Design

The immersive display technology can be used for creating virtual prototypes of products and processes. Integrated production system engineering environments can provide functions to specify, design, engineer, simulate, analyze, and evaluate a production system. Some examples of the functions which might be included in an integrated production system engineering environment are:

1. Identification of product specifications and production system requirements
2. Producibility analysis for individual products
3. Modeling and specification of manufacturing processes
4. Measurement and analysis of process capabilities
5. Modification of product designs to address manufacturability issues
6. Plant layout and facilities planning
7. Simulation and analysis of system performance
8. Consideration of various economic/cost tradeoffs of different manufacturing processes, systems, tools, and materials
9. Analysis supporting selection of systems/vendors
10. Procurement of manufacturing equipment and support systems
11. Specification of interfaces and the integration of information systems
12. Task and workplace design
13. Management, scheduling, and tracking of projects.

The interoperability of the commercial engineering tools that are available today is a challenge. Examples of production systems which may eventually be engineered using this type of integrated environment include: transfer lines, group technology cells, automated or manually operated workstations, customized multipurpose equipment, and entire plants.

15.2.2 Material Handling and Manufacturing Systems

An important outcome of virtual manufacturing is virtual factories of the future. When a single factory may cost over a billion US dollars (as is the case, for instance, in the semiconductor industry), it is evident that manufacturing decision-makers need tools that support good decision making about their design, deployment, and operation. However, in the case of manufacturing models, there is usually no testbed but the factory itself; development of models of manufacturing operations is very likely to disrupt factory operations while the models are being developed and tested.

Virtual factories are based on sophisticated computer simulations for a distributed, integrated, computer-based composite model of a total manufacturing environment, incorporating all the tasks and resources necessary to accomplish the operation of designing, producing, and delivering a product. With virtual factories capable of accurately simulating factory operations over time scales of months, managers would be able to explore many potential production configurations and schedules or different control and organizational schemes at significant savings of cost and time in order to determine how best to improve performance. A virtual factory model involves a comprehensive model or structure for integrating a set of heterogeneous and hierarchical submodels at various levels of abstraction. An ultimate goal would be the creation of a demonstration platform that would compare the results of real factory operations with the results of simulated factory operations. This demonstration platform would use a computer-based model of an existing factory and would compare its performance with that of a similarly equipped factory running the same product line, but using, for example, a new layout of equipment, a better scheduling system, a paperless product and process description, or fewer or more human operators. The entire factory would have to be represented in sufficient detail so that any model user, from factory manager to equipment operator, would be able to extract useful results. To accomplish this, two broad areas need to be addressed:

1. Hardware and software technology to handle sophisticated graphics and data-oriented models in a useful and timely manner
2. Representation of manufacturing expertise in models in such a way that the results of model operation satisfy manufacturing experts' needs for accurate responses.

Some of the research considerations related to virtual modeling technology are as follows:

- *User interfaces.* Since factory personnel in the 21st century will be interacting with many applications, it will not suffice for each application to have its own set of interfaces, no matter how good any individual one is. Much thought has to be given both to the nature of the specific interfaces and to the integration of interfaces in a system designed so as not to confuse the user. Interface tools that allow the user to filter and abstract large volumes of data will be particularly important.

- *Model consistency.* Models will be used to perform a variety of geographically dispersed functions over a short time scale during which the model information must be globally accurate. In addition, pieces of the model may themselves be widely distributed. As a result, a method must be devised to ensure model consistency and concurrency, perhaps for extended time periods. The accuracy of models used concurrently for different purposes is a key determinant of the benefits of using such models.

- *Testing and validation of model concepts.* Because a major use of models is to make predictions about matters that are not intuitively obvious to decision-makers, testing and validation of models and their use are very difficult. For factory operations and design alike, there are many potential *right answers* to important questions, and none of these is *provably correct* (e.g., what is the *right* schedule?). As a result, models have to be validated by being tested against understandable conditions, and in many cases, common sense must be used to judge if a model is *correct*. Because models must be tested under stochastic factory conditions, which are hard to duplicate or emulate, outside the factory environment, an important area for research involves developing tools for use in both testing and validating model operation and behavior. Tools for automating sensitivity analysis in the testing of simulation models would help to overcome model validation problems inherent in a stochastic environment. For more details, please refer to [15.16].

Figure 15.2 shows a schematic user interface for remote facility management using virtual manufacturing concepts. The foreground represents a factory floor layout that identifies important regions of the plant. The walls separating different regions have been removed to obtain a complete bird's eye view of the plant. Transmitted streaming motion from video capture of important regions can be provided as a background. For this purpose one needs to identify and specify the objects for motion streaming. A combination of video and VR is a powerful tool in this context. The advantage of streaming motion coordinates to animate the factory floor would be to give the floor manager an overview of the floor. A suitable set of symbols can be designed to designate the status modes of various regions, namely running state, breakdown state, shutdown state, waiting state (e.g., for parts, worker), scheduled maintenance state, etc. An avataric representation of the floor manager can

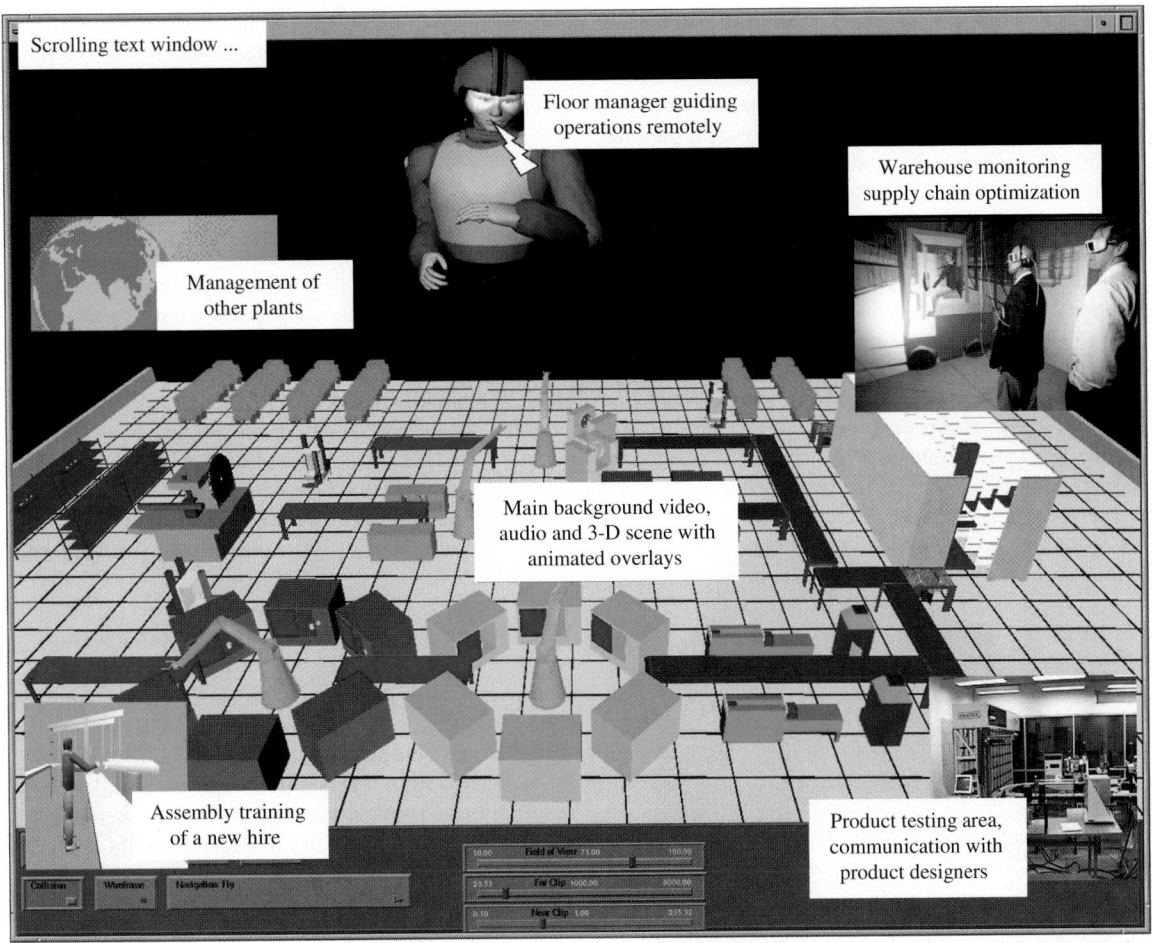

Fig. 15.2 Conceptual demonstration of remote facility management user interface showing many telecollaborative activities

be used to guide operations. A window for the product testing area and communication with product designers can be used to efficiently streamline the product design process. A video camera device can be integrated to stream motion coordinates after the objects in the product design area and their layouts have been specified. Another window showing the assembly training of a new employee can illustrate the use of this con-

cept for another important area of the plant. A similar feature can be designed for warehouse monitoring and supply chain management. A window for management of other plants can take one to different plants, located anywhere in the world. Finally a scrolling text window can pop up any important or emergency messages such as suddenly scheduled meetings, changes of deadline, sudden shifts in strategy, etc.

15.3 Medical Applications

Medical simulations using virtual automation have been a recent area of growth. Some of our activities in this area are highlighted in this Section. Using hap-

tics and VR in high-fidelity open-surgical simulation, certain principles of design and operational prototyping can be highlighted. A philosophy of designing device prototypes by carefully analyzing contact based on open surgical simulation requirements and mapping them with open-source software and off-the-shelf hardware components is presented as an example. Applications in neurosurgery (ventriculostomy), ophthalmology (capsulorrhexis), and dentistry (periodontics) are illustrated. First, a prototype device known as ImmersiveTouch is described, followed by three applications.

ImmersiveTouch is a patent-pending [15.17] next-generation augmented VR technology invented by us, and is the first system that integrates a haptic device with a head and hand tracking system, and a high-resolution high-pixel-density stereoscopic display (Fig. 15.3). Its ergonomic design provides a comfortable working volume in the space of a standard desktop. The haptic device is collocated with the 3-D

graphics, giving the user a more realistic and natural means to manipulate and modify 3-D data in real time. The high-performance, multisensorial computer interface allows relatively easy development of VR simulation and training applications that appeal to many stimuli: audio, visual, tactile, and kinesthetic.

ImmersiveTouch represents an integrated hardware and software solution. The hardware integrates 3-D stereo visualization, force feedback, head and hand tracking, and 3-D audio. The software provides a unified applications programming interface (API) to handle volume processing, graphics rendering, haptics rendering, 3-D audio feedback, and interactive menus and buttons. ImmersiveTouch is an evolutionary virtual-reality system resulting from the integration of a series of hardware solutions to 3-D display issues *and* the development of a unique VR software platform (API) drawing heavily on open-source software and databases.

15.3.1 Neurosurgical Virtual Automation

Neurosurgical procedures, particularly cranial applications, lend themselves to VR simulation. The working space around the cranium is limited. Anatomical relationships within the skull are generally fixed and respiratory or somatic movements do not significantly impair imaging or rendering. The same issues that make cranial procedures so suitable for intraoperative navigation also apply to virtual operative simulation. The complexity of so many cerebral structures also allows little room for error, making the need for skill-set acquisition prior to the procedure that much more significant. The emergence of realistic neurosurgical simulators has been predicted since the mid-1990s when the explosion in computer processing speed seemed to presage future developments [15.18]. VR techniques have been attempted to simulate several types of spinal procedures such as lumbar puncture, in addition to craniotomy-type procedures.

Ventriculostomy is a high-frequency surgical intervention commonly employed for treatment of head injury and stroke. This extensively studied procedure is a good example problem for augmented VR simulators. We investigated the ability of neurosurgery residents at different levels of training to cannulate the ventricle, using computer tomography data on an ImmersiveTouch workstation. After a small incision is made and a bur hole is drilled on a strategically chosen spot in the patient's skull, a ventriculostomy catheter is inserted,

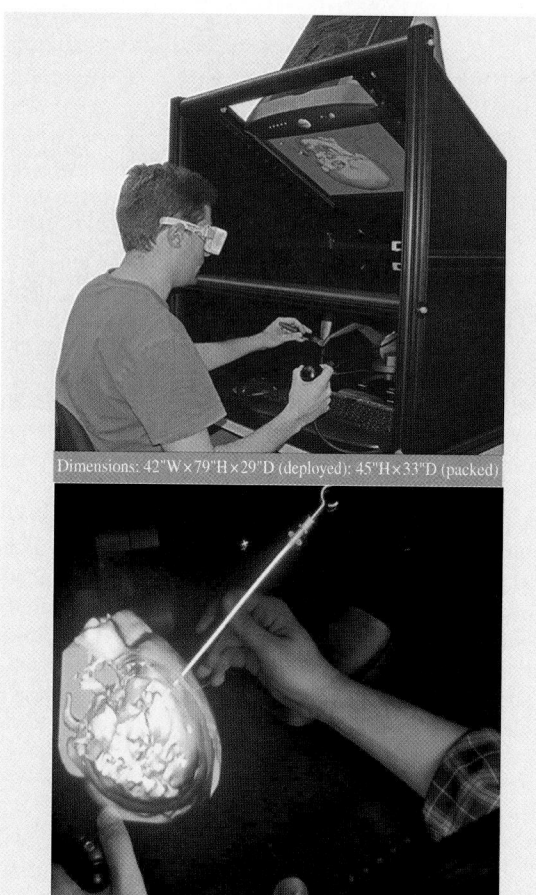

Dimensions: 42"W×79"H×29"D (deployed); 45"H×33"D (packed)

Fig. 15.3 Immersive touch and its neurosurgical application

aiming for the ventricles. A distinct *popping* or puncturing sensation is felt as the catheter enters the frontal horn of the lateral ventricle [15.19].

The performance evaluation from the Immersive-Touch simulator based on a large sample of neurosurgeons was comparable to clinical study findings, in which the average distance and standard deviation of the catheter tip to the foramen of Monroe in both cases were close [15.18].

15.3.2 Ophthalmic Virtual Automation

Cataract surgery is the most common surgical procedure performed by ophthalmic residents-in-training and general ophthalmologists in the USA. With over 1.5 million procedures performed every year, it represents the largest single surgical care provided under Medicare in the USA.

Cataract surgery involves the removal of the opacified crystalline lens by phacoemulsification and replacement of the lens with an intraocular lens implant. Over the years cataract surgery has become a highly technical procedure requiring a number of very fine skills, under an operating microscope. It involves the use and understanding of sophisticated technology, as well as fine hand–eye coordination movements. The training of this highly skilled procedure to residents has become a challenging task for teaching physicians because of its highly technical nature. Current ways of training residents include:

- Practising surgical procedures in a *wet lab* situation, where surgery is practised on animal eyes. Though this procedure has some similarity to human surgery, there are many problems associated with this technique, which include: (a) Porcine and bovine models that do not provide a realistic sense of tissue tension, nor do they mimic the accurate dimensions of the human eye; (b) Any assessment of skills acquired during such animal laboratory procedures with an observing physician is often subjective and repetition is difficult.
- Another technique that has been used extracts eyes from a postmortem human eye bank to perform phacoemulsification. These tissues are very difficult to obtain, and are fairly expensive to procure.
- A common technique, the *apprenticeship* technique, is a *see one, do one, teach one* approach. However, apprenticeship is not the best approach, especially for a highly skilled procedure such as phacoemulsification.

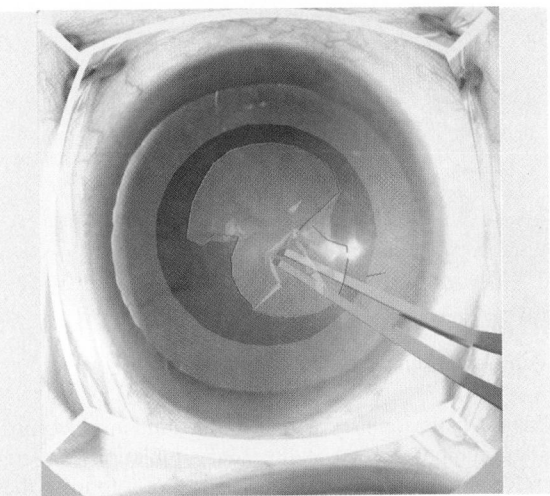

Fig. 15.4 Capsulorrhexis simulation in cataract surgery using virtual reality and haptics

In summary, all of the current paradigms available are suboptimal from the perspective of surgical education as well as patient care, and developments in virtual-reality simulation will allow the surgeon to gain realistic operative experience before their first cataract surgery.

The ophthalmic development of virtual-reality procedures has been minimal. The EYESI ophthalmic surgical simulator (VRmagic GmbH, Mannheim, Germany) simulates intraocular surgery but they have a heavy reliance on physical prototyping that is hard to reconfigure.

The various steps in cataract surgery can be broken down into many steps, including incision, performing capsulorrhexis, phacoemulsification of the cataract and removal, removal of cortex, and placement of the intraocular lens. Many of these steps are being simulated by us using virtual reality and haptics (Fig. 15.4).

15.3.3 Dental Virtual Automation

Dentistry provides a rich collection of clinical procedures which require dexterity in tactile feedback and thus provide a number of very interesting simulation problems. The rationale for using dental simulators is that, by using them, preclinical dental students learn procedure skills faster and more efficiently. The majority of current dental simulators use realistic manikins along with dentiform models (Kavo, Adec or Nevin) incorporated into a simulated dental operatory [15.19]. A number of dental schools also use the DentSim simulator (DenX Ltd., Jerusalem, Israel).

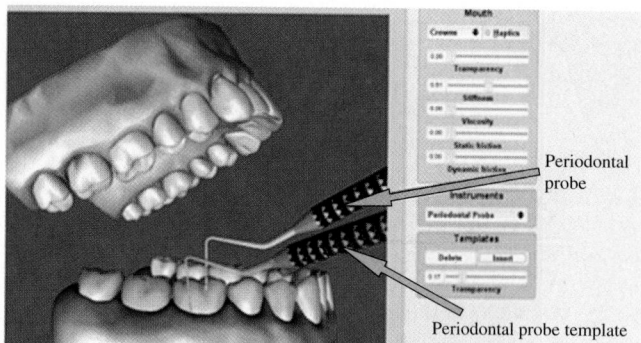

Fig. 15.5 Periodontal simulations

This is a more sophisticated manikin simulator incorporating computer-aided audiovisual simulations having a VR component. It uses a tracking system to trace the movements of a handpiece and scores the accuracy of a student's cavity preparation in a manikin's synthetic tooth.

The addition of haptics to dental simulators is extremely important because sensory motor skills must be developed by dental and hygiene trainees in order to be successful in their profession. This is especially true in the ability to perform a variety of periodontal procedures (i.e., scaling and root planning, periodontal probing, and the use of a periodontal explorer). These tactile skills are currently acquired by having trainees observe instructor demonstrations of a specific procedure and then having them practise on manikins or animal heads. After sufficient training and practice on models, the students proceed to patients. This time-consuming teaching process requires excessive one-on-one instructor–student interaction without students actually feeling what the instructor feels or being physically guided by the instructor while performing a procedure on a manikin or a patient. With the current critical shortage of dental faculty, the training problem has been compounded.

Dental simulators that have been developed include those from Novint Technologies (Novint, Albuquerque,

NM) and Simulife Systems (Simulife Systems, Paris, France). Our work has led to PerioSim at the University of Illinois at Chicago, College of Dentistry [15.19, 20]. The PerioSim prototype system for periodontal probing is designed to provide a haptic force feedback and guidance system for interaction with an on-screen display. The display is a 3-D VR model of a periodontal probe and a human upper and lower dental arch, which includes teeth, their supportive structures, and other oral tissues. The fidelity of the haptic-based system is currently sufficiently sophisticated to differentiate between hard and soft, and normal and pathological tissues. While the development is taking place the other goals are to determine how realistic the PerioSim simulator is and the time required for learning how to use it.

The steps followed by a trainee are as follows: Using a control panel, one of three periodontal instruments can be selected for on-screen use: a periodontal probe, a periodontal explorer or a Gracy scaler. The monitor graphic display is used in conjunction with a haptic device (PHANToM from SensAble Corp., Woburn, MA) for force feedback and perceiving the textural feel of the gingival crevice/pocket area. The 3-D VR periodontal probe can be used to locate and measure crevice or pocket depths around the gingival margins of the teeth. These sites can be identified via the haptic feedback obtained from the PHANToM stylus with or without the actual instrument attached to the stylus. A trainee will be able to differentiate the textural feel of pocket areas and locate regions of subgingival calculus. Since the root surface is covered by gingiva, the trainee cannot see the area being probed or the underlying calculus and must depend totally on haptic feedback to identify these areas. This situation corresponds to conditions encountered clinically. The control panel, which can be made to appear or disappear as needed, has a variety of controls, including adjusting the haptic feel and the degree of transparency of the gingiva, roots, crowns, bone or calculus. The control panel also permits the instructor to insert a variety of templates to guide student instrument positioning in a 3-D display environment (Fig. 15.5).

15.4 Conclusions and Emerging Trends

Virtual-reality and automation technologies are aimed at reducing the time spent in providing service and desired training. At present almost all major companies operate globally. Management, control, and service of overseas facilities and products deployed overseas is

a major challenge. Both processes and factories need to be simulated, although the current thrust seems to be on individual process simulation. One of the important emerging themes is constant evolution of enabling technologies.

A review of ongoing evolution of enabling technologies is available in *Burdea* and *Coiffet*'s book on VR technologies [15.21]. The two main areas in this regard are hardware and software technologies (e.g., [15.22]). The hardware can be broken down into input and output devices and computing hardware architectures. Input devices consist of various trackers; mechanical, electromagnetic, ultrasonic, optical, and hybrid inertial trackers are covered. Tracking is used to navigate and manipulate in a VR environment. There are many more three-dimensional navigation devices that are available, including hand and finger movement tracking. Output devices address graphics displays, sound, and haptic feedback. The graphics displays include head-mounted displays, binocular-type hand-supported displays, floor-supported displays, desktop displays, and large displays based on large monitors and projectors supporting multiple participants simultaneously. The haptics tac-

tile feedback covers tactile mouse, touch-based glove, temperature-feedback glove, force-feedback joysticks, and haptic robotic arms. Hardware architectures include the two major rendering pipelines: graphics and haptics. Personal computer (PC) graphics accelerator cards such as those from nVidia are currently in vogue. Various distributed VR architectures addressing issues such as graphics and haptics pipelines synchronization, PC clusters for tiled visual displays, and multiuser shared virtual environments are equally important. Software challenges include those in modeling and VR programming. Modeling addresses geometric modeling, kinematics modeling, physical modeling, and behavior modeling. VR programming will continue to be aided by evolution of more powerful toolkits. The enabling technologies outlined above can provide answers to many of these challenges in virtual reality and automation.

References

15.1   B. Delaney: *The Market for Visual Simulation/Virtual Reality Systems*, 6th edn. (Cyberedge Information Services, Mountain View 2004), www.cyber-edge.com

15.2   H.Y. Kan, V.G. Duffy, C.-J. Su: An Internet virtual reality collaborative environment for effective product design, Comput. Ind. **45**(2), 197–213 (2001)

15.3   M. Pouliquen, A. Bernard, J. Marsot, L. Chodorge: Virtual hands and virtual reality multimodal platform to design safer industrial systems, Comput. Ind. **58**(1), 46–56 (2007)

15.4   B. Stone, G. Pegman: Robots and virtual reality in the nuclear industry, Serv. Robot **1**(2), 24–27 (1995)

15.5   P.R. Chakraborty, C.J. Bise: Virtual-reality-based model for task-training of equipment operators in the mining industry, Miner. Res. Eng. **9**(4), 437–449 (2000)

15.6   S. Ottosson: Virtual reality in the product development process, J. Eng. Des. **13**(2), 159–172 (2002)

15.7   Y. Jun, J. Liu, R. Ning, Y. Zhang: Assembly process modeling for virtual assembly process planning, Int. J. Comput. Integr. Manuf. **18**(6), 442–451 (2005)

15.8   D. Lee, M. Woo, D. Vredevoe, J. Kimmick, W.J. Karplus, D.J. Valentino: Ophthalmoscopic examination training using virtual reality, Virtual Real. **4**(3), 184–191 (1999)

15.9   C.H. Park, G. Jang, Y.H. Chai: Development of a virtual reality training system for live-line workers, Int. J. Human-Comput. Interact. **20**(3), 285–303 (2006)

15.10   V.S. Pantelidis: Virtual reality and engineering education, Comput. Appl. Eng. Educ. **5**(1), 3–12 (1997)

15.11   Y.S. Shin: Virtual reality simulations in Web-based science education, Comput. Appl. Eng. Educ. **10**(1), 18–25 (2002)

15.12   T.M. Rhyne: Going virtual with geographic information and scientific visualization, Comput. Geosci. **23**(4), 489–491 (1997)

15.13   O.N. Kwon, S.H. Kim, Y. Kim: Enhancing spatial visualization through virtual reality (VR) on the Web: software design and impact analysis, J. Comput. Math. Sci. Teach. **21**(1), 17–31 (2002)

15.14   P. Queau: Televirtuality: the merging of telecommunications and virtual reality, Comput. Graph. **17**(6), 691–693 (1993)

15.15   M. Torabi: Mobile virtual reality services, Bell Labs Tech. J. **7**(2), 185–194 (2002)

15.16   P. Banerjee, D. Zetu: *Virtual Manufacturing* (Wiley, New York 2001)

15.17   P. Banerjee, C. Luciano, L. Florea, G. Dawe: Compact haptic and augmented virtual reality system, US Patent Appl. No. 11/338434 (2006), (previous version: C. Luciano, P. Banerjee, L. Florea, G. Dawe: Design of the ImmersiveTouch: A High-Performance Haptic Augmented Virtual Reality System, CD ROM Proc. Human-Comput. Interact. (HCI) Int. Conf. (Las Vegas 2005))

15.18   P.P. Banerjee, C. Luciano, G.M. Lemole Jr, F.T. Charbel, M.Y. Oh: Accuracy of ventriculostomy catheter placement on computer tomography data using head and hand tracked high resolution virtual reality, J. Neurosurg. **107**(3), 515–521 (2007)

15.19   C. Luciano: Haptics–based virtual reality Periodontal Training Simulator. Ph.D. Thesis (University of
Illinois, Chicago 2006)

15.20   A.D. Steinberg, P. Banerjee, J. Drummond, M. Ze-
fran: Progress in the development of a hap-
tic/virtual reality simulation program for scaling
and root planing, J. Dent. Educ. **67**(2), 161 (2003)

15.21   G. Burdea, P. Coiffet: *Virtual Reality Technology*,
2nd edn. (Wiley Interscience, New York 2003)

15.22   C. Luciano, P. Banerjee, G.M. Lemole, J. Charbel, F. Charbel: Second generation haptic ventriculostomy simulator using the immersivetouch
system, Proc. 14th Med. Meets Virtual Real. (2006)
pp. 343–348

16. Automation of Mobility and Navigation

Anibal Ollero, Ángel R. Castaño

This chapter deals with general concepts on the automation of mobility and autonomous navigation. The emphasis is on the control and navigation of autonomous vehicles. Thus, after an introduction with historical background and basic concepts, the chapter briefly reviews general concepts on vehicle motion control by using models of the vehicle, as well as other approaches based on the information provided by humans. Autonomous navigation is also studied, involving not only motion planning and trajectory generation but also interaction with the environment to provide reactivity and adaptation in the autonomous navigation. These interactions are represented by means of nested loops closed at different frequencies with different bandwidth requirements. The human interactions at different levels are also analyzed, taking into account transmission of control commands and feedback of sensory information. Finally, the

chapter studies multiple mobile systems by analyzing coordinated navigation of multiple autonomous vehicles and cooperation paradigms for autonomous mission execution.

16.1 Historical Background

The ambition to emulate the motion of living things has been sustained throughout history, from old automata to humanoid robots today. The Ancient Greeks and also other cultures from around the world created many automata and managed to make animated statues of mechanical people, animals, and objects (e.g., the moving stone statues made by Daedalus in 520 BC, the flying magpie created by King-shu in 500 BC, the mechanical pigeon made by Archytas of Tarentum in 400 BC, the singing blackbird and other figures that drank and moved created by Ctesibius in 280 BC, etc.). In the 14th century, remarkable automata were produced in Europe (by Johannes Müller and Leonardo da Vinci). The descriptions of mechanical automata in the 16th, 17th, and 18th centuries are also well known (Gianello Della Tour, Salomon de Caus, Christiaan Huygens, Jacques

de Vauçanson, etc.). They had complex mechanisms and produced sounds and music synchronized with the motions. Many interesting automata were made in the 19th century. This was the time when production techniques decreased in cost. The robots (*R.U.R.* (Rossum's Universal Robots), 1921) gave their name to many mechanical men built in the 1920 and 1930s to be shown in films or fairs or to demonstrate remote control techniques, such as the Westinghouse robots in the 1930s.

By the end of the 19th and beginning of the 20th century, mass industrial production required the development of automation. Mobility automation also played an important role. Then, for example, primitive conveyor belts were used in the 19th century and the technology was introduced into assembly lines by 1913 in Ford Motor Company's factory. Automation of mobility in general became a key issue in factory automation. The first programmable paint-spraying mechanism was designed in 1938. Industrial automation met robotics with the design in 1954 of the first programmable robot by Devol, who coined the term *universal automation*. General Motors applied the first industrial robot on a production line in 1962. From 1966 through 1972, the Artificial Intelligence Center at SRI International (then the Stanford Research Institute) conducted research on the mobile robot system *Shakey*.

Control and automation in the transportation of goods and people has also been the target for many centuries. Rudimentary elevators, operated by animal and human power or by water-driven mechanisms, were in use during the Middle Ages and can be traced back to the third century BC. The elevator as we know it today was first developed during the 19th century and relied on steam or hydraulic plungers for lifting capability under manual control. Thus, the valves governing the water flow were manipulated by passengers using ropes running through the cab, a system later enhanced with the incorporation of lever controls and pilot valves to regulate cab speed.

Control has also been a key concept in the development of vehicles. Watt's steam engines controlled by ball governor in 1787 played a central role in the railway developed by 1804 in Great Britain. Control also played a decisive role in the manned airplane flights of the Wright brothers in 1903. Automatic feedback control for flight guidance was possible by 1903 thanks to the gyroscope. Spinning gyroscopes were first mounted in torpedoes. The gyroscope resisted any change in direction by controlling the rudder, automatically correcting any deviation from a straight course. By 1910 gyroscopic stabilizing devices had been mounted in ships, and even in an airplane.

16.2 Basic Concepts

Automation of mobility can be examined by considering the degree of flexibility. Railways are mechanically constrained by the rails. Thus, railway automation has been accomplished since the 1980s in France, where automated subway lines have been in operation since the 1990s for mass transit. The same happens with elevators and industrial automated warehouse systems based on overhead trolleys, as shown in Fig. 16.1 [16.1]. This is an effective solution in many warehouse systems and industry transportation in general. In other cases the path of motion is also physically predetermined by different type of guides, such as inductive guide wires embedded in the floor, which are used in industrial automation to guide so-called *automated guided vehicles* (AGVs). Other AGVs use guideways painted in the floor that require maintenance to be detectable by AGV optical sensors. Installation of new AGVs or changing the pathways is time consuming and expensive in these systems. A more flexible solution consists of using beacons or marker arrangements installed in the factory. These systems are more flexible but require line-of-sight communication between devices to be mounted on the vehicles and a certain number of devices in the industrial environment. Increasing flexibility in industrial transportation has been an objective for many years. The application of industrial autonomous vehicles (Fig. 16.2) and mobile robots with onboard environment sensing for autonomous navigation provides higher flexibility but poses reliability problems. In general the trade-off between flexibility and reliability continues to be a critical aspect in many factory implementations.

Flexible automation of transportation in outdoor environments poses significant challenges because of the difficulties in their conditioning and dynamic characteristics. Automation of cars for transportation has also been a research and development subject since the 1990s. The autonomous navigation of cars and vans has been a testbed for perception, planning, and

Fig. 16.1 Car seat covers industrial automated warehouse system based on overhead trolleys [16.1]

control techniques for almost 20 years. Thus, for example, autonomous visual-based car navigation was demonstrated in Germany by the mid 1980s, and the NavLab project was running in Carnegie Mellon University by the end of the 1980s [16.2] (Fig. 16.3). Many demonstrations in different environments have also been presented. However, the daily operation of these systems is still constrained and only in operation at low speed and in short trips in restricted areas. In [16.3] the state of the technology and future directions of these so-called *cybercars* are analyzed. The use of dedicated infrastructures and the gradual transition from driver assistance to full automation are highlighted as realistic paths toward fully autonomous cars. Some applications require equipped infrastructures, such as guideways for automatic vans with magnetic tracks integrated in the road pavement. The majority of automated highways systems need an equipped road with an adapted architecture, i. e., the PATH (partners for advanced transit and highways) project [16.4].

So-called unmanned vehicles serve as means of carrying or transporting something, but explicitly do not carry a human being. Thus, unmanned ground vehicles in the broader sense includes any machine that moves across the surface of the ground, such as legged machines and machines with onboard tools and robot manipulators (Fig. 16.4). In this chapter, we will concentrate on the control and navigation aspects of these vehicles.

Unmanned aerial vehicles (UAVs) are self-propelled air vehicles that are either remotely controlled by a human operator (remotely piloted vehicles (RPV)) or are capable of conducting autonomous operations. During recent decades significant efforts have been devoted to

Fig. 16.2 Packmobile AGV, Egemin Automation

increase the flight endurance, flight range, and payload of UAVs. Today, UAVs with several thousands of kilometers of flight range, more than 24 h of flight endurance, and more than 1000 kg of payload (Fig. 16.5) are in operation. Furthermore, autonomous airships, helicopters of different size (Fig. 16.6), and other vertical take-off and landing UAVs have also been developed. UAV technology has also evolved to increase the onboard computation and communication capabilities.

Fig. 16.3 Navlab I at Carnegie Mellon University

Fig. 16.4 RAM 2 Mobile robot with manipulator at the University of Málaga

Fig. 16.5 Global Hawk, Northrop Grumman Corporation

Fig. 16.6 Helicopter UAV at the University of Seville

The development of new navigation sensors, actuators, embedded control, and communication systems and the trend towards miniaturization point to mini- and micro-UAVs with increasing capabilities. In [16.5] many UAVs are presented and compared.

Autonomous underwater vehicles (AUV) and underwater robotics is also a very active field of research and development in which many new developments have been presented in recent years. These vehicles are natural extensions of the well-known remotely operated underwater vehicles (ROV) that are controlled and powered from the surface by an operator/pilot

via an umbilical cable and have been used in many applications.

Finally, it should be noted that in many cases mobility is strongly constrained by the particular characteristics of the environment. Thus, for example, the internal inspection [16.6] and eventual repairing of pipes impose constraints on the design of the robots that should navigate inside the pipe, carrying cameras and other sensors and devices. However, these applications will be not considered in this chapter.

The autonomy of all of the above-mentioned vehicles is based on automatic motion control. The next section will summarize general concepts of vehicle

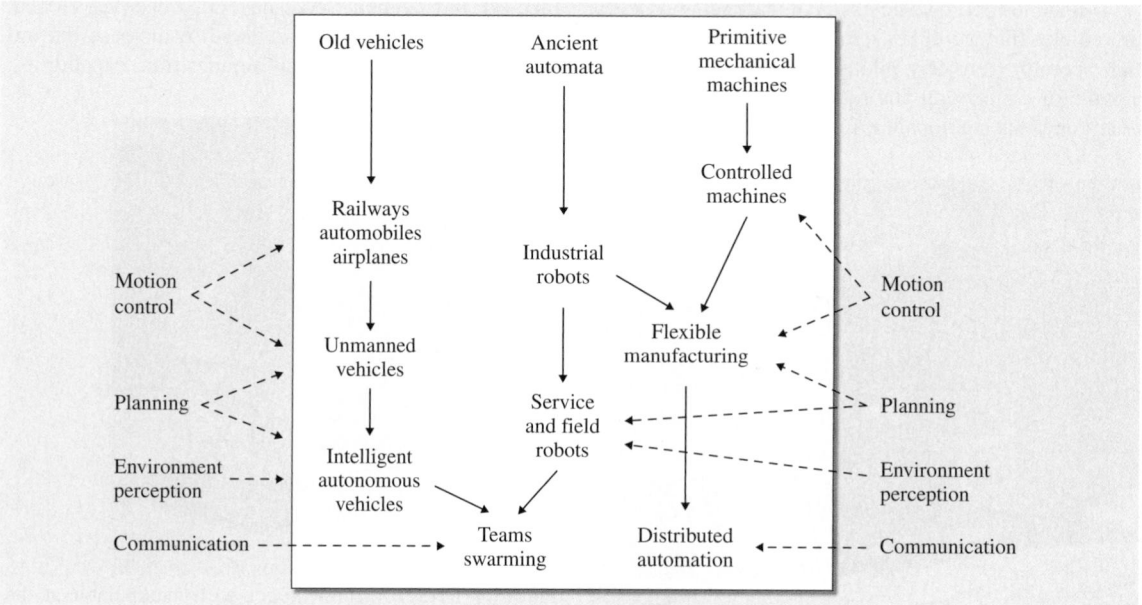

Fig. 16.7 Evolution of automation in mobility and navigation

motion control. Autonomous operation also requires environment perception. General concepts on environment perception and reactivity will be examined in the fourth section of this chapter. Another approach related to mobility enhancement is human augmentation, the objective of which is to augment human capabilities by means of motion-controlled devices interfacing with the human. Two different approaches can be followed. The first is to provide teleoperation capabilities to control the motion of remote vehicles or devices for transportation. The second approach is to augment the person's physical abilities by means of wearable devices or exoskeletons. The idea is not new, but recent progress in sensing human body signals and embedded control systems makes it possible to build exoskeletons to carry heavy loads and march faster and longer.

The applications are numerous and include handicapped people, military, and others. Section 16.5 will be devoted to the human–machine interaction for mobility and particularly for vehicle guidance. Finally, in recent years, a general emerging trend is the development of fleets or teams of autonomous vehicles. This trend involves ground, aerial, and aquatic fleets of vehicles. Section 16.6 is devoted to the coordination of fleets of vehicles. In all the following sections the background and significance of each topic, as well as the existing methods and trends are described. Figure 16.7 shows the evolution of vehicles, robotics, and industrial automation related to the automation of mobility and navigation, illustrating the impact of motion control, motion planning, environment perception, and communication technology.

16.3 Vehicle Motion Control

The lowest-level vehicle control typically consists of the vehicle motion axis in which conventional linear proportional–integral-derivative (PID) motion controllers are usually applied. Other techniques such as fuzzy logic [16.7] are less commonly used. The reference signals of these servo controllers are provided by navigation control loops in which the objective is that the vehicle follows previously defined trajectories. In these control loops navigation sensors such as gyroscopes, accelerometers, compass, and global positioning systems (GPS) are used. The objectives are perturbation rejection and improving of dynamic response. The kinematics and dynamic behavior of the vehicle play an important role in these control loops. The usual formulation is given by the equation

$$\dot{x} = f(x, u) \, , \qquad (16.1)$$

where $x \in \mathbb{R}^n$ is the vehicle state vector, usually consisting of the position, the Euler angles, the linear velocities, and the angular velocities, and $u \in \mathbb{R}^m$ are the control variables. In general the above equation involves the rigid-body kinematics and dynamics, the force and moment generation, and the actuator dynamics. Kinematics is usually considered by means of nonlinear transformations between the reference systems associated to the vehicle body and the global reference system. The nonholonomic constraints, $\phi(\dot{x}, x) = 0$, where ϕ is a nonintegral function, restrict the vehicle's admissible directions of motion and make the control problem more difficult. Thus, it has been shown [16.8] that it is not possible to stabilize a nonholonomic system to

a given set-point by a continuous and time-invariant feedback control. Moreover, in many cases, there are underactuated control systems in which the number of control variables m is lower than the number of degrees of freedom. The dynamic model involves the relation between forces and torques, generated by the propulsion system and the environment, and the accelerations of the vehicle. These relations can be described by means of the Newton–Euler equations. Furthermore, in aerial vehicles, aerodynamics plays a significant role and should be considered.

In ground vehicles the tires–terrain interactions create significant complexity. Several models have been developed to consider these interactions [16.9, 10]. The complexity of the complete dynamic models means that there are a lot of vehicle, road, and tire parameters to estimate and tune. Besides, these models change with tire pressure, road surface and conditions, vehicle weight, etc. so they are more commonly used for controller analysis and simulation, braking controller research or tire failure detection [16.11, 12].

In many formulations only simplified dynamic models are considered and then only the position and angles are included in the state vector of the above equation. Furthermore, if the vehicles move in a plane and the interactions with the terrain are neglected, then only the position in the plane and the orientation angle are considered as state variables. Sometimes a simple actuator dynamic model [16.13] is added to these simplified models. Control theory has been extensively applied to vehicle control. Many linear and nonlinear control systems have been proposed to maintain stability and tracking paths or time trajectories [16.14]. The limitation in the stabilization of nonholonomic system has been avoided by means of time-varying state feedback and discontinuous feedback. The trajectory tracking can be formulated by defining the reference model

$$\dot{x}_r = f(x_r, u_r) \, . \tag{16.2}$$

Then, if the model of the vehicle is (16.1), the problem is to find a control law $u = \varphi(x, x_r, u, u_r)$ such that

$$\lim_{t \to \infty} |x(t) - x_r(t)| = 0 \, . \tag{16.3}$$

The path tracking problem consists of the tracking of a defined path, taking into account the vehicle motion constraints defined by the equation above. Different trajectory and path tracking methods for ground vehicles have been formulated by implementing linear and nonlinear control laws, as shown and summarized in [16.14, 15]. Many practical implementations of path tracking are based on geometric methods [16.16–18]; for example, the simple *pure-pursuit method* is a linear proportional control law of the error of the vehicle position with respect to a goal point. The gain is given by the distance to this goal point defined in the path. The appropriate selection of this gain is a critical issue. Then, gain self-scheduling approaches can be applied. The stability of the path tracking control loop is analyzed in [16.19]. Even though these geometric methods are not optimal, they are usually easy to tune and have a good trade-off between performance and simplicity in real-time implementations as shown in the Defense Advanced Research Projects Agency (DARPA) Grand Challenge 2005 competition [16.20], where most of the teams used simple geometric approaches. Also traditional automatic control techniques such as generalized predictive control (GPC), robust control [16.21] or LQG/LQR (linear quadratic Gaussian/linear quadratic regulator) [16.22] have been successfully applied. These methods require a linear vehicle model so a simplified linearized version of more complex models, such as the previously commented ones, are typically employed.

The steering of ground articulated vehicles and tractor–trailer systems (see, for example, [16.23, 24]) also presents interesting control problems. Tractor–trailer systems (Fig. 16.8) have as an additional state variable the angle between the tractor and the trailer. They are usually underactuated and the saturation in the actuators can play a significant role, particularly when maneuvering backwards because the vehicle tends to

Fig. 16.8 Romeo 4R with trailer in a parking manoeuvre

reach a so-called *jack-knife* when the angle between the tractor and the trailer is greater than a certain value due to perturbations coming from the vehicle–terrain interaction.

In aerial and underwater vehicles the navigation problem is typically formulated in three-dimensional (3-D) space and different control levels involving the dynamic behavior of the vehicle can be identified. The objective of the lower levels is to keep the vehicle in a given attitude by maintaining the stability. The linear and angular velocities in the vehicle body axis, and orientation angles are usually considered as state variables. The higher-level motion control loop consists of the path or trajectory tracking (which includes precision timing). In this case, the course angle error and the cross-track distance can be considered as error signals in the guidance loop.

In [16.25] control techniques for autonomous aerial vehicles are reviewed. The navigation of these UAVs is based on GPS positioning but visual-based position estimation have also been applied [16.26], particularly in case of GPS signal loss (see the next section). In [16.27] model-based control techniques for autonomous helicopters are reviewed and different experiments involving dynamic nonlinear behaviors are presented. The position and orientation of an helicopter is usually controlled by means of five control inputs: the main rotor collective pitch, which has a direct effect on the helicopter height; the longitudinal cyclic, which modifies the helicopter pitch angle and the longitudinal translation; the lateral cyclic, which affects the helicopter roll angle and the lateral translation; the tail rotor, which controls the heading of the helicopter and compensates the antitorque generated by the main rotor; and the throttle control. It is a multivariable nonlinear system with strong coupling in some control loops. Autonomous helicopter control has been a classical control benchmark and many model-based control techniques have been applied, including multi-PID controllers, robust control, predictive control, and nonlinear control. The significance of each of these methods for practical implementations is still an open question. Safe landing on mobile platforms and transportation of loads by means of several UAVs overcoming the payload limitation of individual UAVs are open challenges. Very recently the joint transportation of a load by several helicopters has been demonstrated in the AWARE (acronym of the project *platform for autonomous self-deploying and operation of wireless sensor-actuator networks cooperating with aerial objects*) project [16.28].

There are also approaches in which learning from skilled human pilots or teleoperators plays the most significant role. In these approaches, fuzzy logic [16.29], neural networks [16.30], neurofuzzy techniques, and other artificial intelligence (AI) techniques are applied. Control theory and AI techniques have also been combined. Thus, for example, Takagi–Sugeno fuzzy systems have been applied to learn from human drivers by generating closed-loop control systems that can be analyzed and tuned by means of stability theory. These methods have been applied to drive autonomously trucks (Fig. 16.9) and heavy machines at high speed [16.31] by estimating the position of the vehicle by means of the fusion of GPS and dead-reckoning with simple vehicle models. The system has

Fig. 16.9 Autonomous 16 t Scania truck

been applied to test the tires of vehicles navigating autonomously in testing tracks.

The general challenges are the analysis and design of reliable control techniques that could be implemented in real time at high frequency in the onboard processors, providing reactivity to perturbations while maintaining acceptable performance. This is particularly challenging when considering small or very small vehicles, such as micro-UAVs with important limitations in onboard processing. The application of micro-electromechanical systems (MEMS) to implement these control systems is an emerging technology trend. The practical application in real time of fault-detection techniques and fault-tolerant control systems to improve reliability is another emerging trend.

16.4 Navigation Control and Interaction with the Environment

The consideration of interactions with the environment is also an important problem in mobility and navigation automation. These interactions can also be represented by means of loops closed at different frequencies, as shown in Fig. 16.10. The vehicle control described in the above paragraphs is also embedded in this figure. Reactivity dominates the higher-frequency loops, which also require higher bandwidth in communication channels (inner loops, towards the right in the figure), while deliberation is the main component of the lower-frequency loops, which typically have lower bandwidth requirements in communication channels (outer loops, towards the left in the figure). The inner loops can be considered as the lower lever in the control hierarchy, while the outer loops are the higher levels in this hierarchy. This chapter will not provide details on mobile

robot control architectures, but will merely describe the main interactions.

Environment perception is based on the use of sensors such as cameras, radars, lasers, ultrasonic, and other range sensors. Thus, cameras and radars have been applied extensively for the guidance of autonomous cars. The processing of the images leads to the computation of relevant environment features that can be used to guide the vehicle by means of visual servoing techniques (image-based visual servoing). Alternatively, the features can be used to compute the position/orientation of the vehicle and then apply position-based visual servoing. The stability of the visual control loop in the guidance of vehicles has been studied by several authors (see, for example, [16.32]). The main drawbacks of these methods are robustness to

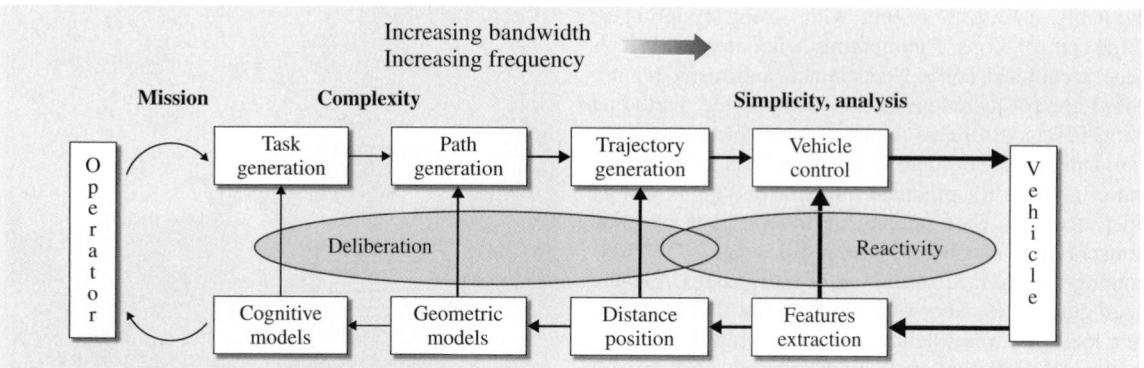

Fig. 16.10 Vehicle control and decision loops. The width of the arrows indicates the frequency of the loops and required bandwidth in communication. The inner loops (*right in the figure*) correspond to higher-frequency loops with higher bandwidth requirements, while the outer loops (*left in the figure*) correspond to lower-frequency loops with lower bandwidth requirements

illumination changes and real-time constraints. Laser-based environment perception techniques are also used in navigation [16.33], as shown by the 2005 DARPA Grand Challenge (DGC) [16.34]. However, even laser measurements have some drawbacks in outdoor environments; for example, dust clouds could be treated as transient obstacles or weeds and large rocks cannot be differentiated.

Compensation for such different environmental conditions plays an important role. The use of several sensors and the application of sensor data-fusion methods significantly improve robustness against changes in these conditions. Thus, for example, most autonomous vehicles in the DGC (see Fig. 16.11 for the 2007 Urban Grand Challenge) applied sensor data-fusion techniques for autonomous navigation. In some cases environment perception can substitute or complement GPS position-

ing, overcoming problems related to the visibility of satellites and degradation of the GPS signal.

The above-mentioned techniques to compute the position of the vehicle with respect to the environment can also be applied to generate trajectories in these environments. Trajectory generation methods can eventually consider the kinematics and dynamic constraints of the vehicles in order to obtain trajectories that can be realistically executed with the vehicle control system described in the previous section. The relevance of these techniques depends on the characteristics of the vehicles. Thus, for example, they are very relevant in the navigation of ground wheeled vehicles with conventional car-like locomotion systems or in fixed-wing airplanes, but could be ignored in omnidirectional ground vehicles navigating at low speeds.

The computation of distances and positions of the vehicle with respect to the environment can also be used to obtain geometric models of the environment. Particularly mapping techniques can be applied. Moreover, many probabilistic simultaneous localization and mapping techniques (SLAM) have been proposed and successfully applied in robotics in the last decade [16.36]. Most implementations have been carried out by using lasers in two-dimensional (2-D) environments. However, the methods for the application of SLAM in 3-D environments are also promising.

The results of vehicle position and environment mapping can be used for the planning of vehicle motion. The planning problem consists of the computation of a path for the vehicle from a starting configuration (position/orientation) to a goal position/orientation configuration, avoiding obstacles and minimizing a cost

Fig. 16.11 Winners of the Urban Grand Challenge, November 2007 [16.35]

index usually related to the length of the path [16.37]. Today, many different techniques for automatic path planning can be applied. In the basic problem a geometric model of the vehicle is assumed and the model of the environment is assumed to be completely known (map known) and static, without other vehicles and moving obstacles. Many path planning methods are solved in the configuration space C, defined by the configuration variables $q \in \mathbb{R}^n$ that completely specify the position and orientation of the vehicle. Then, given $q_{\text{start}}, q_{\text{goal}} \in C$, the problem is to find a sequence of configurations in the obstacle-free configuration space $q_i \in C_{\text{free}}$ connecting q_{start} and q_{goal}.

The problem can be solved by searching in the discretized space. Well-known methods are visibility graphs and Voronoi diagrams. In these methods the connectivity of the free space is represented by means of a network of one-dimensional (1-D) curves. These methods have been extensively applied in 2-D environments. The consideration of 3-D models adds significant complexity for execution in real time. Another well-known strategy consists of searching in an adjacency graph of the free-space cells, obtained by discretization of the environment model into occupancy cells. These methods need the implementation of graph searching algorithms, such as A*, to find a solution and they are usually quite time consuming. The application of multiresolution techniques greatly improves the computational efficiency of these methods. On the other hand, in recent years, randomized methods and particularly so-called rapidly exploring random trees (RRT) [16.38], have been used to explore the free space, obtaining good results. This method makes it possible to explore high-dimensional configuration spaces, and even to include different constraints. Several of these techniques have been applied and extensively tested in the DGC 2005 competition.

There are also methods based on the optimization of potential functions attracting the vehicle to the goal (global minimum) by considering at the same time the effect of repulsive forces exerted by the obstacles. The obvious difficulty of these latter methods, initially proposed for real-time local collision avoidance [16.39], is the existence of local minima. The potential-based methods have also been applied for motion planning, eventually combined with other planning methods such as the space-cell decomposition mentioned above. Different potential functions have been proposed to avoid the local minima problem in these methods [16.40].

The extensions of the basic motion planning problem described above include the consideration of the nonholonomic and dynamic constraints of the vehicle [16.41]. Other extensions include the uncertainties in the models of the vehicle and the environment, as well as the motion of other vehicles and obstacles in the environment.

The stability of reactive navigation is studied in [16.42], where Lyapunov techniques, input/output stability (conicity criterion), and frequency response methods are applied to study the stability of the navigation of an autonomous ground vehicle by using ultrasonic sensors. The stability is related to the parameters of the reactive navigation such as the sensor range and the velocity. The influence of the time delay, due to communication and computation, on the stability of the reactive navigation is also considered. The analysis is based on the definition of the perception function $p = \psi(d, \theta)$, where d and θ are, respectively, the distance and angle at which an obstacle is detected. The values of p are provided to the closed-loop controller. Then, the feedback controller $u = \varphi(p)$ is applied to the vehicle with model given by (16.1).

Planning under uncertainty has been a research topic for many years. In the classical planning methods the world is assumed to be deterministic, and the state observable. The uncertainty can be taken into account by considering stochastic Markov decision processes (MDPs) with observable states, which leads to stochas-

Fig. 16.12 The Aurora robot spraying a greenhouse

tic accurate models. If the state is assumed to be only partially observable, then partially observable Markov decision processes (POMDPs) can be used to consider stochastic inaccurate models.

The outer loop (the upper level in the control architecture) in Fig. 16.10 is based on the consideration of cognitive models of the environment obtained from the geometric representations, knowledge extracted from the sensorial information, i. e., identification of particular objects in the environment (i. e., traffic signals), and previous human knowledge on the existing relations between these objects and the vehicle navigation.

A key architectural issue is the appropriated combination of reactivity and planning. This is related to the interaction between the different levels or control loops in Fig. 16.10 and has a significant impact on the methods to be applied. Thus, some control architectures are based on the application of very simple motion strategies, without considering uncertainties or mobile objects in the environment, and providing reactivity based on real-time sensing of the environment. Reac-

tive techniques can be implemented in behavior-based control architectures to navigate in natural environments without models or with a minimal model of the environment. Figure 16.12 shows the Aurora mobile robot [16.43] which uses a behavior-based architecture to navigate in greenhouses using ultrasonic sensors. Other architectures are based on dynamic planning by incorporating environment information in real time and producing a new plan that reacts appropriately to the new information. In the architecture presented in [16.42] planning techniques based on the kinematics model of the vehicle are used to generate parking maneuvers for articulated vehicles (Fig. 16.8).

One of the main general trends is the integration of control and perception components into embedded systems that can be networked using wired or wireless technologies, leading to *cooperating objects*, with sensing and/or actuation capabilities based on sensor fusion methods that allow full interaction with the environment. Open challenges are related to the development of tools for the analysis and design of these systems.

16.5 Human Interaction

In practice mobility automation requires the intervention of humans at a certain level. The key point is the level of interaction. Thus, with regards to Fig. 16.10, the human can be provided with information from the cognitive models, which encode expertise from other operators, and use this information to decompose a mission into tasks to be planned by the task generator module in the Fig. 16.10.

However, if the task planner does not exist, the human operator can interact with the second outer loop. In this case the human operator can use the map of the environment displayed on a suitable display to generate a sequence of waypoints for the vehicle using an appropriate interface to specify these waypoints by means of a joystick or a simple mouse. The human operator can also generate in the next inner loop a suitable trajectory assisted by computer tools to visualize in an appropriate way the distances to the surrounding obstacles and check the suitability of these trajectories to be executed by the particular vehicle being commanded. Going to the right in the control loops of Fig. 16.10, the operator could directly provide commands to the vehicle control loop by observing significant environment features in the images provided by the onboard camera or cameras.

The above-mentioned interactions involve hardware and software technologies to provide appropriate senso-

rial feedback (visual, audio . . .) to the human pilots and generate actions at different levels from direct guidance to waypoint and task specification.

At this point it is necessary to distinguish between human interventions onboard the vehicle and operation in a remote teleoperation station or from suitable remote devices such as personal digital assistants (PDAs) or even mobile phones involving the communication system, as shown in Fig. 16.13. The first approach can be considered as a compromise solution between vehicle full driving automation, which removes the driver from the control loop, to assisted driving to improve efficiency and reduce accidents, as mentioned in the Introduction. Integration of automatic functions in conventional cars has been a trend in the last years. Autonomous parking of conventional vehicles is an example of this trend. On the other hand, the development of mixed autonomous/manual driving cars seems a suitable approach for the gradual integration of autonomous vehicles on regular roads. Furthermore, the automation of functions in aircraft navigation is also well known.

In the following, the remote teleoperation of vehicles is considered. Figure 16.13 illustrates teleoperation schemes.

The low-level classical teleoperation approach consists of the presentation to the remote teleoperator of images from a camera or cameras mounted on the front of the vehicle. Then, the vehicle is guided manually using joysticks, pedals or similar interfaces to the ones existing in the driving position onboard the vehicles. This approach has many problems; for example, the images may be degraded due to bandwidth instability, leading to poor spatial resolution and variable update rates, degrading the perception of the motion. Furthermore, an important problem is the presence of delays in both the images and sensor data sent to the operator and the operator commands sent back to the vehicle. These delays may generate instability of the teleoperation control loop.

Various technologies have been proposed to overcome these problems. Thus, instead of showing in the display the data and images from the vehicles, it is possible to process these data to extract relevant features to be displayed. Note that this approach can be included in the inner (faster) control loop of Fig. 16.10. The computation of distances in the second loop could be relevant too. Moreover, the displaying of geometric representations of the environment around the vehicle (see also Fig. 16.13) combined with the images (augmented-reality technologies) could also be very useful.

A classical method to reduce the harmful effect of delays in the transmission of images is the use of predictive displays with a synthetic graphic of the vehicle obtained by means of a simulation model. The graphic is overlaid on the real delayed images from the vehicle. This representation leads the operator to perceive in advance the effect of his commands, which can be used to compensate for delays. This can be easily combined with the augmented reality mentioned above. Thus, real-world imagery is embedded within a display of computer-generated landmarks or objects representing the same scene. The computer-generated component of a display can be updated immediately in response to control inputs from the human operator, providing rapid feedback to the operator. If a model of the environment is known, it can be stored in databases and rendered based, for example, on the current GPS position of the vehicle.

Obviously the navigation conditions may have a significant effect on the operators. Thus, it has been pointed out that UAV operators may not modify their visual scanning methods to compensate for the non-recreated multisensory cues. In order to improve the perception of the operators, haptic and multimodal interfaces (e.g., tactile and auditory) have been proposed. Multimodal interfaces may be used not just to compensate for the teleoperator's sensory environment, but more generally to reduce cognitive-perceptual workload levels. Thus, for example, [16.44] have found that audio and tactile messages can improve many aspects of flight control and overall awareness of the situation in UAV teleoperation.

The teleoperation methods presented above greatly depend on communication between the vehicle and the teleoperation station. The development of mobile communication in the last decade has changed the sit-

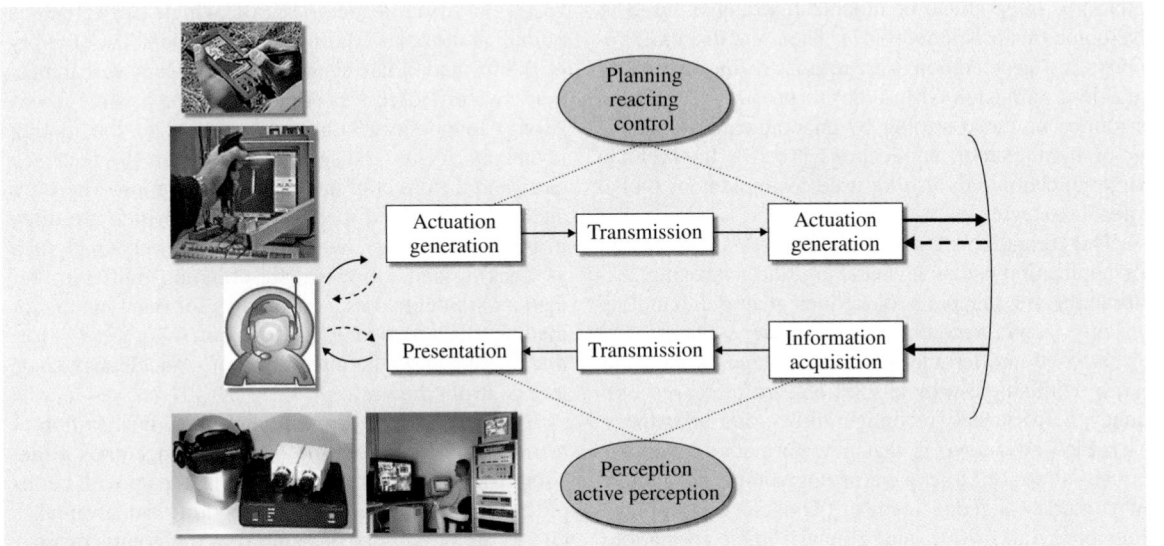

Fig. 16.13 Human interaction loops

uation when compared with the existing technologies when the first autonomous vehicles where developed in the 1980s. Obviously, the communication technologies to be applied greatly depend of the level of intervention of the human teleoperator in the architecture of Fig. 16.13. Thus, high-bandwidth communication is essential at the lower- and high-frequency loops, where vehicle onboard autonomy is low and generation of teleoperation commands depends on observation of the images and information from other sensors onboard the vehicle. However, if the vehicle has onboard autonomy, the communication with the user does not require high bandwidth. Thus, for example, GSM (global system for mobile communications) and GPRS (general packet radio service) have been used for communicat-

ing the vehicles with the users through their mobile phones and PDAs. Furthermore, Wi-Fi (IEEE 802.11) has been applied for communication with the vehicle at low velocity and short range. An emerging trend is the application of ad hoc mobile networks to take into account the particular mobility of vehicles.

The development of models of the loops in Fig. 16.13, as well as new analysis and design tools using these models, are important challenges to be addressed. These models should involve not only the vehicles, the control devices, and communication channels, but also suitable models of the human perception and action mechanisms, which typically require significant experimentation efforts to cope with the behavior of operators under different working conditions.

16.6 Multiple Mobile Systems

The automation of multiple vehicles offers many application possibilities. The interest in transportation is obvious. A basic configuration of multiple vehicles consists of a leader followed by vehicles in a single row. This is usually known as *platooning* [16.45]. The control of a platoon can be implemented by means of a local strategy, i.e., each vehicle is controlled from the unique data received from the vehicle at the front [16.46]. This approach relies mainly on the single-vehicle control problem considered above. The main drawback is that the regulation errors introduced by sensors noises grow from the first vehicle to the last one, leading to oscillations. Intervehicle communication can be used to overcome this problem [16.47]. Then, the distance, velocity, and acceleration with respect to the preceding vehicle are transmitted in order to predict the position and improve the controller by guaranteeing the stability of tight platoon applications [16.48]. Intervehicle communication can also be used to implement global control strategies.

The formation of multiple vehicles is also useful for applications such as searching and surveying, exploration and mapping, hazardous material handling systems, active reconfigurable sensing systems, and space-based interferometry. The advantages when comparing with single-vehicle solutions are increased efficiency, performance, reconfigurability, and robustness. An added advantage is that new formation members can be introduced to expand or upgrade the formation, or to replace a failed member. Thus several applications of aerial, marine, and ground vehicle formations have been proposed. In these formations, the members

of the group of vehicles should keep user-defined distances from other group members. The control problem consists of maintaining these user-defined distances. Formation control involves the design of distributed control laws with limited and disrupted communication, uncertainty, and imperfect or partial measurements. The most common approach is the leader-follower [16.49]. This approach has limitations when considering the reliability of the leaders and the lack of explicit feedback from the follower to the leader. Then, if the follower is perturbed by some disturbances, the formation cannot be maintained. There are also alternative approaches based on virtual leaders [16.50], which is a reference point that moves according to the mission. The stability of the formation has been studied by many researchers that have proposed robust controllers to provide insensitivity to possibly large uncertainties in the motion of nearby agents, transmission delays in the feedback path, and the effect of quantized information. There are also behavior-based methods [16.51], which are often inspired by biology, where formation behaviors such as flocking and following are common. Different behaviors are defined as control laws for reaching and/or maintaining a particular goal. An emerging trend in formation control is the integration of obstacle avoidance into control schemes.

Other approaches are based on the consideration of teams of robots describing different trajectories to accomplish tasks. Furthermore, having a team with multiple heterogeneous vehicles offers additional advantages due to the possibility of exploiting the complementarities of vehicles with different mobility attributes and also different sensors with different perception functionalities. The vehicles need to be coordinated in time (synchronization) to accomplish missions such as monitoring. Spatial coordination is required to ensure that each vehicle will be able to perform its plan safely and coherently, regarding the plans of the others. Assuming that multiple robots share the same world, a path should be computed for each one that avoids collisions with obstacles and with other robots. Some formulations are based on the extension of single-robot path planning concepts such as the configuration space. If there are nr robots and each robot has a configuration space $C^i, i = 1, \ldots, nr$, the state space is defined as the Cartesian product $X = C^1 \times C^2 \times \cdots \times C^{nr}$ and the obstacle region in X is

$$X_{\text{obs}} = \left(\bigcup_{i=1}^{nr} X_{\text{obs}}^i \right) \cup \left(\bigcup_{ij, i \neq j}^{nr} X_{\text{obs}}^{ij} \right), \tag{16.4}$$

where X_{obs}^i and X_{obs}^{ij} are the robot–obstacle and the robot–robot collision states. The problem is to find a continuous path in the free space from the initial state to the goal state, avoiding the obstacle region defined by (16.4). The classical planning algorithms for a single robot with multiple bodies [16.37] could be applied without adaptation in case of a centralized planning that takes into account all robots. The main concern, however, is that the dimension of the state space grows linearly in the number of robots. Complete algorithms require time that is at least exponential in dimension. Sampling-based algorithms are more likely to scale well in practice when there many robots, but the resulting dimension might still be too high. There are also decoupled path planning approaches such as the prioritized planning that considers one robot at a time according to a global priority.

On the other hand, cooperation is defined in the robotic literature as a *joint collaborative behavior that is directed toward some goal in which there is a common interest or reward*. According to [16.54], given some task specified by a designer, a multiple-robot system displays cooperative behavior if, due to some underlying mechanism, there is an increase in the total utility of the system. Cooperative perception can be defined as the task of creating and maintaining a consistent view of a world containing dynamic objects by a group of agents, each equipped with one or more sensors. Cooperative vision perception has become a relevant topic in the multirobot domain, mainly in structured environments [16.55, 56]. In [16.57] cooperative perception methods for multi-UAV system are proposed.

Fig. 16.14 Coordinated flights in the COMETS project [16.52]

These methods have been implemented in the architecture designed in the COMETS project (acronym of the project *real-time coordination and control of multiple heterogeneous unmanned aerial vehicles*) [16.58] (Fig. 16.14). Cooperative perception requires integration of the results of individual perception. Each robot extracts knowledge by applying individual perception techniques, and the overall cooperative perception is performed by merging the individual results. This approach requires knowing the relative position and orientation of the robots. If the GPS signal is not available, position estimation based on environment perception should be applied [16.59, 60]. The cooperation of mobile entities also involves the generation of appropriated motion of the involved entities. In [16.61] the coopera-

Fig. 16.15 Experiment of the CROMAT system for the cooperation of aerial and ground robots [16.53], http://grvc.us.es/cromat

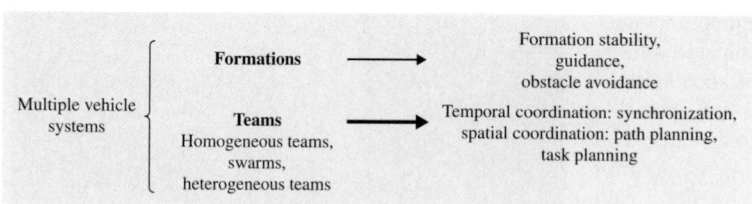

Fig. 16.16 Methods in multiple vehicle systems

tion is categorized into: swarm type, dealing with a large number of homogeneous robots, usually involving numerous repetitions of the same activity over a relatively large area; and intentional cooperation, usually requiring a smaller number of possibly heterogeneous robots (Fig. 16.15) performing several distinct tasks. In these systems the multirobot task allocation problem [16.62] is applied to maximize the efficiency of the team and ensure proper coordination among team members to allow them to complete their mission successfully. Recently, a very popular approach to multirobot task allocation has been the application of market-based negotiation rules by means of the contract net protocol [16.63, 64].

Figure 16.16 shows the different types of multiple-vehicle systems and the methods that are applied in each type.

Communication and networking also play an important role in the implementation of control systems for multiple unmanned vehicles. The star-shaped network configuration with all the vehicles linked to the control station with an unshared link only works well with small teams. When the number of vehicles grows it could be necessary to apply wireless heterogeneous networks with radio nodes mounted at fixed ground stations, on ground vehicles, and in UAVs, and the routing techniques allow any two nodes to communicate either directly or through an arbitrary number of other nodes which act as relays. Furthermore, when there is little or no infrastructure, networks could be formed in an ad hoc fashion and information exchanges occur only via the wireless networking equipment carried by the individual UAVs.

Finally, it should be noted that the wireless networking of teams of robots with sensors and actuators embedded in the infrastructure is a new research and development trend with many potential applications. The AWARE project (http://www.aware-project.net) is developing a new platform for the cooperation of autonomous aerial vehicles with ground wireless sensor–actuator networks. This platform will have self-deployment and self-configuration features for operation in sites without sensing and communication infrastructure.

16.7 Conclusions

Mobility and navigation have been very relevant topics in automation. Thus, automation of mobility plays an important role in factory automation. The automation of the transportation of people and goods in noncontrolled environments is more difficult and its complexity depends on the flexibility. This chapter has analyzed automation of mobility and navigation by focusing on autonomous vehicles. Then, vehicle motion control has been examined, and the main problems in navigation control and interaction of the vehicle with the environment were also studied. Moreover, taking into account that practical applications usually require some degree of human intervention, human interaction and related technologies were reviewed. The last part of the chapter was devoted to systems of multiple autonomous vehicles, including formations and fleets of homogeneous vehicles and also teaming of heterogeneous vehicles. The control and cooperation of these autonomous vehicles to accomplish tasks is an emerging trend that poses different challenges.

References

16.1   R. Marín, J. Garrido, J.L. Trillo, J. Sáez, J. Armesto: An industrial automated warehouse based on overhead trolleys, MCPL'97 IFAC/IFIP Conf. Manag. Control Prod. Logist. (Campinas, 1997) pp. 137–142

16.2 C.E. Thorpe (Ed.): *Vision and Navigation: The Carnegie Mellon Navlab* (Kluwer, Boston 1990)

16.3 M. Parent, A. de La Fortelle: Cybercars: past, present and future of the technology, Proc. ITS World Congr. (2005)

16.4 R. Horowitz, P. Varaiya: Control design of an automated highway system, Proc. IEEE **88**(7), 913–925 (2000)

16.5 UAV Forum: http://www.uavforum.com/ (last accessed March 5, 2009)

16.6 J. Moraleda, A. Ollero, M. Orte: A robotic system for internal inspection of water pipelines, IEEE Robot. Autom. Mag. **6**(3), 30–41 (1999)

16.7 H.M. Kim, J. Dickerson, B. Kosko: Fuzzy throttle and brake control for platoons of smart cars, Fuzzy Sets Syst. **84**, 209–234 (1996)

16.8 R.W. Brockett: Asymptotic stability and feedback stabilization. In: *Differential Geometric Control Theory*, ed. by R.S. Millman, R.W. Brockett, H.H. Sussmann (Birkhauser, Boston 1983)

16.9 C.Y. Chan, H.S. Tan: Feasibility analysis of steering control as a driver-assistance function in collision situations, IEEE Trans. Intell. Transp. Syst. **2**(1), 1–9 (2001)

16.10 J.H. Hahn, R. Rajamani, L. Alexander: GPS-based real-time identification of tire–road friction coefficient, IEEE Trans. Control Syst. Technol. **10**(3), 331–343 (2002)

16.11 B. Samadi, R. Kazemi, K.Y. Nikravesh, M. Kabganian: Real-time estimation of vehicle state and tire–road friction forces, Proc. Am. Control Conf. (Arlington 2001) pp. 3318–3323

16.12 J. Huang, J. Ahmed, A. Kojic, J.P. Hathout: Control oriented modeling for enhanced yaw stability and vehicle steerability, Proc. Am. Control Conf. (Boston 2004) pp. 3405–3410

16.13 A. Kamga, A. Rachid: Speed, steering angle and path tracking controls for a tricycle robot, Proc. IEEE Int. Symp. Computer-Aided Control Syst. Des. (Dearborn 1996) pp. 56–61

16.14 C. deWit, B. Siciliano, G. Bastin: *Theory of Robot Control* (Springer, Berlin Heidelberg 1997)

16.15 A. Ollero: *Robótica. Manipuladores y Robots Móviles* (Marcombo, Spain 2001), in Spanish

16.16 J. Wit, C.D. Crane, D. Armstrong: Autonomous ground vehicle path tracking, J. Robot. Syst. **21**(8), 439–449 (2004)

16.17 A. Rodríguez-Castaño, A. Ollero, B.M. Vinagre, Y.Q. Chen: Setup of a spatial lookahead path tracking controller, Proc. 16th IFAC World Congr. (Prague 2005)

16.18 T. Hellström, T. Johansson, O. Ringdahl: *Development of an autonomous forest machine for path tracking*, Springer Tracts Adv. Robot., Vol. 25 (Springer, Berlin Heidelberg 2006) pp. 603–614

16.19 G. Heredia, A. Ollero: Stability of autonomous vehicle path tracking with pure delays in the control loop, Adv. Robot. **21**(1), 23–50 (2007)

16.20 DARPA Grand Challenge: Special issue, J. Field Robot. **23**(8/9), 461–835 (2006)

16.21 J.Y. Wang, M. Tomizuka: Robust H∞ lateral control for heavy-duty vehicles in automated highway systems, Proc. Am. Control Conf. (San Diego 1999) pp. 3671–3675

16.22 G.H. Elkaim, M. O'Connor, T. Bell, B. Parkinson: System identification and robust control of farm vehicles using CDGPS, Proc. ION GPS-97 (Kansas City 1997) pp. 1415–1424

16.23 A. González-Cantos, A. Ollero: Backing-up maneuvers of autonomous tractor-trailer vehicles using the qualitative theory of nonlinear dynamical systems, Int. J. Robot. Res. **28**(1), 49–65 (2009)

16.24 A. Astolfi, P. Bolzern, A. Locatelli: Path-tracking of a tractor-trailer vehicle along rectilinear and circular paths: a Lyapunov-based approach, IEEE Trans. Robot. Autom. **20**(1), 154–160 (2004)

16.25 A. Ollero, L. Merino: Control and perception techniques for aerial robotics, Annu. Rev. Control **28**, 167–178 (2004)

16.26 O. Amidi, T. Kanade, K. Fujita: A visual odometer for autonomous helicopter flight, Robot. Auton. Syst. **28**, 185–193 (1999)

16.27 M. Bejar, A. Ollero, F. Cuesta: Modeling and control of autonomous helicopters. In: *Advances in Control Theory and Application*, Lect. Notes Control Inf. Sci., Vol. 353, ed. by C. Bonivento, A. Isidori, L. Marconi, C. Rossi (Springer, Berlin Heidelberg 2007) pp. 1–27

16.28 AWARE Project: http://www.aware-project.net (last accessed March 5, 2009)

16.29 A. Ollero, A. García-Cerezo, J.L. Martínez, A. Mandow: Fuzzy tracking methods for mobile robots. In: *Applications of Fuzzy Logic: Towards High Machine Intelligence Quotient Systems*, Vol. 9, ed. by M. Jamshidi, L. Zadeh, A. Titli, S. Boverie, (Prentice Hall, Upper Saddle River 1997) pp. 347–364, Chap. 17

16.30 G. Buskey, G. Wyeth, J. Roberts: Autonomous helicopter hover using an artificial neural network, Proc. IEEE Int. Conf. Robot. Autom. (2001) pp. 1635–1640

16.31 A. Ollero, A. Rodríguez-Castaño, G. Heredia: Analysis of a GPS-based fuzzy supervised path tracking system for large unmanned vehicles, Proc. 4th IFAC Int. Symp. Intell. Compon. Instrum. Control Appl. (SICICA) (Buenos Aires 2000) pp. 141–146

16.32 F. Conticelli, D. Prattichizzo, F. Guidi, A. Bicchi: Vision-based dynamic estimation and set-point stabilization of nonholonomic vehicles, Proc. 2000 IEEE Int. Conf. Robot. Autom. (San Francisco 2000) pp. 2771–2776

16.33 J. González, A. Stenz, A. Ollero: A mobile robot iconic position estimator using a radial laser scanner, J. Intell. Robot. Syst. **13**, 161–179 (1995)

16.34 M. Buehler, K. Iaguemma, S. Singh: *The 2005 DARPA Grand Challenge*, Springer Tracts Adv. Robot., Vol. 36 (Springer, Berlin Heidelberg 2007)

16.35 DARPA Urban Challenge: http://www.darpa. mil/grandchallenge/images/photos/11_4_07/D2X_ 1328.jpg (last accessed March 5, 2009)

16.36 S. Thrun, W. Burgard, D. Fox: *Probabilistic Robotics, Intelligent Robotics and Autonomous Agents* (MIT Press, Cambridge 2005)

16.37 R.C. Latombe: *Robot Motion Planning* (Kluwer, Boston 1991)

16.38 S.M. LaValle: *Rapidly-exploring random trees: A new tool for path planning TR 98-11* (Iowa Univ., Iowa 1998)

16.39 O. Khatib: Real-time obstacle avoidance for manipulators and mobile robots, Int. J. Robot. Res. **5**(1), 90–98 (1986)

16.40 S.A. Masoud, A.A. Masoud: Motion planning in the presence of directional and regional avoidance constraints using nonlinear, anisotropic, harmonic potential fields: a physical metaphor, IEEE Trans. Syst. Man Cybern. Part A, **32**(6), 705–723 (2002)

16.41 V.F. Muñoz, A. Ollero, M. Prado, A. Simón: Mobile robot trajectory planning with dynamic and kinematic constraints, Proc. IEEE Int. Conf. Robot. Autom., San Diego (1994) pp. 2802–2807

16.42 F. Cuesta, A. Ollero: *Intelligent mobile robot navigation*, Springer Tracts Adv. Robot., Vol. 16 (Springer, Berlin Heidelberg 2005)

16.43 A. Mandow, J. Gomez de Gabriel, J.L. Martinez, V.F. Muñoz, A. Ollero, A. García-Cerezo: The autonomous mobile robot aurora for greenhouse operation, IEEE Robot. Autom. Mag. **3**(4), 18–28 (1996)

16.44 G.L. Calhoun, M.H. Draper, H.A. Ruff, J.V. Fontejon: Utility of a tactile display for cueing faults, Proc. Hum. Factors Ergon. Soc. 46th Annu. Meet. (2002) pp. 2144–2148

16.45 P. Daviet, M. Parent: Platooning for small public urban vehicles, 4th Int. Symp. Exp. Robot. (ISER'95) (Stanford 1995) pp. 345–354

16.46 J. Bom, B. Thuilot, F. Marmoiton, P. Martinet: Nonlinear control for urban vehicles platooning, relying upon a unique kinematic GPS, 22nd Int. Conf. Robot. Autom. (ICRA'05) (Barcelona 2005) pp. 4149–4154

16.47 Y. Zhang, E.B. Kosmatopoulos, P.A. Ioannou, C.C. Chien: Autonomous intelligent cruise control using front and back information for tight vehicle following maneuvers, IEEE Trans. Veh. Technol. **48**(1), 319–328 (1999)

16.48 T.S. No, K.-T. Chong, D.-H. Roh: A Lyapunov function approach to longitudinal control of vehicles in a platoon, IEEE Trans. Veh. Technol. **50**(1), 116–124 (2001)

16.49 J.P. Desai, J.P. Ostrowski, V. Kumar: Modeling and control of formations of nonholonomic mobile robots, IEEE Trans. Robot. Autom. **17**(6), 905–908 (2001)

16.50 M. Egerstedt, X. Hu, A. Stotsky: Control of mobile platforms using a virtual vehicle approach, IEEE Trans. Autom. Control **46**, 1777–1782 (2001)

16.51 T. Balch, R.C. Arkin: Behavior-based formation control for multi-robot teams, IEEE Trans. Robot. Autom. **14**, 926–939 (1998)

16.52 A. Ollero, I. Maza: *Multiple Heterogeneous Aerial Vehicles*, Springer Tracts Adv. Robot., Vol. 37 (Springer, Berlin Heidelberg 2007)

16.53 I. Maza, A. Viguria, A. Ollero: Aerial and ground robots networked with the environment, Proc. Workshop Netw. Robot Syst. IEEE Int. Conf. Robot. Autom. (2005) pp. 1–10

16.54 Y.U. Cao, A.S. Fukunaga, A. Kahng: Cooperative mobile robotics: Antecedents and directions, Auton. Robots **4**(1), 7–27 (1997)

16.55 T. Schmitt, R. Hanek, M. Beetz, S. Buck, B. Radig: Cooperative probabilistic state estimation for vision-based autonomous mobile robots, IEEE Trans. Robot. Autom. **18**(5), 670–684 (2002)

16.56 S. Thrun: A probabilistic online mapping algorithm for teams of mobile robots, Int. J. Rob. Res. **20**(5), 335–363 (2001)

16.57 L. Merino, F. Caballero, J.R. Martínez-de Dios, J. Ferruz, A. Ollero: A cooperative perception system for multiple UAVs: application to automatic detection of forest fires, J. Field Robot. **23**(3), 165–184 (2006)

16.58 A. Ollero, S. Lacroix, L. Merino, J. Gancet, J. Wiklund, V. Remuss, I.V. Perez, L.G. Gutiérrez, D.X. Viegas, M.A. González, A. Mallet, R. Alami, R. Chatila, G. Hommel, F.J. Colmenero, B.C. Arrue, J. Ferruz, J.R. Martinez-de Dios, F. Caballero: Multiple eyes in the skies, IEEE Robot. Autom. **12**(2), 46–57 (2005)

16.59 K. Konolige, D. Fox, B. Limketkai, J. Ko, B. Stewart: Map merging for distributed robot navigation, IEEE Int. Conf. Intell. Robot. Syst. (2003) pp. 212–217

16.60 L. Merino, F. Caballero, J. Wiklund, A. Moe, J.R. Martínez-de Dios, P.-E. Forssen, K. Nordberg, A. Ollero: Vision-based multi-UAV position estimation, Robot. Autom. Mag. **13**(3), 53–62 (2006)

16.61 L.E. Parker: Alliance: An architecture for fault-tolerant multi-robot cooperation, IEEE Trans. Robot. Autom. **14**(2), 220–240 (1998)

16.62 B.P. Gerkey, M.J. Mataric: A formal analysis and taxonomy of task allocation in multi-robot systems, Int. J. Robot. Res. **23**(9), 939–954 (2004)

16.63 S.C. Botelho, R. Alami: M+: a scheme for multi-robot cooperation through negotiated task allocation and achievement, Proc. IEEE Int. Conf. Robot. Autom. (Detroit 1999)

16.64 B. Gerkey, M. Mataric: Sold: Auction methods for multi-robot coordination, IEEE Trans. Robot. Autom. **18**(5), 758–768 (2002)

17. The Human Role in Automation

Daniel W. Repperger, Chandler A. Phillips

A survey of the history of how humans have interacted with automation is presented. Starting with the early introduction of automation into the Industrial Revolution to the modern applications that occur in unmanned air vehicle systems, many issues are brought to light. Levels of automation are quantified and a preliminary list delineating what tasks humans can perform better than machines is presented. A number of application areas are surveyed that have or are currently dealing with positive and negative issues as humans interact with machines. The application areas where humans specifically interact with automation include agriculture, communications systems, inspection systems, manufacturing, medical and diagnostic applications, robotics, and teaching. The benefits and disadvantages of how humans interact with modern automation systems are presented in a trade-off space discussion. The modern problems relating to how humans have to deal with automation include trust, social acceptance, loss of authority, safety concerns, adaptivity of automation leading to unplanned unexpectancy, cost advantages, and possible performance gained.

The modern use of the term *automation* can be traced to a 1952 *Scientific American* article; today it is widely employed to define the interaction of humans with machines. Automation (machines) may be electrical, mechanical, require interaction with computers, involve informatics variables, or possible relate to parameters in the environment. As noted [17.1], the first actual automation was the mechanization of manual labor during the Industrial Revolution. As machines became increasingly useful in reducing the drudgery and dan-

ger of manual labor tasks, questions begin to arise concerning how best to proportion tasks between humans and machines. Present applications still address the delineation of tasks between humans and machines [17.2], where, e.g., in chemistry and laboratory tasks, the rule of thumb is to use automation to eliminate much of the *3-D tasks* (dull, dirty, and dangerous). In an effort to be more quantitative in the allocation of work and responsibility between humans and machines, *Fitts* [17.3] proposed a list to identify tasks

Table 17.1 Fitts' list [17.3]

Tasks humans are better at	Tasks machines are better at
Detecting small amounts of visual, auditory, or chemical energy	Responding quickly to control signals
Perceiving patterns of light or sound	Applying great force smoothly and precisely
Improvising and using flexible procedures	Storing information briefly, erasing it completely
Storing information for long periods of time and recalling appropriate parts	Reasoning deductively
Reasoning inductively	
Exercising judgment	

that are better performed by humans or machines (Table 17.1).

This list initially raised some concern that humans and machines were being considered equivalent in some sense and that human could easily be replaced by a mechanical counterpart. However, the important point that Fitts raised was that we should consider some proper allocation of tasks between humans and machines (functional allocation [17.4, 5]) which may be very specific to the skill sets of the human and those the machine may possess. This task-sharing concept has been discussed by numerous authors [17.6]. Ideas of this type have nowadays been generalized into how humans interact with computers, the Internet, and a host of other modern apparatus. It should be clarified, however, that present-day thinking has now moved away from this early list concept [17.7].

17.1 Some Basics of Human Interaction with Automation

In an attempt to be more objective in delineating the interaction of humans with machines, *Parasuraman* et al. [17.8] defined a simple four-stage model of human information processing interacting with computers with the various levels of automation possible delineated in Table 17.2. Note how the various degrees of auto- mation affect decision and action selection. This list differs from other lists generated, e.g., using the concept of supervisory control [17.9, p. 26] or on a scale of *degrees of automation* [17.1, p. 62], but are closely related. As the human gradually allocates more work and responsibility to the machine, the human's role then

Table 17.2 Levels of automation of decision and action selection by the computer

Level	
High = 10	The computer decides everything, acts autonomously, ignoring the human
9	Informs the human only if the computer decides to
8	Informs the human only if asked
7	Executes automatically, then necessarily informs the human
6	Allows the human a restricted time to veto before automatic execution
5	Executes the suggestion if the human approves
4	Suggests one alternative
3	Narrows the selection down to a few alternatives
2	The computer offers a complete set of decision/action alternatives
Low = 1	The computer offers no help; the human must take all actions and make decisions

becomes as the supervisor of the mechanical process. Thus the level of automation selected in Table 17.2 also defines the roles and duties of the human acting in the position of supervisor. With these basics in mind, it is appropriate to examine a brief sample of modern application areas involving humans dealing with automation. After these applications are discussed, the current and most pertinent issues concerning how humans interact with automation will be brought to light.

17.2 Various Application Areas

Besides the early work by Fitts, automation with humans was also viewed as a topic of concern in the automatic control literature. In the early 1960s, *Grabbe* et al. [17.10] viewed the human operator as a component of an electrical servomechanism system. Many advantages were discovered at that time in terms of replacing some human function via automated means; for example, a machine does not have the same temperature requirements as humans, performance advantages may result (improved speed, strength, information processing, power, etc.), the economics of operation are significantly different, fatigue is not an issue, and the accuracy and repeatability of a response may have reduced variability [17.1, p. 163]. In more recent applications (for example, [17.11]) the issues of automation and humans extend into the realm of controlling multiple unmanned air vehicle systems. Such complex (unmanned) systems have the additional advantage that the aircraft does not need a life-support system (absence of oxygen supply, temperature, pressure control or even the requirement for a transparent windshield) if no humans are onboard. Hence the overall aircraft has lower weight, less expense, and improved reliability. Also there are political advantages in the situation of the aircraft being shot down. In this case, there would not be people involved in possible hostage situations. The mitigation of the political cost nowadays is so important that modern military systems see significant advantages in becoming increasingly autonomous (lacking a human onboard).

Concurrent with military applications is the desire to study automation in air-traffic control and issues in cockpit automation [17.12], which have been topics of wide interest [17.13–15]. In air-traffic control, software can help predict an aircraft's future position, including wind data, and help reduce near collisions. This form of predictive assistance has to be accepted by the air-traffic controller and should have low levels of uncertainty. *Billings* [17.13] lists seven principles of human-centered automation in the application domain of air-traffic control. The human-centered (user-centered) design is widely popularized as the proper approach to integrate humans with machines [17.14, 16, 17]. Automation also extends to the office and other workspace situations [17.18] and to every venue where people have to perform jobs.

Modern applications are also implicitly related to the revolution in information technology (IT) [17.19] noting that Fitts' list was preceded by Watson's (IBM's founder in the 1930s) concept that *Machines should work. People should think.* In the early stages of IT, *Licklider* envisioned (from the iron, mainframe computer) the potential for *man–computer symbiosis* [17.20], interacting directly with the brain. IT concepts also prevail in the military. *Rouse* and *Boff* [17.21] equate military success via the IT analogies whereas bits and bytes have become strategically equated with bombs and bullets and advances in networks and communications technologies which have dramatically changed the nature of conflict. A brief list of some current applications are now reviewed to give a flavor of some modern areas that are presently wrestling with future issues of human interaction with automation.

17.2.1 Agriculture Applications

Agriculture applications have been ubiquitous for hundreds of years as humans have interacted with various mechanical devices to improve the production of food [17.22, 23]. The advent of modern farm equipment has been fundamental to reducing some of the drudgery in this occupation. The term *shared control* occurs in situations where the display of the level of automation is rendered [17.9]. In related fields such as fish farming (aquaculture), there are mechanized devices that significantly improve production efficiency. Almost completely automated systems that segment fish, measure them, and process them for the food industry have been described [17.24]. In all cases examples prevail on humans dealing with various levels of automation to improve the quality and quantity of agriculture goods produced. Chapter 63 discusses in more detail automation in agriculture.

17.2.2 Communications Applications

The cellphone and other mobile remote devices have significantly changed how people deal with communications in today's world. The concept of *Bluetooth* [17.25] explores an important means of obviating some of the problems induced when humans have to deal with these communication devices. The Bluetooth idea is that the system will recognize the user when in reasonably close proximity and readjust its settings and parameters so as to yield seamless integration to the specific human operator in question. This helps mitigate the human–automation interaction problems by programming the device to be tailored to the user's specifications. Other mobile devices include remote controls that are associated with all types of new household and other communication devices [17.26]. Chapter 13 discusses in more detail communication in automation including networking and wireless.

17.2.3 Inspection Systems Applications

In [17.27], they discuss 100% automated inspection systems rather than having humans sample a process, e.g., in an assembly-line task. Prevailing thinking is that humans still outperform machines in most attribute-inspection tasks. In the cited evaluation, three types of inspection systems were considered, with various levels of automation and interaction with the human operator. For vision tasks [17.28] pattern irregularity is key to identification of a vector of features that may be untoward in the inspection process.

17.2.4 Manufacturing Applications

Applications in automation are well known to be complex [17.29], such as in factories where production and manufacturing provide a venue to study these issues. In [17.30] a human-centered approach is presented for several new technologies. Some of the negative issues of the use of automation discussed are increased need for worker training, concerns of reliability, maintenance, and upgrades of software issues, etc. In [17.31], manufacturing issues are discussed within the concept of the Internet and how distributed manufacturing can be controlled and managed via this paradigm. More and more modern systems are viewed within this framework of a machine with which the human has to deal by interfacing via a computer terminal connected to a network involving a number of distributed users. Chapters 49,

50, 51 presents different aspects of manufacturing automation.

17.2.5 Medical and Diagnostic Applications

Automation in medicine is pervasive. In the area of anesthesiology, it is analogous to piloting a commercial aircraft, ("hours of boredom interspersed by moments of terror" [17.32]). Medical applications (both treatment and diagnosis) now abound where the care-giver may have to be remote from the actual site [17.33]. An important example occurs in modern robotic heart surgery where it is now only required to make two small incisions in the patient's chest for the robotic end-effectors. This minimally invasive insult to the patient results in reduced recovery time of less than 1 week compared with typically 7 weeks for open heart surgery without using the robotic system. As noted by the surgeon, the greatest advantage gained is [17.34]:

> Without the robot, we must make a large chest incision. The only practical reason for this action is because of the 'size of the surgeon's hands'. However, using the robot now obviates the need for the large chest incision.

The accompanying reduction of medical expenses, decreased risk of infection, and faster recovery time are important advantages gained. The automation in this case is the robotic system. Again, as with Fitts' list, certain tasks of the surgery should be delegated to the robot, yet the other critically (medically) important tasks must be under the control and responsibility of the doctor. From a diagnostics perspective, the concept of remote diagnostic methods are explored [17.35]. As in the medical application, the operator (supervisor) must have an improved sense of presence about an environment remote from his immediate viewpoint and has to deal with a number of reduced control actions. Also, it is necessary to attempt to monitor and remotely diagnose situations when full information may not be typically available. Thus automation may have the disadvantage of limiting the quality of information received by the operator. Chapters 77, 78, 80, 82 provide more insights into automation in medical and healthcare systems.

17.2.6 Robotic Applications

Robotic devices, when they interact with humans, provide a rich venue involving problems of safety, task delegation, authority, and a host of other issues. In [17.36] an application is discussed which stresses the concept of *learning from humans*. This involves teaching the robotic device certain motion trajectories that emulate successful applications gleaned from humans performing similar tasks. In robotic training, it is common to use these *teach pendant* paradigms, in which the robotic path trajectory is stored in a computer after having a human move the end-effector through an appropriate motion field environment. In [17.37], the term *biomimicking motion* is commonly employed for service robots that directly interact with humans. Such assistive aids are commonly used, for example, in a hospital setting, where a robotic helper will facilitate transfer of patients, removing this task from nurses or other care-givers who are normally burdened with this responsibility. Chapter 21 provides an insightful discussion on industrial robots.

17.2.7 Teaching Applications

In recent years, there has been an explosive growth of teaching at universities involving long-distance learning classes. Many colleges and professional organizations now offer online courses with the advantage to the student of having the freedom to take classes at any time during the day, mitigating conflicts, travel, and other costs [17.38]. The subject areas abound, including pharmacy [17.39], software engineering [17.40], and for students from industry in various fields [17.41]. The problem of humans interacting with automation that occurs is when the student has to take a class and deal directly with a computer display and a possibly *not so user-friendly* web-based system. Having the ability to interrelate (student with professor) has now changed and there is a tradeoff space that occurs between the convenience of not having to attend class versus the loss of ability to fully interact, such as in a classroom setting. In [17.42], a variety of multimedia methods are examined to understand and help obviate the loss of full interactive ability that occurs in the classroom. The issue pertinent to this example is that taking the course online (a form of automation) has numerous advantages but incurs a cost of introducing certain constraints into the human–computer interaction.

These examples represent a small sample of present interactions of humans with automation. Derived from these and other applications, projections for some present and future issues that are currently of concern will be discussed in further detail in the next section.

17.3 Modern Key Issues to Consider as Humans Interact with Automation

17.3.1 Trust in Automation

Historically, as automation became more prevalent, concern was initially raised about the changing roles of human operators and automation. *Muir* [17.43] performed a literature review on trust in automated systems since there were alternative theories of trust at that time. More modern thinking clarifies these issues via a definition [17.44]:

> *Automation is often problematic because people fail to rely upon it appropriately. Because people respond to technology socially, trust influences reliance on automation. In particular, trust guides reliance when complexity and unanticipated situations make a complete understanding of the automation impractical. People tend to rely on automation they trust and tend to reject automation they do not.*

Taking levels of trust by pilots as an example, it was found [17.45] that, when trust in the automated system is too low and an alarm is presented, pilots spend extra time verifying the problem or will ignore the alarm entirely. These monitoring problems are found in systems with a high propensity for false alarms, which leads to a reduced level of trust in the automation [17.46]. From the other extreme, if too much trust is placed on the automation, a false sense of security results and other forms of data are discounted, much to the peril of the pilot. From a trust perspective, *Wickens* and colleagues tested heterogeneous and homogeneous crews based on flight experience [17.47] and found little difference in flying proficiency for various levels of automation. However, the homogeneous crews obtained increased benefit from automation, which may be due to, and interpreted in terms of, having a different authority gradient. In considering the level of trust (overtrust or undertrust), *Lee* and *Moray* [17.48] later noted that, as more and more false alarms occur, the operator will decrease their level of trust accordingly even if the automation is adapted. It is as if the decision-aiding system must first prove itself before trust is developed. In more recent work [17.49], a quantitative approach to the trust issue is discussed, showing that overall trust in a system has a significantly inverse relationship with the uncertainty of a system. Three levels of uncertainty were examined using National Institute of Standards (NIST) guidelines, and users rated their trust at each level through questionnaires. The bottom line was that performance of a hybrid system can be improved by decreasing uncertainty.

17.3.2 Cost of Automation

As mentioned previously, cost saving is one of the significant advantages of the use of automation, e.g., in an unmanned air vehicle, for which the removal of the requirement to have a life-support system makes such devices economically advantageous over alternatives. In [17.50] low-cost/cost-effective automation is discussed. By delegating some of the task responsibilities to the instrumentation, the overall system has an improved response. In a cost-effective sense, this is a better design. In [17.51] the application of a fuzzy-logic adaptive Kalman filter which has the ability to provide improved real-time control is discussed. This allows more intelligence at the sensor level and unloads the control requirements at the operator level. The costs of such devices thus drops significantly and they become easier to operate. See Chap. 41 for more details on cost-oriented automation.

17.3.3 Adaptive Versus Nonadaptive Automation

One could argue that, if automation could adapt, it could optimally couple the level of operator engagement to the level of mechanism [17.52]. To implement such an approach, one may engage biopyschometric measures for when to trigger the level of automation. This would seem to make the level of automation related to the workload on the human operator. One can view this concept as adaptive aiding [17.53]. For workload consideration, the well-known National Aeronautics and Space Administration (NASA) task load index (TLX) scale provides some of these objective measures of workload [17.54]. As a more recent example, in [17.55], adaptive function allocation as the dynamic control of tasks which involves responsibility and authority shifts between humans and machines is addressed. How this affects operator performance and workload is analyzed. In [17.56] the emphasis focuses on human-centered design for adaptive automation. The human is viewed

as an active information processor in complex system control loops and supports situational awareness and effective performance. The issues of workload overload and situational awareness are researched.

17.3.4 Safety in Automation

Safety is a two-edged sword in the interaction of humans with automation. The human public demands safety in automation. As mentioned in regard to robotic surgery applications, the patient has a significantly reduced recovery time with the use of a robotic device, however, all such machines may fail at some time. When humans interact with robotic devices, they are always at risk and there are a number of documented cases in which humans have been killed by being close to a robotic device. In factories, safety when humans interact with machines has always been a key concern. In [17.57], a way to approach the safety assessment in a factory in terms of a safety matrix is introduced. This differs from the typical probability method. The elements of the matrix can be integrated to provide a quantitative safety scale. For traffic safety [17.58] human-centered automation is a key component to a successful system and is recommended to be multilayered. The prevailing philosophy is that it is acceptable for a human to give up some authority in return for a reduction in some mundane drudgery.

Remote control presents an excellent case for safety and the benefit of automation. Keeping the human out of harm's way through teleoperation and other means but providing a machine interface allows many more human interactions with external environments which may be radioactive, chemically adverse or have other dangers. See Chap. 39 for a detailed discussion on safety warnings for automation.

17.3.5 Authority in Automation

The problem of trading off authority between the human and computer is unsettling and risky, for example, in a transportation system, such as a car. Events that are unplanned or produce unexpectancy may degrade performance. Familiarity with antibrake systems can be a lifesaver for the novice driver first experiencing a skidding event on an icy or wet road. However, this could also work to the detriment of the driver in other situations. As mentioned earlier, automation has a tradeoff space with respect to authority. In Table 17.2 it is noted that, the higher the level of automation, the greater the loss of authority. With more automation, the effort from the operator is proportionally reduced; however, with loss of authority, the risk of a catastrophic disaster increases. This tradeoff space is always under debate. In [17.59] it is shown that some flexibility can be maintained. Applications where the tradeoff space exists between automation and authority include aircraft, nuclear power plants, etc., *Billings* [17.13] calls for the operator to maintain an active role in a system's operation regardless of whether the automation might be able to perform the particular function in question better than the operator. *Norman* [17.60] stresses the need for feedback to be provided to the operator, and this has been overlooked in many recent systems.

17.3.6 Performance of Automation Systems

There are many ways to evaluate human−machine system performance. Based on the signal detection theory framework, various methods have been introduced [17.61–63]. The concept of likelihood displays shows certain types of statistical optimality in reducing type 1 and type 2 errors and provides a powerful approach to measure the efficacy of any human−machine interaction. Other popular methods to quantify human−machine performance include the information-theoretic models [17.64] with measures such as baud rate, reaction time, accuracy, etc. Performance measurement is always complex since it is well known that three human attributes interact [17.65] to produce a desired result. The requisite attributes include skill, and rule- and knowledge-based behaviors of the human operator.

17.3.7 When Should the Human Override the Automation?

One way to deal with the adaptive nature of automation is to consider *when* the human should intervene [17.66].

It is well known that automation may not work to the advantage of the operator in certain situations [17.67]. Rules can be obtained, e.g., it is well known that it is easier to modify the automation rather than modify the human. A case in point is when high-workload situations may require the automation level to increase to alleviate some of the stress accumulated by the human in a critical task [17.68] in situations such as flight-deck management.

17.3.8 Social Issues and Automation

There are a number of issues of the use of automation and social acceptance. In [17.69] the discussion centers on robots that are socially acceptable. There must be a balance between a design which is human-centered versus the alternative of being more socially acceptable. Tradeoff spaces exist in which these designs have to be evaluated as to their potential efficacy versus agreement in the venue in which they were designed. Also humans are hedonistic in their actions, that is, they tend to favor decisions and actions that benefit themselves or their favored parochial classes. Machines, on the other hand, may not have to deal with these biases. Another major point deals with the disadvantages of automation in terms of replacing human workers. If a human is replaced, alienation may result [17.70], resulting in a disservice to society. Not only do people become unemployed, but they also suffer from loss of identity and reduced self-esteem. Also alienation allows people to abandon responsibility for their own actions, which can lead to reckless behavior. As evidence of this effect, the Luddites in 19th century England smashed the knitting machines that had destroyed their livelihood. Turning anger against an automation object offers, at best, only temporary relief of a long-term problem.

17.4 Future Directions of Defining Human−Machine Interactions

It is seen that the following parameters strongly affect how the level of automation should be modified with respect to humans:

- Trust
- Social acceptance
- Authority
- Safety
- Possible unplanned unexpectancy with adaptive automation
- Application-specific performance that may be gained

17.5 Conclusions

Modern application still wrestle with the benefit and degree of automation that may be appropriate for a task [17.71]. Emerging trends include how to define those jobs that are dangerous, mundane, and simple where there are benefits of automation. The challenges include defining a multidimensional tradeoff space that must take into consideration what would work best in a mission-effectiveness sense as humans have to deal with constantly changing machines that obviate some of the danger and drudgery of certain jobs. For further discussion of the human role in automation refer to the chapters on *Human Factors in Automation Design* Chap. 25 and *Integrated Human and Automation Systems, Including Automation Usability, Human Interaction and Work Design in (Semi)-Automated Systems* Chap. 34 later in this Handbook.

References

17.1   T.B. Sheridan: *Humans and Automation – System Design and Research Issues* (Wiley, New York 2002), pp. 62, 163

17.2   G.E. Hoffmann: Concepts for the third generation of laboratory systems, Clinica Chimica Acta **278**, 203–216 (1998)

17.3   P.M. Fitts (ed.): *Human Engineering for an Effective Air Navigation and Traffic Control System* (National Research Council, Washington 1951)

17.4   A. Chapanis: On the allocation of functions between men and machines, Occup. Psychol. **39**, 1–11 (1965)

17.5   A. Bye, E. Hollnagel, T.S. Brendeford: Human-machine function allocation: a functional modeling approach, Reliab. Eng. Syst. Saf. **64**, 291–300 (1999)

17.6   B.H. Kantowitz, R.D. Sorkin: Allocation of functions. In: *Handbook of Human Factors*, ed. by G. Salvendy (Wiley, New York 1987) pp. 355–369

17.7   N. Moray: Humans and machines: allocation of function. In: *People in Control – Human Factors in Control Room Design*, IEEE Control Eng., Vol. 60, ed. by J. Noyes, M. Bransby (IEEE, New York 2001) pp. 101–113

17.8   R. Parasuraman, T.B. Sheridan, C.D. Wickens: A model for types and levels of human interaction with automation, IEEE Trans. Syst. Man Cybern. A: Syst. Hum. **30**(3), 286–297 (2000)

17.9   T.B. Sheridan: *Telerobotics, Automation, and Human Supervisory Control* (MIT Press, Cambridge 1992), p. 26

17.10   L.J. Fogel, J. Lyman: The human component. In: *Handbook of Automation, Computation, and Control*, Vol. 3, ed. by E.M. Grabbe, S. Ramo, D.E. Wooldridge (Wiley, New York 1961)

17.11   M.L. Cummings, P.J. Mitchell: Operator scheduling strategies in supervisory control of multiple UAVs, Aerosp. Sci. Technol. **11**, 339–348 (2007)

17.12   N.B. Sarter: Cockpit automation: from quantity to quality, for individual pilot to multiple agents. In: *Automation and Human Performance*, ed. by R. Parasuraman, M. Mouloua (Lawrence Erlbaum, Mahwah 1996)

17.13   C.E. Billings: Toward a human-centered aircraft automation philosophy, Int. J. Aviat. Psychol. **1**(4), 261–270 (1991)

17.14   C.E. Billings: *Aviation Automation: The Search for a Human-Centered Approach* (Lawrence Erlbaum Associates, Mahwah 1997)

17.15   D.O. Weitzman: Human-centered automation for air traffic control: the very idea. In: *Human/Technology Interaction in Complex Systems*, ed. by E. Salas (JAI Press, Stamford 1999)

17.16   W.B. Rouse, J.M. Hammer: Assessing the impact of modeling limits on intelligent systems, IEEE Trans. Syst. Man Cybern. **21**(6), 1549–1559 (1991)

17.17   C.D. Wickens: Designing for situational awareness and trust in automation, Proc. Int. Fed. Autom. Control Conf. Integr. Syst. Eng. (Pergamon, Elmsford 1994)

17.18   G. Salvendy: Research issues in the ergonomics, behavioral, organizational and management aspects of office automation. In: *Human Aspects in Office Automation*, ed. by B.G.F. Cohen (Elsevier, Amsterdam 1984) pp. 115–126

17.19   K.R. Boff: Revolutions and shifting paradigms in human factors and ergonomics, Appl. Ergon. **37**, 391–399 (2006)

17.20   J.C.R. Licklider: Man–computer symbiosis. *IRE Trans. Hum. Factors Electron.* In: *Digital Center Research Reports*, Vol. 61, ed. by J.C.R. Licklider, R.W. Taylor (Human Factors Accociation, Palo Alto 1990), reprinted in Memoriam (1960)

17.21   W.B. Rouse, K.R. Boff: Impacts of next-generation concepts of military operations on human effectiveness, Inf. Knowl. Syst. Manag. **2**, 1–11 (2001)

17.22   M. Kassler: Agricultural automation in the new Millennium, Comput. Electron. Agric. **30**(1–3), 237–240 (2001)

17.23   N. Sigrimis, P. Antsaklis, P. Groumpos: Advances in control of agriculture and the environment, IEEE Control Syst. Mag. **21**(5), 8–12 (2001)

17.24   J.R. Martinez-de Dios, C. Serna, A. Ollero: Computer vision and robotics techniques in fish farms, Robotica **21**, 233–243 (2003)

17.25 O. Diegel, G. Bright, J. Potgieter: Bluetooth ubiquitous networks: seamlessly integrating humans and machines, Assem. Autom. **24**(2), 168–176 (2004)

17.26 L. Tarrini, R.B. Bandinelli, V. Miori, G. Bertini: *Remote Control of Home Automation Systems with Mobile Devices*, Lecture Notes in Computer Science (Springer, Berlin Heidelberg 2002)

17.27 X. Jiang, A.K. Gramopadhye, B.J. Melloy, L.W. Grimes: Evaluation of best system performance: human, automated, and hybrid inspection systems, Hum. Factor Ergon. Manuf. **13**(2), 137–152 (2003)

17.28 D. Chetverikov: Pattern regularity as a visual key, Image Vis. Comput. **18**(12), 975–985 (2000)

17.29 D.D. Woods: Decomposing automation: apparent simplicity, real complexity. In: *Automation and Human Performance – Theory and Applications*, ed. by R. Parasuraman, M. Mouloua (Lawrence Erlbaum, Mahwah 1996) pp. 3–17

17.30 A. Mital, A. Pennathur: Advanced technologies and humans in manufacturing workplaces: an interdependent relationship, Int. J. Ind. Ergon. **33**, 295–313 (2004)

17.31 S.P. Layne, T.J. Beugelsdijk: Mass customized testing and manufacturing via the Internet, Robot. Comput.-Integr. Manuf. **14**, 377–387 (1998)

17.32 M.B. Weinger: Automation in anesthesiology: Perspectives and considerations. In: *Human–Automation Interaction – Research and Practice*, ed. by M. Mouloua, J.M. Koonce (Lawrence Erlbaum, Mahwah 1996) pp. 233–240

17.33 D.W. Repperger: Human factors in medical devices. In: *Encyclopedia of Medical Devices and Instrumentation*, ed. by J.G. Webster (Wiley, New York 2006) pp. 536–547

17.34 D.B. Camarillo, T.M. Krummel, J.K. Salisbury: Robotic technology in surgery: past, present and future, Am. J. Surgery **188**(4), 2–15 (2004), Supplement 1

17.35 E. Dummermuth: Advanced diagnostic methods in process control, ISA Trans. **37**(2), 79–85 (1998)

17.36 V. Potkonjak, S. Tzafestas, D. Kostic: Concerning the primary and secondary objectives in robot task definition – the learn from humans principle, Math. Comput. Simul. **54**, 145–157 (2000)

17.37 A. Halme, T. Luksch, S. Ylonen: Biomimicing motion control of the WorkPartner robot, Ind. Robot **31**(2), 209–217 (2004)

17.38 D.G. Perrin: It's all about learning, Int. J. Instruct. Technol. Dist. Learn. **1**(7), 1–2 (2004)

17.39 K.M.G. Taylor, G. Harding: Teaching, learning and research in McSchools of Pharmacy, Pharm. Educ. **2**(2), 43–49 (2002)

17.40 N.E. Gibbs: The SEI education program: the challenge of teaching future software engineers, Commun. ACM **32**(5), 594–605 (1989)

17.41 C.D. Grant, B.R. Dickson: New approaches to teaching and learning for industry-based engineering professionals, Proc. 2002 ASEE Annu. Conf. Expo., session 2213 (2002)

17.42 Z. Turk: Multimedia: providing students with real world experiences, Autom. Constr. **10**, 247–255 (2001)

17.43 B. Muir: Trust between humans and machines, and the design of decision aids, Int. J. Man–Mach. Stud. **27**, 527–539 (1987)

17.44 J. Lee, K. See: Trust in automation: designing for appropriate reliance, Hum. Factors **46**(1), 50–80 (2004)

17.45 J. Lee, N. Moray: Trust and the allocation of function in the control of automatic systems, Ergonomics **35**, 1243–1270 (1992)

17.46 E.L. Wiener, R.E. Curry: Flight deck automation: promises and problems, Ergonomics **23**(10), 995–1011 (1980)

17.47 C.D. Wickens, R. Marsh, M. Raby, S. Straus, R. Cooper, C.L. Hulin, F. Switzer: Aircrew performance as a function of automation and crew composition: a simulator study, Proc. Hum. Factors Soc. 33rd Annu. Meet., Santa Monica (Human Factors Society, 1989) pp. 792–796

17.48 J. Lee, N. Moray: Trust, self-confidence, and operators' adaptation to automation, Int. J. Hum.-Comput. Stud. **40**, 153–184 (1994)

17.49 A. Uggirala, A.K. Gramopadhye, B.J. Melloy, J.E. Toler: Measurement of trust in complex and dynamic systems using a quantitative approach, Int. J. Ind. Ergon. **34**, 175–186 (2004)

17.50 H.-H. Erbe: Introduction to low cost/cost effective automation, Robotica **21**, 219–221 (2003)

17.51 J.Z. Sasiadek, Q. Wang: Low cost automation using INS/GPS data fusion for accurate positioning, Robotica **21**, 255–260 (2003)

17.52 J.G. Morrison, J.P. Gluckman: Definitions and prospective guidelines for the application of adaptive automation. In: *Human Performance in Automated Systems: Current Research and Trends*, ed. by M. Mouloua, R. Parasuraman (Lawrence Erlbaum, Hillsdale 1994), pp. 256–263

17.53 W.B. Rouse: Adaptive aiding for human/computer control, Hum. Factors **30**, 431–443 (1988)

17.54 S.G. Hart, L.E. Staveland: Development of the NASA-TLX (task load index): results of empirical and theoretical research. In: *Human Mental Workload*, ed. by P.A. Hancock, N. Meshkati (Elsevier, Amsterdam 1988)

17.55 S.F. Scallen, P.A. Hancock: Implementing adaptive function allocation, Int. J. Aviat. Psychol. **11**(2), 197–221 (2001)

17.56 D.B. Kaber, J.M. Riley, K.-W. Tan, M. Endsley: On the design of adaptive automation for complex systems, Int. J. Cogn. Ergon. **5**(1), 37–57 (2001)

17.57 Y. Mineo, Y. Suzuki, T. Niinomi, K. Iwatani, H. Sekiguchi: Safety assessment of factory automation systems, Electron. Commun. Jap. Part 3, **83**(2), 96–109 (2000)

17.58 T. Inagaki: Design of human–machine interactions in light of domain-dependence of human-centered automation, Cogn. Tech. Work **8**, 161–167 (2006)

17.59 T. Inagaki: Automation and the cost of authority, Int. J. Ind. Ergon. **31**, 169–174 (2003)

17.60 D. Norman: The problem with automation: inappropriate feedback and interaction, not over-automation, Proc. R. Soc. Lond. B237 (1990) pp. 585–593

17.61 R.D. Sorkin, D.D. Woods: Systems with human monitors: a signal detection analysis, Hum.-Comput. Interact. **1**, 49–75 (1985)

17.62 R.D. Sorkin: Why are people turning off our alarms?, J. Acoust. Soc. Am. **84**, 1107–1108 (1988)

17.63 R.D. Sorkin, B.H. Kantowitz, S.C. Kantowitz: Likelihood alarm displays, Hum. Factors **30**, 445–459 (1988)

17.64 C.A. Phillips: *Human Factors Engineering* (Wiley, New York 2000)

17.65 J. Rasmussen: *Information Processing and Human–Machine Interaction* (North-Holland, New York 1986)

17.66 J.M. Haight, V. Kecojevic: Automation vs. human intervention: What is the best fit for the best performance?, Process. Saf. Prog. **24**(1), 45–51 (2005)

17.67 R. Parasuraman, V. Riley: Humans and automation: use, misuse, disuse, and abuse, Hum. Factors **39**(2), 230–253 (1997)

17.68 B.H. Kantowitz, J.L. Campbell: Pilot workload and flight deck automation. In: *Automation and Human Performance – Theory and Applications*, ed. by R. Parasuraman, M. Mouloua (Lawrence Erlbaum Associates, Mahwah 1996) pp. 117–136

17.69 J. Cernetic: On cost-effectiveness of human-centered and socially acceptable robot and automation systems, Robotica **21**, 223–232 (2003)

17.70 S. Zuboff: *In the Age of the Smart Machine: The Future of Work and Power* (Basic Books, New York 1988)

17.71 S. Sastry: A SmartSpace for automation, Assem. Autom. **24**(2), 201–209 (2004)

18. What Can Be Automated? What Cannot Be Automated?

Richard D. Patton, Peter C. Patton

The question of what can and what cannot be automated challenged engineers, scientists, and philosophers even before the term *automation* was defined. While this question may also raise ethical and educational issues, the focus here is scientific. In this chapter the limits of automation and mechanization are explored and explained in an effort to address this fundamental conundrum. The evolution of computer languages to provide domain–specific solutions to automation design problems is reviewed as an illustration and a model of the limitations of mechanization. The

current state of the art and a general automation principle are also provided.

18.1 The Limits of Automation

The recent (1948) neologism *automation* comes from autom(atic oper)ation. The term *automatic* means self-moving or self-dictating (Greek *autómatos*) [18.1]. The *Oxford English Dictionary* (OED) defines it as

> *Automatic control of the manufacture of a product through a number of successive stages; the application of automatic control to any branch of industry or science; by extension, the use of electronic or other mechanical devices to replace human labor.*

Thus, the primary theoretical limit of automation is built into the denotative meaning of the word itself. Automation is all about self-moving or self-dictating as opposed to self-organizing. Another way of saying this is that the very notion of automation is based upon, and thus limited by, its own mechanical metaphor. In fact, most people would agree that, if an automated process began to self-organize into something else, then it would not be a very good piece of automation per se. It would perhaps be a brilliant act of creating a new form of *life* (i. e., a self-organizing system), but that is certainly not what automation is all about.

Automation is fundamentally about taking some process that itself was created by a life process and making it more mechanical or in the modern computing metaphor *hard-wired*, such that it can be executed without any volitional or further expenditure of *life-process* energy. This phenomenon can even be seen in living organisms themselves. The whole point of skill-building in humans is to drive brain and other neural processes to become more nearly *hard-wired*. It is well known that, as the brain executes some particular circuit more and more, it *hard-wires* itself by growing more and more synaptic connections. This, in effect, automates a volitional process by making it more mechanical and thus more highly automated. Such *nerve training* is driven by practice and repetition to achieve greater performance levels in athletes and musicians.

This is also what happens in our more conscious processes of automation. We create some new process and then seek to *automate* it by making it more mechanical. An objection to this notion might be to say that the brain itself is a mechanism, albeit a very complex one, and since the brain is capable of this self-organizing or volitional behavior how can we make this distinction? I think there are primarily two and possibly a third answer to this, depending on one's belief system.

1. For those who believe that there is life after death it seems to me conclusive to say that, if we live on then, there must be a distinction between our mind and our brain.
2. For almost everyone else there is good research evidence for a distinction between the mind and the brain. This can especially be seen in people with obsessive compulsive disorder and, in fact, *The Mind and the Brain* is very persuasive in showing that a person has a mind which is capable of programming their brain or overcoming some faulting programming of their brain – even when the brain has been hard-wired in some particularly un-useful way [18.2].
3. However, the utter materialist would further argue that the mind is merely an artifact of the elec-

trochemical function of the brain as an ensemble of neurons and synapses and does not exist as a separate entity in spite of its phenomenological differences in functionality.

In any case, as long as we acknowledge that there is some operational degree of distinction between the mind as analogous to software and the brain as analogous to hardware, then we simply cannot determine to what extent the mind and the brain together are operating as a mechanism and to what extent they are operating as a self-organizing open system. The ethical and educational issues related to what can and cannot be automated, though important, are outside of the scope of this chapter. Readers are invited to read Chap. 47 on *Ethical Issues of Automation Management*, Chap. 17 on *The Human Role in Automation*, and Chap. 44 on *Education and Qualification*.

18.2 The Limits of Mechanization

So, another way of asking *what are the limits of automation* is to ask *what are the limits of mechanization?* or, *what are machines ultimately capable of doing autonomously?* But what does *mechanical* mean? Fundamentally it means linear or stepwise, i.e., able to carry out an algorithmically defined process having clear inputs and clear outputs. The well-known Carnot circle used by the military engineer and founder of thermodynamics L. N. S. Carnot (1796–1832) to describe the heat engine and other mechanical systems (Fig. 18.1) separates the engine or system under study from the rest of the universe by a circle with an arrow in for input and an arrow out for output. This is a simple but powerful philosophical (and graphical) tool

for understanding the functions and limits of systems and familiar to every mechanical and systems engineer. More complex mechanical processes may have to go beyond strictly linear algorithmic control to include negative feedback, but still operate in such a way that they satisfy a clear objective function or goal. When the goal requires extremely precise positioning of the automation, a subprocess or correction function called *dither* is added to force the feedback loop to hunt for the precise solution or positioning in a heuristic manner but still subsidiary to and controlled by the algorithm itself. In any case, *mechanical* means non-context-sensitive and discrete, even if it involves dither. Machine theory is basically the opposite of general system theory. And by a *general system* we mean an open system, i.e., a system that is capable of locally overcoming entropy and is self-organizing. Today such systems are typically referred to as complex adaptive systems (CAS).

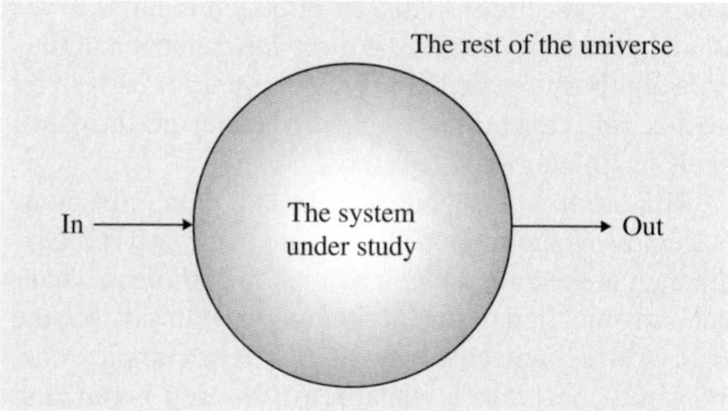

An open system is fundamentally different from a machine. The core difference is that an open system is nonlinear. That is, in general, everything within the system is sensitive to its entire context, i.e., to everything else in the system, not just the previous step in an algorithm. This is most easily seen in quantum theory with the two-slit experiment. Somehow the single electron is aware that there are two slits rather than one and thus behaves like a wave rather than a particle.

Fig. 18.1 The Carnot circle as a system definition tool

In essence, the single electron's behavior is sensitive to the entire experimental context. Now imagine a human brain with 100 billion neurons interconnected with some 100 trillion synapses, where even local synaptic structures make their own locally complex circuits. The human brain is also bathed in chemicals that influence the operation of this vast network of neurons whose resultant behavior then influences the mix of chemicals. And then there are the hypothesized quantum effects, giving rise to potential nonlocal effects, within any particular neuron itself. This is clearly a vastly different sort of thing than a machine: a difference in degree of context sensitivity that amounts to a difference in kind.

One way to illustrate this mathematically is to define a system of simultaneous differential equations. Denoting some measure of elements, $p_i(i = 1, 2, \ldots, n)$, by Q_i, these, for a finite number of elements and in the simplest case, will be of the form

$$dQ_1/dt = f_1(Q_1, Q_2, \ldots, Q_n)$$
$$dQ_2/dt = f_2(Q_1, Q_2, \ldots, Q_n)$$
$$\ldots$$
$$dQ_n/dt = f_n(Q_1, Q_2, \ldots, Q_n) \, .$$

Change of any measure Q_i is therefore a function of all Q, from Q_1 to Q_n; conversely, change of any Q_i entails change of all other measures and of the system as a whole [18.3].

What has been missing from the Strong Artificial Intelligence (AI) debate and indeed from most of Western thought is the nature of open systems and their propensity to generate emergent behavior. We are still mostly caught up in the mechanical metaphor believing that it is possible to mechanize or *model* any system using machines or algorithms, but this is exactly the opposite of what is actually happening. It is systems themselves that use a mechanizing process to improve their performance and move onto higher levels of emergent behavior.

However, it is no wonder that we are so caught up in the mechanical metaphor, as it is the very basis of the industrial revolution and is thus responsible for much of our wealth today. In the field of business application software this peripheral blindness (or, rather, intense focus) has led to huge failures in building complex business application software systems. The presumption is that building a software system is like building a bridge; that it is like a construction project; that it is fundamentally an engineering problem where there is a design process and then a construction process where the design process can fully specify something that can be constructed using engineering principles; more specifically, that there is some design process done by designers and that the construction process is the process of programming done by programmers. However, it is not. It is really a design problem through and through and, moreover, a design problem that does not have general-purpose engineering-like principles and solutions that have already been *reduced to code* by years of practice. It is a design problem of pure logic where the logic can only be fully specified in a completed program where the *construction* process is then simply running a compiler against the program to generate machine instructions that carry out the logic.

1st Generation	Machine code	
2nd Generation	Assembler Machine specific symbolic languages	≈10× over 1st generation Allowed complex operating systems to be built – cost was built into hardware price.
3rd Generation	High level language C, COBOL, FORTRAN	≈10× over 2nd generation Allowed for independent software companies to become profitable.
4th Generation	High-level language integrated with a (virtual) machine environment Visual Basic, Powerbuilder, Java	≈10× over 3rd generation Greatly increased the scale of how large a software company could grow to.
"5th Generation"	Non-existent as a general purpose language must be domain-specific Lawson Landmark (Business Applications)	10–20× over 4th generation

Fig. 18.2 Software language generations

There are no general-purpose solutions to all logic design problems. There are only special-purpose domain-specific solutions. This can be seen in the evolution, or lately lack thereof, of computer languages. We went from first-generation languages (machine code) to second-generation languages (assembler) to third-generation languages (FORTRAN, COBOL, C) to fourth-generation languages (Java, Progress) where each generation represented some tenfold productivity improvement in our ability to produce systems (Fig. 18.2). This impressive progression however slowed down and essentially stopped dead in its tracks some 20 years ago in its ability to significantly improve productivity. Since then there have been many attempts to come up with the next (fifth) generation language or general-purpose framework for dramatically improving productivity. In the 1980s computer-aided software engineering (CASE) tools were the hoped-for solution. In the 1990s object-oriented frameworks looked promising as the ultimate solution and now the industry as a whole seems primarily focused on the notion of service-oriented architecture (SOA) as our best way forward. While these attempts have helped boost productivity to some extent they have all failed to produce an order of magnitude shift in the productivity of building software systems (Fig. 18.3). What they all have in common is the continuation of the mechanical metaphor. They are all attempts at general-purpose engineering-like solutions to this problem. General-purpose languages can only go so far in solving the productivity problem. However, special-purpose languages or domain-specific languages (DSL) take us much farther. There are multi-

ple categories of DSLs. There are *horizontal* DSLs that address some functional layer, like SQL does for data access or UML does for conceptual design (as opposed to the detail design which is the actual complete program in this metaphor). And there are vertical DSLs which address some topical area such as mathematical analysis, molecular modeling or business process applications. There is also the distinction between a DSL and a domain-specific design language (DSDL). All DSDLs are DSLs but not all DSLs are DSDLs (a *prescission* distinction in Peirce's nomenclature [18.4]). A design language is distinct from a programming or implementation language in that in a design language everything is defined relative to everything else. This is like the distinction between a design within a CAD system and the actually built component. In the CAD system one can move a line and everything changes around it. In the actually built component that *line* is a physical entity in some hardened structural relationship. Then there is the further prescission distinction of a pattern language versus a DSDL. A pattern language is a DSDL that has built-in syntax that allows for specific design solutions to be applied to specific domain problems. This tends to be the most vertically domain-specific and also the most powerful sort of language for improving the productivity of building systems. Lawson Landmark is one example of a pattern language that delivers a 20 times improvement over fourth-generation languages in its particular domain.

It is not possible to mechanize or model an entire complex system. We can only mechanize aspects of any particular system. To mechanize the entire system is to destroy the system as such; i. e., it is no longer a system. Analysis of a system into its constituent parts loses its essence as a system, if a system is truly *more than the sum of its parts*. That is, the mechanistic model lacks the system's inherent capability for self-organization and thus adaptability, and thus becomes more rigid and more limited in its capabilities, although, perhaps, much faster. And the limit to what aspects of a system can be mechanized is really the limit of our cognitive ability to model those aspects. In other words it is a design problem. In general the theoretical limit of mechanization is always the current limit of our capacity to design linear stepwise models in any particular solution domain. Thus the key to what can be automated is the depth of our knowledge and understanding within any specific problem domain, the depth required being that depth necessary to be able to design automatons that *fit*, where fit is determined both by the automaton actually producing what is expected as well as fitting appropriately

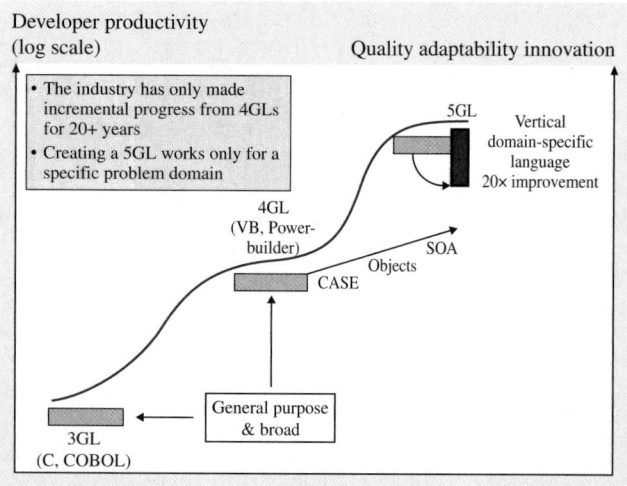

Fig. 18.3 Productivity progression of languages

within the system as a whole; i. e., if the automation disturbs the system that it is within beyond some threshold then the very system that gives rise to the useful automaton will change so dramatically that the automation is no longer useful. An example of this might be a software program for paying employees. If this program does pay employees correctly but requires dramatic changes to the manner in which time card information is gathered such that it becomes too great a burden on some part of the system then the program will likely soon be surrounded by organizational *T cells* and rejected.

Organisms *are* not machines, but they can to a certain extent *become* machines, or perhaps congeal into machines. Never completely, however, for a thoroughly mechanized organism would be incapable of reacting to the incessantly changing conditions of the outside world. The *principle of progressive mechanization* expresses the transition from undifferentiated wholeness to higher function, made possible by specialization and *division of labor*; this principle also implies loss of potentialities in the components and of regularity in the whole [18.5].

18.3 Expanding the Limit

Human business processes (buying, selling, manufacturing, invoicing, and so on), like actual organisms, operate fundamentally as open systems or rather complex adaptive system (CAS). Other examples of CAS are cities, insect colonies, and so on. The architecture of an African termite mound requires more bits to describe than the number of neurons in a termite brain. So, where do they store it? In fact, they do not, because each termite is programmed as an automaton to instinctively perform certain functions under certain stimuli and the combined activity of the colony produces and maintains the mound. Thus, the entire world is fundamentally made up of vastly complex interacting systems. When we automate we are ourselves engaging in the principle of progressive mechanization. We are mechanizing some portion or aspect of a system. We can only mechanize those aspects that are themselves already highly specialized within the system itself or that we can make highly specialized; for example, it is possible to make a mechanical heart but it is impossible to make a mechanical stem cell or a mechanical human brain. It is possible to make a mechanical payroll program but impossible to make a mechanical company that responds to dynamic market forces and makes a profit. The result of this is that any mechanization requires specific understanding of some particular specialization within some particular system. In other words, there are innumerable specific domains of specialization within all of the complex interacting systems in the world. And thus any mechanization is inherently domain- pecific and requires detailed domain knowledge. The process of mechanization is therefore fundamentally a domain-specific design problem. *Thus, the limit of mechanization is our ability to comprehend some particular aspect of a system well enough to be able to extract some portion or aspect of that system in which it is possible to define a boundary or Carnot circle with clear inputs and clear outputs along with a transfer function mapping those inputs to those outputs.* If we cannot enumerate the inputs and outputs and then describe what inputs result in what outputs then we simply cannot automate the process. This is the fundamental requirement of mechanization.

This does not, however, mean that we are limited to our current simple mechanical metaphor of a machine as a set of inputs into a box that transforms those inputs into their associated outputs. Indeed the next step of mechanization under way is to try to mimic aspects of how a system itself works. One aspect of this is the object-oriented revolution in software, which itself grew out of the work done in complex adaptive systems research. Unfortunately, the prevailing simple mechanical metaphor and lack of understanding of the nature of CAS has blunted the evolution of object-oriented technology and limited its impact.

Object-oriented concepts have taken us from a simplistic view of a machine as a *black box* process or function F as shown in Fig. 18.4, in which ($outputs = F(inputs)$), to conceptualizing a machine as an interacting set of agents or objects. These concepts have allowed us to manage higher levels of complexity in the machines we build but they have not taken us to a higher order of magnitude in our ability to mecha-

Fig. 18.4 The engineer's black box

nize complex systems. In the realm of business process software what has been missing is the following:

1. The full realization that what we are modeling with business process software is really a portion or aspect of a complex adaptive system
2. That this is fundamentally a logic design problem rather than an engineering problem
3. That, in this case, the problem domain is primarily about semiotics and pragmatics, i.e., the nature of sign processes and the impact on those who use them

Fortunately there is a depth of research in these areas that provides guidance toward more powerful techniques and tools for making further progress. John Holland, the designer of the Holland Machine, one of the first parallel computers, has done extensive research into CAS and has discovered many basic principles that all CAS seem to have in common. His work led to object-oriented concepts; however, the object-oriented community has not continued to seek to implement his further discoveries and principles regarding CAS into object-oriented computing languages. One very useful concept is how CAS systems use rule-block formations in driving their behavior and the nature of how adaptation continues to build rule-block hierarchies on top of existing rule-block structures [18.6].

Christopher Alexander [18.7] is a building architect who discovered the notion of a *pattern language* for designing and constructing buildings and cities [18.8]. In general, a pattern language is a set of patterns or design solutions to some particular design problem in some particular domain. The key insight here is that design is a domain-specific problem that takes deep understanding of the problem domain and that, once good design solutions are found to any specific problem, we can codify them into reusable patterns. A simple example of a pattern language and the nature of domain specificity is perhaps how farmers go about building barns. They do not hire architects but rather get together and, through a set of *rules of thumb* based on how much livestock they have and storage and processing considerations, build a barn with such and such dimensions. One simple pattern is that, when the barn gets to be too long for just doors on the ends, they put a door in the middle. Now, can someone use these *rules of thumb* or this pattern language to build a skyscraper in New York?

Charles Sanders Peirce was a 19th century American philosopher and semiotician. He is one of the founders of the quintessential American philosophy known as pragmatism, and came up with a triadic semiotics based on his metaphysical categories of firstness, secondness, and thirdness [18.9]. Firstness is the category of idea or possibility. Secondness is the category of brute fact or instance, and thirdness is the category of laws or behavior. Peirce argues that existence itself requires these categories; that they are in essence a unity and that there is no existence without these three categories. Using this understanding one could argue that this demystifies to some extent the Christian notion of God being a unity and yet also being triune by understanding the Father as firstness, the Son as secondness, and the Holy Spirit as thirdness. Thus the trinity of God is essentially a metaphysical statement about the nature of existence. Peirce goes on to build a system of signs or semiotics based on this triadic structure and in essence argues that reality is ultimately the stuff of signs; e.g., God *spoke* the universe into existence. At first one could seek to use this notion to support the Strong AI view that all intelligence is symbol representation and symbol processing and thus a computer can ultimately model this reality. However, a key concept of Peirce is that the meaning of a sign requires an *interpretant*, which itself is a sign and thus also requires a further interpretant to give it meaning in a never-ending process of meaning creation; that is, it becomes an eternally recursive process. Another way of viewing this is to say that this process of making meaning is ultimately an open system. And while machines can model and hence automate closed systems they cannot fully model open systems.

Peirce's semiotics and his notion of firstness, secondness, and thirdness provide the key insights required for building robust ontology models of any particular domain. The key to a robust ontology model is not that it is *right* once and for all but rather that it can be done using a design language and that it has perfect fidelity with the resulting execution model. An ontology model is never *right*; it is just more and more useful. This is because of the notion of the *relativity of categories*.

Perception is universally human, determined by man's psychophysical equipment. Conceptualization is culture-bound because it depends on the symbolic systems we apply. These symbolic systems are largely determined by linguistic factors, the structure of the language applied. Technical language, including the symbolism of mathematics, is, in the last resort, an efflorescence of everyday language, and so will not be independent of the structure of the latter. This, of course, does not mean that the content of mathematics is *true* only within a certain culture. It is a tautological system of a hypothetico-deductive nature, and hence any rational being accepting the premises must agree to all its deductions [18.10].

The most critical aspect of achieving the theoretical limit of automation is the ability to continue to make execution models that are more and more useful. And this is true for two reasons:

1. CAS are so complex that we cannot possibly understand large portions of them perfectly ab initio. We need to model them, explore them, and then improve them.
2. CAS are continually adapting. They are continually changing and thus, even if we perfectly modeled some aspect of a CAS, it will change and our model will become less and less useful.

In order to do this we need a DSDL capable of quickly building high-fidelity models. That is, models which are themselves the drivers of the actual execution of automation.

In order for the model of the CAS to ultimately actually drive the execution it is critical that the ontology model be in perfect fidelity with the execution model. Today there is a big disconnect between what one can build in an *analysis model* and the resultant executing code or execution model. This is analogous to having a design of a two-storey home on a blueprint and then having the result become a three-storey office building. What is required is a design language that can both model the ontology in its full richness as the analysis model but also be able to model the full execution in perfect fidelity with the ontology model. In essence this means a single model that fully incorporates firstness, secondness, and thirdness in all their richness that can then either be fully interpreted on a machine or fully generated to some other machine language or a combination of both.

Only with such a powerful design or pattern language can we overcome the inherent limitations we have in our ability to comprehend the workings of some particular CAS, model, and then automate those portions that can be mechanized and continue to keep pace with the continually evolving CAS that we are seeking to *progressively mechanize*.

18.4 The Current State of the Art

The computing industry in general is still very much caught up in the mechanical metaphor. While object-oriented design and programming technology is more than decades old, and there is a *patterns movement* in the industry that is just over a decade old, there are very few examples of this new DSDL technology in use today. However, where it has been used the results have been dramatic, i. e., on the order of a 20-fold reduction of complexity as measured in *lines of code*. Lawson Landmark^{TM} is such a pattern language intended for the precise functional specification of business enterprise computer applications by domain specialists rather than programmers [18.11]. The result of the domain expert's work is then run through a metacompiler to produce Java code which can be loaded and tested against a transaction script also prepared by the same or another domain expert (i. e., accountant, supply-chain expert, etc.). The traditional application programmer does not appear in this modern business application development scenario because his or her function has been *automated*. It has taken nearly 50 years for specification-based programming to arrive at this level of development and utility and several other aggressive attempts at automation of intellectual functions have taken as long.

Automation of human intellectual functions is basically an aspect of AI, and the potential of AI is basically a philosophical rather than a technical question. The two major protagonists in this debate are the philosopher *Hubert Dreyfus* [18.12, 13] and *Ray Kurzweil* [18.14–16], an accomplished engineer. Dreyfus argues that computers will never ever achieve anything like human intelligence simply because they do not have consciousness. He gives the argument from what he calls associative or peripheral consciousness in the human brain. Kurzweil has written extensively as a futurist arguing that computers will soon exceed human intelligence and even develop spiritual capabilities.

Early in the computer era *Herbert Simon* hypothesized that a computer program that was able to play master-level chess would be exhibiting intelligence [18.17]. Dreyfus and other Strong AI opponents argue that it is not, because the chess program does not automatically play chess like a human does and therefore does not exhibit AI at all. They go on to show that is impossible for a program to play an even simpler game like Go well using this same technology. Playing chess was the straw man set up by Herbert Simon in the mid-1950s to be the benchmark of AI, i.e., if a computer could beat the best human chess player, then it would show true intelligence, but it actually does not. It took nearly 50 years to develop a special-purpose computer able to beat the leading human chess master, Gary Kasparov, but it does not automate or mechanize human intelligence to do so. In fact, it just came up with a different design for a computational algorithm that could play chess. This is analogous to the problem of flying. We did not copy how a bird flies but rather came up with a special-purpose design that suited our particular requirements. The Strong AI hypothesis appears to assume that all human thought, or at least intelligent thought, can be reduced to computation, and since computers can compute orders of magnitude faster than humans they will soon, says Ray Kurzweil, exhibit human-like intelligence, and eventually intelligence even superior to that of humans. Of course, the philosophers who gainsay the Strong AI hypothesis argue that not all intelligent human thought and its consequent behavior can be reduced to simple computation, or even logic.

An early goal of AI was to mechanize the function of an autonomous vehicle, that is, to give the vehicle a goal or set of objectives and the algorithms needed to accomplish them and let it function. The annual autonomous vehicle race in the Nevada desert shows that at least simple goals can be met autonomously or at least as well as a mildly intoxicated human driver. The semi-autonomous Mars rovers show off this technology even better but still fall far short of being *intelligent*.

Another of the early intrinsically human activities that AI researchers tried to automate was the typewriter. Prof. Marvin Minsky started his AI career as a graduate student at MIT in the mid-1950s working on a voice typewriter to type military correspondence for the Department of Defense (DoD). Now, more than 50 years

later, the technology is modestly successful as a software program.

At the same time Prof. Anthony Oettenger did his dissertation at the Harvard Computation Laboratory on automatic language translation with the *Automatic Russian–English Dictionary*. It was modestly successful for a narrow genre of Russian technical literature and created the whole new field of computational linguistics. Today, more than 50 years later, a similar technology is available as a software program for Russian and a few other languages with a library of genre-specific and technical-area-specific vocabulary plug-ins sold separately. The best success story on automatic language translation today is the European Economic Community (EEC), which writes its memos in French in the Brussels headquarters and converts them into the 11 languages of the EEC and then sends them out to member nations. French bureaucratese is a very narrow language genre and probably the easiest case for autotranslation automation due not only to the genre but the source language. No one has ever suggested that *Eugene Onegin* will ever be translated automatically from Russian to English, since so far it has even resisted human efforts to do so. Amazing as it may seem, Shakespeare's poetry translates beautifully to Russian but Pushkin's does not translate easily to English.

Linguists are divided, like philosophers, on whether computational linguistics technology will ever really achieve true automation, but we think that it probably will, subject to human pre- and post-editing in actual volume production practice, and then only for very narrow subject areas in a few restricted genres. Technical prose in which the translator is only copying one fact at a time from one language to another will be possible, but poetry will always be too complex, since poetry always violates the rules of grammar of its own source language.

18.5 A General Principle

What we need is a general principle which will cleanly divide all the things that can be done into those which can be automated and those which cannot be automated. You cannot automate what you cannot do manually, but the converse it not true, since you cannot always automate everything you *can* do manually [18.4, 18, 19]. However, this principle is much too blunt. In his book *Darwin's Black Box* Professor *Michael Behe* argued

from the principle of irreducible complexity that neo-Darwinism was inadequate to explain the development of the eye or of the endocrine system because too many mutations and their immediate useful adaptations had to happen at once, since each of 20 or more required mutations would have no competitive advantage in adaptation individually [18.20]. In his second book *The Edge of Evolution* he sharpens his principle significantly to divide biological adaptations into those which can be explained by neo-Darwinism (single mutations) and those which cannot (multiple mutations). The refined principle, while sharper, still leaves a ragged edge between the two classes of adaptive biological systems, according to *Behe* [18.21].

In our case we postulate the principle of design. Anything that can be copied can be copied automatically. However, any process involving design cannot be automated and any sufficiently (i.e., irreducibly) complex adaptive system cannot be automated (as Behe shows), however simple adaptive systems can be modeled to some extent mechanistically. Malaria can become resistant to a new antibiotic in weeks by Darwin's *black box* if it requires only a one-gene change, however in 10 000 years malaria has not been able to overcome the cell-cycle mutation in humans because it would require too many concurrent mutations.

We conclude that anything that can be reduced to an algorithm or computational process can be automated, but that some things, like most human thought and most functions of complex adaptive systems, are not reducible to a logical algorithm or a computational process and therefore cannot be automated.

References

18.1 *American Machinist*, 21 Oct. 1948. Creation of the term *automation* is usually attributed to Delmar S. Harder

18.2 J.M. Schwartz, S. Begley: *The Mind and the Brain: Neuroplasticity and the Power of Mental Force* (Harper, New York 2002)

18.3 K.L. von Bertalanffy: *General System Theory: Foundations, Development, Applications* (George Braziller, New York 1976) p. 56

18.4 D.D. Spencer: *What Computers Can Do* (Macmillan, New York 1984)

18.5 K.L. von Bertalanffy: *General System Theory: Foundations, Development, Applications* (George Braziller, New York 1976) p. 213, revised edition

18.6 J. Holland: *Hidden Order, How Adaptation Builds Complexity* (Addison-Wesley, Reading 1995)

18.7 C. Alexander: *The Timeless Way of Building* (Oxford, New York 1979)

18.8 C. Alexander: *A Pattern Language* (Oxford, New York 1981)

18.9 C. S. Peirce: *Collected Papers of Charles Sanders Peirce*, Vols. 1–6 ed. by C. Hartshorne, P. Weiss, 1931–1935; Vols. 7–8 ed. by A. W. Burks, (Harvard University Press, Cambridge 1958)

18.10 K.L. von Bertalanffy: *General System Theory: Foundations, Development, Applications* (George Braziller, New York 1976) p. 237

18.11 B.K. Jyaswal, P.C. Patton: *Design for Trustworthy Software: Tools, Techniques and Methodology of Producing Robust Software* (Prentice-Hall, Upper Saddle River 2006) p. 501

18.12 H. Dreyfus: *What Computers Can't Do: A Critique of Artificial Intelligence* (Harper Collins, New York 1978)

18.13 H. Dreyfus: *What Computers Still Can't Do: A Critique of Artificial Reason* (MIT Press, Cambridge 1992)

18.14 R. Kurzweil: *The Age of Intelligent Machines* (MIT Press, Cambridge 1992)

18.15 R. Kurzweil: *The Age of Spiritual Machines: When Computers Exceed Human Intelligence* (Viking, New York 1999)

18.16 R. Kurzweil: *The Singularity is Near: When Humans Transcend Biology* (Viking, New York 2005)

18.17 H. Simon: Perception in chess, Cognitive Psychol. **4**, 11–27 (1973)

18.18 B. W. Arden (Ed.): *What Can Be Automated?*, The Computer Science and Engineering Research Study (COSERS) Ser., (MIT Press, Cambridge 1980)

18.19 I.Q. Wilson, M.E. Wilson: *What Computers Cannot Do* (Vertex, New York 1970)

18.20 M. Behe: *Darwin's Black Box: The Biochemical Challenge to Evolution* (Free, New York 2006)

18.21 M. Behe: *The Edge of Evolution: The Search for the Limits of Darwinism* (Free, New York 2007)

Part C
Automat

Part C Automation Design: Theory, Elements, and Methods

Automation Design: Theory, Elements, and Methods Part C From theory to building automation machines, systems, and systems-of-systems this part explains the fundamental elements of mechatronics, sensors, robots, and other components useful for automation, and how they are combined with control and automation software, including models and techniques for automation software engineering, and the automation of the design process itself. Design theories and methods cover also soft automation, automation modeling and programming languages, real-time and autonomic techniques, and emerging networking and service grids for automation. Human factors engineering and science in the design of automation, including interaction and interface design, and issues of trust and collaboration focus on systems and infrastructures integrating people with decision-support and with teleoperated, remote automatic equipment. Also in this part are advanced design methods and tools of distributed agents, evolutionary techniques and computing algorithms for automation, and design of eight key automation functions to prevent or recover from errors and conflicts, to assure automation reliability and sustainability.

19. Mechatronic Systems – A Short Introduction

Rolf Isermann

Many technical processes and products in the area of mechanical and electrical engineering show increasing integration of mechanics with digital electronics and information processing. This integration is between the components (hardware) and the information-driven functions (software), resulting in integrated systems called mechatronic systems. Their development involves finding an optimal balance between the basic mechanical structure, sensor and actuator implementation, and automatic information processing and overall control. Frequently formerly mechanical functions are replaced by electronically controlled functions, resulting in simpler mechanical structures and increased functionality. The development of mechatronic systems opens the door to many innovative solutions and synergetic effects which are not possible with mechanics or electronics alone. This technical progress has a very strong influence on a multitude of products in the areas of mechanical, electrical, and electronic engineering and is increasingly changing the design, for example, of conventional electromechanical components, machines, vehicles, and precision mechanical devices.

19.1 From Mechanical to Mechatronic Systems

Mechanical systems generate certain motions or transfer forces or torques. For the oriented command of, e.g., displacements, velocities or forces, feedforward and feedback control systems have been applied for many years. The control systems operate either without auxiliary energy (e.g., a fly-ball governor), or with electrical, hydraulic or pneumatic auxiliary energy, to manipulate the commanded variables directly or with a power amplifier. A realization with added fixed wired (analog) devices turns out to enable only relatively simple and limited control functions. If these analog devices are replaced with digital computers in the form of, e.g., online coupled microcomputers, the information processing can be designed to be considerably more flexible and more comprehensive.

Figure 19.1 shows the example of a machine set, consisting of a power-generating machine (DC motor) and a power-consuming machine (circulation pump): (a) a scheme of the components, (b) the resulting sig-

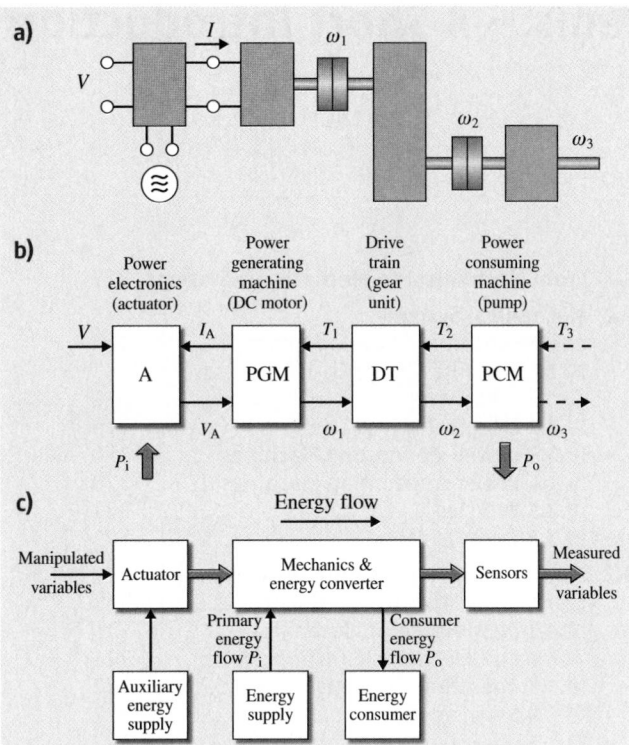

Fig. 19.1a–c Schematic representation of a machine set: **(a)** scheme of the components; **(b)** signal flow diagram (two-port representation); **(c)** open-loop process. V – voltage; V_A – armature voltage; I_A – armature current; T – torque; ω – angular frequency; P_i – drive power; P_o – consumer power◄

nal flow diagram in two-port representation, and (c) the open-loop process with one or several manipulated variables as input variables and several measured variables as output variables. This process is characterized by different controllable energy flows (electrical, mechanical, and hydraulic). The first and last flow can be manipulated by a manipulated variable of low power (auxiliary power), e.g., through a power electronics device and a flow valve actuator. Several sensors yield measurable variables. For a mechanical–electronic system, a digital electronic system is added to the process. This electronic system acts on the process based on the measurements or external command variables in a feedforward or feedback manner (Fig. 19.2). If then the electronic and the mechanical system are merged to an autonomous overall system, an integrated mechanical–electronic system results. The electronics processes information, and such a system is characterized at least by a mechanical energy flow and an information flow.

These integrated mechanical–electronic systems are increasingly called mechatronic systems. Thus, *mecha*nics and elec*tronics* are joined. The word *mechatronics* was probably first created by a Japanese engineer in 1969 [19.1] and had a trademark by a Japanese company until 1972 [19.2]. Several definitions can be found in [19.3–7]. All definitions agree that mechatronics is an interdisciplinary field, in which the following disciplines act together (Fig. 19.3):

- Mechanical systems (mechanical elements, machines, precision mechanics)
- Electronic systems (microelectronics, power electronics, sensor and actuator technology)
- Information technology (systems theory, control and automation, software engineering, artificial intelligence).

The *solution of tasks* to design mechatronic systems is performed on the mechanical as well as on the digital-electronic side. Thus, interrelations during design play an important role; because the mechan-

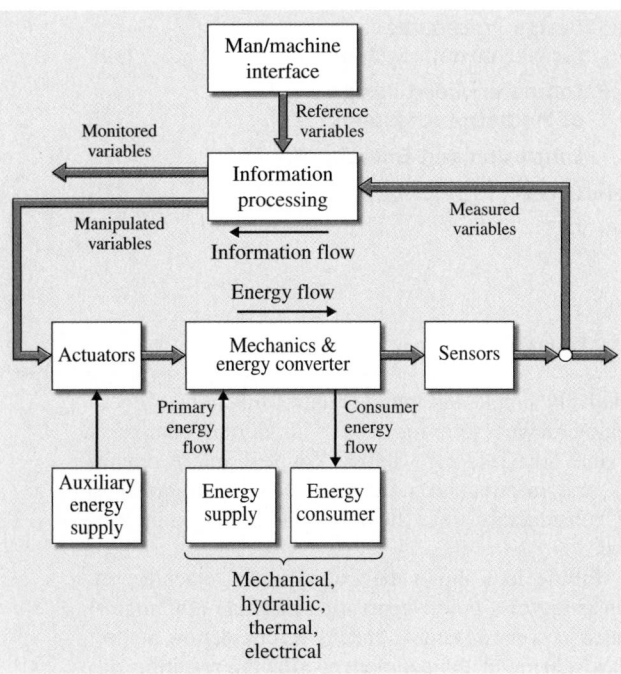

Fig. 19.2 Mechanical process and information processing develop towards a mechatronic system◄

Fig. 19.3 Mechatronics: synergetic integration of different disciplines▶

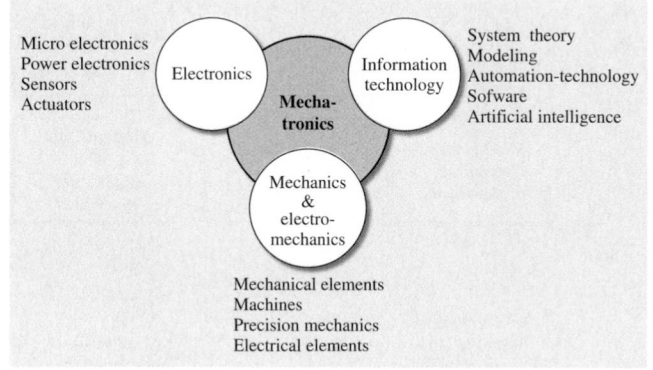

ical system influences the electronic system, and vice versa, the electronic system influences the design of the mechanical system (Fig. 19.4). This means that simultaneous engineering has to take place, with the goal of designing an overall integrated system (an *organic system*) and also creating synergetic effects.

A further feature of mechatronic systems is integrated digital information processing. As well as basic control functions, more sophisticated control functions may be realized, e.g., calculation of nonmeasurable variables, adaptation of controller parameters, detection and diagnosis of faults and, in the case of failures, reconfiguration to redundant components. Hence, mechatronic systems are developing with adaptive or even learning behavior, which can also be called intelligent mechatronic systems.

The developments to date can be found in [19.2, 7–11]. An insight into general aspects are given editorially in journals [19.5, 6], conference proceedings such as [19.12–17], journal articles by [19.18–21], and books [19.22–27]. A summary of research projects at the Darmstadt University of Technology can be found in [19.28].

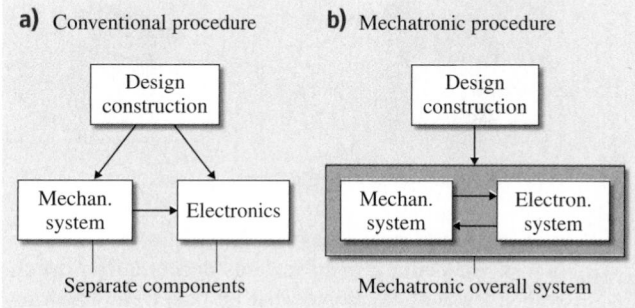

Fig. 19.4a,b Interrelations during the design and construction of mechatronic systems

19.2 Mechanical Systems and Mechatronic Developments

Mechanical systems can be applied to a large area of mechanical engineering. According to their construction, they can be subdivided into mechanical components, machines, vehicles, precision mechanical devices, and micromechanical components.

The design of mechanical products is influenced by the interplay of energy, matter, and information. With regard to the basic problem and its solution, frequently either the energy, matter or information flow is dominant. Therefore, one main flow and at least one side flow can be distinguished [19.29].

In the following some examples of mechatronic developments are given. The area of mechanical components, machines, and vehicles is covered by Fig. 19.5.

19.2.1 Machine Elements, Mechanical Components

Machine elements are usually purely mechanical. Figure 19.5 shows some examples. Properties that can be

improved by electronics are, for example, self-adaptive stiffness and damping, self-adaptive free motion or pretension, automatic operating functions such as coupling or gear shifting, and supervisory functions. Some examples of mechatronic approaches are hydrobearings for combustion engines with electronic control of damping, magnetic bearings with position control [19.30], automatic electronic–hydraulic gears [19.31], and adaptive shock absorbers for wheel suspensions [19.32].

19.2.2 Electrical Drives and Servo Systems

Electrical drives with direct-current, universal, asynchronous, and synchronous motors have used integration with gears, speed sensors or position sensors and power electronics for many years. Especially the development of transistor-based voltage supplies and cheaper power electronics on the basis of transistors and thyristors with variable-frequency three-phase cur-

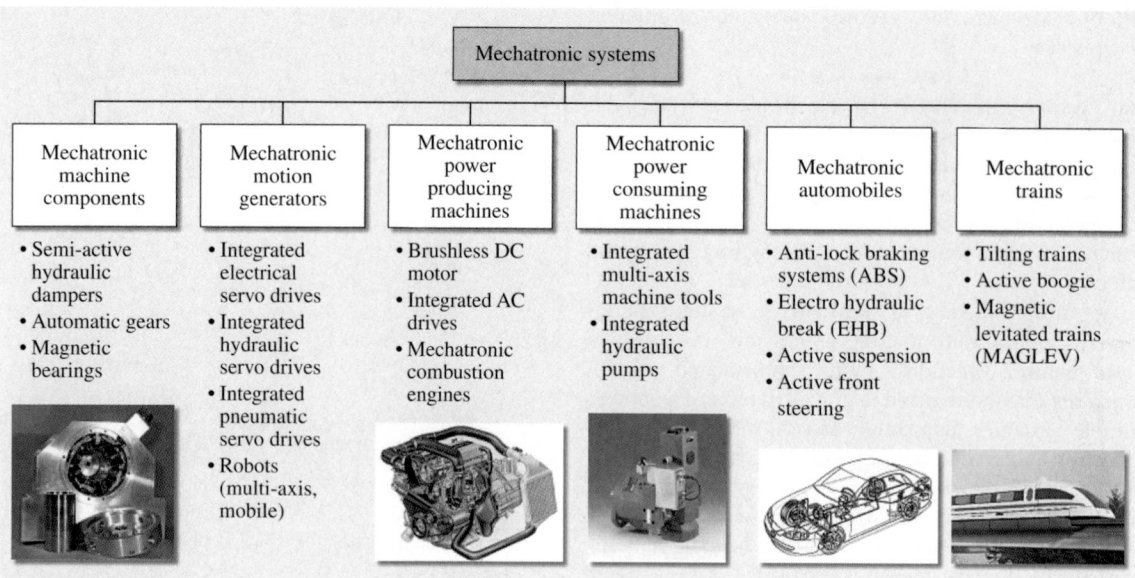

Fig. 19.5 Examples of mechatronic systems

rent supported speed control drives also for smaller power. Herewith, a trend towards decentralized drives with integrated electronics can be observed. The way of integration or attachment depends, e.g., on space requirement, cooling, contamination, vibrations, and accessibility for maintenance. Electrical servo drives require special designs for positioning.

Hydraulic and pneumatic servo drives for linear and rotary positioning show increasingly integrated sensors and control electronics. Motivations are requirements for easy-to-assemble drives, small space, fast change, and increased functions [19.33].

Multiaxis robots and mobile robots show mechatronic properties from the beginning of their design.

19.2.3 Power-Generating Machines

Machines show an especially broad variability. Power-producing machines are characterized by the conversion of hydraulic, thermodynamic or electrical energy and delivery of power. Power-consuming machines convert mechanical energy to another form, thereby absorbing energy. Vehicles transfer mechanical energy into movement, thereby consuming power.

Examples of mechatronic electrical power-generating machines are brushless DC motors with electronic commutation or speed-controlled asynchronous and synchronous motors with variable-frequency power converters.

Combustion engines increasingly contain mechatronic components, especially in the area of actuators. Gasoline engines showed, for example, the following steps of development: microelectronic-controlled injection and ignition (1979), electrical throttle (1991), direct injection with electromechanical (1999) and piezoelectric injection valves (2003), variable valve control (2004); see, for example, [19.34].

Diesel engines first had mechanical injection pumps (1927), then analog-electronic-controlled axial piston pumps (1986), and digital-electronic-controlled high-pressure pumps, since 1997 with common-rail systems [19.35]. Further developments are the exhaust turbochargers with wastegate or controllable vanes (variable turbine geometry, VTG), since about 1993.

19.2.4 Power-Consuming Machines

Examples of mechatronic power-consuming machines are multiaxis machine tools with trajectory control, force control, tools with integrated sensors, and robot transport of the products; see, e.g., [19.36]. In addition to these machine tools with open kinematic chains between basic frame and tools and linear or rotary axes with one degree of freedom, machines with parallel kinematics will be developed. Machine tools show a tendency towards magnetic bearings if ball bearings cannot be applied for high speeds, e.g., for high-speed milling of aluminum, and also for ultracentrifuges [19.37]. Within the area of manufacturing, many machinery, sorting, and transportation devices are characterized by integration with electronics, but as yet they are mostly not fully hardware-integrated. For hydraulic piston pumps the control electronics is now attached to the casing [19.33]. Further examples are packing machines with decentralized drives and trajectory control or offset-printing machines with replacement of the mechanical synchronization axis through decentralized drives with digital electronic synchronization and high precision.

19.2.5 Vehicles

Many mechatronic components have been introduced, especially in the area of vehicles, or are in development: antilock braking control (ABS) [19.38], controllable shock absorbers [19.39], controlled adaptive suspensions [19.40], active suspensions [19.41, 42], drive dynamic control through individual braking (electronic

stability program, ESP) [19.43, 44], electrohydraulic brakes (2001), and active front steering (AFS) (2003). Of the innovations for vehicles 80–90% are based on electronic/mechatronic developments. Here, the value of electronics/electrics of vehicles increases to about 30% or more.

19.2.6 Trains

Trains with steam, diesel or electrical locomotives have followed a very long development. For wagons the design with two boogies with two axes are standard. ABS braking control can be seen as the first mechatronic influence in this area [19.45, 46]. The high-speed trains (TGV, ICE) contain modern asynchronous motors with power electronic control. The trolleys are supplied with electronic force and position control. Tilting trains show a mechatronic design (1997) and actively damped and steerable boogies also [19.47]. Further, magnetically levitated trains are based on mechatronic construction; see, e.g., [19.47].

19.3 Functions of Mechatronic Systems

Mechatronic systems enable, after the integration of the components, many improved and also new functions. This will be discussed by using examples.

19.3.1 Basic Mechanical Design

The basic mechanical construction first has to satisfy the task of transferring the mechanical energy flow (force, torque) to generate motions or special movements, etc. Known traditional methods are applied, such as material selection, calculation of strengths, manufacturing, production costs, etc. By attaching sensors, actuators, and mechanical controllers, in earlier times, simple control functions were realized, e.g., the fly-ball governor. Then gradually pneumatic, hydraulic, and electrical analog controllers were introduced. After the advent of digital control systems, especially with the development of microprocessors around 1975, the information processing part could be designed to be much more sophisticated. These digitally controlled systems were first added to the basic mechanical construction and were limited by the properties of the sensors, actuators, and electronics, i.e., they frequently did not satisfy reliability and lifetime requirements under rough environmental conditions (temperature, vibrations, contamination)

and had a relatively large space requirement and cable connections, and low computational speed. However, many of these initial drawbacks were removed with time, and since about 1980 electronic hardware has become greatly miniaturized, robust, and powerful, and has been connected by field bus systems. Based on this, the emphasis on the electronic side could be increased and the mechanical construction could be designed as a mechanical–electronic system from the very beginning. The aim was to result in more autonomy, for example, by decentralized control, field bus connections, plug-and-play approaches, and distributed energy supply, such that self-contained units emerge.

19.3.2 Distribution of Mechanical and Electronic Functions

In the design of mechatronic systems, interplay for the realization of functions in the mechanical and electronic parts is crucial. Compared with pure mechanical realizations, the use of amplifiers and actuators with electrical auxiliary energy has already led to considerable simplifications, as can be seen in watches, electronic typewriters, and cameras. A further considerable simplification in the mechanics resulted from the introduction of microcomputers in connection with decentralized electrical drives, e.g., for electronic typewriters, sewing machines, multiaxis handling systems, and automatic gears.

The design of lightweight constructions leads to elastic systems that are weakly damped through the material itself. Electronic damping through position, speed or vibration sensors and electronic feedback can be realized with the additional advantage of adjustable damping through algorithms. Examples are elastic drive trains of vehicles with damping algorithms in the engine electronics, elastic robots, hydraulic systems, far-reaching cranes, and space constructions (e.g., with flywheels).

The addition of closed-loop control, e.g., for position, speed or force, does not only result in precise tracking of reference variables, but also an approximate linear overall behavior, even though mechanical systems may show nonlinear behavior. By omitting the constraint of linearization on the mechanical side, the effort for construction and manufacturing may be reduced. Examples are simple mechanical pneumatic and electromechanical actuators and flow valves with electronic control.

With the aid of freely programmable reference variable generation, the adaptation of nonlinear mechanical systems to the operator can be improved. This is already used for driving-pedal characteristics within engine electronics for automobiles, telemanipulation of vehicles and aircraft, and in the development of hydraulically actuated excavators and electric power steering.

However, with increasing number of sensors, actuators, switches, and control units, the cables and electrical connections also increase, such that reliability, cost, weight, and required space are major concerns. Therefore, the development of suitable bus systems, plug systems, and fault-tolerant and reconfigurable electronic systems are challenges for the designer.

19.3.3 Operating Properties

By applying active feedback control, the precision of, e.g., a position is reached by comparison of a programmed reference variable with a measured control variable and not only through the high mechanical precision of a passively feedforward-controlled mechanical element. Therefore, the mechanical precision in design and manufacturing may be reduced somewhat and simpler constructions for bearings or slideways can be used. An important aspect in this regard is compensation

of larger and time-variant friction by adaptive friction compensation. Larger friction at the cost of backlash may also be intended (e.g., gears with pretension), because it is usually easier to compensate for friction than for backlash. Model-based and adaptive control allow operation at more operating points (wide-range operation) compared with fixed control with unsatisfactory performance (danger of instability or sluggish behavior). A combination of robust and adaptive control enables wide-range operation, e.g., for flow, force, and speed control, and for processes involving engines, vehicles, and aircraft. Better control performance allows the reference variables to be moved closer to constraints with improved efficiencies and yields (e.g., higher temperatures, pressures for combustion engines and turbines, compressors at stalling limits, and higher tensions and higher speed for paper machines and steel mills).

19.3.4 New Functions

Mechatronic systems also enable functions that could not be performed without digital electronics. Firstly, nonmeasurable quantities can be calculated on the basis of measured signals and influenced by feedforward or feedback control. Examples are time-dependent variables such as the slip for tires, internal tensions, temperatures, the slip angle and ground speed for steering control of vehicles or parameters such as damping and stiffness coefficients, and resistances. The automatic adaptation of parameters, such as damping and stiffness for oscillating systems based on measurements of displacements or accelerations, is another example. Integrated supervision and fault diagnosis becomes increasingly important with more automatic functions, increasing complexity, and higher demands on reliability and safety. Then, fault tolerance by triggering of redundant components and system reconfiguration, maintenance on request, and any kind of teleservice makes the system more *intelligent*.

19.3.5 Other Developments

Mechatronic systems frequently allow flexible adaptation to boundary conditions. A part of the functions and also precision becomes programmable and rapidly changeable. Advanced simulations enable the reduction of experimental investigations with many parameter variations. Also, shorter time to market is possible if the basic elements are developed in parallel and the functional integration results from the software.

Table 19.1 Some properties of conventional and mechatronic designed systems

Conventional design	Mechatronic design
Added components	**Integration of components (hardware)**
Bulky	Compact
Complex	Simple mechanisms
Cable problems	Bus or wireless communication
Connected components	Autonomous units
Simple control	**Integration by information processing (software)**
Stiff construction	Elastic construction with damping by electronic feedback
Feedforward control, linear (analog) control	Programmable feedback (nonlinear) digital control
Precision through narrow tolerances	Precision through measurement and feedback control
Nonmeasurable quantities change arbitrarily	Control of nonmeasurable estimated quantities
Simple monitoring	Supervision with fault diagnosis
Fixed abilities	Adaptive and learning abilities

A far-reaching integration of the process and the electronics is much easier if the customer obtains the functioning system from one manufacturer. Usually, this is the manufacturer of the machine, the device or the apparatus. Although these manufacturers have to invest a lot of effort in coping with the electronics and the information processing, they gain the chance to add to the value of the product. For small devices and machines with large production numbers, this is obvious. In the case of larger machines and apparatus, the process and its automation frequently comes from different manufacturers. Then, special effort is needed to produce integrated solutions.

Table 19.1 summarizes some properties of mechatronic systems compared with conventional electromechanical systems.

19.4 Integration Forms of Processes with Electronics

Figure 19.6a shows a general scheme of a classical mechanical–electronic system. Such systems resulted from adding available sensors and actuators and analog or digital controllers to the mechanical components. The limits of this approach were the lack of suitable sensors and actuators, unsatisfactory lifetime under rough operating conditions (acceleration, temperature, and contamination), large space requirements, the required cables, and relatively slow data processing. With increasing improvements in the miniaturization, robustness, and computing power of microelectronic components, one can now try to place more emphasis on the electronic side and design the mechanical part from the beginning with a view to a mechatronic overall system. Then, more autonomous systems can be envisaged, e.g., in the form of encapsulated units with noncontacting signal transfer or bus connections and robust microelectronics.

Integration within a mechatronic system can be performed mainly in two ways: through the integration of components and through integration by information processing (see also Table 19.1).

The integration of components (hardware integration) results from designing the mechatronic system as an overall system and embedding the sensors, actuators, and microcomputers into the mechanical process (Fig. 19.6b). This spatial integration may be limited to the process and sensor or the process and actuator. The microcomputers can be integrated with the actuator, the process or sensor, or be arranged at several places. Integrated sensors and microcomputers lead to smart sensors, and integrated actuators and microcomputers develop into smart actuators. For larger systems, bus connections will replace the many cables. Hence, there are several possibilities for building an integrated overall system by proper integration of the hardware.

Integration by information processing (software integration) is mostly based on advanced control functions. Besides basic feedforward and feedback control, an additional influence may take place through process knowledge and corresponding online information processing (Fig. 19.6c). This means processing of available signals at higher levels, as will be discussed in the next Section. This includes the solution of tasks

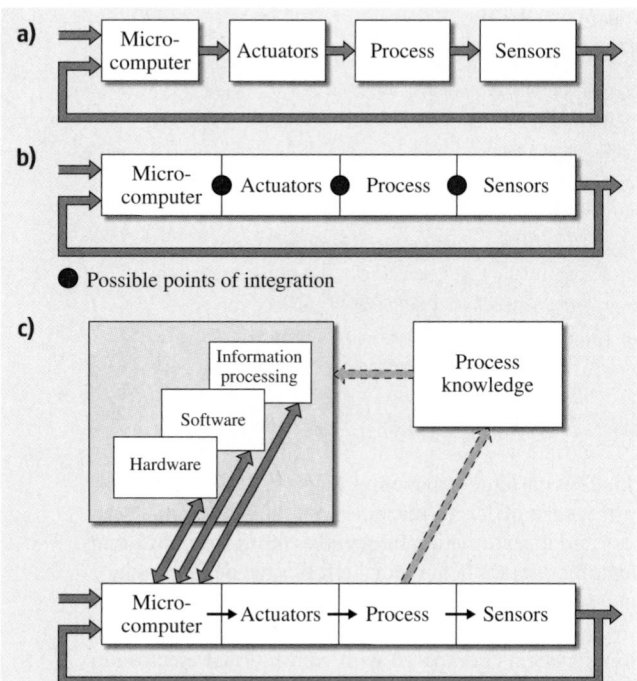

Fig. 19.6a–c Integration of mechatronic systems: (**a**) general scheme of a (classical) mechanical–electronic system; (**b**) integration through components (hardware integration); (**c**) integration through functions (software integration)

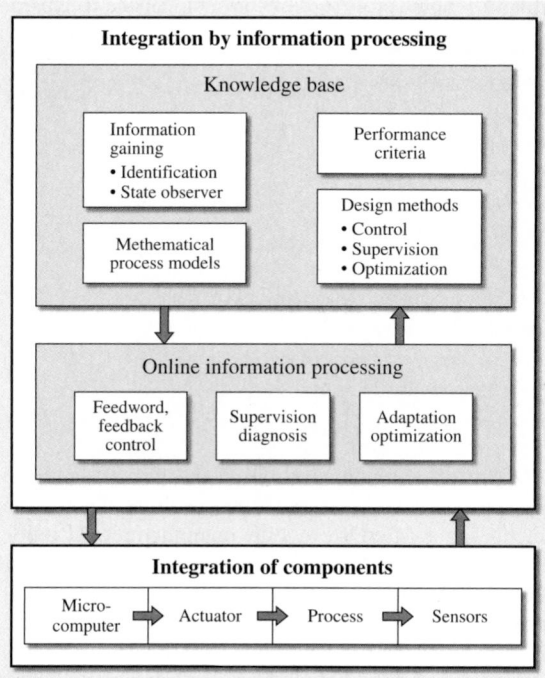

Fig. 19.7 Integration of mechatronic systems: integration of components (hardware integration); integration by information processing (software integration)

such as supervision with fault diagnosis, optimization, and general process management. The corresponding problem solutions result in online information processing, especially using real-time algorithms, which must be adapted to the properties of the mechanical process, e.g., expressed by mathematical models in the

form of static characteristics, differential equations, etc. (Fig. 19.7). Therefore, a knowledge base is required, comprising methods for design and information gain, process models, and performance criteria. In this way, the mechanical parts are governed in various ways through higher-level information processing with intelligent properties, possibly including learning, thus resulting in integration with process-adapted software. Both types of integration are summarized in Fig. 19.7. In the following, mainly integration through information processing will be considered.

Recent approaches for mechatronic systems mostly use signal processing at lower levels, e.g., damping or control of motions or simple supervision. Digital information processing, however, allows the solutions of many more tasks, such as adaptive control, learning control, supervision with fault diagnosis, decisions for maintenance or even fault-tolerance actions, economic optimization, and coordination. These

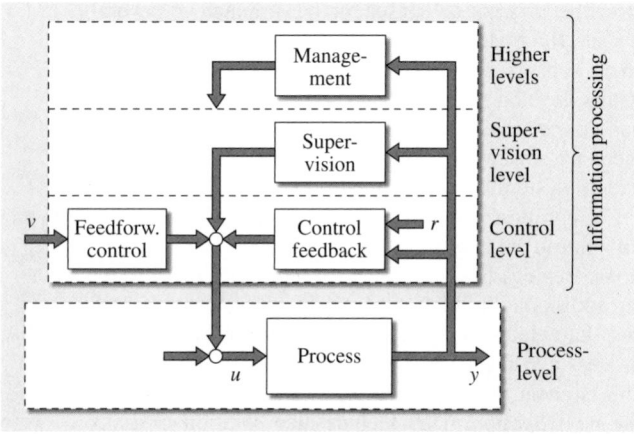

Fig. 19.8 Different levels of information processing for process automation. u: manipulated variables; y: measured variables; v: input variables; r: reference variables ◄

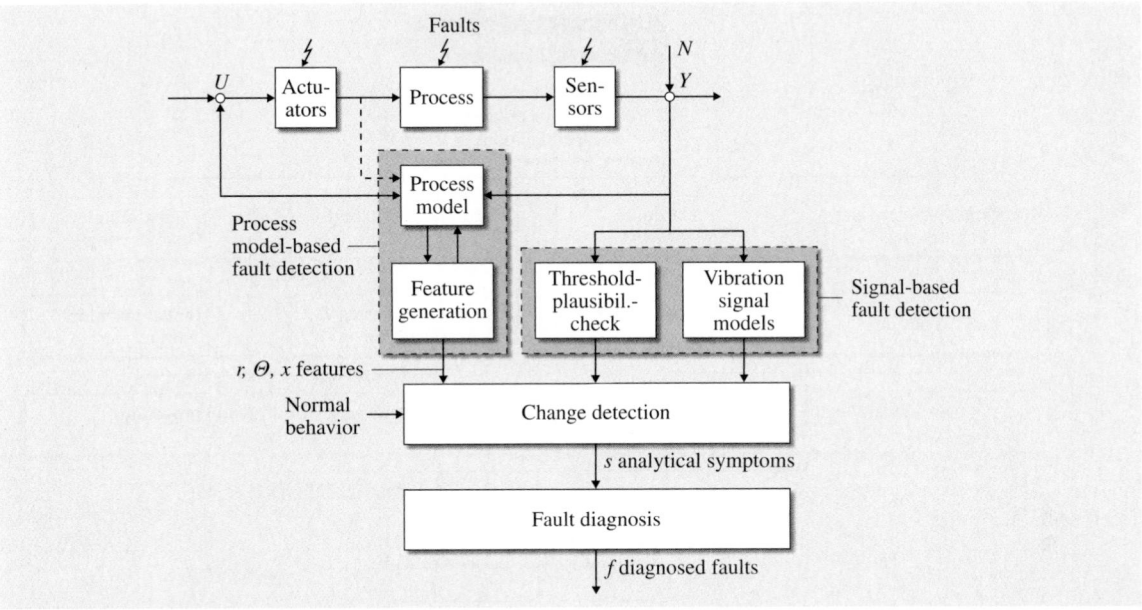

Fig. 19.9 Scheme for model-based fault detection

higher-level tasks are sometimes summarized as *process management*. Information processing at several levels under real-time condition is typical for extensive *process automation* (Fig. 19.8).

With the increasing number of automatic functions (autonomy) including electronic components, sensors, and actuators, increasing complexity, and increasing demands on reliability and safety, integrated supervision with fault diagnosis becomes increasingly important. This is, therefore, a significant natural feature of an intelligent mechatronic system. Figure 19.9 shows a process influenced by faults. These faults indicate unpermitted deviations from normal states and can be generated either externally or internally. External faults are, e.g., caused by the power supply, contam-

ination or collision, internal faults by wear, missing lubrication, and actuator or sensor faults. The classic methods for fault detection are limit-value checking and plausibility checks of a few measurable variables. However, incipient and intermittent faults cannot usually be detected, and in-depth fault diagnosis is not possible with this simple approach. Therefore, model-based fault detection and diagnosis methods have been developed in recent years, allowing early detection of small faults with normally measured signals, also in closed loops [19.48–51]. Based on measured input signals $U(t)$, output signals $Y(t)$, and process models, features are generated by, e.g., parameter estimation, state and output observers, and parity equations (Fig. 19.9).

19.5 Design Procedures for Mechatronic Systems

The design of mechatronic systems requires systematic development and use of modern software design tools. As with any design, mechatronic design is also an iterative procedure. However, it is much more involved than for pure mechanical or electrical systems. Figure 19.10 shows that, in addition to traditional domain-specific engineering, integrated simultaneous (concurrent) engineering is required due to the inte-

gration of engineering across traditional boundaries that is typical of the development of mechatronic systems.

Hence, the mechatronic design requires a simultaneous procedure in broad engineering areas. Traditionally, the design of mechanics, electrics and electronics, control, and human–machine interface were performed in different departments with only occasionally contact,

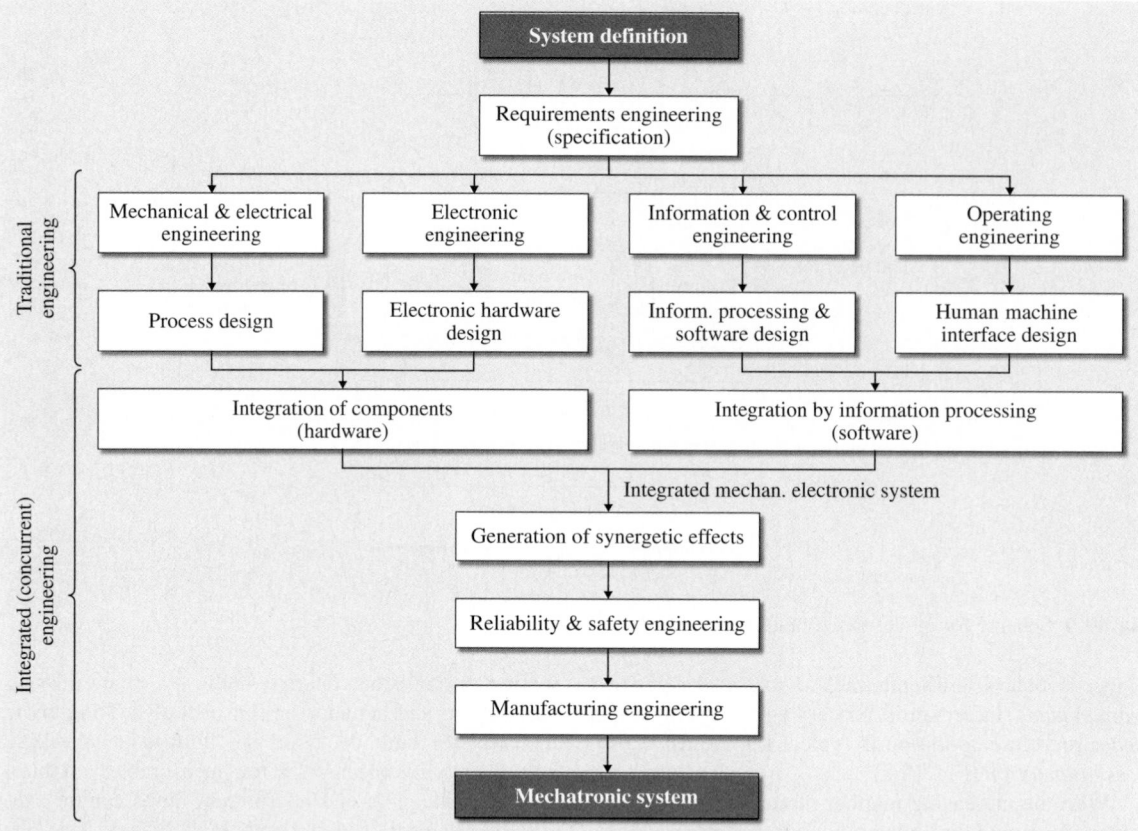

Fig. 19.10 From domain-specific traditional engineering to integrated, simultaneous engineering (iteration steps are not indicated)

sometimes sequentially (bottom-up design). Because of the requirements for integration of hardware and software functions, these areas have to work together and the products have to be developed more or less simultaneously to an overall optimum (concurrent engineering, top-down design). Usually, this can only be realized with suitable teams.

The principle procedure for the design of mechatronic systems is, e.g., described in the VDI-Richtlinie (guideline) 2206 [19.11]. A flexible procedural model is described, consisting of the following elements:

1. Cycles of problem solutions at microscale:

- Search for solutions by analysis and synthesis of basis steps
- Comparison of requirements and reality
- Performance and decisions
- Planning

2. Macroscale cycles in the form of a V-model:

- Logical sequence of steps
- Requirements
- System design
- Domain-specific design
- System integration
- Verification and validation
- Modeling (supporting)
- Products: laboratory model, functional model, pre-series product

3. Process elements for repeating working steps:

- Repeating process elements
- System design, modeling, element design, integration, ...

The V-model, according to [19.11, 52], is distinguished with regard to system design and system

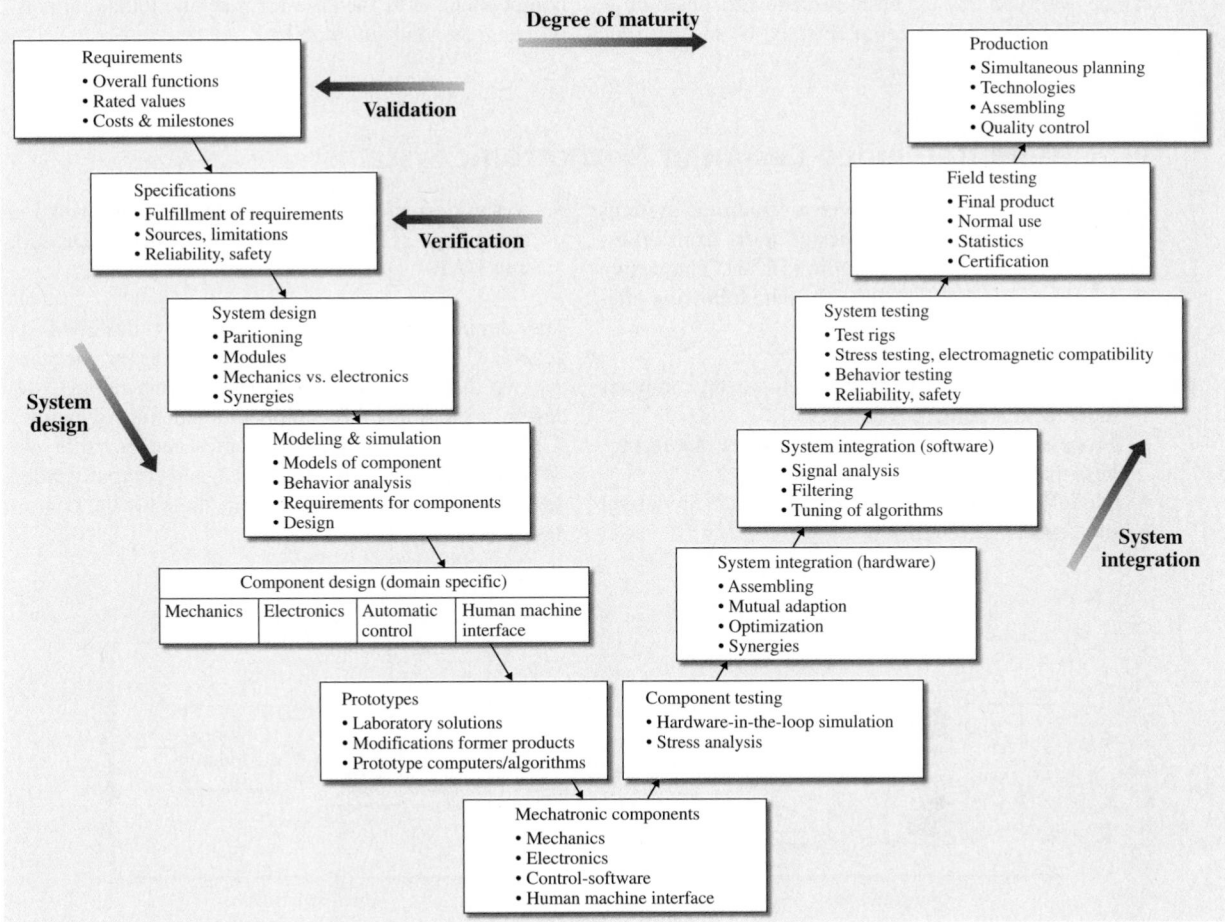

Fig. 19.11 A "V" development scheme for mechatronic systems

integration with domain-specific design in mechanical engineering, electrical engineering, and information processing. Usually, several design cycles are required, resulting, e.g., in the following intermediate products:

- Laboratory model: first functions and solutions, rough design, first function-specific investigations
- Functional model: further development, fine-tuning, integration of distributed components, power measurements, standard interfaces
- Pre-series product: consideration of manufacturing, standardization, further modular integration steps, encapsulation, field tests.

The V-model originates most likely from software development [19.53]. Some important design steps for mechatronic systems are shown in Fig. 19.11 in the form of an extended V-model, where the following are distinguished: system design up to laboratory model, system integration up to functional model, and system tests up to pre-series product.

The maturity of the product increases as the individual steps of the V-model are followed. However, several iterations have to be performed, which is not illustrated in the figure.

Depending on the type of product, the degree of mechatronic design is different. For precision mechanic devices the integration is already well developed. In the case of mechanical components one can use as a basis well-proven constructions. Sensors, actuators, and electronics can be integrated by corresponding changes, as can be seen, e.g., in adaptive shock absorbers, hydraulic brakes, and fluidic actuators. In machines and vehicles it

19.6 Computer–Aided Design of Mechatronic Systems

The general goal in the design of mechatronic systems is the use of computer-aided design tools from different domains. A survey is given in [19.11]. The design model given in [19.52] distinguishes the following integration levels:

- Basic level: specific product development, computer-aided engineering (CAE) tools
- Process-oriented level: design packages, status, process management, data management
- Model-oriented level: common product model for data exchange (STEP)

- System-oriented level: coupling of information technology (IT) tools with, e.g., CORBA, DCOM, and JAVA

The domain-specific design is usually designed on general CASE tools, such as CAD/CAE for mechanics, two-dimensional (2-D) and three-dimensional (3-D) design with AutoCAD, computational fluid dynamics (CFD) tools for fluidics, electronics and board-layout (PADS), microelectronics (VHDL), and computer aided design of control systems CADCS tools for the control design (see, e.g., [19.52]).

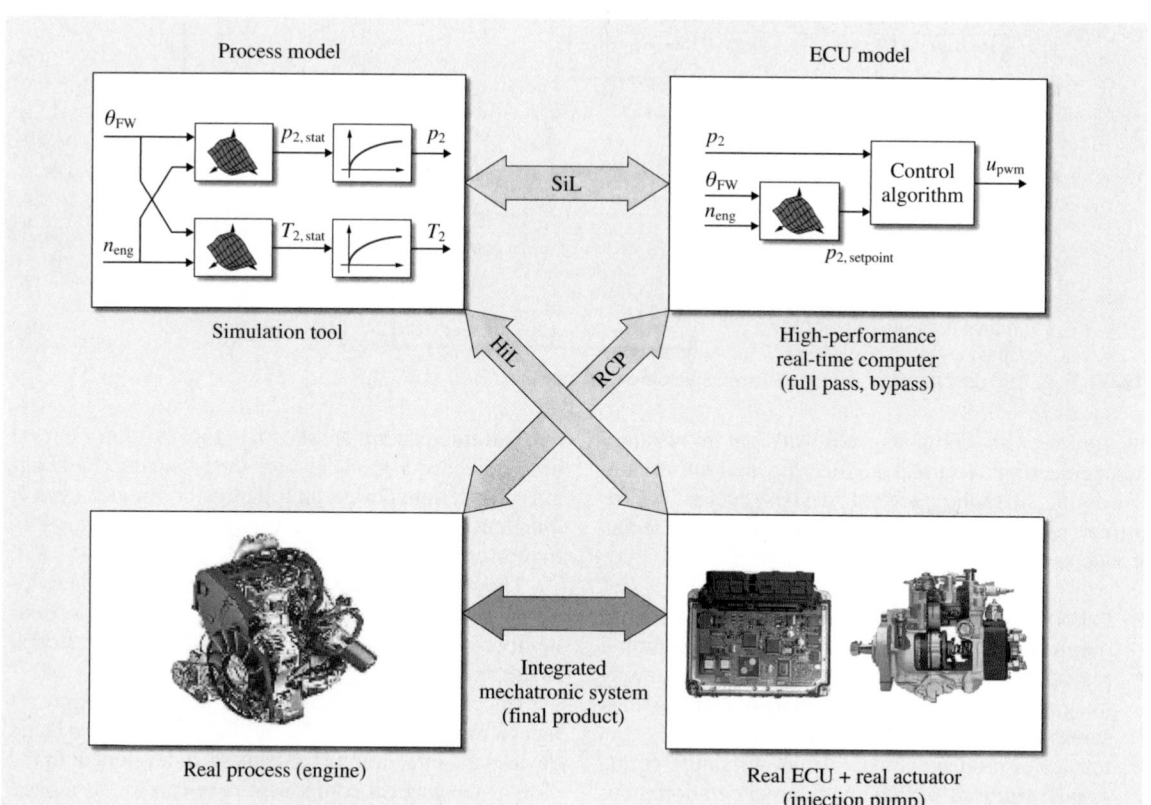

Fig. 19.12 Different couplings of process and electronics for a mechatronic design. SiL: software-in-the-loop; RCP: rapid control prototyping; HiL: hardware-in-the-loop

For overall modeling, object-oriented software is especially of interest based on the use of general model-building laws. The models are first formulated as noncausal objects installed in libraries. They are then coupled with graphical support (object diagrams) by using methods of herit and reusability.

Examples are MODELICA, MOBILE, VHDL-AMS, 20 SIM; see, e.g., [19.54–59]. A broadly used tool for simulation and dynamics design is MAT-LAB/SIMULINK.

To design mechatronic systems various simulation environment models have been developed, as shown in the V-model (Fig. 19.11). In the case of software-in-the-loop (SiL) simulation the process and its control is simulated in a higher language to carry out basic investigations (Fig. 19.12). This does not require real-time simulation and is directed towards consideration of general process behavior and control structure in an earlier stage to avoid too many prototypes.

If first mechatronic prototypes exist but the final hardware of the control is missing, the rapid-control-prototyping (RCP) procedure can be used. In this, a mechatronic prototype operates as a real system with a simulated control on a test rig in order to test, e.g., control algorithms under real conditions. The prototyping computer is a powerful real-time computer with higher-language programming.

Hardware-in-the-loop (HiL) simulation is used to perform various tests in a laboratory with final hardware (electronic control unit: ECU) and the final software together with the simulated process on a powerful computer. Through HiL simulation, extreme operational and environmental conditions can also be investigated, along with faults and failures that cannot be realized with a real process on a test rig or a real vehicle, because the situations would be either too dangerous or too expensive. HiL simulation requires special electronics for reconstruction of the sensor signals and usually includes real actuators (e.g., hydraulics, pneumatics or injection pumps). Through these simulation methods the development of mechatronic systems can be performed without synchronous development on the side of the process, the electronics, or the software.

When designing mechatronic systems the tradiional borders of various disciplines have to be crossed. For the classical mechanical engineer this frequently means that knowledge of electronic components, information processing, and systems theory has to be deepened, and for the electrical/electronic engineer that knowledge on thermodynamics, fluid mechanics, and engineering mechanics has to be enlarged. For both, more knowledge on modern control principles, software engineering, and information technology may be necessary (see also [19.60]).

19.7 Conclusion and Emerging Trends

This Chapter could only give a brief overview of mechatronic systems. As outlined, mechatronic systems cover a very broad area of engineering disciplines. Advanced mechatronic components and systems are realized in many products such as automobiles, combustion engines, aircraft, electrical drives, actuators, robots, and precision mechanics and micromechanics. However, the integration aspects of mechanics and electronics include increasingly more components and systems in the wide areas of mechanical and electrical engineering.

As the development towards the integration of computer-based information processing into products and their manufacturing comprises large areas of engineering, suitable education in modern engineering and also training is fundamental for technological progress. This means, among others, to take multidisciplinary solutions and method-oriented procedures into account. The development of curricula for mechatronics as a proper combination of electrical and mechanical engineering and computer science during the last decade shows this tendency.

References

19.1 N. Kyura, H. Oho: Mechatronics – An industrial perspective, IEEE/ASME Trans. Mechatron. **1**, 10–15 (1996)

19.2 F. Harashima, M. Tomizuka: Mechatronics – "What it is, why and how?", IEEE/ASME Trans. Mechatron. **1**, 1–2 (1996)

19.3   P.A. MacConaill, P. Drews, K.-H. Robrock: *Mechatronics and Robotics I* (ICS, Amsterdam 1991)

19.4   S.J. Ovaska: Electronics and information technology in high range elevator systems, Mechatronics **2**, 88–99 (1992)

19.5   IEEE/ASME Trans. Mechatron., **1**(1) (IEEE, Piscataway 1996), (scope)

19.6   Mechatronics. An International Journal. Aims and Scope. (Pergamon Press, Oxford 1991)

19.7   G. Schweitzer: Mechatronics – a concept with examples in active magnetic bearings, Mechatronics **2**, 65–74 (1992)

19.8   J. Gausemeier, D. Brexel, T. Frank, A. Humpert: Integrated product development, 3rd Conf. Mechatron. Robot. (Teubner, Paderborn, Stuttgart 1996)

19.9   R. Isermann: Modeling and design methodology of mechatronic systems, IEEE/ASME Trans. Mechatron. **1**, 16–28 (1996)

19.10   M. Tomizuka: Mechatronics: from the 20th to the 21st century, 1st IFAC Conf. Mechatron. Syst. (Elsevier, Oxford, Darmstadt 2000) pp. 1–10

19.11   VDI 2206: Entwicklungsmethodik für mechatronische Systeme (Design methodology for mechatronic systems) (Beuth, Berlin 2004), in German

19.12   UK Mechatronics Forum. Conferences in Cambridge (1990), Dundee (1992), Budapest (1994), Guimaraes (1996), Skovde (1998), Atlanta (2000), Twente (2002). IEE & ImechE (1990–2002).

19.13   R. Isermann (Ed.): IMES. Integrated Mechanical Electronic Systems Conference (in German) TU Darmstadt, March 2–3, Fortschr.-Ber. VDI Series 12, 179. (VDI, Düsseldorf 1993)

19.14   DUIS. Mechatronics and Robotics. M. Hiller, B. Fink (Eds). 2nd Conf., Duisburg/Moers, Sept 27–29. (IMECH, Moers 1993)

19.15   O. Kaynak, M. Özkan, N. Bekiroglu, I. Tunay (Eds.): Recent Advances in Mechatronics, Proceedings of International Conference ICRAM'95 (Istanbul, 1995)

19.16   AIM (1999, 2001, 2003). IEEE/ASME Conference on Advanced Intelligent Mechatronics. Atlanta (1999), Como (2001), Kobe (2003), Monterey (2005), Zürich (2007). (IEEE, Piscataway 1999–2007)

19.17   IFAC-Symposium on Mechatronic Systems: Darmstadt (2000), Berkeley (2002), Sydney (2004), Heidelberg (2006). (Elsevier, Oxford 2000–2006)

19.18   M. Hiller: Modelling, simulation and control design for large and heavy manipulators, Int. Conf. Recent Adv. Mechatron. (Istanbul, 1995) pp. 78–85

19.19   J. Lückel (Ed.): 3rd Conf. Mechatron. Robot. (Teubner, Paderborn, Stuttgart 1995)

19.20   J. van Amerongen: Mechatronic design, Mechatronics **13**, 1045–1066 (2003)

19.21   R. Isermann: Mechatronic systems – Innovative products with embedded control. Control Eng. Pract. **16**, 14–29 (2008)

19.22   K. Kitaura: *Industrial Mechatronics* (New East Business Ltd., 1986), in Japanese

19.23   D. Bradley, D. Dawson, D. Burd, A. Loader: *Mechatronics–electronics in Products and Processes* (Chapman Hall, London 1991)

19.24   P. McConaill, P. Drews, K.-H. Robrock: *Mechatronics and Robotics* (ICS, Amsterdam 1991)

19.25   B. Heimann, W. Gerth, K. Popp: *Mechatronik (Mechatronics)* (Fachbuchverlag Leipzig, Leipzig 2001), in German

19.26   R. Isermann: *Mechatronic Systems* (Springer, Berlin 2003), German edition: 1999

19.27   C. Bishop: *The Mechatronics Handbook* (CRC, Boca Raton 2002)

19.28   R. Isermann, B. Breuer, H. Hartnagel (Eds.): *Mechatronische Systeme für den Maschinenbau (Mechatronic Systems for Mechanical Engineering.* (Wiley, Weinheim 2002) results of the special research project 241 IMES in German

19.29   G. Pahl, W. Beitz, J. Feldhusen, K.-H. Grote: *Engineering Design*, 3rd edn. (Springer, London 2007)

19.30   R. Nordmann, M. Aenis, E. Knopf, S. Straßburger: Active magnetic bearings, 7th Int. Conf. Vib. Rotating Mach. (IMechE) (Nottingham, 2000)

19.31   R. Ingenbleek, R. Glaser, K. H. Mayr: Von der Komponentenentwicklung zur integrierten Funktionsentwicklung am Beispiel der Aktuatorik und Sensorik für Pkw-Automatengetriebe (From the design of components to the development of integrated functions). In: *VDI-Conf. Mechatronik 2005 – Innovative Produktentwicklung*, VDI Bericht Ser., Vol. 1892 Wiesloch, Germany. (VDI, Düsseldorf 2005) pp. 575–592, in German

19.32   R. Kallenbach, D. Kunz, W. Schramm: *Optimierung des Fahrzeugverhaltens mit semiaktiven Fahrwerkregelungen (Optimization of the vehicle behavior with semiactive chassis control)* (VDI, Düsseldorf 1988), in German

19.33   A. Feuser: Zukunftstechnologie Mechatronik (Future technology mechatronics), Ölhydraul. Pneum. **46**(9), 436 (2002), in German

19.34   R. Bosch: *Handbook for Gasoline Engine Management* (Wiley, New York 2006)

19.35   R. Bosch: *Diesel Engine Management* (Wiley, New York 2006)

19.36   D.A. Stephenson, J.S. Agapiou: *Metal Cutting Theory and Practice*, 2nd edn. (CRC, Boca Raton 2005)

19.37   S. Kern, M. Roth, E. Abele, R. Nordmann: Active damping of chatter vibrations in high speed milling using an integrated active magnetic bearing, Adaptronic Congress 2006; Conference Proceedings (Göttingen 2006)

19.38   M. Mitschke, H. Wallentowitz: *Dynamik der Kraftfahrzeuge (Vehicle Dynamics)*, 4th edn. (Springer, Berlin 2004), in German

19.39   P. Causemann: *Kraftfahrzeugstoßdämpfer (Shockabsorbers)* (Verlag Moderne Industrie, Landsberg/Lech 2001), in German

19.40 J. Bußhardt, R. Isermann: Parameter adaptive semi-active shock absorbers, ECC Eur. Control Conf., Vol. 4 (Groningen, 1993) pp. 2254–2259

19.41 D. Metz, J. Maddock: Optimal ride height and pitch control for championship race cars, Automatica **22**(5), 509–520 (1986)

19.42 W. Schramm, K. Landesfeind, R. Kallenbach: Ein Hochleistungskonzept zur aktiven Fahrwerkregelung mit reduziertem Energiebedarf, Automobiltech. Z. **94**(7/8), 392–405 (1992), in German

19.43 A.T. van Zanten, R. Erhardt, G. Pfaff: FDR – Die Fahrdynamik-Regelung von Bosch, Automobiltech. Z. **96**(11), 674–689 (1994), in German

19.44 P. Rieth, S. Drumm, M. Harnischfeger: *Elektronisches Stabilitätsprogramm (Electronical Stability Program)* (Verlag Moderne Industrie, Landsberg/Lech 2001), in German

19.45 B. Breuer, K.H. Bill: *Bremsenhandbuch (Handbook of Brakes)*, 2nd edn. (Vieweg, Wiesbaden 2006), in German

19.46 H.-J. Schwartz: *Regelung der Radsatzdrehzahl zur maximalen Kraftschlussausnutzung bei elektrischen Triebfahrzeugen*, Dissertation (TH Darmstadt 1992) in German

19.47 R. Goodall, W. Kortüm: Mechatronics developments for railway vehicles of the future, IFAC Conf. Mechatron. Syst. (Elsevier, Darmstadt, London 2000)

19.48 R. Isermann: Supervision, fault-detection and fault-diagnosis methods – an introduction, Control Eng. Pract. **5**(5), 639–652 (1997)

19.49 R. Isermann: *Fault-Diagnosis Systems – An Introduction from Fault Detection to Fault Tolerance* (Springer, Berlin, Heidelberg 2006)

19.50 J. Gertler: *Fault Detection and Diagnosis in Engineering Systems* (Marcel Dekker, New York 1998)

19.51 J. Chen, R.J. Patton: *Robust Model-Based Fault Diagnosis for Dynamic Systems* (Kluwer, Boston 1999)

19.52 J. Gausemeier, M. Grasmann, H.D. Kespohl: *Verfahren zur Integration von Gestaltungs- und Berechnungssystemen*. VDI-Berichte Nr. 1487. (VDI, Düsseldorf 1999), in German

19.53 STARTS Guide: *The STARTS Purchases Handbook: Soft-Ware Tools for Application to Large Real-Time Systems*, 2nd edn. (National Computing Centre Publications, Manchester 1989)

19.54 J. James, F. Cellier, G. Pang, J. Gray, S.E. Mattson: The state of computer-aided control system design (CACSD), IEEE Control Syst. Mag. **15**(2), 6–7 (1995)

19.55 M. Otter, C. Cellier: Software for modeling and simulating control systems. In: *The Control Handbook*, ed. by W.S. Levine (CRC, Boca Raton 1996) pp. 415–428

19.56 H. Elmqvist: *Object-Oriented Modeling and Automatic Formula Manipulation in Dymola* (Scand. Simul. Soc. SIMS, Kongsberg 1993)

19.57 M. Hiller: Modelling, simulation and control design for large and heavy manipulators, International Conference on Recent Advances in Mechatronics (Istanbul, 1995) pp. 78–85

19.58 M. Otter, E. Elmqvist: Modelica – language, libraries, tools, Workshop and EU-Project RealSim, Simul. News Eur. **29/30**, 3–8 (2000)

19.59 M. Otter, C. Schweiger: Modellierung mechatronischer Systeme mit MODELICA (Modelling of mechatronic systems with MODELICA). *Mechatronischer Systementwurf: Methoden – Werkzeuge – Erfahrungen – Anwendungen*, Darmstadt 2004. VDI Ber. 1842, 39–50. (VDI, Düsseldorf 2004)

19.60 J. van Amerongen: Mechatronic education and research – 15 years of experience, 3rd IFAC Symp. Mechatron. Syst. (Sydney, 2004) pp. 595–607

20. Sensors and Sensor Networks

Wootae Jeong

Sensors are essential devices in many industrial applications such as factory automation, digital appliances, aircraft/automotive applications, environmental monitoring, and system diagnostics. The main role of those sensors is to measure changes of physical quantities of surroundings. In general, sensors are embedded into sensory devices with a circuitry as a part of a system. In this chapter, various types of sensors and their working principles are briefly explained as well as their technical advancement to recent smart microsensors is introduced. Specifically, the individual sensor issue is also extended to emerging networked sensors and their applications from recent research activities. Through this chapter, readers can also understand how multisensors or networked sensors can be configured and how they can collaborate with each other to provide higher performance and reliability within networked sensor systems.

20.1 Sensors

A sensor is an instrument that responds to a specific physical stimulus and produces a measurable corresponding electrical signal. A sensor can be mechanical, electrical, electromechanical, magnetic or optical. Any devices that are directly altered in a predictable, measurable way by changes in a real-world parameter can be a sensor for that parameter. Sensors have an important role in daily life because of the need to gather information and process it conveniently for specific tasks. Recent advances in microdevice technology, microfabrication, chemical processes, and digital signal processing have enabled the development of micro/nanosized, low-cost, and low-power sensors called *microsensors*. Microsensors have been successfully ap-

plied to many practical areas, including medical and space devices, military equipment, telecommunication, and manufacturing applications [20.1, 2]. When compared with conventional sensors, microsensors have certain advantages, such as interfering less with the environment they measure, requiring less manufacturing cost, being used in narrow spaces and harsh environments, etc. The successful application of microsensors depends on sensor capability, cost, and reliability

20.1.1 Sensing Principles

Sensors can be technically classified into various types according to their working principle, as listed

Table 20.1 Technical classification of sensors according to their working principle

Sensing principle	Sensors
Resistance change	Strain gage, potentiometer, potentiometric throttle position sensor (TPS), resistance temperature detector (RTD), thermistor, piezoresistive sensor, magnetoresistive sensor, photoresistive sensor
Capacitance change	Capacitive-type torque meter, capacitance level sensor
Inductance change	Linear variable differential transformer (LVDT), inductive angular position sensor (magnetic pick-up), inductive torque meter
Electromagnetic induction	Electromagnetic flow meter
Thermoelectric effect	Thermocouple
Piezoelectric effect	Piezoelectric accelerometer, sound navigation and ranging (SONAR)
Photoelectric effect	Photodiode, phototransistor, photo-interrupter (optical encoder)
Hall effect	Hall sensor

in Table 20.1. That is, sensors can measure physical phenomena by capturing resistance change, capacitance change, inductance change, thermoelectric effect, piezoelectric effect, photoelectric effect, Hall effect, and so on [20.3, 4]. Among these effects, most sensors utilize the resistance change of a conductor, i.e., *resistivity*. As long as the current density is uniform in the insulator, the resistance R of a conductor of cross-sectional area A can be computed as

$$R = \frac{\rho l}{A} , \qquad (20.1)$$

where l is the length of the conductor, A is the cross-sectional area, and ρ is the electrical resistivity of the material. Resistivity is a measure of the material's ability to oppose electric current. Therefore, change of resistance can be measured by detecting physical deformation (l or A) of conductive materials or by sensing resistivity (ρ) of conductor. As an example, a strain gage is a sensor that measures resistance by deformation of length or cross-sectional area, and a thermometer is a sensor that measures resistance by examining the resistivity change of a material. By expanding (20.1) in a Taylor series and then simplifying the equation, resistance change can be expressed as

$$\Delta R = \Delta \rho \frac{l}{A} + \Delta l \frac{\rho}{A} - \Delta A \frac{\rho l}{A^2} . \qquad (20.2)$$

By dividing both side of (20.2) by the resistance R, the resistance change rate can be expressed as

$$\frac{\Delta R}{R} = \frac{\Delta \rho}{\rho} + \frac{\Delta l}{l} - \frac{\Delta A}{A} = \frac{\Delta \rho}{\rho} + \varepsilon + 2\nu\varepsilon , \qquad (20.3)$$

where ε and ν are the strain and the Poisson's ratio of the material, respectively. When the resistivity (ρ) of a sensing material is close to constant, the resistance can be determined from the values of strain (ε) and Poisson's ratio (ν) of the material (e.g., strain gage). When the resistivity of a sensing material is sensitive to the measuring targets and values of ε, ν can be neglected; the resistance can be measured from the resistivity change ($\Delta \rho / \rho$) (e.g., resistance temperature detector (RTD)).

In capacitance-based sensors, the sensor measures the amount of electric charge stored between two plates of capacitors. The capacitance C can be calculated as

$$C = \frac{\varepsilon A}{d} , \qquad (20.4)$$

where A is the area of each plate, d is the separation between the plates, and ε is the dielectric constant (or permittivity) of the material between the plates. The dielectric constant for a number of very useful dielectrics changes as a function of the applied electrical field. Thus, capacitance-based sensors utilize capacitance change by measuring the dielectric constant, the area (A) of, or the separation (d) between the plates. A capacitive-type torque meter is an example of a capacitance-based sensor.

Inductance-based sensors measure the ratio of the magnetic flux to the current. Linear variable differential transformers (LVDT; Fig. 20.1) and magnetic pick-up sensors are representative inductance-based sensors.

Electromagnetic induction-based sensors are based on Faraday's law of induction, which is involved in the operation of transformers, inductors, and many forms of electrical generators. The law states that *the induced*

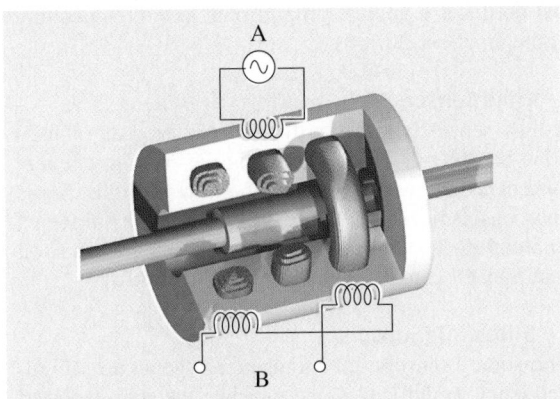

Fig. 20.1 Cutaway view of an LVDT. Current is driven through the primary coil at A, causing an induction current to be generated through the secondary coils at B

electromotive force (EMF) in any closed circuit is equal to the time rate of change of the magnetic flux through the circuit. Quantitatively, the law takes the following form

$$E = -\frac{d\Phi_B}{dt} \, , \tag{20.5}$$

where E is the electromotive force (EMF) and Φ_B is the magnetic flux through the circuit.

Besides these types of sensors, thermocouples measure the temperature difference between two points rather than absolute temperature. In traditional applications, one of the junctions (the cold junction) is maintained at a reference temperature, while the other end is attached to a probe. Having available a cold junction at a known temperature is simply not convenient for most directly connected control instruments. They incorporate into their circuits an artificial cold junction using some other thermally sensitive device, such as a thermistor or diode, to measure the temperature of the input connections at the instrument, with special care being taken to minimize any temperature gradient between terminals. Hence, the voltage from a known cold junction can be simulated, and the appropriate correction applied. Photodiodes, phototransistors, and photo-interrupters are sensors that use the photoelectric effect. Other types of sensors are listed in Table 20.1.

20.1.2 Position, Velocity, and Acceleration Sensors

Sensors can also be classified by the physical phenomena measured, such as position, velocity, acceleration, heat, pressure, flow rate, sound, etc. This classification of sensors is briefly explained below.

Position Sensors

A position sensor is any device that enables position measurement. Position sensors include limit switches or proximity sensors that detect whether or not something is close to or has reached a limit of travel. Position sensors also include potentiometers that measures rotary or linear position. The linear variable differential transformer (LVDT) is an example of the potentiometers for measuring linear displacement, while resolvers and optical encoders measure the rotary position of a rotating shaft. The LVDT and resolver function much like a transformer. The optical encoder produces digital signals that convert motion into a sequence of digital pulses. In fact, there also exist optical encoders for measuring linear motion.

Some position sensors are classified by their measuring techniques. Sonars measure distance with sonic/ultrasonic waves and radar utilizes electronic/radio to detect or measure the distance between two objects. Many other sensors are used to measure position or distance.

Velocity Sensors

Speed measurement can be obtained by taking consecutive position measurements at known time intervals and computing the derivative of the position values. A tachometer is an example of a velocity sensor that does this for a rotating shaft. The typical dynamic time constant of a tachometer is in the range $10-100\,\mu s$. A tachometer is a passive analog sensor that provides an output voltage proportional to the velocity of a shaft. There is no need for an external reference or excitation voltage. Traditionally tachometers have been used for velocity measurement and control only, but all modern tachometers have quadratic outputs which are used for velocity, position, and direction measurements, making them effectively functional as position sensors.

Acceleration Sensors

An acceleration sensor or accelerometer is a sensor designed to measure continuous mechanical vibration such as aerodynamic flutter and transitory vibration such as shock waves, blasts or impacts. Accelerometers are normally mechanically attached or bonded to an object or structure for which acceleration is to be measured. The accelerometer detects acceleration along one axis and is insensitive to motion in orthogonal directions. Strain gages or piezoelectric elements constitute the sensing element of an accelerometer, converting vibration into a voltage signal. The design of an accelerometer is based on the inertial effects associated with a mass connected to a moving object.

Detailed information and technical working processes about position, velocity, and acceleration sensors can be found in many references [20.5, 6].

20.1.3 Miscellaneous Sensors

There are other groups of sensors to measure physical quantities such as force, strain, temperature, pressure, and flow. Force sensors are represented by a load cell that is used to measure a force. The load cell consists of several strain gages connected to a bridge circuit to yield a voltage proportional to the load. Temperature sensors are devices that indirectly measure quantities such as pressure, volume, electrical resistance, and strain and then convert the values using the physical relationship between the quantity and temperature; for example: (a) a bimetallic strip composed of two metal layers with different coefficients of thermal expansion utilizes the difference in the thermal expansion of the two metal layers, (b) a resistance temperature sensor constructed of metallic wire wound around a ceramic or glass core and hermetically sealed utilizes the resistance change of the metallic wire with temperature, and (c) a thermocouple constructed by connecting two dissimilar metals in con-

tact produces a voltage proportional to the temperature of the junction [20.7, 8].

Flow Sensors

A flow sensor is a device for sensing the rate of fluid flow. In general, a flow sensor is the sensing element used in a flow meter to record the flow of fluids. Some flow sensors have a vane that is pushed by the fluid (e.g., a potentiometer), while other flow sensors are based on heat transfer caused by the moving medium.

Ultrasonic Sensors

Ultrasonic sensors or transducer generates high-frequency sound waves and evaluate the echo received back by the sensor. An ultrasonic sensor computes the time interval between sending the signal and receiving the echo to determine the distance to an object. Radar or sonar works on a principal similar to that of the ultrasonic sensor. Some sensors are depicted in Fig. 20.2. However, there are many other groups of sensors not listed in this section. With the advent of semiconductor electronics and manufacturing technology, sensors have become miniaturized and accurate, and brought into existence micro/nanosensors.

Vision Sensors

Another widely used sensor is a vision sensor. A vision sensor is typically used embedded in a vision system. A vision system can be used to measure shape, orientation, area, defects, differences between parts, etc. Vision technology has improved significantly over the last decade in that they have become rather standard smart sensing components in most factory automation systems for part inspection and location detection. In general, a vision system consists of a vision camera, an image processing computer, and a lighting system. The basic principle of operation of a vision system is that it forms an image by measuring the light reflected from objects, and the sensor head analyzes the output voltage from the light intensity received. The sensor head consists of an array of photosensitive, photodiodes or charge-coupled devices (CCD). Currently, various signal-processing techniques for the reflected signals are applied for many industrial applications to provide accurate outputs, as illustrated in Fig. 20.3.

20.1.4 Micro- and Nanosensors

A microsensor is a miniature electronic device functioning similar to existing large-scale sensors. With recent micro-electromechanical system (MEMS) technology,

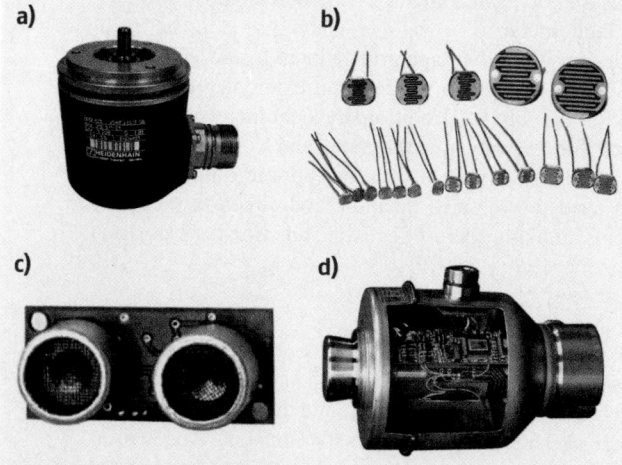

Fig. 20.2a–d Various sensors: (**a**) absolute encoder, (**b**) photoresistor, (**c**) sonar, (**d**) digital load cell cutaway (courtesy of Society of Robots)

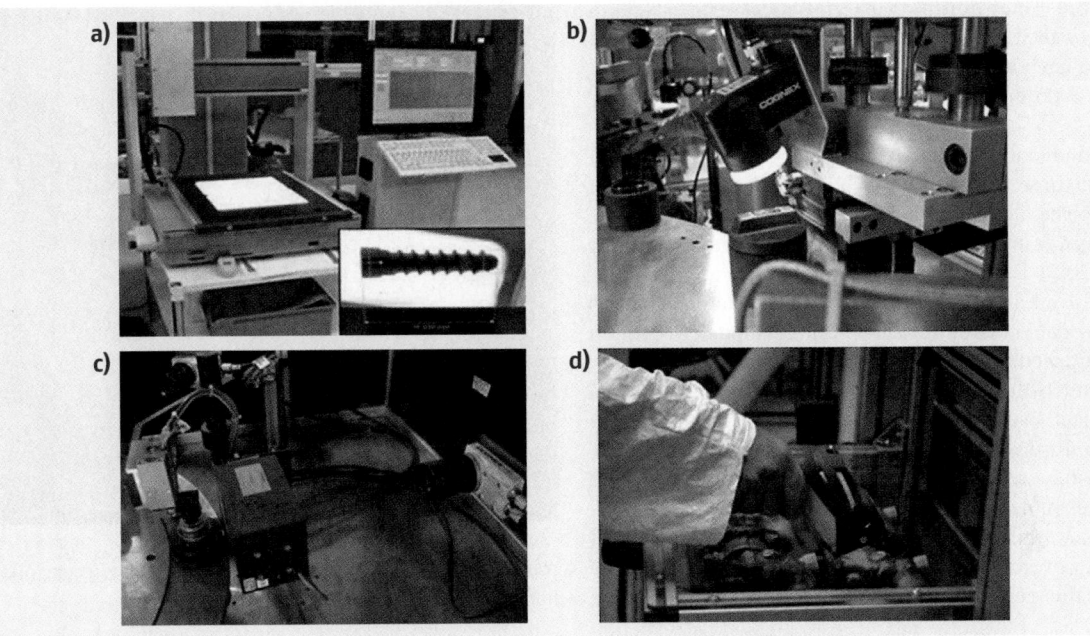

Fig. 20.3a–d Types of vision sensor applications: (**a**) automated low-volume/high-variety production, (**b**) vision sensors for error-proof oil cap assembly, (**c**) defect-free parts with 360° inspection, (**d**) inspection of two-dimensional (2-D) matrix-marked codes (courtesy of Cognex Corp.)

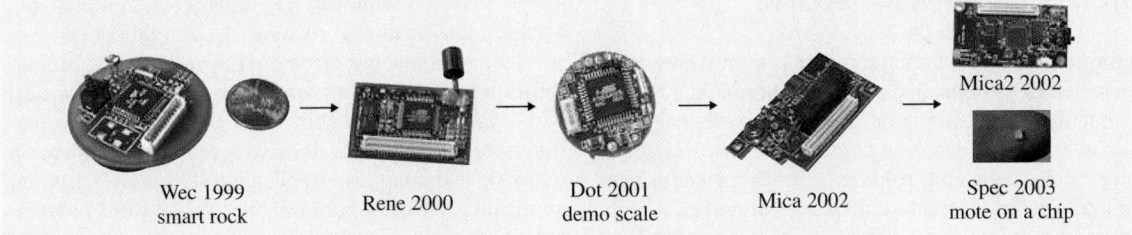

Wec 1999 smart rock   Rene 2000   Dot 2001 demo scale   Mica 2002   Mica2 2002   Spec 2003 mote on a chip

Fig. 20.4 Evolution of smart wireless microsensors (courtesy of Crossbow Technology Inc.)

microsensors are integrated with signal-processing circuits, analog-to-digital (A/D) converters, programmable memory, and a microprocessor, a so-called *smart microsensor* [20.9, 10]. Current smart microsensors contain an antenna for radio signal transmission. Wireless microsensors are now commercially available and are evolving with more powerful functionalities, as illustrated in Fig. 20.4.

In general, a wireless microsensor consists of a sensing unit, a processing unit, a power unit, and communication elements. The sensing unit is an electrical part detecting the physical variable from the environment. The processing unit (a tiny microprocessor) performs signal-processing functions, i.e., integrating

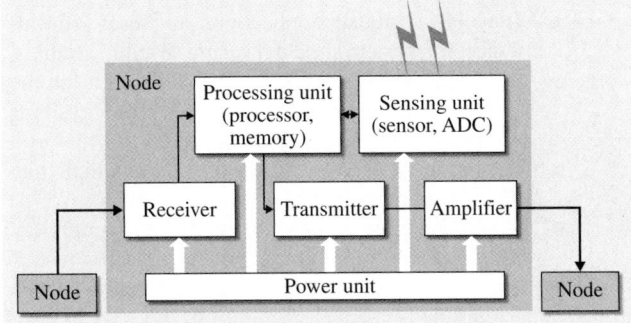

Fig. 20.5 Wireless micronode model. Each node has a sensing module (analog-to-digital converter (ADC)), processing unit, and communication elements

Fig. 20.6 Three-dimensional (3-D) model of three types of single-walled carbon nanotubes, like those used to make certain nanosensors (created by Michael Ströck on February 1, 2006)▶

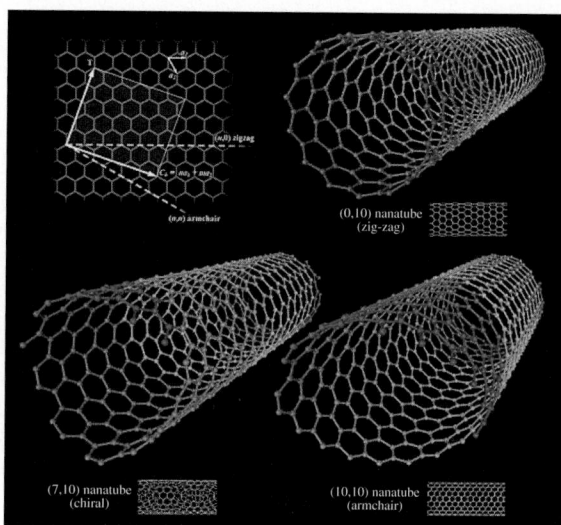

data and computation required in the processing of information. The communication elements consist of a receiver, a transmitter, and an amplifier if needed. The power unit provides energy source with other units (Fig. 20.5). Basically, all individual sensor nodes are operated by a limited battery, but a base-station node as a final data collecting center can be modeled with an unlimited energy source.

Under the microscale, nanosensors are used in chemical and biological sensory applications to deliver information about nanoparticles. As an example, nanotubes are used to sense various properties of gaseous molecules, as depicted in Fig. 20.6. In developing and commercializing nanosensors, developers still need to overcome high costs of production and reliability challenges. In the near future, there is tremendous

room to enhance the technology and implement various nanosensors in real-life applications.

20.2 Sensor Networks

20.2.1 Sensor Network Systems

Before the advent of microminiaturization technology, single-sensor systems had an important role in a variety of practical applications because they were relatively easy to construct and analyze. Single-sensor systems, however, were the only solution when there was critical limitation of implementation space. Moreover, single-sensor systems for recently emerging applications have various limitations and disadvantages:

- They have limited applications and uses; for instance, if a system should measure several variables, e.g., temperature, pressure, and flow rate, at the same time in the application, single-sensor systems are insufficient.
- They cannot tolerate a variety of failures which may take place unexpectedly.
- A single sensor cannot guarantee timely delivery of accurate information all of the time because it is inevitably affected by noise and other uncertain disruptions.

These limitations are critical when a system requires highly reliable and timely information. Therefore,

single-sensor systems are not suitable when robust and accurate information is required in the application.

To overcome the critical disadvantages of single-sensor systems in most applications, multisensor network systems which require replicated sensory information have been studied, along with their communication network technologies. Replicated sensor systems are applicable not only because microfabrication technology enables production of various microsensors at low manufacturing cost, but also because microsensors can be embedded in a system with replicated deployment. These redundantly deployed sensors enable a system to improve accuracy and tolerate sensor failure, i. e., distributed microsensor arrays and networks (DMSA/DMSN) are built from collections of spatially scattered microsensor nodes. Each node has the ability to measure the local physical variable within its accuracy limit, process the raw sensory data, and cooperate with its neighboring nodes.

Sensors incorporated with dedicated signal-processing functions are called intelligent, or smart, sensors. The main roles of dedicated signal processing functions are to enhance design flexibility and realize new sensing functions. Additional roles are to reduce loads on central processing units and signal transmission lines by distributing information processing to the lower layers of the system [20.10].

A set of microsensors deployed close to each other to measure the same physical quantity of interest is called a cluster. Sensors in a cluster can be either of the same or different type to form a distributed sensor network (DSN). A DSN can be utilized in a widely distributed sensor system and implemented as a locally concentrated configuration with a high density.

20.2.2 Multisensor Data Fusion Methods

There are three major ways in which multiple sensors interact [20.11, 12]: (1) *complementary*, when sensors do not depend on each other directly, but are combined to give a more complete image of the phenomena being studied; (2) *competitive*, when sensors provide independent measurement of the same information regarding a physical phenomenon; and (3) *cooperative*, when sensors combine data from independent sensors to derive information that would be unavailable from the individual sensors.

In order to combine information collected from each sensor, various multisensory data fusion methods can be applied. Multisensor data fusion is the process of combining observations from a number of different sensors to provide a robust and complete description of an environment or process of interest. Most current data fusion methods employ probabilistic descriptions of observations and processes and use Bayes' rule to combine this information [20.13, 14].

Bayes' Rule

Bayes' rule lies at the heart of most data fusion methods. In general, Bayes' rule provides a means to make inferences about an object or environment of interest described by a state x, given an observation z. Based on the rule of conditional probabilities, Bayes' rule is obtained as

$$P(x|z) = \frac{P(z|x)P(x)}{P(z)} . \qquad (20.6)$$

The conditional probability $P(z|x)$ serves the role of a sensor model. The probability is constructed by fixing the value of $x = x$ and then asking what probability density $P(z|x = x)$ on x is inferred. The multisensory form of Bayes' rule requires conditional independence

$$P(z_1, \ldots, z_n|x) = P(z_1|x) \ldots P(z_n|x)$$

$$= \prod_{i=1}^{n} P(z_i|x_i) . \qquad (20.7)$$

The recursive form of Bayes' rule is

$$P(x|Z^k) = \frac{P(z_k|x)P(x|Z^{k-1})}{P(z_k|Z^{k-1})} . \qquad (20.8)$$

From this equation, one needs to compute and store only the posterior density $P(x|Z^{k-1})$, which contains a complete summary of all past information.

Probabilistic Grids

Probabilistic grids are the means to implement the Bayesian data fusion technique to problems in mapping [20.15] and tracking [20.16]. Practically, a grid of likelihoods on the states x_{ij} is produced in the form $P(z = z|x_{ij}) = \Lambda(x_{ij})$. It is then trivial to apply Bayes' rule to update the property value at each grid cell as

$$P^+(x_{ij}) = C\Lambda(x_{ij})P(x_{ij}) \quad \forall i, j , \qquad (20.9)$$

where C is a normalizing constant obtained by summing posterior probabilities to 1 at node ij only. Computationally, this is a simple pointwise multiplication of two grids. Grid-based fusion is appropriate to situations where the domain size and dimension are modest. In such cases, grid-based methods provide straightforward and effective fusion algorithms. Monte Carlo and particle filtering methods can be considered as grid-based methods, where the grid cells themselves are samples of the underlying probability density for the state.

The Kalman Filter

The Kalman filter is a recursive linear estimator that successively calculates an estimate for a continuous-valued state on the basis of periodic observations of the state. The Kalman filter may be considered a specific instance of the recursive Bayesian filter [20.17] for the case where the probability densities on states are Gaussian.

The Kalman filter algorithm produces estimates that minimize mean-squared estimation error conditioned on a given observation sequence and so is the conditional mean

$$\hat{x}(i|j) \triangleq E[x(i)|z(1), \ldots, z(j)] \triangleq E[x(i)|Z^j] . \qquad (20.10)$$

The estimate variance is defined as the mean-squared error in this estimate

$$P(i|j) \triangleq E\{[x(i) - \hat{x}(i|j)][x(i) - \hat{x}(i|j)]^\top |Z^j\} . \qquad (20.11)$$

The estimate of the state at a time k, given all information up to time k, is written $\hat{x}(k|k)$. The estimate of the state at a time k given only information up to time $k-1$ is called a one-step-ahead prediction and is written $\hat{x}(k|k-1)$.

The Kalman filter is appropriate to data fusion problems where the entity of interest is well defined by a continuous parametric state. Thus, it would be useful to estimate position, attitude, and velocity of an object, or the tracking of a simple geometric feature. Kalman filters, however, are inappropriate for estimating properties such as spatial occupancy, discrete labels or processes whose error characteristics are not easily parameterized.

Sequential Monte Carlo Methods

The sequential Monte Carlo (SMC) filtering method is a simulation of the recursive Bayes update equations using sample support values and weights to describe the underlying probability distributions. SMC recursion begins with an a posterior probability density represented by a set of support values and weights $\{x_{k-1}^i, w_{k-1|k-1}^i\}_{i=1}^{N_{k-1}}$ in the form

$$P\left(x_{k-1}|Z^{k-1}\right) = \sum_{i=1}^{N_{k-1}} w_{k-1}^i \delta\left(x_{k-1} - x_{k-1}^i\right) .$$

$$(20.12)$$

Leaving the weights unchanged $w_k^i = w_{k-1}^i$ and allowing the new support value x_k^i to be drawn on the basis of old support value x_{k-1}^i, the prediction becomes

$$P\left(x_k|Z^{k-1}\right) = \sum_{i=1}^{N_{k-1}} w_{k-1}^i \delta\left(x_k - x_k^i\right) .$$ $$(20.13)$$

The SMC observation update step is relatively straightforward and described in [20.13, 18].

SMC methods are well suited to problems where state-transition models and observation models are highly nonlinear. However, they are inappropriate for problems where the state space is of high dimension. In addition, the number of samples required to model a given density faithfully increases exponentially with state-space dimension.

Interval Calculus

Interval representation of uncertainty has a number of potential advantages over probabilistic techniques. An interval to bound true parameter values provides a good measure of uncertainty in situations where there is a lack of probabilistic information, but in which sensor and parameter error is known to be bounded. In this

technique, the uncertainty in a parameter x is simply described by a statement that the true value of the state x is known to be bounded between a and b, i.e., $x \in [a, b]$. There is no other additional probabilistic structure implied. With $a, b, c, d \in \mathbb{R}$, interval arithmetic is also possible as

$$d([a, b], [c, d]) = \max(|a-c|, |b-d|) . \quad (20.14)$$

Interval calculus methods are sometimes used for detection, but are not generally used in data fusion problems because of the difficulties to get results converged to anything value, and to encode dependencies between variables.

Fuzzy Logic

Fuzzy logic has achieved widespread popularity for representing uncertainty in high-level data fusion tasks. Fuzzy logic provides an ideal tool for inexact reasoning, particularly in rule-based systems. In the conventional logic system, a membership function $\mu_A(x)$ (also called the characteristic function) is defined. Then the fuzzy membership function assigns a value between 0 and 1, indicating the degree of membership of every x to the set A. Composition rules for fuzzy sets follow the composition processes for normal crisp sets as

$$A \cap B \rightleftharpoons \mu_{A \cap B}(x) = \min[\mu_A(x), \mu_B(x)] , \quad (20.15)$$
$$A \cup B \rightleftharpoons \mu_{A \cup B}(x) = \max[\mu_A(x), \mu_B(x)] . \quad (20.16)$$

The relationship between fuzzy set theory and probability is, however, still debated.

Evidential Reasoning

Evidential reasoning methods are qualitatively different from either probabilistic methods or fuzzy set theory. In evidential reasoning, belief mass cannot only be placed on elements and sets, but also sets of sets, while in probability theory a belief mass may be placed on any element $x_r \in \chi$ and on any subset $A \subseteq \chi$. The domain of evidential reasoning is the power set 2^χ. Evidential reasoning methods play an important role in discrete data fusion, attribute fusion, and situation assessment, where information may be unknown or ambiguous.

Multisensory fusion methods and their models are summarized in Table 20.2, and details can be founded in [20.13, 14].

In addition, multisensor integration or fusion is not only the process of combining inputs from sensors with information from other sensors, but also the logical procedure of inducing optimal output from multiple inputs with one representative format [20.19]. In the fusion of

Table 20.2 Multisensor data fusion methods [20.13]

Approach	Method	Fusion model and rule
Probabilistic modeling	Bayes' rule	$P(x\|Z^k) = \frac{P(z_k\|x)P(x\|Z^{k-1})}{P(z_k\|Z^{k-1})}$
	Probabilistic grids	$P^+(x_{ij}) = C\Lambda(x_{ij})P(x_{ij})$
	The Kalman filter	$P(i\|j)$ $\triangleq E\{[x(i) - \hat{x}(i\|j)][x(i) - \hat{x}(i\|j)]^\top \|Z^j\}$
	Sequential Monte Carlo methods	$P(x_k\|Z^k)$ $= C\sum_{i=1}^{N_k} w_{k-1}^i P(z_k = z_k\|x_k = x_k^i)\delta(x_k - x_k^i)$
Nonprobabilistic modeling	Interval calculus	$d([a, b], [c, d]) = \max(\|a - c\|, \|b - d\|)$
	Fuzzy logic	$A \cap B \rightleftharpoons \mu_{A \cap B}(x) = \min[\mu_A(x), \mu_B(x)]$ $A \cup B \rightleftharpoons \mu_{A \cup B}(x) = \max[\mu_A(x), \mu_B(x)]$
	Evidential reasoning	$2^X = \{\{\text{occupied, empty}\}, \ldots$ $\{\text{occupied}\}, \{\text{empty}\}, 0\}$

large-size distributed sensor networks, an additional advantage of multisensor integration (MSI) is the ability to obtain more fault-tolerant information. This fault tolerance is based on redundant sensory information that compensates for faulty or erroneous readings of sensors. There are several types of multisensor fusion and integration methods, depending on the types of sensors and their deployment [20.20]. This topic has received increasing interest in recent years because of the sensibility of networks built with many low-cost micro- and nanosensors. As an example, a recent improvement of the fault-tolerance sensor integration algorithm (FTSIA) by *Liu* and *Nof* [20.21, 22] enables it not only to detect possibly faulty sensors and widely faulty sensors, but also to generate a final data interval estimate from the correct sensors after removing the readings of those faulty sensors.

20.2.3 Sensor Network Design Considerations

Sensor networks are somewhat different from traditional operating networks because sensor nodes, especially microsensors, are highly prone to failure over time. As sensor nodes weaken or even die, the topology of the active sensor networks changes frequently. Especially when mobility is introduced into the sensor nodes, maintaining the robustness and discovering topology consistently become challenging. Therefore, the algorithms developed for sensor network communication and task administration should be flexible and stable against changes of network topology and work properly under unexpected failure of sensors.

In addition, in order to be used in most applications, DSN systems should be designed with application-specific communication algorithms and task administration protocols because microsensors and their networking systems are extremely resource constrained. Therefore, most research efforts have focused on application-specific protocols with respect to energy consumption and network parameters such as node density, radio transmission range, network coverage, latency, and distribution. Current network protocols also use broadcasting for communication, while traditional and ad hoc networks use point-to-point communication. Hence, the routing protocols, in general, should be designed by considering crucial sensor network features as follows:

1. *Fault tolerance*: Over time, sensor nodes may fail or be blocked due to lack of power, physical damage or environmental interference. The failure of sensor nodes, however, should not affect the overall operation of the sensor network. Thus, fault tolerance or reliability is the ability to sustain sensor network functionality despite likely problems.
2. *Accuracy improvement*: Redundancy of information can reduce overall uncertainty and increase the accuracy with which events are perceived. Since nodes located close to each other are combining information about the same event, fused data improve the quality of the event information.
3. *Timeliness*: DSN can provide the processing parallelism that may be needed to achieve an effective integration process, either at the actual speed that a single sensor could provide, or at even faster operation speed.
4. *Network topology*: A large number of nodes deployed throughout the sensory field should be maintained by carefully designed topology because any changes in sensor nodes and their deployments

affect the overall performance of DSN. Therefore, a flexible and simple topology is usually preferred.

5. *Energy consumption*: Since each wireless sensor node is working with a limited power source, the design of power-saving protocols and algorithms is a significant issue for providing longer lifetime of sensor network systems.

6. *Lower cost*: Despite the use of redundancy, a distributed microsensor system obtains information at lower cost than the equivalent information expected from a single sensor because it does not require the additional cost of functions to obtain the same reliability and accuracy. The current cost of a microsensor node, e.g., dust mote [20.9], is still expensive (US$ 25–172), but it is expected to be less than US$ 1 in the near future, so that sensor networks can be justified.

7. *Scalability*: The coverage area of a sensor network system depends on the transmission range of each node and the density of the deployed sensors. The density of the deployed nodes should be carefully designed to provide a topology appropriate for the specific application.

To provide the optimal solution to meet these design criteria in the sensor network, researchers have considered various protocols and algorithms. However, none of these studies has been developed to improve all

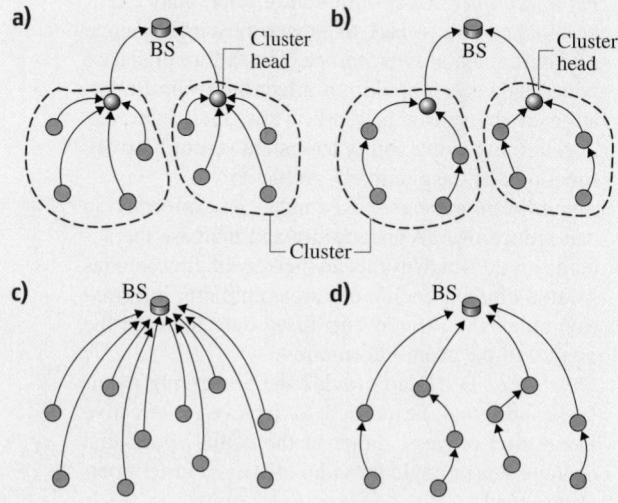

Fig. 20.7a–d Four different configurations of wireless sensor networks: (**a**) single hop with clustering, (**b**) multihop with clustering, (**c**) single hop without clustering, and (**d**) multihop without clustering. BS – base station

design factors because the design of a sensor network system has typically been application specific.

A distributed network of microsensor arrays (MSA) can yield more accurate and reliable results based on built-in redundancy. Recent developments of flexible and robust protocols with improved fault tolerance will not only meet essential requirements in distributed systems but will also provide advanced features needed in specific applications. While MEMS sensor technology has advanced significantly in recent years, scientists now realize the need for design of effective MEMS sensor communication networks and task administration.

20.2.4 Sensor Network Architectures

A well-designed distributed network of microsensor arrays can yield more accurate and reliable results based on built-in redundancy. Recent developments of flexible and robust protocols with improved fault tolerance will not only meet essential requirements in distributed systems but will also provide advanced features needed in specific applications. They can produce widely accessible, reliable, and accurate information about physical environments.

Various architectures have been proposed and developed to improve the performance of systems and fault-tolerance functionality of complex networks depending on their applications. General DSN structures for multisensor systems were first discussed by *Wesson* et al. [20.23]. *Iyengar* et al. [20.24] and *Nadig* et al. [20.25] improved and developed new architectures for distributed sensor integration.

A network is a general graph $G = (V, L)$, where V is a set of nodes (or vertices) and L is a set of communicating links (or edges) between nodes. For a DSN, a node means an intelligent sensing node consisting of a computational processor and associated sensors, and an edge is the connectivity of nodes. As shown in Fig. 20.7, a DSN consists of a set of sensor nodes, a set of cluster-head (CH) nodes, and a communication network interconnecting the nodes [20.20, 24]. In general, one sensor node communicates with more than one CH, and a set of nodes communicating with a CH is called a cluster. A clustering architecture can increase system capacity and enable better resource allocation [20.26, 27]. Data are integrated in CH by receiving required information from associated sensors of the cluster. In the cluster, CHs can interact not only with other CHs, but also with higher-level CHs or a base station. A number of network configurations have been developed to prolong network lifetime and reduce energy consumption in forwarding data. In order to minimize energy consumption, routing schemes can be broadly classified into two categories: (1) clustering-based data forwarding scheme (Fig. 20.7a,b) and (2) multihop data forwarding scheme without clustering (Fig. 20.7b,d).

In recent years, with the advancement of wireless mobile communication technologies, ad hoc wireless sensor networks (AWSNs) have become important. With this advancement, the above wired (microwired) architectures remain relevant only where wireless communication is physically prohibited; otherwise, wireless architectures are considered superior. The architecture of AWSN is fully flexible and dynamic, that is, a mobile ad hoc network represents a system of wireless nodes that can freely reorganize into temporary networks as needed, allowing nodes to communicate in areas with no existing infrastructure. Thus, interconnection between nodes can be dynamically changed, and the network is set up only for a short period of communication [20.28]. Now the AWSN with an optimal ad hoc routing scheme has become an important design concern.

In applications where there is no given pattern of sensor deployment, such as battlefield surveillance or environmental monitoring, the AWSN approach can provide efficient sensor networking. Especially in dynamic network environments such as AWSN, three main distributed services, i.e., lookup service, composition service, and dynamic adaptation service by self-organizing sensor networks, are also studied to control the system (see, for instance, [20.29]).

In order to route information in an energy-efficient way, directed diffusion routing protocols based on the localized computation model [20.30, 31] have been studied for robust communication. The data consumer will initiate requests for data with certain attributes. Nodes will then diffuse the requests towards producers via a sequence of local interactions. This process sets up gradients in the network which channel the delivery of data. Even though the network status is dynamic, the impact of dynamics can be localized.

A mobile-agent-based DSN (MADSN) [20.32] utilizes a formal concept of agent to reduce network bandwidth requirements. A mobile agent is a floating processor migrating from node to node in the DSN and performing data processing autonomously. Each mobile agent carries partially integrated data which will be fused at the final CH with other agents' information. To save time and energy consumption, as soon as certain requirements of a network are satisfied in the

progress of its tour, the mobile agent returns to the base station without having to visit other nodes on its route. This logic reduces network load, overcoming network latency, and improves fault-tolerance performance.

20.2.5 Sensor Network Protocols

Communication protocols for distributed microsensor networks provide systems with better network capability and performance by creating efficient paths and accomplishing effective communication between the sensor nodes [20.29, 33, 34].

The point-to-point protocol (PTP) is the simplest communication protocol and transmits data to only one of its neighbors, as illustrated in Fig. 20.8a. However, PTP is not appropriate for a DSN because there is no communication path in case of failure of nodes or links.

In the flooding protocol (FP), the information sent out by the sender node is addressed to all of its neighbors, as shown in Fig. 20.8b. This disseminates data quickly in a network where bandwidth is not limited and links are not loss-prone. However, since a node always sends data to its neighbors, regardless of whether or not the neighbor has already received the data from another source, it leads to the implosion problem and wastes resources by sending duplicate copies of data to the same node.

The gossiping protocol (GP) [20.35, 36] is an alternative to the classic flooding protocol in which, instead of indiscriminately sending information to all its neighboring nodes, each sensor node only forwards the data to one randomly selected neighbor, as depicted in Fig. 20.8c. While the GP distributes information more slowly than FP, it dissipates resources, such as energy, at a relatively lower rate. In addition, it is not as robust relative to link failures as a broadcasting protocol (BP), because a node can only rely on one other node to resend the information for it in the case of link failure.

In order to solve the problem of implosion and overlap, *Heinzelman* et al. [20.37] proposed the sensor

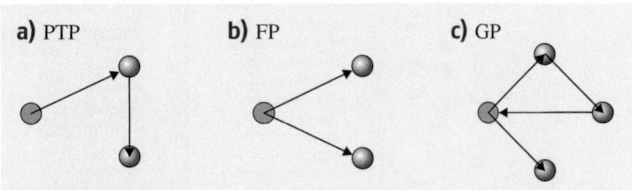

Fig. 20.8a–c Three basic communication protocols: (**a**) point-to-point protocol (PTP), (**b**) flooding protocol (FP), and (**c**) gossiping protocol (GP)

protocol for information via negotiation (SPIN). SPIN nodes negotiate with each other before transmitting data, which helps ensure that only useful transmission of information will be executed. Nodes in the SPIN protocol use three types of messages to communicate: ADV (new data advertisement), REQ (request for data), and DATA (data message). Thus, SPIN protocol works in three stages: ADV–REQ–DATA. The protocol begins when a node advertises the new data that is ready to be disseminated. It advertises by sending an ADV message to its neighbors, naming the new data (ADV stage). Upon receiving an ADV, the neighboring node checks to see whether it has already received or requested the advertised data to avoid implosion and the overlap problem. If not, it responds by sending a REQ message for the missing data back to the sender (REQ stage). The protocol completes when the initiator of the protocol responds to the REQ with a DATA message, containing the missing data (DATA stage).

In a relatively large sensor network, a clustering architecture with a local cluster-head (CH) is necessary. *Heinzelman* et al. [20.38] proposed the low-energy adaptive clustering hierarchy (LEACH), which is a clustering-based protocol that utilizes randomized rotation of local cluster base stations to evenly distribute the energy load of sensors in DSN. Energy-minimizing routing protocols have also been developed to extend the lifetime of the sensing nodes in a wireless network; for example, a minimum transmission energy (MTE) routing protocol [20.39] chooses intermediate nodes such that the sum of squared distances is minimized by assuming a square-of-distance power loss between two nodes. This protocol, however, results in unbalanced termination of nodes with respect to the entire network.

In recent years, a time-based network protocol has been developed. The objective of a time-based protocol is to ensure that, when any tasks keep the resource idle for too long, their exclusive service by the resource is disabled; that is, the time-based control protocol is intended to provide a rational collaboration rule among tasks and resources in the networked system [20.40]. Here, slow sensors will delay timely response and other sensors may need to consume extra energy. The patented fault tolerant time-out protocol (FTTP) uses the basic concept of a time-out scheme effectively in a microsensor communication control (FTTP is a patent-pending protocol of the PRISM Center at Purdue University, USA).

The design of industrial open protocols for mostly wired communication known as fieldbuses, such as DeviceNet and ControlNet, have also been evolved to provide open data exchange and a messaging framework [20.41]. Further development for wireless has been investigated in asset monitoring and maintenance using an open communication protocol such as ZigBee [20.42].

Wireless sensor network application designers also require a middleware to deliver a general runtime environment that inherits common requirements under application specifications. In order to provide robust functions to industrial applications under somewhat limited resource constraints and the dynamics of the environment, appropriate middleware is also required to contain embedded trade-offs between essential quality-of-service (QoS) requirements from applications. Typically, a sensor network middleware has a layered architecture that is distributed among the networked sensor systems. Based on traditional

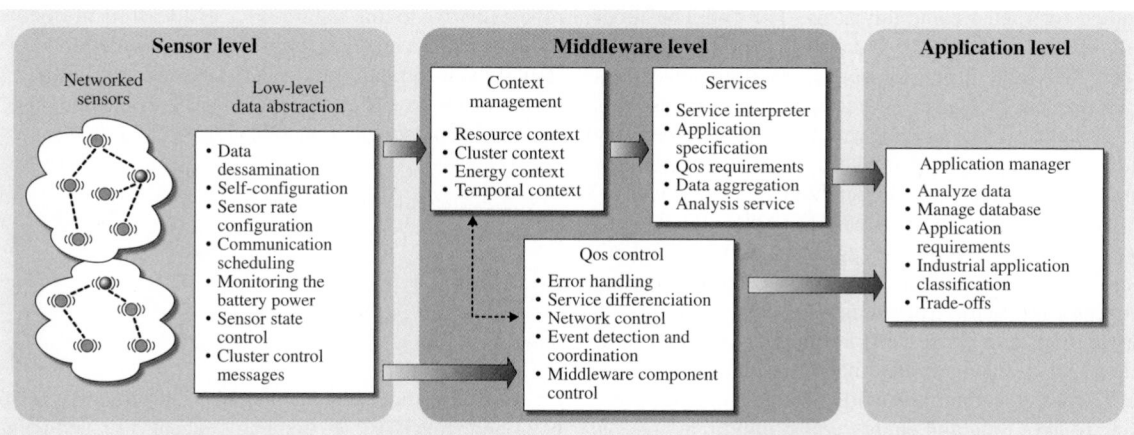

Fig. 20.9 Middleware architecture for facility sensor network applications (MidFSN) (after [20.36])

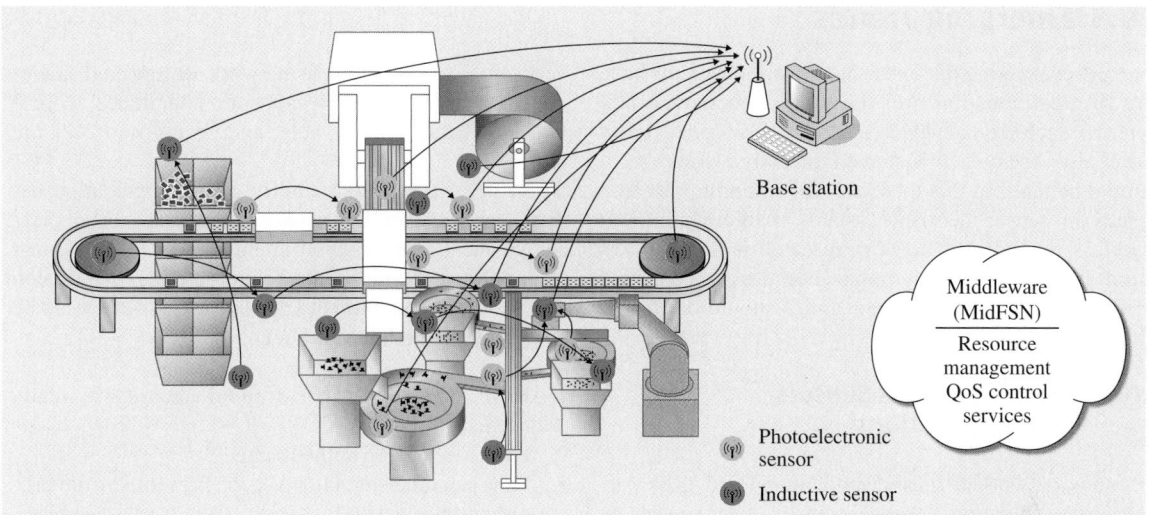

Fig. 20.10 Wireless microsensor network system with two different types of microsensor nodes in an industrial automation application (after [20.43])

architectures, researchers [20.43] have recently been developing a middleware for facility sensor network (MidFSN) whose layered structure is classified into three layers as depicted in Fig. 20.9.

20.2.6 Sensor Network Applications

Distributed microsensor networks have mostly been applied to military applications. However, a recent trend in sensor networks has been to apply the technology to various industrial applications. Figure 20.10 illustrates a networked sensor application used in factory automation. Environment applications such as examination of flowing water and detection of air contaminants require flexible and dynamic topology for the sensor network. Biomedical applications of collecting information from internal human body are based on bio/nanotechnology. For these applications, geometrical and dynamical characteristics of the system must be considered at the design step of network architecture [20.44]. It is also essential to use fault-tolerant network protocols to aggregate very weak signals without losing any critical signals. Specifically designed sensor network systems can also be applicable for intelligent transportation systems, monitoring material flow, and home/office network systems.

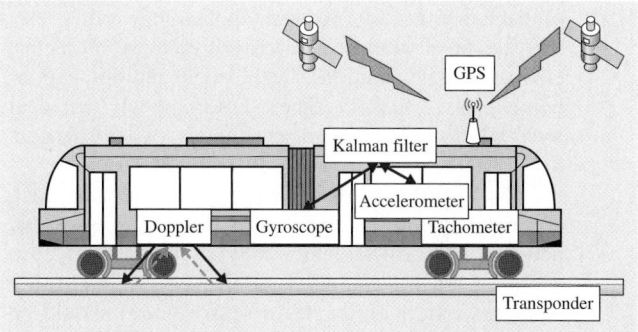

Fig. 20.11 Networked sensors for train tracking and tracing (global positioning system (GPS))

Public transportation systems are another example of sensor network applications. Sensor networks have been successfully implemented into highway systems for vehicle and traffic monitoring, providing key technology for intelligent transportation systems (ITS). Recently various networked sensors have been applied to railway system to monitor the location of rolling stocks and detect objects or obstacles on the rail in advance (Fig. 20.11). Networked sensors can cover a wide monitoring area and deliver more accurate information.

20.3 Emerging Trends

Energy-conserving microsensor network protocols have drawn great attention over the years. Other important metrics such as latency, scalability, and connectivity have also been deeply studied recently. However, it should be realized that there are still emerging research issues in sensor network systems. Although current wireless microsensor network research is moving forward to more practical application areas, emerging research on the following new topics should be further examined and will surface.

20.3.1 Heterogeneous Sensors and Applications

In many networked sensor applications and their performance evaluation, homogeneous or identical sensors were most commonly considered; therefore network performance was mainly determined by the geometrical distances between sensors and the remaining energy of each sensor. In practical applications other factors can also influence coverage, such as obstacles, environmental conditions, and noise. In addition to non-homogeneous sensors, other sensor models can deal with nonisotropic sensor sensitivities, where sensors have different sensitivities in different directions.

The integration of multiple types of sensors such as seismic, acoustic, and optical sensors in a specific network platform and the study of the overall coverage of the system also present several interesting challenges. In addition, when sensor nodes should be shared by multiple applications with differing goals, protocols must efficiently serve multiple applications simultaneously. Therefore, for heterogeneous sensors and their network application development, several research questions should be considered:

1. How should resources be utilized optimally in heterogeneous sensor networks?
2. How should heterogeneous data be handled efficiently?
3. How much and what type of data should be processed to meet quality-of-service (QoS) goals while minimizing energy usage?

20.3.2 Security

Another emerging issue in wireless sensor networks is related to network security. Since the sensor network may operate in a hostile environment, security needs to be built into the network design and not as an afterthought. That is, network techniques to provide low-latency, survivable, and secure networks are required.

In general, low probability of communication detection is needed for networks because sensors are envisioned for use behind enemy lines. For the same reasons, the network should be protected again intrusion and spoofing. For the network security, some research questions should be examined:

1. How much and what type of security is really needed?
2. How can data be authenticated?
3. How can misbehaving nodes be prevented from providing false data?
4. Can energy and security be traded-off such that the level of network security can be easily adapted?

20.3.3 Appropriate Quality-of-Service (QoS) Model

Research in QoS has received considerable attention over the years. QoS has to be supported at media access control (MAC), routing, and transport layers. Most existing ad hoc routing protocols do not support QoS. The routing metric used in current work still refers to the shortest path or minimum hop. However, bandwidth, delay, jitter, and packet loss (reliability or data delivery ratio) are other important QoS parameters. Hence, mechanisms of current ad hoc routing protocols should allow for route selection based on both QoS requirements and QoS availability. In addition to establishing QoS routes, QoS assurance during route reconfiguration has to be supported too. QoS considerations need to be made to ensure that end-to-end QoS requirements continue to be supported. Hence, there is still significant room for research in this area.

20.3.4 Integration with Other Networks

In the near future, sensor networks may interface with other networks, such as a Wi-Fi network, a cellular network, or the Internet. Therefore, to find the best way to interface these networks will be a big issue. Sensor network protocols should support (or at least not compete with) the protocols of the other networks; otherwise sensors could have dual network interface capabilities.

References

20.1 R.C. Luo: Sensor technologies and microsensor issues for mechatronics systems, IEEE/ASME Trans. Mechatron. **1**(1), 39–49 (1996)

20.2 R.C. Luo, C. Yih, K.L. Su: Multisensor fusion and integration: approaches, applications, and future research directions, IEEE Sens. J. **2**(2), 107–119 (2002)

20.3 J.S. Wilson: *Sensor Technology Handbook* (Newnes Elsevier, Amsterdam 2004)

20.4 J. Fraden: *Handbook of Modern Sensors: Physics, Designs, and Applications*, 3rd edn. (Springer, Berlin, Heidelberg 2003)

20.5 J.G. Webster (Ed.): *The Measurement, Instrumentation and Sensors Handbook* (CRC, Boca Raton 1998)

20.6 C.W. de Silva (Ed.): *Mechatronic Systems: Devices, Design, Control, Operation and Monitoring* (CRC, Boca Raton 2008)

20.7 S. Cetinkunt: *Mechatronics* (Wiley, New York 2007)

20.8 D.G. Alciatore, M.B. Histand: *Introduction to Mechatronics and Measurement Systems* (McGraw-Hill, New York 2007)

20.9 B. Warneke, M. Last, B. Liebowitz, K.S.J. Pister: Smart dust: communicating with a cubic-millimeter computer, Computer **34**(1), 44–51 (2001)

20.10 H. Yamasaki (Ed.): *Intelligent Sensors*, Handbook of Sensors and Actuators, Vol. 3 (Elsevier, Amsterdam 1996)

20.11 R. Brooks, S. Iyengar: *Multi-Sensor Fusion* (Prentice Hall, New York 1998)

20.12 K. Faceli, C.P.L.F. Andre de Carvalho, S.O. Rezende: Combining intelligent techniques for sensor fusion, Appl. Intell. **20**, 199–213 (2004)

20.13 D.-W. Hugh, T.C. Henderson: Multisensor data fusion. In: *Springer Handbook of Robotics*, ed. by B. Siciliano, O. Khatib (Springer, Berlin, Heidelberg 2008)

20.14 H.B. Mitchell: *Multi-Sensor Data Fusion* (Springer, Berlin, Heidelberg 2007)

20.15 A. Elfes: Sonar-based real-world mapping and navigation, IEEE Trans. Robot. Autom. **3**(3), 249–265 (1987)

20.16 L.D. Stone, C.A. Barlow, T.L. Corwin: *Bayesian Multiple Target Tracking* (Artech House, Norwood 1999)

20.17 P.S. Mayback: *Stochastic Models, Estimation and Control*, Vol. I (Academic, New York 1979)

20.18 Y. Bar-Shamon, T.E. Fortmann: *Tracking and Data Association* (Academic, New York 1998)

20.19 R.C. Luo, M.G. Kay: Multisensor integration and fusion in intelligent systems, IEEE Trans. Syst. Man Cybern. **19**(5), 901–931 (1989)

20.20 S.S. Iyengar, L. Prasad, H. Min: *Advances In Distributed Sensor Integration: Application and Theory*, Environmental and Intelligent Manufacturing Systems, Vol. 7 (Prentice Hall, Upper Saddle River 1995)

20.21 Y. Liu, S.Y. Nof: Distributed micro flow-sensor arrays and networks: Design of architecture and communication protocols, Int. J. Prod. Res. **42**(15), 3101–3115 (2004)

20.22 S.Y. Nof, Y. Liu, W. Jeong: Fault-tolerant time-out communication protocol and sensor apparatus for using same, patent pending (Purdue University, 2003)

20.23 R. Wesson, F. Hayes-Roth, J.W. Burge, C. Stasz, C.A. Sunshine: Network structures for distributed situation assessment, IEEE Trans. Syst. Man Cybern. **11**(1), 5–23 (1981)

20.24 S.S. Iyengar, D.N. Jayasimha, D. Nadig: A versatile architecture for the distributed sensor integration problem, IEEE Trans. Comput. **43**(2), 175–185 (1994)

20.25 D. Nadig, S.S. Iyengar, D.N. Jayasimha: A new architecture for distributed sensor integration, Proc. IEEE Southeastcon'93E (1993)

20.26 S. Ghiasi, A. Srivastava, X. Yang, M. Sarrafzadeh: Optimal energy aware clustering in sensor networks, Sensors **2**, 258–269 (2002)

20.27 C.R. Lin, M. Gerla: Adaptive clustering for mobile wireless networks, IEEE J. Sel. Areas Commun. **15**(7), 1265–1275 (1997)

20.28 M. Ilyas: *The Handbook of Ad Hoc Wireless Networks* (CRC, Boca Raton 2002)

20.29 A. Lim: Distributed services for information dissemination in self-organizing sensor networks, J. Franklin Inst. **338**(6), 707–727 (2001)

20.30 D. Estrin, J. Heidemann, R. Govindan, S. Kumar: Next century challenges: Scalable coordination in sensor networks, Proc. 5th Annu. Int. Conf. Mobile Comput. Netw. (MobiCOM '99) (1999)

20.31 C. Intanagonwiwat, R. Govindan, D. Estrin, J. Heidemann, F. Silva: Directed diffusion for wireless sensor networking, IEEE/ACM Trans. Netw. **11**(1), 2–16 (2003)

20.32 Y.A. Chau, E. Geraniotis: Multisensor correlation and quantization in distributed detection systems, Proc. 29th IEEE Conf. Decis. Control, Vol. 5 (1990) pp. 2692–2697

20.33 W.B. Heinzelman, A.P. Chandrakasan, H. Balakrishnan: An application-specific protocol architecture for wireless microsensor networks, IEEE Trans. Wirel. Commun. **1**(4), 660–670 (2002)

20.34 S.S. Iyengar, M.B. Sharma, R.L. Kashyap: Information routing and reliability issues in distributed sensor networks, IEEE Trans. Signal Process. **40**(12), 3012–3021 (1992)

20.35 A.S. Ween, N. Tomecko, D.E. Gossink: Communications architectures and capability improvement evaluation methodology, 21st Century Military Commun. Conf. Proc. (MILCOM 2000), Vol. 2 (2000) pp. 988–993

Part C | 20

20.36   S.M. Hedetniemi, S.T. Hedetniemi, A.L. Liest-man: A survey of gossipingand broadcasting in communication networks, Networks **18**, 319–349 (1988)

20.37   W.R. Heinzelman, J. Kulik, H. Balakrishnan: Adaptive protocols for information dissemination in wireless sensor networks, Proc 5th Annu. ACM/IEEE Int. Conf. Mobile Comput. Netw. (MobiCom'99) (1999) pp. 174–185

20.38   W.R. Heinzelman, A. Chandrakasan, H. Balakrishnan: Energy-efficient communication protocol for wireless microsensor networks, Proc. 33rd Annu. Hawaii Int. Conf. Syst. Sci. (2000) pp. 3005–3014

20.39   M. Ettus: System capacity, latency, and power consumption in multihop-routed SS-CDMA wireless networks, Radio Wirel. Conf. (RAWCON'98) (1998) pp. 55–58

20.40   Y. Liu, S.Y. Nof: Distributed micro flow-sensor arrays and networks: Design of architecture and communication protocols, Int. J. Prod. Res. **42**(15), 3101–3115 (2004)

20.41   OPC HAD specifications (2003). Version 1.20.1.00 edition, http://www.opcfoundation.org.

20.42   ZigBee Alliance. Network Specification, Version 1.0, 2004.

20.43   W. Jeong, S.Y. Nof: A collaborative sensor network middleware for automated production systems, Comput. Ind. Eng. (2008), in press

20.44   W. Jeong, S.Y. Nof: Performance evaluation of wireless sensor network protocols for industrial applications, J. Intell. Manuf. **19**(3), 335–345 (2008)

21. Industrial Intelligent Robots

Yoshiharu Inaba, Shinsuke Sakakibara

It has been believed for a long time since the birth of the industrial robot that the only task it could perform was to play back simple motions that had been taught in advance. At the beginning of the 21st century, the industrial robot was born again as the industrial intelligent robot, which performs highly complicated tasks like skilled workers on a production site, mainly due to the rapid advancement in vision and force sensors. The industrial intelligent robot has recently been a key technology to solve issues that today's manufacturing industry is faced with, including the decreasing number of skilled workers and demands for reducing manufacturing costs and delivery time. In this chapter, the latest technology trends in its element technologies such as vision and force sensors are introduced with some of its applications such as the robot cell, which has succeeded in drastically reducing machining costs.

21.1 Current Status of the Industrial Robot Market

Industrial robots are now taking active part in various fields, including automotive and general industries. Today, industrial robots are used in many industries in many countries. The operational stock of industrial robots in major industrialized countries is shown in Table 21.1. Current industrial robots are typically used in such applications as spot welding, arc welding, spray painting, and material handling, as shown in Figs. 21.1–21.4, respectively.

Devol started the history of the industrial robot by filing the patent of its basic idea in 1954. A teaching-playback type industrial robot was delivered as a product for the first time in the USA in 1961. The robot featured two basic operations, teaching and playback, which are adopted in almost all robots on current factory floors.

Part C | 21.2

Table 21.1 Shipments and operational stock of multipurpose industrial robots in 2005 and 2006 and forecasts for 2007–2010. Number of units (source: World Robotics 2007)

Country	Yearly installations				Operational stock at year-end			
	2005	2006	2007	2010	2005	2006	2007	2010
America	21 986	17 910	21 400	24 400	143 634	154 680	167 100	209 000
North America (Canada, Mexico, USA)	21 567	17 417	20 500	23 000	139 984	150 725	162 400	200 900
Central and South America	419	493	900	1400	3650	3955	4700	8100
Asia/Australia	76 047	61 748	66 000	75 000	481 652	479 027	500 500	579 900
China	4461	5770	6600	7900	11 557	17 327	23 900	47 000
India	450	836	1600	4500	1069	1905	3500	14 100
Japan	50 501	37 393	39 900	42 300	373 481	351 658	355 000	362 900
Republic of Korea	13 005	10 756	10 700	11 800	61 576	68 420	73 600	94 000
Taiwan, Province of China	4096	4307			15 464	19 204		
Thailand	1458	1102			2472	3574		
Other Asia	1163	812			11 095	11 385		
Australia/New Zealand	913	772			4938	5554		
Europe	28 432	31 536	35 000	39 000	296 918	315 624	329 800	380 000
Austria	485	498			4148	4382		
Benelux	1097	1459			9362	10 128		
Denmark	354	417			2661	3013		
Finland	556	321			4159	4349		
France	3077	3071	3300	3200	30 236	32 110	34 000	38 800
Germany	10 075	11 425	12 700	13 000	126 294	132 594	137 900	147 400
Italy	5425	6259	6900	6400	56 198	60 049	63 800	72 000
Norway	115	181			811	960		
Portugal	144	268			1542	1710		
Spain	2709	2409			24 141	26 008		
Sweden	939	865			8028	8245		
Switzerland	442	458			3732	3940		
Turkey	207	368			403	771		
United Kingdom	1363	1220	1000	800	14 948	15 082	15 300	13 800
Central/Eastern European countries	1287	1322			9446	10 781		
other Europe	157	995			809	1502		
Africa	204	426	700	900	634	1060	1700	4400
Total	126 669	112 203	123 100	139 300	922 838	950 974	999 100	1 173 300

Source: IFR, national robot associations and UNECE (up to 2004)

21.2 Background of the Emergence of Intelligent Robots

The use of industrial robots on the production site started a rapid expansion in the 1980s because it came to be known that they had the possibility of improving productivity and making the quality of products stable. However, this led to a situation where, for example, dedicated equipment had to be prepared to supply work-

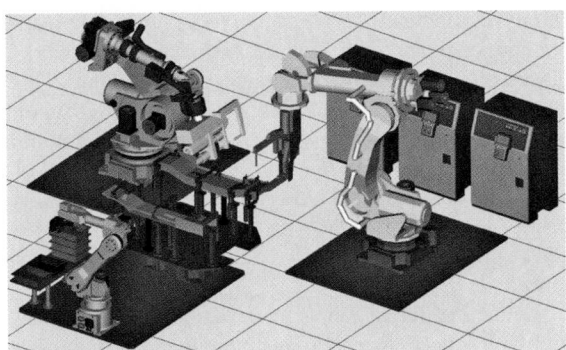

Fig. 21.1 Spot welding

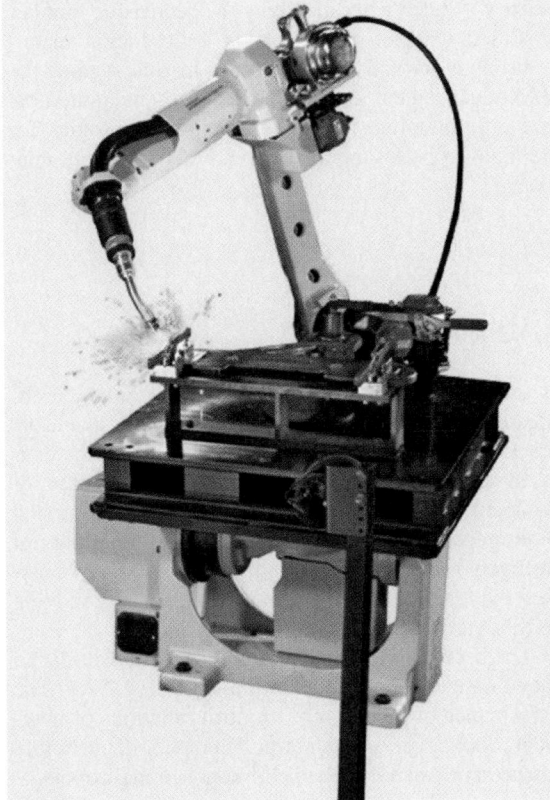

Fig. 21.2 Arc welding

Fig. 21.3 Spray painting

Fig. 21.4 Material handling

pieces to a robot. Also, human operators had to prepare workpieces in alignment for the dedicated equipment before the robot loaded a workpiece to such a machine tool as a lathe. In 2001, the industrial intelligent robots (hereafter *intelligent robots*) appeared on the industrial scene mainly to automate loading workpieces to the fixtures of machine tools such as machining centers.

Before discussing specifics, it is necessary to look back at the history of machining process automation. The first private sector numerical control (NC) was developed in the 1950s, followed by the dramatic enhancement of the NC machine tool market. The machining itself was almost completely automated by the NC; however, loading and unloading of workpieces to and from machine tools were still done by human operators even in the 1990s. The intelligent robot appeared in 2001 for the first time. The term *intelligent robot* does not mean a humanoid robot that walks and talks like a human being, but rather one that performs highly complicated tasks like a skilled worker on the

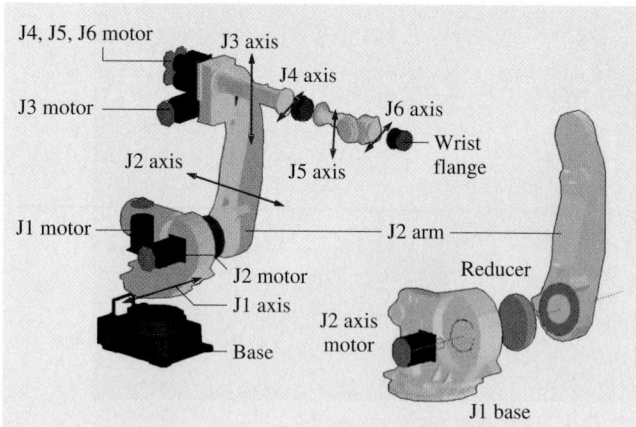

Fig. 21.5 Mechanical structure of an industrial robot

supply equipment, as the robot picks up workpieces one by one using its vision sensor once the workpieces are delivered to a basket near the robot. This also eliminated a burdensome process imposed upon human operators; that of arraying workpieces for the dedicated equipment. In addition, the automation of several tasks, which follow the machining, such as deburring, most of which had been difficult for conventional robots, was also realized by the intelligent robot.

Thus, intelligent robots have recently been increasingly introduced in production, mainly due to their high potential for enhancing global competitiveness as a key technology to solve issues that today's manufacturing industry is faced with, including the decreasing number of skilled workers and demands for reducing manufacturing costs and delivery time. In this regard, the rapid advancement in vision and force sensors and offline programming, and the element technologies for intelligent robots support the trend of robotic automation.

production site by utilizing vision sensors and force sensors. It has first enabled the automatic precision loading of workpieces to the fixture of the machining center and eliminated the need for dedicated parts

21.3 Intelligent Robots

21.3.1 Mechanical Structure

Figure 21.5 shows a typical configuration of a vertical articulated type six-axis robot. There are no big differences between the mechanical structure of an intelligent robot and that of a conventional robot. It comprises several servomotors, reducers, bearings, arm castings, etc.

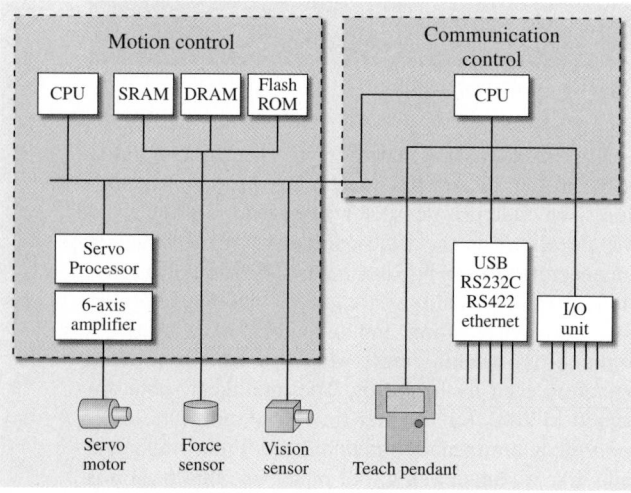

Fig. 21.6 Control system

21.3.2 Control System

As shown in Fig. 21.6, the sensor interface, which enables the connection of sensors such as vision sensors and/or force sensors to the controller, features the control system of an intelligent robot compared to that of a conventional robot. High-speed microprocessors and communication interface also features it. Operators of intelligent robots make robot motion programs by operating the teach pendant shown in Fig. 21.7 and moving the robot arm step by step.

The servo control of intelligent robots is similar to that of conventional robots as shown in Fig. 21.8. The performance of several servo control functions of intelligent robots, such as the interpolation period, are highly enhanced compared with that of conventional robots.

21.3.3 Vision Sensors

Two types of vision sensors are often used on the factory floor: two-dimensional (2-D) and three-dimensional (3-D) vision sensors. The 2-D vision sensors acquire two-dimensional images of an object by irradiating natural light or artificial light on the object and taking the image of the reflected light with a CCD or other type of camera. This enables obtaining a two-dimensional po-

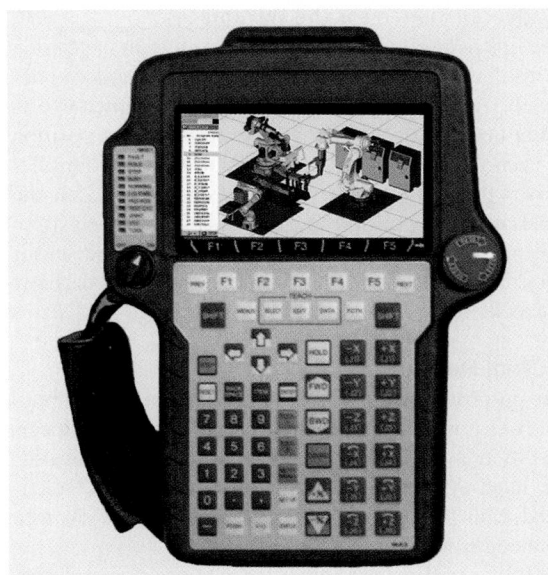

Fig. 21.7 Teach pendant

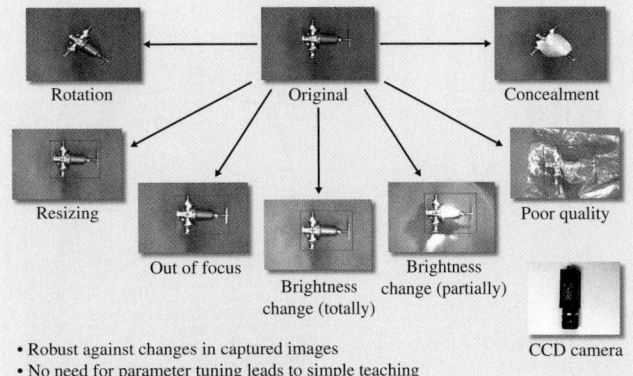

• Robust against changes in captured images
• No need for parameter tuning leads to simple teaching

Fig. 21.9 Images captured by 2-D vision sensor

sition and rotation angle of the object. Recently, 2-D vision sensors have been made useful under the severe production environment at the factory floor due to enhanced tolerance to change in brightness and image degradation based on improved processing algorithms and increased processing speed. Figure 21.9 shows

several images captured and processed by 2-D vision sensors.

There are two major methods for 3-D vision sensors, the structured light method and the stereo method. Of these, the structured light method irradiates the structured light such as slit light or pattern light on an object, and takes images of the reflected light by a CCD or other type of camera to obtain images of the object. The 3-D position and posture of the object are calculated with high accuracy from these images. Additional information on sensors is also provided in Chap. 20.

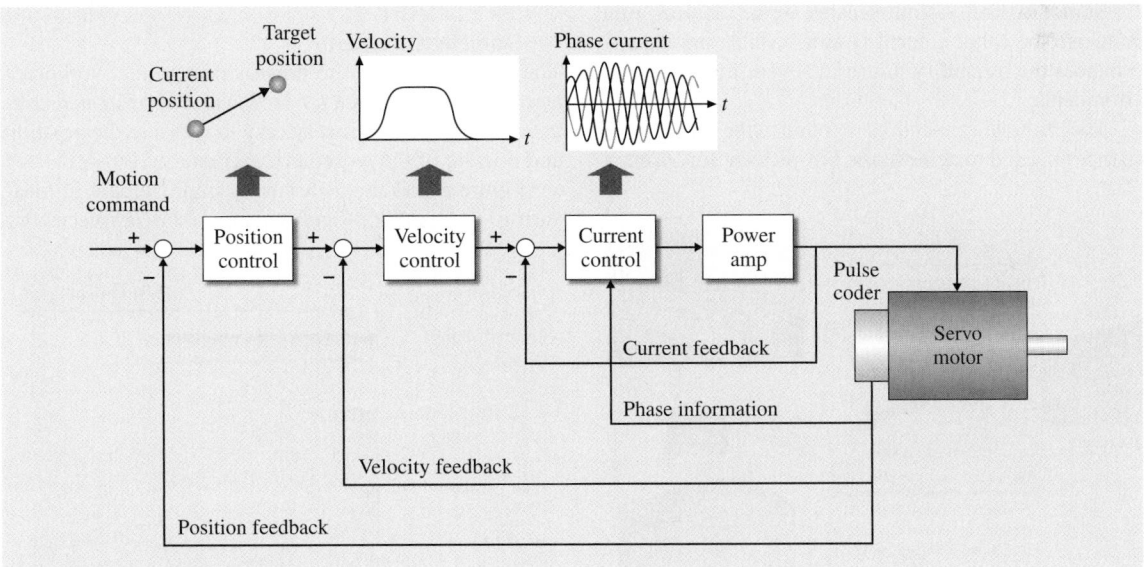

Fig. 21.8 Servo control

Part C | 21.3

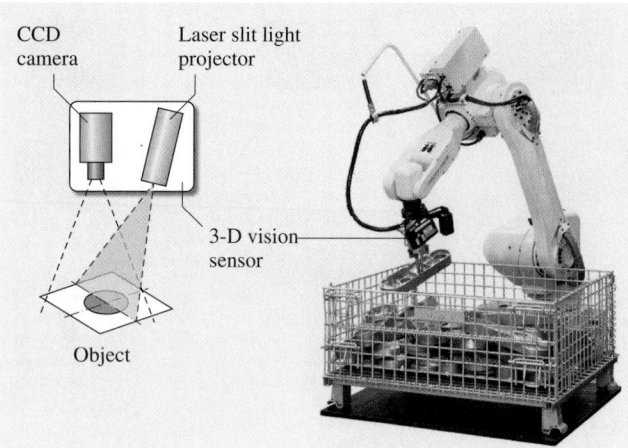

Fig. 21.10 Bin picking

Figure 21.10 shows the bin-picking function realized by using the 3-D vision sensor with the structured light method. The robot can pick up workpieces one by one from those randomly piled in a basket using the 3-D vision sensor.

With respect to the economic effect of the 3-D vision sensor, this reduces the capital investment expense by simplifying the peripheral equipment such as the workpiece feeder, and relieves the operator from the daily burden of arraying workpieces as shown in Fig. 21.11.

Some of the vision sensors have control units built into the robot control system, which significantly enhances the reliability in use in severe production environments.

The following section explains the typical sequences needed to achieve the bin-picking function.

Overall Search for the Workpieces

The image of piled workpieces is taken from upper side by the far-sighted-eye, and the pretaught model is detected from within the image. This far-sighted-eye is calibrated in advance. Then the approximate position of each detected workpiece is acquired. The controller moves the hand-eye (a vision sensor mounted on the hand) near the work-piece based upon this 3-D position, and measurement instructions are issued to the sensor controller, then the trial measurement explained below is executed.

Trial Measurement

The purpose of the trial measurement is to acquire both the 3-D position and posture of the workpiece specified by the overall search. This trial measurement uses the hand-eye and its measurement position is calculated with the approximate position of the workpiece obtained from the overall search.

In the trial measurement the 3-D position and posture of the workpiece are obtained by projecting the structured light by the hand-eye sensor. Because the approximate position of the workpiece obtained from the overall search does not include the posture of the workpiece, the trial measurement position and posture are not always appropriate to the workpiece. This may cause unsuitable accuracy for workpiece handling. To compensate this condition, fine measurement is performed in the next step.

Fine Measurement

The hand-eye is able to come closer to the workpiece after the *overall search for the bin* and the *trial measurement*; thus it becomes very easy to measure the position and posture of the workpiece accurately.

Figure 21.12 shows accuracy improvement in measuring the position and posture of workpieces by

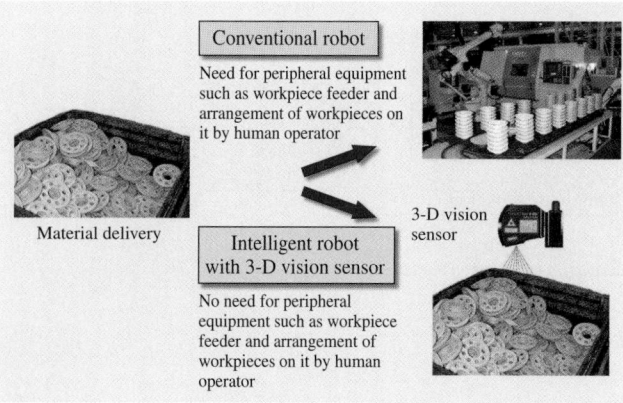

Fig. 21.11 Economic effect of the vision sensor

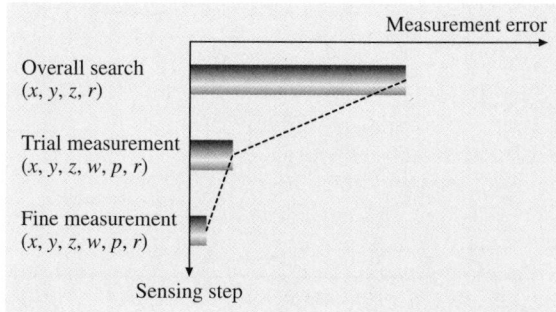

Fig. 21.12 Accuracy improvement by repetitive measurement

this repetitive measurement using the 3-D vision sensor.

Error Recovery

When building an automation system, it is desirable to prevent errors that might stop the system beforehand, but complete provisions to prevent errors are impossible, and the occurrence of unexpected errors is inevitable. For this reason, intelligent robots incorporate exception-handling functions that work in the event of an error occurring. Exception handling analyzes the cause of the error to recover the corresponding error state. This prevents the fall in the operation rate of the bin-picking system.

For instance, when a workpiece is taken out from among those in the basket, the interference of the robot and peripheral equipment might happen occasionally. Then the robot controller with its high sensitivity collision detection function, the details of which are shown in Fig. 21.13, detects the load suddenly added to the robot by the interference, and stops the movement of the robot instantaneously. The position and the posture of the robot are retrieved, and it becomes possible for the robot to continue to move afterwards. This function prevents robot and peripheral equipment from being damaged beforehand, and is useful for improving the operation rate of the system.

The other 3-D vision method, the stereovision method, uses images taken by two cameras, by matching corresponding points on the left and right images to calculate the objects' 3-D position and posture (Fig. 21.14). Although it takes some time to match corresponding points, this method may be advantageous in such a case where a mobile robot has to recognize its surroundings, as there is no need to irradiate auxiliary light as in the structured light method.

21.3.4 Force Sensors

A six-axis force sensor is used to give robots dexterity by mounting it on the wrist of the robot arm. A force sensor generally detects x-, y-, and z-direction forces and each axis' moment (a total of six degrees of freedom). The force sensor usually has a strain gauge on the distortion part, which becomes distorted when applied with force or moment and makes it possible to determine its value. Figure 21.15 shows an example of the force sensor. The use of force sensors has enabled the robot to do such tasks as shaft fitting and gear phase matching for precision machine parts assembly, as well as deburring and polishing, which require

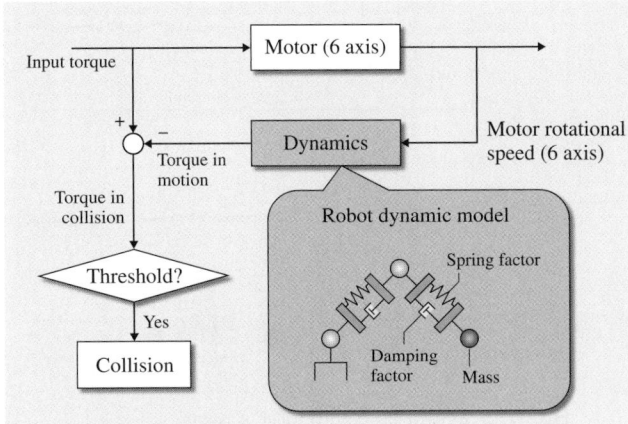

Fig. 21.13 Collision detection

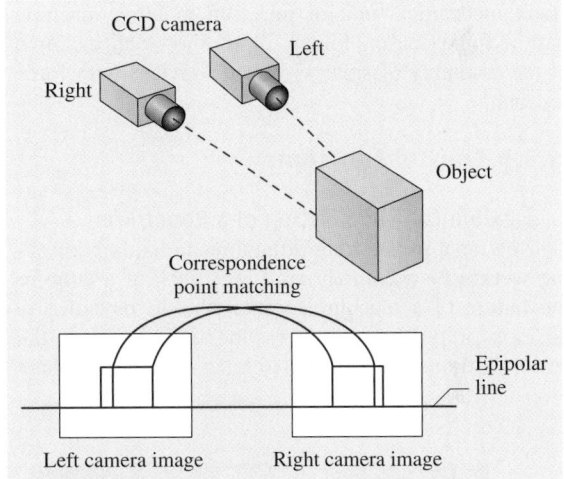

Fig. 21.14 Stereovision method

a certain pressure. Figure 21.16 shows the change of the force and the moment in a peg-in-hole operation.

Fig. 21.15 Force sensor

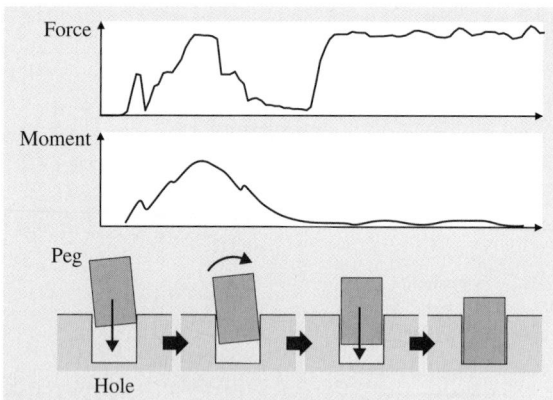

Fig. 21.16 Peg-in-hole with force sensor

Figure 21.17 shows an example of the assembly of the crank mechanical unit of injection molding machine with a force sensor. Figure 21.18 shows an example of the assembly of small electronic devices with force sensors.

21.3.5 Control Functions

Flexible Control Function of a Robot Arm
Conventional robots have difficulties in loading a casting workpiece accurately onto the chuck of a lathe or the fixture of a machining center due to deviance in position or posture of the casting deriving from the casting's dimensional dispersion. In die-casting, there

Fig. 21.17 Assembly of a crank mechanical unit with a force sensor

is also the risk that the conventional robot arm with a workpiece does not comply with the ejector motion when the workpiece is drawn out from the die-mold by the ejector. The robot arm's flexible control function enables the robot to softly control its arm in the

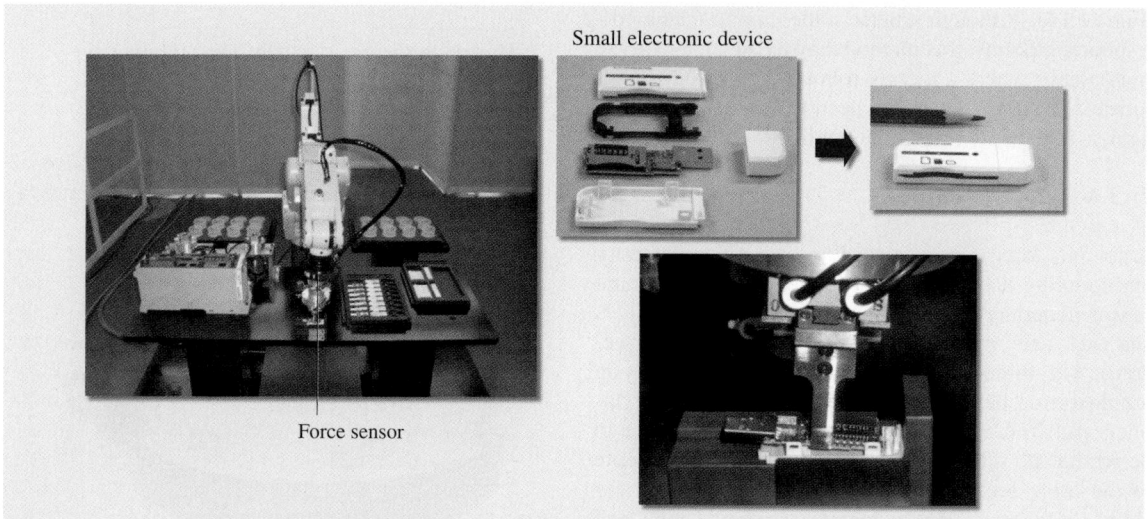

Fig. 21.18 Assembly of small electronic device with a force sensor

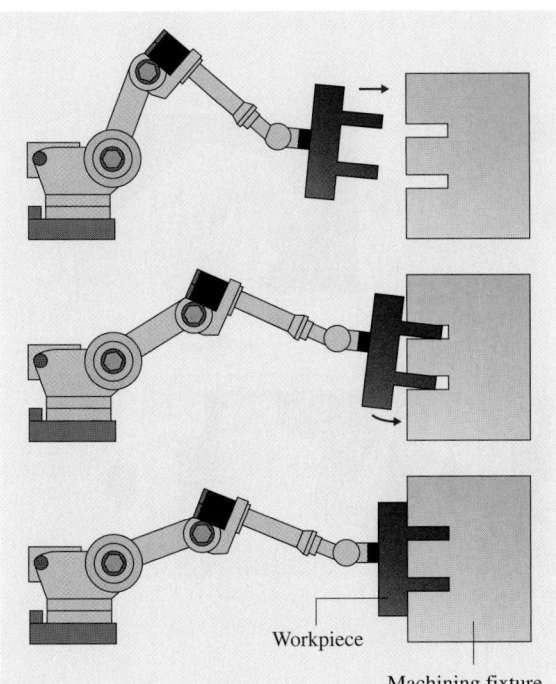

Fig. 21.19 Robot arm and workpiece comply with the machining fixture so as to fit the workpiece to the fixture

designated direction in orthogonal coordinate systems. It can thus accurately load a workpiece to the machine tool and draw out a workpiece from the die-mold without any disturbing force in the die-casting operation. Figure 21.19 shows the flexible control function of robot arm. It does not use any force sensors but rather the current of the servomotors that drive the robot arm.

Coordinated Control Function of Multiple Robots

Coordinated operation by multiple robots has been made possible by synchronizing robots connected to each other via the Ethernet. This function, for example, enables multiple robots to carry a heavy workpiece such as a car body in coordination, where the weight of the workpiece exceeds the payload capacity of one robot. Also, a flexible system can be configured as shown in Fig. 21.20, in which two robots rotate a workpiece while gripping it, where the two robots are used as rotating fixtures, and another robot arc-welds the workpiece in coordination with other robots' motion.

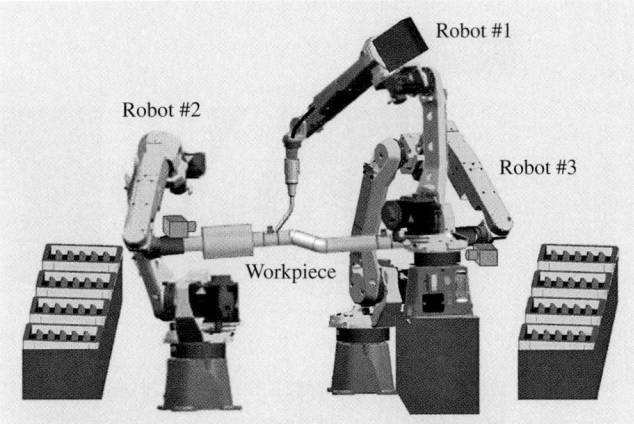

Fig. 21.20 Coordinated arc welding system with three robots

Collision Detection Function

Humans may have a sense of affinity with robots. However, robots are a highly rigid machine by nature, and may cause heavy damage to human beings or other machines at the time of collision. In the latest robot control technology, the drive power supply is immediately cut when a robot collides with an object because it detects the change in the servomotor current that drives the robot arm. This minimizes the damage if a robot collides with machines. Human beings must be separated from the robot in space and/or time by such safety means as fences.

21.3.6 Offline Programming System

Offline programming systems have greatly contributed to decreasing the robot's programming hours as PC performance has improved. There is usually a library of robot models in the offline programming system. The data of workpiece shapes, which has recently been compiled by using 3-D CAD systems, is read into the offline programming system. Peripheral equipment is often defined by using an easy shape generating function of the offline programming system, or by using 3-D CAD data. As shown in Fig. 21.21, after loading the workpiece's 3-D CAD data into the offline programming system, the robot motion program can easily be generated automatically, just by designating the deburring path and the posture of the deburring tool on the PC display.

However, the robot motion program automatically generated on the PC cannot immediately operate the robot on the production floor, as the positional relation of the robot and the workpiece on the PC display is

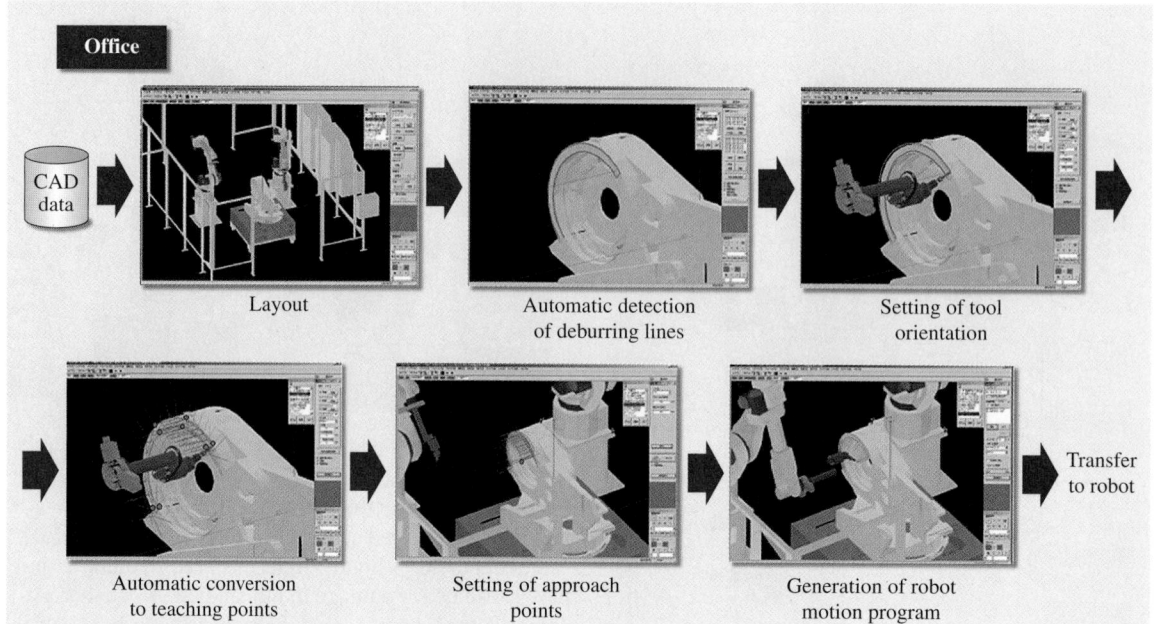

Fig. 21.21 Deburring program generation by offline programming

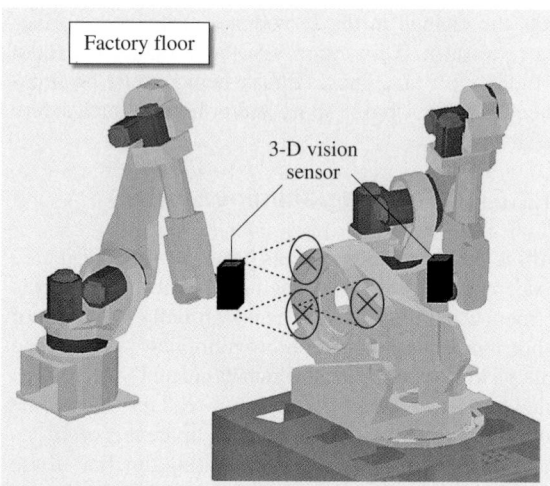

Fig. 21.22 Measurement of workpiece position and posture by the vision sensor

slightly different from that of the real robot and workpiece on the production site. The latest technology in robot vision sensors calculates the real workpiece's position and posture on the production floor, enabling an automatically generated program to be corrected automatically, as shown in Fig. 21.22, thus significantly decreasing the programming time for deburring.

21.3.7 Real-Time Supervisory and Control System

The real-time supervisory and control system is a software package to monitor and control the production system's real time, which is generally called SCADA (supervisory control and data acquisition). As shown in Fig. 21.23, the system enables an operator to generate screens interactively, by making use of modularized functions of data collection, graphical drawing and data analysis software, as well as standard communication network and common database. Various production data including machining data and cycle time can be controlled and analyzed from an office via the Ethernet, which links the office computers with the CNC machine tools and robots on the production floor. The plant manager can directly access the production outcome and operational status of machine tools and robots in the plant he/she is responsible for via internet or by using a mobile phone even during his/her travels abroad, to give precise and timely directions. For a related discussion see Chap. 23 on programming real-time systems for automation.

There is also a system available with a function to strongly support finding causes of temporal system stoppages during the system operation on the production floor.

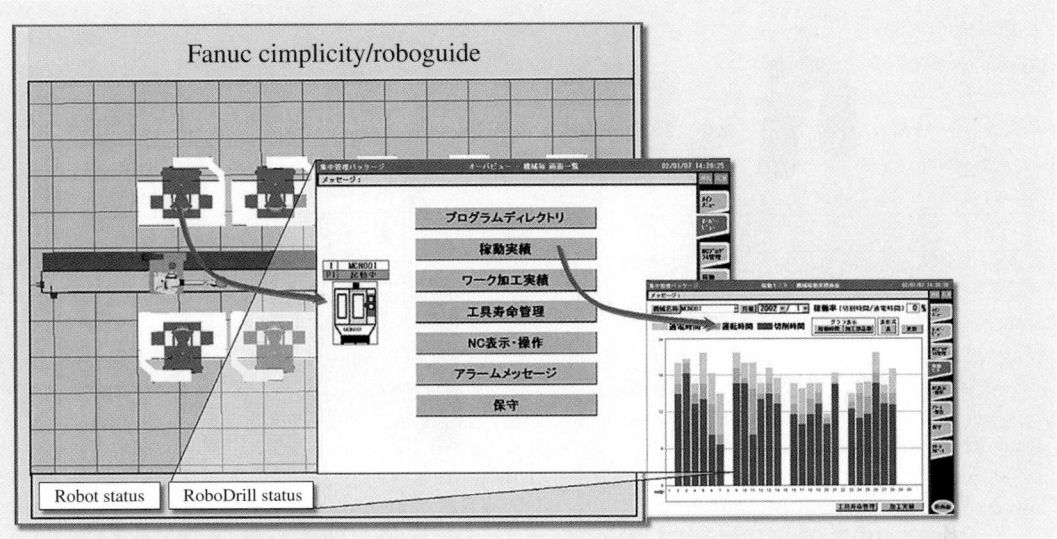

Fig. 21.23 Real-time supervisory and control system

21.4 Application of Intelligent Robots

21.4.1 High-Speed Handling Robot

In the food and pharmaceutical handling field, there has been a delay in robotic automation compared with heavy workpiece handling, as the goods handled in this field are relatively lighter, and as there is a stronger requirement for higher-speed and continuous handling operation. Figure 21.24 shows an example of the high-speed handling by an intelligent robot, which can operate continuously at high speeds, and is clean, washable, and chemical proof. This robot adopts the dual drive and torque tandem control method as shown in Fig. 21.25, the first case in robots, to achieve high-speed and continuous operation. Each basic axis has two motors, and by optimally controlling these motors, high acceleration/deceleration and continuous operation for high duty performance are achieved.

Using the visual tracking function, which combines a vision sensor and a tracking function, can increase handling efficiency. With the vision function built into the robot controller, the vision system has become highly reliable for use on the production floor. With its vision sensor the robot recognizes the position of

a workpiece coming on the conveyer and compensates the robot motion trajectory by comparing the workpiece's position with the conveyor speed data received from the pulse coder. Also, the system allocates handling operations among multiple robots in order to increase handling efficiency. Figure 21.24 shows the visual tracking function by three high-speed handling robots, which can handle 300 parts/min.

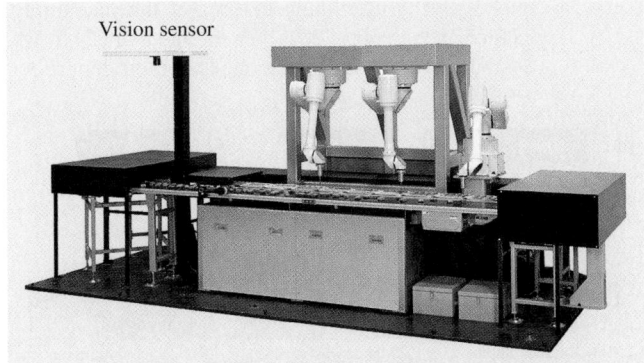

Fig. 21.24 Visual tracking system by high-speed handling robots

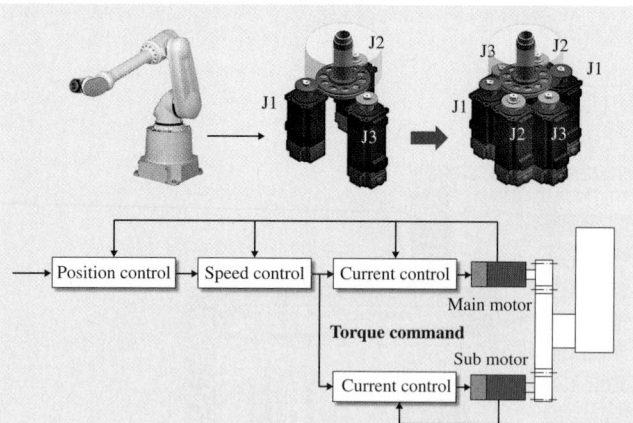

Fig. 21.25 Dual drive and torque tandem control

21.4.2 Machining Robot Cell – Integration of Intelligent Robots and Machine Tools

Generally, the machining system can reduce machining costs by operating long hours continuously. The history of machining system is summarized in Fig. 21.26.

1. In the 1980s, a 24 h continuous operation was achieved by introducing a CNC machine tool system equipped with pallet magazines (Fig. 21.26a).
2. In the 1990s, a 72 h continuous operation including weekends was achieved by introducing a machining system equipped with a large-scale multilayer pallet stocker (Fig. 21.26b).
3. In the 2000s, a robot cell system was developed in which an intelligent robot directly loads workpieces to the machining fixtures of the machining center, achieving 720 h per month, or 24 h for 30 days, continuous operation. The intelligent robot

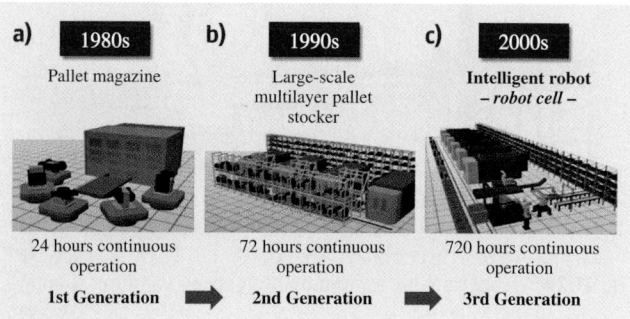

Fig. 21.26a–c A trend in systems for long hours of continuous machining

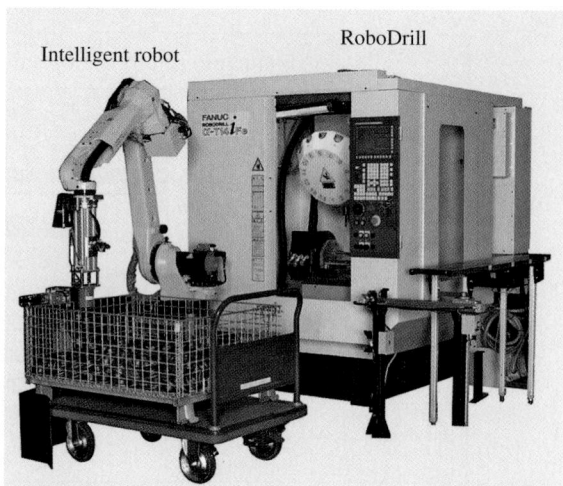

Fig. 21.27 Mini robot cell. Decreases machining cost of long hours of continuous machining drastically

compensates the deviation in its gripping of the workpiece, resulting from the dimensional variation of the casting, with its 3-D vision sensor. Also, the workpiece is loaded to the machining fixture with precision, because the workpiece is pressed softly to the surface of the fixture by controlling the robot arm softly as shown in Fig. 21.19. The robot cell can substantially reduce labor and machining costs, as well as initial capital investment (Fig. 21.26c).

Figure 21.27 shows a mini robot cell comprised of a CNC drill and an intelligent robot, in which the intelligent robot loads and unloads workpieces to and from the CNC drill, and measures the dimensions of the ma-

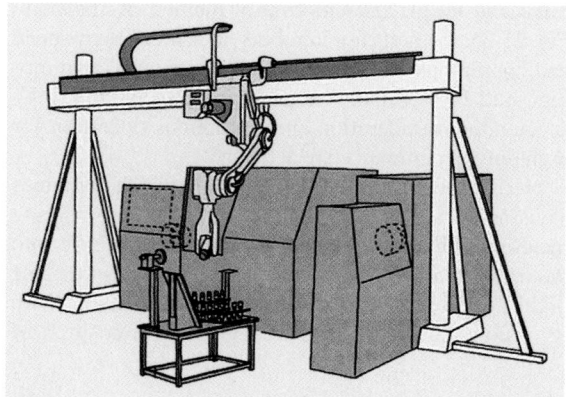

Fig. 21.28 Top mount loader robot

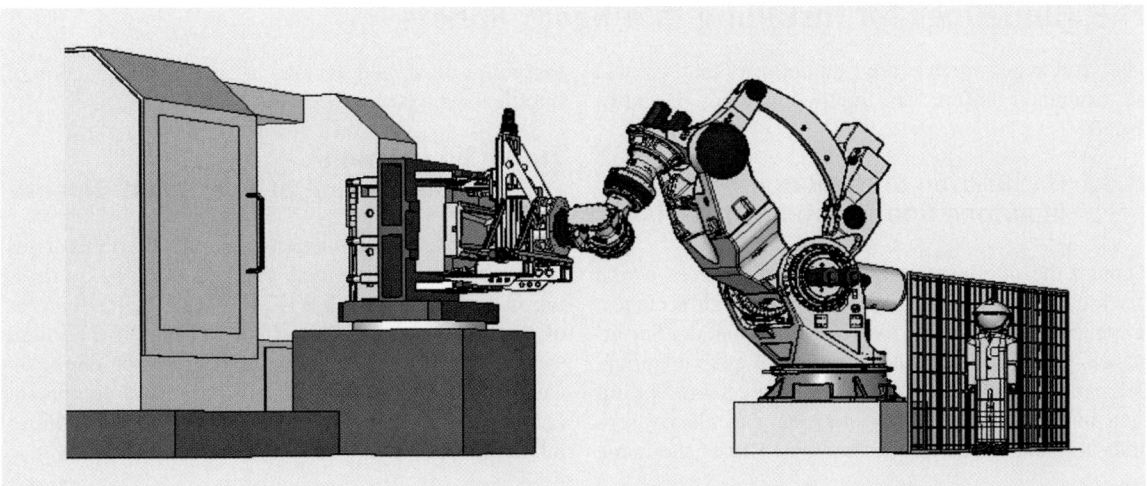

Fig. 21.29 Big robot

chined workpieces. An operator carries a basket filled with workpieces in front of the robot. The intelligent robot with a vision sensor picks up a workpiece from among those randomly placed in the basket and loads it to the drill. This robotic system eliminates dedicated workpiece supply equipment, as well as the manual work by human operators to array workpieces on it.

Mini robot cells are available from the minimum configuration of one machine tool and one intelligent robot to a system with multiple machine tools and robots, meeting the various requirements of users. In the case where multiple machine tools work on a workpiece for different machining processes, there is also a system in which a mobile intelligent robot runs between the machine tools to transfer the workpiece as shown in Fig. 21.28.

Until now, a heavy workpiece of about 1000 kg had to be handled by the coordination of more than two robots, and workers had to use the crane to handle the heavy workpiece. A big intelligent robot, whose payload is about 1000 kg appeared recently. This robot can simplify the robot cell as shown in Fig. 21.29.

21.4.3 Assembly Robot Cell

Intelligent robots are also expected to take an active role in the assembly job, which comprises as large

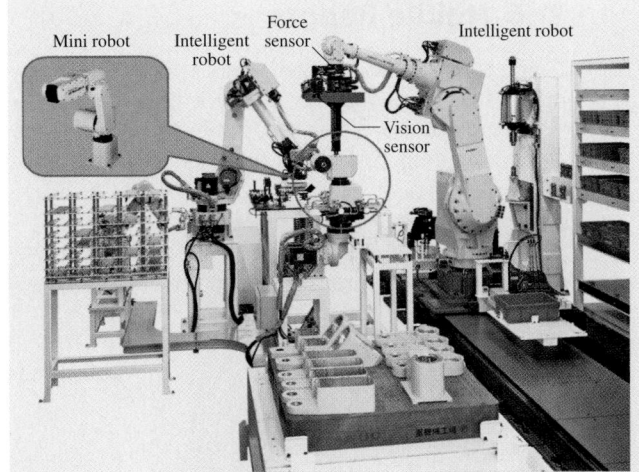

Fig. 21.30 Assembly robot cell

a part of the machine industry as the machining job. The intelligent robot can perform highly accurate assembly jobs, picking up a workpiece from randomly piled workpieces on a tray, assembling it with the fitting precision of 10 μm or less clearance with its force sensor. Figure 21.30 shows an assembly robot cell in which intelligent robots are assembling mini robots.

21.5 Guidelines for Installing Intelligent Robots

The following items are guidelines that should be examined before an intelligent robot is introduced.

21.5.1 Clarification of the Range of Automation by Intelligent Robots

Though the intelligent robot loads workpieces to the machine tool and assembles parts with high accuracy, it cannot do everything a skilled worker can do. For instance, the task such as the assembly of a flexible thing belongs to a field in which humans are more skillful than intelligent robots. It is necessary to clearly separate the range that can be automated from the range that relies on skilled workers before the introduction of intelligent robots.

21.5.2 Suppression of Initial Capital Investment Expense

The effect of the introduction of intelligent robots arises if the peripheral equipment can be simplified by making use of flexibility that is one of the major features of intelligent robots. This effect weakens if the initial capital investment expense is not suppressed due to an easy increase of the expense for peripheral equipment compared with that of the automation that does not use intelligent robots.

21.6 Mobile Robots

Figure 21.31 shows an autonomous mobile cleaning robot. The robot is mainly used for floor-cleaning in skyscrapers. It moves between stories by operating the elevator by itself, and cleans the floors at night. It au-tonomously returns to the start position after it has worked. More than 10 cleaning systems using the robots have already been introduced into several skyscrapers in Japan.

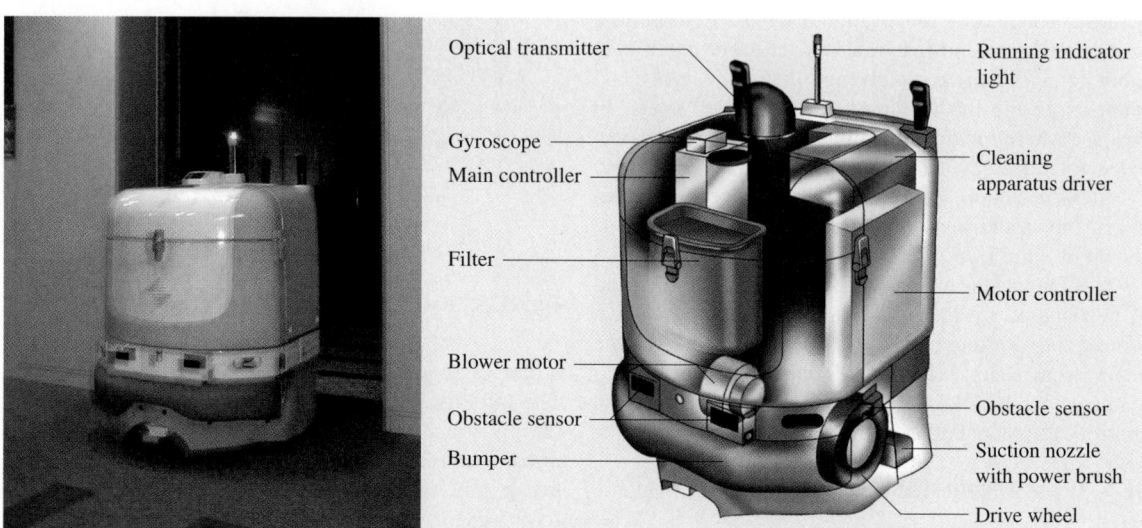

Fig. 21.31 Autonomous mobile cleaning robot (source: Fuji Heavy Industries Ltd.)

21.7 Conclusion

The intelligent robot appeared on the factory floor at the beginning of 2000 and has vision and force sensors. It can work unattended at night and during holidays because it reduces manual preparations such as arrangement of workpieces and/or the necessity of system monitoring compared with the conventional robot. Thus, the use efficiency of equipment rises, machining and assembly costs can be reduced, and the global competitiveness of a product can be improved.

The industrial intelligent robots still have tasks in which they cannot compete with skilled workers, though they have a high level of skills, as has been explained so far. The assembly task of flexible objects such as wire-harnesses is one such task. There are several on-going research and development activities in the world to solve these challenges. One idea is to automate such a task completely, and another is to do it partially. In the latter case, robots and skilled workers work together; robots assemble mechanical parts and skilled workers assemble flexible parts, for example. In any case, the degree of cooperation between humans and robots will increase in the near future.

21.8 Further Reading

- Y. Bar-Cohen, C. Breazeal: *Biologically Inspired Intelligent Robots* (SPIE Press, Bellingham 2003)
- G.A. Bekey: *Autonomous Robots: From Biological Inspiration to Implementation and Control* (MIT Press, Cambridge 2005)
- J.M. Holland: *Designing Autonomous Mobile Robots: Inside the Mind of an Intelligent Machine* (Newnes, Amsterdam 2003)
- S.C. Mukhopadhyay, G.S. Gupta: *Autonomous Robots and Agents* (Springer, New York 2007)

- S.Y. Nof (Ed.): *Handbook of Industrial Robotics* (Wiley, New York 1999)
- R. Siegwart, I.R. Nourbakhsh: *Introduction to Autonomous Mobile Robots* (MIT Press, Cambridge 2004)
- P. Stone: *Intelligent Autonomous Robotics: A Robot Soccer Case Study* (Morgan Claypool, San Rafael 2007)

References

21.1 D.E. Whitney: Historical perspective and state of the art in robot force control, Proc. IEEE Int. Conf. Robot. Autom. ICRA (1985)

21.2 T. Suehiro, K. Takase: Skill based manipulation system, J. Robot. Soc. Jap. **8**(5), 551–562 (1990)

21.3 M.T. Mason: Compliance and force control for computer controlled manipulators, IEEE Trans. Syst. Man Cybern. **11**(6), 418–432 (1981)

21.4 N. Hogan: Impedance control, Part 1–3, Trans. ASME J. Dyn. Syst. Meas. Control **107**, 1–24 (1985)

21.5 J.J. Craig, P. Hsu, S. Sastry: Adaptive control of mechanical manipulators, Int. J. Robot. Res. **6**(2), 16–28 (1987)

21.6 B. Yao, M. Tomizuka: Adaptive coordinated control of multiple manipulators handling a constrained object, Proc. IEEE Conf. Robot. Autom. (1993)

21.7 S. Inaba: Assembly of robots by AI robot, Proc. 22nd IEEE IECON (Int. Conf. Ind. Electron. Control Instrum.), Vol.1 (1996) pp. xxxvii–xl

21.8 S. Sakakibara, A. Terada, K. Ban: An innovative automatic assembly system where a two-armed intelligent robot builds mini robots, Proc. 27th ISIR (1996) pp. 937–942

21.9 S. Sakakibara: Intelligent assembly system suited for module assembly, Proc. 30th ISR (1999) pp. 385–390

21.10 S. Sakakibara: The role of intelligent robot in manufacturing system of the 21st century, Proc. 32nd ISR (2001)

21.11 K. Hariki, K. Yamanashi, K. Otsuka, M. Oda: Intelligent robot cell, FANUC Tech. Rev. **16**(1), 37–42 (2003)

Part C | 21

22. Modeling and Software for Automation

Alessandro Pasetti, Walter Schaufelberger (△)

Automation is in most cases done through the use of hardware and software. Software-related costs account for a growing share of total development costs for automation systems. In the automation field, containment of software costs can be done either through the use of model-based tools (e.g., Matlab) or through a higher level of reuse. This chapter argues that both technologies have their place. The first strategy can be used for the design of software for a large number of identical installations or for the implementation of only part of the software (i. e., the control algorithms). The second strategy is advantageous in the case of industrial automation systems targeting niche markets where systems tend to be one-of-a-kind and where they can be organized in *families* of related applications. In many applications, a combination of the two approaches will produce the best results. Both approaches are treated in the paper. The main focus of the chapter is on developing software for automation. As such software will often be implemented for slightly different processes, it is highly appropriate that the production process is at least partly automated. As space is limited, this chapter can not cover all aspects of the design and implementation of software for automation, but we claim that the methods discussed here integrate very well with traditional methods of software and systems engineering.

Experience from contributions to various research projects recently performed at Swiss Federal Technical university (ETH) are also summarized.

Software plays an increasing role in the design and implementation of automation systems. The efficient construction of reliable software is therefore a general goal in many automation projects. Automation projects are demanding and multifunctional, and nonfunctional (timing, reliability, testability etc.) requirements have to be satisfied. From the early days of the development of computers, their use in automation was investigated, even if the cost of the computer was in many cases much higher than the cost of the plant in the early 1970s.

From these early days, two competing developments emerged. One line developed from the imperative programming languages used at the time (Fortran, later C and Pascal) and the other from the control engineer-

```
C        PROGRAMM ZUR DIGITALEN REGELUNG MIT EINEM P-I-REGLER (DIGREG)
         EXTERNAL PIREG
         COMMON R,S
         CALL STRT
         CALL IGNORE (17)
         CALL SMOD (5,IND)
         PAUSE 1
         CALL SMOD (3,IND)
         CALL DLAYS (1)
         CALL CNCT (18,PIREG,IND)
         READ (4,100) R
         WRITE (4,200) R
         READ (4,101) K
         WRITE (4,201) K
         S = 0
         PAUSE 2
         CALL SMOD (1.IND)
         CALL EARM
         CALL EENA
         CALL SPIG (K,3,2,IND)
1        CONTINUE
         GO TO 1
100      FORMAT (F7.5)
200      FORMAT (10X, 5HR = ,F7.5)
101      FORMAT (I3)
201      FORMAT (10X,5HDT = ,I3)
         STOP
         END


         SUBROUTINE PIREG
         DIMENSION X(1),U(1)
         COMMON R,S
         CALL RIDC (1,1,X,IND)
         S = S + R-X(1)
         U(I) = R-X(1) + .0075*S
         CALL SDIC (1,1,U,IND)
         RETURN
         END
```

Fig. 22.1 Fortran program for a PI controller in 1970

ing approach of describing systems in a declarative way by block diagrams or similar means. Both of these ways were continued and led to solutions widely accepted today in industry.

A Fortran program for a proportional–integral (PI) controller on the hybrid computer available at the time to test ideas of digital control is shown in Fig. 22.1.

This is clearly an imperative style. In a text of the mid 1980s on real-time programming of automation systems [22.1], conventional programming was still the only way to program real-time systems, and the real-time industrial language PORTAL (similar to Pascal) was used for practical experiments, implementing PI and adaptive controllers.

A program we developed in the late 1980s for educational purposes with a graphics environment on a personal computer (PC) [22.2, 3] is shown next in Fig. 22.2 as controller for a small heating system. Simulink was not yet available at the time. At the time, Gem by Digital Research was found to be better suited than Windows for such tasks, the program was therefore written in Gem.

A block diagram is drawn interactively on screen as the declarative description of the control algorithm, which can be directly executed to produce experimental results. This is declarative in nature; the order of execution is not evident from the drawing. A major problem when realizing such programs at the time in Modula-2 was the fact that the language is strongly typed and that lists of function blocks had to be kept. Obviously, these function blocks are of differing types. Programming tricks had to be used to overcome this. A redesign in the object-oriented language Oberon [22.4] demonstrates how easily such programs can be designed with the appropriate mechanisms such as subclassing. In the object–process methodology (OPM) function, structure, and behavior are integrated into a single, unifying model, OPM significantly extends the system modeling capabilities of current object-oriented methods [22.5].

These early attempts to efficient program design and implementation clearly aimed at producing a program for a given task. Graphical and textual libraries were created, but essentially an individual program was the goal of the design. Though quite different in terms of the approach taken, these early attempts were all model based in the sense that a modeling or programming environment was available and that models were used in all stages of the development process.

This changed with the introduction of object-based and object-oriented techniques into software engineering. It became possible to design adaptive programs or programs for families of systems with better orientation on reuse. This is an issue we want to look at in more detail in the following sections.

First versions of Matlab became available in the late 1970s, and Matlab quickly became a de facto standard for control engineers, especially for system design.

22.1 Model-Driven Versus Reuse-Driven Software Development

The fundamental problem of software engineering is to translate a set of user requirements into executable code that implements them in a manner that is as cost effective as possible. There are several ways to solve this problem which, at least conceptually, can be arranged along a continuous spectrum that has at its two extremes the so-called *model-driven* and *reuse-driven approaches*. These two approaches are illustrated in Fig. 22.3.

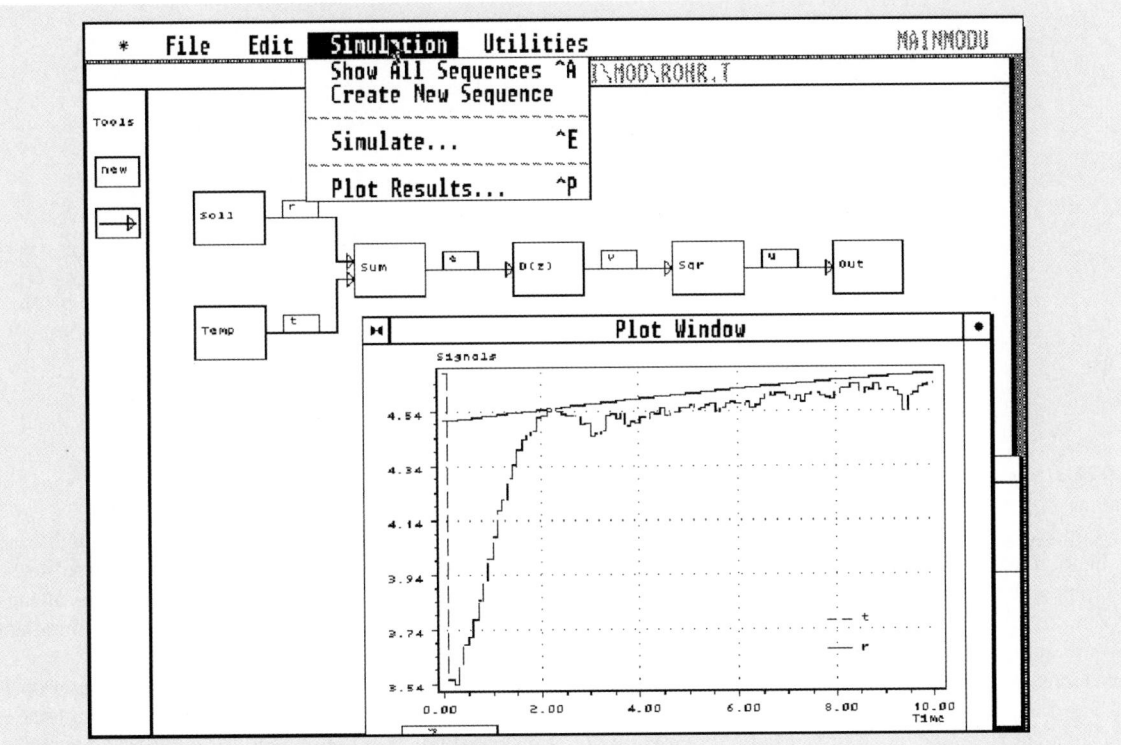

Fig. 22.2 An early programming and experiment environment

Although most projects will take an intermediate or mixed approach, it is useful to consider briefly the two extreme approaches in isolation from each other as a way to clarify their distinctive features.

In the model-driven approach (left-hand side of Fig. 22.3), the user requirements are expressed in a modeling environment that is capable of automatically generating the application code. The prototypical example of such an environment is the Matlab tool suite. The cost and schedule savings arise from the fact that the software design and implementation phases are fully automated.

In the reuse-driven approach (right-hand side of Fig. 22.3), the user requirements are implemented by configuring and composing a set of predefined software building blocks. The cost and schedule savings in this case originate from the possibility of reusing existing software artifacts (modules, components, code fragments, etc.). Traditionally, the reuse-driven approach was implemented by developing libraries of reusable modules. More recently, software frameworks have emerged as a more effective alternative to achieve the

same aim of minimizing software development costs by leveraging reuse.

Each approach has its strengths and weaknesses. The model-driven approach holds the promise of completely automating the software development process, but is in reality limited by the expressive power of the selected modeling language. Thus, for instance, Matlab provides powerful modeling facilities for describing transfer functions and state machines but it does not support well modeling of other functionalities that are equally important in embedded applications (such as management of external units, generation of housekeeping data, processing of operator commands, management and implementation of failure detection and recovery mechanisms, etc.).

The reuse approach can be more flexible both because reusable building blocks can, in principle, be provided to cover as wide a range of functionalities as desired, and because it can be applied in an incremental way with repositories of reusable building blocks being built up over time. The main drawback of this approach is that the selection, configuration, and composition of

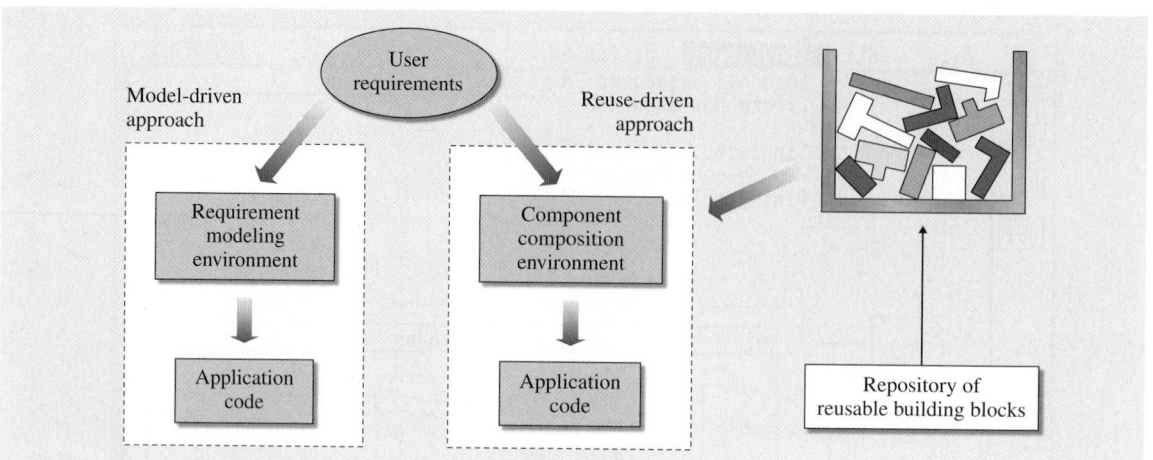

Fig. 22.3 Model-driven versus reuse-driven approach

the building blocks is difficult to automate and, if done by hand, remains a tedious and error-prone task.

In practice, no application is entirely reuse or model driven. Instead, the two approaches are complementary and are applied to different parts of the same application. Consider, for instance, the case of a typical satellite on-board application. A model-driven approach is ideally suited to cover the modeling and implementation of the control algorithms of the application. This is appropriate because there are very good modeling techniques for expressing control algorithms and there are very powerful tools for translating models into code-level implementations. The existence and high level of maturity of the modeling techniques and of the tools is in turn a consequence of the wide range of applications that need control algorithms.

A model-driven approach would, however, be unsuitable to cover functionalities such as the management of satellite sensors or the processing commands from the ground station. These functionalities are entirely specific to the satellite domain and this domain is too

narrow to justify the effort required to develop dedicated modeling techniques and support tools. In these cases, a reuse-driven approach offers the most effective means to improve the efficiency of the software development process.

In general, the reuse-driven approach is appropriate wherever there is a need to model and implement functionalities that are specific to a particular domain. This is due to the high cost of developing industrial-quality model-driven environments. This type of tools only makes economic sense for functionalities that are widely used, namely for functionalities where there is a sufficiently large base of potential users to justify their development costs. In other cases, reuse-driven approaches remain the only viable option.

The model- and reuse-driven paradigms are considered in greater detail in the next two sections (Sects. 22.2 and 22.3). The model- and reuse-driven approaches represent the state of the practice. Section 22.4 extends the discussion to consider some current research trends.

22.2 Model-Driven Software Development

22.2.1 The Matlab Suite (Matlab/Simulink/Stateflow, Real-Time Workshop Embedded Coder)

The prototypical example of model-driven development in the control and automation field is the

Matlab tool suite. This consists of Simulink and Stateflow as toolboxes for programming of continuous time, discrete-time systems, and state machines in graphical form and of the Real-Time Workshop to generate C code from Simulink and Stateflow models. This is a well-known solution, well covered on the World Wide Web (WWW) with many ex-

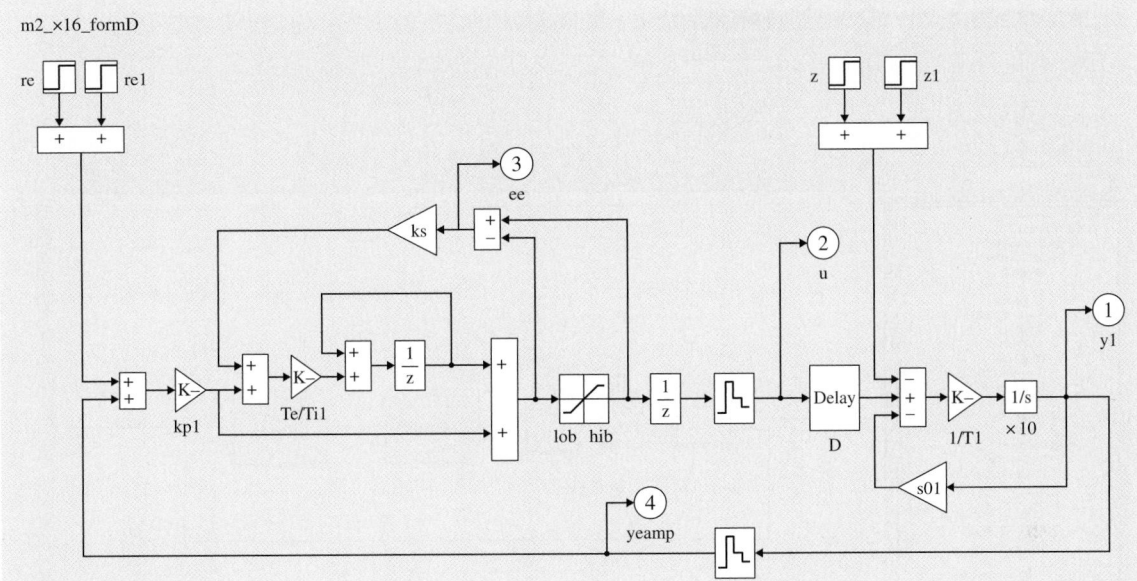

Fig. 22.4 Simulink diagram

amples and will for this reason not be treated in any detail here (see http://www.mathworks.com/ and http://www.mathworks.com/products/rtwembedded/).

A very useful fact worth mentioning here is that Simulink/Stateflow diagrams are stored in textual form and can be analyzed independently of the interpreter, which can also be replaced by other compilers or interpreters. One such case will be mentioned later.

Figure 22.4 shows a typical Simulink diagram of a nonlinear sampled data control system from [22.6].

22.2.2 Synchronous and Related Languages

Behind any model-driven environment there is a language that provides high-level abstractions allowing the designer to express his design. In the case of the Matlab tool suite, this language is hidden from the user, who only interacts with a graphical interface. In other cases, the designer is instead expected to make direct use of the modeling language.

Synchronous languages are one of the most successful families of modeling languages for control and automation systems. Their main strength is that the synchronous paradigm makes it easier to build verification engines that can automatically verify that a model satisfies certain functional properties. The development process then becomes as follows. The designer first

builds a model that is intended to satisfy certain requirements. He then translates his initial requirements into functional properties expressed in a suitable formalism. The verification engine can then verify that the model does indeed implement the requirements. After successful verification, a code generator is used to translate the model into source code in a general-purpose language (typically C).

Several imperative and declarative synchronous languages have been developed following a data flow model of computation [22.7]. This is well adapted to the way in which control engineers model their systems by block diagrams. All the languages must have special constructs for time, concurrency, sequences of values (signals, flows), shift operators etc.

The language Lustre and its industrial version SCADE use data flow models well known to control engineers and all calculations are based on flows (signals). Esterel is an imperative language in the same domain, and Signal is another language more suitable for the design of systems.

Efforts are under way to generate Lustre programs automatically from subsets of Simulink and Stateflow descriptions [22.8]. This makes programs in Simulink/Stateflow accessible to formal analysis.

SCADE from Esterel Technologies [22.9] offers tools for the design and verification in the area of safety critical systems. A very simple example of a standard

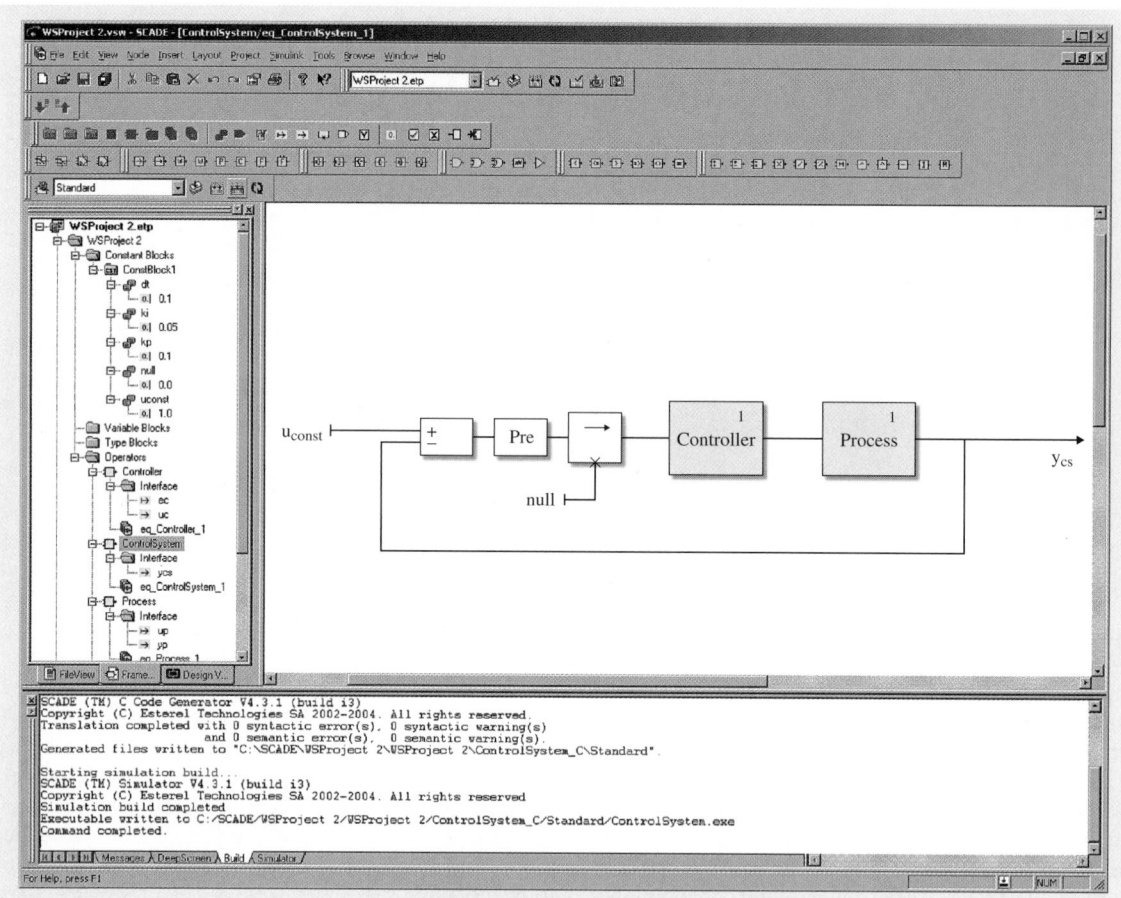

Fig. 22.5 SCADE environment with a very simple example

control loop in a SCADE development environment is shown in Fig. 22.5.

A slightly different approach is taken by Giotto, a time-triggered language for embedded programming [22.10]. Here, the focus is to specify exactly the real-time interactions between the software components and the physical world.

22.2.3 Other Domain-Specific Languages

A more general version of the model-driven approach is the following: Given a problem area, design and implement a specific modeling language for this area, including an environment for editing, compilation etc. This procedure has special relevance at ETH, where Wirth designed and implemented many languages of general (Pascal, Modula, Oberon) and also of a more specific nature (Lola, a logic language). Worth men-

tioning here is Active Oberon by Jürg Gutknecht, an object-oriented language where every object can have an integrated thread of control. This high parallelism is of special interest for reactive systems. The next example will provide more information on this approach.

The IEC 61131 languages have been designed for use in automation. IEC 61131 is an International Electrotechnical Commission (IEC) standard for programmable logic controllers (PLCs). IEC 61131 specifies the syntax and semantics of a unified suite of programming languages for programmable controllers (PCs). These consist of two textual languages, IL (instruction list) and ST (structured text), and two graphical languages, LD (ladder diagram) and FBD (function block diagram). Sequential and parallel processing are possible. The standard is developed at http://www.plcopen.org/, where more information can be found.

22.2.4 Example: Software for Autonomous Helicopter Project

The autonomous helicopter project of ETHZ [22.11] is a typical example of a system where special languages were developed by Wirth and Sanvido for several tasks of the onboard processor. The operating system uses threads instead of tasks [22.12] for speed-up, the logic language Lola [22.13] is used for the design of field-programmable gate arrays (FPGAs), and the mission control language is used for missions. An excerpt from a program in the mission control language is easily readable, as shown in Fig. 22.6.

The controller for the helicopter has also been implemented in Giotto [22.14].

22.3 Reuse–Driven Software Development

22.3.1 The Product Family Approach

Within the reuse-driven paradigm, software product families have emerged as the most successful form of software reuse. A *software product family* [22.15] is a set of applications that can be constructed from a set of shared software assets. The shared assets can be seen as generic *building blocks* from which applications in the family can be built. Usually, a product family is aimed at facilitating the instantiation of applications within a narrow domain. Figure 22.7 illustrates the concept of product family. On the left-hand side, the building blocks offered by the product family are shown. These building blocks are used during the *family instantiation process* to construct a particular application within the family domain.

Product families are characterized by two distinct development processes (Fig. 22.8). In the *family creation process*, the family's reusable assets are designed and developed. In the *family instantiation process*, the reusable assets offered by the family are used to construct a specific application within the family domain.

The family creation process is in turn divided into three phases. In the *domain analysis phase*, the set of applications that must be covered by the family are identified and characterized. The output of this phase is a domain model. In the *domain design phase*, the reusable assets that are to support the instantiation of applications within the family are designed. The output of this phase is one or more models of the family assets. The models express various aspects of the domain design (e.g., there may be functional models, timing models, etc.) In the *domain implementation phase*, the family assets are implemented as concrete building blocks that can be used towards the construction of family applications.

Often, the implementation of the family assets is done automatically by processing the models defined in the domain design phase.

Three matching phases can be identified in the family instantiation process (bottom half of Fig. 22.8). In the *requirement definition phase*, the family domain model is used to verify whether the target application falls within the family domain. This decides whether the family assets can be used to help build the application. If this is the case, a sizable proportion of the application requirements can be expressed in terms of the domain model, for instance by identifying the features in the family domain that are needed by the target application. In the *tailoring phase*, the software assets required for the target application are selected from among those

```
PLAN Example;

VAR phi, theta, psi: REAL; LandOK*: BOOLEAN;

PROCEDURE LiftOff; (* TakeOff Procedure *)
BEGIN
  ACCELERATE(1.0, 0.0, 0.0, 0.0, -0.5);
  TRAVEL(9.0, 0.0, 0.0, 0.0, -0.5);
  ACCELERATE(1.0, 0.0, 0.0, 0.0, 0.5);
END LiftOff;

PROCEDURE Hovering(time: REAL); (* Hovering Procedure *)
BEGIN
  TRAVEL(time, 0.0, 0.0, 0.0, 0.0);
END Hovering;

PROCEDURE Landing; (* Landing Procedure *)
BEGIN
  ACCELERATE(1.0, 0.0, 0.0, 0.0, 0.25);
  TRAVEL(19.0, 0.0, 0.0, 0.0, 0.25);
  ACCELERATE(1.0, 0.0, 0.0, 0.0, -0.25)
END Landing;

BEGIN
  GETATTITUDE(phi, theta, psi);
  LiftOff;
  Hovering(3.0);
  BROADCAST("READYTOLAND");
  LandOK := FALSE;
  WHILE ~LandOK DO SLEEP() END;
  Landing
END Example.
```

Fig. 22.6 Code for helicopter control

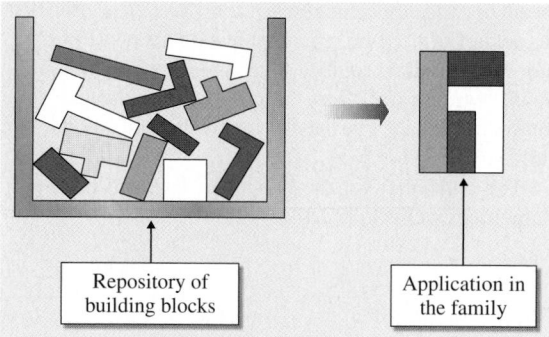

Fig. 22.7 Software product families

offered by the family. They are then adapted and configured to match the needs of the target application. Depending on how the family assets are organized and implemented, adaptation and configuration can be done either at the level of the asset models, or at the level of the implemented assets. Finally, in the *integration and testing phase*, the target application is constructed by assembling the configured and adapted building blocks offered by the framework. Usually, some integration with building blocks that are external to the family is also required in this phase.

22.3.2 The Software Framework Approach

The product family concept is very general and does not imply any assumptions about the nature of the building blocks (Are they components? Procedures? Code fragments? Design models?) or about their mutual relationships (Can they be used independently of each

other or are they embedded within a higher-level structure?). The software framework concept particularizes the family concept. A *software framework* [22.16] is a kind of product family where the reusable building blocks consist of software components embedded within an architecture optimized for the target domain of the framework. Thus, a software framework is a particular way of organizing the shared assets of a software product family in the sense that it defines the type of building blocks that can be provided by the product family and it defines an architecture within which these building blocks are to be used.

Figure 22.9 recasts Fig. 22.7 to illustrate the concept of software framework. The figure highlights the fact that the family reusable assets (the building blocks in the repository) are now organized as a set of interacting entities embedded within an architecture that is itself reusable. The framework approach, in other words, allows the reuse not only of the individual items but also of their mutual interconnections (the latter being an important and often neglected added value).

Frameworks are component based in the sense that the reusable building blocks consist of software components. The term *component* is used to designate a software entity with the following characteristics:

- It can be deployed as a stand-alone unit (hence it owns a clear specification of its *required interface*).
- It provides an implementation for one or more interfaces (hence it owns a clear specification of its *provided interface*).
- It interacts with other components exclusively through these (required and provided) interfaces.

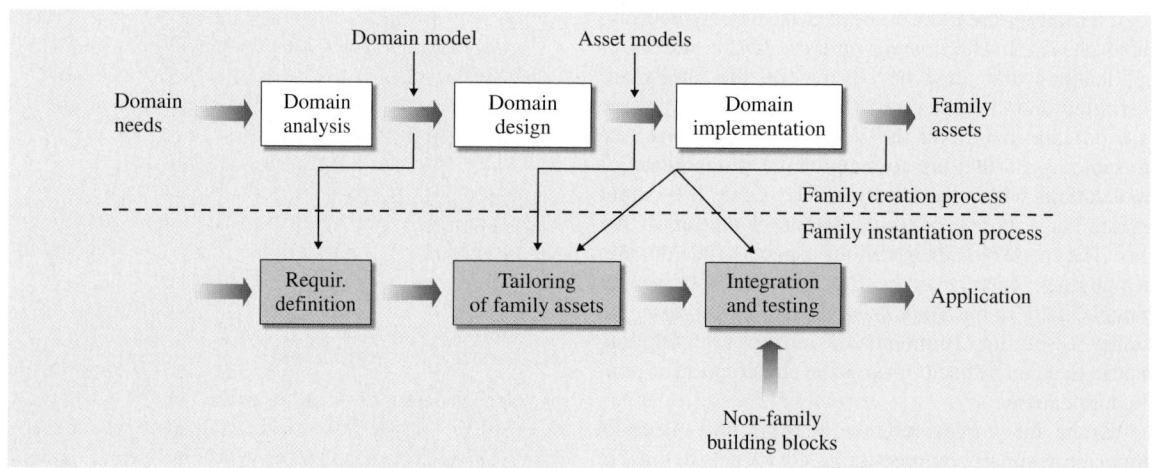

Fig. 22.8 Development process for software product families

The architecture predefined by the framework is defined by the set of interfaces that are implemented by the framework components. Thus, a software framework could also be defined as a product family whose reusable building blocks consist of components and interfaces. The interfaces define the architecture within which the components are to be used. The components encapsulate behavior that is factored out from all (or at least a sizable proportion of) applications in the framework domain.

Still another view is possible of what constitutes a building block of a software framework. This view sees a building block as a *unit of reuse*. At its most basic, the unit of reuse of a component-based framework is a component. This follows from the fact that one of the distinctive features of a component is that it is deployable as a stand-alone unit. The components provided by a software framework, however, are embedded within an architecture (which is also defined by the framework). Hence, users of the framework are likely to focus their attention not on individual components but on groups of cooperating components that, taken together, support the implementation of some function that is important within the framework domain. In fact, well-designed frameworks encourage this higher granularity of reuse by being organized as a bundle of functionalities that users (the application developers) can choose to include in their applications. Inclusion of such functionality implies that a whole set of cooperating components and interfaces are imported into the application. The true unit of reuse – and hence, according to this view, the true building block – is precisely such a set of components and interfaces.

An example may help clarify the above concept. Consider a software framework for satellite onboard applications. One typical functionality that is often found in such systems is the storage of key housekeeping data on a mass-memory device. Accordingly, the software framework would implement default mechanisms for managing such devices. This would probably be done through a set of cooperating components and interfaces. Application developers who need the mass-memory functionality for their target application and who decide to implement it with the help of the assets provided by the framework will import the entire set of components and interfaces. Use of individual components or interfaces is unlikely to make sense because the components and interfaces are specifically designed to work together within a certain architecture. The building block in this case is the set of components and interfaces that support the implementation of the mass-memory functionality.

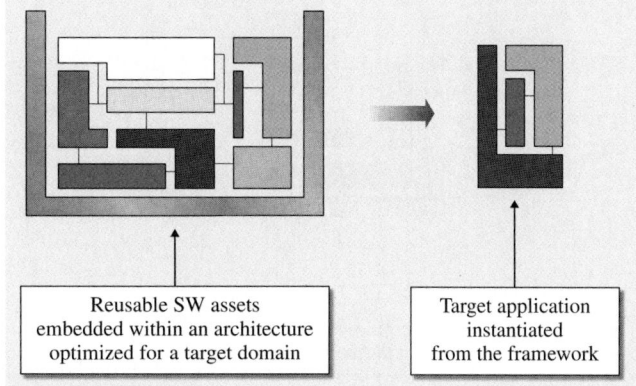

Reusable SW assets embedded within an architecture optimized for a target domain

Target application instantiated from the framework

Fig. 22.9 The software framework concept

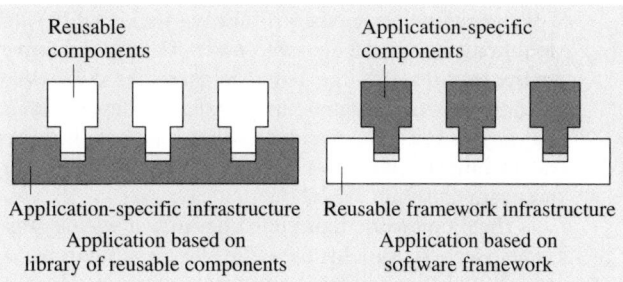

Reusable components

Application-specific components

Application-specific infrastructure

Reusable framework infrastructure

Application based on library of reusable components

Application based on software framework

Fig. 22.10 Software frameworks and libraries of reusable components

Figure 22.10 provides another view of the software framework concept that illustrates the difference from the more traditional forms of software reuse based on libraries of reusable software components. An application instantiated from a framework consists of a reusable architectural skeleton which has been customized with application-specific components (right-hand side in the figure). An application constructed with the help of items from libraries of reusable components (left-hand side in the figure) consists instead of an application-specific architectural skeleton that uses (calls) the services offered by the reusable components. The framework approach thus places emphasis on the reuse of entire architectures.

22.3.3 Software Frameworks and Adaptability

Software frameworks fall within the reuse-driven paradigm. To reuse a software asset (a component, a fragment of code, a design model, etc.) means to use it in different operational contexts. In practice, differ-

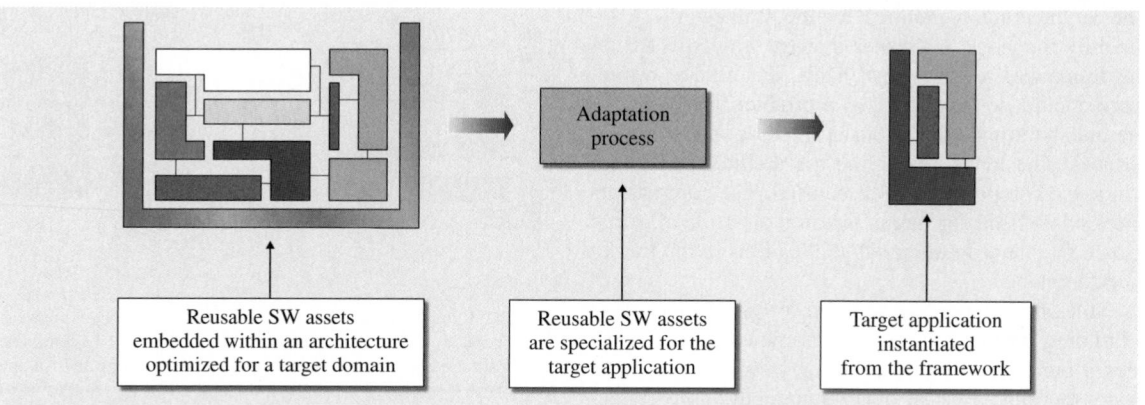

Fig. 22.11 Software frameworks and adaptability

ent operational contexts will always impose differing requirements on the reusable assets. Hence, effective reuse requires that the reusable assets be *adaptable* to different requirements. In this sense, adaptability is the key to reusability and the availability of software adaptability techniques is the necessary precondition for software reusability [22.17].

The framework representation shown in the previous section should therefore be modified as in Fig. 22.11. The items that are selected from the repository are passed through an adaptation or tailoring stage before being integrated to build the target application. In the tailoring stage, the characteristics of the reusable assets are modified to make them match the requirements of the target application.

Software reuse is perhaps the oldest approach to the reduction of software costs and has often been tried in the past. Past attempts however had only mixed success primarily because they either ignored the adaptation phase shown in Fig. 22.11 or because the state-of-the-practice adaptation techniques available at the time were not sufficiently powerful to model the extent of variability in the target domain.

A software framework is defined as a repository of reusable building blocks embedded within an architecture optimized for applications within a certain domain. The quality of a software framework largely depends on the ease with which the artifacts it offers – components and interfaces – can be adapted to the requirements of its users. Software frameworks are therefore categorized on the basis of the adaptation technology they use. Virtually all frameworks built in recent years are object oriented in the sense that they use *inheritance* and *object composi-*

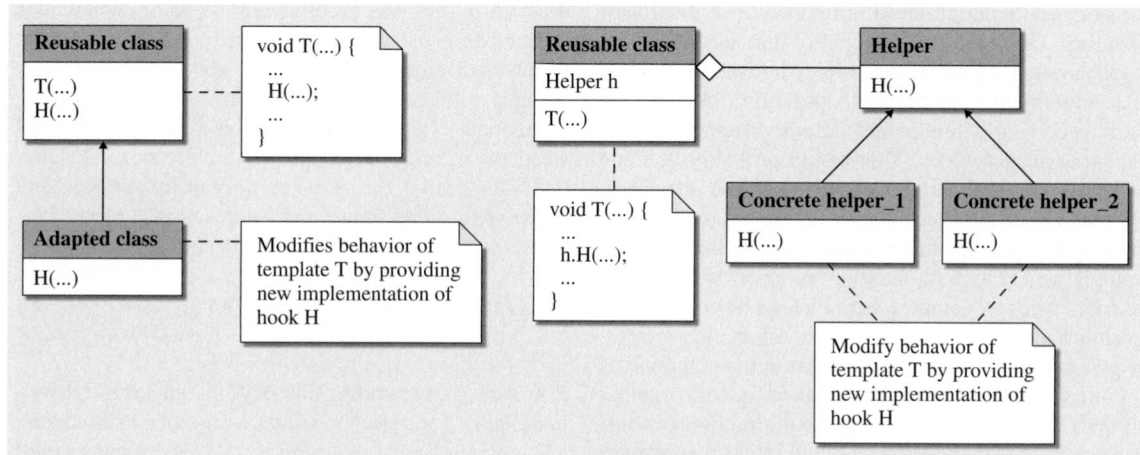

Fig. 22.12 Adaptability through inheritance (*left*) and object composition (*right*)

tion through abstract coupling as their chief adaptation techniques.

These two adaptation techniques are briefly illustrated in Fig. 22.12. The framework components are implemented as reusable classes. Inheritance (left-hand class diagram in the figure) allows a user to modify only a subset of the reusable class behavior. Object composition (right-hand class diagram in the figure) lets the reusable class delegate part of its behavior to an external class that is characterized through an abstract interface. It is notable that, in both cases, the behavior of the reusable class is adapted without touching its source code. This is important because it means that the reusable class can be qualified once at framework level and can then be adapted to different operational situations without having to undergo a full requalification process because its source code was not changed. Some delta qualification effort will always be required because of the different operational context but its extent will be more limited than would be the case if the adaptation had required manual modifications to the source code.

Object-oriented techniques provide adaptability with respect to functional requirements only. Control applications however are characterized by the presence of nonfunctional requirements covering issues such as timing, reliability, observability, testability, and so forth. The lack of techniques to model nonfunctional adaptability was one of the prime causes of the low level of reuse in the control domain. Recently, aspect-oriented programming (AOP) has emerged as a remedy for this problem.

Aspect-oriented programming [22.18] is a software paradigm that allows cross-cutting concerns to be expressed and implemented in a modular manner. At the most basic level, aspect-oriented techniques can be seen as a means to perform automatic transformations of some base source code. An aspect-oriented environment consists of two primary items: an *aspect language* and an *aspect weaver*. The aspect language allows the cross-cutting concerns to be specified and encapsulated in self-contained modules. The aspect weaver is a compiler-like tool that reads an aspect program and projects the changes it specifies onto some base code. This is illustrated in Fig. 22.13.

Current software engineering practice privileges the modeling of the functional aspects of an application. Most software modeling tools are accordingly designed to decompose a software system into functional units (which, depending on the implementation technology, can be modules, classes, objects, etc.).

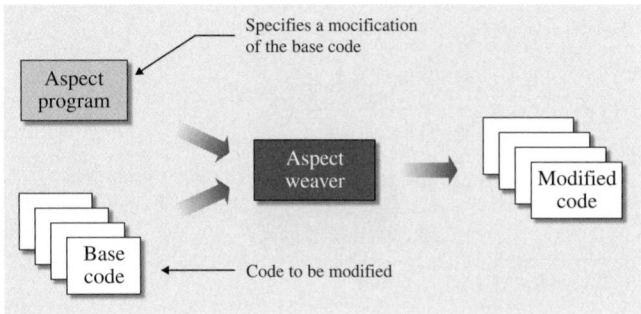

Fig. 22.13 Adaptability through aspect-oriented techniques

Nonfunctional issues tend to cross-cut such functional models and cannot therefore be easily dealt with. Aspect-oriented techniques can be used to encapsulate them and to facilitate their modeling and implementation. The use of aspect-oriented techniques, in other words, allows the application of the principle of separation of concerns to the nonfunctional aspects of a software system. As such, aspect-oriented techniques provide the means to achieve adaptability to nonfunctional requirements. Aspect-oriented techniques are very powerful but they are comparatively new and tool support is weak. One AOP tool that deserves mention because it is specifically targeted at critical systems is XWeaver [22.19]. XWeaver is a source-level weaver that allows source code to be modified in a controlled way that is designed to mimic the effect of manual modifications and to allow code inspections and other quality checks to be performed on the modified code.

22.3.4 An Example: The OBS Framework

The on-board software (OBS) framework was partly developed at ETH and partly at P&P Software GmbH (a research spin-off of ETH) to investigate the application of advanced software engineering techniques to the development of embedded control systems. The framework currently exists only as a research prototype that has been instantiated in laboratory experiments.

The OBS framework was intended to demonstrate the industrial maturity of object-oriented techniques and generative technique [22.19, 20] for software frameworks for embedded control systems [22.20, 21].

The OBS framework was designed in 2002. The OBS framework is a software framework that aims to cover the onboard satellite application domain and in particular the attitude and orbit control system (AOCS)

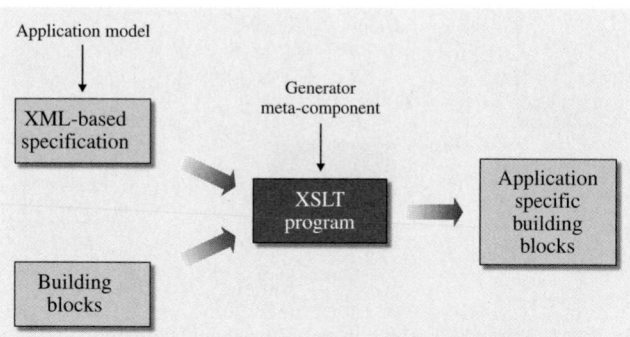

Fig. 22.14 Conceptual structure of a framework-based generative system

and data handling subsystems. The OBS framework is designed to offer reusable assets of four different types:

- *Design patterns* that describe high-level design solutions to recurring design problems in the framework domain
- *Abstract interfaces* that define abstract services that have to be provided by the framework
- *Concrete components* that provide default implementations for some of the services defined by the framework interfaces
- *Generator metacomponents* that encapsulate programs to generate application-specific implementations for some of the framework interfaces.

The design patterns are the vehicle through which the architecture predefined by the framework is captured. The architecture of a target application to be instantiated from the framework is derived by instantiating one or more framework design patterns. The abstract interfaces and the concrete components support the instantiation of the design patterns and will normally directly appear in the target application. The metacomponents do not directly enter the final application but are instead used to generate components that do, or to modify existing components so as to make them compatible with the requirements of a particular target application.

Generator metacomponents are perhaps the most innovative element of the OBS framework. They were introduced to partially automate the adaptation process whereby the assets provided by the framework are modified to match the needs of a target application. In the OBS framework, generator metacomponents are implemented as extensible stylesheet language transformation (XSLT) programs that process a specification written in extensible markup language (XML) and generate either code for the application-specific component or the configuration code for clusters of components. Their mode of operation is illustrated in Fig. 22.14.

The code generation process is driven by a set of specifications that are expressed in an XML document. This is the so-called application model which is a specification of the target application expressed in an XML-based language that is specifically tailored to the needs of the OBS framework.

One concrete example of generator metacomponents provided by the OBS framework is a set of XSLT programs that can automatically generate a wrapper for the code generated by the Matlab tool. The wrapper transforms the C routines generated by Matlab into components that are suitable for integration with other OBS framework components.

The architecture of the OBS framework is fully object oriented. The basic principles of this architecture are summarized in [22.21, 22]. Implementation is done using a restricted subset of C/C++. The OBS framework is packaged as a website which gives access to all the items provided by the OBS framework and to their documentation (www.pnp-software.com/ObsFramework).

22.4 Current Research Directions

As already indicated, the reuse- and model-driven approaches are complementary and are often used to realize different parts of the same application. Current research strengthens this complementarity because it tries to find ways to merge them. There are several ways in which such a merge can be done and the next two sections describe two ways that have been explored at ETH.

22.4.1 Automated Instantiation Environments

The effectiveness of the product family approach derives from the fact that the level of design abstraction is raised from that of individual applications to that of domains of related applications. This allows investment in the design and development of software assets to be reused across applications. Current research attempts to extend the effectiveness of product families further by automating their instantiation process. The objective is to arrive at a generative environment of the kind shown in Fig. 22.15. The environment automatically translates a specification of an application in the family domain into a configuration of the family assets that implements it.

An environment of the type shown in Fig. 22.15 would represent a synthesis of the model- and reuse-driven approach because it would allow the target application to be constructed automatically from its specification while at the same time taking advantage of the existence of predefined building blocks that implement part of the application functionality.

No practical generative environment is yet available, but some work has already been done in developing graphical user interface (GUI)-based tools where predefined components can be configured and linked together to form a complete application. A research prototype has been realized at ETH [22.23] and is described in [22.24, 25]. This tool is based on a modified version of a standard JavaBeans composition environment.

The JavaBeans standard [22.22] supports the definition of components to create GUI-based applications. Several commercial environments are available in which users can create GUI-based applications by composing JavaBean components. At their most basic, these environments offer a palette where the predefined components are shown, a canvas where the components can be linked together, and a property editor where the components can be configured. The user selects the components required for the application from the palette and pulls them down onto the canvas, where they can be linked through graphical or semigraphical operations. The attributes of the component are set in a wizard-like property editor.

In our project, we modified a JavaBeans composition environment to handle the components provided by the AOCS framework (a predecessor of the OBS framework described in Sect. 22.3.4, see also [22.26]). The objective was to allow users to instantiate the framework without writing any code. The user selects the components required for the target application from a palette and composes and configures them in a GUI-based canvas and editor. When the task is completed, the framework instantiation code is automatically generated by the environment. The code is designed to be legible to allow manual modifications if required.

22.4.2 Model–Level Reuse

Automated instantiation environments represent one way in which the model- and reuse-driven paradigms can be merged. A second attempt to achieve the same goal of merging the two approaches and thus reaping the benefits of both is model-based reuse. In virtually all industrial applications of the framework approach, reuse takes place at code level: the reused entities are the framework components and the framework interfaces expressed as source code (or, sometimes, as binary entities). In reality, reuse could also take place at model level.

With reference to Fig. 22.8, in a framework development process, the output of the domain design phase is a set of design models that describe the framework interfaces and components. In practice, such design models are often expressed as generic universal modeling language (UML) design models [22.27, 28]. The semantics of UML is in many respects ambiguous. This means that the models can, at most, have a descriptive/informative role. Reusability can only take place at the code level because it is only the code that unambiguously describes the reusable assets provided by the framework.

The latest version of UML (UML2) includes profiling facilities that allow users to create their own version of the language with a precise semantics [22.29, 30]. This in turns means that the design models can be made as precise as code. In fact, it becomes possible to fully generate the implementation of the framework assets from their models. In this case, reuse can take place at the model level since the models contain the same information as the code.

Such a model-driven approach is being explored at ETH in the ASSERT project [22.31] and is described in [22.32]. It is being applied to industrial applications by P&P Software in the currently ongoing CORDET project (www.pnp-software.com/cordet). The approach

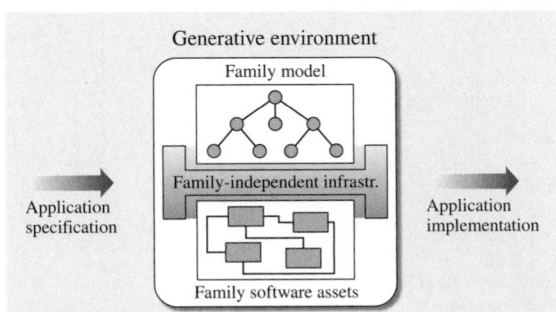

Fig. 22.15 Automated instantiation environment for software product families

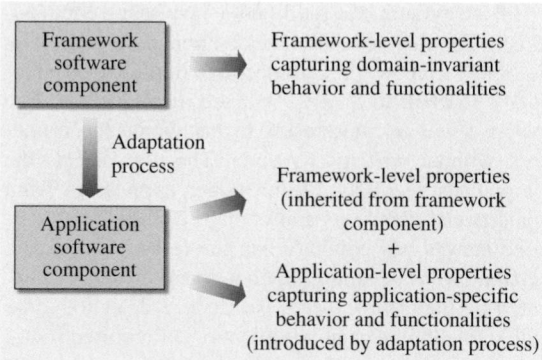

Fig. 22.16 Property-preserving framework adaptation

is based on a UML2 profile, the framework *(FW) profile*, that is specifically targeted at framework development [22.33]. This profile uses UML2 class diagrams to describe the interfaces of the framework components and UML2 state machines to describe their internal behavior. The profile also defines adaptation mechanisms that are based on both class and state machine extension.

A framework is conceptualized as a repository of *models* (not code) of adaptable components. The framework-level models guarantee certain *properties* that represent functional invariants in the framework domain. Since the properties are associated to the design models, they can, if desired, be formally verified on the models.

The FW profile constrains the adaptation mechanisms of the models to ensure that the properties that are defined at framework level still hold at application level. This ensures that all applications that are instantiated from the framework will satisfy the properties defined at framework level. At application level, developers are of course free to add new properties that encapsulate application-specific behavior (Fig. 22.16).

Space does not permit to explain in detail how property invariance can be combined with extensibility and adaptation, but Fig. 22.17 gives a flavor of the approach. The top half of the figure represents the framework-level models which, as mentioned above, consist of class diagrams and state diagrams describing the internal behavior of the framework classes.

The framework-level properties express logical relationships among the variables that define the state of one or more framework components. Formally, functional properties may be expressed as formulas in linear temporal logic. Physically, they encapsulate functional invariants in the framework domain. In practice, the framework properties are defined as properties on the behavior of the state machines associated to the framework classes.

During the framework instantiation process, the framework-level models are extended to capture application-specific behavior. Both the framework classes and their state machines must be extended. However, since it is desired to ensure that applications that use the extended models still satisfy the properties defined at framework level, the extension mechanism must be such that the new classes and their state machines still satisfy the properties that were defined at framework level.

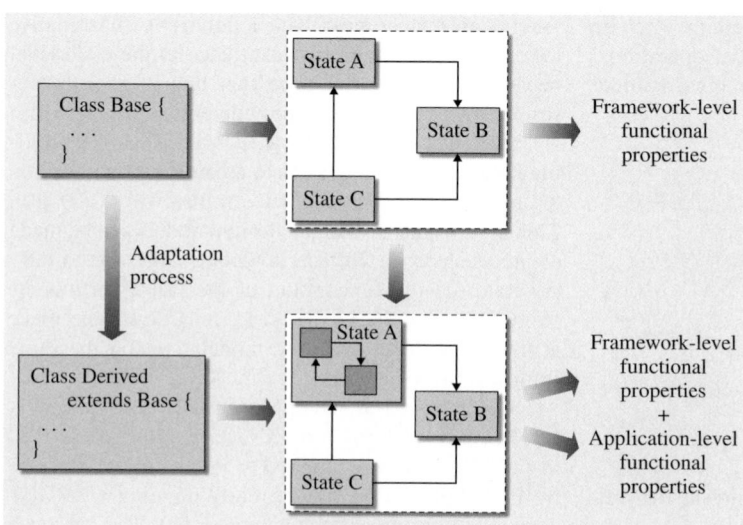

Fig. 22.17 Class extension and state machine extension

The properties defined on the base state machine capture aspects of the state machine topology and of its state transition logic. Hence, the simplest way of preserving them during the extension process is to constrain the extension process to define the internal behavior of one or more of the states of the base state machine without altering its topology and transition logic. This is illustrated in Fig. 22.17, where the derived state machine differs from the base state machine only in having an embedded state machine added to one of the base states. The derived state machine, in other words, defines the internal behavior of a state that was initially defined as being a simple state.

The FW profile adopts the extension approach of Fig. 22.17 and forbids all other kinds of state machine extensions that are allowed by UML2 (redefinition of transitions, definition of new transitions between existing states, definition of new states or regions, etc.).

The extension mechanism sketched above, though very simple, corresponds to a realistic situation that often arises in framework design. This is the case that is described by the well-known template design pattern of [22.34]. This design pattern describes the case of a class that defines some skeleton behavior that offers hooks where application-specific behavior can be added by overriding virtual methods or by providing implementation for abstract methods. The behavior encapsulated in the skeleton is intended to be invariant.

In terms of the model underlying the FW profile, the invariant skeleton behavior is encapsulated by the base state machine, whereas the variable hook behavior is encapsulated by the nested state machines added by the derived class.

The modeling and adaptation concept sketched above is formally captured by the FW profile. The profile can also be seen as a domain-specific language that is targeted at the definition of the functional behavior of software frameworks. This approach thus has all the basic elements of both the model-based approach (design expressed through models with unambiguous semantics and use of domain-specific languages to express the models) and of the reuse-driven approach (definition of domain-specific, reusable, and adaptable software assets).

22.5 Conclusions and Emerging Trends

Software for automation has some inherent difficulties, such as real-time conditions, distributed and parallel operation, easy-to-use interfaces, data storage and reporting, etc.. The software often has to be implemented on one-of-a-kind processes or plants requiring a highly automated development. Two possible routes, which are often combined in practical solutions, have been presented in this chapter. One consists of automatically generating code from a textual or graphical description of the software (model driven) and the other consists of the use of frameworks (reuse driven). Both have their application domains; the framework approach seem to be better suited for the development of software for families of systems because of its inherent possibility of adaptation. Examples of both approaches from our work at ETH are given. Design of software is a challenging task; easy recipes and cookbooks are in our view not appropriate, and no attempt has been made in this direction.

While consulting tasks with industry are proprietary, a considerable shift has been noted over the last few years in the area of software development towards outsourcing and consulting services. Many small companies offer solutions, which often compete favorably with in-house solutions. In such circumstances, it may be advisable to order a framework instead of a program. To use existing frameworks provided under free or open software licences it may also be advisable to organize some training or consulting to speed up the development. The management of such software projects with several contributing teams is, however, still a challenging task.

References

22.1 W. Schaufelberger, P. Sprecher, P. Wegmann: *Echtzeit-Programmierung bei Automatisierungssystemen* (Teubner, Stuttgart 1985), in German

22.2 G.E. Maier, W. Schaufelberger: Simulation and implementation of discrete-time control systems on IBM-compatible PCs by FPU, 11th IFAC World Congr. (Pergamon, Tallinn 1990)

22.3   P. Kolb, M. Rickli, W. Schaufelberger, G.E. Maier: Discrete time simulation and experiments with FPU and block–sim on IBM PC's, IFAC ACE (Pergamon, Boston 1991)

22.4   M. Kottmann, X. Qiu, W. Schaufelberger: *Simulation and Computer Aided Control System Design using Object-Orientation* (vdf ETH, Zürich, 2000)

22.5   D. Dori: *Object-Process Methodology* (Springer, Berlin, Heidelberg 2002)

22.6   A.H. Glattfelder, W. Schaufelberger: *Control Systems with Input and Output Constraints* (Springer, Berlin, Heidelberg 2003)

22.7   A. Benveniste, P. Caspi, S.A. Edwards, N. Halbwachs, P. Le Guernic, R. de Simone: The synchronous languages 12 years later, Proc. IEEE **91**(1), 64–83 (2003)

22.8   P. Caspi, A. Curic, A. Maignan, C. Sofronis, S. Tripakis: *Translating Discrete-Time Simulink to Lustre.* In: ACM Transactions on Embedded Computing Systems (TECS) **4**(4) (New York 2005)

22.9   Esterel Technologies: http://www.esterel-technologies.com/ (last accessed February 6, 2009)

22.10   T.A. Henzinger, B. Horowitz, C.M. Kirsch: Giotto: a time-triggered language for embedded programming, Proc. IEEE **91**(1), 84–99 (2003)

22.11   J. Chapuis, C. Eck, M. Kottmann, M.A.A. Sanvido, O. Tanner: Control of helicopters. In: *Control of Complex Systems*, ed. by A. Åström, P. Albertos, M. Blanke, A. Isidori, W. Schaufelberger, R. Sanz (Springer, Berlin, Heidelberg 2001) pp. 359–392

22.12   N. Wirth: Tasks versus threads: An alternative multiprocessing paradigm, Softw.–Concepts Tools **17**(1), 6–12 (1996)

22.13   N. Wirth: *Digital Circuit Design* (Springer, Berlin, Heidelberg 1995)

22.14   T.A. Henzinger, M.C. Kirsch, M.A.A. Sanvido, W. Pree: From Control Models to Real-Time Code Using Giotto, IEEE Control Syst. Mag. **23**(1), 50–64 (2003)

22.15   P. Donohoe (ed): *Software Product Lines – Experience and Research Directions* (Kluwer, Dordrecht 2000)

22.16   M. Fayad, D. Schmidt, R. Johnson (Eds.): *Building Application Frameworks – Object Oriented Foundations of Framework Design* (Wiley, New York 1995)

22.17   V. Cechticky, A. Pasetti, W. Schaufelberger: The adaptability challenge for embedded system software, IFAC World Congr. Prague (Elsevier, 2005)

22.18   G. Kiczales, J. Lamping, A. Mendhekar, C. Maeda, C. Videira Lopes, J. Loingtier, J. Irwin: Aspect-Oriented Programming, Eur. Conf. Object-Oriented Program. ECOOP '97 (Springer, 1997)

22.19   I. Birrer, P. Chevalley, A. Pasetti, O. Rohlik: An aspect weaver for qualifiable applications, Proc. 15th Data Syst. Aerosp. (DASIA) Conf. (2004)

22.20   J. Cleaveland: *Program Generators with XML and Java* (Prentice Hall, Upper Saddle River 2001)

22.21   K. Czarnecki, U. Eisenecker: *Generative Programming* (Addison-Wesley, Reading 2000)

22.22   R. Englander: *Developing JavaBeans (Java Series)* (O'Reilly and Associated, Köln 1997)

22.23   A. Pasetti: http://control.ee.ethz.ch/~ceg/AutomatedFrameworkInstantiation/index.html, (last accessed February 6, 2009)

22.24   V. Cechticky, A. Pasetti: *Generative programming for space applications*, Proc. 14th Data Syst. Aerosp. (DASIA) Conf. (Prague 2003)

22.25   V. Cechticky, A. Pasetti, W. Schaufelberger: *A generative approach to framework instantiation.* In: *Generative Programming and Component Engineering (GPCE)*, Lecture Notes in Computer Science, Vol. 2830, ed. by F. Pfenning, Y. Smaragdakis (Springer, Berlin, Heidelberg 2003)

22.26   A. Blum, V. Cechticky, A. Pasetti: *A Java-based framework for real-time control systems*, Proc. 9th IEEE Int. Conf. Emerg. Technol. Fact. Autom. (ETFA) (Lisbon, 2003)

22.27   M.R. Blaha, J.R. Rumbaugh: *Object-oriented Modeling and Design with UML* (Prentice Hall, Upper Saddle River 2004)

22.28   D. Rosenberg, M. Stephens: *Use Case Driven Object Modeling with UML: Theory and Practice* (Apress, Berkeley 2007)

22.29   S.W. Ambler: *The Elements of UML 2.0 Style* (Cambridge University Press, Cambridge 2005)

22.30   R. Miles, K. Hamilton: *Learning UML 2.0* (O'Reilly Media, Köln 2006)

22.31   European Space Agency: http://www.assert-project.net/assert.html (last accessed February 6, 2009)

22.32   M. Egli, A. Pasetti, O. Rohlik, T. Vardanega: A UML2 profile for reusable and verifiable real-time components. In: *Reuse of Off-The-Shelf Components (ICSR)*, Lecture Notes in Computer Science, Vol. 4039, ed. by M. Morisio (Springer, Berlin, Heidelberg 2006)

22.33   P&P Software GmbH: http://www.pnp-software.com/fwprofile/ (last accessed February 6, 2009)

22.34   E. Gamma, R. Helm, R. Johnson, J. Vlissides: *Design Patterns, Elements of Reusable Object-Oriented Software* (Addison-Wesley, Reading 1995)

23. Real-Time Autonomic Automation

Christian Dannegger

The world is becoming increasingly linked, integrated, and complex. Globalization, which arrives in waves of increasing and decreasing usage, results in a permanently dynamic environment. Global supply networks, logistics processes, and production facilities try to follow these trends and – if possible – anticipate the volatile demand of the market. This challenging world can no longer be mastered with static, monolithic, and inert information technology (IT) solutions; instead it needs autonomic, adaptive, and agile systems – living systems. To achieve that, new systems not only need faster processors, more communication bandwidth, and modern software tools – more than ever they have to be built following a new design paradigm. As determined by the industrial and business environment, systems have to mirror and implement the real-world distribution of data and responsibility, the market (money)-driven decision basis for all stakeholders in the (real or virtual) market, and the goal orientation of people, which leads to on-demand, loosely coupled, communication with relevant partners (other players, roles) based on reactive or proactive activities.

This chapter provides an insight into the core challenges of today's dynamics and complexity, briefly describes the ideas and goals of the new concept of software agents, and then presents and discusses industry-proven solutions in real-time environments based on this distributed solution design.

Part C | 23

The following examples are discussed in detail in this chapter, covering the solution approach, challenges, and customer demand as well as relevant pros and cons:

- An autonomic machine control system applied to the adaptive control of a modular soldering machine. The particular case is concerned with the creation of a novel modular production machine with an integrated distributed agent control system, which has been sold worldwide since the middle of 2008. The agent model is described in terms of the specific customer requirements and the advantages of the approach.

- A solution to real-time road freight transportation optimization using a commercial multiagent-based system, LS/ATN (living systems adaptive transportation networks), which has been proven through real-world deployment to reduce transportation costs for both small and large fleets. After describing the challenges in this business domain and the real-time optimization approach, we discuss how

the platform is currently evolving to accept live data from vehicles in the fleet in order to improve optimization accuracy. A selection of the predominant pervasive technologies available today for enhancing intelligent route optimization is described.

Both examples reflect their specific history and background, which motivated the customer and the developers to apply an autonomous automation approach.

Although software agents are a core principle of the autonomous automation examples in this chapter we only touch this field slightly, as other chapters in this book focus and elaborate on agent-based automation.

23.1 Theory

23.1.1 Dig into the Subject

First let us briefly delve deeper into the title of this chapter *Real-Time Autonomic Automation*, from right to left.

Automation always has been – and still is – the basic starting point to relieve workers of routine tasks and increase the utilization of resources, because of production machinery, transportation capacity, and limited material stock. Automation means using any kind of mechanical device or program to carry out a repetitive job faster, more reliably, and with higher quality than human beings can normally achieve. According to Wikipedia [23.1]:

> *Automation (ancient Greek: self dictated) ... is the use of control systems such as computers to control industrial machinery and processes, reducing the need for human intervention.*

Autonomic is already a term very close or even similar to the term *agent-based*. In many aspects both describe the same characteristics. Autonomic systems firstly have sensing capabilities to keep in touch constantly with their environment and *know what's going on* (belief). Then they know what they want – their purpose – their goals (desire). And finally autonomic systems are able to derive and decide what to do (intention) based on the current situation and predefined goals. Putting together these terms results in belief–desire–intention (BDI), which is a software model developed for programming intelligent agents. To again cite Wikipedia:

An autonomic system is a system that operates and serves its purpose by managing its own self without external intervention even in case of environmental changes.

The last part of this definition gives the important hint that autonomic systems in particular are able to adapt to changes in the environment (making use of their sensors) in order to decide constantly on the best action according to the current situation.

Real time is a critical term, as it is used ambiguously and greatly depends on the environment in which it is applied. The main differentiation made is between hard and soft real-time systems. *Hard* real-time systems guarantee a configured time from an event to system response, since otherwise the whole system will fail, e.g., in brake control for cars or tail plane fin control for a jet fighter. However, hard real-time control could also apply to comparably *slow* systems as long as the response is guaranteed, such as a heating control where the reaction may take several minutes. *Soft* real-time control means that a system is typically fast enough to react in due time. Deadlines are normally met, but if not kept the system will not fail, but at most lose some quality; for example, in dispatching systems real-time reaction is important, but if a decision (to reshuffle order allocations) is delayed there might be a loss of efficiency and increased costs, but the system does not fail as a whole.

Since distributed systems such as those discussed specifically in this chapter require solutions supporting heterogeneous environments, developers of such systems are attracted by the platform independency of Java. Since its invention, Java has become increasingly fast, but does not have built-in real-time capabilities. Thus the real-time specification for Java (RTSJ, JSR-001 (Java specification request)) has been developed and implemented by Sun and others as an add-on to standard Java, and offered to the market as *real-time Java* or the Java real-time system (Java RTS). On their website Sun gives another short and concise definition of *real time*, which again emphasizes the difference between speed and predictability: "Real-time in the RTSJ context means the ability to reliably and predictably respond to a real-world event. So real-time is more about timing than speed." Additionally to a real-time programming environment a real-time operating system is needed, like Solaris 10, SUSE Linux Enterprise Real-Time 10 (SP2) or Red Hat Enterprise MRG 1.0.1 Errata releases.

Both examples described in this chapter do not fulfill and do not need to fulfill the hard real-time specification. Why? The second example is a decision support system for transport optimization where the dispatchers can still decide and become active independent of the systems, as with a navigation system. The first example has to fulfill hard real-time requirements, but the system design succeeded in keeping the critical real-time parts within the local controller of each module of the machine and thus outside the Java-based control logic. With this approach the control system where the system is normally fast enough to trigger the actuators and the reaction guarantee is given by the lowest-level controllers, the machine manufacturer could save significant costs by being able to use standard hardware, a standard operating system, and a standard runtime environment (Java SE).

Real Time through Autonomy

If hard real-time support will be needed in the future, this does not conflict with the system layout – quite the contrary: autonomous agent-based systems are initially designed to build real-time control systems. The major design principle to enhance real-time capabilities is the natural and elementary separation of responsibilities, and thereby the distribution of tasks. This design for scalability allows making use of all (local) processing power available in a solution and system environment. Keeping local tasks and local decisions entirely local results in high reactivity, independent of the overall system load.

23.1.2 Optimization: Linear Programming Versus Software Agents

To make things even more complex the real world has given us not only the challenge of real-time reactivity but also the parallel goal of optimal decisions (or at least decision support). This means that a software system not only has to cope with very complex, normally NP-hard (nondeterministic polynomial-time hard), optimization problems, but also should solve them again and again, building on the second by second changes in the environment. Unfortunately new design paradigms, such as software agents, are always compared with traditional approaches in terms of solution quality, despite the fact that traditional mathematical operation research (OR) methods are not designed to handle real-time events. For this reason the following descriptions summarize the major differences and objectives of both approaches; a more detailed discussion can be found in [23.2].

Linear Programming

Pro. Current optimizers, which means traditional OR methods such as linear programming, are designed to find *the* optimum for a problem, independent of the processing duration.

Con. However, it is very hard and time consuming to cover the full real-world complexity and map all different aspects into a linear equation system, not to mention changes of requirements, processes or business goals. Despite many traditional optimizers try to achieve real-time response, their basic intention to find the absolute optimum works against it. The optimization interval, even combined with tricks and workarounds, is normally too long to be considered real-time.

Result. The optimum can be found, but too late. The world might already have changed dramatically. The optimum was only valid for the past, when the optimization started. The difference between the calculated and the current optimum is large, and the accumulated error between the optimal curve and the found (to be optimal) curve increases significantly over time.

Software Agents

Con. The distributed negotiation approach with its bottom-up search space expansion is not designed to find the optimum.

Part C | 23.1

Pro. Rather, it permanently strives for the optimum, as the optimization interval is very short and the approach maps the real-world complexity in all its details and can be customized to nearly any specific need without touching the core optimization principle.

Result. A close-to-optimal result is found every second or faster. The difference from the theoretical optimum is relatively small and the total error over time is kept to a minimum.

Conclusion

The optimum should not be seen and understood as a single point in time, but instead as the difference between the continuously calculated optimization curve and the real-world volatile optimum. This short excursion to real-time optimization and its application in automated systems is properly summarized in the following two statements: *How does it help knowing what would have been the optimum one hour ago?* Or: *Better be roughly right instead of precisely wrong.*

23.1.3 Classification of Agent-Based Solutions

To understand and correctly apply agent-based solutions it is important to follow a clear classification of software systems. Based on the short agent definition as a *sense–decide–act loop* it is straightforward to classify agent-based solutions (Fig. 23.1) depending on the existence of real-world or artificial interfaces for the sensors and actuators.

Simulation System

If you input only simulated, recorded historical data or forecasted data into an agent solution and only use the system output for analysis but not for direct decisions, then it is a simulation system. This applies, e.g., if real-time captured order data for a dispatching systems are fed into the dispatching system again, mostly for the purpose of verifying the system configuration (i. e., the cost model, see below). The result of the simulation is stored within databases, data warehouses or presented in business graphs, but not used directly to control or trigger any actions. It is important to understand that the

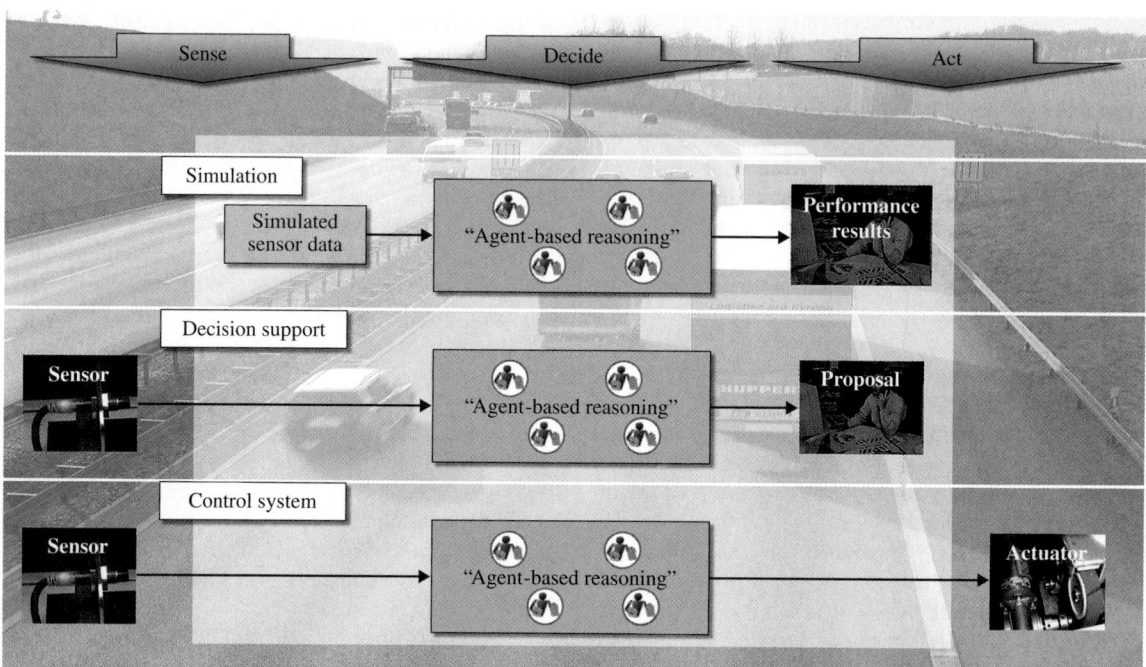

Fig. 23.1 Solution classification: depending on the existence of real-world or simulated sensors and actuators you can distinguish between three system types: simulation, decision support or control system

core of the system – the optimization and decision algorithm – is not simulated, but instead the sensor input and actuator behavior, and hence the real world. Very often it needs even more effort to create a realistic simulation of the world model than *only* implementing the sensor and actuator interfaces.

Decision Support System

If the input is directly linked to real-world sensors (e.g., a telematics system) and the output is only used to support and inform the dispatcher, then we talk about a decision support system. A navigation system is a good example of a decision support system, as its sensors, the GPS (global positioning system) antenna, is directly linked to the real world, the GPS satellites. On the output side, the actions resulting from the best route are not directly executed; the navigation system does not turn the steering wheel. Instead it only suggests to the driver what he/she should do. However, the driver has the final decision.

Control System

If the sensors are real and the output directly executes decisions without human interaction, then it is a closed control loop and thus a control system. One typical representative is the antilock brake system of a car. It automatically – fully autonomic – releases the brake if needed without having the driver in the decision loop. This autonomic behavior is based on at least two different conflicting goals: first to reduce the speed of the car and second to keep the wheels turning.

In each mode of operation the core agent solution is the same; only the real-world interfaces differ. The transportation optimization system described in this chapter is mainly used as a decision support system, but is also used to analyze history and forecasts as a simulation. The discussed machine control system is, as the name implies, a control system, where the results of the decision algorithm directly influence, e.g., the drive speed of transportation belts, heating power, and pump strength.

23.1.4 Self-Management

The ever-accelerating complexity and dynamics of IT systems makes their administration and optimization with only human resources no longer feasible or at times even impossible. Hence, it is reasonable and necessary to equip IT systems with capabilities that increasingly allow them to administrate, monitor, and maintain themselves. These self-management properties of so-called autonomic solutions according to [23.3] are:

Self-configuration: The system automatically changes its operating parameters to adapt to mutable external conditions, some of which may even be unpredictable at the time of a system's development.

Self-optimization: The system continuously assesses its own performance, explores possible courses of actions that would result in performance improvements, and adopts the ones that are most promising.

Self-healing: The system has abilities to recover from certain unfavorable conditions that may result in malfunctions. It autonomously attempts to determine compensation actions and performs them.

Self-protection: The system detects threats against its functioning and takes preventive and corrective measures to ensure correct operation.

This group of properties are often also referred to as *self-properties* or self-* properties.

23.2 Application Example: Modular Production Machine Control

23.2.1 Motivation

Efforts to increase the flexibility of production lines are a pivotal trend in manufacturing. Whole assembly lines as well as individual machines are increasingly subdivided into modules in order to adapt precisely and just in time to constantly changing specifications (quasi make-to-order). An instance of this trend can be also found in microproduction.

However, the centralized, *hardwired* design of traditional control software imposes limits on dealing successfully with unpredictability. It is thus necessary to choose a novel approach to the development of control software, such that it is capable of managing modular machines dynamically and with minimal manual intervention while automatically maximizing the throughput and thereby optimize investment into production resources.

This allows a production line to adapt continuously to changing boundary conditions and order specifications. Such a control system stands out due to its superior flexibility and adaptivity, and drives the automation and optimization of modern production lines further, while at the same time embracing the increasing complexity and dynamics of its environment.

An innovative offering in this area is a key differentiator for all vendors and users of modular production lines. Whitestein's product living systems autonomic machine control (LS/AMC) makes use of these principles and applies them in the modular machine control market.

The particular case discussed is a concrete industrial application that entered live production in mid-2008 and is being offered as a solution to the general market.

23.2.2 Case Environment

The particular application case of the LS/AMC control system is a modular soldering machine wherein each module is governed by an independent local agent controller. Coordination of the individual module operational parameters and the transition of boards from one module to another are the key control aspects.

Machine Setup

After many years of successful soldering using a conventional monolithic machine, the project team decided to prepare for the future by initiating a redesign of the centrally controlled machine (Fig. 23.2) as a novel modular approach employing distributed control (Fig. 23.3).

The modular setup of the new design not only allows configuration of the machine according to the customer's needs, but also has a separate local control within each module. The single drive for the one and only conveyor belt also has been replaced by one conveyor and one drive per module. This gives broad processing flexibility, as the target market for this machine typically requires changing production programs in real time.

A typical machine setup is composed of a feeder, a fluxer, one to three heaters, a soldering wave module, and a cooler module.

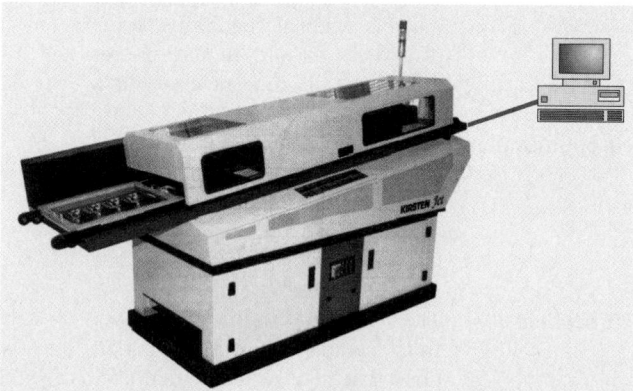

Fig. 23.2 Old machine design: the processing units (modules) are contained within one static monolithic block and centrally controlled

Sensors and Actuators

Each machine module has several sensors and actuators connected to the local controller board. There are digital switch sensors such as *end-of-belt*, *zero-position*, *emergency-stop*, and *liquid level* as well as linear sensors including *temperature* and *encoder* of step motors. Actuators comprise motors, pumps, and heaters as well as fans and signal lights. Overall a small standard configuration with five modules already contains around 40 sensors and 50 actuators, which have to be managed and coordinated.

Customer Requirements

The goal of using agent technology in this project was to minimize the complexity of development, operation, and maintenance of machines, without reducing

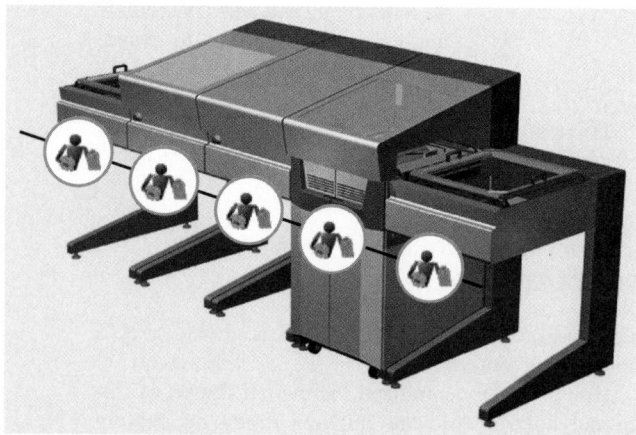

Fig. 23.3 New machine design: each module of the machine is controlled by its own agent

the degrees of freedom for future application scenarios. Specifically, this implies:

Autonomic Equipment Adaptation. The control software of a modern production line must autonomously adapt to the ideal equipment configuration for each order. This effectively eliminates the need for manual reconfiguration. It also ensures that future enhancements of the system remain possible with only minor outlay.

Dynamically Varying Solder Programs. Typically this machine is used for batch-size-one tasks, which means that each and every board is processed with different soldering parameters and the boards are processed in parallel, i. e., pipelined.

Dynamic Performance Optimization. The capability to optimize capacities dynamically with changing configurations and target values is a top priority. This ensures maximum throughput and minimizes idle capacities and quality failures.

Seamless Integration Capability. At the macrolevel it is required that the control software for modular production lines such as this offer standard interfaces to integrate into a total production control system.

Intuitive User Interface. Not least, such an advanced solution also needs to provide an intuitive user interface, which automatically adapts to the actual machine setup (Fig. 23.4). It offers simple controls for the machine operator, extended functionalities for specialists and technicians, and comprehensive remote maintenance capabilities via the Web.

23.2.3 Solution Design

Existing Solutions for Agent–Based Control

Whitestein Technologies has applied agent-based distributed control in many related domains throughout recent years. Before describing the path from monolithic to modular agent-based control we give three examples from other areas where distributed optimization is applied. The following examples all make use of multilateral negotiation algorithms to continuously seek optimal solutions.

Production Scheduling. The resources in a production environment including personnel, machines, and materials are represented by software agents that use negotiation algorithms (e.g., auctions) to offer and *sell*

Fig. 23.4 The modularity of the machine is reflected on the graphical user interface

their capacity to bidding orders, which are also represented by agents. One of the prominent industry examples is described in [23.4].

Road Logistics. To automate the creation of dispatching plans for transportation logistics systems each resource (vehicle) is represented by an agent, which coordinates and exchanges loads with others by making use of bilateral negotiations [23.5]. (See also the next application example.)

Supply Networks. All the players in a supply network continuously need to coordinate their demand forecasts and capacity availability. Agents can assist in this time-consuming and time-sensitive task perfectly. Monitoring agents along the supply chain fire an alarm and trigger activities if reality deviates too much from the plan [23.6].

From Monolithic to Modular Control

As in all previous examples, the LS/AMC-based soldering machine solution uses modular control principles because each module not only needs coordination with neighboring modules but also needs local, autonomic control to optimize the overall process; for example, the heater module must maintain the temperature within tolerance limits irrespective of environmental changes caused by a board running through the module or a user opening a lid. Each module must thus combine its local control tasks with overall process coordination.

Part C | 23.2

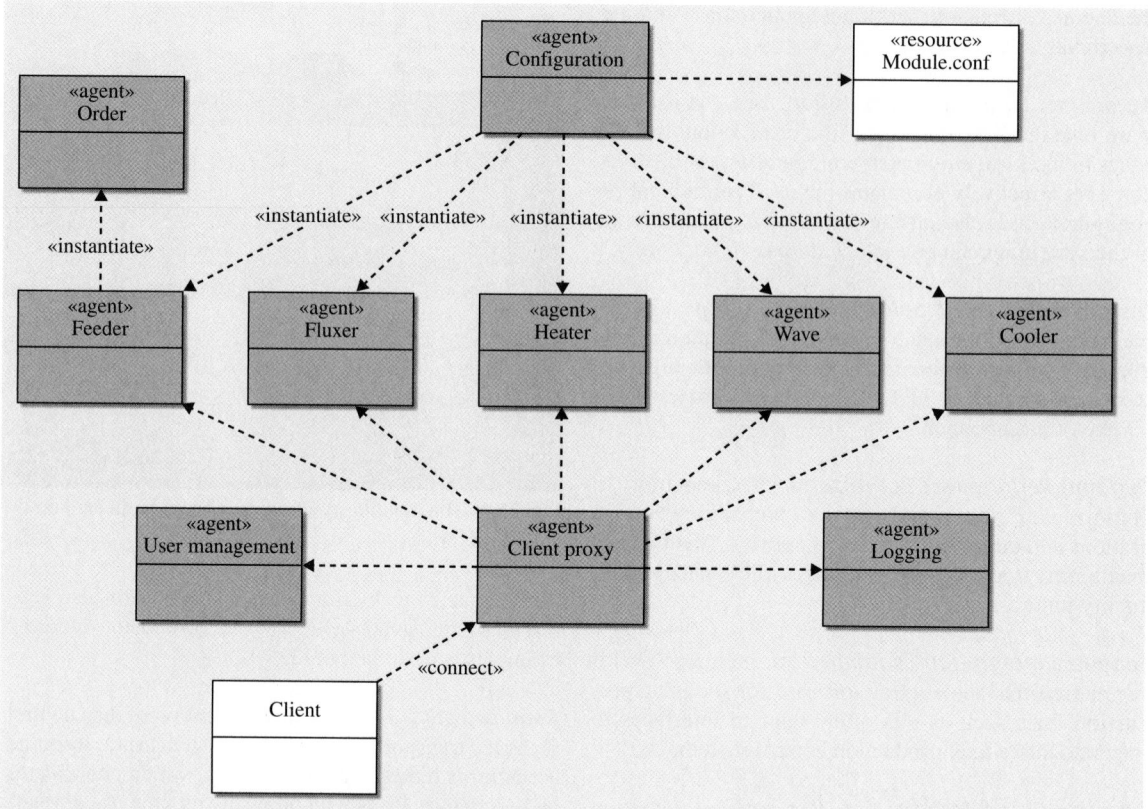

Fig. 23.5 The agent model in AML (agent modeling language)

Modular control also means that each module holds its own production schedule and is able to give a production forecast in a backward-chain manner to enable the feeder module to estimate when best to start a new board. The module agents combine this production planning part with the real-time control when a board physically appears and when target temperatures are reached in reality.

Agent Model

Besides an agent type per physically available module type, the agent model (Fig. 23.5) comprises one agent per order (printed circuit board (PCB) to be soldered) and some administrative agents for user management, configuration management, and client communication. To be precise, an agent in the agent model is an agent type, analogous to a class in object orientation. In a running application an agent is an instance of an agent type and thus correspondent to an object, as an instance of a class. The agent types used within this solution are:

- The *configuration agent* is responsible for detecting the attached modules of a concrete customer machine configuration via the CANopen (CAN: controller area network) bus. It then instantiates the corresponding module agents, where, depending on the detection, several agents of one type might be started, e.g., if two or more heater modules are used.
- One *module agent* type per physical module type, as there are:
  - the *feeder* module
  - the *fluxer* module
  - the *heater* module
  - the *wave* modules, one for the oil and one for the nitrogen version
  - the *cooler* module.

Each of these module agents control their module, e.g., heat up the tin, ensure the needed tin level or keep the temperature stable, and they communicate with neighboring modules for the preliminary and real-time scheduling of the soldering process. New module types will be developed and added to the configuration as needed, e.g., a lift module to bring a board back to the first module of the machine.

The feeder agent has the additional task of instantiating the *order agent* when it detects a new board and the user presses the start button. The order agent then prepares its processing schedule by *talking* to each module agent, and then supervises and logs its soldering process in detail.

The *client-proxy agent* collects and holds all information needed to keep the connected client(s) up to date.

Besides the *logging agent* and the *user-management agent* there are more administrative agents, which are not shown in the agent model diagram for reasons of clarity.

The following are the core features of the implemented agent model.

Autonomic Module Control. Every machine module is represented by a specifically adapted software agent that optimizes the module's operations and capacity utilization.

Superordinate Coordination. Through permanent bilateral negotiation and coordination between neighboring modules (i. e., of their software agents) the system constantly reaches a state of superordinate coordination. This eliminates the need for a central control instance.

Self-Managing Orders. As for every module (resource), software agents are also responsible for the control of each production unit (order). They self-manage the order's progress through the machine(s) autonomously and ensure that all requirements relating to (cost-)efficiency, speed, and quality are optimally satisfied.

Distributed Communication. The decentralized approach based on bilateral communication allows for virtually unlimited scaling possibilities, while at the same time increasing robustness against malfunctions and various external influences.

Standards Compliance. At the controller level the software provides full support for the CANopen industry-standard machine control and communication interface.

Interaction Model

One of the core principles of this solution is to dynamically create one agent per module detected on the CANbus and establish a communication link to the two neighbor agents. Consequently there is no global communication among the agents but only the one on the *left* and the *right* side. This bilateral communication model is very lean but still powerful enough to drive the backward scheduling and real-time synchronization between the modules.

Here is one example of this synchronization task. As each board (production job) and each module has different processing parameters the conveyor belts typically run at different speeds. To ensure clean handover from one to the next module, LS/AMC has implemented a communication protocol (Fig. 23.6) following a notify-and-pull principle, where the *sender* stops and notifies the *receiver* and, as soon as it is ready, the *receiver* sets the receiving speed and then grants the *sender* permission to send at this speed.

23.2.4 Advantages and Benefits

The following advantages are only qualitative. Detailed metrics are not yet available and proven comparisons with other (monolithic) approaches have not been conducted, as this is an ongoing project in its final deployment phase. However, during the course of the development we experienced many of the advantages in real life, and even unexpected ones.

Especially we found the modular design to be extremely helpful in a project like this with moving targets over more than 2 years. The moving target was caused by the learning curve while designing the machine – the hardware itself. Even though sensors, actuators, and their behavior changed every week, the core of the solution has been stable and unchanged since its initial design.

We received more feedback from real life just before the publishing of this Handbook. The machine has been extended for a new customer by two lift modules, two more transportation modules, and a barcode reader. The agent-based design of the solution has shown that it can schedule and optimize the throughput and performance of the machine without any change of the algorithm. The additionally instantiated module agents naturally latched into the processing chain. They coordinated with the *older* module agents to control the soldering process as expected.

At least some of the following – theoretical obvious – advantages have thus been materialized.

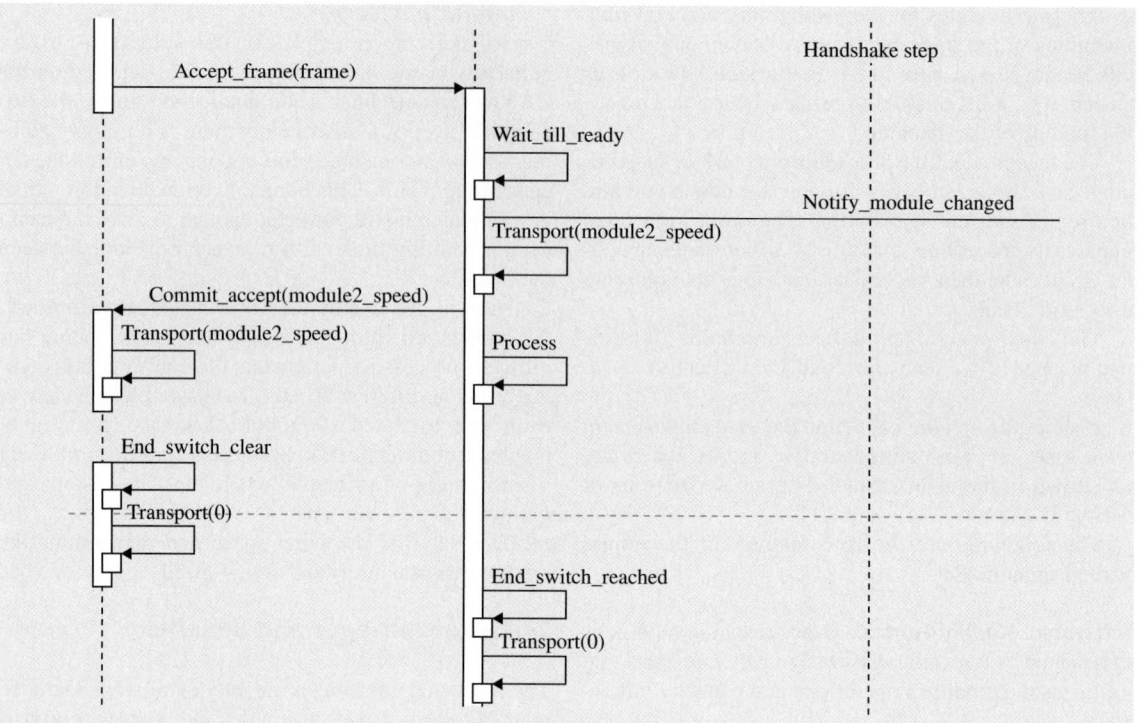

Fig. 23.6 Extract of the agent communication protocol: handshake step

Flexibility

The modular and distributed architecture of the LS/AMCs agent system allows for easy addition of new modules, without causing fundamental changes to the existing system architecture. The introduction of new kinds of machine modules only requires the development of a new module agent, which can be integrated into the current system with minimal effort.

Autonomic Adaptivity

New modules or modules that are failing or in need of maintenance can be exchanged while the system is running. Moreover thanks, to the LS/AMC distributed system architecture and intrinsic feedback-based adaptivity, machine control is updated autonomically at runtime without requiring restart of the control software.

Maintainability

Compared with traditional procedural or purely object-oriented approaches, the agent-oriented design of LS/AMC offers the advantage of intuitively mapping the real-world production line and order structure one-to-one. This makes the system better to understand and use, increases its durability, and improves its maintainability. An agent system also supports the easy and targeted customization of logging routines at the process level. This ensures the availability of more helpful and efficient methods of error monitoring and analysis.

Simulation

Complex simulation scenarios are easy to develop with LS/AMC, since a realistic mirror of a production line is more straightforward to simulate than an abstract model. Many different machine states and process flows can be recreated quickly and realistically. This significantly reduces the cost of quality control and improves personnel training and product demonstrations.

Goal Orientation

The software agents employed in this solution explicitly represent their behavior using partially conflicting logical goals. Order agents, for example, pursue the minimization of throughput time, and module agents have the goal of optimizing the modules' resource consumption. With LS/AMC these goals do not block one another but rather dynamically coordinate toward achieving optimal overall performance.

Dynamic Optimization

A production optimization program is coupled to each work item (board) and transitions together with it through the machine modules. Each module adapts to the particular program and dynamically anticipates parameter and control adjustments when appropriate. The program is linked to the individual order or batch and not tied to a central, fixed setting for the entire machine.

23.2.5 Future Developments and Open Issues

- Integration into preceding and successive processing machines, e.g., automated optical inspection (AOI), cleaning or packaging
- Machine-controlled board loaded through new combined lift/feeder modules. This allows throughput improvement by allowing agents to influence the sequence of production, which is not the case when boards are loaded manually
- Making use of optional surface temperature sensors to improve the control of the temperature curve directly on the processed board.

23.2.6 Reusability

The foundation for the customer- and machine-specific solution is a reusable and generic product kernel providing the following features and functionality:

- The standard agent platform, Living Systems Technology Suite (LS/TS) [23.7], is used as the runtime environment for the agent-based solution.
- The agent principle implies distributed autonomic control for each resource or entity in the system.
- The agent-type framework allows a jump-start for the solution building as it provides all general agents and templates for the application-specific agents.
- User, roles, and rights management is needed in every multiuser environment. New functions can easily be put under the generic access control.
- The built-in standard CANopen interface allows fast integration of every CANopen-compliant controller devices. LS/AMC has implemented a generic interface to CANopen to give each agent transparent access to its sensors and actuators.

23.3 Application Example: Dynamic Transportation Optimization

23.3.1 Motivation

Across Europe and worldwide, road freight transportation is a demanding high-pressure environment. Competition is fierce, margins are slender, and coordination is both distributed and often intensely complex. As a result many companies are seeking methods to control costs by enhancing their traditional dispatching methods with technology capable of intelligent, real-time freight capacity and route optimization. The former ensures that transport capacity is maximally used, while the latter ensures that trucks take the most efficient calculated route between order pickups and deliveries. These are tractable, yet complex, optimization problems because plans can effectively become obsolete the moment a truck leaves the loading dock due to unforeseen real-world events. It thus becomes mission-critical to assist human dispatchers with the computational tools to quickly replan capacity and routing.

A considerable volume of research exists concerning the domain of automatic planning and scheduling, but many real-world scheduling problems, and especially that of transportation logistics, remain difficult to solve. In particular, this domain demands sched-ule optimization for every vehicle in a transportation fleet where pickup and delivery of customer orders is distributed across multiple geographic locations, while satisfying time-window constraints on pickup and delivery per location.

Living systems adaptive transportation networks (LS/ATN) is a novel software agent-based resource management and decision support system designed to address this highly dynamic and complex domain in commercial settings. It makes use of agent cooperation algorithms to derive truck schedules that optimize the use of available resources, leading to significant cost savings. The solution is designed to support, rather than replace, the day-to-day activities of human dispatchers. The agent design chosen for optimization directly reflects the manner in which logistics companies actively manage the complexity of this domain. The global business is divided into regional business entities, which are usually dispatched via distributed dispatching centers. Interacting software agents represent this distribution.

While one of the largest customers of LS/ATN has demonstrated a reduction of 11.7% in costs compared with the manual dispatching solution, we typically guarantee a reduction of at least 4–6%. This improvement

Part C | 23.3

is significant for transportation companies with large numbers of orders to manage, significant costs, and small profit margins.

The achievements made thus far have been attained using only traditional manual communication (mobile phone) between the driver and dispatcher. Using this data LS/ATN generates global dispatching suggestions and improves the communication among the distributed dispatching centers. Incorporating sensor data on, for example, traffic conditions and vehicle status allows more accurate continuous estimation of vehicle estimated time of arrival (ETA), thus presenting yet further opportunities for cost savings and reduced fuel consumption. One key to this is the integration of real-time track-and-trace data feeds from en route vehicles, which act as feedback measures to an optimizer engine. This allows continuous adaptation and regeneration of dynamic route plans based on the real-world environment.

Close integration with key pervasive technologies such as GPS and reliable multinetwork communication offers the capability of enhancing core system intelligence with fast, timely, and accurate measures of the live environment [23.8]. Continuous transmission of vehicle state and location information provides live feedback metrics for the optimization platform, allowing human dispatchers to improve the efficiency of entire fleets. This flexibility enables logistics providers to react quickly to new customer requirements, altering transport routes at very short notice in order to accommodate unexpected events and new orders.

There can be little doubt that the future of freight transportation in Europe and beyond lies with the widespread adoption of pervasive technologies and intelligent transportation systems. One of the few questions remaining is simply how rapidly firms will adapt.

The remainder of this chapter examines the business domain characterizing the identified problems and then presents an industry-proven solution to these problems, LS/ATN (Fig. 23.7). It has been developed in close collaboration with worldwide logistics providers such as DHL, and has been proven through real-world deployment to reduce transportation costs through the optimized route solving for both small and large truck fleets. The primary aspects of our agent-based solution approach are discussed, followed by the presentation of benefits and savings, which are then continued with emerging options for incorporating state-of-the-art mobile technologies and pervasive computing into the solution.

23.3.2 Business Domain

Today most logistics companies use computational tools, collectively known as transport management sys-

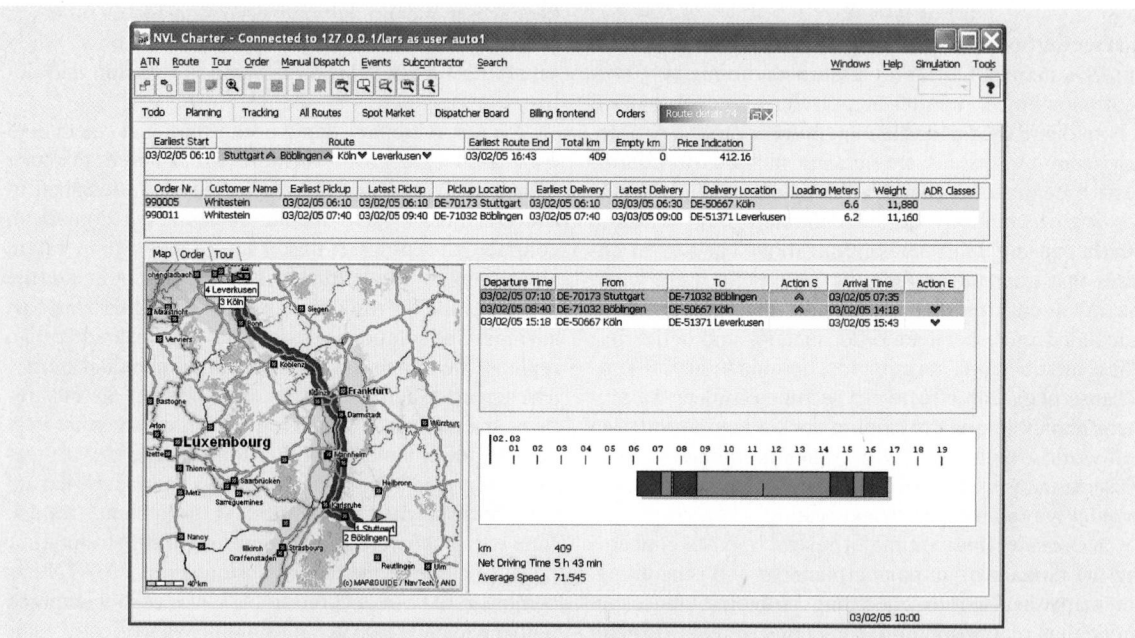

Fig. 23.7 Details of a route in the LS/ATN dispatcher control center, as suggested by an optimizer agent

tems (TMS), such as Transportation Planner from i2 Logistics, AxsFreight from Transaxiom, Cargobase, Elit, and Transflow, to plan their transportation network from a strategic level all the way through to sub-daily route schedules. However, many TMS are unable to handle unexpected events adequately and generate plan alterations in real time. When dealing with large numbers of distributed customers, limited fleet size, last-minute changes to orders, or unexpected unavailability of vehicles due to traffic jams, breakdowns or accidents, static planning systems suffer from limited effectiveness. Significant human effort is required to manually adapt plans and control their execution.

In addition, vehicles can be of different types and capacities, are usually available at different locations, and drivers must observe regulated drive-time restrictions. To cope with all this, new intelligent approaches to route planning are emerging that are capable of continuously determining optimal routes in response to transportation requests arriving simultaneously from many customers. The key challenge lies in allocating a finite number of vehicles of varying capacity and available at different locations such that transportation time and costs are minimized, while the number of on-time pickups and deliveries, and therefore customer satisfaction, is maximized.

Road Freight Transportation

Road freight transportation is a very heterogeneous business environment serving a wide variety of customers with many different types of transportation, each configurable in many ways. In addition, large companies add the challenge of different business structures regarding processes, culture, and information technology.

One of the most significant challenges is the permanent handling of unexpected events such as traffic jams or other reasons for delays and new, changed or canceled customer orders. While new orders are an expected component of everyday business, their precise characteristics and appearance time are highly variable. A good solution must address the decentralized responsibilities of dispatchers working across the world with potentially overlapping geographical responsibilities, and supporting individual strategies and local approaches to dispatching.

To survive in an environment of significant cost pressure with margins of only 1–3%, logistics providers must address how to structure strong interaction between regional or organizational logistics networks and effectively manage the increasing complexity.

Core Challenge

The ongoing challenge for a logistics dispatcher is to find the best balance between:

- His reaction speed (time, effectiveness)
- The quality of a solution (schedule)
- The cost (efficiency) of a solution.

A comprehensive solution not only requires a core real-time optimization algorithm, but also a cooperative process bringing together all involved people.

Load Constraints

In a linear programming approach first of all you have to cover and configure the following load constraints:

- Precedence (pickup before delivery)
- Pairing (pickup and delivery by the same truck)
- Capacity limitation (dependent on truck type)
- Weight limitation (dependent on truck type)
- Order–truck compatibility (type, equipment)
- Order–order compatibility (dangerous goods)
- Last-in first-out (LIFO) loading of orders (optional).

Additionally it is important at least to take into account the following time constraints:

- Order-dependent load and unload durations
- Earliest and latest pickup
- Earliest and latest delivery
- Opening hours for pickup and delivery
- Legal drive-time restrictions
- Maximum allowed tour duration
- Lead time for ordering spot market trucks.

Problem Classification

One approach to tackle this optimization problem is by considering it as a multiple pick up and delivery problem with time windows (mPDPTW) [23.9], which concerns the computation of the optimal set of routes for a fleet of vehicles in order to satisfy a collection of transportation orders while complying with available time windows at customer locations. To solve the real-world challenge to an acceptable degree it is necessary to add another two aspects: first the capability to react in real time, and second to deal with time constraints in a flexible manner, using penalty costs to decide between a new vehicle or being late. This results in the even more complex multiple pick up and delivery problem with soft time windows in real time (R/T mPDPSTW) [23.10–12].

Thus, in addition to a pickup and delivery location, each order includes the time windows within which the order must be picked up and delivered. Vehicles are dispatched from selected starting locations and routes are computed such that each request can be successfully transferred from origin to destination. The goal of R/T mPDPSTW is to provide feasible schedules that satisfy the time window constraints for each vehicle to deliver to a set of customers with known demands on minimum-cost vehicle routes. Another aspect is the capability to suggest charter trucks (dynamically add resources) when appropriate, i.e., when charter trucks are cheaper than the company's own existing or fixed-contract trucks.

Further Challenges

A further significant challenge is managing opening hours, meaning to support multiple time windows during a day (e.g., lunch breaks). One of the major topics outside the optimization core problem is the ability to combine global dispatching suggestions automatically created by the system with local individual dispatcher decisions. Not forgetting the difficulty of combining continuous planning (perpetual with rolling horizon) with discrete decisions, track and trace, and billing processes. Then there is also the recurrent decision to transport direct or indirect (via a hub or depot) and to consider the limited docking or handling capacity at a hub.

Finally customer requests to parallelize the optimization of the three main resources, truck/tractor, trailer/swap body, and driver(s), must also be handled. Each may take a different route due to the pulling unit (truck/tractor), with drivers also potentially changing during a tour.

23.3.3 Solution Concept

The centralized, batch-oriented nature of traditional IT systems imposes intrinsic limits on dealing successfully with unpredictability and dynamic change. Multiagent systems are not restricted in this way because collaborating agents quickly adapt to changing circumstances and operational constraints. For real-time route optimization, it is simply not feasible to rerun a batch optimizer to adjust a transport plan every time a new event is received. Reality has shown that events such as order changes occur, on average, 1.3 times per order. Distributed, collaborating software processes, i. e., agents, can however work together by partitioning the optimization problem and following the bottom-up approach, thereby solving the optimization in near-real time.

Software Agents

To solve the domain challenges described above it is necessary and advantageous to apply a new software design concept: software agents. This technology offers an ideal approach to allow real-time system response and assessment in a distributed heterogeneous environment. Software agents are grounded by the notion of communication between independent active objects, each of which may have its own goal objectives and role assignments. These capabilities inherently mirror typical business structures and processes. Technically, software agents operate using sense–decide–act loops, which can be either purely reactive or proactively goal oriented.

In the transportation business domain an agent could be a packet, a pallet, a truck, a driver, an order or a dispatcher. They follow a reverse, bottom-up optimization principle with decentralized solution discovery and escalation strategies: first a dispatcher mentally optimizes within his domain of responsibility (e.g., 20 trucks), then in steps expands the search space to his office, his subsidiary, the region, the country, and finally tries to improve a solution globally.

Bilateral Order Trade

As mentioned, the agent design principle is based on communication and interaction among autonomous objects mirroring the real world. This optimization model closely follows cooperation in reality, where all trucks are driven and managed by self-employed drivers (and truck owners). They first accept each new order they get from any customer and then start to search, and negotiate with, other truck drivers in order to exchange or transfer orders looking for a win–win situation for both sides. This is triggered by each order event, where an order exchange also counts as an event. Each truck negotiates with other trucks in sequence with a tight restriction to bilateral order trades. However multiple trades can take place in parallel, always between a pair of trucks. This solution design allows fully distributed and parallel solution discovery, which scales very well and allows individual goals and strategies per truck (agent).

Agent Model and Strategy

To solve the R/T mPDPSTW problem dynamically, the LS/ATN transportation optimizer [23.13], used by DHL throughout Europe, segments and distributes the problem across a population of goal-directed software agents. Each agent represents a dispatcher, who manages one or more vehicles (resources). This is slightly different to exactly one agent per one truck, but the principle is the same, even closer to reality where a dispatcher manages more than one truck. The reason was technical performance optimization while keeping the core principle of bilateral negotiations.

The system is completely event driven: a new order, a changed order, a delay or a successful order exchange triggers a local activity. The dispatcher of the affected vehicle becomes active and tries to optimize by negotiating with *neighbor* trucks, trying to exchange or move loads and by checking and calculating all reasonable combinations and selecting the cheapest. The global optimum is striven for through a kind of *snowball effect*, which stops when there is no more optimization found. A threshold savings value, which avoids an order exchange for too little saving, reduces plan perturbation.

To find the optimal allocation the agents work on a strict cost basis. Each possible route is checked against a configurable, individual, and fully detailed cost model. This market-based approach, the money to be spent, is the common denominator to make the multiple conflicting goals comparable, which are:

- Reduction of empty driven distance
- Reduction of waiting times
- Increase of capacity utilization.

For R/T mPDPSTW optimization, an agent represents each geographical region, or business unit, with freight movement modeled as information flow between the agents (Fig. 23.8). Incoming transportation requests are distributed by an AgentRegionBroker (not shown) to the AgentRegionManager governing the region containing the pickup location. The number of such agents depends on the customer's setup of (regional) business units and varies between 6 and 60 for current deployments. In the larger case, 10 000 vehicles and up to 40 000 order requests are processed daily. This implies that no more than a few seconds are available to reoptimize a transportation plan when, for example, a new order must be integrated. Each AgentRegionManager generates a transportation plan specifying which orders to combine into which routes and which vehicles should be assigned to those routes. Agents exchange information using a negotiation protocol to insert transportation requests sequentially, while continually verifying vehicle availability, capacity, and costs.

While the optimization function is 100% cost based, other objectives must be satisfied in parallel when calculating routes. Some of these constraints are compulsory (hard), such as capacity and weight limitations of the vehicle, customer opening hours, that pickup date is before delivery date, and that pickup and delivery are

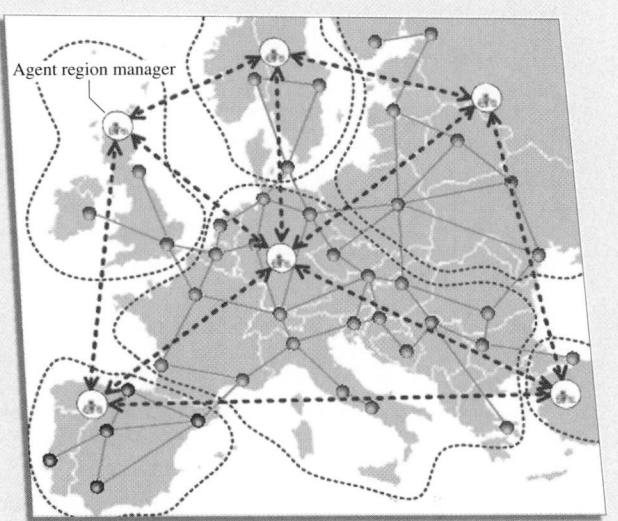

Fig. 23.8 Illustration of freight transportation in Europe partitioned into six regions, each with its own agent region manager. *Blue circles* represent major transport hubs and *red lines* indicate example routes connecting hubs

performed by the same vehicle. Other soft constraints can be violated with a cost penalty, such as missing the latest possible pickup time or delivery time.

Experiments and First Findings

In the course of the software development we evaluated the effect of certain key parameters. One of these is the number of orders being negotiated and transferred between trucks (k). Our experiments showed that the runtime increases linearly with increasing k, but the

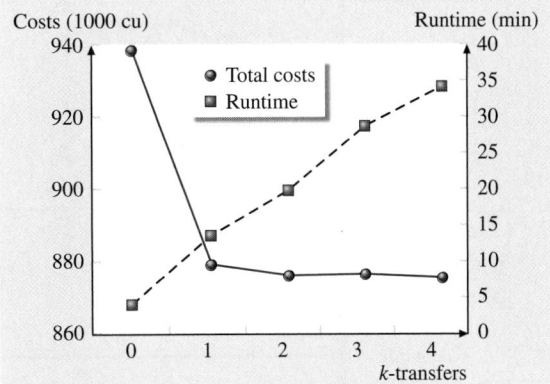

Fig. 23.9 Optimization results with increasing k (number of orders exchanged)

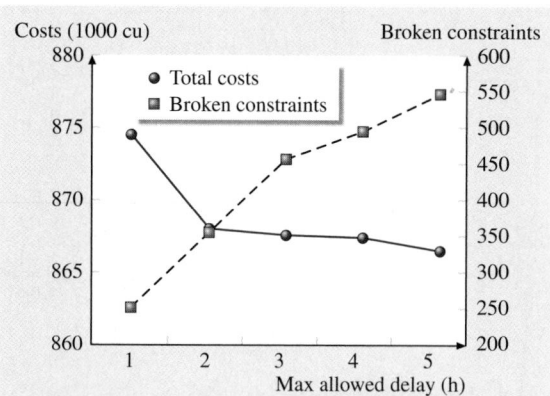

Fig. 23.10 Optimization results with increased soft time windows

costs decrease only marginally (Fig. 23.9). $k = 0$ means that there is no order exchange taking place, but only the first-time allocation.

Another experiment is the effect of the maximum allowed delay for soft time windows. The graph (Fig. 23.10) shows that a maximum delay above 2 h decreases the quality (number of broken constraints increase) while not reducing costs significantly.

The Decision Support Process of LS/ATN
To integrate the globally generated optimization recommendations with the distributed dispatchers performing manual optimization we identify the following decision support process (Fig. 23.11): Orders arriving from an external system (ERP (enterprise ressource planning), TMS (tranportation management system) or other) flow into the core agent system, which generates dispatching recommendations in the form of a globally optimal matching proposal of orders to trucks. There is no time limit into the future; LS/ATN has an endless planning horizon. A certain lead time before the orders are to be checked and released, the system transfers them to the *to-do* board of the responsible dispatchers. They approve and adjust the suggested plan if needed prior to fixing the tour and purchasing the required transportation capacity. This automatically sends a confirmation to the subcontracted carrier and can issue a message to the truck driver, if equipped accordingly. The released routes then switch to tracking mode where the agents take over responsibility for monitoring incoming tracking messages, verifying whether they indicate an existing or upcoming time violation. The dispatcher is informed only if needed. As a final step the dispatcher releases a finished tour for billing.

This *ideal flow* covers the standard case, but reality often intercedes to force alterations in real time. In any situation a dispatcher can put a tour or an order into his manual dispatcher board and adjust the plan. This might be needed if, for example, an actual order differs from the booking only when loading it at the customer site.

23.3.4 Benefits and Savings

Higher Service Level at Reduced Cost
LS/ATN agent-based optimization guarantees a higher service level in terms of results quality. The high solution quality corresponds to a reduced number of

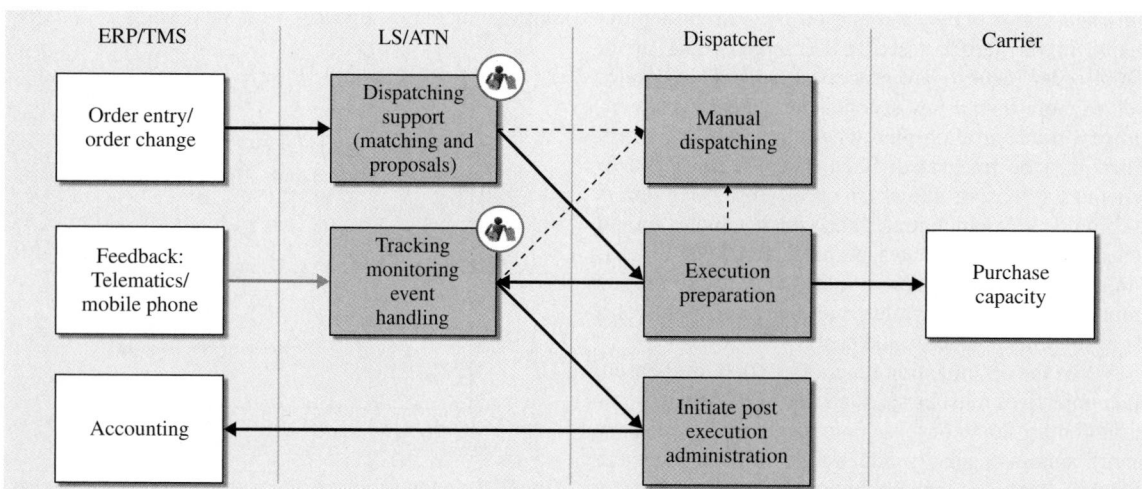

Fig. 23.11 Decision support process of LS/ATN

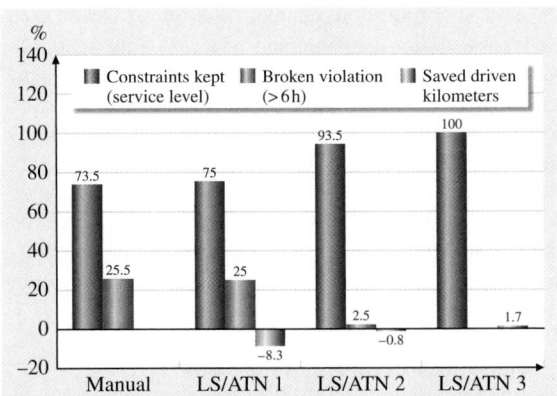

Fig. 23.12 Improvements obtained with LS/ATN over manual dispatching. Higher service level at reduced cost. Saved driven kilometers are compared with the manual figure

violated constraints. The system allows the desired level of service quality to be fine-tuned. Figure 23.12 presents results obtained from LS/ATN relative to the manual dispatching solution of very experienced dispatchers (manual).

The first proposed solution (LS/ATN 1), using relaxed soft constraints as the manual dispatchers do, provides a reduction of 8.3% in driven kilometers at the same service level with no more than 25% violated constraints.

The second solution (LS/ATN 2), configured with higher penalty costs for late delivery, shows a reduction in driven kilometers of 0.8% relative to the manual dispatching solutions, while providing a significant higher

service level: only 2.5% violated constraints with more than 6 h delay.

The third solution (LS/ATN 3), not allowing any time window violation, shows an increase of only 1.7% in terms of driven kilometers, while meeting all the constraints to 100%.

In this analysis we compared the system results with real-world (manual) results, as the customers regard these metrics as optimal. Furthermore, there is the cost and resource problem if one would like to compare the results with the real optimum. This would require setup and development of a parallel solution based on linear programming, which is – according to our experience – not capable of covering all detailed requirements and constraints. Customers do not pay for such a comparison, as it is far too expensive and, even if one could develop a parallel solution fulfilling all requirements in the same way, there is no guarantee that the (one-time) optimal solution will end in due time.

Significantly Increased Process Efficiency

Through the use of automatic optimization a lower process cost is achieved. This is due to automatic handling of plan deviations and evaluation of solution options in real time. Moreover, through automation, the communication costs in terms of dispatcher's time and material is reduced. Better customer support can be guaranteed through fast, comprehensive, and up-to-date information about order execution. Automation also allows processing of a higher number of orders than with manual dispatching only. This is an important issue as the volume of data to be managed is constantly increasing.

Without LS/ATN					
Count	VA*	Trucks	Driven km	€/km	Cost in €
261	**Z**	261	96 231	1.06	102 005
50	**D**	50	19 266	0.85	16 376
100	**R planned**	100	23 836	0.80	19 069
0	**R new**	0	0	0.85	0
76	**B planned**	153	70 803	0.80	56 642
0	**B new**	0	0	0.85	0
Total		564	210 136		194 092

With LS/ATN					
Count	VA*	Trucks	Driven km	€/km	Cost in €
104	**Z**	104	36 204	1.06	38 376
41	**D**	41	18 337	0.85	15 586
89	**R planned**	89	21 450	0.80	17 160
42	**R new**	42	11 798	0.85	10 028
80	**B planned**	160	74 110	0.80	59 280
49	**B new**	98	47 496	0.85	40 372
Total		534	209 385		180 803

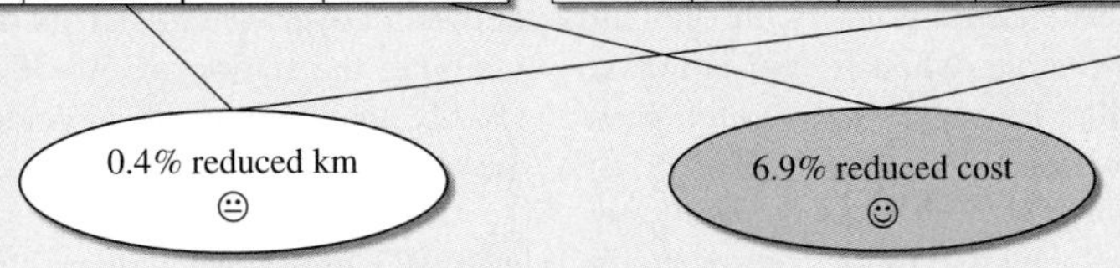

0.4% reduced km ☺

6.9% reduced cost ☺

Fig. 23.13 Significant cost savings through optimized capacity utilization

Significant Savings through Optimized Capacity Utilization

Cost savings cannot only be achieved by avoiding empty trips and reducing driven kilometers; an important aspect of cost reduction is the optimal use and allocation of the company's own and chartered trucks to a mixture of one-way, back-, and round-trips. Without the ability to reduce the driven kilometers significantly there is still a saving potential of up to 7%, as shown in Fig. 23.13.

Savings Potential in Numbers

A partial dataset from our major customer, DHL Freight, contains around 3500 real business transportation requests. In terms of the optimization results, obtained by comparing the solution of manual dispatching of these requests against processing the same orders with LS/ATN, a total 11.7% cost saving was achieved, where 4.2% of the cost savings stem from an equal reduction in driven kilometers. An additional achievement is that the number of vehicles used is 25.5% lower compared with the manual solution. The cost savings would be even higher if fixed costs for the vehicles were included, which is not the case in the charter business, but possibly in other transportation settings.

Combined with other real-world comparisons we can estimate an overall transportation cost saving of 5–10%, which are variable costs (subcontractor payment) and thus have an immediate effect. Fixed cost savings of 50–100% resulting from process and communication improvements are long-term effects, which only pay back when the resources are reallocated.

23.3.5 Emerging Trends: Pervasive Technologies

Although capacity and route optimization tools are proven to produce significant reductions in operating costs, many in the transportation industry are acutely aware that one key and often missing component of the optimization strategy is the provision of real-time feedback from en route vehicles. The objective is an intelligent transportation management system with every vehicle providing up-to-date information of progress through a pickup/delivery schedule and with onboard sensors detecting, for example, when freight is loaded and unloaded, and whether its condition (e.g., temperature) is within tolerance limits.

The *intelligent transportation management systems* model [23.14] developed within the transportation industry is grounded in the principle of vehicle track-

ing and incorporation of real-time information into the transportation management process using available pervasive technologies. The emerging approaches to realizing this model involve various combinations of pervasive technologies, some of which are all highlighted in the following section. This section highlights some of the most relevant technologies in use today, or in the early phases of adoption. LS/ATN is able to make use of data sourced from, manipulated by, or transmitted by any of these technologies to enhance the route optimization process.

Global Positioning System (GPS)

Automatic vehicle location (positional awareness) uses GPS signals for real-time persistent location monitoring of vehicles. Both human dispatchers and route planners such as LS/ATN can then track vehicles continuously as they move between pickup and delivery locations. Active GPS systems allow automatic location identification of a mobile vehicle; at selected time intervals the mobile unit sends out its latitude and longitude, as well its speed and other technical information. Passive GPS uses the onboard units (OBU) to log location and other GPS information for later upload. Accuracy can vary, typically between 2 and 20 m, according to the availability of enhancement technologies such as the wide-area augmentation system (WAAS), available in the USA. The European Galileo system will augment GPS to provide open-use accuracies in the region of 4–8 m within the European region.

The adoption of GPS is growing quickly as the technology becomes commoditized, but some transportation companies remain reliant on legacy equipment for measuring vehicle location. Some of the alternatives to GPS in use today include dead-reckoning, which uses a magnetic compass and wheel odometers to track distance and direction from a known starting point, and the long-range navigational (LORANC) system, which determines a vehicle's location using in-vehicle receivers and processors that measure the angles of synchronized radio pulses transmitted from at least two towers with predetermined position. Another system in use by some transportation companies is cellphone signal triangulation, which estimates vehicle location by movement between coverage cells. This only offers accuracy typically in the region of 50–350 m, but is a cheap and readily available means of determining location.

Onboard Units (OBU)

An OBU, otherwise known as a *black box*, is a vehicle-mounted module with a processor and local memory that is capable of integrating other onboard technologies such as load-status sensors, digital tachographs, toll collection units, onboard and fleet management systems, and remote communications facilities. The majority of OBUs in use today, such as the VDO FM Onboard series from Siemens, the CarrierWeb logistics platform, and EFAS from Delphi Grundig, are typically used to record vehicle location, calculate toll charges, and store vehicle-specific information such as identity, class, weight, and configuration. Some emerging OBUs will have increased processing capabilities allowing them to correlate and preprocess collected data locally prior to transmission. This offers the possibility of more computational intelligence installed within the vehicle, enabling in situ diagnostics and dynamic coordination with the remote planning optimizer such that the vehicle becomes an active participant in the planning process, rather than simply a passive provider and recipient of data.

Vehicle data, in its most common form, relates to the state of the vehicle itself, including, for example, tire pressure, engine condition, and emissions data. Automatic acquisition of this data by onboard sensors and its transmission to a remote system has been available within the automotive industry from some years and is now gaining substantial interest in the freight transportation business. The OBU gathers information from sensors with embedded processors capable of detecting unusual or deviant conditions, and informs a central control center if a problem is detected. Sensors also measure the status of a shipment while en route, such as detecting whether the internal temperature of refrigerated containers is within acceptable tolerance limits or whether a door is open or closed.

RFID

Many assets, including freight containers, swap-bodies, and transport vehicles, are now being fitted with transponders not only to identify themselves, but also to detect shipment contents and maintain real-time inventories. In the latter case, units are equipped with radiofrequency identification (RFID) readers tuned to detect RFID tags within the confined range of the container. Some tags, such as the Intermec Intellitag with an operating range of 4 m, are specifically designed for pallet and container tracking, where tags are attached to every item and automatically scanned whenever cargo is loaded or unloaded. The live inventory serves as both local information for the driver and as real-time feedback to the TMS, which uses it for record keeping and as input to the real-time route planner.

In addition, e-Seals, whether electronic or mechanical, are now often placed on shipments or structures to detect unauthorized entry and send remote alerts via the OBU. E-Seals on a container door can also store information about the container, the declaration of its contents, and its intended route through the system. They document when the seal was opened and, in combination with digital certificates and signatures, identify whether the people accessing the container are authorized to do so.

Mobile Communications

Electronic communication is the key enabler of pervasive technologies. In transportation the most basic form in use is the short message service (SMS), which is commonly used to communicate job status such as when a driver has delivered an order. Technology is already in place to automatically process SMSs and input the data into the route planner.

Also now in relatively widespread use is dedicated short-range communications (DSRC) operating in the short-range 5.8–5.9 GHz microwave band for use between vehicles and roadside transponders. Its primary use in Europe and Japan is for electronic toll collection. DSRC is also used for applications such as verifying whether a passing vehicle has a correctly operating OBU.

Currently, the technology with the greatest utility is machine-to-machine (M2M) [23.15] communication, which is the collective term for enabling direct connectivity between machines (e.g., a vehicle's OBU and the remote planning engine) using widespread wireless technologies. Legacy second-generation (2G) infrastructure is most commonly used as third-generation (3G) technologies enter the mainstream for day-to-day human telecommunications. M2M is quickly emerging as a principle enabler of networked embedded intelligence, the cornerstone of pervasive computing. It can eliminate the barriers of distance, time, and location, and as prices for the use of 2G continue to drop due to continued rollout of 3G technologies, many transportation companies are taking advantage and adopting M2M as their primary means of electronic communication.

Emerging solutions take M2M to another level by enabling always-on and highly reliable communication through automatic selection of connection technology, e.g., general packet radio service (GPRS), enhanced data rates for GSM evolution (EDGE), universal mobile telecommunications system (UMTS), satellite services, and WiFi according to availability. The LS/ATN route optimizer, for example, can be augmented with a remote connection agent module [23.16] installed in vehicles that offers seamless M2M over cellular technologies, wireless local-area network (LAN), and even short-range ad hoc connections if available. The selection of a particular communication technology can be made either manually or automatically, depending on several metrics including location, connection availability, transmission cost, and service type or task; for example, a fleet operator may prefer the use of satellite to communicate directly with a driver, but then a combination of cellular technologies for remote monitoring, trailer tracking, and diagnostics. Low-cost GPRS might be selected to download position coordinates from an onboard GPS; whereas, a higher-bandwidth (and cost) option such as UMTS/WCDMA (wideband code division multiple access) might be preferred for an over-the-air update to the OBU or onboard sensors.

The position information of a single vehicle is used to adjust dispatching plans immediately in the case of deviations (described below in more detail). If the number and density of vehicles in a region is high enough, this *floating vehicle data* may be integrated into a map containing real-time traffic flow information [23.17].

Opportunities of Using Pervasive Technologies

Transportation route optimizers can take advantage of real-time data sourced from vehicles equipped with pervasive technologies by incorporating information relating to vehicle location, state, and activity into their planning processes.

Figure 23.14 shows the major difference and step from the current deployment of the agent-based real-time dispatching system, only making use of traditional communication and track-and-trace capabilities with many manual human activities involved, toward a full real-time control loop leaving the dispatcher in a purely supervisor role. In the future, the sensor and actuator interfaces will be increasingly automated while the decision core system is already in place.

Pervasive communication provides a permanent bilateral link between the vehicles and the dispatching system. Onboard preprocessing is available to calculate continuously the estimated time of arrival (ETA) at the next node, which is periodically sent to the server in order to check immediately the impact on the dispatching plan. Taking speed and other local knowledge into account the local preprocessor is able to deduce traffic conditions and forward this (only temporarily valuable)

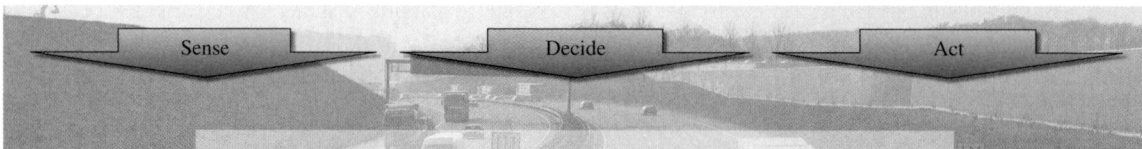

Existing: state-of-the-art real-time dispatching combined with traditional track&trace and communication

Future: M2M real-time track&trace and re-scheduling combined with instant driver update

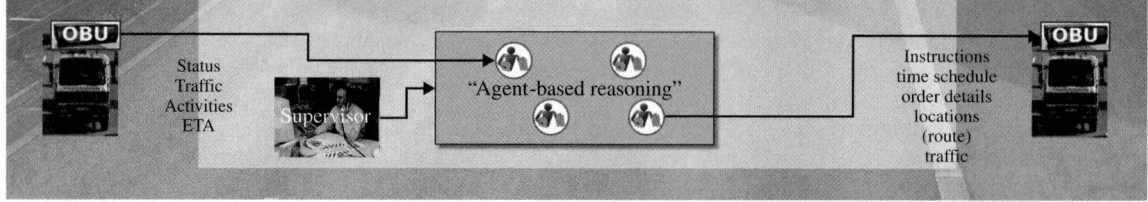

Fig. 23.14 Analysis of the sense–decide–act loop: despite a state-of-the-art reasoning engine, existing applications still involve many media breaks with human involvement, which are increasingly being replaced by integrated pervasive technologies

knowledge to other vehicles via the dispatching system. The combination of sensed speed with the current location can trigger an automatic status message if the truck is waiting for loading/unloading at a customer site or is idle in a traffic jam. A similar functionality is so-called *geo-fencing*, which issues a status messages when entering or leaving a destination. A further automation is the local communication between truck and a *smart container* for docking and undocking messages.

All of the above simplify and speed up execution tracking and dramatically increase status frequency, quality, and accuracy. Real-time plan adjustments ensure that a co-loading opportunity is never missed, or that time is never lost when informing the customer about short-term changes.

In particular, during the decision phase, route optimization and the derivation of schedules can directly use both information relating to vehicles movements as they proceed through delivery schedules and feedback from RFID transponders notifying when orders have been added to or removed. This real-time component implies that time windows can be more finely tuned according to current events, resulting in alternative schedules that can either compensate for delays or take advantage of time saved. Preliminary results with a prototype demonstrate that employing real-time data in the optimization process can further reduce transportation operating costs by up to 3% beyond the 5–10% achieved from the standard optimization process described earlier, depending on the particular case and system configuration.

23.3.6 Future Developments and Open Issues

There remain many scientific and practical challenges related to the design and use of real-time dispatching system. A selection of these that we consider relevant to LS/ATN and for consideration by the community at large are described below.

A major challenge is the effective handling of intercompany, interregion, and intermodal transportation. Transportation intrinsically involves multiple carriers operating both within and across sectors (i.e., road, air, and shipping) and across geographical boundaries. Each carrier has its own, often proprietary, systems that do not necessarily integrate easily with one another. Addressing this integration problem is a significant engineering issue to be faced as the technologies addressed in this chapter come into more widespread use.

The integration of transportation planners into supply chain and production systems is also important. As previously mentioned, freight is now often delivered directly to manufacturing plants without passing through transitional storage. Integration of these systems thus becomes a priority when shaping dynamic supply chains, and supply networks.

OBUs in use today typically consist of a simple processor, memory, and communication interfaces. Installed software is often designed solely for reading data from sensors and transmitting it to the TMS. One method of improving on this design is the integration of an autonomous software controller into the OBU to assist with the manipulation and coordination of collected onboard data. Example uses include assisting in the selection of M2M connection type in a multi-provider environment according to the type and volume of data to be transmitted and caching data locally if connections are temporarily unavailable. The controller can be further extended with a software agent that extends the distributed intelligence offered by the route optimizer. This agent essentially acts as a remote extension of the optimization platform, allowing the agent to act as a proxy representative of the vehicle itself within the context of route scheduling. Vehicles can thus become active participants in the planning process, forming a network overlay of communicating data processors.

Further research is required on so-called *smart* freight containers capable of announcing their presence and even negotiating with external devices; for example, a simple OBU fixed to a container will allow it to communicate with vehicles, customs checks, and equipment at freight consolidation centers. Many major transportation companies use such centers, distributed at strategic locations, with the primary goal of consolidating freight onto as few vehicles as possible to maximize use of available capacity. With the installation of RFID readers, incoming freight with RFID tags can be traced as it moves through a facility, providing TMS optimizers with complete coverage of freight location throughout its entire lifecycle within the business chain.

In addition, external factors also favor early adoption of pervasive technologies, such as the ongoing escalation of fuel prices, new regulations for pollution reduction, and constant increases in demand for fast, high-volume freight shipping. This is recognized in the European Union white paper *European Transport Policy for 2010* [23.18] which discusses the use of intelligent information services integrated with route planning systems and mobile communications to provide real-time, intelligent end-to-end freight and vehicle tracking and tracing.

There can be little doubt that the adoption of intelligent transportation planners capable of using real-time data sourced from pervasive technologies such as those discussed in this chapter is a major objective of many freight transportation operators both in Europe and other areas of the world. With these techniques now widely recognized as an important means of reducing operating costs, many companies are already well advanced on the path to adoption.

23.4 How to Design Agent-Oriented Solutions for Autonomic Automation

In the past decade a lot of work has been done on agent-oriented analysis and design. The works presented in [23.19] and [23.20] are only two examples, but very good starting points to dig into the agent world. More details about agent organization, agent platforms, tools and development can be found in [23.7, 21–23].

Other chapters in this Handbook also cover agent-based solutions, the following gives just a brief outline on how to start thinking in an agent-oriented way in the form of a questionnaire. A detailed discussion would be far beyond the scope of this Handbook.

Questions to structure the overall solution:

- What are the processes?
- Who/what drives the processes?
- Which roles do the process drivers play?

Questions to ask for each agent/role:

- What is the responsibility of the agent?
- Which goals does the agent aim at?
- What is the strategy to reach the goals?
- What knowledge does the agent need to follow this strategy?
- With which agents does he need to communicate?
- Which sensors and actuators are needed or available?

These questions have been discussed and answered to design and implement the two application examples contained in this chapter.

Two main aspects of an agent-based solution should be considered in order to analyze and prove the quality of the design. These aspects cannot be given as a concrete metric, but should be understood as general indicators relative to the size and type of the intended system:

- *Local knowledge*: A good agent solution has been achieved (or is possible) if the local knowledge needed by an agent to achieve its goals can be kept to a minimum. If a solution requires an agent to hold a large amount of data, or – in an extreme case – each agent needs to know *everything*, either the design should be rethought or an agent-based solution is not appropriate. Consider, for example, a sales representative (agent) for a car manufacturer, whose goal is to sell as many cars at the highest price possible. For that he does not need to know all the details about car production or supply chain organization. He only needs some extract of the whole business knowledge.
- *Communication*: Although message exchange and service-oriented architectures can accompany agent-based ideas, a good solution keeps communication to a minimum and makes careful use of resource *bandwidth*. If a design requires too much messaging among the participants then the role, and therefore the goal assignment, is not distinct enough. Agent-oriented design means to define a good level of responsibility and assign it to a software entity, which allows it to pursue its goals and to decide actions based on local knowledge. Cooperation with the environment is needed to sense *what's going on* but should not be needed to draw conclusions.

23.5 Emerging Trends and Challenges

23.5.1 Virtual Production and the Digital Factory

The automation industry, which includes at least all machine manufacturers, has the vision that all components of a production facility will be accompanied by a full digital description in a standardized format. Besides easy and straightforward integration into factory simulation tools, the goal is also to let modules carry their own electronic description to enable them to plug in and integrate automatically into a production system in the sense of self-configuration. Testing and putting into operation would become as easy as attaching a new mouse to a computer. Even though it is obvious that this will not work for all machines and that it is very hard to achieve, it is a very worthwhile goal to work towards. If the modules additionally come along with their own agents, they can also dynamically negotiate with the environment in which they are placed and (self-)optimize their activities.

23.5.2 Modularization

There is a clear trend and motivation in the industry to modularize machines and production facilities, yielding many advantages.

The customer (user) of a machine gets much greater flexibility, as he can order, configure, and dynamically adapt his production lines according to market needs. A common keyword in this regard is the selling argument *grow as you need*. On the maintenance side more cost reductions are possible as fewer spare parts are needed for modules built on the same framework, and if a defective module needs to be replaced this is normally easier than replacing a whole machine. Some production machines even offer a *shopping-cart*-like system, where a component can be exchanged without a screwdriver.

The manufacturer has smaller components to produce, which needs less space, at least for each production cell. In the same way quality tests become easier and faster because only a single module has to be tested. Last but not least, smaller modules are easier and cheaper to transport and deliver. The increased number of common parts leads to cheaper production because fewer tools are needed, and less space for many different parts and higher purchasing discounts can be achieved.

Overall, modularization is a win–win concept for all parties.

23.5.3 More RFID, More Sensors, Data Flooding

As RFID technology increasingly finds its way into industrial usage and other sensors based on video cameras, induction loops or microwaves we increase the amount of generated data day by day. Many companies are hungry for data, but have not clearly defined what to do with this new data flood. Admittedly the data and its accuracy have a high value, but one has to be aware that all this new data keeps its high value for only a very limited time. In other words: you have to process and gain the value from fresh data immediately. Because of the huge volumes involved, this can only be done by automated processes, which handle sensor input where it appears and drive activities without too much data transfer through the network. One can therefore conclude that RFID and other sensors increase the need for software agent concepts.

23.5.4 Pervasive Technologies

Limitations: Onboard Agents

As discussed in Sect. 23.5.1, one vision is to equip machine components and modules directly with their self-* logic (Sect. 23.1.4) and software representative. However, since the processing power of most controller boards is still not sufficient, and more importantly since such a huge variety of controller boards exist, it is not the first goal to support all of these directly. Instead it is much more convenient, faster, and not to forget cheaper to let the agent logic run on dedicated computers and *just* implement interfaces to the different controllers. This approach is not at odds with the general distributed solution design. It is just a special deployment decision. The very specific controller boards are not loaded with additional computing tasks, but only used as interfaces to the attached sensors and actuators. The software agents are deployed to one or more standard computers installed in the field as needed. A solution architect could theoretically use one dedicated personal computer (PC) per agent (per controller), which would directly reflect the distributiveness of the solution, but – again for cost reasons – several controllers, located in the same module, machine, area, room or building, can easily host the agents of many attached controllers. The agents still somehow work *locally*, close to the physical installation and thus the overall solutions provides redundancy, reduced latency, and near-real-time responsiveness. Step by step, as controller boards become more powerful in the coming years, the agents can be run directly on the board. This will be a smooth transition without changing the solution's core algorithms and allows one to take advantage of autonomous concepts already today.

References

23.1   Wikipedia: Automation, http://en.wikipedia.org/wiki/Automation (last accessed February 2009)

23.2   P. Davidsson, S. Johansson, J. Persson, F. Wernstedt: Agent-based approaches and classical optimization techniques for dynamic distributed resource allocation: a preliminary study, AAMAS'03 workshop on Representations and Approaches for Time-Critical Decentralized Resource/Role/Task Allocation (2003)

23.3   J.O. Kephart, D.M. Chess: The vision of autonomic computing, IEEE Comput. Mag. **36**(1), 41–50 (2003)

23.4   S. Bussmann, K. Schild: Self-organizing manufacturing control: an industrial application of agent technology, Proc. 4th Int. Conf. Multi-Agent Syst. (2000) pp. 87–94

23.5   D. Greenwood, C. Dannegger: An industry-proven multi-agent systems approach to real-time plan optimization, 5th Workshop Logist. Supply Chain Manag. (2007)

23.6   R. Zimmermann: *Agent-based supply network event management*. Whitestein Series in Software Agent Technologies and Autonomic Computing. (Birkhäuser, Basel 2006)

23.7   G. Rimassa, D. Greenwood, M.E. Kernland: The living systems technology suite: an autonomous middleware for autonomic computing, Int. Conf. Auton. Auton. Syst. ICAS (2006)

23.8   S. Kim, M.E. Lewis, C.C. White: Optimal vehicle routing with real-time traffic information, IEEE Trans. Intell. Transp. Syst. **6**(2), 178–188 (2005)

23.9   M.W.P. Savelsbergh, M. Sol: The general pickup and delivery problem, Transp. Sci. **29**(1), 17–29 (1995)

23.10   K. Dorer, M. Calisti: An adaptive solution to dynamic transport optimization, Proc. 4th Int. Jt. Conf. Auton. Agents Multiagent Syst. (ACM, New York 2005) pp. 45–51

23.11   S. Mitrovic-Minic: *Pickup and delivery problem with time windows: A survey*. Technical Report TR 1998–12 (Simon Fraser University, Burnaby 1998)

23.12   W.P. Nanry, J.W. Barnes: Solving the pickup and delivery problem with time windows using reactive tabu search, Transp. Res. B **34**, 107–121 (2000)

23.13   M. Pěchouček, S. Thompson, J. Baxter, G. Horn, K. Kok, C. Warmer, R. Kamphuis, V. Mařík, P. Vrba, K. Hall, F. Maturana, K. Dorer, M. Calisti: Agents in industry: the best from the AAMAS 2005 industry track, IEEE Intell. Syst. **21**(2), 86–95 (2006)

23.14   General Datacom: *Transportation and Wireless Connections*, http://www.gdc.com/inotes/pdf/transportation.pdf (Naugatuck 2004)

23.15   G. Lawton: Machine-to-machine technology gears up for growth, IEEE Computer **37**(9), 12–15 (2004)

23.16   D. Greenwood, M. Calisti: The living systems connection agent: seamless mobility at work, Proc. Communication in Distributed Systems (KiVS) (Berne 2007) pp. 13–14

23.17   A. Gühnemann, R. Schäfer, K. Thiessenhusen, P. Wagner: New approaches to traffic monitoring and management by floating car data, Proc. 10th World Conf. Transp. Res. (Istanbul 2004)

23.18   The European Commission: *European Transport Policy for 2010: Time to Decide*, (2001) http://ec.europa.eu/transport/white_paper/documents/index_en.htm

23.19   M. Wooldridge, N.R. Jennings, D. Kinny: *The Gaia methodology for agent-oriented analysis and design*, J. Auton. Agents Multi-Agent Syst. **3**(3), 285–312 (2000)

23.20   F. Zambonelli, N.R. Jennings, M. Wooldridge: Developing Multiagent Systems: The Gaia Methodology, ACM Trans. Softw. Eng. Methodol. **12**(3), 317–370 (2003)

23.21   C. van Aart: *Organizational Principles for Multi-Agent Architectures*. Whitestein Series in Software Agent Technologies and Autonomic Computing (Birkhäuser, Basel 2005)

23.22   R. Unland, M. Klusch, M. Calisti (Eds.): *Software Agent-Based Applications, Platforms and Development Kits*. Whitestein Series in Software Agent Technologies and Autonomic Computing (Birkhäuser, Basel 2005)

23.23   R. Červenka, I. Trenčanský, M. Calisti, D. Greenwood: *AML: Agent Modeling Language Toward Industry-Grade Agent-Based Modeling*, Lecture Notes in Computer Science (Springer, Berlin, Heidelberg 2005)

24. Automation Under Service-Oriented Grids

Jackson He, Enrique Castro-Leon

For some companies, information technology (IT) services constitute a fundamental function without which the company could not exist. Think of UPS without the ability to electronically track every package in its system or any large bank managing millions of customer accounts without computers. IT can be capital and labor intensive, representing anywhere between 1% and 5% of a company's gross expenditures, and keeping costs commensurate with the size of the organization is a constant concern for the chief information officer (CIO)s in charge of IT.

A common strategy to keep labor costs in check today is through a deliberate *sourcing* or service procurement strategy, which may include in-sourcing using in-house resources or outsourcing, which involve the delegation of certain standardized business processes such as payroll to service companies such as ADP.

Yet another way of keeping labor costs in check with a long tradition is through the use of automation, that is, the integrated use of technology, machines, computers, and processes to reduce the cost of labor.

The convergence of three technology domains, namely virtualization, service orientation, and grid computing promises to bring the automation of provision and delivery of IT services to levels never seen before. An IT environment where these three technology domains coexist is said to be a *virtual service-oriented environment* or *VSG* environment. The cost savings are accrued through systemic reuse of resources and the ability to quickly integrate resources not just within one department, but across the whole company and beyond.

In this Chapter we will review each of the constituent technologies for a virtual service-oriented grid and examine how each contributes to the automation of delivery of IT services.

The increasing adoption of service-oriented architectures (SOAs) represents the increasing recognition by IT organizations of the need for business and technology alignment. In fact, under SOA there is no difference between the two. The unit of delivery for SOA is a service, which is usually defined in business terms.

In other words, SOA represents the up-leveling of IT, empowering IT organizations to meet the business needs of the community they serve. This up-leveling creates a gap, because for IT business requirements eventually need to be translated into technology-based solutions.

Our research indicates that this gap is being fulfilled by the resurgence of two very old technologies, namely virtualization and grid computing.

To begin with, SOA allowed the decoupling of data from applications through the magic of extensible mark-up language (XML).

A lot of work that used to be done by application developers and integrators now gets done by computers. When most data centers run at 5–10% utilization, growing and deploying more data centers is not a good solution. Virtualization technology came in very handy to address this situation, allowing the decoupling of applications from the platforms on which they run. It acts as the gearbox in a car, ensuring efficient transmission of power from the engine to the wheels.

The net effect of virtualization is that it allows utilization factors to increase to 60–70%. The technique has been applied to mainframes for decades. Deploying virtualization to tens of thousands of servers has not been easy.

Finally, grid technology has allowed very fast, on-the-fly resource management, where resources are allocated not when a physical server is provisioned, but for each instance that a program is run.

24.1 Emergence of Virtual Service-Oriented Grids

Legacy systems in many cases represent a substantial investment and the fruits of many years of refinement. To the extent that legacy applications bring business value with relatively little cost in operations and maintenance there is no reason to replace them. The adoption of virtual service-oriented grids does not imply wholesale replacement of legacy systems by any means. Newer virtual service-oriented applications will coexist with legacy systems for the foreseeable future. If anything, the adoption of a virtual service-oriented environment will create opportunities for legacy integration with the new environment through the use of web-service-based exportable interfaces. Additional value will be created for legacy systems through extended life cycles and new revenue streams through repurposing older applications.

These goals are attained through the increasing use of machine-to-machine communications during setup and operation.

To understand how the different components of virtual service-oriented grids came to be, we have to look at the evolution of its three constituent technologies: virtualization, service orientation, and grids. We also need to examine how they become integrated and interact with each other to form the core of a virtual service-oriented grid environment. This section also addresses the tools and overall architecture components that keep

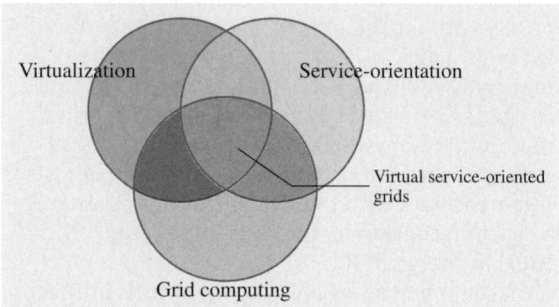

Fig. 24.1 Virtual service-oriented grids represent the confluence of three key technology domains

virtual service-oriented functions as integral business services that deliver ultimate value to businesses. Figure 24.1 depicts the abstract relationship between the three constituent technologies. Each item represents a complex technical domain of its own.

In the following sections, we describe how key technology components in these domains define the foundation for a virtual service-oriented grid environment, elements such as billing and metering tools, service-level agreement (SLA) management, as well as security, data integrity, etc. Further down, we discuss architecture considerations to put all these components together to deliver tangible business solutions.

24.2 Virtualization

Alan M. Turing, in his seminal 1950 article for the British psychology journal *Mind*, proposed a test for whether machines were capable of thinking by having a machine and a human behind a curtain typing text messages to a human judge [24.1]. A thinking machine would pass the test if the judge could not reliably determine whether answers from the judge's question came from the human or from the machine.

The emergence of virtualization technology poses a similar test, perhaps not as momentous as attempting to distinguish a human from a machine. Every computation result, such as a web page retrieval, a weather report, a record pulled from a database or a spreadsheet result can be ultimately traced to a series of state changes. These state changes can be represented by monkeys typing at random at a keyboard, humans scribbling numbers on a note pad, or a computer running a program. One of the drawbacks of the monkey method is the time it takes to arrive at a solution [24.2]. The *human* or manual method is also too slow for most problems of practical size today, which involve millions or billions of records.

Only machines are capable of addressing the scale of complexity of most enterprise computational problems today. In fact, their performance has progressed to such an extent that, even with large computational tasks, they are idle most of the time. This is one of the reasons for the low utilization rates for servers in data centers today. Meanwhile, this infrastructure represents a sunk cost, whether fully utilized or not.

Modern microprocessor-based computers have so much reserve capacity that they can be used to simulate other computers. This is the essence of virtualization: the use of computers to run programs that simulate computers of the same or even different architectures. In fact, machines today can be used to simulate many computers, anywhere between 1 and 30 for practical situations. If a certain machine shows a load factor of 5% when running a certain application, that machine can easily run ten virtualized instances of the same application.

Likewise, three machines of a three-tier e-commerce application can be run in a single physical machine, including a simulation of the network linking the three machines.

Virtual computers, when compared with *real* physical computers, can pass the Turing test much more easily than when humans are compared to a machine. There is essentially no difference between results of computations in a physical machine versus the computation in a virtual machine. It may take a little longer, but the results will be identical to the last bit. Whether running in a physical or a virtualized host, an application program goes through the same state transitions, and eventually presents the same results.

As we saw, *virtualization* is the creation of substitutes for real resources. These substitutes have the same functions and external interfaces as their counterparts, but differ in attributes, such as size, performance, and cost. These substitutes are called *virtual resources*. Because the computational results are identical, users are typically unaware of the substitution. As mentioned, with virtualization we can make one physical resource look like multiple virtual resources; we can also make multiple physical resources into shared pools of virtual resources, providing a convenient way of divvying up a physical resource into multiple logical resources.

In fact, the concept of virtualization has been around for a long time. Back in the mainframe days, we used to have virtual processes, virtual devices, and virtual memory [24.3–5]. We use virtual memory in most operating systems today. With virtual memory, computer software gains access to more memory than is physically installed, via the background swapping of data to disk storage. Similarly, virtualization concepts can be applied to other IT infrastructure layers including networks, storage, laptop or server hardware, operating systems, and applications. Even the notion of process is essentially an abstraction for a virtual central processing unit (CPU) running a single application.

Virtualization on x86 microprocessor-based systems is a more recent development in the long history of virtualization. This entire sector owes its existence to a single company, VMware; and in particular, to founder *Rosenblum* [24.6], a professor of operating systems at Stanford University. Rosenblum devised an intricate series of software workarounds to overcome certain intrinsic limitations of the x86 instruction set architecture in the support of virtual machines. These workarounds became the basis for VMware's early products. More recently, native support for virtualization hypervisors and virtual machines has been developed to improve the performance and stability of virtualization. An example is Intel's virtualization technology (VTx) [24.7].

To look further into the impact of virtualization on a particular platform, Fig. 24.2 illustrates a typical configuration of a single operating system (OS) platform without virtual machines (VMs) and a configuration of multiple virtual machines with virtualization. As indicated in the chart on the right, a new layer of abstraction is added, the virtual machine monitor (VMM), between physical resources and virtual resources. A VMM presents each VM on top of its virtual resources and maps virtual machine operations to physical resources. VMMs can be designed to be tightly coupled with operating systems or can be agnostic to operating systems. The latter approach provides customers with the capability to implement an OS-neutral management infrastructure.

24.2.1 Virtualization Usage Models

Virtualization is not just about increasing load factors; it brings a new level of operational flexibility and convenience to the hardware that was previously associated with software only. Virtualization allows running

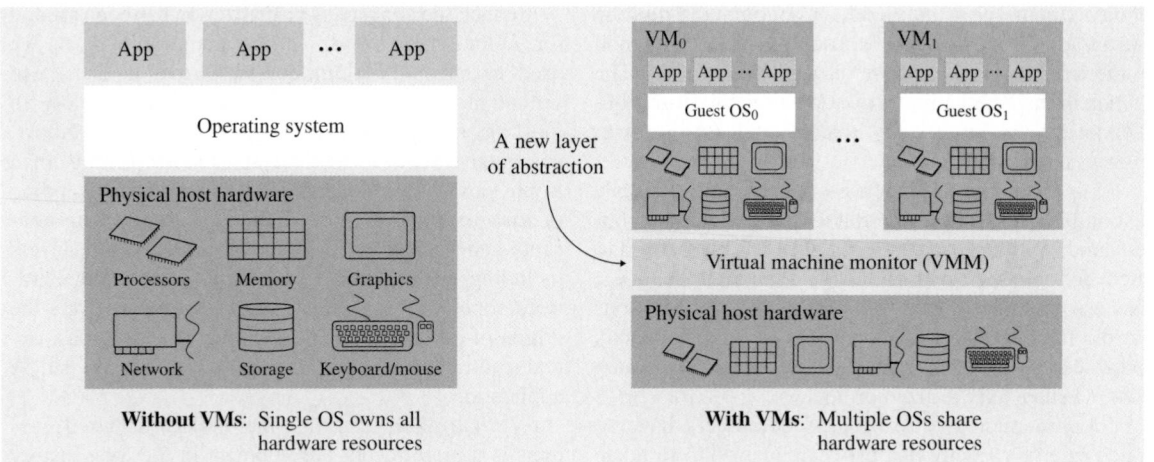

Fig. 24.2 A platform with and without virtualization

instances of virtualized machines as if they were applications. Hence, programmers can run multiple VMs with different operating systems, and test code across all configurations simultaneously. Once systems administrators started running hypervisors in test laboratories, they found a treasure trove of valuable use cases. Multiple physical servers can be consolidated onto a single, more powerful machine. The big new box still draws less energy and is easier to manage on a per-machine basis. This server consolidation model provides a good solution to *server sprawl*, the proliferation of physical servers, a side effect of deploying servers supporting a single application.

There exist additional *helper* technologies complementing and amplifying the benefits brought by virtualization; for example, extended manageability technologies will be needed to automatically track and manage the hundreds and thousands of virtual machines in the environment, especially when many of them are

created and terminated dynamically in a data center or even across data centers.

Virtual resources, even though they act in lieu of physical resources, are in actuality software entities that can be scheduled and managed automatically under program control, replacing the very onerous process of physically procuring machines. The capability exists to deploy these resources on the fly instead of the weeks or months that it would take to go through physical procurement.

Grid computing technologies are needed to transparently and automatically orchestrate disparate types of physical systems to become a pool of virtual resources across the global network.

In this environment, standards-based web services become the fabric of communicate among heterogeneous systems and applications coming from different manufacturers, insourced and outsourced, legacy and new alike.

24.3 Service Orientation

Service orientation represents the natural evolution of current development and deployment models. The movement started as a programming paradigm and evolved into an application and system integration methodology with many different standards behind it. The evolution of service orientation can be traced back to the object-oriented models in the 1980s and even earlier and the component-based development model in the 1990s. As the evolution continues, ser-

vice orientation retains the benefits of component-based development (self-description, encapsulation, dynamic discovery, and loading).

Service orientation brings a shift in paradigm from remotely invoking methods on objects, to one of passing messages between services. In short, service orientation can be defined as a design paradigm that specifies the creation of automation logic in the form of services based on standard messaging schemas.

Messaging schemas describe not only the structure of messages, but also behavior and semantics of acceptable message exchange patterns and policies. Service orientation promotes interoperability among heterogeneous systems, and thus becomes the fabric for systems integration, as messages can be sent from one service to another without consideration of how the service handling those messages has been implemented.

Service orientation provides an evolutionary approach to building distributed systems that facilitate loosely coupled integration and resilience to change. With the arrival of web services and WS* standards (WS* denotes the multiple standards that govern web service protocols), service-oriented architectures (SOAs) have made service orientation a feasible paradigm for the development and integration of software and hardware services.

Some advantages yielded by service orientation are described below.

Progressive and Based on Proven Principles

Service orientation is *evolutionary* (not revolutionary) and grounded in well-known information technology principles taking into account decades of experience in building real-world distributed applications. Service orientation incorporates concepts such as self-describing applications, explicit encapsulation, and dynamic loading of functionality at runtime-principles first introduced in the 1980s and 1990s through object-oriented and component-based development.

Product-Independent and Facilitating Innovation

Service orientation is a set of architectural principles supported by open industry standards independent of any particular product.

Easy to Adopt and Nondisruptive

Service orientation can and should be adopted through an incremental process. Because it follows some of the same information technology principles, service orientation is not disruptive to current IT infrastructures and can often be achieved through wrappers or adapters to legacy applications without completely redesigning the applications.

Adapt to Changes and Improved Business Agility

Service orientation promotes a loosely coupled architecture. Each of the services that compose a business solution can be developed and evolved independently.

24.3.1 Service-Oriented Architectural Tenets

The fundamental building block of service orientation is a *service*. A service is a program interacting through well-defined message exchange interfaces. The interfaces are self-describing, relatively stable over time, and versioning resilient, and shield the service from implementation details. Services are built to last while service configurations and aggregations are built for change. Some basic architectural tenets must be followed for service orientation to accomplish a successful service design and minimize the need for human intervention.

Designed for Loose Coupling

Loose coupling is a primary enabler for reuse and automatic integration. It is the key to making service-oriented solutions resilient to change. Service-oriented application architects need to spend extra effort to define clear boundaries of services and assure that services are autonomous (have fewer dependencies), to ensure that each component in a business solution is loosely coupled and easy to compose (reuse).

Encapsulation. Functional encapsulation, the process of hiding the internal workings of a service to the outside world, is a fundamental feature of service orientation.

Standard Interfaces. We need to force ubiquity at the edge of the services. Web services provide a collection of standards such as simple object access protocol (SOAP), web service definition language (WSDL), and the WS* specifications, which take anything not functionally encapsulated for conversion into reusable components. At the risk of oversimplification, in the same way the browser became the universal graphic user interface (GUI) during the emergence of the First Web in the 1990s, web services became the universal machine-to-machine interface in the Second Web of the 2000s, with the potential of automatically integrating self-describing software components without human intervention [24.8, 9].

Unified Messaging Model. By definition, service orientation enables systems to be loosely bound for both the composition of services, as well as the integration of a set of services into a business solution. Standard messaging and interfaces should be used for both service integration and composition. The use of a unified messaging model blurs the distinction between integration and composition.

Designed for Connected Virtual Environments

With advances in virtualization as discussed in the previous section, service orientation is not limited to a single physical execution environment, but rather can be applied using interconnected virtual machines (VMs). Related content on virtual network of machines is usually referred to as a *service grid*. Although the original service orientation paradigm does not mandate full resource virtualization, the combination of service orientation and a grid of virtual resources hint at the enormous potential business benefits brought by the autonomic and on-demand compute models.

Service Registration and Discovery. As services are created, and their interfaces and policies are registered and advertised, using ubiquitous formats. As more services are created and registered, a shareable service network is created.

Shared Messaging Service Fabric. In addition to networking the services, a secure and robust messaging service fabric is essential for service sharing and communication among the virtual resources.

Resource Orchestration and Resolution. Once a service is discovered, we need to have effective ways to allocate sufficient resources for the services to meet the service consumer's needs. This needs to happen dynamically and be resolved at runtime. This means special attention is placed on discovering resources at runtime and using these resources in a way that they can be automatically released and regained based on resource orchestration policies.

Designed for Manageability

The solutions and associated services should be built to be managed with sufficient interfaces to expose information for an independent management system, which by itself could be composed of a set of services, to verify that the entire loosely bound system will work as designed.

Designed for Scalability

Services are meant to scale. They should facilitate hundreds or thousands of different service consumers.

Designed for Federated Solutions

Service orientation breaks application silos. It spans traditional enterprise computing boundaries, such as network administrative boundaries, organizational and operational boundaries, and the boundaries of time and space. There are no technical barriers for crossing corporate or transnational boundaries. This means services need a high degree of built-in security, trust, and internal identity, so that they can negotiate and establish federated service relationships with other services following given policies administrated by the management system. Obviously, cohesiveness across services based on standards in a network of service is essential for service federation and to facilitate automated interactions.

24.3.2 Services Needed for Virtual Service–Oriented Grids

A set of foundation services is essential to make service-oriented grids work. These services, carried out through machine-to-machine communication, are the building blocks for a solution stack at different layers providing some of the automated functions supported by VSGs.

Fundamental services can be divided into the following categories similar to the open grid services architecture (OGSA) approach, as outlined in Fig. 24.3 and in applications in Fig. 24.4. The items are provided for conceptual clarity without an attempt to provide an exhaustive list of all the services. We highlight some example services, along with a high-level description.

Resource Management Services

Management of the physical and virtual resources (computing, storage, and networking) in a grid environment.

Asset Discovery and Management

Maintaining an automatic inventory of all connected devices, always accurate and updated on a timely basis.

Provisioning

Enabling bare metal provisioning, coordinating the configuration between server, network, and storage in a synchronous and automatic manner, making sure software gets loaded on the right physical machines, taking platforms in and out of service as required for testing, maintenance, repair or capacity expansion; remote booting a system from another system, and managing the licenses associated with software deployment.

Monitoring and Problem Diagnosis

Verifying that virtual platforms are operational, detecting error conditions and network attacks, and responding by running diagnostics, deprovisioning platforms and reprovisioning affected services, or isolating network segments to prevent the spread of malware.

Infrastructure Services

Services to manage the common infrastructure of a virtual grid environment and offering the foundation for service orientation to operate.

QoS Management

In a shared virtualized environment, making sure that sufficient resources are available and system utilization is managed at a specific quality-of-service (QoS) level as outlined in the service-level agreement (SLA).

Load Balancing

Dynamically reassigning physical devices to applications to ensure adherence to specified service (performance) levels and optimized utilization of all resources as workloads change.

Capacity Planning

Measuring and tracking the consumption of virtual resources to be able to plan when to reserve resources for certain workloads or when new equipment needs to be brought on line.

Utilization Metering

Tracking the use of particular resources as designated by management policy and SLA. The metering service could be used for chargeback and billing by higher-level software.

Execution Management Services

Execution management services are concerned with the problems of instantiating and managing to completion units of work or an application.

Business Processes Execution

Setting up generic procedure as building blocks to standardize business processes and enabling interoperability across heterogeneous system management products.

Workflow Automation

Managing a seamless flow of data as part of the business process to move from application to application. Tracking the completion of workflow and managing exceptions.

Execution Resource Allocation

In a virtualized environment, selecting optimal resources for a particular application or task to execute.

Execution Environment Provisioning

Once an execution environment is selected, dynamically provision the environment as required by the application, so that a new instance of the application can be created.

Managing Application Lifecycle

Initiate, track status of execution, and administer the end-of-life phase of a particular application

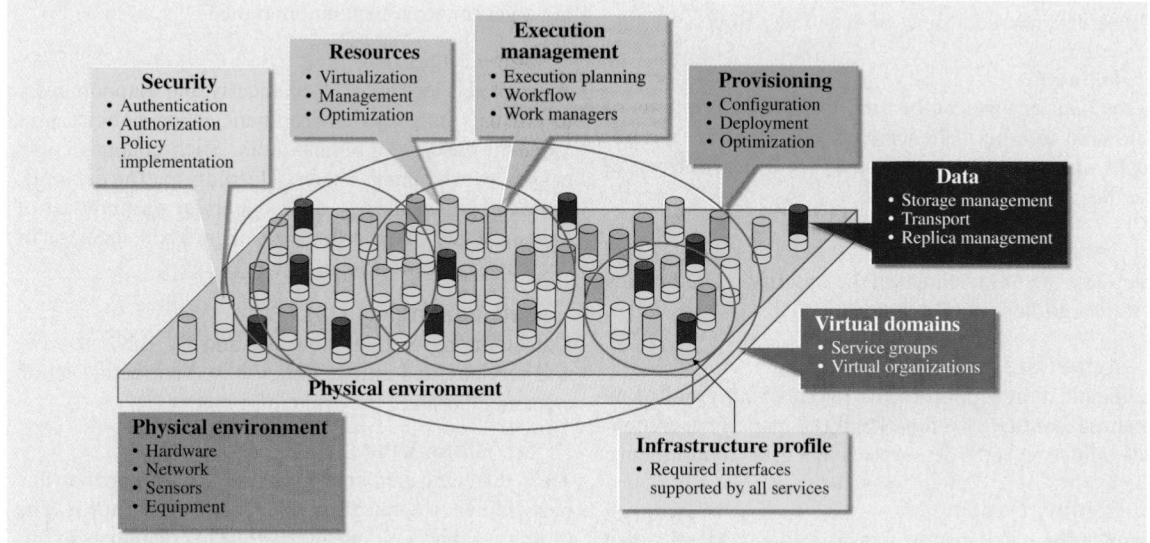

Fig. 24.3 OGSA service model from OGSA Spec 1.5, July 2006

Part C | 24.3

and release virtual resources back to the resource pool.

Data Services
Moving data as required, such as data replication and updates; managing metadata, queries, and federated data resources.

Remote Access
Access remote data resources across the grid environment. The services hide the communication mechanism from the service consumer. They can also hide the exact location of the remote data.

Staging
When jobs are executed on a remote resource, the data services are often used to stage input data to that resource ready for the job to run, and then to move the result to an appropriate place.

Replication
To improve availability and to reduce latency, the same data can be stored in multiple locations across a grid environment.

Federation
Data services can integrate data from multiple data sources that are created and maintained separately.

Derivation
Data services should support the automatic generation of one data resource from another data source.

Metadata
Some data service can be used to store descriptions of data held in other data services. For example, a replicated file system may choose to store descriptions of the files in a central catalogue.

Security Services
Facilitate the enforcement of the security-related policy within a grid environment.

Authentication
Authentication is concerned with verifying proof of an asserted identity. This functionality is part of the credential validation and trust services in a grid environment.

Identity Mapping
Provide the capability of transforming an identity that exists in one identity domain into an identity within another identity domain.

Authorization
The authorization service is to resolve a policy-based access-control decision. For the resource that the service requestor requests, it resolves, based on policy, whether or not the service requestor is authorized to access the resource.

Credential Conversion
Provide credential conversion from one type of credential to another type or form of credential. This may include such tasks as reconciling group membership, privileges, attributes, and assertions associated with entities (service consumers and service providers).

Audit and Secure Logging
The audit service, similarly to the identity mapping and authorization services, is policy driven.

Security Policy Enforcement
Enforcing automatic device and software load authentication; tracing identity, access, and trust mechanisms within and across corporate boundaries to provide secure services across firewalls.

Logical Isolation and Privacy Enforcement
Ensuring that a fault in a virtual platform does not propagate to another platform in the same physical machine, and that there are no data leaks across virtual platforms which could belong to different accounts.

Self-Management Services
Reduce the cost and complexity of owning and operating a grid environment autonomously.

Self-Configuring
A set of services adapt dynamically and autonomously to changes in a grid environment, using policies provided by the grid administrators. Such changes could trigger provisioning requests leading to, for example, the deployment of new components or the removal of existing ones, maybe due to a significant increase or decrease in the workload.

Self-Healing
Detect improper operations of and by the resources and services, and initiate policy-based corrective action without disrupting the grid environment.

Self-Optimizing
Tune different elements in a grid environment to the best efficiency to meet end-user and business needs. The tuning actions could mean reallocating resources to improve overall utilization or optimization by enforcing an SLA.

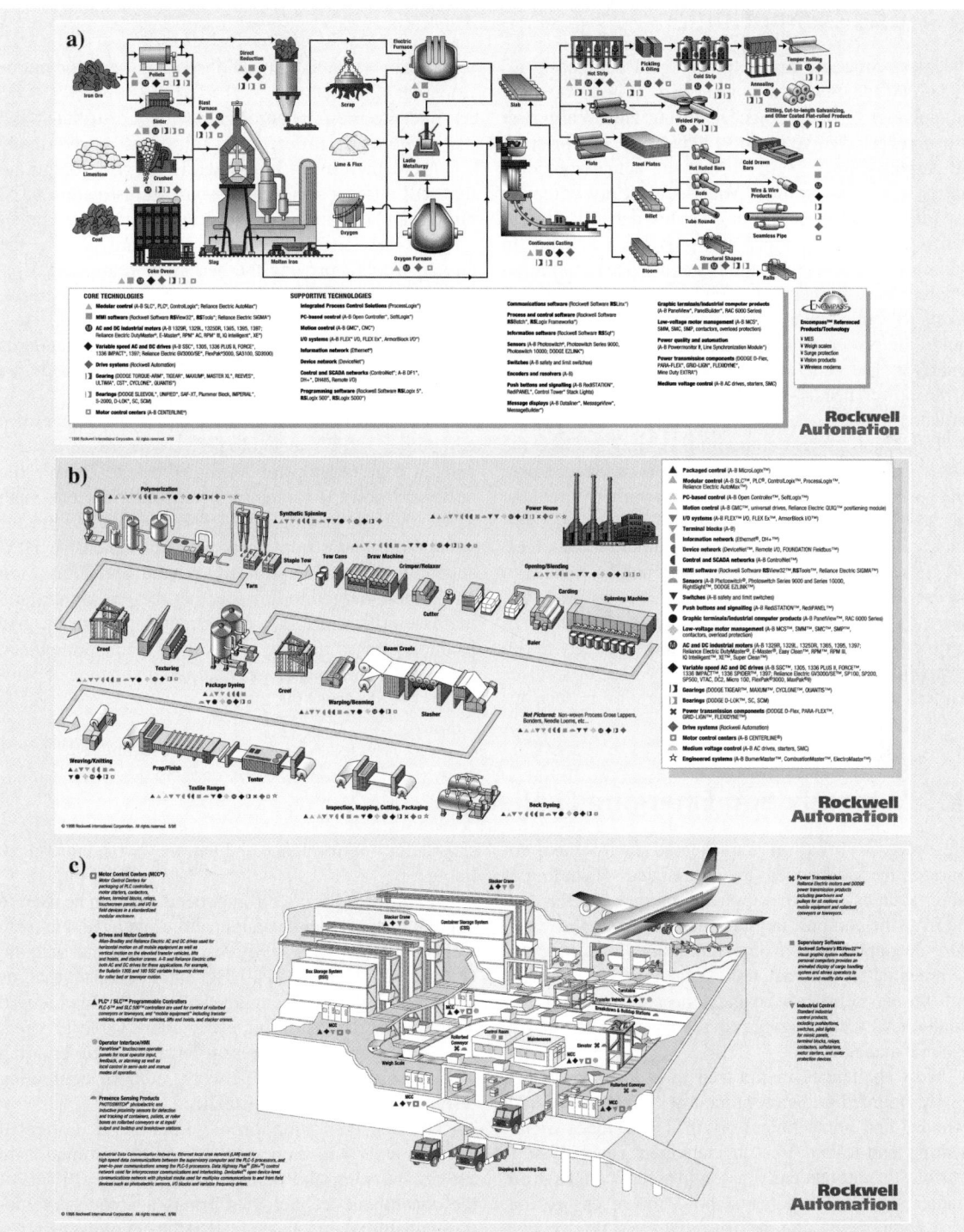

Fig. 24.4a–c Examples of applications that can benefit from virtual services as shown in Fig. 24.3. (**a**) Steel industry; (**b**) textile industry; (**c**) material handling at a cargo airport (courtesy of Rockwell Automation, Inc.)

24.4 Grid Computing

The most common description of grid computing includes an analogy to a power grid. When you plug an appliance or other object requiring electrical power into a receptacle, you expect that there is power of the correct voltage available, but the actual source of that power is not known. Your local utility company provides the interface into a complex network of generators and power sources and provides you with (in most cases) an acceptable quality of service for your energy demands. Rather than each house or neighborhood having to obtain and maintain its own generator of electricity, the power grid infrastructure provides a virtual generator. The generator is highly reliable and adapts to the power needs of the consumers based on their demand.

The vision of grid computing is similar. Once the proper grid computing infrastructure is in place, a user will have access to a *virtual computer* that is reliable and adaptable to the user's needs. This virtual computer will consist of many diverse computing resources. But these individual resources will not be visible to the user, just as the consumer of electric power is unaware of how their electricity is being generated. In a grid environment, computers are used not only to run the applications but to secure the allocation of services that will run the application. This operation is done automatically to maintain the abstraction of anonymous resources [24.10].

Because these resources are widely distributed and may belong to different organizations across many countries, there must be standards for grid computing that will allow a secure and robust infrastructure to be built. Standards such as the open grid services architecture (OGSA) and tools such as those provided by the Globus Toolkit provide the necessary framework. Initially, businesses will build their own infrastructures, but over time, these grids will become interconnected. This interconnection will be made possible by standards such as OGSA and the analogy of grid computing to the power grid will become real.

The ancestry of grids is rooted in high-performance computing (HPC) technologies, where resources are ganged together toward a single task to deliver the necessary power for intensive computing projects such as weather forecasting, oil exploration, nuclear reactor simulation, and so on. In addition to expensive HPC supercomputer centers, mostly government funded, an HPC grid emerged to link together these resources and increase utilization. In a concurrent development, grid technology was used to join not just supercomputers, but literally millions of workstations and personal computers (PCs) across the globe.

24.5 Summary and Emerging Challenges

In essence, a virtual service-oriented environment encourages the independent and automated scheduling of data resources from the applications that use the data and from the compute engines that run the applications. SOA decouples data from applications and provides the potential for automated mechanisms for aligning IT with business through business process management. Finally, grid technologies provide dynamic, on-the-fly resource management.

Most challenges in the transition to a virtualized service-oriented grid environment will likely be of both technical and nontechnical origin, for instance implementing end-to-end trust management: even if it is possible to automatically assemble applications from simpler service components, how do we ensure that these components can be trusted? How do we also ensure that, even if the applications that these components support function correctly, that they will provide satisfactory performance and that they will function reliably?

A number of service components that can be used to assemble more complex applications are available from well-known providers: Microsoft Live, Amazon.com, Google, eBay, and PayPal. The authors expect that, as technology progresses, smaller players worldwide will enter the market, fulfilling every conceivable IT need. These resources may represent business logic building blocks, storage over the network, or even computing resources in the form of virtualized servers.

The expected adoption of virtual service-oriented environments will increase the level of automation in the provisioning and delivery of IT services. Each of the constituent technologies brings a unique automation capability into the mix. Grid technology enables the automatic harnessing of geographically distributed, anonymous computing resources. Service orientation enables on-the-fly, automatic integration of the distributed resources through the use of standards-based interfaces enabled by the use of web services and XML technology. Finally, virtualization enables carving out a physical resource into a multiplicity of virtual resources, allowing the transparent (automatic) matching of the demand for a resource to the physical manifestation of the resource.

24.6 Further Reading

- L. Camarinha-Matos, H. Afsarmanesh, M. Ollus: *Methods and Tools for Collaborative Network Organizations* (Springer, Berlin, Heidelberg 2008)
- E. Castro-Leon, J. He, M. Chang: Scaling down SOA to small businesses, IEEE Int. Conf. Serv.-Oriented Comput. Applic. (SOCA), Newport Beach (2007)
- E. Castro-Leon, J. He, M. Chang, J. Hahn-Steichen, J. Hobbs, G. Yohanan: Service orchestration of Intel-based platforms under a service-oriented infrastructure, Intel Technol. J. **10**(04), 265–273 (2006), http://www.intel.com/technology/itj/2006/v10i4/2-service/1-abstract.htm
- E. Castro-Leon: Using SOA to lower legacy costs and free up manpower, CIO Update (June 2006), http://www.cioupdate.com/trends/article.php/3612206
- E. Castro-Leon: Enterprise grid computing, seven part series, Enterprise Sys. J. (2006), http://www.esj.com/News/article.aspx?editorialsid=1616
- E. Castro-Leon, K. King, M. Linesch, Y. Benvenisti, P. Lee: The missing link, a virtual roundtable interview on grid computing, Busin. Man. Mag. 156–162 (Nov–Dec 2005), http://www.busmanagement.com/pastissue/article.asp?art=25245&issue=138
- E. Castro-Leon: An introduction to web services, Ziff Davis Channel Zone (November 2003), http://channelzone.ziffdavis.com/article2/0,3973,1399287,00.asp
- T. Erl: *Service Oriented Architecture, Concepts, Technology and Design* (Prentice Hall, Englewood Cliffs 2005)
- R. Fogel, E. Castro-Leon, W. Fellows, S. Wallage, A. Mulholland, A. Sinha, R.B. Cohen, T. Gibbs, K. Vizzini, M. Linesch, W. Mougayar, E. Stokes, M.P. Haynos, D. Becker, R. Subramaniam, J. Pike, T. Abels, S. Brewer, R. Vrablik, H.J. Schwarz, N. Devireddy, M. Brunridge, S. Zhou, A. Shum, V. Livschitz, P. Chavez, R. Schecterle, Z. Mahmood, A. Fernandez, D. Kusnetzky, P. Peiravi, L. Schubert, B. Rangarajan, D. Stimson: *The Emergence of Grid and Service Oriented IT: An Industry Vision for Business Success* (Tabor Communications, San Diego 2006)
- I. Foster, C. Kesselman: *The Grid 2: Blueprint for a New Computing Infrastructure* (Morgan Kaufmann, New York 2003)
- L. Grandinetti: *Grid Computing: The New Frontier of High Performance Computing* (Elsevier Science, Amsterdam 2005)
- B. Goldworm, A. Skamarock: *Blade Servers and Virtualization: Transforming Enterprise Computing While Cutting Costs* (Wiley, New York 2007)
- J. Joseph, C. Fellenstein: *Grid Computing* (IBM Press, Prentice Hall 2004)
- V. Moreno, K. Reddy: *Network Virtualization* (Cisco Press, Indianapolis 2006)
- A. Sharp, P. McDermott: *Workflow Modeling: Tools for Process Improvement and Application Development* (Artech House, London 2001)
- W. vand der Aalst, K. van Hee: *Workflow Management: Models, Methods and Systems* (MIT Press, Cambridge 2004)
- B. Woolf: *Exploring IBM SOA Technology & Practice* (Clear Horizon, 2008)
- T. G. Robertazzi: *Networks and Grids: Technology and Theory* (Springer, Berlin, Heidelberg 2007)
- F. Travostino, J. Mambretti, G. Karmous-Edwards: *Grid Networks: Enabling Grids with Advanced Communication Technology* (Wiley, New York 2006)
- G. Papakonstantinou, M. P. Bekakos, G. A. Gravvanis, H. R. Arabnia: *Grid Technologies: Emerging Distributed Architectures to Virtual Organizations (Advances in Management Information)* (WIT Press, 2006)
- A. Chakrabarti: *Grid Computing Security* (Springer, Berlin, Heidelberg 2007)
- T. Priol: *Towards Next Generation Grids: Proceedings of the CoreGRID Symposium 2007* (Springer, Berlin, Heidelberg 2007)
- S. Gorlatch, P. Fragopoulou, T. Priol: *Grid Computing: Achievements and Prospects* (Springer, Berlin, Heidelberg 2008)

References

24.1   A.M. Turing: Computing machinery and intelligence, Mind **59**(236), 433–4600 (1950)

24.2   The Internet Society: The Infinite Monkey Protocol Suite (IMPS), RFC 2795 (2000)

24.3   The IBM CP-40 Project, http://en.wikipedia.org/wiki/IBM_CP-40

24.4   J. Fotheringham: Dynamic storage allocation in the atlas computer, Commun. ACM **4**(10), 435–436 (1961)

24.5   Burroughs Large Systems, http://en.wikipedia.org/wiki/Burroughs_large_systems

24.6   M. Rosenblum, E. Bugnion, S. Devine, S.A. Herrod: Using the SimOS machine simulator to study complex computer systems, ACM TOMACS, Special Issue on Computer Simulation (1997)

24.7   G. Neiger, A. Santoni, F. Leung, D. Rodgers, R. Uhlig: Intel virtualization technology: Hardware support for efficient processor virtualization, Intel Technol. J. **10**(3), 167–177 (2006)

24.8   E. Castro-Leon: Web services readiness, WebServices.org (February 2002), http://www.mywebservices.org/index.php/article/articleview/113/1/61/

24.9   E. Castro-Leon: The web within the web, IEEE Spectrum **41**(2), 42–46 (2004)

24.10   E. Castro-Leon, J. Munter: Grid computing looking forward, Technology@Intel Mag. (May 2005), http://www.intel.com/technology/magazine/computing/grid-computing-0605.htm

25. Human Factors in Automation Design

John D. Lee, Bobbie D. Seppelt

Designers frequently look toward automation as a way to increase system efficiency and safety by reducing human involvement. This approach often leads to disappointment because the role of people becomes more, not less, important as automation becomes more powerfull and prevalent. Developing automation without consideration of the human operator leads to new and more catastrophic failures. For automation to fulfill its promise, designers must avoid a technology-centered approach and adopt an approach that considers the joint operator–automation system. Automation-related problems arise because introducing automation changes the type and extent of feedback that operators receive, as well as the nature and structure of tasks. In addition, operators' behavioral, cognitive, and emotional responses to these changes can leave the system vulnerable to failure. Automation is not a homogenous technology. There are many types of automation and each poses different design challenges. This chapter describes how different types of automation place different demands on operators. It also presents strategies that can help designers achieve the promise of automation. The chapter concludes with future challenges in automation design.

Designers often view automation as the path toward greater efficiency and safety. In many cases, automation does deliver these benefits. In the case of the control of cargo ships and oil tankers, automation has made it possible to operate a vessel with as few as 8–12 crew members, compared with the 30–40 that were required 40 years ago [25.1]. In the case of aviation, automation has reduced flight times and increased fuel efficiency [25.2]. Similarly, automation in the form of decision-support systems has been credited with saving millions of dollars in guiding policy and production decisions [25.3]. Automation promises greater efficiency, lower workload, and fewer human errors; however, these promises are not always fulfilled.

A common fallacy is that automation can improve system performance by eliminating human variability and errors. This fallacy often leads to mishaps that surprise operators, managers, and designers. As an example, the cruise ship Royal Majesty ran aground because the global positioning system (GPS) signal was lost and the position estimation reverted to position extrapolation based on speed and heading (dead reckoning). For over 24 h the crew followed the compelling electronic chart display and did not notice that the GPS signal had been lost or that the position error had been accumulating. The crew failed to heed indications from boats in the area, lights on the shore, and even salient changes in water color that signal shoals. The surprise of the GPS failure was only discovered when the ship ran aground [25.4, 5]. As this example shows, automation does not guarantee improved efficiency and error-free performance.

For automation to fulfill its promise, designers must focus not on the design of the automation, but on the design of the joint human–automation system. Automation often fails to provide expected benefits because it does not simply replace the human in performing a task, but also transforms the job and introduces a new set of tasks [25.6].

One way to view the automation failure that led to the grounding of the Royal Majesty is that it was simply a malfunction of an otherwise well-designed system – a problem with the technical implementation. Another view is that the grounding occurred because the interface design failed to support the new navigation task

and failed to counteract a general tendency for people to overly on generally reliable automation – a problem with human–technology integration. Although it is often easiest to blame automation failures on technical problems or on human errors, many problems result from a failure to consider the challenges of designing not just automation, but a joint human–automation system.

Automation fails because the role of the person performing the task is often underestimated, particularly the need for people to compensate for the unexpected. Although automation can handle typical cases it often lacks the flexibility of humans to handle unanticipated situations. Avoiding these failures requires a design process with a focus on the joint human–automation system. In most applications, neither the human nor the automation can accommodate all situations – each has limits. Successful automation design must empower the operator to compensate for the limits of the automation and help the operator capitalize on the capabilities of the automation.

This chapter provides an overview of some of the problems frequently encountered with automation. It then describes how these problems relate to types of automation and what design strategies can help designers achieve the promise of automation. The chapter concludes with future challenges in automation design.

25.1 Automation Problems

Automation is often designed and implemented with a focus on the technical aspects of sensors, algorithms, and actuators. These are necessary but not sufficient design considerations to ensure that automation enhances system performance. Such a technology-centered approach often confronts the human operator with challenges that lead to system failures. Because automation often dramatically extends the influence of operators on the system (e.g., automation makes it possible for one person to do the work of ten), the consequences of these failures can be catastrophic. The factors underlying these failures are complex and interacting. Failures arise because introducing automation changes the type and extent of feedback that operators receive, as well as the nature and structure of tasks. In addition, operators' behavioral, cognitive, and emotional response to these changes can leave the system vulnerable to failure. A technology-centered approach to au-

tomation design often ignores these challenges, and as a consequence, fails to realize the promise of automation.

25.1.1 Problems Due to Changes in Feedback

Feedback is central to control. One reason why automation fails is that automation often dramatically changes the type and extend of the feedback the operator receives. In the context of driving a car, the driver keeps the car in the center of the lane by adjusting the steering wheel according to visual feedback regarding the position of the car on the road and haptic feedback from the forces on the steering wheel. Emerging vehicle technology may automate lane keeping. Such automation may leave the driver with the visual cues, but may remove the haptic cues. Diminished or eliminated feedback is a common occurrence with automation and it can leave people less prepared to intervene if manual control is required [25.7, 8].

Automation can replace the feedback available in manual control with qualitatively different feedback. As an example, introducing automation into paper-making plants moved operators from the plant floor and placed them in control rooms. This move distanced them from the physical process and eliminated the informal feedback associated with vibrations, sounds, and smells that many operators relied upon [25.9]. At best, this change in cues requires operators to relearn how to control the plant. At worst, the instrumentation and associated displays may not have the information needed for operators to diagnose automation failures and intervene appropriately. Automation can also qualitatively shift the feedback from raw system data to processed, integrated information. Although such integrated data can be simple and easily understood, particularly during routine situations, it may also lack the detail necessary to detect and understand system failures. As an example, the bridge of the cruise ship Royal Majesty had an electronic chart that automatically integrated inertial and GPS navigation data to show operators their position relative to their intended path. This high-level representation of the ship's position remained on the intended course even when the underlying GPS data were no longer used and the ship's actual position drifted many miles off the intended route. In this case, the lack of low-level data and of any indication of the integrated data quality left operators without the feedback they needed to diagnose and respond to the failures of the automation.

The diminished feedback that accompanies automation often has a direct influence on a mishap, as illustrated by the case of the Royal Majesty. However, diminished feedback can also act over a longer time period to undermine operators' ability to perform tasks. In situations in which the automation takes on the tasks previously assigned to the operator, the operator's skills may atrophy as they go unexercised [25.10]. Operators with substantial previous experience and well-developed mental models detect disturbances more rapidly than operators without this experience, but extended periods of monitoring automatic control may undermine skills and diminish operators' ability to generate expectations of correct behavior [25.8]. Such deskilling leaves operators without the skills to accommodate the demands of the job if they need to detect failures and assume manual control. This is a particular concern in aviation, where pilots' aircraft handling skills may degrade when they rely on the autopilot. In

response, some pilots disengage the autopilot and fly the aircraft manually to maintain their skills [25.11].

Automation design requires the specification of sensor, algorithm, and actuator characteristics and their interactions. A technology-centered approach might stop there; however, automation that works effectively requires specification of the feedback to the operators. Without careful design, implementing automation can eliminate and change feedback in a way that can undermine the ability of automation to enhance system performance.

25.1.2 Problems Due to Changes in Tasks and Task Structure

One reason for automation is that it can relieve operators of labor-intensive and error-prone tasks. Frequently, however, the situation becomes more complex in that automation does not simply relieve the operator of tasks, it changes the nature of tasks that must be performed. In most instances, this means that automation requires new skills of operators. Often automation eliminates simple physical tasks, and leaves complex cognitive tasks that appear easy. These complex, yet superficially easy, tasks often lead organizations to place less emphasis on training. On ships, training and certification unmatched to the demands of the automation have led to accidents because of the operators' misunderstanding of new radar and collision avoidance systems [25.12]. For example, on the exam used by the US Coast Guard to certify radar operators, 75% of the items assess skills that have been automated and are not required by the new technology [25.13]. The new technology makes it possible to monitor a greater number of ships, thereby enhancing the need for interpretive skills such as understanding the rules of the road that govern maritime navigation and the automation. These are the very skills that are underrepresented on the Coast Guard exam. Though automation may relieve the operator of some tasks, it often leads to new and more complex tasks that require more, not less, training.

Automation can also change the nature and structure of tasks so that easy tasks are made easier and hard tasks harder – a phenomenon referred to as clumsy automation [25.14]. As *Bainbridge* [25.15] notes, designers often leave the operator with the most difficult tasks because the designers found them difficult to automate. Because the easy tasks have been automated, the operator has less experience and an impoverished context for responding to the difficult tasks. In this situation, automation has the effect of both reducing workload during already low-workload periods and increasing it during high-workload periods; for example, a flight management system tends to make the low-workload phases of flight (e.g., straight and level flight or a routine climb) easier, but high-workload phases (e.g., the maneuvers in preparation for landing) more difficult as pilots have to share their time between landing procedures, communication, and programming the flight management system. Such effects are seen not only in aviation but also in many other domains, such as the operating room [25.16, 17].

The effects of clumsy automation often occur at the level of individual operators and over the span of several minutes, but such effects can also occur across teams of operators over hours or days of operation. Such macrolevel clumsy automation is evident in maritime operations, where automation used for open-ocean sailing reduces the task requirements of the crew, prompting reductions in the crew size. In this situation, automation can have the consequence of making the easy part of the voyage (e.g., open-ocean sailing) easier and the hard part (e.g., port activities) harder [25.18]. Avoiding clumsy automation requires a broad consideration of how automation affects the task structure of operators.

Because automation changes the task structure, new forms of human error often emerge. Ironically, managers and system designers introduce automation to eliminate human error, but new and more disastrous errors often result, in part because automation extends the scope of and reduces the redundancy of human actions. As a consequence, human errors may be more likely to go undetected and do more damage; for example, a flight-planning system for pilots can induce dramatically poor decisions because the automation assumes weather forecasts represent reality and lacks the flexibility to consider situations in which the actual weather might deviate from the forecast [25.19].

Automation-induced errors also occur because the task structure changes in a way that undermines collaboration between operators. Effective system performance involves performing both formal and informal tasks. Informal tasks enable operators to compensate for the limits of the formal task structure; for example, with paper charts mariners will check each others' work, share uncertainties, and informally train each other as positions are plotted [25.20]. Eliminating these informal tasks can make it more difficult to detect and recover from errors, such as the one that led to the grounding of the Royal Majesty. Automation can also disrupt the cooperation between operators reflected in these informal tasks. Cooperation occurs when a person acts in a way that is in the best interests of the group even when it is contrary to his or her own best interests. Most complex, multiperson systems depend on cooperation. Automation can disrupt interactions between people and undermine the ability and willingness of one operator to compensate for another. Because automation also acts on behalf of people, it can undermine cooperation by giving one operator the impression that another operator is acting in a competitive manner, even though the automation's behavior may be due to a malfunction [25.21].

Automation does not simply eliminate tasks once performed by the operator. It changes the task structure and creates new tasks that need to be supported, thereby opening the door to new types of error. Contrary to the expectations of a technology-centered approach to automation design, introducing automation makes it more rather than less important to consider the operators' tasks and role.

25.1.3 Problems Due to Operators' Cognitive and Emotional Response to Changes

Automation sometimes causes problems because it changes operators' feedback and tasks. Operators' cognitive and emotional responses to these changes can amplify these problems; for example, as automation changes the operator's task from direct control to monitoring, the operator may be more prone to direct attention away from the monitoring task, further diminishing the feedback the operator receives from the system. The tendency to trust and complacently rely on automation, particularly during multitask situations, may underlie this tendency to disengage from the monitoring task [25.22–24].

People are not passive recipients of the changes to the task structure that automation makes. Instead, people adapt to automation and this adaptation leads to a new task structure. One element of this adaptation is captured by the ideas of reliance and compliance [25.25]. Reliance refers to the degree to which operators depend on the automation to perform a function. Compliance refers to the degree to which automation changes the operators' response to a situation. Inappropriate reliance and compliance are common automation problems that occur when people rely on or comply with automation in situations where it performs poorly, or when people fail to capitalize on its capabilities [25.26].

Maladaptive adaptation generally, and inappropriate reliance specifically, depends, in part, on operators' attitudes, such as trust and self-confidence [25.27, 28]. In the context of operator reliance on automation, trust has been defined as an attitude that the automation will help achieve an operator's goals in a situation characterized by uncertainty and vulnerability [25.29]. Several studies have shown that trust is a useful concept in describing human–automation interaction, both in naturalistic [25.9] and in laboratory settings [25.30–33].

These and other studies show that people tend to rely on automation they trust and to reject automation they do not trust [25.29]. As an example, the difference in operators' trust in a route-planning aid and their self-confidence in their own ability was highly predictive of reliance on the aid [25.34]. People respond socially to technology in a way that is similar to how they respond to other people [25.35]. *Sheridan* had a similar insight, and suggested that, just as trust mediates relationships between people, it may also mediate relationships between people and automation [25.36, 37]. Because trust often has a powerful effect on mediating relationships between people, trust might exert a similarly strong effect on mediating reliance and compliance with automation [25.38–42].

Inappropriate reliance often stems from a failure of trust to match the true capabilities of the automation. Calibration refers to the correspondence between a person's trust in the automation and the automation's capabilities [25.29]. Overtrust is poor calibration in which trust exceeds system capabilities; with distrust, trust falls short of automation capabilities. Trust often responds to automation as one might expect; it increases over time as automation performs well and declines when automation fails. Importantly, however, trust does not always follow the changes in automation performance. Often, it is poorly calibrated. Trust displays inertia and changes gradually over time rather than responding immediately to changes in automation performance. After a period of unreliable performance, trust is often slow to recover, remaining low even when the automation performs well [25.43]. More surprisingly, trust sometimes depends on surface features of the system that seem unrelated to its capabilities, such as the colors and layout of the interface [25.44–46].

Attitudes such as trust and the associated influence on reliance can exacerbate automation problems such as clumsy automation. As noted earlier, clumsy automation occurs when automation makes easy tasks easier and hard tasks harder. Inappropriate trust can make automation more clumsy because it leads operators to be more willing to delegate tasks to the automation during periods of low workload, compared with periods of high workload [25.15]. This observation demonstrates that clumsy automation is not simply a problem of task structure, but one that depends on operator adaptation that is mediated by attitudes, such as trust.

The automation-related problems associated with inappropriate trust often stem from operators' shift from being a direct controller to a monitor of the automation. This shift also changes how operators receive feedback. Automation shifts people from direct involvement in the action–perception loop to supervisory control [25.47, 48]. Passive observation associated with supervisory control is qualitatively different than active monitoring associated with manual control [25.49, 50]. In manual control, perception directly supports control, and control actions guide perception [25.51]. Monitoring automation disconnects the operators' actions from actions on the system. Such disconnects can undermine the operator's mental model (i. e., their working knowledge of system dynamics, structure, and causal relationships between components), leaving the mental model inadequate to guide expectations and control [25.52, 53].

The shift from direct controller to supervisory controller can also have subtle but important effects on behavior as operators adapt to the automation. Over time automation can unexpectedly shift operators' safety norms and behavior relative to safety boundaries. Behavioral adaptation describes this effect and refers to the tendency of operators to adapt to the new capabilities of the automation in which they change their behavior so that the potential safety benefits of the technology are not realized. Automation intended by designers to enhance safety may instead lead operators to reduce effort and leave safety unaffected or even diminished. Behavioral adaptation occurs at the individual [25.54–56], organizational [25.57], and societal levels [25.58].

Antilock brake systems (ABS) for cars demonstrate behavioral adaptation. ABS automatically modulates brake pressure to maintain maximum brake force without skidding. This automation makes it possible for drivers to maintain control in extreme crash avoidance maneuvers, which should enhance safety. However, ABS has not produced the expected safety benefits. One reason is that drivers of cars with ABS tend to drive less conservatively, adopting higher speeds and shorter following distances [25.59]. Vision enhancement systems provide another example of behavioral adaption. These systems make it possible for drivers to see more at night – a potential safety enhancement; however, drivers tend to adapt to the vision systems by increasing their speed [25.60].

A related form of behavioral adaptation that undermines the benefits of automation is the phenomenon in which the presence of the automation causes a diffusion of responsibility and a tendency to exert less effort when the automation is available [25.61,62]. As a result, people tend to commit more omission errors (failing to detect events not detected by the automation) and more commission errors (incorrectly concurring with erroneous detection of events by the automation) when they work with automation. This effect parallels the adaptation of people when they work in groups; diffusion of responsibility leads people to perform more poorly when they are part of a group compared with individually [25.63].

The issues noted above have primarily addressed the direct performance problems associated with automation. Job satisfaction is another human–automation interaction issue that goes well beyond performance to consider the morale and moral implications of the worker whose job is being changed by automation [25.64]. Automation that is introduced merely because it increases the profit of the company may not necessarily be well received. Automation often has the effect of deskilling a job, making skills that operators worked for years to perfect suddenly obsolete. Properly implemented, automation should reskill workers and make it possible for them to leverage their old skills into new ones that are extended by the support of the automation. Many operators are highly skilled and proud of their craft; automation can either empower or demoralize them [25.9]. Demoralized operators may fail to capitalize on the potential of an automated system.

The cognitive and emotional response of operators to automation can also compromise operators' health. If automation creates an environment in which the demands of the work increase, but the decision latitude decreases, it may then lead to problems ranging from increased heart disease to increased incidents of depression [25.65]. However, if automation extends the capability of the operator and gives him or her greater decision latitude, job satisfaction and health can improve. As an example of improved satisfaction, night-shift operators who had greater decision latitude than day-shift operators leveraged their increased latitude to learn how to manage the automation more effectively [25.9].

Automation problems can be described independently, but they often reflect an interacting and dynamic process [25.66]. One problem can lead to another through positive feedback and vicious cycles. As an example, inadequate training may lead the operator to disengage from the monitoring task. This disengagement leads to poorly calibrated trust and overreliance, which in turn leads to skill loss and further disengagement. A similar dynamic exists between clumsy automation and automation-induced errors. Clumsy automation produces workload peaks, which increase the chance of mode and configuration errors. Recovering from these errors can further increase workload, and so on. Designing and implementing automation without regard for human capabilities and defining the human role as a byproduct is likely to initiate these negative dynamics.

25.2 Characteristics of the System and the Automation

The likelihood and consequences of automation-related problems depend on the characteristics of the automation and the system being controlled. Automation is not a homogenous technology. Instead, there are many types of automation and each poses different design challenges. As an example, automation can highlight, alert, filter, interpret, decide, and act for the operator. It can assume different degrees of control and can operate over timescales that range from milliseconds to months. The type of automation and the operating environment interact with the human to produce the problems just discussed. As an example, if only a single person manages the system then diminished cooperation and collaboration are not a concern. Some important system and automation characteristics include:

- Automation as information processing stages
- Automation authority and autonomy
- Complexity and observability
- Time-scale and multitasking demands
- Agent interdependencies
- Interaction with environment.

25.2.1 Automation as Information Processing Stages

Defining automation in terms of information processing stages describes it according to the information processing functions of the person that it supports or replaces. Automation can sense the world, analyze information, identify appropriate responses to states of the world or control actuators to change those states [25.67]. Information acquisition automation refers to technology that replaces the process of human perception. Such automation highlights targets [25.68, 69], provides alerts and warnings [25.70, 71], organizes, prioritizes, and filters information. Information analysis automation refers to technology that supplants the interpretation of a situation. An example of this type of automation is a system that critiques a diagnosis generated by the operator [25.72]. Action selection automation refers to technology that combines information in order to make decisions on behalf of the operator. Unlike information acquisition and analysis, action selection automation suggests or decides on actions using assumptions about the state of the world and the costs and values of the possible options [25.73]. Action implementation automation supplants the operators' activity in executing a response. The types of automation at each of these four stages of information process can differ according to degree of authority and autonomy.

Automation authority and autonomy concern the degree to which the automation can influence the system [25.74]. Authority reflects the extent to which the automation amplifies the influence of operators' actions and overrides the actions of other agents. One facet of authority concerns whether or not operators interact with automation by switching between manual and automatic control. With some automation, such as cruise control in cars, drivers simply engage or disengage the automation, whereas automation on the flight deck involves managing a complex network of modes that are appropriate for some situations and not for others. Interacting with such flight-deck automation requires the operator to coordinate multiple goals and strategies to select the mode of operation that fits the situation [25.75]. With such multilevel automation the idea of manual control may not be relevant, and so the issues of skill loss and other challenges with manual intervention may be of less concern. The problems with high-authority, multilevel automation are more likely to be those associated with mode confusion and configuration errors.

Autonomy reflects the degree to which automation acts without operator knowledge or opportunity to intervene. *Billings* [25.11] describes two levels of autonomy: *management by consent*, in which the automation acts only with the consent of the operator, and *management by exception*, in which automation initiates activities autonomously. As another example, automation can either highlight targets [25.68, 69], filter information, or provide alerts and warnings [25.70, 71]. Highlighting targets exemplifies a relatively low degree of autonomy because it preserves the underlying data and allows operators to guide their attention to the information they believe to be most critical. Filtering exemplifies a higher degree of autonomy because operators are forced to attend to the information the automation deems relevant. Alerts and warnings similarly exemplify a relatively high level of autonomy because they guide the operator's attention to automation-dictated information and environmental states. High levels of authority and autonomy make automation appear to act as an independent agent, even if the designers had not intended operators to perceive it as such [25.76]. High levels of these two automation characteristics are an important cause of clumsy automation and mode error and can also undermine cooperation between people [25.77].

25.2.2 Complexity and Observability

Complexity and observability refer to the degrees of freedom of the automation algorithms and how directly that complexity is revealed to the operator [25.74]. As automation becomes increasingly complex it can transition from what operators might consider a tool that they use to act on the environment to an agent that acts as a semiautonomous partner. According to the agent metaphor, the operator no longer acts directly on the environment, but acts through an intermediary agent [25.78] or intelligent associate [25.79]. As an agent, automation initiates actions that are not in direct response to operators' commands. Automation that acts as an agent is typically very complex and may or may not be observable. One of the greatest challenges with automated agents is that of mutual intelligibility. Instructing the agent to perform even simple tasks can be onerous, and agents that try to infer operators' intent and act autonomously can surprise operators who might lack accurate mental models of agent behavior. One approach is for the agents to learn and adapt to the characteristics of the operator through a process of remembering what they have been being told to do in similar situations [25.80]. After the agent completes a task it can be equally challenging to make the results observable and meaningful to the operator [25.78]. Because of these characteristics, agents are most useful for highly repetitive and simple activities, where the cost of failure is limited. In high-risk situations, constructing effective management strategies and providing feedback to clarify agent intent and communicate behavior becomes critical [25.75, 81]. The challenges associated with agents reflect a general tradeoff with automation design: more complex automation is often more capable, but less understandable. As a consequence, even though more complex automation may appear superior, the performance of resulting human–automation system may be inferior to that of a simpler, less capable version of the automation.

25.2.3 Time–Scale and Multitasking Demands

This distinction concerns the tempo of the interactions with the automation. The timescale of automation varies dramatically, from decision-support systems that guide corporate strategies over months and years to antilock brake systems that modulate brake pressure over milliseconds. These distinctions can be described in terms of strategic, tactical, and operational automation. Strategic automation concerns balancing values and costs, as well as defining goals; tactical automation, on the other hand, involves setting priorities and coordinating tasks. In contrast, operational automation concerns the moment-to-moment perception of system state and adjustment. With operational automation, operators can experience substantial time pressure as the tempo of activity, on the order of milliseconds to seconds, exceeds their capacity to monitor the automation and still respond in a timely manner to its limits [25.82, 83].

25.2.4 Agent Interdependencies

Agent interdependencies describe how tightly coupled the work of one operator or element of automation is with another [25.6, 57]. In some situations, automation might directly support work of a team of people and in other situations automation might support the activity of a person that has little interaction with others. An important source of automation-related problems is the assumption that automation affects only one person or one set of tasks, causing important interactions with other operators to be neglected. Often seemingly independent tasks may actually be coupled, and automation has a tendency to tighten this coupling. As an example, on the surface, adaptive cruise control affects only the individual driver who is using the system. Because adaptive cruise control responds to the behavior of the vehicle ahead, however, its behavior cannot be considered without taking into account the surrounding traffic dynamics. Failing to consider these interactions of intervehicle velocity changes can lead to oscillations and instabilities in the traffic speed, potentially compromising driver safety [25.84, 85]. Similar failures occur in supply chains, as well as in petrochemical processes where people and automation sometimes fail to coordinate their activities [25.86]. Designing for such situations requires a change in perspective from one centered on a single operator and a single element of automation to one that considers multi-operator–multi-automation interactions [25.87, 88].

25.2.5 Environment Interactions

Interaction with the environment refers to the degree to which the automation system is isolated from or interactive with the surrounding environment. The environmental context can affect the reliability and behavior of the automation, the operator's perception of the automation, and thus the overall effectiveness of the human–automation partnership [25.89–92]. An explicit environmental representation is necessary to understand the joint human–automation performance [25.89].

25.3 Application Examples and Approaches to Automation Design

The previous section described some important characteristics of automation and systems that contribute to automation-related problems. These distinctions help identify design approaches to minimize these problems. This section describes specific strategies for designing effective automation, which include:

- Function allocation with Fitts' list
- Operator–automation simulation and analysis
- Representation aiding and enhanced feedback
- Expectation matching and automation simplification.

25.3.1 Fitts' List and Function Allocation

Function allocation with the Fitt's list is a long-standing technique for identifying the role of operators and automation. This approach assesses each function and whether a person or automation might be best suited to performing it [25.93, 94]. Functions better performed by automation are automated and the operator remains responsible for the rest, and for compensating for the limits of the automation. The relative capability of the automation and human depend on the stage of automation [25.95].

Applying a Fitts' list to determine an appropriate allocation of function has, however, substantial weaknesses. One weakness is that any description of functions is a somewhat arbitrary decomposition of activities that can mask complex interdependencies. As a consequence, automating functions as if they were independent has the tendency to fractionate the operator's role, leaving the operator with an incoherent collection of functions that were too difficult to automate [25.15]. Another weakness is that this approach neglects the tendency for operators to use automation in unanticipated ways because automation often makes new functions possible [25.96]. Another challenge with this general approach is that it often carries the implicit assumption that automation can substitute for functions previously performed by operators and that operators do not need to be supported in performing functions allocated to the automation [25.97]. This substitution-based function allocation fails to consider the qualitative change automation can bring to the operators' work, and the adaptive nature of the operator.

As a consequence of these challenges, the Fitts' list provides only general guidance for automation design and has been widely recognized as problematic [25.73, 95, 97]. Ideally, the function allocation process should not focus on what functions should be allocated to the automation or to the human, but should identify how the human and the automation can complement each other in jointly satisfying the functions required for system success [25.98].

Although imperfect, the Fitts' list approach has some general considerations that can improve design. People tend to be effective in perceiving patterns and relationships amongst data and less so with tasks requiring precise repetition [25.64]. Human memory tends to organize large amounts of related information in a network of associations that can support effective judgments. People also adapt, improvise, and accommodate unexpected variability. For these reasons it is important to leave the *big picture* to the human and the *details* to the automation [25.64].

25.3.2 Operator–Automation Simulation

Operator–automation simulation refers to computer-based techniques that explore the space of operator–automation interaction to identify potential problems. Discrete event simulation tools commonly used to evaluate manufacturing processes are well-suited to operator–automation analysis. Such techniques provide a rough estimate of some of the consequences of introducing automation into complex dynamic systems. As an example, simulation of a supervisory control situation made it possible to assess how characteristics of the automation interacted with the operating environment to govern system performance [25.99]. This analysis showed that the time taken to engage the automation interacted with the dynamics of the environment to undermine the value of the automation such that manual control was more appropriate than engaging the automation.

Although discrete event simulation tools can incorporate cognitive mechanisms and performance constraints, developing this capacity requires substantial effort. For automation analysis that requires a detailed cognitive representation, cognitive architectures, such as adaptive control of thought-rational (ACT-R), offer a promising approach [25.100]. ACT-R is a useful tool for approximating the costs and benefits of various automation alternatives when a simple discrete event simulation does not provide a sufficiently detailed representation of the operator [25.101].

Simulation tools can be used to explore the potential behavior of the joint human–automation system, but may not be the most efficient way of identifying potential human–automation mismatches associated with inadequate mental models and automation-related errors. Network analysis techniques offer an alternative. State-transition networks can describe operator–automation behavior in terms of a finite number of states, transitions between those states, and actions. Figure 25.1 provides an example presentation, defining at a high level the behavior of adaptive cruise control (ACC). This formal modeling language makes it possible to identify automation problems that occur when the interface or the operator's mental model is inadequate to manage the automation [25.102]. Figure 25.2 shows how combining the concurrent processes of the ACC model with its internal states and transitions with the associated driver model of the ACC's behavior reveals

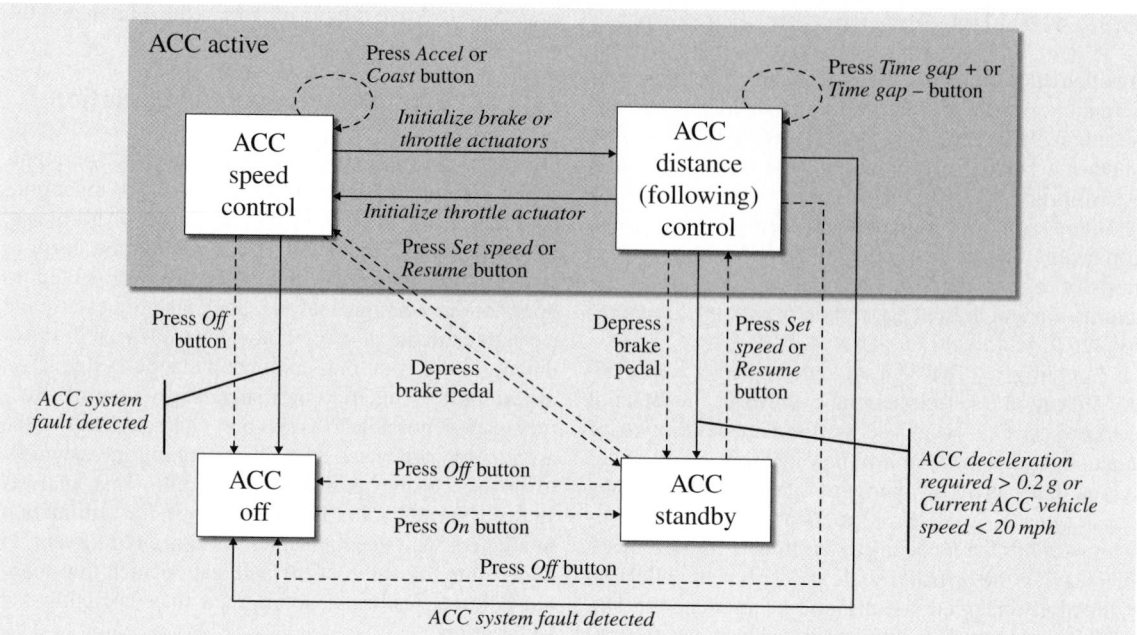

Fig. 25.1 ACC states and transitions. *Dashed lines* represent driver-triggered transitions. *Solid lines* represent ACC-triggered transitions

mismatches. These mismatches can cause automation-related errors and surprises to occur. More specifically, when the automation model enters a particular state and the operator's model does not include this state then the analysis predicts that the associated ambiguity will surprise operators and lead to errors [25.103]. Such ambiguities have been discovered in actual aircraft autopilot systems, and network analysis can identify how to avoid them with improvements to the interface and training materials [25.103].

25.3.3 Enhanced Feedback and Representation Aiding

Enhanced feedback and representation aiding can help prevent problems associated with inadequate feedback that range from developing appropriate trust and clumsy automation to the out-of-the-loop phenomenon. Automation typically lacks adequate feedback [25.104]. Providing sufficient feedback without overwhelming the operator is a critical design challenge. Poorly presented or excessive feedback can increase operator workload and undermine the benefits of the automation [25.105].

A promising approach to avoid overloading the operator is to provide feedback through sensory channels that are not otherwise used (e.g., haptic, tactile, and auditory) to prevent overload of the more commonly used visual channel. Haptic feedback (i.e., vibration on the wrist) has proven more effective in alerting pilots to mode changes of cockpit automation than visual cues [25.106]. Pilots receiving visual alerts only detected 83% of the mode changes, but those with haptic warnings detected 100% of the changes. Importantly, the haptic warnings did not interfere with performance of concurrent visual tasks. Even within the visual modality, presenting feedback in the periphery helped pilots detect uncommanded mode transitions and such feedback did not interfere with concurrent visual tasks any more than currently available automation feedback [25.107]. Similarly, *Seppelt* and *Lee* [25.108] combined a more complex array of variables in a peripheral visual display for ACC. Figure 25.3 shows how this display includes relevant variables for headway control (i.e., time headway, time-to-collision, and range rate) relative to the operating limits of the ACC. This display promoted faster failure detection and more appropriate engagement strategies compared with the standard ACC interface. Although promising, haptic, auditory and peripheral visual displays cannot convey the detail possible in visual displays, making it difficult to convey the complex relationships that some-

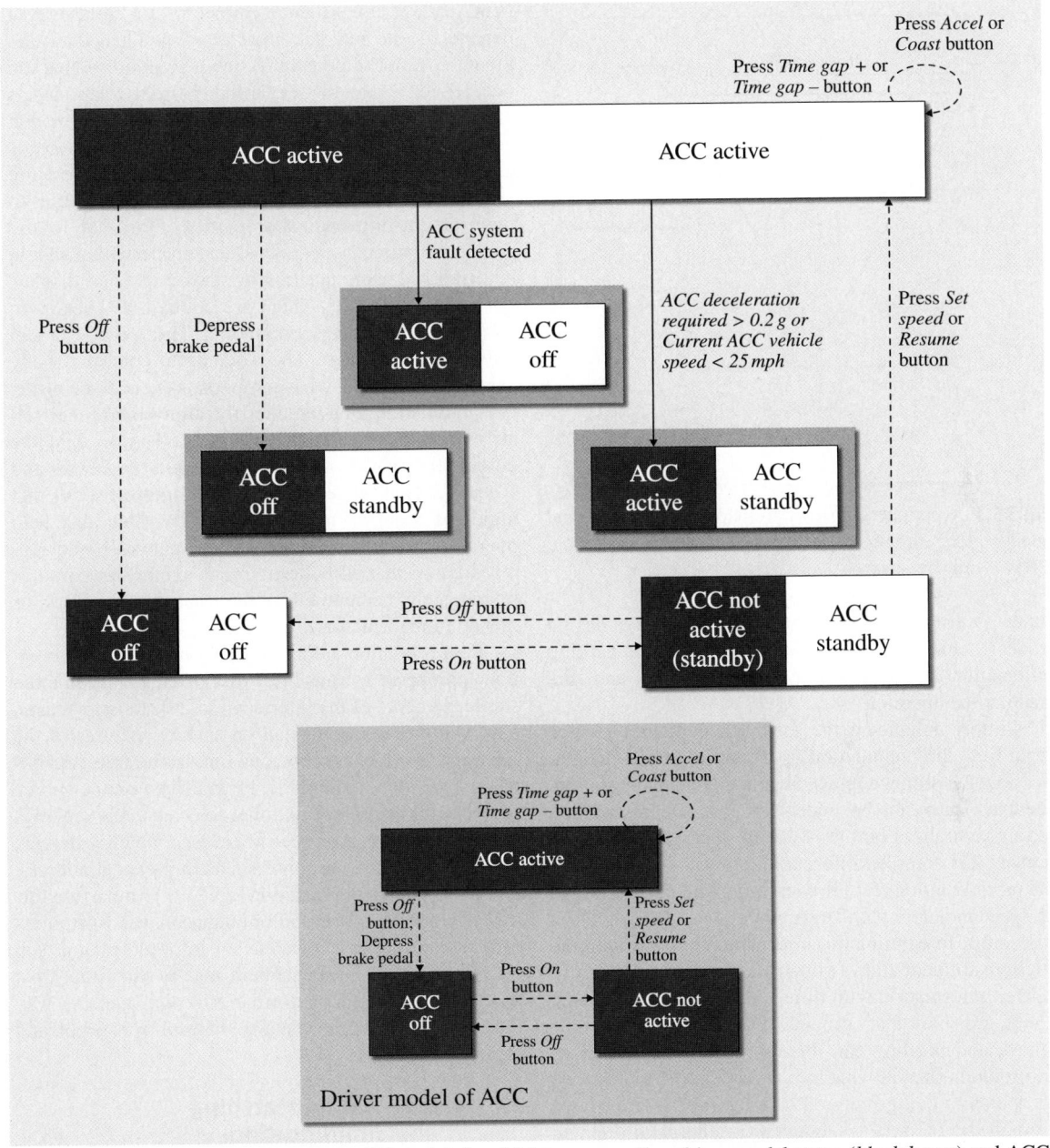

Fig. 25.2 Composite of the driver and ACC models in which corresponding driver model states (*black boxes*) and ACC model states (*white boxes*) are combined into state pairs. Error states, or model mismatches, occur when a particular transition leads to discrepant states. Composite states *ACC not active (standby)/ACC off*, *ACC off/ACC standby*, and *ACC active/ACC standby* are error states. The driver is unaware of the shift of the ACC system into standby when deceleration and vehicle speed limits are reached, and of the ACC system disengaging when system faults are detected, as neither state change is clearly communicated to the driver. The state change that results from the driver depressing the brake pedal is similarly ambiguous

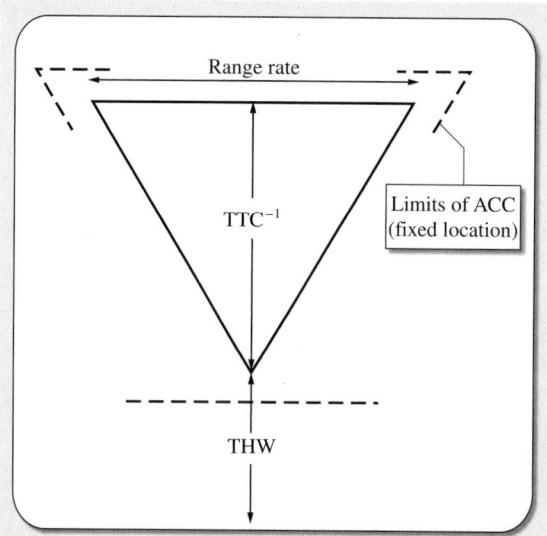

Fig. 25.3 A peripheral display to help drivers understand adaptive cruise control [25.108] (TTC – time-to-collision; THW – time headway

times govern automation behavior. An important design tradeoff emerges: provide sufficient detail regarding automation behavior, but avoid overloading and distracting the operator.

Simply enhancing the feedback operators receive regarding the automation is sometimes insufficient. Without the proper context, abstraction, and integration, feedback may not be understandable. Representation aiding capitalizes on the power of visual perception to convey this complex information; for example, graphical representations for pilots can augment the traditional airspeed indicator with target airspeeds and acceleration indicators. Integrating this information into a traditional flight instrument allows pilots to assimilate automation-related information with little additional effort [25.87]. Using a display that combines pitch, roll, altitude, airspeed, and heading can directly specify task-relevant information such as what is *too low* [25.109] as opposed to operators being required to infer such relationships from the set of variables. Integrating automation-related information with traditional displays and combining low-level data into meaningful information can help operators understand automation behavior.

In the context of process control, *Guerlain* and colleagues [25.110] identified three specific strategies for visual representation of complex process control algorithms. First, create visual forms whose emergent features correspond to higher-order relationships. Emergent features are salient symmetries or patterns that depend on the interaction of the individual data elements. A simple emergent feature is *parallelism* that can occur with a pair of lines. Higher-order relationships are combinations of the individual data elements that govern system behavior. The boiling point of water is a higher-order relationship that depends on temperature and pressure. Second, use appropriate visual features to represent the dimensional properties of the data; for example, magnitude is a dimensional property that should be displayed using position or size on a visual display, not color or texture, which are ambiguous cues as to an increase or decrease in amount. Third, place data in a meaningful context. The meaningful context for any variable depends on what comparisons need to be made. For automation, this includes the allowable ranges relative to the current control variable setting, and the output relative to its desired level. Similarly, *Dekker* and *Woods* [25.97] suggest event-based representations that highlight changes, historical representations that help operators project future states, and pattern-based representations that allow operators to synthesize complex relationships perceptually rather than through arduous mental transformations.

Representation aiding helps operators trust automation appropriately. However, trust also depends on more subtle elements of the interface [25.29]. In many cases, trust and credibility depend on surface features of the interface that have no obvious link to the true capabilities of the automation [25.111, 112]. An online survey of over 1400 people found that for web sites, credibility depends heavily on *real-world feel*, which is defined by factors such as response speed, a physical address, and photos of the organization [25.113]. Similarly, a formal photograph of the author enhanced trustworthiness of a research article, whereas an informal photograph decreased trust [25.114]. These results show that trust tends to increase when information is displayed in a way that provides concrete details that are consistent and clearly organized.

25.3.4 Expectation Matching and Simplification

Expectation matching and simplification help operators understand automation by using algorithms that are more comprehensible. One strategy is to simplify the automation by reducing the number of functions, modes, and contingencies [25.115]. Another is to match its algorithms to the operators' mental model [25.116]. Automation designed to perform in a manner consistent with operators' mental model, preferences, and expectations can make it easier for operators to recognize failures and intervene. Expectation matching and simplification are particularly effective when a technology-centered approach has created an overly complex array of modes and features.

ACC is a specific example of where matching the mental model of an operator to the automation's algorithms may be quite effective. Because ACC can only apply moderate levels of braking, drivers must intervene if the car ahead brakes heavily. If drivers must intervene, they must quickly enter the control loop because fractions of a second can make the difference in avoiding a collision. If the automation behaves in a manner consistent with drivers' expectations, drivers will be more likely to detect and respond to the operational limits of the automation quickly [25.116]. *Goodrich* and *Boer* [25.116] designed an ACC algorithm consistent with drivers' mental models such that ACC behavior was partitioned according to perceptually relevant variables of inverse time-to-collision and time headway. Inverse time-to-collision is the relative velocity divided by the distance between the vehicles. Time headway is the distance between the vehicles divided by the velocity of the driver's vehicle. Using these variables it is possible to identify a perceptually salient boundary that separates routine speed regulation and headway maintenance from active braking associated with collision avoidance.

For situations in which the metaphor for automation is an agent, the mental model people may adopt to understand the automation is that of a human collaborator. Specifically, *Miller* [25.117] suggests that computer etiquette may have an important influence on human–automation interaction. Etiquette may influence trust because category membership associated with adherence to a particular etiquette helps people to infer how the automation will perform. Some examples of automation etiquette are for the automation to make it

easy for operators to override and recover from errors, to enable interaction features only when and if necessary, to explain what is being done and why, to interrupt operators only in emergency situations, and to provide information that is unique to the information known by the operator.

Developing automation etiquette could promote appropriate trust, but also has the potential to lead to inappropriate trust if people infer inappropriate category memberships and develop distorted expectations regarding the capability of the automation. Even in simple interactions with technology, people often respond as they would to another person [25.35, 118]. If anticipated, this tendency could help operators develop appropriate expectations regarding the behavior of the automation; however, unanticipated anthropomorphism could lead to surprising misunderstandings of the automation.

An important prerequisite for designing automation according to the mental model of the operator is the existence of a consistent mental model. Individual differences may lead to many different mental models and expectations. This is particularly true for automation that acts as an agent, in which a mental-model-based design must conform to complex social and cultural expectations. In addition, the mental model must be consistent with the physical constraints of the system if the automation is to work properly [25.119]. Mental models often contain misconceptions, and transferring these to the automation could lead to serious misunderstandings and automation failures. Even if an operator's mental model is consistent with the system constraints, automation based on such a mental model may not achieve the same benefits as automation based on more sophisticated algorithms. In this case, designers must consider the tradeoff between the benefits of a complex control algorithm and the costs of an operator not understanding that algorithm. Enhanced feedback and representation aiding can mitigate this tradeoff.

25.4 Future Challenges in Automation Design

The previous section outlined strategies that can make the operator–automation partnership more effective. As illustrated by the challenges in applying the Fitts' list, the application of these strategies, either individually or collectively, does not guarantee effective automation. In fact, the rapid advances in software and hardware development, combined with an ever expanding range

of applications, make future problems with automation likely. The following sections highlight some of these emerging challenges. The first concerns the demands of managing swarm automation, in which many semiautonomous agents work together. The second concerns large, interconnected networks of people and automation, in which issues of cooperation and competition become critical. These examples represent some emerging challenges facing automation design.

25.4.1 Swarm Automation

Swarm automation consists of many simple, semiautonomous entities whose emergent behavior provides a robust response to environmental variability. Swarm automation has important applications in a wide range of domains, including planetary exploration, unmanned aerial vehicle reconnaissance, land-mine neutralization, and intelligence gathering; in short, it is applicable in any situation in which hundreds of simple agents might be more effective than a single, complex agent. Biology-inspired robotics provides a specific example of swarm automation. Instead of the traditional approach of relying on one or two larger robots, they employ swarms of insect robots [25.120, 121]. The swarm robot concept assumes that small robots with simple behaviors can perform important functions more reliably and with lower power and mass requirements than can larger robots [25.122–124]. Typically, the simple algorithms controlling the individual entity can elicit desirable emergent behaviors in the swarm [25.125, 126]. As an example, the collective foraging behavior of honeybees shows that agents can act as a coordinated group to locate and exploit resources without a complex central controller.

In addition to physical examples of swarm automation, swarm automation has potential in searching large complex data sets for useful information. Current approaches to searching such data sources are limited. People miss important documents, disregard data that is a significant departure from initial assumptions, misinterpret data that conflicts with an emerging understanding, and disregard more recent data that could revise interpretation [25.127]. The parameters that govern discovery and exploitation of food sources for ants might also apply to the control of software agents in their discovery and exploitation of information. Just as swarm automation might help explore physical spaces, it might also help explore information spaces [25.128].

The concept of hortatory control describes some of the challenges of controlling swarm automation. Hortatory control describes situations where the system being controlled retains a high degree of autonomy and operators must exert indirect rather than direct control [25.129]. Interacting with swarm automation requires people to consider swarm dynamics and not just the behavior of the individual agents. In these situations, it is most useful for the operator to control parameters affecting group rather than individual agents and to

receive feedback about group rather than individual behavior. Parameters for control might include the degree to which each agent tends to follow successful agents (positive feedback), the degree to which they follow the emergent structure of their own behavior (stigmergy), and the amount of random variation that guides their paths [25.130]. In exploration, a greater amount of random variation will lead to a more complete search, and a greater tendency to follow successful agents will speed search and exploitation [25.131]. Swarm automation has great potential to extend human capabilities, but only if a thorough empirical and analytic investigation identifies the display requirements, viable control mechanisms, and the range of swarm dynamics that can be comprehended and controlled by humans [25.132].

25.4.2 Operator–Automation Networks

Complex operator–automation networks emerge as automation becomes more pervasive. In this situation, the appropriate unit of analysis shifts from a single operator interacting with a single element of automation to that of multiple operators interacting with multiple elements of automation. Important dynamics can only be explained with this more complex unit of analysis. The factors affecting microlevel behavior may have unexpected effects on macrolevel behavior [25.133]. As the degree of coupling increases, poor coordination between operators and inappropriate reliance on automation has greater consequences for system performance [25.6].

Supply chains represent an increasingly important example of multi-operator–multi-automation systems. A supply chain is composed of a network of suppliers, transporters, and purchasers who work together, usually as a decentralized virtual company, to convert raw materials into products. The growing popularity of supply chains reflects the general trend of companies to move away from vertical integration, where a single company converts raw materials into products. Increasingly, manufacturers rely on supply chains [25.134] and attempt to manage them with automation [25.86].

Supply chains suffer from serious problems that erode their promised benefits. One is the bullwhip effect, in which small variations in end-item demand induces large-order oscillations, excess inventory, and back-orders [25.135]. The bullwhip effect can undermine a company's efficiency and value. Automation that forecasts demands can moderate these oscillations [25.136, 137]. However, people must trust and rely on that automation, and substantial cooperation between supply-chain members must exist to share such information.

Vicious cycles also undermine supply-chain performance, through an escalating series of conflicts between members [25.138]. Vicious cycles can have dramatic negative consequences for supply chains; for example, a strategic alliance between Office Max and Ryder International Logistics devolved into a legal fight in which Office Max sued Ryder for US $21.4 million and then Ryder sued Office Max for US $75 million [25.139]. Beyond the legal costs, these breakdowns threaten competitiveness and undermine the market value of the companies involved [25.134]. Vicious cycles also undermine information sharing, which can exacerbate the bullwhip effect. Even with the substantial benefits of cooperation, supply chains frequently fall into a vicious cycle of diminishing cooperation.

Inappropriate use of automation can contribute to both vicious cycles and the bullwhip effect, but has received little attention. A recent study used a simulation model to examine how reliance on automation influences cooperation and how sharing two types of automation-related information influences cooperation between operators in the context of a two-manufacturer one-retailer supply chain [25.21]. This study used a decision field-theoretic model of the human operator [25.140, 141] to assess the effects of automation failures on cooperation and the benefit of sharing automation-related information in promoting cooperation. Sharing information regarding automation performance improved operators' reliance on automation, and the more appropriate reliance promoted cooperation by avoiding unintended competitive behaviors caused by inappropriate use of automation. Sharing information regarding the reliance on automation increased willingness to cooperate even when the other occasionally engaged in competitive behavior. Sharing information regarding the operators' reliance on automation led to a more charitable interpretation of the other's intent and therefore increased trust in the other operator. The consequence of enhanced trust is an increased chance of cooperation. Figure 25.4 shows that these two types of information sharing influence cooperation and result in an additive improvement in cooperation. This preliminary simulation study showed that cooperation depends on the appropriate use of automation and that sharing automation-related information can have a profound effect on cooperation, a result that merits verification with experiments with human subjects.

The interaction between automation, cooperation, and performance seen with supply-chain management

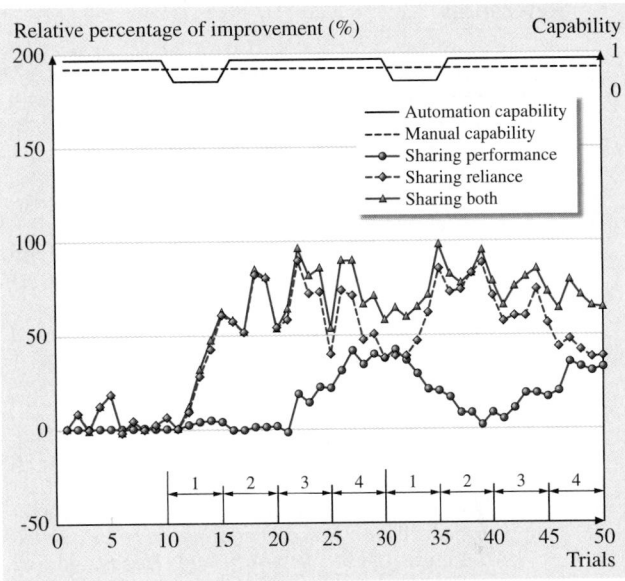

Fig. 25.4 The effect of sharing information regarding the performance of the automation and reliance on the automation [25.21]

may also apply to other domains; for example, power-grid management involves a decentralized network that makes it possible to efficiently supply the USA with power, but it can fail catastrophically when cooperation and information-sharing breaks down [25.142]. Similarly, datalink-enabled air-traffic control makes it possible for pilots to negotiate flight paths efficiently, but it can fail when pilots do not cooperate or have trouble anticipating the complex dynamics of the system [25.143, 144]. Overall, technology is creating many highly interconnected networks that have great potential, but also raise important concerns. Resolving these concerns partially depends on designing effective multi-operator–multi-automation interactions.

Swarm automation and complex operator–automation networks pose challenges beyond those of traditional systems and require new design strategies. The automation design strategies described earlier, such as function allocation, operator–automation simulation, representation aiding, and expectation matching are somewhat limited in addressing the new challenges of swarm automation and complex operator–automation networks. A particular challenge in automation design is developing analytic tools, interface designs, and interaction concepts that consider issues of cooperation and coordination in operator–automation interactions. For further discussion on the automation interactions and interface design refer to Chap. 34.

References

25.1 M.R. Grabowski, H. Hendrick: How low can we go?: Validation and verification of a decision support system for safe shipboard manning, IEEE Trans. Eng. Manag. **40**(1), 41–53 (1993)

25.2 D.C. Nagel: Human error in aviation operations. In: *Human Factors in Aviation*, ed. by E. Weiner, D. Nagel (Academic, New York 1988) pp. 263–303

25.3 D.T. Singh, P.P. Singh: Aiding DSS users in the use of complex OR models, Ann. Oper. Res. **72**, 5–27 (1997)

25.4 NTSB: *Marine accident report – Grounding of the Panamanian Passenger Ship ROYAL MAJESTY on Rose and Crown Shoal Near Nantucket, Massachusetts June 10, 1995* (NTSB, Washington 1997)

25.5 M.H. Lutzhoft, S.W.A. Dekker: On your watch: Automation on the bridge, J. Navig. **55**(1), 83–96 (2002)

25.6 D.D. Woods: Automation: Apparent simplicity, real complexity. In: *Human Performance in Automated Systems: Current Research and Trends*, ed. by M. Mouloua, R. Parasuraman (Lawrence Erlbaum, Hillsdale 1994) pp. 1–7

25.7 S. McFadden, A. Vimalachandran, E. Blackmore: Factors affecting performance on a target monitoring task employing an automatic tracker, Ergonomics **47**(3), 257–280 (2003)

25.8 C.D. Wickens, C. Kessel: Failure detection in dynamic systems. In: *Human Detection and Diagnosis of System Failures*, ed. by J. Rasmussen, W.B. Rouse (Plenum, New York 1981) pp. 155–169

25.9 S. Zuboff: *In the Age of Smart Machines: The Future of Work Technology and Power* (Basic Books, New York 1988)

25.10 M.R. Endsley, E.O. Kiris: The out-of-the-loop performance problem and level of control in automation, Hum. Factors **37**(2), 381–394 (1995)

25.11 C.E. Billings: *Aviation Automation: The Search for a Human-Centered Approach* (Erlbaum, Mahwah 1997)

25.12 NTSB: *Marine accident report – Grounding of the US Tankship Exxon Valdez on Bligh Reef, Prince William Sound, near Valdez, Alaska, March 24, 1989* (NTSB, Washington 1990)

25.13 J.D. Lee, T.F. Sanquist: Augmenting the operator function model with cognitive operations: Assessing the cognitive demands of technological innovation in ship navigation, IEEE Trans. Syst. Man Cybern.- Part A: Syst. Hum. **30**(3), 273–285 (2000)

25.14 E. L. Wiener: *Human Factors of Advanced Technology ("Glass Cockpit") Transport Aircraft*, NASA Contractor Report 177528 (NASA Ames Research Center, 1989)

25.15 L. Bainbridge: Ironies of automation, Automatica **19**(6), 775–779 (1983)

25.16 R.I. Cook, D.D. Woods, E. McColligan, M.B. Howie: Cognitive consequences of 'clumsy' automation on high workload, high consequence human performance, SOAR 90, Space Oper. Appl. Res. Symp. (NASA Johnson Space Center 1990)

25.17 D.D. Woods, L. Johannesen, S.S. Potter: *Human Interaction with Intelligent Systems: Trends, Problems, new Directions* (The Ohio State University, Columbus 1991)

25.18 J.D. Lee, J. Morgan: Identifying clumsy automation at the macro level: development of a tool to estimate ship staffing requirements, Proc. Hum. Factors Ergon. Soc. 38th Annu. Meet., Vol. 2 (1994) pp. 878–882

25.19 P.J. Smith, E. McCoy, C. Layton: Brittleness in the design of cooperative problem-solving systems: the effects on user performance, IEEE Trans. Syst. Man Cybern. – Part A: Syst. Hum. **27**(3), 360–371 (1997)

25.20 E. Hutchins: *Cognition in the Wild* (MIT Press, Cambridge 1995) p. 381

25.21 J. Gao, J.D. Lee: A dynamic model of interaction between reliance on automation and cooperation in multi-operator multi-automation situations, Int. J. Ind. Ergon. **36**(5), 512–526 (2006)

25.22 R. Parasuraman, M. Mouloua, R. Molloy: Monitoring automation failures in human-machine systems. In: *Human Performance in Automated Systems: Current Research and Trends*, ed. by M. Mouloua, R. Parasuraman (Lawrence Erlbaum, Hillsdale 1994) pp. 45–49

25.23 R. Parasuraman, R. Molloy, I. Singh: Performance consequences of automation-induced "complacency", Int. J. Aviat. Psychol. **3**(1), 1–23 (1993)

25.24 U. Metzger, R. Parasuraman: The role of the air traffic controller in future air traffic management: an empirical study of active control versus passive monitoring, Hum. Factors **43**(4), 519–528 (2001)

25.25 J. Meyer: Effects of warning validity and proximity on responses to warnings, Hum. Factors **43**(4), 563–572 (2001)

25.26 R. Parasuraman, V. Riley: Humans and automation: use, misuse, disuse, abuse, Hum. Factors **39**(2), 230–253 (1997)

25.27 M.T. Dzindolet, L.G. Pierce, H.P. Beck, L.A. Dawe, B.W. Anderson: Predicting misuse and disuse of combat identification systems, Mil. Psychol. **13**(3), 147–164 (2001)

25.28 J.D. Lee, N. Moray: Trust, self-confidence, and operators' adaptation to automation, Int. J. Hum.-Comput. Stud. **40**, 153–184 (1994)

25.29 J.D. Lee, K.A. See: Trust in technology: designing for appropriate reliance, Hum. Factors **46**(1), 50–80 (2004)

25.30 S. Halprin, E. Johnson, J. Thornburry: Cognitive reliability in manned systems, IEEE Trans. Reliab. **R-22**, 165–169 (1973)

25.31 J. Lee, N. Moray: Trust, control strategies and allocation of function in human-machine systems, Ergonomics **35**(10), 1243–1270 (1992)

25.32 B.M. Muir, N. Moray: Trust in automation 2: experimental studies of trust and human intervention in a process control simulation, Ergonomics **39**(3), 429–460 (1996)

25.33 S. Lewandowsky, M. Mundy, G. Tan: The dynamics of trust: comparing humans to automation, J. Exp. Psychol.-Appl. **6**(2), 104–123 (2000)

25.34 P. de Vries, C. Midden, D. Bouwhuis: The effects of errors on system trust, self-confidence, and the allocation of control in route planning, Int. J. Hum.-Comput. Stud. **58**(6), 719–735 (2003)

25.35 B. Reeves, C. Nass: *The Media Equation: How People Treat Computers, Television, and New Media Like Real People and Places* (Cambridge University Press, New York 1996)

25.36 T.B. Sheridan, R.T. Hennessy: *Research and Modeling of Supervisory Control Behavior* (National Academy Press, Washington 1984)

25.37 T.B. Sheridan, W.R. Ferrell: *Man-machine Systems: Information, Control, and Decision Models of Human Performance* (MIT Press, Cambridge 1974)

25.38 M. Deutsch: Trust and suspicion, J. Confl. Resolut. **2**(4), 265–279 (1958)

25.39 M. Deutsch: The effect of motivational orientation upon trust and suspicion, Hum. Relat. **13**, 123–139 (1960)

25.40 J.B. Rotter: A new scale for the measurement of interpersonal trust, J. Pers. **35**(4), 651–665 (1967)

25.41 J.K. Rempel, J.G. Holmes, M.P. Zanna: Trust in close relationships, J. Pers. Soc. Psychol. **49**(1), 95–112 (1985)

25.42 W. Ross, J. LaCroix: Multiple meanings of trust in negotiation theory and research: A literature review and integrative model, Int. J. Confl. Manag. **7**(4), 314–360 (1996)

25.43 S. Lewandowsky, M. Mundy, G.P.A. Tan: The dynamics of trust: comparing humans to automation, J. Exp. Psychol.-Appl. **6**(2), 104–123 (2000)

25.44 Y.D. Wang, H.H. Emurian: An overview of online trust: Concepts, elements, and implications, Comput. Hum. Behav. **21**(1), 105–125 (2005)

25.45 D. Gefen, E. Karahanna, D.W. Straub: Trust and TAM in online shopping: an integrated model, Manag. Inf. Syst. Q. **27**(1), 51–90 (2003)

25.46 J. Kim, J.Y. Moon: Designing towards emotional usability in customer interfaces – trustworthiness of cyber-banking system interfaces, Interact. Comput. **10**(1), 1–29 (1998)

25.47 T.B. Sheridan: *Telerobotics, Automation, and Human Supervisory Control* (MIT Press, Cambridge 1992)

25.48 T.B. Sheridan: *Supervisory control.* In: *Handbook of Human Factors*, ed. by G. Salvendy (Wiley, New York 1987) pp. 1243–1268

25.49 J.J. Gibson: Observations on active touch, Psychol. Rev. **69**, 477–491 (1962)

25.50 A.R. Ephrath, L.R. Young: Monitoring vs. man-in-the-loop detection of aircraft control failures. In: *Human Detection and Diagnosis of System Failures*, ed. by J. Rasmussen, W.B. Rouse (Plenum, New York 1981) pp. 143–154

25.51 J.M. Flach, R.J. Jagacinski: *Control Theory for Humans* (Lawrence Erlbaum, Mahwah 2002)

25.52 L. Bainbridge: Mathematical equations of processing routines. In: *Human Detection and Diagnosis of System Failures*, ed. by J. Rasmussen, W.B. Rouse (Plenum, New York 1981) pp. 259–286

25.53 N. Moray: Human factors in process control. In: *The Handbook of Human Factors and Ergonomics*, ed. by G. Salvendy (Wiley, New York 1997)

25.54 G.J.S. Wilde: Risk homeostasis theory and traffic accidents: propositions, deductions and discussion of dissension in recent reactions, Ergonomics **31**(4), 441–468 (1988)

25.55 G.J.S. Wilde: Accident countermeasures and behavioral compensation: the position of risk homeostasis theory, J. Occup. Accid. **10**(4), 267–292 (1989)

25.56 L. Evans: *Traffic Safety and the Driver* (Van Nostrand Reinhold, New York 1991)

25.57 C. Perrow: *Normal Accidents* (Basic Books, New York 1984) p. 386

25.58 E. Tenner: *Why Things Bite Back: Technology and the Revenge of Unanticipated Consequences* (Knopf, New York 1996)

25.59 F. Sagberg, S. Fosser, I.A.F. Saetermo: An investigation of behavioural adaptation to airbags and antilock brakes among taxi drivers, Accid. Anal. Prev. **29**(3), 293–302 (1997)

25.60 N.A. Stanton, M. Pinto: Behavioural compensation by drivers of a simulator when using a vision enhancement system, Ergonomics **43**(9), 1359–1370 (2000)

25.61 K.L. Mosier, L.J. Skitka, S. Heers, M. Burdick: Automation bias: decision making and performance in high-tech cockpits, Int. J. Aviat. Psychol. **8**(1), 47–63 (1998)

25.62 L.J. Skitka, K. Mosier, M.D. Burdick: Accountability and automation bias, Int. J. Hum.-Comput. Stud. **52**(4), 701–717 (2000)

25.63 L.J. Skitka, K.L. Mosier, M. Burdick: Does automation bias decision-making?, Int. J. Human-Comput. Stud. **51**(5), 991–1006 (1999)

25.64 T.B. Sheridan: *Humans and Automation* (Wiley, New York 2002)

25.65 K.J. Vicente: *Cognitive Work Analysis: Towards Safe, Productive, and Healthy Computer-based Work* (Lawrence Erlbaum Associates, Mahwah 1999)

Part C | 25

25.66   J.D. Lee: Human factors and ergonomics in automation design. In: *Handbook of Human Factors and Ergonomics*, ed. by G. Salvendy (Wiley, Hoboken 2006) pp. 1570–1596

25.67   J.D. Lee, T.F. Sanquist: Maritime automation. In: *Automation and Human Performance*, ed. by R. Parasuraman, M. Mouloua (Lawrence Erlbaum, Mahwah 1996) pp. 365–384

25.68   M.T. Dzindolet, L.G. Pierce, H.P. Beck, L.A. Dawe: The perceived utility of human and automated aids in a visual detection task, Hum. Factors **44**(1), 79–94 (2002)

25.69   M. Yeh, C.D. Wickens: Display signaling in augmented reality: effects of cue reliability and image realism on attention allocation and trust calibration, Hum. Factors **43**, 355–365 (2001)

25.70   J.P. Bliss: Alarm reaction patterns by pilots as a function of reaction modality, Int. J. Aviat. Psychol. **7**(1), 1–14 (1997)

25.71   J.P. Bliss, S.A. Acton: Alarm mistrust in automobiles: how collision alarm reliability affects driving, Appl. Ergonom. **34**, 499–509 (2003)

25.72   S. Guerlain, P.J. Smith, J.H. Obradovich, S. Rudmann, P. Strohm, J.W. Smith, J. Svirbely: Dealing with brittleness in the design of expert systems for immunohematology, Immunohematology **12**(3), 101–107 (1996)

25.73   R. Parasuraman, T.B. Sheridan, C.D. Wickens: A model for types and levels of human interaction with automation, IEEE Trans. Syst. Man Cybern. -Part A: Syst. Hum. **30**(3), 286–297 (2000)

25.74   N.B. Sarter, D.D. Woods: Decomposing automation: autonomy, authority, observability and perceived animacy. In: *Human Performance in Automated Systems: Current Research and Trends*, ed. by M. Mouloua, R. Parasuraman (Lawrence Erlbaum, Hillsdale 1994) pp. 22–27

25.75   W.A. Olson, N.B. Sarter: Automation management strategies: pilot preferences and operational experiences, Int. J. Aviat. Psychol. **10**(4), 327–341 (2000)

25.76   N.B. Sarter, D.D. Woods: Team play with a powerful and independent agent: operational experiences and automation surprises on the Airbus A-320, Hum. Factors **39**(4), 553–569 (1997)

25.77   N.B. Sarter, D.D. Woods: Team play with a powerful and independent agent: a full-mission simulation study, Hum. Factors **42**(3), 390–402 (2000)

25.78   M. Lewis: Designing for human-agent interaction, Artif. Intell. Mag. **19**(2), 67–78 (1998)

25.79   P.M. Jones, J.L. Jacobs: Cooperative problem solving in human-machine systems: theory, models, and intelligent associate systems, IEEE Trans. Syst. Man Cybern. – Part C: Appl. Rev. **30**(4), 397–407 (2000)

25.80   S.R. Bocionek: Agent systems that negotiate and learn, Int. J. Hum.-Comput. Stud. **42**(3), 265–288 (1995)

25.81   N.B. Sarter: The need for multisensory interfaces in support of effective attention allocation in highly dynamic event-driven domains: the. case of cockpit automation, Int. J. Aviat. Psychol. **10**(3), 231–245 (2000)

25.82   N. Moray, T. Inagaki, M. Itoh: Adaptive automation, trust, and self-confidence in fault management of time-critical tasks, J. Exp. Psychol.-Appl. **6**(1), 44–58 (2000)

25.83   T. Inagaki: Automation and the cost of authority, Int. J. Ind. Ergon. **31**(3), 169–174 (2003)

25.84   C.Y. Liang, H. Peng: Optimal adaptive cruise control with guaranteed string stability, Veh. Syst. Dyn. **32**(4-5), 313–330 (1999)

25.85   C.Y. Liang, H. Peng: String stability analysis of adaptive cruise controlled vehicles, JSME Int. J. Ser. C: Mech. Syst. Mach. Elem. Manuf. **43**(3), 671–677 (2000)

25.86   J.D. Lee, J. Gao: Trust, automation, and cooperation in supply chains, Supply Chain Forum: Int. J. **6**(2), 82–89 (2006)

25.87   J. Hollan, E. Hutchins, D. Kirsh: Distributed cognition: Toward a new foundation for human-computer interaction research, ACM Trans. Comput.-Hum. Interact. **7**(2), 174–196 (2000)

25.88   J. Gao, J.D. Lee: Information sharing, trust, and reliance – a dynamic model of multi-operator multi-automation interaction, Proc. 5th Conf. Hum. Perform. Situat. Aware. Autom. Technol., ed. by D.A. Vincenzi, M. Mouloua, P.A. Hancock (Lawrence Erlbaum, Mahwah 2004) pp. 34–39

25.89   A. Kirlik, R.A. Miller, R.J. Jagacinsky: Supervisory control in a dynamic and uncertain environment: a process model of skilled human-environment interaction, IEEE Trans. Syst. Man Cybern. **23**(4), 929–952 (1993)

25.90   J.M. Flach: The ecology of human-machine systems I: Introduction, Ecol. Psychol. **2**(3), 191–205 (1990)

25.91   K.J. Vicente, J. Rasmussen: The ecology of human-machine systems II: Mediating "direct perception" in complex work domains, Ecol. Psychol. **2**(3), 207–249 (1990)

25.92   J.D. Lee, K.A. See: Trust in technology: Design for appropriate reliance, Hum. Factors **46**(1), 50–80 (2004)

25.93   B.H. Kantowitz, R.D. Sorkin: Allocation of functions. In: *Handbook of Human Factors*, ed. by G. Salvendy (Wiley, New York 1987) pp. 355–369

25.94   J. Sharit: Perspectives on computer aiding in cognitive work domains: toward predictions of effectiveness and use, Ergonomics **46**(1-3), 126–140 (2003)

25.95   T.B. Sheridan: Function allocation: algorithm, alchemy or apostasy?, Int. J. Hum.-Comput. Stud. **52**(2), 203–216 (2000)

25.96   A. Dearden, M. Harrison, P. Wright: Allocation of function: scenarios, context and the economics of effort, Int. J. Hum.-Comput. Stud. **52**(2), 289–318 (2000)

25.97 S.W.A. Dekker, D.D. Woods: MABA-MABA or abracadabra? Progress on human-automation coordination, Cogn. Technol. Work **4**, 240–244 (2002)

25.98 E. Hollnagel, A. Bye: Principles for modelling function allocation, Int. J. Hum.-Comput. Stud. **52**(2), 253–265 (2000)

25.99 A. Kirlik: Modeling strategic behavior in human-automation interaction: Why an "aid" can (and should) go unused, Hum. Factors **35**(2), 221–242 (1993)

25.100 J.R. Anderson, C. Libiere: *Atomic Components of Thought* (Lawrence Erlbaum, Hillsdale 1998)

25.101 M. D. Byrne, A. Kirlik: Using computational cognitive modeling to diagnose possible sources of aviation error, Int. J. Aviat. Psychol. **12**(2), 135–155

25.102 A. Degani, A. Kirlik: Modes in human–automation interaction: initial observations about a modeling approach, IEEE-Syst. Man Cybern. **4**, 3443–3450 (1995)

25.103 A. Degani, M. Heymann: Formal verification of human–automation interaction, Hum. Factors **44**(1), 28–43 (2002)

25.104 D.A. Norman: The 'problem' with automation: Inappropriate feedback and interaction, not 'overautomation', Philos. Trans. R. Soc. Lond. Ser. B, Biol. Sci. **327**(1241), 585–593 (1990)

25.105 E.B. Entin, E.E. Entin, D. Serfaty: Optimizing aided target-recognition performance. In: *Proc. Hum. Factors Ergon. Soc.* (Human Factors and Ergonomics Society, Santa Monica 1996) pp. 233–237

25.106 A.E. Sklar, N.B. Sarter: Good vibrations: Tactile feedback in support of attention allocation and human-automation coordination in event-driven domains, Hum. Factors **41**(4), 543–552 (1999)

25.107 M.I. Nikolic, N.B. Sarter: Peripheral visual feedback: a powerful means of supporting effective attention allocation in event-driven, data-rich environments, Hum. Factors **43**(1), 30–38 (2001)

25.108 B.D. Seppelt: Making the limits of adaptive cruise control visible, Int. J. Hum.-Comput. Stud. **65**, 192–205 (2007)

25.109 J.M. Flach: Ready, fire, aim: a "meaning-processing" approach to display design. In: *Attention and Performance XVII: Cognitive Regulation of Performance: Interaction of Theory and Application*, ed. by D. Gopher, A. Koriat (MIT Press, Cambridge 1999) pp. 197–221

25.110 S.A. Guerlain, G.A. Jamieson, P. Bullemer, R. Blair: The MPC elucidator: a case study in the design for human–automation interaction, IEEE Trans. Syst. Man Cybern. Part A: Syst. Hum. **32**(1), 25–40 (2002)

25.111 S. Tseng, B.J. Fogg: Credibility and computing technology, Commun. ACM. **42**(5), 39–44 (1999)

25.112 P. Briggs, B. Burford, C. Dracup: Modeling self-confidence in users of a computer-based system

showing unrepresentative design, Int. J. Hum.-Comput. Stud. **49**(5), 717–742 (1998)

25.113 B. Fogg, J. Marshall, O. Laraki, A. Osipovich, N. Fang: What makes web sites credible? A report on a large quantitative study, Proc. Chi Conf. Hum. Fact. Comput. Syst. 2001 (ACM, Seattle 2001)

25.114 B. Fogg, J. Marshall, T. Kameda, J. Solomon, A. Rangnekar, J. Boyd, B. Brown: Web credibility research: a method for online experiments and early study results, Chi Conf. Hum. Fact. Comput. Syst. (2001) pp. 293–294

25.115 V. Riley: A new language for pilot interfaces, Ergon. Des. **9**(2), 21–27 (2001)

25.116 M.A. Goodrich, E.R. Boer: Model-based human-centered task automation: a case study in ACC system design, IEEE Trans. Syst. Man Cybern. – Part A: Syst. Hum. **33**(3), 325–336 (2003)

25.117 C.A. Miller: Definitions and dimensions of etiquette. In: *Etiquette for Human-Computer Work: Technical Report FS-02-02*, ed. by C. Miller (American Association for Artificial Intelligence, Menlo Park 2002) pp. 1–7

25.118 C. Nass, K.N. Lee: Does computer-synthesized speech manifest personality? Experimental tests of recognition, similarity-attraction, and consistency-attraction, J. Exp. Psychol.-Appl. **7**(3), 171–181 (2001)

25.119 K.J. Vicente: Coherence- and correspondence-driven work domains: implications for systems design, Behav. Inf. Technol. **9**, 493–502 (1990)

25.120 R.A. Brooks, P. Maes, M.J. Mataric, G. More: Lunar base construction robots, Proc. 1990 Int. Workshop Intell. Robots Syst. (1990) pp. 389–392

25.121 P.J. Johnson, J.S. Bay: Distributed control of simulated autonomous mobile robot collectives in payload transportation, Auton. Robots **2**(1), 43–63 (1995)

25.122 R.A. Brooks, A.M. Flynn: A robot being. In: *Robots and Biological Systems: Towards a New Bionics*, ed. by P. Dario, G. Sansini, P. Aebischer (Springer, Berlin 1993)

25.123 G. Beni, J. Wang: Swarm intelligence in cellular robotic systems. In: *Robots and Biological Systems: Towards a New Bionics*, ed. by P. Dario, G. Sansini, P. Aebischer (Springer, Berlin 1993)

25.124 T. Fukuda, D. Funato, K. Sekiyama, F. Arai: Evaluation on flexibility of swarm intelligent system, Proc. 1998 IEEE Int. Conf. Robotics Autom. (1998) pp. 3210–3215

25.125 K. Sugihara, I. Suzuki: Distributed motion coordination of multiple mobile robots, 5th IEEE Int. Symp. Intell. Control (1990) pp. 138–143

25.126 T.W. Min, H.K. Yin: A decentralized approach for cooperative sweeping by multiple mobile robots, Proc. 1998 IEEE/RSJ Int. Conf. Intell. Robots Syst. (1998)

25.127 E.S. Patterson: A simulation study of computer-supported inferential analysis under data overload, Proc. Hum. Factors Ergon. 43rd Annu. Meet., Vol. 1 (1999) pp. 363–368

25.128 P. Pirolli, S. Card: Information foraging, Psychol. Rev. **106**(4), 643–675 (1999)

25.129 J. Murray, Y. Liu: Hortatory operations in highway traffic management, IEEE Trans. Syst. Man Cybern. – Part A: Syst. Hum. **27**(3), 340–350 (1997)

25.130 T.R. Stickland, N.F. Britton, N.R. Franks: Complex trails and simple algorithms in ant foraging, **260**(1357), 53–58 (1995)

25.131 M. Resnick: *Turtles, Termites, and Traffic Jams: Explorations in Massively Parallel Microworlds* (MIT Press, Cambridge 1991)

25.132 J.D. Lee: Emerging challenges in cognitive ergonomics: Managing swarms of self-organizing agent-based automation, Theor. Issues Ergon. Sci. **2**(3), 238–250 (2001)

25.133 T.C. Schelling: *Micro Motives and Macro Behavior* (Norton, New York 1978)

25.134 J.H. Dyer, H. Singh: The relational view: cooperative strategy and sources of interorganizational competitive advantage, Acad. Manag. Rev. **23**(4), 660–679 (1998)

25.135 J.D. Sterman: Modeling managerial behavior: misperceptions of feedback in a dynamic decision-making experiment, Manag. Sci. **35**(3), 321–339 (1989)

25.136 X.D. Zhao, J.X. Xie: Forecasting errors and the value of information sharing in a supply chain, Int. J. Prod. Res. **40**(2), 311–335 (2002)

25.137 H.L. Lee, S.J. Whang: Information sharing in a supply chain, Int. J. Technol. Manag. **20**(3-4), 373–387 (2000)

25.138 H. Akkermans, K. van Helden: Vicious and virtuous cycles in ERP implementation: a case study of interrelations between critical success factors, Eur. J. Inf. Syst. **11**(1), 35–46 (2002)

25.139 R.B. Handfield, C. Bechtel: The role of trust and relationship structure in improving supply chain responsiveness, Ind. Mark. Manag. **31**(4), 367–382 (2002)

25.140 J. Gao, J.D. Lee: Extending the decision field theory to model operators' reliance on automation in supervisory control situations, IEEE Syst. Man Cybern. **36**(5), 943–959 (2006)

25.141 J.R. Busemeyer, J.T. Townsend: Decision field theory: A dynamic cognitive approach to decision making in an uncertain environment, Psychol. Rev. **100**(3), 432–459 (1993)

25.142 T.S. Zhou, J.H. Lu, L.N. Chen, Z.J. Jing, Y. Tang: On the optimal solutions for power flow equations, Int. J. Electr. Power Energy Syst. **25**(7), 533–541 (2003)

25.143 T. Mulkerin: Free flight is in the future – large-scale controller pilot data link communications emulation testbed, IEEE Aerosp. Electron. Syst. Mag. **18**(9), 23–27 (2003)

25.144 W.A. Olson, N.B. Sarter: Management by consent in human-machine systems: when and why it breaks down, Hum. Factors **43**(2), 255–266 (2001)

26. Collaborative Human–Automation Decision Making

Mary L. Cummings, Sylvain Bruni

The development of a comprehensive collaborative human–computer decision-making model is needed that demonstrates not only what decision-making functions should or could be assigned to humans or computers, but how many functions can best be served in a mutually supportive environment in which the human and computer collaborate to arrive at a solution superior to that which either would have come to independently. To this end, we present the human–automation collaboration taxonomy (HACT), which builds on previous research by expanding the *Parasuraman* information processing model [26.1], specifically the decision-making component. Instead of defining a simple level of automation for decision making, we deconstruct the process to include three distinct roles: the moderator, generator, and decider. We propose five levels of collaboration (LOCs) for

each of these roles, which form a three-tuple that can be analyzed to evaluate system collaboration, and possibly identify areas for design intervention. A resource allocation mission planning case study is presented using this framework to illustrate the benefit for system designers.

In developing any complex supervisory control system that involves the integration of human decision making with automation, the question often arises as to where, how, and how much humans should be in the decision-making loop. Allocating roles and functions between the human and the computer is critical in defining efficient and effective system architectures. However, role allocation does not necessarily need to be mutually exclusive, and instead of systems that clearly define specific roles for either human or automation, it is possible that humans and computers can collaborate in a mutually supportive decision-making environment. This is especially true for aspects of supervisory control that include planning and resource allocation (e.g., how should multiple aircraft be routed to avoid bad weather, or how to allocate ambulances in a disaster), which is the focus of this chapter. For discussion purposes, we define collaboration as *the mutual engagement of agents in a coordinated and synchronous effort to solve a prob-lem based on a shared conception of it* [26.2, 3]. We define agents as either humans or some form of automation/computer that provides some level of interaction.

For planning and resource allocation supervisory control tasks in complex systems, the problem spaces are large with significant uncertainty, so the use of automation is clearly warranted in attempting to solve a particular problem; for example, if bad weather prevents multiple aircraft from landing at an airport, air-traffic controllers need to know right away which alternate airports are within fuel range, and of these, which have the ability to service the different aircraft types, the predicted traffic volume, routing conflicts, etc. While automation could be used to provide optimized routing recommendations quickly, computer-generated solutions are unfortunately not always the best solutions. While fast and able to handle complex computation far better than humans, computer optimization algorithms are notoriously *brittle* in that they

Part C | 26

can only take into account those quantifiable variables identified in the design stages that were deemed to be critical [26.4]. In supervisory control systems with inherent uncertainties (weather impacts, enemy movement, etc.), it is not possible to include a priori every single variable that could impact the final solution. Moreover, it is not clear exactly what characterizes an *optimal* solution in uncertain such scenarios. Often, in these domains, the need to generate an *optimal* solution should be weighed against a *satisficing* [26.5] solution. Because constraints and variables are often dynamic in complex supervisory control environments, the definition of optimal is also a constantly changing concept. In those cases of time pressure, having a solution that is good enough, robust, and quickly reached is often preferable to one that requires complex computation and extended periods of times, which may not be accurate due to incorrect assumptions.

Recognizing the need for automation to help navigate complex and large supervisory control problem spaces, it is equally important to recognize the critical role that humans play in these decision-making tasks. Optimization is a word typically associated with computers but humans are natural optimizers as well, although not necessarily in the same linear vein as computers. Because humans can reason inductively and generate conceptual representations based on both abstract and factual information, they also have the ability to optimize based on qualitative and quantitative information [26.6]. In addition, allowing operators active participation in decision-making processes provides not only safety benefits, but promotes situation awareness and also allows a human operator, and thus a system, to respond more flexibly to uncertain and unexpected events. Thus, decision support systems that leverage the collaborative strength of humans and automation in supervisory control planning and resource allocation tasks could provide substantial benefits, both in terms of human and system performance,

Unfortunately, little formal guidance exists to aid designers and engineers in the development of collaborative human–computer decision support systems. While many frameworks have been proposed that detail levels of human–automation role allocation, there has been no focus on what specifically constitutes collaboration in terms of role allocation and how this can be quantified to allow for specific system analysis as well as design guidance. Therefore, to better describe human-collaborative decision support systems in order to provide more detailed design guidance, we present the human–automation collaboration taxonomy (HACT) [26.7].

26.1 Background

There is little previous literature that attempts to classify, describe, or provide design guidance on human–automation (or computer) collaboration. Most previous efforts have generally focused on developing application-specific decision support tools that promote some open-ended form of human–computer interaction (e.g., [26.8–10]). In an attempt to categorize human–computer collaboration more formally, *Silverman* [26.11] proposed categories of human–computer interaction in terms of critiquing, although this is a relatively narrow field of human–computer collaboration. *Terveen* [26.12] attempted to seek some unified approach and more broadly define and categorize human–computer collaboration in terms of human emulation and human "complementary" [sic]. Beyond these broad definitions and categorizations of human–computer collaboration and narrow applications of specific algorithms and visualizations, there has been no underlying theory addressing how collaboration with an automated agent supports operator decision making at the most fundamental information processing level.

So while the literature on human–automation *collaboration* in decision making is sparse, the converse is true in terms of scales and taxonomies of automation levels that describe *interactions* between a human operator and a computer/automation. These levels of automation (LOAs) generally refer to the role allocation between automation and the human, particularly in the analysis and decision phases of a simplified information processing model of acquisition, analysis, decision, and action phases [26.1, 13, 14]. The originators of the concept of levels of automation, *Sheridan* and *Verplank* (SV), initially proposed that automation could range from a fully manual system with no computer intervention to a fully automated system where the human is kept completely out of the loop [26.15]. *Parasuraman* [26.1] expanded the original SV LOA to include ten levels (Table 26.1).

Table 26.1 Levels of automation (after [26.1, 15])

Automation level	Automation description
1	The computer offers no assistance: human must take all decision and actions
2	The computer offers a complete set of decision/action alternatives, or
3	Narrows the selection down to a few, or
4	Suggests one alternative, and
5	Executes that suggestion if the human approves, or
6	Allows the human a restricted time to veto before automatic execution, or
7	Executes automatically, then necessarily informs humans, and
8	Informs the human only if asked, or
9	Informs the human only if it, the computer, decides to
10	The computer decides everything and acts autonomously, ignoring the human

At the lower levels, LOAs 1–4, the human is actively involved in the decision-making process. At level 5, the automation takes on a more active role in executing decisions, while still requiring consent from the operator before doing so (known as management-by-consent). Level 6, typically referred to as management-by-exception, allows the automation a more active role in decisions, executing solutions unless vetoed by the human. For levels 7–10, humans are only allowed to accept or veto solutions presented to them. Thus, as levels increase, the human is increasingly removed from the decision-making loop, and the automation is increasingly allocated additional authority. This scale addresses primarily authority allocation, i.e., who is given the authority to make the final decision, although only to a much smaller and limited degree does it address the solution-generation aspect of decision making, which is a critical aspect of human–computer collaboration.

The solution-generation process in supervisory control planning and resource allocation tasks is critical because this is the aspect of the human–computer interaction where the variables and constraints can be manipulated to determine solution alternatives. This access creates a sensitivity analysis trade space that allows human operators the ability to cope with uncertainty and apply judgment and experience that are unavailable to computer algorithms. While the LOAs in Table 26.1

provide some indirect guidance as to how the solution-generation process can be allocated either to the human or computer, it is only tangentially inferred, and there is no level that allows for joint construction or modification of solutions.

Other LOA taxonomies have addressed the need to examine authority and solution generation LOAs, although none have addressed them in an integrated fashion; for example, *Endsley* [26.16] incorporated artificial intelligence into a five-point LOA scale, thus addressing some aspects of solution generation and authority. *Riley* [26.17] investigated the use of the level of information attribute in addition to the automation authority attribute, creating a two-dimensional scale. Another ten-point scale was created by *Endsley* and *Kaber* [26.16] where each level corresponds to a specific task behavior of the automation, going from *manual control* to *full automation*, through intermediate levels such as *blended decision making* or *supervisory control*. While all of these scales acknowledge that there are possible collaborative processes between humans and automated agents, none specifically detail how this interaction can occur, and how different attributes of a collaborative system can each have a different LOA. To address this shortcoming in the literature, we developed the human–automation collaboration taxonomy (HACT), which is detailed in the next section.

26.2 The Human–Automation Collaboration Taxonomy (HACT)

In order to better understand how human operators and automation collaborate, the four-stage information-

processing flow diagram of *Parasuraman* [26.1] (with stages: information acquisition, information analysis,

Part C | 26.2

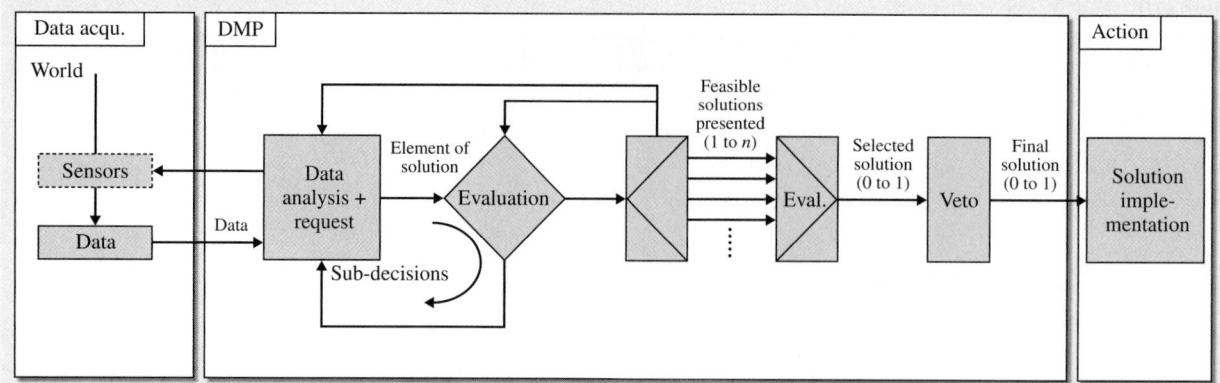

Fig. 26.1 The HACT collaborative information-processing model

decision selection, and action implementation) was modified to focus specifically on collaborative decision making. This new model, shown in Fig. 26.1, features three steps: data acquisition, decision making, and action taking. The data acquisition step is similar to that proposed by *Parasuraman* [26.1] in that sensors retrieve information from the outside world or environment, and transform it into working data. The collaborative aspect of this model occurs in the next stage, the decision-making process, which corresponds to the integration of the analysis and decision phases of the *Parasuraman* [26.1] model.

First, the data from the acquisition step is analyzed, possibly in an iterative way where requests for more data can be sent to the sensors. The data analysis outputs some elements of a solution to the problem at hand. The evaluation block estimates the appropriateness of these elements of solutions for a potential final solution. This block may initiate a recursive loop with the data analysis block; for instance, operators may request more analysis of the domain space or part thereof. At this level, subdecisions are made to orient the search and analysis process. Once the evaluation step is validated, i. e., subdecisions are made, the results are assembled to constitute one or more feasible solutions to the problem. In order to generate feasible solutions, it is possible to loop back to the previous evaluation phase, or even to the data analysis step. At some point, one or more feasible solutions are presented in a second evaluation step.

The operator or automation (depending on the level of automation) will then select one solution (or none) out of the pool of feasible solutions. After this selection procedure, a veto step is added, since it is possible for one or more of the collaborating agents to veto the solu-

tion selected (such as in management-by-exception). An agent may be a human operator or an automated computer system, also called automation. If the proposed solution is vetoed, the output of the veto step is empty, and the decision-making process starts again. If the selected solution is not vetoed, it is considered the *final solution* and is transferred to the action mechanism for implementation.

26.2.1 Three Basic Roles

Given the decision-making process (DMP) shown in Fig. 26.1, three key roles have been identified: moderator, generator, and decider. In the context of collaborative human–computer decision making, these three roles are fulfilled either by the human operator, by automation, or by a combination of both. Figure 26.2 displays how these three basic roles fit into the HACT collaborative information-processing model. The generator and the decider roles are mutually exclusive in that the domain of competency of the generator (as outlined in Fig. 26.2) does not overlap with that of the decider. However, the moderator's role subsumes the entire decision-making process. As will be discussed, each of the three roles has its own possible LOA scale.

The Moderator

The moderator is the agent(s) that keeps the decision-making process moving forward, and ensures that the various phases are executed; for instance, the moderator may initiate the decision-making process and interaction between the human and automation. The moderator may prompt or suggest that subdecisions need to be made, or evaluations need to be considered. It could also be involved keeping the decision processing within

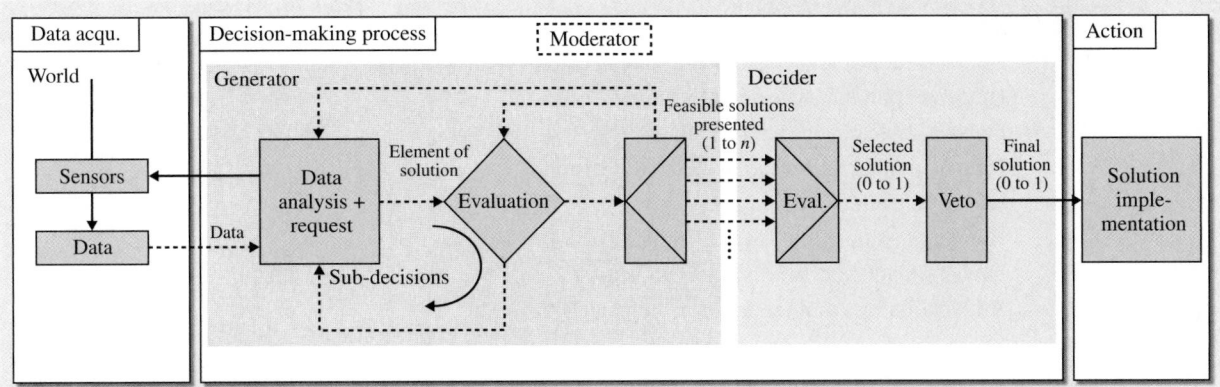

Fig. 26.2 The three collaborative decision-making process roles: moderator, generator, and decider

prespecified limits when time pressure is a concern. In relation to the ten-level SV LOA scale (Table 26.1), the step between LOA 4 and 5 implies this role, but does not address the fact that moderation can occur across multiple segments of the decision-making process and separate from the tasks of solution generation and selection.

The Generator

The generator is the agent(s) that generates feasible solutions from the data. Typically, the generator role involves searching, identifying, and creating solution(s) or parts thereof. Most of the previously discussed LOAs (e.g., [26.1, 16]) address the role of a solution generator. However, instead of focusing on only the actual solution (e.g., automation generating one or many solutions), we expand in detail the notion of the generator to include other aspects of solution generation, i.e., all the other steps within the generator box (Fig. 26.2), such as the automation analyzing data, which makes the solution generation easier for the human operator. Additionally, the role allocation for generator may not be mutually exclusive but could be shared to varying degrees between

Table 26.2 Moderator and generator levels

Level	Who assumes the role of generator and/or moderator?
2	Human
1	Mixed, but more human
0	Equally shared
−1	Mixed, but more automation
−2	Automation

the human operator and the automation; for example, in one system the human could define multiple constraints and the automation searches for a set of possible solutions bounded by these constraints. In another system, the automation could propose a set of possible solutions and then the human operator narrows down these solutions.

For both the moderator and generator roles, the general LOAs can be seen in Table 26.2, which we recharacterize as LOCs (levels of collaboration). While the levels could be parsed into more specific levels, as seen in previously discussed LOAs, these five levels were chosen to reflect degrees of collaboration with the center scale reflecting balanced collaboration. At either end of the LOC scale (2 or −2), the system, in terms of moderation and generation, is not collaborative. The negative sign should not be interpreted as a critical reflection on the use of automation; it simply reflects scaling in the opposite direction. A system at LOC 0, however, is a balanced collaborative system for either the moderator and/or generator.

The Decider

The third role within the HACT collaborative decision-making process is the decider. The decider is the agent(s) that *makes the final decision*, i.e., that selects the potentially final solution out of the set of feasible solutions presented by the generator, and who has veto power over this selection decision. Veto power is a non-negotiable attribute: once an agent vetoes a decision, the other agent cannot supersede it. This veto power is also an important attribute in other LOA scales [26.1, 16], but we have added more resolution to the possible role allocations in keeping with our collaborative approach, listed in Table 26.3. As in Table 26.2, the most balanced

Level	Who assumes the role of decider?
2	Human makes final decision, automation cannot veto
1	Human or automation can make final decision, human can veto, automation cannot veto
0	Human or automation can make final decision, human can veto, automation can veto
−1	Human or automation can make final decision, human cannot veto, automation can veto
−2	Automation makes final decision, human cannot veto

Table 26.3 Decider levels

collaboration between the human and the automation is seen at the midpoint, with the greatest lack of collaboration at the extreme levels.

The three roles, moderator, generator and decider, focus on the tasks or actions that are undertaken by the human operator, the automation, or the combination of both within the collaborative decision-making process.

26.2.2 Characterizing Human Supervisory Control System Collaboration

Given the scales outlined above, decision support systems can be categorized by the collaboration across the three different roles (moderator, generator, and decider) in the form of a three-tuple, e.g., $(2, 1, 2)$ or $(-2, -2, 1)$. In the first example of $(2, 1, 2)$, this system includes the human as both the moderator and the decider, as well as generating most of the solution, but leverages some automation for the solution generation. An example of such a system would be one where an operator needs to plan a mission route but must select not just the start and goal state, but all intermediate points in order to avoid all restricted zones and possible hazards. Automation is used to ensure fuel limits

are not exceeded and to alert the operator in the case of any area violations.

This is in contrast to the highly automated $(-2, -2, 1)$ example, which is the characterization of the Patriot missile system. This antimissile missile system notifies the operator that a target has been detected, allows the operator approximately 15 s to veto the automation's solution, and then fires if the human does not intervene. Thus the automation moderates the flow, analyzes the solution space, presents a single solution, and then allows the human to veto this. Note that under the ten LOAs in Table 26.1, this system would be characterized at LOA 6, but the HACT three-tuple provides much more information. It demonstrates that the system is highly automated at the moderator and generator levels, while the human has more authority than the automation for the final decision. However, a low decider level does not guarantee a human-centered system in that the Patriot system has accidentally killed three North Atlantic Treaty Organization (NATO) airmen because operators were not able to determine in the 15 s window that the targets were actually friendly aircraft and not enemy missiles. This example illustrates that all three entries in the HACT taxonomy are important for understanding a system's collaborative potential.

26.3 HACT Application and Guidelines

In order to illustrate the application and utility of HACT, a case study is presented. Given the increased complexity, uncertainty, and time pressure of mission planning and resource allocation in command and control settings, increased automation is an obvious choice for system improvement. However just what level of automation/collaboration should be used in such an application is not so obvious. As previously mentioned,

too much automation can induce complacency and loss of situation awareness, and coupled with the inherent inability of automated algorithms to be perfectly correct in dynamic command and control settings, high levels of automation are not advisable. However, low levels of automation can cause unacceptable operator workload as well as suboptimal, very inefficient solutions. Thus the resource allocation aspect of mission planning

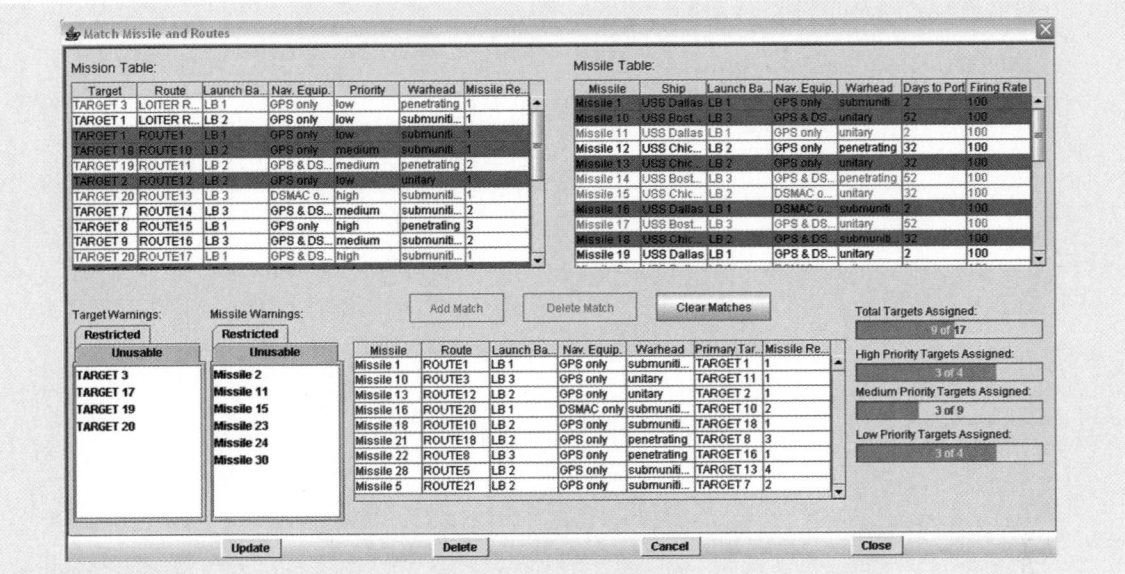

Fig. 26.3 Interface 1

is well suited for some kind of collaborative human–computer environment. To investigate this issue, three interfaces were designed for a representative system, each with a different LOA/LOC detailed in the next section.

The general objective of this resource allocation problem is for an operator to match a set of military missions with a set of available resources, in this case Tomahawk missiles aboard ship and submarine launch platforms.

Interface 1 (Fig. 26.3) was designed to support manual matching of the missiles to the missions at a low level of collaboration. This interface provides raw data tables with all the characteristics of missions and missiles that must be matched, but only provides very limited automated support, such as basic data sorting, mission/missile assignment summaries by categories, and feedback on mission–missile incompatibility and current assignment status. Therefore, this interface mostly involves manual problem solving. As a result, interface 1 is assigned a level 2 moderator because the human operator fully controls the process. Because interface 1 only features basic automation support, the generator role is at level 1. The decider is at level 2 since only the human operator can validate a solution for further implementation, with no possible automation veto.

Interface 2 (Fig. 26.4) was designed to offer the human operator the choice to either solve the mission–

missile assignment task manually as in interface 1 (note in Fig. 26.4 that the top part of interface 2 is a replica of interface 1 shown in Fig. 26.3), or to leverage automation and collaborate with the computer to generate solutions. In the latter instance, termed Automatch, the human operator can steer the search of the automated solution in the domain space by selecting and prioritizing search criteria. Then, the automation's fast computing capabilities perform a heuristic search based on the criteria defined by the human. The operator can either keep the solution output or modify it manually. The operator can also elect to modify the search criteria to get a new solution.

Therefore, for interface 2, the moderator remains at level 2 because the human operator is still in full control of the process, including which tasks are completed, at what pace, and in which order. Because of the flexibility in obtaining a solution in that the human can define the search criteria, thus orienting the automation which does the bulk of the computation, the generator is labeled 0. The decider is at level 2 since only the human operator can validate a final solution, which the automation cannot veto.

While interfaces 1 and 2 are both based on the use of raw data, interface 3 (Fig. 26.5) is completely graphical, and allows the operator to only have access to postsolution sensitivity analysis tools. For interface 2, the automated solution process is guided by the human, who also can conduct sensitivity analysis via an Au-

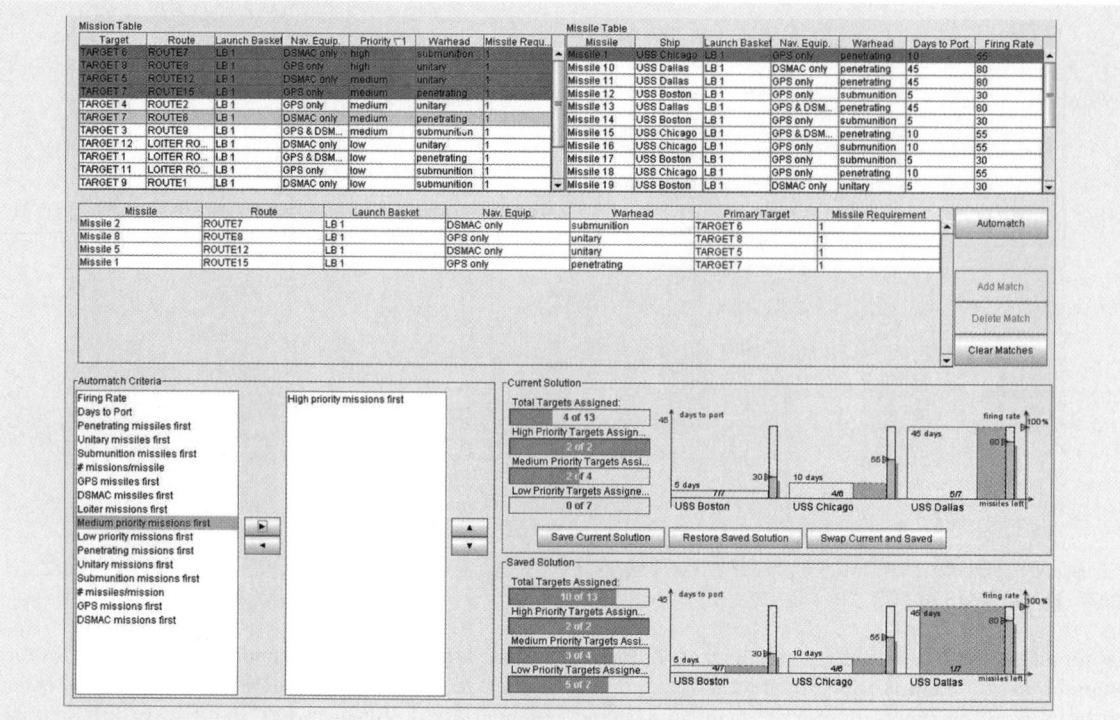

Fig. 26.4 Interface 2

tomatch function; the Automatch button at the top of interface 3 is similar to that in interface 2. However, the user can only select a limited subset of information criteria by which to orient the algorithmic search, causing the operator to rely more on the automation than in interface 2. Thus the HACT three-tuple in this case

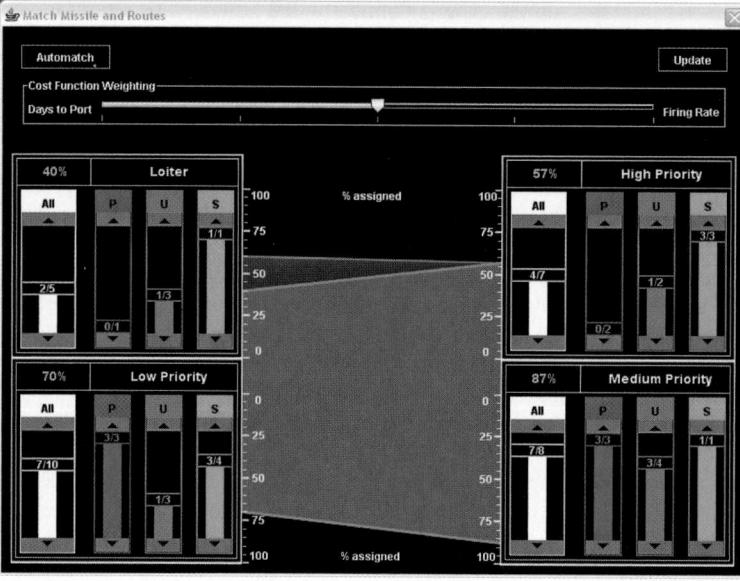

Fig. 26.5 Interface 3

is $(2, -1, 2)$ as neither the moderator nor decider roles changed from interface 2, although the generator's did.

The three interfaces were evaluated with 20 US Navy personnel who would use such a tool in an operational setting. While the full experimental details can be found elsewhere [26.18], in terms of overall performance, operators performed the best with interfaces 1 and 2, which were not statistically different from each other ($p = 0.119$). Interface 3, the one with the predominantly automation-led collaboration, produced statistically worse performance compared with both interfaces 1 and 2 ($p = 0.011$ and 0.031 respectively). Table 26.4 summarizes the HACT categorization for the three interfaces, along with their relative performance rankings. The results indicate that, because the moderator and decider roles were held constant, the degraded performance for those operators using interface 3 was a result of the differences in the generator aspect of the decision-making process. Furthermore, the decline in performance occurred when the LOC was weighted towards the automation. When the solution process was either human-led or of equal contribution, operators performed no differently. However, when the solution generation was automation led, operators struggled.

While there are many other factors that likely affect these results (trust, visualization design, etc.), the HACT taxonomy is helpful in first deconstructing the automation components of the decision-making process. This allows for more specific analyses across different *collaboration levels* of humans and automation, which has not been articulated in other LOA scales. In addition, as demonstrated in the previous example, when comparing systems, such a categorization will also pinpoint which LOCs are helpful, or at the very least, not detrimental. In addition, while not explicitly illustrated here, the HACT taxonomy can also provides

Table 26.4 Interface performance and HACT three-tuples; M – moderator; G – generator; D – decider

	HACT three-tuple (M, G, D)	Performance
Interface 1	(2, 1, 2)	Best
Interface 2	(2, 0, 2)	Best
Interface 3	(2, -1, 2)	Worst

designers with some guidance on system design, i. e., to improve performance for a system; for example, in interface 3, it may be better to increase the moderator LOC instead of lowering the generator LOC.

In summary, application of HACT is meant to elucidate human–computer collaboration in terms of an information processing theoretic framework. By deconstructing either a single or competing decision support systems using the HACT framework, a designer can better understand how humans and computers are collaborating across different dimensions, in order to identify possible problem areas in need of redesign; for example, in the case of the Patriot missile system with a $(-2, -2, 1)$ three-tuple and its demonstrated poor performance, designers could change the decider role to a 2 (only the human makes the final decision, automation cannot veto), as well as move towards a more truly collaborative solution generation LOC. Because missile intercept is a time-pressured task, it is important that the automation moderate the task, but because of the inability of the automation to always correctly make recommendations, more collaboration is needed across the solution-generation role, with no automation authority in the decider role. Used in this manner, HACT aids designers in the understanding of the multiagent roles in human–computer collaboration tasks, as well as identifying areas for possible improvement across these roles.

26.4 Conclusion and Open Challenges

The human–automation collaboration taxonomy (HACT) presented here builds on previous research by expanding the *Parasuraman* [26.1] information processing model, specifically the decision-making component. Instead of defining a simple level of automation for decision making, we deconstruct the process to include three distinct roles, that of the moderator (the agent that ensures the decision-making process moves forward), the generator (the agent that is primarily responsible for generating a solution or set of

possible solutions), and the decider (the agent that decides the final solution along with veto authority). These three distinct (but not necessarily mutually exclusive) roles can each be scaled across five levels indicating degrees of collaboration, with the center value of 0 in each scale representing balanced collaboration. These levels of collaboration (LOCs) form a three-tuple that can be analyzed to evaluate system collaboration, and possibly identify areas for design intervention.

As with all such levels, scales, taxonomies, etc., there are limitations. First, HACT as outlined here does not address all aspects of collaboration that could be considered when evaluating the collaborative nature of a system, such as the type and possible latencies in communication, whether or not the LOCs should be dynamic, the transparency of the automation, the type of information used (i.e., low-level detail as opposed to higher, more abstract concepts), and finally how adaptable the system is across all of these attributes. While this has been discussed in earlier work [26.7], more work is needed to incorporate this into a comprehensive yet useful application.

In addition, HACT is descriptive versus prescriptive, which means that it can describe a system and identify post hoc where designs may be problematic, but cannot indicative how the system should be designed to achieve some predicted outcome. To this end, more research is needed in the application of HACT and the interrelation of the entries within each three-tuple, as well as more general relationships across three-tuples. Regarding the within three-tuples issue, more research is needed to determine the impact and relative importance of each of the three roles; for example, if the moderator is at a high LOC but the generator is at a low LOC, are there generalizable principles that can be seen across different decision support systems? In terms of the between three-tuple issue, more research is needed to determine under what conditions certain three-tuples produce consistently poor (or superior) performance, and whether these are generalizable under particular contexts; for example, in high-risk time-critical supervisory control domains such as nuclear power plant operations, a three-tuple of $(-2, -2, -2)$ may be necessary. However, even in this case, given flawed automated algorithms such as those seen in the Patriot missile, the question could be raised of whether it is ever feasible to design a safe $(-2, -2, -2)$ system.

Despite these limitations, HACT provides more detailed information about the collaborative nature of systems than did previous level-of-automation scales, and given the increasing presence of *intelligent* automation both in complex supervisory control systems and everyday life, such as global positioning system (GPS) navigation, this sort of taxonomy can provide for more in-depth analysis and a common point of comparison across competing systems. Other future areas of research that could prove useful would be the determination of how levels of collaboration apply in the other data acquisition and action implementation information processing stages, and what the impact on human performance would be if different collaboration levels were mixed across the stages. Lastly, one area often overlooked that deserves much more attention is the ethical and social impact of human–computer collaboration. Higher levels of automation authority can reduce an operator's awareness of critical events [26.19] as well as reduce their sense of accountability [26.20]. Systems that promote collaboration with an automated agent could possibly alleviate the offloading of attention and accountability to the automation, or collaboration may further distance operators from their tasks and actions and promote these biases. There has been very little research in this area, and given the vital nature of many time-critical systems that have some degree of human–computer collaboration (e.g., air-traffic control and military command and control), the importance of the social impact of such systems should not be overlooked.

References

26.1   R. Parasuraman, T.B. Sheridan, C.D. Wickens: A model for types and levels of human interaction with automation, IEEE Trans. Syst. Man Cybern. – Part A: System and Humans **30**(3), 286–297 (2000)

26.2   P. Dillenbourg, M. Baker, A. Blaye, C. O'Malley: The evolution of research on collaborative learning. In: *Learning in Humans and Machines. Towards an Interdisciplinary Learning Science*, ed. by P. Reimann, H. Spada (Pergamon, London 1995) pp. 189–211

26.3   J. Roschelle, S. Teasley: The construction of shared knowledge in collaborative problem solving. In: *Computer Supported Collaborative Learning*, ed. by C. O'Malley (Springer, Berlin 1995) pp. 69–97

26.4   P.J. Smith, E. McCoy, C. Layton: Brittleness in the design of cooperative problem-solving systems: the effects on user performance, IEEE Trans. Syst. Man Cybern. **27**(3), 360–370 (1997)

26.5   H.A. Simon, G.B. Dantzig, R. Hogarth, C.R. Plott, H. Raiffa, T.C. Schelling, R. Thaler, K.A. Shepsle, A. Tversky, S. Winter: Decision making and problem solving, Paper presented at the Research Briefings 1986: Report of the Research Briefing Panel on Decision Making and Problem Solving, Washington D.C. (1986)

26.6   P.M. Fitts (ed.).: *Human Engineering for an Effective Air Navigation and Traffic Control system* (National Research Council, Washington D.C. 1951)

26.7 S. Bruni, J.J. Marquez, A. Brzezinski, C. Nehme, Y. Boussemart: Introducing a human–automation collaboration taxonomy (HACT) in command and control decision-support systems, Paper presented at the 12th Int. Command Control Res. Technol. Symp., Newport (2007)

26.8 M.P. Linegang, H.A. Stoner, M.J. Patterson, B.D. Seppelt, J.D. Hoffman, Z.B. Crittendon, J.D. Lee: Human–automation collaboration in dynamic mission planning: a challenge requiring an ecological approach, Paper presented at the Human Factors and Ergonomics Society 50th Ann. Meet., San Francisco (2006)

26.9 Y. Qinghai, Y. Juanqi, G. Feng: Human–computer collaboration control in the optimal decision of FMS scheduling, Paper presented at the IEEE Int. Conf. Ind. Technol. (ICIT '96), Shanghai (1996)

26.10 R.E. Valdés-Pérez: Principles of human–computer collaboration for knowledge discovery in science, Artif. Intell. **107**(2), 335–346 (1999)

26.11 B.G. Silverman: Human–computer collaboration, Hum.–Comput. Interact. **7**(2), 165–196 (1992)

26.12 L.G. Terveen: An overview of human–computer collaboration, Knowl.-Based Syst. **8**(2–3), 67–81 (1995)

26.13 G. Johannsen: *Mensch-Maschine-Systeme (Human–Machine Systems)* (Springer, Berlin 1993), in German

26.14 J. Rasmussen: Skills, rules, and knowledge; signals, signs, and symbols, and other distractions in human performance models, IEEE Trans. Syst. Man Cybern. **13**(3), 257–266 (1983)

26.15 T.B. Sheridan, W. Verplank: *Human and Computer Control of Undersea Teleoperators* (MIT, Cambridge 1978)

26.16 M.R. Endsley, D.B. Kaber: Level of automation effects on performance, situation awarness and workload in a dynamic control task, Ergonomics **42**(3), 462–492 (1999)

26.17 V. Riley: A general model of mixed-initiative human–machine systems, Paper presented at the Human Factors Society 33rd Ann. Meet., Denver (1989)

26.18 S. Bruni, M.L. Cummings: Tracking resource allocation cognitive strategies for strike planning, Paper presented at the COGIS 2006, Paris (2006)

26.19 K.L. Mosier, L.J. Skitka: Human decision makers and automated decision aids: made for each other? In: *Automation and Human Performance: Theory and Applications*, ed. by R. Parasuraman, M. Mouloua (Lawrence Erlbaum, Mahwah 1996) pp. 201–220

26.20 M.L. Cummings: Automation and accountability in decision support system interface design, J. Technol. Stud. **32**(1), 23–31 (2006)

27. Teleoperation

Luis Basañez, Raúl Suárez

This chapter presents an overview of the teleoperation of robotics systems, starting with a historical background, and including the description of an up-to-date specific teleoperation scheme as a representative example to illustrate the typical components and functional modules of these systems. Some specific topics in the field are particularly discussed, for instance, control algorithms, communications channels, the use of graphical simulation and task planning, the usefulness of virtual and augmented reality, and the problem of dexterous grasping. The second part of the chapter includes a description of the most typical application fields, such as industry and construction, mining, underwater, space, surgery, assistance, humanitarian demining, and education, where some of the pioneering, significant, and latest contributions are briefly presented. Finally, some conclusions and the trends in the field close the chapter.

The topics of this chapter are closely related to the contents of other chapters such as those on *Communication in Automation, Including Networking and Wireless* (Chap. 13), *Virtual Reality and Automation* (Chap. 15), and *Collaborative Human–Automation Decision Making* (Chap. 26).

The term *teleoperation* is formed as a combination of the Greek word $\tau\eta\lambda\varepsilon$-, (*tele-*, offsite or remote), and the Latin word *operatĭo, -ōnis* (*operation*, something done). So, *teleoperation* means performing some work or action from some distance away. Although in this sense teleoperation could be applied to any operation performed at a distance, this term is most commonly associated with robotics and mobile robots and indicates the driving of one of these machines from a place far from the machine location.

There are of lot of topics involved in a *teleoperated robotic system*, including human–machine interaction, distributed control laws, communications, graphic simulation, task planning, virtual and augmented reality, and dexterous grasping and manipulation. Also the fields of application of these systems are very wide and teleoperation offers great possibilities for profitable applications. All these topics and applications are dealt with in some detail in this chapter.

27.1 Historical Background and Motivation

Since a long time ago, human beings have used a range of tools to increase their manipulation capabilities. In the beginning these tools were simple tree branches, which evolved to long poles with tweezers, such as blacksmith's tools that help to handle hot pieces of iron. These developments were the ancestors of master–slave robotic systems, where the slave robot reproduces the master motions controlled by a human operator. Tele-operated robotic systems allow humans to interact with robotic manipulators and vehicles and to handle objects located in a remote environment, extending human manipulation capabilities to far-off locations, allowing the execution of quite complex tasks and avoiding dangerous situations.

The beginnings of teleoperation can be traced back to the beginnings of radio communication when Nikola Tesla developed what can be considered the first tele-operated apparatus, dated 8 November 1898. This development has been reported under the US patent 613 809, *Method of and Apparatus for Controlling Mechanism of Moving Vessels or Vehicles*. However, bilateral teleoperation systems did not appear until the late 1940s. The first bilateral manipulators were developed for handling radioactive materials. Outstanding pioneers were Raymond Goertz and his colleagues at the Argonne National Laboratory outside of Chicago, and Jean Vertut at a counterpart nuclear engineering laboratory near Paris. The first mechanisms were mechanically coupled and the slave manipulator mimicked the master motions, both being very similar mechanisms (Fig. 27.1). It was not until the mid 1950s that Goertz presented the first electrically coupled master–slave manipulator (Fig. 27.2) [27.1].

In the 1960s applications were extended to underwater teleoperation, where submersible devices carried cameras and the operator could watch the remote robot and its interaction with the submerged environment. The beginnings of space teleoperation dates form the 1970s, and in this application the presence of time delay started to cause instability problems.

Technology has evolved with giant steps, resulting in better robotic manipulators and, in particular, increasing the communication means, from mechanical to

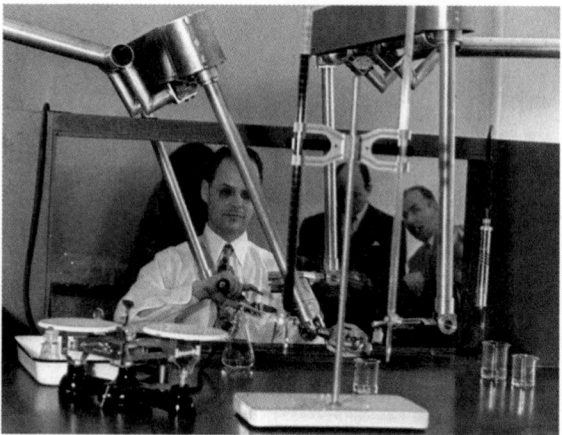

Fig. 27.1 Raymond Goertz with the first mechanically coupled teleoperator (Source: Argonne National Labs)

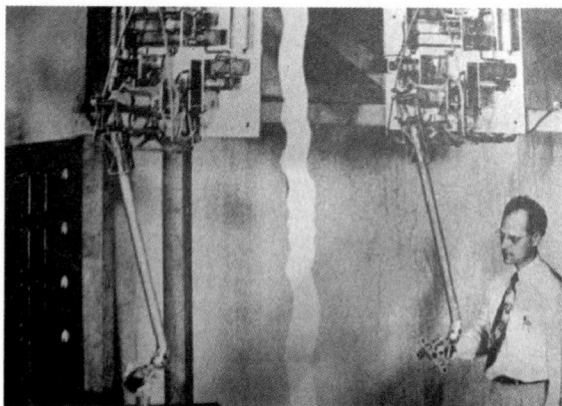

Fig. 27.2 Raymond Goertz with an electrically coupled teleoperator (Source: Argonne National Labs)

electrical transmission, using optic wires, radio signals, and the Internet which practically removes any distance limitation.

Today, the applications of teleoperation systems are found in a large number of fields. The most illustrative are space, underwater, medicine, and hazardous environments, which are described amongst others in Sect. 27.4

27.2 General Scheme and Components

A modern teleoperation system is composed of several functional modules according to the aim of the system. As a paradigm of an up-to-date teleoperated robotic system, the one developed at the Robotics Laboratory of the Institute of Industrial and Control Engineering (IOC), Technical University of Catalonia (UPC), Spain, will be described below [27.2].

The outline of the IOC teleoperation system is represented in Fig. 27.3. The diagram contains two large blocks that correspond to the local station, where the human operator and master robots (haptic devices) are located, and the remote station, which includes two industrial manipulators as slave robots. The system contains the following system modules.

Relational positioning module: This module provides the operator with a means to define geometric relationships that should be satisfied by the part manipulated by the robots with respect to the objects in the environment. These relationships can completely define the position of the manipulated part and then fix all the

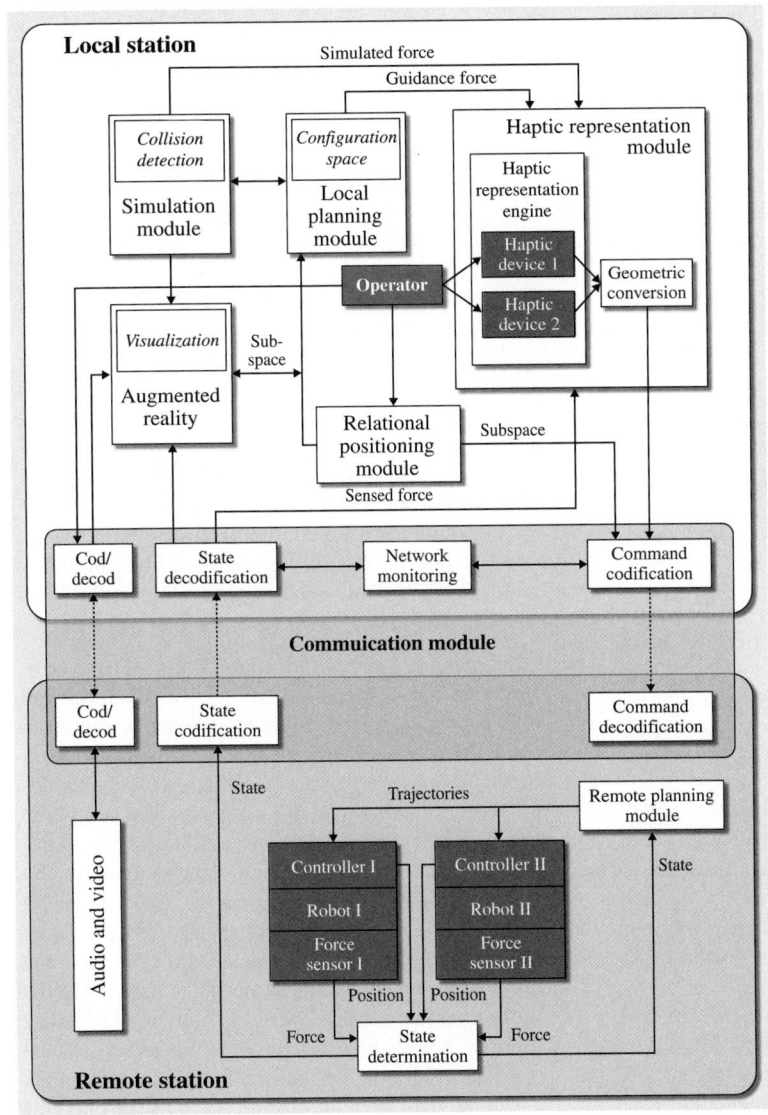

Fig. 27.3 A general scheme of a teleoperation system (courtesy of IOC-UPC)

robots' degrees of freedom (DOFs) or they can partially determine the position and orientation and therefore fix only some DOFs. In the latter case, the remaining degrees of freedom are those that the operator will be able to control by means of one or more haptic devices (master robots). Then, the output of this module is the solution subspace in which the constraints imposed by the relationships are satisfied. This output is sent to the modules of *augmented reality* (for visualization), *command codification* (to define the possible motions in the solution subspace), and *planning* (to incorporate the motion constraints to the haptic devices).

Haptic representation module: This module consists of the *haptic representation engine* and the *geometric conversion submodule*. The haptic representation engine is responsible for calculating the force to be fed back to the operator as a combination of the following forces:

- *Restriction force*: This is calculated by the *planning* module to assure that, during the manipulation of the haptic device by the operator, the motion constraints determined by the *relational positioning* module are satisfied.
- *Simulated force*: This is calculated by the *simulation* module as a reaction to the detection of potential collision situations.
- *Reflected force*: This is the force signal sent from the remote station through the *communication* module to the local station corresponding to the robots' actuators forces and those measured by the force and torque sensors in the wrist of the robots produced by the environmental interaction.

The *geometric conversion submodule* is in charge of the conversion between the coordinates of the haptic devices and those of the robots.

Augmented-reality module: This module is in charge of displaying to the user the view of the remote station, to which is added the following information:

- *Motion restrictions imposed by the operator.* This information provides the operator with the understanding and control of the unrestricted degrees of freedom can be commanded by means of the haptic device (for example, it can visualize a plane on which the motions of the robot end-effector are restricted).
- *Graphical models of the robots in their last configuration received from the cell.* This allows the operator to receive visual feedback of the robots' state from the remote station at a frequency faster

than that allowed by the transmission of the whole image, since it is possible to update the robots' graphical models locally from the values of their six joint variables.

This module receives as inputs: (1) the image of the cell, (2) the state (pose) of the robots, (3) the model of the cell, and (4) the motion constraints imposed by the operator. This module is responsible for maintaining the coherence of the data and for updating the model of the cell.

Simulation module: This module is used to detect possible collisions of the robots and the manipulated pieces with the environment, and to provide feedback to the operator with the corresponding force in order to allow him to react quickly when faced with these possible collision situations.

Local planning module: The planning module of the local station computes the forces that should guide the operator to a position where the geometric relationships he has defined are satisfied, as well as the necessary forces to prevent the operator from violating the corresponding restrictions.

Remote planning module: The planning module of the remote station is in charge of reconstructing the trajectories traced by the operator with the haptic device. This module includes a feedback loop for position and force that allows safe execution of motions with compliance.

Communication module: This module is in charge of communications between the local and the remote stations through the used communication channel (e.g., Internet or Internet2). This consists of the following submodules for the information processing in the local and remote stations:

- *Command codification/decodification*: These submodules are responsible for the codification and decodification of the motion commands sent from the local station and the remote station. These commands should contain the information of the degrees of freedom constrained to satisfy the geometric relationships and the motion variables on the unrestricted ones, following the movements specified by the operator by means of the haptic devices (for instance, if the motion is constrained to be on a plane, this information will be transferred and then the commands will be the three variables that define the motion on that plane). For each robot, the following three qualitatively different situations are possible:
  - The motion subspace satisfying the constraints defined by the relationships fixed by the operator

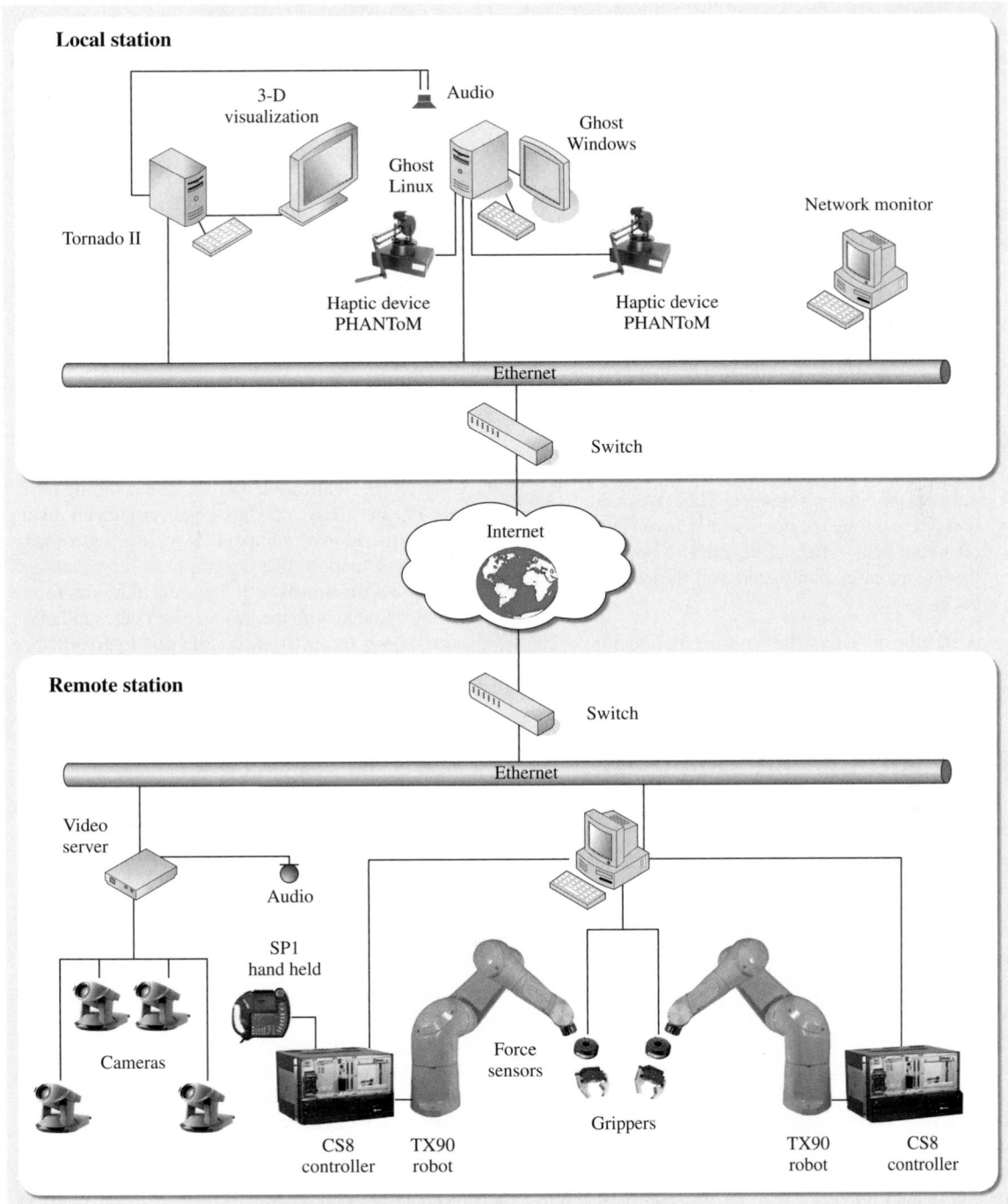

Fig. 27.4 Physical architecture of a teleoperation system (courtesy of IOC-UPC)

has dimension zero. This means that the constraints completely determine the position and

orientation (pose) of the manipulated object. In this case the command is this pose.

- The motion subspace has dimension six, i. e., the operator does not have any relationship fixed. In this case the operator can manipulate the six degrees of freedom of the haptic device and the command sent to the remote station is composed of the values of the six joint variables.
- The motion subspace has dimension from one to five. In this case the commands are composed of the information of this subspace and the variables that describe the motion inside it, calculated from the coordinates introduced by the operator through the haptic device or determined by the local planning module.

- *State codification/decodification*: These submodules generate and interpret the messages between the remote and the local stations. The robot state is coded as the combination of the position and force information.
- *Network monitoring system*: This submodule analyzes in real time the quality of service (QoS) of the communication channel in order to properly adapt the teleoperation parameters and the sensorial feedback.

A scheme depicting the physical architecture of the whole teleoperation system is shown in Fig. 27.4.

27.3 Challenges and Solutions

During the development of modern teleoperation systems, such as the one described in Sect. 27.2, a lot of challenges have to be faced. Most of these challenges now have a partial or total solution and the main ones are reviewed in the following subsections.

27.3.1 Control Algorithms

A control algorithm for a teleoperation system has two main objectives: telepresence and stability. Obviously, the minimum requirement for a control scheme is to preserve stability despite the existence of time delay and the behavior of the operator and the environment. Telepresence means that the information about the remote environment is displayed to the operator in a natural manner, which implies a feeling of presence at the remote site (immersion). Good telepresence increases the feasibility of the remote manipulation task. The degree of telepresence associated to a teleoperation system is called *transparency*.

27.2.1 Operation Principle

In order to perform a robotized task with the described teleoperation system, the operator should carry out the following steps:

- Define the motion constraints for each phase of the task, specifying the relative position of the manipulated objects or tools with respect to the environment.
- Move the haptic devices to control the motions of the robots in the subspace that satisfies the imposed constraints. The haptic devices, by means of the force feedback applied to the operator, are capable of:
  - guiding the operator motions so that they satisfy the imposed constraints
  - detecting collision situations and trying to avoid undesired impacts
- Control the realization of the task availing himself of an image of the scene visualized using three-dimensional augmented reality with additional information (like the graphical representation of the motion subspace, the graphical model of the robots updated with the last received data, and other outstanding information for the good performance of the task).

Scattering-based control has always dominated the control field in teleoperation systems since it was first proposed by *Anderson* and *Spong* [27.3], creating the basis of modern teleoperation system control. Their approach was to render the communications passive using the analogy of a lossless transmission line with scattering theory. They showed that the scattering transformation ensures passivity of the communications despite any constant time delay. Following the former scattering approach, it was proved [27.4] that, by matching the impedances of the local and remote robot controllers with the impedance of the virtual transmission line, wave reflections are avoided. These were the beginnings of a series of developments for bilateral teleoperators. The reader may refer to [27.5, 6] for two advanced surveys on this topic.

Various control schemes for teleoperated robotic systems have been proposed in the literature. A brief description of the most representative approaches is presented below.

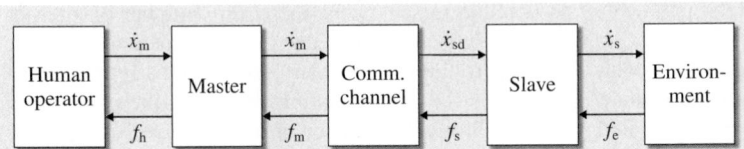

Fig. 27.5 Traditional force reflection

Traditional force reflection. This is probably the most studied and reported scheme. In this approach, the master sends position information to the slave and receives force feedback from the remote interaction of the slave with the environment (Fig. 27.5). However, it was shown that stability is compromised in systems with high time delay [27.3].

Shared compliance control. This scheme is similar to the traditional force reflection, except that on the slave side a compliance term is inserted to modify the behavior of the slave manipulator according to the interaction with the environment.

Scattering-based teleoperation. The scattering transformation (wave variables) used in the transmission of power information makes the communication channel passive even if a time delay T affects the system (Fig. 27.6). However, the scattering transformation presents a tradeoff between stability and performance. In an attempt to improve performance using the scattering transformation, several approaches have been reported, for instance, transmitting wave integrals [27.7, 8] and wave filtering and wave prediction [27.9].

Four-channel control. Velocity and force information is sent to the other side in both directions, thereby defining four channels. In both controllers a linear combination of the available force and velocity information is used to fit the specifications of the control design [27.10].

Proportional (P) and proportional–derivative (PD) controllers. It is widely known that use of the classic scattering transformation may give raise to position drift. In [27.11] position tracking is achieved by sending the local position to the remote station, and adding a proportional term to the position error in the remote controller. Following this approach, [27.12] proposed a symmetric scheme by matching the impedances and adding a proportional error term to the local and remote robots, such that the resulting control laws became simple PD-like controllers. Stability of PD-like controllers, without the scattering transformation, has been proved in [27.13] under the assumption that the human interaction with the local manipulator is passive. In [27.14] it is shown that, when the human operator applies a constant force on the local manipulator,

a teleoperation system controlled with PD-like laws is stable.

Variable-time-delay schemes. In the presence of variable time delays, the basic scattering transformation cannot provide the passivity needed in the communications [27.15]. In order to solve this issue, the use of a time-varying gain that is a function of the rate of change of the time delay has been proposed [27.16]. Recently it has been shown [27.17] that, under an appropriate dissipation strategy, the communications can dissipate an amount of energy equal to the generated energy. Applying the strategy of [27.15], in [27.18] it was proven that, under power scaling factors for microteleoperation, the resulting communications remain passive.

27.3.2 Communication Channels

Communication channels can be classified in terms of two aspects: their physical nature and their mode of operation. According to the first aspect, two groups can be defined: physically connected (mechanically, electrically, optically wired, pneumatically, and hydraulically) and physically disconnected (radiofrequency and optically coupled such as via infrared). The second aspect entails the following three groups:

- *Time delay free.* The communication channel connecting the local and the remote stations does not affect the stability of the overall teleoperation system. In general this is the kind of channel present when the two stations are near to each other. Examples of these communication channels are some surgical systems, where the master and slave are lo-

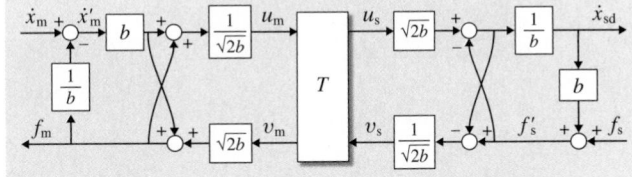

Fig. 27.6 Scattering transformation with impedance adaptation

cated in the same room and connected through wires or radio.

- *Constant time delay*. These are often associated with communications in space, underwater teleoperation using sound signals, and systems with dedicated wires across large distances.
- *Variable time delay*. This is the case, for instance, of packet-switched networks where variable time delays are caused by many reasons such as routing, acknowledge response, and packing and unpacking data.

One of the most promising teleoperation communication channels is the Internet, which is a packet-switched network, i.e., it uses protocols that divide the messages into packets before transmission. Each packet is then transmitted individually and can follow a different route to its destination. Once all packets forming a message have arrived at the destination, they are recompiled into the original message. The transmission control protocol (TCP) and user datagram protocol (UDP) work in this way and they are the Internet protocols most suitable for use in teleoperation systems.

In order to improve the performance of teleoperation systems, quality of service (QoS)-based schemes have been used to provide priorities on the communication channel. The main drawback of today's best-effort Internet service is due to network congestion. The use of high-speed networks with recently created protocols, such as the Internet protocol version 6 (IPv6), improves the performance of the whole teleoperation system [27.19].

Besides QoS, IPv6 presents other important improvements. The current 32 bit address space of IPv4

is not able to satisfy the increasing number of internet users. IPv6 quadruples this address space to 128 bits, which provides more than enough globally unique IP addresses for every network device on the planet. See Fig. 27.7 for a comparison of these protocols.

When using packet-switched networks for real-time teleoperation systems, besides bandwidth, three effects can result in decreased performance of the communication channel: packet loss, variable time delay, and in some cases, loss of order in packet arrival.

27.3.3 Sensory Interaction and Immersion

Human beings are able to perceive information from the real world in order to interact with it. However, sometimes, for engineering purposes, there is a need to interact with systems that are difficult to build in reality or that, due to their physical behavior, present unknown features or limitations. Hence, in order to allow better human interaction with such systems, as well as their evaluation and understanding, the concepts of *virtual reality* and *augmented reality* have been researched and applied to improve development cycles in engineering.

In virtual reality a nonexistent world can be simulated with a compelling sense of realism for a specific environment. So, the real world is replaced by a computer-generated world that uses input devices to interact with and obtain information from the user and capture data from the real world (e.g., using trackers and transducers), and uses output displays that represent the responses of the virtual world by means of visual, touch, aural or taste displays [e.g., haptic devices, head-mounted displays (HMD), and headphones] in order to be perceived by any of the human senses. In this context,

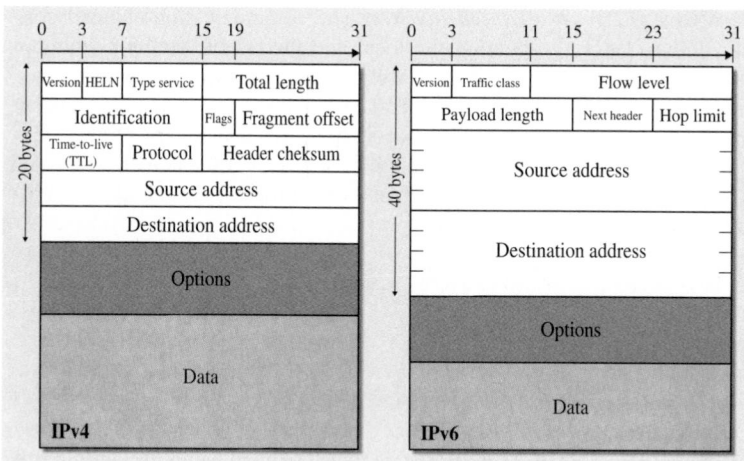

Fig. 27.7 Comparison of IPv4 and IPv6 protocols

immersion is the sensation of being in an environment that actually does not exist and that can be a purely mental state or can be accomplished through physical elements [27.20].

Augmented reality is a form of human–computer interaction (HCI) that superimposes information created by computers over a real environment. Augmented reality enriches the surrounding environment instead of replacing it as in the case of virtual reality, and it can also be applied to any of the human senses. Although some authors put attention on hearing and touch [27.21], the main augmentation route is through visual data addition. Furthermore augmented reality can remove real objects or change their appearance [27.22], operations known as diminished or mediated reality. In this case, the information that is shown and superposed depends on the context, i.e., on the observed objects.

Augmented reality can improve task performance by increasing the degree of reliability and speed of the operator due to the addition or reduction of specific information. Reality augmentation can be of two types: *modal* or *multimodal*. In the modal type, augmentation is referred to the enrichment of a particular sense (normally sight), whereas in the multimodal type augmentation includes several senses. Research done to date has focused mainly on modal systems [27.21, 23].

In teleoperation environments, augmented reality has been used to complement human sensorial perception in order to help the operator perform teleoperated tasks. In this context, augmented reality can reduce or eliminate the factors that break true perception of the remote station, such as time delays in the communication channel, poor visibility of the remote scene, and poor perception of the interaction with the remote environment.

Amongst the applications of augmented reality it is worthwhile to mention interaction between the operator and the remote site for better visualization [27.24, 25], better collaboration capacity [27.26], better path or motion planning for robots [27.27, 28], addition of specific virtual tools [27.29], and multisensorial perception enrichment [27.30].

27.3.4 Teleoperation Aids

Some of the problems arising in teleoperated systems, such as an unstructured environment, communication delays, human operator uncertainty, and safety at the remote site, amongst others, can be reduced using teleoperation aids.

Amongst the teleoperation aids aimed to diminish human operator uncertainty one can highlight virtual fixtures for guiding motion, which have recently been added in surgical teleoperation in order to improve the surgeon's repeatability and reduce his fatigue.

The trajectories to be described by a robot end-effector – either in free space or in contact with other objects – strongly depend on the task to be performed and on the topology of the environment with which it is interacting; for instance, peg-in-hole insertions require alignment between the peg and the hole, spray-painting tasks require maintenance of the nozzle at a fixed distance and orientation with respect to the surface to be painted, and assembly tasks often involve alignment or coincidence of faces, sides, and vertices of the parts to be assembled. For all these examples, virtual guides can be defined and can help the operator to perform the task.

Artificial fixtures or motion guidance can be divided into two groups, depending on how the motion constraints are created, either by software or by hardware. To the first group belong the methods that implement geometric constraints for the operator motions: points, lines, planes, spheres, and cylinders [27.2], which can usually be changed without stopping the teleoperation. An often-used method is to provide obstacles with a repulsive force field, avoiding in this way that the operator makes the robot collide with the obstacles. In the second group, specific hardware is used to guide the motion, for example, guide rails and sliders with circled rails. Figure 27.8 shows a teleoperated painting task restricted to a plane.

An example of a motion constraints generator is the PMF (positioning mobile with respect to fixed) solver [27.31]. PMF has been designed to assist execution of teleoperated tasks featuring precise or repetitive motions. By formulating an object positioning problem in terms of symbolic geometric constraints, the motion

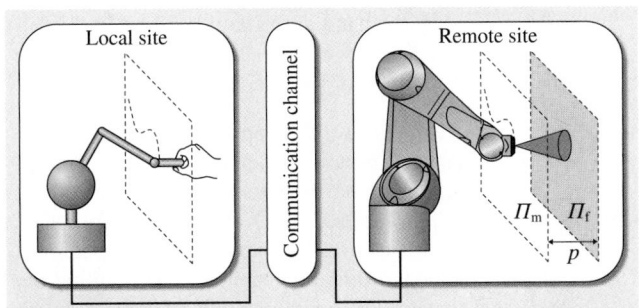

Fig. 27.8 A painting teleoperation task with a plane constraint on the local and the remote sites

of an object can be totally or partially restricted, independently of its initial configuration. PMF exploits the fact that, in geometric constraint sets, the rotational component can often be decoupled from the translational one and solved independently. Once the solution is obtained, the resulting restriction forces are fed to the operator via a haptic interface in order to guide its motions inside this subspace.

27.3.5 Dexterous Telemanipulation

A common action in robotics applications is grasping of an object, and teleoperated robotics is no exception. Grasping actions can often be found in telemanipulation tasks such as handling of dangerous material, rescue, assistance, and exploration, amongst others.

In planning a grasping action, two fundamental aspects must be considered:

1. *How to grasp the object*. This means the determination of the contact points of the grasping device on the object, or at a higher level, the determination of the relative position of the grasping device with respect to the object (e.g., [27.32–34]).
2. *The grasping forces*. This means the determination of the forces to be applied by the grasping device actuators in order to properly constrain the object (e.g., [27.35]).

These two aspects can be very simple or extremely complex depending on the type of object to be grasped, the type of grasping device, and the requirements of the task. In a teleoperated grasping system, besides the general problems associated with a teleoperation system mentioned in the previous sections, the following particular topics must be considered.

Sensing information in the local station. In telemanipulation using complex dexterous grasping devices, such as mechanical anthropomorphic hands directly commanded by the hand of the human operator, the following approaches have been used in order to capture the pose information of the operator hand:

- *Sensorized gloves*. The operator wears a glove with sensors (usually strain gauges) that identify the position of the fingers and the flexion of the palm [27.36]. These gloves allow the performance of tasks in a natural manner, but they are delicate devices and it is difficult to achieve good calibration.
- *Exoskeletons*. The operator wears over the hand an exoskeleton equipped with encoders that identify the position of the fingers [27.37]. Exoskeletons are

more robust in terms of noise, but they are rather uncomfortable and reduce the accessibility of the hand in certain tasks.
- *Vision systems*. Computer vision is used to identify hand motions [27.38]. The operator does not need to wear any particular device and is therefore completely free, but some parts of the hand may easily fall outside of the field of vision of the system and recognition of hand pose from images is a difficult task.

Capturing the forces applied by the operator is a much more complex task, and only some tests using pressure sensors at the fingertips have been proposed [27.39].

Feedback information from the remote station. This can be basically of two types:

- *Visual information*. This kind of information can help the operator to realize how good (robust or stable) the remote grasp is, but only in a very simple grasp can the operator conclude if it is actually a successful grasp.
- *Haptic information*. Haptic devices allow the operator to feel the contact constraints during the grasp in the remote station. Current approaches include gloves with vibratory systems that provides a kind of tactile feeling [27.40], and exoskeletons that attached to the hand and fingers and generate constraints to their motion and provide the feeling of a contact force [27.37]. Nevertheless, these devices have limited performance and the development of more efficient haptic devices with the required num-

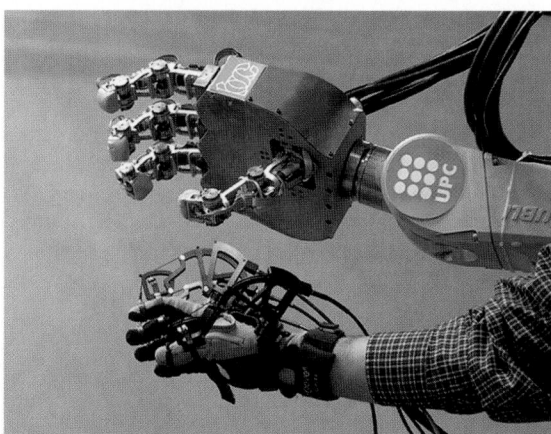

Fig. 27.9 Operator hand wearing a sensorized glove and an exoskeleton, and the anthropomorphic mechanical hand MA-I (courtesy of IOC-UPC)

ber of degrees of freedom and the configuration of the human hand is still an open problem.

Need for kinematics mapping. In real situations, the mechanical gripper or hand in the remote station will not have the same kinematics as the operator hand, even when an anthropomorphic mechanical hand is used. This means that in general the motions of the operator cannot be directly replicated by the remote grasping device, and they have to be interpreted and then adapted from one kinematics to the other, which may be computationally expensive [27.41].

Use of assistance tools. The tools developed with the aim of performing grasps in an autonomous way can be used as assistance tools in telemanipulation;

for instance, grasp planners used to determine optimal grasping points automatically on different types of objects can be run considering the object to be telemanipulated and then, using augmented reality, highlight the grasping points on the object so the operator can move the fingers directly to those points. Of still greater assistance in this regard is the computation and display of independent grasping regions on the object surface [27.42] such that placing a finger on any point within each of these regions will achieve a grasp with a controlled quality [27.43].

Figure 27.9 shows an example where the operator is wearing a commercial sensorized glove and an exoskeleton in order to interact with the anthropomorphic mechanical hand MA-I [27.44].

27.4 Application Fields

The following subsections present several application fields where teleoperation plays a significant role, describing their main particular aspects and some relevant works.

27.4.1 Industry and Construction

Teleoperation in industry-related applications covers a wide range of fields. One of them is mostly oriented towards inspection, repair, and maintenance operations in places with difficult or dangerous access, particularly in power plants [27.45], as well as to manage toxic wastes [27.46]. In the nuclear industry the main reason to avoid the exposure of human workers is the existence of a continuous radioactive environment, which results in international regulations to limit the number of hours that humans can work in these conditions. This application was actually the motivation for early real telemanipulation developments, as stated in Sect. 27.1. Some typical teleoperated actions in nuclear plants are the maintenance of nuclear reactors, decommissioning and dismantling of nuclear facilities, and emergency interventions. The challenges in these tasks include operation in confined areas with high radiation levels, risk of contamination, unforeseen accidents, and manipulation of materials that can be liquid, solid or have a muddy consistency.

Another kind of application is the maintenance of electrical power lines, which require operations such as replacement of ceramic insulators or opening and reclosing bridges, which are very risky for human operators due to the height of the lines and the possibility

of electric shocks, specially under poor weather conditions [27.47]. That is why electric power companies are interested in the use of robotic teleoperated systems for live-line power maintenance. Examples of these robots are the TOMCAT [27.48] and the ROBTET (Fig. 27.10) [27.49].

Another interesting application field is construction, where teleoperation can improve productivity, reliability, and safety. Typical tasks in this field are earth-moving, compaction, road construction and maintenance, and trenchless technologies [27.50]. In general, applications in this field are based on direct visual feedback. One example is radio operation of construc-

Fig. 27.10 Robot ROBTET for maintenance of electrical power lines (courtesy of DISAM, Technical University of Madrid – UPM)

tion machinery, such as bulldozers, hydraulic shovels, and crawler dump trucks, to build contention barriers against volcanic eruptions [27.51]. Another example is the use of an experimental robotized crane with a six-DOF parallel kinematic structure, to study techniques and technologies to reduce the time required to erect steel structures [27.52].

Since the tasks to be done are quite different in the different applications, the particular hardware and devices used in each case can vary a lot, ranging from a fixed remote station in the dangerous area of a nuclear plant, to a mobile remote station assembled on a truck that has to move along a electrical power line or a heavy vehicle in construction. See also Chap. 61 on *Construction Automation* and Chap. 62 on *Smart Buildings*.

27.4.2 Mining

Another interesting field of application for teleoperation is mining. The reason is quite clear: operation of a drill underground is very dangerous, and sometimes mines themselves are almost inaccessible. One of the first applications started in 1985, when the thin-seam continuous mining Jeffrey model 102HP was extensively modified by the US Bureau of Mines to be adapted for teleoperation. Communication was achieved using 0.6 inch wires, and the desired entry orientation was controlled using a laser beam [27.53]. Later, in 1991, a semiautomated haulage truck was used underground, and since then has hauled 1.5 million tons of ore without failure. The truck has an on-board personal computer (PC) and video cameras and the operator can stay on the surface and teleoperate the vehicle using an interface that simulates the dashboard of the truck [27.54]. The most common devices used for teleoperation in mining are load–haul–dump (LHD) machines, and thin-seam continuous mining (TSCM) machines, which can work in a semiautonomous and teleoperated way.

Position measurement, needed for control, is not easy to obtain when the vehicle is beneath the surface, and interference can be a problem, depending on the mine material. Moreover, for the same reason, video feedback has very poor quality. In order to overcome these problems, the use of gyroscopes, magnetic electronic compasses, and radar to locate the position of vehicles while underground has been considered [27.55]. The problems with visual feedback could be solved by integrating, for instance, data from live video, computer-aided design (CAD) mine models, and process control parameters, and presenting the

operator a view of the environment with augmented reality [27.56]. In this field, in addition to information directly related to the teleoperation, the operator has to know other measurements for safety reasons, for instance, the volatile gas (like methane) concentration, to avoid explosions produced due to sparks generated by the drilling action.

Teleoperated mining is not only considered on Earth. If it is too expensive and dangerous to have a man underground operating a mining system, it is much more so for the performance of mining tasks on the Moon. As stated in Sect. 27.4.4, for space applications, in addition to the particularities of mining, the long transmission delay between the local and remote stations is a significant problem. So, the degree of autonomy has to be increased to perform the simplest tasks locally while allowing a human teleoperator to perform the complex tasks at a higher level [27.57]. When the machines in the remote station are performing automated actions, the operator can teleoperate some other machinery, thus productivity can be improved by using a multiuser schema at the local station to operate multiple mining systems at the remote station [27.58]. See also Chap. 57 on *Automation in Mining and Mineral Processing*.

27.4.3 Underwater

Underwater teleoperation is motivated by the fact that the oceans are attractive due to the abundance of living and nonliving resources, combined with the difficulty for human beings to operate in this environment. The most common applications are related to rescue missions and underwater engineering works, among other scientific and military applications. Typical tasks are: pipeline welding, seafloor mapping, inspection and reparation of underwater structures, collection of underwater objects, ship hull inspection, laying of submarine cables, sample collection from the ocean bed, and study of marine creatures.

A pioneering application was the cable-controlled undersea recovery vehicle (CURV) used by the US Army in 1966 to recover, in the Mediterranean sea south of Spain, the bombs lost due to a bomber accident [27.59]. More recent relevant applications are related to the inspection and object collection from famous sunken vessels, such as the Titanic with the ARGO robot [27.60], and to ecological disasters, such as the sealing of crevices in the hull of the oil tanker Prestige, which sank in the Atlantic in 2002 [27.61].

Fig. 27.11 Underwater robot Garbi III AUV (courtesy of University of Girona – UdG)

Specific problems in deep underwater environments are the high pressure, quite frequently poor visibility, and corrosion. Technological issues that must be considered include robust underwater communication, the power source, and sensors for navigation. A particular problem in several underwater applications is the position and force control of the remote actuator when it is floating without a fixed holding point.

Most common unmanned underwater robots are remotely operated vehicles (ROVs) (Fig. 27.11), which are typically commanded from a ship by an operator using joysticks. Communication between the local and remote stations is frequently achieved using an umbilical cable with coaxial cables or optic fiber, and also the power is supplied by cables. Most of these underwater vehicles carry a robotic arm manipulator (usually with hydraulic actuators), which may have negligible effects on a large vehicle, but that introduce significant

Fig. 27.12 Canadarm 2 (courtesy of NASA)

perturbation on the system dynamics of a small one. Moreover, there are several sources of uncertainties, mainly due to buoyancy, inertial effects, hydrodynamic effects (of waves and currents), and drag forces [27.62], which has motivated the development of several specific control schemes to deal with these effects [27.63, 64]. The operational cost of these vehicles is very high, and their performance largely depends on the skills of the operator, because it is difficult to operate them accurately as they are always subject to undesired motion. In the oil industry, for instance, it is common to use two arms: one to provide stability by gripping a nearby structure and another to perform the assigned task.

A new use of underwater robots is as a practice tool to prepare and test exploration robots for remote planets and moons [27.65].

27.4.4 Space

The main motivation for the development of space teleoperation is that, nowadays, sending a human into the space is difficult, risky, and quite expensive, while the interest in having some devices in space is continuously growing, from the practical (communications satellites) as well as the scientific point of view.

The first explorations of space were carried out by robotic spacecrafts, such as the Surveyor probes that landed on the lunar surface between 1966 and 1968. The probes transmitted to Earth images and analysis data of soil samples gathered with an extensible claw. Since then, several other ROVs have been used in space exploration, such as in the Voyager missions [27.66].

Various manipulation systems have been used in space missions. The remote manipulator system, named Canadarm after the country that built it, was installed aboard the space shuttle Columbia in 1981, and since then has been employed in a variety of tasks, mainly focused on the capture and redeployment of defective satellites, besides providing support for other crew activities. In 2001, the Canadarm 2 (Fig. 27.12) was added to the International Space Station (ISS), with more load capacity and maneuverability, to help in more sensitive tasks such as inspection and fault detection of the ISS structure itself. In 2009, the European Robotic Arm (ERA) is expected to be installed at the ISS, primarily to be used outside the ISS in service tasks requiring precise handling of components [27.67].

Control algorithms are among the main issues in this type of applications, basically due to the significant delay between the transmission of information from the local station on the Earth and the reception of the response from remote station in space (Sect. 27.3.1). A number of experimental ground-based platforms for telemanipulation such as the Ranger [27.68], the Robonaut [27.69], and the space experiment ROTEX [27.70] have demonstrated sufficient dexterity in a variety of operations such as plug/unplug tasks and tools manipulation. Another interesting experiment under development is the Autonomous Extravehicular Activity Robotic Camera Sprint (AERCam) [27.71], a teleoperated free-flying sphere to be used for remote inspection tasks. An experiment in bilateral teleoperation was developed by the National Space Development Agency of Japan (NASDA) [27.72] with the Engineering Test Satellite (ETS-VII), overcoming the significant time delay (up to 7 s was reported) in the communication channel between the robot and the ground-based control station.

Currently, most effort in planetary surface exploration is focused on Mars, and several remotely operated rovers have been sent to this planet [27.73]. In these experiments the long time delays in the control signals between Earth-based commands and Mars-based rovers is especially relevant. The aim is to avoid the effect of these delays by providing more autonomy to the rovers. So, only high-level control signals are provided by the controllers on Earth, while the rover solves low-level planning of the commanded tasks. Another possible scenario to minimize the effect of delays is teleoperation of the rovers with humans closer to them (perhaps in orbit around Mars) to guarantee a short time delay that will allow the operator to have real-time control of the rover, allowing more efficient exploration of the surface of the planet [27.74]. See also Chap. 69 on *Space and Exploration Automation* and Chap. 93 on *Collaborative Analytics for Astrophysics Explorations*.

27.4.5 Surgery

There are two reasons for using teleoperation in the surgical field. The first is the improvement or extension of the surgeon's abilities when his/her actions are mapped to the remote station, increasing, for instance, the range of position and motion of the surgical tool (motion scaling), or applying very precise small forces without oscillations; this has greatly contributed to the development of major advances in the field of microsurgery, as well as in the development of minimally invasive surgery (MIS) techniques. Using teleoperated systems, surgeries are quicker and patients suffer less than with the normal approach, also allowing faster recovery. The second reason is to exploit the expertise of

very good surgeons around the world without requiring them to travel, which could waste time and fatigue these surgeons.

A basic initial step preceding teleoperation in surgical applications was telediagnostics, i. e., the motion of a device, acting as the remote station, to obtain information without working on the patient. A simple endoscope could be considered as a basic initial application in this regard, since the position of a camera is teleoperated to obtain an appropriate view inside the human body. A relevant application for telediagnostic is an endoscopic system with 3-D stereo viewing, force reflection, and aural feedback [27.75].

It is worth to highlight the first real remote telesurgery [27.76]. The scenario was as follows: the local station, i. e., the surgeon, was located in New York City, USA, and the remote station, i. e., the patient, was in Strasbourg, France. The performed surgery was a laparoscopic cholecystectomy done to a 68-year-old female, and it was called *operation Lindbergh*, based on the last name of the patient. This surgery was possible thanks to the availability of a very secure high-speed communication line, allowing a mean total time delay between the local and remote stations of 155 ms. The time needed to set up the robotic system, in this case the Zeus system [27.77], was 16 min, and the operation was done in 54 min without complications. The patient was discharged 48 h later without any particular postoperative problems.

A key problem in this application field is that someone's life is at risk, and this affects the way in which information is processed, how the system is designed, the amount of redundancy used, and any other factors that may increase safety. Also, the surgical tool design must integrate sensing and actuation on the millimeter scale.

Normally, the instruments used in MIS do not have more than four degrees of freedom, losing therefore the ability to orient the instrument tip arbitrarily, although specialized equipment such as the Da Vinci system [27.78] already incorporates a three-DOF wrist close to the instrument tip that makes the whole system benefit from seven degrees of freedom. In order to perform an operation, at least three surgical instruments are required (the usual number is four): one is an endoscope that provides the video feedback and the other two are grippers or scissors with electric scalpel functions, which should provide some tactile and/or force feedback (Fig. 27.13).

The trend now is to extend the application field of the current surgical devices so that they can be used in different types of surgical procedures, particularly including tactile feedback and virtual fixtures to minimize the effect of any imprecise motion of the surgeon [27.79]. So far, there are more than 25 surgical procedures in at least six medical fields that have been successfully performed with telerobotic techniques [27.80]. See Chap. 78 on *Medical Automation and Robotics*.

27.4.6 Assistance

The main motivation in this field is to give independence to disabled and elderly people in their daily domestic activities, increasing in this way quality of life. One of the first relevant applications in this line was seen in 1987, with the development of the Handy 1 [27.81], to enable an 11-year-old boy with cerebral palsy to gain independence at mealtimes. The main components of Handy 1 were a robotic arm, a microcomputer (used as a controller for the system), and an expanded keyboard for human–machine interface (HMI).

The most difficult part in developing assistance applications is the HMI, as it must be intuitive and appropriate for people that do not have full capabilities. In this regard different approaches are considered, such as tactile, voice recognition, joystick/haptic interfaces, buttons, and gesture recognition, among others [27.82]. Another very important issue, which is a significant difference with respect to most teleoperation scenarios, is that the local and the remote stations share the same space, i.e., the teleoperator is not isolated from the working area; on the contrary, actually he is part of it. This leads to consider the safety of the teleoperator as one of the main topics.

The remote station is quite frequently composed of a mobile platform and an arm installed on it, and the whole system should be adaptable to unstructured and/or unknown environments (different houses), as it is desirable to perform actions such as going up and down stairs, opening various kinds of doors, grasping and manipulating different kind of objects, and so on. Improvements of the HMI to include different and more friendly ways of use is one of the main current challenges: the interfaces must be even more intuitive and must achieve a higher level of abstraction in terms of user commands. A typical example is understanding of an order when a voice recognition system is used [27.83].

Various physical systems are considered for teleoperation in this field, for instance, fixed devices (the disabled person has to get into the device workspace), or devices based on wheelchairs or mobile robots [27.84]; the latest are the most flexible and versatile, and therefore the most used in recently developed assistance robots, such as RobChair [27.85], ARPH [27.86], Pearl NurseBot [27.87], and ASIBOT [27.82].

27.4.7 Humanitarian Demining

This particular application is included in a separate subsection due to its relevance from the humanitarian point of view. Land mines are very easy to place but very hard to be removed. Specific robots have been developed to help in the removal of land mines, especially to reduce the high risk that exists when this task is performed by humans. Humanitarian demining differs from the mil-

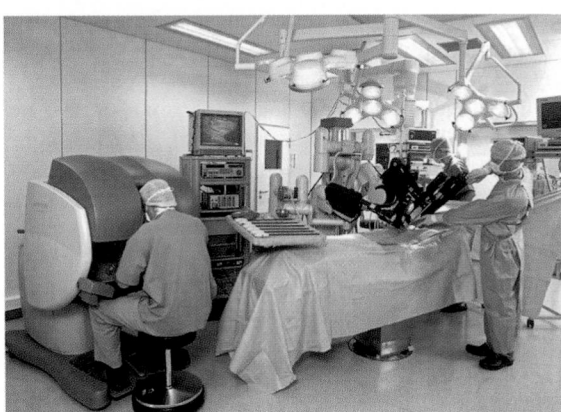

Fig. 27.13 Robotics surgery at Dresden Hospital (with permission from Intuitive Surgical, Inc. 2007)

Fig. 27.14 SILO6: A six-legged robot for humanitarian demining tasks (courtesy of IAI, Spanish Council for Scientific Research – CSIC)

itary approach. In the latter it is only required to find a path through a minefield in the minimum time, while the aim in humanitarian demining is to cover the whole area to detect mines, mark them, and remove/destroy all of them. The time involved may affect the cost of the procedure, but should not affect its efficiency. One key aspect in the design of teleoperated devices for demining is that the remote station has to be robust enough to resist a mine explosion, or cheap enough to minimize the loss when the manipulation fails and the mine explodes.

The removal of a mine is quite a complex task, which is why demining tools include not only teleoperated robotic arms, but also teleoperated robotic hands [27.88]. Some proposals are based on walking machines, such as TITAN-IX [27.89] and SILO6 [27.90] (Fig. 27.14). A different method includes the use of machines to mechanically activate the mine, like the Mini Flail, Bozena 4, Tempest or Dervish, among others; many of these robotic systems have been tested and used in the removal of mines in countries such as Japan, Croatia, and Vietnam [27.91, 92].

27.4.8 Education

Recently, teleoperation has been introduced in education, and can be collated into two main types. In one of these, the professor uses teleoperation to illustrate the (theoretical) concepts to the students during the a lecture by means of the operation of a remote real plant, which obviously cannot be brought to the classroom and that would require a special visit, which would probably be expensive and time consuming. The second type of educational application is the availability of remote experimental plants where the students can carry out experiments and training, working at common facilities at the school or in their own homes at different times. In this regard, during the last 5 years, a number of remote laboratory projects have been developed to teach fundamental concepts of various engineering fields, thanks to remote operation and control of scientific facilities via the Internet. The development of e-Laboratory platforms, designed to enable distance training of students in real scenarios of robot programming, has proven useful in engineering training for mechatronic systems [27.93]. Experiments performed in these laboratories are very varied; they may go from a single user testing control algorithms in a remote real plant [27.94] to multiple users simulating and teleoperating multiple virtual and real robots in a whole production cell [27.95].

The main feature in this type of applications is the almost exclusive use of the Internet as the communication channel between the local and remote stations. Due to its ubiquitous characteristic these applications are becoming increasingly frequent.

27.5 Conclusion and Trends

Teleoperation is a highly topical subject with great potential for expansion in its scientific and technical development as well as in its applications.

The development of new wireless communication systems and the diffusion of global communication networks, such as the Internet, can tremendously facilitate the implementation of teleoperation systems. Nevertheless, at the same time, these developments give rise to new problems such as real-time requirements, delays in signal transmission, and loss of information. Research into new control algorithms that guarantee stability even with variable delays constitutes an answer to some of these problems. On the other hand, the creation of new networks, such as the Internet2, that can guarantee a quality of service can help considerably to solve the real-time necessities of teleoperated systems.

The information that the human operator receives about what is happening at the remote station is es-

sential for good execution of teleoperated tasks. In this regard, new techniques and devices are necessary in order to facilitate immersion of the human operator in the task that he/she is carrying out. Virtual-reality, augmented-reality, haptics, and 3-D vision systems are key elements for this immersion.

The function of the human operator can also be greatly facilitated by aids to teleoperation. These aids, such as relational positioning, virtual guides, collision avoidance methods, and operation planning, can help the construction of efficient teleoperation systems.

An outstanding challenge is dexterous telemanipulation, which requires the coordination of multiple degrees of freedom and the availability of complete sensorial information.

The fields of application of teleoperation are multiple nowadays, and will become even more vast in the future, as research continues to outline new solutions to the aforementioned challenges.

References

27.1 T. Sheridan: *Telerobotics, Automation and Human Supervisory Control* (MIT Press, Cambridge 1992)

27.2 E. Nuño, A. Rodríguez, L. Basañez: Force reflecting teleoperation via IPv6 protocol with geometric constraints haptic guidance. In: *Advances in Telerobotics*, STAR, Vol. 31 (Springer, Berlin, Heidelberg 2007) pp. 445–458

27.3 R.J. Anderson, M.W. Spong: Bilateral control of teleoperators with time delay, IEEE Trans. Autom. Control **34**(5), 494–501 (1989)

27.4 G. Niemeyer, J.J.E. Slotine: Stable adaptive teleoperation, IEEE J. Ocean. Eng. **16**(1), 152–162 (1991)

27.5 P. Arcara, C. Melchiorri: Control schemes for teleoperation with time delay: A comparative study, Robot. Auton. Syst. **38**, 49–64 (2002)

27.6 P.F. Hokayem, M.W. Spong: Bilateral teleoperation: An historical survey, Automatica **42**, 2035–2057 (2006)

27.7 R. Ortega, N. Chopra, M.W. Spong: A new passivity formulation for bilateral teleoperation with time delays, Proc. CNRS–NSF Workshop: Advances in time-delay systems (Paris, 2003)

27.8 E. Nuño, L. Basañez, R. Ortega: Passive bilateral teleoperation framework for assisted robotic tasks, Proc. IEEE Int. Conf. Robot. Autom. (Rome, 2007) pp. 1645–1650

27.9 S. Munir, W.J. Book: Control techniques and programming issues for time delayed internet based teleoperation, ASME J. Dyn. Syst. Meas. Control **125**(2), 205–214 (2004)

27.10 S.E. Salcudean, M. Zhu, W.-H. Zhu, K. Hashtrudi-Zaad: Transparent bilateral teleoperation under position and rate control, Int. J. Robot. Res. **19**(12), 1185–1202 (2000)

27.11 N. Chopra, M.W. Spong, R. Ortega, N. Barbanov: On tracking performance in bilateral teleoperation, IEEE Trans. Robot. **22**(4), 844–847 (2006)

27.12 T. Namerikawa, H. Kawada: Symmetric impedance matched teleoperation with position tracking, Proc. 45th IEEE Conf. Decis. Control (San Diego, 2006) pp. 4496–4501

27.13 E. Nuño, R. Ortega, N. Barabanov, L. Basañez: A globally stable proportional plus derivative controller for bilateral teleoperators, IEEE Trans. Robot. **24**(3), 753–758 (2008)

27.14 R. Lozano, N. Chopra, M.W. Spong: Convergence analysis of bilateral teleoperation with constant human input, Proc. Am. Control Conf. (New York 2007) pp. 1443–1448

27.15 R. Lozano, N. Chopra, M.W. Spong: Passivation of force reflecting bilateral teleoperators with time varying delay, Proc. Mechatron. Conf. (Entschede 2002)

27.16 N. Chopra, M.W. Spong: Adaptive synchronization of bilateral teleoperators with time delay. In:

Advances in Telerobotics, STAR, Vol. 31 (Springer, Berlin, Heidelberg 2007) pp. 257–270

27.17 C. Secchi, S. Stramigioli, C. Fantuzzi: Variable delay in scaled port-Hamiltonian telemanipulation, Proc. 8th Int. IFAC Symp. Robot Control (Bologna 2006)

27.18 M. Boukhnifer, A. Ferreira: Wave-based passive control for transparent micro-teleoperation system, Robot. Auton. Syst. **54**(7), 601–615 (2006)

27.19 P. Loshin: *IPv6, Theory, Protocol, and Practice*, 2nd edn. (Morgan Kaufmann, San Francisco 2003)

27.20 W.R. Sherman, A. Craig: *Understanding Virtual Reality. Interface, Application and Design* (Morgan Kaufmann, San Francisco 2003)

27.21 R. Azuma: A survey of augmented reality, Presence Teleoper. Virtual Environ. **6**(4), 355–385 (1997)

27.22 M. Inami, N. kawakami, S. Tachi: Optical camouflage using retro-reflective projection technology, Proc. Int. Symp. Mixed Augment. Real. (Tokyo 2003) pp. 18–22

27.23 R. Azuma, Y. Baillot, R. Behringer, S. Feiner, S. Julien, B. MacIntyre: Recent advances in augmented reality, IEEE Comp. Graphics Appl. **21**(6), 34–47 (2001)

27.24 A. Rastogi, P. Milgram, J. Grodski: Augmented telerobotic control: A visual interface for unstructured environments, Proc. KBS/Robot. Conf. (Montreal 1995)

27.25 A. Kron, G. Schimdt, B. Petzold, M. Zäh, P. Hinterseer, E. Steinbach: Disposal of explosive ordnances by use of a bimanual haptic system, Proc. IEEE Int. Conf. Robot. Autom. (New Orleans 2004) pp. 1968–1973

27.26 A. Ansar, D. Rodrigues, J. Desai, K. Daniilidis, V. Kumar, M. Campos: Visual and haptic collaborative tele-presence, Comput. Graph. **25**(5), 789–798 (2001)

27.27 B. Dejong, E. Faulring, E. Colgate, M. Peshkin, H. Kang, Y. Park, T. Erwing: Lessons learned from a novel teleoperation testbed, Ind. Robot: Int. J. **33**(3), 187–193 (2006)

27.28 Y. Xiong, S. Li, M. Xie: Predictive display and interaction of telerobots based on augmented reality, Robotica **24**, 447–453 (2006)

27.29 S. Otmane, M. Mallem, A. Kheddar, F. Chavand: Active virtual guides as an apparatus for augmented reality based telemanipulation system on the internet, Proc. 33rd Annu. Simul. Symp. (Washington 2000) pp. 185–191

27.30 J. Gu, E. Auguirre, P. Cohen: An augmented reality interface for telerobotic applications, Proc. Workshop Appl. Comput. Vis. (Orlando 2002) pp. 220–224

27.31 A. Rodríguez, L. Basañez, E. Celaya: A Relational Positioning Methodology for Robot Task Specification and Execution, IEEE Trans. Robot. **24**(3), 600–611 (2008)

27.32   Y.H. Liu: Computing *n*-finger form-closure grasps on polygonal objects, Int. J. Robot. Res. **19**(2), 149–158 (2000)

27.33   D. Ding, Y. Liu, S. Wang: Computation of 3-D form-closure grasps, IEEE Trans. Robot. Autom. **17**(4), 515–522 (2001)

27.34   M. Roa, R. Suárez: Finding locally optimum force-closure grasps, Robot. Comput.-Integr. Manuf. **25**, 536–544 (2009)

27.35   J. Cornellà, R. Suárez, R. Carloni, C. Melchiorri: Dual programming based approach for optimal grasping force distribution, Mechatronics **18**(7), 348–356 (2008)

27.36   T.B. Martin, R.O. Ambrose, M.A. Diftler, R. Platt, M.J. Butzer: Tactile gloves for autonomous grasping with the NASA/DARPA Robonaut, Proc. IEEE Int. Conf. Robot. Autom. (New Orleans 2004) pp. 1713–1718

27.37   M. Bergamasco, A. Frisoli, C. A. Avizzano: Exoskeletons as man–machine interface systems for teleoperation and interaction in virtual environments. In: *Advances in Telerobotics* STAR Ser., Vol. 31 (Springer, New York 2007) pp. 61–76

27.38   S.B. Kang, K. Ikeuchi: Grasp recognition using the contact web, Proc. IEEE/RSJ Int. Conf. Intell. Robots Syst. (Raleigh 1992) pp. 194–201

27.39   R.L. Feller, C.K.L. Lau, C.R. Wagner, D.P. Pemn, R.D. Howe: The effect of force feedback on remote palpation, Proc. IEEE Int. Conf. Robot. Autom. (New Orleans 2004) pp. 782–788

27.40   M. Benali-Khoudja, M. Hafez, J.M. Alexandre, A. Kheddar: Tactile interfaces: a state-of-the-art survey, Proc. 35th Int. Symp. Robot. (Paris 2004) pp. 721–726

27.41   T. Wotjara, K. Nonami: Hand posture detection by neural network and grasp mapping for a master-slave hand system, Proc. IEEE/RSJ Int. Conf. Intell. Robots Syst. (Sendai 2004) pp. 866–871

27.42   M. Roa, R. Suárez: Independent contact regions for frictional grasps on 3-D objects, Proc. IEEE Int. Conf. Robot. Autom. (Pasadena 2008)

27.43   K.B. Shimoga: Robot grasp synthesis algorithms: a survey, Int. J. Robot. Res. **15**(3), 230–266 (1996)

27.44   R. Suárez, P. Grosch: Mechanical hand MA-I as experimental system for grasping and manipulation, Video Proc. IEEE Int. Conf. Robot. Autom. (Barcelona 2005)

27.45   A. Iborra, J.A. Pastor, B. Alvarez, C. Fernandez, J.M. Fernandez: Robotics in radioactive environments, IEEE Robot. Autom. Mag. **10**(4), 12–22 (2003)

27.46   W. Book, L. Love: Teleoperation telerobotics telepresence. In: *Handbook of Industrial Robotics*, 2nd edn. (Wiley, New York 1999) pp. 167–186

27.47   R. Aracil, M. Ferre: Telerobotics for aerial live power line maintenance. In: *Advances in Telerobotics*, STAR Ser., Vol. 31, (Springer, Berlin, Heidelberg 2007) 459–469

27.48   J.H. Dunlap, J.M. Van Name, J.A. Henkener: Robotic maintenance of overhead transmission lines, IEEE Trans. Power Deliv. **1**(3), 280–284 (1986)

27.49   R. Aracil, M. Ferre, M. Hernando, E. Pinto, J.M. Sebastian: Telerobotic system for live-power line maintenance: ROBTET, Control Eng. Prac. **10**(11), 1271–1281 (2002)

27.50   C.T. Haas, Y.S. Ki: Automation in infrastructure construction, Constr. Innov. **2**, 191–210 (2002)

27.51   Y. Hiramatsu, T. Aono, M. Nishio: Disaster restoration work for the eruption of Mt Usuzan using an unmanned construction system, Adv. Robot. **16**(6), 505–508 (2002)

27.52   A.M. Lytle, K.S. Saidi, R.V. Bostelman, W.C. Stones, N.A. Scott: Adapting a teleoperated device for autonomous control using three-dimensional positioning sensors: experiences with the NIST RoboCrane, Autom. Constr. **13**, 101–118 (2004)

27.53   A.J. Kwitowski, W.D. Mayercheck, A.L. Brautigam: Teleoperation for continuous miners and haulage equipment, IEEE Trans. Ind. Appl. **28**(5), 1118–1125 (1992)

27.54   G. Baiden, M. Scoble, S. Flewelling: Robotic systems development for mining automation, Bull. Can. Inst. Min. Metall. **86.972**, 75–77 (1993)

27.55   J.C. Ralston, D.W. Hainsworth, D.C. Reid, D.L. Anderson, R.J. McPhee: Recent advances in remote coal mining machine sensing, guidance, and teleoperation, Robotica **19**(4), 513–526 (2001)

27.56   A.J. Park, R.N. Kazman: Augmented reality for mining teleoperation, Proc. of SPIE Int. Symp. Intell. Syst. Adv. Manuf. – Telemanip. Telepresence Technol. (1995) pp. 119–129

27.57   T.J. Nelson, M.R. Olson: Long delay telecontrol of lunar mining equipment, Proc. 6th Int. Conf. Expo. Eng., Constr., Oper. Space, ed. by R.G. Galloway, S. Lokaj (Am. Soc. Civ. Eng., Reston 1998) pp. 477–484

27.58   N. Wilkinson: Cooperative control in teleoperated mining environments, 55th Int. Astronaut. Congr. Int. Astronaut. Fed., Int. Acad. Astronaut. Int. Inst. Space Law (Vancouver 2004)

27.59   P. Ridao, M. Carreras, E. Hernandez, N. Palomeras: Underwater telerobotics for collaborative research. In: *Advances in Telerobotics*, STAR Ser., Vol. 31 (Springer, Berlin, Heidelberg 2007) pp. 347–359

27.60   S. Harris, R. Ballard: ARGO: Capabilities for deep ocean exploration, Oceans **18**, 6–8 (1986)

27.61   M. Fontolan: Prestige oil recovery from the sunken part of the Wreck, PAJ Oil Spill Symp. (Petroleum Association of Japan, Tokyo 2005)

27.62   G. Antonelli (Ed.): *Underwater Robots: Motion and Force Control of Vehicle–Manipulator Systems* (Springer, Berlin 2003)

27.63   C. Canudas-de-Wit, E.O. Diaz, M. Perrier: Robust nonlinear control of an underwater vehicle/manipulator system with composite dynamics,

Proc. IEEE Int. Conf. Robot. Autom. (Leuven 1998) pp. 452–457

27.64 M. Lee, H-S. Choi: A robust neural controller for underwater robot manipulators, IEEE Trans. Neural Netw. **11**(6), 1465–1470 (2000)

27.65 J. Kumagai: Swimming to Europe, IEEE Spectrum **44**(9), 33–40 (2007)

27.66 L. Pedersen, D. Kortenkamp, D. Wettergreen, I. Nourbakhsh: A survey of space robotics, Proc. 7th Int. Symp. Artif. Intell., Robot. Autom. Space (Nara 2003)

27.67 F. Doctor, A Glas, Z. Pronk: Mission preparation support of the European Robotic Arm (ERA), National Aerospace Laboratory report NLR-TP-2002-650 (Netherlands 2003)

27.68 S. Roderick, B. Roberts, E. Atkins, D. Akin: The ranger robotic satellite servicer and its autonomous software-based safety system, IEEE Intell. Syst. **19**(5), 12–19 (2004)

27.69 W. Bluethmann, R. Ambrose, M. Diftler, S. Askew, E. Huber, M. Goza, F. Rehnmark, C. Lovchik, D. Magruder: Robonaut: A robot designed to work with humans in spaces, Auton. Robots **14**, 179–197 (2003)

27.70 G. Hirzinger, B. Brunner, K. Landzettel, N. Sporer, J. Butterfass, M. Schedl: Space Robotics – DLR's telerobotic concepts, lightweight arms and articulated hands, Auton. Robots **14**, 127–145 (2003)

27.71 S.E. Fredrickson, S. Duran, J.D. Mitchell: Mini AER-Cam inspection robot for human space missions, AIAA Space 2004 Conf. Exhib. (San Diego 2004)

27.72 T. Imaida, Y. Yokokohji, T. Doi, M. Oda, T. Yoshikawa: Ground–space bilateral teleoperation of ETS-VII robot arm by direct bilateral coupling under 7-s time delay condition, IEEE Trans. Robot. Autom. **20**(3), 499–511 (2004)

27.73 R.A. Lindemann, D.B. Bickler, B.D. Harrington, G.M. Ortiz, C.J. Voorhees: Mars exploration rover mobility development, IEEE Robot. Autom. Mag. **13**(2), 19–26 (2006)

27.74 G.A. Landis: Robots and humans: synergy in planetary exploration, Acta Astronaut. **55**(12), 985–990 (2004)

27.75 J.W. Hill, P.S. Green, J.F. Jensen, Y. Gorfu, A.S. Shah: Telepresence surgery demonstration system, Proc. IEEE Int. Conf. Robot. Autom. (IEEE Computer Society, San Diego 1994) pp. 2302–2307

27.76 J. Marescaux, J. Leroy, M. Gagner, F. Rubino, D. Mutter, M. Vix, S.E. Butner, M. Smith: Transatlantic robot-assisted telesurgery, Nature **413**, 379–380 (2001)

27.77 S.E. Butner, M. Ghodoussi: A real-time system for tele-surgery, Proc. 21st Int. Conf. Distrib. Comput. Syst. (IEEE Computer Society, Washington 2001) pp. 236–243

27.78 G.S. Guthart, J.K. Jr. Salisbury: The Intuitive telesurgery system: overview and application, Proc. IEEE Int. Conf. Robot. Autom. (San Francisco 2000) pp. 618–621

27.79 M. Li, A. Kapoor, R.H. Taylor: Telerobotic control by virtual fixtures for surgical applications. In: *Advances in Telerobotics*, STAR, Vol. 31 (Springer, Berlin, Heidelberg 2007) pp. 381–401

27.80 A. Smith, J. Smith, D.G. Jayne: Telerobotics: surgery for the 21st century, Surgery **24**(2), 74–78 (2006)

27.81 M. Topping: An Overview of the development of Handy 1, a rehabilitation robot to assist the severely disabled, J. Intell. Robot. Syst. **34**(3), 253–263 (2002)

27.82 C. Balaguer, A. Giménez, A. Jardón, R. Correal, S. Martínez, A.M. Sabatini, V. Genovese: Proprio, teleoperation of a robotic system for disabled persons' assistance in domestic environments. In: *Advances in Telerobotics*, STAR, Vol. 31 (Springer, Berlin, Heidelberg 2007) pp. 415–427

27.83 O. Reinoso, C. Fernández, R. Ñeco: User voice assistance tool for teleoperation. In: *Advances in Telerobotics*, STAR Ser., Vol. 31 (Springer, Berlin, Heidelberg 2007) pp. 107–120

27.84 K. Kawamura, M. Iskarous: Trends in service robots for the disabled and the elderly, Proc. IEEE/RSJ/GI Int. Conf. Intell. Robots Syst. (Munich 1994) pp. 1647–1654

27.85 G. Pires, U. Nunes: A wheelchair steered through voice commands and assisted by a reactive fuzzy logic controller, J. Intell. Robot. Syst. **34**(3), 301–314 (2002)

27.86 P. Hoppenot, E. Colle: Human-like behavior robot-application to disabled people assistance, Proc. IEEE Int. Conf. Syst., Man, Cybern. (Soc. Syst. Man Cyber., Nashville 2000) pp. 155–160, .

27.87 M.E. Pollack, S. Engberg, J.T. Matthews, S. Thrun, L. Brown, D. Colbry, C. Orosz, B. Peintner, S. Ramakrishnan, J. Dunbar-Jacob, C. McCarthy, M. Montemerlo, J. Pineau, N. Roy: Pearl: A mobile robotic assistant for the elderly, AAAI Workshop Autom. Eldercare (Alberta 2002)

27.88 T. Wojtara, K. Nonami, H. Shao, R. Yuasa, S. Amano, D. Waterman, Y. Nobumoto: Hydraulic master-slave land mine clearance robot hand controlled by pulse modulation, Mechatronics **15**, 589–609 (2005)

27.89 K. Kato, S. Hirose: Development of the quadruped walking robot, TITAN-IX–mechanical design concept and application for the humanitarian demining robot, Adv. Robot. **15**(2), 191–204 (2001)

27.90 P. Gonzalez de Santos, E. Garcia, J.A. Cobano, A. Ramirez: SILO6: A six-legged robot for humanitarian demining tasks, Proc. 10th Int. Symp. Robot. Appl. World Autom. Congr. (2004)

27.91 J.-D. Nicoud: Vehicles and robots for humanitarian demining, Ind. Robot **24**(2), 164–168 (1997)

27.92 M.K. Habib: Humanitarian demining: reality and the challenge of technology – the state of the arts, Int. J. Adv. Robot. Syst. **4**(2), 151–172 (2007)

27.93 C.S. Tzafestas, N. Palaiologou, M. Alifragis: Virtual and remote robotic laboratory: comparative experimental evaluation, IEEE Trans. Educ. **49**(3), 360–369 (2006)

27.94   X. Giralt, D. Jofre, R. Costa, L. Basañez: Proyecto de Laboratorio Remoto de Automática: Objetivos y Arquitectura Propuesta, III Jornadas de Trabajo EIWISA 02, Enseñanza vía Internet/Web de la In-

geniería de Sistemas y Automática (Alicante, 2002) pp. 93–98, in Spanish

27.95   M. Alencastre, L. Munoz, I. Rudomon: Teleoperating robots in multiuser virtual environments, Proc. 4th Mexican Int. Conf. Comp. Sci. (Tlaxcala 2003) pp. 314–321

28. Distributed Agent Software for Automation

Francisco P. Maturana, Dan L. Carnahan, Kenwood H. Hall

Agent-based software and hardware technologies have emerged as a major approach to organize and integrate distributed elements of complex automation. As an example, this chapter focuses on a particular situation. Composite curing, a rapidly developing industry process, generates high costs when not properly controlled. Curing autoclaves require tight control over temperature that must be uniform throughout the curing vessel. This chapter discusses how agent-based software is being implemented into the curing process by pushing control logic down to the lowest level of the control hierarchy into the process controller, i. e., the programmable logic controller (PLC). The chapter also discusses how the benefits of process survivability, diagnostics, and dynamic reconfiguration are achieved through the use of autoclave and thermocouple intelligent agents.

A general introduction and overview of agent-based automation can be found in the additional reading listed. Autoclave curing is vital to the production and support of many industries. In fact, there are so many industries that are dependent upon autoclave curing that improvements to curing production controls would lend a competitive advantage to the corporation that could deliver these improvements. While the focus of this particular work is directed predominately towards the production of composite materials, there are many other processes that could benefit from this work, with almost no differences in terms of the autoclave curing controls. This is especially true for the following products that require autoclave curing: composite materials, polymers, rubber, concrete products (e.g., sand-lime brick, asbestos, hydrous calcium silicate, cement (steam curing)), tobacco, textiles, electrical and electronics products (e.g., printed circuit boards (PCBs), ceramic substrates) chemical, medical/pharmaceutical products, wood and building products, metallurgical products, tire rethreading, glass laminating, and aircraft/aerospace products.

Much work has recently been reported for new curing methods and autoclave curing devices. This is one indication that there is a trend towards researching and developing new and better ways of curing materials. This research and development (R&D) has increased to meet market demands and technical papers regarding materials curing is proliferating. Market research studies also support the notion that there is increased demand for composite materials and that this trend will continue in the future. With increased demand comes a need for increased efficiencies in the curing production process. Composite material manufacturers, aerospace manufacturers, and manufacturers of other materials that require curing are increasing their capital expenditure budgets for the development and/or acquisition of improved autoclave manufacturing capabilities to meet this demand head-on.

Composite curing is accomplished through the proper application of heating and cooling to composite material inside autoclaves or automated ovens. Depending upon the type of composite material or the application of the composite material, a different set of curing parameters is used to control the curing process. Composite materials are cured under very stringent specifications, especially for materials that are used for aerospace applications. If composites are cured at a temperature that is too high, the material could become brittle and will be susceptible to breaking. If cured at a temperature that is too low, the material may not bond correctly and will eventually come apart.

The specifications that govern the curing process for composite materials are called recipes or profiles. Recipes/profiles differ slightly for an autoclave that has convection heaters versus an autoclave that has gas heaters. Likewise, an autoclave that has cryogenic coolers versus an autoclave that is air-cooled or fan-cooled would have slightly different control parameters within its control recipe. In the case of autoclave operations there are, generally, three profiles used to control the composite curing process: temperature, pressure, and vacuum. Of these three, temperature is the most important and the most difficult to control. As seen in Fig. 28.1, six separate periods control the temperature within the curing autoclave. Temperature increases and decreases are ramped up or down according to a particular rate of change (i. e., a slope) that is also regulated within the specification.

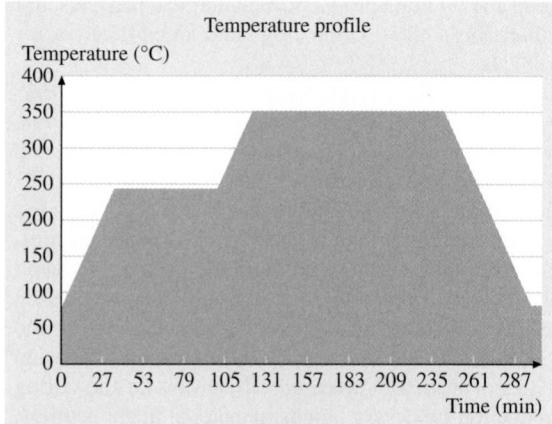

Fig. 28.1 Typical autoclave temperature profile

State-of-the-art systems use centralized material management control loops to guide the curing activity through a prescribed set of thermal, pressure, and vacuum profiles. Controlling autoclave operations so that they conform to thermal profiles is the main concern of the system and, hence, dictates the requirements for system control. For a composite material to bond properly, temperature throughout an autoclave must be maintained at a very specific level. Arrays of thermocouples provide temperature readings back to the control program through output data cards and are used to control the operation of heaters and, sometimes, coolers. One of the challenges of controlling temperature within the autoclave chamber is to keep it within a range of temperature values even when the temperature must be modified several times during the production process (as shown in Fig. 28.1). The nonlinear behaviors of a temperature profile are monitored by a *lead* thermocouple that is designated based upon its position in the autoclave, the type of material (or materials) being cured, chamber idiosyncrasies, and other production variables.

Whichever thermocouple is selected as the *lead* sensor, the feedback temperature data from that sensor will not be allowed to exceed the temperature profile temperature during heating operations. During the period of time when the thermal profile specifies a decrease in temperature (i. e., cooling operations), the *lead* thermocouple is usually used to determine the minimum temperature that will be allowed to occur. Naturally, for this type of control to be effective, network connectivity is critical to maintain high-level material state control. If a *lead* thermocouple disconnects from the composite material while the composite is being cured, the disconnection must be detected as soon as possible, and another thermocouple must be selected by the software to act as the lead. The former lead thermocouple is ignored from that point forward while the curing process continues. For cooling operations, the thermocouple that is used to control the cooling is actually called the *lag* thermocouple and temperatures are not allowed to go below that of the *lag* sensor.

On the surface of this control application, it does not seem that the requirements would be difficult to accomplish through ordinary ladder logic routines. However, there is a commercial issue that takes precedence over the control issue. Millions of dollars of composite materials could be lost during one imperfect curing process run. The survivability of the curing session must be ensured with greater probability despite perturbations in the production process.

Classical control methods suffice to monitor and control the thermal process within the autoclaves. Although the curing activity is performed in a controlled environment, there are dynamic perturbations that affect the thermocouples, which could generate unsatisfactory results, provoking complete rejection of an expensive piece of composite material. Perturbations of interest relate to potential malfunctioning in the thermocouple themselves. A malfunctioning thermocouple may appear healthy on visual inspection but its internal operations may be faulty. This is the case when thermocouples generate bogus readings. This type of problem is very difficult to detect offline and is not detected until the curing process has undergone several steps. Classical control programs that reside in the controller have limitations in terms of enabling enough reasoning to detect such problems early.

Another type of problem occurs when thermocouples detach from the material during curing. The autoclave is a sealed controlled environment that cannot be easily interrupted to reattach the sensors. The controller must be capable of reacting to the failing reader by performing corrective actions on the fly without disrupting operation.

A viable solution to the problem above is to augment the reasoning capability of the control system with more sophisticated reasoning algorithms. These augmented algorithms follow the process to generate a model from it. Condition monitoring rules can then be added to detect malfunctioning sensors. Typically, in industrial implementations, such advanced capabilities are placed in the personal computer (PC) level. A PC workstation is added to supervise the control system. This approach converts the solution into a centralized system. A centralized system suffers from other problems such as single point of failure and connectivity

issues, which exacerbate the problem of maintaining a robust system for the whole duration of the process. Current trends in industry and industrial techniques show a strong tendency to move away from centralized supervisory models.

The endeavor in this work is to show a new technique to cope with the issues above. A programmable logic controller (PLC) can be designed with agent capabilities to enable for advanced reasoning at the controller level. Agents are software components that encapsulate physical equipment knowledge (rules) and properties in the form of capabilities, behaviors, and procedures [28.1–3]. The capabilities express the type of functions that an agent contributes to the well-being of the system. Each capability is a construct of behaviors. Moreover, each behavior is made of sequentially organized procedures. In the example, the agents are seamless integrated with the control algorithm inside the PLC.

The fundamental technical milestone is to eliminate the connectivity problem, and to augment the reconfiguration of the control system by moving the condition monitoring capabilities closer to the physical process, into the PLC. The PLC is the core of the control operation and therefore these devices have been widely adopted in industrial automation with the necessary redundancy to prevent losses.

The intention is to show fundamental aspects of the agent-based control technology to solve a particular composite curing application. The technology discussed is intended for general use; it is possible to use the same infrastructure to solve different applications.

The discussion will begin with a little bit of background about curing technology. Then, industrial agent technology will be discussed. The chapter concludes with an example and discussion of results on how to model the curing application with intelligent agents.

28.1 Composite Curing Background

There are several markets in which large growth in demand for autoclave cured products and materials is forecast. The carbon fiber composites market has been experiencing a growth of 12% during the last 23 years and is expected to grow to US$ 12.2 billion by 2011 [28.4, 5]. Carbon fiber reinforced thermoplastics will be improved by 37% between 2006 and 2010 [28.6]. The opportunities for continuous fiber reinforced thermoplastic composites are likewise increasing

at a strong rate of growth. During 2002 alone, the growth rate was 93% [28.7, 8]. The aircraft maker Airbus has projected increases of thermoplastic composite use of 20% per year.

The natural rubber market is also seeing higher demands for its products. Rubber products have the second largest need for autoclave curing production improvements. Higher demands for natural rubber products are largely due to the spectacular economic growth in the People's Republic of China (PRC) and greater demand for consumer as well as industrial products. Production levels just within Thailand have risen from 1.805 million tons in 1995 to 2.990 million tons in 2005. Actual production exceeded 1995 demand projections by 6.25% in 2000 and 40.05% in 2005. By 2010, it is expected that 3.148 million tons of natural rubber will have to be produced. This will be more than 63% than was originally thought to be needed in 1995.

There have been many technical and scientific papers published recently that address many different methods of curing within an autoclave. In 2004, *Salagnac* et al. sought to improve the curing process through heat exchanges that were mainly free convection and radiation within a narrow-diameter autoclave [28.9, 10]. Transient thermal modeling of in situ curing during the tape winding of composite cylinders was attempted and successfully executed by *Kim* et al. at the University of Texas at Austin in 2002 [28.11]. *Chang* et al. studied the optimal design of the cure cycle for the consolidation of thick composite laminates in 1996 at the National Cheng-Kung University in Taiwan [28.12]. Smart autoclave processing of thermoset resin matrix composites that were based upon temperature and internal strain monitoring was studied by *Jinno* et al. in 2003 [28.13]. Their work looked at the control of temperature ramp rate so that the peak temperature predicted by Springer's thermochemical model is kept below an allowable value. Cure completion was determined by a cure rate equation, and internal strain monitoring with embedded optical-fiber sensors [28.13]. In 2003, a report was performed by *Mawardi* and *Pitchumani* at the University of Connecticut [28.14]. Their work was depicted as the development of optimal temperature and current cycles for the curing of composites using embedded resistive heating elements. Heating and cooling was controlled internal to the materials rather than through external heating and cooling equipment.

To get to a better understanding of the curing process *Thomas* et al. performed experimental characterizations of autoclave-cured glass–epoxy composite laminates at Washington University in St. Louis, MO [28.15]. From the results of over 100 experimental autoclave curing runs in 1996, the study team sought to verify shrinking horizon model predictive control (SHMPC) to predict and control the thickness and void content of their composite materials. At the Korea Advanced Institute of Science and Technology, *Kim* and *Lee* reported the results from their experiments with composites to reduce fabrication thermal residue stress of hybrid co-cured structures through the use of dielectrometry [28.16]. *Ghasemi-Nejhad* reported in 2005 his method of manufacturing and testing of active composite panels with embedded piezoelectric sensors and actuators [28.17]. The embedded sensors and actuators were used to minimize the problems associated with layering composites such as voids and weak bonds. *Pantelelis* et al. of the National Technical University of Athens presented a computer simulation tool in 2004 that was coupled with a numerical optimization method developed for use in the optimal design of cure cycles in the production of thermoset-matrix composite parts [28.18]. The presenters, *Georg* et al. from Boeing, *Madsen* and *Teng* from Northrup Grumman, and *Courdji* from Convergent Manufacturing, discussed the exploration of composites processing and producibility by analysis [28.19]. The discussion centered predominately on accelerated insertion of materials-composites (AIM-C) processing and producibility modules and the results that could be achieve by linking it with the robust design computation system (RDCS). AIM-C was being conducted jointly by the Naval Air Systems Command (NASC) and the Defense Advance Research Projects Agency (DARPA). In particular, AIM-C sought to significantly reduce the time and cost of inserting new materials.

For more than 10 years, the University of Delaware has been formulating advanced curing controls in its Department of Chemical Engineering and Center for Composite Materials. A team consisting of *Pillai, Beris,* and *Dhurjati* developed an expert system tool in 1997 to operate an autoclave for the intelligent curing of composite materials [28.20]. At the same facility in 2002, *Michaud* et al. developed a robust, simulation-based optimization and control method methodology to identify and implement the optimal curing conditions for thick-sectioned resin transfer molding (RTM) composite manufacturing [28.21].

To get better validation results for the curing process, *Arms* et al. developed a technique for remote powering and bidirectional communication with a network of micro-miniature, multichannel addressable sensing modules [28.22]. Use of the embedded sensors sought to eliminate the damage that is commonly incurred when removing thermocouples from composite materials in the post-curing phase of production. *Uryasev* and *Trindade* also studied methods to achieve better validation results and settled on a method combining an analytical model and experimental test data for optimal determination of failure tolerance limits [28.23].

Alternative curing schemes are also being developed. Electron beam (EB) curing is a technique that researchers at Acsion in Manitoba, Canada say could have a massive impact in the aerospace and defense industries, especially for the composite materials used in aircraft wings [28.24]. In electron beam curing, the incoming high-speed electrons knock off other electrons in the polymer resin, causing the material to cure. X-rays generated from the high-energy beams can penetrate even deeper into the composite, yielding a uniform cure in material several centimeters thick. The process takes only minutes compared with hours for conventional autoclave curing. Eight independent studies have shown potential manufacturing cost savings of 26–65% for prototyping alone. This could rise to as much as 90–95% [28.24]. The National Aeronautics and Space Administration (NASA) encouraged the advancement of the EB curing process through a small business innovation research (SBIR) grant awarded to Science Research Laboratory (SRL), Inc. As a result of the work conducted, a new curing method was produced that used automated tape placement with the electron beam. The new method is known as in situ electron beam curing and the SRL-devised system is being utilized at the NASA Marshall Space Flight Center [28.25].

Commercial interests have been developing other methods and products to supplement the curing controls that have already been put into use in the factory. The Blair Rubber Company of Seville, OH has developed very precise process specifications for vulcanizing rubber products in curing autoclaves. Multichannel communications modules are in common use with thermocouples. In particular, Data Translation, Inc. has universal serial bus (USB) thermocouple measurement modules that can monitor several thermocouple devices simultaneously. Flexible thermocouple management programs have been developed to minimize the costs associated with thermocouple wear and tear. Thermocouples are reworked and/or recalibrated and returned to service through this type of program. Other thermocouple management programs are involved with manual and automatic device slaving, intelligent thermocouples that protect against ground fault current, automatic compensation, and automatic swapping of lead thermocouples [28.26–30].

Even the design and manufacture of thermocouples has become specialized. Customized wiring is provided within thermocouples by manufacturers depending upon the industrial application and/or the environment in which the sensors will be placed. Within autoclave-based composite applications, there are at least seven types of thermocouples.

28.2 Industrial Agent Architecture

Industrial agents have been designed to execute on programmable logic controllers (PLCs). The agents are built on workstations, compiled, and then downloaded to the PLCs [28.31–33]. The downloading of the agents is directed by central management software that executes an agent assignment script. A decision was made to expand the PLC firmware to enable for the programming of advanced decision-making engines.

The agents have three main parts:

1. Reasoning
2. Data table interface
3. Execution control.

The first part is the actual brain of the agent. The brain is composed of behaviors, which execute as needed, as prescribed in the process operations. Dynamically generated events trigger the execution of behaviors in the agent to follow a particular sequence of

procedures. An industrial agent is essentially reactive, but its software infrastructure permits the incorporation of proactive behaviors. The data table interface serves as the repository of control and agent data. The reasoning part sets values in the data table to estimable the control loops. The control part is the control execution level. This part is in charge of executing control loops in a periodic or continuous fashion to maintain steady-state conditions during the execution of the operations.

As shown in Fig. 28.2, a hierarchical library of interrelated agents represents the composite curing application. Each node of the library corresponds to an agent of a given type such as a thermocouple or autoclave. Each agent will expose a set of user-configurable attributes to give it a personality and operational ranges.

The interagent communication is based on the FIPA (Foundation for Intelligent Physical Agents) [28.34] language specification. The content of the agent message is written in a job description language

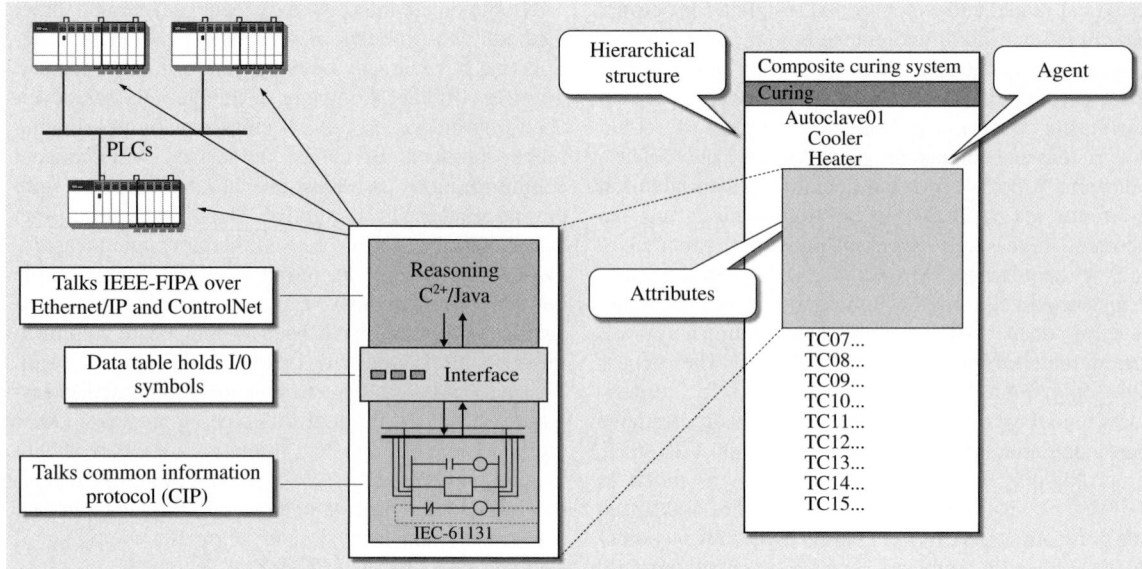

Fig. 28.2 Industrial agent components. CIP – common information protocol

(JDL) [28.31–33]. The industrial platform is based on Rockwell Automation's PLCs and NetLinx networking.

28.2.1 Agent Design and Partitioning Perspective

The benefits of using agents can be observed in three main technological aspects:

1. System design
2. Upgrading
3. Runtime.

For an appropriate system design technique, the effects of changing components (adding, removing, and altering) must be measured. The agent technology provides new techniques to design truly distributed systems by using libraries of functionality where the designers specify the components and their corresponding behaviors in generic terms. The libraries of functionality augment the system design capabilities by eliminating the need to redesign the whole system when changes are made.

System upgradability is traditionally a difficult task in industrial systems due to the hard-coupled dependencies among the controlled components (tightly coupled systems). In a nonagent system, the logical dependencies among the components must be hard-coded into the software during design time. Therefore, the effect of changing one component cascades throughout

the system, forcing the modification of other components, and so on. In an agent-based system, one attempts to eliminate this cascading effect by emphasizing loosely coupled relationships among the controlled components. Agents generate dynamic interconnections with counterparts via agent messages. The dynamically emerging interconnectivity effects the agents' behaviors by locally changing their world view.

The runtime perspective of agent-based systems is helped by the agent's capability (or service) of infrastructure discovery. Using the dynamic discovery infrastructure, an agent can initiate the discovery of other agents than can help in responding to a particular event. As opposed to having a specific set of predefined plans or agents to talk to under particular circumstances, the agents opt to discover their associates that can currently supply a solution. This distributed nature of agent technology opens the door to creating a more survivable system by eliminating single points of failure.

A very significant aspect of the agent-based control technology is system scalability. An agent solution can be scaled up and down to fit different system sizes.

A powerful feature of agents is the ability for interagent communications to detect, isolate, and accommodate component failures. In the case of a multithermocouple system, agent-to-agent communication is directed at validating normative temperature readings. Any variations from the group's tendency are quickly detected and assessed by the individual agents that are affected by such variations. Inconsistency between observed temperatures provides the basis to suspect, detect, and further isolate the faulty element. Fault detection and diagnosis are key enabling features of the agent infrastructure that drive the processes of configuration and planning. Dynamic reconfiguration may include changing the operating state of a group of system devices in a coordinated manner as well as dynamically changing the control for system elements.

28.2.2 Agent Tool

The technological advantage provided by the agents is the ability to create agent components in industrial commercial off-the-shelf PLCs without having to create specialized hardware and programming tools. The advantage of using PLC devices as the primary control and agent-hosting platform is the reliability of its operating system, hardware components, and packaging, which are readily available for industrial use.

By embedding the agents directly into the controller, it is possible for the agent to interoperate with the control programs faster. This embodiment eliminates the potential loss of controllability due to a lost connection between a controlling device and the supervising computer. Control programs ensure the hard, real-time response required by many control applications and an agent part ensures high-level coordination and flexibility. This synergy of agent and control inside the PLC and in combination with smaller PLCs has made it possible to transform a physical machine into an intelligent machine by localizing the intelligence.

The changes made to the PLC firmware include software extensions to the tasking model of the controller, object instantiation and retention during power cycles, and a more general communication layer. Great

effort took place initially to make the PLC an agent-hosting device with the capability to absorb executable agent code via object downloading. The extensions to the communication layer covered the generation and parsing of FIPA-based messages. With this feature, the control-based agents talk to other agents that comply with the FIPA specifications, even those that may have been implemented using a different infrastructure.

To make the agent system scalable and manageable in terms of software for agent development, it was necessary to create a development environment for the programming of agent libraries [28.35]. There are five main phases of agent development:

1. Template library design
2. Facility editor
3. Control system editor
4. Assignment wizard
5. Control code generator.

In the template library editor, a collection of components to represent the physical system is identified and built with control and reasoning parts. These components are generic, without referring to specific instances or devices. The library of components is then instantiated and arranged in a specific tree to represent the hierarchical order and attributes of the components that will be part of a specific system. This is where parameters of controlled devices and locations are added. The control system editor helps in the selection of the control and communication network equipment. In addition, it helps in the assignment of the input and output (I/O) points.

Once the pieces above are defined, the control and reasoning components defined within the application (in the facility editor) are assigned to specific controllers (PLCs). After the assignment, the executable code is generated and downloaded to the controllers.

28.3 Building Agents for the Curing System

The purpose of modeling the composite process using industrial agents is to increase the degree of survivability, diagnostics, and reconfigurability of the system [28.36]. Survivability of the control system is very important since it is directly related to safety and the creation of scrap material.

Curing is a long process and therefore there is a greater time span during which disruption may oc-

cur. Situations such as power shutdowns and network discontinuity are of critical importance and these must be handled in a timely way to sustain high process integrity. Since the curing process consists of applying heat and cooling to composite material inside a sealed autoclave, a reliable system to operate on top of the temperature sensors is fundamental. The thermocouples read the temperature of the composite material at differ-

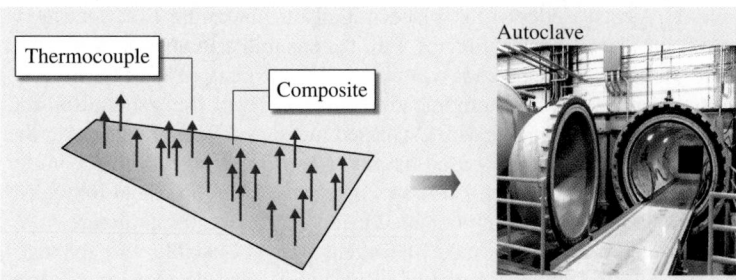

Fig. 28.3 Composite curing

ent locations. Moreover, since these constitute a group of logically interconnected sensors, the thermocouples are treated as a multisensor problem. Figure 28.3 illustrates how a composite material part is populated with the multisensor array prior to its placement inside the autoclave furnace. Thermocouple agents will be created to represent the thermocouple devices in the controller. The agents will use their social skills to organize the array into a smart and highly reconfigurable logical structure.

Given that the process operates in a controlled vacuum, it is not possible to stop the process to adjust a loose thermocouple, for example. Therefore, the autodiagnosability of the process is a critical factor that is made a part of the decision-making roles of the agents. As a consequence, both the supervisory and control roles must be able to detect problems and reconfigure the physical system to cope with such eventualities in order to enable continuous, uninterrupted activity.

Factors such as nonlinearities of the process and material (exothermal reactions provoking thermal spikes) must be considered in the modeling of the supervisory intelligence. The agent technology discussed herein allows for an incremental implementation of the necessary rules. The agent's decision power resides in rules, which are encapsulated in the behaviors. A validation framework based on simulation helps to create a virtual model for testing and enhancing the curing control rules and loops of the agents and control algorithms.

The agent software that is downloaded to the PLC is associated with a control-level driver (IEC-61131 control programs). The agent directs the control system using a lightweight supervisory strategy. Agents do not perform direct control. The device drivers carry out direct control upon the equipment. Thus, to avoid decision collisions and misinterpretation of the world, the sphere of influence of each agent must be defined with respect to its associated physical device.

The distribution of the agents throughout specific sections of the application is a necessary partitioning of responsibility that needs to take place before programming the rules. In the current case, the application is made of a single machine (autoclave) with a multisensor array (thermocouples).

There are three partitioning criteria to model the system:

1. One-to-one
2. One-to-many
3. Many-to-one.

One-to-one portioning exists in physical devices that can be isolated and classified as independent work units. Typical examples of such a case are a valve, switch, or breaker. In this case, one agent per physical device is enough. A one-to-many case appears in complex systems, where there is still an easy classification of the work units. This type of partitioning happens in equipment grouping situations, whereby one agent can be associated with a group of related physical devices. A rarer case is many-to-one partitioning, in which multiple agents are associated with a single or grouped

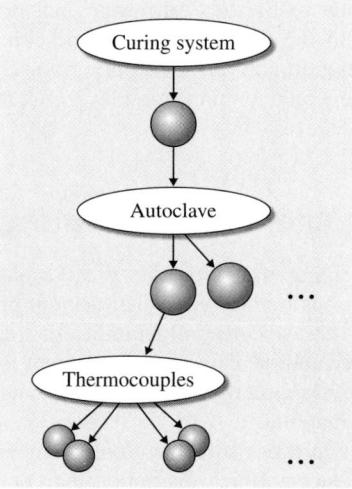

Fig. 28.4 Agent classification for curing system

physical device. A typical situation of this type occurs when a single agent cannot cope with too many concurrent processes. Efficiency of the decision-making process is then measured in terms of communication latencies and decision-making convergence. A system with too many agents and processes has many more messages on the network. The network may become overwhelmed with too many packets and communication latencies may increase. Since agents collaborate with their peers through the exchange of information, they heavily depend on timely arrival of information during the group decision process. Thus, it may occur that the agents begin dropping out of the conversation without concluding a decision process. In this case, agents would retry communication, causing more messages and therefore problems in convergence. Thus, special care must be taken to obtain a good partitioning of agent functionality.

The composite curing system corresponds to a three-tier one-to-one system. As shown in Fig. 28.4, the top-level tier has one agent to represent the overall curing system. The second tier represents the autoclave machinery that will be part of the curing system. There is one autoclave agent per autoclave machine. The third tier represents the thermocouple sensor level. There is one agent per thermocouple.

28.4 Autoclave and Thermocouple Agents

A bottom-up design (from the simplest part of operation to the more complex behavior) of the components takes place to build the curing system. Each component is associated with a piece of equipment or a particular process as an agent. Figure 28.5 shows a group of curing components. Most of the components are control-oriented drivers. However, autoclave and thermocouple components are agents.

The left-hand side of Fig. 28.5 shows the component list. The upper-right part shows the low-level programming blocks of the component. The lower-right part shows the list of control parameters of the component, such as input and output (I/O) variables.

28.4.1 Autoclave Agent

The autoclave agent contains curing artifacts such as cooler, heater, pressure compressor, and thermocouples. Thermocouple agents have their own temperature monitoring system and personality. Depending on the stage of the curing process, an autoclave agent periodically requests a reorganization of the thermocouples to se-

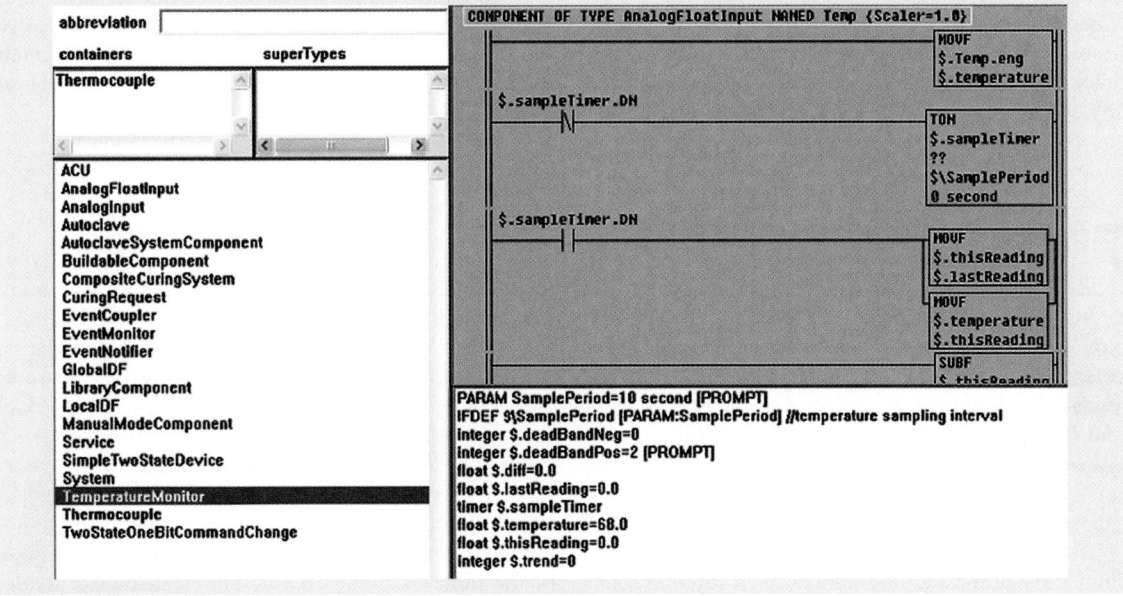

Fig. 28.5 Component list

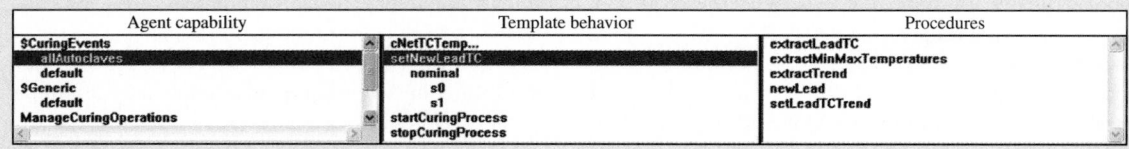

Fig. 28.6 Agent capabilities, behaviors, and procedures

lect a leading and a lagging thermocouple. The nominal behaviors given to an autoclave agent consist of the following:

- *cNetTCTemp*: This behavior generates a multicast message to all dependent thermocouples (TC1–TC16 in Fig. 28.2 to have them summarize their current temperature trends. Asynchronous responses are sent by the thermocouple agents to report their temperature information.
- *SetNewLeaTC*: In this behavior, the autoclave agent applies process-specific criteria to generate a list of leading and lagging thermocouples. The autoclave agent uses a min–max selection mechanism. The process recipe and the stage of curing determine the temperature boundaries (upper and lower).
- *startCuringProcess*: This behavior contains the startup sequence of the autoclave.
- *stopCuringProcess*: This behavior contains the shutdown sequence of the autoclave.

The aggregation of capabilities into the agent behavior creates the rules of the agent incrementally. In Fig. 28.6 all autoclave agents have the *CuringEvent* capability, which implements a set of template behaviors. Each behavior is associated with a set of procedures.

28.4.2 Thermocouple Agent

A thermocouple agent has two essential behaviors to monitor the temperature of the process and the sensor condition and trending using the following behaviors

- *getTemperature*: This behavior uses a sampling algorithm to calculate temperature average and standard deviations. This behavior allows for continuous adjustment of temperature measurement to compensate for natural jitter.
- *provideTemperatureTrend*: This behavior enables the detection of temperature variations outside the nominal ranges. The feasible range allows for the detection of extreme deviations from nominal trends. These events are reported to the autoclave agent as events.

Each thermocouple carries out continuous sampling of the temperature. This information is processed statistically. Temperature sampling is an internal-to-thermocouple action, which corresponds to an interagent communication. The thermocouple agents receive requests from the autoclave agent to transmit their temperature trends in the form of agent messages in a periodic fashion.

28.5 Agent–Based Simulation

Simulation is made up of two parts: the machine and the material. Machine simulation covers the autoclave machine itself with its mechanical devices (such as heater, cooler, and compressor) and thermocouples. Material simulation mimics the behavior of the composite material. A finite-difference mesh simulates the thermal dissipation throughout the material. The model represents a single piece of composite material of square shape with arbitrary sides and thickness.

The control process takes place in a PLC. However, for the purpose of the simulation, a soft controller PLC (Rockwell Automation, SoftLogix 5800) with

firmware extensions to support the agents was used. The Simulink/Matlab engine was used as the simulation toolset. To mimic the physical process, a simulation of the material, thermocouples, and material was created. The agents use the simulation as the process model instead of the actual machine and material.

In Fig. 28.7 there are horizontally and vertically arranged gray rectangular boxes. These boxes contain heat transfer differential equations to calculate the accumulation and dissipation of heat on a specific node of the finite-difference mesh. The heat transfer model is subjected to anisotropic properties to mimic the heat

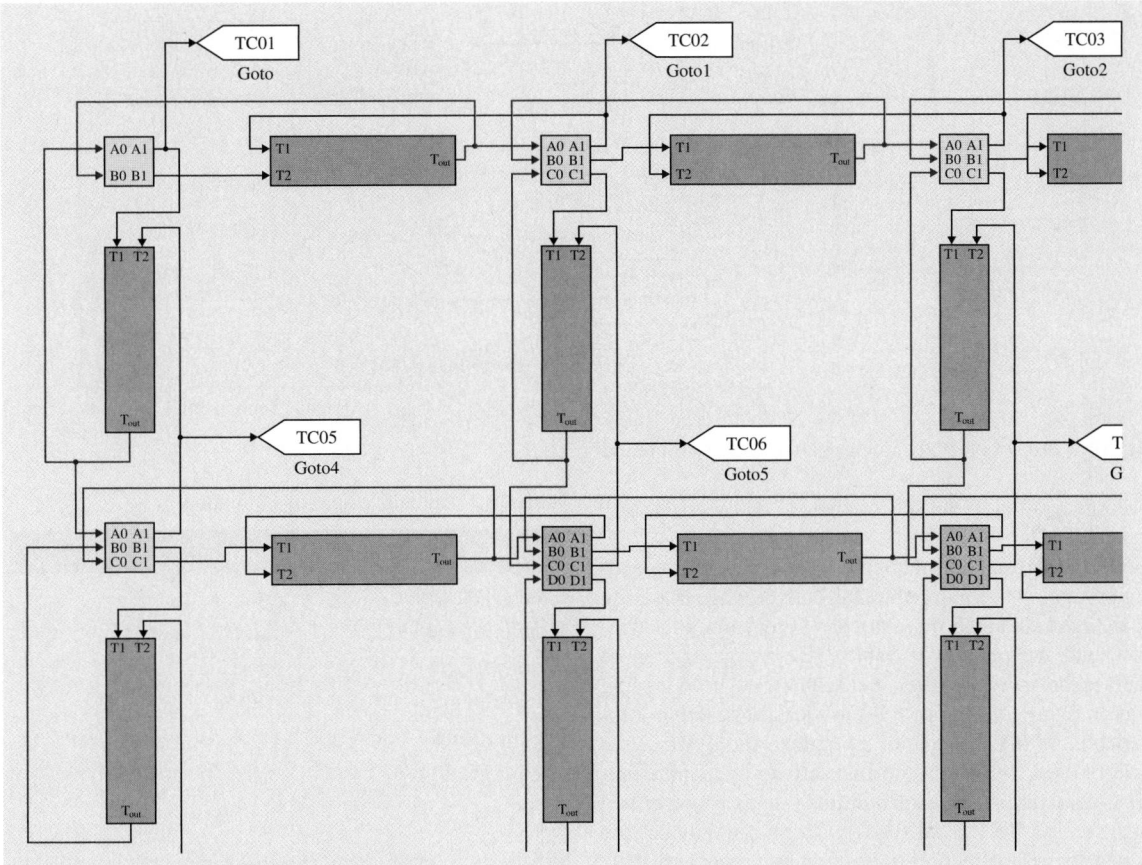

Fig. 28.7 Composite material simulation

dissipation rates throughout the material. The chemical composition of the resin affects the heat transfer, making it vary with location and direction of propagation.

The square boxes represent finite-difference nodes (concentration nodes) to calculate the average temperature of the material at the specific location of measurement. Figure 28.7 shows just a small section of the finite-difference mesh. In the actual model, there are 16 concentration points and 16 thermocouples. The thermocouple sensors are scattered throughout the material to measure the temperature of the composite at each node. The temperature readings are sent to the PLC as an input array of temperatures to be further interpreted by the thermocouple control drivers.

Figure 28.8 shows the simulation blocks that contain the autoclave and autoclave's inner atmosphere. The autoclave agent connects to the simulation throughout six inputs parameters (heaterIsOn, coolerIsOn, coolerIsEnabled, heaterIsEnabled, heatin-

gRate, coolingRate) to control the autoclave. The maximum and minimum heating and cooling rates are provided in the curing recipe. Heating and cooling rates are used as the control variables to control the amount of heat that is applied to the material. Table 28.1 shows an example of the command that initiates a heating-up phase in the simulation.

Control commands drive the simulation in desired directions to achieve a particular curing profile. All 16 thermocouple readings must be maintained within the curing envelope (refer to Fig. 28.1 for the profile shape).

Deviations from the curing envelop are regulated by the cooperative actions of the agents. The agents continuously monitor the health of the thermocouples using temperature trending and standard deviations. When a thermocouple fails, the agent that represents that thermocouple decides to dismiss itself from the monitoring array. This decision is taken autonomously at the lowest level. The dismissal cascades up the hierarchy, inducing

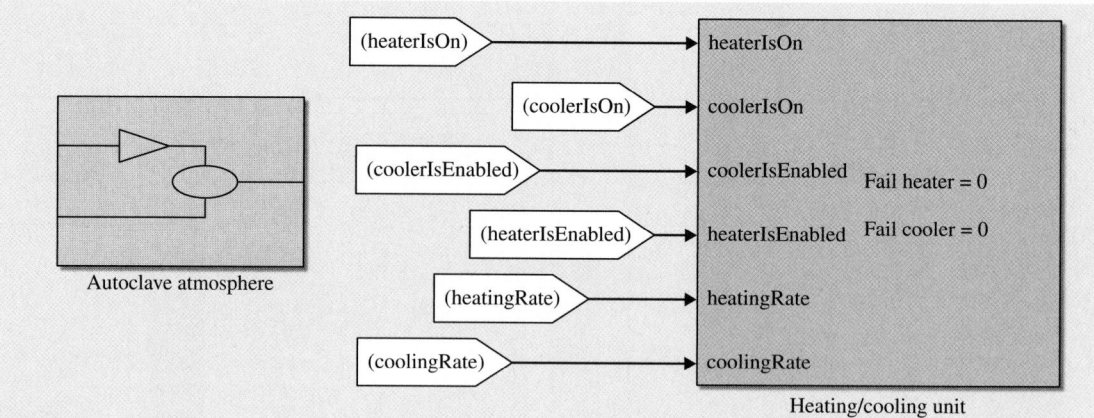

Fig. 28.8 Autoclave and autoclave atmosphere simulation

a reorganization of the sensor array to adjust to the new reading set.

If a departing thermocouple was operating as a leading or lagging thermocouple, the current stage of curing is at high risk since the autoclave agent loses its window into the process variable. The reorganization of the sensor arrays implies a discovery of new leading and/or lagging thermocouples to reestablish the process variable. While the agents reorganize the system, the control must continue its normal activity by maintaining all temperatures within the nominal ranges as prescribed in the recipe, but this fall-back position can only be sustained for a short period until the new process variable is reestablished. This reorganization activity is the core

Table 28.1 Autoclave heat-up command

Variable name	Value
heaterIsEnabled	true
coolerIsEnabled	false
coolerIsOn	false
heaterIsOn	true
heatingRate	%.%
coolingRate	%.%

benefit of the multiagent system. Agents are intended to handle these dynamic decisions.

28.6 Composite Curing Results and Recommendations

The composite curing process begins with downloading of the agents and device drivers into the PLC. Next, the process simulation is started and connected to the soft controller throughout and automation proxy. The role of the automation proxy is to exchange I/O data between the controller and the simulation, as well as the synchronization of their clocks.

There are several aspects of interest that can be extracted from this technology. First, there is a permanent learning process associated with the encapsulation of the behaviors into particular agents, as well as the partitioning of the domain into regions.

28.6.1 Designing the Validation System

The system designer carries out preliminary verifications to test the agent behaviors. It is natural to see some chaotic results in the initial iterations. The designer adjusts the agent and control functions to make the prototype stable and observable using an interactive process.

The design of the agent is based on templates [28.35]; for instance, only one thermocouple agent template is built for all 16 thermocouples. Likewise, in object-oriented programming, agent programming allows for the creation of agent instances from a single agent template. Thus, all thermocouple agents share the same reasoning rules but vary in personality (name, location, ranges, etc.).

Another consideration is the speed of the simulation. The real curing process takes hours to complete. However, during the modeling and validation phase it is not affordable to spend so much time waiting for events and changes in the process, so the simulation has to be accelerated to compress the total time. This requirement adds complexity to the model since the controller cycles must be in real time to mimic actual control. Time compression is factored into the control timers to make events observable in shorter time spans. In this curing model, ideal time spans during the validation phase d should not exceed 30 min/cycle. Once the overall model is ready, tests begin to check the control/agent responses against the system dynamics.

28.6.2 Modeling Process Dynamics

A critical requirement in agent-based engineering is the interdisciplinary attitude of the engineers. The creation of these sophisticated models requires more than one design perspective to create valid results. The designer transits from sole software engineer into a control engineer who understands the dynamics of the process.

Figure 28.9 shows the distribution of temperature for one of the test runs. The colored regions represent isothermals for heating and cooling. The isothermals change in shape and size as the curing process evolves. The curing profile is shown in the lower half of Fig. 28.9. The temperature distribution is not uniform due to the properties of the material. The thermal reaction of the composite makes hot and cold spots change location and size over time. The control system must maintain all temperatures within the upper and lower

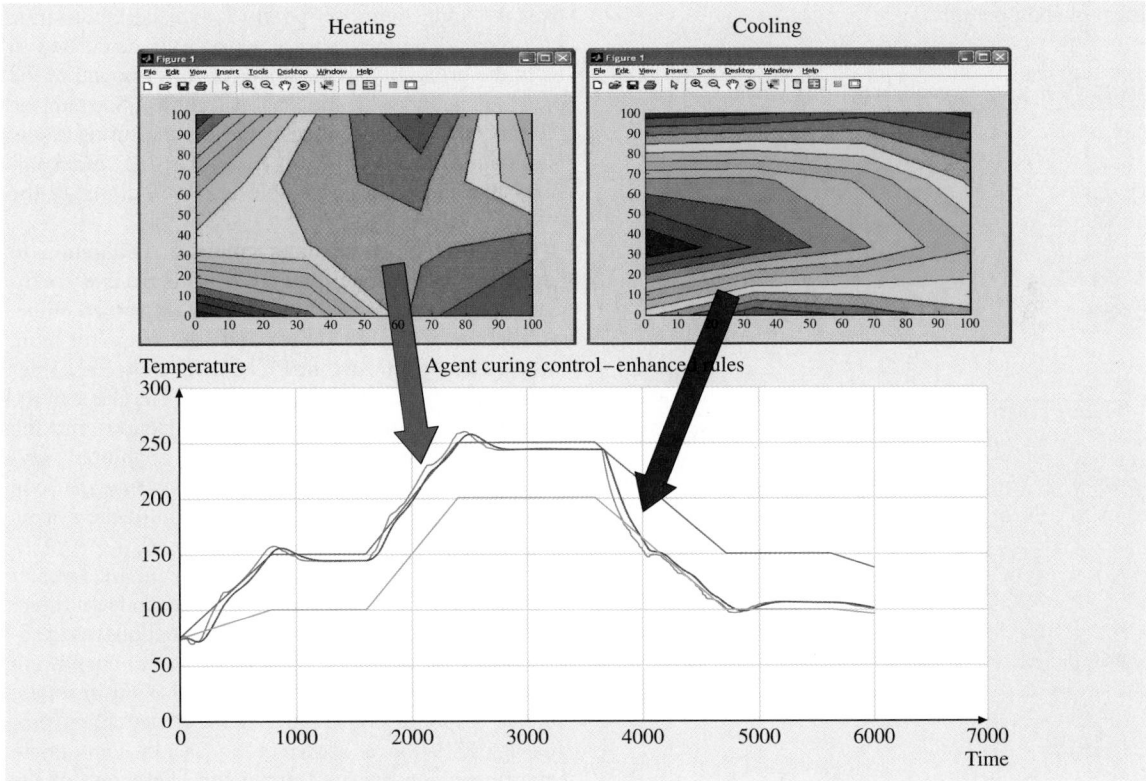

Fig. 28.9 Heating and cooling simulation

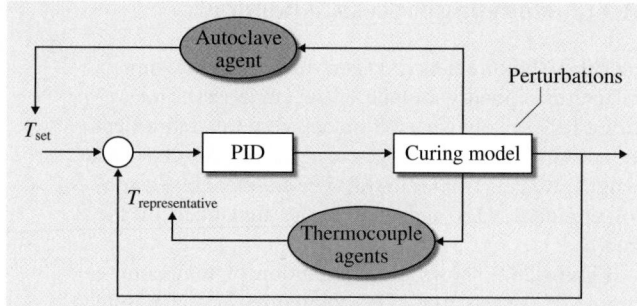

Fig. 28.10 Agents and control cooperation

bound of the curing envelope. The profile introduces abrupt changes in the temperature slopes when moving from a neutral stage into a heating or cooling stage. The control system must regulate the overshooting amplitude while maintaining a good configuration of the thermocouple. There is overshooting in the process but the recipe allows for some operation outside the boundaries for short periods.

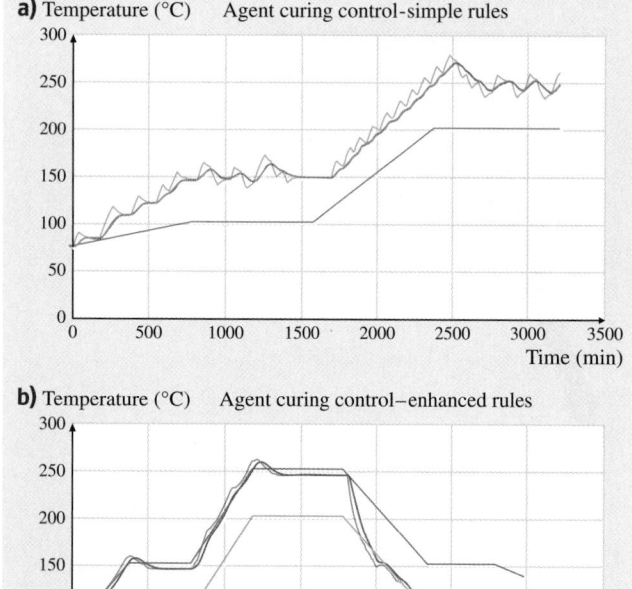

Fig. 28.11 (a) Control response for large look-ahead factor. **(b)** Control response for small look-ahead factor

The core activity is to observe and diagnose the thermocouple readings to select the representative temperature of the process as the process variable. Cooperation among the thermocouples helps in determining this information autonomously. It is very important that the agents maintain close inspection of the temperature sensors and ensure correct selection of the representative temperatures to be able to set the heating and cooling rates, as the material is very susceptible to these rates.

The control system has been implemented as a combination of a proportional–integral–derivative (PID) block with agents, as shown in Fig. 28.10. The combined operations of PID and agents constitute the main control loop to generate heating and cooling rates as the control response. The agents interact with the PID in two main ways. First, to execute an accurate control response, the agents select the lead temperature for the particular stage of the process, which could be heating, cooling or neutral. The autoclave agent calculates the temperature set-point (T_{set}) as an input to the PID block. The temperature set-point is calculated to be in the proximity of the upper or lower bound of the profile depending on the stage of curing. Second, the thermocouple agents collect data from the curing model (simulation) to select $T_{representative}$. The PID block ensures that the process variable is driven closer to the set-point value.

The PID block requires empirical calibration to adjust the proportional, integral, and derivative coefficients of the equation. The tuning is performed online while running the simulation and controller (for more information on how to tune PID coefficients please refer to the control system design literature). The shape of the curing profile introduces too drastic inflections, making it harder to maintain the steady-state condition. Nevertheless, the introduction of the agents into the loop allows for quicker and better utilization of the system variables to cope with such changes.

The ability to look ahead into the future (greater than 60 s) was introduced into the agents to learn future inflections in the curing profile. The intention was to adjust the set-point to begin compensation early to prevent overshoot. However, it was learned that the resultant control response constrained the PID action by provoking oscillation, as shown in Fig. 28.11a. Throughout experimentation, it was learned that the extent of the look-ahead factor played a critical role in the instability observations. The look-ahead factor was reduced to a less than 30 s horizon to make the PID filtering more dominant in the control of the overshooting, as shown in Fig. 28.11b. Nonetheless, the inflexion points are still a difficult aspect to handle using the discussed agent control algorithm. Without agent support, the inflexion points that are introduced by the curing profile will be confined to a PID loop fine-tuning task. By expanding and contracting the look-ahead factor of the agents, it is possible to perceive the effect of the agent reasoning on the PID loop response. Ideally, agents will act rapidly with a very short look-ahead factor to provide better input to the PID loop. However, agents are not intended to carry out control; they are designed to influence the control algorithm in a beneficial direction. An important lesson from this experiment is the notion of role separation: that which belongs to control must be in the control layer. How to define this boundary is still an art rather than a science.

Figure 28.11b shows an improved response, where the temperature profile follows the recipe with no oscillations. According to specifications, the resultant cured material from this experimental trial would be considered of high quality. To obtain an optimum solution further adjustments to the PID coefficients and agent actions are required and perhaps an expansion of the agent

intelligence. Nevertheless, these improvements fall outside the scope of this chapter.

During the agent cooperation, the agents talk to each other using the FIPA/JDL messaging protocol. Figure 28.12 shows an example agent communication. The observer can trace the agent transactions to learn what decisions have been made. The transactions show the information that the agents exchanged to arrive at a control decision. In Fig. 28.12 a multicasting example is shown, where a single agent contacts multiple agents in a two-way transaction. In multiagent systems, interagent communication consists of request/inform transactions, where an agent transmits a request for information to one or a group of agents. The receiving agents process the request locally but may initiate consultation with other agents as well by using similar request/inform transactions. Once the receiving agent has finished preparing its response, it uses an inform message to send the information back to the requester. The requester then receives multiple responses from the agents. Each response provides a different perspective of the situation under consultation. The requester must select the most valuable information or a combination of it to conclude its own decision-making cycle.

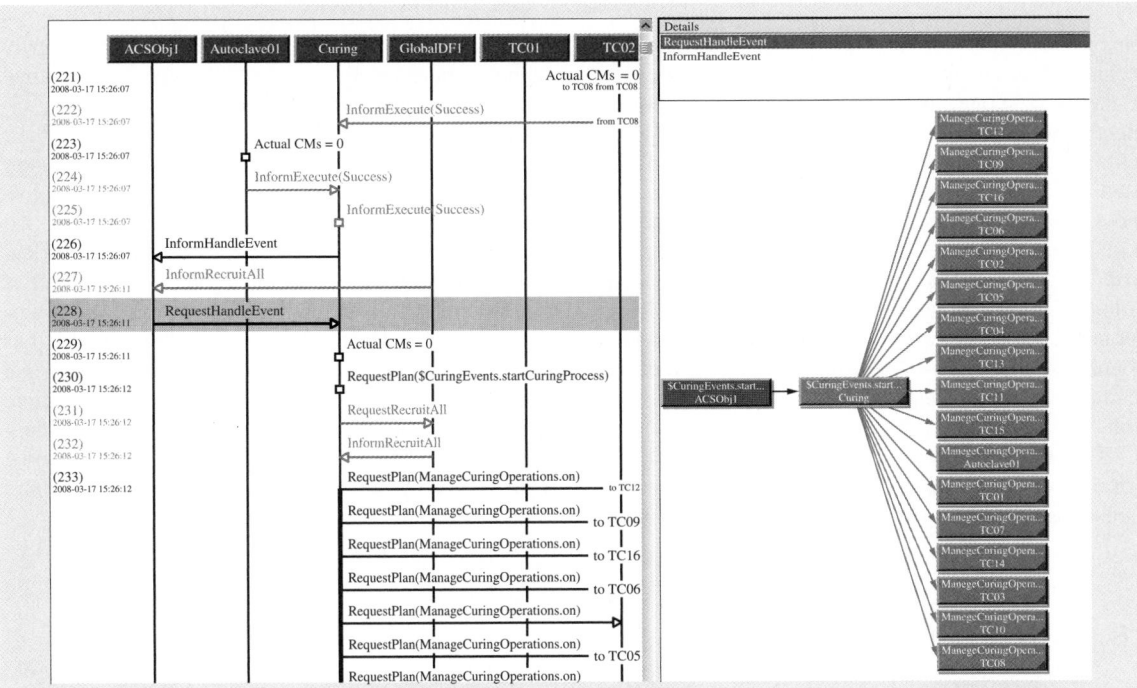

Fig. 28.12 Agent communication

The distributed nature of agent messaging allows for a realistic expansion of the system to add/remove functionality as desired.

In the current implementation, all agents reside within one controller, but the agents can be distributed in any desired topology. The distribution topology is a design decision that must comply with the application's characteristics and requirements. A rule of thumb to decide on the agent distribution is to weigh the desired degree of survivability, diagnosability, and reconfigurability of the system.

In a distributed agent implementation, agents reside in separate platforms. The distribution of the agents depends on the partitioning criteria selected for the application. The partitioning criteria takes under considerations the number of messages used in negotiation. The idea is to use a reduced number of messages. A poor division of functionality among the agents will result in more messaging among them. Message traffic and bandwidth utilization are good metrics to define optimum partitioning.

28.6.3 Timing and Stability Criteria

The multiagent system guarantees an exhaust search for an optimum configuration within a specific time horizon. For small systems (< 100 agents), it is fairly simply to establish the time horizon required by the agents to accomplish a complete search of configurations. The time horizon estimation is established by injecting various conditions into the simulation model so the system designer can observe what time is required to search for and converge to a solution. Once this time horizon is estimated, each agent is given a factored time limit to participate in a particular negotiation process. An infinite time horizon is never given to the agents to avoid deadlocks. Unknown singularities could generate deadlocks in the system. It is guaranteed that the agents will find a solution within the predefined time horizon. However, the value of the solution will fluctuate within optimum and-near optimum performances, depending on the complexity of the emergent search space.

28.7 Conclusions

This chapter has shown how to use industrial agent technology to solve a composite curing application. It has also shown some of the technology requirements and implications of extending industrial PLCs with agent firmware. An example application was used to show the design and modeling phases and the validation of the solution using simulation. Important results and recommendations can be extracted from this to help the modernization of curing technology. Using interagent messaging and the ability to cooperate on the creation and assessment of thermocouple configurations, the agents enable a dynamic learning system, always adapting to current conditions.

The PLC firmware provides for a multitasking priority-based executive control cycle. Agents on built as user level tasks to execute at a low priority. Fundamental tasks in the PLC that manage the PLC's integrity are never interrupted by the agent tasks. User-defined tasks such as stand-alone periodic and continuous control loops compete with agent tasks, but based on their priority levels. Among these, periodic tasks are higher order than agents, but never higher than the executive functions.

The extent of learning depends on the sophistication of the rules given to the agents. This aspect is a user-defined characteristic of the system. Agent programming provides for this type of flexibility. It has been learned that loosely coupled agents and control components ease the scalability of the control solution. Finally, this chapter has shown how survivability, diagnostics, and reconfiguration of a control system can be achieved with component-level intelligence in a more compact manner.

28.8 Further Reading

1. F. Bergenti, M.P. Gleizes, F. Zambonelli (Eds.): *Methodologies and Software Engineering for Agent* *Systems: The Agent-Oriented Software Engineering Handbook* (Springer, New York 2004)

2. M. D'Inverno, M. Luck, M. Fisher, C. Preist (Eds.): *Foundations and Applications of Multi-Agent Systems: UKMAS Workshops* (Springer, New York 2002)
3. S.M. Deen: *Agent Based Manufacturing* (Springer, New York 2003)
4. C.Y. Huang, S.Y. Nof: Formation of autonomous agent networks for manufacturing systems, Int. J. Prod. Res. **38**(3), 607–624 (2000)
5. N.R. Jennings, M.J. Wooldridge (Eds.): *Agents Technology Foundation, Applications and Markets* (Springer, New York 1998)
6. M. Klusch, S.P. Bergamaschi Edwards, P. Petta (Eds.): *Intelligent Information Agents* (Springer, New York 2003)
7. P. Kopacek (Ed.): *Multi-Agent Systems in Production: A Proceedings volume of the IFAC workshop* (Pergamon, Oxford 2000)
8. R.S.T. Lee, V. Loia: *Computational Intelligence for Agent-based Systems* (Springer, New York 2007)
9. P. Maes: Modeling adaptive, autonomous agents, Artif. Life **1**(1/2), 135–162 (1994)
10. G. Moro, M. Koubarakis (Eds.): *Agents and Peer-to-Peer Computing: First International Workshop* (Springer, New York 2002)
11. S.Y. Nof: Intelligent collaborative agents. In: *McGraw Hill Yearbook of Science and Technology* (McGraw Hill, Columbus 2000) pp. 219–222
12. S.Y. Nof: *Handbook of Industrial Robotics* (Wiley, New York 1999)
13. L. Steels: When are robots intelligent, autonomous agents?, Robot. Auton. Syst. **15**, 3–9 (1995)
14. Z. Zhang: *Agent-based Hybrid Intelligent Systems: An Agent-based Framework for Complex Problem Solving* (Springer, New York 2004)

Part C | 28

References

28.1 R.G. Smith: The Contract Net Protocol: high-level communication and control in a distributed problem solver, IEEE Trans. Comput. **29**(12), 1104–1113 (1980)

28.2 N.R. Jennings, M.J. Wooldridge: Co-operating agents: concepts and applications. In: *Agent Technology: Foundations, Applications, and Markets*, ed. by H. Haugeneder, D. Steiner (Springer, Berlin Heidelberg 1998) pp. 175–201

28.3 W. Shen, D. Norrie, J.P. Bartès: *Multi-Agent Systems for Concurrent Intelligent Design and Manufacturing* (Taylor & Francis, London 2001)

28.4 Commercial Technology for Maintenance Activities – CTMA (2008), http://ctma.ncms.org/default.htm

28.5 Research and Markets: *Growth Opportunities in Carbon Fiber Composites Market 2006–2011* (Research and Markets, Dublin 2006), http://www.researchandmarkets.com/reports/362300/growth_opportunities_in_carbon_fiber_composites

28.6 T. Roberts: Rapid growth forecast for carbon fibre market, Reinf. Plast. **51**(2), 10–13 (2007)

28.7 Research and Markets: *Opportunities in Continuous Fiber Reinforced Thermoplastic Composites 2003–2008* (Research and Markets, Dublin 2003) p. 240

28.8 K. Burger: The changing outlook for natural rubber, Natuurrubber **40**(4), 1–4 (2005)

28.9 P. Salagnac, P. Dutournié, P. Glouannec: Curing of composites by radiation and natural convection in an autoclave, Am. Inst. Chem. Eng. **50**(12), 3149–3159 (2004)

28.10 P. Salagnac, P. Dutournié, P. Glouannec: Simulations of heat transfers in an autoclave. Applications to the curing of composite material parts, J. Phys. IV France **120**, 467–472 (2004)

28.11 J. Kim, T.J. Moon, J.R. Howell: Transient thermal modeling of in-situ curing during tape winding of composite cylinders, J. Heat Transf. **125**(1), 137–146 (2003)

28.12 M.-H. Chang, C.-L. Chen, W.-B. Young: Optimal design of the cure cycle for consolidation of thick composite laminants, Polym. Compos. **17**(5), 743–750 (2004)

28.13 M. Jinno, S. Sakai, K. Osaka, T. Fukuda: Smart autoclave processing of thermoset resin matrix composites based on temperature and internal strain monitoring, Adv. Compos. Mater. **12**(1), 57–72 (2003)

28.14 A. Mawardi, R. Pitchumani: Optimal temperature and current cycles for curing composites using embedded resistive heating elements, J. Heat Transf. **125**(1), 126–136 (2003)

28.15 M.M. Thomas, B. Joseph, J.L. Kardos: Experimental characterization of autoclave-cured glass-epoxy composite laminates: cure cycle effects upon thickness, void content and related phenomena, Polym. Compos. **18**(3), 283–299 (2004)

28.16 H.S. Kim, D.G. Lee: Reduction of fabricational thermal residual stress of the hybrid co-cured structure using a dielectrometry, Compos. Sci. Technol. **67**(1), 29–44 (2007)

28.17 M.N. Ghasemi-Nejhad, R. Russ, S. Pourjalali: Manufacturing and testing of active composite panels with embedded piezoelectric sensors and actuators, J. Intell. Mater. Syst. Struct. **16**(4), 319–333 (2005)

28.18   N. Pantelelis, T. Vrouvakis, K. Spentzas: Cure cycle design for composite materials using computer simulation and optimisation tools, Forsch. Ingenieurwes. **67**(6), 254–262 (2007)

28.19   P. George, J. Griffith, G. Orient, J. Madsen, C. Teng, R. Courdji: Exploration of composites processing and producibility by analysis, Proc. 34th Int. SAMPE Tech. Conf. (Baltimore 2002)

28.20   V. Pillai, A.N. Beris, P. Dhurjati: Intelligent curing of thick composites using a knowledge-based system, J. Compos. Mater. **31**(1), 22–51 (1997)

28.21   D.J. Michaud, A.N. Beris, P.S. Dhurjati: Thick-sectioned RTM composite manufacturing, Part II. Robust cure optimization and control, J. Compos. Mater. **36**(10), 1201–1231 (2002)

28.22   S.W. Arms, C.P. Townsend, M.J. Hamel: Validation of remotely powered and interrogated sensing networks for composite cure monitoring (2003), http://www.microstrain.com/white/Validationof RemotelyPoweredandInterrogatedSensingNetworks.pdf

28.23   S. Uryasev, A.A. Trindade: Combining Analytical Model and Experimental Test Data for Optimal Determination of Failure Tolerance Limits (2004). http://www.arpa.mil/dso/thrusts/matdev/aim/ AIM%20PDFs/presentation_2004/Case2328_ attach.pdf

28.24   S. Hill: Electron Beam Curing, Mater. World **7**(7), 398–400 (1999)

28.25   NASA: A New Kind of Curing (2007), http://command. fuentek.com/matrix/success-story.cfm? successid=154

28.26   Blair Rubber Company: Rubber Lining Manual, Section 14: Curing Instructions, Engineering and Application Manual (2005), http://www.blairrubber.com/ manual/PDF_Docs/Sec14_Curing/CURING_ INSTRUCTIONS_Rev3.pdf

28.27   DT9805 Series USB Thermocouple Measurement Modules, Datatranslation product datasheet (2008), http://www.datx.com/docs/datasheets/ dt9805.pdf

28.28   Flexible Thermocouple Management Programs (2008), http://www.vulcanelectric.com/pdf/ calibration.pdf

28.29   The Hot Runner Manager (2008). http://www.fast-heat.com/hotrunner/HR_Specialized.html (thermocouple management)

28.30   TE Wire & Cable: Industrial Application Guide (2008). http://www.tewire.com/10-10.html

28.31   P. Vrba, V. Marík: Simulation in agent-based manufacturing control systems, Proc. IEEE Int. Conf. Syst. Man Cybern. (Hawaii 2005) pp. 1718–1723

28.32   F.P. Maturana, P. Tichý, P. Šlechta, R. Staron: Using dynamically created decision-making organizations (holarchies) to plan, commit, and execute control tasks in a chiller water system, Proc. 13th Int. Workshop Database Expert Syst. Appl. (DEXA 2002), HoloMAS 2002 (Aix-en-Provence 2002) pp. 613–622

28.33   F.P. Maturana, R.J. Staron: *Integration of Collaborating Agent-based Sub-systems – Final Report*, prepared for Johns Hopkins University Applied Physics Laboratory under Subcontract 864123 and the Office of Naval Research under Prime Contract N00014-02-C-0526 (2004)

28.34   IEEE: *The Foundation for Intelligent Physical Agents* (FIPA, IEEE Standards 2008), http://www.fipa.org

28.35   R.J. Staron, F.P. Maturana, P. Tichý, P. Šlechta: Use of an agent type library for the design and implementation of highly flexible control systems, 8th World Multiconf. Syst. Cybern. Inf. (SCI2004) (Orlando 2004)

28.36   F.M. Discenzo, F.P. Maturana, R.J. Staron, P. Tichý, P. Šlechta, V. Marík: Prognostics and control integration with dynamic reconfigurable agents, 1st WSEAS Int. Conf. Electrosci. Technol. Nav. Eng. All-electr. Ship (Vouliagmeni, Athens 2004)

29. Evolutionary Techniques for Automation

Evolutionary

Mitsuo Gen, Lin Lin

In this chapter, evolutionary techniques (ETs) will be introduced for treating automation problems in factory, manufacturing, planning and scheduling, and logistics and transportation systems. ET is the most popular metaheuristic method for solving *NP*-hard optimization problems. In the past few years, ETs have been exploited to solve design automation problems. Concurrently, the field of ET reveals a significant interest in evolvable hardware and problems such as routing, placement or test pattern generation.

The rest of this chapter is organized as follows. First the background developments of evolutionary techniques are described. Then basic schemes and working mechanism of genetic algorithms (GAs) will be given, and multiobjective evolutionary algorithms for treating optimization problems with multiple and conflicting objectives are presented. Lastly, automation and the challenges for applying evolutionary techniques are specified.

Next, the various applications based on ETs for solving factory automation (FA) problems will be surveyed, covering planning and scheduling problems, nonlinear optimization problems in manufacturing systems, and optimal design problems in logistics and transportation systems.

Finally, among those applications based on ETs, detailed case studies will be introduced. The first case study covers dispatching of automated guided vehicles (AGV) and machine scheduling in a flexible manufacturing system (FMS). The second ET case study for treating automation problems is the robot-based assembly line balancing (ALB) problem. Numerical experiments for various scales of AGV dispatching problems and robot-based ALB

problems will be described to show the effectiveness of the proposed approaches with greater search capability that improves the quality of solutions and enhances the rate of convergence over existing approaches.

29.1 Evolutionary Techniques

Evolutionary techniques (ET) is a subfield of artificial intelligence (AI), and refers to a synthesis of methodologies from fuzzy logic (FL), neural networks, genetic algorithm (GA), and other evolutionary algorithms (EAs). ET uses the evolutionary process within a computer to provide a means for addressing complex engineering problems involving chaotic disturbances, randomness, and complex nonlinear dynamics that traditional algorithms have been unable to conquer (Fig. 29.1).

Computer simulations of evolution started as early as in 1954; however most publications were not widely noticed. From these beginnings, computer simulation of evolution by biologists became more common in the early 1960s. Evolution strategies (ES) was introduced by *Rechenberg* in the 1960s and early 1970s, and it was able to solve complex engineering problems [29.2]. Another approach was the evolutionary programming (EP) of *Fogel* [29.3], which was proposed for generating artificial intelligence. EP originally used finite state machines for predicting environments, and used variation and selection to optimize the predictive logic. Genetic algorithms (GAs) in particular became popular through the work of *Holland* in the early

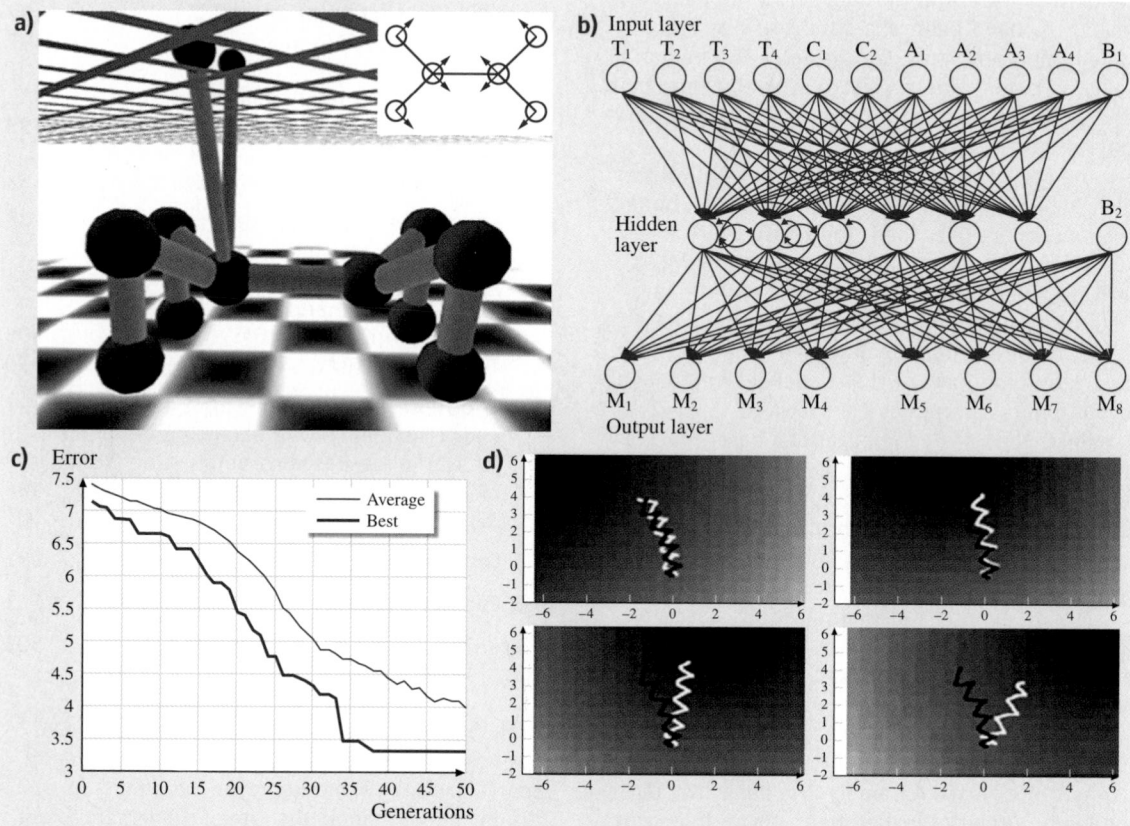

Fig. 29.1a–d Evolving a controller for a fixed morphology: (**a**) The morphology of the machine contains four legs actuated with eight motors, four ground-touch sensors, four angle sensors, and two chemical sensors. (**b**) The machine is controlled by a recurrent neural net whose inputs are connected to the sensors and whose outputs are connected to the motors. (**c**) Evolutionary progress shows how the target misalignment error reduces over generations. (**d**) *White trails* show the motion of the machine towards high concentration (*darker area*). *Black trails* show tracks when the chemical sensors are turned off (after [29.1])

Table 29.1 Classification of evolutionary techniques [29.12]

Optimization algorithms		Other approaches	
Evolutionary algorithms	Genetic algorithms Evolutionary programming Evolution strategy	Self-organization	Self-organizing maps Growing neural gas Competetive learning demo applet
Swarm intelligence	Genetic programming Learning classifier systems Ant colony optimization Particle swarm optimization	Differential evolution Artificial life Cultural algorithms Harmony search algorithm Artificial immune systems Learnable evolution model	Digital organism

1970s [29.4]. His work originated with studies of cellular automata, conducted by Holland and his students at the University of Michigan [29.5]. Holland introduced a formalized framework for predicting the quality of the next generation, known as Holland's schema theorem. Research in GAs remained largely theoretical until the mid-1980s. Genetic programming (GP) is an extended technique of GA popularized by *Koza* in which computer programs, rather than function parameters, are optimized [29.6]. Genetic programming often uses tree-based internal data structures to represent the computer programs for adaptation instead of the list structures typical of genetic algorithms. As academic interest grew, the dramatic increase in desktop computational power allowed for practical application of the new technique [29.7–11]. Evolutionary techniques are generic population-based metaheuristic optimization algorithms, summarized in Table 29.1.

Several conferences and workshops have been held to provide an international forum for exchanging new ideas, progress or experience on ETs and to promote better understanding and collaborations between the theorists and practitioners in this field. The major meetings are the Genetic and Evolutionary Computation Conference (GECCO), the IEEE Congress on Evolutionary Computation (CEC), Parallel Problem Solving from Nature (PPSN), the Foundations of Genetic Algorithms (FOGA) workshop, the Workshop on Ant Colony optimization and Swarm Intelligence (ANTS) and the Evo* and EuroGP workshops etc. The major journals are *Evolutionary Computation*, *IEEE Transactions on Evolutionary Computation*, *Genetic Programming*, and *Evolvable Machines*.

29.1.1 Genetic Algorithm

Among evolutionary techniques (ETs), genetic algorithms (GA) are the most widely known type of ETs today. GA includes the common essential elements of ETs, and has wide real-world applications. The original form of GA was described by *Goldberg* [29.5]. GA is a stochastic search technique based on the mechanism of natural selection and natural genetics. The central theme of research on GA is to keep a balance between exploitation and exploration in its search to the optimal solution for survival in many different environments. Features for self-repair, self-guidance, and reproduction are the rules in biological systems, whereas they barely exist in the most sophisticated artificial systems. GA has been theoretically and empirically proven to provide a robust search in complex search spaces.

GA, differing from conventional search techniques, starts with an initial set of random solutions called the *population*. Each individual in the population is called a *chromosome*, representing a solution to the problem at hand. A chromosome is a string of symbols, usually but not necessarily, a binary bit string. The chromosomes *evolve* through successive iterations, called *generations*. During each generation, the chromosomes are *evaluated*, using some measures of *fitness*. To create the next generation, new chromosomes, called *offspring*, are generated by either merging two chromosomes from the current generation using a *crossover* operator and/or modifying a chromosome using a *mutation* operator. A new generation is formed by selecting some of the parents, according to the fitness values, and offspring, and rejecting others so as to keep the *population size* constant. Fitter chromosomes have higher probabilities of being selected. After several generations, the algorithms converge to the best chromosome, which hopefully represents the optimum or suboptimal solution to the problem. In general, GA has five basic components, as summarized by *Michalewicz* [29.8]:

1. A genetic representation of potential solutions to the problem
2. A way to create a population (an initial set of potential solutions)
3. An evaluation function rating solutions in terms of their fitness
4. Genetic operators that alter the genetic composition of offspring (crossover, mutation, selection, etc.)
5. Parameter values that genetic algorithms use (population size, probabilities of applying genetic operators, etc.).

Figure 29.2 shows a general structure of GA, where $P(t)$ and $C(t)$ are parents and offspring in the current generation t.

29.1.2 Multiobjective Evolutionary Algorithm

Multiple objective problems arise in the design, modeling, and planning of many complex real systems in the areas of industrial production, urban transportation, capital budgeting, forest management, reservoir management, layout and landscaping of new cities, energy distribution, etc. It is easy to find that almost every important real-world decision problem involves multiple and conflicting objectives which need to be tackled while respecting various constraints, leading to overwhelming problem complexity. Since the 1990s, EAs have been received considerable attention as a novel approach to multiobjective optimization problems, resulting in a fresh body of research and applications known as evolutionary multiobjective optimization (EMO).

Features of Genetic Search
The inherent characteristics of EAs demonstrate why genetic search is well suited for multiple-objective op-

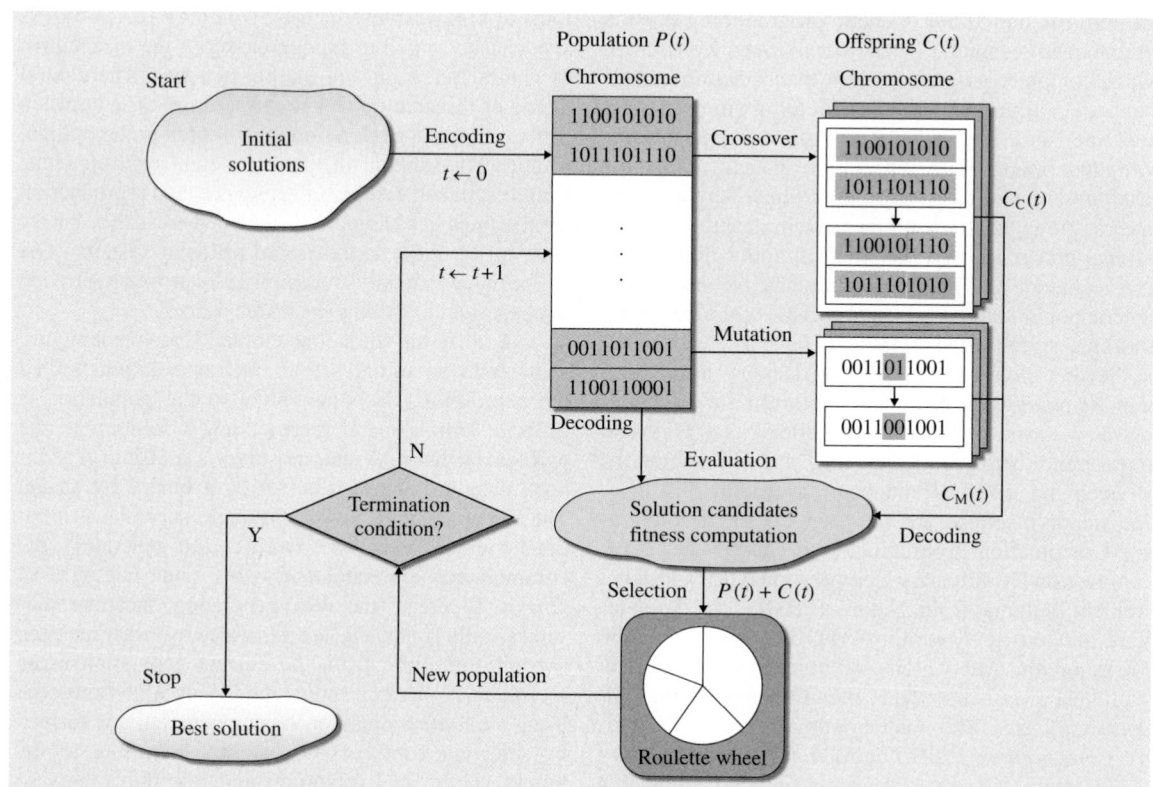

Fig. 29.2 The general structure of genetic algorithms

timization problems. The basic feature of the EAs is multiple-directional and global search by maintaining a population of potential solutions from generation to generation. It is hoped that the population-to-population approach will explore all Pareto solutions.

EAs do not have much mathematical requirements about the problems, and can handle any kind of objective functions and constraints. Due to their evolutionary nature, EAs can search for solutions without regard to the specific internal workings of the problem. Therefore, it is hopeful to solve complex problem by using the evolutionary algorithms.

Because evolutionary algorithms, as a kind of meta-heuristics, provide us with great flexibility to hybridize conventional methods into their main framework, we can take advantage of both evolutionary algorithms and conventional methods to make much more efficient implementations. The growing research on applying EAs to multiple-objective optimization problems presents a formidable theoretical and practical challenge to the mathematical community [29.10].

Fitness Assignment Mechanism

A special issue in the multiobjective optimization problem is the fitness assignment mechanism. Since the 1980s, several fitness assignment mechanisms have been proposed and applied to multiobjective optimization problems. Although most fitness assignment mechanisms are just different approaches and are applicable to different cases of multiobjective optimization problems. In order to understanding the development of EMO, we classify the fitness assignment mechanisms by according to the published year.

Type 1: Vector evaluation approach. The vector evaluated genetic algorithm (veGA) was the first notable work to solve multiobjective problems in which a vector fitness measure is used to create the next generation [29.13].

Type 2: Pareto ranking + diversity: *Fonseca* and *Fleming* proposed a multiobjective genetic algorithm (moGA) in which the rank of a certain individual corresponds to the number of individuals in the current population that dominate it [29.14]. *Srinivas* and *Deb* also developed a Pareto-ranking-based fitness assignment and called it the nondominated sorting genetic algorithm (nsGA) [29.15]. In each method, the nondominated solutions constituting a nondominated front are assigned the same dummy fitness value.

Type 3: Weighted sum + elitist preserve: *Ishibuchi* and *Murata* proposed a weighted-sum-based fitness assignment method, called the random-weight genetic

algorithm (rwGA), to obtain a variable search direction toward the Pareto frontier [29.16]. The weighted-sum approach can be viewed as an extension of methods used in the multiobjective optimizations to GAs. It assigns weights to each objective function and combines the weighted objectives into a single objective function. *Gen* et al. proposed another weight-sum-based fitness assignment method called the adaptive-weight genetic algorithm (awGA), which readjusts weights for each objective based on the values of nondominated solutions in the current population to obtain fitness values combined with the weights toward the Pareto frontier [29.11]. *Zitzler* and *Thiele* proposed the strength Pareto evolutionary algorithm (spEA) [29.16] and an extended version spEA II [29.17, 18] that combines several features of previous multiobjective genetic algorithms (moGA) in a unique manner. *Deb* suggested a nondominated sorting-based approach called the nondominated sorting genetic algorithm II (nsGA II) [29.10], which alleviates three difficulties: computational complexity, nonelitism approach, and the need to specify a sharing parameter. nsGA II was advanced from its origin, nsGA. *Gen* et al. proposed an interactive adaptive-weight genetic algorithm (i-awGA), which is an improved adaptive-weight fitness assignment approach with consideration of the disadvantages of the weighted-sum and Pareto-ranking-based approaches [29.11].

29.1.3 Evolutionary Design Automation

Automation is the use of control systems such as computers to control industrial machinery and processes, replacing human operators. In the scope of industrialization, it is a step beyond mechanization. Whereas mechanization provided human operators with machinery to assist them with the physical requirements of work, automation greatly reduces the need for human sensory and mental requirements as well. Processes and systems can also be automated.

Automation plays an increasingly important role in the global economy and in daily experience. Engineers strive to combine automated devices with mathematical and organizational tools to create complex systems for a rapidly expanding range of applications and human activities.

Evolutionary algorithms (EAs) have received considerable attention regarding their potential as novel optimization techniques. There are three major advantages when applying EA to design automation:

1. *Adaptability*: EAs do not have many mathematical requirements regarding the optimization problem. Due to their evolutionary nature, EAs will search for solutions without regard to the specific internal workings of the problem. EAs can handle any kind of objective functions and any kind of constraints (i. e., linear or nonlinear, defined on discrete, continuous or mixed search spaces).
2. *Robustness*: The use of evolution operators makes EA very effective in performing global search (in probability), while most conventional heuristics usually perform local search. It has been proven by many studies that EA is more efficient and more robust at locating optimal solution and reducing computational effort than other conventional heuristics.
3. *Flexibility*: EAs provide great flexibility to hybridize with domain-dependent heuristics to make an efficient implementation for a specific problem.

However, to exploit the benefits of an effective EA to solve design automation problems, it is usually necessary to examine whether we can build an effective genetic search with the encoding. Several principles were proposed to evaluate effectiveness [29.11]:

Property 1 (*Space*): Chromosomes should not require extravagant amounts of memory.

Property 2 (*Time*): The time required for executing evaluation, recombination, and mutation on chromosomes should not be great.

Property 3 (*Feasibility*): A chromosome corresponds to a feasible solution.

Property 4 (*Legality*): Any permutation of a chromosome corresponds to a solution.

Property 5 (*Completeness*): Any solution has a corresponding chromosome.

Property 6 (*Uniqueness*): The mapping from chromosomes to solutions (decoding) may belong to one of the following three cases (Fig. 29.5): 1-to-1 mapping, n-to-1 mapping, and 1-to-n mapping. The 1-to-1 mapping is the best among the three cases, and 1-to-n mapping is the most undesirable.

Property 7 (*Heritability*): Offspring of simple crossover (i. e., one-cut-point crossover) should correspond to solutions which combine the basic features of their parents.

Property 8 (*Locality*): A small change in a chromosome should imply a small change in its corresponding solution.

29.2 Evolutionary Techniques for Industrial Automation

Currently, for manufacturing, the purpose of automation has shifted from increasing productivity and reducing costs to broader issues, such as increasing quality and flexibility in the manufacturing process. For example, automobile and truck pistons used to be installed into engines manually. This is rapidly being transitioned to automated machine installation, because the error rate for manual installment was around 1–1.5%, but has been reduced to 0.00001% with automation. Hazardous operations, such as oil refining, manufacturing of industrial chemicals, and all forms of metal working, were always early contenders for automation.

However, many applications of automation, such as optimizations and automatic controls, are formulated with complex structures, complex constraints, and multiple objects simultaneously, which makes the problem intractable to traditional approaches. In recent years, the evolutionary techniques community has turned much of its attention toward applications in industrial automation.

29.2.1 Factory Automation

In a manufacturing system, layout design is concerned with the optimum arrangement of physical facilities, such as departments or machines, in a certain area. Usually the design criterion is considered to be minimizing material-handling costs. Because of the combinatorial nature of the facility layout problem, the heuristic technique is the most promising approach for solving practical-size layout problems. The interest in application of ETs to facility layout design has been growing rapidly. *Tate* and *Smith* applied GA to the shape-constrained unequal-area facility layout problem [29.19]. *Cohoon* et al. proposed a distributed GA for the floorplan design problem [29.20]. *Tam* reported his experiences of applying genetic algorithms to the facility layout problem [29.21].

A flexible machining system is one of the forms of factory automation. The design of a flexible machining system (FMS) involves the layout of machine and workstations. *Kusiak* and *Heragu* wrote a survey paper on the machine layout problem [29.22]. The layout of machines in a FMS is typically determined by the type of material-handling devices used. The most used material-handling devices are: material-handling robot, automated guided vehicle (AGV), and gantry robot. *Gen* and *Cheng* provide various GA approaches for the machine layout and facility layout problems [29.9].

29.2.2 Planning and Scheduling Automation

The planning and scheduling of manufacturing systems always require resource capacity constraints, disjunctive constraints, and precedence constraints, owing to the tight due dates, multiple customer-specific orders, and flexible process strategies. Here, some hot topics of applications of ETs in advanced planning and scheduling (APS) are introduced. These models mainly support the integrated, constraint-based planning of the manufacturing system to reduce lead times, lower inventories, increase throughput, etc.

The flexible jobshop problem (fJSP) is a generalization of the jobshop and parallel machine environment [29.23], which provides a closer approximation to a wide range of real manufacturing systems. *Kacem* et al. proposed the operations machine-based GA approach [29.24], which is based on a traditional representation called the schemata theorem representation. *Zhang* and *Gen* proposed a multistage operation-based encoding for fJSP [29.25].

The objective of the resource-constrained project scheduling problem (rcPSP) is to schedule activities such that precedence and resource constraints are obeyed and the makespan of the project is to be minimized. *Gen* and *Cheng* adopted priority-based encoding for this rcPSP [29.9]. In order to improve the effectiveness of priority-based GA approach for an extended resource constrained multiple project scheduling problem, *Kim* et al. combined priority dispatching rules in priority-based encoding process [29.26].

The advanced planning and scheduling (APS) model includes a range of capabilities from finite capacity planning at the plant floor level through constraint-based planning to the latest applications of advanced logic for supply-chain planning and collaboration [29.27]. Several related works by *Moon* et al. [29.28] and *Moon* and *Seo* [29.29] have reported a GA approach especially for solving such kinds of APS problems.

29.2.3 Manufacturing Automation

Assembly-line balancing problems (ALB) consist of distributing work required to assemble a product in mass or series production on an assembly line among a set of workstations. Several constraints and different objectives may be considered. The simple assembly-line balancing problem consists of assigning tasks to workstations such that precedence relations between tasks and zoning or other constraints are met. The objective is to make the work content at each station most balanced. GAs have been applied to solve various assembly-line balancing problems [29.30–32]. *Gao* et al. proposed an innovative GA hybridized with local search for a robotic-based ALB problem [29.33]. Based on different neighborhood structures, five local search procedures are developed to enhance the search ability of GA.

An automated guided vehicle (AGV) is a mobile robot used widely in industrial applications to move materials from point to point. AGVs help to reduce costs of manufacturing and increase efficiency in a manufacturing system. For a recent review of AGV problems and issues the reader is referred to [29.34–37]. *Lin* et al. adopted priority-based encoding for solving the AGV dispatching problem in an FMS [29.38].

29.2.4 Logistics and Transportation Automation

Logistics is the last frontier for cost reduction and the third profit source of enterprises [29.39]. The interest in developing effective logistics system design models and efficient optimization methods has been stimulated by high costs of logistics and the potential for securing considerable savings. One of the most important topics of logistics and transportation automation is automated container terminals.

Handling equipment scheduling (HES) is important for scheduling different types of handing equipment in order to improve the productivity of automated container terminals. Lau and Zhao give a mixed-integer programming model, which considers various constraints related to the integrated operations between different types of handling equipment. This study proposes a multilayer GA to obtain the near-optimal solution of the integrated scheduling problem [29.39].

Berth allocation planning (BAP) is to allocate space along the quayside to incoming ships at a container terminal in order to minimize some objective function.

Imai et al. introduce a formulation for the simultaneous berth and crane allocation problem and employed GA to find an approximate solution for the problem [29.40].

The aim of storage locations planning (SLP) is to determine the optimal storage strategy for various container-handling schedules. *Preston* and *Kozan* developed a container location model and proposed a GA approach with analyses of different resource levels and a comparison with current practice at the Port of Brisbane [29.41].

29.3 AGV Dispatching in Manufacturing System

Automated material handling has been called the key to integrated manufacturing. An integrated system is useless without a fully integrated, automated material handling system. In the manufacturing environment, there are many automated material handling possibilities. Currently, automated guided vehicles systems (AGV systems), which include automated guided vehicles (AGVs), are the state of the art, and are often used to facilitate automatic storage and retrieval systems (AS/RS).

Traditionally, AGV systems were mostly used in manufacturing systems. In manufacturing areas, AGVs are used to transport all types of materials related to the manufacturing process. The transportation network connects all stationary installations (e.g., machines) in the center. At stations, pickup and delivery points are installed that operate as interfaces between the production/storage system and the transportation system of the center. At these points a load is transferred by, for example, a conveyor from the station to the AGV and vice versa. AGVs travel from one pickup and delivery point to another on fixed or free paths. Guide paths are determined by, for example, wires in the ground or markings on the floor. More recent technologies allow AGVs to operate without physical guide paths.

29.3.1 Network Modeling for AGV Dispatching

In this Subsection, we introduce simultaneous scheduling and routing of AGVs in a flexible manufacturing system (FMS) [29.38]. An FMS environment requires a flexible and adaptable material handling system. AGVs provide such a system. An AGV is a material-handling equipment that travels on a network of guide paths. The FMS is composed of various cells, also called workstations (or machines), each with a specific operation such as milling, washing or assembly. Each cell is connected to the guide path network by a pickup/delivery (P/D) point where pallets are transferred from/to the AGVs. Pallets of products are moved between the cells by the AGVs. Assumptions considered are as follows:

1. AGVs only carry one kind of products at a time.
2. A network of guide paths is defined in advance, and the guide paths have to pass through all pickup/delivery points.
3. The vehicles are assumed to travel at a constant speed.
4. The vehicles can just travel forward, not backward.
5. As many vehicles travel on the guide path simultaneously, collisions are avoided by hardware and are not considered herein.
6. At each workstation, there is pickup space to store the operated material and delivery space to store the material for the next operation.
7. The operation can be started any time after an AGV took the material to come. And also the AGV can transport the operated material from the pickup point to the next delivery point any time.

Definition 1: A node is defined as task T_{ij}, which represents a transition task of the j-th process of job J_i for moving from the pickup point of machine $M_{i,j-1}$ to the delivering point of machine M_{ij}.

Definition 2: An arc can be defined as many decision variables, such as, capacity of AGVs, precedence constraints among the tasks, or costs of movement. Lin et al. defined an arc as a precedence constraint, and give a transition time $c_{jj'}$ from the delivery point of machine M_{ij} to the pickup point of machine $M_{i'j'}$ on the arc.

Definition 3: We define the task precedence for each job; for example, task precedence for three jobs is shown in Fig. 29.3.

The notation used in this Chapter is summarized as follows. Indices: i, i': index of jobs, $i, i' = 1, 2, \ldots, n$; j, j': index of processes, $j, j' = 1, 2, \ldots, n$. Parameters: n: total number of jobs, m: total number of machines, n_i: total number of operations of job i, o_{ij}: the j-th operation of job i, p_{ij}: processing time of operation o_{ij}, M_{ij}: machine assigned for operation o_{ij}, T_{ij}:

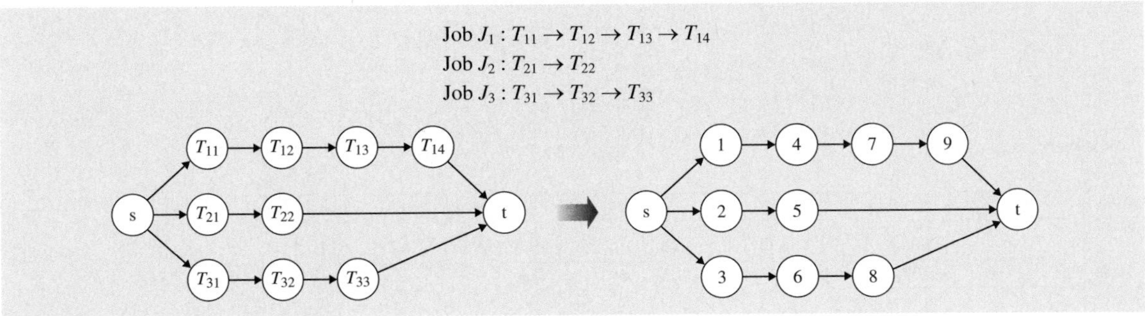

Job J_1 : $T_{11} \to T_{12} \to T_{13} \to T_{14}$
Job J_2 : $T_{21} \to T_{22}$
Job J_3 : $T_{31} \to T_{32} \to T_{33}$

Fig. 29.3 Illustration of the network structure of the example

transition task for operation o_{ij}, t_{ij}: transition time from $M_{i,j-1}$ to M_{ij}. Decision variables: x_{ij}: assigned AGV number for task T_{ij}, t_{ij}^{S}: starting time of task T_{ij}, c_{ij}^{S}: starting time of operation o_{ij}.

The objective functions are to minimize the time required to complete all jobs (i. e., the makespan) t_{MS} and the number of AGVs n_{AGV}, and the problem can be formulated as follows:

$$\min t_{MS} = \max_{i} \left\{ t_{i,n_i}^{S} + t_{M_{i,n_i},0} \right\} , \tag{29.1}$$

$$\min n_{AGV} = \max_{i,j} \left\{ x_{ij} \right\} , \tag{29.2}$$

$$\text{s.t.} \ c_{ij}^{S} - c_{i,j-1}^{S} \geq p_{i,j-1} + t_{ij} , \quad \forall i, j = 2, \ldots, n_i , \tag{29.3}$$

$$\left(c_{ij}^{S} - c_{i'j'}^{S} - p_{i'j'} + \Gamma \left| M_{ij} - M_{i'j'} \right| \geq 0 \right)$$
$$\vee \left(c_{i'j'}^{S} - c_{ij}^{S} - p_{ij} + \Gamma \left| M_{ij} - M_{i'j'} \right| \geq 0 \right) ,$$
$$\forall (i, j), (i', j') , \tag{29.4}$$

$$\left(t_{ij}^{S} - t_{i'j'}^{S} - t_{i'j'} + \Gamma \left| x_{ij} - x_{i'j'} \right| \geq 0 \right)$$
$$\vee \left(t_{i'j'}^{S} - t_{ij}^{S} - t_{ij} + \Gamma \left| x_{ij} - x_{i'j'} \right| \geq 0 \right) ,$$
$$\forall (i, j), (i', j') , \tag{29.5}$$

$$\left(t_{i,n_i}^{S} - t_{i'j'}^{S} - t_{i'j'} + \Gamma \left| x_{ij} - x_{i'j'} \right| \geq 0 \right)$$
$$\vee \left(t_{i'j'}^{S} - t_{i,n_i}^{S} - t_i + \Gamma \left| x_{ij} - x_{i'j'} \right| \geq 0 \right) ,$$
$$\forall (i, n_i), (i', j') , \tag{29.6}$$

$$c_{ij}^{S} \geq t_{i,j+1}^{S} - p_{ij} , \tag{29.7}$$

$$x_{ij} \geq 0 , \quad \forall i, j , \tag{29.8}$$

$$t_{ij}^{S} \geq 0 , \quad \forall i, j , \tag{29.9}$$

where Γ is a very large number, and t_i is the transition time from the pickup point of machine M_{in} to the

delivery point of loading/unloading. Constraint (29.3) describes the operation precedence constraints. In (29.4)–(29.6), since one or the other constraint must hold, it is called disjunctive constraint. This represents the operation nonoverlapping constraint (29.4) and the AGV nonoverlapping constraint (29.5, 29.6).

29.3.2 Evolutionary Approach: Priority-Based GA

For solving the AGV dispatching problem in FMS, the special difficulty arises from (1) that the task sequencing is *NP-hard problem*, and (2) that a random sequence of AGV dispatching usually does not satisfy the operation precedence constraint and routing constraint.

Firstly, we give a priority-based encoding method that is an indirect approach, encoding some guiding information to construct a sequence of all tasks. As is known, a gene in a chromosome is characterized by two factors: the locus, i. e., the position of the gene within the structure of the chromosome, and the allele, i. e., the value the gene takes. In this encoding method, the position of a gene is used to represent the ID which mapping the task in Fig. 29.3 and its value is used to represent the priority of the task for constructing a sequence among candidates. A feasible sequence can be uniquely determined from this encoding with consideration of the operation precedence constraint. An example of a gen-

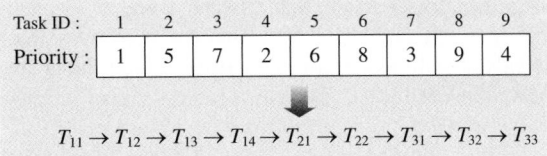

Fig. 29.4 Example generated chromosome and its decoded task sequence

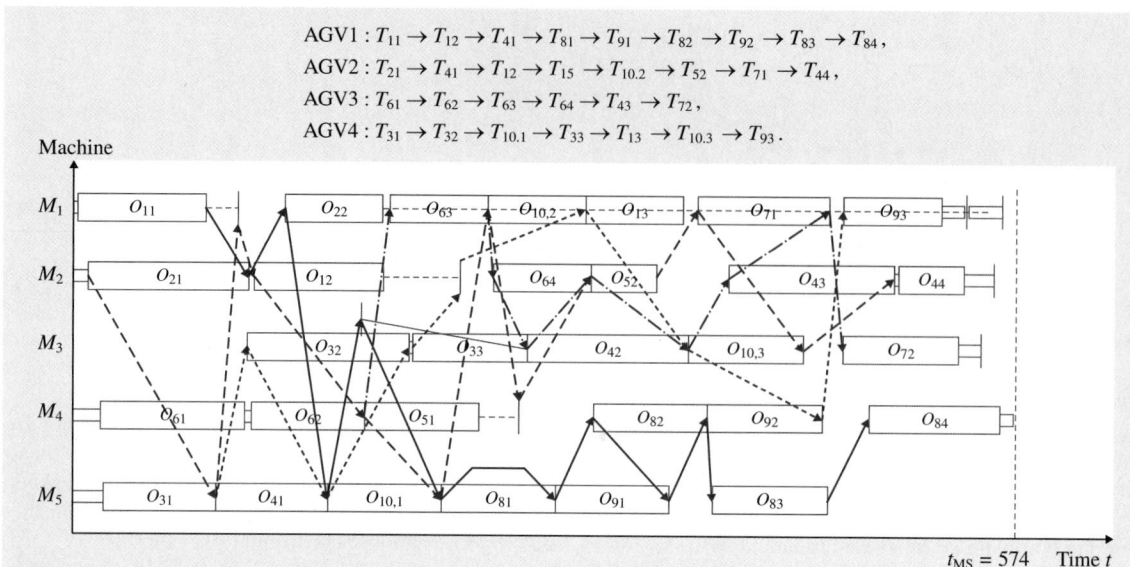

$$\text{AGV1} : T_{11} \rightarrow T_{12} \rightarrow T_{41} \rightarrow T_{81} \rightarrow T_{91} \rightarrow T_{82} \rightarrow T_{92} \rightarrow T_{83} \rightarrow T_{84},$$
$$\text{AGV2} : T_{21} \rightarrow T_{41} \rightarrow T_{12} \rightarrow T_{15} \rightarrow T_{10.2} \rightarrow T_{52} \rightarrow T_{71} \rightarrow T_{44},$$
$$\text{AGV3} : T_{61} \rightarrow T_{62} \rightarrow T_{63} \rightarrow T_{64} \rightarrow T_{43} \rightarrow T_{72},$$
$$\text{AGV4} : T_{31} \rightarrow T_{32} \rightarrow T_{10.1} \rightarrow T_{33} \rightarrow T_{13} \rightarrow T_{10.3} \rightarrow T_{93}.$$

Fig. 29.5 Gantt chart of the schedule of the case study considering AGVs routing (after [29.1], by permission of Macmillan Nature 2004)

erated chromosome and its decoded path is shown in Fig. 29.4 for the network structure of Fig. 29.3.

After generating the task sequence, we separate tasks into several groups for assigning different AGVs. We find the breakpoints, which the tasks are the final transport of job i from pickup point of operation O_{in} to delivery point of loading/unloading. Then we separate the part of tasks sequence by the breakpoints. An example of grouping is shown as follows, using the chromosome (Fig. 29.4):

$$\text{AGV1} : T_{11} \rightarrow T_{12} \rightarrow T_{13} \rightarrow T_{14}$$
$$\text{AGV2} : T_{21} \rightarrow T_{22}$$
$$\text{AGV3} : T_{31} \rightarrow T_{32} \rightarrow T_{33} \,.$$

As genetic operators, we combine a weight mapping crossover (WMX), insertion mutation, and immigration operator based on the characteristic of this representation, and adopt an interactive adaptive-weight fitness assignment mechanism that assigns weights to each objective and combines the weighted objectives into a single objective function. The detailed procedures are showed in [29.38].

29.3.3 Case Study

For evaluating the efficiency of the AGV dispatching algorithm suggested in the case study, a simulation program was developed using Java on a Pentium IV processor (3.2 GHz clock). The detailed test data is given by *Yang* [29.42] and *Kim* et al. [29.43]. GA parameter settings were taken as follows: population size, *pop-Size* = 20; crossover probability, $p_C = 0.70$; mutation probability, $p_M = 0.50$; immigration rate, $\mu = 0.15$. In an FMS case study, ten jobs are to be scheduled on five machines. The maximum number of processes for the operations is four. The detailed date sets are shown in [29.44].

We can draw a network depended on the precedence constraints among tasks $\{T_{ij}\}$ of the case study. The best result is shown in Fig. 29.5. The final time required to complete all jobs (i. e., the makespan) is 574, and four AGVs are used. Figure 29.5 shows the result on a Gantt chart. As discussed above, the AGV dispatching problem is a difficult problem to solve by conventional heuristics. Adaptability, robustness, and flexibility make EA very effective for such automation problems.

29.4 Robot-Based Assembly-Line System

Assembly lines are flow-oriented production systems which are still typical in the industrial production of high-quantity standardized commodities and are even gaining importance in low-volume production of customized products. Usually, specific tooling is developed to perform the activities needed at each station. Such tooling is attached to the robot at the station. In order to avoid the wasted time required for tool change, the design of the tooling can take place only after the line has been balanced. Different robot types may exist at the assembly facility. Each robot type may have different capabilities and efficiencies for various elements of the assembly tasks. Hence, allocating the most appropriate robot for each station is critical for the performance of robotic assembly lines.

29.4.1 Assembly-Line Balancing Problems

This problem concerns how to assign the tasks to stations and how to allocate the available robots for each station in order to minimize cycle time under the constraint of precedence relationships. Let us consider a simple example to describe the problem, in which ten tasks are to be assigned to four workstations, and

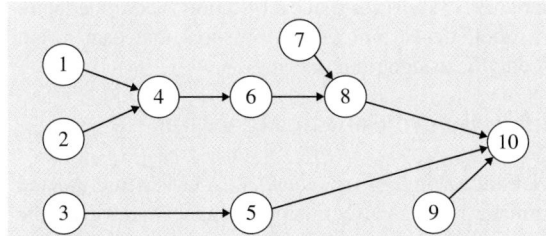

Fig. 29.6 Precedence graph of the example problem

Table 29.2 Data for the example

i	Suc(i)	R_1	R_2	R_3	R_4
1	4	17	22	19	13
2	4	21	22	16	20
3	5	12	25	27	15
4	6	29	21	19	16
5	10	31	25	26	22
6	8	28	18	20	21
7	8	42	28	23	34
8	10	27	33	40	25
9	10	19	13	17	34
10	–	26	27	35	26

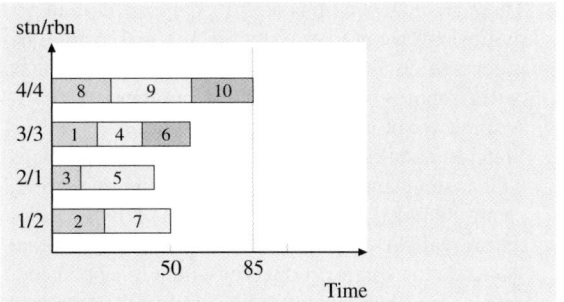

Fig. 29.7 The feasible solution for the example (stn – workstation number, rbn – robot number)

four robots are to be equipped on the four stations. Figure 29.6 shows the precedence constraints for the ten tasks, and Table 29.2 gives the processing time for each of the tasks processed by each robot. We show a feasible solution for this example in Fig. 29.7.

The balancing chart for the solution can be drawn to analyze the solution. Figure 29.7 shows that the idle time of stations 1–3 is very large, which means that this line is not balanced for production. In the real world, an assembly line is not just used for producing one unit of the product; it should produce several units. So we give the Gantt chart for three units to analyze the solution, as shown in Fig. 29.8.

29.4.2 Robot-Based Assembly-Line Model

The following assumptions are stated to clarify the setting in which the problem arises:

1. The precedence relationship among assembly activities is known and invariable.
2. The duration of an activity is deterministic. Activities cannot be subdivided.

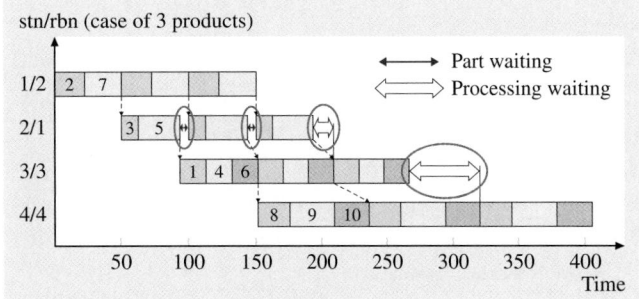

Fig. 29.8 Gantt chart for producing three units

3. The duration of an activity depends on the assigned robot.
4. There are no limitations on the assignment of an activities or a robot to any station. If a task cannot be processed on a robot, the assembly time of the task on the robot is set to a very large number.
5. A single robot is assigned to each station.
6. Material handling, loading, and unloading times, as well as setup and tool changing times are negligible, or are included in the activity times. This assumption is realistic on a single model assembly line that works on the single product for which it is balanced. Tooling on such a robotic line is usually designed such that tool changes are minimized within a station. If tool change or other type of setup activity is necessary, it can be included in the activity time, since the transfer lot size on such line is of a single product.
7. The number of workstations is determined by the number of robots, since the problem aims to maximize the productivity by using all robots at hand.
8. The line is balanced for a single product.

The notation used in this section can be summarized as follows. Indices: i, j: index of assembly tasks, $i, j = 1, 2, \ldots, n$; k: index of workstations, $k = 1, 2, \ldots, m$; l: index of robots, $l = 1, 2, \ldots, m$. Parameters: n: total number of assembly tasks, m: total number of workstations (robots), t_{il}: processing time of the i-th task by robot l, pre(i): the set of predecessor of task i in the precedence diagram. Decision variables

$$x_{jk} = \begin{cases} 1 & \text{if task } j \text{ is assigned to workstation } k \\ 0 & \text{otherwise ,} \end{cases}$$

$$y_{kl} = \begin{cases} 1 & \text{if robot } l \text{ is allocated to workstation } k \\ 0 & \text{otherwise .} \end{cases}$$

Problem formulation

$$\min C_T = \max_{1 \le k \le m} \left\{ \sum_{i=1}^{n} \sum_{l=1}^{m} t_{il} x_{ik} y_{kl} \right\} , \tag{29.10}$$

$$\text{s.t.} \sum_{k=1}^{m} k x_{jk} - \sum_{k=1}^{m} k x_{ik} \ge 0, \ \forall i, j \in \text{pre}(i) , \tag{29.11}$$

$$\sum_{k=1}^{m} x_{ik} = 1 , \quad \forall i , \tag{29.12}$$

$$\sum_{l=1}^{m} y_{kl} = 1 , \quad \forall k , \tag{29.13}$$

$$\sum_{k=1}^{m} y_{kl} = 1 , \quad \forall l , \tag{29.14}$$

$$x_{ik} \in \{0, 1\} \quad \forall k, i , \tag{29.15}$$

$$y_{kl} \in \{0, 1\} \quad \forall l, k . \tag{29.16}$$

The objective (29.10) is to minimize the cycle time, C_T. Constraint (29.11) represents the precedence constraints. It ensures that, for each pair of assembly activities, the precedent cannot be assigned to a station before the station of the successor if there is precedence between the two activities. Constraint (29.12) ensures that each task has to be assigned to one station. Constraint (29.13) ensures that each station is equipped with one robot. Constraint (29.14) ensures that each robot can only be assigned to one station.

29.4.3 Hybrid Genetic Algorithm

Order encoding for task sequence: A GA's structure and parameter settings affect its performance. However, the primary determinants of a GA's success or failure are the coding by which its genotypes represent candidate solutions, and the interaction of the coding with the GA's recombination and mutation operators.

A solution of the robot-based assembly line balancing (rALB) problem can be represented by two integer vectors: the task sequence vector, v_1, which contains a permutation of assembly tasks ordered according to their technological precedence sequence, and the robot assignment vector, v_2. The solution representation method is illustrated in Fig. 29.9. The detailed processes of decoding the task sequence and assigning robots to workstations are shown in [29.34].

In the real world, an assembly line is not just for producing one unit of the product; tt should produce several units. So we give the Gantt chart

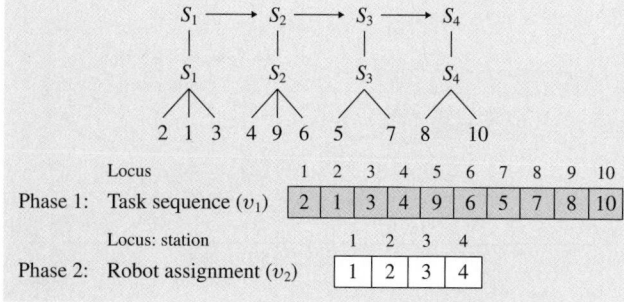

Fig. 29.9 Solution representation of a sample problem

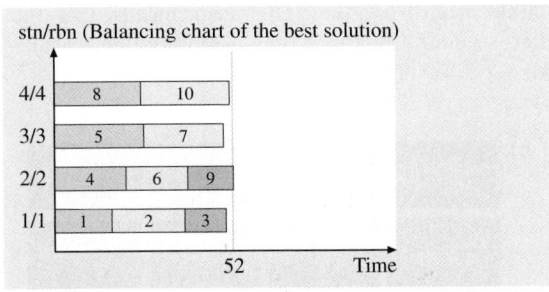

Fig. 29.10 The balancing chart of the best solution (stn – workstation number, rbn – robot number)

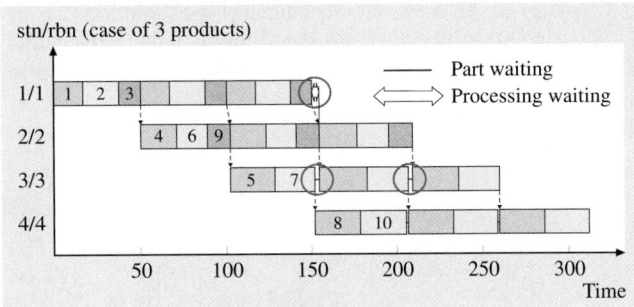

Fig. 29.11 Gantt chart for producing three units

Table 29.3 Performance of the proposed algorithm

Test problem No. of tasks	No. of stations	WEST ratio	Cycle time (C_T) Levitin et al. recursive	Levitin et al. consecutive	Proposed approach
25	3	8.33	518	503	503
	4	6.25	351	330	327
	6	4.17	343	234	213
	9	2.78	138	125	123
35	4	8.75	551	450	449
	5	7.00	385	352	244
	7	5.00	250	222	222
	12	2.92	178	120	113
53	5	10.60	903	565	554
	7	7.57	390	342	320
	10	5.30	35	251	230
	14	3.79	243	166	162
70	7	10.00	546	490	449
	10	7.00	313	287	272
	14	5.00	231	213	204
	19	3.68	198	167	154
89	8	11.13	638	505	494
	12	7.42	455	371	370
	16	5.56	292	246	236
	21	4.24	277	209	205
111	9	12.33	695	586	557
	13	8.54	401	339	319
	17	6.53	322	257	257
	22	5.05	265	209	192
148	10	14.80	708	638	600
	14	10.57	537	441	427
	21	7.05	404	325	300
	29	5.10	249	210	202
297	19	15.63	1129	674	646
	29	10.24	571	444	430
	38	7.82	442	348	344
	50	5.94	363	275	256

with three units for analyzing the solution as in Fig. 29.11. We can see the solution reduce the waiting time for the line by comparing with the feasible solution from Fig. 29.8. This also means that the better solution improved the assembly-line balancing.

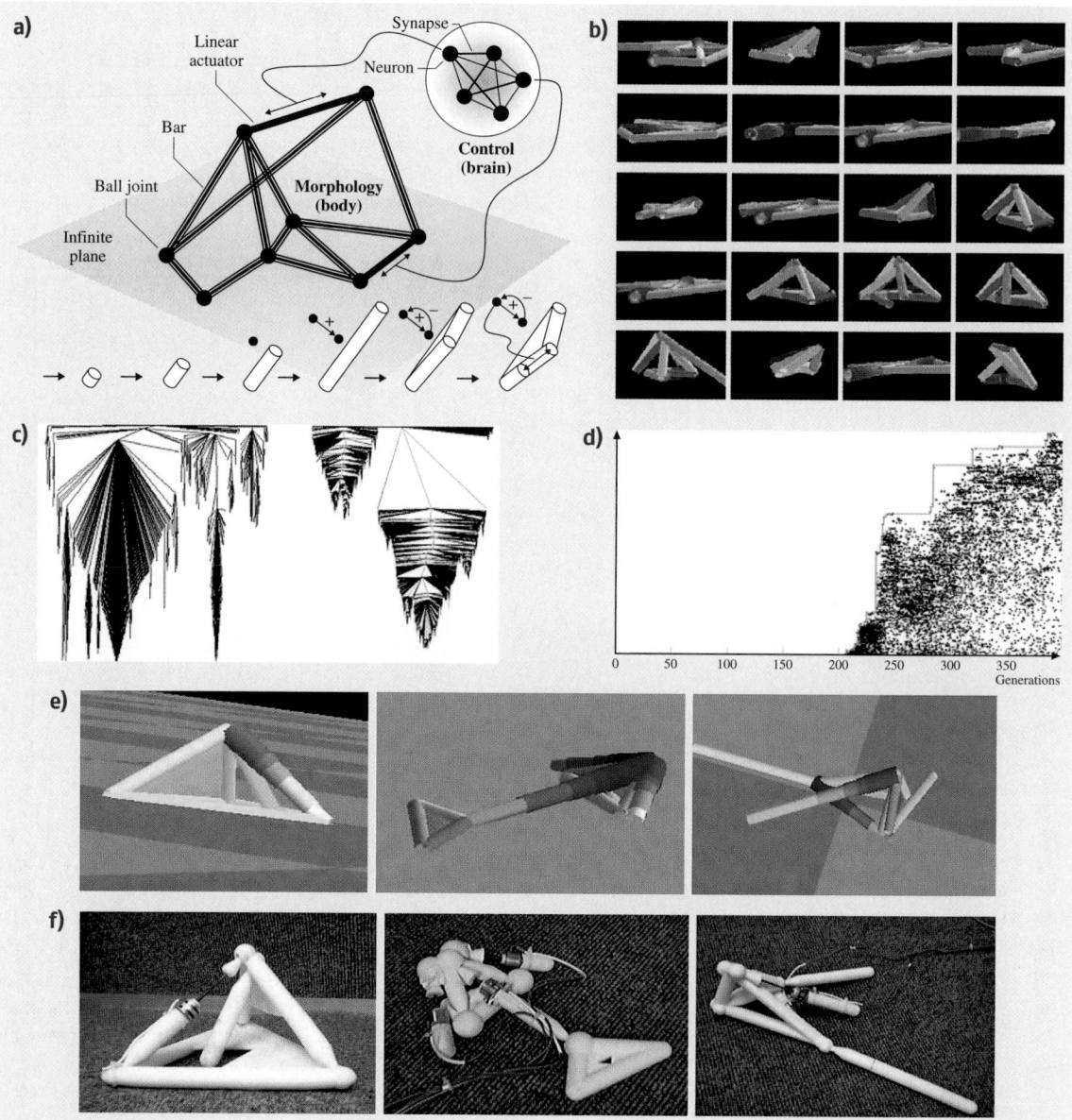

Fig. 29.12a–f Evolving bodies and brains. (**a**) Schematic illustration of an evolvable robot. (**b**) An arbitrarily sampled instance of an entire generation, thinned down to show only significantly different individuals. (**c**) Phylogenetic trees of two different evolutionary runs, showing instances of speciation and massive extinctions. (**d**) Progress of fitness versus generation for one of the runs. Each *dot* represents a robot (morphology and control). (**e**) Three evolved robots, in simulation. (**f**) The three robots from (**c**) reproduced in physical reality using rapid prototyping (after *Lipson* and *Pollack* (2000))

29.4.4 Case Study

In order to evaluate the performance of the proposed method, a large set of problems were tested. In the literature, no benchmark data sets are available for rALB. There are eight representative precedence graphs [29.45], used in the simple assembly line balancing (sALB) literature [29.46]. These precedence graphs contain 25–297 tasks.

Table 29.3 shows the experiment results of Levitin et al.'s two algorithms and proposed approach for 32 different scale test problems [29.47]. As depicted in Table 29.3, all of the results by the proposed approach are better than Levitin et al.'s recursive assignment method, and most results of the proposed approach are better than Levitin et al.'s consecutive assignment method, except for tests 25-3, 35-7, and 111-17. Depending on the experiment results, the way in which a solution of the automation problem is encoded into a chromosome is a key issue for applying the evolutionary algorithm.

29.5 Conclusions and Emerging Trends

The techniques presented in this Chapter are bioinspired and their influence on factory, planning and scheduling, manufacturing, logistics and transportation is discussed. Emerging trends will continue to follow bioinspired control and automation, such as the development of evolvable robots to better meet the needs of evolving and flexible lines (Fig. 29.12). For a related discussion refer to Chap. 14 in an earlier section of this Handbook.

29.6 Further Reading

- T. Gomi (Ed.): *Evolutionary Robotics: From Intelligent Robotics to Artificial Intelligence* (Springer, Tokyo 2001)
- H. Lipson: Evolutionary robotics and open-ended design automation. In: *Biomemetics: Biologically Inspired Technologies*, ed. by Y. Bar-Cohen (CRC Press, Boca Raton 2006) pp. 129–156
- H. Lipson, J.B. Pollack: Automatic design and manufacture of artificial lifeforms, Nature **406**, 974–978 (2000)
- S. Nolfi, D. Floreano: *Evolutionary Robotics: The Biology, Intelligence, and Technology of Self-Organizing Machines* (MIT Press, Cambridge 2000)

References

29.1 J. Bongard, H. Lipson: Integrated Design, Deployment and Inference for Robot Ecologies, Proc. Robosphere 2004 (NASA Ames Research Center 2004)

29.2 I. Rechenberg: *Evolution Strategie: Optimierung Technischer Systeme nach Prinzipien der Biologischen Evolution* (Frommann–Holzboog, Stuttgart 1973)

29.3 L.A. Fogel, M. Walsh: *Artificial Intelligence Through Simulated Evolution* (Wiley, New York 1966)

29.4 J. Holland: *Adaptation in Natural and Artificial Systems* (University of Michigan Press, Ann Arbor 1975), (MIT Press, Cambridge 1992)

29.5 D. Goldberg: *Genetic Algorithms in Search, Optimization and Machine Learning* (Addison-Wesley, Reading 1989)

29.6 J.R. Koza: *Genetic Programming* (MIT Press, Cambridge 1992)

29.7 H. Schwefel: *Evolution and Optimum Seeking*, 2nd edn. (Wiley, New York 1995)

29.8 Z. Michalewicz: *Genetic Algorithm + Data Structures = Evolution Programs*, 3rd edn. (Springer, New York 1996)

29.9 M. Gen, R. Cheng: *Genetic Algorithms and Engineering Design* (Wiley, New York 1997)

29.10 K. Deb: *Multi-Objective Optimization Using Evolutionary Algorithms* (Wiley, New York 2001)

29.11 M. Gen, R. Cheng, L. Lin: *Network Models and Optimization: Multiobjective Genetic Algorithm Approach* (Springer, London 2008)

29.12 Wikipedia: *Evolutionary Computation*, http://en.wikipedia.org/wiki/Evolutionary_computation

29.13 J.D. Schaffer: Multiple Objective Optimization with Vector Evaluated Genetic Algorithms, Proc. 1st Int. Conf. on Genet. Algorithms (1985) pp. 93–100

29.14 C. Fonseca, P. Fleming: An overview of evolutionary algorithms in multiobjective optimization, Evolut. Comput. **3**(1), 1–16 (1995)

29.15   N. Srinivas, K. Deb: Multiobjective function optimization using nondominated sorting genetic algorithms, Evolut. Comput. **3**, 221–248 (1995)

29.16   H. Ishibuchi, T. Murata: A multiobjective genetic local search algorithm and its application to flow-shop scheduling, IEEE Trans. Syst. Man Cybern. **28**(3), 392–403 (1998)

29.17   E. Zitzler, L. Thiele: Multiobjective evolutionary algorithms: a comparative case study and the strength Pareto approach, IEEE Trans. Evolut. Comput. **3**(4), 257–271 (1999)

29.18   E. Zitzler, L. Thiele: SPEA2: Improving the Strength Pareto Evolutionary Algorithm, *Technical Report 103*, (Computer Engineering and Communication Networks Lab, Zurich 2001)

29.19   D. Tate, A. Smith: Unequal-area facility layout by genetic search, IIE Trans. **27**, 465–472 (1995)

29.20   J. Cohoon, S. Hegde, N. Martin: Distributed genetic algorithms for the floor-plan design problem, IEEE Trans. Comput.-Aided Des. **10**, 483–491 (1991)

29.21   K. Tam: Genetic algorithms, function optimization, facility layout design, Eur. J. Oper. Res. **63**, 322–346 (1992)

29.22   A. Kusiak, S. Heragu: The facility layout problem, Eur. J. Oper. Res. **29**, 229–251 (1987)

29.23   M. Pinedo: *Scheduling Theory, Algorithms and Systems* (Prentice-Hall, Upper Saddle River 2002)

29.24   I. Kacem, S. Hammadi, P. Borne: Approach by localization and multiobjective evolutionary optimization for flexible job-shop scheduling problems, IEEE Trans. Syst. Man Cybern. Part C **32**(1), 408–419 (2002)

29.25   H. Zhang, M. Gen: Multistage-based genetic algorithm for flexible job-shop scheduling problem, J. Complex. Int. **11**, 223–232 (2005)

29.26   K.W. Kim, Y.S. Yun, J.M. Yoon, M. Gen, G. Yamazaki: Hybrid genetic algorithm with adaptive abilities for resource-constrained multiple project scheduling, Comput. Ind. **56**(2), 143–160 (2005)

29.27   D. Turbide: Advanced planning and scheduling (APS) systems, Midrange ERP Mag. (1998)

29.28   C. Moon, J.S. Kim, M. Gen: Advanced planning and scheduling based on precedence and resource constraints for e-Plant chains, Int. J. Prod. Res. **42**(15), 2941–2955 (2004)

29.29   C. Moon, Y. Seo: Evolutionary algorithm for advanced process planning and scheduling in a multi-plant, Comput. Ind. Eng. **48**(2), 311–325 (2005)

29.30   Y. Tsujimura, M. Gen, E. Kubota: Solving fuzzy assembly-line balancing problem with genetic algorithms, Comput. Ind. Eng. **29**(1/4), 543–547 (1995)

29.31   M. Gen, Y. Tsujimura, Y. Li: Fuzzy assembly line balancing using genetic algorithms, Comput. Ind. Eng. **31**(3/4), 631–634 (1996)

29.32   J. Rubinovitz, G. Levitin: Genetic algorithm for line balancing, Int. J. Prod. Econ. **41**, 343–354 (1995)

29.33   J. Gao, G. Chen, L. Sun, M. Gen, An efficient approach for type II robotic assembly line balancing problems, Comput. Ind. Eng., in press (2007)

29.34   L. Qiu, W. Hsu, S. Huang, H. Wang: Scheduling and routing algorithms for AGVs: a survey, Int. J. Prod. Res. **40**(3), 745–760 (2002)

29.35   I.F.A. Vis: Survey of research in the design and control of automated guided vehicle systems, Eur. J. Oper. Res. **170**(3), 677–709 (2006)

29.36   T. Le-Anh, D. Koster: A review of design and control of automated guided vehicle systems, Eur. J. Oper. Res. **171**(1), 1–23 (2006)

29.37   J.K. Lin: Study on guide path design and path planning in automated guided vehicle system. Ph.D. Thesis (Waseda University, Japan 2004)

29.38   L. Lin, S.W. Shinn, M. Gen, H. Hwang: Network model and effective evolutionary approach for AGV dispatching in manufacturing system, J. Intell. Manuf. **17**(4), 465–477 (2006)

29.39   Y.K. Lau, Y. Zhao: Integrated scheduling of handling equipment at automated container terminals, Ann. Operat. Res. **159**(1), 373–394 (2008)

29.40   A. Imai, H.C. Chen, E. Nishimura, S. Papadimitriou: The simultaneous berth and quay crane allocation problem, Transp. Res. Part E: Logist. Transp. Rev. **44**(5), 900–920 (2008)

29.41   P. Preston, E. Kozan: An approach to determine storage locations of containers at seaport terminals, Comput. Oper. Res. **28**(10), 983–995 (2001)

29.42   J.B. Yang: GA-based discrete dynamic programming approach for scheduling in FMS environment, IEEE Trans. Syst. Man Cybern. B **31**(5), 824–835 (2001)

29.43   K. Kim, G. Yamazaki, L. Lin, M. Gen: Network-based hybrid genetic algorithm to the scheduling in FMS environments, J. Artif. Life Robot. **8**(1), 67–76 (2004)

29.44   S.H. Kim, H. Hwang: An adaptive dispatching algorithm for automated guided vehicles based on an evolutionary process, Int. J. Prod. Econ. **60/61**, 465–472 (1999)

29.45   A. Scholl, N. Boysen, M. Fliedner, R. Klein: Homepage for assembly line optimization research, http://www.assembly-line-balancing.de/

29.46   A. Scholl: *Data of Assembly Line Balancing Problems*. Schriften zur Quantitativen Betriebswirtschaftslehre 16/93, (TH Darmstadt, Darmstadt 1993)

29.47   G. Levitin, J. Rubinovitz, B. Shnits: A genetic algorithm for robotic assembly balancing, Eur. J. Oper. Res. **168**, 811–825 (2006)

30. Automating Errors and Conflicts Prognostics and Prevention

Xin W. Chen, Shimon Y. Nof

Errors and conflicts exist in many systems. A fundamental question from industries is How can errors and conflicts in systems be eliminated by automation, or can we at least use automation to minimize their damage? The purpose of this chapter is to illustrate a theoretical background and applications of how to automatically prevent errors and conflicts with various devices, technologies, methods, and systems. Eight key functions to prevent errors and conflicts are identified and their theoretical background and applications in both production and service are explained with examples. As systems and networks become larger and more complex, such as global enterprises and the Internet, error and conflict prognostics and prevention become more important and challenging; the focus is shifting from passive response to proactive prognostics and prevention. Additional theoretical developments and implementation efforts are needed to advance the prognostics and prevention of errors and conflicts in many real-world applications.

Part C | 30

30.1 Definitions

All humans commit errors ("To err is human") and encounter conflicts. In the context of automation, there are two main questions: (1) Does automation commit errors and encounter conflicts? (2) Can automation help humans prevent errors and eliminate conflicts? All human-made automation includes human-committed errors and conflicts, for example, human programming errors, design errors, and conflicts between

Table 30.1 Examples of errors and conflicts in production automation

Error	Conflict
• A robot drops a circuit board while moving it between two locations • A machine punches two holes on a metal sheet while only one is needed, because the size of the metal sheet is recognized incorrectly by the vision system • A lathe stops processing a shaft due to power outage • The server of a computer-integrated manufacturing system crashes due to high temperature • A facility layout generated by a software program cannot be implemented due to irregular shapes	• Two numerically controlled machines request help from the same operator at the same time • Three different software packages are used to generate optimal schedule of jobs for a production facility; the schedules generated are totally different • Two automated guided vehicles collide • A DWG (drawing) file prepared by an engineer with AutoCAD cannot be opened by another engineer with the same software • Overlapping workspace defined by two cooperating robots

human planners. Two automation systems, designed separately by different human teams, will encounter conflicts when they are expected to collaborate, for instance, the need for communication protocol standards to enable computers to interact automatically. Some errors and conflicts are inherent to automation, similar to all human-made creations, for instance, a robot mechanical structure that collapses under weight overload.

An error is any input, output or intermediate result that has occurred or will occur in a system and does not meet system specification, expectation or comparison objective. A conflict is an inconsistency between different units' goals, plans, tasks or other activities in a system. A system usually has multiple units, some of which collaborate, cooperate, and/or coordinate to complete tasks. The most important difference between an error and a conflict is that an error can involve only one

unit, whereas a conflict involves two or more units in a system. An error at a unit may cause other errors or conflicts, for instance, a workstation that cannot provide the required number of products to an assembly line (a conflict) because one machine at the workstation breaks down (an error). Similarly, a conflict may cause other errors and conflicts, for instance, a machine that did not receive required products (an error) because the automated guided vehicles that carry the products collided when they were moving toward each other on the same path (a conflict). These phenomena, errors leading to other errors or conflicts, and conflicts leading to other errors or conflicts, are called *error and conflict propagation*.

Errors and conflicts are different but related. The definition of the two terms is often subject to the understanding and modeling of a system and its units. Mathematical equations can help define errors and con-

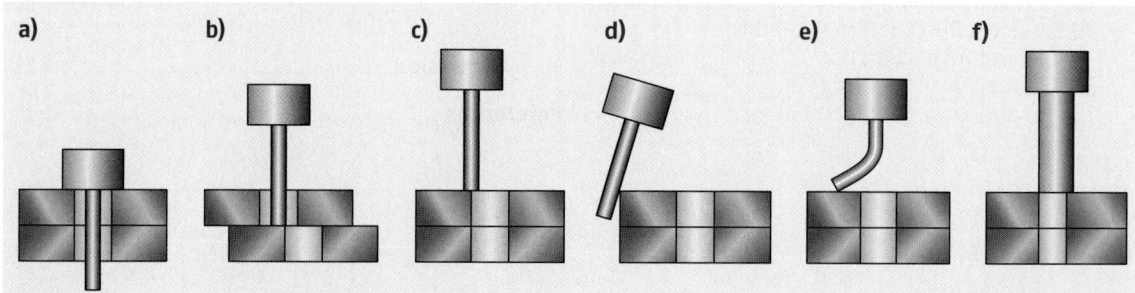

Fig. 30.1a–f Errors and conflicts in a pin insertion task: (**a**) successful insertion; (**b–f**) are unsuccessful insertion with (1) errors if the pin and the two other components are considered as one unit in a system, or (2) conflicts if the pin is a unit and the two other components are considered as another unit in a system [30.1]

Table 30.2 Examples of errors and conflicts in service automation

Error	Conflict
• The engine of an airplane shuts down unexpectedly during the flight • A patient's electronic medical records are accidently deleted during system recovery • A pacemaker stops working • Traffic lights go off due to lightening • A vending machine does not deliver drinks or snacks after the payment • Automatic doors do not open • An elevator stops between two floors • A cellphone automatically initiates phone calls due to a software glitch	• The time between two flights in an itinerary generated by an online booking system is too short for transition from one flight to the other • A ticket machine sells more tickets than the number of available seats • An ATM machine dispenses $250 when a customer withdraws $260 • A translation software incorrectly interprets text • Two surgeries are scheduled in the same room due to a glitch in a sensor that determines if the room is empty

flicts. An error is defined as

$$\exists E\left[u_{r,i}(t)\right] , \quad \text{if } \vartheta_i(t) \xrightarrow{\text{Dissatisfy}} \text{con}_r(t) . \quad (30.1)$$

$E\left[u_{r,i}(t)\right]$ is an error, $u_i(t)$ is unit i in a system at time t, $\vartheta_i(t)$ is unit i's state at time t that describes what has occurred with unit i by time t, $\text{con}_r(t)$ denotes constraint r in the system at time t, and $\xrightarrow{\text{Dissatisfy}}$ denotes that a constraint is not satisfied. Similarly, a conflict is defined as

$$\exists C\left[n_r(t)\right] , \quad \text{if } \theta_i(t) \xrightarrow{\text{Dissatisfy}} \text{con}_r(t) . \quad (30.2)$$

$C\left[n_r(t)\right]$ is a conflict and $n_r(t)$ is a network of units that need to satisfy $\text{con}_r(t)$ at time t. The use of constraints helps define errors and conflicts unambiguously. A constraint is the system specification, expectation, comparison objective or acceptable difference between different units' goals, plans, tasks or other activities. Tables 30.1 and 30.2 illustrate errors and conflicts in automation with some typical examples. There are also human errors and conflicts that exist in automation systems. Figure 30.1 describes the difference between errors and conflicts in pin insertion.

This Chapter provides a theoretical background and illustrates applications of how to prevent errors and conflicts automatically in production and service. Different terms have been used to describe the concept of errors and conflicts, for instance, failure (e.g., [30.2–5]), fault (e.g., [30.4, 6]), exception (e.g., [30.7]), and flaw (e.g., [30.8]). *Error* and *conflict* are the most popular terms appearing in literature (e.g., [30.3,4,6,9–15]). The

related terms listed here are also useful descriptions of errors and conflicts. Depending on the context, some of these terms are interchangeable with error; some are interchangeable with conflict; and the rest refer to both error and conflict.

Eight key functions have been identified as useful to prevent errors and conflicts automatically as described below [30.16–19]. Functions 5–8 prevent errors and conflicts with the support of functions 1–4. Functions 6–8 prevent errors and conflicts by managing those that have already occurred. Function 5, *prognostics*, is the only function that actively determines which errors and conflicts will occur, and prevents them. All other seven functions are designed to manage errors and conflicts that have already occurred, although as a result they can prevent future errors and conflicts directly or indirectly. Figure 30.2 describes error and conflict propagation and their relationship with the eight functions:

1. Detection is a procedure to determine if an error or a conflict has occurred.
2. Identification is a procedure to identify the observation variables most relevant to diagnosing an error or conflict; it answers the question: Which of them has already occurred?
3. Isolation is a procedure to determine the exact location of an error or conflict. Isolation provides more information than identification function, in which only the observation variables associated with the error or conflict are determined. Isolation does not provide as much information as the diagnostics function, however, in which the type, magnitude,

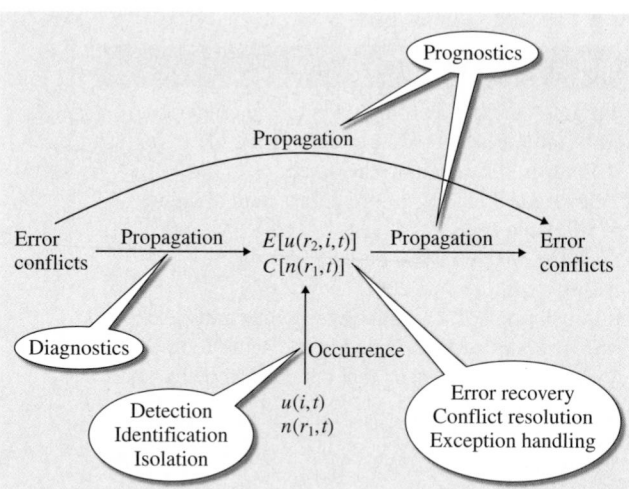

Fig. 30.2 Error and conflict propagation and eight functions to prevent errors and conflicts

and time of the error or conflict are determined. Isolation answers the question: Where has an error or conflict occurred?

4. Diagnostics is a procedure to determine which error or conflict has occurred, what their specific characteristics are, or the cause of the observed out-of-control status.

5. Prognostics is a procedure to prevent errors and conflicts through analysis and prediction of error and conflict propagation.

6. Error recovery is a procedure to remove or mitigate the effect of an error.

7. Conflict resolution is a procedure to resolve a conflict.

8. Exception handling is a procedure to manage exceptions. Exceptions are deviations from an ideal process that uses the available resources to achieve the task requirement (goal) in an optimal way.

There has been extensive research on the eight functions, except prognostics. Various models, methods, tools, and algorithms have been developed to automate the management of errors and conflicts in production and service. Their main limitation is that most of them are designed for a specific application area, or even a specific error or conflict. The main challenge of automating the management of errors and conflicts is how to prevent them through prognostics, which is supported by the other seven functions and requires substantial research and developments.

30.2 Error Prognostics and Prevention Applications

30.2.1 Error Detection in Assembly and Inspection

As the first step to prevent errors, error detection has attracted much attention, especially in assembly and inspection; for instance, researchers [30.3] have studied an integrated sensor-based control system for a flexible assembly cell which includes error detection function. An error knowledge base has been developed to store information about previous errors that had occurred in assembly operations, and corresponding recovery programs which had been used to correct them. The knowledge base provides support for both error detection and recovery. In addition, a similar machine-learning approach to error detection and recovery in assembly has been discussed. To realize error recovery, failure diagnostics has been emphasized as a necessary step after the detection and before the recovery. It is noted that, in assembly, error detection and recovery are often integrated.

Automatic inspection has been applied in various manufacturing processes to detect, identify, and isolate errors or defects with computer vision. It is mostly used to detect defects on printed circuit board [30.20–22] and dirt in paper pulps [30.23, 24]. The use of robots has enabled automatic inspection of hazardous materials (e.g., [30.25]) and in environments that human operators cannot access, e.g., pipelines [30.26]. Automatic inspection has also been adopted to detect errors in many other products such as fuel pellets [30.27], printing the contents of soft drink cans [30.28], oranges [30.29], aircraft components [30.30], and microdrills [30.31]. The key technologies involved in automatic inspection include but are not limited to computer or machine vision, feature extraction, and pattern recognition [30.32–34].

30.2.2 Process Monitoring and Error Management

Process monitoring, or fault detection and diagnostics in industrial systems, has become a new subdiscipline within the broad subject of control and signal processing [30.35]. Three approaches to manage faults for process monitoring are summarized in Fig. 30.3. The analytical approach generates features using detailed mathematical models. Faults can be detected and diagnosed by comparing the observed features with the features associated with normal operating conditions directly or after some transformation [30.19]. The data-driven approach applies statistical tools on large amount of data obtained from complex systems. Many quality control methods are examples of the data-driven approach. The knowledge-based approach uses qualitative models to detect and analyze faults. It is especially suited for systems in which detailed mathematical models are not available. Among these three approaches, the data-driven approach is considered most promising because of its solid theoretical foundation compared with the knowledge-based approach and its ability to deal with large amount of data compared with the analytical approach. The knowledge-based approach, however, has gained much attention recently. Many errors and conflicts can be detected and diagnosed only by experts who have extensive knowledge and experience, which need to be modeled and captured to automate error and conflict prognostics and prevention.

30.2.3 Hardware Testing Algorithms

The three fault management approaches discussed in Sect. 30.2.2 can also be classified according to the way that a system is modeled. In the analytical approach, quantitative models are used which require the complete specification of system components, state variables, observed variables, and functional relationships among them for the purpose of fault management. The data-driven approach can be considered as the effort to develop qualitative models in which previous and current data obtained from a system are used. Qualitative models usually require less information about a system than do quantitative models. The knowledge-based approach uses qualitative models and other types of models; for instance, pattern recognition techniques use multivariate statistical tools and employ qualitative models, whereas the signed directed graph is a typical dependence model which represents the cause–effect relationships in the form of a directed graph [30.36].

Similar to algorithms used in quantitative and qualitative models, optimal and near-optimal test sequences have been developed to diagnose faults in hardware [30.36–45]. The goal of the test sequencing problem is to design a test algorithm that is able to unambiguously identify the occurrence of any system state (faulty or fault-free state) using the test in the test set and minimizes the expected testing cost [30.37].

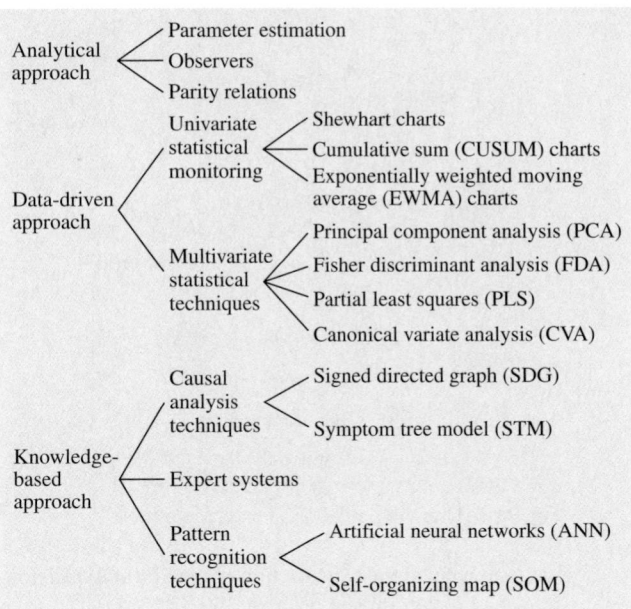

Fig. 30.3 Techniques of fault management in process monitoring

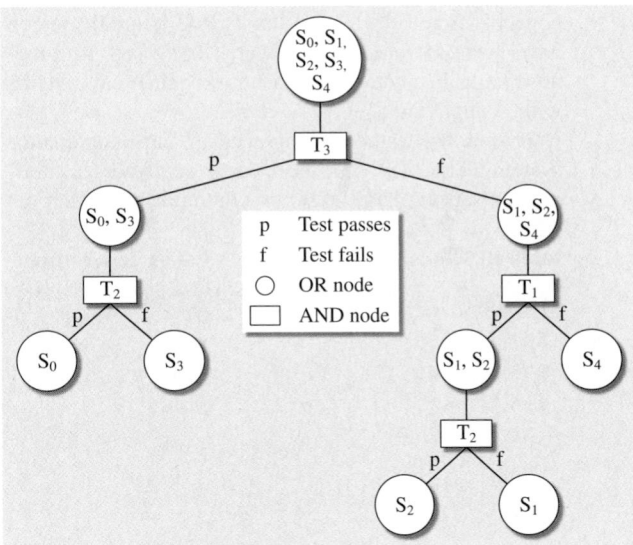

Fig. 30.4 Single-fault test strategy

The test sequencing problem belongs to the general class of binary identification problems. The problem to diagnose single fault is a perfectly observed Markov decision problem (MDP). The solution to the MDP is a deterministic AND/OR binary decision tree with OR nodes labeled by the suspect set of system states and AND nodes denoting tests (decisions) (Fig. 30.4). It is

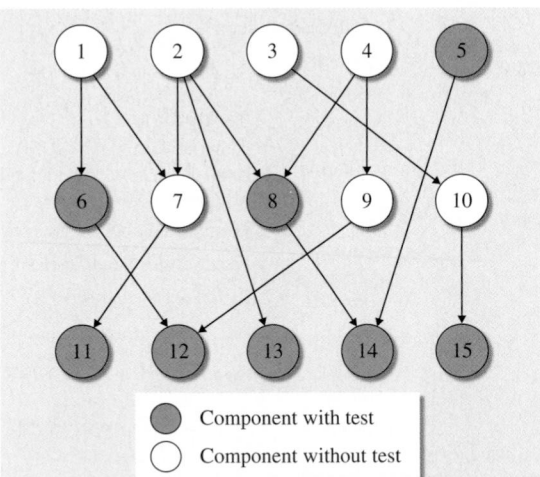

Fig. 30.5 Digraph model of an example system

Part C | 30.2

50 000 faults and 45 000 test points [30.36]. TEAMS can be used to model individual systems and generate near-optimal diagnostic procedures. Research on test sequencing then expanded to diagnose multiple faults [30.41–45] in various real-world systems including the Space Shuttle's main propulsion system. Test sequencing algorithms with unreliable tests [30.40] and multivalued tests [30.45] have also been studied.

To diagnose a single fault in a system, the relationship between the faulty states and tests can be modeled by directed graph (digraph model) (Fig. 30.5). Once a system is described in a diagraph model, the full order dependences among failure states and tests can be captured by a binary test matrix, also called a dependency matrix (D-matrix, Table 30.3). Other researchers have used digraph model to diagnose faults in hypercube microprocessors [30.46]. The directed graph is a powerful tool to describe dependences among system components and tests.

Three important issues have been brought to light by extensive research on test sequencing problem and should be considered when diagnosing faults for hardware:

1. The order of dependences. The first-order cause–effect dependence between two nodes, i.e., how a faulty node affects another node directly, is the simplest dependence relationship between two nodes. Earlier research did not consider the dependences among nodes [30.37, 38], whereas in

well known that the construction of the optimal decision tree is an *NP*-complete problem [30.37].

To subdue the computational explosion of the optimal test sequencing problem, algorithms that integrate concepts from information theory and heuristic search have been developed and were first used to diagnose faults in electronic and electromechanical systems with a single fault [30.37]. An X-Windows-based software tool, the testability engineering and maintenance system (TEAMS), has been developed for testability analysis of large systems containing as many as

Table 30.3 D-matrix of the example system derived from Fig. 30.5

State/test	$T_1(5)$	$T_2(6)$	$T_3(8)$	$T_4(11)$	$T_5(12)$	$T_6(13)$	$T_7(14)$	$T_8(15)$
$S_1(1)$	0	1	0	1	1	0	0	0
$S_2(2)$	0	0	1	1	0	1	1	0
$S_3(3)$	0	0	0	0	0	0	0	1
$S_4(4)$	0	0	1	0	1	0	1	0
$S_5(5)$	1	0	0	0	0	0	1	0
$S_6(6)$	0	1	0	0	1	0	0	0
$S_7(7)$	0	0	0	1	0	0	0	0
$S_8(8)$	0	0	1	0	0	0	1	0
$S_9(9)$	0	0	0	0	1	0	0	0
$S_{10}(10)$	0	0	0	0	0	0	0	1
$S_{11}(11)$	0	0	0	1	0	0	0	0
$S_{12}(12)$	0	0	0	0	1	0	0	0
$S_{13}(13)$	0	0	0	0	0	1	0	0
$S_{14}(14)$	0	0	0	0	0	0	1	0
$S_{15}(15)$	0	0	0	0	0	0	0	1

most recent research different algorithms and test strategies have been developed with the consideration of not only the first-order, but also high-order dependences among nodes [30.43–45]. The high-order dependences describe relationships between nodes that are related to each other through other nodes.

2. Types of faults. Faults can be classified into two categories: functional faults and general faults. A component or unit in a complex system may have more than one function. Each function may become faulty. A component may therefore have one or more functional faults, each of which involves only one function of the component. General faults are those faults that cause faults in all functions of a component. If a component has a general fault, all its functions are faulty. Models that describe only general faults are often called *worst-case* models [30.36] because of their poor diagnosing ability;

3. Fault propagation time. Systems can be classified into two categories: zero-time and nonzero-time systems [30.45]. Fault propagation in zero-time systems is instantaneous to an observer, whereas in nonzero-time systems it is several orders of magnitude slower than the response time of the observer. Zero-time systems can be abstracted by taking the propagation times to be zero.

Another interesting aspect of the test sequencing problem is the list of assumptions that have been discussed in several articles, which are useful guidelines for the development of algorithms for hardware testing:

1. At most one faulty state (component or unit) in a system at any time [30.37]. This may be achieved if the system is tested frequently enough [30.42].
2. All faults are permanent faults [30.37].
3. Tests can identify system states unambiguously [30.37]. In other words, a faulty state is either identified or not identified. There is not a situation such as: There is a 60% probability that a faulty state has occurred.
4. Tests are 100% reliable [30.40,45]. Both false positive and false negative rates are zero.
5. Tests do not have common setup operations [30.42]. This assumption has been proposed to simplify the cost comparison among tests.
6. Faults are independent [30.42].
7. Failure states that are replaced/repaired are 100% functional [30.42].
8. Systems are zero-time systems [30.45].

Note the critical difference between assumptions 3 and 4. Assumption 3 is related to diagnostics ability. When an unambiguous test detects a fault, the conclusion is that the fault has definitely occurred with 100% probability. Nevertheless, this conclusion could be wrong if the false positive rate is not zero. This is the test (diagnostics) reliability described in assumption 4. When an unambiguous test does not detect a fault, the conclusion is that the fault has not occurred with 100% probability. Similarly, this conclusion could be wrong if the false negative rate is not zero. Unambiguous tests have better diagnostics ability than ambiguous tests. If a fault has occurred, ambiguous tests conclude that the fault has occurred with a probability less than one. Similarly, if the fault has not occurred, ambiguous tests conclude that the fault has not occurred with a probability less than one. In summary, if assumption 3 is true, a test gives only two results: a fault has occurred or has not occurred, always with probability 1. If both assumptions 3 and 4 are true, (1) a fault must have occurred if the test concludes that it has occurred, and (2) a fault must have not occurred if the test concludes that it has not occurred.

30.2.4 Error Detection in Software Design

The most prevalent method to detect errors in software is model checking. As *Clarke* et al. [30.47] state, *model checking* is a method to verify algorithmically if the model of software or hardware design satisfies given requirements and specifications through exhaustive enumeration of all the states reachable by the system and the behaviors that traverse them. Model checking has been successfully applied to identify incorrect hardware and protocol designs, and recently there has been a surge in work on applying it to reason about a wide variety of software artifacts; for example, model checking frameworks have been applied to reason about software process models, (e.g., [30.48]), different families of software requirements models (e.g., [30.49]), architectural frameworks (e.g., [30.50]), design models (e.g., [30.51]), and system implementations (e.g., [30.52–55]). The potential of model checking technology for (1) detecting coding errors that are hard to detect using existing quality assurance methods, e.g., bugs that arise from unanticipated interleavings in concurrent programs, and (2) verifying that system models and implementations satisfy crucial temporal properties and other lightweight specifications has led a number of international corporations and government research laboratories such as Microsoft,

IBM, Lucent, NEC, the National Aeronautics and Space Administration (NASA), and the Jet Propulsion Laboratory (JPL) to fund their own software model checking projects.

A drawback of model checking is the state-explosion problem. Software tends to be less structured than hardware and is considered as a concurrent but asynchronous system. In other words, two independent processes in software executing concurrently in either order result in the same global state [30.47]. Failing to execute checking because of too many states is a particularly serious problem for software. Several methods, including symbolic representation, partial order reduction, compositional reasoning, abstraction, symmetry, and induction, have been developed either to decrease the number of states in the model or to accommodate more states, although none of them has been able to solve the problem by allowing a general number of states in the system.

Based on the observation that software model checking has been particularly successful when it can be optimized by taking into account properties of a specific application domain, *Hatcliff* and colleagues have developed Bogor [30.56], which is a highly modular model-checking framework that can be tailored to specific domains. Bogor's extensible modeling language allows new modeling primitives that correspond to domain properties to be incorporated into the modeling language as first-class citizens. Bogor's modular architecture enables its core model-checking algorithms to be replaced by optimized domain-specific algorithms. Bogor has been incorporated into Cadena and tailored to checking avionics designs in the common object request broker architecture (CORBA) component model (CCM), yielding orders of magnitude reduction in verification costs. Specifically, Bogor's modeling language has been extended with primitives to capture CCM interfaces and a real-time CORBA (RT-CORBA) event channel interface, and Bogor's scheduling and state-space exploration algorithms were replaced with a scheduling algorithm that captures the particular scheduling strategy of the RT-CORBA event channel and a customized state-space storage strategy that takes advantage of the periodic computation of avionics software.

Despite this successful customizable strategy, there are additional issues that need to be addressed when incorporating model checking into an overall design/development methodology. A basic problem concerns incorrect or incomplete specifications: before verification, specifications in some logical formalism (usually temporal logic) need to be extracted from design requirements (properties). Model checking can verify if a model of the design satisfies a given specification. It is impossible, however, to determine if the derived specifications are consistent with or cover all design properties that the system should satisfy. That is, it is unknown if the design satisfies any unspecified properties, which are often assumed by designers. Even if all necessary properties are verified through model checking, code generated to implement the design is not guaranteed to meet design specifications, or more importantly, design properties. Model-based software testing is being studied to connect the two ends in software design: requirements and code.

The detection of design errors in software engineering has received much attention. In addition to model checking and software testing, for instance, *Miceli* et al. [30.8] has proposed a metric-based technique for design flaw detection and correction. In parallel computing, synchronization errors are major problems and a nonintrusive detection method for synchronization errors using execution replay has been developed [30.14]. Besides, concurrent error detection (CED) is well known for detecting errors in distributed computing systems and its use of duplications [30.9, 57], which is sometimes considered a drawback.

30.2.5 Error Detection and Diagnostics in Discrete-Event Systems

Recently, Petri nets have been applied in fault detection and diagnostics [30.58–60] and fault analysis [30.61–63]. Petri nets are formal modeling and analysis tool for discrete-event or asynchronous systems. For hybrid systems that have both event-driven and time-driven (synchronous) elements, Petri nets can be extended to global Petri nets to model both discrete-time and event elements. To detect and diagnose faults in discrete-event systems (DES), Petri nets can be used together with finite-state machines (FSM) [30.64, 65]. The notion of diagnosability and a construction procedure for the diagnoser have been developed to detect faults in diagnosable systems [30.64]. A summary of the use of Petri nets in error detection and recovery before the 1990s can be found in the work of *Zhou* and *DiCesare* [30.66].

To detect and diagnose faults with Petri nets, some of the places in a Petri net are assumed observable and others are not. All transitions in the Petri net are also unobservable. Unobservable places, i.e., faults, indicate that the number of tokens in those places is not observable, whereas unobservable transitions indicate that their occurrences cannot be observed [30.58, 60]. The objective of the detection and diagnostics is to identify the occurrence and type of a fault based on observable places within finite steps of observation after the occurrence of the fault. It is clear that to detect and diagnose faults with Petri nets, system modeling is complex and time consuming because faulty transitions and places must be included in a model. Research on this subject has mainly involved the extension of previous work using FSM and has made limited progress.

Faults in discrete-event systems can be diagnosed with the decentralized approach [30.67]. Distributed diagnostics can be performed by either diagnosers communicating with each other directly or through a co-ordinator. Alternatively, diagnostics decisions can be made completely locally without combining the information gathered [30.67]. The decentralized approach is a viable direction for error detection and diagnostics in large and complex systems.

30.2.6 Error Detection in Service and Healthcare Industries

Errors tend to occur frequently in certain service industries that involve intensive human operations. As the use of computers and other automation devices, e.g., handwriting recognition and sorting machines in postal service, becomes increasingly popular, errors can be effectively and automatically prevented and reduced to minimum in many service industries including delivery, transportation, e-Business, and e-Commerce. In some other service industries, especially in healthcare systems, error detection is critical and limited research has been conducted to help develop systems that can automatically detect human errors and other types of errors [30.68–72]. Several systems and modeling tools have been studied and applied to detect errors in health industries with the help of automation devices (e.g., [30.73–76]). Much more research needs to be conducted to advance the development of automated error detection in service industries.

30.2.7 Error Detection and Prevention Algorithms for Production and Service Automation

The fundamental work system has evolved from manual power, human–machine system, computer-

aided and computer-integrated systems, and then to e-Work [30.77], which enables distributed and decentralized operations where errors and conflicts propagate and affect not only the local workstation, but the entire production/service network. Agent-based algorithms, e.g., (30.3), have been developed to detect and prevent errors in the process of providing a single product/service in a sequential production/service line [30.78, 79]. Q_i is the performance of unit i. U'_m and L'_m are the upper limit and lower limit, respectively, of the acceptable performance of unit m. U_m and L_m are the upper limit and lower limit, respectively, of the acceptable level of the quality of a product/service after the operation of unit m. Units 1 through $m-1$ complete their operation on a product/service before unit m starts its operation on the same product/service. An agent deployed at unit m executes (30.3) to prevent errors

$$\exists E(u_m), \quad \text{if} \left\{ U_m - L'_m < \sum_{i=1}^{m-1} Q_i \right\}$$
$$\cup \left\{ L_m - U'_m > \sum_{i=1}^{m-1} Q_i \right\}. \quad (30.3)$$

In the process of providing multiple products/services, traditionally, the centralized algorithm (30.4) is used to predict errors in a sequential production/service line. $I_i(0)$ is the quantity of available raw materials for unit i at time 0. η_i is the probability a product/service is within specifications after being operated by unit i, assuming the product/service is within specifications before being operated by unit i. $\varphi_m(t)$ is the needed number of qualified products/services after the operation of unit m at time t. Equation (30.4) predicts at time 0 the potential errors that may occur at unit m at time t. Equation (30.4) is executed by a central control unit that is aware of $I_i(0)$ and η_i of all units. Equation (30.4) often has low reliability, i.e., high false positive rates (errors are predicted but do not occur), or low preventability, i.e., high false negative rate (errors occur but are not predicted), because it is difficult to obtain accurate η_i when there are many units in the system.

$$\exists E[u_m(t)], \quad \text{if} \min_{i=1}^{m} \left\{ I_i(0) \times \prod_{i}^{m} \eta_i \right\} < \varphi_m(t)$$
$$(30.4)$$

To improve reliability and preventability, agent-based error prevention algorithms, e.g., (30.5), have been de-

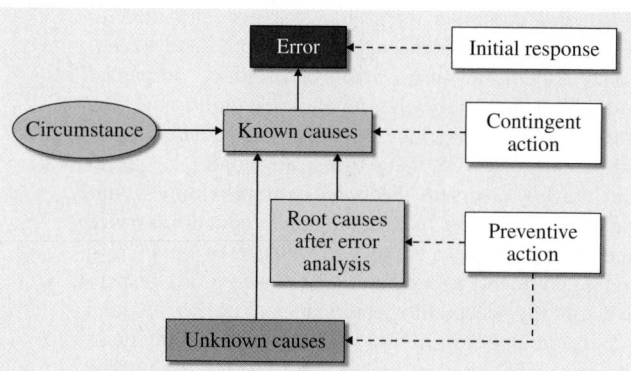

Fig. 30.6 Incident mapping

veloped to prevent errors in the process of providing multiple products/services [30.80]. $C_m(t')$ is the number of cumulative conformities produced by unit m by time t'. $N_m(t')$ is the number of cumulative nonconformities produced by unit m by time t'. An agent deployed at unit m executes (30.5) by using information about unit $m-1$, i.e., $I_{m-1}(t')$, η_{m-1}, and $C_{m-1}(t')$ to prevent errors that may occur at time t, $t' < t$. Multiple agents deployed at different units can execute (30.5) simultaneously to prevent errors. Each agent can have its own attitude, i.e., optimistic or pessimistic, toward the possible occurrence of errors. Additional details about agent-based error prevention algorithms can be found in the work by *Chen* and *Nof* [30.80]:

$$\exists E[u_m(t)] \quad \text{if } \min\big[I_m(t'), I_{m-1}(t') \times \eta_{m-1} \\ + C_{m-1}(t') - N_m(t') \\ - C_m(t')\big] \times \eta_m + C_m(t') \\ < \varphi_m(t), \quad t' < t. \quad (30.5)$$

30.2.8 Error-Prevention Culture (EPC)

To prevent errors effectively, an organization is expected to cultivate an enduring error-prevention culture (EPC) [30.81], i.e., the organization knows what to do to prevent errors when no one is telling it what to do. The EPC model has five components [30.81]:

1. Performance management: the human performance system helps manage valuable assets and involves five key areas: (a) an environment to minimize errors, (b) human resources that are capable of performing tasks, (c) task monitoring to audit work, (d) feedback provided by individuals or teams through collaboration, and (e) consequences provided to encourage or discourage people for their behaviors.
2. System alignment: an organization's operating systems must be aligned to get work done with discipline, routines, and best practices.
3. Technical excellence: an organization must promote shared technical and operational understanding of how a process, system or asset should technically perform.
4. Standardization: standardization supports error prevention with a balanced combination of good manufacturing practices.
5. Problem-resolution skills: an organization needs people with effective statistical diagnostics and issue-resolution skills to address operational process challenges.

Not all errors can be prevented manually and/or by automation systems. When an error does occur, incident mapping (Fig. 30.6) [30.81] as one of the exception-handling tools can be used to analyze the error and proactively prevent future errors.

30.3 Conflict Prognostics and Prevention

Conflicts can be categorized into three classes [30.82]: goal conflicts, plan conflicts, and belief conflicts. Goals of an agent are modeled with an intended goal structure (IGS; e.g., Fig. 30.7), which is extended from a goal structure tree [30.83]. Plans of an agent are modeled with the extended project estimation and review technique (E-PERT) diagram (e.g., Fig. 30.8). An agent has (1) a set of goals which are represented by circles (Fig. 30.7), or circles containing a number (Fig. 30.8), (2) activities such as Act 1 and Act 2 to

achieve the goals, (3) the time needed to complete an activity, e.g., T1, and (4) resources, e.g., R1 and R2 (Fig. 30.8). Goal conflicts are detected by comparing goals by agents. Each agent has a PERT diagram and plan conflicts are detected if agents fail to merge PERT diagrams or the merged PERT diagrams violate certain rules [30.82].

The three classes of conflicts can also be modeled by Petri nets with the help of four basic modules [30.84]: sequence, parallel, decision, and decision-free, to detect conflicts in a multiagent system. Each agent's goal and plan are modeled by separate Petri nets [30.85], and many Petri nets are integrated using a bottom-up approach [30.66, 84] with three types of operations [30.85]: AND, OR, and precedence. The synthesized Petri net is analyzed to detect conflicts. Only normal transitions and places are modeled in Petri nets for conflict detection. The Petri-net-based approach for conflict detection developed so far has been rather limited. It has emphasized more the modeling of a system and its agents than the analysis process through which conflicts are detected.

The three common characteristics of available conflict detection approaches are: (1) they use the agent concept because a conflict involves at least two units in a system; (2) an agent is modeled for multiple times because each agent has at least two distinct attributes: goal and plan; and (3) they not only detect, but mainly prevent conflicts because goals and plans are determined before agents start any activities to achieve them. The main difference between the IGS and PERT approach, and the Petri net approach is that agents communicate with each other to detect conflicts in the former approach whereas a centralized control unit analyzes the integrated Petri net to detect conflicts in the latter approach [30.85]. The Petri net approach does not detect conflicts using agents, although systems are modeled with agent technology. Conflict detection has been mostly applied in collaborative design [30.86–88]. The ability to detect conflicts in distributed design activities is vital to their success because multiple designers tend to pursue individual (local) goals prior to considering common (global) goals.

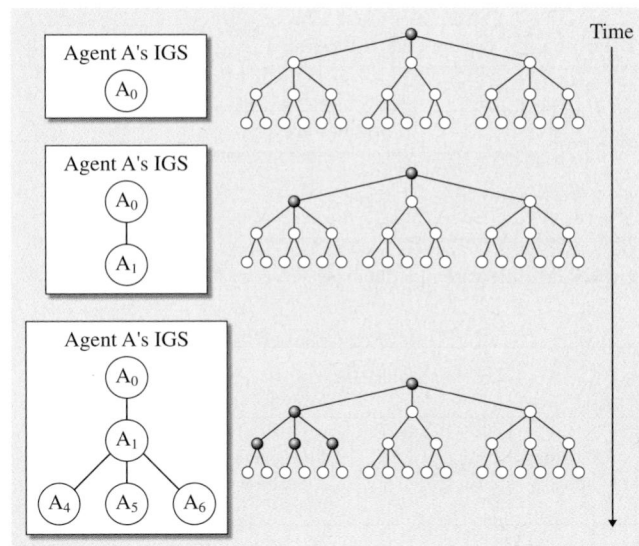

Fig. 30.7 Development of agent A's intended goal structure (IGS) over time

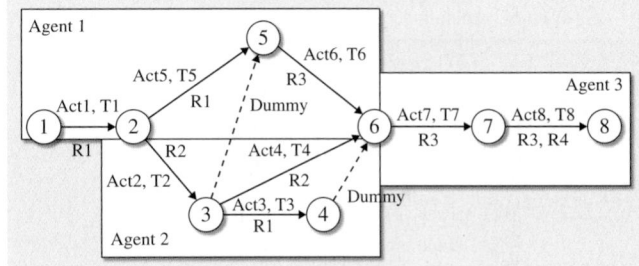

Fig. 30.8 Merged project estimation and review technique (PERT) diagram

30.4 Integrated Error and Conflict Prognostics and Prevention

30.4.1 Active Middleware

Middleware was originally defined as software that connects two separate applications or separate products and serves as the glue between two applications; for example, in Fig. 30.9, middleware can link several different database systems to several different web servers. The middleware allows users to request data from any database system that is connected to the middleware using the form displayed on the web browser of one of the web servers.

Active middleware is one of the four circles of the "e-" in e-Work as defined by the PRISM

Center (Production, Robotics, and Integration Software for Manufacturing & Management) at Purdue University [30.77]. Six major components in active middleware have been identified [30.89, 90]: modeling tool, workflows, task/activity database, decision support system (DSS), multiagent system (MAS), and collaborative work protocols. Active middleware has been developed to optimize the performance of interactions in heterogeneous, autonomous, and distributed (HAD) environments, and is able to provide an e-Work platform and enables a universal model for error and conflict prognostics and prevention in a distributed environment. Figure 30.10 shows the structure

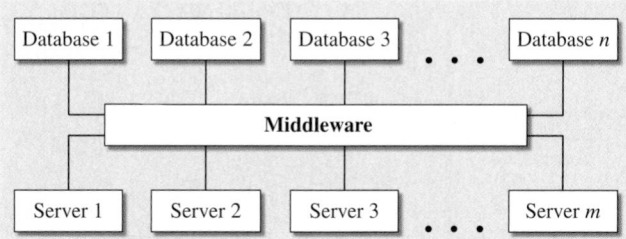

Fig. 30.9 Middleware in a database server system

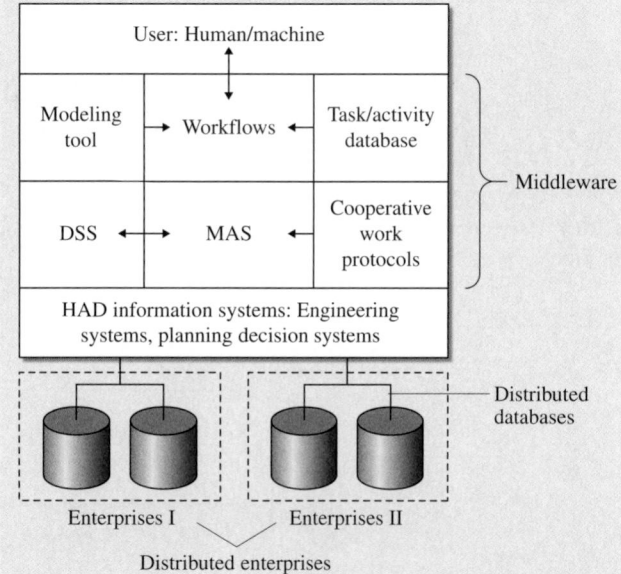

Fig. 30.10 Active middleware architecture (after [30.89])

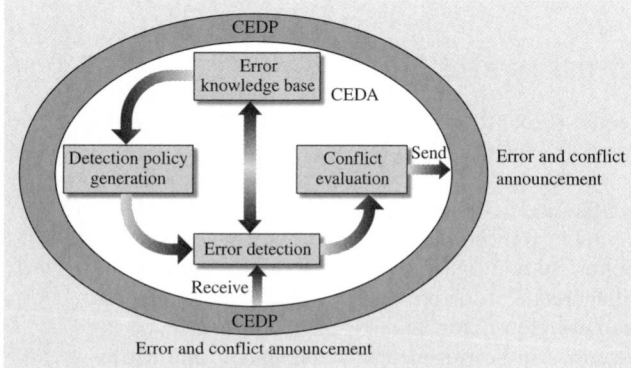

Fig. 30.11 Conflict and error detection model (CEDM)

of the active middleware; each component is described below:

1. *Modeling tool*: The goal of a modeling tool is to create a representation model for a multiagent system. The model can be transformed to next-level models, which will be the base of the system implementation.
2. *Workflows*: Workflows describe the sequence and relations of tasks in a system. Workflows store the answer to two questions: (1) Which agent will benefit from the task when it is completed by one or more given agents? (2) Which task must be finished before other tasks can begin? The workflows are specific to the given system, and can be managed by a workflow management system (WFMS).
3. *Task/activity database*: This database is used to record and help allocate tasks. There are many tasks in a large system such as those applied in automotive industries. Certain tasks are performed by several agents, and others are performed by one agent. The database records all task information and the progress of tasks (activity) and helps allocate and reallocate tasks if required.
4. *Decision support system* (DSS): DSS for the active middleware is like the operating system for a computer. In addition, DSS already has programs running for monitoring, analysis, and optimization. It can allocate/delete/create tasks, bring in or take off agents, and change workflows.
5. *Multiagent system* (MAS): MAS includes all agents in a system. It stores information about each agent, for example, capacity and number of agents, functions of an agent, working time, and effective date and expiry date of the agent.
6. *Cooperative work protocols*: Cooperative work protocols define communication and interaction protocols between components of active middleware. It is noted that communication between agents also includes communication between components because active middleware includes all agents in a system.

30.4.2 Conflict and Error Detection Model

A conflict and error detection model (CEDM; Fig. 30.11) that is supported by the conflict and error detection protocol (CEDP, part of collaborative work protocols) and conflict and error detection agents (CEDAs, part of MAS) has been developed [30.91] to detect errors and conflicts in different network topologies. The CEDM integrates CEDP, CEDAs, and four error and conflict detection components (Fig. 30.11). A CEDA is deployed at each unit of a system to (1) detect errors and conflicts by three components (detection policy generation, error detection, and conflict evaluation), which interact with and are supported by error knowledge base, and (2) communicate with other CEDAs to send and receive error and conflict announcements with the support of CEDP. The CEDM has been applied to four different network topologies and the results show that the performance of CEDM is sometimes counterintuitive, i. e., it performs better on networks that seem more complex. To be able to detect both errors and conflicts is desired when they exist in the same system. Because errors are different from conflicts, the activities to detect them are often different and need to be integrated.

30.4.3 Performance Measures

Performance measures are necessary for the evaluation and comparison of various error and conflict prognostics and prevention methods. Several measures have already been defined and developed in previous research:

1. Detection latency: The time between the instant that an error occurs and the instant that the error is detected [30.10, 91].
2. Error coverage: The percentage of detected errors with respect to the total number of errors [30.10].
3. Cost: The overhead caused by including error detection capability with respect to the system without the capability [30.10].
4. Conflict severity: The severity of a conflict. It is the sum of the severity caused by the conflict at each involving unit [30.91].
5. Detectability: The ability of a detection method. It is a function of detection accuracy, cost, and time [30.92].
6. Preventability: The ratio of the number of errors prevented divided by the total number of errors [30.80].
7. Reliability: The ratio of the number of errors prevented divided by the number of errors identified or predicted, or the ratio of the number of errors detected divided by the total number of errors [30.40, 45, 80].

Other performance measures, e.g., total damage and cost–benefit ratio, can be developed to compare different methods. Appropriate performance measures help determine how a specific method performs in different situations and are often required when there are multiple methods available.

30.5 Error Recovery and Conflict Resolution

When an error or a conflict occurs and is detected, identified, isolated or diagnosed, there are three possible consequences: (1) other errors or conflicts that are caused by the error or conflict have occurred; (2) other errors or conflicts that are caused by the error or conflict will (probably) occur; (3) other errors or conflicts, or the same error or conflict, will (probably) occur if the current error or conflict is not recovered or resolved, respectively. One of the objectives of error recovery and conflict resolution is to avoid the third consequence when an error or a conflict occurs. They are therefore part of error and conflict prognostics and prevention.

There has been extensive research on automated error recovery and conflict resolution, which are often domain specific. Many methods have been developed and applied in various real-world applications in which the main objective of error recovery and conflict resolution is to keep the production or service flowing;

for instance, Fig. 30.12 shows a recovery tree for rheostat pick-up and insertion, which is programmed for automatic error recovery. Traditionally, error recovery and conflict resolution are not considered as an approach to prevent errors and conflicts. In the next two Sections, we describe two examples, error recovery in robotics [30.93] and conflict resolution in collaborative facility design [30.88, 94], to illustrate how to perform these two functions automatically.

30.5.1 Error Recovery

Error recovery cannot be avoided when using robots because errors are an inherent characteristic of robotic applications [30.95] that are often not fault tolerant. Most error recovery applications implement preprogrammed nonintelligent corrective actions [30.95–98]. Due to the large number of possible errors and the

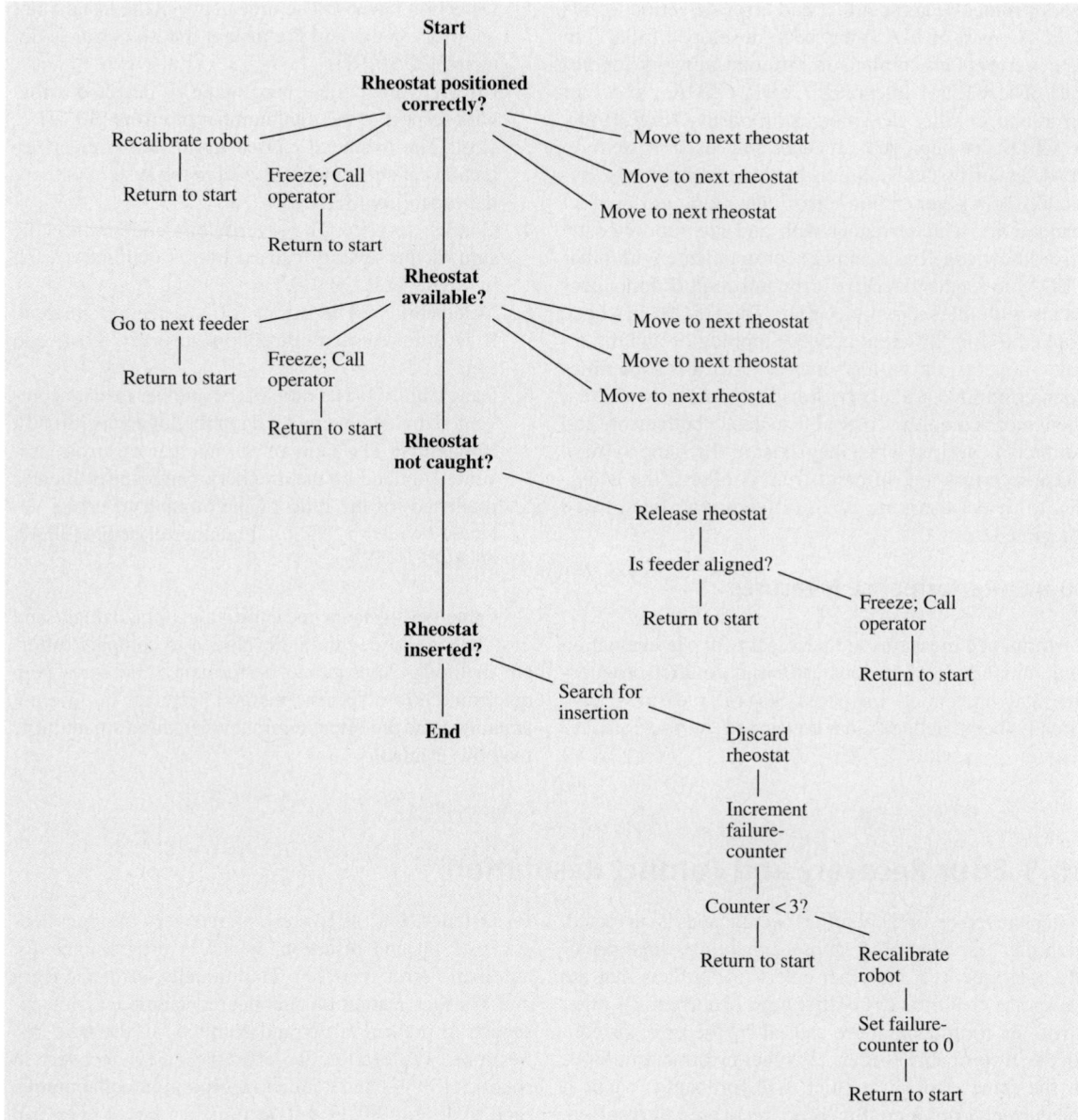

Fig. 30.12 Recovery tree for rheostat pick-up and insertion recovery. A branch may only be entered once; on success branch downward; on failure branch to right if possible, otherwise branch left; when the end of a branch is reached, unless otherwise specified return to last sensing position; "?" signifies sensing position where sensors or variables are evaluated (after [30.1])

inherent complexity of recovery actions, to automate error recovery fully without human interventions is difficult. The emerging trend in error recovery is to equip

Table 30.4 Multiapproach conflict resolution in collaborative design (Mcr) structure [30.88, 94] (after [30.94], courtesy Elsevier, 2008)▶

Stage	Strategy	Steps to achieve conflict resolution	Methodologies and tools
Mcr(1)	Direct negotiation	1. Agent prepares resolution proposal and sends to counterparts 2. Counterpart agents evaluate proposal. If they accept it, go to step 5; otherwise go to step 3 3. Counterpart agents prepare a counteroffer and send it back to originating agent 4. Agent evaluates the counteroffer. If accepted go to step 5; otherwise go to Mcr(2) 5. End of the conflict resolution process	Heuristics; Knowledge-based interactions; Multiagent systems
Mcr(2)	Third-party mediation	1. Third-party agent prepares resolution proposal and sends to counterparts 2. Counterpart agents evaluate the proposal. If accepted, go to step 5; otherwise go to step 3 3. Counterpart agents prepare a counteroffer and send it back to the third-party agent 4. Third-party agent evaluates counteroffer. If accepted go to step 5; otherwise go to Mcr(3) 5. End of the conflict resolution process	Heuristics; Knowledge-based interactions; Multiagent systems; PESUADER [30.99]
Mcr(3)	Incorporation of additional parties	1. Specialized agent prepares resolution proposal and sends to counterparts 2. Counterpart agents evaluate the proposal. If accepted, go to step 5; otherwise go to step 3 3. Counterpart agents prepare a counteroffer and send it back to the specialized agent 4. Specialized agent evaluates counteroffer. If accepted go to step 5; otherwise go to Mcr(4) 5. End of the conflict resolution process	Heuristics; Knowledge-based interactions; Expert systems
Mcr(4)	Persuasion	1. Third-party agent prepares persuasive arguments and sends to counterparts 2. Counterpart agents evaluate the arguments 3. If the arguments are effective, go to step 4; otherwise go to Mcr(5) 4. End of the conflict resolution process	PERSUADER [30.99]; Case-based reasoning
Mcr(5)	Arbitration	1. If conflict management and analysis results in common proposals (X), conflict resolution is achieved through management and analysis 2. If conflict management and analysis results in mutually exclusive proposals (Y), conflict resolution is achieved though conflict confrontation 3. If conflict management and analysis results in no conflict resolution proposals (Z), conflict resolution must be used	Graph model for conflict resolution (GMCR) [30.100] for conflict management and analysis; Adaptive neural-fuzzy inference system (ANFIS) [30.101] for conflict confrontation; Dependency analysis [30.102] and product flow analysis for conflict resolution

Table 30.5 Summary of error and conflict prognostics and prevention theories, applications, and open challenges

Applications	Assembly and inspection	Process monitoring			Hardware testing	Software testing
Methods/ technologies	Control theory; Knowledge base; Computer/machine vision; Robotics; Feature extraction; Pattern recognition	Analytical	Data-driven	Knowledge-based	Information theory; Heuristic search	Model checking; Bogor; Cadena; Concurrent error detection (CED)
Functions						
Detection	×	×	×	×	×	×
Diagnostics	×	×	×	×		×
Identification	×	×	×	×	×	×
Isolation	×	×	×	×	×	×
Error recovery	×					
Conflict resolution						
Prognostics		×	×	×		
Exception handling	×					
Errors/conflicts	E	E	E	E	E	E
Centralized/ decentralized	C	C	C	C	C	C
Strengths	Integration of error detection and recovery	Accurate and reliable	Can process large amount of data	Does not require detailed system information	Accurate and reliable	Thorough verification with formal methods
Weaknesses	Domain specific; Lack of general methods	Require mathematical models that are often not available	Rely on the quantity, quality, and timeliness of data	Results are subjective and may not be reliable	Difficult to derive optimal algorithms to minimize cost; Time consuming for large systems	State explosion; Duplications needed in CED; Cannot deal with incorrect or incomplete specifications
	[30.3, 20–34]	[30.17, 19, 35]			[30.36–45]	[30.8, 9, 14, 47–57]

systems with human intelligence so that they can correct errors through reasoning and high-level decision making. An example of an intelligent error recovery system is the neural-fuzzy system for error recovery (NEFUSER) [30.93]).

The NEFUSER is both an intelligent system and a design tool of fuzzy logic and neural-fuzzy models for error detection and recovery. The NEFUSER has

been applied to a single robot working in an assembly cell. The NEFUSER enables interactions among the robot, the operator, and computer-supported applications. It interprets data and information collected by the robot and provided by the operator, analyzes data and information with fuzzy logic and/or neural-fuzzy models, and makes appropriate error recovery decisions. The NEFUSER has learning ability to im-

Table 30.5 (cont.)

Applications	Discrete event system	Collaborative design		Production and service		
Methods/ technologies	Petri net; Finite-state machine (FSM)	Intended goal structure (IGS); Project evaluation and review technique (PERT); Petri net; Conflict detection and management system (CDMS)	Facility description language (FDL); Mcr; CDMS	Detection and prevention algorithms; Reliability theory; Process modeling; Workflow	Conflict and error detection model (CEDM); Active middleware	Fuzzy logic; Artificial intelligence
Functions						
Detection	×	×		×	×	
Diagnostics						
Identification	×	×		×	×	
Isolation	×	×		×	×	
Error recovery						×
Conflict resolution			×			
Prognostics	×	×		×		
Exception handling			×			×
Errors/conflicts	E	C	C	E	E/C	E
Centralized/ decentralized	C/D	C/D	C/D	C/D	D	C/D
Strengths	Formal method applicable to various systems	Modeling of systems with agent-based technology	Integration of traditional human conflict resolutions and computer-based learning	Reliable; Easy to apply	Short detection time	Correct errors through reasoning and high-level decision making
Weaknesses	State explosion for large systems; System modeling is complex and time-consuming	An agent may be modeled for multiple times due to many conflicts it is involved	The adaptability of the methods to other design activities has not been validated	Limited to sequential production and service lines; Domain specific	Needs further development and validation	Needs further development for various applications
	[30.58–67]	[30.66, 82–88]	[30.77, 86, 88, 94] [30.104–114]	[30.68–80]	[30.77, 89–91]	[30.93, 95–98, 103]

prove corrective actions and adapt to different errors. The NEFUSER therefore increases the level of automation by decreasing the number of times that the robot has to stop and the operator has to intervene due to errors.

Figure 30.13 shows the interactions between the robot, the operator, and computer-supported applications. The NEFUSER is the error recovery brain and is programmed and run on MATLAB, which provides a friendly windows-oriented fuzzy inference system (FIS) that incorporates the graphical user interface tools of the fuzzy logic toolbox [30.103]. The example in Fig. 30.13 includes a robot and an operator in an assembly cell. In general, the NEFUSER design for error

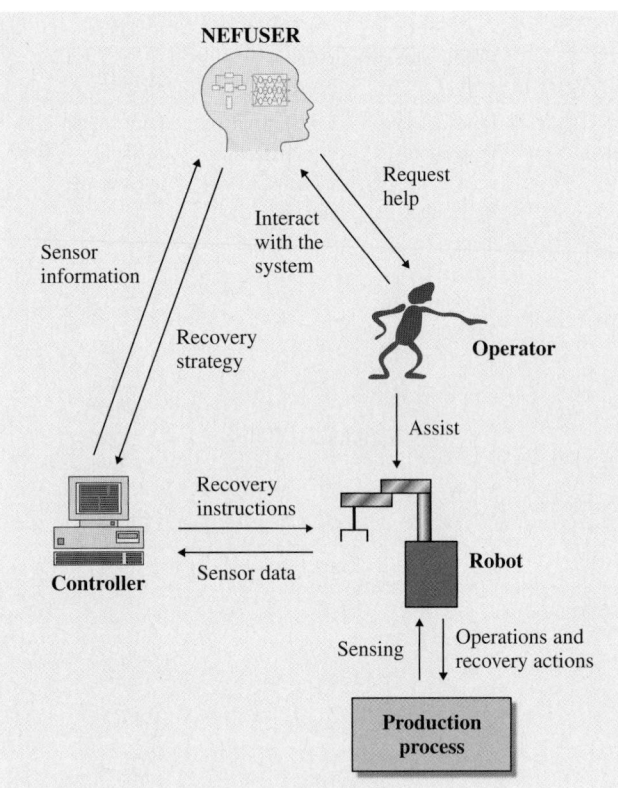

Fig. 30.13 Interactions with NEFUSER (after [30.93])

recovery includes three main tasks: (1) design the FIS, (2) manage and evaluate information, and (3) train the FIS with real data and information.

30.5.2 Conflict Resolution

There is a growing demand for knowledge-intensive collaboration in distributed design [30.94, 113, 114].

Conflict detection has been studied extensively in collaborative design, as has conflict resolution, which is often the next step after a conflict is detected. There has been extensive research on conflict resolution (e.g., [30.105–110]). Recently, a multiapproach method to conflict resolution in collaborative design has been introduced with the development of the facility description language–conflict resolution (FDL-CR) [30.88]. The critical role of computer-supported conflict resolution in distributed organizations has been discussed in great detail [30.77, 104, 111, 112]. In addition, *Ceroni* and *Velasquez* [30.86] have developed the conflict detection and management system (CDMS) and their work shows that both product complexity and number of participating designers have a statistically significant effect on the ratio of conflicts resolved to those detected, but that only complexity had a statistically significant effect on design duration.

Based on the previous work, most recently, a new method, Mcr (Table 30.4), has been developed to automatically resolve conflict situations common in collaborative facility design using computer-support tools [30.88, 94]. The method uses both traditional human conflict-resolution approaches that have been used successfully by others and principles of conflict prevention to improve design performance and apply computer-based learning to improve usefulness. A graph model for conflict resolution is used to facilitate conflict modeling and analysis. The performance of the new method has been validated by implementing its conflict resolution capabilities in the FDL, a computer tool for collaborative facility design, and by applying FDL-CR, to resolve typical conflict situations. Table 30.4 describes the Mcr structure.

Table 30.5 summarizes error and conflict prognostics and prevention methods and technologies in various production and service applications.

30.6 Emerging Trends

30.6.1 Decentralized and Agent-Based Error and Conflict Prognostics and Prevention

Most error and conflict prognostics and prevention methods developed so far are centralized approaches (Table 30.5) in which a central control unit controls data and information and executes some or all eight functions to prevent errors and conflicts. The centralized approach often requires substantial time to execute

various functions and the central control unit often possesses incomplete or incorrect data and information [30.80]. These disadvantages become apparent when a system has many units that need to be examined for errors and conflicts.

To overcome the disadvantages of the centralized approach, the decentralized approach that takes advantage of the parallel activities of multiple agents has been developed [30.16, 67, 79, 80, 91]. In the decentralized approach, distributed agents detect, identify or isolate errors and conflicts at individual units of a system, and communicate with each other to diagnose and prevent errors and conflicts. The main challenge of the decentralized approach is to develop robust protocols that can ensure effective communications between agents. Further research is needed to develop and improve decentralized approaches for implementation in various applications.

30.6.2 Intelligent Error and Conflict Prognostics and Prevention

Compared with humans, automation systems perform better when they are used to prevent errors and conflicts through the violation of specifications or violation in comparisons [30.13]. Humans, however, have the ability to prevent errors and conflicts through the violation of expectations, i.e., with tacit knowledge and high-level decision making. To increase the effectiveness degree of automation of error and conflict prognostics and prevention, it is necessary to equip automation systems with human intelligence through appropriate modeling techniques such as fuzzy logic, pattern recognition, and artificial neural networks. There has been some preliminary work to incorporate high-level human intelligence in error detection and recovery (e.g., [30.3, 93]) and conflict resolution [30.88, 94]. Additional work is needed to develop self-learning, self-improving artificial intelligence systems for error and conflict prognostics and prevention.

30.6.3 Graph and Network Theories

The performance of an error and conflict prognostics and prevention method is significantly influenced by the number of units in a system and their relationship. A system can be viewed as a graph or a network with many nodes, each of which represents a unit in the system. The relationship between units is represented by the link between nodes. The study of network topologies has a long history stretching back at least to the 1730s. The classic model of a network, the random network, was first discussed in the early 1950s [30.115] and was rediscovered and analyzed in a series of papers

published in the late 1950s and early 1960s [30.116–118]. Most recently, several network models have been discovered and extensively studied, for instance, the small-world network (e.g., [30.119]), the scale-free network (e.g., [30.120–123]), and the Bose–Einstein condensation network [30.124]. Bioinspired network models for collaborative control have recently been studied by *Nof* [30.125] (see also Chap. 75 for more details).

Because the same prognostics and prevention method may perform quite differently on networks with different topologies and attributes, or with the same network topology and attributes but with different parameters, it is imperative to study the performance of prognostics and prevention methods with respect to different networks for the best match between methods and networks. There is ample room for research, development, and implementation of error and conflict prognostics and prevention methods supported by graph and network theories.

30.6.4 Financial Models for Prognostics Economy

Most errors and conflicts must be detected, isolated, identified, diagnosed or prevented. Certain errors and conflicts, however, may be tolerable in certain systems, i.e., fault-tolerant systems. Also, the cost of automating some or all eight functions of error and conflict prognostics and prevention may far exceed the damages caused by certain errors and conflicts. In both situations, cost–benefit analyses can be used to determine if an error or a conflict needs to be dealt with. In general, financial models are used to analyze the economy of prognostics and prevention methods for specific errors and conflicts, to help decide which of the eight functions will be executed and how they will be executed, e.g., the frequency. There has been limited research on how to use financial models to help justify the automation of error and conflict prognostics and prevention [30.92, 126]. One of the challenges is how to appropriately evaluate or assess the damage of errors and conflicts, e.g., short-term damage, long-term damage, and intangible damage. Additional research is needed to address these economical decisions.

30.7 Conclusion

In this Chapter we have discussed the eight functions that automate error and conflict prognostics and prevention and their applications in various production and service areas. Prognostics and prevention methods for errors and conflicts are developed based on extensive theoretical advancements in many science and engineering domains, and have been successfully applied to various real-world problems. As systems and networks become larger and more complex, such as global enterprises and the Internet, error and conflict prognostics and prevention become more important and the focus is shifting from passive response to active prognostics and prevention.

References

30.1 S.Y. Nof, W.E. Wilhelm, H.-J. Warnecke: *Industrial Assembly* (Chapman Hall, New York 1997)

30.2 L.S. Lopes, L.M. Camarinha-Matos: A machine learning approach to error detection and recovery in assembly, Proc. IEEE/RSJ Int. Conf. Intell. Robot. Syst. 95, 'Human Robot Interaction and Cooperative Robots', Vol. 3 (1995) pp. 197–203

30.3 H. Najjari, S.J. Steiner: Integrated sensor-based control system for a flexible assembly, Mechatronics 7(3), 231–262 (1997)

30.4 A. Steininger, C. Scherrer: On finding an optimal combination of error detection mechanisms based on results of fault injection experiments, Proc. 27th Annu. Int. Symp. Fault-Toler. Comput., FTCS-27, Digest of Papers (1997) pp. 238–247

30.5 K.A. Toguyeni, E. Craye, J.C. Gentina: Framework to design a distributed diagnosis in FMS, Proc. IEEE Int. Conf. Syst. Man. Cybern. 4, 2774–2779 (1996)

30.6 J.F. Kao: Optimal recovery strategies for manufacturing systems, Eur. J. Oper. Res. 80(2), 252–263 (1995)

30.7 M. Bruccoleri, Z.J. Pasek: Operational issues in reconfigurable manufacturing systems: exception handling, Proc. 5th Biannu. World Autom. Congr. (2002)

30.8 T. Miceli, H.A. Sahraoui, R. Godin: A metric based technique for design flaws detection and correction, Proc. 14th IEEE Int. Conf. Autom. Softw. Eng. (1999) pp. 307–310

30.9 C. Bolchini, W. Fornaciari, F. Salice, D. Sciuto: Concurrent error detection at architectural level, Proc. 11th Int. Symp. Syst. Synth. (1998) pp. 72–75

30.10 C. Bolchini, L. Pomante, F. Salice, D. Sciuto: Reliability properties assessment at system level: a co-design framework, J. Electron. Test. 18(3), 351–356 (2002)

30.11 M.D. Jeng: Petri nets for modeling automated manufacturing systems with error recovery, IEEE Trans. Robot. Autom. 13(5), 752–760 (1997)

30.12 G.A. Kanawati, V.S.S. Nair, N. Krishnamurthy, J.A. Abraham: Evaluation of integrated system-level checks for on-line error detection, Proc. IEEE Int. Comput. Perform. Dependability Symp. (1996) pp. 292–301

30.13 B.D. Klein: How do actuaries use data containing errors?: models of error detection and error correction, Inf. Resour. Manag. J. 10(4), 27–36 (1997)

30.14 M. Ronsse, K. Bosschere: Non-intrusive detection of synchronization errors using execution replay, Autom. Softw. Eng. 9(1), 95–121 (2002)

30.15 O. Svenson, I. Salo: Latency and mode of error detection in a process industry, Reliab. Eng. Syst. Saf. 73(1), 83–90 (2001)

30.16 X.W. Chen, S.Y. Nof: Prognostics and diagnostics of conflicts and errors over e-Work networks, Proc. 19th Int. Conf. Production Research (2007)

30.17 J. Gertler: *Fault Detection and Diagnosis in Engineering Systems* (Marcel Dekker, New York 1998)

30.18 M. Klein, C. Dellarocas: A knowledge-based approach to handling exceptions in workflow systems, Comput. Support. Coop. Work 9, 399–412 (2000)

30.19 A. Raich, A. Cinar: Statistical process monitoring and disturbance diagnosis in multivariable continuous processes, AIChE Journal 42(4), 995–1009 (1996)

30.20 C.-Y. Chang, J.-W. Chang, M.D. Jeng: An unsupervised self-organizing neural network for automatic semiconductor wafer defect inspection, IEEE Int. Conf. Robot. Autom. ICRA (2005) pp. 3000–3005

30.21 M. Moganti, F. Ercal: Automatic PCB inspection systems, IEEE Potentials 14(3), 6–10 (1995)

30.22 H. Rau, C.-H. Wu: Automatic optical inspection for detecting defects on printed circuit board inner layers, Int. J. Adv. Manuf. Technol. 25(9–10), 940–946 (2005)

30.23 J.A. Calderon-Martinez, P. Campoy-Cervera: An application of convolutional neural networks for automatic inspection, IEEE Conf. Cybern. Intell. Syst. (2006) pp. 1–6

30.24 F. Duarte, H. Arauio, A. Dourado: Automatic system for dirt in pulp inspection using hierarchical image segmentation, Comput. Ind. Eng. 37(1–2), 343–346 (1999)

30.25 J.C. Wilson, P.A. Berardo: Automatic inspection of hazardous materials by mobile robot, Proc. IEEE Int. Conf. Syst. Man. Cybern. 4, 3280–3285 (1995)

30.26 J.Y. Choi, H. Lim, B.-J. Yi: Semi-automatic pipeline inspection robot systems, SICE-ICASE Int. Jt. Conf. (2006) pp. 2266–2269

30.27 L.V. Finogenoy, A.V. Beloborodov, V.I. Ladygin, Y.V. Chugui, N.G. Zagoruiko, S.Y. Gulvaevskii, Y.S. Shul'man, P.I. Lavrenyuk, Y.V. Pimenov: An optoelectronic system for automatic inspection of the external view of fuel pellets, Russ. J. Nondestr. Test. **43**(10), 692–699 (2007)

30.28 C.W. Ni: Automatic inspection of the printing contents of soft drink cans by image processing analysis, Proc. SPIE **3652**, 86–93 (2004)

30.29 J. Cai, G. Zhang, Z. Zhou: The application of area-reconstruction operator in automatic visual inspection of quality control, Proc. World Congr. Intell. Control Autom. (WCICA), Vol. 2 (2006) pp. 10111–10115

30.30 O. Erne, T. Walz, A. Ettemeyer: Automatic shearography inspection systems for aircraft components in production, Proc. SPIE **3824**, 326–328 (1999)

30.31 C.K. Huang, L.G. Wang, H.C. Tang, Y.S. Tarng: Automatic laser inspection of outer diameter, run-out taper of micro-drills, J. Mater. Process. Technol. **171**(2), 306–313 (2006)

30.32 L. Chen, X. Wang, M. Suzuki, N. Yoshimura: Optimizing the lighting in automatic inspection system using Monte Carlo method, Jpn. J. Appl. Phys., Part 1 **38**(10), 6123–6129 (1999)

30.33 W.C. Godoi, R.R. da Silva, V. Swinka-Filho: Pattern recognition in the automatic inspection of flaws in polymeric insulators, Insight Nondestr. Test. Cond. Monit. **47**(10), 608–614 (2005)

30.34 U.S. Khan, J. Iqbal, M.A. Khan: Automatic inspection system using machine vision, Proc. 34th Appl. Imag. Pattern Recognit. Workshop (2005) pp. 210–215

30.35 L.H. Chiang, R.D. Braatz, E. Russell: *Fault Detection and Diagnosis in Industrial Systems* (Springer, London New York 2001)

30.36 S. Deb, K.R. Pattipati, V. Raghavan, M. Shakeri, R. Shrestha: Multi-signal flow graphs: a novel approach for system testability analysis and fault diagnosis, IEEE Aerosp. Electron. Syst. Mag. **10**(5), 14–25 (1995)

30.37 K.R. Pattipati, M.G. Alexandridis: Application of heuristic search and information theory to sequential fault diagnosis, IEEE Trans. Syst. Man. Cybern. **20**(4), 872–887 (1990)

30.38 K.R. Pattipati, M. Dontamsetty: On a generalized test sequencing problem, IEEE Trans. Syst. Man. Cybern. **22**(2), 392–396 (1992)

30.39 V. Raghavan, M. Shakeri, K. Pattipati: Optimal and near-optimal test sequencing algorithms with realistic test models, IEEE Trans. Syst. Man. Cybern. A **29**(1), 11–26 (1999)

30.40 V. Raghavan, M. Shakeri, K. Pattipati: Test sequencing algorithms with unreliable tests, IEEE Trans. Syst. Man. Cybern. A **29**(4), 347–357 (1999)

30.41 M. Shakeri, K.R. Pattipati, V. Raghavan, A. Patterson-Hine, T. Kell: *Sequential Test Strategies for Multiple Fault Isolation* (IEEE, Atlanta 1995)

30.42 M. Shakeri, V. Raghavan, K.R. Pattipati, A. Patterson-Hine: Sequential testing algorithms for multiple fault diagnosis, IEEE Trans. Syst. Man. Cybern. A **30**(1), 1–14 (2000)

30.43 F. Tu, K. Pattipati, S. Deb, V.N. Malepati: *Multiple Fault Diagnosis in Graph-Based Systems* (International Society for Optical Engineering, Orlando 2002)

30.44 F. Tu, K.R. Pattipati: Rollout strategies for sequential fault diagnosis, IEEE Trans. Syst. Man. Cybern. A **33**(1), 86–99 (2003)

30.45 F. Tu, K.R. Pattipati, S. Deb, V.N. Malepati: Computationally efficient algorithms for multiple fault diagnosis in large graph-based systems, IEEE Trans. Syst. Man. Cybern. A **33**(1), 73–85 (2003)

30.46 C. Feng, L.N. Bhuyan, F. Lombardi: Adaptive system-level diagnosis for hypercube multiprocessors, IEEE Trans. Comput. **45**(10), 1157–1170 (1996)

30.47 E.M. Clarke, O. Grumberg, D.A. Peled: *Model Checking* (MIT Press, Cambridge 2000)

30.48 C. Karamanolis, D. Giannakopolou, J. Magee, S. Wheather: Model checking of workflow schemas, 4th Int. Enterp. Distrib. Object Comp. Conf. (2000) pp. 170–181

30.49 W. Chan, R.J. Anderson, P. Beame, D. Notkin, D.H. Jones, W.E. Warner: Optimizing symbolic model checking for state charts, IEEE Trans. Softw. Eng. **27**(2), 170–190 (2001)

30.50 D. Garlan, S. Khersonsky, J.S. Kim: Model checking publish-subscribe systems, Proc. 10th Int. SPIN Workshop Model Checking Softw. (2003)

30.51 J. Hatcliff, W. Deng, M. Dwyer, G. Jung, V.P. Ranganath: Cadena: An integrated development, analysis, and verification environment for component-based systems, Proc. 2003 Int. Conf. Softw. Eng. (ICSE 2003) (Portland 2003)

30.52 T. Ball, S. Rajamani: Bebop: a symbolic modelchecker for Boolean programs, Proc. 7th Int. SPIN Workshop, Lect. Notes Comput. Sci. **1885**, 113–130 (2000)

30.53 G. Brat, K. Havelund, S. Park, W. Visser: Java PathFinder – a second generation of a Java modelchecker, Proc. Workshop Adv. Verif. (2000)

30.54 J.C. Corbett, M.B. Dwyer, J. Hatcliff, S. Laubach, C.S. Pasareanu, Robby, H. Zheng: Bandera: Extracting finite-state models from Java source code, Proc. 22nd Int. Conf. Softw. Eng. (2000)

30.55 P. Godefroid: Model-checking for programming languages using VeriSoft, Proc. 24th ACM Symp. Princ. Program. Lang. (POPL'97) (1997) pp. 174–186

30.56 Robby, M.B. Dwyer, J. Hatcliff: Bogor: An extensible and highly-modular model checking framework, Proc. 9th European Softw. Eng. Conf. held jointly with the 11th ACM SIGSOFT Symp. Found. Softw. Eng. (2003)

30.57 S. Mitra, E.J. McCluskey: Diversity techniques for concurrent error detection, Proc. IEEE 2nd Int. Symp. Qual. Electron. Des. IEEE Comput. Soc., 249–250 (2001)

30.58 S.-L. Chung, C.-C. Wu, M. Jeng: *Failure Diagnosis: A Case Study on Modeling and Analysis by Petri Nets* (IEEE, Washington 2003)

30.59　P.S. Georgilakis, J.A. Katsigiannis, K.P. Valavanis, A.T. Souflaris: A systematic stochastic Petri net based methodology for transformer fault diagnosis and repair actions, J. Intell. Robot. Syst. Theory Appl. **45**(2), 181–201 (2006)

30.60　T. Ushio, I. Onishi, K. Okuda: *Fault Detection Based on Petri Net Models with Faulty Behaviors* (IEEE, San Diego 1998)

30.61　M. Rezai, M.R. Ito, P.D. Lawrence: *Modeling and Simulation of Hybrid Control Systems by Global Petri Nets* (IEEE, Seattle 1995)

30.62　M. Rezai, P.D. Lawrence, M.R. Ito: *Analysis of Faults in Hybrid Systems by Global Petri Nets* (IEEE, Vancouver 1995)

30.63　M. Rezai, P.D. Lawrence, M.B. Ito: *Hybrid Modeling and Simulation of Manufacturing Systems* (IEEE, Los Angeles 1997)

30.64　M. Sampath, R. Sengupta, S. Lafortune, K. Sinnamohideen, D. Teneketzis: Diagnosability of discrete-event systems, IEEE Trans. Autom. Control **40**(9), 1555–1575 (1995)

30.65　S.H. Zad, R.H. Kwong, W.M. Wonham: Fault diagnosis in discrete-event systems: framework and model reduction, IEEE Trans. Autom. Control **48**(7), 1199–1212 (2003)

30.66　M. Zhou, F. DiCesare: *Petri Net Synthesis for Discrete Event Control of Manufacturing Systems* (Kluwer, Boston 1993)

30.67　Q. Wenbin, R. Kumar: Decentralized failure diagnosis of discrete event systems, IEEE Trans. Syst. Man. Cybern. A **36**(2), 384–395 (2006)

30.68　A. Brall: Human reliability issues in medical care – a customer viewpoint, Proc. Annu. Reliab. Maint. Symp. (2006) pp. 46–50

30.69　H. Furukawa: Challenge for preventing medication errors–learn from errors–: what is the most effective label display to prevent medication error for injectable drug? Proc. 12th Int. Conf. Hum.-Comput. Interact.: HCI Intell. Multimodal Interact. Environ., Lect. Notes Comput. Sci. **4553**, 437–442 (2007)

30.70　G. Huang, G. Medlam, J. Lee, S. Billingsley, J.-P. Bissonnette, J. Ringash, G. Kane, D.C. Hodgson: Error in the delivery of radiation therapy: results of a quality assurance review, Int. J. Radiat. Oncol. Biol. Phys. **61**(5), 1590–1595 (2005)

30.71　A.-S. Nyssen, A. Blavier: A study in anesthesia, Ergonomics **49**(5/6), 517–525 (2006)

30.72　K.T. Unruh, W. Pratt: Patients as actors: the patient's role in detecting, preventing, and recovering from medical errors, Int. J. Med. Inform. **76**(1), 236–244 (2007)

30.73　C.C. Chao, W.Y. Jen, M.C. Hung, Y.C. Li, Y.P. Chi: An innovative mobile approach for patient safety services: the case of a Taiwan health care provider, Technovation **27**(6–7), 342–361 (2007)

30.74　S. Malhotra, D. Jordan, E. Shortliffe, V.L. Patel: Workflow modeling in critical care: piecing to-gether your own puzzle, J. Biomed. Inform. **40**(2), 81–92 (2007)

30.75　T.J. Morris, J. Pajak, F. Havlik, J. Kenyon, D. Calcagni: Battlefield medical information system-tactical (BMIST): The application of mobile computing technologies to support health surveillance in the Department of Defense, Telemed. J. e-Health **12**(4), 409–416 (2006)

30.76　M. Rajendran, B.S. Dhillon: Human error in health care systems: bibliography, Int. J. Reliab. Qual. Saf. Eng. **10**(1), 99–117 (2003)

30.77　S.Y. Nof: Design of effective e-Work: review of models, tools, and emerging challenges, Product. Plan. Control **14**(8), 681–703 (2003)

30.78　X. Chen: Error detection and prediction agents and their algorithms. M.S. Thesis (School of Industrial Engineering, Purdue University, West Lafayette 2005)

30.79　X.W. Chen, S.Y. Nof: Error detection and prediction algorithms: application in robotics, J. Intell. Robot. Syst. **48**(2), 225–252 (2007)

30.80　X.W. Chen, S.Y. Nof: Agent-based error prevention algorithms, submitted to the IEEE Trans. Autom. Sci. Eng. (2008)

30.81　K. Duffy: Safety for profit: Building an error-prevention culture, Ind. Eng. Mag. **9**, 41–45 (2008)

30.82　K.S. Barber, T.H. Liu, S. Ramaswamy: Conflict detection during plan integration for multi-agent systems, IEEE Trans. Syst. Man. Cybern. B **31**(4), 616–628 (2001)

30.83　G.M.P. O'Hare, N. Jennings: *Foundations of Distributed Artificial Intelligence* (Wiley, New York 1996)

30.84　M. Zhou, F. DiCesare, A.A. Desrochers: A hybrid methodology for synthesis of Petri net models for manufacturing systems, IEEE Trans. Robot. Autom. **8**(3), 350–361 (1992)

30.85　J.-Y. Shiau: A formalism for conflict detection and resolution in a multi-agent system. Ph.D. Thesis (Arizona State University, Arizona 2002)

30.86　J.A. Ceroni, A.A. Velásquez: Conflict detection and resolution in distributed design, Prod. Plan. Control **14**(8), 734–742 (2003)

30.87　T. Jiang, G.E. Nevill Jr: Conflict cause identification in web-based concurrent engineering design system, Concurr. Eng. Res. Appl. **10**(1), 15–26 (2002)

30.88　M.A. Lara, S.Y. Nof: Computer-supported conflict resolution for collaborative facility designers, Int. J. Prod. Res. **41**(2), 207–233 (2003)

30.89　P. Anussornnitisarn, S.Y. Nof: The design of active middleware for e-Work interactions, PRISM Res. Memorandum (School of Industrial Engineering, Purdue University, West Lafayette 2001)

30.90　P. Anussornnitisarn, S.Y. Nof: e-Work: the challenge of the next generation ERP systems, Prod. Plan. Control **14**(8), 753–765 (2003)

30.91 X.W. Chen, S.Y. Nof: An agent-based conflict and error detection model, submitted to Int. J. Prod. Res. (2008)

30.92 C.L. Yang, S.Y. Nof: Analysis, detection policy, and performance measures of detection task planning errors and conflicts, PRISM Res. Memorandum, 2004-P2 (School of Industrial Engineering, Purdue University, West Lafayette 2004)

30.93 J. Avila-Soria: Interactive Error Recovery for Robotic Assembly Using a Neural-Fuzzy Approach. Master Thesis (School of Industrial Engineering, Purdue University, West Lafayette 1999)

30.94 J.D. Velásquez, M.A. Lara, S.Y. Nof: Systematic resolution of conflict situation in collaborative facility design, Int. J. Prod. Econ. **116**(1), 139–153 (2008), (2008)

30.95 S.Y. Nof, O.Z. Maimon, R.G. Wilhelm: Experiments for Planning Error-Recovery Programs in Robotic Work, Proc. Int. Comput. Eng. Conf. Exhib. **2**, 253–264 (1987)

30.96 M. Imai, K. Hiraki, Y. Anzai: Human-robot interface with attention, Syst. Comput. Jpn. **26**(12), 83–95 (1995)

30.97 T.C. Lueth, U.M. Nassal, U. Rembold: Reliability and integrated capabilities of locomotion and manipulation for autonomous robot assembly, Robot. Auton. Syst. **14**, 185–198 (1995)

30.98 H.-J. Wu, S.B. Joshi: Error recovery in MPSG-based controllers for shop floor control, Proc. IEEE Int. Conf. Robot. Autom. ICRA **2**, 1374–1379 (1994)

30.99 K. Sycara: Negotiation planning: An AI approach, Eur. J. Oper. Res. **46**(2), 216–234 (1990)

30.100 L. Fang, K.W. Hipel, D.M. Kilgour: *Interactive Decision Making* (Wiley, New York 1993)

30.101 J.-S.R. Jang: ANFIS: Adaptive-network-based fuzzy inference systems, IEEE Trans. Syst. Man. Cybern. **23**, 665–685 (1993)

30.102 A. Kusiak, J. Wang: Dependency analysis in constraint negotiation, IEEE Trans. Syst. Man. Cybern. **25**(9), 1301–1313 (1995)

30.103 J.-S.R. Jang, N. Gulley: *Fuzzy Systems Toolbox for Use with MATLAB* (The Math Works Inc., 1997)

30.104 C.Y. Huang, J.A. Ceroni, S.Y. Nof: Agility of networked enterprises: parallelism, error recovery and conflict resolution, Comput. Ind. **42**, 73–78 (2000)

30.105 M. Klein, S.C.-Y. Lu: Conflict resolution in cooperative design, Artif. Intell. Eng. **4**(4), 168–180 (1989)

30.106 M. Klein: Supporting conflict resolution in cooperative design systems, IEEE Trans. Syst. Man. Cybern. **21**(6), 1379–1390 (1991)

30.107 M. Klein: Capturing design rationale in concurrent engineering teams, IEEE Computer **26**(1), 39–47 (1993)

30.108 M. Klein: Conflict management as part of an integrated exception handling approach, Artif. Intell. Eng. Des. Anal. Manuf. **9**, 259–267 (1995)

30.109 X. Li, X.H. Zhou, X.Y. Ruan: Study on conflict management for collaborative design system, J. Shanghai Jiaotong University (English ed.) **5**(2), 88–93 (2000)

30.110 X. Li, X.H. Zhou, X.Y. Ruan: Conflict management in closely coupled collaborative design system, Int. J. Comput. Integr. Manuf. **15**(4), 345–352 (2000)

30.111 S.Y. Nof: Tools and models of e-Work, Proc. 5th Int. Conf. Simul. AI (Mexico City 2000) pp. 249–258

30.112 S.Y. Nof: Collaborative e-Work and e-Manufacturing: challenges for production and logistics managers, J. Intell. Manuf. **17**(6), 689–701 (2006)

30.113 X.F. Zha, H. Du: Knowledge-intensive collaborative design modeling and support part I: review, distributed models and framework, Comput. Ind. **57**, 39–55 (2006)

30.114 X.F. Zha, H. Du: Knowledge-intensive collaborative design modeling and support part II: system implementation and application, Comput. Ind. **57**, 56–71 (2006)

30.115 R. Solomonoff, A. Rapoport: Connectivity of random nets, Bull. Mater. Biophys. **13**, 107–117 (1951)

30.116 P. Erdos, A. Renyi: On random graphs, Publ. Math. Debr. **6**, 290–291 (1959)

30.117 P. Erdos, A. Renyi: On the evolution of random graphs, Magy. Tud. Akad. Mat. Kutato Int. Kozl. **5**, 17–61 (1960)

30.118 P. Erdos, A. Renyi: On the strenth of connectedness of a random graph, Acta Mater. Acad. Sci. Hung. **12**, 261–267 (1961)

30.119 D.J. Watts, S.H. Strogatz: Collective dynamics of 'small-world' networks, Nature **393**(6684), 440–442 (1998)

30.120 R. Albert, H. Jeong, A.L. Barabasi: Internet: Diameter of the World-Wide Web, Nature **401**(6749), 130–131 (1999)

30.121 A.L. Barabasi, R. Albert: Emergence of scaling in random networks, Science **286**(5439), 509–512 (1999)

30.122 A. Broder, R. Kumar, F. Maghoul, P. Raghavan, S. Rajagopalan, R. Stata, A. Tomkins, J. Wiener: Graph structure in the Web, Comput. Netw. **33**(1), 309–320 (2000)

30.123 D.J. de Solla Price: Networks of scientific papers, Science **149**, 510–515 (1965)

30.124 G. Bianconi, A.L. Barabasi: Bose-Einstein condensation in complex networks, Phys. Rev. Lett. **86**(24), 5632–5635 (2001)

30.125 S.Y. Nof: Collaborative control theory for e-Work, e-Production, and e-Service, Annu. Rev. Control **31**(2), 281–292 (2007)

30.126 C.L. Yang, X. Chen, S.Y. Nof: Design of a production conflict and error detection model with active protocols and agents, Proc. 18th Int. Conf. Prod. Res. (2005)

Part C | 30

Auto Part D mati

Part D Automation Design: Theory and Methods for Integration

Automation Design: Theory and Methods for Integration. Part D After focusing in the previous part on the details, principles and practices of the methodologies for automation design, the chapters in this part cover the basic design requirements for the automation and illustrate examples of how the challenging issues can be solved for the deign and integration of automation with respect to its main purpose: Continuous and discrete processes and industries, such as chemicals, refineries, machinery, and instruments; process automation safety; automation products such as circuit breakers, motors, drives, robots, and other components for consumer products; and services, such as maintenance, logistics, upgrade and repair, remote support operations, and tools for service personnel; design issues and criteria when integrating humans with the automation; design techniques, criteria and algorithms for flow lines, such as assembly lines, transfer lines, machining lines; and the design of complex, large-scale, integrated automation. Another view of automation integration is with computer-aided design (CAD) and computer-aided engineering (CAE), which are themselves fine examples of integrated automation, and are required for the design of any automation and non-automation components, products, microelectronics, and services. Concluding this part is the design for safety of automation, and of automation for safety, as they have become an obligatory and mandatory concern of automation integrators.

31. Process Automation

Thomas F. Edgar, Juergen Hahn

The field of process automation is concerned with the analysis of dynamic behavior of chemical processes, design of automatic controllers, and associated instrumentations. Process automation as practised in the process industries has undergone significant changes since it was first introduced in the 1940s. Perhaps the most significant influence on the changes in process control technology has been the introduction of inexpensive digital computers and instruments with greater capabilities than their analog predecessors. During the past 20 years automatic control has assumed increased importance in the process industries, which has led to the application of more sophisticated techniques.

31.1 Enterprise View of Process Automation

Process automation is used in order to maximize production while maintaining a desired level of product quality and safety and making the process more economical. Because these goals apply to a variety of industries, process control systems are used in facilities for the production of chemicals, pulp and paper, metals, food, and pharmaceuticals. While the methods of production vary from industry to industry, the principles of automatic control are generic in nature and can be universally applied, regardless of the size of the plant.

In Fig. 31.1 the process automation activities are organized in the form of a hierarchy with required functions at the lower levels and desirable functions at the higher levels. The time scale for each activity is shown on the left side of Fig. 31.1. Note that the frequency of execution is much lower for the higher-level functions.

31.1.1 Measurement and Actuation (Level 1)

Measurement devices (sensors and transmitters) and actuation equipment (for example, control valves) are used to measure process variables and implement the calculated control actions. These devices are interfaced to the control system, usually digital control equipment such as a digital computer. Clearly, the measurement and actuation functions are an indispensable part of any control system.

31.1.2 Safety and Environmental/ Equipment Protection (Level 2)

The level 2 functions play a critical role by ensuring that the process is operating safely and satisfies environ-

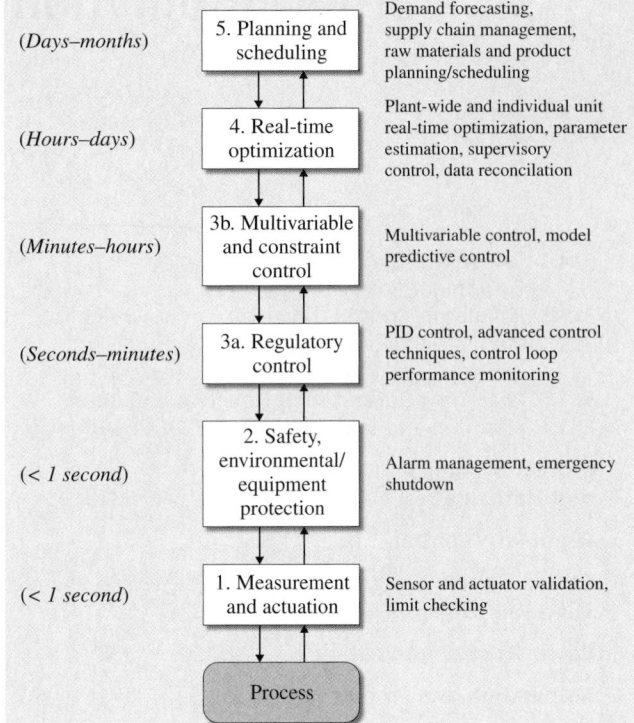

Fig. 31.1 The five levels of process control and optimization in manufacturing. Time scales are shown for each level [31.1]

mental regulations. Process safety relies on the principle of multiple protection layers that involve groupings of equipment and human actions. One layer includes process control functions, such as alarm management during abnormal situations, and safety instrumented systems for emergency shutdowns. The safety equipment (including sensors and control valves) operates independently of the regular instrumentation used for regulatory control in level 3a. Sensor validation techniques can be employed to confirm that the sensors are functioning properly.

31.1.3 Regulatory Control (Level 3a)

Successful operation of a process requires that key process variables such as flow rates, temperatures, pressures, and compositions be operated at, or close to, their set points. This level 3a activity, regulatory control, is achieved by applying standard feedback and feedforward control techniques. If the standard control techniques are not satisfactory, a variety of advanced control techniques are available. In recent years, there

has been increased interest in monitoring control system performance.

31.1.4 Multivariable and Constraint Control (Level 3b)

Many difficult process control problems have two distinguishing characteristics: (1) significant interactions occur among key process variables, and (2) inequality constraints exist for manipulated and controlled variables. The inequality constraints include upper and lower limits; for example, each manipulated flow rate has an upper limit determined by the pump and control valve characteristics. The lower limit may be zero or a small positive value based on safety considerations. Limits on controlled variables reflect equipment constraints (for example, metallurgical limits) and the operating objectives for the process; for example, a reactor temperature may have an upper limit to avoid undesired side reactions or catalyst degradation, and a lower limit to ensure that the reaction(s) proceed.

The ability to operate a process close to a limiting constraint is an important objective for advanced process control. For many industrial processes, the optimum operating condition occurs at a constraint limit, for example, the maximum allowed impurity level in a product stream. For these situations, the set point should not be the constraint value because a process disturbance could force the controlled variable beyond the limit. Thus, the set point should be set conservatively, based on the ability of the control system to reduce the effects of disturbances.

The standard process control techniques of level 3a may not be adequate for difficult control problems that have serious process interactions and inequality constraints. For these situations, the advanced control techniques of level 3b, multivariable control and constraint control, should be considered. In particular, the model predictive control (MPC) strategy was developed to deal with both process interactions and inequality constraints.

31.1.5 Real-Time Optimization (Level 4)

The optimum operating conditions for a plant are determined as part of the process design, but during plant operations, the optimum conditions can change frequently owing to changes in equipment availability, process disturbances, and economic conditions (for example, raw materials costs and product prices). Consequently, it can be very profitable to recalculate the optimum operating conditions on a regular basis. The new optimum conditions are then implemented as set points for controlled variables.

Real-time optimization (RTO) calculations are based on a steady-state model of the plant and economic data such as costs and product values. A typical objective for the optimization is to minimize operating cost or maximize the operating profit. The RTO calculations can be performed for a single process unit and/or on a plant-wide basis.

The level 4 activities also include data analysis to ensure that the process model used in the RTO calculations is accurate for the current conditions. Thus, data reconciliation techniques can be used to ensure that steady-state mass and energy balances are satisfied. Also, the process model can be updated using parameter estimation techniques and recent plant data.

31.1.6 Planning and Scheduling (Level 5)

The highest level of the process control hierarchy is concerned with planning and scheduling operations for the entire plant. For continuous processes, the production rates of all products and intermediates must be planned and coordinated, based on equipment constraints, stor-

age capacity, sales projections, and the operation of other plants, sometimes on a global basis. For the intermittent operation of batch and semibatch processes, the production control problem becomes a batch scheduling problem based on similar considerations. Thus, planning and scheduling activities pose large-scale optimization problems that are based on both engineering considerations and business projections.

The activities of levels 1–3a in Fig. 31.1 are required for all manufacturing plants, while the activities in levels 3b–5 are optional but can be very profitable. The decision to implement one or more of these higher-level activities depends very much on the application and the company. The decision hinges strongly on economic considerations (for example, a cost–benefit analysis), and company priorities for their limited resources, both human and financial. The immediacy of the activity decreases from level 1 to level 5 in the hierarchy. However, the amount of analysis and the computational requirements increase from the lowest to the highest level. The process control activities at different levels should be carefully coordinated and require information transfer from one level to the next. The successful implementation of these process control activities is a critical factor in making plant operation as profitable as possible.

31.2 Process Dynamics and Mathematical Models

Development of dynamic models forms a key component for process automation, as controller design and tuning is often performed by using a mathematical representation of the process. A model can be derived either from first-principles knowledge about the system or from past plant data. Once a dynamic model has been developed, it can be solved for a variety of conditions that include changes in the input variables or variations in the model parameters. The transient responses of the output variables are calculated by numerical integration after specifying both the initial conditions and the inputs as functions of time.

A large number of numerical integration techniques are available, ranging from simple techniques (e.g., the Euler and Runge–Kutta methods) to more complicated ones (e.g., the implicit Euler and Gear methods). All of these techniques represent some compromise between computational effort (computing time) and accuracy. Although a dynamic model can always be solved in principle, for some situations it may be difficult to generate useful numerical solutions. Dynamic models that exhibit a wide range of time scales (stiff equations)

are quite difficult to solve accurately in a reasonable amount of computation time. Software for integrating ordinary and partial differential equations is readily available. Popular software packages include MATLAB, Mathematica, ACSL, IMSL, Mathcad, and GNU Octave.

For dynamic models that contain large numbers of algebraic and ordinary differential equations, generation of solutions using standard programs has been developed to assist in this task. A graphical user interface (GUI) allows the user to enter the algebraic and ordinary differential equations and related information such as the total integration period, error tolerances, the variables to be plotted, and so on. The simulation program then assumes responsibility for:

1. Checking to ensure that the set of equations is exactly specified
2. Sorting the equations into an appropriate sequence for iterative solution
3. Integrating the equations
4. Providing numerical and graphical output.

Examples of equation-oriented simulators used in the process industries include DASSL, ACSL, gPROMS, and Aspen Custom Modeler.

One disadvantage of equation-oriented packages is the amount of time and effort required to develop all of the equations for a complex process. An alternative approach is to use modular simulation where prewritten subroutines provide models of individual process units such as distillation columns or chemical reactors. Consequently, this type of simulator has a direct correspondence to the process flowsheet. The modular approach has the significant advantage that plant-scale simulations only require the user to identify the appropriate modules and to supply the numerical values of model parameters and initial conditions. This activity requires much less effort than writing all of the equations. Furthermore, the software is responsible for all aspects of the solution. Because each module is rather general in form, the user can simulate alternative flowsheets for a complex process, for example, different configurations of distillation towers and heat exchangers, or different types of chemical reactors. Similarly, alternative process control strategies can be quickly evaluated. Some software packages allow the user to add custom modules for novel applications.

Modular dynamic simulators have been available since the early 1970s. Several commercial products are available from Aspen Technology and Honeywell. Modelica is an example of a collaborative effort that provides modeling capability for a number of application areas. These packages also offer equation-oriented capabilities. Modular dynamic simulators are achieving a high degree of acceptance in process engineering and control studies because they allow plant dynamics, real-time optimization, and alternative control configurations to be evaluated for an existing or a new plant. They also can be used for operator training. This feature allows dynamic simulators to be integrated with software for other applications such as control system design and optimization.

While most processes can be accurately represented by a set of nonlinear differential equations, a process is usually operated within a certain neighborhood of its normal operating point (steady state), thus the process model can be closely approximated by a linearized version of the model. A linear model is beneficial because it permits the use of more convenient and compact methods for representing process dynamics, namely Laplace transforms. The main advantage of Laplace transforms is that they provide a compact representation of a dynamic system that is especially useful for the analysis of feedback control systems. The Laplace transform of a set of linear ordinary differential equations is a set of algebraic equations in the new variable s, called the Laplace variable. The Laplace transform is given by

$$F(s) = L[f(t)] = \int_0^\infty f(t) e^{-st} \, dt , \qquad (31.1)$$

where $F(s)$ is the symbol for the Laplace transform, $f(t)$ is some function of time, and L is the Laplace operator, defined by the integral. Tables of Laplace transforms are well documented for common functions [31.1]. A linear differential equation with a single input u and single output y can be converted into a transfer function using Laplace transforms as follows

$$Y(s) = G(s)U(s) , \qquad (31.2)$$

where $U(s)$ is the Laplace transform of the input variable $u(t)$, $Y(s)$ is the Laplace transform of the output variable $y(t)$, and $G(s)$ is the transfer function, obtained from transforming the differential equation.

The transfer function $G(s)$ describes the dynamic characteristic of the process. For linear systems it is independent of the input variable and so it can readily be applied to any time-dependent input signal. As an example, the first-order differential equation

$$\tau \frac{dy(t)}{dt} + y(t) = Ku(t) \qquad (31.3)$$

can be Laplace-transformed to

$$Y(s) = \frac{K}{\tau s + 1} U(s) . \qquad (31.4)$$

Note that the parameters K and τ, known as the process gain and time constant, respectively, map into the transfer function as unspecified parameters. Numerical values for parameters such as K and τ have to be determined for controller design or for simulation purposes. Several different methods for the identification of model parameters in transfer functions are available. The most common approach is to perform a step test on the process and collect the data along the trajectory until it reaches steady state. In order to identify the parameters, the form of the transfer function model needs to be postulated and the parameters of the transfer function can be estimated by using nonlinear regression. For more details on the development of various transfer functions, see [31.1].

31.3 Regulatory Control

When the components of a control system are connected, their overall dynamic behavior can be described by combining the transfer functions for each component. Each block describes how changes in the input variables of the block will affect the output variables of the block. One example is the feedback/feedforward control block diagram shown in Fig. 31.2, which contains the important components of a typical control system, namely process, controller, sensor, and final control element. Regulatory control deals with treatment of disturbances that enter the system, as shown in Fig. 31.2. These components are discussed in more detail below.

Most modern control equipment require a digital signal for displays and control algorithms, thus the analog-to-digital converter (ADC) transforms the transmitter analog signal to a digital format. Because ADCs may be relatively expensive if adequate digital resolution is required, incoming digital signals are usually multiplexed. Prior to sending the desired control action, which is often in a digital format, to the final control element in the field, the desired control action is usually transformed by a digital-to-analog (DAC) converter to an analog signal for transmission. DACs are relatively inexpensive and are not normally multiplexed. Widespread use of digital control technologies has made ADCs and DACs standard parts of the control system.

Sensors

The hardware components of a typical modern digital control loop shown in Fig. 31.2 are discussed next. The function of the process measurement device is to sense the values, or changes in values, of process variables. The actual sensing device may generate, e.g., a physical movement, a pressure signal, or a millivolt signal. A transducer transforms the measurement signal from one physical or chemical quantity to another, e.g., pressure to milliamps. The transduced signal is then transmitted to a control room through the transmission line. The transmitter is therefore a signal generator and a line driver. Often the transducer and the transmitter are contained in the same device.

The most commonly measured process variables are temperature, flow, pressure, level, and composition. When appropriate, other physical properties are also measured. The selection of the proper instrumentation for a particular application is dependent on factors such as: the type and nature of the fluid or solid involved; relevant process conditions; range, accuracy, and repeatability required; response time; installed cost; and maintainability and reliability. Various handbooks are available that can assist in selecting sensors for particular applications (e.g., [31.2]). Sensors are discussed in detail on Chap. 20.

Control Valves

Material and energy flow rates are the most commonly selected manipulated variables for control schemes. Thus, good control valve performance is an essential ingredient for achieving good control performance. A control valve consists of two principal assemblies: a valve body and an actuator. Good control valve performance requires consideration of the process characteristics and requirements such as fluid characteristics, range, shut-off, and safety, as well as control

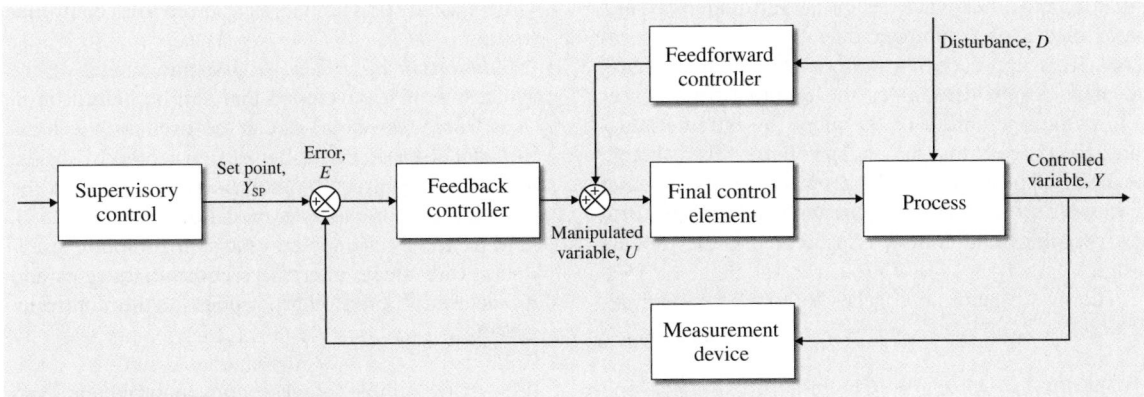

Fig. 31.2 Block diagram of a process

requirements, e.g., installed control valve characteristics and response time. The proper selection and sizing of control valves and actuators is an extensive topic in its own right [31.2].

Controllers

The most commonly employed feedback controller in the process industry is the proportional–integral (PI) controller, which can be described by the following equation

$$u(t) = \bar{u} + K_C \left(e(t) + \frac{1}{\tau_I} \int_0^t e(t') \, dt' \right) . \tag{31.5}$$

Note that the controller includes proportional as well as integrating action. The controller has two tuning parameters: the proportional constant K_C and the integral time constant τ_I. The integral action will eliminate offset for constant load disturbances but it can potentially lead to a phenomenon known as reset windup. When there is a sustained error, the large integral term in (31.5) causes the controller output to saturate. This can occur during start-up of batch processes, or after large set point changes or large sustained disturbances.

PI controllers make up the vast majority of controllers that are currently used in the chemical process industries. If it is important to achieve a faster response that is offset-free, a PID (D = derivative) controller can be utilized, described by the following expression

$$u(t) = \bar{u} + K_C \left(e(t) + \frac{1}{\tau_I} \int_0^t e(t') \, dt' + \tau_D \frac{de(t)}{dt} \right) . \tag{31.6}$$

The PID controller of (31.6) contains three tuning parameters because the derivative mode adds a third adjustable parameter τ_D. However, if the process measurement is noisy, the value of the derivative of the error may change rapidly and derivative action will amplify the noise, as a filter on the error signal can be employed. In the 21st century, digital control systems are ubiquitous in process plants, mostly employing a discrete (finite-difference) form of the PID controller equation given by

$$u_k = \bar{u} + K_C \left(e_k + \frac{\Delta t}{\tau_I} k \sum_{i=0}^{k} e_i + \tau_D \frac{e_k - e_{k-1}}{\Delta t} \right) , \tag{31.7}$$

where Δt is the sampling period for the control calculations and k represents the current sampling time. If the process and the measurements permit to chose the sampling period Δt to be small then the behavior of the digital PID controller will essentially be the same as for an analog PID controller.

31.4 Control System Design

Traditionally, process design and control system design have been separate engineering activities. Thus, in the traditional approach, control system design is not initiated until after plant design is well underway and major pieces of equipment may even have been ordered. This approach has serious limitations because the plant design determines the process dynamics as well as the operability of the plant. In extreme situations, the process may be uncontrollable, even though the design appears satisfactory from a steady-state point of view. A more desirable approach is to consider process dynamics and control issues early in the process design.

The two general approaches to control system design are:

1. *Traditional approach.* The control strategy and control system hardware are selected based on knowledge of the process, experience, and insight. After the control system is installed in the plant, the controller settings (such as in a PID controller) are adjusted. This activity is referred to as controller tuning.

2. *Model-based approach.* A dynamic model of the process is first developed that can be helpful in at least three ways: (a) it can be used as the basis for model-based controller design methods, (b) the dynamic model can be incorporated directly in the control law (for example, model predictive control), and (c) the model can be used in a computer simulation to evaluate alternative control strategies and to determine preliminary values of the controller settings.

For many simple process control problems controller specification is relatively straightforward and a detailed analysis or an explicit model is not required. However, for complex processes, a process model is invaluable both for control system design and for an improved understanding of the process.

The major steps involved in designing and installing a control system using the model-based approach are shown in the flowchart of Fig. 31.3. The first step, formulation of the control objectives, is a critical decision. The formulation is based on the operating objectives for the plants and the process constraints; for example, in the distillation column control problem, the objective might be to regulate a key component in the distillate stream, the bottoms stream, or key components in both streams. An alternative would be to minimize energy consumption (e.g., heat input to the reboiler) while meeting product quality specifications on one or both product streams. The inequality constraints should include upper and lower limits on manipulated variables, conditions that lead to flooding or weeping in the column, and product impurity levels.

After the control objectives have been formulated, a dynamic model of the process is developed. The dynamic model can have a theoretical basis, for example, physical and chemical principles such as conservation laws and rates of reactions, or the model can be developed empirically from experimental data. If experimental data are available, the dynamic model should be validated, with the data and the model accuracy characterized. This latter information is useful for control system design and tuning.

The next step in the control system design is to devise an appropriate control strategy that will meet the control objectives while satisfying process constraints. As indicated in Fig. 31.3, this design activity is based on models and plant data. Finally the control system can be installed, with final adjustments performed once the plant is operating.

31.4.1 Multivariable Control

In most industrial processes, there are a number of variables that must be controlled, and a number of variables can be manipulated. These problems are referred to as multiple-input multiple-output (MIMO) control problems. For almost all important processes, at least two variables must be controlled: product quality and throughput. Several examples of processes with two controlled variables and two manipulated variables are shown in Fig. 31.4. These examples illustrate a characteristic feature of MIMO control problems, namely, the presence of process interactions; that is,

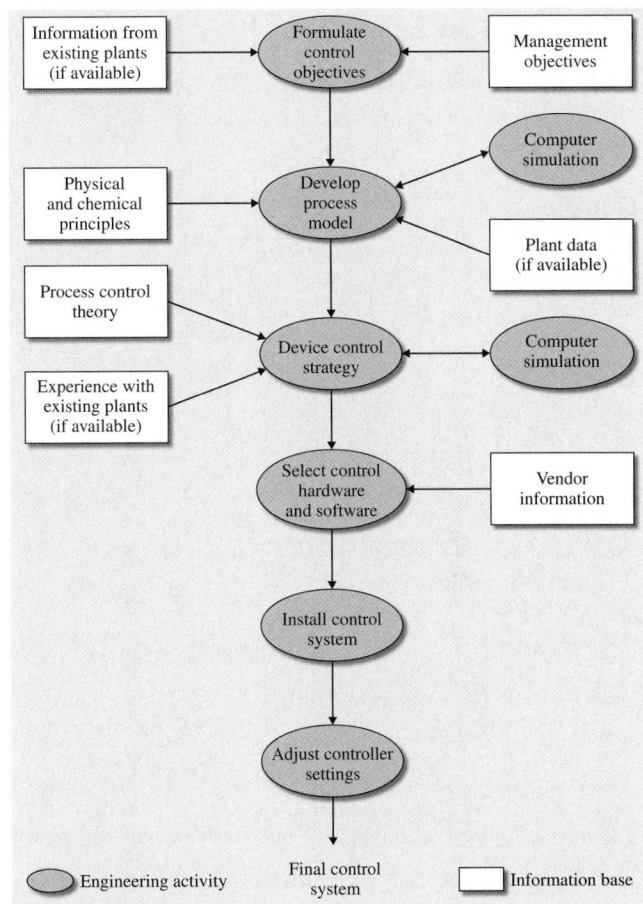

Fig. 31.3 Major steps in control system development [31.1]

each manipulated variable can affect both controlled variables. Consider the inline blending system shown in Fig. 31.4a. Two streams containing species A and B, respectively, are to be blended to produce a produce stream with mass flow rate w and composition x, the mass fraction of A. Adjusting either manipulated flow rate, w_A or w_B, affects both w and x.

Similarly, for the distillation column in Fig. 31.4b, adjusting either reflux flow rate R or steam flow S will affect both distillate composition x_D and bottoms composition x_B. For the gas–liquid separator in Fig. 31.4c, adjusting the gas flow rate G will have a direct effect on pressure P and a slower, indirect effect on liquid level h because changing the pressure in the vessel will tend to change the liquid flow rate L and thus affect h. In contrast, adjusting the other manipulated variable L directly affects h but has only a relatively small and indirect effect on P.

a) Inline blending system

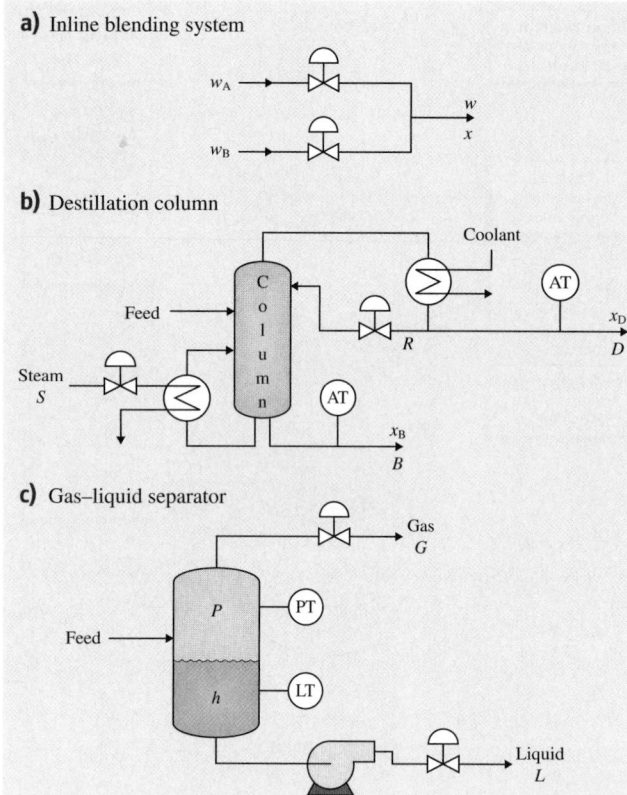

b) Destillation column

c) Gas–liquid separator

Fig. 31.4a–c Physical examples of multivariable control problems [31.1]

Model predictive control offers several important advantages: (1) the process model captures the dynamic and static interactions between input, output, and disturbance variables, (2) constraints on inputs and outputs are considered in a systematic manner, (3) the control calculations can be coordinated with the calculation of optimum set points, and (4) accurate model predictions can provide early warnings of potential problems. Clearly, the success of MPC (or any other model-based approach) depends on the accuracy of the process model. Inaccurate predictions can make matters worse, instead of better.

First-generation MPC systems were developed independently in the 1970s by two pioneering industrial research groups. Dynamic matrix control (DMC) was devised by Shell Oil [31.3], and a related approach was developed by ADERSA [31.4]. Model predictive control has had a major impact on industrial practice; for example, an MPC survey by *Qin* and *Badgwell* [31.5] reported that there were over 4500 applications worldwide by the end of 1999, primarily in oil refineries and petrochemical plants. In these industries, MPC has become the method of choice for difficult multivariable control problems that include inequality constraints.

The overall objectives of an MPC controller are as follows:

1. Prevent violations of input and output constraints
2. Drive some output variables to their optimal set points, while maintaining other outputs within specified ranges
3. Prevent excessive movement of the manipulated variables
4. Control as many process variables as possible when a sensor or actuator is not available.

A block diagram of a model predictive control system is shown in Fig. 31.5. A process model is used to predict the current values of the output variables. The residuals (the differences between the actual and predicted outputs) serve as the feedback signal to a prediction block. The predictions are used in two types of MPC calculations that are performed at each sampling instant: set-point calculations and control calculations. Inequality constraints on the input and output variables, such as upper and lower limits, can be included in either type of calculation.

The model acts in parallel with the process and the residual serves as a feedback signal, however, it should be noted that the coordination of the control and set-point calculation is a unique feature of MPC.

Pairing of a single controlled variable and a single manipulated variable via a PID feedback controller is possible, if the number of manipulated variables is equal to the number of controlled variables. On the other hand, more general multivariable control strategies do not make such restrictions. MIMO control problems are inherently more complex than single-input single-output (SISO) control problems because process interactions occur between controlled and manipulated variables. In general, a change in a manipulated variable, say u_1, will affect all of the controlled variables $y_1, y_2, \ldots, y_n$. Because of process interactions, selection of the best pairing of controlled and manipulated variables for a multiloop control scheme can be a difficult task. In particular, for a control problem with n controlled variables and n manipulated variables, there are $n!$ possible multiloop control configurations. Hence there is a growing trend to use multivariable control, in particular an approach called model predictive control (MPC).

Furthermore, MPC has had a significant impact on industrial practice because it is more suitable for constrained MIMO control problems.

The set points for the control calculations, also called targets, are calculated from an economic optimization based on a steady-state model of the process, traditionally, a linear steady-state model. Typical optimization objectives include maximizing a profit function, minimizing a cost function, or maximizing a production rate. The optimum values of set points are changed frequently owing to varying process conditions, especially changes in the inequality constraints. The constraint changes are due to variations in process conditions, equipment, and instrumentation, as well as economic data such as prices and costs. In MPC the set points are typically calculated each time the control calculations are performed.

The control calculations are based on current measurements and predictions of the future values of the outputs. The predictions are made using a dynamic model, typically a linear empirical model such as a multivariable version of the step response models that were discussed in Sect. 31.2. Alternatively, transfer function or state-space models can be employed. For very nonlinear processes, it can be advantageous to predict future output values using a nonlinear dynamic model. Both physical models and empirical models, such as neural networks, have been used in nonlinear MPC [31.5].

The objective of the MPC control calculations is to determine a sequence of control moves (that is, manipulated input changes) so that the predicted response moves to the set point in an optimal manner. The actual output y, predicted output $\hat{y}$, and manipulated input u are shown in Fig. 31.6. At the current sampling instant, denoted by k, the MPC strategy calculates a set of M values of the input $\{u(k+i-1), i = 1, 2, \ldots, M\}$. The set consists of the current input $u(k)$ and $M-1$ future inputs. The input is held constant after the M control moves. The inputs are calculated so that a set of P predicted outputs $\{\hat{y}(k+i), i = 1, 2, \ldots, P\}$ reaches the set point in an optimal manner. The control calculations are based on optimizing an objective function. The number of predictions P is referred to as the prediction horizon while the number of control moves M is called the control horizon.

A distinguishing feature of MPC is its receding horizon approach. Although a sequence of M control moves is calculated at each sampling instant, only the first move is actually implemented. Then a new sequence is calculated at the next sampling instant, after new

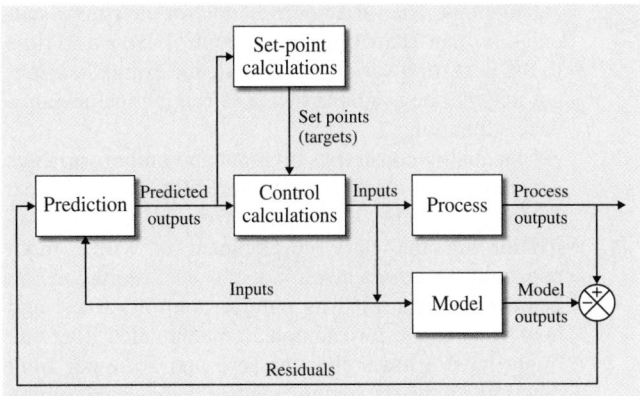

Fig. 31.5 Block diagram for model predictive control [31.1]

measurements become available; again, only the first input move is implemented. This procedure is repeated at each sampling instant.

In MPC applications, the calculated input moves are usually implemented as set points for regulatory control loops at the distributed control system (DCS) level, such as flow control loops. If a DCS control loop has been disabled or placed in manual mode, the input variable is no longer available for control. In this situation, the control degrees of freedom are reduced by one. Even though an input variable is unavailable for control, it can serve as a disturbance variable if it is still measured.

Before each control execution, it is necessary to determine which outputs (controlled variables (CV)), inputs (manipulated variables (MV)), and disturbance variables (DVs) are currently available for the MPC

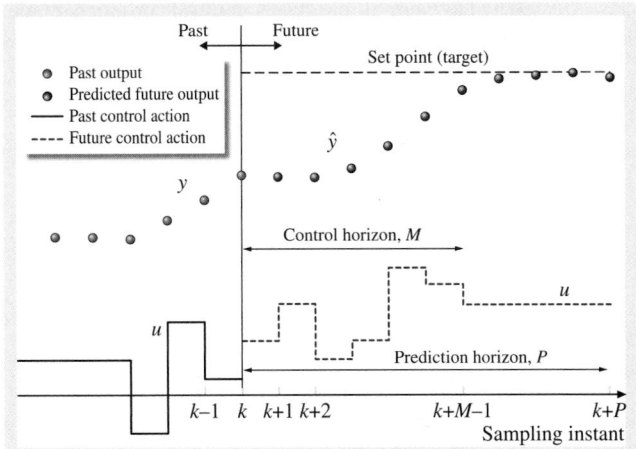

Fig. 31.6 Basic concept for model predictive control

calculations. The variables available for the control calculations can change from one control execution time to the next for a variety of reasons; for example, a sensor may not be available owing to routine maintenance or recalibration.

Inequality constraints on input and output variables are important characteristics for MPC applications. In fact, inequality constraints were a primary motivation for the early development of MPC. Input constraints occur as a result of physical limitations on plant equipment such as pumps, control valves, and heat exchangers; for example, a manipulated flow rate might have a lower limit of zero and an upper limit determined by the pump, control valve, and piping characteristics. The dynamics associated with large control valves impose rate-of-change limits on manipulated flow rates.

Constraints on output variables are a key component of the plant operating strategy; for example, a common distillation column control objective is to maximize the production rate while satisfying constraints on product quality and avoiding undesirable operating regimes such as flooding or weeping. It is convenient to make a distinction between hard and soft constraints. As the name implies, a hard constraint cannot be violated at any time. By contrast, a soft constraint can be violated, but the amount of violation is penalized by a modification of the cost function. This approach allows small constraint violations to be tolerated for short periods of time [31.1].

31.5 Batch Process Automation

Batch processing is an alternative to continuous processing. In batch processing, a sequence of one or more steps, either in a single vessel or in multiple vessels, is performed in a defined order, yielding a specific quantity of a finished product. Because the volume of product is normally small, large production runs are achieved by repeating the process steps on a predetermined schedule. In batch processing, the production amounts are usually smaller than for continuous processing; hence, it is usually not economically feasible to dedicate processing equipment to the manufacture of a single product. Instead, batch processing units are organized so that a range of products (from a few to possibly hundreds) can be manufactured with a given set of process equipment. Batch processing can be complicated by having multiple stages, multiple products made from the same equipment, or parallel processing lines. The key challenge for batch plants is to consistently manufacture each product in accordance with its specifications while maximizing the utilization of available equipment. Benefits include reduced inventories and shortened response times to make a specialty product compared with continuous processing plants. Typically, it is not possible to use blending of multiple batches in order to obtain the desired product quality, so product quality specifications must be satisfied by each batch.

Batch processing is widely used to manufacture specialty chemicals, metals, electronic materials, ceramics, polymers, food and agricultural materials, biochemicals and pharmaceuticals, multiphase materials/blends, coatings, and composites – an extremely broad range of processes and products. The unit operations in batch processing are also quite diverse, and some are analogous to operations for continuous processing.

In analogy with the different levels of plant control depicted in Fig. 31.1, batch control systems operate at various levels:

- Batch sequencing and logic controls (levels 1 and 2)
- Control during the batch (level 3)
- Run-to-run control (levels 4 and 5)
- Batch production management (level 5).

Figure 31.7 shows the interconnections of the different types of control used in a typical batch process. Run-to-run control is a type of supervisory control that resides principally in the production management block. In contrast to continuous processing, the focus of control shifts from regulation to set-point changes, and sequencing of batches and equipment takes on a much greater role.

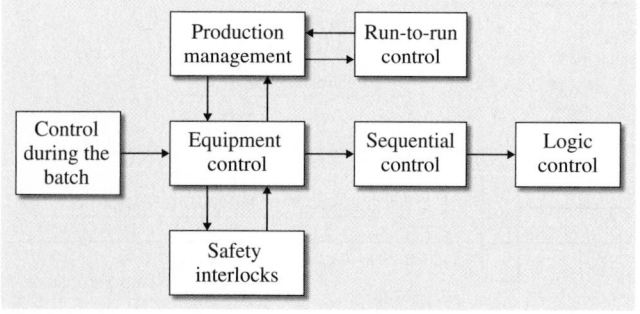

Fig. 31.7 Overview of a batch control system

Batch control systems must be very versatile to be able to handle pulse inputs and discrete input/output (I/O) as well as analog signals for sensors and actuators. Functional control activities are summarized as follows:

1. *Batch sequencing and logic control:* Sequencing of control steps that follow a recipe involves, for example, mixing of ingredients, heating, waiting for a reaction to complete, cooling, and discharging the resulting products. Transfer of materials to and from batch tanks or reactors includes metering of materials as they are charged (as specified by the recipe), as well as transfer of materials at the completion of the process operation. In addition to discrete logic for the control steps, logic is needed for safety interlocks to protect personnel, equipment, and the environment from unsafe conditions. Process interlocks ensure that process operations can only occur in the correct time sequence.

2. *Control during the batch:* Feedback control of flow rate, temperature, pressure, composition, and level, including advanced control strategies, falls in this category, which is also called *within-the-batch* control [31.6]. In sophisticated applications, this requires specification of an operating trajectory for the batch (that is, temperature or flow rate as a function of time). In simpler cases, it involves tracking of set points of the controlled variables, which includes ramping the controlled variables up and down and/or holding them constant for a prescribed period of time. Detection of when the batch operations should be terminated (end point) may be performed by inferential measurements of product quality, if direct measurement is not feasible.

3. *Run-to-run control:* Also called batch-to-batch control, this supervisory function is based on offline product quality measurements at the end of a run. Operating conditions and profiles for the batch are adjusted between runs to improve the product quality using tools such as optimization.

4. *Batch production management:* This activity entails advising the plant operator of process status and how to interact with the recipes and the sequential, regulatory, and discrete controls. Complete information (recipes) is maintained for manufacturing each product grade, including the names and amounts of ingredients, process variable set points, ramp rates, processing times, and sampling procedures. Other database information includes batches produced on a shift, daily, or weekly basis, as well as material and energy balances. Scheduling of process units

is based on availability of raw materials and equipment and customer demand.

Recipe modifications from one run to the next are common in many batch processes. Typical examples are modifying the reaction time, feed stoichiometry, or reactor temperature. When such modifications are done at the beginning of a run (rather than during a run), the control strategy is called run-to-run control. Run-to-run control is frequently motivated by the lack of online measurements of the product quality during a batch run. In batch chemical production, online measurements are often not available during the run, but the product can be analyzed by laboratory samples at the end of the run. The process engineer must specify a recipe that contains the values of the inputs (which may be time-varying) that will meet the product requirements. The task of the run-to-run controller is to adjust the recipe after each run to reduce variability in the output product from the stated specifications.

Batch run-to-run control is particularly useful to compensate for processes where the controlled variable drifts over time; for example, in a chemical vapor deposition process the reactor walls may become fouled owing to byproduct deposition. This slow drift in the reactor chamber condition requires occasional changes to the batch recipe in order to ensure that the controlled variables remain on target. Eventually, the reactor chamber must be cleaned to remove the wall deposits, effectively causing a step disturbance to the process outputs when the inputs are held constant. Just as the run-to-run controller compensates for the drifting process, it can also return the process to target after a step disturbance change [31.7, 8].

The Instrument Society of America (ISP) SP-88 standard deals with the terminology involved in batch control [31.9]. There are a hierarchy of activities that take place in a batch processing system [31.10]. At the highest level, procedures identify how the products are made, that is, the actions to be performed (and their order) as well as the associated control requirements for these actions. Operations are equivalent to unit operations in continuous processing and include such steps as charging, reacting, separating, and discharging. Within each operation are logical points called *phases*, where processing can be interrupted by operator or computer interaction. Examples of different phases include the sequential addition of ingredients, heating a batch to a prescribed temperature, mixing, and so on. Control steps involve direct commands to final control elements, specified by individual control instructions in software.

As an example, for {operation = charge reactant} and {phase = add ingredient B}, the control steps would be: (1) open the B supply valve, (2) total the flow of B over a period of time until the prescribed amount has been added, and (3) close the B supply valve.

The term *recipe* has a range of definitions in batch processing, but in general a recipe is a procedure with the set of data, operations, and control steps required to manufacture a particular grade of product. A *formula* is the list of recipe parameters, which includes the raw materials, processing parameters, and product outputs. A recipe *procedure* has operations for both normal and abnormal conditions. Each operation contains resource requests for certain ingredients (and their amounts). The *operations* in the recipe can adjust set points and turn equipment on and off. The complete production run for a specific recipe is called a *campaign* (multiple batches).

In multigrade batch processing, the instructions remain the same from batch to batch, but the formula can be changed to yield modest variations in the product; for example, in emulsion polymerization, different grades of polymers are manufactured by changing the formula. In flexible batch processing, both the formula (recipe parameters) and the processing instructions can change from batch to batch. The recipe for each product must specify both the raw materials required and how conditions within the reactor are to be sequenced in order to make the desired product.

Many batch plants, especially those used to manufacture pharmaceuticals, are certified by the International Standards Organization (ISO). ISO 9000 (and the related ISO standards 9001–9004) state that every manufactured product should have an established, documented procedure, and the manufacturer should be able to document that the procedure was followed. Companies must pass periodic audits to main ISO 9000 status. Both ISO 9000 and the US Food and Drug Administration (FDA) require that only a certified recipe be used. Thus, if the operation of a batch becomes *abnormal*, performing any unusual corrective action to bring it back within the normal limits is not an option. In addition, if a slight change in the recipe apparently produces superior batches, the improvement cannot be implemented unless the entire recipe is recertified. The FDA typically requires product and raw materials tracking, so that product abnormalities can be traced back to their sources.

Recently, in an effort to increase the safety, efficiency, and affordability of medicines, the FDA has proposed a new framework for the regulation of pharmaceutical development, manufacturing, and quality assurance. The primary focus of the initiative is to reduce variability through a better understanding of processes than can be obtained by the traditional approach. Process analytical technology (PAT) has become an acronym in the pharmaceutical industry for designing, analyzing, and controlling manufacturing through timely measurements (i.e., during processing) of critical quality and performance attributes of raw and in-process materials and processes, with the goal of ensuring final product quality. Process variations that could possibly contribute to patient risk are determined through modeling and timely measurements of critical quality attributes, which are then addressed by process control. In this manner processes can be developed and controlled in such a way that quality of product is guaranteed.

Semiconductor manufacturing is an example of a large-volume batch process [31.7]. In semiconductor manufacturing an integrated circuit consists of several layers of carefully patterned thin films, each chemically altered to achieve desired electrical characteristics. These devices are manufactured through a series of physical and/or chemical batch unit operations similar to the way in which speciality chemicals are made. From 30 to 300 process steps are typically required to construct a set of circuits on a single-crystalline substrate called a wafer. The wafers are 4 to 4–12 inch (100–300 mm) in diameter, 400–700 μm thick, and serve as the substrate upon which microelectronic circuits (devices) are built. Circuits are constructed by depositing the thin films (0.01–10 μm) of material of carefully controlled composition in specific patterns and then etching these films to exacting geometries (0.35–10 μm).

The main unit operations in semiconductor manufacturing are crystal growth, oxidation, deposition (dielectrics, silicon, metals), physical vapor deposition, dopant diffusion, dopant-ion implantation, photolithography, etch, and chemical–mechanical polishing. Most processes in semiconductor manufacturing are semibatch; for example, in a single-wafer processing tool the following steps are carried out:

1. A robotic arm loads the boat of wafers
2. The machine transfers a single wafer into the processing chamber
3. Gases flow continuously and reaction occurs
4. The machine removes the wafer
5. The next wafer is processed.

When all wafers are finished processing, the operator takes the boat of wafers to the next machine. All of these steps are carried out in a clean room designed to minimize device damage by particulate matter.

For a given tool or unit operation a specified number of wafers are processed together in a lot, which is carried in a boat. There is usually an extra slot in the boat for a pilot wafer, which is used for metrology reasons. A cluster tool refers to equipment which has several single-wafer processing chambers. The chambers may carry out the same process or different processes; some vendors base their chamber designs on series operation, while others utilize parallel processing schemes.

The recipe for the batch consists of the regulatory set points and parameters for the real-time controllers on the equipment. The equipment controllers are normally not capable of receiving a continuous set-point trajectory. Only furnaces and rapid thermal processing tools are able to ramp up, hold, and ramp down their temperature or power supply. A recipe can consist of several steps; each step processes a different film based on specific chemistry. The same recipe on the same type of chamber may produce different results, due to different processes used in the chamber previously. This lack of repeatability across chambers is a big problem with cluster tools or when a fabrication plant (fab) has multiple machines of the same type, because it requires that a fab keep track of different recipes for each chamber. The controller translates the desired specs into a machine recipe. Thus, the fab supervisory controller only keeps track of the product specifications.

Factory automation in semiconductor manufacturing integrates the individual equipment into higher levels of automation, in order to reduce the total cycle time, increase fab productivity, and increase product yield [31.8]. The major functions provided by the automation system include:

1. Planning of factory operation from order entry through wafer production
2. Scheduling of factory resources to meet the production plan
3. Modeling and simulation of factory operation
4. Generation and maintenance of process and product specification and recipes
5. Tracking of work-in-progress (WIP)
6. Monitoring of factory performance
7. Machine monitoring, control, and diagnosis
8. Process monitoring, control, and diagnosis.

Automation of semiconductor manufacturing in the future will consist of meeting a range of technological challenges. These include the need for faster yield ramp, increasing cost pressures that compel productivity improvements, environmental safety and health concerns, and shrinking device dimensions and chip size. The development of 300 mm platforms in the last few years has spawned equipment with new software systems and capabilities. These systems will allow smart data collection, storage, and processing on the equipment, and transfer of data and information in a more efficient manner. Smart data management implies that data are collected as needed and based upon events and metrology results. As a result of immediate and automatic processing of data, a larger fraction of data can be analyzed, and more decisions are data driven. New software platforms provide the biggest opportunity for a control paradigm shift seen in the industry since the introduction of statistical process control.

31.6 Automation and Process Safety

In modern chemical plants, process safety relies on the principle of multiple protection layers. A typical configuration is shown in Fig. 31.8. Each layer of protection consists of a grouping of equipment and/or human actions. The protection layers are shown in the order of activation that occurs as a plant incident develops. In the inner layer, the process design itself provides the first level of protection. The next two layers consist of the basic process control system (BPCS) augmented with two levels of alarms and operator supervision or intervention. An alarm indicates that a measurement has exceeded its specified limits and may require operator action.

The fourth layer consists of a safety interlock system (SIS), which is also referred to as a safety instrumented system or as an emergency shutdown (ESD) system. The SIS automatically takes corrective action when the process and BPCS layers are unable to handle an emergency; for example, the SIS could automatically turn off the reactant pumps after a high-temperature alarm occurs for a chemical reactor. Relief devices such as rupture discs and relief valves provide physical protection by venting a gas or vapor if overpressurization occurs. As a last resort, dikes are located around process units and storage tanks to contain liquid spills. Emergency response plans are used

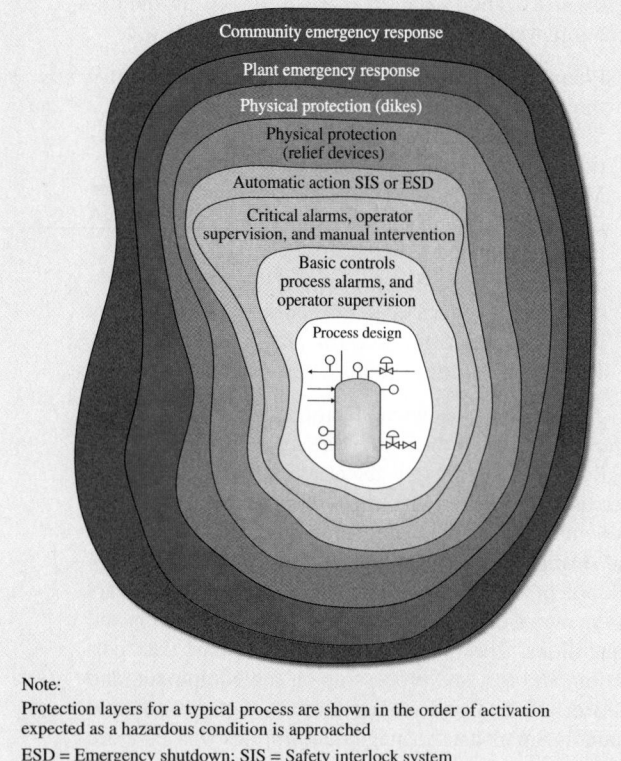

Note:
Protection layers for a typical process are shown in the order of activation expected as a hazardous condition is approached

ESD = Emergency shutdown; SIS = Safety interlock system

Fig. 31.8 Typical layers of protection in a modern chemical plant [31.11]

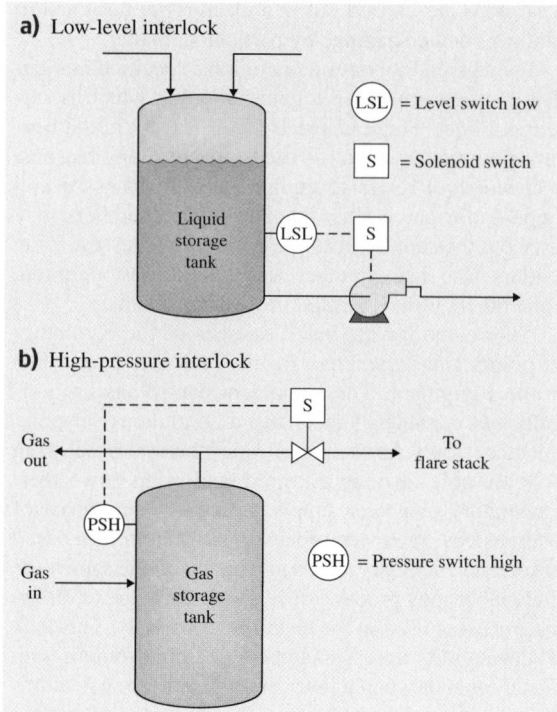

Fig. 31.9a,b Two interlock configurations [31.1]

to address emergency situations and to inform the community.

The functioning of the multiple layer protection system can be summarized as follows [31.11]:

> *Most failures in well-designed and operated chemical processes are contained by the first one or two protection layers. The middle levels guard against major releases and the outermost layers provide mitigation response to very unlikely major events. For major hazard potential, even more layers may be necessary.*

It is evident from Fig. 31.8 that automation plays an important role in ensuring process safety. In particular, many of the protection layers in Fig. 31.8 involve instrumentation and control equipment. The SIS operation is designed to provide automatic responses after alarms indicate potentially hazardous situations. The objective is to have the process reach a safe condition. The automatic responses are implemented via interlocks and automatic shutdown and start-up systems. Distinc-

tions are sometimes made between safety interlocks and process interlocks; process interlocks are used for less critical situations to provide protection against minor equipment damage and undesirable process conditions such as the production of off-specification product.

Two simple interlock systems are shown in Fig. 31.9. For the liquid storage system, the liquid level must stay above a minimum value in order to avoid pump damage such as cavitation. If the level drops below the specified limit, the low-level switch (LSL) triggers both an alarm and a solenoid (S), which acts as a relay and turns the pump off. For the gas storage system in Fig. 31.9b, the solenoid-operated valve is normally closed. However, if the pressure of the hydrocarbon gas in the storage tank exceeds a specified limit, the high-pressure switch (PSH) activates an alarm and causes the valve to open fully, thus reducing the pressure in the tank. For interlock and other safety systems, a switch can be replaced by a transmitter if the measurement is required. Also, transmitters tend to be more reliable.

The SIS in Fig. 31.9 serves as an emergency backup system for the BPCS. The SIS automatically starts when a critical process variable exceeds specified alarm limits that define the allowable operating region. Its initiation results in a drastic action such as starting or stopping a pump or shutting down a process unit. Consequently, it is used only as a last resort to prevent injury to people or equipment.

It is very important that the SIS function independently of the BPCS; otherwise emergency protection will be unavailable during periods when the BPCS is not operating (e.g., due to a malfunction or power failure). Thus, the SIS should be physically separated from the BPCS and have its own sensors and actuators. Sometimes redundant sensors and actuators are utilized; for example, triply redundant sensors are used for critical measurements, with SIS actions based on the median of the three measurements. This strategy prevents a single sensor failure from crippling SIS operation. The SIS also has a separate set of alarms so that the operator can be notified when the SIS initiates an action (e.g., turning on an emergency cooling pump), even if the BPCS is not operational.

31.7 Emerging Trends

Main emerging trends in process automation have to do with process integration, information integration, and engineering integration. The theme across all of these is the need for a greater degree of integration of all components of the process. A thorough discussion on the implications of these trends and their challenges on process automation can be found in this Handbook in Chap. 8.

31.8 Further Reading

- A. Cichocki, H.A. Ansari, M. Rusinkiewicz, D. Woelk: *Workflow and Process Automation: Concepts and Technology*, 1st edn. (Springer, London 1997)
- P. Cleveland: Process automation systems Control Eng. **55**(2), 65–74 (2008)

- S.-L. Jämsä-Jounela: Future trends in process automation Annu. Rev. Control, **31**(2), 211–220 (2007)
- J. Love: *Process Automation Handbook: A Guide to Theory and Practice*, 1st edn. (Springer, London 2007)

References

31.1   D.E. Seborg, T.F. Edgar, D.A. Mellichamp: *Process Dynamics and Control*, 2nd edn. (Wiley, New York 2004)

31.2   T.F. Edgar, C.L. Smith, F.G. Shinskey, G.W. Gassman, P.J. Schafbuch, T.J. McAvoy, D.E. Seborg: Process control. In: *Perry's Chemical Engineering Handbook*, ed. by R.H. Perry, D.W. Green, J.O. Maloney (McGraw-Hill, New York 2008)

31.3   C.R. Cutler, B.L. Ramaker: Dynamic matrix control – a computer control algorithm, Proc. Jt. Auto. Control Conf., paper WP5-B (San Francisco 1980)

31.4   J. Richalet, A. Rault, J.L. Testud, J. Papon: Model predictive heuristic control: applications to industrial processes, Automatica **14**, 413–428 (1978)

31.5   S.J. Qin, T.A. Badgwell: A survey of industrial model predictive control technology, Control Eng. Pract. **11**, 733–764 (2003)

31.6   D. Bonvin: Optimal operation of batch reactors – a personal view, J. Process Control **8**, 355–368 (1998)

31.7   T.G. Fisher: *Batch Control Systems: Design, Application, and Implementation* (ISA, Research Triangle Park 1990)

31.8   J. Parshall, L. Lamb: *Applying S88: Batch Control from a User's Perspective* (ISA, Research Triangle Park 2000)

31.9   T.F. Edgar, S.W. Butler, W.J. Campbell, C. Pfeiffer, C. Bode, S.B. Hwang, K.S. Balakrishnan, J. Hahn: Automatic control in microelectronics manufacturing: practices, challenges and possibilities, Automatica **36**, 1567–1603 (2000)

31.10   J. Moyne, E. del Castillo, A.M. Hurwitz (Eds.): *Run to Run Control in Semiconductor Manufacturing* (CRC, Boca Raton 2001)

31.11   AIChE Center for Chemical Process Safety: *Guidelines for Safe Automation of Chemical Processes* (AIChE, New York 1993)

32. Product Automation

Friedrich Pinnekamp

The combined effects of rapidly growing computational power, and the shrinking of the associated hardware in recent decades, mean that almost all products used in industry have acquired some form of *intelligence*, and can perform at least part of their functions automatically.

The influence of this development on global society is breathtaking. Today, only 50 years after the first indication of automation, the life of individuals and the way industries work has been transformed fundamentally.

The automation of a product requires the ability to achieve unsupervised interaction between the device's various sensors and actuators, and ultimately the ability to communicate and interact with other units.

This chapter gives an overview of the requirements to be fulfilled in the automation of products, and gives a flavor of today's state of the art by presenting typical examples of automated products from a wide range of industrial applications. These examples cover automation in instrumentation, motors, circuit breakers, drives, robots, and embedded systems.

32.1 Historical Background

Some 50 years ago the term *product automation* was not even known. The door opener to the automation of individual devices, components or *products* was a tiny electronic circuit, commercially introduced by Fairchild Semiconductors and Texas Instruments in 1961: the microprocessor.

Already 5 years later the trend towards progressive miniaturization, known as Moore's law, was initiated and is still ongoing.

In the 1970s the development of the microprocessors geared up and led to the RISC (reduced instruction set computer) processors and very-large-scale integration (VLSI) in the 1980s. In the 1980s the step from 32 to 64 bit processors was taken.

In 1992, DEC introduced the Alpha 21064 at a speed of 200 MHz. The superscalar, superpipelined 64 bit processor design was pure RISC, but it outperformed the other chips and was referred to by DEC as the world's fastest processor.

Today in 2007 the flagships of microprocessors are built in 65 nm technology, have about 170 million transistors, and tact frequencies of 2000 MHz.

With such a powerful *brain* embedded in a device, automation has almost unlimited potential.

The enormous impact of this revolutionary technical development on society is obvious. It changed the daily life of almost all people in the world. The personal computer (PC) is based on microprocessors, transforming the way of working in offices and factories fundamentally. Microprocessors *crept* into almost all devices and made the use of machines much more convenient (at least after all the manuals themselves have developed into embedded systems).

Production of goods boosted to new productivity levels with microprocessors and there is hardly a single niche in our life that is not covered by automation devices today.

32.2 Definition of Product Automation

Product automation is a notation that can easily be confused with *production automation* or *automation products*, all names used in industry. To add to the confusion, the individual terms *product* and *automation* have a wide range of meanings themselves. Before we discuss *product automation* it seems to be appropriate to clearly describe what we understand by this.

Production automation is the automation of individual steps or the whole chain of steps necessary to produce. The otherwise manual part of the *manufacturing* is therefore carried out or supported by tools, machines or other devices.

Industries that produce such tools or machines call their devices *products*. Products used to automate production can be called *automation products*. Examples are numerical controlled machines, robots or sorting devices. These automation products serve individual steps or support the infrastructure of an automated production line.

To be able to do this, these products or devices must possess a certain degree of automation themselves. A motor that drives the arm of a robot, for example, must be able to receive signals for its operation and must have some mechanism to start its operation on request. Thus a motor, to use this example, must itself be automated.

When we talk about *product automation* we have in mind the automation of devices that fulfill various tasks in industry, not necessarily only tasks in production processes. A device with automation capabilities can also be used as a stand-alone unit to serve individual functions.

Thus *product automation* is the attempt to equip products with functionality so that they can fulfill their tasks fully or partly in an automated way.

In this chapter we want to describe the state of the art and the trends in automating products.

The functions required to transform a simple tool, say a hammer, into an automated product – in this case it may be a robot executing the same movement with the hammer as we would do with our arm – are very different, both in nature and complexity.

From the large variety of combinations of these functions we select a dozen typical applications or product examples to give a feeling of the status of implementation of automation on the product level.

Further examples can be found in [32.1]

32.3 The Functions of Product Automation

The hammer mentioned above is a good example of a whole class of *products* (or devices) with the task of providing a mechanical impulse to another object (in most cases a nail). This hammer is useless if no one is taking it and using it as a tool.

To make a simple hammer an automated hammer, we need an additional system that provides at least the functions a person would apply to the hammer in order to make it useful.

We have to give the hammer a target (hit that specific nail), we need a force (an actuator) to move the hammer, and we have to control the movement in various aspects: acceleration, direction, speed, angle of impact, precision of the path. We have to inform the

Fig. 32.1 The functional blocks required for automating a device

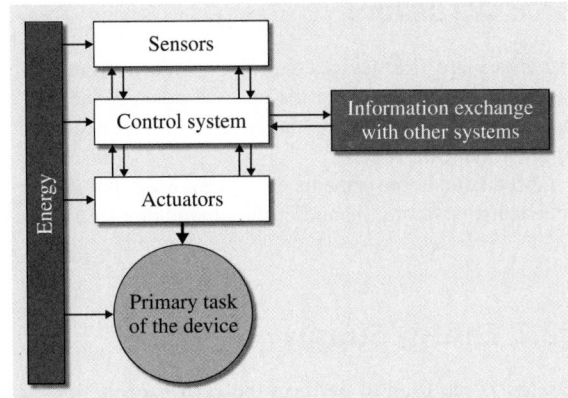

control system (the brain) that the task is executed and we have to put the hammer back into a default position after the hit. In addition, we need an energy source to provide the power for the operation and we need some sensors (eyes, ears, fingers) to inform the control system about the status of the hammer and its position during the action.

In a more schematic way the requirements for automation look like the following.

32.4 Sensors

To automate a device, sensors are required to inspect the environment and provide information about the subsequent reaction of the device.

It depends very much on the task to be carried out with the *product* which type of sensor is adequate (see Chap. 20 in this handbook and [32.2, 3]).

Table 32.1 gives an overview on the physical parameters to be measured by the sensor when specific properties are applicable.

Table 32.1 Physical parameters to be measured by a sensor

Mechanical properties	Distance, speed, acceleration, position, angle, mass flow, level, tension, movement, vibration
Thermodynamic properties	Temperature, pressure, composition, density, energy content
Electrical properties	Voltage, current, phase, frequency, phase angle, conductivity
Magnetic properties	Magnetic field
Electromagnetic properties	Radiation intensity, light fluctuations, parameters of light propagation
Other properties	Radioactivity

32.5 Control Systems

The signals from the sensors, in either analog or digital form, have to be interpreted and compared with a model of the device task in order to initiate the actuators. The control system therefore contains all algorithms to provide a proper operation of the device.

Simple logical controllers are used for on–off functions. Sequence controllers are used for time-dependent operations of the devices (Chap. 9 on control theory and [32.4–6]).

Proportional–integral–derivative (PID) controllers are widely used for parameter control, and the even more sophisticated model-based control systems see use in device control. For example, the starter of a motor to close a valve may need to adjust its current ramp-up curve according to the behavior of the overall system in which this valve and its automated motor are embedded.

32.6 Actuators

Actuators are the devices through which a controller acts on the system. As in the case of sensors, the specific type of the actuator depends very much on the application [32.7, 8].

Mechanical movements are introduced by spring mechanisms, hydraulic and/or pneumatic devices, magnetic forces, valves or thermal energy. Changes of thermodynamic properties are introduced through heating or cooling or pressure variations, for example.

Electrical properties are modified by charging or discharging capacitors or application of voltage and current, just to mention a few.

32.7 Energy Supply

Energy is required to perform the primary function of the device. It can be supplied in various forms such as mechanical, electrical or thermal.

Energy is also required to operate the sensors, actuators, and control systems as well as the communication channels within or to the outside of the device.

Energy (in most cases electrical energy) can be supplied from remote sources, either via cables or without a wired connection (for example, transmitted through electromagnetic fields), or taken from device internal storage systems, mostly batteries (in rare cases also fuel cells).

In the long term, the energy supply for systems that must operate in remote locations for long periods without intervention will develop into a design issue – for example, through suitable storage mechanisms.

32.8 Information Exchange with Other Systems

To operate in an automatic way, a device often has to communicate with its environment. For stand-alone devices, this information exchange is provided through sensors that observe the external parameters. In the majority of cases, however, the automated device must communicate with peer devices or with a superordinate control system.

Basic information for the function of the device has to be transferred and, if necessary, updated. Such updates can affect anything from operating data to the master program for the operation.

It may be necessary for an operator to communicate with the device, thus a man–machine interface is required.

Other devices or higher control systems may require information exchange in both directions, to activate the device, get a status report or synchronize with other devices in a larger system.

A language for the communication has to be defined and here several standards have been developed over the years as automation was spreading across industry.

32.9 Elements for Product Automation

The overview above covers the broad spectrum of aspects and applications in product automation. In the following, some prominent examples of state-of-the-art implementation of the related technologies are given.

32.9.1 Sensors and Instrumentation

Instrumentation is a crucial element in product automation. The information required to effectively automate a product has to be gathered in an adequate way. The state of the art in sensor technology shall be demonstrated with a few examples only, characterizing the level of technology required in modern sensors.

Figure 32.1 shows both a pressure and a temperature sensor. In both cases the probe itself is the crucial but is by far a minor part of the whole system. Most of the technology is located in the electronic part that serves as data evaluation and signal transmission.

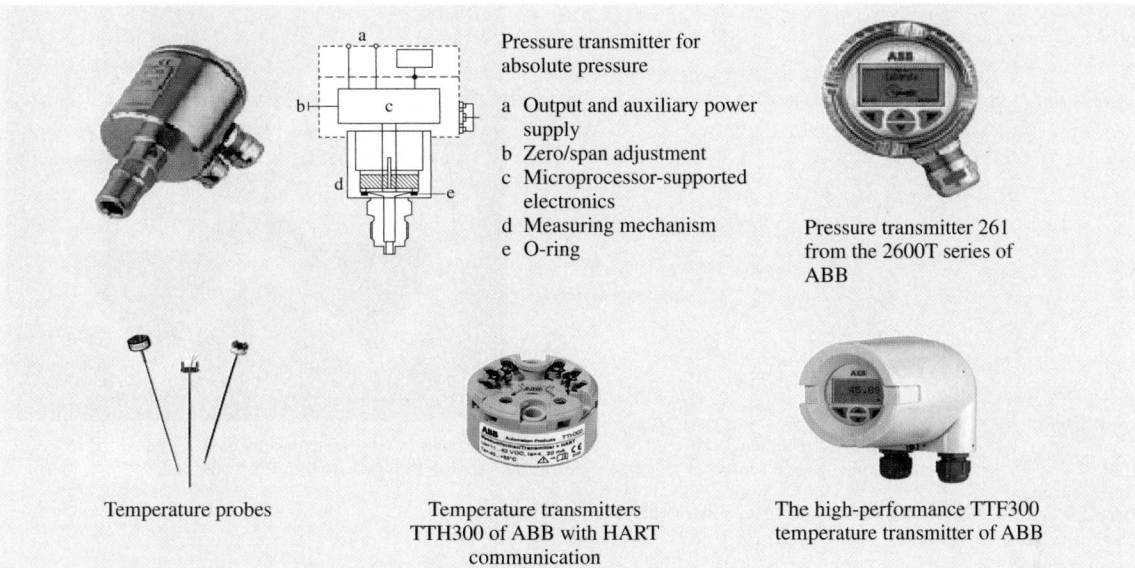

Fig. 32.1 Examples of pressure and temperature sensors (after [32.9, 10])

Signal transmission takes place via a communication protocol.

A further critical element is the human–machine interface, which is becoming more intuitive in its use. The user is provided only a few *simple* buttons to check or modify the settings of the sensors.

A high degree of automation is put into the peripheral aspects of the sensors after the primary measurement system of the physical parameters have been developed to perfection.

Other sensors are built on new sensing technology, perfecting physical effects such as the influence of a magnetic field on the polarization of light waves propagating in this field. Figure 32.3 shows such a sensor for measuring the direct-current (DC) in a foundry.

The sensor as such consists of an optical fiber wound around the current-carrying bar. The magnetic field generated by the current influences the light propagation in the fiber, from which the amplitude of the current can be derived.

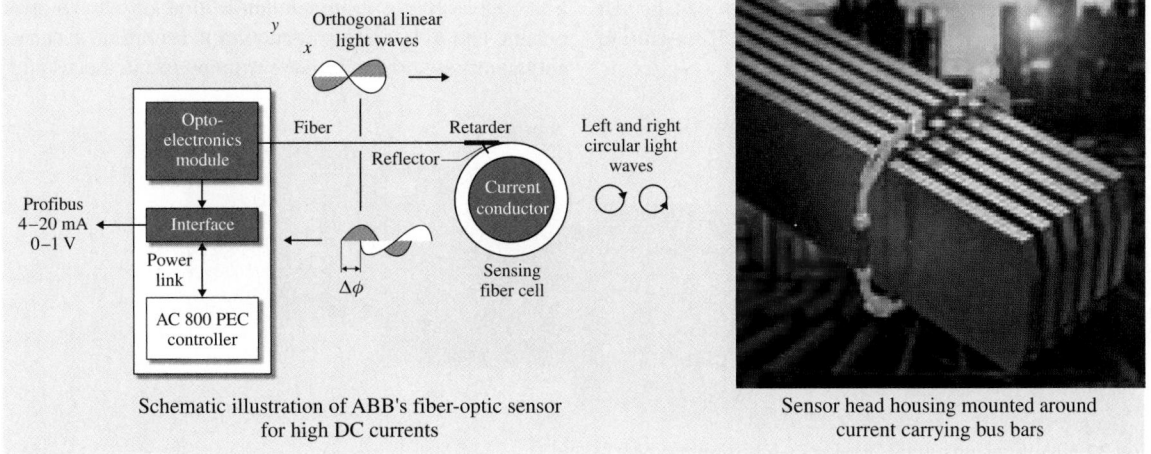

Fig. 32.3 Fiber-optic current sensor (after [32.11])

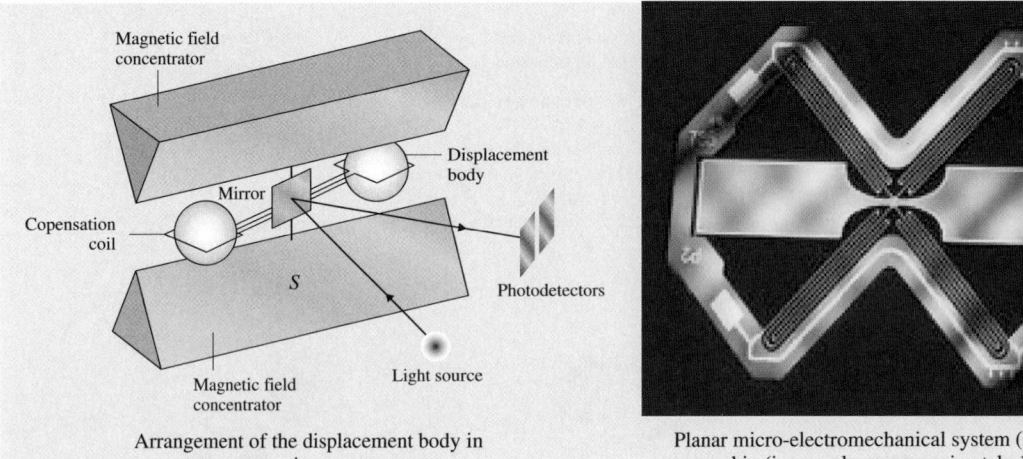

| Arrangement of the displacement body in a paramagnetic oxygen sensor | Planar micro-electromechanical system (MEMS) sensor chip (inner volume approximately 100 mm³) |

Fig. 32.4 Micro-electromechanical oxygen sensor (after [32.12])

The electronic processing of the transformed light signal occupies the central part of the sensor.

Making use of micro-electromechanical systems (MEMS) technology, the sensors themselves become more accurate and easier to manufacture [32.13, 14].

An example of this development towards silicon-based systems is the oxygen sensor head shown in Fig. 32.4.

Compared with other gases, oxygen shows a high magnetic susceptibility, and this is used to detect it in a gas flow. The sensor consists of a displacement body located in a strong magnetic field gradient. The torque of this body is a measure of the concentration of the magnetic oxygen gas. With silicon manufacturing technology these delicate mechanical devices can be cut out on chips and integrated into the overall measuring system.

32.9.2 Circuit Breakers

Circuit breakers are devices that connect or disconnect electrical energy to a user and, in addition, interrupt a short-circuit current in case of emergency (Fig. 32.5) [32.15].

While the primary mechanism for this function is the mechanical movement of metal contacts, modern versions of this product show a very high degree of automation.

State-of-the-art circuit breakers therefore have a quick and user-friendly way of setting trip parameters – preferably a method that can be instantiated offline and before installation of the circuit breaker. In addition, they offer a complete indication of why a trip occurs, and a data-logger function to record all electrical quantities surrounding the tripping event. Accessing

Fig. 32.5 Automated circuit breakers for different power levels

this information quickly and from anywhere, without the need to plug a direct physical connections between the trip unit and the PC or personal digital assistant (PDA) are also features of modern products.

An important element in any circuit breaker is a current sensor. In low-voltage circuit breakers, which are fitted with electronic trip units, these sensors are not only used for current measurement, but they must also provide sufficient energy to power the electronics.

A commonly used sensor is a Rogowsky coil that provides a signal proportional to the derivative of the current – this signal needs to be integrated. This is done digitally with a powerful digital signal processor (DSP), which is part of the overall multiprocessor architecture, and essentially the heart of the trip unit. In fact, this DSP is used to carry out other functions, for example, communications, that in previous designs required separate hardware components.

The elimination of these hardware components combined with a simplified trip unit input stage means that a single printed circuit board (PCB) is all that is required for the unit's electronics. This is a vast improvement on previous designs where four PCBs were needed to provide the same functionality.

The circuit breaker shown in Fig. 32.6 also has an integrated human–machine interface (HMI) [32.15]. A high-definition, low-power-consuming graphical display makes data easier to read. And, because of an energy-storing capacitor, a description of the alarms can be displayed for up to 48 h without the need for an auxiliary power supply. Nevertheless, these alarm descriptions are saved and can be viewed long after this 48 h period has elapsed by simply powering the trip unit.

A wireless link, based on Bluetooth technology, connects the trip unit to a portable PC, PDA or laptop. This enables users to operate in a desktop environment familiar to them. From this environment operators can

use electric network dimensioning support programs to ensure optimal adjustment of the protection functions. In addition, users can print reports, save data on different media, or send data by e-Mail to other parties from the comfort of their own desk.

The use of fieldbus plug devices allows a choice of fieldbus (e.g., DeviceNet or Profibus DP) that best suits the user's specific needs when connecting the circuit breakers to an overall system.

32.9.3 Motors

Motors are the most common devices in production automation, being part of production lines or of robots [32.16]. The need to automate the motors themselves is therefore very high, but also motors in stand-alone application have automation features built in for various reasons.

To avoid negative influence on the electrical supply system when motors are switched on, the most common automation device is the *soft starter*: when a certain motor power output is required, the optimal solution (depending on the power network conditions and requirements) becomes a frequency converter start, also called a soft start. This allows the motor to be started at high torque without causing any voltage drop on the power network.

The converter brings the motor up to speed. Upon reaching nominal speed and after being synchronized to the network, the circuit breaker between the converter and the power network is opened. The breaker between the motor and the network is then closed. Finally, the breaker between the motor and the converter is opened.

The control systems to manage the motor come in two basic varieties. Closed-loop control systems have encoders in the motor to report its status. This is used as feedback information for the control algorithm. Open-loop systems are simpler because these encoders are omitted, but at the price of a lower control accuracy. With model-based control, however, the accuracy of a closed-loop is achieved without encoders. ABB's direct torque control technology is an example for this modern approach: it uses mathematical functions to predict the motor status. The accuracy and repeatability delivered is comparable to closed-loop systems, but with the added bonus of a higher responsiveness (up to ten times as fast).

Direct torque control (DTC) is a control method that gives electronic variable-speed motor controllers [alternating-current (AC) drives] an excellent torque response time. For AC induction machines, it deliv-

Fig. 32.6 HMI of a circuit breaker

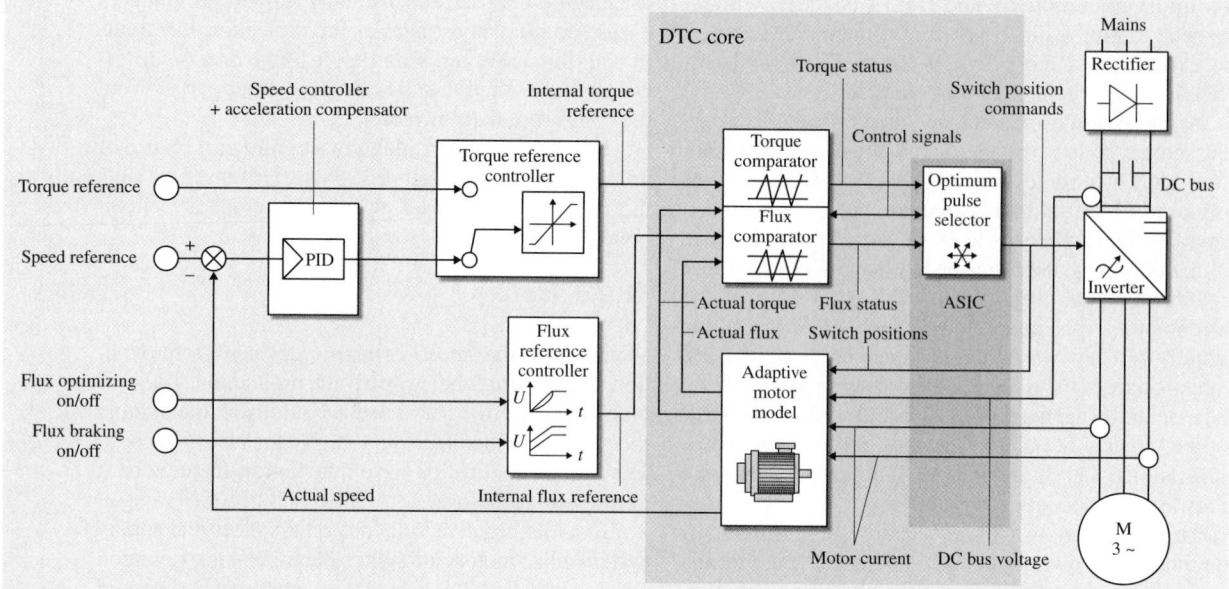

Fig. 32.7 Block diagram of DTC (after [32.17])

ers levels of performance and responsiveness reaching the machine's theoretical limits in terms of torque and speed control (Fig. 32.7).

DTC uses a control algorithm that is implemented on a microcontroller embedded in the drive. The technology was first used commercially in 1995, and rapidly became the preferred control scheme for AC drives, especially for demanding or critical applications, where the quality of the control system could not be compromised.

32.9.4 Drives

In the above example of motor automation, soft starters for smaller motors are often integrated into the motor itself.

The control systems to manage larger motors become large in themselves – like their smaller counterparts, they are equipped with power electronic components as well as control processors and peripheral systems, but the size of these rises with their power capability.

Such units are called drives. Today drive units can power motors of up to 50 MW. The continuing decrease in size and increase in power in such units is driven by improvements in power electronics and microelectronics.

When we look at a drive device as a product in itself, the automation of this product is mainly by two growing requirements: a simplified man–machine interface and communication issues within a system of which the drive is part.

Intelligent drives are certain to benefit from the growth of Ethernet communications by becoming an integral part of control, maintenance, and monitoring systems. Decentralized control systems will be created in which multiple drives share control functions, with one taking over in the event of a fault or error in another drive. The advantage of this is that reliance on costly

Fig. 32.8 Assistant control panels allow easy programming of standard drives (after [32.18])

programmable logic controllers (PLC) is greatly reduced and automation reliability improves dramatically.

The modern drive is programmed via a control panel that is similar in look, feel, and functionality to a mobile phone. A large graphical display and soft keys make it extremely easy to navigate. This detachable, multilingual alphanumeric control panel, as shown in Fig. 32.8, allows access to various *assistants* and a built-in help function to guide the user during start-up, maintenance, and diagnostics. For example, a real-time clock assists in rapid fault diagnostics. When a fault is detected, a diagnostic assistant will suggest ways to fix the problem. Drive setup and configuration data can be copied from one motor controller to another to ensure that, in the event of a drive failure, there is no need to start the setup process from the beginning.

Today, mid-range drives can store five times the amount of information compared with typical drives from the 1980s. In addition, drives with increased processing power and memory enable configurations that are better suited to an application.

In industry today, software modifications are the most useful and cost-effective way of modifying a drive. This is because developments in software have given drives increased capability with less hardware; for example, a drive controlling a conveyor belt in a biscuit factory can be programmed to operate in many different ways, such as starting and stopping at certain intervals or advancing a certain distance. The same drive used in a ventilation system can be programmed to maintain constant air pressure in a ventilation duct.

Software developments are also leading to drives with adaptive programming. Adaptive programming enables the user to freely program the drive with a set of predefined software blocks with predefined functions.

32.9.5 Robots

Robots are the prototype of product automation as they present the impression of a self-determined machine in the most visible way.

The classical application of robots is found in car manufacturing, where hundreds of devices perform movements, grinding, welding, painting, assembly, and other operations in the flow production (see Chap. 21 and [32.19, 20]).

Other applications of robotic support are less well known but are entering the market on a broad base.

The FlexPicker from ABB (Fig. 32.9) is one such example: It is a parallel kinematics robot that offers a great combination of speed and flexibility [32.21].

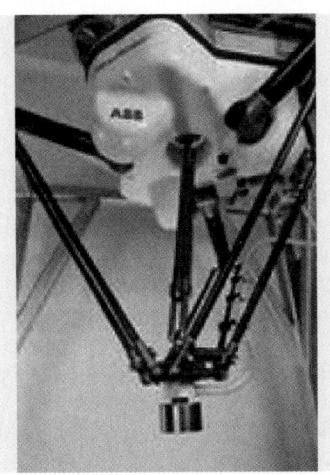

Fig. 32.9 The FlexPicker robot achieves accelerations of 10 g and handles up to 120 item/min

The gripper can pick items with picking rates exceeding 120 items per minute. The products can be picked and placed one by one. Since all the motors and gears are fixed on the base of the robot, the mass of the moving arms is limited to a few kilograms. This means that accelerations above 10 g can be achieved. Automation of this fast movement puts different requirements on the automation system of the device than, for example, the slow welding of a car roof.

While the primary automation of the robot movement is fully developed (the world's first electrically driven robot was introduced as early as 1974), the

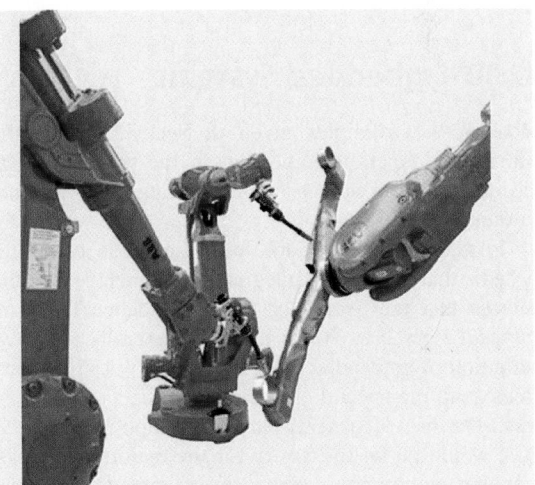

Fig. 32.10 The MultiMove function, which is embedded into the IRC5 software, allows up to four robots and their work-positioners or other devices to work in full coordination

further improvements come, such as in the drives examples, from better coordination of several robots,
a simplified man–machine interface, and easier programming methods for the robot systems.

The recently launched robot controller IRC5 from
ABB, presenting the fifth generation of robot controllers, introduces MultiMove (Fig. 32.10). MultiMove
is a function embedded into the IRC5 software that allows up to four robots and their work-positioners or
other devices to work in full coordination. This advanced functionality has been made possible by the
processing power and modularity of the IRC5 control
module, which is capable of calculating the paths of up
to 36 servo axes [32.22].

The FlexPendant (Fig. 32.11) [32.22] is a handheld
robot–user interface for use with the IRC5 controller.
It is used for manually controlling or programming
a robot, or for making modifications or changing settings during operation. The FlexPendant can jog a robot
through its program, or jog any of its axes to drive the
robot to a desired position, and save and recall learnt
positions and actions.

The ergonomically designed unit weighs less than
1.3 kg. It has a large color touchscreen, eight keys, a joystick, and an emergency stop button. The user-friendly
interface is based on Microsoft's CE.NET system. It
can display information in 12 languages of which 3
can be made active simultaneously, meaning languages
can be changed during operation. This is useful when
staff from different countries work in the same fac-

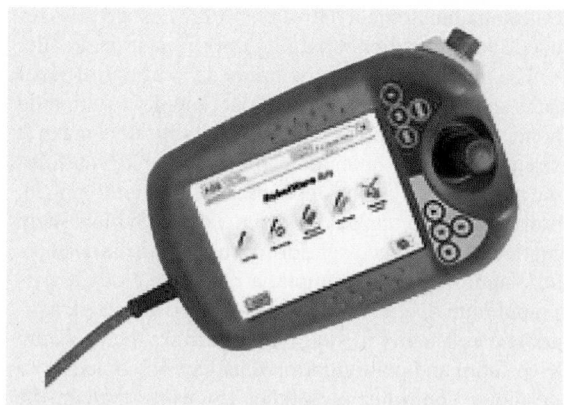

Fig. 32.11 Handheld programming

tory. The display can be flipped by 180° to make the
FlexPendant suitable for left- or right-handed operators. The options can be customized to suit the robot
application.

Programming of robots to fulfill their sometimes
quite complicated task requires a lot of time and can
in most cases be done by specialists only. However,
considerable progress has been made by virtual robot
programming that allows offline programming with
models. Recently these systems can also be directly
connected to the robots, which is especially useful to set
up a new robot installation. Background software programs and the use of templates significantly facilitate
this programming work.

32.10 Embedded Systems

Most of the examples given in Sect. 32.9 have the
automation function embedded in the product, making embedded system a common feature for product
automation.

Embedded systems are special-purpose computer
systems that are totally integrated and enclosed by the
devices that they serve or control – hence the term
embedded systems. While this is a generally accepted
definition of embedded systems, it does not give many
clues as to the special characteristics the systems possess. The use of general-purpose computers, such as
PCs, would be far too costly for the majority of products that incorporate some form of embedded system
technology. A general-purpose solution might also fail
to meet a number of functional or performance requirements such as constraints in power consumption, size
limitations, reliability or real-time performance.

Most present product automation could not have
been conceived without embedded system technology.
Examples are distributed control systems (DCS) that
can safely automate and control large and complex industrial plants, such as oil refineries, power plants, and
paper mills. In the early days of industrial automation,
relay logic was used to perform simple control functions. With the advent of integrated circuits and the first
commercial microcontrollers in the 1970s and 1980s,
programmable industrial controllers were introduced to
perform more complex control logic.

Industrial requirements vary enormously from application to application, but special industrial requirements typically include:

- Availability and reliability
- Safety

- Real-time, deterministic response
- Power consumption
- Lifetime

Automated products as well as automation and power systems must have very high availability and be extremely reliable. Economic security and personal safety depend on high-integrity systems. Embedded systems play a critical role in such mission-critical configurations.

Real time is a term often associated with embedded systems. Because these systems are used to control or monitor real-time processes, they must be able to perform certain tasks reliably within a given time. The response time associated with this *real time* varies with the application and can range from seconds to microseconds. Embedded systems must operate with as little power consumption as possible. For this reason various *power-harvesting* technologies are often included in the design.

Yet another requirement that is frequently imposed on industrial embedded systems is a long lifetime of the product itself and the lifecycle of the product family. While modern consumer electronics may be expected to last for less than 5 years, most industrial devices are expected to work in the field for 20 years or more. This imposes challenges not only on the robustness of the electronics, but also on how the product should be handled throughout its lifecycle: Hardware components, operating systems, and development tools are constantly evolving and individual products eventually become obsolete.

The key issues in developing embedded systems are complexity, connectivity, and usability. While the steadily increasing transistor density and speeds of integrated circuits offer tremendous opportunities, these improvements also present huge challenges to handle the complexity: a modern embedded system can consist of hundreds of thousands of lines of software code.

Before the widespread deployment of digital communication, most embedded systems operated in a stand-alone mode (Fig. 32.12). They may have had some capabilities for remote supervision and control, but, by and large, most functions were performed autonomously. This is changing rapidly. Embedded systems are now often part of sophisticated distributed networks. Simple sensors with basic transmitter electronics have been replaced by complex, intelligent field devices. As a consequence, individual products can no longer be automated in isolation; they must have common components. Communication has gone from being a small part of a system to being a significant function. Where serial peer-to-peer communication was once the only way to connect a device to a control system, fieldbuses are now able to integrate large numbers of complex devices. The need to connect different applications within a system to information and services in field devices drives the introduction of standard information and communication technologies (ICT) such as Ethernet and web services.

Complex field devices are often programmable or configurable. Today's pressure transmitters can contain several hundred parameters. The interaction with a device – either from a built-in panel or from a software

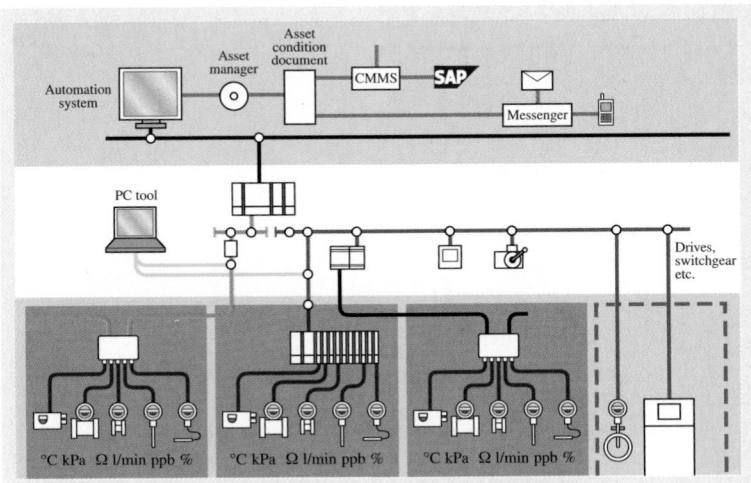

Fig. 32.12 Automated products like instruments as part of a control network (after [32.23])

application in the system – has become more complex. The task of hiding this complexity from the user through the creation of a user-friendly device has sometimes been underestimated. Most other requirements are easily quantifiable or absolute, but *usability* is somewhat harder to define.

The emergence of system-on-chip (SoC) technology has enabled extremely powerful systems to run on configurable platforms that contain all the building blocks of an embedded system: microprocessors, digital signal processors (DSP), programmable hardware logic, memory, communication processors, and display drivers, to give but a few examples. A further important aspect of the evolution of embedded systems is the trend towards networking of embedded nodes using specialized network technologies, frequently referred to as networked embedded systems (NES).

SoC can be defined as a complex integrated circuit, or integrated chipset, that combines the main functional elements or subsystems of a complete end product in a single entity (Fig. 32.13). Nowadays, the

most challenging SoC designs include at least one programmable processor, and very often a combination of at least one RISC (reduced instruction set computing) control processor and one DSP. They also include on-chip communications structures – processor bus(es), peripheral bus(es), and sometimes a high-speed system bus. A hierarchy of on-chip memory units, as well as links to off-chip memory, is especially important for SoC processors. For most signal-processing applications, some degree of hardware-based accelerating functional unit is provided, offering higher performance and lower energy consumption. For interfacing to the external world, SoC design includes a number of peripheral processing blocks consisting of analogue components as well as digital interfaces (for example, to system buses at board or backplane level). Future SoC may incorporate MEMS-based (microelectromechanical system) sensors and actuators or chemical processing (lab-on-a-chip).

Recently, the scope of SoC has broadened. From implementations using custom integrated circuits (ICs),

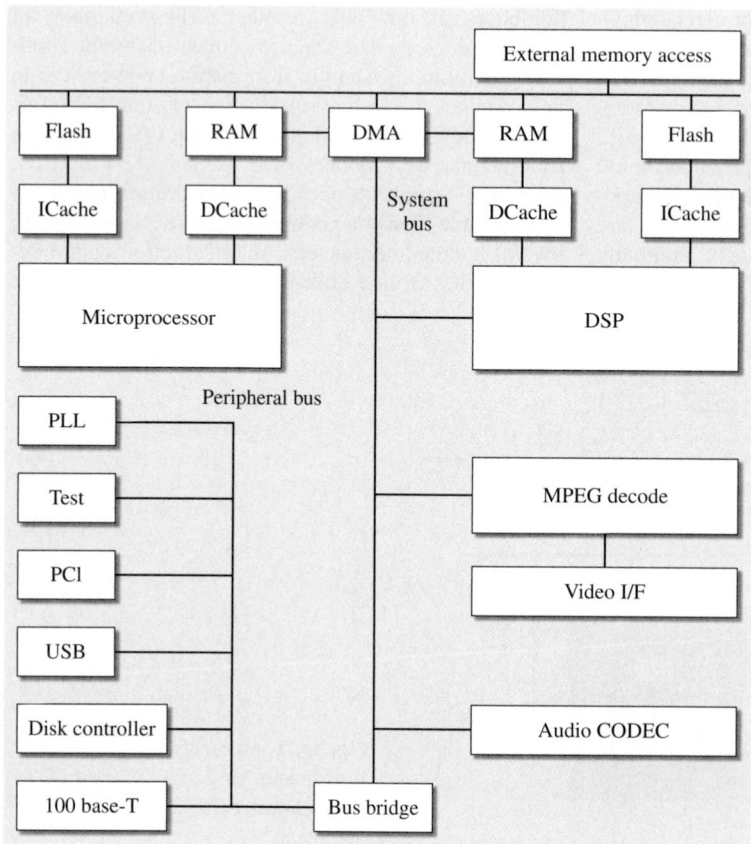

Fig. 32.13 Typical SoC system for consumer applications (after [32.24]). USB – universal serial bus, PLL – phase locked loop, PCI – peripherical component interconnect, DMA – direct memory access, I/F – interface

Fig. 32.14 Fully fledged board for the automation of a substation in an electric grid: (a) EPROM, (b) signal preprocessing FPGA, (c) device internal 100 bit/s serial communication, (d) power supply, (e) multiport Ethernet switch with optical and electrical 100 Mbit/s Ethernet media access, (f) 18–300 V binary inputs, (g) binary input processing ASIC, (h) RAM, (i) PowerPC microcontroller (after [32.25])

application-specific IC (ASIC) or application-specific standard part (ASSP), the approach now includes the design and use of complex reconfigurable logic parts with embedded processors. In addition other application-oriented blocks of intellectual property, such as processors, memories or special-purpose functions from third parties are incorporated into unique designs.

These complex field-programmable gate arrays (FPGAs) are offered by several vendors. The guiding principle behind this approach to SoC is to combine large amounts of reconfigurable logic with embedded RISC processors, in order to enable highly flexible and tailorable combinations of hardware and software processing to be applied to a design problem. Algorithms that contain significant amounts of control logic, plus large quantities of dataflow processing, can be

partitioned into the control RISC processor with reconfigurable logic for hardware acceleration.

Another important facet of the evolution of embedded systems is the emergence of distributed embedded systems, frequently termed networked embedded systems, where the word *networked* signifies the importance of the networking infrastructure and communication protocol. A networked embedded system is a collection of spatially and functionally distributed embedded nodes, interconnected by means of wired and/or wireless communication infrastructure and protocols, and interacting with the environment (via sensor/actuator elements) and with each other. Within the system, a master node can also be included to coordinate computing and communication, in order to achieve specific objectives.

Early implementation of numerical power system protection and control devices used specialized digital signal processing (DSP) units. Today's implementations are leveraging the vast computing power available in general-purpose central processing units (CPU). As such, PowerPC microcontrollers deliver high computing power at low power consumption and, therefore, low power dissipation. Random-access memory (RAM) is utilized for the program execution memory and erasable read-only memory (EPROM) stores program and configuration information. A typical configuration can include a 400 MHz PowerPC, 64 MB of EPROM, and 64 MB of RAM. The CPU can be complemented with field-programmable gate arrays (FPGAs) that integrate logic and signal preprocessing functionality. An automation device usually includes a number of printed circuit board assemblies (PCBA), accommodating requirements for the diversity and number of different input and output circuitry. High-speed serial communication is built in for intermodule communication that enables the CPU to send and acquire data from the input and output modules. Application-specific circuits are designed to optimize overall technical and economical objectives. Figure 32.14 shows a sample of a high-performance CPU module, connected to a binary input and Ethernet communication module.

32.11 Summary and Emerging Trends

Product automation is a highly dynamic and interdisciplinary field of technology.

The attempt to automate almost all individual devices from consumer products to large machines and systems is an ongoing trend.

It is driven by the advances in miniaturization, cost, and performance of electronic components and the standardization of cross-device functions such as communication.

The broad application range of product automation requires both general understanding of the development of electronics and software as well as specific knowledge of the tasks of the products to be automated.

With a higher degree of standardization and modularization of embedded systems the implementation of product automation will continue with high speed.

To facilitate an efficient implementation of embedded systems as the heart of product automation future research will focus on three areas:

- Reference designs and architecture
- Seamless connectivity and middleware
- System design methods and tools

The target will be a generic platform of abstract components with high reusability for the applications. This platform shall facilitate a standardized interface to the environment and allow for the addition of application-specific modules. An overriding feature here is the lowest possible power consumption of the embedded systems.

Further steps will deal with the self-configuration and self-organization of components that form the product automation system. There is a clear trend towards ubiquitous connectivity schemes and networks for those systems.

Last not least, design methods and tools must be further developed addressing the various levels of the complex systems. This set of methods will include open interface standards, automatic validation and testing as well as simulation. We will also see progress in the rapid design and prototyping of complex systems.

References

32.1 V.L. Trevathan (Ed.): *A Guide to the Automation Body of Knowledge*, 2nd edn. (ISA, Durham 2006)

32.2 J. Fraden: *Handbook of Modern Sensors: Physics, Designs, and Applications* (Springer, New York 2003)

32.3 J.S. Wilson: *Sensor Technology Handbook* (Elsevier, Amsterdam 2005)

32.4 G.K. McMillan, D.M. Considine: *Process/Instruments and Controls Handbook* (McGraw Hill, New York 1999)

32.5 N.S. Nise: *Control Systems Engineering* (Wiley, New York 2007)

32.6 R.C. Dorf, R.H. Bishop: *Modern Control Systems* (Prentice Hall, Upper Saddle River 2007)

32.7 H. Janocha: *Actuators: Basics and Applications* (Springer, New York 2004)

32.8 B. Nesbitt: *Handbook of Valves and Actuators* (Butterworth–Heinemann, Oxford 2007)

32.9 R. Huck: Feeling the heat, ABB Rev. Spec. Rep. Instrum. Anal. (2006) pp. 17–19

32.10 W. Scholz: Performing under pressure, ABB Rev. Spec. Rep. Instrum. Anal. (2006) pp. 14–16

32.11 K. Bohnert, P. Guggenbach: A revolution in DC high current measurement, ABB Rev. **1**, 6–10 (2005)

32.12 P. Krippner, B. Andres, P. Szasz, T. Bauer, M. Wetzko: Microsystems at work, ABB Rev. **4**, 68–73 (2006)

32.13 C. Liu: *Foundations of MEMS* (Prentice Hall, Upper Saddle River 2005)

32.14 J.W. Gardner, V. Varadan, O.O. Awadelkarim: *Microsensors, MEMS and Smart Devices* (Wiley, New York 2001)

32.15 F. Viaro: Breaking news, ABB Rev. **3**, 27–31 (2004)

32.16 A. Hughes: *Electric Motors and Drives* (Elsevier, Amsterdam 2005)

32.17 I. Ruohonen: Drivers of change, ABB Rev. **2**, 23–25 (2006)

32.18 M. Paakkonen: Simplicity at your fingertips, ABB Rev. Spec. Rep. Motors Drives (2004) pp. 55–57

32.19 S.Y. Nof: *Handbook of Industrial Robotics* (Wiley, New York 1999)

32.20 F.L. Lewis, D.M. Dawson, C.T. Abdallah: *Robot Manipulator Control: Theory and Practice* (Marcel Dekker, New York 2004)

32.21 H.J. Andersson: Picking pizza picker, ABB Rev. Spec. Rep. Robot. (2005) pp. 31–34

32.22 C. Bredin: Team-mates, ABB Rev. Spec. Rep. Robot. (2005) pp. 53–56

32.23 S. Keeping: The future of instrumentation, ABB Rev. Spec. Rep. Instrum. Anal. (2006) pp. 40–43

32.24 G. Martin, R. Zurawski: Trends in embedded systems, ABB Rev. **2**, 9–13 (2006)

32.25 K. Scherrer: Embedded power protection, ABB Rev. **2**, 18–22 (2006)

33. Service Automation

Friedrich Pinnekamp

A fast and effective industrial service, supporting plants with a broad spectrum of assistance from preventive maintenance to emergency repair, rests on two legs: the physical transport of people and equipment, and the provision of the vast variety of information required by service personnel. While the automation of physical movement is limited, data management for efficient servicing, including optimized logistics for transport, is increasingly expanding throughout the service industry.

This chapter discusses the basic requirements for the automation of service and gives examples of how the challenging issues involved can be solved.

Part D | 33

33.1 Definition of Service Automation

To be able to describe the automation of service, we must first define what we mean by *service*. While it is obvious that in the context of this Handbook we are not talking about public religious worship or the act of putting a ball into play in a tennis match, the wide application of the word *service* still requires a stricter definition.

Service in general is an act of helpful activity, but in the context of this chapter we shall restrict our-selves to considering the provision of the activities that industry requires. In particular, we want to address organized systems of apparatus, appliances, employees, etc. for providing supportive activities to industrial operations.

In this narrower sense we will discuss how industrial service is automated. Other types of service automation are discussed in more detail in Chaps. 62, 65, 66, 68 and 71–74 of this Handbook.

33.2 Life Cycle of a Plant

Every physical object in use shows signs of wear once in operation. This is true for cars, airplanes, ships, computers, and combining many individual devices, for plants.

Operators of an industrial plant are well aware of this *natural* behavior of equipment and react accordingly to avoid deterioration of plant performance.

The value of the installation is decreasing, the reason for the depreciation, and there are different strategies for countermeasures.

Figure 33.1 shows an overview of the main strategies to extend the useful life cycle of an industrial plant.

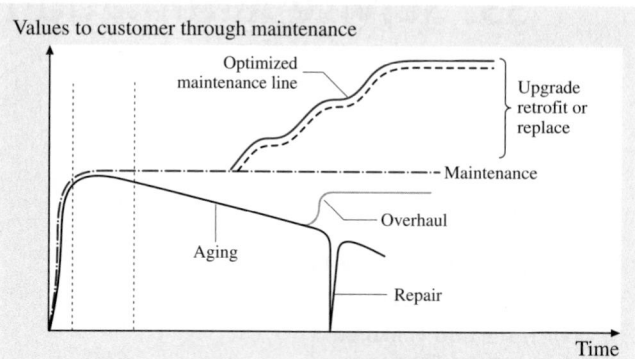

Fig. 33.1 Various approaches to maintenance during the lifecycle of an industrial plant [33.5]

Continuous maintenance is a way to keep performance at a high level, while overhaul is done at regular intervals and repair in case of need (see Chap. 42 on *Reliability, Maintenance and Safety*, and [33.1–4]). When retrofit and replacement is combined with a performance upgrade, the value of the plant can even be increased.

Performance of maintenance, overhaul, repair or upgrade has an associated cost and which strategy is most beneficial depends very much on the plant concerned. A car manufacturer with a robotized workflow and high production rate cannot afford the outage of a line due to a malfunctioning robot, and the cost of a standing line can easily be calculated. However, a factory in batch operation with long delays in production may be less sensitive to an outage.

In any case the cost of service shall be as low as possible to provide a positive balance to the economic equation. Cost of service has three main elements: the parts that have to be replaced, the man-hours for performing the service, and last but not least, the cost of interrupted production, the latter being proportional to time of outage.

Thus time and cost efficiency are the critical factors of any service and are the main driving forces to automate the task to the highest possible extent. In this sense service is following the trend of implementing information technology (IT) to speed up and rationalize its performance, as is done in office work or production.

33.3 Key Tasks and Features of Industrial Service

Before we discuss the automation of industrial services, we must describe the aspects and tasks of industrial service in order to understand the different approaches to automation in this area. Service comes onto the stage after an industrial plant or subsystems of plants or even individual products in an industrial operation have been installed and placed into operation.

For the operator of an industrial plant, it is of utmost important that the equipment is available with optimum performance at any time that it is needed. For some systems in a plant this means 24 h a day and 365 days a year, whereas for others this may apply for only a few weeks spread over the year.

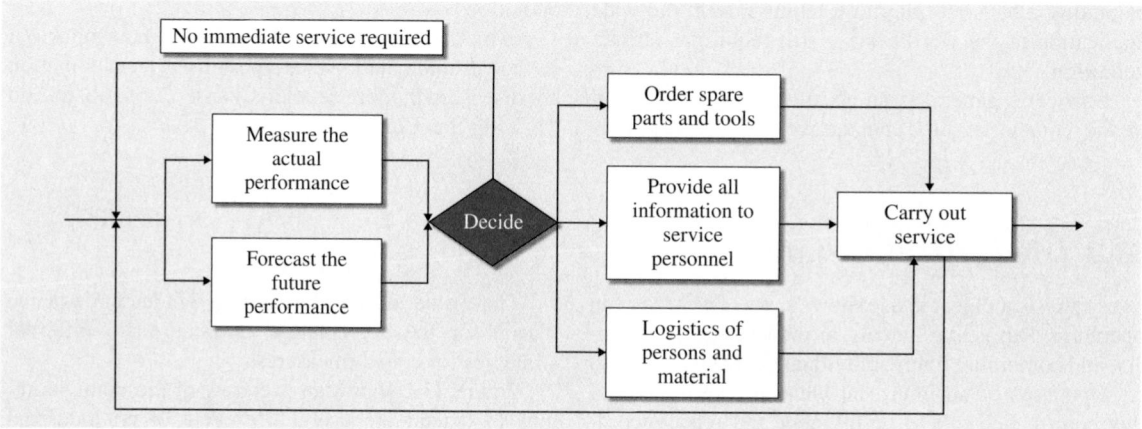

Fig. 33.2 Major steps in industrial service

Even though industrial plants prefer to have equipment that does not require any service, maintenance or repair, planned outage for overhaul is generally accepted and sometimes unavoidable. A large rotating kiln in a cement plant, for example, must usually be fitted with a new liner every year, stopping production for a short time. Mostly unacceptable, however, are unplanned outages due to malfunction of equipment.

The better prepared a supplier of this equipment is to react to these unplanned events, the more appreciated this service will be.

Taking this into account, industrial service must strive to:

- Keep equipment at maximum performance all the time
- Provide well-planned service at times of planned outage
- Restore optimum performance as quickly and effectively as possible after unplanned outages.

Figure 33.2 shows the major steps to be taken for any service action. As we can see in this figure, there is a layer of information gathering, analyzing, and exchange and a logistic element, including the physical transport. Both elements can be automated and it is ob-vious that the information-related aspect has attracted the highest degree of automation to date.

In a more detailed view, the following aspects have to be considered:

- The actual performance of the equipment must be known.
- A forecast of possible changes in the performance must be available.
- Detailed knowledge of the scheduled use of the equipment is necessary.
- Efficient, real-time communication about disturbances must exist.
- Optimum logistics must enable fast and effective service action.

In more technical terms, an optimum service is based on:

- Real-time performance monitoring of the equipment
- Knowledge-based extrapolation of future performance
- Knowledge about use of the equipment in the industrial environment

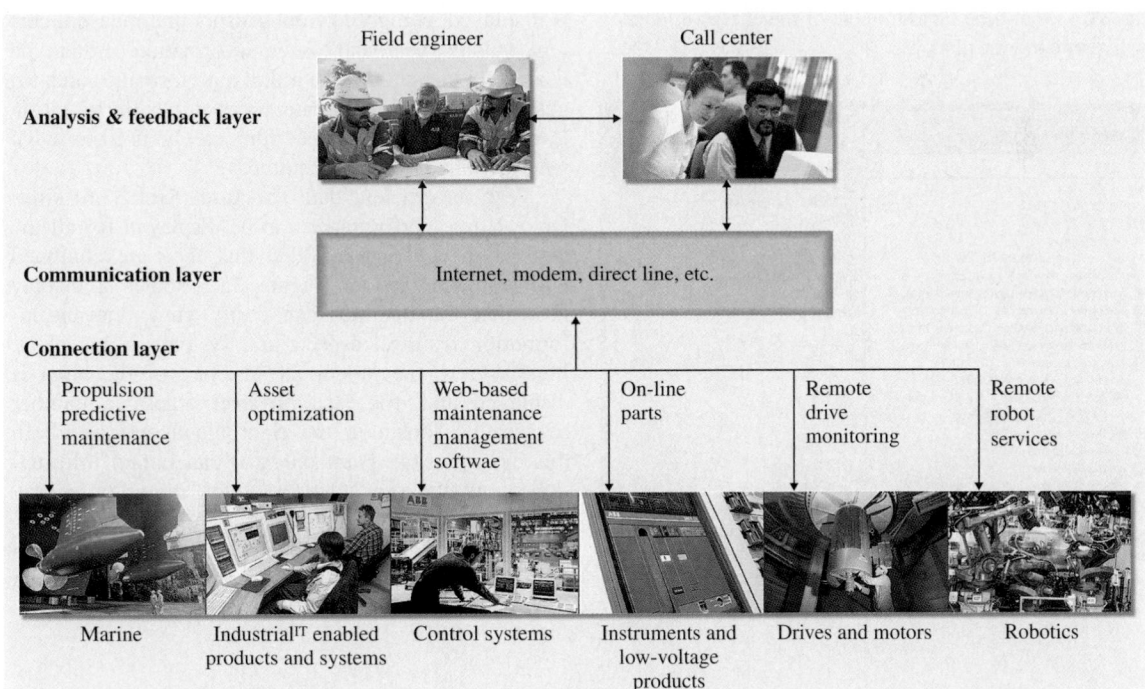

Fig. 33.3 Tasks and communication channels of a service organization [33.6]

- Adequate communication channels between equipment and service provider
- Access to relevant equipment data and metadata for the service personnel
- Access to hardware and tools to restore equipment performance efficiently.

Figure 33.3 shows the tasks and communication channels for a service organization that takes care of a variety of products. Each of these tasks can be automated.

The following sections describe in more detail the critical aspects of service and how they can be automated.

33.4 Real-Time Performance Monitoring

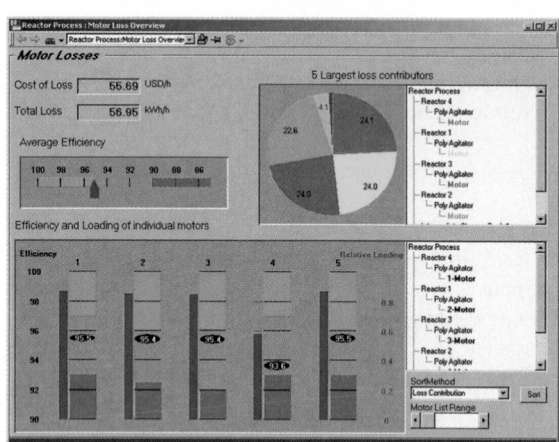

Fig. 33.4 Real-time measurement of the energy efficiency of five motors in a plant [33.7]

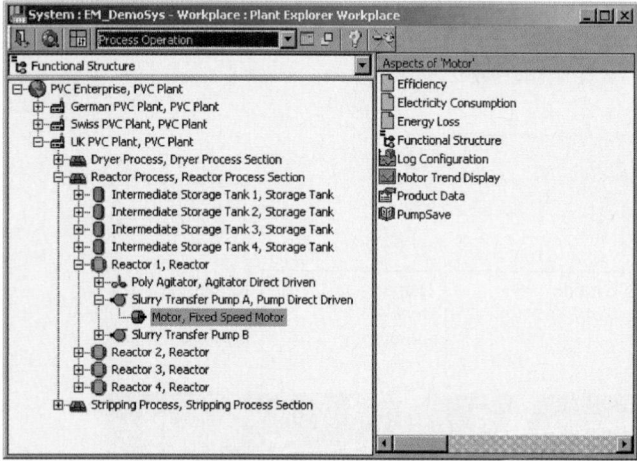

Fig. 33.5 Structured view of devices in a plant with some aspects related to them [33.7]

In order to keep a plant in optimum operation, all devices must work properly; this means that their performance must be monitored. This can be done by regular inspection by service personnel, or in an automated way, i. e., by equipping each device with adequate sensors [33.8–10].

In Chap. 20 the sensors suitable for the automation of the functions of a product are described in more detail. As an illustrative example of the automation of service, we have chosen a motor, which is a typical example in industry with motors accounting for more than 60% of the electrical energy consumed by industry.

Figure 33.4 shows a real-time measurement of five motors in a plant. Here the specific aspect of efficiency is displayed, composed from sensors that measure current, voltage, heat, and torque. Information of this type is a useful indicator of the motor's performance and, together with other motor data, for example, the vibration level and its frequency spectrum, can be used to judge whether intervention is required.

With an efficient data structure (Sect. 33.6) similar real-time performance can be displayed for all the devices in a plant, provided that they are equipped with suitable sensors. Figure 33.5 shows a display in which the operator can easily view relevant information from all aspects and systems in his plant. In this view, the functional structure of the plant is displayed and, for each physical object, a number of aspects, shown in the right menu, are listed. In this example, the plant operator can obtain information about the product data of the motor or energy losses.

All these data are valuable for the service task, based on which predictions about future performance of the parts can be made.

33.5 Analysis of Performance

An experienced plant operator or service engineer can use device performance information measured in real time to determine further actions. Modern IT systems support this task. The various data are analyzed and knowledge-based systems propose further actions, which may for example include the replacement of a component, a scheduled service within a certain interval or some other preventive intervention. When more background knowledge is considered in the programming, these systems can also include the financial and environmental consequences of a replacement, as shown in Fig. 33.6.

The example given in Fig. 33.6 also shows that information from different functions, here engineering, service, and finances, are increasingly being combined to provide a consistent view of the whole industrial operation.

This accumulated knowledge and experience helps engineers to decide what equipment needs to be inspected and when, as well as to establish where failure would be least acceptable and cause the most problems. This makes it easier to see where effort has to be focused in order to maximize return. It facilitates the optimization of examination intervals whilst at the same time identifying equipment for which noninvasive examinations would be equally effective.

More innovative automation in this direction can be expected in the future.

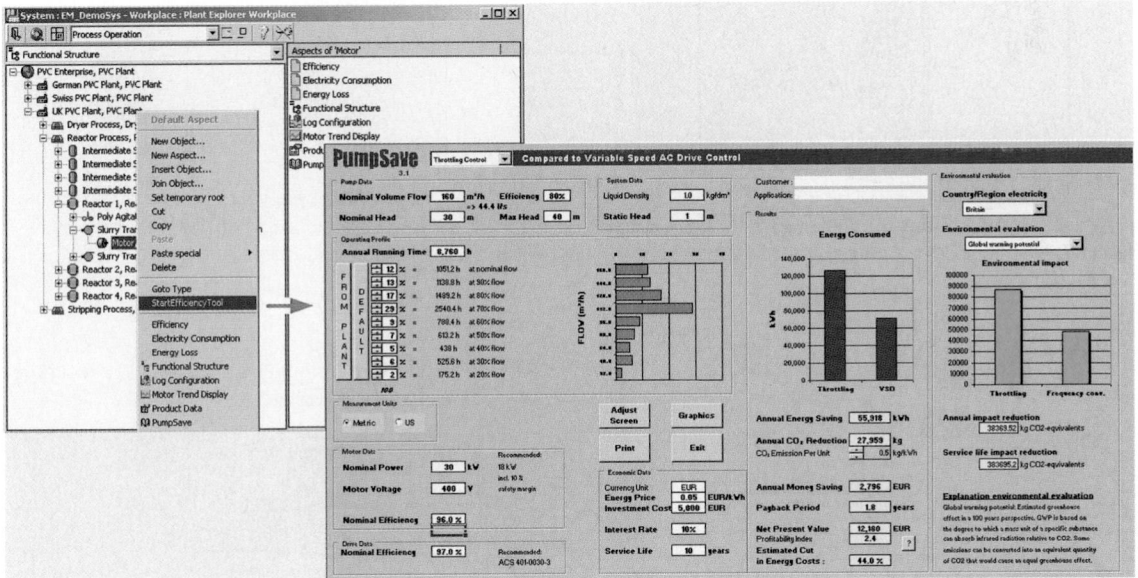

Fig. 33.6 Result of online real-time analysis of motor performance based on evaluation of various parameters [33.7]

33.6 Information Required for Effective and Efficient Service

The amount of data and background information that a service engineer requires to enable a fast reaction is enormous; the list below describes the major aspects:

- Location of the equipment and description of how to reach it, including local access procedures
- Technical data, including drawings and part lists

- Purpose of the equipment and expected performance
- Connection of equipment to other parts of the installation
- Security issues in handling the equipment
- Historical data for the equipment, including previous service actions

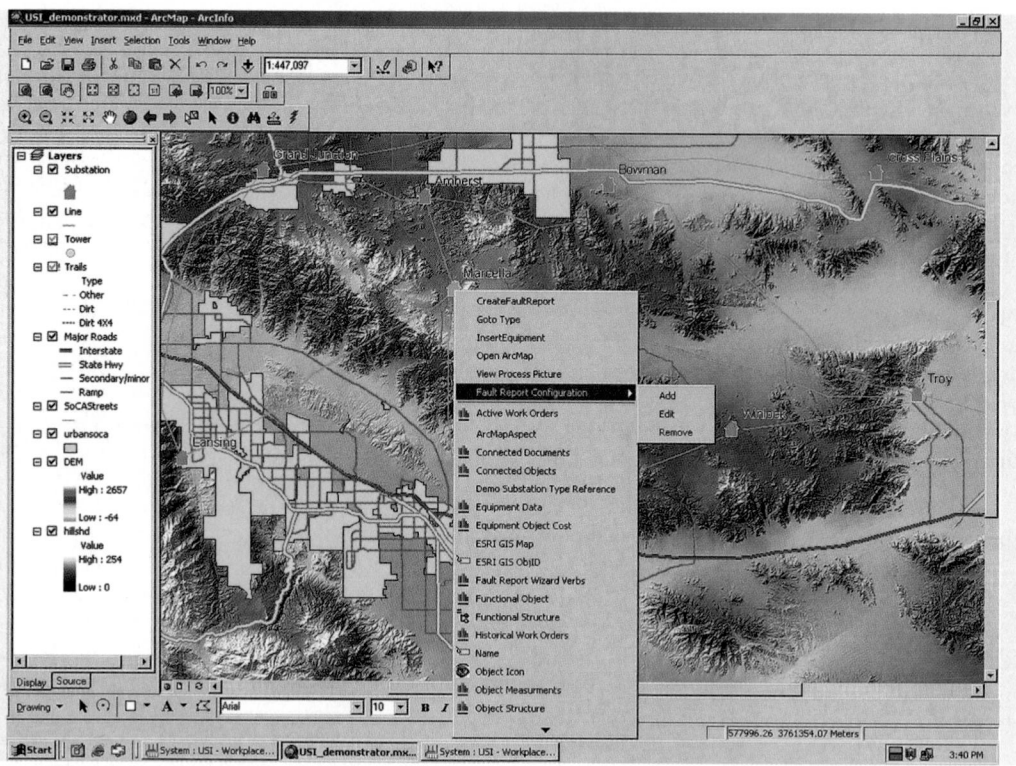

Fig. 33.7 Display of geographical information in relation to a fault in a transmission grid [33.11]

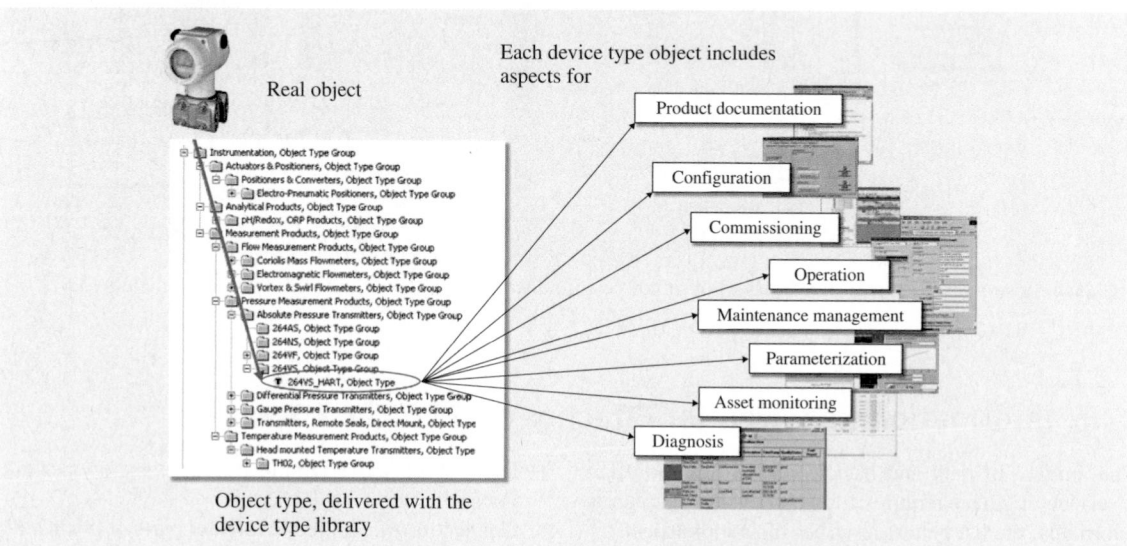

Fig. 33.8 Multidimensional structure connecting objects with their various aspects [33.12]

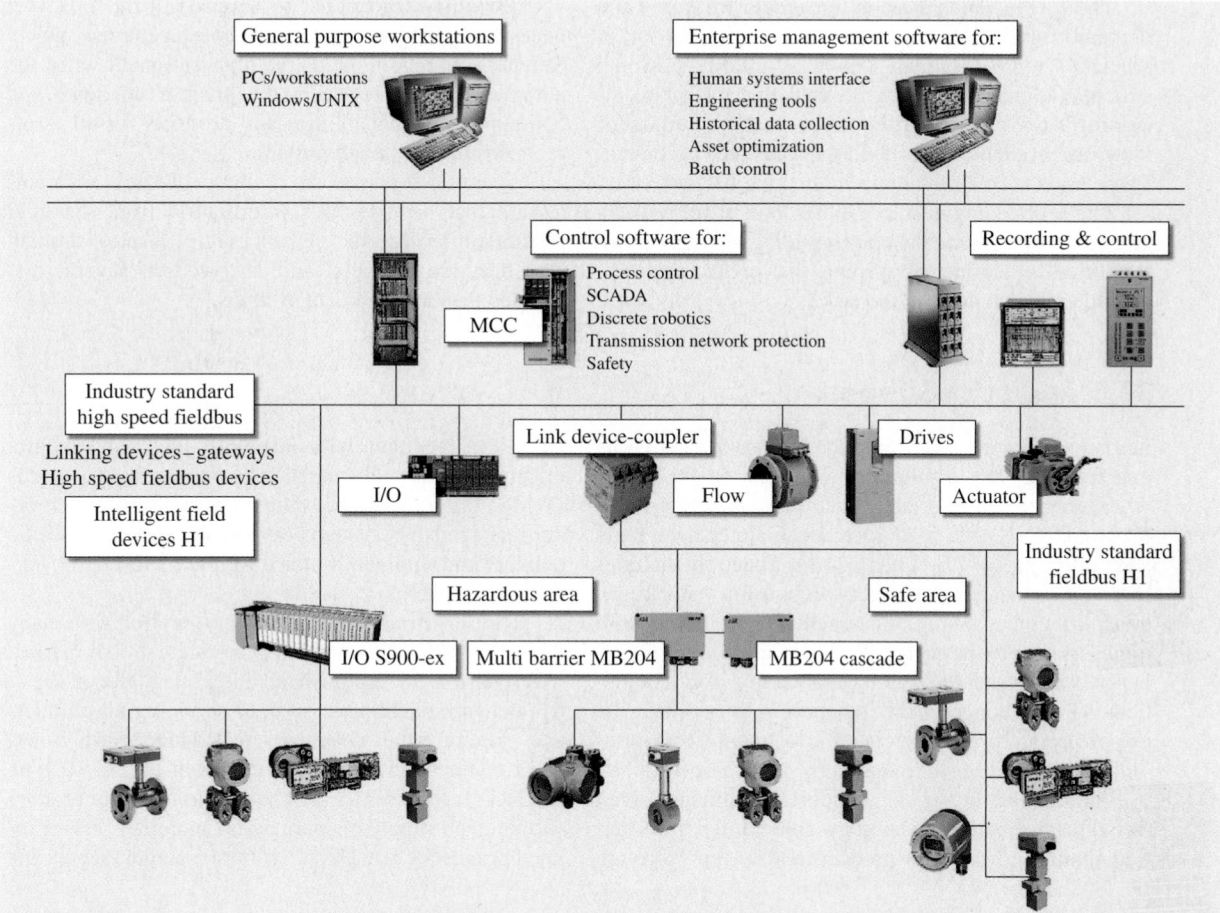

Fig. 33.9 Architecture of a plant with control and information flow down to the field device level [33.13] (MCC: motor control center; I/O: input/output; SCADA: supervisory control and data acquisition)

- Failure reports
- Proposed service actions
- Detailed advice of how to carry out these actions, including required tools
- Information about spare parts, and their availability and location
- Information about urgency and time constraints
- Data for administrative procedures.

Most of this information can be supplied automatically to service personnel if it is available in a form that allows digital data transfer.

This availability is the major bottleneck for the automation of service. Drawings of equipment may be difficult to come by and/or the original supplier of this equipment may no longer be available.

Drawings, for example, may be available but not scanned, so that only physical copies can be used by service personnel.

Gradually, suppliers of industrial equipment are transferring their data to platforms allowing various forms of access, a trend already established for many consumer goods such as cars or washing machines, for example.

Figure 33.7 shows the example of a data set used in the service of transmission systems, where a failure in the grid has to be located and identified, and the service crew has to be sent to the correct location for repair. In this multilayer view, geographical data, taken from global position system (GPS) systems are combined with detailed technical information, failure reports, and historical aspects of the devices to be serviced.

The way in which these different data are stored and prepared for fast and easy retrieval is important for an efficient service. Figure 33.8 shows a multidimensional structure, in which, for each physical object of an installation – in this example a flow meter – a large number of views are available: so called aspects of the equipment. These aspects cover the items in the long list above.

The service engineer can either look at the aspects of interest by opening the corresponding view or can be supported by automated systems that collect and prepare the relevant data for the task.

Obviously, for full use to be made of the data, it is necessary that the data are available in the first place. Service automation yields the highest benefit when the automation architecture in the plant is adequate and communication capabilities and products for all asset-related information are possible.

Figure 33.9 shows an example of such an architecture that provides this information from the field instrument level to the highest control level of a plant. The data are recorded and analyzed in several distributed data management systems.

33.7 Logistics Support

Logistics is a support function whose objective is guaranteeing the availability of required hardware when service is performed. Such hardware is primarily spare parts, but also includes maintenance parts and tools [33.14–17]. The logistics function includes inventory management and warehousing (stocking), transportation planning and execution, including both transportation to the customer site (or site where service is performed), and the return of material (reverse logistics). The challenge of this function is to optimize the use of inventory and associated investment, while providing the service level required by service operations.

Automating the logistics support function is enabled through information technology, integrating: information about spare parts for products and systems through

an online catalogue with inventory status, availability information through warehouse management systems (WMS) ordering, order status information through order managements systems (OMS) [33.18], and specific delivery and shipment status information through track-and-trace systems.

In the environment of a large corporation with many different products and systems serving a global market, it is possible to establish an integrated global logistics network enabling support of servicing all products and systems to all customers worldwide. Such a network could be organized as shown in Fig. 33.10. The product services center (PSC) is responsible for product support, and supply of spare parts (and other service related products). The DC (distribution center) represents

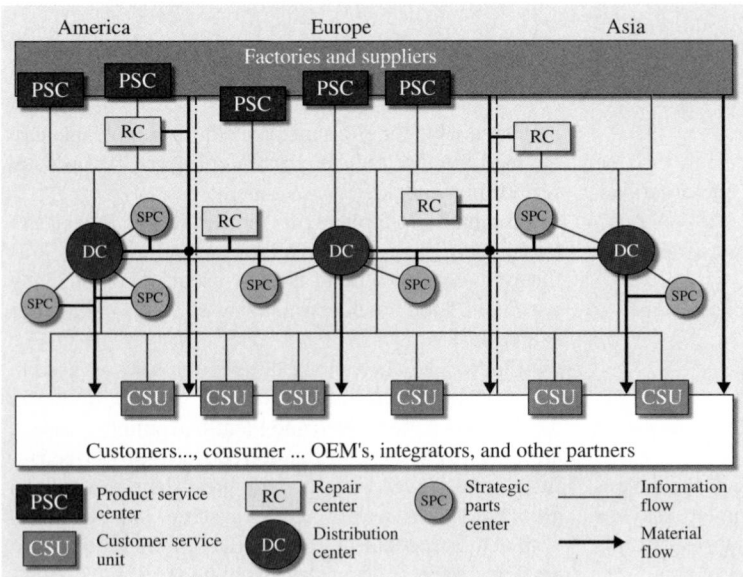

Fig. 33.10 A network to deliver fast and effective service to a widely distributed customer base. OEM, original equipment manufacturer

cost-based operations focused on distribution, providing logistics services to different units. The DCs provide all logistics services, including warehousing and shipping, and are strategically located. In an efficient network most spare part orders are directly shipped to end customers from the central DCs. The DC will handle delivery of parts in a specified time window depending on where the customer is located in relation to the DC.

Other approaches to efficient logistics may have critical parts stocked at a customer site, with all movements of parts tracked and automatic replenishment done based on parameter settings. When a part has to be removed, the integrated network will generate the replacement order automatically.

Customer support units (CSUs) are close to the sites at which service may be required. The CSU works with customers to ensure they have access to the right spare parts. Proactive support might enable customers to directly access the CSU to order parts online, for example, or ask for technical and order information.

The repair center (RC) is another important function in the service network. Customers are not only interested in replacement of faulty parts but could require repair or reconditioning as an alternative. The suppliers of this service are the RCs, providing the repair function itself, and the *reverse logistics*. Administration of product warranty often requires return of faulty parts, which is an element of the reverse logistics flow.

33.8 Remote Service

Remote service [33.19] is an umbrella term for a variety of technologies that have one concept in common: bringing the problem to the expert rather than bringing the expert to the problem.

Remote services use existing and cutting-edge technologies to support field engineers, irrespective of location, in ways only dreamt of as little as 5 years ago. The Internet, together with advances in communications and encryption techniques, has contributed enormously to this end. Remote service developments are a direct result of the changing needs of plant operators expecting more support at lower costs. Remote services are designed to maximize knowledge bases in the most cost-effective manner. The result ensures that the best knowledge is in the right place, at the right time. With a large number of different types of devices in a plant, this can be a complex undertaking.

Generally equipment (control systems, sensors, motors etc.) is accessible from a remote location primarily via the Internet or a telephone line. Several advantages are obvious:

- Support can be given in a shorter period of time.
- The right experts support the local service personnel.
- Cost may be reduced by avoiding travel.

One of many examples where remote service is introduced is robots. Robots play a crucial role in the high productivity and availability of a production line. Any problem or reduced performance of a robot has a direct negative influence on the output of the line. The operator's expectation is to avoid delays and disturbances during production.

Recently, Asea Brown Boveri (ABB) developed a communication module that can easily be plugged into the robot controller for both old and new robot generations [33.20]. This module reads the data from the controller and sends them directly to a remote service center, where the data are automatically analyzed. This is another example of the ever-growing application of machine-to-machine technology, which has now been applied to the world of robots. By accessing all relevant information on the conditions, the support expert can remotely identify the cause of a failure and provide fast support to the end user to restart the system. Many issues can hence be solved without a field intervention. In a case where a field intervention is necessary, the resolution at the site will be rapid and minimal, supported by the preceding remote diagnostics.

This automatic analysis not only provides an alert when a failure with the robot occurs but also predicts a difficulty that may present itself in the future. To achieve this, performance of the robot is regularly analyzed and the support team is automatically notified of any condition deviation.

The degree of automation of a remote service is again limited by the fact that personnel have to be present at the location where the service is required.

Even though the replacement of people by service robots is possible in principle, practical implementation of this approach is hardly feasible.

33.9 Tools for Service Personnel

The service task is, like almost all industrial operations, an administrative process with many steps that need to be organized, executed, and monitored. To support this administrative work, a number of software solutions are available on the market. These tools help to oversee and order the steps to be taken by analyzing the situation down to execution of the service on site.

Besides this administrative element of service there is another level of organization, which is more difficult to manage. When service personnel arrive at the location where the service is needed, they should already have all necessary information about the repair or maintenance. If this is not possible (because the automation level was too low) the information has to be collected on site – a costly and time-consuming process.

Devices with a data port and a self-analysis system can be connected and may help service personnel to find the optimum approach for the task. In connection to the back office, the service person is supplied with documentation such as handbooks, drawings, guidelines for maintenance etc. via various communication channels. While the availability of communication channels and portable devices to handle the data are

not a major bottleneck today, the access to all the information, preferably in digital form, is the limiting factor.

To alleviate this, increasing amounts of data related to the specific device are stored together with the hardware, for example, in form of radiofrequency identification (RFID) [33.21, 22] or built-in direct Internet connection.

Future tools could integrate technologies of *augmented reality*. These tools combine the real picture of a device with displayed information about the object inspected. Looking at a valve in a plant, for example, the system would display the data of last inspection, the present performance status, analyzed by a background system, the scheduled maintenance interval, etc. Drawings of the part can be displayed and maintenance instructions can be given.

In a number of very specific applications such as surgery, the defense industry or aircraft, for example, head-up devices are in use. Due to the need for integrated data from many different sources in the service application, it will take some time until this technology will get wider use in service.

33.10 Emerging Trends: Towards a Fully Automated Service

While industrial service has various branches, from preventive maintenance to scheduled repair, we take emergency repair as an example to paint the future of service automation.

What is currently state of the art in many consumer products and some industrial devices will spread through the whole industry in the form of products and systems with self-monitoring capability and self-diagnostic features to detect a performance problem and analyze the probable cause.

These devices will be equipped with communication functions to report their status to a higher-level data management system, which will then automatically initiate the necessary steps: further analysis of the case and comparison with previous events, thereby implementing a knowledge-based repair strategy. It will collect the necessary documents for efficient repair according to the chosen strategy and provide these to the service personnel, who will have been alerted automatically. It will provide access information to the service staff and

inform the personnel at the location about the details of the subsequent repair.

The system will furthermore automatically order the spare parts or any special tools needed for the service and initiate their delivery to the site.

Once the service personnel is on site the system will give automated advice related to the service and, if needed, connect to a back-up engineer who will have full access to all data and the actual situation on site.

In the far future robots installed or transported to the site of service may take over the manual work of service engineers: as a first step with direct remote control, but in the future as independent service robots.

While the automated functions can be carried out with almost no time delay, the transport of people and hardware required to execute the repair will be the time-limiting factor. In this regard improvements in warehousing and transportation logistics, for example, with the help of GPS systems, will drive future development.

Service automation is one of the fastest-growing disciplines in industry. Ideally, it makes utmost use of information and communication technology. It will however take some time for the valuable tools already installed in some parts of industry to penetrate fully into industrial service. Service automation requires close cooperation of many experts working in the areas of embedded systems, sensors, knowledge management, logistics, communication, data architecture, control systems, etc.

We can expect major steps in the future towards more effective and faster industrial service.

References

33.1　J. Levitt: *Complete Guide to Predictive and Preventive Maintenance* (Industrial Press, New York 2003)

33.2　L.R. Higgins, K. Mobley: *Maintenance Engineering Handbook* (McGraw Hill, New York 2001)

33.3　B.W. Niebel: *Engineering Maintenance Management* (Marcel Dekker, New York 1994)

33.4　R.K. Mobley: *Maintenance Fundamentals* (Butterworth–Heinemann, Burlington 2004)

33.5　K. Ola: Lifecycle management for improved product and system availability, ABB Rev. Spec. Rep. Ind. Serv. **29**, 36–37 (2004)

33.6　G. Cheever: High tech means high service performance, ABB Rev. Spec. Rep. Ind. Serv., 26–28 (2004)

33.7　T. Haugen, J. Wiik, E. Jellum, V. Hegre, O.J. Sørdalen, G. Bennstam: Real time energy performance management of industrial plants, ABB Rev. Spec. Rep. Ind. Serv., 10–15 (2004)

33.8　J.J.P. Tsai, Y. Bi, S.J.H. Yang, R.A.W. Smith: *Distributed Real-Time Systems: Monitoring, Visualization, Debugging and Analysis* (Wiley, New York 1996)

33.9　I. Lee, J.Y.-T. Leung, S.H. Son: *Handbook of Real-Time and Embedded Systems* (Taylor & Francis, Boca Raton 2007)

33.10　S.A. Reveliotis: *Real-Time Management of Resource Allocation Systems: A Discrete Event Systems Approach* (Springer, New York 2004)

33.11　J. Bugge, D. Julian, L. Gundersen, M. Garnett: Map of the future, ABB Rev. **2**, 30–33 (2004)

33.12　A. Kahn, S. Bollmeyer, F. Harbach: The challenge of device integration, ABB Rev. Special Rep. Autom. Syst., 79–81 (2007)

33.13　H. Wuttig: Asset optimization solutions, ABB Rev. Spec. Rep. Ind. Serv., 19–22 (2004)

33.14　M.S. Stroh: *A Practical Guide to Transportation and Logistics* (Logistics Network Dumont, New Jersey 2006)

33.15　M. Christopher: *Logistics and Supply Chain Management: Creating Value-Adding Networks* (FT, London 2005)

33.16　J.V. Jones: *Integrated Logistics Support Handbook* (McGraw Hill, New York 2006)

33.17　A. Rushton, P. Croucher, P. Baker: *The Handbook of Logistics and Distribution Management* (Kogan Page, Philadelphia 2006)

33.18　E.H. Sabri, A. Gupta, M. Beitler: *Purchase Order Management Best Practices: Process, Technology and Change* (J. Ross Pub., Fort Lauderdale 2006)

33.19　O. Zimmermann, M.R. Tomlinson, S. Peuser: *Perspectives on Web Services: Applying SOAP, WSDL and UDDI to Real-World Projects* (Springer, New York 2005)

33.20　D. Blanc, J. Schroeder: Wellness for your profit line, ABB Rev. **4**, 42–44 (2007)

33.21　B. Glover, H. Bhatt: *RFID Essentials* (O'Reilly Media, Sebastopol 2006)

33.22　D. Brown: *RFID Implementation* (McGraw Hill, New York 2006)

Part D | 33

34. Integrated Human and Automation Systems

Dieter Spath, Martin Braun, Wilhelm Bauer

Over the last few decades, automation has developed into a central technological strategy. Automation technologies augment human life in many different fields. However, after having an unrealistic vision of fully automated production, we came to the realization that automation would never be able to replace man completely, but rather support him in his work. A contemporary model is the human-oriented design of an automated man–machine system. Here, the technology helps man to accomplish his tasks and enables him at the same time to expand his capacities.

In addition to traditional usage in industrial process automation nowadays automation technology supports man through the help of *smarter*, so to speak, better linked, efficient, miniaturized systems. In order to facilitate the interaction taking place between man and machine, functionality and usability are stressed.

In addition to basic knowledge, examples of use, and development prospects, this chapter will present strategies, procedures, methods, and rules regarding human-oriented and integrative design of automated man–machine systems.

Within the last decades automation has become a central technological strategy. Automation technologies have penetrated into several fields of application, for example, industrial production of goods, selling of tickets in the field of public transport, and light control in the domestic environment, where it has shaped our daily life and work situation.

Work automatons have liberated man from processing dangerous or inadequate work, on the one hand. Since automatons continue to accomplish more demanding functions, which had previously been accomplished by man himself, the meaning of human work is challenged, on the other. This question has already been elaborated on at the beginning of the endeavor of industrial automation. Nowadays we hardly discuss the necessity of the application of automated systems anymore. We rather discuss their human-oriented design.

Progress within the field of information technology is one reason for increased automation. A characteristic of modern automation technologies is the miniaturization of components and the decentralization of systems. In addition to applications of industrial process automation, which are strongly influenced by mechanical engineering, information and automation technology allows for the design of *smarter*, more efficient assisting systems – such as personal digital assisting systems or ambient intelligent systems – whose technical components the user often does not see anymore.

In the past, technical progress became a synonym for replacement of man by a technical system. When working systems were developed rationally, the role of human work and its contribution to the overall result often only found partial consideration.

Following the rationalization and automation of work with the aim of increasing production, we risk increasing dissolution of the relation between individual and work. However, we have to realize that the human contribution will always influence the performance and quality of any work system.

After visionary illusions of an entirely automated and deserted factory based on mass production, the notion that automation technologies could not entirely replace man became increasingly dominant. An increasing number of individualized products demand a high degree of production flexibility and dependability, a requirement that an entirely automated system cannot fulfill. These demands can be accomplished by hybrid automation, which appropriately integrates the specific strengths of man and technology.

The contemporary approach is the human-oriented design of a man–machine system. Instead of subordinating man to the technical-organizational conditions of any work process, this approach contributes to humane conditions, which helps man to accomplish his tasks and supports the expansion of his capacities. The special meaning of a human-oriented work environment results from the fact that man is the agent of his own manpower. Consequently, individual, work process, and work result are closely linked with each other. Only if the process of automation – in addition to every necessary rationalization – also serves the humanization of the working conditions can it fulfill functional and economical expectations.

The interaction of man and technology is the focus when developing a human-oriented automated man–machine system.

In order to apply human resources in an optimal way and to use them synergetically, we need an integrative approach for the development of a man–machine system.

According to an integrated procedure regarding the development of (partly) automated man–machine systems, the present chapter will first elaborate some basics and definitions (Sect. 34.1). In Sect. 34.2 we present practical knowledge of the technology of automation regarding its usage in different life circumstances and fields of work and establish the relation between automation and man.

In Sect. 34.3 we present successful rules of development and processes of automated systems. The presentation will end with an outlook of future development and its implication for man (Sect. 34.4).

The main focus will be on the interaction of man and technology. Specific methods and instruments for the development of technical and technological aspects of automated work systems will be presented in this Handbook in the corresponding chapters.

34.1 Basics and Definitions

The development of an automated man–machine system demands relevant basic knowledge as well as well-chosen definitions of some individual terms.

34.1.1 Work Design

The goal of work design is adjustment of work and man so that functional organization with regard to hu-

man effectiveness and needs can be realized. Thus, the goal is to achieve good interaction between the working man, the technical object, and the work tools (see Sect. 34.1.8). Goals of the work design are:

- *Humanization*, which means human-friendly design of a work system regarding its demands and effects on the working human being.

- *Rationalization*, which means the increase of effectiveness or efficiency of the human or technical work in relation to the work product, for example, the amount, quality, dependability, and security, or avoiding of failure. The goal is to gain the same effect with less means, or a better effect with the same means.
- *Cost effectiveness*, expressing the relation between cost and gain; it will be codetermined by the two other target areas.

The work design encompasses ergonomic, organizational, and technical conditions of work systems and work processes in order to achieve the main design requirements.

Ergonomic Work Design

The object of ergonomic design is adjustment of work to the characteristics and capacities of man [34.1]. With the help of human-oriented design of the workplace and working conditions we want to achieve the following for the worker to enable productive and efficient working processes:

- Harmless, accomplishable, tolerable, and undisturbed working conditions
- Standards of social adequacy according to work content, work task, work environment as well as salary and cooperation
- Capacities in order to fulfill learning tasks that can support and develop the worker's personality [34.2].

The basis of ergonomic design of the workplace is the *anthropometry*, which defines the doctrine of dimensions, proportions, and measurements in relation to the human body. The goal of the anthropometrical design of the workplace is the adjustment of the workplace according to human dimensions. This can be realized by including spatial dimensions and functions of the human body (Sect. 34.3.4).

This physiological design of the workplace takes into consideration human factors engineering as well as the work plan and work process, which are adapted to the physiological demand of the worker (Sect. 34.1.8).

Software usability engineering aims for optimization of the various elements of the man–machine interface and communication between man and machine. The term *usability engineering* implies the development, analyses, and evaluation of information systems, so that man with his demands and capacities is the center of interest. The adjustment of software–

technical application and user is supposed to increase productivity, flexibility, and quality within the work system.

Usability aims at the optimization of the different procedures, which enables the user to accomplish a certain task with the help of a technical product. The main goals are: easy handling, learning ability, and optimal usage. Usability is not only a characteristic of a product, but rather an attribute of the interaction between a group of users and a product within a certain context [34.3].

Organizational Work Design

Organizational work design aims at the coordination of division of labor, meaning the appropriate segmentation of a task into subtasks and their goal-oriented adjustment. Organizational work can pursue different goals, for example:

- Addressing economic problems (deficient flexibility, poor capacity utilization, inappropriate quality)
- Addressing personal problems (for example, dissatisfaction, high fluctuation)
- Reshaping of the technical system.

Regarding the notion of humanization, organizational development contributes to good matching of work content and conditions to the capacities and interests of each individual worker [34.4].

From an economical point of view, organizational development aims for efficient application of scant resources, so that the final goal can be achieved. When competing for scant resources, the form of organization that provides smooth handling of the division of labor prevails.

34.1.2 Technical and Technological Work Design

The *technological design* of a work system is based on the selection of a certain class of technologies. It refers to the work procedure, that is to say, the basic decision of how to achieve a change of the work object [34.5]. From a technological point of view we have to increase the reliability and efficiency of the work system. The tasks of the technological work design are the constructive design of the technical tangible means (for example, equipment and facility) and the design of the man–machine interface (see Sect. 34.1.3).

Further technical development will modify the technological work design. The *technical work design* will define the functional separation of man and technical

Table 34.1 Level of technologies and functional division between man and technical system [34.5]

Technical level	Energy supply	Process control
Manual realization	Man	Man
Mechanical realization	Technical system	Man
Automated realization	Technical system	Technical system

tangible means, as shown in the level of technology of the particular work system [34.1]. Table 34.1 presents schematically the relation between the different levels of technology and the functional separation between man and technical system.

34.1.3 Work System

We understand human work within the realm of a work system, in which the worker functions with a goal in mind. A work system consists of the three elements: *man*, *technical tangible means* (for example, machines), and the *environment*, and is characterized by a task. Tasks are understood as either a change in the configuration of the work object (for example, to process material, to change energy, to inform men) or in place (for example, transportation of goods, energy, information or men). These elements of the work system are all connected by the time of activity.

Man does not always have a direct influence on the work object; very often man exercises the influence indirectly through a means he uses at work, such as a tool, machine, vehicle, and computer. The influence of man on the object is then characterized by the mechanization of this applied means [34.6].

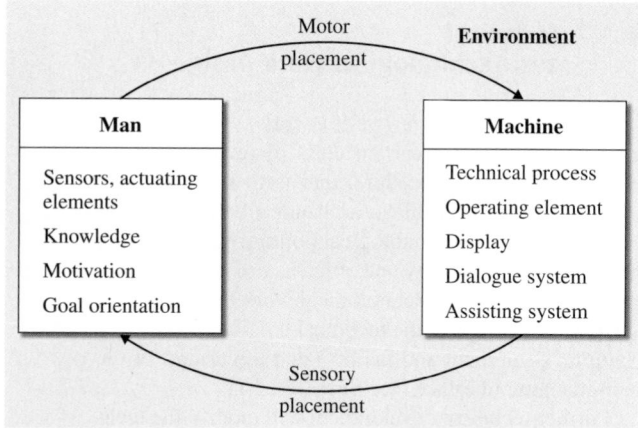

Fig. 34.1 General structure of a man–machine system

34.1.4 Man–Machine System

A man–machine system (MMS) is one specific form of a work system. It is understood as a functional abstraction when analyzing, designing, and evaluating the many forms of goal-oriented exchange of information between man and the technical system in order to fulfill the work task. All man–machine systems comprise man, interface, and the according technical system. The term *machine* is used for all technical shapes. Important components of a machine are display and control units, automated subunits, and computerized assistant systems [34.7].

Processes of human or physical-technical data processing characterize the conduct of a man–machine system. The general structure of a MMS (Fig. 34.1) is a feedback control system, where man, according to his goal, the information he has received, and the work task, decides and, thus, exercises control of the technical system.

We can differentiate between man–machine-systems that show goal-oriented dialog systems and dynamic systems. Further subcategorizations are shown in Fig. 34.2.

The spectrum of goal-oriented dialog systems comprises end devices belonging to information technology, mobile communication technology, consumer electronics, medical technology, domestic appliance technology, service automatons (Fig. 34.3), and process control workstations.

Dialogue systems are interactive, goal-oriented systems from information technology, which react on external input.

Dynamic systems are characterized by a continuously changeable variable. The condition of a manual control system is called *man in the loop*, while the condition of an automated control system is called *man out of the loop*.

We find manual control systems in all kinds of vehicles, on the water, in the air, and on the ground. Man has the following tasks in a manually control system [34.8]:

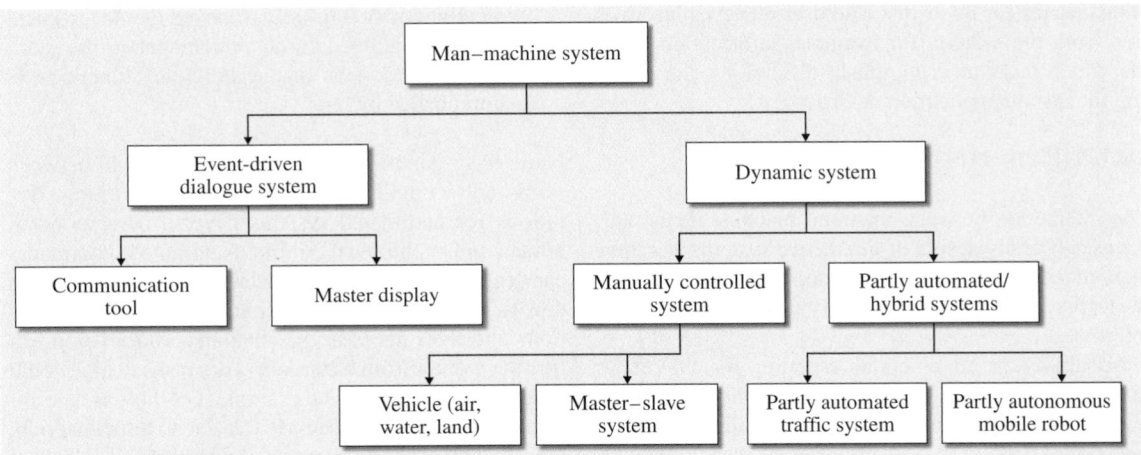

Fig. 34.2 Types of man–machine systems [34.8]

- *Communication:* Creation of communication interfaces as well as sending and reception of information
- *Scanning and evaluation of the situation:* Scanning of state variables of the system and the environment, directly via the natural senses or indirectly via technical sensors and displays
- *Planning:* Determination of distance from start to destination
- *Navigation:* Compliance with the planned distance and estimation
- *Stabilization:* Maintenance of necessary positions (for example, machine direction control on the street)
- *System management:* Utilization and supply of subsystems as well as error diagnosis and elimination.

Fig. 34.3 Ticket machine as an event-driven dialogue system

Another application range with manual control is the master–slave system used for telepresence, i. e., the enhancement of man's sensorial and manipulation capacities in order to work further away. The teleoperator benefits from sensors, actuating elements, and multimodal channels of communication to and from the human user. This creates a telepresence from a tele-workplace for the operation in a nonadmissible physical environment – for example, over large distances or in microworlds [34.9].

We sometimes find automatic regulation of state variables (Sect. 34.1.5) in *partly automated systems*. During automated processing, the worker is placed outside of the closed loop, but nevertheless supervises the automated process and handles disturbances.

Hybrid systems characterize a simultaneous situation- and function-oriented combination of manual or automated processing of the task within a collective work system. Due to flexible adaptation of the degree of automation, we aim for optimal usage of the specific and supplementary capacities or characteristics of man and machine [34.10].

Planes and cars are representatives of partly automated systems. As we do not want to rely solely on automaton when using *partly automated transportation systems*, a human observer is assigned to the system. In this way intervention is guaranteed in case of breakdown.

Another category of man–machine systems is partly automated robots. Here, the human being undertakes mission management by transmitting goals to a removed, partly automated robot, controls its actions, and compensates for its errors. The user retrieves information from the system, for example, information about the which tasks to accomplish, difficulties, the robot, and the distant application environment.

34.1.5 Man–Machine Interaction

The interaction between man and machine deals with the user-friendly design of interactive systems and their man–machine interface in general. Man interacts with the technical system (Sect. 34.1.3) via the man–machine interface.

Usability is an essential criterion for the man–machine interaction. The design of a man–machine interaction takes account of aspects of usability engineering, context analyses, and information design [34.11]. To provide the user with programs, which the untrained worker is able to learn quickly and the professional can use in a productive and accurate way, is the goal of software ergonomic technologies.

Computer-based word-processing systems are a good example of a man–machine interaction. In the past, computer systems often used text-based man–machine interfaces. In the meantime, the graphical desktop has become the main interface; even language and gesture identification are becoming increasingly important.

34.1.6 Automation

To assign functions to a machine, which once were accomplished by man, is the goal of automation. The degree of automation is determined by how many subfunctions are done by man or by the machine. Automation is also characterized by process control, which beyond mechanization is also a result of the technical system [34.12]. Depending on the complexity of the control tasks, we differentiate between complete and semiautomatic functions:

- *A complete automated system* of a machine does not need any human support. The machine completely relieves man from work. Complete automation is appropriate or functional when the worker cannot complete his work precisely enough, when he is not able to complete it at all, or when the working task, for example, is too dangerous for him.
- *Semiautomation* is a work characteristic of a machine that only needs some degree of support from man. In contrast to a completely automated system, semiautomation does not achieve complete relief from work for the worker. The control of the individual functions is usually achieved by the technical system. Program control, which means the start, end, and succession of the individual functions, is accomplished by man.

Numerical controlled machines can switch between semi- and complete automation while working. Solutions for automated systems have to correspond to human and economical criteria. Automated workplaces can lead to work relief and decrease of physical strain due to the elimination or limitation of certain situations and their necessary adaptations. Automation can free the worker from hazardous work tasks, which could influence his health. Good examples for this case are automatic handling machines. In relation to humanization, the following criteria promote automation:

- Abolition of monotone work
- Abolition of difficult physical strain due to unfavorable body position and exertion, for example, lifting heavy parts
- Abolition of unfavorable environmental impact, for example, provoked by heat, dirt, and noise
- Reduction of risk of accident [34.4].

The design of complete work tasks creates better work conditions. To reach this goal we need qualitative job enrichment based on new combinations of work tasks as well as the realization of new work functions.

Problem shifting can occur as well. Automation changes the job requirement in terms of attention, concentration, reaction rate and reliability, combination capacity, and distinct optical and acoustical perceptual capacity [34.13].

Due to the combination of manual and automated functions, we aim for adequate and practical application of specific human/engine capacity resources in hybrid systems [34.14].

34.1.7 Automation Technology

Automation technology refers to the use of strategies, methods, and appliances (hardware and software), which are able to fulfill predetermined goals mostly automatically and without constant human interference [34.15]. Automation technology addresses the conceptualization and development of automatons or other automatically elapsing and technical processes in the following areas:

- Engine construction, automotive engineering, air and space technology, robotics

- Automation of factories and buildings
- Computer process control of chemical and procedural machines
- Traffic control.

Process automation (during continuous processes such as power generation) and *production automation* (for discrete processes such as assembly of machines or controlling of machine tools) are main application areas in the field of automation technology.

34.1.8 Assisting Systems

Assisting systems are components of man–machine communication (Sect. 34.3.4). Assisting systems do not substitute a human being, but support him occasionally while accomplishing tasks that overburden or do not challenge him enough. An assisting system should:

- Reduce the subjectively felt complexity of a technical system
- Facilitate the spontaneous use of a technical system
- Enable fast learning of the functions or handling of the system
- Make the use of the system more reliable and secure.

Assistance creates a connection between the demands, capabilities, and capacities of the user on the one hand, and the functions of the interactive system on the other. The following assisting functions seem relevant [34.16]:

- Motivation, activation, and goal orientation (i. e., activation, orientation, and warning assistance)
- Information reception, and perception of signals of interactive systems and of environmental information (i. e., display, amplifier, and repetition assistance)
- Information integration and production of situational consciousness (i. e., presentation, translation, and explanation assistance)
- Decisiveness to take action, to decide and choose a course of action (i. e., offer, filter, proposition, and acceptance assistance; informative and silent construction assistance)
- To take action or carry out an operation (i. e., power and limit assistance; input assistance)
- Processing of system feedback or of a situation (i. e., feedback assistance).

As an example, we use assisting systems through personal computer (PC) software, automatic copilots in planes and vehicles, in the form of personal digital assistants (PDAs), or as part of smart-home concepts.

Assisting systems can be differentiated according to their adaptation of different user profiles, tasks, and situational conditions [34.17]:

- *Constant assisting systems* always show the same conduct, independent of the operator or situation. Their advantage is consistency and transparency. However, these systems are inflexible; the provided support does not always suit the user nor the situation.
- *Assisting systems designed according to user specification* adjusted to the needs of certain users and their tasks in specific contexts. This kind of adjustment proves to be problematic when, for example, the context of usage changes.
- *Adaptable assisting systems* can be adjusted by the user to specific needs, tasks, and situations of use. The calibration of assisting systems occurs via selection or adjustment of parameters. The user takes the initiative to adjust the system.
- Changes of *adaptive assisting systems* do not occur on the basis of explicit guidelines given by the user, but through the system's evaluation of actual and saved context characteristics. Adaptive systems autonomously adjust assistance to the user, his preferences, and needs in certain situations.

34.1.9 The Working Man

The development and evaluation of human-oriented work (Sect. 34.1.1) requires knowledge about the human factors. Selected work-scientific concepts are described below.

Concept of Stress and Strain. All human requirements, which evolve from workplace, work object, work organization, and environmental influences, are part of the notion of *stress and strain*, which concept describes the defined reaction of the individual body to external stress (or workload). Individual capability is the factor connecting impact and stress [34.1]. The workload and individual capabilities are decisive for the strain factor.

The higher the impact of the workload, the more the worker has to use his individual capabilities in order to fulfill the task effectively. Figure 34.4 shows an ergonomic *stress–strain* model. This *stress–strain* concept has been primarily described for the field of physical labor, but can also be used when talking about psychological stress.

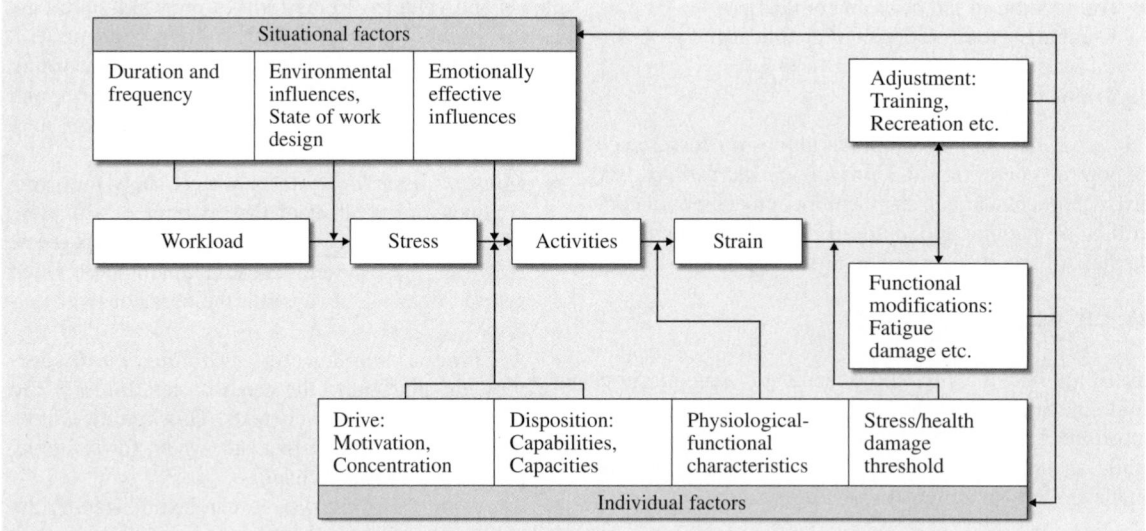

Fig. 34.4 Stress–strain model (according to [34.18])

Prerequisites for Performance. The entirety of information processing and energy transformation, which leads to the achievement of a goal, is defined as work performance. In order to be work efficient, we need human and objective prerequisites of performance (i.e., work organization and technical facilities). Human prerequisites for performance refer to capability and motivation (Fig. 34.5).

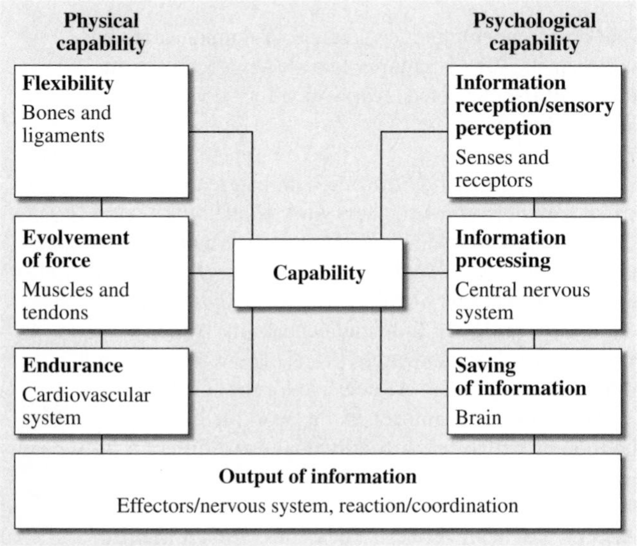

The term *physical motivation* comprises the sum of all biological body activity. It is limited to physical aspects. Work performance is not a constant factor, but undergoes changes. In order to be efficient, physical performance requires several psychological performance prerequisites, such as motivation, willingness of effort, and cognitive understanding of the task. These factors are called psychological motivation.

Stress. *Task-related stress* results mainly from the comprehension and processing of information, the movement of the body, and the release of muscle power when using work equipment. Comprehension and processing of information occurs through sensor and discriminatory work. Perception stresses our visual, auditory, haptics/tactile, and proprioceptive sense organs. Discriminating work leads to stress, based on recognition and identification of signals. The intensity of stress, which results from the work function, depends on:

- The duration and frequency of the task
- The complexity of the task itself and of different work processes
- The dynamics of the processes that need to be controlled
- The expected precision of the accomplishment of the task
- The level of concentration required while working

Fig. 34.5 Capability factors

- The specific characteristics of appearing signals
- Flexibility of comprehension of the information.

Stress that results from the *work environment* can result from a physical, chemical or social cause that may affect factors such as lighting, sound, climate, and mechanical vibration. Bodily stress can be provoked by manual handling of heavy loads or work equipment due to inappropriate body movement and enforced body position. Workplace conditions that demand a static body position are very stressful for the body as the circulation of blood is negatively affected.

Stress from *work organization* can result from regulation of the work schedule (for example, shift work), operating speed, the succession of tasks, inappropriate amounts of work during peak times, lack of influence on one's work, strict control, and uniform tasks. We also speak of stress when a task demands constant willingness of action, even though human interference is only necessary in exceptional cases [34.19].

The operation of *information* (for example, complex information or information deficits) can also provoke further situations of stress.

Strain. We differentiate between physical and psychological strain. Consequences of strain present themselves on a muscular–vegetative or cognitive–emotional level. Disturbances of the psychophysical balance are provoked by either excessive or unchallenging situations. Both of these situations imply that the individual prerequisite of performance does not correspond to the corresponding performance condition. In the case of an excessively challenging situation operational demands exceed the individual's capability and motivation. Unchallenging situations are characterized by the fact that individual capabilities and needs are not sufficiently taken into consideration. Both situations can lead to reversible disturbance of the individual performance prerequisite, such as tiredness, monotony, stress, and psychological saturation. On the level of performance, they can cause changes in work processing (e.g., output and quality). These disturbances can be eliminated by a change of physical and psychological performance functions, as well as by implementing phases of recovery time. Nevertheless, disaccord between strain and recovery can only be tolerated for a limited period of time [34.20].

34.2 Use of Automation Technology

Automation technology can be found in many fields at work as well as in public and private life. Examples of automation technologies are:

- Assembly automation (factory automation)
- Process automation in chemical and procedural facilities
- Automation of office work
- Building automation
- Traffic control
- Vehicle and aeronautical technology.

Increasing functional and economical demands as well as technical development will lead to ongoing automation of technical systems in many different walks of life and work – from the office to factory level. Regarding automation effort we can identify two developing trends [34.21]:

- Increase of *complexity* of automation solutions via enhancement and process-oriented integration of functionalities: linked and standardized control systems are increasingly replacing often proprietary isolated applications. The continuous automation of

processes and work systems leads to an increase of effectiveness and quality as well as to a decrease of costs in industrial processes of value creation.

- The increasing effectiveness of automation and control techniques and miniaturization of components leads to decentralization of control systems and their integration on-site. Next to process automation, more and more often, we also find product automation.

Information technology, which continues to produce more powerful hardware components, is a driving force for this development. Selected areas of application, which are characterized by a high degree of interaction between man and technology, will be presented in the following sections. Partly automated systems and assisting systems will be one focus of the following elaboration.

34.2.1 Production Automation

Production automation is a discipline within the field of automation technology that aims at the automation of discontinuous processes (i. e., discrete parts manufacturing) using technical automatons. Production automation is engaged in the entirety of control, closed-loop control devices, and optimization equipment in the space of production facilities [34.22].

Through history, production automation shows several levels of development, starting with automation related to the working process and ending with complex assembly automation. In the early days of automation, extensive complete processing of specific work objects was the focus. This development started with numerically controlled (NC) machine tools, following by mechanical workstations, reaching the level of flexible processing systems.

Flexible production systems are based on the material and informational linking of automated machines. They are characterized by the integration of the functions of transportation, storing, handling, and operating, and comprise the following subsystems:

- Technical system
- Flow of material, maintenance, and disposal system
- Storage system and application system
- Information and energy system
- Equipment system, machine system, and inspection equipment system
- Maintenance system.

Flexible production systems are designed for an assortment of geometrical and technological work equipment.

Due to the marginal effort required for their adjustment they can be fitted to changing production tasks as well as to fluctuating capacity loads. Control of the corresponding subsystems can be realized by a linked computer system, which takes into account machining situations, storage, and transport systems among others, and which receives requirements from a central processing control. The computer system controls usage of the appropriate machines, appropriation of the technical control programs, availability and progress controls, logging of production, and information from the maintenance service. An agent can influence the system in case of disturbance. Figure 34.6 summarizes the different levels of the respective machines up to the flexible production system.

A main technical component of production automation is the industrial robot. An industrial robot is a universally usable automaton with at least three axes, which have claws or tools, and whose movements are programmable without mechanical interaction. The main application areas of industrial robots are welding, assembling, and handling of tools.

The present level of development of production automation exhibits process-related integration of computer-based construction (computer-aided design, CAD) and production (computer-aided manufacturing, CAM). Process-oriented CAD/CAM solutions result in a continuous information processing system during the preparation and execution of production, in

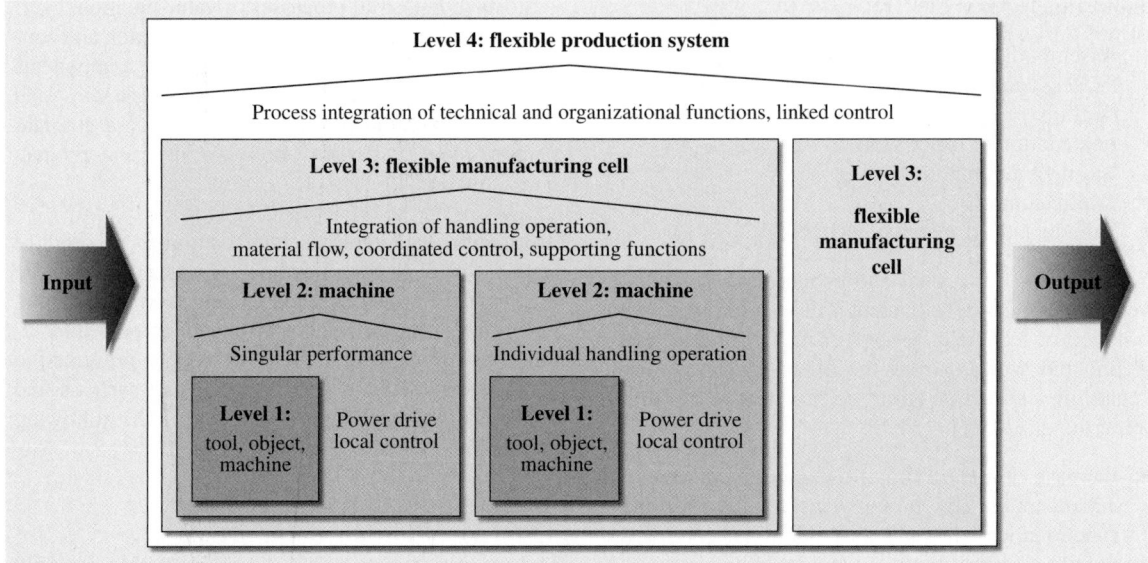

Fig. 34.6 Shell model of production automation (according to [34.14])

which data is produced and managed automatically and is circulated to other fields of operation (for example, systems of merchandise management). During the complex and continuous process of automation, development and production as well as the corresponding business fields are informationally and materially linked with each other. Continuous solutions are the basis of an automated plant, which at first only captures selected production areas, but then entire companies [34.23].

An extreme degree of automation is not always the perfect solution for manufacturing technology. If fewer pieces and complex tasks determine the production, the usage of less automated systems is a better solution [34.24]. Robots reach their limit if the execution of the task demands a high degree of perception, skill, or decisiveness, which cannot be realized in a robust or cost-effective way.

Unpredictable production range and volume as well as higher costs and quality demands increase the area of conflict between flexibility and automation in the production. Due to progress in the field of man–machine interaction and robotics, the field of hybrid systems has become established, in which mobile or stationary assisting robots represent the most economical form of production [34.25]. Assisting robots support flexible manual positions by accomplishing tasks together with the human being.

Man's sensory capabilities, knowledge, and skill are thus combined with the advantages of a robot (e.g., power, endurance, speed, and preciseness). Assisting robots can now not only handle special tasks, but also cover a broad spectrum of assistance of widely different tasks. Figure 34.7 shows an exemplary utilization of an assisting robot while assembling.

An assisting robot can either be installed stationary at the workplace or used in a mobile way at different locations. In both cases the movement and workplace of man and robot overlap. In order to make it possible for man and robot to cooperate efficiently in complex situations, it is necessary that the robot system has a sensory survey of the environment and an understanding of the job definition.

34.2.2 Process Automation

Chemical and procedural industries use automation, first and foremost, for monitoring and control of autonomous processes. This can usually be realized by the application of process computers, which are directly connected to the technical process. They collect situa-

Fig. 34.7 Hybrid system during assembly (photo was taken with permission of Fraunhofer IPA)

tional data, analyze errors, and control and optimize the process.

Interaction of man and technology is limited to man–machine communication when using process computers and control equipment. Section 34.3.4 will elaborate on these aspects in more detail.

34.2.3 Automation of Office Work

In this regard, we can differentiate between *informational* and *manual office work*. Manual office work can especially be found in the infrastructure sector. The use of decentralized computer technologies close to the workplace accelerates the automation of algorithmic office work with manual and informational character.

Automation of office processes aims to increase work efficiency. Some business processes can only be realized by use of technology. The comprehensive information offered by the Internet is, for example, only usable by computer-based search algorithms.

Workflow systems are important for the automation of *infrastructural office* work. They use optimized structures of organization for the automation of work processes. They influence the work process of each individual worker [34.26] by allocating individual procedure steps and forwarding them after processing.

Workflow systems support among others the following functions:

- Classification, excavation, and removal of information carriers in an archive
- Physical transport of information carriers
- Recognition of documents and transmission to the appropriate agent
- Connecting of incoming documents with previous information (for example, an incoming document will be connected with previously saved data about the client)
- Deadline monitoring and resubmission, capacity exchange in case of employer absence (for example, forwarding systems)
- Updating of data in the course of the process of individual processing steps (e.g., verification of the inventory when sending out orders).

Moreover, the use of automated work tools in the form of copy technology, microfilm technology, word-processing systems, and information transfer technology contributes to an increase of efficiency in the field of office work.

A characteristic of *informational office work* is tight connection of creative and processing work phases. The human being creates ideas, processes the task, and evaluates the results of the work. As organizational task are algorithmatized and delegated to the computer, the machine can support the working man [34.27]. Man has the possibility to intervene in the process by correcting, modifying, evaluating, and controlling it. Computer-based work connects the advantages of a computer, for example, high-speed operations and manipulation of extended data, with man's decision-making ability in an optimal way. Computer-based work does not eliminate man's creativity, but rather reinforces them. New solutions will also be bound to man's creativity.

34.2.4 Building Automation

The term *building automation* defines the entirety of monitoring, control, and optimization systems in buildings. It is part of technical facility management. This includes the integration of building-specific processes, functions, and components in the fields of heating, ventilation, climate, lighting, security, or accession control. The continuous cross-linking of all components and functions in the building as well as their decentralized control are the characteristics of building automation [34.28].

Building automation aims to reduce costs of buildings via a methodological approach to planning, design, construction, and operation. For this, operational sequences are conducted in an independent way or simplified in their handling or controlling. Functions can, for example, be aligned according to changing operating conditions (season, time of day, weather, etc.) and activities can be combined into scenarios.

Due to construction automation the technical and organizational degree of complexity increases as well as the demand according to a functional integration. Here, we have to differentiate between the demands and needs of different user groups (i. e., users, operators, and service staff):

- *User*: Functions of the MMS have to be reduced to a necessary minimum. They have to be intuitively comprehensible and easily manageable. Interference in operation has to be possible at any time (with the exception of safety functions).
- *Operator*: MMS has to provide optimal support for maintenance and support as well as optimizing operation of the construction technology. Depending on the object's size and complexity, this can comprise a spectrum of simple fault indication up to teleguided control systems.
- *Service staff:* Operating functions, which are unnecessary in a common process, have to be available exclusively to service staff.

34.2.5 Traffic Control

Traffic control means active control of the flow of traffic by traffic management systems. Traffic telemetry is a main application of traffic management systems, comprising all electronic control and assisting systems that coordinate the flow of traffic automatically and support driver routing [34.29]. Traffic telemetry has the following goals [34.30]:

- Increase of efficiency of existing traffic infrastructure for a high volume of traffic
- Avoidance of traffic jams as well as of empty cars and *look-around* drives
- Combination of advantages of the individual carriers (that is to say, railway, street, water, air) and integration into one general concept
- Increase of traffic safety: decrease of accidents and traffic jams
- Decrease of environmental burden due to traffic control

Traffic control depends on the availability of appropriate traffic data, which is collected by optical or inductive methods, amongst others. The collected data is processed in a primary traffic control unit and transformed into traffic information, on the basis of which traffic scenarios can be developed and traffic streams can be processed. Manipulation of traffic streams occurs, for example, through warning notices, speed limitations, or rerouting recommendations.

Information on the infrastructure is carried out from vehicle to driver through intercommunication signs, light signals, navigation systems or radio.

Satellite-navigation systems are widely used in vehicles; they carry out supported route calculation and vehicle-specific traffic control or goal orientation. Telemetric systems are also used for mileage measurement in regard to road charging.

Traffic telemetry appliances can avoid or lessen disturbances and optimize traffic flow in a timely, geographical, and modal way [34.31]. Even though traffic telemetry systems have great potential to ameliorate the entire traffic situation, their usage is still limited. One reason is the insufficient quality of collected data regarding traffic and street status for traffic control. Another reason is that the possibility of intervention in daily traffic is limited.

In the future, oriented adaptive traffic control systems will have to be able to dynamically link individual and public vehicles as well as traffic data. In addition to stationary detectors, vehicle-generated traffic announcements are also included in the area-wide extension of databases regarding traffic situation and traffic prognosis. Mobil communication systems contribute to the exchange of information between vehicle and infrastructure.

34.2.6 Vehicle Automation

Automated systems are used in road vehicles and air transportation. They support and control the driver or pilot in his task.

Assisting systems in a car stabilize the vehicle (for example, antiblock system, stability programs, brake assistant, and emergency brake) and help the car to remain in lane, maintain the correct distance to the car in front, control lighting (for example, through an adaptive lighting system), assist in routing (for example, navigation systems and destination guides), as well as accomplish driving maneuvers and parking [34.32].

Moreover, assisting systems facilitate the usage of numerous comfort, communication, and entertain-

ment functions in the vehicle. Figure 34.8 presents an overview of assisting systems in a car. Assisting systems, which are only barely noticed, are for example, a radio that automatically regulates the volume according to the surrounding noise, a display that optimizes its brightness according to the external light level, and automatic windscreen wipers that regulate their cleaning intervals according to the force of rain. Nowadays we combine several assisting functions into adaptive driver assisting systems [34.33].

In order to support steering, numerous development ideas have been discussed. The final goal is to create a system that supports steering in order to stay in lane and maintain the distance to the preceding cars [34.34]. In this regard, a video camera records the forward lane structure. With the help of some dynamic driving factors and evaluation of the captured images, the system can determine the current position of the vehicle relative to the lane markings. If the vehicle diverges from the required lane, the driver will feel small but continuous correcting forces through the steering wheel. Using the same technology we can also produce an alert in case of strong divergence from the road, using a synthetically generated noise or vibration feeling. These stimuli reduce the possibility of unintended divergence from the road and increase the likelihood that an error will be noticed and corrected early enough.

Another system that helps lane control and maintenance of the correct distance to preceding cars is called the *electronic drawbar*. This is a nontactile coupling be-

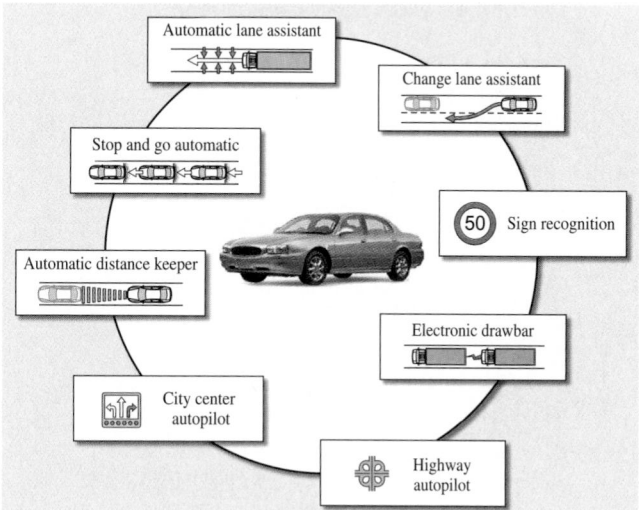

Fig. 34.8 Overview of an assistance system in a car

Fig. 34.9 Adaptive, multifunctional driver-assistance system

tween (usually commercial) vehicles based on sensors and computer technology. The following vehicles follow the leading vehicle automatically, as if they were connected to the preceding vehicle with a drawbar. The *electronic drawbar* should make it possible for just one driver to lead and control a line of cars following each other, all within a short distance. Due to the resulting low aerodynamic residence, gas consumption should be decreased.

Figure 34.9 shows a prototype adaptive, multifunctional driver-assistance system in a car.

Advising navigation systems that react to the current traffic situation are in serial production and show increasingly better results. Integrated diagnosis systems refer to disturbances and maintenance intervals. Table 34.2 presents the assisting functions according to the realized assistant type. Moreover, this

Table 34.2 Assisting systems in cars [34.8]

Modality/ assistance type	Visual	Acoustic	Haptic	Without information
Intervention			Antiblock system (ABS), servotronics, reflex control, collision assistant	Crash prevention, belt pretensioner, seat conditioning, sunroof
Command		Lane control, problem report, navigation order	Lane control, distance keeping, intersection assistant, curve assistant	
Consultation	Navigation systems, traffic jam alarm			
Information	Diagnoses systems			

Table 34.3 Assisting systems in aircrafts [34.8]

Modality/ assisting type	Visual	Acoustic	Haptic
Intervention			Maneuver delimiter, quickening
Command		Ground proximity warning system (GPWS)	Flight vector display, stick-shaker, callouts
Consultation	Flight management system (FMS), electronically centralized aircraft monitoring system (ECAM)		
Information	Navigation databank, plane state variables and maneuver borders		

table shows which dialog system addresses with sense modality.

The idea of the self-controlling vehicle has been abandoned. Although technical control systems are generally available, public opinion has determined that the driver is responsible for driving and not the lane control or tailgate protection system. Possible legal consequences in the case of failure of the technology are difficult to assess.

The application area of *aviation* is traditionally characterized by progressive technologies. As a result, many assisting systems exist in the field. Table 34.3 presents an assistant type and the used modality for some assisting functions in a plane.

On the level of plane stabilization, we find intervening systems, including maneuver delimiter and commanding systems such as light vector displays, a stick-shaker, and callouts. Steering is supported by collision-prevention commands and ground-proximity warnings.

For the planning of the flight path, the flight management system is supported by an informational system that leads the pilot from the start to the destination via previously entered travel points. The electronically centralized aircraft monitoring (ECAM) system is a consultation system that supports the pilot actively when managing system resources and correcting errors. Error messages are determined by individual phases of the flight. Furthermore, a plan possesses dialog systems through which the pilot executes the flight path, the management of subsystems, and the management of errors. More assisting systems concern information exchange with air-traffic control, communication with the operating airline (company regulations), and execution of braking maneuvers on the runway (break-to-vacate).

34.3 Design Rules for Automation

The design of automated work systems are usually geared to systematic work design. It is subject to specific requirements and methods, which will be presented below.

34.3.1 Goal System for Work Design

The decision for a specific kind of design of a work system – including the option of automation – is mainly determined by the following three goals [34.35]:

1. *Functional goals:* Optimal accomplishment of functions (for example, dependability, endurance, precision, and reproduction). The limited capabilities of man regarding sensors (for example, no adequate receptors for voltage, operations being fast, objects being small) or motor functions (for example, regarding the height and endurance of body power) often demand the use of technologies. Man's broad capabilities and flexibility, which allows him to diverge, if necessary, from fixed algorithms and to react to changed situations, demand the inclusion of a human being [34.36].
2. *Human-oriented goals:* Goals that concern human health, as well as performance prerequisites (for example, work safety, level of task demands, and qualification).
3. *Economic goals:* Given function fulfillment with preferably low costs, or best function fulfillment with fixed costs.

It is not possible to derive a priority for a single design goal in the goal system, as usually one has to choose an optimum between opposing subgoals. In individual cases anthropocentric or technocentric approaches of design result from the emphasis of different goals. Here, we favor a anthropocentric (that is to say, human-oriented) approach to design.

34.3.2 Approach to the Development and Design of Automated Systems

Consideration of Life Cycles
The development of a man–machine system occurs based on a formalized plan. The development of a system is subdivided into individual phases (Fig. 34.10):

- The process of development and design begins with a *prerun phase*, during which a system concept is processed. This phase aims to identify the current problem, and to evaluate whether it is really necessary to build the intended system and where it should be used.

- During the *definition phase*, the system is deconstructed into subfunctions and subsystems. Specifications are developed, in which all characteristics and performance features as well as demands for man and technology have to be included. Performance is related to the system's different phases of use, e.g., operation and maintenance [34.37]. With regard to the predetermined aspects, the layout and evaluation of alternative solutions are developed for each subsystem.
- During the *development phase*, a detailed design of the subsystem's hardware and software is realized. The developed solutions are integrated into a complete functional system.
- The *acquisition phase* comprises production of the system's hard- and software, testing of the technical subsystem's functionality, integration of mechanical components, and production of user and maintenance instructions. Furthermore, staff will be included in this phase.
- During the *use phase*, usage of the system is realized. Operational experiences, which include a normal, disturbed, and maintained operation, form the basis for improvement, modernization, and re-design of similar systems.

A review at the end of each phase decides upon the start of the next phase, when the development project has achieved a definite level of maturity.

Design Projects
The following design projects are exercised for each phase of a man–machine system (Fig. 34.9)

- With the help of system analysis the general goals for the complete system are determined, then the problem is firmly established and detailed.

- Within the framework of the *function analysis*, the system functions necessary to achieve the predetermined goals are established. The analyses of individual functions and their allocation to the machine or to man is the task of the technological work design.
- The tasks for the human result from the assignment of functions. Through a *task analysis* we have to evaluate whether the function and its related task can generally be conducted by the available staff. A group of users has to be selected and trained.
- *Technical, organizational, and ergonomic work design* aims for optimal development of the system components and work conditions through alignment of the conditions to the user (Sect. 34.1.1).
- An evaluation helps to verify the efficiency of the enforced design measurements for the implemented system. In order to achieve system optimization, the design methods have to be adjusted in case of goal divergence.

Emphasis on Human–Oriented Design
Within the realm of the human-oriented design of a man–machine system, the functions and components with a high degree of interaction between men and machine [34.38] are the most important. When designing partly automated systems or assisting systems, the following tasks are mainly affected:

- Tasks for function division between man and machine as well as for structuring the task given to man (i. e., technical/technological or organizational work design)
- Tasks for optimization of integration of communication at the interface of man and machine (i. e., ergonomic design)

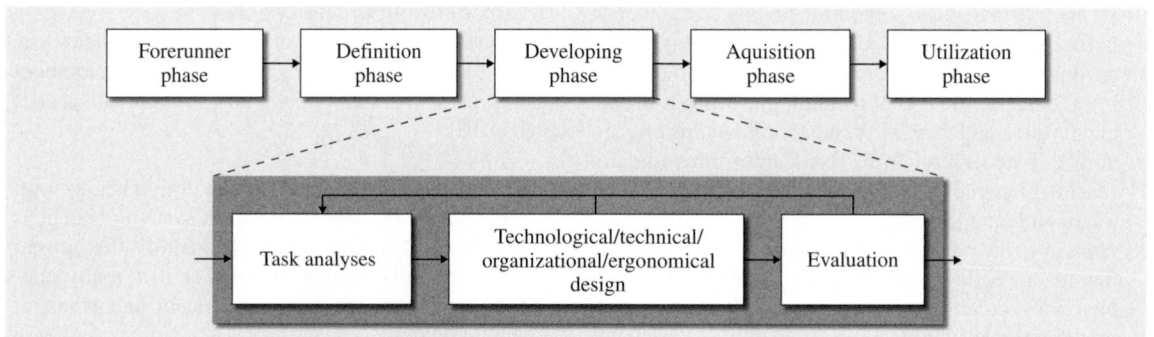

Fig. 34.10 Phases of an automated system and their development project

- Tasks to increase job or machine safety (i. e., technical design) by including hazardous factors of the work environment (e.g., noise, climate, vibrations).

Task-specific criteria and requirements of human-oriented work design as well as methodical approaches will be discussed in the following sections.

34.3.3 Function Division and Work Structuring

Subfunctions of a work system can be achieved by man and by a machine. If man is the center of interest, the design of the work system is geared towards the following levels of evaluation of human work [34.39]:

- *Feasibility:* anthropometrical, psychophysical, biomechanical thresholds for brief workload duration in order to prevent health damage
- *Tolerability:* physiological and medical thresholds for a long workload duration
- *Reasonability:* sociological, group specific, and individual thresholds for a long workload duration
- *Satisfaction:* individual sociopsychological thresholds with long and short validity.

In manual assembly, for example the joining of parts, certain thresholds result for the worker due to the required speed or accuracy of motions. Figure 34.11 presents this situation schematically.

Ergonomic optimization is desirable, if the combination of stress parameters resulting from the work task leads to tolerable work conditions (i. e., ergonomic

design, see Sect. 34.1.1). If work can be done but is not tolerable, mainly the content of the working task has to be changed with the help of measurements regarding work structuring (i. e., organizational design, see Sect. 34.1.1). If the work task cannot be fulfilled, automation (i. e., a technical/technological design, see Sect. 34.1.1) of the work system is recommended, which implies extensive transfer of functions to a machine.

Further influences on the design of systems result from technical and economical requirements. Due to the numerous influences on the design, it is impossible to distinguish specific design dimensions from each other. In factories we can find work systems with different degrees of automation among which are obvious combinations of manual and automated systems (hybrid systems).

When assigning functions within the man–machine system, the psychological and physiological performance requirements of the working man have to be taken into consideration (Sect. 34.1.8). In this way overextension of the worker can be avoided and health damage can be averted. Appropriate work structuring also takes into consideration not giving the man solely leftover functions, where he compensates or conducts nonautomated functions. Function division and work structuring therefore have to be developed in such a way that interesting, motivating, and diversified task arise [34.40] and optimal system effectiveness is guaranteed.

34.3.4 Designing a Man–Machine Interface

A central task when developing human-oriented automated work systems is the design of the man–machine interface. Ergonomically designed workplaces and machines – both in regard to hardware and software with the corresponding desktop – contribute to user-friendly and efficient task accomplishment.

Workplace
The workplace is the place where the task will be accomplished. At workplaces on the production level the following problems often occur:

- Inappropriate posture, forced posture due to limited free space or awkward positioning of appliances
- One-sided, repetitive movements
- Inappropriate movement of arms and legs
- Appliance of body forces.

In terms of ergonomic work design the demand arises to adjust the workplace in regard to its dimensions, visual

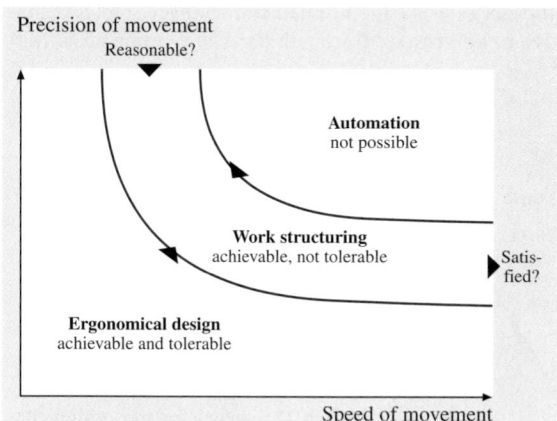

Fig. 34.11 Dimensions of work design from an ergonomic point of view [34.4]

and active area, and the forces that have to be applied to the conditions of the working man. The starting point for workplace design is the working task. Depending on the working task, specific demands arise in regard to motor functions and visual perception. The demands of the working task have to match man's performance prerequisites. Figure 34.12 shows factors that have to be taken into consideration when designing a workplace. The technologies or materials that are used can lead to further demands in regard to the configuration of the workplace, for example, lighting, ventilation, and storage.

Anthropometric Design of Workplaces. The starting point for anthropometric design of workplaces is the size of the human body. In order to prevent forced posture, e.g., intense bending forward or to the side, workplaces have to be aligned with the body size of the workers who work at that particular place.

When discussing the anthropometric design of the workplace, it has to be clarified whether work will take place sitting or standing (Fig. 34.13). A balanced position can be achieved by a change of posture. Basically it is important to set the human body size in relation to the correct distance between floor or feet height, working height, and seat height. The working height has to be adjusted to the conditions of the work task. In addition it is important to have sufficient horizontal and vertical space.

If several persons work at different times at the same workplace (for example, shift work), the workplace has to be equipped with appropriate adjustability in order to be adjusted to different body shapes and heights.

Physiological Design of Workplaces. Physical strength also has to be taken into consideration at automated

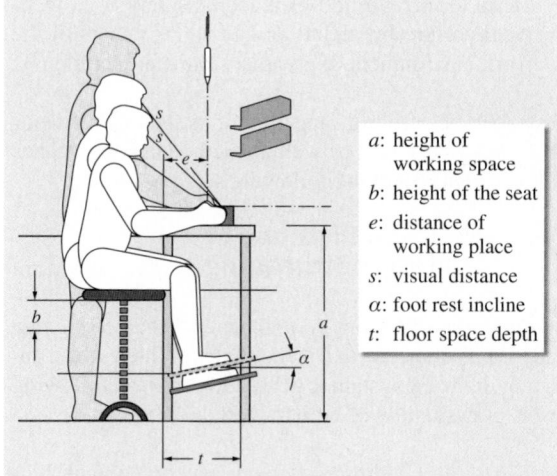

a: height of working space
b: height of the seat
e: distance of working place
s: visual distance
α: foot rest incline
t: floor space depth

Fig. 34.13 Blueprint of a possible seat-standing workplace

workplaces, as the basis for lifting and carrying of loads and operation of the system elements. Tasks that require intensive body movements and a large amount of physical strength should be conducted in a standing position. Tasks that require a high degree of preciseness (e.g., subtle assembly) and little physical strength can be conducted while sitting. Information about the dimensions of working tasks and their required physical strength is given in ergonomic literature [34.1, 5, 41].

Anatomically Favorable Alignment of Working Tools. Anatomically favorable alignment of appliances and working tools avoids unnecessary movement and assists balanced physical strain on the human body. Figure 34.14 presents an example of how to align appliances in a partly automated workplace. Objects that have to be grasped (bin with parts) are arranged within

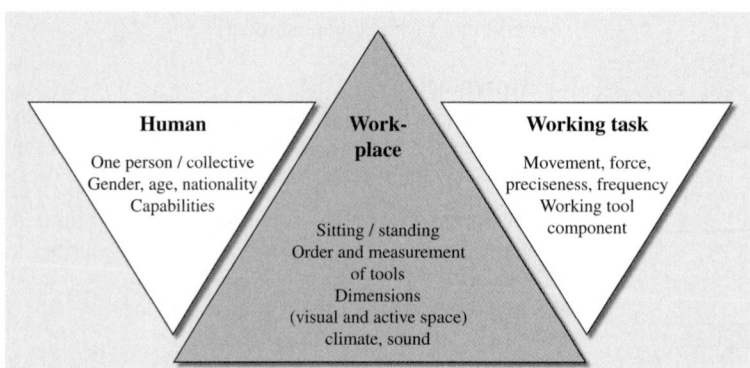

Human
One person / collective
Gender, age, nationality
Capabilities

Work-place

Working task
Movement, force, preciseness, frequency
Working tool component

Sitting / standing
Order and measurement of tools
Dimensions
(visual and active space)
climate, sound

Fig. 34.12 Factors for the design of a workplace

Fig. 34.14 Partly automated manufacturing workplace

a reachable distance and the machine is positioned in the center.

Design of VDU Workplaces. Most of the time automated work in an office is performed at visual display units (VDUs). In addition to the working postures already mentioned above, the following factors have to be taken into account when working at a VDU [34.43]:

- *Strain of the supporting apparatus or musculoskeletal system:* Continuous, static sitting strains the supporting apparatus and the musculoskeletal system extremely and can lead to muscle tensions and degeneration effects. Repetitive tasks such as data entry can lead to pain in the wrist (for example, repetitive strain injury) and chronic back pain.
- *Eyestrain:* Frequent change of eye contact between screen, draft, and keyboard (that is to say, accommodation) and the resulting adjustment to the changing light level (that is to say, adaptation) provoke enormous strain on the eyes and can cause eye damage.
- *Psycho-mental strain:* Man is stressed while carrying out demanding informational tasks when being exposed to (unintentional) interruptions, time pressure, and fragmentation of the work task.

The demand to configure the daily work schedule with a change of activities or regular breaks is very important for the human-oriented design of a workplace. VDU tasks and other activities should be alternated in order to avoid burden on the eyes and encourage movement. If a change of activity cannot be realized, regularly short breaks from work at the screen should be included [34.27].

Man–Machine Communication

In the course of automation, interaction between man and machine becomes an information exchange, i.e., man–machine communication. This information exchange is realized by man's sensors (i.e., senses) and actuating elements (i.e., effectors) on the one hand, and input and display systems of the machine on the other. This exchange of information is controlled by the human or technical processing of the information. Software ergonomic work design has the function of designing input and display systems and the machine's processing of information according to ergonomic goals (Sect. 34.3.3). While so doing, the task and conditions of the working environment are taken into account. Additional tasks result from the selection and education of users who operate these systems.

Sensory, cognitive, and motor characteristics have to be taken into account when designing man–machine communication. The visual channel (the eye) can be addressed via optical displays, the auditory channel (the ear) via acoustic displays, and the tactile channel (sense of touch) via haptic displays. As the hand and foot motor functions are available for the mechanical input after processing of information by the brain (cognition), so language is available for linguistic input. Now input via body movement (gestures) also plays an important role, for example, through the use of the mouse or via the measurement of hand, head, and eye movement (movement tracking). According to Fig. 34.15, the

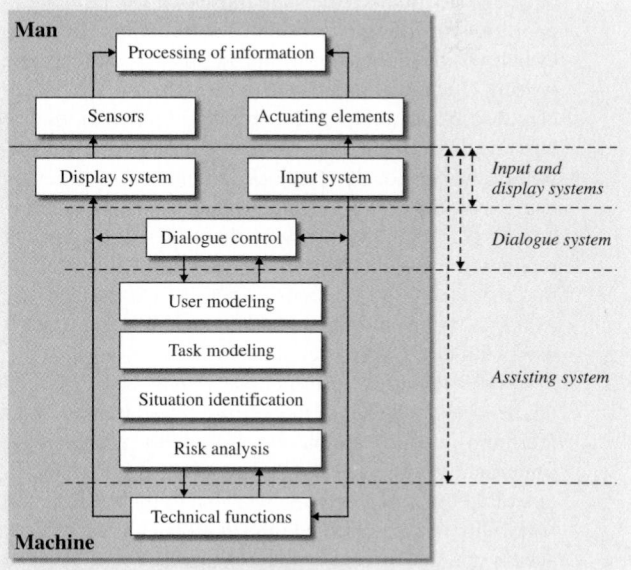

Fig. 34.15 Levels of man–machine communication [34.42]

design of man–machine communication can be represented at three levels:

- The highest level shows the interface between man and machine, the input and display systems, which are often defined as the *desktop*.
- The next level is the *dialogue system*, which creates the connection between the input and display, and which causes the machine's control of information flow.
- The third level refers to the application of assisting systems, which support the user when developing processing strategies and system control.

The design of these levels has the goal of tuning the technical system for the acceptance, processing, and display of information for man and his tasks.

Input and Display Systems. The design of an input and display system has to resolve three tasks [34.44]:

- It has to choose appropriate design parameters according to human conditions when adjusting the system to the human's motor function and sensors.
- It has to display the information codes that are necessary for exchange of information between man and machine; in this case, compatibility has to be taken into account.
- The organization of information designs the input or output of connected information.

Appropriate input systems are those that use human capabilities for the transmission of information: exercise of body force on objects, gestures, mimics, talking, and writing. Technical input systems are also useful if they can absorb and interpret the provided information. The best available technology allows the extended utilization of the previously mentioned human capabilities by using switches, levers, hand wheels, dials, keys, keyboards (for example, work at the VDU), etc. Speech recognition systems have seen immense progress, so that the use of speech signal input can be realized, for example, when entering numeric codes. Writing input is used with electronic notebooks.

When releasing information, visual, auditory, and tactile kinesthetic sense modalities are addressed. Furthermore, speech output is becoming increasingly important. Haptic displays are mainly used to facilitate blind operation of switches. Due to the amount and variability of the presented information and the optional access options, optical displays – mostly in the form of screens – play a dominant role. Multimedia forms of communication connect the different forms of coding: number, text, speech, and picture.

Compatibility of the design of (complex) input and output display systems is extremely important. Compatibility can also be defined as the decoding process, which the human has to achieve when evaluating the different forms of information. Inner compatibility is achieved when compatibility exists between man's periphery and the inner models (i. e., associations and stereotypes). A good example is the fact that one expects an increase in value when turning the adjusting knob to the right. An outer compatibility is in place when a display turns to the right according to the movement of the operating element.

Dialogue Systems. Dialogue represents the interactive exchange of information between man and machine in order to achieve a task. Depending on the user's previous knowledge, different dialogue forms are possible: question–answer dialogue, form dialogue, menu dialogue, key dialogue, command dialogue, and natural conversational dialogue. Dialogue forms as well as a direct manipulation comprise concrete actions (for example, display or sketching), pictures, and speech. These demand less abstraction from the user than text-based dialogues. There is always the possibility to adjust the dialogue system to a user-specific demand.

Continuous technical progress leads to rapid introduction of new program versions with dialogue alternatives. In order to guarantee reliable utilization, the user has to learn continuously. As printed instruction sheets are often ignored, integrated guidance of the user in the dialogue is practical. This guidance should be in line with the user's learning progress and the assisting and support systems, which the user can use in the dialogue.

Assisting Systems. Man–machine communication is generally designed as a dialogue, i. e., interaction between information input and output. Assisting systems (or dialogue assistants) support the user during goal-oriented use of a dialogue system by explaining the system's functions and user directions (Sect. 34.1.7). In order to guarantee progress of the dialogue, data is taken from the user (that is to say, user modeling), from the task to be solved (that is to say, task modeling), and from the present state of the system (that is to say, situation identification). In so doing, the field of man–machine communication comprises increasingly higher levels of technical processing of information. Whereas at the beginning only the desktop was affected in the form of display and input elements, nowadays the entire design of informational–technical systems are directed to the user.

Principles of Design for Man–Machine Communication

The design of man–machine communication mainly comprises aspects of dialogue development and control hierarchy.

Dialogue Development. Software ergonomic development of man–machine communication is geared towards the following demands for appropriate dialogue development [34.45]:

- *Usability* characterizes the degree to which a product can be used effective, efficiently, and in a satisfying way.
- *Effectiveness* determines the precision and completeness with which the user can achieve a certain task goal. Errors can have a negative impact on effectiveness. This is the case when the user is incapable of achieving the defined goal correctly or completely.
- *Efficiency* describes the cost-related effectiveness, referring to the relation between the achieved precision and completeness of the effort used by the user while attaining the goal. Deficiencies affect efficiency of utilization (*fitness for use*). The result is correct, but the effort is inappropriate, as the user, for example, can hardly avoid making mistakes.
- *Satisfaction* is an indicator of the acceptance of utilization.

Central criteria for dialogue design are *functionality* (i.e., the suitability of the task) and *usability* (i.e., suitability for learning). The ISO 9241 standard [34.3] defines established principles for the design of man–machine communication (Fig. 34.16):

- *Suitability for the task:* While working, the user should experience support instead of interference or unnecessary demands.
- *Self-descriptiveness:* The dialogue should make it clear what the user should do next.
- *Controllability:* The user should be able to control the pace and sequence of the interaction.
- *Compatibility, conformity with user expectations:* General experiences, schooling, experiences with work processes or similar software are effectively applicable. In terms of the expectation conformity, boundaries of the technical system have to be transparent (for example, safety functions).
- *Error tolerance:* The dialogue should be forgiving. In spite of defective input, the result will be achieved with only a few corrections.
- *Suitability for individualization:* The dialogue system allows adjustments to the demands of the working task, to the user's individual preferences, and to the utilization capability.
- *Suitability for learning:* The user will be supported and instructed while learning the dialogue system.

Burmester [34.46] recommends a simple and clear presentation of information, direct and unambiguous language, direct feedback, as well as user-oriented terminology for the implementation of design principles. In order to design the utilization capability of the man–

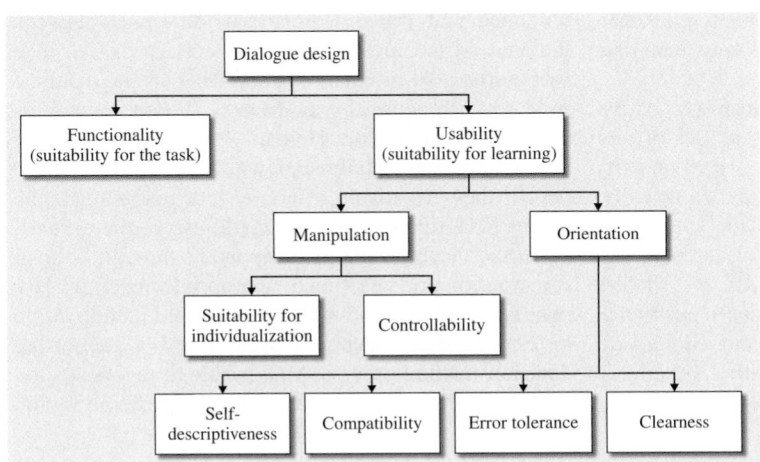

Fig. 34.16 Design principles of the ISO 9241 standard for man–machine communication

machine system more intuitively, known objects, for example, office folders, are displayed on the software's desktop.

Control Hierarchy. The embedding of the dialogue system into the control hierarchy of a man–machine system is as important as the dialogue design itself. Here we have to clarify, if and how far the technical system is allowed to limit man's freedom to take decisions and carry out actions. This question can be illustrated with the help of the example of a driving assisting system, which leads to increased safety for the driver when maintaining an inappropriate distance to the preceding vehicle and a reduction in speed.

Man's capability to make a decision is dependent on his position in the MMS. *Decker* [34.47] classifies three different performance types:

- *Restitution performances:* Here, the machine enables a disabled human being to achieve standard performances. Examples are prosthesis for amputated body limbs.
- *Expansion performances:* Here, man achieves together with the machine a higher level of performance than a standard performance. A good example is a databank system, which simplifies the data management and data organization.
- *Substitution performances:* The performance, which so far has been accomplished by man, is now achieved by a machine supervised by man. An example is the autopilot in a plane.

Experts are of the opinion that, as long as it is not possible to completely substitute man by a machine, man should have the highest level of decision making in a man–machine environment [34.48]. This implies that man will have the final decision. In this way, man can correct the MMS in case of errors.

The following problem appears however with partly automated MMS: During long working phases of uninterrupted automated production, man is permanently unchallenged and easily fatigued. If man then actually has to interfere due to a breakdown of the automated system, the once unchallenging situations transforms into a too challenging one. Consequently, we should aim for a continuous activity level, in order to guarantee manual takeover of the controlling system. Moreover, redundancy should assure that the entire system is reliable.

34.3.5 Increase of Occupational Safety

Automation of working systems needs to take into account occupational safety. Occupational safety is a necessary requirement for work processing. Safe working conditions should minimize or even prevent health damage to the working man. Generally speaking, the work result cannot be guaranteed if unsafe work conditions exist. A high level of safety is guaranteed by measures of machine safety. Occupational safety comprises safety measures and rules of conduct, which should provide the user of technical systems with the highest protection possible.

Automation measures basically reduce the hazardous risks to the worker. However, when breakdowns occur, the untrained worker has to interfere under time pressure, which can cause new danger situations.

The design of a man–machine interface with an abstract representation of information between user and system, and that also screens mechanical noises and vibrations, results in the risk that man's contact with reality is weakened. It has been observed that some drivers compensate for safety gained through assisting systems by changing to a risky driving style [34.49]. As a result we can state that the demand for increased occupational safety always has to be seen in the context of system reliability.

System Reliability

The term *system reliability* defines the probability that the system works without any errors in a defined timeframe. An error can be defined as an intolerable divergence from a defined level of quality. Human reliability describes man's capability to accomplish a task in an acceptable way under fixed conditions within a defined time frame. Here a basic difference between human and technical reliability becomes visible: man works in a goal-oriented manner, whereas the machine works in a functional way. Man is able to control his actions autonomously. Human errors do not usually lead to an immediate breakdown, but can be corrected before they negatively affect system operations. The technical design of systems should assure the reliability of the machine and its components through optimal construction and selection of appropriate materials. Human reliability can be increased by rapid identification of errors by the technical system and a supporting decision-making process. Man is able to recognize and correct for his own errors prior to their interference with the system. He can also stop processes that seem to be becoming problematic [34.50].

Design for Safety

Design Strategies. Adequate actions can contribute to an increase in the safety level of automated working systems and minimization of hazardous risks. The following kinds of risks have to be taken into account:

- Hazardous risks linked to the utilization of the working tool
- Hazardous risks that emerge at the workplace due to interaction of the working tools with each other
- Hazardous risks caused by the working materials or the working environment.

Here we have to conduct a risk analysis of the working tool at the predetermined workplace. We have to set this analysis in relation to other working mediums and take all operating conditions (i.e., normal mode, malfunction mode, and reconnection) into consideration.

The following rules have to be followed when designing for safety [34.43]:

- Elimination or minimization of hazardous risks (i.e., indirect measures, for example, integration of a safety concept when developing or constructing a machine)
- Implementing safety measures against risks that cannot be eliminated (i.e., indirect measures)
- Supply of information to users about remaining risks due to the incomplete effectiveness of the implemented safety measures; indication of eventual

Fig. 34.17 Disjunctive safety device at a palletizing automaton

necessary special training and personal protection equipment (i.e., information measures).

Risk Minimization of Mechanical Hazards. Mechanical hazards are, for example, objects that fall or slide out, dangerous surfaces and forms, moveable parts, instability or material failure. Appropriate protective measures have to been chosen for risk minimization of mechanical hazards. Here we have to differentiate between inherently safe construction and technical safety measures.

Inherently safe construction minimizes or eliminates risks through an adequate design. Examples of inherently safe constructions are the following:

- Avoidance of sharp edges
- Consideration of geometric and physical factors (for example, limitation of amount, speed, energy, and noise)
- Electrical supply of energy at extra-low voltage.

Technical safety measures should be used when the inherently safe constructions are not sufficient enough. Technical safety measures differ between disjunctive and nondisjunctive appliances [34.51].

A separating safety device is part of a machine which is used as a special kind of bodily shield (Fig. 34.17). Depending on its construction, a separating safety measure can be a cabinet, cover, shield, door, casing, etc. The following two devices are examples of separating safety devices [34.40]:

- *Fixed separating safety device:* These are either lasting (for example welded) or fixed to the machine with the help of elements (for example, a bolt). They block access to a dangerous area.
- *Movable separating safety device:* These are mostly mechanical in nature and connected to the machine (by hinges, for example). They can be opened even without the utilization of tools. A touch-sensitive switch prevents the execution of dangerous machine functions while the safety device is open.

The physical barrier between man and the hazardous machine function is absent in the case of a nonseparating safety device. The worker will be recognized as soon as he enters or reaches into the danger area. As soon as the worker is recognized, the existing risk is minimized or eliminated. We can differentiate between the following nonseparable safety devices:

- *Control device with automatic reset device:* Control devices that start and maintain operation of mechan-

Fig. 34.18 Safety measure with approximation function (light barrier) at a partly automated assembly area

ical parts only as long as a hand control or operating element is actuated. Examples are hand drills and an edge grinder:

- *Two-hand coupling:* Control devices that demands the use of both hands at the same time in order to start or maintain the machine's functions.
- *Protective device with approximation function:* Devices that stop hazardous mechanical parts as soon as a person or a part of the human body crosses a well-defined boundary (for example, light barrier, see Fig. 34.18).

Man–Machine Interface. Switches for the operation mode and the removal of the energy supply are applied when designing a safe man–machine-interface (Sect. 34.1.3). Consequently, we find the following specific design features:

- *Switch for operation mode:* Machines have to be reliably stopped in all functions or safety levels (for example, maintenance, inspection).
- *Energy supply:* The spontaneous restart of a machine after a breakdown of energy supply has to be avoided, if this implies danger to the worker.

When designing a man–machine interface for safety, displays have to be adjusted to human perceptive senses so that relevant signals, warnings or information can be quickly and reliably noticed. In order to evaluate a situation, man allocates the role of key elements to a multiplicity of sensations. These key elements are incorporated into superior strategies. If complex information is arranged in functional groups, decision-making processes will be accomplished in a better way. Man, for example, recognizes qualitative changes of process parameters faster and is also able to assess them better if they are presented in graphical patterns. The combination of graphical elements produces a pattern in which changes can be recognized right away. In the case of a change in the pattern, man can selectively change to the next deeper level of information processing, in which detailed information about the situation and other characteristics relevant to the decision such as quantitative measures is available. A superior graphical model functions as an early warning system.

34.4 Emerging Trends and Prospects for Automation

34.4.1 Innovative Systems and Their Application

In the light of technical innovation it can be expected that automation of the value-creation processes will increase in all branches and application areas. When used adequately, automation can contribute to the improvement of product quality, the increase of productivity, and the enhancement of quality of work [34.52]. Without automation technology, some fields of human work would not be accomplishable. The operation of complex control systems in modern aircrafts may exemplify this circumstance.

With the help of efficient methods and procedures of automation technology, new systems and products can be obtained. Flexible automation, rather then the highest degree of automation, is the aim. Characteristics of these systems and products are increasing complexity and decentralization, a higher degree of cross-linking, and dynamic behavior. However, they are only controllable with the help of automated appliances. Example products can be found in the field of mechatronics. As well as in the traditional application areas in the field of process and production automation, automated solutions are also frequently used in the service industries, maintenance areas, and the field of leisure activities

Fig. 34.19 Care-O-bot is a supporting system for the care of elderly people (photo taken with permission of Fraunhofer IPA)

(Fig. 34.19). Appliances can be found in the following fields:

- Health, e.g., intelligent prosthesis
- Service industry, e.g., fueling, cleaning, and household robot
- Building industry, e.g., energy optimizing systems
- Biotechnology, e.g., monitoring systems
- Technology of mobile systems, e.g., digital assistants.

An automated product has to fulfill strong requirements regarding precision and reliability, utilization characteristics, handling, and cost–value ratio [34.53]. Further development of all technical innovations demands the central inclusion of man. All technology should be geared to the human being, and his demands and performance prerequisites. The development of methods and procedures for ambient intelligence [34.54] is associated with human-oriented design (ambient intelligence is a technological paradigm which, first of all, is connected to the European research program on Information Society Technologies. It is related to the more hardware-oriented approach of the US-American research project on Ubiquitous Computing, as well as the industrial concept of pervasive computing). Ambient intelligence includes the vision of the information society, which stresses the factors of usability and efficiency of user support with the help of intelligent systems, in order to facilitate interaction between men or between man and computer.

One example application area is the *intelligent house*, whose entire functions and appliances (e.g., heat, kitchen appliances, and shutters) can be operated by a computer and be adjusted to the inhabitants' needs.

Natural and demand-oriented forms of interaction with an intelligent environment should lead to the situation in which utilization of the computer does not require more attention than the fulfillment of other daily activities, such as walking, eating, or reading.

The technical core of ambient intelligence is the omnipresence of information technology and consequently unlimited access to information and performance measures at any time, from any location. Man will thus be surrounded by a multitude of artifacts with intelligent and intuitive usable interfaces – from household objects used daily to facilities in public spaces. They will be able to recognize people and react actively to their informational needs in a discrete and fast way [34.55]. It is necessary that the ambient intelligence system registers the users' presence and situation and reacts to their needs, habits, gestures, and emotions in a sensitive and adaptive way.

Hence, ambient intelligence systems differ from present computer-based man–machine systems, whose technology is mostly focused on the realization of the task and forces the user to adjust to these requirements.

Ambient intelligence unifies a multiplicity of complex technologies and methods from different disciplines, such as: multisensor ad hoc networks, social user interfaces, dynamic integration of components for speech, and gesture recognition, and invisible computing. Further development of ambient intelligence systems requires intensive cooperation of these disciplines. Ambient intelligence always puts the human being with all his capabilities and needs at the center of the technologies that need to be developed.

34.4.2 Change of Human Life and Work Conditions

Continuous automation of different fields of life – first and foremost at work – leads to significant changes of human habits, task contents, and qualification requirements.

In production, workplaces that have been replaced by automation technology now demand higher and different qualifications from its workers. On the one hand, we have to assume that repetitious use, charging, disposal, and control functions, as well as administration and formalization of control tasks will continue to decrease. On the other hand, it is likely that maintenance work, demanding control functions, goal-oriented tasks, programming, and analytical tasks will increase [34.56].

As the importance of intellectually demanding tasks at work increases, so will demands on the workers, and stress will change. In the future, man will take a more social, and communicative role.

As the level of automation increases, so does its responsibility for business economics. The amount of instrumentation necessary for production will increase.

From an economical point of view it is important to achieve multilayered efficiency for this productive technology. As a result, the habits of workers doing shift work and their rhythm of life have to change. The demands of the level of organization and the peculiarity of mutual responsibility will also increase correspondingly.

Different forms of automation development can create long-lasting effects that will diminish the separation between *white-collar* and *blue-collar* work that has existed to date. In so doing, they overcome the fixed and sometimes rigid separation of functions when working. The resulting changes regarding the content and execution of work comprise an individual chance for development for the working human.

References

34.1   H.-J. Bullinger: *Ergonomie – Produkt- und Arbeitsplatzgestaltung* (Teubner, Stuttgart 1994), in German

34.2   H. Luczak, W. Volpert, A. Raeithel, W. Schwier: *Arbeitswissenschaft, Kerndefinition – Gegenstandsbereich – Forschungsgebiete* (RKW Eschborn, Edingen-Neckarhausen 1987), in German

34.3   ISO 9241: Design ergonomischer Benutzerschnittstellen (Ergonomics of human-system interaction) in German

34.4   H.-J. Bullinger, M. Braun: Arbeitswissenschaft in der sich wandelnden Arbeitswelt. In: *Erträge der interdisziplinären Technikforschung*, ed. by G. Ropohl (Schmidt, Berlin 2001) pp. 109–124, in German

34.5   H. Luczak: *Arbeitswissenschaft*, 2nd edn. (Springer, Berlin 1998), in German

34.6   J.-H. Kirchner: Das Arbeitssystem. In: *Handbuch Arbeitswissenschaft*, ed. by H. Luczak, W. Volpert (Schäffer-Poeschel, Stuttgart 1997) pp. 606–608, in German

34.7   K.-P. Timpe: Mensch-Maschine-System. In: *Handbuch Arbeitswissenschaft*, ed. by H. Luczak, W. Volpert (Schäffer-Poeschel, Stuttgart 1997) pp. 609–612, in German

34.8   K.-F. Kraiss: *Anthropotechnik in der Fahrzeug- und Prozessführung* (Rheinisch-Westfälische Technische Universität, Aachen 2005), in German

34.9   T. Sheridan: *Telerobotics, Automation, and Human Supervisory Control* (MIT Press, Cambridge 1992)

34.10 B. Lotter: Hybride Montagesysteme. In: *Montage in der industriellen Produktion*, ed. by B. Lotter, H.-P. Wiendahl (Springer, Heidelberg 2006) pp. 193–217, in German

34.11 H. Charwat: *Lexikon der Mensch-Maschine-Kommunikation* (Oldenbourg, München 1994), in German

34.12 K.-F. Kraiss: Benutzergerechte Automatisierung. Grundlagen und Realisierungskonzepte, Automatisierungstechnik **46**(10), 457–467 (1998), in German

34.13 H.-J. Bullinger: Technologische Ansätze. In: *Handbuch Arbeitswissenschaft*, ed. by H. Luczak, W. Volpert (Schäffer-Poeschel, Stuttgart 1997) pp. 82–86, in German

34.14 D. Spath, M. Weck, G. Seliger: Produktionssysteme. In: *Betriebshütte Produktion und Management*, 7th edn., ed. by W. Eversheim, G. Schuh (Springer, Berlin 1996) pp. 10.1–10.36, in German

34.15 G. Bretthauer, W. Richter, H. Töpfer: *Automatisierung und Messtechnik* (Mittag, Maria Rain 1996), in German

34.16 Y. Hauss, K.-P. Timpe: Automatisierung und Unterstützung im Mensch-Maschine-System. In: *Mensch Maschine Systemtechnik – Konzepte, Modellierung, Gestaltung, Evaluation*, ed. by K.-P. Timpe, T. Juergensohn, H. Kolrep (Symposion, Düsseldorf 2000) pp. 41–62, in German

34.17 H. Wandke, E. Wetzenstein-Ollenschläger: Assistenzsysteme: woher und wohin?, Proc. 1st Annu. GC-UPA Track (Stuttgart 2003), in German

34.18 W. Rohmert: Das Belastungs-Beanspruchungs-Konzept, Z. Arbeitswiss. **38**(4), 193–200 (1984), in German

34.19 R. Bokranz, K. Landau: *Einführung in die Arbeitswissenschaft* (Ulmer, Stuttgart 1991), in German

34.20 W. Quaas: Ermüdung und Erholung. In: *Handbuch Arbeitswissenschaft*, ed. by H. Luczak, W. Volpert (Schäffer-Poeschel, Stuttgart 1997) pp. 347–353, in German

34.21 R.D. Schraft, W. Schäfer: Die Automatisierungstechnik fordert miniaturisierte Systeme, Innov. Tech. neue Anwend. **6**(19), 4–5 (2001), in German

34.22 R.D. Schraft, R. Kaun: *Automatisierung der Produktion* (Springer, Berlin 1998), in German

34.23 R.D. Schraft: Neue Trends in der Automatisierungstechnik, Total. Integr. Autom. **3**(4), 32–35 (2004), in German

34.24 G. Lay, E. Schirrmeister: *Sackgasse Hochautomatisierung? Praxis des Abbaus von Overengineering in der Produktion* (Fraunhofer ISI, Karlsruhe 2000), in German

34.25 E. Helms, R.D. Schraft, M. Hägele: rob@work: Robot Assistant in Industrial Environments, Proc. 11th IEEE Int. Workshop Robot Hum. Interact. Commun. (ROMAN, Berlin 2002)

34.26 F. Lehmann, E. Ortner: Workflow-Systeme – ein interdisziplinäres Forschungs- und Anwendungsgebiet, Inform. **5**(2), 2–10 (1998), in German

34.27 D. Spath, M. Braun, P. Grunewald: *Gesundheits- und leistungsförderliche Gestaltung geistiger Arbeit* (Schmidt, Berlin 2003), in German

34.28 S. Baumgarth, E. Elmar Bollin, M. Büchel: *Digitale Gebäudeautomation* (Springer, Berlin 2004), in German

34.29 R. Schmidt-Clausen: *Verkehrstelematik im internationalen Vergleich. Folgerungen für die deutsche Verkehrspolitik* (Lang, Frankfurt 2004), in German

34.30 R. Gassner, A. Keilinghaus, R. Nolte: *Telematik und Verkehr. Elektronische Wege aus dem Stau?* (Nomos, Baden-Baden 1994), in German

34.31 Acatech (ed.): *Mobilität 2020. Perspektiven für den Verkehr von morgen* (Fraunhofer IRB, Stuttgart 2006), in German

34.32 C. Albus, B. Friede, F. Nicklisch, H. Schulze: Intelligente Transport-Systeme. Fahrer-Assistenz-Systeme, Z. Verkehrssicherh. **45**, 98–104 (1999), in German

34.33 C. Marberger: Adaptive Mensch-Maschine-Schnittstellen in Fahrzeugen, 52. Kongr. Ges. Arbeitswiss. (GfA, Dortmund 2006) pp. 79–82

34.34 K. Schattenberg: *Fahrzeugführung und gleichzeitige Nutzung von Fahrerassistenz- und Fahrerinformationssystemen*. Dissertation (Rheinisch-Westfälische Technische Hochschule, Aachen 2002) in German

34.35 T. Müller: Technologische und technische Arbeitsgestaltung. In: *Handbuch Arbeitswissenschaft*, ed. by H. Luczak, W. Volpert (Schäffer-Poeschel, Stuttgart 1997) pp. 579–583, in German

34.36 J. Lee: Human Factors and Ergonomics in Automation Design. In: *Handbook of Human Factors and Ergonomics*, 3rd edn., ed. by G. Salvendy (Wiley, New York 2006) pp. 1570–1596

34.37 M. Braun, L. Wienhold: Systematisierung betrieblicher Anforderungen an Arbeits- und Gesundheitsschutzinformationen. In: *Sicherheit und Gesundheit bei betrieblichen Entwicklungs- und Planungsprozessen*, ed. by H. Gebhardt, K.-H. Lang, B.H. Müller, M. Stein, R. Tielsch (Wirtschaftsverlag, Bremerhaven 2003) pp. 49–71, in German

34.38 J.-M. Hoc: From human-machine interaction to human-machine cooperation, Ergonomics **43**(7), 833–843 (2000)

34.39 W. Rohmert: Der Beitrag der Ergonomie zur Arbeitssicherheit, Werkstattstechnik, Z. Ind. Fert. **66**(1), 345–350 (1976), in German

34.40 P. Nicolaisen: *Sicherheitseinrichtungen für automatisierte Fertigungssysteme* (Hanser, München 1993), in German

34.41 G. Salvendy (ed.): *Handbook of Human Factors and Ergonomics*, 3rd edn. (Wiley, New York 2006)

34.42 G. Geiser: Informationstechnische Arbeitsgestaltung,. In: *Handbuch Arbeitswissenschaft*, ed. by H. Luczak, W. Volpert (Schäffer-Poeschel, Stuttgart 1997) pp. 589–594, in German

34.43 P. Kern, M. Schmauder, M. Braun: *Einführung in den Arbeitsschutz* (Hanser, München 2005), in German

34.44 B. Shneiderman, C. Plaisant: *Designing the User Interface: Strategies for Effective Human-Computer Interaction*, 4th edn. (Addison-Wesley, Boston 2005)

34.45 N. Bevan, M. MacLeod: Usability measurement in context, Behaviour & Information Technology **13**(1/2), 132–145 (1993)

34.46 M. Burmester: *Guidelines and Rules for Design of User Interfaces for Electronic Home Devices* (Fraunhofer IRB, Stuttgart 1997), ESPRIT-Project 6984

34.47 M. Decker: *Perspektiven der Robotik. Überlegungen zur Ersetzbarkeit des Menschen*, 2nd edn. (Europäische Akademie zur Erforschung von Folgen wissenschaftlich-technischer Entwicklungen, Bad Neuenahr-Ahrweiler 2001), in German

34.48 T. Sheridan: *Humans and Automation* (Wiley, New York 2002)

34.49 E. Assmann: *Untersuchung über den Einfluss einer Bremsweganzeige auf das Fahrverhalten*. Dissertation (Technische Universität, München 1985), in German

34.50 M. Braun: Leistungskompensation bei betrieblichen Störungen, Werkstattstech. online **94**(1/2), 2–6 (2004), in German

34.51 R. Skiba: *Taschenbuch Arbeitssicherheit* (Schmidt, Bielefeld 2000), in German

34.52 J. Rech, K.-D. Althoff: Artificial intelligence and software engineering: Status and future trends, Künstliche Intell. **18**(3), 5–11 (2004)

34.53 G. Bretthauer: Automatisierungstechnik – Quo vadis? Neun Thesen zur zukünftigen Entwicklung, Automatisierungstechnik **53**(4/5), 155–157 (2005), in German

Part D | 34

34.54 C. Ressel, J. Ziegler, E. Naroska: An approach towards personalized user interfaces for ambient intelligent home environments, 2nd IET Int. Conf. Intell. Environ. IE 06, Vol. 1 (Institution of Engineering and Technology, London 2006)

34.55 M. Friedewald, O. Da Costa, Y. Punie: Perspectives of ambient intelligence in the home environment, Telemat. Inform. **22**(3), 221–238 (2005)

34.56 D. Spath, M. Braun, L. Hagenmeyer: Human factors in manufacturing and process control. In: *Handbook of Human Factors and Ergonomics*, 3rd edn., ed. by G. Salvendy (Wiley, New York 2006) pp. 1597–1625

35. Machining Lines Automation

Xavier Delorme, Alexandre Dolgui, Mohamed Essafi, Laurent Linxe, Damien Poyard

This chapter deals with automation of machining lines, sometimes called transfer lines, which are serial machining systems dedicated to the production of large series. They are composed of a set of workstations and an automatic handling system. Each workstation carries out one identical set of operations every cycle time. The design of transfer lines is comprised of several steps: product analysis, process planning, line configuration, transport system design, and line implementation. In this chapter, we deal with line configuration. Its design performance is crucial for companies to compete in the market. The main problem at this step is to assign the operations necessary to manufacture a product to different workstations while respecting all constraints (i. e., the line balancing problem). The aim is to minimize the cost of this line while ensuring a desired production rate. After a review of the existing types of automated machining lines, an illustration of a developed

methodology for line configuration is given using an industrial case study of a flexible and reconfigurable transfer line.

Part D | 35

Manufacturers are increasingly interested in the optimization of their production systems. The objective is to optimize some criteria such as total investment cost, floor area, number of workstations, production rate, etc.

The automatic serial line, often called a transfer line, is a widely used production system in machining environments [35.1–6]. Transfer lines also exist in the assembly industry. Their properties are defined, for example, by *Nof* et al. [35.7]. In such a line, a repeatable set of operations is executed each cycle. The line is composed of sequentially arranged workstations and a transport system which ensures a constant flow of parts along the workstations. This automatic handling system is generally composed of conveyors fixed on rails that transfer the part from one station to the next with holder robots for part loading and unloading at sta-

tions. The transfer machining lines produce large series of identical or similar items.

Automation of a machining line for a given product family (or reconfiguration of an existing line for a new product family) is a significant investment, and requires a long period for its design (often 18 months). Manufacturers have to invest heavily when installing these lines or for their reconfiguration. This investment influences to a large extent the cost of the finished products over the lifetime of the line. Therefore, profitability depends directly on the success of the line design or reconfiguration. Investment cost should be minimized and the configuration obtained should be as efficient as possible. Thus, optimization is a crucial issue at the transfer line design or reconfiguration stage.

The design of transfer lines is comprised of several steps: product analysis, process planning, line configuration, transport system design, and line implementation. In this chapter, we deal with line configuration. Its design performance is crucial for companies to compete in the market.

As a rule, the configuration of a transfer line involves two principle steps:

1. choice of line type
2. logical synthesis of the manufacturing process, which consists of grouping the operations into stations (i. e., *line balancing*).

In this chapter, we focus on the second step of this procedure, because the decisions made there define the principal characteristics of the line. An error at this time is too costly to rectify. A brief description of this problem is in order.

Automated machining lines are composed of a set of serial workstations. The stations are visited in a given order. The line investment cost depends on the number of stations and equipment of each station. Both are defined via an assignment of operations to workstations. Usually, each task is characterized by: (1) its time, (2) a set of operations which must be assigned

before (precedence constraints), (3) a set of operations which must be executed on the same workstation (inclusion constraints), and (4) a set of operations which cannot be executed on the same workstation (exclusion constraints). Of course, in actual industrial problems, various additional specific constraints may have to be taken into account as well. Thus, at the line configuration stage, it is necessary to solve the line balancing problem, which consists of assigning the operations to workstations, minimizing the line investment cost while respecting the objective production rate as well as the aforementioned constraints.

This chapter is organized as follows. In Sect. 35.1, the fundamental assumptions and existing types of automated machining lines are introduced. In Sect. 35.2, some challenges and a general methodology for design and reconfiguration of these lines are explained. In Sect. 35.3, the role and importance of line balancing at the design or reconfiguration stage are presented. Section 35.4 illustrates our approach and models on an industrial case study. Moreover, in this section, a novel and promising exact resolution method is suggested for balancing of machining lines with parallel machines and setup times.

35.1 Machining Lines

A machining line is a production system composed of several sequential workstations; each workstation contains various machining equipment. A given set of operations is performed at each station to obtain the final product. The most frequent machining operations are:

- *Drilling*, to fabricate holes in parts
- *Milling* of shapes or removal of material with various milling cutters to form concave or convex grooves, etc.
- *Tapping*, which involves cutting internal screw threads in holes
- *Boring* to enlarge a hole that has already been drilled to precise dimensions.

A combination of these operations is usually needed to manufacture complex parts such as cylinder heads, cylinder blocks; see, for example, Fig. 35.1.

The machining process has numerous specific properties that directly influence the organization of the automated machining lines. Because of the complex-

ity involved, special studies and decision-aid tools are required for competitive design and reconfiguration of these lines.

There are three principal types of automated machining lines for large series, namely, dedicated, flexible, and reconfigurable transfer lines. Each of these has its own characteristics and assumptions, which will be briefly detailed in the following.

35.1.1 Dedicated Transfer Lines

A dedicated transfer line (DTL) is the most economic form of machining systems with a large productivity and profitability if there is enough volume. DTLs are used for the production of a single type of product (or close variants) in large series: a large quantity of identical products is manufactured with the same sequences of operations on stations. The stations are arranged serially. Each station is equipped with multispindle heads. Each multispindle head executes several operations simultaneously. Depending on the architecture of the line, spindle heads can be activated at each station in parallel

Fig. 35.2 A multispindle head for a dedicated transfer line (property of PCI/SCEMM)

- Simplifying handling operations reduces the needs for flow management and production control.

  Disadvantages of these lines are:

- The dedicated transfer lines demand large investments and must have a long lifetime to be profitable, the *ramp up* of the production is relatively long (2–4 weeks).
- Taking into account specific aspects of the product during the line design stage is possible, but once the line is defined it is very difficult to modify (line reconfiguration is costly).
- Breakdowns are a crucial problem; when a breakdown occurs in a single station, the entire line is stopped (in addition, if the corresponding operation did not end on time because of this breakdown, the product is automatically defective).

Note that for these lines the criterion to be optimized is easy to identify and calculate: minimizing the investment cost. Moreover, the interest in studying DTLs lies in the fact that their structure represents a basic form of organization for other machining systems. Indeed, all the problems that appear during the optimization of a dedicated transfer line are present in the design of other automated machining lines.

35.1.2 Flexible Transfer Lines

The flexible transfer line (FTL) is a special case of a flexible manufacturing system (FMS). The flexibility of a FMS is ensured thanks to the utilization of computer numerical control (CNC) machines (machining centers), automated transport, and warehousing systems with sophisticated control software [35.8]. An exam-

Fig. 35.1 Examples of parts produced by automated machining lines

or in sequence. An example of such a multispindle head is shown in Fig. 35.2.

When customer demand is significant and stable for a number of years, this type of machining line is the most profitable solution.

One of the design principles for the dedicated lines is the reduction of the cycle time and minimizing the amount of equipment (machines, spindle heads, tools, etc.) which, as mentioned above, has a direct influence on the reduction of the unit production cost.

Principal *advantages* of dedicated transfer lines are:

- High precision: these lines are designed to maximize the accuracy when machining of the part.
- Quality: there are no tool changes; therefore, once quality is established, it is stable.
- Mass production: the annual production can be in the millions.

Part D | 35.1

ple of a flexible machining center with the devices for change of tools is shown in Fig. 35.3.

FMS can produce several types of products, belonging to a broad family. By family we mean products having comparable dimensions and similar geometric characteristics, as well as the same tolerances. These related products can be manufactured by the same equipment. Software takes care of possible changes by reprogramming the machining or rescheduling the products to be manufactured.

There are three basic types of FMS [35.7]:

- Flexible lines: these consist generally of sequentially arranged workstations with programmable CNC machines (machining centers) and are used especially for products with several product variations and a short lifetime for each variation.
- Flexible cells: such a system is composed of disconnected programmable cells, where each cell consists of one or several machining centers and carries out

processes that comprise complete or almost complete tasks. The number of distinct parts in such a cell is often restricted, from eight to ten. This is due to a limited capacity of the cells.
- Flexible systems: are composed of linked flexible cells. There are two types of linkage: (1) with a rigid sequence of linking (cells are connected in a given invariable order); (2) with linkages that can be adapted to any particular production and/or assembly process.

The main objective of flexible transfer lines (FTLs) is to be able to produce several variations of the same product in large series. These lines assure a quick passage from one variation to another. FTLs are also able to change production volumes, if necessary, within a given range. The *ramp up* of production is short (1–2 days). However, they present a certain number of drawbacks:

1. These systems are very expensive mainly because they are composed of CNC machines. These machines are designed for a forecasted family of parts and produced without optimal process planning for each actual part. At the line design stage, the machining specifications are not accurately known. Therefore, the designer tends to insert more functions than necessary. Obviously, this increases the cost.
2. The development of software for the line control system is also very expensive because this flexible equipment requires sophisticated rules of management for each machine as well as for the entire line.
3. Contrary to dedicated machines, which contain multispindle heads with fixed tools, CNC machines use single spindle heads with frequent tool changes. Therefore, it is often difficult to maintain the level of precision in machining operations equal to that of dedicated lines.
4. Owing to rapid technological advances, these sophisticated and costly machines are quickly subject to obsolescence.
5. Because of their complexity, these flexible transfer lines tend to be less reliable.

35.1.3 Reconfigurable Transfer Lines

The concept of reconfigurable manufacturing systems (RMS) was introduced in *Koren* et al. [35.9]. The authors highlight the industry's new requirements for machining systems given the increasingly shorter product runs and the need for more customization. From the beginning, an RMS is designed to be able to make

Fig. 35.3 Machining center Meteor ML (property of PCI/SCEMM)

changes in its physical configuration to answer market fluctuations in both volume and type of product.

For RMS, and especially for reconfigurable transfer lines (RTL), the principal characteristics are:

- Modularity: in a reconfigurable manufacturing system, all the major components are modular (system, software, control, machines, and process). Selection of basic modules and the way they can be connected provide systems that can be easily integrated, diagnosed, customized, and converted.
- Integrability: to aid in designing reconfigurable systems, a set of system configurations and their integration rules must be established. Initially, such rules were developed for configurable computing [35.10]. In the machining domain, these rules should allow designers to relate clusters of part features and their corresponding machining operations to workstations and machine modules, thereby enabling product–process integration.
- Customization: this characteristic distinguishes RMS from FMS and DTL, and can reduce system and machine costs. This type of system provides customized flexibility for a particular part family, and is open ended.
- Convertibility: rapid changeover between members of the existing part family and quick system adaptability for future products.

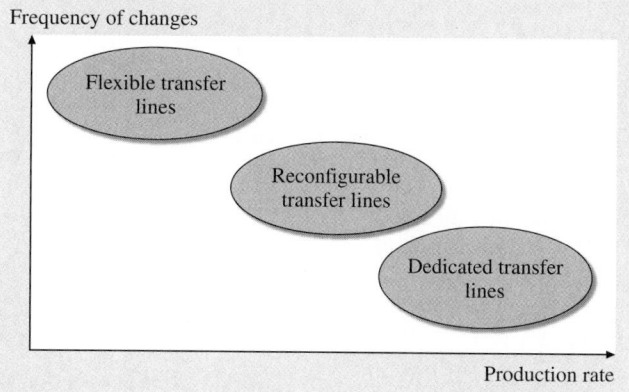

Fig. 35.4 Trade-off between production rate and frequency of market changes

- Diagnosability: detects machine failure and identifies causes of unacceptable part quality.

RTLs are usually conceived when there is little or no knowledge of future production volume or product changes. In this sense, they can be viewed as a compromise between the DTL and FTL (see Fig. 35.4). On the other hand, as they allow hardware reconfiguration in addition to software reconfiguration of FTL, some authors judge them more flexible.

35.2 Machining Line Design

We will now consider the preliminary design of transfer lines: corresponding challenges and general methodology.

35.2.1 Challenges

Usually for this type of project, the procedure is as follows: a company (client) contacts the transfer line manufacturer. The client gives the parts properties (part plans, characteristics, etc.) and the required output (production rate). Then comes the critical phase: the manufacturer should quickly offer a complete preliminary design solution for the corresponding line in terms of line architecture, number of machines, etc., and an approximate line cost. The acceptance of this solution by the client, and consequently the continuation of the negotiation and further development of the project, depends on the quality of this early solution. The temporal

progress of the negotiation process and its critical phase are illustrated in Fig. 35.5.

The manufacturer's objective is to reduce the preliminary design time while minimizing the cost of the potential line. This is decisive, due to strong competition among manufacturers in this domain. Moreover, these lines are technically very complex and require huge investments.

If a preliminary solution is more expensive than those of the competitors, then the contract (several hundred million euros) may be lost. If it is cheaper, then the manufacturer increases the chances of obtaining the contract. However, if the proposal is not feasible, because some constraints were not considered due to the lack of time, then this can generate additional costs for the manufacturer in correcting the solution. Therefore, this contract may be not profitable. Thus, the manufacturer is under a deadline to produce an initial feasible

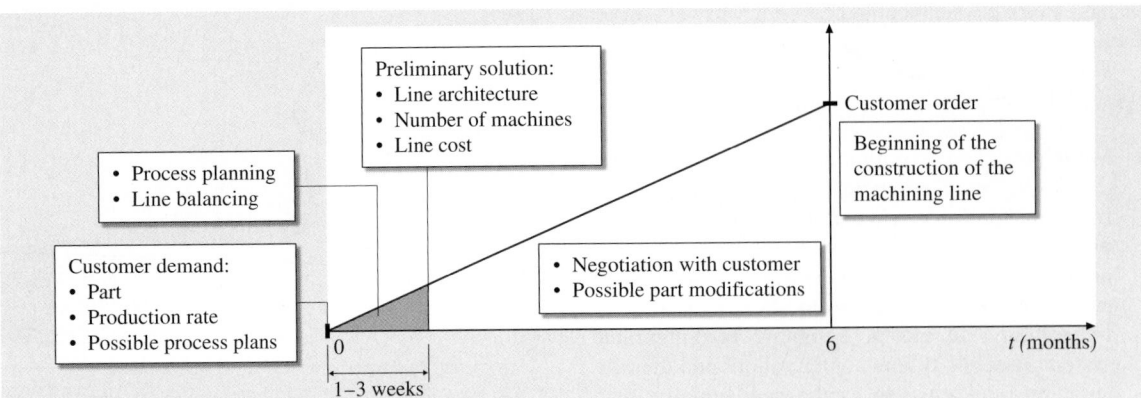

Fig. 35.5 Negotiation process: the critical phase

solution at the lowest possible cost within a very short time period.

In addition, after the preliminary design, almost always the product to be manufactured undergoes some modifications during the stage of detailed design of the line. The line manufacturer must continuously take into account these modifications. Furthermore, modifications of the design solution are difficult and time consuming. Therefore, decision-aid models are eminently useful for the preliminary design and to take into account the modifications during the detailed design. We will now present the methodology as applied to one such decision-making process.

35.2.2 General Methodology

Independent of the type of transfer line considered (dedicated, flexible or reconfigurable), its design demands an overall approach requiring the resolution of several interconnected problems [35.2]. Ideally, decisions relating to all these problems must be considered simultaneously. However, the total problem is very complex. Therefore, it is necessary to decompose this problem into several subproblems, each engendering less complex decisions [35.3].

Note that only the preliminary design stage is considered in this chapter, i.e., when all principal decisions are made concerning line architecture and its elements. Usually, this is followed by a detailed design (specifications for mechanical elements, tools, spindle heads, etc.), which is outside the scope of this chapter (see [35.4, 5] for a presentation).

The following general steps can summarize the preliminary design process. Note that the importance of each step depends on the type of transfer line considered. Some steps can be omitted.

- Product analysis: this gives a complete description of the operations that have to be executed for the future products.
- Process planning: covers the selection of processes required to transform raw parts into finished products. Here, technological constraints are defined. For instance, during process planning, partial order between operations, inclusion and exclusion constraints are established. This requires an accurate understanding of the functional specifications for the products and technological conditions for the operations.
- Configuration design and balancing problem: selection of the type of machining line and the resolution of the balancing problem, i.e., the allocation of operations to workstations in order to obtain the necessary production rate meeting demand while achieving the quality required. It is imperative to consider here all the constraints, particularly those of precedence.
- Dynamic flow analysis and transport system design: simulation is used to study the flow of products taking into account random events as well as variability in production. The objective is to analyze the dynamic flows and choose the material handling system as well as optimize the facilities layout, i.e., placement of machines. The decisions must be coherent with those defined at the previous steps.
- Detailed design and implementation of the line.

In addition, for flexible lines, a scheduling step also has to be considered [35.11]. After the implementation, if product and/or volume change, a similar analysis should be performed for the optimal reconfiguration of the transfer line (note that this is rarely considered for dedicated lines of mass production; more precisely, a reconfiguration of a dedicated line deals with specific engineering approaches). As illustrated in Fig. 35.6, these steps are executed sequentially. Of course, the designer can return to the previous steps as often as necessary (i. e., the decision-making process is iterative).

Such a methodology was already implemented in a decision-aid software tool for the preliminary design of dedicated transfer lines [35.12]. The developed software includes a database of parameterized features for product analysis. The product analysis provides a set of features which will be used at the process planning step. The process planning generates several process plans with the best one chosen for each feature. A set of operations and constraints are obtained. Then, a type of transfer line is selected by considering the process plans, part dimensions, required productiveness, cost of equipment, variability and longevity of market demand, etc. For the obtained process plans and production system type, the corresponding line balancing problem is solved. Finally, an estimation of the cost of the production system is made. If the solution or the cost is unsatisfactory, the designer can modify the data and constraints and restart the procedure.

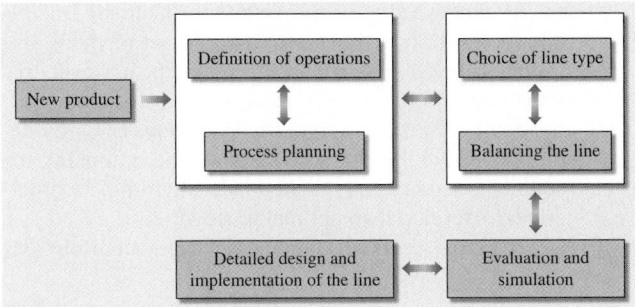

Fig. 35.6 Design of transfer lines

While this software was initially developed to design dedicated transfer lines, the general methodology is valid for all transfer lines (dedicated, flexible or reconfigurable). The difference is that for the dedicated transfer line it is applied once (at the preliminary design stage). In the case of flexible transfer lines, this tool should be used in real time, each time a new product is launched. For reconfigurable transfer lines, it is useful for each physical reconfiguration of the line.

This approach is based on a set of engineering procedures, knowledge-based constraints, and some optimization techniques for transfer line balancing. The optimization techniques are the core (and originality) of this methodology, which is why, the rest of this chapter will consider this aspect.

35.3 Line Balancing

As aforementioned, the line balancing (assignment of operation to workstations) is the key problem in the design of transfer lines.

Historically, the line balancing problem was first stated for assembly lines. As far as we know, the earliest publication on assembly line balancing (ALBP) was presented by *Salveson* [35.13]. Furthermore, exhaustive studies were made by several researchers in the last 50 years, with many interesting applications covered. One comprehensive state of the art has been presented in a special issue [35.14]. Several articles provide broad surveys of this problem; see, for example [35.15–20]. To summarize, the ALBP is NP-hard; see, for example [35.21]. Much research has been generated to solve the problem by developing approximate or exact methods [35.22–32].

The problem of machining line balancing is rather recent. This problem was mentioned in [35.33]. In *Dol-*

gui et al. [35.34], it was defined for dedicated transfer lines and first called transfer line balancing problem (TLBP).

Industry favors solving TLBP because the machining lines become too expensive otherwise. The TLBP consists of answering the following questions:

1. Which machining units are to be chosen to execute the required operations?
2. How many workstations are necessary?
3. How should the machining units be assigned to the stations?

These questions can be answered by an intelligent assignment of operations and machining units to workstations, minimizing the line cost while satisfying the objective production rate as well as respecting all other constraints.

Several exact and approximate (or heuristic) methods for TLBP have been proposed. Exact methods are useful to better understand the problem, however for large-scale problems they require excessive computation time. Contrarily, approximate methods can provide quicker results but do not guarantee the optimality of solutions. Additionally, a heuristic algorithm is often easier to develop than optimal procedures.

The most significant methods for an exact resolution of the TLBP are:

- Linear programming in mixed variables: the problem is modeled as a mixed integer program and solved with an optimization tool such as ILOG Cplex [35.35–37].
- Dynamic programming: a recursive method used for the resolution of problems having an additive objective function. Examples of this approach for TLBP are given in [35.33, 34, 38, 39], where the initial problems were transformed into constrained shortest-path problems and solved with appropriate algorithms.
- Branch and bound: an implicit enumerative procedure which avoids verifying all solutions. Several works use this approach for the resolution of the TLBP; see, for example, [35.40, 41].

Also, the column generation method can be used for TLBP. Indeed, it was already successfully used for assembly line balancing; see, for example, [35.42].

For large-scale problems, or when the allocated computing time is severely limited (e.g., for flexible transfer lines), several approximate methods have been designed. We classify these methods into two categories:

1. Heuristics based on priority rules derived from the methods for ALBP. There are several heuristic algorithms, which differ in the rule(s) used:
   - Ranked positioned weight (RPW) [35.22]): based on the weights of the operations calculated from their execution time and the operational times of their successors [35.43].
   - Computer method of sequencing operations for assembly lines (COMSOAL) [35.24]): solutions are generated by assigning operations randomly to the stations [35.44–47].
2. Metaheuristics, i.e., solving strategies applicable to a wide range of combinatorial optimization problems:
   - A multistart decomposition approach was suggested in [35.43, 48].

An example of machining line balancing via simulation can be found in [35.49].

Note that most of these methods were developed for dedicated transfer lines. In the next section, we will show how this approach can be applied to flexible and reconfigurable transfer lines. To illustrate, an industrial case study will be presented with a mixed integer programming model.

35.4 Industrial Case Study

35.4.1 Description of the Case Study

In Fig. 35.7, the machining line considered in this case study is presented. This line is designed to manufacture automotive cylinder heads. It is equipped with CNC machines (machining centers) for the output of 1250 parts per day. All the machines are identical (line modularity principle), with some exceptions. In contrast to dedicated transfer lines with multispindle machines, here, each machine contains one spindle and a magazine for tools. For each machine, to pass from one operation to the next it is necessary to consider an additional time due to tool changes and displacements or/and the rotation of the part (setup time). Taking into account the fact that a part is held at a machine with some fixtures in a given position (part fixing and clamping), some faces

and elements of the part are not accessible for machining even after part displacement or rotation. Whatever positioning and clamping are chosen some areas on the part will be hidden or covered. Therefore, the choice of a part position for part fixing should be also considered in the optimization procedure.

In Fig. 35.7, lines (1) represent the transport system composed of conveyors. Robots are used for part loading and unloading. The boxes (2) represent the CNC machines. Machines in a group aligned vertically represent a workstation. Then a workstation can comprise more than one machine; in this case, the same operations are duplicated and executed on different machines. With the parallel machines at each station, the line is easily reconfigurable. The line cycle time can be modified, if necessary, and even be shorter than the

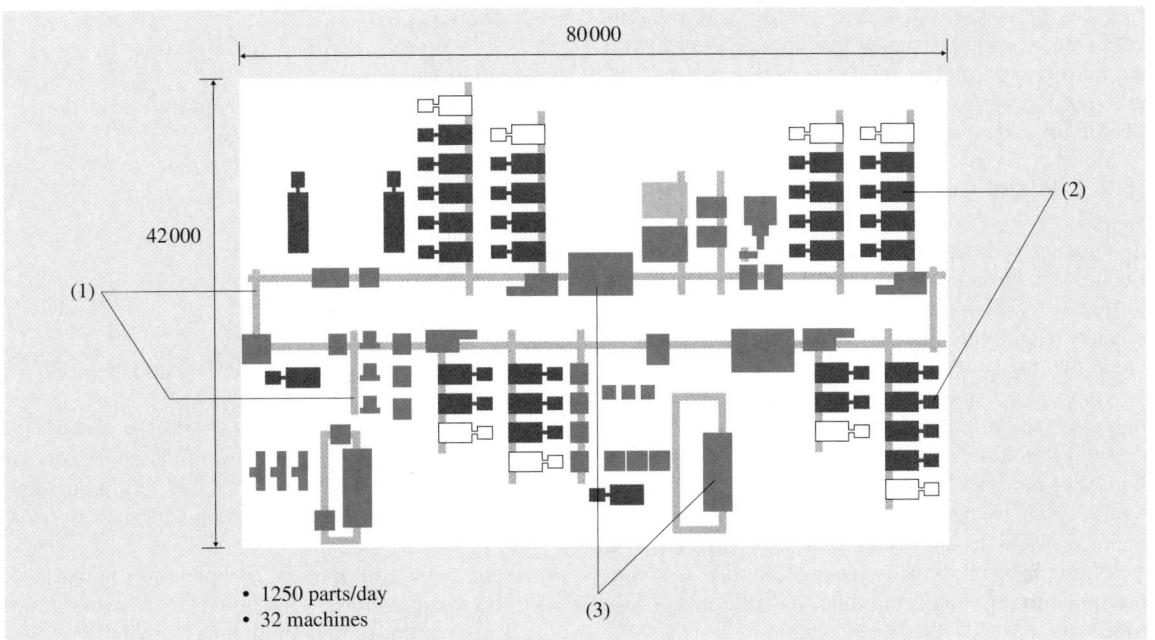

Fig. 35.7 Schema of a line for machining cylinder heads (PCI/SCEMM)

time of an operation. The boxes (3) represent dedicated stations for specific operations such as assembly or washing.

To help the designer of this line, we developed a model for line balancing. The input data used were:

- Cycle time (*takt* time) imposed by the objective production rate: one part is produced at each cycle.
- Precedence constraints: relations of order between operations. These relations define feasible sequences of operations.
- Inclusion constraints: the need to carry out fixed groups of operations on the same workstation.
- Exclusion constraints: the impossibility of carrying out certain subsets of operations at the same workstation.
- Accessibility constraints: these are related to the positioning of the part; indeed, for a position some part sides are not accessible, and thus operations on these sides cannot be carried out without repositioning. In the considered machining line, only one part fixing position is defined for each workstation (part repositioning occurs between two stations).
- Sequence-dependent setup times: the time required for the execution of two sequential operations is not

equal to the sum of their times but also depends on the order in which they are done, because the time needed for the displacement/change of tool and part rotation are not negligible.

- Parallel machines: at each workstation several identical CNC machines are installed. Thus, the local cycle time of the workstation is equal to the number of parallel machines multiplied by the line cycle time (*takt* time). The machines of the same workstation execute the same operations (in parallel on different product units).

Hence, here, we have a special case of line balancing with a sequential execution of operations, setup times, parallel machines, as well as accessibility, exclusion, and inclusions constraints.

The line of the case study can be regarded as reconfigurable. Indeed, while designed for the production of a single product, if there are changes on the product characteristics, the reconfiguration of this line is possible and easy thanks to:

- The use of standard and identical CNC machining centers, which simplifies the reallocation of operations to the workstations.
- At each station, machining centers can be added or eliminated as needed thanks to this modularity.

Now, we present a mixed integer programming (MIP) model for the design of this line for a given product. Furthermore, at the end of this section, we will give an extension of this model which can be used when reconfiguring the line for another product.

35.4.2 Mixed Integer Programming (MIP)

To summarize the optimization problem, we will enumerate its main assumptions.

The set of all operations N to be executed at the line is determined by the process plans for the product for which the line is designed. A part to be machined will pass through a sequence of workstations in the order of their installation. Each workstation is provided with at least one machine which carries out operations during the line cycle time. In the case where workload time of a workstation exceeds the line cycle time, parallel and identical machines are installed. In this case, the local cycle time is equal to the number of parallel machines multiplied by the line cycle time. All machines of the same station execute the same operations.

There are four types of additional constraints on the assignment of the operations (as detailed earlier), namely:

- Precedence constraints
- Exclusion constraints
- Inclusion constraints
- Accessibility constraints.

The time required for the execution of two operations is not equal to the sum of their times but depends on the sequence in which they are executed (Fig. 35.8).

The optimization problem consists of assigning operations to workstations to minimize the total number of machines on the line while respecting the given constraints.

Mathematical Model
We will introduce the following notations.

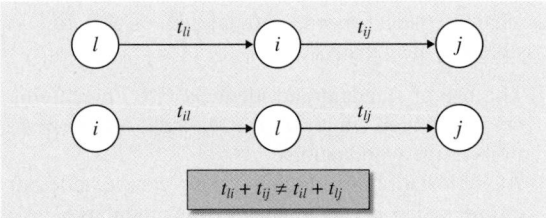

Fig. 35.8 Sequence-dependent setup times

Indexes:

- i, j for operations
- q for the place (order) of an operation in the sequence of assigned operations
- n for the number of parallel machines at a workstation
- k for the workstations
- a for the part fixing positions

Parameters:

- N, the set of operations to be assigned ($i, j = 1, \ldots, |N|$)
- A, the set of possible part positions for part fixing in a machining center; only one of these positions is chosen for each workstation; a part fixing position defines the accessibility constraints for the part ($a = 1, \ldots, |A|$)
- l_0, the maximum number of operations authorized to be assigned to a workstation: each workstation created cannot contain more than l_0 operations
- n_0, the maximum number of machines on a workstation
- m_0, the maximum number of workstations
- $q_0 = l_0 \cdot m_0$, the maximum number of possible assignments (places) for operations
- t_i, the operational time for operation i ($i = 1, \ldots, |N|$)
- t_{ij}, the setup time when operation j is processed directly after operation i at the same workstation
- T_0, the objective line cycle time (*takt* time)
- P_i, the set of direct predecessors of operation i
- P_i^*, the set of all predecessors of i (direct and indirect predecessors)
- F_i^*, the set of all successors of i (direct and indirect successors)
- ES, the collection of subsets e ($e \subset N$) of operations which must be imperatively assigned to the same workstation
- $\overline{ES}$, the set of pairs of operations (i, j) which cannot be assigned to the same workstation
- $A(i)$, the set of the possible part fixing positions for which the execution of operation i is possible
- $S(k)$, the set of possible places for operations at workstation k; this set is given by an interval of indexes; the maximum possible interval is $S(k) = \{l_0(k - 1) + 1, l_0(k - 1) + 2, \ldots, l_0 k\}$; $\forall k = 1, 2, \ldots, m_0$
- $K(i)$, the set of workstations on which operation i can be processed: $K(i) \subseteq \{1, 2, \ldots, m_0\}$

- $Q(i)$, the set of possible places for operation i in the sequence of all operations: $Q(i) \subseteq \{1, 2, \ldots, l_0 m_0\}$
- $N(k)$, the set of operations which can be processed at workstation k
- $M(q)$, the set of operations which can be assigned to the place q in the sequence
- E_i, the earliest workstation to which operation i can be assigned
- L_i, the last workstation to which operation i can be assigned

Variables:

- $x_{iq} = 1$, if operation i is in qth place (q is its order in the overall assignment sequence), otherwise $x_{iq} = 0$;
- τ_q, the setup time required between operations assigned to the same workstation in place q and $q+1$ (Fig. 35.9);
- $y_{nk} = 1$, if there are n parallel machines at the workstation k, 0 otherwise;
- $z_{ka} = 1$, if for the part of the workstation k the fixing position a is used, 0 otherwise.

Note that, if an operation is assigned to place q, it is the $q - (\lceil q/l_0 \rceil - 1) \cdot l_0$th operation of the workstation $\lceil q/l_0 \rceil$.

The optimization model is as follows:

- The objective function (35.1) minimizes the total number of machines

$$\text{Minimize} \sum_{k=1}^{m_0} \sum_{n=1}^{n_0} n \cdot y_{nk} . \tag{35.1}$$

- Equation (35.2) verifies that there is only one value for the number of parallel machines on each workstation

$$\sum_{n=1}^{n_0} y_{nk} \leq 1, \forall k = 1, 2, \ldots, m_0 . \tag{35.2}$$

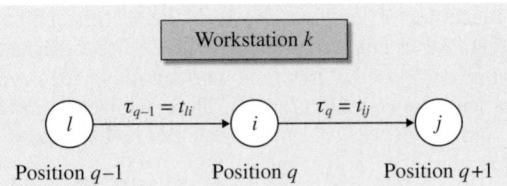

Fig. 35.9 Definition of the parameter τ_q

- Equation (35.3) assures that a workstation is open only if the preceding workstation is also open

$$\sum_{n=1}^{n_0} y_{nk} \geq \sum_{n=1}^{n_0} y_{n(k+1)} , \quad \forall k = 1, 2, \ldots, m_0 - 1 . \tag{35.3}$$

- Equation (35.4) assures that each operation i is assigned once and only once

$$\sum_{q \in Q(i)} x_{iq} = 1, \quad \forall i \in N . \tag{35.4}$$

- The constraints (35.5) assure that a place in the sequence is occupied by only one operation

$$\sum_{i \in M(q)} x_{iq} \leq 1, \quad \forall q = 1, 2, \ldots, q_0 . \tag{35.5}$$

- Equation (35.6) assures that an operation is assigned to a place only if another operation is assigned to the preceding place of the sequence (there is no empty place in the sequence of assigned operations)

$$\sum_{i \in M(q-1)} x_{i(q-1)} \geq \sum_{i \in M(q)} x_{iq} ,$$
$$\forall q \in S(k) \setminus \min\{S(k)\} , \quad \forall k = 1, 2, \ldots, m_0 . \tag{35.6}$$

- Equation (35.7) verifies that only one part fixing position is chosen for each workstation

$$\sum_{a \in A} z_{ka} \leq 1, \quad \forall k = 1, 2, \ldots, m_0 . \tag{35.7}$$

- Equation (35.8) assures that accessibility constraints are respected (the part fixing position chosen for a workstation authorizes the execution of every operation assigned to this station)

$$\sum_{q \in S(k)} x_{iq} \leq \sum_{a \in A(i)} z_{ka} , \quad \forall k = 1, 2, \ldots, m_0, \forall i \in N . \tag{35.8}$$

- Equation (35.9) calculates the additional time between operation i and operation j when operation j is processed directly after operation i at the same workstation

$$\tau_q \geq t_{ij} \cdot (x_{iq} + x_{j(q+1)} - 1) ,$$
$$\forall i \in M(q) , \quad \forall j \in M(q+1) ,$$
$$\forall q \in S(k) \setminus \max\{S(k)\} ,$$
$$\forall k = 1, 2, \ldots, m_0 . \tag{35.9}$$

- Equation (35.10) assures that the workload time of every workstation does not exceed the local cycle time, which corresponds to the number of installed parallel machines at this workstation multiplied by the objective cycle time of the line

$$\sum_{q \in S(k) \setminus \max\{S(k)\}} \tau_q + \sum_{i \in N(k)} \sum_{q \in S(k)} t_i \cdot x_{iq}$$
$$\leq T_0 \cdot \sum_{n=1}^{n_0} n \cdot y_{nk}, \quad \forall k = 1, 2, \ldots, m_0.$$
(35.10)

- Equation (35.11) defines the precedence constraints between operations

$$\sum_{q \in Q(j)} q \cdot x_{jq} \leq \sum_{q \in Q(i)} q \cdot x_{iq}, \forall i \in N, \quad \forall j \in P_i.$$
(35.11)

- Equation (35.12) represents the inclusion constraints

$$\sum_{q \in S(k) \cap Q(i)} x_{iq} = \sum_{q \in S(k) \cap Q(j)} x_{jq},$$
$$\forall i, j \in e, \quad \forall e \in ES, \quad \forall k \in K(i).$$
(35.12)

- Equation (35.13) represents the exclusion constraints

$$\sum_{q \in S(k)} (x_{iq} + x_{jq}) \leq 1,$$
$$\forall (i, j) \in \overline{ES}, \quad \forall k \in K(i) \cap K(j).$$
(35.13)

- Equations (35.14)–(35.17) provide additional constraints on the possible values of variables

$$\tau_q \geq 0, \quad \forall q = 1, 2, \ldots, q_0,$$
(35.14)
$$x_{iq} \in \{0, 1\}, \quad \forall i \in N, \quad \forall q \in Q(i),$$
(35.15)
$$y_{nk} \in \{0, 1\},$$
$$\forall n = 1, 2, \ldots, n_0, \quad \forall k = 1, 2, \ldots, m_0,$$
(35.16)
$$z_{ka} \in \{0, 1\}, \quad \forall k = 1, 2, \ldots, m_0, \quad \forall a \in A.$$
(35.17)

35.4.3 Computing Ranges for Variables

The model (35.1–35.17) can be solved using a standard operational research solver, for example, ILOG Cplex. Nevertheless, the calculation time is prohibitive. The resolution time for the model (35.1–35.17) can be greatly decreased using efficient techniques to reduce the number of variables (the size of the model) and

consequently to accelerate the search for an optimal solution.

We propose a technique for calculating bounds for the possible indexes for the variables of the mathematical model. This can simplify the problem and thus reduce the calculation time.

Taking into account the different constraints between operations, we can calculate the sets $K(i)$, $N(k)$, $S(k)$, $Q(i)$, and $M(q)$ more precisely. Note that these sets give intervals of possible values for the corresponding indexes.

The following additional notations can be defined:

- $E_i[r]$ is a recursive variable for the step by step calculation of the value of E_i taking into account setup times between operations, $r = 0, 1$.
- $L_i[r]$ is a recursive variable for the step by step calculation of the value of L_i taking into account setup times between operations, $r = 0, 1$.

With P_i^*, which is the set of all predecessors of operation i, and F_i^*, which is the set of all successors of operation i, we can also introduce:

- $Sp_i[r]$: the sum of the $(|P_i^*| - E_i[r] + 1)$ shortest setup times between the operations of the set $P_i^* \cup \{i\}$ composed of operation i and all its predecessors, $i \in N$
- $Sf_i[r]$: the sum of the $(|F_i^*| - m_0 + L_i[r])$ shortest setup times between the operations of the set $F_i^* \cup \{i\}$ composed of operation i and all its successors, $i \in N$
- $d[i, j]$: a parameter (distance) which has the following property: if (i, j) or $(j, i) \in \overline{ES}$, then $d[i, j] = 1$, else $d[i, j] = 0$.

The total operational time T_{sum} without considering the setup times between operations is calculated as follows

$$T_{\text{sum}} = \sum_{i \in N} t_i.$$

A lower bound on the number of workstations can be calculated by supposing that each workstation contains n_0 machines. Therefore, the local cycle time of each workstation is equal to $(T_0 \cdot n_0)$. The line becomes a serial line composed of identical workstations with a cycle time which is equal to $(T_0 \cdot n_0)$. Then, a lower bound on the number of workstations LB_{ws} can be calculated as follows

$$LB_{\text{ws}} = \lceil T_{\text{sum}} / (T_0 \cdot n_0) \rceil,$$

where the notation $\lceil x \rceil$ indicates the lowest integer value higher than or equal to x.

In the same way, a lower bound on the number of machines in the line (LB_m) can be determined by the following expression

$$LB_m = \lceil T_{sum}/T_0 \rceil .$$

Thus, the following procedure calculates the sets $K(i)$, $Q(i)$, $M(q)$, $N(k)$, and $S(k)$. Note that the operations are numbered in order of precedence graph ranks (in topological order). Some lines are annotated with comments. The symbol "//" is used to mark the beginning and the end of these comments.

Algorithm

Step 1 // step-by-step calculation of E_i and L_i, taking into account precedence constraints and setup times//

for all $i \in N$ **do**

  begin

  // calculate the earliest workstation $E_i[0]$ on which operation i can be processed taking into account the precedence constraints; note that an operation cannot be processed before its predecessors //

$$E_i[0] \leftarrow \left\lceil (t_i + \sum_{j \in P_i^*} t_j)/(n_0 \cdot T_0) \right\rceil ;$$

// calculate the latest workstation $L_i[0]$ on which operation i can be processed considering the precedence constraints, note that an operation cannot be processed after its successors //

$$L_i[0] \leftarrow m_0 - \left\lceil (t_i + \sum_{j \in F_i^*} t_j)/(n_0 \cdot T_0) \right\rceil + 1;$$

// calculate $E_i[1]$, which are new values of E_i obtained by taking into account in addition setup times between operations //

$$E_i[1] \leftarrow \left\lceil (t_i + Sp_i[0] + \sum_{j \in P_i^*} t_j)/(n_0 \cdot T_0) \right\rceil ;$$

// calculate $L_i[1]$, which are new values of L_i obtained by taking into account in addition setup times between operations //

$$L_i[1] \leftarrow m_0 - \left\lceil (t_i + Sf_i[0] + \sum_{j \in F_i^*} t_j)/(n_0 \cdot T_0) \right\rceil + 1;$$

// updating the values of E_i //

if $E_i[1] \neq E_i[0]$ **then** $E_i \leftarrow \max \Big(E_i[0] + 1,$

$$\left\lceil (t_i + Sp_i[1] + \sum_{j \in P_i^*} (t_j))/(n_0 \cdot T_0) \right\rceil \Big)$$

else $E_i \leftarrow E_i[1];$

// updating the values of L_i //

if $L_i[1] \neq L_i[0]$ **then** $L_i \leftarrow \min \Big(L_i[0] - 1,$

$$m_0 - \left\lceil (t_i + Sf_i[1] + \sum_{j \in F_i^*} t_j)/(n_0 \cdot T_0) \right\rceil + 1 \Big)$$

else $L_i \leftarrow L_i[1];$

end

Step 2 // step-by-step calculation of E_i, taking into account exclusion and inclusion constraints//

$j_{cur} \leftarrow 1;$

do

  $j_{min} \leftarrow j_{cur};$

  $j_{cur} \leftarrow |N|;$

  // new values of E_i are calculated by considering exclusion constraints //

  for $j \leftarrow j_{min} + 1, \ldots, |N|$ **do**

    $E_j \leftarrow \max \Big(\max_{i \in P_j^*}\{E_i + d[i, j]\}, E_j \Big);$

  for each $e \in ES$

    begin

    $E_e \leftarrow \max_{j \in e}(E_j);$

    for each $j \in e$ **if** $E_j < E_e$ **then**

      begin

      // new value of E_i is calculated, now taking into account an inclusion constraint//

      $E_j \leftarrow E_e;$

      $j_{cur} \leftarrow \min\{j_{cur}, j\};$

      end

    end

until $j_{cur} = |N|.$

Step 3 // step-by-step calculation of L_i, taking into account inclusion and exclusion constraints //

$j_{cur} \leftarrow |N|;$

do

  $j_{max} \leftarrow j_{cur};$

  $j_{cur} \leftarrow 1;$

  // new values of L_i are calculated by considering exclusion constraints //

  for $j \leftarrow j_{max} - 1, \ldots, 1$ **do**

    $L_j \leftarrow \min \big(\min_{i \in F_j^*}\{L_i - d[j, i]\}, L_j \big);$

  for each $e \in ES$

    begin

    $L_e \leftarrow \min_{j \in e}(L_j);$

    for each $j \in e$ **if** $L_j > L_e$ **then**

      begin

      // new values of L_i are again calculated, now taking into account inclusion constraints //

      $L_j \leftarrow L_e;$

      $j_{cur} \leftarrow \max\{j_{cur}, j\};$

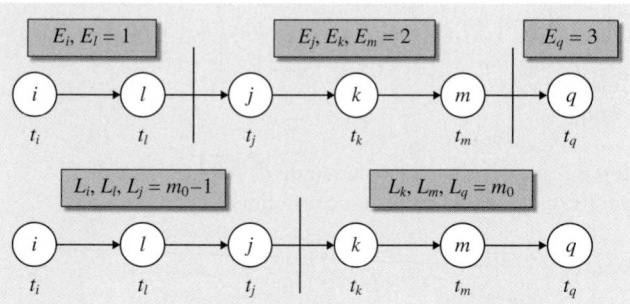

Fig. 35.10 An example of initial values for E_i and L_i

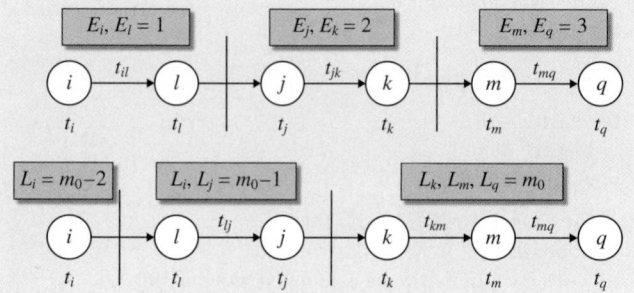

Fig. 35.11 Modified values of E_i and L_i taking into account setup times

           end
         end
   until $j_{\text{cur}} = 1$.

Step 4 // calculation of the sets $K(i)$, $N(k)$, $S(k)$, $Q(i)$, and $M(q)$ //
   for all $i \in N$ **do**
      $K(i) \leftarrow [E_i, L_i]$;
   for $k \leftarrow 1, 2, \ldots, m_0$ **do**

begin
   $N(k) \leftarrow \{i | i \in N, k \in K(i)\}$;
$$S(k) \leftarrow \left[1 + \sum_{k'=1}^{k-1} |S(k')|, \ \min\left(|N(k)|, l_0\right) \right.$$
$$\left. + \sum_{k'=1}^{k-1} |S(k')| \right];$$
end
for all $i \in N$ **do**
   $Q(i) \leftarrow \left[\min\{S(E_i)\}, \max\{S(L_i)\} \right]$;
for $q \leftarrow 1, 2, \ldots, \max\{S(m_0)\}$ **do**
   $M(q) \leftarrow \{i | q \in Q(i)\}$;
End of algorithm.

Some illustrations of the algorithm rules are presented in Figs. 35.10–35.13.

Numerical Example

In order to better explain the suggested algorithm, we present a numerical example with ten operations. Figure 35.14 shows the precedence graph and operational times.

The objective line cycle time is: $T_0 = 16$ units of time; the maximum number of stations $m_0 = 6$; the maximum number of machines to be installed on a station $n_0 = 3$; the maximum number of operations to be assigned to a station $l_0 = 8$.

The inclusion constraints are: $ES = \{(2, 4); (8, 9); (5, 6)\}$.

The exclusion constraints are: $\overline{ES} = \{(2, 7); (3, 4)\}$.

The setup times are reported in Table 35.1. For example, the setup time $t_{4,5} = 3$ corresponds to the time that is required to perform operation 5 immediately after operation 4.

The total operational time, $T_{\text{sum}} = \sum_{i \in N} t_i = 161$ units of time.

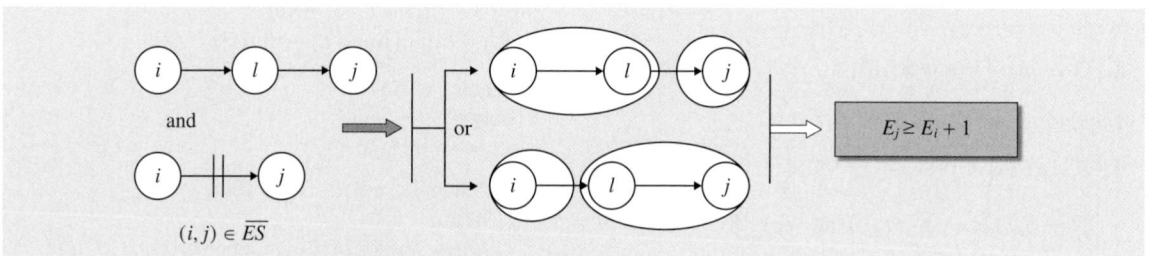

Fig. 35.12 An example of modifications of E_i by considering an exclusion constraint

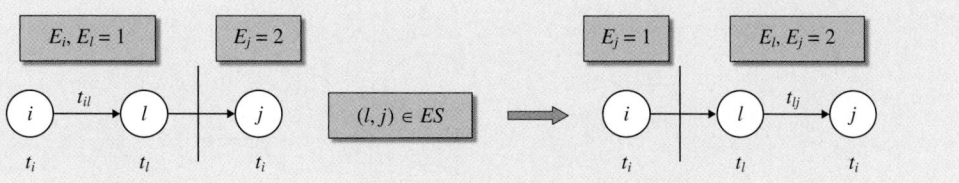

Fig. 35.13 An example of modification for E_i by taking into account an inclusion constraint

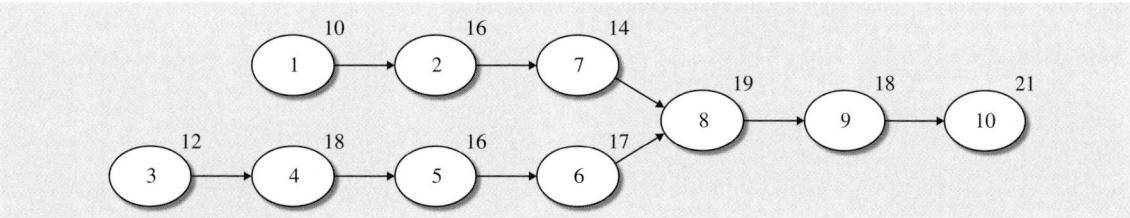

Fig. 35.14 Precedence graph

Table 35.1 Setup times

i \ j	1	2	3	4	5	6	7	8	9	10
1	–	4	4	3	2	1	1	2	1.5	1.5
2	5	–	1.5	1	1	2.5	3	3	3	3.5
3	1.5	3.5	–	2.5	1.2	3.4	4	4.2	3	2.2
4	4.5	4	3	–	3	1.5	3	2	4.5	1.8
5	4	4	5	5	–	2.5	2	2	4.5	1
6	25	1.5	3	1.2	1	–	2	2	3	4
7	4	3.8	3	1.8	4	3	–	4.7	4	1.4
8	3	2.5	4	3.4	3.2	2.4	1.6	–	2	4
9	3	4.9	4	2.3	1.6	3.6	3	1.2	–	3
10	1.5	4	1.5	4.6	3.7	2.2	2.7	1.2	4.8	–

A lower bound on the number of workstations is:
$LB_{ws} = \lceil T_{sum}/(T_0 \cdot n_0) \rceil = \lceil 161/(16 \cdot 3) \rceil = 4$.

Thus, the optimal solution cannot have fewer than four workstations.

A lower bound on the number of machines is:
$LB_m = \lceil T_{sum}/T_0 \rceil = \lceil 161/16 \rceil = 11$.

Then, the optimal solution cannot have fewer than 11 machines.

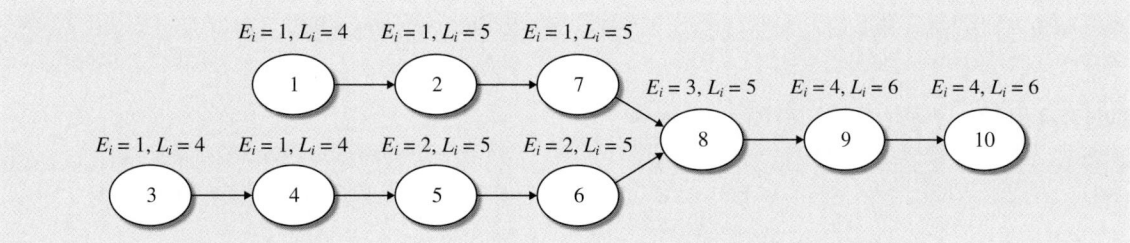

Fig. 35.15 Values of E_i and L_i obtained considering precedence constraints and setup times

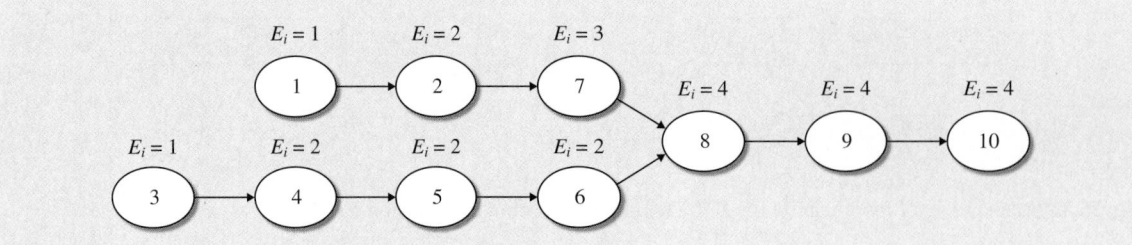

Fig. 35.16 Values of E_i considering exclusion and inclusion constraints

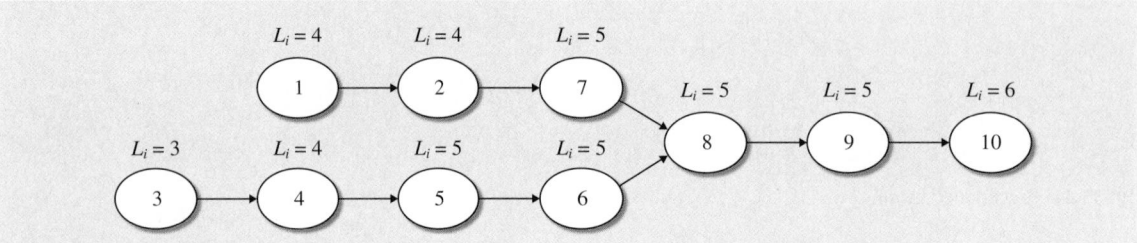

Fig. 35.17 Values of L_i considering exclusion and inclusion constraints

Now, the procedure of range calculation for indexes is applied:

- *Step 1* The initial values of E_i and L_i for each operation are calculated considering set-up times between operations and precedence constraints (Fig. 35.15).
- *Step 2* The new values of E_i are obtained by considering the exclusion and inclusion constraints (Fig. 35.16).
- *Step 3* The new values of L_i are calculated by considering the exclusion and inclusion constraints (Fig. 35.17).
- *Step 4*
  - Sets $K(i)$ for operations $i \in N$ are obtained (Table 35.2)

- Sets of operations $N(k)$ for stations $k = 1, 2, \ldots, m_0$ are defined (Table 35.3)
- Range of places $S(k)$ for operations of station k is calculated, $k = 1, 2, \ldots, m_0$ (Table 35.4)
- Finally, the range of places $Q(i)$ for operation i is found, for all $i \in N$ (Table 35.5).

35.4.4 Reconfiguration of the Line

As indicated at the beginning of this section, the studied line is reconfigurable. After the implementation of the line, if there are changes in the product characteristics or if there is a new product to be machined, the line can be reconfigured. Such a reconfiguration problem consists of reassigning operations to the stations

Table 35.2 The ranges $K(i)$ for the operations

Operation i	1	2	3	4	5	6	7	8	9	10
$K(i)$	[1,4]	[2,4]	[1,3]	[2,4]	[2,5]	[2,5]	[3,5]	[4,5]	[4,5]	[4,6]

Table 35.3 The set of operations $N(k)$ for each station

Station k	1	2	3	4	5	6
$N(k)$	{1,3}	{1,2,3,4,5,6}	{1,2,3,4,5,6,7}	{1,2,4,5,6,7,8,9,10}	{5,6,7,8,9,10}	{10}

Table 35.4 The ranges of places $S(k)$ for stations

Station k	1	2	3	4	5	6
$S(k)$	[1,2]	[3,8]	[9,15]	[16,23]	[24,29]	[30,30]

Table 35.5 The ranges of places $Q(i)$ for operations

Operation i	1	2	3	4	5	6	7	8	9	10
$Q(i)$	[1,23]	[3,23]	[1,15]	[3,23]	[3,29]	[3,29]	[9,29]	[16,29]	[16,29]	[16,30]

while minimizing the number of additional machines and/or those that move from a workstation to another in order to rebalance the line. This problem is similar to the design problem considered in the previous sections. Therefore, the proposed MIP model (35.1–35.17) can be easily adapted for this new problem.

The following modifications are made in the model (35.1–35.17): a new objective function is considered (35.1') along with additional constraints (35.18) and a new set of variables (35.19) which represents the gap between the number of machines for each station in the line before reconfiguration (y_{nk}^{Old}), and their number in the line reconfigured.

$$\text{Minimize} \sum_{k=1}^{m_0} \left(\delta_k^+ + \delta_k^- \right), \tag{35.1'}$$

$$\sum_{n=1}^{n_0} n \cdot y_{nk} - \sum_{n=1}^{n_0} n \cdot y_{nk}^{\text{Old}} = \delta_k^+ - \delta_k^-,$$
$$\forall k = 1, 2, \ldots, m_0, \tag{35.18}$$

$$\delta_k^+, \delta_k^- \geq 0, \quad \forall k = 1, 2, \ldots, m_0 \tag{35.19}$$

35.5 Conclusion and Perspectives

Transfer lines are used in many manufacturing domains, especially in machining systems, to efficiently effectuate high-quality and economical production. In today's competitive business environment, several manufacturers have opted for transfer lines to benefit from their advantages, namely precision, quality, productivity, reduction of handling cost, etc. However, transfer lines also present some drawbacks, such as requiring a large investment. Normally, transfer lines are highly automated, but the level of automation depends on the type of customer demand. Three types of transfer lines exist: dedicated, flexible, and reconfigurable. Dedicated lines are composed of workstations with multispindle heads. Flexible transfer lines have several types of CNC machines. Reconfigurable lines offer a mix of different types of machines (special machines, CNC machines, machining units, etc.) and can have different architectures (simple line, U-line, parallel stations, etc.).

With increasing technological progress and development of ever more sophisticated and efficient machining equipment, the problem of automated machining line design is exceptionally pertinent. Indeed, the concepts for machining lines are continuously improved through the development of new types of architectures and machines. Unfortunately, there is a gap between industrial cases and research problems treated. In contrast with assembly systems, in the domain of machining lines, the gap is often due to the lack of collaboration between the industrial and academic worlds.

In this chapter, written jointly by academic (Ecole des Mines de Saint Etienne) and industrial (PCI/SCEMM) partners, a general overview of the automated machining lines is presented. The principal characteristics of these lines and a general methodology for their design are introduced. This methodology is valid independent of the type of the line: dedicated, flexible or reconfigurable. The goal is to help machining line manufacturers to design efficient lines and become more competitive.

On receipt of a customer demand for a line which includes a part description (plans, characteristics, etc.) and the required output, the machining line manufacturer must be able to propose a complete solution within a very short time interval. This preliminary solution concerns the line architecture, number of machines, and equipment, with a line cost evaluation. A major difficulty deals with line balancing, which is a hard combinatorial optimization problem. All types of transfer lines are concerned. In this chapter, a short survey of general line balancing approaches is given. The methods developed for the balancing of the automated machining lines are enumerated and commented. Then, an industrial case study is presented. It illustrates and highlights the importance of the line balancing problem. Afterwards, a mixed integer program for the considered case is proposed. The presented model and approach are useful from a practical perspective. They generate more appropriate preliminary solutions to customer needs within a very short timeframe. From an academic point of view, this is a new formulation of the line balancing problem with sequence-dependent setup times and parallel machines.

For future research work, beside the improvement of the models and resolution algorithms, numerous research perspectives have yet to be studied in this field. Among them, the combination of optimization methods

and discrete-event simulation seems promising. Simulation is a powerful method to illustrate and study the flow of material on the line and to determine the effect of its architecture on line reliability and performance. Also, the development of interactive and iterative software could provide useful decision-aid systems for industry.

References

35.1   M.P. Groover: *Automation, Production Systems and Computer Integrated Manufacturing* (Prentice Hall, Eaglewood Cliffs 1987)

35.2   R.G. Askin, C.R. Standridge: *Modeling and Analysis of Manufacturing Systems* (Wiley, New York 1993)

35.3   K. Hitomi: *Manufacturing System Engineering* (Taylor & Francis, London 1996)

35.4   A.I. Dashchenko (Ed.): *Manufacturing Technologies for Machines of the Future: 21st Century Technologies* (Springer, Berlin, Heidelberg 2003)

35.5   A.I. Dashchenko (Ed.): *Reconfigurable Manufacturing Systems and Transformable Factories* (Springer, Berlin, Heidelberg 2006)

35.6   A. Dolgui, J.M. Proth: *Les systèmes de production modernes* (Hermes-Science, Paris 2006)

35.7   S.Y. Nof, W.E. Wilhelm, H.J. Warnecke: *Industrial Assembly* (Chapman Hall, London 1997)

35.8   A. Kusiak: *Modelling and Design of Flexible Manufacturing Systems* (Elsevier, Amsterdam 1986)

35.9   Y. Koren, U. Heisel, F. Javane, T. Moriwaki, G. Pritchow, H. Van Brussel, A.G. Ulsoy: Reconfigurable manufacturing systems, CIRP Ann. **48**(2), 527–598 (1999)

35.10   J. Villasenor, W.H. Mangione-Smith: Configurable computing, Sci. Am. **276**(6), 66–71 (1997)

35.11   G.W. Zhang, S.C. Zhang, Y.S. Xu: Research on flexible transfer line schematic design using hierarchical process planning, J. Mater. Process. Technol. **129**, 629–633 (2002)

35.12   A. Dolgui, O. Guschinskaya, N. Guschinsky, G. Levin: Decision making and support tools for design of machining systems. In: *Encyclopedia of Decision Making and Decision Support Technologies*, Vol. 1, ed. by F. Adam, P. Humphreys, (Idea Group, Hershey 2008) pp. 155–164

35.13   M.E. Salveson: The assembly line balancing problem, J. Ind. Eng. **6**(4), 18–25 (1955)

35.14   A. Dolgui (Ed.): Feature cluster on the balancing of assembly and transfer lines, Eur. J. Oper. Res. **168**(3), 663–951 (2006)

35.15   I. Baybars: A survey of exact algorithms for the simple assembly line balancing problem, Manag. Sci. **32**(8), 909–932 (1986)

35.16   S. Ghosh, R.J. Gagnon: A comprehensive literature review and analysis of the design, balancing and

scheduling of assembly line systems, Int. J. Prod. Res. **27**, 637–670 (1989)

35.17   E. Erel, S.C. Sarin: A survey of the assembly line balancing procedures, Prod. Plan. Control **9**(5), 414–434 (1998)

35.18   B. Rekiek, A. Dolgui, A. Delchambre, A. Bratcu: State of the art of assembly lines design optimisation, Annu. Rev. Control **26**(2), 163–174 (2002)

35.19   N. Boysen, M. Fliedner, A. Scholl: A classification of assembly line balancing problems, Eur. J. Oper. Res. **183**(2), 674–693 (2007)

35.20   N. Boysen, M. Fliedner, A. Scholl: Assembly line balancing: which model to use when?, Int. J. Prod. Econ. **111**, 509–528 (2008)

35.21   T.K. Bhattachajee, S. Sahu: Complexity of single model assembly line balancing problems, Eng. Costs Prod. Econ. **18**, 203–214 (1990)

35.22   W.P. Helgeson, D.P. Birnie: Assembly line balancing using the ranked positional weight technique, J. Ind. Eng. **12**, 394–398 (1961)

35.23   C.L. Moodie, H.H. Young: A heuristic method for assembly line balancing for assumption of constant or variable elements time, J. Ind. Eng. **16**, 23–29 (1965)

35.24   A.L. Arcus: COMSOAL: a computer method of sequencing operations for assembly lines, Int. J. Prod. Res. **4**(4), 259–277 (1966)

35.25   F.F. Boctor: A multiple-rule heuristic for assembly line balancing, J. Oper. Res. Soc. **46**, 62–69 (1995)

35.26   B. Rekiek, P. De Lit, A. Delchambre: Designing mixed-product assembly lines, IEEE Trans. Robot. Autom. **16**(3), 414–434 (1998)

35.27   A. Scholl: *Balancing and Sequencing of Assembly Lines* (Physica, Heidelberg 1999)

35.28   M. Amen: Heuristic methods for cost-oriented assembly line balancing, A comparison on solution quality and computing time, Int. J. Prod. Econ. **69**, 255–264 (2001)

35.29   M. Amen: Heuristic methods for cost oriented assembly line balancing, a survey, Int. J. Prod. Econ. **68**, 1–14 (2000)

35.30   J. Bukchin, M. Tsur: Design of flexible assembly line to minimize equipment cost, IIE Trans. **32**, 585–598 (2000)

35.31   J. Bukchin, A. Rubinovitz: A weighted approach for assembly line design with station paralleling and equipment selection, IIE Trans. **35**, 73–85 (2002)

35.32   C. Andrés, C. Miralles, R. Pastor: Balancing and scheduling tasks in assembly lines with sequence-dependent setup times, Eur. J. Oper. Res. **187**(3), 1212–1223 (2008)

35.33   J. Szadkowski: Critical path concept for multi-tool cutting processes optimization. In: *Manufacturing Systems Modeling, Management and Control: Proceedings of the IFAC Workshop*, ed. by P. Kopacek (Elsevier, Vienna 1997) pp. 393–398

35.34   A. Dolgui, N. Guschinski, G. Levin: On problem of optimal design of transfer lines with parallel and sequential operations, Proc. 7th IEEE Int. Conf. Emerg. Technol. Fact. Autom. (ETFA'99), Vol. 1, ed. by J.M. Fuertes (IEEE, Barcelona 1999) pp. 329–334

35.35   S. Belmokhtar: Lignes d'usinage avec équipements standards: modélisation, configuration et optimisation. Ph.D. Thesis (Ecole des Mines de Saint Etienne, Saint Etienne 2006), in French

35.36   S. Belmokhtar, A. Dolgui, N. Guschinsky, G. Levin: An integer programming model for logical layout design of modular machining lines, Comput. Ind. Eng. **51**(3), 502–518 (2006)

35.37   A. Dolgui, B. Finel, N. Guschinsky, G. Levin, F. Vernadat: MIP approach to balancing transfer lines with blocks of parallel operations, IIE Trans. **38**, 869–882 (2006)

35.38   A. Dolgui, N. Guschinsky, G. Levin: A Special case of transfer lines balancing by graph approach, Eur. J. Oper. Res. **168**(3), 732–746 (2006)

35.39   A. Dolgui, N. Guschinsky, G. Levin, J.M. Proth: Optimisation of multi-position machines and transfer lines, Eur. J. Oper. Res. **185**(3), 1375–1389 (2008)

35.40   A. Dolgui, I. Ihnatsenka: Branch and bound algorithm for a transfer line design problem: sta-

tions with sequentially activated multi-spindle heads, Eur. J. Op. Res. (2007) available online, doi:10.1016/j.ejor.2008.03.028, (in press)

35.41   A. Dolgui, I. Ihnatsenka: Balancing modular transfer lines with serial-parallel activation of spindle heads at stations, Discret. Appl. Math. **157**(1), 68–89 (2009)

35.42   W.E. Wilhelm: A column-generation approach for the assembly system design problem with tool changes, Int. J. Flex. Manuf. Syst. **11**, 177–205 (1999)

35.43   O. Guschinskaya, A. Dolgui, N. Guschinsky, G. Levin: A heuristic multi-start decomposition approach for optimal design of serial machining lines, Eur. J. Oper. Res. **189**(3), 902–913 (2008)

35.44   B. Finel: Structuration de lignes d'usinage: méthodes exactes et heuristiques. Ph.D. Thesis (Université de Metz, Metz 2004), in French

35.45   A. Dolgui, B. Finel, N. Guschinsky, G. Levin, F. Vernadat: An heuristic approach for transfer lines balancing, J. Intell. Manuf. **16**(2), 159–171 (2005)

35.46   B. Finel, A. Dolgui, F. Vernadat: A random search and backtracking procedure for transfer line balancing, Int. J. Comput. Integr. Manuf. **21**(4), 376–387 (2008)

35.47   M. Essafi, X. Delorme, A. Dolgui: A heuristic method for balancing machining lines with paralleling of stations and sequence-dependent setup times, Proc. Int. Workshop LT'2007 (Sousse 2007) pp. 349–354

35.48   O. Guschinskaya: Outils d'aide à la décision pour la conception en avant-projet des systèmes d'usinage à boîtiers multibroches. Ph.D. Thesis (Ecole des Mines de Saint Etienne, Saint Etienne 2007), in French

35.49   S. Masood: Line balancing and simulation of an automated production transfer line, Assem. Autom. **26**(1), 69–74 (2006)

Part D | 35

36. Large-Scale Complex Systems

Florin-Gheorghe Filip, Kauko Leiviskä

Large-scale complex systems (LSS) have traditionally been characterized by large numbers of variables, structure of interconnected subsystems, and other features that complicate the control models such as nonlinearities, time delays, and uncertainties. The decomposition of LSS into smaller, more manageable subsystems allowed for implementing effective decentralization and coordination mechanisms. The last decade revealed new characteristic features of LSS such as the networked structure, enhanced geographical distribution and increased cooperation of subsystems, evolutionary development, and higher risk sensitivity. This chapter aims to present a balanced review of several traditional well-established methods and new approaches together with typical applications. First the hierarchical systems approach is described and the transition from coordinated control to collaborative schemes is highlighted. Three subclasses of methods that are widely utilized in LSS – decentralized control, simulation-based, and artificial-intelligence-based schemes – are then reviewed. Several basic aspects of decision support systems (DSS) that are meant to enable effective cooperation between man and machine and among the humans in charge with LSS management and control are briefly exposed. The chapter concludes by presenting several technology trends in LSS.

There is not yet a universally accepted definition of the *large-scale complex systems* (LSS) though the *LSS movement* started more than 40 years ago. However, by convention, one may say that a particular system is a large and complex one if it possesses one or several characteristic features. For example, according to *Tomovic* [36.1], the set of LSS characteristics includes the structure of interconnected subsystems and the presence of multiple objectives, which, sometimes, are vague and even conflicting. A similar viewpoint is proposed by *Mahmoud*, who describes a LSS as [36.2]:

A system which is composed of a number of smaller constituents, which serve particular functions, share common resources, are governed by interrelated goals and constraints and, consequently, require more than one controllers.

Šiljak [36.3] states that a LSS is characterized by its high dimensions (large number of variables), constraints in the information infrastructure, and the presence of uncertainties. At present there are software products on the market which can be utilized to solve optimization problems with thousands of variables. A good example is Solver.com [36.4]. Complications may still be caused by system non-linearities, time delays, and different time constants, and, especially over recent years, risk sensitivity aspects.

36.1 Background and Scope

In real life one can encounter lots of natural, man-made, and social entities that can be viewed as LSS. From the early years of the *LSS movement*, the LSS class has included several particular subclasses such as: steelworks, petrochemical plants, power systems, transportation networks, water systems, and societal organizations [36.5–7]. Interest in designing effective control schemes for such systems was primarily motivated by the fact that even small improvements in the LSS operations could lead to large savings and important economic effects.

The structure of interconnected subsystems has apparently been the characteristic feature of LSS to be found in the vast majority of definitions. Several subclasses of interconnections can be noticed (Fig. 36.1).

First there are the *resource sharing* interconnections described by *Findeisen* [36.8], which can be identified at the *system level* as remarked by *Takatsu* [36.9]. Also, at the *system level*, subsystems may be interconnected through their common objectives [36.8]. Subsystems may also be interconnected through buffer units (tanks), which are meant to attenuate the effects of possible differences in the operation regimes of plants which feed or drain the stock in the buffer. This type of *flexible interconnection* can frequently be met in large industrial and related systems such as refineries, steelworks, and water systems [36.10]. The dynamics of the stock value s in the buffer unit can be modeled by a differential equation. In some cases buffering units are not allowed because of technological reasons; for example, electric power cannot be stocked at all and reheated ingots in steelworks must go immediately to rolling mills to be processed. When there are no buffer units, the subsystems are coupled through *direct interconnections*, at the *process level* [36.9].

In the 1990s, integration of systems continued and new paradigms such as the *extended/networked/virtual enterprise* were articulated to reflect real-life developments. In this context, *Mårtenson* [36.11] remarked that complex systems became even more complex. She provided several arguments to support her remark: first, the ever larger number of interacting subsystems that perform various functions and utilize technologies belonging to different domains such as mechanics, electronics, and *information and communication technologies* (ICT); second, that experts from different domains can encounter hard-to-solve communication problems; and also, that people in charge of control and maintenance tasks, who have to treat both routine and emergence situations, possess uneven levels of skills and training and might even belong to different cultures.

Nowadays, *Nof* et al. show that [36.12]:

There is the need to create the next generation manufacturing systems with higher levels of flexibility, allowing these systems to respond as a component of enterprise networks in a timely manner to highly dynamic supply-and-demand networked markets.

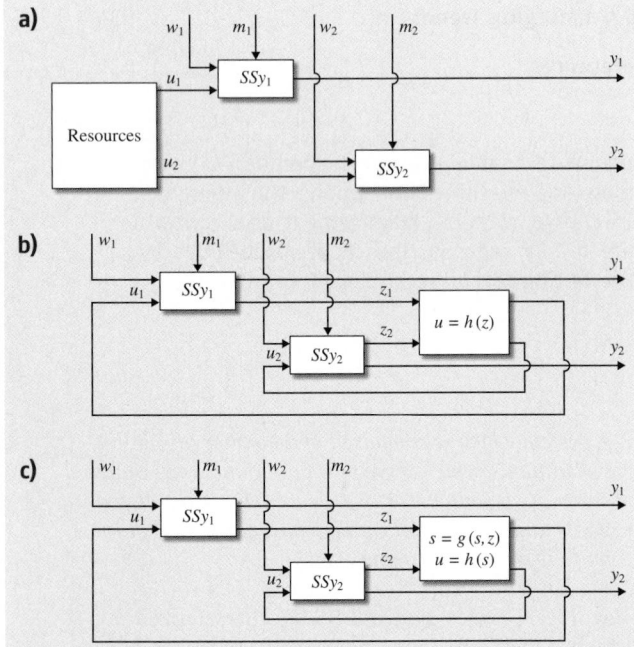

Fig. 36.1a–c Interconnection patterns: (**a**) resource sharing, (**b**) direct interconnection, (**c**) flexible interconnection; SSy = subsystem, m = control variable, y = output variable, w = disturbance, u = interconnection input, z = interconnection output, $g(\cdot)$ = stock dynamics function, $h(\cdot)$ = interconnection function

Table 36.1 Summary of methods described in this chapter

Decomposition-coordination-based methods	*Mesarovic, Macko, Takahara* [36.7]; *Findeisen* et al. [36.16]; *Titli* [36.17]; *Jamshidi* [36.6]; *Brdys, Tatjewski* [36.18]
Optimization-based methods	*Dourado* [36.19]; *Filip*, et al. [36.20]; *Filip* et al. [36.21]; *Guran* et al. [36.22]; *Peterson* [36.23]; *Tamura* [36.24]
Decentralized control	*Aybar* et al. [36.25]; *Bakule* [36.26]; *Borrelli* et al. [36.27]; *Inalhan* et al. [36.28]; *Krishnamurthy* et al. [36.29]; *Langbort* et al. [36.30]; *Šiljak* et al. [36.31]; *Šiljak* et al. [36.32]
Simulation-based methods	*Arisha* and *Yong* [36.33]; *Chong* et al. [36.34]; *Filip* et al. [36.20]; *Gupta* et al. [36.35]; *Julia* and *Valette* [36.36]; *Lee* et al. [36.37]; *Leiviskä* et al. [36.38, 39]; *Liu* et al. [36.40]; *Ramakrishnan* et al. [36.41]; *Ramakrishnan* and *Thakur* [36.42]; *Taylor* [36.43]
Intelligent methods	
• Fuzzy logic	*Ichtev* [36.44]; *Leiviskä* [36.45]; *Leiviskä* and *Yliniemi* [36.46];
• Neural networks	*Arisha* and *Yong* [36.33]; *Azhar* et al. [36.47]; *Hussain* [36.48]; *Liu* et al. [36.49]
• Genetic algorithms	*Dehghani* et al. [36.50]; *El Mdbouly* et al. [36.51]; *Liu* et al. [36.40]
• Agent-based methods	*Akkiraju* et al. [36.52]; *Hadeli* et al. [36.53]; *Heo* and *Lee* [36.54]; *Mařík* and *Lažanský* [36.55]; *Park* and *Lim* [36.56]; *Parunak* [36.57]

They also emphasize that *e-Manufacturing is highly dependent on the efficiency of collaborative human–human and human–machine e-Work*. See Chap. 88 on *Collaborative e-Work, e-Business, and e-Service*. In general, there is a growing trend to understand the design, management, and control aspects of complex *supersystems* or *systems of systems* (SoS). Systems of systems can be met in space exploration, military and civil applications such as computer networks, integrated education systems, and air transportation systems. There are several definitions of SoS, most of them being articulated in the context of particular applications; for example, *Sage* and *Cuppan* [36.13] state that a SoS is not a monolithic entity and possesses the majority of the following characteristics: geographic distribution, operational and management independence of its subsystems, emergent behavior, and evolutionary development. All these developments obviously imply ever more complex control and decision problems. A particular case which has received a lot of attention over recent years is large-scale *critical infrastructures* (communication networks, the Internet, highways, water systems, and power systems) that serve not only the business sector but society in general [36.14, 15]. All

of these recent developments are likely to provide fresh strong stimuli for new research in the LSS domain.

36.1.1 Approaches

The progresses made in information and communication technologies have enabled the designer to overcome several difficulties he might have encountered when approaching a LSS, in particular those caused by a large number of variables and the low performance (with respect to throughput and reliability) of communication links. However, as *Cassandras* points out [36.58]:

The complexity of systems designed nowadays is mainly defined by the fact that computational power alone does not suffice to overcome all difficulties encountered in analyzing, planning and decision-making in the presence of uncertainties.

A plethora of methods have been proposed over the last four decades for managing and controlling large-scale complex systems such as: decomposition, hierarchical control and optimization, decentralized control, model reduction, robust control, perturbation-based techniques, usage of artificial-intelligence-based techniques, integrated problems of system optimization and parameter estimation [36.59], and so on. Two common ideas can be found in the vast majority of approaches proposed so far:

a) Replacing the original problem with a set of simpler ones which can be solved with the available tools and accepting the satisfactory, near optimal solutions
b) Exploiting the particular structure of each system to the extent possible.

Table 36.1 presents a summary of the main methods to be described in this chapter.

36.1.2 History

Though several ideas and methods for controlling LSSs were proposed in the 1960s and even earlier, it is accepted by many authors that the book of *Mesarovic* et al. published in 1970 [36.7] triggered the *LSS move-ment*. The concepts revealed in that book, even though they were strongly criticized in 1972 by *Varaiya* [36.60] (an authority among the pioneers of the *LSS move-ment*), have inspired many academics and practitioners. A series of books including those of *Wismer* [36.61], *Titli* [36.17], *Ho* and *Mitter* [36.62], *Sage* [36.63], *Šiljak* [36.3, 64], *Singh* [36.65], *Findeisen* et al. [36.16], *Jamshidi* [36.6], *Lunze* [36.66], and *Brdys* and *Tat-jewski* [36.18] followed on and contributed to the consolidation of the LSS domain of research and paved the way for practical applications.

In 1976, the first International Federation of Automatic Control (IFAC) conference on *Large-Scale Systems: Theory and Applications* was held in Udine, Italy. This was followed by a series of symposia which were organized by the specialized Technical Committee of IFAC and took place in various cities in Europe and Asia (Toulouse, Warsaw, Zurich, Berlin, Beijing, London, Patras, Bucharest, Osaka, and Gdansk). The scientific journal *Large Scale Systems* published by North Holland played an important role in the development of LSS domain, especially in the 1980s.

36.2 Methods and Applications

36.2.1 Hierarchical Systems Approach

The central idea of the *hierarchical multilevel systems* (HMS) approach to LSS consists of replacing the original system (and the associated control problem) with a multilevel structure of *smaller* subsystems (and associated less complicated problems). The subproblems at the bottom of the hierarchy are defined by the *interventions* made by the higher-level subproblems, which in turn utilize the *feedback information* they receive from the solutions of the lower-level subproblems.

There are three main subclasses of hierarchies which can be obtained in accordance with the complexity of *description*, *control task*, and *organization* [36.7].

Levels of Description
The first step in analyzing an LSS and designing the corresponding control scheme consists of model building. As *Steward* [36.67] points out, practical experience witnessed there is a *paradoxical law of systems*. If the description of the plant is too complicated, then the designer is tempted to consider only a part of the system or a limited sets of aspects which characterize its behavior. In this case it is very likely that the very ignored parts and aspects have a crucial importance. Consequently it emerges that more aspects should be considered, but this may lead to a problem which is too complex to be solved in due time. To solve the conflict between the necessary *simplicity* (to allow for the usage of existing methods and tools with a reasonable consumption of time and other computer resources) and the acceptable *precision* (to avoid obtaining wrong or unreliable results), the LSS can be represented by a family of models. These models reflect the behavior of the LSS as viewed from various perspectives, called [36.7] *levels of description* or *strata*, or *levels of influence* [36.63, 68]. The description levels are governed by independent laws and principles and use different sets of descriptive variables. The lower the level is, the more detailed the description of a certain entity is. A unit placed on the n-th level may be viewed as a subsystem at level $n - 1$. For example, the same manufacturing system can be described from the top stratum in terms of economic and financial models, and, at the same time, by control variables (states, controls, and disturbances) as viewed from the middle stratum, or by physical and chemical variables as viewed from the bottom description level (Fig. 36.2).

Levels of Control

In order to act in due time even in emergency situations, when the available data are uncertain and the decision consequences are not fully explored and evaluated, a hierarchy of specialized control functions can be an effective solution as shown by *Eckman* and *Lefkowitz* [36.70]. Several examples of sets of levels of control are:

a) Regulation, optimization, and organization [36.71]
b) Direct control, supervisory control, optimization, and coordination [36.72]
c) Stabilization, dynamic coordination, static optimization, and dynamic optimization [36.8]
d) Measurement and regulation, production planning and scheduling, and business planning [36.73].

The levels of control, also called *layers* by *Mesarovic* et al. [36.7], can be the result of a time-scale decomposition. They can be defined on the basis of time horizons taken into consideration, or the frequency of disturbances which may show up in process variables, operation conditions, parameters, and structure of the plant as stated by *Schoeffler* [36.68], as shown in Fig. 36.2.

Levels of Organization

The hierarchies based on the complexity of organization were proposed in mid 1960s by *Brosilow* et al. [36.74] and *Lasdon* and *Schoeffler* [36.75] and were formalized in detail by *Mesarovic* et al. [36.7]. The hierarchy with several levels of organization, also called *echelons* by *Mesarovic* et al. [36.7], has been, for many years, a natural solution for management of large-scale military, industrial, and social systems, which are made up of several interconnected subsystems when a centralized scheme cannot be either technically possible or economically acceptable.

The central idea of the multiechelon hierarchy is to place the control/decision units, which might have different objectives and information bases, on several levels of a management and control pyramid. While the multilayer systems implement the vertical division of the control effort, the multiechelon systems include also a horizontal division of work. Thus, on the n-th organization level the i-th control unit, CU_i^n, has limited autonomy. It sends *coordination* signals downwards to a well-defined subset of control units which are placed at the level $n-1$ and it receives coordination signals from the corresponding unit placed

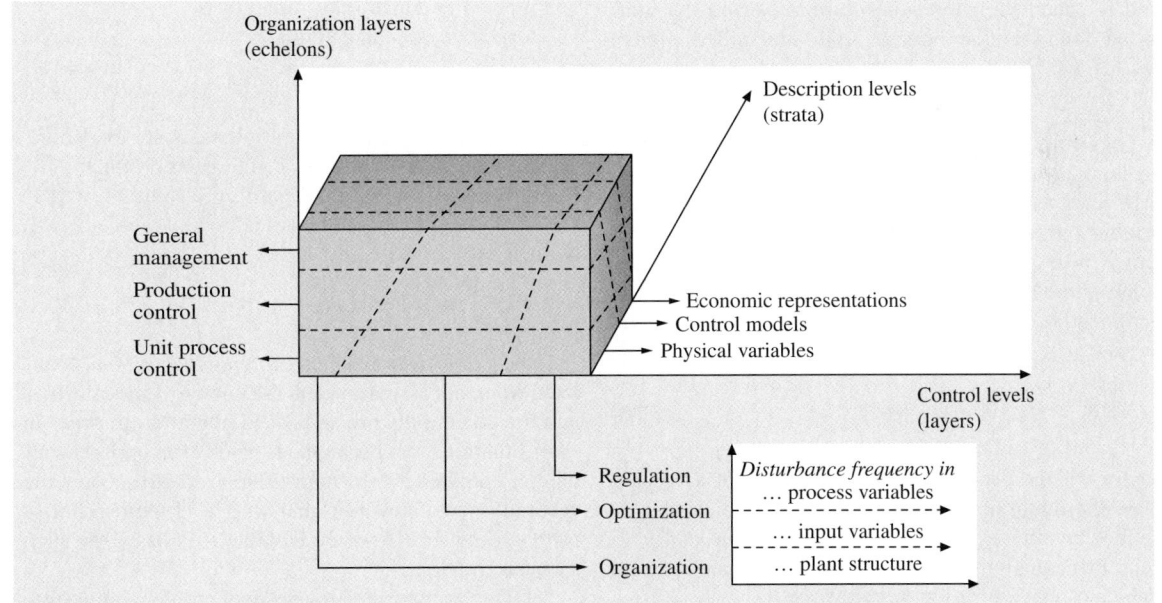

Fig. 36.2 A hierarchical system approach applied to an industrial plant (after [36.69])

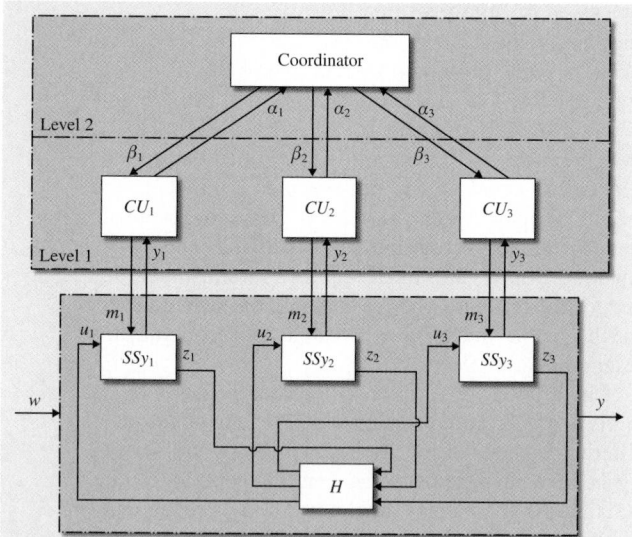

Fig. 36.3 A simple two-level multilevel control system ($CU =$ control unit, $SSy =$ controlled subsystem, $H =$ interconnection function, $m =$ control variable, $u =$ input interconnection variable, $z =$ output interconnection variable, $y =$ output variable, $w =$ disturbance)

on level $n + 1$. The unit on the top of the pyramid is called the *supremal coordinator* and the units to be found at the bottom level are called *infimal units*.

Manipulation of Complex Mathematical Problems

To take advantage of possible benefits of hierarchical multilevel systems a systematic decomposition of the original large-scale system and associated control problem is necessary. There are many situations when the control problem may be formulated as (or reduced to) an *optimization problem* (**P1**), which is, defined in general terms as

$$(\mathbf{P1}): \quad \underset{v}{\mathrm{extr}}\, J(v)\,; \quad v \in V\,, \tag{36.1}$$

where v is the decision variable (a scalar, or a vector), V is the admissible variation domain (which can be defined by differential or difference equations and/or algebraic inequations), and J is the performance measure (which can be a function or a functional).

The decomposition methods are based on various combinations of several *elementary manipulations* [36.76]. There are two main subsets of *elementary manipulations*:

a) Transformations, which are meant to substitute the original large-scale complex problem by a more manipulable one

b) *Decompositions*, which are meant to replace a large-scale problem by a number of smaller subproblems.

In the sequel several elementary manipulations will be reviewed following the lines exposed by *Wilson* [36.76].

The *variable transformation* replaces the original problem (**P1**) by an equivalent one (**P2**) through the utilization of a new variable $y = f(x)$ and a new performance measure $Q(y)$ and admissible domain Y, so that there is the inverse function $v = f^{-1}(v)$. The new problem is defined as

$$(\mathbf{P2}): \quad \underset{y}{\mathrm{extr}}\, Q(y)\,;$$
$$(y \in Y)\,, \ (\forall y) = f(v)[Q(y) = J(v)]\,. \tag{36.2}$$

The *Lagrange transformation* can simplify the admissible domain; for example, let the domain V be defined by complicated equalities and inequalities

$$V = \{v : (v \in V_1)\,, \ (g_0(v)\,, \ g_-(v) \leq 0)\}\,, \tag{36.3}$$

where V_1 is a certain set, g_0 and g_- represent equality and inequality constraints, respectively.

A Lagrangian can be defined

$$L(v, \pi, \gamma) = J(v) + \langle \pi, g_0(v)\rangle - \langle \gamma, g_-(v)\rangle\,, \tag{36.4}$$

where π are the Lagrange multipliers, γ are the Kuhn–Tucker multipliers, and $\langle \cdot, \cdot \rangle$ is the scalar product.

If L possesses a saddle point, the solution of (**P1**) is also the solution of the transformed problem (**P2**) defined as

$$(\mathbf{P2}): \quad \underset{\pi\gamma}{\max}\, \underset{v}{\min}[L(v, \pi, \gamma)]\,; \ v \in V_1\,. \tag{36.5}$$

The manipulation called *evolving the problem* is utilized when not all parameters are known or the priorities and the constraints are subject to alteration in time. In such situations, the problem is solved even under uncertainties and then is reformulated to take into account the accumulation of new information. The *repetitive control* proposed by *Findeisen* et al. [36.16] is based on such a transformation.

Having transformed the original problem into a convenient form, a subset of *smaller* subproblems can be obtained through *decomposition* as shown in the sequel.

The partitioning of the large-scale problem can be applied if several subsets of independent variables can be identified; for example, let (**P1**) be defined as

$$(\textbf{P1}): \quad \text{extr}[J^1(v^1) + J^2(v^2)] \; ; \; v^1 \in V^1 \; ; \; v^2 \in V^2 \; ,$$

(36.6)

then, two independent subproblems (**P2**^{1}) and (**P2**^{2}) can be obtained

$$(\textbf{P2}^1): \quad \text{extr}[J^1(v^1)] \; ; \quad v^1 \in V^1 \; , \tag{36.7}$$

$$(\textbf{P2}^2): \quad \text{extr}[J^2(v^2)] \; ; \quad v^2 \in V^2 \; . \tag{36.8}$$

This decomposition is utilized in assigning separate subproblems to the controllers which are situated at the same level of a hierarchical pyramid or in decentralized control schemes where the controllers act independently.

The parametric decomposition divides the large-scale problem into a pair of subproblems by setting temporary values to a set of *coupling parameters*. While in one problem of the pair the coupling parameters are fixed and all other variables are free, in the second subproblem they are free and the remaining variables are fixed as solutions of the first subproblem. The two subproblems are solved through an iterative scheme which starts with a set of guessed values of the coupling parameters; for example, let the large-scale problem be defined as follows

$$(\textbf{P1}): \quad \underset{v}{\text{extr}}[J(v)] \; ; \; v = (\alpha, \beta) \; ; \; (\alpha \in A) \; , \; (\beta \in B) \; ;$$

$$\alpha R \beta \; ,$$

where α and β are the components of v, A and B are two admissible sets, β is the coupling parameter, and R is a relation between α and β. The problem (**P1**) can be divided into the pair of subproblems (**P2**) and (**P3**)

$$(\textbf{P2}): \quad \underset{\alpha}{\text{extr}}\left[J\left(\alpha, \overset{*}{\beta}\right)\right] \; ; \; (\alpha \in A) \; , \; (\alpha R \beta) \; ,$$

$$(\textbf{P3}): \quad \underset{\beta}{\text{extr}}\left[J\left(\overset{*}{\alpha}(\beta), \beta\right)\right] \; ; \; (\beta \in B) \; ,$$

$$\left\{ \exists \overset{*}{\alpha}\left[\left(\overset{*}{\alpha} \in A\right) \; , \; \left(\overset{*}{\alpha} R \beta\right)\right]\right\} \; ,$$

where $\overset{*}{\alpha}(\beta)$ is the solution of (**P2**) for the given value $\beta = \overset{*}{\beta}$ and $\overset{*}{\beta}$ is the solution of (**P3**) for the given value $\alpha = \overset{*}{\alpha}$.

The parametric decomposition is utilized to divide the effort between a coordinating unit and the subset of coordinated units situated at lower organization level (echelon).

The *structural decomposition* divides the large-scale problem into a pair of subproblems through

modifying the performance measure and/or constraints. While one subproblem consists in setting the best/satisfactory formulation of the performance measure and/or admissible domain, the second one is to find the solution of the modified problem. This manipulation is utilized to divide the control effort between two levels of control (layers).

From Coordination to Cooperation

The traditional multilevel systems proposed in the 1970s to be used for the management and control of large-scale systems can be viewed as *pure hierarchies* [36.77]. They are characterized by the circulation of *feedback* and *intervention* signals only along the vertical axis, up and down, respectively, in accordance with traditional concepts of the command and control systems. They constituted a theoretical basis for various industrial distributed control systems which possess at highest level a powerful minicomputer. Also the *multilayer* and *multiechelon* hierarchies served in the 1980s as a conceptual reference model for the efforts to design *computer-integrated manufacturing* (CIM) systems [36.78, 79].

Several new schemes have been proposed over the last 25 years to overcome the drawbacks and limits of the practical management and control systems designed in accordance with the concepts of pure hierarchies such as: inflexibility, difficult maintenance, and limited robustness to major disturbances. The more recent solutions exhibit ever more increased communication and cooperation capabilities of the management and control units. This trend has been supported by the advances in communication technology and artificial intelligence; for example, even in 1977, *Binder* [36.80] introduced the concept of *decentralized coordinated control with cooperation*, which allowed limited communication among the control unit placed at the same level. Several years later, *Hatvany* [36.81] proposed the *heterarchical* organization, which allows for exchange of information among the units placed at various levels of the hierarchy.

The term *holon* was first proposed by *Koestler* in 1967 [36.82] with a view to describing a general organization scheme able to explain the evolution and life of biological and social systems. A holon cooperates with other holons to build up a larger structure (or to solve a complex problem) and, at the same time, it works toward attaining its own objectives and treats the various situations it faces without waiting for any instructions from the entities placed at higher levels. A *holarchy* is a hierarchy made up of holons. It is characterized by several features as follows [36.83]:

- It has a tendency to continuously grow up by attracting new holons.
- The structure of the holarchy may permanently change.
- There are various patterns of interactions among holons such as: communication messages, negotiations, and even aggressions.
- A holon may belong to more than one holarchy if it observes their operation rules.
- Some holarchies may work as pure hierarchies and others may behave as heterarchical organizations.

Figure 36.4 shows an object-oriented representation of a holarchy. The rectangles represent various classes of objects such as pure hierarchies, heterarchical systems, channels, and holons. This shows that the class of holarchies may have particular subclasses such as *pure hierarchies* and *heterarchical systems*. Also a holarchy is composed of several constituents (subclasses) such as: holons (at least one *coordinator* unit and two *infimal/coordinated units* in the care of pure hierarchies) and *channels for coordination* (in the case of pure hierarchies) or *channels for cooperation* (in the case of pure heterarchies). Coordination channels link the supremal unit to, at least, two infimal units. While there are, at least, two such coordination links in the case of pure hierarchies, a heterarchical system may have no such link. While, at least, one cooperation channel is present in a heterarchical system, no such a link is allowed in a pure hierarchy.

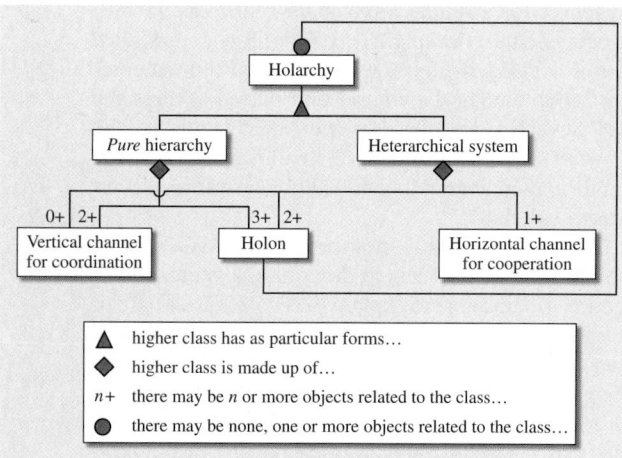

Fig. 36.4 Holarchies: an object-oriented description

Management and control structures based on holarchy concepts were proposed by *Van Brussel* et al. [36.84], *Valckenaers* et al. [36.85] for implementation in complex discrete-part manufacturing systems.

To increase the autonomy of the decision and control units and their cooperation the multiagent technology is recommended by *Parunak* [36.57] and *Hadeli* et al. [36.53]. An *intelligent software agent* encapsulates its code and data, is able to act in a proactive way, and cooperates with other agents to achieve a common goal [36.86]. The control structures which utilize the agent technology have the advantage of simplifying industrial transfer by incorporating existing legacy systems, which can be encapsulated in specific agents. *Mařík* and *Lažanský* [36.55] make a survey of industrial applications of agent technologies which also considers pros and cons of agent-based systems. They also present two applications:

a) A shipboard automation system which provides flexible and distributed control of a ship's equipment
b) A production planning and scheduling system which is designed for a factory with the possibility of influencing the developed schedules by customers and suppliers.

36.2.2 Other Methods and Applications

Decentralized Control

Feedback control of large-scale systems poses the standard control problem: *to find a controller for a given system with control input and control output ensuring closed-loop systems stability and reach a suitable input–output behavior*. The fundamental difference between small and large systems is usually described by a pragmatic view: a system is large if it is conceptually or computationally attractive to decompose it into interconnected subsystems. Such subsystems are typically of small size and can be solved easier than the original system. The subsystem solutions can be combined in some manner to obtain a satisfactory solution for the overall system [36.87].

Decentralized control has consistently been a control of choice for large-scale systems. The prominent reason for adopting this approach is its capability to solve effectively the particular problems of dimensionality, uncertainty, information structure constraints, and time delays. It also attenuates the problems that communication lines may cause. While in the hierarchical control schemes, as shown above, the control units are coordinated through intervention signals and may be allowed to exchange cooperation messages, in decentralized control, the units are completely independent or at least almost independent. This means that the information flow network among the control units can be divided into completely independent partitions. The units that belong to different subnetworks are completely separate from each other. Only restricted communication at certain time moments or intervals or limited to small part of information among the units is allowed. Decentralized structures are often used but their performance is worse compared with the centralized case. The basic decentralized control schemes are as follows:

- *Multichannel system.* The global system is considered as one whole. The control inputs and the control outputs operate only locally. This means that each channel has available only local information about the system and influences only a local part of the system.
- *Interconnected systems.* The overall system is decomposed according to a selected criterion. Then local controllers are designed for each subsystems. Finally, the local closed-loop subsystems and interconnections are tested to satisfy the desired overall system requirements.

At present a serious problem is the lack of relevant theoretic and methodological tools to support the scalable solution of new networked complex large-scale problems including asynchronous issues. The recent accomplishments are aimed at broadening the scope of decentralized control design methods using *linear matrix inequalities* (LMIs) [36.31], dynamic interaction coordinator design to ensure the desired level of interconnections [36.32], advanced decentralized control strategies for complex switching systems [36.26], hybrid large-scale systems [36.27], Petri nets [36.25], large-scale supply chain decentralized coordination [36.28, 29], and distributed control systems with network communication [36.30].

Simulation-Based Scheduling and Control in LSS

In continuous, large-scale industrial plants such as in chemical, power, and paper industries and waste-water treatment plants, simulation-based scheduling starts from creating scenarios for production and comparing these scenarios for optimality and availability. Problems can vary from order allocation between multiple production lines to optimal storage usage and detection and compensating for bottlenecks. Heuristic rules are usually connected to simulation, making it possible to adjust the production to varying customer needs, minimize the use of raw materials and energy, decrease the environmental load, stabilize or improve the quality, etc. Early applications in the paper industry are given by *Leiviskä* et al. [36.38, 39]. The main problem is to balance the production and several intermediate storages in (multiple) production lines, and give room for maintenance shutdowns and coordinate production rate changes. The model is based on the state model with storage capacities as the state variables and production rates as the control variables. Heuristics and bottleneck considerations are connected to these systems. A newer, agent-based solution has also been proposed [36.52]. There are also several classical optimization-based solutions for this problem [36.19, 21–24].

Modern chemical batch processes are large scale, complex, serial/parallel, multipurpose processes. They are especially common in the food and fine chemicals industries. They resemble flexible manufacturing systems common in electronics production. From the scheduling and control point of view complexity brings along also difficult interactions and uncertainty that are difficult to tackle with conventional tools. Simulation-based scheduling can include as much complexity as needed, and it is a largely used tool in the evaluation of the performance of different optimizing systems. Connecting heuristics or rule-based systems to simulation makes it also a flexible tool for batch process scheduling. Modeling approaches differ, e.g., real-time simulation using Petri nets [36.36] and the combination of discrete event simulation with genetic algorithms for the steel annealing shop have been proposed [36.40].

Flexible manufacturing systems, e.g., for components assembly, offer several difficulties for production scheduling and control. Dynamic, random nature is one main concern in operation control. Also quickly changing products and production environments, especially in electronics production, lead to a great variability in requirements for production control. In real cases, it is also typical that several scenarios must be created and evaluated. The handling of uncertain and vague information itself causes also problems in real-world applications. Uncertain data has to be extracted from data sources avoiding noise, or at least avoiding increasing it.

Discrete event simulation models the system as it propagates over time, describing the changes as separate discrete events. This approach also found a lot of applications in manufacturing industries, queuing systems, and so on. An early application to jobshop scheduling is presented by *Filip* et al. [36.20], who utilize various combinations of several dispatching rules to create the list of future events. *Taylor* [36.17] reported on an application of discrete event simulation, combined with heuristics, to the scheduling of the printed circuit board (PCB) assembly line. The situation is complicated by the fact that the production control must operate on three levels: at the system level concerning production mix problems, at the cell level for routing problems, and at the machine level to solve sequencing problems. Discrete event simulation is also the key element in the shop floor scheduling system proposed by *Gupta* et al. [36.35]. The procedure starts by creating feasible schedules for the telephone terminals plant, helps in taking other requirements into account and in tackling uncertainties, and makes rescheduling possible. A system integrating simulation and neural networks has been used in photolithography toolset scheduling in wafer production [36.33]. The system uses the weighted-score approach, and the role of the neural network is to update the weights set to different selection criteria. Fuzzy logic provides the arsenal of methods for dealing with uncertainties. Several examples for PCB production are given by *Leiviskä* [36.45].

Two-stage approaches have been used in bottleneck-based approaches [36.34]. The first-pass simulation recognizes the bottlenecks, and their operation are optimized during the second-pass simulation. Better control of work in bottlenecks improves the performance of the whole system. The main dispatching rule is to group together the lots that need the same setups. The system also reveals the non-bottleneck machines and makes it possible to apply different dispatching rules according to the process state. The example is from semiconductor production.

In practice, scheduling is a part of the decision hierarchy starting from the enterprise-level strategic decisions and going down to machine-level order or tools scheduling. Simulation is used at different levels of this hierarchy to provide interactive means for guaranteeing the overall optimality or at least the feasibility of the decisions made at different levels. Such integrated and interactive approaches exist also in supply-chain management systems. In large-scale manufacturing systems, supply-chain control must take four interacting factors into account: suppliers, manufacturing, distribution, network, and customers. To control all these interactions successfully, various operating factors and constraints – processing times, production capacities, availability of raw materials, inventory levels, and transportation times – must be considered.

Discrete event simulation is also one possibility to create an object-oriented, scalable, simulation-based control architecture for supply-chain control [36.41]. Requirements for modularity and maintainability also lead to distributed simulation models, especially when a simulation-based control architecture is controlling supply chain interactions. This means a modeling technique including a federation of simulation models that are solved in a coordinated manner. The system architecture is presented in [36.42]. Each supply-chain entity has two simulation models associated with it – one running in real time and the other as a lookahead simulation. The lookahead model is capable of predicting the impact of a disturbance observed by the real-time model. A *federation object coordinator* (FOC) coordinates the real-time simulation models. In this case, a *master event calendar* allocates interprocess events to all simulation models and resynchronizes all simulations at the end of every activity [36.37].

In simulation-based control the controller makes decisions based both on the current state of the system and future scenarios, usually produced by simulation. Here, the techniques for calculation of these scenarios play the main role. *Ramakrishnan* and *Thakur* [36.42] proposed the extension *sequential dynamic systems* (SDS) that they call *input–output SDS* to model and analyze distributed control systems and to compensate for the weaknesses of automata-based models. They use the discrete-part production plant as an example.

Artificial Intelligence–Based Control in LSS

Artificial intelligence (AI)-based control in large-scale systems uses, in practice, all the usual methods of intelligent control: fuzzy logic, neural networks, and genetic algorithms together with different kinds of hybrid solutions [36.88]. The complex nature of applications makes the use of intelligent systems advantageous. Dealing with this complexity is also the biggest challenge for the methodological development: the large-scale process structures, complicated interconnections, nonlinearity, and multiple time scales make the systems difficult to model and control. *Fuzzy logic control* (FLC) has found most of its applications in cases which are difficult to model, suffer from uncertainty or imprecision, and where a skilful operator is superior to conventional automation systems. *Artificial neural networks* (ANN) contribute to modeling and forecasting tasks and combined with fuzzy logic in neuro-fuzzy systems combine the benefits of both approaches. *Genetic algorithms* (GA), which are basically optimization systems, are used in tuning models and controllers. See Chap. 14 on *Artificial Intelligence and Automation* for additional content.

As shown above, the control of large-scale industrial plants have usually been based on distributed hardware and hierarchical design of control functions [36.89, 90]. The supervisory and local control levels lay under enterprise and mill-wide control levels. Supervisory control provides the local controls with the set points that fulfil the quality and schedule requirements coming from the mill-wide level and help in optimizing the operation of the whole plant. This *optimization* leaves room for versatile application of intelligent methods. Local units on the other hand, control the actual process variables according to the set points given by the supervisory control level. Even though the proportional–integral–differential (PID) controller is by far the most important tool, intelligent control plays an increasing role also at the local control level. Intelligent methods have been useful in tuning local PID controllers. In practice, fuzzy controllers must have adaptive capabilities. Gain scheduling is a typical approach for large-scale systems, but applications of model reference adaptive control and self-tuning adaptive control exist. Self-tuning has been used in controlling a pilot-scale rotary drum where the disturbances are due to long and varying time delays and changes in the raw materials [36.46].

Model-based control techniques, e.g., *model predictive control* (MPC), have been applied for the control of processes with a long delay or dead time. In MPC, the controller based on a plant model determines a manipulated variable profile that optimizes some performance objectives over the time in question. ANN are used in replacing the mathematical models in optimization as shown in a survey made by *Hussain* [36.48]. Also Takagi–Sugeno fuzzy models are used in connection with *model-based predictive control* [36.44]. Hybrid systems include both continuous- and/or discrete-time dynamics together with discrete events. So their state consists of real-valued, discrete-valued, and/or logical variables. *Support vector machines* have been used as a part of MPC strategy for hybrid systems [36.91].

Power systems have been an important application field for intelligent control since 1990s [36.92]. Design of centralized controllers is difficult for many obvious reasons: power systems are large scale and decentralized by nature. They are also nonlinear and have multiple dynamics and considerable time delays. Decentralized local control can apply linear models and purely local measurements. *Available transfer capability* (ATC) is a real-time index used in monitoring and controlling the power transactions and avoiding overloading of the transmission lines [36.47]. There are difficulties in calculating it accurately online for large-scale systems. Decreasing the number of input variables to only three and using fuzzy modeling helps in this. Simulations show that neural-networks-based local excitation controls can take care of interactions between generators and dampen oscillations effectively. Neural networks are used in approximating unknown dynamics and interconnections [36.40]. The designing of the controller for two-area hydrothermal power systems based on genetic algorithm improves the rise time and settling time, and simulations show that the proposed technique is superior to the traditional methods [36.51]. A local Kalman filter and genetic algorithms estimate all local states and interactions between subsystems in a large-scale power system. The controller uses these estimates, optimizes a given performance index, and then regulates the system states [36.50].

Agent-based technologies have been used in complex, distributed systems. Good examples come from intelligent control of highly distributed systems in the chemical industry and in the area of utility distribution (power, gas, and waste-water treatment). As shown above, holonic agents take care of machine or cell-level (local) controls, sometimes even integrated with machines. Intelligent agents can be associated with each manufacturing unit and they communicate, coordinate their activities, and cooperate with each other.

Fault detection and diagnosis (FDD) may be tackled by decomposing the large-scale problem into smaller subtasks and performing control and FDD locally [36.93]. Large-scale complex power systems need systematic tools for protection and control. The supervisory control technique and a design procedure of a supervisor that coordinates the behavior of relay agents to isolate fault areas are presented in [36.56]. Multiagent systems have also been used in identification and control of a 600 MW boiler–turbine–generator unit [36.54]. In this case, online identifiers are used for control and offline identifiers for fault diagnosis. Event-based approaches are used for building large-scale distributed systems and applications, especially in a networked environment. A hybrid approach of event-based communications for real-time manufacturing supervisory control is applied for large-scale warehouse management [36.94]. See Chap. 30 on *Automating Error and Conflict Prognostics and Prevention* for additional content.

Computer-Supported Decision Making in Large-Scale Complex Systems

As shown above, a possible solution to many LSS control problems is the use of artificial-intelligence methods. However, in the field, due to strange combinations of external influences and circumstances, rare or new situations may show up that were not taken into consideration at design time. Already in 1990, *Martin* et al. remarked that [36.95]:

although AI and expert systems were successful in solving problems that resisted to classical numerical methods, their role remains confined to support functions, whereas the belief that evaluation by man of the computerized solutions may become superfluous is a very dangerous error.

Based on this observation, *Martin* et al. [36.95] recommended *appropriate automation*, which integrates technical, human, organizational, economical, and cultural factors.

The *decision support system concept* (DSS) appeared in the early 1970s. As with any new term, the significance of DSS was in the beginning rather vague and controversial. While some people viewed it as a new redundant term used to describe a subset of management information systems (MIS), some other argued it was a new label abusively used by some vendors to take advantage of a new fashion. Since then many research and development activities and applications have witnessed that the DSS concept definitely meets a real need and there is a market for it even in the context of real-time applications in the industrial milieu [36.96, 97].

The Nobel Prize winner *H. Simon* [36.98] identified three steps of the *decision-making* (DM) process, namely:

a) *Intelligence*, consisting of activities such as data collection and analysis in order to recognize a decision problem
b) *Design*, including activities such as model statement and identification/production and evaluation of various potential solutions to the problem
c) *Choice*, or selection of a feasible alternative for implementation.

Later, he added a fourth step – implementation and result evaluation – which may correspond to supervisory control in industrial milieu. If a decision problem cannot be entirely clarified and all possible decision alternatives cannot be fully explored and evaluated before a choice is made, then the problem is said to be *unstructured* or *semistructured*. If the problem were completely structured, an automatic device could have solved the problem without any human intervention. On the other hand, if the problem has no structure at all, nothing but hazard can help. If the problem is semistructured a computer-aided decision can be envisaged.

Most of the developments in the DSS domain have initially addressed business applications not involving any real-time control. However, even in the early 1980s DSS were reported to be used in manufacturing control [36.20, 99]. In 1987, *Bosman* [36.100] stated that control problems could be looked upon as a *natural extension* and as a *distinct element* of planning *decision-making processes* (DMP). Almost 20 years later, *Nof* et al. state [36.12]:

... the development and application of intelligent decision support systems can help enterprises cope with problems of uncertainty and complexity, to increase efficiency, join competitively in production networks, and improve the scope and quality of their customer relations management (CRM).

Real-time decision-making processes (RT DMPs) for control applications are characterized by several particular aspects such as:

a) They involve continuous monitoring of a dynamic environment.
b) They are short time horizon oriented and are carried out on a repetitive basis.
c) They normally occur under time pressure.
d) Long-term effects are difficult to predict [36.101].

It is quite unlikely that an *econological* (*economically logic*) approach, involving optimization, be technically possible for genuine RT DMPs. *Satisficing* approaches, which reduce the search space at the expense of the decision quality, or fully automated DM systems, if taken separately, cannot be accepted either, but for some exceptions. At the same time, one can notice that genuine RT DMP can show up in *crisis* situations only; for example, if a process unit must be shut down due to an unexpected event, the production schedule of the entire plant might become obsolete. The right decision will be to take the most appropriate compensation measures to *manage the crisis* over the time period needed to recomputed a new schedule or update the current one. In this case, a *satisficing decision* may be appropriate. If the crisis situation has been met previously and successfully surpassed, an almost automated solution based on past decisions stored in the

Table 36.2 Possible task assignment in DSS

Decision steps and activities	EU	NU	NM	ES	ANN	CBR	IA
Intelligence							
• *Setting objectives*	I	M				I/M	P/M
• *Perception of DM situation*	I	M		P			M/I
• *Problem recognition*	I	M			P	M/I	
Design							
• *Model selection*	E	M		I		I	
• *Model building*	M	P		I/M	P		
• *Model validation*	I	M					
• *Setting alternatives*	P	M	P				
Choice							
• *Model experimenting*				M/I			
– *Model solving*		E	E		I		
– *Result interpreting*	I				P	P	
– *Parameter changing*	E			M/I			
• *Solution adotiing*	E						
• *Sensitivity analysis*	M		I				
Release for implementation	E	E					P

EU – expert user, NU – novice user, NM – numerical model, ES – rule-based expert system, ANN – artificial neural network, CBR – case-based reasoning, GA – genetic algorithm, IA – intelligent agent, P – possible, M – moderate, I – intensive, E – essential

information system can be accepted and validated by the human operator. On the other hand, the minimization of the probability of occurrences of crisis situations should be considered as one of the inputs (expressed as a set of constraints or/and objectives) in the scheduling problem [36.96, 102].

In many problems, decisions are made by a group of persons instead of an individual. Because the *group decision* is either a combination of individual decisions or a result of the selection of one individual decision, this may not be *rational* in Simon's acceptance. The group decision is not necessarily the best choice or a combination of individual decisions, even though those might be optimal, because various individuals might have various perspectives, goals, information bases, and criteria of choice. Therefore, group decisions show a high *social* nature, including possible conflicts of interest, different visions, influences, and relations [36.103]. Consequently, a *group* (or *multiparticipant*) DSS needs an important communication facility.

The generic framework of a DSS, proposed by *Bonczek* et al. in 1980 [36.104] and refined later by

Holsapple and *Whinston* [36.105] is quite general and can accommodate the most recent technologies and architectural solutions. It is based on three essential components. The first one is the *language* (and *communications*) *subsystem* (LS). This is used for:

a) Directing data retrieval, allowing the user to invoke one out of a number of report generators
b) Directing numerical or symbolic computation, enabling the user either to invoke the models by names or construct model and perform some computation at his/her free will
c) Maintaining knowledge and information in the system
d) Allowing communication among people in case of a group DM
e) Personalizing the user interface.

The *knowledge subsystem* (KS) normally contains:

a) *Empirical knowledge* about the state of the application environment in which the DSS operates

b) *Modeling knowledge*, including basic modeling blocks and computerized simulation and optimization algorithms to use for deriving new knowledge from the existing knowledge

c) *Derived knowledge* containing the constructed models and the results of various computations

d) *Meta-knowledge* (knowledge about knowledge) supporting model building and experimentation and result evaluation

e) *Linguistic knowledge* allowing the adaptation of system vocabulary to a specific application

f) *Presentation knowledge* to allow for the most appropriate information presentation to the user.

The third essential component of a DSS is the *problem processing subsystem* (PPS), which enables combinations of abilities and functions such as information acquisition, model formulation, analysis, evaluation, etc.

It has been noticed that some DSS are *oriented* towards the left hemisphere of the human brain and some others are oriented towards the right hemisphere. While in the first case quantitative and computational aspects are important, in the second pattern recognition and reasoning based on analogy prevail. In this context, there is a significant trend towards combin-

ing numerical models and models that emulate the human reasoning to build advanced DSS [36.106]. A great number of optimization algorithms have been developed and carefully tested so far. However, their effectiveness in decision making has been limited. Over the last three decades traditional numerical methods have, along with databases, been essential ingredients of DSS. From an information technology perspective, their main advantages [36.107] are: compactness, computational efficiency (if the model is correctly formulated), and the market availability of software products. On the other hand, they present several disadvantages. Because they are the result of intellectual processes of abstraction and idealization, they can be applied to problems which possess a certain structure, which is hardly the case in many real-life problems. In addition, the use of numerical models requires that the user possesses certain skills to formulate and experiment the model. As was shown in the previous section, AI-based methods supporting decision making are already promising alternatives and possible complements to numerical models. New terms such as *tandem systems*, or *expert DSS (XDSS)* have been proposed for systems that combine numerical models with AI-based techniques. An ideal task assignment is given in Table 36.2 [36.97].

36.3 Case Studies

The following case studies illustrate how combinations of methods may be utilized to solve large-scale complex problems.

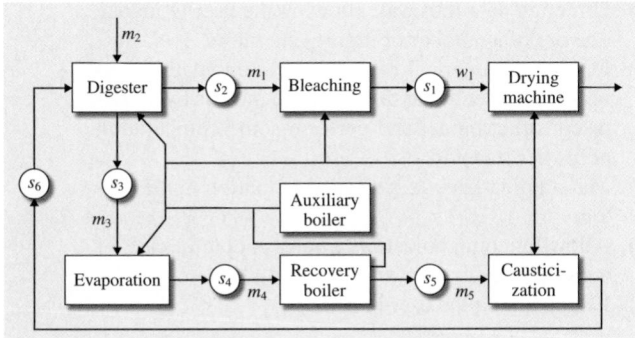

Fig. 36.5 Pulp mill model

36.3.1 Case Study 1: Pulp Mill Production Scheduling

Figure 36.5 shows the pulp mill modeled as a common state-space system. The state of the system $s(t)$ is described by the amount of material in each storage tank. The production rates of the processes are chosen as control variables forming the control vector $m(t)$. The required pulp production is usually taken as a deterministic known disturbance vector $w(t)$.

The operation of the plant presented in Fig. 36.5 is described by the vector–matrix differential equation

$$\frac{ds(t)}{dt} = Bm(t) + Cw(t) \,,$$

where B and C are coefficient matrices describing the relationships between the model flows (transfer ratios).

Since the most storage tanks have only one input flow and one output flow, most elements in **B** and **C** matrices equal zero.

If the steam balance (dashed line in Fig. 36.5) is included in scheduling, an additional variable describing the steam development in the auxiliary boiler is required. It is a scalar variable denoted by S. Accordingly, the steam balance is

$$S(t) = \mathbf{D}\boldsymbol{m}(t) + \mathbf{E}\boldsymbol{w}(t) .$$

Note that the right-hand side of the balance includes both consumption and generation terms. The variables in the model are constrained by the capacity limits of tanks and processes in the following way

$$\boldsymbol{s}^{\min} \leq \boldsymbol{s}(t) \leq \boldsymbol{s}^{\max} ,$$
$$\boldsymbol{m}^{\min} \leq \boldsymbol{m}(t) \leq \boldsymbol{m}^{\max} ,$$
$$S^{\min} \leq S(t) \leq S^{\max} .$$

Due to the fact that scheduling is concerned with relatively long time intervals, no complete and complicated process models are necessary. If all the small storage tanks are included in the model, the system dimensions increase and it becomes difficult to deal with. These tanks also have no meaning from the control point of view. Simpler model follows by combining small storage tanks.

There are several ways to solve the scheduling problem as shown before. Optimization can benefit from decomposition and solving of smaller problems as described in Sect. 36.2.1. A review of methods is presented in [36.39]. It seems, however, that no approach alone can deal with this problem successfully. Hybrid systems, consisting of algorithmic, rule-based, and intelligent parts integrated with each other, and also agent-based systems, could be the best possible answer [36.97, 110].

36.3.2 Case Study 2: Decision Support in Complex Disassembly Lines

In [36.108], the control of a complex industrial disassembly process of out-of-use manufactured products is studied. The disassembly processes are subject to uncertainties. The most difficult problem in such systems is that a disassembly operation can fail at any moment because of the product or component degradation. In this case one has to choose between applying an alternative disassembly destructive operation (dismantling), and aborting the disassembly procedure. This decision must be taken in real time because in a used product the components states are not known from the beginning of the

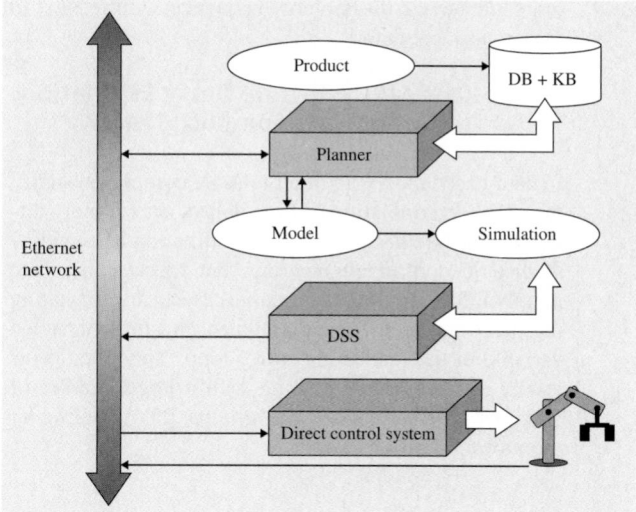

Fig. 36.6 DSS integration in the multilayer control system (after [36.108])

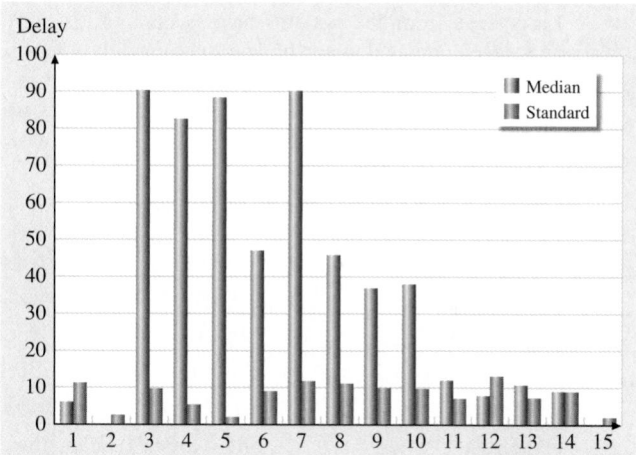

Fig. 36.7 The results of time delay estimation for one group (after [36.109])

process. The solution is to integrate a decision support system (DSS) in the architecture of a multilayer system. As shown in Fig. 36.6, the control and decision tasks are distributed among three levels: planning, decision support, and direct control. The disassembly planner gives the sequence of the components that must be separated to achieve the target component. The planner fuses the information from the artificial vision system with that contained in the database for each component or subassembly. A model of the product is generated. The DSS integrates the model and performs the simulation to recommend a *good* disassembly sequence with respect to the economical criteria.

36.3.3 Case Study 3: Time Delay Estimation in Large-Scale Complex Systems

In data-based modeling of large-scale complex systems, the exact determination of time delays is extremely difficult. The methods for delay estimation are widely studied in control engineering, but these studies are mainly limited to the two-variable cases, i. e., estimating the delay between the manipulated and the controlled variable in the feedback control loop. The situation is totally different when dealing with a large number of variables grouped in several groups for modeling or monitoring purposes.

Mäyrä et al. [36.109] discuss a delay estimation scheme combining genetic algorithms and principal component analysis (PCA). Delays are optimized with genetic algorithms with objective functions based on PCA. Typically, a genetic algorithm maximizes the variance explained by the first or two first principal components. The paper gives an example using simulation data of the paper machine, which includes over 50 variables. The variables were first grouped based on the cross-correlation and graphical analysis into five groups, and delays were estimated both for the variables inside the groups and between the groups. The results for one group of 15 variables are given in Fig. 36.7. The estimation was repeated 60 times and the figure shows the median and standard variance of these simulations.

36.4 Emerging Trends

Large-scale complex systems have become a research and development domain of automation with a series of rather established method and technologies and industrial application. Table 36.3 contains a summary of references to basic concepts.

At present academics and industrial practitioners are working to adapt the methods and practical solutions in the LSS field to modern information and communication technologies and new enterprise paradigms. Several significant trends which can be noticed or forecast are:

Table 36.3 Key to references on basic concepts

Basic books	*Mesarovic, Macko, Takahara* [36.7]; *Wismer* [36.61]; *Titli* [36.17]; *Ho* and *Mitter* [36.62]; *Sage* [36.63]; *Šiljak* [36.3, 64]; *Singh* [36.65]; *Findeisen* et al. [36.16]; *Jamshidi* [36.6]; *Lunze* [36.66]; *Brdys* and *Tatjewski* [36.18]
Hierarchies	*Mesarovic, Macko, Takahara* [36.7]
Strata	*Sage* [36.63]; *Schoeffler* [36.68]
Layers	*Findeisen* [36.8]; *Havlena* and *Lu* [36.73]; *Isermann* [36.72]; *Lefkowitz* [36.111]; *Schoeffler* [36.68]; *Brdys* and *Ulanicki* [36.112]
Echelons	*Brosilow, Lasdon* and *Pearson* [36.74]; *Lasdon* and *Schoeffler* [36.75]
Heterarchy	*Hatvany* [36.81]
Holarchy	*Hop* and *Schaeffer* [36.83]; *Koestler* [36.82]; *Van Brussel* et al. [36.84]; *Valckenaers* et al. [36.85]
Decision support systems	*Bonczek, Holsapple* and *Whinston* [36.104]; *Bosman* [36.100]; *Chaturverdi* et al. [36.101]; *De Michelis* [36.103]; *Dutta* [36.107]; *Filip* [36.96]; *Filip* et al. [36.21]; *Filip, Donciulescu* and *Filip* [36.97]; *Holsapple* and *Whinston* [36.105]; *Kusiak* [36.106]; *Martin* et al. [36.95]; *Nof* [36.99]; *Nof* et al. [36.12]; *Simon* [36.98]

- A promising modern form to coordinate the actions of the intelligent agents is *stigmergy*. This is inspired by the behavior of social insects which use a form of indirect communication mediated by an active environment to coordinate their actions [36.53].
- Advanced decentralized control strategies for large-scale complex systems have recently been extended into new applied areas, such as flexible structures [36.113, 114], Internet congestion control [36.115], aerial vehicles [36.116], or traffic control [36.117], to mention a few of them.
- Recent theoretic achievements in decentralized control can be progressively extended into the areas

of integrated/embedded control, distributed control (over communication networks), hybrid/discrete-event systems and networks, and autonomous systems to serve as a very efficient tool to solve various large-scale control problems.
- Incorporation and combination of newly developed numeric optimization and simulation models and symbolic/and connectionist or agent-based will continue in an effort to reach the *unification* of humans, numerical models, and AI-based tools.
- Mobile communications and web technology will be ever more considered in LSS management and control applications. In multiparticipant DSS, people will make co-decisions in *virtual teams*, no matter where they are temporarily located.

References

36.1 R. Tomovic: Control of large systems. In: *Simulation of Control Systems*, ed. by I. Troch (North Holland, Amsterdam 1972) pp. 3–6

36.2 M.S. Mahmoud: Multilevel systems control and applications, IEEE Trans. Syst. Man Cybern. **SMC-7**, 125–143 (1977)

36.3 D.D. Šiljak: *Large Scale Dynamical Systems: Stability and Structure* (North Holland, Amsterdam 1978)

36.4 Solver.com: *Premium Solver Platform for Excel* www.solver.com (2007)

36.5 M. Athans: Advances and open problems in the control of large-scale systems, Plenary paper, Proc. 7th IFAC Congr. (Pergamon, Oxford 1978), 871–2382

36.6 M. Jamshidi: *Large Scale Systems: Modeling and Control* (North Holland, New York 1983) 2nd edn. (Prentice Hall, Upper Saddle River 1997)

36.7 M.D. Mesarovic, D. Macko, Y. Takahara: *Theory of Hierarchical Multilevel Systems* (Academic, New York 1970)

36.8 W. Findeisen: Decentralized and hierarchical control under consistence or disagreements of interests, Automatica **18**(6), 647–664 (1982)

36.9 S. Takatsu: Coordination principles for two-level satisfactory decision-making systems, Syst. Sci. **7**(3/4), 266–284 (1982)

36.10 F.G. Filip, D.A. Donciulescu: On an online dynamic coordination method in process industry, IFAC J. Autom. **19**(3), 317–320 (1983)

36.11 L. Mårtenson: Are operators in control of complex systems?, Proc. 13th IFAC World Congr., Vol. B (Pergamon, Oxford 1990) pp. 259–270

36.12 S.Y. Nof, G. Morel, L. Monostori, A. Molina, F.G. Filip: From plant and logistics control to multienterprise collaboration, Annu. Rev. Control **30**(1), 55–68 (2006)

36.13 A.P. Sage, C.D. Cuppan: On the system engineering of systems of systems and federations of systems, Inf. Knowl. Syst. Manage. **2**(4), 325–349 (2001)

36.14 A.V. Gheorghe: Risks, vulnerability, maintainability and governance: a new landscape for critical infrastructures, Int. J. Crit. Infrastruct. **1**(1), 118–124 (2004)

36.15 A.V. Gheorghe: *Integrated Risk and Vulnerability Management Assisted by Decision Support Systems. Relevance and Impact on Governance* (Springer, Dordrecht 2005)

36.16 W. Findeisen, M. Brdys, K. Malinowski, P. Tatjewski, A. Wozniak: *Control and Coordination in Hierarchical Systems* (Wiley, Chichester 1980)

36.17 A. Titli: *Commande hierarchisée des systemes complexes* (Dunod Automatique, Paris 1975)

36.18 M. Brdys, P. Tatjewski: *Iterative Algorithms for Multilayer Optimizing Control* (Imperial College, London 2001)

36.19 A. Dourado Correia: Optimal scheduling and energy management in industrial complexes: some new results and proposals, Preprints CIM Process Manufact. Ind. IFAC Workshop (Pergamon Press, Espoo 1992) pp. 139–145

36.20 F.G. Filip, G. Neagu, D. Donciulescu: Jobshop scheduling optimization in real-time production control, Comput. Ind. **4**(3), 395–403 (1983)

36.21 F.G. Filip, D. Donciulescu, R. Gaspar, M. Muratcea, L. Orasanu: Multilevel optimisation algorithms in computer aided production control in the process industries, Comput. Ind. **6**(1), 47–57 (1985)

36.22 M. Guran, F.G. Filip, D.A. Donciulescu, L. Orasanu: Hierarchical optimisation in computer dispatcher systems in process industry, Large Scale Syst. **8**, 157–167 (1985)

36.23 J. Pettersson, U. Persson, T. Lindberg, L. Ledung, X. Zhang: Online pulp mill production optimization, Proc. 16th IFAC World Congr. (Prague 2005), on CD ROM

36.24 H. Tamura: Decentralised optimization for distributed-lag models of discrete systems, Automatica **11**, 593–602 (1975)

36.25 A. Aybar, A. Iftar, H. Apaydin-Özkan: Centralized and decentralized supervisory controller design to enforce boundedness, liveness, and reversibility in Petri nets, Int. J. Control **78**, 537–553 (2005)

36.26 L. Bakule: Stabilization of uncertain switched symmetric composite systems, Nonlinear Anal.: Hybrid Syst. **1**, 188–197 (2007)

36.27 F. Borrelli, T. Keviczky, G.J. Balas, G. Steward, K. Fregene, D. Godbole: *Hybrid Decentralized Control of Large Scale Systems*, Hybrid Systems: Computation and Control (Springer, Heidelberg 2005) pp. 168–183

36.28 G. Inalham, J. How: Decentralized inventory control for large-scale supply chains, Proc. Am. Control Conf. (Minneapolis 2006) pp. 568–575

36.29 P. Krishnamurthy, F. Khorrami, D. Schoenwald: Computationally tractable inventory control for large-scale reverse supply chains, Proc. Am. Control Conf. (Minneapolis 2006) pp. 550–555

36.30 C. Langbort, V. Gupta, R.M. Murray: Distributed control over falling channels. In: *Networked Embedded Sensing and Control*, ed. by P. Antsaklis, P. Tabuada (Springer, Berlin 2006) pp. 325–342

36.31 D.D. Šiljak, A.I. Zečević: Control of large-scale systems: beyond decentralized feedback, Annu. Rev. Control **29**, 169–179 (2005)

36.32 D.D. Šiljak: *Dynamic Graphs. Plenary paper. The International Conference on Hybrid Systems and Applications* (University of Louisiana, Lafayette 2006)

36.33 A. Arisha, P. Young: Intelligent simulation-based lot scheduling of photolithography toolsets in a wafer fabrication facility, Proc. 2004 Winter Simul. Conf. (Washington 2004) pp. 1935–1942

36.34 C.S. Chong, A.I. Sivakumar, R. Gay: Simulation based scheduling using a two-pass approach, Proc. 2003 Winter Simul. Conf. (New Orleans 2003) pp. 1433–1439

36.35 A.K. Gupta, A.I. Sivakumar, S. Sarawgi: Shopfloor scheduling with simulation based proactive decision support, Proc. Winter Simul. Conf. (San Diego 2002) pp. 1897–1902

36.36 S. Julia, R. Valette: Real-time scheduling of batch systems, Simul. Pract. Theory **8**, 307–319 (2000)

36.37 S. Lee, S. Ramakrishnan, R.A. Wysk: A federation object coordinator for simulation based control and analysis, Proc. Winter Simul. Conf. (San Diego 2002) pp. 1986–1994

36.38 K. Leiviskä, P. Uronen, H. Komokallio, H. Aurasmaa: Heuristic algorithm for production control of an integrated pulp and paper mill, Large Scale Syst. **3**, 13–25 (1982)

36.39 K. Leiviskä: Benefits of intelligent production scheduling methods in pulp mills, Proc. CESA'96 IMACS Multiconf. Comput. Eng. Syst. Appl. Symp. Control Optim. Supervis., Vol. 2 (Lille 1996) pp. 1246–1251

36.40 Q.L. Liu, W. Wang, H.R. Zhan, D.G. Wang, R.G. Liu: Optimal scheduling method for a bell-type batch annealing shop and its application, Control Eng. Pract. **13**, 1315–1325 (2005)

36.41 S. Ramakrishnan, S. Lee, R.A. Wysk: Implementation of a simulation-based control architecture for supply chain interactions, Proc. Winter Simul. Conf. (San Diego 2002) pp. 1667–1674

36.42 S. Ramakrishnan, M. Thakur: An SDS modeling approach for simulation-based control, Proc. Winter Simul. Conf. (Orlando 2005) pp. 1473–1482

36.43 G.D. Taylor Jr: A flexible simulation framework for evaluating multilevel, heuristic-based production control strategies, Proc. Winter Simul. Conf. (New Orleans 1990) pp. 567–569

36.44 A. Ichtev, J. Hellendoom, R. Babuska, S. Mollov: Fault-tolerant model-based predictive control using multiple Takagi–Sugeno fuzzy models, Proc. IEEE Int. Conf. Fuzzy Syst. FUZZ-IEEE'02, Vol. 1 (Honolulu 2002) pp. 346–351

36.45 K. Leiviskä: Applications of intelligent systems in electronics manufacturing, Proc. 2nd Conf. Manag. Control Prod. Logist. MCPL'2000 (Grenoble 2000), on CD-ROM

36.46 K. Leiviskä, L. Yliniemi: Design of adaptive fuzzy controllers. In: *Do Smart Adaptive Systems Exist?*, ed. by B. Gabrys, K. Leiviskä, J. Strackeljan (Springer, Berlin 2005) pp. 251–266

36.47 B. Azhar, A.B. Khairuddin, S.S. Ahmed, M.W. Mustafa, A. Zin, H. Ahmad: A novel method for ATC computations in a large-scale power system, IEEE Trans. Power Syst. **19**(2), 1150–1158 (2004)

36.48 M.A. Hussain: Review of the applications of neural networks in chemical process control-simulation and online implementation, Artif. Intell. Eng. **13**, 55–68 (1999)

36.49 W. Liu, J. Sarangapani, G.K. Venayagamoorthy, D.C. Wunsch, D.A. Cartes: Neural network based decentralized excitation control of large scale power systems, Proc. Int. Jt. Conf. Neural Netw. (Vancouver 2006)

36.50 M. Dehghani, A. Afshar, S.K. Nikravesh: Decentralized stochastic control of power systems using genetic algorithms for interaction estimation, Proc. 16th IFAC World Congr. (Prague 2005), on CD ROM

36.51 E.E. El Mdbouly, A.A. Ibrahim, G.Z. El-Far, M. El Nassef: Multilevel optimization control for large-scale systems using genetic algorithms, Proc. 2004 Int. Conf. Electr., Electron. Comput. Eng. ICEEC '04 (Cairo 2004) pp. 193–197

36.52 R. Akkiraju, P. Keskinocak, S. Murthy, F. Wu: An agent-based approach for scheduling multiple machines, Appl. Intell. **14**(2), 135–144 (2001)

36.53 H. Hadeli, P. Valckenaers, C.B. Zamfirescu, H. Van Brussel, B.S. Germain: Self-organising in multi-agent coordination and control using stigmergy. In: *Self-Organising Applications: Issues, challenges and trends. Lecture Notes in Artificial Intelligence*, Vol. 2977, ed. by H. Hadeli (Springer, Heidelberg 2004) pp. 325–340

36.54 J.S. Heo, K.Y. Lee: A multi-agent system-based intelligent identification system for power plant control and fault-diagnosis, Proc. IEEE Power Eng. Soc. Gen. Meet. (Montreal 2006) pp. 1–6

36.55 V. Mařík, J. Lažanský: Industrial applications of agent technologies, Control Eng. Pract. **15**(11), 1364–1380 (2007)

36.56 S.J. Park, J.T. Lim: Modelling and control of agent-based power protection systems using supervisors, IEEE Proc. Control Theory Appl. **153**, 92–99 (2006)

36.57 H.V.D. Parunak: *Practical and Industrial Applications of Agent-Based Systems* (Industrial Technology Institute, Ann Arbor 1998)

36.58 C.G. Cassandras: Complexity made simple – at small price, Proc. 9th IFAC Symp. Large Scale Systems: Theory and Applications 2001, ed. by F.G. Filip, I. Dumitrache, S. Iliescu (Elsevier, Oxford 2001) pp. 1–5

36.59 P.D. Roberts: An algorithm for steady-state system optimization and parameter estimation, Int. J. Syst. Sci. **10**(7), 719–734 (1979)

36.60 P.P. Varaiya: Review of the book *Theory of Hierarchical Multilevel Systems*, IEEE Trans. Autom. Control **17**, 280–281 (1972)

36.61 D. Wismer: *Optimization Methods for Large Scale Systems* (Mc. Graw Gill, New York 1971)

36.62 Y.C. Ho, S.K. Mitter: *Directions in Large-Scale Systems* (Plenum, New York 1976)

36.63 A.P. Sage: *Methodology for Large Scale Systems* (McGraw Hill, New York 1977)

36.64 D.D. Šiljak: *Decentralized Control of Complex Systems* (Academic, Cambridge 1990)

36.65 M.G. Singh: *Dynamic Hierarchical Control* (North Holland, Amsterdam 1978)

36.66 J. Lunze: *Feedback Control of Large-Scale Systems* (Prentice Hall, New York 1992)

36.67 D. Steward: *Systems Analysis and Management: Structure, Strategy and Design* (Petrocelli Book, New York 1981)

36.68 J. Schoeffler: Online multilevel systems. In: *Optimization Methods for Large Scale Systems*, ed. by D. Wismer (McGraw Hill, New York 1971) pp. 291–330

36.69 J. Minsker, S. Piggot, G. Freidson: Hierarchical automation control systems for large-scale systems and applications, Proc. 5th IFAC World Congr. (Paris 1972)

36.70 D.P. Eckman, I. Lefkowitz: Principles of model technique in optimizing control, Proc. 1st IFAC World Congr. (Moscow 1960) pp. 970–974

36.71 I. Lefkowitz: Multilevel approach to control system design, Proc. JACC (1965) pp. 100–109

36.72 R. Isermann: Advanced methods of process computer control for industrial, Int. J. Comput. Ind. **2**(1), 59–72 (1981)

36.73 V. Havlena, J. Lu: A distributed automation framework for plant-wide control, optimisation, scheduling and planning, selected plenaries, semiplenaries, milestones and surveys, Proc. 16th IFAC World Congr., ed. by P. Horacet, M. Simandl, P. Zitek (2005) pp. 80–94

36.74 C.B. Brosilow, L. Ladson, J.D. Pearson: Feasible optimization methods for interconnected systems, Proc. Joint Autom. Control Conf. – JACC (Rensselaer Polytechnic Institute, Troy, New York 1965) pp. 79–84

36.75 L.S. Lasdon, J.D. Schoeffer: A multilevel technique for optimization, Proc. JACC (1965) pp. 85–92

36.76 I.D. Wilson: Foundations of hierarchical control, Int. J. Control **29**(6), 899 (1979)

36.77 F.G. Vernadat: *Enterprise Modelling and Integration Principles and Applications* (Chapman Hall, London 1996)

36.78 T.J. Williams: *Analysis and Design of Hierarchical Control Systems with Special Reference to Steel Plant Operations* (Elsevier, Amsterdam 1985)

36.79 T.J. Williams: *A Reference Model for Computer Integrated Manufacturing* (Instrument Society of America, Research Triangle Park 1989)

36.80 Z. Binder: *Sur l'organisation et la conduite des systemes complexes*, These de Docteur (LAG, Grenoble 1977) in French

36.81 J. Hatvany: Intelligence and cooperation in heterarchic manufacturing systems, Robot. Comput. Integr. Manuf. **2**(2), 101–104 (1985)

36.82 A. Koestler: *The Ghost in the Machine* (Hutchinson, London 1967)

36.83 M. Hopf, C.F. Schoeffer: *Holonic Manufacturing Systems, Information Infrastructure Systems for Manufacturing* (Chapmann Hall, London 1997) pp. 431–438

36.84 H. Van Brussel, P. Valckenaers, J. Wyns: HMS – holonic manufacturing system test case (IMS Project). In: *Enterprise Engineering and Integration: Building International Consensus*, ed. by K. Kosanke, J.G. Nell (Springer, Berlin 1997) pp. 284–292

36.85 P. Valckenaers, H. Van Brussel, K. Hadeli, O. Bochmann, B.S. Germain, C. Zamfirescu: On the design of emergent systems an investigation of integration and interoperability issues, Eng. Appl. Artif. Intell. **16**, 377–393 (2003)

36.86 G. Tecuci: *Building Intelligent Agents: An Apprenticeship Multistrategy Learning Theory, Methodol-*

ogy, Tool and Case Studies (Academic, New York 1998)

36.87   L. Bakule: Complexity-reduced guaranteed cost control design for delayed uncertain symmetrically connected systems, Proc. 2005 Am. Control Conf. (Portland 2005) pp. 2500–2505

36.88   P.P. Groumpos: Complex systems and intelligent control: issues and challenges, Proc. 9th IFAC Symp. Large Scale Syst.: Theory Appl. 2001, ed. by F.G. Filip, I. Dumitrache, S. Iliescu (Elsevier, 2001) pp. 29–36

36.89   A. Kamiya, S.J. Ovaska, R. Roy, S. Kobayashi: Fusion of soft computing and hard computing for large-scale plants: a general model, Appl. Soft Comput. J. **5**, 265–279 (2005)

36.90   K. Leiviskä: Control systems. In: *Process Control. Papermaking Science and Technology, Book 14*, ed. by K. Leiviskä (Fapet Oy, Jyväskylä 1999) pp. 13–17

36.91   B.M. Åkesson, M.J. Nikus, H.T. Toivonen: Explicit model predictive control of a hybrid system using support vector machines, Proc. 1st IFAC Workshop Appl. Large Scale Ind. Syst. ALSIS '06 (Helsinki/Stockholm 2006), on CD ROM

36.92   K. Kawai: Knowledge engineering in power-plant control and operation, Control Eng. Pract. **4**, 1199–1208 (1996)

36.93   G. Stephanopoulos, J. Romagnoli, E.S. Yoon: *Online Fault Detection and Supervision in the Chemical Process Industries 2001* (Jejudo Island, Korea 2004)

36.94   D.H. Zhang, J.B. Zhang, Y.Z. Zhao, M.M. Wong: Event-based communications for equipment supervisory control, Proc. 10th IEEE Conf. Emerg. Technol. Fact. Autom. (Catania 2005) pp. 341–347

36.95   T. Martin, J. Kivinen, J.E. Rinjdorp, M.G. Rodd, W.B. Rouse: Appropriate automation integrating human, organisation and culture factors, Preprints IFAC 11th World Congr., Vol. 1 (1990) pp. 47–65

36.96   F.G. Filip: Towards more humanized real-time decision support systems. In: *Balanced Automation Systems; Architectures and Design Methods*, ed. by L.M. Camarinha-Matos, H. Afsarmanesh (Chapman Hall, London 1995) pp. 230–240

36.97   F.G. Filip, D. Donciulescu, C.I. Filip: Towards intelligent real-time decision support systems, Stud. Inf. Control SIC **11**(4), 303–312 (2002)

36.98   H. Simon: *The New Science of Management Decisions* (Harper & Row, New York 1960)

36.99   S.Y. Nof: Theory and practice in decision support for manufacturing control. In: *Data Base Management*, ed. by C.W. Holsapple, A.B. Whinston (Reidel, Dordrecht 1981) pp. 325–348

36.100 A. Bosman: Relations between specific DSS, Decis. Support Syst. **3**, 213–224 (1987)

36.101 A.R. Charturverdi, G.K. Hutchinson, D.L. Nazareth: Supporting real-time decision-making through machine learning, Decis. Support Syst. **10**, 213–233 (1997)

36.102 M. Cioca, L.I. Cioca, S.C. Buraga: Spatial [elements] decision support system used in disaster management, Digit. EcoSyst. Technol. Conf. 2007. DEST '07. Inaugural IEEE-IES (Cairn 2006) pp. 607–612

36.103 G. De Michelis: Coordination with cooperative processes. In: *Implementing Systems for Support Management Decisions*, ed. by P. Humphrey, L. Bannon, A. McCosh, P. Migliarese, J.C. Pomerol (1996) pp. 108–123

36.104 R.H. Bonczek, C.W. Holsapple, A.B. Whinston: *Foundations of Decision Support Systems* (Academic, New York 1980)

36.105 C.W. Holsapple, A.B. Whinston: *Decision Support System: a Knowledge-Based Approach* (West, Mineapolis 1996)

36.106 A. Kusiak: *Intelligent Management Systems* (Prentice Hall, Englewood Cliffs 1990)

36.107 A. Dutta: Integrated AI and optimization for decision support: a survey, Decis. Support Syst. **18**, 213–226 (1996)

36.108 L. Duta, J.M. Henrioud, F.G. Filip: Applying equal piles approach to disassembly line balancing problem, Proc. 16th IFAC World Congr. (Session Industrial Assembly and Disassembly) (Prague 2005), on CD ROM

36.109 O. Mäyrä, T. Ahola, K. Leiviskä: Time delay estimation in large data bases, IFAC LSSTA Symp. (Gdansk 2007), on CD ROM

36.110 F.G. Filip: System analysis and expert systems techniques for operative decision making, J. Syst. Anal. Model. Simul. **8**(2), 296–404 (1990)

36.111 I. Lefkowitz: Hierarchical control in large-scale industrial systems. In: *Large Scale Systems*, ed. by D.D. Haimes (North Holland, Amsterdam 1982) pp. 65–98

36.112 M. Brdys, B. Ulanicki: *Operatioal Control of Water Systems* (Prentice Hall, New York 1994)

36.113 L. Bakule, F. Paulet-Crainiceanu, J. Rodellar, J.M. Rossell: Overlapping reliable control for a cable-stayed bridge benchmark, IEEE Trans. Control Syst. Technol. **4**, 663–669 (2006)

36.114 L. Bakule, J. Rodellar, J.M. Rossell: Robust overlapping guaranteed cost control of uncertain steady-state discrete-time systems, IEEE Trans. Autom. Control **12**, 1943–1950 (2006)

36.115 R. Srikant: *The Mathematics of Internet Congestion Control* (Birkhäuser, Boston 2004)

36.116 D.M. Stipanovic, G. Inalham, R. Teo, C.J. Tomlin: Decentralized overlapping control of a formation of unmanned aerial vehicles, Automatica **40**(1), 1285–1296 (2004)

36.117 S.S. Stankovic, M.J. Stanojevic, D.D. Šiljak: Decentralized overlapping control of a platoon of vehicles, IEEE Trans. Control Syst. Technol. **8**, 816–832 (2000)

37. Computer-Aided Design, Computer-Aided Engineering, and Visualization

Gary R. Bertoline, Nathan Hartman, Nicoletta Adamo-Villani

This chapter is an overview of computer-aided design (CAD) and computer-aided engineering and includes elements of computer graphics, animation, and visualization. Commercial brands of three-dimensional (3-D) modeling tools are dimension driven, parametric, feature based, and constraint based all at the same time. The term *constraint-based* is intended to include all of these many facets. This means that, when geometry is created, the user specifies numerical values and requisite geometric conditions for the elemental dimensional and geometric constraints that define the object. Many of today's modern CAD tools also operate on similar interfaces with similar geometry-creation command sequences [37.1] that operate interdependently to control the modeling process. Core modules include the sketcher, the solid modeling system itself, the dimensional constraint engine, the feature manager, and the assembly manager [37.2]. In most cases, there is also a drawing tool, and other modules that interface with analysis, manufacturing process planning, and machining. The 3-D animation production process can be divided into three main phases:

- Concept development and preproduction
- Production
- Postproduction and delivery.

These processes can begin with the 3-D geometry generated by CAD systems in the design process or 3-D models can be created as a separate process. The second half of the chapter explains the process commonly used to create animations and visualizations.

37.1 Modern CAD Tools

Today's commercial brands of 3-D modeling tools essentially contain many of the same types of functions across the various vendor offerings. They are dimension driven, parametric, feature based, and constraint based all at the same time, and these terms have come to be synonymous when describing modern CAD systems [37.3]. For the purposes of this chapter, the term *constraint-based* will be intended to include all of these many facets. Generally this means that, when geometry is created, the user specifies numerical values and requisite geometric conditions for the elemental dimensional and geometric constraints that define the object; for example, a rectangular prism would be defined by parameter dimensions that control its height, width, and depth. In addition, many of today's modern CAD tools also operate on similar interfaces with similar geometry-creation command sequences [37.1]. Generally, most constraint-based CAD tools consist of software modules that operate interdependently to control the 3-D modeling process. They include core modules such as the sketcher, the solid modeling system itself, the dimensional constraint engine, the feature manager, and the assembly manager [37.2]. In most cases, there is also a drawing tool, and other modules that interface with analysis, manufacturing process planning, and machining. The core modules are used in conjunction with each other (or separately as necessary) to develop a 3-D model of the desired product. In so doing, most modern CAD systems will produce the same kinds of geometry, irrespective of the software interface they possess. Many of the modern 3-D CAD tools combine constructive solid geometry (CSG) and boundary representation (B-rep) modeling functionality to form hybrid 3-D modeling packages [37.2, 3]. Traditionally, CSG used mathematical primitives to create 3-D models. They were efficient for the storage of the database, but they had difficulty with sculpted surfaces and editing the finished model. B-rep modelers use surfaces directly to represent the object three-dimensionally, so they tend to be very accurate. However, they also tend to have large database structures, hence the development of hybrids to capture the best characteristics of both B-rep and CSG.

Constraint-based CAD tools create a solid model as a series of features that correspond to operations that would be used to create the physical object. Features can be created dependently or independently of each other with respect to the effects of modifications made to the geometry. If features are dependent, then an update to the parent feature will affect the child feature. This is known as a *parent–child reference*, and these references are typically at the heart of most modeling processes performed by the user [37.2]. The geometry of each feature is controlled by the use of modifiable constraints that allow for the dynamic update of model geometry

as the design criteria change. When a parent feature is modified, it typically creates a ripple effect that yields changes in the child features. This is one example of *associativity* – the fact that design changes propagate through the geometric database and associated derivatives of the model due to the interrelationships between model features. This dynamic editing capability is also reflected in assembly models that are used to document the manner in which components of a product interact with each other. Modifications to features contained in a part will be displayed in the parent part as well as in the assembly that contains the part. Any working drawings of the part or assembly will also update to reflect the changes. This is another example of *associativity*.

A critical issue in the use of constraint-based CAD tools is the planning that happens prior to the creation of the model [37.3]. This is known as design intent. Much of the power and utility of constraint-based CAD tools is derived from the fact that users can edit and redefine part geometry as opposed to deleting and recreating it. This requires a certain amount of thought with respect to the relationships that will be established between and within features of a part and between components in an assembly. The ways in which the model will be used in the future and how it could potentially be manipulated during design changes are both factors to consider when building the model. The manner in which the user expects the CAD model to behave under a given set of circumstances, and the effects of that behavior on other portions of the same model or on other models within the assembly, is known as design intent [37.2, 3]. The eventual use and reuse of the model will have a profound effect on the relationships that are established within the model as well as the types of features that are used to create it, and vice versa.

37.2 Geometry Creation Process

Geometry is created in modern constraint-based CAD systems using the modules and functionality described above, especially the sketcher, the dimensional constraint engine, the solid modeling system, and the feature manager. Modern CAD systems create many different kinds of geometry, which generally fall into one of three categories: wireframe, surface or solid. Most users work towards creating solid geometry. In doing so, the user often employs the larger functionality of the CAD system described in the previous section. To create solid geometry, the user considers their de-

sign intent and proceeds to make the first feature of the model. The most common way to create feature geometry within a part file is to sketch the feature's cross-section on a datum plane (or flat planar surface already existing in the part file), dimension and constrain the sketched profile, and then apply a feature form to the cross-section. Due to the inherent inaccuracies of sketching geometric entities on a computer screen with a mouse, CAD systems typically employ a constraint solver. This portion of the software is responsible for resolving the geometric relationships and general proportions between the sketched entities and the dimensions that the user applies to them, which is another example of automation in the geometric modeling process. The final stage of geometry creation is typically the application of a feature form, which is what gives a sketch its depth element. This model creation process is illustrated in Fig. 37.1. This automated process of capturing dimensional and parametric information as part of the geometry creation process is what gives modern CAD systems their advantage over traditional engineering drawing techniques in terms of return on investment and efficiency of work. Without this level of automation, CAD systems would be nothing more than an electronic drawing board, with the user being required to recreate a design from scratch each time.

As the user continues to use the feature creation functions in the CAD system, the feature list continues to grow. It lists all of the features used to create a model in chronological order. The creation of features in a particular order also captures design intent from the user, since the order in which geometry is created will have a final bearing on the look (and possibly the function) of the object. In most cases, the feature tree is also the location where the user would go to consider modifying the order in which the model's features were created (and rebuilt whenever a change is made to the topology of the model).

As users become more proficient at using a constraint-based CAD system to create geometry, they adopt their own mental model for interfacing with the software [37.4]. This mental model typically evolves to match the software interface metaphor of the CAD system. In so doing, they are able to leverage their expertise regarding the operation of the software to devise highly sophisticated methods for using the CAD systems. This level of sophistication and automation by the user is due in some part to the nature of the constraint-based CAD tools. It is also what enables the user to dissect geometric models created by others (or themselves at a prior time) and reuse them to develop new or modified designs. Effective use of the tools requires that the user's own knowledge base comprised of the conceptual relationships regarding the capture of design intent in the geometric model and the specific software skills necessary to create geometry be used. This requires the use of an object–action interface model and metaphor on the part of the user in order to be effective [37.1]. This interface model correlates the objects and actions used in the software with those used in the physical construction of the object being modeled. If a person is to

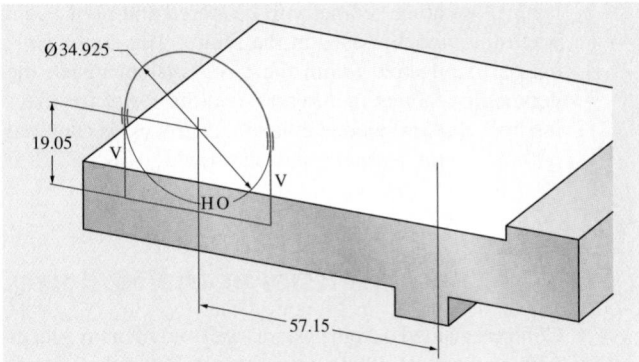

Fig. 37.1 Sketch geometry created on a plane and extruded for depth

use the CAD tool effectively, these two sets of models should be similar. In relation to the object–action interface model is the idea of a user's mental model of the software tool [37.1]. This mental model is comprised of semantic knowledge of how the CAD system operates, the relationships between the different modules and commands, and syntactic knowledge that is comprised of specific knowledge about commands and the interface (Fig. 37.2).

The process of creating 3-D geometry in this fashion allows the user to automate the capture of their design intent. Semantic and syntactic knowledge are combined once in the initial creation of the model to develop the intended shape of the object being modeled. This encoding of design knowledge allows the

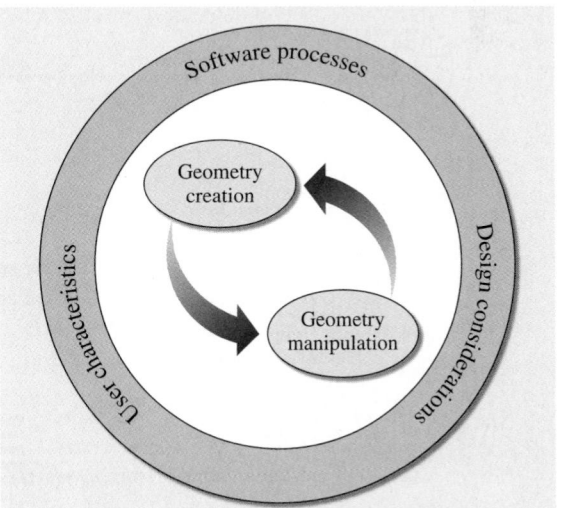

Fig. 37.2 Expert mental model of modern CAD system operation

labor of creating geometry to be stored and used again when the model is used in the future. This *labor storage* is manifested within the CAD system inside the geometric features themselves, and the *script* for playing back that knowledge-embedding process is captured within the feature manager as described in Sect. 37.2. It

is generally common knowledge within the modern 3-D modeling environment that a user will likely work with models created by other people and vice versa. As such, having a predictable means to include design intent in the geometric model is critical for the reuse of existing CAD models within an organization.

37.3 Characteristics of the Modern CAD Environment

Computer-aided design systems are used in many places within a product design environment, but each scenario tends to have a common element: the need to accurately define the geometry which represents an object. This could be in the design engineering phase to depict a product, or during the manufacturing planning stage for the design if a fixture to hold a workpiece. Recently, these CAD systems have been coupled with *product data management* (PDM) *systems* to track the ongoing changes through the lifecycle of a product. By so doing, the inherent use of the CAD system can be tracked, knowledge about the design can be stored, and permissions can be granted to appropriate users of the system. While the concept of concurrent engineering is not new, contemporary depictions of that model typically show a CAD system (and often a PDM system) at the center of the conceptual model, disseminating embedded information for use by the entire product development team throughout the product lifecycle [37.3].

To use a modern CAD system effectively, one must understand the common inputs and outputs of the system, typically in light of a concurrent and distributed design and manufacturing environment. Input usually takes the form of numerical information regarding size, shape, and orientation of geometry during the product model creation process. This information generally comes directly from the user responsible for developing the product; however, it is not uncommon to get CAD input data from laser scanning devices used for quality control and inspection, automated scripts for generating seed geometry, or translated files from other systems. As with other types of systems, the quality of the information put into the system greatly affects the quality of the data coming out of the system. In today's geographically dispersed product development environment, CAD geometry is often exported from the CAD system is a neutral file format (e.g., IGES or STEP) to be shared with other users up

```
# collect the list of fours
Collect
    Reverse(  _GetPointsBySide:(1.          xdc_StartingPt_TE: - 1, $lineNum, $Rstk, $Phi, $slnt)) +
    Reverse(  _GetPointsBySide:(  xdc_StartingPt_TE: - 1, RR_NPT:, $lineNum, $Rstk, $Phi, $slnt));
    #
    }; # end Loop Through Sections
}; # end List


(List)    _Points_Radial:
    If (   PtOrder: = Sides) Then
    Loop
    {
        For $eachSection In   _Points_Section::
        Collect first(nth(1,$eachSection)) Into $LE_SS;
        Collect last(nth(1,$eachSection)) Into $SS_TE;
        Collect first(nth(3,$eachSection)) Into $TE_PS;
        Collect last(nth(3,$eachSection)) Into $PS_LE;
        Return Is {$LE_SS, $SS_TE, $TE_PS, $PS_LE};
    }
    Else If (   PtOrder: = Wrap) Then
    Loop
    {
        For $eachSection In   _Points_Wrap::
        Collect nth(  _NSP: + (2 *  _NLE:) + 1,       $eachSection) Into $PSLEpts;
        Collect nth(  _NSP:,                          $eachSection) Into $SSLEpts;
        Collect nth((2 *   _NSP:) + (2 *   _NLE:),    $eachSection) Into $PSTEpts;
        Collect nth(1,                               $eachSection) Into $SSTEpts;
        Return Is {$SSLEpts,$SSTEpts,$PSTEpts,$PSLEpts};
    }
    Else
        {};
```

Fig. 37.3 User script for automatic geometry generation

and down the supply chain. Detailed two-dimensional drawings are often derived from the 3-D model in a semi-automated fashion to document the product and to communicate with suppliers. In addition, 3-D CAD data is generally shared in an automated way (due to integration between digital systems) with structural and manufacturing analysts for testing and process planning.

Geometry creation within CAD systems is also automated for certain tasks, especially those of a repetitive nature. The use of geometry duplication functions often involves copying, manipulating, or moving selected entities from one area to another on a model. This reduces the amount of time that it takes a user to create their finished model. However, it is critical that the user be mindful of parent–child references as described previously. While these references are elemental to the very nature of modern CAD systems, they can make the modification and reuse of design geometry tenuous at a later date, thereby negating any positive effects of a user having copied geometry in an effort to save time. Geometry automation also exists in the form of using scripting and programming functionality in modern CAD systems to generate geometry based on common templates. This scenario is particularly helpful when it is necessary to produce variations of objects with high degrees of accuracy and around which exists a fair amount of tribal knowledge and corporate practice. A set of parameters are created that represents corporate knowledge to be embedded into the geometry to control its shape and behavior and then the CAD system generates the desired geometry based on user inputs (Figs. 37.3 and 37.4). In the example of the airfoil, aerodynamic data has been captured by an engineering analyst and input into a CAD system us-

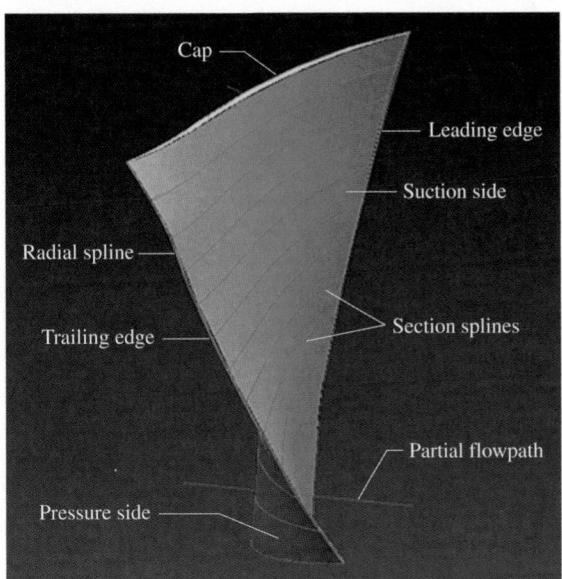

Fig. 37.4 Airfoil geometry generated from script (labels not generated as part of script)

ing a knowledge-capture module of the software. These types of modules allow a user to configure the behavior of the CAD system when it is supplied with a certain type of data in the requisite format. This data represents the work of the analyst, which is then used to automate the creation of the 3-D geometry to represent the airfoil. Such techniques are beginning to replace the manual geometric modeling tasks performed by users on designs that require a direct tie to engineering analysis data, or on those designs where a common geometry is shared among various design options.

37.4 User Characteristics Related to CAD Systems

Contemporary CAD systems require a technological knowledge base independent of (yet complementary to) normal engineering fundamentals. An understanding of design intent related to product function and how that is manifested in the creation of geometry to represent the product is critical [37.4, 5]. Users require the knowledge of how the various modules of a CAD systems work and the impact of their command choices on the usability of geometry downstream in the design and manufacturing process. In order to enable users to accomplish their tasks when using CAD tools, training in how to use the system is critical. Not just at a basic level for understanding the commands themselves, but the development of a community of practice to support the ongoing integration of user knowledge into organizational culture and best practices is critical.

Complementary to user training, and one of the reasons for why relevant training is important in the use of CAD tools, is for users to develop *strategic knowledge* in the use of the design systems. Strategic knowledge is the application of procedural and factual knowledge in the use of CAD systems directed towards a goal within a specific context [37.6–8]. It is through the development of strategic knowledge that users are able to effectively utilize the myriad functionality within modern CAD systems. Nearly all commercial CAD systems have similar user interfaces, similar geometry creation techniques, and similar required inputs and optional outputs. Productivity in the use of CAD systems requires that users employ their knowledge of engineering fundamentals and the tacit knowledge gathered from their environment, in conjunction with technological and strategic knowledge of the CAD system's capabilities, to generate a solution to the design problem at hand.

37.5 Visualization

Visualization information can be presented in visual formats such as text, graphics, charts, etc. This visualization makes applications simpler to understand by human users. Visualization is useful in automation not only for supervision, control, and decision support, but also for training. A variety of visualization methods and software are available, including geographic information systems and virtual reality (see examples in Chaps. 15, 16, 26, 27, 34, 38, and 73).

A geographic information system (GIS) is a computer-based system for capturing, manipulating, and displaying information using digitized maps. Its key characteristic is that every digital record has an identified geographical location. This process, called geocoding, enables automation applications for planning and decision making by mapping visualized information.

Virtual reality is interactive, computer-generated, three-dimensional imagery displayed to human users through a head-mounted display. In virtual reality, the visualization is artificially created. Virtual reality can be a powerful medium for communication and collaboration, as well as entertainment and learning. Table 37.1 lists examples of visualization applications (see also Chap. 15 on *Virtual Reality and Automation*).

Table 37.1 Examples of visualization applications

Application domain	Examples of visualization applications
Manufacturing	Virtual prototyping and engineering analysis
	Training and experimenting
	Ergonomics and virtual simulation
Design	Design of buildings
	Design of bridges
	Design of tools, furniture
Business	Advertising and marketing
	Presentation in e-Commerce, e-Business
	Presentation of financial information
Medicine	Physical therapy and recovery
	Interpretation of medical information and planning surgeries
	Training surgeons
Research and development	Virtual laboratories
	Representation of complex math and statistical models
	Spatial configurations
Learning and entertainment	Virtual explorations: art and science
	Virtual-reality games
	Learning and educational simulators

37.6 3-D Animation Production Process

Computer animations and simulations are commonly used in the engineering design process to visualize movement of parts, determine possible interferences of parts, and to simulate design analysis attributes such as fluid and thermal dynamics. The 3-D model files of most CAD systems can be converted into a format that can be used as input into popular animation software programs, such as Maya and 3ds Max. Once the files have been input into the animation software program, the animation process can begin. The animation process can be quite complex, depending on the level of realism necessary for the design visualization. This section will describe the steps necessary to create design animations from CAD models.

The 3-D animation production process can be divided into three main phases:

- Concept development and preproduction (Sect. 37.6.1)
- Production (Sect. 37.6.2)
- Postproduction and delivery (Sect. 37.6.3).

37.6.1 Concept Development and Preproduction

Several key activities take place during this phase, including story development and visual design, production planning, storyboarding, soundtrack recording, animation timing, and production of the animatic.

Every animation tells a *story*, "... you need not to have characters to have a story ..." [37.9]; for example

an architectural walkthrough or a medical visualization has a story in the sense that the events progress in a logical and effectively developed way. Story development begins with a *premise* – an idea in written form [37.10]. When the premise is approved, it is expanded into an *outline* or *treatment* – a scene-by-scene description of the animation – and the treatment is fleshed out into a full *script*. Visual design is carried out concurrently to story development. It is during this conceptual stage that the visual style of the animation is defined, character/object/environment design is finalized and approved for production, and the story idea is translated into a visual representation – the *storyboard*.

A *storyboard* is a sequence of images – *panels* – and textual descriptions describing the story, design, action, pacing, sound track, effects, camera angles/moves, and editing of the animation. Figure 37.5 shows an example of a preliminary storyboard. *Animation timing* is the process of pacing the action on the storyboard panels in order to tell the story in a clear and effective manner. The most common method of timing is the creation of a *story reel* or *animatic*. The creation of the animatic [37.11]: "... is essentially the process of combining the sound track with the storyboard to pace out the sequence". In addition to scanned-in storyboard panels with digitized soundtrack, the animatic can include simulated camera moves and rough motion of characters and objects. The main purpose of the animatic is to show the flow of the story by blocking the timing of the individual shots and defining the transitions between them. It provides an opportunity to

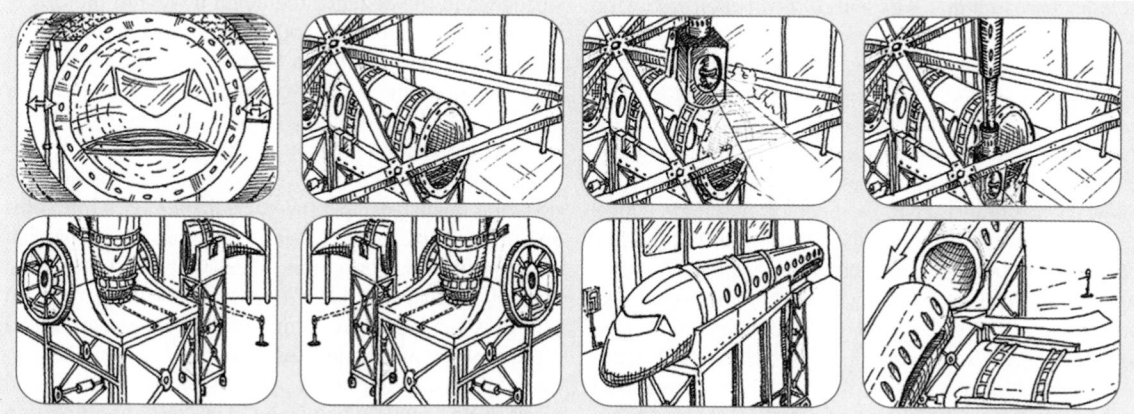

Fig. 37.5 An example of preliminary storyboard illustrating the futuristic assembly process of a Boeing 787 (courtesy of Purdue University, with permission from *N. Adamo-Villani, C. Miller*)

experiment with different cinematic solutions and to visualize whether the final animation makes sense as a filmic narrative.

37.6.2 Production

The production phase includes the following activities: 3-D modeling, texturing, rigging, animation, camera setup, lighting, and rendering.

Modeling

"Modeling is the spatial description and placement of objects, characters, environments and scenes with a computer system" [37.12]. In general, 3-D models for animation are produced using one of four approaches: *surface modeling*, *particle-system modeling*, *procedural modeling* or *digitizing techniques*.

In *surface modeling* surfaces are created using *spline*, *polygon* or *subdivision surfaces* modeling methods. A *spline* model consists of one or several *patches*, i.e., surfaces generated from two spline curves. Different splines generate different types of patches; the majority of spline models used in 3-D animation consist of nonuniform rational B-splines (NURBS) patches, which are generated from NURBS curves. A *polygonal model* consists of flat polygons, i.e., multisided objects composed of *edges*, *vertices*, and *faces*; a *subdivision surface* results from repeatedly refining a polygonal mesh to create a progressively finer mesh. Each subdivision step refines a submesh into a supermesh by inserting more vertices. In this way several levels of detail are created, allowing highly detailed modeling in isolated areas.

Common techniques used to create surface models include: *lathe*, *extrude*, *loft*, and *Boolean operations*; for instance, a polygonal mesh or a NURBS surface can be created by drawing a curve in space and rotating it around an axis (*lathe* or *revolve*); or by drawing a curve and pushing it straight back in space (*extrude*); or by connecting a series of contour curves (*loft*). *Boolean operators* allow for combination of surfaces in various ways to produce a single piece of geometry. Three Boolean operations are commonly used in 3-D modeling for animation: *addition* or *union*, *subtraction*, and *intersection*. The *addition* operation combines two surfaces into a single, unified surface; the *subtraction* operation takes away from one object the space occupied by another object, and the *intersection* operation produces an object consisting of only those parts shared by two objects, for instance, overlapping parts.

Particle-system modeling is an approach used to represent phenomena as fire, snow, clouds, smoke, etc. which do not have a stable and well-defined shape. Such phenomena would be very difficult to model with surface or solid modeling techniques because they are composed of large amounts of molecule-sized *particles* rather than discernible surfaces. In particle-system modeling, the animator creates a system of particles, i.e., graphical primitives such as points or lines, and defines the particles' physical attributes. These attributes control how the particles move, how they interact with the environment, and how they are rendered. Dynamics fields can also be used to control the particles' motion.

Procedural modeling includes a number of techniques to create 3-D models from sets of rules. *L-systems*, *fractals*, and *generative modeling* are examples of procedural modeling techniques since they apply algorithms for producing scenes; for instance, a terrain model can be produced by plotting an equation of fractal mathematics that recursively subdivides and displaces a patch.

When a physical model of an object already exists, it is possible to create a corresponding 3-D model using various *digitizing methods*. Examples of digitizing tools include *3-D digitizing pens* and *laser contour scanners*. Each time the tip of a 3-D pen touches the surface of the object to be digitized, the location of a point is recorded. In this way it is possible to compile a list of 3-D coordinates that represent key points on the surface. The 3-D modeling software uses these points to build the corresponding digital mesh, which is often a polygonal surface. In *laser contour scanning* the physical object is placed on a turntable, a laser beam is projected onto its surface, and the distance the beam travels to the object is recorded. After each 360° rotation a contour curve is produced and the beam is lowered a bit. When all contour curves have been generated, the 3-D software builds a lofted surface.

Surface models can be saved to a variety of formats. Some file formats are exclusive to specific software packages (*proprietary formats*), while others are *portable*, which means they can be exchanged among different programs. The two most common portable formats are ".obj" (short for object) introduced by Alias for high-end computer animation and visual effects productions, and the drawing interchange format (DXF) developed by Autodesk and widely used to exchange models between CAD and 3-D animation programs.

Texturing

Texturing is the process of defining certain characteristics of a 3-D surface such as color, shininess, reflectivity, transparency, incandescence, translucence, and smoothness. Frequently, these characteristics or *parameters* are treated as a *single set* called a *shader* or a *computer graphics (CG) material*. A *shader parameter* can be assigned a single value (for, example the color parameter can be assigned an RGB value of 255,0,0; in this case the entire surface is red), or the value can vary across the surface. *Two-dimensional (2-D) texture mapping* is a method of varying the texture parameter values across the surface using 2-D images. For example a 2-D picture can be applied to a 3-D surface to produce a certain color or transparency pattern. The 2-D picture can be a digital photo, a scanned image, an image produced with a 2-D paint program, or it can be generated procedurally. In general, the application of the 2-D image to the 3-D surface can be implemented in two ways: by projecting the image onto the surface (*projection mapping*) or by stretching it across the surface (*parameterized mapping*). In certain situations 2-D texture mapping does not produce realistic texture effects; for instance, a virtual block of marble rendered using 2-D texture techniques will appear to be wrapped in marble-patterned paper, rather than made of marble. To solve this problem it is possible to use another technique called *solid texture mapping*. The idea behind this technique is that *you create a virtual volume of texture and you immerse your object in that volume* [37.9]. Many 3-D software packages allow for creation of procedural 3-D textures. Figure 37.6 shows a rendering produced using a variety of texturing techniques.

Fig. 37.6 A 3-D image produced using a variety of texture maps: color, transparency, bump, reflection, and translucence (courtesy of Purdue University, with permission from *N. Adamo-Villani*)

Rigging

Rigging is the process of setting up a 3-D object or character for animation. Common rigging techniques include: *forward kinematics* (FK) and *inverse kinematics* (IK), *hierarchical models*, *skeletal systems*, *limits*, and *constraints*. When the 3-D object/character to be animated is made of multiple segments, the segments (or *nodes*) can be organized in an FK *hierarchical model*. In an FK hierarchy a node just below another node is called a *child*, while the node just above is called a *parent*, and the flow of transformations goes from parent to child. In order to animate an FK hierarchical model, each node needs to be selected and transformed individually to attain a certain pose. This process can become complicated and tedious when the model is very elaborate; for example, imagine a situation in which the animator needs to place the hand of a 3-D human-like character on a particular object. With an FK model, the animator has to first rotate the shoulder, then the lower arm, then the wrist, hand and fingers, working from the top of the hierarchy down. He cannot select the hand and place it on the object because the other parts of the arm will not follow, as they are parents of the hand. This problem can be solved by creating an *inverse kinematics* model in which the transformations travel upward through the hierarchy, for instance, from the hand to the shoulder. In this case the animator can place the hand on the object and have the other segments (lower arm and upper arm) follow the motion. An IK model is also called an *IK chain* and each node is referred to as a *link*. The first link of the chain is called the *root* of the chain and the end point of the last link is called the *effector* (as it *affects* the positions of all the other links, i.e., it effects the *IK solution*). In principle, each link in a chain can rotate any number of degrees around any axis. While these unrestricted rotations may be appropriate in certain situations, they are not likely to produce realistic results when the IK system is applied, for instance, to a human or animal model. This is due to the fact that human and animal joints have rotational limits; for example, the knee joint cannot bend beyond $\approx 180°$. To solve this problem it is common to set up *limits* and *constraints*. Limits can be defined for any of a model's three basic transformations (translation, rotation and scale); in addition, the transformations can be constrained (i.e., associated) to the transformations of other objects. *Position* or *point*, *rotation* or *orient*, and *direction* or *aim* are common types of constraints used in 3-D animation.

Many 3-D animation packages allow for creation of both FK and IK models. In general, to rig complex characters or objects the animator creates a *skeletal system*, i. e., a hierarchical model composed of joints connected by bones. The segments that make up the 3-D character/object are parented to the joints of the skeleton and the skeleton can function as an FK or IK model, with the possibility to switch between the two modes of operation during the animation process. If the 3-D object/character is supposed to deform during motion, the 3-D geometry can be attached (or *skinned*) to the skeletal joints; in this case the skeleton functions as a *deformation system*.

Animation

Common 3-D animation techniques include: *keyframe*, *motion path*, *physically-based*, and *motion capture animation*. In *keyframe animation*, the animator sets *key* values to various objects' parameters and saves these values at particular points in the timeframe; this process is called *setting keyframes*. After the animator has defined the keyframes, the 3-D software interpolates the values of the object's parameters between the keyframes. To gain more control of the interpolation, a parameter curve editor is available in the majority of 3-D animation packages. The parameter curve editor shows a graphical representation of the variation of a parameter's value over time (the *animation curve*. The animation curves are Bézier curves whose control points are the actual keyframes. Two tangent handles (vectors) are available at each control point (or keyframe) and allow the animator to manipulate the shape of the curve and thus the interpolation.

Motion path animation is used when the object to be animated needs to follow a well-defined path (for instance, a train moving along the tracks). In this case the animator draws the path, attaches the object to the path, and defines the number of frames required to reach the end of the path. In addition, the animator has the ability to control the rate of motion along the path, and the orientation of the object. The main advantage of motion path over keyframe animation is that, if the path is modified, the animation of the object that follows it updates to the path's changes.

Physically based animation is based on dynamics methods and is used to generate physically accurate simulations. Several steps are required to set up a dynamic simulation including definition of the objects' physical attributes (i. e., mass, initial velocity, elasticity, etc.); modeling of dynamic forces acting upon objects; and definition of collisions. A dynamic simulation can be *baked* before the animation is rendered; *baking* is the process of generating an animation curve for each parameter of an object whose change over time is caused by the dynamic simulation.

Motion capture animation, also referred to as *performance animation* or *digital puppetry* [37.13]:

> ... *involves measuring an actor's (or object) position and orientation in physical space and recording that information in a computer-usable form.*

In general the position or orientation of the actor is measured by a collection of input devices (*optical markers* or *sensors*) attached to the actor's body. Each input device has three degrees of freedom (DOF) and produces 3-D rotational or translational data which are channeled to the joints of a virtual character. As the actor moves, the input devices send data to the computer model. These data are used to control the movements of the character in real time, and to generate the animation curves. Motion capture animation is often used when the animation of the 3-D character needs to match the performance of the actor very precisely.

Camera Setup

The point of view from which a scene is observed is defined by the CG camera. The point of view is determined by two components or nodes: the *location of the camera* and the camera *center of interest*. The location is a point in space, while the center of interest can be specified as a location in space (i. e., a triplet of *XYZ* coordinate values) or as camera direction (i. e., a triplet of *XYZ* rotational values). In addition to location and center of interest, important attributes of a CG camera include the *zoom* parameter [whose value determines the width of the field of view (FOV) angle], *depth of field*, and *near and far clipping planes*, used to *clip* the viewable (and therefore renderable) 3-D space in the *Z* direction.

Lighting

The process of lighting a CG scene involves selecting the types of light to be used, defining their attributes, and placing them in the virtual environment. Common types of light used in CG lighting include *ambient lights*, *spotlights*, *point lights*, and *directional lights*. An *ambient light* simulates the widely distributed, indirect light that has bounced off objects in the 3-D scene and provides a uniform level of illumination; a *point light* emanates light in all directions from a specific location in space (simulating a light bulb); a *spotlight* is defined by location and direction and emits light in a cone-shaped beam of variable width; and a *directional light (or infinite light)* is assumed to be located infinitely far away and simulates the light coming from the sun. Common parameters of CG lights are: *intensity, color, falloff* (i. e., decrease of intensity with distance from the light source), and *shadow characteristics* such as shadow color, resolution, and density. In general, CG shadows are calculated using two popular techniques: *ray-tracing* and *shadow depth-maps*. *Ray-tracing* traces the path of a ray of light from the light source and determines whether objects in the scenes would block the ray to create a shadow; *depth-mapped shadows* use a pre-calculated *depth map* to determine the location of the shadows in the scene. Each pixel in the depth map represents the distance from the light source to the nearest shadow-casting surface in a specific direction. During rendering the light is cut off at the distances specified by the depth map with the result of making the light appear to be blocked by the objects. The shadows in Fig. 37.7 were generated using this technique.

Rendering

Rendering is the process of producing images from 3-D data. Most rendering algorithms included in 3-D animation software packages use an approach called *scan-line rendering*. A *scan line* is a row of pixels in a digital image; in scan-line rendering the program calculates the color of each pixel one after the other, scan line by scan line. The calculation of the pixels' colors can be done using different algorithms such as *ray-casting, ray-tracing*, and *radiosity*. The idea behind

ray-casting is to cast rays from the camera location, one per pixel, and find the closest object blocking the path of that ray. Using the material properties and the effect of the lights on the object, the ray-casting algorithm can determine the shading of this object. Ray-casting algorithms do not render reflections and refractions because they render the shading of each surface in the scene as if it existed in isolation, in other words other objects in the scene do not have any effect on the object being rendered. *Ray-tracing* addresses this limitation by considering all surfaces in the scene simultaneously. As each ray per pixel is cast from the camera location, it is tested for intersection with objects within the scene. In the event of a collision, the pixel's color is updated, and the ray is either recast or terminated based on material properties (such as reflectivity and refraction) and maximum recursion allowed. Although the ray-tracing algorithm can represent optical effects in a fairly realistic way, it does not render the diffuse reflection of light from one surface to another. This effect happens, for instance, when a blue object is close to a white wall. Even if the wall has very low specularity and reflectivity it will take a bluish hue because light bounces in a very diffuse way from the surface of the object to the surface of the wall. It is possible to render this phenomenon using a *radiosity* algorithm, which divides the surfaces in the scene into smaller subsurfaces or patches. A *form factor* is computed for each pair of subsurfaces; form factors are a coefficient describing how well the patches can see each other. Patches that are far away from each other, or oriented at oblique angles relative to one an-

Lights – 91 spotlights
Intensity – 0.2
Falloff – 52
Depth map resolution – 600
Shadow density – 1.35

All geometry resides within
the blue circle

Fig. 37.7 A 3-D image rendered using a lighting setup composed of a dome of spotlights with depth-mapped shadows (courtesy of Purdue University, with permission from *N. Adamo-Villani, C. Miller*)

other, will have small form factors while patches that are close to each other and facing each other 100% will have a form factor close to 1. The form factors are used as coefficients in a linearized form of the rendering equation, which yields a linear system of equations. Solving this system yields the radiosity, or brightness, of each patch, taking into account diffuse interreflections and soft shadows [37.14].

Key activities of the preproduction and production phases are illustrated in Fig. 37.8.

37.6.3 Postproduction

The postproduction phase includes two main activities: digital compositing and digital output.

Digital Compositing
In general an animated sequence includes images from multiple sources that are integrated into a single, seamless whole. *Digital compositing* is the process of digitally manipulating and combining at least two source images to produce an integrated result [37.15];

for example, Fig. 37.9 is a composite created from three different original images: the buildings, roads, and grass areas are a CG rendering produced from a three-dimensional model; the trees are also computer generated 3-D imagery rendered in a different pass as paint effects strokes; and the sky is a digital photograph projected onto a 3-D dome and used as a backdrop. In addition many of the elements in the image had some additional processing performed on them as they were added to the scene (for example, color and size adjustments).

Digital Output
Computer animation sequences can be output in the form of digital files, video or film. Some of the most popular digital file formats for saving animation sequences include: QuickTime, Motion Pictures Expert Group (MPEG), audio video interleaved (AVI), and Windows Media. The QuickTime format stores both video and audio data. It is a cross-platform format that supports different spatial and temporal resolutions and provides a variety of compression options. MPEG

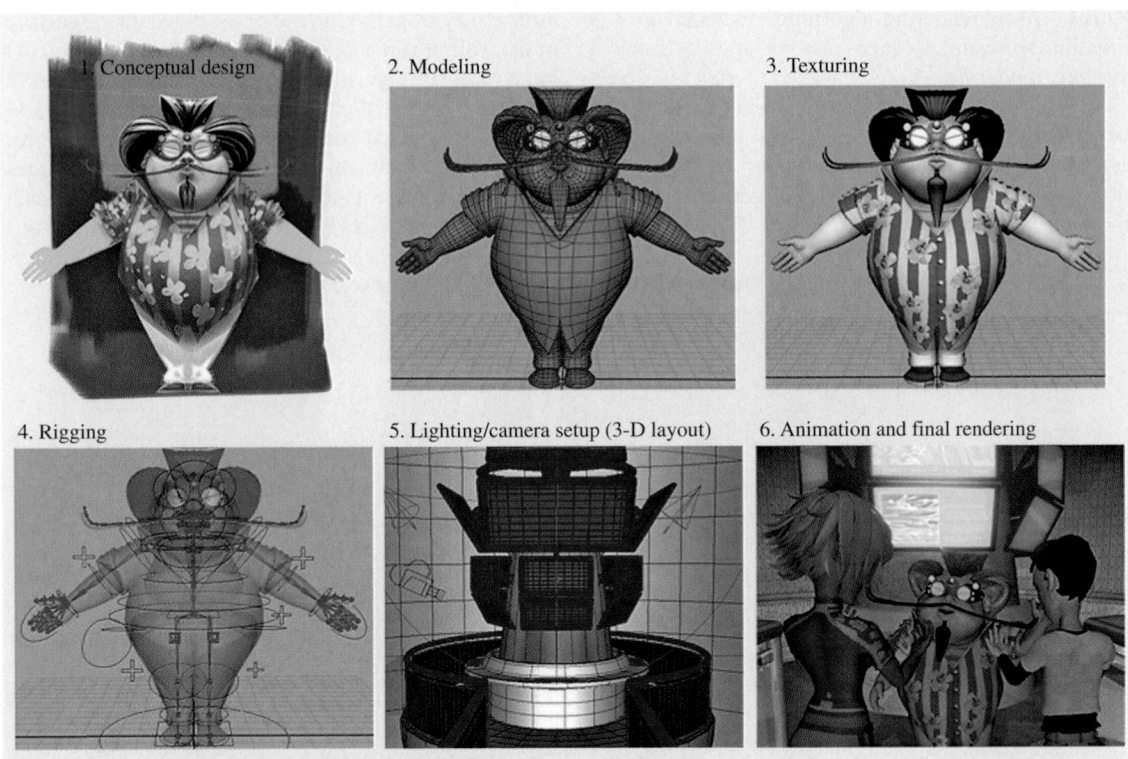

Fig. 37.8 Image showing key activities of the 3-D animation production process (courtesy of Purdue University and Educate for Tomorrow, Inc., with permission from *N. Adamo-Villani, R. Giasolli*)

(developed by the Motion Pictures Expert Group) is a popular format for compressing animations; the data compression is based on the removal of data that are identical or similar not just within the frame, but also between frames [37.12]. The AVI format (introduced by Microsoft in 1992) is a generic Windows-only format for moving images; a more recent version of this format is *Windows Media*, which offers more efficient compression for streaming high-resolution images.

Output on video can be done on a variety of video formats (both analog and digital). Commonly used digital formats include *high-definition* formats such as D6, HD-D5, HDCAM, DVCPRO and *standard-definition* formats such as D1, Digital Betacam, IMX, DV(NTSC), D-VHS, and DVD.

Fig. 37.9 An example composite image (courtesy of Purdue University, with permission from *N. Adamo-Villani*, *G. Bertoline* and *M. Sozen*)

Conclusions and Emerging Trends

This chapter provided an overview of computer-aided design and computer-aided engineering and included elements of computer graphics, animation, and visualization. Today's commercial brands of 3-D modeling tools essentially contain many of the same functions, irrespective of which software vendor is chosen. CAD software programs are dimension driven, parametric, feature based, and constraint based all at the same time, and these terms have come to be synonymous when describing modern CAD systems. Computer animations and simulations are commonly used in the engineering design process to visualize movement of parts, determine possible interferences of parts, and to simulate design analysis attributes such as fluid and thermal dynamics. Today most CAD systems 3-D model files can be converted into a format that can be used as input into popular animation software programs, such as Maya and 3ds Max. However, in the future it is anticipated that there will be a tighter integration between CAD and animation programs. CAD vendors will be under greater pressure to partner large enterprise software companies and become more product life cycle management (PLM) centric. This will result in CAD and animation becoming a part of a larger suite of software products used in industry. The rapid development of information technology and computer graphics technology will impact the hardware platforms and software development related to CAD and animation. This will result in even more feature-rich CAD software programs and capabilities. Faster screen refresh rates of large CAD models, the ability to collaborate at great distances in real time, shorter animation rendering times, and higher-resolution images are a few improvements that will result from the rapid development of information technology and computer graphics technology. Overall, there is an exciting future for CAD and animation in the future that will result in many positive impacts and changes for industries and businesses that depend on CAD and animation as a part of their day-to-day business.

References

37.1   E.N. Wiebe: 3-D constraint-based modeling: finding common themes, Eng. Des. Graph. J. **63**(3), 15–31 (1999)

37.2   P.J. Hanratty: Parametric/relational solid modeling. In: *Handbook of Solid Modeling*, ed. by D.E. Lacourse (McGraw-Hill, New York 1995) pp. 8.1–8.25

37.3   G.R. Bertoline, E.N. Wiebe: *Fundamentals of Graphic Communications*, 5th edn. (McGraw-Hill, Boston 2006)

37.4   N.W. Hartman: Defining expertise in the use of constraint-based CAD tools by examining practicing professional, Eng. Des. Graph. J. **69**(1), 6–15 (2005)

37.5   N.W. Hartman: The development of expertise in the use of constraint-based CAD tools: examining practicing professionals, Eng. Des. Graph. J. **68**(2), 14–25 (2004)

37.6   S.K. Bhavnani, B.E. John: Exploring the unrealized potential of computer-aided drafting, Proc. CHI'96 (1996) pp. 332–339

37.7   S.K. Bhavnani, B.E. John: From sufficient to efficient usage: An analysis of strategic knowledge, Proc. CHI'97 (1997) pp. 91–98

37.8   S.K. Bhavnani, B.E. John, U. Flemming: The strategic use of CAD: An empirically inspired, theory-based course, Proc. CHI'99 (1999) pp. 183–190

37.9   M. O'Rourke: *Principles of Three-Dimensional Computer Animation*, 3rd edn. (Norton, New York 2003)

37.10   J.A. Wright: *Animation Writing and Development* (Focal, Oxford 2005)

37.11   C. Winder, Z. Dowlatabadi: *Producing Animation* (Focal, Oxford 2001)

37.12   I.V. Kerlow: *The Art of 3-D: Computer Animation and Effects*, 2nd edn. (Wiley, Indianapolis 2000)

37.13   S. Dyer, J. Martin, J. Zulauf: Motion capture white paper (1995)

37.14   C. Goral, K.E. Torrance, D.P. Greenberg, B. Battaile: Modeling the interaction of light between diffuse surfaces, Comput. Graph. **18**(3), 213–222 (1984)

37.15   R. Brinkmann: *The Art and Science of Digital Compositing* (Morgan Kaufmann, Sand Diego 1999)

38. Design Automation for Microelectronics

Deming Chen

Design automation or computer–aided design (CAD) for microelectronic circuits has emerged since the creation of integrated circuits (IC). It has played a crucial role to enable the rapid development of hardware and software systems in the past several decades. CAD techniques are the key driving forces behind the reduction of circuit design time and the optimization of circuit quality. Meanwhile, the exponential growth of circuit capacity driven by Moore's law prompts new and critical challenges for CAD techniques. Moore's law describes an important trend in the history of the semiconductor industry: that the number of transistors per unit chip area would be doubled approximately every 2 years. This observation was first made by Intel co–founder Gordon E. Moore in a paper in 1965. Moore's law has held true for the past four decades, and many people believe that it will continue to apply for at least another decade before reaching the fundamental physical limits of device fabrication.

In this chapter we will introduce the fundamentals of design automation as an engineering field. We begin with several important processor

technologies and several existing IC technologies. We then present a typical CAD flow covering all the major steps in the design cycle. We also cover some important topics such as verification and technology computer–aided design (TCAD). Finally, we introduce some new trends in design automation.

38.1 Overview

Microelectronic circuits are ubiquitous nowadays. We can find them not only in desktop computers, laptops, and workstations, but also in consumer electronics, home and office appliances, automobiles, military applications, telecommunication applications, etc. Due to different requirements of these applications, circuits are designed differently to pursue unique features suitable for the specific application. In general, these devices are built with two fundamental and orthogonal technologies: *processor technology* and *IC technology*.

38.1.1 Background on Microelectronic Circuits

Processor technology refers to the architecture of the computation engine used to implement the desired functionality of an electronic circuit. It is categorized into three main branches: *general-purpose processors*, *application-specific instruction set processors*, and *single-purpose processors* [38.1]. A general-purpose processor, or microprocessor, is a device that executes software through instruction codes. Therefore, they are

Part D | 38

software-programmable. This processor has a program memory that holds the instructions and a general datapath that executes the instructions. The general datapath consists of one or several general-purpose arithmetic logic units (ALUs). An application-specific instruction set processor (ASIP) is a software-programmable processor optimized for a particular class or domain of applications, such as signal processing, telecommunication, or gaming applications. To fit the application, the datapath and instructions of such a processor are customized; for example, one type of ASIP, the digital signal processor (DSP), may have special-purpose datapath components such as a multiply–accumulate unit, which can perform multiply-and-add operations using only one instruction. Finally, a single-purpose processor is a digital circuit designed to serve a single purpose – executing exactly one program. It represents an exact fit of the desired functionality and is not software-programmable. Its datapath contains only the essential components for this program and there is no program memory. A general-purpose processor offers maximum flexibility in terms of the variety of applications it can support but with the least efficiency in terms of performance and power consumption. On the contrary, single-purpose processors offer the maximum performance and power efficiency but with the least flexibility in terms of the applications they can support. The ASIP offers a compromise between these two extremes.

IC technology refers to the specific implementation method or the design style of the processing engine on an IC. It is categorized into three main branches: *full custom*, *semicustom*, and *programmable logic device* (PLD) [38.2]. Full custom refers to the design style where the functional and physical designs are hand-crafted. This would provide the best design quality but also requires the extensive effort of a design team to optimize each detailed feature of the circuit. Since the design effort and cost are high, this design style is usually used in high-volume (thus cost can be amortized) or high-performance applications. Semicustom design is also called application-specific integrated circuit (ASIC). It tries to reduce the design complexity by restricting the circuit primitives to a limited number. Such a restriction allows the designer to use well-designed circuit primitives (gates or functional blocks) and focus on their efficient interconnection. Such a restriction also makes it easier to develop computer-aided design tools for circuit design and optimization and reduce the design time and cost. Today the number of semicustom designs outnumbers custom designs significantly, and some high-performance microprocessors have been designed partially using semicustom style, especially for the control logic (e.g., IBM's POWER-series processors and SUN Microsystems' UltraSPARC T1 processors).

Semicustom designs can be further partitioned into several major classes. Figure 38.1a shows such a partition. *Cell-based design* generally refers to *standard cell design*, where the fundamental cells are stored in a library. Cells often are simple gates and latches, but can be complex gates, memories, and logic blocks as well. These cells are pretested and precharacterized. The maintenance of the library is not a trivial task since each cell needs to be characterized in terms of area, delay, and power over ranges of temperatures and supply voltages [38.2]. Companies offer standard cell libraries for use with their fabrication and design technologies, and amortize the effort of designing the

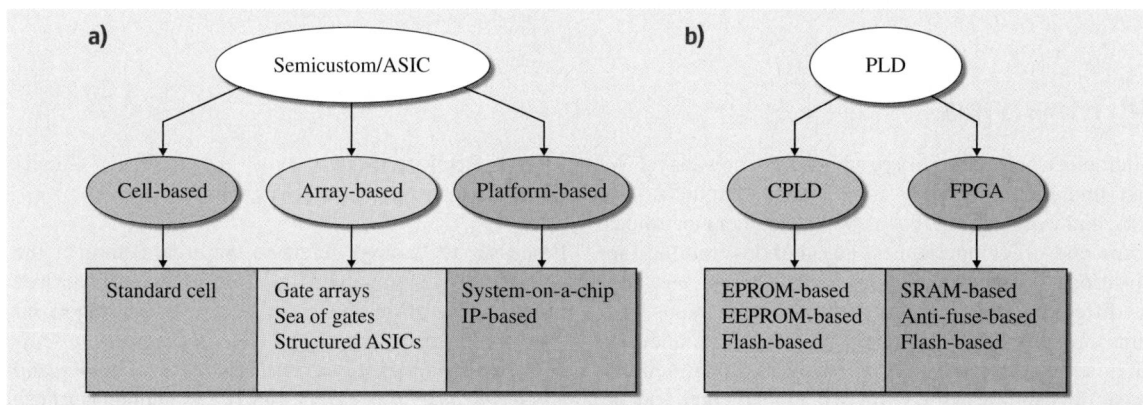

Fig. 38.1a,b Classification of (**a**) semicustom design and (**b**) PLD design (SRAM: static random access memory; EPROM: erasable programmable read-only memory; EEPROM: electrically erasable programmable read-only memory)

Table 38.1 Comparison of IC technologies

Metrics	Full custom	Semicustom	PLD
Nonrecurring engineering (NRE) cost	Very high	Medium-high	Low
Unit cost (low volume)	Very high	Medium-high	Low
Unit cost (high volume)	Low	Low	High
Design time	Very long	Medium-long	Short
Logic density	Very high	High	Low-medium
Circuit performance	Very high	High	Low-medium
Circuit power consumption	Low	Medium	High
Flexibility	Low	Medium	High

cell library over all the designs that use it. *Array-based design* in general refers to the design style of constructing a common base array of transistors and personalizing the chip by altering the metallization (the wiring between the transistors) that is placed on top of the transistors. It mainly consists of *gate arrays*, *sea of gates*, and *structured ASICs* (refer to Chap. 8 in [38.3] for details). *Platform-based design* [38.4–6] refers to the design style that heavily reuses hardware and software intellectual property (IP), which provide preprogrammed and verified design elements. Rather than looking at IP reuse in a block-by-block manner, platform-based design aggregates groups of components into a reusable platform architecture. There is a slight difference between *system-on-a-chip* and *IP-based* design. A system-on-a-chip approach usually incorporates at least one software-programmable processor, on-chip memory, and accelerating function units. IP-based design is more general and may not contain any software-programmable processors. Nonetheless, both styles heavily reuse IPs.

The third major IC technology is programmable logic device (PLD) technology (Fig. 38.1b). In PLD, both transistor and metallization are already fabricated but they are hardware-programmable. Such programming is achieved by creating and destroying wires that connect logic blocks either by making an *antifuse*, which is an open circuit device that becomes a short device when traversed by an appropriate current pulse, or setting a bit in a programmable switch that is controlled by a memory cell. There are two major PLD types: complex programmable logic devices (CPLDs) and field-programmable gate arrays (FPGA). The main difference between these two types of PLDs is that the basic programmable logic element in a CPLD is the PLA (programmable logic array) (two-level AND/OR array), and the basic element in an FPGA is the look-

up table (LUT). The PLAs are programmed through a mapping of logic functions in a two-level representation onto the AND/OR logic array, and the LUTs are programmed by setting bits in the LUT memory cells that store the truth table of logic functions. In general, CPLDs' routing structures are simpler than those of FPGAs. Therefore, the interconnect delay of CPLD is more predictable compared with that of FPGAs. FPGAs usually offer much larger logic capacity than CPLDs, mainly because LUTs offer finer logic granularity than PLAs so they are suitable to be replicated massively to help achieve complex logic designs. Nowadays, a high-end commercial FPGA, such as Altera Stratix III and Xilinx Virtex-5, can contain more than 300 K LUTs. Hardware programmability has significant advantages of short design time, low design cost, and fast time to market, which become more important when the design is complex. However, PLDs offer less logic density compared with semicustom designs, mainly because they occupy a significant amount of circuit area to add in the programming bits [38.7]. Nonetheless, the number of new design starts using PLDs significantly outnumbers the new semicustom design starts. According to research firm Gartner/Dataquest, in the year 2007, there were nearly 89 000 FPGA design starts, and this number will swell to 112 000 in 2010 – some 25 times that of semicustom/ASIC designs [38.8]. Figure 38.1b provides further characterization for the implementation styles of CPLDs and FPGAs.

An important fact is that different IC technologies have different advantages and disadvantages in terms of circuit characteristics. Table 38.1 lists the comparison of these technologies using some key metrics: nonrecurring engineering (NRE) cost (a one time charge for design and implementation of a specific product), unit cost, design time, logic density, circuit performance, power consumption, and flexibility (referring

Table 38.2 Combinations between processor technologies and IC technologies

IC technology types	Processor types		
	General purpose	Single purpose	ASIP
Full custom	Intel Core 2 Quad	Intel 3965ABG	TI TMS320C6000
	AMD Opteron	(802.11 a/b/g wireless chip)	(DSP)
		Analog ADV202	
		(JPEG 2000 Video CODEC)	
Semicustom	ARM9	ATMEL AT83SND1C	Infineon C166
	PowerPC	(MP3 Decoder)	(Microcontroller)
PLD	Altera NIOS II	Altera Viterbi Decoder	AllianceCORE C32025
	Xilinx MicroBlaze	(Error detection)	(DSP)

to the ease of changing the hardware implementation corresponding to design changes). Another important fact is that processor technologies and IC technologies are orthogonal to each other, which means that each of the three processor technologies can be implemented in any of the three IC technologies. Table 38.2 lists some representative combinations between these two technologies; for instance, general-purpose processors can be implemented using full-custom (Intel Core 2), semicustom (ARM9), or PLD (NIOS II). Each of the nine combinations has the combined features of the two corresponding technologies; for instance, Intel Core 2 represents the top but costly implementation of a general-purpose processor(s), and the Altera Viterbi Decoder represents the fast-time-to-market version of a single-purpose processor.

38.1.2 History of Electronic Design Automation

Electronic design automation (EDA) creates software tools for computer-aided design (CAD) of electronic systems ranging from printed circuit boards (PCBs) to integrated circuits. We describe a brief history of EDA next. General CAD information was provided in Chap. 37 of this Handbook. Integrated circuits were designed by hand and manually laid out before EDA. This design method obviously could not handle large and complex chips. By the mid 1970s, designers started to automate the design process using placement and routing tools. In 1986 and 1987 respectively, Verilog and VHDL (very high speed integrated circuit hardware description language) were introduced as hardware description languages. Circuit simulators quickly followed these inventions, allowing direct simulation of

IC designs. Later, logic synthesis was developed, which would produce circuit netlists for downstream placement and routing tools. The earliest EDA tools were produced academically, and were in the public domain. One of the most famous was the *Berkeley VLSI Tools Tarball*, a set of UNIX utilities used to design early VLSI (very large scale integration) systems. Meanwhile, the larger electronic companies had pursued EDA internally, where the seminal work was done at IBM and Bell Labs. In the early 1980s, managers and developers spun out of these big companies to start up EDA as a separate industry. Within a few years there were many companies specializing in EDA, each with a slightly different emphasis. Many of these EDA companies merged with one another through the years. Currently, the major EDA companies include Cadence, Magma, Mentor Graphics, and Synopsys. The total annual revenue of EDA is close to six billion US dollars.

According to the International Technology Roadmap for Semiconductors, the IC technology scaling driven by the Moore's law will continue to evolve and dominate the semiconductor industry for at least another 10 years. This will lead to over 14 billion transistors integrated on a single chip in the 18 nm technology by the year 2018 [38.9]. Such a scaling, however, has already created a large design productivity gap due to inherent design complexities and deep-submicron issues. The study by the research consortium SEMATECH shows that, although the level of on-chip integration, expressed in terms of the number of transistors per chip, increases at an approximate 58% annual compound growth rate, the design productivity, measured in terms of the number of transistors per staff-month, grows only at a 21% annual compound rate. Such a widening gap between IC capacity and design productivity presents critical challenges and also opportunities for the CAD community.

Better and new design methodologies are needed to bridge this gap.

38.2 Techniques of Electronic Design Automation

EDA can work on digital circuits and analog circuits. In this article, we will focus on EDA tools for digital integrated circuits because they are more prominent in the current EDA industry and occupy the major portion of the EDA market. For analog and mixed-signal circuit design automation, readers are referred to [38.10] and [38.11] for more details. Note that we can only briefly introduce the key techniques in EDA. Interested readers can refer to [38.12–14] for more details.

38.2.1 System–Level Design

Modern system-on-a-chip or FPGA designs contain embedded processors (hard or soft), busses, memory, and hardware accelerators on a single device. These embedded processors are software-programmable IP cores. Hard processors are built with full-custom or semicustom technologies, and soft processors are built with PLD implementations (Table 38.2). On the one hand, these types of circuits provide opportunities and flexibilities for system designers to develop high-performance systems targeting various applications. On the other hand, they also immediately increase the design complexity considerably, as mentioned in Sect. 38.1. To realize the promise of large system integration, a complete tool chain from concept to implementation is required. System- and behavior-level synthesis techniques are the building blocks for this automated system design flow. System-level synthesis compiles a complex application in a system-level description (such as in C or SystemC) into a set of tasks to be executed on various software-programmable processors (referred to as *software*), or a set of functions to be implemented in single-purpose processors (referred to as *customized hardware* or simply *hardware*), together with the communication protocols and the interface logic connecting different components. Such capabilities are part of the *electronic system-level* (ESL) design automation that has emerged recently to deal with the design complexity and improve design productivity. The design challenges in ESL are mainly on effective hardware/software partitioning and co-design, system integration, and related issues such as standardization of IP integration, system modeling,

performance/power estimation, and system verification, etc.

Figure 38.2 illustrates a global view of the ESL design flow. The essential task is the hardware/software co-design, which requires hardware/software partitioning and incorporates three key synthesis tasks: *processor synthesis*, *interface synthesis*, and *behavioral synthesis*. Hardware/software partitioning defines the parts of the application that would be executed in software or hardware. Processor synthesis for software-programmable processors usually involves instantiation of processor IP cores or generation of processors with customized features (customized cache size, datapath, bitwidth, or pipeline stages, etc.). Behavioral synthesis is also called high-level synthesis. It is a process that takes a given behavioral description of a hardware circuit and produces an RTL (register transfer level) design automatically. We will introduce more details about behavioral synthesis in the next section. Every time the designer explores a different system architecture with hardware/software partitioning, the system interfaces must be redesigned. Interface synthesis is the process of automatic derivation of both the hardware and software interfaces to bind hardware/software elements together and permit them to communicate correctly and efficiently. Interface synthesis results need to meet bandwidth and performance requirements. The end product of ESL is an integrated system-level IC (e.g., system on a chip,

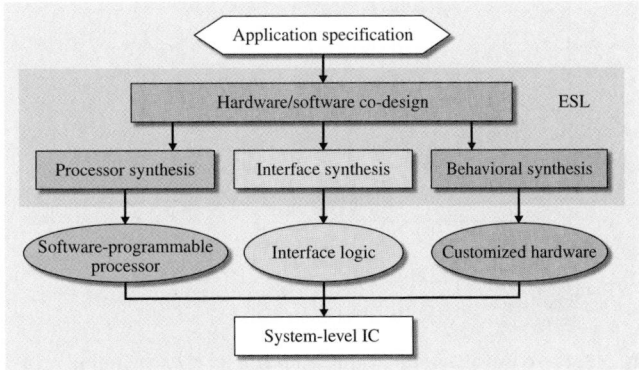

Fig. 38.2 Electronic system-level (ESL) design flow

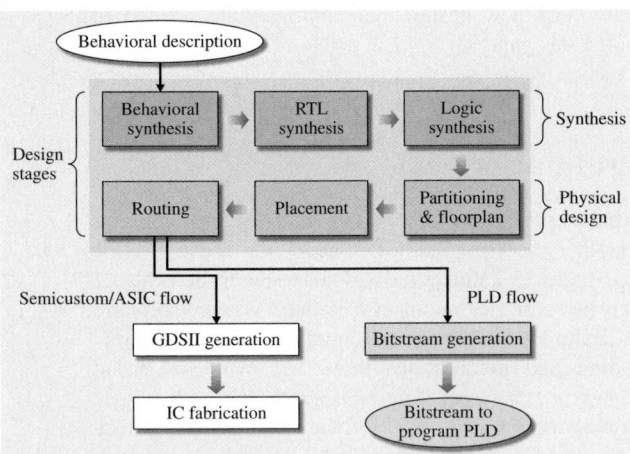

Fig. 38.3 A typical design flow

system in an FPGA, etc.) that aggregates software-programmable processors, customized hardware, and interface logic to satisfy the overall area, delay, and power constraints of the design. Interested readers can refer to [38.1, 12, 13] and [38.15–23] for further study.

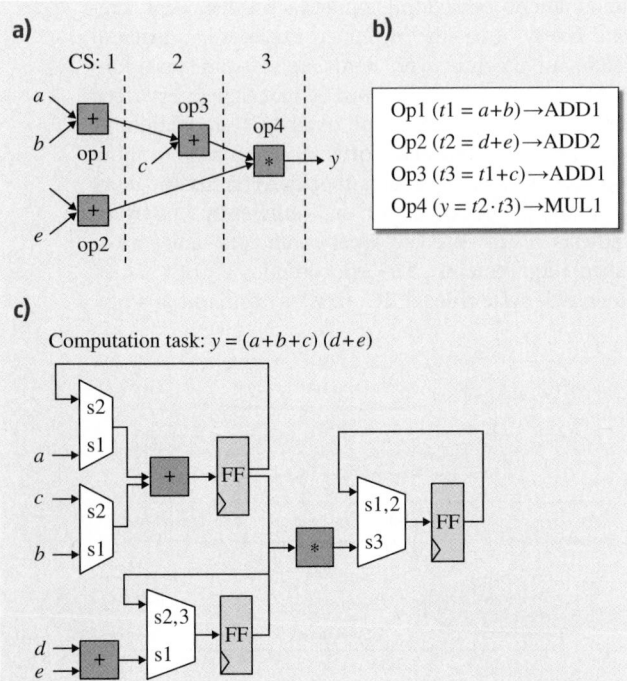

Fig. 38.4a–c A behavioral synthesis example: (**a**) scheduling solution, (**b**) binding solution, and (**c**) final datapath (after [38.24])

38.2.2 Typical Design Flow

The majority of the development effort for CAD techniques is devoted to the design of single-purpose processors using semicustom or PLD IC technologies. We will introduce a typical design flow step by step as shown in Fig. 38.3.

Behavior Synthesis

The basic problem of behavioral synthesis or high-level synthesis is the mapping of a behavioral description of a circuit into a cycle-accurate RTL design consisting of a *datapath* and a *control unit*. Designers can skip behavioral synthesis and directly write RTL codes for circuit design. This design style is facing increasing challenges due to the growing complexity of circuit design. A datapath is composed of three types of components: *functional units* (e.g., ALUs, multipliers, and shifters), *storage units* (e.g., registers and memory), and *interconnection units* (e.g., buses and multiplexers). The control unit is specified as a finite-state machine which controls the set of operations for the datapath to perform during every control step (clock cycle). The behavioral synthesis process mainly consists of three tasks: *scheduling*, *allocation*, and *binding*. Scheduling determines when a computational operation will be executed; allocation determines how many instances of resources (functional units, registers, or interconnection units) are needed; binding binds operations, variables, or data transfers to these resources. In general, it has been shown that the code density and simulation time can be improved by tenfold and hundredfold, respectively, when moving to behavior-level synthesis from RTL synthesis [38.23]. Such an improvement in efficiency is much needed for design in the deep-submicron era. Figure 38.4 shows the scheduling and the binding solution for a computation $y = (a+b+c) \times (d+e)$. Figure 38.4a shows the scheduling result, where CS means control step or clock cycle number. Figure 38.4b shows the binding solution for operations, which is a mapping between operations and functional units ($t1$, $t2$, $t3$ are temporary values). Figure 38.4c shows the final datapath. Note that the marks s1, s2, and s3 in the multiplexers indicate how the operands are selected for control steps 1, 2, and 3, respectively. A controller will be generated accordingly (not shown in the figure) to control the data movement in the datapath.

Behavior synthesis is a well-studied problem [38.2, 24–27]. Most of the behavioral synthesis problems are NP-hard problems due to various constraints, including latency and resource constraints. The subtasks of behavioral synthesis are highly interrelated with one another; for example, the scheduling of operations is directly constrained by resource allocation. Behavioral synthesis also faces challenges on how to connect better to the physical reality. Without physical layout information, the interconnect delay cannot be accurately estimated. In addition, there is a need of powerful data-dependence analysis tools to analyze the operational parallelism available in the design before one can allocate proper amount of resources to carry out the computation in parallel. In addition, how to carry out memory partitioning, bitwidth optimization, and memory access pattern optimization, together with behavioral synthesis for different application domains are important problems. Given all these challenges, much research is still needed in this area. Some recent representative works are presented in [38.28–35].

RTL Synthesis
The next step after behavioral synthesis is RTL synthesis. RTL synthesis performs optimizations on the register-transfer-level design. Input to an RTL synthesis tool is a Verilog or VHDL design that includes the number of datapath components, the binding of operations/variables/transfers to datapath components, and a controller that contains the detailed schedule of computational, input/output (I/O), and memory operations. In general, an RTL synthesis tool would use a front-end parser to parse the design and generate an intermediate representation of the design. Then, the tool can traverse the intermediate representation and create a netlist that consists of typical circuit substructures, including memory blocks, *if* and *case* blocks, arithmetic operations, registers, etc. Next, synthesis and optimization can be performed on this netlist, which can include examining adders and multipliers for constants, operation sharing, expression optimization, collapsing multiplexers, re-encoding finite-state machines for controllers, etc. Finally, an inferencing stage can be invoked to search for structures in the design that could be mapped to specific arithmetic units, memory blocks, registers, and other types of logic blocks from an RTL library. The output of the RTL synthesis provides such a mapped netlist. For the controller and glue logic, generic Boolean networks can be generated. RTL synthesis may need to consider the target IC technologies; for example, if the target IC technology is PLD, the regularity of PLD logic fabric offers opportunities for directly mapping datapath components to PLD logic blocks, producing regular layout, and reducing chip delay and synthesis runtime [38.37]. There are interesting

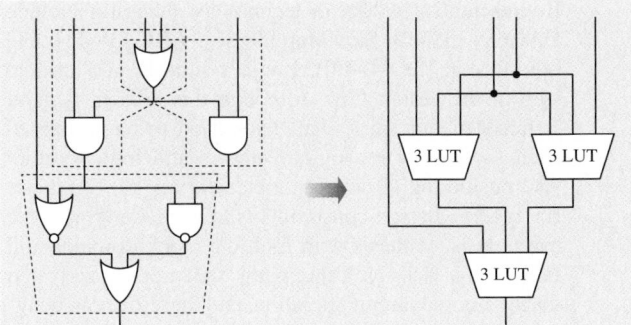

Fig. 38.5 An example of technology mapping for FPGAs (after [38.36])

research topics for further study in RTL synthesis, such as retiming for glitch power reduction, resource sharing for multiplexer optimization, and layout-driven RTL synthesis, to name just a few.

Logic Synthesis
Logic synthesis is the task of generating a structural view of the logic-level implementation of the design. It can take the generic Boolean network generated from the RTL synthesis and perform logic optimization on top of it. Such optimizations include both sequential logic optimization and combinational logic optimization. Typical sequential optimization includes finite-state machine encoding/minimization and retiming for the controller, and typical combinational optimization includes constant propagation, redundancy removal, logic network restructuring, and optimization, and don't-care-based optimizations. Such optimizations can also be carried out either in a general sense or targeting a specific IC technology. General optimization is also called *technology-independent optimization* with objectives such as minimizing the total amount of gates or reducing the logic depth of the Boolean network. Famous examples include the two-level logic minimizer ESPRESSO [38.38], the sequential circuit optimization system SIS [38.39], binary decision diagram (BDD)-based optimizations [38.40], and satisfiability (SAT)-based optimizations [38.41]. Logic optimization targeting a specific IC technology is also called *technology-dependent optimization*. The main task in this type of optimization is *technology mapping*, which transforms a Boolean network into an interconnection of logic cells provided from a cell library. Figure 38.5 demonstrates an example of mapping a Boolean network into an FPGA. In Fig. 38.5, each subcircuit in the dotted box is mapped into a three-input LUT.

Representative works in technology mapping include DAGON [38.42], FlowMap [38.36], ABC [38.43], and others (e.g. [38.44–46]). Logic synthesis is a critical step in the design flow. Although this area in general is fairly mature, new challenges need to be addressed such as fault-aware logic synthesis and logic synthesis considering circuit parameter variations. Synthesis for specific design constraints is also challenging. One example is synthesis with multiple clock domains and false paths [38.47]. False paths will not be activated during normal circuit operation, and therefore can be ignored. Multicycle paths refer to signal paths that carry a valid signal every few clock cycles, and therefore have a relaxed timing requirement.

Partitioning and Floorplan

We now get into the domain of physical design (Fig. 38.3). The input of physical design is a circuit netlist, and the output is the layout of the circuit. Physical design includes several stages such as *partitioning*, *floorplan*, *placement*, and *routing*. Partitioning is usually required for multimillion-gate designs. For such a large design, it is not feasible to layout the entire chip in one step due to the limitation of memory and computation resources. Instead, the circuit will be first partitioned into subcircuits (blocks), and then these blocks can go through a process called floorplan to set up the foundation of a good layout. A disadvantage of the partitioning process, however, is that it may degrade the performance of the final design if the components on a critical path are distributed into different blocks in the design [38.48]. Therefore, setting timing constraints is important for partitioning. Meanwhile, partitioning should also work to minimize the total number of connections between the blocks to reduce global wire usage and interconnect delay. Represen-

tative partitioning works include *Fiduccia–Mattheyses* (FM) partitioning [38.49] and hMETIS [38.50]. Other works (e.g. [38.51, 52]) are also well known.

Floorplan will select a good layout alternative for each block and for the entire chip as well. Floorplan will consider the area and the aspect ratio of the blocks, which can be estimated after partitioning. The number of terminals (pins) required by each block and the nets used to connect the blocks are also known after partitioning. The net is an important concept in physical design. It represents one wire (or a group of connected wires) that connect a set of terminals (pins) so these terminals will be made electrically equivalent.

In order to complete the layout, we need to determine the shape and orientation of each block and place them on the surface of the layout. These blocks should be placed in such a way as to reduce the total area of the circuit. Meanwhile, the pin-to-pin wire delay needs to be optimized. Floorplan also needs to consider whether there is sufficient routing area between the blocks so that the routing algorithms can complete the routing task without experiencing routing congestions. Partitioning and floorplan are optional design stages. They are required only when the circuit is highly complex. Usually, physical design for PLDs can skip these two stages. The PLD design flow, especially for hierarchical-structured FPGAs, would require a clustering design stage after the technology mapping stage. The clustering stage would gather groups of LUTs into logic blocks (e.g., each logic block contains ten LUTs). The netlist of logic blocks is then fed to a placement engine to determine the locations of the logic blocks on the chip. Some floorplanning works are [38.53–56].

Placement

Placement is a key step in the physical design flow. It deals with the similar problem as floorplan – determining the positions of physical objects (logic blocks and/or logic cells) on the layout surface. The difference is that, in placement, we can deal with a large number of objects (up to millions of objects) and the shape of each object is predetermined and fixed. Therefore, placement is a scaled and restricted version of the floorplan problem and is usually applied within regions created during floorplanning. Placement has a significant impact on the performance and routability of a circuit in nanometer design because a placement solution, to a large extent, defines the amount of interconnects, which have become the bottleneck of circuit performance. Figure 38.6 shows a simple example of a placement problem [38.48]. It shows

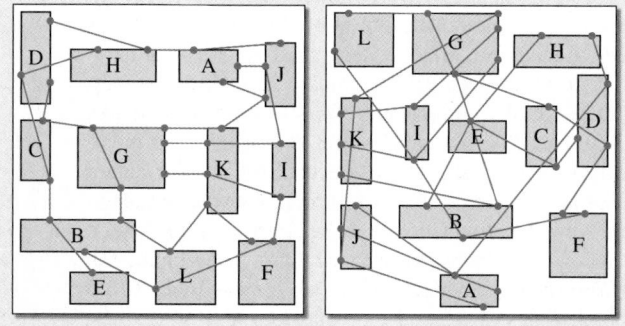

Fig. 38.6 Two different placements of the same problem (after [38.48])

two different placements for the same problem. The wire congestion in Fig. 38.6a is much less than that in Fig. 38.6b. Thus, the solution in Fig. 38.6a can be considered more easily routable than that in Fig. 38.6b. In placement, several optimization objectives may contradict each other; for example, minimizing layout area may lead to increased critical path delay and vice versa. Placement, like most of other physical design tasks, is an *NP*-hard problem and hence the algorithms used are generally heuristic in nature. Because of the importance of placement, an extensive amount of research has been carried out in the CAD community. Placement algorithms can be mainly categorized into simulated-annealing-based (e.g., [38.57, 58]), partitioning-based (e.g., CAPO [38.59]), analytical placement (e.g., BonnPlace [38.60]), and multilevel placement (e.g., mPL [38.61]). Some other well known placers include FastPlace [38.62], grid warping [38.63], Dragon [38.64], NTUplace [38.65], and APlace [38.66]. In general, for small placement instances (<50 K movable objects), simulated annealing is successful. For routability-driven placement of midsize instances (up to 100–200 K movable objects), partitioning-based placement does well. However, for large instances (above 200 K and into the millions), especially with many fixed pins/IP blocks, analytical and multilevel placement methods are the most successful.

It is worth mentioning that there is research on placement/floorplan-driven synthesis (also called *physical synthesis*). Once placement is available, interconnects are defined and may become a performance bottleneck. These interconnect delay values however can be fed back to the synthesis stages so that further optimization can be carried out in the presence of interconnect delays, including operations rebinding, logic restructuring, and remapping, etc. After such operations, an incremental placement step is needed to finalize the placement again given the new synthesis results. Placement-driven optimization is optional, but may improve design performance considerably.

Routing

After placement, the routing stage determines the geometric layouts of the nets to connect all the logic blocks and/or logic cells together. Routing is the last step in the design flow before either creating the GDSII (graphic data system II) file for fabrication in the semicustom/ASIC design style or generating the bitstream to program the PLD (Fig. 38.3). GDSII is a database file format used as the industry standard for IC layout

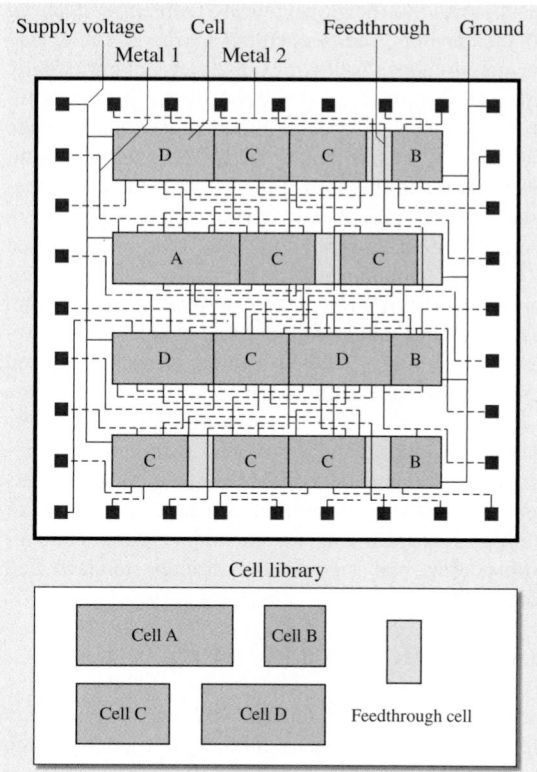

Fig. 38.7 The final layout of a simple standard cell-based design (metal 1: the first metal layer; metal 2: the second metal layer; feedthrough cell, dummy cells making space for routing wires to go through) (after [38.48])

data exchange. The objective of routing can be reducing the total wire length, minimizing the critical path delay, minimizing power consumption, or improving manufacturability, etc. A deep-submicron VLSI chip may contain tens of millions of gates. As a result, millions of nets have to be routed while each net may have hundreds of possible routes. This makes the routing problem computationally expensive and hard. To deal with such complexity, the current solution is to divide routing into two phases: *global routing* and *detailed routing*. Global routing will generate a coarse route for each net, which basically assigns routing regions to each net without specifying the actual geometric layout of the net. Detailed routing then finds the actual geometric layout of each net within the assigned regions.

The global routing problem is typically studied as a graph problem with different graph models, including the grid model, checkerboard model, and channel intersection model. Also, there are two kinds of approaches to solve the global routing problem: *sequential* and *concurrent*. The sequential approach routes the nets one by one, following an order determined by some criteria such as the nets' criticality or number of terminals. Some important algorithms include maze routing [38.67], line-probe [38.68], shortest-path-based, negotiation-based [38.69], and Steiner tree-based [38.70] algorithms. The concurrent approach avoids the ordering problem by considering routing of all the nets simultaneously. It usually follows a hierarchical partitioning of the problem instance into smaller subinstances, which can be solved by integer programming. Detailed routing is usually solved incrementally by routing a net one region at a time in a predefined order considering number of terminals, net width and type, pin locations, via restrictions, number of metal layers, etc. There has been an extensive amount of research published for routing algorithms [38.12, 48, 71]. Figure 38.7 illustrates the final layout of a simple standard cell design.

38.2.3 Verification and Testing

The major design steps in Sect. 38.2.2 focus on design implementation. Implementation is a transformation process that converts a design from a more abstract level into a lower, more detailed level. Verification, on the other hand, is the task of verifying that such a transformation is done correctly. Also, due to manufacturing imperfections, each fabricated chip needs to go through a testing procedure to make sure it is functioning as desired. Verification and testing are essential for timely delivery of correct ICs. These steps can occupy more than half of the total design time. We briefly introduce these topics here. Interested readers can refer to [38.3, 12, 13] for further details.

Verification
Ideally, verification should be carried out after each implementation step to catch any design errors. Otherwise, errors can propagate to the lower design levels and may eventually lead to faulty manufacturing masks, which then requires a design respin and generates a large cost overhead. Verification can be carried out in several different ways, namely *simulation*, *formal verification*, *emulation*, and *post-silicon validation*.

Simulation. Simulation uses mathematical models to simulate the behavior of an actual electronic device or circuit. In general, people obtain typical input vectors

(stimuli), track the propagation of these input values through the circuit, and check whether the simulated outputs are identical to the intended outputs of the circuit. Once mismatches are identified, the errors need to be localized and fixed. Simulations can be carried out at different levels (e.g., system level, RT (register transfer) level, or gate level). Usually, one cannot afford exhaustive simulation using all the input vectors because it would be very slow (the number of input vectors is an exponential function of the number of total inputs). Thus, in practice, only a subset of all the input vectors is used. As a result, how to select such a subset becomes extremely important; only the most relevant vectors should be selected for the maximum simulation coverage, which then leads to high fault coverage. Fault coverage is defined as a percentage that reports the ratio of output ports actually toggling between 1 and 0 during simulation, compared with the total number of output ports present in the circuit. However, the downside of this compromise is that some corner case design errors may remain undetected if testing vectors are not properly selected.

Simulation is also widely used to analyze the timing of the circuit, especially for analog and mixed-signal circuits. Simulation-based timing analysis is able to consider the correlation among the circuits' inputs and avoid timing analysis hurdles due to false paths. However, the main concern of this approach is its runtime complexity, especially for large and complex digital circuits. A popular replacement is static timing analysis (STA), which is carried out in an input-independent manner and tries to find the worst-case delay of the circuit over all possible input combinations. The computational complexity is liner in the number of edges in the circuit netlist. However, due to the *static* feature of STA, it is vulnerable to false paths, which the STA may treat as critical, but in reality they may never be sensitized. A series of works have been dedicated to dealing with this problem [38.72–74].

Formal Verification. Instead of simulation, formal verification strives to prove the correctness of a circuit implementation using formal methods of mathematics. Used correctly, this method can decrease the verification time as well as guarantee the correctness. However, due to the intrinsic difficulty of the problem, this proof-based method cannot scale to very large designs. There are two main techniques for formal verification, namely *model checking* and *equivalence checking*. model checking verifies that a design satisfies certain properties. It generally requires that the designer knows what desired properties the design should have. Equivalence checking compares two implementations (one of which is known to be correct) and proves that they are functionally equivalent; for example, two Boolean functions can be compared using their reduced-ordered binary decision diagram (ROBDD) representations, and two finite-state machines can be compared using their state diagrams.

Emulation and Post-silicon Validation. Emulation is the implementation of a prototype of the design using a PLD (Sect. 38.1.1). Once the prototype is completed, the verification process applies input vectors to the PLD and compares its outputs with the intended outputs. This is similar to simulation in principle, but is much faster due to the significant hardware acceleration effect provided by the hardware resources in the PLD. Thus, designers can afford to operate on more input vectors for better verification results. A drawback of this approach is that, for each design, designers need to spend time and effort to come up with its prototype upfront, before emulation starts (system- and behavior-level design automation can speed up this process). Post-silicon validation uses actual fabricated chips and tests them at full speed. This method is obviously the fastest but is also the most expensive among all the verification methods.

Testing
Verification works on the functional or timing aspect of the design before manufacturing. Even if the chip is without design errors, hardware defects may occur during the manufacturing process. A number of factors such as optical proximity effects, airborne impurities, and processing material flaws during fabrication can result in defective transistors and interconnect. There are also hard errors caused by sophisticated mechanisms, such as those caused by antenna, thermal, and inductive effects. Thus, testing after manufacturing is essential. Since each individual chip must be tested, testing can be very time consuming and expensive. Testing can be also challenging because the behavior of a million-gate VLSI chip has to be tested using only a small number of pads (e.g., < 100).

People have identified fault models for testing purposes. The best known is the model of *stuck-at* faults, where a fault causes the signal of a wire to be fixed at the value 0 or at the value 1. With this model, one tries to observe the faults' effect on the circuit behavior in-

stead of directly detecting a physical defect. The main issues and techniques involved with testing include the following:

1. Automatic test pattern generation (ATPG): This technique automatically selects high-quality test vectors to minimize the testing time of each chip on the tester.
2. Controllability and observability of signals: *Controllability* means ease of forcing an internal logic gate to 0 or 1 by driving input pins of the chip, whereas *observability* is the ease of observing an internal logic gate by watching external output pins of the chip. Testing ideally would check every gate in the circuit to prove it is not stuck. Therefore, it is desirable to design the chip to increase gate observability and controllability.
3. Scan chain: Convert each flip–flop to a scan register, which has normal mode and scan mode. In the scan mode, specific values can be scanned in for testing purposes.
4. Built-in self-test (BIST): Circuits test themselves. The circuit contains extra testing circuitry, which in the testing mode generates input vectors to test the circuit by itself.

38.2.4 Technology CAD

Technology computer-aided design (TCAD) is an important branch in CAD which carries out numeric simulations of semiconductor processes and devices. Process TCAD takes a process flow, including essential steps such as ion implantation, diffusion, etching, deposition, lithography, oxidation, and silicidation, and simulates the active dopant distribution, the stress distribution, and the device geometry. Mask layout is also an input for process simulation. The layout can be selected as a linear cut in a full layout for a two-dimensional simulation or a rectangular cut from the layout for a three-dimensional simulation. Process TCAD produces a final cross-sectional structure. Such a structure is then provided to device TCAD for modeling the device electrical characteristics. The device characteristics can be used to either generate the coefficients of compact device models or develop the compact models themselves. These models are then used in circuit simulators, such as the SPICE (simulation program with integrated circuit emphasis) simulator, to model the circuit behavior. Because of the detailed physical modeling involved, TCAD is mostly used to aid the design of single devices.

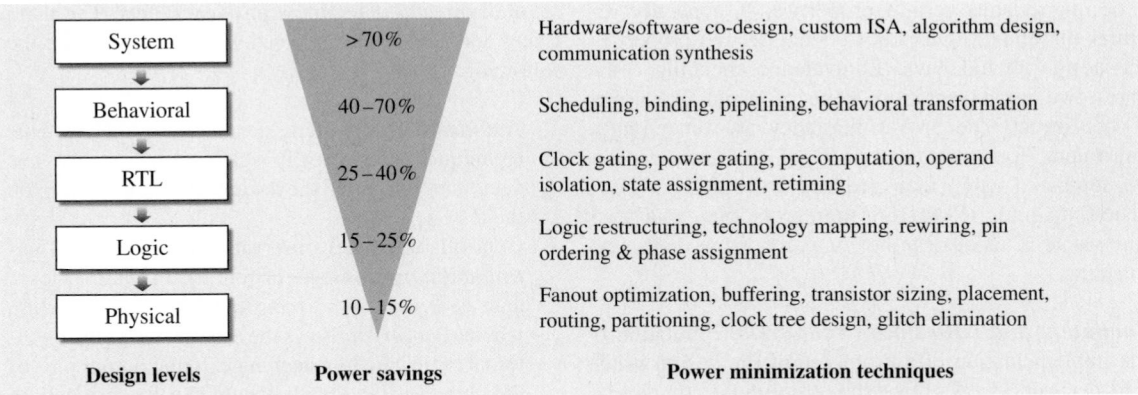

Design levels	Power savings	Power minimization techniques
System	>70%	Hardware/software co-design, custom ISA, algorithm design, communication synthesis
Behavioral	40–70%	Scheduling, binding, pipelining, behavioral transformation
RTL	25–40%	Clock gating, power gating, precomputation, operand isolation, state assignment, retiming
Logic	15–25%	Logic restructuring, technology mapping, rewiring, pin ordering & phase assignment
Physical	10–15%	Fanout optimization, buffering, transistor sizing, placement, routing, partitioning, clock tree design, glitch elimination

Fig. 38.8 Power-saving opportunities at different design levels (after [38.76]) (ISA: instruction set architecture)

TCAD has become a critical tool in the development of next-generation IC processes and devices. The reference [38.75] summarizes applications of TCAD in four areas:

1. Technology selection: TCAD tools can be used to eliminate or narrow technology development options prior to starting experiments.
2. Process optimization: Tune process variables and design rules to optimize performance, reliability, cost, and manufacturability.
3. Process control: Aid the transfer of a process from one facility to another (including from development to manufacturing) and serve as reference models for diagnosing yield issues and aiding process control in manufacturing.
4. Design optimization: Optimize the circuits for cost, power, performance, and reliability.

The challenge for TCAD is that the physics and chemistry of fabrication processes are still not well understood. Therefore, TCAD cannot replace experiments except in very limited applications so far. It is worth mentioning that electromagnetic field solvers are considered part of TCAD as well. These solvers solve Maxwell's equations, which govern the electromagnetic behavior, for the benefits of IC and PCB design; for example, one objective is to help account accurately for parasitic effects of complicated interconnect structures.

38.2.5 Design for Low Power

With the exponential growth of the performance and capacity of integrated circuits, power consumption has become one of the most constraining factors in the IC design flow. There are three power sources in a circuit: *switching power*, *short-circuit power*, and *static* or *leakage power*. The first two types of power can only occur when a signal transition takes place at the gate output; together they are called *dynamic power*. There are two types of signal transitions: one is the signal transition necessary to perform the required logic functions; the other is the unnecessary signal transition due to the unbalanced path delays to the inputs of a gate (called *spurious transition* or *glitch*). Static power is the power consumption when there is no signal transition for a gate. As technology advances to feature sizes of 90 nm and below, static power starts to become a dominating factor in the total chip power dissipation. Design for low power is a vast research topic involving low-power device/circuit/system architecture design, device/circuit/system power estimation, and various CAD techniques for power minimization [38.77–80]. Power minimization can be performed in any of the design stages. Figure 38.8 shows the power-saving techniques and power saving potentials during each design level [38.76]. Note that some techniques are not unique to only one design level; for example, glitch elimination and retiming can be applied to logic-level design as well.

38.3 New Trends and Conclusion

Due to technology scaling, nanoscale process technologies are fraught with nonidealities such as process variations, noise, soft errors, leakage, and others. Designers are also facing unprecedented design complexity due to these issues. CAD techniques need new innovations to continue to deliver high-quality IC designs in a short period of time. Under this vision, we introduce some new trends in CAD below.

Design for Manufacturing (DFM)

Nanometer IC designs are deeply challenged by manufacturing variations. The industry is currently using 193 nm photolithography for fabrication of ICs in 130 nm and down (to 32 nm or even 22 nm). Therefore, it is challenging for the photolithography process to precisely control the manufacturing quality of the circuit features. There are other manufacturing/process challenges, such as topography variations, random defects due to missing/extra material, via void/failure, etc. DFM will take the manufacturing issues into the design process to improve circuit manufacturability and yield. The essential task in DFM is the development of resolution enhancement techniques (RETs), such as tools for optical proximity correction (OPC) and phase-shift mask (PSM) [38.81–85]. As an example, Fig. 38.9 shows the OPC optimization for a layout, which manipulates mask geometry to compensate for image distortions. Another area is to develop efficient engineering change order (ECO) tools, so that when some changes need to be made, as few layers as possible need to be modified [38.86, 87]. Meanwhile, post-silicon debug and repair techniques are gaining importance as well [38.88, 89].

Statistical Static Timing Analysis (SSTA)

Large variation in process parameters makes worst-case design too expensive in terms of power and delay. Meanwhile, nominal case design will result in a loss in yield as performance specifications may not be met for a large percentage of chips. SSTA is an effort to specifically improve performance yield to combat manufacturing variations. SSTA treats the delay of each gate as a random variable and propagates gates' probability density functions (PDFs) through the circuit to create a PDF of the output delay random variable. Spatial correlations among the circuit components need to be considered. A vast amount of research has been reported (e.g., [38.90–98]) in the past 5 years. SSTA is critical to guide statistical design methodologies. An important application is SSTA-driven placement and routing to improve performance yield.

Design for Nanotechnology

Sustained exponential growth of complex electronic systems will require new breakthroughs in fabrication and assembly with controlled engineering of nanoscale components. *Bottom-up* approaches, in which integrated functional device structures are assembled from chemically synthesized nanoscale building blocks (so-called nanomaterials), such as *carbon nanotubes*, *nanowires*, and other molecular electronic devices, have the potential to revolutionize the fabrication of electronic systems. Nanoelectronic circuits always have a certain percentage of defects as well as nanomaterial-specific variations over and above process variations introduced by lithography. Using simplified nanode-

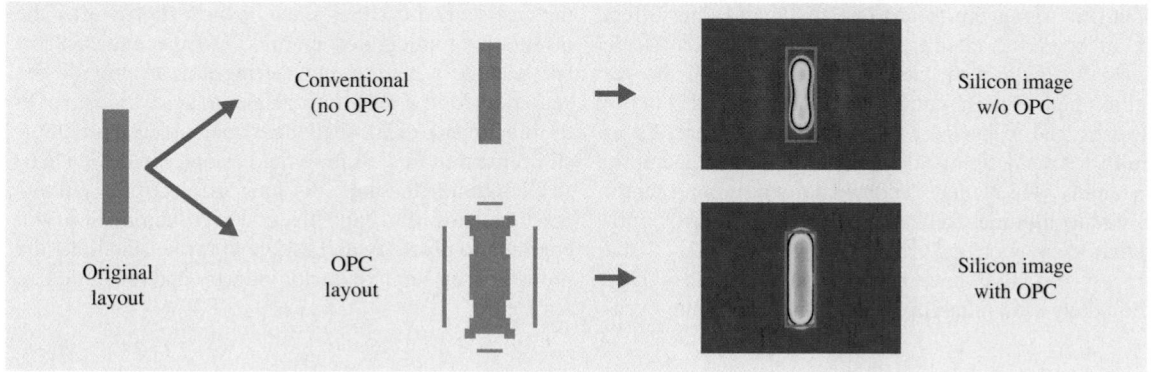

Fig. 38.9 An illustration of optical proximity correction (after [38.81])

vice assumptions and traditional scaled design flows will lead to suboptimal and impractical nanocircuit designs and inaccurate system evaluation results. For nanotechnology to fulfill its promise, there is a need to understand and incorporate nano-specific design techniques, such as nanosystems modeling, statistical approaches, and fault-tolerant design, systematically from devices all the way up to systems. Initial effort has been made in this important area [38.99–107], but much more research has to be done to enable the large integration capability of nanosystems. Chapter 53 provides more information on micro and nano manipulation related to design of nanotechnology.

Design for 3-D ICs

One promising way to improve circuit performance, logic density or power efficiency is to develop three-dimensional integration, which increases the number of active die layers and optimizes the interconnect network vertically [38.108–115]. Potentially, three-dimensional (3-D) IC provides improved bit bandwidth with reduced wire length, delay, and power. There are different bonding technologies for 3-D ICs, including die-to-die, die-to-wafer, and wafer-to-wafer, and the two parties can be bonded face-to-face or face-to-back. One disadvantage of the 3-D IC is its thermal penalty. The 3-D stacks will increase heat density, leading to degraded chip performance and reliability if not handled properly.

Design for Reliability

Besides fabrication defects, soft errors and aging errors have emerged as the new sources of circuit unreliability for nanometer circuit designs. A soft error occurs when a cosmic particle, such as a neutron, strikes a portion of the circuit, upsetting the state of a bit. Aging errors are due to the wearing effect of an operating circuit. As device dimensions scale down faster than the supply voltage [38.9], the resulting high electric fields combined with temperature stresses lead to device aging and hence failure. Especially, transistor aging due to negative-bias temperature instability (NBTI) has become the determining factor in circuit lifetime. Reliability analysis and error mitigation techniques under soft errors, aging effects, and process variations have been proposed [38.116–126]. Ultimately, chip reliability would need to become a crit-

ical design metric incorporated into mainstream CAD methodologies.

Design with Parallel Computing

An important way to deal with design complexity is to take advantage of the latest advances of parallel computing with multicore computer systems so that computation can be carried out in parallel for acceleration. Although there are some studies on parallel CAD algorithms (e.g., [38.127, 128]), much more work is needed to come up with parallel CAD algorithms to improve design productivity.

Design for Network on Chip (NoC)

The increasing complexity and heterogeneity of future SoCs (system-on-a-chip) prompt significant system scalability challenge using conventional on-chip communication schemes, such as the point-to-point (P2P) and bus-based communication architectures. NoC emerged recently as a promising solution for the future [38.129–133]. In a NoC system, modules such as processor cores, memories, and other IP blocks exchange data using a network on a single chip. NoC communication is constructed from a network of data links interconnected by switches (or routers) such that messages can be relayed from any source module to any destination module. Because all links in the NoC can operate simultaneously on different data packets, a high level of parallelism can be achieved with a great scaling capability. However, many challenging research problems remain to be solved for NoC, from the design of the physical link through the network-level structure, all the way up to the system architecture and application software.

Electronic design automation or computer-aided design as an engineering field has been evolving through the past several decades since its birth shortly after the invention of integrated circuits. On the one hand, it has become a mature engineering area to provide design tools for the electronic semiconductor industry. On the other hand, many challenges and unsolved problems still remain in this exciting field as on-chip device density continues to scale. As long as electronic circuits are impacting our daily lives, design automation will continue to diversify and evolve to further facilitate the growth of the semiconductor industry and revolutionize our future.

References

38.1   F. Vahid, T. Givargis: *Embedded System Design: A Unified Hardware/Software Introduction* (Wiley, New York 2002)

38.2   G. De Micheli: *Synthesis and Optimization of Digital Circuits* (McGraw-Hill, Upper Saddle River 1994)

38.3   N. Weste, D. Harris: *CMOS VLSI Design: A Circuits and Systems Perspective*, 3rd edn. (Addison Wesley, Indianapolis 2004)

38.4   K. Keutzer, S. Malik, R. Newton, J. Rabaey, A. Sangiovanni-Vincentelli: System level design: orthogonalization of concerns and platform-based design, IEEE Trans. CAD Integr. Circuits Syst. **19**(12), 1523–1543 (2000)

38.5   A. Sangiovanni-Vincentelli, G. Martin: A vision for embedded systems: platform-based design and software methodology, IEEE Des. Test Comput. **18**(6), 23–33 (2001)

38.6   G. Martin, H. Chang: *Winning the SoC Revolution: Experiences in Real Design* (Kluwer, Dordrecht 2003)

38.7   I. Kuon, J. Rose: Measuring the gap between FPGAs and ASICs, IEEE Trans. CAD Integr. Circuits Syst. **26**(2), 203–215 (2007)

38.8   D. Orecchio: FPGA explosion will test EDA, Electronic Design Update (2007), http://electronicdesign.com/Articles/ArticleID/15910/15910.html. Accessed 18 June 2007

38.9   L. Wilson: International Technology Roadmap for Semiconductors. http://www.itrs.net/ (2008)

38.10 G.G.E. Gielen, R.A. Rutenbar: Computer-aided design of analog and mixed-signal integrated circuits, Proc. IEEE **88**(12), 1825–1854 (2000)

38.11 P. Wambacq, G. Vandersteen, J. Phillips, J. Roychowdhury, W. Eberle, B. Yang, D. Long, A. Demir: CAD for RF circuits, Proc. Des. Autom. Test Eur. (2001)

38.12 L. Scheffer, L. Lavagno, G. Martin (eds.): *Electronic Design Automation for Integrated Circuits Handbook* (CRC, Boca Raton 2006)

38.13 D. Jansen (ed.): *The Electronic Design Automation Handbook* (Springer, Norwell 2003)

38.14 C.J. Alpert, D.P. Mehta, S.S. Sapatnekar (eds.): *The Handbook of Algorithms for VLSI Physical Design Automation* (CRC, Boca Raton 2007)

38.15 S. Kumar, J. Aylor, B.W. Johnson, W.A. Wulf: *The Codesign of Embedded Systems: A Unified Hardware/Software Representation* (Kluwer, Dordrecht 1996)

38.16 Y.T. Li, S. Malik: *Performance Analysis of Real-Time Embedded Software* (Kluwer, Dordrecht 1999)

38.17 G. De Micheli, R. Ernst, W. Wolf (eds.): *Readings in Hardware/Software Codesign* (Morgan Kaufmann, New York 2001)

38.18 F. Balarin, H. Hsieh, L. Lavagno, C. Passerone, A. Pinto, A. Sangiovanni-Vincentelli, Y. Watanabe, G. Yang: Metropolis: a design environment for heterogeneous systems. In: *Multiprocessor Systems-on-Chips*, ed. by A. Jerraya, W. Wolf (Morgan Kaufmann, New York 2004), Chap. 16

38.19 R.P. Dick, N.K. Jha: MOCSYN: multi-objective core-based single-chip system synthesis, Proc. IEEE Des. Autom. Test Eur. (1999)

38.20 P. Petrov, A. Orailoglu: Tag compression for low power in dynamically customizable embedded processors, IEEE Trans. CAD Integr. Circuits Syst. **23**(7), 1031–1047 (2004)

38.21 S.P. Levitan, R.R. Hoare: *Structural Level SoC Design Course* (The Technology Collaborative, Pittsburgh 1991)

38.22 B. Bailey, G. Martin, A. Piziali: *ESL Design and Verification: A Prescription for Electronic System Level Methodology* (Elsevier, Amsterdam 2007)

38.23 K. Wakabayashi, T. Okamoto: C-based SoC design flow and EDA tools: an ASIC and system vendor perspective, IEEE Trans. CAD Integr. Circuits Syst. **19**(12), 1507–1522 (2000)

38.24 J.P. Elliott: *Understanding Behavioral Synthesis: A Practical Guide to High-Level Design* (Kluwer, Dordrecht 1999)

38.25 D. Gajski, N. Dutt, A. Wu: *High-Level Synthesis: Introduction to Chip and System Design* (Kluwer, Dordrecht 1992)

38.26 A. Raghunathan, N.K. Jha, S. Dey: *High-Level Power Analysis and Optimization* (Kluwer, Dordrecht 1998)

38.27 R. Camposano, W. Wolf: *High-Level VLSI Synthesis* (Springer, New York 2001)

38.28 J. Chang, M. Pedram: *Power Optimization and Synthesis at Behavioral and System Levels Using Formal Methods* (Kluwer, Boston 1999)

38.29 A. Chandrakasan, M. Potkonjak, J. Rabaey, R. Brodersen: Hyper-LP: a system for power minimization using architectural transformations. In: *The Best of ICCAD, 20 Years of Excellence in Computer-Aided Design*, ed. by A. Kuehlman (Kluwer, Boston 2003)

38.30 S. Gupta, N.D. Dutt, R. Gupta, A. Nicolau: *SPARK: A Parallelizing Approach to the High-Level Synthesis of Digital Circuits* (Kluwer, Norwell 2004)

38.31 S. Memik, E. Bozorgzadeh, R. Kastner, M. Sarrafzadeh: A scheduling algorithm for optimization and early planning in high-level synthesis, ACM Trans. Des. Autom. Electron. Syst. **10**(1), 33–57 (2005)

38.32 J. Jeon, D. Kim, D. Shin, K. Choi: High-level synthesis under multi-cycle interconnect delay, Proc. Asia South Pac. Des. Autom. Conf. (2001)

38.33 P. Brisk, A. Verma, P. Ienne: Optimal polynomial-time interprocedural register allocation for high-level synthesis and ASIP design, Proc. Int. Conf. Comput.-Aided Des. (2007)

38.34   D. Chen, J. Cong, Y. Fan, G. Han, W. Jiang, Z. Zhang: xPilot: a platform–based behavioral synthesis system, Proc. SRC Techcon Conf. (2005)

38.35   F. Wang, X. Wu, Y. Xie: Variability-driven module selection with joint design time optimization and post-silicon tuning, Proc. Asia South Pac. Des. Autom. Conf. (2008)

38.36   J. Cong, Y. Ding: FlowMap: an optimal technology mapping algorithm for delay optimization in lookup-table based FPGA designs, IEEE Trans. CAD Integr. Circuits Syst. **13**(1), 1–12 (1994)

38.37   T.J. Callahan, P. Chong, A. DeHon, J. Wawrzynek: Fast module mapping and placement for datapaths in FPGAs, Proc. Int. Symp. FPGAs (1998)

38.38   R. Brayton, G. Hachtel, C. McMullen, A. Sangiovanni-Vincentelli: *Logic Minimization Algorithms for VLSI Synthesis* (Kluwer, Boston 1984)

38.39   E. Sentovich, K. Singh, L. Lavagno, C. Moon, R. Murgai, A. Saldanha, H. Savoj, P. Stephan, R. Brayton, A. Sangiovanni-Vincentelli: *SIS: A System for Sequential Circuit Synthesis, Memo. UCB/ERL M92/41* (Univ. of California, Berkeley 1992)

38.40   R. Bryant: Graph-Based Algorithms for Boolean Function Manipulation, IEEE Trans. Comput. **35**(8), 677–691 (1986)

38.41   J. Marques Silva, K. Sakallah: Boolean satisfiability in electronic design automation, Proc. Des. Autom. Conf. (2000)

38.42   K. Keutzer: DAGON: technology mapping and local optimization, Proc. IEEE/ACM Des. Autom. Conf. (1987)

38.43   Berkeley ABC: A system for sequential synthesis and verification. http://www.eecs.berkeley.edu/~alanmi/abc/ (2005)

38.44   C.-W. Kang, A. Iranli, M. Pedram: A synthesis approach for coarse-grained, antifuse-based FPGAs, IEEE Trans. CAD Integr. Circuits Syst. **26**(9), 1564–1575 (2007)

38.45   A. Ling, D. Singh, S. Brown: FPGA PLB architecture evaluation and area optimization techniques using boolean satisfiability, IEEE Trans. CAD Integr. Circuits Syst. **26**(7), 1196 (2007)

38.46   A.K. Singh, M. Mani, R. Puri, M. Orshansky: Gain-based technology mapping for minimum runtime leakage under input vector uncertainty, Proc. Des. Autom. Conf. (2006)

38.47   L. Cheng, D. Chen, D.F. Wong, M. Hutton, J. Govig: Timing constraint-driven technology mapping for FPGAs considering false paths and multi-clock domains, Proc. Int. Conf. Comput.-Aided Des. (2007)

38.48   N. Sherwani: *Algorithms for VLSI Physical Design Automation* (Kluwer, Dordrecht 1999)

38.49   C.M. Fiduccia, R.M. Matheysses: A linear-time heuristic for improving network partitions, Proc. IEEE/ACM Des. Autom. Conf. (1982) pp. 175–181

38.50   G. Karypis, R. Aggarwal, V. Kumar, S. Shekhar: Multilevel hypergraph partitioning: application in VLSI domain, Proc. IEEE/ACM Des. Autom. Conf. (1997)

38.51   Y.C. Wei, C.K. Cheng: Ratio cut partitioning for hierarchical designs, IEEE Trans. CAD Integr. Circuits Syst. **10**, 911–921 (1991)

38.52   H. Liu, D.F. Wong: Network-flow-based multiway partitioning with area and pin constraints, IEEE Trans. CAD Integr. Circuits Syst. **17**(1), 50–59 (1998)

38.53   L. Stockmeyer: Optimal orientation of cells in slicing floorplan designs, Inf. Control **57**(2–3), 91–101 (1984)

38.54   D.F. Wong, C.L. Liu: A new algorithm for floorplan design, Proc. Des. Autom. Conf. (1986)

38.55   P. Sarkar, C.K. Koh: Routability-driven repeater block planning for interconnect-centric floorplanning, IEEE Trans. CAD Integr. Circuits Syst. **20**(5), 660–671 (2001)

38.56   M. Healy, M. Vittes, M. Ekpanyapong, C. Ballapuram, S.K. Lim, H. Lee, G. Loh: Multi-objective microarchitectural floorplanning for 2-D and 3-D ICs, IEEE Trans. CAD Integr. Circuits Syst. **26**(1), 38–52 (2007)

38.57   W. Sun, C. Sechen: Efficient and effective placement for very large circuits, IEEE Trans. CAD Integr. Circuits Syst. **14**(3), 349–359 (1995)

38.58   V. Betz, J. Rose, A. Marquardt: *Architecture and CAD for Deep-Submicron FPGAs* (Kluwer, Dordrecht 1999)

38.59   A. Caldwell, A.B. Kahng, I. Markov: Can recursive bisection produce routable placements?, Proc. IEEE/ACM Des. Autom. Conf. (2000) pp. 477–482

38.60   U. Brenner, A. Rohe: An effective congestion-driven placement framework, Proc. Int. Symp. Phys. Des. (2002)

38.61   T. Chan, J. Cong, T. Kong, J. Shinnerl: Multilevel circuit placement. In: *Multilevel Optimization in VLSICAD*, ed. by J. Cong, J. Shinnerl (Kluwer, Boston 2003), Chap. 4

38.62   C. Chu, N. Viswanathan: FastPlace: efficient analytical placement using cell shifting, iterative local refinement, and a hybrid net model, Proc. Int. Symp. Phys. Des. (2004) pp. 26–33

38.63   Z. Xiu, J. Ma, S. Fowler, R. Rutenbar: Large-scale placement by grid warping, Proc. Des. Autom. Conf. (2004)

38.64   T. Taghavi, X. Yang, B. Choi, M. Wang, M. Sarrafzadeh: Dragon2006: blockage-aware congestion-controlling mixed-size placer, Proc. Int. Symp. Phys. Des. (2006)

38.65   T. Chen, Z. Jiang, T. Hsu, H. Chen, Y. Chang: NTUplace2: a hybrid placer using partitioning and analytical techniques, Proc. Int. Symp. Phys. Des. (2006)

38.66   A.B. Kahng, S. Reda, Q. Wang: APlace: a general analytic placement framework, Proc. Int. Symp. Phys. Des. (2005)

38.67   C.Y. Lee: An algorithm for path connections and its applications, Proc. IRE Trans. Electron. Comput. (1961)

38.68 D.W. Hightower: A solution to the line routing problem on a continuous plane, Proc. Des. Autom. Workshop (1969)

38.69 L. McMurchie, C. Ebeling: PathFinder: a negotiation-based performance-driven router for FPGAs, Proc. Int. Symp. FPGAs (1995)

38.70 C. Chu, Y.C. Wong: FLUTE: fast lookup table based rectilinear steiner minimal tree algorithm for VLSI design, IEEE Trans. CAD Integr. Circuits Syst. **27**(1), 70–83 (2008)

38.71 J. Hu, S. Sapatnekar: A survey on multi-net global routing for integrated circuits, Integration: VLSI J. **31**, 1–49 (2001)

38.72 P. McGeer, R. Brayton: Efficient algorithms for computing the longest viable path in a combinational network, Proc. Des. Autom. Conf. (1989)

38.73 P. Ashar, S. Dey, S. Malik: Exploiting multicycle false paths in the performance optimization of sequential logic circuits, IEEE Trans. CAD Integr. Circuits Syst. **14**(9), 1067–1075 (1995)

38.74 S. Zhou, B. Yao, H. Chen, Y. Zhu, M. Hutton, T. Collins, S. Srinivasan, N. Chou, P. Suaris, C.K. Cheng: Efficient timing analysis with known false paths using biclique covering, IEEE Trans. CAD Integr. Circuits Syst. **26**(5), 959–969 (2006)

38.75 J. Mar: The application of TCAD in industry, Proc. Int. Conf. Simul. Semiconduct. Process. Dev. (1996)

38.76 M. Pedram: Low power design methodologies and techniques: an overview. http://atrak.usc.edu/~massoud/ (1999)

38.77 W. Nebel, J. Mermet (eds.): *Low Power Design in Deep Submicron Electronics* (Springer, New York 1997)

38.78 K. Roy, S. Prasad: *Low-Power CMOS VLSI Circuit Design* (Wiley, New York 2000)

38.79 M. Pedram, J.M. Rabaey: *Power Aware Design Methodologies* (Springer, New York 2002)

38.80 R. Puri, L. Stok, J. Cohn, D. Kung, D. Pan, D. Sylvester, A. Srivastava, S. Kulkarni: Pushing ASIC performance in a power envelope, Proc. Des. Autom. Conf. (2003)

38.81 L. Huang, D.F. Wong: Optical proximity correction (OPC): friendly maze routing, Proc. Des. Autom. Conf. (2004)

38.82 P. Yu, S.X. Shi, D.Z. Pan: True process variation aware optical proximity correction with variational lithography modeling and model calibration, J. Micro/Nanolith. MEMS MOEMS **6**, 031004 (2007)

38.83 P. Berman, A.B. Kahng, D. Vidhani, H. Wang, A. Zelikovsky: Optimal phase conflict removal for layout of dark field alternating phase shifting masks, Proc. Int. Symp. Phys. Des. (1999)

38.84 L.W. Liebmann, G.A. Northrop, J. Culp, M.A. Lavin: Layout optimization at the pinnacle of optical lithography, Proc. SPIE Des. Process Integr. Electron. Manuf. (2003)

38.85 F.M. Schellenberg, L. Capodieci: Impact of RET on physical layouts, Proc. Int. Symp. Phys. Des. (2001)

38.86 Y.M. Kuo, Y.T. Chang, S.C. Chang, M. Marek-Sadowska: Engineering change using spare cells with constant insertion, Proc. Int. Conf. Comput.-Aided Des. (2007)

38.87 S. Ghiasi: Incremental component implementation selection: enabling ECO in compositional system synthesis, Proc. Int. Conf. Comput.-Aided Des. (2007)

38.88 A. Krstic, L.C. Wang, K.T. Cheng, T.M. Mak: Diagnosis-based post-silicon timing validation using statistical tools and methodologies, Proc. Int. Test Conf. (2003)

38.89 K.H. Chang, I. Markov, V. Bertacco: Automating post-silicon debugging and repair, Proc. Int. Conf. Comput.-Aided Des. (2007)

38.90 S. Sapatnekar: *Timing* (Springer, New York 2004)

38.91 A. Srivastava, D. Sylvester, D. Blaauw: *Statistical Analysis and Optimization for VLSI: Timing and Power* (Springer, New York 2005)

38.92 V. Mehrotra, S. Sam, D. Boning, A. Chandrakasan, R. Vallishayee, S. Nassif: A methodology for modeling the effects of systematic within-die interconnect and device variation on circuit performance, Proc. Des. Autom. Conf. (2000)

38.93 M. Guthaus, N. Venkateswaran, C. Visweswariah, V. Zolotov: Gate sizing using incremental parameterized statistical timing analysis, Proc. Int. Conf. Comput.-Aided Des. (2005)

38.94 J. Le, X. Li, L.T. Pileggi: STAC: statistical timing analysis with correlation, Proc. IEEE/ACM Des. Autom. Conf. (2004)

38.95 M. Orshansky, A. Bandyopadhyay: Fast statistical timing analysis handling arbitrary delay correlations, Proc. IEEE/ACM Des. Autom. Conf. (2004)

38.96 V. Khandelwal, A. Srivastava: A general framework for accurate statistical timing analysis considering correlations, Proc. IEEE/ACM Des. Autom. Conf. (2005)

38.97 A. Ramalingam, A.K. Singh, S.R. Nassif, M. Orshansky, D.Z. Pan: Accurate waveform modeling using singular value decomposition with applications to timing analysis, Proc. Des. Autom. Conf. (2007)

38.98 J. Xiong, V. Zolotov, L. He: Robust extraction of spatial correlation, Proc. Int. Symp. Phys. Des. (2006)

38.99 J. Heath, P. Kuekes, G. Snider, S. Williams: A defect-tolerant computer architecture: opportunities for nanotechnology, Science **280**, 1716–1721 (1998)

38.100 A. DeHon, H. Naeimi: Seven strategies for tolerating highly defective fabrication, IEEE Des. Test Comput. **22**(4), 306–315 (2005)

38.101 J. Deng, N. Patil, K. Ryu, A. Badmaev, C. Zhou, S. Mitra, H.S. Wong: Carbon nanotube transistor circuits: circuit-level performance benchmarking and design options for living with imperfections, Proc. IEEE Int. Solid-State Circuits Conf. (2007)

38.102 S.C. Goldstein, M. Budiu: NanoFabric: spatial computing using molecular electronics, Proc. Int. Symp. Comput. Archit. (2001) pp. 178–189

38.103 D.B. Strukov, K.K. Likharev: A reconfigurable architecture for hybrid CMOS/nanodevice circuits, Proc. Int. Symp. FPGAs (2006)

38.104 A. Raychowdhury, A. Keshavarzi, J. Kurtin, V. De, K. Roy: Analysis of carbon nanotube field effect transistors for high performance digital logic – modeling and DC simulations, IEEE Trans. Electron. Dev. **53**(11) (2006)

38.105 W. Zhang, N.K. Jha, L. Shang: NATURE: a CMOS/nanotube hybrid reconfigurable architecture, Proc. Des. Autom. Conf. (2006)

38.106 M. Ben Jamaa, K. Moselund, D. Atienza, D. Bouvet, A. Ionescu, Y. Leblebici, G. De Micheli: Fault-tolerant multi-level logic decoder for nanoscale crossbar memory arrays, Proc. Int. Conf. Comput.-Aided Des. (2007)

38.107 A. Nieuwoudt, M. Mondal, Y. Massoud: Predicting the performance and reliability of carbon nanotube bundles for on-chip interconnect, Proc. Asian South Pac. Des. Autom. Conf. (2007)

38.108 C. Ababei, P. Maidee, K. Bazargan: Exploring potential benefits of 3-D FPGA integration, Proc. Field Programmable Logic and Application, Vol. 3203 (Springer, Berlin Heidelberg 2004) pp. 874–880

38.109 K. Banerjee, S.J. Souri, P. Kapur, K.C. Saraswat: 3-D ICs: a novel chip design for improving deep-submicrometer interconnect performance and systems-on-chip integration, Proc. IEEE **89**(5), 602–633 (2001)

38.110 M. Lin, A. El Gamal, Y.C. Lu, S. Wong: Performance benefits of monolithically stacked 3-D-FPGA, Proc. Int. Symp. FPGAs (2006)

38.111 W.R. Davis, J. Wilson, S. Mick, J. Xu, H. Hua, C. Mineo, A.M. Sule, M. Steer, P.D. Franzon: Demystifying 3-D ICs: the pros and cons of going vertical, IEEE Des. Test Comput. **22**(6), 498–510 (2005)

38.112 C. Dong, D. Chen, S. Haruehanroengra, W. Wang: 3-D nFPGA: a reconfigurable architecture for 3-D CMOS/nanomaterial hybrid digital circuits, IEEE Trans. Circuits Syst. **54**(11), 2489–2501 (2007)

38.113 Y. Xie, G. Loh, B. Black, K. Bernstein: Design space exploration for 3-D architecture, ACM J. Emerg. Technol. Comput. Syst. **2**(2), 65–103 (2006)

38.114 J. Cong, Y. Ma, Y. Liu, E. Kursun, G. Reinman: 3-D architecture modeling and exploration, Proc. Int. VLSI/ULSI Multilevel Interconnect. Conf. (2007)

38.115 M. Pathak, S.K. Lim: Thermal-aware steiner routing for 3-D stacked ICs, Proc. Int. Conf. Comput.-Aided Des. (2007)

38.116 M. Zhang, N.R. Shanbhag: Soft error-rate analysis (SERA) methodology, IEEE Trans. CAD Integr. Circuits Syst. **25**(10), 2140–2155 (2006)

38.117 N. Miskov-Zivanov, D. Marculescu: MARS-C: modeling and reduction of soft errors in combinational circuits, Proc. Des. Autom. Conf. (2006)

38.118 R.R. Rao, K. Chopra, D. Blaauw, D. Sylvester: An efficient static algorithm for computing the soft error rates of combinational circuits, Proc. Des. Autom. Test Eur. (2006)

38.119 S. Mitra, N. Seifert, M. Zhang, Q. Shi, K.S. Kim: Robust system design with built-in soft error resilience, IEEE Comput. **38**(2), 43–52 (2005)

38.120 B. Paul, K. Kang, H. Kufluoglu, A. Alam, K. Roy: Impact of NBTI on the temporal performance degradation of digital circuits, IEEE Electron. Dev. Lett. **26**(8), 560–562 (2005)

38.121 D. Marculescu: Energy bounds for fault-tolerant nanoscale designs, Proc. Des. Autom. Test Eur. (2005)

38.122 W. Wang, S. Yang, S. Bhardwaj, R. Vattikonda, S. Vrudhula, F. Liu, Y. Cao: The impact of NBTI on the performance of combinational and sequential circuits, Proc. IEEE/ACM Des. Autom. Conf. (2007)

38.123 W. Wu, J. Yang, S.X.D. Tan, S.L. Lu: Improving the reliability of on-chip caches under process variations, Proc. Int. Conf. Comput. Des. (2007)

38.124 A. Mitev, D. Canesan, D. Shammgasundaram, Y. Cao, J.M. Wang: A robust finite-point based gate model considering process variations, Proc. Int. Conf. Comput.-Aided Des. (2007)

38.125 S. Sarangi, B. Greskamp, J. Torrellas: A model for timing errors in processors with parameter variation, Proc. Int. Symp. Qual. Electron. Des. (2007)

38.126 L. Cheng, Y. Lin, L. He, Y. Cao: Trace-based framework for concurrent development of process and FPGA architecture considering process variation and reliability, Proc. Int. Symp. FPGAs (2008)

38.127 P. Banerjee: *Parallel Algorithms for VLSI Computer-Aided Design* (Prentice-Hall, Englewood Cliffs 1994)

38.128 A. Ludwin, V. Betz, K. Padalia: High-quality, deterministic parallel placement for FPGAs on commodity hardware, Proc. Int. Symp. FPGAs (2008)

38.129 G. De Micheli, L. Benini: *Networks on Chips: Technology and Tools* (Morgan Kaufmann, New York 2006)

38.130 A. Jantsch, H. Tenhunen (eds.): *Networks on Chip* (Kluwer, Dordrecht 2003)

38.131 A. Hemani, A. Jantsch, S. Kumar, A. Postula, J. Öberg, M. Millberg, D. Lindqvist: Network on a chip: an architecture for billion transistor era, Proc. IEEE NorChip Conf. (2000)

38.132 H.G. Lee, N. Chang, U.Y. Ogras, R. Marculescu: On-chip communication architecture exploration: a quantitative evaluation of point-to-point, bus, and network-on-chip approaches, ACM Trans. Des. Autom. Electron. Syst. **12**(3) (2007)

38.133 H. Wang, L.S. Peh, S. Malik: Power-driven design of router microarchitectures in on-chip networks, Proc. Int. Symp. Microarchit. (2003)

39. Safety Warnings for Automation

Mark R. Lehto, Mary F. Lesch, William J. Horrey

Automated systems can provide tremendous benefits to users; however, there are also potential hazards that users must be aware of to safely operate and interact with them. To address this need, safety warnings are often provided to operators and others who might be placed at risk by the system. This chapter discusses some of the roles safety warnings can play in automated systems, from both the traditional perspective of warnings as a form of hazard control and the perspective of warnings as a form of automation. During this discussion, the chapter addresses some of the types of warnings that might be used, along with issues and challenges related to warning effectiveness. Design recommendations and guidelines are also presented.

Automated systems have become increasingly prevalent in our society, in both our work and personal lives. Automation involves the execution by a computer (or machine) of a task that was formerly executed by human operators [39.1]; for example, automation may be applied to a particular function in order to complete tasks that humans cannot perform or do not want to perform, to complete tasks that humans perform poorly or that incur high workload demands, or to augment the capabilities and performance of the human operator [39.2]. The potential benefits of automation include increased productivity and quality, greater system safety and reliability, and fewer human errors, injuries or occupational illnesses. These benefits follow because some demanding or dangerous tasks previously performed by the operator can be completely eliminated through automa-

tion, and many others can be made easier. On the other hand, automation can create new hazards and increase the potential for catastrophic human errors [39.3]; for example, in advanced manufacturing settings, the use of robots and other forms of automation has reduced the need to expose workers to potentially hazardous materials in welding, painting, and other operations, but in turn has created a more complex set of maintenance, repair, and setup tasks, for which human errors can have serious consequences, such as damaging expensive equipment, long periods of system downtime, production of multiple runs of defective parts, and even injury or death.

As implied by the above example, a key issue is that automation increases the complexity of systems [39.4]. A second issue is that the introduction of automation into a system or task does not necessarily remove the human operator from the task or system. Instead, the role and responsibilities of the operator change. One common result of automation is that operators may go from active participants in a task to passive monitors of the system function [39.5, 6]. This shift in roles from active participation to passive monitoring can reduce the operator's situation awareness and ability to respond appropriately to automation failures [39.7]. Part of the problem is that the operator may have few opportunities to practise their skills because automation failures tend to be rare events. Further complicating the issue, system monitoring might be done from a remote location using one or more displays that show the status of many different subsystems. This also can reduce situation awareness, for many different reasons. Another common problem is that workload may be too low during routine operation of the automated system, causing the operator to become complacent and easily distracted. Furthermore, designers may assign additional unrelated tasks to operators to make up for the

reduced workload due to automation. This again can impair situation awareness, as performing these unrelated tasks can draw the operator's attention away from the automated system. The need to perform these additional tasks can also contribute to a potentially disastrous increase in workload in nonroutine situations in which the operator has to take over control from the automated system.

Many other aspects of automated systems can make it difficult for operators and others to be adequately aware of the hazards they face and how to respond to them [39.4, 8]. To address this issue, safety warnings are often employed in such systems. This chapter discusses some of the roles safety warnings can play in automated systems, from both the traditional perspective of warnings as a form of hazard control and the perspective of warnings as a form of automation. During this discussion, the chapter addresses some of the types of warnings that might be used. We also discuss issues related to the effectiveness of warnings and provide design recommendations and guidelines.

39.1 Warning Roles

The role of warnings in automated systems can be viewed from two overlapping perspectives: (1) warnings as a method of hazard control, and (2) warnings as a form of automation.

39.1.1 Warning as a Method of Hazard Control

Warnings are sometimes viewed as a method of last resort to be relied upon when more fundamental solutions to safety problems are infeasible. This view corresponds to the so-called hierarchy of hazard control, which can be thought of as a simple model that prioritizes control methods from most to least effective. One version of this model proposes the following sequence: (1) eliminate the hazard, (2) contain or reduce the hazard, (3) contain or control people, (4) train or educate people, and (5) warn people [39.9]. The basic idea is that designers should first consider design solutions that completely eliminate the hazard. If such solutions are technically or economically infeasible, solutions that reduce but do not eliminate the hazard should then be considered. Warnings and other means of changing human behavior, such as training, education, and supervision, fall in this latter category for obvious reasons. Sim-

ply put, these behavior-oriented approaches will never completely eliminate human errors and violations. On the other hand, this is also true for most design solutions. Consequently, warnings are often a necessary supplement to other methods of hazard control [39.10].

There are many ways warnings can be used as a supplement to other methods of hazard control; for example, warnings can be included in safety training materials, hazard communication programs, and within various forms of safety propaganda, including safety posters and campaigns, to educate workers about risks and persuade them to behave safely. Particularly critical procedures include start-up and shut-down procedures, setup procedures, lock-out and tag-out procedures during maintenance, testing procedures, diagnosis procedures, programming and teaching procedures, and numerous procedures specific to particular applications. The focus here is to reduce errors and intentional violations of safety rules by improving worker knowledge of what the hazards are and their severity, how to identify and avoid them, and what to do after exposure. Inexperienced workers are often the target audience at this stage. Warnings can also be included in manuals or job performance aids (JPAs), such as written procedures, checklists, and instructions. Such warnings usually consist of brief statements that either instruct less-skilled workers or remind skilled workers to take necessary precautions when performing infrequent maintenance or repair tasks. This approach can prevent workers from omitting precautions or other critical steps in a task. To increase their effectiveness, such warnings are often embedded at the appropriate stage within step-by-step instructions describing how to perform a task. Warning signs, barriers, or markings at appropriate locations, can play a similar role; for example, a warning sign placed on a safety barrier or fence surrounding a robot installation might state that no one except properly authorized personnel is allowed to enter the area. Placing a label on a guard to warn that removing the guard creates a hazard also illustrates this approach.

Warning signals can also serve as a supplement to other safety devices such as interlocks or emergency

Fig. 39.2 Safety laser scanner application (courtesy of Sick Inc., Minneapolis)

braking systems; for example, presence sensing and interlock devices are sometimes used in installations of robots to sense and react to potentially dangerous workplace conditions. Sensors used in such systems include (1) pressure-sensitive floor mats, (2) light curtains, (3) end-effector sensors, (4) ultrasound, capacitive, infrared, and microwave sensing systems, and (5) computer vision. Floor mats and light curtains are used to determine whether someone has crossed the safety boundary surrounding the perimeter of the robot. Perimeter penetration will trigger a warning signal and in some cases will cause the robot to stop. End-effector sensors detect the beginning of a collision and trigger emergency stops. Ultrasound, capacitive, infrared, and microwave sensing systems are used to detect intrusions. Computer vision theoretically can play a similar role in detecting safety problems.

Figure 39.1a and b illustrate how a presence sensing system, in this case a safety light curtain device in an

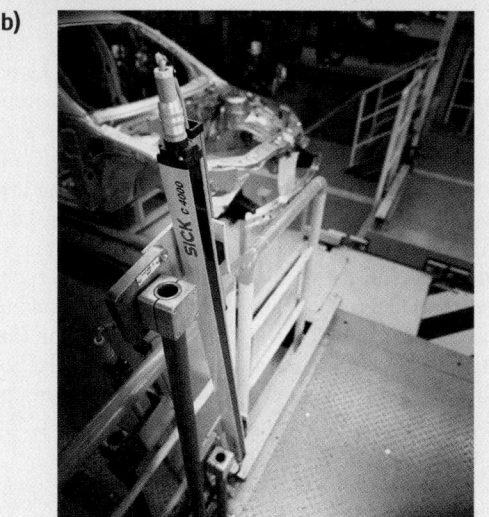

Fig. 39.1a,b Auto-body assembly line safety light curtain system (courtesy of Sick Inc., Minneapolis)

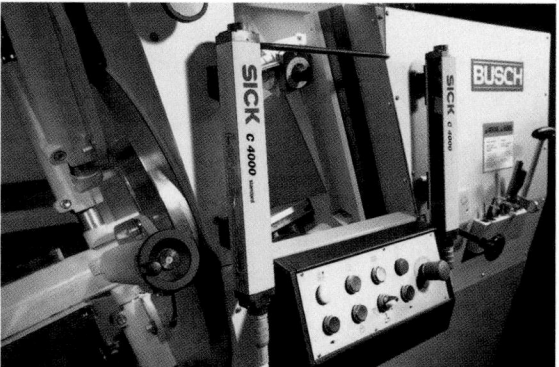

Fig. 39.3 C4000 safety light curtain hazardous point protection (courtesy of Sick Inc., Minneapolis)

Fig. 39.4 Safety laser scanner AGV (automated guided vehicle) application (courtesy of Sick Inc., Minneapolis)

auto-body assembly line, might be installed for point of operation, area, perimeter, or entry/exit safeguarding. Figure 39.2 illustrates a safety laser scanner. By reducing or eliminating the need for physical barriers, such systems make it easier to access the robot system during setup and maintenance. By providing early warnings prior to entry of the operator into the safety zone, such systems can also reduce the prevalence of nuisance machine shut-downs (Fig. 39.3). Furthermore, such systems can prevent the number of accidents by providing warning signals and noises to alert personnel on the floor (Fig. 39.4).

39.1.2 Warning as a Form of Automation

Automated warning systems have been implemented in many domains, including aviation, medicine, process control and manufacturing, automobiles and other surface transportation, military applications, and weather forecasting, among others. Some specific examples of automated systems include collision warning systems and ground proximity warning systems in automobiles and aircraft, respectively. These systems will alert drivers or pilots when a collision with another vehicle or the ground is likely, so that they can take evasive action. In medicine, anesthesiologists and medical care

workers must monitor patients' vitals, sometimes remotely. Similarly, in process control, such as nuclear power plants, workers must continuously monitor multiple subsystems to ensure that they are at safe and tolerable levels. In these situations, automated alerts can be used to inform operators of any significant departures from normal and acceptable levels, whether in the patients' condition or in plant operation and safety. Automation may be particularly important for complex systems, which may involve too much information (sometimes referred to as *raw data*), creating difficulties for operators in finding relevant information at the appropriate times. In addition to simply informing or alerting the human operator, automation can play many different roles, from guiding human information gathering to taking full control of the system.

As implied by the above examples, automated warning systems come in many forms, across a wide variety of domain applications. *Parasuraman* et al. [39.11] propose a taxonomy of human–automation interaction that provides a useful way of categorizing the function of these systems according to the psychological process they are intended to replace or supplement. As shown in Fig. 39.5, automation can be applied at any of four stages: (1) information acquisition, (2) information analysis (3) decision selection, and (4) action implementation. These four stages are based on a simple model of human information processing (sensory processing, cognition/working memory, decision making, response execution). The model proposed by *Parasuraman* et al. [39.11] also maps onto *Endsley*'s [39.12] model of situation awareness (SA), with early stages of automation contributing to the establishment and maintenance of SA (as also shown in Fig. 39.5). Good situation awareness is an important precursor to accurate decision making and action selection.

For any given automated system, the level of automation at each stage of the model can vary from low to high and this level will dictate how much control the human is afforded in the operation of the system. As expanded upon in the following discussion, the functions performed by automated warning systems tend to fall into the second and third stages of automation (information analysis and decision selection, respectively), depending on whether they simply provide human operators with alerts or whether they indicate also the appropriate course of action.

Stage 1: Information Acquisition
At the first stage, automation involves the acquisition and registration of multiple sources of input data. Au-

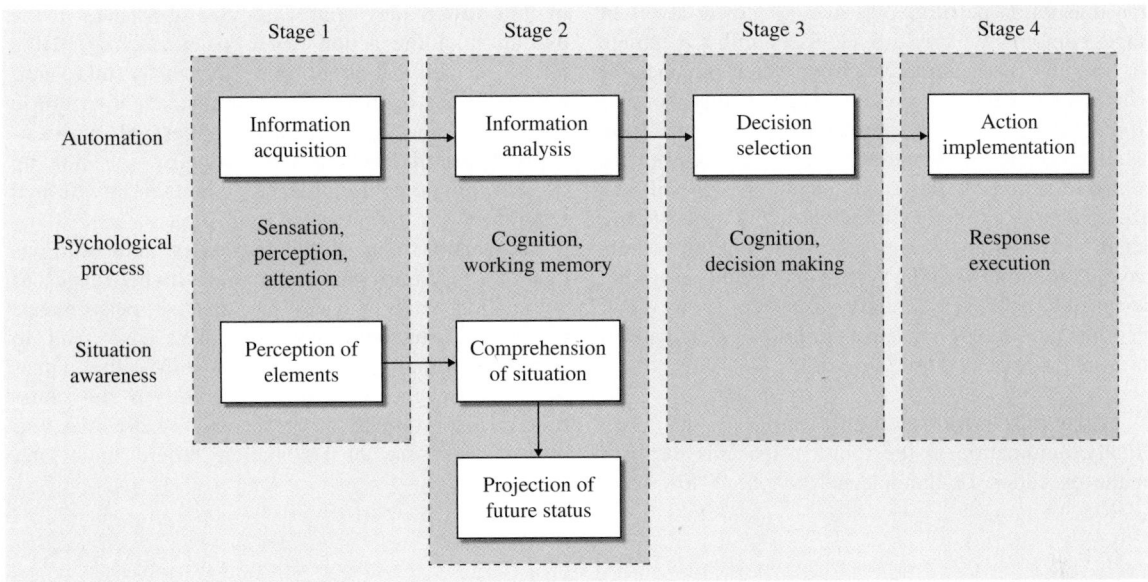

Fig. 39.5 Stages of automation [39.13] and the corresponding psychological processes and level of situation awareness [39.12] (after *Horrey* et al. [39.14])

tomation in this stage acts to support human sensory and attentional processes (e.g., detection of input data). A high level of automation at this stage may filter out all information deemed to be irrelevant or less critical to the current task, presenting only the most critical information to the operator. Thus, only relevant cues and reports (at least those deemed relevant by the system) would pass through the filters, allowing operators the capacity to make more effective decisions, especially when under time duress. Systems using a lower level of automation, on the other hand, may present all of the available input data but guide attention to what the automation infers to be the most relevant features (e.g., target cueing; information highlighting).

There has been extensive research into the effects of stage 1 automation (attention guidance) in target detection tasks. Basic research has reliably demonstrated the capacity for visual cues to reduce search times in target search tasks [39.15]. Applied research has also demonstrated these benefits in military situations [39.16, 17], helicopter hazard detection [39.18], aviation and air-traffic control [39.6, 19], and a number of other domains. These generally positive results support the potential value of stage 1 automation in applications where operators can receive an excessive number of warnings. Example applications might filter the warnings in terms of urgency or limit the warnings to relevant subsystems.

Stage 2: Information Analysis

Information analysis involves higher cognitive functions such as working memory, information integration, and cognitive inference. Automation at stage 2 may help operators by integrating the raw data, drawing inferences, and/or generating predictions. In this stage, lower levels of automation may extrapolate current information and predict future status (e.g., cockpit predictor displays [39.20]). Higher levels of automation at this stage may reduce information from a number of sources into a single hypothesis regarding the state of the world; for example, collision warning systems in automobiles will use information regarding the speed of the vehicle ahead, the intervehicle separation, and the driver's own velocity (among other potential information) to indicate to the driver when a forward collision is likely [39.21–24]. In general, operators are quicker to respond to the relevant event when provided with these alerts. Studies of automated alerts have been performed in many different domains, including aviation [39.25], process control [39.26], unmanned aerial vehicle operation [39.27], medicine [39.28], air-traffic control [39.6], and battlefield operations [39.14].

Stage 3: Decision Selection

The third stage involves the selection, from among many alternatives, of the appropriate decision or action. This will typically follow from some form of information integration performed at stage 2. Lower levels of stage 3 automation may provide users with a complete set (or subset) of alternatives from which the operator will select which one to execute (whether correct or no). Higher levels may only present the *optimal* decision or action or may automatically select the appropriate course of action. At this stage the automation will utilize implicit or explicit assumptions about the costs and benefits of different decision outcomes; for example, the ground proximity warning system in aviation – a system designed to help avoid aircraft–ground collisions – will recommend a single maneuver to pilots when a given threshold is exceeded (*pull up*).

Stage 4: Action Implementation
Finally, automation in the fourth stage aids the user in the execution of the selected action. A low level of automation may simply provide assistance in the execution of the action (e.g., power steering). High levels of automation at this stage may take control from the operator; for example, adaptive cruise control (ACC) systems in automobiles will automatically adjust the vehicle's headway by speeding up or slowing down in order to maintain the desired separation.

In general, one of the ironies of automation is that those systems that incorporate higher stages of automation tend to yield the greatest performance in normal situations; however, these also tend to come with the greatest costs in off-normal situations, where the automated response to the situation is inappropriate or erroneous [39.29]. We will discuss the issue of automation failure in a later section.

39.2 Types of Warnings

Warning systems serve many functions. Typically, they provide the user/operator with information on the status of the system. This information source aids the user in maintaining situational awareness, and does not necessarily require that specific action be taken (stage 2 automation). Alternatively, other warning systems signal the user that a specific response needs to be made at a specific time in order to reduce associated risks (stage 3 automation). Below, we will review the different methods and modes of presenting warning information that can be used to accomplish these goals.

39.2.1 Static Versus Dynamic Warnings

Perhaps the most familiar types of warnings are the visually based signs and labels that we encounter everyday whether on the road (e.g., *slippery when wet*), in the workplace (industrial warnings such as *entanglement hazard – keep clear of moving gears*), or on consumer products (e.g., *do not take this medication if you might be pregnant*). These signs and labels indicate the presence of a hazard and may also indicate required or prohibited actions to reduce the associated risk, as well as the potential consequences of failing to comply with the warning. This type of warning is *static* in the sense that its status does not change over time [39.30]. However, as noted by *Lehto* [39.31], even these static displays have a dynamic component in that they are no-

ticed at particular points in time. In order to increase the likelihood that a static warning, such as a sign or label, is *received* (i. e., noticed, perceived, and understood) by the user at the appropriate moment, it should be physically as well as temporally placed such that using the product requires interaction with the label prior to the introduction of the hazard to the situation; for example, *Duffy* et al. [39.32] examined the effectiveness of a label on an extension cord which stated: "Warning. Electric shock and fire. Do not plug more than two items into this cord." Interactive labels in which the label was affixed to the outlet cover on the female receptacle were found to produce greater compliance than a no-label control condition, and a tag condition in which the warning label was attached to the extension cord 5 cm above the female receptacle. Other studies [39.33, 34] have found that a warning that interrupts a user's script for interacting with a product increases compliance. A *script* consists of a series of temporally ordered actions or events which are typical of a user's interactions with a class of objects [39.35].

Additionally, varying the warning's physical characteristics can also increase its conspicuity or noticeability [39.36]; for example, larger objects are more likely to capture attention than smaller objects. Brightness and contrast are also important in determining whether an object is discernible from a background. As a specific form of contrast, highlighting can be used to emphasize different portions of a warning label or sign [39.37]. Ad-

Table 39.1 Contrast between the fundamental properties of visual, auditory, and haptic modalities of information processing (after *Sanderson* [39.38])

Visual	Auditory	Haptic
Persistent – signal typically persistent in time so that information about past has same sensory status as information about present	*Transitory* – signal happens in time and recedes into past: creating persistent information or information about past are design challenges	*Transitory* – signal happens in time and recedes into past: creating persistent information or information about past are design challenges
Localized – can be sensed only from specific locations such as monitors or other projections (eyeballs needed)	*Ubiquitous* – can be sensed from any location unless technology is used to create localized qualities	*Personal* – can be sensed only by the person whom the display is directed (unless network or shared)
Optional – there are proximal physical means for completely eliminating signal (eyeballs, eyelids, turn)	*Obligatory* – there are no proximal physical means for completely eliminating signal (unless earplugs block signal)	*Obligatory* – there are no proximal physical means for completely eliminating signal (unless remove device)
Moderately socially inclusive – others aware of signal but need not look at screen	*Socially inclusive* – others always receive signal unless signal is sent only to an earpiece	*Not socially inclusive* – others probably unaware of signal
Sampling-based monitoring – temporal sampling process needed for coverage of all needed variables	*Peripheral monitoring* – temporal properties of process locked into temporal properties of display	*Interrupt-based monitoring* – monitoring based on interrupts
High information density – many variables and relationships can be simultaneously presented	*Moderate information density* – several variables and relationships can be simultaneously presented	*Low information density* – few variables and relationships can be simultaneously presented

ditionally, lighting conditions influence detectability of signs and labels (i. e., reduced contrast).

In contrast to static warnings, dynamic warning systems produce different messages or *alerts* based on input received from a sensing system – therefore, they indicate the presence of a hazard that is not normally present [39.30, 39]. Environmental variables are monitored by a sensor, and an alert is produced if the monitored variables exceed some threshold. The threshold can be changed based on the criticality of potential consequences – the more trivial the consequences, the higher the threshold and, for more serious consequences, the threshold would be set lower so as reduce the likelihood of *missing* the critical event. However, the greater the system's sensitivity, the greater the

likelihood that an alert will be generated when there is no hazard present. Ideally, an alert should always be produced when there is a hazard present, but never be produced in the absence of a hazard. Implications of system departures from this ideal will be discussed later in this chapter.

The remainder of this chapter will focus on the implementation and functioning of dynamic warning systems. These warnings serve to alert users to the presence of a hazard and its associated risks. Accordingly, they must readily capture attention and be easily/quickly understood. The ability of the warning to capture attention is especially important in the case of complex systems, in which an abundance of available information may overload the operator's limited attentional resources. When designing a warning system, it is critical to take into account the context in which the warning will appear; for example, in a noisy construction environment, in which workers may be wearing hearing protection, an auditory warning is not likely to be effective. Whatever the context, the warning should be designed to *stand out* against any background information (i. e., visual clutter, ambient noise). *Sanderson* [39.38] has provided a taxonomy/terminology for thinking about sensory modality in terms of whether information is persistent in time; whether information delivery is localized, ubiquitous, or personal; whether sensing the information is optional or obligatory; whether the information is socially inclusive; whether monitoring occurs through sampling, peripheral awareness, or is interrupt based; and information density (Table 39.1). Next, advantages and disadvantages of different modes of warning presentation will be discussed in related terms.

39.2.2 Warning Sensory Modality

Visual Warnings

The primary challenge in using visual warnings is that the user/operator needs to be looking at a specific location in order to be alerted, or the warning needs to be sufficiently salient to cause the operator to reorient their focus towards the warning. As discussed earlier in the section on static warnings, the conspicuity of visually based signals can be maximized by increasing size, brightness, and contrast [39.36]. Additionally, flashing lights attract attention better than continuous indicator-type lights (e.g., traffic signals incorporating a flashing light into the red phase) [39.36]. Since flash rates should not be greater than the critical flicker fusion frequency (≈ 24 Hz; resulting in the perception of a continuous light), or so slow that the *on* time might be missed, *Sanders* and *McCormick* [39.40] recommend flash rates of around 10 Hz.

Auditory Warnings

Auditory stimuli have a naturally alerting quality and, unlike visual warnings, the user/operator does not have to be oriented towards an auditory warning in order to be alerted, that is, auditory warnings are omnidirectional (or *ubiquitous* in Sanderson's terminology) [39.41, 42]. Additionally, localization is possible based on cues provided by the difference in time and intensity of the sound waves arriving at the two ears. To maximize the likelihood that the auditory warning is effective, the signal should be within the range of

about 800–5000 Hz (the human auditory system is most sensitive to frequencies within this range – frequencies contained in speech; e.g., *Coren* and *Ward* [39.43]) – and should have a tonal quality that is distinct from that of expected environmental sounds – to help reduce the possibility that it will be masked by those sounds (see *Edworthy* and *Hellier* [39.44] for an in-depth discussion of auditory warning signals).

Verbal Versus Nonverbal

Any auditory stimulus, from a simple tone to speech, can serve as an alert as long as it easily attracts attention. However, the human auditory system is most sensitive to sound frequencies contained within human speech. Speech warnings have the further advantage of being composed of signals (i. e., words) which have already been well learned by the user/operator. There is a redundancy in the speech signal such that, if part of the signal is lost, it can be *filled in* based on the context provided by the remaining sounds [39.45, 46]. However, since speech is a temporally based code that unfolds over time, it is only physically available for a very limited duration. Therefore, earlier portions of a warning message must be held in working memory while the remainder of the message continues to be processed. As a result, working memory may become overloaded and portions of the warning message may be lost. With visually based verbal warnings, on the other hand, there is the option of returning to, and rereading, earlier portions of the warning; that is, they persist over time. However, since the eyes must be directed towards the warning source, the placement of visually based warnings is critical so as to minimize the loss of other potentially critical information (i. e., such that other *signals* can be processed in peripheral vision).

While verbal signals have the obvious advantage that their meaning is already established, speech warnings require the use of recorded, digitized, or synthesized speech which will be produced within a noisy background – therefore, intelligibility is a major issue [39.44]. Additionally, as indicated earlier, the speech signal unfolds over time and may take longer to produce/receive than a simpler nonverbal warning signal. However, nonverbal signals must somehow encode the urgency of the situation – that is, how quickly a response is required by the user/operator. Extensive research in the auditory domain indicates that higher-frequency sounds have a higher perceived urgency than lower-frequency sounds, that increasing the modulation of the amplitude or frequency of a pulse decreases urgency, that increases in number of harmonics increases perceived urgency, and that spectral shape also impacts perceived urgency [39.44, 47, 48]. *Edworthy* et al. [39.48] found that faster, more repetitive bursts are judged to be more urgent; regular rhythms are perceived as more urgent than syncopated rhythms; bursts that are speeded up are perceived as more urgent than those that stay the same or slow down; and the larger the difference between the highest and lowest pitched pulse in a burst, the higher the perceived urgency.

Haptic/Tactile Warnings

While the visual and auditory channels are most often used to present warnings, haptic or tactile warnings are also sometimes employed; for example, the improper maneuvering of a jet will cause tactile vibrations to be delivered through the pilot's control stick – this *alert* serves to signal the need to reorient the control. In the domain of vehicle collision warnings, *Lee* et al. [39.49] examined driver preferences for auditory or haptic warning systems as supplements to a visual warning system. Visual warnings were presented on a head-down display in conjunction with either an auditory warning or a haptic warning, in the form of a vibrating seat. Preference data indicated that drivers found auditory warnings to be more annoying and that they would be more likely to purchase a haptic warning system.

In the domain of patient monitoring, *Ng* et al. [39.50] reported that a vibrotactile wristband results in a higher identification rate for heart rate alarms than does an auditory display. In *Sanderson*'s [39.38] terminology, a haptic alert is discrete, transitory, has low precision, has obligatory properties, and allows the visual and auditory modalities to continue to monitor other information sources. Therefore, patient monitoring represents a good use of haptic alarms.

Another example of a tactile or haptic warning, though not technologically based, is the use of rumble strips on the side of the highway. When a vehicle crosses these strips, vibration and noise is created within the vehicle. Some studies have reported that the installation of these strips reduced drift-off road accidents by about 70% [39.51]. A similar application is the tactile ground surface indicators that have become more common as a result of the Americans with Disabilities Act (ADA). These surfaces are composed of truncated cones which provide a distinctive pattern which can be felt underfoot or through use of a cane. It is intended to alert individuals with visual impairments of hazards associated with blending pedestrian and vehicular traffic (i. e., on curb ramps on the approach to the street surface).

Multimodal Warnings

For the most part, we have focused our discussion on the use of individual sensory channels for the presentation of automated warnings. However, research suggests that multimodal presentation results in significantly improved warning processing. *Selcon* et al. [39.52] examined multimodal cockpit warnings. These warnings must convey the nature of the problem to the pilot as quickly as possible so that immediate action can be taken. The warnings studied were visual (presented using pictorials), auditory (presented by voice), or both (incorporating visual and auditory components) and described real aircraft *warning* (high priority/threat) and *caution* (low priority/threat) situations. Participants were asked to classify each situation as either *warning* or *caution* and then to rate the threat associated with it. Response times were measured. Depth of understanding was assessed using a measure of situational awareness [39.53, 54]. Performance was faster in the condition incorporating both visual and auditory components (1.55 s) than in the visual (1.74 s) and auditory (3.77 s) conditions and there was some indication that this condition was less demanding and resulted in improved depth of understanding as well.

Sklar and *Sarter* [39.55] examined the effectiveness of visual, tactile, and redundant visual and tactile cues for indicating unexpected changes in the status of an automated cockpit system. The tactile conditions produced higher detection rates and faster response times. Furthermore, provision of tactile feedback and performance of concurrent visual tasks did not result in any interference suggesting that tactile feedback may better support human–machine communication in information-rich domains (see also *Ng* et al.'s [39.50] findings reviewed earlier).

Research also indicates that multimodal presentation provides comprehension and memory benefits for verbal warnings. Using a laboratory task in which participants measured and mixed *chemicals*, *Wogalter* and *Young* [39.42] (see also *Wogalter* et al. [39.41]) observed higher compliance rates in conditions in which warnings to wear a mask and gloves were presented using both text and voice (74% of participants complied) than in conditions in which the warnings were presented via voice (59%) or text alone (41%). A similar pattern was observed in a field experiment in which text and voice warnings warned of a wet floor in a shopping center. Voice warnings are more likely to capture attention. However, most of the participants in the text warning condition also reported awareness of the warning. Therefore, awareness alone cannot account for the higher compliance rates. Additionally, the combined text and voice condition resulted in higher compliance than the voice-alone condition. The combination of voice and print appears more persuasive than print or voice alone.

In another study, *Conzola* and *Wogalter* [39.56] used voice and print warnings to supplement product manual instructions during the unpacking of a computer disk drive. The supplemental voice warnings were presented via digitized voice while the text was presented via printed placard. Compliance was higher in conditions with the supplemental messages than in the manual-only condition, but there was no significant difference in compliance between the voice and print conditions. As regards to memory, the voice message resulted in greater recall than the print or manual-only conditions. This finding is consistent with what is known as the *modality effect* in working memory

research – since verbal information is stored in memory in an auditory/acoustic format, memory for verbal information is better when the information is presented auditorily (i.e., by voice) than when that same information is presented visually (i.e., as text) (see *Penney* [39.57] for a review). Performance suffers when verbal information is presented visually as translation into an auditory code is required for storage in memory. Translation is unnecessary with voice presentation.

To summarize, by providing redundant delivery channels, a multimodal approach helps to ensure that the warning attracts attention, is received (i.e., understood) by the user/operator, and is remembered. Future research should focus on further developing relatively underused channels for warning delivery (i.e., haptic) and using multiple modalities in parallel in order to increase the attentional and performance capacity of the user/operator [39.38].

39.3 Models of Warning Effectiveness

Over the past 20 years, much research has been conducted on warnings and their effectiveness. The following discussion will first introduce some commonly used measures of effectiveness. Attention will then shift to modeling perspectives and related research findings which provide guidance as to when, where, and why warnings will be effective.

39.3.1 Warning Effectiveness Measures

The performance of a warning can be measured in many different ways [39.39]. The ultimate measure of effectiveness is whether a warning reduces the frequency and severity of human errors and accidents in the real world. However, such data is generally unavailable, forcing effectiveness to be evaluated using other measures, sometimes obtained in controlled settings that simulate the conditions under which people receive the warning; for example, the effectiveness of a collision avoidance warning might be assessed by comparing how quickly subjects using a driving simulator notice and respond to obstacles with and without the warning system. For the most part, such measures can be derived from models of human information processing that describe what must happen after exposure to the warning, for the warning to be effective [39.31, 39]. That is, the human operator must notice the warning, correctly comprehend its meaning, decide on the ap-

propriate action, and perform it correctly. Analysis of these intervening events can provide substantial guidance into factors influencing the overall effectiveness of a particular warning.

A complicating issue is that the design of warnings is a polycentric problem [39.58], that is, the designer will have to balance several conflicting objectives when designing a warning. The most noticeable warning will not necessarily be the easiest to understand, and so on. As argued by *Lehto* [39.31], this dilemma can be partially resolved by focusing on decision quality as the primary criterion for evaluating applications of warnings; that is, warnings should be evaluated in terms of their effect on the overall quality of the judgments and decisions made by the targeted population in their naturalistic environment. This perspective assumes that decision quality can be measured by comparing people's choices and judgments to those prescribed by some objective *gold standard*. In some cases, the gold standard might be prescribed by a normative mathematical model, as expanded upon below.

39.3.2 The Warning Compliance Hypothesis

The warning compliance hypothesis [39.59] states that people's choices should approximate those obtained by applying the following optimality criterion:

If the expected cost of complying with the warning is greater than the expected cost of not complying, then it is optimal to ignore the warning; otherwise, the warning should be followed.

The warning compliance hypothesis is based on statistical decision theory which holds that a rational decision-maker should make choices that maximize expected value or utility [39.60, 61]. The expected value of choosing an action A_i is calculated by weighting its consequences C_{ik} over all events k, by the probability P_{ik} that the event will occur. More generally, the decision-maker's preference for a given consequence C_{ik} might be defined by a value or utility function $V(C_{ik})$, which transforms consequences into preference values. The preference values are then weighted using the same equation. The expected value of a given action A_i becomes

$$EV[A_i] = \sum_k P_{ik} V(C_{ik}) . \qquad (39.1)$$

From this perspective, people who decide to ignore the warning feel that avoiding the typically small cost of compliance outweighs the large, but relatively unlikely cost of an accident or other potential consequence of not complying with the warning. The warning compliance hypothesis clearly implies that the effectiveness of warnings might be improved by

1. Reducing the expected cost of compliance, or
2. Increasing the expected cost of ignoring the warning.

Many strategies might be followed to attain these objectives; for example, the expected cost of compliance might be reduced by modifying the task or equipment so the required precautionary behavior is easier to perform. The benefit of following this strategy is supported by numerous studies showing that even a small cost of compliance (i.e., a short delay or inconvenience) can encourage people to ignore warnings (for reviews see *Lehto* and *Papastavrou* [39.62], *Wogalter* et al. [39.63], and *Miller* and *Lehto* [39.64]). Some strategies for reducing the cost of compliance include providing the warning at the time it is most relevant or more convenient to respond to.

Increasing the expected cost of ignoring the warning is another strategy for increasing warning effectiveness suggested by the warning compliance hypothesis. The potential value of this approach is supported by studies indicating that people will be more likely to take precautions when they believe the danger is present and perceive a significant benefit to taking the precaution. This might be done through supervision and enforcement or other methods of increasing the cost of ignoring the warning. Also, assuming the warning is sometimes given when it does not need to be followed (i.e., a false alarm), the expected cost of ignoring the warning will increase if the warning is modified in a way that reduces the number of false alarms. This point leads us to the topic of information quality.

39.3.3 Information Quality

In a perfect world, people would be given warnings if, and only if, a hazard is present that they are not already aware of. When warning systems are perfect, the receiver can optimize performance by simply following the warning when it is provided [39.65]. Imperfect warning systems, on the other hand, force the receiver to decide whether to consult and comply with the provided warning.

The problem with imperfect warning systems is that they sometimes provide false alarms or fail to detect the hazard. From a short-term perspective, false alarms are often merely a nuisance to the operator. However, there are also some important long-run costs, because repeated false alarms shape people's attitudes and influence their actions. One problem is the *cry-wolf* effect which encourages people to ignore (or mistrust) warnings [39.66, 67]. Even worse, people may decide to completely eliminate the nuisance by disconnecting the warning system [39.68]. Misses are also an important issue, because people may be exposed to hazards if they are relying on the warning system to detect the hazard. Another concern is that misses might reduce operator trust in the system [39.18].

Due to the potentially severe consequences of a miss, misses are often viewed as automation failures. The designers of warning systems consequently tend to focus heavily on designing systems that reliably provide a warning when a hazard is present. One concern, based on studies of operator overreliance upon imperfect automation [39.1, 19, 69], is that this tendency may encourage overreliance on warning systems. Another issue is that this focus on avoiding misses causes warning systems to provide many false alarms for each correct identification of the hazard. This tendency has been found for warning systems across a wide range of application areas [39.65, 67].

39.3.4 Information Integration

As discussed by *Edworthy* [39.70] and many others, in real-life situations people are occasionally faced with choices where they must combine what they already know about a hazard with information they obtain from hazard cues and a warning of some kind. In some cases, this might be a warning sign or label. In others, it might be a warning signal or alarm that indicates a hazard is present that normally is not there. A starting point for analyzing how people might integrate the information from the warning with what they already know or have determined from other sources of information is given by Bayes' rule, which describes how to infer the probability of a given event from one or more pieces of evidence [39.61]. Bayes' rule states that the posterior probability of hypothesis H_i given that evidence E_j is present, or $P(H_i|E_j)$, is given by

$$P(H_i|E_j) = \frac{P(E_j|H_i)}{P(E_j)} P(H_i) , \qquad (39.2)$$

where $P(H_i)$ is the probability of the hypothesis being true prior to obtaining the evidence E_j, and $P(E_j|H_i)$ is the probability of obtaining the evidence E_j given that the hypothesis H_i is true. When a receiver is given an imperfect warning, we can replace $P(E_j|H_i)$ in the above equation with $P(W|H)$ to calculate the probability that the hazard is present after receiving a warning. That is,

$$P(H|W) = \frac{P(W|H)}{P(W)} P(H) , \qquad (39.3)$$

where $P(H)$ is the prior probability of the hazard, $P(W)$ is the probability of sending a warning, $P(W|H)$ is the probability of sending a warning given the hazard is present, and $P(H|W)$ is the probability that the hazard is present after receiving the warning.

A number of other models have been developed in psychology that describe mathematically how people combine sources of information. Some examples include social judgment theory, policy capturing, multiple cue probability learning models, information integration theory, and conjoint measurement approaches [39.31]. From the perspective of warnings design, these approaches can be used to check which cues are actually used by people when they make safety-related decisions and how this information is integrated. A potential problem is that research on judgment and decision making clearly shows that people integrate information inconsistently with the prescriptions of Bayes' rule in some settings [39.71]; for example, sev-

eral studies show that people are more likely to attend to highly salient stimuli. This effect can explain the tendency for people to overestimate the likelihood of highly salient events.

One overall conclusion is that significant deviations from Bayes' rule become more likely when people must combine evidence in artificial settings where they are not able to fully exploit their knowledge and cues found in their naturalistic environment [39.72–74]. This does not mean that people are unable to make accurate inferences, as emphasized by both Simon and researchers embracing the ecological [39.75, 76] and naturalistic [39.77] models of decision making. In fact, the use of simple heuristics in rich environments can lead to inferences that are in many cases more accurate than those made using naive Bayes, or linear, regression [39.75]. Unfortunately, many applications of automation place the operator in a situation which removes them from the rich set of naturalistic cues available in less automated settings, and forces the operator to make inferences from information provided on displays. In such situations, it may become difficult for operators to make accurate inferences since they no longer can rely on simple heuristics or decision rules that are adapted to particular environments.

39.3.5 The Value of Warning Information

As mentioned earlier, the designers of warning systems tend to focus heavily on designing systems that reliably provide a warning when a hazard is present, which results in many false alarms. From a theoretical perspective it might be better to design the warning system so that it is less conservative. That is, a system that occasionally fails to detect the hazard but provides fewer false alarms might improve operator performance. From the perspective of warning design, the critical question is to determine how selective the warning should be to minimize the expected cost to the user as a function of the number of false alarms and correct identifications made by the warning [39.68, 78, 79].

Given that costs can be assigned to false alarms and misses, an optimal warning threshold can be calculated that maximizes the expected value of the provided information. If it is assumed that people will simply follow the recommendation of a warning system (i. e., the warning system is the sole decision-maker) the optimal warning threshold can be calculated using classical signal detection theory (SDT) [39.80, 81]. That is, a warning should be given when the likelihood ratio $P(E|S)/P(E|N)$ exceeds the optimal warning threshold β, calculated as shown below.

$$\beta = \frac{1 - P_S}{P_S} \times \frac{c_r - c_a}{c_i - c_m} , \qquad (39.4)$$

where

P_S the a priori probability of a signal (hazard) being present,

$P(E|S)$ the conditional probability of the evidence given a signal (hazard) is present,

$P(E|N)$ the conditional probability of the evidence given a signal (hazard) is not present,

c_r cost of a correct rejection,

c_a cost of a false alarm,

c_i cost of a correct identification,

c_m cost of a missed signal.

In reality, the problem is more complicated because, as mentioned earlier, people might consider other sources in addition to a warning when making decisions. The latter situation corresponds to a team decision made by a person and warning system working together, as expanded upon later.

39.3.6 Team Decision Making

The distributed signal detection theoretic (DSDT) model focuses on how to determine the *optimal* decision thresholds of both the warning system and the human operator when they work together to make the best possible decision [39.65]. The proposed approach is based on the distributed signal detection model [39.82–86]. The key insight is that a warning system and human operator are *both* decision-makers who *jointly* try to make an optimal team decision.

The DSDT model has many interesting implications and applications. One is that the warning system and human decision-maker should adjust their decision thresholds in a way that depends upon what the other is doing. If the warning system uses a low threshold and provides a warning even when there is not much evidence of the hazard, the DSDT model shows that the human decision-maker should adjust their own threshold in the opposite direction. That is, the rational human decision-maker will require more evidence from the environment or other source before complying with the warning. At some point, as the threshold for providing the warning gets lower, the rational decision-maker will ignore the warning completely.

The DSDT model also implies that the warning system should use different thresholds depending upon how the receiver is performing. If the receiver is dis-

abled or unable to take their observation from the environment, the warning should take the role of primary decision-maker, and set its threshold accordingly to that prescribed in traditional signal detection theory. Along the same lines, if the decision-maker is not responding in an optimal manner when the warning is or not given, the DSDT model prescribes ways of modifying the warning systems threshold; for example, if the decision-maker is too willing to take the precaution when the warning is provided, the warning system should use a stricter warning threshold. That is, the warning system should require more evidence before sending a warning.

Research has been performed that addresses predictions of the DSDT model [39.65, 87, 88]. One study showed that the performance of subjects on a simple inference task changed dramatically, depending on the warning threshold value [39.65]. The optimal warning threshold varied between subjects, and subjects changed their own decision thresholds, consistently with the DSDT model's predictions, when the warning threshold was modified. A second study, using a similar inference task, also showed that the optimal threshold varied between subjects [39.87]. Performance was significantly improved by adjusting the warning threshold to the optimal DSDT value calculated for each particular subject based on their earlier performance. A third study, in a more realistic environment, compared the driving performance of licensed drivers on a driving simulator [39.88]. Overall, use of the DSDT threshold improved passing decisions significantly over the SDT warning threshold. Drivers also changed their own decision thresholds, in the way the DSDT model predicted they should, when the warning threshold changed. Another interesting result was that use of the DSDT threshold resulted in either *risk-neutral* or *risk-averse* behavior, while, on the other hand, use of the SDT threshold resulted in some *risk-seeking* behavior; that is, people were more likely to ignore the warning and visual cues indicating that a car might be coming. Overall, these results suggest that, for familiar decisions such as choosing when to pass, people can behave nearly optimally.

One of the more interesting aspects of the DSDT model is that it suggests ways of adjusting the warning threshold in response to how the operator is performing.

39.3.7 Time Pressure and Stress

Time pressure and stress is another important issue in many applications of warnings. Reviews of the literature suggest that time pressure often results in poorer task performance and that it can cause shifts between the cognitive strategies used in judgment and decision-making situations [39.89, 90]. One change is that people show a tendency to shift to noncompensatory decision rules. This finding is consistent with contingency theories of strategy selection. In other words, this shift may be justified when little time is available, because a non-compensatory rule can be applied more quickly. *Maule* and *Hockey* [39.89] also note that people tend to filter out low-priority types of information, omit processing information, and accelerate mental activity when they are under time pressure.

The general findings above indicate that warnings may be useful when people are under time pressure and stress. People in such situations are especially likely to make mistakes. Consequently, warnings that alert people after they make mistakes may be useful. A second issue is that people under time pressure will not have

a lot of extra time available, so it will become especially important to avoid false alarms. A limited amount of research addresses the impact of time stress on warnings compliance. In particular, a study by *Wogalter* et al. [39.63] showed that time pressure reduced compliance with warnings. Interestingly, subjects performed better in both low- and high-stress conditions when the warnings were placed in the task instructions than on a sign posted nearby. The latter result supports the conclusion that warnings which efficiently and quickly transmit their information may be better when people are under time stress or pressure. In some situations, this may force the designer to carefully consider the tradeoff between the amount of information provided and the need for brevity. Providing more detailed information might improve understanding of the message but actually reduce effectiveness if processing the message requires too much time and attentional effort on the part of the receiver.

39.4 Design Guidelines and Requirements

Safety warnings can vary greatly in behavioral objectives, intended audiences, content, level of detail, format, and mode of presentation. Accordingly, the design of adequate warnings will often require extensive investigations and development activities involving significant resources and time [39.91], well beyond the scope of this chapter. The following discussion will briefly address methods of identifying hazards for applications of automation, along with some legal requirements and design specifications found in warning standards.

39.4.1 Hazard Identification

The first step in the development of warnings is to identify the hazards to be warned against. This process is guided by past experience, codes and regulations, checklists, and other sources, and is often organized by separately considering systems and subsystems, and potential malfunctions at each stage in their life cycles. Numerous complementary hazard analysis methods, which also guide the process of hazard identification, are available [39.92, 93]. Commonly used methods include work safety analysis, human error analysis, failure modes and effects analysis, and fault tree analysis.

Work safety analysis (WSA) [39.94] and human error analysis (HEA) [39.95] are related approaches that organize the analysis around tasks rather than system

components. This process involves the initial division of tasks into subtasks. For each subtask, potential effects of product malfunctions and human errors are then documented, along with the implemented and the potential countermeasures. In automation applications, the tasks that would be analyzed fall into the categories of normal operation, programming, and maintenance.

Failure modes and effects analysis (FMEA) is a systematic procedure for documenting the effects of system malfunctions on reliability and safety [39.93]. Variants of this approach include preliminary hazard analysis (PHA), and failure modes effects and criticality analysis (FMECA). In all of these approaches, worksheets are prepared which list the components of a system, their potential failure modes, the likelihood and effects of each failure, and both the implemented and the potential countermeasures that might be taken to prevent the failure or its effects. Each failure may have multiple effects. More than one countermeasure may also be relevant for each failure. Identification of failure modes and potential countermeasures is guided by past experience, standards, checklists, and other sources. This process can be further organized by separately considering each step in the operation of a system; for example, the effects of a power supply failure for a welding robot might be separately considered for each step performed in the welding process.

Fault tree analysis (FTA) is a closely related approach used to develop fault trees. The approach is top-down in that the analysis begins with a malfunction or accident and works downwards to basic events at the bottom of the tree [39.93]. Computer tools which calculate minimal cut-sets and failure probabilities and also perform sensitivity analysis [39.96] make such analysis more convenient. Certain programs help analysts draw fault trees [39.97,98]. Human reliability analysis (HRA) event trees are a classic example of this approach.

FTA and FMEA are complementary tools for documenting sources of reliability and safety problems, and also help organize efforts to control these problems. The primary shortcoming of both approaches is that the large number of components in many automated systems imposes a significant practical limitation on the analysis, that is, the number of event combinations that might occur is an exponential function of the large number of components. Applications for complex forms of automation consequently are normally confined to the analysis of single-event failures [39.100].

FTA and FMEA both provide a way of estimating the system reliability from the reliability of its components. *Dhillon* [39.101] provides a comprehensive overview of documents, data banks, and organizations for obtaining failure data to use in robot reliability analysis. Component reliabilities used in such analysis can be obtained from sources such as handbooks [39.102], data provided by manufacturers (Table 39.2), or past experience. Limited data is also available that documents error rates of personnel performing reliability-related tasks, such as maintenance (Table 39.3). Methods for estimating human error rates have been developed [39.103], such as technique for human error rate prediction (THERP) [39.104] and success likelihood index method-multiattribute utility decomposition (SLIM-MAUD) [39.105]. THERP follows an approach analogous to fault tree analysis, to estimate human error probabilities (HEPs). In SLIM-MAUD expert ratings are used to estimate HEPs as a function of performance shaping factors (PSFs).

Given that system component failure rates or probabilities are known, the next step in quantitative analysis is to develop a model of the system showing how system reliability is functionally determined by each component. The two most commonly used models are: (1) the systems block diagram, and (2) the system fault tree. Fault trees and system block diagrams are both useful for describing the effect of component configurations on system reliability. The most commonly considered configurations in such analysis are: (1) serial systems, (2) parallel systems, and (3) mixed serial and parallel systems.

39.4.2 Legal Requirements

In most industrialized countries, governmental regulations require that certain warnings be provided to

Table 39.2 Estimates of fire protection systems operational reliability (probability of success) [39.99]

Protection system	Warrington Delphi UK (Delphi Group)		Fire eng guidelines Australia (expert surveys)		Japanese studies (incident data)	
	Smoldering	Flaming	Smoldering	Flaming	Tokyo FD	Watanabe
Heat detector	0	89	0	90/95	94	89
Home smoke system	76	79	65	75/74	NA	NA
System smoke detector	86	90	70	80/85	94	89
Beam smoke detectors	86	88	70	80/85	94	89
Aspirated smoke detectors	86	NA	90	95/95	NA	NA
Sprinklers operate	95		50	95/99	97	NA
Sprinklers control but do not extinguish	64			NA	NA	NA
Sprinklers extinguish	48			NA	96	NA
Masonry construction	81					
	29% probability an opening will be fixed open		95% if no opening 90% if opening with auto closer		NA	NA
Gypsum partitions	69					
	29% probability an opening will be fixed open		95% if no opening 90% if opening with auto closer		NA	NA

Table 39.3 Summary of predicted human error probabilities in offshore platform musters (reprinted from [39.108], with permission from Elsevier). MO – man overboard, GR – gas release, F&R – fire and explosion, TSR – temporary safe refuge, OIM – offshore installation manager, PA – public announcement

No.	Action	HEP			Phase	Loss of defences
		MO	GR	F&E		
1	Detect alarm	0.00499	0.0308	0.396	Awareness	Do not hear alarm. Do not properly
2	Identify alarm	0.00398	0.0293	0.386		identify alarm. Do not maintain
3	Act accordingly	0.00547	0.0535	0.448		composure (panic)
4	Ascertain if danger is imminent	0.00741	0.0765	0.465	Evaluation	Misinterpret muster initiator seriousness and fail to muster in
5	Muster if in imminent danger	0.00589	0.0706	0.416		a timely fashion. Do not return process to safe state. Leave workplace
6	Return process equipment to safe state	0.00866	0.0782	0.474		in a condition that escalates initiator or impedes others' egress
7	Make workplace as safe as possible in limited time	0.00903	0.0835	0.489		
8	Listen and follow PA announcements	0.00507	0.0605	0.42	Egress	Misinterpret or do not hear PA announcements. Misinterpret tenability
9	Evaluate potential egress paths and choose route	0.00718	0.0805	0.476		of egress path. Fail to follow a path which leads to TSR; decide to follow
10	Move along egress route	0.00453	0.0726	0.405		a different egress path with lower
11	Assess quality of egress route while moving to TSR	0.00677	0.0788	0.439		tenability. Fail to assist others. Provide incorrect assistance which
12	Choose alternate route if egress path is not tenable	0.00869	0.1	0.5		delays or prevents egress
13	Assist others if needed or as directed	0.0101	0.0649	0.358		
14	Register at TSR	0.00126	0.01	0.2	Recovery	Fail to register while in the TSR.
15	Provide pertinent feedback attained while en route to TSR	0.00781	0.0413	0.289		Fail to provide pertinent feedback. Provide incorrect feedback. Do not don personal survival suit in an
16	Don personal survival suit or TSR survival suit if instructed to abandon	0.00517	0.026	0.199		adequate time for evacuation. Misinterpret OIM's instructions or do not follow OIM's instructions
17	Follow OIM's instructions	0.0057	0.0208	0.21		
18	Follow OIM's instructions	0.0057	0.0208	0.21		

workers and others who might be exposed to hazards. For example, in the USA, the Environmental Protection Agency (EPA) has developed several labeling requirements for toxic chemicals. The Department of Transportation (DOT) makes specific provisions regarding the labeling of transported hazardous materials. The most well-known governmental standards in the USA applicable to applications of automation are the general industry standards specified by the Occupation Safety and Health Administration (OSHA). The OSHA has also promulgated a hazard communication standard that applies to workplaces where toxic or hazardous materials are in use. Training, container labeling and

other forms of warnings, and material data safety sheets are all required elements of the OSHA hazard communication standard. Other relevant OSHA publications addressing automation are the *Guidelines for Robotics Safety* [39.106] and the *Occupational Safety and Health Technical Manual* [39.107].

In the USA, the failure to warn also can be grounds for litigation holding manufacturers and others liable for injuries incurred by workers. In establishing liability, the *theory of negligence* considers whether the failure to adequately warn is unreasonable conduct based on (1) the foreseeability of the danger to the manufacturer, (2) the reasonableness of the assumption that a user would realize the danger and, (3) the degree of care that the manufacturer took to inform the user of the danger. The *theory of strict liability* only requires that the failure to warn caused the injury or loss.

39.4.3 Voluntary Standards

A large set of existing standards provide voluntary recommendations regarding the use and design of safety information. These standards have been developed by both: (1) international groups, such as the United Nations, the European Economic Community (EURONORM), the International Organization for Standardization (ISO), and the International Electrotechnical Commission (IEC), and (2) national groups, such as the American National Standards Institute (ANSI), the British Standards Institute, the Canadian Standards Association, the German Institute for Normalization (DIN), and the Japanese Industrial Standards Committee.

Among consensus standards, those developed by ANSI in the USA are of special significance. Over the last decade or so, five new ANSI standards focusing on safety signs and labels have been developed and one significant standard has been revised. The new standards are: (1) ANSI Z535.1 *Safety Color Code*, (2) ANSI Z535.2 *Environmental and Facility Safety Signs*, (3) ANSI Z535.3 *Criteria for Safety Symbols*, (4) ANSI Z535.4 *Product Safety Signs and Labels*, and (5) ANSI Z535.5 *Accident Prevention Tags*. The recently revised standard is ANSI Z129.1-1988, *Hazardous Industrial Chemicals – Precautionary Labeling*. Furthermore, ANSI has published a *Guide for Developing Product Information*.

Warning requirements for automated equipment can also be found in many other standards. The most well-known standard in the USA that addresses automation safety is ANSI/RIA R15.06. This standard was first published in 1986 by the Robotics Industries Association (RIA) and the American National Standards Institute (ANSI) as ANSI/RIA R15.06, the *American National Standard for Industrial Robots and Robot Systems – Safety Requirements* [39.109]. A revised version of the standard was published in 1992 and the standard is currently undergoing revisions once again. Several other standards developed by the ANSI are potentially important in automation applications. The latter standards address a wide variety of topics such as machine tool safety, machine guarding, lock-out/tagout procedures, mechanical power transmission, chemical labeling, material safety data sheets, personal protective equipment, safety markings, workplace signs, and product labels. Other potentially relevant standards developed by nongovernmental groups include the National Electric Code, the Life Safety Code, and the proposed UL1740 safety standard for industrial robots and robotics equipment. Literally thousands of consensus standards contain safety provisions. Also, many companies that use or manufacture automated systems will develop their own guidelines [39.110]. Companies often will start with the ANSI/RIA R15.06 robot safety standard, and then add detailed information that is relevant to their particular situation.

39.4.4 Design Specifications

Design specifications can be found in consensus and governmental safety standards specifying how to design (1) material safety data sheets (MSDS), (2) instructional labels and manuals, (3) safety symbols, and (4) warning signs, labels, and tags.

Material Safety Data Sheets
The OSHA hazard communication standard specifies that employers must have a MSDS in the workplace for each hazardous chemical used. The standard requires that each sheet be written in English, list its date of preparation, and provide the chemical and common name of hazardous chemicals contained. It also requires the MSDS to describe (1) physical and chemical characteristics of the hazardous chemical, (2) physical hazards, including potential for fire, explosion, and reactivity, (3) health hazards, including signs and symptoms of exposure, and health conditions potentially aggravated by the chemical, (4) the primary route of entry, (5) the OSHA permissible exposure limit, the American Conference of Governmental Industrial Hygienists (ACGIH) threshold limit value, or other recommended limits, (6) carcinogenic properties, (7) generally applicable precautions, (8) generally applicable control measures, (9) emergency and first-aid procedures, and (10) the name, address, and telephone number of a party able to provide, if necessary, additional information on the hazardous chemical and emergency procedures.

Instructional Labels and Manuals
Few consensus standards currently specify how to design instructional labels and manuals. This situation is, however, quickly changing. The ANSI *Guide for Developing User Product Information*, was recently published in 1990, and several other consensus or-

Table 39.4 Summary of recommendations in selected warning systems (after *Lehto* and *Miller* [39.39], and *Lehto* and *Clark* [39.9])

System	Signal words	Color coding	Typography	Symbols	Arrangement
ANSI Z129.1 Precautionary labeling of hazardous chemicals	Danger Warning Caution Poison optional words for *delayed* hazards	Not specified	Not specified	Skull-and-crossbones as supplement to words. Acceptable symbols for three other hazard types	Label arrangement not specified; examples given
ANSI Z535.2 Environmental and facility safety signs	Danger Warning Caution Notice [General safety] [Arrows]	Red Orange Yellow Blue Green as above; B&W otherwise per ANSI Z535.1	Sans serif, Upper case, Acceptable typefaces, Letter heights	Symbols and pictographs per ANSI Z535.3	Defines signal word, word message, symbol panels in 1–3 panel designs. Four shapes for special use. Can use ANSI Z535.4 for uniformity
ANSI Z535.4 Product safety signs and labels	Danger Warning Caution	Red Orange Yellow per ANSI Z535.1	Sans serif, Upper case, Suggested typefaces, Letter heights	Symbols and pictographs per ANSI Z535.3; also Society of Automotive Engineers (SAE) J284 Safety Alert Symbol	Defines signal word, message, pictorial panels in order of general to specific. Can use ANSI Z535.2 for uniformity. Use ANSI Z129.1 for chemical hazards
National Electrical Manufacturers Association (NEMA) guidelines: NEMA 260	Danger Warning	Red Red	Not specified	Electric shock symbol	Defines signal word, hazard, consequences, instructions, symbol. Does not specify order
SAE J115 Safety signs	Danger Warning Caution	Red Yellow Yellow	Sans serif, Typeface, Upper case	Layout to accommodate symbols; specific symbols/pictographs not prescribed	Defines 3 areas: signal word panel, pictorial panel, message panel. Arrange in order of general to specific
ISO standard: ISO R557, 3864	None. Three kinds of labels: Stop/prohibition Mandatory action Warning	Red Blue Yellow	Message panel is added below if necessary	Symbols and pictographs	Pictograph or symbol is placed inside appropriate shape with message panel below if necessary

ganizations are working on draft documents. Without an overly scientific foundation, the ANSI Consumer Interest Council, which is responsible for the above guidelines, has provided a reasonable outline to manufacturers regarding what to consider in producing instruction/operator manuals. They have included sec-

Table 39.4 (cont.)

System	Signal Words	Color Coding	Typography	Symbols	Arrangement
OSHA 1910.145 Specification for accident prevention signs and tags	Danger Warning (tags only) Caution Biological Hazard, BIOHAZARD, or symbol [Safety instruction] [Slow-moving vehicle]	Red Yellow Yellow Fluorescent Orange/ Orange-Red Green Fluorescent Yellow-Orange & Dark Red per ANSI Z535.1	Readable at 5 ft or as required by task	Biological hazard symbol. Major message can be supplied by pictograph (tags only). Slow-moving vehicle (SAE J943)	Signal word and major message (tags only)
OSHA 1910.1200 [Chemical] Hazard communication	Per applicable requirements of Environmental Protection Agency (EPA), Food and Drug Administration (FDA), and Consumer Product Safety Commission (CPSC)		In English		Only as material safety data sheet
Westinghouse handbook; FMC guidelines	Danger Warning Caution Notice	Red Orange Yellow Blue	Helvetica bold and regular weights, Upper/lower case	Symbols and pictographs	Recommends 5 components: signal word, symbol/ pictograph, hazard, result of ignoring warning, avoiding hazard

tions covering *organizational elements*, *illustrations*, *instructions*, *warnings*, *standards*, *how to use language*, and *an instructions development checklist*. While the guideline is brief, the document represents a useful initial effort in this area.

Safety Symbols

Numerous standards throughout the world contain provisions regarding safety symbols. Among such standards, the ANSI Z535.3 standard, *Criteria for Safety Symbols*, is particularly relevant for industrial practitioners. The standard presents a significant set of selected symbols shown in previous studies to be well understood by workers in the USA. Perhaps more importantly, the standard also specifies methods for designing and evaluating safety symbols. Important

provisions include: (1) new symbols must be correctly identified during testing by at least 85% of 50 or more representative subjects, (2) symbols which do not meet the understandability criteria should only be used when equivalent word messages are also provided, and (3) employers and product manufacturers should train users regarding the intended meaning of the symbols. The standard also makes new symbols developed under these guidelines eligible to be considered for inclusion in future revisions of the standard.

Warning Signs, Labels, and Tags

ANSI and other standards organizations provide very specific recommendations about how to design warning signs, labels, and tags. These include, among other factors, particular signal words and text, color coding schemes, typography, symbols, arrangement, and hazard identification (Table 39.4).

Among the most popular signal words recommended are: *danger*, to indicate the highest level of hazard; *warning*, to represent an intermediate hazard; and *caution*, to indicate the lowest level of hazard. Color coding methods, also referred to as a *color system*, consistently associate colors with particular levels of hazard; for example, red is used in all of the standards to represent the highest level of danger. Explicit recommendations regarding typography are given in nearly all the systems. The most general commonality between the systems is the recommended use of sans-serif typefaces. Varied recommendations are given regarding the use of symbols and pictographs.

The FMC and the Westinghouse systems advocate the use of symbols to define the hazard and to convey the level of hazard. Other standards recommend symbols only as a supplement to words. Another area of substantial variation shown in Table 39.4 pertains to the recommended label arrangements. The proposed arrangements generally include elements from the above discussion and specify the image's graphic content and color, the background's shape and color, the enclosure's shape and color, and the surround's shape and color. Many of the systems also precisely describe the arrangement of the written text and provide guidance regarding methods of hazard identification.

Certain standards also specify the content and wording of warning signs or labels in some detail; for example, ANSI Z129.1 specifies that chemical warning labels include (1) identification of the chemical product or its hazardous component(s), (2) signal word, (3) statement of hazard(s), (4) precautionary measures, (5) instructions in case of contact or exposure, (6) an-

tidotes, (7) notes to physicians, (8) instructions in case of fire and spill or leak, and (9) instructions for container handling and storage. This standard also specifies a general format for chemical labels that incorporate these items and recommended wordings for particular messages.

There are also standards which specifically address the design of automated systems and alerts; for example, the Department of Transportation, Federal Aviation Administration's (FAA) *Human Factors Design Standard* (HFDS) (2003) indicates that alarm systems should alert the user to the existence of a problem, inform of the priority and nature of the problem, guide the user's initial response, and confirm whether the user's response corrected the problem. Furthermore, it should be possible to identify the first event in a series of alarm events (information valuable in determining the cause of a problem). Consistent with our earlier discussion, the standard suggests that information should be provided in multiple formats (e.g., visual and auditory) to improve communication and reduce mental workload, that auditory signals be used to draw attention to the location of a visual display, and that false alarms should not occur so frequently so as to undermine user trust in the system. Additionally, users should be informed of the inevitability of false alarms (especially in the case of low base rates). A more in-depth discussion of the FAA's recommendations is beyond the scope of this chapter – for additional information, see the *Human Factors Design Standard* [39.111]. Recommendations also exist for numerous other applications including automated cruise control/collision warning systems for commercial trucks greater than 10 000 pounds [39.112] as well as for horns, backup alarms, and automatic warning devices for mobile equipment [39.113].

39.5 Challenges and Emerging Trends

The preceding sections of this chapter reveal that warnings can certainly play an important role in automated systems. One of the more encouraging results is that people often tend to behave consistently with the predictions of normative models of decision making. From a general perspective, this is true both if people comply with warnings because they believe the hazard is more serious, and if people ignore warnings with little diagnostic value or when the cost of compliance is believed to be high. This result supports the conclusion that normative models can play an important role in suggesting and evaluating design solutions that address

issues such as operator mistrust of warnings, complacency, and overreliance on warnings. Perhaps the most fundamental design challenge is that of increasing the value of the information provided by automated warning systems. Doing so would directly address the issue of operator mistrust.

The most fundamental method of addressing this issue is to develop improved sensor systems that more accurately measure important variables that are strongly related to hazards or other warned against events. Successful implementations of this approach could increase the diagnostic value of the warnings provided by automated warning systems by reducing either false alarms or misses, and hopefully both. Given the significant improvements and reduced costs of sensor technology that have been observed in recent years, this strategy seems quite promising. Another promising strategy for increasing the diagnostic value of the warnings is to develop better algorithms for both integrating information from multiple sensors and deciding upon when to provide a warning. Such algorithms might include methods of adaptive automation that monitor the operator's behavior, and respond accordingly; for example, if the system detects evidence that the user is ignoring the provided warnings, a secondary, more urgent warning that requires a confirmatory response might be given to determine if the user is disabled (i.e., unable to respond because they are distracted or even sleeping). Other algorithms might track the performance of particular operators over a longer period and use this data to estimate the skill of the operator or determine the types of information the operator uses to make decisions. Such tracking might reveal the degree to which the operator relies on the warning system. It also might reveal the extent to which the other sources of information used by operator are redundant or independent of the warning.

Another challenge that has been barely, if it all, addressed in most current applications of warning systems is related to the tacit assumption that the perceived costs and benefits of correct detections, misses, false alarms, and correct rejections are constant across operators and situations. This assumption is clearly false, because operators will differ in their attitudes toward risk. Furthermore, the costs and benefits are also likely to change greatly between situations; for example, the expected severity of an automobile accident changes, depending on the speed of the vehicle. This issue might be addressed by algorithms based on normative models which treat the costs and benefits of correct detections, misses, false alarms, and correct rejections as random variables which are a function of particular operators and situations.

Many other challenges and areas of opportunity exist for improving automated warnings; for example, more focus might be placed on developing warning systems that are easier for operators to understand. Such systems might include capabilities of explaining why the warning is being provided and how strong the evidence is. Other systems might give the user more control over how the warning operates.

References

39.1   R. Parasuraman, V.A. Riley: Humans and automation: use, misuse, disuse, abuse, Hum. Factors **39**, 230–253 (1997)

39.2   C.D. Wickens, J.G. Hollands: *Engineering Psychology and Human Performance*, 3rd edn. (Prentice Hall, Upper Saddle River 2000)

39.3   N.B. Sarter, D.D. Woods: How in the world did we get into that mode? Mode error and awareness in supervisory control, Hum. Factors **37**(1), 5–19 (1995)

39.4   C. Perrow: *Normal Accidents: Living with High-Risk Technologies* (Basic Books, New York 1984)

39.5   M.R. Endsley: Automation and situation awareness. In: *Automation and Human Performance*, ed. by R. Parasuraman, M. Mouloua (Lawrence Erlbaum, Mahwah 1996) pp. 163–181

39.6   U. Metzger, R. Parasuraman: The role of the air traffic controller in future air traffic management: an empirical study of active control versus passive monitoring, Hum. Factors **43**(4), 519–528 (2001)

39.7   M.R. Endsley, E.O. Kiris: The out-of-the-loop performance problem and level of control in automation, Hum. Factors **37**(2), 381–394 (1995)

39.8   J. Reason: *Human Error* (Cambridge Univ. Press, Cambridge 1990)

39.9   M.R. Lehto, D.R. Clark: Warning signs and labels in the workplace. In: *Workspace, Equipment and Tool Design*, ed. by W. Karwowski, A. Mital (Elsevier, Amsterdam 1990) pp. 303–344

39.10   M.R. Lehto, G. Salvendy: Warnings: a supplement not a substitute for other approaches to safety, Ergonomics **38**(11), 2155–2163 (1995)

39.11   R. Parasuraman, T.B. Sheridan, C.D. Wickens: A model for types and levels of human interaction with automation, IEEE Trans. Syst. Man Cyber. Part A: Syst. Hum. **30**(3), 286–297 (2000)

39.12   M.R. Endsley: Toward a theory of situation awareness in dynamic systems, Hum. Factors **37**(1), 32–64 (1995)

39.13   R. Parasuraman: Designing automation for human use: empirical studies and quantitative models, Ergonomics **43**(7), 931–951 (2000)

39.14   W.J. Horrey, C.D. Wickens, R. Strauss, A. Kirlik, T.R. Stewart: Supporting situation assessment through attention guidance and diagnostic aiding: the benefits and costs of display enhancement on judgment skill. In: *Adaptive Perspectives on Human-Technology Interaction: Methods and Models for Cognitive Engineering and Human-Computer Interaction*, ed. by A. Kirlik (Oxford Univ. Press, New York 2006) pp. 55–70

39.15   P. Flanagan, K.I. McAnally, R.L. Martin, J.W. Meehan, S.R. Oldfield: Aurally and visually guided visual search in a virtual environment, Hum. Factors **40**(3), 461–468 (1998)

39.16   M. Yeh, C.D. Wickens, F.J. Seagull: Target cuing in visual search: the effects of conformality and display location on the allocation of visual attention, Hum. Factors **41**(4), 524–542 (1999)

39.17   S.E. Graham, M.D. Matthews: *Infantry Situation Awareness: Papers from the 1998 Infantry Situation Awareness Workshop* (US Army Research Institute, Alexandria 1999)

39.18   H. Davison, C.D. Wickens: Rotorcraft hazard cueing: the effects on attention and trust, Proc. 11th Int. Symp. Aviat. Psychol. Columbus (The Ohio State Univ., Columbus 2001)

39.19   K.L. Mosier, L.J. Skitka, S. Heers, M. Burdick: Automation bias: decision making and performance in high-tech cockpits, Int. J. Aviat. Psychol. **8**(1), 47–63 (1998)

39.20   C.D. Wickens, K. Gempler, M.E. Morphew: Workload and reliability of predictor displays in aircraft traffic avoidance, Transp. Hum. Factors **2**(2), 99–126 (2000)

39.21   J.D. Lee, D.V. McGehee, T.L. Brown, M.L. Reyes: Collision warning timing, driver distraction, and driver response to imminent rear-end collisions in a high-fidelity driving simulator, Hum. Factors **44**(2), 314–334 (2001)

39.22   A.F. Kramer, N. Cassavaugh, W.J. Horrey, E. Becic, J.L. Mayhugh: Influence of age and proximity warning devices on collision avoidance in simulated driving, Hum. Factors **49**(5), 935–949 (2007)

39.23   M. Maltz, D. Shinar: Imperfect in-vehicle collision avoidance warning systems can aid drivers, Hum. Factors **46**(2), 357–366 (2004)

39.24   T.A. Dingus, D.V. McGehee, N. Manakkal, S.K. Jahns, C. Carney, J.M. Hankey: Human factors field evaluation of automotive headway maintenance/collision warning devices, Hum. Factors **39**(2), 216–229 (1997)

39.25   N.B. Sarter, B. Schroeder: Supporting decision making and action selection under time pressure and uncertainty: the case of in-flight icing, Hum. Factors **43**(4), 573–583 (2001)

39.26   D.A. Wiegmann, A. Rich, H. Zhang: Automated diagnostic aids: the effects of aid reliability on users' trust and reliance, Theor. Issues Ergon. Sci. **2**(4), 352–367 (2001)

39.27   S.R. Dixon, C.D. Wickens: Automation reliability in unmanned aerial vehicle control: a reliance-compliance model of automation dependence in high workload, Hum. Factors **48**(3), 474–486 (2006)

39.28   F.J. Seagull, P.M. Sanderson: Anesthesia alarms in context: an observational study, Hum. Factors **43**(1), 66–78 (2001)

39.29   L. Bainbridge: Ironies of automation, Automatica **19**, 775–779 (1983)

39.30   J. Meyer: Responses to dynamic warnings. In: *Handbook of Warnings*, ed. by M.S. Wogalter (Lawrence Erlbaum, Mahwah 2006) pp. 89–108

39.31   M.R. Lehto: Optimal warnings: an information and decision theoretic perspective. In: *Handbook of Warnings*, ed. by M.S. Wogalter (Lawrence Erlbaum, Mahwah 2006) pp. 89–108

39.32   R.R. Duffy, M.J. Kalsher, M.S. Wogalter: Increased effectiveness of an interactive warning in a realistic incidental product-use situation, Int. J. Ind. Ergon. **15**, 159–166 (1995)

39.33   J.P. Frantz, J.M. Rhoades: A task-analytic approach to the temporal and spatial placement of product warnings, Hum. Factors **35**, 719–730 (1993)

39.34   M.S. Wogalter, T. Barlow, S.A. Murphy: Compliance to owner's manual warnings: influence of familiarity and the placement of a supplemental directive, Ergonomics **38**, 1081–1091 (1995)

39.35   R.C. Schank, R. Abelson: *Scripts, plans, goals, and understanding* (Lawrence Erlbaum, Hillsdale 1977)

39.36   M.S. Wogalter, W.J. Vigilante: Attention switch and maintenance. In: *Handbook of Warnings*, ed. by M.S. Wogalter (Lawrence Erlbaum, Mahwah 2006) pp. 89–108

39.37   S.L. Young, M.S. Wogalter: Effects of conspicuous print and pictorial icons on comprehension and memory of instruction manual warnings, Hum. Factors **32**, 637–649 (1990)

39.38   P. Sanderson: The multimodal world of medical monitoring displays, Appl. Ergon. **37**, 501–512 (2006)

39.39   M.R. Lehto, J.M. Miller: *Warnings, Volume 1: Fundamentals, Design and Evaluation Methodologies* (Fuller Technical, Ann Arbor 1986)

39.40   M.S. Sanders, E.J. McCormick: *Human Factors in Engineering and Design* (McGraw-Hill, New York 1993), 7th edn.

39.41   M.S. Wogalter, M.J. Kalser, B.M. Racicot: Behavioral compliance with warnings: effects of voice, context, and location, Saf. Sci. **16**, 637–654 (1993)

39.42   M.S. Wogalter, S.L. Young: Behavioural compliance to voice and print warnings, Ergonomics **34**, 79–89 (1991)

39.43   S. Coren, L.M. Ward: *Sensation and Perception*, 3rd edn. (Harcourt Brace Jovanovich, San Diego 1989)

39.44   J. Edworthy, E. Hellier: Complex nonverbal auditory signals and speech warnings. In: *Handbook of Warnings*, ed. by M.S. Wogalter (Lawrence Erlbaum, Mahwah 2006) pp. 89–108

39.45   R.M. Warren, R.P. Warren: Auditory illusions and confusions, Sci. Am. **223**, 30–36 (1970)

39.46   A.G. Samuel: Phonemic restoration: insights from a new methodology, J. Exp. Psychol. Gen. **110**, 474–494 (1981)

39.47   K.L. Momtahan: *Mapping of psychoacoustic parameters to the perceived urgency of auditory warning signals* (Carleton Univ., Ottawa Ontario 1990), unpublished master's thesis

39.48 J. Edworthy, S.L. Loxley, I.D. Dennis: Improving auditory warning design: relationship between warning sound parameters and perceived urgency, Hum. Factors **33**, 205–231 (1991)

39.49 J.D. Lee, J.D. Hoffman, E. Hayes: Collision warning design to mitigate driver distraction. In: *Proceedings of the SIGCHI Conference on Human Factors in Computing Systems* (ACM, New York 2004) pp. 65–72

39.50 J.Y.C. Ng, J.C.F. Man, S. Fels, G. Dumont, J.M. Ansermino: An evaluation of a vibro-tactile display prototype for physiological monitoring, Anesth. Analg. **101**, 1719–1724 (2005)

39.51 J.J. Hickey: *Shoulder Rumble Strip Effectiveness: Drift-off-road Accident Reductions on the Pennsylvania Turnpike (Transportation Research Record 1573)* (National Research Council, Washington 1997) pp. 105–109

39.52 S.J. Selcon, R.M. Taylor, R.A. Shadrake: Multimodal cockpit warnings: pictures, words or both?, Proc. Hum. Factors Soc. 36th Annu. Meet. (Human Factor Society, Santa Monica 1992) pp. 57–61

39.53 S.J. Selcon, R.M. Taylor: Evaluation of the situational awareness rating technique (SART) as a tool for aircrew systems design, Proc. AGARD AMP Symp. Situational Awareness in Aerospace Operations (Copenhagen 1989)

39.54 R.M. Taylor: Situational awareness rating technique (SART): the development of a tool for aircrew systems design, Proc. AGARD AMP Symp. Situational Awareness in Aerospace Operations (Copenhagen 1989)

39.55 A.E. Sklar, N.B. Sarter: Good vibrations: tactile feedback in support of attention allocation and human-automation coordination in event-driven domains, Hum. Factors **41**, 543–552 (1999)

39.56 V.C. Conzola, M.S. Wogalter: Using voice and print directives and warnings to supplement product manual instructions – conspicuous print and pictorial icons, Int. J. Ind. Ergon. **23**, 549–556 (1999)

39.57 C.G. Penney: Modality effects in short-term verbal memory, Psychol. Bull. **82**, 68–84 (1975)

39.58 J.A. Henderson, A.D. Twerski: Doctrinal collapse in products liability: the empty shell of failure to warn, New York Univ. Law Rev. **65**(2), 265–327 (1990)

39.59 J. Papastavrou, M.R. Lehto: Improving the effectiveness of warnings by increasing the appropriateness of their information content: some hypotheses about human compliance, Saf. Sci. **21**, 175–189 (1996)

39.60 J. von Neumann, O. Morgenstern: *Theory of Games and Economic Behavior* (Princeton Univ. Press, Princeton 1947)

39.61 L.J. Savage: *The Foundations of Statistics* (Dover, New York 1954)

39.62 M.R. Lehto, J. Papastavrou: Models of the warning process: important implications towards effectiveness, Saf. Sci. **16**, 569–595 (1993)

39.63 M.S. Wogalter, D.M. Dejoy, K.R. Laughery: Organizing theoretical framework: a consolidated-human information processing (C-HIP) model. In: *Warnings and Risk Communication*, ed. by M. Wogalter, D. DeJoy, K. Laughery (Taylor and Francis, London 1999)

39.64 J.M. Miller, M.R. Lehto: *Warnings and Safety Instructions: The Annotated Bibliography*, 4th edn. (Fuller Technical, Ann Arbor 2001)

39.65 J. Papastavrou, M.R. Lehto: A distributed signal detection theory model: implications for the design of warnings, Int. J. Occup. Saf. Ergon. **1**(3), 215–234 (1995)

39.66 S. Breznitz: *Cry-Wolf: The Psychology of False Alarms* (Lawrence Erlbaum, Hillsdale 1984)

39.67 J.P. Bliss, C.K. Fallon: Active warnings: false alarms. In: *Handbook of Warnings*, ed. by M.S. Wogalter (Lawrence Erlbaum, Mahwah 2006) pp. 231–242

39.68 R.D. Sorkin, B.H. Kantowitz, S.C. Kantowitz: Likelihood alarm displays, Hum. Factors **30**, 445–459 (1988)

39.69 N. Moray: Monitoring, complacency, scepticism and eutactic behaviour, Int. J. Ind. Ergon. **31**(3), 175–178 (2003)

39.70 J. Edworthy: Warnings and hazards: an integrative approach to warnings research, Int. J. Cogn. Ergon. **2**, 2–18 (1998)

39.71 M.R. Lehto, H. Nah: Decision making and decision support. In: *Handbook of Human Factors and Ergonomics*, 3rd edn., ed. by G. Salvendy (Wiley, New York 2006) pp. 191–242

39.72 M.S. Cohen: The naturalistic basis of decision biases. In: *Decision Making in Action: Models and Methods*, ed. by G.A. Klein, J. Orasanu, R. Calderwood, E. Zsambok (Ablex, Norwood 1993) pp. 51–99

39.73 H.A. Simon: A behavioral model of rational choice, Q. J. Econ. **69**, 99–118 (1955)

39.74 H.A. Simon: Alternative visions of rationality. In: *Reason in Human Affairs*, ed. by H.A. Simon (Stanford Univ. Press, Stanford 1983)

39.75 G. Gigerenzer, P. Todd, ABC Research Group: *Simple Heuristics That Make Us Smart* (Oxford University Press, New York 1999)

39.76 K.R. Hammond: *Human Judgment and Social Policy: Irreducible Uncertain, Inevitable Error, Unavoidable Injustice* (Oxford Univ. Press, New York 1996)

39.77 G.A. Klein, J. Orasanu, R. Calderwood, E. Zsambok: *Decision Making in Action: Models and Methods* (Ablex, Norwood 1993)

39.78 D.A. Owens, G. Helmers, M. Sivak: Intelligent vehicle highway systems: a call for user-centred design, Ergonomics **36**(4), 363–369 (1993)

39.79 P.A. Hancock, R. Parasuraman: Human factors and safety in the design of intelligent vehicle-highway systems (IVHS), J. Saf. Res. **23**, 181–198 (1992)

39.80 D.M. Green, J.A. Swets: *Signal Detection Theory and Psychophysics* (Wiley, New York 1966)

39.81 C.D. Wickens: *Engineering Psychology and Human Performance*, 2nd edn. (Harper Collins, New York 1992)

39.82 R.R. Tenney, N.R. Sandell Jr.: Detection with distributed sensors, IEEE Trans. Aerosp. Electron. Syst. **17**(4), 501–510 (1981)

39.83 L.K. Ekchian, R.R. Tenney: Detection networks, Proc. 21st IEEE Conf. Decis. Control (1982) pp. 686–691

39.84 J.D. Papastavrou, M. Athans: On optimal distributed decision architectures in a hypothesis testing environment, IEEE Trans. Autom. Control **37**(8), 1154–1169 (1992)

39.85 J.D. Papastavrou, M. Athans: The team ROC curve in a binary hypothesis testing environment, IEEE Trans. Aerosp. Electron. Syst. **31**(1), 96–105 (1995)

39.86 J.N. Tsitsiklis: Decentralized detection, Adv. Stat. Signal Proc. **2**, 297–344 (1993)

39.87 M.R. Lehto, J.P. Papastavrou, W. Giffen: An empirical study of adaptive warnings: human vs. computer adjusted warning thresholds, Int. J. Cogn. Ergon. **2**(1/2), 19–33 (1998)

39.88 M.R. Lehto, J.P. Papastavrou, T.A. Ranney, L. Simmons: An experimental comparison of conservative versus optimal collision avoidance system thresholds, Saf. Sci. **36**(3), 185–209 (2000)

39.89 A.J. Maule, G.R.J. Hockey: State, stress, and time pressure. In: *Time Pressure and Stress in Human Judgment and Decision Making*, ed. by O. Svenson, A.J. Maule (Plenum, New York 1993) pp. 83–102

39.90 E. Edland, O. Svenson: Judgment and decision making under time pressure. In: *Time Pressure and Stress in Human Judgment and Decision Making*, ed. by O. Svenson, A.J. Maule (Plenum, New York 1993) pp. 27–40

39.91 J.P. Frantz, T.P. Rhoades, M.R. Lehto: Warnings and risk communication. In: *Warnings and Risk Communication*, ed. by M.S. Wogalter, D.M. DeJoy, K. Laughery (Taylor and Francis, London 1999) pp. 291–312

39.92 M.R. Lehto, G. Salvendy: Models of accident causation and their application: review and reappraisal, J. Eng. Technol. Manag. **8**, 173–205 (1991)

39.93 W. Hammer: *Product Safety Management and Engineering*, 2nd edn. (American Society of Safety Engineers (ASSE), Des Plaines 1993)

39.94 J. Suoakas, V. Rouhiainen: *Work Safety Analysis: Method Description and User's Guide*, Research Report 314 (Technical Research Center of Finland, Tempere 1984)

39.95 J. Suoakas, P. Pyy: *Evaluation of the Validity of Four Hazard Identification Methods with Event Descriptions*, Research Report 516 (Technical Research Center of Finland, Tempere 1988)

39.96 S. Contini: Fault tree and event tree analysis, Conf. Adv. Inf. Tools Saf. Reliab. Anal. ISPA (1988) pp. 24–28

39.97 S. Ruthberg: DORISK – a system for documentation and analysis of fault trees, SRE Symp. (Trondheim 1985)

39.98 M. Knochenhauer: ABB Atom's SUPER NET programme package for reliability and risk analysis, Conf. Adv. Inf. Tools Saf. Reliab. Anal. ISPA (1988)

39.99 R.W. Bukowski, E.K. Budnick, C.F. Schemel: Estimates of the operational of fire protection systems, Proc. Soc. Fire Prot. Eng. Am. Inst. Archit. (2002) pp. 111–124

39.100 M.L. Visinsky, J.R. Cavallaro, I.D. Walker: Robotic fault detection and fault tolerance: a survey, Reliab. Eng. Syst. Saf. **46**:2, 139–158 (1994)

39.101 B.S. Dhillon: *Robot Reliability and Safety* (Springer, Berlin, Heidelberg 1991)

39.102 Department of Defense: *MIL-HBDK-217F: Reliability Prediction of Electronic Equipment* (Rome Laboratory, Griffiss Air Force Base, NY 1990)

39.103 D.I. Gertman, H.S. Blackman: *Human Reliability, Safety Analysis Data Handbook* (Wiley, New York 1994)

39.104 A.D. Swain, H. Guttman: *Handbook for Human Reliability Analysis with Emphasis on Nuclear Power Plant Applications* (US Nuclear Regulatory Commission, Washington 1983), NUREG/CR-1278

39.105 D.E. Embrey: *SLIM-MAUD: An Approach to Assessing Human Error Probabilities Using Structured Expert Judgment* (US Nuclear Regulatory Commission, Washington 1984), NUREG/CR-3518, Vols. 1 and 2

39.106 OSHA: *Guidelines for Robotics Safety* (US Department of Labor, Washington 1987)

39.107 OSHA: Industrial robots and robot system safety. In: *Occupational Safety and Health Technical Manual* (US Department of Labor, Washington 1996)

39.108 D.G. DiMattia, F.I. Khan, P.R. Amyotte: Determination of human error probabilities for offshore platform musters, J. Loss Prev. Proc. Ind. **18**, 488–501 (2005)

39.109 American National Standard for Industrial Robots and Robot Systems: Safety Requirements, ANSI/RIA R15.06, (Robotic Industries Association, Ann Arbor, MI, and the American National Standards Institute, New York 1992)

39.110 K.M. Blache: Industrial practices for robotic safety. In: *Safety, Reliability, and Human Factors in Robotic Systems*, ed. by J.H. Graham (Van Nostrand Reinhold, Dordrecht 1991), Chap. 3

39.111 Department of Transportation, Federal Aviation Administration: *Human Factors Design Standard* (DTFAA, Springfield 2003), Report No. DOT/FAA/CT-03/05 HF-STD-001

39.112 A. Houser, J. Pierowicz, R. McClellan: *Concept of operations and voluntary operational requirements for automated cruise control/collision warning systems (ACC/CWS) on-board commercial motor vehicles* (Federal Motor Carrier Safety Administration, Washington 2005), report No. FMCSA-MCRR-05-007, retrieved from http://www.fmcsa.dot.gov/facts-research/research-technology/report/forward-collision-warning-systems.htm

39.113 US Department of Labor, Mine Safety and Health Administration, (n.d.): Horns, backup alarms, and automatic warning devices, Title 30 Code of Federal Regulations (30 CFR 56.14132, 57.1413230, 77.410, 77.1605), retrieved from http://www.msha.gov/STATS/Top20Viols/tips/14132.htm

Part E

Automati

Part E Automation Management

Automation Management. Part E The main aspects of automation management are covered by the chapters in this part: Cost effectiveness and economic reasons for the design, feasibility analysis, implementation, rationalization, use, and maintenance of particular automation; performance and functionality measures and criteria, such as quality of service, energy cost, reliability, safety, usability, and other criteria; issues involved with managing automation over its life cycles, and the use of embedded automation for this management; how to best prepare the next generation of automation engineers, practitioners, inventors, developers, scientists, operators, and general users, and how to best prepare and qualify automation professionals. Related also to the above topics are the issues of how to manage automatically and control the increasingly complex and rapidly evolving software assets, their maintenance, replacement, and upgrading; how to simplify the specification of increasingly more integrated, inter-dependent and complex automation; and what are some of the ethical concerns with automation and solutions being developed to prevent abuse of automation and with automation.

This part concludes the first portion of the Handbook, which is devoted to the science, theory, design, and management aspects and challenges of automation. The next portion is devoted to the main functional areas of automation, demonstrating the role of the previously discussed theories, techniques, models, tools and guidelines, as they are applied and implemented specifically and successfully and the obstacles they face in those functional areas.

40. Economic Rationalization of Automation Projects

José A. Ceroni

The future of any investment project is undeniably linked to its economic rationalization. The chance that a project is realized depends on our ability to demonstrate the benefits that it can convey to a company. However, traditional investment evaluation must be enhanced and used carefully in the context of rationalization to reflect adequately the characteristics of modern automation systems. Nowadays automation systems often take the form of complex, strongly related autonomous systems that are able to operate in a coordinated fashion in distributed environments. Reconfigurability is a key factor affecting automation systems' economic evaluation due to the reusability of equipment and software for the manufacturing of several products. A new method based on an analytical hierarchy process for project selection is reviewed. A brief discussion on risk and salvage consideration is included, as are aspects needing further development in future rationalization techniques.

Worldwide adoption of automation advanced technologies such as robotics, flexible manufacturing, and computer-integrated manufacturing systems has been key to continued improvement of competitiveness in present global markets [40.1]. Adequate definition and selection of automation technology offers substantial potential for cost savings, increased flexibility, better product consistency, and higher throughput. However, justifying automation technology based only on traditional economic criteria is at least biased and often wrong. Lack of consideration of automation strategic and long-term benefits has often led to failure in adopting it [40.2–5]. Ignoring automation's long-term benefits and impact on company strategy leads to poor decision-making on technologies to implement. Usually nowadays, the long-range cost of not automating can turn out to be considerably greater than the short-term cost of acquiring automation technology [40.6].

A primary objective of any automation project must be to develop a new integrated system able to provide financial, operational, and strategic benefits. Thus the automation project should avoid replicating current operational methods and support systems. In this way, the automation project should make clear the four differences from any other capital equipment project:

- Automation provides flexibility in production capability, enabling companies to respond effectively to market changes, an aspect with clear economic value.
- Automation solutions force users to rethink and systematically define and integrate the functions of

their operations. This reengineering process creates major economic benefits.

- Modern automation solutions are reprogrammable and reusable, with components often having lifecycles longer than the planned production facility.
- Using automation significantly reduces requirements for services and related facilities.

These differences lead to operational benefits that include:

- Increased flexibility
- Increased productivity
- Reduced operating costs
- Increased product quality
- Elimination of health and safety hazards
- Higher precision
- Ability to run longer shifts
- Reduced floor space.

Time-based competition and mass customization of global markets are key competitive strategies of present-day manufacturing companies [40.7]. Average product lifecycle in marketplaces has changed from years to months for products based on rapidly evolving technologies. This demands agile automated systems, created through the concept of common manufacturing processes organized modularly to allow rapid deployment in alternative configurations. These reconfigurable automated systems represent the cornerstone in dealing with time-based competition and mass customization.

Justification of reconfigurable automation systems must necessarily include strategic aspects when comparing them with traditional manufacturing systems developed under the product-centric paradigm. Product-specific systems generally lack the economical reconfiguration ability that would allow them to meet the needs of additional products. Consequently traditional systems are typically decommissioned well before their capital cost can be recovered, and are then held in storage until fully depreciated for tax purposes before being sold at salvage value. However, it must be kept in mind that reconfigurability claims for additional investment in design, implementation, and operation of the system. Quick-change tooling are an example of reconfigurability equipment which allows rapid product changeover. It is estimated that generic system capabilities can increase the cost of reconfigurable system hardware by as much as 25% over that of a comparable dedicated system. On the other hand, software required to configure and run reconfigurable automation systems is often much more expensive to develop than simple part-specific programs. Traditional economic evaluation methods fail to consider benefits from capital reutilization over multiple projects and also disregard strategic benefits of technology. Upgrade of traditional economic evaluation methods is required to account for the short-term economic and long-term strategic value of investing in reconfigurable automation technologies to support the evolving production requirements of a family of products. In this chapter the traditional economic justification approach to automation system justification is first addressed. A related discussion on economic aspects of automation not discussed in this chapter can be found in Chap. 7. New approaches to automated system justification based on strategic considerations are presented next. Finally, a discussion on justification approaches currently being researched for reconfigurable systems is presented.

40.1 General Economic Rationalization Procedure

In general terms, an economic rationalization enables us to compare the financial benefits expected from a given investment project with alternative use of investment capital. Economic evaluation measures capital cost plus operating expenses against cash-flow benefits estimated for the project. This section describes a general approach to economic rationalization and justification of automation system projects.

40.1.1 General Procedure for Automation Systems Project Rationalization

The general procedure for rationalization and analysis of automation projects presented here consists of a pre-cost-analysis phase, followed by a cost-analysis phase. Figure 40.1 presents the procedure steps and their sequence. The sequence of steps in Fig. 40.1 is reviewed in detail in the rest of this section and an example cost-analysis phase is described.

40.1.2 Pre-Cost-Analysis Phase

The pre-cost-analysis phase evaluates the feasibility of the automation project. Feasibility is evaluated in terms of the technical capability to achieve production capacity and utilization as estimated in production schedules. The first six steps of the procedure include determining the most suitable manufacturing method, selecting the tasks to automate, and the feasibility of these options (Fig. 40.1). Noneconomic considerations must be studied and all data pertinent to product volumes and operation times gathered.

Alternative Automated Manufacturing Methods

Production unit cost at varying production volumes for three main alternative manufacturing methods (manual labor, flexible automation, and hard automation) are compared in Fig. 40.2 [40.8]. Manual labor is usually the most cost-effective method for low production volumes; however, reconfigurable assembly is changing this situation drastically. Flexible, programmable automation is most effective for medium production volumes, ranging from a few tens or hundreds of products per year per part type to hundreds of thousands of products per year. Finally, annual production volumes of 500 000 or above seem to justify the utilization of hard automation systems.

Boothroyd et al. [40.9] have derived specific formulas for assembly cost (Table 40.1). By using these formulas they compare alternative assembly systems such as the one-operator assembly line, assembly center with two arms, universal assembly center, free-transfer machine with programmable workheads, and dedicated machine. The last three systems are robotics-based automated systems. The general expression derived by [40.9] is

$$C_{pr} = t_{pr}\left(W_t + \frac{WM'}{SQ}\right),$$

where C_{pr} = unitary assembly cost, t_{pr} = average assembly time per part, W_t = labor cost per time, W = operator's rate in dollars per second, M' = assembly equipment cost per time, S = number of shifts, and Q = operator cost in terms of capital equivalent.

Parameters and variables in this expression present alternative relationships, depending on the type of assembly system.

Figures 40.3 and 40.4 show the unitary cost for the assembly systems at varying annual production volumes. It can be seen that production of multiple products increases costs, by approximately 100%, for the assembly center with two arms and free-transfer ma-

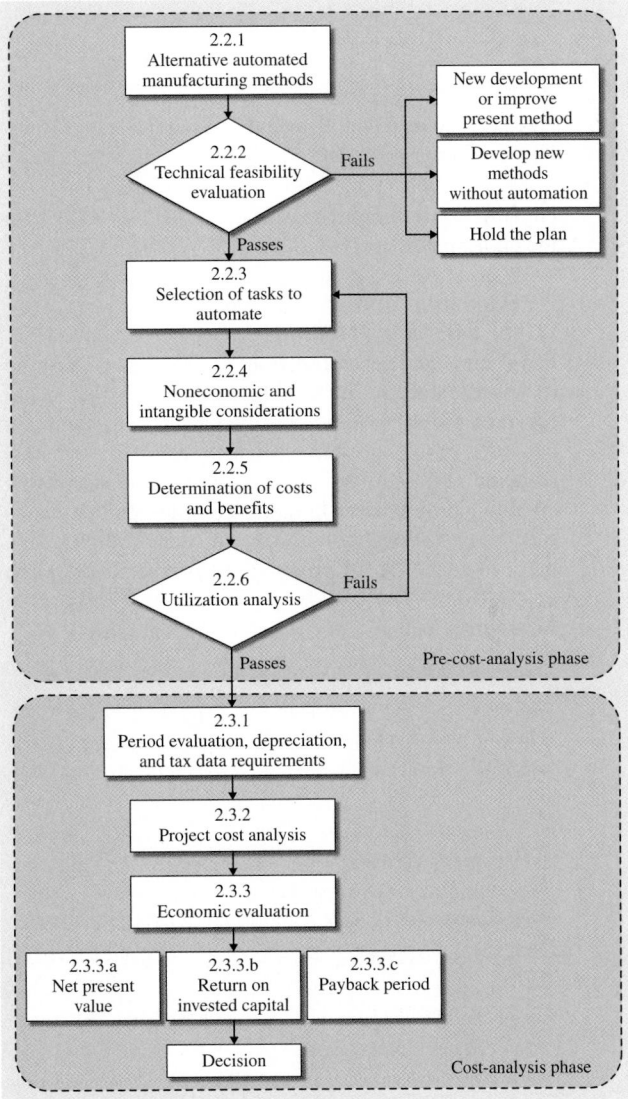

Fig. 40.1 Automation project economic evaluation procedure

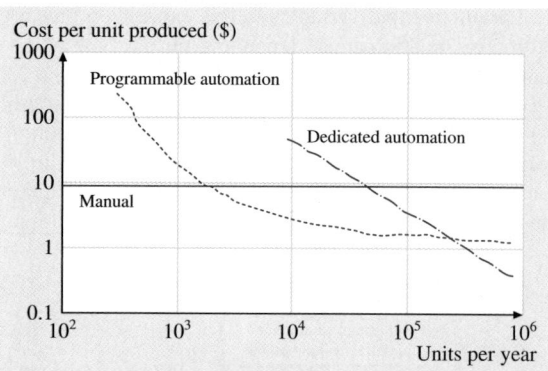

Fig. 40.2 Comparison of manufacturing methods for different production volumes

chine with programmable workheads, and by 1000% for the dedicated hybrid machine.

Evaluation of Technical Feasibility for Alternative Methods

Feasibility of the automation system plan must be reviewed carefully. It is perfectly possible for an automation project to have a positive economic evaluation but have problems with its feasibility. Although this situation may seem strange it must be considered that an automation project is rather complex and demands specific operational conditions, far more complex than those in conventional production systems. A thorough

feasibility review must consider aspects such as the answers to the following questions in case of automated assembly:

- Is the product designed for automated assembly?
- Is it possible to do the job with the planned procedure and within the given cycle time?
- Can reliability be ensured as a component of the total system?
- Is the system sufficiently staffed and operated by assigned engineers and operators?
- Is it possible to maintain safety and the designed quality level?
- Can inventory and material handling be reduced in the plant?
- Are the material-handling systems adequate?
- Can the product be routed in a smooth batch-lot flow operation?

Following the feasibility analysis, alternatives are considered further in the evaluation. If the plan fails due to lack of feasibility, a search for other type of solutions is in order. Alternative solutions may involve the development of new equipment, improvement of proposed equipment or development of other alternatives.

Selection of Tasks to Automate

Selection of tasks for automation is a difficult process. The following five job grouping strategies may assist the determination of tasks to automate:

Table 40.1 Comparison of assembly systems cost

Assembly system	t_{pr}	W_t	M'
Operator assembly line no feeders	$Kt_0(1+x)$	nW/k	$(n/k)(2C_B + N_pC_C)$
Operator assembly line with feeders	$Kt_0'(1+x)$	nW/k	$(n/k)(2C_B + N_pC_C) + N_p(ny + N_dC_F)$
Dedicated hybrid machine	$t + xT$	$3W$	$[nyC_T + (T/t)(ny)C_B] + N_p\{(ny + N_d)(C_F + C_w)$
			$+[ny + (T/2t)(ny)]C_C\}$
Free-transfer machine	$k(t + xT)$	$3W$	$(n/k)[C_{dA} + (T/t + 1)C_B + N_p[(ny + N_d)C_M + nC_g$
with programmable workheads			$+(n/k)(T/2t + 0.5)C_C]$
Assembly center with two arms	$n(t/2 + xT)$	$3W$	$2C_{dA} + N_p[C_C + nC_g + (ny + N_d)C_M]$
Universal assembly center	$n(t/2 + xT)$	$3W$	$(2C_{dA} + nyC_{PF} + 2C_{ug}) + N_pC_C$

where: C_{pr} = product unit assembly cost, S = number of shifts, Q = equivalent cost of operator in terms of capital equivalent, W = operator's rate in dollars per second, d = degrees of freedom, k = number of parts assembled by each operator or programmable workhead, n = total number of parts, N_p = number of products, N_d = number of design changes, C_{dA} = cost of a programmable robot or workhead, C_B = cost of a transfer device per workstation, C_C = cost of a work carrier, C_F = cost of an automatic feeding device, C_g = cost of a gripper per part, C_M = cost of a manually loaded magazine, C_{PF} = cost of a programmable feeder, C_S = cost of a workstation for a single station assembly, C_t = cost of a transfer device per workstation, C_{ug} = cost of a universal gripper, C_W = cost of a dedicated workhead, T = machine downtime for defective parts, t_0, t_0' = machine downtime due to defective parts, t = mean time of assembly for one part, x = ratio of faulty parts to acceptable parts, and y = product styles

- Components of products of the same family
- Products presently being manufactured in proximity
- Products consisting of similar components that could share part-feeding devices
- Products of similar size, dimensions, weight, and number of components
- Products with simple design possible to manufacture within a short cycle time.

Noneconomic and Intangible Considerations

Issues related to specific company characteristics, company policy, social responsibility, and management policy need to be addressed both quantitatively and qualitative in the automation project. Adequate justification of automation systems needs to consider aspects such as:

- Compliance with the general direction of the company's automation
- Satisfaction of equipment and facilities standardization policies
- Adequate accommodation of future product model changes or production plans
- Improvement of working life quality and workers morale
- Positive impact on company reputation
- Promotion of technical progress at the company.

Special differences among automated solutions (e.g., robots) and other case-specific capitalization equipment also provide numerous intangible benefits, as the following list illustrates:

- Robots are reusable.
- Robots are multipurpose and can be reprogrammed for many different tasks.
- Because of reprogrammability, robotic systems service life can often be three or more times longer than that of fixed (hard) automation devices.
- Tooling costs for robotic systems also tend to be lower owing to the programming capability around certain physical constraints.
- Production startup occurs sooner because of less construction and tooling constraints.
- Plant modernization can be implemented by eliminating discontinued automation systems.

Determination of Costs and Benefits

Although costs and benefits expected from automation projects vary according to each particular case being analyzed, a general classification of costs can

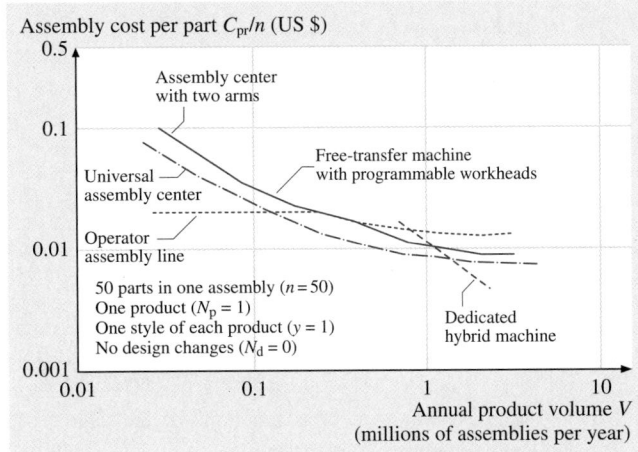

Fig. 40.3 Comparison of alternative assembly systems (one product)

include operators wages, capital, maintenance, design, and power costs. However, it must be noted that, while usually wages decrease at higher levels of automation, the rest of the cost tend to increase. Figure 40.5 shows the behavior of assembly costs at different automation levels [40.10]. Consequently, it would be possible to determine the optimal degree of automation based on the minimum total operational cost of the system.

Questions regarding benefits of automation systems often arise concerning long-range, unmeasurable effects on economic issues. A few such issues include the impact of the automation system on:

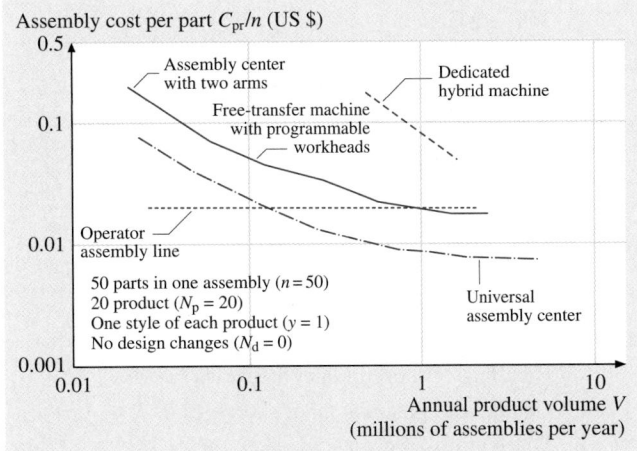

Fig. 40.4 Comparison of alternative assembly systems (20 products)

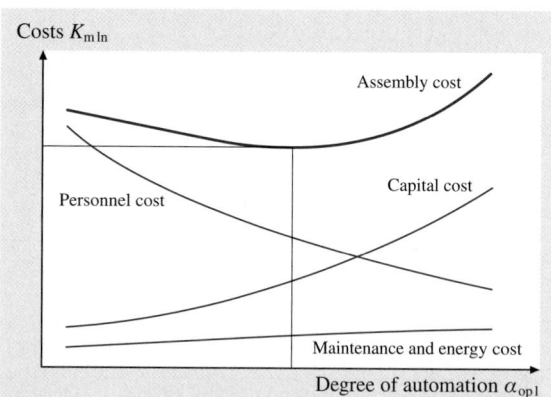

Fig. 40.5 Assembly costs as functions of the degree of automation in the assembly case

- Product value and price
- Increase of sales volume
- Decrease of production cost
- Decrease of initial investment requirements
- Reduction of products lead time
- Decrease of manufacturing costs
- Decrease of inventory costs

- Decrease of direct and indirect labor costs
- Decrease of overhead rate
- Full utilization of automated equipment
- Decrease of setup time and cost
- Decrease of material-handling cost
- Decrease of damage and scrap costs.

Table 40.2 lists additional difficult-to-quantify benefits usually associated with automation projects.

Utilization Analysis

Underutilized automated systems usually cannot be cost-justified, mainly due to the high initial startup expenses and low labor savings they result in. Consideration of additional applications or planned future growth are required to drive the potential cost-effectiveness up; however, there are also additional costs to consider, for example, tooling and feeder costs associated with new applications.

40.1.3 Cost–Analysis Phase

This phase of the methodology focuses on detailed cost analysis for investment justification and includes five

Table 40.2 Difficult-to-quantify benefits of automation. Analyzing the amount of change in each of these categories in response to automation and assigning quantitative values to these intangible factors is necessary if they are to be included in the financial analysis. Otherwise they can only be used as weighting factors when determining the best alternative

Automation can improve	Automation can reduce or eliminate
Flexibility	Hazardous, tedious jobs
Plant modernization	Safety violations and accidents
Labor skills of employees	Personnel costs for training
Job satisfaction	Clerical costs
Methods and operations	Cafeteria costs
Manufacturing productivity capacity	Need for restrooms, need for parking spaces
Reaction to market fluctuations	Burden, direct, and other overhead costs
Product quality	Manual material handling
Business opportunities	Inventory levels
Share of market	Scrap and errors
Profitability	New product launch time
Competitive position	
Growth opportunities	
Handling of short product lifecycles	
Handling of potential labor shortages	
Space utility of plant	
Level of management	

Table 40.3 Example data

Project costs	
Machine cost	$ 80 000
Tooling cost	$ 13 000
Software integration	$ 30 000
Part feeders	$ 20 000
Installation cost	$ 25 000
Total	$ 168 000
Actual realizable salvage	$ 12 000

Table 40.4 MACRS percentages

Year	Percentage
1	20.00
2	32.00
3	19.20
4	11.52
5	11.52
6	5.76

steps (Fig. 40.1). To evaluate economically the automation project installation, the following data are required:

- Capital investment of the project
- Estimated changes in gross incomes (revenue, sales, savings) and costs expected from the project.

To illustrate the remaining steps of the methodology, an example will be developed. Installation cost, operation costs, and salvage for the example are as given in Table 40.3.

Period Evaluation, Depreciation, and Tax Data Requirements

Before proceeding with the economic evaluation, the evaluation period, tax rates, and tax depreciation method must be specified. We will consider in the example an evaluation period of 6 years. We will use the US Internal Revenue Service's Modified Accelerated Cost Recovery System (MACRS) for 5 years (Table 40.4). Tax rate considered is 40%. These values are not fixed and can be changed if deemed appropriate.

Project Cost Analysis

The project cost is as given in Table 40.3 (US$ 168 000). To continue it is necessary to determine (estimate) the yearly changes in operational cost and cost savings (benefits). For the example these are as shown in Table 40.5.

Economic Rationalization

Techniques used for the economic analysis of automation applications are similar to those for any manufacturing equipment purchase. They are usually based on net present value, rate of return or payback (payout) methods. All of these methods require the determination of the yearly net cash flows, which are defined as

$$X_j = (G - C)_j - (G - C - D)_j(T) - K + L_j ,$$

where X_j = net cash flow in year j, G_j = gross income (savings, revenues) for year j, C_j = total costs for year j, D_j = tax depreciation for year j, T = tax rate (as-

Table 40.5 Costs and savings (dollars per year)

Year	1	2	3	4	5	6
Labor savings	70 000	70 000	88 000	88 000	88 000	88 000
Quality savings	22 000	22 000	28 000	28 000	28 000	28 000
Operating costs (increase)	(25 000)	(25 000)	(19 000)	(12 000)	(12 000)	(12 000)

Table 40.6 Net cash flow

End of year	K&L	Total G[a]	C	D[b]	X
0	168 000	–	–	–	−168 000
1		92 000	25 000	33 600	53 640
2		92 000	25 000	53 760	61 704
3		116 000	19 000	32 256	71 102
4		116 000	12 000	19 354	70 142
5		116 000	12 000	19 354	70 142
6	L = 12 000	116 000	12 000	9677	78 271

[a]: These are the sums of labor and quality savings (Table 40.5)
[b]: Computed with the MACRS for each year

Table 40.7 Pairwise comparison of criteria and weights

Criteria	Price	Weight	Power	Spindle	Diameter	Stroke
Price	1.00	0.52	0.62	0.41	0.55	0.59
Weight	1.93	1.00	2.60	1.21	1.30	2.39
Power	1.61	0.38	1.00	0.40	0.41	1.64
Spindle	2.44	0.83	2.47	1.00	0.40	0.46
Diameter	1.80	0.77	2.40	2.49	1.00	2.40
Stroke	1.69	0.42	0.61	2.17	0.42	1.00

sumed constant), K = project cost (capital expenditure), and L_j = salvage value in year j.

The net cash flows are given in Table 40.6.

Net Present Value (NPV). Once the cash flows have been determined, the net present value (NPV) is determined using the equation

$$\text{NPV} = \sum_{j=0}^{n} \frac{X_j}{(1+k)^j} = \sum_{j=0}^{n} X_j(P/F, k, j),$$

where X_j = net cash flow for year j, n = number of years of cash flow, k = minimum acceptable rate of return (MARR), and $1/(1+k)^j$ = discount factor, usually designated as $(P/F, k, j)$.

With the cash flows of Table 40.6 and $k = 25\%$, the NPV is

$$\text{NPV} = -168\,000 + 53\,640(P/F, 25, 1)$$
$$+ 61\,704(P/F, 25, 2) + \cdots$$
$$+ 78\,271(P/F, 25, 6) = \$\,18\,431\,.$$

The project is economically acceptable if its NPV is positive. Also, a positive NPV indicates that the rate of return is greater than k.

Return on Invested Capital (ROIC). The ROIC or rate of return is the interest rate that makes the NPV = 0. It is sometimes also referred to as the internal rate of return (IRR). Mathematically, the ROIC is defined as

$$0 = \sum_{j=0}^{n} \frac{X_j}{(1+i)^j} = \sum_{j=0}^{n} X_j(P/F, i, j),$$

where i = ROIC.

For this example the ROIC is determined from the following expression

$$0 = -168\,000 + 53\,640(P/F, i, 1)$$
$$+ 61\,704(P/F, i, 2) + \cdots + 78\,271(P/F, i, 6)\,.$$

To solve the previous expression for i, a trial-and-error approach is needed. Assuming 25%, the right-hand side gives $\$\,18\,431$ (NPV calculation) and with 35% it is $\$\,-11\,719$. Therefore the ROIC using linear interpolation is approximately 31%. This ROIC is now compared with the minimum acceptable rate of return (MARR). In this example the MARR is that used for calculating the NPV. If ROIC $\geq$ MARR the project is acceptable; otherwise it is unacceptable. Consequently the NPV and the rate-of-return methods will give the same decision regarding the economic desirability of a project (investment). It is pointed out that the definitions of cash flow and MARR are not independent. Also, the omission of debt interest in the cash-flow equation does not necessarily imply that the initial project cost (capital expenditure) is not being financed by some combination of debt and equity capital. When total cash flows are used, the debt interest is included (approximately) in the definition of MARR as

$$\text{MARR} = k_e(1 - c) + k_d(1 - T)c\,,$$

where k_e = required return for equity capital, k_d = required return for debt capital, T = tax rate, and c = debt ratio of the pool of capital used for current capital investments.

It is not uncommon in practice to adjust (increase) k_e and k_d to account for project risk and uncertainties in economic conditions.

The effects of automation on ROIC (Fig. 40.6) are documented elsewhere in the literature [40.11]. The main effects of automation can be classified into reduction of capital or increased profits or, more desirably, both simultaneously. Automation may generate investment capital savings in project engineering, procurement costs, purchase price, installation, configuration, calibration or project execution. Working capital requirements may be lowered by reducing raw material (quantity or price), product inventories, spares parts for equipment, reduced energy and utilities utilization or

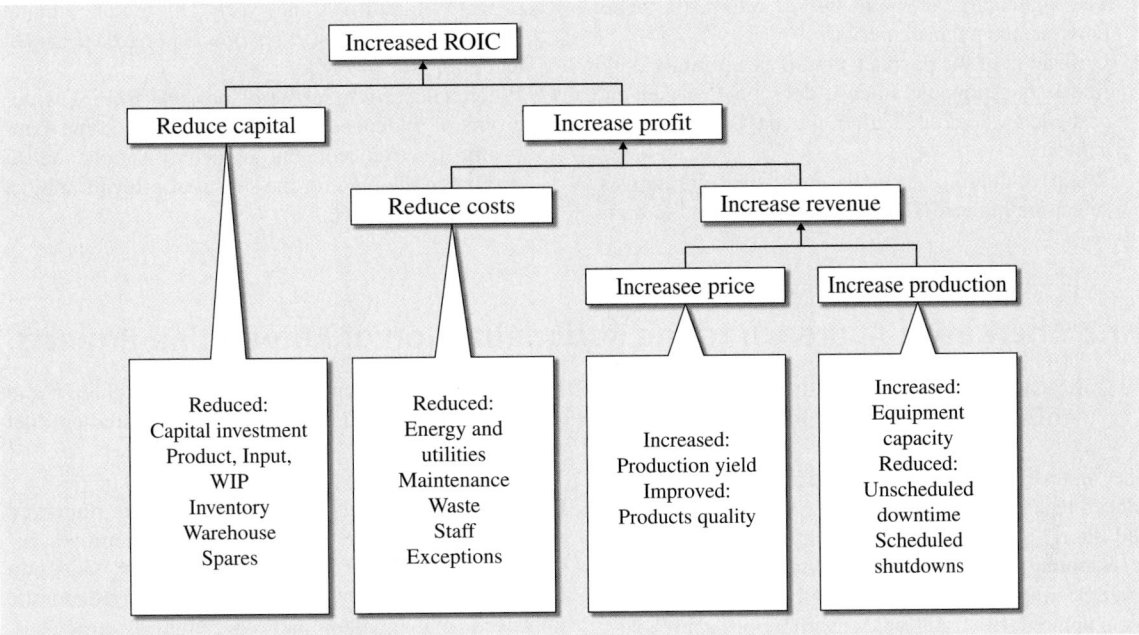

Fig. 40.6 Effects of automation on ROIC

increased product yields. Maintenance cost are diminished in automation solutions by reducing unscheduled maintenance, number of routine checks, time required for maintenance tasks, materials purchase, and number and cost of scheduled shutdown tasks. Automation also contributes to reduce impacts (often hard to quantify) due to health, safety, and environmental issues in production systems.

Profits could increase due to automation by increasing the yield of more valuable products. Reduced work-in-process inventory and waste result in higher revenue per unitary input to the system. Although higher production yield will be meaningful only if the additional products can be sold, today's global markets will surely respond positively to added production capacity.

Payback (Payout) Period. An alternative method used for economic evaluation of a project is the payback period (or payout period). The payback period is the number of years required for incoming cash flows to balance the cash outflows. The payback period (p) is obtained from the expression

$$0 = \sum_{j=0}^{p} X_j \,.$$

This is one definition of the payback period, although an alternative definition that employs a discounting procedure is most often used in practice. Using the cash flow given in Table 40.6, the payback equations for 2 years gives

$$-168\,000 + 53\,640 + 61\,704 = \$-52\,656$$

and for 3 years it is

$$-168\,000 + 53\,640 + 61\,704 + 71\,102 = \$18\,446 \,.$$

Therefore, using linear interpolation, the payback period is

$$p = 2 + \frac{52\,656}{71\,102} = 2.74 \text{ years} \,.$$

40.1.4 Additional Considerations

The following aspects must be kept in mind when applying the general procedure and related techniques described in this section. Careful review of these issues will ensure that the evaluation is proven correct:

● Cash-flow equation component values are incremental. They represent increases or decreases resulting directly from the project (investment) under consideration.

- The higher the NPV and rate of return, the better (shorter) the payback period.
- Utilization of the payback period as a primary criterion is questionable since it does not consider the cash flows generated after the payback period is achieved.
- When evaluating mutually exclusive alternatives, select the highest NPV alternative. Using the highest rate of return is incorrect. This point is made clear by *Stevens* [40.12], *Blank* [40.13], *Thuesen* and *Fabrycky* [40.14].
- When selecting a subset of projects from a larger group of independent projects due to some constraint (restriction), the objective should be to maximize the NPV of the subset of projects subject to the constraint(s).

40.2 Alternative Approach to the Rationalization of Automation Projects

40.2.1 Issues in Strategic Justification of Advanced Technologies

The analysis typically performed for justifying advanced technology as outlined in Sect. 40.1 is financial and short term in nature. This has caused difficulty in adopting systems, and technology having both strategic implications and intangible benefits usually not captured by traditional justification approaches. Elsewhere in the literature a number of other issues that make justification and adoption of strategic technologies difficult can be found, including high capital costs and risks, difficulty in quantifying indirect and intangible benefits, inappropriate capital budgeting procedures, and technological uncertainties [40.15]. Another complicating factor in the justification of integrated technologies is the cultural and organizational issues involved. The impact of implementing a flexible manufacturing system crosses many organizational boundaries. The success or failure of this implementation depends on the buy-in of all organizations and individuals involved. Traditional methods of justification often do not consider these organizational impacts and are not designed for group consensus building.

The literature discusses many of the intangible and nonquantifiable benefits of implementing automated systems; the most often mentioned is flexibility. *Zald* [40.16] discusses four kinds of flexibility provided by automated systems:

- Mix flexibility: the ability to have multiple products in the same product process at the same time
- Volume flexibility: the ability to change the process so that additional or less throughput is achieved
- Multifunction flexibility: the ability to have the same device do different tasks by changing tools on the device
- New product flexibility: the ability to change and reprogram the process as the manufactured product changes.

Other frequently mentioned benefits include improved product quality, better customer service, improved response time, improved product consistency, reduction in inventories, improved safety, better employee morale, improved management and operation of processes, shorter cycle times and setups, and support for continuous improvement and just-in-time (JIT) efforts [40.3–6].

40.2.2 Analytical Hierarchy Process (AHP)

Rarely do automation systems comprise out-of-the-box solutions. In fact, most automated systems nowadays comprise a collection of equipment properly integrated into an effective solution. This integration process makes evaluation of alternative solutions more complex, due to the many combinations possible for the configuration of all available components. To assist the equipment selection process AHP has been implemented in the form of decision support systems (DSSs) [40.17]. AHP was developed by *Saaty* [40.18] as a way to convey the relative importance of a set of activities in a quantitative and qualitative multicriteria decision problem. The AHP method is based on three principles: the structure of the model [40.19], comparative judgment of alternatives and criteria [40.20], and synthesis of priorities. Despite the wide utilization of AHP, selection of casting process [40.21], improvement of human performance in decision-making [40.19], and improvement of quality-based investments [40.22], the method has shortcomings related to its inability to handle decision-maker's uncertainty and imprecision in determining values for the pairwise comparison process involved. Another difficulty with AHP lies in the fact

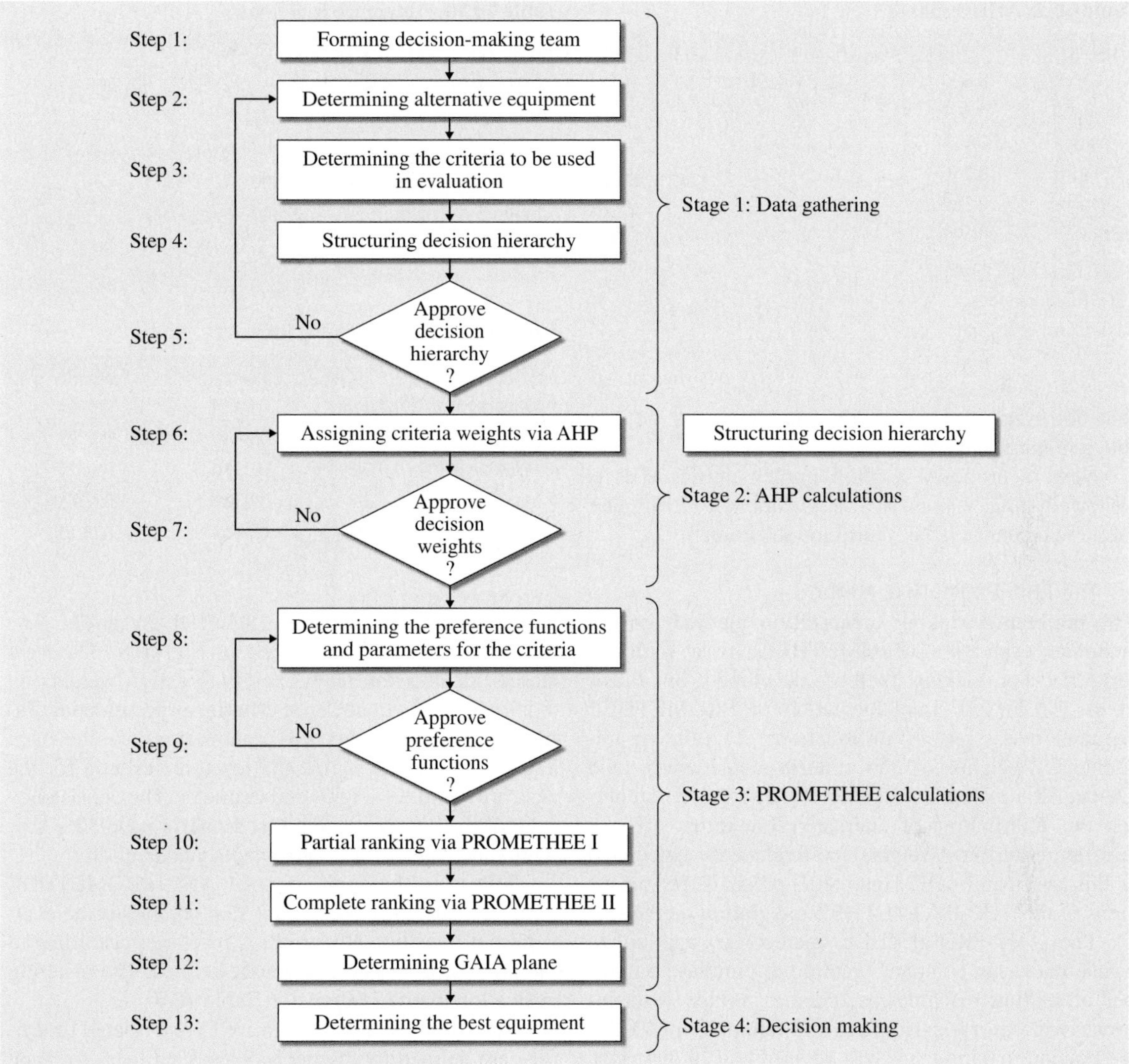

Economic Rationalization of Automation Projects | 40.2 Alternative Approach to the Rationalization of Automation Projects 709

Part E | 40.2

Step 1: Forming decision-making team

Step 2: Determining alternative equipment

Step 3: Determining the criteria to be used in evaluation

Step 4: Structuring decision hierarchy

Step 5: Approve decision hierarchy ? No

⎫ Stage 1: Data gathering

Step 6: Assigning criteria weights via AHP

Structuring decision hierarchy

Step 7: Approve decision weights ? No

⎫ Stage 2: AHP calculations

Step 8: Determining the preference functions and parameters for the criteria

Step 9: Approve preference functions ? No

Step 10: Partial ranking via PROMETHEE I

Step 11: Complete ranking via PROMETHEE II

Step 12: Determining GAIA plane

⎫ Stage 3: PROMETHEE calculations

Step 13: Determining the best equipment

⎫ Stage 4: Decision making

Fig. 40.7 Steps of AHP-PROMETHEE method

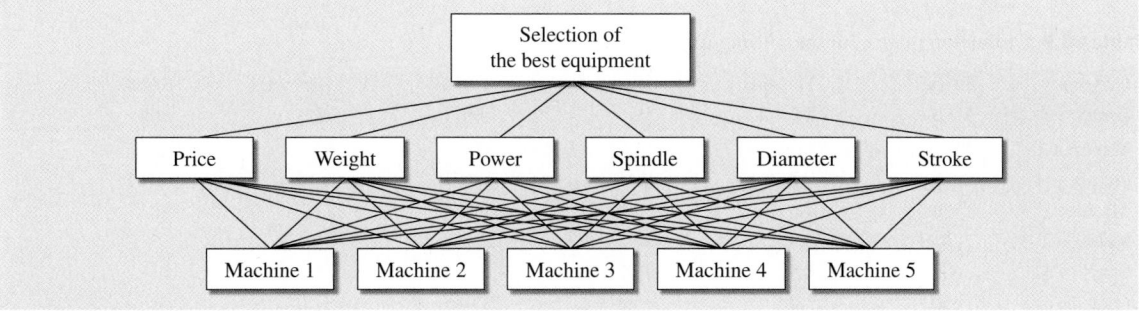

Fig. 40.8 Decision structure

Table 40.8 AHP results

Criteria	Weights (ω)	$\lambda_{\max}$, CI, RI	CR
Price	0.090	$\lambda_{\max} = 6.201$	0.032
Weight	0.244		
Power	0.113	CI = 0.040	
Spindle	0.266		
Diameter	0.186	RI = 1.24	
Stroke	0.101		

CR: Consistency ratio
RI: Random index
CI: Consistency index

Table 40.10 Preference functions

Criteria	PF	Thresholds		
		q	p	s
Price	Level	600	800	–
Weight	Gaussian	–	–	4
Power	Level	800	1200	–
Spindle	Level	20 000	23 000	–
Diameter	Gaussian	–	–	6
Stroke	V-shape	–	50	–

Table 40.11 PROMETHEE flows

Alternatives	Φ^+	Φ^-	Φ
Machine 1	0.0199	0.0139	0.0061
Machine 2	0.0553	0.0480	0.0073
Machine 3	0.0192	0.0810	−0.0618
Machine 4	0.0298	0.0130	0.0168
Machine 5	0.0478	0.0163	0.0315

that not every decision-making problem may be cast into a hierarchical structure.

Next, a proposed method implementing AHP is reviewed using a numerical application for computer numerical control (CNC) machine selection.

The AHP–PROMETHEE Method

The preference ranking organization method for enrichment evaluation (PROMETHEE) is a multicriteria decision-making method developed by *Brans* et al. [40.23, 24]. Implementation of PROMETHEE requires two types of information: (1) relative importance (weights of the criteria considered), and (2) decision-maker's preference function (for comparing the contribution of alternatives in terms of each separate criterion). Weights coefficients are calculated in this case using AHP. Figure 40.7 presents the various steps of the AHP-PROMETHEE integration method.

The AHP-PROMETHEE method is applied to a manufacturing company wanting to purchase a number of milling machines in order to reduce work-in-process inventory and replace old equipment [40.17]. A decision-making team was devised and its first task was to determine the five milling machines candidates for the purchasing and six evaluation criteria:

price, weight, power, spindle, diameter, and stroke. The decision structure is depicted in Fig. 40.8. The next step is for decision team experts to assign weights on a pairwise basis to decision criteria, as presented in Table 40.7. Results from AHP calculations are shown in Table 40.8 and show that the top three criteria for the case are spindle, weight, and diameter. The consistency ratio of the pairwise comparison matrix is 0.032 < 0.1, which indicates weights consistency and validity.

Following the application of AHP, PROMETHEE steps are carried out. The first step comprises the evaluation of five alternative milling machines according to the evaluation criteria previously defined. The resulting evaluation matrix is shown in Table 40.9.

Next, a preference function (PF) and related thresholds are defined by the decision-making team for each criterion. PF and thresholds consider features of the milling machines and the company's purchasing pol-

Table 40.9 Evaluation matrix for the milling machine case

Criteria	Price	Weight	Power	Spindle	Diameter	Stroke
Unit	US $	kg	W	rpm	mm	mm
Max/min	Min	Min	Max	Max	Max	Max
Weight	0.090	0.244	0.113	0.266	0.186	0.101
Machine 1	936	4.8	1300	24 000	12.7	58
Machine 2	1265	6.0	2000	21 000	12.7	65
Machine 3	680	3.5	900	24 000	8.0	50
Machine 4	650	5.2	1600	22 000	12.0	62
Machine 5	580	3.5	1050	25 000	12.0	62

Economic Rationalization of Automation Projects | 40.3 Future Challenges and Emerging Trends in Automation Rationalization 711

Part E | 40.3

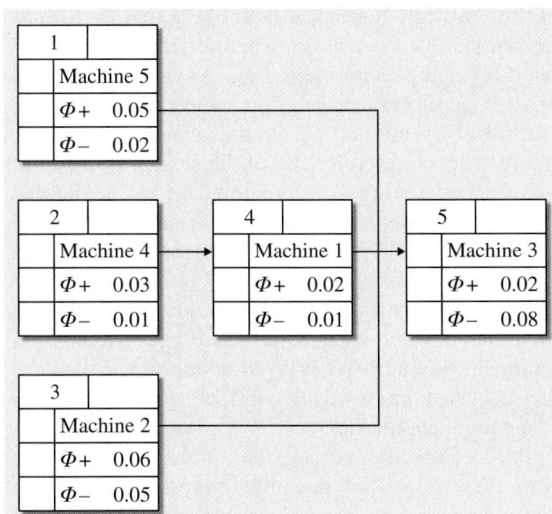

Fig. 40.9 PROMETHEE I partial ranking

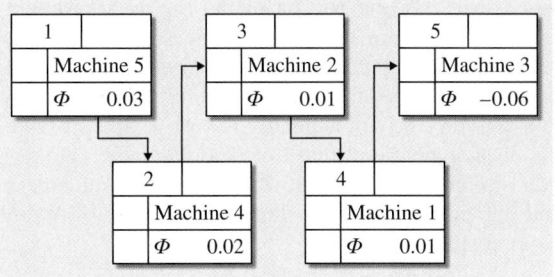

Fig. 40.10 PROMETHEE II complete ranking

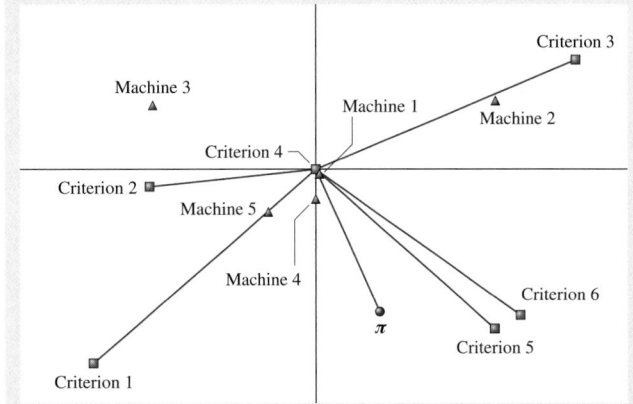

Fig. 40.11 GAIA decision plane

icy. Table 40.10 shows preference functions and their thresholds.

The partial ranking of alternatives is determined according to PROMETHEE I, based on the positive and negative flows shown in Table 40.11. The resulting partial ranking is shown in Fig. 40.9 and reveals that machine 5, machine 2, machine 4, and machine 1 are preferred over machine 3, and machine 4 is preferred over machine 1. The partial ranking also shows that machine 5, machine 4, and machine 2 are not comparable, as well as machine 5 and machine 1, and machine 2 and machine 1.

PROMETHEE II uses the net flow in Table 40.11 to compute a complete ranking and identify the best alternative. According to the complete ranking, machine 5 is selected as the best alternative, while the other machines are ranked accordingly as machine 4, machine 2, machine 1, then machine 3 (Fig. 40.10).

The geometrical analytic for interactive aid (GAIA) plane [40.25] representing the decision (Fig. 40.11) shows that: price has great differentiation power, criteria 1 (price) and 3 (power) are conflicting, machine 2 is very good in terms of criterion 3 (power), and machine 3 is very good in terms of criteria 2 and 4. The vector π (decision axis) represents the compromise solution (selection must be in this direction).

40.3 Future Challenges and Emerging Trends in Automation Rationalization

40.3.1 Adjustment of Minimum Acceptable Rate of Return in Proportion to Perceived Risk

Since capital investments involve particular levels of risk, it is common practice for management to increase the MARR for automation projects involving higher risk. Assigning a higher MARR forces the proposed projects to generate greater return on their capital investment. A similar strategy is proposed to recognize the fact that capital equipment that can be reused for additional projects is less likely to experience a decline in its value due to unforeseen reductions in a given product's demand and market life.

Traditionally, more stringent MARR requirements are applied to the entire capital investment. It is proposed that only the portion of the capital investment that is at risk due to sudden changes in market demand or operating conditions should be forced to meet these more demanding rate of return. This approach to assigning risk will explicitly recognize and promote the development and reuse of reconfigurable automation by compensating for its higher initial development and implementation costs through lower rate-of-return requirements.

40.3.2 Depreciation and Salvage Value Profiles

Mechanisms for capital depreciation and estimated project salvage values also have a significant effect on the financial justification of automation. Two mechanisms of depreciation must be considered: tax and book depreciation. Tax depreciation methods provide a systematic mechanism to acknowledge the reduction of capital asset value over time. Since depreciation is tax-deductible, it is generally in the best interests of the company to depreciate the asset as quickly as possible. Allowable depreciation schedules are determined by the country tax code. Tax depreciation can become a factor when comparing product-specific automation for reconfigurable systems when the projected product life is less than the legislated tax depreciation life. Under these circumstances the product-specific systems can be sold on the open market and the remaining tax depreciation would be forfeited. A second alternative is that the system would be decommissioned and stored by the company until it is fully depreciated and then sold for salvage. Instead, it is unlikely that the life of a reconfigurable system would be shorter than the established tax depreciation period due to its redeployment for a new application project. Tax depreciation terms are determined by tax and accounting conventions rather than the expected service life of the asset. Book depreciation schedules are therefore developed to predict the realizable salvage value at the end of an asset's useful life. Two approaches for determining realizable salvage value can be utilized: the internal asset value in the organization (when it is used in an additional project) and the asset value when sold for salvage. Automation equipment profitably redeployed within the same company is clearly more valuable to that organization than to an equipment reseller.

Product-specific automation, when sold off for scrap, has value for potential buyers only due to their interest in key system components. Product- and process-specific tooling will represent very little value for equipment resellers and, most likely, all the custom engineering and software development required to field the system will be lost. On the other hand, if the system has been developed based on modular components, well understood by the user and other manufacturing organizations, it may represent more value than a completely new set of system components. A redeployed system may result in lower time and cost to provide useful automation resources. Investments made in the development of process technology may be of value in subsequent projects [40.26]. Application software, if developed in a modular fashion, also has the potential for reutilization.

40.4 Conclusions

Although computation of economic performance indicators for automation projects is often straightforward, rationalization of automation technology is fraught with difficulty and many opportunities for long-term improvements are lost because purely economic evaluation apparently showed no direct economic benefit. Modern methods, taking into account risks involved in technology implementation or comparison of complex projects, are emerging to avoid such high-impact mistakes. In this chapter, in addition to providing the traditional economic rationalization methodology, strategic considerations are included plus a discussion on current trends of automation systems towards agility and reconfigurability.

References

40.1   S. Brown, B. Squire, K. Blackmon: The contribution of manufacturing strategy involvement and alignment to world-class manufacturing performance, Int. J. Oper. Prod. Manag. **27**(3–4), 282–302 (2007)

40.2   A.M.A. Al-Ahmari: Evaluation of CIM technologies in Saudi industries using AHP, Int. J. Adv. Manuf. Technol. **34**(7–8), 736–747 (2007)

40.3   K. Feldmann, S. Slama: Highly flexible assembly – scope and justification, CIRP Ann. Manuf. Technol. **50**(2), 489–498 (2001)

40.4   D. Dhavale: Justifying manufacturing cells, Manuf. Eng. **115**(6), 31–37 (1995)

40.5   J.F. Kao, J.L. Sanders: Analysis of operating policies for manufacturing cells, Int. J. Prod. Res. **33**(8), 2223–2239 (1995)

40.6   R. Quaile: What does automation cost – calculating total life-cycle costs of automated production equipment such as automotive component manufacturing systems isn't straightforward, Manuf. Eng. **138**(5), 175 (2007)

40.7   D. Vazquez-Bustelo, L. Avella, E. Fernandez: Agility drivers, enablers and outcomes – empirical test of an integrated agile manufacturing model, Int. J. Oper. Prod. Manag. **27**(12), 1303–1332 (2007)

40.8   J.J. Mills, G.T. Stevens, B. Huff, A. Presley: Justification of robotics systems. In: *Handbook of Industrial Robotics*, ed. by S.Y. Nof (Wiley, New York 1999) pp. 675–694

40.9   G. Boothroyd, C. Poli, L.E. Murch: *Automatic Assembly* (Marcel Dekker, New York 1982)

40.10 S.Y. Nof, W.E. Wilhelm, H.-J. Warnecke: *Industrial Assembly* (Chapman Hall, London 1997)

40.11 D.C. White: Calculating ROI for automation projects, Emerson Process Manag. (2007), available at www.EmersonProcess.com/solutions/ Advanced Automation (Last access date: March 20, 2009)

40.12 G.T. Stevens Jr.: *The Economic Analysis of Capital Expenditures for Managers and Engineers* (Ginn, Needham Heights 1993)

40.13 L.T. Blank, A.J. Tarquin: *Engineering Economy*, 6th edn. (McGraw-Hill, New York 2004)

40.14 G.J. Thuesen, W.J. Fabrycky: *Engineering Economy*, 9th edn. (Prentice Hall, Englewood Cliffs 2000)

40.15 O. Kuzgunkaya, H.A. ElMaraghy: Economic and strategic perspectives on investing in RMS and FMS, Int. J. Flex. Manuf. Syst. **19**(3), 217–246 (2007)

40.16 R. Zald: Using flexibility to justify robotics automation costs, Ind. Manag. **36**(6), 8–9 (1994)

40.17 M. Dağdeviren: Decision making in equipment selection: an integrated approach with AHP and PROMETHEE, J. Intell. Manuf. **19**, 397–406 (2008)

40.18 T.L. Saaty: *The Analytic Hierarchy Process* (McGraw-Hill, New York 1980)

40.19 E. Albayrak, Y.C. Erensal: Using analytic hierarchy process (AHP) to improve human performance: an application of multiple criteria decision making problem, J. Intell. Manuf. **15**, 491–503 (2004)

40.20 J.J. Wang, D.L. Yang: Using hybrid multi-criteria decision aid method for information systems outsourcing, Comput. Oper. Res. **34**, 3691–3700 (2007)

40.21 M.K. Tiwari, R. Banerjee: A decision support system for the selection of a casting process using analytic hierarchy process, Prod. Plan. Control **12**, 689–694 (2001)

40.22 Z. Güngör, F. Arikan: Using fuzzy decision making system to improve quality-based investment, J. Intell. Manuf. **18**, 197–207 (2007)

40.23 J.P. Brans, P.H. Vincke: A preference ranking organization method, Manag. Sci. **31**, 647–656 (1985)

40.24 J.P. Brans, P.H. Vincke, B. Mareschall: How to select and how to rank projects: the PROMETHEE method, Eur. J. Oper. Res. **14**, 228–238 (1986)

40.25 A. Albadvi, S.K. Chaharsooghi, A. Esfahanipour: Decision making in stock trading: an application of PROMETHEE, Eur. J. Oper. Res. **177**, 673–683 (2007)

40.26 S.L. Jämsä-Jounela: Future trends in process automation, Annu. Rev. Control **31**(2), 211–220 (2007)

41. Quality of Service (QoS) of Automation

Heinz-Hermann Erbe (△)

Quality of service (QoS) of automation involves issues of cost, affordability, energy, maintenance, and dependability. This chapter focuses on cost, affordability, and energy. (The next chapter addresses the other aspects.) Cost-effective or cost-oriented automation is part of a strategy called *low-cost automation*. It considers the life cycle of an automation system with respect to their owners: design, production, operating, and maintenance, refitting or recycling. *Affordable automation* is another part of the strategy. It considers automation or automatic control in small enterprises to enhance their competitiveness in manufacturing and service. Despite relative expensive components the automation system can be cheap with respect to operation and maintenance. As examples are discussed: numerical controls of machine tools; shop floor control with distributed information processing; programmable logic controllers (PLCs) shifting to general-purpose (PC); smart devices, i. e. information processing integrated in sensors and actuators; and distributed manufacturing, and maintenance.

Energy saving can be supported by automatic control of consumption in households, office buildings, plants, and transport. *Energy intensity* is decreasing in most developing countries, caused by changing habits of people and by new control strategies. Centralized generation of electrical energy has advantages in terms of economies of scale, but also wastes energy. Decentralized generation of electricity and heat in regional or local units are of advantage. A combination of wind energy, solar energy, hydropower, energy from biomass, and fossil fuel in small units could provide electrical energy and heat in regions isolated from grids. These hybrid energy concepts are demanding advanced, but low-cost, controls.

The term *low-cost automation* was born in 1986 at a symposium in Valencia sponsored by the International Federation of Automatic Control [41.1]. However, the use of this term led to a misunderstanding of low-cost automation as a technology with poor performance, although the intention was to promote affordable automation devices and to reduce the life cycle cost or cost of ownership of automation systems. The intention was also to bridge the gap between control theory and control engineering practice by applications using low-cost techniques. *Ortega* [41.2] pointed out that the transfer of knowledge between the academic community and the industrial user is far from satisfactory, and it still is, particular regarding small- and medium-sized industry. However, developments in control theory are based on principles that can appeal to concepts with which the practical engineer is familiar. Despite the successfully demonstration of advances of modern control methods over classical ones, actual implementations of automatic control in manufacturing plants show a preference for classical proportional–integral–derivative (PID) control. Two key factors are considered [41.3]:

1. Return on investment (better, cheaper, and faster)
2. Ease of application.

To be utilized broadly, a new technology must demonstrate tangible benefits, be easier to implement and maintain, and/or substantially improve performance and efficiency. Sometimes a new control method is not pursued due to poor usability during operation and troubleshooting in an industrial environment. The PID, due to its simplicity, whether implemented analog or digital, provides advantages in its application. However, manufacturing systems are becoming more complex. Control needs include single-input/single-output SISO and multiple-input/multiple-output MIMO controllers. Therefore control techniques beyond PID control become necessary. State-space methods are well developed and provide many advantages if well understood [41.3].

Low-cost automation is now established as a strategy to achieve the same performance as sophisticated automation but with lower costs. The designers of automation systems have a cost frame within which they have to find solutions. This is a challenge to theory and technology of automatic control, as the main parts of automation. Low-cost automation is not an oxymoron, like *military intelligence* or *jumbo shrimps*. It opposes the rising cost of sophisticated automation and propagates the use of innovative and intelligent solutions at affordable cost. The concept can be regarded as a collection of methodologies aiming at the exploitation of tolerance to imprecision or uncertainties to achieve tractability, robustness, and finally low-cost solutions. Mathematically elegant designs of automation systems are often not feasible because of their neglect of real-world problems, and in addition are often very expensive for their owners.

Cost aspects are mostly considered when designing automation systems. However, in the end, industry is looking for intelligent solutions and engineering strategies for saving cost but that nevertheless have secure,

high performance. Field robots in several domains such as manufacturing plants, buildings, offices, agriculture, and mining are candidates for reducing operation cost. Enterprise integration and support for networked enterprises are considered as cost-saving strategies. Human–machine collaboration is a new technological challenge, and promises more than cooperation. Last but not least, condition monitoring of machines to reduce maintenance cost and avoid downtime of machines and equipment, if possible, is also a new challenge, and promotes e-Maintenance [41.4] and e-Service [41.5].

The reliability of low-cost automation is independent of the grade of automation, i.e., it covers all possible circumstances in its field of application. Often it is more suitable to reduce the grade of automation and involve human experience and capabilities to bridge the gap between theoretical findings and practical requirements [41.6]. On the other hand, theoretical findings in control theory and practice foster intelligent solutions with respect to saving costs. Anyway, reliability is a must for all automation systems, although this requirement has no one-to-one relation to cost.

As an example one may consider computer-integrated manufacturing (CIM). The original concept of CIM connected automatically the design of parts to machines at the workshop via shop floor planning and scheduling software, and therefore used a lot of costly components and instruments. After a while this kind of automation turned out not to be cost effective, because the centralized control system had to fight against uncertainties and unexpected events. Decentralization of control and the involvement of human experience and knowledge along the added-value chain of the production process required less sophisticated hardware and software and reduced manufacturing cost, which allowed CIM to break through even in small and medium-sized enterprises [41.7].

Low-cost automation also concerns the implementation of automation systems. This should be as easy as possible and also facilitate maintenance. Maintenance is very often the crucial point and an important cost factor to be considered. Standardization of components of automation systems could also be very helpful to reduce cost, because it fosters usability, distribution, and innovation in new applications, for example fieldbus technology in manufacturing and building automation.

The components of an automation/control system, such as sensors, actuators and the controller itself, can incorporate advances in information technology. Local information processing allows for integration at the level of components, reducing total cost and providing new features in control lines. They may be wireless connected because of low data stream rates. Smart sensors and actuators are new developments for industrial automation [41.8]. The cost of wireless links will fall while the cost of wired connections will remain about constant, making wireless increasingly a logical choice for communication links. However, wireless links will not be applicable in all situations, notably those cases where high reliability and low latencies are required. The implication for factory automation systems is that processing and storage will become cheap: every sensor, actuator, and network node can be economically provided with unlimited processing power. If processing and storage systems become inexpensive relative to wiring costs, then the trend will be to locate processing power near where it is needed in order to reduce wiring costs. The trend will be to apply more processing and storage systems when and where they will reduce the cost of interconnections. The cost of radio and networking technology has fallen to the point where a wireless connection is already less expensive than many wired connections. New technology promises to further reduce the cost of wireless connections [41.9].

Distributed collaborative engineering, i. e., the control of common work over remote sites, is now an emerging topic in cost-oriented automation. Integrated product and process development as a cost-saving strategy has been partly introduced in industry. However, as *Nnaji* et al. [41.10] mention, lack of information from suppliers and working partners, incompleteness and inconsistency of product information/knowledge within the collaborating group, and incapability of processing information/data from other parties due to the problem of interoperability hamper effective use. Hence, collaborative design tools are needed to improve collaboration among distributed design groups, enhance knowledge sharing, and assist in better decision making. (See also Chaps. 26, 88.)

Mixed-reality concepts could be useful for collaborative distributed work because they address two major issues: seamlessness and enhancing reality. In mixed-reality distributed environments information flow can cross the border between reality and virtuality in an arbitrary bidirectional way. Reality may be the continuation of virtuality, or vice versa [41.11], which provides a seamless connection between the two worlds. This bridging or mixing of reality and virtuality opens up some new perspectives not only

for work environments but also for learning or training environments [41.12, 13]. (See also Chaps. 15, 86.)

The changing global context is having an impact on local and regional economies, particularly on small and medium-sized enterprises. Global integration and international competitive pressures are intensifying at a time when some of the traditional competitive advantages – such as relatively low labor costs – enjoyed by certain countries are vanishing.

One can see growing emphasis on strategies for encouraging supply-chain (vertical) and horizontal networking. These could be means for facilitating agile manufacturing. Agile manufacturing is built around the synthesis of a number of independent enterprises forming a network to join their core skills, competencies, and capacities to be able to operate profitably in a competitive environment characterized by unpredictable and continually changing customer demands. Agile manufacturing is underpinned by collaborative design and manufacturing in networks of legally separate and spatially distributed companies. Such networks are useful in optimizing current processes and mastering sporadic change in demand, material, and technology [41.14]. Distributed collaborative automation in manufacturing networks is therefore an emerging area for automatic control.

Energy-saving strategies and also individual solutions for reducing energy consumption are challenging politicians, the public, and researchers. The cost aspect is very important. Components of controllers such as sensors and actuators may be expensive, but one has to calculate savings in energy costs over a certain period or life cycle of a building or plant.

Energy provision and consumption based on finite resources can cause conflicts between customers, and between customers and suppliers. Energy savings are necessary due to these finite resources. Effective use of available energy can be supported by automatic control of consumption in households, office buildings, plants, and transportation. Energy intensity is decreasing in most developing countries. This is caused by changing habits of people but also by new control strategies. The use of energy in a country, in the residential sector, the commercial building sector, the transportation sector, and the industrial sector, influences the competitiveness of the economy, the environment, and the comfort of the inhabitants.

Building automation is generally a highly cost-oriented business. The goal is to provide acceptable comfort conditions at the lowest possible cost in terms of implementation, operation, and maintenance. Energy saving is one of the most important goals in building automation [41.15]. (See also Chap. 62.)

In Sect. 41.1 the strategy of cost-oriented automation will be explained with the examples of cost of ownership and robotics. Section 41.2 continues, covering affordable automation with subsections on smart devices, i. e., information processing embedded in sensors, and actuators, programmable logic controllers as important components of an affordable automation, and examples of low-cost production technology. Section 41.3 covers energy saving with automatic control using the aforementioned strategies. (See also Chap. 40.)

41.1 Cost-Oriented Automation

Cost-oriented automation as part of the strategy of low-cost automation considers the *cost of ownership* with respect to the life cycle of the system:

- Designing
- Implementing
- Operating
- Reconfiguring
- Maintenance
- Recycling.

Components and instruments could be expensive if life cycle costs are to decrease. An example is enterprise integration or networked enterprises as production systems that are vertically (supply chain) or horizontally and vertically (network) organized.

Cost-effective product and process realization has to consider several aspects regarding automatic control [41.13]:

- Virtual manufacturing supporting integrated product and process development
- Tele- or web-based maintenance (cost reduction with e-Maintenance systems in manufacturing)
- Small and medium-sized (SME)-oriented agile manufacturing.

Agile manufacturing here is understood as the synthesis of a number of independent enterprises forming a network to join their core skills.

As mentioned above, life cycle management of automation systems is important regarding cost of ownership. The complete production process has to be considered with respect to its performance, where maintenance is the most important driver of cost. *Nof* et al. [41.16] consider the performance of the complete automation system, which interests the owner in terms of cost, rather than only the performance of the control system; i. e., a compromise between cost of maintenance and cost of downtime of the automation system has to be found.

41.1.1 Cost of Ownership

A large amount of the cost of a manufacturing plant over its lifetime is spent on implementation, ramp up, maintenance, and reconfiguration. Cost-of-ownership analysis makes *life cycle costs* transparent regarding purchase of equipment, implementation, operation costs, energy consumption, maintenance, and reconfiguration. It can be used to support acquisition and planning decisions for a wide range of assets that have significant maintenance or operating costs throughout their usable life. Cost of ownership is used to support decisions involving computing systems, vehicles, laboratory and test equipment, manufacturing equipment, etc.

It brings out the *hidden* or nonobvious ownership costs that might otherwise be overlooked in making purchase decisions or planning budgets. The analysis is not a complete cost–benefit analysis. It pays no attention to benefits other than cost savings when different scenarios are compared. When this approach is used in decision support, it is assumed that the benefits from all alternatives are more or less equal, and that choices differ only on the cost side.

In highly industrialized countries the type and amount of automation mostly depends on the cost of labor. However, automation in general is not always effective [41.17]. A recent study carried out by the Fraunhofer Institute of Innovation and System Technology regarding the grade of automation within the German industry [41.6] found that companies are refraining from implementing highly automated systems. The costs of maintenance and reconfiguration are considered too high; it would be more cost effective to involve well-qualified operators. The same considerations were discussed by *Blasi* and *Puig* [41.18]:

Automation is not useful by itself: there are a lot of additional requirements to accomplish its func-

tion as needed, not only putting automation into the factories, putting automation just where it is needed and economically justified.

Blasi and *Puig* [41.18] consider the challenges of manufacturing processes of consumer goods. They stress that automation helps in providing homogeneous operation, but for many operations humans can sometimes carry out the task better than automation systems. However, humans are failure prone, emotionally driven, and do not constantly operate at the same level, which also has to be considered.

Morel et al. [41.19], as mentioned in the Introduction, stress the consideration of the whole performance of a plant, rather than the control performance only, which is what interests the owner regarding cost; i. e., a compromise between cost of maintenance and cost of downtime of the automation systems in a plant regarding commitments to customers or market must be considered. Scheduled maintenance can be shifted, but it is of course better to implement condition-based maintenance, in which degradation of parts can be observed and the decision to run a machine or piece of equipment for a certain extra time to fulfill a relevant customer order can be made reasonably based on collected data.

Blasi and *Puig* [41.18] consider the conditions for successful automation in industrial applications. Based on their experiences, a manufacturing engineer dealing with plant automation should bear in mind that proper automation is much more than a matter of machines and equipment. The *organization of work* in the whole plant involving the human experience at all stages and always considering possible improvements with automation or not is the challenge for engineers and management to save costs. Automation helps in providing homogeneous manufacturing processes. In many operations, humans can sometimes carry out the task better without automation. Humans may perform better than any automatic machine, for the time being, but it is impossible for them to assure constant quality. Where reasonable cost is critical, inspecting the image quality of a screen, for example, is a task reserved, for the moment, to humans, because computer vision is too expensive.

To summarize this section Fig. 41.1 shows the main cost contributions which are of interest to the owner of the machine and manufacturing equipment.

41.1.2 Robotics

Robots were created in order to automate, optimize, and ameliorate work processes that had up to then been

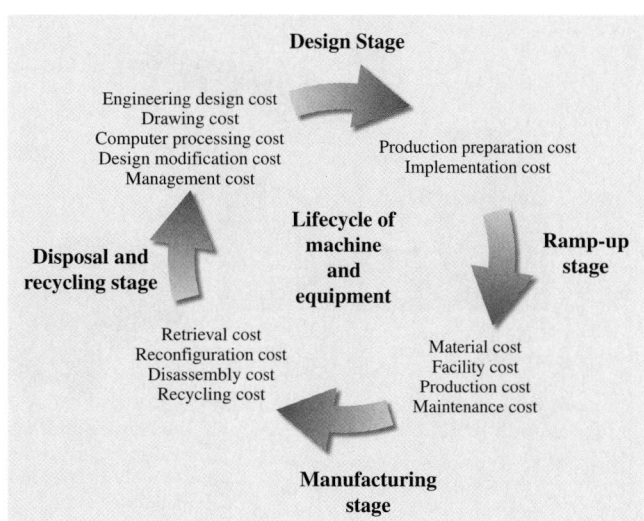

Fig. 41.1 Cost contributions of the life cycle of machine and equipment

carried out by humans alone. For reasons of safety, humans may not enter the working space of the robot. Recently it has been shown, however, that robots cannot come close to matching the abilities or intelligence of humans. Therefore, new systems which enable collaboration between humans and robots are becoming increasingly important. The advantages of using innovative robots can especially be seen in medical or other service-oriented areas as well as in the emerging area of humanoid robots.

Although robot technology is mostly regarded as costly, one can see today applications not only in customized mass production (automobile industry) but also in small and medium-sized enterprises (SMEs), manufacturing small lots of complex parts. Increasing product variety and customer pressure for short delivery times put robots in the focus. Robots can be used for loading and unloading of machine tools or other equipment such as casting machines, injection molding machines, etc. Here mostly the robots are stationary, and their control is coordinated with the control of the machine or equipment they serve.

Even *field robots* are applicable with affordable cost. *Ollero* et al. [41.20] describe as an example (among other applications) an autonomous truck, which could also be an unmanned vehicle in a manufacturing site. Low cost should be referred to the components to be used as well as to system design and maintenance. Sometimes a trade-off between general-purpose components and components tailored to the application has

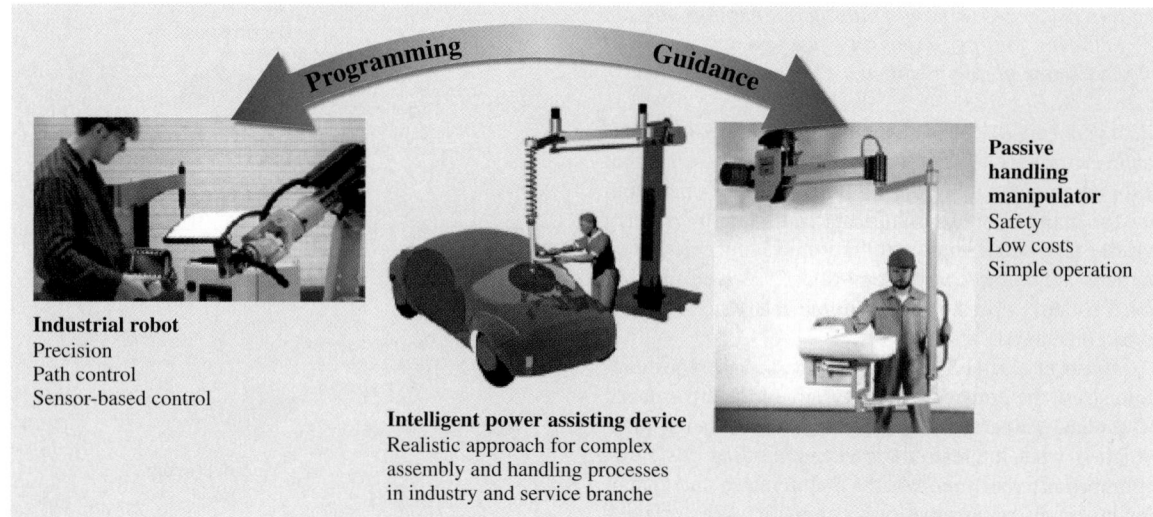

Fig. 41.2 Cobots, a new class of handling devices, which combine the characteristics of robots and hand-guided manipulators [41.24]

to be considered. Given the high cost of components, the design of the field robot should consider modularity, and simple assembly, reliability maintenance, and fault-detection properties are important for reducing cost in terms of the life cycle. *Sasiadek* and *Wang* [41.21] report on low-cost positioning of autonomous robots fusing, data sensed by global positioning system (GPS) and inertial navigation system (INS). Automation in deep mining with autonomous guided vehicles (AGV) saves cost in terms of healthcare of workers, provision of energy (using fuel cells), and compressed air required to maintain working conditions. For navigation, GPS is not available for this application, therefore radio beacons together with INS can be used. The vehicles and their control are discussed by *Dragt* et al. [41.22] and *Sasiadek* and *Lu* [41.23].

Wang et al. [41.25] developed a low-cost robot platform for control education.

Applications of robots in automated assembly and disassembly fail mostly because the environment cannot be sufficiently structured. This was the reason for the development of *collaborative robots* (cobots) or *intelligent (power) assisting devices* (I(P)AD) (Fig. 41.2). It was also motivated by ergonomic problems in assembly of parts, where parts weight endangered the

human body. Cobots offer a cost-effective solution for material handling. Complete automation of assembly processes is complicated, if not impossible. Reconfiguration of robots performing assembly tasks could be very costly. Humans, on the other hand, have capabilities that are difficult to automate, such as parts-picking from unstructured environments, identifying defective parts, fitting parts together despite minor shape variations, etc. Cobots are passive systems which are set in motion and guided by humans. The cobot concept supposes that shared control, rather than amplification of human power, is the key enabler [41.26]. The main task of the cobot is to convert a virtual environment, defined in software, into physical effect on the motion of a real payload, and thus also on the motion of the worker. Overhead gantry-style rail systems used in many shops can be considered cobots but without the virtual surface. Virtual surfaces separate the region where the worker can freely move the payload from the region that cannot be penetrated. These surfaces or walls have an effect on the payload like a ruler guiding a pencil. Technically cobots are based on *continuously variable transmissions* (CVT). *Peshkin* et al. [41.26] developed spherical CVTs, and *Surdilovic* et al. [41.24] developed CVTs based on a differential transmission.

41.2 Affordable Automation

This strategy of low-cost automation focuses on making systems affordable for their owners with respect to the problem to be solved. An example is a manufacturing system in a small or medium-sized enterprise, where automation increases productivity and therefore competitiveness. Although small enterprises agree that at least routine work can be done better when automated, there is still a fear that automation will be sophisticated, failure prone, need experts for maintenance and reconfiguration, and therefore would be costly.

Soloman [41.27] points to shortening product life cycles that need more intelligent, faster, and more adaptable assembly and manufacturing processes with reduced setup, reconfiguration, and maintenance time. Machine vision, despite costly components, can reduce manufacturing cost when properly applied [41.28]. In order to survive in a competitive market it is essential that manufactures have the capability to deploy rapidly affordable automation to adapt to a changing manufacturing environment with increased productivity but reduced production costs.

41.2.1 Smart Devices

Smart devices (sensors, actuators) with local information processing in connection with data fusion are in steady development, achieving cost reduction of components in several application fields such as automotive, robotics, mechatronics, and manufacturing [41.29, 30]. One of the first discussions on smart devices regarding cost aspects was given by *Boettcher* and *Traenkler* [41.31].

The developments allow for computer-based automation system to evolve from centralized architectures to distributed ones. The first level of distribution in order to reduce the wiring cost consisted of exchanging inputs and outputs through a fieldbus as a communication support. The second level integrates data processing in a modular setting as close as possible to sensors and actuators. These smart sensors and actuators can communicate, self diagnose or make decisions [41.32]. One step to realize smart sensors is to add electronics *intelligence* for postprocessing of outputs of conventional sensors prior to use by the control system. The same can be applied to actuators. Advantages are tighter tolerance, improved performance, automatic actuator calibration, etc.

Smart devices used in continuous systems benefit from the addition of microelectronics and software that runs inside the device to perform control and diagnostic functions. Very small components such as inputs/outputs blocks and overload relays are too small to integrate data processing for technical/economic reason. However, it is possible to develop embedded intelligence and control for the smallest factory floor devices. It is not always possible to implement data storage or processing on each sensor or actuator. The alternative solution is to implement data-processing units connected to some sensors or actuators, connected together by communication links in order to obtain remote inputs/outputs. Current trends are for the development of smart equipment associated to the fieldbus, which leads to a distributed architecture. Automation systems have evolved from a centralized to a distributed architecture, yielding an automation system with an intelligent distributed architecture. Robots are following this evolution, and increasingly becoming decomposed into divided subsystems, each of which realizes an elementary function. Distributed automation systems yield several advantages, such as greater flexibility, simplicity of operation, and better commissioning and maintenance. Today's smart field devices consist of two essential parts: sensor or actuator modules, and elec-

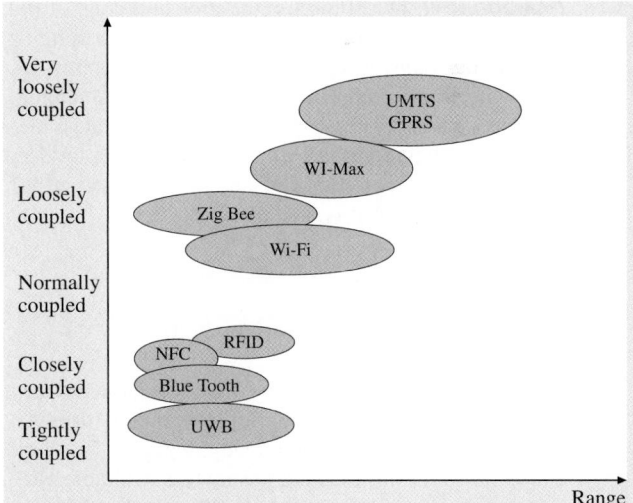

Fig. 41.3 Range and coupling modes of wireless technologies [41.9] (UMTS – universal mobile telecommunications system, GPRS – general packet radio service, WI-Max – worldwide interoperability for microwave access, Wi-Fi – wireless interoperability fidelity, RFID – radio frequency identification, NFC – near field communication, UWB – ultra wire band)

tronic modules. Microcomputers, initially responsible for PID control, communication, etc., now have diagnosis functions included. Advanced diagnosis addresses fault detection, fault isolation, and root analysis. The early detection of anomalies, either process or device related, is a key to improving plant availability and reducing production costs.

When smart devices come together with wireless communication a great cost saving can be achieved. This infrastructure will be flexible for reconfiguration. The reconfiguration of existing software for the new configuration is sometimes very costly and has to be considered. However, wireless communication eases the problem of physically inflexible communication infrastructures.

In mobile devices *wireless connections* are mandatory. Distributed sensors in a wide area do not need wireless communication, unless wiring is cost prohibitive. Without cables, cost-intensive wiring plans are not necessary. The freedom to place wireless sensors and actuators anywhere in a plant or a building becomes limited if the devices need a main power source, in which case power cables become necessary. It depends on the sensors if they can use internal batteries or can harvest energy from the environment. Several technologies for wireless communication are available at the market, with different standards and ranges (Fig. 41.3). *Cardeira* et al. [41.9] discuss the pros and cons of the available technologies. They conclude that, in spite of some initial skepticism, wireless communication is imposing itself as a complement to wired communication. Location awareness is a new feature of wireless devices. This feature may have a strong impact on service, where the physical location of a device is important for tracking, safety, security, and maintenance.

41.2.2 Programmable Logic Controllers as Components for Affordable Automation

Programmable logic controllers (PLCs) can be regarded as the classic components of affordable automation. An affordable application was already reported by *Jörgl* and *Höld* [41.33]. PLCs are meanwhile available with full capabilities for less than US$ 100. A PLC can be defined as a microprocessor-based control device, with the original purpose of supplementing relay logic. Early PLCs were only able to perform logical operations. PLCs can now perform more complex sequential control algorithms with the increase in microprocessor performance. They can admit analog inputs and out-

puts. The main difference from other computers is the special input/output arrangements, which connect the PLC to smart devices as sensors and actuators. PLCs read, for example, limit switches, dual-level devices, temperature indicators, and the positions of complex positioning systems. On the actuator side, PLCs can drive any kind of electric motor, pneumatic or hydraulic cylinders or diaphragms, magnetic relays or solenoids. The input/output arrangements may be built into a simple PLC, or the PLC may have external I/O modules attached to a proprietary computer network that plugs into the PLC.

PLCs were invented as less expensive replacements for older automated systems that used hundreds or thousands of relays. Programmable controllers were initially adopted by the automotive manufacturing industry, where software revision replaced rewiring of hard-wired control panels. The functionality of the PLC has evolved over the years to include typical relay control, sophisticated motion control, process control, distributed control systems, and complex networking.

There are other ways for automating machines, such as a custom microcontroller-based design, but there are differences between the two approaches: PLCs contain everything needed to handle high-power loads, while a microcontroller would need an electronics engineer to design power supplies, power modules, etc. Also a microcontroller-based design would not have the flexibility of in-field programmability of a PLC, which is why PLCs are used in production lines. Typically they are highly customized systems, so the cost of a PLC is low compared with the cost of contacting a designer for a specific one-time-only design.

The earliest PLCs expressed all decision-making logic in simple ladder diagrams (LDs) inspired by electrical connection diagrams. Electricians were quite able to trace out circuit problems with schematic diagrams using ladder logic. This was chosen mainly to address the apprehension of technicians. Today, the line between a personal computer and a PLC is thinning. PLCs have connections to personal computers (PCs) and Windows-based software-programming packages allow for easy programming and simulation. With the IEC 61131-3 standard, it is now possible to program using structured programming languages and logic elementary operations. A graphical programming notation called *sequential function charts* is available on certain programmable controllers. IEC 61131-3 currently defines five programming languages for programmable control systems: function block diagram (FBD), ladder diagram (LD), structured text (ST; similar to the Pascal programming language), instruction list (IL; similar to assembly language), and sequential function chart (SFC). These techniques emphasize logical organization of operations. *Susta* [41.34] presents a method to convert a PLC program in either of the languages mentioned above into programming statements, which can be used either for low-cost emulation of the program or as an auxiliary tool when a debugged PLC program is moved to another cheaper hardware, for instance, to a PC.

Continuous processes cannot be accomplished fast enough by PLC on–off control. The control system most often used in continuous processes is PID control. PID control can be accomplished by mechanical, pneumatic, hydraulic, or electronic control systems, as well as by PLCs. PLCs, including low-cost ones, have PID control functions included, which are able to accomplish process control effectively. To program control functions IL, ST or derived function block diagrams (DFBD) are used.

41.2.3 Production Technology

Within the last years so-called *shop-floor-oriented technologies* have been developed [41.17], and have achieved success at least but not only in small and medium-sized enterprises (SMEs). These technologies are focused on agile manufacturing, which involves using intelligent automation combined with human skill and experience on the shop floor. With shop-floor-oriented production support, human skill and automation create synergetic effects to master the manufacturing process.

Automation provides the necessary support to execute tasks and rationalize decisions. This represents a strand of low-cost automation. Running the manufacturing process effectively is not only a question of technology, though this is essential. Together with adequate work organization in which human skills can be developed, it establishes the framework for cost-effective, competitive manufacturing in SMEs.

Recent achievements for manufacturing are:

- Shop floor planning and control based on operators experience
- Low-cost numerical controls for machine tools and manufacturing systems (job-shop controls).

Shop floor control is the link between the administrative and planning section of an enterprise and the manufacturing process at the shop floor. It is the information backbone of the entire production process. What at least small and medium-sized enterprises need is shop floor control support to use the skills and experience of the workforce effectively. However, it should be stressed that this is support, not determination of what to do based on centralized automatic decision making.

In small-batch or single production (molds, tools, spare parts), devices for dynamic planning are desired. Checking all solutions to the problems arising on the shop floor while taking into account all relevant restrictions with short-term scheduling outside the shop floor either by manual or automatic means is not very effec-

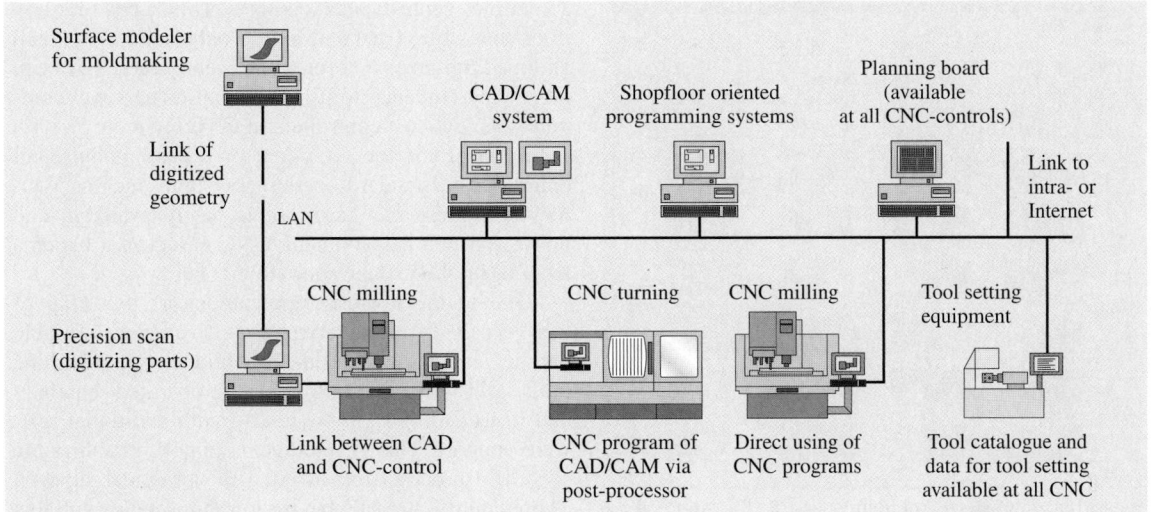

Fig. 41.4 Modules of a shop floor network

tive. It is senseless to schedule manufacturing processes exactly for weeks ahead. However, devices that are capable of calculating time corridors are of advantage. The shop floor can do the fine planning with respect to actual circumstances much better then central planning.

Human experience regarding solutions, changing parameters, and interdependencies is the very basis of shop floor decisions and needs to be supported rather than replaced.

Figure 41.4 shows a network system linking all relevant modules to be used by skilled workers. Apart from necessary devices such as tool-setting, an electronic planning board is integrated. The screen of the board is available at all computer numerical control (CNC) controls to be used at least to obtain information on tasks to be done at certain workplaces to a certain schedule. As all skilled workers in a group are responsible for the manufacturing process they have, beyond access to information, the task of fine planning of orders they received with frame data from the management. They use electronic planning boards at the CNC controls or alternatively at PCs besides the machines.

Planning and scheduling support with soft constraints was developed by [41.35], called Job-Dispo (Fig. 41.5). It consists mainly of an *intelligent planning board* with drag-and-drop functionality, a graphic editor, and a structured query language (SQL) database running on PCs under Windows. In SMEs, manufacturing groups are empowered to regulate their tasks themselves based on frame data from the management. The operators receive only rough data for orders

from central management, concerning delivery time, material, required quality, supply parts, etc. Using this support the workers decide on the sequences of tasks to be done at the different parts of the manufacturing system. The electronic planning board simulates the effects of their decisions. Normally more than one task needs the same resource at the same time. The operators are enabled to change the resource limits of workplaces and machines (working time, adding or changing shift work, etc.) until the simulation results in acceptable work practice and fulfills customers' demands. If this cannot be achieved, central management has to be involved in the decision. Job-Dispo is partly automatic but allows for soft constraints to make use of the experience of human operators.

Machine tools have a key position in manufacturing. Their functionality, attractiveness, and acceptance are determined by the efficiency of their control systems. Typical control tasks can be divided into the numerical controller, programmable logic control, controls for drives, and auxiliary equipment. Manufacturers of computer numerical control (CNC) systems are increasing effort on developing new concepts permitting flexible control functions with a broad scope of freedom to adapt the functions to the specific requirements of the planned application in order to increase the use of standardized components. This simplifies the integration of CNC of different manufactures into the same workshop environment. Even today, CNC systems are mainly offered as closed manufacturer-specific solutions and some machine tool builders develop their own, sometimes sophisticated, controls. This keeps the shop floor inflexible, but flexibility is only needed in small and medium-sized enterprises. Recently there has been a growing tendency to use industrial or personal computers as low-cost controllers in CNC technology, with a numerical control (NC) core on a card module, but connected to standard operating systems such as Windows. Because PC hardware is easily available and under constant development, CNC based on it benefits from technological advances in this field.

One of the low-cost applications of this kind of CNC is called *job-shop control*, which are now available on the market (e.g., Siemens, Heidenhain). Machine tools with job-shop control can be operated manually and if needed or appropriate also with additional software support. This is intelligent support, enabling the switch in process from manual to numerical support. These controls are suitable for job shops with small lots and are easy to handle manually (conventional turning

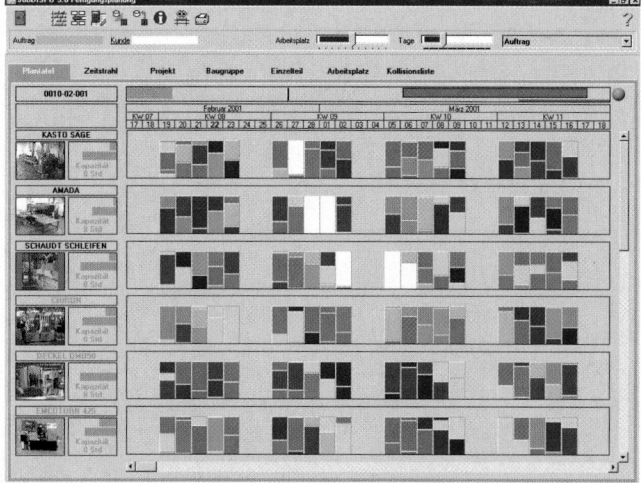

Fig. 41.5 Electronic, PC-based planning board for shop floor use [41.35]

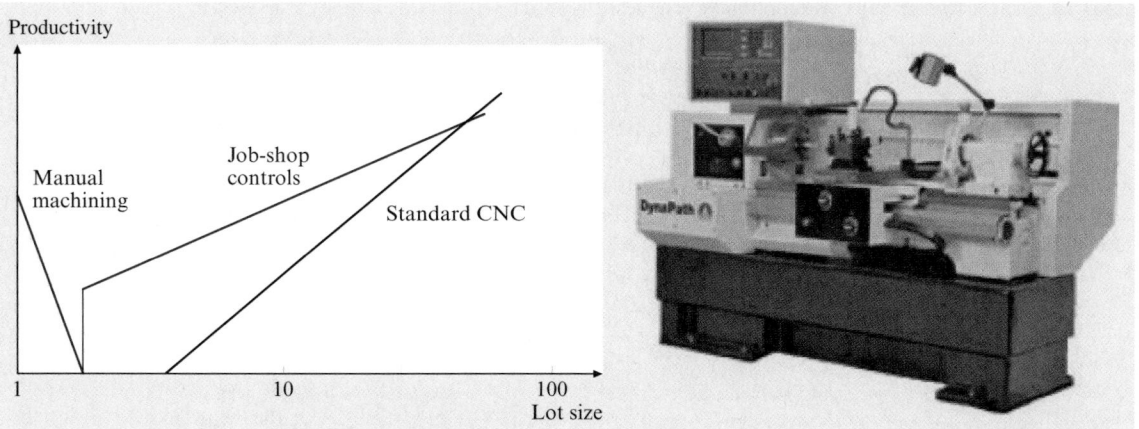

Fig. 41.6 Advantage of job-shop controls. Lathe with job-shop control

or milling) or programmable with interactive graphic support. Therefore the knowledge and experience of skilled workers can be challenged and division of work between programming and operating is unnecessary, saving costs while avoiding organizational effort and increasing flexibility. Job shop controls with PC operation systems can be integrated into an enterprise network, allowing for flexible manufacturing. It provides not only easy programming at the machines but also archiving of programs and loading of programs from other locations. Moreover connection to tool data management systems allows for quick search and ordering of the right tool at the workplace. Figure 41.6 demonstrates the range of job-shop controls with respect to lot size and productivity.

As an example of affordable automation in small enterprises, consider how to enhance the productivity and flexibility of a manufacturing process to acceptable costs. The main points in this respect are work organization and the technology used. Both of these aspects have to be considered together because they affect one another. Investing in a new, or at least better, technology is connected to decisions for machines with enhanced productivity and also to producing better quality that customers will be willing to pay for.

Considering machine tools or manufacturing cells or systems it is not always necessary to replace them completely. In many enterprises one can find conventional machines in a very good state but meanwhile not suitable to produce parts of high quality in an adequate time. Conventional machine tools such as lathes or conventional milling machines usually have a machine bed with good quality and stiffness; they should not be thrown onto the scrapheap. These machines could be equipped at least with electronic measurement devices as linear rules to improve the manufactured quality with respect to required tolerances. The next improvement could be refitting with numerical control. This certainly requires servo drives for each controllable axis while the drive of the spindle will be controlled using a frequency converter. Numerically controlled machines of the first generation sometimes only need a new control to bring them up to today's standards, a process called *upgrading*. Sometimes it is desired to retain conventional handling of machines despite the retrofit, i.e., moving tables and saddles mechanically with hand wheels in addition to numerically controlled servo drives. This facilitates the manufacturing of simple parts while using the advantages of CNC control to manufacture geometrically complex work pieces.

41.3 Energy-Saving Automation

41.3.1 Energy Generation

With energy *waste* levels in the process of electricity generation running at 66%, this sector has great potential for improvement. Using standard technology, only 25–60% of the fuel used is converted into electrical power. Combined-cycle gas turbines (CCGT) are among the most efficient plants now available, as compared with old thermal solid-fuel plants, some of which were commissioned in the 1950s.

Table 41.1 Energy savings and consumption (TW h/year) [41.36]

	Electricity savings achieved in the period 1992–2008	Consumption in 2003	Consumption in 2010 (with current policies)	Consumption in 2010, available potential (with additional policies)
Washing machines	10–11	26	23	14
Refrigerators, freezers	12–13	103	96	80
Electrical ovens	–	17	17	15.5
Standby	1–2	44	66	46
Lighting	1–5	85	94	79
Dryers	–	13.8	15	12
Domestic electrical storage water heaters	–	67	66	64
Air-conditioners	–	5.8	8.4	6.9
Dishwashers	0.5	16.2	16.5	15.7
Total	**24.5–31.5**	**377.8**	**401.9**	**333.1**

The biggest waste in the electricity supply chain (generation–transmission–distribution–supply) is the unused heat which escapes in the form of steam, mostly by heating the water needed for cooling in the generation process. The supply chain is still largely characterized by *central generation of electricity* in large power plants, followed by costly transport of the electricity to final consumers via cables. This transport generates further losses, mainly in distribution. Thus, centralized generation has advantages in terms of economies of scale, but also wastes energy. *Decentralized generation of electricity* and heat in regional or local units (including single buildings) could be of advantage. Such units are under development based on gas, or fuel cells in buildings. A combination of wind energy, solar energy, hydropower, energy from biomass, and fossil fuels in small units could provide electrical energy and heat in regions isolated from grids. These hybrid energy concepts are demanding advanced, but affordable, controls.

41.3.2 Residential Sector

Households accounted for 17% of the estimated gross energy consumption of 1725 Mtoe in 2005 in the European Union (EU), according to Eurostat energy balances. This amount could be reduced if people would change their habits to:

- Switch off appliances that are not in use, as standby consumes energy
- Select energy-efficient domestic appliances
- Use low-energy light bulbs
- Increase levels of recycling

- Monitor energy consumption
- Ensure systems are operating correctly
- Adjust the central heating set-point, and ensure correct distribution of sensors
- Double glazing of windows against heat and cold
- Insulation of walls
- etc.

However, people are lazy; automatic control can give support by managing the consumption of household appliances. The so-called *intelligent home* could provide solutions, but anyway people have to be aware of unnecessary energy consumption. Table 41.1 below shows possible savings of electrical energy in households of the EU.

41.3.3 Commercial Building Sector

Energy and therefore costs can be saved with suitable and intelligent automation. *Energy management control systems* (EMCS) are centralized computer control systems intended to operate a facility's equipment efficiently. These systems are still evolving rapidly, and they are controversial. Some applications are appropriate for computer control systems, and many are not. A range of simpler alternatives are available.

Advantage of *building automation systems* (BAS) includes monitoring, report generation, and remote control of equipment. Pitfalls are system cost, skilled staffing requirements, software limitations, vendor support, maintenance, rapid obsolescence, and lack of standardization.

BAS are also known by a variety of other names, including *energy management systems* (EMS) and *smart*

Table 41.2 Three levels of observing and improving energy consumption [41.37]

	Policy	Structure	Resources
Improving performance →	Formal energy policy and implementation plan. Commitment and active involvement of top management	Energy management fully integrated into management structure and systems from board level down	Full time staff and budget resources related to energy spend at recommended levels	← **Increase commitment**
	Energy policy set and reviewed by middle management	A management structure exists but there is no direct reporting to top management	Staff and budget resources not linked to energy spend
	Technical staff have developed their own guidelines	Informal and unplanned	Informal allocation of staff time and no specific energy budget

building controls. A system typically has a central computer, distributed microprocessor controllers (called *local panels, slave panels, terminal equipment controllers,* and other names), and a digital communication system. The communication system may carry signals directly between the computer and the controlled equipment, or there may be tiers of communications.

Building automation can be a very effective way to reduce building operational costs and improve overall comfort and efficiency of a building. There are many definitions and examples of building automation. Simply put, building automation uses software to connect and control electrical functions in a building. Those functions usually include but are not limited to the heating, ventilation, air-conditioning (HVAC) and lighting systems. For further reading see [41.15] and Chap. 62.

41.3.4 Transportation Sector

Transport accounts of 20% of the estimated gross energy consumption of 1725 Mtoe in the European Union in 2005, according to Eurostat energy balances. Reducing this consumption is not only a technical problem, but mainly a political one. Road transport of goods receives direct or indirect subventions. The mostly state-owned railway organizations in Europe are not able to install a common system to reduce inefficient road transport. Low-cost air travel is increasing air pollution. Public transport does not always offer an acceptable alternative to individual car use.

To get people or goods from A to B there are various means of transport, via air, road, rail, or water. Today no interoperable information system is available. however,

this would be an assumption for an efficient transport management system. These problems of information control could be solved.

Of course, achievements of automatic control in motor management and electrical drives reducing energy consumption can be seen. Hybrid drives in cars with automatic control of power management are favored. However, more effort is necessary to find acceptable and cost-effective solutions.

41.3.5 Industrial Sector

An enterprise should always observe and improve its energy consumption at all levels: the energy improvement cycle. A precondition for entering the cycle is commitment at all levels of the enterprise. Table 41.2 illustrated an uncommitted enterprise on the bottom row and an enterprise that is highly committed to energy improvement on the top row. It is important that the enterprise demonstrates commitment to sustained energy improvement before the process of delivering that improvement can begin.

The *energy improvement strategy* (Fig. 41.7) need not be long or complex but it is vitally important, as it will set the direction of all efforts to manage and improve energy consumption. Essentially there are two ways to cut energy bills (Fig. 41.7):

- Pay less for energy
- Consume less energy.

To reduce energy consumption it is necessary to analyze all possible sources of wasted energy: buildings, machines and equipment, production processes of goods, recycling of wasted energy (heat, etc.), and transport of

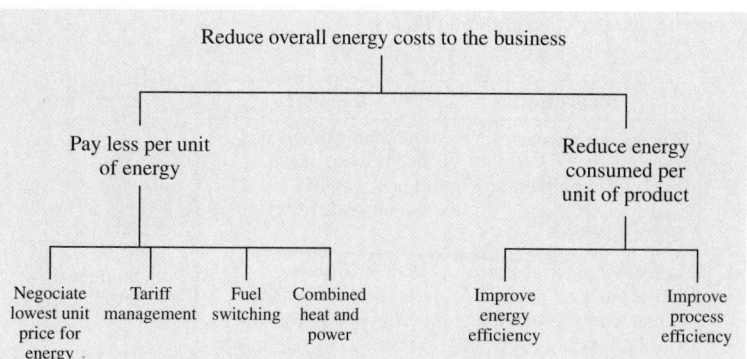

Fig. 41.7 Energy improvement strategy in an enterprise [41.37]

raw material, premanufactured parts, and manufactured products.

Developed automatic control concepts can help (less energy-consuming drives), but control concepts for an enterprise as a whole may be of advantage in terms of energy consumption reduction.

The use of energy in a country, in the residential sector, the commercial building sector, the transportation sector, and the industrial sector, influences the competitiveness of the economy, the environment, and the comfort of the inhabitants.

While energy efficiency measures energy inputs for a given level of service, *energy intensity* measures the efficient use of energy. It is defined as the ratio of energy consumption to a measure of the demand for services (e.g., number of buildings, total floor space, number of employees), or more generally the energy required to generate US$ 1000 of gross domestic product (GDP) (Fig. 41.8).

High energy intensity indicates a high price or cost of converting energy into GDP, while low energy intensity indicates a lower price or cost of converting energy into GDP.

Many factors influence an economy's overall energy intensity. It may reflect requirements for general standards of living and weather conditions in an economy. It is not untypical for particularly cold or hot climates to require greater energy consumption in homes and

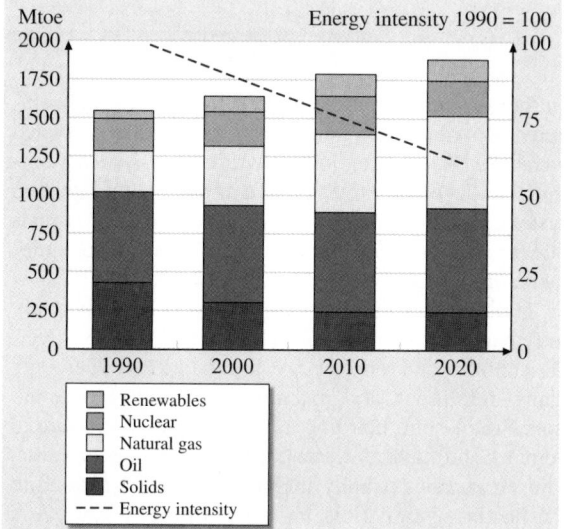

Fig. 41.8 Estimated total energy consumption by fuel and energy intensity 1990–2020 of the 25 EU member states [41.38]

workplaces for heating or cooling. A country with an advanced standard of living is more likely to have a wider prevalence of consumer goods and thereby be impacted in its energy intensity compared with a country with a lower standard of living.

41.4 Emerging Trends

41.4.1 Distributed Collaborative Engineering

Supporting cost-effective *human–human collaboration* in networked enterprises is a challenge for cost-oriented

automation. With the trend to extend the design and processing of products over different and remotely located factories, the problem arises of how to secure effective collaboration of the involved workforce. Usual face-to-face work will be replaced at least partly if not totally by computer-mediated collaboration. The development and implementation of information and communication technology, suitably adapted to the needs of the workforce and facilitating remotely distributed collaborative work, is a challenge to engineers. Information mediated only via vision and sound is insufficient for collaboration. In designing and manufacturing it is often necessary to have the parts in your hands. To grasp a part at a remote site requires force (haptic) feedback in addition to vision and sound. Consider, for example, remote service for maintenance.

Bruns et al. [41.11, 39, 40] developed low-cost devices based on *mixed-reality concepts* for web-based learning as a first step to distributed collaborative work.

Mixed-reality environments as defined in [41.41] are those in which real-world and virtual-world objects are presented together on a single display. Mixed-reality techniques have proven valuable in single-user applications. Meanwhile, research has been done on applications for collaborative users. Mixed reality could be useful for collaborative distributed work because it addresses two major issues: seamlessness and enhancing reality [41.42].

Bruns [41.11] notes that most existing collaborative workspaces strictly separate reality and virtuality; for example, when controlling a remote process, one can sense and view specific system behavior, control the system by changing parameters, and observe the process by video cameras. The process, as a flow of energy – controlled by signals and information – is either real or completely modeled in virtuality and simulated. In mixed-reality distributed environments, information flow can cross the border between reality and virtuality in an arbitrary bidirectional way. Reality may be the continuation of virtuality, or vice versa, which provides a seamless connection between the two worlds. This bridging or mixing of reality and virtuality opens up some new perspectives for learning or training environments [41.12] as well as for distributed work environments.

The connection between the real and the virtual world is mediated through an *energy interface* called a *hyperbond*. In dynamic systems the system components are connected through energy (or power) transfer.

If a real dynamic system is to be extended digitally with software running on a computer (virtual dynamic system) analog sensor signals have to be converted into digital values for the software side of the interface. Flow in the opposite direction requires digital values from the software to be converted into analog signals that generate effort and flow to the real dynamic system.

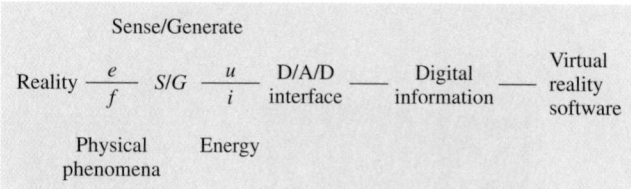

Fig. 41.9 Architecture of a hyperbond connection

Figure 41.9 describes in general the interface between the real and the virtual world: Effort (e) and flow (f) are sensed (S) (or generated (G)), providing voltage (u) and current (i) (or effort and flow in the opposite), an analog-to-digital (A/D) converter provides digital information for the software, or a digital-to-analog (D/A) converter converts digital information into analog signals to drive a generating mechanism for effort and flow to the real world.

The interface generates or dissipates energy (or power). The power, provided through the real system, has to be dissipated, because the virtual continuation with software requires nearly negligible power. In the opposite direction, the digital information provided through the software has to generate the power necessary to connect to the real system.

Application for a Discrete-Time Event System

Electropneumatic components installed at a real workbench are connected to a virtual workbench with electropneumatic components stored in a library. Because the valves can only be open or closed, the cylinders only on or off, one has only discrete events. The energy interface, called the hyperbond, receives digital information from the virtual workbench and generates air pressure and air flow as well as voltage and current for the solenoid valves and cylinders at the real workbench (Fig. 41.10). In the opposite direction the energy interface senses air pressure and air flow as well as electrical signals to convert into digital signals. No feedback control within the energy interface is necessary unless it is desired.

The *virtual and real workbench* can be located either at a local site or at remote sites connected via the Internet. Also virtual workbenches may be distributed at different sites connected via the Internet to the (only) real workbench. The software of the virtual workbench allows access by many users at the same time. Therefore students or workers distributed at different locations can solve tasks together at their virtual workbenches and export it on the real workbench to test their common solution in reality.

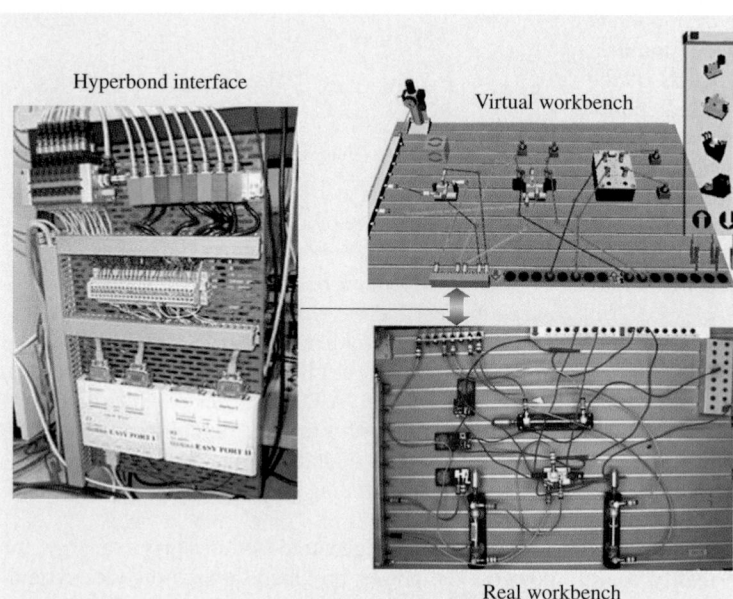

Fig. 41.10 Real and virtual pneumatic workbench connected with an energy interface (hyperbond)

The connected workbenches are located in computer animated virtual environment (CAVE)-like constructions. CAVEs consist of scaffoldings with canvases onto which the images of other workspaces with the people working on them are beamed (Fig. 41.11).

The example presents distributed workbenches for discrete events (valves, cylinders, on/off). *Yoo* and *Burns* [41.39] enhanced this application with a generally described energy interface for the connection of real (continuous-time dynamic) and virtual (discrete-time dynamic) systems. *Yoo* [41.43] developed this interface with low-cost components for examples with haptic feedback.

41.4.2 e-Maintenance and e-Service

Among the main cost factors responsible for the performance of a plant is the availability of machines and equipment. Therefore maintenance takes an important part in plant automation. Avoiding downtime, or at least minimizing it, is the goal. *Advanced maintenance strategies and services* are candidates for cost savings. These integrate information processing and the processing of physical objects. e-Maintenance is an emerging concept, generally defined as a maintenance

Fig. 41.11 Workspace with the real workbench and the virtual one in front, and the remote collaborators projected onto the left and right canvases

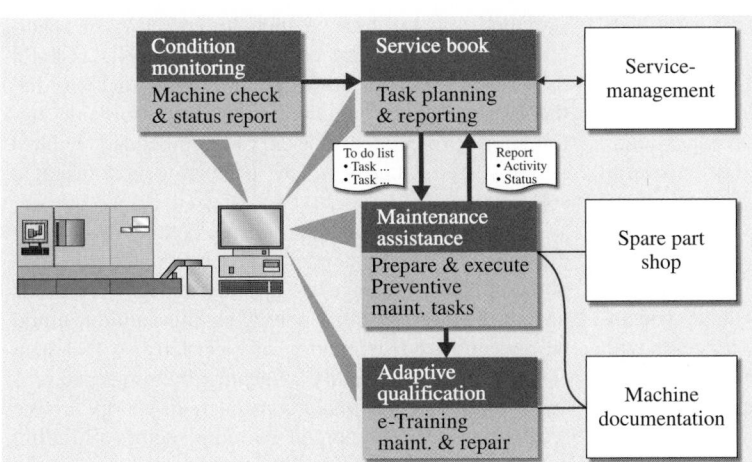

Fig. 41.12 Platform for web-based maintenance service [41.5]

management strategy in which assets are monitored (conditioned-based monitoring) and action is synchronized with the business process through the use of web-enabled and wireless infotronics technologies [41.4,44]. Despite the development of new maintenance strategies (Fig. 41.12) and the efforts of service providers one should not underestimate the knowledge and experiences of the operators of automation systems, and use their expertise if possible. Here one has to find a compromise between the cost of service from outside or specialists from inside and the cost of qualified operators and their permanent training. To empower people effectively depends on confidence between management and shop floor. It is not easy for the management, but this problem is solvable if it reduces costs and everybody understands it. *e-Maintenance* strategies were developed for the automotive industry. Many small enterprises still rely on preventive maintenance or even on breakdown maintenance. The challenge is to make e-Maintenance available for these enterprises for affordable cost.

Tele-service was developed to support small enterprises to enhance their productivity at affordable cost. The principal goal is to minimize trouble-shooting costs by allowing service personnel the flexibility to work from a distance. The main advantage for the machine operator lies in reducing machine downtime. e-Services

go beyond such conventional concepts. *e-Service* is considered to include support of customer service via information and communication components and services. e-Service makes the installation and start-up of machines and plants possible, as well as trouble-shooting, the transfer of new software versions, the provision of replacement parts, and ordering spare parts on time. In the future, e-Service will also find application in process support and customer consultation [41.5].

Two emerging areas have been described with respect to their influence of cost savings. Some others that should be mentioned are:

1. *Semi-automated disassembly* of electronic waste, such as mobile phones, will gain more importance in the near future: a flexible, modular system for the development of semi-automatized, intelligent disassembly cells including stationary robots and especially a *low-cost*, hierarchical control structure [41.45].
2. *Harrison* and *Colombo* [41.46] propose *collaborative automation* in manufacturing. Their approach is to define a suitable set of basic production functions and then to combine these functions in different arrangements. This approach creates more complex production activities and saves costs of reconfiguration of rigid automation in manufacturing.

41.5 Conclusions

Low-cost automation was considered under two aspects: cost-oriented automation as a strategy to reduce the cost of ownership of an automation system, and af-

fordable automation focused on the needs of small and medium-sized enterprises to enhance productivity and thereby competitiveness.

In the first case the life cycle of a production system was discussed, and design and maintenance/reconfiguration were identified as the main cost drivers. Recent developments of e-Maintenance as emerging trends are promising for cost saving. Maintenance was discussed holistically regarding the overall performance of a production system. Life cycle engineering with its different facets for cost savings was described, and further developments were suggested. Also human–human collaboration (including support for human–machine collaboration) has been considered with respect to cost. Mixed-reality concepts with application in learning environments to train for cost-effective collaborative work over remote sites has also been presented.

In the second case, examples of affordable automation for old and new developments of automatic control such as smart devices were presented. Web based e-Services to solve maintenance problems and to avoid downtime of machines have recently been provided by machine manufacturers. It is a challenge to the automatic control community to transfer their developments in modern control design and theory to make them applicable for affordable automation projects; for example, embedded control systems (see Chap. 43), are promising cost-oriented solutions.

Energy saving and its automatic control was briefly discussed for electrical energy generation in large-scale power plants. Wastage of primary energy occurs as unused heat. A possible energy-saving solution could be decentralized generation of electricity so that heat could be used more easily for heating buildings and processes in plants. The main consumers of energy are the residential and commercial building sectors. Building automation systems are being developed to save energy. The challenge is to make a possible sophisticated control understandable for the operators. To summarize, energy saving is not only a matter of technology but also depends on people's habits.

References

41.1   P. Albertos, J.A. de la Puente (Eds.): *Proc. LCA'86 IFAC Symp. Compon. Instrum. Tech. Low Cost. Appl. Valencia* (Pergamon, Oxford 1986)

41.2   R. Ortega: New techniques to improve performance of simple control loops, Proc. Low Cost Autom., ed. by P. Kopacek, P. Albertos (Pergamon, Oxford 1989) pp. 1–5

41.3   S. Chand: From electrical motors to flexible manufacturing: control technology drives industrial automation, Selected Plenaries, Milestones and Surveys, Proc. 16th IFAC World Congr., ed. by P. Horacek, M. Simandl, P. Zitek (Elsevier, Oxford 2005) pp. 40–45

41.4   Z. Chen, J. Lee, H. Qui: Infotronics technologies and predictive tools for next-generation maintenance systems, Proc. 11th Symp. Inf. Control Probl. Manuf. (Elsevier 2004)

41.5   R. Berger, E. Hohwieler: Service platform for web-based services, Proc. 36th CIRP Int. Semin. Manuf. Syst. Produktionstech., Vol. 29 (2003) pp. 209–213

41.6   G. Lay: Is high automation a dead end? Cutbacks in production overengineering in the German industry, Proc. 6th IFAC Symp. Cost Oriented Autom. (Elsevier, Oxford 2002)

41.7   P. Kopacek: Low cost factory automation, Proc. Low Cost Autom., ed. by P. Kopacek, P. Albertos (Pergamon, Oxford 1992)

41.8   M. Starosviecki, M. Bayart: Models and languages for the interoperability of smart instruments, Automatica **32**(6), 859–873 (1996)

41.9   C. Cardeira, A.W. Colombo, R. Schoop: Wireless solutions for automation requirements, ATP Int. – Autom. Technol. Pract. **4**(2), 51–58 (2006)

41.10   B.O. Nnaji, Y. Wang, K.Y. Kim: Cost effective product realization, Proc. 7th IFAC Symp. Cost Oriented Autom. (Elsevier, Oxford 2005)

41.11   F.W. Bruns: Hyper-bonds – distributed collaboration in mixed reality, Annu. Rev. Control **29**(1), 117–123 (2005)

41.12   D. Müller: Designing learning spaces for mechatronics. In: *MARVEL – Mechatronics Training in Real and Virtual Environments*, Impuls, Vol. 18, ed. by D. Müller (NA/Bundesinstitut für Berufsbildung, Bremen 2005)

41.13   H.H. Erbe: Introduction to low cost/cost effective automation, Robotica **21**(3), 219–221 (2003)

41.14   E.O. Adeleye, Y.Y. Yusuf: Towards agile manufacturing: models of competition and performance outcomes, Int. J. Agil. Syst. Manag. **1**, 93–110 (2006)

41.15   T. Salsbury: A survey of control technologies in the building automation industry, Selected Plenaries, Milestones and Surveys, Proc. 16th IFAC World Congr., ed. by P. Horacek, M. Simandl, P. Zitek (Elsevier, Oxford 2005) pp. 331–341

41.16   S.Y. Nof, G. Morel, L. Monostori, A. Molina, F. Filip: From plant and logistics control to multi-enterprise collaboration, Annu. Rev. Control **30**(1), 55–68 (2006)

41.17   H.-H. Erbe: Technology and human skills in manufacturing, Balanced Automation Systems II,

BASYS '96 (Chapman Hall, London 1996) pp. 483–490

41.18   A. Blasi, V. Puig: Conditions for successful automation in industrial applications. A point of view, Proc. 15th IFAC World Congr. (Elsevier, Oxford 2005)

41.19   G. Morel, B. Iung, M.C. Suhner, J.B. Leger: Maintenance holistic framework for optimizing the cost/availability compromise of manufacturing systems, Proc. 6th IFAC Symp. Cost Oriented Autom. (Elsevier, Oxford 2002)

41.20   A. Ollero, G. Morel, P. Bernus, S.Y. Nof, J. Sasiadek, S. Boverie, H. Erbe, R. Goodall: Milestone report of the IFAC manufacturing and instrumentation coordinating committee: from MEMS to enterprise systems, Annu. Rev. Control **26**, 151–162 (2002)

41.21   J. Sasiadek, Q. Wang: Low cost automation using INS/GPS data fusion for accurate positioning, Robotica **21**, 255–260 (2003)

41.22   B.J. Dragt, F.R. Camisani-Calzolari, I.K. Craig: An overwiew of the automation of load-haul-dump vehicles in an underground mining environment, Proc. 16th IFAC World Congr. (Elsevier, Oxford 2005)

41.23   J. Sasidak, Y. Lu: Path tracking of an autonomous LHD articulated vehicle, Proc. 16th IFAC World Congr. (Elsevier, Oxford 2005)

41.24   D. Šurdilovic, R. Bernhardt, L. Zhang: New intelligent power-assist systems based on differential transmission, Robotica **21**, 295–302 (2003)

41.25   W. Wang, Y. Zhuang, W. Yun: Innovative control education using a low cost intelligent robot platform, Robotica **21**, 283–288 (2003)

41.26   M.A. Peshkin, E. Colgate, W. Wannasuphoprasit, C.A. Moore, R.B. Gillespie, P. Akella: Cobot architecture, IEEE Trans. Robot. Autom. **17**(4), 377–390 (2001)

41.27   S. Soloman: *Affordable Automation* (McGraw-Hill, New York 1996)

41.28   F. Lange, G. Hirzinger: Is vision the appropriate sensor for cost oriented automation?, Proc. Cost Oriented Autom., ed. by R. Bernhardt, H.-H. Erbe (Elsevier, Oxford 2002) pp. 129–134

41.29   M. Bayart: LARII: development tool for smart sensors and actuators, Proc. Cost Oriented Autom., ed. by R. Bernhardt, H.-H. Erbe (Elsevier, Oxford 2002) pp. 83–88

41.30   M. Bayart: Smart devices for manufacturing equipment, Robotica **21**, 325–333 (2003)

41.31   J. Boettcher, H.R. Traenkler: Trends in intelligent instrumentation, Proc. Low Cost Autom., ed. by A. de Carli (Pergamon Press, Oxford 1989) pp. 241–248

41.32   M.K. Masten: Electronics: the intelligence in intelligent control, Proc. 3rd IFAC Symp. Intell. Compon. Instrum. Control Appl. SICICA'97, Annecy, France (1997)

41.33   H.P. Jörgl, G. Höld: Low cost PLC design and application, Proc. Low Cost Autom., ed. by P. Kopacek, P. Albertos (Pergamon, Oxford 1992) pp. 7–12

41.34   R. Susta: Low cost simulation of PLC programs, Proc. 7th IFAC Symp. Cost Oriented Autom. (Elsevier, Oxford 2005)

41.35   http://www.Fauser.de

41.36   European Commission: *Directorate General for Energy and Transport* (European Commission, 2005)

41.37   M. Pattison: Energy prices are rising and continue to rise. In: *Memorias de Primera Jornadas de Mantenimiento*, ed. by Escuela Politecnica Nacional, ASEA (Brown Boveri, Quito 2006)

41.38   European Commission: *PRIMES baseline, European Energy and Transport – Scenarios on Key Drivers* (European Commission, 2004)

41.39   Y. Yoo, F.W. Bruns: Energy interfaces for mixed reality, Proc. 12th Symp. Inf. Control Probl. Manuf. (Elsevier, Oxford 2006)

41.40   F.W. Bruns, H.-H. Erbe: Mixed reality with hyperbonds – a means for remote labs, Control Eng. Pract. **15**(11), 1435–1444 (2006)

41.41   P. Milgram, F. Kishino: A taxonomy of mixed reality visual displays, IEICE Trans. Inf. Syst. **E77-D**(12), 1321–1329 (1994)

41.42   H. Ishii, B. Ullmer: Tangible bits: towards seamless interfaces between people, bits and atoms, Proc. CHI'97 (1997) pp. 234–241

41.43   Y.H. Yoo: Mixed Reality Design using unified Energy Interfaces (Univ. Bremen). Ph.D. Thesis (Shaker, Aachen 2007)

41.44   A. Müller, M.C. Suhner, B. Iung: Formalization of a new prognosis model for supporting proactive maintenance implementation on industrial system, Reliab. Eng. Syst. Saf. **93**(2), 234–253 (2008)

41.45   B. Kopacek, P. Kopacek: Semi-automatised disassembly, Proc. 10th Int. Workshop Robotics in Alpe Adria Danube Region RAAD'01, Vienna (2001) pp. 363–370

41.46   R. Harrison, A.W. Colombo: Collaborative automation – from rigid coupling towards dynamic reconfigurable production systems, Proc. 16th IFAC World Congr. (Elsevier, Oxford 2005)

Part E | 41

42. Reliability, Maintainability, and Safety

Gérard Morel, Jean-François Pétin, Timothy L. Johnson

Within the last 20 years, digital automation has increasingly taken over manual control functions in manufacturing plants, as well as in products. With this shift, reliability, maintainability, and safety responsibilities formerly delegated to skilled human operators have increasingly shifted to automation systems that now *close the loop*. In order to design highly *dependable* automation systems, the original concept of design for reliability has been refined and greatly expanded to include new engineering concepts such as availability, safety, maintainability, and survivability. Technical definitions for these terms are provided in this chapter, as well as an overview of engineering methods that have been used to achieve these properties. Current standards and industrial practice in the design of dependable systems are noted. The integration of dependable automation systems in multilevel architectures has also evolved greatly, and new concepts of control and monitoring, remote diagnostics, software safety, and automated reconfigurability are described. An extended example of the role of dependable automation

systems at the *enterprise level* is also provided. Finally, recent research trends, such as automated verification, are cited, and many citations from the extensive literature on this topic are provided.

Industrial automation systems are intensively embedding infotronics and mechatronics technology (IMT) in order to fulfil complex applications required by the increasing customization of both services and goods [42.2–6]. The resulting behavior of these IMT-based automation systems is shifting system dependability responsibility [42.7] from the human operator to the automation software.

Management, engineering, and maintenance personnel have a primary responsibility to assure reliability [42.8, 9], maintainability, and safety of all automated systems, and manufacturing systems in particular. Therefore, safety, reliability, and availability as performance attributes to access the dependability of a system are threatened by a rapid growth in software

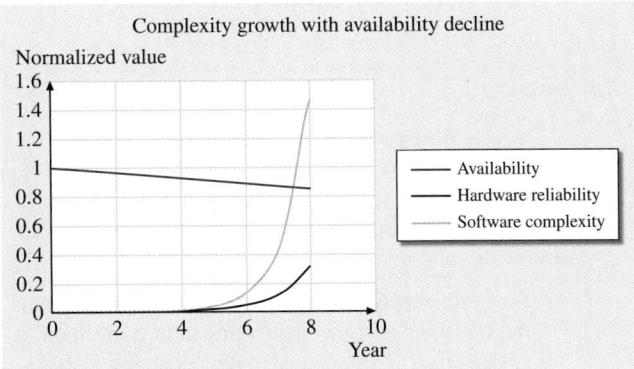

Fig. 42.1 Growth of software complexity and its impact on system availability (after [42.1])

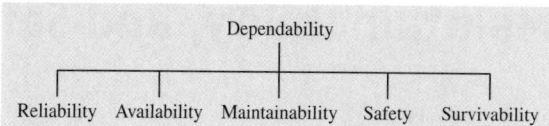

Fig. 42.2 The dependability tree (after [42.10])

complexity that could limit further automation progress (Fig. 42.1).

Section 42.1 provides definitions of dependability key concepts (Fig. 42.2) that enlarge reliability, maintainability, and safety (RMS) concepts [42.11, 12] by characterizing the ability of a device or system to deliver the correct service that can justifiably be trusted by all stakeholders in the automated process.

42.1 Definitions

Dependability is an integrative concept that encompasses required attributes (qualities) of a system assessed by quantitative measures (reliability, maintainability) or qualitative ones (safety) in order to cope with the chain of fault–error–failure threats of an operational system, by combining a set of means related to fault prevention, fault tolerance, fault removal, and fault forecasting [42.14].

Reliability is the ability of a device or system to perform a required function under stated conditions for a specified period of time. This property is often measured by the probability $R(t)$ that a system will operate without failure before time t, often defined according to the failure rate ($\lambda(t)$) as

$$R(t) = \exp\left(-\int_0^t \lambda(u)\,du\right),$$

meaning

$$R(t) = \Pr(\mathrm{TTF} > t),$$

where TTF is the time to failure.

This definition of reliability is concerned with the following four key elements:

1. First, reliability is a probability. This means that there is always some chance for failure. Reliability engineering is concerned with achieving a specified probability of success, at a specified statistical confidence level.

Then, methods for design of highly dependable automation systems are outlined in Sect. 42.2. Section 42.3 discusses the methods for achieving long-term dependable operation for an existing system.

Finally, dependability has evolved from reliability/availability concerns to information control concerns, as an outgrowth of the technological deployment of information-intensive systems and the economical pressure for cost-effective automation [42.13]. Section 42.4 concludes with challenges, trends, and open issues related to system resilience, aiming to cope with system dependability in the presence of active faults, i.e., system survivability. Chapter 39 of this handbook contains information related to the concepts covered in this Chapter.

2. Second, reliability is predicated on *intended function*. The system requirements specification is the criterion against which reliability is measured.
3. Third, reliability applies to a specified period of time. In practical terms, this means that a system has a specified chance that it will operate without failure before a final time (e.g., $0 < t < T$).
4. Fourth, reliability is restricted to operation under stated conditions. This constraint is necessary because it is impossible to design a system for unlimited conditions. Both normal and abnormal operating environments must be addressed during design and testing.

Maintainability is the ease with which a device or system can be repaired or modified to correct and prevent faults, anticipate degradation, improve performance or adapt to a changed environment. Beyond simple physical accessibility, it is the ability to reach a component to perform the required maintenance task: maintainability should be described [42.15] as the characteristic of material design and installation that determines the requirements for maintenance expenditures, including time, manpower, personnel skill, test equipment, technical data, and facilities, to accomplish operational objectives in the user's operational environment. Like reliability, maintainability can be expressed as a probability $M(t)$ based on the repair rate ($\mu(t)$) as

$$M(t) = 1 - \exp\left(-\int_0^t \mu(u)\,du\right),$$

meaning

$$M(t) = \Pr(\text{TTR} < t) \,,$$

where TTR is the time to repair.

Availability characterizes the degree to which a system or equipment is operable and in a committable state at the start of a mission, when the mission lasts for an unknown, i. e., random, time. A simple representation for availability is the proportion of time a system is in a functioning condition, and this can be expressed mathematically [42.17] by

$$A(t) = \frac{\mu}{\mu + \lambda} + \frac{\lambda}{\mu + \lambda} e^{-(\mu+\lambda)t} \,,$$

where λ is the constant failure rate and μ the constant repair rate, meaning

$$A(t) \equiv \Pr(Z(t) = 1) \,,$$

with

$$Z(t) \equiv \begin{cases} 1 & \text{if the system is up at time } t \\ 0 & \text{if the system is down at time } t \,. \end{cases}$$

System availability is important in achieving production rate goals, but additional processes must be invoked to assure a high level of product quality. Historically (before 1960), a *quality laboratory* would draw samples from the production line and subject them to a battery of material, dimensional, and/or functional tests, with the objective of verifying that quality was being attained *for a typical part*. In recent years, the focus has shifted from assurance of average quality to assurance of quality of *every part produced*, driven by consumer product safety concerns. *Deming* [42.18] and others were instrumental in developing methods for statistical process control, which focused on the use of quality control data to adjust process parameters in a *quality feedback loop* that assured consistently high product quality; these techniques were developed and perfected in the 1970s and 1980s. Still more recently, sensors to measure critical quality variables online have been developed, and the *quality feedback loop* is now often automated (algorithmic statistical process control). At the same time, the standards for product quality have moved up from about two sigma (1 defective product in 100) to five or six sigma (about 1 defective product in 100 000).

Increasing availability consists of reducing the number of failures (reliability) and reducing the time to repair (maintainability) according to the following for-

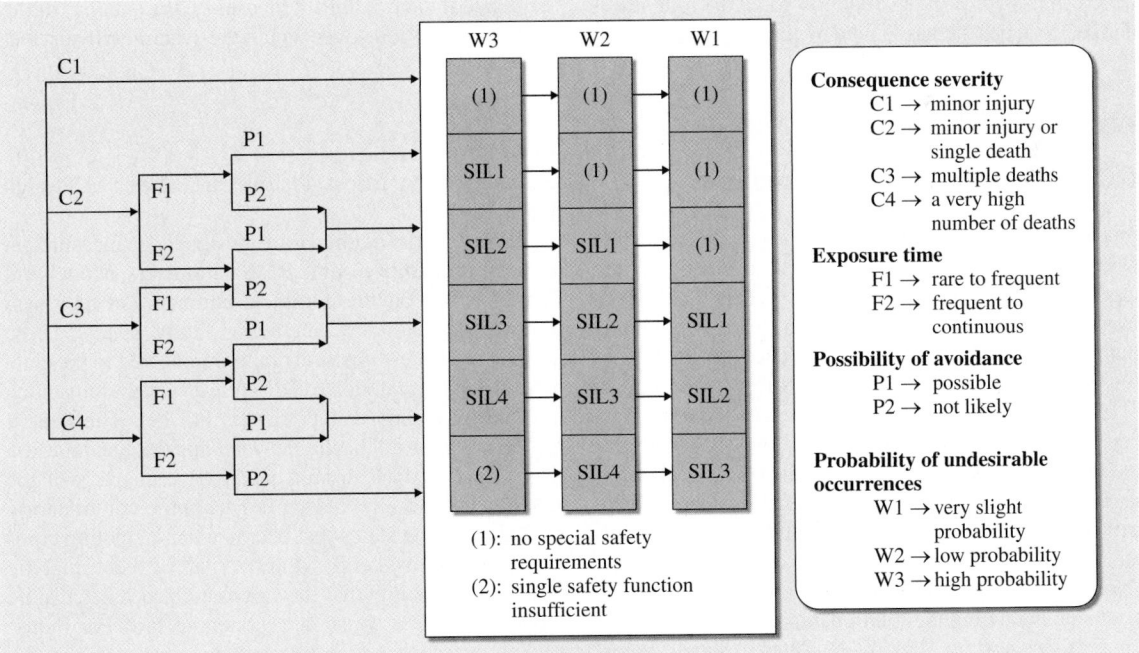

(1): no special safety
     requirements
(2): single safety function
     insufficient

Consequence severity
C1 → minor injury
C2 → minor injury or
     single death
C3 → multiple deaths
C4 → a very high
     number of deaths

Exposure time
F1 → rare to frequent
F2 → frequent to
     continuous

Possibility of avoidance
P1 → possible
P2 → not likely

**Probability of undesirable
occurrences**
W1 → very slight
     probability
W2 → low probability
W3 → high probability

Fig. 42.3 Determining safety integrity level according to IEC [42.16]

mula

$$A(\infty) = \frac{\text{MTBF}}{\text{MTBF} + \text{MTTR}}$$

as the asymptotic value of $A(t)$, where MTBF is the mean time between failures and the MTTR is the mean time to repair.

Safety is the state of being safe, the condition of the automated system being protected against catastrophic consequences to the user(s) and the environment due to component failure, equipment damage, operator error, accidents or any other undesirable abnormal event. Safety hazard mitigation can take the form of being protected from the event or from exposure to something that causes health or economical losses. It can include protection of people and limitation of environmental impact.

Industrial automation standards (Fig. 42.3), introduce engineering and design requirements that vary according to the safety integrity levels (SIL). SIL specifies the target level of safety integrity that can be determined by a risk-based approach to quantify the desired average probability of failure of a designed function, probability of a dangerous failure per hour, and the consequent severity of the failure. Combining these criteria for a given function leads to four levels of SILs that can be associated with specific engineer-

ing guidelines and architecture recommendations; for example, SIL 4 is the most critical level and the use of formal methods is strongly recommended to handle the complexity of software-intensive applications and to prove safety properties. To achieve RMS properties over the lifecycle of an automated system, two complementary activities must be undertaken:

- During the system development and design phase, the occurrence of faults should be prevented by using appropriate models and methods: quantitative approaches based on stochastic models can be used to perform a predictive RMS analysis, and qualitative approaches focusing on engineering process (e.g., Six Sigma) can be used to improve the quality of the automated system and its products.
- During the operational life of the automated system, personnel should avoid or react to undesired situations by deploying appropriate safety architectures, maintenance procedures, and management methods.

Survivability is the quantified ability of a system to continue to fulfil its mission during and after a natural or manmade disturbance. In contrast to dependability studies, which focus on analysis of system dysfunction, resilience for survivability focuses on the analysis of the range of conditions over which the system will survive.

42.2 RMS Engineering

42.2.1 Predictive RMS Assessment

To evaluate and measure the various parameters that characterize system dependability, many methods and approaches have been developed. Their goal is to provide a structured framework to represent failures qualitatively and/or quantitatively. They are mainly of two types: declarative and probabilistic.

Declarative methods are designed to identify, classify, and bracket the failures and provide methods and techniques to avoid them. Most classical models use graphical classification of failure, causes, and criticality (failure mode, effects and criticality analysis (FMECA), hazardous operation (Hazop), etc.), block diagrams, and fault trees to provide a graphical means of evaluating the relationships between different parts of the system (Fig. 42.4). These models incorporate predictions based on parts-count failure rates taken from historical data. While the predictions are often not very accurate in an absolute sense, they

are valuable to assess relative differences in design alternatives.

Probabilistic methods are designed to measure, in terms of probability, some RMS parameters. Models are mainly based on the complete enumeration of a system's possible states, including faulty states. These models use state-transition notation involved in the classical stochastic models of discrete event systems such as Markov chains and Petri nets [42.19]. The benefit of Markov and stochastic Petri net approaches relies on their capability to support quantitative analysis of the models, but these models suffer from the combinatoric explosion of the states that occurs when modeling complex industrial systems. Moreover, all of these analytic approaches assume that the stochastic processes can be modeled using a constant exponential law. For industrial processes that do not fit with this strong Markovian hypothesis, the definition of simulation models, such as Monte Carlo simulation, remains the only way to evaluate the RMS parameters.

a)

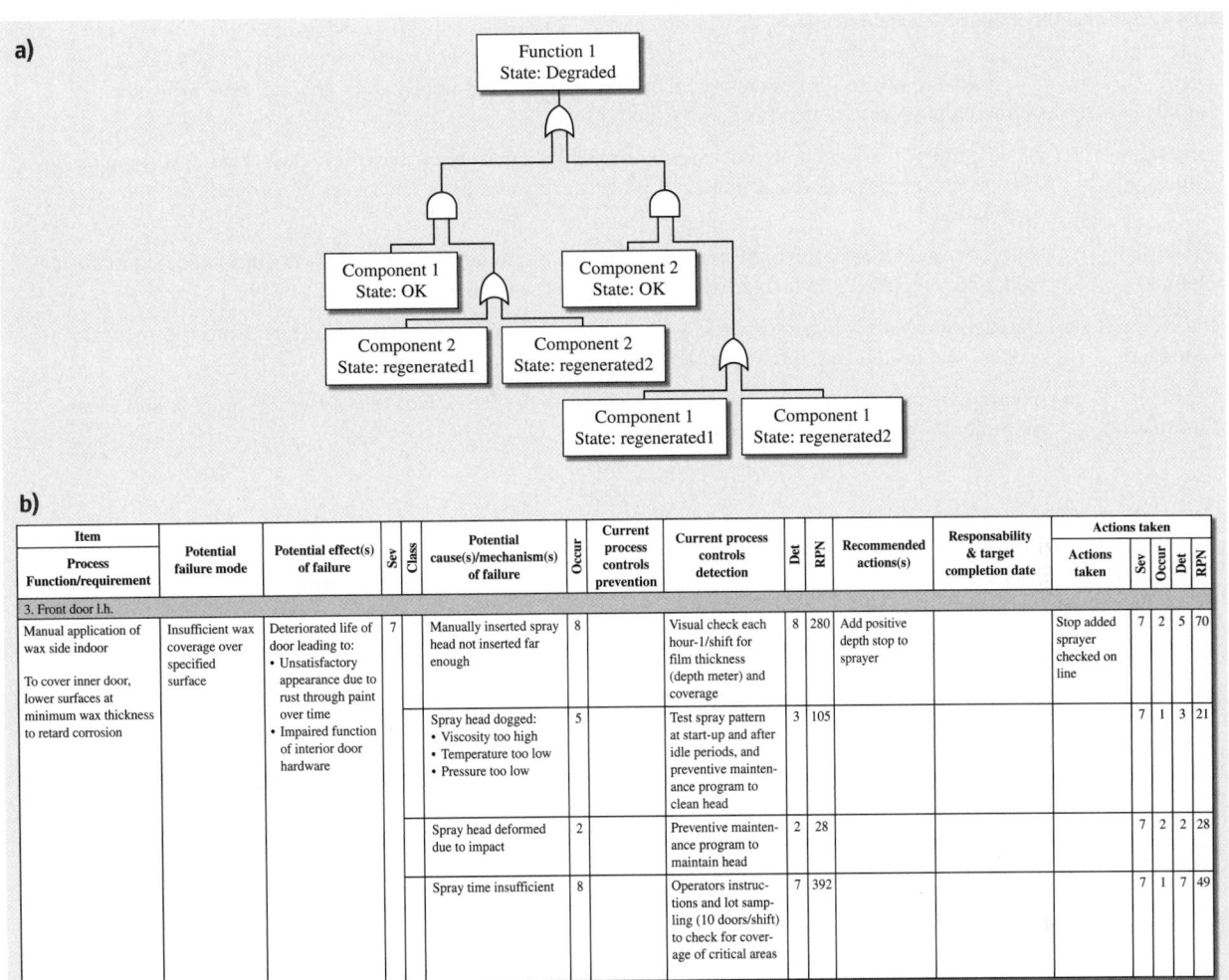

b)

Item	Potential failure mode	Potential effect(s) of failure	Sev	Class	Potential cause(s)/mechanism(s) of failure	Occur	Current process controls prevention	Current process controls detection	Det	RPN	Recommended actions(s)	Responsability & target completion date	Actions taken				
Process Function/requirement													Actions taken	Sev	Occur	Det	RPN
3. Front door l.h.																	
Manual application of wax side indoor To cover inner door, lower surfaces at minimum wax thickness to retard corrosion	Insufficient wax coverage over specified surface	Deteriorated life of door leading to: • Unsatisfactory appearance due to rust through paint over time • Impaired function of interior door hardware	7		Manually inserted spray head not inserted far enough	8		Visual check each hour-1/shift for film thickness (depth meter) and coverage	8	280	Add positive depth stop to sprayer		Stop added sprayer checked on line	7	2	5	70
					Spray head dogged: • Viscosity too high • Temperature too low • Pressure too low	5		Test spray pattern at start-up and after idle periods, and preventive maintenance program to clean head	3	105				7	1	3	21
					Spray head deformed due to impact	2		Preventive maintenance program to maintain head	2	28				7	2	2	28
					Spray time insufficient	8		Operators instructions and lot sampling (10 doors/shift) to check for coverage of critical areas	7	392				7	1	7	49

Fig. 42.4a,b Example of declarative models. (**a**) Fault tree. (**b**) FMECA (RPN – risk priority number, Sev – severity, Occur – occurence, Det – detectability (high detectability implies lower risk))

Whatever the kind of used approaches, models for predictive RMS evaluation rely upon system data collection that does not always reflect the system reality due to a gap between real and estimated states. This limitation reinforces the need to establish reliable gates between RMS engineering and system deployment to update the RMS model data with real-time information provided by the automated system.

42.2.2 Towards a Safe Engineering Process for RMS

Automation techniques have proven their effectiveness in controlling the behavior of complex systems, based on the use of suitable mathematical relationships involving feedback system dynamics during the design process. Nevertheless, the process of automating a system, as addressed by *system theory* for *automatic control*, also deals with qualitative phases [42.19] that require intuitive modeling of real phenomena (a quantity of material, energy, information, a robot, a cell, a plant, etc.) to be controlled for achieving end-user goals. The modeler's intuition remains important [42.20, 21] to build the model as an abstraction of the real system by identifying the appropriate input, output, and state variables in order to logically define the required system behavior. The main difficulty is to handle the quality of the automation engineering pro-

Table 42.1 Capability maturity model [42.24]

Level 1: Initial	The software process is characterized as ad hoc and occasionally even chaotic. Few processes are defined and success depends on individual effort.
Level 2: Repeatable	Basic project management processes are established to track cost, schedule, and functionality. The necessary process discipline is in place to repeat earlier successes on projects with similar applications.
Level 3: Defined	The software process for both management and engineering activities is documented, standardized, and integrated into a standard software process for the organization.
Level 4: Managed	Detailed measures of the software process and product quality are collected. Both the software process and product are quantitatively understood and controlled.
Level 5: Optimizing	Continuous process improvement is enabled by quantitative feedback from the process and from piloting innovative ideas and technologies.

cesses from definition and development to deployment and operation of the target system by standardization and use of best practices that are generic to well-identified problem classes and whose quality has been established by experience. *Capability maturity models* (CMM) [42.22], and *validation–verification* methods, guide engineers to combine prescriptive and descriptive models in order to meet system requirements such as RMS, but without any formal proof of accuracy of the resulting system model. Finally, the present trend to compose automation logic by assembling standardized, configurable, off-the-shelf components [42.23] strengthens the need to first better relate the modeling process and the system goals and then to preserve them through the transformation of models of the automation engineering chain. The CMM, was developed as a means of rating the thoroughness of a software development process, by the Carnegie Melon University Software Engineering Institute in the 1990s.

To pave the way toward CMM level 5, there is a growing demand for formalized methods for assuring dependability in industrial automation engineering, in order to compensate for the increasing complexity of software-intensive applications [42.25]. In particular, high levels of safety integrity, as addressed by the International Electrotechnical Commission (IEC) 61508 standard, should be formally checked and proven by mathematically sound techniques in order to verify the required completeness, consistency, unambiguity, and finally correctness of the system models throughout the definition, development, and deployment phases of the engineering lifecycle [42.26, 27].

The conformance measure of system models with regards to the requirements, and especially RMS features, can be obtained using:

- Assertion methods that include the properties to be checked in the system models proceed to an a posteriori verification using automatic techniques such as model checking [42.28].
- Refinement methods that start with the formalization of a requirement model and progressively enrich this model until a concrete model of the system that fulfils, by construction, the identified requirements is obtained. They can be based on:
  - Semiformal mechanisms that identify and classify RMS requirements and then allocate those requirements to the function, components, and equipment of the automated system. In this case, classical models combine computer-science approaches such as unified modeling language (UML) with discrete-event analysis models.
  - Formal mechanisms [42.29] that allow a sequence of formal models to be systematically derived while preserving the link between formal models and required properties (goals): an extension of the spiral method for software engineering.

All of these techniques may be combined to contribute to RMS issues [42.30, 31], but the emphasis on correct system definition is then shifted to earlier requirements analysis and elicitation phases.

42.3 Operational Organization and Architecture for RMS

Taking advantage of technological advances in the field of communications (web services embedded in programmable logic controllers) or in the field of electronics and information technology (radiofrequency identification (RFID), sensor networks, software embedded components, etc.), automated systems now include an increasing part of information technology and communication distributed at the very heart of production processes and products. However, this automation comes at a price: the complexity of the control system in terms of both heterogeneous material (dedicated computers, communications networks, supply chain operations and capture, etc.) and software functions (scheduling, control, supervisory control, monitoring, diagnosis, reconfiguration, etc.) that it houses (Fig. 42.5).

This section deals with the operational architectures and organizations required to enable active dependability of the automated system by providing information processing, storage, and communication capabilities to anticipate undesired situations or to react as effectively as possible to fault occurrences.

42.3.1 Integrated Control and Monitoring Systems

In order to maintain an acceptable quality of service, dependability should no longer be considered redundant, but should be integrated with production systems in order to be an asset in the business competitive environment. This leads to integration of additional monitoring functions with the classical control functions of an automated system in order to provide the system with the ability to reconfigure itself to continue some or all of its missions. The main idea is to avoid a complete shutdown of the system when a failure (with a consequent reduction in the productive potential of the system) occurs. Considering the system's intrinsic flexibilities, the aim is to promote system reconfiguration using a reflex loop including:

- Failure detection reports about the normal or abnormal behavior of the system. These are mainly based on a theoretical model of the functional and dysfunctional behavior of the devices involved in the automated system.

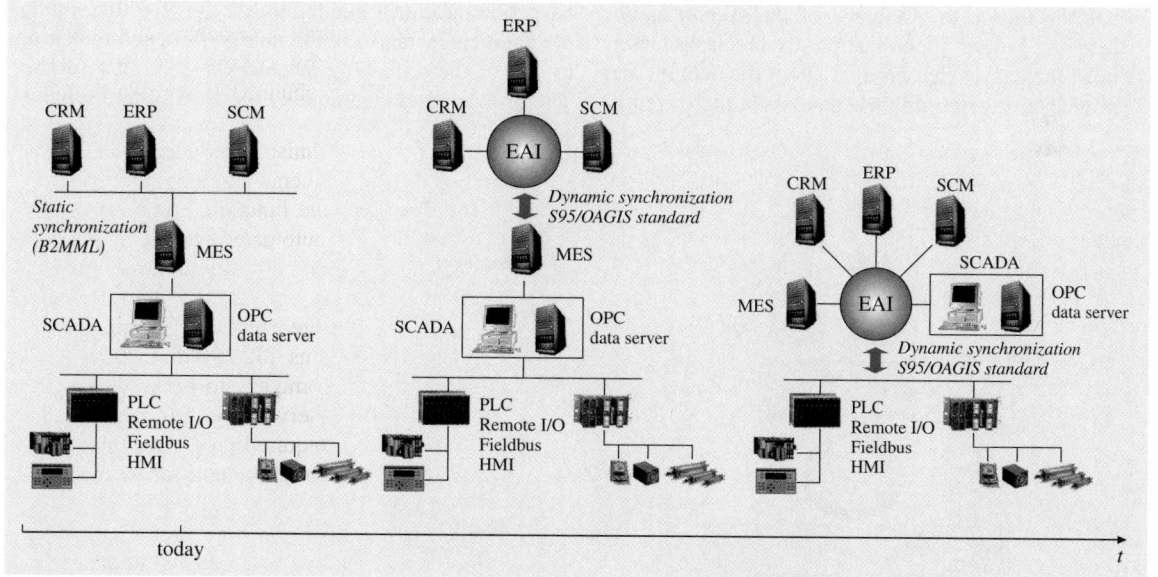

Fig. 42.5 Evolution of automated system architecture (CRM – customer relationship management, ERP – enterprise resource planning, SCM – system configuration maintenance, MES – manufacturing execution system, OPC – online process control, PLC – programmable logic controller, SCADA – supervisory control and data acquisition, EAI – enterprise architecture interface, HMI – human machine interface, OAGIS – open applications group integration specification)

Part E | 42.3

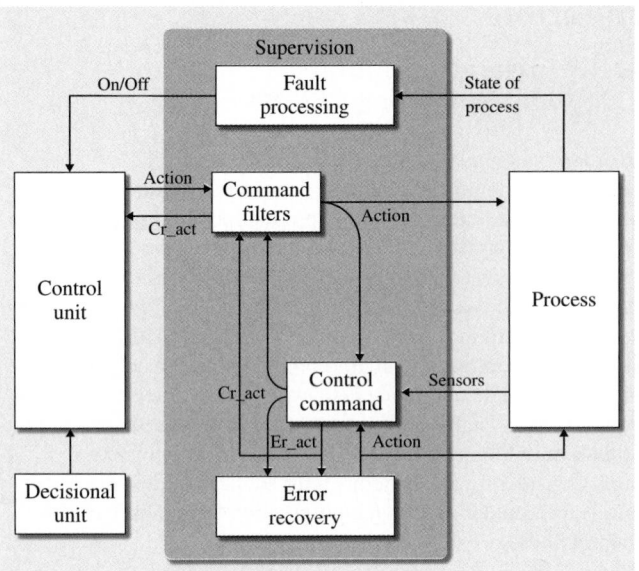

Fig. 42.6 Integrated control and monitoring systems (after [42.32])

- Diagnosis is mandated to establish a causal connection between an observed symptom and the failure that occurred, its causes, and its consequences. This function involves failure localization to isolate the failure to a subarea of the system and/or devices, failure identification to precisely determine the causes that brought about the default, and prognosis to determine whether or not there are im-

mediate consequences of the failure on the plant's future operation.
- Reconfiguration concerns reorganization of hardware and/or software of a control system to ensure production within a timeframe compatible with the specifications. This function involves decision-making activities to define the most appropriate control policy and operational activities to implement the reconfigured control actions.

Integration of monitoring [42.33, 34], diagnosis [42.35] or even prognosis into control for manufacturing systems have been widely explored for discrete-event systems (Fig. 42.6) and today provide material for identifying degradation or failure modes where control reconfiguration may be required [42.36].

Reconfiguration exploits the various flexibilities of the automated system (functional and/or material redundancies). In this way, it aims to satisfy fault-tolerance properties that characterize the ability of a system (often computer-based) to continue operating properly in the event of the failure of some of its components. One of the most important design techniques is replication – providing multiple identical instances of the same system or subsystem, directing tasks or requests to all of them in parallel, and choosing the correct result on the basis of a quorum – and redundancy – providing multiple identical instances of the same system and switching to one of the remaining instances in case of a failure. These techniques significantly increase system reliabil-

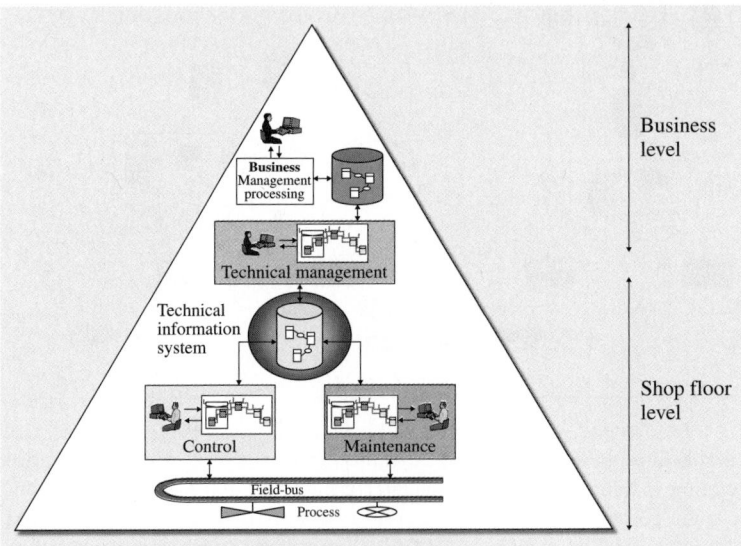

Fig. 42.7 Integrated control, maintenance, and technical management: layers of automation

ity, and are often the only viable means of doing so. However, they are difficult to design and expensive to implement, and are therefore limited to critical parts of the system.

While automation of these functions is obviously necessary for ensuring the best reactivity of the industrial production system to failure occurrence, it is nevertheless true that system stoppage is often performed by the human operator, who must act manually to put it back into a admissible state. This justifies the use of supervision and supervisory control and data-acquisition (SCADA) systems that help human operators for plant monitoring and decision-making related to the various corrective actions to be performed in order to get back to a normal functioning situation (reconfiguration, management of operating mode). Given the ever-increasing complexity of industrial processes, the burden itself tends to become difficult or even impossible. For these reasons, much research is aimed at developing and proposing solutions aimed at assisting the human operator in the phases of reconfiguration.

42.3.2 Integrated Control, Maintenance, and Technical Management Systems

Further developments of integrated control and monitoring systems have lead European projects in intelligent actuation and measurement [42.37–40] to demonstrate the benefit of integrating control, maintenance, and technical management (CMTM) activities [42.41]:

- To optimize control activities by exploiting the plant as efficiently as possible and taking into account real-time information about process status (device and function availability) provided by monitoring and maintenance activities
- To optimize the scheduling of the maintenance activities by taking into account production constraints and objectives
- To optimize, by technical management based on validated information, the operation phase by modifying control or maintenance procedures, tools, and materials

Applying this principle at the shop-floor level of the production system consists of integrating the operational activities of the CMM *agents* responsible for the plant and its lower-level interfaces with the system devices. They are also linked with the business level of the enterprise (enterprise resource planning, etc.) for business-to-manufacturing integration issues (manufacturing execution system (MES)). These oper-

ational activities are based on collaboration between human stakeholders and technical resources that support schedule management, quality management, etc., but also process management and maintenance management, which are more dependent on the e-*Connectivity* of the supporting devices.

The expected integrated organization for shop-floor activities requires that information is made available for use by all the operational activities (MES or CMM). In this way, *intelligence* embedded in field devices (e.g., devices such as actuators, sensors, PLCs (programmable logic controllers), etc.) and digital communication provide a solution to an informational representation of the production process as efficiently as possible: the system provides the right information at the right time and at the right place. In other words, the closer the data representation (e.g., in an object-oriented system) to the physical and material flows, the better the semantics of its informational representation for integration purposes (Fig. 42.7).

At the shop-floor level, local intelligence (software) allows distribution of information processing, information storage, and communication capabilities in field devices and adds to their classical roles new services related to monitoring, validation, evaluation, decision making, etc., with regard to their own operations (increased degree of autonomy) but also their application context (increased degree of component interaction).

42.3.3 Remote and e-Maintenance

Modern production equipment (manufactured by original equipment manufacturers, OEMs) is highly specialized; for example, a semiconductor manufacturing plant may have over 200 specialized production stages and over 100 equipment suppliers. In a serial process of this type, all 200 steps must operate within specification to produce an operational semiconductor at the end of the line. This type of process requires extraordinarily high reliability (and availability) of the OEM production equipment. When such equipment must be taken out of service, it is not uncommon to incur production loss rates of over 100 000 $/h, and therefore accurate diagnosis and rapid repair of equipment are essential. Since the year 2000, OEMs have increasingly provided network-capable diagnostic interfaces to equipment, so that experts do not have to come to the site to make a diagnosis or repair, but can guide plant personnel in doing this, and can order and ship parts overnight. This is often termed *e-Diagnostics*, and is crucial to maintaining high availability of production equipment. Using e-Diagnostics, a manufacturer may maintain remote service contracts with dozens of OEM suppliers to assure reliable operation of an entire production process.

In some processes, where production equipment is subject to wear or usage that is predictably related, for example, to the number of parts produced, it is possible to forecast the need for inspection, repair, or periodic replacement of critical parts, a process called *prognostics*. Although some statistical methods for prognostics (such as Weibull analysis) are well known, the ability to accurately predict the need for *service of an individual part* is still not well developed, and is not yet widely accepted. One goal of this type of analysis is *condition-based maintenance* (CBM), the practice of maintaining equipment based on its condition rather than on the basis of a fixed schedule [42.43].

Proactive maintenance is a new maintenance policy [42.44] based on prognostics, and improves on condition-based maintenance (CBM). CBM acquires real-time information in order to propose actions and to repair only when maintenance is necessary. CBM consists of equipment health monitoring to determine the equipment state; CBM is a kind of *just-in-time maintenance*. CBM is not able to predict the future state of equipment. The prognostic capability of the proactive maintenance is based on the history of the equipment operation, its current state, and its future operating conditions. The objective of proactive maintenance is to know if the system is able to accomplish its function for a given time (for example, until the next plant maintenance shutdown).

Information from control systems (distributed or not), automation, data-acquisition systems, and sensors makes it possible to measure variables continuously in order to produce symptoms or indicators of malfunction, to acquire the number of cycles of production, the time of production, the energies consumed, etc., in order to correlate this information with the diagnosis and assess the probabilities of root cause. Based on these monitoring and diagnosis functions, proactive maintenance, thanks to prognosis, propagates the drift of system behavior through time, taking into account the future exploitation conditions. Based on this

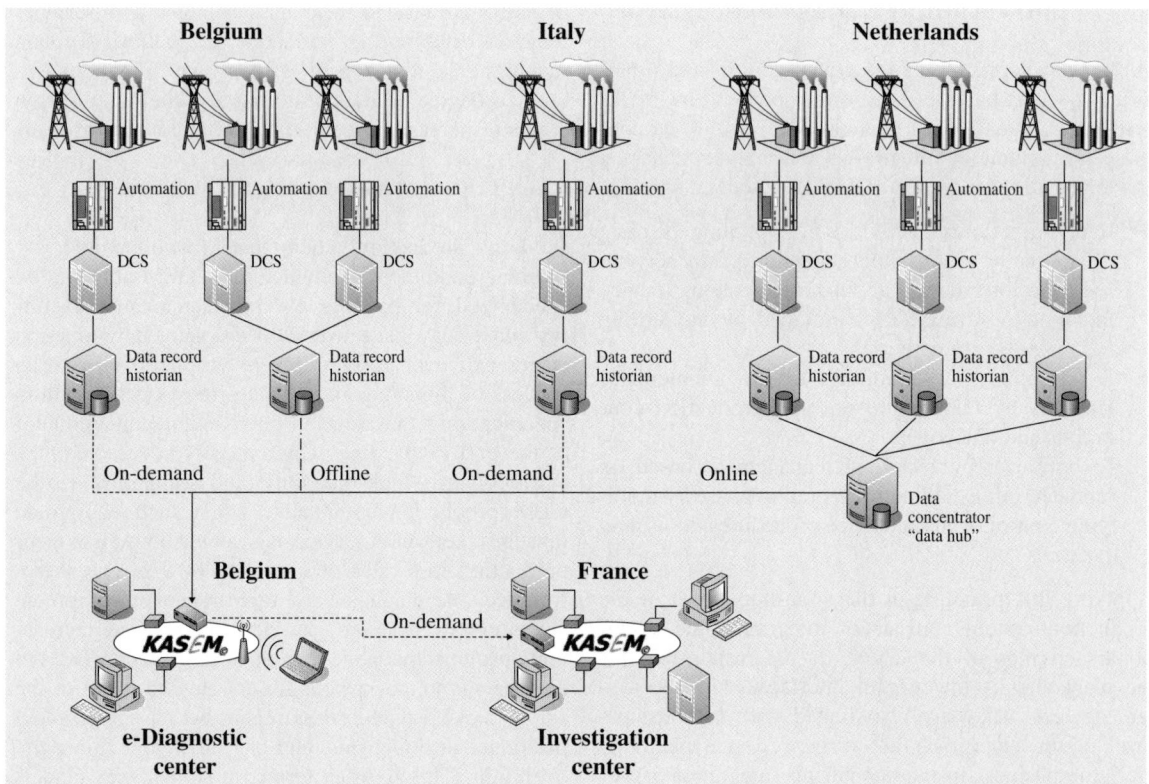

Fig. 42.8 Distributed e-Maintenance infrastructure in a power energy plant [42.42] (DCS – distributed control system)

extrapolation, prognostics can be used to evaluate the time when the drift will exceed a threshold and to propose a time before next potential failure. In this way, proactive maintenance can optimize maintenance actions and planning in order to minimize production downtime.

Proactive maintenance allows a maintenance action improvement (mean availability), to follow the degradation tendency (quality of service), to avoid the occurrence of dangerous situations (safety), and finally to support the operator with knowledge oriented to the degradation cause and effect (maintainability).

e-Maintenance is an organizational point of view of maintenance. The concept of e-Maintenance comes from remote maintenance capabilities coupled with information and communication capabilities. Remote maintenance was first a concept of remote data acquisition or consultation. Data are accessible during a limited time. In order to realize e-Maintenance objectives data storage must be organized to allow flexible access to historical data.

In order to improve remote maintenance, a new concept of e-Maintenance emerged at the end of the 1990s. The e-Maintenance concept integrates cooperation, collaboration, and knowledge-sharing capabilities in order to evolve the existing maintenance processes and to try to tend towards new enterprise concepts: extended enterprise, supply-chain management, lean maintenance, distributed support and expert centers, etc. Based on web technologies, the e-Maintenance concept is nowadays available and industrial e-Maintenance software platforms exist. e-Maintenance platforms (sometimes termed asset management systems) manage the whole of the maintenance processes throughout the system lifecycle from engineering, maintenance, logistic, experience feedback, maintenance knowledge capitalization, optimization, etc. to reengineering and revamping.

e-Maintenance is not based on software functions but on maintenance services that are well-defined, self-contained, and do not depend on the context or state of other services. So, with the advent of service-oriented architectures (SOA) and enterprise service-bus technologies [42.45], e-Maintenance platforms are easy to evolve and can provide interoperability, flexibility, adaptability, and agility. e-Maintenance platforms are a kind of *hub* for maintenance services based on existing, new, and future applications.

42.3.4 Industrial Applications

Industrial software platforms have been developed during the 1990s in order to provide the proof of concept of this RMS modeling framework before marketing off-the-shelf products. The first applications have appeared since 2000 in various sectors such as power energy, steel factory, petrochemical process, navy logistics and maintenance support, nuclear fuel manufacturing and waste treatment, etc.

A common objective of these multisector applications is to reduce operation costs by increasing the availability, maintainability, and reliability of plants and systems, and to facilitate their compliance with regulation laws. Another common objective is to elicit and save the implicit knowledge acquired by skilled operators as well as by skilled engineers when performing their tasks. Others objectives are specific to an industrial sector; for example, understanding complex phenomena to anticipate maintenance operations is critical to optimize the impact of shutdown and startup operations in process plants [42.46].

Return on investment is estimated to be at most 1 year from these industrial experiments, and leads to a distributed service-oriented e-Maintenance infrastructure to warrantee by contract a level of availability in plant operation (Fig. 42.8).

42.4 Challenges, Trends, and Open Issues

All aspects of dependability such as reliability, maintainability, and safety should be viewed in a broader context depending on both management and technical processes within the enterprise system to ensure the necessary resilience to intrinsic and extrinsic complex phenomena occurring when systems are operating in changing environments. For example, MTBF is a measure of the random nature of an event and does not predict when something will fail but only predicts the probability that a system will fail within a certain time

boundary. Contrary to conventional wisdom, accidents often result from interactions between perfectly functioning components, i.e., before a system has reached its expected life as predicted by RMS analysis.

Such considerations underscore that other advanced concepts are beyond traditional RMS analyses and the individual mind-set of each engineering discipline to cope with emergent behavior as one of the results of complexity. In other words, dependability assumes that cause–effect relationships can be ordered in known and knowable ways, while resilience [42.47] should confine the contextual emergence of complex relationships within the system and between the system and its environment in unordered ways [42.48].

An initial challenge is to understand that the concept of system as the unique result of normal emergence within a collaborative systems engineering process leads to an ad hoc solution based on heuristics and normative process-driven guidelines [42.49].

A second challenge relies on weak emergence [42.50] to perceive, model, and check added behaviors

due to the interactions between the component systems. This should be led by extensive model-driven requirements analysis adding more details than current practices and complementary experiments such as multiagent simulation to track self-organizing patterns in order to improve component systems' adaptability.

A third open challenge deals with the quality of the engineering process to determine whether a system can survive a strongly emergent event, as well as the adaptability of the whole enterprise to come into play in facing an inevitable systemic instability.

References

42.1   T.L. Johnson: Improving automation software dependability: a role for formal methods?, Control Eng. Pract. **15**(11), 1403–1415 (2007)

42.2   J. Stark: *Handbook of Manufacturing Automation and Integration* (Auerbach, Boston 1989)

42.3   R.S. Dorf, A. Kusiak: *Handbook of Design, Manufacturing and Automation* (Wiley, New York 1994)

42.4   A. Ollero, G. Morel, P. Bernus, S.Y. Nof, J. Sasiadek, S. Boverie, H. Erbe, R. Goodall: From MEMS to enterprise systems, IFAC Annu. Rev. Control **26**(2), 151–162 (2002)

42.5   S.Y. Nof, G. Morel, L. Monostori, A. Molina, F. Filip: From plant and logistics control to multi-enterprise collaboration, IFAC Annu. Rev. Control **30**(1), 55–68 (2006)

42.6   G. Morel, P. Valckenaers, J.M. Faure, C.E. Pereira, C. Diedrich: Manufacturing plant control challenges and issues, IFAC Control Eng. Pract. **15**(11), 1321–1331 (2007)

42.7   A. Avizienis, J.C. Laprie, B. Randell, C. Landwehr: Basic Concepts and Taxonomy of Dependable and Secure Computing, IEEE Trans. Dependable Secur. Comput. **1**(1), 11–33 (2004)

42.8   S.E. Rigdon, A.P. Basu: *Statistical Methods for the Reliability of Repairable Systems* (Lavoisier, Paris 2000)

42.9   J. Moubray: *Reliability-Centered Maintenance* (Industrial, New York 1997)

42.10   A. Avizienis, J.C. Laprie, B. Randell: Fundamental concepts of dependability, LAAS Techn. Rep. **1145**, 1–19 (2001), http://www.laas.fr

42.11   J.W. Foster, D.T. Philips, T.R. Rogers: *Reliability Availability and Maintainability: The Assurance Technologies Applied to the Procurement of Production Systems* (MA Press, 1979)

42.12   M. Pecht: *Product Reliability, Maintainability and Supportability Handbook* (CRC, New York 1995)

42.13   H. Erbe: Technologies for cost-effective automation in manufacturing, IFAC Professional Briefs (2003) pp. 1–32

42.14   IEEE: *IEEE Standard Computer Dictionary: A Compilation of IEEE Standard Computer Glossaries* (IEEE, 1990), http://ieeexplore.ieee.org/xpls/abs_all.jsp?tp=&isnumber=4683&arnumber=182763&punumber=2267

42.15   D. Kumar, J. Crocker, J. Knezevic, M. El-Haram: *Reliability, Maintenance and Logistic Support. A life Cycle Approach* (Springer, Berlin, Heidelberg 2000)

42.16   IEC 61508: Functional safety of electrical/electronic/programmable electronic (E/E/PE) safety-related systems

42.17   T. Nakagawa: *Maintenance Theory of Reliability* (Springer, London 2005)

42.18   W.E. Deming: *Out of the Crisis: For Industry, Government, Education* (MIT Press, Cambridge 2000)

42.19   C.G. Cassandras, S. Lafortune: *Introduction to Discrete Event Systems* (Kluwer Academic, Norwell 1999)

42.20   F. Lhote, P. Chazelet, M. Dulmet: The extension of principles of cybernetics towards engineering and manufacturing, Annu. Rev. Control **23**(1), 139–148 (1999)

42.21   N. Viswanadham, Y. Narahari: *Performance Modeling of Automated Manufacturing Systems* (Prentice-Hall, Englewood Cliffs 1992)

42.22   http://www.sei.cmu.edu/cmmi

42.23   http://www.oooneida.info

42.24   M.C. Paulk: How ISO 9001 compares with the CMM, IEEE Softw. **12**(1), 74–83 (1995)

42.25   K. Polzer: Ease of use in engineering – availability and safety during runtime, Autom. Technol. Pract. **1**, 49–60 (2004)

42.26   T. Shell: Systems functions implementation and behavioural modelling: system theoretic approach, Int. J. Syst. Eng. **4**(1), 58–75 (2001)

42.27   A. Moik: Engineering-related formal method for the development of safe industrial automation systems, Autom. Technol. Pract. **1**, 45–53 (2003)

42.28   E.M. Clarke, O. Grunberg, D.A. Peled: *Model Checking* (MIT Press, Cambridge 2000)

42.29 J.R. Abrial: *The B Book: Assigning Programs to Meanings* (Cambridge Univ. Press, Cambridge 1996)

42.30 T. Kim, D. Stringer-Calvert, S. Cha: Formal verification of functional properties of a SCR-style software requirements specification using PVS, Reliab. Eng. Syst. Saf. **87**, 351–363 (2005)

42.31 J. Yoo, T. Kim, S. Cha, J.-S. Lee, H.S. Son: A formal software requirements specification method for digital nuclear plant protection systems, Syst. Softw. **74**(1), 73–83 (2005)

42.32 S. Elkhattabi, D. Corbeel, J.C. Gentina: Integration of dependability in the conception of FMS, 7th IFAC Symp. on Inf. Control Probl. Manuf. Technol., Toronto (1992) pp. 169–174

42.33 R. Vogrig, P. Baracos, P. Lhoste, G. Morel, B. Salzemann: Flexible manufacturing shop, Manuf. Syst. **16**(3), 43–55 (1987)

42.34 E. Zamaï, A. Chaillet-Subias, M. Combacau: An architecture for control and monitoring of discrete events systems, Comput. Ind. **36**(1–2), 95–100 (1998)

42.35 A.K.A. Toguyeni, E. Craye, L. Sekhri: Study of the diagnosability of automated production systems based on functional graphs, Math. Comput. Simul. **70**(5–6), 377–393 (2006)

42.36 M.G. Mehrabi, A.G. Ulsoy, Y. Koren: Reconfigurable manufacturing systems: key to future manufacturing, J. Intell. Manuf. **11**(4), 403–419 (2000)

42.37 ESPRIT II-2172 DIAS Distributed Intelligent Actuators and Sensors

42.38 ESPRIT III-6188 PRIAM Pre-normative Requirements for Intelligent Actuation and Measurement

42.39 ESPRIT III-6244 EIAMUG European Intelligent Actuation and Measurement User Group

42.40 ESPRIT IV-23525 IAM-PILOT Intelligent Actuation and Measurement Pilot

42.41 J.F. Pétin, B. Iung, G. Morel: Distributed intelligent actuation and measurement system within an integrated shop-floor organisation, Comput. Ind. J. **37**, 197–211 (1998)

42.42 http://www.predict.fr

42.43 http://www.openoandm.org

42.44 B. Iung, G. Morel, J.-B. Léger: Proactive maintenance strategy for harbour crane operation improvement, Robotica **21**, 313–324 (2003)

42.45 F.B. Vernadat: Interoperable enterprise systems: Principles, concepts and methods, IFAC Annu. Rev. Control. **31**(1), 137–145 (2007)

42.46 D. Galara: Roadmap to master the complexity of process operation to help operators improve safety, productivity and reduce environmental impact, Annu. Rev. Control **30**, 215–222 (2006)

42.47 http://www.resilience-engineering.org

42.48 C.F. Kurtz, D.J. Snowden: The new dynamics of strategy: sense-making in a complex and complicated world, IBM Syst. J. **42**(3), 462–483 (2003)

42.49 ISO/IEC 15288, http://www.incose.org

42.50 M. Bedau: *Weak Emergence, Philosophical Perspectives: Mind, Causation and World*, Vol. 11 (Blackwell, Oxford 1997)

Part E | 42

43. Product Lifecycle Management and Embedded Information Devices

Dimitris Kiritsis

The closed-loop product lifecycle management (PLM) system focuses on tracking and managing the information of the whole product lifecycle, with possible feedback of information to product lifecycle phases. It provides opportunities to reduce the inefficiency of lifecycle operations and gain competitiveness. Thanks to the advent of hardware and software related to product identification technologies, e.g., radiofrequency identification (RFID) technology, recently closed-loop PLM has been highlighted as a tool of companies to enhance the performance of their business models. However, implementing the PLM system requires a high level of coordination and integration. In this chapter we present the background methodologies and techniques and the main components for closed-loop PLM and how they are related to each other. We start with the concept of closed-loop PLM and a system architecture in Sect. 43.1. In Sect. 43.2 we describe the necessary components for closed-loop PLM and how to integrate and coordinate them with respect to business models, hardware, and software. In Sect. 43.3 we propose a development guide based on experiences gathered from prototype applications developed to date. In Sect. 43.4 we introduce a real case example that implements a closed-loop PLM solution focusing on

Part E | 43

end-of-life (EOL) of vehicles (ELV). Finally, Sect. 43.5 discusses some challenging issues and emerging trends in the implementation of closed-loop PLM.

43.1 The Concept of Closed-Loop PLM

Product lifecycle management (PLM) is a new strategic approach to manage product-related information efficiently over the whole product lifecycle. Conceived as an extension to product data management (PDM), its vision is to provide more product-related information to the extended enterprise over the whole product lifecycle. Its concept appeared in the late 1990s, moving

beyond the engineering aspects of a product and providing a shared platform for creation, organization, and dissemination of product-related knowledge across the extended enterprise [43.1]. PLM facilitates the innovation of enterprise operations by integrating people, processes, business systems, and information throughout product lifecycle and across extended enterprise. It aims to derive the advantages of horizontally connecting functional silos in organizations, enhancing information sharing, efficient change management, use of past knowledge, and so on [43.2]. To meet this end, a PLM system should be able to monitor the progress of a product at any stage in its lifecycle, to analyze issues that might arise at any product lifecycle phase, to make suitable decisions to address problems, and to execute and enforce these decisions. In spite of its vision, PLM as defined above has not received much attention so far from industry because there are no efficient tools to gather product data over the entire product lifecycle. However, recent applications of product identification technologies in various PLM aspects [43.3–9] demonstrate that a sound technological framework is now available for PLM to implement its vision. Product identification technologies enable products to have embedded information devices (e.g., RFID tags and on-board computers), which makes it possible to gather the whole lifecycle data of products at any time and at any place. A new generation of PLM systems based on product identification technologies will make the whole product lifecycle totally visible and will allow all actors involved in the product lifecycle to access, manage, and control product-related information, especially information after product delivery to customers and up to its final destiny, without temporal or spatial constraints. During the whole product lifecycle, we can now have visibility of not only forward but also backward information flow; for example, beginning of life (BOL) information related to product design and production can be used to streamline operations of middle of life (MOL) and end of life (EOL). Furthermore, MOL and EOL information can also go back to designers and production engineers for the improvement of BOL decisions. This indicates that information flow is horizontally closed over the whole product lifecycle. In addition, based on data gathered by product embedded information devices (PEID), we can analyze product-related information and take some decisions on the behavior of products, which will affect data gathering again [43.10]. This means that information flow is also vertically closed. We call this concept and relevant systems the *closed-loop PLM*. The concept of closed-loop PLM can be defined as follows: *a strategic business approach for the effective management of product lifecycle activities by using product data/information/knowledge which can compensate PLM to realize product lifecycle optimization dynamically in closed loops with the support of PEIDs and product data and knowledge management (PDKM) system.*

The objective of closed-loop PLM is to optimize the performance of product lifecycle operations over the whole product lifecycle, based on seamless product information flow through a local wireless network of PEIDs and associated devices and through remote Internet connection [43.11] to knowledge repositories in PDKM. In addition to PEIDs, sensors can be built in products and linked to PEIDs for gathering status data [43.12]. During product lifecycle, each lifecycle actor can have access to PEIDs locally with PEID controllers (e.g., RFID readers) or to a remote PLM system for getting necessary information. Furthermore, in closed-loop PLM, decision support systems (DSS) integrated to PDKM systems may provide lifecycle actors with suitable advice or decision support at any time.

In the closed-loop PLM, all business activities performed along the product lifecycle must be coordinated and efficiently managed. Although there are a lot of information flows and interorganizational workflows, the business operations in closed-loop PLM are based on the interactions among three organizations: the *PLM agent*, *PLM system*, and *product*. The PLM agent can gather product lifecycle information quickly from each product with a mobile device such as a personal digital assistant (PDA) or a fixed reader with built-in antenna. He sends information gathered at each site (e.g., retail sites, distribution sites, and disposal plants) to a PLM system, as illustrated in Fig. 43.1.

A PLM system provides lifecycle information or knowledge generated by PLM agents through product lifecycle activities realized through the three main product lifecycle phases: BOL, MOL, and EOL.

BOL is the phase where the product concept is generated and subsequently physically realized. In the closed-loop PLM, designers and production engineers will receive feedback about detailed product information from distributors, maintenance/service engineers, customers or remanufacturers on product status, product usage, product service, conditions of retirement, and disposal of their products. The feedback information is extremely valuable for product design and production because designers and production engineers are able to exploit expertise and knowhow of other actors in the product lifecycle. Hence, closed-loop PLM can improve the quality of product design and the efficiency of production.

MOL is the phase where products are distributed, used, maintained, and serviced by customers or engineers. In the closed-loop PLM, a PEID can log the product history related to distributing routes, usage conditions, failure, maintenance or service events, and

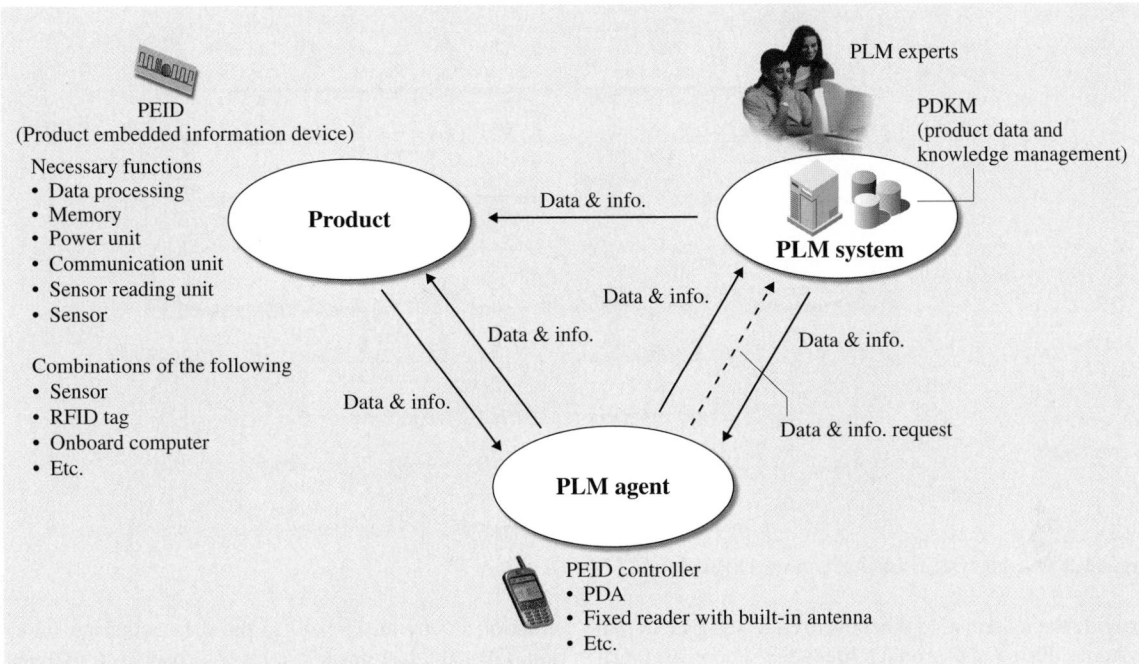

Fig. 43.1 Basic framework for PEID applications in PLM

so on. This information is later gathered into a PLM system for analysis and sharing. Thus, during MOL, an up-to-date report about the status of products and real-time assistance can be obtained from this system through the Internet or wireless mobile technology. Based on these feedbacks, predictive maintenance can be done by maintenance engineers [43.13]. Furthermore, optimizing logistics operations for maintenance and service can be facilitated.

EOL is the phase where EOL products are collected, disassembled, refurbished, recycled, reassembled, reused or disposed. It can be said that EOL starts from the time when the product no longer satisfies an initial purchaser [43.14]. In the closed-loop PLM, the use of PEIDs can greatly increase the effectiveness of EOL management; for example, material recycling can be significantly improved because recyclers and reusers can obtain accurate information about *valuable parts and materials* arriving via EOL routes: what materials they contain, who manufactured them, and other knowledge that facilitates material reuse [43.15].

43.2 The Components of a Closed-Loop PLM System

The components of a closed-loop PLM system and their relations are presented in the five layers of the system architecture schema shown in Fig. 43.2 [43.16].

These layers are mainly classified into business process, software, and hardware. PEID is an important hardware component for facilitating the closed-loop PLM concept. Furthermore, software related to applications and middleware layers, and their interfaces play important roles in closed-loop PLM. Following are some more details about the components of whole

closed-loop PLM system: PEID, middleware, DSS, and PDKM.

43.2.1 Product Embedded Information Device (PEID)

PEID stands for *product embedded information device*. It is defined as a device embedded in (or attached to) a product, which contains information about the product [e.g., product identity (ID), and which is able to

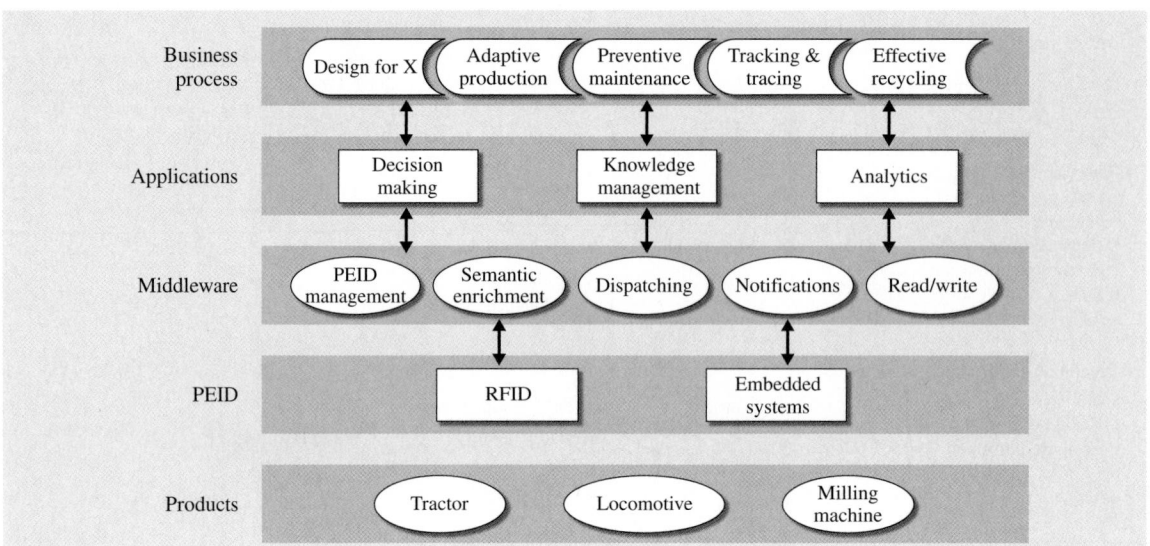

Fig. 43.2 Overall system architecture for closed-loop PLM

provide the information whenever requested by external systems during the product lifecycle. There are various kinds of information devices built in products to gather and manage product information, for example, various types of RFID tags and onboard computers. A PEID has a unique ID and provides data gathering, processing, and data-storage functions. Power management of a PEID is important to allow it to provide its functionality along the product lifecycle. Particular attention is obviously paid to the data gathering function. Over the last decade, a lot of sensor technologies have been developed to gather environmental status data of products related to mechanical, thermal, electrical, magnetic, radiant, and chemical data [43.12, 17]. These sensor technologies can be incorporated into the PEID to gather the history of product status with the data gathering function. Eventually, these functions enable a PEID to gather data from several sensors, to retain or

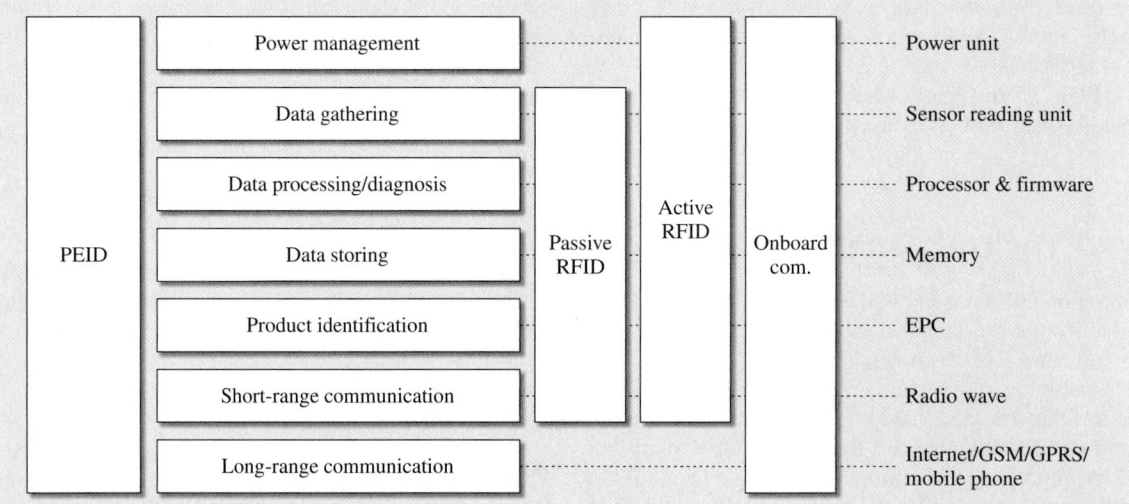

Fig. 43.3 PEID functions and types (EPC – electronic product code, GSM – global system for mobile communications, GPRS – general packet radio service)

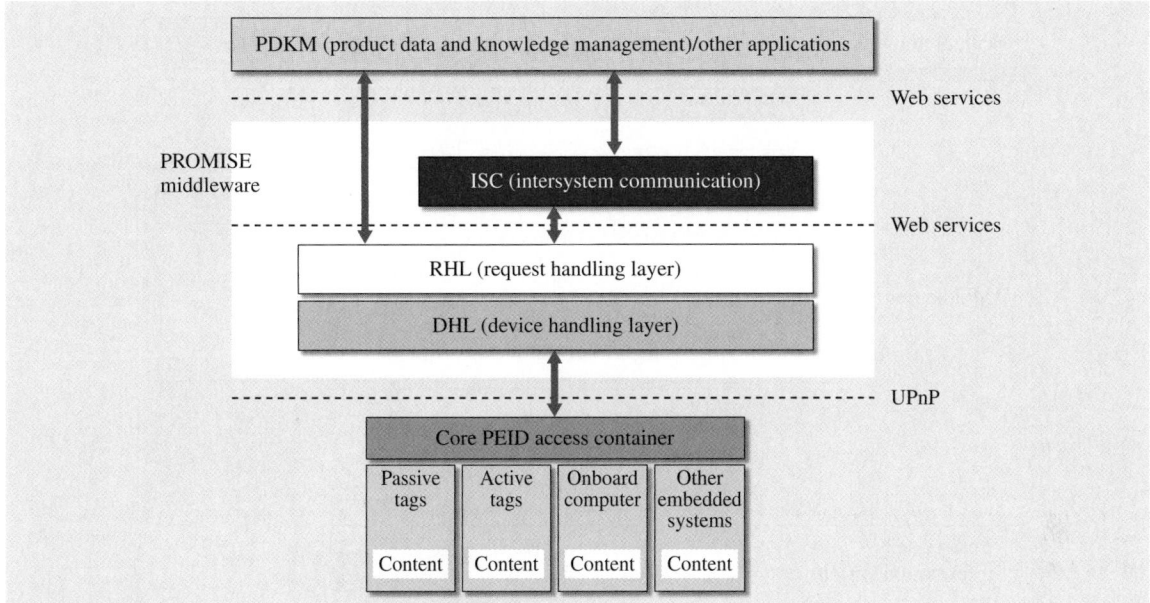

Fig. 43.4 Middleware architecture (UPnP – universal plug and play)

store them, and if necessary to analyze them or support associated decision making. In addition, it should have a communication function with external environments for exchanging data. For this, a PEID should have a processing unit, communication unit, sensor reader, data processor, and memory. Depending on the combinations of these functions, the PEID has several types such as passive RFID tag, active RFID tag, and onboard computer. In particular the manufacturing cost of the PEID is greatly affected by power management and data function specification. Hence, the PEID should be carefully designed considering application characteristics. The overall architecture of PEID is depicted in Fig. 43.3.

43.2.2 Middleware

Middleware can be considered as intermediate software between different applications. Developing middleware is one of the most challenging areas in the closed-loop PLM since it is the core technology to efficiently gather and distribute PEID data. It plays a role as the interface between different software layers, e.g., between PEIDs and PDKM, as shown in Fig. 43.4. It is used to support complex and distributed applications, e.g., applications between RFID tags and business information systems, to communicate, coordinate, and manage data by converting the data in a proper way. In the closed-loop PLM, it has a role to map the low-level data gathered from PEID readers to more meaningful data of other high-level application such as field DB/PDKM and PLM business applications. There are several issues to be resolved: data security, consistency, synchronization of data, tracking and tracing, exception handling, and so on. Figure 43.4 shows the overall architecture of the middleware developed in the PROMISE (product lifecycle management and information tracking using smart embedded systems) project (www.promise-plm.com).

43.2.3 Decision Support System (DSS)

Decision support system (DSS) software provides lifecycle actors with the ability to transform gathered data into necessary information and knowledge for specific applications. To this end, diagnosis/analysis tools for gathered data and data transformer are required. There are a lot of decision support areas which are highlighted in the closed-loop PLM, mainly transforming lifecycle information of other lifecycle phases into streamlining current lifecycle operations; for example, main areas in decision support for MOL include assisting efficient maintenance diagnosis and prognosis, whereas in EOL this includes efficient waste management. Figure 43.5 shows the overall architecture of the middleware developed in the PROMISE project (see also Chap. 87 on *Decision Support Systems*).

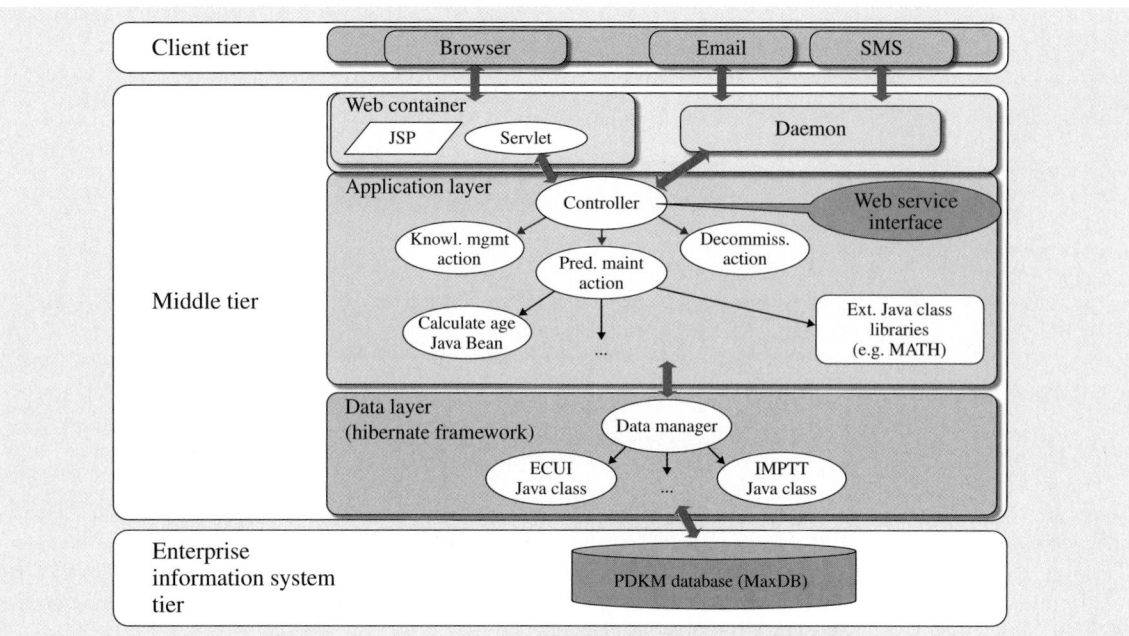

Fig. 43.5 Decision support system architecture (JSP – JavaServer pages)

43.2.4 Product Knowledge and Management System (PDKM)

The PDKM manages information and knowledge generated during the product lifecycle. It is generally linked with decision support systems and data transformation software. PDKM is a process and the associated technology to acquire, store, share, and secure understandings, insights, and core distinctions. PDKM should link not only product design and development such

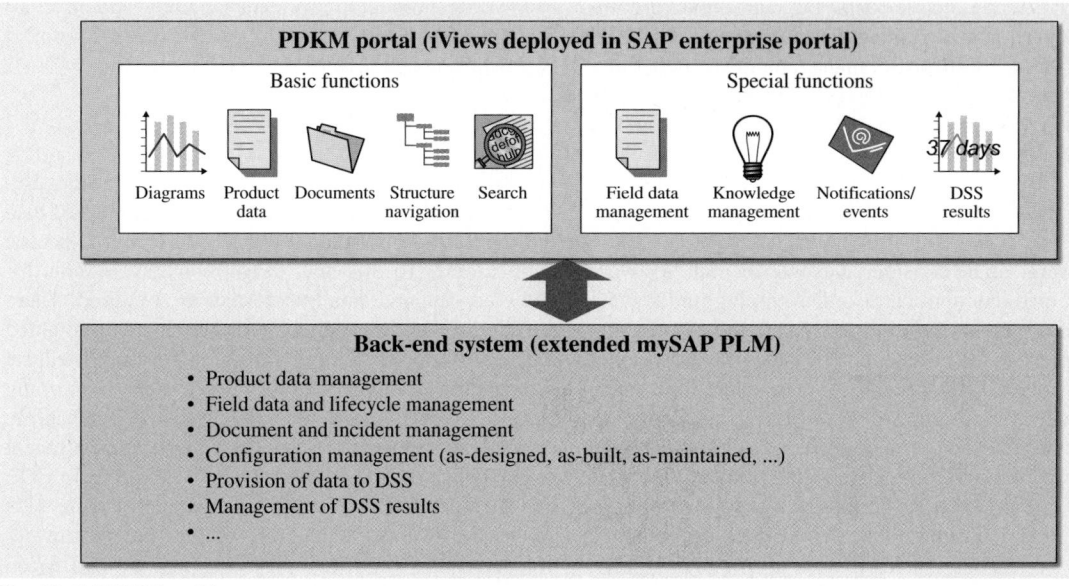

Fig. 43.6 PDKM architecture

as computer-aided design/manufacture (CAD/CAM) but also other back-end software to achieve interoperability of all activities that affect a product and its lifecycle. Figure 43.6 shows the overall architecture of the PDKM developed in the PROMISE project.

43.3 A Development Guide for Your Closed-Loop PLM Solution

In this section we describe the main elements of the development of a closed-loop PLM solution: modeling, selection of PEID system, data and data flow definition, PDKM, DSS and middleware.

43.3.1 Modeling

PLM has specific objectives at each phase of the lifecycle: BOL, MOL or EOL; for example, at BOL, improving product design and production quality are main concerns. During MOL, improving reliability, availability, and maintainability of products are the most interesting issues. In EOL, optimizing EOL product recovery operations is one of the most challenging issues.

It is advisable to begin the development of a closed-loop PLM solution by modeling the various characteristics of the solutions we want to develop. If, for example, we consider the EOL phase of a product, a use-case diagram such as the one shown in Fig. 43.7 below will help to identify the main actors and activities of the solution. This model shows how a PLM system,

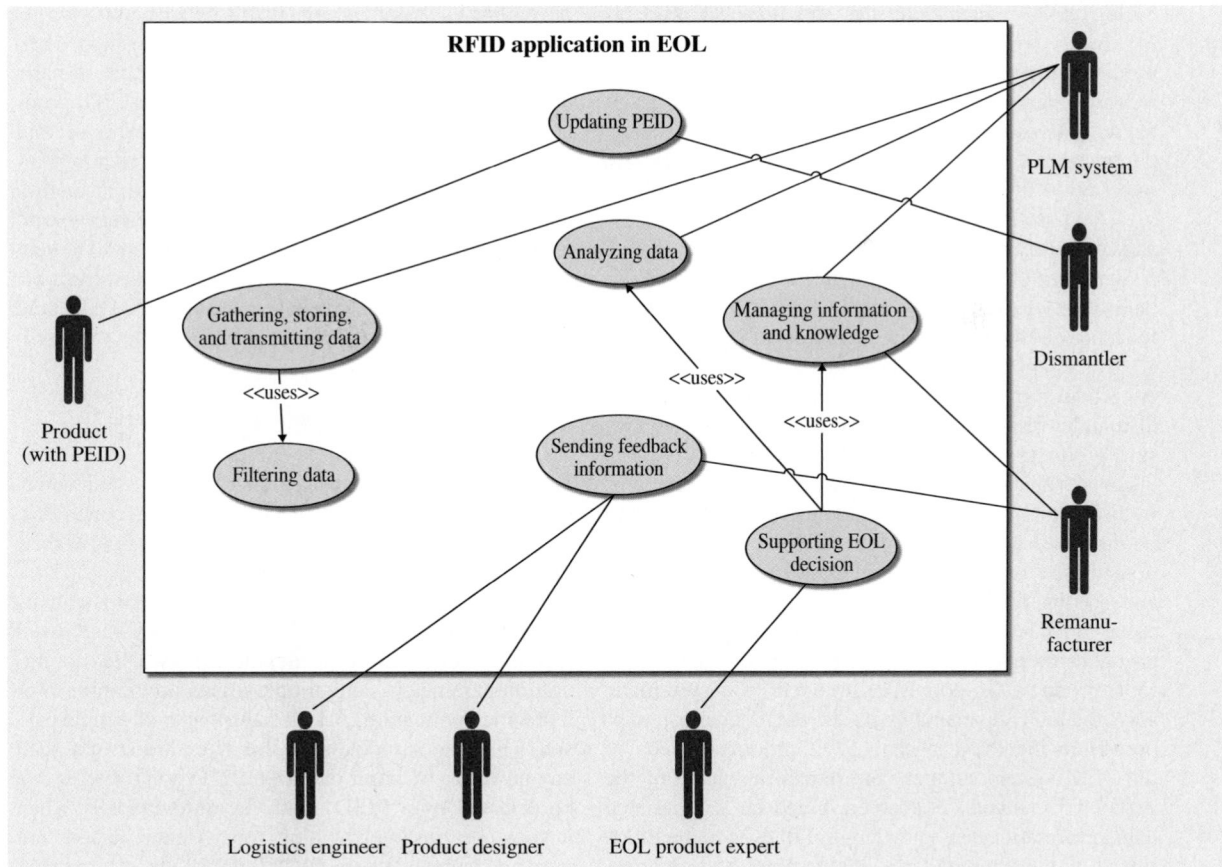

Fig. 43.7 Use case for PEID application at EOL

using PEID technology, can gather accurate data related to product lifecycle history at the collecting and dismantling phase of EOL products, e.g., which components they consist of, what materials they contain, who manufactured them, and other data that facilitate reuse of materials, components, and parts. Based on gathered data, EOL product experts in the PLM system can predict degradation status and remaining lifetime of parts or components. With this information, at the inspection phase, the dismantler can implement *EOL product recovery optimization*, in other words, deciding on suitable EOL recovery options such as recycle, reuse, remanufacturing, and disposal, with the objective of maximizing values of EOL products considering product status. This decision also provides useful information to remanufacturers for making an efficient remanufacturing plan in advance. Furthermore, logistics engineers can improve logistics at EOL (*reverse logistics*) from collecting to remanufacturing, reuse or disposal. They can obtain supply volume data for recycle, reuse, remanufacturing, and disposal products in advance from the EOL decision. In addition, EOL product recovery decision data and product status at EOL dismantling can give useful information to product designers for improving product design with several purposes, e.g., design for reliability, reuse, recycle, and so on.

The next step of modeling concerns the process and events of the solution. This is well achieved with a swim-lane chart. Figure 43.8 below shows the swim-lane chart of closed-loop application at EOL, mainly focusing on EOL product recovery optimization.

In this application, at first, the EOL collector gathers products that have lost their values. Then, the EOL dismantler inspects collected products visually. As a result, products can be simply classified into two parts: disposal, and disassembly for more detailed inspections. In disassembly, the concerned components or parts will be inspected in detail and sorted into several EOL options based on some criteria. During the inspection and sorting process, if necessary, the dismantler accesses PEIDs of the parts or components concerned to gather necessary data for inspecting and sorting the EOL products. To sort EOL products in a systematic way, the EOL dismantler asks for EOL decision support from the PLM system. EOL product experts in the PLM system estimate the remaining value of the parts or components concerned, based on accumulated data, information, and knowledge at PDKM in the PLM. Based on the estimated remaining values and other information such as costs and benefits of recycle, reuse,

remanufacture, disposal, and so on, EOL product experts decide on an adequate EOL option for each part or component, i.e., which parts or components should be recycled, reused, remanufactured or disposed, under some constraints related to environmental regulation and product quality. This information will be stored in the PDKM and transmitted to dismantlers. If necessary, product designer and logistics engineer receive this information from the PLM system to improve their operations.

Based on the proposed EOL decision, dismantlers sort the parts or components. When the EOL dismantler sorts products, depending on the sorting results of EOL products, operations related to PEIDs may be different. They may be removed and replaced with new ones; or its data contents can be reset or updated without replacement for the second life of parts or components; for example, in the recycle case, PEIDs will usually be detached from products. Then, recyclable products will be sent to specific lots that have similar materials features. For each lot, a new PEID will be used for its management. Each lot will be sent to recycling companies. In the reuse case, after quality data of parts or components are updated to an existing PEID, products will be sent to a remanufacturing site or second market. In the remanufacturing case, after required information for remanufacturing such as current quality data, required quality data, product specification, and production instruction are updated, products will be sent to remanufacturing sites. In the case of disposal, after updating disposal-relevant data to each PEID and PLM system by disposal engineers, products are sent to disposal companies.

43.3.2 Selection of PEID System

Table 43.1 shows the basic functions and their corresponding components of a PEID with its specifications. Here, PEIDs can be classified into four types by their functions and specifications.

Type A is for simple applications. It contains only its own identification function. For this, it has a small-sized read-only microchip that includes its own ID, configuration parameters, and simple logics programmed for a specific application. A 1 bit transponder or simple passive RFID tag is included in this type. It does not need any power to transmit data to a PEID controller. It can be detected by a PEID controller automatically when it goes into the interrogation zone. Hence, it does not need any battery. When lifecycle actors just want to read a small amount of product identification data without

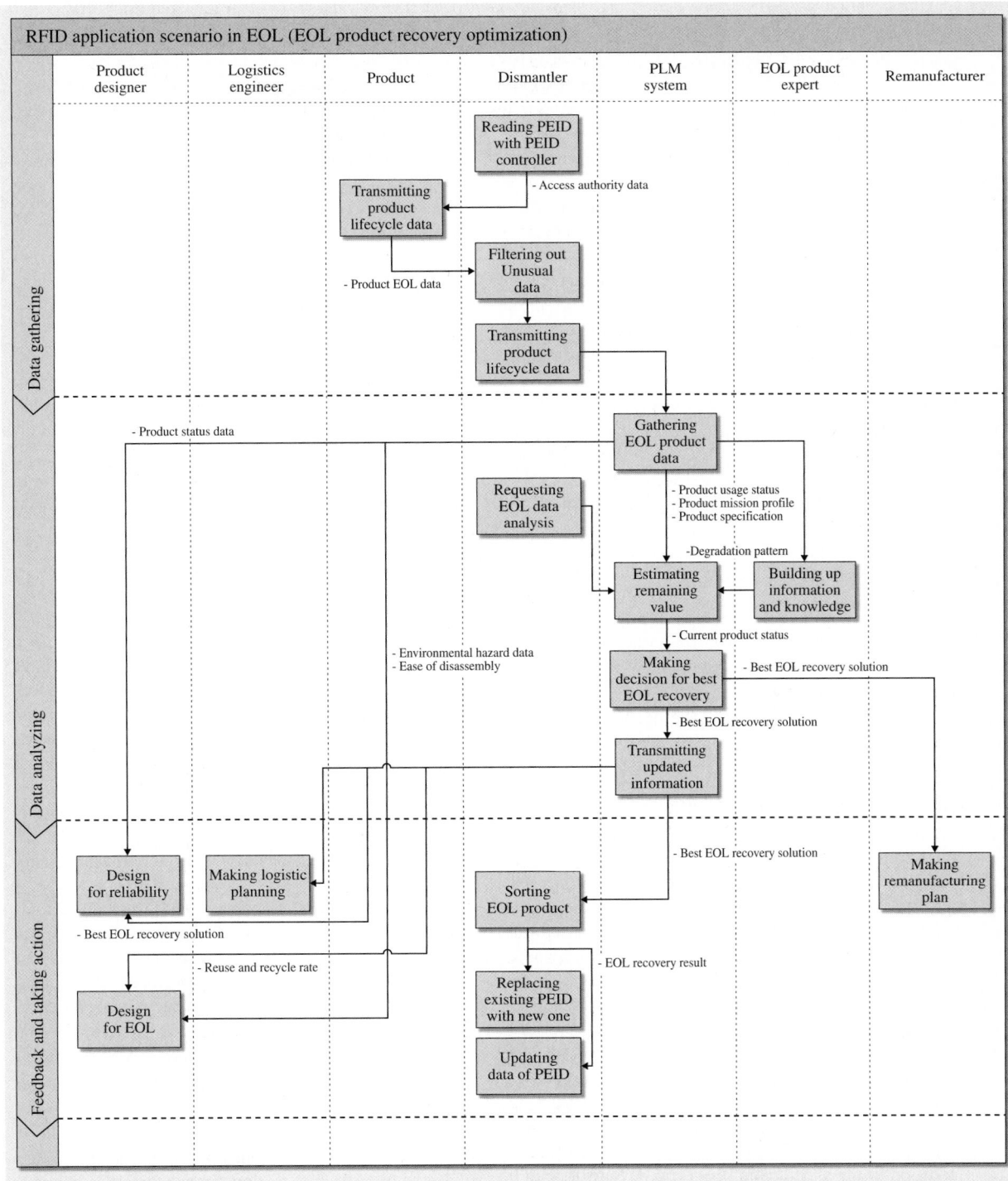

Fig. 43.8 EOL swim-lane chart

Table 43.1 Classification of PEIDs (*filled bullet*: high capacity, *empty bullet*: low capacity) (LF – low frequency, HF – high frequency, UHF – ultra high frequency)

Category / Our classification	Type A (1 bit transponder)	Type B (passive or semi-passive type with memory)	Type C (active type with memory and sensor)	Type D (device for smart product)
Function (corresponding component)				
Product identification (simple serial number in built-in chip)	•	•	•	•
Sensing (sensor)	–	–	•	•
Data processing (microprocessor)	–	○	○	•
Data storage (memory)	–	•	•	•
Power management (battery)	–	○	•	•
Communication (communication module)	–	–	–	•
Specification				
Memory type (RO[a], WORM[b], RW[c])	RO	WORM, RW	RW	RW
Reading distance (L[d], M[e], H[f])	L, M	M	M, H	M, H
Data rate (L, M, H)	L	L, M	M, H	M, H
Processing ability (L, M, H)	–	L	L, M	M, H
Frequency of operation (LF, HF, UHF)	LF, HF	LF, HF	HF, UHF	HF, UHF
Application level				
	Component/ item level	Component/ item or lot level	Assembly level or lot level	Product level

[a] Read only, [b] Write once and read many, [c] Read/write, [d] up to 1 cm, [e] Up to 1 m, [f] Over 1 m (L = low, M = medium, H = high)

storing additional data to a product itself during product lifecycle, this type of PEID is suitable.

Compared with type A, type B has additionally storage capability. Hence, it enables storage of necessary data in a product itself during its lifecycle. In other words, a PEID controller can update new data or information to this type of PEID, if necessary. This requires a read/write type of memory. Depending on applications, some may need processing ability to filter gathered data. Furthermore, some may need a battery because data storage requires a large amount of power. To keep not only static but also dynamic data about products (but a small amount, such as product history data within a product itself) this type is preferable. A semipassive or active tag can be used for this type of application. This type can be used in production lines or warehouses, or in supply chains for item management applications, e.g., checking item status, classifying items, tracing item history, and so on.

Type C has sensing and power management functions to gather environmental data of a product, in addition to the specifications of type B. Sensors can be installed into a RFID tag or separately, independent of a RFID tag. It should have its own battery since sensors require a large amount of power. It may also have a communication module, depending on the application. Through the communication module, it is able to transmit gathered data from sensors to back-end systems by itself. Its size is larger and its reading distance is longer than those of the previously described types. Predictive maintenance domain is a major application of this type of device. Depending on the types of sensors used, the application areas are huge, from food to machinery products.

Type D is the most complex PEID, which has additionally communication and processing ability. It can keep some amount of product status data gathered from sensors in its own memory. Furthermore, it can analyze gathered data and make some decisional processes based on them autonomously. This reduces the amount of data to be handled by the back-end systems. In addition, it can communicate with a PLM system directly without the help of PLM agents.

43.3.3 Data and Data Flow Definition

Table 43.2 describes the main data in several information flows in PLM.

Table 43.2 Main data of information flows in PLM

Information flow	Category	Main data
BOL to MOL	BOM information	Product ID, product structure, part ID, component ID product/part/component design specification, etc.
	Information for maintenance/service	Spare part ID list, price of spare part, maintenance/service instructions, etc.
	Production information	Assemble/disassemble instruction, production specifications production history data, production routing data, production plan, inventory status, etc.
BOL to EOL	Product information	Material information, BOM, part/component cost, disassemble instruction, assembly information for remanufacturing, etc.
	Production information	Production date, lot ID, production location, etc.
MOL to EOL	Maintenance history information	Number of breakdowns, parts/components' IDs in problem, installed date, maintenance engineers' IDs, list of replaced parts, aging statistics after substitution, maintenance cost, etc.
	Product status information	Degree of quality of each component, performance definition, etc.
	Usage environment information	Usage condition (e.g., average humidity, internal/external temperature), user mission profile, usage time, etc.
	Updated BOM	Updated BOM by repairing or changing parts and components, etc.
MOL to BOL	Maintenance and failure information for design improvement	Ease of maintenance/service, reliability problems, maintenance date, frequency of maintenance, $MTBF^{1}$, $MTTR^{2}$, failure rate, critical component list, root causes, etc.
	Technical customer support information	Customer complaints, customer profiles, response, etc.
	Usage environment information	Usage condition (e.g., average humidity, internal/external temperature), user mission profile, usage time, etc.
EOL to MOL	Recycling/reusing part or component information	Reuse part or component, remanufacturing information, quality of remanufacturing part or component, etc.
EOL to BOL	EOL product status information	Product/part/component lifetime, recycling/reuse rate of each component or part, etc.
	Dismantling information	Ease to disassemble, reuse or recycling value, disassembly cost, remanufacturing cost, disposal cost, etc.
	Environmental effects information	Material recycle rate, environmental hazard information, etc.

[1] Mean time between failures, [2] Mean time to repair

Part E | 43.3

43.3.4 PDKM, DSS and Middleware

PLM has emerged as an enterprise solution. Thus, all software tools/systems/databases used by various departments and suppliers throughout the whole product lifecycle have to be integrated so that the information contained in their systems can be shared promptly and correctly between people and applications [43.2]. Hence, it is important to understand how application software in a PLM fits with others

Table 43.3 Functions and specifications of main software components

Classification	Middleware	PDKM	DSS
Function	• Request-driven reading • Event-driven reading • Filter data • Data transformation • Write • PEID management • Service management • Data transition	• Document management • Field data management • User requirement management • Data transformation • Communication requirement management • Information requirement management	• Making decision • Decision support • Data analysis and transformation
Specification	• Location (within product, outside of product) • Reading distance • Reading rate • Data format • Interface protocol with PEID • Type of controller	• Location • Main user • Types of knowledge management • Data format • Amount of data	• Location • Purpose of decision support • Decision-maker • Data format • Expected output type • Types of DSS • Types of decision model

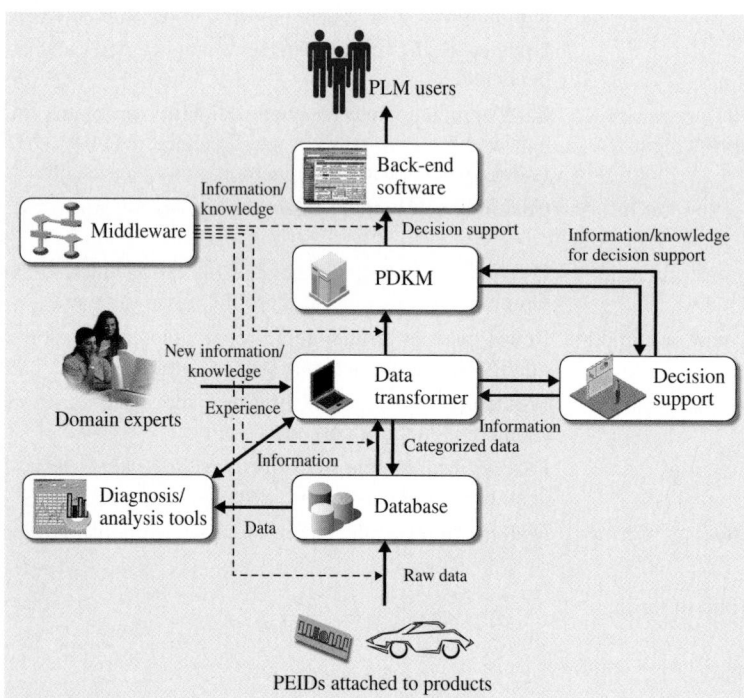

Fig. 43.9 Software architecture for closed-loop PLM

in order to manage product information and operations [43.18]. For this, a software architecture is required. Software architecture is the high-level structure of a software system concerned with how to design software components and make them work together.

Figure 43.9 shows a software architecture for closed-loop PLM. It takes a vertical approach in the sense that its structure represents a hierarchy of software of closed-loop PLM, from gathering raw data to business applications. Embedded software (called firmware) built into PEIDs plays the role of controlling and managing PEID data. The embedded software can have the ability to filter raw data gathered by various sensors, if necessary (this function can also be done in middleware). This can resolve the problem of memory size in PEIDs by removing duplicate and unnecessary data. Furthermore, firmware can do simple analyses based on the gathered data, or this function can also be implemented in other parts such as middleware, diagnosis, and analysis tools, and PDKM.

Database (DB) software is required to store processed data and manage them efficiently. A DB can be distributed or located on a central server. Regarding the format of the database, relational and object-oriented databases have been considered in the relevant research community. The configuration of the DB should be determined considering a trade-off between cost and efficiency of data management, which is different for each case.

PDKM, decision support, and middleware software components must be designed and implemented according to their description in Sect. 43.2.

Finally, back-end software can be defined as the part of a software system that processes the input from the front-end system that interacts with the user. These usually involve legacy systems of an enterprise, e.g., enterprise resource planning (ERP), supply chain management (SCM), and customer relationship management (CRM). The back-end software will support PLM users in implementing several business processes (see Chap. 90 on *Business Process Automation: CRM, SM and ERP*).

Table 43.3 presents the functions and specifications of the main software components such as middleware, PDKM, and DSS.

43.4 Closed-Loop PLM Application

The domain of application presented here is the end-of-life (EOL) phase of the product lifecycle. It specifically deals with the take back of end-of-life vehicles (ELVs) by dismantlers so that they can be reprocessed: this strategy allows for both the feedback of vital information (design information, usage statistics on components, etc.), and the materials/components themselves to the beginning-of-life (BOL) stage of the product lifecycle; as well as the take back of selected components into the middle-of-life (MOL) phase of the product lifecycle as secondhand parts.

This industrial application was developed and implemented by the Research Center of Fiat (CRF) and its partners in PROMISE, and is reported in detail in [43.19] and in the case study description A1 in [43.20]. It focuses specifically on the dismantler and the operations performed to achieve the correct removal decision [i.e., removal for reuse (BOL or MOL), removal for remanufacturing (BOL), disposal, etc.]; and the correct categorization and analysis of various environmental usage statistics associated with specific components from the ELV.

The dismantler decides on the ELV's recycling/recovery path and converts the ELV into components for reuse, remanufacturing or recycling. The dismantler's role is critical for returning ELV components and information from EOL to BOL. He retrieves from external databases the list of standard components to be removed from the car and checks if the components in the onboard diary are included in the standard list. At the same time he retrieves models (algorithms and costs) and thresholds from PDKM in order to compute wear-out level for each component and analyze the economic value of parts.

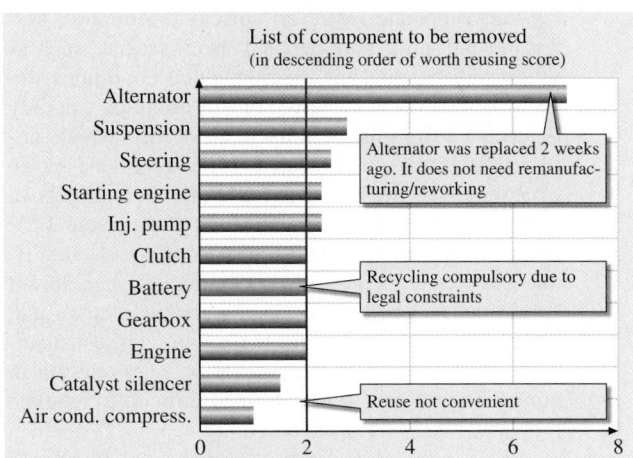

Fig. 43.10 List of EOL components of a car

In particular he decides which parts should be removed from the vehicle, how to recover (reuse or remanufacture) the removed parts, to which customers the parts should be delivered, and where to store the parts.

In the first stage, the system generates a bill of materials (BOM) of the car automatically based on the car model or identity number inputted in the background database; this BOM is used as the basis for developing a list of potentially valuable parts to remove, which also takes into account the requirements of legislation.

Using the dealer back-end system, the list of components to be removed from the car is computed. An example is shown in Fig. 43.10.

43.4.1 ELV Information and PEID Technology

In order to make these decisions properly and accurately, large amounts of information are required by the ELV dismantlers, which may be classified into six categories:

(1) Product-related information
(2) Location-related information
(3) Utilization-related information
(4) Legislative information
(5) Market information
(6) Process information

Generally the information in categories (1), (4), and (6) above is relatively easy to acquire, because the product-related information is usually obtained from the automotive producers and from other relatively static legislative bodies. However, current information systems cannot give more detailed specifications, such as usage statistics and the environmental conditions under which the ELV was used; this information can only be obtained from the ELV itself, a situation which, until recently, was only deemed to be resolvable by an experienced dismantler that could use their subjective judgement to make decisions about the present ELV based upon their knowledge of past ELVs. Naturally, this was seen as an unsatisfactory situation: the relevant knowledge required was subjective and qualitative, and, more important, linked to the personality of the dismantler. It was difficult to write down or communicate in numeric terms, and so was deemed difficult to regulate properly.

PEID technology, used as an enabler of PLM, can help to remove many of the unsatisfactory elements of the dismantling problem. By exploiting the capabilities of PEIDs, sensors embedded in particular vehicle components can collect and record relevant information about the vehicle's lifecycle, including production, usage, maintenance, and dismantling data. The dismantlers only need to read the data from the ELVs' PEID system, and can thereby obtain all the required location- and utilization-related information.

Thus, the solution suggested here removes the qualitative, subjective elements of the dismantling problem by emphasizing the use of a PEID technology infrastructure that cumulatively develops an information store of usage statistics as the ELV moves through its product lifecycle. More importantly, the use of PEIDs allows developers to remove the decision making at EOL from the experienced dismantler's hands, and allows an EOL-dedicated DSS to be developed based upon numeric usage statistics from the PEIDs in the ELV.

43.4.2 Decision Flow

The decision support for ELVs consists of two web-based process stages (Fig. 43.11). In the first stage, (1) remove decision, the system automatically generates a bill of materials (BOM) of the car based on the car model or identity number inputted and the background database; this BOM is used as the basis for developing a list of potentially valuable parts to remove, which also takes into account the requirements of legislation [that is, if there are hazardous parts, such as batteries, that must be removed by European Union (EU) law, whether valuable or not]. Once this list of parts to be removed has been generated, the user moves into the second part of the web interface: (2) recovery path, to determine what on the projected list of car parts should actually be removed and what should not be removed owing to actual damage, abnormal wear and tear or other factors that reduce the parts' value.

There are two key removal decisions involved at this point for each part in the ELV under consideration:

(1) Remove part from the vehicle for further treatment or
(2) Leave part on the vehicle to be shredded

If a part is (1) removed, it is because it is worth it: the quality of the part, the cost of labor to remove, the market conditions, and the present stock levels are assessed; if the part passes all of these thresholds then it is removed. If a part is (2) left on the ELV, it is because it is not worth removing: the value of the part does not cover its quality, the cost to remove, or the market may

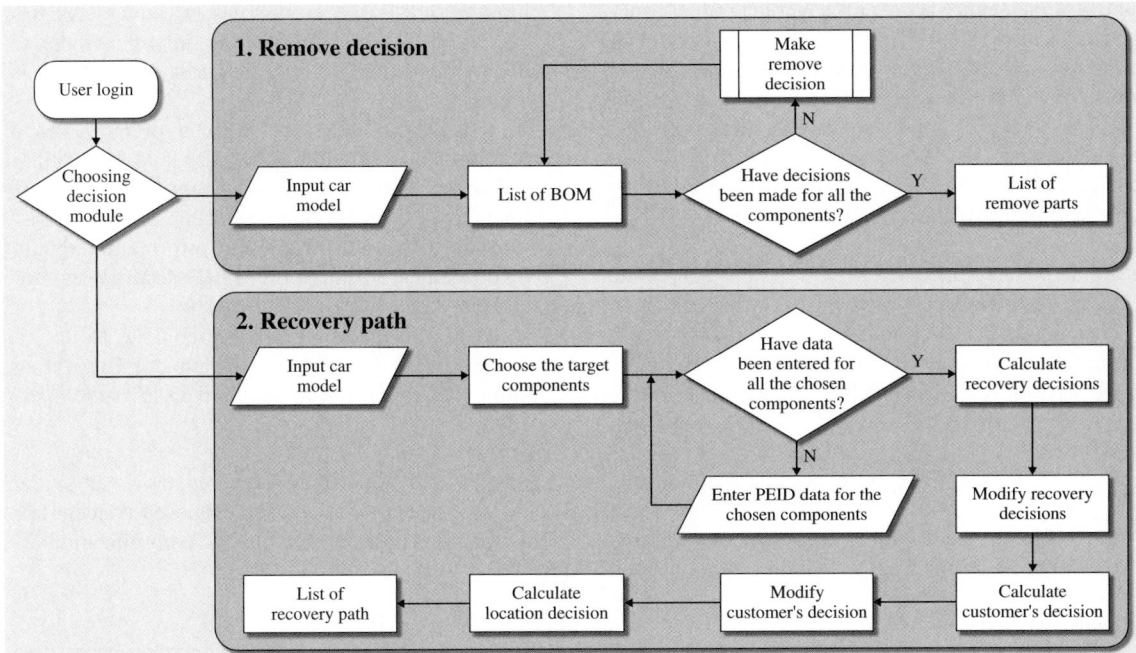

Fig. 43.11 A schematic view of DSS for ELVs

be unfavorable or the dismantler overstocked; the part is left on the ELV to be shredded as base material for recycling.

In the second stage, (2) recovery path in Fig. 43.11, the system assumes the recovery of a number of parts from the first stage and focuses on the *recovery path* required for these removed components. The two main recovery paths that any component can now take are: remanufacturing (i. e., retooling of a part to original quality levels; normally performed at the BOL phase) or reuse (i. e., use of the part in the secondary market; the part flow path is to the MOL if this is the case). Using the information derived from the PEIDs located on the recovered parts, the DSS can direct the user to the optimal recovery path for each of the re-

moved parts; this is performed by a set of algorithms that use the usage statistics on the PEIDs of the recovered parts to determine the correct recovery path for each component (the particular algorithms are not detailed as they are beyond the scope of this paper). Once the recovery method is issued, the system will cooperate with its back-end system to suggest a potential downstream customer and potential storage position for the component. Again, at each decision-making stage, the decision-maker has the authority to change the decision based on their judgement. When all of the decisions relating to the ELV are settled, the system records all the necessary information to the PDKM system, which is available to the BOL designers of the vehicle in question for examination.

43.5 Emerging Trends and Open Challenges

Total management of the product lifecycle is critical to innovatively meet customer needs throughout its entire lifecycle without driving up costs, sacrificing quality or delaying product delivery. For this, it is necessary to develop a PLM system in which information flow is horizontally and vertically closed, i. e., closed-loop PLM.

The closed-loop PLM system provides opportunities to reduce the inefficiency of lifecycle operations and gain competitiveness.

In this chapter, we have also discussed a system architecture for product lifecycle management where information flows are closed due to emerging product identification technology over the whole product lifecycle (closed-loop PLM). To gather product lifecycle data during all product lifecycle phases, the concept and architecture of PEID has been introduced. Furthermore, necessary software components and their relations have been addressed.

The following is a list of issues to be resolved for implementing the closed-loop PLM concept:

- In the business model aspect, it is necessary to develop a good business model to apply the concept of closed-loop PLM for optimizing the profit of a company. For this, trade-off analysis for cost and effect are prerequisite. Depending on each case, partial implementation of closed-loop PLM may be cost-effective.
- Regarding the PEID, it is necessary to develop a generic concept of a PEID that can be used over the whole product lifecycle. For this, however, first, the lifecycle of the PEID, including reuse, should

be modeled. Based on this design, suitable PEIDs should be designed, because the great bottleneck to deployment of PEID into business applications is their cost.

- In terms of middleware, it is a prerequisite to develop a method for managing and controlling enormous amounts of PEID event data. Methods for filtering huge amounts of event data and transforming them into meaningful information should be developed. Furthermore, PEID security and authority problems should be resolved.
- In terms of PDKM, it is a prerequisite to design the product lifecycle data schema for integrating all relevant data objects required in lifecycle operations.

Finally, the case studies developed so far in the PROMISE project show that the proposed concept can yield great benefit to product lifecycle optimization efforts.

References

43.1   F. Ameri, D. Dutta: Product life cycle management: needs, concepts and components, Technical Report (Product Lifecycle Management Development Consortium PLMDC-TR3-2004, 2004)

43.2   M. Macchi, M. Garetti, S. Terzi: Using the PLM approach in the implementation of globally scaled manufacturing, Proc. Int. IMS Forum 2004: Global Challenges in Manufacturing (2004)

43.3   H.B. Jun, D. Kiritsis, P. Xirouchakis: Closed-loop PLM. In: *Advanced Manufacturing – An ICT and Systems Perspective*, ed. by M. Taisch, K.-D. Thoben, M. Montorio (Taylor & Francis, London 2007) pp. 90–101

43.4   H. B Jun, J. H Shin, D. Kiritsis, P. Xirouchakis: System architecture for closed-loop product lifecycle management, Int. J. Comput. Integr. Manuf. **20**(7), 684–698 (2007)

43.5   H.B. Jun, J.H. Shin, Y.S. Kim, D. Kiritsis, P. Xirouchakis: A framework for RFID applications in product lifecycle management, Int. J. Comput. Integr. Manuf. (2007), DOI: 10.1080/09511920701501753

43.6   D. Kiritsis, A. Bufardi, P. Xirouchakis: Research issues on product life cycle management and information tracking using smart embedded systems, Adv. Eng. Inform. **17**, 189–202 (2003)

43.7   D. Kiritsis, A. Rolstadås: PROMISE – a closed-loop product life cycle management approach, Proc. IFIP 5.7 Adv. Prod. Manag. Syst.: Model. Implement. Integr. Enterp. (2005)

43.8   A.K. Parlikad, D. McFarlane, E. Fleisch, S. Gross: The role of product identity in end-of-life decision

making, Technical Report (Auto-ID Center, Institute of Manufacturing, Cambridge 2003)

43.9   M. Schneider: Radio frequency identification (RFID) technology and its application in the commercial construction industry, Technical Report (University of Kentucky, 2003)

43.10   S.S. Chawathe, V. Krishnamurthy, S. Ramachandran, S. Sarma: Managing RFID data, Proc. 30th VLDB Conf. (2004) pp. 1189–1195

43.11   T. Nieva: Remote data acquisition of embedded systems using Internet technologies: a role based generic system specification. Ph.D. Thesis (EPFL, Lausanne 2001)

43.12   Z. Gsottberger, X. Shi, G. Stromberg, T.F. Sturm, W. Weber: Embedding low-cost wireless sensors into universal plug and play environments, Proc. 1st Eur. Workshop Wirel. Sens. Netw. (EWSN 04) (2004) pp. 291–306

43.13   J. Lee, H. Qiu, J. Ni, D. Djurdjanovic: Infotronics technologies and predictive tools for next-generation maintenance systems, Proc. 11th Symp. Inf. Control Probl. Manuf. (Elsevier, 2004)

43.14   C.M. Rose, A. Stevels, K. Ishii: A new approach to end-of-life design advisor (ELDA), Proc. 2000 IEEE Int. Symp. Electr. Environ. (ISEE 2000) (2000)

43.15   PROMISE: PROMISE – integrated project: annex I – description of work, Project proposal (2004)

43.16   G. Hackenbroich, Z. Nochta: A process oriented software architecture for product life cycle management, Proc. 18th Int. Conf. Prod. Res. (2005)

43.17   I. Inasaki, H.K. Tönshoff: Roles of sensors in man-
        ufacturing and application ranges. In: *Sensors in
        Manufacturing*, ed. by H.K. Tönshoff, I. Inasaki
        (Wiley, New York 2001)
43.18   CIMdata: Product life cycle management – em-
        powering the future of business, Technical Report
        (CIMdata, 2002)

43.19   H. Cao, P. Folan, L. Zheng Lu, J. Mascolo,
        N. Frantone, J. Browne: Design of an end-of-life
        decision support system using product embed-
        ded information device technology, ICE Conf. Proc.
        (2006)
43.20   PROMISE: PROMISE case studies, public version
        available at www.promise-plm.com

44. Education and Qualification for Control and Automation

Bozenna Pasik-Duncan, Matthew Verleger

Engineering education has seen an explosion of interest in recent years, fueled simultaneously by reports from both industry and academia. Automatic control education has recently become a core issue for the international control community. This has occurred in tandem with the explosion of interest in engineering education as a whole. The applications of control are growing rapidly. There is an increasing interest in control from researchers from outside of traditionally control-based fields such as aeronautics, chemical, mechanical, and electrical engineering. Recently control and systems theory have had much to offer to nontraditional control fields such as biology, biomedicine, finance, actuarial science, and the social sciences as well as transportation and telecommunications networks. Complementary, innovative developments of control and systems theory have been motivated and inspired by complex real-world problems. These new developments present huge challenges in control education. Meeting these challenges will require a multifaceted approach by the control community that includes new approaches to teaching, new preparations for facing new theoretical control and systems theory problems, and a critical review of the status quo. This chapter discusses these new challenges as well as new approaches to education and outreach. This chapter starts by presenting an argument towards the future of controls as the application of control theory expands into new and unique disciplines. It provides two case studies of nontraditional areas where control theory has been applied: finance and biomedicine. These two case studies show a high potential for using powerful fundamental principles and tools of automatic control in research with an inter-

disciplinary nature. The chapter then outlines current and future pedagogical approaches being employed in control education, particularly introductory courses, around the world. It concludes with a discussion about the role of scholarship, teaching, and learning in control education both now and in the coming years.

Part E | 44

44.1 The Importance of Automatic Control in the 21st Century

The field of automatic control has a rich history (Fig. 44.1), well presented in a special issue of the *European Journal of Control* [44.1]. *Fleming* [44.2] provides an assessment of its status and needs of control theory through the 1980s. A broader and more updated report is provided in [44.3–5].

At its core, control systems engineering involves a variety of tasks including modeling, identification, estimation, simulation, planning, decision making, optimization, and deterministic and stochastic adaptation. While the overarching purpose of any control system is to assist with the automation of an event, the successful application of control principles involves the integration of various tools from related disciplines such as signal processing, filtering, stochastic analysis, electronics, communication, software, algorithms, real-time computing, sensors and actuators, as well as application-specific knowledge.

The applications of control automation range from transportation, telecommunications networks, manufacturing, communications, aerospace, process industries through commercial products reaching as diverse fields of study as biology [44.6], medicine [44.7–10], and finance [44.11, 12]. New issues are already appearing in the next generation of transportation [44.13] and telecommunications network [44.14] problems, as well as in the next generation of sensor networks (particularly for applications such as weather prediction), emergency response systems, and medical devices. These are new challenges that can be solved through the careful application of control and systems theory. The impact of systems and control in the changing world is described in the 2007 Chinese Control Conference (CCC) plenary talk *Systems and Control Impact in a Changing World* delivered by Ted Djaferis, the 2007 President of the Control Systems Society [44.15]. A more thorough description of control theory for automation can be found in Chaps. 9–11.

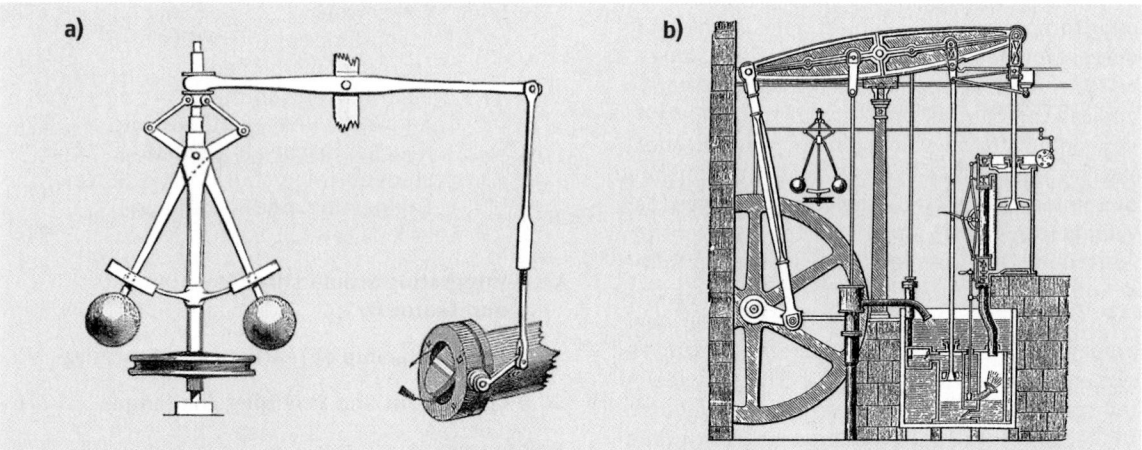

Fig. 44.1a,b The centrifugal governor (**a**) is widely considered to be the first practical control system, dating back to the 1780s. It was used in the Boulton and Watt steam engine (**b**) to regulate the amount of steam allowed into the cylinders to maintain a constant engine speed

44.2 New Challenges for Education

Marketplace pressures and advances in technology are driving a need in modern industry for well-trained control and systems scientists. With its cross-boundary nature and ever-growing application base, helping students in all disciplines of science, technology, engineering, and mathematics (STEM) to understand the power and complexity of control systems is becoming an even more critical component to the future of technical education. The need to train all STEM graduates to be comfortable with control theory generates many new challenges in control education which are extensively discussed in [44.16] as well as in the report of the National Science Foundation (NSF) and the Control Systems Society (CSS) panel on an assessment of the field [44.4] with its summary given in [44.5].

While the skills necessary for students to become successful practitioners of their craft are changing, so too is the background of our students. They are better prepared to work with modern computing technologies. The ability to interact with and manipulate a computer is second nature to today's *connected* student [44.18]. Thus the time is ripe for major renovations in control education as it applies to STEM disciplines. The first step in the renovation process is to develop cross-disciplinary examples, demonstrations, and laboratory exercises that illustrate systems and control across the entire spectrum of STEM education [44.3–5, 19].

The recent National Academy of Engineering (NAE) report [44.17] *identified the attributes and abilities engineers will need to perform well in a world driven by rapid technological advancements, national security needs, aging infrastructure in developed countries (Table 44.1), environmental challenges brought about by population growth and diminishing resources, and creation of new disciplines at the interfaces between engineering and science.* The systems and control community has been actively involved and engaged in taking a leading role in shaping the future automatic control engineering cur-

Table 44.1 America's infrastructure is in dire need of repair: just one of the many problems today's engineering students will be facing as they enter the workforce (after [44.17])

Area	Grade	Trend (since 2001)
Roads	D$_+$	↓
Bridges	C	↔
Transit	C$_-$	↓
Aviation	D	↔
Schools	D$_-$	↔
Drinking water	D	↓
Watewater	D	↓
Dams	D	↓
Solid waste	C$_+$	↔
Hazardous waste	D$_+$	↔
Navigable waterways	D$_+$	↓
Energy	D$_+$	↓
America's infrastructure GPA[a] total	D$_+$	
investement		US $ 1.6 trillion (estimated 5-year need)

[a] grade point average

riculum [44.20], as well as in educating and making nonengineering communities aware of the benefits and the power of the systems and control approaches and tools.

44.3 Interdisciplinary Nature of Stochastic Control

Stochastic adaptive control, whereby the unknown parameters of a control system are modeled as random variable or random processes [44.21, 22], can be used to illustrate the interdisciplinary nature of control. The general approach to adaptive control (Fig. 44.2) involves a splitting, or separation, of parameter identification and adaptive control. A system's behavior depends on some set of parameters, and the fact that the values of the parameters are unknown makes the system unknown. Some crucial information concerning the system is not available to the controller, and this information can be learned during the system's performance. Using that information, the system's performance can then be altered to respond. This altered response may in turn alter the previously unknown parameters. This process is repeated in a recursive manner until the system is shut down.

The described problem is the basic problem of adaptive control. A stochastic control system can be described using a stochastic differential equation or a stochastic partial differential equation. The solution to the adaptive control problem consists of showing the strong consistency of the family of estimators of an unknown parameter and the self-optimality of an adaptive control that uses the family of estimates. The disturbance, or noise, in the system is modeled by a Brownian motion or more generally by a fractional Brownian motion (more accurate for recent problems in telecommunication, finance or biomedicine) [44.11]. Industrial operation models are often described by stochastic controlled systems [44.23].

Let us describe a simple adaptive control using a simple investment model. Consider a model where an investor has a choice in investing in two assets, a simple

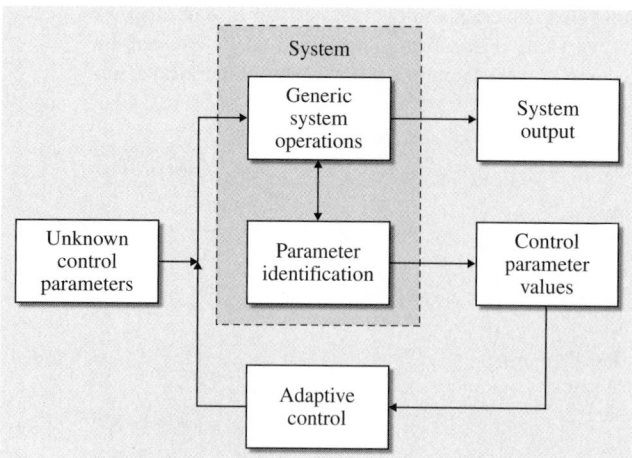

Fig. 44.2 The general structure of an adaptive control system is one where the critical input control parameters are outputs of the system. As the system operates over time, the parameters are adapted to control how the system functions in the future◄

stock and transfer the money to the savings account, but when the stock is expected to go up to repurchase the stock before it goes up. The control variable is the total amount of money transferred from the stock to the savings account and can be positive (stock has just been sold) or negative (stock is about to be purchased). The savings account is governed by a traditional differential equation and the stock is governed by a stochastic differential equation. The goal is to find the optimal control so that the expected rate of growth is maximized. The identification problem is to estimate the unknown parameters, the drift and variance of the stock's randomness, based on the available observations. The adaptive control problem is to construct the (certainty equivalence) adaptive control as a function of the current state and the current estimate of what is expected to happen.

savings account with a fixed rate of growth (say 3.5%) or shares of Microsoft stock whose growth is governed by a Brownian motion with unknown drift and variance. The investor controls his asset by transferring money between the stock and the savings account. The goal is, when the stock is expected to go down, to sell the

44.4 New Applications of Systems and Control Theory

The use of systems and control theory has seen an expansion in recent years into new and emerging fields of study. This expansion functions as a control system in itself, with the unknown parameter being the number of STEM students receiving controls training. The expansion into new disciplines demonstrates the necessity of controls education for all students in STEM disciplines by showing that controls theory can be applied in a wide variety of unique places for producing solutions to some of today's most complex problems. This demonstration in turn acts as a driver for the curricular change necessary to result in more STEM students experiencing control education. As those students graduate, they too will push the boundaries of what control theory can be applied to and in turn feed the expansion cycle.

44.4.1 Financial Engineering and Financial Mathematics

Research in control methods in finance has experienced tremendous progress in recent years. The rapid progress has necessitated communication and networking between researchers in different disciplines. This area brings together researchers from mathematical sciences, finance, economics, and engineering. Together, these people work on advances and future directions of control methods in portfolio management, stochastic models of markets and pricing, and hedging of options. The finance area is highly interdisciplinary. This interdisciplinary nature comes from the necessity for a wide variety of skills and knowledge bases.

As a major impetus for the development of financial management and economics, the research in financial engineering has had a major impact on the global economy. For instance, the Black–Scholes model [44.24] and its various extensions for pricing of options [44.25] using stochastic calculus have become a standard practice nowadays and have led to a revolution in the financial industry. Incidentally, it also won Scholes and Merton the Nobel Prize for Economics in 1997 (Black had unfortunately passed away 2 years prior).

Powerful techniques of stochastic analysis and stochastic control have been brought to almost all aspects of finance and resulted in a number of important advances [44.12]. To name just a few, they include the studies of valuation of contingent claims in complete and incomplete markets, consumption–investment models with or without constraints, portfolio management for institutional investors such as pension funds and banks, and risk assessment and management using financial derivatives. At the same time, these applications require and stimulate many new and exciting theoretical discoveries within the systems and control field. Take for instance the study of arbitrage theory, risk assessment, and portfolio management, which have collectively led to new developments in martingale theory and stochastic control. Moreover, the development of financial engineering has created a large demand for graduates at both Master and Ph.D. levels in industry, resulting in the introduction of the curriculum in many universities including Kent State University, Princeton University, Columbia University, and the University of California at Berkeley.

Another contribution to the control community from financial engineering and financial mathematics was the identification and control of stochastic systems with noise modeled by a fractional Brownian motion [44.11], a process that can possess a long-range dependence. Motivated by the need for this type of process in telecommunications for models of Ethernet and asynchronous transfer mode (ATM) traffic, a new stochastic calculus for fractional Brownian motion was developed. This work in financial engineering and financial mathematics has since been successfully used in other fields such as telecommunications and medicine, in particular epilepsy analysis of brain waves [44.8, 9].

44.4.2 Biomedical Models: Epilepsy Model

Epilepsy is a condition where a person has unprovoked seizures at two or more separate times in her/his life. A seizure is an abnormal electrical discharge within the brain resulting in involuntary changes in movement, sensation, perception, behavior, and/or level of consciousness. It is estimated that 1% of the populations of industrialized countries have epilepsy whereas 5–10% of the populations of nonindustrialized countries have epilepsy. One link has been found between epilepsy and malnutrition [44.26]. In the USA the number of epilepsy cases is significantly larger than the number of cases of people who have Parkinson's disease, muscular dystrophy, multiple sclerosis, acquired immunodeficiency syndrome (AIDS) or Alzheimer's disease [44.10]. The organizers of the Third International Workshop on Seizure Prediction in Epilepsy held in Freiburg, Germany stated in the Welcome to the Workshop that:

The great interest of participants from all over the world and the high number of original contributions presented ... instills confidence in us that seizure prediction is a promising field for the years to come.

Epilepsy models, with their complexity, can serve as an example of interdisciplinary and multidisciplinary research for which systems and control approaches are showing considerable promise. The methods pioneered in the financial mathematics and engineering sector described above have been successfully used for detection and prediction of epileptic seizures. There is published evidence that the seizure periods of brain waves of some patients have long-range dependencies with nonseizure periods. Based on some initial work, it seems that the estimates of the Hurst parameter, which is a characterization of the long-range dependence in fractional Brownian motion, have a noticeable change prior to and during a seizure. Some of the algorithms [44.8] used in identifying the Hurst parameter use stochastic calculus for fractional Brownian motion that was developed within the financial engineering and financial mathematics sector. A new application for the Hurst parameter, real-time event detection, has recently been identified [44.9]. The high sensitivity to brain state changes, ability to operate in real time, and small computational requirements make Hurst parameter estimation well suited for implementation into miniature implantable devices for contingent delivery of antiseizure therapies.

This innovative interdisciplinary research has developed a new technology for future automated therapeutic intervention devices to lessen, abort or prevent seizures, opening the possibility of creating a brain pacemaker. The goal is, by eliminating the unpredictability of seizures, to minimize or prevent the disability caused by epilepsy and hardship it imposes on patients and their families and communities. It brings a hope to improve productivity and better quality life for those afflicted with epilepsy and their families as well as for care-givers and healthcare providers. The creation of a seizure warning device will minimize risks of injury, the degrading experience associated with having seizures in public, and the unpredictable disruption of normal daily-life activities.

Only collaborative work of engineers, computer scientists, physicists, physicians, biologists, and mathematicians can be successful in solving this type of complex problem. Based partly on the success thus far, there is a strong desire for a partnership with control engineers. The University of Kansas (KU) Stochastic Adaptive Control Group (SACG) has a long history of established collaboration in control education and research with the KU Medical Center Comprehensive Epilepsy Center and with Flint Hills Scientific, LLC (FHS). FHS has one of the largest recorded collections of long-term patient seizure data and it has used this data to develop real-time seizure prediction algorithms which have outperformed other prediction algorithms. Recently FHS has initiated electrical stimulation control with a seizure prediction algorithm to prevent the occurrence of seizures.

44.5 Pedagogical Approaches

The field of control is currently suffering from a fundamental design flaw. Because of a push from both those external to the control community (e.g., the NAE's *Engineer of 2020* reports) as well as those internal to the control community, control has become an essential component for STEM education. This has resulted in a flood of "current-state" papers describing methods and mechanisms that are currently being employed for teaching control.

The beauty of control as an area of study lies in its use of many different areas of mathematics: from functional analysis through stochastic processes, stochastic analysis, stochastic calculus of a fractional Brownian motion, stochastic partial differential equations, stochastic optimal control to methods of mathematical statistics, as well as current computational methods in stochastic differential and partial differential equations. The curse of control as an area of study lies in its use of many different areas of mathematics. Because of the broad spectrum of mathematical tools for approaching systems and control problems, it is not a subject that allows for practical understanding without some amount of deep coverage. Likewise, it is a topic that almost demands some hands-on experimentation, which can be both costly and time consuming. Several approaches are described for building adequate and appropriate control coursework into an already packed curriculum.

44.5.1 Coursework

A special pair of issues of the *Control Systems Magazine* on innovations in undergraduate education [44.28, 29] presents a variety of articles related to undergraduate controls and systems education broken down into six broad categories; Kindergarten-12th grade (K-12) education, course and curriculum development, experiment development, special projects, software and laboratory development, and lecture material.

Djaferis [44.27] describes a three-pronged approach to introducing systems and controls in a first-year engineering course. A combination of lectures, simulations, and a hands-on experiment where student develop a collision avoidance system for a model car (Fig. 44.3) are used to indoctrinate students into how control theory can be practically applied within the engineering design process. Because the only prerequisite requirement for the course is first-semester calculus, it can be offered during the spring term of the first year, allowing students the opportunity to experience control early in their education. Additionally, the author found that the early introduction to control resulted in a number of students choosing to take a more advanced course in control as well as for some to pursue graduate studies of systems and control.

An example of a new introductory control course for both engineers and general scientists is described in [44.30]. The course was taught for over 10 years at Sweden's Lund University. The goals for the course include:

- Demonstrating the benefits of control and the power of feedback
- Describing the role of control in the design process and the importance of integrated systems

Fig. 44.3 Control laboratory experiment for first-year engineering students (after [44.27])

- Introducing the language of systems and control
- Introducing the basic ideas and concepts of control
- Explaining fundamental limitations
- Explaining how to formulate, interpret, and test specifications
- Exploring computational tools such as MATLAB and Simulink [44.32] to compute time and frequency responses
- Developing practical experience of sketching Bode diagrams and performing calculations using the associated plots.

The challenge is to teach students sufficient information for understanding the validity of the information contained in computer plots and to appreciate how to achieve a satisfactory design that meets the specifications and necessary criteria. One of the curricular design ideas that fueled the course's creation was to provide a course that functioned as both an introduction for those that elected to continue their control studies as well as a functionally complete overview for students who would only take one course on control.

One approach for demonstrating and appreciating the importance of the dynamics of a system is the use of the bicycle discussed in detail in [44.33] and proposed for use in an introductory course. Because of the universal familiarity of the bicycle, it represents a highly approachable problem context. It also presents a wide variety of unique control- and dynamics-related problems; for example, the fact that when at rest it is an unstable system but when in motion it exhibits aspects of stability presents a unique problem situation for students to understand. The problem can also be extended into issues of rear-wheeled steering and other assorted complexity inducing issues. Finally, the scope of the problem is such that other non-control-related problems can be introduced, such as discussions of the elemen-

tal physics of motion and the theoretical aspects of design.

In response to the National Academy of Engineering report [44.17, 20] in which interdisciplinary system engineering is cited as an increasingly important aspect of modern engineering and of the education of future engineers projects that integrate elements of mechanical design, modeling, control system design, and software implementation were developed at the University of Illinois at Urbana-Champaign [44.34]. In these projects, control design becomes an integral part of the larger systems engineering problems and must be carried out in conjunction with design and optimization of structural members, choice and placement of sensors and actuators, electronics, power considerations, modeling, simulations, identification, and software developments.

44.5.2 Laboratories as Interactive Learning Environments

Understanding systems and control demands hands-on experience. It is nearly impossible to appreciate the complexities and interactions of a physical system without witnessing it. Therefore a good control laboratory is important for control education. Advances in technology have reduced the cost of developing laboratories significantly. Numerous courses currently utilize the Lego Mindstorms kits which include a variety of sensors and a programmable computer capable of interpreting the sensor's raw value [44.35–39]. While the majority of those courses utilize the Lego system as a vehicle for learning programming concepts or mechanical design principles, the sensor capabilities open the door for use in a systems and control course.

The Intelligent Control Systems Laboratory [44.16] at Australia's Griffith University demonstrates con-

Fig. 44.4 Autonomous vehicles competing in the DARPA Grand Challenge [44.31]

trol through cooperative driverless vehicles. The idea of intelligent vehicles has brought with it promises of heightened safety, reliability, and efficiency. In 2004, the Defense Advanced Research Projects Agency (DARPA) held the first of three DARPA Grand Challenge prize competitions for driverless vehicles [44.31] (Fig. 44.4). The first year, the best vehicle drove only 7.36 miles of the 150 mile desert course. In 2006 at the second challenge, five vehicles completed the course and all but one of the 23 participating teams achieved a distance greater than the maximum 7.36 miles obtained during the first challenge. In 2007, the third challenge required participants to navigate a more urban setting. Six of the 11 participating teams rose to the challenge and successfully drove through the 60 mile urban course.

Finally, the number of interactive tools for education in automatic control is growing rapidly. Many examples are provided [44.40–43]. A sample plain talk presented below provides other examples of creative, interesting, and important modern control engineering education laboratories.

44.5.3 Plain Talk on Control for a Wide Range of the Public

The IEEE Control Systems Society Technical Committee on Control Education has been in the process of developing a series of short presentations prepared for a wide range of the public demonstrating the power, beauty, and excitement of systems and control. These presentations were given at the workshops for high-school teachers and students, sponsored by the National Science Foundation and the Control Systems Society. They are presented in [44.44, 45].

The following is a sample of such talks that were presented at the 2007 Control and Decision Conference in New Orleans:

- T. Djaferis: *The Power of Feedback* (University of Massachusetts Amherst)
- C. G. Cassandras: *Joys and Perils of Automation* (Dept. of Manufacturing Engineering and Center for Information and Systems Engineering Boston University)
- R. M. Murray: *Control Education and the DARPA Grand Challenge* (Control and Dynamical Systems, California Institute of Technology)
- P. R. Kumar: *The Next Phase of the Information Technology Revolution* (University of Illinois, Urbana-Champaign)

- C. Tomlin: *Controlling Air Traffic* (University of California, Berkeley)
- M. Spong: *Control in Mechatronics and Robotics* (University of Illinois, Urbana-Champaign)
- W. S. Levine, D. Hristu-Varsakelis: *Some Uses for Computer-Aided Control System Design Software in Control Education* (University of Maryland)
- I. Osorio: *Application of Control Theory to the Problem of Epilepsy* (University of Kansas Medical Center and Mark Frei Flint Hills Scientific, LLC)
- D. Duncan, T. Duncan, B. Pasik-Duncan: *Random Walk around Some Problems in Stochastic Systems and Control* (Yale University, University of Kansas)
- K. Furuta: *Understanding Phenomena through Real Physical Objects–Controlling Pendulum* (Tokyo Denki University, Japan)
- J. Baillieul: *Risk Engineering–Past Successes and Future Challenges* (Intelligent Mechatronics Laboratory, Boston University)

Developing plain talks that can be used by members of the control community for noncontrol communities in different settings is very important and has become a major goal for the control education committees. It is through these talks that the general population, particularly children and young adults, can learn to appreciate control and potentially pursue further study of the topic.

44.5.4 New Approaches to Cultivating Students Interest in Math, Science, Engineering, and Technology at K–12 Level

The IEEE CSS Technical Committee on Control Education together with the American Automatic Control Council (AACC) and the International Federation of Automatic Control (IFAC) Committees on Control Education has organized a series of workshops for middle- and high-school students and teachers on *The Power, Beauty, and Excitement of Control* at all major control conferences sponsored by CSS, AACC, and IFAC in the USA and around the world since 2000 [44.46]. The model for these workshops was created and developed at the University of Kansas (KU) 15 years ago. The university organized semiannual half-day workshops for fifth- and sixth-graders from local schools to promote STEM disciplines. They became very successful and played a major role in establishing an important partnership between K-12 and KU [44.47]. Another important outreach activity for encouraging young people to consider a career in control was the workshop organized for girls at the International Symposium on Intelligent Control/Mediterranean Conference on Control and Automation (ISIC/MED) conference [44.48].

The purpose of NSF/CSS workshops is to inspire interest from youth towards studies in control systems and to assist high-school teachers in promoting the discipline of control systems among their students. It is composed of several short but effective presentations, as listed in Sect. 44.5.3, on various problems from the real world that have been solved by using control engineering methods, techniques, and technologies. The workshops bring together all-star teams of some of the most eminent senior control researchers and some of the most prominent younger researchers involved in new technologies to present control systems as an exciting and intellectually stimulating field. The attractiveness and excitement of choosing a career in control engineering has been addressed at workshops. Live interaction between the presenters and the audience has been an important feature of the workshops. The workshops have become very popular, with the last one in San Diego bringing over 650 students and teachers. Additional information on the workshops can be found in [44.49,50].

44.6 Integrating Scholarship, Teaching, and Learning

The integration of scholarship, teaching, and learning into the classroom can be subdivided in many ways. An important subdivision is *horizontal integration* versus *vertical integration* [44.51]. Horizontal integration deals with institutional integration, drawing perspectives from across the institution, whereas vertical integration is programmatic, drawing perspectives from throughout a specific school, department, or program. Both are important for success.

To increase horizontal integration, faculty from different disciplines work together in teaching a single course. Faculty in different disciplines often work on similar topics and each one can provide his or her particular insights for understanding the material and assimilating the information. Their combined efforts result in a variety of advantages. First, they develop cross-disciplinary ties which often boost collaboration. Second, students learn a topic from multiple perspectives, which increases their potential ability to understand the topics, as they are more likely to hear at least one explanation that they will understand, as well as increasing their interactions with a wider variety of faculty members.

Vertical integration incorporates students and researchers at different levels in the teaching activity, that is, there is involvement of senior high-school students, undergraduates, graduates, and postdoctoral students. Often a student is more likely to discuss various questions with someone near to his or her educational level, and these educationally close people can often identify more easily the causes of difficulty. The Stochastic Adaptive Control Group (SACG) research group at the University of Kansas has successfully implemented the approach [44.46,49,52].

44.7 The Scholarship of Teaching and Learning

No modern educational discussion would be complete without some discussion of the scholarship of teaching and learning (SoTL). While the wording of SoTL is similar to that of Sect. 44.6, there is a careful distinction. Integrating scholarship, teaching, and learning is about using research (scholarship) in the classroom to enhance teaching and learning. SoTL is about analyzing and reflecting on teaching and learning using the same scholarly processes (literature review, testing, data collection, statistics, as well as more qualitative measures such as systematically coded interviews) used in other scholarly work. Illinois State University defines SoTL to be "systematic reflection on teaching and learning made public" [44.53].

What makes aspects of SoTL interesting for control educators is that the process of *systematically reflecting* can itself be a type of control problem. We learn about our students each time we teach and we adapt the course and methods of teaching to this particular class. It can be considered a stochastic system; stochastic because there is a lot of noise in the system, but typically also having some degree of consistency. Sometimes referred to as *action research*, the process involves an iterative cycle of *planning, action, and fact-finding about the result of the action* [44.54]. With each new set of facts, a new plan is formed, upon which action is taken and new facts emerge. The process then begins anew. Over time, a lesson is refined, with the best approaches and attributes of the lesson being used, while the weakest elements are systematically removed or replaced. *Boyer* refers to this as "scholarship in teaching" [44.55].

Education should be regarded as a stochastic process that changes over time, a process with several components such as vision, design, data collection, and data analysis. As instructors, we should collect information, build a portfolio, analyze our reports and data after every class, and want to do better each time. We should want to apply the same rigor to our teaching that we do to our other scholarly endeavors.

44.8 Conclusions and Emerging Challenges

The control field is an exciting field. It has a cross-boundary nature. Many nontraditional disciplines recognize the power and the need for systems and control approaches, principles, and technologies. Control education of tomorrow is a collaborative effort integrating scholarship, teaching, and learning. This collaborative effort needs to include K-12 teachers, students, and scholars. They need to work together as partners who are learners in the process of control education. A classroom should be treated as a scientific laboratory, actively engaged in SoTL.

Similarly, instructors must learn to integrate their scholarship into their teaching and learning by building ties both across a discipline and across an institution. It is important for control instructors to build bridges with mathematics, computer science, and science. A future control engineer has to integrate engineering with computations, communications, mathematics, and science.

This is an extraordinary time for control, with extraordinary opportunities. Several special sessions on control educations have been organized at major control conferences, bringing together leading control scholars from academia and industry. Those important discussions have been well documented [44.16]. Every issue of the *IEEE Control Systems Magazine* is either devoted fully or has an article on control education. The biggest challenge is to attract young people to engineering and, in particular, systems and control engineering. It is important for all control communities to be involved in

outreach programs. It is important to build new bridges with other disciplines. It is important to focus on good communication and writing and it is most important to be passionate and enthusiastic about systems and control, and pass on this passion and enthusiasm to young people, in particular women and minorities, encouraging them to pursue and stay in the control profession, which has so much to offer, both now and as control expands into more and more new and unique disciplines.

Future engineers must be well prepared for the complex technical, social, and ethical questions raised by emerging disciplines and technologies. To be successful, students will need to be broadly educated. Teachers and students need to understand the full range of problems covering requirement, specifications, implementation, commissioning, and operation that is reliable, efficient, and robust. They need to understand that new materials and devices are made possible through advanced control of manufacturing processes. They need to recognize that control theories can be used for achieving breakthroughs in highly diverse settings, including biomedicine and finance. As the NSF/CSS panel report summarizes [44.3]:

> *... perhaps most important is the continued development of individuals who embrace a system perspective and provide technical leadership in modeling, analysis, design and testing of complex engineering systems.*

References

44.1   S. Bittanti, M. Gevers (Eds.): On the dawn and development of control science in the XX-th century (Special Issue), Eur. J. Control **13**, 1–81 (2007)

44.2   W.H. Fleming: *Future Directions in Control Theory. A mathematical Perspective* (Society for Industrial and Applied Mathematics, Philadelphia 1988)

44.3   P. Antsaklis, T. Basar, R. DeCarlo, N.H. McClamroch, M. Spong, S. Yurkovich: Report on the NSF/CSS workshop on new directions in control engineering education, IEEE Control Syst. Mag. **19**, 53–58 (1999)

44.4   R.M. Murray: *Control in an Information Rich World* (Society for Industrial and Applied Mathematics, Philadelphia 2003)

44.5   R.M. Murray, K.J. Åström, S.P. Boyd, R.W. Brockett, G. Stein: Future directions in control in an information rich world, IEEE Control Syst. Mag. **23**, 20–23 (2003)

44.6   E.D. Sontag: Molecular systems biology and control, Eur. J. Control **11**, 396–436 (2005)

44.7   J.M. Bailey, W.M. Haddad: Paradigms, benefits, and challenges. Drug dosing control in clinical pharmacology, IEEE Control Syst. Mag. **25**, 35–51 (2005)

44.8   S.H. Haas, M.G. Frei, I. Osorio, B. Pasik-Duncan, J. Radel: EEG ocular artifact removal through ARMAX model system identification using extended least squares, Commun. Inf. Syst. **3**, 19–40 (2003)

44.9   I. Osorio, M.G. Frei: Hurst parameter estimation for epileptic seizure detection, Commun. Inf. Syst. **7**, 167–176 (2007)

44.10   I. Osorio, M.G. Frei, S.B. Wilkinson: Real time automated detection and quantitative analysis of seizures and short term prediction of clinical onset, Epilepsia **39**, 615–627 (1998)

44.11   B. Pasik-Duncan: Random walk around some problems in identification and stochastic adaptive control with applications to finance. In: *AMS-IMS-SIAM Joint Summer Research Conference on Mathematics of Finance* (2003) pp. 273–287

44.12   B. Pasik-Duncan: Special issue on the stochastic control methods in financial engineering, IEEE Trans Automatic Control (2004)

44.13   P. Varaiya: Reducing highway congestion: an empirical approach, Eur. J. Control **11**, 301–310 (2005)

44.14   P.R. Kumar: New technological vistas for systems and control: the example of wireless networks. What does the future hold for the design of wireless networks?, IEEE Control Syst. Mag. **21**, 24038 (2001)

44.15   T.E. Djaferis: Systems and control impact in a changing world. In: *Chinese Control Conference (CCC)* (Hunan, China 2007) pp. 21–28

44.16   B. Pasik-Duncan, R. Patton, K. Schilling, E.F. Camacho: Four focused forums. Math, science and technology in control engineering education, IEEE Control Syst. Mag. **26**, 93–98 (2006)

44.17   G. Clough: *The Engineer of 2020: Visions of Engineering in the New Century* (National Academy, Washington 2004)

44.18   D. Oblinger: Boomers, Gen-Xers, and Millennials: Understanding the "New Students", EDUCAUSE Rev. **38**, 36–45 (2003)

44.19   B.S. Heck, D.S. Dorato, D.S. Bernstein, C.C. Bissell, N.H. McClamroch, J. Fishstrom: Future directions in control education, IEEE Control Syst. Mag. **19**, 36–53 (1999)

44.20   G. Clough: *Educating the Engineer of 2020: Adapting Engineering Education to the New Century* (National Academy, Washington 2005)

44.21   T.E. Duncan, B. Pasik-Duncan: Stochastic adaptive control. In: *The Control Handbook*, ed. by W.S. Levine (CRC, Boca Raton 1995) pp. 1127–1136

44.22   T.E. Duncan, B. Pasik-Duncan: Adaptive control of continuous time stochastic systems, J. Adapt. Control Signal Process. **16**, 327–340 (2002)

44.23   B. Pasik-Duncan: Stochastic systems. In: *Wiley Encyclopedia of IEEE*, Vol. 20, ed. by J.G. Webster (Wiley, New York 1999) pp. 543–555

44.24   F. Black, M. Scholes: The pricing of options and corporate liabilities, J. Polit. Econ. **81**, 637–654 (1973)

44.25   R.C. Merton: Theory of rational option pricing, Bell J. Econ. Manag. Sci. **4**, 141–183 (1973)

44.26   S. Crepin, D. Houinato, B. Nawana, G.D. Avode, P. Preux, J. Desport: Link between epilepsy and malnutrition in a rural area of Benin, Epilepsia **48**, 1926–1933 (2007)

44.27   T.E. Djaferis: Automatic control in first-year engineering study, IEEE Control Syst. Mag. **24**, 35–37 (2004)

44.28   D.S. Bernstein: Innovations in undergraduate education, Special Issue, IEEE Control Syst. Mag. **24**, 1–101 (2004)

44.29   D.S. Bernstein: Innovations in undergraduate education: Part II, Special Issue, IEEE Control Syst. Mag. **25**, 1–106 (2005)

44.30   K.J. Åström: Challenges in control education, 7th IFAC Symposium on Advances in Control Education (Universidad Politecnica de Madrid, 2006)

44.31   DARPA: http://www.darpa.mil/grandchallenge/ Vol. 2008: Defense Advanced Research Projects Agency (DARPA, 2007)

44.32   The MathWorks Inc.: *Using MATLAB. The Language of Technical Computing* (The MathWorks Inc., Natick 2002)

44.33   K.J. Åström, R.E. Klein, A. Lennartson: Bicycle dynamics and control, IEEE Control Syst. Mag. **25**, 26–47 (2005)

44.34   M.W. Spong: Project based control education, 7th IFAC Symposium on Advances in Control Education (Universidad Politecnica de Madrid, 2006)

44.35   P.J. Gawthrop, E. McGookin: Using LEGO in control education, 7th IFAC Symposium on Advances in Control Education (Universidad Politecnica de Madrid, 2006)

44.36   J. LaCombe, C. Rogers, E. Wang: *Using Lego Bricks To Conduct Engineering Experiments* (American Society for Engineering Education, Salt Lake City 2004)

44.37   N. Jaksic, D. Spencer: *An Introduction To Mechatronics Experiment: Lego Mindstorms Next Urban Challenge* (American Society for Engineering Education, Honolulu 2007)

44.38   D. Hansen, B. Self, B. Self, J. Wood: *Teaching Undergraduate Kinetics Using A Lego Mindstorms Race*

Car Competition (American Society for Engineering Education, Salt Lake City, 2004)

44.39   B.S. Heck, N.S. Clements, A.A. Ferri: A LEGO experiment for embedded control system design, IEEE Control Syst. Mag. **24**, 43–56 (2004)

44.40   D.S. Bernstein: The quanser DC motor control trainer – individual or team learning for hands-on education, IEEE Control Syst. Mag. **25**, 90–93 (2005)

44.41   S. Dormido: The role of interactivity in control learning, 6th IFAC Symposium on Advances in Control Education (Oulu, 2003)

44.42   J.L. Guzman, K.J. Åström, S. Dormido, T. Hagglund, Y. Piguet: Interactive learning modules for PID control, 7th IFAC Symposium on Advances in Control Education (Universidad Politecnica de Madrid, 2006)

44.43   M. Johansson, M. Galvert, K.J. Åström: Interactive tools for education in automatic control, IEEE Control Syst. Mag. **18**, 33–40 (1998)

44.44   B. Pasik-Duncan: Workshop for Czech high school students and teachers at the 16th IFAC world congress, IEEE Control Syst. Mag. **26**, 110–111 (2006)

44.45   M.H. Shor, F.B. Hanson: Bringing control to students and teachers, IEEE Control Syst. Mag. **24**, 20–30 (2004)

44.46   B. Pasik-Duncan: Mathematics education of tomorrow, Assoc. Women Math. (AWM) Newslett. **34**, 6–11 (2004)

44.47   D. Duncan, B. Pasik-Duncan: Undergraduates' partnership with K-12, American Control Conference (Anchorage, 2002) pp. 1103–1107

44.48   M.K. Michael: Encouraging young women toward engineering and applied sciences: ISIC/MED special preconference workshop, IEEE Control Syst. Mag. **26**, 100–101 (2006)

44.49   B. Pasik-Duncan, T. Duncan: KU stochastical adaptive control undergraduate success stories, Am. Control Conf. (ACC) (Anchorage 2002) pp. 1085–1086

44.50   IFAC Technical Committee on Control Education (EDCOM): http://www.griffith.edu.au/centre/icsl/edcom/ (last accessed February 19, 2009)

44.51   D. Purkerson Hammer, S.M. Paulsen: Strategies and processes to design an integrated, longitudinal professional skills development course sequence, Am. J. Pharmaceut. Educ. **65**, 77–85 (2001)

44.52   B. Pasik-Duncan, T.E. Duncan: Research experience at all levels, American Control Conference (Maui, 2003) pp. 3036–3038

44.53   Illinois State University: http://www.sotl.ilstu.edu/ (last accessed February 19, 2009)

44.54   K. Lewin: Action research and minority problems, J. Social Issues **2**, 34–46 (1946)

44.55   E.L. Boyer: *Scholarship Reconsidered: Priorities of the Professoriate* (Princeton Univ. Press, Lawrenceville 1990)

45. Software Management

Peter C. Patton, Bijay K. Jayaswal

This chapter is an introduction to software management in the context of automation. It recognizes how software and automation are intertwined and have been mutually enabling in enhancing the reach and seemingly unimaginable applications of these two disciplines. It further identifies software engineering as application of various tools, techniques, methodologies, and disciplines to produce and maintain an automated solution to a problem and how software management plays a central role in making it possible. We recognize that software must be managed like any other corporate or organizational resource albeit as a virtual rather than an actual or tangible entity. In this chapter we restrict ourselves to three crucial issues of software management in the context of software as a component in automation and how it enhances its value and availability by effective software distribution, asset management, and cost estimation. It presents current best practices in software automation, distribution, asset management, and cost estimation.

Part E | 45

45.1 Automation and Software Management

Automation and software are remarkably intertwined. As organized disciplines, they have independent origins but are deeply interlocked now. Software, as an integral component of computers, is an essential tool for automation of a wide range of industrial, enterprise, educational, scientific, military, medical, home, entertainment, and personal devices, processes, and equipment. On the other hand, automation software, compilers, and a range of computer-assisted software engineering (CASE) tools are crucial for estimating, developing, testing, maintaining, and integrating all kinds of software, especially large and complex ones. CASE

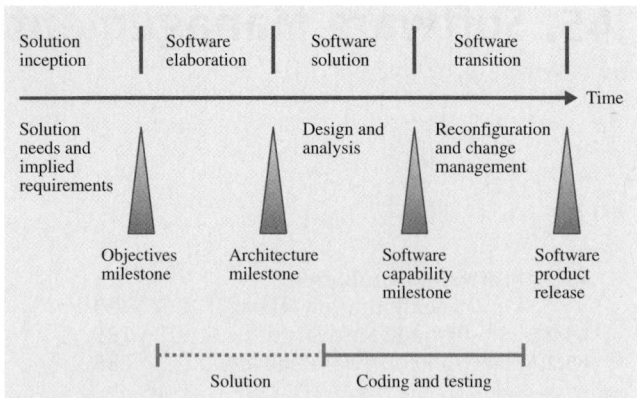

Fig. 45.1 Phases and milestones of software development

automation tools are nothing but computer programs that assist software analysts, engineers, and coders during all phases of the software and system development lifecycle (Fig. 45.1). They are used to automate software development process and to ease the task of coordinating various events in the development cycle. CASE tools are usually divided into two main groups: those that deal with the upstream parts of the system development lifecycle (preliminary estimating, investigation, analysis, and design) are referred to as front-end CASE tools, and those that deal mainly with the implementation and installation are referred to as back-end CASE tools [45.1]. Thus software is indispensable for automation and automation software tools are essential for developing reliable and cost-effective software.

In its earliest applications, automation was mechanization of certain operation or a series of mechanical tasks and operations that would be done manually otherwise. While automation developed initially to meet the challenges to automate mechanical and manual operations in industries, software's origins could be traced to solving (computing) military and subsequently, scientific and business-related data-based problems. It is amazing what a crucial role software has come to play in computing, automation, and other numerous applications, given that it was actually an afterthought to hardware. The term software was not even used until 1958 almost two decades after the invention of the ENIAC computer in 1940s.

45.1.1 Software Engineering and Software Management

Software as a tool in automation has enabled a quantum leap in improving quality, reliability, cost, precision,

safety, and security of a large number of products and systems and in expanding automation to seemingly unimaginable applications in the last several decades. Software has enabled applications such as the Internet and mobile platforms that have transformed our lives. However, there are huge intellectual, resource, and research challenges to deliver and control complex systems involving systems of systems, networks of networks, and agents of agents. They create huge intellectual control and software management problems [45.2]. Automation and component technology would play a critical role in meeting these software management challenges.

Software consists of computer programs, procedures, and (possibly) associated documentation and data pertaining to the operation of a computer system [45.3]. It is important to note that software is not just codes or even programs. Further, software runs on hardware and has been, relatively speaking, a laggard as regards quality, delivery, and cost relative to hardware part of computers. Software is thus the crucial element to be addressed in determining cost and performance of a large number of computer applications. Often computers operate as part of communication networks, increasingly the web and the Internet, that provide connectivity for importing and exporting data for a wide variety of commercial, military, educational, entertainment, and public services. The limits of software availability and trustworthiness are similarly determining factors as regards effectiveness and viability of an increasingly large number of automation applications.

Software engineering, on the other hand, could be described as the application of tools, techniques, methodologies, and disciplines to produce and maintain an *automated* solution to a problem. Designing and delivering trustworthy software is one of the great technological challenges of our times. It would require, among others, the implementation of a lifecycle approach to software development that addresses software trustworthiness at upstream phases of the software development process [45.3].

That brings us to the subject of software management. What exactly is software management? An inclusive view of a software business should include organizational, strategic, and competitive contexts of an enterprise processes and technologies that are used in developing software. That is a vast area and beyond the scope of this Handbook and indeed this chapter. Even the software development process is a large discipline consisting of tools, techniques, and methodologies that aid planning, estimating, staffing, organizing, and controlling the software development process [45.3, 4]. In this chapter we would restrict ourselves to three crucial issues of software management in the context of software as a component in automation and how it enhances its value and availability by effective *software distribution*, *asset management*, and *cost estimation*.

45.2 Software Distribution

45.2.1 Overview of Software Distribution/Software Delivery

Software distribution is a generic term used to describe automated or semiautomated distribution or delivery of software, usually on a network including the Internet. Such facilities enable software providers to distribute, fix, and update software packages to clients and servers on an organization's network. It is usually configured to install software with no user intervention and as such can be used to keep organization's network up to date with minimum disruptions. Enterprises face the challenges of maintaining security and compliance requirements while keeping pace with meeting regulatory and technological changes. The information technology (IT) staff has to constantly manage new security threats and patches, updates as well as innovations. Business success depends on maintaining control in the face of all these changes. A smart configuration management suite must meet several requirements (Fig. 45.2) [45.5]:

1. Increase efficiency and save time with a complete hardware and software management solution for all the systems and users in your complex network environment.
2. Reduce costs and the demands on help desk resources with tools that help you securely and easily support users in any networked environment.
3. Protect user productivity and reduce resource needs by easily keeping up with patches and updates and maintaining system-level security.
4. Save time and network bandwidth with patented, ultra-efficient, and fault-tolerant software distribution technologies.
5. Decrease software licensing costs and quickly respond to audits with comprehensive software licence monitoring capabilities.
6. Increase efficiency and save time by easily migrating users and their profiles to new operating systems.

Software distribution is one of the important tools of configuration management tools. The rate and scale of change is increasing. Demands such as Microsoft Vista migrations, data center consolidation initiatives, and stringent time-to-market requirements have placed a new emphasis on the ability to execute changes effectively and efficiently. At the same time, IT is faced with resource constraints, a geographically dispersed and mobile workforce, and ongoing security threats.

HP configuration management solutions enable IT to respond to these demands through automated deployment and continuous management of software, including operating systems, applications, patches, content, and configuration settings, on the widest breadth and largest volume of devices throughout the lifecycle for:

- IT efficiency to control management costs
- Agility to bring services to customers and users faster.

Software distribution is essentially the process of making the software products and services available to users in a manner that meets their cost, quality, and delivery expectations. It means that the software provider successfully delivers the product that the users need, when they and in the form need it, and at the price they are willing to pay for it. It also implies that the

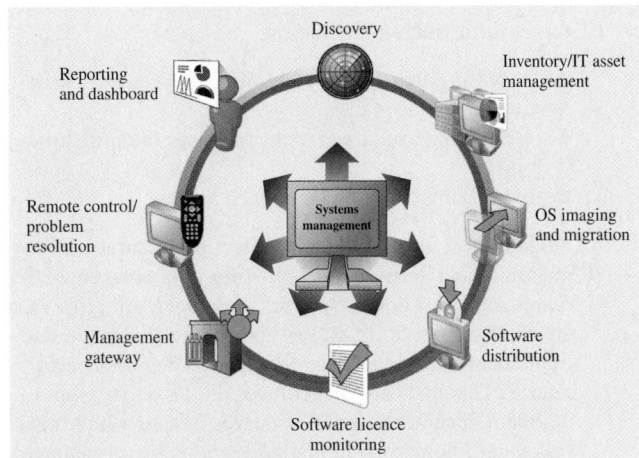

Fig. 45.2 A software management console (after [45.5] with permission)

software provider is able to do so in a cost-effective manner that is economically and competitively viable. While for many physical products we take this process for granted, for software, especially, large and complex ones, it is quite a challenge. The challenge comes often from software's complexity and the very nature of software design and use.

In addition to what was stated above regarding software distribution benefits it also ensures ubiquity and currency of software updates, and allows for certain compliance with all licensing provisions, as well as economy as just licences currently in use will be paid for.

45.2.2 Software Distribution in MS Configuration Manager 2007

Microsoft's System Center Configuration Manager (MSCCM) is a state-of-the-art toolset for software distribution and management in a networked organization. We cite this product extensively as an exemplar of technology available today without loss of generality, even though it is a particular proprietary product [45.6]. Other quality software distribution products in this general area include LANDesk [45.7], and Akamai [45.8]. Microsoft System Center is a suite of IT management solutions that help IT departments proactively plan, deploy, manage, and optimize the software lifecycle of a networked IT environment.

Planning Software Distribution
Planning is the first step in effectively upgrading the network server infrastructure. During this phase, the IT department must collect critical information about server infrastructure, including:

- Assessing the current state of the server infrastructure and datacenter
- Identifying each asset that comprises the infrastructure
- Identifying the purpose of each asset.

For most organizations, collecting accurate information about server assets within the datacenter is easier said than done. Networked datacenters grow increasingly complex daily as companies introduce and implement new technology to enhance business performance. This makes it difficult for the IT department to maintain accurate records of server assets, which also makes it difficult to plan upgrades and enhancements to server infrastructure. Microsoft System Center (MSCC) delivers capabilities that make it easier for the IT or-

ganization to collect information needed for in-depth knowledge of existing infrastructure.

The first step of planning a server upgrade is to identify all the assets that make up the network. The IT department needs a centralized management solution that automatically identifies software assets. Microsoft System Center Configuration Manager 2007 (MSCCM 2007) simplifies this task with hardware and software inventory capabilities that identify hardware and software assets, catalog who is using those assets, and understand where they are located. Through asset intelligence, MSCCM 2007 presents a clear picture of IT assets by providing identification and categorization of the servers, desktops, laptops, mobile devices, and software installed across both physical and virtual environments. Within the datacenter, this provides a fast method for understanding what server devices are in use today and who is using them. A new feature available in the first service pack for MSCCM 2007 also enables asset intelligence to identify new and changing systems and notify IT administrators of changes. This can reduce time spent identifying and tracking assets during and after an upgrade project.

As IT organizations move through the phases of the infrastructure optimization model, planning a server upgrade presents an opportunity to cut both medium- and long-term costs by optimizing the use of server resources within the datacenter. Virtualization is one of the most important trends that can impact server resource optimization by changing how an IT department manages servers and workloads. Virtual machine technology decouples the physical hardware from software so that the IT department can run multiple virtual machines on a single physical server.

Microsoft System Center Operations Manger 2007 and Microsoft System Center Virtual Machine Manager 2007 help the IT department identify how servers are being used, how each server is performing, and how each server can be used to its fullest potential. System Center Operations Manager 2007 monitors server health and stores vital performance information in a database that System Center Virtual Machine Manager 2007 can access and analyze. Virtual Machine Manager 2007 then generates a consolidation report that provides an easy-to-understand summary of the long-term performance of a workload. This information helps project teams make informed decisions about which servers would be ideal candidates for consolidation. Also, information about the performance of the hardware running virtualized applications provides data that decision-makers need to smartly move those applications off one server onto another, re-image the server, and then return the applications while maintaining full availability of the datacenter resources.

Microsoft System Center Data Protection Manager 2007 helps companies plan a server upgrade with confidence by enabling the IT department to back up existing data. System Center Data Protection Manager 2007 was built to protect and recover:

- Microsoft SQL Server
- Microsoft Exchange Server
- Microsoft Office SharePoint Server
- Microsoft Virtual Server
- Microsoft Active Directory directory service
- Windows file services.

Configuring Software Distribution

When the IT department has created an accurate inventory of server assets, it can then design the datacenter and determine which changes should be made to guarantee the most cost-efficient infrastructure. Then further steps will enable the department to successfully deploy the Windows Server 2008 operating system, Microsoft SQL Server 2008, and Exchange Server 2007 SP1 and transform the datacenter into a strategic asset. During the configuration or build phase, the IT department must create server images, convert physical servers to virtual servers, create a disaster recovery plan, and monitor the testing process. The build phase offers an opportunity for IT departments to identify areas for reducing costs, improving efficiency, and supporting compliance efforts. One way to accomplish this is by creating standardized server images for all server components, for both physical and virtual machines. System Center Operations Manager 2007 and System Center Virtual Machine Manager 2007 facilitate this process.

The task sequencer, driver packages, and dynamic driver catalog included with Configuration Manager 2007 significantly reduce the number of server images that the IT organization must create and be deployed to either physical or virtual machines. IT administrators can create a simple generic image and dynamically add the necessary drivers during the build. In addition, by integrating vendor-provided tools, Configuration Manager 2007 can automate the setup of redundant array of independent disks (RAID), storage area network (SAN), and Internet small computer system interface (iSCSI) hard-drive configurations as part of the task sequence. This can have a favorable impact on the amount of work required later as upgrades are issued. Upon creation of the server images for physical machines,

Virtual Machine Manager 2007 converts the appropriate images for virtual machines. Traditionally, this task can be slow and disrupt business operations, but Virtual Machine Manager 2007 uses the volume shadow copy service, which helps administrators create virtual machines without interrupting the source physical server. Virtual Machine Manager 2007 also simplifies this whole process by providing a task-based wizard that helps guide administrators. Once images are created, Virtual Machine Manager 2007 supports a complete library that organizes and manages all the *building blocks* of the virtual datacenter within a single interface.

Data Protection Manager 2007 helps prevent the IT department from losing critical business data when upgrading server infrastructure. By integrating a point-in-time database restore with existing application logs, Data Protection Manager can deliver nearly zero data loss recovery for Microsoft Exchange Server, SQL Server, and SharePoint Server, eliminating the need to replicate or synchronize data. Data Protection Manager also uses both disk and tape media to enable fast restore from disk and supports long-term data retention and off-site portability with disks.

Before deploying upgrades to the server environment, the IT department must perform tests to ensure business continuity when the new server products go live. System Center Operations Manager 2007 makes it easy to access the results of these tests, much in the same way that it monitors the overall health of the server infrastructure. An IT department can also create scenarios that act like the end-user of a specific service to monitor success and failure rates and performance statistics results that can help identify potential deployment issues. In addition, administrator-simulated end-users can access Virtual Machine Manager by way of a web portal that is designed for user self-service. This portal enables test users and development users to quickly provision new virtual machines for themselves, according to the controls set by the administrator. Not only can IT personnel quickly test new configurations, but they can also uncover problems before deployment.

Managing Software Distribution

During deployment, IT departments must quickly roll out new products while remaining agile so they can respond to changes. Costs must also be kept to a minimum and business operations must not be disrupted. In the past, deploying new server software required someone to sit down at each server and complete the upgrade. This manual process took significant resources and did not guarantee that servers were deployed with consistent configurations. Determining which virtual and physical machines to link together was also difficult because companies did not have the data, such as workloads, performance metrics, and network capacity, to create optimal arrangements. Companies often risked losing vital company data during the migration process. System Center helps alleviate these challenges.

With Configuration Manager 2007, IT administrators can roll out new servers rapidly and consistently by automating operating system deployments and task sequences. IT administrators can fully deploy and configure servers from previous states, either by updating or replacing original equipment manufacturer (OEM) builds, or by installing the operating system and applications on new computers. Preboot execution environment protocol and Windows deployment services also make it easier to deploy servers that have no operating system installed: just plug in the server and turn it on.

The task sequencer in Configuration Manager 2007 fully automates the end-to-end deployment process, enabling zero-touch to near-zero-touch deployments. This means that the process of building servers, which can include more than 80 steps, including image loads, driver loads, update loads, and multiple reboots, can be handled by Configuration Manager automatically.

IT departments can also maintain visibility of the state of the infrastructure throughout the entire datacenter deployment and management process. Configuration Manager 2007 generates detailed reports about the deployments and provides information about those that have failed. This information helps the IT department resolve problems quickly, easily, and proactively. To maximize server utilization, it is critical that IT administrators select the appropriate virtual machine host for a given workload. Virtual Machine Manager 2007 helps IT departments with this complex task of *intelligent placement*. Virtual Machine Manager 2007 uses a holistic approach to selecting the appropriate hosts based on four factors:

1. The resource consumption characteristics of the workload
2. Minimum central processing unit (CPU), disk, random-access memory (RAM), and network capacity requirements
3. Performance data from virtual machine hosts
4. Preselected business rules and models associated with each workflow that contain knowledge from the entire lifecycle of the workload.

After the analysis, Virtual Machine Manager 2007 produces an intelligent placement report that helps the

IT department select the appropriate host for a given workload.

As IT administrators migrate information to an updated server platform, it is critical that data is not lost or corrupted. Once the new platform is in place, Data Protection Manager 2007 will identify the new server environment and enable customers to quickly and easily restore the data where it needs to go. Administrative delays associated with restores are also reduced by using a restore user interface that is based on the calendar, robust media management functionality, and disk-based end-user recovery. With Data Protection Manager 2007, restoring information takes seconds and involves simply browsing a share and copying directly from Data Protection Manager to the production server. By enabling customers to restore data from disk, Data Protection Manager significantly shortens the amount of time it takes to recover data, allowing customers to recover data in minutes versus the hours it takes to recover from tape. Data Protection Manager also minimizes the risk of failure that is associated with recovering data from tape.

Monitoring Software Distribution

After successfully upgrading the server infrastructure with next-generation server technology from Microsoft, the IT department must continue to monitor the infrastructure to ensure technology and licences are up to date, the network is secure, and commitments to meet service level agreements for performance and availability are met. In addition, the IT department must ensure consistency within server configurations, for example, guaranteeing that every exchange server has the same configuration and that server resources are being used with maximum efficiency to derive the most value from existing resources. Meeting these goals was once a challenge because the IT department did not have a solution that enabled the management of the entire server infrastructure from a central location.

System Center Server Management Suite Enterprise not only simplifies and speeds the deployment of new server software, it also eases the ongoing task of managing the entire server infrastructure on a day-to-day basis. Centralized Management of Server Networks System Center offers many ways for IT departments to proactively manage the state of IT infrastructure regardless of its complexity; for example, System Center Operations Manager 2007 provides an easy-to-use management environment that can oversee thousands of servers and applications, delivering a comprehensive view of the health of the datacenter. System Center Operations

Manager 2007 also comes with over 60 management packs, which extend management capabilities to the operating systems, applications, and other technology components that make up the datacenter. With these management packs, IT departments have access to best-practice knowledge about specific Microsoft products and can more easily discover, monitor, trouble shoot, report on, and resolve problems for a specific technology component. Consequently, they can keep their datacenter running smoothly and efficiently. System Center Operation Manager also has a high-availability architecture that can leverage the latest network load-balancing and clustering capabilities to help ensure the datacenter is managed day and night.

To help guarantee that the infrastructure has the right configurations across all required server components, IT administrators can use System Center Configuration Manager 2007. The *desired configuration management* feature in Configuration Manager 2007 allows IT administrators to automatically assess how computers comply with predefined configurations; for example, an IT department can monitor the health of a configuration implemented for Microsoft Exchange Server or Windows Server and are alerted when a server's configuration drifts from the standard configuration.

Configuration Manager also ships with configuration packs, which provide predefined, optimized configurations for a range of servers. In addition, one of the most time-consuming aspects of ongoing management of the datacenter can be automated and managed by using Configuration Manager. Updating servers with patches, drivers, etc. within enforced maintenance windows remains a key challenge for IT departments. The desired configuration management feature can automate this process, ensuring that servers are maintained, available, and compliant with organizational standards.

Improve Disaster Recovery Capabilities. The IT department cannot prevent organizational disasters, but can take the appropriate steps to ensure that data is protected by developing and implementing a well-planned backup and recovery strategy for network outages. Data Protection Manager 2007 delivers the best possible recovery experience because it features continuous data protection with traditional backup, disk-based recovery, tape-based storage, database synchronizations, and log shipping. Consequently, with just a few mouse clicks the IT administrator can restore a SQL Server database directly back to the original server, restore data to a re-

covery database on the original server, or copy database files to an alternate server or disk.

Software Management Lifecycle
As the IT department updates and maintains datacenter server infrastructure and transitions to a dynamic IT infrastructure, Microsoft System Center can play a major role at each step. Because System Center is an integrated solution for the datacenter, IT departments can derive the most value in the shortest time. Every capability is built on a common framework and design, so IT departments can smoothly transition from one phase of the lifecycle to the next. Some examples of these transitions include:

- The ability to configure, deploy, and monitor server images, automatically, and then patch or update these images as required
- The ability to monitor datacenter applications and servers (such as Microsoft SQL Server 2008), be alerted to failures, and then recover from backup data
- The ability to report server performance, identify problem servers, backup servers, and convert to a virtual form to allow uninterrupted service while switching to new hardware.

System Center delivers the capabilities the IT department needs for the complete distributed software management lifecycle, and even offers specific licensing to support the evolution of the datacenter with the Server Management Suite Enterprise.

45.2.3 On-Demand Software

Microsoft's chairman Bill Gates recently provided a glimpse into the software giant's strategy for the future. In a widely circulated memo, Gates indicated that Microsoft will shift its focus from packaged software to the software as a service (SaaS), or *on-demand*, model. Instead of buying software outright and installing it on desktops or servers, businesses would rent applications on a per-user, per-month basis. Other enterprise software vendors have long been testing similar SaaS offerings as well. For small businesses, the on-demand model promises to reduce costs and complexity while at the same time increase the level of software functionality available to them. There are no packaged applications or hardware to buy upfront, and no dedicated personnel needed to install and maintain software. What's more SaaS will unleash a wave of service applications and that will dramatically change the nature and cost of solutions deliverable to enterprises or small businesses.

In this case as with the web a decade ago the revolution started without Microsoft, since other companies have been selling on-demand software for years. However, a push by the world's most powerful software maker will certainly accelerate the trend, and it is noteworthy that the immediate target of the new Microsoft Office Live on-demand platform is small business. Prerelease or beta versions of e-Mail, project management, collaboration, website design, and analytics programs will be available in early 2006. Microsoft aims to offer some services to small businesses for free; the basic version of Microsoft Office Live, will be supported by advertisers and will allow small businesses to establish a domain name, complete with a hosted web site with 30 MB of storage, among other services.

45.2.4 Electronic Software Delivery

Distributing software over the Internet is becoming an increasingly popular means to distribute software to users. Electronic software delivery (ESD) refers to such distribution and particularly the practice of enabling users to download software from the Internet. In fact, ESD is the future of software distribution, sales, invoicing, payment, maintenance, and customer service and support. It is already an efficient and preferred way for vendors and buyers to distribute/acquire a wide range of software including music, videos, productivity soft-

ware, and increasingly texts and books. Compared with physical distribution involving compact disks (CDs), digital video disks or digital versatile disks (DVDs), and paper formats, ESD offers dramatic improvements in cost, delivery time, and convenience.

The trend toward ESD is irreversible. Nearly all software vendors offer electronic delivery options, and many new companies choose online downloads as their only vehicle for software delivery. An increasingly large proportion of consumer software and enterprise software are being delivered electronically.

Major Challenges
of Electronic Software Delivery System

ESD however has a long way to go as regards quality, reliability, and security of delivery. The following constitute the major challenges that ESD faces [45.8]:

a) *Ensuring robust download performance* to promote user satisfaction, avoid costly physical fulfillment, and prevent lost sales
b) *Providing high-performance infrastructure* that meets highly unpredictable demands in a cost-effective manner
c) *Measuring and understanding download results* and completion rates to gain insight into customer base and ESD efficacy
d) *Identifying specific geographic regions* where end-users are downloading from and controlling/restricting access accordingly.

45.3 Asset Management

45.3.1 Software Asset Management and Optimization

Typically the investment in software amounts to 20% of an organization's IT budget. The portfolio of software development and maintenance tools and its enterprise applications may be licensed from third-party vendors, developed in-house, or in some cases be open-source system software such as Linux, or open-source applications such as Apache or EMACS. In any case the IT organization must take full systems responsibility including currency, interoperability, and local support for all of the software in its portfolio, managing these as the organization would any other asset. In the case of in-house developed application packages the organization has conventional ownership of the asset; in the case of third-party systems or application software the or-

ganization has a lease-like licence for the use of the software and perhaps a maintenance service subscription; for open-source software the situation is less clear but can be sufficiently unambiguous that any risk of use is vastly outweighed by the technological and economic benefits of use.

Depending on the size of the organization and its IT commitment, the human and financial resources required to manage software as an asset may vary widely. At an IBM conference for IT directors in Atlanta recently one of the authors sat next to a senior technical person from Bell-South. As the presenter went around the room of 30 IT directors asking for their titles and affiliation, he got to my seat mate who declared he was a Unix system administrator. The others looked at him in disgust, wondering how a low-level *grunt* like him got into this august gathering of senior managers. The fellow went on to finish his sentence saying that he managed 6700 Unix servers for Bell-South. Their expressions quickly turned from disgust to admiration as he probably had the biggest and most critical IT responsibility in the group. The complexity that the software asset manager must deal with comes in two flavors: first the inherent complexity of the software functionality itself, but second the number of desktop or laptop workstations that must be dealt with. Two decades ago most chief information officers (CIOs) avidly avoided taking responsibility for small computers outside their immediate purview, but since then the proliferation of client–server computing organization has long insisted that they take responsibility for them without the mandate of *ownership*. This introduces the complexity of numbers or volume, similarity to the complexity of a 1000 piece jigsaw puzzle, which is more than twice as difficult to assemble as a 500 piece puzzle. Few IT administrators are responsible for 6700 servers, but many are responsible for the effective operation of 10 000 or more client workstations remotely attached to the mainframe(s) or servers for which they do have *ownership*. As client–server technology as grown along with the third-party software vendor industry, so has the need for automation in software asset management. A glance at the Internet will reveal dozens of packages designed to aid the software asset manager in his or her task. We will not attempt to catalog this dynamic market here but rather only mention the availability and capability of such support software packages and describe what you might expect from one.

45.3.2 The Applications Inventory or Asset Portfolio

The director of development in the IT organization usually takes responsibility for what is called the applications portfolio of all the software systems, application packages, and tools licensed, owned, or used to fulfill the organization's functions. In larger organizations the responsibility may be divided between a director of systems for system software and a director of applications for applications software. An organization supporting thousands of client platforms may have a third person responsible for the management of client hardware asset leasing, software asset licensing, and end-user support. As in any other IT function these managers must walk a tightrope to avoid too slow incorporation and support of new technology and too rapid incorporation of new technology. In the former case their end-users lose

competitive advantage because they do not have access to current software features and functions, and in the latter they lose competitive advantage due to unreliability of early release software, and perhaps even end-user training problems.

The software asset portfolio should include all environmental, legal, functional, and resource requirements information about any program system, application, or tool, whether licensed from the hardware vendor, a third-party software vendor, open source or locally developed. During the preparation for addressing year 2000 (Y2K) computer compliance one of the most often asked questions about a perfectly working but not Y2K compliant COBOL75 program to the IT manager was: Do you have current source code? If the answer was "yes" the second question was: Do you still have an archived copy of your COBOL75 compiler? All too often the answer to the second question was that it had long ago been discarded *because it was no longer supported by the vendor*. It was very easy (and relatively inexpensive) to help those IT managers who saved everything and knew exactly where it all was. As the prophet Isaiah once said, "my people go into exile (or captivity) for lack of knowledge" (Isaiah 5:13). The requirements for a software portfolio are relatively simple:

- Know what software you are using and its source and environmental requirements
- Know which of your users needs it, how often, and what their applications are
- If still supported by a vendor, know who to contact at the vendor for emergency support
- If not supported, maintain a current copy of the source code and a compiler that will compile it
- Know the legal or licensing constraints of the software and any sublicensed components that may be used within it
- Maintain a current source code file if possible; if not, insure that one is escrowed for you by the vendor
- Archive software that is no longer used; it is amazing how often someone very high up in the organization will need it one more time.

45.3.3 Software Asset Management Tools

The issues that software asset management (SAM) tools offer to handle for an organization concern IT managers, financial managers, and corporate legal staff. For IT managers the issues of concern are:

- Manage site licences for software by user and work-station
- Get more out of underutilized assets by moving un-used software to new users rather than purchasing more licences
- Manage unlicensed and open-source software any-where in the system
- Summarize entitlement information
- Reconcile installations to entitlements
- Provide user services high-level access to vendor support as required
- Monitor lifecycles and proposed upgrades of all li-censed software
- Ensure uniform corporate-wide installation of up-grades and support.

For financial managers:

- Know whether all the software in firm is licensed and legal
- Avoid compliance fines and fees
- Ensure that site licences are sufficient but also at lowest cost
- Maximize underused software assets by redistribut-ing existing licences to new users
- Minimize overall cost of licensed software.

For corporate legal staff in support of IT and financial management:

- Ensure absolute compliance with site licence provi-sions
- Inform all staff and department heads about the risks of using unlicensed software
- Inhibit maverick departments and personnel from downloading or sharing unlicensed software
- Limit corporate liability by ensuring that unau-thorized or illegal (e.g., off-shore gaming, porno-graphic, child pornography, etc.) software is never installed on corporate desktops or laptops.

While managing these issues calls for little more than the application of common sense, doing so in a large corporation, university, or government orga-nization with several hundred servers and 15 000 or more workstations and laptops is best done in a rig-orous and disciplined way. One of the most pervasive problems is easiest solved; when CIO at a research uni-versity one of the authors had difficulty discouraging some faculty, research staff, and department personnel from downloading and sharing nonlicensed, non-open-source software that they felt entitled to use without licence. A short conversation with the general counsel solved the problem with a single joint memo inform-ing such users and department that, in the event they were sued for such noncompliance, they would have to pay the legal fees for their defense and any fines out of personal or departmental budgets, since the uni-versity would not be doing so. As for the latter, more sensitive and even higher risk issue, making sure ev-eryone in the organization knows the consequences of noncompliance with corporate standards is usually sufficient.

A somewhat larger issue is that software licensing agreements are not all the same. Site licences for col-leges and universities are often more lenient, and may for example allow a faculty person to use two copies of the software, one on a desktop and the other on a laptop, providing that only one of them is being used at a time. Provisions such as this may require invoking a bit of the honor system but are usually manageable without undue risk.

45.3.4 Managing Corporate Laptops and Their Software

Laptops have a special set of problems due to their mo-bility and susceptibility to shock-related damage. Many organizations support both personal computer (PC) and Mac laptops and thus may have to license the same soft-ware in more than one way. As laptop hard drives are more subject to shock damage and consequent loss of both system and application software as well as data, an inventory of the software licensed to each laptop may be critical. The mobility of laptops introduces haz-ards beyond shock in terms of theft or just loss of the whole machine. Corporate data may be seriously com-promised in these situations so security and backup is especially important for laptops. If an organization has thousands of laptop users it may require a full-time per-son to maintain currency and security of both laptop hardware and software. Now that many organizations issue only laptops and not a desktop computer as well, a special focus on laptop computing assets beyond just SAM generally is worthy of consideration.

45.3.5 Licence Compliance Issues and Benefits

As every personal computer user well knows, if you do not agree to the license agreement, the software will not install and open for you. Thus, the first benefit from li-censing compliance is access. At the beginning the user was forced to actually read the licence before the installer would accept their agreement, but users never read these things (licences and contracts are written by lawyers and therefore only read by lawyers). When the vendors realized that everyone (except lawyers) was just scrolling to the end and clicking on the accept radio button, they began to allow the user to just hit the button and install. The licence is a form of contract and thus protects both the vendor and the user. Software is no longer bought and sold; rather the use of it or access to it is guaranteed by the vendor to purchaser of a licence. This means that, if it does not work as delivered, you can insist on certain fixes and upgrades, although you may have to pay an additional sum for the upgrades. After spending a great deal of creative effort building a system of application, it is some comfort to know that it will not be arbitrarily dropped or made unavailable to you. In such a case the end-user may need to finally read the licence agreement to determine his or her rights and what remedies may be available. If you build a complex application without benefit of licence and something goes wrong then you have no recourse. In a large organization this is very important and careful management of site licences may be an important business continuity consideration if a real disaster happens.

45.3.6 Emerging Trends and Future Challenges

The trend in personal computing and information sharing is definitely towards mobility. As WiFi wireless networks proliferate beyond today's local-area networks (LANs) and small digital assistants, including enhanced cell phones, become more popular, one may expect to see the corporations computing, software, and information assets spread ever more widely. As in any other management situation it is necessary to know where everything is, how it is being used, and to minimize risk.

45.4 Cost Estimation

45.4.1 Estimating Project Scope

One of the major problems in computer systems development is estimating the time and cost of a development project. It has often been said that the next major development in hardware technology forecast to be available in 10 years arrives in about half that time, but the time forecast to build a new software system takes twice as long and costs at least twice as much as the initial forecast. The field of software development is strewn with disasters and abandoned projects because the scope of the project was not understood from the beginning or was expanded beyond the time and resources available during development. A few tragic examples are given in the book *Design for Trustworthy Software* [45.3, p. 536].

One of the earliest and best contributions to the solution of this problem was *Prof. Barry Boehm*'s magisterial work entitled *Software Engineering Economics* [45.9]. The constructive cost model (COCOMO) which he developed for estimating the cost and development time for large mainframe software projects has been updated for the current client–server era and will be discussed below. While there are other similar development models and methodologies, COCOMO has long since become the gold standard for such methods and is available in several commercial versions. Here we will briefly review the COCOMO suite and a recent product, SEER-SEM, which is based on it. These are typical of the software estimation packages available to the software developer today.

45.4.2 Cost Estimating for Large Projects

Large software development projects have an unfortunate history of time and cost overruns and the largest and more complex of them have often been abandoned after several years and many millions of dollars invested. It is also true that some large software projects do finish, on time, within budget, and satisfy their users. *Capers Jones*, the dean of software estimation, did a study of both 59 manual and 50 automated estimates for large projects (e.g., in the 5000 function-point range) [45.10]. The manual estimates were created by managers using conventional methodologies and the automated ones by commercially available estimating tools. The manual estimates yielded lower costs and shorter times than actually experienced and only 4 of the 50 were within 10% of actual. Only one of the automated estimates was optimistic, most were much too conservative, but 22 were within 10% of actual experience. Curiously, both were fairly accurate estimating the actual coding involved, but the manual estimates were much too optimistic for the nonprogram-

Table 45.1 Relative software development activities and effort by genre (after [45.10, p. 3])

Activity	WWW %	MIS %	Outsource %	Commercial %	System %	Military %
Requirements	5	7.5	9	4	4	7
Prototype	10	2	2.5	1	2	2
Architecture		0.5	1	2	1.5	1
Planning		1	1.5	1	2	1
Func design		8	7	6	7	6
Tech design		7	8	5	6	7
Design review			0.5	1.5	2.5	1
Coding	30	20	16	23	20	16
Reuseability	5		2	2	2	2
Package purchase		1	1		1	1
Code inspection				1.5	1.5	1
Verification and validation						1
Config management		3	3	1	1	1.5
Integration		2	2	1.5	2	1.5
Documentation	10	7	9	12	10	10
Unit testing	30	4	3.5	2.5	5	3
Functional testing		6	5	6	5	5
Integration testing		5	5	4	5	5
System testing		7	5	7	5	6
Field testing				6	1.5	3
Acceptance testing		5	3		1	3
Independent testing						1
Quality testing			1	2	2	1
Training		2	3		1	1
Project management	10	12	12	11	12	13

ming parts of the project such as design, documentation, rework, testing, project management, etc. Jones notes that, for projects with fewer than 1000 function points (or about 125 000 C source LOC), programming is the major cost driver, but for projects above 10 000 function points (or about 1 250 000 C source LOC), testing and bug removal, plus documentation are much more costly than the programming or actual implementation cost. The dynamic of software development cost estimation is that developers tend to be conservative, especially those who have survived one or more unsuccessful project *death marches*, but their customers, i. e., the senior managers who need the application tend to be optimistic about both cost and time and crowd the developers toward more optimistic estimates. If the developer is too cautious and conservative the manager will simple not *buy* the project. Any estimate, whether it turns out to be optimistic or conservative in the end, must be defensible and is best based on experience with previous projects of the same size and complexity.

As software development projects get larger and the specter of later abandonment becomes more threatening, the need for more accurate estimates has prompted the development of a whole new software development service industry to provide quality tools. Some of the estimating tools on the market today are COCOMO II, CoStar, CostModeler, CostXpert, KnowledgePlan, PriceS, SEER, SLIM, and SoftCost [45.10, p. 2]. Some of the first generation of tools are still in use but are no longer being supported or sold. Jones reports that the basic kernel of functions supported by such tools includes:

- Sizing logic for specifications, source code, and test cases
- Phase-level, activity-level, and task-level estimation
- Schedule adjustments for holidays, vacations, and overtime
- Local salary and burden rate adjustments
- Adjustments for military, systems, and commercial standards

- Metrics support for function points and/or LOC
- Support for maintenance and enhancement.

More advanced features available in some estimating systems include quality and reliability estimates, risk and value analysis, return on investment (ROI), models to collect historical data, statistical analysis, and currency conversion for overseas development. Table 45.1 summarizes Jones experience with six different application genres. While he styles the table as merely illustrative it certainly fits one of the author's more than 50 years developing software in all six of these genres.

The new class of automatic software estimating tools may not be perfect but they are very good and may prompt the user to include factors he or she might otherwise overlook. As always in any complex human endeavor, there is no substitute for experience.

45.4.3 Requirements-Based a priori Estimates

Of course it is easier to estimate a large software project if this is the tenth one you have done just like it, but that is rarely the case. Developers who managed to do two or three projects fairly well are often promoted to chief software architect or director of MIS. What can be done when you need to create a requirements-based estimate and the requirements are not completely known yet? Amazing as it may seem to an engineer, this case occurs very often in software development, and can be tolerated because software is not manufactured like other systems and products designed by engineers. A software development project is the result of a negotiation and, like any other negotiated decision, can be renegotiated. Software is simply designed, redesigned, and redesigned again as it is implemented. The only software implementation analog to hard goods manufacturing is the almost completely error-free process of copying a computer file on to one or more CD-ROMs.

The hardware designer would think that the software engineer would follow the logical process of requirements discovery, functional design, technical design, implementation, testing, documentation, and delivery; however, the more common case in software development is that the first truly known and understood parameter is the delivery date to the customer. Hence, the software developer's commonly employed process, known as *date-driven estimation* [45.11]. So at this point we now when the project is to be completed but as yet do not know what is to be delivered. In traditional engineering one makes the estimate after design but in software engineering one must make the estimate *before* design since, for software, design replaces manufacturing. This places a tremendous burden on requirements discovery [45.12] and on the subsequent specification process. It has long been conventional wisdom in software development that one could write a precise functional specification (i.e., precise to *both* the implementers *and* the end users). Efforts to do so have greatly improved in the current era of object-oriented programming analysis and design, and have given new hope for the decades-long search for the holy grail of software development: specification-based programming [45.3, p. 502–506]. New technologies coming on to the market today such as Lawson Software's Landmark and Wescraft's Java refactoring system, inter alia promise a high degree of programming automation for Java-based software [45.3, p. 501]. Moreover, since the precise functional specification is written by domain specialist(s) and the Java code produced automatically by a metacompiler, what you as domain expert or allele for the end-user specify is truly what you get. This innovation in programming will dramatically change the way software is designed, implemented, and maintained, but it also will put even more burden on the requirements discovery and specification part of the project, which according to *Dekkers* are already the source of 60–99% of the defects delivered into implementation [45.12]. We have long known that defects are written into software, just like villains are written into melodramas, and with similar high drama and anxiety we see as we approach the end of the project, i.e., the well-known and very precise date of delivery.

In our opinion the best way to deal with this commonly occurring situation is the analytic hierarchy process (AHP) developed by *Saaty* [45.13], which naturally prioritizes the user requirements as they are identified. Clearly, the most critical and time-consuming requirements come first to the user's mind and are the most important to the user's ultimate satisfaction with the end result of the software development project. A description and overview of AHP together with examples of its application is given in [45.3, Chap. 8].

45.4.4 Training Developers to Make Good Estimates

The need to train developers to make precise estimates is an artifact of the conventional software design process, in which requirements and functional designers write in one *language*, technical or detail designers in another, testers in yet another, and the end-user in even yet another, but a bit more similar to the first language in the sequence. Understanding software design documents (including the actual source code) going from one stage to the next in the process seems to involve an inherent *language translation* effort which can be the source of many defects in the end result; for example, functional specification writers tend to think in terms of effort units, placing the project in a cost and resource framework. Developers however, tend to describe size and complexity in terms of the things they will have to make to implement the requirements [45.14]. Commercial estimation packages need either concrete measures such as source LOC or abstract measures such as function points as input, but are *encoded* with sufficient industry experience to translate back and forth between such measures.

Putnam and his associates, the developers of SLIM, have developed a process for mapping *units of need* into *units of work* which is able to train estimators to take best advantage of the new software estimation tools coming onto the market. As an alternative to the conventional software development process, one determines the size of the software components by analyzing them into common low-level software implementation units; creating a model-based first-cut estimate using known experiential productivity assumptions, project size, and critical components; performing what-if modeling until an agreed-upon (i.e., negotiated) estimate has been developed; and finally creating a detailed plan for the project [45.14]. The what-if modeling part of this alternative employs the estimation tool (SLIM in his example) in a feedback loop as well as a feedforward process and provides a powerful learning experience for the estimator. The intermediate units, for example, would include forms, new reports, changed reports, table changes, job control language (JCL) changes, and SQL procedures, and for each such unit of work a qualitative measure of effort, e.g., simple, average or complex. This is a more analytical and finer-grained approach than the conventional method of estimating the number of reports, estimating the number of forms, then the number of transactions per component, and then the number of database accesses, etc. Highly experienced COBOL programmers of yesteryear could use the time-honored conventional approach to estimate a project's source LOC to within a few percent using conventional methods and experience, but object-oriented analysis, design and programming (OOAPD)-based design begs for a more analytical and finer-grained modeling as proposed and implemented by Putnam.

45.4.5 Software Tools for Software Development Estimates

In the late 1970s and early 1980s Barry Boehm developed the original COCOMO system mentioned above to support the first generation of true software engineers in estimating the cost of mainframe software development. By 2000 the need for a version this tool able to handle object-oriented programming (OOP) and client–server software development led to COCOMO II. There is also a large suite of collateral but genre or methodology specific variations and derivatives of CO-COMO models and packages available today [45.15, p. 2]. The major variations are COCOMO II, COIN-COMO, and DBA COCOMO, which are fundamentally the same model but tailored for different development situations. Commercial version of COCOMO such as Costar (http://www.softstarsystems.com) and CostXpert (http://www.CostXpert.com) provide even further cost estimation capability (see also http://sunset.usc.edu/).

The basic logic of COCOMO models is based on the general model or formula

$$ PM = A \times \left(\sum Size \right)^{\sum B} \times \prod EM \,, $$

where PM is person months, A is a calibration factor, Size is a measure of functional size for a software model that has an additive effect on the software development effort, B are scale factors that have an exponential of nonlinear effect on software development effort, and EM are effort multipliers that influence software development effort.

Clearly this is an experience-based model or formula, not merely an equation to be evaluated. Each factor in the equation can be represented by either a single or multiple values depending on the purpose of the factor; for example, Size may be measured in either source LOC or function points, and EM may be used to describe development environment factors such as software complexity or software reuse. COCOMO II has 1 additive, 5 exponential, and 17 multiplicative factors, but other models in the suite have a different number depending on the scope of the effort and software genre being estimated by that model [45.15]. Boehm's current research at the University of Southern California (USC) is directed toward a unification of the COCOMO suite of models in order to provide a comprehensive estimate for the development of a software system to help developers make more precise cost and time estimates for their projects.

SEER-SEM is a cousin of COCOMO since both followed the early work of both Jensen method and Halstead's software science. As in most other models the development environment is characterized by parameters. The architecture of a software cost estimation model is characterized by the way it answers the following questions [45.16]:

- How large is the project?
- How productive are the developers?
- How much effort and time are required to complete the project?
- How does the outcome change under resource constraints?
- How much will the project cost?
- What is the expected quality of the outcome?
- How much effort will be required to maintain and upgrade the system in the field?

While the model can estimate program size in either LOC or function points we will display only the source LOC formula here

$$S_e = \text{NewSize} + \text{ExistingSize}$$
$$\times (0.4 \times \text{Redesign} + 0.25 \times \text{Reimpl}$$
$$+ 0.35 \times \text{Retest}),$$

where S_e increases in direct proportion to the amount of new code being added, but by a lesser amount for the redesign, implementation, and retesting of existing code [45.16, p. 1]. This pragmatic approach reflects the fact that today's programmer rarely starts with a blank piece of paper (or a blank screen, as it were), but rather is redesigning or upgrading some existing system or application package to run in a new or enhanced environment.

The formula for effort in SEER-SEM is

$$K = D^{0.4}(S_e/C_{te})^{1.2},$$

where S_e is effective size, and C_{te} is effective technology, a composite metric capturing factors relating to development efficiency or productivity, based on an extensive set of experiential people, process, and product parameters. D is staffing complexity, which depends on how quickly qualified staff can be added to the project [45.16]. Once effort is obtained, the time to implement can be found from

$$t_d = D^{-0.2}(S_e/C_{te})^{0.4}.$$

The exponent on the critical ratio reflects the fact that as the project size increases so does the time to complete, however at a lesser rate.

45.4.6 Emerging Trends and Future Challenges

The future of software development may be a very interesting revolutionary one rather than the evolutionary one we have experienced for the past 50 years. The emergence of very high-quality open-source software such as Linux and Apache raises the natural question as to why management should pay large sums and wait long times to develop custom proprietary software when open-source alternatives are available. Of course the answer to this question is they do not need to: if a time- and industry-tested alternative is available it should be added to the portfolio, but managed in a somewhat different way. We should only write software when we have to, and this principle is leading to a restructuring of the software industry from a vertical to a horizontal model. In the future most software vendors will sell their products to other software vendors, similar to the structure of today's automobile industry. Henry Ford began his company with an extremely vertically integrated business model like those of European manufacturers: he had his own taconite mines in Minnesota to feed his own steel mills, his own sand mines in New Mexico for producing glass windows, his own sheep ranches in Montana to produce the wool for upholstery, etc. Today, every small engineering and machining firm in Windsor, Ontario produces part for Ford. We think the same development is happening in software, aided by OOP and the functional componentization and consequent reusability of software. Many firms will write software but only a few of them will deliver it to end-users.

The computer revolution has produced a hardware performance gain of more than 10^{10} in the 60 years since the ENIAC was announced in 1946; however, the productivity gain in software development has been much less impressive. Countess Ada Lovelace invented the idea of the program loop in 1844 while watching Jacquard loom operate; Betty Holberton invented the sort–merge generator at the Harvard Computation Laboratory 100 years later; a decade later Dr. Grace Hopper left Harvard to build the first compilers, Math-Matic and FlowMatic at Univac, while Mandaly Grems at Boeing was creating sophisticated scientific programming interpretive systems for the IBM 701.26. The progression of programmer-oriented languages such as COBOL, FORTAN, and ALGOL in the second software generation led to a factor of 10–20 in programming productivity. Third- and fourth-generation software added additional increase by at least an order of magnitude each, but three or four orders of magnitude overall is much smaller than ten orders of magnitude.

The future of software development belongs to pattern languages. Christopher Alexander is a building architect who discovered the notion of a pattern language for designing and constructing buildings and cities. In general, a pattern language is a set of patterns or design solutions to some particular design problem in some particular domain. The key insight here is that design is a domain-specific problem that takes deep understanding of the problem domain and that, once good design solutions are found to any specific problem, we can codify them into reusable patterns. A simple example of a pattern language and the nature of domain specificity is perhaps how farmers go about building barns. They do not hire architects but rather get together and, through a set of *rules of thumb* based on how

much livestock they have and storage and processing considerations, build a barn with certain dimensions. One simple pattern is that, when the barn gets to be too long for doors only at the ends, they put a double door in the middle. Only with a powerful design or pattern language – and in general a domain-specific design language (DSDL) – can we overcome the inherent limitations we have in our ability to comprehend the working of some complex system, to model it, and then automate those portions which can be mechanized [45.17]. Lawson Software's experience with Richard Patton and Richard Lawson's Landmark [45.3, p. 501] DSDL has not only shown more than another order of magnitude in business enterprise software development, but in the first year of delivering accounting, supply chain, and human resources (HR) applications only one bug has been reported. This is the future of software development.

45.5 Further Reading

- J. Bosch: *Design and Use of Software Architectures: Adopting and Evolving a Product Line Approach* (Addison-Wesley, Reading 2000)
- A. Cockburn: *Agile Software Development* (Longman, Boston 2002)
- D. Dikel, D. Kane, S. Ornburn, W. Loftus, J. Wilson: Applying software product-line architecture, IEEE Comput. **30**(8), 49–55 (1997)
- D. Dori: *Object-Process Methodology* (Springer, Berlin, Heidelberg 2002)
- A. De Lucia, F. Ferrucci, G. Tortora, M. Tucci: *Emerging Methods, Technologies, and Process Management in Software Engineering* (Wiley, IEEE Computer Society, New York 2008)
- R. Fantina: *Practical Software Process Improvement* (Artech House, Norwood 2005)
- P. Hall, J. Fernandez-Ramil: *Managing the Software Enterprise: Software Engineering and Information*

Systems in Context (Cengage Learning Business, London 2007)
- I. Jacobson, G. Booch, J. Rumbaugh: *The Unified Software Development Process* (Addison-Wesley, Reading 1999)
- D. Leffingwell: *Scaling Software Agility: Best Practices for Large Enterprises* (Addison-Wesley, Reading 2007)
- D. Leffingwell, D. Widrig: *Managing Software Requirements: A Unified Approach* (Addison-Wesley, Reading 1999)
- A. Mathur: *Foundations of Software Testing* (Addison-Wesley, Reading 2008)
- P. Robillard, P. Kruchten, P. d'Astous: *Software Engineering Processes: With UPEDU* (Addison-Wesley, Reading 2002)
- W. Royce: *Software Project Management: A Unified Framework* (Addison-Wesley, Reading 1998)

References

45.1   S. Barclay, S. Padusenko: *Case Tools History*, Curriculum Methods Class CURR 309, Computer Science, Faculty of Education (Queen's University, Kingston 2008), http://educ.queensu.ca/~compsci/units/casetools.html#HIST

45.2   B. Boehm: Foreword. In: *Software Management*, ed. by D.J. Reifer (Wiley Interscience, New York 2006) p. ix

45.3   B.K. Jayaswal, P.C. Patton: *Design for Trustworthy Software: Tools Techniques of Developing Robust*

Software (Prentice Hall, Upper Saddle River 2006) p. 58

45.4   D.J. Reifer (Ed.): *Software Management* (Wiley Interscience, New York 2006)

45.5   LANDesk Software Ltd.: http://www.networkd.com/pdf/LDMS8/Data_Sheets/ds_mgtsuite_en-US.pdf (South Jordan, 2007)

45.6   Microsoft System Center: *Controlling Costs and Driving Agility in the Datacenter*, MS Whitepaper (Microsoft, Redmond 2007)

45.7 LANDesk Software Ltd.: Solutions Brief (LANDesk, South Jordan 2007), http://www.landesk.com

45.8 Akamai Electronic Software Delivery: *Business Benefits and Best Practices* (Akamai, Cambridge 2007), http://www.akamai.com/dl/whitepapers Akamai_ESD_Whitepaper.pdf

45.9 B.W. Boehm: *Software Engineering Economics* (Prentice Hall, Upper Saddle River 1981)

45.10 C. Jones: Software cost estimating methods for large projects, Crosstalk (2005), http://www.stsc.hill.af.mil/crosstalk/2005/04/0504Jones.html

45.11 S. McConnell: After the Gold Rush, 2004 Systems and Software Technology Conference (Salt Lake City 2004)

45.12 C.A. Dekkers: Creating requirements-based estimates before requirements are complete, Crosstalk (2005), http://www.stsc.hill.af.mil/crosstalk/2005/04/0504Dekkers.html

45.13 T.L. Saaty: *The Analytic Hierarch Planning Process: Planning, Priority Setting, Resource Allocation* (McGraw-Hill, New York 1980)

45.14 L.H. Putnam, D.T. Putnam, D.H. Beckett: A method for improving developer's software size estimates, Crosstalk (2005), http://www.stsc.hill.af.mil/crosstalk/2005/04/0504Putnam.html

45.15 B.W. Boehm, R. Valerdi, J.A. Lane, A.W. Brown: COCOMO suite methodology and evolution, Crosstalk (2005), http://www.stsc.hill.af.mil/crosstalk/2005/04/0504Boehm.html

45.16 L. Fischman, K. McRitchie, D.D. Galorath: Inside SEER-SEM, Crosstalk (2005), http://www.stsc.hill.af.mil/crosstalk/2005/04/0504Fischman.html

45.17 R.D. Patton: What can be automated? What cannot be automated?. In: *Springer Handbook of Automation*, ed. by S.Y. Nof (Springer, Berlin, Heidelberg 2009), Chap. 18

46. Practical Automation Specification

Wolfgang Mann

This chapter specifies equipment-based control system structures for complex and integrated systems and describes the approach to and implementation of an equipment-based control strategy. Based on a view of subsystems in a production, process or a single machine the control system has to abstract the subunits in an object-oriented manner to obtain their methods and properties. The base subunits will run as separate state machines (either on centralized or decentralized control devices) representing themselves to the next control hierarchy level only by said methods and properties. These base subunits form functional subsystems in the same way.

Advantages of such a modular specification are: easy replacement of different base units with the same functionality to the next hierarchy level, high efficiency in construction kit engineering of systems, easy integration of systems to vertical integration attempts – especially in the field of networking and data concentration. The challenge is the implementation on standard industrial programmable logic controller (PLC) systems with a standard industrial-like programming language (e.g., EN 61131). An example demonstrates the implementation in a modern test stand for heat meters for the German Physikalisch-Technische Bundesanstalt (PTB) institute, a system with about

1000 physical input/output (I/O) and measurement points.

46.1 Overview

Complex systems with many dozens of subsystems become increasingly challenging in terms of programming, interfacing, and commissioning, in both operation and maintenance.

The reasons for this intricacy are multifarious and include [46.1]:

1. Product lifecycles getting shorter, and even technological cycles getting shorter.
2. Orders to stock are replaced by short orders to delivery.
3. Change form self-production to integration of sub-suppliers.

4. Decentralized stock and service.
5. High price pressure.
6. Change form manually process integration to integrated processes.
7. Change from product supplier to system supplier.
8. Production processes control multiple enterprises.
9. e-Commerce and quality management systems additionally produce enormous quantity of data and force the introduction of data management throughout the company.

These requests can be covered by a horizontal and vertical integration of process and production systems and subsystems within the company and extending the network to suppliers and even customers. So the systems become more complex, open, distributed, and heterogeneous. Dataflow often has to be implemented throughout the complete enterprise hierarchy form the shop floor area to enterprise management level.

To address these attempts, new software concepts have to be implemented throughout the complete information chain within an enterprise, starting at the I/O level of the production line.

Although there are well-developed methods and strategies to produce well-structured, reusable, and sophisticated code in the information technology (IT) personal computer (PC)-based environment, industrial control at its base level is still done using standard PLC systems with EN 61131-like programming language using functional blocks, structured text or instruction lists at large scale [46.2].

As long as the projects are small and centralized a programming technique limited to the I/Os of industrial controllers and their periphery is applicable, with some restrictions.

However, with the above-stated imperative demands in mind, conventional I/O-based engineering attempts leads to unstructured, error-prone code, resulting in cost overruns during installation and operation.

We consider the traditional PLC programming style more as an expression of old-fashioned thinking, having its tradition as far back as relay-based times, rather than an exigency coming from existing hardware and in particular software systems.

Before the implementation strategy is described, we look for the *must-haves* to overcome the shortcomings of traditional industrial control programming.

46.2 Intention

Many of the following topics have been implemented for years on IT-based systems using C++, Java or other object-oriented (OO) languages. Some of these have found their way into the International Electrotechnical Commission (IEC) 61499 function blocks (with a lack of practical implementations untill now), but actually the bulk of industrial control applications is still done on EN 61131-based systems. So, in the following, focus is placed on the implementation on these existing systems [46.3].

46.2.1 Encapsulation

The most important part of changing the thinking in the traditional control programming approach is to leave the physical I/O connections as the primary way of composition and interaction of systems.

Normally an object in the real world is experienced by humans through its properties, attributes, features (however named), and the possibilities it offers or what we can do with this object (called methods). In general we realize this object only in a certain abstracted layer.

The level of this abstraction is chosen according to our actual demands.

As an example we take a human being (a car mechanic). If our car is broken, we abstract him as an object: human with the attributes educated, skilled on certain brand, available, etc., and his methods: takes order, repairs car, renders account, etc. We are not interested in how the blood flows through his veins, how his heart functions, etc. His doctor on the other hand is interested in exactly these methods and attributes. He will abstract him as object: human with the attributes: sex, age, blood pressure, etc. and methods: sees, smells, tastes, etc.

Coming back to automation. We are interested in automated systems for production and processes, control and testing, etc. Usually complex systems are built as a network of subsystems (objects) based again on a network of subsystems.

In our namespace these subsystems are called equipment. The term *equipment* could lead to some unaesthetic grammar constructs in the following, but we have chosen it intentionally: firstly, equipment expresses things we need for some activities, especially in production, and secondly it is already a plural expression, although it could be a single device or the whole factory. So it represents the modularity of the method down to the smallest base modules (that in general again consist of parts).

We differentiate between base equipment and abstracted equipment. A motor, a valve or a pump is a single base equipment. A pump station, for example, is a single abstracted equipment.

What is the definition of base equipment in contrast to abstracted equipment? A base equipment implements interface access via physical I/Os to the real world, whereas an abstracted equipment implements interface access to physical I/Os only via base equipment. Nevertheless base equipment can of cause host other base equipment (see Sect. 46.2.2 as well).

The intention for building equipment is to encapsulate the actual physical implementation of one subsystem from the collaboration with other subsystems, as well as to support the abstraction of these systems for different service subscribers (with different interests, e.g., production, service, and management) to its methods and/or properties.

The encapsulation consists of three layers: the interface layer (providing abstracted methods and properties), the implementation layer (running the algorithms, the event routines, and the exception routines), and the base layer. The base layer is the interface to the physical I/Os and the access to the implementation layer of subequipment for base equipment and the same without direct access to physical I/Os for abstract equipment.

Access from one single equipment to another is allowed only by use of its interface layer. This strict access method allows one to change the implementation layer as well as the base layer of equipment without affecting how it is accessed.

46.2.2 Generalization (Inheritance)

In the implementation of object-oriented languages the method of inheritance is an important feature. It describes the deployment from more generic objects to more specialized ones. As in the following implementation we have no special software tools in standard EN 61131-based systems for constructing strict inheritance (as is possible in C++), we talk about generalization and bear in mind the underlying ideas. So, for our needs, generalization allows hierarchical structures and establishes a relationship between a more

general equipment and a more specific one, without redefining all the specific methods and properties, although if required it is possible to override the generic methods and properties of the generic equipment with more specific ones.

To follow our first example, a woman inherits all the methods and properties of the object human, but has of course more specific attributes and methods, most importantly to give life to new humans. We will give more technical examples in the implementation section.

46.2.3 Reusability

Probably it is the basic idea of every engineer to built up a system (whichever) out of already existing, used, and tested parts coming from a toolbox (as children do when playing Lego) to create new systems and projects. By doing this an engineer mostly thinks about saving time, lowering risks, and reducing overall costs.

The basis for doing this is encapsulation and the concept of instances. As a single equipment is the implementation of methods and at the same time its abstraction, and the base layer is the interface to the physical world, it is only a general construct. The base equipment, e.g., is brought to life when it is connected via its interface to defined physical I/O points and placed as a defined instance in the state machine call (see the implementation section); for example, we define an equipment pump with all its methods and properties, but only when we have certain pumps (FreshwaterPump01, WasteWaterPump01, . . .) instantiated in our project will water flow.

So with these concepts we are able to forget the internal details of the equipment for all future implementations of the same sort of pump and can concentrate on the method and property interface.

46.2.4 Interchangeability

Interchangeability has a strong relation to reusability as it is based on the same concepts, although it describes another topic.

Reusability refers to the ability to use the same equipment without major efforts again in the same or another project. Interchangeability applies when we want to replace existing equipment with another type of equipment, but with the same functionality. According to our encapsulation approach, equipment with the same functionality should have the same interface layer, although the implementation layer and the base layer are different. We advise strict application of this rule, at least for base equipment.

Consider, for example, that one has to change a pump with all its control and monitoring periphery for any reason (customer demand, unavailability, etc.) with one from another manufacturer. It is evident that often the physical implementation of the machine will be different, even though it has the same functionality.

Having two equipments describing the two pump systems with the same interface layer makes software changes possible in a matter of minutes.

46.2.5 Interoperability

As long as the defined equipment runs on a centralized system, communication between them leads to no trouble. As we pointed out in the Introduction systems nowadays are highly complex and distributed, so we need implementation concepts to simplify access between equipment even in high-grade networked and heterogeneous environments, starting at low control levels with existing equipment.

We will not describe all the systems and attempts in hardware and software that exist or are in development, which could fill books. We only want to focus on the way we structure the issue.

As pointed out in the paragraph on encapsulation, the only way of accessing other equipment is by using the interface layer. As a fetch-ahead of the implementation, we will see that this interface is not part of the equipment functions, but is a separate, generally available data structure. For the time being, this data structure is only composed of basic data types.

As a paradigm we do not allow the implementation layer of a single equipment to be spread over several controllers. This seems like a strong restriction on first reflection, but it enforces well-elaborated encapsulation, and good and fine-structured systems. As an additional advantage, it simplifies equipment communication, which is one of our aims.

46.3 Strategy

46.3.1 Device Drivers

As we defined in the paragraph on encapsulation the basis of every system is its base equipment – the equipment connected directly to the physical world. We consider this equipment like device drivers in the PC environment. In the same way that a program in Windows does not care about the actual physical mouse type connected to the computer, but only the abstracted events, other base or abstracted equipment does not care about the physical implementation of the accessed equipment.

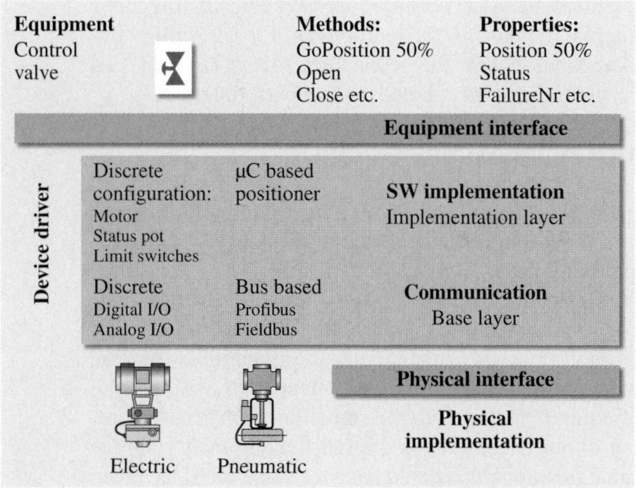

All base equipment, at least, is implemented in the implementation layer as a state machine, executing the functions necessary to the actual state and checking all allowed transitions to switch to another. Typically another cycle is used to catch all possible exceptions.

Figure 46.1 shows the example of a control valve (base equipment) with two possible physical and software implementations. The left column in the implementation part depicts a discrete built-on electric valve with a motor, a potentiometer providing feedback, and limit switches for the end positions. In this case the implementation layer has to build up the complete functionality of the valve, such as the control of the motor, the control loop with the feedback potentiometer, etc. In the right column a micro controller (μC) positioner-based pneumatic valve is represented. Here the implementation layer has to manage mainly the communication to the μC positioner, as the control itself is done by the external logic.

Additionally the independency of the physical implementation is visible. As the electrical valve is

Fig. 46.1 Example of a single base equipment (control valve)

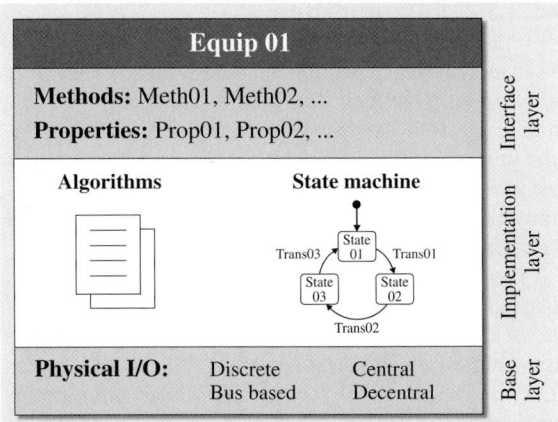

Fig. 46.2 Example of a specifying single base equipment (equipment block diagram); Equip01 does not use other equipment, accessing only physical I/O

connected via the base interface to discrete I/O, the pneumatic valve is running on a bus system.

Nevertheless the interface layer is the same, and we can control both valve types from higher-level equipment with a handful of methods and properties. So we fulfill at least the demands for encapsulation, reusability, and interchangeability.

46.3.2 Equipment Blocks

To model a complex system a graphical representation of its equipment is helpful. We choose a unified modeling language (UML) class-like representation, although the diagram has an additional area for the base layer [46.4].

Figure 46.2 shows an equipment block for a single base equipment interfacing the base layer singly to

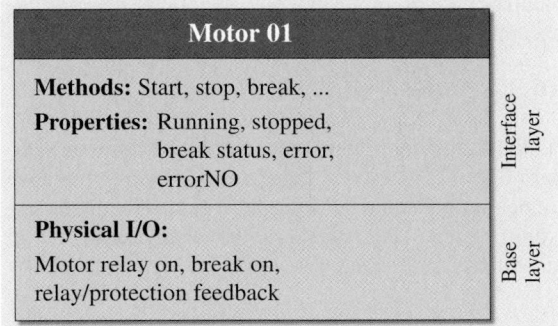

Fig. 46.3 Specifying implementation of single base equipment (relay/motor protection controlled motor)

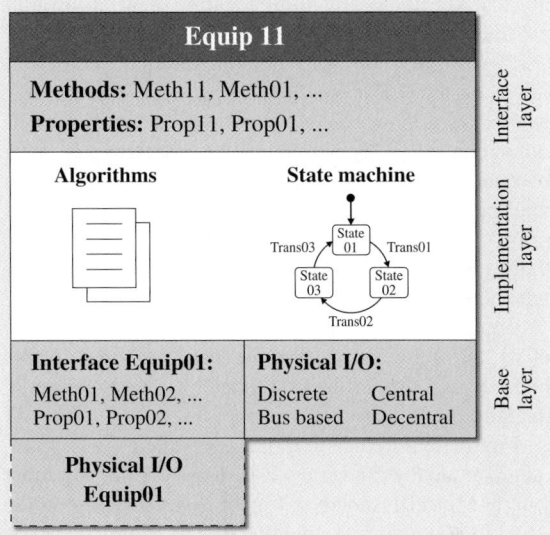

Fig. 46.4 Specifying generalization of single base equipment

physical I/O and not to other equipment. The implementation layer is not a must for the diagram representation, but can be used for clarification of the actual execution of the equipment.

Figure 46.3 depicts the diagram for a simple motor with a brake controlled only by a relay–motor protection switch combination with feedback contacts.

Figure 46.4 shows a more specific equipment *inherited* from a general one. Equip11 will use the methods and properties of Equip01, adding some specific new methods and properties by implementing additional

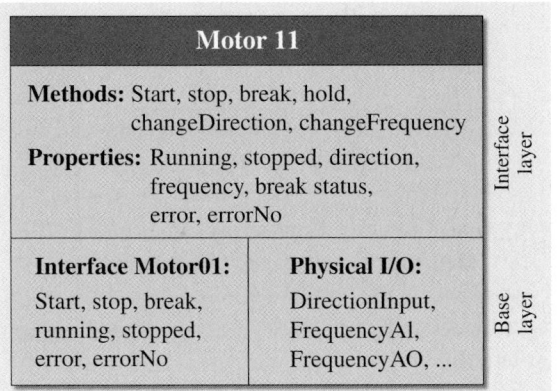

Fig. 46.5 Specifying implementation of single base equipment based on generalized base equipment (motor controlled by frequency converter)

physical IO and/or by implementing new algorithms and transitions in the interface of Equip01.

Figures 46.4 and 46.5 demonstrate the generalization (inheritance) of equipment in general based on the example of a more sophisticated implementation of the motor controlled by a frequency converter unit. Bear in mind that there are no automatic mechanisms for inheritance in EN 61131 as in object-oriented programming (OOP) languages. Nevertheless, we implement the underlying idea and call the method generalization to indicate the difference.

Equip11 can forward the interface layer of Equip01 to its own interface layer or can override certain methods and properties of Equip01 as its own implementation.

Figure 46.5 depicts a motor controlled via an intelligent frequency converter. It is derived from the more general Motor01 shown in Fig. 46.3. Depending on the actual implementation, the interface of Motor01 can be forwarded directly to the interface layer of Motor11, or some methods and/or properties have to be overridden.

Figure 46.6 represents a single abstract equipment, built up on two sublevels, as Equip02 accesses another equipment. Euip21 is abstract, as it does not access physical I/O directly.

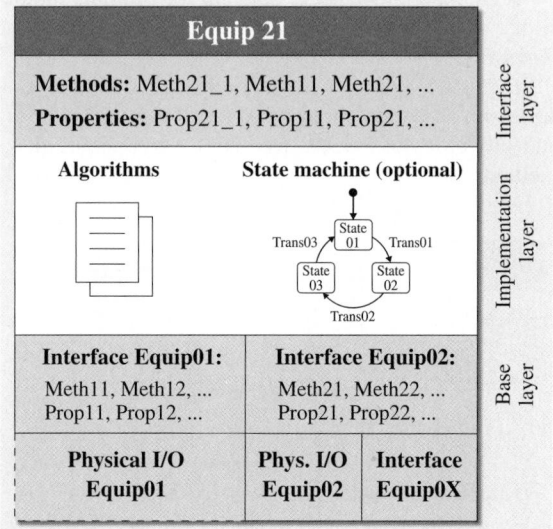

Fig. 46.6 Abstract equipment (does not access physical I/O directly)

46.3.3 Communication

As stated above, communication between equipment is allowed only through use of the interface level. For communication between equipment on the same control unit, shared data blocks are used in general. For homogenous distributed systems all open or proprietary transfer mechanisms are allowed, ensuring coherent data transfer.

For data communication between heterogeneous distributed systems a standardized communication protocol has to be used. As object linking and embedding (OLE) for process control (OPC) data access (DA) is widespread, we normally use this protocol. As version 2 of this protocol does not support well-structured data, the interface layer data has to built up by basic data types only when using OPC DA 2.

With broad distribution of OPC DA 3 or OPC extensible markup language (XML) data access (OPC XML-DA) this constriction will vanish in the foreseeable future, reducing implementation time for the interface communication [46.5].

Special attention has to paid to access to the interface layer of a single equipment by another equipment. As indicated in Figs. 46.4 and 46.6, in general, a single subequipment is embedded into another equipment only via connecting the interface layer of subequipment to the base layer of the accessing equipment. In this way we will obtain a well-structured and strictly hierarchical system.

An exception to this rule has been implemented for accessing data of interest to a group of equipment, bypassing this strict hierarchical system. A separate handler extracts filtered data from the interface layer of defined equipment, stores it in separated data blocks, and present these to a certain interface of the implementation layer. In general we use this outlier only for exception handling.

46.3.4 Rules

As the following implementation will be done on standard EN 61131-based systems, the following rules have to be implemented as a design guide, as these existing systems do not force the engineer to act in an equipment-oriented way. On the other hand, only by using a C++ compiler, nobody is barred from writing unstructured code in an objectless method. The rules are:

1. Encapsulation is a must: no direct access to the physical level other than by distinctive base equipment.
2. Abstraction has to be forced to the highest level possible, which makes especially interchangeability simple.
3. Interequipment communication has to be done only by use of the interface level (methods and properties).
4. Generalization has to be used whenever possible.
5. The use of abstract equipment should be done in the lowest possible level.
6. Implementation of one single equipment on a single controller (no implementation of algorithms or state machines in the implementation layer via more then one controlling unit).

46.4 Implementation

As pointed out at the beginning, in this chapter we want to focus on the implementation of the topics elaborated so far for EN 61131-based PLCs, since this poses a particularly challenge.

We have chosen the implementation in instruction list form, as the flexibility of this method in general is the largest for most control systems. Nevertheless one can use structured text or function block diagrams; ladder diagrams or sequential function charts are not adequate [46.2].

All equipment is implemented in function blocks; each instance of the function block represents a physical unit for base equipment or a certain virtual or composed unit for abstract equipment.

The base layer is represented by the function block's own encapsulated data block, whereas the interface layer is represented by a separate, general data block, unique for each implementation of the equipment. This allows simple data communication and easy multiple access for different higher-level equipment.

For base equipment a state-machine implementation is a must, although abstract equipment should also use this method if it makes sense.

All base equipment function block are called cyclically by a service routine and run their state-machine functions. Implementing this approach, at least error handling is done for all devices, independently of whether the device is actually used or not. This is a major advantages for early failure detection. The way of connecting physical I/O to base equipment is not limited and could be either discrete on a centralized unit or via fieldbuses, Ethernet or even wireless for decentralized peripherals.

As mentioned earlier, one homogenous control unit is responsible for implementing the algorithms and state-machine functions for a single equipment, i.e., such a control unit can of course host a large number of equipments, but the implementation layer of one single equipment cannot be executed by several control devices.

So, considering the communication aspect, even in heterogeneous networked systems different communication methods simply have to map the interface data blocks to the participating control units in a coherent manner.

Practically we use system-specific communication methods, as long we are working in homogenous platforms, because they are in general more efficient. If we leave the homogenous domain, we normally use OPC DA and OPC AE (alarms and events) [46.5].

Because we do not mix data with data communication in our strategy, functional engineering and data communication setup can be handled separately.

46.5 Additional Impacts

46.5.1 Vertical Integration and Views

In the Introduction of this chapter we found that vertical integration of processes is an answer to many of the demands of modern business [46.6]. Figure 46.7 represents the business as well as the communication pyramid in a contemporary enterprise. Nowadays the control level is the domain of PLCs, whereas

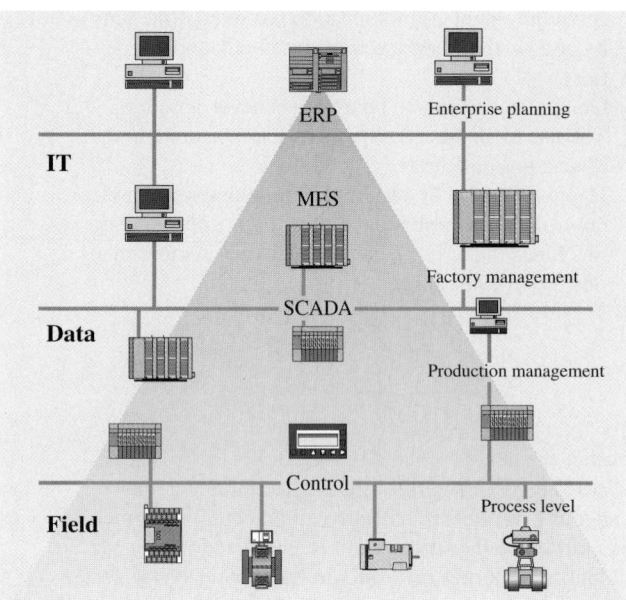

Fig. 46.7 Business and communication pyramid in a modern enterprise

supervisory control and data acquisition (SCADA), manufacturing execution systems (MES), and enterprise resource planning (ERP) are the domains of PC-based devices [46.7].

With the implementation of an equipment-based programming structure at the control level we overcome the existing gap between the field and the data/IT level.

As many enterprise MES and ERP solutions are already object oriented, equipment-based control structures can easily be mapped to objects in higher-level applications, as the equipment constructs are built up in a class-like manner [46.7].

The hierarchical structure of the equipment implementation follows the natural business pyramid, and the encapsulation and abstraction support the inherent enforcement of data reduction in the lower field level.

46.5.2 Testing

A major part of resources during the development of a system is spent on testing its functionality and exception handling. By using software strategies with a clear hierarchical structure, a modular concept with a strictly defined interface layer architecture, these resources can be dramatically reduced, concurrently increasing performance and stability.

46.5.3 Simulation

Another attempt to reduce testing resources is the simulation of systems or subsystems of a project. If a simulation is connected to real hardware, we speak of *hardware in the loop* simulation. In testing developed equipment, one can differentiate between two cases. The higher-level equipment accessing the part to be tested is simulated, or a lower-level part we want to access is simulated.

In both cases the concept of equipment-based structures simplifies the simulation part. Firstly, most simulation tools are object oriented, so we are able to map our equipment easily to their classes. Second the simulation environment has to model only the abstracted interface layer of the simulated subsystem, rather than simulating all the necessary I/Os.

46.6 Example

Profactor, amongst others, has been a supplier of test stands for heat and flow meters for 20 years; additionally the automation group of ARC-sr delivers innovative automation solutions for the process and production industry and also develops new automation processes for a couple of sectors [46.8].

In 2000 Profactor was put in charge of building one of the largest and most modern test stands for heat meters in Berlin. The customer was the well-known Physikalisch Technische Bundesanstalt (PTB) [46.9].

In addition to building up the mechanics and measurement system, challenging the frontiers of today's realizability, the test stand had to be embedded into an overall test administration system. This started with the management of customers, test orders, and devices under test, and ended with management of test reports and quality management for the test stand with its own subsystems.

As the system for the PTB was the most complex test stand ever, incorporating about 1000 physical I/O and measurement points, including dozen of subsystems, and has to be vertically integrated into the testing hierarchy of the PTB, Profactor decided to implement the described system for the actual and future system as well as for other complex systems in their field of activities [46.10].

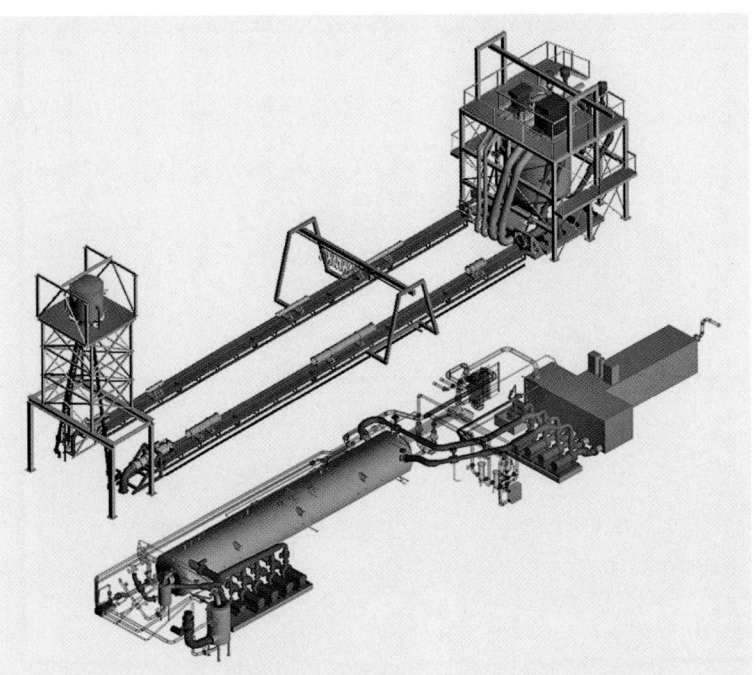

Fig. 46.8 Layout of the test stand for flow and heat meters

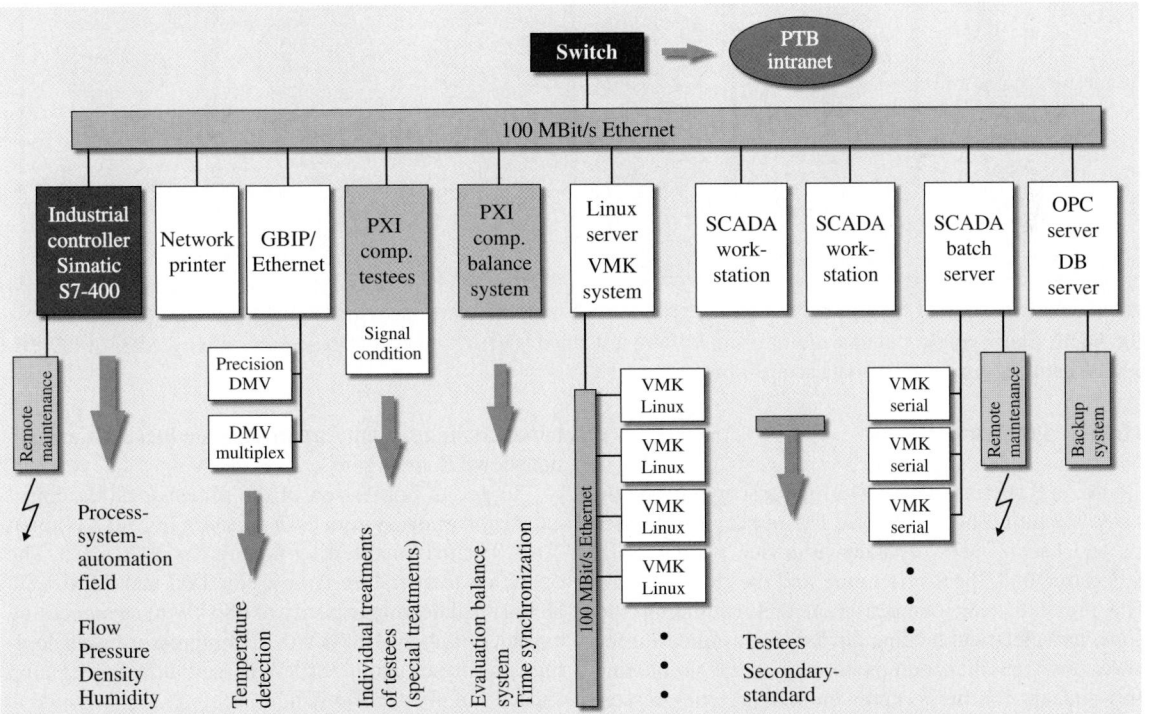

Fig. 46.9 Specifying subsystems and their communication implementation (Seibersdorf Research) (VMK – measurement interface, GBIP – general purpose interface bus, PXI – PCI extensions for instrumentation)

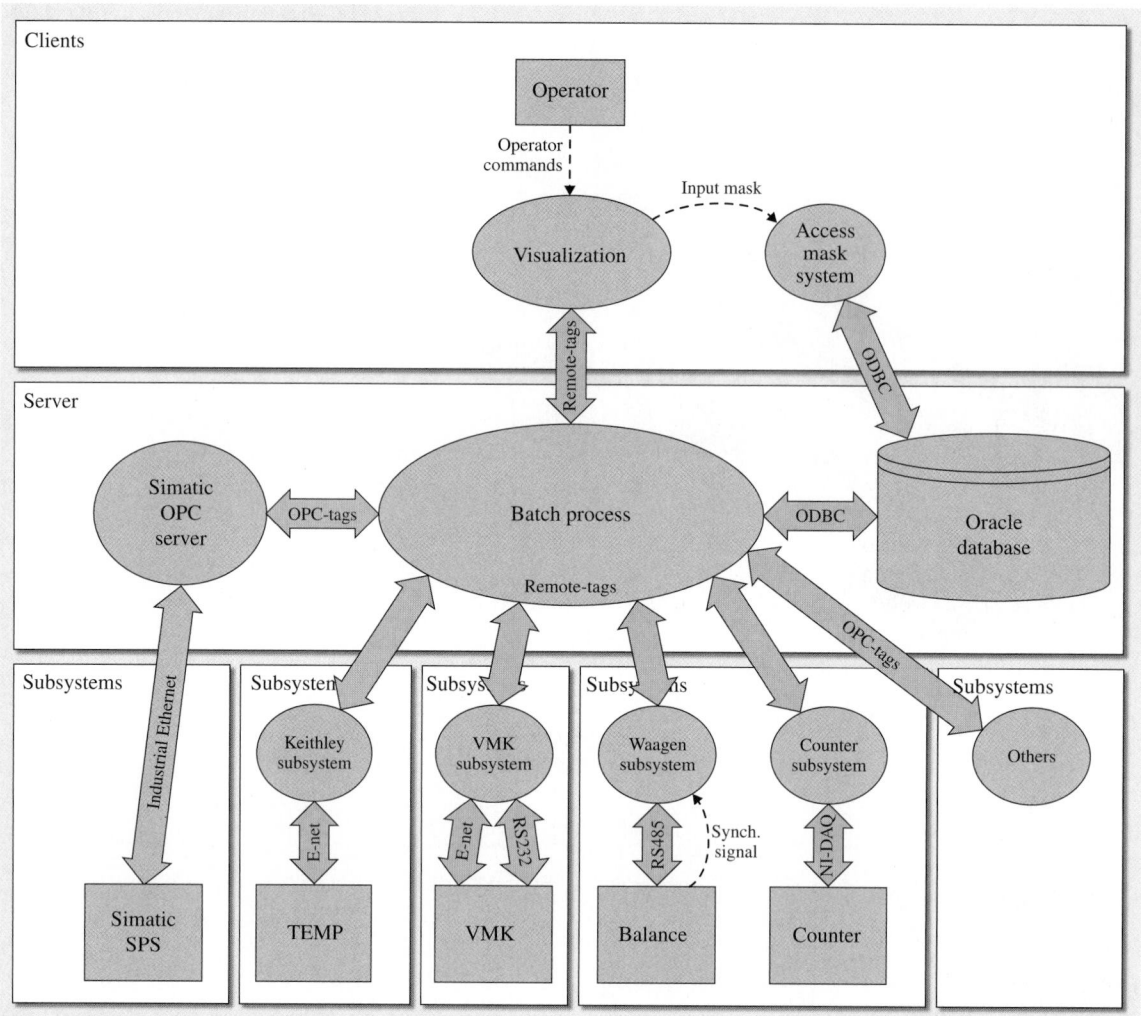

Fig. 46.10 Specifying communication pyramid of the test stand (ODBC – open database connectivity, SPS – PLC, NI-DAQ – national instruments – data acquisition)

46.6.1 System

Figure 46.8 depicts the layout of the test stand. The facility is built up on two floors. The upper floor houses the test bench itself, the balance device with its self-calibration unit, the diverter unit, and the elevated tank. The pressure tank, compensation tank, pumping stations, and electrical heating device are installed on the lower floor. Auxiliary equipment such as the gas heating unit and air cooling systems as well as compressor-based cooling systems are in separate locations and are not shown in the figure.

To get an impression of the dimensions, the linear expansion of the system as depicted in Fig. 46.8 is about 40 m. The maximal testing flow rate is $1000 \, \mathrm{m^3/h}$. The operation temperature is between $3\,°\mathrm{C}$ and $90\,°\mathrm{C}$ with an installed heating capacity of $680\,\mathrm{kW\,h}$, an air cooling capacity of about $1000\,\mathrm{kW\,h}$, a compressor-based cooling capacity of about $380\,\mathrm{kW\,h}$, and an installed pump capacity of about $350\,\mathrm{kW\,h}$.

The technical requests for the commissioning were at the limit of technical feasibility:

- Stability of temperature in all operating points < 0.1 K
- Stability of flow (up to $1000\,\mathrm{m^3/h}$) $< 1\%$
- Stability of pressure (in pressure mode) < 3 mbar
- Accuracy of balance unit < 50 g (on 20 t of load)
- Overall measurement uncertainty $< 4 \times 10^{-4}$.

These demands resulted in highest efforts to the control and measurement system.

The key factor for the integration request was the integration of the test stand management database (Oracle or MS SQL server (Microsoft structured query language)) with its user front-end in the production database that organizes the test routines and is responsible for the storage of real-time trend data from the testing process [46.11].

In these databases not only the test and test management data is stored, but also the complete setup and control parameters as well as the calibration data of the test stands measurement devices.

For exact repetition of tests all setup and control parameters of the complete facility can be reloaded to the subsystems automatically by recalling a certain date or test order from the database. In additionally providing the necessary automatic documents for quality control the customer owns not only one of the most modern test stands in measuring technology, but also in terms of data management.

Figure 46.9 shows the communication of the subsystems. The control of the test stand itself and its auxiliary systems is realized by an EN 61131-like PLC system. This PLC system alone manages about 800 physical I/O points on a centralized and decentralized basis. The programming strictly follows the described equipment structure.

46.6.2 Impacts

The first PLC implementation of this concept required more effort and expense compared with traditionally programming, but major benefits were observed already during start of operation and integration into the vertical environment.

46.6.3 Succession

Subsequently Profactor was put in charge of building another, smaller test stand for a major German power authority. Although Profactor was forced to use a lot of different components compared with the PTB installation, the decrease of time, efforts, cost, and failures was significant.

46.7 Conclusion

By following the described methods of implementing an equipment-based control system structure, the system designer is able to deliver software with a clear hierarchical structure and a modular concept with strictly defined interface layer architecture, simultaneously decreasing the required resources and increasing performance, stability, and openness.

It is possible to implement a lot of concept features normally only supported by object-oriented languages. Most importantly, the engineer can use existing software and hardware platforms that have been available for many years and anywhere.

46.8 Further Reading

1. M. Frappier, H. Habrias: *Software Specification Methods: An Overview Using a Case Study* (Springer, Berlin Heidelberg 2000)

2. H. Ehrig: *Integration of Software Specification Techniques for Application in Engineering: Introduction and Overview of Results* (Springer, Berlin Heidelberg 2004)

References

46.1   G. Strohrmann: *Automatisierungstechnik 1* (Olden-burg, Munich 1998), in German

46.2   K.H. John, M. Tiegelkamp: *SPS Programmierung mit IEC 61131-3* (Springer, Berlin Heidelberg 2000), in German

46.3   R.W. Lewis: *Modelling Control Systems Using IEC 61499: Applying Function Blocks to Distributed Systems* (Inst. Engineering and Technology, London 2001)

46.4   H.E. Eriksson: *UML Toolkit* (Wiley, New York 2001)

46.5   OPC Task Force: *OPC Overview* (OPC Foundation, Scottsdale 1998)

46.6   Arbeitskreis Systemaspekte des ZVEI Fachverbandes AUTOMATION: Die Prozessleittechnik im Spannungs-feld neuer Standards und Technologien, J. Appl. Test. Technol. **43**, 53–60 (2001), in German

46.7   A. Dedinak, G. Kronreif, C. Wögerer: Vertical inte-gration of production systems, IEEE Int. Conf. Ind. Technol. ICIT'03 (Maribor 2003)

46.8   A. Dedinak, C. Wögerer, H. Haslinger, P. Hadinger: Vertical integration of mechatronic systems demonstrated on industrial examples – theory and implementation examples, BASYS'04, Proc. 6th IFIP Int. Conf. Inf. Technol. Autom. Syst. Manuf. Serv. (Vienna 2004)

46.9   A. Dedinak, C. Wögerer: *Automatisierung von Großprüfanlagen am Beispiel eines Wärmezäh-lerprüfstandes für die PTB*, White Paper (ARC Seibersdorf Research, Vienna 2002), in German

46.10   A. Dedinak, W. Studecker, A. Witt: Fully automated test-plant for calibration of flow-/heat-meters, Int. Fed. Autom. Control (IFAC), 16th World Congr. (Prag 2005)

46.11   A. Dedinak, S. Koetterl, C. Wögerer, H. Haslinger: *Integrated vertical software solutions for industrial used manufacturing and testing systems for re-search and development* (Advanced Manufacturing Technology, London 2004)

47. Automation and Ethics

Srinivasan Ramaswamy, Hemant Joshi

Should we trust automation? Can automation cause harm to individuals and to society? Can individuals apply automation to harm other individuals? The answers are yes; hence, ethical issues are deeply associated with automation. The purpose of this chapter is to provide some ethical background and guidance to automation professionals and students. Governmental action and economic factors are increasingly resulting in more global interactions and competition for jobs requiring lower-end skills as well as those that are higher-end endeavors such as research. Moreover, as the Internet continually eliminates geographic boundaries, the concept of doing business within a single country is giving way to companies and organizations focusing on serving and competing in international frameworks and a global marketplace. Coupled with the superfluous nature of an Internet-driven social culture, the globally-distributed digitalization of work, services and products, and the reorganization of work processes across many organizations have resulted in ethically challenging questions that are not just economically, or socially sensitive, but also highly culturally sensitive. Like the shifting of commodity manufacturing jobs in the late 1900s, standardization of information technology and engineering jobs have also accelerated the prospect of services and jobs more easily moved across the globe, thereby driving a need for innovation in design, and in the creation of higher-skill jobs. In this chapter, we review the fundamental concepts of ethics as it relates to automation, and then focus on the impacts of automation and their significance in both education and research.

47.1 Background

To educate a man in mind and not in morals is to educate a menace to society. (Theodore Roosevelt)

In this chapter we attempt to address a key issue facing people from industry and academia, especially with the rapid pace of globalization and technological advancement related to automation. Why is ethics, and what makes studying and understanding ethics and its link to automation important; both the inculcation of it among our present and future colleagues, employees, and public services, and understanding it within the context of academic, government, and corporate research. After describing the ethical issues related to automation, we focus our presentation on two specific areas, education and research, respectively. In the section on education, we present a mechanism whereby the inculcation of ethics can, and should, be integrated within a student's curricular program and learning experience, instead of the simpler one-course approach that is taken by educational institutions today, in response to the mandatory requirement of *teaching* ethics as sought by employers and accreditation agencies such as ABET. The section on research could have been written on many levels – from ethics in workplace, personal ethics, to social and professional perspectives of what can be considered ethical behavior in research. Since these topics are widely covered elsewhere (references are given below), we have chosen to illustrate and explore the critically emerging issues of user profiling by logging user activities on a network (the Internet and automation networking in general). This illustration is important because this issue is beginning to assume a greater degree of significance in today's world, with the ability of people and organizations to use advanced automation to gather, store, mine and analyze enormous amounts of data, very cheaply. Hence, addressing this issue will likely prompt ethical questions (not just limited to what we present here) across all the above different perspectives.

47.2 What Is Ethics, and How Is It Related to Automation?

New and emerging automation technologies and solutions pose significant new challenges for ethical individuals, organizations, and policy-makers. (Automation Scholars)

Ethics is a set of principles of right and wrong that individuals apply when making decisions influencing their behavior. Many decisions can clearly be recognized by most people as being wrong or immoral, including violations of the law, dishonesty, and any other behaviors that conflict with common behavioral norms and societal values. The role of ethics, ethical thinking, is important especially when there are no clear-cut guidelines, for example, when individuals encounter conflicts between objectives and their principles, and as often happens with the emergence of new technology, including automation technology [47.1–7]. As new choices and new experiences become available to individuals and organizations, they face dilemmas between risks and benefits, short-term benefits against long-term risks, risks to individuals versus benefits to a group, and so on. A major challenge to ethical behavior is the fact that not only changes in technological abilities over time pose new ethical dilemmas, but that ethics is deeply rooted in local and domain cultures, hence, it requires adjustments and calibration in the interfaces and exchanges. This dual challenge for inter-cultural ethical behavior over time and location has been evident throughout history, and is particularly sharp at the edges during our age of tremendous automation innovations coupled with intensifying global exchanges (Fig. 47.1).

Automation has several particular impacts on ethics:

1. Automation enables unethical behavior, e.g., applying automatic imaging to monitor private situations violates privacy rights, but may be necessary for security and prevention of theft.
2. Automation simplifies unethical behavior by obscuring its source, e.g., people blaming automation for mistakes, delays, inefficiency, and other weaknesses (*It's not me; it's this dumb computer*).
3. Automation increasingly enables unethical behavior related to information and communication, e.g., recording conversations and proprietary knowledge; maintaining and visiting web-sites with illegal, violent, or hateful contents.
4. Automation enables replacement of labor, e.g., by robots, automated sorting, and automatic inspection.
5. Automation affords anonymous access over and to private or restricted property.

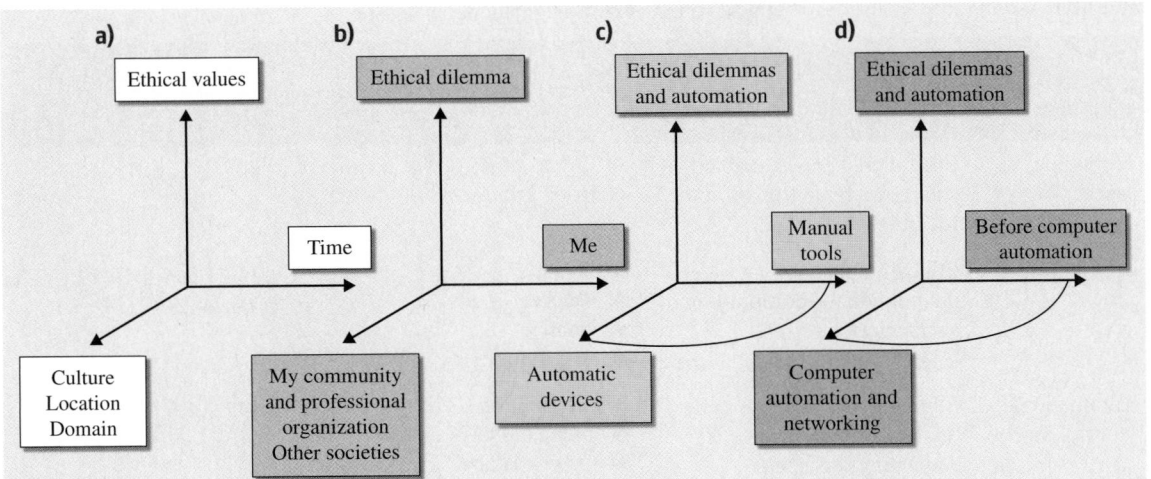

Fig. 47.1a–d Ethics values and dilemmas: (**a**) Ethics of today may not be the same as ethics of yesteryears due to changes in cultural and technological evolution (changes) around the globe. (**b**) Ethical dilemmas are conflicts between individual's or groups of individuals' rights, benefits, and rewards versus community, organization, and society at large gains and sustainability. (**c**) Major ethical dilemmas emerge when changing from manual tools and procedures to automated and automatic devices, e.g., remote imaging, banking automation, and Internetworking. (**d**) Major ethical dilemmas further emerge, more frequently and with farther impact when automation evolves and with computers and worldwide network communications advancements

6. Automation enables cyber-crime, cyber-terrorism, information hiding or obscuring, forgery, identity theft, or identity hiding.

Some of these examples overlap with criminal and other illegal behavior [47.6, 8, 9]. But there are many examples where the situations are ambiguous, or ambivalent. When society realizes the severity and damage caused by some such cases, laws are developed and implemented. Often, however, ethical issues emerge and require urgent individual and organizational responses in the face of far-reaching ethical dilemmas.

47.3 Dimensions of Ethics

Dimensions of ethics can be considered in multiple aspects, which are inter-related (Table 47.1): From the *aspect of automation technology*, how and what it enables in challenging ethical behaviors, e.g., financial crimes through banking automation; from the *aspect of impacts on individuals, on communities, and on society*, e.g., hate crimes through the Internet. From the *aspect of automation security*, how automation's own security can be breached with unethical schemes and outcomes, e.g., by intentionally or unintentionally disabling software safety functions. In all dimensions, however, it is clear that people are responsible, directly or indirectly, intentionally or unintentionally, for their ethical decisions, behaviors, and the outcomes; furthermore,

people, not automation, are the potential misusers and abusers of automation in the context of ethics.

Ethics and automation can also generally be divided into ethical issues involving information-focused automation [47.1, 2, 4, 10], e.g., information security and privacy; and automatic device/systems ethic [47.7, 9, 11–13], e.g., ethics of robotics (sometimes called robo-ethics), for instance, trust in tele-surgery by robots. There are, of course, overlapping ethical dimensions, for instance, when information systems are hacked (security breach) to disrupt automatic traffic and aviation control (impact dimension), or to dysfunction automatic power distribution (technology and impact dimensions) [47.5, 14–16].

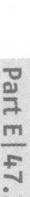

Table 47.1 Aspects and dimensions of ethical concerns with automation

Aspect of ethical dimensions	Scope	Main ethical dimensions	Sample references
Technology aspect	Ethical challenges enabled and raised by automation functions and abilities	● Cyber-ethics (a.k.a. e-Ethics) ● Robo-ethics	[47.3, 7, 8, 11–15, 17–20]
Impact aspect	Ethical impacts on individuals, communities, and society	● Privacy ● Property ● Quality ● Accessibility	[47.1, 2, 4, 6, 10, 14, 16, 19, 21–23]
Automation security aspect	Ethical issues of man-made (malicious and erroneous) and natural disasters causing security threats through automation, and vulnerabilities caused by automation	● Information security ● Technical failures ● Cyber-crime ● Cyber-terrorism ● Cyber-warfare; robo-warfare	[47.2, 4, 6, 7, 12–15, 20, 24–30]

Consider the four main automation areas (Chap. 3, Fig. 3.2):

1. Automation with just computers – Data processing and decision support, e.g., enterprise resource planning, accounting services
2. Automation with various automation platforms and applications, but without robots, meaning automation with devices, sensors, and communication, e.g., weather forecasting, air-traffic control
3. Automation applying also robotics, e.g., fire safety including alarms and robotic sprinklers
4. Automation with robotics, e.g., robot painting, robotics in microelectronics fabrication and assembly

Each of these automation areas involves ethical decisions and behaviors, along the dimensions indicated in Table 47.1, by managers, operators, maintenance personnel, and designers, who have to adhere to ethical values to enable sustainable services and viable society. Additional examples follow below.

Another common view of ethics and automation has been the view from the aspect of impacts on individuals, on communities, and on society. Four main dimensions of ethics in this context, as related to automation, are privacy, property, quality, and accessibility.

Privacy: Privacy issues are related to gathering, maintaining, distributing, analyzing, and mining information about individuals. For example:

- What rights do individuals have to their own information and its protection?
- What information about themselves do individuals have to share with others?
- Who is responsible for the security of private information about individuals when it is maintained in a database?
- What rights to surveillance over individuals do organizations and government services have?

Property: Issues involving ownership of physical and intellectual property. For example:

- Can corporate automation equipment be used for personal purposes?
- How should software, music, and other media piracy be handled?
- Who is accountable and liable for damage caused by automation?
- How will intellectual property be traced and accounted for, when automation enables its easy and rapid copying and transfer?
- Who is responsible and accountable for backup records?

Quality: Quality of automation implies its integrity and safety of functions, fidelity, authenticity and accuracy. For example:

- Can an individual trust an automatic device, e.g., in medical diagnostics and treatment?

- What quality standards are needed to protect society's and individuals' safety and health, also including long-term environmental concerns?
- What quality standards and protocols concerning automation and information are required to protect individuals' rights?
- Who is responsible, accountable, and liable for the accuracy and authenticity of information and of automatic functions?
- Who is responsible, accountable, and liable when functions relying on automation fail?

Accessibility: Accessibility issues involve the right to access and benefit from automation, the authority of who can and cannot access certain automation assets and resources, and the increasing dependency on automation. For example:

- What skills and what values should be preserved and maintained in a society increasingly relying on automation?
- What about loss of judgment due to such reliance?
- Who is authorized to use automation and access automation resources?
- How can such access be managed and controlled?
- Can employees or clients with disabilities be provided with access to automation, for their work, healthcare, learning, and entertainment?
- How and under what conditions should access to automation be priced and charged?
- Can automation limit political freedom?
- Does automation cause addiction and isolation from family and community?

47.3.1 Automation Security

Automation security involves security of computer and controller software and hardware; of information and knowledge stored, maintained, and collected by automation, e.g., the Internet, imaging satellites, and sensor networks; and of automation devices, appliances, systems, networks, and other platforms. Most of the ethical concerns in automation security overlap the previous aspects and dimensions, but have certain unique security related dimensions. Some of the ethical issues associated with automation security are:

- What are the vulnerabilities of automation security that impact on privacy, property, quality, and accessibility ethical dimensions, and who is responsible for overcoming them? For recovering from them?
- With increasing automatic interconnections and automatic interactions between various automation systems and devices, how can security levels be maintained, shared, and warranted over entire services? Who is responsible for tracking, tracing, and blocking the instigators and initiators of the security shortcomings causing unsatisfactory service?
- Who is responsible and who is liable in the case of harmful and damaging security breaches, such as trespassing, espionage, sabotage, information extortion, data acquisition attacks, cyber-terrorism, cyber-crime, compromised intellectual property, private information theft, and so on? What would be the difference between breaches caused by unintentional human error versus malicious, unethical acts?
- Who is responsible and who is liable when there are automation software attacks, e.g., software viruses, worms, Trojan horses, denial of service, phishing, spamware, and spyware attacks?

Governments, national and international organizations, and companies have already advanced various measures of defenses and protection mechanisms against security breaches. Examples are the Business Software Alliance (www.bsa.org), the cyber consequences unit in the US Department of Homeland Security, computer and information security enterprises such as www.cybertrust.com, and university centers such as CERIAS (Center of Education and Research in Information Assurance and Security, www.cerias.purdue.edu), and CERT (Computer Emergency Response Team, www.cert.org). Yet, automation security poses complex and difficult challenges because of the high cost of preventing hazards, associated with the difficulty to justify such controls, the difficulty to protect automation networks that cross platforms, organizations, countries, and continents, and the rapid automation advances, which render new security measures obsolete. More about automation security can be found in [47.25–30].

The ethical dilemmas discussed above and their dimensions illustrate some of the ethical questions raised by developing and applying automation, and by its rapid advancement and influence over our society, from automatic control devices, robots and instruments, to the computing, information, communication, and Internet applications.

47.3.2 Ethics Case Studies

Ethics is best taught and explained through case studies and examples [47.2, 3, 5, 14, 21, 31, 32]. In Table 47.2, examples of ethical issues related to automation are described. Some examples are clear ethical dilemmas.

Some of them are more subtle ethical problems. As often is the case in ethical dilemmas, solutions are usually not simple.

Table 47.2 Examples of ethical issues in automation and their dimensions ▶

47.4 Ethical Analysis and Evaluation Steps

What is hateful to you, do not do to your fellow human. (Hillel the Elder, Talmud, Shabbat 31a)

How can one rationalize situations and decisions involving ethical conflicts? And how can automation systems be designed and operated with assurance that intended ethical imperatives and decisions would indeed be followed?

Some of the earliest thinking about ethics and automation, in the area of robotics, is attributed to Isaac Asimov, a prominent scientist and science fiction author, who wrote in his book *I, Robot,* [47.33] and also in *Looking Ahead* [47.34] *The Three Laws of Robotics*:

1. A robot may not injure a human being, or, through inaction, allow a human being to come to harm.
2. A robot must obey the orders given it by human beings except where such orders would conflict with the First Law.
3. A robot must protect its own existence as long as such protection does not conflict with the First or Second Law.

Substituting *an automation system* for *a robot* in the three laws above would still make a lot of sense in any context of automation, including the threat of automation singularity (see Chap. 3). But a critical issue is how to implement it during the design and activation of any automation functions. While this challenge is still open to research and discoveries, ethics educators and scholars recommend a five-step approach, as follows [47.1–8, 18, 19]:

Step 1 *Characterize and Specify the Facts*
Establish the stakeholders and events involved, including the *6 Ws*, who, what, when, why, to whom, and where.
Notes:
a) Sometimes, just clarifying the facts results in simplifying the resolution and the decision.
b) Often getting multiple parties, even those in conflict, to agree on the facts may help resolve the ethical conflict.

c) Often the clarification of facts sharpens and simplifies the realization of an ethical imperative, leading one or more of the participants to share the facts with authorities (known as the *whistle blower*), thus leading to a resolution of the ethical dilemma.

Step 2 *Formulate the Dilemma and Conflict (or Conflicts), and Find the Involved Values*
Ethical issues are always linked with values; the parties in conflict usually claim their motivation as the pursuit of high values, such as fairness, freedom, protection of privacy and property, saving resources and the environment, and increasing quality.

Step 3 *Clarify Who Would Benefit and Who Would Be Harmed by the Given Ethical Issue*
Beyond the facts established in Step 1, including the stakeholders, analyzing and finding who may benefit and who may be harmed can be useful in clarifying and understanding which solution, or solutions, may be effective and feasible and practicable.

Step 4 *Weigh and Balance the Resolution Options*
Ethical dilemmas and conflicts are characterized by having complex variables and dependencies, and rarely present a simple solution. Usually, not every one of the stakeholders and other involved individuals, organizations, and society members can be satisfied. Moreover, the thorny realization is that there would almost always be some who may suffer or would consider themselves harmed under any given decision. In some cases, there may not be any optional strategies that could balance the consequences to all the involved parties.

Step 5 *Analyze and Clarify the Potential Outcome of the Ethical Decision*
Certain options of ethical strategy and policy to resolve a given ethical dilemma may satisfy our principles and values, yet they may be harmful from other aspects. For example, a policy

Ethical case	Dimensions of ethical issues			
	Privacy	Property	Quality	Accessibility
1. Company database highlights employees' personal attributes, e.g., nearing retirement, potentially being discriminatory. Furthermore, who needs to know this information? Who is authorized to access it?	✓			✓
2. Service providers monitor employees' access to certain websites. Employees cannot prevent being monitored while using company computers; employers may abuse the gathered private information.	✓	✓		✓
3. Organization audits individuals' use of unauthorized software, either to create policies, protect itself from property lawsuits, or monitor individual private behavior.	✓	✓		
4. Company is using automated imaging technology to replace employees. This case illustrates a typical conflict between economy, accuracy, and efficiency goals achievable with automation, and the loyalty to dedicated employees (who will lose access to work with automation).			✓	✓
5. New automatic sorter has hidden design deficiencies that are too costly to repair after deployment of thousands of such devices. This case is common, as evident by some ethical companies occasionally recalling defective automation equipment for upgrade and repair. If there is no recall in such cases, clients and users are denied access to better quality and safer equipment.			✓	✓
6. A robot controller, under certain undisclosed conditions, will cause substantial chemical waste and pollution. This multi-dimensional ethical problem, involving issues of significant potential damages to life, life quality and property, possibly denying access to inflicted areas and properties, and potentially costing also in major remedial and recovery efforts, is illustrated by cases of whistle blowers, ethical individuals who risked their employment to warn about imminent hazards.		✓	✓	✓
7. Company has superior medical automation technology but will not produce it for several years till it recovers all previous investments in the inferior product currently being marketed to hospitals. This case is similar to Case 6, except it is a different scenario.		✓	✓	✓
8. A vending machine delivers (a) the right item, but returns too much change; (b) the wrong item, and no change. Ethical dilemmas are caused by automation's dysfunctional quality.			✓	
9. A manager blames automation for faulty packaging. Is automation to blame, or is it its designer/implementer/user?			✓	
10. A student blames the school's computer for lost homework. Ethical challenges concerning work quality (and computer automation quality) are posed to both the student and the instructor.			✓	
11. (Think of an ethical dilemma with your home automation.)				
12. (Think of an ethical dilemma unique to your organization's use of automation.)				

Table 47.3 Examples of conflicts between ethical values and principles

Values/Principles in conflict	Illustration
1. Short versus long term	Software patch solving security problems now but causing hazards later
2. Individual versus community	Wasteful exploitation of resources harming later generations
3. Justice versus mercy	Charging for mass email to prevent spamming
4. Privacy versus convenience	Reading the fine details of use-contracts for each downloaded software
5. Loyalty versus truth	Divulging harmful private or proprietary information gained in confidence
6. Loyalty to present versus former organization (employer)	Sharing knowledge about relative advantages or shortcomings of design or applications
7. Efficiency versus safety	Higher speed limits and lower weight versus automobile accidents' severity

that works well for some situations may not work well, or may work only partially well for the same situations under different conditions (it is a conditional solution); or not work well at another time period (it is a time-dependent solution). In analyzing potential outcomes, one may consider the conflicts arising between wrong and right solutions or decisions, between two wrong solutions or decisions, and between two right solutions or decisions. Examples of such ethical conflict analyses are illustrated in Table 47.3.

In analyzing ethical conflicts, usually the conflicts between two or more right solutions, or between two or more right values, pose the most complicated dilemmas.

For example, decisions in examples 1 and 5, which may be relatively simple when the potential hazard is enormous. The situations are also relatively more complex when multiple conflicts combine, e.g., individual versus community for short versus long term implications. Additional guidance is offered by ethics principles.

47.4.1 Ethics Principles

Numerous ethics principles have evolved since ancient times, and have been suggested by ethics philosophers and scholars. Seven of the well known principles are listed in Table 47.4.

Principles (1) and (2) are considered the *individual fairness* principles. Principle (3) is similar to (2), but stated from a group aspect. Principle (4) represents the

Table 47.4 Seven ethics principles

Principle's name	Ethical principle imperative/lesson
1. Hillel the Elder's principle	Do not do to others what you do not want to be done to you.
2. *The Golden rule*	Do to others what you would accept if done to you.
3. Immanuel Kant's categorical imperative	If an action is not right for everyone (in a team, or group, or community) to take, then it is not right for anyone.
4. Descartes' rule of change	If an action cannot be taken repeatedly (e.g., a small action that may snowball out of control), it is wrong to take it at any time.
5. Utilitarian principle	Decide on the action that leads to the higher, or greater, or more significant value (if values can be prioritized, and if consequences can be predicted).
6. Risk aversion principle	Decide on the action that leads to the least damage, or the smallest hazard.
7. *No Free Lunch* rule	Respect the ownership of tangible and intangible assets, and if ownership is unknown to you, assume somebody owns assets that do not belong to you.

impact of time and changes over time (at least those changes that are predictable. Principle (5) addresses the issue of conflict between several objectives and principles, and maximizing the value of consequences. Principle (6) is similar to (5) but from the aspect of minimizing damage. Finally, principle (7) addresses the value and concern for intellectual property protection, on par with physical property protection, as a fair principle for a globally sustainable society.

The above principles provide some guidance for initial analysis. Often, however, they may point to conflicting strategies, and individuals still need to carefully weigh their decisions and take responsibility for each of their decisions. On the other hand, these basic principles offer clear tests for actions and decisions that should not be followed if they fail these tests.

47.4.2 Codes of Ethics

To address the complexities of ethical issues, corporations and organizations define, accept, and publicize their code of ethics [47.10, 21–23]. Such a code prescribes values to which the organization or corporation members are supposed to adhere. The typical structure of a code of ethics closely related to automation is illustrated in the Appendix. Examples of codes of ethics in different countries are included on the USA National Academy of Engineering site ([47.21], http://onlineethics.org). Values that may be incorporated in a code of ethics include:

- Care for others
- Compliance with the law
- Consideration of cultural differences
- Courtesy
- Fairness
- Honesty
- Integrity
- Loyalty
- Reliability
- Respect for sustainable environment
- Trustworthiness
- Waste avoidance and elimination.

General moral imperatives that are included in a code of ethics are listed as follows:

- Follow fairness principles
- Contribute to society and human well-being and sustainability
- Avoid harming others
- Be honest and trustworthy
- Honor property rights including copyrights and patents
- Give proper credit to intellectual property
- Access automation resources only when authorized
- Respect the privacy, diversity, and rights of others.

47.5 Ethics and STEM Education

I didn't know I was a slave until I found out I couldn't do the things I wanted. (Frederick Douglass)

Rapid advancements in automation have led to significant challenges, as indicated earlier in this chapter. Automation has also influenced changes to demographics, and the creeping of problems associated with student and employee recruitment, retention, and *focused* funding. Good educational preparation in the science, technology, engineering, and mathematics (STEM) disciplines is one of the primary means available to prepare the workforce to compete globally for highly skilled technology-based and automation-based jobs. In their current work environments, not only do students need to understand and deal with the increased knowledge expectations from the workforce, but they need to also understand and deal with the pervasive and dominant role of automation technology within their chosen fields, and operate effectively in an increasingly multi-cultural and multi-ethnic, global environment. In these jobs, softer skills, which relate to *how we go about getting things done*, being language, society, and culture-sensitive, are becoming equally important as the hard functional skills (e.g., programming, problem solving, techniques selection, modeling) that have traditionally defined what it means to be competent in a chosen professional field. The widespread globalization of the job market calls for future employees to be adaptive, curious, and nurturing so as to work effectively in a team, which may be either co-located or geographically separated.

47.5.1 Preparing the Future Workforce and Service-Force

I cannot tell anybody anything; I can only make them think. (Socrates)

Many organizations, for profit and not-for-profit, now realize that hiring workers who have been trained to understand international issues, specifically from an ethical and cultural perspective, will provide their businesses and services the necessary competitive-sustainable advantage in a global society and global market. For example, conducting transactions in another country can be riddled with cultural issues that require deft personal touch, such as demonstrating appropriate hospitality, and respecting cultural and religious diversity. Thus, the future in many professional disciplines is not in merely a collective ability to prepare and graduate good designers, programmers, practitioners, managers, and technologists – these skills have now become commodities that can be outsourced. It lies is in the ability to prepare entry-level employees and continuing education employees who are highly comfortable with the theory, can appropriately blend it with necessary practice and possess an understanding of both the business culture and the social issues involved, while being able to effectively share, communicate, articulate, and advance their ideas for an innovative product or solution. Hence, how we educate students to become such successful employees and entrepreneurs, while acting ethically in the global economy and society is an important consideration, often better taught through the use of appropriate case studies.

The rapidly emerging and evolving, and highly sensitive global economy is profoundly affecting the employment patterns and the professional lives of graduates. Thus educating the future workforce to understand such issues in a global context is becoming a highly sought-after experience and a critical differentiator in their employability, often testing their ability to bridge discipline-specific theoretical research issues with real-world practice, including addressing and resolving ethical dilemmas, as reflected by the inundation of single and multi-semester capstone projects in many disciplines. While it has been widely reported that despite intensifying competition, off shoring between developed and developing countries can benefit both parties, many students from western countries have shunned STEM careers because they fear that job opportunities and salaries in these fields will decline. Thus, education is confronted with needing to provide

students with higher-order technological skills aptly blended with the consideration of emerging social needs across the globe to provide much needed experiences to thrive in the future, as well as be frontline contributors to the technologically and ethically savvy workforce.

A fundamental change in the education of future workforce and service-force is necessary to assure that we are well prepared for the increasingly more professionally demanding roles. These demands relate to success in the job market, responsibilities toward employers, customers, clients, community and society, and responsibilities as developers of powerful and pervasive automation technologies. In addition to strong technical and management skills, future software and automation designers need the skills to design customized products and integrated services that meet the diverse needs of a multi-cultural, multi-ethnic, and increasingly smaller world united by rapid scientific and technological advances, and facing globally and tightly inter-related hazards and challenges. These trends come with unforeseen social and ethical challenges and tremendous opportunities.

47.5.2 Integrating Social Responsibility and Sensitivity into Education

Effectively integrating social-responsibility, sensitivity and sustainability into our educational curricula has become essential for employers and organization leaders [47.32, 35–37]. See also, for example, the IEEE and ACM model curricula in the context of automation (IEEE.org, ACM.org). Students, trainees, and employees need the diverse exposure to problems and ideas to develop a broad, yet pragmatic vision of the technologically-shifting employment and business landscape. A case study-based approach to teaching, training, and inculcating ethical behavior can provide adequate opportunities to develop the necessary soft skills for being successful in the global service and workplace. Such an exposure can vastly benefit those who may very well be charged with developing policies, priorities, and making investments that can help regions and nations to remain competitive and integrated in the global automation systems and services industry.

Many STEM curricula, in response to these growing industry needs, have placed emphasis on team-based projects and problem-based instruction styles. However, these projects have their own bag of pit falls; for example, in project-based software development classes students often epitomize software development as building the best solution to address customers' requirements. In the following section, a dilemma-based case study approach that goes beyond a project-based curriculum is described. It encourages students to reflect upon the social and ethical ramifications of technology, expanding the narrow, functional-focused tunnel vision that currently (subliminally) exists across many computing and automation curricula, and the automation and software industry, in particular. This is an attempt to address some specific concerns that arise out of such a problem/project focused curricula. With respect to automation and software-related issues, some of these concerns include:

1. In today's post-scandal business climate, additional scrutiny, public condemnation, and possible legal consequences could result if individuals and companies continue to violate accepted ethics and fairness standards. While it is often difficult, if not impossible to predict the future, or the negative consequences of a creation, is ignoring such possible consequences on individuals not ethically questionable?
2. Is it responsible automation development practice, when creating a new technology, and is it ethically sound enough with regard to any possible negative consequences of the new creation and its effects on society?

47.5.3 Dilemma-Based Learning

Education is what remains after one has forgotten what one has learned in school. (John Dryden)

Case-based learning has long been used in management and business schools [47.38, 39]. It has also proved to be highly effective in other disciplines [47.40–42]. According to [47.43], "Students change profoundly in their ability to undertake critical analysis and discuss issues intelligently". Case-based instruction offers a number of advantages and is effective for increasing student motivation [47.40, 41]. In summary, it is thought to be more effective than didactic teaching methods because real-world cases:

1. More accurately represent the complexity and ambiguity of problems
2. Provide a framework for making explicit the problem-solving processes of both novices and experts
3. Provide a means for helping students develop the kind of problem-solving strategies that practicing professionals need [47.44].

Problem-based learning, a case-based derivative, is also widely used, where students are required to learn and apply assimilated knowledge [47.45]. It is reported to broaden students' views and causes a new awareness of their own ideologies and capabilities, and effects growth, questioning, or affirmation [47.42].

In *dilemma-based learning* [47.31, 37], another case-based derivative, a story or game is used to communicate the feeling of real-life dilemmas, while challenging its users to learn from the results of their actions. Dilemmas are chosen for their relevancy to complex and costly situations that are difficult for people to comprehend. For example, dilemmas may reflect the complexities of network implementations or the impact of blame on team productivity and project costs. Dilemmas in the classroom challenge learners to balance trade-offs between short-term rewards and long-term results [47.37]. In prior work, it has been noticed that discussions on real-world topics through dilemma-based case studies that couple logical investigative thinking of the problem-based approaches with strategic needs assessments – cost, performance metrics, etc., make appropriate sense in motivating CS students [47.32, 35, 37].

The use of enthusiasm, empathy, and role-play by students has also been shown to be beneficial in improving overall student attitude and encouraging more participation by women students and minorities [47.36, 46]. It helps develop learning communities and other forms of peer support structures, while emphasizing the positive social benefits of automation and computing. It instills a good feeling among students and motivates them to be participative [47.47, 48]. Hence, a secondary effect of this approach is to help student retention efforts, as they explore related technology issues and interests in the various domains based upon their own personal analogical contexts and experiences. Thus, a recurring dilemma-based approach integrated into multiple automation and computing classes could help increase retention of acceptable ethical standards among students regarding automation technology and help them better understand different ethical issues and perspectives.

Dilemma-based learning, by adopting and building upon themes that dominate our everyday lives, in introductory level classes can not only have the greatest impact on subsequent classes, but also help correct the bad blame-driven rapport that the engineering and computing disciplines have received since the 2001 market crash. Progressive refinement of knowledge gained through more dilemma-based cases in different classes throughout the curriculum provide the natural progression necessary for the retention of ethical issues, while allowing for reinforcement learning through similar dilemmas, but with increasing technical content of cases.

Currently, for interested educators, there are several archival case resources (this is a partial set of references to such material) on ethics with appropriate real-world cases that can be adapted to the needs of a particular class (e.g., [47.1–7, 21, 50, 51]). They can serve as resources to start the building of dilemma-based case studies across several core classes in automation-related curricula.

47.5.4 Model-Based Approach to Teaching Ethics and Automation (Learning)

Several *model-based approaches to teaching* ethics and automation have been developed and implemented effectively. For example, in [47.21] a model for teaching information assurance ethics is presented. The model is composed of four dimensions:

1. The moral development dimension
2. The ethical dimension
3. The security dimension
4. The solutions dimension.

The ethical dimension explores the ethical ramifications of a topic from a variety of perspectives. The security dimension includes ways in which an informa-

tion assurance topic manifests to information assurance professionals. The solutions dimension focuses on remedies that individuals, groups of individuals, and society have created to address security problems and associated ethical dilemmas. The moral development dimension describes the stages and transitions that humans experience as they develop morally, and as they develop their own personal beliefs and behaviors about right and wrong.

Another model-based approach [47.49] is the IDEA model, described next.

The IDEA model presents how dilemma-based learning can be accomplished. There are two primary players and four steps to the IDEA model (Fig. 47.2). The players include the teachers involved in teaching the courses and the participating students. The four steps are, in turn, specific to these players. The four steps are explained in more detail and illustrated next.

IDEA Step 1: Involve and Identify

From the teacher's perspective, the 'I' in IDEA stands for *involve* and from the students' perspective it stands for *identify*. The teacher begins by engaging in a discussion of specific cases that are related to the topic being discussed in the class. For example, in an introductory programming course the discussion may be based on a case that is related to the issue of outsourcing. The teacher presents various concerns with respect to the case in question while at the same time engaging the students' interest through discussions (several societal

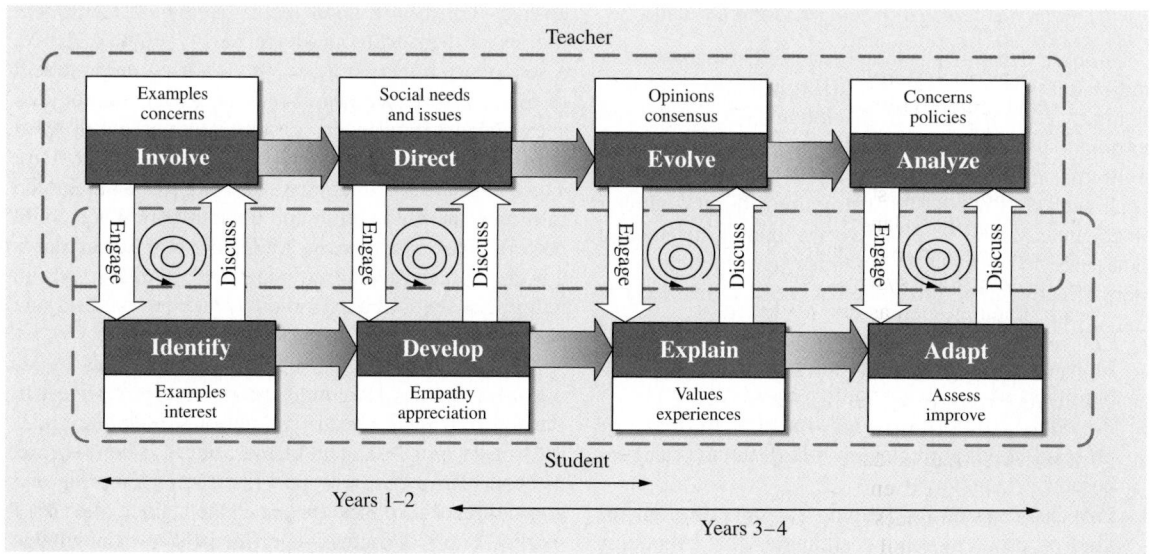

Fig. 47.2 The IDEA Model (after [47.49])

issues can be discussed here: job loss, immigration issues, changing business culture, companies relocating to other countries, etc.).

By engaging the students in the identification of appropriately interesting cases, they become active participants in the class discussions and hence are more likely to engage in investigating the case study further from various socially-interesting perspectives.

A case study on outsourcing provides the ideal opportunity to dispel some of the pervasive myths that students seem to be swayed by, in their choice of automation and computing as a career choice. Current world news information is critical to involving students in the topic of discussion.

For example, at the time of writing this chapter, in the current state of the economy (August 2008), according to the CIO magazine, the unemployment rate for people in the IT industry is less than 3%, while that for the entire USA is 5.9%. Such information opens up the classroom for engaging discussions on IT-driven outsourcing myths and realities. In the rest of this section, we use outsourcing as an engaging example to illustrate the IDEA model. However, this example is by no means meant to be restrictive; other relevant examples may be issues of poor GUI design, issues with electronic voting machines (especially in years of national elections), issues of multi-language support in browsers, issues of robots in tele-surgery, issues with automation for earthquake rescue and refugee survival, automation innovations for energy production, distribution, and delivery, etc.

IDEA Step 2: Direct and Develop

In step 2, most possibly in a follow up class, the student is directed (guided) by the teacher to explore some specific issues of the case further to develop a deeper understanding of the various issues involved. Following the outsourcing case study identified earlier, say an automation assembly language programming class, students can be engaged in a discussion of software outsourcing for embedded systems, say the development of software modules such as drivers that are further integrated into everyday automation systems. Issues of security and privacy that are affected by these low-level software modules, which may be produced in any part of the world, can be discussed and articulated. It has been observed that students participate in such engaging topics with great enthusiasm. This enthusiasm allows learners and trainees to develop a mental model of the entire issue, as well as understand some of the subtle issues in the globalized system of automation

software development, as well as appreciate the finer details of even studying a subject such as assembly language programming and its need within an automation and computing-based curriculum.

Often, students tend to develop a *follow the herd* mentality and are swayed by what they see and hear as requisite *job skills*. Students may often espouse the clouded view that they need to spend most of their time in the program learning *marketable* skills – such as the next hot programming language or system. By association, they may believe they should not spend time learning issues that may not be directly related to their immediate future jobs. This learning misconception has indeed been the observation of instructors and trainers in many disciplines.

Hence, although highly relevant to learning the fundamentals of automation, or computer science, courses such as assembly language programming, evoke less interest among current-day students. Integrating such a dilemma-oriented case study driven discussion can help assure the students of the need for focusing on such fundamental courses as well as understanding its high relevance to societal needs – for example, helping build privacy and security in I/O drivers and embedded automation devices and systems.

IDEA Step 3: Evolve and Explain

The mediocre teacher tells. The good teacher explains. The superior teacher demonstrates. The great teacher inspires. (William Ward)

In step 3, the 'E' in IDEA stands for *evolve* from the teacher's perspective, and *explain* from the students' perspective. The student, in the same (automation programming/digital design/assembly language) or a follow up class (say a database systems class that normally appears in a junior/senior year of the curriculum) is guided by the teacher to explore more details of the case to understand the magnitude and implications of the various issues involved. Again, on the issue of outsourcing, the teacher can engage the students in cases such as *credit card sales and marketing* (or cellular communication devices, etc.), whereby the jobs of identifying and seeking *likely* customers are outsourced to BPO companies (business process outsourcing).

Foreign governments are offering significant fiscal and non-fiscal incentives to attract such foreign direct investments into their respective countries and hence it is difficult for a business to ignore such compelling benefits. Experts who see the growing global demand for BPO (estimated to be at US$180 billion in 2010) indicate a shift from cost-effectiveness to issues of skills, quality, and competence. Issues of personal, professional, and business ethics would definitely be factored as we move towards meeting such expectations, often driven by concerned citizens whose personal data is at stake as part of such BPO decision processes in multinational organizations.

In a course such as database systems, the teacher can guide discussion on how such practices effect the compilation, sharing, and administration of the data contained in large-scale distributed databases in question, their effect on issues of an individual's privacy, which possibly is no longer within the geographical confines of the source country, and issues of checks and bounds verifications that need to occur for such business arrangements between business operating in across different countries that are culturally different.

How is an individual's right to privacy different across cultures and what does *privacy* mean in a different society? What are the issues a business needs, or service needs, to be concerned with respect to the laws of the country? How can the business or service contain and secure the assimilation and sharing of such data? Instructors can promote discussions that can actually engage the student in understanding core values that may be viewed differently across cultures and grow by discussing cases that involve such experiences.

IDEA Step 4: Analyze and Adapt

Through the use of the three earlier steps, students would have incrementally developed the mental and subject-level maturity needed to understand the various issues, their interrelatedness and the socio-cultural effects of the various aspects of automation and computing. In an appropriate junior/senior level course, say systems analysis and design, software engineering or a capstone automation course, where students normally develop large-scale projects to demonstrate their deep understanding of their career subject, students can focus on better understanding the design and development, or process; issues that need to be enforced for guaranteeing globally standardized automation development practices when dealing with data and signals that can be potentially misused.

Students will also be better prepared to understand and discuss issues of professional codes of ethics, since they would have been exposed to and have developed a deeper understanding of the need for them in a globalized sense. In addition, they may have actually gained the necessary skills to analyze and assess ethical dilemmas and conflicts, good versus evil ideas and policies, and issues of sensitivity to social and global sustainability concerning the design and enforcement of such policies for globally-distributed services and businesses.

47.6 Ethics and Research

Collectively this book provides a wealth of automation-related research topics: sensor networks, cybernetics, communication, automatic control, soft computing, artificial intelligence, evolutionary automation, etc. All these automation research topics may serve as valid, timely topics for ethical concerns related to research; highly appropriate for this section. For the purposes of demonstration of emerging issues that can be ethically-sensitive vis-á-vis research, we focus specifically on the ethical issues related to research aided by the exponential growth of the World Wide Web and the information it could offer research about Internet users [47.52].

In order to advance research and serve the users of their products, many Internet companies keep web access logs, search history logs, or transaction logs. Why is this perspective of *logs* important? On bulletin boards, peer-to-peer and social networks, e-Commerce sites, and the Internet in general, individuals can behave and operate with certain anonymity in the absence of

the presentation of self. Individuals online have a sense of complete autonomy and anonymity. Often the learnt social norm from such interactions is that there is little incentive to feel responsible for one's own actions or sensitivity to the open public and community, in general, if *the community* does not provide some kind of *instantaneous* visible reward or tangible penalty.

47.6.1 Internet–Based Research

The scaling up of web content as well as users has resulted in increased difficulty in searching for information over the web. The ever increasing number of pages that match any given set of query words compel users to modify their queries a number of times before obtaining the required information. This repeated, inefficient search results in increased traffic on the network and in a spiraling effect, which in turn results in higher resource consumption and overload. Search engines have made it possible for anyone to look up information from any corner of the world on the Internet.

In an unprecedented decision statement a judge in New Zealand banned online media from publishing the names of two people accused of murder [47.53]. All other news media such as TV, printed media, etc. were allowed to publish the names except the Internet media. This distinction was based on the concern that information about the accused is available on Internet for a long time even after the trial is over.

This case poses a dilemma about the information available on the Internet and in search engine logs much longer after the validity of the information has expired. The availability of query log datasets such as AOL has opened up the doors for carrying out exploratory research on searching user query logs and coming up with possible solutions to make the user search sessions more productive with the intent to provide better search experience for users. While AOL seems to have not taken adequate measures to hide personally identifiable information, the availability of the data set itself poses interesting ethical questions.

Several related developments can be summarized about research and internet-based search, which may shed light on ethical concerns and conflicts in this domain:

- In order to encourage research with user search query logs, Microsoft announced that it would avail its dataset to selected research organizations upon signing agreements. Such safeguards are necessary to protect user privacy and advance research, while developing better tools to help search engine users. Users' *opt-in* and *opt-out*, meaning personally selective, optional acceptance or rejection of sharing their personal information, have become common as part of *codes of privacy* [47.4].
- Users' web searching behavior has been an interesting research area for some time now. Researchers have studied the overall nature of *information behavior*, including information seeking behavior [47.54], *information retrieval* (IR) with *hidden behavioral patterns* and semantically *super-concepts* [47.55,56]. Sometimes, the thirst for information and convenience influence human searchers (as well as purchasers on the web researching available options) to willingly compromise, at least in part, their principled sensitivity to protect their privacy.
- Privacy rights and protection privileges are also associated with the availability of user search query

log datasets. Included are multi-faceted logs coupled with relevant information such as time spent on the web page clicked on, web pages opened, printed and/or bookmarked, and whether the user's true or at least intended information needs are satisfied. Potential ramifications or lack of such query logs dataset vis-á-vis user privacy issues are outlined in [47.57–59] and are subject to further research.

Addressing privacy rights and issues and Internet-based research requires review boards, as described also in the next section. Ethical thinking in this direction includes, for example:

- Setting up a review board for release of query log data for research purposes, while adhering to certain guidelines of ethical practices [47.60].
- Classifying sensitive queries in the query log dataset from a privacy perspective; for example, by partial anonymization of queries [47.61].
- Specific methods of anonymizing sensitive queries in the AOL and similar such datasets [47.62, 63]. For instance:
  a) applying threshold cryptography systems that eliminate highly identifying queries in real time, and
  b) dealing with a set of aggregated queries that are overly identifying, and addressing issues of tradeoff between privacy and utility of the query log data.

47.6.2 More on Research Ethics and User Privacy Issues

While internet users' search session data availability for research and other exploitation illustrates serious ethical issues, some of which are described above, other privacy, property, quality, and accessibility ethical concerns need to be addressed. These concerns need to be appropriately handled, including: fair use of information, ethics of anonymity, and critical need for carefully enabling selective access to private information and behavioral research for specific goals of information for safety, health, security, and other essential public needs [47.64].

Privacy and Accessibility Rights Versus Significant Public Service

What about limiting research on behavior patterns, which may result in losing the opportunity to obtain unique results for targeted services that are significantly beneficial to society, even critical for sustainability? For example, health, safety, and security related issues may need Internet-based and mobile phone-based research. An emerging optional way is to have informed consent from the users at appropriate instances, to enable the fair use of behavior information for agreed upon and selectively chosen research activities. This area is being addressed already by different industry segments and is handled by various legal means.

Policies for Conducting Research Based on Automation

Policies for research based on knowledge obtained by automation have been developed and are still emerging to address ethical concerns, e.g., [47.65]. Initiatives have emerged and need to be strengthened and widened to satisfy World Wide Web media related issues, as such data may become increasingly available for organizations to mine for gaining competitive advantage in the market place, e.g., [47.33]. A consortium for university researchers, industries, government agencies, and other concerned organizations to discuss policy and other related issues of conducting such research is being developed.

The Myth About User Privacy with Automation

One myth about privacy of automation-users is that protecting privacy rights is the onus of the user. In today's world, where information systems security management is a discipline that is fast emerging, its peripheries are yet to be well defined. Who are the gatekeepers?

- Internet service provider (ISP) are burdened with the responsibility of being gatekeepers of their users' privacy; they have to regularly compromise with governmental agencies trying to gain access to ISP user data in order to prevent crime or conduct data forensics.
- Search engine services have a similar responsibility, though they are not burdened with the bulk of keeping the identity of their users private (exceptions being Google or Yahoo! users who may opt to log in before conducting a web-based search).

Neither of these entities, the ISP and the search engine service, would like to be burdened with the bulk of the responsibility of protecting the identity of a user, when the user performs web-based searches. But the fact is that the necessary interface for Internet access is provided to the user by the ISP. This fact lays the primary responsibility of user identity obfuscation squarely on the ISP. ISP employees may be able to gain access to searches conducted by their users and may be able to exploit these details in various unethical or ethical ways. This risk is higher in smaller communities that have populations less than 50 000 and are typically serviced by a few local ISPs.

Policies on Data Mining for Efficiency

Automation data preservation, analysis and indexing are important for web-based search engines and other Internet companies and automation services to perform efficiently, since correlating diverse user searches and interactions are the modus operandi of enhancing performance results. This information can be useful for automation design and architecture evolution. However, this data mining can also be misused by the automation service provider. Self regulations should be supported by clearly defined policies on how the data is collected, accessed, and distributed even for research purposes.

Institutional Review Boards

As common with any research involving human subject, universities and research organizations need to follow strict review board scrutiny. The Internet data research initiatives undertaken by universities and research organizations should also go through institutional review boards' (IRB) formal approval process to make sure human interests, rights, and privacy are protected. Since the review, scrutiny and approval procedures can also be automated, Internet service providers and companies should set clearly defined guidelines and policies for its researchers and users. Many companies already focus on establishing a working group of individuals from privacy, legal, IRB, and security teams to discuss various aspects of the problem and proposed solutions. Such working groups study problems on a case by case basis ensuring a company's competitive advantage without compromising on the ethical issues (if any) involved in the research.

The ethical issues about research and automation will undoubtedly be addressed as organizations and society learn the pitfalls and find methods to resolve the ethical dilemmas that have been mentioned. At the same time, it is clear (as indicated in Fig. 47.1) that newly developed and far reaching automation functions will continue to pose tremendous ethical challenges to individuals, organizations, and society at large.

47.7 Challenges and Emerging Trends

In this chapter, ethical challenges, dilemmas, and conflicts related to automation and enabled or introduced by automation have been highlighted. The context of automation and internationalization, or globalization of services and businesses bring further need for rational, acceptable, and sustainable ethical sensitivities and behaviors. These should continue to be the responsibility of individuals, and of individuals within organizations, but should also be supported, monitored, and maintained by automation mechanisms. Therefore, the increasing attention being paid to ethics in the context of automation, specifically from the perspective of education and research, has been explained. Challenging ethical issues have been presented and illustrated relative to the dimensions of technology, security, privacy, property, quality, and accessibility.

For education and training, the model-based approach for integrating ethics and socially responsible automation/computing into the undergraduate curricula, as well as training courses, has been presented. Examples from automation and computing curricular perspective have been used, and can be adapted to other science and technological disciplines, and commercial and service organizations. For the effective application of this approach, or similar programs, one needs the participation of several multi-disciplinary members, instructors or trainers. However, the attractiveness of such an approach is in its ability to engage the students and trainees meaningfully while still undertaking the primary task of learning the skills and techniques they would need to be successful upon graduation or completion.

For research, certain open challenges in gathering, mining, and observing user information-seeking behavior, while maintaining individuals' privacy rights have been highlighted. Policies, including review boards, have been and are being developed to address these ethical concerns. In such situations a strong rational balance between advanced research and user privacy must be maintained at all times. While the research community at large would come up with the solutions, privacy, anonymity, and fair use issues need to be effectively addressed to demonstrate the innumerable benefits that such research work can yield for the great benefits of individuals and of society. Marketing and information dissemination in a digital world represent an emerging area of research that can be timely and exciting for students, for users, for organizations, and for the public

– for example, issues such as cookies leaving digital trail mixes on people's machines, in light of protecting society, while also protecting individual freedom and individual rights.

47.7.1 Trends and Challenges

Ethical issues, dilemmas, and conflicts, and unethical behaviors, some of which are horrendous and tragic, are unfortunately an integral part of the proliferation of computers and automation in our lives. Major concerns range from privacy, copyrights, and cyber crime issues, to the global impact of computers and communication, online communities and social networks, and effects of virtual reality. Articles, books, conferences, on-line resources, social and political processes have evolved and continue to grow in importance and influence, contributing to ethics expertise in diverse disciplines. The breadth of multi-disciplinary scope allows students and professionals to learn, understand, and evaluate the individual, social, and ethical issues brought about by computer and automation technologies.

Some specific trends to consider:

International Policies
Impact of digitized information on individuals, communities, organizations, and societies, including continued discussions and necessary development of international policies on:

- Privacy
- Automation quality and reliability
- Automation security
- Copyrights and intellectual property
- Collaborative protocols for rational automation control, equality of access under authorization procedures, and trust and authentication agreements

Frameworks and Regulations
Development of ethical frameworks and regulatory processes are needed for substantial treatment of the interrelated automation issues of cyber-ethics: accessibility, free speech and expression, property, privacy, and security.

Self-Repair and Self-Recovery
Research and development of automatic self-repair and self-recovery are needed to address the risks as-

sociated with unexpected computer and automation break-downs, disasters, and failures that open up vulnerabilities to unethical, unsustainable scenarios.

Ethical Automation

Research and development are needed of *inherently* ethical software and ethical automation (including ethical robotics) able to automatically handle and automatically help resolve issues such as media copying, file sharing, infringement of intellectual property, security risks and threats, Internet-based crime, automation-assisted forgery, identity theft, unethical employee surveillance, individual privacy, and compliance with ethical and professional codes.

Ethics of Robotic Automation

Major advancements are needed in robo-ethics to address the fact that (1) robots and robotic automation are increasingly more capable, and (2) there are humans that will increasingly abuse these powerful capabilities, deploying them in ethically questionable situations and environments (e.g., in schools, hospitals, etc.) where ethically wrong robotic automation conduct could have disastrous impacts on humans.

- We must develop ways to ensure that automation, without robots and with robots, will always behave in an ethically correct manner.
- We need to be able to trust that automation, through software-inherent ethics-rationale reflecting ethical human logic (preferably specified in natural languages) will always behave under strict ethical constraints. These constraints must follow previously defined ethical codes, and be able to limit their actions and behavior under these constraints, always reflecting ethical humans' instructions, even without human supervision.

The dual challenge in front of us is that as we develop more powerful, intelligent, and autonomous automation, we must also be careful it is not and cannot be abused by unethical people against us and against other people; we must also be careful that this powerful automation does not assume independence to, on its own, hurt people and inflict damage. The challenge for automation scientists, designers, and managers is that we need to consider how to ethically control the behavior of automation and how to ethically restrict its autonomy – because automation is all around us and because we are so dependent on it.

47.8 Additional Online Resources

Source materials relevant to ethics and automation are available from the ACM and IEEE model curricula, national societies such as ACM, IEEE, AAAS, ASEE, AAES, AIS, and others, for guidelines on ethics; groups such as ACM SIGCAS, CERIAS, CPSR, EFF, EPIC, and other professional organizations that promote responsible behavior. Their conferences, journals and materials provide rich, additional topics on automation and ethics. In addition, the following are several online resources relevant for ethics and automation:

http://www.bsa.org
http://catless.ncl.ac.uk/risks
http://www.cerias.purdue.edu/
http://computingcases.org/index.html
http://www.cpsr.org/ethics/eei
http://csethics.uis.edu/dolce
http://www.cyberlawclinic.org/casestudy.htm

http://www.dhs.gov/dhspublic
(on strategy to secure the cyberspace)
http://ethics.iit.edu/resources/onlineresources.html
http://ethics.iit.edu/codes/engineer.html
http://ethics.iit.edu/emerging/index.html
http://ethics.sandiego.edu/resources/cases/
HomeOverview.asp
http://ethics.tamu.edu/1995nsf.htm
http://www.georgetown.edu/research/nrcbl/nrc
http://microsoft.com/piracy
http://onlineethics.org
http://privacyrights.org
http://www.rbs2.com/ethics.htm
http://repo-nt.tcc.virginia.edu/ethics/index.htm
http://seeri.etsu.edu/Ethics.htm
http://government.zdnet.com/?p=3935
(Modern Wars: Cyber assisted warfare).

47.A Appendix: Code of Ethics Example

ACM (Association for Computing Machinery) Code of Ethics and Professional Conduct

Adopted by ACM Council 10/16/92.

Preamble

Commitment to ethical professional conduct is expected of every member (voting members, associate members, and student members) of the Association for Computing Machinery (ACM).

This Code, consisting of 24 imperatives formulated as statements of personal responsibility, identifies the elements of such a commitment. It contains many, but not all, issues professionals are likely to face. Section 47.A.1 outlines fundamental ethical considerations, while Sect. 47.A.2 addresses additional, more specific considerations of professional conduct. Statements in Sect. 47.A.3 pertain more specifically to individuals who have a leadership role, whether in the workplace or in a volunteer capacity such as with organizations like ACM. Principles involving compliance with this Code are given in Sect. 47.A.4.

The Code shall be supplemented by a set of Guidelines, which provide explanation to assist members in dealing with the various issues contained in the Code. It is expected that the Guidelines will be changed more frequently than the Code.

The Code and its supplemented Guidelines are intended to serve as a basis for ethical decision making in the conduct of professional work. Secondarily, they may serve as a basis for judging the merit of a formal complaint pertaining to violation of professional ethical standards.

It should be noted that although computing is not mentioned in the imperatives of Sect. 47.A.1, the Code is concerned with how these fundamental imperatives apply to one's conduct as a computing professional. These imperatives are expressed in a general form to emphasize that ethical principles which apply to computer ethics are derived from more general ethical principles.

It is understood that some words and phrases in a code of ethics are subject to varying interpretations, and that any ethical principle may conflict with other ethical principles in specific situations. Questions related to ethical conflicts can best be answered by thoughtful consideration of fundamental principles, rather than reliance on detailed regulations.

47.A.1 General Moral Imperatives

As an ACM member I will...

Contribute to Society and Human Well-Being

This principle concerning the quality of life of all people affirms an obligation to protect fundamental human rights and to respect the diversity of all cultures. An essential aim of computing professionals is to minimize negative consequences of computing systems, including threats to health and safety. When designing or implementing systems, computing professionals must attempt to ensure that the products of their efforts will be used in socially responsible ways, will meet social needs, and will avoid harmful effects to health and welfare.

In addition to a safe social environment, human well-being includes a safe natural environment. Therefore, computing professionals who design and develop systems must be alert to, and make others aware of, any potential damage to the local or global environment.

Avoid Harm to Others

Harm means injury or negative consequences, such as undesirable loss of information, loss of property, property damage, or unwanted environmental impacts. This principle prohibits use of computing technology in ways that result in harm to any of the following: users, the general public, employees, employers. Harmful actions include intentional destruction or modification of files and programs leading to serious loss of resources or unnecessary expenditure of human resources such as the time and effort required to purge systems of *computer viruses*.

Well-intended actions, including those that accomplish assigned duties, may lead to harm unexpectedly. In such an event the responsible person or persons are obligated to undo or mitigate the negative consequences as much as possible. One way to avoid unintentional harm is to carefully consider potential impacts on all those affected by decisions made during design and implementation.

To minimize the possibility of indirectly harming others, computing professionals must minimize malfunctions by following generally accepted standards for system design and testing. Furthermore, it is often necessary to assess the social consequences of systems to project the likelihood of any serious harm to others. If system features are misrepresented to users, coworkers, or supervisors, the individual computing professional is responsible for any resulting injury.

In the work environment the computing professional has the additional obligation to report any signs of system dangers that might result in serious personal or social damage. If one's superiors do not act to curtail or mitigate such dangers, it may be necessary to *blow the whistle* to help correct the problem or reduce the risk. However, capricious or misguided reporting of violations can, itself, be harmful. Before reporting violations, all relevant aspects of the incident must be thoroughly assessed. In particular, the assessment of risk and responsibility must be credible. It is suggested that advice be sought from other computing professionals. See principle 2.5 regarding thorough evaluations.

Be Honest and Trustworthy

Honesty is an essential component of trust. Without trust an organization cannot function effectively. The honest computing professional will not make deliberately false or deceptive claims about a system or system design, but will instead provide full disclosure of all pertinent system limitations and problems.

A computer professional has a duty to be honest about his or her own qualifications, and about any circumstances that might lead to conflicts of interest.

Membership in volunteer organizations such as ACM may at times place individuals in situations where their statements or actions could be interpreted as carrying the *weight* of a larger group of professionals. An ACM member will exercise care to not misrepresent ACM or positions and policies of ACM or any ACM units.

Be Fair and Take Action not to Discriminate

The values of equality, tolerance, respect for others, and the principles of equal justice govern this imperative. Discrimination on the basis of race, sex, religion, age, disability, national origin, or other such factors is an explicit violation of ACM policy and will not be tolerated.

Inequities between different groups of people may result from the use or misuse of information and technology. In a fair society, all individuals would have equal opportunity to participate in, or benefit from, the use of computer resources regardless of race, sex, religion, age, disability, national origin or other such similar factors. However, these ideals do not justify unauthorized use of computer resources nor do they provide an adequate basis for violation of any other ethical imperatives of this code.

Honor Property Rights Including Copyrights and Patent

Violation of copyrights, patents, trade secrets and the terms of license agreements is prohibited by law in most circumstances. Even when software is not so protected, such violations are contrary to professional behavior. Copies of software should be made only with proper authorization. Unauthorized duplication of materials must not be condoned.

Give Proper Credit for Intellectual Property

Computing professionals are obligated to protect the integrity of intellectual property. Specifically, one must not take credit for other's ideas or work, even in cases where the work has not been explicitly protected by copyright, patent, etc.

Respect the Privacy of Others

Computing and communication technology enables the collection and exchange of personal information on a scale unprecedented in the history of civilization. Thus there is increased potential for violating the privacy of individuals and groups. It is the responsibility of professionals to maintain the privacy and integrity of data describing individuals. This includes taking precautions to ensure the accuracy of data, as well as protecting it from unauthorized access or accidental disclosure to inappropriate individuals. Furthermore, procedures must be established to allow individuals to review their records and correct inaccuracies.

This imperative implies that only the necessary amount of personal information be collected in a system, that retention and disposal periods for that information be clearly defined and enforced, and that personal information gathered for a specific purpose not be used for other purposes without consent of the individual(s). These principles apply to electronic communications, including electronic mail, and prohibit procedures that capture or monitor electronic user data, including messages, without the permission of users or bona fide authorization related to system operation and maintenance. User data observed during the normal duties of system operation and maintenance must be treated with strictest confidentiality, except in cases where it is evidence for the violation of law, organizational regulations, or this Code. In these cases, the nature or contents of that information must be disclosed only to proper authorities.

Honor Confidentiality

The principle of honesty extends to issues of confidentiality of information whenever one has made an explicit promise to honor confidentiality or, implicitly, when private information not directly related to the performance of one's duties becomes available. The ethical concern is to respect all obligations of confidentiality to employers, clients, and users unless discharged from such obligations by requirements of the law or other principles of this Code.

47.A.2 More Specific Professional Responsibilities

As an ACM computing professional I will...

Strive to Achieve the Highest Quality, Effectiveness and Dignity in Both the Process and Products of Professional Work

Excellence is perhaps the most important obligation of a professional. The computing professional must strive to achieve quality and to be cognizant of the serious negative consequences that may result from poor quality in a system.

Acquire and Maintain Professional Competence

Excellence depends on individuals who take responsibility for acquiring and maintaining professional competence. A professional must participate in setting standards for appropriate levels of competence, and strive to achieve those standards. Upgrading technical knowledge and competence can be achieved in several ways: doing independent study; attending seminars, conferences, or courses; and being involved in professional organizations.

Know and Respect Existing Laws Pertaining to Professional Work

ACM members must obey existing local, state, province, national, and international laws unless there is a compelling ethical basis not to do so. Policies and procedures of the organizations in which one participates must also be obeyed. But compliance must be balanced with the recognition that sometimes existing laws and rules may be immoral or inappropriate and, therefore, must be challenged. Violation of a law or regulation may be ethical when that law or rule has inadequate moral basis or when it conflicts with another law judged to be more important. If one decides to violate a law or rule because it is viewed as unethical, or for any other reason, one must fully accept responsibility for one's actions and for the consequences.

Accept and Provide Appropriate Professional Review

Quality professional work, especially in the computing profession, depends on professional reviewing and critiquing. Whenever appropriate, individual members should seek and utilize peer review as well as provide critical review of the work of others.

Give Comprehensive and Thorough Evaluations of Computer Systems and Their Impacts, Including Analysis of Possible Risks

Computer professionals must strive to be perceptive, thorough, and objective when evaluating, recommending, and presenting system descriptions and alternatives. Computer professionals are in a position of special trust, and therefore have a special responsibility to provide objective, credible evaluations to employers, clients, users, and the public. When providing evaluations the professional must also identify any relevant conflicts of interest, as stated in imperative 1.3.

As noted in the discussion of principle 1.2 on avoiding harm, any signs of danger from systems must be reported to those who have opportunity and/or responsibility to resolve them. See the guidelines for imperative 1.2 for more details concerning harm, including the reporting of professional violations.

Honor Contracts, Agreements, and Assigned Responsibilities

Honoring one's commitments is a matter of integrity and honesty. For the computer professional this includes ensuring that system elements perform as intended. Also, when one contracts for work with another party, one has an obligation to keep that party properly informed about progress toward completing that work.

A computing professional has a responsibility to request a change in any assignment that he or she feels cannot be completed as defined. Only after serious consideration and with full disclosure of risks and concerns to the employer or client, should one accept the assignment. The major underlying principle here is the obligation to accept personal accountability for professional work. On some occasions other ethical principles may take greater priority.

A judgment that a specific assignment should not be performed may not be accepted. Having clearly identified one's concerns and reasons for that judgment, but failing to procure a change in that assignment, one may yet be obligated, by contract or by law, to proceed as directed. The computing professional's ethical judgment should be the final guide in deciding whether or not to proceed. Regardless of the decision, one must accept the responsibility for the consequences.

However, performing assignments *against one's own judgment* does not relieve the professional of responsibility for any negative consequences.

Improve Public Understanding of Computing and Its Consequences

Computing professionals have a responsibility to share technical knowledge with the public by encouraging understanding of computing, including the impacts of computer systems and their limitations. This imperative implies an obligation to counter any false views related to computing.

Access Computing and Communication Resources only when Authorized to Do so

Theft or destruction of tangible and electronic property is prohibited by imperative 1.2–*Avoid harm to others*. Trespassing and unauthorized use of a computer or communication system is addressed by this imperative. Trespassing includes accessing communication networks and computer systems, or accounts and/or files associated with those systems, without explicit authorization to do so. Individuals and organizations have the right to restrict access to their systems so long as they do not violate the discrimination principle (see 1.4). No one should enter or use another's computer system, software, or data files without permission. One must always have appropriate approval before using system resources, including communication ports, file space, other system peripherals, and computer time.

47.A.3 Organizational Leadership Imperatives

As an ACM member and an organizational leader, I will...

Background Note: This section draws extensively from the draft IFIP Code of Ethics, especially its sections on organizational ethics and international concerns. The ethical obligations of organizations tend to be neglected in most codes of professional conduct, perhaps because these codes are written from the perspective of the individual member. This dilemma is addressed by stating these imperatives from the perspective of the organizational leader. In this context *leader* is viewed as any organizational member who has leadership or educational responsibilities. These imperatives generally may apply to organizations as well as their leaders. In this context *organizations* are corporations, government agencies, and other *employers*, as well as volunteer professional organizations.

Articulate Social Responsibilities of Members of an Organizational Unit and Encourage Full Acceptance of Those Responsibilities

Because organizations of all kinds have impacts on the public, they must accept responsibilities to society. Organizational procedures and attitudes oriented toward quality and the welfare of society will reduce harm to members of the public, thereby serving public interest and fulfilling social responsibility. Therefore, organizational leaders must encourage full participation in meeting social responsibilities as well as quality performance.

Manage Personnel and Resources to Design and Build Information Systems that Enhance the Quality of Working Life

Organizational leaders are responsible for ensuring that computer systems enhance, not degrade, the quality of working life. When implementing a computer system, organizations must consider the personal and professional development, physical safety, and human dignity of all workers. Appropriate human-computer ergonomic standards should be considered in system design and in the workplace.

Acknowledge and Support Proper and Authorized Uses of an Organization's Computing and Communication Resources

Because computer systems can become tools to harm as well as to benefit an organization, the leadership has the responsibility to clearly define appropriate and inappropriate uses of organizational computing resources. While the number and scope of such rules should be minimal, they should be fully enforced when established.

Ensure that Users and Those Who Will Be Affected by a System Have Their Needs Clearly Articulated During the Assessment and Design of Requirements; Later the System Must Be Validated to Meet Requirements

Current system users, potential users and other persons whose lives may be affected by a system must have their needs assessed and incorporated in the statement of requirements. System validation should ensure compliance with those requirements.

Articulate and Support Policies that Protect the Dignity of Users and Others Affected by a Computing System

Designing or implementing systems that deliberately or inadvertently demean individuals or groups is ethically unacceptable. Computer professionals who are in decision making positions should verify that systems are designed and implemented to protect personal privacy and enhance personal dignity.

Create Opportunities for Members of the Organization to Learn the Principles and Limitations of Computer Systems

This complements the imperative on public understanding (2.7). Educational opportunities are essential to facilitate optimal participation of all organizational members. Opportunities must be available to all members to help them improve their knowledge and skills in computing, including courses that familiarize them with the consequences and limitations of particular types of systems. In particular, professionals must be made aware of the dangers of building systems around oversimplified models, the improbability of anticipating and designing for every possible operating condition, and other issues related to the complexity of this profession.

47.A.4 Compliance with the Code

As an ACM member I will...

Uphold and Promote the Principles of this Code

The future of the computing profession depends on both technical and ethical excellence. Not only is it important for ACM computing professionals to adhere to the principles expressed in this Code, each member should encourage and support adherence by other members.

Treat Violations of this Code as Inconsistent with Membership in the ACM

Adherence of professionals to a code of ethics is largely a voluntary matter. However, if a member does not follow this code by engaging in gross misconduct, membership in ACM may be terminated.

This Code and the supplemental Guidelines were developed by the Task Force for the Revision of the ACM Code of Ethics and Professional Conduct: Ronald E. Anderson, Chair, Gerald Engel, Donald Gotterbarn, Grace C. Hertlein, Alex Hoffman, Bruce Jawer, Deborah G. Johnson, Doris K. Lidtke, Joyce Currie Little, Dianne Martin, Donn B. Parker, Judith A. Perrolle, and Richard S. Rosenberg. The Task Force was organized by ACM/SIGCAS and funding was provided by the ACM SIG Discretionary Fund. This Code and the supplemental Guidelines were adopted by the ACM Council on October 16, 1992.

This Code may be published without permission as long as it is not changed in any way and it carries the copyright notice. Copyright 1997, Association for Computing Machinery, Inc.

References

47.1   G. Reynolds: *Ethics in Information Technology*, 2nd edn. (Cengage Learning, Florence 2006)

47.2   H.T. Tavani: *Ethics and Technology: Ethical Issues in an Age of Information and Communication Technology* (Wiley, New York 2006)

47.3   D.M. Hester, P.J. Ford: *Computers and Ethics in the Cyberage* (Prentice Hall, Englewood Cliffs 2000)

47.4   R.K. Rainer, E. Turban: Ethics, privacy, and information security. In: *Introduction to Information Systems*, 2nd edn. (Wiley, New York 2009), Chap. 3

47.5   D.G. Johnson: *Computer Ethics*, 3rd edn. (Prentice Hall, Englewood Cliffs 2000)

47.6   S. Baase: *A Gift of Fire: Social, Legal, and Ethical Issues for Computers and the Internet*, 2nd edn. (Prentice Hall, Englewood Cliffs 2002)

47.7   J.M. Kizza: *Computer Network Security and Cyber Ethics*, 2nd edn. (McFarland, Jefferson 2006)

47.8   R. Spinello: *Cyber Ethics: Morality and Law in Cyberspace*, 3rd edn. (Jones Bartlett, Boston 2006)

47.9   P.M. Asaro: Robots and responsibility from a legal perspective, Proc. IEEE ICRA (2007)

47.10   J.M. Kizza: *Ethical and Social Issues in the Information Age*, 3rd edn. (McFarland, Jefferson 2007)

47.11   S. Bringsjord, K. Arkoudas, P. Bello: Toward a general logicist methodology for engineering ethically correct robots, IEEE Intell. Syst. **21**, 38–44 (2006)

47.12   D. Dennett: When HAL kills, who's to blame?. In: *HAL's Legacy: 2001's Computer as Dream and Reality*, ed. by D. Stork (MIT Press, Cambridge 1996)

47.13   G. Veruggio, F. Operto: Roboethics: Social and ethical implications of robotics. In: *Springer Handbook of Robotics*, ed. by B. Siciliano, O. Khatib (Springer, Berlin Heidelberg 2007), Chap. 64

47.14   R. Spinello: *Readings in CyberEthics*, 2nd edn. (Jones Bartlett, Boston 2004)

47.15   S. Helmers, U. Hoffmann, J.-B. Stamos-Kaschke: (How) Can software agents become good net citizens?, CMC Magazine **3**, 1–9 (1997)

47.16   V. Perri: Ethics, regulation and the new artificial intelligence, part II: autonomy and liability, Inf. Commun. Soc. **4**, 406–434 (2001)

47.17   J. Sullins: When is a robot a moral agent?, Int. J. Inf. Ethics **6**, 12 (2006)

47.18   K. Himma: Artificial agency, consciousness, and the criteria for moral agency: what properties must an artificial agent have to be a moral agent?, 7th Int. Comput. Ethics Conf. (San Diego 2007)

47.19   R. Lucas: Moral theories for autonomous software agents, ACM SIGCAS Comput. Soc. Arch. **34**, 4 (2004)

47.20   R.C. Arkin: Governing lethal behavior: embedding ethics in a hybrid deliberative/reactive robot architecture, Proc. 3rd ACM/IEEE Int. Conf. Human–Robot Interact. (Amsterdam 2008) pp. 121–128

47.21   Online Ethics Center, National Academy of Engineering, Washington (2009) http://www.onlineethics.org/

47.22   T.W. Bynum, S. Rogerson (Eds.): *Computer Ethics and Professional Responsibility* (Blackwell, New York 2004)

47.23   G. Stamatellos: *Computer Ethics: A Global Perspective* (Jones Bartlett, Sudbury 2007)

47.24   P. Lucas: Why bother? Ethical computers – that's why!, ACM Int. Conf. Proc. Ser. **7**, 33–38 (2000)

47.25   J. Berge: *Software for Automation: Architecture, Integration, and Security* (ISA, New York 2005)

47.26   T. Macaulay: *Industrial Automation and Process Control Security: SCADA, DCS, PLC, HMI* (Auerbach, London 2009)

47.27   E. Cayirci, A. Levi, C. Rong: *Security in Wireless Ad Hoc and Sensor Networks* (Wiley, New York 2009)

47.28   J. Dubin: *The Little Black Book of Computer Security*, 2nd edn. (Penton Media, New York 2008)

47.29   C. Easttom: *Computer Security Fundamentals* (Prentice Hall, Englewood Cliffs 2005)

47.30   D. Gollmann: *Computer Security* (Wiley, New York 2006)

47.31   M. Dark: *Learning and Teaching with Case Studies*, Center for Education and Research in Information Assurance and Security (Purdue University, Purdue 2008) http://www.cerias.purdue.edu/education/ post_secondary_education/undergrad _and_grad/ curriculum_development/ethical_social _prof_ issues/learning_and_teaching.php

47.32   T. Brignall, S. Ramaswamy: A Framework for Integrating Ethical and Values-Based Instruction into the ACM Computing Curricula 2001, Workshop Prot. Inf. Comput. Beyond (CERIAS, Purdue 2001)

47.33   I. Asimov: *I, Robot* (Gnome Press, New York 1950)

47.34   S.Y. Nof (Ed.): *Handbook of Industrial Robotics* (Wiley, New York 1985)

47.35   M. J. Dark, A. Ghafarian, M. A. Saheb, J. Yu: Information ethics and social issues in the undergraduate computer science curriculum: a curriculum development and implementation report, (Redmond 2002) http://www.cerias.purdue.edu/education/ post_secondary_education/undergrad _and_grad/ curriculum_development/ethical_social _ prof_issues/framework.php

47.36   P. James: Transformative learning: promoting change across cultural worlds, J. Vocat. Educ. Train. **49**, 197–219 (1997)

47.37   M. Dark, N. Harter, L. Morales, M.A. Garcia: An information security ethics education model, J. Comput. Sci. Coll. **23**, 82–88 (2008)

47.38   T.L. Beauchamp: *Case Studies in Business, Society and Ethics* (Prentice Hall, Englewood Cliffs 1998)

47.39   R. Spinello: *Case Studies in Information Technology Ethics*, 2nd edn. (Prentice Hall, New York 2003)

47.40   B.B. Levin: Using the case method in teacher education: the role of discussion and experience in teachers' thinking about cases, J. Teach. Teach. Educ. **10**, 2 (1996)

47.41   J. Kleinfeld: Changes in problem solving abilities of students taught through case methods, 1991 Annu. Meet. Am. Educ. Res. Assoc. (Chicago 1991)

47.42   J. Cossom: Teaching from cases: Education for critical thinking, J. Teach. Soc. Work **5**, 139–155 (1991)

47.43   D.W. Ewing: *Inside the Harvard Business School* (Random House, New York 1990)

47.44   M.J. Julian, M.B. Kinzie, V.A. Larsen: Compelling case experiences: performance, practice, and application for emerging instructional designers, Perform. Improv. Q. **13**, 164–201 (2000)

47.45   L. Wilkerson, W.H. Gijselaers: *Bringing Problem-Based Learning to Higher Education: Theory and Practice* (Jossey-Bass, New York 1996)

47.46   J. Lave: The culture of acquisition and the practice of understanding. In: *Cultural Psychology: Essays on Comparative Human Development*, ed. by J.W. Stigler, R.A. Schweder, G. Herdt (Cambridge Univ. Press, Boston 1990) pp. 309–327

47.47   J.M. Cohoon: Recruiting and retaining women in undergraduate computing majors, ACME SIGCSE Bulletin **34**, 48–52 (2002)

47.48   K. Treu, A. Skinner: Ten Suggestions for a gender-equitable CS classroom, ACME SIGCSE Bulletin **34**, 165–167 (2002)

47.49   S. Ramaswamy: Societal-consciousness in the computing curricula: a time for serious introspection, Int. Symp. Technol. Soc.: Risk, Vulnerability, Uncertainty, Technology and Society (Las Vegas 2007)

47.50   Privacy Rights Clearinghouse: San Diego, CA 92103, USA, http://privacyrights.org (last accessed January 2009)

47.51   ComputingCases.org: http://computingcases.org/index.html (last accessed January 2009)

47.52   P. Lyman, H.R. Varian, P. Charles, N. Good, L.L. Jordan, J. Pal: *How much Information?* (SIMS Lab, University of Berkeley 2003), http://www.sims.berkeley.edu/research/ projects/how-much-info-2003/ (last accessed January 2009)

47.53   E. Gay: Judge restricts online reporting of case, The New Zealand Herald, Aug 25 (2008)

http://www.nzherald.co.nz/nz/news/article.cfm?c_id=1&objectid=10528866 (last accessed January 8, 2009)

47.54 T.D. Wilson, D. Ellis, N. Ford, A. Foster: *Uncertainty in Information Seeking*, Final Report, LIC Res. Rep. **59** (Department of Information Studies, University of Sheffield 1999) http://informationr.net/tdw/publ/unis/report.html (Dec 1, 2006)

47.55 L.D. Catledge, J.E. Pitkow: *Characterizing browsing strategies in the world wide web*, Comput. Netw. ISDN Syst. **27** (Elsevier Science, Amsterdam 1995)

47.56 H. Joshi, S. Ito, S. Kanala, S. Hebbar, C. Bayrak: Concept set extraction with user session context, Proc. 45th ACM South East Conf. (Winston-Salem 2007)

47.57 C. Silverstein, H. Marais, M. Henzinger, M. Moricz: Analysis of a very large web search engine query log, ACM Special Interest Group on Information Retrieval (SIGIR) Conference (1999) pp. 6–12

47.58 A. Goker, D. He: Analyzing web search log to determine session boundaries for user-oriented learning, Proc. Adapt. Hypermedia Adapt. Web Based Syst. Int. Conf. AH 2000 (Trento 2000) pp. 319–322

47.59 D. Downey, S. Dumais, E. Horvitz: Models of searching and browsing: languages, studies, and application, Int. Conf. Artificial Intell. IJCAI (2007)

47.60 J. Bar-Ilan: Access to query logs – An academic researcher's point of view, Position paper at Query Log Anal. Workshop at WWW 2007 Conf. (Banff 2007)

47.61 X. Li, E. Agichtein: Towards privacy-preserving query log publishing, Position paper at Query Log Anal. Workshop at WWW 2007 Conf. (Banff 2007)

47.62 E. Adar: User 4XXXXX9: Anonymizing query logs, Query Log Anal. workshop at WWW 2007 Conf. (Banff 2007)

47.63 L. Weinstein: Search engine privacy dilemmas – and paths towards solution, on line blog entry at http://lauren.vortex.com/archive/000188.html (February 12, 2007)

47.64 J. Waldo, H.S. Lin, L.I. Millett (Eds.): *Engaging Privacy and Information Technology in a Digital Age* (The National Academies Press, Washington 2007)

47.65 Committee on Technical and Privacy Dimensions of Information for Terrorism Prevention and Other National Goals, National Research Council: *Protecting Individual Privacy in the Struggle Against Terrorists: A Framework for Program Assessment* (The National Academies Press, Washington 2008)

Part F

Industrial

Part F Industrial Automation

Industrial Automation. Part F Industrial automation is well known and fascinating to all of us who were born in the 20th century, from visits to plants and factories to watching movies highlighting the automation marvels of industrial operations and assembly lines. This part begins with explanation of machine tool automation, including various types of numerical control (NC), flexible, and precision machinery for production, manufacturing, and assembly, digital and virtual industrial production, to detailed design, guidelines and application of automation in the principal industries, from aerospace and automotive to semiconductor, mining, food, paper and wood industries. Chapters are also devoted to the design, control and operation of functions common to all industrial automation, including materials handling, supply, logistics, warehousing, distribution, and communication protocols, and the most advanced digital manufacturing, RFID-based automation, and emerging micro-automation and nano-manipulation. Industrial automation represents a major growth and advancement opportunity, because as explained in this part, it can provide significant innovative solutions to the grand challenges of our generation, including the production and distribution capacity of needed goods and equipment, as well as food, medical and other essential sustenance supplies for the quality of life around the globe.

48. Machine Tool Automation

Keiichi Shirase, Susumu Fujii

Numerical control (NC) is the greatest innovation in the achievement of machine tool
automation in manufacturing. In this chapter, first a history of the development up to
the advent of NC machine tools is briefly reviewed (Sect. 48.1). Then the machining centers
and the turning centers are described with
their key modules and integration into flexible manufacturing systems (FMS) and flexible
manufacturing cells (FMC) in Sect. 48.2. NC part
programming is described from manual programming to the computer-aided manufacturing
(CAM) system in Sect. 48.3. In Sect. 48.4 and
Sect. 48.5, following the technical innovations in the advanced hardware and software
systems of NC machine tools, future control
systems for intelligent CNC machine tools are
presented.

Numerical control (NC) is the greatest innovation in the
achievement of machine tool automation in manufacturing. Machine tools have expanded their performance
and ability since the era of the Industrial Revolution;
however all machine tools were operated manually
until the birth of the NC machine tool in 1952. Numerical control enabled control of the motion and
sequence of machining operations with high accuracy and repeatability. In the 1960s, computers added
even greater flexibility and reliability of machining
operations. These machine tools which had computer
numerical control were called CNC machine tools.
A machining center, which is a highly automated NC

milling machine performing multiple milling operations, was developed to realize process integration as
well as machining automation in 1958. A turning center, which is a highly automated NC lathe performing
multiple turning operations, was also developed. These
machine tools contributed to realize the flexible manufacturing system (FMS), which had been proposed
during the mid-1960s. FMS aims to perform automatic
machining operations unaided by human operators to
machine various parts.

The automatically programmed tool (APT) is the
most important computer-assisted part programming
language and was first used to generate part programs

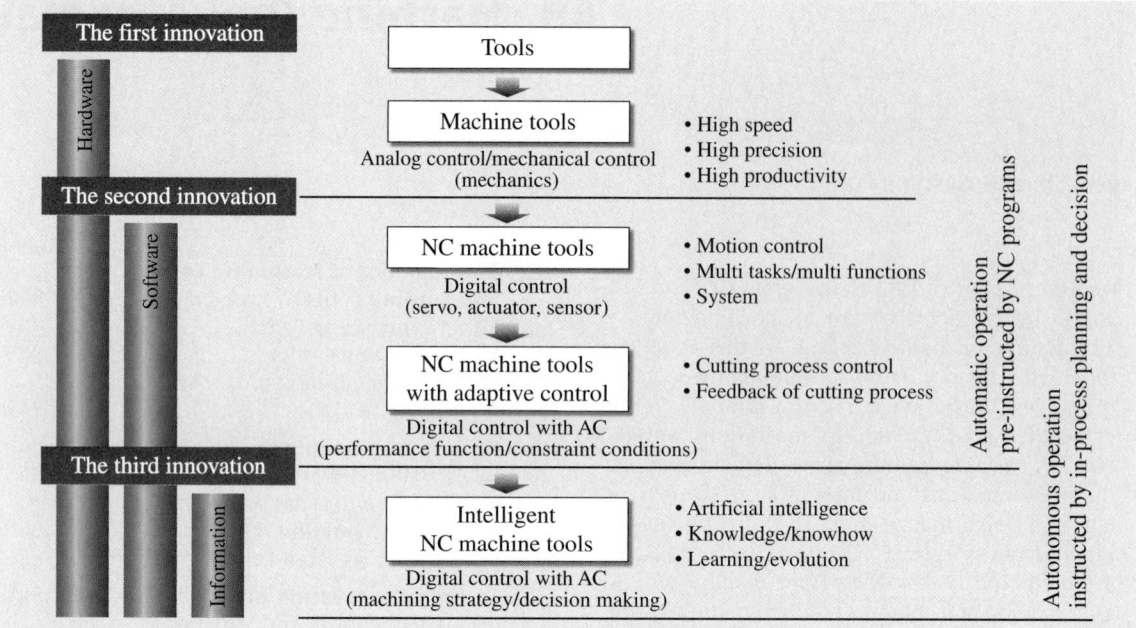

Fig. 48.1 Evolution of machine tools toward the intelligent machine for the future

in production around 1960. The extended subset of APT (EXAPT) was developed to add functions such as setting of cutting conditions, selection of cutting tool, and operation planning besides the functions of APT. Another pioneering NC programming language, COMPACT II, was developed by Manufacturing Data Systems Inc. (MDSI) in 1967. Technologies developed beyond APT and EXAPT were succeeded by computer-aided manufacturing (CAM). CAM provides interactive part programming with a visual and graphical environment and saves significant programming time and effort for part programming. For example, COMPACT II has evolved into open CNC software, which enables integration of off-the-shelf hardware and software technologies. NC languages have also been integrated with computer-aided design (CAD) and CAM systems.

In the past five decades, NC machine tools have become more sophisticated to achieve higher accuracy and faster machining operation with greater flexibility. Certainly, the conventional NC control system can perform sophisticated motion control, but not cutting process control. This means that further intelligence of NC control system is still required to achieve more sophisticated process control. In the near future, all machine tools will have advanced functions for process planning, tool-path generation, cutting process monitoring, cutting process prediction, self-monitoring, failure

prediction, etc. Information technology (IT) will be the key issue to realize these advanced functions. The paradigm is evolving from the concept of autonomy to yield next-generation NC machine tools for sophisticated manufacturing systems.

Machine tools have expanded their performance and abilities as shown in Fig. 48.1. The first innovation took place during the era of the Industrial Revolution. Most conventional machine tools, such as lathes and milling machines, have been developed since the Industrial Revolution. High-speed machining, high-precision machining, and high productivity have been achieved by these modern machine tools to realize mass production.

The second innovation was numerical control (NC). A prototype machine was demonstrated at MIT in 1952. The accuracy and repeatability of NC machine tools became far better than those of manually operated machine tools. NC is a key concept to realize programmable automation. The principle of NC is to control the motion and sequence of machining operations. Computer numerical control (CNC) was introduced, and computer technology replaced the hardware control circuit boards on NC, greatly increasing the reliability and functionality of NC. The most important functionality to be realized was adaptive control (AC). In order to improve the productivity of the machining process and the quality of machined surfaces, several AC systems for real-time adjustment of cutting parameters have been proposed and developed [48.1].

As mentioned above, machine tools have evolved through advances in hardware and control technologies. However, the machining operations are fully dominated by the predetermined NC commands, and conventional machine tools are not generally allowed to change the machining sequence or the cutting conditions during machining operations. This means that conventional NC machine tools are allowed to perform only automatic machining operations that are pre-instructed by NC programs.

In order to realize an *intelligent* machine tool for the future, some innovative technical breakthroughs are required. An intelligent machine tool should be good at learning, understanding, and thinking in a logical way about the cutting process and machining operation, and no NC commands will be required to instruct machining operations as an intelligent machine tool thinks about machining operations, and adapts the cutting processes itself. This means that an intelligent machine tool can perform autonomous operations that are instructed by in-process planning made by the tool itself. Information technology (IT) will be the key issue to realize this third innovation.

48.1 The Advent of the NC Machine Tool

48.1.1 From Hand Tool to Powered Machine

It is well known that John Wilkinson's boring machine (Fig. 48.2) was able to machine a high-accuracy cylinder to build Watt's steam engine. The perfor-

Fig. 48.2 Wilkinson's boring machine (1775)

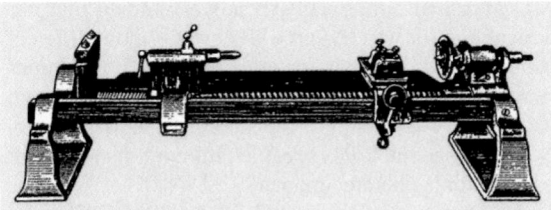

Fig. 48.3 Maudsley's screw-cutting lathe with mechanized tool carriage (1800)

mance of steam engines was improved drastically by the high-accuracy cylinder. With the spread of steam engines, machine tools changed from hand tools to powered machines, and metal cutting became widespread to achieve modern industrialization. During the era of the Industrial Revolution, most conventional machine tools, such as lathes and milling machines, were developed.

Maudsley's screw-cutting lathe with mechanized tool carriage (Fig. 48.3) was a great invention which was able to machine high-accuracy screw threads. The screw-cutting lathe was developed to machine screw threads accurately; however the mechanical tool carriage equipped with a screw allowed precise repetition of machined shapes. Precise repetition of machined shape is an important requirement to produce many of the component parts for mass production. Therefore, Maudsley's screw-cutting lathe became a prototype of lathes.

Whitney's milling machine (Fig. 48.4) is believed to be the first successful milling machine used for cutting plane of metal parts. However, it appears that Whitney's milling machine was made after Whitney's death. Whitney's milling machine was designed to manufacture interchangeable musket parts. Interchangeable parts require high-precision machine tools to make exact shapes.

Fitch's turret lathe (Fig. 48.5) was the first automatic turret lathe. Turret lathes were used to produce complex-shaped cylindrical parts that required several operating sequences and tools. Also, turret lathes can perform automatic machining with a single setup and can achieve high productivity. High productivity is an

Fig. 48.4 Whitney's milling machine (1818)

Fig. 48.6 A copy milling machine

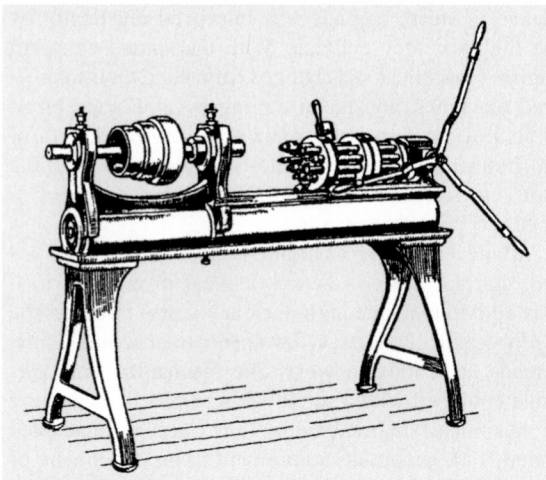

Fig. 48.5 Fitch's turret lathe (1845)

important requirement to produce many of the component parts for mass production.

As mentioned above, the most important modern machine tools required to realize mass production were developed during the era of the Industrial Revolution. Also high-speed machining, high-precision machining, and high productivity had been achieved by these modern machine tools.

48.1.2 Copy Milling Machine

A copy milling machine, also called a tracer milling machine or a profiling milling machine, can duplicate freeform geometry represented by a master model

for making molds, dies, and other shaped cavities and forms. A prove tracing the model contour is controlled to follow a three-dimensional master model, and the cutting tool follows the path taken by the tracer to machine the desired shape. Usually, a tracing prove is fed by a human operator, and the motion of the tracer is converted to the motion of the tool by hydraulic or electronic mechanisms. The motion in Fig. 48.6 shows an example of copy milling. In this case, three spindle heads or three cutting tools follow the path taken by the tracer simultaneously. In some copy milling machines the ratio between the motion of the tracer and the cutting tool can be changed to machine shapes that are similar to the master model.

Copy milling machines were widely used to machine molds and dies which were difficult to generate with simple tool paths, until CAD/CAM systems became widespread to generate NC programs freely for machining three-dimensional freeform shapes.

48.1.3 NC Machine Tools

The first prototype NC machine tool, shown in Fig. 48.7, was demonstrated at the MIT in 1952. The name *numerical control* was given to the machine tool, as it was controlled numerically. It is well known that numerical control was required to develop more efficient manufacturing methods for modern aircraft, as aircraft components became more complex and required more machining. The accuracy, repeatability, and productivity of NC machine tools became far better than those of machine tools operated manually.

The concept of numerical control is very important and innovative for programmable automation, in which the motions of machine tools are controlled or

Fig. 48.7 The first NC machine tool, which was demonstrated at MIT in 1952

later required the prior development of numerical control.

A program to control NC machine tools is called a part program, and the importance of a part program was recognized from the beginning of NC machine tools. In particular, the definition for machining shapes of more complex parts is difficult by manual operation. Therefore, a part programming language, APT, was developed at MIT to realize computer-assisted part programming.

Recently, CNC has become widespread, and in most cases the term NC is used synonymously with CNC. Originally, CNC corresponded to an NC system operated by an internal computer, which realized storage of part programs, editing of part programs, manual data input (MDI), and so on. The latest CNC tools allow generation of a part program interactively by a machine operator, and avoid machine crash caused by a missing part program. High-speed and high-accuracy control of machine tools to realize highly automated machining operation requires the latest central processing unit (CPU) to perform high-speed data processing for several functions.

instructed by a program containing coded alphanumeric data. According to the concept of numerical control, machining operation becomes programmable and machining shape is changeable. The concept of the flexible manufacturing system (FMS) mentioned

48.2 Development of Machining Center and Turning Center

48.2.1 Machining Center

A machining center is a highly automated NC milling machine that performs multiple machining operations such as end milling, drilling, and tapping. It was developed to realize process integration as well as machining automation, in 1958. Figure 48.8 shows an early machining center equipped with an automatic tool changer (ATC). Most machining centers are equipped with an ATC and an automatic pallet changer (APC) to perform multiple cutting operations in a single machine setup and to reduce nonproductive time in the whole machining cycle.

Machining centers are classified into horizontal and vertical types according to the orientation of the spindle axis. Figures 48.9 and 48.10 show typical horizontal and vertical machining centers, respectively. Most horizontal machining centers have a rotary table to index the machined part at some specific angle relative to the cutting tool. A horizontal machining center which has a rotary table can machine the four vertical faces of boxed workpieces in single setup with minimal human assistance. Therefore, a horizontal ma-

chining center is widely used in an automated shop floor with a loading and unloading system for workpieces to realize machining automation. On the other

Fig. 48.8 Machining center, equipped with an ATC (courtesy of Makino Milling Machine Co. Ltd.)

Fig. 48.9 Horizontal machining center (courtesy of Yamazaki Mazak Corp.)

Fig. 48.10 Vertical machining center (courtesy of Yamazaki Mazak Corp.)

hand, a vertical machining center is widely used in a die and mold machine shop. In a vertical machining center, the cutting tool can machine only the top surface of boxed workpieces, but it is easy for human operators to understand tool motion relative to the machined part.

Automatic Tool Changer (ATC)

ATC stands for automatic tool changer, which permits loading and unloading of cutting tools from one machining operation to the next. The ATC is designed to exchange cutting tools between the spindle and a tool magazine, which can store more than 20 tools. The

Fig. 48.11 ATC: automatic tool changer (courtesy of Yamazaki Mazak Corp.)

large capacity of the tool magazine allows a variety of workpieces to be machined. Additionally, higher tool-change speed and reliability are required to achieve a fast machining cycle. Figure 48.11 shows an example of a twin-arm-type ATC driven by a cam mechanism to ensure reliable high-speed tool change.

Automatic Pallet Changer (APC)

APC stands for automatic pallet changer, which permits loading and unloading of workpieces for machining automation. Most horizontal machining centers have two pallet tables to exchange the parts before and after ma-

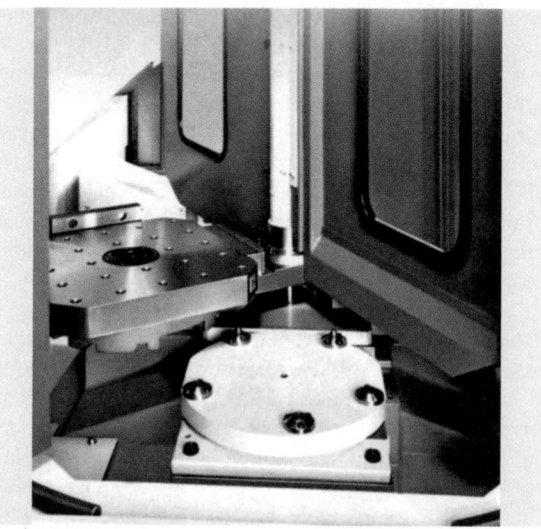

Fig. 48.12 APC: automatic pallet changer (courtesy of Yamazaki Mazak Corp.)

Fig. 48.13 Turning center or CNC lathe (courtesy of Yamazaki Mazak Corp.) ▶

chining automatically. Figure 48.12 shows an example of an APC. The operator can be unloading the finished part and loading the next part on one pallet while the machining center is processing the current part on another pallet.

48.2.2 Turning Center

A turning center is a highly automated NC lathe to perform multiple turning operations. Figure 48.13 shows a typical turning center. Changing of cutting tools is performed by a turret tool changer which can hold about ten turning and milling tools. Therefore, a turning center enables not only turning operations but also milling operations such as end milling, drilling, and tapping in a single machine setup. Some turning centers have two spindles and two or more turret tool changers to complete all machining operations of cylindrical parts in a single machine setup. In this case, the first half of the machining operations of the workpiece are carried out on one spindle, then the second half of the machining operations are carried out on another spindle, without unloading and loading of the workpiece. This reduces production time.

Turret Tool Changer
Figure 48.14 shows a tool turret with 12 cutting tools. A suitable cutting tool for the target machining operation is indexed automatically under numerical control for continuous machining operations. The most sophisticated turning centers have tool monitoring systems which check tool length and diameter for automatic tool alignment and sense tool wear for automatic tool changing.

48.2.3 Fully Automated Machining: FMS and FMC

Flexible Manufacturing System (FMS)
The concept of the flexible manufacturing system (FMS) was proposed during the mid-1960s. It aims to perform automatic machining operations unaided by human operators to machine various parts. Machining centers are key components of the FMS for flexible machining operations. Figure 48.15 shows a typical FMS, which consists of five machining centers, one conveyor, one load/unload station, and a central computer that controls and manages the components of the FMS.

Fig. 48.14 Tool turret in turning center (courtesy of Yamazaki Mazak Corp.)

Fig. 48.15 Flexible manufacturing system (courtesy of Yamazaki Mazak Corp.)

No manufacturing system can be completely flexible. FMSs are typically used for mid-volume and mid-variety production. An FMS is designed to machine parts within a range of style, sizes, and processes, and its degree of flexibility is limited. Additionally, the machining shape is changeable through the part programs that control the NC machine tools, and the part programs required for every shape to be machined have to be prepared before the machining operation. Therefore a new shape that needs a part program is not acceptable in conventional FMSs, which is why a third innovation of machine tools is required to achieve autonomous machining operations instead of automatic machining operations to achieve true FMS.

An FMS consists of several NC machine tools such as machining centers and turning centers, material-handling or loading/unloading systems such as industrial robots and pallets changer, conveyer systems such as conveyors and automated guided vehicles (AGV), and storage systems. Additionally, an FMS has a central computer to coordinate all of the activities of the FMS, and all hardware components of the FMS generally have their own microcomputer for control. The central computer downloads NC part programs, and controls the material-handling system, conveyer system, storage system, and management of materials and cutting tools, etc.

Human operators play important roles in FMSs, performing the following tasks:

1. Loading/unloading parts at loading/unloading stations
2. Changing and setting of cutting tools
3. NC part programming
4. Maintenance of hardware components
5. Operation of the computer system.

These tasks are indispensable to manage the FMS successfully.

Flexible Manufacturing Cell (FMC)

Basically, FMSs are large systems to realize manufacturing automation for mid-volume and mid-variety production. In some cases, small systems are applicable to realize manufacturing automation. The term flexible manufacturing cell (FMC) is used to represent small systems or compact cells of FMSs. Usually, the number of machine tools included in a FMC is three or fewer. One can consider that an FMS is a large manufacturing system composed of several FMCs.

48.3 NC Part Programming

The task of programming to operate machine tools automatically is called NC part programming because the program is prepared for a part to be machined. NC part programming requires the programmer to be familiar with both the cutting processes and programming procedures. The NC part program includes the detailed commands to control the positions and motion of the machine tool. In numerical control, the three linear axes (x, y, z) of the Cartesian coordinate system are used

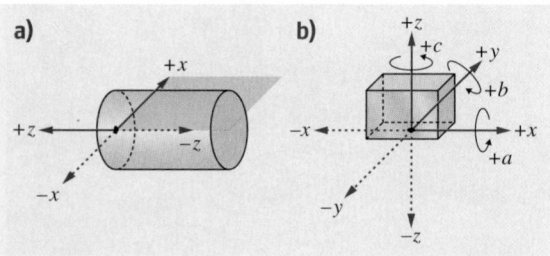

Fig. 48.16a,b Coordinate systems in numerical control. **(a)** Cylindrical part for turning; **(b)** cuboid part for milling

to specify cutting tool positions, and three rotational axes (a, b, c) are used to specify the cutting tool postures. In turning operations, the position of the cutting tool is defined in the x–z plane for cylindrical parts, as shown in Fig. 48.16a. In milling operations, the position of the cutting tool is defined by the x-, y-, and z-axes for cuboid parts, as shown in Fig. 48.16b.

Numerical control realizes programmable automation of machining. The mechanical actions or motions of the cutting tool relative to the workpiece and the control sequence of the machine tool equipments are coded by alphanumerical data in a program. NC part programming requires a programmer who is familiar with the metal cutting process to define the points, lines, and surfaces of the workpiece, and to generate the alphanumerical data. The most important NC part programming techniques are summarized as follows:

1. Manual part programming
2. Computer-assisted part programming – APT and EXAPT
3. CAM-assisted part programming.

48.3.1 Manual Part Programming

This is the simplest way to generate a part program. Basic numeric data and alphanumeric codes are entered manually into the NC controller. The simplest commands example is shown as follows:

N0010	M03	S1000	F100	EOB
N0020	G00	X20.000	Y50.000	EOB
N0030	Z20.000	EOB		
N0040	G01	Z − 20.000	EOB	

Each code in the statement has a meaning to define a machining operation. The "N" code shows the sequence number of the statement. The "M" code and the following two-digit number define miscellaneous functions; "M03" means to spindle on with clockwise rotation. The "S" code defines the spindle speed; "S1000" means that the spindle speed is 1000 rpm. The "F" code defines the feed speed; "F100" means that the feed is 100 mm/min. "EOB" stands for "end of block" and shows the end of the statement. The "G" code and the following two-digit number define preparatory functions; "G00" means rapid positioning by point-to-point control. The "X" and "Y" codes indicate the x- and y-coordinates. The cutting tool moves rapidly to the position $x = 20$ mm and $y = 50$ mm with the second statement. Then, the cutting tool moves rapidly again to the position $z = 20$ mm with the third statement. "G01" means linear positioning at controlled feed speed. Then the cutting tool moves with the feed speed, defined by "F100" in this example, to position $z = -20$ mm.

The positioning control can be classified into two types, (1) point-to-point control and (2) continuous path control. "G00" is a positioning command for point-to-point control. This command only identifies the next position required at which a subsequent machining operation such as drilling is performed. The path to get to the position is not considered in point-to-point control. On the other hand, the path to get to the position is controlled simultaneously in more than one axis to follow a line or circle in continuous path control. "G01" is a positioning command for linear interpolation. "G02" and "G03" are positioning commands for circular interpolation. These commands permit the generation of two-dimensional curves or three-dimensional surfaces by turning or milling.

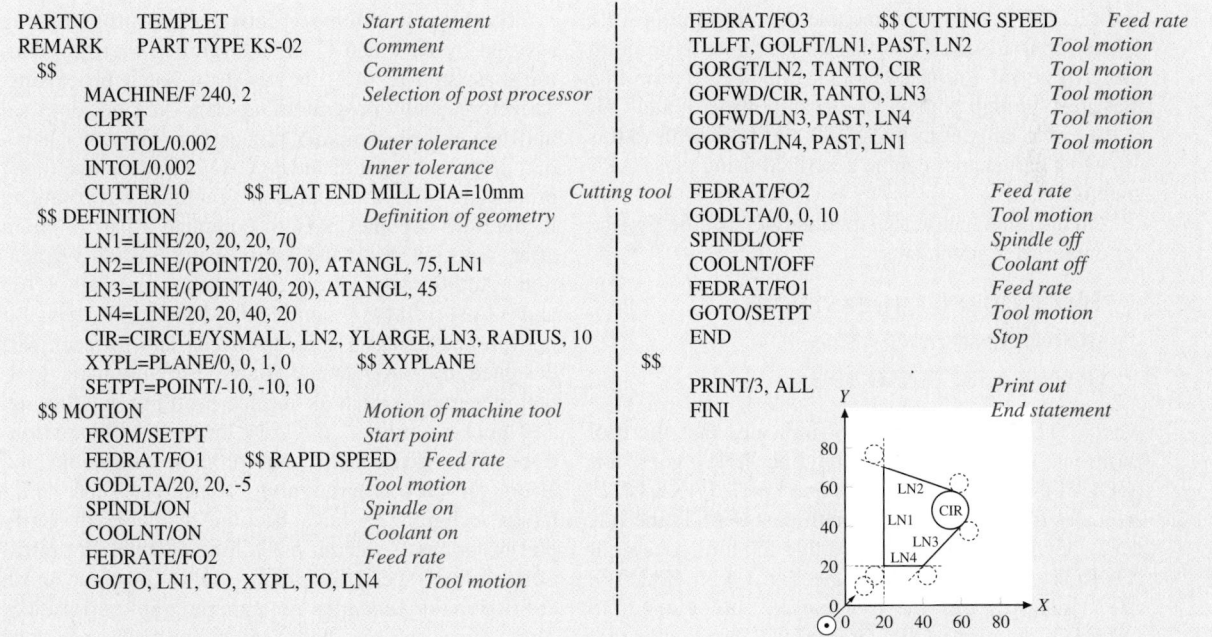

PARTNO	TEMPLET	*Start statement*
REMARK	PART TYPE KS-02	*Comment*
$$		*Comment*
	MACHINE/F 240, 2	*Selection of post processor*
	CLPRT	
	OUTTOL/0.002	*Outer tolerance*
	INTOL/0.002	*Inner tolerance*
	CUTTER/10 $$ FLAT END MILL DIA=10mm	*Cutting tool*
$$ DEFINITION		*Definition of geometry*
	LN1=LINE/20, 20, 20, 70	
	LN2=LINE/(POINT/20, 70), ATANGL, 75, LN1	
	LN3=LINE/(POINT/40, 20), ATANGL, 45	
	LN4=LINE/20, 20, 40, 20	
	CIR=CIRCLE/YSMALL, LN2, YLARGE, LN3, RADIUS, 10	
	XYPL=PLANE/0, 0, 1, 0 $$ XYPLANE	
	SETPT=POINT/-10, -10, 10	
$$ MOTION		*Motion of machine tool*
	FROM/SETPT	*Start point*
	FEDRAT/FO1 $$ RAPID SPEED	*Feed rate*
	GODLTA/20, 20, -5	*Tool motion*
	SPINDL/ON	*Spindle on*
	COOLNT/ON	*Coolant on*
	FEDRATE/FO2	*Feed rate*
	GO/TO, LN1, TO, XYPL, TO, LN4	*Tool motion*

FEDRAT/FO3 $$ CUTTING SPEED		*Feed rate*
TLLFT, GOLFT/LN1, PAST, LN2		*Tool motion*
GORGT/LN2, TANTO, CIR		*Tool motion*
GOFWD/CIR, TANTO, LN3		*Tool motion*
GOFWD/LN3, PAST, LN4		*Tool motion*
GORGT/LN4, PAST, LN1		*Tool motion*
FEDRAT/FO2		*Feed rate*
GODLTA/0, 0, 10		*Tool motion*
SPINDL/OFF		*Spindle off*
COOLNT/OFF		*Coolant off*
FEDRAT/FO1		*Feed rate*
GOTO/SETPT		*Tool motion*
END		*Stop*
PRINT/3, ALL		*Print out*
FINI		*End statement*

Fig. 48.17 Example program list in APT

48.3.2 Computer-Assisted Part Programming: APT and EXAPT

Automatically programmed tools is the most important computer-assisted part programming language and was first used to generate part programs in production around 1960. EXAPT contains additional functions such as setting of cutting conditions, selection of cutting tool, and operation planning besides the functions of APT. APT provides two steps to generate part programs: (1) definition of part geometry, and (2) specification of tool motion and operation sequence. An example program list is shown in Fig. 48.17. The following APT statements define the contour of the part geometry based on basic geometric elements such as points, lines, and circles:

LN1 =LINE/20, 20, 20, 70

LN2 =LINE/(POINT/20, 70), ATANGL, 75, LN1

LN3 =LINE/(POINT/40, 20), ATANGL, 45

LN4 =LINE/20, 20, 40, 20

CIR =CIRCLE/YSMALL, LN2,

　　　YLARGE, LN3, RADIUS, 10

where LN1 is the line that goes through points (20, 20) and (20, 70); LN2 is the line that goes from point (20, 70) at 75° to LN1; LN3 is the line that goes from point (40, 20) at 45° to the horizontal line; LN4 is the line that goes through points (20, 20) and (40, 20); and CIR is the circle tangent to lines LN2 and LN3 with radius 10. Most part shapes can be described using these APT statements.

On the other hands, tool motions are specified by the following APT statements:

TLLFT, GOLFT/LN1, PAST, LN2

GORGT/LN2, TANTO, CIR

GOFWD/CIR, TANTO, LN3

where "TLLFT, GOLFT/LN1" indicates that the tool positions left (TLLFT) of the line LN1, goes left (GOLFT), and moves along the line LN1. "PAST, LN2" indicates that the tool moves until past (PAST) the line LN2. "GORGT/LN2" indicates that the tool goes right (GORGT) and moves along the line LN2. "TANTO, CIR" indicates that the tool moves until tangent to (TANTO) the circle CIR. GOFWD/CIR indicates that the tool goes forward (GOFWD) and moves along the circle CIR. "TANTO, LN3" indicates that the tool moves until tangent to the line LN3.

Additional APT statements are prepared to define feed speed, spindle speed, tool size, and tolerances of tool paths. The APT program completed by the part programmer is translated by the computer to the cutter location (CL) data, which consists of all the geometry and cutter location information required to machine the part. This process is called main processing or preprocessing to generate NC commands. The CL data is converted to the part program, which is understood by the NC machine tool controller. This process is called postprocessing to add NC commands to specify feed speed, spindle speed, and auxiliary functions for the machining operation.

48.3.3 CAM-Assisted Part Programming

CAM systems grew based on technologies relating to APT and EXAPT. Originally, CAM stood for computer-aided manufacturing and was used as a general term for computer software to assist all operations while realizing manufacturing. However, CAM is now used to indicate computer software to assist part programming in a narrow sense.

The biggest difference between part programming assisted by APT and CAM is usability. Part programming assisted by APT is based on batch processing. Therefore, many programming errors are not detected until the end of computer processing. The other hand, part programming assisted by CAM is interactive-mode processing with a visual and graphical environment. It therefore becomes easy to complete a part program after repeated trial and error using visual verification. Additionally, close cooperation between CAD and CAM offers a significant benefit in terms of part programming. The geometrical data for each part designed by CAD are available for automatic tool-path generation, such as surface profiling, contouring, and pocket milling, in CAM through software routines. This saves significant programming time and effort for part programming. Recently, some simulation technologies have become available to verify part programs free from machining trouble. Optimization of feed speed and detection of machine crash are two major functions for part program verification. These functions also save significant production lead time.

48.4 Technical Innovation in NC Machine Tools

48.4.1 Functional and Structural Innovation by Multitasking and Multiaxis

Turning and Milling Integrated Machine Tool
Recently, a turning and milling integrated machine tool has been developed as a sophisticated turning center. It also has a rotating cutting tool which can perform milling operation besides turning operation, as shown in Fig. 48.18. The benefits of the use of turning and milling integrated machine tools are

1. Reduction of production time
2. Improved machining accuracy
3. Reduction of floor space and initial cost.

As the high performance of these machine tools was accepted, the configuration became more and more complicated. Multispindles and multiturrets are integrated to perform multitasks simultaneously. The machine tool shown in Fig. 48.18 has two spindles: one milling spindle with four axes and one turret with two axes. Increasing the complexity of these machine tools causes the risk of machine crashes during machining operation, and requires careful part programming to avoid machine crashes.

Five-Axis Machining Center
Multiaxis machining centers are expanding in practical applications rapidly. The multiaxis machining center is applied to generate a workpiece with complex geometry with a single machine setup. In particular, five-axis machining centers have become popular for machining aircraft parts and complicated surfaces such as dies and molds. A typical five-axis machining center is shown in Fig. 48.19. Benefits to the use of multiaxis machining centers are

1. Reduction of preparation time
2. Reduction of production time
3. Improved machining accuracy.

Parallel Kinematic Machine Tool
A parallel kinematic machine tool is classified as a multiaxis machine tool. In the past years, parallel kinematic machine tools (PKM) have been studied with interest for their advantages of high stiffness, low inertia, high accuracy, and high-speed capability. Okuma Corporation in Japan developed the parallel mechanism machine tool COSMO CENTER PM-600 shown in

Fig. 48.20. This machine tool achieves high-speed and high-degrees-of-freedom machining operation for practical products. Also, high-speed milling of a free surface is shown in Fig. 48.20.

Ultraprecision Machine Tool
Recently, ultraprecision machining technology has experienced major advances in machine design, performance, and productivity. Ultraprecision machining was successfully adopted for the manufacture of computer memory discs used in hard-disk drives (HDD), and also

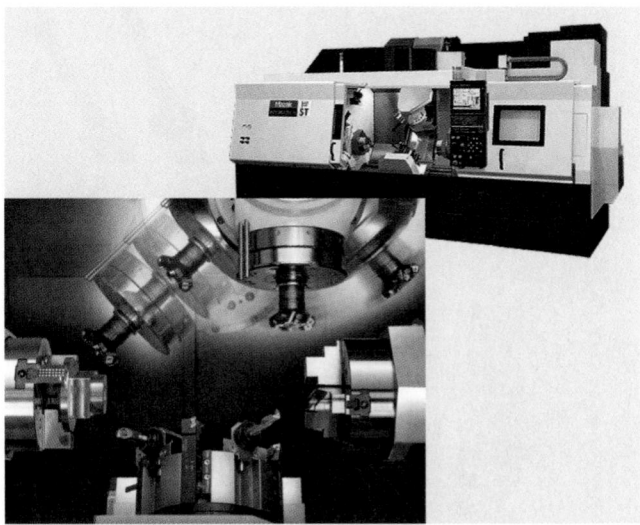

Fig. 48.18 Milling and turning integrated machine tool (courtesy of Yamazaki Mazak Corp.)

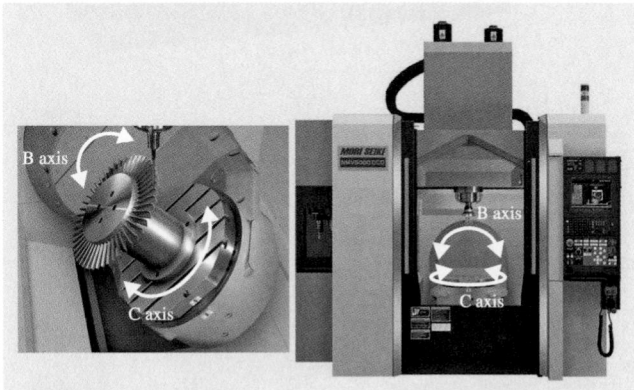

Fig. 48.19 Five-axis machining center (courtesy of Mori Seiki Co. Ltd.)

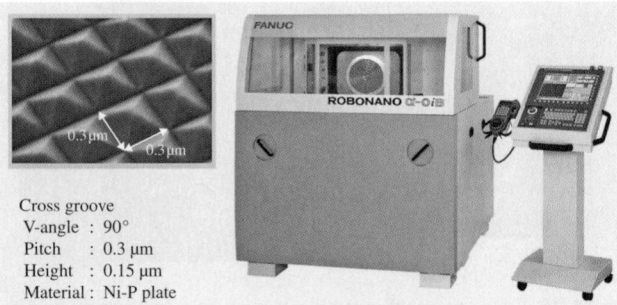

Cross groove
V-angle : 90°
Pitch   : 0.3 μm
Height   : 0.15 μm
Material : Ni-P plate

Fig. 48.21 Ultraprecision machine tool (courtesy of Fanuc Ltd.)

Fig. 48.20 Parallel kinematic machining center (courtesy of OKUMA Corp.) ◄

photoreflector components used in photocopiers and printers. These applications require extremely high geometrical accuracies and form deviations in combination with supersmooth surfaces.

The FANUC ROBONANO α-0iB is shown in Fig. 48.21 as an example of a five-axis ultraprecision machine tool. Nanometer servo-control technologies and air-bearing technologies are combined to realize an ultraprecision machine tool. This machine provides various machining methods for mass production with nanometer precision in the fields of optical electronics, semiconductor, medical, and biotechnology.

48.4.2 Innovation in Control Systems Toward Intelligent CNC Machine Tools

The framework of future intelligent CNC machine tools is summarized in Fig. 48.22. A conventional CNC control system has two major levels: the servo control (level 1 in Fig. 48.22) and the interpolator (level 2) for the axial motion control of machine tools. Certainly, the conventional CNC control system can achieve highly sophisticated motion control, but it cannot achieve sophisticated cutting process control. Two additional levels of control hierarchy, levels 3 and 4 in Fig. 48.22, are required for a future intelligent CNC control system to achieve more sophisticated process control.

Machining operations by conventional CNC machine tools are generally dominated by NC programs, and only feed speed can be adapted. For sophisticated cutting process control, dynamic adaptation of cutting parameters is indispensable. The adaptive control (AC) scheme is assigned at a higher level (level 3) of the control hierarchy, enabling intelligent process monitoring, which can detect machining state independently of cutting conditions and machining operation.

Level 4 in Fig. 48.22 is usually regarded as a supervisory level that receives feedback from measurements of the finished part. A reasonable index to evaluate the cutting results and a reasonable strategy to improve cutting results are required at this level. For this purpose, the utilization of knowledge, knowhow, and skill related to machining operations has to be considered. Effective utilization of feedback information regarding the cutting results is very important.

Additionally, an autonomous process planning strategy, which can generate a flexible and adaptive working

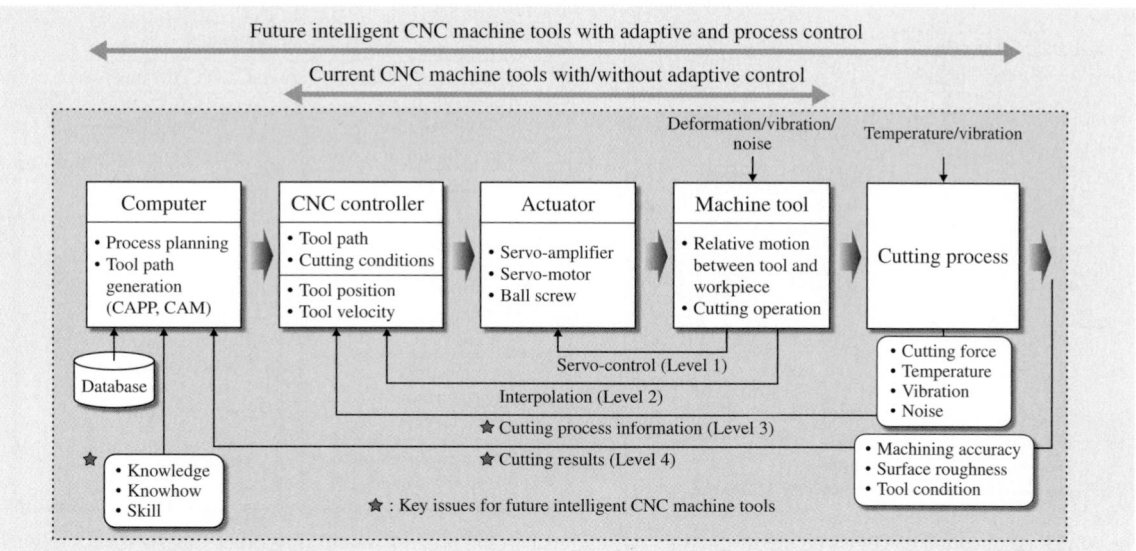

Fig. 48.22 Framework of intelligent machine tools (CAPP – computer aided process planning)

plan, is required as a function of intelligent CNC machine tools. It must be responsive and adaptive to unpredictable changes, such as job delay, job insertion, and machine breakdown on machining shop floors. In order to generate the operation plan autonomously, several planning and information processing functions are needed. Operation planning, cutting tool selection, cutting parameters assignment, and tool-path generation for each machining operation are required at the machine level. Product data analysis and machining feature recognition are important issues as part of information processing.

48.4.3 Current Technologies of Advanced CNC Machine Tools

Open Architecture Control

The concept of open architecture control (OAC) was proposed in the early 1990s. The main aim of OAC was easy implementation and integration of customer-specific controls by means of open interfaces and configuration methods in a vender-neutral standardized environment [48.2]. It provides the methods and utilities for integrating user-specific requirements, and it is required to implement several intelligent control applications for process monitoring and control.

Altintas has developed a user-friendly, reconfigurable, and modular toolkit called the open real-time operating system (ORTS). ORTS has several intelligent machining modules, as shown in Fig. 48.23. It can be used for the development of real-time signal processing, motion, and process control applications. A sample tool-path generation using quintic spline interpolation for high-speed machining is described as an application, and a sample cutting force control has also been demonstrated [48.3].

Mori and *Yamazaki* developed an open servo-control system for an intelligent CNC machine tool to minimize the engineering task required for implementing custom intelligent control functions. The conceptual design of this system is shown in Fig. 48.24. The software model reference adaptive control was implemented as a custom intelligent function, and a feasibility study was conducted to show the effectiveness of the open servo control [48.4]. Open architecture control will reach the level of maturity required to replace current CNC controllers in the near future. The custom intelligent control functions required for an intelligent machine tool will be easy to implement with the CNC controller. Machining performance in terms of higher accuracy and productivity will thereby be enhanced.

Feedback of Cutting Information

Yamazaki proposed TRUE-CNC as a future-oriented CNC controller. (TRUE-CNC was named after the following key words. T: transparent, transportable, transplantable, R: revivable, U: user-reconfigurable, and E: evolving.) The system consists of an information service, quality control and diagnosis, monitoring, control, analysis, and planning sections, as shown in

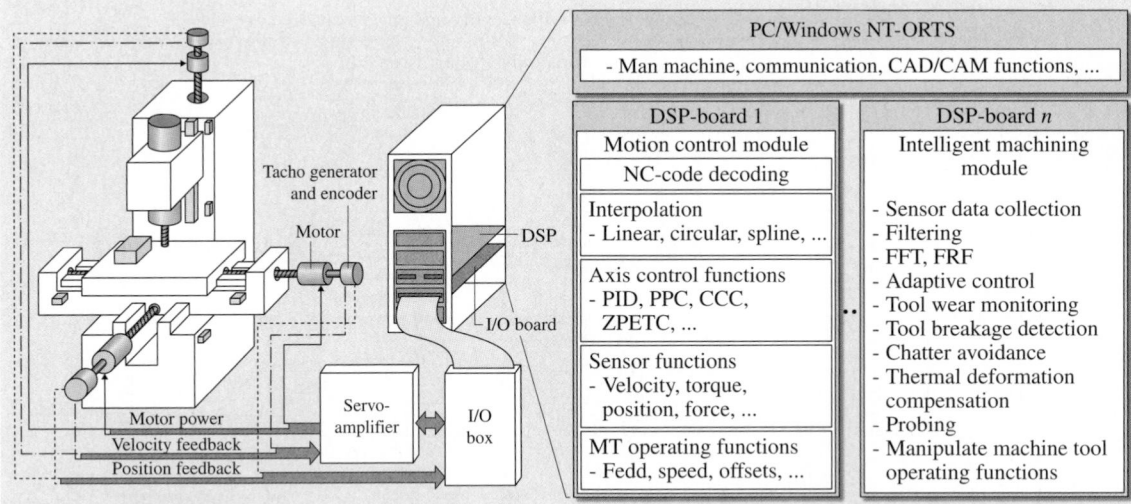

Fig. 48.23 Application of ORTS on the design of CNC and machining process monitoring (after [48.3]) (DSP – digital signal processor, FFT – fast Fourier transform, FRF – frequency response function, PID – proportional–integral– derivative controller, PPC – pole placement controller, CCC – cross coupling controller, ZPETC – zero phase error tracking controller, I/O – input/output, MT – machine tool)

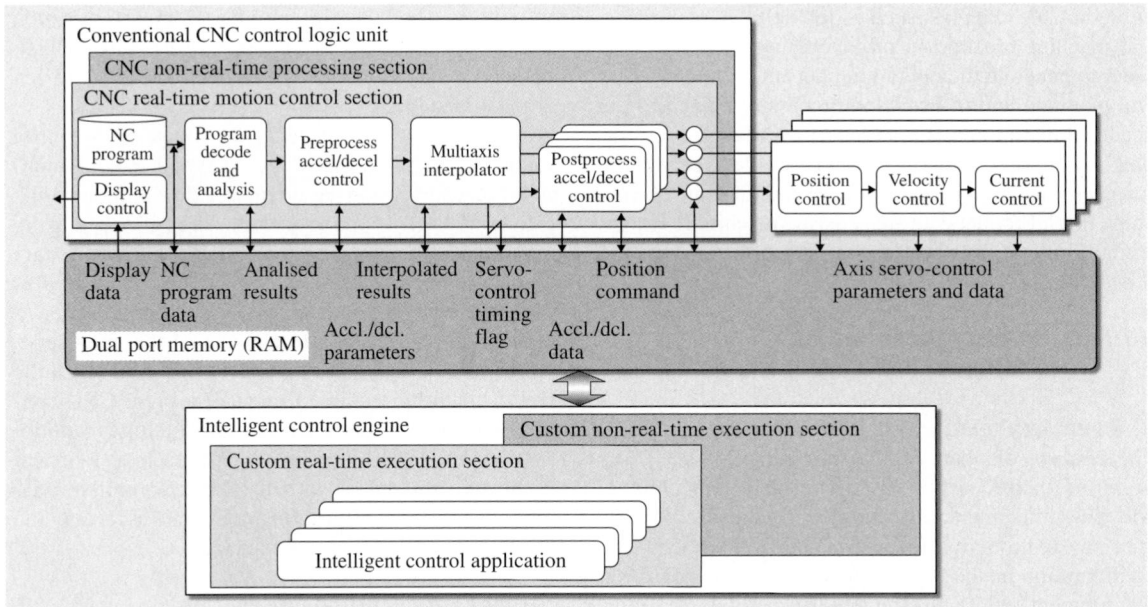

Fig. 48.24 Conceptual design of open servo system (after [48.4])

Fig. 48.25 [48.5]. TRUE-CNC allows the operator to achieve maximum productivity and highest quality for machined parts in a given environment with autonomous capture of machining operation proficiency and machining knowhow. The autonomous coordinate

measurement planning (ACMP) system is a component of TRUE-CNC, and enhances the operability of coordinate measuring machines (CMMs). The ACMP generates probe paths autonomously for inline measurement of machined parts [48.6]. Inspection results

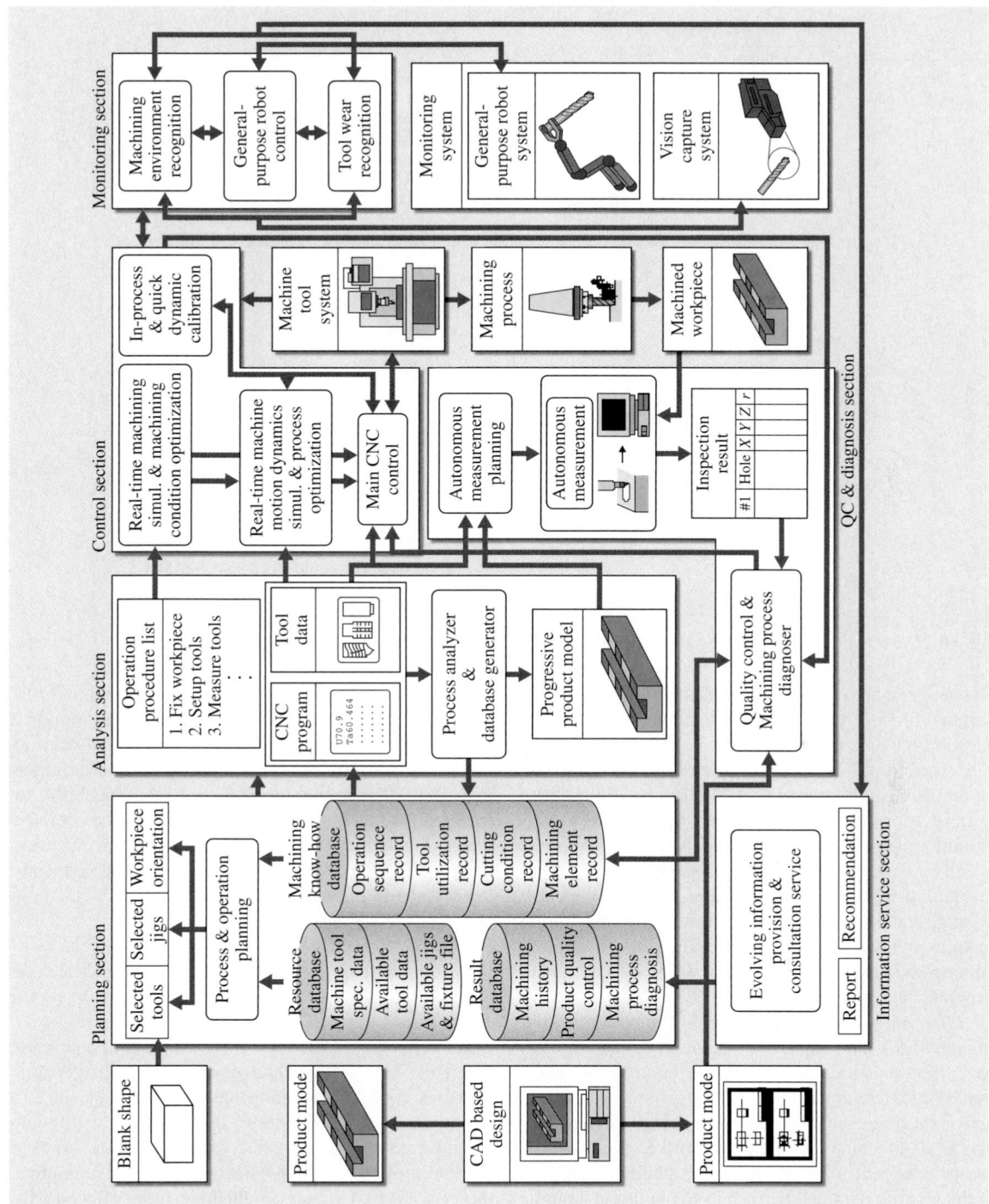

Fig. 48.25 Architecture of TRUE-CNC (after [48.5])

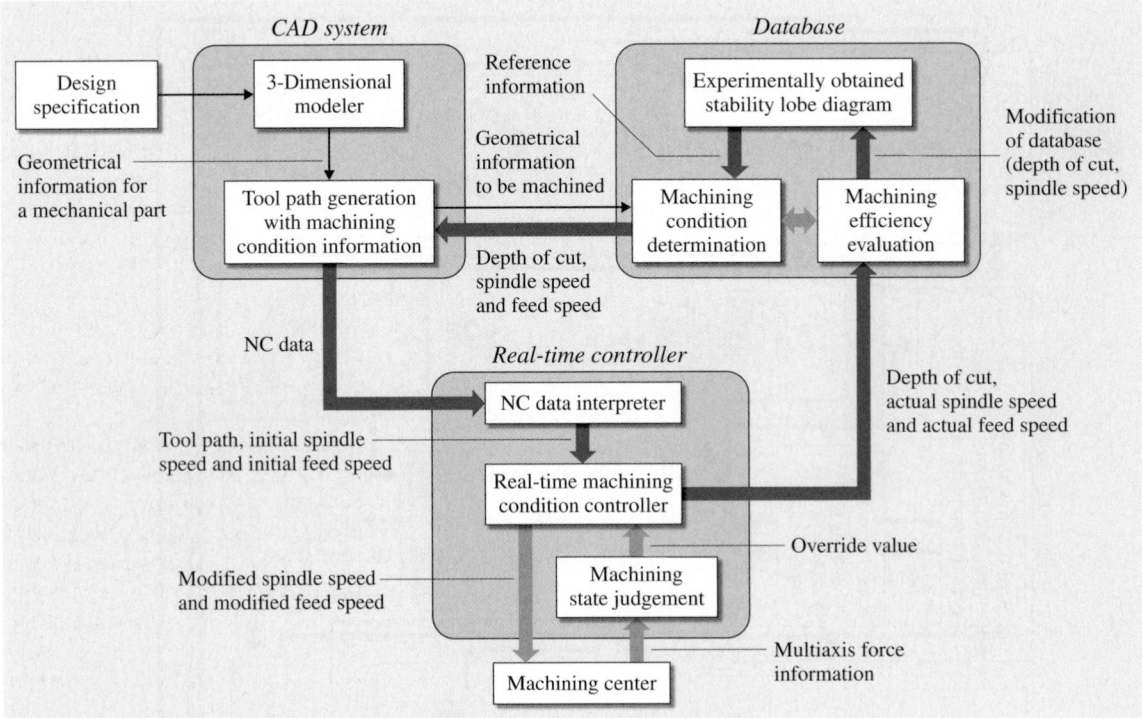

Fig. 48.26 Software system configuration of open architecture CNC (after [48.8])

or measurement data are utilized to evaluate the machining process to be finished and to assist in the decision-making process for new operation planning. The autonomous machining process analyzer (AMPA) system is also a component of TRUE-CNC. In order to retrieve knowledge, knowhow, and skill related to machining operations, the AMPA analyzes NC programs coded by experienced machining operators and gathers machining information. Machining process sequence, cutting conditions, machining time, and machining features are detected automatically and stored in the machining knowhow database [48.7], whihc is then used to generate new operation plans.

Mitsuishi developed a CAD/CAM mutual information feedback machining system which has capabilities for cutting state monitoring, adaptive control, and learning. The system consists of a CAD system, a database, and a real-time controller, as shown in Fig. 48.26 [48.8]. The CNC machine tool equipped with a six-axis force sensor was controlled to obtain the stability lobe diagram. Cutting parameters, such as depth of cut, spindle speed, and feed speed, are modified dynamically according to the sequence for finding stable cutting states, and the stability lobe diagram is obtained au-

tonomously. The stability lobe diagram is then used to determine chatter-free cutting conditions. Furthermore, *Mitsuishi* proposed a networked remote manufacturing system which provides remote operating and monitoring [48.9]. The system demonstrated the capability to transmit the machining state in real time to the operator who is located far from the machine tool. The operator can modify the cutting conditions in real time depending on the machining state monitored.

Five-Axis Control
Most commercial CAM systems are not sufficient to generate suitable cutter location (CL) data for five-axis control machining. The CL data must be adequately generated and verified to avoid tool collision with the workpiece, fixture, and machine tool. In general, five-axis control machining has the advantage of enabling arbitrary tool posture, but it makes it difficult to find a suitable tool posture for a machining strategy without tool collision. *Morishige* and *Takeuchi* applied the concept of C-space to generate tool-collision-free CL data for five-axis control [48.10, 11]. The two-dimensional C-space is used to represent the relation between the tool posture and the collision area, as

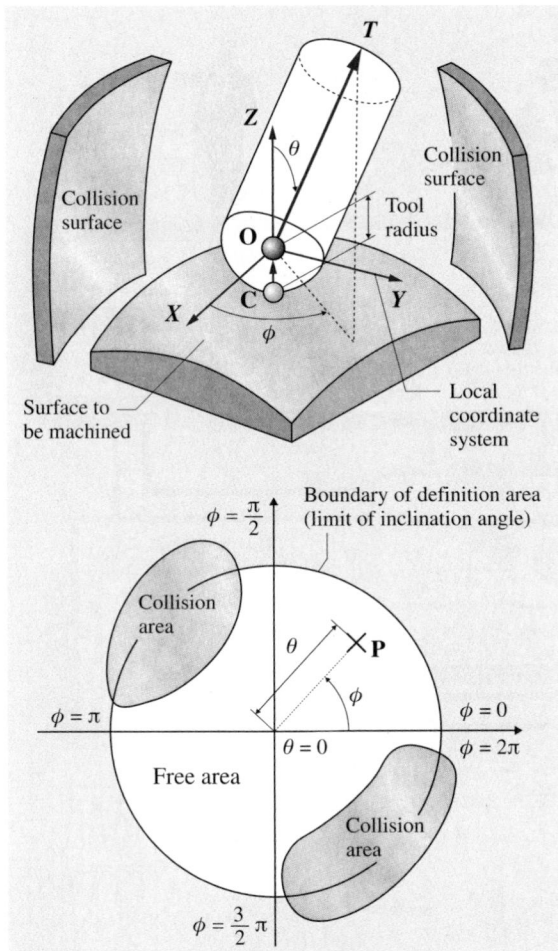

Fig. 48.27 Configuration space to define tool posture (after [48.11])

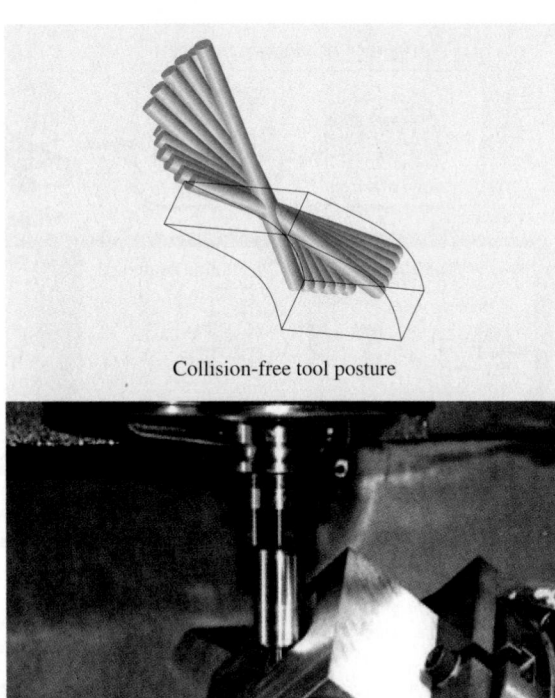

Collision-free tool posture

Fig. 48.28 Five-axis control machining (after [48.11])

shown in Fig. 48.27. Also, three-dimensional C-space is used to generate the most suitable CL data which satisfy the machining strategy, smooth tool movement, good surface roughness, and so on. Experimental five-axis-control collision-free machining was performed successfully, as shown in Fig. 48.28.

48.4.4 Autonomous and Intelligent Machine Tool

The whole machining operation of conventional CNC machine tools is predetermined by NC programs. Once the cutting conditions, such as depth of cut and stepover, are given by the machining commands in the NC programs, they are not generally allowed to be changed during machining operations. Therefore NC programs must be adequately prepared and verified in advance, which requires extensive amounts of time and effort. Moreover, NC programs with fixed commands are not responsive to unpredictable changes, such as job delay, job insertion, and machine breakdown found on machining shop floors.

Shirase proposed a new architecture to control the cutting process autonomously without NC programs. Figure 48.29 shows the conceptual structure of autonomous and intelligent machine tools (AIMac). AIMac consists of four functional modules called management, strategy, prediction, and observation. All functional modules are connected with each other to share cutting information.

Digital Copy Milling for Real-Time Tool-Path Generation

A technique called digital copy milling has been developed to control a CNC machine tool directly. The

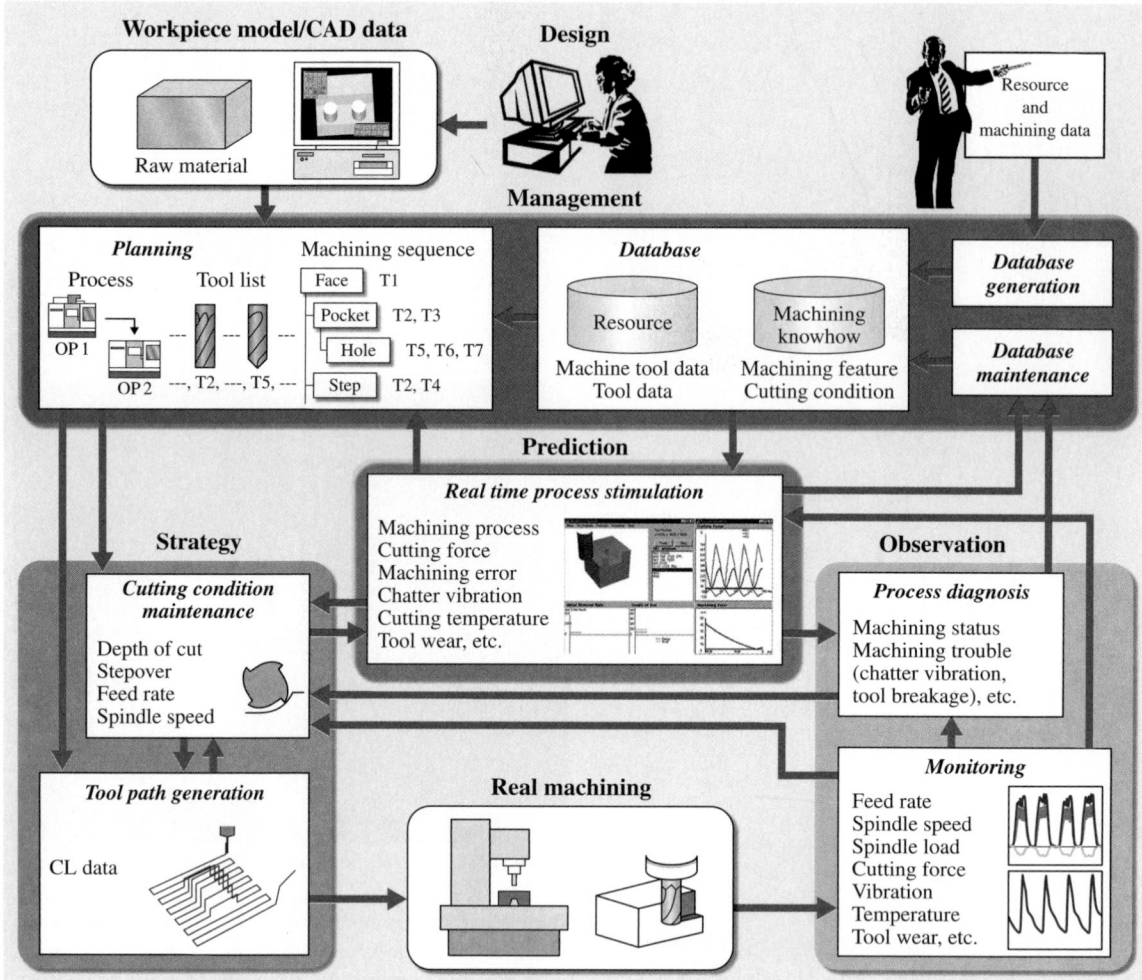

Fig. 48.29 Conceptual structure of AIMac

digital copy milling system can generate tool paths in real time based on the principle of traditional copy milling. In digital copy milling, a tracing probe and a master model in traditional copy milling are represented by three-dimensional (3-D) virtual models in a computer. A virtual tracing probe is simulated to follow a virtual master model, and cutter locations are generated dynamically according to the motion of the virtual tracing probe in real time. In the digital copy milling, cutter locations are generated autonomously, and an NC machine tool can be instructed to perform milling operation without NC programs. Additionally, not only stepover, but also radial and axial depths of

cut can be modified, as shown in Fig. 48.30. Also, digital copy milling can generate new tool paths to avoid cutting problems and change the machining sequence during operation [48.12].

Furthermore, the capability for in-process cutting parameters modification was demonstrated, as shown in Fig. 48.31 [48.13]. Real-time tool-path generation and the monitored actual milling are shown in the lower-left corner and the upper-right corner of this figure. The monitored cutting torque, adapted feed rate, and radial and axial depths of cut are shown in the lower-right corner of this figure. The cutting parameters can be modified dynamically to maintain the cutting load.

Fig. 48.30a–d Example of real-time tool-path generation. (**a**) Bilateral zigzag paths; (**b**) contouring paths; (**c**) change of stepover; (**d**) change of cutting depth ▶

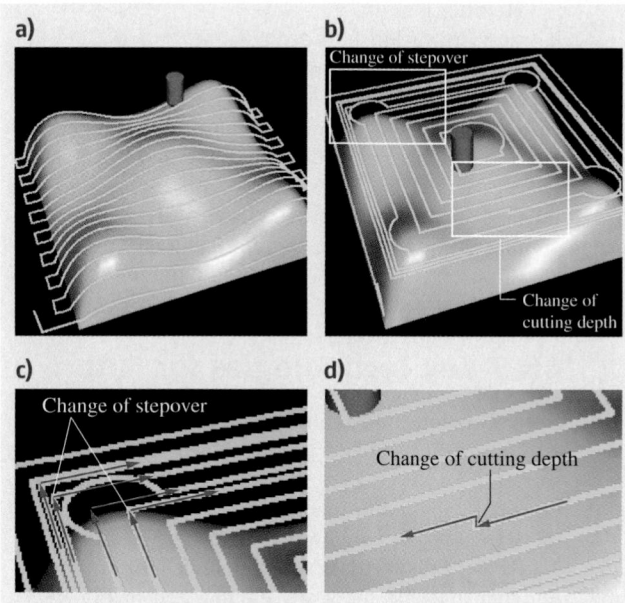

a)

b) Change of stepover

Change of cutting depth

c) Change of stepover

d) Change of cutting depth

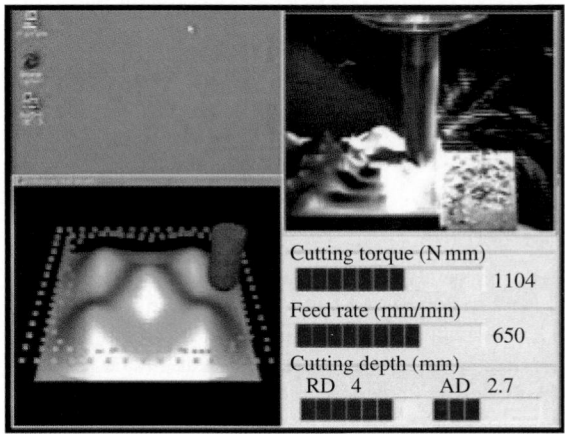

Cutting torque (N mm)

1104

Feed rate (mm/min)

650

Cutting depth (mm)
RD 4   AD 2.7

Fig. 48.31 Adaptive milling on AIMac

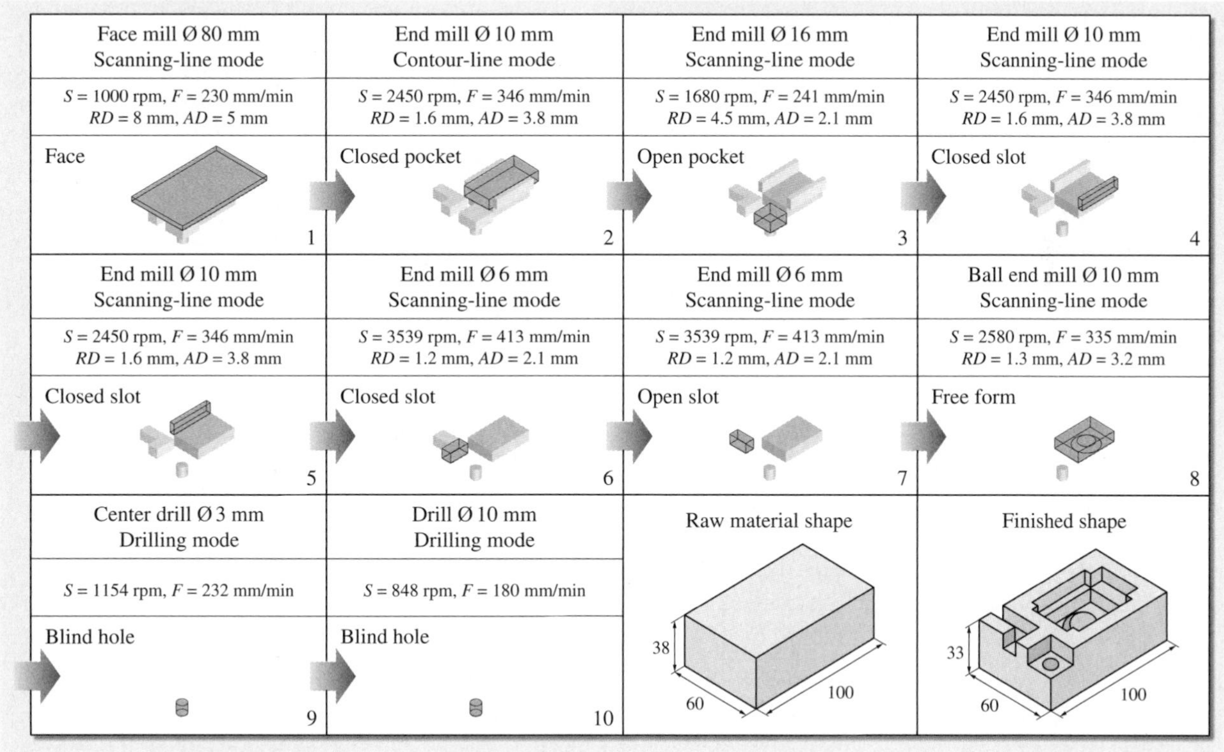

Face mill Ø 80 mm Scanning-line mode	End mill Ø 10 mm Contour-line mode	End mill Ø 16 mm Scanning-line mode	End mill Ø 10 mm Scanning-line mode
$S = 1000$ rpm, $F = 230$ mm/min $RD = 8$ mm, $AD = 5$ mm	$S = 2450$ rpm, $F = 346$ mm/min $RD = 1.6$ mm, $AD = 3.8$ mm	$S = 1680$ rpm, $F = 241$ mm/min $RD = 4.5$ mm, $AD = 2.1$ mm	$S = 2450$ rpm, $F = 346$ mm/min $RD = 1.6$ mm, $AD = 3.8$ mm
Face 1	Closed pocket 2	Open pocket 3	Closed slot 4
End mill Ø 10 mm Scanning-line mode	End mill Ø 6 mm Scanning-line mode	End mill Ø 6 mm Scanning-line mode	Ball end mill Ø 10 mm Scanning-line mode
$S = 2450$ rpm, $F = 346$ mm/min $RD = 1.6$ mm, $AD = 3.8$ mm	$S = 3539$ rpm, $F = 413$ mm/min $RD = 1.2$ mm, $AD = 2.1$ mm	$S = 3539$ rpm, $F = 413$ mm/min $RD = 1.2$ mm, $AD = 2.1$ mm	$S = 2580$ rpm, $F = 335$ mm/min $RD = 1.3$ mm, $AD = 3.2$ mm
Closed slot 5	Closed slot 6	Open slot 7	Free form 8
Center drill Ø 3 mm Drilling mode	Drill Ø 10 mm Drilling mode	Raw material shape	Finished shape
$S = 1154$ rpm, $F = 232$ mm/min	$S = 848$ rpm, $F = 180$ mm/min		
Blind hole 9	Blind hole 10	38 60 100	33 60 100

Fig. 48.32 Results of machining process planning on AIMac

Flexible Process and Operation Planning System

A flexible process and operation planning system has been developed to generate cutting parameters dynamically for machining operation. The system can generate the production plan from the total removal volume (TRV). The TRV is extracted from the initial and finished shapes of the product and is divided into machining primitives or machining features. The flexible process and operation planning system can generate cutting parameters according to the machining features detected. Figure 48.32 shows the operation sequence and cutting tools to be used. Cutting parameters are determined for the experimental machining shape. The digital copy milling system can generate the tool paths or CL data dynamically according to these results and perform the autonomous milling operation without requiring any NC program.

48.5 Key Technologies for Future Intelligent Machine Tool

Several architectures and technologies have been proposed and investigated as mentioned in the previous sections. However, they are not yet mature enough to be widely applied in practice, and the achievements of these technologies are limited to specific cases.

Achievements of key technologies for future intelligent machine tools are summarized in Fig. 48.33. Process and machining quality control will become more important than adaptive control. Dynamic tool-path generation and in-process cutting parameters modification are required to realize flexible machining operation for process and machining quality control. Additionally, intelligent process monitoring is needed to evaluate the cutting process and machining quality

for process and machining quality control. A reasonable strategy to control the cutting process and a reasonable index to evaluate machining quality are required. It is therefore necessary to consider utilization and learning of knowledge, knowhow, and skill regarding machining operations.

A process planning strategy with which one can generate flexible and adaptive working plans is required. An operation planning strategy is also required to determine the cutting tool and parameters. Product data analysis and machining feature recognition are important issues in order to generate operation plans autonomously.

Sections 48.4.2–48.5 are quoted from [48.14].

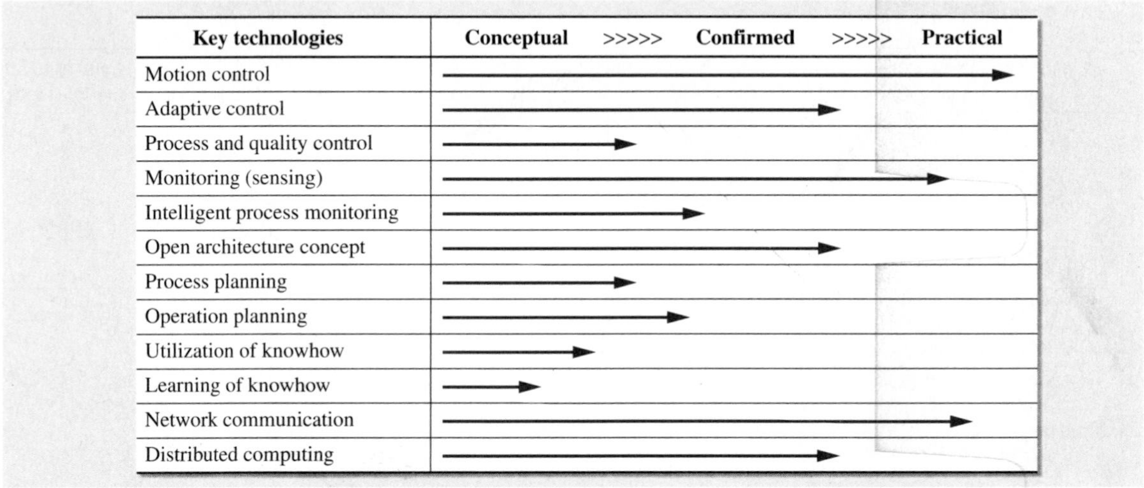

Key technologies	Conceptual >>>>> Confirmed >>>>> Practical
Motion control	
Adaptive control	
Process and quality control	
Monitoring (sensing)	
Intelligent process monitoring	
Open architecture concept	
Process planning	
Operation planning	
Utilization of knowhow	
Learning of knowhow	
Network communication	
Distributed computing	

Fig. 48.33 Achievements of key technologies for future intelligent machine tools

48.6 Further Reading

- Y. Altintas: *Manufacturing Automation: Metal Cutting Mechanics, Machine Tool Vibrations, and CNC Design* (Cambridge Univ. Press, Cambridge 2000)
- J.G. Bollinger, N.A. Duffie: *Computer Control of Machines and Processes* (Addison-Wesley, Boston 1988)
- E.P. DeGarmo, J.T. Black, R.A. Kosher: *DeGarmo's Materials and Processes in Manufacturing* (Wiley, New York 2007)
- K. Evans: *Programming of CNC Machines* (Industrial, New York 2007)
- Y. Ito: *Modular Design for Machine Tools* (McGraw-Hill, New York 2008)
- K.-H. John, M. Tiegelkamp: *IEC 61131-3: Programming Industrial Automation Systems* (Springer, Berlin Heidelberg 2001)

- S. Krar, A. Gill, P. Smid, P. Wanner: *Machine Tool Technology Basics* (Industrial, New York 2003)
- I.D. Marinescu, C. Ispas, D. Boboc: *Handbook of Machine Tool Analysis* (CRC, Boca Raton 2002)
- B.W. Niebel, A.B. Draper, R.A. Wysk: *Modern Manufacturing Process Engineering*, (McGraw-Hill, London 1989)
- G.E. Thyer: *Computer Numerical Control of Machine Tools*, (Butterworth-Heinemann, London 1991)
- K.-H. Wionzek: *Numerically Controlled Machine Tools as a Special Case of Automation* (Didaktischer Dienst, Berlin 1982)

References

48.1   Y. Koren: Control of machine tools, ASME J. Manuf. Sci. Eng. **119**, 749–755 (1997)

48.2   G. Pritschow, Y. Altintas, F. Jovane, Y. Koren, M. Mitsuishi, S. Takata, H. Brussel, M. Weck, K. Yamazaki: Open controller architecture – past, present and future, Ann. CIRP **50**(2), 463–470 (2001)

48.3   Y. Altintas, N.A. Erol: Open architecture modular tool kit for motion and machining process control, Ann. CIRP **47**(1), 295–300 (1998)

48.4   M. Mori, K. Yamazaki, M. Fujishima, J. Liu, N. Furukawa: A study on development of an open servo system for intelligent control of a CNC machine tool, Ann. CIRP **50**(1), 247–250 (2001)

48.5   K. Yamazaki, Y. Hanaki, Y. Mori, K. Tezuka: Autonomously proficient CNC controller for high performance machine tool based on an open architecture concept, Ann. CIRP **46**(1), 275–278 (1997)

48.6   H. Ng, J. Liu, K. Yamazaki, K. Nakanishi, K. Tezuka, S. Lee: Autonomous coordinate measurement planning with work-in-process measurement for TRUE-CNC, Ann. CIRP **47**(1), 455–458 (1998)

48.7   X. Yan, K. Yamazaki, J. Liu: Extraction of milling know-how from NC programs through reverse engineering, Int. J. Prod. Res. **38**(11), 2443–2457 (2000)

48.8   M. Mitsuishi, T. Nagao, H. Okabe, M. Hashiguchi, K. Tanaka: An open architecture CNC CAD-CAM ma-

chining system with data-base sharing and mutual information feedback, Ann. CIRP **46**(1), 269–274 (1997)

48.9   M. Mitsuishi, T. Nagao: Networked manufacturing with reality sensation for technology transfer, Ann. CIRP **48**(1), 409–412 (1999)

48.10   K. Morishige, Y. Takeuchi, K. Kase: Tool path generation using C-space for 5-axis control machining, ASME J. Manuf. Sci. Eng. **121**, 144–149 (1999)

48.11   K. Morishige, Y. Takeuchi: Strategic tool attitude determination for five-axis control machining based on configuration space, CIRP J. Manuf. Syst. **31**(3), 247–252 (2003)

48.12   K. Shirase, T. Kondo, M. Okamoto, H. Wakamatsu, E. Arai: Trial of NC programless milling for a basic autonomous CNC machine tool, Proc. 2000 JAPAN-USA Symp. Flex. Autom. (JUSFA2000) (2000) pp. 507–513

48.13   K. Shirase, K. Nakamoto, E. Arai, T. Moriwaki: Digital copy milling – autonomous milling process control without an NC program, Robot. Comput. Integr. Manuf. **21**(4-5), 312–317 (2005)

48.14   T. Moriwaki, K. Shirase: Intelligent machine tools: current status and evolutional architecture, Int. J. Manuf. Technol. Manag. **9**(3/4), 204–218 (2006)

49. Digital Manufacturing and RFID-Based Automation

Wing B. Lee, Benny C.F. Cheung, Siu K. Kwok

Advances in the Internet, communication technologies, and computation power have accelerated the cycle of new product development as well as supply chain efficiency in an unprecedented manner. Digital technology provides not only an important means for the optimization of production efficiency through simulations prior to the start of actual operations but also facilitates manufacturing process automation through efficient and effective automatic tracking of production data from the flow of materials, finished goods, and people, to the movement of equipment and assets in the value chain. There are two major applications of digital technology in manufacturing. The first deals with the modeling, simulation, and visualization of manufacturing systems and the second deals with the automatic acquisition, retrieval, and processing of manufacturing data used in the supply chain. This chapter summarizes the state of the art of digital manufacturing which is based on virtual manufacturing (VM) simulation and radio frequency identification (RFID)-based automation. The associated technologies, their key techniques, and current research work are highlighted. In addition, the social and technological obstacles to the development of a VM system and in an RFID-based manufacturing process automation system, and some practical application case studies of digital manufacturing based on VM and RFID-based automation, are also discussed.

49.1 Overview

The industrial world is undergoing profound changes as the information age unfolds [49.1]. The competitive advantage in manufacturing has shifted from the mass-production paradigm to one that is based on fast responsiveness and on flexibility [49.2]. One of the important issues in manufacturing is related to the integration of engineering and production activities. This includes integration of developers, suppliers, and customers through the entire production cycle, involving design, production, testing, servicing, and marketing. The scope of digital manufacturing includes manifesting physical parts directly from three dimensional (3-D) computer-aided design (CAD) files or data, using additive fabrication techniques such as 3-D printing, rapid prototyping from virtual manufacturing (VM) models, and the use of radiofrequency identification (RFID) for supporting manufacturing process optimization and resources planning. Examples can be found in *RedEye* [49.3], *Stratasys* [49.4], and *Autodesk* [49.5]. To achieve the integration, digital manufacturing, which covers all the engineering functions, information flow, and the precise characteristics of a manufacturing system are needed. Manufacturing enterprises are now forced to digitize manufacturing information and accelerate their manufacturing innovation in order to improve their competitive edge in the global market.

There are two major applications of digital technology in manufacturing. One is based on virtual manufacturing, which deals with the modeling, simulation, and visualization of manufacturing systems. The second is based on the automation of the manufacturing process and deals with the automatic *acquisition, retrieval, and processing of manufacturing data encountered in the supply chain*. In this chapter, the state of the art of digital manufacturing based on recent advances in VM and RFID-based automation is summarized. The concept and benefits of VM and RFID are presented, while the associated technologies, their key techniques, and current research work are also highlighted. The social and technological obstacles in the development of a VM system and an RFID-based manufacturing process automation system together with some case studies are discussed at the end of the chapter.

49.2 Digital Manufacturing Based on Virtual Manufacturing (VM)

49.2.1 Concept of VM

Digital manufacturing based on VM integrates manufacturing activities dealing with models and simulations, instead of objects and their operations in the real world. This provides a digital tool for the

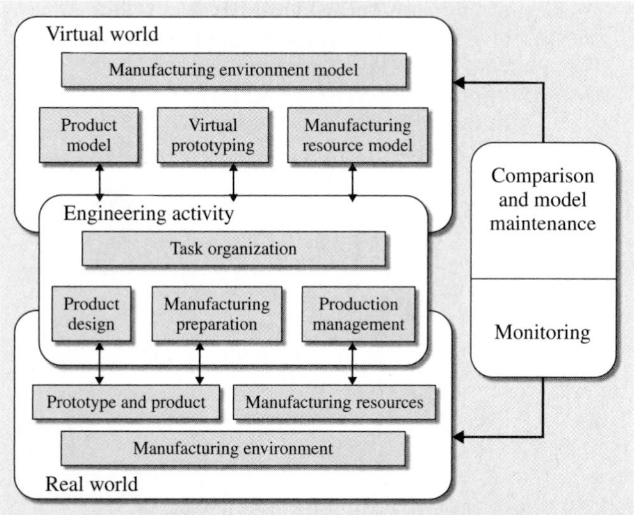

Fig. 49.1 Conceptual view of a virtual manufacturing system (after *Kimura* et al. [49.6])

optimization of the efficiency and effectiveness of the manufacturing process through simulations prior to actual operation and production. A VM system can produce digital information to facilitate physical manufacturing processes. The concept, significance, and related key techniques of VM were addressed by *Lawrence Associate Inc.* [49.7] while the contribution and achievements of VM were reviewed by *Shukla* [49.8]. As mentioned by *Kimura* [49.6], a typical VM system consists of a manufacturing resource model, a manufacturing environment model, a product model, and a virtual prototyping model. Some active research work is found in the study of both conceptual and constructive VM systems. *Onosato* and *Iwata* [49.9] developed the concept of a VM system, and *Kimura* [49.6] described the product and process model of a VM system. Based on the concept and the model, *Iwata* et al. [49.10] proposed a general modeling and simulation architecture for a VM system. *Gausemeier* et al. [49.11] has developed a cyberbike VM system for the real-time simulation of an enterprise that produces bicycles. With the use of a VM system, people can observe the information in terms of the structure, states, and behaviors equivalent to a real manufacturing environment, in a virtual environment [49.12, 13]. Various manufacturing processes can be integrated and realized in one system so that manufacturing cost and time to market can be reduced and productivity can be significantly improved.

A conceptual view of a VM system according to *Kimura* [49.6] is shown in Fig. 49.1. The manufacturing activities and processes are modeled before, and sometimes in parallel with, the operations in the real world. Interaction between the virtual and real worlds is accomplished by continuous monitoring of the performance of the VM system. Since a VM model is based on real manufacturing facilities and processes, it provides realistic information about the product and its manufacturing processes, and also allows for their evaluation and validation. Since no physical conversion of materials into products is involved in VM, this helps to enhance production flexibility and reduce the cost of production as the cost for making the physical prototypes can be reduced.

Basically, the classification of a VM system is based on the type of system integration, the product and process design, and the functional applications. According to the definitions proposed by *Onosato* and *Iwata* [49.9], every manufacturing system can be decomposed into four different subsystems: a real and physical system (RPS), a real information system (RIS), a virtual physical system (VPS), and a virtual information system (VIS). A RPS consists of substantial entities such as materials, parts, and machines that exist in the real world, while a RIS involves the activities of information processing and decision making such as design, scheduling, controlling, and prediction, etc. On the other hand, a computer system that simulates the responses of a real physical system is a virtual physical system, which can be represented by a factory model, product model, and a production process model. The production process models are used to determine the interactions between the factory model and each of the product models. A VIS is a computer system which simulates a RIS and generates control commands for the RPS.

VM can be subdivided into product-design-centered VM, production-centered VM, and control-centered VM according to the product design and process design functions. The product-design-centered VM makes use of different virtual designs to produce the production prototype. The relevant information about a new product (product features, tooling, manufacturability, etc.) is provided to the designer and to the manufacturing system designers to support decision making in the product design process. Production-centered VM is based on the RPS or VPS to simulate the activities in process development and alternative process plans

while the control-centered VM aims at optimizing the production cycles based on dynamic control of process parameters.

Production-centered VM, on the other hand, is based on the functional use so as to provide interactive simulation of various manufacturing or business processes such as virtual prototyping, virtual operational system, virtual inspection, virtual machining, virtual assembly, etc. Virtual prototyping (VP) mainly deals with the processes, tooling, and equipment such as injection molding processes, while virtual machining mainly deals with cutting processes such as turning, milling, drilling and grinding, etc. Finally, control-centered VM technology is used to study the factors affecting the quality, machining time, and costs based on the modeling and simulation of the material removal process as well as on the relative motion between the tool and the workpiece. Virtual inspection makes use of VM technology to model and simulate the inspection process, and the physical and mechanical properties of the inspection equipment. The aim of this is to study the inspection methodologies, inspection plans, and the factors affecting the accuracy of the inspection process, etc. In assembly work, VM is mainly used to investigate the assembly processes, the mechanical and physical characteristics of the equipment and tooling, and the interrelationship between different parts so as to predict the quality of an assembly or product cycle. It will also examine the costs as well as evaluating the feasibility of the plan of the assembly process. VM can also be used for virtual operational control, the aim of which is to evaluate the design and operational performance of the material flow and information flow system, etc. VM technology is also used for investigating the human behavior of the different workers who handle the various tasks such as assembling parts, queuing, handling documents, etc. The human factors affecting the operation of a manufacturing or business system can be predicted and evaluated.

49.2.2 Key Technologies Involved in VM

The development of VM demands multidisciplinary knowledge and technologies related to computer hardware and software, information technology, microelectronics, manufacturing, and mathematical computation. The key technological areas related to VM are:

1. *Visualization technologies*
   VM makes use of direct graphic interfaces to display highly accurate, easily understandable, and accept-

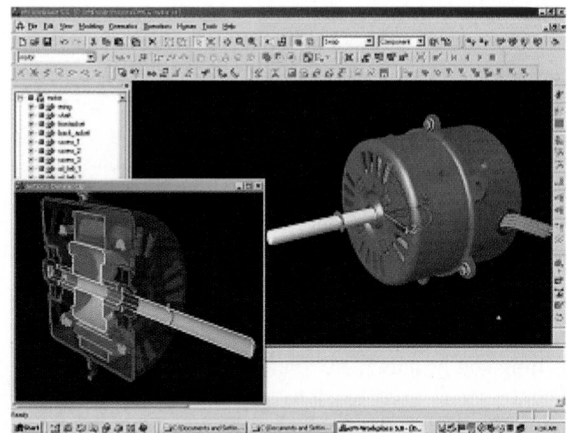

Fig. 49.2 Evaluation of product design

able input and output information for the user. This demands advanced visualization technologies such as image processing, virtual reality (VR), multimedia, design of graphic interfaces, animation, etc.

2. *Techniques for the establishment of a virtual manufacturing environment*
   A computerized environment for VM operations is vital. This includes the hardware and software for the computer, modeling and simulation of the information flow, support of the interface between the real and the virtual environment, etc. This needs the research and development of devices for VM operational environment, interface and control between the VM system and the real manufacturing (RM) system, information and knowledge integration and acquisition, etc.

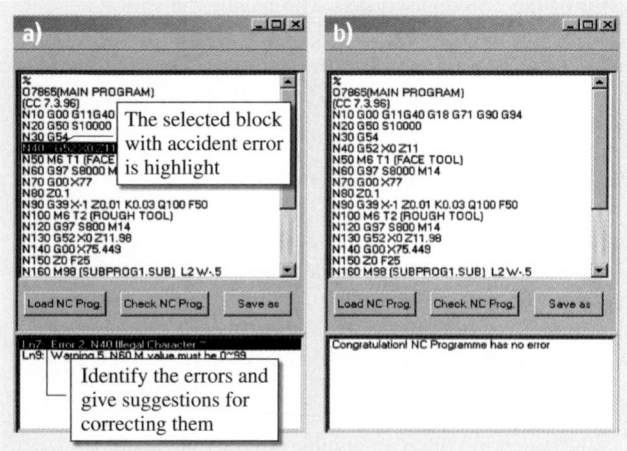

Fig. 49.3a,b NC program validation

3. *Information-integrated infrastructure*
   This refers to the hardware and software development for supporting the models and the sharing of resources, i.e., information and communication technologies (ICT) among dispersed enterprises.

4. *Methods of information presentation*
   The information about product design and manufacturing processes and related solid objects are represented using different data formats, languages, and data structure so as to achieve the data sharing in the information system. There is a need for research into advanced technologies for 3-D geometrical representation, knowledge-based system description, rule-based system description, customer-orientated expert systems, feature-based modeling, physical process description, etc.

5. *Model formulation and reengineering techniques*
   In order to define, develop, and establish methods and techniques which are capable of realizing the functions and interrelationships among various models in the VM system, various techniques are employed, including model exchange, model management, model structure, data exchange, etc.

6. *Modeling and simulation techniques*
   These refer to processes and methods used to mimic the real world, on the computer. Further research on the technologies related to dispersed network modeling, continuous system modeling, model databases and their management optimization analysis, validation of simulation results, development of simulation tools, and software packaging technique are much needed.

7. *Verification and evaluation*
   To ensure that the output from the VM system is equivalent to that from the RM system, related technologies such as standards of evaluation, decision tools, and evaluation methods are needed. These methods are useful for verifying and evaluating the performance and reliability of different models in VM systems and their outputs.

Some of these technologies are comparatively mature. However, most of them have to be further developed before they can be used to form an integrated VM platform.

49.2.3 Some Typical Applications of VM

1. VM can be used in the evaluation of the feasibility of a product design, validation of a production and business plan, and optimization of the product

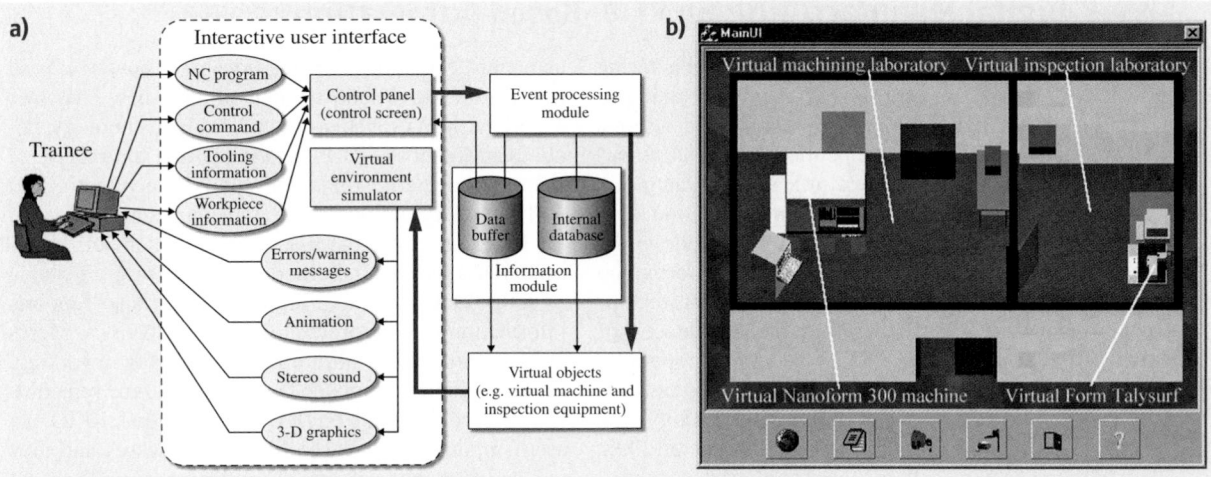

Fig. 49.4a,b Virtual training workshop (VTW) for ultraprecision machining center

design and of the business processes. Fine-tuning of these will result in reduction of the cost of the product throughout its lifecycle.

2. As shown in Figs. 49.2 and 49.3, VM can be used to test and validate the accuracy of the product and business process designs, for example, the design of the appearance of a product, analysis of its dynamic characteristics, checking for the tool path during machining process, numerical control (NC) program validation, checking for possible collision problems in machining [49.14] and assembly [49.15].

3. With the use of VM, it is possible to conduct training (Fig. 49.4) in a distributed virtual environment for the operators, technicians, and management people on the use of manufacturing facilities. The costs of training and production can thus be reduced.

4. As a knowledge-acquisition vehicle, VM can be used to acquire continuously the manufacturing or business knowhow, traditional manufacturing or business processes, production data, etc.

49.2.4 Benefits Derived from VM

1. VM can be used not only to predict the costs of product or process development but can also be used to provide the information about the process capability [49.16]. It allows for the modeling and simulation of the activities involved in process development and also for alternative process plans. It also enables the rapid evaluation of a production plan, the evaluation of the operational status

of a manufacturing system as well as of the objectives of the design of the physical system such as the degree of optimization of manufacturing resources and facilities, etc. The information generated from VM system is useful for improving the accuracy of the decisions made by the designer and the management. Possible problems in product or process development can be predicted and resolved prior to the actual operation.

2. VM can support VP, which simulates the materials, processes, tooling, and equipment in the fabrication of prototypes. The factors affecting the process, product quality, and hence the material properties, processing time, and manufacturing costs can be analyzed with the use of modeling and simulation techniques. As more computer-based product models are developed and prototyped upstream in the product development process using VM, the number of downstream physical prototypes traditionally made to validate the product models and new designs can be reduced. Hence, the product development time can be shortened.

3. With the virtual environment provided by VM [49.17], customers can take part in the product development process. Design engineers can respond more quickly to customer queries and hence provide better solutions to the customers. Discussion, manipulation, and modification of the product data model directly among personnel with different technical backgrounds are also facilitated. As a result the competitive edge of an enterprise in the market can be enhanced.

49.3 Digital Manufacturing by RFID-Based Automation

Another major application of digital manufacturing deals with the automatic acquisition and processing of manufacturing data in the supply chain. This is due to the fact that the keen competition in global manufacturing has rekindled interest in lean manufacturing, reducing inventory, and efficiency in production control. There has been growing interest worldwide in the use of RFID to digitalize the manufacturing information so as to automate the manufacturing process. As technical problems are slowly being overcome and the cost of using RFID is decreasing, RFID is becoming popular in manufacturing industries. According to industrial statistics, the worldwide market for RFID technology was US$ 1.49 billion in 2004. In the first 6 months of 2008, 6.8 billion tags were sold for these applications as well as 15.3 billion tags for pallets and cases. There is great market demand, which is ever increasing, for various RFID applications. It is predicted that RFID industry figures will increase from US$ 1.95 billion in 2005 to US$ 26.9 billion in 2015 [49.18].

The rapid increase in use of RFID technology in the retail industry has been driven by major players such as Gillette, Tesco, Wal-Mart, and Metro AG in Germany.

Wal-Mart, the world's largest retailer, has started deploying RFID applications and implementing new procedures in some of its distribution centers and stores. By January 1st, 2005, Wal-Mart required its top 100 suppliers to put RFID tags on shipping crates and pallets, and by January 1st, 2006, this was expanded to its next 200 largest suppliers. The aim of the applications is to reduce out-of-stock occurrences by providing visibility of the location of goods by using RFID tags. Out-of-stock items that are RFID-tagged have been found to be replenished three times faster than before, and the amount of out-of-stock items that have to be manually filled has been cut by 10%.

Gillette and Tesco implemented an item-level RFID project in the UK. Gillette razor blade cartridges were tagged with RFID tags and Tesco, the retailer, used an RFID reader embedded smart shelf system to search for items in the field, in order to take on-shelf inventories. Metro AG has implemented an item-level RFID trial for use in its future stores. RFID tags are attached to each pack of Gillette razor blades, Proctor and Gamble (P&G) shampoos, Kraft cream cheese, and digital versatile disks (DVDs). In addition to enhancing stock replenishment operations, consumers also benefit. Shopping trolleys which automatically update shopping lists and self-check-out systems have also been implemented using this technology. This demonstrates how RFID can revamp the retail industry and provide new customer experiences. With better tag and reader technology, declines in the cost of RFID tagging, and the release of information-sharing platforms, it is likely that RFID will be widely adopted across the entire supply chain.

In recent years, the use of RFID has enabled real-time visibility and increased the processing efficiency of shop-floor manufacturing data. RFID also supports information flow in process-linked applications. Moreover, it can help to minimize the need for reworking, improve efficiency, reduce line stoppages, and replenish just-in-time materials on the production line. RFID can assist in automating assembly-line processes and thus reduce labor and cost, and minimize errors on the plant floor. The integration of RFID with various manufacturing systems is still a challenge to many corporations. As most large retailers will gradually demand the use of RFID in the goods from their suppliers, this creates both pressure and the opportunity for small and medium sized enterprises (SMEs) to adopt this technology in their logistics operations and extend it to the control of their manufacturing processes. Some previous work [49.19] has discussed the point that RFID can be more cost effective in bridging the gap between automation and information flow by providing better traceability and reliability on the shop floor.

Traditional shop-floor control in a production environment, although computerized, still requires manual input of shop-floor data to various systems such as the enterprise resources planning (ERP) for production planning and scheduling. Such data includes product characteristics, labor, machinery, equipment utilization, and inspection records. Companies such as Lockheed Martin, Raytheon, Boeing [49.20], and Bell Helicopter have installed lean data-capture software and technologies and are in the process of converting barcodes to RFID. Honeywell was using barcodes to collect data related to part histories, inventories, and billing, and sharing data with its clients, and is accelerating its plan to switch from barcodes to RFID. The use of RFID has several advantages over barcodes, as tags that contain microchips that can store as well as transmit dynamic data have a fast response, do not require line of sight, and possess high security. RFID offers greater scope for the automation of data capture for manufacturing process control. In recent years, the cost of RFID tags has continuously decreased [49.21, 22] while their data capability has increased, which makes practical applications of RFID technology in manufacturing automation economically feasible.

For manufacturers, it is becoming increasing important to design and integrate RFID information into various enterprise application software packages and to solve connectivity issues relating to plant floor and warehousing. Real-time manufacturing process automation is dependent on the principle of closed-loop automation that senses, decides, and responds from automation to plant and enterprise operations; for example, a pharmaceuticals manufacturer [49.23] makes use of RFID to trace the route or history taken by an individual product at multiple locations along the production line. This allows the pharmaceuticals manufacturer easily to trace all final products that might have been affected by any production miscarriage. An aerospace company named Nordam Group uses RFID to track its high-cost molds. Through the use of RFID tags, they save the cost of real-time tool tracking. With growing emphasis on real-time responsiveness, manufacturers are seeking to control more effectively the production processes in real time in order to eliminate waste and boost throughput. The desire to extend supply-chain execution dispatching within a plant makes closed-loop automation an imperative. The inability to achieve true closed-loop manufacturing process automation presents one of the greatest barriers to successful real-time operation strategies. Critical elements that support manufacturing process automation include:

- Dynamic information technology (IT) systems to support high-mix, variable environments
- Dynamic modeling, monitoring, and management of manufacturing resources
- Real-time synchronization between activities in the offices, manufacturing plant, and supply chain
- Visibility of the real-time status of production resources

Although some previous studies [49.24–26] have shown that RFID technology has the potential to address these problems and has great potential for supporting manufacturing automation, critical deficiencies in the current systems in keeping track of processes under changing conditions include:

- Lack of integration of manufacturing with the supply chain
- Lack of real-time shop-floor data for predictive analysis and for decision support
- Lack of a common data model for all operational applications

- Inability to sense and analyze inputs ranging from the factory floor to the supply chain
- Ineffective integration links between manufacturing process automation and enterprise IT applications
- Inability to provide intelligent recommendations and directives to targeted decision points for quick action.

In light of these, this chapter presents a review of the RFID-based manufacturing process automation system (MPAS), which embraces heterogeneous technologies that can be applied in the manufacturing process environment with the objective of enhancing digital manufacturing at the level of automation across the enterprise and throughout the value chain. The proposed RFID-based manufacturing process automation system aims to address these deficiencies.

49.3.1 Key RFID Technologies

RFID is an advanced automatic identification technology, which uses radiofrequency signals to capture data remotely from tags within reading range [49.27, 28]. The basic principle of RFID, i.e., the reflection of power as the method of communication, was first described in 1948. One of the first applications of RFID technology was *identify friend or foe* (IFF) detection deployed by the British Royal Air Force during World War II. The IFF system allowed radar operators and pilots to distinguish automatically between friendly and enemy aircraft using radiofrequency (RF) signals. The main objective was to prevent *friendly fire* and to aid the effective interception of enemy aircraft. The radiofrequency used is the critical factor for the type of application for which an RFID system is best suited. Basically, the radiofrequencies can be classified as shown in Table 49.1.

A typical RFID system contains several components, including an RFID tag, which is the identification device attached to the item to be tracked, and an RFID reader and antenna, which are devices that can recognize the presence of RFID tags and read the information stored on them. After receiving the information, in order to process the transmission of information between the reader and other applications, RFID middleware is needed, which is software that facilitates the communication between the system and the RFID devices. Figure 49.5 shows a typical RFID system to illustrate how the pieces fit together.

As shown in Fig. 49.5, radio frequency (RF) tags are devices that contain identification and other information

Table 49.1 Comparison of RFID frequency bands and their respective applications

Frequency	Approximate read range	Data speed	Cost of tags	Applications
Low frequency (LF) (125 kHz)	< 5 cm (passive)	Low	High	Animal identification Access control
High frequency (HF) (13.56 MHz)	10 cm – 1 m (passive)	Low to moderate	Medium to low	Smart cards Payment
Ultrahigh frequency (UHF) (433, 868–928 MHz)	3–7 m (passive)	Moderate to high	Low	Logistics and supply chain Pallet and case tracking Baggage tracking
Microwave (2.45 and 5.8 GHz)	10–15 m (passive) 20–40 m (active)	High	High	Electronic toll collection (ETC) Container tracking

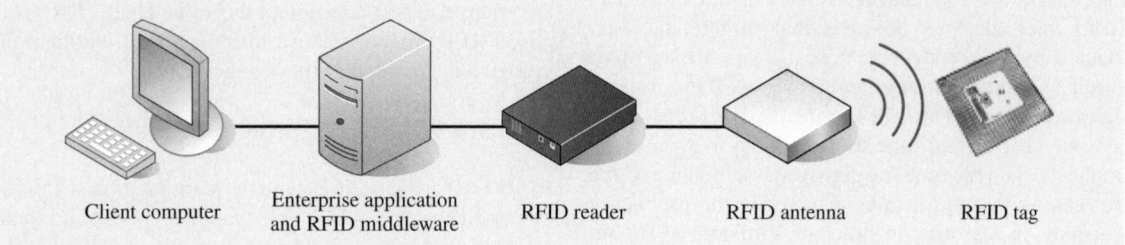

Client computer   Enterprise application   RFID reader   RFID antenna   RFID tag
                     and RFID middleware

Fig. 49.5 Infrastructure for an RFID system

that can be communicated to a reader from a distance. The tag comprises a simple silicon microchip attached to a small flat aerial which is mounted on a substrate. RFID tags (Fig. 49.6) can be divided into three main types with respect to the source of energy used to power them:

1. *Active tags*: use a battery to power the tag transmitter and receiver to broadcast their own signals to readers within the life of batteries. This allows them to communicate over distances of several meters.

2. *Semipassive tags*: have built-in batteries to power the chip's circuitry, resist interference, and circumvent a lack of power from the reader signal due to long distance. They are different from active tags in that they only transmit data at the time a response is received.

3. *Passive tags*: derive their power from the field generated by the reader without having an active transmitter to transfer the information stored.

The reader is called an interrogator, and it sends and receives RF data to and from the tag via antennas.

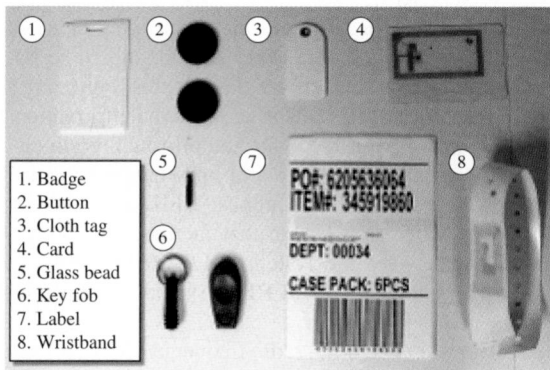

1. Badge
2. Button
3. Cloth tag
4. Card
5. Glass bead
6. Key fob
7. Label
8. Wristband

Fig. 49.6 Samples of different types of tags

Fig. 49.7 Example RFID reader

As shown in Fig. 49.7, the reader contains a transmitter, a receiver, and a microprocessor. The reader unit also contains an antenna as part of the entire system. The antennas broadcast the RF signals generated by the reader and receive responses from tags within range. The data acquired by the readers is then passed to a host computer, which may run specialist RFID software or middleware to filter the data and route the data to the correct application to be processed into useful information. Middleware refers to software that lies between two interfaces; RFID middleware is software that lies between the readers and the data collection systems. It is used to collect and filter data from readers. It can transfer those useful data, which are filtered, to the data collection system.

49.3.2 Applications of RFID–Based Automation in Digital Manufacturing

Manufacturing process automation relies heavily on efficient and effective automatic tracking of assets during manufacturing. The tracked assets include physical assets such as equipment, raw materials, work-in-progress (WIP), finished goods, etc. Efficient automatic tracking of assets is beneficial for manufacturers. One of the ma-

jor challenges for manufacturers today is management of their movable assets. Managing assets includes activities such as locating assets, tracking status, and keeping a history of flow information. If the location of an asset is not effectively located, workers are required to spend much time searching for it, which results in increased process costs. As a result, manufacturers are turning to RFID to investigate whether they can use it to bring the benefits of managing assets individually, real-time location tracking of assets, and better information accuracy and automation into their business operations.

Examples can be found in the garment industry [49.23]. The manufacturer ties the RFID tags to the bundles and sends them to the sewing workstation. At each process station, the data stored in the RFID tag is captured. The real-time status of the WIP can be automatically captured during the manufacturing process. It is also interesting to note that Boeing and Airbus are moving forward and have created an RFID strategy for airplane manufacturing [49.20, 28], as a component to meet US Department of Defense mandates. The RFID tags are attached to removable parts of the airplane to aid the control of maintenance programs. In the present study, four major areas of RFID in manufacturing process automation are introduced.

49.4 Case Studies of Digital Manufacturing and RFID–Based Automation

49.4.1 Design of Assembly Line and Processes for Motor Assembly

Kaz (Far East) Limited, formerly Honeywell Consumer Product (HK) Ltd., is a multinational manufacturing

company with about 45 employees in Hong Kong and around 2650 in China. Its corporate headquarters, design, sales offices, and production plants are widely dispersed in Europe, North America, and Asia, as are its key suppliers and customers. As shown in Fig. 49.8,

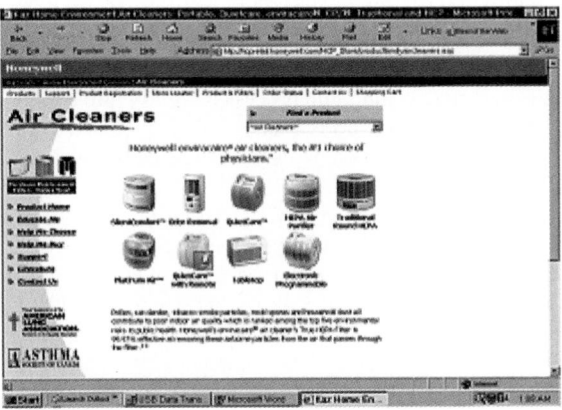

Fig. 49.8 Home comfort products from Kaz

Fig. 49.9 Virtual assembly station for assembling a motor

the company has been involved in the seasonal products business which includes home comfort products such as portable air cleaners, humidifiers, heaters, fans, etc. It delivers a portfolio of innovative products and trusted brands to serve customers worldwide, though most of

their products are sold to customers in North America and Europe.

The company possesses a number of assembly lines and clusters for assembling the motors for the home comfort products. For new assembly line planning, there are many factors which affect the efficiency and throughput rate. These include the design of the workplace, the investigation of the assembly process, the allocation of the resources (e.g., number of workers), and the scheduling of the assembly lines. Conventionally, the production team was essentially empowered to use their own judgment in deciding *what, when, and how* to setup the assembly lines, based on their own experience. The optimal efficiency, throughput rate, and quality for new assembly lines can only be obtained through iterative trial operations and process refinement. This is not only time consuming but also costly. With the use of the VM approach, Kaz can investigate the assembly processes and optimum utilization of resources through modeling and simulation. The feasibility of the assembly process plan (Fig. 49.9) can be evaluated prior to the actual production. As a result, the time and cost of setting up new assembly operations can be significantly reduced.

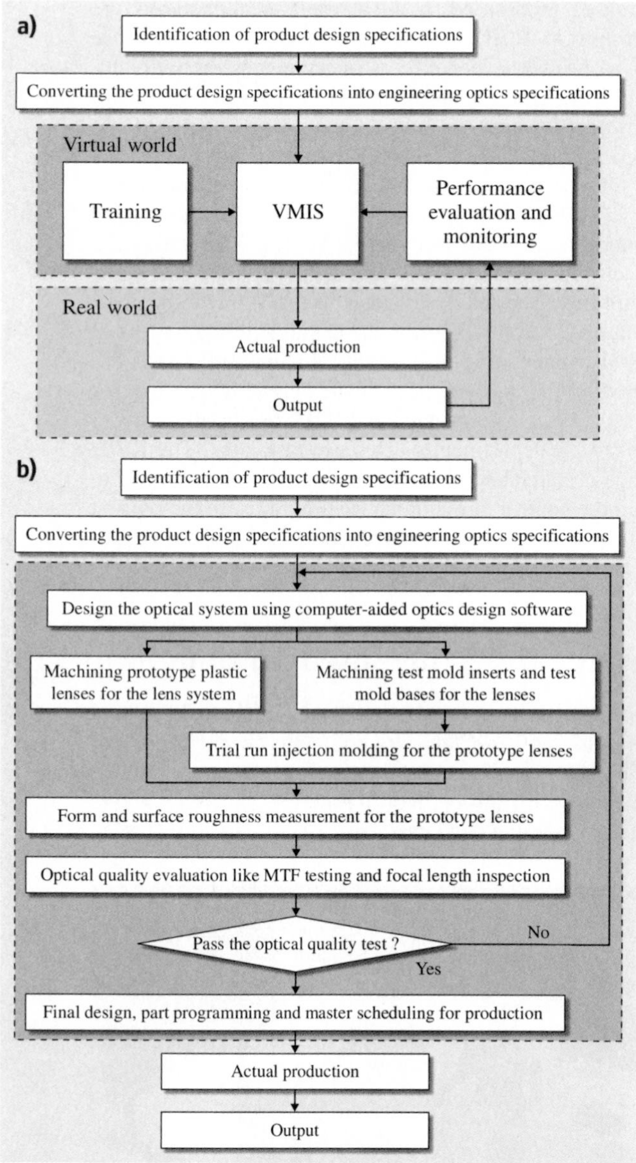

Fig. 49.10a,b A comparison between (**a**) the conventional approach and (**b**) the virtual manufacturing approach for the design and manufacture of precision optics (MTF – modular transfer function)

49.4.2 A VM System for the Design and the Manufacture of Precision Optical Products

The conventional approach to the design and the manufacture of precision optical products is based on a trial-and-error method. As shown in Fig. 49.10a, the optical product is designed using computer-aided optics design software (Fig. 49.11b). Then a lens prototype is made by either direct machining or injection-molding from a test mold insert machined by ultraprecision machining (Fig. 49.12). Quality tests will then be conducted on the prototype lenses or mold inserts (Fig. 49.13). It should be noted that the design, prototyping, and evaluation processes are iterative until a satisfactory mock-up is found. This is expensive and time consuming; it also creates a bottleneck for the overall process flow.

Using the VM approach, as shown in Fig. 49.10b, the iterative design, prototyping, and testing processes are accomplished by a virtual machining and inspection system (VMIS) [49.29–33]. As shown in Fig. 49.14, the VMIS has been developed with the aim of creating a virtual manufacturing environment. This is done by electronically representing the activities of optic design,

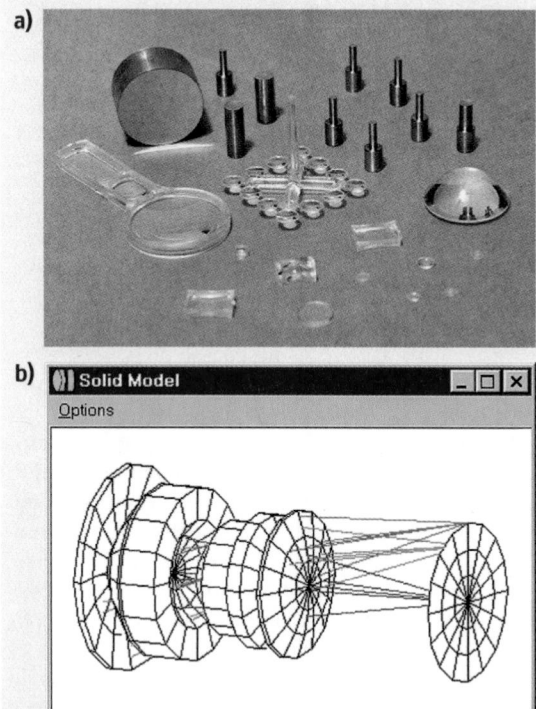

Fig. 49.11 (a) Precision optical products and **(b)** their computer-aided optics design

Fig. 49.12a,b Comparison between **(a)** actual and **(b)** virtual machining process

prototyping ultraprecision machining and inspection of the design, and manufacture of precision optical products. Interaction between the virtual and real worlds is accomplished by continuous training and by monitoring the performance of the VMIS system by comparing the simulation results and the real cutting test results (Fig. 49.15).

The feasibility of a product design and the optimal resources needed to turn a design into a real product can be determined prior to any manufacturing resources being committed and before any costly scrap is generated. The VMIS allows the precision optics manufacturers to evaluate the feasibility of an optical product design and a manufacturing process plan prior to actual production. This avoids the need for conducting expensive production trials and physical prototyping.

49.4.3 Physical Asset Management (PAM)

A physical asset management (PAM) program involves the tracking of physical assets such as expensive ma-

chinery and equipment, as well as returnable items such as containers and pallets. Across industries and regions, PAM is a major concern in the manufacturing industry. PAM includes engineering activities, as well as the adoption of a variety of approaches to maintenance such as reliability-centered maintenance, multiskills, total productive maintenance, and hazard and operability studies [49.34]. Manufacturing enterprises that have better PAM systems usually excel when compared with their counterparts in terms of utilization of assets in manufacturing. However, many organizations today still face significant challenges in relation to tracking the location, quantity, condition, and maintenance and depreciation status of their fixed assets.

Traditionally, a popular approach is to track fixed assets by utilizing serially numbered asset tags, such as barcodes, for easy and accurate reading. This process is often performed by scanning a barcode on a physical asset identification (ID) tag that has been affixed to the asset. However, PAM may also involve a physical asset ID tag with a human-readable number only, which

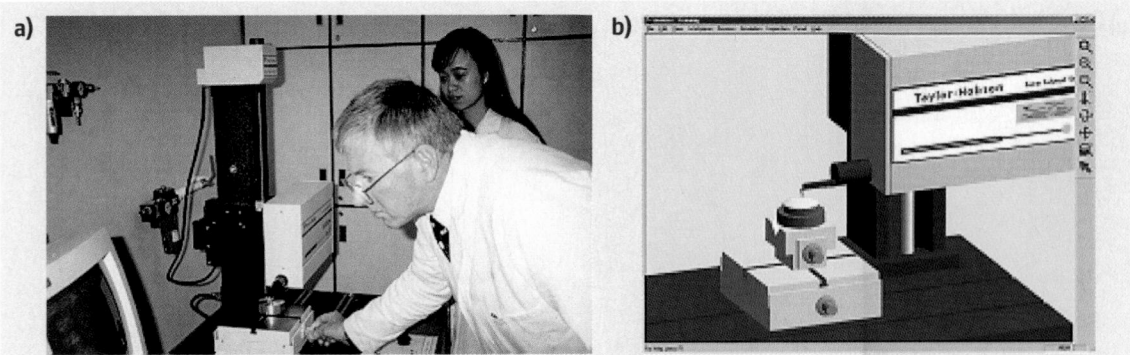

Fig. 49.13a,b Comparison between (**a**) actual and (**b**) virtual inspection of precision optics

is documented manually. The barcode approach has its limitations though, e.g., lack of automation and inability to provide real-time tracking.

To increase the utilization rate of movable equipment, manufacturing enterprises usually adopt a fluid approach to the composition of production lines. Movable equipment is shared across lines for some nonbottleneck procedures. Previously, applications for transfer of physical assets required tedious manual operations. Since most manufacturing processes were related to the requested equipment, late arrival of equipment lowered the overall efficiency of the company, which in turn impacted on the company's profit margin. Also, as the transfer records were entered into a database after a long verification process, the timeliness of the information was questionable. Worse, due to inevitable human errors, the database records were incomplete and inaccurate. Common discrepancies included asset locations, utilization, and history of maintenance and repair. As a result, equipment could not be repaired in a timely fashion and some was even lost.

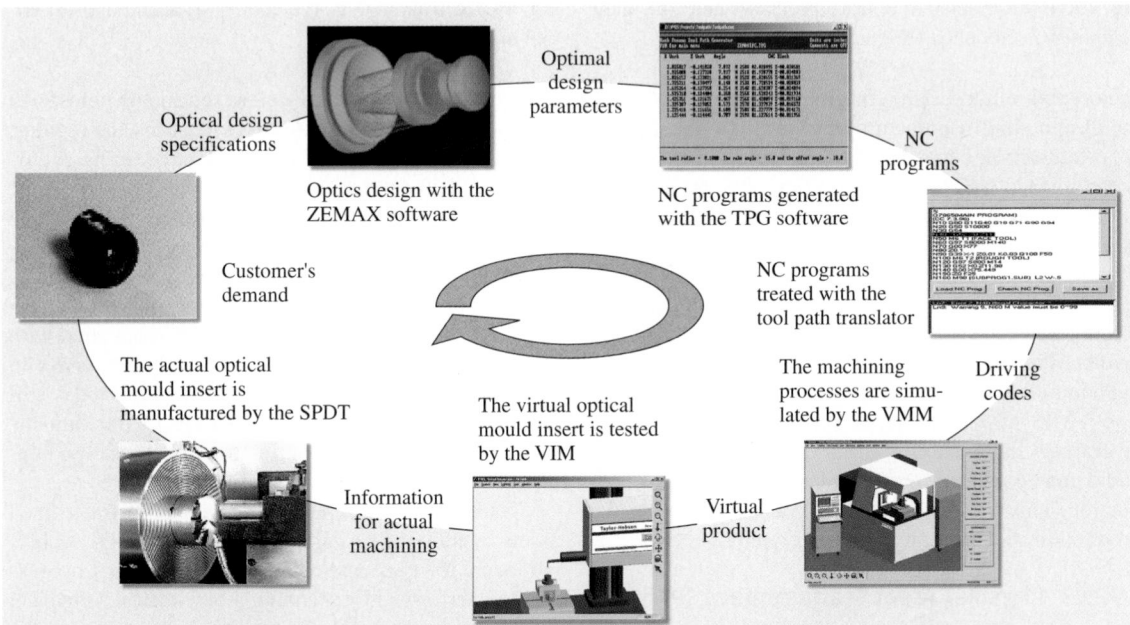

Fig. 49.14 Graphical illustration of the process flow for the design and manufacture of precision optics (SPDT – single-point diamond turning, VMM – virtual machining module, VIM – virtual inspection module)

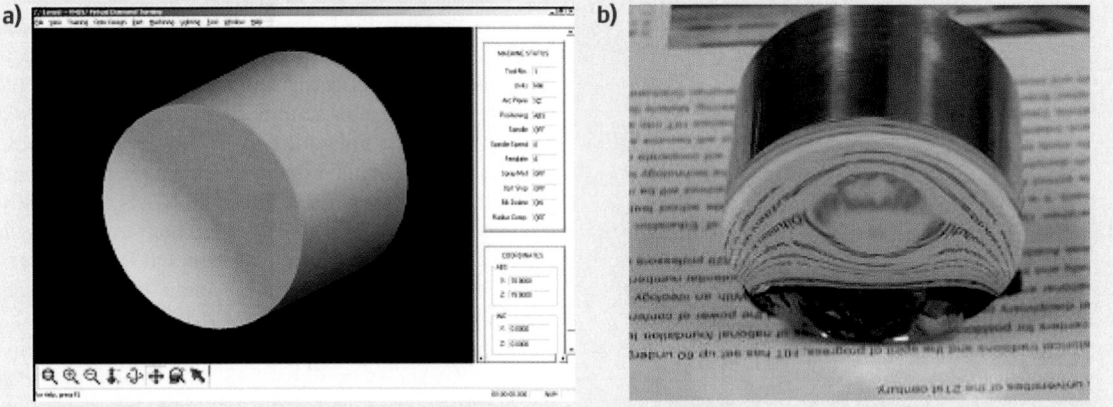

Fig. 49.15 (**a**) Virtual and (**b**) actual mold inserts produced based on the VMIS

To improve the overall efficiency and increase the visibility of physical assets, an RFID-enabled solution is used to track the movement of physical assets easily when they pass through antenna detection points without interfering with normal operations. It was therefore proposed that RFID antennas should be set up near the door of each production site so that the moving in and out of physical assets would be automatically recorded, as illustrated in Figs. 49.16 and 49.17.

RF signals are affected by metal because of interference and reflection from it, which affects the readability of RFID tags. Since most equipment is made of metal, this is one of the main challenges of implementing an RFID-enabled solution in the plant. To address this problem, a feasibility study was carried out to investigate the necessary settings for satisfactory RFID-enabled tracking performance. Several experiments were performed and the results are shown in Fig. 49.18. Special attention was given to the height and angle of the antennas in order to provide optimal results for autotracking.

To evaluate the feasibility of the proposed RFID-enabled PAM system, a case study was carried out in a selected company named SAE Magnetics (HK) Ltd., which is one of the world's largest independent manufacturers of magnetic recording heads for hard-disk drives used in computers and, increasingly, in consumer electronics such as digital video recorders, MP3 players, and even mobile phones. While the bulk of SAE's operations are based in two modern facilities located in mainland China, the company needs to manage large-scale warehouses for its sophisticated logistics. Differ-

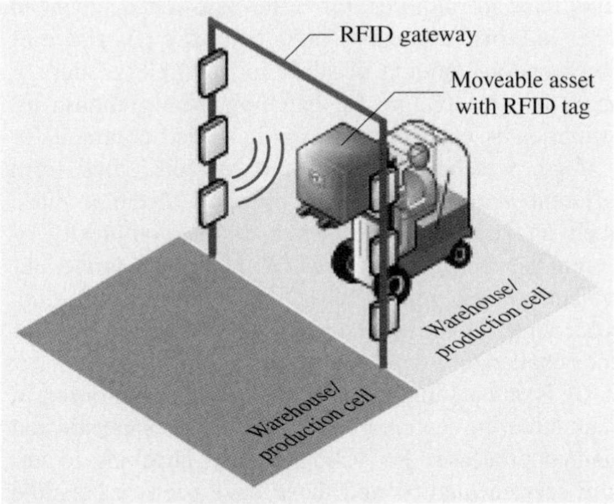

Fig. 49.16 Tracking the movement of equipment by RFID technology

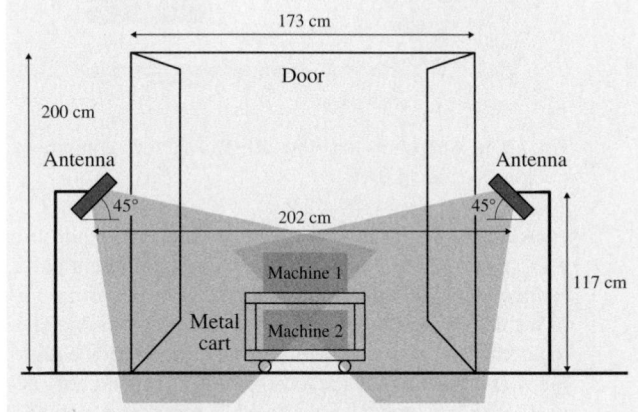

Fig. 49.17 Deployment setup of RFID gateway

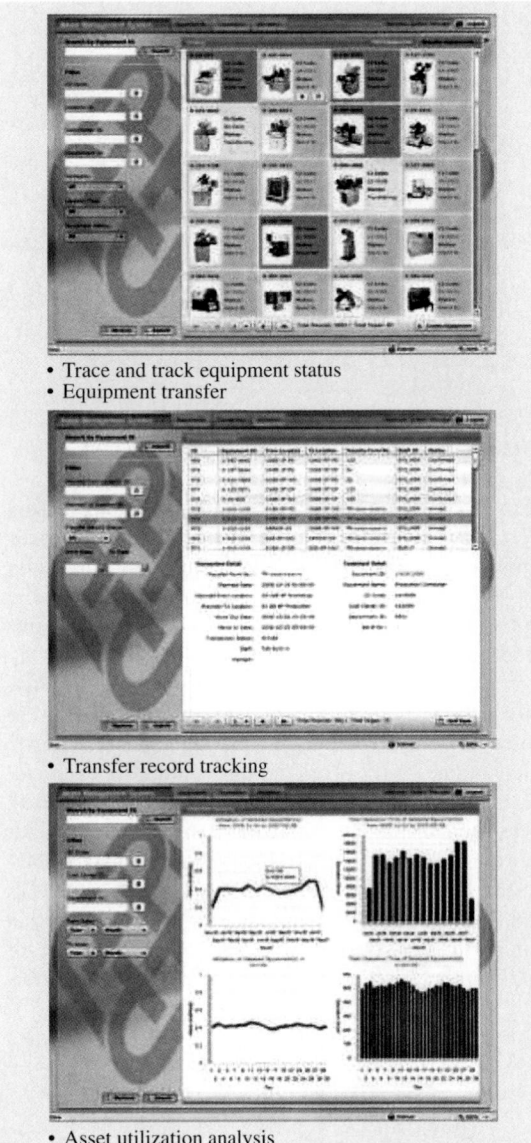

- Trace and track equipment status
- Equipment transfer

- Transfer record tracking

- Asset utilization analysis

Fig. 49.18 Snapshots of the RFID-enabled equipment tracking system in SAE

ent kinds of equipment are stored in these warehouses, e.g., semiproducts, subassemblies, and equipment parts (collectively known as physical assets). It is normal for movable physical assets to be transferred from one site to another to maximize their use, e.g., production equipment and assets that are costly and advanced. In the past, SAE's warehouses adopted a pen-and-paper approach to record the transfer of physical assets. In time,

this became a complicated process that generated piles of forms and involved many approval processes, which ultimately limited the company's operational efficiency.

This new method has enabled equipment or required parts and subassemblies to arrive promptly and has increased the utilization of important and valuable physical assets for the company. This system automatically tracks the transfer of physical assets and greatly reduces the processing time of equipment transfer. With the data transfer between the system and RFID devices being in real time, information about physical assets can be captured. These data are then presented in the RFID-enabled equipment tracking system in a systematic way, as shown in Fig. 49.18. The system provides the ability to track and trace the status of equipment, issue equipment transfer orders, and provides reports for asset utilization analysis. As a result, location and utilization of equipment can be easily identified using visual diagrams that help in maintenance planning and repair scheduling.

The introduction of an RFID-enabled equipment tracking system at SAE enabled the operational inefficiencies associated with the PAM approach to be targeted and allowed the equipment transfer operation and production planning at SAE to be streamlined. The system has generated great benefits for the company, and has ultimately achieved cost savings while enhancing the effectiveness of asset management.

49.4.4 Warehouse Management

Warehouse management in manufacturing enterprises usually requires much more complex and accurate systems than are required for warehouse management in other industries. Since a single company takes care of a tremendous amount of stock for multiple vendors at one time, the loading for handling the warehouse information is especially heavy. It is also common for a single warehouse to be used to store stock from different vendors so as to enhance utilization rates. Such an approach also increases the complexity of warehouse storage structures. Without an accurate, automated, and comprehensive stock transfer recording procedure, warehouse management becomes a tremendous challenge.

It is interesting to note that many warehouses in manufacturing enterprises still rely on sophisticated manual processes for validating and checking to ensure service quality, and these have been responsible for limiting the efficiency of the operation; for example, updating inventory lists in the warehouse, checking of replenishment requirements, and entering updated data into the Kerry warehouse management system (KWMS) are all manual processes. Not surprisingly, these paper-based processes are excessively time consuming and susceptible to manual errors.

Low warehouse visibility is another main area of concern. Warehouse visibility can be evaluated by measuring the discrepancy between the actual status and the system warehouse inventory status. Since the system is such that warehouse inventory status is only updated after office staff have entered all the latest information into the KWMS, there might be a discrepancy between the actual status and the system warehouse inventory status. As a result, staff might not know the real-time warehouse inventory status from the KWMS. Such information gaps could result in wrong decisions in inbound, relocation, and outbound processes. This is especially important for dialysis solutions since they have a limited period of validity. Extensive warehouse storage time causes product expiry and ultimately financial loss to the company.

To address these problems, RFID is used to increase the visibility of the warehouse and to streamline warehouse operations. Through the implementation of RFID-enabled solutions in warehouse management, instant and accurate inventory records with automatic real-time record update and physical stocktaking can be achieved. However, most of the existing RFID-enabled solutions for warehouse management are lacking in automation. In the present study, an RFID-enabled intelligent forklift is proposed to achieve a higher degree of automation for the warehouse management via RFID. Two RFID antennas were placed at the front of the forklift, as shown in Fig. 49.19, one for location scanning and the other for pallet scanning. A tailor-made application was installed in the mobile computer placed in the forklift, which is used to show the information captured by the antennas and the information stored in the database system. Forklift drivers can view the information on the vehicle-mounted computer screen.

A tag containing product storage information, including particular stock keeping unit (SKU) and lot numbers, is pasted to each pallet. At the same time, RFID tags that store unique location information are placed on the racks. When a driver has loaded a pallet onto the forklift, the pallet information is scanned automatically and that product information and its planned location are shown on the screen attached to the forklift (Fig. 49.20). After the pallet has been placed in the storage position, the forklift scans the location tag on the rack, and completes the stock-in process (Fig. 49.21).

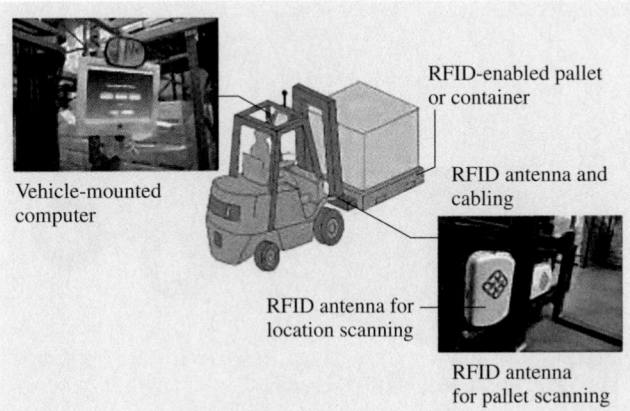

Fig. 49.19 The RFID-enabled intelligent forklift

With such modifications in warehouse operation, numerous benefits can be achieved. Since most paper processes are eliminated by RFID-enabled automation, both the speed and accuracy of the stock transaction process are greatly enhanced; for example, warehouse staff are no longer required to fill in put-away lists and relocation lists, neither do they need to input data manually.

To realize the capability of the RFID-enabled intelligent forklift, a case study was carried out in a selected company named Kerry Logistics (Hong Kong) Limited. Kerry Logistics (Hong Kong) Limited is part of the diversified international conglomerate, the Kuok Group, and is a subsidiary of the Hong-Kong-listed Kerry Properties Limited. Kerry's Asian logistics business has been built on more than two decades of warehousing and logistics operations in Hong Kong. Its third-party logistics (3PL) business has dramatically accelerated since 1998. Kerry Logistics now encompasses contract logistics,

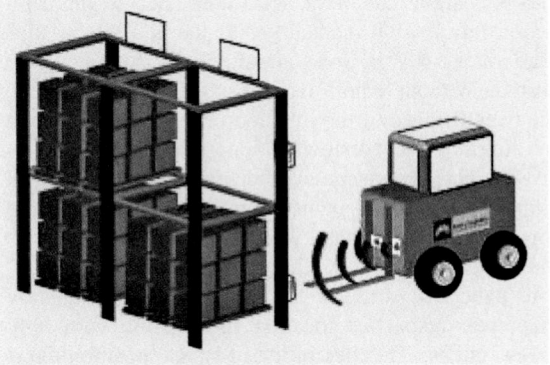

Fig. 49.20 Location RFID scanning

Fig. 49.21 Pallet RFID scanning

distribution centers, international air and sea freight forwarding, transportation, distribution, and value-added services.

Like most third-party logistics (3PL) service providers around the world, Kerry Logistics has recently recognized some inefficiency in the warehouse management of one of its major logistics centers located in Kwai Chung. Because of the complexity of taking care of multiple companies and products at the same time, 3PL service providers require considerable documentation to record the mass of information generated in the course of their work. One warehouse inside the logistics center stores healthcare products, especially dialysis solutions, for renal patients. To reduce human error and streamline the operation flow, an RFID-enabled intelligent forklift was developed.

With such modifications in warehouse operation, numerous benefits can be achieved. Since most paper processes are eliminated by RFID-enabled automation, both the speed and accuracy of the stock transaction process are greatly enhanced. For example, warehouse staffs are no longer required to fill in put-away lists and relocation lists, neither do they need to input location information manually into the KWMS, as this kind of activity is done automatically by the RFID system. Without tedious paper-based location mapping, the overall speed of the inbound process has improved by almost 15%, not to mention the cost saving factors of increased accuracy and reduced manpower. In addition, the RFID-enabled intelligent forklift allows for unstructured storage, which completely eliminates the relocation process. Forklift drivers can freely locate pallets of stock on the racks, and subsequently map their respective location information with only a few clicks. The flexibility of stock positioning is thus improved, resulting in better utilization of ware-

house space and further shortened inbound processing time.

49.4.5 Information Interchange in Global Production Networks

The trend towards the establishment of a global production network in the manufacturing industry poses some new challenges to information interchange; for example, some components of a product may be produced in one country, some part of the product may be outsourced and made in a second country, and the final assembly of the finished product may be done a third country, before being ultimately sold somewhere else. However, apart from the obvious need for collaboration, the problems of data integration and information sharing within the supply chain are challenging. Therefore, an RFID-based intra-supply-chain information system (ISCIS) is much needed in order to streamline the supply-chain activities and form an intra-supply-chain network.

Figure 49.22 shows the architecture of a proposed RFID-based ISCIS which is basically divided into platform tiers. The system consists of the application of RFID tags to generate the data within the supply-chain network. Data from the RFID tags is scanned by readers and synchronized with the internal information systems. The RFID-based ISCIS enables the different partners in the supply chain to share real-time information.

1. Data acquisition tier: The data from the operational level on the textile supply chain are captured by RFID technology. RFID readers are installed in the warehouses of the textile and apparel manufacturers and at the retailers. Tags are placed either at the pallet level or the item level, depending on the stage in the supply chain. At the level of the individual item of merchandise it is difficult to achieve 100% reading accuracy with RFID tags. However, it is already technically feasible to track RFID tags at the pallet or carton level. As the raw materials progress through the stages of WIP to the finished goods, the tagging level changes from the pallet level to the carton level and finally to the item level.
2. Information systems tier: This integrates the RFID-based information systems with the internal information systems of the different supply-chain partners, such as enterprise resources planning (ERP) systems. The most successful new business models are probably those that can integrate information technology into all activities of the enterprise-wide value chain.

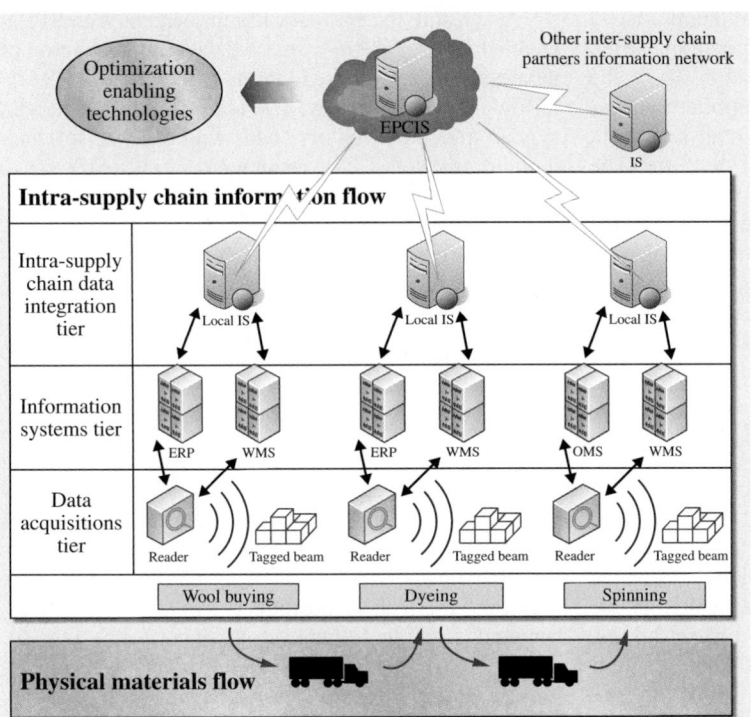

Fig. 49.22 Architecture of RFID-based ISCIS (IS – information system, WMS – warehouse management system)

3. Intra-supply-chain data integration platform: Figure 49.23 shows how the supply-chain information is shared and integrated with the different supply-chain parties. On this platform, an integrated information system is built to enable the supply-chain partners to download the information from the chain. As discussed before, the RFID gateway can act as a check-in and check-out system and synchronize the internal information systems without delay. Supply-chain partners can select the information to be shared, which will be stored in a centralized database.

4. Web-based platform: On this platform, a web-based information-sharing portal is built so that supply-

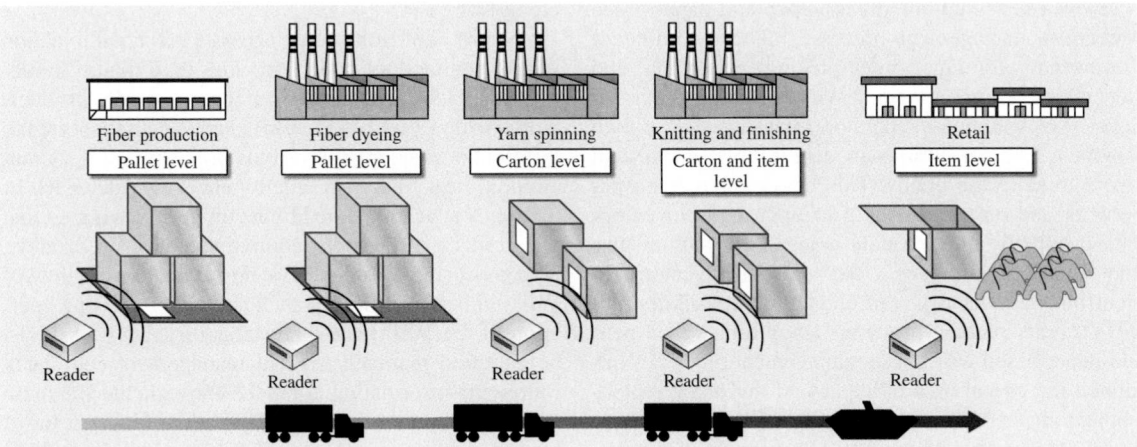

Fig. 49.23 Information sharing along the supply chain

chain partners can retrieve related supply-chain information. Intra-supply-chain partners can use the web browser to search for information about the chain. The production status, the supplier's inventory level, the delivery status, and retail sales data can be displayed on the web-based platform. The above platforms help an intra-supply-chain company to obtain data on their operations and provide an interface permitting them to share information. To facilitate the intra-supply-chain collaboration, different optimization technologies, such as intelligent agent and data mining, can be used to analyze these data.

To realize the capability of the RFID-based ISCIS, a case study was conducted in a reference site in the textile industry named Novetex. Novetex Spinners Limited was established in 1976 and, employing over 2000 people, is now recognized as the world's largest single-site woolen spinner. As part of the Novel Group, Novetex has its headquarters in Hong Kong and a factory located in Zhuhai, Southern China. The factory houses four spinning mills, a dyeing mill, and 51 production lines, resulting in an annual capacity of 7500 t of high-quality yarn. Novetex has recognized the need to be a truly global supplier and has invested in offices and agents worldwide to offer customers the most efficient service possible. The trend towards the establishment of a global production network in the textile industry poses some new challenges to Novetex; for example, fibres might be produced in one country, spun to yield yarns in a second country, woven into fabrics in a third, and sewn into clothing in yet another one, before being ultimately sold somewhere else.

Novetex has traditionally used pen-and-paper-based warehouse management processes. The recording of the inventory, storage, order picking, packaging, and stocktaking processes were heavily dependent on the efficiency of warehouse operators, who used their own judgment and perceptions to decide how, when, and where to store the goods. This heavy reliance on paperwork and on human input to update the inventory data inevitably leads to data inaccuracy. This in turn may affect the company's decisions on inventory replenishment, inventory control, and the first-in first-out (FIFO) order-picking process. Heavy reliance on pen-and-paper-based warehouse management processes has limited the operational efficiency in inventory replenishment, inventory control, and the FIFO order picking process of Novetex. These issues become more complex in the era of the global production network.

As a result, the company has implemented an RFID-enabled ISCIS. The outcomes of the implementation of the system were positive in terms of increased stock visibility and data accuracy. The textile manufacturer can synchronize the inventory status with the ERP software without any delay. The attachment of an RFID tag to each bundle of goods facilitates identification and visualization of items. The storage location of the items can be easily identified and retrieved from the ERP system. Order picking and stock relocation is easier for the warehouse.

49.4.6 WIP Tracking

RFID-based solutions have been used for a decade in manufacturing. Their applications in a manufacturing enterprise are usually tracking of parts during manufacture and tracking of assembled items, i. e., WIP. As a result, efficiency can be increased while entry errors and manpower are saved. For WIP tracking, the tags can be attached to products as they are being assembled or created along the production line [49.35]. The status of the product can be updated as it progresses along the production line via RFID readers placed above and below its path [49.36]. According to *Hedgepeth* [49.37], the history or route taken by an individual product can be ascertained from data stored on tags attached to the item, by installing RFID readers at single or multiple locations along the production line. As a result, the location as well as the flow history of any item can be recorded in the manufacturer's database. To realize the advantages of RFID-enabled WIP tracking, a case study was conducted in a precision mold manufacturing company named Nypro Tool Hong Kong Ltd.

Nypro Tool Hong Kong Ltd. is a precision injection mold manufacturer which provides high-quality molding tools globally. Since the company provides products with a wide variety, WIP items need to be transported to different workshops for various processes such as machining, heat treatment, quality checking, etc. Each of the items is accompanied by its unique drawing so that operators can refer to its required operations. Currently, Nypro is using a barcode system to monitor the flow of WIP items within the plants. For every processing operation of the WIP items, the status of the process must be captured manually so that management can check follow-up information for WIP through the database system. However, the manual process induces a lot of human errors and is time consuming. The application of RFID in WIP tracking is believed to be able to au-

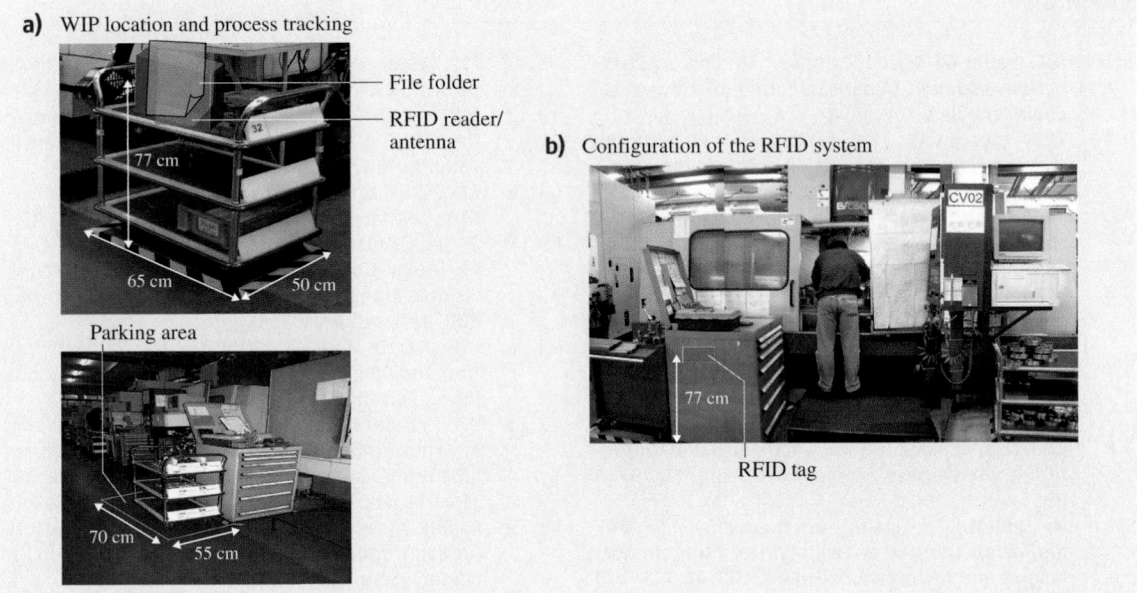

Fig. 49.24a,b RFID-based location tracking and process status tracking for WIP application (**a**) WIP location and process tracking (**b**) Configuration of the RFID system

tomate the tracking process and reduce the number of errors.

As shown in Fig. 49.24a, which shows the location tracking process, RFID gateways are set up at the entrance of the workshops. WIP items and drawings transported to the workshop are required to pass through the gateway and the tag information for the arriving item is automatically tracked when it is pulled through the gateway. Figure 49.24b shows the setup for the RFID-based process. The process is designed in such a way as to capture the machining status of the WIP items accurately and automatically. The result of the trial implementation shows that the management of the company can accurately plan job allocation during the production process through automatic and accurate capture and tracking of the information related to the flow, location, and processing status of the WIP items. In other words, visibility of WIP information can be significantly enhanced and this makes manufacturing process automation possible.

49.5 Conclusions

Digital manufacturing through virtual manufacturing simulation, and real-time tracking of production data in the supply chain, have led to improved accuracy of information, reduction of human errors, and automation of business operations. This chapter presents a study of digital manufacturing which covers computer simulation of manufacturing processes, and systems for manufacturing process automation based on real-time online capture of RFID data from the movement of materials, people, equipment, and assets in the production value chain. Various case studies are described which involve virtual assembly, VP, PAM, warehouse management, WIP management, and the management of a global production network.

Virtual manufacturing has gone beyond the graphical simulation of product and process design, has accelerated the product development cycle, and deepens the level of customer interaction in the preproduction phase of products. Combined with the growing demand for RFID technology for automatic capture, tracking, and processing of the huge amount of data available in the production of goods and services, digital manufacturing is going to revolutionize the way in which the supply chain is managed. It will also greatly change the behavior of global producers and consumers.

References

49.1   H.C. Crabb: *The Virtual Engineer: 21st Century Product Development* (American Society of Mechanical Engineers, New York 1998)

49.2   W.B. Lee, H.C.W. Lau: Factory on demand: the shaping of an agile production network, Int. J. Agil. Manag. Syst. **1/2**, 83–87 (1999)

49.3   RedEye: http://www.redeyerpm.com (2008)

49.4   Stratasys: http://www.stratsys.com (2008)

49.5   Autodesk: http://usa.autodestl.com (2008)

49.6   F. Kimura: Product and process modelling as a kernel for virtual manufacturing environment, Ann. CIRP **42**, 147–150 (1993)

49.7   Lawrence Associates Inc. (Ed.): *Virtual Manufacturing User Workshop*, Tech. Rep. (Lawrence Associates, Wellesley 1994)

49.8   C. Shukla, M. Vazquez, F.F. Chen: Virtual manufacturing: an overview, Comput. Ind. Eng. **13**, 79–82 (1996)

49.9   M. Onosato, K. Iwata: Development of a virtual manufacturing system by integrating product models and factory models, Ann. CIRP **42**, 475–478 (1993)

49.10   K. Iwata, M. Onosato, K. Teramoto, S. Osaki: A modelling and simulation architecture for virtual manufacturing systems, Ann. CIRP **44**, 399–402 (1995)

49.11   J. Gausemeier, O.V. Bohuszewicz, P. Ebbesmeyer, M. Grafe: Cyberbikes-interactive visualization of manufacturing processes in a virtual environment. In: *Globalization of Manufacturing in the Digital Communications Era of the 21 Century-Innovation, Agility, and the Virtual Enterprise*, ed. by G. Jacucci, G.J. Olling, K. Preiss, M.J. Wozny (Kluwer Academic, Dordrecht 1998) pp. 413–424

49.12   K. Iwata, M. Onosato, K. Teramoto, S. Osaki: Virtual manufacturing systems as advanced information infrastructure for integrated manufacturing resources and activities, Ann. CIRP **46**, 335–338 (1997)

49.13   K.I. Lee, S.D. Noh: Virtual manufacturing system - a test-bed of engineering activities, Ann. CIRP **46**, 347–350 (1997)

49.14   S. Jayaram, H. Connacher, K. Lyons: Virtual assembly using virtual reality techniques, Comput. Aided Des. **28**, 575–584 (1997)

49.15   R. Tesic, P. Banerjee: Exact collision detection using virtual objects in virtual reality modelling of a manufacturing process, J. Manuf. Syst. **18**, 367–376 (1999)

49.16   U. Jasnoch, R. Dohms, F.B. Schenke: Virtual engineering in investment goods industry-potentials and application concept. In: *Globalization of Manufacturing in the Digital Communications Era of the 21 Century-Innovation, Agility, and the Virtual Enterprise*, ed. by G. Jacucci, G.J. Olling, K. Preiss,

49.17   M.J. Wozny (Kluwer Academic, Dordrecht 1998) pp. 487–498

49.17   M. Weyrish, P. Drew: An interactive environment for virtual manufacturing: the virtual workbench, Comput. Ind. **38**, 5–15 (1999)

49.18   RNCOS: RFID Industry – A Market Update, http://www.rncos.com/Report/COM16.htm (2005)

49.19   R. Qiu, Q. Xu: Standardized shop floor automation: An integration perspective, Proc. 14th Int. Conf. Flexible Automation and Intelligent Manufacturing (NRC Research Press 2004) pp. 1004–1012

49.20   C. Poirier, D. Mccollum: *RFID Strategic Implementation and ROI: A Practical Roadmap to Success* (Book News, Portland 2006)

49.21   D. McFarlane, S. Sarma, J. Chirn, C. Wong, K. Ashton: Auto-ID systems and intelligent manufacturing control, Eng. Appl. Artif. Intell. **16**, 365–376 (2003)

49.22   R. Qiu: A service-oriented integration framework for semiconductor manufacturing systems, Int. J. Manuf. Technol. Manage. **10**, 177–191 (2007)

49.23   Kali Laboratories: RFID Implementation, http://www.icegen.net/implement.htm (2004)

49.24   A. Kambil, J. Brooks: Auto-ID across the value chain: from dramatic potential to greater efficiency and profit, White Paper of MIT Auto-ID Center (2002)

49.25   R. Qiu: RFID-enabled automation in support of factory integration, Robot. Comput. Integr. Manuf. **23**, 677–683 (2007)

49.26   R. Moroz Ltd.: Understanding radio frequency identification, http://www.rmoroz.com/pdfs/UNDERSTANDING%20RFID_November22_2004.pdf (2004)

49.27   F. Klaus: *RFID Handbook*, 2nd edn. (Wiley, New York 2003)

49.28   B. Bacheldor: Aircraft Parts Maker Adds Tags to Molds, http://www.rfidjournal.com/articleview/2411/1/1 (2006)

49.29   C.F. Cheung, W.B. Lee: A framework of a virtual machining and inspection system for diamond turning of precision optics, J. Mater. Proc. Technol. **119**, 27–40 (2001)

49.30   W.B. Lee, J.G. Li, C.F. Cheung: Research on the development of a virtual precision machining system, Chin. J. Mech. Eng. **37**, 68–73 (2001)

49.31   W.B. Lee, C.F. Cheung, J.G. Li: Applications of virtual manufacturing in materials processing, J. Mater. Proc. Technol. **113**, 416–423 (2001)

49.32   Y.X. Yai, J.G. Li, W.B. Lee, C.F. Cheung, Z.J. Yuan: VMMC: a test-bed for machining, Comput. Ind. **47**, 255–268 (2002)

49.33   W.B. Lee, J.G. Li, C.F. Cheung: Development of a virtual training workshop in ultra-precision machining, Int. J. Eng. Educ. **18**, 584–596 (2002)

49.34 I. Hipkin: Knowledge and IS implementation: case studies in physical asset management, Int. J. Oper. Prod. Manage. **21**, 1358–1380 (2001)

49.35 Datamonitor: *RFID in Manufacturing: The Race to Radio-Tag is Heating Up in Manufacturing* (Datamonitor, New York 2005)

49.36 D.E. Brown: *RFID Implementation* (McCraw–Hill, New York 2007)

49.37 W.O. Hedgepeth: *RFID Metrics: Decision Making Tools for Today's Supply Chains* (CRC, Boca Raton 2006)

50. Flexible and Precision Assembly

Brian Carlisle

Flexible assembly refers to an assembly system that can build multiple similar products with little or no reconfiguration of the assembly system. It can serve as a case study for some of the emerging applications in flexible automation. A truly flexible assembly system should include flexible part feeding, grasping, and fixturing as well as a variety of mating and fastening processes that can be quickly added or deleted without costly engineering. There is a limited science base for how to design flexible assembly systems in a manner that will yield predictable and reliable throughputs. The emergence of geometric modeling systems (computer-aided design, CAD) has enabled work in geometric reasoning in the last few years. Geometric models have been applied in areas such as machine vision for object recognition, design and throughput analysis of flexible part feeders, and dynamic simulation of assembly stations and assembly lines. Still lacking are useful techniques for automatic model generation, planning, error representation, and error recovery. Future software architectures for flexible automation should include geometric modeling and reasoning capabilities to support autonomous, sensor-driven systems.

Part F | 50

50.1 Flexible Assembly Automation

Flexible assembly automation is used in many industries such as electronics, medical products, and automotive, ranging in scale from disk drive to automotive body assembly. The ubiquitous yardstick for a flexible assembly system is the human. While dedicated machines can far exceed the speed of a human for certain applications, for example, circuit board assembly, there is no automation system than can assemble a hard disk drive one day and an automobile fuel injector the next.

A flexible assembly system must present, grasp, mate, and attach parts to each other. It may also perform quality tests during these processes. In order to be useful, it must compare favorably in terms of development cost and product assembly cost with manual alternatives. To date, automatic assembly has only been justified for applications where volumes exceed tens of thousands of assemblies per year.

Key capabilities in sensing, modeling, reasoning, manipulation, and planning are still in their infancy for automated systems. The following discussion of flexible automation for assembly will illustrate the current state of the art. Further advances in flexible automation will require machine control software architectures that can integrate complex motion control, real-time sensing, three-dimensional (3-D) modeling, 3-D model generation from sensor data, automatic motion planning from task-level goals, and means to represent and recover from errors.

50.1.1 Feeding Parts

Assembly requires picking up parts, orienting them, and fastening them together. In almost all automatic assembly systems in production today, parts are oriented and located by some means *before* the assembly system picks them up. In fact, it should be recognized that orientation has value; if part orientation is lost, it costs money to restore it. Due to this requirement for pre-oriented parts, most automatic assembly systems only deal with small parts, typically less than $10\,\mathrm{cm}^3$ in volume. Traditional small part feeders include indexing tape feeders (Fig. 50.1), tray feeders, tube feeders, Gel Pack (a sticky film in a round frame for semiconductor parts) feeders, and vibratory bowl feeders (Fig. 50.2). Each of these feeders is typically limited to a fairly narrow class of parts; for example, tape feeders are sized to the width of a part; a 3 mm-wide tape cannot feed a 25 mm-wide part. Tape and tray systems have several disadvantages in addition to their cost: they are not space efficient for shipping as parts are stored at a low density, and the packing material can add several cents to the cost of each part. Larger parts are almost univer-

sally transported on pallets, or in boxes or bins, and are at best only partially oriented.

The development of machine vision systems has begun to change this situation (Fig. 50.3). Useful machine vision systems began to emerge in the 1980s. Early systems that could recognize the silhouette of a part in two dimensions cost US$ 50 000–100 000. Today, you can buy two-dimensional (2-D) vision systems for a few thousand dollars, and the price continues to drop rapidly. At the time of writing, 3-D vision systems are beginning to emerge, and in a few cases, are being installed in factories to guide robots in the acquisition of large heavy parts from pallets and even bins [50.1].

Vision systems are allowing the development of part feeders that can separate parts from bulk, inspect the parts for certain critical dimensions, and guide a robot or other automatic assembly machine to acquire the part reliably.

Several such part feeders are shown in Fig. 50.4.

These feeders utilize either conveyors or vibratory plates to separate parts from bulk and advance them under a machine vision system, which then determines if the part is in an orientation that can be picked up by the assembly machine. If not, parts are usually recirculated, since it takes a lot of time to pick them up and hand them off to a part-reorienting mechanism, and then regrasp them. These feeders can handle a wide variety of part sizes and geometries without changing the feeder. In fact, this class of feeder is increasingly being accepted in applications where parts are changed daily or several times a day.

Predicting the throughput of this type of feeder requires predicting the probability that a part will come

Fig. 50.1 Sticky-tape part feeder (courtesy of Pelican Packaging Inc.)

Fig. 50.2 Bowl part feeder (courtesy of Pelican Packaging Inc.)

to rest in a desired stable state (will not topple or roll) when it is separated from the other parts. The author worked with *Ken Goldberg* and others to develop a method for predicting the distribution of stable states from part geometry [50.2]. Part geometries can be complex, and the distribution of stable states may not be obvious. The part illustrated in Fig. 50.5, from a plastic camera, has 12 stable states, of which four are shown. Goldberg et al. developed an algorithm that could predict the distribution of these 12 stable states from a CAD model of the part.

The ability to simulate feeder throughput from an analysis of part geometry is an example of using modeling and geometric reasoning to help design a flexible assembly system.

It is desirable to be able to generate both 2-D and 3-D vision recognition algorithms directly from CAD models of parts. While this sounds simple for the case of 2-D parts, to date the author is not aware of any commercially available vision systems that offer this capability. Real-world challenges from lighting, reflections, shadowing, lens and parallax distortion, camera and lens geometry, etc. combine to make this a challenging task. The task becomes more challenging when 3-D vision is considered. Robust models that take into account these factors need to be developed and extensive algorithm testing done to ensure reliable real-world vision system performance from CAD-generated algorithms.

In summary, recent advances in machine vision are allowing large parts to be picked directly from bins and small parts to be picked from mechanisms that separate them under a camera. These advances allow assembly systems to be changed over quickly to new products without designing and installing new part feeders.

50.1.2 Grasping Parts

In today's commercial assembly systems almost 100% of part grasping is done by either a vacuum cup or a two-finger gripper with custom fingers designed for the particular part to be grasped. Even for systems with the vision-based flexible part feeders described above, if we change parts, we must change the part gripper. Part pick-up strategies are generated by explicit manual training or programming. Obstacle avoidance of other parts generally also requires part-specific programming. This approach works fine for systems that only have to handle one or two types of parts for an extended period of time.

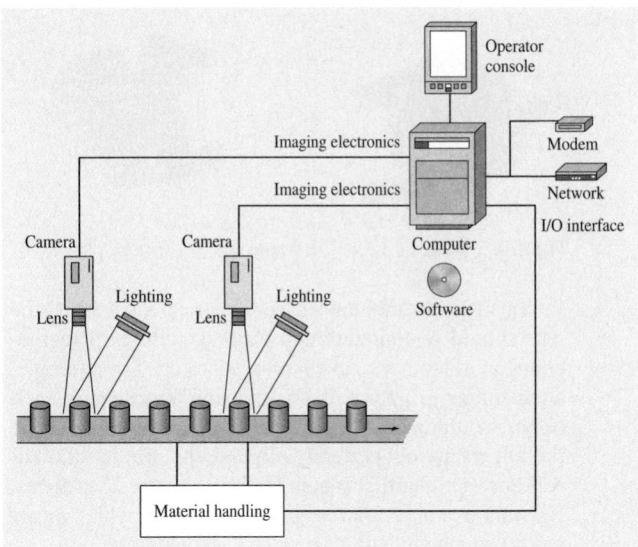

Fig. 50.3 Machine vision system (courtesy of High-Tech Digital, Inc.)

As part dimensions shrink, new considerations enter into gripper design. As part dimensions approach 1 mm, chemical attraction, static electricity or other forces may exceed the force of gravity on the part. So, while it may be possible to pick up a small part with a vacuum needle, it may be necessary to use air pressure or other means to release it. With tiny parts, it can be very difficult to grasp a part with the precision needed to place it at the desired location. Therefore, for most very small parts, machine vision is used to refine the position of the part in the gripper after it has been grasped.

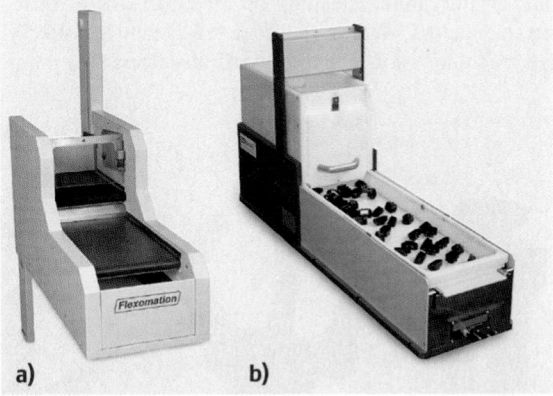

Fig. 50.4a,b Flexible part feeders. (**a**) Adept Technology [50.3], (**b**) Flexfactory [50.4]

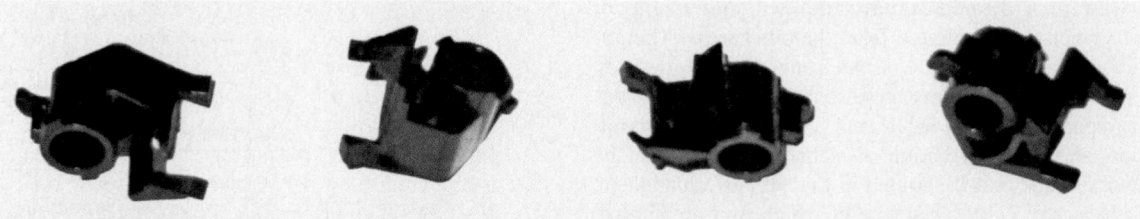

Fig. 50.5 Four of 12 stable states for a complex part

In order to maintain a stable grasp of a part, the part should be uniquely constrained, with gripper mechanical tolerances taken into account; for example, a two-finger gripper with an opposing V groove in each finger would appear to locate a round part. However, if the fingers are not perfectly aligned, the part in fact will roll between parallel plates of the opposite V grooves. For this example, one V groove opposite a flat finger would be a better solution.

In a similar manner, attempting to grasp a prismatic part with two flat fingers is a bad idea. Any slight variation in parallelism of the part or fingers will result in a poor grasp that may not resist acceleration or part mating forces. In this case three pads to define a plane on one finger and a single pad to press the part against this plane on the opposing finger is a better solution and can handle draft angle on parts, finger splay, and other tolerances.

These simple examples illustrate the more general concept of planning and achieving stable grasp in multiple dimensions, and resisting slip and torque in the dimensions important to the process.

In the popular image of a service robot, the robot performs a wide range of menial tasks, perhaps cooking and serving dinner, cleaning up after the kids or dusting the shelves. Within the next few years it is likely that machine vision technology will develop to the point

where, given explicit models of objects, a robot could pick up a large variety of objects from an unstructured environment. However, this will be possible only if it can grasp them.

Multifinger prehensile hands were developed in research laboratories in the 1980s. Several companies now offer commercial versions [50.6, 7]. However, to date, we lack the sensing and control technologies to reorient a part within the hand once it is grasped, except for ad hoc preprogrammed strategies developed manually with a good deal of trial and error. Even the automatic generation of stable grasping strategies for these complex hands is still a topic for research and has not found commercial acceptance.

What is often not realized by the casual observer of these humanoid hands is that the control of these hands is far more difficult than the control of a six-axis robot arm. Planning stable grasps with humanoid hands from CAD models of objects has been the topic of numerous research papers for many years now [50.5, 8] (Fig. 50.6). *Kragic* et al. [50.8] present a typical approach to this issue and point out that having a CAD model and grasp plan are not sufficient to achieve a stable grasp. The robot must also determine the actual orientation of the object. Kragic et al. used simple objects with strong markers so that a simple 3-D vision system could determine the object orientation for the

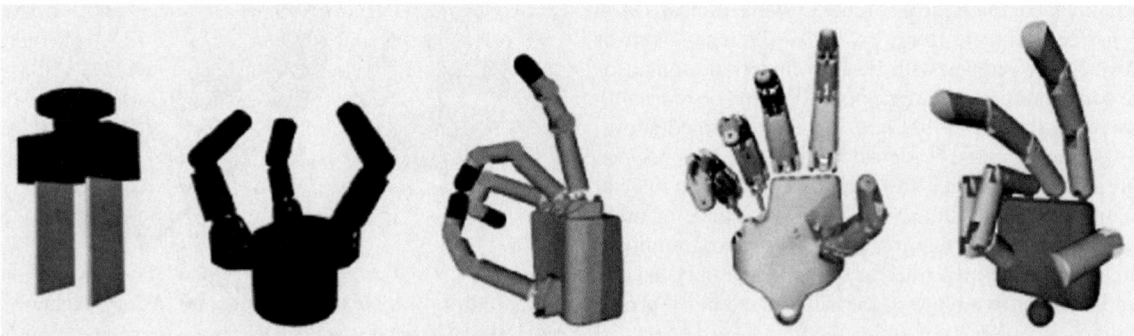

Fig. 50.6 A typical industrial gripper and several humanoid grippers [50.5]

planner to plan a grasp. Further, they point out that not all their grasp plans were stable and discuss the need for further work.

This work further emphasizes the need for a 3-D modeler integrated with the motion control system to support both the grasp planning and 3-D object recognition with machine vision.

50.1.3 Flexible Fixturing

Most assemblies are based around a larger part to which smaller parts are affixed. The larger part typically moves through several assembly stations by means of a material handling system. Each station feeds and attaches one or two smaller parts. Some sort of fixture is usually designed to hold the larger part in a desired orientation.

Current industry practice is for this fixture to be some sort of pallet, or a two- or three-jaw clamping system. In some cases fixtured parts must withstand significant assembly forces from press fits or secondary machining operations.

There is interest in reducing the time and cost to design and fabricate fixtures, and being able to use them again for a different product rather than scrap them when a product changes. This has led to various approaches to flexible part fixturing.

The simplest approach is based on a modular series of adjustable blocks and clamps that can be used to locate and support parts. This approach has been used in machining centers for some time and can provide a high degree of rigidity.

A more general approach was proposed and tested by Sandia Labs in 1996. This was a planar clamp-ing scheme in which a geometric planner analyzed the perimeter of a part, and placed some pins in a location on a grid such that a single clamp could uniquely restrain the part from translation and rotation [50.10].

More recently, a team at Sandia extended this work to a 3-D planner [50.9] (Fig. 50.7). Inputs to this planner include a 3-D (ACIS) model of the part, a fixture kit specification detailing the fixture building tools and clamps, friction data for the fixture components and the workpiece, and disturbance forces to be applied to the fixtured part. The tool then outputs a series of fixture designs with a quality score for each.

For larger workpieces, one or more multiaxis robots are now being used to hold and reposition workpieces while other robots add parts to the assembly. Parallel link structures are being offered commercially for orienting larger workpieces (Fig. 50.8). In November 2007, at the International Japan Robot Exhibition in Tokyo, Yaskawa showed a two-armed robot performing assembly, with one arm holding a workpiece and a second arm inserting components. All programming for this demonstration was done manually.

Fixture design is essentially the same problem as grasping design except there is a strong commercial need for fixtures to be low cost, as for many systems there may be a large number of fixtures in the system. Generating reliable fixture designs quickly from CAD geometry and knowledge of assembly forces remains an interesting challenge. With the advent of rapid

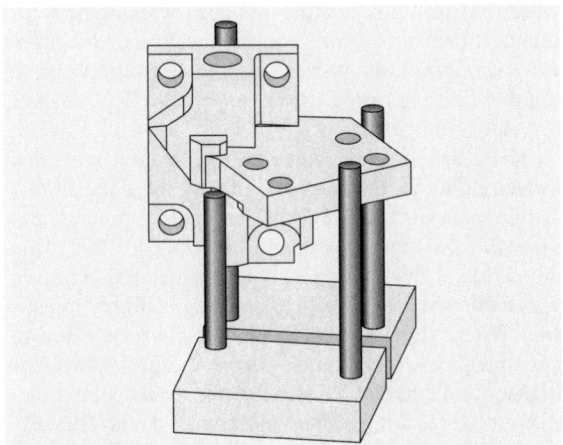

Fig. 50.7 A computer-generated fixture (after [50.9])

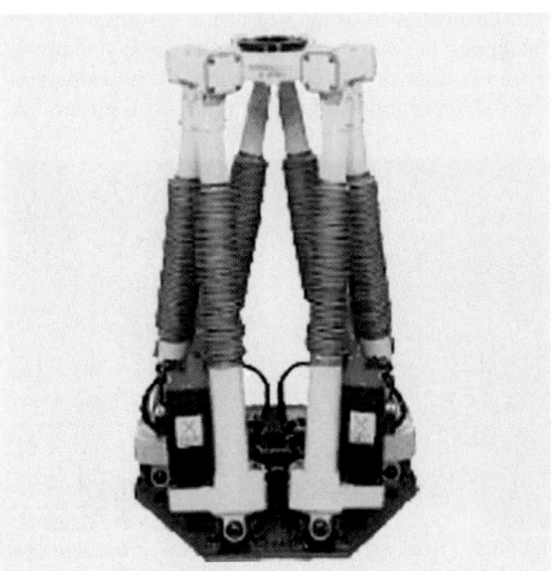

Fig. 50.8 A robot fixture

prototyping machines, it is now possible to quickly fabricate fairly complex fixture geometry. It is important that fixtures provide kinematic grasps that can withstand assembly (or machining) forces.

50.2 Small Parts

50.2.1 Aligning Small Parts

For many industries part dimensions are shrinking to a point where humans can no longer handle or assemble parts. Parts with submillimeter dimensions often have manufacturing tolerances that can be a substantial percentage of the overall part size, and may not be registered to any part dimension that can be grasped; for example, laser diodes emit light with a Gaussian intensity profile whose peak must be measured optically to position the laser diode in an optical assembly. Cancer cells in a fluid must be located optically so they can be sampled by means of a microliter pipette.

While semiautomated systems employing people looking through microscopes are currently used for many such applications, advances in integrating machine vision with motion control now allow parts to be actively steered into position, using vision to take a series of pictures to measure alignment error. The workpiece is moved into position until the alignment error falls below a threshold. Commercial applications use 2-D machine vision for this [50.11] (Fig. 50.9).

In the next several years it is likely that this work will be extended to 3-D vision. There are some significant challenges in using 3-D vision for aligning very small parts. For these applications the parts typically fill a large portion of the field of view of the camera(s). For 3-D applications this means that, as a gripper approaches a part, the part image will change dramatically in size and perspective. For applications where a number of pictures will be taken and processed, the vision system must be able to recognize these changing images without explicit programming for each image. In addition, for the case where small parts are being actively mated to each other, one part may obscure the second part so that only a few features on the second part may be visible. Lighting, reflections, and depth of field all contribute to make this work challenging.

Automatically generating 3-D vision recognition and part-mating algorithms is another area where the integration of a 3-D modeling system and a 3-D motion planning system with the actual online motion control system will be necessary.

50.2.2 Fastening Small Parts

There are many well-developed techniques for fastening large parts together. However, when part geometries shrink to submillimeter levels, fastening techniques are largely confined to adhesive bonding, eutectic bonding such as soldering, and welding.

Since all of these material-based fastening techniques involve physical changes in the material, they all tend to introduce dimensional changes during the fastening process; for example, in aligning laser diodes in fiber-optic transceivers, the desired optical alignment is within $10\,\mu\text{m}$. Epoxy curing and laser welding both introduce dimensional shifts equal to the desired assembly tolerance. It took the photonics industry over 5 years to improve first-pass production yields from 30% to over 90% due to this issue.

There are several approaches to dealing with this problem. One is to model and predict the dimensional change that will occur during the bonding process and offset the assembly positions to account for this. This only works if the amount of bonding material is highly repeatable and the assembly geometry is highly repeatable. Where the assembly geometry must vary due to part tolerances, for example where the peak power of a laser diode moves relative to the package outline, a fixed offset is not possible. In this case a real-time offset would need to be computed based on the alignment geometry and bonding material properties.

Fig. 50.9 Visual servoing: steering parts into alignment using vision in motion control loop (courtesy of Precise Automation)

A second approach is to bond the parts, then measure the resulting geometry, and deform the resulting bonded structure to achieve the desired alignment. This technique is employed by the photonics industry to align laser diodes, and is referred to as *bend align*. Techniques must be used to stress-relieve the assembly after deforming so that residual stress does not affect alignment over time.

A third approach is to create a separate, rigid, kinematic support structure for the parts, which controls part alignment, and use a parallel bonding process around this support structure to place it in compression. A properly designed support structure can then resist the forces from the bonding material. The author suggested the kinematic structure in Fig. 50.10, in which a fiber is mounted in either a solid ferrule or a V-block ferrule. The ferrule rests on a hollow ceramic tube with a V notch. A solder preform is placed inside this tube. The assembly system aligns the fiber and the preform is heated, bonding the ferrule to the mounting substrate, with the ceramic tube resisting the compression forces when the solder solidifies. A sixth degree of freedom can be added by using a half-sphere instead of a cylinder for the fiber ferrule.

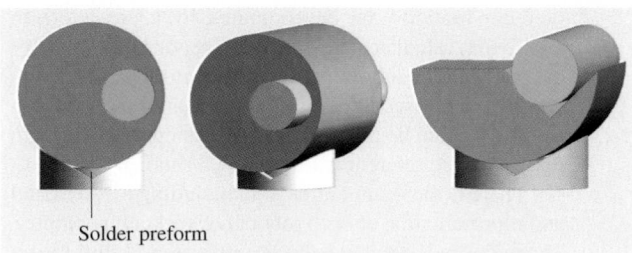

Solder preform

Fig. 50.10 A five-degree-of-freedom kinematic mount for optical fibers (courtesy of B. Carlisle)

A fourth approach proposed in the photonics industry is to bond parts to microservo mechanisms which can be actively controlled after bonding to align parts [50.12]. However, this approach is relatively expensive and may suffer from reliability problems over extended service periods.

In general, the stable alignment of small parts to micron or submicron tolerances remains a challenge, which will benefit from advances in modeling stresses and bonding material deformation in systems that can be linked to real-time motion systems.

50.3 Automation Software Architecture

The foregoing discussion points out the need for some new features in a flexible automation programming language. Some of these features are available in today's robot languages; some of them (3-D modeling) are available in separate packages; and some are not available yet, other than in research environments. Surprisingly, some robot vendors still do not offer general-purpose programming languages, preferring to offer application-specific programming tools. For a current state-of-the-art robotic programming language, see [50.13]. Also see Chap. 22 on *Modeling and Software for Automation*.

A general-purpose automation programming language should include the features described below.

50.3.1 Basic Control and Procedural Features

In addition to common control features such as looping, branching, multithreading, mathematical functions, and data structures, an automation language should be able to coordinate multiple mechanisms on a common time base. It should be possible to do this over a network, with a structure that allows master–slave or peer-to-peer communication and control. It should be possible to use sensor data to alter motion in real time.

50.3.2 Coordinate System Manipulation

Robots and other automation equipment are usually programmed in a Cartesian coordinate system, which may be different from the joint coordinates of the robot. Mathematical models, loosely referred to as *kinematics*, transform joint coordinates into Cartesian coordinates via matrix algebra. Modern robot languages store Cartesian positions in a format known as a homogeneous coordinate transformation. These coordinate transformations can be multiplied together to offset positions relative to a pallet, for example, or a tool. Sensor data must be transformed from the sensor coordinate system, for example, a camera frame, to the robot coordinate system. It is useful to be able to compute elements of these coordinate transformations in real time at the application level; for example, circular, elliptical, spline or other motions can be computed by a procedure in a loop that calculates the destination coordinates in a location variable and then calls a trajectory generator to compute and execute an incremental motion. Tracking a conveyor can be accomplished by using an encoder to update a base reference frame in real time.

There is increasing interest in including a 3-D model and representation of both robot and workcell geometry. Today's robot languages do not offer the ability to define the surfaces of the robot or workcell in 3-D space, other than some simple planes and cylinders. Accurate representation of 3-D surfaces would be very useful for collision avoidance, path planning, and safety systems.

There is also increasing interest in improving the absolute accuracy of robots and other automatic machines. Today's kinematic models of six-axis robots assume that the physical robot is manufactured perfectly, i.e., that right angles are perfect, link lengths are perfect, there are no deflections from gravity or loads, etc. For offline programming of robots working in complex environments, for example, spot-welding of an automobile, it is desirable for the robot motion to be one to two orders of magnitude more accurate than typical manufacturing tolerances permit. Recently, more complete models of robots have been developed in offline simulators. Actual robots can be programmed to go through a calibration routine and a more accurate model of the robot is built in the simulator. Then the simulator commands an offset path which is downloaded so the real robot will make accurate motions. This is currently an offline function. Improved robot controls will include this capability as an online function.

50.3.3 Sensor Interfaces and Sensor Processing

Machine vision, force sensors, and other sensors have been integrated with commercial robot control systems since the early 1980s. However, what is needed today is the ability for sensors to deal with 3-D geometry, and for sensor programs to be generated automatically from 3-D geometry.

In fact, it would be very useful if sensors could be used to create 3-D models when no explicit models exist. A mobile robot entering a new environment should be able to build up a model of the environment from sensor data. Extensive work in the research community has shown success building 3-D models from a series of 2-D images [50.14] as well as fusing sensor data from stereo vision and laser range finders for complex navigation tasks such as the Defense

Advanced Research Projects Agency (DARPA) Grand Challenge competitions in 2006 and 2007, in which multiple teams of researchers developed vehicles that could build models and navigate through unknown outdoor environments.

Sensor-generated models must then be available to the motion planning system so the robot can plan moves, update progress through the environment, avoid collisions, and detect and recover from errors.

50.3.4 Communications Support and Messaging

Many commercially available robot controllers currently support only limited-bandwidth messaging and file handling. With the rapid development of distributed motion control where motion can be coordinated at frequencies of 1 kHz or more over a network, the technology now exists for higher-bandwidth, time-synchronized communication; for example, a mobile camera could broadcast an image to a group of mobile devices (imagine soccer-playing robots) that also exchange strategy and planning information. The automation language and operating system should be capable of handling high-bandwidth communications without interfering with other deterministic tasks such as trajectory planning, image processing, and servo loops. The ability to set time slots and task priorities in the operating system becomes very important as communication loads increase. Personal computers remain disappointing in this regard, with large, unpredictable communication delays.

50.3.5 Geometric Modeling

From much of the foregoing discussion it should be clear that the author believes that automation languages should be extended with 3-D modeling systems. One important difference though, is that robotic and other automation systems are dynamic, while many modeling systems are static, or updated at a low rate. In order to be useful, an online modeling system needs to be capable of being updated at rates similar to those in 3-D video games, as robots can make large motions in a few hundred milliseconds. It is likely that some of the simplifications and data-compression methods used in video games may be useful for real-time geometric modeling for motion control.

Eventually, it may be useful to include dynamic models as well as structural deflection models in automation languages. Today's robots are large, rigid, heavy structures with masses that exceed their rated payloads by a factor of ten or more. Lighter, more flexible robots would waste less energy, but will be harder to control and harder to program without some capability to predict their trajectories under load.

50.3.6 Application Error Monitoring and Branching

In most industrial robot applications, over 50% of the programming effort is devoted to anticipating, sensing, and recovering from possible errors. Today, much of this programming is done on site, in an ad hoc manner, and takes a long time to develop and debug. As more sensors are used, this problem becomes larger, as sensors can introduce new errors. It may not be obvious to the programmer why a sensor-driven system is not reliable. To aid debugging, data-logging features, time-stamping of data and communication messages, and single stepping of motion programs are becoming common automation language features. In assembly systems, many errors are due to poorly understood or poorly modeled part tolerances. Often weeks or even months of testing is required for a production system to meet reliability standards. This testing is used to find software bugs, make sensors reliable, and make the system robust within a statistical range of part tolerances.

However, we still do not have general methods to analyze assembly systems for errors, represent errors or generate error monitoring and recovery strategies. Work by Deming and others in statistical process control contributed greatly to understanding manufacturing process tolerances and designing products and processes for reliable production with known tolerances. However, these techniques, while used in the metal-forming and semiconductor industries, seem to be largely absent in automated assembly systems. In 1997 *Carlisle* and *Craig* developed a simulation tool [50.15] for assembly tolerance process analysis that was used by Nokia to analyze and improve cellphone production yields. However, since assembly systems vary so dramatically, there are few, if any, generally accepted practices for modeling assembly processes and part tolerances and predicting and improving yields.

It is not clear if it is productive to try to predict errors. In general, an error is a deviation from a plan. It may be more productive to detect errors quickly and for the system to have enough geometric and sensor data to make a new plan quickly to try to recover from the error. Assembly systems with this capability have yet to be demonstrated.

50.3.7 Safety Features

The cost of assembly robots and sensors is coming down quickly. At the same time the speed of these devices is increasing. Motions of 1 m in a fraction of a second are now common. As a result, robots can present a substantial danger to humans who may enter the robot workspace.

The industry approach to this issue is to create walls around the robot with sensors and interlocks to prevent people from entering the robot workspace when it is moving under computer control. This approach is both expensive and inefficient. A US$ 15 000 robot may be surrounded by US$ 5000 of screens, light curtains or safety mats. Creating cells with walls tends to require more floor space for each cell.

More generally, there are more and more applications where it is desirable to have robots work with, and in some cases touch, people. To address this issue in a general way we need control systems that can model the robot's structure as well as the environment, and sensors that can detect people entering the robot's workspace. We need motion control systems that can respond dynamically to space intrusion and modify the motion appropriately. An operator should be able to walk up to a robot workcell and load a new tray of parts in the workspace without fear of injury in the same manner that he or she would interact with a human assembler.

50.3.8 Simulation and Planning

It is time for robot simulation to move from an offline capability to an online capability, where the control system contains a real-time geometric simulation of the complete assembly system. Simulation systems are now widely used for programming robots for spot-welding, arc-welding, and some material handling tasks. However, they could be more widely used for many applications described here, including programming flexible part feeders, programming 2-D and 3-D vision systems, optimizing sensor-driven motions through workcells, detecting and recovering from errors, and allowing robots to interact safely with people.

Online simulation offers the opportunity to develop high-level representations of common tasks; for example, a task-level command such as "Drive a screw at location hole 1 to torque X" is much easier for an application programmer to work with than many lines of detailed programming code. However for task-level instructions to be fairly general, they should be able to access geometric and process information from databases, and a simulation system with motion planning ability and knowledge of the robot and workcell should generate the motion plan for the task. If an error occurs, the planning system should generate a new plan online.

More generally, simulation and task-level planning are necessary tools to address the broader issue of robot interoperability, and program and data sharing. Users would like to be able to move programs from one brand of robot to another. This will remain an elusive goal until robots can be instructed at a very high level of abstraction, in almost the same manner as a human is given high-level, abstract directions along with some data, and figures out how to perform the task. The instruction "Bolt down the manifold cover", given a CAD model, is far easier to translate than today's explicit low-level programs.

50.3.9 Pooling Resources and Knowledge

In the last few years it has become technically possible for people to share knowledge and collaborate remotely over the Internet. This has created the potential for open-architecture systems where many people pool knowledge to create complex systems efficiently. The Linux operating system is one example that has been widely accepted by an enthusiastic user base.

An ideal robot programming system would have a core set of functions that could support a wide range of applications, yet it would be open enough and extensible enough that users could create new capabilities that could be shared or integrated. There is an interesting balance between providing enough structure that programs from different developers can be integrated while allowing enough flexibility that users can add new features easily.

50.4 Conclusions and Future Challenges

In general, we need to raise the abstraction level of robot programming if robots are going to be able to perform increasingly complex tasks in increasingly less structured environments. Programming languages for robots need to incorporate modeling, sensing, and planning capabilities to allow libraries of tasks and actions to be compiled. These high-level tasks need to be robust and

safe. We also need to think about how to pool ideas and resources from many different developers and locations, to build up the large knowledge base that will be needed for robots to move from simple, highly structured tasks to complex, unstructured tasks. A more thorough discussion on the future of flexible and precise automation is provided in Chap. Sect. 21.3 of the handbook.

50.5 Further Reading

1. S. Y. Nof (Ed.): *Handbook of Industrial Robotics* (Wiley, New York 1999)

2. B. Siciliano, O. Khatib (Eds.): *Springer Handbook of Robotics* (Springer, Berlin, Heidelberg 2008)

References

50.1   A. Shafi: *Bin Picking Axle Shafts* (Shafi Inc., Michigan 2008), http://www.shafiinc.com/solutions/sol36/test_1.htm (last accessed 2008)

50.2   K. Goldberg, B. Mirtich, Y. Zhuang, J. Craig, B. Carlisle, J. Canny: Part pose statistics: estimators and experiments, IEEE Trans. Robot. Autom. **15**(5), 849–857 (1999)

50.3   B. Carlisle: Feeder developed by at Adept Technology and licensed to Flexomation Inc. http://www.flexomation.com/ (last accessed 2008)

50.4   Flexfactory: *Feeder developed by Flexfactory AG* (Flexfactory, Dieticon 2008), http://www.flexfactory.com/ (last accessed 2008)

50.5   A. Miller, P. Allen: From robotic hands to human hands: a visualization and simulation engine for grasping research, Ind. Robot **32**(1), 55–63 (2005)

50.6   Barret Technology: http://www.barrett.com/robot/products-hand.htm (2008) (last accessed 2008)

50.7   Schunk Gripper: http://www.schunk.co/ (2008) (last accessed 2008)

50.8   D. Krajic, A. Miller, P. Allen: Real time tracking meets online grasp planning, Proc. 2001 ICRA IEEE Int. Conf. on Robotics and Automation, Vol. 3 (2001) pp. 2460–2465

50.9   R. Brown, R. Brost: A 3-d Modular Gripper Design Tool, Sandia Rep. SAND97-0063, UC-705 (1997)

50.10   R. Brost, K. Goldberg: A complete algorithm for designing planar fixtures using modular components, IEEE Trans. Robot. Autom. **12**(1), 31–46 (1996)

50.11   J. Shimano: *Visual Servoing Taylor Made for Robotics* (Motion Systems Design, 2008), (last accessed 2008)

50.12   A. Hirschberg: Active alignment photonics assembly, US Patent 6295266 (2001)

50.13   B. Shimano: *Guidance Programming Language* (Precise Automation, 2004–2008), www.preciseautomation.com (last accessed 2008)

50.14   M. Lin, C. Tomasi: Surface Occlusion from Layered Stereo. Ph.D. Thesis (Stanford Univ., Stanford 2003)

50.15   J. Craig: Simulation-based robot cell design in AdeptRapid, Proc. 1997 ICRA IEEE Int. Conf. on Robotics and Automation, Vol. 4 (1997) pp. 3214–3219

Part F | 50

51. Aircraft Manufacturing and Assembly

Branko Sarh, James Buttrick, Clayton Munk, Richard Bossi

Increasingly the manufacturing of complex products and component parts involves significant automation functions. This chapter describes a cross section of automated manufacturing systems used to fabricate, inspect, and assemble aircraft. Aircraft manufacturing cost reductions were made possible by development of advanced technologies and applied automation to produce high-quality products, make air transportation affordable, and improve the standard of living for people around the globe. Fabrication and assembly of a commercial aircraft involve a variety of detail part fabrication and assembly operations. Fuselage assembly involves riveting/fastening operations at five major assembly levels. The wing has three major levels of assembly. The propulsion systems, landing gear, interiors, and several other electrical, hydraulic, and pneumatic systems are installed to complete the aircraft structurally and, after functional tests, it normally gets painted and goes to the flight ramp for final customer acceptance checks and delivery. Aircraft manufacturing techniques are well developed, fabrication and assembly processes follow a defined sequence, and process parameters for manual and mechanized/automated manufacturing are precisely controlled. Process steps are inspected and documented to meet the established Federal Aviation Administration quality requirements, ensuring reliable functions of components, structures, and systems, which result in dependable aircraft performance.

Part F | 51

The emergence of the industrial age brought significant changes and impacts on living conditions of human societies. Innovative product development in a free market environment, driven by a desire to improve standards of living, led to revolutionary products in computing, communications, and transportation that impacted all levels of human activities. The key enabler of this progress was improved industrial productivity, which climaxed with automation technologies, starting with hard automation (suitable for mass production of consumer goods) and evolved into intelligent automation (selectively used for batch-type single-component fabrication) using computers and software to precisely control complex processes.

Automation has successfully met the need to improve quality, reduce cost, and improve ergonomics of aircraft fabrication and assembly. The conversion to digitally defined aircraft and advancements in ma-

chine tools have enabled widespread use of automation. The trend to machine/fabricate components accurately with automated machines started with a Massachusetts Institute of Technology (MIT) demonstration of the first-ever developed numerically controlled (N/C) machine in 1952. This quickly led to a huge machine tool market that enabled rapid production of precision machined parts and accurate assembly of large aircraft structures.

51.1 Aircraft Manufacturing and Assembly Background

Aircraft manufacturing cost reductions were made possible by development of advanced technologies and applied automation to produce high-quality products, make air transportation affordable, and improve the standard of living for people around the globe [51.1–3].

Fabrication and assembly of a commercial aircraft, such as the one depicted in Fig. 51.1, involves a variety of detail part fabrication and assembly operations. A number of raw materials are machined and fabricated into detail parts, which are then assembled into various levels of structural configurations. Starting with basic assembly of detail parts into simple panels, they are then combined into super panels and higher-level assemblies to produce the fuselage, wings, and finally the complete aircraft. Integral designs and efficient production of new aircraft involve tradeoffs to optimize materials, the number of parts and size of structures, use of innovative processes, and adaptations of appropriate existing equipment and facilities.

Fuselage assembly involves riveting/fastening operations at five major assembly levels. At the first (lowest) level, skins, doublers, longerons, and shear ties are joined to form a single panel. The size and complexity of single panels is primarily driven by aerodynamics, load requirements, and function in operation of the aircraft. At the second assembly level, several single panels are joined using additional detail parts along longitudinal and radial joints into super panels. Frames are usually attached to shear ties during this operation. Assembly of the floor grid structure (also second-level assembly) joins the floor beams and seat tracks. At the third assembly level, half shells are created by joining super panels and the floor grid is often attached to the upper or lower half-shell of the fuselage. Fourth-level assembly involves longitudinal joining of barrel halves to complete the individual barrel structures. Fifth-level (highest) assembly does the 360° radial joins which fasten the nose, forward, center, and aft fuselage barrels together. Inside these structures, multiple radial doublers, couplings, and fittings are installed to complete the aircraft's fuselage structure.

The wing has three major levels of assembly. The first (lowest) assembly level joins the upper and lower skin panels, spars, and bulkheads (consisting of N/C-machined skins, stringers, and stiffeners). At the second assembly level, spars and bulkheads are joined to form the wing grid, to which skin panels are attached to create the wing box. The third (highest) assembly level joins leading and trailing edge components to the wing box to complete the wing structure. The wing is joined to the fuselage in the appropriate sequence to complete the airframe.

The propulsion systems, landing gear, interiors, and several other electrical, hydraulic, and pneumatic systems are installed to complete the aircraft structurally and, after functional tests, it normally gets painted and

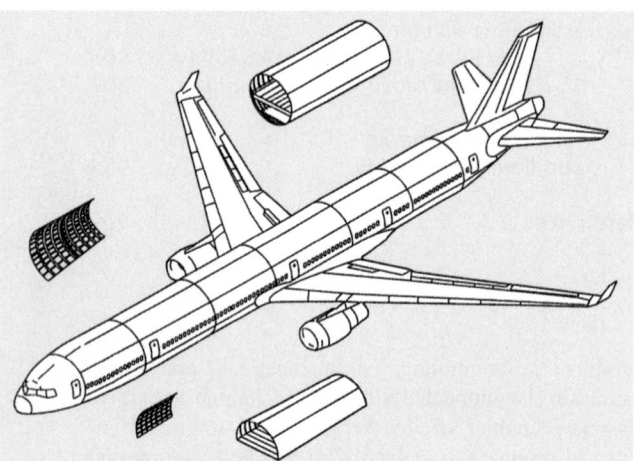

Fig. 51.1 Major aircraft fuselage and wing components layout

goes to the flight ramp for final customer acceptance checks and delivery.

Aircraft manufacturing techniques are well developed fabrication and assembly processes that follow a defined sequence, and process parameters for manual and mechanized/automated manufacturing are precisely controlled. Process steps are inspected and documented to meet the established Federal Aviation Administration quality requirements, ensuring reliable functions of components, structures, and systems, which result in dependable aircraft performance.

All activities at aircraft factories are organized around flow of materials, parts, and structures to the final assembly line. In the early stages of aircraft manufacturing, detail parts are fabricated (involving machining, heat treatment, stretch forming, superplastic forming, chemical treatment, composite material layup, curing, trimming, etc.), followed by part inspection (x-ray, ultrasonic, etc.). The next manufacturing steps focus on the assembly of detail parts into subassemblies and larger structures using both manual assembly tasks and automated machinery (C-frame, ring riveters, etc.), and also moving lines for final aircraft assembly.

Due to economic pressures and ergonomic necessities, the majority of manual aircraft manufacturing has been replaced during past decades by mechanized and/or automated processes and systems, yielding significant process and cost saving improvements; for example, the productivity of machining processes has generally improved by a factor of ten, with some highly automated assembly processes enjoying improvements in excess of a factor of 15.

51.2 Automated Part Fabrication Systems: Examples

Automated aircraft part fabrication involves a variety of manufacturing techniques and systems, all tailored to processing specific materials and part configurations, ranging from aluminum and titanium alloys to carbon-fiber epoxy materials, using intelligent automation to produce strong, lightweight parts at affordable/competitive costs.

51.2.1 N/C Machining of Metallic Components

Process Description

Since the advent of metallic airframe construction, airplane manufacture has been machining intensive, largely because the starting material forms (i.e., plate, extrusion, die forging, etc.) were not available in near-net shapes. To minimize airplane fly-weight and ensure good fatigue life, most metallic surfaces are machined to obtain the final component configuration and achieve a specified surface finish. An important expression in the aerospace machining industry is the *buy-to-fly ratio*, which indicates the ratio of excess material removed during a given machining operation versus the remaining material that flies away on the airplane. For an average commercial aircraft N/C-machined part, this ratio is about 8 : 1.

The machining process usually employs a cutter mounted in a rotating spindle, where the spindle or part can be moved relative to one another by a numerical controller (N/C) using servo motors. The spindle revo-lutions per minute (RPM) may vary from 0 to 40 000, depending on the material, with the largest wing skin mills employing multiple spindles with power ratings up to 200 horsepower (HP). N/C machine tools commonly used within the aerospace industry are some of the largest machine tools in the world, with skin mill bed sizes ranging up to 24 ft wide by 270 ft long [51.4–8].

Wing Skin Mills

The wing skin mill shown in Fig. 51.2 is capable of machining two wing skins simultaneously. The aluminum plate from which the wing skins are produced is held down to the skin mill bed using a vacuum. Typically the aerodynamic outer wing surface, known as the outside mold line (OML), is machined first. Once completed, the wing skin is flipped over, the vacuum is reapplied, and machining on the inside mold line (IML) is completed. The IML contains pads and other features which mate to other wing structure components such as wing ribs and stringers. In addition, the wing skins, which are thickest near the fuselage, taper down to approximately 0.25 in thick at the outboard wing tip. Thickness tolerances for these flight critical components are typically held to ±0.005 in Mammoth gantries (weighing nearly 30 t) carrying two 200 HP machining spindles over a wing skin move with precision, while holding the necessary tolerances. Face mill cutters up to 1 ft in diameter are employed to quickly cover the vast expanse of the wing skins, generating a large volume of

Fig. 51.2 Cincinnati milacron wing skin mill and wing skin after machining

aluminum chips, which are collected by a vacuum chip collection system requiring a 100 HP motor.

Customized Long-Bed Gantry-Style Milling Machines

Dedicated long-bed purpose-designed gantry-style milling machines have been the traditional choice for machining major wing structural components (i.e., stringers, spar chords, and channel vents). These machines employ multiple high-power (112 kW), low-RPM spindles, each capable of high material removal rates (MRR). Each wing component (upper chord, lower chord, stringers, and channel vents) for each airplane model has a dedicated part holding fixture. Large-diameter cutters mounted in steep-taper tool holders experience high cutting forces and bending mo-

ments. This technology is still appropriate when high MRR is required for long extruded parts of adequate stiffness. These machines have been regarded as the aerospace machining standard for 50 years.

The right side of Fig. 51.3 shows a typical 110 ft-long completed part after machining and a typical cross section of the extrusion that makes this part. The cross-sectional area reduction is evident in this illustration. Approximately 12 kg of material is machined away for each (1 kg) of flyaway part. When buy-to-fly ratios for various manufacturing methods are compounded by the sheer size of commercial airplanes, it becomes obvious how an airframe manufacturer can produce over 35 million pounds (16 Gg) of aluminum chips annually. This is roughly equivalent to the airframe weight of 100 Boeing 747 aircraft.

Fig. 51.3 Dedicated spar mill gantry with typical cross section and machined part

High-Performance Machining

As future airframe designs are considered, more emphasis will be placed on component producibility and manufacturing costs. Traditional built-up assemblies are being replaced by monolithic designs. Contrary to popular belief, the value for the airframe producer to utilize high-speed machining techniques does not lie solely in reduced machining cycle times. Economies of scale are gained not by machining cycle time reduction alone, but by conversion of multiple-piece assemblies to monolithic components. This ensures a more accurate final part and the elimination of assembly tooling, labor, equipment, and facilities previously required by built-up designs. High-performance machining provides designers a manufacturing process to produce thin-wall monolithic parts quickly with minimal distortion.

However, monolithic designs are placing new burdens on high-performance machining technology. Such designs often require more material be removed during the machining process than previously required for built-up or assembled sheet metal components. It is difficult to use extrusions or die forgings for monolithic airplane components that are long with significant cross-sectional changes. Components up to 110 ft (35 m) with a tapering cross-section must either be produced from multiple die forgings that are joined or machined from plate stock (if they are designed as a monolithic part). This requires removing large amounts of material, as depicted in Fig. 51.3. Increased buy-to-fly ratios will necessitate that more emphasis be placed on maximizing material removal rates in the future.

51.2.2 Stretch Forming Machine for Aluminum Skins

Machine Description

The machine depicted in Fig. 51.4 is capable of stretch-forming large contoured fuselage skins. The major machine components are:

1. A die table for supporting and moving stretch form dies (dies contain the configuration of the final skin and are placed on the table) vertically during the stretch forming process.
2. Two articulating jaws (one on each side of the table), consisting of multiple jaw segments, enabling articulation of each jaw around longitudinal and transversal axis to accommodate the contoured skin geometry. Jaw segments have built-in hydraulic clamps which firmly clamp the metallic sheets prior to the stretch-forming process. During the forming process each jaw moves longitudinally away from the die table.
3. A computerized numerical control (CNC) machine controller, which executes stretch-forming programs by activating/moving machine components [51.1–3].

Heat-Treat and Stretch-Forming Processes

Common structural materials used for fuselage skins, fuselage frames, and stringers are the high-performance aluminum alloys, namely the 2000 series and 7000 series aluminum alloys (i. e., 2024, 7075). Both of these aluminum alloys are heat-treatable for strength, toughness, and corrosion resistance. To achieve high strength,

Fig. 51.4 Stretch-forming machine for skins

Technical data
- Stretch forming of large contoured fuselage skins
- Machine size: $L = 20$ m; $W = 4$ m; $H = 3$ m
- Sheet size max. $t = 6$ mm; $W = 2.5$ m; $L = 12$ m
- Articulating jaws stretch force = 1500 tons
- Articulating jaws min. radius: longitudinal axis = 10 m
- Articulating jaws min. radius: transverse axis = 2 m
- Stretch die table max force = 1000 tons
- CNC controller
- All machine motions can be actuated manually or controlled by program

the 2024 alloy, commonly used for fuselage skins, is heat-treated in furnaces up to 496 °C. A quenching process from these elevated temperatures dissolves the alloy constituents (i. e., such as copper) in a solid state in the solid aluminum alloy. After quenching, room-temperature aging occurs, causing copper constituents to *precipitate* along the grain boundaries and along slip planes of the alloy. This action distorts the crystal lattice interferes with any smooth slip process, resulting in increased strength of the material. Immediately after quenching, the material is relatively soft and can be formed as in as-quenched (AQ) tempering. However after about 20 min the alloy will start *room*-temperature aging and strengthen to the T4 temper condition. Room-temperature aging is also referred to as *natural* aging. The final T4 strength obtained is about 96 h.

Alloy sheets are moved from the quenching system/equipment to the stretch-forming machine and laid up on the stretch die, then both ends of the sheets are pushed into the jaws and jaw segment clamps are activated, clamping onto sheet ends. Jaw segments are configured around the longitudinal axis to accommodate skin shape, dictated by the stretch die geometry, and longitudinal forces are applied to the sheet by driving jaws away from the table. At a certain point, the table is activated, moving the stretch die vertically while the jaws rotate around the transversal axis, pulling the sheet and forcing it to comply to the stretch die geometry. After sufficient stretch (plastic deformation) is achieved, the jaws and table reverse direction, relieving tension on the sheet once it has attained the desired skin geometry (and allowing for some spring-back). Ideally, aluminum skin stretch form dies are built with spring-back compensation. This is especially important for large contoured skins. Software for stretch die design tools are readily available and have proven very useful.

For many decades, the stretch forming process was (and partially still is) a black art. It requires very experienced personnel to drive machine elements (table, jaws) in linear and rotational axes while observing the skin during stretch-forming operations. Variations in skin and stretching behavior result from changes in material properties during the incubation time and make it difficult to establish precise process parameters. After years of experimentation, and collecting empirical data (required degree of stretch, etc.), the majority of skin form operations can be computer controlled or at least semi-automated, whereby the operator observing the skin behavior can make slight adjustments to the degree of stretch to compensate for material property variations. Programs can be generated offline, or

recorded/stored in the teach mode on the machine during the stretch-forming process.

51.2.3 Chemical Milling and Trimming Systems for Aluminum Skins

System Description

Chemical milling is a material removing process using chemical reaction to dissolve material in certain locations to produce contoured skins with variation in cross sections (thickness), accommodating changing design load conditions along the fuselage. This process involves several subsystems:

1. Galvanic treatment tanks for cleaning and surface preparation of stretch-formed skins
2. A five-degree-of-freedom (DOF) robotic system applying a mask to the skin surface
3. A five-DOF gantry robotic system and flexible pogo fixture using a carbon-dioxide laser to scribe the mask, enabling mask removal in certain skin locations
4. Chemical milling and galvanic treatment tanks to perform metal removal (chemical milling)
5. A five-DOF CNC gantry system with a flexible pogo table to trim and drill skins [51.1–3].

Process Description

Stretch-formed skins have to be cleaned and surface prepared for the application of the chemical milling mask. A robotic system under program control automatically applies the mask of defined thickness to both

Fig. 51.5 Skin with scribed mask

sides of the skin, moving a spray nozzle along optimized patterns to achieve homogeneous mask coverage thickness and to minimize overspray. After the mask is sufficiently dry, skins are positioned onto the flexible pogo table, stabilized with suction cups to pogos. The skin surface pointing to the outside of the aircraft is positioned onto pogos and does not require any chemical milling. A gantry robot moves the head with a carbon-dioxide laser in five DOFs across the inner surface of the skin, scribing (cutting) mask along certain patterns to facilitate peeling off of the mask. After all patterns are cut (Fig. 51.5) skins are processed using computer-controlled cranes to move skins through the chemical milling tanks. First, the mask is removed in areas which will lead to the thinnest skin cross-sections. Skin is dipped into the chemical milling tank, allowing NaOH chemical (kept at elevated temperatures) to etch exposed aluminum surface long enough to achieve the desired remaining cross section. Etching velocity depends on the NaOH concentration (which is controlled daily). The etching velocity is used as an input parameter to the crane controller, which pulls the skin out of the tank at a predetermined time automatically. If measurement of the desired aluminum skin thickness is verified, the next mask area is peeled off, and the next chemical milling process is repeated, until all required skin areas are processed. The last skin processing step involves trimming of boundaries and drilling of tooling and determinant assembly holes. These tasks are accomplished on the five-axis trimming and drilling machine (Fig. 51.6). Pogos of the flexible machine table are driven in three linear DOFs to positions dictated by the skin configuration, and swiveling suction cups on top of pogos stabilize the skin during the trimming and drilling operations. Mechanical cutters and chip extraction nozzles surrounding the cutter are used. Net trimmed skins with tooling and determinant assembly holes are now ready for assembly.

All crane movements in the galvanic process line and the robotic chemical mask application process are controlled by a simple program. For mask scribing, the laser gantry robot, and the trimming/drilling robotic system, computer-aided design (CAD) skin geometry data is imported into the process simulation and a semi-automated program creation system generates processing programs.

51.2.4 Superplastic Forming (SPF) and Superplastic Forming/ Diffusion Bonding (SPF/DB)

Process Description

The SPF process is an elevated-temperature process where fine-grain material, such as alpha-beta titanium (Ti-6Al-4V is most common) and certain aluminum alloys (7475, 2004, and 5083), can be formed into complex shapes using gas pressure. The process temperature depends on the material and alloy being formed: titanium ($774-927\,°C$ aluminum ($454-510\,°C$). SPF parts, such as those shown in Fig. 51.7, are produced with typical elongations up to 300% [51.4–8].

The basic principles of the superplastic forming process are illustrated in Fig. 51.8, typically using heat and gas pressure to fully form aluminum or titanium part blanks to match the tool's contour.

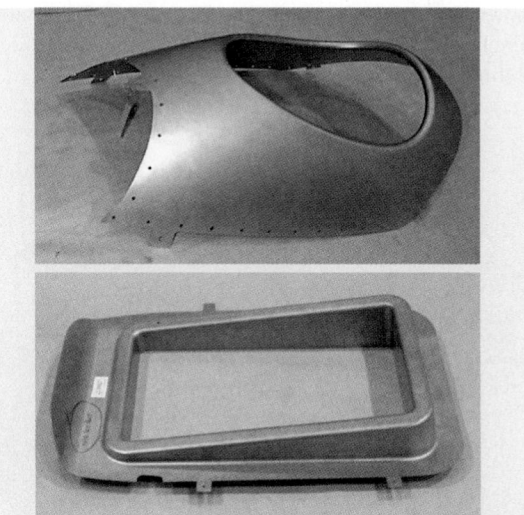

Fig. 51.7 Parts fabricated using super plastic forming (SPF)

Fig. 51.6 CNC trimming and drilling system

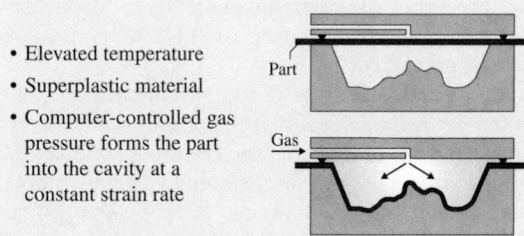

- Elevated temperature
- Superplastic material
- Computer-controlled gas pressure forms the part into the cavity at a constant strain rate

Fig. 51.8 SPF process principles

The SPF/DB process combines an SPF operation with a diffusion bonding (DB) process, whereby two or more sheets of titanium are used to create an integrally stiffened panel structure. For DB to occur, the titanium sheets must contact each other in an inert atmosphere at controlled temperatures and pressures for a specified time. Several methods can be used to achieve these conditions. The one shown in Fig. 51.9 uses a heated press and tool pressure to bring the sheets together. Another method uses gas pressure inside the welded titanium pack to force the sheets into contact with each other. Diffusion bonding is a solid-state process and no melting occurs at the bond line. Once the individual grains on the surface touch each other, they start growing across the interface of the two sheets. This process continues until the sheets are completely diffusion bonded to each other and there is no evidence microscopically of there having been two, or more, pieces of material.

The typical hot (up to 982 °C) shuttle table press, shown in Fig. 51.9, produces SPF and SPF/DB aluminum and titanium parts and has computer-controlled heating, pressure, and gas systems.

Fig. 51.9 Hot press, tool, and SPF part

SPF Benefits. The benefits of the SPF process are that:

- It replaces multipiece assemblies with one monolithic component, saving cost, weight, and tooling.
- It can produce complex geometry and sharp radii.
- Components contain very little, if any, residual stress (no spring-back).
- Less assembly is required (lower cost, lighter weight, and better dimensional accuracy).
- Titanium parts are corrosion resistant.

SPF/DB Benefit. The benefit of the SPF/DB process is that:

- It reduces assembly, producing an integral structure, with no fasteners needed to attach inner structure to outer skin.

51.2.5 Automated Composite Cutting Systems

Ultrasonic Cutting Machine

Uncured unidirectional and fabric carbon fiber, glass fiber, Kevlar, prepreg, and honeycomb materials can be cut into various shapes or forms (preforms) prior to hand placement. Automated computer-controlled ultrasonic cutting machines precisely perform this task with minimal waste of these expensive materials due to advanced computer nesting programs. Up to ten plies of prepreg can be cut at the same time by a carbide ultrasonic knife, which translates up and down at 30 000 strokes per second to provide a clean cut.

The prepreg material, which includes a backing film, is pulled off a roll at the end of the cutting machine bed (Fig. 51.10) and is laid down on a rubber table. A disposable bag is then placed over the prepreg material and any wrinkles are smoothed out. Vacuum is turned on in the table, which pulls the bag down on top of the prepreg material and stabilizes it during the cutting operation. The two-axis N/C machine accurately positions the ultrasonic knife along a preprogrammed path to achieve the desired shape. Once the cutting operation is complete, the vacuum is released and the vacuum bag is removed. Then the preforms and scrap material are manually removed from the machine bed [51.9–15].

Abrasive Water Jet

Cured graphite–epoxy composite structure has very strong fibers in a softer matrix, so trimming them with

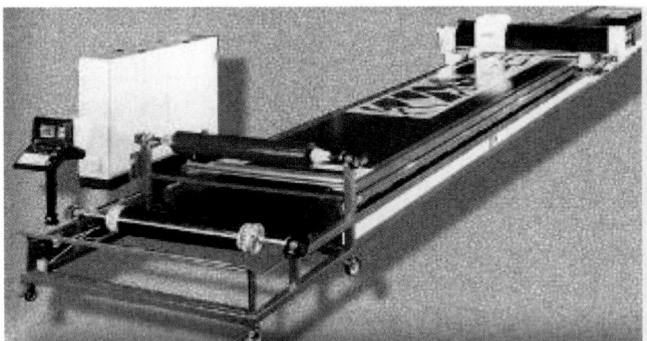

Technical data
- Typical configuration: flat bed two-axis gantry
- Work zone: up to 10 ft × 100 ft
- Cutting speed: up to 1000 in/min
- Rapid traverse speed: 2000 in/min
- Positioning accuracy: 0.002 in/ft–0.015 in overall
- Knife: 20 000–30 000 strokes/s

Fig. 51.10 Two-dimensional ultrasonic cutting machine

Technical data
- Precision multiaxis overhead gantry, 5 up to 11 axes
- Work zone: up to 20 ft × 50 ft × 5 ft
- Cutting speed: up to 250 in/min
- Rapid traverse speed: 1200 in/min
- Positioning accuracy: 0.002 in/ft–0.015 in overall
- Thin stream: 0.020 to 0.050 in dia.

← Insert shows a 60 000 psi waterjet stream cutting a part

Fig. 51.11 Gantry abrasive water jet with *pogostick* tooling

conventional machining techniques does not work very well, since heat and the abrasive nature of composites tend to wear out cutters rapidly and delaminate or pull fibers out of the adhesive matrix. Abrasive water jets are being used extensively to trim, and in some cases drill, holes in these materials due to the clean cut, low fiber pull out, and little or no delamination. The majority of water-jet systems used in aerospace are high-pressure (60 000–87 000 psi) garnet or aluminum-oxide grain abrasive water-jet units (Fig. 51.11).

An N/C machine positions a high-pressure abrasive water jet near the periphery of a composite part that is held in the correct contour by a series of headers or pogos (Fig. 51.11). The computer program drives the N/C machine in a very accurate path at various speeds to cut off the excess material and provide the finished edge of the part.

51.2.6 Automated Tape Layup Machine

Automated tape layup (ATL) is an additive process used to construct large structures from composite prepreg tape material. It is primarily used in the aerospace industry. Machines typically have a five-axis precision overhead gantry and an application head that is suspended from a cross rail. The machine motion and head functions are controlled by a computer with specialized programming. Prepreg tape is typically more than 3 in wide, which is suited to flat or mild contours (Fig. 51.12).

The highly specialized head can precisely lay any number of plies of composite filament tape, in any desired orientation, assuring consistent part shape, thickness, and quality. A typical machine draws from supply reels, then deposits 3, 6 or 12 in tape on flat or mild-contour layup tools. The layup heads can heat the

Technical data
- Precision five-axis overhead gantry with multifunction head
- Work zone: up to 20 ft × 100 ft
- Feed rate: up to 1200 in/min
- Traverse speed: 2200 in/min
- Positioning accuracy: 0.002 in/ft−0.015 in/overall
- Layup rate: up to 50 unknown unit lbs/h

Fig. 51.12 Overhead gantry with work piece below and multifunction tape application head

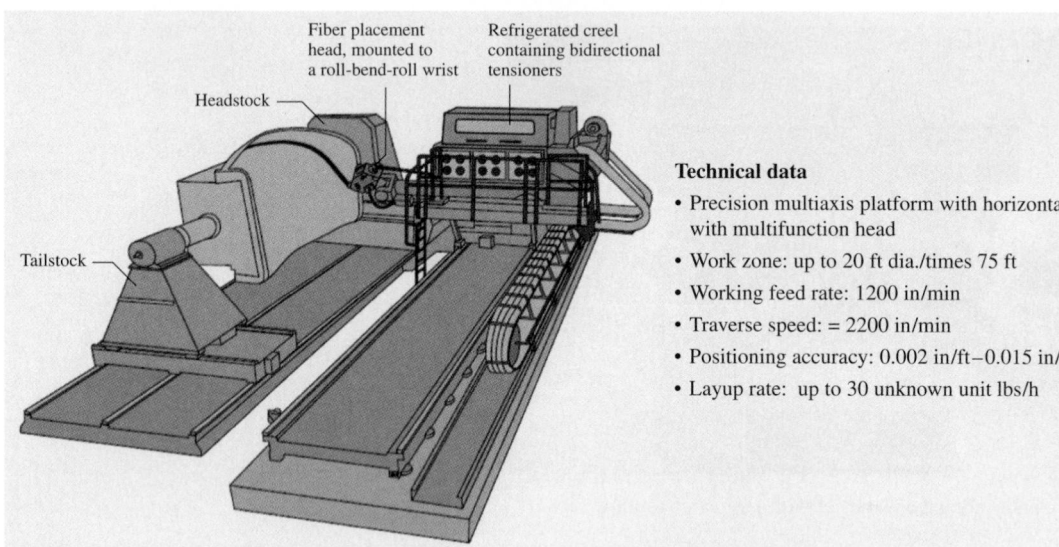

Technical data
- Precision multiaxis platform with horizontal ram with multifunction head
- Work zone: up to 20 ft dia./times 75 ft
- Working feed rate: 1200 in/min
- Traverse speed: = 2200 in/min
- Positioning accuracy: 0.002 in/ft−0.015 in/overall
- Layup rate: up to 30 unknown unit lbs/h

Fig. 51.13 Automated fiber placement machine

tape prior to laying it down, then compact or compress the tape after it is placed on the layup tool. Each layer or ply of tape can be oriented in a direction that optimizes the specific desired part characteristics [51.9–15].

51.2.7 Automated Fiber Placement Machine

Automated fiber placement (AFP) machines, such as that shown in Fig. 51.13, combine two technologies widely used in industry: automated tape layup (ATL) and filament winding (FW). The AFP process is used by the aerospace industry to construct large-circumference and complex structures such as fuselage barrels, ducts, and pressure vessels from composite prepreg materials. The Boeing 787 fuselage barrel in Fig. 51.14 is a primary example [51.9–15].

Fig. 51.14 Boeing 787 fuselage barrel

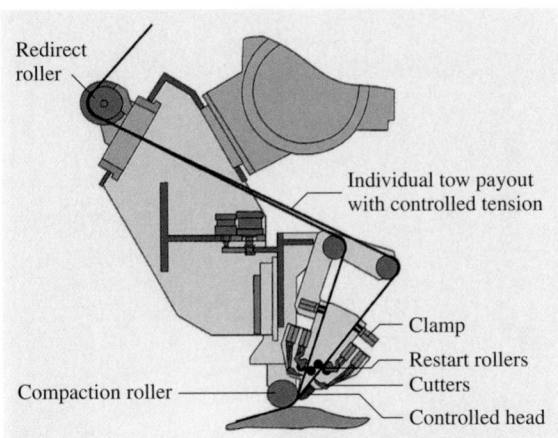

Fig. 51.15 Automated tape layup process head

This additive process utilizes relatively narrow strips of unidirectional composite prepreg tape, commonly called tow, which have unidirectional fibers preimpregnated with a thermoset resin that is later cured. Central to the process is the fiber placement machine, basically a seven-axis manipulator with a head (Fig. 51.15) that arrays a group of tows side-by-side into a continuous band and compacts them against the surface of a concave, convex, contoured combined layup mandrel. The mandrel is mounted on a trunion system similar to a lathe, so that it can rotate as the manipulator is placing the tow. AFP combines the advantages of both filament winding and automated tape layup. The raw materials used are tow-preg or slit-tape rolls of aramid, fibreglass or carbon fiber, preimpregnated, typically with epoxy resin. The width of tow or slit-tape ranges from 3.2 mm to 6.4 mm with thicknesses ranging from 0.13 mm to 0.35 mm. Typical systems permit the use of 12, 24, or 32 tows simultaneously and can lay up on the top of a honeycomb core without degrading it.

51.3 Automated Part Inspection Systems: Examples

The nondestructive inspection (NDI) of aircraft systems is performed at the very highest level of sensitivity because of the criticality of the components. X-ray radiography is the primary method for the inspection of metallic components, particularly welds in tubes and ducts of titanium and inconel as well as other aerospace welded joints [51.1, 2]. Ultrasonic inspection is the primary method for carbon-fiber polymer composites [51.3, 4]. Both x-ray and ultrasonics benefit in terms of quality and value by the implementation of highly automated systems. Ultrasonic systems have required major developments in robotics for inspection rate and sensitivity requirements for both production [51.5–7] and for field in-service inspection operations [51.8, 9]. Automated x-ray systems have been slower to be implemented, but progress is being made. The future direction in aircraft part inspection will be automated interpretation of the NDI data that is currently being implemented in higher-production-rate industries.

51.3.1 X-ray Inspection Systems

Figure 51.16 shows a diagram and photograph of a seven-axis CNC system for digital radiography (DR) of welds at Boeing Commercial Airplanes Fabrication Division in Auburn, WA, USA. The CNC manipulator, source, and detector are located in a radiation vault. A complex welded duct is positioned by the CNC manipulator at a series of preprogrammed locations between the x-ray source and digital detector, as shown in the right-hand image. The insert in the lower right of the figure shows the DR image from the operator's console display. The system consists of five major components: the Siemens controller-based CNC manipulator, the x-ray source, the digital x-ray detector, the control computer for the CNC manipulator, and the image display and analysis system. The system requirement is for x-ray image quality indicator sensitivity of 1-1T (1% part thickness with visible hole of 1% part thickness diameter) in the radiographic image for 100% coverage of the part. To achieve this image quality an x-ray spot size of 20 μm nominal and a magnification of 4.5× is used to create images with greater than ten line-pairs per millimeter resolution and better than 1% contrast sensitivity. Position of the weld to be inspected is critical to achieving the required image quality and this is accomplished by automated control of the CNC manipulator. Table 51.1 lists the critical characteristics of the CNC manipulator system. The CNC manipulator is programmed to begin a testing session by positioning and imaging a test standard at the same geometric factors, exposure parameters, and image display settings as will be used for the part to be inspected. Once the operator approves the quality of the inspection for the

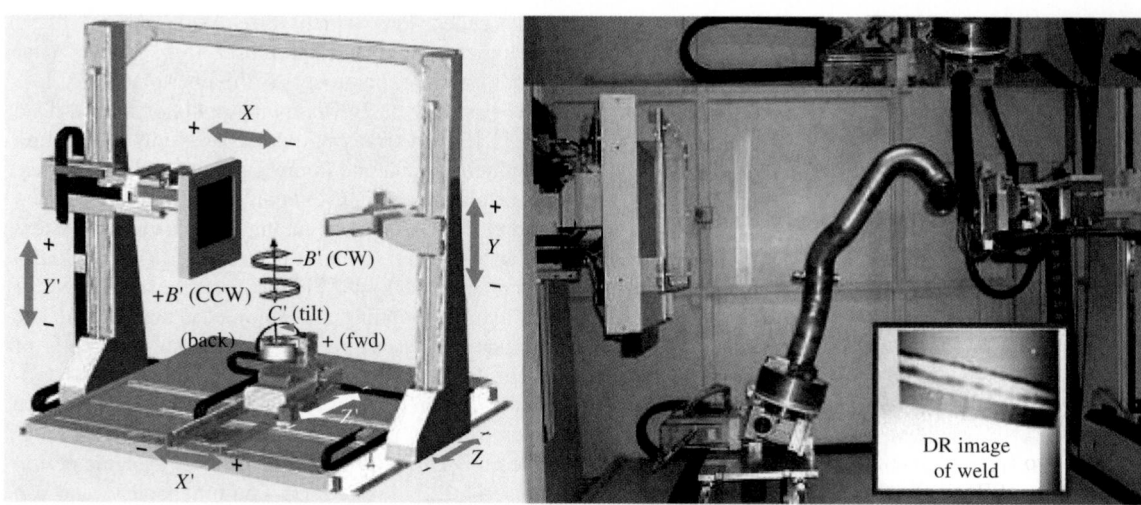

Fig. 51.16 Diagram and photograph of robotic digital radiography system (courtesy of Boeing)

standard, inspection of the part begins. Each part configuration is programmed in the CNC manipulator to allow the weld to be 100% inspected by a series of radiographic views. The CNC manipulator positions the part according to the program for an exposure at the first location. Following operator review of the resulting radiographic image, the CNC manipulator advances to the next position in the sequence and the process repeats until the entire part is inspected. Typical inspection sequences for the CNC manipulator include 20–50 views, taking approximately 10–20 min. Images are reviewed by automated sequence of viewing parameters followed by preset adjustments of the image display for enhancement of areas of interest for detail review. Enhancement and measurement features include different window/level parameters, digital magnification, and contrast enhancement [51.16–28].

51.3.2 Ultrasonic Inspection Systems

For carbon-fiber polymer composite materials including laminate, sandwich structure, and bonds, ultrasound is the principle inspection method and requires highly automated scanners to keep up with the production rate. Sophisticated automation is also needed to handle the complex contoured geometry of the large composite assemblies. Ultrasonic scanning systems can be constructed in a variety of forms from small portable units to large gantry systems. Ultrasonic inspection is most commonly performed with some type of water coupling of the ultrasonic energy between the piezoelectric transducer and the test part. Methods include: immersion where the part is submerged in a water bath with the transducer; bubbler systems, where the transducer rides on the surface in a shoe that also has a flow of water;

Table 51.1 Boeing commercial airplane robotic x-ray system characteristics

Characteristic		Values
Robot	Type	Siemens Simotion seven axis
	Range of motion	Magnification axis: ≈ 1.5 m, loading axis: 0.3 m, rotation axis: 360°, tilting table axis: $\pm 60°$
	Positional accuracy	0.5 mm, angular axis $< 1°$
	Load-carrying ability	70 kg
X-ray source	Model	FEINFOCUS, Model: FXE-225, 320 W (225 kV at 1400 mA)
	Spot size	Less than 20 μm
Detector	Model	Perkin Elmer 1620
	Pixel size/bit depth	200 μm/16 bit
Weld radiography quality	Image quality	1-1T, with > 10 line pairs/mm at 4.5×
	Inspection time	10 s per view, 30 s to 1 min per weld

Fig. 51.17 Two types of automated ultrasonic inspection systems: independent tower system (*right*) and gantry system (*left*) (courtesy of Boeing)

and squirter systems, where the transducer is located in a nozzle that shoots water at a part from a short distance. The ultrasonic inspection can use one or more transducers in a variety of combinations. The typical inspection uses either one transducer in pulse echo (PE) mode, or two transducers, each aligned on opposite sides of the part, for through transmission ultrasound (TTU). Pitch-catch mode uses two transducers on the same side of the part. For high throughputs using automated systems, the scanning robotics may handle arrays of transducers that provide coverage of large areas of material [51.16–27].

As aerospace structures become larger and more complicated, overhead gantry systems or tower gantry systems such as that illustrated in Fig. 51.17 are used. The overhead bridge scanning system can be used with transducer manipulators with up to five axes of motion: X, Y, and Z translations, rotation, and tilt. Using motion control software, the transducer can be oriented normal to complex curved part surface during scanning. Two transducer manipulators are employed for through transmission imaging. The computer software can keep the squirter transducers aligned and normal to the part surface for complex geometric configurations. The part geometries are taught by manual selection of a few data points along a scan or using CAD surface data. In some cases the systems will take pulse echo data from each

transducer on each side of the part and through transmission data simultaneously. The tower scanner uses independent machines on each side of the part to be tested. The advantages of this configuration include the independent surface following from each side, the improved *reach in* capability stiffness, and reduced ceiling height. The ultrasonic testing (UT) data from a quality surface following scanner includes 1-to-1 flaw sizing on complex curvature objects. The test part shown in the tower system in Fig. 51.17 is a landing gear pod fairing. The UT C-scan image data shown in the lower-right side of the figure is the two-dimensional (2-D) ultrasonic C-scan representation of the three-dimensional (3-D) object, in which light areas indicate laminate and darker areas indicate honeycomb core. Laminate inspections are usually performed at 5 MHz, while honeycomb is commonly inspected with 1 or 2.25 MHz ultrasound.

Inspection speeds depend on the data acquisition rates. The data acquisition rate in X and the step size in Y are determined by the minimum defect size that is to be detected. Three data points are required for the minimum defect size, such that 0.08 in data space is used for 0.25 in defect sensitivity. Many scanners and the associated acquisition electronics can scan at up to 40 in/s while maintaining 0.04 in data spacing in the scan direction. Coverage of 25–50 ft^2/h is possible on many parts.

51.4 Automated Assembly Systems/Examples

Conversion to digitally defined parts is a key factor that has enabled the widespread use of automation.

Aircraft assembly machines are custom-designed to meet specific requirements, where the combination of tight engineering tolerances, the need to reduce part variation, and large machine envelope drives large and expensive machines. Despite the high initial cost, the use of automation has been very successful in addressing the needs to improve quality, reduce costs, and improve ergonomics of aircraft fabrication and assembly. Assembly systems are designed around the type of structures to be assembled. For single and very large panels, C-frame riveting/fastening machines are commonly used. The most suitable system to assemble half-shells (fuselage barrels) is a ring riveter, and final aircraft assembly is performed manually using advanced hand tools and, more recently, newly developed, flexible, adaptable, portable assembly systems.

51.4.1 C-Frame Fastening Machine

Automated wing panel fastening machines are used to build stringer stiffened wing panels by riveting stringer-to-wing skins and fastening adjacent wing panels together. Structurally, they are large C-frames mounted on rails that travel the length of the wing. Floor-mounted header tooling supports the temporarily tacked together wing panel to ensure proper panel positioning for permanent fastener installation. The C-frame wraps around the wing skin and applies drilling and fastener installation tools to both sides of the panel. For accurate positioning, wing riveters obtain positional accuracy from the stringers mounted on the underside of the wing panel. A typical rivet installation cycle involves positioning the machine to the proper location, clamping the skin and stringer together, drilling a hole, inserting a rivet, hydraulically squeezing a rivet, and shaving the formed head. Then, an electronic vision system verifies if the shaved rivet meets the flushness re-

quirement. Bolts are applied in a similar sequence, with sealant being applied prior to bolt insertion, and then the nuts are installed and torqued.

A Boeing wing panel fastening system is shown in production in Fig. 51.18. Typical production wing lines have multiple wing assembly machines capable of operating on multiple parallel rail systems with turntables and interconnecting tracks allowing a machine to move to work stations within the track network [51.4–8].

Machine Features
- 50 000 lbs rivet upset capability to accommodate 7/16 in-diameter 7050 rivets
- Full servo programmable control including:
  - High-speed upper head transfer
  - High-speed drill and shave spindles
  - High-speed servo buck
  - High-speed servo lower head
  - Servo clamp
- Up to six upper head positions
- Statistical process control (SPC) of fastener installation processes
- Automatic fastener selection
- Fastest machine cycle time in the industry
- Standard slug squeeze process
- Squeeze/squeeze III slug installation process
- Vibratory insertion process for two-piece fasteners in high interference fit conditions
- Torque-controlled nut runner
- Lockbolt swage collar tooling

51.4.2 Ring Riveter for Fuselage Half-Shell Assembly

System Description
This computer-controlled riveting/fastening machine is being used to automatically join fuselage half-shells made from single panels along longitudinal and radial joints as shown in Fig. 51.19. It consists of:

1. Outer ring, moving on longitudinal rails, carrying a multifunction end-effector, which moves radially on the inside of the machine ring structure
2. A robotic arm, moving on the floor inside the ring on rails longitudinally, carrying a second multifunction end-effector
3. A flexible fixture to support the half-shell during the assembly operations
4. Machine controller, executing assembly sequence programs generated using an offline programming system [51.1–3].

Fig. 51.18 Gemcor automated wing fastening system

Technical data
- Assembly of 180° half shells
- Machine size: $L = 10-80$ m; $W = 6-12$ m; $H = 6-10$ m
- Fully automated process, CNC control
- Automatic work piece transfer
- Precision rivet/fastener head position
- Solid work piece clamping during run time
- Number of rivet/fastener cassettes = 16 or more
- Drill spindle rpm = 0 to 20 000
- Spindle feed = 0 to 0.006 ipr
- Pneumatic hammer for interference fastener insertion
- Rivet installation rate = 4 to 10 rivets/min
- Fastener installation rate = 2 to 6 fasteners/min
- Off-line or teach-in programming

Fig. 51.19 Ring riveter assembly system (courtesy of Broetje)

Assembly Process

Single panels and all parts building the half-shell are tacked together and half-shell stabilized on the flexible fixture is moved into the ring's working envelope. The multifunction end-effector moving on the outside of the half-shell, and the internal multifunction end-effector, perform synchronous riveting/fastening operations through the skin. The outside multifunction end-effector uses a drilling module, a rivet fastener feeding module, and a rivet upsetting tool; and the internal multifunction end-effector uses a clamping tool and rivet upsetting or sleeve installation tool module, if two-piece fasteners are being installed. Process parameters such as the clamping force generated by internal and external end-effector bushings at the drilling location are adjusted for structural material (aluminum, titanium, composite) and stiffness. The drill unit's RPM and feed force are selected as a function of hole diameter, required hole tolerance (e.g., 0.001 in) and cutter material (e.g., carbide, PCD). The position of all machine components (ring, robot, end-effectors) is CNC controlled and a vision system built into the outside multifunction end-effector provides the operator with visual control and enables precision position adjustments if required. The internal multifunction end-effector can be quickly decoupled from the robotic arm and replaced with the C-frame type of multifunction end-effector for shear tie-to-frame riveting operations. This multifunction end-effector locates a feature (hole or rivet) on the frame, using a vision system, and communicates information to the robot controller, which then positions the end-effector to the proper location for clamping, drilling, and rivet insertion/upsetting tasks. All process parameters are monitored and saved to a quality-assurance data bank to verify and document process and part quality.

51.4.3 Airplane Moving Line Assembly

System Description

In 1913 Henry Ford put the first moving assembly line ever used for large-scale manufacturing into use.

Fig. 51.20 Ford Motor Company moving assembly line

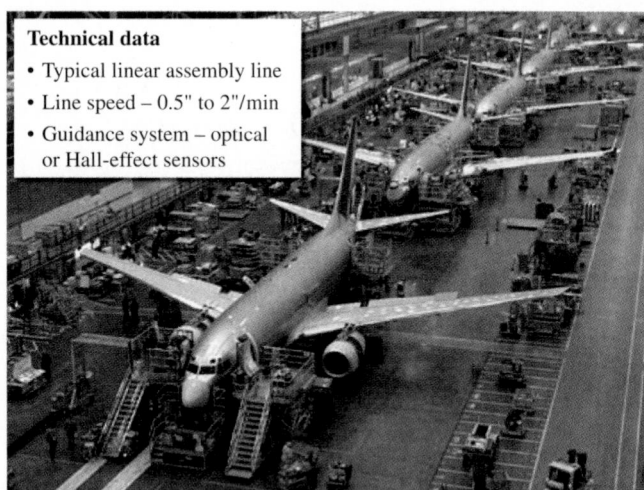

Technical data
- Typical linear assembly line
- Line speed – 0.5" to 2"/min
- Guidance system – optical or Hall-effect sensors

Fig. 51.21 Boeing 737 final assembly line

Assembly Process

Major sections of commercial aircraft are assembled in stationary fixtures until they are structurally stable and need little or no support from external tools. At this point the aircraft is placed on a barge or carrier and begins its final assembly as it is towed by a motorized automated tug. The tug attaches to the front of the barge and pulls it forward under the power of a motor that is computer controlled. Steering is accomplished by an optical sensor that follows a white line along the floor.

Major subassemblies and components such as the landing gear, interior systems, and passenger seats are installed by mechanics as the airplane moves down the assembly line. In addition, functional testing is performed on the various systems in the airplane and the engines are attached. The use of moving assembly line can typically reduce the final assembly time by 50% and significantly reduce the number of days for this task. The assembly time reduction is due to the application of lean manufacturing techniques, which were introduced into the aerospace industry in late 1999. Moving lines help companies achieve higher efficiencies because they create a sense of urgency as well as streamlining and standardizing assembly processes to eliminate waste and nonvalue operations. Computer-controlled tugs set the pace or *takt time* (the manufacturing time needed to accomplish certain predetermined tasks). As the airplane moves past visible marks on the floor, teams of mechanics install prekitted parts and tools using standard processes within the allotted time so that the next team can continue adding value to the airplane as assembly progresses [51.9–15].

Ford lowered the price of a car by producing them at record-breaking rates with this new assembly process. Today almost all automobile manufacturers use moving lines to assemble their products for cost, quality, and flow time reasons. In the past 5 years some aerospace companies have moved away from traditional stationary dock assembly systems in favor of the more efficient moving assembly lines similar to those used in the automobile industry. State-of-the-art examples are the Ford production line at Flatrock, MI, USA, shown in Fig. 51.20 and the Boeing 737 final assembly line in Renton, WA, USA, shown in Fig. 51.21.

51.5 Concluding Remarks and Emerging Trends

Aircraft fabrication and assembly technologies are undergoing significant changes driven by the need to achieve performance and economic targets and only a few predominant trends are discussed here. There is a never-ending search for higher aircraft performance, coupled with the desire to reduce the production labor content, processing time, and cost. This requires progressiveness and innovative use of exotic materials with improved specific mechanical properties (such as carbon-fiber composites), and development of more efficient fabrication and assembly technologies for metallic and nonmetallic structures. The move to more composite aircraft mitigates some typical problems with metallic parts (i. e., the shorter fatigue life and galvanic corrosion inherent to aluminum alloys compared with equivalent composite parts).

A majority of today's manual high-level assembly operations, such as joining fuselage barrels and wing boxes, could be replaced in the future with flexible, adaptable, and affordable, semi-automated assembly systems to perform clamping of parts using electromagnets, drilling, and countersinking, and fastener installation tasks.

One example offering potentially significant improvements in assembly efficiency, is the friction stir welding (FSW) process, developed for space launch vehicle fabrication (for joining Delta II and IV aluminum panels to fuel tanks). This technology, once thoroughly tested and approved, can be introduced into commercial aircraft structural assembly, replacing today's time-consuming mechanical joining techniques.

The potential capability to produce high-level assemblies with composite structures helps to eliminate several lower-level assembly tasks. This has been achieved, as evidenced by the redesign of multipanel aluminum fuselage barrels into a one-piece composite barrel. Such moves call for development of innovative structural configurations, and require mastering engineering challenges associated with tooling, equipment, processes, and inspection.

The emergence and growth of rapid prototyping/fabrication technology could revolutionize the fabrication/manufacturing of parts and assemblies. Parts will be *grown* in a system that requires only the electronic part geometry information and raw material in powder form as inputs. Parts are currently created (grown) by layered material deposition and particle fusion using lasers. These plastic net-shape, or near-net-shape, metallic parts eliminate the need for a majority of the material removal processes and reduce the material buy-to-fly ratios and manufacturing cost. This trend could evolve into *growing substructures* or even complete aircraft segments, eliminating all part fabrication and assembly tasks.

The trend toward monolithic metallic and large-scale integral composite structures will probably continue in the future, and will require the development of advanced automated fabrication and assembly systems to meet demands for improved aircraft performance at minimal cost. The current trend is to machine or fabricate components accurately with automated machines and then use accurately machined features in the detail parts as references to build larger assemblies. At some point, as the structure increases in size, it becomes cost prohibitive to use conventional automation to assemble large parts, so the fall-back position has been to join or splice larger assemblies manually. To capture the benefits of automation with larger assemblies in the future, a new generation of flexible portable automation is being developed. These lightweight portable systems use the aircraft structure as their foundation and will produce quality parts at an affordable cost.

References

51.1 C. Wick, J.T. Benedict, R.F. Veilleux: *Tool and Manufacturing Engineering Handbook – Vol. 2: Forming* (SME, Dearborn 1984)

51.2 E.H. Zimmerman: *Getting Factory Automation Right: The First Time* (SME, Dearborn 2001)

51.3 J.A. Schey: *Introduction to Manufacturing Processes* (McGraw Hill, New York 1987)

51.4 M. Watts: High performance machining in aerospace, Proc. 4th Int. Conf. Metal Cutt. High Speed Mach. (Boeing, Seattle 2002)

51.5 M. Watts: Evolving aerospace machining processes, 4th Int. Conf. High Speed Mach. – Ind. Tool. Conf. (Southampton 2001)

51.6 L. Hefti: Innovations in fabricating superplastically formed components, First and Second Int. Symp. Superplast. Superplast. Form. Technol. (ASM, Materials Park 2003) pp. 124–130

51.7 D. Sanders: A production system using ceramic die technology for superplastic forming, Superplast. Adv. Mat. ICSAM 2003 (Trans Tech, 2004) pp. 177–182

51.8 GEMCOR: http://www.gemcor.com (2009)

51.9 PASER: Abrasive waterjet helps make composites affordable for Boeing, http://www.flowcorp.com/waterjet-resources.cfm?id=251 (2008)

51.10 Boeing completes first 787 composite fuselage section, http://www.boeing.com/companyoffices/gallery/images/commercial/787/k63211-1.html (2005)

51.11 R.A. Kisch: *Automated Fiber Placement Historical Perspective* (Boeing, Seattle 2006), http://www.ingersoll.com/ind/tapelayer.htm

51.12 Boeing reduces 737 airplane's final-assembly time by 50 percent, http://www.boeing.com/news/releases/2005/q1/nr_050127g.html (2005)

51.13 T.G. Gutowski: *Advanced Composites Manufacturing* (Wiley, New York 1997)

51.14 S. Mazumdar: *Composites Manufacturing: Materials, Product, and Process Engineering* (CRC Press, Boca Raton 2002)

51.15 F. Campbell Jr.: *Manufacturing Processes for Advanced Composites* (Elsevier, Amsterdam 2004)

51.16 R. Bossi, F. Iddings, G. Wheeler (Eds.): *Nondestructive Testing Handbook, Vol. 4 – Radiographic Testing*, 3rd edn. (American Society for Nondestructive Testing, Columbus 2002)

51.17 R. Halsmshaw: *Nondestructive Testing*, 2nd edn. (Edward Arnold, London 1991)

51.18 G.L. Workman, D. Kishoni (Eds.): *Nondestructive Testing Handbook, Vol. 7 – Ultrasonic Testing*, 3rd edn. (American Society for Nondestructive Testing, Columbus 2007)

51.19 ASM: *ASM Handbook, Vol. 21 – Composites, Quality Assurance* (ASM, Metals Park 2001)

51.20 J. Summerscales (Ed.): *Nondestructive Testing of Fibre-Reinforced Plastics Composites*, Vol. 2 (Elsevier, New York 1990), pp. 107–111

51.21   G.L. Workman: Robotics and nondestructive test-
         ing – a primer, World Conf. Nondestruct. Test. (NDT)
         (1985) pp. 1822–1829

51.22   P. Walkden, P. Wright, S. Melton, G. Field: Auto-
         mated ultrasonic systems, World Conf. NDT (1985)
         pp. 1822–1829

51.23   T.S. Jones: Inspection of composites using the au-
         tomated ultrasonic scanning system (AUSS), Mater.
         Eval. **43**(6), 746–753 (1985)

51.24   M.K. Reighard, T.W. Van Oordt, N.L. Wood: Rapid
         ultrasonic scanning of aircraft structures, Mater.
         Eval. **49**(12), 1506–1514 (1991)

51.25   Y. Bar-Cohen, P.G. Backes: Scanning aircraft struc-
         tures using open-architecture robotic crawlers as
         platforms with NDT boards and sensors, Mater.
         Eval. **57**(3), 361–366 (1999)

51.26   J.J. Gallar: Modular robotic manipulation in radio-
         graphic inspection, Mater. Eval. **46**(11), 1397–1399
         (1988)

51.27   D. Mery: Automated radioscopic testing of alu-
         minum die castings, Mater. Eval. **64**(2), 135–143
         (2006)

51.28   ASTM: ASTM E 1025-84, Standard Practice for Hole-
         Type Image Quality Indicators Used for Radiography

52. Semiconductor Manufacturing Automation

Tae-Eog Lee

We review automation requirements and technologies for semiconductor manufacturing. We first discuss equipment integration architectures and control to meet automation requirements for modern fabs. We explain tool architectures and operational issues for modern integrated tools such as cluster tools, which combine several processing modules with wafer-handling robots. We then review recent progress in tool science for scheduling and control of integrated tools and discuss control software architecture, design, and development for integrated tools. Next, we discuss requirements and technologies in fab integration architectures and operation such as modern fab architectures and automated material-handling systems, communication architecture and networking, fab control application integration, and fab control and management.

52.1 Historical Background

The world semiconductor market has been growing fast and amounted to US$ 270 billion in 2007. The semiconductor manufacturing industry has kept making innovations in circuit design and manufacturing technology. Some key innovations include circuit width reductions from 1.0 μm in 1985 to 60 nm in 2005, 40 nm in 2007, and even down to 14 nm by 2020, and wafer size increase from 200 mm to 300 mm wafers, and even to 450 mm or larger in the near future. Some fabs are producing 1 Gb random-access memory (RAM) by using 50 nm technology, which reduces the cost by about 50% compared with 60 nm technology. Such technology innovations have led to higher circuit density, increased circuit speed, and remarkable price reduction, which also have created new demand and expanded the market.

In 2007, 35 new wafer fabs began to ramp up world monthly fab capacity by two million 200 mm wafers, that is, a 17% increase. A modern fab construction costs about US$ 2 billion. On the other hand, the semiconductor manufacturing industry has suffered strong competition due to excessive capacity. Therefore, the industry has tried to reduce costs, improve quality, and shorten the manufacturing cycle time. Automation has been the key for such manufacturing improvement and business success. Consequently, there have been many aggressive technology innovations and standardizations for fab automation. We therefore need to review those

efforts, the state of art, and the future challenges for fab automation.

In this chapter, we briefly introduce semiconductor manufacturing systems and automation requirements, architecture and control for processing equipment and material-handling systems, communication architecture and networking, and software architecture for process control, equipment control, and fab-wide control. We explain academic research works as well as industrial technologies and practices.

52.2 Semiconductor Manufacturing Systems and Automation Requirements

52.2.1 Wafer Fabrication and Assembly Processes

The semiconductor manufacturing process consists of wafer fabrication and assembly. In the wafer fabrication process, multiple circuit layers (up to 30 or more) are laid out on a wafer surface through the repetition of identical sequences of process steps. Most fabrication process steps are chemical processes that oxidize a wafer surface, coat photosensitive chemicals onto the surface, expose it to a circuit image from a light source, develop and etch the circuit pattern, deposit other chemicals onto it, diffuse and implant additional chemicals on the etched pattern, and so on. Once a circuit layer is formed, the wafer reenters the fabrication line to form the next circuit layer. The total number of process steps may amount to 480 or more. A wafer has several hundreds of formed circuit devices. For strict quality control, the formed circuits are measured by metrology equipment frequently after some key process steps. Based on the metrology results, some devices in a wafer may be repaired, reworked or scrapped. Wafter yield may be rather low, especially during the ramp-up stage for the initial 3–6 months. Wafers are transported and loaded into processing tools using a carrier called a *cassette* or *pod* that loads 25 wafers. A typical fab produces 40 000 wafers each month. The fabrication cycle time is several weeks or even a few months, depending on the fab management performance. About 20 000–100 000 wafers may be in progress at any given time.

Once a wafer completes the fabrication processes, devices on a wafer undergo intensive circuit tests called electronic die sorting (EDS). Depending on the test results, the devices are classified into different final products with specifications on clock speed, number of

effective transistors, and so on. A device that fails to satisfy the specification of a high-grade product is classified into a lower-grade product. Such a sorting process is also called *binning*. Some devices may be defective. Due to the yield problem and binning, it is difficult to predict the number of final products of each grade or type.

Wafers that complete EDS are sent to an assembly or packaging plant. The fabrication processes leading to EDS and the assembly processes after EDS are called *front-end* and *back-end processes*, respectively. In the back-end processes, a wafer is sliced into individual devices. The sliced devices undergo packaging processes that include tape mounting, wire bonding, molding, and laser marking. The packaged devices take final tests, where additional binning is carried out. The back-end processes have been regarded as relatively low technology with low value added and tend to be subcontracted. However, multichip packages (MCP) that combine several chips together into a single package are becoming increasingly popular due to growing demand from the mobile-device industry. MCP or other advanced packaging technologies such as wafer-scale packaging and flip chips increase the value and importance of the back-end processes. Hence, a number of back-end processes still involve manual material handling while the front-end processes have become highly automated. Figure 52.1 summarizes the overall semiconductor manufacturing processes.

A process step is performed by a number of similar or identical wafer processing tools. Due to strict quality requirements, some wafer lots should be processed only with a restricted set of tools. Different types of wafer lots flow concurrently through the fab. Therefore, the fab can be viewed as a hybrid flow shop. Reentrant job

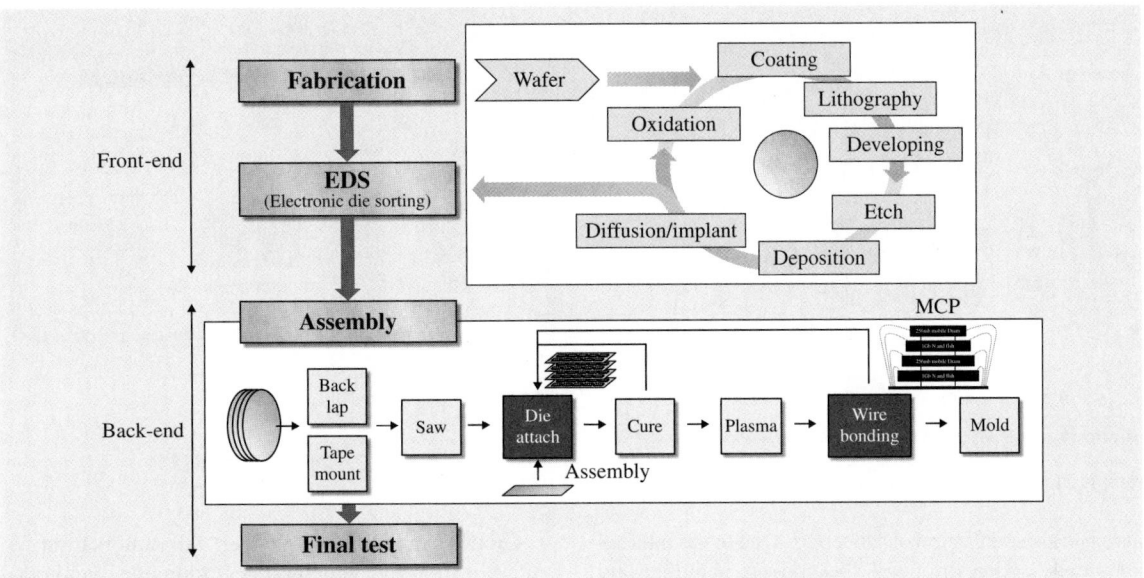

Fig. 52.1 Overall manufacturing processes

flows for processing multiple circuit layers and random yield make planning and scheduling complicated. A fab consists of several hundreds of processing and inspection tools. The tools are grouped into bays, where each bay consists of 10–20 processing tools. Each bay has a stocker, where wafer cassettes are waiting for processing or moving to the next bay.

52.2.2 Automation Requirements for Modern Fabs

There are several drivers for fab automation. The material-handling tasks in a fab are very large; for instance, a fab that processes 40 000 wafers a month requires 200 operators per shift just for moving wafer cassettes [52.1]. Therefore, automated material-handling systems (AMHSs) are used to reduce such high human operator requirements. Other drivers for material-handling automation include prevention of human's handling errors such as wafer dropping, and better tool utilization and reduced manufacturing cycle time by fast and reliable material transfer [52.1]. The key technological innovations in the front-end processes during the past decades are the continuing reduction of circuit features for higher density and functionality, and wafer size increase to 300 mm for higher throughput. These have led to significant fab automation. Extreme circuit shrinkage requires strict quality control and higher-class clean rooms to reduce increased risk of

particle contamination. As human operators are a significant source of particle generation, the number of operators needs to be reduced. Wafer size increase leads to significantly heavier weight of a wafer cassette beyond human operator's adequate workload. Therefore, in recent 300 mm fabs, wafer-handling operations have been mostly automated. Control applications for equipment and AMHSs from many different vendors should be easily integrated. Design, scheduling, and control of fully automated fabs are highly complicated and require new concepts and ideas (Fig. 52.2).

Traditionally, wafers in a cassette have been processed in batch mode for most chemical processes such as etching, deposition, etc. However, as the wafer size increases and quality requirements become stricter due to circuit shrinkage, it becomes difficult to control gas or chemical diffusion on all wafer surfaces within a large processing chamber to be uniform enough for strict quality requirements. Therefore, single-wafer processing (SWP) technology that processes wafers one by one has been extensively introduced for most processes. In order to reduce excessive moving tasks between SWP chambers, several SWP chambers are integrated within a closed environment together with a wafer-handling robot. Such a system is called *cluster tools*. An integrated system of SWP chambers with multiple handling robots is often called a *track equipment* or *track system*. It can be considered as a combination of multiple cluster tools. Cluster tools or track equipment have been

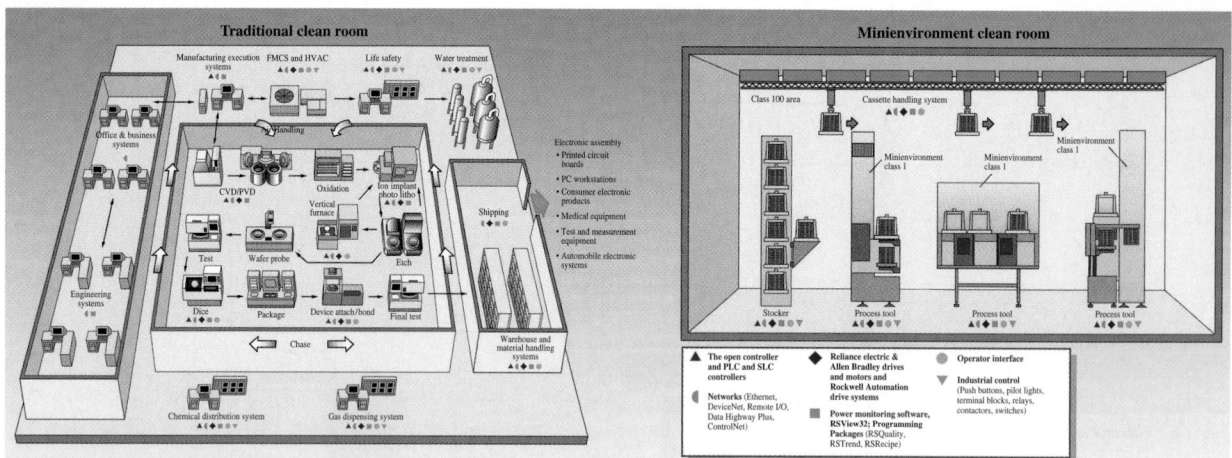

Fig. 52.2 Semiconductor fabrication clean rooms (courtesy of Rockwell Automation, Inc.)

increasingly used for most processes. Due to the internal complexity and restrictions, they pose scheduling and control challenges. First, their operations should be optimized to maximize throughput. Second, wafer delays within a processing chamber after processing should be controlled because residual gases and heat affect wafer quality significantly. Third, the tool controller should be reliable and easily adaptable for different tool configurations and changing wafer flow patterns or recipes. Scheduling and control, and tool application integration are not trivial.

Another important issue for fab automation is standardization for reducing integration effort and performance risk. Semiconductor Equipment and Material International (SEMI), an international organization, has developed extensive standards on architectural and interface standards of material-handling hardware, communication, and control software for fab automation. The standards themselves are based on state-of-the-art automation technologies; however, they should be continuously improved for higher operational goals and changing automation requirements.

52.3 Equipment Integration Architecture and Control

52.3.1 Tool Architectures and Operational Requirements

In a cluster tool, there is no intermediate buffer between the process modules (PMs). A wafer, once unloaded from a loadlock, can return to the loadlock only after it completes all required process steps and is often cooled down at a cooler module, if any. This is because a hot wafer returned to the wafer cassette at the loadlock may damage other wafers there and a hot wafer in progress should not be excessively cooled down before processing at the next PM. A wafer loaded into a PM immediately starts processing since the PM's chamber already has gases and heat. There are different cluster tool architectures, as illustrated in Fig. 52.3. Most tools have radial configurations of chambers, where robot move times between chambers are minimized. Linear

configurations are also considered to add or remove chambers flexibly. The robot has a single arm or dual arms. The dual arms keep opposite positions. Dual-armed tools are known to have higher throughput than single-armed tools [52.2]. There are also tools with intermediate vacuuming buffers between chambers and loadlocks [52.3] in order to save vacuuming and venting times at the chambers. Some new cluster tools use multiple wafer slots in a chamber in order to improve throughput above that of SWP tools by processing several wafers together [52.4]. However, those new tool architectures tend to increase scheduling complexity significantly.

Track equipment or systems are also widely used for integrating several process steps. Photolithography processes use track systems that supply steppers with wafers coated with photosensitive chemicals and de-

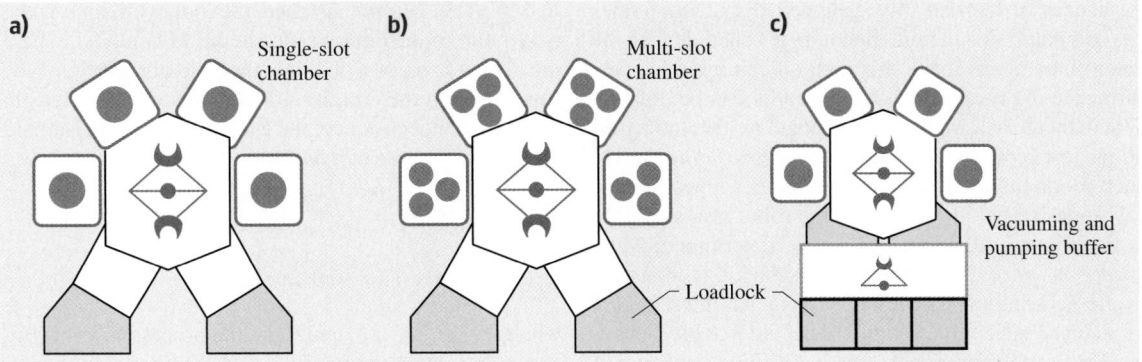

Fig. 52.3a–c Tool architectures: (**a**) single-slot cluster tool, (**b**) multi-slot cluster tool, (**c**) tool with intermediate buffers

velop the circuit patterns on the wafers that are formed by exposures to circuit pattern picture images at the steppers. Process modules for coating and developing, and accompanying baking and cooling modules are combined into a track tool with several robots, as illustrated in Fig. 52.4. Each process step has five to ten parallel modules [52.5, 6]. An automated wet station also has a series of chemical and rinsing baths for cleaning wafer surfaces, which are combined by several robots moving on a rail [52.7]. Recently, EDS processes for testing devices on wafers are automated to form a kind of track system. A number of testing tools for wafer burn-in (WBI) test, hot pretest, cold pretest, laser repair, and posttest are configured in series–parallel by several robots moving on a rail. EDS systems and wet stations can process several different wafers concurrently while most cluster tools or track tools for coating and developing repeatedly process identical wafers.

Wafers mostly go through a sequence of process steps in series. For some processes, wafers visit some process steps again; for instance, unlike conventional chemical vapor deposition, atomic-layer deposition process controls the deposition thickness by repeating extremely thin deposition multiple times. Therefore, a wafer reenters the chambers many times. In track systems, wafer reentrance can be achieved, depending on the chamber configuration and process recipe. In some processes, a chamber should be cleaned after a specified number of wafers have been processed or when sensors within the chamber detect significant contamination. If a wafer remains in a chamber after processing, this can lead to quality problems. This idle time, called *wafer delay*, must be bounded, reduced or regulated. Process times or tasks times are rather constant, but can be subject to random variation, mostly within a few percent. There can be exceptional delay, even if only rare, due to

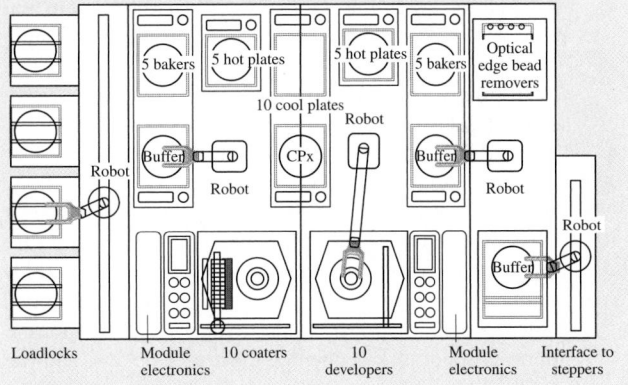

Fig. 52.4 A track system

abnormal process conditions. A wafer alignment task, which correctly locates a wafer unloaded from a loadlock onto a robot arm by using a laser pointing system, sometimes fails and needs to be retried. Integrated tools mostly limit intermediate buffers. Therefore, blocking and waiting are common and even deadlocks can occur. Reentrance, wafer delays, cleaning cycles, and uncertainty all increase scheduling complexity significantly. Tool productivity by intelligent scheduling and control is critical for maximizing fab productivity and even affects wafer quality significantly.

52.3.2 Tool Science: Scheduling and Control

Scheduling Strategies

There can be alternative scheduling strategies for cluster tools. First, a dispatching rule determines the next robot task depending on the tool state. It can be considered dynamic and real time. However, it is hard to optimize the rule. We are only able to compare performances of heuristically designed dispatching rules by computer simulation. Second, a schedule can be determined in advance. This method can optimize performance if a proper scheduling model can be defined. When there is a significant change in the tool situation, rescheduling is done. *Cyclic scheduling* makes each robot and each processing chamber repeat identical work cycles [52.7, 8]. Once the robot task sequence is determined, all work cycles are determined. Most academic works on cluster tool scheduling consider cyclic scheduling. Cyclic scheduling has merits such as reduced scheduling complexity, predictable behavior, improved throughput, steady or periodical timing patterns, and regulated or bounded task delays or wafer delays and work in progress [52.7–10]. In cyclic scheduling, the timings of tasks can be controlled in real time while the sequence or work cycle is predetermined.

A cluster tool that repeats identical work cycles can be formally modeled and analyzed by a timed event graph (TEG), a class of Petri nets [52.12]. Transitions, places, arcs, and tokens usually represent activities or events, conditions or activities, precedence relations between transitions and places, and entities or conditions, respectively. They are represented graphically by rectangles, circles, arrows, and dots, respectively. Figure 52.5 is an example of a TEG model for cluster tools. Once a TEG model is made, the tool cycle time, the optimal robot task sequence, the wafer delays, and the optimal timing schedules can be systematically identi-

fied [52.7, 9, 10]; for instance, the tool cycle time is the maximum of the circuit ratios in the TEG model, where the *circuit ratio* of a circuit is the ratio of the sum of the total times in the circuit to the number of the tokens in the circuit. For instance, the cycle time of a dual-armed cluster tool can be derived from the ratio as

$$\max\left[\max_{i=1,\dots,n}\frac{p_i+2u+2l+3v}{m_i},\right.$$
$$\left.(n+1)(u+l+2v)\right],$$

where p_i, m_i, u, l, v, and n are the process time of process step i, the number of parallel chambers for process step i, the unloading time, the loading time, the move time between the chambers, and the number of process steps, respectively [52.13].

Schedule Quality

For a cluster tool with a given cyclic sequence, there can be different classes of schedules, each of which corresponds to a firing schedule of the TEG model. A periodic schedule repeats an identical timing pattern for each d work cycles. When $d = 1$, the schedule is called *steady*. In a steady schedule, task delays such as wafer delays are all constant. In a d-periodic schedule, the wafer delays have d different values, while the average is the same as that of a steady schedule. The period d is determined from the TEG model. A schedule that starts each task as soon as the pre-

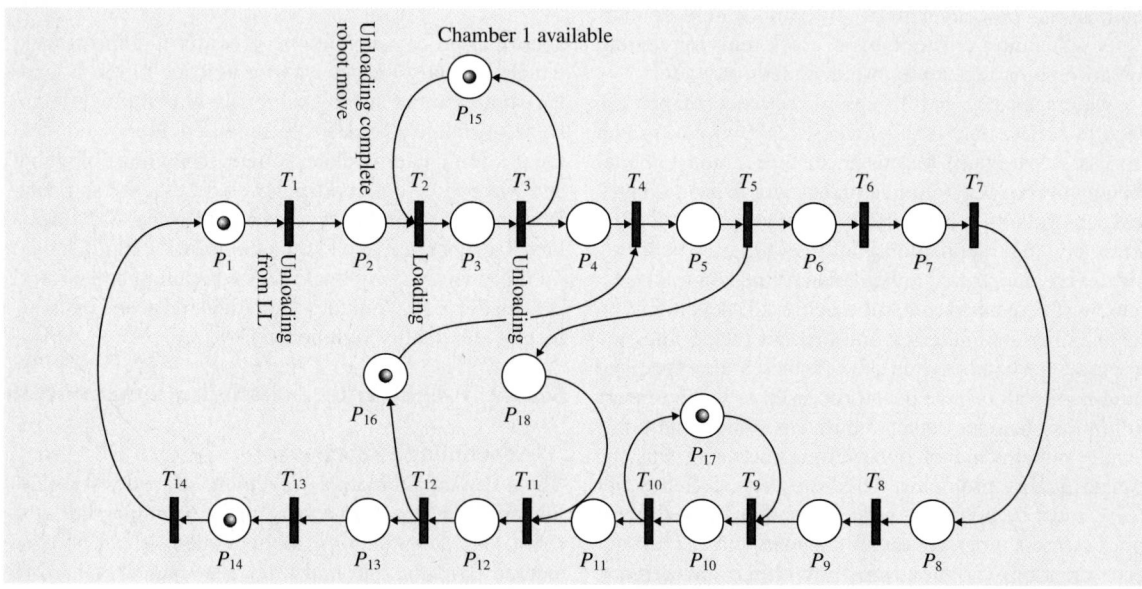

Fig. 52.5 A timed event graph model for a dual-armed cluster tool [52.11]

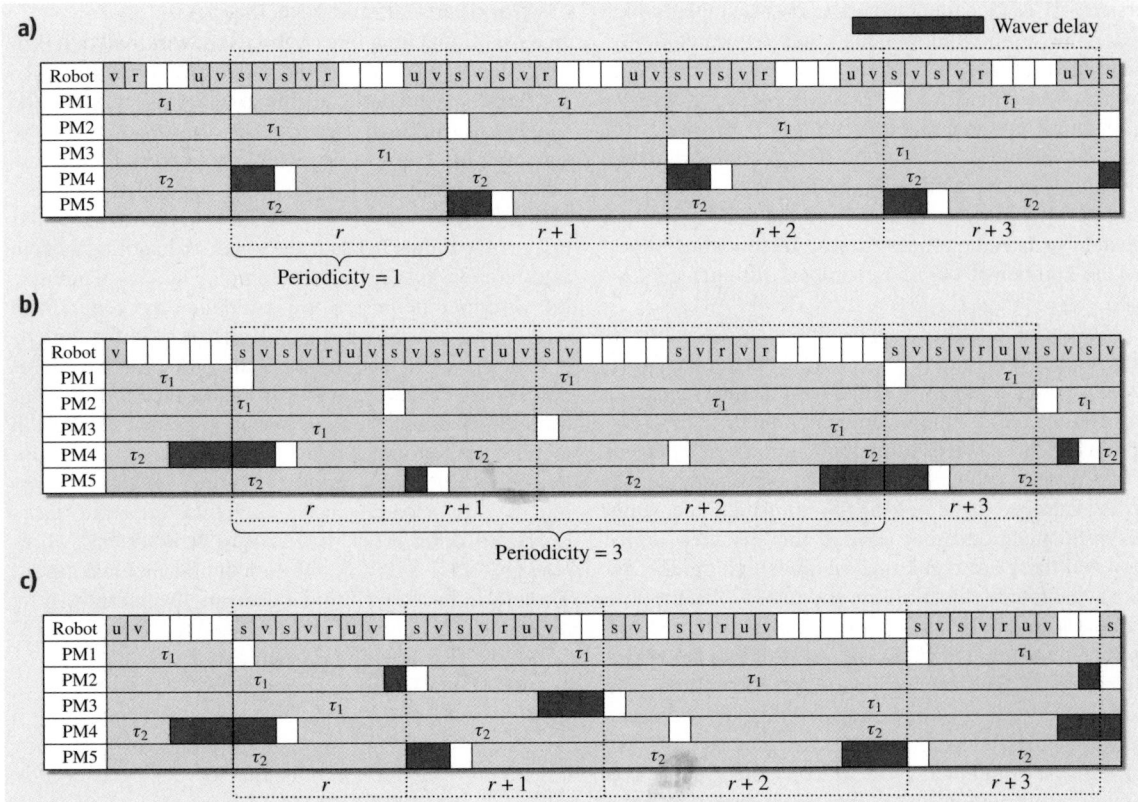

Fig. 52.6a–c Examples of schedules. (**a**) Steady schedule: a SESS, (**b**) 3-periodic schedule, (**c**) irregular schedule

ceding ones complete is called *earliest*. An earliest schedule can be generated by the earliest firing rule of the TEG model that fires each transition as soon as it is enabled. In other words, an earliest starting schedule need not be generated and stored in advance. The TEG model with the earliest firing rule can be used as a real-time scheduler or controller for the tool. Therefore, an earliest schedule can be implemented by an event-based control, which initiates a task when an appropriate event, for instance, a task completion, occurs. Therefore, an earliest starting schedule based on such event-based control has merits. First, potential logical errors due to message sequence changes can be prevented. When a tool is controlled by a predetermined timing schedule, communication or computing delays may cause a change in a message sequence and a critical logical error; for instance, a robot may try to unload a wafer at a chamber before processing at the chamber has been completed and hence when the wafer slot is still closed. Second, the earliest schedule minimizes the average tool cycle time, which is the same as the

maximum circuit ratio of the TEG model. Therefore, the most desirable schedule is a *steady and earliest starting schedule* (SESS). For a cluster tool with cyclic operation, there always exists a SESS. Figure 52.6a is an example of SESS for the TEG model. A SESS can be computed in advance using the max-plus algebra or a kind of longest-path algorithm [52.9] and implemented by an event-based controller based on the TEG model [52.10, 13].

Controlling Wafer Delays

When a tool has a strict constraint on the maximum wafer delay, as in low-pressure chemical vapor deposition, coating processes or chemical cleaning processes, it is important to know whether there exists a feasible schedule that satisfies the constraint. There have been works on the schedulability of a cluster tool, that is, the existence of a feasible SESS [52.11, 14]. *Lee* and *Park* [52.14] propose a necessary and sufficient condition for schedulability, that is, the existence of a feasible SESS, based on circuits in an extended

..ion of TEG called negative event graph, which models the time-window constraints on wafer delays by negative places and tokens. In fact, schedulability can also be verified by the existence of a feasible solution in an associated linear program. However, the necessary sufficient condition identifies why the time constraints are violated, and often gives a closed-form schedulability condition based on the scheduling parameters such as the process times, the robot task times, and the number of parallel chambers for each process step.

Most schedulability analyses assume deterministic process and task times. When a cluster tool is operated by a SESS, the wafer delays are kept constant. However, in reality, there can be sporadic random disruptions such as wafer alignment failures and retrials or exceptional process times. In this case, the schedule is disturbed to a non-SESS, in which the wafer delays fluctuate and may exceed the specified limits. However, there are regulating methods that quickly restore a disrupted schedule. *Kim* and *Lee* [52.15] propose a *schedule stability* condition for which a disrupted earliest firing schedule of a TEG or a cluster tool converges to the original SESS regardless of the disruptions, and a simple way of enforcing such stability by adding an appropriate delay to some selected tasks. Therefore, we can regulate wafer delays to be constant. Such a stability control method has been proven to be effective even when there are persistent time variations of a few percent [52.15]. Even when the process times or the robot task times vary significantly, but only if they are within a bounded range, schedulability against wafer delay constraints can be verified by an efficient algorithm on an associated graph [52.15]. When the initial timings are not appropriately controlled or a SESS is disrupted, the earliest schedule converges to a periodic schedule whose period is determined from the TEG. Therefore, the wafer delays can be much larger than the constant value for a SESS. For a given wafer delay constraint, even if the schedulability condition is satisfied, that is, a feasible SESS exists, a periodic schedule may have wafer delays that exceed the limit. Therefore, we are concerned with whether such a periodic schedule with fluctuating wafer delays can satisfy the wafer delay constraint. *Lee* et al. [52.10] proposed a systematic method for identifying exact values of task delays of a TEG or wafer delays of a cluster tool for each type of schedule: steady or periodic, earliest or not. From the method, the schedulability of periodic schedules, which occurs when timings are not well controlled, can be verified.

Workload Balancing for Tools

In a traditional flow line or shop, the workload of a process step is the sum of the process times of all jobs for the step. The bottleneck is the process step with the maximum workload. Imbalance in the workloads of the process steps causes waiting of jobs or work in progress before the bottleneck. However, in automated manufacturing systems such as cluster tools, the workload is not easy to define because the material-handling system interferes with the job processing cycle. To generalize the workload definition, we can define the generalized workload for a resource as the circuit ratio for the circuit in the TEG that corresponds to the work cycle of the resource [52.10, 16]; for instance, the workload for a chamber at process step i with m_i parallel chambers in a single-armed tool is $(p_i + 2l + 2u + 3v)/m_i$, because each work cycle of a chamber requires a wafer processing (p_i), two loading tasks ($2l$), two unloading tasks ($2u$), and three robot moves ($3v$). A robot has workload $(n+1)(u+l+2v)$, the sum of all robot task times. Therefore, the overall tool cycle time is determined by the bottleneck resource as

$$\max\left[\max_{k=1,2,\ldots,n}\frac{p_k+2l+2u+3v}{m_k},\right.$$
$$\left.(n+1)(u+l+2v)\right].$$

Imbalance between the workloads or circuit ratios causes task delays such as wafer delays. In a single-armed tool, the workload imbalance between process step i's cycle and the whole tool cycle is

$$\max\left[\max_{k=1,2,\ldots,n}\frac{p_k+2l+2u+3v}{m_k},\right.$$
$$\left.(n+1)(u+l+2v)\right]-\frac{(p_i+2l+2u+3v)}{m_i}.$$

Notice that each chamber at process step i has cycle time $(p_i+2l+2u+3v)$, while the overall cycle time at the process step is $(p_i+2l+2u+3v)/m_i$. Therefore, the delay in each cycle of a chamber at process step i is m_i times as long as the workload imbalance at the process step. Consequently, the average wafer delay at a chamber at process step i is [52.10]

$$m_i\max\left[\max_{k=1,2,\ldots,n}\frac{p_k+2l+2u+3v}{m_k},\right.$$
$$\left.(n+1)(u+l+2v)\right]-(p_i+2l+2u+3v).$$

We note from the well-known queueing formula, Little's law, that the average delay is proportional to the average work in progress. In a cluster tool, wafer delays are more important than the number of waiting wafers because of extreme limitation on the wafer waiting space. Wafer delays can be reduced or eliminated by balancing the circuit ratios. Such *generalized workload balancing* can be done by adding parallel chambers to a bottleneck process step, accommodating the process times within technologically feasible ranges or intentionally delaying some robot tasks [52.10, 16]. *Lee* et al. [52.10, 16] proposed a linear programming model that optimizes such workload balancing decisions under given restrictions. Workload balancing is essential for cluster tool engineering.

Additional Works

Cluster tools with cleaning cycles, multi-slots, and reentrance present more challenging scheduling problems. There are some works on using cyclic scheduling for these problems [52.4, 17, 18]. For a tool controlled by a dispatching rule, we cannot optimize the rule and identify or control wafer delays. Wafer delays are unexpected and can be excessively long. Nonetheless, dispatching rules are inevitable when the scheduling problem is too complex or involves uncontrollable significant uncertainty. Reentrance, cleaning cycles, and multi-slots contribute significantly to scheduling complexity. In general, process times and robot task times in cluster tools and track equipment are relatively well regulated and have variations within a few percent, because most processes are designed to terminate within a specified time. However, modern adaptive process control that adapts process control parameters based on real-time sensor information may cause significant time variation. Cleaning based on chamber conditions may occur randomly and hence increase uncertainty significantly. There are some works on dispatching rules for cluster tools with cleaning and multi-slots [52.19].

52.3.3 Control Software Architecture, Design, and Development

In a cluster tool, each processing module or chamber is controlled by a process module controller (PMC). The robot, loadlocks, and slot valves at the chamber are controlled by a transport module controller (TMC). A module controller receives data from the sensors in a chamber, and issues control commands to the actuators such as gas valves, pumps, and heaters. The module controllers use bus-type control networks called *fieldbuses* such as process field bus–decentralized peripherals (PROFIBUS-DP) and control area networks (CANs) for communication and control with sensors and actuators. The module controllers are also coordinated by a system controller, called the *cluster tool controller* (CTC). A CTC has a module manager and a real-time scheduler. A module manager receives essential event messages from the PMCs, manages the states of the process modules, and sends the PMCs detailed control commands to perform a scheduling command from the scheduler. Communication between the PMCs, TMC, and the CTC usually uses transmission control protocol/Internet protocol (TCP/IP) based on Ethernet because they are well-known and accepted universal standards.

A real-time scheduler monitors the key events from each PMC and the TMC through the module manager. The events include starts and completions of wafer processing or robot tasks, which are essential for scheduling. Then, the scheduler determines the states of the modules and scheduling decisions as specified by the scheduling logic or rules, and issues the scheduling commands to the module manager. Since the wafer flow pattern can change, the scheduling logic should be easily changed without much programming work. The modules are often configured by a tool vendor to fulfill a specific cluster tool order. For large liquid-crystal display (LCD) fabrication, the modules are often integrated at a fab to assemble a large-scale cluster tool. Therefore, the scheduler should implement the scheduling logic in a modular way for flexibility when changing logic. To do this, the scheduling logic can be implemented by an extended finite state machine (EFSM) [52.13]. An EFSM models state change of each module and embeds a short programming code for the scheduling logic or procedure. The scheduling logic also includes procedures for handling exceptions such as wafer alignment failures, processing chamber failures, robot arm failures, etc. Figure 52.7 illustrates a typical architecture for communication and control in a cluster tool. A track system has a similar communication and control architecture.

A SEMI standard, cluster tool module communication (CTMC), specifies a model of distributed application objects for module controllers and a CTC, and a messaging standard between the objects [52.20]. *Lee* et al. [52.21] also propose an object-oriented application integration framework based on a high-level fieldbus communication protocol and service standard, PROFIBUS-field message specification (FMS), which defines a messaging standard between manufacturing equipment based on their object models. They sug-

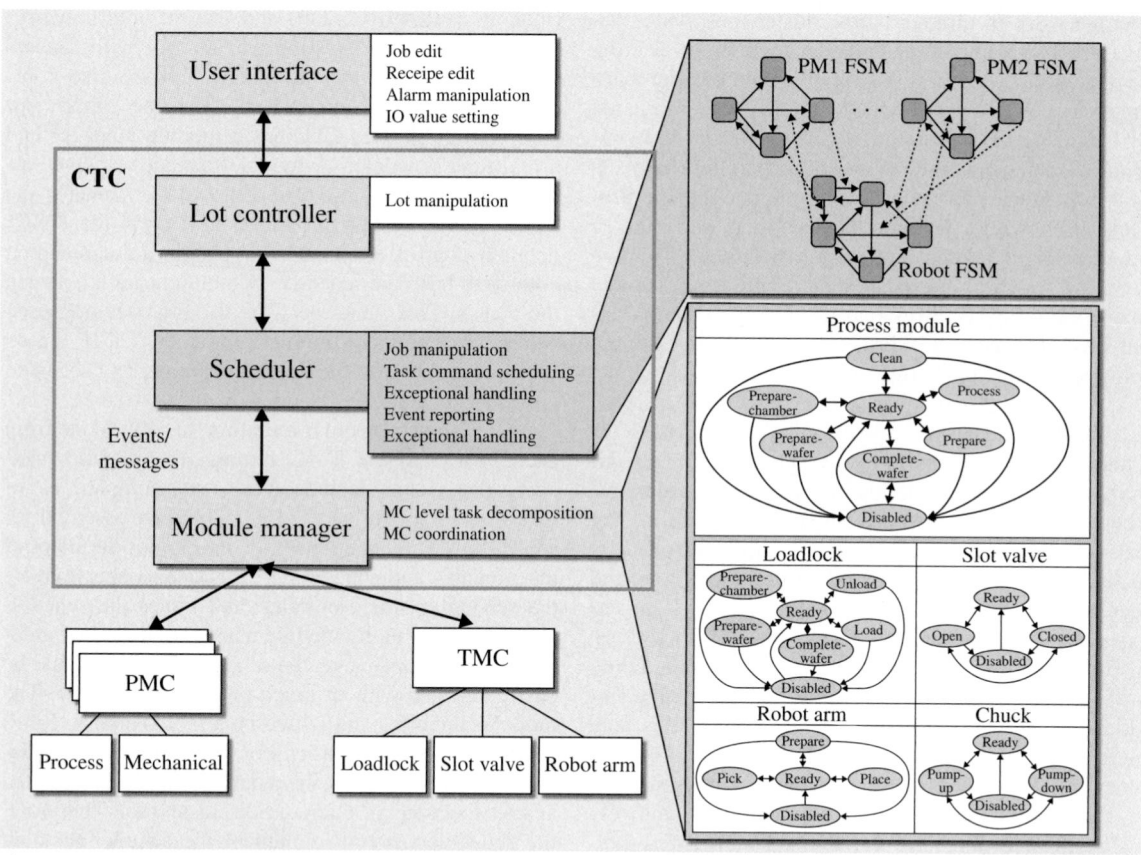

Fig. 52.7 A cluster tool controller architecture [52.22]

Part F | 52.3

gest that some object models in CTMC, which were defined based on a traditional object model for material-handling systems, need to be modified to handle the robot tasks in a cluster tool.

Each time a new cluster tool is developed, the scheduling logic and a CTC application should be integrated and extensively tested. However, tool testing and verification involve difficulties. First, a real tool is expensive and hence cannot be tied up for extensive testing. Second, testing with a real tool can be hazardous due to mechanical or space restrictions. Third, since the dynamics of a real cluster tool is slow, it takes significant time to test the system. Finally, it is often difficult to recognize subtle logical errors by observing operational behavior of a real tool. Therefore, the CTC and scheduler need to be tested in a virtual environment such as a *virtual cluster tool* (VCT), in which the process modules and the transport modules are replaced by their emulators [52.22]. The emulators receive control commands from the scheduler through the module man-

ager and/or the module controllers, and create messages for events such as process completions or robot task completions at appropriate times. The process times can be accelerated for initial rough-cut testing. Tool engineers examine the sequence of the events generated at the CTC or module controllers, and detect an anomaly. Such verification takes several days or weeks and is tedious. Some errors are hard to recognize and are often missed. *Joo* and *Lee* [52.22] propose the use of event sequence finite state machines for automatic error detection, which is basically identical to a finite state machine except that, when an event other than allowed ones at a state occurs, an error is assumed. They detected several unexpected logical errors, including logical errors caused by message sequence changes due to communication delay. Most tool simulators, such as ToolSim by Brooks Automation, focus on performance evaluation of a configured tool rather than high-fidelity modeling and verification of tool operation and messaging between a CTC and module controllers.

52.4 Fab Integration Architectures and Operation

52.4.1 Fab Architecture and Automated Material-Handling Systems

In modern 300 mm fabs, wafer cassette-handling tasks for interbay moves as well as intrabay ones are automated. In order to save the footprint and secure human operator access for equipment maintenance or exception handling, overhead transport (OHT) systems are mostly used. Traditional automated guided vehicle (AGV) or rail-guided vehicle (RGV) systems have been replaced by OHTs. In order to reduce particle contamination risk, tasks of loading and unloading wafer at tools are automated by using a new wafer carrier, the front open unified pod (FOUP), and a standard mechanical interface (SMIF). Processing tools are often enclosed in a minienvironment with extreme cleanness. Design and operation of the architecture and AMHSs of such fully automated fabs should be optimized to maximize throughput and reduce the cycle time while minimizing capital investment. An AMHS itself can be a bottleneck due to a limited number of vehicles and congestion on transport rails. Transport routes are not so flexible and should be considered as a limited resource. Therefore, in some 300 mm fabs, even critical metrology steps are skipped in order to reduce excessive vehicle traffic and the cycle time. Scheduling and dispatching systems are not yet well designed to handle such fully automated fabs. Control software such as manufacturing execution systems (MESs), material control systems (MCSs), equipment controllers, and schedulers as well as the AMHS architecture are not yet as intelligent and flexible as human operators, who make adaptive and intelligent decisions depending on the situation. There still remain many challenges to smart and efficient fully automated fabs. Figure 52.8 illustrates a typical OHT system layout, which consists of intrabay and interbay loops. There are works on optimal design of OHT networks, optimal number of OHTs, and performance analysis [52.23, 24].

Automated material-handling systems mostly have limited handling capacity and flexibility due to restricted paths and limited number of vehicles. Therefore, stockers or waiting places have been mandatory solutions for such problems. Stocking wafer cassettes at a bay involves significant delay due to prior waiting cassettes and handling operations. Therefore, in some 300 mm fabs, a desire to minimize the delivery cycle time led to attempts to combine several bays into a larger cell by eliminating bay-stockers in order to enforce direct delivery. However, this may cause significant OHT congestion and blocking, and hence throughput degradation. Nonetheless, direct delivery is one of the key technological challenges for next-generation 450 mm fabs [52.25]. To achieve the goal of direct delivery, we need quite different architectures of fabs and material-transfer systems. A solution might be to mimic a transfer line or a conveyor system, where wafer cassettes go through a significant number of process tools without intermediate stocking. Such a system is called an *inline* system. One of the most serious disadvantages of inline systems is lack of flexibility. In future fabs, lot sizes will continue to shrink. Therefore, con-

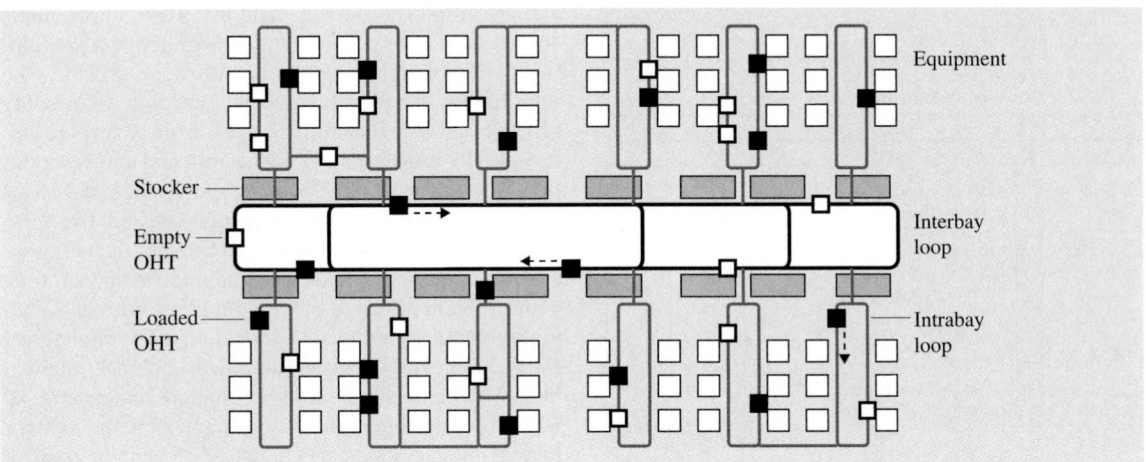

Fig. 52.8 An overhead transport system

Stocker
Empty OHT
Loaded OHT

Equipment
Interbay loop
Intrabay loop

flicting goals of flexibility and direct delivery should be resolved. LCD fabs, for which material transfer has been fully automated from the early stage due to manual handling difficulties, tend to introduce inline systems for more process steps as the panel size increases continually. A future 450 mm fab may also resemble an LCD line [52.25]. Stocker racks may be extensively located in parallel to the inline system [52.25]. Several alternatives for future fab and material-handling system architectures are now being discussed [52.26].

Traditionally, AMHSs have been scheduled and controlled separately from job scheduling. That is, wafer processing jobs are scheduled disregarding the limited capacity of the AMHS, and then material-transfer tasks, which are requested from the job schedule executor such as a real-time dispatcher, are separately planned and controlled by a material control system (MCS), that is, the AMHS controller. However, such decoupling is not so effective for modern integrated systems where job scheduling is significantly restricted by the AMHS, and vice versa. Interaction between job schedules and material transfer control should be considered, or they should be simultaneously scheduled as in cluster tools. MCSs have been engineered by AMHS vendors and are managed by automation engineers in fabs. However, job scheduling has been done by production management or control staffs. In the future, the two staff groups should better collaborate to tightly couple job scheduling and AMHS control. As fab technologies evolve, material-handling requirements become more challenging. SEMI has updated a roadmap for AMHSs for future fabs [52.27].

52.4.2 Communication Architecture and Networking

SEMI communication standards have been widely used in fabs to reduce system integration efforts [52.28]. While old tools are connected only by RS-232 ports, modern tools have Ethernet connections. The semiconductor equipment communication standard I (SECS-I) and high-speed message standard (HSMS) define data standards on RS-232-based serial communication and TCP/IP communication over Ethernet connection, respectively. SECS-II defines messaging standards. The generic equipment model (GEM) and virtual factory equipment interfaces (VFEI) are object-based application interface standards for equipments and factory control applications, respectively. The overall communication architecture is summarized in Fig. 52.9. AMHSs use fieldbus or control networks, either open or proprietary.

As advanced process control (APC) technology for real-time process sensing and real-time adaptive control becomes widespread, there is increasing demand on high-speed real-time communication technology, beyond the current communication architecture, in order to process massive process sensing data in real time.

52.4.3 Fab Control Application Integration

The most critical application for factory integration is a manufacturing execution system (MES). Its basic functions are to monitor equipment, send recipes, and keep track of wafers or other auxiliary materials such as photomasks. Quality monitoring and scheduling functions tend to be performed by separate applications from specialized vendors. MES applications should be easily and reliably integrated with equipment control applications. Traditionally, MESs used middleware based on message queueing to reliably process massive event messages from many equipments. No messages should be lost and the response time should be controlled. Therefore, such messages from many different tools are queued and the message queues are served by reasonable queueing or service policies for load balancing and response time control. Such message-based communication and integration require significant application work to integrate MES applications with equipment control applications. An application designer should understand all low-level messages and their required sequence for logical interaction between the MES and equipment controllers. Debugging, verification, and modification

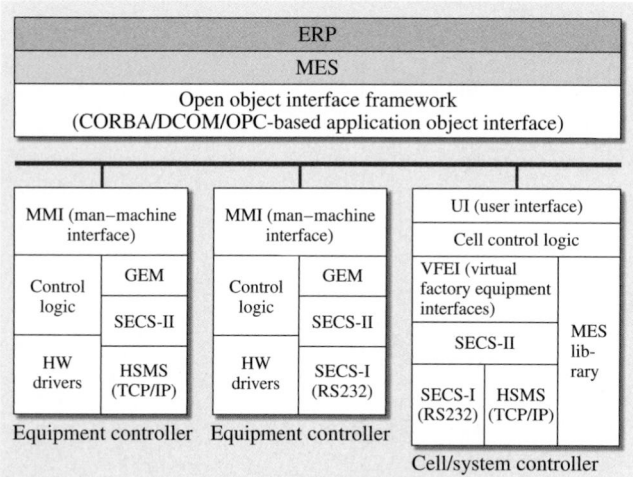

Fig. 52.9 Communication architecture for fab automation

are not easy. An alternative approach is object-based application integration. Each equipment and an MES application have a model of constituent objects, which specify the functions and informational states. Then, interactions between an MES and equipment are implemented by method calls or service requests between their corresponding objects. The common object request broker architecture (CORBA) is a middleware solution for facilitating application integration and interaction between such distributed objects and managing objects and services. MES application designers can conveniently make use of the high-level services of the objects in equipment control applications as well as common MES application objects. Detailed messaging sequences are handled by the methods of the objects that provide the relevant services. SEMI proposed an object-based MES application design standard, called the computer-integrated manufacturing (CIM) framework. SEMI also developed a standard object model for control applications of process equipment, called the object-based equipment model (OBEM). There have been concerns about whether CORBA can work reliably and fast enough for modern fab environments that generate massive amounts of real-time data. However, MES vendors have successfully implemented CORBA-based MES solutions, for example, IBM's SiView and AIM Sys-

tem's NanoMES. Figure 52.10 illustrates object-based interaction.

Recently, the service-oriented architecture (SOA) has been increasingly popular for business and enterprise applications [52.29]. Business processes tend to change frequently to cope with business requirement changes, and to be distributed over the Internet. Therefore, more flexibly composable *services* are defined and called as needed to form a new business process. Objects are considered to have too small granularity to be used for business processes [52.29]. Further, distributed objects technology such as CORBA and the distributed component object model (DCOM) are not easy standards to work with, because it is difficult to integrate object applications that were developed by different people at different places on different platforms at different times. Furthermore, CORBA and DCOM are not widely understood by software engineers and control and automation engineers. Web services have been open standards for easily integrating applications distributed on the Internet by using extensible markup language (XML)-based open standards such as simple object access protocol (SOAP), web services description languages (WSDLs), and universal description, discovery, and integration (UDDI), and standard web protocols such as XML, hypertext transfer protocol (HTTP), and transmission control protocol/Internet

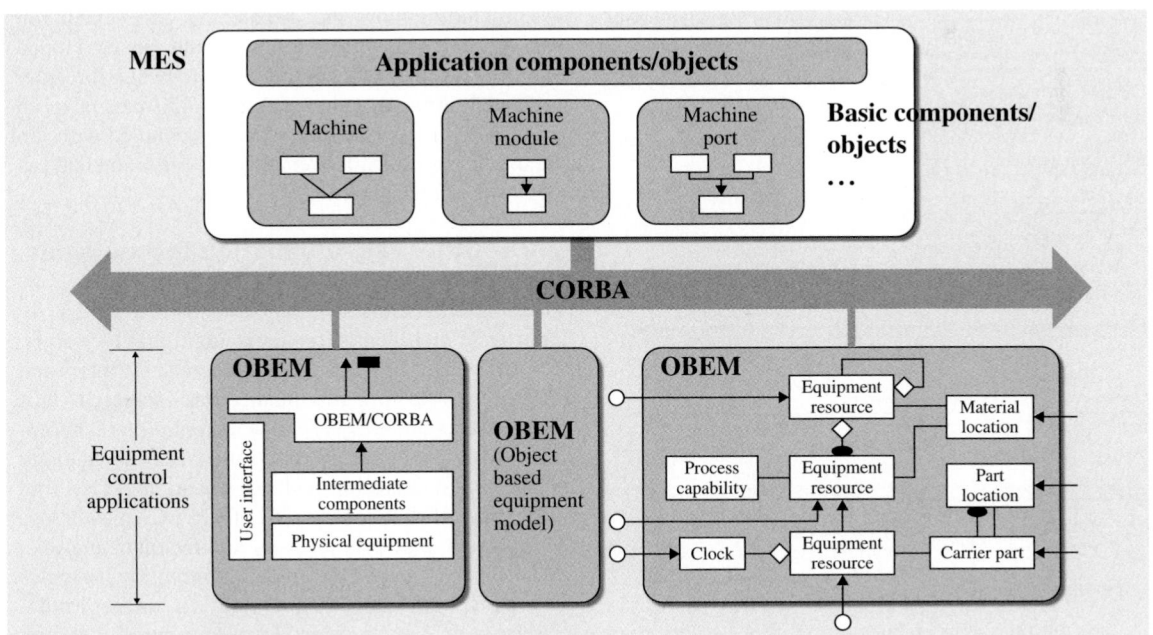

Fig. 52.10 Object-based interaction for MES and equipment control applications

protocol (TCP/IP). Therefore, SOA based on web services can provide open standards for easily integrating distributed factory applications with proper granularity. Therefore, some fabs or vendors for MESs or fab management applications are also now considering SOA-based design. However, it should be studied more whether SOA really makes sense for factory applications in terms of reliability and real-time performance.

52.4.4 Fab Control and Management

Fab operation is highly complicated due to the complex process steps and the massive number of lots in progress. One of the most crucial fab control applications is a real-time dispatcher that keeps track of the lots and equipment states, and determines which lots will be processed at which tools. It uses dispatching or scheduling rules that are proven to be effective for fab operation. The rules may be developed and tested for each fab through extensive simulation in advance. The essential function of a dispatcher is to process massive amounts of job and equipment data reliably and quickly, and compute a dispatch list quickly. A dispatcher sends a scheduling command to the MCS and the process equipment directly in an automated fab, whereas in a manual fab human operators perform the job of loading tasks as specified in the dispatch list.

An alternative scheduling approach to the dispatching rules is to have a separate scheduler that determines an appropriate work-in-progress level for each process step by using a dynamic lot flow model and then determines an optimal schedule for each process step separately under the restriction of the ready times and the due dates that are imposed by the schedule of other process steps. Frequent rescheduling is needed to cope with changes in fabs. Even in this case, the dispatcher retains the basic functions except for the scheduling function, and may change the schedule from the scheduler by local rules depending on the fab state. This approach has potential for further improving fab performance. However, there should be more experimental studies on which approach is more effective for different fab management environments.

A production planning system or supply-chain planning system determines daily production requirements for key process stages to meet order due dates or demand forecasts while minimizing inventory level. The system also considers binning due to random yields and capacity constraints. Other important fab control applications include yield management systems and advanced planning and scheduling (APS) systems. An overall fab control application architecture is summarized in Fig. 52.11.

In spite of extensive literature on fab scheduling, control, and management, there still remain many issues, including how the dispatching and scheduling systems, and scheduling rules should be developed to fulfill complex scheduling requirements for fully automated 300 mm fabs or future 450 mm fabs, in which AMHSs will be more strongly coupled with job scheduling for direct delivery, and lot definition and job flows will change significantly.

52.4.5 Other Fab Automation Technologies

Fab automation aims at an autonomous factory that reliably and intelligently produces high-quality wafers. As quality requirements have become stricter and the cost of attaining this quality has increased, fabs have developed *quality-sensitive automation* technologies. Advanced process control (APC) technology includes fault detection and classification (FDC) and run-to-run (R2R) control [52.30]. FDC makes use of statistical methods such as multivariate analysis, or intelligent computing or data-mining technologies such as neural networks or rules, in order to detect early any anomaly in process control that will cause significant quality problems, classify the prob-

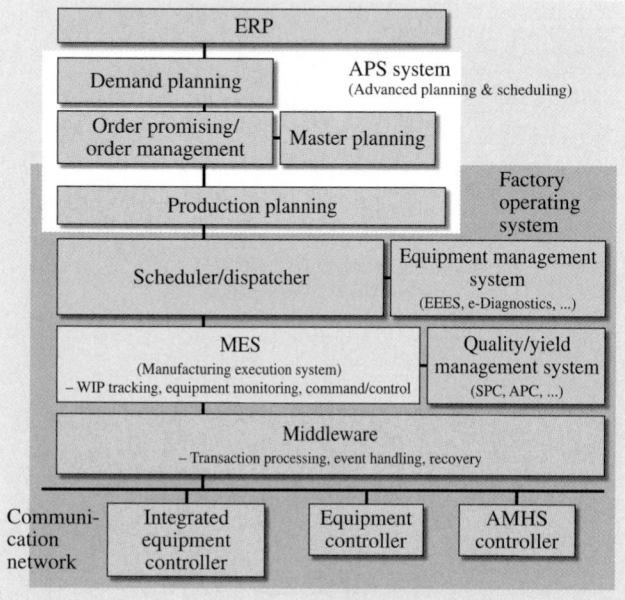

Fig. 52.11 A fab control system architecture (EEES – engineering equity extension service, SPC – statistical process control)

lems, and report them to the quality engineers. R2R control intelligently adapts process control parameters based on in situ measurements from process sensors. The response models between the measurement and control parameters are dynamic, nonlinear, multiple-input multiple-output (MIMO), and uncertain [52.30]. Therefore, advanced stochastic or statistical functional models and algorithms, or neural networks are used. An equipment engineering system (EES) is for tool vendors to remotely monitor process control of tools at fabs and tune process parameters. It is intended to reduce the initial ramp-up period and cost. Tool vendors cannot keep high-class engineers at customer sites for long periods, for instance, even more than 6 months. Another automation technology for quality is e-Diagnostics, which enables tool vendors in remote locations to detect an anomaly in tools in production at fabs quickly. It can prevent or reduce production of defective wafers and reduce the lead time to dispatch tool engineers to customer sites. Tool vendors and SEMI have developed EES and e-Diagnostics technologies and standards, including data standards, security control, remote control or manipulation, etc. [52.31].

52.5 Conclusion

Semiconductor manufacturing fabs have extensively developed and implemented state-of-the-art industrial automation technologies. We have briefly reviewed them in this Chapter. There remain many challenges for the future, such as for 450 mm fabs. Future fabs for manufacturing nanodevices may require quite new concepts of equipment and material handling, and hence new automation technologies. Concepts, technologies, and practices of semiconductor manufacturing automation can give insights into automation of other manufacturing industries or service systems.

References

52.1 C. Haris: Automated material handling system. In: *Semiconductor Manufacturing Handbook*, ed. by H. Geng (McGraw-Hill, New York 2005) pp. 32.1–32.11

52.2 S. Venkatesh, R. Davenport, P. Foxhoven, J. Nulman: A steady-state throughput analysis of cluster tools: Dual-blade versus single-blade robots, IEEE Trans. Semicond. Manuf. **10**(4), 418–424 (1997)

52.3 J.-H. Paek, T.-E. Lee: Operating strategies of cluster tools with intermediate buffers, Proc. 7th Annu. Int. Conf. Ind. Eng. (2002) pp. 1–5

52.4 C. Jung: Stedy State Scheduling and Modeling of Multi-Slot Cluster Tools. M. Sc. Thesis (Department of Industrial Engineering, KAIST 2006)

52.5 H.L. Oh: Conflict resolving algorithm to improve productivity in single-wafer processing, Proc. Int. Conf. Model. Anal. Semicond. Manuf. (MASM) (2000) pp. 55–60

52.6 H.J. Yoon, D.Y. Lee: Real-time scheduling of wafer fabrication with multiple product types, Proc. IEEE Int. Conf. Syst. Man Cybern. (1999) pp. 835–840

52.7 T.-E. Lee, H.-Y. Lee, S.-J. Lee: Scheduling a wet station for wafer cleaning with multiple job flows and multiple wafer-handling robots, Int. J. Prod. Res. **45**(3), 487–507 (2007)

52.8 T.-E. Lee, M.E. Posner: Performance measures and schedules in periodic job shops, Oper. Res. **45**(1), 72–91 (1998)

52.9 T.-E. Lee: Stable earliest starting schedules for periodic job shops: a linear system approach, Int. J. Flex. Manuf. Syst. **12**(1), 59–80 (2000)

52.10 T.-E. Lee, R. Sreenivas, H.-Y. Lee: Workload balancing for timed event graphs with application to cluster tool operation, Proc. IEEE Int. Conf. Autom. Sci. Eng. (2006) pp. 1–6

52.11 J.-H. Kim, T.-E. Lee, H.-Y. Lee, D.-B. Park: Scheduling of dual-armed cluster tools with time constraints, IEEE Trans. Semicond. Manuf. **16**(3), 521–534 (2003)

52.12 T. Murata: Petri nets: properties, analysis and applications, Proc. IEEE **77**(4), 541–580 (1989)

52.13 Y.-H. Shin, T.-E. Lee, J.-H. Kim, H.-Y. Lee: Modeling and implementating a real-time scheduler for dual-armed cluster tools, Comput. Ind. **45**(1), 13–27 (2001)

52.14 T.-E. Lee, S.-H. Park: An extended event graph with negative places and negative tokens for time window constraints, IEEE Trans. Autom. Sci. Eng. **2**(4), 319–332 (2005)

52.15 J.-H. Kim, T.-E. Lee: Schedule stabilization and robust timing control for time-constrained cluster tools, Proc. IEEE Conf. Robot. Autom. (2003) pp. 1039–1044

52.16   T.-E. Lee, H.-Y. Lee, Y.-H. Shin: Workload balancing and scheduling of a single-armed cluster tools, Proc. Asian-Pac. Ind. Eng. Manag. Syst. Conf. (2004) pp. 1–6

52.17   H.J. Kim: Scheduling and Control of Dual-Armed Cluster Tools With Post Processes. M. Sc. Thesis (Department of Industrial Engineering, KAIST 2006)

52.18   H.-Y. Lee, T.-E. Lee: Scheduling single-armed cluster tools with reentrant wafer flows, IEEE Trans. Semicond. Manuf. **19**(2), 224–240 (2006)

52.19   J.-S. Lee: Scheduling Rules for Dual-Armed Cluster Tools With Cleaning Processes. M. Sc. Thesis (Department of Industrial Engineering, KAIST 2008)

52.20   SEMI E38.1-95: Cluster tool module communication(CTMC), *SEMI International Standards* (2007)

52.21   J.-H. Lee, T.-E. Lee, J.-H. Park: Cluster tool module communication based on a high-level fieldbus, Int. J. Comput. Integr. Manuf. **17**(2), 151–170 (2004)

52.22   Y.-J. Joo, T.-E. Lee: A virtual cluster tool for testing and verifying a cluster tool controller and a scheduler, IEEE Robot. Autom. Mag. **11**(3), 33–49 (2004)

52.23   D.-Y. Liao, H.-S. Fu: A simulation-based, two-phased approach for dynamic OHT allocation and

dispatching in large-scaled 300 mm AMHS management, Proc. IEEE Int. Conf. Robot. Autom. **4**, 3630–3635 (2002)

52.24   D.-Y. Liao, H.-S. Fu: Speedy delivery-dynamic OHT allocation and dispatching in large-scale, 300 mm AMHS management, IEEE Robot. Autom. Mag. **11**(3), 22–32 (2004)

52.25   J.S. Pettinato, D. Pillai: Technology decisions to minimize 450-mm wafer size transition risk, IEEE Tans. Semicond. Manuf. **18**(4), 501–509 (2005)

52.26   D. Pillai: The future of semiconductor manufacturing, IEEE Robot. Autom. Mag. **13**(4), 16–24 (2006)

52.27   SEMI The international technology roadmap for semiconductors (ITRS): an update, SEMI Eur. Stand. Autumn Conf. (2006)

52.28   SEMI International Standards, SEMI (2007), CD-ROM

52.29   D. Krafzig, K. Banke, D. Slama: *Enterprise SOA: Service-Oriented Architecture Best Practices* (Prentice Hall, Upper Saddle River 2005)

52.30   J. Moyne, E. del Castillo, A.M. Hurwitz: *Run-to-Run Control in Semiconductor Manufacturing* (CRC, New York 2001)

52.31   H. Wohlwend: *e-Diagnostics Guidebook: Revision 2.1* (Int. SEMATECH Manuf. Initiative, 2005), http://www.sematech.org/docubase/abstracts/4153deng.htm

53. Nanomanufacturing Automation

Ning Xi, King Wai Chiu Lai, Heping Chen

This chapter reports the key developments for nanomanufacturing automation. Automated CAD guided nanoassembly can be performed by an improved atomic force microscopy (AFM). Although CAD guided automated manufacturing has been widely studied in the macro-world, nanomanufacturing is challenging. In nanoenvironments, the nanoobjects are usually distributed on a substrate randomly, so the nanoenvironment and the available nanoobjects have to be modeled in order to design a feasible nanostructure. Because of the positioning errors due to the random drift, the actual position of each nanoobject has to be identified by our local scanning method. The advancement of AFM increases the efficiency and accuracy to manipulate and assemble nanoobjects. Besides, the manufacturing process of carbon nanotube (CNT) based nanodevices is discussed. A novel automated manufacturing system has been especially designed for manufacturing nanodevices. The system integrates a new dielectrophoretic (DEP) microchamber into a robotic based deposition workstation and increases the yield to form semi-conducting CNTs for manufacturing nanodevices. Therefore, by using the proposed CNT separation and deposition system, CNT based nanodevices with specific and consistent electronic properties can be manufactured automatically and effectively.

Part F | 53

53.1 Overview

Nanoscale materials with unique mechanical, electronic, optical, and chemical properties have a variety of potential applications such as nanoelectromechanical systems (NEMS) and nanosensors. The development of nanoassembly technologies will potentially lead to breakthroughs in manufacturing new revolutionary industrial products. The techniques for nanoassembly can be generally classified into *bottom-up* and *top-down* methods. Self-assembly in nanoscale is reported as the most promising *bottom-up* technique, which is applied to make regular, symmetric patterns of nanoentities. However, many potential nanostructures and nanodevices are asymmetric, which cannot be manufactured using self-assembly only. A *top-down* method would be desirable to fabricate complex nanostructures.

The semiconductor fabrication technique is a matured *top-down* method, which has been used in the fabrication of microelectromechanical systems (MEMS). However, it is difficult to build nanostructures using this method due to limitations of the traditional lithography. Although smaller features can be made by electron beam nanolithography, it is practically very difficult to position the feature precisely using e-Beam nanolithography. The high cost of the scanning electron microscopy (SEM), ultrahigh vacuum condition, and space limitation inside the SEM vacuum capsule also impede its wide application.

Atomic force microscopy (AFM) [53.1] has proven to be a powerful technique to study sample surfaces down to the nanoscale. It can work with both conductive and insulating materials and in many conditions, such as air and liquid. Not only can it characterize sample surfaces, it can also modify them through nanolithography [53.2, 3] and nanomanipulation [53.3, 4], which is a promising nanofabrication technique that combines *top-down* and *bottom-up* advantages. In recent years, many kinds of AFM-based nanolithographies have been implemented on a variety of surfaces such as semiconductors, metals, and soft materials [53.5–8]. A variety of AFM-based nanomanipulation schemes have been developed to position and manipulate nanoobjects [53.9–13]. However, nanolithography itself can hardly be considered as sufficient for fabrication of a complete device. Thus, manipulation of nanoobjects has to be involved in order to manufacture nanostructures and nanodevices. The AFM-based nanomanipulation is much more complicated and difficult than the AFM-based nanolithography because nanoobjects have to be manipulated from one place to another by the AFM tip, and sometimes it is necessary to relocate the nanoobjects during nanomanipulation while nanolithography can only draw patterns. Since the AFM tip as the manipulation end-effector can only apply a point force on a nanoobject, the pushing point on the nanoobject has to be precisely controlled in order to manipulate the nanoobjects to their desired positions. In the most recently available AFM-based manipulation methods, the manipulation paths are obtained either manually using haptic devices [53.9, 10] or in an interactive way between the users and the atomic force microscope (AFM) images [53.11, 12]. The main problem of these schemes is their lack of real-time visual feedback, so an augmented reality interface has been developed [53.14, 15]. But positioning errors due to deformation of the cantilever and random drift such as thermal drift cause the nanoobjects to be easily lost or manipulated to

wrong places during manipulation; the result of each operation has to be verified by a new image scan before the next operation starts. This scan-design-manipulation-scan cycle is usually time consuming and inefficient.

In order to increase the efficiency and accuracy of AFM-based nanoassembly, automated CAD guided nanoassembly is desirable [53.16]. In the macroworld, CAD guided automated manufacturing has been widely studied [53.17]. However, it is not a trivial extension from the macroworld to the nanoworld. In the nanoenvironments, the nanoobjects, which include nanoparticles, nanowires, nanotubes, etc., are usually distributed on a substrate randomly. Therefore, the nanoenvironment and the available nanoobjects have to be modeled in order to design a feasible nanostructure. Because manipulation of nanoparticles only requires translation, while manipulation of other nanoobjects such as nanowires involve both translation and rotation, manipulation of nanowires is more challenging than that of nanoparticles. To generate a feasible path to manipulate nanoobjects, obstacle avoidance must also be considered. Turns around obstacles should also be avoided since they may cause the failure of the manipulation. Because of the positioning errors due to the random drift, the actual position of each nanoobject must be identified before each operation.

Beside, the deformation of the cantilever caused by manipulation force is one of the most major nonlinearities and uncertainties. It causes difficulties in accurately controlling the tip position, and results in missing the position of the object. The softness of the conventional cantilevers also causes the failure of manipulation of sticky nanoobjects because the tip can easily slip over the nanoobjects. An active atomic force microscopy probe is used as an adaptable end effector to solve these problems by actively controlling the cantilever's flexibility or rigidity during nanomanipulation. Thus, the adaptable end effector is controlled to maintain straight shape during manipulation [53.18].

Apart from nanoassembly, manufacturing process of nanodevices is important. Carbon nanotube (CNT) has been investigated as one of the most promising candidates to be used for making different nanodevices. CNTs have been shown to exhibit remarkable electronic properties, such as ballistic transport and semiconducting behavior, which depend on their diameters and chiralities. Recently, it was demonstrated that CNTs can be used to build various types of devices such as nanotransistors [53.19], logic devices [53.20], infrared detectors [53.21, 22], light emitting devices [53.23], chemical sensors [53.24, 25], etc. The general man-ufacturing processes of CNT-based devices is shown in Fig. 53.1. The most challenging parts include CNT selection, deposition, and assembly. Basically, CNT assembly can be done by our AFM-based nanoma-nipulation system [53.26]. With the advancement of our automated local scanning method for AFM sys-tems [53.27], automated assembly for CNT-based devices can be done effectively. However, electronic properties of CNTs vary and CNTs can be classified into two types: semiconducting CNTs and metallic CNTs. Therefore, an automatic method for the selection and deposition of a single CNT with a specific electronic property should be established [53.28, 29].

Selection of a CNT with the desired electronic property is crucial to its application. Basically, sev-eral approaches have been pursued to separate different electronic types of CNTs. *Arnold* et al. demonstrated that semiconducting CNTs and metallic CNTs were separated by using some encapsulating agents or sur-factants [53.30]. Besides, *Avouris* et al. demonstrated turning a metallic CNT into a semiconducting CNT after removal of metallic carbon shells by an electri-cal breakdown process [53.31]. *Krupke* et al. reported a technique to enrich metallic CNT thin film. They demonstrated that metallic CNTs were concentrated on a substrate by using dielectrophoresis [53.32,33]. Based on the review of these CNT separation techniques, we develop a microchamber to filter different types of CNTs effectively.

Various methods have been proposed to move and deposit a CNT to the metal microelectrodes; this ad-vances the manufacturing process of the CNT-based nanodevices. A nanorobotic technique uses nanomanip-ulators inside a scanning electron microscopy (SEM) to perform the nanomanipulation. Since a sample cham-ber of an SEM is spacious, it is possible to put some custom-design nanomanipulators inside the chamber. *Yu* et al. put a custom piezoelectric vacuum manipula-tor inside the chamber of an SEM, and they visually observed the manipulation process of CNTs [53.34]. *Dong* et al. also developed a 16-degree-of-freedom nanorobotic manipulator to characterize CNTs inside an SEM system [53.35]. The idea of nanoassembly inside SEM is promising, but it needs a vacuum en-vironment for proper operation. Alternatively, electric field assisted methods have been proposed to manipu-late and deposit CNTs directly. *Green* et al. introduced AC electrokinetics forces to manipulate sub-micrometer particles on microelectrode structures [53.36]. Bundled CNTs have also been manipulated by dielectrophoretic

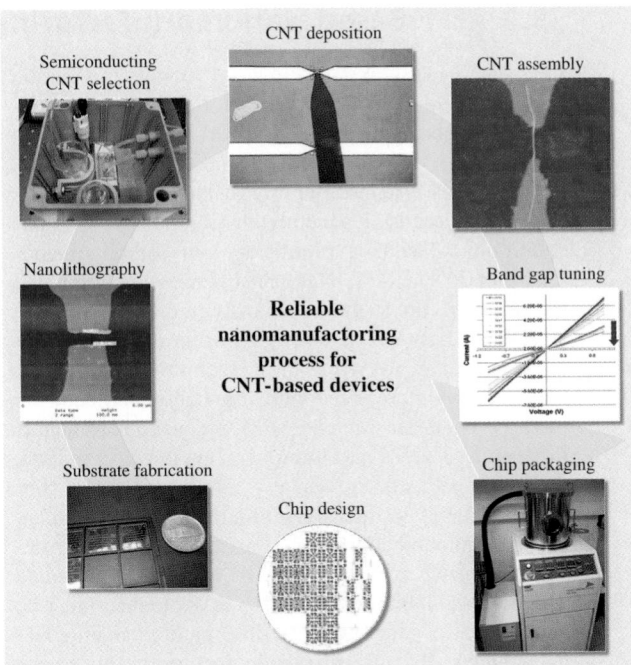

Fig. 53.1 Flow chart of nanomanufacturing of CNT-based devices

(DEP) force [53.37, 38]. Moreover, *Dong* and *Nel-son* reported the batch fabrication of CNT bearings and transistors by assembling CNTs on a silicon chip using DEP force [53.39, 40]. A fabricated chip was immersed in a reservoir that contained CNT suspen-sion, and CNTs were deposited on the microchip by applying a composite AC/DC electric field. This elec-tric field manipulation technique is an effective and feasible method to batch manipulate CNTs manually. However, an automated robotic system for mass pro-duction of consistent CNT-based devices has not been archived.

An automated nanomanipulation system is dis-cussed in Sect. 53.2. The collision-free paths are generated based on the CAD model, the environment model, and the model of the nanoobjects. A local scanning method is developed to obtain the actual po-sition of each nanoobject to compensate for the random drift. Moveover, automatic nanoassembly of nanostruc-tures using the designed CAD models is presented. The nanomanufacturing process of CNT-based device is discussed in Sect. 53.3. The process includes the de-velopment of a novel CNT separation system and an automated deposition processes for both single-walled carbon nanotubes (SWCNTs) and multi-walled carbon nanotubes (MWCNTs).

53.2 AFM-Based Nanomanufacturing

In recent years, many kinds of nanomanipulation schemes have been developed to manipulate nanoobjects. A nanorobotic technique uses nanomanipulators inside an SEM to perform the nanomanipulation [53.34, 35]. The idea of nanoassembly inside SEM is promising, but it needs a vacuum environment for proper operation. AFM is a promising tool for nanomanufacturing [53.12, 41]. Nanoobjects were manipulated by an AFM tip to build nanostructures and devices effectively because of its high resolution. Besides, it does not need to work in a vacuum environment, which allows more freedom in the nanomanufacturing process. In order to use AFM for nanomanufacturing, further studies and improvements have been done. Since nanoobjects are usually distributed on a substrate randomly in the nanoworld, the nanoenvironment and the nanoobjects must be modeled in order to design a feasible nanostructure. In order to manipulate nanoobjects automatically, obstacle avoidance must be considered to generate a feasible path to manipulate nanoobjects. Because of positioning errors due to the random drift, the actual position of each nanoobject must be identified before each operation; this correction can be done by our local scanning method. In order to increase the efficiency and accuracy of AFM-based nanoassembly, automated CAD guided nanoassembly is desirable.

53.2.1 Modeling of the Nanoenvironments

Because the nanoobjects are randomly distributed on a surface, the position of each nanoobject must be determined in order to perform automatic manipulation. Also the nanoobjects have different shapes, such as nanoparticles and nanowires, as shown in Fig. 53.2. They must be categorized before manipulation because the manipulation algorithms for these nanoobjects are different.

After an AFM image is obtained, the nanoobjects can be identified and categorized. The X, Y coordinates and the height information of each pixel can be obtained from the AFM scanning data. Because the height, shape, and size of nanoobjects are known, they are used as criteria to identify nanoobjects and obstacles based on a fuzzy method as follows. Firstly, all pixels higher than a threshold height are identified. The shapes of clustered pixels are categorized and compared with the ideal shapes of nanoobjects. If the shape of clustered pixels is close to the ideal shape, the pixels are assigned a higher probability (p_1). Secondly, if the height of a pixel is close to the ideal height of nanoobjects, a higher probability (p_2) is assigned to it. Thirdly, the neighboring pixels with higher probability ($p_1 p_2$) are counted and the area of the pixels is identified. If the area is close to the size of nanoobjects, the pixels are assigned a higher probability (p_3). If the probability ($p_1 p_2 p_3$) of a pixel is higher than a threshold, it is in a nanoobject. Using the neighboring relationship of pixels, objects can then be identified. The length of a nanoobject can be calculated by finding the long and short axes using a least squares fitting algorithm. If the length/width ratio is larger than a set value, it is considered as a nanowire, otherwise, as a nanoparticle.

53.2.2 Methods of Nanomanipulation Automation

Since the AFM tip can only apply force to a point on a nanoobject in AFM-based nanomanipulation, it is very challenging to generate manipulation paths to manipulate nanoobjects to a desired location, especially for nanowires, because manipulation of nanoparticles only requires translation, while that of nanowires involves translation as well as rotation. Turns around obstacles should be avoided since they may cause manipulation failure. In the following sections, automated manipulation of nanoparticles and nanowires, respectively, will be discussed.

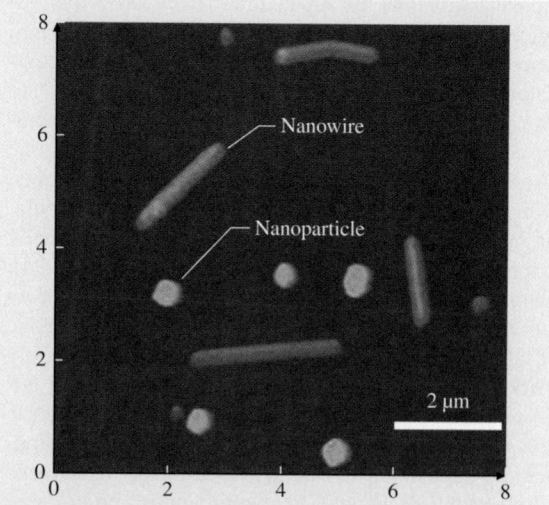

Fig. 53.2 Nanoobjects obtained from AFM scanning. The scanning area is $8\,\mu m \times 8\,\mu m$

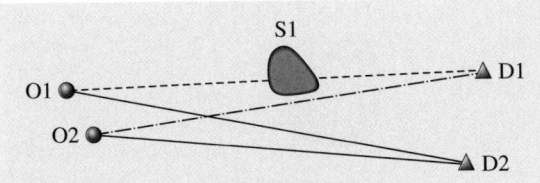

Fig. 53.3 The straight line connection between an object and a destination. O1 and O2 are objects, D1 and D2 are destinations, S1 is an obstacle

Automated Manipulation of Nanoparticles

Once the destinations, objects, and obstacles are determined, a collision-free path can be generated by the tip path planner.

A direct path (straight path) is a connection from an object to a destination using a straight line without any obstacles or potential obstacles in between. Figure 53.3 shows the connections between objects and destinations. The paths from O2 to D2 and from O1 to D2 are direct paths, and the path between O1 and D1 is not a direct path due to collision.

Due to the van der Waals force between an object and an obstacle, the object may be attracted to the obstacle if the distance between them is too small. Therefore, the minimum distance has to be determined first to avoid the attraction. Figure 53.4a shows a particle object and a particle obstacle, and Fig. 53.4b a particle object and a nanowire obstacle, respectively.

In the first case, all objects and obstacles are assumed to be spheres; the van der Waals force can be expressed as [53.42]

$$F_w = \frac{-A}{6D} \frac{R_1 R_2}{R_1 + R_2} , \tag{53.1}$$

where F_w is the van der Waals force, A is the Hamaker constant, D is the distance between the two spheres, and R_1 and R_2 are the radii of the two spheres. For the second case that the obstacle is a nanowire, the nanowire can be considered as separated nanoparticles, so the van der Waals force between a nanoparticle and each separated nanoparticles can be calculated using (53.1).

Different materials have different Hamaker constants. Nevertheless, the Hamaker constants are found to lie in the range $(0.4 - 4) \times 10^{-19}$ J [53.42]. If an object is not attracted to an obstacle, the van der Waals force between an object and a destination has to be balanced by the friction force between the object and the surface. The friction force between the object and the surface can be formulated as [53.43]

$$F_c = \mu_{os} F_{os}^r + \nu F_{os}^a , \tag{53.2}$$

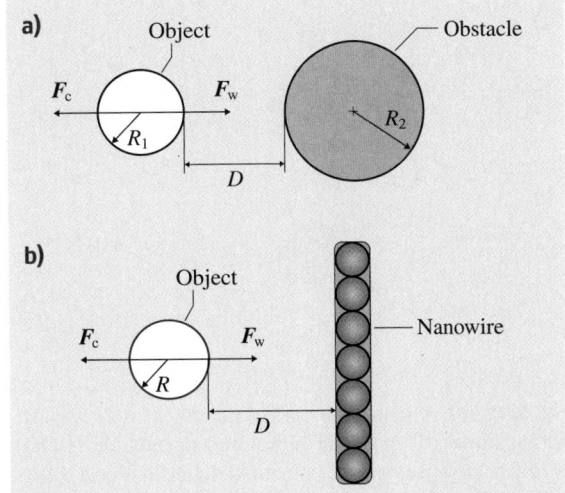

Fig. 53.4a,b The van der Waals force between objects and obstacles. The objects are nanoparticles. (**a**) The obstacle is a nanoparticle. R_1 and R_2 are the radius of the two spheres respectively, D is the distance between the two spheres, F_w is the van der Waals force, and F_c the friction force. (**b**) The obstacle is a nanowire. The nanowire can be considered as a line of nanoparticles. R is the radius of the sphere

where F_c is the friction force, μ_{os} is the sliding friction coefficient between an object and the substrate surface, ν is the shear coefficient, F_{os}^r is the repulsive force, and F_{os}^a is the adhesive force. When pushing an object, the minimum repulsive force equals the adhesive force. Then (53.2) becomes

$$F_c = (\mu_{os} + \nu) F_{os}^a . \tag{53.3}$$

The adhesive force F_{os}^a can be estimated by [53.43]

$$F_{os}^a = \frac{A_{os}}{A_{ts}} F_{ts}^a , \tag{53.4}$$

where A_{os} is the nominal contact area between an object and a substrate surface, A_{ts} is the nominal contact area between the AFM tip and the substrate surface, and F_{ts}^a is the measured adhesive force between the AFM tip and the surface.

Since the van der Waals force must be balanced by the friction force during manipulation, the minimum distance D_{min} can be calculated using (53.1), (53.3) and (53.4)

$$D_{min} = \frac{A}{6} \frac{R_1 R_2}{R_1 + R_2} \frac{A_{ts}}{(\mu_{os} + \nu) A_{os} F_{ts}^a} . \tag{53.5}$$

The distance between an object and a nanowire must be larger than D_{min} during manipulation. If there is an

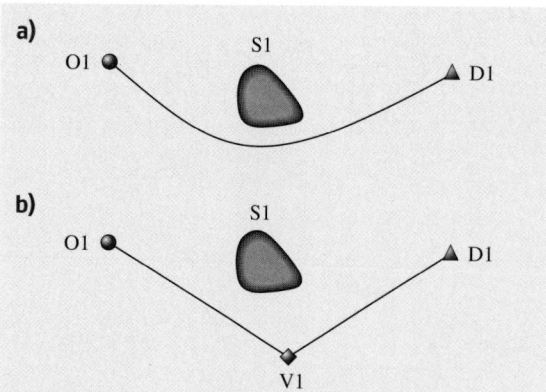

Fig. 53.5 (a) A path with turns. An object may be lost during turns. **(b)** A virtual object and destination (VOD) connects an object and a destination. O1 is an object, D1 is a destination, S1 is an obstacle, and V1 is a VOD

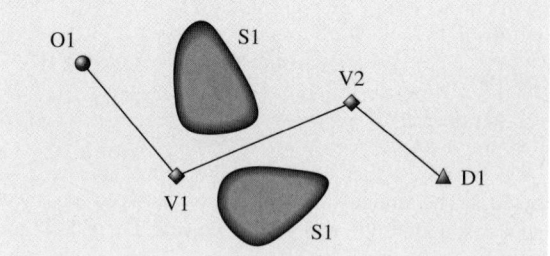

Fig. 53.6 Two VODs connect an object with a destination. O1 is an object, D1 is the destination, and S1 and S2 are obstacles

obstacle that is close or on the straight line, the path formed by the straight line is not considered as a direct path. For example, the path between O2 and D1 in Fig. 53.3 is not a direct path due to the attraction.

After the direct paths are generated, objects are assigned to the destinations one to one. There are some destinations that may not have any objects assigned to them. This is because there are no direct paths to some destinations. Therefore, indirect paths (curved paths) to avoid the obstacles must be generated. In general, it is possible to lose particles during nanomanipulation in both direct paths or curved paths, but a curved path as shown in Fig. 53.5 has a much higher risk of losing objects than a direct path. The AFM-based manipulation system can use force feedback to detect the lost particle during the manipulation. A surface must be scanned again if an object is lost during manipulation. Because the scanning time is much longer than the manipulation time, turns should be avoided during nanomanipulation. To solve the problem, a virtual-object-destination algorithm has been developed. Figure 53.6 shows a virtual-object-destination (VOD).

An object and a destination are connected using direct paths through a VOD. Since there are many possible VODs to connect an object and a destination, a minimum distance criterion is applied to find a VOD. The total distance to connect an object and a destination through a VOD is

$$d = \sqrt{(x_2 - x_0)^2 + (y_2 - y_0)^2}$$
$$+ \sqrt{(x_2 - x_1)^2 + (y_2 - y_1)^2} \,, \qquad (53.6)$$

where x_2, y_2 are the coordinates of the center of a VOD, x_0, y_0 are the coordinates of the center of an object, and x_1, y_1 are the coordinates of the center of a destination.

The connections between the VOD, object and destination have to avoid the obstacles, i. e.,

$$\sqrt{(x - x_s)^2 + (y - y_s)^2} \geq D_{\min} + R' \,, \qquad (53.7)$$

where x, y are the coordinates of the object center along the path and x_s, y_s are the coordinates of the center of the obstacle. R' is defined as

$$R' = R_1 + R_2 \,. \qquad (53.8)$$

Then a constrained optimization problem is formulated

$$\min_{x_2, y_2} d = \sqrt{(x_2 - x_0)^2 + (y_2 - y_0)^2}$$
$$+ \sqrt{(x_2 - x_1)^2 + (y_2 - y_1)^2} \,,$$
$$\text{subject to: } \sqrt{(x - x_s)^2 + (y - y_s)^2} \geq D_{\min} + R' \,.$$
$$(53.9)$$

This is a single objective constrained optimization problem. A quadratic loss penalty function method [53.44] is adopted to deal with the constrained optimization problem by formulating a new function $G(x)$

$$\min_{x_2, y_2} G(x) = \min_{x_2, y_2} d + \beta (\min[0, g])^2 \,, \qquad (53.10)$$

where β is a big scalar and g is formulated using the given constraint, i. e.

$$g = \sqrt{(x - x_s)^2 + (y - y_s)^2} - (D_{\min} + R') \,. \qquad (53.11)$$

Then the constrained optimization problem is transferred into an unconstrained one using the quadratic loss penalty function method. The pattern search method [53.45] is adopted here to optimize the unconstrained optimization problem to obtain the VOD.

If one virtual object and destination cannot reach an unassigned destination, two or more VODs must be found to connect an object and a destination. Figure 53.6 illustrates the process. Similarly, the total distance to connect the object and destination can be calculated. The constraint is the same as (53.9). Then, a single objective constrained optimization problem can be formulated to obtain the VODs.

Automated Manipulation of Nanowires

The manipulation of a nanowire is much more complicated than that of a nanoparticle because there is only translation during manipulation of a nanoparticle, while there are both translation and rotation during manipulation of a nanowire. A nanowire can only be manipulated to a desired position by applying force alternatively close to its ends. From an AFM image, nanowires can be identified and represented by their radius and two end points. Each end point on a nanowire must be assigned to the corresponding point on the destination. The starting pushing point is important since it determines the direction along which the object moves. By choosing a suitable step size, an AFM tip path can be generated. Therefore, the steps of automated manipulation of nanowire are: find the initial position and destination of a nanowire, find the corresponding points, find starting pushing point, and calculate the pushing step and plan the tip trajectory. The details of the steps are given below.

To automatically manipulate a nanowire, its behavior under a pushing force has to be modeled. When a pushing force is applied to a nanowire, the nanowire starts to rotate around a pivot if the pushing force is larger than the friction force. Figure 53.7 shows the applied pushing force and the pivot. The nanowire rotates around point D when it is pushed at point C by the AFM tip. The nanowire under pushing may have different kinds of behavior, which depend on its own geometry property. If the aspect ratio of a nanowire is defined as

$$\sigma = d/L . \tag{53.12}$$

A nanowire with the aspect ratio of $\sigma > 25$ usually behaves like a wire, which will deform or bend under pressure. The rotation behavior was observed for nanowires with aspect ratio of $\sigma < 15$. In this case, the pushing force $\boldsymbol{F}$ from the tip causes the friction and shear force $\boldsymbol{F}' = \mu_{ot}\boldsymbol{F} + \nu\boldsymbol{F}_{ot}^a$ along the rod axis direction when the pushing direction is not perpendicular to the rod axis, where μ_{ot} and ν are the friction and shear coefficients between the tip and the nanowire,

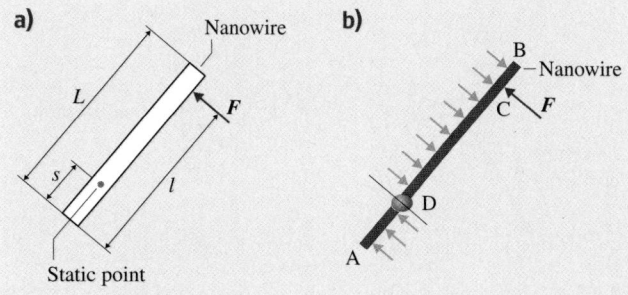

Fig. 53.7a,b The behavior of a nanowire under a pushing force: **(a)** F is the applied external force, L is the length of the nanowire **(b)** the detailed force model. D is the pivot where the nanowire rotates, C is the pushing point

which depend on the material properties and the environment, $\boldsymbol{F}_{ot}^a$ is the adhesion force between the tip and the nanowire.

Fortunately, it is easy to prove that the force $\boldsymbol{F}'$ hardly causes the rod to move along the rod axis direction. Assuming that the shear forces between rod and surface are equal along all directions during moving,

$$fL = f'_{max}d . \tag{53.13}$$

Because the shear force is usually proportional to the contact area, and the contact area between a nanowire and surface is much greater than that between the tip and the nanowire,

$$\nu\boldsymbol{F}_{ot}^a \ll f'd . \tag{53.14}$$

Also note that

$$\boldsymbol{F} \le fL = f'_{max}d , \tag{53.15}$$

and because μ is usually very small, finally it is reasonable to assume that

$$f'd = \boldsymbol{F}'\mu\boldsymbol{F} + \nu\boldsymbol{F}_{ot}^a < f'_{max}d . \tag{53.16}$$

This means that the rod will have no motion along the axis direction and, therefore, the static point D must be on the axis of the nanowire. Considering the above analysis, the nanowire can be simplified as a rigid line segment. The external forces applied on the nanowire in surface plane can be modeled as shown in Fig. 53.7. The pivot D can be either inside the nanowire or outside the nanowire. First assume that D is inside the nanowire. In this case, all the torques around D are self-balanced during smooth motion.

$$\boldsymbol{F}(l-s) = \frac{1}{2}f(L-s)^2 + \frac{1}{2}fs^2 , \tag{53.17}$$

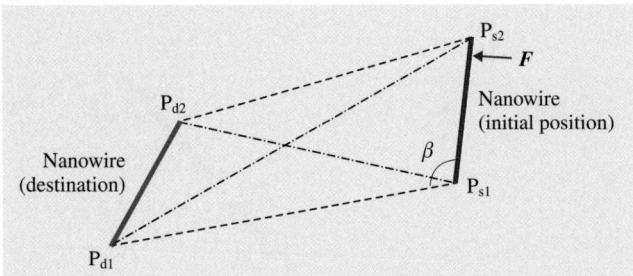

Fig. 53.8 The initial position of the nanowire and the destination where it is manipulated

where F is the applied external force, f is the evenly distributed friction and shear force density on the nanowire, L is the length of the nanowire, s is the distance from one end of the nanowire (point A in Fig. 53.7) to the pivot D, and l is the distance from A to C, where the external force is applied. Equation (53.17) can be written as

$$F = \frac{f(L-s)^2 + fs^2}{2(l-s)}. \tag{53.18}$$

The pivot can be found by minimizing F with respect to s, i.e.

$$\frac{\mathrm{d}F}{\mathrm{d}s} = 0 \;\Rightarrow\; s^2 - 2ls + lL - L^2/2 = 0. \tag{53.19}$$

Since we have assumed that $0 < s < L$, a unique solution of the pivot for any $0 < l < L$ except $l = L/2$ can be determined by

$$s = \begin{cases} l + \sqrt{l^2 - lL + L^2/2} & l < L/2 \\ l - \sqrt{l^2 - lL + L^2/2} & l > L/2 \end{cases}. \tag{53.20}$$

When $l = L/2$, there is no unique solution. A detailed analysis will show that s can be any value when the

force F is applied in the exact middle of the rod, the point T becomes a bifurcation point. Now, assume the static point D is outside of rod and on the left side. Noting that $s < 0$ now, the self-balanced torque equation becomes

$$f(l - s) = fL(L/2 - s), \tag{53.21}$$

namely

$$F = \frac{fL(L/2 - s)}{2(l - s)}. \tag{53.22}$$

It can be seen that F can only be minimized at $l = L/2$

$$\frac{\mathrm{d}F}{\mathrm{d}s} = 0, \qquad \text{for} \quad l = L/2. \tag{53.23}$$

Similarly, if static point D is on the right side ($s > 0$), the analysis results should be the same. Practically, it is hard to keep T at this bifurcation point ($l = L/2$). Therefore, during manipulation, it is better to avoid pushing the exact middle of the rod because it is hard to predict the behavior of the rod in this case.

The corresponding points between a nanowire and its destination have to be matched in order to plan a manipulation path. Figure 53.8 shows the initial position and the destination of a nanowire. P_{s1} and P_{s2} are the initial positions, and P_{d1} and P_{d2} are the destinations.

The nanowire rotates anti-clockwise and moves downward if the starting pushing point is close to P_{s2}. Similarly, the nanowire rotates clockwise and moves upward if the starting pushing point is close to P_{s1}. The starting pushing point can be determined by the angle β as shown in Fig. 53.8. If $\beta > 90°$, the starting pushing point should be close to P_{s2}. Otherwise, P_{s1}. Figure 53.9 shows the process to manipulate a nanowire from its initial position to its destination. The manipulation scheme of a nanowire has to go through a zigzag strategy in order to position the nanowire with specified orientation.

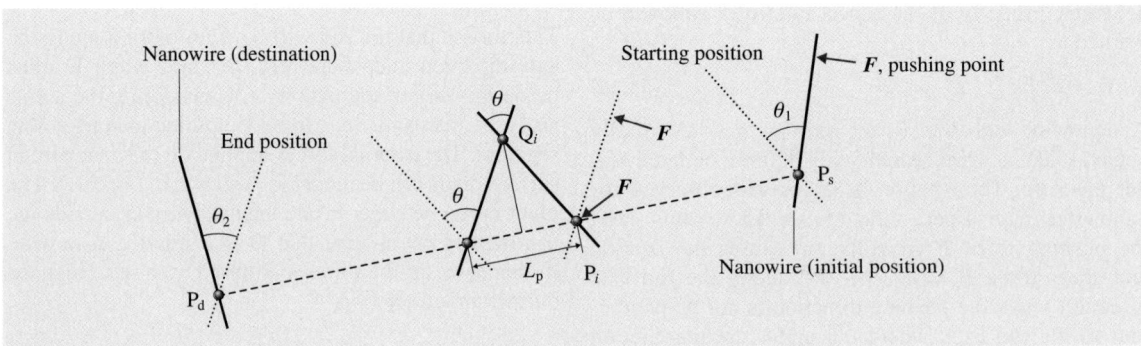

Fig. 53.9 The manipulation of a nanowire from an initial position to its destination

When the alternating pushing forces at two points on the nanowire are applied, a nanowire rotates around two pivots P_i and Q_i. The distance L_1 between P_i and Q_i can be calculated if the pushing points are determined. The two pivots P_s and P_d are connected to form a straight line, and the distance d between the two points is calculated. d is then divided into N small segments (the number of manipulations). Then L_p in Fig. 53.9 can be obtained

$$L_p = \frac{d}{N}, \qquad 0 < L_p < 2L_1 .$$ (53.24)

During manipulation, the pivot P_i is always on the line generated by the two points P_s and P_d. Then the rotation angle for each step can be obtained

$$\theta = 2a \cos\left(\frac{L_p}{2L_1}\right) .$$ (53.25)

The rotation angle θ stays the same during manipulation. The initial pushing angle θ_1 and the final pushing angle θ_2 in Fig. 53.9 can be calculated by finding the starting position and the ending position. After θ is determined, the pivots P_i and Q_i ($i = 1, \ldots, N$) can be calculated. The pushing points can then be determined. Here we show how to determine the pushing points when a nanowire rotates around the pivot P_i as an example. Figure 53.10 shows the frames used to determine the tip position.

The following transformation matrix can be easily calculated. The transformation matrix of the frame originated at P_s relative to the original frame is $\mathbf{T}_s$. The transformation matrix of the frame originated at P_i relative to the frame originated at P_s is $\mathbf{T}_i$. Supposing the rotation angle is $\beta (0 < \beta \leq \theta)$, the transformation matrix relative to the frame originated at P_i is $\mathbf{T}_{\beta i}$. β can be obtained by setting a manipulation step size. The coordinates of the pushing point can then be calculated

$$\begin{pmatrix} X_F \\ Y_F \\ 1 \end{pmatrix} = \mathbf{T}_s \mathbf{T}_i \mathbf{T}_{\beta i} \begin{pmatrix} 0 \\ L - 2s \\ 1 \end{pmatrix} .$$ (53.26)

Similar steps can be followed to determine the pushing position for a nanowire rotating around the pivot Q_i, $i \in [1, N]$. After the coordinates of the pushing point are obtained, the manipulation path for a nanowire can be generated.

53.2.3 Automated Local Scanning Method for Nanomanipulation Automation

The random drift due to thermal extension or contraction causes a major problem during nanomanipulation,

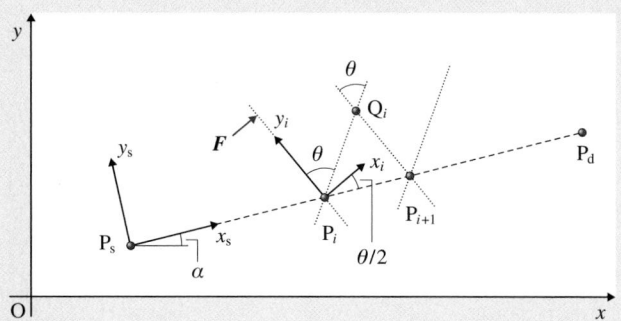

Fig. 53.10 The coordinates used to compute the AFM tip pushing position at each step

because the object may be easily lost or manipulated to wrong destinations. Before the manipulation, the objects on the surface are identified and their positions are labeled. However, the labeled positions of the nanoobjects have errors due to random drift. To compensate for the random drift, the actual position of each nanoobject must be identified before each operation. Because the time to scan a big area is quite long, a quick local scanning mechanism is developed to obtain the actual position of each nanoobject in a short time. Nanomanipulation is then performed immediately after the local scan. Figure 53.11 shows the local scanning method.

From the path data, the original position of a nanoobject is obtained. Also, the nanoobject is categorized into two groups: the nanoparticle and nanowire. A scanning pattern is generated for the nanoobject according to its group. The scanning pattern is fed to the imaging interface to scan the surface. If the nanoob-

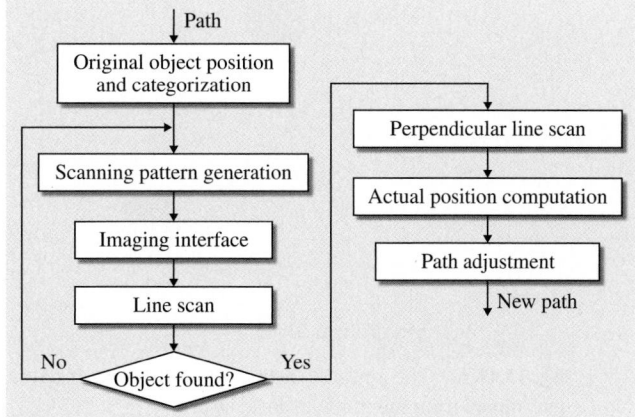

Fig. 53.11 The local scanning strategy to obtain the actual positions of nanoobjects

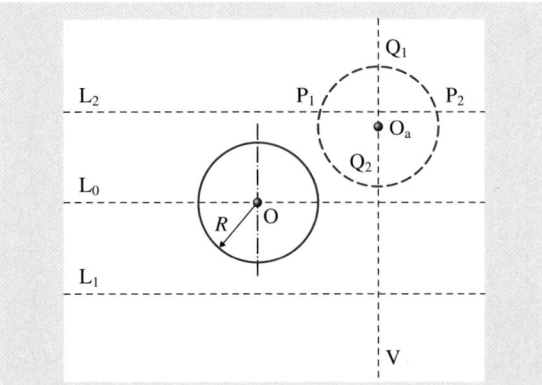

Fig. 53.12 Local scan pattern to search the actual position of a nanoparticle. O is the original center of the particle, R is the radius of the particle, O_a is the actual center of the particle, L_0, L_1, and L_2 are the horizontal scan lines, V is the vertical scan line. P_1 and P_2 are the interactions between the particle edge and a horizontal scan line, Q_1 and Q_2 are the intersections between the particle edge and the vertical scan line

For example, the location of a nanoparticle can be represented by its center and radius. The radius of each particle R has been identified before the manipulation starts. The actual center of a nanoparticle can be relocated by two lines, a lateral and a cross line as shown in Fig. 53.12. First, the nanoparticle is scanned using line L_0, which passes the original center of the particle in the image. If the particle is not found, then the scanning line moves up and down alternatively by a distance of $3/2R$. Once the particle has been found, two intersection points, P_1 and P_2 between the particle edge and the lateral line are located. A cross line scan V, which goes through the middle point between P_1 and P_2, is used to locate the center of the particle. The cross scan line has two intersection points, Q_1 and Q_2, with the particle edge. The middle point between Q_1 and Q_2 is the actual center of the nanoparticle. The local scanning range (the length of the scanning line) l can be determine by the maximum random drift such that $l > R + r_{max}$, where r_{max} is the estimated maximum random drift distance.

After the center of a nanoobject has been identified, the drifts in the XY directions are calculated. The drifts in the XY directions are then used to update the destination position as shown in Fig. 53.13a. Finally a new path is generated to manipulate the nanoparticle.

After the local scan of the first nanoparticle, the direction and size of the drift can be estimated. The information can be used to generate the scanning pattern for the next nanoobject as shown in Fig. 53.13b.

ject is not found, a new scanning pattern is generated. The process continues until the nanoobject is discovered. The actual position of the nanoobject can then be computed. The manipulation path is then adjusted based on the actual position. For nanoparticles and nanoobjects, different scanning patterns must be used in order to obtain their actual position.

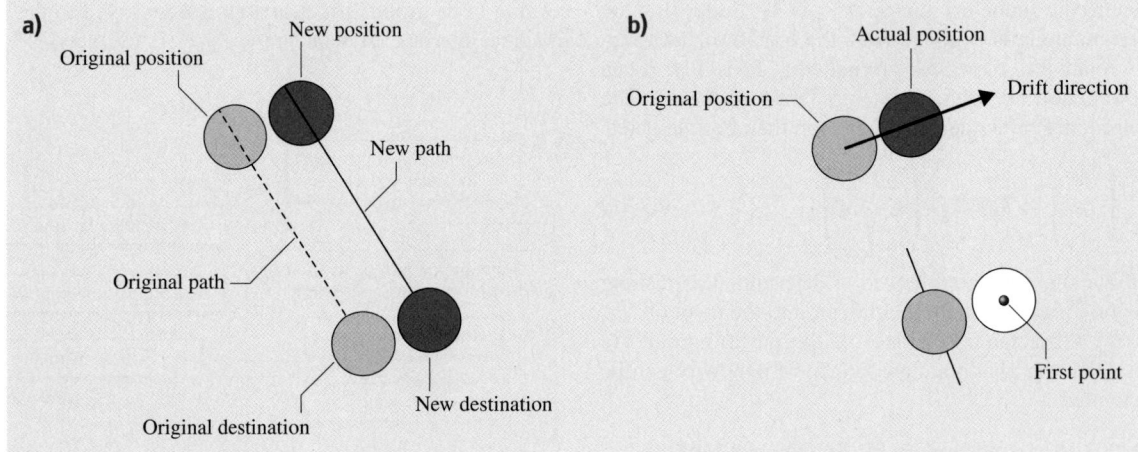

Fig. 53.13 (a) The updated path after drift compensation. **(b)** The local scan after the drift direction and size are determined from the previous local scan

53.2.4 CAD Guided Automated Nanoassembly

In order to increase the efficiency and accuracy of AFM-based nanoassembly, automated CAD guided nanoassembly is desirable. A general framework for automated nanoassembly is developed to manufacture nanostructures and nanodevices, as illustrated in Fig. 53.14. Based on the CAD model of a nanostructure and the distribution of nanoobjects on a surface from an AFM image, the tip path planner generates manipulation paths to manipulate the nanoobjects. The paths are fed to a user interface to simulate the manufacturing process and then to the AFM system to perform the nanoassembly process.

The AFM tip path planner is the core of the general framework. Figure 53.15 shows the architecture of the tip path planner. Nanoobjects on a surface are first identified based on the AFM image. A nanostructure is then designed using the available nanoobjects. Initial collision-free manipulation paths are then generated based on the CAD model of a designed nanostructure. In order to overcome the random drift, a local scanning method is applied to identify the actual po-

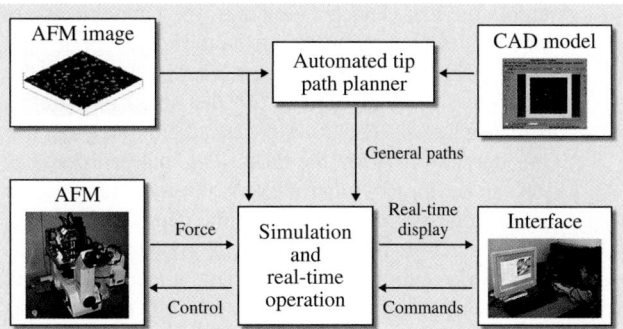

Fig. 53.14 The general framework for automated path generation system. The *bottom left* is the AFM system and the *bottom right* is the augmented reality interface used for simulation and real-time operation

sition of a nanoobject before its manipulation. Each manipulation path of the nanoobject is adjusted accordingly based on its actual position. The regenerated path is then sent to the AFM system to manipulate the nanoobject. The process continues until all nanoobjects are processed. A nanostructure is finally fabricated.

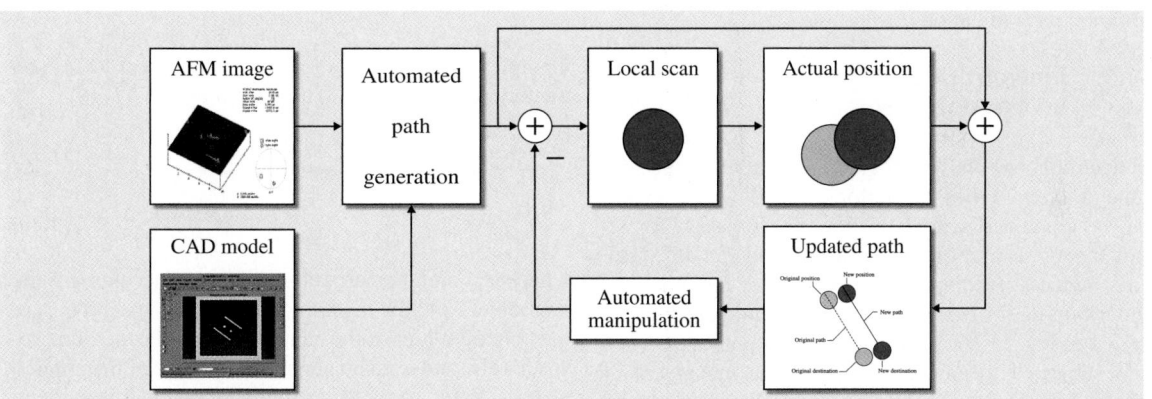

Fig. 53.15 Automated tip path planner. Initial paths are generated based on the CAD model of a designed nanostructure and the randomly distributed nanoobjects on a surface. The manipulation path of each nanoobject is adjusted accordingly based on the local scanning result

53.3 Nanomanufacturing Processes

The nanomanufacturing process for nanodevices is not straightforward, especially in nanomaterial preparation, selection, and deposition processes. To prepare the nanomaterial, nanoobjects are usually dissolved into solution, then the nanoobject suspension is put in an ultrasonicator or a centrifuge for dispersing the nanoobjects. Afterwards, specific properties of the nanoobjects should be selected. Finally, they are delivered to assemble the nanodevices. However, the nanoobjects are too small to be manipulated by traditional robotic systems, novel devices and systems must be developed for this. Since the nanoobjects are dissolved into fluids, dielectrophoresis and microfluidic technology can be considered to perform the tasks. The material preparation can be done by micromixers; the selection process can be done by microfilters; and the deposition process can be done by integrating the microchannel and microactive nozzle to deposit the nanoobject suspension. CNT is one of the most common nanoobjects and it has some promising properties that are useful for generation nanodevices. In this chapter, the development of a novel automated CNT separation system to classify the electronic types of CNTs will be described, which involves the analysis for DEP force on CNTs and fabrication of a DEP microchamber. Moreover, this DEP microchamber was successfully integrated into an automated deposition workstation to manipulate a single CNT to multiple pairs of microelectrodes repeatedly. The automated deposition processes for both SWCNTs and MWCNTs will be presented. As a result, CNT-based nanodevices with specific and consistent electronic properties can be manufactured automatically. The resulting devices can potentially be used in commercial applications.

53.3.1 Dielectrophoretic Force on Nanoobjects

Dielectrophoresis has been used to manipulate and separate different types of biological cells. DEP forces can be combined with field-flow fractionation for simultaneous separation and measurement [53.46]. DEP force induces movement of a particle or a nanoobject under non-uniform electric fields in liquid medium as shown in Fig. 53.16. The nanoobject is polarized when it is subjected to an electric field. The movement of the nanoobject depends on its polarization with respect to the surrounding medium [53.47]. When the nanoobject is more polarizable than the medium, a net dipole is induced parallel to the electric field in the nanoobject and, therefore, the nanoobject is attracted to the high electric field region. On the contrary, an opposite net dipole is induced when the nanoobject is less polarizable than the medium, and the nanoobject is repelled by the high electric field region. The direction of the DEP force on the particle is given by the Clausius–Mossotti factor (CM factor, K). It is defined as a complex factor, describing a relaxation in the effective permittivity of the particle with a relaxation time

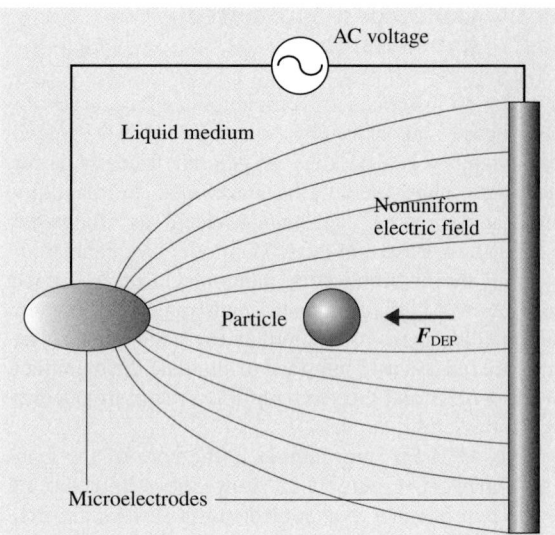

Fig. 53.16 Illustration of the dielectrophoretic manipulation

described by [53.47, 48]

$$K(\varepsilon_p^*, \varepsilon_m^*) = \frac{\varepsilon_p^* - \varepsilon_m^*}{\varepsilon_p^* + 2\varepsilon_m^*} . \tag{53.27}$$

Complex permittivities of the nanoobject (ε_p^*) and medium (ε_m^*) are defined and given by [53.47, 48]

$$\varepsilon_p^* = \varepsilon_p - \mathrm{i}\frac{\sigma_p}{\omega} , \tag{53.28}$$

$$\varepsilon_m^* = \varepsilon_m - \mathrm{i}\frac{\sigma_m}{\omega} , \tag{53.29}$$

where ε_p and ε_m are the real permittivities of the nanoobject and the medium, respectively, σ_p and σ_m are the conductivities of the nanoobject and the medium, respectively, and ω is the angular frequency of the applied electric field; the CM factor is frequency-dependent. The time-averaged DEP force acting on the particle is given by [53.47, 48]

$$F_{\mathrm{DEP}} = \frac{1}{2}V\varepsilon_m\mathrm{Re}(K)\nabla \mid E \mid^2 , \tag{53.30}$$

where V is the volume of the nanoobject and $\nabla \mid E \mid^2$ is the root-mean-square of the applied electric field. Based on this equation, the direction of the DEP force is determined by the real part of the CM factor K. When $\mathrm{Re}[K] > 0$, the DEP force is positive, and therefore the CNT is moved toward the microelectrode in the high electric field region. When $\mathrm{Re}[K] < 0$, the DEP force is negative, the particle is repelled away from the microelectrode. Moreover, we know that the magnitude and direction of DEP forces depends on the size and material properties of the nanoobjects, so separation of nanoobjects can be done.

53.3.2 Separating CNTs by an Electronic Property Using the Dielectrophoretic Effect

A theoretical analysis of DEP manipulation on a CNT was performed, and CM factors were calculated for a metallic SWCNT (m-SWCNT) and a semiconducting MWCNTs (s-SWCNT), respectively. In the analysis, semiconducting and metallic CNT mixtures are dispersed in the alcohol medium assuming the permittivities of a s-SWCNT and a m-SWCNT are $5\varepsilon_0$ [53.32] and $10^4\varepsilon_0$ [53.37], respectively, where ε_0 is the permittivity of free space ($\varepsilon_0 = 8.854188 \times 10^{-12}$ F/m). The conductivities of a s-SWCNT and a m-SWCNT are 10^5 S/m and 10^8 S/m [53.37], respectively. The permittivity and conductivity of the alcohol are $20\varepsilon_0$ and $0.13\,\mu$S/m, respectively. Based on these parameters and (53.27), plots of Re[K] for different CNTs are obtained and shown in Fig. 53.17. The result indicated that s-SWCNTs undergo a positive DEP force at low frequencies (< 1 MHz) while the DEP force is negative when the applied frequency is larger than 10 MHz. However, m-SWCNTs always undergo a positive DEP force at the applied frequency from 10 to 10^9 Hz. The result also matched the experimental result from [53.33], which showed that the positive DEP effect on SWCNTs reduced as the frequency of applied electric field increased. In addition, the theoretical result provides a better understanding of DEP manipulation on different types of CNTs. DEP force can be used to separate and identify different electronic types of CNTs (metallic and semiconducting). Based on the result shown in Fig. 53.17, metallic CNTs can be se-

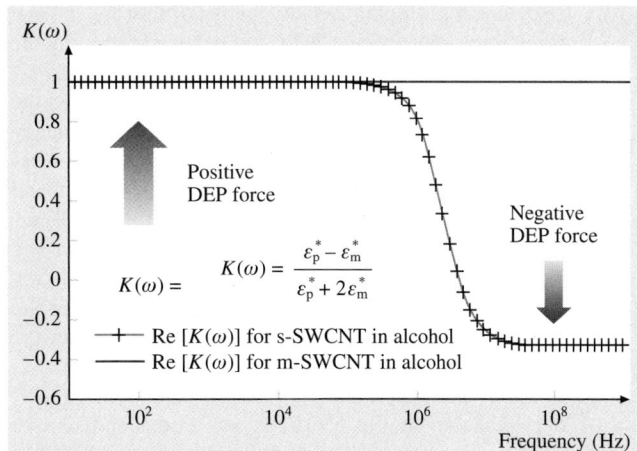

Fig. 53.17 Plots of Re[$K(\omega)$] that indicated positive and negative DEP forces on different CNTs

lectively attracted to the microelectrodes by applying AC voltage in the high frequency range (> 10 MHz). However, semiconducting CNTs cannot be attracted by using the same frequency range; this makes the selection of semiconducting CNTs difficult. In order to select semiconducting CNTs to make nanodevices, we fabricated a microchamber (DEP chamber) with arrays of microelectrodes to filter metallic CNTs in the medium. Design and fabrication of the DEP chamber will be discussed in the next section.

53.3.3 DEP Microchamber for Separating CNTs

A DEP microchamber was designed and fabricated to filter metallic CNTs in CNT suspension as shown in Fig. 53.18. Many finger-like gold microelectrodes were first fabricated inside the chamber. The performance of the filtering process was affected by the design of these finger-like microelectrodes; the microelectrode

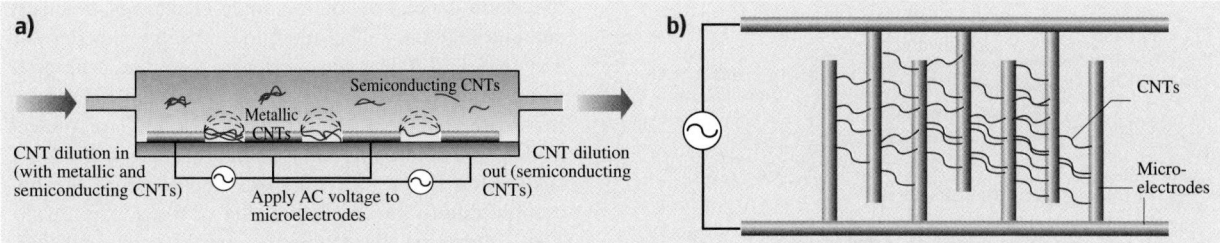

Fig. 53.18a,b DEP microchamber to filter metallic CNT. Metallic CNTs are attracted on microelectrodes. Only semiconducting CNTs flow to the outlet. (**a**) Side view; (**b**) top view

structure with higher density induced a stronger DEP force such that more CNTs could be attracted to the microelectrodes. The gap distance between these microelectrodes is $5-10\,\mu m$. The micropump pumped the CNT suspension to the DEP chamber; a high frequency AC voltage was applied to the finger-like microelectrodes so that metallic CNTs were attracted to them and stayed in the DEP chamber. Semiconducting CNTs remained in the suspension and flowed out of the chamber. Finally, the filtered suspension (with semiconducting CNTs only) was transferred to an active nozzle for the CNT deposition process. This will be described in the next section.

The fabrication process of the DEP microchamber is shown in Fig. 53.19. It was composed of two different substrates. Polymethylmethacrylate (PMMA) was used as the top substrate because it is electrically and thermally insulating, optically transparent, and biocompatible. By using a hot embossing technique, the PMMA substrate was patterned with a microchannel ($5\,mm\,L \times 1\,mm\,W \times 500\,\mu m\,H$) and a microchamber ($1\,cm\,L \times 5\,mm\,W \times 500\,\mu m\,H$) by replicating from a fabricated metal mold. In order to protect the PMMA

substrate from the CNTs-alcohol suspension, a parylene C thin film layer was coated on the substrate, because parylene resists chemical attack and is insoluble in all organic solvents. Alternatively, quartz was used as the bottom substrate, and arrays of the gold microelectrodes were fabricated on the substrate by using a standard photolithography process. A layer of AZ5214E photoresist with thickness of $1.5\,\mu m$ was first spun onto the $2'' \times 1''$ quartz substrate. It was then patterned by AB-M mask aligner and developed in an AZ300 developer. A layer of titanium with a thickness of $3\,nm$ was deposited by thermal evaporator followed by depositing a layer of gold with thickness of $30\,nm$. The titanium provided a better adhesion between gold and quartz. Afterwards, photoresist was removed in acetone solution, and arrays of microelectrodes were formed on the substrate. Finally, PMMA and quartz substrate were bonded together by UV-glue to form a close chamber. The fittings were connected at the ends of the channel to form an inlet and an outlet for the DEP chamber.

The separation performance of the DEP chamber should be optimized for different nanoobjects. Several parameters should be considered in the process: the concentration of nanoobjects in the suspension, the strength of the DEP force, the flow rate of the suspension in the DEP chamber, the structure of the channel, and the microelectrodes of the DEP chamber.

53.3.4 Automated Robotic CNT Deposition Workstation

In order to manipulate a specific type of CNTs precisely and fabricate the CNT-based nanodevices effectively, a new CNT deposition workstation has been developed as shown in Fig. 53.20. The system consists of a microactive nozzle, a DEP microchamber, a DC microdiaphragm pump, and three micromanipulators. By integrating these components into the deposition workstation, a specific type of CNT can be deposited to the desired position of the microelectrodes precisely and automatically. The micron-sized active nozzle with a diameter of $10\,\mu m$ was fabricated from a micropipette using a mechanical puller and is shown in Fig. 53.21. It transferred the CNT suspension to the microelectrodes on a microchip, and a small droplet of the CNT suspension (about $400\,\mu m$) was deposited on the microchip due to the small diameter of the active nozzle. The volume of the droplet is critical because excessive CNT suspension easily causes the formation of multiple CNTs. The microactive nozzle was then con-

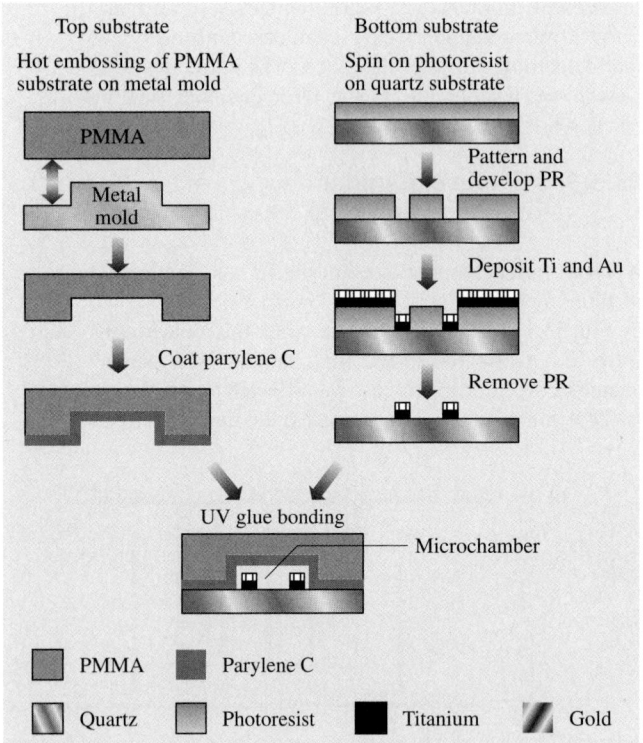

Fig. 53.19 The fabrication process of a DEP microchamber

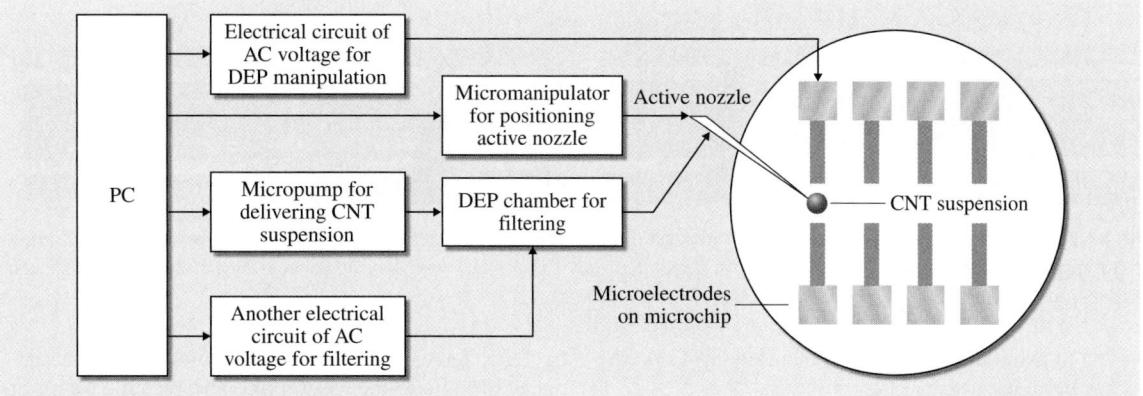

Fig. 53.20 Illustration of the CNT deposition workstation

nected to the DEP chamber, which was designed to filter metallic CNTs and select semiconducting CNTs in the CNT suspension. The raw CNT suspension was firstly pumped to the DEP chamber through a DC microdiaphragm pump (NF10, KNF Neuberger, Inc.). After the filtering process, the CNT suspension from the DEP chamber was delivered to the active nozzle for CNT deposition. By mounting the active nozzle to one of the computer controllable micromanipulators (CAP945, Signatone Corp.), the active nozzle could be moved to the desired position of the microelectrodes automatically. In order to apply the electric field to the microelectrodes during the deposition process, the other pair of micromanipulators was connected to an electrical circuit and moved to the desired location of the microelectrodes; therefore, AC voltage with different magnitudes and frequencies could be applied.

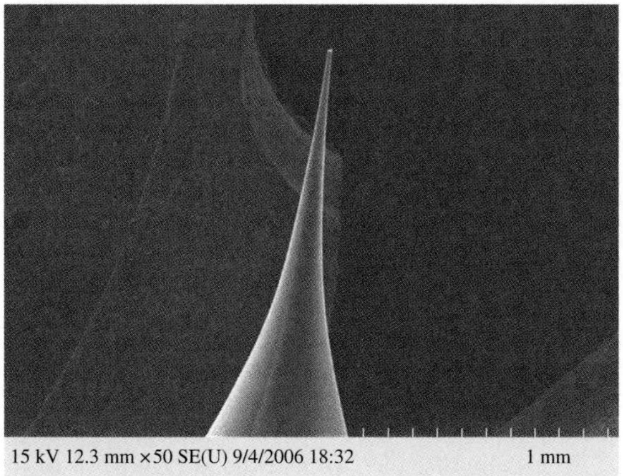

The micromanipulators, DC microdiaphragm pump, and electrical circuit were connected to the computer and controlled simultaneously during the deposition process. By controlling the position of the micromanipulators, the magnitude and frequency of the applied AC voltage, and the flow rate of the micropump, the CNT suspension can be handled automatically and deposited to the desired position.

In the deposition process, AC voltage of 1.5 V peak-to-peak with frequency of 1 kHz was applied; a positive DEP force was induced to attract CNTs to the microelectrodes. CNT deposition on multiple pairs of microelectrodes was implemented by controlling the movement of the micromanipulator, which was connected with the active nozzle. Since the position of each pair of the microelectrodes was known from the design CAD file, distances (along x and y axes) between each pair of microelectrodes were then calculated and recorded in the deposition system. At the start, the active nozzle was aligned to the first pair of the microelectrodes as shown in Fig. 53.22a. The position of the active nozzle tip was 2 mm above the microchip. When the deposition process started, the active nozzle moved down 2 mm and a droplet of CNT suspension was deposited on the first pair of microelectrodes as shown in Fig. 53.22b. Afterwards, the active nozzle moved up 2 mm and traveled to the next pair of microelectrodes as shown in Fig. 53.22c. The micromanipulator moved down again to deposit the CNT suspension on the second pair of microelectrodes as shown in Fig. 53.22d. This process repeated continuously until CNT suspension was deposited on each pair of microelectrodes on the microchip. By activating the AC voltage simultaneously, a CNT was attracted and connected between each pair of microelectrodes. The activation time was short

15 kV 12.3 mm ×50 SE(U) 9/4/2006 18:32   1 mm

Fig. 53.21 SEM image of the micro active nozzle with 10 μm tip diameter

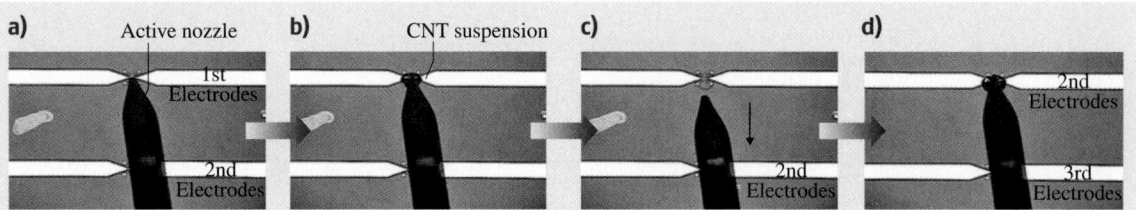

Fig. 53.22a–d CNT deposition process flow observed under the optical microscope. (**a**) The active nozzle tip aligned to the initial electrodes, (**b**) CNT suspension deposited, (**c**) the nozzle was moving to next electrodes, and (**d**) CNT suspension deposited on the second electrodes

(≈ 2 s) to avoid the formation of bundled CNTs on the microelectrodes.

After the deposition process, AFM was used to check the CNT formation as shown in Fig. 53.23. Sometimes, there are some impurities or more than one CNT trapped between the microelectrodes as shown in Fig. 53.23f. Therefore, it is necessary to take another step to clean up the microelectrodes gap area and adjust the position of the CNT to make the connection. This final step is very critical and is termed CNT assembly; it can be done by our AFM-based nanomanipulation system. I–V characteristics of the CNT-based devices were also obtained as shown in the inset images of Fig. 53.23. Based on the results, this indicates that both SWCNTs and MWCNTs could be repeatedly and automatically manipulated between the microelectrodes by using the deposition system. This CNT deposition workstation integrates all essential components to manipulate the specific type of CNTs to desired positions precisely

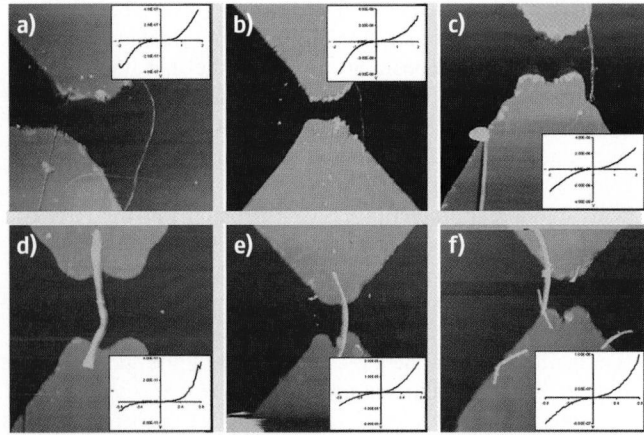

Fig. 53.23a–f AFM images showing the individual CNTs was deposited on the microelectrodes. Panels (**a–c**) are SWCNTs, panels (**d–f**) are MWCNTs. The *inset images* are the corresponding I–V curves of each CNT

by DEP force. The development of this system produces benefits to the assembling and manufacturing of CNT-based devices. The yield of depositing CNTs on the microelectrodes is very high after optimizing the following factors: the concentration of the CNT suspension, the volume of the CNT suspension droplet, and the activation time, magnitude and frequency of the applied electric field.

In order to validate the separation performance of different CNTs by the DEP chamber, experiments for both raw CNT suspension (before passing through the DEP chamber) and filtered CNT suspension were conducted, respectively. The procedure for preparing the CNT suspension was the same as the process presented in the previous section. SWCNT powder (BU-203, Bucky USA, Nanotex Corp.) was dispersed in an alcohol liquid medium, and the CNT suspension was put in the ultrasonicator for 15 min. The length of the SWCNT is 0.5–$4\,\mu$m. Finally, the raw SWCNT suspension was prepared and the concentration was about $1.1\,\mu$g/ml. Afterwards, the separation process was performed on the raw SWCNT suspension as illustrated in Fig. 53.24. During the process, the raw SWCNT suspension was pumped to the DEP chamber through the micropump. The flow rate was about 0.03 l/min. A high frequency AC voltage (1.5 Vpp, 40 MHz) was applied to the microelectrodes in the DEP chamber; it induced a positive DEP force on metallic SWCNTs in the suspension but a negative DEP force on semiconducting SWCNTs. Since the metallic SWCNTs were attracted to the microelectrodes and stayed in the DEP chamber, it was predicted that only semiconducting SWCNTs remained in the filtered SWCNT suspension. The filtered SWCNT suspension was then collected at the outlet of the chamber for later CNT disposition process.

After preparing the raw and filtered SWCNT suspension, the deposition process was performed by using our CNT deposition workstation, which was intro-

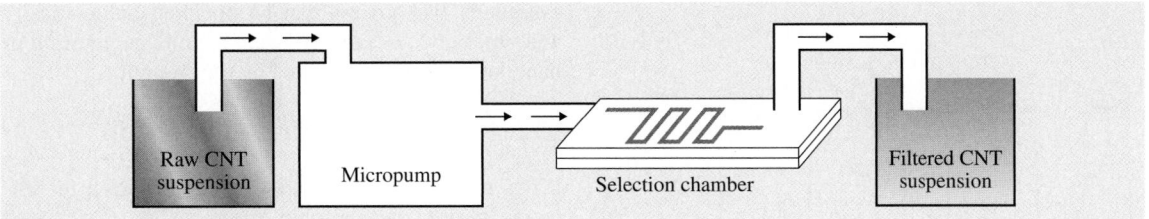

Fig. 53.24 The CNT filtering process

duced in the previous section. In the experiment, the raw SWCNT suspension and the filtered SWCNT suspension were deposited on the microchip 20 times, respectively. The electronic properties of CNTs in both suspensions were then studied by measuring the I–V curves. The yields to obtain semiconducting CNTs from the raw CNT suspension and filtered CNT suspension were also compared. Based on the preliminary results, the yield of depositing semiconducting SWCNTs (from the raw SWCNT suspension) was about 33% as shown in Fig. 53.25; the yield of depositing semiconducting SWCNTs (from the filtered SWCNT suspension) was about 65% as shown in Fig. 53.25. The yield to form semiconducting CNTs is very important because many devices require materials with semiconducting properties. The results indicated that there was significant improvement in forming semiconducting CNTs on the microelectrodes by using our DEP chamber. The yield

in forming semiconducting CNTs changed from 33% (before the filtering process) to 65% (after the filtering process). The yield should be improved by optimizing the concentration of CNT suspension, the strength of the DEP force, the flow rate of the suspension in the DEP chamber, the structure of the channel, and the microelectrodes of the DEP chamber.

The yield to form semiconducting CNTs is very important because it affects the successful rate to fabricate nanodevices. The yield to form semiconducting CNTs was increased by using our system. Although there is a synthesis that produces nearly 90% of semiconducting CNTs by PECVD [53.49], both CNT synthesis methods and post-processing separation methods are important and can be combined for different applications. Our separation system is a post-processing method, which can be used together with different CNT synthesis methods. Since our system used electrical signal to control the

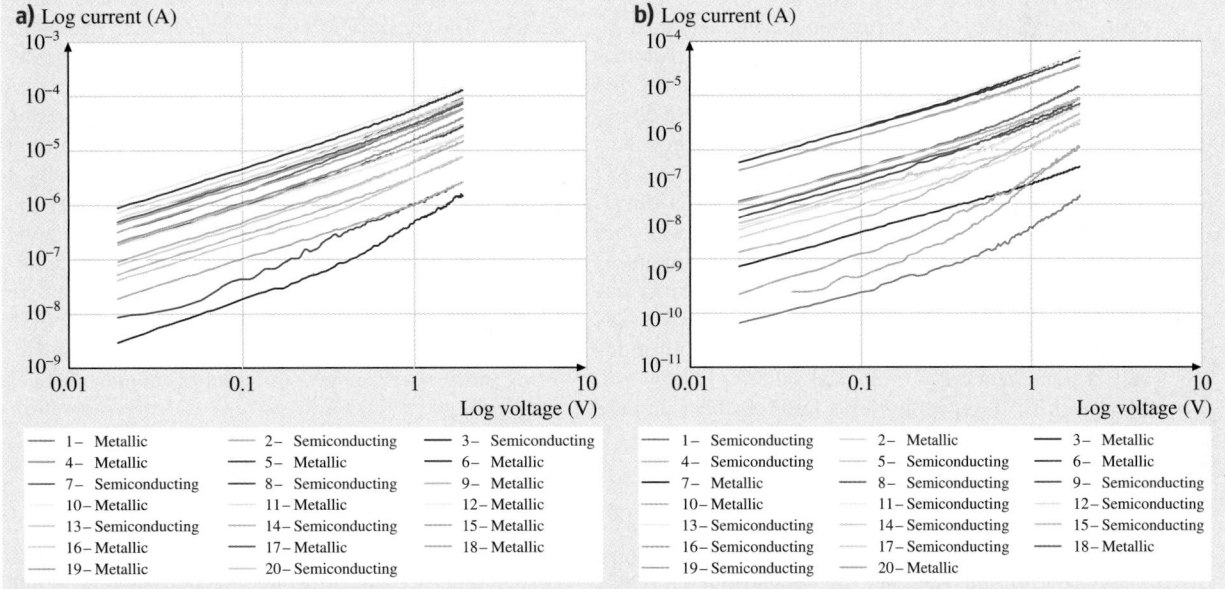

Fig. 53.25a,b I–V characteristics of SWCNTs. (**a**) For the raw SWCNT suspension, (**b**) for the filtered SWCNT suspension

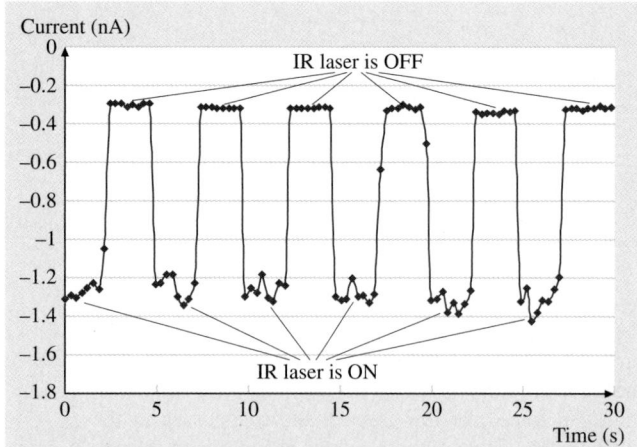

Fig. 53.26 Temporal photoresponses of a CNT-based IR detector

DEP manipulation and separation, it can be integrated with current robotic manufacturing systems easily, and

eventually the process can be operated automatically and precisely. As a result, batch nanomanufacturing of nanodevices can be achieved by this system.

53.3.5 CNT-Based Infrared Detector

When semiconducting CNTs are deposited on the microelectrodes, the photonic effects of the CNT-based nanodevice can be studied. For example, a CNT device was put under the infrared (IR) laser source (UH5-30G-830-PV, World Star Tech, optical power: 30 mW; wavelength: 830 nm), and the photocurrent from the CNT-based nanodevice was measured. The laser source was configured to switch on and off in several cycles; the temporal photoresponses of the device are shown in Fig. 53.26. The experimental result showed the CNT-based device was sensitive to the IR laser, so CNTs can be used to make novel IR detectors. More detail design and fabrication of CNT-based IR detectors are given in [53.50–52]

53.4 Conclusions

Automated nanomanipulation is desirable to increase the efficiency and accuracy of nanoassembly. Automated nanoassembly of nanostructures is very challenging because of the manipulation path generation for different nanoobjects, position errors due to random drift, and cantilever deformation during nanomanipulation. This chapter discussed automated nanomanipulation technology for nanoassembly. Automated nanomanipulation methods of nanoobjects were developed, and an automated local scanning method was presented to compensate for the random drift. A CAD guided automated nanoassembly method was developed. CAD guided automated nanoassembly was able to open a door to assembly of complex nanostructures and nanodevices. The effectiveness of the system has also been verified by inscribing nano features on soft surface, manipulating nanoparticles DNA molecules and characterization of biological samples [53.53–56]. Moreover, CNT separation by a DEP chamber and the

development of an automated CNT deposition workstation that applies DEP manipulation on CNTs were presented. The system assembles semiconducting CNTs to the microelectrodes effectively and, therefore, it is possible to improve the success rate to fabricate nanodevices. The separation method developed in this paper is a post-processing method, which can be used together with different CNT synthesis methods. Since our system used electrical signals to control CNT separation and DEP manipulation, it can be integrated into current robotic manufacturing systems easily, and eventually it will be possible to operate the process automatically and precisely. It opens the possibility of batch fabricating CNT-based devices. Furthermore, the nanomanufacturing process is not limited to CNTs, but it can be used on other nano materials such as ZnO and InSb nanowires etc. The development of the nanomanufacturing process will achieve different novel nano devices effectively.

References

53.1   G. Binning, C.F. Quate, C. Gerber: Atomic force microscope, Phys. Rev. Lett. **56**(9), 930–933 (1986)

53.2   D. Wang, L. Tsau, K.L. Wang, P. Chow: Nanofabrication of thin chromium film deposited on Si(100) surfaces by tip induced anodization in atomic force microscopy, Appl. Phys. Lett. **67**, 1295–1297 (1995)

53.3   D.M. Schaefer, R. Reifenberger, A. Patil, R.P. Andres: Fabrication of two-dimensional arrays of nanometer-size clusters with the atomic force microscope, Appl. Phys. Lett. **66**, 1012–1014 (1995)

53.4   T. Junno, K. Deppert, L. Montelius, L. Samuelson: Controlled manipulation of nanoparticles with an atomic force microscope, Appl. Phys. Lett. **66**(26), 3627–3629 (1995)

53.5   P. Avouris, T. Hertel, R. Martel: Atomic force microscope tip-induced local oxidation of silicon: kinetics, mechanism, and nanofabrication, Appl. Phys. Lett. **71**, 285–287 (1997)

53.6   R. Nemutudi, N. Curson, N. Appleyard, D. Ritchie, G. Jones: Modification of a shallow 2DEG by AFM lithography, Solid-State Electron. **57/58**, 967–973 (2001)

53.7   S.J. Ahn, Y.K. Jang, S.A. Kim, H. Lee, H. Lee: AFM nanolithography on a mixed LB film of hexadecylamine and palmitic acid, Ultramicroscopy **91**, 171–176 (2002)

53.8   E. Dubois, J.-L. Bubbendorff: Nanometer scale lithography on silicon, titanium and PMMA resist using scanning probe microscopy, Solid-State Electron. **43**, 1085–1089 (1999)

53.9   M. Sitti, H. Hashimoto: Tele-nanorobotics using atomic force microscope, Proc. IEEE Int. Conf. Intell. Robot. Syst. (Victoria 1998) pp. 1739–1746

53.10   M. Guthold, M.R. Falvo, W.G. Matthews, S. Paulson, S. Washburn, D.A. Erie, R. Superfine, F.P. Brooks Jr., R.M. Taylor II: Controlled manipulation of molecular samples with the nanomanipulator, IEEE/ASME Trans. Mechatron. **5**(2), 189–198 (2000)

53.11   A.A.G. Requicha, C. Baur, A. Bugacov, B.C. Gazen, B. Koel, A. Madhukar, T.R. Ramachandran, R. Resch, P. Will: Nanorobotic assembly of two-dimensional structures, Proc. IEEE Int. Conf. Robot. Autom. (Leuven 1998) pp. 3368–3374

53.12   L.T. Hansen, A. Kühle, A.H. Sørensen, J. Bohr, P.E. Lindelof: A technique for positioning nanoparticles using an atomic force microscope, Nanotechnology **9**, 337–342 (1998)

53.13   G.Y. Li, N. Xi, M. Yu, W.K. Fung: 3-D nanomanipulation using atomic force microscope, Proc. IEEE Int. Conf. Robot. Autom. (Taipei 2003)

53.14   G. Li, N. Xi, H. Chen, C. Pomeroy, M. Prokos: Videolized atomic force microscopy for interactive nanomanipulation and nanoassembly, IEEE Trans. Nanotechnol. **4**(5), 605–615 (2005)

53.15   G. Li, N. Xi, M. Yu, W.-K. Fung: Development of augmented reality system for AFM-based nanomanipulation, IEEE/ASME Trans. Mechatron. **9**(2), 358–365 (2004)

53.16   H. Chen, N. Xi, G. Li: CAD-guided automated nanoassembly using atomic force microscopy-based nonrobotics, IEEE Trans. Autom. Sci. Eng. **3**(3), 208–217 (2006)

53.17   H. Chen, N. Xi, W. Sheng, Y. Chen: General framework of optimal tool trajectory planning for free-form surfaces in surface manufacturing, J. Manuf. Sci. Eng. **127**(1), 49–59 (2005)

53.18   J. Zhang, N. Xi, G. Li, H.-Y. Chan, U.C. Wejinya: Adaptable end effector for atomic force microscopy based nanomanipulation, IEEE Trans. Nanotechnol. **5**(6), 628–642 (2006)

53.19   A. Javey, J. Guo, D.B. Farmer, Q. Wang, D. Wang, R.G. Gordon, M. Lundstrom, H. Dai: Carbon nanotube field-effect transistors with integrated ohmic contacts and high-k gate dielectrics, Nano Lett. **4**(3), 447–450 (2004)

53.20   A. Bachtold, P. Hadley, T. Nakanishi, C. Dekker: Logic circuits with carbon nanotube transistors, Science **294**, 1317–1320 (2001)

53.21   I.A. Levitsky, W.B. Euler: Photoconductivity of single-walled carbon nanotubes under CW illumination, Appl. Phys. Lett. **83**, 1857–1859 (2003)

53.22   L. Liu, Y. Zhang: Multi-wall carbon nanotube as a new infrared detected material, Sens. Actuators A **116**, 394–397 (2004)

53.23   J.A. Misewich, R. Martel, P. Avouris, J.C. Sang, S. Heinze, J. Tersoff: Electrically induced optical emission from a carbon nanotube FET, Science **300**, 783–786 (2003)

53.24   L. Valentini, I. Armentano, J.M. Kenny, C. Cantalini, L. Lozzi, S. Santucci: Sensors for sub-ppm NO_2 gas detection based on carbon nanotube thin films, Appl. Phys. Lett. **82**, 4623–4625 (2003)

53.25   J. Kong, N.R. Franklin, C. Zhou, M.G. Chapline, S. Peng, K. Cho, H. Dai: Nanotube molecular wires as chemical sensors, Science **287**, 622–625 (2000)

53.26   G.Y. Li, N. Xi, M. Yu, W.K. Fung: Augmented reality system for real-time nanomanipulation, Proc. IEEE Int. Conf. Nanotechnol. (San Francisco 2003)

53.27   L. Liu, Y. Luo, N. Xi, Y. Wang, J. Zhang, G. Li: Sensor referenced real-time videolization of atomic force microscopy for nanomanipulations, IEEE/ASME Trans. Mechatron. **13**(1), 76–85 (2008)

53.28   K.W.C. Lai, N. Xi, C.K.M. Fung, J. Zhang, H. Chen, Y. Luo, U.C. Wejinya: Automated nanomanufacturing system to assemble carbon nanotube based devices, Int. J. Robot. Res. (IJRR) **28**(4), 523–536 (2009)

53.29   K.W.C. Lai, N. Xi, U.C. Wejinya: Automated process for selection of carbon nanotube by electronic property using dielectrophoretic manipulation, J. Micro-Nano Mechatron. **4**(1), 37–48 (2008)

53.30   M.S. Arnold, A.A. Green, J.F. Hulvat, S.I. Stupp, M.C. Hersam: Sorting carbon nanotubes by electronic structure using density differentiation, Nat. Nanotechnol. **1**, 60–65 (2006)

53.31   P.G. Collins, M.S. Arnold, P. Avouris: Engineering carbon nanotubes and nanotube circuits using electrical breakdown, Science **292**, 706–709 (2001)

Part F | 53

53.32   R. Krupke, F. Hennrich, H. von Lohneysen, M.M. Kappes: Separation of metallic from semiconducting single-walled carbon nanotubes, Science **301**, 344–347 (2003)

53.33   R. Krupke, S. Linden, M. Rapp, F. Hennrich: Thin films of metallic carbon nanotubes prepared by dielectrophoresis, Adv. Mater. **18**, 1468–1470 (2006)

53.34   M. Yu, M.J. Dyer, G.D. Skidmore, H.W. Rohrs, X. Lu, K.D. Ausman, J.R.V. Ehr, R.S. Ruoff: Three-dimensional manipulation of carbon nanotubes under a scanning electron microscope, Nanotechnology **10**, 244–252 (1999)

53.35   T. Fukuda, F. Arai, L. Dong: Assembly of nanodevices with carbon nanotubes through nanorobotic manipulations, Proc. IEEE **91**, 1803–1818 (2003)

53.36   N.G. Green, A. Ramos, H. Morgan: AC electrokinetics: a survey of sub-micrometre particle dynamics, J. Phys. D: Appl. Phys. **33**, 632–641 (2000)

53.37   M. Dimaki, P. Boggild: Dielectrophoresis of carbon nanotubes using microelectrodes: a numerical study, Nanotechnology **15**, 1095–1102 (2004)

53.38   J. Li, Q. Zhang, N. Peng, Q. Zhu: Manipulation of carbon nanotubes using AC dielectrophoresis, Appl. Phys. Lett. **86**, 153116–153118 (2005)

53.39   A. Subramanian, L.X. Dong, J. Tharian, U. Sennhauser, B.J. Nelson: Batch fabrication of carbon nanotube bearings, Nanotechnology **18**, 075703 (2007)

53.40   A. Subramanian, T. Choi, L.X. Dong, D. Poulikakos, B.J. Nelson: Batch fabrication of nanotube transducers, Proc. 7th IEEE Conf. Nanotechnol. (IEEE-NANO2007) (Hong Kong 2007)

53.41   R. Resch, C. Baur, A. Bugacov, B.E. Koel, A. Madhukar, A.A.G. Requicha, P. Will: Building and manipulating three-dimensional and linked two-dimensional structures of nanoparticles using scanning force microscopy, Langmuir **14**(23), 6613–6616 (1998)

53.42   J. Israelachvili: *Intermolecular and surface forces* (Academic Press, London 1991)

53.43   G.Y. Li, N. Xi, M. Yu, W.K. Fung: Modeling of 3-D interactive forces in nanomanipulation, IEEE/RSJ Int. Conf. Intell. Robot. Syst. (Las Vegas 2003)

53.44   P.M. Garth: *Nonlinear programming: Theory, algorithm and applications* (Wiley, New York 1983)

53.45   M. Avriel: *Nonlinear Programming: Analysis and Methods* (Prentice Hall, Englewood Cliffs 1976)

53.46   J.C. Giddings: Field-Flow Fractionation: Analysis of macromolecular, colloidal, and particulate materials, Science **260**, 1456–1465 (1993)

53.47   H. Morgan, N. G. Green: *AC Electrokinetics: colloids and nanoparticles* (Research Studies Press Ltd. Hertfordshire 2003)

53.48   T.B. Jones: *Electromechanics of Particles* (Cambridge Univ. Press, Cambridge 1995)

53.49   Y. Li, D. Mann, M. Rolandi, W. Kim, A. Ural, S. Hung, A. Javey, J. Cao, D. Wang, E. Yenilmez, Q. Wang, J.F. Gibbons, Y. Nishi, H. Dai: Preferential growth of semiconducting single-walled carbon nanotubes by a plasma enhanced CVD method, Nano Lett. **4**(2), 317–321 (2004)

53.50   J. Zhang, N. Xi, H. Chen, K.W.C. Lai, G. Li: Design, manufacturing and testing of single carbon nanotube based infrared sensors, IEEE Trans. Nanotechnol. **8**(2), 245–251 (2009)

53.51   J. Zhang, N. Xi, H. Chen, K.W.C. Lai, G. Li: Photovoltaic effect in single carbon nanotube based Schottky diodes, Int. J. Nanopart. **1**(2), 108–118 (2008)

53.52   J. Zhang, N. Xi, K. Lai: Fabrication and testing of a nano infrared detector using a single carbon nanotube (CNT), SPIE Newsroom (2007), online at: http://spie.org/x8489.xml

53.53   J. Zhang, N. Xi, L. Liu, H. Chen, K.W.C. Lai, G. Li: Atomic force yields a master nanomanipulator, IEEE Nanotechnol. Mag. **2**(2), 13–17 (2008)

53.54   G. Li, N. Xi, D.H. Wang: Probing membrane proteins using atomic force microscopy, J. Cell. Biochem. **97**, 1191–1197 (2006)

53.55   G. Li, N. Xi, D.H. Wang: Investigation of angiotensin II type 1 receptor by atomic force microscopy with functionalized probe, Nanomed. Nanotechnol. Biol. Med. **1**(4), 302–312 (2005)

53.56   G. Li, N. Xi, D.H. Wang: In situ sensing and manipulation of molecules in biological samples using a nano robotic system, Nanomed. Nanotechnol. Biol. Med. **1**(1), 31–40 (2005)

54. Production, Supply, Logistics and Distribution

Rodrigo J. Cruz Di Palma, Manuel Scavarda Basaldúa

To effectively manage a supply chain it is necessary to coordinate the flow of materials and information both within and among companies. This flow goes from suppliers to consumers, as it passes through manufacturers, wholesalers, and retailers. While materials and information move through the supply chain, automation is used in a variety of forms and levels as a way to raise productivity, enhance product quality, decrease labor costs, improve safety, and even to perform tasks that go beyond the precision and reliability of humans. A rapid development in information technology has transformed not only the way people work and interact with each other; electronic media enable enterprises to collaborate on their work and missions within each organization and with other independent enterprises, including suppliers and customers.

Within this chapter, the focus is on the main benefits of automation in production, supply, logistics, and distribution environments. The first Section centers on machines and equipment automation for production. The second section focuses on computing/communication automation for planning and operations decisions. Finally, the last section highlights some considerations regarding economics, productivity, and flexibility important to bear in mind while designing an automation strategy.

54.1 Historical Background

Automation is any technique, method or system of operating or controlling a process without continuous input from an operator, thereby reducing human intervention to a minimum. Many believe that automation of supply chain networks began with the use of personal computers in the late 1970s, while others date it back to the use of electricity in the early 1900s. The fact is that,

regardless of when we place the beginning of automation, it has changed the way we work, think, and live our lives. Children are now in contact with automation from the day they are born, such as automated machines that monitor the vital signs of premature infants. As people grow older, they continually have contact with automation via automatic teller machines (ATMs),

Table 54.1 Supply chain evolution at a food-products production and service company

	Before supply chain	After supply chain design	Advantages
Forecasting/ ordering	The company determines the amount of nuts that a customer will expect in its food products.	The company and its customer share sales forecasts based on current point-of-sale data, past demand patterns, and upcoming promotions, and agree on an amount and schedule to supply.	Forecasting accuracy, collaborative replenishment planning
Procurement	The company phones its Brazilian office and employees deliver the orders in person to local farmers, who load the raw nuts on trucks and deliver them to the port.	The company contacts its Brazilian office by email, but employees still must contact local farmers personally.	Enhanced communication
Transportation	The shipping company notifies the company when the nuts have sailed. When the nuts arrive in a US port, a freight-forwarder processes the paperwork to clear the shipment through customs, locates a truck to deliver them to the company plants, and delivers the nuts to the company's manufacturing plant, although it may be only half-full and return empty, costing the company extra money.	Shippers and truckers share up-to-date data online via a collaborative global logistics system that connects multiple manufacturers and transportation companies and handles the customs' process. The system matches orders with carriers to assure that trucks travel with full loads.	Online tracking, transportation cost minimization
Manufacturing	The nuts are cleaned, roasted, and integrated with various food products manufactured by the company according to original production forecast.	Production forecast is updated based on the current demand for product mix, and products are manufactured under lean manufacturing and just-in-time principles.	Capacity utilization, production efficiency
Distribution	The products are packed, and trucks take them to the company's multiple warehouses across the country, from where they are ready to be shipped to stores. However, they may not be near the store where the customer needs them because local demand has not been considered.	After food products are inspected and packed at the plant, the company sends the products to a third-party distributor, which relieves the company of a supply chain activity not among its core competencies. The distributor consolidates the products on trucks with other products, resulting in full loads and better service.	Enhanced distribution planning, inventory management and control.
Customer	If the company ordered too many nuts, they will turn soft in the warehouses, and if they ordered too few, the customer will buy food products with nuts elsewhere.	The company correctly knows the customer's needs so there is neither a shortage nor an oversupply of the products. Transportation, distribution, warehousing, and inventory costs drop, and product and service quality improve.	Increased customer satisfaction. Cost minimization, profit maximization

self-operated airplanes, and self-parking cars. In the industrial realm, automation now can be seen from such simple tasks as milking a cow to complex repetitive task such as building a new car. Computing and communications have also transformed production and service organizations over the past 50 years. While working in parallel, error recovery, and conflict resolution have been addressed by human workers since early days of industry, they have recently been transformed by computer into integrated functions [54.1].

Automation is the foundation of many of society's advances. Through productivity advances and reductions in costs, automation allows more complex and sophisticated products and services to be available to larger portions of the world's population. The evolution of supply chain management at a company as an automation example of the production, supply, logistics, and distribution is described in Table 54.1. An example of such evolution for a specific company is discussed in [54.2].

54.2 Machines and Equipment Automation for Production

54.2.1 Production Equipment and Machinery

Automation has been used for many years in manufacturing as a way to increase speed of production, enhance product quality, decrease labor costs, reduce routine labor-intensive work, improve safety, and even to perform tasks that go beyond the precision and reliability of normal human abilities [54.3, 4]. Examples of au-

Fig. 54.1 Industrial robotics in car production (courtesy of KUKA Robotics)

tomation range from the employment of robots to install a windshield on a car, machine tools that process parts to build a computer processor, to inspection and testing systems for quality control to make certain the precise amount of cereal is in each box of cornflakes. An example of industrial robots for car production is shown in Fig. 54.1.

There are many reasons that explain why automation has become increasingly important in today's production systems. The main justifications arise from the relative strengths that automation provide in comparison with humans within a production environment. Automated machines are able to perform systematic and repetitive tasks, requiring precision, storing large amounts of information, executing commands fast and accurately, handling multiple tasks simultaneously, and conducting dangerous/hazardous work [54.5, 6].

However, in spite of the numerous benefits and advantages automation may offer, it is not always the best solution and in some cases it is not even a feasible one. In situations with elevated levels of variation, short product lifecycles, or highly customized production, a solution involving automation can be unnecessarily complex and expensive when compared with a traditional manual process. Under these circumstances, the benefits of a manual process, such as the flexibility and lower capital requirements, begin to gain relative advantage when compared with an automated process.

54.2.2 Material Handling and Storage for Production and Distribution

Material handling equipment plays an important role in the automation of production, storage, and distribution systems, by interconnecting the fixed or flexible work-

Fig. 54.2 Inertial guidance automatic guided vehicle (courtesy of the Jervis B. Webb Company)

stations that compose them. Typical automated material handling systems include conveyors for moving product in a general direction inside a facility, sorters and carousels for distributing products to specific locations, and automated storage and retrieval systems (ASRS) for storage and automated guided vehicles (AGVs) for transporting materials between work stations [54.7]. For instance, AGVs play an important role in the paper industry, where moving roles quickly and efficiently is critical. The key is not to damage rolls during this handling process. An example of an AGV in the paper industry is shown in Fig. 54.2.

Conveyors are believed to be the most common material handling system used in production and distribution processes. These systems are used when materials must be transferred in relatively large quantities between specific machines or workstations following a predetermined path or route. Most of these systems use belts or wheels to transfer materials horizontally or gravity to move materials between points at different heights. Frequently within production or distribution systems several types of conveyors are utilized in a combined manner, constituting conveyor networks or integrated systems.

In more sophisticated conveyor networks, sorters are utilized. A sorter consists of an array of closely coupled, high-density diverters used to sort units (materials, products, parts, etc.) to specific lanes for further consolidation. Consequently, in addition to the transfer functionality supported by regular conveyor systems, sorter mechanisms provide, by means of sensors, a classification capability that makes them especially attractive for highly variable, small-size shipments. Examples of operations that utilize sorters for their shipments are Federal Express, United Parcel Services, and Amazon.

Kimberly-Clark claims that a sorter mechanism implemented at one of its distribution facilities in Latin America has improved the truck loading operation time from 1–3 h to 20 min. With capacity of 200 cases per minute and fully customizable logic (i. e., it can be programmed to follow a balanced sorting sequence per dock or a sorting sequence by customer orders), this sorter has significantly increased truck rotation at the Kimberly-Clark distribution center versus the previous manual system.

54.2.3 Process Control Systems in Production

Production systems can be designed with different levels of automation. However in all automated production systems, even at the lowest levels of sophistication constituted by automated devices such as valves and actuators, a control system of some kind is required.

At the individual machine level, the automatic control is executed by computer systems. An example is computer numerical control (CNC), which reads instructions from an operator in order to drive a machine tool. At the automatic process level there are two main types of control systems, the programmable logic controller (PLC) and the distributed control system (DCS). These systems were initially developed to support distinctive process control functions. PLCs began replacing conventional relay/solid-state logic in machine control while DCSs were merely a digital replacement of analog controllers and panel-board displays. However, the technology of both types of process control systems has evolved over the years and the differences in their functionalities have become less straightforward. Further up in the hierarchy, at the production cell level, cell controllers provide coordination among individual workstations by interacting with multiple PLCs, DCSs, and other automated devices. At the highest production automation level, control systems such as manufacturing control systems and/or integrated plant systems initiate, coordinate, and provide visibility of the manufacturing operation of all other lower control levels [54.6, 8].

54.3 Computing and Communication Automation for Planning and Operations Decisions

54.3.1 Supply Chain Planning

To manage the supply chain effectively it is necessary to coordinate the flow of materials and information both within and between companies (e.g., [54.9]). The focus of the supply chain planning process is to synchronize activities from raw materials to the final customer. Supply chain planning processes strive to find an integrated solution for the strategic, tactical, and operational activities in order to allow companies to balance supply and demand for the movement of goods and services.

Information and communication technology (ICT) plays a vital role in supply chain planning by facilitating the flow of information and enhancing the cooperation between customers, suppliers and third party partners. As show in Fig. 54.3, intranets can be used to integrate information from isolated business processes within the firm to help them manage their internal supply chains.

Access to these private intranets can also be extended to authorized supplies, distributors, logistics services, and to retail customers to improve coordination of external supply chain processes [54.10].

Electronic data interchange (EDI) and other similar technologies not only save money by reducing or eliminating human intervention and data entry, but also pave the way for collaboration initiatives between organizations. An example would be vendor-managed inventory (VMI) where customers send inventory and consumption information to suppliers, who schedule deliveries to keep customer inventory within agreed upon ranges. VMI not only provides benefits for the customer but also increases demand visibility for the supplier and leads to cost savings and reduced inventory investment for the supplier. As shown in Fig. 54.4, EDI can significantly improve productivity. A typical illustration is the case of Warner-Lambert that increased its prod-

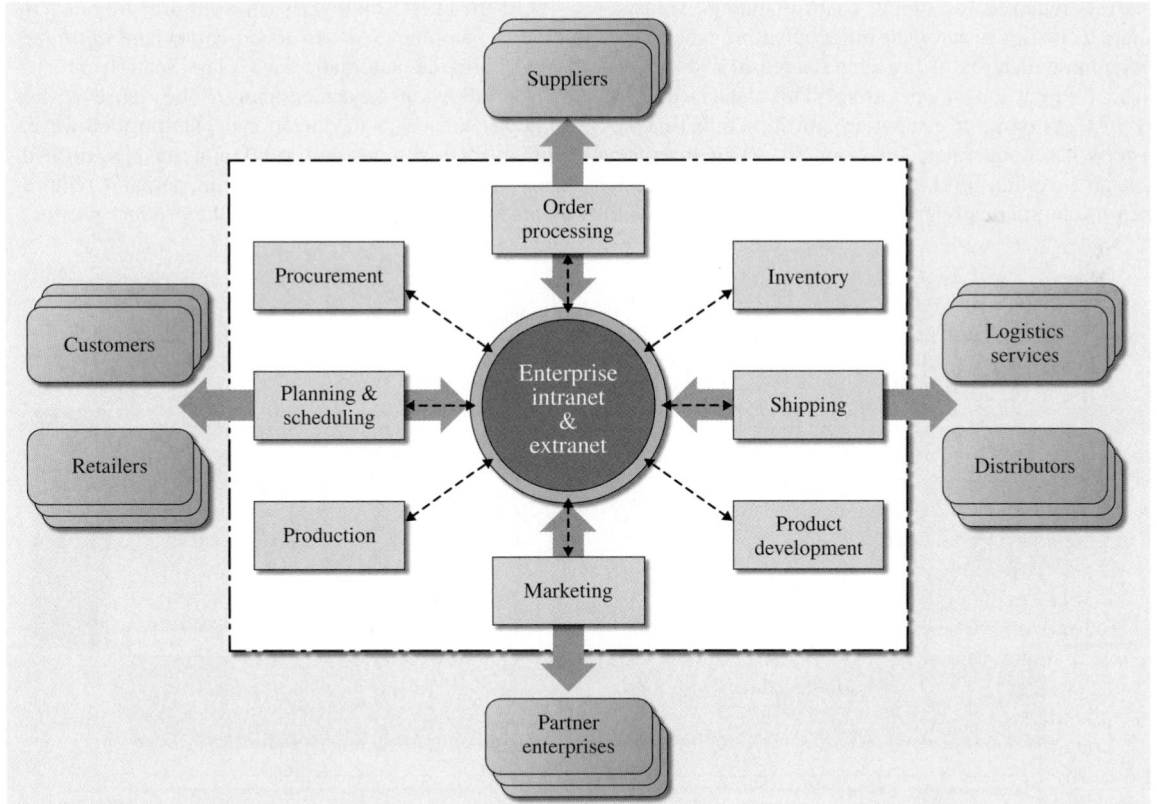

Fig. 54.3 Intranet and extranet for supply chain planning

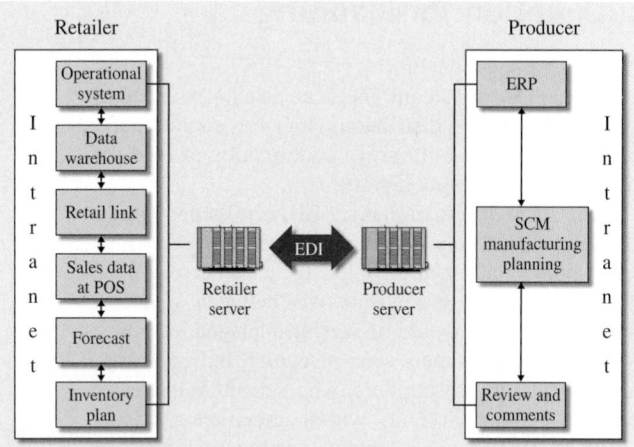

Fig. 54.4 Producer and retailer EDI application

ICT systems can also be used for simulating and optimizing a system. An example could be where ICT data is used to simulate the impact on a process with different levels of work in process on total productivity.

54.3.2 Production Planning and Programming

As the rate of technological innovation increases, maintaining a competitive cost structure may rely heavily on production efficiency generated by an effective production planning and programming process. For large-scale, global enterprises where traditional MRPI (material resource planning (1st generation)) and MRPII (material resource planning (2nd generation)) approaches may not be sufficient, new enterprise resource planning (ERP) solutions are being deployed that not only link all aspects from the bill of materials to suppliers to customer orders, but also utilize different algorithms in order to efficiently solve the scheduling conundrum. A general framework of ERP implementation is shown in Fig. 54.5.

Off-the-shelf solutions such as those provided by software supplier SAP advance planner and optimizer (APO) provide automatic data classification and retrieval, allowing key measures to be retrieved for strategic, tactical, and operational planning. However, typical ERP systems are rigid and have a difficult time adjusting to the complexity in demand requirements and constant innovation of the product portfolio

ucts' shelf-fill rate at its retailer Wal-Mart from 87% to 98% by EDI application, earning the company about US $8 million a year in additional sales [54.11].

ICT is key to the strategic, tactical, and operational analysis required for supply chain planning. Transactional ICT helps to automate the acquisition, processing, and communication of raw data related to historic and current supply chain operations. This data is utilized by ICT systems for evaluating and disseminating decisions based on rules; for example, when inventory reaches a certain level, a purchase order for replenishment is automatically generated [54.12]. The data from

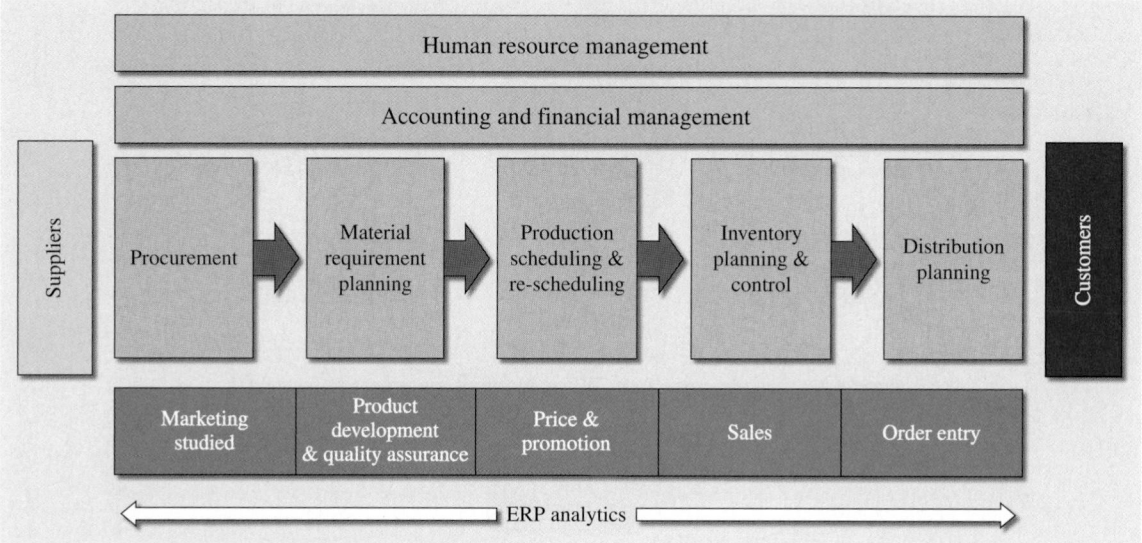

Fig. 54.5 A general framework of ERP implementation

mix. Global companies that compete in diverse marketplaces may choose to address these issues by building their own large-scale optimization models. With a solid database structure, these models are able to adapt to continuous changes in portfolios and can incorporate external influences on demand, such as market trends in promotions and advertisement. Over the past decade, there has been a shift in focus from business functions to business processes and further into value chains. Nowadays, enterprises focus on the effectiveness of the operation that requires functions to be combined to achieve end-to-end business processes [54.13].

54.3.3 Logistic Execution Systems

Logistic execution systems (LES), seek to consolidate all logistic functions, such as receiving, storage, inventory management, order processing, order preparation, yard management and shipping into an integrated process. The LES can be designed to communicate with the firms enterprise network and external entities such as customers and suppliers.

The LES might be part of the ERP system or otherwise should communicate with it through interfaces in order to interact with other modules such as finance, purchasing, administration, and production planning. Usually it also communicates with customers, suppli-

ers, and carriers via EDI or web-based systems with the purpose of sharing logistic information relevant to all parties such as order status, trucks availability, confirmation of order reception, etc.

A LES is usually composed of a warehouse management system (WMS) complemented with a labor management system (LMS) and a transportation management system (TMS). Essentially WMS issues, manages, and monitors tasks related to warehouse operation performance. A WMS improves efficiency and productivity of warehousing operations usually supported by (1) barcode and (2) radiofrequency data communications technologies. Both of these technologies provide a WMS with instantaneous visibility of warehouse operations, facilitating precise inventory control as well as accurate knowledge of labor and equipment resources availability. The addition of labor and transportation management in a LES has expanded the span of WMS functionality to the point of practically embracing every logistic function from receiving to the actual shipping of goods. An example of a WMS is shown in Fig. 54.6.

54.3.4 Customer–Oriented Systems

Many times, automation is thought to only apply to the production environment. However, there are a myriad of examples where other critical non-production-related processes have been automated, such as order entry,

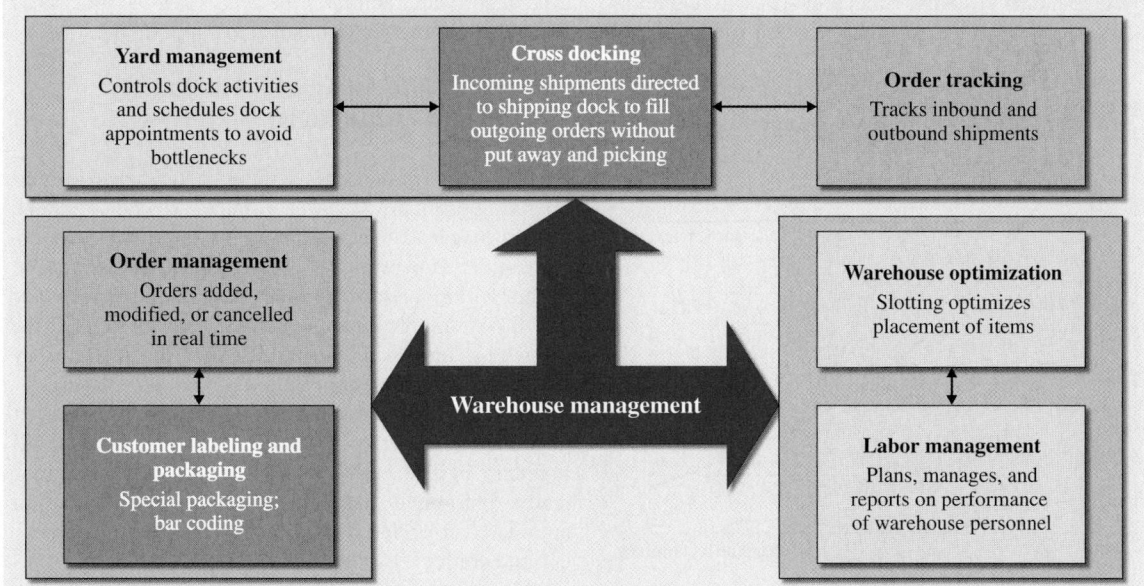

Fig. 54.6 A warehouse management system with decision support systems

inventory management, customer service, and product portfolio management. The goal of these applications is to improve upon the customers' experience with suppliers.

For many companies, the customer experience is a very tangible and measurable effect that should be *viewed through the eyes of the customer*. The goal for

the supplier is to be *easy to do business with*. For this reason, many companies have decided to automate how the customer exchanges information related to sales and logistics functions. One example of this automation is the use of cellular phone messaging technology to coordinate in real time to the customer the status of their order throughout the delivery process.

54.4 Automation Design Strategy

54.4.1 Labor Costs and Automation Economics

Competitive dynamics and consumer needs in today's marketplace demonstrate the need for manufacturer flexibility in response to the speed of change. Modern facilities must be aligned with the frequent creation of new products and processes and be prepared to manage these changes and resulting technological shifts. Therefore, one of the most difficult issues now facing companies is identifying a manufacturing strategy that includes the optimal degree of automation for a given competitive environment.

As shown in Fig. 54.7, increased production costs, including the mitigation of possible labor shortages, is one of the initial reasons companies look towards automation. However, production costs alone are usually not sufficient to justify the cost of investment. Productivity, customer response time, and speed to market are many times key factors to the success of automating

a production process and must be measured in order to effectively determine the impact on costs and expected revenues in the future.

54.4.2 The Role of Simulation Software

Determining the benefits of automation may prove to be a challenge, especially when there are complex relationships among processes. Will there be improvement in throughput and will this be sufficient to justify the required investment? With the availability of software such as ARENA by Rockwell Automation, simulation can be used as a tool to test different possible scenarios without having to make any physical changes to an existing system. This tool can be especially useful in cases where the required investment for automation is high and the expected benefits are not easily measured. In addition, the visual simulation capabilities related to this type of software facilitates the analysis of the process as well as obtaining top management support.

54.4.3 Balancing Agility, Flexibility, and Productivity

Typical business concerns such as increasing sales and operating profit are always considered key performance drivers that lead the investment strategy within an organization. However, with the complexity that can be found within certain marketplaces, a local niche or a well-developed global market, the need for sustainable growth has increasingly become one of the most important aspects when determining a competitive strategy.

A major competitive concern in the global market is agility in order to react to a dynamic market. To maintain agility between autonomous and geographically distributed functions significant investment in automated error detection is required to facilitate recovery and conflict resolution [54.1].

This complexity has adjusted the traditional return-on-investment (ROI) approach to automation invest-

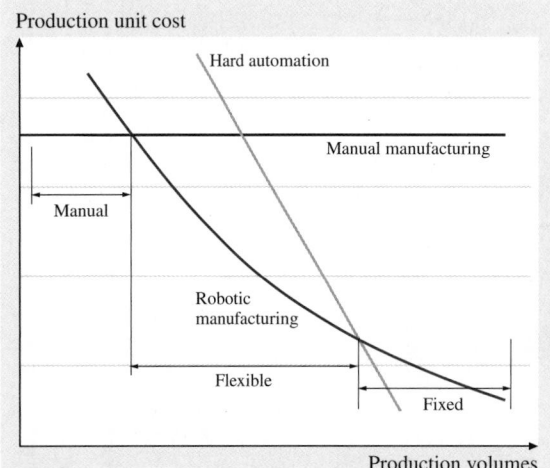

Fig. 54.7 Automation tradeoffs based on production volumes

ment justification by determining success on how well the solution is able to create value for customers through services or products. What are customers looking for? Is it on-shelf availability at a slightly higher price? A highly computerized ASRS (automated storage and retrieval system) may provide increased throughput, less errors, and speed in distribution. What if the customers change how and where products are required? An AS/RS may not be flexible enough to adjust quickly to changes in demand patterns or shifts in the demand network nodes. The correct solution should address customers' expectations regarding product quality, order response times, and service costs, seeking to obtain and sustain customer loyalty.

54.5 Emerging Trends and Challenges

The use of automation in production and supply chain processes has expanded dramatically in recent years. As globalization advances along with product and process innovation, it would seem that the importance of automation will continue to intensify into the future.

The landscape of global manufacturing is changing. More and more production plants are being built in India, Brazil, China, Indonesia, Mexico, and other developing countries. The playing field is rapidly being leveled. However, by crossing borders, companies also increase the complexity of operations and supply chains. Virtually seamless horizontal and vertical integration of information, communication, and automation technology throughout the organization is needed by companies such as Wal-Mart, Kimberly-Clark, Procter and Gamble, Clorox, and Nestle, in order to address the dynamics of today's manufacturing environment. Collaborative design, manufacturing, planning, commerce, plant-to-business connectivity, and digital manufacturing are just some of the many models that seem to be on the horizon for leading manufacturing companies in order to induce further integration of processes. Several technologies and systems are reaching the required level of maturity to support these models, and thereby accelerate the adoption of automation in production and supply chains. Examples of these technologies would be:

- Supply chain planning systems
- Supply chain security
- Manufacturing operations management solutions
- Active radiofrequency identification (RFID)
- Sensor-based supply functions
- Industrial process automation.

Furthermore other technologies such as manufacturing process management frameworks, supplier relationship management suites, supply chain execution systems, passive RFID technology, Six-Sigma IT (information technology), and lean manufacturing systems are more on an emerging or adolescence level of maturity [54.3, 4].

In addition, collaborative e-Work theory and techniques are emerging as powerful automation support for production, supply, logistics, and distribution [54.14, 15] (see also Chap. 88).

54.5.1 RFID Technology in Supply Chain and Networks

An emerging technology with great promise is the use of radiofrequency identification (RFID) to automate the collection of data. A RFID tag is a small computer chip that can send via radiofrequency a small amount of information a short distance. The signal is captured by a RFID antenna and then transferred to a computer network for data processing (Fig. 54.8). The general RFID system architecture, applications, frequencies, and standards are shown in Fig. 54.9 and Table 54.2.

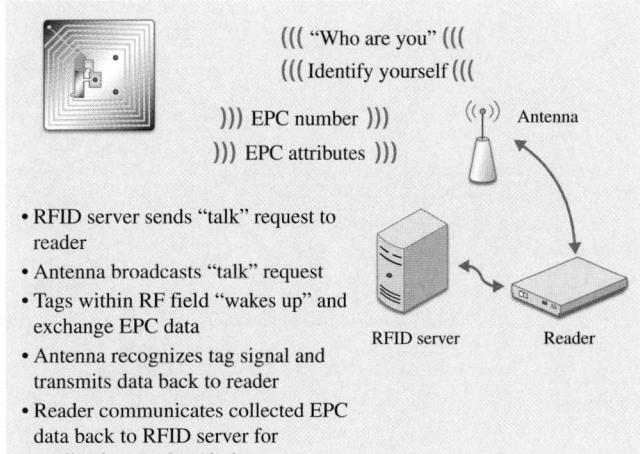

- RFID server sends "talk" request to reader
- Antenna broadcasts "talk" request
- Tags within RF field "wakes up" and exchange EPC data
- Antenna recognizes tag signal and transmits data back to reader
- Reader communicates collected EPC data back to RFID server for applications and analytics

Fig. 54.8 How RFID works

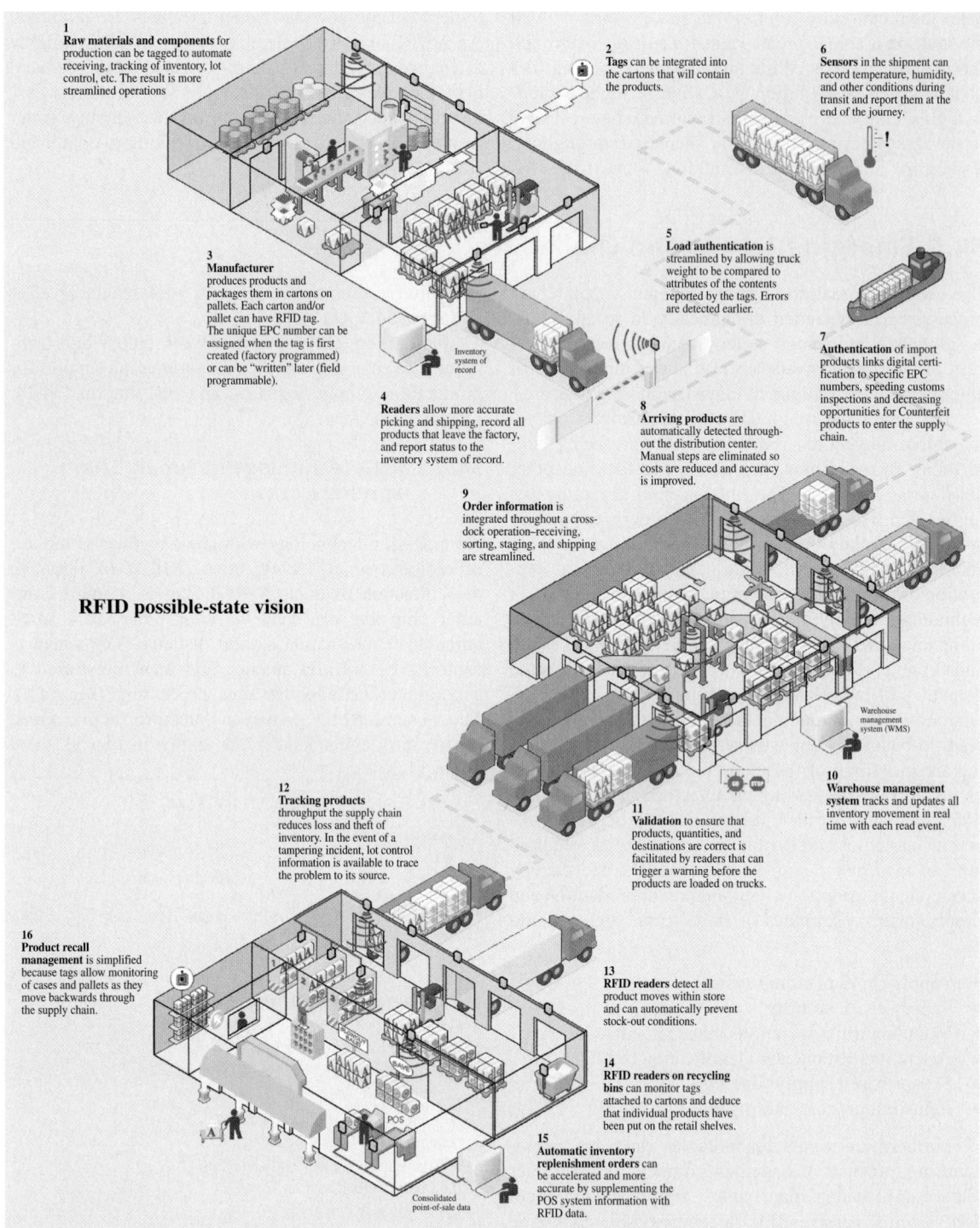

1
Raw materials and components for production can be tagged to automate receiving, tracking of inventory, lot control, etc. The result is more streamlined operations

2
Tags can be integrated into the cartons that will contain the products.

6
Sensors in the shipment can record temperature, humidity, and other conditions during transit and report them at the end of the journey.

3
Manufacturer
produces products and packages them in cartons on pallets. Each carton and/or pallet can have RFID tag. The unique EPC number can be assigned when the tag is first created (factory programmed) or can be "written" later (field programmable).

5
Load authentication is streamlined by allowing truck weight to be compared to attributes of the contents reported by the tags. Errors are detected earlier.

7
Authentication of import products links digital certification to specific EPC numbers, speeding customs inspections and decreasing opportunities for Counterfeit products to enter the supply chain.

4
Readers allow more accurate picking and shipping, record all products that leave the factory, and report status to the inventory system of record.

Inventory system of record

8
Arriving products are automatically detected throughout the distribution center. Manual steps are eliminated so costs are reduced and accuracy is improved.

9
Order information is integrated throughout a cross-dock operation–receiving, sorting, staging, and shipping are streamlined.

RFID possible-state vision

Warehouse management system (WMS)

12
Tracking products
throughput the supply chain reduces loss and theft of inventory. In the event of a tampering incident, lot control information is available to trace the problem to its source.

11
Validation to ensure that products, quantities, and destinations are correct is facilitated by readers that can trigger a warning before the products are loaded on trucks.

10
Warehouse management system tracks and updates all inventory movement in real time with each read event.

16
Product recall management is simplified because tags allow monitoring of cases and pallets as they move backwards through the supply chain.

13
RFID readers detect all product moves within store and can automatically prevent stock-out conditions.

14
RFID readers on recycling bins can monitor tags attached to cartons and deduce that individual products have been put on the retail shelves.

15
Automatic inventory replenishment orders can be accelerated and more accurate by supplementing the POS system information with RFID data.

Consolidated point-of-sale data

POS

Fig. 54.9a,b RFID in the supply chain (courtesy of BearingPoint's RFID solution)

Table 54.2 RFID functions, frequencies, and standards [54.16]

Applications	Frequencies	Standards
Animal, identification dogs, cats, cattle	< 135 kHz	ISO 18000–2 ISO 11784 ISO 11785 ISO 14223
Smart cards, passport, books at library	13.553–13.567 MHz	ISO 18000–3 ISO 7618 ISO 14443 ISO 15693 13.56 MHz ISM band class 1
Supply chain for retail	868–928 MHz	EPC global class 1 Gen-2 ISO 18000–6

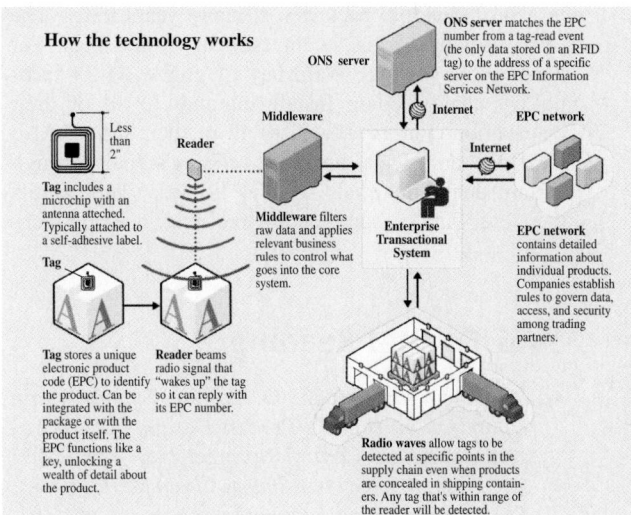

Fig. 54.9 (b)

RFID is becoming increasingly prevalent as the price of the technology decreases. Some RFID applications are summarized in Table 54.3. In supply chain applications, current uses of RFID technology are focused on location identification of products. These are as varied as identifying trailers and ocean freight trailers in a trailer yard, stopping of shoplifting of small but higher-priced fast-moving consumer goods such as razor blades, as well as an alternative to a WMS. In broader uses Wal-Mart is discovering RFID technology can help it increase sales by making sure inventory at the store's loading dock is actually placed on the shelf.

The barriers to implementing RFID technology are cost, effectiveness, and fears of a loss of personal privacy. The cost of a RFID tag has declined 90% in the past few years, but is still expensive and their use is usually limited to pallets and not on individual cases, products or boxes. It is believed that the uses of RFID

Table 54.3 RFID applications and examples

Application	Examples
Documents (e.g., passports)	Year 2000: Malaysia, Year 2005: New Zealand, The Netherlands, Norway, Year 2006: Ireland, Japan, Pakistan, Germany, Portugal, Poland, Year 2007: UK, Australia, and USA
Transportation payments	Electronic Road Pricing (Canada) T-Money (Korea) Octopus Card (Hong Kong) Super Urban Intelligent Card (Japan) Chicago Card and the Chicago Card Plus, PayPass, CharlieCard (USA)
Product tracking	Cattle tracking Jewelry tracking Library book or bookstore tracking Truck and trailer tracking
Supply chain network	Wal-Mart inventory system Boeing 787 Dreamliner maintenance and inventory system Promotion tracking

tags on individual packages is many years away. The other cost barrier is the investment in antennas. For a company such as Wal-Mart to utilize RFID technology, they need to install antennas in all of their distribution centers as well as all of their stores. Also RFID technology has problems sending signals through certain dense materials such as liquids, which limits their use. Finally, some people feel that, if RFID technology improves in terms of the distance a signal can be sent, then people will be able to determine which products are in people's homes, and thoughts of *Big Brother* come to mind. Currently the technology is not capable of fulfilling these privacy concerns, but the concern will continue to slow the acceptance of the technology. See also Chap. 49 on *Digital Manufacturing and RFID-Base Automation*.

54.6 Further Reading

- A. Dolgui, J. Soldek, O. Zaikin: *Supply Chain Optimization: Product/Process Design, Facility Location and Flow Control* (Springer, New York 2005)
- A.G. Kok, S.C. Graves: *Supply Chain Management; Design, Coordination, and Operation*, 1st edn. (Elsevier, Amsterdam Boston 2003)
- A. Rushton, P. Croucher, P. Baker: *The Handbook of Logistics and Distribution Management* (Kogan Page 2006)
- B. Kim: *Supply Chain Management* (Wiley (Asia), Hoboken 2005)
- C.E. Heinrich: *RFID and Beyond: Growing Your Business through Real World Awareness* (Wiley, Indianapolis 2005)
- D.E. Mulcahy: *Warehouse Distribution and Operations Handbook* (McGraw Hill, New York 1993)
- D.F. Ross: *Distribution: Planning and Control* (Springer, London 1995)
- D.J. Bowersox, D.J. Closs, M.B. Cooper: *Supply Chain Logistics Management*, 2nd edn. (McGraw-Hill/Irwin, Boston 2007)
- E.W. Schuster, S.J. Allen, D.L. Brock: *Global RFID: The Value of the EPC Global Network for Supply Chain Management* (Springer, London 2007)
- G. Simson: *RFID: Applications, Security, and Privacy* (Addison-Wesley, Upper Saddle River 2006)
- H. Chen, P.B. Luth: Scheduling and coordination in manufacturing enterprise automation, Proc. 2000 IEEE international Conference on Robotics and Automation (2000), pp. 389–394
- Harvard Business School Press: *Harvard Business Review on Supply Chain Management* (Harvard Business Review, Boston 2006)
- I. Bose, R. Pal: Auto-ID: Managing anything, anywhere, anytime in the supply chain, Commun. ACM **48**(8), 100–106 (2005)
- J. Berger, J.L. Gattorna: *Supply chain cybermastery: building high performance supply chains of the future* (Gower Publishing Company 2001)
- J.J. Coyle, E.J. Bardi, C.J. Langley: *The Management of Business Logistics: A Supply Chain Perspective* (South-Western/Thomson Learning, Mason 2003)
- J.-S. Song: *Supply Chain Structures: Coordination, Information and Optimization* (Springer, London 2001)
- N. Nicosia, N.Y. Moore: *Implementing Purchasing and Supply Chain Management: Practices in Market Research* (RAND, Santa Monica 2006)
- N. Viswanadham: Supply chain engineering and automation, Proc. 2000 IEEE international Conference on Robotics and Automation (2000) pp. 408–413
- N. Viswanadham: The past, present and future of supply-chain automation, IEEE Robot. Automat. Mag. **9**(22), 48–56 (2002)
- T.E. Vollmann, W.L. Berry, D. Clay: *Manufacturing Planning and Control System for Supply Management*, 5th edn. (McGraw-Hill, New York 2005)
- S. Chopra, P. Meindl: *Supply Chain Management: Strategy, Planning, and Operation*, 2nd edn. (Prentice Hall, Upper Saddle River 2004)

References

54.1 C.Y. Huang, J.A. Ceroni, S.Y. Nof: Agility of networked enterprises – parallelism, error recovery and conflict resolution, Comput Ind. **42**, 275–287 (2000)

54.2 F. Keenan: Logistics Gets a Little Respect, Bus. Week, 112–115 (2007)

54.3 Gartner, Inc.: Hype Cycle for Manufacturing 2005, Gartner's Hype Cycle Special Report (2005)

54.4 S.-L. Jämsä-Jounela: Future trends in process automation, Annu. Rev. Control **31**(2), 211–220 (2007)

54.5 H. Jack: Integration and Automation of Manufacturing Systems (2001)

54.6 R.A. LeMaster: Lectures on Automated Production Systems (Department of Engineering, University of Tennessee at Martin)

54.7 F. Gómez-Estern: Cintas Transportadoras en Automatización de la Producción, in Spanish

54.8 M. Piszczalski: Plant control evolution – technology update information, Automot. Des. Prod. (2002), http://www.thefreelibrary.com/Plant+control+evolution.+(Technology+Update+Information).-a084237484

54.9 R.S. Russell, B.W. Taylor III: *Operations Management*, 4th edn. (Prentice Hall, Upper Saddle River 2003)

54.10 K.C. Laudon, J.P. Laudon: *Management Information Systems*, 9th edn. (Prentice Hall, Upper Saddle River 2006)

54.11 E. Turban, D. Leidner, E. Mclean, J. Wetherbe: *Information Technology for Management*, 6th edn. (Wiley, New York 2008)

54.12 J.F. Shapiro: Business Process Expansion to Exploit Optimization Models For Supply Chain Planning (2002)

54.13 P. Anussornnitisarn, S.Y. Nof: e-Work: the challenge of the next generation ERP systems, Prod. Plan. Control **14**(8), 753–765 (2003)

54.14 S.Y. Nof: Collaborative control theory for e-Work, e-Production, and e-Service, Annu. Rev. Control **31**, 281–292 (2007)

54.15 S.Y. Nof, F.G. Filip, A. Molina, L. Monostori, C.E. Pereira: Advances in e-Manufacturing, e-Logistics, and e-Service systems. Milestone report, Proc. IFAC Congress'08 (Seoul 2008)

54.16 EPCglobal: http://www.epcglobalinc.org

55. Material Handling Automation in Production and Warehouse Systems

Jaewoo Chung, Jose M.A. Tanchoco

This chapter presents material handling automation for production and warehouse management systems that process: receipt of parts from vendors, handling of parts in production lines, and storing and shipping in warehouses or distribution centers. With recent advancements in information interface technology, innovative system design technology, and intelligent system control technology, more sophisticated systems are being adopted to enhance the productivity of material handling systems. Information interface technology utilizing wireless devices such as radiofrequency identification (RFID) tags and mobile personal computers significantly simplifies information tracking, and provides more accurate data, which enables the development of more reliable systems for material handling automation. Highly flexible and efficient automated material handling systems have been newly designed for various applications in many industries. Recently these systems have been connected into large-scale integrated automated material handling systems (IAMHS) that create synergy with material handling automation by proving speedy and robust infrastructures. As a benefit of high-level material handling automation, the modern supply chain management (SCM) successfully synchronizes sales, procurement, and production in enterprises.

In today's competitive environment, suppliers must be equipped with more cost-effective and faster supply chain systems to remain in the market. Companies are investing in material handling automation (MHA) not only to reduce labor cost, delivery time, and product damage, but also to increase throughput, transparency, and integratability in production and warehouse management systems. The material handling industry has grown consistently over many years. The Material Handling Industry of America (MHIA) estimates that, in 2006, new orders of material handling equipment machines (MHEM) grew 10% compared with 2005 and set a new record high at US\$ 26.3 billion in the USA [55.1].

In the past, labor cost was the most important element for estimating the return on investment (ROI) of a stand-alone automated material handling system (AMHS), and the system was a relatively small part of the production or warehouse facility. Nowadays, the impact of the system throughout the supply chain is becoming larger and more complicated; for example, a radiofrequency identification (RFID) system enhances customer satisfaction by providing convenience in data tracking as well as reducing order picking times and shipping errors in warehouse. AMHSs are not alternatives selected after prudent economic analysis, but are rather major components in a production and warehouse facility. Also, the sizes of systems and the complexities of their operations are increasing. Multiple AMHSs consisting of RFID systems, automated guided vehicles (AGVs) systems, and au-

a) Warehouse system for pharmaceutical industry

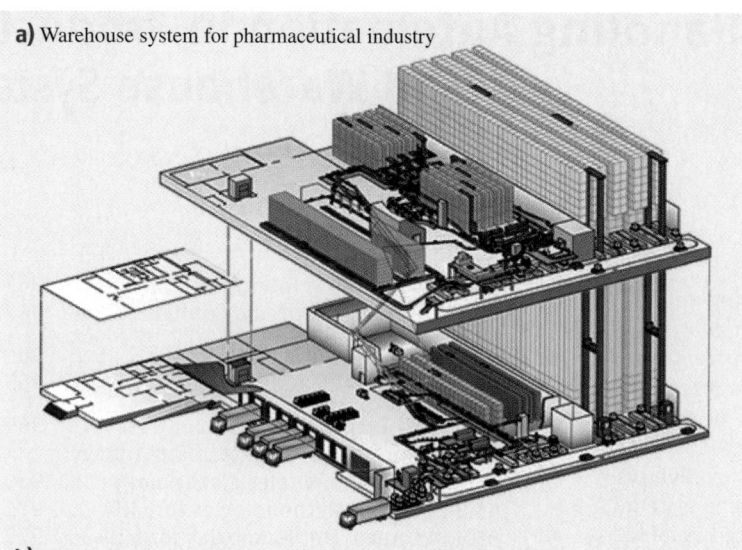

Fig. 55.1a,b IAMHS for pharmaceutical industry (courtesy of Murata Machinery). **(a)** Warehouse system for pharmaceutical industry, **(b)** material flows in warehouse system above

b) Material flows in warehouse system above

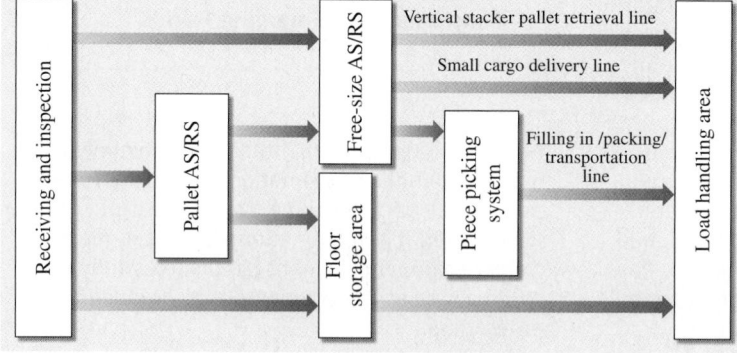

tomated storage and retrieval systems (AS/RSs) are typically installed in a production and warehouse facility as a connected system. As its complexity has increased, optimization of the design and operation of these systems has become of interest to both AMHS vendor companies and their customers. Many examples of these integrated systems can be observed in the semiconductor [55.2], automotive [55.3], and freight industries [55.4, 5].

This chapter introduces practical applications of MHA for production and warehouse systems. It starts by introducing a concept of the IAMHS that uses several types of the AMHS in a single integrated system (Figs. 55.1 and 55.2). The focus is particularly on what an IAMHS consists of and how it collaborates with other systems in SCM. Based on this introduction, components of the IAMHS and their recent technology advancement in the MHA will be reviewed.

55.1 Material Handling Integration

55.1.1 Basic Concept and Configuration

An IAMHS integrates different types of automated material handling equipment in a single control en-

vironment. A simple type of IAMHS was used for seaport or airport cargo terminals, which are served by stacker cranes and AGV systems [55.5]. The main issue of the simple IAMHS is how to reduce wait-

Finally, the IAMHS is operated by handheld terminals providing many applications in the warehouse. It is equipped with an RFID or barcode reader that allows flexible adaptation to changes in distribution quantity.

The semiconductor industry is equipped with one of the most complex IAMHSs for wafer fabrication (fab) lines (Fig. 55.3). A fab line may consist of more than 300 steps and 500 process tools. Material transportation between tools in a wafer fab line is fully automated by the overhead hoist transporter (OHT) system, which is a type of rail-guided vehicle (RGV) system, an AS/RS called the *stocker*, a lifting system that transfers wafer carriers between different floors, and a mini-environment that is used for a standard interface of machines with the AMHS. For the next generation of IAMHS in a fab line, Middlesex has proposed a new concept using conveyor systems (Fig. 55.2) instead of OHT systems and stockers, which guarantees larger-capacity transfers and quick response times for deliveries. Middlesex has focused on high-end conveyor systems for many years. More reviews of the IAMHS in the semiconductor industry are provided by *Montoya-Torres* [55.2].

These IAMHSs are generally highly flexible in design for customized usage, and some of them are even unique and revolutionary. *Kempfer* [55.7] introduced an order picking system utilizing a voice recognition system and RFID system in a large-scale automated distribution center. The article reports that the average order picking performance was improved from 150 cases per man-hour to 220 cases per man-hour by reducing operators' information handling time. A few companies also achieved similar

Fig. 55.2 New IAMHS design for next-generation semiconductor fab (courtesy of Middlesex)

ing time during job transition between two different AMHSs to increase throughput; for instance, an AGV has to wait after arriving at the load position if the crane is not ready to unload a container to the AGV. If their jobs are poorly synchronized, the waiting time will be longer, and as a consequence throughput will drop.

Recently IAMHSs with more complicated component systems have been implemented for many companies in different industries. Figure 55.1 shows an example IAMHS used in a warehouse system in the pharmaceutical industry [55.6]. In this configuration, there are five different types of the AMHS. First, a pallet AS/RS is installed and the temperatures of each shelf in the AS/RS can be controlled according to the characteristics of the products stored to maintain product quality. Second, a free-size AS/RS is used for storage of individual orders and items that are frequently replenished. It can store items regardless of their size, shape, or weight since it uses a hoisting carriage that can handle a wide range of products. Third, an automated overhead traveling vehicle is installed to replenish items with minimum labor cost and waiting time. It uses overhead space to increase space efficiency. Another AMHS used is a digital picking system, providing convenience for picking tasks by displaying directions on a digital panel installed on the shelves.

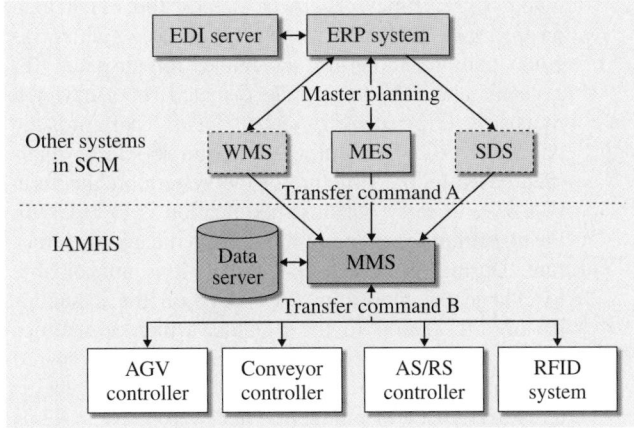

Fig. 55.3 IAMHS in hierarchical system architecture

improvements by adopting an integrated RFID and voice recognition system [55.8, 9]. *Chang* et al. [55.9] proposed an integrated multilevel conveying device for an automated order picking system that transfers articles between two different levels of a multi-

storey building to improve the operational and spatial efficiency of the warehouse system. The system employs a specially designed device comprised of a stacker crane, a vehicle-based transporter, and conveyor system.

55.2 System Architecture

In the design of a large-scale IAMHS, a well-structured system reduces the redundancies of functions in different modules, unnecessary transactions between modules, and system errors caused by large and complex individual functions of these modules. An algorithm ignoring the system architecture sometimes tends to create many problems during implementation, mainly because of the lack of necessary information and difficulty in interacting with existing systems [55.10]. Examples of this limitation can be found in the literature. An AGV scheduling algorithm under an FMS environment determines a sequence of the AGV route within a certain time horizon, considering the information from both the work centers and AGV systems on a shop floor. However, under this system architecture, it is very difficult for an AGV controller to take into consideration complex constraints of work centers such as machine status, processing times, and setup times because of the long calculation time. Therefore, generally, job sequencing and scheduling are performed independently by the scheduling and dispatching system, which is then connected to the AGV controller using a sequence of protocols. An AGV controller only takes care of requested transfer commands, which specify source and destination locations, priorities, and command trigger times. There is already too much load on the AGV controller in its original tasks, which include path planning for a vehicle, job dispatch for a newly idle vehicle, vehicle dispatch for a new job requested, error recovery, etc. [55.11]. Therefore, the developed AGV scheduling algorithm should be modified based on the structure of the system architecture. One way to carry out this modification is to break up the algorithm for different modules in the system structure. During this break-up process, it is unavoidable to change the algorithm depending on the availability of information to the module, which sometimes causes significant performance degradation compared with the original algorithm. As the number of subsystems being used in production and warehouse systems continues to increase, a well-structured system will be

beneficial for facilitating collaborations between different departments as well as these systems. However, it is an open challenge to construct a well-designed system structure that accommodates all the different types of AMHS regardless of the size of the system and the type of business on which the IAMHS is centered.

Various types of system architecture can be used to design an IAMHS with other application systems, depending on the manufacturing type of the shop floor, the size of the total system, the number of transactions per second, etc. Figure 55.3 illustrates a design example of the system architecture for the IAMHS presented in Fig. 55.1. The focus of this figure is on software modularity. Each AMHS has its own controller (the four controllers at the bottom of Fig. 55.3), which is responsible for its own tasks and communication with the material management system (MMS), which is a high-level integrating system that will be explained later in more detail; for example, an AGV controller addresses job allocation, path planning, and collision avoidance, receives a transfer command from the MMS (transfer command B in Fig. 55.3), and reports necessary activities such as vehicle allocation and job completion to the MMS so that data are kept for tracking in the future. Each controller also has to process error-recovery routines for robustness of the system control. The MMS manages multiple controllers of different AMHSs, and has a database server to store all transactions of the subsystems in the IAMHS. It receives transfer commands or short-term scheduling results of processing machines from the scheduling module in a higher-level system in the SCM (transfer command A in Fig. 55.3). In this structure, long-term optimization of processing machines is responsible for the higher-level system, and the MMS focuses on efficiencies during the transportation of unit loads within production and warehouse facilities. Details of the MMS are explained in the next section. As shown in Fig. 55.3, the higher-level systems of the MMS can be a manufacturing execution system (MES), warehouse management system (WMS), and enterprise resource planning (ERP) system. The WMS can be substituted by the MMS if the warehouse is composed of relatively simple systems.

Understanding high-level decision-support systems in SCM helps to understand the scope of control tasks performed by the IAMHS. The advanced planning and scheduling (APS) system generally consists of planning and scheduling modules. Sometimes the scheduling module is again broken down into scheduling and dispatching modules (SDS in Fig. 55.3). The ERP system generally includes the planning module; however, the scheduling and dispatching modules can be included in any other systems such as the MES and WMS. In Fig. 55.3, it is assumed that the modules are running in a stand-alone system called the SDS that communicates with the ERP system, MES, and MMS. The planning module makes a long-term production or procurement plan based on customer orders, demand forecasting results, and capacity constraints. Its time horizon varies from weeks to months. Practically, it hardly optimizes complex factors of resources on the shop floor because of the long computation time, but constructs highly aggregated planning. Its results include production quantities for each product type and time bucket, or production due dates for each product type or product group. Detailed resource requirement plans are not specified by the planning module due to the uncertainties and complexities of operations. The scheduling module is responsible for delineating more concrete plans for the shop floor to meet the target production plan from the planning module. It typically tries to optimize various resource constraints with several objectives such as due-date satisfaction and throughput maximization. Detailed resource requirement plans over time buckets within a time horizon are created by the scheduling module. It sometimes takes into account constraints in the AMHS for more robust scheduling. The time horizon of the scheduling module varies from a few hours to days. The dispatching module determines the best unit load for a machine in real time following a trigger event from the machine or unit load. It tries to follow up closely the scheduling results, which are globally optimized. The MMS in the IAMHS receives transfer commands from either the scheduling module or dispatching module based on its system architecture. These transfer commands are the result of the scheduling, machine assignment or job sequencing on processing machines. The MMS manages the process of the given transfer command by creating more detailed transfer commands to the AMSHs in the IAMHS.

The dispatching module is sometimes included in the MMS, and creates the transfer commands based on the scheduling results and its own dispatching rules for the real-time status of the shop floor. The IAMHS takes charge of the final execution of the SCM in an enterprise and also provides useful information as described above.

The higher-level systems of the IAMHS automate information processing throughout an enterprise. The MES in Fig. 55.3 is a tracking system that collects important data from processing machines and stores them in well-structured database tables for analysis of quality and process controls; however, it has expanded its role into many other areas based on a powerful open architecture. It has been popularly used in the electronic industry such as in semiconductor fabs and surface-mounting technology (SMT) lines, and has recently spread into other industries. The warehouse management system (WMS) is generally used in mid- or large-size warehouse facilities, similar to MES for a production shop floor; it tracks every movement of materials and support operations for material handling in the warehouse. Its focus is on information processing automation. The objective of implementing an ERP system in a company is information sharing for rapid and correct decision-making, and implementation throughout the enterprise by using an integrated database system [55.12]. Chapter 90 provides a more thorough discussion of ERP and related concepts. The whole procedures of order entry, production planning, material procurement, order delivery, and corresponding cash flow are managed by the system. All the data from different applications in an enterprise or between different enterprises are exchanged by an electronic data interchange (EDI) server, which allows automated exchange of data between applications. Based on the EDI technology, applications freely exchange purchase orders, invoices, advance ship notices, and other business documents directly from one business system to the other without human support. Figure 55.4 illustrates the connectivity of IAMHS to other systems in SCM, which is used in an actual industry.

55.2.1 Material Management System

The role of the MMS is very important in a complex IAMHS for high-level automation. The main functions of the MMS are summarized below, in increasing order of importance. This summary does not discuss the dispatching module that assigns unit loads to process-

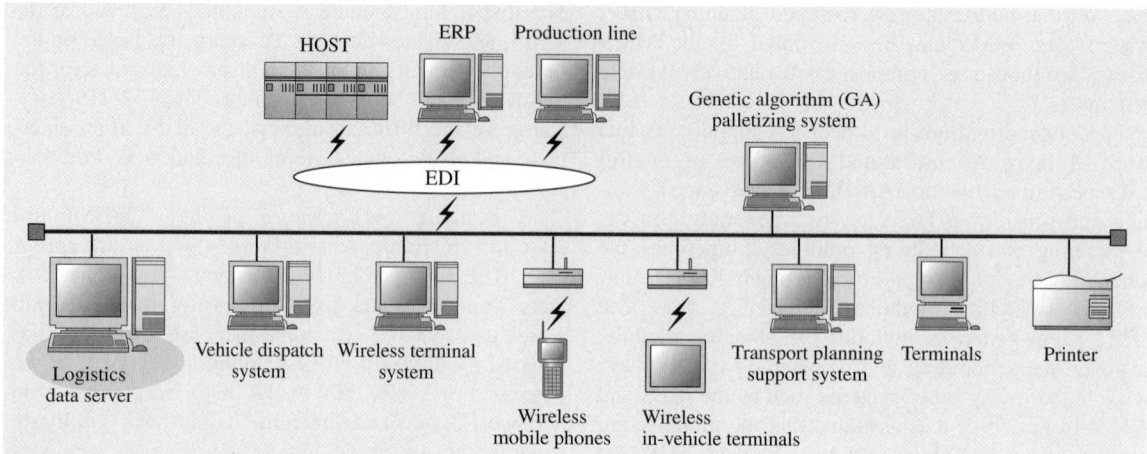

Fig. 55.4 Connectivity of IAMHS to other systems in SCM (courtesy of Murata Machinery)

ing machines because this involves so many topics; however, the dispatching functions bounded to AMHSs (i.e., dispatching unit loads while not considering process machines) will be discussed here.

The roles of the MMS in a complex IAMHS are:

1. Determining the best destination among several possible AMHS alternatives
2. Determining the best route to get to the destination from a source location via several AMHSs
3. Determining a proper priority for the transfer command
4. Storing and reporting various data using a database server
5. Transfer command management between different AMHSs
6. Error detection and recovery for the transfer command
7. Providing a user interface for control, monitoring, and reporting.

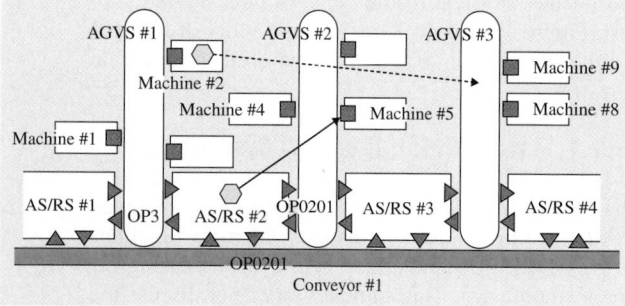

Fig. 55.5 Example of IAMHS

For a unit load to be transferred, its source and destination locations are mainly determined by the dispatching or scheduling module; however, when the candidate destinations are AMHSs, it is sometimes more efficient for the MMS to determine the final destination than for the scheduling or dispatching module to do so. Consequently, the dispatching and scheduling system provides a destination group to the MMS. Figure 55.5 illustrates the execution process of a transfer command. In the figure, if a unit load from Machine #2 has just finished its processing and has to be transferred to an AS/RS to be processed next, on one of the machines connected to AGVS #3 (dashed arrow in the figure), there are two candidate AS/RSs connected to AGVS #3: AS/RS #3 and AS/RS #4. The AS/RS group connected to AGVS #3 is named AS_Group #3. The MMS will receive a transfer command from the dispatching module, specifying the source location as Machine #2 and the final destination as AS_Group #3. Since there are two alternative destinations, the MMS may consider the product type of the unit load, the full rate of each AS/RS, the load port status of each AS/RS, the shortest distances from the unit load to the AS/RSs considering current active jobs in each system, and so on. It will determine the best AS/RS amongst the two alternatives and trigger a transfer command to AGVS #1, which will first move the load to AS/RS #1 from Machine #2.

The MMS is also responsible for determining the destination subsystem in an AMHS, such as the load/unload port (or pickup/drop-off port) in an AS/RS, because there are generally multiple load/unload ports with different types and numbers of buffers. The ports may differ, being load only, unload only or of unified type. Assume that a unit load in AS/RS #2 in Fig. 55.5 has to be moved to Machine #5, connected to AGVS #2 (solid arrow in Fig. 55.5). First, the unit load has to be moved to one of the output ports in the AS/RS. The AS/RS controller may not know which output port will be the best among the three possible ones in the figure because it does not know the next destination of the unit load. The MMS may determine a load port connected to AGVS #2, OP0201 in Fig. 55.5. In practical application, the problems are generally much more complicated than this illustration due to the increased instances in the system. Few studies have addressed this type of problem. *Sun* et al. [55.13] and *Jimenez* et al. [55.14] stress the importance of this problem in the literature and introduce a few ideas being used in practical applications; however, their methods leave much room for improvement in that they use static approaches and consider limited factors.

Obtaining the best route to get to the destination is another important task of the MMS. In a complex IAMHS, there are many possible routes consisting of different AMHS types. The IAMHS in Fig. 55.5 is represented by the graph in Fig. 55.6. A graph can be encoded in database tables by using an adjacency matrix or incidence matrix for use by a computer program. The adjacency matrix is a simple from–to chart between a pair of vertices, in which the value of an edge is the distance between the vertex pair, being zero if the pair are not connected. The incidence matrix rep-

resents the connectivity of vertices by edges. Using a graphical representation of the IAMHS, many predefined properties and algorithms of graph theory can be applied to develop algorithms for the MMS; for example, Dijkstra's algorithm can be used to determine the shortest path from a source to destination location.

The time intervals between the arrivals of transfer commands are sometimes completely random in that there are significant fluctuations in the number of arrivals during different time periods. When the queue size increases on AMHSs, use of different priorities for transfer commands often provides a very useful solution to improve overall system performance. It is reported that a good priority algorithm can improve the throughput of a production facility [55.15].

There are two types of tables in the database of the MMS. One type of table stores parameters for control algorithms and status user interfaces (UIs). These need a minimum number of entities to achieve a shorter transaction time when they are queried. The other type of table stores data for movement histories based on communication messages between component systems of the IAMHS. The accuracy of these historical data have been significantly improved by material handling automation with advanced information interface technology (IIT) by using RFID technology or barcode systems. A large amount of information can be extracted from the historical data, including the standard operating time of a machine, the processing routes of

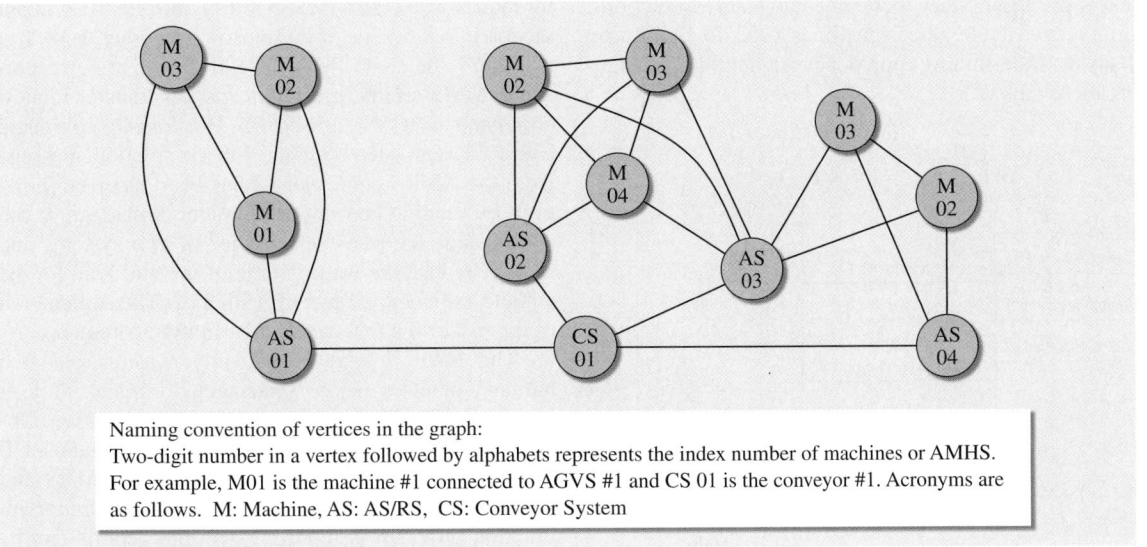

Naming convention of vertices in the graph:
Two-digit number in a vertex followed by alphabets represents the index number of machines or AMHS.
For example, M01 is the machine #1 connected to AGVS #1 and CS 01 is the conveyor #1. Acronyms are as follows. M: Machine, AS: AS/RS, CS: Conveyor System

Fig. 55.6 Graphical representation of the IAMHS in Fig. 55.5

a unit load over machines in different process stages, the lead time of the unit load from start to finish, etc.

These data provide useful information for various purposes. Most of all, without accurate data from production and warehouse systems, it is hard to achieve successful realization of the enterprise-level decision support systems explained above. Good planning strongly depends on accurate data. In practice, many companies have invested in expensive ERP systems capable of automated production planning for their shop floors; however, many of them do not use the module because of poor planning and scheduling quality from the system. One of the main reasons for this poor quality is sometimes due to the lack of good data from the material handling system, which relies on manual jobs and operator paperwork. Accurate data from AMHSs also helps to achieve lean manufacturing on the shop floor by providing precise measures. For a complicated shop floor, it is often difficult to define a bottleneck stage or performance measures of the bottleneck machines. Lean manufacturing starts from well-defined and accurate performance measures. Many details of the machines can be analyzed by data relating to material movements from machine to machine, examples of which include machine throughput, product lead time, and the work-in-process (WIP) for each processing stage. Sometimes they also provide benefits for engineering analysis for the improvement of quality control. The performance of the IAMHS can also be measured and improved by using these historical data. A new algorithm under test can be easily tracked to assess how it performs in an actual application. Since there are many data transactions, summarized tables are sometimes used for long-term analysis. Data-mining approaches are helpful in designing these tables.

Another important task of the MMS is the path management function, which controls a sequence of transportation jobs. Let us consider the following simple transfer request as an example. A transfer request is sent to the MMS from the dispatching module to move a unit load from a rack in AS/RS #2 to AS/RS #3 through AGVS #2 in Fig. 55.5. Figure 55.7 shows a message sequence illustrating communication between the MMS and AS/RS controllers involved in this transfer, and between the MMS and AGV controller. A few more messages might be used in actual systems. As seen in the figure, although this is a relatively simple transferring task, more than 13 messages are used to complete the task. First, transfer request #1 is triggered by the MMS (it can be triggered by either the dispatching module or a procedure of the MMS itself). This request message transmits the source location as AS/RS #2, the destination as the unload port of the AS/RS, and the unit load identity to the AS/RS controller #2. If there are other high-level systems such as a WMS or MES, the MMS will send additional messages to these systems. In this case, the status of the unit load possibly needs to be updated from *Waiting* to *Busy* or *Transferring* for the WMS and MES. To send this message to AS/RS controller #2, the MMS has to make at least two major decisions: it has to select an unload port among several idle ports, and to determine to which among the many other AMHSs this message should be sent. For the former decision, the closest idle port to the next destination (i. e., AGVS #2) is selected based on the MMS algorithm. After receiving the transfer request from the MMS, AS/RS controller #2 will put the job into its queue, if it is performing other tasks. If it is its turn, the controller will send the job assign report to the MMS so that it triggers another transfer request command to AGV controller #2. This transfer command could be sent later after the job completion message from the AS/RS controller #2 has been received; however, by sending before the completion message, it can synchronize the transfer activities of two systems and thereby reduce the waiting time of the unit load for the vehicle at the unload port of AS/RS #2. The explanation of the rest of the messages in the figure is omitted.

The MMS integrates not only systems but also human operators in the system environment. A user interface (UI) plays a major role in this integration. Operators can monitor the number of AMHSs using the UI. Also, parameters to control an individual AMHS and IAMHS are changed through the UI. Another important function provided is reporting. Various reports can be queried directly from the database of the MMS.

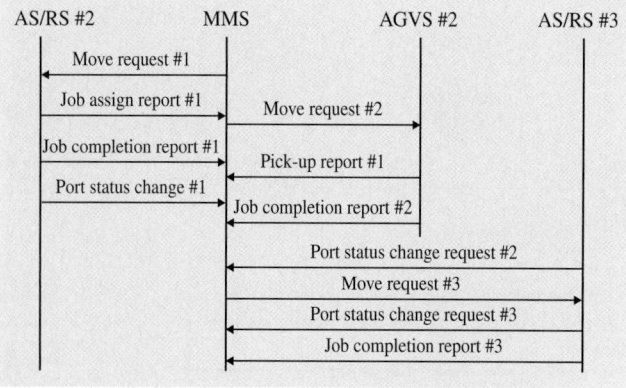

Fig. 55.7 Message sequence for a simple transfer command

55.3 Advanced Technologies

This section surveys advanced technologies enabling the IAMHS to achieve the high-level MHA. First, the IIT utilizing wireless devices will be reviewed, then the focus will move onto design and control issues of the MHA. A wide range of methodologies across artificial intelligence (AI) and operations research (OR) techniques have been adopted to solve challenging problems in design and control of MHA. The design and control issues with AMHS types will be briefly described including different points of interest. The review focuses on the technical issues of the MMS, which is the most important element of the IAMHS. Finally, AI and OR techniques are compared according to several criteria in MHA.

55.3.1 Information Interface Technology (IIT) with Wireless Technology

Benefits of wireless communication systems include mobility, installation flexibility, and scalability. Applications of the wireless communication used for MHA are radiofrequency identification (RFID), wireless local-area network (LAN) (i.e., Ethernet), and wireless input/output (I/O). The wireless sensor network has also great potential for many applications of the MHA to collect data or form a closed-loop control system.

Radiofrequency Identification (RFID)
RFID enhances information tracking with a wide variety of applications for material handling [55.16]; for example, it prevents loss of boxes and incorrect shipping in a distribution center and reduces time for reading tags in boxes or carriers on a manufacturing shop floor. Its greatest advantages over barcode systems are its long read range, flexibility of locating tags in boxes, multitasking for reading many tags at the same time, and robustness against damage. Finally, RFID systems increase the accuracy of data from material handling systems and reduce time for data collection. With more reliable and faster information tracking, more sequential operations can be automated and integrated without affecting system performance or requiring human interventions. It also enables the development of higher-level MHA in production and warehouse systems.

An RFID system consists of tags and readers. An RFID tag has two components, a semiconductor chip and antenna, and there are basically two types of RFID tags, *passive and active tags*, based on the source of the power. A passive tag does not have a battery and is powered by the backscattered RF signal from the reader, while an active tag has a battery and is therefore more reliable. Although the read range of a tag depends strongly on its power level, antenna, and frequency, and the environment in which it is used, an active tag can have a range of up to 30 m or more while a passive tag can be read reliably over a few meters. In between these two types, there is a semiactive tag that is powered by RF from the reader and consumes power from a battery while communicating with the reader. The lifetime of the battery is about 7 years or more. Another classification of RFID tags is based on the ability to write information to tags. Some tags are classified as *read-only* and can be written only once but read many times; these are generally passive tags. Information can be written by both users and producers. There are also rewritable passive tags in which the program can be rewritten by users. Most active tags are rewritable. RFID readers send RF signals to tags, receive signals from tags, and communicate with a central system. Their functions varies from a simple on/off check for data collection to control of a large system. Popularly used tags are as large as an electronic card, being installed in a larger computer system with network capability; however, they can be as small as $0.05 \times 0.05 \, \text{mm}^2$, as shown in Fig. 55.10. On the right-hand side of Fig. 55.8, powder-type RFID chips developed by Hitachi are compared with a human hair.

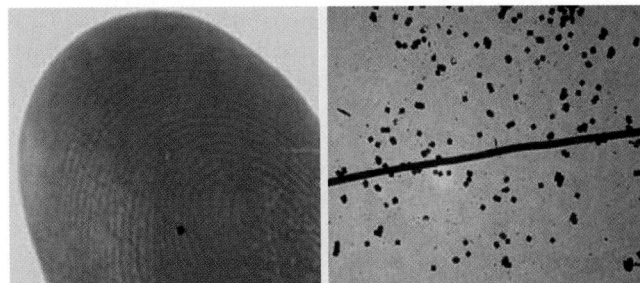

Fig. 55.8 Mu-chips and powder type RFID chips (courtesy of Hitachi)

These powder-type RFID tags are 64 times smaller than those in current use ($0.4 \times 0.4\,\mathrm{mm}^2$ mu-chips, on the left, produced by the same company), which can already be embedded into paper currency, gift certificates, and identification documents. For more information in RFID see Chap. 49.

Wireless LAN

A wireless LAN establishes a network environment by using wireless devices instead of wired ones within a limited space. One popular application that adopts wireless LAN in MHA is AGV systems, which use it for communication between vehicles and controllers. Each vehicle has a network interface card (NIC) that is connected to the wireless LAN. An access point is a gateway to connect to a wired LAN and similar to a LAN hub, connecting 25–50 vehicles within a range of 20–150 m. The infrastructure network is always connected to an access point, which connects the wired LAN with the wireless LAN. In the infrastructure network, the basic service set (BSS) is formed and acts as a base station connecting all vehicles in the cell to the LAN. BSSs that use nonoverlapping channels can be part of an extended service set (ESS). The vehicles within the ESS but in different BSSs are connected through roaming. *Lee* and *Lee* [55.17] develop an integrated communication system that connects Profibus and IEEE 802.11, which are wired and wireless LAN communication protocols, for a container terminal automated by an AGV system. Using this protocol converter, the wireless LAN can be connected to the existing wired fieldbus for soft real-time data exchange that loses some of its usefulness after a time limit.

Wireless I/O

A wireless I/O device is a small circuit card with an antenna installed in a material handling system or its controller; it can be used for both data-acquisition and closed-loop control applications. It receives microwave radio data from I/O points, and sends those data to a central processing device such as a programmable logic controller (PLC), data loggers, supervisory control, and data-acquisition system (SCADA), or a general PC [55.18]. Since it does not use wireless LAN or a fieldbus, implementation is much easier than afor wireless LAN. It can be simply regarded as removing the necessity for wires; however, by itself, it offers many advantages such as broader connectivity, increased mobility and flexibility, reduced installation time, and reduced points of failures. One of the disadvantages of wireless I/O is that, since it uses a relatively narrow range of wireless signals, a smaller number of wireless I/Os can be used in a certain area. Therefore, as the number of points in an area grows, a wireless or wired LAN will become more appropriate.

Wireless Sensor Networks

Sensor networks [55.19] are currently limited to novel systems. Many sensors, distributed in a system or area, can be used to build a network for monitoring a space shuttle, military equipment unit or nuclear power plant. Wireless sensor networks conceptually use small, smart, cheap sensors that consist of a sensing module, a data-processing module, and communication components; however, conventional sensors can also be used. The network is mainly used for monitoring systems that requires highly autonomous and intelligent decision-making in a dynamic and uncertain environment. They have a great deal of potential to be adopted in MHA even though few researchers have studied these applications. There are two areas of wireless sensor network applications for the MHA.

First, reliability is often very important for the MHA because, in a highly automated system, the failure of an AMHS causes the breakdown of multiple machines or a whole area operated by the system. This may be more critical than the failure of an individual processing tool in production systems. Therefore, monitoring and diagnosing the AMHS lead to some important issues; for instance, vibration sensors and optical sensors attached to the crane of an AS/RS collaborate to detect a potential problem that might cause positioning or more critical errors. By detecting the problem before the AS/RS actually breaks down, engineers can recognize the problem more precisely and prepare required parts and tools in advance; hence, repair time can be significantly reduced.

Second, most AMHSs use a closed feedback system that controls the system based on feedback from component systems or sensors. *Walker* et al. [55.20] studied a method to control an industrial robot that handles flexible materials such as wires and rubber hoses. It utilizes feedback from sensor network cameras to predict the motion of the robot with the better vision. Since the feedback can be created from many different points such as grasps, paths, and goal points, it reduces blind spots of unpredictable motions and greatly enhances control precision. Chapter 20 provides additional information on sensor networks.

Table 55.1 Design issues and related studies on MHA

Reference	AMHS type	Design issue	Criteria	Solution approach
Cho and Egbelu [55.21]	IAMHS	MHS equipment selection problem	Qualitative factors, equipment variety (minimizing)	Fuzzy logic and knowledge-based rule
Nadoli and Rangaswami [55.22]	IAMHS	Design and modeling for a new semiconductor fab	Design lead time	Expert system, computer simulation
Jimenez et al. [55.23]	IAMHS	Performance evaluation of AMHS	Delivery time, transport time, throughput	Computer simulation
Huang et al. [55.24]	General MHS	Location of MHS	Total distance, fixed cost of MHS	Lagrangian relaxation and heuristic method
Jang et al. [55.25]	AS/RS	Estimation of AS/RS performance	Delivery rate, in-process inventory	Queuing network model
Lee et al. [55.26]	AS/RS	Optimal design of rack structure with various sized cells	Space utilization (lost space)	Modular cells, heuristic
Ting and Tanchoco [55.27]	AGV	Location of the central path	Total rectilinear distance	MIP
Gaskins and Tanchoco [55.28]	AGV	Guide path design: direction of path segments	Total flow distance	Integer programming, heuristic
Tanchoco and Sinriech [55.29]	AGV	Guide path design: optimal design of a single-loop	Total flow distance	Integer programming
Bozer and Srinivasan [55.30]	AGV	Guide path design: tandem guide path	Balanced workload	Integer programming, set partition
Caricato and Grieco [55.31]	AGV	Guide path design	Flow distance, computation time	Simulated annealing
Nazzal and McGinnis [55.32]	AGV	Estimation of performance measures	Vehicle utilization, blocking time, empty vehicle interarrival time	Queuing network model
Vis et al. [55.33]	AGV	Estimation of the number of vehicles	Service level (waiting time)	Network flow

55.3.2 Design Methodologies for MHA

MHA design studies largely deal with strategic decision-making, which includes optimal selection of automated material handling equipment, locating storage and vehicle paths for new facility planning, rack design for AS/RSs, flow path design for AGV sys-

tems, and capacity estimation of the system. Table 55.1 briefly summarizes studies related to design issues. The MHA design problem is sometimes closely related to the layout design problem in that both consider issues at a very early stage of system implementation. Also, they share performance measures in many areas. *Peters* and *Yang* [55.34] integrate these two methods into a single procedure using the space-filling curve (SFC) method. *Ting* and *Tanchoco* [55.27] propose a new layout design method for a semiconductor fab. They use an integer programming model to determine the optimal location of the AGV track. *Chung* and *Jang* [55.35] also suggest a new layout alternative called *integrated room layout* for better material handling in a semiconductor fab and scrutinize the benefits of the layout compared with existing layout alternatives in the industry using qualitative and quantitative analysis.

One of the difficulties in design of large-scale IAMHSs is estimation of system capacity. Although computer simulation has been used, its feedback cycle from modeling to results analysis is very slow for a large problem, which is an issue as timing of the solution is sometimes very important. Also, a simple deterministic analysis using from–to charts of material flows cannot provide a precise estimation of variances in the system. As an alternative approach studied for capacity analysis, the queuing network approach shows good performance [55.32]. *Rembold* and *Tanchoco* [55.36] explore a framework that evaluates and improves a sequence of modeling tasks for material flow systems. They aim to develop a more fundamental solution to the problems while encountered while designing an IAMHS. The framework addresses the following questions of designers: selection of the software application for solving a problem, organizing the data sets required for the design, incorporation of the design into parts that cannot be automated, and diagnosing problems in material flow systems. Those authors use an open architecture for the framework, since advance identification of all factors and cases for evaluation and redesign of the material flow processes are limited. With the open architecture for the framework, users can easily find their own methods by incorporating ad hoc situations into the framework.

55.3.3 Control Methodologies for MHA

Extensive research has been performed on the control of the AMHS. Especially, AGV control problems have benefited from strong research streams in academia and the MHA industry, since AGVs have been popular for use in many industries. Figure 55.9 shows an interesting AGV design with many storage racks that is used in a hospital. Recently, two well-organized literature surveys on the AGV system were published by *Vis* [55.37], and by *Le-Anh* and *De Koster* [55.38]. One of the characteristics of control algorithms of MHA is that minimizing flow distance in time is a dom-

Fig. 55.9 AGV used in a hospital (courtesy of Egemin)

inant criterion, among others. Other criteria such as resource utilization, throughput, and load balance have frequently been subgoals to achieve the minimum flow time. Necessity for a very short response time is another characteristic of control algorithms for the MHA; for example, a vehicle dispatch algorithm for the AGV controller should respond within a few seconds or less, otherwise the vehicle will have to wait for a job command on the path. For a short response time, the time horizon of the control algorithms is zero or very short, because a longer time horizon often causes an explosion of the search space. The minimum control horizon also helps to yield a reliable solution because uncertain parameters will be used less. If a control algorithm malfunctions, the result will be more serious than just a performance drop. It sometimes causes a detrimental failure in the shop floor. Hence, a conservative approach tends to be used in real applications.

A big challenge in AGV control problems is that users want to use a larger loop with many vehicles in order to reduce transportation time and investment. AGV systems implemented earlier generally used a modular structure to avoid heavy load on one AGV loop and had many loops, with a maximum of about five vehicles in a loop; however, these days, a large loop with a maximum about 40 vehicles is used. Therefore, the vehicle dispatch, scheduling, routing, and deadlock avoidance problems are becoming more complicated and important. Table 55.2 summarizes control issues and their studies in the MHA.

Table 55.2 Control issues and related studies on MHA

Researchers	AMHS type	Control issue	Criteria	Solution approach
Dotoli and *Fanti* [55.39]	IAMHS	Integrated AS/RS and RGV control	Throughput, computation time	Colored Petri nets
Mahajan et al. [55.40]	AS/RS	Job sequencing	Throughput	Heuristic (nearest neighborhood)
Lin and *Tsao* [55.8]	AS/RS	Crane scheduling for batch job in CIM environment	Total fulfillment time of batch	Heuristic (dynamic availability oriented controller)
Lee et al. [55.41]	AS/RS	Rack assignment for cargo terminals stochastic demand	Expected travel time	Heuristic (storage reservation policy), stochastic
Chetty and *Reddy* [55.42]	AS/RS	Job sequencing	10 criteria (mean flow time, mean waiting time, min/max completion time, etc.)	Genetic algorithm
Sinriech and *Palni* [55.43]	AGV	Vehicle scheduling	Optimality of scheduling solution	MIP, heuristic (branch and bound)
Correa et al. [55.44]	AGV	Vehicle scheduling	Solution time, job processing time	MIP and CP hybrid method
Jang et al. [55.45]	AGV	Vehicle routing in clean bay	AGV utilization, WIP level	Heuristic, look-ahead control procedure
Koo et al. [55.46]	AGV	Vehicle dispatching	Production throughput, lead time	Heuristic, bottleneck-machine first
Kim et al. [55.47]	AGV	Vehicle dispatching in floor shop	Production throughput	Heuristic (balanced work load)
Jeong and *Randhawa* [55.48]	AGV	Vehicle dispatching	Vehicle travel time, blocking time, WIP	Heuristics, multi-attribute dispatching
Moorthy et al. [55.49]	AGV	Deadlock avoidance in large-scale AGVS (cycle deadlock)	Number of AGVs in a loop, number deadlocks	Heuristic, state prediction
Bruno et al. [55.50]	AGV	Empty vehicle parking	Response time	Heuristic (location model (MIP) and shortest path algorithm)

IAMHS Research

Researchers recently started to study complicated issues of the IAMHS. A major concern is routing strategies from source to destination location in a complicated IAMHS, in which there are multiple routes from one location to the others. The routes consist of not only physical paths such as an AGV path or conveyor track but also AMHS themselves, such as AS/RSs, AGVSs, and buffer stations. Practical applications generally store predetermined static shortest routes in a database for all pairs of source locations and destinations; however, when the number of components increases, maintenance problems for the parameters involved become much more difficult since there are too many combinations of nodes. The shortest-distance algorithm using graph theory with an adjacency matrix might be a better approach.

A new concept called *flow diversion* is proposed to determine dynamic routing based on the load rate of the routes in automated shipment handling systems by *Cheung* et al. [55.51]. The authors utilize the multicommodity flow models using linear programming (LP) to solve this problem. In this model, the transfer time for a route is a function of the loads assigned to all pairs of unit loads in the system, which generates a nonlinear function in the objective function; those authors transform this nonlinear function to a piecewise-linear function to make the problem tractable. *Lau* and *Zhao* [55.4] study a joint job scheduling problem for the automated air cargo terminal at Hong Kong, which is mainly composed of AGV systems, AS/RSs, cargo hoists, and conveyors. In the model, activities between different AMHSs are triggered by communication between the systems. The scheduling algorithm constructs a cooperative sequential job served by different AMHSs, employing the maximum matching algorithm of the bipartite graph. A task for an AGV is assigned or matched to an stacker crane (SC) to reduce the SC delay time. A similar problem is solved by *Meersmans* and *Wagelmans* [55.5]. Their research focuses on the scheduling problem of the IAMHS in seaport terminals employing a local beam search algorithm. The nodes explored in the search algorithm are represented by a sequence of container IDs to be processed by different AMHSs, and the nodes in branches are cut based on the beam width determined by an evaluation function. Those authors prove that there exists an optimal sequence of tasks for one AMHS when the sequence is assigned to the other AMHS.

Sujono and *Lashkari* [55.52] study another integrating method allocating a part type to a processing machine and material handling (MH) equipment type simultaneously in a flexible manufacturing system (FMS). In that research, there are nine different types of the material handling systems in the experimental model. The method improves the algorithms proposed by *Paulo* et al. [55.53] and *Lashkari* et al. [55.54] and uses a 0/1 mixed integer programming model. Two objective functions are modeled: one minimizes operating costs related to machine operations, setup, and MH operations; the other maximizes the compatibility of the part types using MH equipment types. To measure compatibility, parameters are quantified from the subjective factors defined by *Ayres* [55.55]. Some of the constraints are: balance equations between parts and process plans, machines and process plan, processing machines, and MH equipment types. The other important constraint sets are capacity constraints: the total load of the allocated tasks for an MH equipment type cannot exceed its capacity, and a machine cannot be allocated more than its capacity. A test problem consisting of 1356 constraints and 3036 binary variables was solved in about 9.2 s by using LINGO in a Pentium 4 PC. Since this model considers many details of the practical factors in the FMS, and showed a successful calculation result, it can be used for many other practical applications.

In addition to the examples shown above, large-scale optimization problems such as the vehicle routing problem (VRP), vehicle scheduling problem (VSP), and integrated scheduling problem of IAMHS with consideration of processing machines have been modeled to increase MHA efficiency. However, to be used for actual applications and thereby achieve a higher-level MHA, shorter computation times are urgently required. In a complicated IAMHS, integrating software packages such as the MMS need sophisticated algorithms; however, it also needs high reliability in a dynamic environment. For most tasks, real-time decision-making that requires response times within a few seconds is a precondition for IAMHS algorithms.

MMS-related Issues

The MMS is a key component to integrate different AMHSs in an IAMHS. Destination allocation, routing algorithm, and prioritizing algorithm are essential roles of the MMS, among others. Graph theory is popularly used to represent components and relationships in the IAMHS. In Fig. 55.6, nodes represent the AMHSs and their subcomponents, such as load/unload ports. Edges represent the connection and distance between nodes. As mentioned above, this graph is stored in database tables using the adjacency and incidence matrices. The shortest-path algorithm is the most important and fundamental algorithm for an MMS since it is used for several purposes in the system such as destination assignment and best routing determination. Dijkstra's algorithm is popularly used [55.56]. The Bellman–Ford algorithm can be used if there are negative weights of the edges.

To determine the final destination of a unit load, the MMS has to evaluate various factors on the same scale. More specifically, to determine an AS/RS as the final destination among several alternatives, the shortest distance is generally the most important criterion; however, the full rates of the AS/RSs are sometimes also important to make the loads balanced between different AS/RSs. There are two applicable ways to standardize different scales of factors on the same scale. First, different weight values can be applied for each factor to find the best alternatives. Second, a priority and its threshold value can be given to each factor, and the most important alternative is selected if it is within the threshold, otherwise the next alternative will be considered.

Determining the best route from a source to destination location via several AMHSs is relatively simple when compared with the vehicle routing problem (VRP) or vehicle scheduling problem (VSP), because the graph generally has a smaller number of nodes than those of the general VRP or VSP. However, the problem can be complicated when the load level of systems has to be taken into account. A flow cost function determines the weight value of an edge based on the queue size and system processing time for one unit in an AMHS, i. e., the load level is measured by these factors. It converts the weight value to a distance value by using the speed factor of the system. In an actual problem, this task can be considerably more complicated.

Prioritizing for the unit load is sometimes very useful for vehicle-initiated dispatching rules [55.11] when the IAMHS becomes a bottleneck in an FMS for a certain period of time. The priority determined by the MMS can be used by an AGV controller to determine a job priority for a vehicle that has just become idle; for example, the first-come first-served rule picks up the job with the longest waiting time for all unit loads in the queue. If a priority is given to each job from 1 to 5, the priority unit can be treated as a certain time scale, e.g., 10 min, for each unit. Together with the actual waiting time, the controller can prioritize the unit loads; for instance, if a unit load waits for a vehicle assignment for 5 min and its priority given by the MMS is 3, then its final priority can be $10 \times 2 + 5$ min, which is equal to 25 min. The prioritizing methods used by the MMS generally address problems of how to avoid machine starvation. While various MMS prioritizing rules can be used based on the constraints of the shop floor, the importance of considering bottleneck machines to determine transfer priorities of unit loads is emphasized by *Koo* et al. [55.46] and *Li* et al. [55.15].

55.3.4 AI and OR Techniques for MHA

It is worthwhile to compare AI search and OR optimization techniques with respect to several different criteria of logical flexibility, computation time, and application areas in MHA. In general, AI search algorithms define problems with four instances: initial state, successor function, goal test, and path cost function [55.57]. The initial state is a state in which the given problem starts. The successor function receives a state as a parameter and returns a set of actions and successors. And the successors are new states reachable from the given new state. The definition of the state together with the successor function is very important to determine the overall search space of the given problem and information necessary for the solution. The goal test determines whether a given state satisfies all the conditions of the goal state. The path cost function calculates a numerical cost for each path explored by the successor function.

There are two types of AI search algorithms: uninformed and informed search. Uninformed search does not use prior information to explore a solution. Examples of uninformed search algorithms are the depth-first search, breadth-first search, and bidirectional search. Informed search utilizes given information for new states that will be opened during the search. Informed search is also called heuristic search, which includes greedy best-first search, A* search, memory-bounded, and local beam search, which again includes simulated annealing, tabu search, and genetic algorithm. Constraint programming (CP) is one of the AI search methods that uses a standard structured representation consisting of the *problem domain* and *constraint*. Figure 55.10 explains the main procedure used by the ILOG CP solver [55.58]. The domain is a set of possible values of the variable representing the problem, and the constraint is a rule that imposes a limitation on the variable. The most powerful aspect of this method is that it utilizes the concept of *constraint propagation*. It narrows down the search space by imposing a constraint on variables and the constraint imposed further reduces domains of other variables based on the constraints already posted on the variables. Among reduced domains of variables, the method uses a branching process with a backtracking algorithm to find the best solution. Because of high modeling flexibility, AI techniques have been popularly used for a wide variety of control applications such as robotics and automated planning.

The following studies illustrate the use of AI techniques for MHA. *Cho* and *Egbelu* [55.21] use

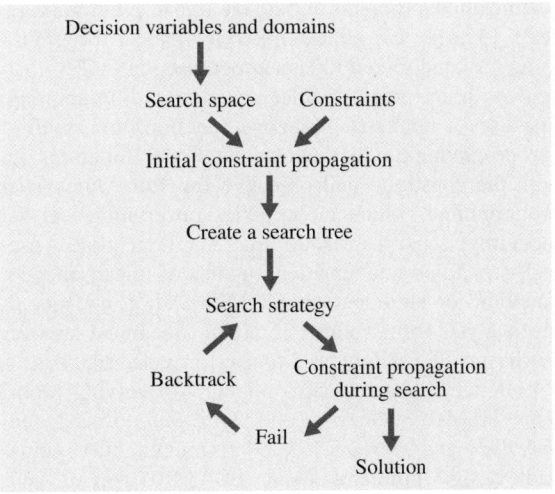

Fig. 55.10 Main solution procedure of CP (courtesy of ILOG)

Table 55.3 Comparison of AI and OR techniques

Comparison items	AI approaches	OR approaches	Hybrid approaches
Modeling flexibility	High	Low	High
Time horizon	Short	Long	Long
Response time	Short	Long	Medium
Problem size	Small	Large	Large
Illustrations	IAMHS design: AMHS equipment type selection [55.21] AS/RS: Job sequencing [55.40], Stacker scheduling [55.8], Rack design [55.26] AGV: Deadlock avoidance [55.49]	IAMHS design: Performance evaluation [55.25, 32], AGV: Guide path design [55.28], AMHS location [55.24, 27]	AGV vehicle scheduling [55.44, 60] IAMHS: Integrated scheduling of AMHS and FMS [55.52]

knowledge-based rules, fuzzy logic, and decision algorithms to address AMHS equipment type selection problems. Their procedures consist of three phases: material handling equipment selections for each material flow connections, redundancy and excess capacity check, and budget constraint consideration. *Chan* et al. [55.59] also solve a similar problem using an expert system. An order picking sequence problem in an AS/RS is addressed by *Mahajan* et al. [55.40] by using an AI technique. In their procedure, the state is represented by a sequence of the orders, and a success function providing a selection criterion of the order sequence is developed by a nearest-neighborhood strategy.

Operations research techniques mainly focus on the optimization problems based on linear programming (LP) [55.61]. LP is extended to integer programming (IP) and mixed integer programming (MIP), that deal with integral variables, quadratic programming that uses a nonlinear objective function, and nonlinear programming that allows nonlinear functions in both the constraint and objective function. Stochastic programming, which incorporates uncertainties in its modeling, is also a variant of the LP. OR techniques use well-structured mathematical models of linear, integer, quadratic or nonlinear models. Simulation and queuing analysis form another important technical area of stochastic OR, mainly used for performance analysis.

OR techniques are also popular for solving problems in MHA. *Gaskins* and *Tanchco* [55.28] and *Tanchoco* and *Sinreich* [55.29] formulate AGV guide path design problems using 0/1 MIP. *Nazzal* and *McGinnis* [55.32] estimate the system capacity requirement of the large-scale AMHS in a semiconductor

fab by utilizing a queuing network model. *Ting* and *Tanchco* [55.27] and *Huang* et al. [55.24] address the location problems of the AMHSs in facility layouts using MIP formulations. Huang et al. further use a heuristic approach employing the Lagrangian relaxation method.

AI and OR have their backgrounds in computer science and industrial engineering, respectively. AI approaches utilize knowledge representation to solve a problem; however, OR techniques use mathematical modeling of the problem. Knowledge representation consists of symbols and mathematical equations with relationships. OR techniques generally use dedicated solvers such as CPLEX and LINDO to solve mathematical models of the problems, whereas AI techniques use their own languages such as list processing (LISP), and programming in logics (Prolog). Constraint programming (CP) uses a solver similarly to the OR solvers but has much greater flexibility in using procedures and algorithms. A widely known CP solver is the ILOG solver. One advantage of AI over OR techniques is their flexibility in expressing problems. Since the AI techniques listed above do not use strict mathematical formulations to represent problems, there is a great deal of flexibility to deal with instances and activities in the problems. On the other hand, OR techniques model problems with strict mathematical procedures that are generally used repeatedly in different problems. While OR techniques find optimal or near-optimal solutions, AI techniques find good solutions for the given problems. OR techniques have focused on large-scale optimization problems for decision support systems while AI techniques are rooted in control problems that have shorter horizons but need reliable solutions. However, it is also true that there has been some overlap between AI and OR techniques, especially for local beam search algorithms. Also, a group of researchers has tried to take advantage of the two techniques by integrating procedures in the techniques [55.44]. For more complete reviews on the history and state of the art comparing AI and OR techniques, refer to *Gomes* [55.61], *Kobbacy* et al. [55.62], and *Marcus* [55.63].

Table 55.3 compares AI, OR, and their hybrid approaches with several different characteristics. An application area that needs great logical flexibility, such as the selection problem of material handling systems, tends to use AI techniques more frequently. Problems with shorter time horizon use AI heuristic search approaches (second row in the table), and prob-

lems with a longer time horizon tend to use the OR approaches. Hybrid approaches focus on reducing computation times. OR techniques are frequently used for AMHS design problems because response time is less important for them and they consider a large number of instances in the system. The AGV dispatching and routing problems tend to use both heuristic and OR approaches to similar degrees. Approaches integrating AI and OR approaches pursue both flexibility and optimality and have been applied to very complicated problems for the MHA [55.44], which deal with AGV scheduling problems and integrated scheduling of IAMHS with FMS. Examples on application areas and their approaches used are listed in the last row of Table 55.3.

55.4 Conclusions and Emerging Trends

Material handling automation (MHA) in production and warehouse management systems provides speedy and reliable infrastructure for information systems in SCM such as ERP system, FMS, WMS, and MES. Most of all, it enhances accurate data tracking during material handling in shop floors and warehouses. Relying on these accurate data, high-level automation such as production and procurement planning, scheduling, and dispatching in the SCM systems can be made much more reliable; consequently, more intelligent functions in their decision-making procedures can be added. Another trend in MHA, the integrated and automated material handling system (IAMHS), has been increasingly implemented in various applications to help ever-complicated material handling operations in large-scale production and warehouse systems. The important issues of the IAMHS reviewed in this chapter can be largely broken down into design and control issues. The design issues cover material handling equipment selection, capacity estimation, innovative equipment design,

and system design optimization. The control issues that have been hard constraints for the higher-level MHA tend to involve domain-specific problems for each component system in the IAMHS such as the AS/RS, AGVS, or MMS.

Several potential routes for further increasing the level of intelligence in the MHA are recognized in this chapter. While the number of components in the IAMHS continues to increase, long response time is regarded as a major limitation during implementations of new control algorithms. Continuous efforts to reduce the computation time of algorithms in the future are desired. It is also pointed out that newly developed algorithms should take into account system architecture for practical applications. As another possibility, a sensor network might be used for diagnosis of AMHSs since their reliability is becoming critical, and also its closed feedback mechanism can potentially be used for more precise controls, as seen in each context.

References

55.1 Material Handling Industry of America http://www.mhia.org/irl, (last accessed February 15, 2009)

55.2 J.R. Montoya-Torres: A literature survey on the design approaches and operational issues of automated wafer-transport systems for wafer fabs, Prod. Plan. Control **17**(7), 648–663 (2006)

55.3 T. Feare: GM runs in top gear with AS/RS sequencing, Mod. Mater. Handl. **53**(9), 50–52 (1998)

55.4 H.Y.K. Lau, Y. Zhao: Joint scheduling of material handling equipment in automated air cargo terminals, Comput. Ind. **57**(5), 398–411 (2006)

55.5 P.J.M. Meersmans, A.P.M. Wagelmans: *Dynamic Scheduling of Handling Equipment at Automated*

55.6 *Container Terminals, Econometric Institute Report EI 2001-33* (Erasmus University, Rotterdam 2001)

55.6 Murata Machinery, http://www.muratec-l-system.com/en/example/deliver/medical.html (last accessed February 15, 2009)

55.7 L. Kempfer: Produce delivered fresh and fast, Mater. Handl. Manag. **March**, 40–42 (2006)

55.8 C.W.R. Lin, Y.Z. Tsao: Dynamic availability-oriented control of the automated storage/retrieval system. A computer integrated manufacturing perspective, Int. J. Adv. Manuf. Technol. **29**(9-10), 948–961 (2006)

55.9 T.H. Chang, H.P. Fu, K.Y. Hu: The innovative conveying device application for transferring articles between two-levels of a multi-story building, Int. J. Adv. Manuf. Technol. **28**(1-2), 197–204 (2006)

55.10 B. Rembold, J.M.A. Tanchoco: Modular framework for the design of material flow systems, Int. J. Prod. Res. **32**(1), 1–21 (1994)

55.11 P.J. Egbelu, J.M.A. Tanchoco: Characterization of automated guided vehicle dispatching rules, Int. J. Prod. Res. **22**(3), 359–374 (1984)

55.12 L. Hossain, J.D. Patrick, M.A. Rashid: *Enterprise Resource Planning: Global Opportunities and Challenges* (Idea Group, Hershey 2002)

55.13 D.S. Sun, N.S. Park, Y.J. Lee, Y.C. Jang, C.S. Ahn, T.E. Lee: Integration of lot dispatching and AMHS control in a 300 mm wafer FAB, IEEE/SEMI Adv. Semiconduc. Manuf. Conf. Workshop – Adv. Semiconduct. Manuf. Excellence (2005) pp. 270–274

55.14 J. Jimenez, B. Kim, J. Fowler, G. Mackulak, Y.I. Choung, D.J. Kim: Operational modeling and simulation of an inter-bay AMHS in semiconductor wafer fabrication, Winter Simul. Conf. Proc. **2**, 1377–1382 (2002)

55.15 B. Li, J. Wu, W. Carriker, R. Giddings: Factory throughput improvements through intelligent integrated delivery in semiconductor fabrication facilities, IEEE Trans. Semiconduct. Manuf. **18**(1), 222–231 (2005)

55.16 S.S. Garfinkel, B. Rosenberg: *RFID Applications, Security, and Privacy* (Addison-Wesley, New York 2006)

55.17 K.C. Lee, S. Lee: Integrated network of Profibus-DP and IEEE 802.11 wireless LAN with hard real-time requirement, IEEE Int. Symp. Ind. Electron. **3**, 1484–1489 (2001)

55.18 A. Herrera: Wireless I/O devices in process control systems, Proc. ISA/IEEE Sensors Ind. Conf. (2004) pp. 146–147

55.19 S. Phoha, T. LaPorta, C. Griffin: *Sensor Network Operations* (Wiley, Piscataway 2006)

55.20 I. Walker, A. Hoover, Y. Liu: Handling unpredicted motion in industrial robot workcells using sensor networks, Ind. Robot. **33**(1), 56–59 (2006)

55.21 C. Cho, P.J. Egbelu: Design of a web-based integrated material handling system for manufacturing applications, Int. J. Prod. Res. **43**(2), 375–403 (2005)

55.22 G. Nadoli, M. Rangaswami: Integrated modeling methodology for material handling systems design, Winter Simul. Conf. Proc. (1993) pp. 785–789

55.23 J.A. Jimenez, G. Mackulak, J. Fowler: Efficient simulations for capacity analysis and automated material handling system design in semiconductor wafer fabs, Winter Simul. Conf. Proc. (2005) pp. 2157–2161

55.24 S. Huang, R. Batta, R. Nagi: Variable capacity sizing and selection of connections in a facility layout, IIE Trans. **35**(1), 49–59 (2003)

55.25 Y.J. Jang, G.H. Choi, S.I. Kim: Modeling and analysis of stocker system in semiconductor and LCD fab, IEEE Int. Symp. Semiconduct. Manuf. Conf. Proc. ISSM 2005 (2005) pp. 273–276

55.26 Y.H. Lee, M.H. Lee, S. Hur: Optimal design of rack structure with modular cell in AS/RS, Int. J. Prod. Econ. **98**(2), 172–178 (2005)

55.27 J.-H. Ting, J.M.A. Tanchoco: Optimal bidirectional spine layout for overhead material handling systems, IEEE Trans. Semiconduct. Manuf. **14**(1), 57–64 (2001)

55.28 R.J. Gaskins, J.M.A. Tanchoco: Flow path design for automated guided vehicle systems, Int. J. Prod. Res. **25**(5), 667–676 (1987)

55.29 J.M.A. Tanchoco, D. Sinriech: OSL – optimal single loop guide paths for AGVS, Int. J. Prod. Res. **30**(3), 665–681 (1992)

55.30 Y.A. Bozer, M.M. Srinivasan: Tandem AGV system: a partitioning algorithm and performance comparison with conventional AGV systems, Eur. J. Oper. Res. **63**, 173–191 (1992)

55.31 P. Caricato, A. Grieco: Using simulated annealing to design a material-handling system, IEEE Intell. Syst. **20**(4), 26–30 (2005)

55.32 D. Nazzal, L.F. McGinnis: Analytical approach to estimating AMHS performance in 300 mm fabs, Int. J. Prod. Res. **45**(3), 571–590 (2007)

55.33 I.F.A. Vis, R. de Koster, K.J. Roodbergen, L.W.P. Peeters: Determination of the number of automated guided vehicles required at a semi-automated container terminal, J. Oper. Res. Soc. **52**(4), 409–417 (2001)

55.34 B.A. Peters, T. Yang: Integrated facility layout and material handling system design in semiconductor fabrication facilities, IEEE Trans. Semiconduct. Manuf. **10**(3), 360–369 (1997)

55.35 J. Chung, J. Jang: The integrated room layout for semiconductor facility plan, IEEE Trans. Semiconduct. Manuf. **20**(4), 517–527 (2007)

55.36 B. Rembold, J.M.A. Tanchoco: Material flow system model evaluation and improvement, Int. J. Prod. Res. **32**(11), 2585–2602 (1994)

55.37 I.F.A. Vis: Survey of research in the design and control of automated guided vehicle systems, Eur. J. Oper. Res. **170**(3), 677–709 (2006)

55.38 T. Le-Anh, M.B.M. De Koster: A review of design and control of automated guided vehicle systems, Eur. J. Oper. Res. **171**(1), 1–23 (2006)

55.39 M. Dotoli, M.P. Fanti: A coloured Petri net model for automated storage and retrieval systems serviced by rail-guided vehicles: a control perspective, Int. J. Comput. Int. Manuf. **18**(2-3), 122–136 (2005)

55.40 S. Mahajan, B.V. Rao, B.A. Peters: A retrieval sequencing heuristics for miniload end-of-aisle automated storage/retrieval system, Int. J. Prod. Res. **36**(6), 1715–1731 (1998)

55.41 C. Lee, B. Liu, H.C. Huang, Z. Xu, P. Goldsman: Reservation storage policy for AS/RS at air cargo terminals, Winter Simul. Conf. Proc. (2005) pp. 1627–1632

55.42 O.V.K. Chetty, M.S. Reddy: Genetic algorithms for studies on AS/RS integrated with machines, Int. J. Adv. Manuf. Technol. **22**(11-12), 932–940 (2003)

55.43 D. Sinriech, L. Palni: Scheduling pickup and deliveries in a multiple-load discrete carrier environment, IIE Trans. Inst. Ind. Eng. **30**(11), 1035–1047 (1998)

55.44 A.I. Corréa, A. Langevin, L.M. Rousseau: Scheduling and routing of automated guided vehicles: a hybrid approach, Comput. Oper. Res. **34**(6), 1688–1707 (2007)

55.45 J. Jang, J. Suh, P.M. Ferreira: An AGV routing policy reflecting the current and future state of semiconductor and LCD production lines, Int. J. Prod. Res. **39**(17), 3901–3921 (2001)

55.46 P.H. Koo, J. Jang, J. Suh: Vehicle dispatching for highly loaded semiconductor production considering bottleneck machines first, Int. J. Flex. Manuf. Syst. **17**(1), 23–38 (2005)

55.47 C.W. Kim, J.M.A. Tanchoco, P.-H. Koo: AGV dispatching based on workload balancing, Int. J. Prod. Res. **37**(17), 4053–4066 (1999)

55.48 B.H. Jeong, S.U. Randhawa: A multi-attribute dispatching rule for automated guided vehicle systems, Int. J. Prod. Res. **39**(13), 2817–2832 (2001)

55.49 R.L. Moorthy, W. Hock-Guan, W.-C. Ng, T. Chung-Piaw: Cycle deadlock prediction and avoidance for zone controlled AGV system, Int. J. Prod. Econ. **83**, 309–324 (2003)

55.50 G. Bruno, G. Ghiani, G. Improta: Dynamic positioning of idle automated guided vehicles, J. Intell. Manuf. **11**(2), 209–215 (2000)

55.51 R. Cheung, A. Lee, D. Mo: Flow diversion approaches for shipment routing in automatic shipment handling systems, Proc. – IEEE Int. Conf. Robot. Autom. (2006) pp. 695–700

55.52 S. Sujono, R.S. Lashkari: A multi-objective model of operation allocation and material handling system selection in FMS design, Int. J. Prod. Econ. **105**(1), 116–133 (2007)

55.53 J. Paulo, R.S. Lashkari, S.P. Dutta: Operation allocation and materials-handling system selection in a flexible manufacturing system: a sequential modeling approach, Int. J. Prod. Res. **40**, 7–35 (2002)

55.54 R.S. Lashkari, R. Boparai, J. Paulo: Towards an integrated model of operation allocation and materials handling selection in cellular manufacturing system, Int. J. Prod. Econ. **87**(2), 115–139 (2004)

55.55 R.U. Ayres: Complexity, reliability and design: manufacturing implications, Manuf. Rev. **1**(1), 26–35 (1988)

55.56 R.K. Ahuja, T.L. Magnanti, J.B. Orlin: *Network flows: theory, algorithms, and applications* (Prentice Hall, Upper Saddle River 1993)

55.57 S. Russell, P. Norvig: *Artificial Intelligence: a Modern Approach* (Prentice Hall, New York 2003)

55.58 ILOG Solver 5.3 user manual

55.59 F.T.S. Chan, R.W.L. Ip, H. Lau: Integration of expert system with analytic hierarchy process for the design of material handling equipment selection system, J. Mater. Process. Technol. **116**(2-3), 137–145 (2001)

55.60 D. Naso, B. Turchiano: Multicriteria meta-heuristics for AGV dispatching control based on computational intelligence, IEEE Trans. Syst. Man Cybern. B **35**(2), 208–226 (2005)

55.61 C.P. Gomes: Artificial intelligence and operations research: challenges and opportunities in planning and scheduling, Knowl. Eng. Rev. **15**(1), 1–10 (2000)

55.62 K.A.H. Kobbacy, S. Vadera, M.H. Rasmy: AI and or in management of operations: history and trends, J. Oper. Res. Soc. **58**(1), 10–28 (2007)

55.63 R. Marcus: Application of artificial intelligence to operations research, Commun. ACM **27**(10), 1044–1047 (1984)

56. Industrial Communication Protocols

Carlos E. Pereira, Peter Neumann

This chapter discusses a very relevant aspect in modern automation systems: the presence of industrial communication networks and their protocols. The introduction of Fieldbus systems has been associated with a change of paradigm to deploy distributed industrial automation systems, emphasizing device autonomy and decentralized decision making and control loops. The chapter presents the main wired and wireless industrial protocols used in industrial automation, manufacturing, and process control applications. In order to help readers to better understand the differences between industrial communication protocols and protocols used in general computer networking, the chapter also discusses the specific requirements of industrial applications. As the trend of future automation systems is to incorporate complex heterogeneous networks, consisting of (partially homogeneous) local and wide area as well as wired and wireless communication systems, the concept of virtual automation networks is presented

56.1 Basic Information

56.1.1 History

Digital communication is now well established in distributed computer control systems both in discrete manufacturing as well as in the process control industries. Proprietary communication systems within SCADA (supervisory control and data acquisition) systems have been supplemented and partially displaced by Fieldbus and sensor bus systems. The introduction of Fieldbus systems has been associated with a change of paradigm to deploy distributed industrial automation systems, emphasizing device autonomy and decentralized decision making and control loops. Nowadays, (wired) Fieldbus systems are standardized and are the most important communication systems used in commercial control installations. At the same time, Ethernet won the battle as the most commonly used communication technology within the office domain, resulting in low component prices caused by mass production. This has led to an increasing interest in adapting Ethernet for industrial applications and several approaches have been proposed (Sect. 56.1.4). Ethernet-based solutions are dominating as a merging technology.

In parallel to advances on Ethernet-based industrial protocols, the use of wireless technologies in the industrial domain has also been increasingly researched. Following the trend to merge automation and office networks, heterogeneous networks (virtual automation networks (VAN)), consisting of local and wide area networks, as well as wired and wireless communication systems, are becoming important [56.1].

56.1.2 Classification

Industrial communication systems can be classified as follows regarding different capabilities:

- *Real-time behavior*: Within the automation domain, real-time requirements are of uttermost importance and are focused on the response time behavior of data packets. Three real-time classes can be identified based on the required temporal behavior:
  - Class 1: soft real-time. Scalable cycle time, used in factory floor and process automation in cases where no severe problems occur when deadlines are not met.
  - Class 2: hard real-time. Typical cycle times from 1 to 10 ms, used for time-critical closed loop control.
  - Class 3: isochronous real-time, cycle times from 250 μs to 1 ms, with tight restrictions on jitter (usually less than 1 μs), used for motion control applications.

  Additionally, there is a class *non real-time*, which means systems without real-time requirements; these are not considered here. It means (regarding industrial automation) exchange of engineering data maintenance, etc.
- *Distribution*: The most important achievement of industrial communication systems are local area communication systems, consisting of sensor/actuator networks (Chap. 20 and Sect. 56.3.2), Fieldbus systems, and Ethernet-based local area networks (LAN). Of increasing importance is the use of wide area networks (WAN) (telecommunication networks, Internet, etc.). Thus, it should be advantageous to consider WANs as part of an industrial communication system (Sect. 56.2), mostly within the upper layers of an enterprise hierarchy.
- *Homogeneity*: There are homogeneous parts (e.g. standardized Fieldbus systems) within an industrial communication system. But in real applications the use of heterogeneous networks is more common, especially when using WANs and when connected with services of network providers.
- *Installations types*: While most of the installed enterprise networks are currently wired, the number of wireless installations is increasing and this trend will continue.

56.1.3 Requirements in Industrial Automation Networks

The main requirements are:

- *Real-time behavior*: Diagnosis, maintenance, commissioning, and slow mobile applications are examples of non real-time applications. Process automation and data acquisition usually present soft real-time requirements. Examples of hard real-time applications are closed-loop control applications, such as in fast mobile applications and machine tools. Motion control is an example of an isochronous hard real-time application.
- *Functional safety*: Protection against hazards caused by incorrect functioning including communication via heterogeneous networks. There are several safety integrity levels (SIL) [56.2]. It includes the influence of noisy environments and the degree of reliability.
- *Security*: This means a common security concept for distributed automation using a heterogeneous network with different security integrity levels (not existent yet).
- *Location awareness*: The desired context awareness leads to the usage of location-based communication services and context-sensitive applications.

56.1.4 Chapter Overview

The remainder of the chapter is structured as follows. Section 56.2 discusses the concept of virtual automation networks (VANs), a key concept in future distributed automation systems, which will be composed of (partially homogeneous) local and wide area as well as wired and wireless communication systems leading to complex heterogenous communication networks.

In Sect. 56.3 the main wired industrial communication protocols are presented and compared, while Sect. 56.4 discusses wireless industrial communication systems. Section 56.5 deals with the use of wide area communications to execute remote automation operations.

56.2 Virtual Automation Networks

56.2.1 Definition, Characterization, Architectures

Future scenarios of distributed automation lead to desired mechanisms for geographically distributed automation functions for various reasons:

- *Centralized* supervision and control of (many) *decentralized* (small) technological plants
- Remote control, commissioning, parameterization, and maintenance of distributed automation systems
- Inclusion of remote experts or external machine-readable knowledge for plant operation and maintenance (for example, asset management, condition monitoring, etc.).

This means that heterogeneous networks, consisting of (partially homogeneous) local and wide areas, as well as wired and wireless communication systems, will play an increasing role. Figure 56.1 depicts the communication environment of a complex automation scenario. Following a unique design concept, regarding the objects to be transmitted between geographically distributed communication end points, the heterogeneous network becomes a *virtual automation network* (VAN) [56.3, 4]. VAN characteristics are defined for domains, where the expression *domain* is widely used to address areas and devices with common properties/behavior, common network technology, or common application purposes.

56.2.2 Domains

Within the overall automation and communication environment, a VAN domain covers all devices that are grouped together on a *logical or virtual* basis to represent a complex application such as an industrial application. Therefore, the encompassed networks may be heterogeneous and devices can be geographically

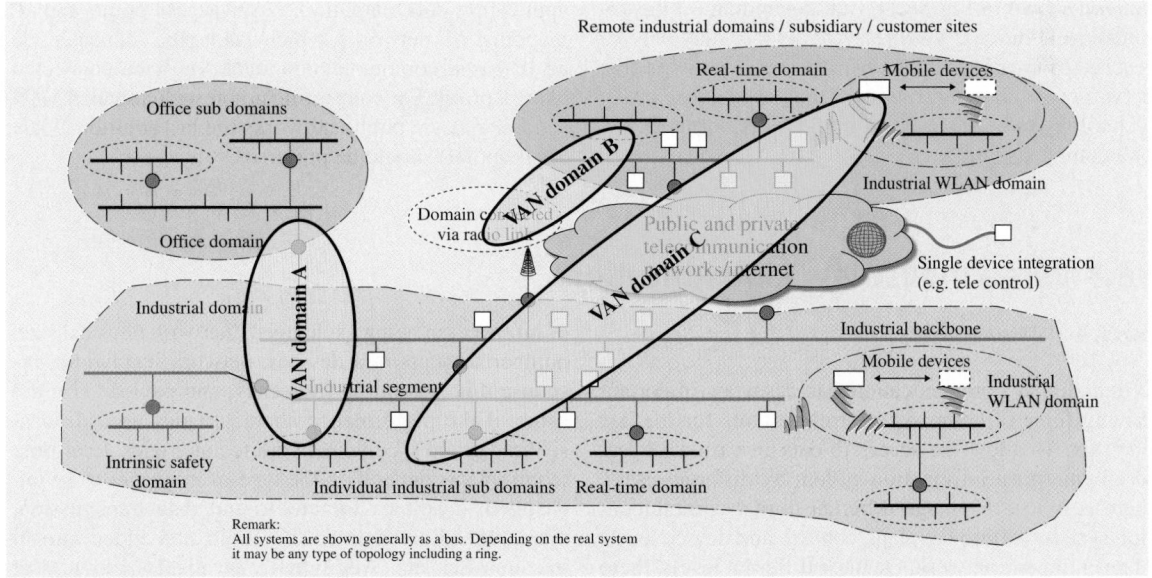

Fig. 56.1 Different VAN domains related to different automation applications [56.3]

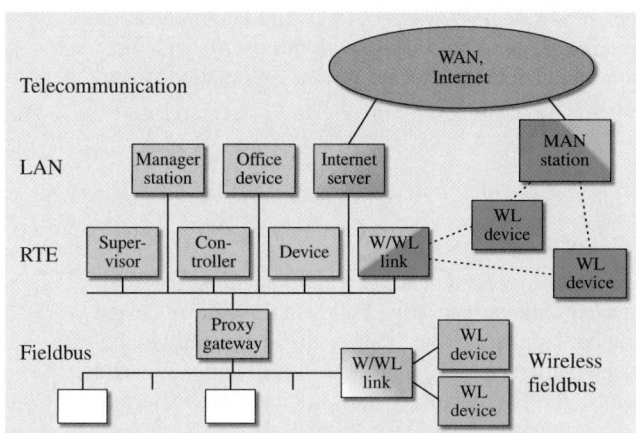

Fig. 56.2 Network transitions (local area networks (LAN), wireless LAN (WL), wired/wireless (W/WL), real-time Ethernet (RTE), metropolitan area network (MAN), wide area network (WAN))

distributed over a physical environment, which shall be covered by the overall application. But all devices that have to exchange information within the scope of the application (equal to a VAN domain) must be VAN aware or *VAN enabled* devices. Otherwise, they are VAN independent and are not a member of a VAN domain. Figure 56.1 depicts VAN domain examples representing three different distributed applications.

Devices related to a VAN domain may reside in a homogeneous network domain (e.g. the *industrial domain* shown in Fig. 56.1). But, depending on the application, additional VAN relevant devices may only be reached by crossing other network types (e.g., wide area network type communication) or they need to use proxy technology to be represented in the VAN domain view of a complex application.

56.2.3 Interfaces, Network Transitions, Transmission Technologies

A VAN network consists of several different communication paths and network transitions. Figure 56.2 depicts the required transitions in heterogeneous networks.

Depending on the network and communication technology of the single path there will be differences in the addressing concept of the connected network segments. Also the communication paths have different communication line properties and capabilities. Therefore, for the path of two connected devices within a VAN domain the following views are possible:

- The logical view: describing the properties/capabilities of the whole communication path
- The physical view: describing the detailed properties/capabilities of the passed technology-dependent communication paths
- The behavioral view: describing the different cyclic/acyclic temporal behavior of the passed segments.

There are different opportunities to achieve a communication path between network segments/devices (or their combinations). These are: Ethernet line (with/without transparent communication devices), wireless path, telecommunication network (1:1), public networks (n:m, provider-oriented), VPN tunnel, gateway (without application data mapping), proxy (with application data mapping), VAN access point, and IP mapping. All networks, which can not be connected via an IP-based communication stack, must be connected using a proxy. For connecting nonnested/cascaded VAN subdomains via public networks the last solution (VAN access point) should be preferred.

56.3 Wired Industrial Communications

56.3.1 Introduction

Wired digital communication has been an important driving force of computer control systems for the last 30 years. To allow the access to data in various layers of an enterprise information system by different users, there is a need to merge different digital communication systems within the plant, control, and device levels of an enterprise network. On these different levels, there are distinct requirements dictated by the nature and type of information being exchanged. Network physical size, number of supported devices, network bandwidth, response time, sampling frequency, and payload size are some of the performance characteristics used to classify and group specific network technologies. Real-time requirements depend on the type of messages to be exchanged: deadlines for end-to-end data transmission, maximum allowed jitter for audio and video stream transmission, etc. Additionally, available resources at the various network levels may vary significantly. At the device level, there are extremely limited resources (hardware, communications), but at the plant level powerful computers allow comfortable software and memory consumption.

Due to the different requirements described above, there are different types of industrial communication systems as part of a hierarchical automation system within an enterprise:

- Sensor/actuator networks: at the field (sensor/actuator) level
- Fieldbus systems: at the field level, collecting/distributing process data from/to sensors/actuators, communication medium between field devices and controllers/PLCs/management consoles
- Controller networks: at the controller level, transmitting data between powerful field devices and controllers as well as between controllers
- Wide area networks: at the enterprise level, connecting networked segments of an enterprise automation system.

Vendors of industrial communication systems offer a set of fitting solutions for these levels of the automation/communication hierarchy.

56.3.2 Sensor/Actuator Networks

At this level, several well established and widely adopted protocols are available:

- HART (HART Communication Foundation): highway addressable remote transducer, coupling analog process devices with engineering tools [56.5]
- ASi (ASi Club): actuator sensor interface, coupling binary sensors in factory automation with control devices [56.6].

Additionally, CAN-based solutions (CAN in automation (CIA)) are used for wide-spread application fields, coupling decentralized devices with centralized devices based on physical and MAC layers of the controller area network [56.7]. Recently, IO Link has been specified for bi-directional digital transmission of parameters between simple sensor/actuator devices in factory automation [56.8,9].

HART

HART Communication [56.5] is a protocol specification, which performs a bi-directional digital transmission of parameters (used for configuration and parameterization of intelligent field instruments by a host system) over analog transmission lines. The host

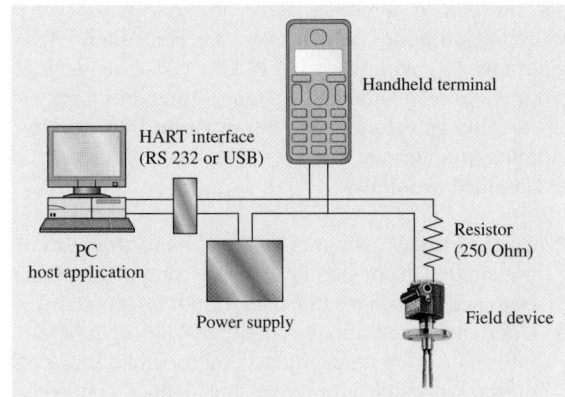

Fig. 56.3 A HART system with two masters (http://www.hartcomm.org)

system may be a distributed control system (DCS), a programmable logic controller (PLC), an asset management system, a safety system, or a handheld device. HART technology is easy to use and very reliable. The HART protocol uses the Bell 202 Frequency Shift Keying (FSK) standard to superimpose digital communication signals at a low level on top of the 4–20 mA analog signal. The HART protocol communicates at 1200 bps without interrupting the 4–20 mA signal and allows a host application (master) to get two or more digital updates per second from a field device. As the digital FSK signal is phase continuous, there is no interference with the 4–20 mA signal. The HART protocol permits all digital communication with field devices in either point-to-point or multidrop network configurations. HART provides for up to two masters (primary and secondary). As depicted in Fig. 56.3, this allows secondary masters (such as handheld communicators) to be used without interfering with communications to/from the primary master (i. e. control/monitoring system).

ASi (IEC 62026-2)
ASi [56.6] is a network of actuators and sensors (optical, inductive, capacitive) with binary input/output signals. An unshielded twisted pair cable for data and power (max. 2 A; max. 100 m) enables the connection of 31 slaves (max. 124 binary signals of sensors and/or actuators). This enables a modular design using any network topology (i. e. bus, star, tree). Each slave can receive any available address and be connected to the cable at any location.

AS-Interface uses the APM method (alternating pulse modulation) for data transfer. The medium access is controlled by a master–slave principle with cyclic polling of all nodes. ASi masters are embedded (ASi) communication controllers of PLCs or PCs, as well as gateways to other Fieldbus systems. To connect legacy sensors and actuators to the transmission line, various coupling modules are used. AS-Interface messages can be classified as follows:

- Single transactions: maximum of 4 bit information transmitted from master to slave (output information) and from slave to master (input information)
- Combined transactions: more than 4 bits of coherent information are transmitted, composed of a series of master calls and slave replies in a defined context.

For more details see www.as-interface.com.

56.3.3 Fieldbus Systems

Nowadays, Fieldbus systems are standardized (though unfortunately not unified) and widely used in industrial automation. The IEC 61158 and 61784 standards [56.11, 12] contain ten different Fieldbus concepts. Seven of these concepts have their own complete protocol suite: PROFIBUS (Siemens, PROFIBUS International); Interbus (Phoenix Contact, Interbus Club); Foundation Fieldbus H1 (Emerson, Fieldbus Foundation); SwiftNet (B. Crowder); P-Net (Process Data); and WorldFIP (Schneider, WorldFIP). Three of them are based on Ethernet functionality: high speed Ethernet (HSE) (Emerson, Fieldbus Foundation); Ethernet/IP (Rockwell, ODVA); PROFINET/CBA: (Siemens, PROFIBUS International). The world-wide

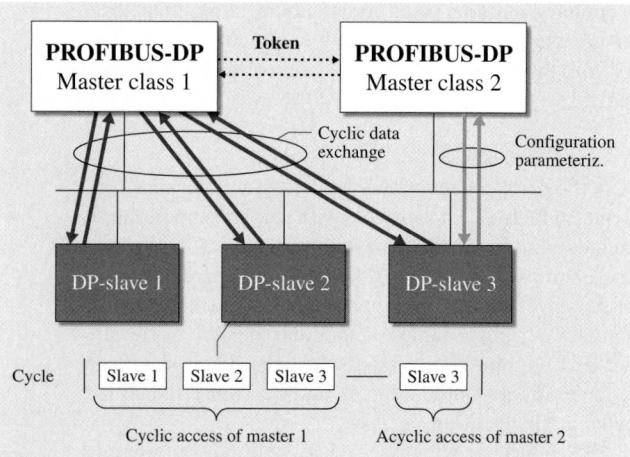

Fig. 56.4 Profibus medium access control (from [56.10])

leading positions within the automation domain regarding the number of installed Fieldbus nodes hold PROFIBUS and Interbus followed by DeviceNet (Rockwell, ODVA), which has not been part of the IEC 61158 standard. For that reason, the basic concepts of PROFIBUS and DeviceNet will be explained very briefly. Readers interested in a more comprehensive description are referred to the related web sites.

PROFIBUS

PROFIBUS is a universal fieldbus for plantwide use across all sectors of the manufacturing and process industries based on the IEC 61158 and IEC 61784 standards. Different transmission technologies are supported [56.10]:

- RS 485: Type of medium attachment unit (MAU) corresponding to [56.13]. Suited mainly for factory automation. Technical details see [56.10, 13]. Number of stations: 32 (master stations, slave stations or repeaters); Data rates: 9.6/19.2/45.45/93.75/187.5/500/1500/3000/6000/12 000 kb/s.
- Manchester bus powered (MBP). Type of MAU suited for process automation: line, tree, and star topology with two wire transmission; 31.25 kBd (preferred), high speed variants w/o bus powering and intrinsic safety; synchronous transmission (Manchester encoding); optional: bus powered devices (≥ 10 mA per device; low power option); optional: intrinsic safety (Ex-i) via additional constraints according to the FISCO model. Intrinsic safety means a type of protection in which a portion of the electrical system contains only intrinsically safe equipment (apparatus, circuits, and wiring) that is incapable of causing ignition in the surrounding atmosphere. No single device or wiring is intrinsically safe by itself (except for battery-operated self-contained apparatus such as portable pagers, transceivers, gas detectors, etc., which are specifically designed as intrinsically safe self-contained devices) but is intrinsically safe only when employed in properly designed intrinsically safe system. There are couplers/link devices to couple MBP and RS485 transmission technologies.
- Fibre optics (not explained here, see [56.10]).

There are two medium access control (MAC) mechanisms (Fig. 56.4):

1. Master–master traffic using token passing
2. Master–slave traffic using polling.

PROFIBUS differentiates between two types of masters:

1. Master class 1, which is basically a central controller that cyclically exchanges information with the distributed stations (slaves) at a specified message cycle.
2. Master class 2, which are engineering, configuration, or operating devices. The slave-to-slave communication is based on the application model *publisher/subscriber* using the same MAC mechanisms.

The dominating PROFIBUS protocol is the application protocol DP (*decentralized periphery*), embedded into the protocol suite (Fig. 56.5).

Depending upon the functionality of the masters, there are different volumes of DP specifications. There are various profiles, which are grouped as follows:

1. Common application profiles (regarding functional safety, synchronization, redundancy etc.)
2. Application field specific profiles (e.g. process automation, semiconductor industries, motion control).

These profiles reflect the broad experience of the PROFIBUS International organization.

DeviceNet

DeviceNet is a digital, multi-drop network that connects and serves as a communication network between industrial controllers and I/O devices. Each device and/or controller is a node on the network. DeviceNet uses a trunk-line/drop-line topology that provides separate twisted pair busses for both signal and power distribution. The possible variants of this topology are shown in [56.14].

Thick or thin cables can be used for either trunklines or droplines. The maximum end-to-end network length varies with data rate and cable thickness. DeviceNet allows transmission of the necessary power on the network. This allows devices with limited power requirements to be powered directly from the network, reducing connection points and physical size. DeviceNet systems can be configured to operate in a master-slave or a distributed control architecture using peer-to-peer communication. At the application layer, DeviceNet uses a producer/consumer application model. DeviceNet systems offer a single point of connection for configuration and control by supporting both I/O and explicit messaging.

DeviceNet uses CAN (controller area network [56.7]) for its data link layer, and CIP (common indus-

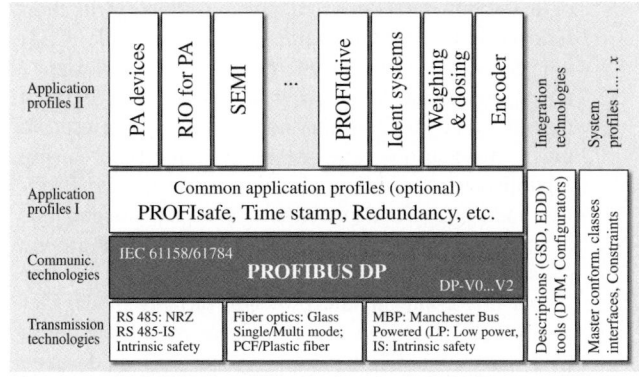

Fig. 56.5 PROFIBUS protocol suite (from [56.10])

trial protocol) for the upper network layers. As with all CIP networks, DeviceNet implements CIP at the session (i. e. data management services) layer and above, and adapts CIP to the specific DeviceNet technology at the network and transport layer, and below. Figure 56.6 depicts the DeviceNet protocol suite.

The data link layer is defined by the CAN specification and by the implementation of CAN controller chips. The CAN specification [56.7] defines two bus states called dominant (logic 0) and recessive (logic 1). Any transmitter can drive the bus to a dominant state. The bus can only be in the recessive state when no transmitter is in the dominant state. A connection with a device must first be established in order to exchange information with that device. To establish a connection, each DeviceNet node will implement either an unconnected message manager (UCMM)

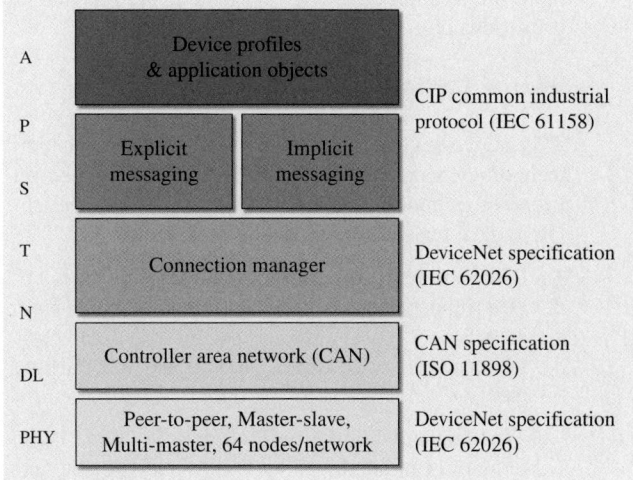

Fig. 56.6 DeviceNet protocol suite (from [56.14])

or a Group 2 unconnected port. Both perform their function by reserving some of the available CAN identifiers. When either the UCMM or the Group 2 unconnected port is selected to establish an explicit messaging connection, that connection is then used to move information from one node to the other (using a publisher/subscriber application model), or to establish additional I/O connections. Once I/O connections have been established, I/O data may be moved among devices on the network.

At this point, all the protocol variants of the DeviceNet I/O message are contained within the 11 bit CAN identifier. CIP is strictly object oriented. Each object has attributes (data), services (commands), and behavior (reaction to events). Two different types of objects are defined in the CIP specification: communication objects and application-specific objects. Vendor-specific objects can also be defined by vendors for situations where a product requires functionality that is not in the specification. For a given device type, a minimum set of common objects will be implemented.

An important advantage of using CIP is that for other CIP-based networks the application data remains the same regardless of which network hosts the device. The application programmer does not even need to know to which network a device is connected.

CIP also defines device profiles, which identifies the minimum set of objects, configuration options, and the I/O data formats for different types of devices. Devices that follow one of the standard profiles will have the same I/O data and configuration options, will respond to all the same commands, and will have the same behavior as other devices that follow that same profile. For more information on DeviceNet readers are referred to www.odva.org.

56.3.4 Controller Networks

This network class requires powerful communication technology. Considering controller networks based on Ethernet technology, one can distinguish between (related to the real-time classes, see Sect. 56.1):

1. Local soft real-time approaches (real-time class 1)
2. Deterministic real-time approaches (real-time class 2)
3. Isochronous real-time approaches (real-time class 3).

The standardization process started in 2004. There were many candidates to become part of the extended Fieldbus standard IEC 61158 (edition 4): high

speed Ethernet HSE (Emerson, Fieldbus Foundation); Ethernet/IP (Rockwell, ODVA); and PROFINET/CBA (Siemens, PROFIBUS International). Nine Ethernet-based solutions have been added. In this section a short survey of the previously mentioned real-time classes will be given, and two practical examples will be examined.

Local Soft Real-Time Approaches (Real-Time Class 1)

These approaches use TCP (UDP)/IP mechanisms over shared and/or switched Ethernet networks. They can be distinguished by different functionalities on top of TCP (UDP)/IP, as well as by their object models and application process mechanisms. Protocols based on Ethernet-TCP/IP offer response times in the lower millisecond range but are not deterministic, since data transmission is based on the best effort principle. Some examples are given below.

MODBUS TCP/IP (Schneider) [56.15]. MODBUS is an application layer messaging protocol for client/server communication between devices connected via different types of buses or networks. Using Ethernet as the transmission technology, the application layer protocol data unit (A-PDU) of MODBUS (function code and data) is encapsulated into an Ethernet frame. The connection management on top of TCP/IP controls the access to TCP.

Ethernet/IP (Rockwell, ControlNet International, Open DeviceNet Vendor Association) uses a common industrial protocol CIP [56.16]. In this context, IP stands for industrial protocol (not for Internet protocol). CIP represents a common application layer for all physical networks of Ethernet/IP, ControlNet and DeviceNet. Data packets are transmitted via a CIP router between the networks. For the real-time I/O data transfer, CIP works on top of UDP/IP. For the explicit messaging, CIP works on top of TCP/IP. The application process is based on a producer/consumer model.

High Speed Ethernet HSE (Fieldbus Foundation) [56.17]. A field device agent represents a specific Fieldbus Foundation application layer function (including Fieldbus message specification). Additionally, there are HSE communication profiles to support the different device categories: host device, linking device, I/O gateway, and field device. These devices share the tasks of the system using distributed function block applications.

PROFINET (PNO PROFIBUS User Organization, Siemens) [56.19]. uses object model CBA (component based architecture) and DCOM wire protocol with the remote procedure call mechanisms (DCE RPC) (OSF C 706) to transmit the soft real-time data. An open source code and various exemplary implementations/portations for different operating systems are available on the PNO web site.

P-Net on IP (Process Data) [56.20]. Based on P-Net Fieldbus standard IEC 61158 Type 4 [56.11], P-Net on IP contains the mechanism to use P-Net in an IP environment. Therefore, P-Net PDUs are wrapped into UDP/IP packages, which can be routed through IP networks. Nodes on the IP network are addressed with two P-Net route elements. P-Net clients (master) can access servers on an IP network without knowing anything about IP addresses.

All of the above mentioned approaches are able to support widely used office domain protocols, such as SMTP, SNMP, and HTTP. Some of the approaches support BOOTP and DHCP for web access and/or for engineering data exchange. But the object models of the approaches differ.

Deterministic Real-Time Approaches (Real-Time Class 2)

These approaches use a middleware on top of the MAC layer to implement scheduling and smoothing functions. The middleware is normally represented by a software implementation. Industrial examples include the following.

PROFINET (PROFIBUS International, Siemens) [56.19]. This variant of the Ethernet-based PROFINET IO system (using the main application model background of the Fieldbus PROFIBUS DP) uses the object model IO (input/output). Figure 56.7 roughly depicts the PROFINET protocol suite, containing the connection establishment for PROFINET/CBA via connection-oriented RPC on the left side, as well as for PROFINET IO via connectionless RPC on the right side. The exchange of (mostly cyclic) productive data uses the real-time functions in the center.

The PROFINET IO service definition and protocol specification [56.21] covers the communication between programmable logical controllers (PLCs), supervisory systems, and field devices or remote input and output devices. The PROFINET IO specification complies with IEC 61158, Parts 5 and 6, specially the Fieldbus application layer (FAL). The PROFINET pro-

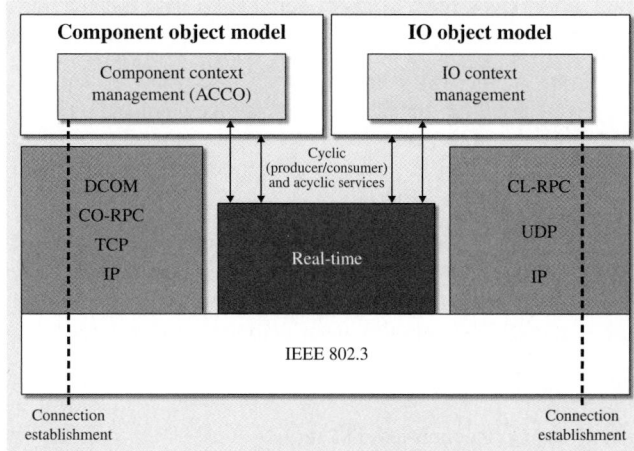

Fig. 56.7 PROFINET protocol suite of PROFINET (from [56.18]) (active control connection object (ACCO), connection-oriented (CO), connectionless (CL), remote procedure call (RPC))

tocol is defined by a set of protocol machines. For more details see [56.22].

Time-Critical Control Network (Tcnet, Toshiba) [56.23]. Tcnet specifies in the application layer a so-called *common memory* for time-critical applications, and uses the same mechanisms as mentioned for PROFINET IO for TCP(UDP)/IP-based non real-time applications. An extended data link layer contains the scheduling functionality. The common memory is a virtual memory globally shared by participating nodes as well as application processes running on each node. It provides a temporal and spatial coherence of data distribution. The common memory is divided into blocks with several memory lengths. Each block is transmitted to member nodes using multicast services, supported by a publisher node. A cyclic broadcast transmission mechanism is responsible for refreshing the data blocks. Therefore, the common memory consists of dedicated areas for the transmitting data to be refreshed in each node. Thus, the application program of a node has quick access to all (distributed) data.

The application layer protocol (FAL) consists of three protocol machines: the FAL service protocol machine (FSPM), the application relationship protocol machine (ARPM), and the data link mapping protocol machine (DMPM). The scheduling mechanism in the data link layer follows a token passing mechanism.

Vnet (Yokogawa) [56.24]. Vnet supports up to 254 subnetworks with up to 254 nodes each. In its application layer, three kinds of application data transfers are supported:

- A one-way communication path used by an end-point for inputs or outputs (conveyance paths)
- A trigger policy
- Data transfer using a buffer model or a queue model (conveyance policy).

The application layer FAL contains three types of protocol machines: the FSPM FAL service protocol machine, ARPMs application relationship protocol machines, and the DMPM data link layer mapping protocol machine. For real-time data transfer, the data link layer offers three services:

1. Connection-less DL service
2. DL-SAP management service
3. DL management service.

Real-time and non real-time traffic scheduling is located on top of the MAC layer. Therefore, one or more time-slots can be used within a macro-cycle (depending on the service subtype). The data can be ordered by four priorities: urgent, high, normal, time-available. Each node has its own synchronized macro-cycle. The data link layer is responsible for clock synchronization.

Isochronous Real-Time Approaches (Real-Time Class 3)

The main examples are as follows.

Powerlink (Ethernet PowerLink Standardization Group (EPSG), Bernecker and Rainer), developed for motion control [56.25]. Powerlink offers two modes: protected mode and open mode. The protected mode uses a proprietary (B&R) real-time protocol on top of the shared Ethernet for protected subnetworks. These subnetworks can be connected to an open standard network via a router. Within the protected subnetwork the nodes cyclically exchange real-time data avoiding collisions. The scheduling mechanism is a time-division scheme. Every node uses its own time slot [slot communication network management (SCNM)] to send its data. The mechanism uses a manager node, which acts comparably with a bus master, and managed nodes act similar to a slave. This mechanism avoids Ethernet collisions. The Powerlink protocol transfers the real-time data isochronously. The open mode can be used for TCP(UDP)/IP based applications. The network normally uses switches. The traffic has to be transmitted within an asynchronous period of the cycle.

EtherCAT [EtherCAT Technology Group (ETG), Beckhoff] developed as a fast backplane communication system [56.26]. EtherCAT distinguishes two modes: direct mode and open mode. Using the *direct mode*, a master device uses a standard Ethernet port between the Ethernet master and an EtherCAT segment. EtherCAT uses a ring topology within the segment. The medium access control adopts the master/slave principle, where the master node (typically the control system) sends the Ethernet frame to the slave nodes (Ethernet device). One single Ethernet device is the head node of an EtherCAT segment consisting of a large number of EtherCAT slaves with their own transmission technology. The Ethernet MAC address of the first node of a segment is used for addressing the EtherCAT segment. For the segment, special hardware can be used. The Ethernet frame passes each node. Each node identifies its subframe and receives/sends the suitable information using that subframe. Within the EtherCAT segment, the EtherCAT slave devices extract data from and insert data into these frames. Using the *open mode*, one or several EtherCAT segments can be connected via switches with one or more master devices and Ethernet-based *basic slave* devices.

PROFINET IO/Isochronous Technology (PROFIBUS User Organization, Siemens) developed for any industrial application [56.27]. PROFINET IO/Isochronous Technology uses a middleware on top of the Ethernet MAC layer to enable high-performance transfers, cyclic data exchange and event-controlled signal transmission. The layer 7 functionality is directly linked to the middleware. The middleware itself contains the scheduling and smoothing functions. This means that TCP/IP does not influence the PDU structure. A special Ethertype is used to identify real-time PDUs (only one PDU type for real-time communication). This enables easy hardware support for the real-time PDUs. The technical background is a 100 Mbps full duplex Ethernet (switched Ethernet). PROFINET IO adds an isochronous real-time channel to the RT channels of real-time class 2 option channels. This channel enables a high-performance transfer of cyclic data in an isochronous mode [56.28]. Time synchronization and node scheduling mechanisms are located within and on top of the Ethernet MAC layer. The offered bandwidth is separated for cyclic hard real-time and soft/non real-time traffic. This means that within a cycle there are separate time domains for cyclic hard real-time, for soft/non real-time over TCP/IP traffic, and for the synchronization mechanism, see also Fig. 56.8.

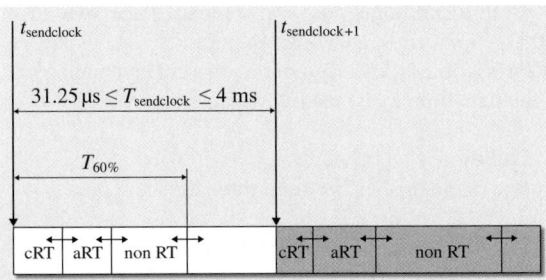

Fig. 56.8 LMPM MAC access used in PROFINET IO [56.22] (cyclic real-time (cRT), acyclic real-time (aRT), nonreal-time (non RT), medium access control (MAC), link layer mapping protocol machine (LMPM))

The cycle time should be in the range of $250\,\mu s$ (35 nodes) up to $1\,ms$ (150 nodes) when simultaneously TCP/IP traffic of about $6\,Mbps$ is transmitted. The jitter will be less than $1\,\mu s$. PROFINET IO/IRT uses switched Ethernet (full duplex). Special four-port and two-port switch ASICs have been developed and allow the integration of the switches into the devices (nodes) substituting the legacy communication controllers of Fieldbus systems. Distances of $100\,m$ per segment (electrical) and $3\,km$ per segment (fiber-optical) can be bridged.

Ethernet/IP with Time Synchronization (ODVA, Rockwell Automation). Ethernet/IP with time synchronization [56.29], an extension of Ethernet/IP, uses the *CIP Synch protocol* to enable the isochronous data trans-fer. Since the CIP Synch protocol is fully compatible with standard Ethernet, additional devices without CIP Synch features can be used in the same Ethernet system. The CIP Synch protocol uses the *precision clock synchronization protocol* [56.30] to synchronize the node clocks using an additional hardware function. CIP Synch can deliver a time-synchronization accuracy of less than $500\,ns$ between devices, which meets the requirements of the most demanding real-time applications. The jitter between master and slave clocks can be less than $200\,ns$.

SERCOS III, (IG SERCOS Interface e.V.). A SERCOS network [56.31], developed for motion control, consists of masters and slaves. Slaves contain integrated repeaters, which have a constant delay time T_{rep} (input/output). The nodes are connected via point-to-point transmission lines. Each node (participant) has two communication ports, which are interchangeable. The topology can be either a ring or a line structure. The ring structure consists of a primary and a secondary channel. All slaves work in forwarding mode. The redundancy provided by the ring structure prevents any downtime caused by a broken cable. The line structure consists of either a primary or a secondary channel. The last physical slave performs the loopback function. All other slaves work in forwarding mode. No redundancy against cable breakage is achieved. It is also possible to insert and remove slaves during operation (hot plug). This is restricted to the last physical slave.

56.4 Wireless Industrial Communications

56.4.1 Basic Standards

Wireless communication networks are increasingly penetrating the application area of wired communication systems. Therefore, they have been faced with the requirements of industrial automation. Wireless technology has been introduced in automation as wireless local area networks (WLAN) and wireless personal area networks (WPAN). Currently, the wireless sensor networks (WSN) are under discussion especially for process automation. For specific application aspects of wireless communications see Chap. 13 on *Communication in Automation, Including Networking and Wireless*, and Chap. 20 on *Sensors and Sensor Networks*. The basic standards are the following:

- Mobile communications standards: GSM, GPRS, and UMTS wireless telephones (DECT)
- Lower layer standards (IEEE 802.11: Wireless LAN [56.32], and 802.15 [56.33]: personal area networks) as a basis of radio-based local networks (WLANs, Pico networks and sensor/actuator networks)
- Higher layer standards (application layers on top of IEEE 802.11 and 802.15.4, e.g. WiFi, bluetooth [56.34], wireless HART, and ZigBee [56.35])
- Proprietary protocols for radio technologies (e.g. wireless interface for sensors and actuators (WISA) [56.36])
- Upcoming radio technologies such as *ultra wide band* (UWB) and WiMedia

For more detailed information and survey see [56.37–41].

56.4.2 Wireless Local Area Networks (WLAN)

The term WLAN refers to a wireless version of the Ethernet used to build computer networks for office and home applications. The original standard (IEEE802.11) specified an infrared, a direct sequence spread spectrum (DSSS) and a frequency hopping spread spectrum (FHSS) physical layer. There is an approval for WLAN to use special frequency bands, however it has to share the medium with other users. The WiFi Alliance was founded to assure interoperability between WLAN clients and access points of different vendors. Therefore, a certification procedure and a WiFi logo are provided. WLANs use a licence-free frequency band and no service provider is necessary.

WLAN is a mature technology and it is implemented in PCs, laptops, and PDAs. Modules for embedded systems development are also available. WLAN can be used almost world wide. Embedded WLAN devices need a powerful microcontroller. WLAN enables wireless access to Ethernet based LANs and is helpful for the vertical integration in an automated manufacturing environment. It offers high speed data transmission that can be used to transmit productive data and management data in parallel. The WLAN propagation characteristics fit into a number of possible automation applications. WLAN enables more flexibility and a cost effective installation in automation associated with mobility and localization. The transition to Ethernet is simple and other gateways are possible. The largest part of the implementation is achieved in hardware; however improvements can be made above the MAC layer.

56.4.3 Wireless Sensor/Actuator Networks

Various wireless sensor network (WSN) concepts are under discussion, especially in the area of industrial automation. Features such as time synchronized operation, frequency hopping, self-organization (with respect to star, tree, and mesh network topologies), redundant routing, and secure data transmission are desired. Interesting surveys on this topic are available in [56.41–44]. Process automation requirements can be generally fulfilled by two mesh network technologies:

- ZigBee (ZigBee Alliance) [56.35]
- Wireless HART [56.45, 46].

Both technologies use the standard IEEE 802.15.4 (2003) *low-rate wireless personal area network* (WPAN) [56.33], specifying the physical layer and parts of the data link layer (medium access control).

ZigBee

ZigBee distinguishes between three device types:

- Coordinator ZC: root of the network tree, storing the network information and security keys. It is responsible for connection the ZigBee network to other networks.
- Router ZR: transmits data of other devices.
- End device ZED: automation device (e.g. sensor), which can communicate with ZR and ZC, but is unable to transmit data of other devices.

An enhanced version allows one to group devices and to store data for neighboring devices. Additionally, to save energy, there are full-function devices and reduced-function devices. The ZigBee application layer (APL) consists of three sublayers: application support layer (APS) (containing the connection lists of the connected devices), an application framework (AF), and Zigbee device objects (ZDO) (definition of devices roles, handling of connection requests, and establishment of communication relations between devices).

For process automation, the ZigBee application model and the ZigBee profiles are very interesting. The application functions are represented by application objects (AO), and the generic device functions by device objects (DO). Each object of a ZigBee profile can contain one or more clusters and attributes, transferred to the target AO (in the target device) directly or to a coordinator, which transfers them to one or more target objects.

WirelessHART

Revision 7 of HART protocol includes the specification of WirelessHART [56.46]. The mesh type network allows the use of redundant communication paths between the radio-based nodes. The temporal behavior is determined by the time synchronized mesh protocol (TSMP) [56.47, 48]. TSMP enables a synchronous operation of the network nodes (called *motes*) based on a time slot mechanism. It uses various radio channels (supported by the MAC layer) for an end-to-end communication between distributed devices. It works comparably with a frequency hopping mechanism missed in the basic standard IEEE 802.15.4.

TSMP supports star, tree, as well as mesh topologies. All nodes have the complete routing function (contrary to ZigBee). A self-organization mechanism enables devices to acquire information of neighboring nodes and to establish connections between them. The messages have their own network identifier. Thus, different networks can work together in the same radio area. Each node has its own list of neighbors, which can be actualized when failures have been recognized.

To support security, TSMP uses mechanisms for encryption (128 bit symmetric key), authentification (32 bit MIC for source address), and integrity (32 bit MIC for message content). Additionally, the frequency hopping mechanism improves the security features. For detailed information see [56.46].

56.5 Wide Area Communications

With the application of remote automation mechanisms (remote supervisory, operation, service) using wide area networks, the stock of existing communication technology becomes broader and includes the following [56.18]:

- All appearances of the Internet (mostly supporting best effort quality of services)
- Public digital wired telecommunication systems: either line-switched [e.g. integrated services digital network (ISDN)] or packet-switched [such as asymmetric/symmetrical digital subscriber line (ADSL, SDSL)]
- Public digital wireless telecommunication systems (GSM-based, GPRS-based, UMTSbased)
- Private wireless telecommunication systems, e.g. trunk radio systems.

The transition between different network technologies can be made easier by using multiprotocol label switching (MPLS) and synchronous digital hierarchy (SDH). There are several private protocols (over leased lines, tunneling mechanisms, etc.) that have been used in the automation domain using these technologies. Most of the wireless radio networks can be used in non real-time applications, some of them in soft real-time applications; however industrial environments and industrial, scientific, and medical (ISM) band limit the applications. Figure 56.9 depicts the necessary remote channels.

The end-to-end connection behavior via these telecommunication systems depends on the recently offered quality of service (QoS). It strongly limits the use of these systems within the automation domains. Therefore, the following application areas have to be distinguished:

Non-Real-Time Communication in Automation
- Non real-time communication (Standard IT: upload/download, SNMP) with lower priority to real-time communication: for configuration, diagnostics, automation-specific up/download.
- Manufacturing-specific functions, context management, establishment of application relationships and connection relationships to configure IO devices, application monitoring to read status data (diagnostics, I&M), read/write services (HMI, application program), open loop control.

The automation domain has the following impact on non real-time WAN connections: addressing between multiple distributed address spaces, and redundant transmission for minimized downtime to ensure its availability for a running distributed application.

Real-Time Communication in Automation
- Cyclic real-time communications (i. e. PROFINET IO data) for closed loop control and acyclic alarms (i. e. PROFINET IO alarms) as major manufacturing-specific services
- Transfer (and addressing) methods for RT data across WAN can be distinguished as follows:

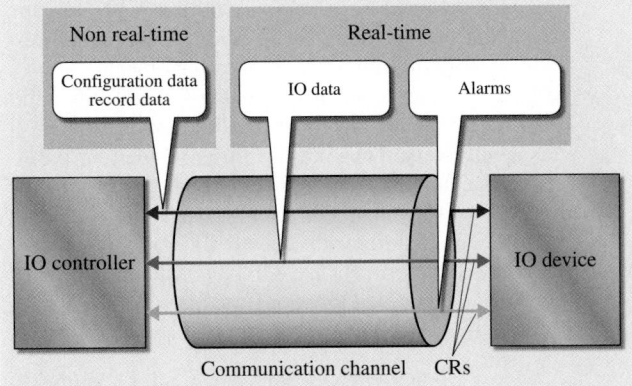

Fig. 56.9 Remote communication channels over WAN (input/output (IO); communication relation (CR))

– *MAC based*: Tunnel (real-time class 1, partially real-time class 2 for longer send intervals, e.g. 512 ms), clock across WAN and reserved send phase for real-time transmission
– *IP based*: Real-time over UDP (routed); web services based [56.49, 50].

The automation domain has the following impact on real-time WAN connections: a constant delay-sensitive and jitter-sensitive real-time base load (e.g. in LAN: up to 50% bandwidth reservation for real-time transmission).

To use a wide area network for geographically distributed automation functions, the following basic design decisions were made following the definitions in Sect. 56.2:

● A virtual automation network (VAN) is an infrastructure for standard LAN-based distributed industrial automation concepts (e.g. PROFINET or other) in an extended environment. The productive automation functions (applications) are described by their object models used in existing industrial communications. The application service elements (ASEs), as they are specified in the IEC 61158 standard, can additionally be used.
● The establishment of the end-to-end connections between distributed objects within a heterogeneous network is based on web services. Once this connection has been established, the runtime channel between these objects is equivalent to the runtime channel within the local area by using PROFINET (or other) runtime mechanisms.
● The VAN addressing scheme is *based on names* to avoid the use of IP and MAC addresses during establishment of the end-to-end path between logically connected applications within a VAN domain. Therefore, the IP and MAC addresses remain transparent to the connected application objects.
● Since there is no new Fieldbus or real-time Ethernet protocol, no new specified application layer is necessary. Thus, the well-tried models of industrial communications (as they are specified in the IEC 61158 standard) can be used. Only the additional requirements caused by the influence of wide area networks have to be considered and they lead to additional functionality following the above-mentioned design guidelines.

Most of the WAN systems that offer quality-of-service (QoS) support cannot provide real guarantees, and this strongly limits the use of these systems within the automation domain. To guarantee a defined QoS for data transmission between two application access points via a wide area network, written agreements between customer and service provider [service level agreements (SLA)] must be contracted. In cases where the provider cannot deliver the promised QoS, an alternative line must be established to hold the connection for the distributed automation function (this operation is fully transparent to the application function). This line should be available from another provider, independent from the currently used provider. The automation devices (so called VAN access points (VAN-APs)) should support functions to switch (either manually or automatically) to an alternative line [56.51].

There are different mechanisms to realize a connection between remote entities:

● *The VAN switching connection*: the *logical* connection between two VAN-APs over a WAN or a public network. One VAN switching connection owns one or more communication paths. VAN switching line is defined as one physical communication path between two VAN-APs over a WAN or a public network. The endpoints of a switching connection are VAN-APs.
● *The VAN switching line*: the *physical* communication path between two VAN-APs over a WAN or a public network. A VAN switching line has its provider and own QoS parameter. If a provider offers connections with different warranted QoS each of these shall be a new VAN switching line.
● *VAN switching endpoint access*: the physical communication path between one VAN-AP and a WAN or a public network. This is a newly introduced class for using the switching application service elements of virtual automation networks for communication via WAN or public networks.

These mechanisms are very important for the concept of VANs using heterogeneous networks for automation. Depending on the priority and importance of the data transmitted between distributed communications partners, the kind of transportation service and communication technology is selected based on economical aspects. The VAN provider switching considers the following alternatives:

● *Use case 1*: For packet-oriented data transmission via public networks a connection from a corresponding VAN-AP to the public network has to be established. The crossover from/to the public network is represented by the VAN switching endpoint access. The requirements made for this line have to be fulfilled by the service level agreements from the chosen provider. Within the public network it is not possible to influence the quality of service. The data package leaves the public network when the VAN switching endpoint access from the faced communication partner is achieved. The connection from the public network to the faced VAN-AP is also provided by the same or an alternative provider and guarantees defined requirements. The data exchange between two communication partners is independent of each other.

- *Use case 2*: For a connection-oriented data transmission (or data packages with high-level priority) the use of manageable data transport technology is needed. The VAN switching line represents a manageable connection. A direct *known* connection between two VAN-APs has to be established and a VAN switching endpoint access is not needed. The chosen provider guarantees the defined requirements for the complete line. When the current line loses the promised requirements it is possible to define the VAN-APs to build up an alternative line and hold on/disconnect the current line automatically.

56.6 Conclusions

As discussed in the above sections, the area of industrial communication protocols has been experiencing a tremendous evolution over the last ten years, being strongly influenced by advances in the area of information technology and hardware/software developments. Existing industrial communication protocols have impacted very positively both in the operation of industrial plants, due to enhanced diagnostic capabilities, which has improved maintenance operations, as well as in the development of complex automation systems.

The chapter has reviewed the main concepts of industrial communication networks and presented the most prominent wired and wireless protocols that are already incorporated in a large number of industrial devices (from thousands to millions).

Within this scope of a multitude of existing protocols and also motivated by the growth of the Internet and increasing possibilities of the World Wide Web network, and the increased demand for geographically distributed automation functions virtual automation networks (VAN), a very interesting approach would appear to be to allow integration via heterogeneous networks. The section presented the concepts of VAN domains and interfaces and the challenges on ensuring timely communication behavior, safety, and security across multiple VAN domains.

56.7 Emerging Trends

The number of commercially available industrial communication protocols has continued to increase, despite some trials to converge to a single and unified protocol, in particular during the definition of the IEC 61178 standard; the automation community has started to accept that no single protocol will be able to meet all different communication requirements from different application areas. This trend will be continued by the emerging wireless sensor networks as well as the integration of wireless communication technologies in all mentioned automation-related communication concepts. Therefore, increasing attention has been given to concepts and techniques to allow integration among heterogeneous networks, and within this context virtual automation networks are playing an increasing role.

With the proliferation of networked devices with increasing computing capabilities, the trend of decentralization in industrial automation systems will increase in the future (Figs. 56.10 and 56.11). This situation will lead to an increased interest in autonomic systems with self-X capabilities, where X stands for alternatives as configuration, organizing, optimizing, healing, etc. The idea is to develop automation systems and devices that are able to manage themselves given high level objectives. Those systems should have sufficient degrees of freedom to allow a self-organized behavior, which will adapt to dynamically changing requirements. The ability to deal with widely varying time and resources demands while still delivering dependable and adaptable services with guaranteed temporal qualities is a key aspect for future automation systems.

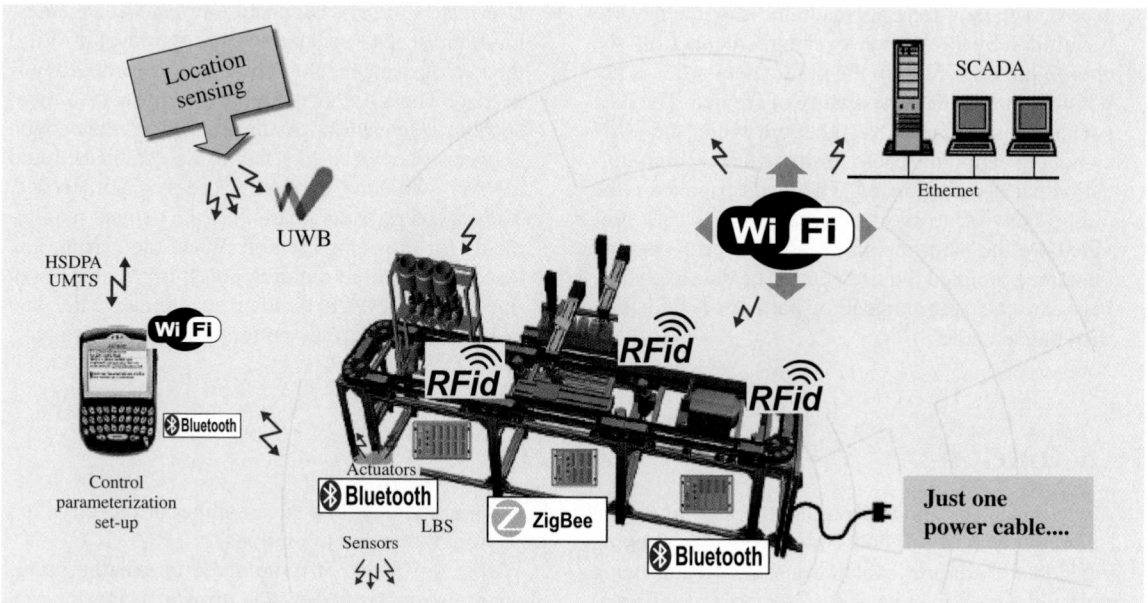

Fig. 56.10 The wireless factory (from [56.52])

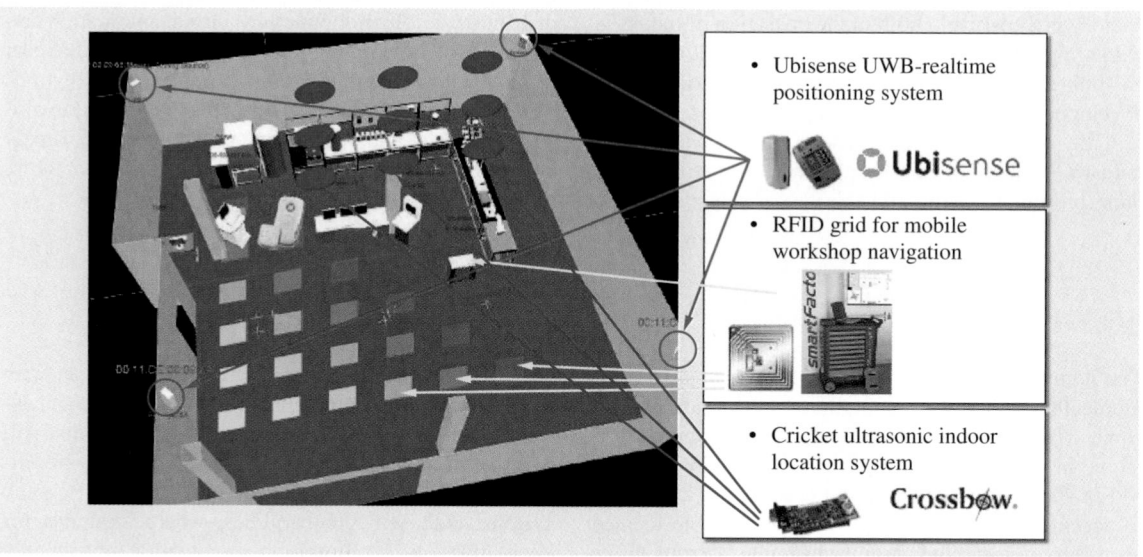

Fig. 56.11 Indoor positioning systems in the SmartFactory (from [56.52])

56.8 Further Reading

56.8.1 Books

- J.W.S. Liu: *Real-Time Systems* (Prentice Hall, Upper Saddle River 2000)
- P.S. Marshall: *Industrial Ethernet* (ISA, 2004)
- P. Pedreiras, L. Almeida: Approaches to enforce real-time behavior in Ethernet. In: *The Industrial Communication Technology Handbook*, ed. by R. Zurawski (CRC, Boca Raton 2005)
- B. Schneider: *Secrets and Lies – Digital Security in a Networked World* (Wiley, New York 2000)
- L.R. Franco, C.E. Pereira: Real-time characteristics of the foundation Fieldbus protocol. In: *Fieldbus Technology: Industrial Network Standards for Real-Time Distributed Control*, ed. by N.P. Mahalik (Springer, Berlin Heidelberg 2003), pp. 57–94

56.8.2 Various Communication Standards

- IEC: IEC 61508: Functional safety of electrical/electronic/programmable electronic safety-related systems (2000)
- IEC: IEC 61158 Ser., Edition 3: Digital data communication for measurement and control – Fieldbus for use in industrial control systems (2003)
- IEC: IEC 61784-1: Digital data communications for measurement and control – Part 1: Profile sets for continuous and discrete manufacturing relative to Fieldbus use in industrial control systems (2003)
- PROFIBUS Guideline: PROFInet Architecture Description and Specification, Version V 2.0 (PNO, Karlsruhe 2003)

56.8.3 Various web Sites of Fieldbus Organizations and Wireless Alliances

- IEEE 802: http://www.ieee802.org (last accessed April 6, 2009)
- HART Communication Foundation: http://hartcomm.org (last accessed April 6, 2009)
- ODVA: http://www.odva.org (last accessed April 6, 2009)
- PROFIBUS Nutzer Organisation/PROFIBUS International: http://www.profibus.com (last accessed April 6, 2009)
- Interbus Club: http://www.interbusclub.com (last accessed April 6, 2009)
- Fieldbus Foundation: http://www.fieldbus.org (last accessed April 6, 2009)
- MODBUS: http://www.modbus.org/ (last accessed April 6, 2009)
- Actor-Sensor-Interface ASi: http://www.as-interface.net (last accessed April 6, 2009)
- Ethernet POWERLINK Standardization Group (EPSG): http://www.ethernet-powerlink.org (last accessed April 6, 2009)
- EtherCAT Technology Group: http://www.ethercat.org (last accessed April 6, 2009)
- Interessengemeinschaft SERCOS interface e.V.: http://www.ig.sercos.de (last accessed April 6, 2009)
- IEEE 802.11TM Wireless Local Area Networks: http://ieee802.org/11/ (last accessed April 6, 2009)
- ZigBee Alliance: http://www.zigbee.org (last accessed April 6, 2009)
- Bluetooth Special Interest Group: http://www.bluetooth.org (last accessed April 6, 2009)
- WISA: Wireless Interface for Sensors and Actuators: http://library.abb.com/global/scot/scot209.nsf/veritydisplay/4e478bd7490a3f8bc12571f100427dcb/$File/2CDC171017K0201.PDF (last accessed April 6, 2009)
- Virtual Automation Networks (VAN): http://www.van-eu.eu/ (last accessed April 6, 2009)

References

56.1   P. Neumann: Communication in industrial automation – what is going on?, INCOM 2004, 11th IFAC Symp. Inf. Control Probl. Manuf., Salvador da Bahia (2004)

56.2   IEC 61508: Functional safety of electrical/electronic/ programmable electronnic safety-related systems (2000)

56.3   P. Neumann, A. Pöschmann, E. Flaschka: Virtual automation networks, heterogeneous networks for industrial automation, atp Int. Autom. Technol. Pract. **2**, 36–46 (2007)

56.4   Virtual Automation Networks: European Integrated Project FP6/2004/IST/NMP/2 016696 VAN. Deliverable D02.2-1: Topology Architecture for the VAN Virtual Automation Domain (2006)

56.5   HART: http://www.hartcomm2.org/ (last accessed April 6, 2009)

56.6   R. Becker, B. Müller, A. Schiff, T. Schinke, H. Walker: *A Compilation of Technology, Functionality and Application* (AS International Association, Gelnhausen 2002)

56.7   CAN: IEC 11898 Controller Area Networks: http://de.wikipedia.org/wiki/Controller_Area _Network (last accessed April 6, 2009), see also http://www.cancia.org/canopen

56.8   IO-Link Communication Specification, V 1.0; PROFIBUS International 2.802 (2008)

56.9   IO-Link Integration, Part 1, V. 1.00; PROFIBUS International 2.812 (2008)

56.10   PROFIBUS: Technical description, http://www.profibus.com/pb/ (last accessed April 6, 2009)

56.11   IEC 61158 Ser., Edition 3: Digital data communication for measurement and control – Fieldbus for use in industrial control systems (2003)

56.12   IEC 61784-1: Digital data communications for measurement and control – Part 1: Profile sets for continuous and discrete manufacturing relative to fieldbus use in industrial control systems (2003)

56.13   ANSI/TIA/EIA-485-A: Electrical characteristics of generators and receivers for use in balanced digital multipoint systems (1998)

56.14   ODVA: http://www.odva.org/portals/ (last accessed April 6, 2009)

56.15   MODBUS TCP/IP: IEC 65C/341/NP: Real-Time Ethernet: MODBUS-RTPS (2004)

56.16   Ethernet IP: Ethernet/IP specification, Release 1.0., (ControlNet International and Open DeviceNet Vendor Association, 2001)

56.17   High Speed Ethernet: HSE Specification documents FF-801, 803, 586, 588, 589, 593, 941 (Fieldbus Foundation, Austin 2001)

56.18   P. Neumann: Communication in industrial automation. What is going on?, Control Eng. Pract. **15**, 1332–1347 (2007)

56.19   PROFIBUS Nutzerorganisation: PROFInet Architecture Description and Specification, Version V 2.0., PROFIBUS Guideline (2003)

56.20   P-Net on IP: IEC 65C/360/NP: Real-Time Ethernet: P-NET on IP (2007)

56.21   PROFINET IO: IEC 65C/359/NP: Real-Time Ethernet: PROFINET IO, Application Layer Service Definition and Application Layer Protocol Specification (2004)

56.22   P. Neumann, A. Pöschmann: Ethernet-based real-time communication with PROFINET IO, WSEAS Trans. Commun. **4**(5), 235–245 (2005)

56.23   Time-critical Control Network: IEC 65C/353/NP: Real-Time Ethernet: Tcnet (2007)

56.24   Vnet/IP: IEC 65C/352/NP: Real-Time Ethernet: Vnet/IP (2007)

56.25   POWERLINK: IEC 65C/356/NP: Real-Time Ethernet: POWERLINK (2007)

56.26   ETHERCAT: IEC 65C/355/NP: Real-Time Ethernet: ETHERCAT (2007)

56.27   W. Manges, P. Fuhr (Eds.): PROFINET IO/Isochronous Technology, IFAC Summer School Control, Computing, Communications, Prague (2005)

56.28   J. Jasperneite, K. Shehab, K. Weber: Enhancements to the time synchronization standard IEEE-1588 for a system of cascaded bridges, 5th IEEE Int. Workshop Fact. Commun. Syst., WFCS2004, Vienna (2004) pp. 239–244

56.29   EtherNet/IP with time synchronization: IEC 65C/361/NP: Real-Time Ethernet: EtherNet/IP with time synchronization (2007)

56.30   IEC 61588: Precision clock synchronization protocol for networked measurement and control systems (2002)

56.31   SERCOS III: IEC 65C/358/NP: Real-Time Ethernet: SERCOS III (2007)

56.32   Wireless LAN: IEEE 802.11: IEEE 802.11 Wireless Local Area Networks Working Group for WLAN Standards, http://ieee802.org/11/ (last accessed April 6, 2009)

56.33   Personal Area Networks: IEEE 802.15: IEEE 802.15 Working Group for Wireless Personal Area Networks, Task Group 1 (TG1), http://www.ieee802.org/15/pub/TG1.html (last accessed April 6, 2009), and: Task Group 4 (TG4), http://www.ieee802.org/15/pub/TG4.html (last accessed April 6, 2009)

56.34   Bluetooth: The Official Bluetooth Membership Site, https://www.bluetooth.org (last accessed April 6, 2009)

56.35   ZigBee: http://www.zigbee.org/ (last accessed April 6, 2009)

56.36   WISA: http://library.abb.com/GLOBAL/SCOT/SCOT209.nsf/VerityDisplay/4E478BD7490A3F8BC12571F100427DCB/$File/2CDC171017K0201.PDF (last accessed April 6, 2009)

56.37 ABI Research Forecast and Information on Emerging Wireless Technologies:
http://www.abiresearch.com (last accessed April 6, 2009)

56.38 W. Stallings: IEEE 802.11. Wireless LANs from a to n, IT Professional **6**(5), 32–37 (2004)

56.39 IEEE 802.11 Tutorial:
www.eecs.berkeley.edu/~ergen/docs/ieee.pdf (last accessed April 6, 2009), see also http://ayman.elsayed.free.fr/msc-student/wlan.tutorial.pdf

56.40 A. Willig, K. Matheus, A. Wolisz: Wireless technology in industrial networks, Proc. IEEE **93**(6), 1130–1151 (2005)

56.41 P. Neumann: Wireless sensor networks in process automation, survey and standardisation, atp **3**(49), 61–67 (2007), (in German)

56.42 I.F. Akyildiz: Key wireless networking technologies in the next decade, IFAC Conf. Fieldbus Technol. FeT 2005, Puebla (2005)

56.43 K. Koumpis, L. Hanna, M. Andersson, M. Johansson: Wireless industrial control and monitoring beyond cable replacement, PROFIBUS Int. Conf. Coombe Abbey, Warwickshire (2005), Paper C1

56.44 Industrial wireless technology for the 21st century, Industrial Wireless Workshop, San Francisco (2002)

56.45 J.-L. Griessmann: HART protocol rev. 7 including WirelessHART, atp Int.-Autom. Technol. Pract. **2**, 21–22 (2007)

56.46 HART 7: HART Protocol Specification, Revision 7 (HART Communication Foundation, 2007), see also http://www.hartcomm.org/ (last accessed April 6, 2009)

56.47 Dust Networks: TSMP Seminar. Online-Präsentation (2006)

56.48 Dust Networks: Technical Overview of Time Synchronized Mesh Protocol (TSMP), White Paper (Dust Networks, 2006)

56.49 IBM: Standards and Web services, http://www-128.ibm.com/developerworks/webservices/standards/ (last accessed April 6, 2009)

56.50 L. Wilkes: The web services protocol stack, Report from CBDI Web Services Roadmap (2005), http://roadmap.cbdiforum.com/reports/protocols/ (last accessed April 6, 2009)

56.51 Virtual Automation Networks: European Integrated Project FP6/2004/IST/NMP/2 016696 VAN. Deliverable D07.2-1: Integration Concept, Architecture Specification (2007)

56.52 D. Zuehlke: SmartFactory from vision to reality in factory technologies, IFAC Congress 2008, Seoul (2008)

57. Automation and Robotics in Mining and Mineral Processing

Sirkka-Liisa Jämsä-Jounela, Greg Baiden

Mines and mineral processing plants need integrated process control systems capable of improving plant-wide efficiency and productivity. Mining automation systems today typically control fixed plant equipment such as pumps, fans, and phone systems. Much work is underway around the world in attempting to create the moveable equivalent of the manufacturing assembly line for mining. This technology has the goals of speeding production, improving safety, and reducing costs. Process automation systems in mineral processing plants provide important plant operational information such as metallurgical accounting, mass balances, production management, process control, and optimization. This chapter discusses robotics and automation for mining and process control in mineral processing. Teleoperation of mining equipment and control

strategies for grinding and flotation serve as examples of current development of field.

57.1 Background

Mining is the act of extracting mineral determined to be ore from the earth to be processed in a mineral processing operation. All mining operations have a least some limited mineral processing available on site. Usually the sophistication of the complex is determined by unit process operations needed to make the product, or distribution and transportation costs (Fig. 57.1). The mineral extraction process can occur using many potential mining methods. Some of the methods include open pit, caving, bulk stoping, and/or selective mining techniques such as cut and fill, as well as *room* and *pillar* [57.1]. Each method and suite of mining equipment has the aim of extracting the mineral at a profit for processing.

The aim of a mineral processing operation is to concentrate a raw ore for the subsequent metal extraction stage. Usually, the valuable minerals are first liberated

from the ore matrix by comminution and size separation processes (crushing, grinding, and size classification), and then separated from the gangue using processes capable of selecting the particles according to their physical or chemical properties, such as surface hydrophobicity, specific gravity, magnetic susceptibility, and color (flotation, magnetic or gravimetric separation, sorting, etc.) [57.2].

Process automation has always played a key role in the mineral process industries and is gaining momentum in mining extraction operations as mobile robotics techniques are being applied. The use of advanced technologies, including modeling, simulation, advanced control strategies, smart equipment, fieldbuses, wireless networks, remote maintenance, etc., is widespread in many sectors (Fig. 57.2). Information-based technologies are responsible for making mineral

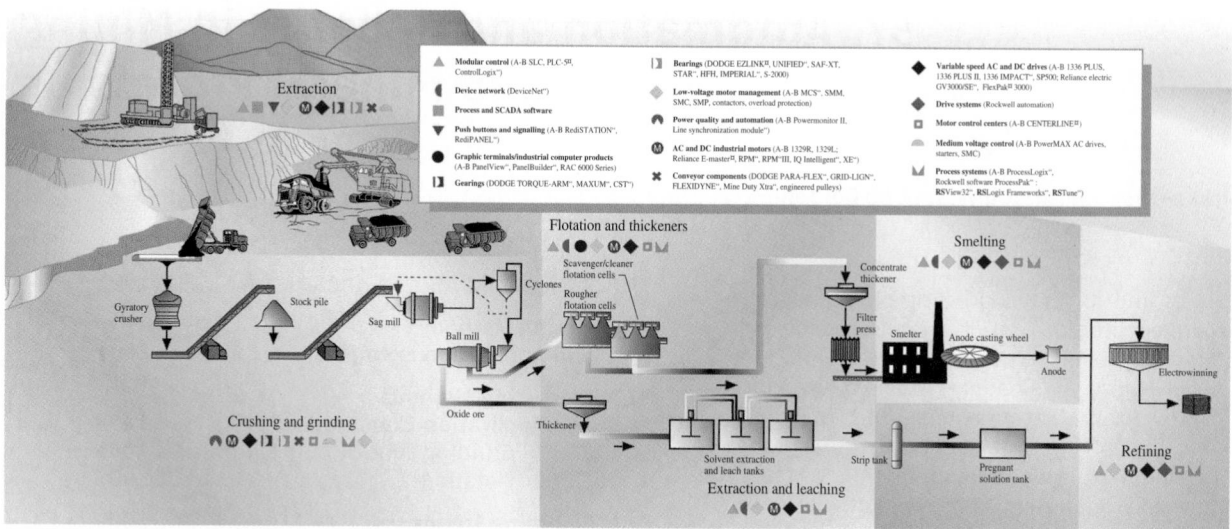

Fig. 57.1 Mineral processing automation (courtesy of Rockwell Automation, Inc)

processing more efficient and reliable, and help the industry to adapt to new competitive environments in a safe and environmentally sound manner. One critical step in achieving these objectives is to develop and apply improved control systems across the full range of applications from mining to processing and utilization.

While mineral processing has had extensive use of many advanced technologies, standard mining applications such as pumping, dewatering, hoist control, and power distribution remain the norm with some individual exceptions in ventilation systems and other mine wide systems. Overall, the complication and scale of mining operations has delayed the wide adoption of advanced technology. Several stand alone technologies have seen successful implementation in mining and pilot projects; full scale mine implementation have been attempted with extremely encouraging results. The Intelligent Mine Program in Scandinavia and the Mining Automation Program [57.3] in Canada were two main projects attempted in the 1990s and early 2000s. The main technology drivers were seen to be: telecommunications, positioning and navigation, integrated software systems, and mobile robotic equipment. The Intelligent Mine Program explored the issues from a rock and process characteristic point of view and the Mining Automation Program from an equipment point of view.

The optimization of the economics of the process operations is the key driver for the application of advanced control. Many successful control strategy implementations in mineral processing have been

reported. The power of model-based control for industrial semi-autogenous grinding circuits was discussed by *Herbst* and *Pate* [57.4]. In the application, they used an expert system and online process models to find the optimum feed rate. A Kalman filter was used to estimate unmeasured variables such as mill filling and ore hardness that were required by the expert system. An 8% improvement in feed rate over manual control was achieved with the control system. In the multivariable control application on a two-stage milling circuit at the East Driefontein Gold Mine in South Africa, the average throughput t/h was increased from 73.1 to 79.2, and the average grain size $\% < 75\,\mu\text{m}$ from 76.5 to 78.5. The standard deviation of the grain size values was reported to decrease from 3 to the 0.9 [57.5]. Successful economic results and benefits from 13 years of computer control in flotation have been reported by *Miettunen* [57.6].

The economics of the application of intelligent robotics for mining was seen as having substantial benefits. These were discussed in *Baiden* [57.7]. This report showed that the fundamental definition of *ore* would be altered by the projected results of cost reduction and mining rate improvements. Further, robotic operation would improve the safety of miners as it would drop exposure levels. Subsequently, the Intelligent Mine Program and Mining Automation Program showed through field feasibility experimentation that these projections were realistic. Several projects around the world now are investigating the opportunity for robotic and teleoperated equipment in particular applications.

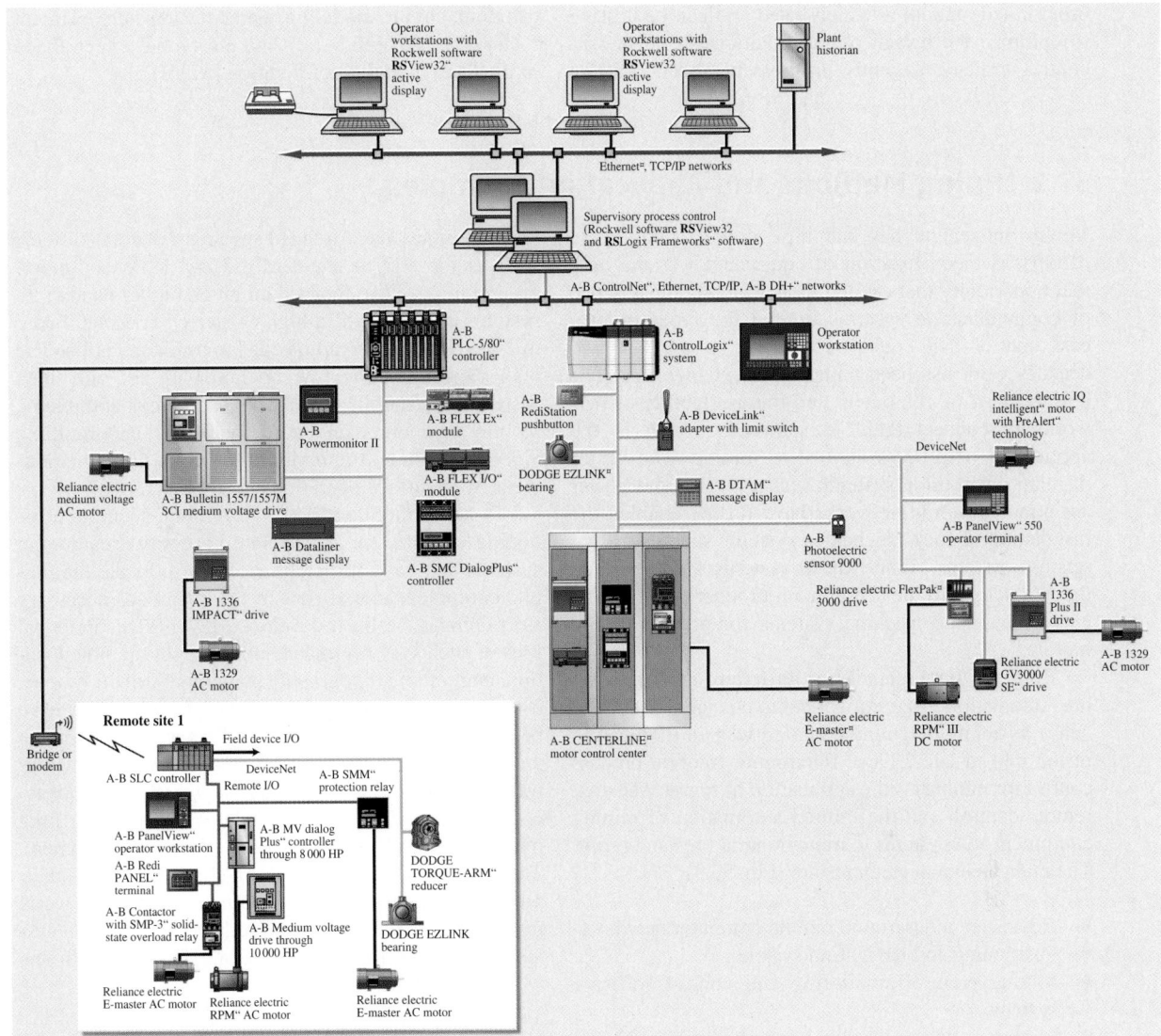

Fig. 57.2 Mining automation architecture (courtesy of Rockwell automation)

However, the control of mineral processes is faced with many challenges. At the present time it is not possible to measure, on a real-time basis, the important physical or chemical properties of the material processes. This is particularly true for the fresh ore feed characteristics (mineral grain size distribution, mineral composition, mineral association, grindability) and the ground material properties (liberation degree, particle composition distribution, particle hydrophobicity). An essential feature of control and optimization strategies is the availability of mathematical models that accurately describe the characteristics of the process. Satisfactory mathematical models are not, however, available for mineral processing unit processes due to the fact that the physics and chemistry of the sub-processes involved are poorly understood. Models for process analysis and optimization for comminution circuits are usually based on population balance models and the use of breakage and selection functions. Numerous empirical and phenomenological models based on various assumptions for flotation have been proposed in the literature. Among the many flotation models, the classical first-order kinetic model is widely used and can be utilized to optimize the design of the flotation circuit and its control strategy. Recently progress has been made in grinding circuit modeling using the discrete element method (DEM) [57.8–12], and efforts have been made in the CFD modeling of flotation [57.13].

57.2 Mining Methods and Application Examples

Mining in general has had little process control capability as mechanization of equipment was the only real opportunity that existed. For example, the absence of communication systems limited the types of process control that could be applied. In the last two decades work has been underway to change this. The portable size of computers and the availability of networks to connect them to spawned growth in the application of process control to mining. The basics of main distribution systems such as water and power are now the norm. Networks have further enabled the installation of rock mechanics systems such as microseismic systems. While these systems are important they do not get to the actual main production technologies because the machine systems for production are mobile.

Both the Intelligent Mine Program and the Mining Automation Program worked to change this and the concepts behind telemining started to gain momentum in the mid to late 1990s. Telemining (mobile process control for mining) is the application of remote sensing, remote control, and the limited automation of mining equipment and systems to mine mineral ores at a profit. The main technical elements are (Fig. 57.3):

- Advanced underground mobile computer networks
- Positioning and navigation systems
- Mining process monitoring and control software systems
- Mining methods designed specifically for telemining
- Advanced mining equipment.

Telemining has the capability to reduce cycle times, improve quality and increase the efficiency of equipment and personnel, resulting in increased revenue and lower costs.

Advanced high capacity mobile computer networks form the foundation of teleremote mining (Fig. 57.4). The mine may be connected via the telecommunication system so mines can be run from operation centers underground or on the surface. Several opportunities exist for communication, depending on the environment.

Surface mines have trended towards network systems such as the 802.11 standard [57.14]. Whereas underground mines have focused on much higher bandwidth systems consisting of a high capacity backbone linked to 2.4 GHz capacity radio cells for communication. The high capacity allows the operation of not only data systems but mobile telephones, handheld computers, mobile computers on board machines, and multiple video channels to run multiple pieces of mining equipment from surface operation centers [57.15, 16].

To apply mobile robotics to mining, accurate positioning systems are an absolute necessity. Positioning systems that have sufficient accuracy to locate the mobile equipment in real-time at the tolerances necessary for mining have been developed [57.17, 18]. Practical uses of such systems include machine set-up, hole location, and remote topographic mapping. Surface systems use GPS for location and several of these systems have been developed. In underground mines, some of the most advanced positioning equipment consists of laser reference positioning, ring-laser-gyro (RLG), and accelerometers. Units are mounted on all types of drilling machines so that operators can position the equipment. These types of systems are just beginning to make their presence known over conventional surveying; several manufacturers offer this new product [57.19]. RLG systems track the location of mobile machinery in the

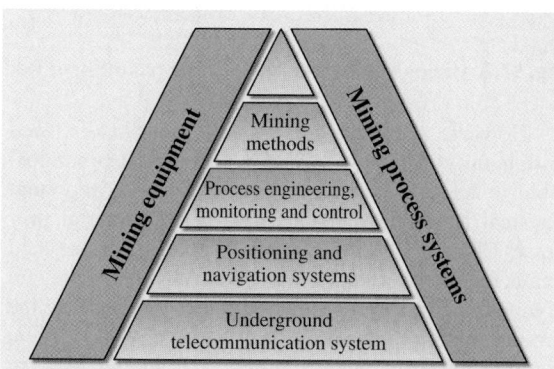

Fig. 57.3 Conceptual representation of the key technological components (after [57.1])

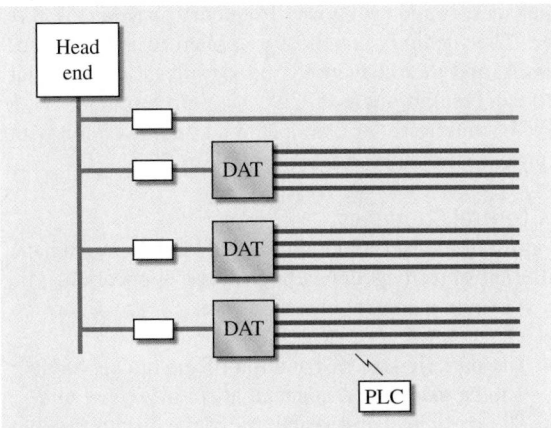

Fig. 57.4 Example of a high capacity cellular network (after [57.1])

mine. Accurate positioning systems mounted on mobile equipment will enable the application of advanced

manufacturing robotics to mining. Usually in advanced manufacturing, robotic equipment is fixed to the floor, allowing very accurate surveying and positioning of the equipment. The positioning systems being used for mining equipment allow accurate positioning of surface equipment using GPS, and inference techniques allow high accuracy positioning of mobile underground equipment.

Mine planning, simulation, and process control systems are growing using the foundations of telecommunications, positioning, and navigation. Linking geology and engineering directly to operations is important for the successful application of these systems. Several systems such as Datamine, Gemcom, Mine 24D, to name a few, are in use around the world today. Further, process control systems for the day-to-day operation of pumping, dewatering, and power distribution are the norm. New systems for ventilation control are starting to emerge as the cost of the overall system infrastructure is reduced.

57.3 Processing Methods and Application Examples

The overall objective of a grinding and flotation unit is to prepare a concentrate, which may be as simple as the net revenue of the plant. In practice, however, the links between the grinding and flotation circuits tuning and the economic objective are not obvious, and the objectives are always broken down into particle size reduction, mineral liberation, and mineral separation objectives. In the following, grinding and flotation areas are briefly discussed as application areas where automation has played an important role in mineral processing. These application areas serve as examples of current developments in the field of automation in mineral processing.

57.3.1 Grinding Control

Grinding ore to the optimum size for mineral extraction by flotation or leaching is an essential but high energy intensive part of most mineral processing operations. The benefits from improved grinding control are substantial, primarily in the areas of improved milling efficiency, more stable operation, higher throughput, and improved downstream processing. Grinding an ore finer than is necessary leads to increased energy costs, reduced throughput, increased mill liner consumption, and increased consumption of grinding media and

reagents. Insufficient ore grinding, on the other hand, reduces the recovery rate of the valuable mineral.

Instrumentation
For grinding instrumentation both basic measurement and advanced indirect instruments are available. The most common measurements are: mass flow rate on a conveyor belt, volume flow rate, pipeline pressure, pulp density, sump level, mill motor power consumption, and mill rotation speed. Online particle size measurement is also a part of the well-instrumented grinding circuit. Indirect instruments are mostly used in mill or hydrocyclone operation monitoring. These measurements are based, for example, on acoustic measurements, vision-based monitoring, mill liner sensors, and mill power frequency analysis.

Mass flow rate measurement on a conveyor belt is mainly performed by nuclear weight gauges. In pipeline flow measurements magnetic instruments are the most typical. Pulp densities can be measured by a nuclear density meter, soft sensors, or alternatively by certain particle size analyzers.

Online particle size analysis can be performed using several techniques. The three most typical online particle size analysis methods in mineral processes are mechanical, ultrasonic, and laser diffraction-based devices. Outokumpu Technology's PSI-200 has been one of the most popular mechanical devices since the 1970s. The measurement is based on a reciprocating caliper with high precision position measurement. The measurement technique limits accurate size measurement to the coarser end of the distribution. The ultrasonic-based measurements were also developed in the 1970s, for example the Svedala Multipoint PSM-400. However, the method requires frequent calibration and is susceptible to air bubbles. The laser diffraction method represents the latest technology in online particle size analysis. The PSI-500 particle size analyzer, manufactured by Outokumpu, uses laser diffraction-based measurement, with automatic sample preparation. The system enables the development of new advanced control employing the full scale of the particle size distribution [57.20].

Some vision-based measurements have recently been developed. Mintek has a product CYCAM for hydrocyclone underflow monitoring. The equipment measures the angle of the discharge and therefore the conditions; for example, roping can be detected. For particle size on-belt monitoring of the grinding and crushing circuit Metso has a product called VisioRock.

Various methods are used for mill charge measurement. These include acoustic measurements [57.9], mill liner sensors and mill power frequency analysis [57.22–24]. The methods are mostly applied to specific processes, and there have been no significant commercial product breakthroughs.

To summarize, an example of a grinding circuit with typical instrumentation is given in Fig. 57.5.

Control Strategies

Control of the wet mineral grinding circuits might have different objectives depending on the application. The most common control objectives are:

- The particle size distribution of the circuit product is to be maintained constant at *constant feed rate*.
- The particle size distribution of the circuit product is to be maintained constant at *maximum feed rate*.
- Both the particle size distribution and solid contents circuit product are to remain *constant*.

The control strategy for the grinding circuit is based on a hierarchical structure. Basic controls mainly consist of traditional PI controllers and ratio controllers. The mill water feed typically has a ratio control with the ore feed. In many cases, sump levels are controlled by changing the pump speed. Furthermore, the pump speed is used to control the hydrocyclone feed pressure.

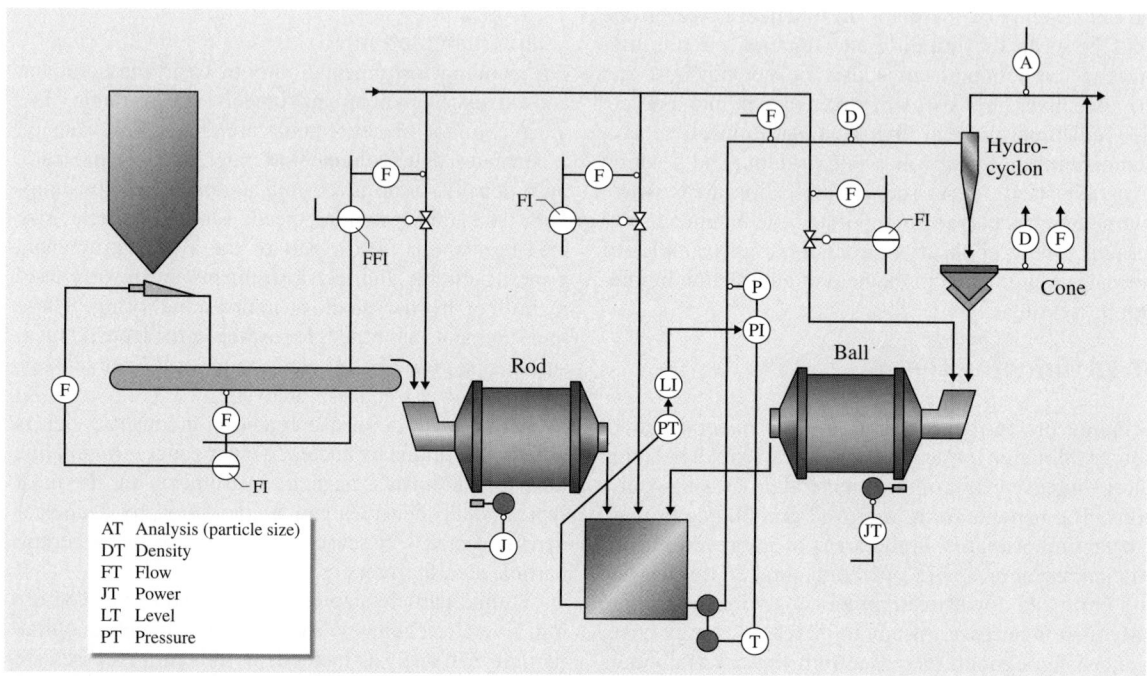

AT Analysis (particle size)
DT Density
FT Flow
JT Power
LT Level
PT Pressure

Fig. 57.5 Typical flowsheet of a grinding circuit (after [57.21])

The cyclone feed density is stabilized by manipulating an additional water feed rate to the sump. Particle size measurement is also currently applied in grinding circuit control. The product particle size measurement can, for example, be used to manipulate the ore feed rate to the primary mill. In addition, higher level optimization methods are typically applied to maximize the throughput with desired constraints.

57.3.2 Flotation

In flotation the aim of the control strategy is adjustment of the operating conditions as a function of the raw ore properties and feed rate, metal market prices, energy, and reagent costs [57.25]. Usually these objectives require a certain amount of trade-off between the concentrate and tonnage, the impurity contents and the operating costs.

Instrumentation

In flotation, instrumentation is available for measuring flow rates, density, cell levels, airflow rate, reagent feed rates, pH, and conductivity. Slurry flow measurement is mainly performed by a magnetic flow meter, and density by a nuclear density meter. The most typical instruments used for measuring the slurry level in a cell are a float with a target plate and ultrasonic level transmitter, a float with angle arms and capacitive angle transmitter, and reflex radar. The instrument for measuring flotation airflow rate contains a thermal gas mass flow sensor or a differential pressure transmitter with a venture tube, pitot tube, or Annubar element. A wide range of different instrumentation solutions for reagent dosing exist. The best choice is to use inductive flow meters and control valves. Electrochemical measurements give important information about the surface chemistry of valuable and gangue minerals in the process. pH is the most commonly measured electrochemical potential, and sometimes pH measurement can be replaced by conductivity measurement, which gives approximately the same information as pH measurement. Recently, other electrochemical potential measurements have also been under study. The use of minerals as working electrodes makes it possible to detect the oxidation state of different minerals and to control their floatability. Stability of the electrodes, however, has been a problem in online use, but some good results have been reported.

X-ray fluorescence is the universal method for online solid composition measurement in flotation. Equipment vendors now offer, however, more efficient, compact, flexible and reliable devices than were available in the 1970s.

To summarize, an example of a flotation circuit with typical basic instrumentation is given in Fig. 57.6. Conductivity and pH are measured in the conditioner. On-stream analyses are taken from the feed, tailings and concentrate, and also from several flows between the flotation sections. Flow rates, levels, and airflow rates are measured at several points. Most of the reagents are added in the grinding circuit, except for frother, which is added in the conditioner and additional sodium cyanide in the cleaner.

Recent developments in instrumentation have provided new instruments, such as image analysis-based devices for froth characteristics measurement. Three different image analysis products have been reported to be available commercially: FrothMaster from Outokumpu, JKFrothCam from JKTech, and VisioFroth from Metso Minerals. Research has been carried out on developing image processing algorithms and on analyzing the correlations between image analysis and process variables, more recently also on flotation control based on image analysis.

A comprehensive description of the flotation plant instrumentation has been reported by *Laurila* et al. [57.26].

Flotation Control

Flotation control designed according to the classical control hierarchy of base level controls, stabilizing control, and optimizing control has been widely accepted as a mature technology since 1970. Basic controls consist of traditional PI controllers for cell levels and airflow rates. A feedforward ratio controller is used for reagent flow rates. For cell levels in series a combination of feedforward and multivariable control strategy has been also widely applied in industrial use [57.27].

Developing flotation control strategies is still an active research topic since the benefits to be gained in terms of improved metallurgical performance are substantial. However, flotation control is becoming more and more difficult due to the emergence of low grade and complex ores. Machine vision technology provides a novel solution to several of the problems encountered in conventional flotation control systems, like the effects of various disturbances appearing in the froth phase. Structural characteristics such as bubble diameter and froth mobility give valuable information for following the trend in metal grade and recovery.

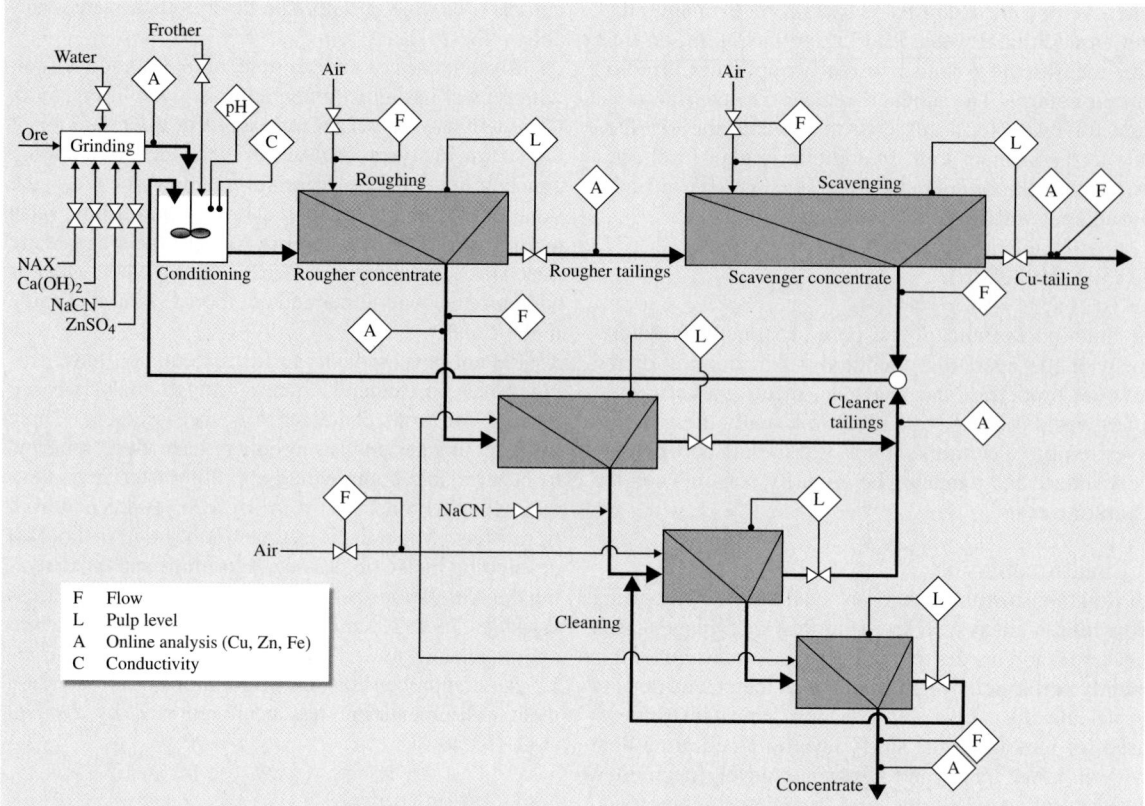

Fig. 57.6 Typical flowsheet of a flotation process (Cu circuit of Pyhäsalmi concentrator) (after [57.28])

In the FrothMaster-based control in the rougher flotation at Cadia Hills Gold Mine in New South Wales, Australia, three FrothMaster units measure froth speed, bubble size, and froth stability. The control strategy contains stabilizing and optimizing options. Stabilizing control strategy is logic based and manipulates the level, frother addition rate, and aeration rate to control the froth speed. Optimizing the control of the grade changes the setpoint values of the froth speed [57.29]. Many industrial implementations of the JKFrothCam system have been reported as well. The control system consists of PID controllers and/or an expert system. Measurements of bubble size, froth structure, and froth velocity are taken and reagent dosages, cell level, and aeration rates are used as manipulated variables [57.30]. VisioFroth is one module in the Metso Minerals CISA optimizing control system, by which froth velocity, bubble size distribution, and froth color can be measured. The largest VisioFroth installations are at Freeport, Indonesia, with 172 cameras and Minera escondida Phase IV, Chile, with 102 cameras. A combination of on-

stream analysis and image analysis technology seems to be the most efficient way to control flotation today. Better concentrate grade consistency, and thus improved plant recovery, have been reported using this combination [57.31].

Flotation, being a time variant and nonlinear process that usually also undergoes large unknown disturbances is, however, difficult to manage optimally by classical linear control theory applications. Operator support systems are needed to overcome these problematic situations. The latest applications of the operator support systems are concentrating on solving the issue of feed type classification. At many mines, changes in the mineralogy of the concentrator feed cause problems in process control. After a change in the feed type a new process control method has to be found. This is usually done by experimentation because the new type is often unknown. These experiments take time and the resulting treatment method might not be optimal. The monitoring system developed by *Jämsä-Jounela* et al. [57.28] uses SOM for online identification of the feed ore type and a knowledge database that contains information about how to handle a specific ore type. A self-learning algorithm scans historical data in order to suggest the best control strategy. The key for successful implementations is the right selection of variables for the ore type determination.

57.4 Emerging Trends

Modern automation systems in plants that have to process ever more complex ore are faced with the challenge of incorporating the increasing capabilities of modern technology in order to be able to succeed in a very competitive and global market, in which product variety and complexity, as well as quality requirements, are increasing, and environmental issues are playing ever more important key roles.

Mines and processing plants need integrated process control systems that can improve plant-wide efficiency and productivity. Advances in information technology have provided the capabilities for sharing information across the globe and, as such, process automation and control have become more directly responsible for assisting in the financial decision making of companies. The future aim of the system approach is to cover the complete value-added chain from the mine to the end product, and to utilize the latest hardware and software technology advances in their systems (Fig. 57.12).

Emerging trends in mining will likely take the form of moving towards advanced manufacturing techniques. Telecommunications systems on the surface and underground have opened the door to a completely new thinking in mining. The three biggest trends will be advances in positioning systems, telerobotic control of machinery, and the techniques that these systems will enable.

57.4.1 Teleremote Equipment

Work continues in the research field in building the equivalent to global positioning systems for underground. This technology development was recently reported at Massmin 2008 in Lulea, Sweden [57.33]. This new development combined with gyro technology will alter current practices in ways not yet comprehended. If this technology is combined with a RLG and a laser scanners mounted on a mobile machine such as shown in Fig. 57.7; the machines can have knowledge of positioning in real time on board the machine. Tasks

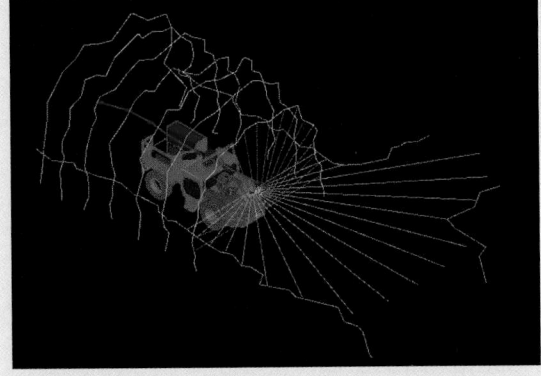

Fig. 57.7 RLG and test-bed (after [57.32])

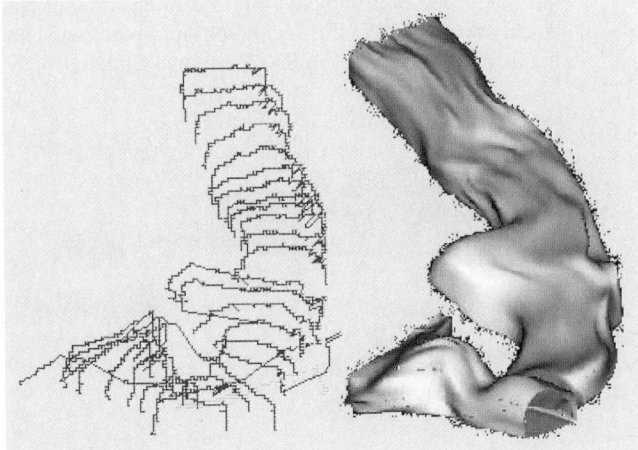

Fig. 57.8 Software generated drift from test-bed machine data (after [57.32])

Fig. 57.9 Teleremote operation chair (after [57.32])

such as mapping, drill setup, and machine guidance systems will become simple to implement. Figure 57.7 shows a RLG and a concept of a machine for surveying. Figure 57.8 shows the actual mapping data collected by this surveying machine. At present, this unit is capable of surveying a 1 km drift (tunnel) in a few hours as opposed to several days using current work practices. The addition of an equivalent to GPS for underground will improve this technology and many more.

Another important trend is enabled by advances being made in communications capacity. The operation of teleremote equipment is possible for all processes and equipment. An operator station as shown in Fig. 57.9 is connected to the machine via the telecommunications system. This allows the operator to run several machines simultaneously, and together with positioning and navigation systems will allow the operator to instantaneously move from machine to machine across multiple mine environments. Several mines around the world are attempting this technology in operation. The list includes Inco, LKAB, Rio Tinto, and Codelco.

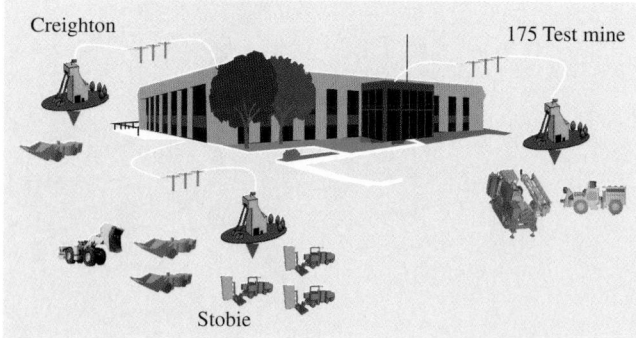

Fig. 57.10 Mines operation center

As the technology becomes more widespread, it will allow mining companies to consider the installation of mine operation centers such as the one shown in Fig. 57.10. Prototypes have been designed and installed around the world. The figure shows a mine operation center (MOC) that connects Stobie Mine, Creighton Mine, and the Research Mine at Inco. As seen in this picture, all are connected to the MOC. Three Tamrock Datasolo drills and five LHDs of various types are working or have worked from the MOC since its inception.

Benefits

Significant benefits of this teleremote style of operation lie in safety, productivity, and value-added time. Operators spend less time underground thus reducing exposure to underground hazards, and productivity is improved from the current one person per machine to one person per three machines. Initial tests indicate that 23 continuous LHD hours of operation in a 24 h period is possible, which is significantly better than the current 15 h. Clearly capital requirements in the latter situation are reduced.

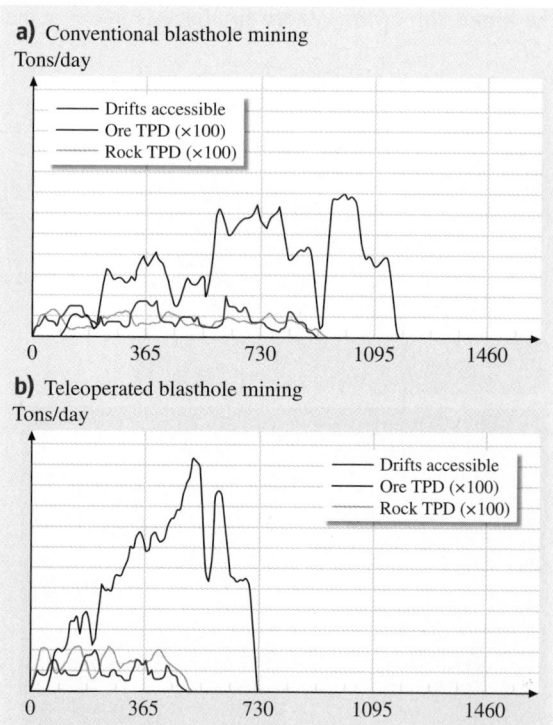

Fig. 57.11a,b Conventional (**a**) and teleremote mine (**b**) life comparisons (after [57.1])

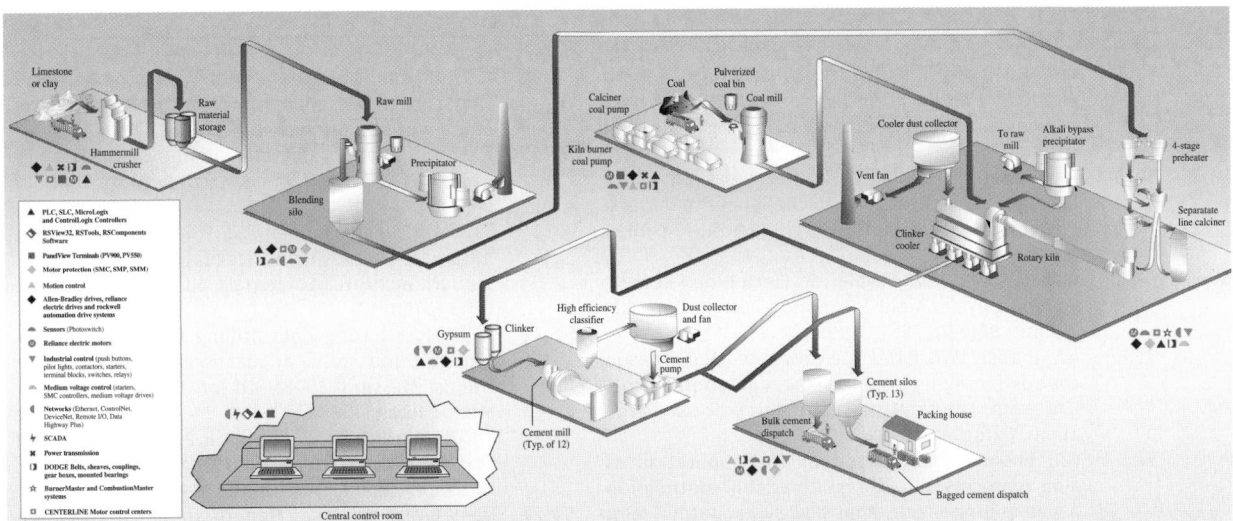

Fig. 57.12 Automation of cement processing by coordinating needed software and hardware (courtesy of Rockwell Automation, Inc)

The effectiveness of teleremote mining may be analyzed in the short term using computer-based simulation systems, which are powerful quantification and visualization tools of technology and operations. The following example shows the impact of the teleoperated mining technology on throughput, mine life, better resource utilization, and increased value generation for the organization.

57.4.2 Evaluation of Teleoperated Mining

A simulation model was used to evaluate the impact of teleremote operations on mine life and provide outputs required to make planning decisions. Teleremote operations have been shown in an operating mine to be capable of 7 to 7.5 h of operation per 8 h shift as compared to 5 h in a conventional mine. Other significant differences between conventional and telerobotic mining are increased flexibility and safety.

The comparative graphs shown in Fig. 57.11 show significant potential results from the application of robotics and automation. Mine life is reduced by 38% using teleremote mining versus conventional mining because of the higher mining rate from improved throughput and face utilization. Moreover, utilization of

LHD equipment is increased by 80% in teleremote mining compared to conventional settings. With a total of two LHDs, high rates of production were achieved.

57.4.3 Future Trends in Grinding and Flotation Control

The optimization methods for grinding control are expected to be developed due to better particle distribution measurement systems, advanced mill condition measurements (e.g. frequency analysis), and the use of efficient grinding simulations based, for instance, on the discrete element method. Flotation is facing a new era in terms of process control and automation. Flotation cells have increased in size dramatically over the past years; flotation circuit design of multiple recycle streams will be replaced by simpler circuits, subsequently leading to a decreasing number of instruments with higher demands on reliability and accuracy. This will set new challenges for the control system design and implementation of these new plants. Process data driven monitoring methods, model predictive control (MPC), and fault tolerant control (FTC) will be among the most favorable methods to be applied together with the recently developed new measurement instruments.

References

57.1   W.A. Hustrulid, R.L. Bullock: *Underground Mining Methods – Engineering Fundamentals and International Case Studies* (Society of Mining Engineers, Littleton 2001)

57.2   D. Hodouin, S.-L. Jämsä-Jounela, M.T. Carvalho, L. Bergh: State of the art and challenges in mineral processing control, Control Eng. Pract. **9**, 995–1005 (2001)

57.3   G.R. Baiden, M.J. Scoble, S. Flewelling: Robotic systems development for mining automation, CIM Bulletin **86**(972), 75–77 (1993)

57.4   J.A. Herbst, W.T. Pate: The power of model based control for mineral processing operations, Proc. IFAC Symp. Autom. Min. Miner. Met. Process. (Pergamon, Oxford 1987) pp. 7–33

57.5   G.I. Gossman, A. Bunconbe: The application of a microprocessor-based multivariable controller to a gold milling circuit, Proc. IFAC Symp. Autom. Min. Miner. Met. Process. (1983)

57.6   J. Miettunen: The Pyhäsalmi concentrator – 13 years of computer control, Proc. IFAC Symp. Autom. Min. Miner. Met. Process. (Pergamon, Oxford 1983) pp. 391–403

57.7   G.R. Baiden: A Study of Underground Automation. Ph.D. Thesis (McGill University, Montreal 1993)

57.8   T. Inoue, K. Okaya: Grinding mechanism of centrifugal mills – A simulation study based on the discrete element method, Int. J. Miner. Process. **44**(5), 425–435 (1996)

57.9   A. Datta, B. Mishra, K. Rajamani: Analysis of power draw in ball mills by discrete element method, Can. Metall. Q. **99**, 133–140 (1999)

57.10 M.M. Bwalya, M.H. Moys, A.L. Hinde: The use of the discrete element method and fracture mechanics to improve grinding rate prediction, Miner. Eng. **14**, 565–573 (2001)

57.11 B.K. Mishra: A review of computer simulation of tumbling mills by the discrete element method: Part I – Contact mechanics, Int. J. Miner. Process. **115**, 290–297 (2001)

57.12 T. Inoue, K. Okaya: Grinding mechanism of centrifugal mills – A simulation study based on the discrete element method, Int. J. Miner. Process. **44**(5), 425–435 (1996)

57.13 P.T.L. Koh, M.P. Schwartz: CFD modeling of collisions and attachments in a flotation cell, Proc. 2nd Int. Flotat. Symp., Flotation'03, Miner. Eng. Int., Helsinki (2003)

57.14 M.G. Lipsett: Information technology in mining: An overview of the Mining IT User Group, CIM Bulletin **1067**, 49–51 (2003)

57.15 G.R. Baiden, M.J. Scoble: Mine-wide information system development, Canadian Institute of Mining and Metallurgy - 93rd Annu. Gen. Meet. Bull., Montreal (1991)

57.16 A. Hulkkonen: Wireless underground communications system, Telemin 1 and 5th Int. Symp. Mine Mech. Autom., Sudbury (1999)

57.17 P. Cunningham: Automatic toping system, Telemin 1 and 5th Int. Symp. Mine Mech. Autom., Sudbury (1999)

57.18 Y. Bissiri, G.R. Baiden, S. Filion, A. Saari: An automated surveying device for underground navigation, Institute of Materials, Minerals and Mining IOM3 (2008)

57.19 A. Zablocki: Long hole drilling trends in Chilean underground mine applications, capacities and trends, Massmin 2008: 5th Int. Conf. Exhib. Mass Min., Lulea (2008)

57.20 T.J. Napier-Munn, S. Morrel, R.D. Kojovic: *Mineral Comminution Circuits: Their Operation and Optimization* (JKMRC, Queensland 1999)

57.21 S.-L. Jämsä-Jounela: Modern approaches to control of mineral processing, Acta Polytech. Scand. Math. Comput. Sci. Ser. **57**, 61 (1990)

57.22 P. Koivistoinen, R. Kalapudas, J. Miettunen: A new method for measuring the volumetric filling of a grinding mill. In: *Comminution Theory and Practice*, ed. by S.K. Kaeatra (Society for Mining, Metallurgy and Exploration, Littleton 1992) pp. 563–574

57.23 J. Järvinen: A volumetric charge measurement for grinding mills, Preprints of the 11th IFAC Symp. Autom. Min. Miner. Met. Process. (1994)

57.24 S.-J. Spencer, J.-J. Campbell, K.-R. Weller, Y. Liu: Acoustic emissions monitoring of SAG mill performance, Proc. 2nd Int. Conf. Intell. Process. Manuf. Mater. IPMM'99 (1999) pp. 939–946

57.25 S.-L. Jämsä-Jounela: Current status and future trends in the automation of mineral and metal processing, Control Eng. Pract. **9**, 1021–1035(2001)

57.26 H. Laurila, J. Karesvuori, O. Tiili: Strategies for instrumentation and control of flotation circuits. In: *Mineral Processing Plant Design, Practice and Control*, ed. by A.L. Mular, D.N. Halbe, D.J. Barratt (Society of Mining, Metallurgy and Exploration, Littleton 2002)

57.27 P. Kampjarvi, S.-L. Jämsä-Jounela: Level control strategies for flotation cells, Miner. Eng. **16**(11), 1061–1068 (2003)

57.28 S.-L. Jämsä-Jounela, S. Laine, E. Ruokonen: Ore type based expert systems in mineral processing plants, Part. Part. Syst. Charact. **15**(4), 200–207 (1998)

57.29 M. van Olst, N. Brown, P. Bourke, S. Ronkainen: Improving flotation plant performance at Cadia by controlling and optimizing the rate of froth recovery using Outokumpu FrothMaster, Proc. 33rd Annu. Oper. Conf. Can. Miner. Process., ed. by M. Smith (Ottawa, 2001) pp. 25–37

57.30 P.N. Holtmam, K.K. Nguyen: On-line analysis of froth surface in coal and mineral flotation using JKFrothCam, Int. J. Miner. Process. **64**, 163–180 (2002)

57.31 Anon.: CICA OCS Expert System, http://www.metso-cisa.com/ (2005)

57.32 P. Koivistoinen, J. Miettunen: Flotation control at Pyhäsalmi. In: *Developments in Mineral Processing, Flotation of Sulphide Minerals*, ed. by K.S.E. Forssberg (Elsevier, Amsterdam 1985) pp. 447–472

57.33 Y. Bissiri, A. Saari, G.R. Baiden: Real time sensing of rock flow in block cave mining, Massmin 2008: 5th Int. Conf. Exhib. Mass Min., Lulea (2008)

58. Automation in the Wood and Paper Industry

Birgit Vogel-Heuser

Plant automation in the *timber (wood) industry* and *paper industry* has many analogies due to the similar process characteristics, i. e., hybrid process. From a plant automation point of view these industries are challenging because of their technical requirements. The USA is the largest producer of paper and paperboard. In a census by the America Census Bureau, the paper industry is placed 7 among 21 different industry groups, with 6% of the total value of product shipments for the years 2005 and 2006 [58.1]. China, as the second largest paper producer after the USA, is expecting a growth rate of 12.4% per annum (for the forecast from 1990 to 2010) [58.2]. The German woodworking machinery sector, with more than 26% of the world market share, had a € 3.4 billion turnover in 2006 and 72% export quota [58.3]. In 2006, the German print and paper industry grew by more than 7%, to € 8.5 billion and more than 84% export quota ([58.4], more details in [58.5]). This chapter will not only highlight the specific requirements from a technical point of view but also from the marketing point of view. Observations from these two points of view will lead to a heterogeneous automation system with

proprietary devices for real-time and machinery-safety-related tasks, and standard devices for the rest. Both industries belong to plant manufacturing industries with their typical business characteristics. Automation in this field is technology driven and its importance is growing because more functionalities are being implemented using automation software to increase systems flexibility. The interface from automation level to enterprise resource planning (ERP) systems is being standardized in international manufacturing companies. Engineering is the key factor for improvement that needs to be considered in the coming years, and therefore also modularity and reusability where applicable.

58.1 Background Development and Theory

Both processes considered in this chapter are characterized by a large number of interrelated but independent processing steps as well as complex control parameters. Despite narrow limits and sophisticated process control, inhomogeneous properties of raw materials will cause variations in product quality. These variations cannot be measured directly at the time of production and can only be determined subsequently by destructive testing.

An overview of the requirements of process automation in plant manufacturing industry is provided in Table 58.1.

The criteria can be structured according to process requirements, automation system architecture, and project. A process automation system controls a *type of process*, e.g., *batch*, *continuous*, or *discrete*. Sometimes processes are composed of different process types and are then known as *hybrid* process. Both processes discussed in this chapter are hybrid process. Hybrid processes require different control strategies and therefore also different modeling notations, e.g., block diagrams (continuous) or state charts (batch). Both processes demand a *hybrid automation system* architecture.

Table 58.1 Overview of the requirements of process automation

Categories/criteria		Functionality/notation aspects
Process (hybrid)	Batch (continuous)	Transfer functions, block diagrams, differential equation
	Discrete	Status model, flow chart, continuous function chart, Petri net
	Real time	Hard real-time and event-controlled system, synchronized drive
	Reliability	Mapping of reliability aspects (failure mode and effects analysis: FMEA, fault tree analysis: FTA)
Automation system (heterogeneous)	Implementation	Programming language (C, IEC 61131-3, PEARL)
		Operation system [proprietary real-time operating system (RTOS proprietary)], Windows CE or XT
	Hardware platform	Programmable logic controller (PLC), distributed control system (DCS), personal computer (PC), microcontroller
	System architecture heterogeneous	Central, decentralized, distributed
	Human–machine interface (HMI)	PC-based standard HMI tools and proprietary tools for fast trending of process data

Specific requirements for both considered processes in terms of process automation are:

1. Wood, as the input material, fluctuates strongly due to the environment.
2. The required mechanical construction relies on heavy machinery.
3. They have hard and fast real-time requirements for the control of central machinery with a large number of control loops, which does not allow the use of one standard automation device, for example, a programmable logic controller (PLC) or a decentralized control system (DCS). As an example, during particle board manufacturing in a hydraulic press, up to 350 closed-loop controllers need to run within 10–20 ms cycle time.
4. Both timber and paper plants can consist of 3000–6000 analogue or digital inputs and outputs that are connected via a fieldbus. A paper plant needs to control up to 3500 control loops using 10–20 central processing units (CPUs) (PLCs or DCSs) that communicate via Ethernet-based bus systems.
5. The huge number of analogue control loops in a machine, in combination with requirements for fast real-time operation, results in the use of specific automation solutions. In the timber industry a versa module eurobus (VMEbus)-based control system is used, whereas in the paper(making) industry, decentralized control systems (DCS) with proprietary automation devices are used.

6. Worldwide environmental conditions lead to different automation architectures, i. e., decentralized and centralized design.
7. Consideration of both cross section and longitudinal section in control is required. The width of a pressed board is up to 2.9 m and in a paper machine up to 11 m, therefore the control and operation task has to deal with these two dimensions.
8. The need for precise synchronization between several drives in the machine, else the paper may be ripped, the timber mat may be ripped or compressed, or the steel belt of the hydraulic press may be damaged, thus causing damage to the surface of the panels. Speed of production depends strongly on product thickness, and ranges from 300 mm/s to 1.5 m/s in the timber industry and up to 25 m/s in the paper industry. Older solutions used vertical shafts to synchronize the drives accurately.
9. Continuous production around the clock, 365 days a year, demands high availability of plant automation system, i. e., about 90% in the timber industry and 99.9% in the paper industry.
10. Plant automation can automate almost 100% of tasks, but some important measurements such as moisture can still not be measured precisely, while other values such as internal bonding and bending strength are only available through laboratory experiments such as destructive testing.
11. Implementation of closed-loop controllers for mat weight or square weight, moisture, temperature, and thickness is required. Quality criteria for the sur-

Table 58.2 Languages of the IEC 61131-3 standard

Name/Functionality	Application
Ladder logic/circuit diagram	On/off, lamps
Instruction list/assembler type	Time-critical modules
Function block language/Boolean operations and functions	Interlocking, controller, reusable functions, communication
Sequential function chart/state diagram	Sequences
Structured text/higher programming language	Controller, technological functions

face structure include color and gloss in the paper industry and varnishing in the timber industry.

12. From the control-theoretic point of view, today both processes still lack a method that can model the whole process and the central machinery, i.e., the continuous hydraulic press in the timber industry and the paper machine in the paper industry. A first attempt to develop a model for the central machine has been made for the timber industry [58.6]. A method to model the process has not yet been developed due to lack of information concerning the process. Support from the operator and technologist is needed to bridge this gap, and to do this they would need, for example, a fast trending tool (sample time 10 ms for 450 analogue inputs) to analyze and optimize control loops in the entire machine or process.

After the discussion of technical requirements the typical market requirements should be mentioned. The company- and/or product-specific technological knowhow for both considered domains are of great value. The market is very competitive, which is why technological knowhow is kept within different production companies. Technological feedback to the machinery supplier is weak and therefore it is difficult to achieve technical control improvement. Suppliers are mostly situated in high-wage countries but the market is global, with growing competition from low-wage countries, therefore engineering efficiency and reduced engineering times are important for suppliers. Additionally, reduced start-up times and improved plant reliability are growing challenges in every part of the world. This means that engineering processes need to be optimized through modularity and reuse, as well as data integration along the engineering lifecycle. Modularity and reusability in modeling and simulation provide the opportunity to increase software quality while reducing start-up time.

During the design of each plant location, various requirements need to be taken into account, regarding: standards, type, and stability of power supply from power companies; ground connection; and the different qualifications of operator and service personnel.

Due to the availability requirement (see item 9 above) and the protection of technological knowhow while optimizing production, there is a need to provide maintainable and adaptable systems for the worldwide market. As far as possible standardized automation systems should be used, i.e., PLCs in the timber industry and DCSs in the paper industry.

The end customer usually specifies the brand of PLC, human–machine interface (HMI), or DCS, and sometimes also the drives used (if technologically possible). PLC and DCS systems are programmed according to the International Electrotechnical Commission (IEC) 61131-3 standards (Table 58.2). Customers in the USA prefer ladder logic as the programming language whenever possible. Technologically complex functions will be encapsulated in function blocks. However, standard PLCs and DCSs are not appropriate for fast real-time requirements and they can be manipulated by end customers; this should not be allowed as they are complex devices and there is the risk of machine hazards. This is why heterogeneous automation systems are standard in both industries. These heterogeneous systems consist of standard devices, standard HMIs, and fieldbus solutions, enriched with domain- or even company-specific solutions for fast real-time control. The important difference between the two industries is that the requirement for intrinsic safety is more relevant to the paper industry compared with the timber industry. The intrinsic safety requirement has a strong impact on the selection of appropriate equipment, especially sensors and actuators as well as fieldbus solutions.

Some green-field plants for the timber industry include glue production and/or paper machines that produce paper necessary for lamination.

Most machinery will be shipped in parts and is only commissioned on site, therefore machines or machine components will only be tested at the supplier's site. The final test is necessary after commissioning on site.

Training of maintenance personnel and operators takes place during commissioning and start-up. *Operator training systems* (OTS), which include simulation of the process, will be developed in the near future.

58.2 Application Example, Guidelines, and Techniques

As discussed, the requirements of the timber and paper industries are similar. This Chapter will first explain the continuous hydraulic press as an application example to show these requirements in more detail and appropriate automation solutions. After that, one solution for a paper machine will be introduced to highlight the similarities in terms of requirements and solutions.

58.2.1 Timber Industry

The production of fiber boards in the timber industry is a *hybrid production process* which generates a product (wooden panels) that has to meet many quality requirements. The thickness and density of the board or the consistency of the material are target values that need to be controlled within the tight intervals. The process

consists of multiple steps where wood as the input material is prepared (Fig. 58.1). In the subsequent steps, the fiber mat is formed through continuous pressing process, heated, hardened, and sawn into discrete boards.

Wood is a natural material that is strongly influenced by temperature and humidity conditions. The input material's characteristics for the process varies widely.

The most costly piece of equipment to acquire and operate is the hydraulic press. It usually determines the maximum capacity of the production line [58.8]. The simplified press principle is described as follows. For the finished board, the material (already mixed with glue) has to be pressed with a specific pressure to obtain a certain thickness that is related to a set value. Wood is a natural material with underlying strong disturbing influences such as temperature or humidity

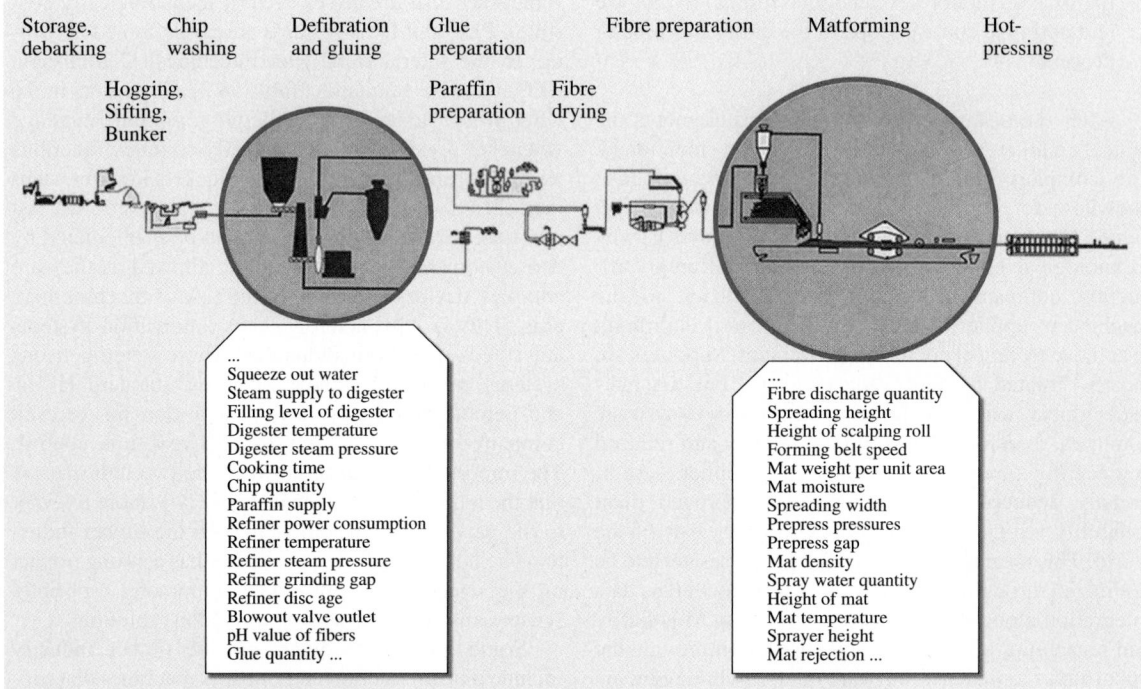

Fig. 58.1 Overview of the fiber board production process (after [58.7])

conditions. Thereby the characteristic of the input material to the press varies widely and cannot be measured with sufficient accuracy for automatic control. Hence the operator should be able to influence the automatic control strongly in order to compensate for such influences, based on his experience. The operator must be able to recognize that the situation is no longer in the range of the selected recipe and that he has to react by changing the distance or pressure parameters in a specific frame or group of frames (Fig. 58.2a). Therefore the operator needs a good overview of the real values of pressure and distance in cross-section and longitudinal directions (Fig. 58.2b).

Inclination of the press, which would produce an improper board, can be monitored on the distance profile (Fig. 58.2b right, section 6 slight inclination).

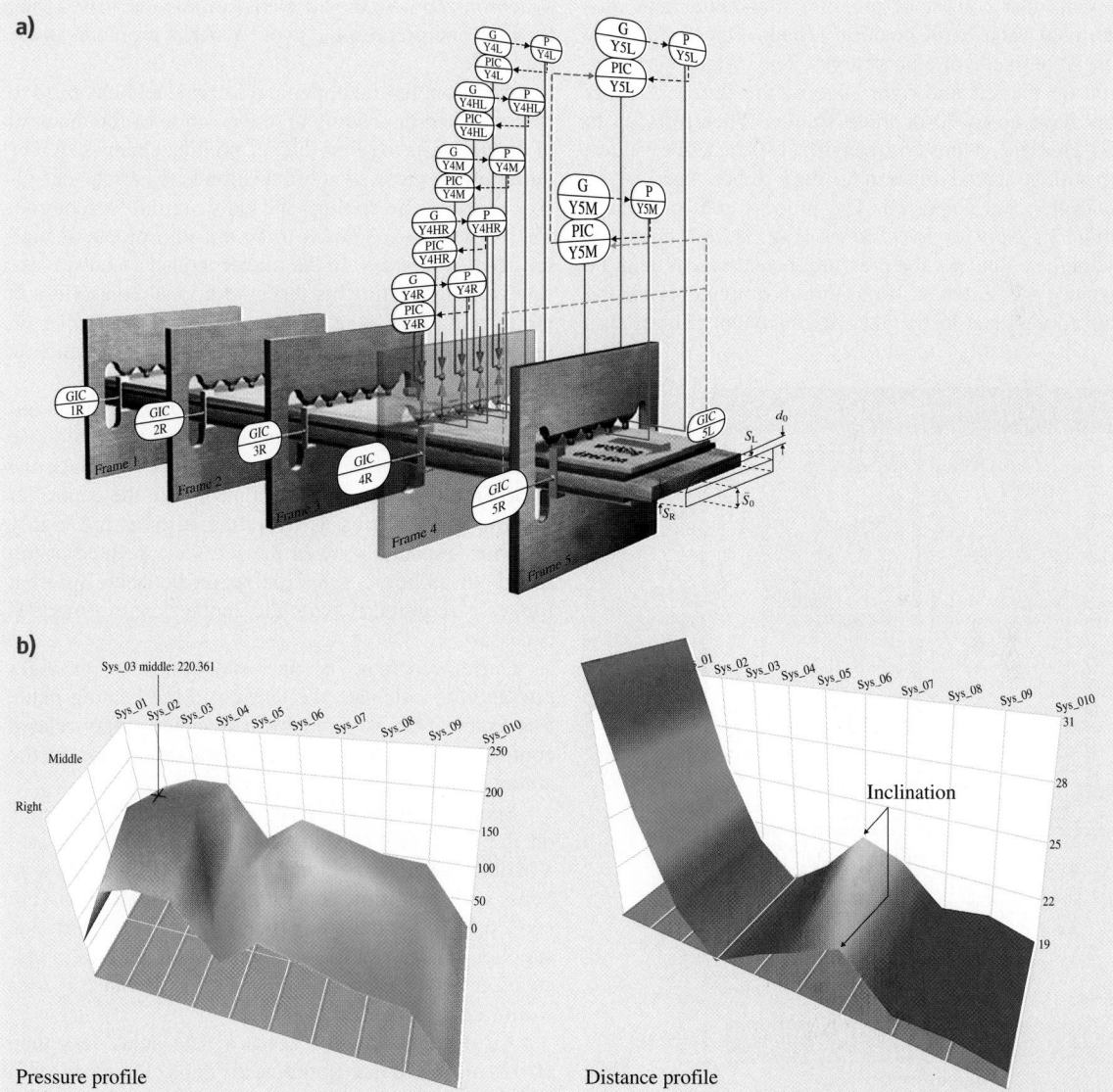

Fig. 58.2 (**a**) Process and instrumentation diagram of a continuous *thermohydraulic press* (only distance and pressure control, working direction from *left* to *right*). (**b**) Pressure and distance profile of a specific hydraulic press in 3-D; *x*-axis shows working direction from *left* to *right*, *y*-axis displays pressure profile (*left*) and distance profile (*right*) (after [58.9])

Torsion of the press table would be identified easily using this presentation. It is possible to identify if the pressure is too high along and/or across the press using the pressure profile (Fig. 58.2b left). The shape allows evaluation of whether the mat will be pressed regularly and symmetrically (across the cross section). Due to technical requirements a maximum pressure is set in order to achieve the thickness of the material. Distance control is realized by several hydraulic systems that consist of pressure transmitter and proportional valve with position sensor. One frame may consist of five hydraulic systems. The distance is measured on the left and right edge of the frame. A press may have up to 70 or more frames. Therefore, up to 350 pressure values and roughly 140 distance values, all with a spatial relation to each other, need to be controlled and displayed. The process and instrumentation diagram for five frames (Fig. 58.2a) shows the hydraulic cylinders, the pressure measurement, and the distance measurement. Real-time requirements for the entire loop need to be taken into account (Fig. 58.2a).

The controllers are more or less simple proportional–integral–differential (PID) controllers with filters or nested PID controllers. Complexity depends on their interdependency. Hydraulic systems are mechanically coupled to a thick steel plate to heat the mat so that the glue will harden. Besides the pressure control and distance control, temperature control and controllers to synchronize different drives at the inlet and the outlet of the press are also included. These drives need to be synchronized with the different forming line drives (mat forming and prepress) to avoid material problems in the press inlet.

Additionally the upper and lower steel belts need to be controlled depending on the position of the material in the press during first inlet or product changes. At the outlet of the press, synchronization between the cut-to-size saw and the cooling and stacking line is required. The endless board needs to be cut into pieces at high speed, e.g., 1.5 m/s. In the timber industry a cut-to-size saw is included for this task. Additional controllers in the forming line or press may be added depending on product requirements and the speed of the production process.

Figure 58.3 shows the usual time delays in the entire control cycle from data measurement to its effective influence on the process. It is necessary to reduce the cycle time for each distance controller in the controller itself to under 20 ms as hydraulic systems are highly dynamic, especially when the press is operated at top speed. In addition, synchronization between different frames is required if switch to another control mode is necessary.

Currently there is one supplier who provides a method to calculate the tension in the heating plate. The calculation is done by a machine-safety-related controller based on a finite-element calculation in the automation system.

Specific process control systems, which depends on the supplier's chosen solution, e.g., Motorola and VMEbus-based or personal computer (PC)-based systems with real-time operating system (RTOS), are needed to fulfill the strict time requirement. Only one supplier delivers a PLC solution, which needs less calculation power for a continuous press because it uses a different mechanical concept.

All distance control needs a data delay less than 10–20 ms in order to optimize the control loops for thin board. In order to optimize a controller, it is necessary to use a data sampling rate that is 5–10 times faster. This results in an optimized data gathering strategy and the development of proprietary trending systems [58.10].

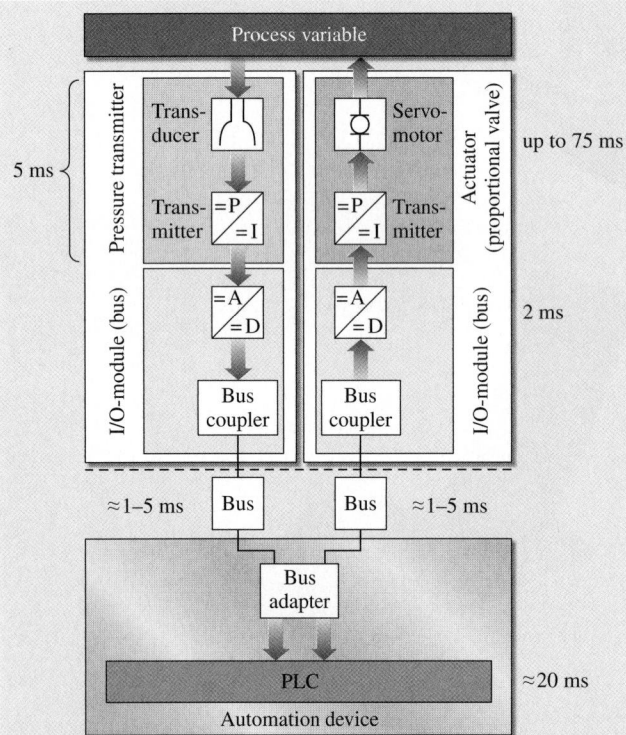

Fig. 58.3 System components of the control loop that needs to be considered for (hydraulic press) cycle time calculation. I/O – input/output

Classically, HMI trending functionality is limited to the fastest sample time of 3 s or even slower because they focus on long-term trending. PLCs are connected with an Ethernet-based network (Fig. 58.4). Sensors and actuators are connected with a fieldbus system to ensure deterministic behavior. The cut-to-size saw is realized on PLCs with controllers for servo drives, or on more or less computer numerical control (CNC)-type devices to enable the required high speed.

Due to customer requirements in plant operation, such as long mean time between failures (MTBF), plant maintenance should be carried out by their plant service personnel. They should have the possibility to manipulate the PLC program within certain limits and to continue the optimization of processes. Therefore PLC is the preferred automation approach. The customer can specify the PLC supplier and bus systems supplier according to the market shares in his company and/or country, the service structure of the PLC supplier, and the skills of his maintenance personnel.

It is a marketing advantage to implement all control functionality on PLC. Due to the growth in system manipulation, a new challenge for PLC is emerging: the ability to track operator input and manipulations made on the PLC so that improper manipulations that may

cause hazards can be detected and tracked. Password protection alone is not sufficient.

Regarding safety requirements, a strategy to overcome loss of power or press stoppage with material in the press is required because the material may start burning after a long time. Uninterruptible power supply (UPS) and emergency power supply are standard equipment.

The *cooling and stacking line* (sanding and trimming of particle boards) is needed to cool down the panel and prepare it for the finishing line with lamination or panel division, sorting, and packing. Intermediate and delivery storages in timber industry often follow a simple but chaotic storage strategy. However, tracking and tracing of boards is becoming increasingly important to the customer. For this reason, ERP and production management are becoming more important and a more precise storage system is needed. Board handling is often realized by using forklift trucks. The boards are chaotically stacked in a building until the truck from the customer arrives. They are identified only by a piece of paper with a printed barcode. Due to the manual handling strategy, tracking of faulty material is nearly impossible. Two strategies will be discussed and evaluated later (Sect. 58.2): radiofrequency identi-

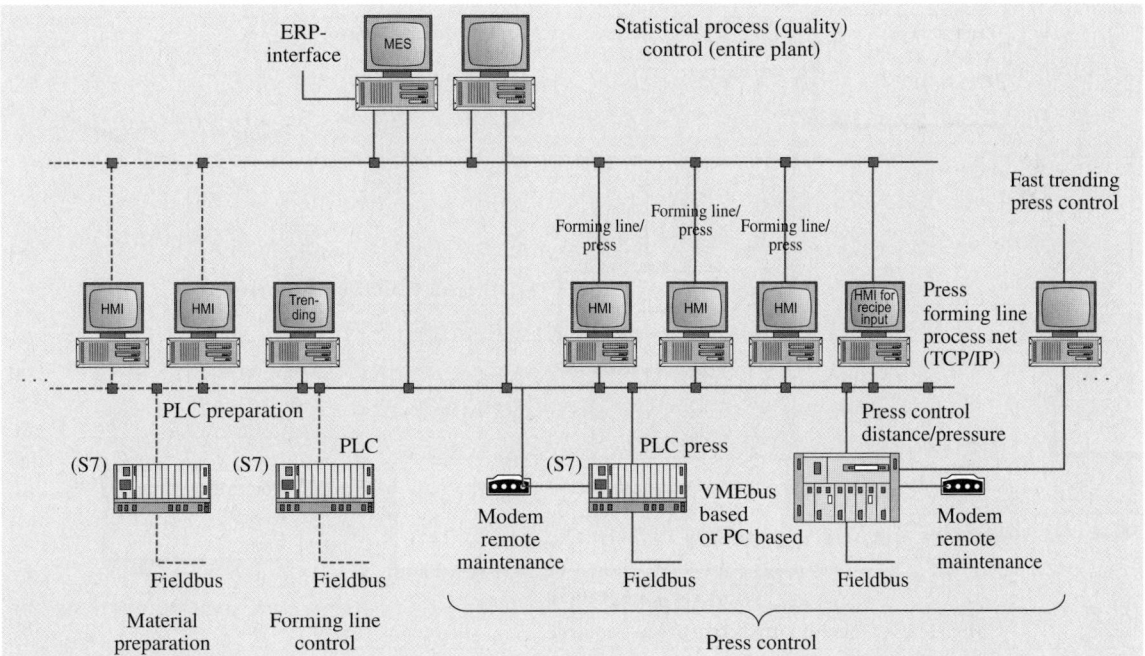

Fig. 58.4 Automation architecture (example from one supplier). TCP/IP – transmission control protocol/Internet protocol (MES – manufacturing execution system)

fication (RFID) of every board and global positioning system (GPS) detection of every stack.

There are also automated intermediate storage systems that allow handling of different board sizes on a mobile electrically powered rail-bound device using steel pallets [58.13]. The material flow is tracked and the product data are available. In production of thin medium-density fiber (MDF), board pallets need to be equipped with top and bottom protection panels to keep the product flat and protect its surface.

The quality of the board is not only influenced by the press but in all sections of the plant, particularly by the properties of the wood mixture and glue used [58.14]. The plant sections involved in the entire manufacturing process – starting with the hogged chip manufacturing, flaking or defibrating, drying, blending, straining, mat forming, up to prepressing and pressing sections – are all interdependent and, like the properties of the raw materials used, are subject to fluctuations. There are more than 100 parameters that have various intensities of effects on a plant's productivity and the quality of the product (Fig. 58.5). These parameters are input values for a process model based on a statistical algorithm [the three-stage least-squares (3SLS) algorithm [58.15, 16]].

Prerequisites for the analysis of data detailing the production history of the product, so-called material tracking and correspondingly time-correlated process data, need to be calculated. First implementations of a model-based quality-prediction and cost-optimization process control are running successfully [58.11, 12].

The quality of relevant process data and control parameters are taken into account to predict the resulting quality. Optimized process settings Y_{opt} for control parameters Y are calculated to ensure that nominal required quality in terms of costs is met

$$\text{cost}(Y) \to \text{Minimum}$$

under the condition

$$Q_{nom}(t+1) < Q_{pred}\big(X(t), Y_{opt}(t+1)\big) + S_{pred}\big(X(t), Y_{opt}(t+1)\big)$$

and

$$Q_{nom}(t+1) > Q_{pred}\big(X(t), Y_{opt}(t+1)\big) - S_{pred}\big(X(t), Y_{opt}(t+1)\big).$$

To close the loop, these optimized process settings are entered as new nominal values for the control parameters.

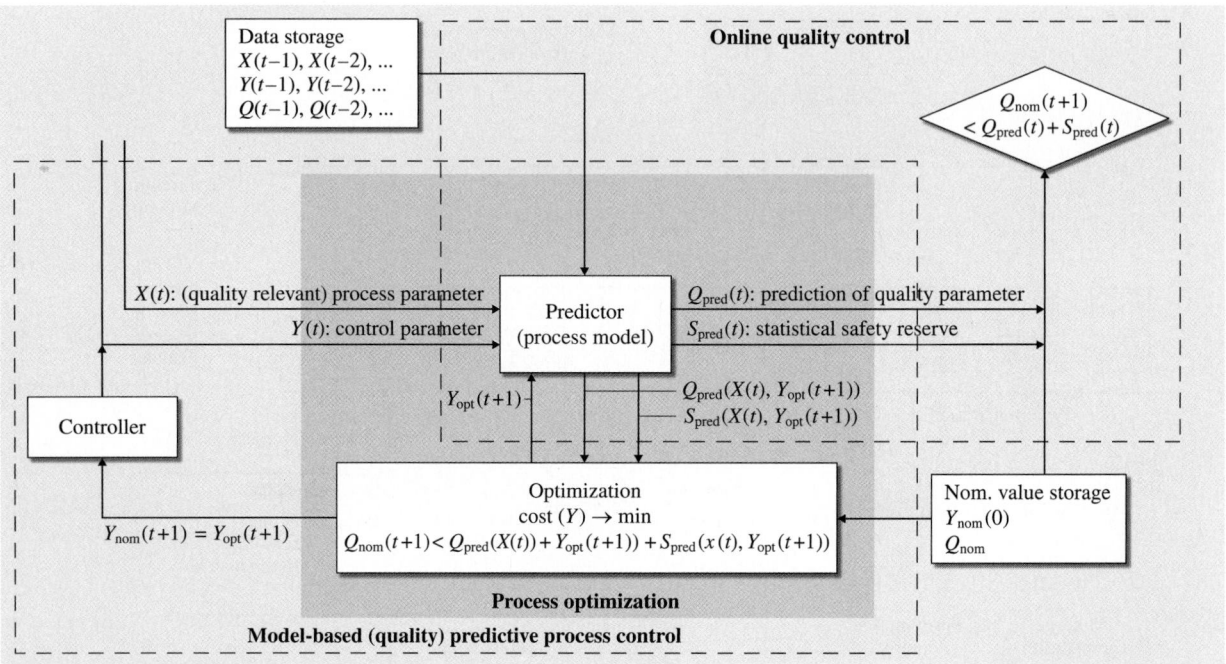

Fig. 58.5 Global control loop using *model-based quality-predictive* and *cost-optimized process* control (after [58.11, 12])

58.2.2 Paper-Making Industry

Heterogeneous automation systems are implemented in the paper industry to control force, torque, temperature, moisture, and vibration. Required reliability is 0.9999 (24 h/365 days per year). An overview of process sections, controllers, and most importantly sensors is given in Fig. 58.6 (working direction: left to right, with the process divided into two figures). The paper machine is a device for continuously forming, dewatering, pressing, and drying a web of paper fibers [58.18].

The automation system is mainly realized using a DCS system. In the given example, the automation system was being realized using PCS 7 Siemens based on the 400 CPUs. The connection to the sensors and actuators is realized using the PROFIBUS DP fieldbus. Specific automation devices are implemented for cross-section control and transmission control. Most measuring devices, e.g., gloss, moisture, caliper and basis weight, and color, are equipped with their own au-

tomation device and will be connected with a controller area network bus (CANbus) to the cross-section control and over a measurement server (PC) to the HMI. Voith, for example, developed its own weight profile control system software named Profilmatic. Profilmatic cross-direction control software continuously and automatically aligns each actuator against its respective measurement position from the downstream scanner. An automapping algorithm monitors the movement of a normal profile control and aligns the cross-direction measurement data boxes against the actuator control zones. Each control output array compares the actual profile change against the expected profile change. A software model continuously updates the process-mapping model using the difference between measured and expected response.

Model-based soft sensors deliver high-quality data for superior longitudinal control of the paper machine. Similar to in the wood industry, a multivariate statistical approach is implemented to forecast, for example, the

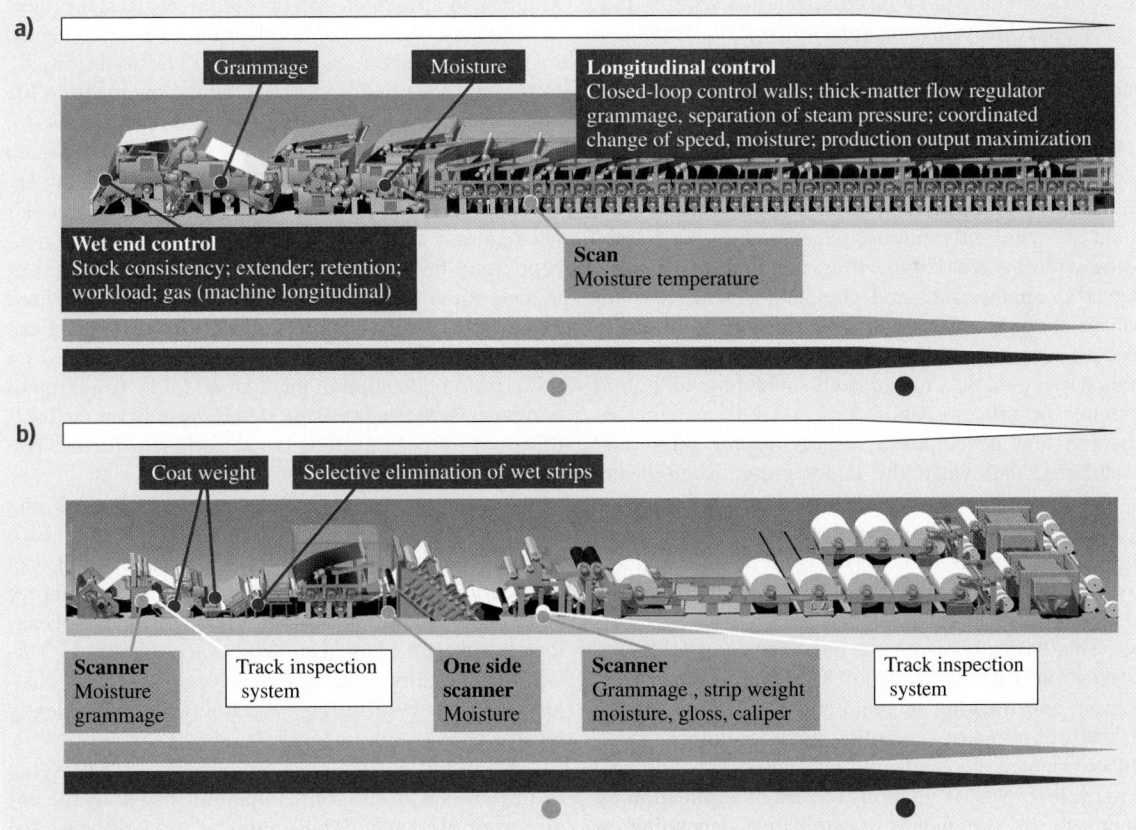

Fig. 58.6 Example of a paper machine: process overview with sensors and actuators (after [58.17])

weight profile at the end of the paper machine. Some basic constraints for the implementation of statistical process (quality) control in paper industry are given in [58.12], e.g.,

> this assumes that all of the important variables are measured in a timely manner … Another point to consider is the sampling frequency. Generally, the more variability in a material, the more often it should be sampled.

Blatzheim gave an example of the benefit of such system: 15–16 km of waste material reduction was achieved after implementing this system, corresponding to an approximate increase of 200 000 of sellable paper per year.

Laukkanen mentioned OTS and appropriate workflow guidelines as being one of the keys to reduce commissioning and start-up, especially abroad [58.3].

Through the engineering life cycle, depending on the project phase, different tools and methods are used. In the worst case, the corresponding data have to be re-entered during the transition from one phase to the next because there are no appropriate interfaces between the individual tools. The ideal is to strive for a higher-level tool that consistently provides all system information in a model and enables the design to be realized both at an abstract level and in a conventional environment (e.g., electrical engineering computer aided engineering (E-CAE), IEC 61131-3).

58.3 Emerging Trends, Open Challenges

There are many open challenges in the timber and paper industries. Some will be discussed in this section, i. e., *engineering lifecycle* and *data integration*, *reduction of complexity* for operators, improvement of operator training, as well as evaluation of new technologies.

Open challenges are to increase engineering efficiency while reducing costs and start-up time by applying modularity and reuse. Communication is a major factor to consider when improving engineering quality because different disciplines such as sales, technology, mechanical engineering (including hydraulic), electrical engineering, and computer science are involved through different phases in the engineering process. Communication could be supported by a comprehensive model. Unfortunately modeling of hybrid systems in process automation is still a field for research and development, especially for reuse and modularity, and when the target group is engineers or technicians coming from different disciplines. One promising approach is to apply a comprehensive modeling notation such as the systems modeling language (SysML) [58.19] that is based on the unified modeling language (UML) and developed for systems engineering. This basically solves the deficiencies of UML for automation, i. e., modeling of hardware aspects, integration, and tracking of requirements. The advantage of UML in terms of supporting modularity through an object-oriented mechanism is fully available. The task is to evaluate SysML in terms of ease of application by engineers and technicians in automation depending on the availability of strong tools. One of the first steps

to integrate modeling into IEC 61131-3 is already in progress in a research and development (R&D) project in Germany. Various UML diagrams are implemented in one of the market leading IEC 61131-3 tools [58.20]. Presenting the modules throughout the engineering life cycle from customers' requirements to operation is still not solved and will not only be a task for research but also for the development department of leading automation companies. Plant manufacturing companies need an easy way to evaluate whether a new plant concept could be realized by reusing existing modules or to determine the number of new modules that need to be developed. Data integration throughout the engineering life cycle is theoretically solved but still far away from realization in the applied tools. Coupling of computer-aided engineering (CAE) system on the basis of a tool-to-tool interface is not sufficient for the two considered industries.

There is also huge potential for cost saving during start-up by integrating engineering data into an ERP system, and offshoring in a global market. Additional elaboration on logistics, e.g., just-in-time machine or component to site, is needed. Today mainly mechanical construction has been subject to offshoring. This will change during the next few years. A lot of challenges for the engineering workflow will consequently be affected.

Operation and maintenance, which is challenging to improve, is also a very important phase in the engineering life cycle. Application of three-dimensional (3-D) visualization can successfully reduce complexity for operators and at the same time enhance their mental model of the process [58.9, 21]. An expert evaluation has proved the benefit of 3-D visualization. It would also be beneficial to implement a sliding mechanism to analyze data offline. Real process data presented in 3-D, e.g., a surface plot (Fig. 58.1b), may be analyzed in slow or fast motion. By selecting the time frame to be analyzed the technologist can view the data like a data player. This is integrated into standard HMI systems and allows analysis of critical situations as well as faulty situations in order to increase understanding of the process. Recent work focuses on application of this approach for operator training.

Regarding new technological trends, Ethernet-based fieldbus systems as well as identification technologies (RFID) need to be evaluated. A first implementation of Profinet in the cooling and stacking line has been realized. The upcoming question and design decision is whether Profinet should also be used for communication between PLC, sensors, and actuators, also in hazardous areas. Market shares outside Europe are hard to predict but this is a prerequisite before coming to a general decision for one bus system based on Ethernet standard.

The use of RFID to optimize material handling from the press outlet to customers is being discussed. There are some constraints that need to be evaluated before a test, e.g., placement of the RFID (how it can be fixed to the board before the press or after the cut-to-size saw); temperature resistance, as the boards are still at about $100\,°C$ at the press outlet; and costs. The idea is to store all relevant data relating to the board during its production process on the RFID tag so that customers have access to the production data on request.

Board-handling satellite data to define the position of a panel stack in a chaotic storage system may also be helpful.

References

58.1 US Census Bureau: http://factfinder.census.gov
58.2 D. He, C. Barr: China's pulp and paper sector: an analysis of supply-demand and medium term projections, Int. For. Rev. **6**(3–4), 254–266 (2004)
58.3 I. Laukkanen: Visions and requirements of automation engineering, interview not published
58.4 http://www.VDMA.org
58.5 VDMA: *Volkswirtschaft und Statistik. Statistisches Handbuch für den Maschinenbau* (Eigenverlag, Frankfurt 2007), in German, Transl.: National economics and statistics: statistical handbook for mechanical engineering
58.6 H. Thoemen, P.E. Humphrey: Modeling the physical processes relevant during hot pressing of wood-based composites – Part I. Heat and mass transfer, Holz Roh- Werkst. **64**(1), 1–10 (2005)
58.7 B. Scherff, G. Bernardy: Prozessmodellierung führt zu Online-Qualitätskontrolle und Prozessoptimierung bei der Span- und Faserplattenproduktion, Holz Roh- Werkst. **55**(3), 133–140 (1997), in German, Transl.: Process optimization leads to online process control and process optimization in particle and fiber board production
58.8 H. Thoemen, C.R. Haselein: Modeling the physical processes relevant during hot pressing of wood-based composites – Part II. Rheology, Holz Roh-Werkst. **64**(2), 125–133 (2005)
58.9 D. Pantförder, B. Vogel-Heuser: Nutzen von 3-D-Pattern in der Prozessführung am Beispiel geeigneter Anwendungsfälle, Autom.-Tech. Praxis **11**, 62–70 (2006), in German, Transl.: Development and application of 3-D pattern in process plant operation – benefit and state of the art
58.10 http://www.siempelkamp.com
58.11 IEEE: *Proceedings of the 24th Annual Conference of IEEE* (Industrial Electronics Society, Aachen 1998)
58.12 G. Bernardy, B. Scherff: SPOC – process modeling provides online quality control and predictive process control in particle and fibreboard production, Proc. 24th Ann. Conf. IEEE (Industrial Electronics Society, Aachen 1998) pp. 1703–1707
58.13 http://www.metsopanelboard.com/panelboard
58.14 H.-J. Deppe, K. Ernst: *Taschenbuch der Spanplattentechnik*, 4th edn. (DRW, Leinfelden-Echterdingen 2000), in German, Transl.: *Paperback of Particle Board Technique*
58.15 G. Bernardy, B. Scherff: Savings potential in chipboard and fibreboard, cost reduction by the integration of process control technology and statistical process optimisation, Asian Timber **1**, 37–40 (1996)
58.16 G. Bernardy, A. Lingen: Prozessdatenbasierte Online-Qualitätskontrolle für die kontinuierliche Überwachung von Prozessen mit zerstörender Stichprobenprüfung, Autom.-Tech. Praxis **9**, 44–51 (2002), in German, Transl.: Process data based online quality control for processes with destructive test of samples
58.17 M. Blatzheim: Automatisierungstechnik in der Papierindustrie, Stand der Technik und besondere Anforderungen, Autom.-Tech. Praxis **2**, 61–63 (2007), in German, Transl.: Automation in the paper industry, State of the Art and other special requirements
58.18 C.J. Biermann: *Handbook of Pulping and Papermaking*, 2nd edn. (Academic, San Diego 1996)
58.19 http://www.sysml.org

58.20   http://www.es.eecs.uni-kassel.de/forschung/
         projekte/uml2iec61131/e_index.html
58.21   B. Vogel-Heuser, K. Schweizer, A. van Burgeler,
         Y. Fuchs, D. Pantförder: Auswirkungen einer drei-
         dimensionalen Prozessdatenvisualisierung auf die
         Fehlererkennung, Z. Arbeitwiss. **1**, 23–34 (2007), in
         German, Transl.: Benefits of 3-D process visualiza-
         tion on fault detection in process plant operation

59. Welding Automation

Anatol Pashkevich

This Chapter focuses on automation of welding processes that are commonly used in industry for joining metals, thermoplastics, and composite materials. It includes a brief review of the most important welding techniques, welding equipment and power sources, sensors, manipulating devices, and controllers. Particular emphasis is given to monitoring and control strategies, seam-tracking methods, integration of welding equipment with robotic manipulators, computer-based control architectures, and offline programming of robotic welding systems. Application examples demonstrating state-of-the-art and recent advances in robot-based welding are also presented. Conclusions define next challenges and future trends in enhancing of welding technology and its automation potential, modeling and control of welding processes, development of welding equipment and dedicated robotic manipulators, automation of robot programming and process planning, human—machine interfaces, and integration of the automated robotic stations within the global production system.

59.1 Principal Definitions

Welding is a manufacturing process by which two pieces of materials (metals or thermoplastics) are joined together through coalescence. This is usually achieved by melting the workpieces and adding a filler material that causes the coalescence and, after cooling, forms a strong joint. Sometimes, pressure is applied in combination with heat, or alone. At present, heat welding is the most common welding process, which is widely used in automotive, airspace, shipbuilding, chemical and petroleum industries, power generating, manufacturing of machinery, and other areas [59.1–3].

For heat welding, many different energy sources can be used, including a gas flame, an electric arc, a laser or an electron beam, friction, etc. Depending on the *mode of energy transfer*, the American Welding Society (ASW) has grouped welding/joining processes and assigned them official letter designations, which are used for identification on drawings and in technological documentation. In particular, the ASW distinguishes *arc welding*, *gas welding*, *resistance welding*, *solid-state welding*, and *other welding processes*. Within each group, processes are distinguished depending on the *influence of capillary attraction* (which is the ability of a substance to draw another substance into it). For instance, the arc welding group includes gas metal arc (GMAW), gas tungsten arc (GTAW), flux cored arc welding (FCAW), and other types of welding. Detailed and complete classification of the welding processes is given in [59.4].

59.2 Welding Processes

59.2.1 Arc Welding

This group uses an electric arc between an electrode and the base material in order to melt metals at the welding point. The arc is created by direct or alternating current using consumable or nonconsumable electrodes. The welding region may also be protected from atmospheric oxidation and contamination by an inert or semi-inert gas (shielding gas). The oldest process of this type, *carbon arc welding* (CAW), uses a carbon electrode and has limited applications today. It has been replaced by *metal arc welding*. A typical example is *shielded metal arc welding* (SMAW), in which a flux-covered metal electrode produces both shielding (CO_2 from decomposition of the covering) and filler metal (from melting of the electrode core). This process is widely used in manual welding and is rather slow, since the consumable electrode rods (or *sticks*) must be frequently replaced.

Automatic arc welding is mainly based on the *gas metal arc welding* (GMAW) process, also known as metal inert gas (MIG) or metal active gas (MAG) welding [59.5]. The process uses a continuous wire feed as a consumable electrode and an inert or semi-inert gas mixture as shielding (Fig. 59.1a). The wire electrode is fed from a spool, through a welding torch. Since the electrode is continuous, this process is faster compared than SMAW. Besides, the smaller arc size allows making overhead joints. However, the GMAW equipment is more complex and expensive, and requires more complex setup. During operation, the process is controlled with respect to arc length and wire feeding speed. GMAW is the most common welding process in industry today; it is suitable for all thicknesses of steels,

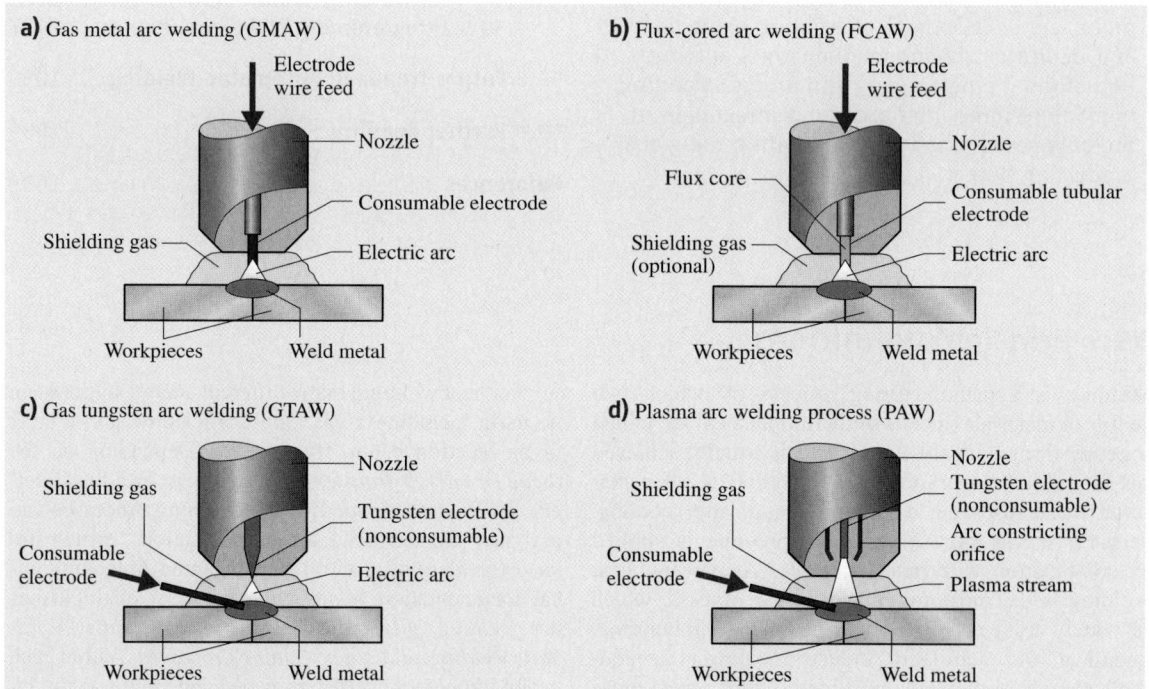

Fig. 59.1a–d Schematics of typical arc welding processes: (after [59.4])

aluminum, nickel, stainless steels, etc. This process has many variations depending on the type of welded metal and shielding gas, and also the metal transfer mode.

A related process, *flux-cored arc welding* (FCAW), uses similar equipment but is based on a continuously fed flux-filled electrode, which consists of a tubular steel wire containing *flux* (a substance which facilitates welding by chemically cleaning the metals to be joined [59.1]) at its core (Fig. 59.1b). The heat of the arc decomposes the electrode core producing gas for shielding and also deoxidizers, ionizers, and purifying agents. Additional shielding may be obtained from externally supplied gas. Obviously, this cored wire is more expensive than the standard solid one, but it enables higher welding speed and greater metal penetration.

Another variation is *submerged arc welding* (SAW) that is also based on the consumable continuously fed electrode (solid or flux cored), but the arc zone is protected by being *submerged* under a covering layer of granular fusible flux. When molten, the flux generates protective gases and provides a current path between the electrode and the base metal. Besides, the flux creates a glass-like slag, which is lighter than the deposited metal from the electrode, so the flax floats on the surface as a protective cover. This increases arc quality, since atmospheric contaminants are blocked by the flux. Also, working conditions are much better because the flux hides the arc, eliminating visible arc light, sparks, smoke, and spatters. However, prior to welding, a thin layer of flux powder must be placed on the welding surfaces.

For nonferrous materials (such as aluminum, magnesium, and copper alloys) and thin sections of stainless steel, welding is performed by the *gas tungsten arc welding* (GTAW) process, also referred to as *tungsten inert gas* (TIG) welding. The process uses a nonconsumable tungsten electrode with high melting temperature, so the arc heat causes melting of the workpiece and additional filling wire only (Fig. 59.1c). As an option, the filling metal may not be used (*autogenous* welding). The weld area is protected from air contamination by a stream of inert gas, usually helium or argon, which is fed through the torch. Because of the smaller heat zone and weld puddle, GTAW yields better quality compared with other arc welding techniques, but is usually slower. The process also allows a precise control, since heat input does not depend on the filler material rate. Another advantage is the wide range of materials that can be welded, so this process is widely used in the airspace, chemical, and nuclear power industries.

A related process, *plasma arc welding* (PAW), uses a slightly different welding torch to produce a more focused welding arc. In this technique, which is also based on a nonconsumable electrode, an electric arc transforms an inert gas into plasma (i. e., an electrically conductive ionized gas of extremely high temperature) that provides a current path between the electrode and the workpiece (Fig. 59.1d). Similar to the GTAW process, the workpiece is melted by the intense heat of the arc, but very high power concentration is achieved. To initiate the plasma arc, a tungsten electrode is located within a copper nozzle. First, a pilot arc is initiated between the electrode and nozzle tip, then it is transferred to the workpiece. Shielding is obtained from the hot ionized gas (normally argon) issuing from the orifice. In addition, a secondary gas is used (argon, argon/hydrogen or helium), which assists in shielding. PAW is characterized by extremely high temperatures (30 000 °F), which enables very high welding speeds and exceptionally high-quality welds; it can be used for welding of most commercial metals of various thicknesses. A variation of PAW is plasma cutting, an efficient steel cutting process.

59.2.2 Resistance Welding

Resistance welding is a group of welding processes in which the heat is generated by high electrical current passing through the contact between two or more metal surfaces under the pressure of copper electrodes. Small pools of molten metal are formed at the contact area, which possess the highest electrical resistance in this circuit. In general, these methods are efficient and produce little pollution, but their applications are limited to relatively thin materials. There are several processes of this type; two of them are briefly described below.

Resistance spot welding (RSW) is used to join overlapping thin metal sheets, typically, of 0.5–3.0 mm thickness. It employs two nonconsumable copper alloy electrodes to apply pressure and deliver current to the welding area (Fig. 59.2a). The electrodes clamp the metal sheets together, creating a temporary electrical circuit through them. This results in rapid heating of the contact area to the melting point, which is transformed into a nugget of welded metal after the current is removed. The amount of heat released in the spot is determined by the amplitude and duration of the current, which are adjusted to match the material and the sheet thickness. The size and shape of the spots also depend on the size and contour of the electrodes. The main advantages of this method are efficient energy use,

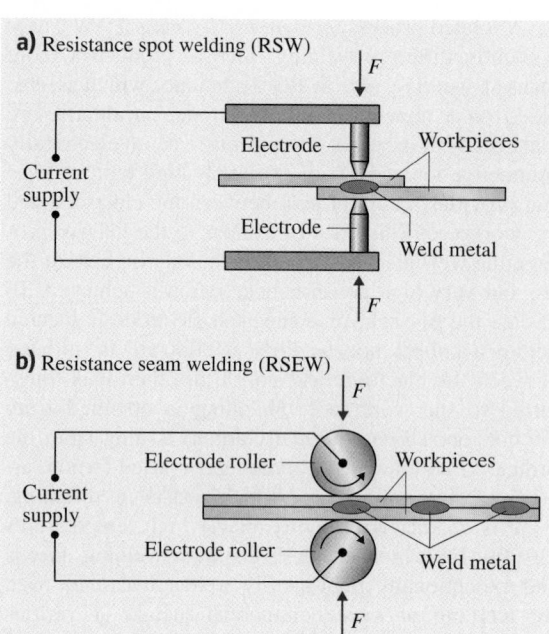

Fig. 59.2a,b Schematics of typical resistance welding processes (after [59.4])

low workpiece deformation, no filler materials, and no requirements for the welding position. Besides, this process allows high production rates and easy automation. However, the weld strength is significantly lower than for other methods, making RSW suitable for certain applications only (it is widely used in the automotive industry where cars can have up to several thousand spot welds).

Resistance seam welding (RSEW) is a modification of spot welding where the bar-shaped electrodes are replaced by rotating copper wheels. The rotating electrodes are moved along the weld line (or vice versa, the workpiece is moved between the electrodes), progressively applying pressure and creating an electrical circuit (Fig. 59.2b). This allows obtaining long continuous welds (for direct current) or series of overlapping spot welds (for alternative or pulsed current). In seam welding, more complicated control is required, involving coordination of the travel speed, applied pressure, and electrical current to provide the overlapping welds. This process may be automated and is quite common for making flange welds, watertight joints for tanks, and metal containers such as beverage cans. There are a number of process variants for specific applications, which include *wide wheel seam*, *narrow wheel seam*, *consumable wire seam welding*, and others.

59.2.3 High–Energy Beam Welding

Energy beam welding is a relatively new technology that has become popular in industry due to its high precision and quality [59.6]. It includes two main processes, laser beam welding and electron beam welding, differing mainly in the source of energy delivered to the welding area. Both processes are very fast, allow for automation, and are attractive for high-volume production.

Laser beam welding (LBW) uses a concentrated coherent light as the heat source to melt metals to be welded. Due to the extremely high energy concentration, it produces very narrow and deep-penetration welds with minimum heat-effective zones. Welds may be fabricated with or without filler metal; the molten pool is protected by an externally supplied shielding gas. It is a versatile process, capable of welding most commercially important metals, including steel, stainless steel, titanium, nickel, copper, and certain dissimilar metal combinations with a wide range of thickness. By using special optical lenses and mirrors, the laser beam can be directed, shaped, and focused on the workpiece surface with great accuracy. Since the light can be transmitted through the air, there is no need for vacuum, which simplifies equipment and lowers operating cost. The beam is usually generated using a gas-based CO_2 solid-state Nd:yttrium–aluminum–garnet (YAG) or semiconductor-based diode lasers, which can operate in pulsed or continuous mode. Furthermore, the beam is delivered to the weld area through fiber optics. For welding, the beam energy is maintained below the vaporization temperature of the workpiece material (higher energy is used for hole drilling or cutting where vaporization is required). Advantages of LBW include high welding speed, high mechanical properties, low distortion, and no slag or spatter. The process is commonly used in the automotive industry.

A derivative of LBW, *dual laser beam welding*, uses two equal power beams obtained by splitting the original one. This leads to a further increase in welding speed and improvement of cooling conditions. Another variation, laser hybrid welding, combines the laser with metal arc welding. This combination also offers advantages, since GMAW supplies molten metal to fill the joint, and a laser increases the welding speed. Weld quality is higher as well, as the potential for undercutting is reduced.

Electron beam welding (EBW) is a welding process in which the heat is obtained from high-velocity electrons bombarding the surfaces to be joined. The electrons are accelerated to a very high velocity (about

50% of the speed of light), so beam penetration is extremely high and the heat-affected zone is small, allowing joining of almost all metals and their combinations. To achieve such a high electron speed and to prevent dispersion, the beam is always generated in high vacuum and then delivered to the workpiece located in a chamber with medium vacuum or even out of vacuum. In the last case, specially designed orifices separate a series of chambers at various vacuum levels. Because of the vacuum, a shielding gas is not used, while a filler metal may be used for some materials (for deoxidizing the melted plain carbon steel that emits gases, to prevent weld porosity). The EBW process provides very narrow and high-quality welds; it is commonly used for joining stainless steels, superalloys, and reactive and refractory metals. The primary disadvantage of the EBW is high equipment cost and high operation price (due to the need for vacuum). Besides, location of the parts with respect to the beam must be very accurate.

59.3 Basic Equipment and Control Parameters

The described welding technologies utilize various types of equipment and control units. However, since arc and resistance spot welding are used in manufacturing most widely, they are expanded upon in more detail.

59.3.1 Arc Welding Equipment

Arc-welding processes employ the basic electrical circuit, where the currents typically vary from 100 to 1000 A, and voltage ranges from 10 to 50 V. The power supply can produce either direct current (DC) or alternating current (AC), and usually can maintain either constant current or constant voltage. Consumable-electrode processes (such as GMAW) generally use direct current, while nonconsumable-electrode processes (GTAW, etc.) can use either direct current (with negative electrode polarity) or alternating current (with square-wave AC pattern) [59.1, 3, 5].

For arc welding processes, the voltage is directly related to the arc length, and the current is related to the amount of heat produced. So, *constant-current* power supplies are most often used for manual welding, because they maintain a relatively constant heat output even if the voltage varies due to imperfect control of electrode position. *Constant-voltage* power supplies are usually utilized for automated welding, since the electrode spatial position (and arc length) is proper controlled and the current sensor can be used for adjusting the electrode position in the feedback loop.

Typical welding equipment for the GMAW process is shown in Fig. 59.3. It includes a power supply, welding cables, a welding gun, a water cooling unit, a shielding gas supplier, wire feed system, and a process control unit. Here, the cathode (negative) cable is connected to the workpiece, and the anode (positive) cable is connected to the welding gun. The consumable welding wire is continuously fed through the gun cable and the contact tube inside the gun, where an electrical connection is made to the power supply. In addition, the shielding gas and cooling water are also fed through the gun cable. The welding gun can be operated either man-

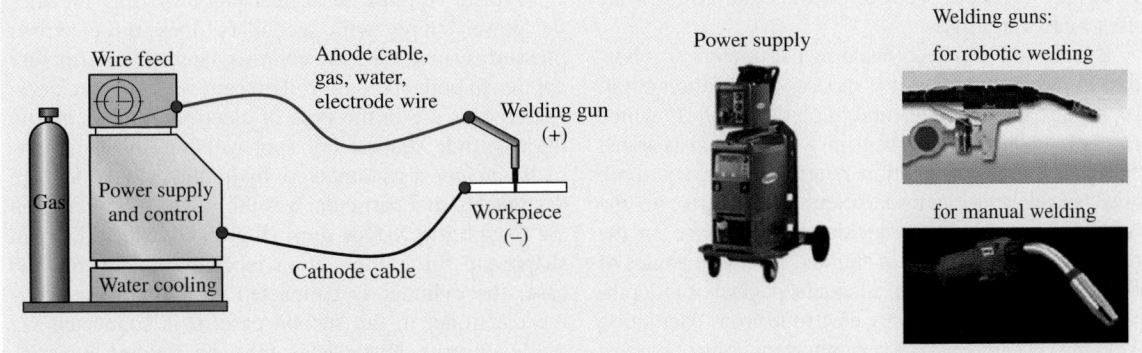

Fig. 59.3 Composition of a typical GMAW machine and its components (http://www.robot-welding.com/welding_torch.htm, http://www.binzel-abicor.com)

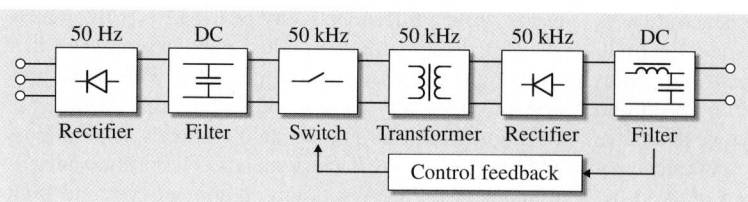

Fig. 59.4 General structure of the inverter-based welding power supply

ually or automatically, by a welding robot or some other automated setup. The gun shape is usually a swan-neck or straight. Guns with low current and light duty cycle are generally gas-cooled whereas those with higher current are water-cooled.

Formerly, welding machines were based on simple transformers with the operational frequency of the main energy source (i. e., 50 or 60 Hz). For DC welding, the transformer was equipped with a rectifier and an additional low-pass filter to suppress the ripples and produce a process-stabilizing effect. In modern *inverter-type* equipment (Fig. 59.4) the main conversion is performed at much higher frequency (approximately 20–50 kHz) allowing to decrease transformer weight, size, and magnetic losses (by about tenfold).

The output stage of the power supply may also include a controlled on/off switch circuit. By varying the on/off period (i. e., the pulse duty factor), the average voltage may be perfectly adjusted. For AC welding, the power source implements additional features such as pulsing the welding current, variable frequencies, variable ratio of positive/negative half-cycles, etc. This allows adjusting the square-wave shape to minimize the electrode thermal stress and the cleaning effect. In some cases, an AC sine wave is combined with high-frequency high voltage in the neighborhood of zero-crossing, to ensure noncontact arc reignition. Other variants use pulsed DC current of low-frequency (1–10 Hz) to reduce weld distortions and compensate cast-to-cast variations.

By relevant settings of welding parameters, it is possible to select three possible modes of operation (short arc mode, spray mode, and globular mode), which are distinguished by the way in which metal is transferred. The weld orientation relative to gravity, torch travel speed, and electrode orientation relative to the welding joint also have considerable influence on the weld formation. For most materials, electrode angles of 60–120° give welds with adequate penetration-depth-to-width ratio. In some cases, electrode cross-oscillation (weaving) is necessary. Other important control parameters are electrode feed speed, distance between the workpiece and contact nozzle, travel motion parameters

(straight or weaving type), composition of shielding gas, and delivery of cooling gas/water.

59.3.2 Resistance Welding Equipment

The implementation of resistance spot welding involves coordinated application of force and current of the proper magnitude and time profile. Typically the current is in the range of 1–100 kA, and the electrode force is 1–20 kN. For the common combination "1.0 + 1.0 mm" sheet steel, the corresponding voltage between the electrodes is only 1.0–1.5 V, however the voltage from the power supply is much higher (5–10 V) because of the very large voltage drop in the electrodes [59.2–4].

The spot welding cycle is divided into four time segments: *squeeze*, *heat* (weld), *cool* (hold), and *off*, as shown in Fig. 59.5. The squeeze segment provides time to bring the electrodes into contact with the workpiece and develop full force. The heat segment is the interval during which the welding current flows through the circuit. The cool segment, during which the force is still held, allows the weld to be solidified, and the off segment is to retract the electrodes, and remove or reposition the workpiece. Typical values for the heat and hold times are 0.1–0.5 s and 0.02–0.10 s, respectively. In industry, the segment duration is often expressed in cycles of the main frequency (50 or 60 Hz).

Typical equipment for resistance welding includes the power supply with secondary lines, the electrode pressure system, and the control system. This structure applies to both spot and roller seam welding machines. Differences are in the type of electrode fittings and in the electrode shapes. For spot welding, the guns normally include a pneumatic or hydraulic cylinder and are designed to fit a particular assembly. The most common are C-type and X-type guns (Fig. 59.6), which differ in shape and force application mechanisms (in the first case, the cylinder is connected directly to the moving electrode; in the second case, it is connected via the lever arm). However, some new welding guns incorporate built-in electromechanical actuators for force generation.

Resistance welding may employ several power supply architectures that differ in the type of the current (AC/DC) and frequency of voltage conversion: *AC power source* based on a low-frequency (50 or 60 Hz) step-down transformer; *DC power source* with a low-frequency (50 or 60 Hz) transformer and rectifier; *impulse capacitive-discharge source*, where the rectified primary current is stored in capacitors and is transformed into high welding currents; *inverter-based power source*, where the primary supply voltage (50 or 60 Hz) is rectified and is converted to a mid-frequency (20–50 kHz) square wave. Similar to arc welding, the inverter-based method gives essential reduction of power supply size and weight. All methods may be used with single- or three-phase mains supply.

From a compositional point of view, there are two main types of resistance welding equipment. In the first type, an AC power unit with electric transformer is built directly into a welding gun. The second type uses a DC power unit with welding cables connected to the gun.

Modern computer-based control units allow programming of all essential process parameters, such as current magnitude, welding cycle times, and electrode force. Some sophisticated controllers also allow regulation of current during welding, control of pre/post-heat operations, or adjustment of the clamping force during the cycle. Particular values of the welding parameters depend on the physical properties and thickness of the

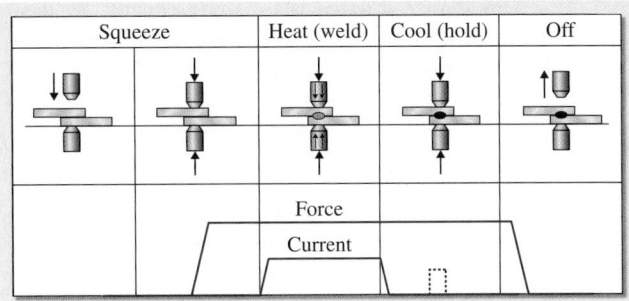

Fig. 59.5 Spot welding cycle

Fig. 59.6 Spot welding guns (X-type and C-type) (http://www.spotco.com)

joining materials, and also on the type of equipment used. Weld current shape is usually rectangular, but can also be trapezoid type with programmed rise/fall times. For some thick materials, several current pulses may be applied.

59.4 Welding Process Sensing, Monitoring, and Control

Automated welding requires accomplishing a number of tasks (such as weld placement, weld joint tracking, weld size control, control of the weld pool, etc.) that are based on real-time monitoring and control of relevant parameters. These actions must be performed in the presence of disturbances caused by inaccurate joint geometry, misalignment of workpiece and welding tool, variations in material properties, etc. The main challenge is that the observable data is indirectly related to final weld quality, so sensing and feedback control relies on a variety of techniques. Basically, they are divided into groups (technological and geometrical), which provide correspondingly control/monitoring of the welding process and positioning of the workpiece relative to the energy source [59.7, 8].

59.4.1 Sensors for Welding Systems

For welding, technological parameters typically include voltage, current, and wire feed speed. The *arc voltage* is usually measured at the contact tube within the weld torch, but the voltage drop between the tube and the wire tip (where the arc starts) must be compensated. Another method is to measure the voltage on the wire inside the feeding system, which provides a more accurate result.

The *welding current* can be measured using two types of sensors, the Hall-effect sensor and current shunt. The former is a noncontact device that responds to the magnetic field induced by the current. The second sensor type employs a contact method where the current flows through a calibrated resistor (shunt) that converts the current into a measured voltage.

The *wire feed speed* is usually estimated by measuring the speed of the drive wheel of the feeder unit. However, this must be complemented with special features of the feeder mechanics, which ensures robustness with respect to wire diameter variations and bending/twisting of the wire conduit.

Sensors for geometrical parameters provide the data for seam tracking during welding and/or seam searching before welding. These capabilities ensure adaptation to the actual (i. e., nonnominal) weld joint geometry and the workpiece position/orientation relative to the torch. The most common geometrical sensors are based on *tactile*, *optical* or *through-arc* sensing principles.

Tactile sensors implement purely mechanical principles, where a spring-loaded guide wheel maintains a fixed relationship between the weld torch and the weld joint. In more sophisticated sensors the signals from the mechanical probe are converted into electrical signals to acquire the geometrical data.

Optical sensors usually use a laser beam, which scans the seam in linear or circular motions, and a charge-coupled device (CCD) array that captures features of the weld joint (Fig. 59.7a). By means of scanning, the sensor acquires a two-dimensional (2-D) image of the joint profile. When the welding torch and sensor are being moved, a full three-dimensional (3-D) description of the weld joint is created. By applying appropriate image processing techniques and the triangulation method, it is possible to compute the gap size and weld location with respect to the welding torch [59.9]. A laser-based optical sensor is typically mounted on the weld torch, ahead of the welding direction, and a one-degree-of-freedom mechanism is required to maintain this configuration during welding. A typical laser scanner provides a sweep frequency of 10–50 Hz and an accuracy of ±0.1 mm, which is sufficient for most of welding processes. However, high price often motivates the use of alternative sensing methods.

Through-arc sensing is based on the measurement of the arc current corresponding to weaving (i. e., scanning) torch motions (Fig. 59.7b). This is a popular and cost-effective method for seam tracking in GMAW and related processes [59.10]. This method employs the relation between variations of the arc current and the electrode/workpiece distance, which is negative proportional for constant arc voltage. Typically, triangular-, sinus- or trapezoid-type motions are used, with a few millimeters of weaving amplitude, to achieve accuracy of about ±0.25 mm. For this method, geometrical information can be retrieved using continuous current measurement or its measurements at the turning and/or center points of the weaving motion. Correspondingly, different control principles are applied based on difference computing or template matching.

In practice, the tracking capability is usually combined with a search function (i. e., preweld sensing of the joint location), where the torch gradually moves in a predefined direction until detecting the weld joint. There are two basic methods for this function, which differ in terms of sensors and search patterns:

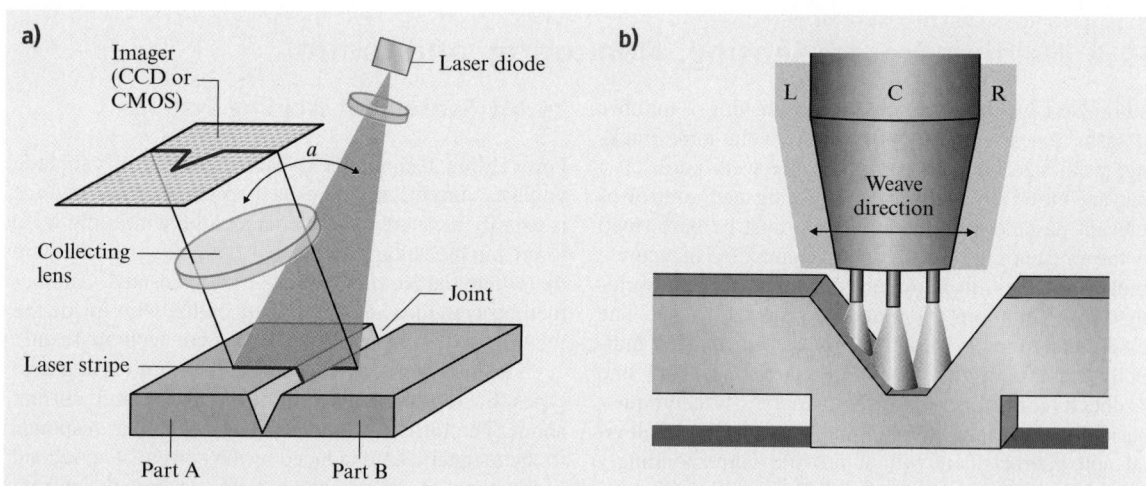

Fig. 59.7a,b Seam tracking using laser scanning (**a**) and through-arc sensing with weaving (**b**) (http://www.roboticsonline.com, http://www.thefabricator.com)

1. Approaching the workpiece without weaving, detecting electrical contact between the electrode and the weld plates, and calculating the starting position from this information
2. Approaching the workpiece with weaving, detecting the arc current, and tracking the seam in a normal way

For welding automation, other sensors can also be used, for example: inductive proximity sensors or eddy-current sensors, infrared sensors, ultrasonic sensors, and also sophisticated computer vision. However, they are not so common in this application, which is characterized by harsh environment with high temperatures, intense light, and high currents.

59.4.2 Monitoring and Control of Welding

Using data obtained from the sensors, it is possible to evaluate the weld quality and detect (or even classify) different weld defects, such as porosity, metal spatter, irregular bead shape, incomplete penetration, etc. These capabilities are implemented in monitoring systems, which usually use high-speed online analysis of welding voltage and/or current that are compared with preset nominal values or time patterns. Based on this analysis, an alarm is triggered if any difference from the preset values exceeds the given threshold. More sophisticated installations use computer-based image processing to evaluate the welding pool geometry and penetration depth [59.11].

To judge weld quality, the monitoring system relies on physical or statistical models, allowing the definition of alarm thresholds correlated with real weld defects or welding process specifications; for instance, for all GMAW processes, the voltage and current shape and mean values allows the detection of the metal transfer mode (short circuit, globular or spray transfer). In pulsed GMAW, the peak current is monitored and compared with preset values. For short-circuit GMAW, the monitoring features includes short-circuit time or frequency, as well as the average short-circuit current and the average arc current. In the general case, the features used for monitoring may be dependent on the specific algorithm and the welding condition.

For process feature analysis, various strategies are applied. The simplest ones employ deterministic decision making based on nominal values and tolerances, where any deviation from these is considered a potential cause of quality decrease. More sophisticated techniques employ template matching or treat the measured features as random variables and apply statistical methods such as control charts or spectrum analysis. However, the user must realize that increasing the detection probability often leads to false alarms that regularity interrupt the process. So, most current commercial monitoring systems utilize simple and robust algorithms, in which process features are averaged within user-defined time segments, filtered, and compared with a predefined threshold corresponding to normal welding conditions.

Welding process deviations detected via monitoring are to be compensated by control actions [59.12]. However, because of process complexity and indirect relevance of the observable data, simple feedback loops cannot be implemented. So, in addition to seam tracking, model-based strategies must be applied to enable adjustment of the welding equipment settings. However, in spite of its tremendous practical significance, this is still an active research area that employs various sophisticated decision-making techniques based on artificial intelligence and knowledge-based modeling.

59.5 Robotic Welding

Most industrial automated welding systems employ robotic manipulators, which are integrated with standard welding equipment that provides energy supply and basic control of welding parameters. The manipulators replace the human operator by handling the welding tool and positioning the workpiece. Usually this leads to an increase in quality and productivity, but poses a number of additional problems related to robot control, programming, calibration, and maintenance [59.13, 14].

59.5.1 Composition of Welding Robotic System

Currently, robots are mainly used for arc and spot welding processes. However, some recent applications deal with laser and plasma welding and also with friction stir welding. Typically, a robotic welding station includes a robot, a robot controller, welding equipment with relevant sensors, and clamping devices (fixture), allowing the workpiece to be held in the desired position in spite

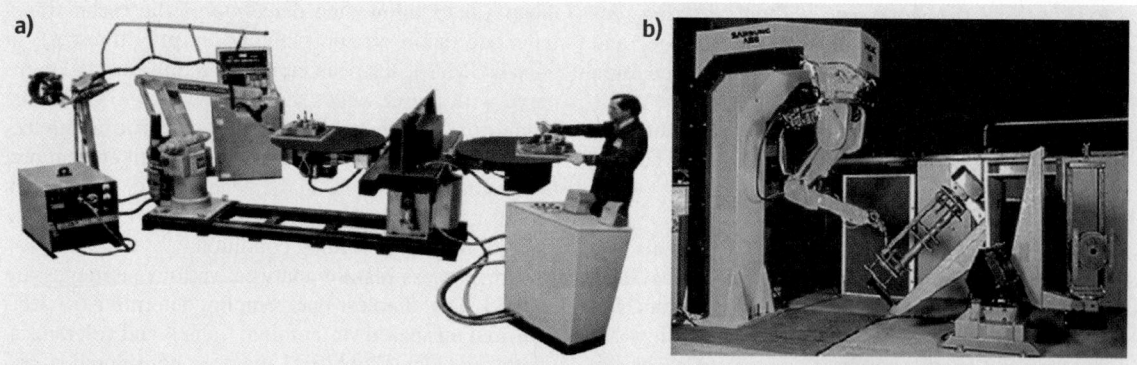

Fig. 59.8a,b Composition of arc welding robotic stations with floor-mounted (**a**) and column-mounted (**b**) robots (after [59.15, pp. 595, 602])

of thermal deformations. In addition, there are a variety of auxiliary mechanisms that provide an increasing in the robot workspace, better weld positioning, safety protection, and workpiece transportation between workstations.

The most common design of an *arc welding robotic station* is shown in Fig. 59.8. Usually, the robot has six actuated axes, so it can access any point within the working range at any orientation of the welding torch. In most cases robots are implemented with a serial architecture with revolute joints, which ensures larger workspaces (Fig. 59.9). Typical arc welding robots have a working envelope of 2000 mm and payload capacity of about 5 kg, which is sufficient for handling welding tools. To extend the working range, robots may be installed in an overhead position. A further extension of the working range can be achieved by installing the robot onto a linear carriage with auxiliary actuated axes (track, gantry or column). The wire feed unit and the spool carriers for the wire electrodes are often fixed to the robot, but can also be placed separately. In many cases the torch is equipped with shock absorption devices (such as springs) to protect it against collisions.

The workpiece positioners allow the location of seams in the best position relative to gravity (i.e., *downhand*) and to provide better weld accessibility. They usually have one or two actuated axes and may handle payload from a few kilograms to several hundred tons. The most common positioners are *turnover, turn–tilt*, and *orbital tables*, but *turning rolls* are also used to rotate the workpiece while making circular seams (in tank manufacturing, for instance). Positioners with orbital design have an advantage for heavy parts, allowing rotation of the workpiece around its center of gravity. In some cases, positioners are implemented with a multitable architecture, in which the operator feeds and removes the welded workpiece on one side, while the robot is welding on the other side. The positioner axes may either turn to certain defined positions (*index-based control*) or be guided by the robot controller and moved synchronically with the internal axes.

For spot welding, the robot payload capacity is essentially higher (about 150 kg), being defined by rather

Fig. 59.9a–d Mechanical components of an arc welding robotic station: (**a**) robotic manipulator, (**b**) positioner; (**c**),(**d**) robots with translational motion units (http://www.kuka.com)

Fig. 59.10 Robotic spot welding line for the automotive industry (http://tal.co.in/solutions/equipments/robotics-automation.html)

heavy equipment mounted on the robotic manipulator arm. Usually, each spot welding station includes several robots working simultaneously to provide the same cycle time along the manufacturing line. An example of a spot welding line for car manufacturing is presented in Fig. 59.10. For such applications, robots usually perform several thousands welds on over a hundred parts with a cycle time of about 1 min. Besides the joining and handling operations, the robots also ensure online measurement and inspection by means of dedicated laser sensors.

To ensure coordination of all components of the automated welding system, a relevant multimicroprocessor control architecture usually comprises two hierarchical levels. At the lower level, the local controllers implement mainly position-based algorithms that can receive a desired trajectory and run it continuously (for each actuated axis separately, but simultaneously and coordinated). High-level controllers ensure trajectory correction in real time, as a function of the observed results of the welding process. Some robot controllers can be connected via the Internet with telediagnostic systems to support service personnel during troubleshooting.

59.5.2 Programming of Welding Robots

To take advantage of robotic welding, especially in small-batch manufacturing, it is necessary to reduce the prewelding phase (or setup time), which includes selection of welding parameters and generation of control program defining motions of the robot, positioner, and other related mechanisms. This process is time con-

suming and may be longer than the actual welding phase [59.16].

For selection of welding parameters, there are at present a number of generally accepted databases. They allow the definition of optimal values of the welding current, voltage, welding speed, wire diameter, and number of weld beads/layers depending on type of weld, welding position, properties of materials, plate thickness, etc. Also, these databases usually provide interface to computer-aided design (CAD) models of the joining components to simplify extraction of geometrical information.

For robot programming, two basic methods exist: *online* (programming at the robot) and *offline* (programming out of the robot cell). The former method, which is also referred to as manual teaching, requires extraction of the robot from the manufacturing process and involves operator-guided implementation of all required motions. The operator uses a dedicated teach-pendant to move the welding torch to notable points of the weld, to store the torch position and orientation, and to create corresponding motion commands with necessary attributes (defining velocity, type of interpolation along the path, weave pattern, welding parameters, etc.). The simplest implementation of the offline programming uses an external computer to create a text file describing the sequence of motions, but the command arguments (i. e., torch position and orientation) are obtained via manual teaching. Nevertheless, this offer essential shortening of programming time because of extensive use of standard macros.

Advanced offline programming systems provide fully autonomous program generation, completely outside of the manufacturing cell. They rely on so-

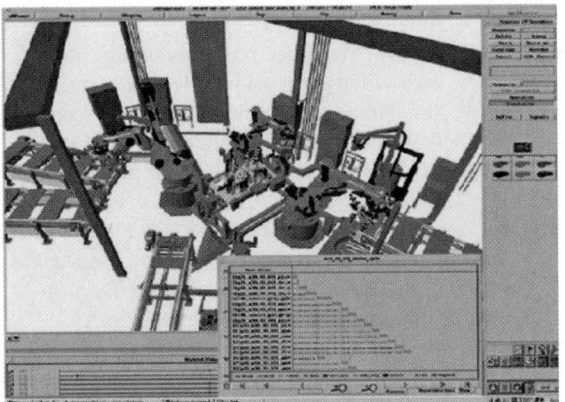

Fig. 59.11 Simulation and programming environment of eM-Workplace (Robcad) (http://www.ugs.com)

phisticated photorealistic 3-D graphical simulation of the robotic system and parts to be welded, allowing the required torch coordinates/orientations to be obtained directly from the models. Moreover, modern CAD-based robotic programming systems [such as em-Workplace (Robcad), IGRIP, CimStation, etc.] provide an interface to all standard 3-D modeling systems and incorporate a number of additional tools for robotic cell design, layout optimization, graphical simulation of the movements, program debugging and verification with respect to collisions and cycle time, program downloading to the robot controller, unloading of existing programs for optimization, etc. An example view of a CAD-based simulation and programming environ-

ment is presented in Fig. 59.11. However, while using the offline programming, it is necessary to ensure good correspondence between the nominal CAD model of the robot and its actual geometrical parameters. In practice, this is not a trivial problem, which is solved via calibration of all geometrical parameters describing the workcell components and their spatial location.

Automation of robot programming is still an active research area, which is targeted to replace movement-oriented program development to task-oriented programming. The final goal is automatic generation of robot programs from CAD drawings and welding databases, similar to programming methods for computer numerical control (CNC) machines.

59.6 Future Trends in Automated Welding

At present, welding automation is on the rise because of stricter customer demands and a shortage of skilled welders. So, equipment manufactures and system integrators are enhancing their production and implementing more advanced technologies. The most important technology-oriented directions defining future trends in welding automation are the following [59.17–20]:

- *Improvement of traditional welding processes* with respect to productivity and environmental issues (including development of better controllable power sources, new electrodes, shielding gases and fluxes; using twin-arc and tandem-arc torches)
- *Industrial implementation of new efficient and environment-friendly processes*, such as laser beam and electron beam welding, friction stir welding, and magnetic pulse welding (including development of new energy sources, relevant control algorithms, manipulating equipment)
- *Creating new knowledge-based welding process models* with ability for online learning and capability for online feedback control of essential process features (such as molten pool geometry, heat distribution, surface temperature profile, thermal deformations)
- *Development of advanced sensors and intelligent seam-tracking control algorithms* (to compensate for parts' mechanical tolerances in 3-D or 6-D space, and making welds in nonflat positions)
- *Development of new process monitoring and non-destructive evaluation methods* (using model-based

condition monitoring and failure analysis techniques, and online ultrasonic and laser-based testing)

Concerning mechanical components (robotic manipulators, positioners, etc.), it is recognized that their current performance satisfies the requirements of most welding processes with respect to ability to reproduce the desired trajectory with given speed and accuracy. Future developments will focus on automation of robot programming and integration with other equipment:

- *Task-oriented offline programming and integration with product design* (using simulation-based methods; simultaneous product and fixture design; implementing 3-D virtual-reality tools and concurrent engineering concepts; developments of human–machine interfaces)
- *Standardization* of mechanical components, control platforms, sensing devices and control architectures (to reduce the system development time/cost and simplify its modification)
- *Monitoring of the welding equipment and robotic manipulators* (to predict or detect machine failures and reduce downtime using predefined exception-handling strategies)

In addition, there are a number of essential issues that are not directly linked with welding technology and equipment. These include marketing aspects, and also networking and collaboration using modern e-Manufacturing concepts.

59.7 Further Reading

- AMS: *ASM Handbook. Vol. 6: Welding, Brazing and Soldering*, 10th edn. (ASM International, Metals Park 1993)
- K. Weman: *Welding Processes Handbook* (Woodhead, Boca Raton 2003)
- J. Pan: *Arc Welding Control* (CRC, Boca Raton 2003)
- R. Hancock, M. Johnsen: Developments in guns and torches, Weld. J. **83**(5), 29–32 (2004)
- F.M. Hassenkhoshnaw, I.A. Hamakhan: Automation capabilities for TIG and MIG welding processes, Weld. Cutt. **5**(3), 154–156 (2006)
- M. Fridenfalk, G. Bolmsjö: Design and validation of a universal 6-D seam tracking system in robotic welding based on laser scanning, Ind. Robot **30**(5), 437–448 (2003)
- A.P. Pashkevich, A. Dolgui: Kinematic control of a robot-positioner system for arc welding application. In: *Industrial Robotics: Programming, Simulation and Applications*, ed. by K.-H. Low (Pro Literatur, Mammendorf 2007) pp. 293–314
- G.E. Cook, R. Crawford, D.E. Clark, A.M. Strauss: Robotic friction stir welding, Ind. Robot **31**(1), 55–63 (2004)
- X. Chen, R. Devanathan, A.M. Fong (eds.): *Advanced Automation Techniques in Adaptive Material Processing* (World Scientific, River Edge 2002)
- U. Dilthey, L. Stein, K. Wöste, F. Reich: Developments in the field of arc and beam welding processes, Weld. Res. Abroad **49**(10), 21–31 (2003)
- G. Bolmsjö, M. Olsson, P. Cederberg: Robotic arc welding – trends and developments for higher autonomy, Ind. Robot **29**(2), 98–104 (2002)
- J. Villafuerte: Advances in robotic welding technology, Weld. J. **84**(1), 28–33 (2005)
- T. Yagi: State-of-the-art welding and de-burring robots, Ind. Robot **31**(1), 48–54 (2004)
- A. Benatar, D.A. Grewell, J.B. Park (eds.): *Plastics and Composites Welding Handbook* (Hanser Gardner, Cincinnati 2003)
- W. Zhang: Recent advances and improvements in the simulation of resistance welding processes, Weld. World **50**(3–4), 29–37 (2006)
- P.G. Ranky: Reconfigurable robot tool designs and integration applications, Ind. Robot **30**(4), 38–344 (2003)
- W.G. Rippey: Network communications for weld cell integration – status of standards development, Ind. Robot **31**(1), 64–70 (2004)
- The Welding Institute, UK. http://www.twi.co.uk
- American Welding Society, USA. http://www.aws.org
- ASM International: Materials Information Society. http://www.asminternational.org
- M. Sciaky: Spot welding and laser welding. In: *Handbook of Industrial Robotics*, ed. by S.Y. Nof (Wiley, New York 1999) pp. 867–886
- J.A. Ceroni: Arc welding. In: *Handbook of Industrial Robotics*, ed. by S.Y. Nof (Wiley, New York 1999) pp. 887–905

References

59.1 L.F. Jeffus: *Welding: Principles and Applications*, 6th edn. (Delmar, New York 2007)

59.2 H.B. Cary, S.C. Helzer: *Modern Welding Technology*, 6th edn. (Prentice Hall, New Jersey 2004)

59.3 W.A. Bowditch, K.E. Bowditch, M.A. Bowditch: *Welding Technology Fundamentals* (Goodheart-Willcox, South Holland 2005)

59.4 AWS: *AWS Welding Handbook. Vol. 1: Welding Science and Technology. Vol. 2: Welding Processes*, 9th edn. (American Welding Society, Miami 2001)

59.5 H.B. Cary: *Arc Welding Automation* (Marcel Dekker, New York 1995)

59.6 J. Norrish: *Advanced Welding Processes* (IOP, London 2006)

59.7 G. Bolmsjö, M. Olsson: Sensors in robotic arc welding to support small series production, Ind. Robot **32**(4), 341–345 (2005)

59.8 Z. Yan, D. Xu, Y. Li, M. Tan, Z. Zhao: A survey of the sensing and control techniques for robotic arc welding, Meas. Control **40**(5), 146–150 (2007)

59.9 S.-M. Yang, M.-H. Cho, H.-Y. Lee, T.-D. Cho: Weld line detection and process control for welding automation, Meas. Sci. Technol. **18**(3), 819–826 (2007)

59.10 U. Dilthey, L. Stein, M. Oster: Through-the-arc sensing – a universal and multipurpose sensor for arc welding automation, Int. J. Join. Mater. **8**(1), 6–12 (1996)

59.11   Y.M. Zhang (ed.): *Real-Time Weld Process Monitoring* (Woodhead, Cambridge 2008)

59.12   J.P.H. Steele, C. Mnich, C. Debrunner, T. Vincent, S. Liu: Development of closed-loop control of robotic welding processes, Ind. Robot **32**(4), 350–355 (2005)

59.13   J.N. Pires, A. Loureiro, G. Bolmsjö: *Welding Robots: Technology, System Issues and Application* (Springer, London 2006)

59.14   R.C. Dorf, S.Y. Nof (eds.): *International Encyclopedia of Robotics: Applications and Automation* (Wiley, New York 1988)

59.15   B.-S. Ryuh, G.R. Pennock: Arc welding robot automation systems. In: *Industrial Robotics: Programming, Simulation and Applications* (Pro Literatur, Mammendorf 2007) pp. 596–608

59.16   J.N. Pires: *Industrial Robots Programming: Building Applications for the Factories of the Future* (Springer, New York 2007)

59.17   T.J. Tarn, S. Chen, C. Zhou (eds.): *Robotic Welding, Intelligence and Automation*, Lecture Notes in Control and Information Sciences, Vol. 362 (Springer, Berlin 2007)

59.18   G.E. Cook: Robotic arc welding: research in robotic feedback control, IEEE Trans. Ind. Electr. **30**(3), 252–268 (2003)

59.19   M. Erickson: Intelligent robotic welding, Tube Pipe J. **17**(3), 34–41 (2006)

59.20   U. Dilthey, L. Stein, C. Berger, K. Million, R. Datta, H. Zimmermann: Future prospects of shape welding, Weld. Cutt. **5**(3), 164–172 (2006)

60. Automation in Food Processing

Darwin G. Caldwell, Steve Davis, René J. Moreno Masey, John O. Gray

Factory-based food production and processing globally forms one of the largest economic and employment sectors. Within it, current automation and engineering practice is highly variable, ranging from completely manual operations to the use of the most advanced manufacturing systems. Yet overall there is a general lag in the use of automation technology compared with other industries. There are many reasons for this lack of uptake and this chapter will initially discuss the factors that make automation of food production so essential and at the same time consider counterinfluences that have prevented this automation uptake.

In particular the chapter will focus on the diversity of an industry covering areas such as bakery, dairy, confectionary, snacks, meat, poultry, seafood, produce, sauce/condiments, frozen, and refrigerated products, which means that generic solutions are often (considered by the industry) difficult or impossible to obtain. However, it will be shown that there are many features in the production process that are almost completely generic, such as labeling, quality/safety automation, and palletization, and others that do in fact require an almost unique approach due to the natural and highly variable features of food products. In considering these needs, this chapter has therefore approached the specific automation requirements of food production from two perspectives. Firstly, it will be shown that in many cases there are generic automation solutions that could be valuably used across the industry ranging from small cottage facilities to large multinational manufacturers. Examples of generic types of automation well suited across the industry will be provided. In addition, for some very specific difficult handling operations, customized solutions will be shown to give opportunities to study the

problems/risks/demands associated with food handling and to provide an insight into the solution, thereby demonstrating that in most instances the difficult/impossible can indeed be achieved.

60.1 The Food Industry

Food and drink manufacturing forms one of the largest global industry sectors. In the European Union (EU), it is in fact the largest manufacturing sector with an annual turnover (in 2006) in excess of € 830 billion and a workforce of 3.8 million people [60.1]. However, unlike in other manufacturing sectors, there is still a very high level of low-skilled, low-paid labor.

Before considering the detailed use of automation in food processing, it is important to understand the nature of the industry. The food industry is not one single sector making a range of broadly similar products. It is in fact wide and diverse both in terms of the products and in structures and is characterized by:

- A very large number of small and medium enterprises (SMEs) operating in a highly competitive environment
- Rapid changes in product lines (often several changes per day)
- Generally low profit margins
- Extensive use of manual labor in often unattractive operating environments
- Low uptake of automation procedures.

It is also very important to note that, except for a few multinationals:

- Engineering research and development activity is low
- Ability to exploit and maintain advanced automation equipment is fairly poor
- Information technology (IT) and e-Commerce infrastructure is generally weak.

Within the industry there is a strong feeling that in the medium term (3–10 years) the number of people willing to work for the current low wages will decline and the industry will have to change to survive. Faced with this problem the industry has identified the application of automation and robotic systems as a major growth area with the aims of:

- Improving production efficiency and impacting on yield margins and profitability
- Reducing waste on all levels: product, energy, pollution, water, etc.
- Enhancing hygiene standards, and conforming to existing and future legislation pertaining to food production, including enhancing hygienic operation and product traceability

- Improving working conditions to improve retention of high-quality, motivated staff
- Improving the consistency of product quality.

However, despite these very significant and compelling driving influences, only a small number of companies are yet making significant use of automation for raw and in-progress product handling. The question therefore arises as to why the food sector should not be making extensive use of automation. There is no single simple answer to this question, but the answers seem to be embedded in a number of technical, financial, and cultural issues [60.2, 3].

Although the limited use of automation is certainly a reflection of a conservative investment policy in a low-margin industry, it is equally clear that in many instances the use of labor-intensive manual techniques is a deliberate policy because of:

i)   The flexibility provided by the human worker. Humans handle manipulative complexities with ease, by combining dextrous handling capabilities (the human hand), advanced sensing, and behavioral models of the product accumulated with experience.
ii) A lack of understanding of the properties of the food product as an *engineering* material. The handling characteristics of many (most) food products cannot be adequately described with geometry-based information, as is usually the case with conventional engineering materials since the geometry of food product is:

- In many/most instances nonrigid, often delicate, and/or perishable
- Variable in texture, color, shape, and size
- Often variable as a function of time and the forces applied
- Affected by environmental conditions including temperature, humidity, and pressure
- Easily bruised and marked when it comes into contact with hard and/or rough surfaces
- Susceptible to bacterial contamination.

iii) The product deforming significantly during handling. Any system developed to handle such food items therefore needs to react accordingly to this deformation and there is a lack of handling strategies and end-effectors designed to cope with the variable characteristics of food products.

iv) The perceived inabilities of current automation systems to cope with the variation in product and production demands.

In addition, the food sector often feels that there are significant issues relating to automation including:

- Robotic systems and application technology have been developed for the engineering manufacturing industry and they cannot transfer across into food manufacture without significant changes.
- Robot manufacturers and system integrators often have a poor understanding of the economics, payback rationale, and operating pressures in the food industry, which are very different from those in other sectors that make more use of automation/robotics. This results in the wrong products being offered for sale at the wrong price and with the wrong sales model.
- Support for complex IT-based systems is largely absent in smaller food manufacturers and there is no cost-effective outsourced support available.
- The space and flexibility requirements of, in particular, the smaller food manufacturer require that any automation fits around, and works with, existing manual operations. This is in contrast to most engineering automation which is physically separated from people.

60.2 Generic Considerations in Automation for Food Processing

When considering automation within the sector it is clear that there are very large differences in the exact nature of the work and the level and form of automation. For instance, unlike in the car industry, which is generally homogenous, making one easily recognizable product, in the food industry this is not true. The sector can be broken down in many ways, e.g., bakery, dairy, confectionary, snacks, meat, poultry, seafood, produce, sauce/condiments, frozen, and refrigerated. Within these areas there are of course many more subdivisions and these subdivisions mean that it is almost impossible to consider the industry as a whole and certainly from the viewpoint of automation this is extremely difficult, although there are several key aspects that have commonality.

60.2.1 Automation and Safety

Issues relating to food safety through accidental or deliberate contamination are of paramount concern in all food manufacturing facilities. To address these concerns and ensure public confidence there are a number of national and international standards. Depending on the country where the food is being manufactured these standards may be compulsory or voluntary and may be more or less strictly enforced. Among the most readily recognized of these standards are

- Hazard analysis and critical control points (HACCP) has been developed as a systematic preventive approach to food safety that addresses physical, chemical, and/or biological hazards during the manufacturing process, rather than through end-of-line and finished product testing/inspection. This goal is achieved by identifying potential food safety risks in the manufacturing process and acting on these critical control points (CCPs), e.g., by cooking, to prevent the hazard being realized. HACCP forms part of the whole manufacturing process including packaging, and distribution [60.4].
- Current good manufacturing practice (cGMP) deals with the control, quality assurance/testing, and management of food, pharmaceutical, and medical products in a manufacturing environment [60.5].
- ISO 22000 aims to bring the structures and benefits of ISO 9000, from which it is derived, to the food and drink processing and manufacturing sectors [60.6].

The introduction of automation and robotic equipment must of course conform to these standards, ideally without introducing any new hazards, but at the same time it is clear that the introduction of automation can have a positive impact since it permits humans, who are the most significant and certainly the most unpredictable contamination source, to be removed from the hazard consideration.

60.2.2 Easy-to-Clean Hygienic Design

Machinery to be used for direct handling of food can be designed following a set of hygienic design guidelines that ensure good standards of hygiene in production [60.7, 8], yet there is often a certain degree of confusion regarding what constitutes hygienic design and how hygienic a particular piece of machinery needs to be. Fundamentally this is product/task specific but it is clear that products such as raw meat, fish, and poultry are highly susceptible to contamination from microorganisms and require very high levels of hygienic design, while for dry foods such as biscuits or cakes lower levels of hygienic design may be more than adequate.

Routine cleaning and disinfection procedures involve the use of acidic, alkaline, and chlorinated cleaning chemicals. The need for frequent wash-downs makes a sealed, waterproof structure essential to enclose and protect internal components. The preferred material for food processing machinery is 304 or 316 grade stainless steel (BS EN 10088:2005), polished to a unidirectional satin polish finish [60.9]. Aluminum is not sufficiently corrosion resistant to commonly used cleaning chemicals and its use should be avoided for food-contact applications. Surfaces should be non-porous and free from cracks, crevices, scratches or pits that could harbor microorganisms after cleaning. Painted or coated surfaces should be avoided on food-contact parts; however, if used, the finish must be resistant to flaking or cracking. All parts likely to come into contact with food should be readily visible for inspection and accessible for cleaning. Joints that are screwed or bolted together inherently have crevices that cannot be adequately cleaned. A rubber seal or gasket should be used between components joined in this way. Exposed threads and fasteners such as screws, bolts, and rivets should be avoided if possible in food contact areas. All corners should be radiused and sharp internal corners should be avoided.

The use of plastics can offer certain advantages over stainless steel in some applications. However, plastics are generally more susceptible to failure from a range of different causes [60.7]. A database of plastics approved for food use by the US Food and Drug Administration (FDA) is available online [60.10]. In Europe, the use of plastics for food-contact applications is regulated by EU Commission Directive 2002/72/EC. One significant disadvantage of plastics is that they are easily scratched through manual cleaning. Surface scratches can accumulate over time and harbor microorganisms. A study by *Midelet* and *Carpentier* [60.11] suggests that microorganisms also attach themselves more strongly to plastics than to stainless steel. Rubber compounds used for seals, gaskets, and suction cups should also be food approved. Nitrile butyl rubber (Buna-N), fluoroelastomers (Viton), and silicone rubber are among those commonly used in the food industry. Likewise lu-

bricants, adhesives, and any other materials that may come into contact with food should be approved for food use [60.10, 12].

60.2.3 Fast Operational Speed (High-Speed Pick and Place)

High speed becomes important when an automation/robotic solution must compete on economic terms with hard automation or human workers. Low profit margins in food manufacturing mean that throughput must be maximized in order to increase profit. Increased production capacity also leads to reduced production costs. Arguably the most common task in food manufacturing is pick-and-place handling, where an object is picked from a conveyor belt and placed into its primary packaging. The pick-and-place speed of industrial robots is based on a standard $25 \times 300 \times 25 \, \text{mm}^3$ cycle and has been steadily rising. Speeds of between 80 and 120 picks/min, which are comparable to that of a human operator, are now becoming commonplace. Conveyor belts used in the food industry are generally no more than 50 or 60 cm wide. This width corresponds to the maximum distance that an operator, standing on one side of the conveyor, can comfortably reach to pick an object on the opposite side.

This need for high-speed handling is most acutely observed in the bakery and confectionary subsectors, where there is a need for high-speed handling of products that do in general have relatively good structural formal and repeatability and can be moved at high speed without disintegrating.

Biscuits/cookies form a particularly good example of this type of product but other breaded products, e.g., croissants and even meat precuts such as pepperoni for a pizza can be considered. In these tasks human operators are required to identify the product (visually), grasp the product, and place it either into a container or on to a secondary product. The operating frequencies are typically high (over 100 picks/min) with motions in the range of 30–50 cm one way. The mass of the objects are typically very low (only a few grams) and when undertaken by humans the user will often pick up several objects at one instance to minimize the movements to and from the conveyor. Recently the ABB IRB 340 FlexPicker robot has been extensively and generally very successfully used in this type of application to pick up multiple products as a group, or one at a time.

With recent advances in vision technology, robotic packing lines can handle varying or irregular products (Fig. 60.1). Automation of this type often integrates with a variety of sensor systems, e.g., checkweighers, vision, and metal detectors to enhance the handling process and combine this with online inspection.

60.2.4 Joints and Seals

Careful attention should be paid to the design of joints in automation/robotic systems to ensure that they are both waterproof and hygienic, and avoid deep, narrow crevices at the joints which are impossible to clean. This is illustrated in Fig. 60.2a. An improved design using a spring-energized polytetrafluoroethylene (PTFE) face seal is shown in Fig. 60.2b. Commercial seals are available where the spring groove is filled with silicone for use in food processing applications. Cover plates, which provide access to the inside of food processing machinery, are sealed using rubber gaskets. The screws securing the cover plates should also be sealed. Small screws can be sealed using a food-grade sealant. The screws should have plain hexagon heads, which are easier to clean.

60.2.5 Actuators

Pneumatic cylinders are low cost and commonly used in the food industry to actuate fixed automation machinery. Accurate position control of pneumatic actuators without the use of mechanical stops is, however, difficult to achieve under either proportional or pulse width modulation control schemes [60.13, 14]. In addition, position control using pulse-width modulation requires rapid cycling of the solenoid valves used to drive the actuator and this wears out the valves extremely quickly. Hydraulic actuators are not used in the food industry, as there is a risk that hydraulic fluid may contaminate the product.

Electric motors are comparatively easy to control and reliable in service. Brushless direct-current (DC) motors, despite their higher initial cost, have a service life many times greater than that of brushed motors, leading to increased reliability, lower maintenance costs, and less down time. This makes brushless motors more economical to use over the lifetime of the automation. The most significant operational issue with motors is ensuring adequate sealing to prevent the ingress of water/solvents during cleaning etc.

60.2.6 Orientation and Positioning

For all forms of automation knowledge of the exact position of a product is vital. This is no less true in the food

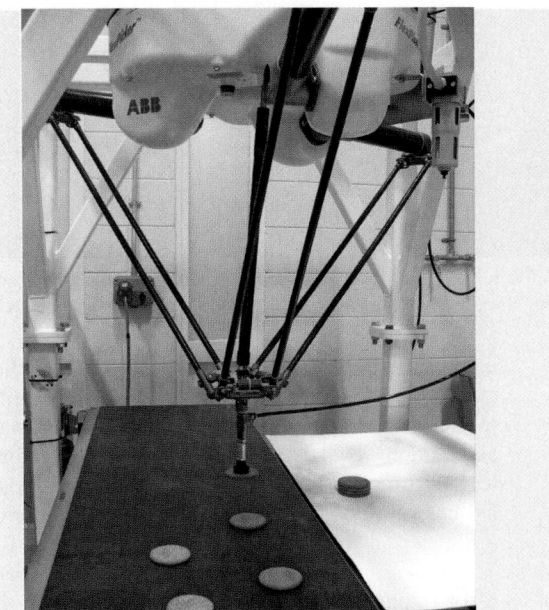

Fig. 60.1 An ABB IRB 340 used to pick and place biscuits

sector than elsewhere but in this sector a number of factors conspire to place a low value on this information.

Figure 60.3 shows a very common example of an ordered layup of product that is taken from the production line and placed in *bins* that have complete disorder. To automate the process of recreating order from this chaos is at best difficult and expensive and at worst currently technically impossible, but humans easily cope with this disorder. In considering examples of this type it is clear that within a food plant many processes that are currently considered difficult or impossible could be automated if greater attention were paid to the retention of position and orientation data. In some instances this will involve changes to the handling process while

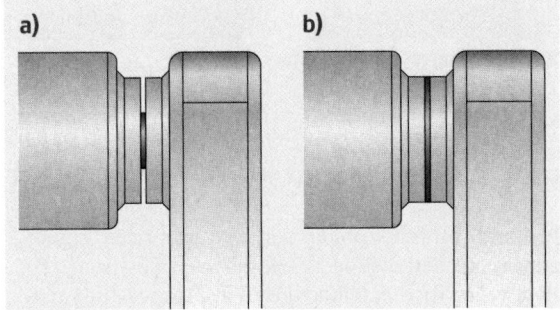

Fig. 60.2a,b Unhygienic and hygienic robot joint design. **(a)** Uncleanable gap, **(b)** PTFE face seal

Fig. 60.3 Food automation will often reverse traditional manufacturing aims that create disorder from the manufactured order

in others simple remedies such as properly adjusting guides, transfer conveyors or feed rates will be sufficient to create the order needed to satisfy the downstream automation. This is a process that has already been learned (often by hard experience) in other industries but has yet to be fully appreciated by the food processing sector.

At the same time it is possible to deal with aspects of orientation and positional inaccuracies and the vision systems that are becoming increasingly common in other industry sectors are finding useful outlets in

food processing also. Indeed in the food sector the relatively low tolerances (due to the product variability) mean that vision-linked handling systems could potentially be even more successful in these applications than elsewhere.

60.2.7 Conveyors

Conveyors are typically belted transport machines to carry products, containers, packs or packaging along a production line or between production centers (Fig. 60.4). There are a very large number of types of conveyor designed to operate with different products, and selection of the correct conveyor system is essential to good automation in the food sector [60.15]. Conveyors can be formed as stand-alone linear or curved units or they can be integrated into complex transportation networks custom-designed for each factory application. In their simplest forms the conveyor merely moves product from point A to B but they can be integrated with control systems, advanced drives, programmable logic controllers (PLCs), sensors etc. and form an integral part of good automation.

Fig. 60.4 Conveying solutions

60.3 Packaging, Palletizing, and Mixed Pallet Automation

One area of food production that has seen significant use of automation is end-of-line operations. For many years, food manufacturers have successfully used traditional hard automation, including wrappers, top loaders, and side loaders, to package easy-to-handle

products, e.g., cartons, boxes, trays, bags, and bottles. In this area of the food production cycle the product has been changed from the highly variable food product into a *containerized* unit. Within the end-of-line packaging operations there are therefore a number of key areas that have wide application across the sector (Fig. 60.5).

These include aspects such as labeling, checkweighing, inspection (visual, metal detection etc.), and palletizing.

60.3.1 Check Weight

The checkweigher is an automatic machine for measuring the weight of packaged commodities and hence ensuring that the product is within specified limits (Fig. 60.6). Any packs that are outside the tolerance are ejected from the line automatically and may be reworked. Although there are many forms of checkweigher they generally follow a fairly common format. From the main production flow line the product is transferred to an accelerating belt that spaces products that are often closely located on the line. This means that individual products can be weighed without interference from neighbors. The weigh station is an instrumented conveyor belt incorporating a high-speed transducer (typically a load cell), with user interface and often data ports for Ethernet etc. At the outflow of the checkweigher there is a reject conveyor to remove out-of-tolerance packs without disrupting the normal flow. The reject mechanism is automatic and may involve a variety of approaches such as air jets and mechanical arms. Checkweighers can have a throughput of up to 750 products per minute. The communication ports ensure that the checkweigher can be integrated into the whole plant operation, communicating production data etc. and forming part of a full SCADA (supervisory control and data acquisition) system. By controlling and monitoring the throughput of the checkweigher it is possible to detect out-of-performance upstream operations and to dynamically change the performance of the upstream operations by adjusting their set-points. Unfortunately this is seldom achieved and machines often run with poor adjustments that increase reject rates or overfill and hence *gives away* product. In addition, by integrating production of several lines or monitoring data over time, it is possible to permit some underweight product, as the overall average is within tolerance and the data from the checkweigher can validate this. This can significantly reduce wastage and has potential enormous savings.

60.3.2 Inspection Systems

Other inspection stations within the typical production line include metal detection and also on occasion

Fig. 60.5 End-of-line automation

x-ray machines (Fig. 60.6). These systems are primarily installed as safety systems to prevent physical contamination of food. In the instance of metal detectors which operate by enclosing the whole of the production conveyor belt, the product passes through the detector, which is tuned to detect small metal shards that may have become located in the food product. If the sensor is triggered the product is automatically rejected but in this instance there is no rework and indeed the product is usually inspected closely to discover the type of contamination and to ensure that this is eliminated. Metals detection is present on almost every production line. X-ray machines, although slightly less common, are used to check for nonmetallic contamination, e.g., glass, plastics, bone, and fibres, and also in some meat products as a quality control system to identify gristle.

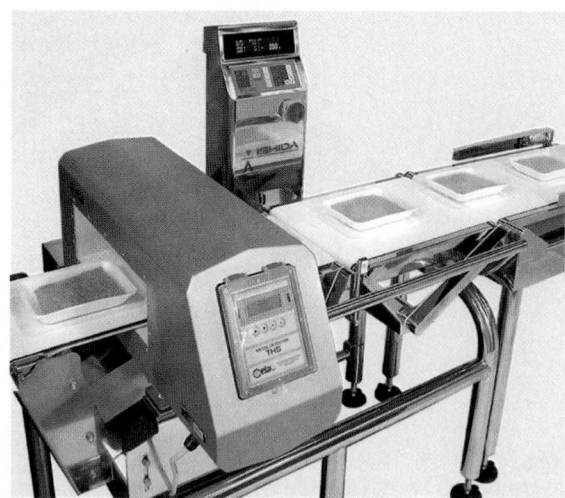

Fig. 60.6 Combined check weight and metal detection station

60.3.3 Labeling

Labels are used on every kind of product to brand, decorate or provide information, and in the food sector it is not uncommon that a label fulfils all three functions simultaneously (Fig. 60.7). Labeling is one of the final aspects of the production process and may be independent or integrated with other systems such as the checkweigher and inspection systems. There are two main types of labeling machine: wet glue and self-adhesive (pressure-sensitive) applicators. For the food

processing industry self-adhesive labelers are by far the more common, using preglued labels that are supplied on a reel of release paper or film. This method of application enables labels to be applied at medium/high speed to soft packages as well as rigid containers. This is very suited to the food sector.

60.3.4 Palletizing

The pallet is the fundamental loading and transportation unit for most food operations. As such, automation of the warehousing and palletizing operations for food companies, as in most industry sectors, is potentially one of the most profitable areas. As with most industry sectors many features influence the selection of automation for palletizing, including line speeds, factory layout, space at the end of production lines, and of course cost, but in food operations there is also the advantage that, by the time the products reach the palletizing stage, the packaging has usually created a fairly repeatable form that is missing in many other upstream areas and this is therefore one of the easiest areas to automate.

While it can be recognized that there are many features in common with other industry sectors, one recent trend that is particularly strongly driven in the food sector is the assembly of mixed product pallets, which reflected demand from retailers for custom pallet loads that suit the store rather than the shipper. The mixed load pallet is therefore emerging as one of the most ef-

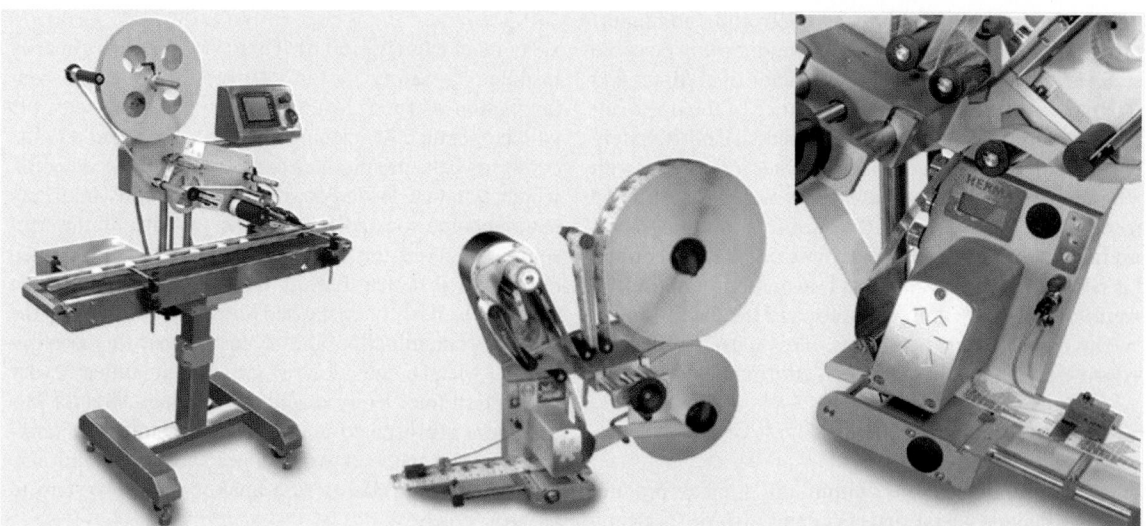

Fig. 60.7 Labeling stations

Fig. 60.8 Robotic palletizing

ficient technologies available in the food supply chain process.

To address these demands and opportunities and the use of multiple feeder lines and rapid pattern changes, the automation industry has focused on the development of software needed to pick the product and *design* the pallet, the hardware to recognize the product (sensors which are often visual based), hardware to manipulate the product (often robots), and the integration of the hardware–software solutions (Fig. 60.8).

Within the robotic community there have been important developments in the software to optimize picking, placement, and overall *construction* of diverse pallets, with many robot manufacturers, software houses, and systems developers introducing dedicated software that will allow online pallet preparation to meet the demands of the manufacturer and, more impor-

tantly, the retailer. These software systems can be fully integrated with external systems, e.g., machine-vision systems and image processing or other sensors to detect the presence of the product and its position and orientation, and this information can be directly communicated to a manipulation system, which is typically robotic, to allow flexibility in programming and motion control. This comprehensive integration of all components into one platform facilitates efficient communication and guarantees reliable robot operation. These software packages typically integrate only with one robot manufacturer's product line and it is therefore necessary to use combined hardware and software solutions. Integration suppliers can often integrate nonstandard units but this has a significant cost implication.

To address the increasing need for and use of robots in the food industry a number of robot manufacturers have or are developing products specifically for these applications. However, to date very few commercial robots have been developed specifically for the food industry. Often, existing models have simply been upgraded for use in food production and this has created a negative impression among sections of the food industry. Examples of industrial robots that are currently used for primary packaging and assembly of foods include the ABB IRB 340 FlexPicker (probably the most common robot in high-speed pick-and-place applications and well suited to handling wrapped and baked products) and Bosch Sigpack Delta robots, the FANUC LR Mate 200iB food robot, Gerhard Schubert's TLM-F4, and more recently the FANUC M-430iA/2F, which has a sleek profile with no food particle retention areas.

60.4 Raw Product Handling and Assembly

While the use of robots and advanced automation for end-of-line operations such as case packing and palletizing is already well established, robotics/automation of primary handling and assembly of foods has so far been limited. However, since financial justification for the installation of an automation/robotic system is typically based on the reduction of labor costs and the bulk of manual labor in a food production line is generally concentrated in primary packaging and assembly operations, this is the area that requires the greatest concentration of effort.

As already noted products handled by traditional automation are usually homogenous in terms of size,

shape, and weight and also tend to be rigid; however, some or all of these conditions do not prevail in food processing. Food is very often fragile and, unless extreme care is taken during handling, products can be damaged and in the worst case this can mean they have to be discarded. This means that the handling techniques used in traditional automation are generally not suited to the handling of *raw* food products and the mechanism of *grasping* the food product (rather than basic motion) is often the key to successful automation.

Taylor [60.15] classified the gripping techniques for nonrigid materials into three separate classes defined by the mechanism of the grasp:

Mechanical techniques – the product is firmly clamped between two or more mechanical fingers and held due to the friction contact. To minimize the grip force the gripper jaws can be compliant or specifically shaped to the particular object. This can only be used where variation between products is relatively small.

Intrusive grippers – pins are fed into the surface or body of the material to be lifted. The pins are precisely located so that when inserted the object becomes locked to the gripper. This technique is generally unsuitable for food products, as it would often cause unacceptable levels of damage.

Surface attraction – adhesives or a vacuum are used to create a bonding force between the gripper and product. Vacuum grippers have been successfully used in the food industry and are well suited to objects with regular or flat surfaces, such as biscuits. However, not all food items can be handled with such grippers due to difficulties in achieving an airtight seal, bruising, and the inflow of particles that could lead to microbial growth unless equipment is sterilized regularly.

As can be seen above designing mechanisms to grasp food products is not straightforward and the techniques used in other industries cannot be directly applied. Different types of food product present different challenges and as a result there are numerous examples of grippers that have been developed for use in the food industry which address these challenges.

60.4.1 Handling Products That Bruise

There are many food products that are easy to bruise. These are typically fruits and vegetables, but other products can also develop unsightly marks if grasped too firmly. For this reason handling techniques that minimize forces and pressures must be developed. One example of a product that is particularly susceptible to bruising is the mushroom. Although not immediately obvious a bruise can appear on a mushroom as long as several days after being handled. This can mean that, while a product may appear acceptable when dispatched from a factory/farm, it can appear damaged by the time it reaches the retailer/customer.

Mushroom harvesting is typically performed manually and, despite the delicate capabilities of the human hand, mushrooms do become bruised during manual harvesting. An automated system for the harvesting of mushrooms was produced by *Reed* et al. [60.16] with the aim of reducing labor but also reducing product damage. The design of the system paid par-

ticular attention to the delicacy of the mushroom contact.

The mushroom harvesting process consists of four main stages: first the position of an individual viable mushroom is obtained, followed by picking and trimming of the mushroom before placing it in a container. The location of the mushroom is obtained from a vision system mounted vertically over the mushroom bed. Image-processing software identifies and numbers each mushroom and then determines how best to pick them. Mushrooms below a certain size threshold are disregarded and are left to be harvested another day. An isolated mushroom is easy to pick, but this is not typically the case and usually mushrooms touch or overlap. The control software must therefore determine the best way to extract each mushroom without disturbing those around it. This is achieved by bending the mushroom away from those that surround it before picking.

The mushrooms are grasped using a vacuum cup mounted through a compliant link to a rack and pinion allowing the cup to be positioned on the surface of the mushroom. The cup is then twisted about the vertical axis to break the mushrooms base and allow it to be removed. A turret mechanism was also included, which allowed the most appropriately sized cup to be used for the particular mushroom being grasped.

The contact between the vacuum cup and the mushroom is the source of potential produce bruising and so determining the optimum vacuum force is critical. Experiments revealed that the force of the vacuum on the mushroom produced a faint mark on the mushroom during grasping but this was not considered by the industry to be unacceptably severe. However, if slip occurred between the mushroom and vacuum cup during rotation this resulted in unacceptable shear damage on the mushroom's surface.

Once a mushroom has been removed from the ground it is placed in a fingered conveyor with the stalk pointing vertically downwards. A blade then removes the lower section of stalk which is discarded and the trimmed mushroom is placed in a plastic tray ready for dispatch. The mushrooms are not dropped as this would result in denting and bruising. The complete system was trialled at a commercial mushroom farm in The Netherlands and by the Horticultural Research International in the UK. The average picking speed of the system was nine mushrooms per minute and in both of these trials the amount of mushroom bruising and damage was found to be significantly lower than when manual picking was used.

60.4.2 Handling Fish and Meat

While less susceptible to bruising than fruits and vegetables, meat and fish present their own grasping challenges. Due to the ease with which such products deform a traditional parallel jaw gripper is typically unable to grasp them with sufficient firmness. Similarly vacuum grippers have had only limited success grasping meats since the fleshy nature of meat means that unsightly peaks can be produced when a vacuum force is applied. Also moisture on the surface of the meat can be drawn into the vacuum system, causing blockages and contamination as well as reducing the moisture content of the product.

For the above reasons a number of alternative approaches have been proposed to address the problem of handling meat. *Khodabandehloo* [60.17] proposed a gripper with similar functionality to the human hand which could use its fingers to grasp a product. A full dexterous hand would be unnecessarily complex and as yet no such system has been demonstrated in an industrial environment, however the principle still appeared promising and a multifingered gripper was developed [60.17], formed from a solid piece of flexible rubber. An internal cavity was created at the finger's knuckle which could be pressurized by an external air supply. As the cavity was filled with air it expanded, causing the rear surface of the finger to elongate. Due to the location of the cavity the front surface of the finger remained unextended. As the finger was formed from one solid piece of rubber this difference in extension caused the knuckle to flex.

A hand consisting of four such fingers was developed and tested at the University of Bristol, UK. It was positioned so that two fingers were located on each side of the piece of meat to be lifted. When activated, the fingers curled around the meat, creating a grasp. Due to the compliant nature of the fingers they did not create damage to the surface of the meat as there was no hard contact. Due to the low number of mechanical parts and lack of moving linkages the gripper was very well suited to the hygiene requirements of the food industry as the gripper could be washed or hosed down without risk of damage.

Whilst proving effective at handling some cuts of meat the Bristol University gripper was unsuited to grasping steaks or thin slices of meat as they deform too much for the fingers to produce a secure grasp.

An alternative approach is the Intelligent Portion Loading Robot produced by AEW Delford Systems Ltd. [60.18]. This system is robot based and is able to handle and manipulate a broad range of meat types including both bone-in and boneless portions, fish, cheese, and sliced products. Meat is fed to the system on a conveyor where a vision system determines the position and orientation of the product to be handled. An ABB IRB 340 FlexPicker robot fitted with a novel end-effector developed by AEW Delford is then used to pick each product and transfer it to packaging or a further processing machine.

The end-effector's design is simple with a low number of parts, making it well suited to the needs of the food industry. The end-effector is essentially a high-speed parallel jawed gripper. Each jaw consists of a very thin plate which, when the gripper activates, is forced under the product as can be seen in Fig. 60.9a,b. The low profile of the jaws means they can be inserted under the product without damaging it. Although the lateral force applied to the product as the jaws are closed is relatively low it is still possible that this might dislodge the product slightly. In order to prevent this, a spring-loaded guide plate rests on the upper surface of the product being lifted whilst the jaws close.

Fish pieces can be particularly difficult to handle as, due to their structure, they can crumble when handled, breaking into many pieces. *Gjersted* [60.19] developed a needle gripper for the picking and packing of pieces of fresh, cooked and uncooked fish.

The gripper operates using a surface hooking principle [60.19] and uses numerous pins which enter the product simultaneously from opposite sides. The pins are angled slightly towards the center of the product

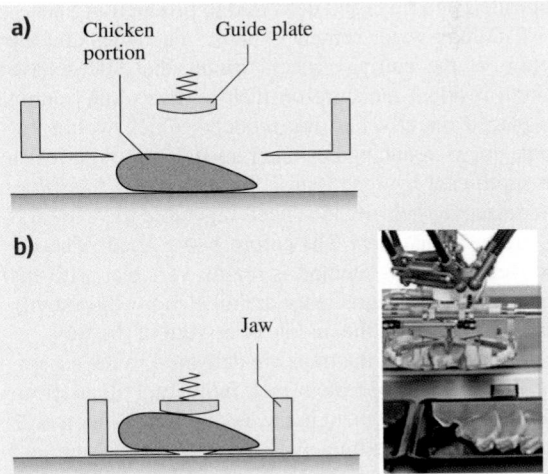

Fig. 60.9a,b AEW Delford gripper raised (**a**) and lowered with jaw closed (**b**)

and as a result when inserted they physically lock the fish firmly in place, which means it can be handled and accelerated rapidly without fear of being dropped. The only way that the product can be dropped whilst the pins are still inserted is if the product breaks apart, but this is unlikely to happen as, whilst the pins are in the product, they form an internal support structure which helps keep the product in one piece.

The gripper was developed according to European Hygienic Engineering and Design Group (EHEDG) principles for hygienic design, meaning it meets with the stringent requirements of the food industry. The gripper has been tested successfully with both salmon and cod and demonstrates excellent holding capability with minimal impact on the product surface and the overall product quality. In fact the impact on product appearance and quality was judged to be less than when conventional human handling was used.

60.4.3 Handling Moist Food Products

Within the food manufacturing industry it is extremely common that the materials to be handled are moist. This can be a result of washing, cooking or cutting or indeed just due to the nature of the product. This moisture can often make traditional grippers ineffective and so a number of novel techniques have been developed.

Sliced tomatoes and cucumbers used in a wide variety of products, e.g., salads and sandwiches, are typically washed and sliced in a secondary part of a factory using large-scale slicing machines capable of processing many kilos per minute. Once sliced the product is deposited into trays and delivered to production lines.

The high water content of most vegetables and the nature of the cutting process means that slices have a high residual moisture on their surfaces and cannot be placed directly into the product, which would become soggy reducing customer appeal, although it has no significant hygiene issues. To reduce this *sogginess* and improve shelf life the sliced vegetable trays are left to drain, for at least 2 h, before being used. The effectiveness of this method is highly variable, with the upper layers of ingredients draining more thoroughly than those towards the middle or bottom of the tray.

After draining, the trays are delivered to the assembly lines, where operators pick individual slices from the trays and place them in the assembled product, e.g., sandwiches. It is extremely difficult to do this without further damaging the slices and as a result it is not uncommon for the center of tomatoes to become detached. Furthermore the moisture causes the slices to stick to-

gether and the operators have to separate them, slowing the overall process. For this reason a production line working at 50 sandwiches per minute can typically have four operators just placing tomato slices and a similar number handling cucumber.

Davis et al. [60.20] proposed an automated system for the handling of sliced tomato and cucumber based on a novel end-effector. The solution involves cutting slices on the actual assembly line for immediate use. A slice would only be cut when required and thus the need to pick an individual slice from a tray is removed. Once cut, each slice is grasped using a noncontact Bernoulli gripper and a robot places it as required (Fig. 60.10).

A Bernoulli gripper operates using compressed air and a flat gripping face. Deflectors on the surface of the gripper direct the supplied air so that it radiates from the center of the gripper across the surface. When the grip-

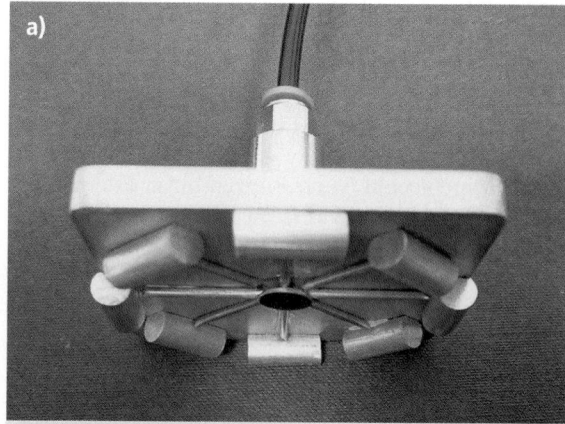

Fig. 60.10 (a) Noncontact Bernoulli gripper. **(b)** Gripper handling tomato

ping surface is brought close to an object to be grasped the gap through which the air travels becomes reduced. To maintain the volumetric air flow through the gripper this results in an increase in air velocity. The rapid flow of air between the object and gripper generates an attractive force in line with Bernoulli's principle. It is this force which allows the object to be grasped.

As well as lifting the products the gripper is also able to remove moisture from the object being handled using the air-knife principle where moisture is atomized by the air and blown off the surface.

Another technique developed for lifting moist products is the cryogenic gripper. *Stephan* and *Seliger* [60.21] developed a *freezing gripper* for use in the textile industry which created a bond between the gripper and products by freezing moisture on the surface of the product using a Peltier element. The reported grip forces were as high as $3.5\,\text{N/cm}^2$ after $3\,\text{s}$ of *freezing* with release after $1\,\text{s}$.

Although this gripper was developed for use in the textile industry the technique appeared to have potential in the food industry. To assess this, the Food Refrigeration and Process Engineering Research Center at the University of Bristol, UK [60.22] carried out tests on cryogenic grippers for the food industry. They were particularly interested to determine whether such techniques could be used to lift sheet-like food materials such as lasagne, sliced fish, cheese, and ham. Similar results to the work of *Stephan* et al. [60.21] regarding grasp times were obtained, however, unforced release times were found to be poor and mechanical release methods were found to produce unacceptable damage to the products' surface. Nonetheless it is suggested that with further development work a viable gripper can probably be developed that has the potential to be very useful in some sectors of the food industry, e.g., frozen foods.

60.4.4 Handling Sticky Products

Many food products are sticky and, whilst there is usually no problem developing automated systems for grasping such objects, releasing them can often present a challenge. The glacé cherry is an example of one such product. When handled with a traditional two-jaw gripper the cherry is found to stick to one of the jaws when released [60.23]. This meant that the cherry could not be positioned accurately. *Reed* et al. [60.23] developed a unique gripper for the production of Bakewell tarts. These small cakes require a decorative cherry to be place at the center of each cake and therefore a method

of picking and reliably releasing a single cherry was developed.

The gripper developed is a two-fingered parallel jaw mechanism as shown in Fig. 60.11a. As with a standard gripper the jaws are closed and an object is held by a frictional grasp. However, the unique feature of this gripper is that the contact surface of each jaw is covered in a polyester film. This film takes the form of a narrow tape which is wound onto spools (Fig. 60.11). When the gripper releases an object a length of tape is wound off the inner spools and onto the outer spools as shown in Fig. 60.11.

To release an object the spools are rotated and the resultant motion of the polyester tape on each jaw causes the object being grasped to be transported downwards. At the tips of the jaws the tape doubles back on itself and this causes it to peel away from the object being held and therefore release it. The sharpness with which the tape doubles back on itself is vital. If insufficiently sharp it would be possible for the object to remain stuck to one of the tapes and be transported along the outside of the jaw. An appropriately tight turn ensures that the contact area between the tape and object is so small that the resulting adhesive force is not large enough to support the weight of the object.

In addition to its ability to handle sticky objects this gripper can be used to position objects in confined spaces as the jaws of the gripper do not need to be opened during product release. This makes the gripper particularly well suited to placing objects into boxes. *Reed* et al. demonstrated how the gripper could be used to place petits fours and fondants into presentation boxes [60.23].

Another sticky product that has a reputation of being particularly difficult to handle is fresh sheets of lasagne. Clamping-type end-effectors cannot be used as

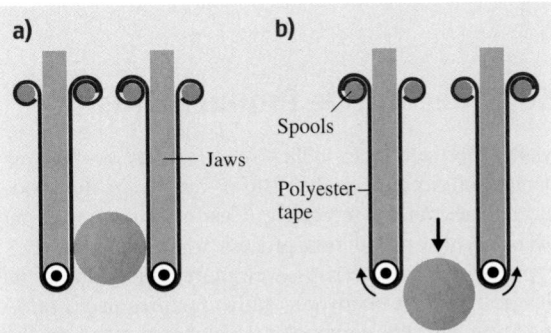

Fig. 60.11a,b Parallel jaw gripper grasping (**a**) and releasing (**b**) a sticky object

Part F | 60.4

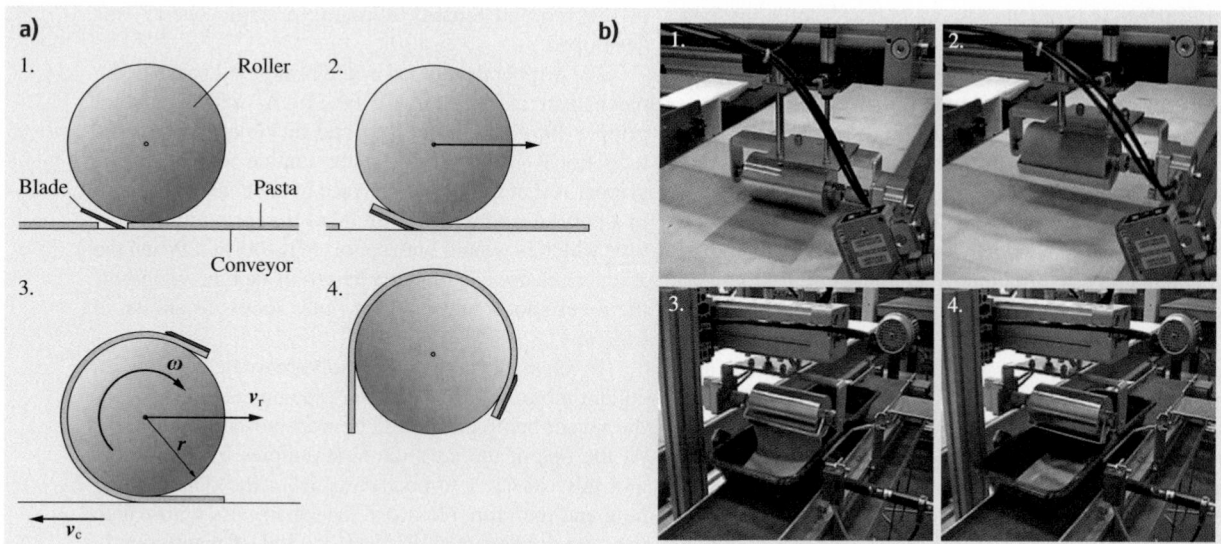

Fig. 60.12 (a) Lasagne lifting. **(b)** Motions in the automated handling of a lasagne pasta sheet

they would damage the surface of the pasta, and for similar reasons vacuum cups are also unsuitable.

Moreno-Masey et al. investigated the possibility of automating the manufacture of lasagne ready (microwave) meals and developed a method based on the rolling action common in making pastry [60.24]. By rolling a sheet of lasagne onto a roller and then gradually unrolling it above a product it was shown that the sheet could be positioned accurately and with little damage. The conceptual design with the envisaged sequence of operations is shown in Fig. 60.12a, with the automation system shown in Fig. 60.12b.

The gripper is initially positioned so that the spatula arm, needed to lift the front edge of the pasta, is located close to the pasta sheet. The gripper then moves horizontally a short distance towards the pasta, forcing the spatula under the leading edge of the sheet. The gripper continues moving in the horizontal direction and simultaneously rotates the roller. By coordinating the two motions the pasta is rolled onto the *gripper* in a controlled manner. To release the pasta and deposit it into a tray the roller is simply rotated in the opposite direction. The weight of the sheet causes the lasagne to peel free of the roller in an equally controlled manner. The machine constructed based on this design is shown in Fig. 60.12b. The motion of all actuators are pneumatically powered with PLC control over the joints using input sensing on the position of the lasagne sheets. The machine is sufficiently simple and low cost that nine identical machines could be used to produce ready meals at a typical production rate of 60 per minute.

60.5 Decorative Product Finishing

Many food products include components or features which add nothing to the taste or quality of the product but are purely decorative. Cake manufacture is an example of a production process where the product's appearance is as (perhaps even more) important as its flavor. There are many decorative features used, ranging from discrete components which are place on the surface of the cake to intricate patterns or text produced using icing.

Park Cakes in the UK is a large bakery producing cakes for special occasions which often include hand-written messages on their upper surface such as "Happy Birthday" or "Congratulations". These messages are produced by skilled staff using icing-filled pastry bags. The operators apply pressure to the bags in order to produce a constant flow of icing with which to *write* the messages. In order to be able to undertake this task to the required high standard requires both experience and training and this increases the overall staff costs.

At certain times of the year (Christmas, Valentines' day, and Easter) the demand for cakes increases dramatically and can mean production rates at the Park Cakes factory can increase by as much as 300%. This presents a significant problem as additional labor is needed but finding skilled staff is difficult and additional training is often required. This is costly and requires detailed planning to ensure labor levels are available at just the right time.

A robot-based solution was proposed and developed [60.25] in which the messages were initially produced in a computer-aided design (CAD) package which converted the written text into a series of coordinate positions used to programme the robot. A four-degree-of-freedom SCARA (selective compliance assembly robot arm) design was sufficient to achieve the icing task. In addition to the basic motions involved in writing, the appearance of the icing is also dependent on the relative distance between the depos-

itor and the surface of the cake. To ensure knowledge and maintenance of correct separation a laser range finder was used to measure the distance to the cake's surface and allowed the robot's height to be adjusted to correspond to the individual profile of each cake. The robot carries the icing head, which was capable of producing a steady stream of chocolate icing. It was essential that the icing depositor could be switched off with a clean cutoff (i. e., when stopped the icing flow would instantly cease without any stringing). This was achieved using a stainless-steel 2200-245-Series KISS Tip Seal Valve. Whilst the robot system operated as intended there were some initial problems with the depositing system caused by the flow characteristics of the icing, and input from a chocolate technologist allowed the icing to be produced more consistently resolving the problems.

The quality of the final products exceeded Park Cakes' expectations and the system removed some of the pressures on the company to find large numbers of skilled laborers during periods of high demand.

60.6 Assembly of Food Products – Making a Sandwich

The instances already studied have shown uses of automation in applications involving containerized food products and individual or discrete food product handling. However, one of the major challenges in the automation of food production is the assembly of food products from a large number of discrete components. The assembly of a sandwich is one such problem.

Until recently sandwich production has been performed almost entirely manually with lines employing up to 80 people. The little automation that is used typically takes the form of slicers or depositors. The high level of labor means that there is a real incentive to look to automation. However, the only successful example of an automated sandwich line is that developed for Uniq Plc. by *Lieder* [60.26]. This system uses industrial robots and an indexing system to automate the entire process from buttering to packing but operates most successfully for products with paste fillings. Due to the use of robots there are significant safety issues and as such the line must be enclosed by guarding. This means it cannot operate alongside humans and also leads to a large machine footprint.

An alternative approach was described by *Davis* et al. based on simpler dedicated yet more flexible concepts [60.27, 28]. In this system the ingredients in-

cluded: two slices of bread, butter, mayonnaise, chicken (diced), lettuce (shredded), four slices of tomato, and four slices of cucumber. This is considered by the industry to represent the most difficult sandwich handling scenario and does not benefit from the binding effects found in the pastes used on the Lieder line. The automated system consisted of a continuously running corded conveyor which transports the bread slices and sandwich assemblies between workstations. This conveyor is separated into two lanes with each lane handling half of the sandwich (top and bottom slice). A piece of bread is placed in each lane and as they progress along the line additional ingredients are added to the bottom (supporting) slice. After all the ingredients are added the second slice must be lifted, inverted, and then placed on the top of the first slice. Studies of current production lines identified a number of individual processes needed to construct the sandwich:

1. Ingredient placement – individual ingredients are placed onto a single slice of bread.
2. Topping – a second slice of bread is placed on top of the first.
3. Cutting – the sandwich is positioned and cut once diagonally to form two triangular sandwiches.

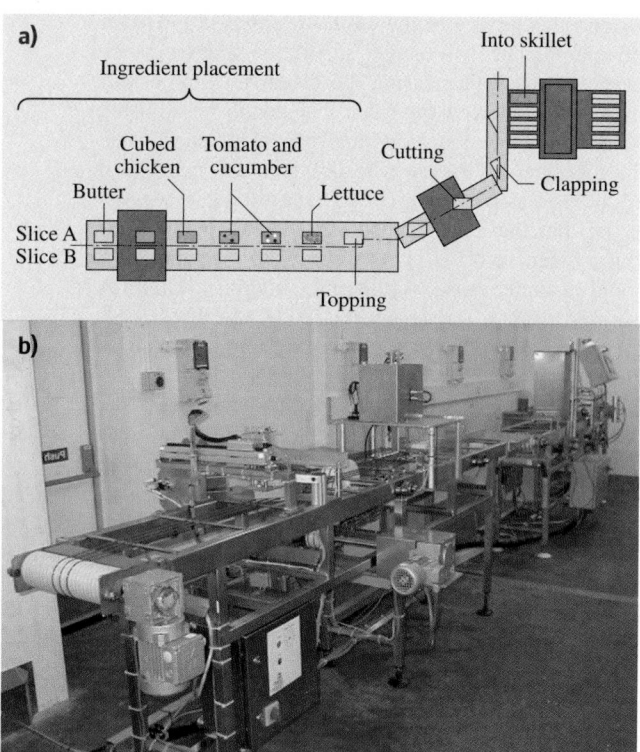

Fig. 60.13a,b Chicken salad sandwich production processes

4. Clapping – one triangular sandwich is placed on top of another prior to top packing.
5. Packing – the two sandwiches are placed into a skillet.

Figure 60.13a shows these five individual operations, while Fig. 60.13b shows the developed system.

An industry-standard programmable logic controller (PLC) is used to control the basic functions of the process. All sensors, variable-speed drives, stepper drives, the vision system and pneumatic control valves are connected to and controlled by this PLC. A touchscreen human–machine interface (HMI) is used to operate the machine and to give the operator feedback about the operating state. The HMI also has an engineer mode that allows qualified personnel to change conveyor speeds and perform basic sequencing checks.

Stepper drives are used where controlled movement of the sandwiches is required. A stepper drive is used at the optimum cut station to precisely align the sandwich before the sandwich is cut. The clapping station uses three stepper drives to place the two halves of the cut sandwich precisely on top of each other prior to being packed. The packing station uses a stepper drive to rotate the assembled sandwich through precisely 90° before being dropped into the awaiting skillet. All the stepper drives are controlled by the PLC and all the sensors are connected to the PLC.

To establish the validity of each principle a series of trials over a 10 week period on 250 000 sandwiches were conducted within a sandwich factory. The aim of the trails was to analyze each process critically and establish whether these principles worked with real products and in a true factory environment. For each of the individual results there were very satisfactory outcomes with high levels of acceptance and reliability.

60.7 Discrete Event Simulation Example

With automation projects being expensive, and the cost of lost production during the installation of equipment being high, it is vital that any potential problems with an automation project be identified as early as possible. Discrete event simulation is a software tool which allows machines, production lines or even entire factories to be produced, tested, and assessed on a computer before any real equipment is purchased or installed.

The simulations provide accurate representations of production processes and product flow. This means an appropriately constructed simulation can be used to identify bottlenecks in existing or planned layouts

(Fig. 60.14). This is particularly useful when installing a discrete piece of automation onto a line as it allows the effect of the machine on the remainder of the line to be assessed.

Simulation can also be used to determine if labor is being used efficiently. The example described below is of a sandwich production line. From video footage of the line it was determined how long each operator took to perform their particular task. The simulation was then created and sandwiches were input onto the line at the appropriate rate. Analysis was then performed to determine what percentage of each operator's time was spent idle. It was shown that each of the three operators tasked

Fig. 60.14 Simulation of sandwich production line

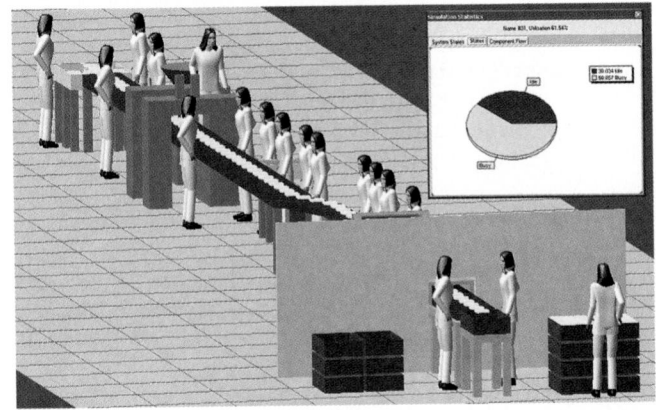

with placing lettuce on the sandwiches was only busy for 60% of their time. This meant that either one of the operators could be removed or the line speed could be increased. The simulation then showed that increasing the line speed produced problems elsewhere as other operators no longer had enough time to perform their tasks.

This is just one example of how simulation can be useful. More complex simulations allow the introduction of variation of task time, variability of product flow, and changes in shift patterns, to mention just three factors.

60.8 Totally Integrated Automation

While production hardware forms the most obvious feature on the food factory landscape, it is only part of a larger automation scheme which can ultimately deliver a totally integrated form of automation encompassing management and strategic functions (logistics, sales, orders, dispatch, traceability, energy, maintenance etc.), production functions including SCADA, networking, HMI, distributed process control (DPC), and (at the machine level) PLC, sensors, actuators etc. A typical format for this type of layout is shown in Fig. 60.15,

and the techniques used in the food sector vary little from those in other sectors.

Although the levels of uptake do vary across the industry, in general, at the managerial and administrative level most food processing companies do make extensive use of at least some aspect of control systems. This is particularly important with regard to production scheduling, logistics, and distribution as in the food sector the concept of *just in time* is taken to perhaps its greatest extreme. In food product manufacture it is not

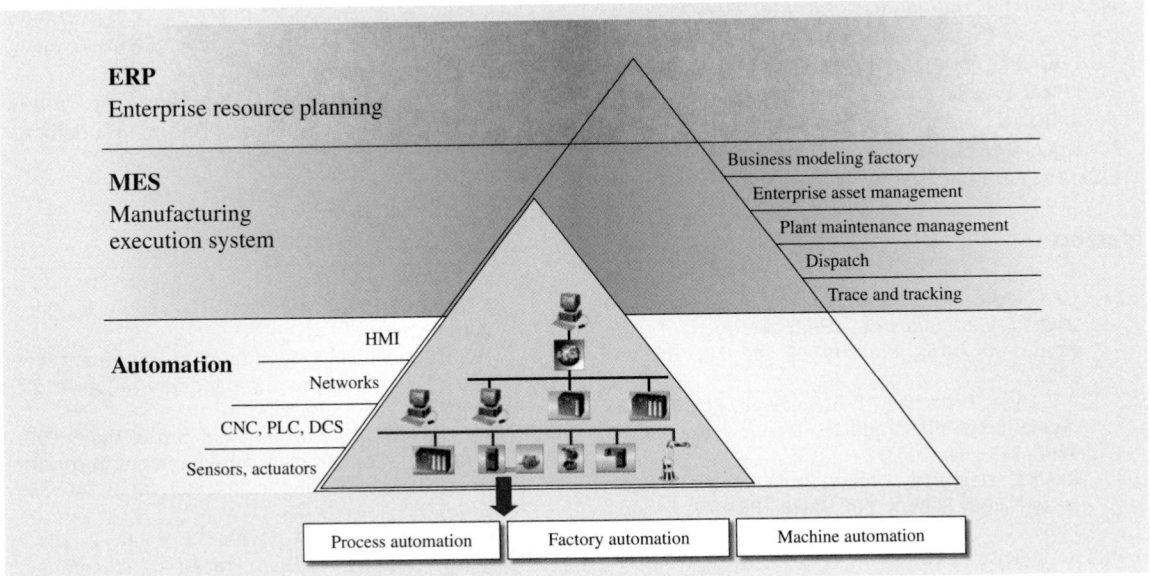

Fig. 60.15 A totally integrated approach to factory and production automation

unusual that orders are placed at midnight for dispatch in under 12 h. These orders can amount to tens or even hundreds of thousands of units. While the manufacturers do try to predict demand it is not unknown for the customer (supermarkets) to vary the orders by 50% or more. Indeed it is not unknown for expected orders to be canceled with only hours of warning. In these instances efficient and effective process control and management is essential.

In terms of process automation there is use of SCADA, DCS, PLC integrated networks, etc. In areas such as batch production, for the beverage sector etc., the use of this technology is common, but in the broader food sector dealing with discrete components the use of process automation is more patchy. Certainly most of the hardware which is developed for the food sector now comes with PLC, DCS, and networking capability and can be integrated into a full SCADA system, however, a fully integrated and networked operation is certainly not the normal practice, although it does represent best practice. There are several reasons for reduced uptake even when automation machinery has been installed that has advanced control functionality.

60.9 Conclusions

The food processing sector is the largest manufacturing industry in many countries, but to date it has been one of the least effective at using automation, and particularly following the newer automation trends. This has been driven by many factors that are both commercial and technical and it is certainly true that in many instances food products pose questions that are not present when handling traditional materials. Yet those within the industry are acutely aware of the pressure to reduce food costs while maintaining quality and it is also true that when there is a genuine desire from the food processing organization there are many ways in which automation can be used to enhance the performance of the business. It is particularly hoped that SMEs that have least experience (and confidence) with advanced automation will find the necessary guidance to reap the benefits.

60.10 Further Reading

- R.G. Moreira: *Automatic Control for Food Processing Systems* (Springer, Berlin, Heidelberg 2000)
- L.M. Cheng: *Food Machinery – For the Production of Cereal Foods, Snack Foods And Confectionery* (Woodhead, Cambridge 1992)
- M. Kutz: *Handbook of Farm, Dairy, and Food Machinery* (William Andrew, Norwich 2007)
- G.D. Saravacos, A.E. Kostaropoulos (Eds.): *Handbook of Food Processing Equipment* (Kluwer Academic Plenum, Norwell 2003)
- B. Siciliano, O. Khatib (Eds.): *Springer Handbook of Robotics* (Springer, Berlin, Heidelberg 2008)

References

60.1   CIAA: *Data and Trends of the European Food and Drink Industry* (Confederation of the Food and Drink Industry of the EU, Brussels 2006)

60.2   P.Y. Chua, T. Ilschner, D.G. Caldwell: Robotic manipulation of food products – a review, Ind. Robot. **30**(4), 345–354 (2003)

60.3   BRA/DTI Technology: *Market Review of the Robotics Sector*, Final Report, 5th March (BRA/DTI, London 1997)

60.4   J. Taylor: *HACCP Made Easy* (Practical HACCP, Manchester 2006)

60.5   J.M. Farber: *Safe Handling of Foods* (CRC, Boca Raton 2000)

60.6   ISO: ISO 22000:2005, Food safety management systems – Requirements for any organization in the food chain (ISO, Geneva 2007)

60.7   Materials of Construction Subgroup of the EHEDG: Materials of construction for equipment in contact with food, Trends Food Sci. Technol. **18**(S1), S40–S50 (2007)

60.8   H.L.M. Lelieveld, M.A. Mostert, J. Holah, B. White (Eds.): *Hygiene in Food Processing* (Woodhead, Cambridge 2003)

60.9 C. Honess: Importance of surface finish in the design of stainless steel. In: *Stainless Steel Ind.* (British Stainless Steel Association, Sheffield 2006) pp. 14–15

60.10 FDA: Indirect food additives: Polymers, Code of Federal Regulations, CFR Title 21, part 177 (FDA, Rockville 2007), Available online from http://www.cfsan.fda.gov/~lrd/FCF177.html

60.11 G. Midelet, B. Carpentier: Transfer of microorganisms, including listeria monocytogenes, from various materials to beef, Appl. Environ. Microbiol. **68**(8), 4015–4024 (2002)

60.12 FDA:. Indirect food additives: Adhesives and components of coatings. Code of Federal Regulations, CFR Title 21, part 175 (FDA, Rockville 2007), Available online from http://www.cfsan.fda.gov/~lrd/FCF175.html

60.13 B.M.Y. Nouri, F. Al-Bender, J. Swevers, P. Vanherck, H. Van Brussel: Modelling a pneumatic servo positioning system with friction, Proc. Am. Control Conf., Vol. 2 (2000) pp. 1067–1071

60.14 R.B. van Varseveld, G.M. Bone: Accurate position control of a pneumatic actuator using on/off solenoid valves, Proc. IEEE Int. Conf. Robotics and Automation ICRA, Vol. 2 (1997) pp. 1196–1201

60.15 P.M. Taylor: Presentation and gripping of flexible materials, Assem. Autom. **15**(3), 33–35 (1995)

60.16 J.N. Reed, S.J. Miles, J. Butler, M. Baldwin, R. Noble: Automatic mushroom harvester development, J. Agric. Eng. Res. **78**(1), 15–23 (2001)

60.17 K. Khodabandehloo: Robotics in food manufacturing. In: *Advanced Robotics and Intelligent Machines*, ed. by J.O. Gray, D.G. Caldwell (IEE, Stevenage 1996) pp. 220–223

60.18 IPL: Marel Food Systems (2008), www.marel.com/company/brands/AEW-Delford/

60.19 T.B. Gjerstad, T.K. Lien: New gripper technology for flexible and efficient fish processing, Proc. Food Factory of the Future 3 (Gothenburg 2006)

60.20 S. Davis, J.O. Gray, D.G. Caldwell: An end effector based on the Bernoulli principle for handling sliced fruit and vegetables, Int. J. Robot. Comput. Integr. Manuf. **24**(2), 249–257 (2008)

60.21 F. Stephan, G. Seliger: Handling with ice – the cryo-gripper, a new approach, Assem. Autom. **19**(4), 332–337 (1999)

60.22 FRPERC: Food Refrigeration and Process Engineering Research Centre, Univ. Bristol, UK (2008) http://www.frperc.bris.ac.uk/

60.23 C. Connolly: Gripping developments at Silsoe, Ind. Robot J. **30**(4), 322–325 (2003)

60.24 R.J. Moreno-Masey, D.G. Caldwell: Design of an automated handling system for limp, flexible sheet lasagna pasta, IEEE Int. Conf. Robot. Autom. ICRA (Rome 2007) pp. 1226–1231

60.25 B. Rooks: The man-machine interface get friendlier at Manufacturing Week, Ind. Robot. J. **25**(2), 112–116 (1998)

60.26 Robot Food Technologies Germany GmbH. Wietze (2008) http://www.robotlieder.de/

60.27 S. Davis, M.G. King, J.W. Casson, J.O. Gray, D.G. Caldwell: Automated handling, assembly and packaging of highly variable compliant food products – Making a sandwich, IEEE Int. Conf. Robot. Autom. ICRA (Rome 2007) pp. 1213–1218

60.28 S. Davis, M.G. King, J.W. Casson, J.O. Gray, D.G. Caldwell: End effector development for automated sandwich assembly, Meas. Control. **40**(7), 202–206 (2007)

Part G Infrastructure and Service Automation

Infrastructure and Service Automation. Part G Without automation, certain infrastructure and services could not even be imagined, as in space exploration and secure electric power distribution. In others, automation is essential in improving them so fundamentally, that their benefit influences tremendous social and economic transformation and even revolution, as with transportation, agriculture and entertainment. Chapters in this part explain how automation is designed, selected, integrated, justified and applied, its challenges and emerging trends in those areas and in the construction of structures, roads and bridges; of smart buildings, smart roads and intelligent vehicles; cleaning of surfaces, tunnels and sewers; land, air, and space transportation; information, knowledge, learning, training, and library services; and in sports and entertainment. With the enormous increase in the importance of the service sector in the global economy, this infrastructure and service automation part clarifies not only how it has evolved and is being enabled, but also how automation will influence future growth and further innovations in these domains.

61. Construction Automation

Daniel Castro-Lacouture

The construction industry is labor intensive, project based, and slow to adopt emerging technologies. Combined, these factors make the construction industry not only one of the most dangerous industries worldwide, but also prone to low productivity and cost overruns due to shortages of skilled labor, unexpected site conditions, design changes, communication problems, constructability challenges, and unsuitability of construction means and techniques. Construction automation emerged to overcome these issues, since it has the potential to capitalize on increasing quality expectations from customers, tighter safety regulations, greater attention to computerized project control, and technological breakthroughs led by equipment manufacturers. Today, many construction operations have incorporated automated equipment, means, and methods into their regular practices.

The Introduction to this chapter provides an overview of construction automation, highlighting the contribution from robotics. Several motivations for automating construction operations are discussed in Sect. 61.1, and a historical background is included in Sect. 61.2. A description of automation in horizontal construction is included in Sect. 61.3, followed by an overview of building construction automation in Sect. 61.4. Some techniques and guidelines for construction management automation are discussed in Sect. 61.5, which also presents several emerging trends. Section 61.6 shows some typical application examples in today's construc-

tion environment. Finally, Sect. 61.7 briefly draws conclusions and points out challenges for the adoption of construction automation.

Construction automation has been continuously redefined throughout the past two decades. In 1988, construction automation was defined as "the work to increase the contribution of machines or tools while decreasing the human input" [61.1]. Another definition states that it is "the technology concerned with the application of electronic, mechanical and computer-based systems to operate and control construction production" [61.2]. Construction automation was further characterized as [61.3]:

the work using construction techniques including equipment to operate and control construction production in order to reduce labor, reduce duration, increase productivity, improve the working environment of labor and decrease the injury of labor during construction process.

From a systemic perspective, construction automation is the technology-driven method of streamlining construction processes with the intention of improving safety, productivity, constructability, scheduling or control, while providing project stakeholders with a tool for prompt and accurate decision making. This method must not be limited to replicating skilled labor or conventional equipment performance. The latter is the purpose of single-task robots, which have been an important component of construction automation. In the year 2000, robotics applications in the construction industry completed 20 years of research, exploration, and prototyping, as documented in the first book on robotics in civil engineering [61.4]. These applications made important contributions to replicating single tasks, which could be completed faster and safer, since

no laborers were operating equipment. However, initial and operating costs have been a problem for the massive deployment of construction robots [61.5, 6]. Later, the distinction between single-task robots and construction automation became more evident. Single-task robots perform a specific job, whereas construction automation uses principles of industrial automation to streamline repetitive tasks, such as just-in-time delivery systems, coded components or computerized information management systems [61.7]. Automation has been associated with repetitive processes, while robotics has targeted single tasks or jobs, imitating skilled labor. Nevertheless, construction processes may not be repetitive as a whole, due to the planning, design, and assembly requirements that must be addressed prior to initiating construction. In addition, site layout and logistics constraints may pose another obstacle so that a theoretical repetitive process must be decomposed into simpler tasks. These simpler tasks may be treated as repetitive in nature. Some examples of repetitive tasks are digging a trench, placing a pipe, backfilling, placing masonry tiles, hauling topsoil, etc.

61.1 Motivations for Automating Construction Operations

Based on a market research questionnaire conducted in 1998 to construction industry respondents from 24 countries, the strongest reasons for robotic construction automation were: productivity improvement, quality and reliability, safety, enhancement of working conditions, savings in labor costs, standardization of components, life cycle cost savings, and simplification of the workforce [61.8].

The project-based nature of the construction industry implies the periodic mobilization of construction equipment, materials, supplies, personnel, and temporary facilities at the start of every construction project. Recent hires, especially field laborers, may not be familiar with the construction practices adopted by the firm on a particular project, making it difficult to engage them in technologically advanced processes from the beginning of the project. Shortages of skilled labor, due to economic fluctuations, immigration policies, or geographic considerations, make the adaptation to project-based means and techniques even more challenging. Therefore, the possibility of automating construction processes would constitute a great opportunity for overcoming the transition to project-based demands. Diffi-

culties in the delivery of supplies and the assembly of materials on site have been alleviated by the adoption of off-site assemblies, manufacturing automation principles, and procurement of premanufactured components.

Another motivation for automating construction tasks is safety in the workplace. Research has found that the causes of accidents can be attributed to factors such as human error, unsafe behavior, and the interaction of humans with materials, tools, and environmental factors [61.9]. Some of the incidents leading to construction injuries and fatalities can be attributed to collisions between workers and equipment, or from workers falling from roofs, scaffolds or trench edges [61.10]. In the USA in 2006, there were 1226 fatalities associated with the construction industry. This accounts for almost 24% of all fatalities in the private sector [61.11]. However, the construction industry accounts for only 5% of the US workforce [61.12]. This high proportion of construction injuries and fatalities may indicate that the industry needs new approaches in order to improve safety environments for workers on construction sites. Some efforts have focused on using machines to complete repetitive tasks that were once performed by workers. This practice has removed workers from hazardous construction environments, but those workers that need to remain in place are still vulnerable to accidents. Automation efforts may further reduce the possibility of accidents or near misses by creating a sensor-based network that tracks the position of workers and equipment, thereby alerting the worker and supervisor when a hazardous condition arises.

Productivity improvement constitutes another reason for automating construction operations. This improvement is critical since traditionally the construction industry has been one of the worst industries with regard to annual increase in productivity [61.13]. This concern, combined with ever-increasing costs, high accident rates, late completions, and poor quality, has been the subject of dedicated studies on construction productivity improvement [61.14]. Construction is one of the largest product-based industries,

and contributes a major portion to the gross domestic product of both developed and developing countries [61.15, 16]. However, when compared with manufacturing, construction is rather slow in terms of technological progress [61.17]. There are also differences associated with purchasing conditions, risk, market environment, sales network, product uniqueness, and project format [61.18]. Construction is often considered an antiquated industry. There have not been dramatic changes in basic construction methods in the last 40 years. The modest attention to research and development by both the public and private construction sectors has undermined possible breakthroughs in construction automation that may impact the overall productivity of the industry. Furthermore, the assumed endless availability of workers and the focus on cost reduction and short-term efficiency have continued to limit the rationale for automation efforts in construction [61.19].

61.2 Background

The historical development of construction automation has been marked by equipment inventions aimed at performing specific tasks originally done by workers, and by ground-breaking methodologies intended for improving the systematic behavior of resources in a construction project setting. Table 61.1 shows the historical development of construction automation, from the early stages of equipment inventions to the latest trends in automated project control and decision support systems. Since construction has been mostly an adopter of innovation from other fields rather than a source of innovation, every development indicated in the table is shown in the period where it had a dramatic impact on construction means, management, and methods, and not necessarily when it was invented.

Table 61.1 includes chronological developments related to technologies, means, and methods that have had a dramatic impact on the way construction operations are performed, managed, and conceived, allowing them to be partially or fully automated. The discipline of construction management studies

the practice of the managerial, technological, and business-oriented characteristics of the construction industry, whereas construction engineering analyzes design, operational, and constructability aspects. This differentiation has been reflected in the way several automation efforts have been focused throughout the last few decades. These efforts and periods have been catalogued in different manners: based on the taxonomy of field operations [61.3, 20, 21]; based on geographical development [61.6, 7]; based on maturity [61.22]; and whether technologies are considered hard or soft [61.23]. This chapter describes automation efforts from the perspective of their purpose in the construction domain, that is, whether the automation effort has been focused primarily on the construction of highway or heavy structures, on buildings, or on computer-supported integrated technologies that facilitate construction management decision making, which can be applied to the design, procurement, assembly, construction, maintenance, and management of any type of facility.

Table 61.1 Historical development of construction automation

Period	Development
1100s	Pulleys, levers
1400s	Cranes
1500s	Pile driver
1800s	Elevators, steam shovels, internal combustion engine, power tools, reinforced concrete
1900s	Slip-form construction
1910s	Gantt charts, work breakdown structures (WBS)
1920s	Dozers, engineering vehicles
1930s	Prefabrication, hydraulic power, concrete pumps
1950s	Project evaluation and review technique (PERT), computers
1960s	Time-lapse studies, critical path method (CPM)
1970s	Robotics, computer-aided design (CAD), discrete-event simulation
1980s	3-D CAD, 4-D CAD, massification of personal computers, spreadsheets, relational databases, geographic information systems (GIS), large-scale manipulators
1990s	Internet, intranets, extranets, personal digital assistants (PDAs), global positioning systems (GPS), barcodes, radiofrequency identification systems (RFID), wireless communications, remote sensing, precision laser radars (LADARS), enterprise resource planning (ERP), object-oriented programming (OOP), concurrent engineering, industry foundation classes (IFC), building information models (BIM), lean construction (LC)
2000s	Web-based project management, e-Work, parametric modeling, Wi-Fi, ultra wide band (UWB) for tracking and positioning, machine vision, mixed augmented reality, nanotechnology

61.3 Horizontal Construction Automation

Horizontal construction has been prone to automation due to the repetitive tasks, intensive labor, and equipment involved in the operation. This type of construction comprises linear projects, such as road construction, paving, drilling, trench excavation, and pipe laying.

One of the first attempts to fully automate the paving process was the Road Robot [61.24]. The aim of this project was to develop a self-navigating, self-steering asphalt paver that would allow road engineers to improve the quality of pavements, while also being more environmentally friendly. The operation of the Road Robot was divided among four subsystems: asphalt materials logistics, traveling mechanism, road surface geometry, and screeding. Although the Road Robot successfully demonstrated the capabilities and advantages of a fully automated asphalt paver, further development appears to have been halted.

The computer-integrated road paving (CIRPAV) prototype was another attempt to automate paving operations [61.25]. The primary functions of the CIRPAV system were to assist the operator in maintaining the paver on its correct trajectory at the correct speed, to automatically adjust the position and cross-slope of the screed, and to record actual work performed by the paver and transmit performance data to a remote ground station, in order to maintain global quality control at the site level. The CIRPAV system consists of three main subsystems: the ground subsystem, the onboard subsystem, and the positioning subsystem. After several trials, the following improvements were achieved, in contrast with the conventional asphalt paving process: the costs of establishing and maintaining references for profile control and equipment operations were reduced from 10% of the total cost of the work to below 5%; the fluctuation of the

Fig. 61.1a,b Slip-form pavers: (**a**) screeding process, (**b**) idle workers during operation [61.26]

layer thickness was decreased, with estimated savings of materials of about 5% of the total cost of the work; and the quality of the final pavement was improved. The CIRPAV prototype was able to place asphalt within ±5 cm in both transversal and longitudinal directions, and within ±0.5 mm for the height component.

The appearance of slip-form pavers semi-automated concrete paving operations. Figure 61.1a shows an automated concrete screeding process.

However, these operations still require several laborers who remain idle most of the process time, as seen in Fig. 61.1b. Furthermore, slip-form machines still depend on visual inspection and manual samples to perform quality control of the concrete mix. In recent years, a prototype design of a fully autonomous robot for concrete paving was developed [61.30]. The Robopaver prototype is a battery-operated robot consisting of several different operations: placing prefabricated steel reinforcement bar cages, placing and distributing concrete, vibrating, screeding, final finishing, and curing. Results from the simulation of prototypical tests yielded productivity improvements for Robopaver of 20% over the traditional slip-form paver, while foreman utilization achieved 99% with no operators involved, as opposed to the traditional operation with 46% foreman utilization and six laborers [61.31]. The construction work zone for Robopaver was less prone to accidents involving construction workers, while being more productive.

Trench excavation and laying buried pipes are among the most dangerous tasks in the construction industry. While heavy construction equipment such as cranes, loaders or backhoe excavators are used to dig, hoist, and lower pieces of pipe into the trench, workers guide the operation from inside the trench to perform final alignment and jointing. Excavating operations, such as trenching, require precise control.

Two controllable struts align pipe

Laser target above laser point

Fig. 61.2a–c Construction excavation automation: (**a**) robotic excavator [61.27], (**b**) Lancaster University Computerized Intelligent Excavator (LUCIE) [61.28], (**c**) PipeMan [61.29]

Previous experiments with robotic excavation have implemented a conventional industrial robot fitted with a bucket as the end-effector [61.32, 33]. More recently, a prototype Komatsu PC05-7 hydraulic mini-excavator was extensively modified to operate as an autonomous robotic excavator [61.27]. Figure 61.2a,b shows modified robotic excavators during trench-forming tasks.

Research in the telerobotic operation of hoisting and placing pipes on trenches has shown promising results for improving safety. Telerobotic systems are mechanical devices that combine human and machine intelligence to perform tasks remotely with the assistance of various sensors, computers, man–machine interface devices, and electronic controls. Building upon a previous-generation pipe manipulator dubbed PipeMan, further improvements were made by adding a laser and video system in order for the operator to control the entire device remotely, becoming a rugged but simple man–machine interface for motion control and control feedback, as shown in Fig. 61.2c. Pipe installation tasks could be initiated and observed by the operator with the help of wireless fidelity (Wi-Fi) interfaces to the electrohydraulic valves mounted on the manipulator, as well as the video images transmitted wireless to the flat screen mounted to the side of the cabin window [61.29].

61.4 Building Construction Automation

In the 1990s, despite numerous attempts to develop highly automated machines and robotics for construction field operations in the previous decade, few practical applications could be found on construction sites [61.21]. In Japan, however, pushed by building corporations and manufacturers, the largest construction firms of the time developed robots for building construction. Among these firms (e.g., Takenaka, Shimizu, Taisei, Kajima, Obayashi, and Kumagai-Gumi), a variety of single-task robots were manufactured for practical construction applications. These applications mainly consisted of concrete floor finishing, exterior wall spray painting, ceiling board installation, and fire proofing. The goals of the deployment of these single-task robots were mostly the improvement of productivity, safety, and quality.

Japanese contractors also developed automated building systems, which consisted of on-site construction factories that used ideas already tested in the manufacturing and automobile industries, such as just-in-time, material tracking or streamlining repetitive operations [61.7].

The WASeda construction robot (WASCOR) project entailed a building system carried out by assembling factory-made interior units installed with frames, boards, papers, and fixtures, and using construction robots [61.37]. The push-up method consisted of assembling the roof floor first, lifting it up with its supporting columns using hydraulic jacks, thereby serving as the working platform for the construction of the lower floors. Every time a lower floor was completed, the roof floor was jacked up [61.7]. Shimizu manufactur-

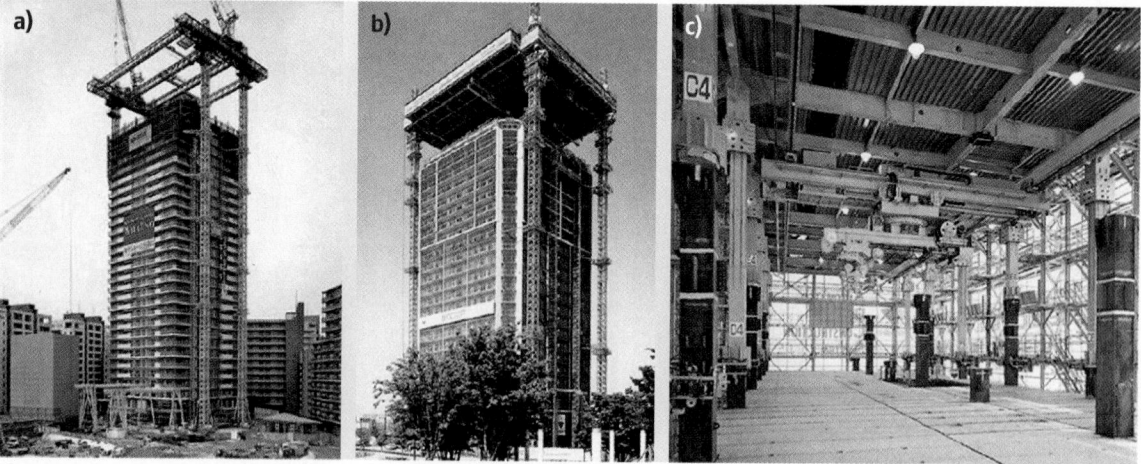

Fig. 61.3a–c Automated building systems in Japan: (**a**) SMART [61.34], (**b**) Big Canopy [61.35], and (**c**) ABCS [61.36]

ing system by advanced robotics technology (SMART) was an integrated system that automated the erection and welding of steel frames, laid concrete floor boards, and installed exterior and interior wall panels and other components. The automated building construction system (ABCS) system consisted of a *construction factory* placed on top of the building, lifting structural members to the lower floors and welding the components with robots. It had the capacity to build two floors at once. Productivity shortcomings and cost considerations associated with the operation of the ABCS prompted the development of the Big Canopy system. This system featured four tower masts and a massive canopy at the top, which lifted prefabricated material to the target floor, where workers controlled the maneuvering of the components with the use of joysticks. The T-up system comprised a support, a base, and a manipulator. The construction process began with the erection of the building core and the base was constructed at the ground

level. Guide columns and hydraulic jacks elevate the base to the target floor. Figure 61.3 shows several examples of automated building systems developed in Japan.

Besides in Japan, there has been significant advancement in robotics research and development for building construction in the past two decades. Some of these research projects have been led by academic institutions and government agencies, as shown in Table 61.2.

Almost 30 years after robotics applications in construction started being researched, explored, and prototyped in the early 1980s, these applications are still considered atypical for the construction industry. However, robotic application efforts are still underway, mainly for performing single tasks that are part of tedious dangerous jobs that demand high quality and productivity. This trend goes along with the faster pace of robotics development in other industries, whose success has not been paralleled by the con-

Table 61.2 Automation and robotics research for building construction, excluding Japan

Project	Institution	Features
TAMIR	Technion – Israel Institute of Technology, Israel	Autonomous multipurpose interior robot
Rebar manufacturing	Technion – Israel Institute of Technology, Israel, and University of Ljubljana, Slovenia	Assembly of rebar cages for beams and columns
Building assembly robot	University of Karlsruhe, Germany	Automatic crane handling system
SHAMIR	Technion – Israel Institute of Technology, Israel	Autonomous multipurpose interior robot
Mobile bricklaying robot	University of Stuttgart, Germany	Brickwork erection
Handling robot	Technion – Israel Institute of Technology, Israel	Conversion of existing cranes into large semi-automatic manipulators
ROCCO	University of Karlsruhe and Technical University of Munich, Germany	Assembly system for computer-integrated construction
Brick laying robot (BLR)	Delft University of Technology, Netherlands	Brick placing from a platform on top of a variable-height telescopic mast
RoboCrane	National Institute of Standards and Technology, USA	Welding system and steel placing
ROMA	University Carlos III de Madrid, Spain	Autonomous climbing for construction inspection
Contour crafting	University of Southern California and Ohio University, USA	Additive fabrication technology for surface-forming troweling to create smooth and accurate planar and free-form surfaces
Freeform construction	Loughborough University, UK	Megascale rapid manufacturing for construction

struction industry. The decision-making and situational analysis complexities, coupled with the project-based nature and short-term focus of the construction industry has hindered success. Nevertheless, for the past decade, construction researchers and practitioners have shifted their efforts toward the development of integrated automated construction systems aimed at providing decision-makers with robust tools for project management. These systems consist of a variety of computer-supported applications that take advantage of the increasing computing power available, as well as accessibility to tracking and positioning technologies, such as global positioning system (GPS), radiofrequency identification (RFID), and Wi-Fi, which have demonstrated dramatic improvements in materials and resources management, progress tracking, cost monitoring, quality control, and equipment operator training.

Automation efforts have been implemented throughout the architecture, engineering, construction, and facility management life cycle. During the design phase, four-dimensional (4-D) computer models, i. e., three-dimensional (3-D) objects plus time dimension, allow project information to be shared among project participants, showing realistic views of objects and activities and their sequence of assembly. Complex designs take advantage of 4-D models by using animations to link computer-aided design (CAD) elements with schedule activities, thereby improving the clarity of design and construction information, allowing early clash detection, and helping track the construction progress.

Building information models (BIM) emerged to formally address these design automation needs, modeling representations of the actual elements and components of a building. BIM is based on industry foundation classes (IFCs) and architecture, engineering and construction extensive markup language (aecXML), which are data structures for representing project information [61.38]. Further implications of BIM to design automation include the possibility of estimating building costs, schedule progress, green building rating, energy performance, safety plans, material availability, and creating a credible baseline for project control, among others.

At the other end of the project life time, facility managers continuously make decisions on whether or not to conduct refurbishments, and prioritize between cost and quality. As the built environment ages, these assessments are applicable to demolition decisions. Automated condition assessment and refurbishment decision support systems are leveraging the complexity of the building systems, such as technical, technological, ecological, social, comfort, aesthetical, etc., where every subsystem influences the total efficiency performance and where the interdependence between subsystems plays a critical role [61.39]. Furthermore, design changes for refurbishment projects appear frequently due to a number of factors such as the lack of suitable design data, insufficient condition data, inadequate information on building condition, and ineffective communication between the client and contractors [61.40].

61.5 Techniques and Guidelines for Construction Management Automation

The management of construction projects deals with planning and scheduling, material procurement, cost control, safety, performance tracking, and design–construction coordination, among other issues.

61.5.1 Planning and Scheduling Automation

Planning and scheduling consists of sequencing processes, activities, and tasks according to time, space, and resources constraints, specifying the duration of such tasks and the relationships between them. Traditionally, scheduling and planning were done through simple bar charts. In the 1960s, the critical path method (CPM) emerged, followed by discrete systems mod-

els [61.23]. Today, several software applications have been developed to automate the CPM, such as Microsoft Project, Primavera SureTrak, and Primavera P3. These applications allow users to link different activities, allocate resources, and optimize the schedule. In addition, they provide a user-friendly representation that can be used in different aggregation levels. A further step in the automation is taken by specialized software applications that include sophisticated methods of resource leveling and scheduling optimization considering uncertainty. Four-dimensional CAD emerged in the 1980s to associate time to spatial elements, thereby allowing visual representation of sequences and communication. To generate a 4-D simulation, a 3-D model is created and a manual or semi-automated process of linking

3-D objects to tasks in time takes place. Visualization of construction sequences is at the task level, which means that the active task is visualized when the associated construction object is highlighted. One drawback of this approach is that activities that are not associated to objects, such as milestones, cannot be visualized on the simulation. In addition, since the links have to be updated every time the 3-D model changes, it is a very time-consuming process. The new trend of scheduling automation is to incorporate BIM into the schedule, including spatial, resource utilization, and productivity information [61.38]. BIM consists of a set of parametric building elements that have associated several properties and relationships between elements, which store information on geometry, material, etc. Using BIM for scheduling, the model is linked to the schedule, not an object is linked to a task. This approach not only saves time and is more accurate, but also allows the evaluation of resources and material consumption over time [61.41].

61.5.2 Construction Cost Management Automation

Construction cost management deals with the relation between the owner's budget, the project estimates, and the actual cost of the project. Several attempts to automate cost estimating started in the 1980s, when software applications emerged to link construction quantities to cost databases or to standard industry databases. Later, this applications evolved to automate quantity take-offs from CAD drawings and to transfer such estimates to job cost management applications that allow automated tracking of contract amounts, subcontracts, purchase orders, quantity totals, billings, and payments. The new trend in cost construction automation is to incorporate a cost estimate application to BIM, becoming helpful in conceptual estimates because those estimates are calculated based on project characteristics and project type (i. e., office building, school, number of floors, parking spaces, number of offices, etc.). Those quantities are not available in CAD models because they do not define object types. On the other hand, BIM allows early quantity extraction and cost estimates. Later in design, BIM allows users to extract the quantities of components, area and volume spaces, and material quantities. In terms of a detailed estimate, the accuracy depends on the detail of the model. An important advantage of using BIM for estimating is that the estimate is automatically updated when the model changes; in addition, it helps reduce bid costs because it lowers uncertainty related to mater-

ial quantities. There are three different ways how BIM can be used to aid the cost management process:

1. To export quantities to estimating software
2. To link BIM objects to estimating software, which allows manual inclusion of costs that are not associated to a specific object
3. As a quantity take-off tool [61.38].

61.5.3 Construction Performance Management Automation

Current research efforts to find appropriate methods for performance monitoring and control mainly fall into three categories: (1) a series of automated techniques for detection of building entities: radiofrequency identification (RFID), global positioning systems (GPS), laser scanners, Wi-Fi, and ultra wideband (UWB) positioning technologies, and visual sensing; (2) a series of visualization techniques used to represent discrepancies between as-planned and as-built; such methods could help in comparing real-world data with the five-dimensional (5-D) as-planned data (i. e., 3-D model plus sequencing and estimated cost described in BIM); for example, it could help automatically detect discrepancies and identify those elements that are ahead of or behind schedule, identify those elements that are on budget or present overruns, and assist in taking corrective control actions; and (3) enterprise resource planning (ERP) and computer-supported frameworks that allow the integration of workflow information for automated project performance control.

Radiofrequency Identification (RFID)
In RFID technology, radiofrequency is used to capture and transmit data from a tag embedded or attached to construction entities. RFID is helpful for material tracking due to its larger data storage capabilities, being more rugged, not requiring line of sight, and being faster in collecting data about batch of components [61.42, 43]. The technology also has the advantage that it can be combined with other tracking technologies (i. e., 3-D laser systems or barcoding) that can complement and enhance the tracking system. This method might be troublesome for tracking numerous and dissimilar types of construction objects and personnel, especially if the RFID chips are not given the appropriate amount of time to be recognized by RFID readers. Other disadvantages include the need for previously identifying and tagging the entity to be tracked. This greatly limits tracking capabilities and makes it unreasonable for

tracking ever-changing entities, such as personnel and cast-in-place materials on a construction site.

Global Positioning System (GPS)

GPS is a system that can provide 3-D position and orientation coordinates. The GPS system can be applied in two ways: differential GPS based on range measurements, and kinematic GPS based on phase measurements. Differential GPS can be used to measure locations at a metric or submetric accuracy. In kinematic GPS, positions are then computed with centimeter accuracy [61.42]. As for differential GPS, the kinematic computation can be performed either in postprocessing or in real-time, which is called real-time kinematic GPS. As a satellite-based technology, standard GPS needs a line of sight between the receiver and the satellite; therefore, it cannot normally operate indoors. Recent developments enable GPS to operate indoors by adding cellular, laser or other technology [61.44]. Research on the use of GPS in the construction industry has been characterized by the positioning of equipment and by construction metrology in field operations [61.25,45,46]. The latter has been complemented with deployment of laser radar imaging of construction sites. The GPS system is limited by several factors. It can only be applied for outdoor use and needs a GPS receiver to be attached to the entity that is being tracked. Since the number of materials involved in a project is usually significantly greater than the number of pieces of equipment involved in installing them, it is, in most cases, unfeasible to attach a GPS receiver to each piece of material.

Laser Scanning

Laser scanning is a terrestrial laser imaging system that creates a highly accurate 3-D image of a surface for use mainly in computer-aided design (CAD) [61.47]. Typical 3-D laser scanners can create highly accurate three-dimensional images of objects from quickly captured 3-D and 2-D data of the objects. Laser scanning is especially applicable in construction sites, where the high amount of objects and materials make it difficult to gather accurate spatial data. Although data acquisition is fast, postprocessing tasks, such as georeferencing and 3-D model generation, can take several times longer, accounting for 80% of the time needed to complete the task, depending on the level of detail of the 3-D model [61.48]. It also has issues with noise, which needs to be removed during the segmentation process. This segmentation process is still not realized automatically and takes considerable processing time. Other disadvantages include the high cost and current lack of automation.

Wi-Fi and UWB

Wireless local area networks (WLAN) featuring Wi-Fi, also technically known as 802.11 for a 2.4 GHz radio band, have been considered as a key opportunity for the construction industry sector [61.49]. The main function of the Wi-Fi system is to integrate hardware components, such as application server, positioning engine server, and finder client, with the CAD drawings of the facility. Integrated through a web interface, the user is allowed to obtain material information from the site and trace it on the finder client's graphical interface in real time. Presently, positioning accuracy is up to a meter with current Wi-Fi technology, although it may become 3 cm with the use of ultra-wide-band [61.50, 51]. Wi-Fi and UWB applications in construction suffer from several disadvantages. The necessity for tagging is one of the deficiencies of this system. Another limitation is the necessity for measurement of infrastructure, which means that , ideally, the use of a total station is required in order to obtain accurate results. This increases the time needed for the setup of the system, as well as the cost of the tracking method.

Vision Tracking

Vision tracking is a method that blends video cameras and computer algorithms to perform a variety of measurement tasks. Traditional vision-based tracking can provide real-time visual information of construction job sites. This technology is unobtrusive, which means that there is no need for tagging or installing sensors to the tracked entities. Additional advantages include the simplicity, commercial availability, and low cost associated with video equipment and the ability to significantly automate the tracking process [61.52]. Disadvantages of this technology are associated with its limitations due to field of view, visibility, and occlusion, and the inability to track indoor elements using peripheral cameras. There are also further complications of differentiating between interesting and irrelevant entities at the site.

e-Work

Research on the automation of preconstruction workflows has been documented with the application of e-Work methods in steel reinforcement [61.53], and in construction materials in general [61.54]. This work is built upon previous research in the manufacturing domain, where e-Work is composed of collaborative, computer-supported activities and communications-supported operations in highly distributed organizations, thereby investigating fundamental design principles for the effectiveness of these activities [61.55, 56]. See Chap. 88 on *Collaborative e-Work, e-Business, and e-Service*.

61.5.4 Design–Construction Coordination Automation

Design–construction coordination consists of synchronizing all designs, field conditions, special conditions, trades, and systems, in order to deliver the project according to the design intent. One of the most important tasks is clash detection. Currently, most clash detec-tion is done manually by overlying drawings in order to try to detect clashes within the different systems, and some contractors use CAD applications to overlay layers to detect conflicts visually; both approaches are very slow and prone to errors. The new trends in clash de-tection are based on BIM tools, which allow automatic clash detection from two perspectives: 3-D geometry conflict detection and object relationship and interac-tion conflict detection [61.38, 57]. As all objects are in the model, clash detection can be performed at different levels of detail as needed. Finally, detected conflicts can be corrected very quickly before the next round of clash detection takes place, making the updating of drawings simpler.

61.6 Application Examples

Four particular application examples have been selected because of their current impact on construction au-tomation. The first three examples are state of the art, whereas the last one is still in the research stage. The first example is related to earthmoving automation; the second example involves planning and scheduling au-tomation; the third example describes the automation of a construction cost estimation process; the last exam-ple, still in the research stage, features the automation of construction progress monitoring.

61.6.1 Grade Control System for Dozers

One of the latest inventions toward the accomplishment of automation in construction is the technology incor-porated into the new line of Caterpillar crawler tractors or bulldozers. Even though these machines are not factory-equipped with all the technical automated op-tions, these units are preinstalled with all the accessories to make them compatible with these new technologies. The Caterpillar factory standard technology Accu-Grade [61.58] is an integrated system that provides accessibility for the implementation of more advanced technologies into these machines; for example, the in-tegration of GPS, laser grading, cross slope, etc. This system has benefits in terms of the productivity, ac-curacy, and overall effectiveness of the equipment use by the operator. It includes the automatic blade con-trol function which ultimately eliminates the operator from the accuracy-demanding task of blade positioning, especially in cases where accuracy is very important.

AccuGrade GPS control system is one of the most effec-tive tools to have been integrated into these machineries, essentially because it creates 3-D model replicas of the ground contours designed by engineers and then auto-matically matches these contours with the actual work done by the machine. This machine eliminates the use of an operator's ability to match, as accurately as pos-sible, the ground's elevation to the design, making it an automatic procedure calculated accurately by a machine and guided by an educated operator. Job site safety is improved, since it reduces the need for personnel to guide the machine, such as flaggers, stakers, checkers, surveyors, etc. Another feature that the AccuGrade sys-tem has integrated is the safety interlock, which locks the blade in a position when the system is not active.

In order to accomplish the overall goal of automat-ing the earthwork activities of a construction project, this system provides a tool to create complex 3-D de-signs and incorporate them into the machine's GPS system. This tool is most commonly used for more com-plex earthwork projects, such as golf courses or specific projects requiring constant slope degrees differences. Also, the ability to create flat, single, and dual-planar designs for less demanding projects, such as parking lots or building pads, is a great advantage that the GPS control system provides in order to achieve greater au-tomation, accuracy, and cost reductions.

Figure 61.4 shows the sloping designs that the GPS grade control system will integrate into the computer of the machine in order to automate the earthmoving necessary to achieve desired specs.

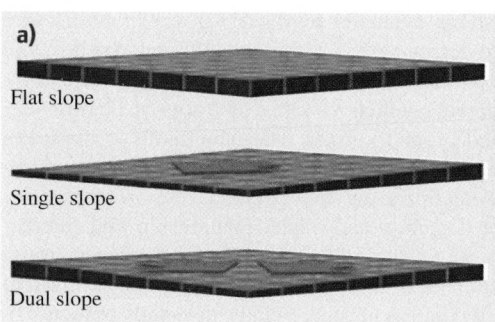 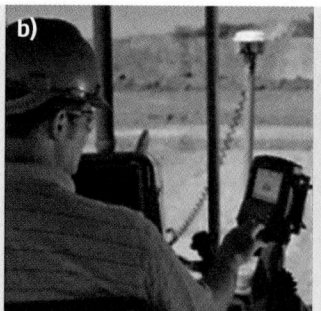 

Fig. 61.4a–c Grade control system for CAT Dozer: (**a**) 3-D sloping designs, (**b**) in-cab computer, (**c**) earthmoving operation [61.58]

Using AccuGrade, the dozer achieves quicker completion times and minimizes the need for staking, string lines, and grade checkers, apart from automatically performing the job without having to worry about accuracy, material waste, and several other time-consuming factors that would regularly be involved in an earthmoving project. According to studies done by Caterpillar, productivity is increased by 50% and surveying costs are reduced by 90% [61.58].

There are some off-board components and some office components involved in the operation of the GPS system. The off-board component includes GPS satellites which send positioning information to the machine, a GPS base station located within radio range of the machine that is used to transfer information from the satellite, and a GPS receiver installed in the machine for information receipt by the machine. In addition, there are office components involved, including 3-D design software and office software used to design and convert any kind of design into a format that the machine can use to create its design models.

61.6.2 Planning and Scheduling Automation

This example shows the use of BIM as an aid in the scheduling process [61.41]. The project was the construction of the Süddeutscher Verlag Corporate Headquarters in Munich, Germany, a 28-storey, high-rise office building with an area of $78\,500\,\text{m}^2$. The construction period was 36 months. In order to manage the schedule of the project, 4-D CAD and BIM were used. The CAD model consisted of 40 000 objects and a schedule that included 800 tasks, 600 of which could be visualized in the 4-D simulation, as shown in Fig. 61.5.

Linking CAD objects and tasks took 4 days. When there was a change in the CAD model, the task linking update took another 4 days. Linking BIM and the time schedule took 0.5 days, while changes on the model just needed to reload the data and apply the relationships, a process that took just a few minutes. After comparing the quantities generated by the BIM tool and the quantities generated manually, the BIM quantities proved to be more accurate.

61.6.3 Construction Cost Estimating Automation

This example demonstrates BIM capabilities as a conceptual cost estimating automation tool [61.38]. The project, Hillwood Commercial in Dallas, USA, is a six-storey, mixed-use building with an area of $12\,500\,\text{m}^2$. The building was constructed under a design–build delivery method. The design team started modeling during the conceptual stage to evaluate different alternatives to be presented to the client, using DProfiler, a parametric BIM tool that assigns cost information to the objects. DProfiler is integrated into a commercial cost database,

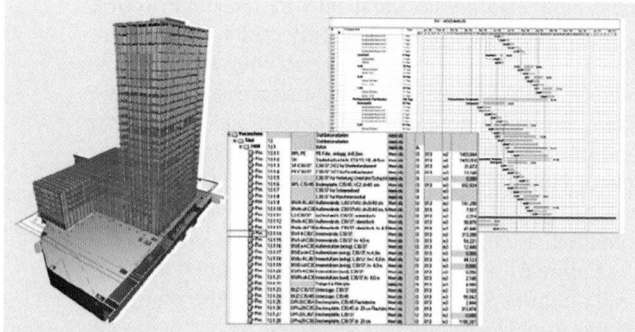

Fig. 61.5 CAD model, bill of quantities, and construction schedule [61.41]

giving the design team and the owner real-time cost data. The use of DProfiler instead of manual-based estimating resulted in a 92% reduction of time used to develop the estimate.

The team developed an initial concept design and created a model in DProfiler with links to cost information. The model took into consideration regional cost factors, building type, and other components using templates that have been developed with experience from similar projects. As the team constructed the building mass, the cost data was updated in real time. Then, as all components and assemblies were associated to cost in a database, when the team included more details the cost was automatically updated. In addition to linking components or assemblies to the cost database, DProfiler also allowed the creation of relationships and rules; for instance, it included one specific component that was not part of the model. In this way, as the model was developed, real-time information linked to the database was presented. The team also used DProfiler to run different scenarios and make informed decisions on floor-to-floor height, square footage, location of components, etc.

The team realized several benefits after using BIM to aid the conceptual estimate process. They found a reduction of labor hours to produce an estimate. DProfiler avoided the take-off process and also had an automatic link to commercial cost databases. The design team was able to produce an accurate estimate in real time, reducing the time spent in verifying the accuracy of the estimate, thereby spending more time in the financial analysis of the different options. Also, a visual representation of the estimate was created, reducing the potential for errors.

61.6.4 Construction Progress Monitoring Automation

This example illustrates a method of automated progress monitoring; the inputs are the as-planned 4-D model and a sequence of on-site time-lapse photographs that represent the as-built model. The objective of this method is to automatically overlay the as-built images with the as-planned model to determine and update the progress status. It consists of putting the real-world camera into the 4-D model; then, the image taken of the

Fig. 61.6 Site photograph, superimposed photograph, and color coding [61.59]

4-D model is overlaid with the site photograph taken at the same time. The 4-D photograph displays the expected progress; therefore, superimposition of images allows comparison of discrepancies between what was intended and what was performed [61.59]. The method uses earned value analysis to present progress indicators and uses a color code to provide a visual representation, where a dark red color represents objects that are behind schedule/cost and dark green represents objects that are ahead of schedule, as shown in Fig. 61.6.

Future trends in this domain include implementing combined progress tracking method that consists of generating a BIM model of the project that contains 3-D geometry, schedule, and cost information with the as-planned data. An as-built model is then generated using the combination of two automated data collection technologies: machine vision and Wi-Fi. To gather real-time object information, that information can be translated using an IFC language to generate a 4-D model with cost information on demand. Both models can be superimposed, detecting graphically in real-time the objects that are on time, the objects that are behind schedule, and the objects that are ahead of schedule. In addition, this application will allow the calculation of earned value indicators with the information stored in both models.

61.7 Conclusions and Challenges

The cyclical nature of the construction industry, featuring longer periods of reduced activity combined with imminent shortages of skilled labor, increasing quality expectations from costumers, tighter safety regulations, greater attention to computerized scheduling and project control, and technological breakthroughs led by equipment manufacturers, has once again put construction automation on the map of researchers and practitioners alike. In spite of challenges, such as cost, lack of governmental incentives, regulatory barriers, disbelief by the majority of the construction sector, or the lack of research and development, much advancement has been possible in construction automation in the past three decades. Presently, the demand for accurate, up-to-date, and timely information to manage construction projects is growing, so automating the construction management process presents enormous benefits. It has been proven that the use of technologies, such as automated data collection: reduces the time required to gather information at the job site; allows the use of real-time information; decreases the response time when corrective actions have to take place, reducing the costs associated with late response; and enhances the visualization and communication interactions within the project team. In addition, the use of BIM to support different project management processes also presents several benefits. It enhances communication with and understanding of stakeholders, and enables constructability decisions to be made during the design phase and not only during construction. It reduces the time spent in redesign and in answering requests for information, and allows a better sense of the time and the cost of the project, since estimates and the schedule are more accurate and do not have delays or overruns because of unexpected change orders.

However, there are still several technological gaps that need further research and development. A significant drawback is the lack of major research in the construction industry. The major challenges faced by the industry in regard to construction management automation are: reduction of the technological gaps and interoperability issues, the cost and availability of such technologies, and the systematic education of construction practitioners on these subjects.

References

61.1   R. Tucker: High payoff areas for automation applications, Proc. 5th Int. Symp. Robotics Constr., Tokyo 1988 (Japan Industrial Robot Association, Tokyo 1988)

61.2   B. Uwakweh: A framework for the management of construction robotics and automation, Proc. 7th Int. Symp. Robotics Constr., Bristol 1990 (Bristol Polytechnic, Bristol 1990)

61.3   J. Hsiao: A Comparison of Construction Automation in Major Constraints and Potential Techniques for Automation in the United States, Japan, and Taiwan. M.Sc. Thesis (MIT, Boston 1994)

61.4   M.J. Skibniewski: *Robotics in Civil Engineering* (Van Nostrand Reinhold, Southampton, Boston, New York 1988)

61.5   R. Kangari: Advanced robotics in civil engineering and construction, Proc. 5th Int. Conf. Adv. Robotics, Pisa 1991 (Institute of Electrical and Electronics Engineers, Los Alamitos 1991)

61.6   R. Best, G. de Valence (Eds.): *Design and Construction: Building in Value* (Butterworth Heinemann, London 2002)

61.7   L. Cousineau, N. Miura: *Construction Robots: The Search for New Building Technology in Japan* (ASCE, Reston 1998)

61.8   D. Cobb: Integrating automation into construction to achieve performance enhancements, Proc. CIB World Build. Congr., Wellington 2001 (International Council for Research and Innovation in Building and Construction, Rotterdam 2001)

61.9   M. Lehto, G. Salvendy: Models of accident causation and their application: review and reappraisal, J. Eng. Technol. Manag. **8**, 173–205 (1991)

61.10   D. Castro-Lacouture, J. Irizarry, C.A. Arboleda: Ultra wideband positioning system and method for safety improvement in building construction sites, Proc. ASCE/CIB Constr. Res. Congr., Grand Bahama Island 2007 (American Society of Civil Engineers, Reston 2007)

61.11   Bureau of Labor Statistics – BLS: *Census of Fatal Occupational Injuries – 2006* (BLS, Washington DC 2007), http://www.bls.gov/iif/oshcfoi1.htm#2006 (last accessed Dec 7, 2007)

61.12   T.S. Abdelhamid, J.G. Everett: Identifying root causes of construction accidents, ASCE J. Constr. Eng. Manag. **126**(1), 52–60 (2000)

61.13   J.J. Adrian: *Construction Productivity Improvement* (Elsevier, Amsterdam 1987)

61.14   C.H. Oglesby, H.W. Parker, G.A. Howell: *Productivity Improvement in Construction* (McGraw Hill, New York 1989)

61.15   D. Crosthwaite: The global construction model: a cross-sectional analysis, Constr. Manag. Econ. **18**, 619–627 (2000)

61.16 J. Lopes, L. Ruddock, L. Ribeiro: Investment in construction and economic growth in developing countries, Build. Res. Inf. **30**(3), 152–159 (2002)

61.17 K.W. Chau: Estimating industry-level productivity trends in the building industry from building cost and price data, Constr. Manag. Econ. **11**, 370–83 (1993)

61.18 D.W. Halpin: *Construction Management*, 3rd edn. (Wiley, Hoboken 2006)

61.19 C. Peterson: A Methodology for Identifying Automation Opportunities in Industrial Construction. M.Sc. Thesis (University of Texas at Austin, Austin 1990)

61.20 R.W. Nielsen: Construction field operations and automated equipment, Autom. Constr. **1**, 35–46 (1992)

61.21 J.G. Everett, A.H. Slocum: CRANIUM: device for improving crane safety and productivity, ASCE J. Constr. Eng. Manag. **119**(1), 1–17 (1994)

61.22 C. Balaguer: Open issues and future possibilities in the EU construction automation, Proc. 17th Int. Symp. Robotics Constr., Taipei 2000 (National Taiwan University, Taipei 2000)

61.23 C. Haas, K. Saidi: Construction automation in North America, Proc. 22nd Int. Symp. Robotics Constr., Ferrara 2005 (University of Ferrara, Ferrara 2005)

61.24 R.D. Schraft, G. Schmierer: *Service Robots: Products, Scenarios, Visions* (A.K. Peters, Natick 2000)

61.25 F. Peyret, J. Jurasz, A. Carrel, E. Zekri, B. Gorham: The computer integrated road construction project, Autom. Constr. **9**, 447–461 (2000)

61.26 Gomaco: *Slipform Pavers* (Gomaco Corporation, Ida Grove 2008), http://www.gomaco.com/Resources/pavers.htm (last accessed Feb 27, 2008)

61.27 Q. Ha, M. Santos, Q. Nguyen, D. Rye, H. Durrant-Whyte: Robotic excavation in construction automation, IEEE Robotics Autom. Mag. **9**(1), 20–28 (2002)

61.28 D.W. Seward: *Control and Instrumentation Research Group* (Lancaster University, Lancaster 2008), http://www.engineering.lancs.ac.uk/REGROUPS/ci/Files/projects/derek.html (last accessed Feb 27, 2008)

61.29 L.E. Bernold: Control schemes for tele-robotic pipe installation, Autom. Constr. **16**, 518–524 (2007)

61.30 C. Maynard, R.L. Williams, P. Bosscher, L.S. Bryson, D. Castro-Lacouture: Autonomous robot for pavement construction in challenging environments, Proc. 10th ASCE Int. Conf. Eng. Constr. Oper. Chall. Environ., League City/Houston 2006 (American Society of Civil Engineers, Reston 2006)

61.31 D. Castro-Lacouture, L.S. Bryson, C. Maynard, R.L. Williams, P. Bosscher: Concrete paving productivity improvement using a multi-task autonomous robot, Proc. 24th Int. Symp. Robotics Constr., Cochi 2007 (Indian Institute of Technology, Madras 2007)

61.32 L.E. Bernold: Motion and path control for robotic excavation, J. Aerosp. Eng. **6**(1), 1–18 (1993)

61.33 D.V. Bradley, D.W. Seward: The development, control and operation of an autonomous robotic excavator, J. Intell. Robotics Syst. **21**, 73–97 (1998)

61.34 Shimizu: *SMART System* (Shimizu Corporation, Tokyo 2008), http://www.shimz.com.sg/techserv/tech_con1.html (last accessed Feb 27, 2008)

61.35 Obayashi: *Appearance of Big Canopy* (Obayashi Corporation, Osaka 2005), http://www.thaiobayashi.co.th/images/obacorp/technology_automate/1n.jpg (last accessed Feb 27, 2008)

61.36 Obayashi: *ABCS Construction Scene* (Obayashi Corporation, Osaka 2005), http://www.thaiobayashi.co.th/images/obacorp/technology_automate/5n.jpg (last accessed Feb 27, 2008)

61.37 M. Handa, Y. Hasegawa, H. Matsuda, K. Tamaki, S. Kojima, K. Matsueda, T. Takakuwa, T. Onoda: Development of interior finishing unit assembly system with robot: WASCOR IV research project report, Autom. Constr. **5**(1), 31–38 (1996)

61.38 C. Eastman, P. Teicholz, R. Sacks, K. Liston: *BIM Handbook: A Guide to Building Information Modeling for Owners, Managers, Designers, Engineers and Contractors* (Wiley, Hoboken 2008)

61.39 A. Kaklauskas, E.K. Zavadskas, S. Raslanas: Multivariant design and multiple criteria analysis of building refurbishments, Eng. Build. **37**, 361–372 (2005)

61.40 Y. Lee, J.D. Gilleard: Collaborative design: a process model for refurbishment, Autom. Constr. **11**(5), 535–544 (2002)

61.41 J. Tulke, J. Hanff: 4-D construction sequence planning – new process and data model, Proc. CIB-W78 24th Int. Conf. Inf. Technol. Constr., Maribor 2007 (Int. Council for Research and Innovation in Building and Construction, Rotterdam 2007)

61.42 R. Navon: Research in automated measurement of project performance indicators, Autom. Constr. **16**(7), 176–188 (2006)

61.43 E. Jaselskis, T. El-Misalami: Implementing radio frequency identification in the construction process, ASCE J. Constr. Eng. Manag. **129**(6), 680–688 (2003)

61.44 S. Kang, D. Tesar: Indoor GPS metrology system with 3-D probe for precision applications, Proc. ASME IMECE Int. Mech. Eng. Congr., Anaheim 2004 (American Society of Mechanical Engineers, New York 2004)

61.45 G. Cheok, W.C. Stone, R. Lipman, C. Witzgall: Ladars for construction assessment and update, Autom. Constr. **9**, 463–477 (2000)

61.46 C. Caldas, D. Grau, C. Haas: Using global positioning systems to improve materials locating processes on industrial projects, ASCE J. Constr. Eng. Manag. **132**(7), 741–749 (2004)

61.47 E. Jaselskis, Z. Gao, R.C. Walters: Improving transportation projects using laser scanning, ASCE J. Constr. Eng. Manag. **131**(3), 377–384 (2005)

61.48   H. Sternberg, T. Kersten, I. Jahn, R. Kinzel: Terrestrial 3-D laser scanning – data acquisition and object modelling for industrial as-built documentation and architectural applications, Proc. 20th ISPRS Congress, Istanbul 2004 (International Society for Photogrammetry and Remote Sensing, Istanbul 2004)

61.49   M. Böhms, C. Lima, G. Storer, J. Wix: Framework for future construction ICT, Int. J. Des. Sci. Technol. **11**(2), 153–162 (2004)

61.50   R.J. Fontana, E. Richley, J. Barney: Commercialization of an ultra wideband precision asset location system, Proc. IEEE Conf. Ultra Wideband Syst. Technol., Reston 2003 (Institute of Electrical and Electronics Engineers, Los Alamitos 2003)

61.51   D. Castro-Lacouture, L.S. Bryson, J. Gonzalez-Joaqui: Real-time positioning network for intelligent construction, Proc. Int. Conf. Comput. Decis. Mak. Civ. Build. Eng., Montreal 2006 (International Society for Computing in Civil and Building Engineering, Montreal 2006)

61.52   Z. Zhu, I. Brilakis: Comparison of civil infrastructure optical-based spatial data acquisition techniques, Proc. ASCE Comput. Civ. Eng., Pittsburgh 2007 (American Society of Civil Engineers, 2007)

61.53   D. Castro-Lacouture, M. Skibniewski: Implementing a B2B e-Work system to the approval process of rebar design and estimation, ASCE J. Comput. Civ. Eng. **20**(1), 28–37 (2006)

61.54   D. Castro-Lacouture, M. Skibniewski: Implementation of e-Work models for the automation of construction materials management systems, Prod. Plan. Control **14**(8), 789–797 (2003)

61.55   S.Y. Nof: Models of e-Work, Proc. IFAC Symp. on Manufacturing, Modelling, Management and Control, Rio, Greece 2000 (Elsevier, Amsterdam 2000)

61.56   P. Anussornnitisarn, S.Y. Nof: e-Work: the challenge of the next generation ERP systems, Prod. Plan. Control **14**(8), 753–765 (2003)

61.57   B. Akinci, M. Fischer, R. Levitt, B. Carlson: Formalization and automation of time-space conflict analysis, ASCE J. Comput. Civ. Eng. **6**(2), 124–135 (2002)

61.58   Caterpillar: *ACCUGRADE GPS Grade Control System* (Caterpillar Inc, Peoria 2008), http://www.cat.com/cda/layout?m=62100&x=7 (last accessed Feb 27, 2008)

61.59   M. Golparvar-Fard, F. Peña-Mora, C. Arboleda, S.H. Lee: Visualization of construction progress monitoring with 4-D simulation model overlaid on time-lapsed photographs, ASCE J. Comput. Civil Eng. (2009), Forthcoming

62. The Smart Building

Timothy I. Salsbury

Buildings account for a large fraction of global energy use and have a correspondingly significant impact on the environment. Buildings are also ubiquitous in virtually every aspect of our lives from where we work, live, learn, govern, heal, and worship, to where we play. The application of control and automation to buildings can lead to significant energy savings, improved health and safety of occupants, and enhance life quality. The aim of this chapter is to describe what makes buildings smart, provide examples of common control strategies, and highlight emerging trends and open challenges. Today, the most prevalent use of automation in buildings is in heating, ventilating, and air-conditioning (HVAC) systems. This chapter reviews common control and automation methods for HVAC, but also describes how automation is being extended to other building processes. The number of controllable and interconnected systems is increasing in modern buildings and this is creating new opportunities for the application of automation to coordinate and manage operation. However, the chapter draws attention to the

fact that the buildings industry is very large, fragmented, and cost-oriented, with significant economic and technical barriers that can, in some cases, impede the adoption and wide-scale deployment of new automation technologies.

62.1 Background

The worldwide energy used to heat, cool, ventilate, light, and deliver basic services to buildings was, on average, approximately 2.4 TW ($= 2.4 \times 10^{12}$ W) in 2004. A further 5.6 TW was attributed to industrial plants, with a large fraction of these housed in buildings such as factories, power plants, and other manufacturing facilities [62.1]. In developed countries, buildings can be responsible for as much as 50% of total energy use. These dramatic statistics coupled with the fact that human beings spend most of their lives inside buildings make this application of critical importance to the well-being of our planet and its global population.

There are many different types of buildings, ranging from simple places of shelter to highly complex ecosystems that provide a range of specialized services to support specific functions. One common purpose of buildings is to create a modified environment that is comfortable for occupants even when outside conditions are unfavorable. Comfortable conditions are maintained through the operation and coordination of mechanical and electrical systems and through the conversion of energy from one type to another. The control of indoor environmental conditions is the most common and widely exploited application of automation technologies in buildings and will be the main focus of this chapter. Factors that affect the indoor environment are air quality, temperature, and humidity, as well as lighting, and safety from fire and security threats.

For a building to be smart it must have some form of automatic control system. Building control systems vary widely in complexity from simple mechanical feedback mechanisms to a network of microprocessor-based digital controllers [62.2]. At the latter end of the spectrum, the network of controllers is often known as the building automation system (BAS). A BAS needs to interface with physical systems in order to effect desired changes on the building, and the interfacing is usually made through means of sensors and actuators. Buildings can contain many types of physical systems, the most common of which are described in more detail below.

HVAC and Plumbing. In terms of energy, the most important systems in buildings are those used to heat, cool, and ventilate the indoor environment. These systems are collectively known as the heating, ventilating, and air-conditioning (HVAC) plant. The HVAC plant is used to condition the psychrometric properties (temperature and humidity) of the indoor environment as well as air quality. HVAC systems range in complexity from simple residential units that may have only heating units to advanced systems for high-performance buildings such as clean rooms and chemical laboratories [62.3]. Plumbing systems are closely associated with the HVAC plant but are often handled by a different group of companies and contractors on a building project. Plumbing serves the HVAC systems with water supplies used for heating and cooling and also handles distribution of potable and waste water. Plumbing and waste disposal systems are critical elements in ensuring occupant health [62.4]. Automatic control is widely used in HVAC systems and ranges from simple feedback loops to complex sequences of operation that manage scheduling and interactions between systems.

Lighting Systems. Lighting systems are responsible for a large fraction of building energy use, close to that used by HVAC for the commercial sector. Most buildings have both artificial and natural lighting and the interaction between these sources is important in creating the right level of illumination. In most buildings, both artificial and natural lighting are operated manually using electric switches and window shades. However, automated lighting systems are available and are starting to be used in modern buildings. For these systems, sensors measure illumination levels in a space and also

whether a space is occupied and this information is used to regulate artificial light levels and control shades on windows.

Fire and Security. Fire systems are found in large buildings and include fire detection and alarming as well as sprinkler systems for abatement. Security systems are used to control access to buildings and internal areas and trigger alarms when unauthorized access is detected. Both of these systems are evolving rapidly largely due to the availability of more advanced sensor technology and imaging devices. The additional and richer sensor information available from these systems is creating new opportunities for intelligent responses to particular situations. For example, having knowledge of where people are located in the case of a fire can be used to manage evacuation routes, improve emergency response team planning, and also provide input to the HVAC system to mitigate the spread of smoke.

Specialized Systems. Many buildings require specialist services to support specific tasks and functions. For example, a hospital might require oxygen supply and distribution and fume hoods for extracting dangerous and toxic chemicals. In common with other building services, these systems require piping or ducting for containment and distribution, and pumps/fans, valves, and dampers for fluid movement and control. Localized power generation and combined heat and power plants that use waste heat from electricity generation to power heating and cooling systems are also found in some buildings or campuses. These systems may include renewable sources of energy such as solar, wind, and geothermal, and other power generation technologies such as fuel cells and microturbines. Specialized systems, and in particular those that are packaged, often have dedicated embedded controls, but in some cases the BAS might be deployed to provide some higher-level control and supervision.

Supporting Infrastructures. Modern buildings contain systems powered by electricity and an increasing number powered by natural gas. Buildings therefore need to have a distribution network for these energy supplies. Key components include wiring, switch panels, circuit breakers, transformers for electricity, and valves, piping, and various safety devices for gas. Another key infrastructure in modern buildings is that associated with information technology (IT). The IT infrastructure in a building is usually considered to be separate from the other systems mentioned so far because it is handled by a different group of companies outside of the construction business. Similar to electricity distribution, IT systems require wires, switch panels, and access points in order to facilitate the transmission of both digital and analog data for devices such as computers and telephones. Specialized cooling systems may also be needed for high-powered computing devices and data centers. Electricity, gas, and IT distribution systems in a building will usually have some control elements, particularly for safety reasons.

The range of systems and processes in buildings is clearly very broad and cannot be covered comprehensively in the space available here. For this reason, this chapter will focus more on the control and automation aspects of a building rather on the systems that are under control. The chapter will also concentrate on the software and algorithmic aspects of control systems, which are the source of the smartness in buildings, rather than the hardware and supporting infrastructures. The outline of the chapter is as follows. The rest of this section provides a discussion on what makes buildings smart and concludes with a historical perspective of building technologies. An overview of common control strategies and their applications is presented in Sect. 62.2, followed by a discussion in Sect. 62.3 on emerging trends that are affecting the buildings industry. Section 62.4 describes open challenges and, in particular, business and technical barriers to the adoption of new technologies. Finally, Sect. 62.5 draws conclusions and reiterates some of the key points identified throughout the chapter.

62.1.1 What is a Smart Building?

A building is made *smart* through the application of intelligence or knowledge to automate the operation of building systems. In modern buildings, the intelligence or *smartness* of building operation is encapsulated in algorithms, which are implemented in software on microprocessor-based computing devices. Many of these computing devices are part of the building automation system, which can be decomposed into the following four main components:

- User interface – allows exchange of information between a human operator and the computer system
- Algorithms – methods or procedures for performing certain tasks such as control and automation
- Network – includes information transmission media (e.g., wiring), routers, and appropriate encoders and decoders for sharing information among devices

- Sensors and actuators – these represent the interfaces between the computing systems and the plant.

The user interface, network, sensors, and actuators are critical components of a BAS, but these are all enabling technologies that only provide the means by which the intelligence inherent in the algorithms can be applied. The algorithms fundamentally determine the operational behavior of the controlled systems and are the source of the smartness. In a typical building, numerous objectives can be defined suitable for the application of control methods. Examples are regulating a room temperature to a set level, turning off systems at a certain time, and controlling access to a room based on information read by a card reader. Controlling a variable, such as temperature, to a set level is probably the most common control objective and is most often carried out using feedback. Feedback is a fundamental building block of control and automation and its application in buildings will be discussed in more detail in Sect. 62.2.

Recent technological advances in information technology, including networking, computing power, and sensor technology, have meant that the number of controllable devices in buildings has proliferated. Not only are there more devices to control, but information can now be shared more easily between disparate systems. Information is more easily accessible both within system groups as well as across different groups. The HVAC group of systems is particularly notable in making information available to the BAS from multiple types of subsystems including boilers, chillers, fans, pumps, cooling towers, and measured by numerous types of sensors. Opportunities abound within just the HVAC group of systems for applying control strategies that take advantage of the available data to improve overall system performance.

The idea of combining information from different systems to implement new and smart control and automation strategies extends easily to system groups that traverse traditional boundaries. The example of combining data from the fire and access control systems to provide improved emergency response information and also more effectively manage evacuation was cited earlier. Another example is in utilizing access control data to estimate the number of people in a building and then using this estimate to operate the HVAC systems more energy-efficiently. Although huge potential exists for coordinated system operation, the reality today is that the handling of interactions is limited and in most cases ad hoc. However, despite the primitive nature of current automation strategies, the sharing of scheduling databases and responses to alarms represents significant progress toward smarter building operation.

Building operation is not the only way in which buildings have been made smarter. The lifecycle of a building includes its planning, design, construction, installation and commissioning, operation, maintenance, retrofit and remodeling, and destruction. Each of these tasks not only consumes energy and resources, but affects subsequent tasks. For example, a building will only operate energy-efficiently if it has been correctly designed and constructed. New and smart technologies are being utilized at each stage in the lifecycle to improve the overall process. A prime example is in being able to simulate the performance of a design before it is built [62.5]. This is a powerful technology that can lead to cost and energy savings for a project. The ability to simulate building systems is also enabling the development of innovative algorithmic smart technologies such as automated design and optimization [62.6].

It is also important to mention the significant energy and resources used during the construction phase of a building. Energy is used at every step; from the production of the building materials, to their transport to site, through to the operation of machines for excavation and assembly of the materials. The businesses devoted to this one phase of the building lifecycle are numerous, diverse, and employ various forms of automation to enhance efficiency. The term *smart* is frequently encountered in various construction technologies including prefabricated components, advanced supply chain and project management, and on-site machinery, to name only a few examples. However, the term *smart buildings* has come to mean smart operation more than anything else, and for this reason the operation phase will be the focus of this chapter. The smart operations, enabled by technological advances, guarantees cost savings in construction and operations, and improved functionality [62.7].

62.1.2 Historical Perspectives

Modification of the environmental conditions inside buildings is not new, with records showing that the ancient Egyptians used aqueducts for cooling as long ago as the second millennium BC [62.8]. Heating systems were also used by the Romans in 100 AD in their northern territories based on underfloor distribution of air heated by furnaces [62.9]. From the perspective of the building plant, evolution had been slow until the advent of commercially viable electrical air-conditioning systems in the first decade of the 1900s. The origin of these systems can be traced to discoveries by Michael Faraday in 1820 on how to create a cooling effect by compressing and liquefying ammonia, but it was the commercial success of the early electrical air-conditioners that triggered a new interest in the indoor environment and its control. Activity in this area was probably at its height in the first half of the 20th century. During this period, the field of heating, ventilating, and air-conditioning (HVAC) would have been considered a high-technology field that posed some of the most interesting engineering challenges of the time [62.10].

The application of automation technology to buildings has paralleled its application to other industries, with the idea of feedback playing a central role. The feedback concept is a fundamental element of automation and has a history of its own dating back to the water clocks of the ancient Greeks and Arabs [62.11]. The Dutch inventor Cornelis Drebbel is credited as creating one of the first feedback temperature controllers for a furnace in the early 1600s [62.12]. The first thermostats for space temperature control using heating plant appeared in the late 1800s. The thermostat feedback system is an implementation of an intelligent human concept, or procedure, to solve the problem of temperature regulation. These early applications of automation thus led to the creation of the first smart buildings. Implementation of feedback and other control methods originally used mechanical transmission of information, such as pneumatics. These systems were replaced by electrical devices with controllers first being implemented using analog circuits. Today, most analog controllers have given way to digital devices where earlier feedback strategies are now implemented as algorithms in software.

The feedback concept has remained central in building automation. It is the most common type of control strategy and is used to control everything from air-conditioning to lighting to fire and security to toilet flushing. Although the common room thermostat principle is still widely used, building automation systems now encompass an evermore sophisticated array of control algorithms that not only provide regulation of individual variables to setpoints, but also provide high-level coordination and management of building assets.

62.2 Application Examples

Possibly excepting certain types of chain stores, every building is different, having a unique mix of structure, geometry, orientation, location, number of people, and types of building plant. The control and automation systems mirror this bespoke nature and are usually specific to each particular building. Finding common elements of control logic that run across all building types is therefore a challenge. This problem is particularly exacerbated at the higher levels of control logic hierarchy that are used to coordinate the operation of different systems. The aim of this section is to identify and review a sample of control strategies that are generic and found in buildings of different types. Although these strategies are outnumbered in practice by ad hoc and rule-based logic, they usually have a better scientific basis and are more easily adapted and scaled across different buildings.

62.2.1 Control without Feedback

A very simple and yet effective form of building automation is to operate systems based on a time clock; for example, a heater in a building could be turned on at a certain time every day and turned off at another time. These times can be determined from some expected behavior that is known to be linked to time, such as people coming in to work, or even the sun rising and setting. This kind of control logic can be used to control everything from building access, lights, and HVAC. Time-based control is also a type of event-triggered logic where the event being monitored is time of day. The strategy does not contain feedback because time is unaffected by the action of the systems being operated. Time-based operational scheduling is commonplace in modern buildings as an overriding or supervisory logic even when more advanced control and building management strategies are employed. The logic is usually implemented as rules such as: IF TIME = 9.00AM TURN ON CHILLER; IF TIME = 5.00PM TURN OFF CHILLER.

Closed-loop control is synonymous with feedback control. *Open loop* refers to the case where no information associated with the controlled variable is used in deciding how to adjust the manipulated variable; the loop is thus open. In the HVAC industry, reference is sometimes made to *open-loop operation*. In most cases this involves having an operator make adjustments to a manipulated variable manually. However, if the operator is making adjustments in response to observations of the controlled variable, this is not truly open loop because the operator is providing the feedback mechanism. An example of this is when a room in a building has a heating device that can be either on or off but only activated by a human operator. When the room is too cold the person will turn on the heater and when it is too hot they will turn it off. Manual operation of this sort is common in buildings and may be carried out by a dedicated building operator who oversees plant operation. In many situations, preexisting automated feedback systems may be overridden by the operator because of lack of confidence or trust. The operator then manually adjusts things such as control valves and dampers to maintain comfort conditions.

62.2.2 Feedback Control

Single Loop

Feedback control is the most widely used automation concept in buildings, with the temperature thermostat being the vanguard of this strategy. Early examples of these devices were based on pneumatics and mechanical transmission of information [62.13]. Modern buildings use thermostats that contain a temperature sensor and a small integrated circuit that determines when the temperature is outside an acceptable range. The thermostat triggers a switch to operate a device which is then expected to bring the temperature back into its acceptable range. Thermostatic temperature devices usually require specification of a setpoint and a control band. A device would be switched either on or off whenever the measured temperature was outside of the control band, which surrounds the setpoint. Figure 62.1 shows an example thermostatic control strategy for a heater in a room with a setpoint of 20 °C and a control band of 1 °C. There is a trade-off between the closeness of control, determined by the control band, and the wear and tear on the equipment resulting from cycling between on and off states. Some equipment types may also have constraints on how long they can remain in an on or off state and these are called minimum on and off times. The maximum cycle frequency is the reciprocal of the sum of the minimum on and off times.

Thermostatic control is an example of single-loop feedback control where the controller is a switch or relay-type device that has infinite gain. The switching of the controlled device causes oscillations to occur in the controlled variable around its setpoint. These oscilla-

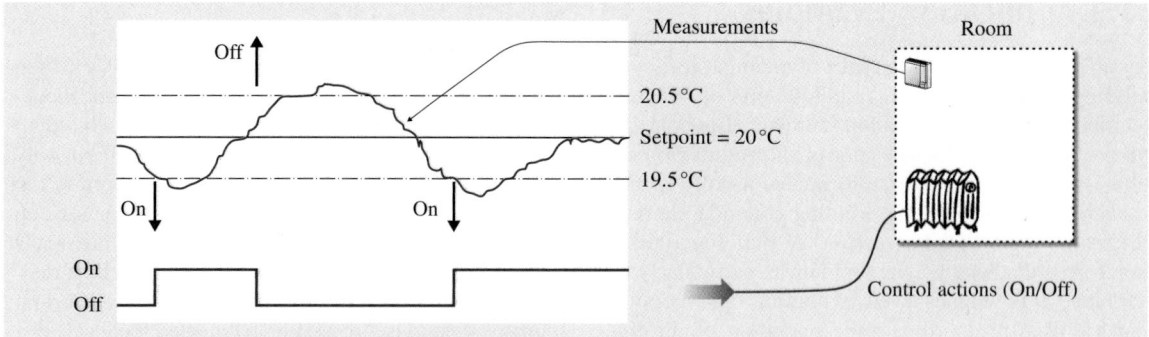

Fig. 62.1 Thermostatic temperature control of a heater

tions are usually undesirable, especially if the amplitude is large due to minimum on and off times being large relative to the dominant time constant of the controlled device. One way to avoid oscillations is to use a physical device whose output can be modulated. Instead of being only on or off, the device output can be modulated by manipulating an input between 0 and 100% of its range. Having a manipulated variable that can be varied opens the way for finite-gain controllers to be used in feedback loops. Modulated systems make up about two-thirds of controlled devices in buildings, with switched systems representing the other third.

Proportional, integral, and derivative (PID) action controllers are the most common type of finite-gain controllers used in building feedback loops. These controllers modulate the input to the controlled device based on the difference between the setpoint and the controlled variable, and the theory behind the algorithm is well established [62.14]. Proportional-only (P) controllers are still common in buildings where control exactly to setpoint is not critical. As is well known, these controllers yield a steady-state offset from setpoint and, when this is not desirable, proportional–integral (PI) controllers are used. This complexity of feedback control works well for most building control loops that are dominated by one time constant. The integral action of the controller ensures that the error signal can be maintained at zero and proper tuning allows good control without oscillations for most loops. Adding the derivate action can yield better results for common applications such as room temperature control, but the drawback is that tuning is more difficult.

PI(D) controllers can also be used in buildings to control switched devices. This is achieved by using an element in the control loop that converts the output signal from the controller (usually between 0 and 100%) to a pulse train (either 0 or 1). Pulse-width-modulation

(PWM) logic can be used to provide this conversion, as illustrated in Fig. 62.2. PWM and other variants have been applied to HVAC system applications with promising results [62.15].

Although the PI algorithm is well established and almost ubiquitously deployed for building control, it has been recognized for some time that control performance is often poor in practice. One reason for this is that many plant items in buildings are nonlinear. When a PI controller is used to control a nonlinear system, its performance will vary with operating point. In severe cases, the control might be too aggressive, causing oscillations at one point and yet be too slow at another. These kinds of performance problems can jeopardize comfort and energy use and also wear out equipment. There are several methods for counteracting nonlinearity that are used in nonbuilding applications, such as gain scheduling and model-based control [62.16]. The research community has investigated using some of these methods in buildings including generalized minimum-variance control [62.17] and model-based control [62.18]. However, the industry has been reluctant to adopt these methods because of the extra time (and money) required for setup and tuning. More successful approaches in the building industry have been those that are self-tuning or automatically adaptive to changes; examples include neural networks [62.19] and pattern-recognition adaptive control [62.20]. The latter method of Seem has been commercialized and is

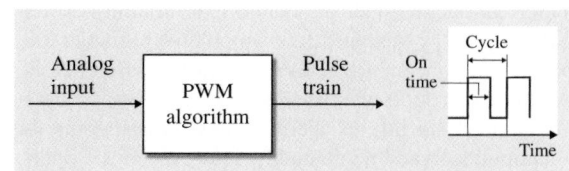

Fig. 62.2 PWM conversion of analog to switched signal

a standard offering from one large controls vendor in the USA.

Multiloop

Multiloop feedback control strategies are used in buildings, with the most common being cascaded control. In these configurations, there is an inner loop that controls a fast variable and an outer loop that controls the setpoint of the inner loop based on the feedback of a slower variable. An example is the control of variable-air-volume (VAV) boxes for regulating room temperature, as illustrated in Fig. 62.3. These boxes are supplied with conditioned air from a central air-handling unit and they can control the flow of conditioned air to a room by means of a damper. In the inner loop, the controlled variable is the airflow rate entering the room. In the outer loop, the controlled variable is the room temperature. The outer loop tries to control the room temperature to its setpoint by modulating the setpoint for the airflow in the inner loop. In turn, the inner loop tries to meet the airflow setpoint by modulating the VAV box damper. Cascaded control allows a control problem to be separated into two parts, one with slow dynamics and the other with fast dynamics. Each type of fast and slow disturbance is then handled by a separate controller, with improved performance compared with having just one controller. A general guideline for cascade control is that the inner loop should be at least three times faster than the outer loop. Further guidelines for implementation can be found in [62.14].

Multivariable control in the sense of having centralized controllers that are multiple-input multiple-output (MIMO) is not typical in building applications. The most common situation is to have multiple feedback controllers that do not share information with each other. For example, a large open-plan office space might have several controlled temperature zones that have separate feedback controllers. An obvious problem with this approach is when there are interactions between zones due to effects such as interzonal airflow. Poor performance in one zone can then propagate to other zones if interactions are strong. Current practice involves trying to minimize interactions by positioning sensors away from each other and making sure that control zones are separated.

There is generally significant interaction amongst building systems and in many cases operation could be improved through application of multivariable control methods. Model-based approaches are one way of handling multivariable systems and these methods have been investigated and applied to buildings, e.g., [62.21].

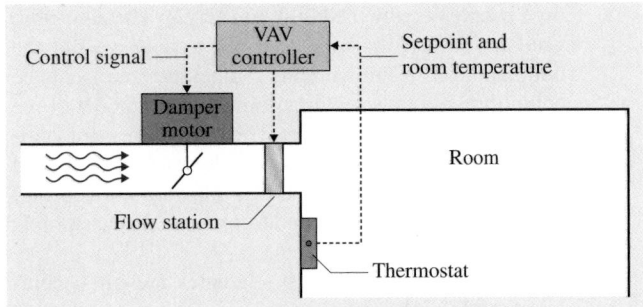

Fig. 62.3 Example cascaded control strategy

However, adoption of these methods by the building industry is impeded by the costs associated with setup and commissioning and of handling the additional complexity. Nevertheless, multivariable control is underexploited in building applications and it appears to be a promising direction for future research.

62.2.3 Energy Management Control Strategies

Strategies for energy management are usually employed at a higher level in the control hierarchy of a building. Many controls vendors offer proprietary algorithms for energy management and these are normally programmed into so-called *supervisory controllers*. In contrast to local controllers, which handle simple tasks such as feedback loops, supervisory controllers have more processing power and memory and consolidate data from multiple nodes on a network. Supervisory controllers perform functions that may affect the operation of several feedback controllers by adjusting things such as setpoint and operation schedules. This section reviews some of the most common energy management control strategies that are found in buildings. The overall idea behind the strategies is explained, but algorithmic details are not provided because no standard implementation exists and many different types of proprietary implementations can be found in practice.

A variable that is commonly used in energy management control strategies is the outside air temperature. This variable is known to affect the thermal loads on a building and its ultimate heating and cooling requirements. One example where the outside air temperature is used as part of feedforward control strategy is the optimum start algorithm. This algorithm uses the outside air temperature to determine when to turn on the heating and cooling systems in a building so that the internal temperatures will be at their respective setpoints by the time people occupy the building [62.22]. The algorithm needs to predict the time required to heat or cool the building from a known start temperature and outside conditions and thus requires some kind of model of the building's thermal response. The complement of optimum start is optimum stop, which determines a time when systems can be turned off, allowing the capacity of the building to hold conditions near their setpoints until people leave.

Optimum start and stop strategies are often combined with *night-setback*. Instead of just turning all systems off during unoccupied periods, night setback control maintains closed-loop control on space temperatures but at different setpoints. During the heating season, for example, temperatures inside the building would be kept lower during the unoccupied period, but only low enough that recovery time would be rapid enough and losses of energy stored in building materials minimized so that this stored energy can be carried over to the next day to reduce overall energy use. Various algorithms have been proposed for predicting the night-setback setpoints and startup and shutdown times based on models and optimization and the use of thermal storage systems, e.g., [62.23, 24]. Integrated optimum start and stop with night-setback algorithms are available from some control vendors, and tuning and setup requirements depend on the particular algorithm that is offered and the systems that are targeted for control.

Another higher-level control strategy that uses feedforward is setpoint reset, where the setpoint of a feedback loop is adjusted based on a feedforward variable; for example, the setpoint of the main supply of conditioned air to a building can be made a function of the outside air temperature. The setpoint would be adjusted to be higher when the outside air temperature is low and made to be lower if the outside temperature is high. The setpoint of cold and hot water supplies generated by building chiller and boiler plant are other variables that are sometimes linked to reset strategies based on outside air temperature. Most of these algorithms adjust setpoints between two specified limits as the outside air temperature varies between upper and lower bounds. Setup and tuning of these reset strategies usually requires specification of the upper and lower values for the setpoints and the outside air temperature.

Feedforward control can also be used to supplement feedback control in situations where disturbances such as changes in airflow and outside temperatures can be measured in systems such as air-handling units [62.25]. The advantage of supplementing feedback with feedforward is that the effect of disturbances can be mitigated when they occur instead of having to wait for their effect to be revealed in the feedback variable. It is unfortunate that many disturbance signals in buildings are in fact measured but not used in the control strategy. The number of potentially useful signals is also increasing as more types of systems become integrated in the BAS; for example, signals from lighting occupancy sensors could be used to improve the control of the HVAC systems by adjusting capacity based on anticipated changes in loads.

Economizer control is a very common energy-saving strategy, particularly in the USA, that is used to reduce cooling loads by switching the air source to the main air-handling systems between recirculated (with a mix of minimum outside air) and 100% outside air. An economizer strategy evaluates the temperature or enthalpy of the two potential air sources and issues control signals to dampers to provide an air supply to cooling heat exchangers with the aim of reducing required cooling capacity. This control method is usually part of sequencing logic that is used to coordinate the operation of heating, cooling, and recirculation systems. The energy saving potential of economizer control has been shown to be as high as 52% [62.26], but results depend on climatic conditions. Both temperature and enthalpy measurements are meant as proxies for the eventual cooling load. Improved results could thus be obtained by using more accurate methods for predicting loads, e.g., by using thermodynamic models of the cooling plant. Today, control vendors normally only offer a choice between temperature or enthalpy economizer strategies. In most cases, enthalpy-based strategies will yield more energy savings, but results can be inconsistent because most humidity sensors, which are used to calculate enthalpy, are notoriously inaccurate.

Peak load management is an aspect of energy management that has received increased attention in recent years. This is mostly due to the fact that utility companies often tie energy tariffs to peak demand statistics, thereby creating incentives to minimize peaks and flatten the load profile. The strategy is commonly termed *demand limiting* and involves shutting down the operation of certain plant items to keep demand below target levels. The concept is illustrated in Fig. 62.4, which shows how building demand is kept below a target level by shedding loads. The objective of this strategy is to control peak demands while at the same time minimizing the disruption to control objectives such as comfort conditions within the building [62.27]. Unresolved research issues are how to set the targets and how to determine the order and time over which equipment

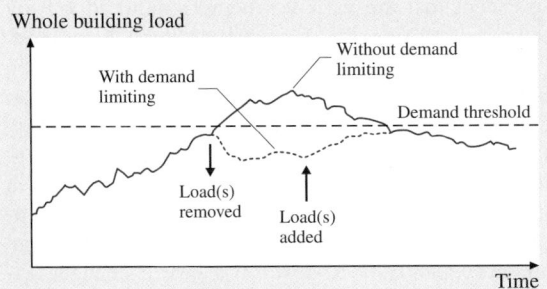

Fig. 62.4 Demand limiting scenario

loads should be added and removed. Currently, the algorithms available from control vendors for demand limiting require the user to set targets based on expert knowledge, which leads to very variable results from these strategies.

The discussion so far has focused on energy management and control of HVAC systems. The reason for this focus is that this group of systems usually represents the only example of where algorithms with more sophistication than simple scheduling and feedback can be found in contemporary buildings. Although more system types are being integrated into the BAS network, the control strategies for these systems are normally primitive and frequently just provide a means for centralized manual operation. However, the potential for more sophisticated control and coordination of systems such as lighting, fire, and security has been recognized for some time and examples of specialized buildings with more sophisticated control strategies can be found. Some of the ideas implemented in these showcase buildings are beginning to trickle down to the rest of the building stock but the rate of adoption is slow, mostly because of cost constraints. Lighting systems are one example of where the use of more sophisticated control can overcome cost constraints by reducing energy use. Traditionally, lights have been controlled manually as a form of open-loop control. However, studies have shown that the implementation of automated feedback signals such as occupancy measurements and outside light sensors can be very effective for control and energy reduction [62.28]. More advanced control is also starting to appear, making use of additional controllable elements such as shading devices that can be used to regulate the inflow of natural light into a space [62.29].

This section has only provided a brief overview of a sample of energy management strategies in buildings. Further examples related to HVAC can be found in [62.30]. The demand for these kinds of strategies is growing in response to higher energy prices and increased environmental concerns, and the computing and networking infrastructure needed to perform this kind of plant-wide control is becoming more widely availably. Most energy management strategies are implemented as supervisory logic on top of lower feedback control loops. In controls terminology the combination of this kind of supervisory logic with low-level feedback control is known as hybrid control. Design of hybrid control strategies in buildings is becoming more formalized due to the availability of new software design tools; for example, the lower-level control strategies might be designed using block diagrams and transfer functions and the supervisory logic designed using state-machine logic diagrams. Modern software programs also allow control logic to be simulated before being implemented, which greatly improves reliability and lowers development costs.

62.2.4 Performance Monitoring and Alarms

Control and operational automation is the most common application of intelligence in buildings. However, recent years have seen a growing interest in intelligent monitoring and analysis of operational performance [62.31]. This has paralleled similar research and application in other industries such as aerospace, chemical processing, and power generation. The most pressing need for monitoring and analysis technologies is to detect faults that could jeopardize performance and, more importantly, safety. The concept of generating an alarm when a critical variable is outside of acceptable bounds is the simplest implementation and is employed quite widely in modern buildings. Although the idea is simple, setting the alarm limits requires intelligence and in some cases may need to be periodically adjusted to cater for changing conditions. A detailed discussion on safety warnings in automation is provided in Chap. 39.

Recent years have seen a demand for more sophisticated performance monitoring and analysis, beyond simple alarming [62.32, 33]. The motivation for this derives from a recognition that control and automation of system operation frequently fall short of expectations. Poor performance can be caused by many things, ranging from faults and deteriorations in the plant, badly tuned controllers, wrongly implemented control logic, to sensor or actuator malfunction. There is also a drive to increase the automation of analysis and operational oversight tasks that were previously carried out manually, not only to reduce costs, but also to help operators deal with the overwhelming amount of data now

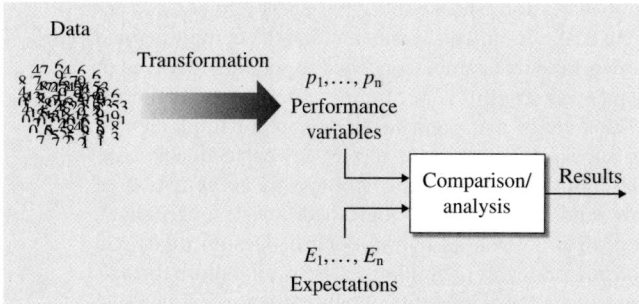

Fig. 62.5 Example performance assessment methodology

available to them through modern building automation systems.

The associated problems of performance analysis and fault detection and diagnosis primarily involve transforming data available from sensor measurements and control actions into variables related to performance that can be compared with expectations, as illustrated in Fig. 62.5. An example of a performance variable is the coefficient of performance (COP) of a chiller, which is calculated from several temperature and power measurements. Fault detection involves comparing performance variables with expectations and making a binary decision as to whether a fault exists or not. Fault detection is therefore very similar to simple alarm-

ing except that the variables being monitored are not raw measurements, but derived from them using models, statistics, expert rules or some other transformation method. Fault diagnosis is more complicated and can be broken down into locating the source of the problem, matching symptoms to a cause, and estimating the magnitude of the fault. In a building, a hierarchical approach might start with the detection of higher than normal energy use, get narrowed down to one air-handling system, and subsequently to a valve on a heat exchanger. Finally the fault might be diagnosed as a valve leakage and estimated to be 50% of maximum flow.

The problem of performance monitoring in buildings is also suitable for the application of single-input single-output (SISO) loop monitoring techniques developed for other industries. Measuring the variance of controlled variables and comparing with theoretical benchmarks for example is one method that has been successfully deployed in many industries [62.34]. However, variance of control variables is normally less important in buildings than in other manufacturing-type industries where product quality and profit is directly correlated with control variable variance. One of the biggest problems in buildings lies in detecting problems with the plant such as leaking valves, stuck dampers, sensor failures, and other more prevalent malfunctions [62.31].

62.3 Emerging Trends

As mentioned at the start of this chapter, buildings can be viewed as large processes with many interacting and diverse systems operating together to achieve various objectives. In this way buildings are broadly similar to other applications such as chemical processing, power generation or oil refineries. A building has numerous

manipulated variables and measured signals like these other systems. The challenge lies in how to connect the variables together and what algorithms and logic to deploy between these linkages. Figure 62.6 illustrates the concept of how algorithms and logic routines link manipulated and measured variables.

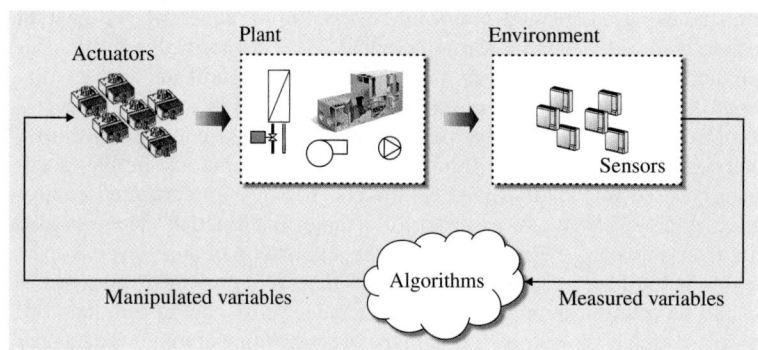

Fig. 62.6 Linking the measurements to the manipulated variables

One trend that is still emerging in some sectors of the buildings industry but is already mature in other sectors is the deployment of IT infrastructure to provide a backbone that allows all sensors actuators and other devices to be visible on a single network. This is part of a more general trend toward the convergence of the building automation and information technology systems in a building. Having an IT infrastructure in place that links disparate building devices facilitates the development of advanced control and operational management algorithms that take advantage of the plant-wide data. Energy management strategies that were mentioned in Sect. 62.2.3 are one example of algorithms that combine information from different subsystems to attain an overall objective. Application of these types of building-wide optimization is likely to increase in the future with the demand driven by rising energy costs and increasing environmental concerns.

IT infrastructure and networking is constantly evolving, making it easier to connect and redistribute devices; for example, wireless networking is becoming more popular in buildings because it reduces wiring costs and makes it easier to reconfigure spaces and move sensors from one location to another. The general trend is also for a greater diversity of devices to be connected to the network and for each device to contain some level of embedded intelligence. These so-called smart devices are part of a trend toward distributed computing. Distribution of computing functions across devices makes a system less prone to catastrophic failure and allows problems to be broken up into smaller pieces and solved using many low-cost devices rather than one high-cost computing engine.

The development of automation algorithms and control techniques has lagged that of hardware and IT infrastructure. Many opportunities therefore now exist for making better use of the information available from a building automation network to improve operation and control and address higher-level objectives such as minimizing energy use. One particularly underdeveloped topic is that of coordinating the operation of different building systems; for example, taking into account the interactions between lighting and HVAC could potentially yield large energy savings for many types of buildings [62.35]. Another example is food retail buildings that use refrigeration systems to keep products cool. Refrigeration systems generate heat and affect the indoor environment in these buildings but their operation is not usually coordinated with the HVAC plant. Combining information from security and access control systems to predict the number of people in

a space for improved climate control is yet another example, mentioned earlier, of how information from one subsystem could be used to enhance the operation of another.

Continuing with the theme of IT infrastructure providing new opportunities for higher level-plant coordination, recent years have also seen the deployment of multibuilding control and operation management [62.27]. The concept here is to tie together several buildings and have operational oversight and alarm management handled in one location rather than in each individual building. This allows for centralized data processing and the possibility for a smaller group of highly qualified operators to spread their expertise over multiple buildings. The approach also opens the way for peer-based benchmarking and performance monitoring that can be particularly advantageous when many buildings are of the same type [62.36].

The discussion so far in this section has identified some potential opportunities for control and operational management that are the result of having more information from multiple diverse systems available on a network. Capitalizing on these opportunities often requires combining ideas and methods from different mathematical disciplines. This has been an emerging trend in recent years with a common example being the combination of statistics and control theory such as when employed in statistical process control [62.37]. Other examples are the use of economic ideas as a way to instigate distributed optimization. The availability of real-time energy prices in certain geographical areas is an example of using economics to encourage distributed optimization of load profiles with the aim of evening out the loads at the power plant [62.38].

The idea of using market-based theories is also beginning to be seen as a viable strategy for optimizing the operation of systems within a building. The way in which building systems are designed and implemented is inherently distributed, with small amounts of processing power usually available locally at each device rather than concentrated in a central location. Furthermore, the number of variables in a building and the complexity of interactions and nonlinear behaviors can make a centralized approach to optimization unviable. The metaphor of each device being an *agent* working to optimize its own objective function according to constraints and general guidelines is an emerging field of research with the aim being that overall building performance would reach some optimum behavior without the need for centralized optimization. This idea is beginning to be explored for building applications [62.39]

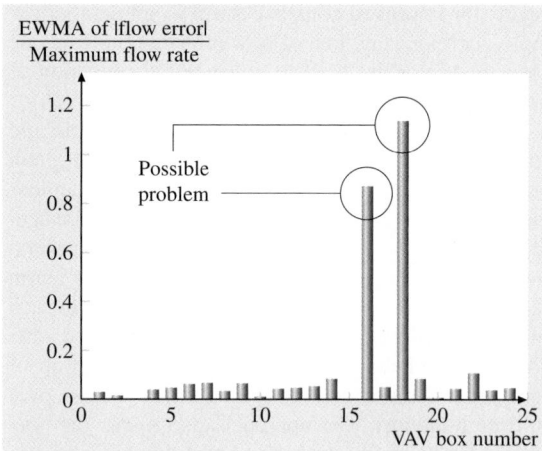

Fig. 62.7 Normalized EWMA statistics for VAV box control performance

and has already been applied to problems in other applications [62.40].

The expansion of control methods and plant operational management ideas that has occurred in other applications is highly relevant to buildings and many of the most promising ideas are beginning to be explored and adopted. The general subject of plant performance assessment and auditing is one example where advances from other applications are being adopted. Some control vendors now offer automatic trending and benchmarking of performance indices that are not just raw measurements, but quantities derived from physics and/or statistics. The methods of analyzing these performance indices and ways in which they can be presented to a user are emerging areas of research and application. The goal of this work is to be able to quickly detect problems and narrow down the cause so that downtimes are reduced, performance is more consistent, and maintenance can be more proactive and targeted. One popular approach is to group together operational statistics from multiple plant items that are of similar type and look for outliers amongst the set as a way to identify

faults. Figure 62.7 illustrates this approach based on actual exponentially weighted moving averages (EWMA) of the setpoint error signal for VAV boxes in a large building [62.41].

An extension of the concepts of detecting and diagnosing problems is to make the control system able to adjust automatically so that the building can become fault-tolerant. Buildings usually have many different types of systems that can be used to compensate for each other; for example, a problem identified with cooling capacity that is affecting the temperature of conditioned air could be compensated for by operating the fans at higher loads to increase airflow. This redundancy is already (inadvertently) used to create fault tolerance in buildings by having multiple closed-loop control strategies that sometimes compete with each other. This creates problems for more conventional methods of fault detection that only check whether setpoints are being maintained because the effects of a fault are masked by the competing closed-loop controllers. Hence, there is increasing awareness that fault detection and diagnosis methods need to share information with the control strategy and be combined for the most effective solution [62.42].

The buildings industry is notoriously fragmented with multiple parties involved in the various lifecycle tasks of, among others: design, construction, occupation, maintenance, and renovation. An emerging area is the application of information management to a building lifecycle. Recent years have seen concerted efforts to standardize various aspects of a building process, ranging from standard data models for building geometry and plant description [62.43] to communication protocols for automation system networks [62.44]. The general goal of these standardization efforts is to improve the efficiency of passing information and also reduce the costs and risk of developing and marketing new products such as software. The proliferation of web-based commerce is testament to how standard (and stable) infrastructure can be an effective stimulant to innovation and technological progress.

62.4 Open Challenges

Although buildings can be considered similar to any other large-scale process, there are some unique issues that can hamper efforts to make buildings smart. Primary systemic issues in the building industry are that it is low cost, fragmented, and risk averse. The

way in which buildings are financed is geared toward minimization of capital costs. Most contracts are awarded on a lowest-cost basis and this means that the operational costs of a building and other lifecycle costs are deemphasized. The result is that operational performance of the building is frequently compromised through poor design and poor-quality installation and maintenance. Several studies have shown that proper commissioning of buildings can lead to large energy savings and improvements in system performance [62.45]. Cost pressures also lead to the installation of low-quality sensors and actuators that detrimentally affect control performance. A related problem is when too few sensors are installed, causing measurements to be inaccurate and control performance and energy use to be affected.

At the control hardware level, cost constraints lead to minimal memory and processing power in local controllers and also low-resolution analog-to-digital and digital-to-analog converters. Eight-bit converters are still commonplace and the level of signal quantization can be severe enough to cause oscillations in feedback loops. Quantization can also corrupt signals, which makes later analysis and diagnosis of problems more difficult. Another source of quantization is logic that is included on a network so that data are only sent when there is a change of value beyond a certain threshold. This strategy provides data compression and network traffic minimization but it severely affects the signals being filtered and poses problems when trying to analyze data for control performance assessment and diagnostics [62.46].

The level of education and training tends to be low for building operators, also because of cost constraints. This means that there is often difficulty in understanding the processes in a building, their interactions, and the role of the control system. It is common therefore for operators to quickly shut-off control logic that they do not properly understand. The situation then arises of having many loops in a building left in override mode, leading to suboptimal and effectively open-loop, or no, control of the building systems. The primary objective of most operators is also to keep the occupants comfortable and respond to hot and cold complaints. The energy efficiency of the building is usually secondary and draws little attention. The lack of consideration for energy use is due to the still relatively cheap cost of energy and also the lack of performance indicators that could help identify energy problems.

The discussion so far has centered on barriers, both industry-systemic and technical, that impede the development, application, and deployment of new smart-building algorithms and technologies. However, many of these barriers can be overcome by adapting smart control and operational management methods to the unique aspects of the buildings industry; for exam-

ple, the problem of too few sensors can be addressed by using models to predict measurements to create virtual sensors based on analytical redundancy. Signal processing methods can be used to reconstruct quantized data, and robust control theory can be used to minimize the effect of slow sampling and potential jitter. Buildings can benefit from the application of methods and algorithms developed in other diverse fields but these methods have to be adapted to the specific issues of the buildings.

Another open challenge is to integrate the diverse array of systems in buildings into a common information-sharing framework. Although this has already started to happen with the advent of open protocols for network communication, the full potential is not yet realized. One reason for this is that buildings contain a very diverse group of systems that are made by many different manufacturers, all with their own embedded controls and electronics. The commoditization of certain building systems such as packaged cooling units has brought down costs, but these systems rarely have external communication interfaces to allow interconnectivity with other systems because of the extra costs this would incur. Although interconnection of building devices is a key to unlocking untapped energy and operational efficiencies, this cannot happen without algorithms and control methods that can take advantage of the newly available information. There is therefore a dilemma because both aspects are needed, and business will be reluctant to invest in development of just one aspect without the other being already available. Possible future avenues that could alleviate some of the barriers to connectivity are solutions such as power line networking, utilization of an existing IT backbone, and general lower-cost solutions for plug-and-play networking.

The issue of energy efficiency has again come to the fore in recent years, and this has led to new legislation and various certification programs being established in several countries around the world to encourage the design and construction of so-called green buildings. Even if financial incentives are lacking, increased concern over the environmental impacts of wasteful energy use within populations is exerting pressure on the construction industry to portray a greener and more energy-efficient image. The most visible sign of a response to these pressures is in the design and construction of modern buildings. There are several examples of new buildings around the world that make such efficient use of sunlight and wind through smart design that they can eliminate the need for mechanical heating/cooling and ventilation and almost eliminate the need for artificial lighting when sunlight is available. The manufacturers of electrical and mechanical systems are also responding to energy concerns by producing more efficient equipment that is often certified by government or other third-party organizations. However, the challenge is for the automation systems and especially the control algorithms to keep pace with the rapid evolution of buildings and inherent equipment. Failure to keep pace will result in buildings operating well below their intended efficiency levels.

62.5 Conclusions

The operation of many aspects of a building is now automated, ranging from temperature control to fire and security. The general trend is for the operation of more systems and appliances to be automated and for these to be connected to a common network. The availability of common networking infrastructure is also stimulating the demand for more advanced control and operational management methods that take into account system interactions and facilitate the optimization of building-wide criteria such as energy use. However, the adoption of IT infrastructure has in many ways outpaced the development of algorithms and automation methods that can capitalize on these advances. There is also the problem of operators not being able to cope with the abundance of data now available on BAS networks. Again, this problem is exacerbated by the lack of algorithms for processing and reducing the data into more manageable statistics or performance reports.

The complex and bespoke nature of building systems makes it difficult to develop and apply generic algorithms and, on the other hand, it is too costly to tailor algorithms to every building system. This issue is difficult to resolve but is being alleviated by standardization and commoditization of building systems. New model-free algorithmic methods for control and optimization have the potential of being able to adapt to different system types without requiring significant engineering effort or tuning. In combination, these developments may help overcome some of the barriers to the adoption of new technology for improved energy management and control in buildings.

References

62.1   IEA: *International Energy Outlook 2007* (United States Department of Energy, Washington 2007), retrieved on 2007-06-06

62.2   H.M. Newman: *Direct Digital Control of Building Systems: Theory and Practice* (Wiley, New York 1994)

62.3   ASHRAE: *HVAC Systems and Equipment* (American Society of Heating, Ventilating, and Air-Conditioning Engineers, Atlanta 2004)

62.4   WHO: *Health Aspects of Plumbing* (World Health Organization and World Plumbing Council, Geneva 2006)

62.5   G. Augenbroe: Trends in building simulation, Build. Environ. **37**(8–9), 891–902 (2002)

62.6   Y. Zhang, J.A. Wright, V.I. Hanby: Energy aspects of HVAC system configurations – problem definition and test cases, HVAC&R Research **12**(3C), 871–888 (2006)

62.7   J. Sinopoli: *Smart Buildings* (Spicewood Publishing, Elk Grove Village 2006)

62.8   I. Shaw: *The Oxford History of Ancient Egypt* (Oxford Univ. Press, Oxford 2003)

62.9   J.W. Humphrey, J.P. Oleson, A.N. Sherwood: *Greek and Roman Technology: A Sourcebook* (Routledge, London 1997)

62.10   W.H. Carrier: *Modern Air-Conditioning, Heating and Ventilating*, 2nd edn. (Pitman Publishing, New York 1950)

62.11   O. Mayr: *The Origins of Feedback Control* (Cambridge MIT Press, Cambridge 1970)

62.12   L.E. Harris: *The Two Netherlanders, Humphrey Bradley and Cornelis Drebbel* (Cambridge Univ. Press, Cambridge 1961)

62.13   W.S. Johnson: Electric Tele-Thermoscope, Patent 281884 (1883)

62.14   K. Åström, T. Hägglund: *PID Controllers: Theory, Design and Tuning*, 2nd edn. (Instrument Society of America, Research Triangle Park 1995)

62.15   T.I. Salsbury: A new pulse modulation adaptive controller (PMAC) applied to HVAC systems, Control Eng. Pract. **10**(12), 1357–1370 (2002)

62.16   T. Marlin: *Process Control*, 2nd edn. (McGraw–Hill Education, New York 2000)

62.17   A.L. Dexter, R.G. Hayes: Self-tuning charge control scheme for domestic stored-energy heating systems, IEE Proc. D: Control Theory Appl. **128**(6), 292–300 (1981)

62.18   X.-C. Xi, A.-N. Poo, S.-K. Chou: Support vector regression model predictive control on a HVAC plant, Control Eng. Pract. **15**(8), 897–908 (2007)

62.19   S.J. Hepworth, A.L. Dexter: Adaptive neural control with stable learning, Math. Comput. Simul. **41**(1–2), 39–51 (1996)

62.20   J.E. Seem: A new pattern recognition adaptive controller with application to HVAC systems, Automatica **34**(8), 969–982 (1998)

62.21   X.-D. He, S. Liu, H.H. Asada: Modeling of vapor compression cycles for multivariable feedback control of HVAC systems, J. Dyn. Syst. Measur. Control Trans. ASME 119 (2) **119**(2), 183–191 (1997)

62.22   A.L. Dexter: Self-tuning optimum start control of heating plant, Automatica **17**(3), 483–492 (1981)

62.23   J.M. House, T.F. Smith: System approach to optimal control for HVAC and building systems, ASHRAE Transactions **101**(2), 647–660 (1995)

62.24   F.B. Morris, J.E. Braun, S.J. Treado: Experimental and simulated performance of optimal control of building thermal storage, ASHRAE Transactions **100**(1), 402–414 (1994)

62.25   T.I. Salsbury: A temperature controller for VAV air-handling units based on simplified physical models, HVAC&R Research **4**(3), 265–279 (1998)

62.26   J.D. Spitler, D.C. Hittle, D.L. Johnson, C.O. Pedersen: A comparative study of the performance of temperature-based and enthalpy-based economy cycles, ASHRAE Transactions **93**(2), 13–22 (1989)

62.27   T.A. Reddy, J.K. Lukes, L.K. Norford, L.G. Spielvogel, L. Norford: Benefits of multi-building electric load aggregation: actual and simulation case studies, ASHRAE Transactions **110**(2), 130–144 (2004)

62.28   B. Von Neida, D. Maniccia, A. Tweed: An analysis of the energy and cost savings potential of occupancy sensors for commercial lighting systems, J. Illum. Eng. Soc. **30**(2), 111–125 (2001)

62.29   A. Guillemin, N. Morel: Innovative lighting controller integrated in a self-adaptive building control system, Energy Build. **33**(5), 477–487 (2001)

62.30   ASHRAE: *HVAC Applications* (American Society of Heating, Ventilating, and Air-Conditioning Engineers, Atlanta 2007)

62.31   J. Hyvarinen, S. Karki: *Final Report Vol 1: Building Optimization and Fault Diagnosis Source Book* (Technical Research Centre of Finland, Espoo 1996)

62.32   S. Katipamula, M.R. Brambley: methods for fault detection, diagnostics, and prognostics for building systems – a review, Part II, HVAC&R Research **11**(2), 169–187 (2005)

62.33   S. Katipamula, M.R. Brambley: Methods for fault detection, diagnostics, and prognostics for building systems – a review, Part I, HVAC&R Research **11**(1), 3–25 (2005)

62.34   T.J. Harris: Assessment of control loop performance, Can. J. Chem. Eng. **67**, 856–861 (1989)

62.35   O. Sezgen, J.G. Koomey: Interactions between lighting and space conditioning energy use in US commercial buildings, Energy **25**(8), 793–805 (2000)

62.36   K.L. Gillespie Jr., P. Haves, R.J. Hitchcock, J. Deringer, K.L. Kinney: Performance monitoring in commercial and institutional buildings, HPAC Engineering **78**(12), 39–45 (2006)

62.37   J. Schein, J.M. House: Application of control charts for detecting faults in variable-air-volume boxes, ASHRAE Transactions **109**(2), 671–682 (2003)

62.38   D.S. Watson, M.A. Piette, O. Sezgen, N. Motegi: Automated demand response, HPAC Engineering **76**, 20–29 (2004)

62.39   P. Davidsson, M. Boman: Distributed monitoring and control of office buildings by embedded agents, Inf. Sci. **171**, 293–307 (2005)

62.40   N.R. Jennings, S. Bussmann: Agent-based control systems, IEEE Control Syst. Mag. **23**(3), 61–73 (2003)

62.41   J.E. Seem, J.M. House, R.H. Monroe: On-line monitoring and fault detection, ASHRAE Journal **41**(7), 21–26 (1999)

62.42   J.E. Seem, J.M. House: Integrated control and fault detection of air-handling units, Proc. IFAC Conf. Energy Sav. Control Plants Build. (Bulgaria 2006)

62.43   T. Froese, M. Fischer, F. Grobler, J. Ritzenthaler, K. Yu, S. Sutherland, S. Staub, J. Kunz: Industry foundation classes for project management – a trial implementation, Electron. J. Inf. Technol. Constr. **4**, 17–36 (1999)

62.44   S.T. Bushby: BACnet: a standard communication infrastructure for intelligent buildings, Autom. Constr. **6**(5–6), 529–540 (1997)

62.45   E. Mills, N. Bourassa, M.A. Piette, H. Friedman, T. Haasl, T. Powell, D. Claridge: The cost-effectiveness of commissioning, HPAC Engineering **77**(10), 20–25 (2005)

62.46   N.F. Thornhill, M. Oettinger, P. Fedenczuk: Refinery-wide control loop performance assessment, J. Process Control **9**, 109–124 (1999)

63. Automation in Agriculture

Yael Edan, Shufeng Han, Naoshi Kondo

The complex agricultural environment combined with intensive production requires development of robust systems with short development time at low cost. The unstructured nature of the external environment increases chances of failure. Moreover, the machines are usually operated by low-tech personnel. Therefore, inherent safety and reliability is an important feature. Food safety is also an issue requiring the automated systems to be sanitized and reliable against leakage of contaminations. This chapter reviews agricultural automation systems including field machinery, irrigation systems, greenhouse automation, animal automation systems, and automation of fruit production systems. Each section describes the different automation systems with many application examples and recent advances in the field.

Agricultural productivity has significantly increased throughout the years through intensification, mechanization, and automation. This includes automated farming equipment for field operations, animal systems, and growing systems (greenhouse climate control, irrigation systems). Introduction of automation into agriculture has lowered production costs, reduced the drudgery of manual labor, raised the quality of fresh produce, and improved environmental control. Unlike industrial applications, which deal with simple, repetitive, well-defined, and a priori known tasks, automation in agriculture requires advanced technologies to deal with the complex and highly variable environment and produce. Agricultural products are natural objects which have a high degree of variability as a result of environmental and genetic variables. The agricultural environment is complex and loosely structured with large variations between fields and even within the same field. Fundamental technologies must be developed to solve difficult problems such as continuously changing conditions, variability in products and environment (size, shape, location, soil properties, and weather), delicate products, and hostile environmental conditions (dust, dirt, and extreme temperature and humidity). Intelligent control systems are necessary for dynamic, real-time interpretation of the environment and the objects. When compared with industrial automation systems, precision requirements in agricultural automation systems may be much lower. Since the product being dealt with is of relative low cost, the cost of the automated system must be low in order for it to be economically justified. The seasonal nature of agriculture makes it difficult to achieve the high utilization found in manufacturing industries.

63.1 Field Machinery

The use of machinery in agriculture has a long history, but the most significant developments occurred during the 20th century with the introduction of tractors. As early as 1903, the first farm tractor powered by an internal combustion engine was built by Hart Parr Company. Using its assembly line techniques, Henry Ford & Son Corporation started mass production of Fordson tractors in 1917. The commercial success of tractors sparked other innovations as well. In 1924, the International Harvester Company introduced a power takeoff device that allowed power from a tractor engine to be transmitted to the attached equipment such as a mechanical reaper. Deere & Company followed in 1927 with a power lift device that raised and lowered hitched implements at the end of each row. Rubber wheels were first designed and used for tractors in 1932 to improve traction and fuel economy. Pulled and powered by tractors, an increasingly wide range of farm implements were developed in the 20th century to mechanize crop production in every step, from tillage, planting, to harvesting. Harvesting equipment trailed only tractors in importance. Early harvesters for small-grain crops were pulled by tractors and powered by tractors' power takeoff (PTO). The development of a self-propelled combine in 1938 by Massey Harris marked a significant progress in increasing productivity. The self-propelled combine incorporated several functions such as vehicle propulsion, grain gathering, and grain threshing into an all-in-one unit for better operation efficiency. The mechanization of harvesting other crops included the developments of mechanical hay balers in the 1930s and mechanical spindle cotton pickers in 1943. Tractors, combines, and other farm machinery were continuously refined during the second half of the 20th century to be more efficient, productive, and user-friendly. The success of agricultural mechanization has built a strong foundation for automation. Automation increases the productivity of agricultural machinery by increasing efficiency, reliability, and precision, and reducing the need of human intervention [63.1]. This is achieved by adding sensors and controls. The blending of sensors with mechanical actuation can be found in many agricultural operations such as automating growing conditions, vision-guided tractors, product grading systems, planters and harvesters, irrigation, and fertilizer applicators. The history of automation for agricultural machinery is almost as old as agricultural mechanization. Two ingenious examples in the early 20th century were the self-leveling system for hillside combines by Holt Co. in 1891 and the implement draft control system by Ferguson in 1925 [63.2]. Early automation systems mainly used mechanical and hydromechanical control devices. Since the 1960s, electronics development for monitoring and control has dominated machine designs, and has led to increased machinery automation and intelligence. Mechatronics technology, a blend of mechanics, electronics, and computing, is often applied to the design of modern automation systems. Automation in contemporary agricultural machines is more complicated than a single control action; for example, the modern combine harvester has automatic control of header height, travel speed, reel speed, rotor speed, concave opening, and sieve opening to optimize the entire harvest process. Farm machinery includes tractors and transport vehicles, tillage and seeding machines, fertilizer applicators and plant protection application equipment, harvesters, and equipment for post-harvest preservation and treatment of produce. Mechanization and automation examples can be found in many of these machines [63.3]. However, the wide variety of agricultural systems and their diversity throughout the world makes it difficult to generalize about the application of automation and control [63.1]. Therefore, only one type of automation – automated navigation of agricultural vehicles – will be presented here. Automated vehicle navigation systems include the operator-assisted steering system, automatic steering system, and autonomous system. These systems can relieve the vehicle operator of the repetitive and monotonous steering operation. Automatic guidance has been the most active research area in the automation history of agricultural machinery. With the introduction of the global positioning system (GPS) to agriculture in the late 1980s, automatic guidance technology has been successfully commercialized. Today, autoguidance is the fastest growing segment in the agricultural machinery industry. The following sections discuss the principles of autoguidance systems, the available technologies, and examples of specific autoguidance systems.

63.1.1 Automatic Guidance of Agricultural Vehicles

For many agricultural operations, an operator is required to perform two basic functions simultaneously: steering the vehicle and operating the equipment. The need to relieve the operator of continuously making steering adjustments has been the main reason for the development of automatic guidance systems. Excellent references to automatic vehicle guidance research in Canada, Japan, Europe, and the USA can be found in *Wilson* [63.4], *Torii* [63.5], *Keicher* and *Seufert* [63.6], and *Reid* et al. [63.7]. Figure 63.1 shows a typical autoguidance system which includes a position sensor, a steering angle sensor, and a steering actuator as the hardware components, and a path planner, a navigation controller, and a steering controller as the software components. The path planner gives the desired (or planned) vehicle position. This desired position is compared with the measured position given by the position sensor. The navigation controller calculates the desired steering control angle based on the difference in the desired and measured positions. Finally, the steering controller uses the difference in the desired and measured steering angles to calculate an implementing steering control signal and sends it to drive the steering actuator. Modern agricultural vehicles often employ electrohydraulic (E/H) steering systems. Developments in each of the system components are described in details below.

Position Sensing

The position sensing system measures vehicle position relative to a reference frame and provides inputs to the navigation controller. Most agricultural guidance applications require position measurement in two-dimensional (2-D) space. In addition, vehicle speed, heading, and rotational movements (roll, pitch, yaw) are often needed by the navigation controller.

Guidance accuracy is the primary factor in selecting a position sensor. *Auernhammer* and *Muhr* [63.8] suggested three levels of accuracy required for different farming operations: 1 m for rough operations (soil sampling, weed scouting), 10 cm for fine operations (pesticide application, soil cultivation), and 1 cm for precise operations (planting, plowing). Different position sensors are selected in a guidance system to meet the accuracy requirements for different farming operations. In general, there are three categories of

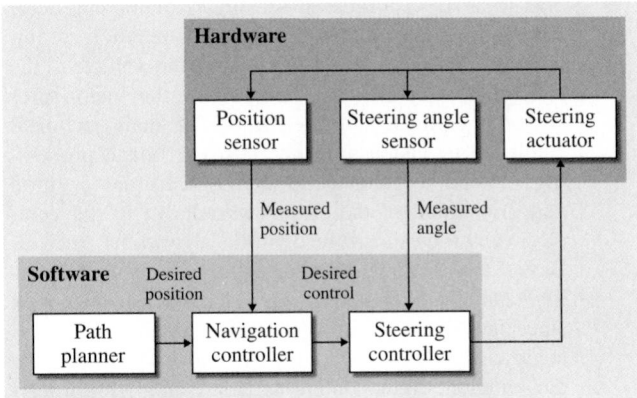

Fig. 63.1 Components of a typical autoguidance system

positioning techniques: absolute positioning, relative positioning, and sensor fusion.

Absolute Positioning. The most common system of absolute positioning is the global navigation satellite system (GNSS). Currently, the NAVSTAR global positioning system (GPS) in the USA is the only fully operational GNSS. A GPS receiver calculates its position by measuring the distance between itself and three or more GPS satellites. The positioning accuracy of an autonomous, mobile GPS receiver is 5–15 m. This accuracy is generally not suitable for vehicle guidance. To improve the accuracy, a differential correction technique is applied. A differential GPS (DGPS) receiver can provide position accuracy within 2–5 m and within 1 m precision in a short time period. The DGPS receiver's accuracy can meet the requirement of positioning accuracy for most guidance applications. Further improvement in GPS accuracy requires carrier-phase enhancement (or a real-time kinematic process), typically using a local base station. The real-time kinematic GPS (RTK GPS) receiver can achieve centimeter accuracy and should meet the positioning accuracy requirements for almost all agricultural field operations. The GPS positioning technique has been successfully implemented for vehicle guidance since its inception [63.9–12]. Other absolute positioning sensors, such as laser [63.13] and geomagnetic direction sensors [63.14], have been developed and applied to vehicle guidance with varying degrees of success. However, currently the GPS receiver remains the only commercially viable choice for absolute positioning systems.

Relative Positioning. The most promising system of relative positioning is computer vision using cameras [63.4]). Vision-based sensing is mainly used for automatic guidance in row crops. Its operation resembles a human operator's steering of the vehicle – the camera is equivalent to the *eye* and the vision processor is equivalent to the *brain*. The main technical challenge to vision guidance is using image processing to find a guidance directrix, i.e., the position and orientation of the crop rows relative to the vehicle. Numerous image recognition algorithms, such as Bayes classification, edge detection, K-means clustering, and the Hough transform, have been developed since the 1980s [63.15–19]. Vision-based system can achieve excellent positioning accuracy under good crop and ambient light conditions; for example, *Billingsley* and *Schoenfisch* [63.20] reported 2 cm accuracy for their vision guidance systems. *Han* et al. [63.19] reported 1.0 cm average root-mean-square (RMS) offset error for soybean images and 2.4 cm for corn images. However, the vision-based system may not be reliable under changing lighting conditions, which are not uncommon in an agricultural environment. Other relative positioning sensors include dead reckoning, odometry, and inertial measurement units (IMU). These sensors are seldom used alone in a vehicle navigation system. Instead, they are integrated with absolute positioning sensors (e.g., GPS) in a sensor fusion approach.

Sensor Fusion. Sensor fusion is the process of combining data from multiple sensors so that the resulting information is better than when these sensors are used individually. No single positioning sensor will work for agricultural vehicle guidance under all conditions; for example, a GPS signal may be blocked by heavy tree shading. Vision sensors may not work under heavy dust conditions. Sensor fusion not only provides a way to automatically switch to a working sensor when one of the sensors quits working, but also blends the outputs from the multiple working sensors to obtain the best results. A good example of sensor fusion is integration of GPS with inertial sensors [63.21]. In this approach, GPS provides the low-frequency absolute position information, and inertial sensors provide the high-frequency relative position information. Inertial sensors can smooth out the short-term GPS errors, and the GPS can correct the bias and scale factor errors of the inertial sensors. If the GPS signals become temporarily unavailable, the inertial sensors can continue to provide position information. Sensor fusion allows the integration of several low-cost sensors to achieve good positioning accuracy [63.22]. Many algorithms are available for sensor fusion [63.23], with the Kalman filtering technique

being the most common approach [63.24]. Adaptive sensor fusion algorithms have also been developed to deal with a priori unknown sensory distributions and asynchronous update of the sensors [63.25]. Terrain compensation is another example of applying sensor fusion to improve guidance accuracy on sloping terrain. A terrain compensation module measures vehicle roll, pitch, and yaw angles, and combines these measurements with the position measurement to compensate for the GPS antenna movement due to side slopes and rough terrain. Many manufactures of autosteering systems now offer terrain compensation features. Additional information on sensor fusion can also be found on Chap. 20.

Path Planning

Path planning is the generation of 2-D sequenced positions or trajectories for the automated vehicle. The sequenced positions account for the vehicle kinematics such as the minimum turn radius and other constraints. Most agricultural operations, such as tillage, planting, spraying, and harvesting, require the vehicle to travel the entire field with parallel paths at a fixed spacing equaling the implement width. Planning such paths is called coverage path planning. Coverage path planning involves two steps. Step one is to decompose a field into subregions. An optimal travel direction is found for each subregion. Step two is to find the optimal coverage pattern within each subregion. Many different algorithms have been developed for coverage path planning [63.26].

Trapezoidal decomposition is a popular technique for subdividing the field. The trapezoids are then merged into larger blocks and the selection is made using certain criteria which take into consideration the area and the route length of the block and the efficiency of driving [63.27]; *Jin* and *Tang* [63.28] used a geometric model to represent the full coverage path planning problem. The algorithm was capable of finding a globally optimal decomposition for a given field and the direction of the boustrophedon paths for each subregion. The search mechanism of the algorithm is guided by a customized cost function that unifies different cost criteria, and a divide-and-conquer strategy is adopted.

A graphical approach is often used to find the optimized coverage pattern within a subregion [63.29–31]. The processes include partitioning the area, building a partition graph, and searching the partition graph. Heuristic functions are used in the searching process to prune the search tree early so the optimized solution can be found within a reasonable time. In the case of multiple vehicles working in the same region, *Gray* [63.32] developed a path planning method.

Navigation and Steering Controllers

The navigation controller takes the desired and measured positions as inputs to compute the desired control variables, typically the lateral and heading corrections. The desired control variables and the measured variables (typically the steering angle) are fed into the steering controller to compute the steering corrections. A typical navigation control algorithm calculates the lateral and heading errors based on a reference point on the vehicle and a target (look-ahead) point on the desired vehicle trajectory. The target point may be dynamically adjusted based on speed to achieve satisfactory path tracking performance [63.33, 34].

Agricultural vehicles frequently operate in challenging conditions such as varying travel speed, operating load, and ground surface conditions. The steering controller design must be robust enough to adapt to these conditions. Several steering controllers, including proportional–integral–differential (PID) controller, feedforward PID (FPID) controller, and fuzzy-logic (FL) controllers, have been developed and implemented in the guidance system [63.35–37]. Additional information on mobility and navigation can be found on Chap. 16.

Commercialization of Autoguidance Systems

Commercial development of autoguidance systems by US manufacturers started in the 1990s soon after the availability of GPS to agricultural applications. Early GPS-based guidance systems used visual aids, commonly referred to as *lightbars*, to show a driver how to steer the vehicle along parallel passes or swaths across a field. The need to improve driving accuracy and repeatability led to the development of the next level of automation – autosteering. The autosteering system steers the vehicle within a path and the driver only needs to turn at the ends. Several preset driving patterns can be used by an autosteering system during field operations. The most popular patterns for ground applications are straight rows and curved rows. The straight-row option allows the operator to follow parallel straight paths separated by a predetermined swath width. An initial path (A–B line) is first defined by the operator and the remaining paths are generated by the guidance system. For the curved-row option, the operator drives the first curved path. The autoguidance system steers the vehicle along the consecutive paths. Other driving patterns such as circles (for center-pivot irrigation

Fig. 63.2 A John Deere 8000 series tractor equipped with the GreenStar AutoTrac assisted steering system

field) and spirals (for field headlands) are also available in some autoguidance systems. Autoguidance systems are now commercially available. Figure 63.2 shows an example of the GreenStar AutoTrac assisted steering system on a John Deere 8000 series tractor. Most autoguidance systems have reported a path-to-path accuracy better than 5 cm with DGPS or RTK under good field conditions.

63.1.2 Autonomous Agricultural Vehicles and Robotic Field Operations

An autonomous vehicle must be able to work without an operator. In addition to steering, it must perform other tasks that a human operator typically does: detecting and avoiding unknown objects, operating at a safe speed, and performing implement tasks while driving. Developing human intelligence for the autonomous vehicle is a challenging job.

Autonomous vehicles working in an unstructured agricultural environment must use sophisticated sensing and control systems to be able to react to any unplanned events. A typical unplanned event is the presence of a human or animal in front of the vehicle. Development of vehicle safeguarding systems is the key to the deployment of the autonomous vehicles. A number of technologies have been investigated for providing vehicle safeguarding. *Guo* et al. [63.38] used two ultrasonic sensors to detect a human being. The reliable detection range was up to 4.6 m for moving objects and 7.5 m for stationary objects under field conditions. *Wei* et al. [63.39] used a binocular stereo camera to detect a person standing in front of a vehicle. The system was able to find the person's relative motion status (speed and heading) relative to the vehicle at a distance range from 3.4 to 13.4 m. *Kise* et al. [63.40] tried a laser rangefinder to estimate the relative motion of a tractor obstacle. In general, ultrasonic sensors are low cost but their detection range is short. Stereo cameras are unreliable under changing lighting conditions. Currently, the most reliable technology is the laser rangefinder, but its use is limited to research vehicle platforms due to the high costs. Multiple levels of system redundancy must be designed into the vehicle, which often requires multiple safeguarding sensors. Development of robotic field operations is an integral part of autonomous vehicles. In order to use an autonomous vehicle, tasks must be automated as well. Over the years, agricultural equipment has evolved to accommodate the automated control of tasks [63.1]. Microprocessor-based electronics control is replacing mechanical control, and electrohydraulically powered actuators are preferred over mechanically powered ones. The adoption of CAN bus standards (SAE J1939, DIN 9684, ISO 11783) in the agricultural equipment industry has allowed networking of multiple control systems.

Task automation examples can be found in many modern agricultural machines. Examples include map-based automatic spraying of fertilizer and chemicals on sprayers, and headland management systems (HMS) for automatic sequencing of tractor functions normally associated with headland turns. *Matsuo* et al. [63.41] described a tilling robot that was able to do tillage, seedling, and soil paddling operations.

Reid [63.42] discussed a number of challenges related to the development of intelligent agricultural machinery and equipment. At present, autonomous agricultural vehicles and robotic field operations are still not reliable and durable enough to meet the requirements of the agricultural industry and its customers. Nevertheless, a number of autonomous vehicle systems have been developed as proof-of-concept machines which may lead to commercialization in the future. Some exemplary systems are briefly introduced below.

Robotic Harvester

A robotic harvester, called Demeter (Fig. 63.3), has been developed by the Carnegie Mellon University Robotics Institute for automated harvesting of windrowed crops. The robot platform was a New Holland 2550 self-propelled windrower equipped with DGPS, inertial navigation system (INS), and two color cameras. The camera system detected the cut/uncut edge of the crop, which gave a relative directrix for the

Fig. 63.3 The robotic harvester (Demeter) (courtesy of Carnegie Mellon University)

harvester to follow. The camera system was also used to detect potential obstacles for vehicle safeguarding. GPS data was fused with vision data for guidance. In addition to steering, speed and header height of the harvester were also automatically controlled. In 1997, the Demeter autonomously harvested 100 acres of alfalfa in a continuous run (excluding stops for refueling). During 1998, the Demeter harvested in excess of 120 acres of crop, cutting in both sudan and alfalfa fields [63.43, 44].

Autonomous Tractor

An autonomous tractor has been jointly developed by John Deere and Autonomous Solutions Inc. for automated spraying, mowing, and tillage in orchards (Fig. 63.4). The robot platform was a John Deere 5000 series tractor with significant modifications. The system components included the vehicle, a mobile control unit, and a base station; all were communicated

Fig. 63.4 A John Deere 5000N autonomous orchard tractor (courtesy of Deere & Company)

by a wireless CAN system. A DGPS and INS were used as positioning sensors. Vehicle controls included steering, brake, clutch, three-point hitch, PTO, and throttle. A long-range obstacle detection system was proposed for vehicle safeguarding. One of the key developments in the project was the path and mission planning, which included dynamic replanning for dynamic service events. The system design followed an industry joint architecture for unmanned ground systems (JAUGS) architecture. A proof-of-concept system was developed and successfully demonstrated, but the production decision was not made, primarily due to safety concerns.

Small Robotic Platforms

In agriculture, small robots can be used for many field tasks such as collection of soil or plant samples and detection of weed, insect or plant stress. When equipped with a larger energy source and appropriate actuators, they can also be used for localized treatments such as spot-spraying of chemicals or mechanical in-row weeding. A number of small robots have been developed, mainly at universities and research institutes [63.45]. *Astrand* and *Baerveldt* [63.46] developed an autonomous robot for mechanical weed control in outdoor environments. The robot employs a grey-level vision system to guide itself along the crop rows and a second, color-based vision system to identify the weed and to control a weeding tool that removes the weed within the row of crops. A plant nursing robot, HortiBot, was developed in Denmark as a tool carrier for precision weeding [63.47–49]. The HortiBot is a radio-controlled slope mower (Spider ILD01, Dvorák Machine Division, Czech Republic) equipped with a robotic accessory kit. A commercial stereo vision system was implemented for automatic guidance within plant rows.

63.1.3 Future Directions and Prospects

Farm productivity has increased significantly during the last century. Today, less than 3% of the US popula-tion works in agriculture, yet they produce more than adequate food for the entire nation. Agricultural mechanization has played a significant role in achieving this miracle. As next steps to mechanization, automation and robotization of farm operations can result in additional productivity improvement.

Autoguidance will continue to be the main focus of future development. The agricultural industry is now developing new systems for automation beyond autosteering of vehicles. Implement guidance and headland management are two examples. An implement guidance system automatically steers both tractor and implement and keeps the implement on the desired path. This helps overcome implement drift on hillsides or contour field conditions. The headland management system automates implement controls (e.g., to raise or lower the implement) and makes automatic turns at headland and interior field boundaries. Other guidance technologies that are close to commercialization include: sensor fusion that employs a multitude of complementary positioning sensors to improve system reliability, path or mission planning that produces the most efficient coverage paths for a single or multiple vehicles, and leader–follower systems for multiple-vehicle navigation and control, as in the case of combine harvester operation. Precision farming has become an area of enormous growth and excitement since the 1980s. The key concept in precision farming is to manage crop production at the subfield level. The labor-intensive nature of precision farming practices brings a great need for automated machines and equipment. Yield mapping and variable-rate application systems are now commercially available. In the future, autonomous field scout vehicles are needed for soil sampling, crop scouting, and real-time data collection. Small robots are desired for individual plant care such as precision weed control and selective crop harvesting. Because precision farming is considered as the future of agriculture, automation and robotics technologies will certainly become a big part of production agriculture in the 21st century.

63.2 Irrigation Systems

Irrigation is the supplemental application of water to the soil for assisting in growing crops. It is used mainly to replace missing rainfall for field crops, and to supply water to crops growing in protected environments such as greenhouses. The main objective is to supply the re-quired amount of water to the plants at the right time. The types of irrigation techniques differ in how the water is distributed within the field. In surface irrigation systems water moves over the land by gravity and infiltrates into the soil. Surface irrigation systems include

Fig. 63.5 Sprinkler irrigation (courtesy of US Fish and Wildlife Service, USFWS/Elkins WV)

Fig. 63.6 Center pivot with drop sprinklers (courtesy of Conversation and Production Res. Lab., Bushland, TX, USDA, ARS)

furrow, border-strip, and basin irrigation. Localized irrigation systems distribute water in piped networks by pressure, and the water is applied locally in the field and to the plant. Localized systems include spray, sprinkler, drip, and bubble systems. Automation provides efficient on-farm use of water and labor for all methods by enabling flexible frequency, rate, and duration of water supply with control of the irrigator at the right application point [63.50].

63.2.1 Types of Irrigation Systems

Flood control automation includes optimal gate operation of irrigation reservoirs [63.51]; surge flooding, which enables release of water at prearranged intervals; telemetering of paddy ponding depth and canal water level [63.52], which can be used to capture runoff and pump it back into the field for reuse; and precision con-

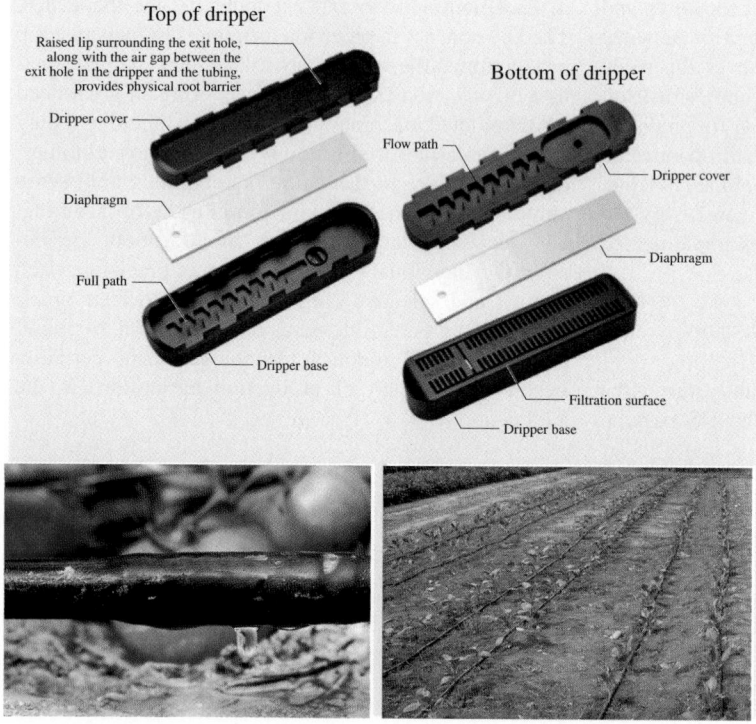

Fig. 63.7 Dripper and drip line irrigation system (courtesy of Netafim)

trol of inflow rate using ground-based remote-sensing feedback control systems [63.53]. Position of the advance of water along the furrow can be determined by contact-type sensors manually positioned in the furrow and recently by imaging systems [63.53, 54].

In sprinkler irrigation, water is piped to several locations in the field and distributed by high-pressure sprinklers or guns (Fig. 63.5). Spatially variable irrigation systems have typically used self-propelled irrigation systems – sprinklers mounted on moving platforms or center pivots [63.56, 57]. Center-pivot irrigation is a sprinkler irrigation system that is composed of several pipe segments joined together that are mounted on wheeled towers with sprinklers positioned along its length (Fig. 63.6). The system moves in a circular pattern.

Drip irrigation systems (Fig. 63.7) were invented in Israel in 1965. Water is applied slowly and directly to the soil, and only where needed. A drip irrigation system consists of valves, back-flow preventers, pressure regulators, filters, emitters, and of course the pipes (the mainline that leads water from the source to the valve, and the subpipe that goes from the valves to the connection point of the drip tubing and the drip tubes). Low-head bubbler irrigation systems are micro-irrigation systems based on gravity flow that operate at low pressure and require no filtration or pumping [63.58]. Their main advantages are simplicity, lower energy requirements, and few mechanical breakdowns [63.58]. However, their application is limited due to the complicated design and installation problems.

63.2.2 Automation in Irrigation Systems

Automation systems include irrigation time clocks – mechanical and electromechanical timers to allow accurate control of water responding to environmental changes and plant demands [63.59], with recent advances in using sensors to measure soil properties such as moisture and salinity using resistance- and capacitance-based sensors, and time-domain reflectometry [63.60, 61]. Sensors for measuring plant stress [63.62] by scanned and spotted canopy temperature measurements have been used in scheduling decisions for center-pivot and subsurface drip irrigation systems [63.63]. Sensors include infrared thermometers, thermal scanner sensors, and multispectral imaging [63.64].

High-resolution data of soil and water dynamics coupled with measurement of crop response to salinity and water stress are important for irrigation management optimization [63.65]. These data are commonly

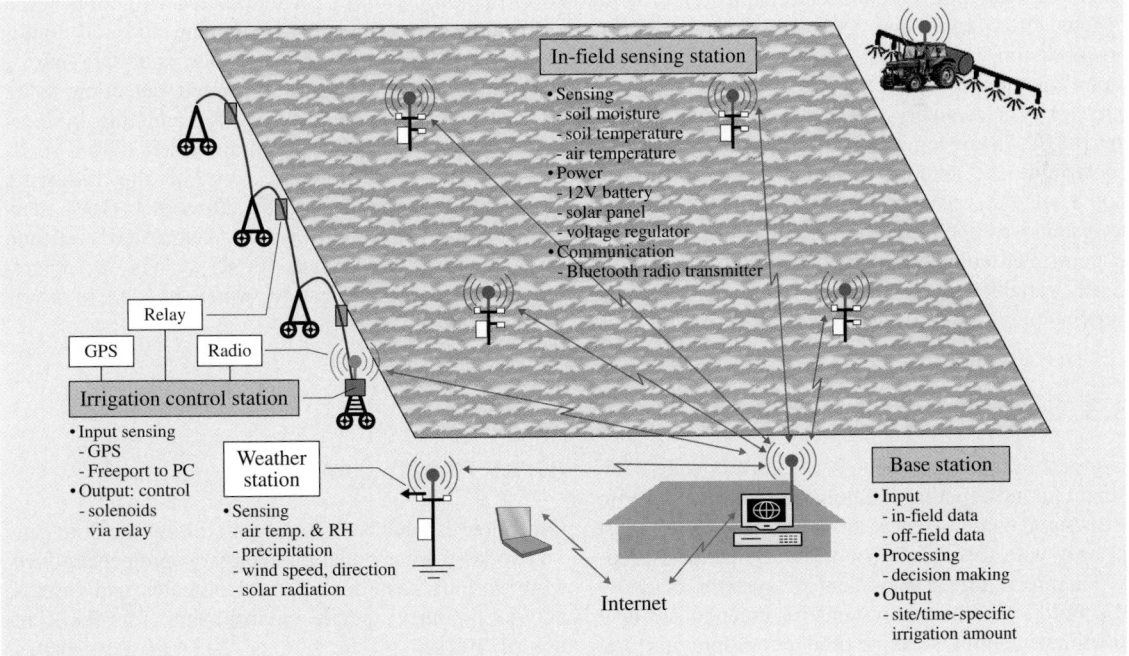

Fig. 63.8 Wireless irrigation system conceptual layout (after [63.55])

provided by weight-based soil lysimeters, with recent development of a volumetric lysimeter system [63.65].

Developments in automated irrigation systems include scheduling programs that use weather data to recommend and control time and amount of irrigation, crop growth stage and water/nutrients needs detected in real time, and commercial yield monitors and remote sensors to map crop production precisely. An example includes a real-time irrigation scheduling program for supplementary irrigation that includes a reference crop evapotranspiration model, an actual evapotranspiration model, a soil water balance model, and an irrigation forecast model, all combined using a mixed linear program [63.67, 68].

Low-cost microprocessor and infrared sensor systems for automating water infiltration measurements [63.69] are important in controlling crop yields and delivering water and agricultural chemicals to soil profile. Control of nutrients with sensors enables optimization of irrigation and fertilization management systems, useful for reducing environmental impact caused by runoff of nutrients into surfaces and groundwater by using ion-sensitive field-effect sensors [63.70].

A wireless in-field sensor-based irrigation management system was developed to provide variable-rate irrigation. Variable-rate irrigation was controlled by a computer that sends control signals to irrigation controllers via real-time wireless communications based on field information and GPS positions of sprinklers [63.55, 66, 71, 72]. A self-propelled linear sprinkler system equipped with a DGPS and a program logic controller was remotely controlled by a base computer [63.72, 73], using a closed-loop irrigation system to determine the amount of irrigation based on distributed soil water measurements (Figs. 63.8 and 63.9). The system was operated by a program logic controller that controlled solenoids to turn sprinkler nozzles on and off. Variable-rate application was implemented by regulating pressure into a group of nozzles.

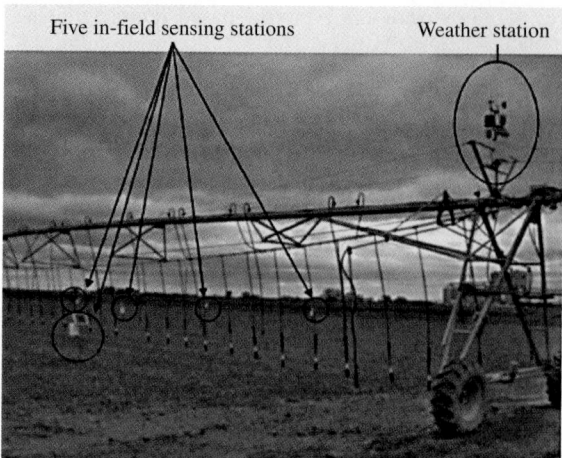

Fig. 63.9 Five in-field sensing stations and weather station mounted on the linear irrigation cart (after [63.66])

To control small areas in field irrigation, solid-set sprinkler and micro-irrigation can be controlled using centralized or distributed irrigation controls [63.74]. Architectures of distributed sensor networks for site-specific irrigation automation combining smart soil moisture sensors and sprinkler valve controllers have been developed [63.75] and are commercially available (e.g., Irriwise, Netafim). This can be expanded to closed-loop control for automated irrigation based on in-field sensing feedback of plant and soil conditions. Further developments include spot-spraying of herbicides based on real-time weed detection using optical sensors. Growers using recirculating systems often choose to sterilize the drain water before sending it back to the plants. One of the following two methods is often used: ultraviolet (UV) sterilization or ozone sterilization. Automated real-time polymerase chain reaction (PCR) system for detecting pathogens in irrigation water has been developed [63.76].

63.3 Greenhouse Automation

The greenhouse environment is a relatively easy environment for introduction of automated machinery due to its structured nature. Hence, the automated system must deal only with the variability of the agricultural product. Therefore, the development of systems is easier and simpler. Automation systems for greenhouses deal with climate control, seedling production, spraying, and harvesting as detailed in the following sections.

63.3.1 Climate Control

Greenhouses have been developed during the 20th century to keep solar radiation energy, to protect products from various hazardous natural climates and insects, and to produce suitable environments for plants by use of $100\,\mu m$ plastic film or $2-3\,mm$ glass plates. Advances in sensors and microcomputers have led to modern greenhouse operations that include control of climate, irrigation, and nutrient supply to plants to produce the best conditions for crop growth in an economical way. Environmental control enables year-round culture and shorter cultivation periods. This section outlines greenhouse environment control and automation.

Parameters and Sensors for Environmental Control

Light. Generally, there are two types of light: sunlight and artificial light from lamps. Visible light (400–700 nm, photosynthetically active radiation) is important for plant growth. Photosynthetic photon flux density (PPFD, measured in $\mu mol\,m^{-2}\,s^{-1}$) from photon sensors is appropriate when light intensity is measured for plant growth, while intensity of illumination (lux) is measured based on human sensitivity [63.77]. Although intensity and color temperature of sunlight vary from time to time and from place to place, artificial lighting devices can change them more drastically. There are several popular lighting devices: incandescent lamp, fluorescent lamp, high-intensity discharge lamp (HID lamp: Hg lamp, Na lamp, metal harido lamp), light-emitting diodes (LED), electroluminescence (EL), hybrid electrode fluorescent lamp (HEFL), and others. It is necessary to use them based on size, shape, efficiency, light intensity, life, color rendering, and color temperature of the lamp [63.78].

Temperature and Humidity. Heating and cooling in greenhouse are important for plant growth. Due to the amount of energy consumed for these operations, their control is critical. Electric heaters are used when it is necessary to specifically heat local sections such as in seedling production. Radiation in greenhouses that use sunlight can cause high air temperatures. Hence, cooling is necessary. To reduce cooling costs, curtain, infrared absorption glass (80% transmittance in visible region and 20% in infrared region), watering on glass roof, whitening the cover material, fan and pad system, fan and mist system, fog and fan system, and other methods are employed [63.79]. The thermocouple sensor is a popular measure of air temperature, while a thermo-camera and other radiation thermometers can measure radiant energy from plant parts or material bodies. Several types of humidity sensors are available: elemental devices whose electrical resistance, capacitance, or impedance is changed with humidity change. The sensors can measure 10–90% relative humidity. Humidity in greenhouses is influenced by air

temperature control, transpiration from plants, water evaporation from soils, and other effects; for example, the fog and fan system can decrease temperature by 2 °C and increase humidity by 20% as compared with external air [63.80]. To reduce humidity, an electric cooling machine is sometimes used, while air ventilation is the simplest method. Thus, humidity control also requires compensation of temperature change. When greenhouse environments are controlled, both heat balance and moisture budget must be considered. PID and adaptive control methods have been developed for temperature and humidity control [63.81–83].

CO_2 Concentration. Plants absorb CO_2 and transform it into sugars and then into new plant tissue [63.84]. Every gram of CO_2 fixated by the plant yields around 10 g of new plant material. This so-called photosynthesis (or CO_2 assimilation) requires good light and suitable growing conditions. Plants consume more CO_2 under more light and also at higher CO_2 level. By CO_2 enrichment the CO_2 uptake can be increased. The effect of CO_2 on yield is proportional to the amount of time of CO_2 enrichment.

CO_2 uptake depends on the crop, the leaf area, and environmental conditions such as soil moisture and atmospheric humidity. It is expressed in gram CO_2 gas per m^2 ground area per hour ($g\,m^{-2}\,h^{-1}$). CO_2 uptake varies from 0 during very poor conditions to about $5\,g\,m^{-2}\,h^{-1}$ under excellent light conditions, and up to $7\,g\,m^{-2}\,h^{-1}$ under excellent light conditions combined with high CO_2 levels. At night no CO_2 is taken up; in contrast, plants produce CO_2 due to respiration. Hence the CO_2 level in a closed greenhouse naturally increases overnight to above-ambient levels.

Ventilation influences the CO_2 level, which has three situations:

1. CO_2 depletion: the CO_2 level is below ambient. Any leakage or ventilation will bring CO_2 into the greenhouse. Ample ventilation can prevent CO_2 depletion.
2. Elevated CO_2 levels due to CO_2 enrichment. CO_2 gas will rapidly be lost during venting, depending on the vent opening, wind speed, and CO_2 level.
3. When the CO_2 level in the greenhouse is equal to the level outside. The influx of fresh air plus the CO_2 supply exactly compensates the CO_2 absorption. In this situation there is no CO_2 loss. The CO_2 demand equals the CO_2 absorption by the plants plus the CO_2 lost by leakage or ventilation.

However, the benefits of CO_2 enrichment should outweigh the costs. This depends on the yield increase due to CO_2, as well as on the price of the produce. Moderate CO_2 enrichment is sometimes more economic than excessive enrichment. CO_2 enrichment should not go beyond 1000 ppm, as it is not beneficial for the plants and is unnecessarily expensive. Sensitive plants (e.g., young or stressed plants, sensitive species) should not be exposed to more than 700 ppm CO_2. Too high CO_2 levels cause partial closing of the pores in the leaves, which leads to low growth. Also, at higher CO_2 concentration, there is higher risk of accumulation of noxious gases that can be present in the CO_2 gas.

Air Flow. It is important to keep uniform temperature, humidity, and CO_2 in the greenhouse for proper plant culture and uniform growth. Air flow in greenhouses is achieved in different ways depending on the greenhouse structure. Natural ventilation is usually used due to its low costs. However, control of airflow with natural ventilation is limited. Therefore, it is necessary to analyze natural ventilation properly and increase ventilation efficiency. Natural ventilation is driven by pressure differences created at the vent openings both by wind and/or temperature differences. Prediction of air exchange rates and optimization of greenhouse design requires complicated models due to the coupling and nonlinearities in the energy balance models. Additional controls of air flow include on/off control of fan ventilation systems, side openings, and water sprayers [63.85] with recent developments in rate control achieved by PID or fuzzy-logic control.

Control Methods
Greenhouse climate control requires consideration of many nonlinear interrelated variables. Control models should take into account weather prediction models, crop growth models, and the greenhouse model. The following methods have been used for control: classical methods (proportional integral derivative control, cascade control), advanced control (nonlinear, predictive, adaptive [63.86]), and artificial intelligence softcomputing techniques (fuzzy control, neural networks, genetic algorithms [63.87, 88]). Control is implemented with programmable logic controllers or microcomputers. Climate controllers that use online measurements of plant temperature, and fruit growth and quality, to estimate actual transpiration and photosynthesis will be the future development. This will enable the development of closed-loop systems that use the *speaking plant* as the feedback for the control system and thereby result in

effective control of the greenhouse climate [63.89, 90]. Effective control of the greenhouse climate must also incorporate long-term management plans to increase profitability and quality [63.91].

63.3.2 Seedling Production

Seedling production is one of the key technologies to grow high-quality products in fruit and vegetable production. Seedling operations such as seed selecting, seeding, irrigating, transplanting, grafting, cutting, and sticking have been mechanized or automated [63.77]. A fully automatic seedling production factory has been reported as a part of a plant factory [63.78], while a precise seeding machine which can seed in the same orientation has also been developed [63.79]. Several grafting robots and robots for transplantation from cell tray to cell tray or to pot have been commercialized. Herein, a grafting robot and a cutting sticking robot will be described as examples.

Grafting Robot
Grafting operations are conducted for better disease resistance, higher yield, and higher-quality products. Opportunities for the grafting operation are recently increasing, because of the agricultural chemical restrictions introduced to improve food safety and sustainable agriculture in the world. As the demand for grafted seedlings increases, a higher-performance model or a fully automatic model of the grafting robot is currently expected, while semiautomatic models have been commercialized since about 20 years ago. Grafting involves the formation of one seedling by uniting two different kinds of seedlings, using the side of the root of one seedling and the side of the seed leaf of the other. The side of the root of a seedling is called a stock and the side of the seed leaf, a scion. In order to graft a watermelon or a cucumber, a pumpkin is frequently used as a stock. The grafting method shown in Fig. 63.10 is called the *single cotyledon grafting method*, and is adopted as the operation process of a grafting robot for cucurbitaceous vegetables. For the stock, one seed leaf and its growing point are cut off. For the scion, the side of the root side is cut off diagonally at the middle of the hypocotyl, and the side of the seed leaf which contains the growing point is used.

Grafting operation of different kinds of plants is carried out by joining the stock and the scion using a special clip as an adhesive. Although stock seedlings and scion seedlings are hung up on spinning discs and supplied synchronously in some robot, mechanical fin-

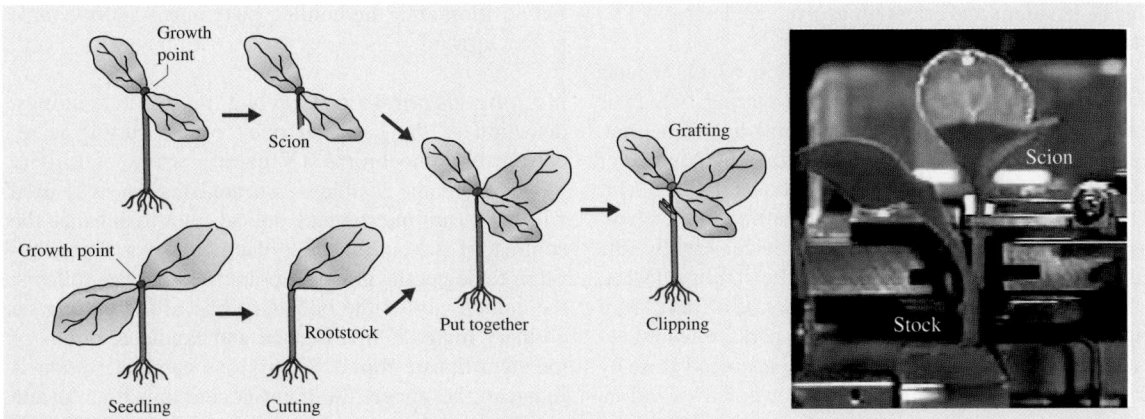

Fig. 63.10 Single cotyledon grafting method and an actual grafted seedling with mechanical fingers in the grafting operation of a robot (after [63.84, 89])

gers handle the seedlings, as shown in Fig. 63.10. Two operators hand the stock and scion to the mechanical fingers individually. At the stock cutting section, the shoot apex which contains one of the seed leaves and growing points are cut off by a spinning cutter which spins in the diagonally upward direction. On the other hand, at the scion cutting section, the side of the root is cut off at the hypocotyl by a spinning cutter which spins in the diagonally downward direction. After removing the useless part, the stock and the scion are gripped by the gripper and sent to the clipping section, where they are joined and fixed by the clip. The most important points in grafting operations are to cut the seedlings at the proper points and fix the stocks and scions precisely. To accomplish a higher success rate, the stock and scion should be properly hung for the spinning cutters when seedlings are handed into the mechanical fingers. The success rate of the grafting robot is 97% and the robot can perform grafting operations ten times faster than human workers [63.77].

Cutting Sticking Robot

Cutting sticking operations are often conducted in flower production in order to enhance productivity by using cuttings obtained from mother plants. Currently, humans manually stick the cuttings, however, the operation is monotonous and requires a lot of time and labor. A semiautomatic and a fully automatic chrysanthemum cutting sticking systems [63.80, 81] have been developed so far. In this section, a fully automatic system for chrysanthemum will be introduced, because it has a function of recognizing complicated-shaped seedlings by machine vision.

Robotic Cutting Sticking System. A prototype robotic cutting sticking system (Fig. 63.11) mainly consists of a cutting-provision system, machine vision, a leaf-removing device, and a planting device. The figure includes the latter three sections. The flow of cutting sticking operation is as follows: first, a bundle of cuttings is put into a water tank for refreshment because the cuttings are usually stored in a refrigerator for about a week until some amount of cuttings need to be prepared through picking from mother plants. The cuttings are floated on the water and spread out by adding vibrations to the tank. After refreshment in water and spread enough in a while, the cuttings are picked up by a manipulator based on information about the cut-

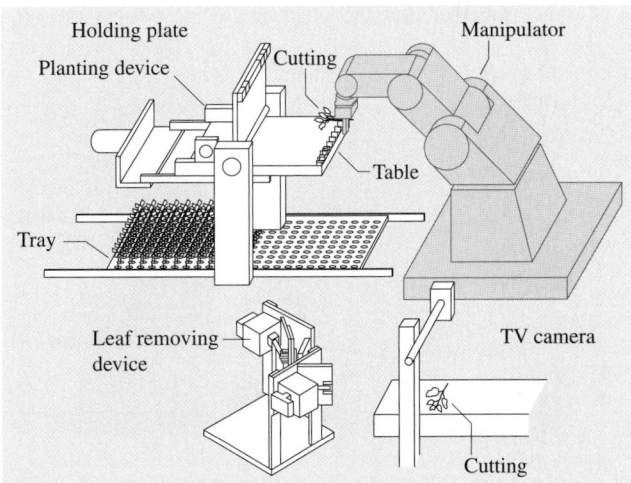

Fig. 63.11 Chrysanthemum cutting sticking system (prototype)

tings – positions and orientations from a television (TV) camera installed above the water tank.

Secondly, another TV camera (Fig. 63.11) detects the position and orientation of the cutting, which is transferred to a table from the water tank by the manipulator. The TV camera indicates the grasping position of the cutting for another manipulator, shown in Fig. 63.11. Thirdly, the manipulator brings the cutting to the planting device via the leaf-removing device. Finally, the cuttings are stuck into a plug tray by the planting device.

The leaf-removing device consists of a frame with cutters, a movable plate with rubber, and a solenoid actuator. The movable plate is driven to open and close by the solenoid actuator in order to cut lower leaves and arrange the shape of upper large leaves by chopping them with the cutters. Two identical devices are placed at an angle of 90° to cut the leaves completely since each leaf emerges at an angle of 144° from the main stem. Parts of the upper large leaves and lower petioles are cut by closing the movable plate. After this operation at the first device, the cuttings are moved to the second device and other leaves desired for removal are cut.

The planting device mainly consists of a table to place the cuttings in a row and a holding plate which opens and closes. The holding plate is driven to open and close by a motor which is mounted on the table. The table and the plate are driven in linear motion by another motor and a screw, and are rotated by a motor. A cell tray is set below the planting device. The holding plate closes after ten cuttings are placed on the table since a row of the tray has ten cells. The table rotates until it is perpendicular to the tray and moves downward. The ten cuttings are stuck into the tray together and the planting device adopts the ini-

tial position after the holding plate opens and the table moves upward.

Machine Vision. To pick up and transfer the cuttings, detection of the grasping point of the cutting is required. A monochrome TV camera whose sensitivity ranges from the visible to infrared regions was used with a 850 nm interference optical filter to enhance the contrast of the cutting on a black conveyor. The algorithm to detect the grasping point [63.82] is as follows: the complexity of the boundary line of the cutting on a binary image is investigated and candidate points of the stem tip are found. If only one candidate point is found in the image, the point is determined as a stem tip. When there are more than two candidate points, the complexity of the boundary line around the candidate points is detailed and points which are not adapted to conditions of the main stem are removed. The condition is that boundary lines around the stem tip have a lot of linearity. If only one candidate point remains after processing, the point is determined as the stem tip. In the case of plural points remaining, the whole boundary line of the cutting is detailed, the region of leaves is detected, and a candidate point which has a certain distance from the region of leaves is determined as the stem tip. When no point meets this condition or when more than two points remain even after the processing, the cutting is transferred back to the first stage, because it is too risky to determine the stem tip in these cases. The grasping point was defined as the position 10 mm above the stem tip. Experimental results indicated that about 95% cuttings are satisfactorily detected with no missed detection, and all remaining cuttings were transferred back.

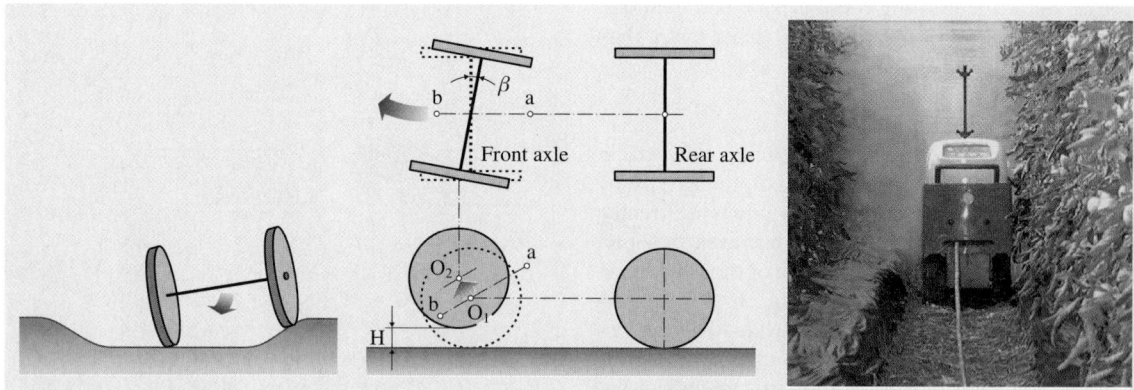

Fig. 63.12 Self-heading-correction mechanism and an unmanned sprayer (courtesy of Maruyama MFg., Co. Inc.) (after [63.84, 89])

63.3.3 Automatic Sprayers

Chemical control is required for crop production in controlled environments, and automation of the chemical spray is desirable to minimize exposure of chemicals. Spraying robots have been commercialized so far [63.83, 92, 93]. A key technology of the robots is autonomous control of the vehicle. Figure 63.12 shows the principle of the self-heading-correction mechanism. The front axle can be turned freely around an axis A–B which is fixed to the body diagonally. Assuming that a front wheel on one side runs on a ridge (it is off course), the center of this front wheel shifts from O1 to O2. At the same time, the other front wheel moves down and back. Consequently, the resulting steering angle β causes the vehicle to descend from the ridge, correcting its moving direction by itself. In the case where both a front and a rear wheel run on a ridge at the same time, the effect of the heading correction will be reduced because the steering angle β may be smaller. To obtain an appropriate steering angle, the rear tread is 35 mm shorter than the front tread. An unmanned sprayer is shown in Fig. 63.12.

Another method is called *electromagnetic induction type*. Induction wires are laid down under ridge aisles and/or headland and a vehicle with an induction sensor detecting the magnetic field created by the wires can automatically travel along the wires. When the vehicles move to the next ridge aisle in narrow headlands in greenhouses, several methods have been reported: a pivot shaft that comes out to make the turn, four-wheel steering, an additional rail system to convey the vehicle to the next aisle, and a manual method. In orchards, automatic speed sprayers using induction wires and induction pipes were developed in 1993 and 1994, respectively. A method that uses a remote-controlled helicopter has been very popular in the fields.

63.3.4 Fruit Harvesting Robots

It can be said that the history of agricultural robots started with a tomato harvesting robot [63.94]. There has been much research on fruit harvesting robots for tomato, cherry tomato, cucumber, eggplant, and strawberry [63.95–100]. Vegetable harvesting robots have also been investigated, but there is no commercial robot yet. The main reasons limiting commercialization of harvesting robots are low success rates due to diversity of plant properties, slow operational speeds, and high costs associated with the seasonal affect. How-

Fig. 63.13 A tomato harvesting robot

ever, practical use of harvesting robots is expected in the future.

Tomato Harvesting Robot

Research on the first tomato harvesting robot started at Kyoto University in 1982 and several different types of tomato harvesting robots and their components have been developed. A cluster harvesting robot is now under development. The main components of many of the tomato harvesting robots are a manipulator, end-effector, machine vision, and traveling device, as shown in Fig. 63.13. The robot automatically travels between ridges and stops in front of a plant using photosensors and reflection plates on the ridges, which can give the location of the robot in the greenhouse. When the traveling device stops, a machine-vision system measures fruit color and location, the manipulator approaches the cluster, and an end-effector picks a fruit. After completing the operation at the location, the robot moves to the next location of the reflection plate.

Phytological Characteristics of Tomato Plant

Most tomato plants for the fresh market are usually grown on a vertical plane with supports or with hanging equipments until many fruit clusters are harvested. However, high-density single-truss tomato production systems (STTPS) have been reported [63.101]. In addition, an attempt was conducted to grow the tomato plant upside down on the tomato production system because of the smaller labor requirement for plant training and ease of mechanical operation. Some varieties are for individual harvesting while others are for cluster harvesting. Some varieties produce round-shaped fruits and longer fruits, depending on the season. There are also many fruit sizes. Fruit clusters are supposed to grow outwards due to a growth rule, but the main stem sometimes twists, which causes random cluster direction so that tomato fruits may sometimes be hidden by leaves and stems. When a robot is introduced to the production system, it should be adaptable to plant diversity.

Manipulator

The basic mechanism of a manipulator depends on the configuration of the plant, the three-dimensional (3-D) positions of its work objects, and the approach paths to the objects. In the first attempt to robotize the tomato harvesting operation, a five-degree-of-freedom (DOF) articulated manipulator was used [63.98], and a seven-DOF manipulator was investigated for harvesting six clusters [63.102]. However, the Dutch-style growing system has been popularly introduced to large-scale greenhouses throughout the world, and target fruit are always located at a similar height. Therefore, a selective compliant robot arm (SCARA)-type manipulator can be used. When the fruit cluster is transferred to a container quickly, cluster swing damping is required.

End-Effector

The fruit cluster has several fruits and their peduncles have joints in many varieties of tomato plants. When a human harvests ripe fruit one by one in the cluster, he/she can pick them off easily by bending them at the joints instead of cutting. To harvest the fruit, several end-effectors have been developed; Fig. 63.14 shows one of them [63.103]. A 10 mm-thick rubber pad is attached to each finger plate to protect the fruit from slipping and damage. The length, width, and thickness of a finger plate are 155, 45, and 10 mm, respectively. The gripping force exerted by the finger plates can be adjusted from 0 to 33.3 N, while these finger plates grip fruits ranging from 50 to 90 mm in diameter. The suction pad was attached to the end of a rack, which is driven back and forth by a DC motor and a pinion between the finger plates. The speed and stroke of the suction pad motion are 38 mm/s and 80 mm, respectively. The suction pad can be moved forward up to 43 mm from the tips of the finger plates. The moving distance and stopping position of the pad can be detected by a rotary-type potentiometer. Two limit switches are attached to both ends of the pad stroke in order to prevent the pad from overrunning.

Machine Vision

A traditional method of detecting 3-D locations of target fruits is feature-based stereo vision. A pair of identical color cameras acquire images and discriminate

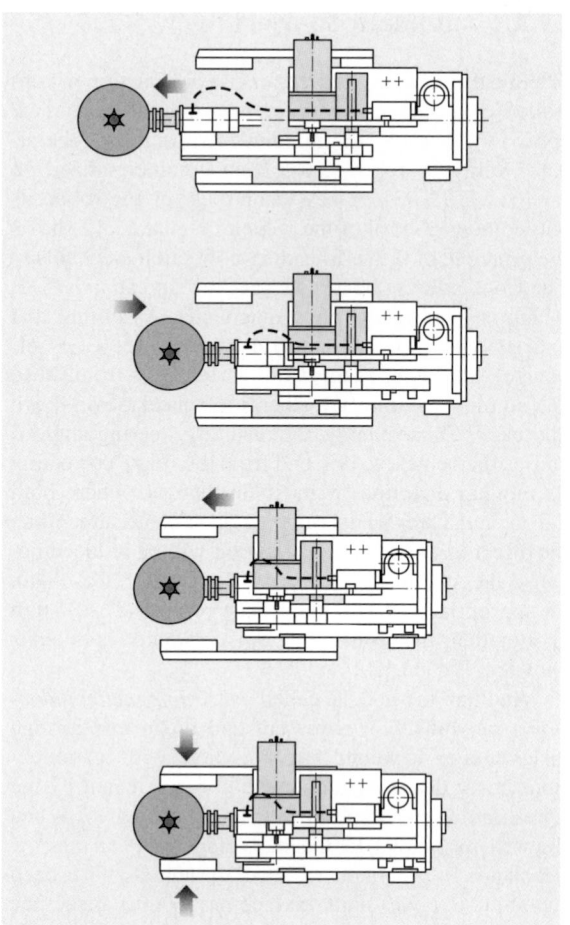

Fig. 63.14 An end-effector

red-colored fruits. Based on the disparity of fruits on both images, the depth of the target fruit can be calculated. Although a small error in the 3-D location occurs because of hidden parts of the fruits, the suction pad can tolerate these error. It is not easy for stereo vision to detect all fruits locations when a corresponding problem happens due to hidden fruits and many fruits in the images. In this case, a 3-D laser sensor or area-based stereo vision may help detect the fruit depths.

Traveling Device

Figure 63.13 shows a four-wheel-type battery car on which a tomato harvesting robot is mounted. The traveling device moves and stops between the ridges and turns at the headlands to go to another ridge. In Dutch-style large-scale greenhouses, two heating pipes are usually used. It is easy to introduce a rail-type traveling device as these pipes can be used as rails. Rail-type traveling devices (manual or self-propelled) are already used for leaf picking, manual harvesting, spraying, and many other operations in greenhouses.

63.4 Animal Automation Systems

Automation of animal husbandry systems includes the development of environmental control systems, automated weighing and monitoring systems, and automated feeding systems.

Climate control of housed animals has an important influence on the productivity and health of the animals and therefore its control is very important [63.104]. However, this is a difficult and complicated task due to the nonlinear effects of the animals on the temperature and humidity conditions inside animal housing buildings [63.104]. In conditions where animals are housed outside, control is further complicated due to changing environmental conditions. Air-quality and environmental monitoring is important for environmental protection aspects and hence is gaining increasing attention and importance.

Devices for electronic animal identification and monitoring became available in the mid 1970s and have enabled implementation of advanced management schemes [63.105], specifically for livestock and swine management. The ISO standardization of injectable electronic transponders in the late 1990s expanded applications to all animal species [63.105]. Several sensors have been developed to provide individual animal parameters such as size, weight, and fat. These parameters are used for management decisions. The current new generation of sensors enable health and production status monitoring, both improving animal welfare and ensuring increased food quality and safety. Recent developments include acoustic passive integrated transponder tags using micro-electromechanical systems (MEMS) technology [63.106]. Tags may be used for tracing animals from growth to final processing for quality control and food security purposes [63.106].

The main expense in animal production systems is food intake. Automated feeding systems decrease production costs while ensuring that animals receive necessary nutrient ingredients. Group and individual feeding systems have been developed to measure and control food intake.

Production, health, and welfare controls are being introduced into modern farms using advanced information systems. Data from multiple sensors at the individual and group levels are taken on a daily basis for advanced monitoring and control. Various systems will be presented in the following sections.

63.4.1 Dairy

The dairy industry is probably the most automated agricultural production system, with almost all processes, from feeding to milking, being completely automated. In the dairy industry, many maintenance routines such as milking, feeding, weighing, and online recording of performance are fully automated on an individual animal basis. Optimal management is defined as producing maximum milk yield while minimizing costs. The computation and data-storage capacity of computers theoretically enable sophisticated decision-making to underpin the automated processes in order to obtain optimal individual and herd performance. These include automated feeders, sensors that measure daily activities of cows, and online automated parlor systems for recording milk production and quality. Reproduction monitoring includes systems for timing of insemination based on oestrus detection. Health care systems include detection of mastitis. The objective is to fully automate every process from feeding to milking to reduce production costs and maximize milk yield.

The physical process of feeding and recording actual feed consumption is based on feed administration of concentrates and roughage, ration composition, and feed calculation for an individual cow or a group of cows. Analysis of performance data indicates that cow performance under a uniform rationing regime is consistent in trend but varies in magnitude, and therefore an optimal feed policy, in terms of *efficient* rationing of concentrates, should be on an individual basis [63.107]. An alternative approach, the sweeping method, is based on average values for the herd. This can cause cows not to reach maximum milk yield because of insufficient concentrate ration or imply that excess feed be consumed since there are cows that would have reached their maximum milk yield with a smaller concentrate ration. Both result in redundant financial expense. Due to the advent of technology, the farmer is able to allocate a different amount to each cow using individual computer-controlled calf feeders [63.108] and

Fig. 63.15 A controlled automatic fodder consumption and feeding system (after [63.109])

integrated real-time control systems for measuring, controlling, and monitoring individual food intake of free-housed dairy cows [63.109]. Individual allocation decisions are made according to each cow's performance, Performance parameters include the individual cow's output (milk yield and composition) and measurements of physiological variables including body composition [63.110], shape, and size. An example of a system consisting of 40 feeding cells is shown in Fig. 63.15. Each cell comprises an identification system, a fodder weight system, and an automatic opening and closing yoke gate [63.109]. Each feeding stall consists of a feeding trough, an electronic weight scale and central processing unit (CPU), identification system, presence sensor, and a cylinder with valve. All components are connected to a programmable logic controller (PLC) which processes the data and activates the electropneumatic actuators. The data is backed up to a management computer. The management computer is also used as a monitoring station and a basic man–machine interface for defining basic operations and preliminary data analysis. The specific yoke design allows the cow's head to enter the yoke gate without enabling access to the fodder. This places the radiofrequency identification tag on the cow's ear close enough to the antenna and simultaneously activates the proximity sensor (by the cow's head). If the cow is allowed to eat according to the predetermined conditions, the PLC records the current scale's weight and the yoke gate bar is lowered by the associated electropneumatic cylinder. The cow may then push its head into the fodder trough and feed. The scale measures and records the weight of the fodder at predefined intervals. Each CPU scale is connected to the PLC directly via bi-

nary code to decimal (BCD) so no time delay is caused by weight transmission. A restriction bar on the fodder trough prevents the cow from pushing its head up, thereby preventing spillage of fodder. The use of presence sensors in the yoke appeared to be very important to determine if the cow had left the yoke station. The feeding troughs were arranged in a row to enable convenient dispersal of fodder (into the containers) by the passage of a semiautomated fodder dispersal wagon.

Several methods have been developed for automatic weighing of cows. Cows are weighted as they exit the milking parlor so as not to interrupt their daily regime. The motion of cows creates measurement problems, including changes along the scale due to applied forces, crowding of cows on the scales, and significant variations between cows and between the same cow at different times of the day or on different days. Dynamic weighing of cows is a common practice in many commercial farms, achieved by filtering the measured signal and averaging it or recording the peak value as the cow transfers its weight [63.111–113] using physical mathematical models that simulate cow walking [63.114].

Milking cows is a complicated task due to the physics combined (teat treatment, control of the milking unit) and variable biological components (milk secretion, udder stimulation) including the risk of infecting the udder with pathogen microbes [63.115]. Although the first proposals for mechanical milking were presented over 100 years ago, milking machinery became common only in the early 1950s, with completely automatic milking systems being introduced in the 1990s [63.116]).

First steps in automating the milking process included detection of end of milking and automatic teat cup detaching [63.115]. Various optical, capacitive, and inductive sensors were developed to detect low milk flow, which indicated end of milking [63.117]. Mechanized stimulation of udder was achieved by using pneumatic and electronic pulsators. Continuous individual variation of vacuum level, pulse rate, and pulse rate for each milking unit was developed. Milk yield recording is implemented using tipping trays and volumetric measuring systems, with many sophisticated measuring systems to separate air from milk to improve accuracy. Automatic milking requires automatic application of teat cups. Ultrasonic sensors, a charge-coupled device (CCD) camera, and a laser are used to locate the teats to control in real time the arm to adapt to the variations in teat positions, spacings, and shape and to the motions of the cow during teat attachment. In most systems a two-stage teat location process has been applied.

First, the approximate teat positions are determined by dead reckoning using body position sensors, ultrasonic proximity sensors or vision systems. The final attachment is achieved by fine-position sensors using arrays of light beams mounted on the robot arm. Automatic checks of udder condition and milk quality include online milk analysis. Milk quality is a critical parameter both from an economic point of view and from health perspectives [63.116]. Measures include conductivity, temperature, and color of milk, integrated with yield information. Biosensors have been used to measure antibiotic residues, mammary infection components, and metabolites including the development of electronic samplers that enable real-time measurements [63.118]. Online inline milk composition sensors measure in real time during milking the concentrations of fat, protein, and lactose, and indicate the presence of blood and somatic cell count (SCC) based on near-infrared analysis [63.119].

Various teat cleaning systems, including brushes and rollers or separate teat-cup-like cleaning devices, have been developed. In addition, systems for cleaning the complete system (circulation cleaning, cleaning with boiling water, cluster flushing [63.116]) are applied.

Robot milking (see Fig. 63.16 for an example), introduced in the early 1990s by several commercial companies (e.g., Lely, DeLaval, GM Zenith, Fullwood Merlin), provides increased yield by increasing the frequency of milking and improved milk quality.

Automatic health measurements during automatic milking include leg health measurement and respiration rate measurement [63.120]. Lameness detection is important due to the important welfare, health, and eco-nomic problems it causes. Leg health is also measured by measuring the dynamic weight or load of each leg while the cows are weighed on scales at the exit of the milking parlor. Several techniques have been developed including pedometers, activity meters worn around the neck, force plates that measure reaction forces on walk-through weighing systems [63.120–123], and increased respiration rate measured using laser distance sensor [63.120]. Ultrasonic back-fat sensor can provide information about the health or growth status of the livestock [63.124].

Another important health measure is mastitis, a main reason for reduced milk yield and early losses in cows, caused by the biological activity of microbes. It can be detected by counting the number of somatic cells in milk. Various methods have been developed to measure it accurately using electrical conductivity measurements, body temperature, and milk temperature. Recently inline near-infrared sensors have been developed to measure milk conductivity and milk temperature of each seperate quarter (a sensor is connected to each udder cup) [63.119].

The primary direct parameter to detect oestrus is concentration of milk hormones (progesterone), indicating the fertility status of the cow. However, it is commonly measured only in laboratories based on samples, although biosensors have been developed for its measurement [63.125]. Several indirect parameters have been developed into automated systems, including electrical conductivity of vaginal secretion, milk temperature, and cow behavior including cow activity measurement using pedometers, heart rate, etc. Improved measurement was achieved by combining information from several parameters (e.g., combining cow activity with milk yield, feed intake, milk temperature).

Behavior measurement has been achieved using different systems: a radar-based automatic local position measurement system for tracking dairy cows in free stall barns [63.126], global positioning systems for measuring grazing behavior (*Turner* et al. [63.127], video measurements [63.128], and automatic tracking systems based on magnetic induction [63.129].

Environmental control systems in the dairy industry are less common since cows are located in barns that are open, shaded or partially shaded. Systems developed include automatic cooling using fans based on online imaging systems that detect crowding (Fig. 63.16) and microclimate and gas emissions in cold uninsulated cattle houses [63.130].

Management information systems that combine herd and individual health and production param-

Fig. 63.16 Lely milking robot in an open barn with fans controlled when cow crowding is detected

eters [63.131, 132] are important to ensure efficient automation. Further advances in design and management of livestock environments will require development of sustainable livestock production systems accounting systematically for the environmental benefits and burdens of the processes using a lifecycle assessment process [63.133]. Strategies will need to be developed to regulate and reduce harmful gas emissions from livestock farms and land application of manure [63.134].

63.4.2 Aquaculture

Physiological rates of cultured species can be regulated by controlling the environmental conditions and system inputs. This yields increased process efficiency, reduced energy and water losses, reduced labor costs, and reduced stress and disease. Automation applications include algae and feed production, feed management, environmental controls such as filtration systems, and automated air-pressure control. An intensive water-quality monitoring program includes routine sampling (twice a week) and 24 h sampling (every 3 months) of

Fig. 63.17 Water channels and air distribution system. Air-flow rate is controlled by regulating air-blower frequency using readings of oxygen concentration in the fish tank

nitrogen (NO_3, NO_2), phosphate, pH, and temperature. Fish and shellfish biomass should be sampled and seaweed should be harvested [63.135, 136].

Automation usually exists in closed systems such as recirculated aquaculture systems, but it can also be applied to pond and offshore aquaculture systems. Intensive recirculating aquaculture systems (RAS) reduce land and water use at the expense of increased energy requirements for operating treatment processes to support high culture densities, often with the addition of pure oxygen (see, e.g., Fig. 63.17). The use of pure oxygen is usually expensive and requires considerable energy for dissolving in the water as well as for stripping off the carbon dioxide created by respiration. In conventional RAS design gas exchange and dissolved waste treatments (e.g., CO_2 stripping and ammonia removal by nitrification) are linked into one water-treatment loop. However, because excretion rates of CO_2 are an order of magnitude greater than ammonia excretion rates this design may result in toxic CO_2 concentrations. In addition, pressurized pumping and pure oxygen addition may increase the risk of gas bubble disease. Hence, low-head recirculating system that separate the gases treatment loop (oxygen and CO_2) from the nitrification and solid filtration treatment loop by using a high-efficiency airlift producing a bubbly flow are used [63.137]. The integrated pond system (IPS) concept suggests a novel solution for environmentally friendly land-based mariculture. The IPS recycles excreted nutrients (valuable nitrogen) through algal biofilters utilizing solar radiation for their photosynthetic processes [63.138].

Accurate size and shape information of wild and cultured fish population is important for managing the growth and harvesting process including feeding regimes, grading times, and optimum harvest time [63.139]. Information on both average weight and distribution is necessary for grading, feeding, and harvesting decisions [63.140]. Machine vision has been used to determine fish size [63.139, 141], mass [63.140]; color [63.141], weight, and activity patterns. The problems with image capture in ponds are the low contrast between fish, the dynamic movement of fish, and changing lighting conditions. Real-time in situ fish behavior quantification and biomass estimation has also been used for management decisions [63.142].

The cost of feed is usually the major operating cost in aquaculture [63.143]. Overfeeding results in leftovers, which leads not only to extra costs but also to poor water quality, causing additional stress and extra loads on mechanical and biofilters and oxygenation

Fig. 63.18 Aquaculture closed system with a feeding station. Feed quantities are calculated on a daily basis to each fish tank according to fish weight, water temperature, and growth rate

devices [63.143]. In addition feeding rhythms affect feed conversion rates and proximal composition of fish flesh. Automated feeding systems (see, e.g., Fig. 63.18) include timer-controlled feeders [63.144], demand feeders, and automated data-acquisition systems to assess fish feeding rhythm, and acoustic, photoelectric sensors to detect the turbidity of the effluent. Hydroacoustic sensors and machine-vision systems have been used to detect left-over pellets.

Future research should be directed towards engineering environmental monitoring and controlling recirculated systems, and the development of sustainable automated systems. Considering that sustainable development is probably the major challenge faced by aquaculture [63.145, 146], one should consider sustainability, which can be considered in three main categories: environmental, economical, and sociological [63.147]. Another perspective of sustainable development relates to resource utilization and external effects that are described by various indicators (mainly in physical terms [63.136]). This should include online reporting of system failures and automation of the final harvesting and grading process [63.148], thereby improving food safety and maintaining product quality.

63.4.3 Poultry

Poultry house controllers include sensors for internal and external temperature measurement, moisture, static pressure, feed lines, water consumption, and gas and vent box status [63.149, 150]. Additional automa-

tion equipment includes feed consumption monitoring equipment, bird weight scales, feed bin load sensors, gas meters, and water meters.

Physiological signals are important for health monitoring and behavior analysis. Several systems have been developed, including an implanted radiotelemetry system for remote monitoring of heart rate and deep body temperature, and multispectral image analysis for real-time disease detection [63.151]. An automated growth and nutrition control system has been developed for broiler production using an online parameter estimation procedure to model the dynamic growth of broiler chickens as a response to feed supply [63.152]. Image based bird behavior analysis can be can be used to develop time profiles of bird activity (movement, response to ventilation, huddling, etc.) as well as to compare activity levels in different portions of the house. Time profiles of bird activity can contribute to improved feeder and water design, and enhanced distribution of ventilation air to provide more uniform bird comfort [63.149, 150, 153].

Several mechanical poultry catching systems [63.154] have led to improvements in bird welfare in addition to manual labor reduction. Systems include [63.154]: rubber paddles that rotate onto the birds from above and then push the birds onto a conveyor belt which carries them back to a loading platform where they are deposited into crates, a hydraulic drive system that advances along the poultry house and picks up the birds with soft rubber-fingered cylinders that gently lift them onto a conveyor that transfers the birds to a caging system, and the Anglia Autoflow (Norfolk, UK) batch-mode catcher that shuttles birds from collection to a separate packing unit.

63.4.4 Sheep and Swine

Robot shearing operations have been developed and commercially applied in Australia [63.155]. The sheep is constrained with straps on a movable platform. Hydraulically position clippers using force feedback control the actual shearing of the wool. Path computations are continuously updated during the shearing process.

Several feeding systems exist in sow farming: commercial electronic feeding systems that feed one at a time by enclosing each sow as it eats, electronic sow feeding systems in loose housing environments that limit the feed ration [63.156], and a computer-controlled system that allows sows to feed from one of two feed formulations to meet their nutritional requirements while satisfying their need for satiety by using bulk ingredients providing automatic body weight and average daily weight gain [63.157]. An important indicator of animal growth and health is the animals' weight, in addition to its importance in determining readiness for market. Weighing has been accomplished using walk-through weighing based on mechanical scales and imaging systems [63.158]. Physiological variables measurements include body shape and size using image analysis [63.159]. Ultrasonic probes have been applied to measure back fat for monitoring animal growth and feeding regimes. A robotic system capable of holding a sensor and placing it on the pig while it is located in the feeding stall has been developed [63.160].

Real-time behavior and control of swine thermal comfort has been achieved using imaging systems [63.161]. Planning individual showering systems for pregnant sows to prevent heat stress [63.162] as been used in automatic shower cages to prevent waste water and improve efficiency. Automatic cleaning systems to reduce infections risks between batches of pigs has been used based on an intelligent sensor for robotic cleaning [63.163].

Recent environmental policies limiting the amount of nitrogen and phosphorus that can be applied in the field have led to the development of online analysis of pig manure systems, including mobile spectroscopy instruments in the visible and near-infrared wavebands [63.164].

63.5 Fruit Production Operations

Fruit production automated systems deal with all stages of production: growing (automated sprayers, weeders), harvesting, and post harvest operation (grading, sorting).

63.5.1 Orchard Automation Systems

Fruit production operations in orchards such as pruning, thinning, harvesting, spraying, and weeding have been mechanized and automated. Even when automation systems have been developed for the same variety of fruit tree, their components differ substantially because plant training systems, cultivation methods, climate conditions, labor conditions, and other conditions and situations differ from country to country. This section describes functions, mechanisms, and important observations of automation systems in orchards.

Fruit Harvesting Robots in Orchards
Several types of shakers are working in orange fruit orchards: trunk shake and catch, mono-boom trunk shake, canopy shake and catch, continuous canopy shake, and others. These shakers are used due to labor shortage, but the harvested fruits are only for processing into juice; they cannot be consumed in the fresh market because of unavoidable damage. Several types of orange harvesting robots that have manipulators with picking end-effectors and machine-vision systems have been reported in the USA, Japan, and European countries [63.165–169]. Figure 63.19 shows an articulated manipulator with three degrees of freedom (DOFs) mounted on the base attached to the boom.

It was developed by Kubota Co., Ltd., Japan. The advantage of the articulated manipulator is its compact size when folded up in a narrow space between trees. Figure 63.20 shows a prismatic arm with three DOFs driven by hydraulic power. Citrus trees have large canopies and many branches, twigs, and leaves. Since these can often be obstacles for fruit harvesting, research on robots with more degrees of freedom has also been reported [63.170, 171]. Color cameras were often used as sensing systems to detect fruit because citrus fruit have orange colors. Fruit locations are calculated by use of stereo vision, differential object size, vision servoing, ultrasonic sensors or a combination of them [63.172–179]. Their end-effectors have the function of rotating semicircular cutters so that they can cut peduncles in various directions.

Fig. 63.19 Orange harvesting robot (Kubota Co., Ltd.)

Fig. 63.20 Orange harvesting robot (University of Florida)

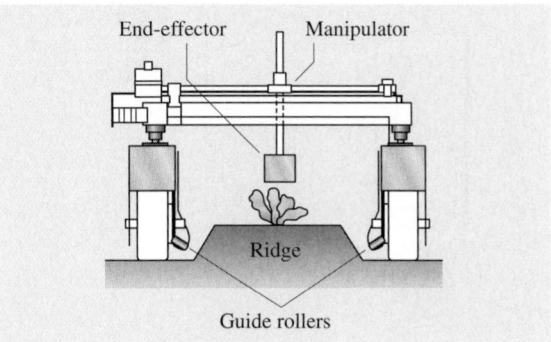

Fig. 63.21 A multioperation robot

Grape [63.180, 181], apple [63.182, 183], melon [63.184], watermelon [63.185], and other fruit harvesting robots [63.186] have also been studied. The basic mechanisms of the manipulators depend on fruit tree canopy size and shape. Grapevines in many of European and American countries are grown in crop rows, but those in Asian countries are grown on a trellis training system due to different climate conditions. Melons and watermelons are grown on the ground. There is research on fence-style training systems for orange trees [63.187] for higher-quality products. This approach of changing the training system takes a horticultural approach to accomplish a higher rate of success for harvesting robots. Since the harvesting operation is usually conducted once in a year in orchards, during a short period, a robot which can only harvest fruits is not economical. Therefore, other operations such as thinning, bagging, and spraying are needed for the orchard robot. Some of these functions are accomplished by replacing end-effectors and software [63.188].

Automation of Spraying and Weeding Operations

Control of disease, insect pests, and weeds is an essential operation to gain a stable high yield of crops and high-quality products. This operation includes biological, physical, and chemical methods. Chemical spraying is widely used in agricultural production environments. Today, technologies with high accuracy to spray only the necessary parts of the plant using a minimum amount of chemicals are required to protect workers as well as the environment.

A nozzle-positioning system for a precision sprayer was studied with a robust crop position detection system at Tohoku Experimental Station, Japan [63.189] under varying field light conditions in rice crop fields. The data from a vision sensor was transmitted to a herbicide applicator that is made up of a microcontroller, with slidable arms coupled with spray nozzles. Nozzles were driven to the optimal positions. The system was tested to evaluate its performance. It had high enough accuracy for use in Japanese rice crop fields. A fluid-handling system to allow on-demand chemical injection was developed for a machine-vision-controlled sprayer. The system was able to provide a wide range of flow rates of chemical solution [63.190].

Wiedemann et al. [63.191] developed a spray boom that could sense mesquite plants. Sprayers were attached to tractors and all-terrain vehicles. Controllers were designed to send fixed-duration voltage pulses to solenoid valves for spray release through flat-fan nozzles when mesquite canopies interrupted the light. The levels of mesquite mortality achieved were equivalent to those achieved by hand-spraying by ground crews.

The speed-sprayer (SS) has been widely used in orchards and its autonomous control is a main theme in the automation of the spraying operation. The *electromagnetic induction type* and *pipe induction type* were commercialized in 1993 and 1994 [63.192, 193]; both types require induction wires or pipes on the ground, underground, or above ground at 150–200 cm height and between tree rows. Induction sensors, safety sensors (ultrasonic sensor, touch sensors), and other internal sensors are installed in the unmanned SS, and autonomous control is conducted with fuzzy theory. Another method of SS control with genetic algorithm and fuzzy theory using GPS has been reported [63.194].

Figure 63.21 shows a multioperation robot with a three-DOF Cartesian coordinate manipulator and an end-effector [63.195]. When an end-effector shown in Fig. 63.22 is attached to the manipulator, it can weed

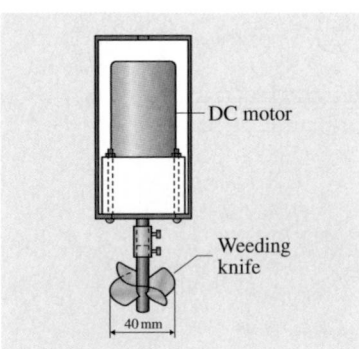

Fig. 63.22
A weeding end-effector

on the ridge between crops. Color images of the weed are fed to a computer from a color camera and the three-dimensional location of the weed is calculated using a binocular stereo method. Weed detection is conducted using color or texture difference between weed and soil or crops [63.196–198]. The end-effector was a weed knife with a spiral shape (4 cm diameter). This robot can also be a leaf-vegetable harvesting robot or a transplanting robot when the end-effectors were replaced [63.186].

63.5.2 Automation of Fruit Grading and Sorting

Because of the ever-growing need to supply high-quality food products within a short time, automated grading of agricultural products is getting special prior-

ity among many farmers associations. The impetus for these trends can be attributed to increased awareness of consumers about their health well-being and a response from producers to provide quality-guaranteed products with consistency. It is in this context that the field of automatic inspection and machine vision comes to play the important role of quality control for agricultural products [63.184–188]. Unlike most industrial products, quality inspection of agricultural products presents specific challenges because nonstandard products must be inspected according to their appearance and internal quality, which are acceptable to customers only for nondestructive methods [63.199]. Several sensors have been developed and applied for internal quality determination, including sugar content, acidity, rind puffing, rotten core, and other internal defects [63.190–194].

Fruit Grading System with Conveyors

Figure 63.23 shows an automated inspection system for quality control of various agricultural products, with fruits and vegetables being the main ones. As a representative of other agricultural products, discussion in this section is focused mainly on the orange fruit, a major agricultural product inspected by this system. The main components of the system for automated inspection and sorting can be outlined as follows:

1. Product reception from supplier
2. Container unpacking and dumping of products
3. Feeding of products to the conveyor line

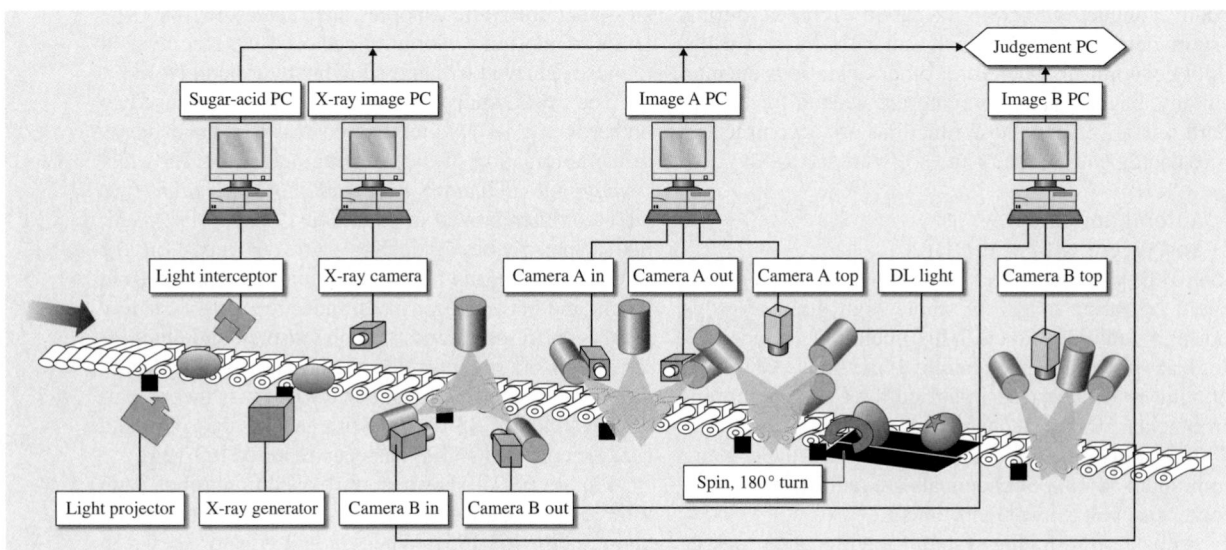

Fig. 63.23 A schematic diagram of the camera and lighting setup

4. Inspection for internal and external conditions and defects followed by assignment of quality rating
5. Weight adjustment and release of the inspected product into packing box
6. Labeling of grade and size using an inkjet printer
7. Box closure and sealing
8. Box transfer onto palette and loading ready for marketing

These features are integrated into an operational line that combines advanced designs, expert fabrications, and automatic mechanical control with the main objective of offering the best visual solutions and stable quality judgment.

Illumination and Image-Capture Devices

Illumination is one of the most important components for the machine-vision system to inspect products, because it determines the quality of images acquired, especially for glossy products whose cuticular layers are thick. Polarizing filters are sometimes used in front of lighting devices and camera lenses to eliminate halation on the acquired images [63.195]. A color CCD TV camera is often employed to sense light through photosensitive semiconductor devices, and the CCD array data is transferred by progressive scan mode to the frame storage area, representing an image of the scene [63.196]. The TV camera is equipped with one chip that transfers red–green–blue (RGB) analogue data and is set at a shutter speed of $1/1\,000$ s during inspection because line speed is usually 1 m/s. Image-capture boards with 8 bit level resolution and spatial resolution of about 512×512 pixels have been used to store and digitize video signals and output the data to computer memory for analysis or display on the monitor. Recently, a special image-acquisition device, a universal serial bus (USB) or a local-area network (LAN) is often used between TV camera and personal computer (PC) instead of the image-capture board, enabling image processing to be performed within $10-30$ ms.

Product Reception and Forwarding

The first step in the inspection procedure starts at the receiving platform, situated on the ground floor. Agricultural products are packed into containers by the farmers and delivered to the inspection factory in trucks. A folk lift is used to unload the containers on one pallet and deliver them to the depalletizer device that separates the containers automatically so that they are fed one by one to the conveyor, which propels them to the upper floor, where the main inspection line is located.

The depalletizer has a capacity to handle $1200-1400$ pallets per hour. After depalletization, the containers are handled by the dumper machine. The dumper is an automated machine that turns and empties the containers gently and then spreads the fruit on a belt conveyor. Using specialized rollers fruits are singulated so that they are fed singly to the roller-pin conveyor before processing. To acquire a complete view of the fruit the roller pins have been designed so that the fruit is always positioned at the center point.

Internal Quality Inspection

The first stage in the inspection sequence is to determine the sugar and acid contents using a near-infrared (NIR) inspection system. A special sensor determines the sugar content (brix equivalent) and acidity level of the fruits from light wavelengths received by specific sensors after light is transmitted through the fruit. The sensor photoelectrically converts the light into signals and sends them to the computer unit, where they are processed and classified. In addition, the internal fruit-quality sensor measures the granulation level of the fruit, which indicates its internal water content. Next, the fruit is conveyed to the x-ray imaging component, which inspects for biological defects such as rind puffing and the granulation status of the juice sacs. X-ray imaging operates by transmitting x-rays released from a generator through the orange fruit. The emitted optical x-ray image is relayed to the x-ray scintillator, an optical device that consists of a thin coat of luminescent materials through which x-rays are converted into the visible light of a normal image. The resulting image is captured by a monochrome CCD camera and copied to computer memory through an image-capture board.

Image Analysis

At the next operation, the fruit is conveyed to the third inspection stage where the main image processing and grading takes place using factory automation computers (Fig. 63.23). Six CCD cameras set in random trigger mode acquire images of the fruits as they are conveyed at constant speed. Firstly, two side cameras on the right side, placed equidistance from the central position of the fruit, capture images of the right side of the fruit. Next, two side cameras on the left side, again placed equidistance from the central position of the fruit, capture images of the left side of the fruit. Next, a top camera acquires the top surface image. Finally, the fruit is then spun through $180°$ around a horizontal axis by mechanically controlled roller pins, which facilitates the acquisition of another image of the lower surface by a sixth camera. Images are captured by CCD cameras after a trigger signal is received through the digital input/output (DIO) board. Sensors, wired to a sequencer box, are used to track the fruit position and the output is relayed to the DIO board from where the program reads the 8 bit output data.

Image processing is then executed and the following features are inspected:

1. Size (maximum and minimum diameter, area, and extrapolated diameter)
2. Color (color space based on hue, saturation, intensity (HSI) values and RGB ratios)
3. Shape (perimeter L, L^2/area, inflection point of outer surface, and center of gravity)
4. Bruise (based on intensity of blue color level and summation of color levels for blue, and R–G derived images)

Results of processing are written to a shared memory area using a memo-link from where a judgment PC makes decision about the quality of the fruit. Grading for quality is assigned by ranks of between four and six. Graded fruits are conveyed to the weight adjustment machine, which controls the total weight of oranges to be packed in a box according to preset values. Between four and eight weight rankings of fruits are used to fill different pack boxes. An automatic barcode labeler then prints the grade and size outputs on the box using an inkjet printer before the packing box is closed by the packing machine and then automatically sealed by the case-sealer machine. At the end of the product inspection line a robot-controlled palletizer completes the process by arranging the boxes onto palettes ready for onward loading onto trucks and transport to consumer markets.

Data Maintenance

An important feature of the grading system design is that it is adaptable to the inspection of different products such as potato, tomato, persimmon, sweet pepper, waxed apple, and kiwi fruits with adjustments only to the processing codes. Several lines for orange fruit inspection combined with high conveyance and high-speed microprocessors enable the system to handle large batches of fruit product at high speeds. Apart from quality inspection another objective of this system is to gather data about product performance. Identification of certain defects and counts can lead to the discovery of the cause and its severity. All data from receipt of fruits at the collection site through the analytical processes, packing, and shipping is stored in an office computer connected through a local-area network (LAN) to the corresponding factory automation PCs. Experienced personnel manage the data, based on which day-to-day operations can be monitored remotely. By making product performance records available on the Internet it will be possible to monitor performance online and provide a useful service to the customers. All in all, the customer of a product is the final judge of its quality. Therefore, keeping internal standards and specifications in line with customer expectations is a priority that is achieved through good relationships and regular communications with the customer.

Fruit Grading Robot

Based on the technologies of the grading system described in the previous section and robotic technologies, a fruit grading robot as shown in Fig. 63.24 was developed in 2002. This robot system has two three-DOF Cartesian manipulators, 16 suction cups as end-effectors, and 16 machine-vision systems consisting of 16 color TV cameras and 36 lighting devices. This can be applied to tomato, peach, pear, apple, and other fruits.

Operation flow with this grading system is as follows:

1. When products are received, four blocks of containers (a block consists of ten containers) are loaded on a pallet.
2. A block of the containers is lifted up to the second floor, where the main parts of the grading system work, and a container separator sends containers one by one to a barcode reader.

Fig. 63.24 A tomato fruit grading robot

3. After obtaining information from each barcode attached to a container, the container is sent to a robot, the fruit providing robot.
4. The robot sucks fruits up by using suction pads and moves them to a halfway stage.
5. Another robot, the grading robot, picks fruits up again from the halfway stage and bottom and side images of fruits are acquired by TV cameras during transferring fruits.
6. The robot transfers them to trays on a conveyor line and a top image of the fruit is acquired in a camera box.
7. After appearance inspection, internal conditions and sugar content are inspected by an infrared analysis sensor.
8. Fruit that pass the internal quality sensor box are packed into a corrugated cardboard box by a packing robot based on their grading results. Grade, size, and name of fruit variety are printed on the box surface by an inkjet printer and the box is closed and sealed.
9. Finally, the boxes are transferred onto palettes and are loaded into a truck for marketing.

Figure 63.25 shows the actions of the two manipulators: the fruit providing and grading robots. A container in which 8×6, 6×5, 6×4 or 5×3 fruits are filled is pushed into the working area of the providing robot by a pusher (1). The providing robot has a three-DOF Cartesian coordinate manipulator and eight suction pads as end-effectors. The robot sucks eight (maximum) fruits up (2) and transfers them to a halfway stage, spacing fruit intervals in the y direction (3). Two providing robots independently work and set 16 (maximum) fruits on a halfway stage. A grading robot which consists of another three-DOF manipulator (two prismatic joints and a rotational joint) and 16 suction pads sucks them up again (4) and moves them to trays on a conveyor line. Bottom images of fruits are acquired as the grading robot moves over 16 color TV cameras. The cameras and lighting devices turn down 90° following the grading robot's motion (5). Before releasing the

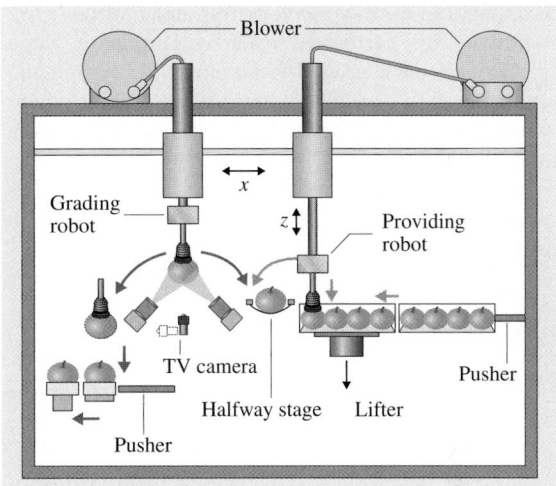

Fig. 63.25 Actions of robots

fruits to trays on the line, the TV cameras acquire four side images of fruits as they rotate through 270° (6). After image acquisition, the robot releases the fruits into trays (7) and a pusher pushes 16 trays to a conveyor line (8).

This grading robot's maximum speed was 1 m/s and its stroke was about 1.2 m. It took 2.7 s for the robot to transfer the 16 fruit to trays, 0.4 s to move down from the initial position, 1 s to move back from releasing fruits, and 0.15 s for waiting. Total time was 4.25 s to move back and forth for the stroke. This makes this robot performance approximately 10 000 fruit/h. In this system, four blowers with specification of 1.4 kW, 3400 rpm, 38 kPa, 1.3 m³/min displacement were used for two providing robots and a grading robot. About 30 kPa vacuum force was suitable for sucking peach fruit, while 45 kPa was used for pear and apple fruits and no damage was observed even after sucking peach fruits twice. The tray has a data carrier (256 byte EE-PROM), and grading information of each fruit is sent from a computer to the data carrier through an antenna after image processing. A conveyor line transfers the trays at 30 m/min [63.197].

63.6 Summary

Despite the problems in introducing automation into agricultural production systems many automation systems have been developed and are commonly applied in agricultural operations. Automation has increased the

efficiency and quality of agricultural production systems.

However, automated or semiautomated farming is far from a reality in many parts of the world. Due to cheap labor in third-world countries, much of the work on farms is still performed manually. Despite the large capital investment needed to purchase the equipment, automation will probably be introduced also into these countries to provide the needs for increased production and land efficiency.

In industrialized countries the production trend is towards large-scale farms and hence automation will be advanced and commercialized to make this feasible. Farmers must produce their food at competitive prices to stay in business and automation of farming technology is the only way forward. With the improvement of sensors and computers, and decrease of automation equipment costs, this is becoming feasible and more systems will be introduced. The current century will probably see significant advances in automation and robotization of farm operations. The future farm will include integration of advanced sensors, controls, and intelligent software to provide viable solutions to the complex agricultural environment.

References

63.1   J.K. Schueller: Automation and control. In: *CIGR Handbook of Agricultural Engineering*, Information Technology, Vol. VI, ed. by A. Munack (CIGR, Tzukuba 2006) pp. 184–195, Chap. 4

63.2   H.G. Ferguson: Apparatus for coupling agricultural implements to tractors and automatically regulating the depth of work, Patent GB 253566 (1925)

63.3   G. Singh: Farm machinery. In: *Agricultural Mechanization & Automation, Encyclopedia of Life Support Systems (EOLSS)*, ed. by P. McNulty, P.M. Grace (EOLSS, Oxford 2002)

63.4   J.N. Wilson: Guidance of agricultural vehicles – a historical perspective, Comput. Electron. Agric. **25**(1), 3–9 (2000)

63.5   T. Torii: Research in autonomous agriculture vehicles in Japan, Comput. Electron. Agric. **25**(1), 133–153 (2000)

63.6   R. Keicher, H. Seufert: Automatic guidance for agricultural vehicles in Europe, Comput. Electron. Agric. **25**(1), 169–194 (2000)

63.7   J.F. Reid, Q. Zhang, N. Noguchi, M. Dickson: Agricultural automatic guidance research in North America, Comput. Electron. Agric. **25**(1), 155–167 (2000)

63.8   H. Auernhammer, T. Muhr: GPS in a basic rule for environment protection in agriculture, Proc. Autom. Agric. **11**(91), 394–402 (1991)

63.9   M. O'Connor, T. Bell, G. Elkaim, B. Parkinson: Automatic steering of farm vehicles using GPS, Proc. 3rd Int. Conf. Precis. Agric. (Minneapolis 1996) pp. 767–778

63.10   T. Stombaugh, E. Benson, J.W. Hummel: Automatic guidance of agricultural vehicles at high field speeds, ASAE Paper No. 983110 (ASAE, St. Joseph 1998)

63.11   T. Bell: Automatic tractor guidance using carrier-phase differential GPS, Comput. Electron. Agric. **25**(1/2), 53–66 (2000)

63.12   N. Noguchi, M. Kise, K. Ishii, H. Terao: Field automation using robot tractor, Automation Technology for Off-road Equipment, Proc. 26–27 July Conf., ed. by Q. Zhang (ASAE, Chicago 2002) pp. 239–245

63.13   G.P. Gordon, R.G. Holmes: Laser positioning system for off-road vehicles, ASAE Paper No. 88-1603 (ASAE, St. Joseph 1988)

63.14   N. Noguchi, K. Ishii, H. Terrao: Development of an agricultural mobile robot using a geomagnetic direction sensor and image sensors, J. Agric. Eng. Res. **67**, 1–15 (1997)

63.15   J.F. Reid, S.W. Searcy, R.J. Babowic: Determining a guidance directrix in row crop images, ASAE Paper No. 85-3549 (ASAE, St. Joseph 1985)

63.16   J.B. Gerrish, G.C. Stockman, L. Mann, G. Hu: Image rocessing for path-finding in agricultural field operations, ASAE Paper No. 853037 (ASAE, St. Joseph 1985)

63.17   J.A. Marchant, R. Brivot: Real time tracking of plant rows using a Hough transform, Real Time Imaging **1**, 363–375 (1995)

63.18   J.A. Marchant: Tracking of row structure in three crops using image analysis, Comput. Electron. Agric. **15**, 161–179 (1996)

63.19   S. Han, Q. Zhang, B. Ni, J.F. Reid: A guidance directrix approach to vision-based vehicle guidance systems, Comput. Electron. Agric. **43**, 179–195 (2004)

63.20   J. Billingsley, M. Schoenfisch: The successful development of a vision guidance system for agriculture, Comput. Electron. Agric. **16**(2), 147–163 (1997)

63.21   J.A. Farrell, T.D. Givargis, M.J. Barth: Real-time differential carrier phase GPS-aided INS, IEEE Trans. Control Sys. Technol. **8**(4), 709–721 (2000)

63.22   L. Guo, Q. Zhang, S. Han: Position estimate of off-road vehicles using a low-cost GPS and IMU, ASAE Paper No. 021157 (ASAE, St. Joseph 2002)

63.23   M.A. Abidi, R.C. Gonzales: Data fusion. In: *Robotics and Machine Intelligence* (Academic, San Diego 1992)

63.24   M.S. Grewal, A.P. Andrews: *Kalman Filter: Theory and Practice Using MATLAB*, 2nd edn. (Wiley, New York 2001)

63.25   O. Cohen, Y. Edan: A new framework for online sensor and algorithm selection, Robot. Auton. Syst. **56**(9), 762–776 (2008)

63.26 H. Choset: Coverage for robotics – a survey of recent results, Ann. Math. Artif. Intell. **31**, 113–126 (2001)

63.27 T. Oksanen, S. Kosonen, A. Visala: Path planning algorithm for field traffic, ASAE Paper No. 053087 (ASAE, St. Joseph 2005)

63.28 J. Jin, L. Tang: Optimal path planning for arable farming, ASAE Paper No. 061158 (ASAE, St. Joseph 2006)

63.29 U. Shani: Filling regions in binary raster images: a graph-theoretic approach, SIGGRAPH '80 Conf. Proc. (ACM, New York 1980) pp. 321–327

63.30 Y.Y. Huang, Z.L. Cao, E.L. Hall: Region filling operations for mobile robot using computer graphics, Proc. of IEEE Int. Conf. Robot. Autom. (1986) pp. 1607–1614

63.31 Z.L. Cao, Y. Huang, E.L. Hall: Region filling operations with random obstacle avoidance for mobile robots, J. Robot. Syst. **5**(2), 87–102 (1988)

63.32 S.A. Gray: Planning and Replanning Events for Autonomous Orchard Tractors. Ph.D. Thesis (Utah State University, Utah 2001)

63.33 J. Park, P.E. Nikravesh: A look-ahead driver model for autonomous cruising on highways, 1996 Future Transport. Technol. Conf. Expo. (Warrendale, 1996)

63.34 Q. Zhang, H. Qiu: A dynamic path search algorithm for tractor automatic navigation, Trans. ASAE. **47**(2), 639–646 (2004)

63.35 D. Wu, Q. Zhang, J.F. Reid, H. Qiu: Adaptive control of electrohydraulic steering system for wheel-type agricultural tractors, ASAE Paper No. 993079 (ASAE, St. Joseph 1999)

63.36 Q. Zhang: Hydraulic linear actuator velocity control using a feedforward-plus-PID control, Int. J. Flex. Autom. Integr. Manuf. **77**, 275–290 (1999)

63.37 H. Qiu, Q. Zhang, J.F. Reid: Fuzzy control of electrohydraulic steering systems for agricultural vehicles, Trans. ASAE **44**(6), 1397–1402 (2001)

63.38 L. Guo, Q. Zhang, S. Han: Agricultural machinery safety alert system using ultrasonic sensors, J. Agric. Saf. Health **8**(4), 385–396 (2002)

63.39 J. Wei, F. Rovira-Mas, J.F. Reid, S. Han: Obstacle detection using stereo vision to enhance safety of autonomous machines, Trans. ASAE **48**(6), 2389–2397 (2005)

63.40 M. Kise, Q. Zhang, N. Noguchi: An obstacle identification algorithm for a laser range finder-based obstacle detector, Trans. ASAE **48**(3), 1269–1278 (2005)

63.41 Y. Matsuo, S. Yamamoto, O. Yukumoto: Development of tilling robot and operation software. In: Autom. Technol. Off-Road Equip. (ATOE) Proc., ed. by Q.Zhang, ASAE Publication No. 701P0509 (2002) pp. 184–189

63.42 J.F. Reid: Mobile intelligent equipment for off-road environments, Proc. ATOE Conf. (ASAE, St. Joseph 2004) pp.1–9

63.43 T. Pilarski, M. Happold, H. Pangels, M. Ollis, K. Fitzpatrick, A. Stentz: The Demeter system for automated harvesting, Proc. 8th Int. Top. Meet. Robot. Remote Syst. (1999)

63.44 T. Pilarski, M. Happold, H. Pangels, M. Ollis, K. Fitzpatrick, A. Stentz: The Demeter system for automated harvesting, Auton. Robot **13**, 19–20 (2002)

63.45 RAS: *Robotics and Automation Society*, Service Robots (IEEE, Piscataway 2008)

63.46 B. Astrand, A.J. Baerveldt: An agricultural mobile robot with vision-based perception for mechanical weed control, Auton. Robot **13**, 21–35 (2002)

63.47 R.N. Jørgensen, C.G. Sørensen, J.M. Pedersen, I. Havn, H.J. Olsen, H.T. Søgaard: Hortibot: An accessory kit transforming a slope mower into a robotic tool carrier for high-tech plant nursing – part I, ASAE Paper No. 63082 (ASAE, St. Joseph 2006)

63.48 C.G. Sørensen, R.N. Jørgensen, M. Nørremark: HortiBot: Application of quality function deployment (QFD) method for horticultural robotic tool carrier design planning – part II, ASAE Paper No. 67021 (ASAE, St. Joseph 2006)

63.49 M. Nørremark, C.G. Sørensen, R.N. Jørgensen: HortiBot: Comparison of potential present and future weeding technologies – part III, ASAE Paper No. 67023 (ASAE, St. Joseph 2006)

63.50 J.L. Merriam, S.W. Styles, B.J. Freeman: Flexible irrigation systems: concept, design, and application, J. Irrig. Drain. Engrg. **133**(1), 2–11 (2007)

63.51 S.J. Kim, P.S. Kim: Optimal gate operation of irrigation reservoir using water management program, ASAE Paper No. 042067 (ASAE, St. Joseph 2004)

63.52 G. Park, M.S. Lee, S.J. Kim: Networking model of paddy irrigation system using archyhydro GIS, ASAE Paper No. 052079 (ASAE, St. Joseph 2005)

63.53 Y. Lam, D.C. Slaughter, W.W. Wallender, S.K. Upadhyaya: Machine vision monitoring for control of water advance in furrow irrigation, Trans. ASAE **50**(2), 371–378 (2007)

63.54 Y. Lam, D.C. Slaughter, S.K. Upadhyaya: Computer vision system for automatic control of precision furrow irrigation system, ASAE Paper No. 062078 (ASAE, St. Joseph 2006)

63.55 Y. Kim, R.G. Evans, W. Iversen, F.P. Pierce, J.L. Chavez: Software design for wireless in-field sensor based irrigation management, ASAE Paper No. 063704 (ASAE, St. Joseph 2006)

63.56 N.L. Klocke, C. Hunter, M. Alam: Application of a linear move sprinkler system for limited irrigation research, ASAE Paper No. 032012 (ASAE, St. Joseph 2003)

63.57 B.A. King, R.W. Wall, L.R. Wall: Distributed control and data acquisition system for closed-loop site-specific irrigation management with center pivots, Appl. Eng. Agric. **21**(5), 871–878 (2005)

63.58 M. Yitayew, K. Didan, C. Reynolds: Microcomputer based low-head gravity-flow bubbler irrigation system design, Comput. Electron. Agric. **22**, 29–39 (1999)

63.59   F.S. Zazueta, A.G. Smajstrla: Microcomputer-based control of irrigation systems, Appl. Eng. in Agric. **8**(5), 593–596 (1992)

63.60   B. Cardenas-Lailhacar, M.D. Dukes, G.L. Miller: Sensor-based control of irrigation in Bermudagrass, ASAE Paper No. 052180 (ASAE, St. Joseph 2005)

63.61   M.B. Haley, M.D. Dukes: Evaluation of sensor based residential irrigation water application, ASAE Paper No. 072251 (ASAE, St. Joseph 2007)

63.62   S.R. Evett, R.T. Peters, T.A. Howell: Controlling water use efficiency with irrigation automation, South. Conserv. Syst. Conf. (Amarillo 2006)

63.63   D.F. Wanjuru, S.J. Maas, J.C. Winslow, D.R. Upchurch: Scanned and spot measured temperatures of cotton and corn, Comput. Electron. Agric. **44**, 33–48 (2004)

63.64   S.R. Herwitz, L.F. Johnson, S.E. Dunagan, R.G. Higgins, D.V. Sullivan, J. Zheng, B.M. Lobitz, J.G. Leung, B.A. Gallmeyer, M. Aoyagi, R.E. Slye, J.A. Brass: Imaging from an unmanned aerial vehicle surveillance and decision support, Comput. Electron. Agric. **44**, 49–61 (2004)

63.65   J.A. Poss, W.B. Russell, P.J. Shouse, R.S. Austin, S.R. Grattan, C.M. Grieve, J.J. Lieth, L. Zheng: A volumetric lysemeter system: an alternative to weighing lysimeters for plant-water relations studies, Comput. Electron. Agric. **43**, 55–68 (2004)

63.66   Y. Kim, R.G. Evans, W. Iversen, F.P. Pierce: Instrumentation and control for wireless sensor network for automated irrigation, ASAE Paper No. 061105 (ASAE, St. Joseph 2006)

63.67   T. Hess: A microcomputer scheduling program for supplementary irrigation, Comput. Electron. Agric. **15**, 233–243 (1996)

63.68   M.J. Upcraft, D.H. Noble, M.K.V. Carr: A mixed linear programme for short-term irrigation scheduling, J. Oper. Res. Soc. **40**(10), 923–931 (1989)

63.69   K. Milla, S. Kish: A low cost microprocessor and infrared sensor system for automating water infiltration measurements, Comput. Electron. Agric. **53**, 122–129 (2006)

63.70   J. Artigas, A. Beltran, C. Jimenez, A. Baldi, R. Mas, C. Dominguez, J. Alonmso: Application of ion sensitive field effect transistor based sensor for soil analysis, Comput. Electron. Agric. **31**(3), 281–293 (2001)

63.71   R.T. Peters, S.R. Evett: Using low-cost GPS receivers for determining field position of mechanized irrigation systems, Appl. Eng. Agric. **21**(5), 841–845 (2005)

63.72   Y. Kim, R.G. Evans, W. Iversen, F.P. Pierce: Evaluation of wireless control for variable rate irrigation, ASAE Paper No. 062164 (ASAE, St. Joseph 2006)

63.73   F.R. Miranda, R. Yoder, J.B. Wilkerson: A site-specific irrigation control system, ASAE Paper No. 031129 (ASAE, St. Joseph 2003)

63.74   F.R. Miranda, R.E. Yoder, J.B. Wilkerson, L.O. Odhiambo: An autonomous controller for site-specific management of fixed irrigation systems, Comput. Electron. Agric. **468**, 183–197 (2005)

63.75   King B.A. W.W. Wall, D.C. Kincaid, D.T. Westermann: Field testing of a variable rate sprinkler and control system for site-specific water and nutrient application. Appl. Eng. Agric. **21**(5), 847–853 (2005)

63.76   A.T. Csordas, M.J. Delwiche, J. Barak: Automated real-time PCR Biosensor for the detection of pathogens in produce irrigation water, ASAE Paper No. 047045 (ASAE, St. Joseph 2004)

63.77   N. Kondo, K.C. Ting (Eds.): *Robotics for Bioproduction Systems* (ASAE, St. Joseph 1998)

63.78   T. Mitsuhashi, A. Yamazaki, T. Shichishima: Automation of plant factory, Proc. 4th SHITA Symp. (Tokyo 1994) pp. 45–57

63.79   N. Kondo, M. Monta, N. Noguchi: *Agri-Robots (II) Mechanisms and Practice* (Corona, Tokyo 2006) pp. 1–223

63.80   W. Simonton: Automatic geranium stock processing in a robotic workcell, Trans. ASAE **33**(6), 2074–2080 (1990)

63.81   N. Kondo, M. Monta: Basic study on chrysanthemum cutting sticking robot, Proc. Int. Symp. Agric. Mech. Autom., Vol. 1 (1997) pp. 93–98

63.82   N. Kondo, M. Monta, Y. Ogawa: Cutting providing system and vision algorithm for robotic chrysanthemum cutting sticking system, Preprints of the International Workshop on Robotics and Automated Machinery for Bioproductions (Valencia 1997) pp. 7–12

63.83   U-shin LTD.: US-500 Users manual (Tokyo, 1993)

63.84   E. Nederhoff: *Energy and CO_2 Enrichment* (Galileo Services Ltd, New Zealand 2007), http://www.redpathaghort.com/bulletins/co2.html

63.85   C. Kittasa, N. Katsoulasa, A. Bailleb: SE-structures and environment: Influence of greenhouse ventilation regime on the microclimate and energy partitioning of a rose canopy during summer conditions, J. Agric. Eng. Res. **79**(3), 349–360 (2001)

63.86   E.J. van Henten: Greenhouse climate management: an optimal control approach. Ph.D. Thesis (Wageningen University, Holland 1994)

63.87   R. Caponetto, L. Fortuna, G. Nunnari, L. Occhipinti, M.G. Xibilia: Soft computing for greenhouse climate control, IEEE Trans. Fuzzy Sys. **8**(6), 1101–1120 (2000)

63.88   T. Morimoto, Y. Hashimoto: An intelligent control for greenhouse automation, orieneted by the concepts of SPA and SFA, Comput. Electron. Agric. **29**, 3–20 (2000)

63.89   L.D. Albright: Controlling greenhouse environments, Acta Horticulturae **578**, 47–54 (2002)

63.90 B. Bailey: Natural and mechanical greenhouse climate control. Acta Horticulturae 710, Int. Symp. Des. Environ. Control Trop. Subtrop. Greenhouses (2006)

63.91 C. Serodio, J. Boaventura Cunha, R. Morais, C. Couto, J. Monteiro: A networked platform for agricultural management systems, Comput. Electron. Agric. 31, 75–90 (2001)

63.92 Maruyama MFG. Co. Inc.: Shuttle spray-car MSC5-100U Users manual (Tokyo 2000)

63.93 Kioritz Corporation: Robotic Spray-car Users manual (Tokyo 2003)

63.94 N. Kawamura, K. Namikawa, T. Fujiura, M. Ura: Study on agricultural robot (Part 1), J. Soc. Agric. Mach. (Japan) 46(3), 353–358 (1984)

63.95 S. Arima, N. Kondo, Y. Shibano, J. Yamashita, T. Fujiura, H. Akiyoshi: Study on cucumber harvesting robot (Part 1), J. Soc. Agric. Mach. (Japan) 56(1), 45–53 (1994)

63.96 S. Arima, N. Kondo, Y. Shibano, T. Fujiura, J. Yamashita, H. Nakamura: Study on cucumber harvesting robot (Part 2), J. Soc. Agric. Mach. (Japan) 56(6), 69–76 (1994)

63.97 T. Fujiura, I.D.M. Subrata, T. Yukawa, S. Nakao, H. Yamada: Cherry tomato harvesting robot, Proc. Int. Symp. Autom. Robot. Bioprod. Process., Vol. 2 (Jpn. Soc. Agric. Mach., Kobe 1995) pp. 175–180

63.98 N. Kondo, M. Monta, Y. Shibano, K. Mohri: Basic mechanism of robot adapted to physical properties of tomato plant, Proc. Int. Conf. Agric. Mach. Process Eng., Vol. 3 (Seoul 1993) pp. 840–849

63.99 N. Kondo, Y. Nishitsuji, P.P. Ling, K.C. Ting: Visual feedback guided robotic cherry tomato harvesting, Trans. ASAE 39(6), 2331–2338 (1996)

63.100 N. Kondo, K.C. Ting: Robotics for Bioproduction Systems (ASAE, St. Joseph 1998)

63.101 K.C. Ting, G.A. Giacomelli, W. Fang: Decision support system for single truss tomato production, Proc. of XXV CIOSTA-CIGR V Congr. (1993) pp. 10–13

63.102 N. Kondo, M. Monta, Y. Shibano, K. Mohri: Basic mechanism of robot adapted to physical properties of tomato plant, Proc. Int. Conf. Agric. Mach. Process Eng. (Seoul 1993) pp. 840–849

63.103 N. Kondo, M. Monta, Y. Shibano, K. Mohri: Two finger harvesting hand with absorptive pad based on physical properties of tomato, Environ. Control Biol. 31(2), 87–92 (1993)

63.104 P.I. Daskalov, K.G. Arvanitis, G.D. Pasgianos, N.A. Sigrimis: Nonlinear adaptive temperature and humidity control in animal buildings, Biosyst. Eng. 93(1), 1–24 (2006)

63.105 W.J. Eradus, M.B. Jansen: Animal identification and monitoring, Comput. Electron. Agric. 24, 91–98 (1999)

63.106 S. Holm, J. Brungot, A. Ronneklein, L. Hoff, V. Jahr, K.M. Kjolerbakken: Acoustic passive integrated transponders for fish tagging and identification, Aquac. Eng. 36(2), 122–126 (2007)

63.107 E. Maltz, S. Devir, O. Kroll, B. Zur, S.L. Spahr, R.D. Shanks: Comparative responses of lactating cows to total mixed rations or computerized individual concentrates feeding, J. Diary Sci. 75(6), 1588–1603 (1992)

63.108 F. Seipelt, A. Bunger, R. Heeren, D. Kähler, M. Lüllmann, G. Pohl: Computer controlled calf rearing, Fifth Int. Dairy Housing Proc. 29–31 January 2003 Conf., Fort Worth, ed. by K. Janni (2003) pp. 356–360, ASAE Publication Number 701P0203

63.109 I. Halachmi, Y. Edan, E. Maltz, U.M. Peiper, U. Moalem, I. Brukental: A real-time control system for individual dairy cow food intake, Comput. Electron. Agric. 20, 131–144 (1998)

63.110 A.V. Fisher: A review of the technique of estimating the composition of livestock using the velocity of ultrasound, Comput. Electron. Agric. 17(2), 217–231 (1997)

63.111 D.E. Filby, M.J.B. Turner, M.J. Street: A walkthrough weigher for dairy cows, J. Agric. Eng. Res. 24, 67–78 (1979)

63.112 J. Ren, N.L. Buck,, S.L. Spahr: A dynamic weight logging system for dairy cows, Trans. ASAE 35, 719–725 (1992)

63.113 U. Peiper, Y. Edan, S. Devir, M. Barak, E. Maltz: Automatic weighing of dairy cows, J. Agric. Eng. Res. 56(1), 13–24 (1993)

63.114 D. Cveticanin, G. Wendl: Dynamic weighing of dairy cows: using a lumped-parameter model of cow walk, Comput. Electron. Agric. 44, 63–69 (2004)

63.115 D. Ordolff: Introduction of electronics into milking technology, Comput. Electron. Agric. 30, 125–149 (2001)

63.116 K. de Koning: Automatic milking lessons from Europe, ASAE Paper No. 044188 (ASAE, St. Joseph 2004)

63.117 D. Ordolff: Introduction of electronics into milking technology, Comput. Electron. Agric. 30(1–3), 125–149 (2001)

63.118 D.M. Jenkins, M.J. Delwiche, R.W. Claycomb: Electrically controlled sampler for milk component sensors, Appl. Eng. Agric. 18(3), 373–378 (2002)

63.119 G. Katz, A. Arazi, N. Pinski, I. Halachmi, Z. Schmilovitz, E. Aizinbud, E. Maltz: Current and Near Term Technologies for Automated Recording of Animal Data for Precision Dairy Farming (ADSA, San Antonio 2007)

63.120 M. Pastell, A. Aisla, M. Hautala, J. Ahokas: Automatic cow health measurement system in a milking robot, ASAE Paper No. 064037 (ASAE, St. Joseph 2006)

63.121 P.G. Rajkondawar, U. Tasch, A.M. Lefcourt, B. Erez, R. Dyer, M.A. Varner: A system for identifying lameness in dairy cattle, Appl. Eng. Agric. 18(1), 87–96 (2002)

63.122 U. Tasch, P.G. Rajkondawar: The development of a SoftSeparator for a lameness diagnostic system, Comput. Electron. Agric. 44(3), 239–245 (2004)

63.123 M. Pastell, H. Takko, H. Gröhn, M. Hautala, V. Poikalainen, J. Praks, I. Veermäe, M. Kujala, J. Ahokas: Assessing cows' welfare: weighing the cow in a milking robot, Biosyst. Eng. **93**(1), 81–87 (2006)

63.124 A.R. Frost, C.P. Schofield, S.A. Beaulah, T.T. Mottram, J.A. Lines, C.M. Wathes: A review of livestock monitoring and the need for integrated systems, Comput. Electron. Agric. **17**(2), 139–159 (1997)

63.125 M. Delwiche, X. Tang, R. Bondurant, C. Munro: Estrus detection with a progesterone biosensor, Transactions ASAE **44**(6), 2003–2008 (2001)

63.126 L. Gygax, G. Neisen, H. Bollhalder: Accuracy and validation of radar based automatic local position measurement system for tracking dairy cows in free-stall barns, Comput. Electron. Agric. **56**, 23–33 (2007)

63.127 L.W. Turner, M. Anderson, B.T. Larson, M.C. Udal: Global positioning systems and grazing behavious in cattle, Livest. Environ. **VI**, 640–650 (2001)

63.128 T.J. DeVries, M.A.G. von Keyserlingk, K.A. Beauchemin: Frequency of feed delivery affects the behavior of lactating dairy cows, J. Dairy Sci. **88**, 3553–3562 (2005)

63.129 L. Gygax, G. Neisen, H. Bollhalder: Accuracy and validation of a radar-based automatic local position measurement system for tracking dairy cows in free-stall barns, Comput. Electron. Agric. **56**(1), 23–33 (2007)

63.130 F. Teye, H. Gröhn, M. Pastell, M. Hautala, A. Pajumägi, J. Praks, V. Poikalainen, T. Kivinen, J. Ahokas: Microclimate and gas emissions in cold uninsulated dairy buildings, 2006 ASABE Annu. Int. Meet. (ASABE, St. Joseph 2006), ASABE Paper No. 064080, pp. 1–8

63.131 R.M.T. Baars, C. Solano, M.T. Baayen, R. Rojas, L. Mannetje: MIS support for pasture and nutrition management of dairy farms in tropical countries, Comput. Electron. Agric. **15**, 27–39 (1996)

63.132 M.A.P.M. van Asseldonk, R.B.M. Hurine, A.A. Didkhuizen, A.J.M. Beulens, A.J. Udink ten Cate: Information needs and information technology on dairy farms, Comput. Electron. Agric. **22**, 97–107 (1999)

63.133 C.M. Wathes, S.M. Abeyesinghe, A.R. Frost: Environmental design and management for livestock in the 21st century: resolving conflicts by integrated solutions, Livest. Environ. VI: Proc. 6th Int. Symp. (2001) pp. 5–14, ASAE Publication No. 701P0201

63.134 J.M. Powell, P.R. Cusick, T.H. Misselbrook, B.J. Holmes: Design and calibration of chambers for mearuing ammonia emissions from tie-stall dairy barns, Trans. ASABE **50**(3), 1045–1051 (2007)

63.135 N. Mozes, O. Zemora, C. Porter, H. Gordin: Marine integrated pond system under desert conditions in southern israel – potential, results and limitations, Aquacult. Eur. 2003 Conf. (Trondheim 2003)

63.136 N. Mozes: Ustainable development of land-based mariculture: integrated system with algal biofilter versus recirculation system with bacterial biofilter, Aquacult. Eur. 2003 Conf. (Trondheim 2003)

63.137 N. Mozes, I. Haddas, D. Conijeski, M. Eshchar: The low-head megaflow air driven recirculating system – minimizing biological and operational risks, Proc. Aquacult. Eur. 2004 Conf. (Barcelona 2004) pp. 598–599

63.138 M. Shpigel, A. Neori, D.M. Popper, H. Gordin: A proposed model for 'environmentally clean' land-based culture of fish, bivalves and seaweeds, Aquaculture **117**(1/2), 115–118 (1993)

63.139 C. Costa, A. Loy, S. Cataudella, D. Davis, M. Scardi: Extracting fish size using dual underwater cameras, Aquacult. Eng. **35**, 218–227 (2006)

63.140 J.A. Lines, R.D. Tillet, L.G. Ross, D. Chan, S. Hockaday, N.J.B. McFarlane: An automatic image base system form estimating mass of free-swimming fish, Comput. Electron. Agric. **31**, 151–168 (2001)

63.141 B. Zion, V. Alchanatis, V. Ostrovsky, A. Barki, I. Karplus: Comput. Electron. Agric. **56**(1), 34–45 (2007)

63.142 P.G. Lee: A review of automated control systems for aquaculture and design criteria for their implementation, Aquacult. Eng. **14**(3), 205–227 (1995)

63.143 C.W. Chang, W.R.C. Fang. Jao, C.Z. Shyu, I.C. Lioa: Development of an intelligent feeding controller for indoor intensive culturing of eel, Aquacult. Eng. **32**, 343–353 (2005)

63.144 N. Papandroulakis, P. Dimitris, D. Pascal: An automated feeding system for intensive hatcheries, Aquacult. Eng. **26**, 13–26 (2002)

63.145 F.J. Muir, C. Brugere Young, A.J.A. Stewart: The solution to pollution? The value and limitations of environmental economics in guiding aquaculture development, Aquacult. Econom. Manag. **3**(1), 43–57 (1999)

63.146 A.W. Wurts: Sustainable aquaculture in the twenty first century, Rev. Fish. Sci. **8**(2), 141–150 (2000)

63.147 R.H Caffey: *Quantifying Sustainability in Aquaculture Production* (Louisiana State University, Baton Rouge 1998)

63.148 F. Wheaton, S. Hall: Research needs for oyster shucking, Aquacult. Eng. **37**, 67–72 (2007)

63.149 G.F. Figueiredo, M.D. Dawson, E.R. Benson, G.L. van Wicklen, N. Gedamu: Development of machine vision based poultry behaviour analysis, ASAE Paper No. 0330834 (ASAE, St. Joseph 2003)

63.150 G.F. Figueiredo, M.D. Dawson, E.R. Benson, G.L. van Wicklen, N. Gedamu: Advancement in whole house machine vision based poultry behaviour analysis, ASAE Paper No. 043084 (ASAE, St. Joseph 2004)

63.151 K. Chao, Y.R. Chen, W.R. Hruschka, B. Park: Chicken heart disease characterization by multispectral imaging, Appl. Eng. Agric. **17**(1), 99–106 (2001)

63.152 K.F. Stacey, D.J. Parsons, A.R. Frost, C. Fisher, D. Filmer, A. Fothergill: An automatic growth and nutrition control system for broiler production, Biosyst. Eng. **89**(3), 363–371 (2004)

63.153 D. Sergeant, R. Boyle, M. Forbes: Computer visual tracking of poultry, Comput. Electron. Agric. **21**, 1–18 (1998)

63.154 S. Jaiswa, E.R. Benson, J.C. Bernard, G.L. van Wicklen: Neural network modeling and sensitivity analysis of mechanical poultry catching system, Biosyst. Eng. **92**(1), 59–68 (2005)

63.155 J.P. Trevelyan: Sensing and control for sheep shearing robots, IEEE Trans. Robot. Autom. **5**(6), 716–727 (1989)

63.156 F. Perez-Munoz, S.J. Hoff, T. Van Hal: A quasi ad-libitum electronic feeding system for gestating sows in loose housing, Comput. Electron. Agric. **19**(3), 277–288 (1998)

63.157 F. Perez-Munoz, S.J. Hoff, T. van Hal: A quasi ad-libitum electronic feeding system for gestating sows in loose housing, Comput. Electron. Agric. **19**, 277–288 (1998)

63.158 Y. Wang, W. Yang, P. Winter, L.T. Walker: Non-contact sensing of hog weights by machine vision, Appl. Eng. Agric. **22**(4), 577–582 (2006)

63.159 C.P. Schofield, C.T. Whittemore, D.M. Green, M.D. Pascual: The determination of beginning and end of period live weights in growing pigs, J. Sci. Food Agric. **82**, 1672–1675 (2002)

63.160 R.D. Tillet, A.R.S. Frost, S.K. Welch: Predicting sensor placement targets on pigs using image analysis, Biosyst. Eng. **81**(4), 453–463 (2002)

63.161 H. Xin, B. Shao: Real-time behaviour-based assessment and control of swine thermal comfort, Proc. 7th Int. Symp. (ASAE, St. Joseph 2005) pp. 694–702, ASAE Publication No. 701P0205

63.162 M. Barbari: Planning individual showering systems for pregnant sows in dynamic groups, Livest. Environ. **VII**, 130–137 (2005), ASAE Publication No. 701P0205

63.163 G. Zhang, J.S. Strom, M. Blanke, I. Braithwaite: Spectral signatures of surface materials in pig buildings, Biosyst. Eng. **94**(4), 495–504 (2006)

63.164 W. Saeys, A.M. Mouazen, H. Ramon: Potential for onsite and online analysis of pig manure using visible and near infrared reflectance spectroscopy, Biosyst. Eng. **91**(4), 393–402 (2005)

63.165 R.C. Harrell, P.D. Adsit, T.A. Pool, R. Hoffman: The Florida Robotic Grove-Lab, Trans. ASAE **33**(2), 391–399 (1990)

63.166 M. Hayashi, Y. Ueda, H. Suzuki: Development of agricultural robot, Proc. 6th Conf. Robot. (Robotics Society of Japan, 1988) pp. 579–580

63.167 T. Fujiura, M. Ura, N. Kawamura, K. Namikawa: Fruit harvesting robot for orchard, J. Soc. Agric. Mach. (Japan) **52**(2), 35–42 (1990)

63.168 F. Juste, I. Fornes: Contributions to robotic harvesting of citrus in Spain, Proc. of the AG-ENG 90 Conf. (Berlin, 1990) pp. 146–147

63.169 G. Rabatel, A. Bourely, F. Sevila, F. Juste: Robotic harvesting of citrus, Proc. Int. Conf. Harvest and Post-harvest Technol. Fresh Fruits and Vegetables (Guanajuato, 1995) pp. 232–239

63.170 N. Kondo, M. Monta, T. Fujuira, Y. Shibano, K. Mohri: Control method for 7 DOF robot to harvest tomato, Proc. Asian Control Conf., Vol. 1 (1994) pp. 1–4

63.171 M.W. Hannan, T. Burks: Current developments in automated citrus harvesting, ASAE Paper No. 043087 (ASAE, St. Joseph 2004)

63.172 E. Molto, F. Pla, F. Juste: Vision systems for the location of citrus fruit in a tree canopy, J. Agric. Eng. Res., **52**, 101–110 (1992)

63.173 N. Kondo, N. Kawamura: Methods of detecting fruit by visual sensor attached to manipulator, J. Soc. Agric. Mach. (Japan) **47**(1), 60–65 (1985)

63.174 N. Kondo, S. Endo: Methods of detecting fruit by visual sensor attached to manipulator (II), J. Soc. Agric. Mach. (Japan) **51**(4), 41–48 (1989)

63.175 N. Kondo, S. Endo: Methods of detecting fruit by visual sensor attached to manipulator (III), J. Soc. Agric. Mach. (Japan) **52**(4), 75–82 (1990)

63.176 T. Fujiura, J. Yamashita, N. Kondo: Agricultural robots (1): Vision sensing system, ASAE Paper No. 92-3517 (ASAE, St. Joseph 1992)

63.177 G. Rabatel: A vision system for the fruit picking robot, Proc. Agric. Eng. '88 Conf. (Paris 1988), AG-ENG Paper No. 88-293

63.178 G. Rabatel, A. Bourely, F. Sevila: Objects detection with machine vision in outdoor complex scenes, Proc. Robot. Syst. Eng. Syst. Intell. (Corfou 1991) pp. 395–403

63.179 D.C. Slaughter, R.C. Harrell: Color vision in robotic fruit harvesting, Trans. ASAE **30**(4), 1144–1148 (1987)

63.180 N. Kondo: Harvesting robot based on physical properties of grapevine, Jpn. Agric. Res. Q. **29**(3), 171–177 (1995)

63.181 A. Sittichareonchai, F. Sevila: A robot to harvest grapes, ASAE Paper No. 89-7074 (ASAE, St. Joseph 1989)

63.182 L. Kassay: Hungarian robotic apple harvester, ASAE Paper No. 92-7042 (ASAE, St. Joseph 1992)

63.183 A. Grand d'Esnon: Robotic harvesting of apples, Proc. Agri-Mation 1st Conf. Expo. (ASAE, Chicago 1885) pp. 210–214

63.184 Y. Edan, G.E. Miles: Design of an agricultural robot for harvesting melons, Trans. ASAE **36**(2), 593–603 (1993)

63.185 M. Iida, K. Furube, K. Namikawa, M. Umeda: Development of watermelon harvesting gripper, J. Soc. Agric. Mach. (Japan), **58**(3), 19–26 (1996)

63.186 N. Kondo, K.C. Ting: *Robotics for Bioproduction Systems* (ASAE, St. Joseph 1998)

63.187 K. Kurokami: Fence type training system of mandarin orange tree, Agric. Hortic. **55**(2), 289–293 (1980)

63.188 M. Monta, N. Kondo, Y. Shibano, K. Mohri: Basic study on robot to work in vineyard (Part 3) – measurement of physical properties for robotization and manufacture of berry thinning hand, J. Soc. Agric. Mach. (Japan) **56**(2), 93–100 (1994)

63.189 K. Nishiwaki, K. Amaha, R. Otani: *Development of Nozzle Positioning System for Precision Sprayer, Automation Technology for Off-Road Equipment* (ASAE, St. Joseph 2004)

63.190 K.P. Gilles, D.K. Giles, D.C. Saughter, D. Downey: Injection and fluid handling system for machine-vision controlled spraying, ASAE Paper No. 011114 (ASAE, St. Joseph 2001)

63.191 H.T. Wiedemann, D. Ueckert, W.A. McGinty: Spray boom for sensing and selectively spraying small mesquite on higway rights-of-way, Appl. Eng. Agric. **18**(6), 661–666 (2002)

63.192 K. Tosaki, S. Miyahara, T. Ichikawa, Y. Mizukura: Development of microcomputer controlled driverless air blast, J. Soc. Agric. Mach. (Japan) **58**(6), 101–110 (1996)

63.193 Japanese Society of Agricultural Machinery: *Handbook of Bioproduction Machinery* (Corona, Tokyo 1996) p. 731

63.194 S.I. Cho, J.H. Lee: Autonomous speed-sprayer using differential GPS system, genetic algorithm and fuzzy control, J. Agric. Eng. Res. **76**, 111–119 (2000)

63.195 M. Dohi: Development of multipurpose robot for vegetable production, Jpn. Agric. Res. Q. **30**(4), 227–232 (1996)

63.196 U. Ahmad, N. Kondo, S. Arima, M. Monta, K. Mohri: Weed detection in lawn field based on gray-scale uniformity, Environ. Control Biol. **36**(4), 227–237 (1998)

63.197 U. Ahmad, N. Kondo, S. Arima, M. Monta, K. Mohri: Weed detection in lawn field using machine vision. utilization of textural features in segmented area, J. Soc. Agric. Mach. (Japan) **61**(2), 61–69 (1999)

63.198 B.L. Steward, L.F. Tian, L. Tang: Distance-based control system for machine vision-based selective spraying, Trans. ASAE **45**(5), 1255–1262 (2002)

63.199 J. Njoroge, K. Ninomiya, N. Kondo, H. Toita: Automated fruit grading system using image processing, Proc. SICE Ann. Conf. (Osaka 2002), MP18-3 on CD-ROM

64. Control System for Automated Feed Plant

Nick A. Ivanescu

Many factories, especially in developing countries, still use old technology and control systems. Some of them are forced to replace at least the control and supervising system in order to increase their productivity. This chapter presents a modern solution for updating the control system of a fodder-producing factory without replacing the field devices or the infrastructure. The automation system was chosen in order to allow correct control of the whole plant, using a single programmable logic controller (PLC). Structure and design of the software project is described. Also, several interesting software solutions for managing special processes such as material extraction and weighing machines calibration are presented. Production quality results and future development are also discussed. In the last part of the chapter some guidelines for automation of a chicken-growing plant are presented.

Part G | 64

64.1 Objectives

One of the recent projects we worked on for was to design and execute a complete control system for a combined fodder-producing plant. From the start it must be specified that we had to find solutions to command and control the existing machines, and the only things that could be replaced or added were sensors and transducers. Most other devices, such as motors, limit sensors, tension transducers, etc., remained in place. Also 90% of the cable system was retained.

This project had several objectives:

1. Design of the command and control system, taking into account the electrical characteristics of the machines, and the number and type of the electrical signals coming from and going to the system

2. Development of a powerful software system to allow manual control of the whole plant, continuous visualization of all signals and commands, and full automatic control of the production process

3. Real-time communication with personal computers in a local network, enabling managers and other personnel to supervise the production flow and results

4. Ensuring that the total quantity of the final product should not exceed or be less than 5% of the programmed quantity. Also the percentage of every component of the resulted material should be less than 5% different from the calculated recipe.

64.2 Problem Description

A combined fodder plant produces food for industrial grown poultry. This food is a mixture of different types of cereals combined with some concentrated products.

Practically, there are three main areas inside the factory:

- The raw material storage area, consisting of several storage bins
- The weighing and mixing area, where the final product is obtained
- The finished-product storage area, consisting of several storage bins, where the resulting product is stored while waiting to be taken to poultry farms.

Part of the plant is presented in Fig. 64.1.

Let us take a brief look at the components of the plant shown in Fig. 64.1:

- There are eight raw-materials storage bins, containing eight different types of cereals.
- From these bins material can be extracted to weighing machine 1, by means of several extractors.
- Weighing machine 1 is used to weigh specific quantities of cereals, with a maximum of 2000 kg supported by the machine. The machine supplies an analog signal in voltage, which we converted to a unified 4–20 mA signal.

- The intermediate tank between weighing machine 1 and the mill is needed because the speed of the mill is lower than the evacuation speed from weighing machine 1.
- The mill is used for milling the cereals.
- The grinding machine is used for grinding the milled cereals.
- The mixing machine is a large tank where the processed cereals are mixed together with some special oil and another component (premix).
- The materials circulate between the machines by means of several transporters and elevators.
- Weighing machines 2 and 3 weigh, respectively, oil and premix extracted from the milling machines. Details will be given later in the chapter.

The production flow can be briefly described as follows:

1. The human operator must establish the quantities for every type of cereal, for oil, and for premix that must be part of the final product. Quantities are determined for only one charge of extraction, because the capacity of the weighing machines is limited. Also the number of charges is specified, in order to produce the whole amount of material wanted.

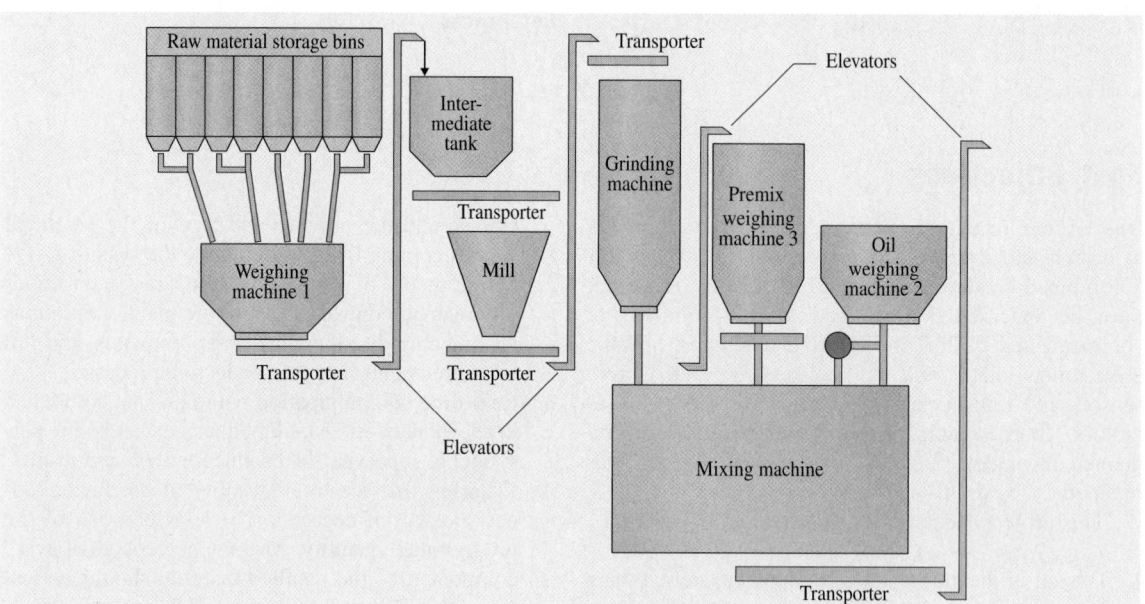

Fig. 64.1 Production part of the plant

2. The process starts with extraction of specified quantities of cereals into weighing machine 1.
3. Weighing machine 1 is emptied into the intermediate tank, if some specific conditions are fulfilled. Immediately a new extraction should begin for the next charge.
4. The milling and grinding processes are done automatically; some sensors signal when the corresponding tanks are empty.
5. When the grinding is over, material must be transferred to the mixing machine. After a short time, oil and premix are also loaded into the mixing machine.

6. The mixing process lasts several minutes (a configurable period of time), at the end of which the machine is opened and the finished product is transported to the finished-product storage bins and the charge can be considered ended.

It is necessary to say that, in order to obtain high productivity, another charge must be in course of processing, even if the previous one is not finished. Of course the detailed process implies several constraints and internal conditions, some of which will be discussed later in the chapter.

64.3 Special Issues To Be Solved

One of the first and most difficult parts of the project was to identify all analog and digital signals that must come from and go to the installations. Without proper documentation or an electrical schematic of the plant, this job proved to be extremely time consuming. After analysis, 192 digital inputs, 126 digital outputs, and 12 analog signals were identified, together with the cables connected to them from the previous control system.

Another issue that had to be taken into account was the improper grounding of most of the machines inside the plant, which can cause undesired variation of analog signals. Most of the work regarding this issue was done in software because electrical solutions were not feasible (as they should have been taken into consideration during the design and build of the factory).

During study of the rest of the machines in the plant, some special situations were determined regard-

ing weighing machine 3. Firstly, the four tensions transducers that measured the weight inside the machine were not of the same type. One transducer had been replaced in the past but with a different type, and had a different output electrical tension for the same measured weight compared with the other three transducers. This resulted in nonlinear variation of the unified analog signal with the weight inside the machine. The solution to this problem is described later in the chapter.

The other special effect discovered was that, when weighing machine 3 was opened and premix was transferred to the mixing machine, an air flow developed and pushed up the weighing machine, resulting in incorrect measurement of weight during this process. Stabilization of the signal appears many seconds after extraction stopped. A software solution was chosen for this practical problem also.

64.4 Choosing the Control System

To choose the main control device for such an industrial system correctly, several factors must be taken into consideration, including:

- The number and type of input/output signals
- The complexity of the software that must be developed
- The communication facilities that are requested and can be fulfilled
- The ease of developing a human–machine interface for supervising and commanding the processes

- The budget allocated to this part of the control system.

The solution chosen in this case was the ThinkIO PLC from the German company Kontron. The ThinkIO device is an innovative concept to integrate high-performance personal computer (PC) functionality, fieldbuses, and input/output (I/O) modules. It has a powerful Pentium processor, two Ethernet interfaces, universal serial bus (USB) for keyboard and mouse, and digital visual interface (DVI) for liquid-crystal display

(LCD) or cathode-ray tube (CRT) displays. For automation applications it can run programs developed in 3S Software Codesys package. Also it can handle hundreds of I/O signals, so it fulfilled the system necessities. Another big advantage of this device is that even the human–machine interface runs on it, so practically you do not need a separate PC to supervise the process or to send manual commands to it. We chose the Linux operating system for this PLC. An object linking and embedding (OLE) process control (OPC) server runs on this PLC so communication with it is easy and reliable. After completing connection of it to the field devices, the next step was to design and develop the software. However, before designing the software solution, there was an important job to do, essential for correct results of the production process.

64.5 Calibrating the Weighing Machines

All three weighing machines have some tension sensors that output an electric signal that is (theoretically) proportional to the weight in the machine. Weighing machines 1 and 3 have four sensors, mounted at the four corners of the weighing tank, that output a voltage (0–100 mV) proportional to the force on each sensor. The sum of the four tensions is converted to a current in the range of 0–20 mA and connected to an analogue input of the PLC.

A potential problem could appear if the sensors are not of the same type (different output tension range) or are not mounted perfectly symmetrical. As already described, such a situation occurs for weighing machine 3. The output of one of the sensors had a different tension range to those of the other three sensors.

The standardization procedure implies the determination of the relation between the weight of the weighing machines and the output current of the measurement sensors, practically being the relation between the weight and the engineering units resulting from the analog-to-digital conversion inside the PLC.

To determine this relation, we used 100 standard 20 kg units. Weighing machines 1 and 3 can support up to 2000 kg load. To determine the specified relation, we loaded the machines with consecutive 200 kg weights, until 2000 kg was reached. After that, the weights were removed, 200 kg at a time, until there was no load on the machine. Fortunately, the hysteresis phenomenon was very small and practically did not matter. The characteristic of the machines is shown in Fig. 64.2.

Machine 1 proved to be linear so we adopted (64.1) for calculating the instantaneous weight value

$$weight1 := (adc\_value - init\_value1)/scale\_1 \, ,$$
$$(64.1)$$

where:

- weight1 is the instant value of the weight (in kg).
- adc_value is the numerical value obtained by converting the analogue signal (0–20 mA) to 12 bits.
- init_value1 is the numerical value obtained from sensors when machine 1 is empty (because the weighing machines have their own weight).
- scale_1 is the approximated slope of the line, calculated using the arithmetic average of all DAC units/kg ratios at all the measured points (Fig. 64.2).

Machine 3 had a different behavior because one of the sensors is of a different type to the other three, outputting a tension in a smaller range than the others. The result is that the resulting graph is more of parabolic type. Anyway because determination of the parabolic equation is very difficult, we chose to piecewise-linearize the graph on ten portions. So the formula for one portion reads

$$weight3 := (adc\_value - init\_value3)/scale\_3i \, ,$$
$$(64.2)$$

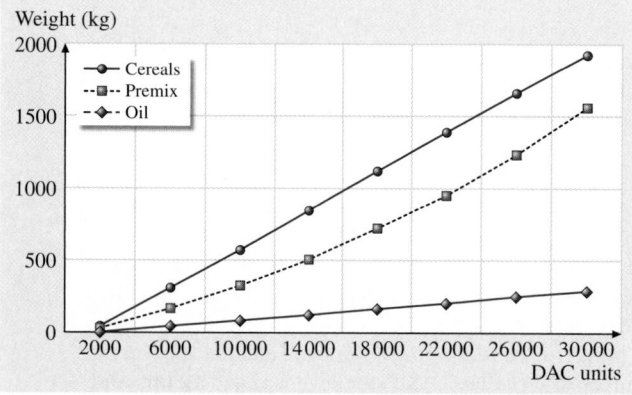

Fig. 64.2 Characteristic of the three weighing machines, for cereals, premix, and oil

where:

- weight3 is the instantaneous value of the weight (in kg).
- adc_value is the numerical value obtained by converting the analogue signal (0–20 mA) on 12 bits.
- init_value3 is the numerical value obtained from sensors when machine 1 is empty (because the weighing machines have their own weight).
- scale_3i is the approximated slope of the approximation line on portion i, calculated from the values

of start point and end point of the corresponding portion (Fig. 64.2).

Weighing machine 2 has only one tension sensor so the relation between the weight value and the output tension is linear, so the same formula was used for the whole weighing range (0–120 kg)

$$weight\_oil := (can\_value - init\_oil)/scale\_oil, \quad (64.3)$$

where the elements of (64.3) have the same meaning as in previous equations.

64.6 Management of the Extraction Process

Another delicate process that had to be correctly controlled is the cereals extraction and premix unloading. The cereals are transported from the storage bins with a system of conveyors and elevators and are loaded into weighing machine 1 using several extractors. However, because the end of the extractors is some distance above the weighing machine, there is still some material in the air that will fall into the weighing machine even after the extractor's motor is stopped.

The quantity falling down in the machine after stopping the extractor depends on various factors:

- Type of cereal
- Quantity desired to be extracted
- Distance from the end of extractor to the weighing machine.

The same considerations apply for unloading the premix in the mixing machine, but the quantity falling after the unloading lid is closed is small. The solution adopted in order to minimize extraction errors was to use some estimated *in-air* quantity to anticipate the quantity loaded into the machine after the extractor is stopped. In this case the extractor should be stopped before the weighing machine measures the desired quantity. So the estimation (for one material) has an initial value but is adjusted at every extraction with

the simple formula

$$new\_estimation = old\_estimation$$
$$+ 0.5(last\_extracted\_quantity - wanted\_quantity), \quad (64.4)$$

where:

- new_estimation is the (*in-air*) estimated quantity that will be used to determine the moment when the extractor must be stopped for the current extraction.
- old_estimation is the estimation used for the previous extraction.
- last_extracted_quantity is the concrete quantity previously extracted.
- wanted_quantity is the ideal quantity that should be extracted.

In this way in three or four extractions the error practically drops below 1 kg. Taking into account that a complete production cycle has tens of extractions, the total extraction error is very small, less than 5%.

After weighing machine 1 is loaded with all types of cereals needed for the current charge, the machine is unloaded completely, when several conditions are fulfilled. Premix and oil quantities are measured somehow differently, i.e., when they are unloaded from the weighing machines into the mixing machine, but the procedure is similar.

64.7 Software Design: Theory and Application

The Codesys logical software package was used to develop the control programs. The software project should have had different functions:

- Graphical user interfaces, for configuring the process, for monitoring all the production stages and all the factory areas, and for manual command of the installations in the factory

- Automatic accomplishment of planned production
- Implementation of constraints and conditions between different machines
- Sorting and mediation of analogue signals
- Network communication with PCs.

First let us briefly discuss some *theoretical* approaches regarding the conception of an automation software project. The easiest and most natural type of programming language that should be used for the control part of a program is the sequential function chart. This programming language is supported by the large majority of PLCs producers. Even if this language is not implemented in a PLC and other languages, such as ladder diagrams, must be used, there are algorithms that help the programmer to *convert* a logical diagram in a ladder diagram. On the other hand, other types of jobs, such as implementation of synchronization and conditioning of events can be done more easily in a ladder diagram type of program. Some examples are shown later in the chapter.

When several industrial processes are be controlled by a single PLC, the best idea is to control every execution device (such as motors, actuators, valves, and so on) *separately* in a separate logical diagram. If related events for several devices are somehow connected, it is better to develop a single logical diagram for their control. Relationship or synchronization between logical diagrams is achieved by *global variables*.

64.7.1 Project Structure and Important Issues

The structure of the software project is shown in Fig. 64.3.

The idea was to separately control almost every subprocess that takes place during the production flow (extractions, mixing, unloading and so on) [64.1]. Most of the programs are written in sequential function chart (SFC) programming language, the *interblocking* program is written in the ladder diagram (LD) language and other smaller programs are written in structure text (ST) language. All of these programs run *simultaneously* in the PLC memory.

Briefly described, the programs have the following functions:

- *Extraction* programs take care of the extraction of each type of cereal, with one program for each type of cereal (written in SFC).

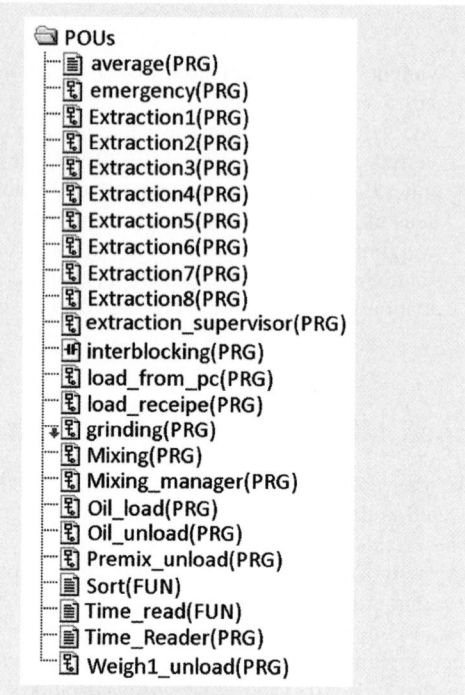

Fig. 64.3 Structure of the software project

- The *interblocking* program takes care of all conditioning between different installations together with automatically stopping some motors when they reach their limits (Fig. 64.4); for example, one elevator cannot be started if a connected transporter was not previously started. The program is written in LD language because it must check all the conditions at every PLC cycle.
- *Loadxxx* programs manage the communication between the PLC and PC (written in SFC).
- The *mixing* program controls the necessary timing needed for the mixing process of the materials.
- The *extraction_supervisor* program supervises the extractions programs. It decides if and when an extraction should start.

Other programs manage the loading and unloading of oil and premix, calculate the average values of analogue signals, and so on. The project can work in two modes:

- *Manual* mode, where only specific commands can be given to some of the machines
- *Automatic* mode, where the production process is automatically managed by the PLC.

One of the major tasks to be efficiently implemented is the correct management of *timing*. The chosen solution was to use only one timer variable for all SFC-type programs that is started when automatic mode is chosen. In all SFC programs that need timing comparisons, the solution adopted was to read this *global timer* as necessary and make the desired calculations of time passed. Other tested solution, using special timing functions, like timer on (off) delay did not give expected results in SFC programs, but performed well in LD diagrams. When the automatic mode ends, the global timer is stopped and reset in order to avoid the situation in which the timer could reach its maximum and overflow. The timer is not needed in the manual mode.

Another special request was to minimize as much as possible the production time, in order to increase productivity. The solution chosen to achieve this goal was to extract the cereals immediately after weighing machine 1 is unloaded into the transporting system. In this way no time is lost; while the milling machine is prepared to receive material, the weighing machine 1 is emptied without any delay. In Fig. 64.4 an example is shown of how the timer-off delay function block is used in a LD program.

Because the impulse duration of variable b_SEL1_op (starts a motor rotating clockwise) is too short, it was extended by 500 ms. In the second line of the diagram you can see that the command is reset when:

- 500 ms has passed or the confirmation from the motor's contactor arrived.
- The motor reached the corresponding limit.
- Confirmation from the counterclockwise contactor is not active (failure situation).

In this way the operator must only start the movement of a motor and the controller automatically stops it when one of the conditions is fulfilled.

Another job to be correctly managed was the emergency status. Because of the nature of the installations of the factory, there are frequent situations when the product process must be immediately interrupted (extractor clogging, transporters stuck, and so on).

When an emergency occurs (that could not be detected automatically) the operator must stop all the processes immediately by simply clicking a button, implemented in the user interface. Inside the programs, almost all the steps verify the status of the emergency button and the status of several sensors. When an emer-

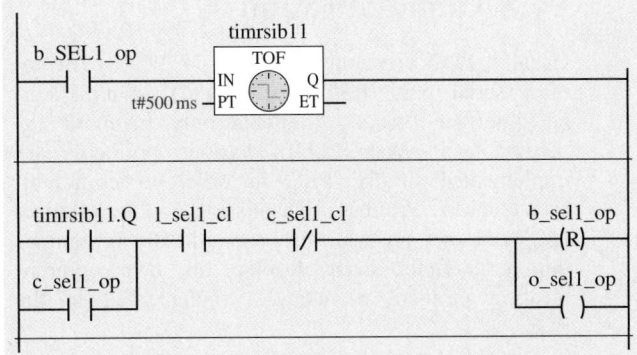

Fig. 64.4 Automatically stopping a motor in an LD program

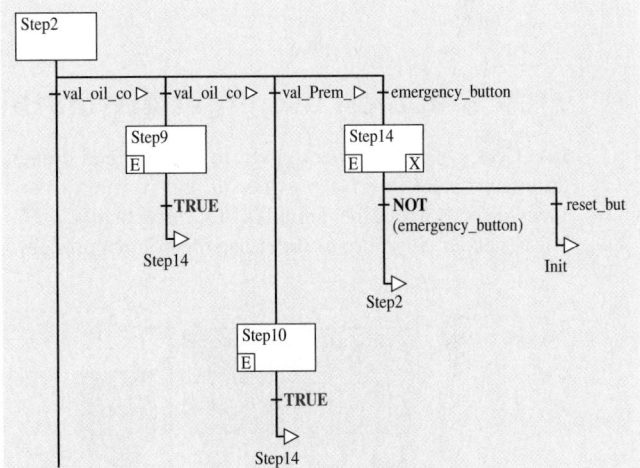

Fig. 64.5 Managing emergency situations in SFC

gency is detected, all SFC programs stop in their current step. Depending on the gravity of the situation, the operator has the possibility to resume the process from the point where it was stopped or to reset all the programs (Fig. 64.5).

Once can see that step 2 has four conditions of transition. The first is the normal behavior of the process, conditions 2 and 3 appear when some material is missing in the bin, and the fourth is true when the operator presses the emergency button on the interface screen. The latter three transitions lead the program to an *emergency* step 14, from where the graph can evolve to step 2 (resuming the program) or the Init step, sending the program back to its initial status.

64.8 Communication

Usually PLCs communicate with a PC by means of a serial port (RS232 or RS485) or an Ethernet interface (usually available only in more expensive and newer PLCs). Various protocols are implemented in the PLC in order to communicate (such as Profibus, transmission control protocol (TCP)/internet protocol (IP) and others). Sometimes the programmer must develop his own communication protocol if none is implemented in the PLC.

ThinkIO controllers have already installed an OPC server that can be used to exchange data between the PLC and other computers connected in the same network. This is a very useful feature because several PCs can communicate with the controller and can be used to monitor all the production process, to configure the recipe, and even to send commands to the installations (if the program allows this).

At this time one PC program was developed, with the following behavior:

- The operator can edit, save, load recipes, and send them to the PLC on the network, acting as an OPC client.
- Reports are printed at the printer after each charge finishes.

64.9 Graphical User Interface on the PLC

In a Codesys development program the user can create as many visual interfaces as needd and run this interfaces directly from the ThinkIO PLC. Practically a PC is not needed anymore to supervise the control process.

For this project five visual screens were developed, each representing a specific area of the factory; for example Fig. 64.6 shows the main area of the factory, with its complicated transport system (formed by conveyors

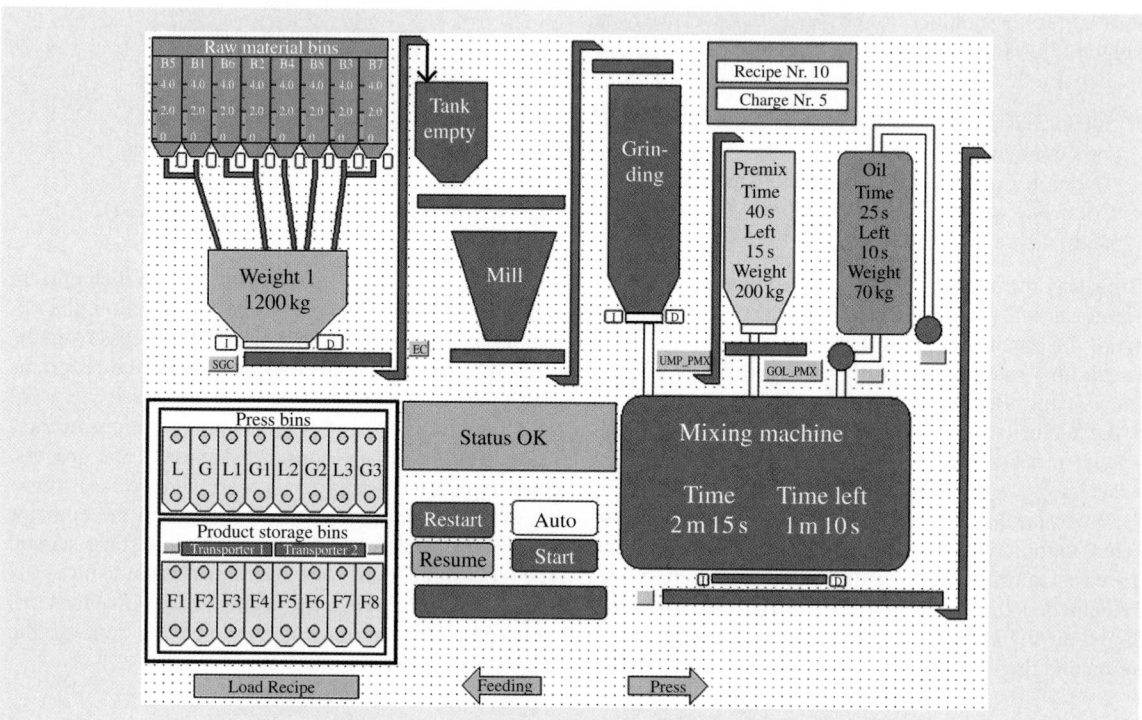

Fig. 64.6 Software view: main production area

and elevators), storage bins for raw material and final product, and other machines. Buttons on the screen can be clicked with the mouse and will start/stop some motors. The color of the objects usually indicates its status (started, stopped, full, empty, and so on), according to the values of associated sensors.

In this screen the operator can see information from all the sensors in the area, even the values from the weighing machines. Also the operator can manually command the execution elements from this area, e.g., can start automatic production or stop it by pressing the emergency button.

64.10 Automatic Feeding of Chicken

All this fodder production obtained in the feed plant is destined for feeding growing chicken in special farms. Chicken growth requires two separate processes: the supply of food and water to them and control of the environment. The chickens are prepared for sacrifice after 40 days, which is the optimum period for the chickens to be large enough and have tender meat. The control system of the factory must be very efficient in order to minimize the cost per kilo of meat. Two factors influence correct growth of chickens: food quality and environmental parameters (temperature and humidity). It is very important that the chicken do not waste energy adapting to the environment, so temperature and humidity control must be very accurate. Environment control strategy is different from summer to winter.

The process of supplying food and drink is not very complicated. There are sensors that check if there is

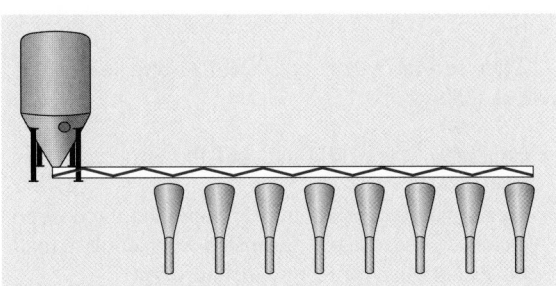

Fig. 64.7 The feeding system

food in the *plate*; when the plate is empty the control system starts the motors of the conveying system and stops them after a specific period of time. There is also an alarm system when the tank reaches a *lower limit* (Fig. 64.7).

64.11 Environment Control in the Chicken Plant

Environment control has to be the most accurate. Temperature and humidity must continually be maintained inside a specific range of values that differs from one day to another. To fulfil this, gas burners and fans are used to ventilate the interior with fresh air. To achieve a good distribution of temperature many small burners are used instead of fewer larger ones. Because the outside temperature differs between day and night, not all capacity of the fans is needed all the time. To adjust this, variable-frequency drive (VFD) speed control is used. A special situation occurs on cold days when, even with no fan speed, the air flow is too strong to maintain the temperature. There are three blowers and normally all of them have to work for a good dispersion of fresh air. Anyway there is the possibility to manually select the operation of each blower. Another option is to use dumpers in each fan to control the opening of the windows. The inside temperature is controlled by a proportional–integral–differential (PID) software al-

gorithm combined with control of the dumpers when needed. This is a continuous process and above it there is a batch process that provides the set point for temperature controller according to a temperature diagram. Usually this system is sufficient also to keep O_2 within prescribed limits, but it is not sufficient for humidity all of the time. The behavior of the indoors environment is for increasing humidity, so controlled opening of windows succeeds in maintaining the desired conditions.

Here the processes are minor, but the combination of a continuous process with a batch one does not result in a simple system. The performance of controlling these parameters combined with the provision of appropriate food for each stage makes the difference between companies. Another important influence on system performance is the positioning of temperature sensors, which should be uniformly distributed throughout the area.

64.12 Results and Conclusions

There were two important requirements to be accomplished by this automation system:

1. The ratio between different components contained in a final product had to be very close to the programmed values (maximum 5% error).
2. The quantity of final product must also be close to the programmed total quantity (2–3% error was requested).

After several weeks of system testing the results were as follows:

- Condition 1 was fulfilled, and the error was very small (2%).
- Final quantity was about 5% more than the desired quantity, because of estimated extraction, small losses in transporting system, and so on.

The solution adopted was to slightly reduce the scales for calculating the weight values for the weighing machines until the resulted quantity was within approximately 2% of the desired value. At that time the automation could be declared finished, with very good results.

We believe that the solution chosen for this automated system is very modern and has several advantages over what one might call classical solutions using other PLC models [64.2]:

- Process control and supervision do not require a PC, as all programs and visual interfaces run on the PLC.
- Fast and reliable network communication with PCs, by OPC server, in contrast to the widely used serial RS232 or RS485 communication in other controllers.
- The short working cycle due to fast processor.
- The extremely large number of I/O modules that can be managed (hundreds of digital signals monitored) by a single controller.

Future work will be to design a supervising user interface program that can access the controller directly the from internet [64.3], so, e.g., the owner of the factory can be informed about production flow even when far from the factory.

64.13 Further Reading

- T. Borangiu, F.D. Anton, S. Tunaru, A. Dogar, N.A. Ivanescu, N. Manu: High speed real-time robot vision design for flexible part feeding, Proc. 14th Int. Conf. Robotics, Alpe-Adria-Danube Region RAAD'05 (2005) pp. 337–345

- N. Ivanescu, S. Brotac, A. Rosu: Modern solution for controlling combined fodder plant, IMS (2007) pp. 236–241

References

64.1   T. Borangiu: *Sequential Controllers and Microprogramming* (Polytechnical, Bucharest 1993)
64.2   N. Ivanescu, S. Brotac, T. Borangiu, A. Rosu: *Modernizing a Fodder Plant*, Proc. IEEE SMC Int. Conf. Distrib. Hum.-Mach. Syst. (Atena 2008) pp. 543–548

64.3   N. Ivanescu, S. Brotac, T. Borangiu: A distributed and configurable architecture for controlling industrial processes, IMS, 132–137 (2001)

65. Securing Electrical Power System Operation

Petr Horacek

Automation in power systems has a very long tradition. Just recall the flyball governor in a steam engine and it becomes clear that power people have been using control principles and instruments for more than a century. There are, however, new challenges in power generation and transmission concerning the security and efficiency of those services that require the attentions of both theoreticians and practitioners. These challenges are the subject of this chapter. The power failures that affect large grids from time to time show that system collapses are not simply a subject of academic debate. Power network operating reliability has become an issue that any country must make a top priority. The reliability of a power system depends for the most part on the quality of the decisions made, both automatically and manually. The term *power system* is very broad, and we focus here on the power system backbone, electric power transmission, and its operational reliability, particularly when automatic control plays a vital role. The content of the chapter remains interdisciplinary, spanning power systems, automation and economy, as changes resulting from the opening of the markets and permanent power system restructuring affect its operation.

Part G | 65

There are several textbooks that survey the power industry – covering a regulated utility structure and the major concepts of deregulation, the history of electricity, the technical aspects, and the business of power – that are recommended for those who would like to get a broader picture of the industry [65.1].

Unlike other commodities, electricity generally cannot be stored for later use; it must be used as it is generated. Therefore, generation must be dispatched instantaneously to respond to real-time changes in consumers' demand for electricity. To support these energy flows, grid operation encompasses scheduling and managing flows over transmission lines and coordinating the operation of the transmission network equipment.

This chapter introduces methods for securing the operation of an electrical power system network, as

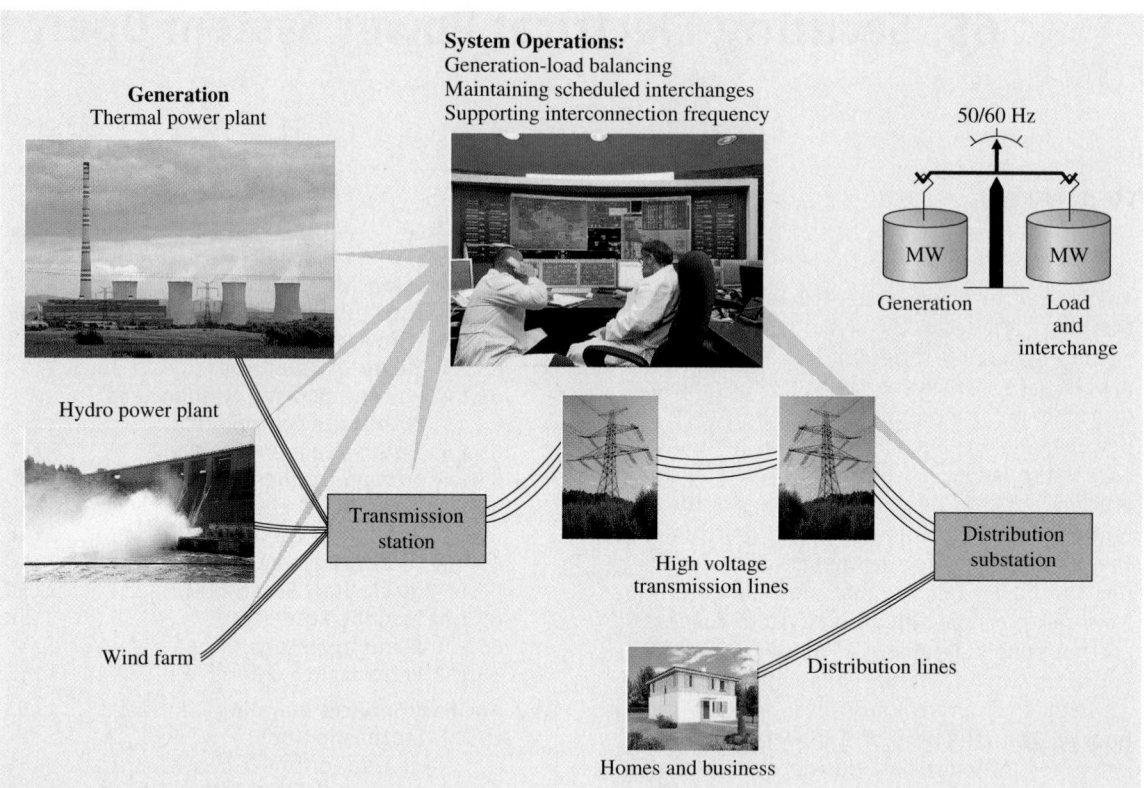

Fig. 65.1 Area system control objectives

shown schematically in Fig. 65.1, in terms of active power balancing through both automatic and manual interventions. Unlike a traditional control problem, where the task is to synthesize the control law for a given plant/process and constraints, our problem is to determine the constraints for a given control law so the control system has the right freedom in taking proper actions. Constraints appear in the form of regulating/balancing power reserves that have to be planned and reserved in advance so that the area control error, the instantaneous difference between the actual and the reference value for the power interchange of a control area (also known as the unintentional deviation), is kept within reasonable bounds. Balancing power is accessible in the form of ancillary services (AS), as provided by power-generating companies and traded in the market. Balancing power reserve activations are variables manipulated by a system operator (SO) which is responsible for power balancing in a geographical area. This is a complex task in a liberalized market environment. Power reserves, or rather the ancillary services providing the reserves, must be carefully planned and

purchased so that the SO is able to use them when required.

The chapter introduces a power-balancing problem, starting from a review of the basic concepts of controlling synchronous generators connected to large grids, such as the technical means for power balancing that are available, how the power balancing is distributed throughout the network, and who is in charge of the task. Next, criteria related to the successful completion of the balancing task are revealed for two large grids: the North American and European grids. Later, the principle of a decision-support tool used by a SO for year-ahead planning and procurement of ancillary services is described. A method originally designed for the area of the UCTE, the *Union for the Co-ordination of Transmission of Electricity*, in which 23 European countries are synchronously interconnected, can be made applicable to other conditions after modification. Some of the challenges associated with continuously evolving power systems, where automation plays an important role, are discussed at the end.

65.1 Power Balancing

In any electrical system, the active power generation and consumption must be balanced to prevent major losses of material and even human lives, as large or long-lasting power deviations may ultimately lead to blackouts if the balance moves out of control. Disturbance to this imbalance, first noticed as a system frequency deviation, will be offset initially by the kinetic energy of the rotating turbo generators and motors connected to the grid. The capacity of kinetic energy storing elements is insufficient to maintain the power equilibrium in real time. Generation units must be manipulated to conduct power balancing so that the network user is not affected by demand changes or generation and transmission outages.

65.1.1 The Problem

Secure operation of a transmission system and prevention of blackouts is an issue of rising importance for countries with deregulated electricity markets. The liberalization of electricity markets is a process that started a while back and is still in progress in many countries [65.2–5]. The basic motivation for this liberalization is to enable more effective power generation as well as investment and expansion planning than would occur in a traditionally vertically integrated electricity supply industry. On the demand side, end users are free to choose their supplier and to negotiate their contracts. On the supply side, producers can sell their electricity to any other market players. It is believed that this could possibly result in lower electricity prices. However, links between physics and business practices must be carefully maintained to ensure that enough balancing power is available when required, so the performance criteria of the system is guaranteed at the lowest cost possible. In another words, there is the need for active power balancing mechanisms, reserve planning and purchasing, which is the task performed by a balancing authority (BA) responsible for power balancing in an area within the electrical grid.

The BA is an electrical power system or combination of electrical power systems bounded by interconnection metering and telemetering. The BA balances the supply and demand within its area, maintains the interchange of power with other balancing authorities, and maintains the frequency of the electrical power system within reasonable limits.

65.1.2 Who Performs Power Balancing?

All grid operators are charged with maintaining the reliability of the systems under their control. Within their footprints, SOs oversee and direct the high-voltage bulk-power system and coordinate electricity generation to maintain a reliable supply of electrical power to electricity users. SOs provide critical reliability services, including outage coordination, generation scheduling, voltage management, ancillary services provision, load forecasting, and more. They improve reliability in part because of their large scope – by consolidating control areas, they reduce the number of decision makers managing the grid, which simplifies coordination and improves reliability.

System operators oversee grid functions and make necessary corrections to ensure reliability on a minute-to-minute basis around the clock. SOs may organize wholesale markets for energy and ancillary services, such as reserves, frequency and voltage regulation, and voltage support. These ancillary services help ensure reliability and help system operators react quickly and effectively to changing conditions on the grid, such as the loss of a generating unit or a transmission line.

Independent system operators (ISOs) and regional transmission organizations (RTOs) in North America and transmission system operators (TSOs) in continental Europe have the size, scope, scale, tools, information and authority to be effective grid managers. ISOs/RTOs/TSOs have near-identical responsibilities for managing the grid over a large geographic scope and operating markets to some extent.

Anyone who would like to get more details on the definitions, terminology, responsibilities and procedures that a particular system operator should use an internet search engine and look for the key word *Grid Code*. The *Grid Code* is an industrial textbook that every system operator must follow every day; it is carefully and regularly updated.

North America

There are three synchronous areas in North America: the Western, Eastern and Texas Interconnections. In each interconnection there are ISOs (BAs) and local balancing authorities (LBAs) that communicate. As part of operating the grid, ISOs/RTOs meet or exceed the reliability standards set by the North American Electric Reliability Corporation (NERC) and its regional councils. Adhering to NERC standards ensures that the entire

Table 65.1 Classification of ancillary services

Generic AS	Description
Continuous regulation	Provided by online resources with automatic controls that respond rapidly to operator requests for up and down movements. Used to track and correct minute-to-minute fluctuations in system load and generator output.
Energy imbalance management	Serves as a bridge between the regulation service and the hourly or half-hourly bid-in energy schedules; similar to but slower than continuous regulation. Also serves a financial (settlement) function in clearing spot markets.
Instantaneous contingency reserve	Provided by online resources equipped with frequency or other controls that can rapidly increase output or decrease consumption in response to a major disturbance or other contingency event.
Replacement reserves	Provided by resources with a slower response time that can be called upon to replace or supplement the instantaneous contingency reserve in restoring system stability.
Voltage control	The injection or absorption of reactive power to maintain transmission system voltages within required ranges.
Black start	Generation is able to start itself without support from the grid and has sufficient real and reactive capability and control to be useful in system restoration.

grid in North America operates at appropriate levels of reliability.

The ISO calculates the area control error (ACE; described later in the chapter) for its balancing authority area and measures the ACE against the applicable balancing standards. There are also local balancing authorities that are responsible for actual interchange.

To summarize, the ISO assumes the role of a BA, planning authority, reliability coordinator, market operator and tariff and interconnection administrator. LBAs take care of load and resource asset management, tie-line checkout and emergency operation coordination. ISO guarantees effective use of assets, including the dispatch of energy and the dispatch of reserves (including contingency reserves). The ISO organizes the energy and operating reserve market (also known as the ancillary services market), regulation and response services, and contingency reserves.

Europe

In continental Europe there is an equivalent to the American ISO/RTO Council, the UCTE. The UCTE also defines standards the members must adhere to, so the UCTE plays the role of the NERC as well. The difference is that members of the UCTE are system operators that are each responsible for a single control area (synchronously interconnected within the UCTE) and own the transmission assets. In many countries, the

transmission system operator (TSO) plays the role of the BA.

65.1.3 Means of Balancing

Power reserves are provided by generators and distributors in the form of ancillary services. By definition, ancillary services are interconnected operations services influencing transfer of electricity between purchasing and selling entities which a balancing authority must include in an open access transmission tariff.

Ancillary services are a collection of secondary services offered to help to insure the reliability and availability of energy to consumers. These services include regulation and contingency reserve (spinning, supplemental – nonspinning). The contingency reserve services are often referred to as operating reserves.

Ancillary services insure that capacity is available when needed to maintain secure power system operations due to a loss of or increase in load and resources.

A contingency is a trip of a transmission line or generator, a loss of load, or some combination of these events. This contingency in turn causes other problems, such as a transmission line overload, an over- or undervoltage in an area, over- or underfrequency, or frequency instability. Contingency reserves are a special percentage of generation capacity resources held back or reserved to meet emergency needs.

AS arrangements vary considerably across electricity markets. Let us introduce six generic AS that are necessary for maintaining system reliability and security in electricity grids. Ancillary services required during normal conditions are continuous regulation and energy imbalance management. Instantaneous contingency reserve and replacement reserve are services that are used during system contingencies. Ancillary services that do not apply to active power balancing directly are voltage support and black start. The generic AS categories are described more precisely in Table 65.1.

Ancillary service markets are critical to power system security and reliability and, given the value that customers place on reliable electrical services, their overall value to society is quite high.

Each market incorporates all six generic ancillary services, although the nomenclature, technical requirements and procurement details vary significantly. This chapter assumes that we are working with a TSO, and an area of the UCTE, in the European system, where transmission ownership and system operation is managed by a TSO as a single authority in the control area.

The role played by the TSO in the real-time power balancing can be understood from the overall block diagram in Fig. 65.2, which shows the essentials of the power balancing task while hiding the complexity of the very large scale power system. The control area in the center is synchronously connected to three adjacent areas A, B and C, making an interconnection.

The task for the TSO is to acquire enough power reserves in the form of ancillary services and activate them in a timely manner in order to guarantee the limits set on performance indices (standards) that evaluate the quality of power balancing. The task must be solved for the lowest cost possible. In this chapter, we define performance indices, describe the methods used in planning, and show how the planned reserves are validated. This chapter considers one year as the horizon for ancillary services planning.

65.1.4 Power Balancing Mechanisms Within the Control Area

What Is a Control Area?
A control area is an electrical power system that is managed under a common automatic control scheme

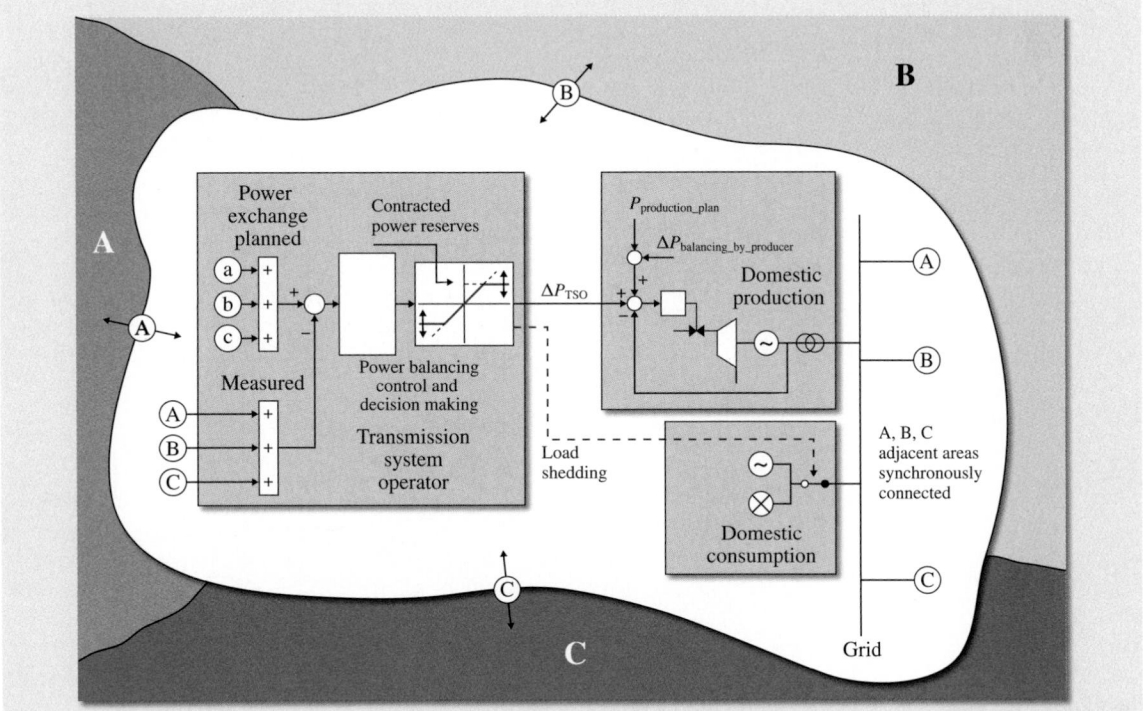

Fig. 65.2 Diagram showing the principle of power balancing in interconnected areas (frequency control loops not shown)

that maintains frequency by balancing load with production. Historically, a utility ran its own control area, regulating frequency, balancing load and generation, purchasing energy and capacity, and maintaining operating reserves as needed. Many large utility control areas served smaller in-area and nearby utilities as well. Today, some control areas exist to manage only generation, but most balance both generation and load. In North America, many control areas that existed historically have been consolidated into ISOs or RTOs and offer system balancing and management services that the control area no longer had to provide itself.

Power Balancing

To operate a large power system and to create suitable conditions for commercial electricity trade, it is necessary to schedule the power to be exchanged at the interconnection borders between the system operators in advance. During daily operation, the schedules are followed by means of the load-frequency control installed in each control area. Despite the functionality of load-frequency control, unintentional deviations occur in energy exchanges. For this reason, unintentional deviations are compensated for through methods of the TSO in every control area.

Real-time power balancing of a control area is usually performed by both automatic control and manual interventions from the TSO's dispatch center. The entire feedback control loop of a particular control area is shown in Fig. 65.3.

From the control theory perspective, the controlled variable is the instantaneous value of the cross-border power exchange $P_{measured}$, which should meet its scheduled value $P_{programmed}$. Manipulated variables are represented by various categories of regulation power available in the form of ancillary services. Disturbance variables influencing cross-border power exchange can be divided into two groups: deterministic, e.g., imbalances caused by trading strategies; and stochastic, mainly random demand fluctuations and generator outages.

The input to the controller is the ACE, the difference between the scheduled and actual cross-border power exchange, corrected for the effect of the primary frequency control so that the central controller does not compensate for outages of generators located in adjacent control areas (for example).

The output of the central controller is to increase or decrease power on the production side or, in the extreme case, on the consumption side as well. Despite the

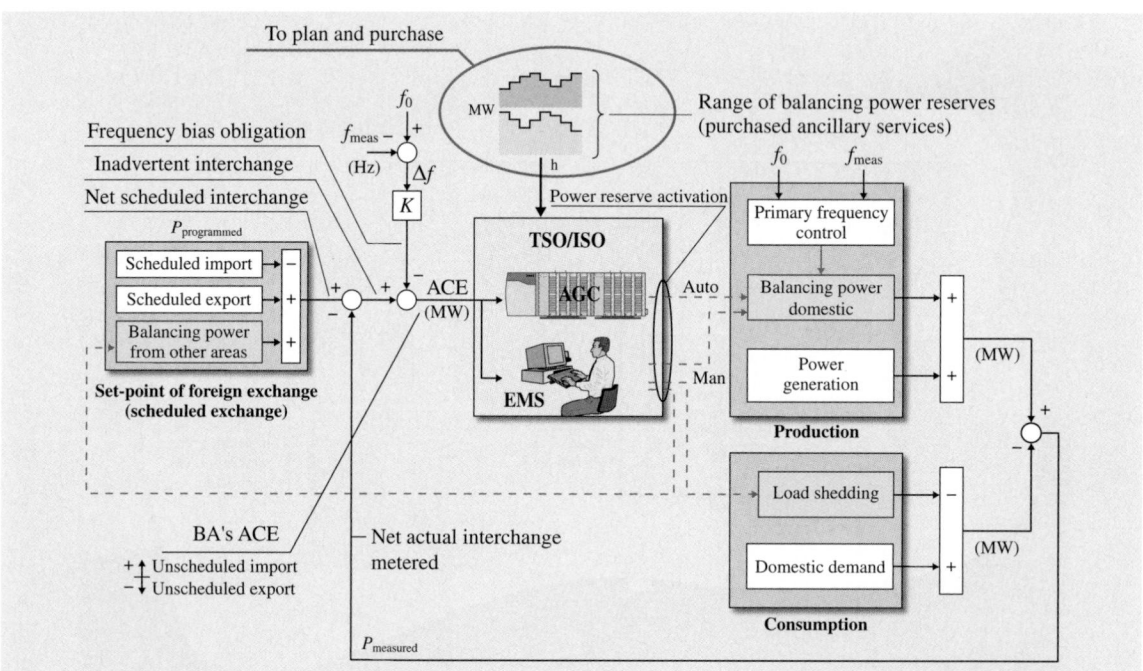

Fig. 65.3 Power balancing in the area as a feedback control system

fact that the manipulated variable is active power, a single variable, we have to deal with a number of outputs due to the fact that there are many actuators with different characteristics spread throughout the power system. Different turbo generators have different control regions and dynamics. Some units are in a standby mode and it takes some time to start them up and reach synchronous speed. There are several modes in which the unit control system can run. Thus, each producer can offer the TSO a certain amount of reserve power in several categories of AS. The TSO defines the portfolio of ancillary services which might differ from country to country, state to state, or area to area, and runs an AS market where a TSO purchases balancing power reserves for potential use. Examples of AS we deal with in this chapter when explaining a method of AS planning are shown in Table 65.2.

Ancillary services that are activated automatically are primary and secondary controls. The functionality of primary frequency control, or rather activation of the active power reserve RZPR, is distributed throughout the control area, and is part of a speed governor control system operating in droop control mode.

Activation of the secondary power reserve is conducted by the central TSO's control system. This scheme is associated with the term *automatic generation control* (AGC), which means the automatic adjustment of a control area's generation from a central location to maintain its interchange schedule plus frequency bias. AGC is also known as a load-frequency controller. Generators under secondary control (UCTE) must reach the demanded increase/decrease in power output within 10 min at a rate of at least 2 MW/min.

The remaining AS, used predominantly for energy imbalance management and during contingencies, are dispatched by the TSO manually.

Let us define basic terms before describing the methodology for AS planning.

65.1.5 System Frequency

The electrical frequency in the network, the system frequency f, is a measure of the rotation speed of the synchronized generators. As the total demand increases, the system frequency will decrease, and vice versa. The turbine speed controller performs an automatic primary control action to balance demand and generation. The frequency deviation Δf is influenced by both the total inertia in the system and the speed of primary control. Under undisturbed conditions, the system frequency must be maintained within strict limits in order to ensure the full and rapid deployment of control facilities in response to a disturbance.

65.1.6 Primary Frequency Control

The objective of primary frequency control (PR) is to maintain a balance between generation and consumption within the synchronous area using turbine speed governors. PR is an automatic function distributed throughout the network. When the generator is part of a large power system, and electrical power generation is shared by two or more machines, the frequency (speed) cannot be controlled such that it remains constant because it would forbid generation sharing between various synchronous generators. Control with speed droop is the solution that allows for fair generation sharing.

PR stabilizes the system frequency in seconds after a disturbance. In principle, the control law is proportional and the controlled system does not have integral character. The result is the steady state error in system frequency and, in addition, scheduled power exchanges between areas are not restored. Through the joint action of all primary controllers, PR ensures the operational reliability of the power system of the synchronous area. It stabilizes the frequency, but cannot drive the system

Part G | 65.1

Table 65.2 Examples of ancillary services and their categories

Generic AS		Preferred use	Area-specific AS identifier	Response time
Continuous regulation	Spinning	Primary frequency control	RZPR	$\approx 30\,\mathrm{s}$
Continuous regulation	Spinning	Secondary power and frequency control	RZSR	$\leq 15\,\mathrm{min}$
Energy imbalance management	Spinning	Load following	RZTR$^+$ RZTR$^-$	$\leq 30\,\mathrm{min}$
Instantaneous contingency reserve	Nonspinning	Operating reserve	RZQS	$\leq 15\,\mathrm{min}$
Replacement reserve	Nonspinning	Dispatch reserve	RZN$_{>30}$	$> 30\,\mathrm{min}$

frequency back to the original setpoint value and cannot restore cross-border power exchanges in the interconnected system, so they will differ from the values agreed between parties.

The overshoot or undershoot of the system frequency is a function of the amplitude and dynamics of the disturbance affecting the balance between power output and consumption, the kinetic energy of rotating machines in the system, the number of generators subject to PR (i.e., the primary control reserve and its distribution between these generators), the dynamic characteristics of the machines and their control systems, and the dynamic characteristics of loads, particularly the self-regulating effect of loads.

The steady state error of the system frequency is a function of the amplitude of the disturbance and the gain of the proportional controller, which is associated with the so-called network power system frequency characteristic.

Starting from undisturbed operation of the UCTE network [65.6], a sudden loss of 3000 MW of generating capacity must be offset by primary control alone, without the need for customer load-shedding in response to a frequency deviation. In addition, when the self-regulating effect of the system load is assumed to be 1%/Hz, the absolute frequency deviation must not exceed 180 mHz. Likewise, the sudden load-shedding of 3000 MW in total must not lead to a frequency deviation exceeding 180 mHz.

The primary control reserve for the entire synchronous area RZPR, also referred to as a continuous regulation reserve, is determined by the required performance described above, taking into account measurements, experience and theoretical considerations.

The share $RZPR_i$ of the control area i is defined by multiplying the calculated reserve for the synchronous area and the contribution coefficient C_i of the control area

$$RZPR_i = RZPR \, C_i \, , \tag{65.1}$$

where the contribution coefficient C_i is calculated on a regular basis for each control area or TSO using

$$C_i = \frac{E_i}{E} \, , \tag{65.2}$$

where E_i is the electricity generated in control area i and E is the total electricity produced in all control areas of the synchronous interconnection.

A deviation in system frequency Δf will, in the steady state, release primary control power throughout the synchronous area ΔP_{PR} as

$$\Delta P_{PR} = \lambda \Delta f \, , \tag{65.3}$$

where the gain λ is called the power system frequency characteristic of the synchronous area and λ is the sum of the power system frequency characteristics of all areas that are synchronously connected.

Each TSO is required to maintain a network power-frequency characteristic λ_i of the control area that is derived from the power-frequency characteristic designed for the overall synchronous area λ and the area contribution coefficient C_i

$$\lambda_i = C_i \lambda \, . \tag{65.4}$$

Each interconnected TSO must activate primary control power that is adequate for the system frequency deviation Δf using

$$\Delta P_{PR\,i} = \lambda_i \Delta f \, , \tag{65.5}$$

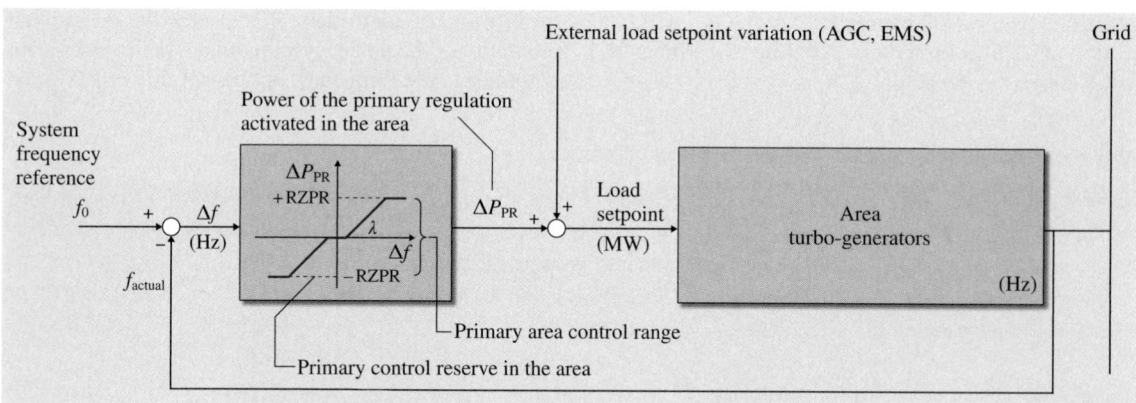

Fig. 65.4 Diagram showing the principle of the primary frequency control loop in the area

where $\Delta P_{\mathrm{PR}\,i}$ is the power variation generated locally in the control area in response to a frequency deviation caused by a disturbance (e.g., generation unit outage).

Figure 65.4 shows the principle of the primary control loop for the area. Note that the index of the area i is omitted from all power variables and λ for simplicity. The block diagram should actually be split further into control loops of individual turbo generators, as shown in Fig. 65.5. Generating units should deliver programmed (contracted) active power and provide power balancing functionality through primary, secondary and tertiary control at the same time. The programmed power is thus modified locally by primary frequency control independently of all other methods of power balancing used centrally by the TSO.

Primary control is performed by the speed governors of the generating units. With primary control, a variation in system frequency that is greater than the dead band will result in a change in unit power generation. Generators are required to participate in this control by setting the droop, which is directly related to the gain of the speed governor in Fig. 65.5, accord-

ing to specifications defined by the TSO. Transients of primary control have a timescale of seconds.

Power engineering is an area where different disciplines use different words for the same function, and many power engineers get frustrated when the literature fails to define or mention the terms *droop*, *isochronous* or *speed/load*, which refer to generator control [65.7]. Here are a few examples.

Rotating equipment people refer to *droop* control. The governor droop characteristic of a generation unit is given by the frequency deviation (% with respect to the nominal frequency) needed to change the generation power output (% with respect to nominal output) multiplied by 100. For example, a 5% droop means that a 5% frequency deviation causes a 100% change in power output. However, controls or instrumentation specialists would call this *proportional* control and might not even recognize the term *droop*. Frequently, utilities or powerhouse people will refer to this as *speed/load* control and may not recognize either of the other terms.

By the same token, people with a rotating equipment background will refer to *isochronous* control. Control engineers will call this PID control, and utilities or pow-

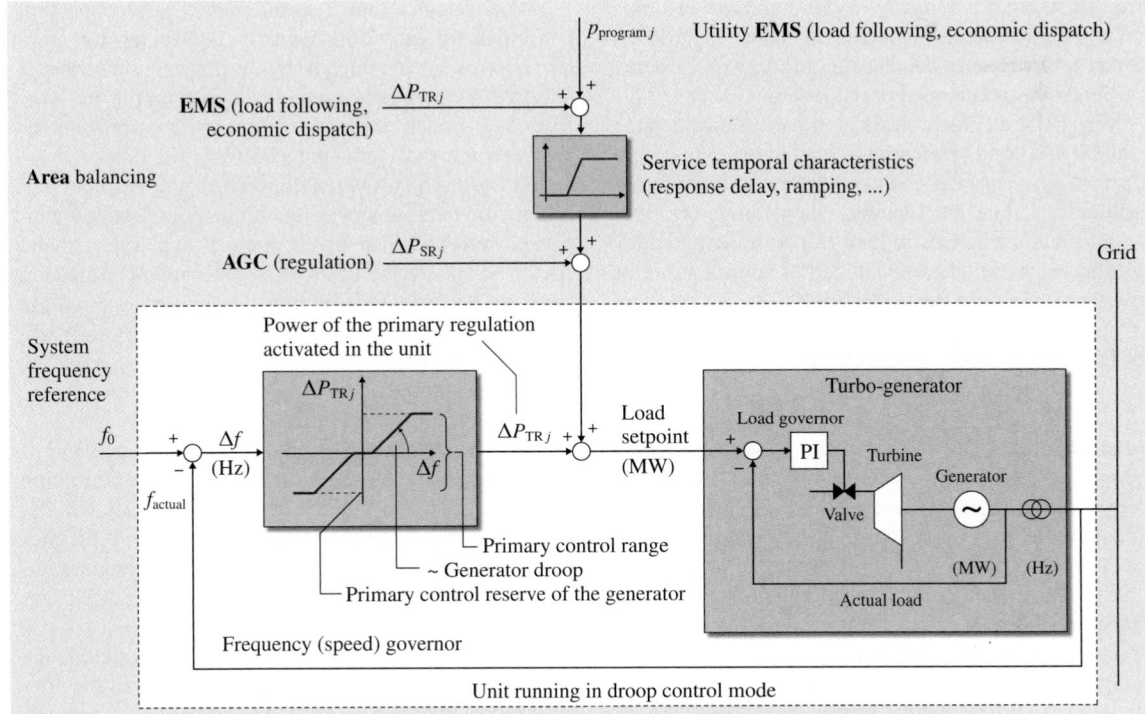

Fig. 65.5 Diagram showing the principle of the primary frequency control loop for a generating unit

erhouse people will use the terms *frequency* or *speed* control to refer to the same thing.

The synchronous generator is operated in frequency (isochronous) control mode when disconnected from the grid and is switched to droop control when synchronized and connected to the grid. On a small electrical grid, one machine is usually operated in isochronous speed control mode, maintaining the system frequency using a PID control loop. Any other (usually smaller) generators that are connected to the grid are operated in droop speed control mode, running under P control. If two prime movers operating in isochronous speed control mode are connected to the same electrical grid, they will usually *fight* to control the frequency, and wild oscillations of the grid frequency usually result. Only one machine can have its governor operating in isochronous speed control mode for stable grid frequency control when multiple units are being operated in parallel.

In very large electrical grids, commonly referred to as *infinite* electrical grids, there is no single machine operating in isochronous speed control mode that is capable of controlling the grid frequency, and all of the prime movers are operated in droop speed control mode. However, there are so many of them and the electrical grid is so large that no single unit can cause the grid frequency to increase or decrease by more than a fraction of a percent as it is loaded or unloaded.

Very large electrical grids require system operators to quickly respond to changes in load in order to control grid frequency properly, since there is no isochronous machine that does so. Usually, when things are operating normally, changes in load can be anticipated and additional generation can be added or subtracted in order to maintain tight frequency control.

65.1.7 Secondary Frequency and Power Control

Secondary control (SR) drives the system frequency and cross-border power exchanges back to the original desired (programmed) values after 15–30 min if the activated power does not reach the saturation limit of the reserve acquired. To prevent saturation and loss of secondary control functionality, additional power should be released to the system by activating reserves other than the primary balancing power reserves.

Transients of secondary control are on the order of minutes. Secondary control is also called AGC. In addition, AGC distributes the imbalance between selected units in an economical way. AGC is usually provided

by the area's balancing authority, the TSO in the case of UCTE interconnection. Generation scheduling and control is an important component of daily power system operation. The overall objective is to control the electrical output of generating units in order to supply the continuously changing customer power demands in an economical manner. Much of this functionality is provided by AGC as run by the balancing authority (TSO) and related functions operating within a utility control center energy management system (EMS).

Figure 65.6 shows a block diagram of the automatic SR control system, the function of which is realized by the TSO.

The output signal of the secondary PI controller ΔP_{SR} is distributed over participating generators with participating factors $\alpha_1, \ldots, \alpha_n$.

The function of SR, also known as load-frequency control or frequency-power control, is to keep or to restore the power balance in each control area, and consequently, to keep or to restore the system frequency f to its set-point value and the power interchanges with adjacent control areas to their programmed, scheduled values. This will ensure that the full reserve of primary control power RZPR will be recovered.

Whereas all control areas of the interconnection participate in providing primary control power, only the control area affected by a power imbalance is required to undertake secondary action for the correction. Parameters for the secondary control schemes are set such that only the controller in the zone affected by the disturbance concerned will respond and initiate the deployment of the requisite secondary control power. Within a given control area, the demand should be covered at all times by electricity produced in that area, including electricity imports (under purchase contracts and/or electricity production from jointly operated plants outside the zone concerned). Secondary control is applied to selected generator units in the power plants comprising the control loop.

Since it is technically impossible to guard against all random variables affecting production, consumption or transmission, the volume of reserve capacity will depend upon the level of acceptable risk. These principles will apply regardless of the distribution of responsibilities between the parties involved in supplying electricity to consumers.

In order to determine whether power interchange deviations are associated with an imbalance in the control area concerned or with the activation of primary control power, a corrective term should be introduced into the power interchange deviation. This is known

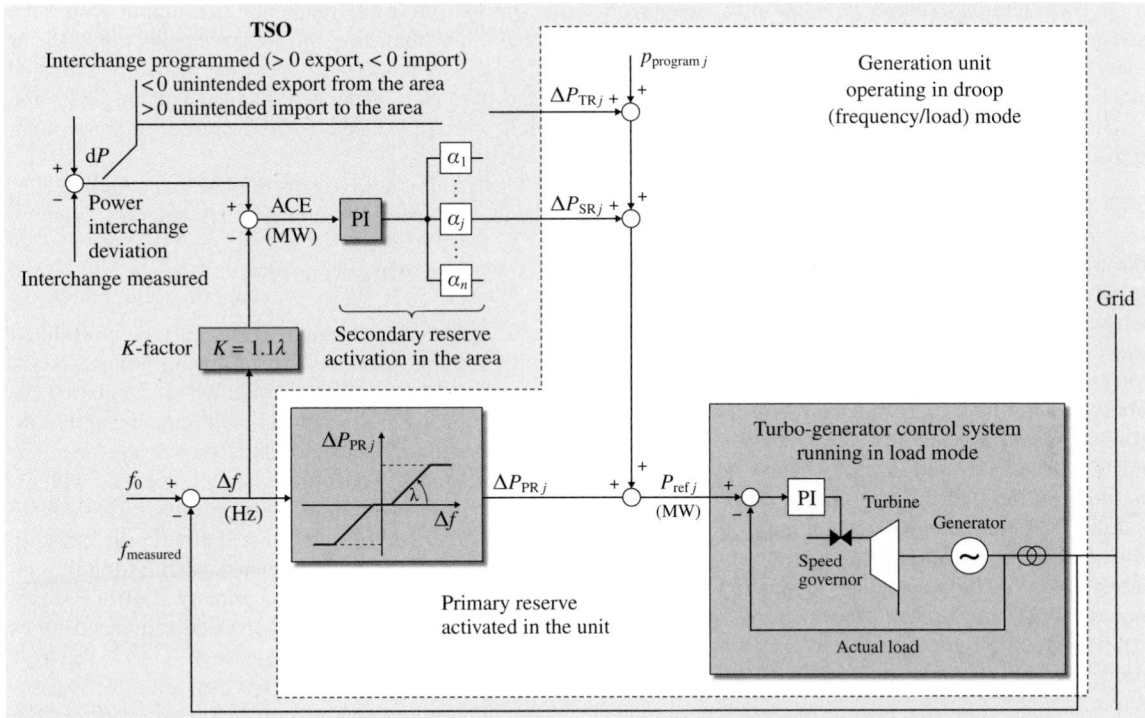

Fig. 65.6 Diagram showing the principle diagram of the secondary frequency and power control loop

as the network characteristic method, and its function is shown through the K-factor link displayed in Fig. 65.6. Due to the uncertainty in the self-regulating effect of the load, the K-factor K_j may be chosen to be slightly higher than the rated value of the power system frequency characteristic λ_j, such that the SR will accentuate the effect of the PR. Each control area is equipped with one secondary controller to minimize the ACE in real time:

$$\begin{aligned} \text{ACE} &= P_{\text{programmed}} - P_{\text{measured}} \\ &\quad - K\,(f_0 - f_{\text{measured}})\,, \end{aligned} \tag{65.6}$$

where P_{measured} is the sum of the instantaneous measured active power transfers on the tie-lines, $P_{\text{programmed}}$ is the exchange program with all the adjacent control areas, K is the K-factor of the control area, a constant (MW/Hz), f_0 is the setpoint, and f_{measured} is the instantaneous measured system frequency.

The ACE is the control area's imbalance $P_{\text{programmed}}$ $- P_{\text{measured}}$ with the hidden effect of the area's primary control. A different sign convention may be used for ACE relative to the respective TSO. In this chapter, the power transits are considered positive for export and negative for import. Hence, a positive (a negative) ACE

requires an increase (a decrease) in the secondary control power. The ACE must be kept close to zero in each control area at all times.

The secondary controller is implemented as a PI controller with additional features such as anti-wind-up and rate limiting.

The secondary control range RZSR is the range of adjustment of the secondary control power ΔP_{SR} within which the secondary controller can operate automatically in both directions (positive and negative).

In control areas of different sizes, load variations of varying magnitude must be corrected within approximately 15 min. The size of the RZSR required depends on the typical load variation, schedule changes, and generating units. The minimum value for the RZSR related to load variations in MW, recommended in UCTE, is obtained using the empirical formula

$$\text{RZSR} = \sqrt{aL_{\text{max}} + b^2} - b\,, \tag{65.7}$$

where L_{max} is the maximum anticipated load in MW for the control area, and the parameters a and b are established empirically, with the following values used by the UCTE: $a = 10\,\text{MW}$ and $b = 150\,\text{MW}$.

If consumption exceeds production for longer periods, immediate action must be applied using tertiary control, standby supplies, or contractual load shedding as a last resort.

Secondary reserve is also referred to as continuous regulation reserve.

65.1.8 Tertiary Reserve

The active power ΔP_{TR} of a tertiary reserve RZTR is usually activated manually by the TSO to free up an exhausted secondary reserve so that it can effectively reject common fluctuations. Tertiary control action involves a signal being sent by the TSO to a producer whose generating unit is included in the tertiary control scheme requesting that the control range of the unit should be changed (e.g., by changing the operating point of the boiler). This usually takes some time, and the frequency of such actions is also limited. The function of the tertiary control can be viewed as a dynamic change in the active power setpoint, which in turn moves the use of the secondary control reserve away from its limit value. The reserve denoted $RZTR^-$ is used to decrease the output and $RZTR^+$ to increase the output.

Tertiary control is an economic dispatch. It is used to drive the system as economically as possible and restore security levels if necessary.

The AS provider must provide the requested portion of the regulating reserve within a specified time limit (e.g., 30 min) and at a specified minimum rate, typically 2 MW/min.

The tertiary reserve is a load-following spinning reserve.

65.1.9 Quick-Start Reserve

The quick-start reserve (RZQS) is usually provided by pumped-storage power stations that are able to start generation within 10 min. The preferred use of RZQS is to compensate for large and sudden deviations due to generator outage. This reserve is activated manually from the TSO's control center. As the capacity of water reservoirs is limited, the use of the quick-start reserve is restricted in time.

The quick-start reserve is also referred to as the contingency nonspinning reserve.

65.1.10 Standby Reserve

The standby reserve (RZDZ) is typically represented by combined-cycle gas-turbine generation units. Depend-

ing on their ability to reach nominal output within the agreed time, they may be in the category RZN_{30}^+ or $RZN_{>30}$. The standby reserves are typically activated in the case of a generator outage or a longer-lasting deficit of energy in the system caused by some other reason.

The standby reserve falls into the category of a contingency replacement (supplemental) reserve.

65.1.11 Planning Reserves

There are lots of constraints that must be considered when planning reserves. The following nonusable capacity must be taken into account when calculating the capacity needed to meet power requirements: units subject to long-term shutdown; units shut down for repair and maintenance; limits on capacity associated with restrictions in fuel supplies; limits on capacity associated with environmental restrictions; limits on the capacity of hydroelectric plants associated with hydraulic and environmental constraints; the primary control reserve; reserves to cover variations in production and consumption (secondary and tertiary reserves).

In addition to these factors, which are directly associated with production, system conditions must also be considered, given that network constraints may limit the transmission of the power produced.

From the TSO point of view, the support system for AS planning should read the reliability requirements for power balance control and provide recommendations for the optimal composition of the AS, which the TSO should then acquire on the market. In particular, the system should be able to answer the following questions:

- What amount of reserves in the form of AS is needed in each category to insure that power balancing is realized with the desired reliability?
- What can providers of AS on the market be expected to offer in terms of their volumes and prices?
- How should the AS be selected based on the offer to insure that the requirements are covered in the optimal way?
- What are the expected costs associated with a certain level of operational reliability of the transmission system?

Similar tasks have emerged in all countries that have gone or are going through the liberalization process. From a global perspective, the safe operation of a transmission system can be viewed as a multicriteria optimization task. Due to a strong interaction between the technical and economic consequences of the planning and operating decisions, the two basic competing criteria – quality of electricity and costs of operating the transmission system – must be considered simultaneously.

Optimal planning of the AS involves market modeling [65.3]. Three main modeling trends can be identified: optimization models, which focus on a profit maximization problem for one of the players competing on the market; equilibrium models, which represent the overall market behavior, taking into account competition amongst all participants; and simulation models, often based on multiagent systems, which are an alternative to equilibrium models in cases where the problem under consideration is too complex. The models use a variety of computational techniques: linear programming, mixed integer programming, dynamic programming, nonlinear programming, heuristic approaches, and others.

Many papers focus on the pricing and optimal procurement of the energy and AS. If the market products (energy and AS) are procured simultaneously through central (usually day-ahead) auctions, the task to solve is to simultaneously co-optimize energy and AS ensuring that costs are minimized but safe operation of the transmission system is retained. This involves addressing issues such as optimizing power flow, unit commitment, congestion management and emission constraints [65.4–8]. The products on the market may be procured sequentially through central auctions managed by the ISO/TSO, usually in the following order: energy market, transmission market to resolve congestions (if needed), and market for each category of the AS (from the fastest-responding to the slowest-responding service). In such cases, the problem of how to optimally purchase AS taking into account the downward substitutability of the AS arises, since the market clearing price in each market depends on the volume required. A rational buyer's algorithm described in papers [65.9, 10] can handle this problem.

However, the complex task of optimally planning AS is still a challenging one. Although methods solving some of the partial subtasks exist, they must be carefully chosen and modified to fit in the context of a specific country or region. The inner structure of the global AS planning problem may also vary from country to country.

Let's assume that the generation portfolio of a particular control area consists of dozens of 110 MW and 200 MW generation units, some 500 MW and 1000 MW

generation units, and that the peak load is approximately 11 GW. Let's also assume a dense network so that no inner flow congestion management is needed; capacity auctions will only be organized for cross-border tie-lines.

This control area is part of the European synchronous interconnection (UCTE), and as such its operation must comply with the UCTE rules [65.6] and relevant legislation, which (among others) states the liability of the TSO for frequency and power balance maintenance. For this purpose, the TSO purchases most of the AS through free-market tenders organized typically for three-year and one-year periods. The TSO only spends money on reserving the AS (purchase); the costs of AS activation are covered later by the market participants that cause the power imbalance. Hence, for the rest of the chapter, when we refer to the cost of AS, we mean the cost of reserving the AS.

Energy markets are separate from AS markets. Most energy is traded in the form of bilateral contracts for base-load products on a year-ahead basis. Peak-load products are mainly associated with short-term energy markets. We assume that the TSO is not allowed to participate in energy markets.

65.1.12 Performance Criteria

Different interconnections may have chosen and use slightly different performance criteria to evaluate the reliability of system services in terms of power balancing.

The North American Electric Reliability Corporation (NERC) defines three control performance standards (CPS) for the assessment of control area generation control performance: CPS1, CPS2 and DCS [65.11]. All control areas in North America implemented CPS in 1998.

CPS1. Each balancing authority operates such that, on a rolling twelve-month basis, the scaled average of clock-minute averages of the ACE of the area multiplied by the corresponding clock-minute averages of the interconnection's frequency error is less than a specific limit. The index of the area i will be omitted for the rest of the chapter for simplicity, as our discussion only concerns a single area of the interconnection.

CPS1 is a statistical measure of ACE variability. CPS1 measures ACE in combination with the interconnection's frequency error. CPS1 requires that the average of the clock-minute averages of a control area's ACE over a given period divided by its K factor multiplied by the corresponding clock-minute averages of the frequency error should be less than a given constant

$$\text{AVG}_{\text{Period}} \left(\frac{\text{ACE}}{K_{\text{area}}} \Delta f \right) \leq \varepsilon_1^2 . \tag{65.8}$$

The constant ε_1 is derived from the targeted frequency bound (the targeted RMS value of one-minute average frequency error based on frequency performance over a given year). For the purpose of control performance evaluation, CPS1 is evaluated monthly for the previous (sliding) twelve-month period.

CPS1 measures control performance by comparing how well a control area's ACE performs in conjunction with the frequency error of the interconnection. Equation 65.8 can be viewed as a correlation between ACE and Δf. Positive correlation means undesirable performance (the area control error contributes to the frequency deviation from the desired value) and is therefore limited by the upper bound ε_1^2. Negative correlation occurs when ACE helps to compensate the total ACE of the interconnection and is helping to offset the system frequency deviation.

CPS2. The other criterion, CPS2, is designed to bound ten-minute ACE averages and provides an oversight function to limit excessive unscheduled power flows that could result from large ACE

$$\text{AVG}_{10\text{-minute}}(\text{ACE}) \leq \varepsilon_{10} \sqrt{K_{\text{area}} K_{\text{interconnection}}} , \tag{65.9}$$

where $K_{\text{interconnection}}$ is the sum of the K factors in the entire interconnection and ε_{10} is the targeted RMS of the ten-minute average frequency error. Each balancing authority operates such that its average ACE for at least 90% of the ten-minute periods (six nonoverlapping periods per hour) during a calendar month is within a specific limit, referred to as L_{10}. Extensive analysis and further discussions of the two performance indices can be found in [65.11–14].

NERC does not have the statutory authority to enforce compliance with its reliability standards. Despite its lack of authority, NERC has made every effort to continuously clarify and upgrade its reliability standards as the electrical power industry has evolved. NERC has also enhanced the monitoring of compliance with its standards by individual entities, independent system operators, and regional reliability councils.

Each control area can meet the CPS standards by any means they wish. A control area can potentially use current control schemes, meet its generating obligations manually, or implement other control schemes.

A control area that does not meet the CPS is not allowed to sell control services to other parties external to its metered boundaries. This impacts the purchasing control services from this control area. This is a significant penalty given new operating environments.

Benefits of CPS include the fact that it is technically defensible, allows smarter control by reducing unit maneuvering, provides more accurate assessment of control area performance, reduces generating unit operating and maintenance costs, provides the ability to monitor and control frequency performance, and allows less restrictive short-term performance requirements in exchange for more restrictive long-term requirements.

Disturbance Control Standard (DCS). The purpose of the disturbance control standard (DCS) is to ensure that the balancing authority is able to utilize its contingency reserve to balance resources and demand and return interconnection frequency within defined limits following a reportable disturbance. Because generator outages are far more common than significant losses of load, and because contingency reserve activation does not typically apply to the loss of load, the application of DCS is limited to the loss of supply and does not apply to the loss of load.

A disturbance is defined as any event that is $\geq 80\%$ of the magnitude of the control area's most severe single contingency. A control area is responsible for recovering from a disturbance within 10 min by recovering the amount of the disturbance or returning ACE to zero. A disturbance is not reportable if it is greater than the control area's most severe contingency.

The control area must comply with the DCS 100% of the time. Any control area that does not comply is required to carry an additional contingency reserve. The extra reserves must be carried for the quarter following the quarter in which the noncompliance occurs.

Europe (UCTE)

According to the UCTE Operation Handbook [65.6], the individual ACE_i needs to be controlled to zero on a continual basis in each control area. In addition, the frequency deviation should decay to the given setpoint in less than 15 min, and any power outage should be compensated for accordingly. Both large and/or long-lasting ACE deviations should be avoided as much as possible.

These general quality requests will be specified in a more rigorous way in Sect. 65.2.2 when working with one-minute and one-hour averages of ACE, $\text{AVG}_{1\text{-min}}(\text{ACE}) = \text{ACE}_1$, $\text{AVG}_{1\text{-h}}(\text{ACE}) = \text{ACE}_{60}$ and other indices, so that it can be used as performance criteria and power balancing quality indices. The soft constraints L_1 and L_{60}, similar to L_{10} for NERC's CPS2, will be used.

Some of the requests can be represented as a mean value of ACE that should be kept at zero and a standard deviation of ACE that should be kept low. The mean and the standard deviation of the ACE is sometimes used to compare the operations of control areas in a single interconnection [65.15].

Setting the limit values of the selected performance indices is a delicate task. In general, if the limits are set too high, larger and/or longer-lasting power imbalances are assumed to be allowed. On the other hand, setting the limits too low may request an amount of balancing power reserve that is simply not available in the respective category in the area, and/or reserving that power will become too expensive.

In the following section, the task and the principle of the algorithm for AS planning is described.

65.2 Ancillary Services Planning

Each ISO/TSO can adopt different AS planning procedures depending on the interconnection and control area specificity. Recent changes in market structures make the problem more difficult, since there is a lack of long-term experience in effective AS planning. The principles, various criteria and constraints adopted in the method described in this section may serve as the basis for a general platform, despite the fact that the approach was developed for one control area in the UCTE interconnection. The method is currently being used by the TSO of the Czech Republic as a support tool and is described in [65.16] in greater detail.

The entire AS planning task can be decomposed into several subtasks (as depicted in Fig. 65.7) that are executed in the order indicated by the respective block numbering.

The control area is viewed as a pool. First, the ACE of the area is modeled as if it would run without AGC and EMS actions used centrally by the TSO (Step 1). Various horizons may be considered. One year is assumed for the reminder of the chapter. The variability of such an *open-loop ACE* is high. Considering the control and disturbance performance standards, the request for AS to decrease ACE variability is calculated

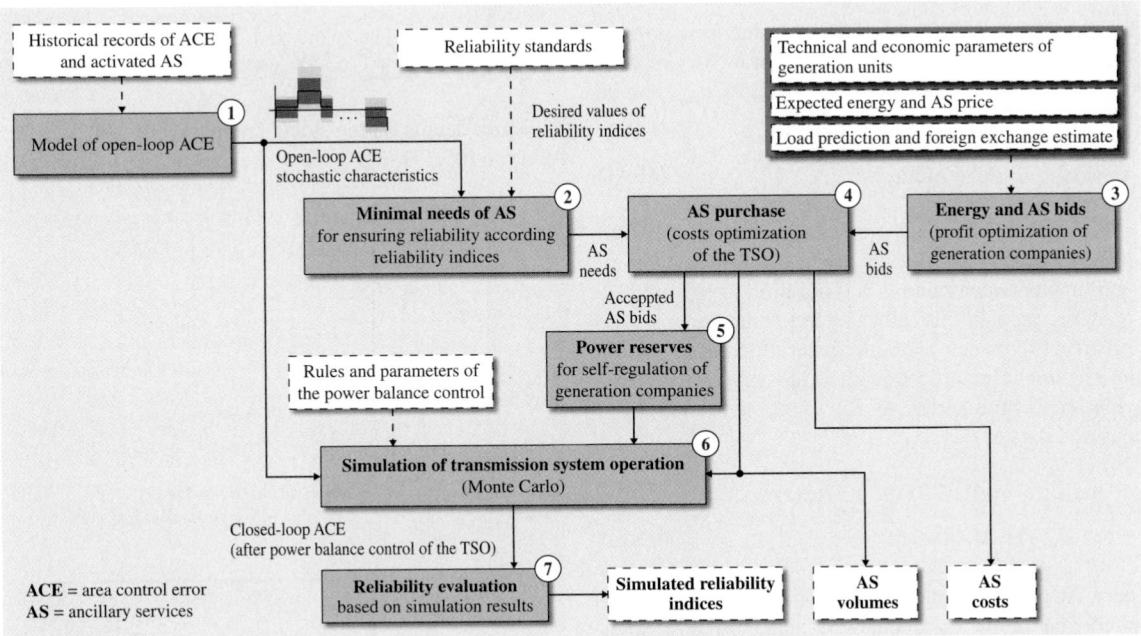

Fig. 65.7 Block diagram of ancillary services planning and validating procedures

(Step 2). If the availability of the AS is not limited and the cost of the AS is not an issue, the algorithm proceeds in running a Monte Carlo simulation of the area (Step 6) that includes TSO's AGC and EMS actions, and gets ACE statistics. Once the ACE time series has been obtained, any performance indices and conditions introduced by respective standards can be evaluated, including NERC's CPS2 and DCS.

In reality, AS is a commodity and the TSO must go to the market, compare the offerings to what is needed to meet the performance standards, and consider the AS price when purchasing. The initial plan for AS may not then be feasible, as some of the services needed may not be offered in sufficient amounts. However, there is some freedom to restructure the needs, as some faster services can be substituted for slower ones. This leads to a linear programming problem with the aim of minimizing the cost of the purchase and meeting the performance standards at the same time. AS bids are calculated (Step 3) and the optimal purchase is made, replacing the originally assumed set of AS by the feasible and cost-optimized one (Step 4).

The next sections briefly review the procedure. More details can be found in [65.16, 17].

65.2.1 Stochastic Model of Area Control Error

The *open-loop ACE* is considered a random process with characteristics related to the recent history of control area operation. The open-loop ACE ACE_{OV} can be decomposed into two components

$$ACE_{OV} = P_V + ACE_O . \tag{65.10}$$

P_V represents forced generation unit outages and will be modeled as a Markov process parameterized by mean time between failures and duration of repairs.

ACE_O accounts for other phenomena, for instance a mismatch between load and generation, and is modeled as a stochastic process with a Gaussian distribution. A historical time series ACE_O^H is reconstructed from measured data as

$$ACE_O^H(t) = ACE^H(t) + \sum AS(t) ,$$
$$\forall t : P_V^H(t) = 0 , \tag{65.11}$$

where ACE^H is the measured ACE, P_V^H are recorded forced generation units outages, and $\sum AS$ denotes the sum of all activated ancillary services at the time

instant t. ACE_O represents open-loop ACE in undisturbed conditions only. Then the statistical parameters of ACE_O are evaluated from ACE_O^H. A convolution between the probability density functions of ACE_O and P_V results in the probability density function of the open-loop ACE, ACE_{OV} (Fig. 65.8), which shows the performance of the control area in terms of how frequently a particular magnitude of the ACE appears in relation to the total number of observations. As shown, the positive part of the function is modulated by the forced generation unit outages.

The component ACE_O can further be decomposed into a *slow component* ACE_{O_slow}, which represents *low-frequency* content with a quasi-period of 10–30 min corresponding to trend changes in ACE_O, and a *fast component* ACE_{O_fast}, which represents *high-frequency* noise with a quasi-period of several minutes caused by common fluctuations in the system

$$ACE_O = ACE_{O\_slow} + ACE_{O\_fast} . \tag{65.12}$$

This decomposition will be used later when determining AS needs, because each of the components is compensated by a different type of AS. The ACE_{O_fast} is calculated from ACE_O^H as a moving hour average, and the remaining *noise* represents ACE_{O_slow}. Finally, the open-loop area control error ACE_{OV} is a sum of slow and fast variations of ACE_O plus forced generator outages

$$ACE_{OV} = ACE_{O\_slow} + ACE_{O\_fast} + P_V . \tag{65.13}$$

More details on the stochastic model of ACE can be found in [65.17].

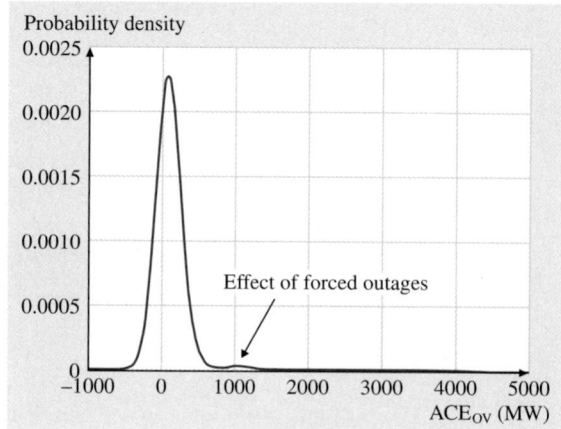

Fig. 65.8 Probability density function of ACE_{OV}

65.2.2 Minimal Needs of AS

In order to determine the power reserves in the form of AS that would technically suffice for proper power balance control with acceptable reliability, it is necessary to refer to reliability standards. Since the UCTE has no specific reliability standards for its members [65.18], seven indices describing reliability are defined in Table 65.3 for this purpose. Each of them represents a probability that ACE exceeds a certain threshold and lasts for a certain amount of time (value at risk). The index $rACE_1$ refers to the one-minute average value of ACE, while the index $rACE_{60}$ refers to an average value of ACE over 1 h.

The indices $rACE_1$ and $rACE_{60}$ are statistical measures of ACE that are related to NERC's CPS2. $rACE_{1t}$ is linked to the UCTE requirement for frequency recovery in less than 15 min after a forced generation unit outage, and thus it is similar to NERC's DCS index. The other indices ($rACE_{60\,tx}$) indicate the possible danger of a longer-lasting power imbalance that may lead to an application of load-shedding mechanisms or to a blackout in the worst case.

The desired values were determined statistically from historical records of ACE. A few consecutive years should be evaluated. The duration of the period assigned for statistical evaluation should correspond to a period when the technical parameters of the transmission system and the market conditions are compatible with the current situation and practices. In addition, no problems related to the area performance should be reported from either TSOs in adjacent areas or from domestic customers. Hence, we can assume that the TSO's performance was satisfactory.

The indices are direct measures of the quality of power balance control in the control area, and indirect measures of the reliability of the power supply to end-consumers.

The AS categories considered throughout the chapter are described in Table 65.2. Analytical computations of the minimal AS needs are based on the stochastic characteristics of ACE and its components, as determined in Step 1. Regarding the fact that some of the AS are mutually substitutable (i.e., faster-responding services such as secondary control can be substituted for slower-responding services such as tertiary control),

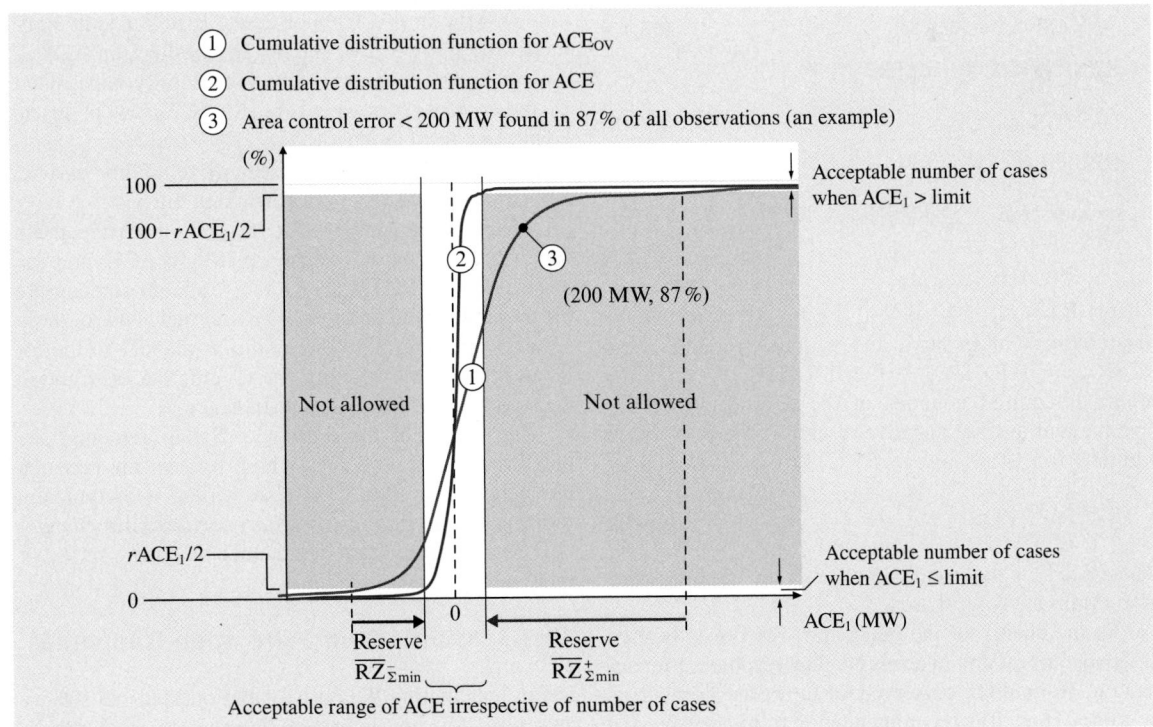

Fig. 65.9 Cumulative distribution function of the open-loop ACE and its use

Table 65.3 Reliability indices based on area control error time series evaluation

Performance index	Limit value (%)	Description
$r\text{ACE}_1$	3.2	Probability that the absolute value of the one-minute average of ACE exceeds $L_1 = 100\,\text{MW}$
$r\text{ACE}_{60}$	4.6	Probability that the absolute value of the one-hour average of ACE exceeds $L_{60} = 20\,\text{MW}$
$r\text{ACE}_{1t}$	0.027	Probability that the one-minute average of ACE is higher/lower than $\pm 100\,\text{MW}$ for more than 15 min
$r\text{ACE}_{60t2}$	0.829	Probability that the one-hour average of ACE is higher/lower than $\pm 20\,\text{MW}$ within two consecutive hours
$r\text{ACE}_{60t3}$	0.140	Probability that the one-hour average of ACE is higher/lower than $\pm 20\,\text{MW}$ within three consecutive hours
$r\text{ACE}_{60t4}$	0.040	Probability that the one-hour average of ACE is higher/lower than $\pm 20\,\text{MW}$ within three consecutive hours

the needs of the AS reserves are expressed in the form of inequalities (65.14)–(65.18) rather than exact recommended values for each AS category

$$\text{RZSR} \geq \text{RZSR}_{\min} , \tag{65.14}$$

$$\text{RZSR} + \text{RZTR}^+ \geq \text{RZSR}_{\min} + \text{RZTR}^+_{\min} , \tag{65.15}$$

$$\text{RZSR} + \text{RZQS} + \text{RZTR}^+ \geq \text{RZ}^+_{\sum \min} , \tag{65.16}$$

$$\text{RZSR} + \text{RZTR}^- \geq \max\left(\text{RZ}^-_{\sum \min}, \text{RZSR}_{\min} + \text{RZTR}^-_{\min}\right) , \tag{65.17}$$

where $\text{RZ}^+_{\sum \min} \geq 0$, $\text{RZ}^-_{\sum \min} \geq 0$ are total minimal volumes of positive and negative reserves, and $\text{RZSR}_{\min} \geq 0$, $\text{RZTR}^+_{\min} \geq 0$ and $\text{RZTR}^-_{\min} \geq 0$ are the minimal required volumes of the secondary, tertiary positive and tertiary negative reserves, respectively. In addition, (65.18) represents a UCTE $N-1$ criterion

$$\text{RZSR} + \text{RZQS} \geq C_{N-1} , \tag{65.18}$$

where C_{N-1} is the installed power of the largest generator within the control area.

As the energy of the quick-start reserve is limited and due to its quality in terms of a fast response after activation, it should be preserved for future use as much as possible. Thus, it is recommended that this fast service should be replaced by a slower one whenever possible, and the same amount of reserve should be planned for such a slower service. This reasoning is expressed by

$$\text{RZN}_{>30} = \text{RZQS} . \tag{65.19}$$

Figure 65.9 illustrates how the total minimal volumes $\text{RZ}^+_{\sum \min} \geq 0$ and $\text{RZ}^-_{\sum \min} \geq 0$ are determined from the cumulative distribution function of ACE_{OV}. According to the reliability standards, the absolute value of the closed-loop ACE is allowed to exceed the threshold $100\,\text{MW}$ in $r\text{ACE}_1\%$ of cases. If this is split symmetrically into positive and negative values, an ACE_{OV} that is higher than $100\,\text{MW}$ should be compensated by the control reserves except in $r\text{ACE}_1/2\%$ of cases. Hence, we need at least $\text{RZ}^+_{\sum \min} \geq 0$ and $\text{RZ}^-_{\sum \min} \geq 0$ reserves to satisfy the standards of reliability derived from area satisfactory performance in the past.

The secondary reserve $\text{RZSR}_{\min}$ should compensate for the fast variations of the open-loop ACE, and the tertiary reserve $\text{RZTR}^+_{\min}$, $\text{RZTR}^-_{\min}$ should compensate for the slow ACE variations. The minimal needs of these services are determined in a similar manner to the total needs, but by utilizing the cumulative distribution functions of $\text{ACE}_{\text{OV\_slow}}$ and $\text{ACE}_{\text{OV\_fast}}$.

The values of the minimal AS requirements take the form of a time series, which reflects the fact that the behavior of the ACE also varies; it is typically more uncertain during *transition periods* with changeable weather and temperature, such as during spring or fall.

65.2.3 AS Bids from Generation Companies

Before calculating the costs of the optimal AS set, we also need to know what energy and AS bids can be expected on the market (Fig. 65.7). The underlying assumption is that under liberalized market conditions, generation companies are only motivated by economic factors, and their aim is to maximize their profit regardless of transmission system needs and reliability issues. A generation company decides how much of its installed capacity will preferably be offered in the form of energy sold to domestic retailers or abroad and what amount will be offered to the TSO in the form of AS. This part of the model solves a mixed integer linear programming problem involving unit commitment. The main model inputs and outputs are summarized in Table 65.4.

The model does not search for a demand-bid equilibrium on the market, because the computational time associated with a year-ahead simulation would be too high. Hence, the expected AS and energy prices are entered as the inputs. In other words, the model tells the TSO the amount of AS reserves that are expected to be available on the market if their prices are given. Thus, for instance, the TSO can enter the maximum acceptable prices of AS and see whether or not the bids meet the minimal AS needs in such a case.

65.2.4 Procurement of Ancillary Services

The key point is to optimize the AS selection covering the minimal AS needs based on the lowest costs for AS reservation. In the case where the minimal needs of AS in a certain category determined in Step 2 are higher than the offer on the market, the optimal purchase algorithm takes this limited offer into account by substituting the AS with a different AS category if possible. The optimal purchase is stated as a linear optimization problem

$$
\min_{\substack{\forall k \\ \text{RZSR}_k \\ \text{RZTR}_k^+ \\ \text{RZTR}_k^- \\ \text{RZQS}_k \\ \text{RZN}_{>30k}}} \sum_{\forall k} \left(C_{\text{RZSR}_k} \text{RZSR}_k + C_{\text{RZTR}_k^+} \text{RZTR}_k^+ \right.
$$
$$
+ C_{\text{RZTR}_k^-} \text{RZTR}_k^- + C_{\text{RZQS}_k} \text{RZQS}_k
$$
$$
\left. + C_{\text{RZN}_{>30k}} \text{RZN}_{>30k} \right) \tag{65.20}
$$

subject to the constraints (65.14)–(65.18). The index k denotes individual AS providers, C_{RZSR_k}, $C_{\text{RZTR}_k^+}$, $C_{\text{RZTR}_k^-}$, C_{RZQS_k}, $C_{\text{RZN}_{>30k}}$ are bid prices of the respective AS, and RZSR_k, RZTR_k^+, RZTR_k^-, RZQS_k, $\text{RZN}_{>30k}$ are the AS volumes purchased from the kth AS provider.

Table 65.4 Main inputs and outputs of the AS market model

Model inputs	
1.	List of power producers acting on the market and portfolio of their generation units
2.	Technical data for generating units (type of technology, minimal and maximal outputs, certified AS ranges, etc.)
3.	Economic characteristics of generating units (variable costs, start-up and shut-down costs)
4.	Planned unit outages
5.	Domestic power consumption and foreign exchange forecast
6.	TSO's expert estimate of future energy and AS prices based on knowledge of the current market situation and possible future trends
Model outputs	
1.	Time series of expected volumes of AS offered by each of the AS providers in the market

65.2.5 Availability of Self-Regulation Power

The transmission system is balanced by not only the TSO but also the area generators. The generation companies are economically motivated to meet their contracted production. Thus, if one or more of their units fail, the remaining capacity of the other units is activated, if possible, to compensate for the forced outage (self-regulation). However, this ability varies in time and depends on instantaneous domestic consumption, foreign exchange, etc. Therefore, an algorithm for computing time-variable available capacity for self-regulation is implemented and conducted as Step 5 in Fig. 65.7. In principle, the remaining available capacity P_{AC} can be determined as

$$
P_{\text{AC}} = P_{\text{IC}} - P_{\text{OUT}} - P_{\text{E}} - P_{\text{AS}} , \tag{65.21}
$$

where P_{IC} is the installed capacity of generation companies, P_{OUT} are planned outages, P_{E} is the sold energy and P_{AS} is sold AS. The self-regulation ability is then taken into account in Monte Carlo simulations where the forced generation unit outages are randomly generated.

65.2.6 Monte Carlo Simulation of Power Balancing

The quality of the transmission system operation must be validated for the intended ancillary services acquired in Step 4 in Fig. 65.7 for several reasons. First, only two of the performance indices $r\text{ACE}_1$ and $r\text{ACE}_{60}$ evaluating the ACE time series were taken into account when calculating technical needs of the AS. Second, purchasing the AS may result in a different composition of AS due to the limited offer from the AS providers. To learn more about the behavior of the transmission system, a power balancing control system simulator is used to simulate the automatic activation of AS and to mimic human operator interventions at the TSO control center [65.19].

The logic used by the TSO operator manually activating the respective categories of AS is modeled as if-then rules. The operator evaluates past and predicted values of ACE, considers the balancing power reserves left for immediate and near-future use and the dynamics of the respective AS (i. e., the delay, ramp up, ramp down), and checks the time taken for the activation or deactivation of the AS to be realized. By means of Monte Carlo simulations, this model generates time series for open-loop ACE [65.19], AS activations/deactivations (both automatic and operator-based) and, consequently, closed-loop ACE during a planning period when using the recommended set of AS. Additional nonguaranteed types of AS, such as emergency assistance from abroad and the purchasing of balancing energy are also used in simulation runs. These types of services are characterized by zero reservation costs and limited availability (i. e., the services might be temporarily unavailable when required by the operator for balance control). The availability of the nonguaranteed types of AS is a parameter that can be set in the simulator.

The master block diagram shown in Fig. 65.10 gives an overall picture of the simulator used in Step 6 in Fig. 65.7. The simulator is directly related to the feedback control scheme from Fig. 65.3.

Simulated ACE_{OV}, ACE with automatic activation of the secondary reserve and manual activation of the tertiary reserve, quick-start and standby reserves, emergency assistance from abroad, and balancing energy from abroad are shown in Fig. 65.11. The simulation shows a record for two days with a large power deviation originating from an outage of a 1000 MW generator. The area control error not compensated by the TSO ACE_{OV} is the curve with the highest magnitude. The producer compensates for the unintended power deviation by activating its own power reserves. As a result, ACE_{OV} tends to be drawn towards zero but the self-regulating action is not efficient enough and so a significant power deviation remains for about 12 h, which is unacceptable.

The situation improves when the TSO activates its reserves. The secondary controller responds to the sudden increase in ACE almost immediately (red). As the secondary power reserve available at that time

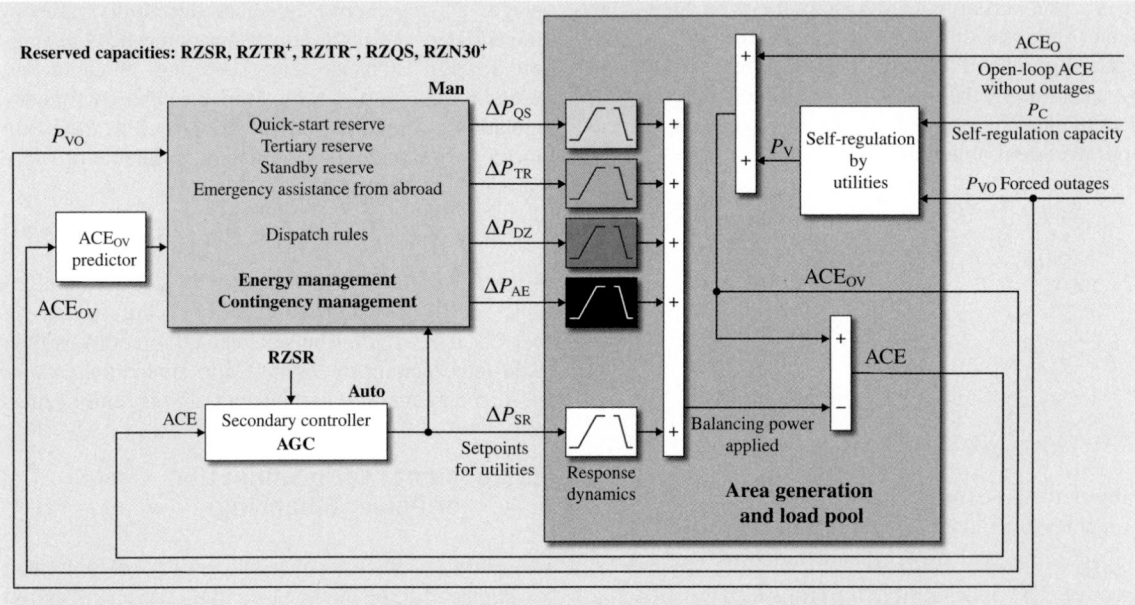

ACE$_{OV}$ = Remaining ACE not compensated by the TSO = P_V: Forced outages compensated by utilities; ΔP_{EA} = Emergency assistance (from abroad)

Fig. 65.10 Block diagram of a power balancing control system simulation for the area

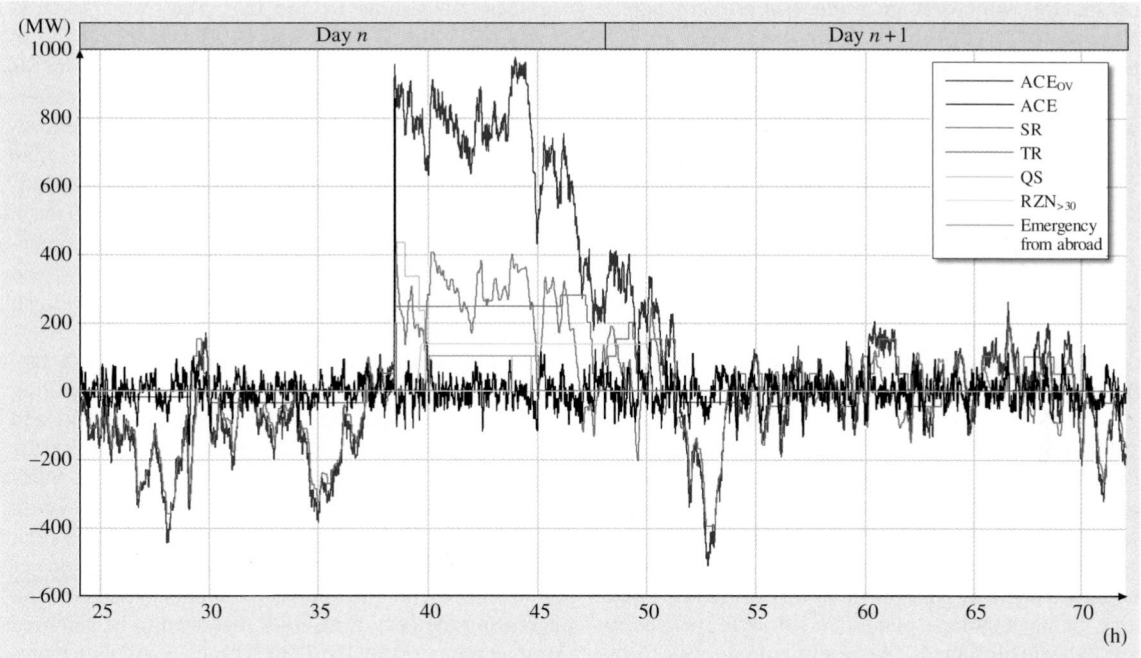

Fig. 65.11 Activation of ancillary services for area power balancing

is about 400 MW, it is not possible to cover the massive outage solely using this reserve, and other services must be called up to assist. The TSO's operator activates the quick-start reserve (cyan) and the tertiary reserve (magenta) at the same time. The pumped-storage quick-start must be saved as its capacity is limited, so the operator tends to replace it as soon as possible with the services with a longer response time, the standby reserve (yellow), and/or tries to acquire services that are not guaranteed: emergency assistance from the adjacent area (green). As a result, the ACE (black) is minimized and the massive outage is well compensated for via a temporary combination of ancillary services, or rather their activation. The TSO's manual intervention also assures that the capacity of the secondary reserve is restored as quickly as possible so the output of the automatic load-frequency controller is driven away from the limits, the power reserve purchased by the TSO in advance.

65.2.7 Control Performance Evaluation

Although the AS needs were determined in Step 2 with the aim of meeting the required reliability in terms of the indices $r\mathrm{ACE}_1$ and $r\mathrm{ACE}_{60}$, not all of the phe-

nomena influencing the reliability could be taken into account in the analytical computations. For instance, the dynamics of AS activation/deactivation and the limited capacity of pumped-storage reservoirs are neglected. Moreover, the AS set recommended for purchase is often different from the minimal set due to limited offers or the substitution of one AS for another. Hence, through the statistical evaluation of the Monte Carlo-simulated ACE, more realistic values of the reliability indices $r\mathrm{ACE}_1$ and $r\mathrm{ACE}_{60}$ can be determined, and the values of the other five indices introduced in Table 65.3 can also be found.

Other performance indices can be evaluated, including NERC's CPS2 and DSC, with the exception of CPS1, as it requires the tie-line frequency deviation to be simulated whereas the simulator [65.17] does not support frequency.

65.2.8 Case Studies

Algorithms of AS planning and the area control system simulator can be used for various studies and analyses supporting TSO's decision-making when planning and purchasing ancillary services.

Technical and economic aspects are closely related in this task, and it is essential to answer the question of how much it will cost to guarantee a commonly agreed and accepted level of reliability in the system services provided by the TSO. The methodology presented here enables the relation between the reliability measured through statistical evaluation of ACE and the cost of purchasing the power reserves to be studied. A typical relation between two reliability indices and the cost of AS is shown in Fig. 65.12.

The plot shows the relative increase or decrease in the AS purchase cost when the request for reliability, measured as the relative number of cases when minute averages and hourly averages of ACE exceed specified levels (e.g., 100 MW for a minute and 20 MW for an hour), is relaxed.

Several studies illustrating the capabilities of the method can be found in [65.16], where the results of year-ahead planning based on specifying the needs for the individual services and the recommended optimal purchase are shown. There can be significant differences between the needs and the purchase, which is due to the fact that RZQS availability is limited to existing pumped-storage plants, and thus influenced by their scheduled outages. A similar rule applies to the standby reserve $RZN_{>30}$. Thus, more positive reserves must be acquired in the category of faster-response

services, according to (65.16). The expert-entered prices of RZSR, $RZTR^+$ and $RZTR^-$ were related as $C_{RZSR} > C_{RZTR^+} + C_{RZTR^-}$, so the symmetrical RZSR service is preferred to tertiary control, which results in low or even zero purchasing of tertiary reserves $RZTR^-$. However, the volume of $RZTR^+$ is higher than the needs during some periods of the year, as it substitutes for the lack of RZQS and $RZN_{>30}$ in order to reach the minimal positive reserve $RZ^+_{\sum min}$. A correlation between greater amounts of faster-response services and improved reliability of power system operations can also be observed.

Self-regulation of power producers is a risky factor. The level to which a generation company is willing to eliminate the difference between the planned and the actual production also depends on the penalty the company will have to pay for the internal error. Studies carried out on the power balancing control system simulator can be used to find out the acceptable relation for internal error charges. Reducing internal power balancing reserves of the power producer will end up increasing the cost of reserves that need to be acquired for the entire area by the TSO. This cost will then be related to the reduction in the producer's balancing power. Simulation studies in [65.16] show that a lack of self-

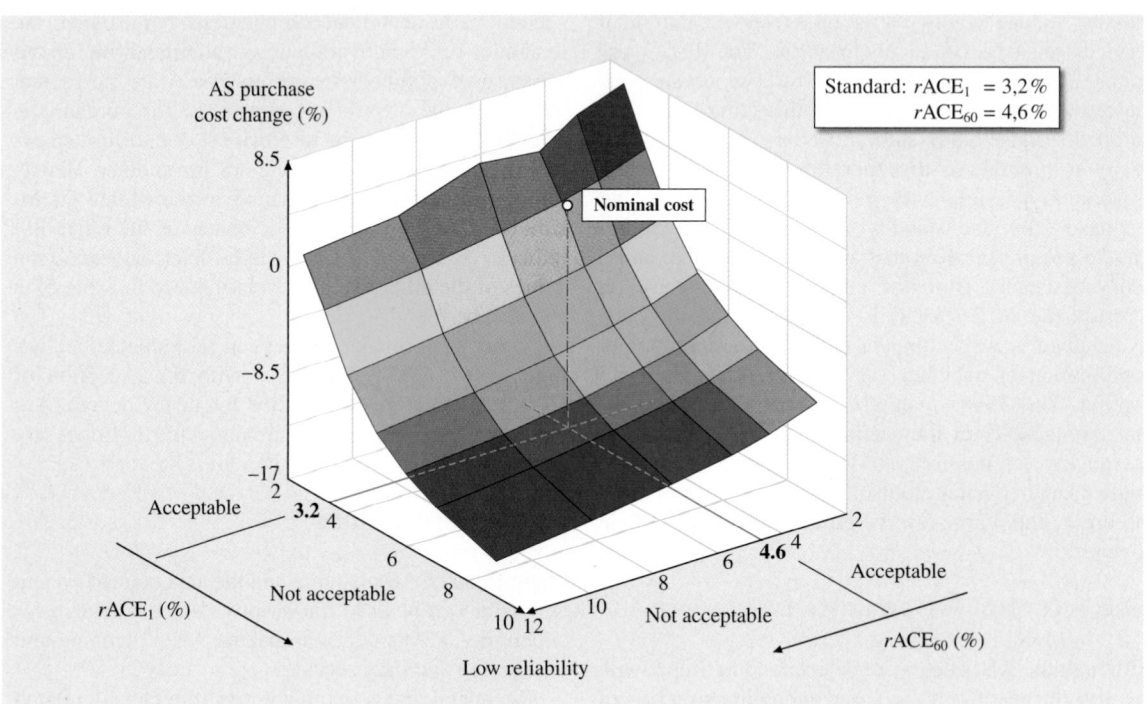

Fig. 65.12 AS costs as a function of required reliability

regulation capability in the utilities could increase the TSO's AS expenses significantly.

The methodology presented deals with the TSO's ancillary services planning and purchasing in order to prevent undesirable situations in the transmission system under changing market conditions and with predicting the influence of these changes on the TSO's expenditure. The AS planning method is shown for year-ahead planning (including various analyses and sensitivity tests, e.g., costs–reliability relation), and the results provide support for the TSO's strategies in the annual AS auction. Despite the fact that the proposed concept is tailored to the needs of a particular control area in the UCTE interconnection, it represents a compact approach to optimal AS planning.

The presented planning and modeling approach is based on historical data. However, the method can incorporate the characteristics of new types of generation units, study what installing wind turbines would mean for ancillary services planning, and evaluate the cost of reaching the same level of reliability.

Price elasticity can be incorporated into the model of energy and AS bids in order to make the entire model more realistic. The costs for AS will then supposedly increase if the (optimal) volumes required in a certain AS category are higher than usual.

The challenging task is to determine the value of unserved energy in the area. The value of unserved energy is the loss to the economy if a MWh of the energy required by consumers cannot be supplied. Consequently, the described methodology allows the optimal balance between the costs of AS and the costs of unserved energy to be found.

65.2.9 Related Topics

Generation from Renewable Resources

Energy sources are undergoing permanent restructuring. Large-scale renewable resource installations like wind power, which is the world's fastest-growing energy technology, represent a new challenge to the power system. The fluctuating nature of wind power affects power system reliability, deviating from the planned power generation, which leads to power balancing problems. Selecting appropriate approaches for calculating the wind power-induced balancing power reserve while respecting the specificity of regulatory and market situations in the region is currently an issue [65.20, 21]. Several detailed technical investigations of the impacts of wind power plants on grid ancillary services needs have recently been performed [65.22]. Although the

approaches vary, three utility timeframes appear to be the most significant issues: regulation, load following and unit commitment. This article describes and compares the analytic frameworks from recent analysis and discusses the implications and cost estimates of wind integration. The findings of these studies indicate that relatively large-scale wind generation will have an impact on power system operation and costs, but that these impacts and costs are relatively low at the penetration rates that are expected over the next few years.

Transmission Congestion Management

Grid management has become notably more complex over the past few decades. The level of grid usage is increasing markedly in every region. Consistently high utilization of critical assets in a network often means that these assets become bottlenecks that limit the use of the network and its ability to serve demands at every level consistently and reliably. These points on the grid create transmission constraints. In some cases they create congestion by limiting the buyers' ability to secure energy from the most economical source, meaning that energy must be purchased from closer, more costly generators. In other cases, transmission constraints can cause a reliability problem when customer demands exceed the delivery capabilities of the transmission system plus local generation.

As the power flows over a grid with rapidly changing electrical topologies increase, the local utility is no longer able to see and manage every factor affecting the flows through its share of the transmission grid. The scope of grid monitoring and control that worked effectively in the days of high capacity margins and limited interutility flows are no longer sufficient. Higher grid usage causes greater reliability challenges, and policy-makers and utilities then come to the opinion that increasing the sizes of grid oversight and control organizations is the only effective way of dealing with these challenges.

This problem is of great importance in countries/areas where, besides energy and ancillary services, transmission capacity is a commodity in a deregulated market for electricity and/or where the transmission capacities are almost/temporarily saturated (as is the case when the transmission systems were not designed for the power flows required).

The problems of ancillary services planning and real-time balancing power reserve activation discussed in this chapter do not address one important issue: the limited capacity of the transmission lines. It can happen that the called-up ancillary service will overload a particular transmission line when there is a scheduled energy exchange. As the approach described so far models the control area as a pool, it does not allow the control area network topology to be taken into consideration. Other approaches must be applied.

Reference [65.23] presents major congestion management methods in the new market environment from the international technical literature, which are presented and classified according to security-constrained generation redispatch, zonal/cluster based management, a network sensitivity factor-based method, congestion management using FACTS devices, congestion pricing and market-based methods, congestion management using demand-side resources, and financial transmission rights.

In a restructured power system, the transmission network is the key mechanism for generators to compete in supplying large users and distribution companies. The role of the competitive electricity markets is described in [65.24]. The lessons learned by ERCOT (Texas ISO) in relation to flow-based zonal redispatch to relieve transmission congestion are discussed in [65.25].

Further Reading

The electrical power industry has undergone dramatic changes in many countries in recent years. Recent deregulation has transformed it from a technology-driven industry into one driven by public policy requirements and the open-access market. Now, just as the utility companies must change to ensure their survival, engineers and other professionals in the industry must acquire new skills, adopt new attitudes, and accommodate other disciplines. The book by *Denny* and *Dismukes* [65.26] provides engineers with a broad overview of most of the topics that should be explored in order to understand and meet the challenges of the new competitive environment. Integrating the business and technical aspects of the restructured power industry, it explains how new methods of power systems operation and energy marketing relate to public policy, regulation, economics, and engineering science. The authors examine the technologies and techniques currently in use and lay the groundwork for the coming era of unbundling, open access, power marketing, self-generation, and regional transmission operations.

References

65.1 L. Philipson, H.L. Willis: *Understanding Electric Utilities and De-regulation* (CRC, Boca Raton 2005)

65.2 A. Orths, A. Schmitt, Z.A. Styczynsky, J. Verstege: Multi-criteria optimization methods for planning and operation of electrical energy systems, Electr. Eng. **83**, 251–258 (2001)

65.3 M. Ventosa, Á. Baíllo, A. Ramos, M. Rivier: Electricity market modeling trends, Energy Policy **33**, 897–913 (2005)

65.4 L. Zuyi, M. Shahidehpour: Security-constrained unit commitment for simultaneous clearing of energy and ancillary services markets, IEEE Trans. Power Syst. **20**(2), 1079–1088 (2005)

65.5 W. Tong, M. Rothleder, Z. Alaywan, A.D. Papalexopoulos: Pricing energy and ancillary services in integrated market systems by an optimal power flow, IEEE Trans. Power Syst. **19**(1), 339–347 (2004)

65.6 UCTE: *UCTE Operation Handbook* (Union for the Coordination of Transmission of Electricity, Brussels 2004), v2.2/20.07.04

65.7 I. Boldea: *The Electric Generators Handbook: Synchronous Generators* (CRC, Boca Raton 2006)

65.8 M.B. Zammit, D.J. Hill, R.J. Kaye: Designing ancillary services markets for power system security, IEEE Trans. Power Syst. **15**(2), 675–680 (2000)

65.9 G. Chicco, G. Gross: Competitive acquisition of prioritizable capacity-based ancillary services, IEEE Trans. Power Syst. **19**(1), 569–576 (2004)

65.10 Y.A. Liu, Z. Alaywan, M. Rothleder, S. Liu, M.A. Assadian: A rational buyer's algorithm used for ancillary service procurement, Proc. IEEE Power Eng. Soc. Winter Meet. (2000) pp. 855–860

65.11 AERC: *Reliability Standards for the Bulk Electric Systems of North America* (North American Electric Reliability Corporation, Princeton 2008)

65.12 G. Gross, J.W. Lee: Analysis of load frequency control performance assessment criteria, IEEE Trans. Power Syst. **16**(3), 520–531 (2001)

65.13 N. Jaleeli, L.S. VanSlyck: NERC's new control performance standards, IEEE Trans. Power Syst. **14**(3), 1092–1096 (1999)

65.14 N. Jaleeli, L.S. VanSlyck: Discussion of analysis of load frequency control performance assessment criteria, IEEE Trans. Power Syst. **17**(2), 530–531 (2002)

65.15 B. Stojkovic: An efficient approach for the load-frequency control and its role under the conditions of deregulated environment, Eur. Trans. Electr. Power **16**, 423–435 (2006)

65.16 P. Havel, P. Horacek, V. Černý, J. Fantík: Optimal planning of ancillary services for reliable power balance control, IEEE Trans. Power Syst. **23**(3), 1375–1382 (2008)

65.17 E. Janeček, V. Černý, A. Fialová, J. Fantík: A new approach to modelling of electricity transmission system operation, Proc. IEEE PES Power Syst. Conf. Expo., Atlanta (2006) pp. 1429–1434

65.18 L. Dale, D. Milborrow, R. Slark, G. Strbac: Total cost estimates for large-scale wind scenarios in UK, Energy Policy **32**(17), 1949–1956 (2004)

65.19 P. Havel, P. Horacek, J. Fantík, E. Janeček: Criteria for evaluation of power balance control performance in UCTE transmission grid, Proc. 17th IFAC World Congr., Seoul (2008)

65.20 V. Černý, P. Janeček, A. Fialová, J. Fantík: Monte-Carlo simulation of electricity transmission system operation, Proc. 17th IFAC World Congr., Seoul (2008)

65.21 C. Ensslin: The influence of modelling accuracy on the determination of wind power capacity effects and balancing needs. In: *Renewable Energies and Energy Efficiency*, Vol. 1, ed. by J. Schmid (Kassel Univ. Press, Kassel 2007)

65.22 B. Parsons, M. Milligan, B. Zavadil, D. Brooks, B. Kirby, K. Dragoon, J. Caldwell: Grid impacts of wind power: a summary of recent studies in the United States, Wind Energy **7**(2), 87–108 (2004)

65.23 L.A. Tuan, K. Bhattacharya, J. Daalder: A review on congestion management methods in deregulated electricity markets, Proc. Power Energy Syst. (Acta 2004)

65.24 L.L. Lai: *Power System Restructuring and Deregulation: Trading, Performance and Information Technology* (Wiley, New York 2001)

65.25 J.M. Griffin, S.L. Puller: *Electricity Deregulation: Choices and Challenges* (Univ. Chicago Press, Chicago 2005)

65.26 F.I. Denny, D.E. Dismukes: *Power Systems Operations and Electricity Markets* (CRC, Boca Raton 2002)

66. Vehicle and Road Automation

Yuko J. Nakanishi

Presently in the USA, Europe, Japan, and in other parts of the world, intelligent transportation system (ITS) technologies are being developed and deployed to increase the *intelligence* of vehicles. The two key benefits of these technologies are enhancement of safety and mobility of the traveling public. The *intelligence* is provided by electronics, communications systems, software, and human–machine interfaces and is assisting drivers with many aspects of the driving task. Drivers may be warned about potential crashes with other cars, about objects that are hidden from their vantage point, and about excessive speeds. Information about real-time traffic conditions including incidents on a driver's preferred route, travel times to specific destinations, and about restaurants, hotels, and other destination points may be provided. In-vehicle navigation systems can tell drivers how to get to a destination on a turn-by-turn basis and may be linked to a central dispatch center to summon help automatically in case of an accident.

The major initiatives and technologies being developed in the USA – integrated vehicle-based safety systems, forward collision warning systems, road departure crash warning systems, vehicle infrastructure integration – are described and discussed in this chapter. In addition, how they

interact with the driver is important in terms of safety, liability, and acceptance of the technologies. The human factors elements that should be considered are presented and discussed in the chapter as well.

66.1 Background

66.1.1 USA – Intelligent Transportation Systems (ITS) Background

In the USA, there were 5 973 000 police-reported motor vehicle traffic accidents in 2006. These crashes resulted in 42 642 deaths and 2 575 000 injuries. During 2006, an average of 117 persons died daily as a result of these crashes. There were another 4 189 000 accidents involving property damage only [66.1]. While the fatality rate has decreased from 1.69 per 100 million vehicle miles of travel (VMT) in 1996 to 1.41 in 2006, these statistics are still staggering and reflect the very serious economic and human costs of these accidents. (National Highway Traffic Safety Administration (NHTSA, http://www.nhtsa.dot.gov/) reports that, in 2000, the economic costs of traffic accidents were US$230.6 billion.)

The latest mobility research by the Texas Transportation Institute concluded that traffic congestion continues to worsen in US cities and causes 4.2 billion hours of delays, depletes US $78 billion from the US economy, and wastes 2.9 billion gallons of fuel. In 2005, congestion in urban areas caused the average peak period traveler to spend an extra 38 h of travel time and consume an additional 26 gallons of fuel, at a cost of US $710 per traveler [66.2].

Intelligent transportation systems (ITS) is the application of electronics, communications, information, and information technology to transportation systems. The primary drivers of these technologies are safety and mobility. Concomitant benefits are efficiency, convenience, and accessibility. Highlights of safety and mobility benefits generated by ITS projects are presented below [66.3].

Safety benefits:

- In Georgia, the Navigator incident management program reduced secondary crashes from an expected 676 to 210 in the 12 months ending April 2004 (2006).
- In North Carolina, a work zone equipped with smart work zone traveler information systems observed fewer crashes compared with other work zones without the technology (2005).
- In Baltimore, a *second train coming* warning system decreased the frequency of the most common risky behavior at crossings (i. e., drivers that crossed the tracks after the protection gates began to ascend from the first train before the protection gates could be redeployed for the second train) by 26% (2002).
- Evaluation indicated that integrating dynamic message signs (DMS) and incident management systems could reduce crashes by 2.8%, and that integrating DMS and arterial traffic control systems could decrease crashes by 2%, in San Antonio, TX (2000).
- In Georgia, call boxes installed on a 39 mile section of I-185 were estimated to eliminate one injury per year, and one fatality every 5 years (2000).
- An advanced curve warning system on an interstate route in northern California caused over 68% of drivers to reduce their speed (2000).
- In a rural area of Virginia, a collision countermeasure system installed on a two-way stop-controlled

intersection reduced vehicle speeds by 2.4 mi/h, and increased the average projected time to collision from 2.5 to 3.5 s (2000).

- An automated enforcement systems in California decreased highway–rail grade crossing violations by up to 92% (1999).
- A dynamic truck downhill speed warning system installed on I-70 in Colorado reduced the average speed of passing trucks by approximately 5.2 mi/h (1999).
- In Colorado, a downhill truck speed warning system installed on I-70 reduced runaway ramp usage by 24% and contributed to a 13% drop in crashes involving trucks and excessive speeds (1997).
- After a ramp rollover warning system was installed at three curved exit ramps on the beltway around Washington, DC, there were no accidents at any of these sites during the 3 year postdeployment test period evaluated (1997).
- Following deployment of the TransGuide freeway management system in San Antonio, TX, crash frequency was reduced by 41% and incident response time decreased by 20% (1997).
- Advanced traffic management systems in Amsterdam and Germany reduced crash rates by 20–23% (1999).
- In Japan, a real-time incident detection and warning system installed on a dangerous curve on the Hanshin Expressway decreased the rate of secondary crashes by 50% (1997).

Mobility benefits [66.3]:

- In Georgia, the Navigator incident management program reduced the average incident duration from 67 min to 21 min, saving 7.25 million vehicle-hours of delay over one year (2006).
- In Georgia, the highway emergency response operator (HERO) motorist assistance patrol program and Navigator incident management activities saved more than 187 million US dollars, yielding a benefit-to-cost ratio of 4.4:1 (2006).
- In Utah, incident management teams in Salt Lake Valley area decreased incident duration by approximately 20 min per incident on three major interstates (2004).
- In 2002, the Maryland coordinated highways action response team (CHART) highway incident management program reduced delay by about 30 million vehicle − hours and saved about 5 million gallons of fuel (2003).

- In Albuquerque, NM, work zone surveillance and response at the *Big I* Interchange reduced average clearance time by 44% (2001).
- During the first year of operations at the *Big I* work zone in Albuquerque, temporary traffic management and motorist assistance patrols reduced the average incident response time to less than 8 min, and no fatalities were reported (2001).
- Modeling performed as part of an evaluation of nine ITS implementation projects in San Antonio, TX, indicated that integrating dynamic message signs (DMS), incident management, and arterial traffic control systems could reduce delay by 5.9% (2000).
- The delay reduction benefits of improved incident management in the Greater Houston area saved motorists approximately US $8 440 000 annually (1997).
- In San Antonio, TX, a freeway management system led to an estimated delay savings of 700 vehicle – hours per major incident (1997).
- In Brooklyn, an incident management system on the Gowanus and Prospect Expressways used closed circuit television (CCTV), highway advisory radio, DMS, and a construction information hotline to improve average incident clearance time by about 1 h, a 66% improvement (1997).

The National ITS Architecture [66.4] provides a common framework for the planning defining and implementation of ITS, and defines the required functions, the physical entities or subsystems where these functions reside, and the information flows and data flows linking these functions and physical subsystems together. The National ITS Architecture consists of three layers, institutional, communications, and transportation:

- *Institutional layer*: represents the existing and emerging institutional constraints and arrangements and addresses policy issues, funding incentives, working arrangements, and jurisdictional structure.
- *Communications layer*: comprises communication equipment such as wireless transmitters and receivers.

- *Transportation layer*: presents relationships among the transportation-related elements, including traveler subsystems, vehicles, transportation management centers field devices, and external system interfaces.

Logical architecture defines ITS processes, process-to-process data flows, and data elements and is not technology specific. Physical architecture is the high-level structure containing major ITS system components. Subsystems are the structural elements of the physical architecture. The subsystems are categorized into center, traveler, field, and vehicles [66.5].

The intelligent vehicles subsystem is comprised of collision avoidance systems, driver assistance systems, and collision notification systems (Fig. 66.1).

Collision avoidance systems (CAS) warn drivers of impending danger. The vehicle warnings need to occur early enough for the driver to take action, and must also be clear and understandable to the driver. Since multiple alarms would be confusing and distracting for the driver, a management system needs to be present to decide which CAS or driver assistance warning to provide when the condition warrants it. CAS systems comprise the following elements:

- *Intersection collision warning* systems detect and warn drivers of approaching traffic at high-speed intersections; these systems use both vehicle- and infrastructure-based technologies. Once vehicle infrastructure integration is in place, other drivers would also be able to receive warnings about impending violations (Fig. 66.2).
- *Obstacle detection* systems use vehicle-mounted sensors to detect obstructions, such as other vehicles, road debris or animals, in a vehicle's path and alert the driver. An example of a more advanced system for obstacle detection and road navigation in intelligent vehicles is presented in Fig. 66.3.
- *Lane-change warning* systems have been deployed to alert bus and truck drivers of vehicles or obstructions in adjacent lanes when the driver prepares to change lanes.
- *Lane departure warning* systems warn drivers that their vehicle is unintentionally drifting out of the lane.
- *Rollover warning* systems notify drivers of heavy trucks when they are traveling too fast for an approaching curve, given their vehicles operating characteristics. This system would also be useful for light trucks (sports utility vehicles (SUVs)) as well.

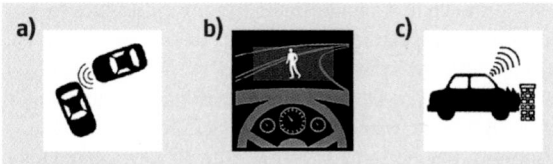

Fig. 66.1 **(a)** Collision avoidance system. **(b)** Driver assistance system. **(c)** Collision notification system

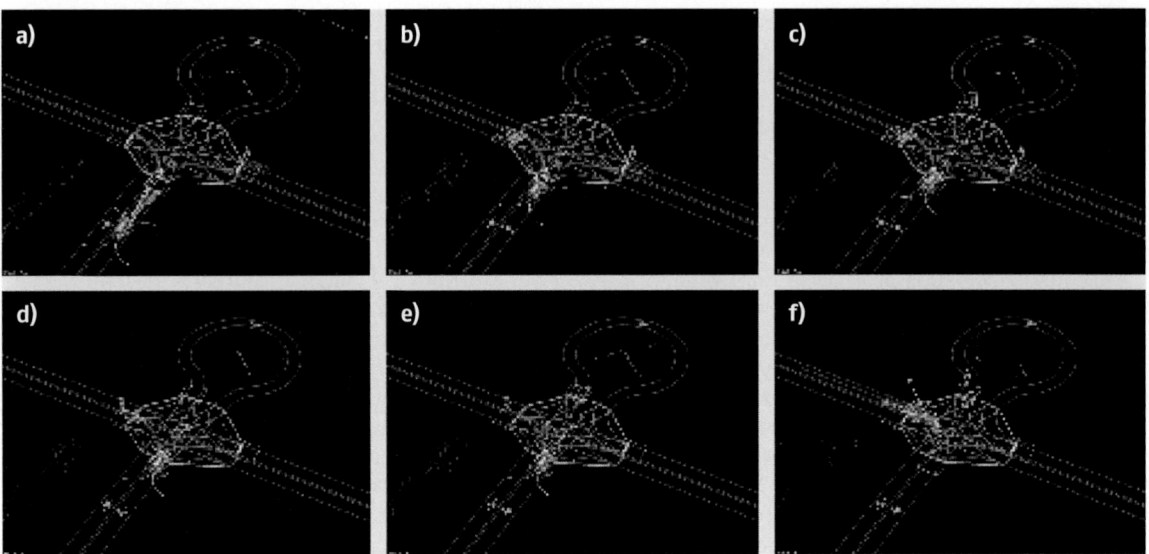

Fig. 66.2 (a) The robot approaches an intersection and identifies a waiting car (*rectangle* marked by a *blue diamond*). As the gets closer (**b**) it detects a second vehicle (*rectangle, left of frame*), and upon coming to a stop (**c**) tracks a third vehicle, which arrives late (*rectangle middle of frame*). (**d**) The first vehicle begins moving through the intersection, and precedence is transferred to the second vehicle, denoted by the *teal diamond*. (**e**) The second vehicle begins traversing the intersection. (**f**) Upon all clear, the robot takes precedence ahead of the third vehicle and safely navigates the intersection (courtesy of Tartan Racing, Carnegie Mellon University)

- *Road departure warning* systems have been tested using machine vision and other in-vehicle systems to detect and alert drivers of potentially unsafe lane-keeping practices and to keep drowsy drivers from running off the road.
- *Forward collision warning* systems use microwave radar and machine vision technology to help de-

tect and avert crashes. Drivers are warned of unsafe conditions via vehicle displays or audible alerts. If a driver does not properly apply brakes in a critical situation, some systems automatically assume control and apply the brakes in an attempt to avoid a collision.

- *Rear-impact warning* systems use radar detection to prevent accidents. A warning sign is activated on the rear of the vehicle to warn tailgating drivers of impending danger.

Driver assistance systems (DAS) assist drivers navigate their vehicles safely, enhance their vision, and control the speed of their vehicles. These systems will

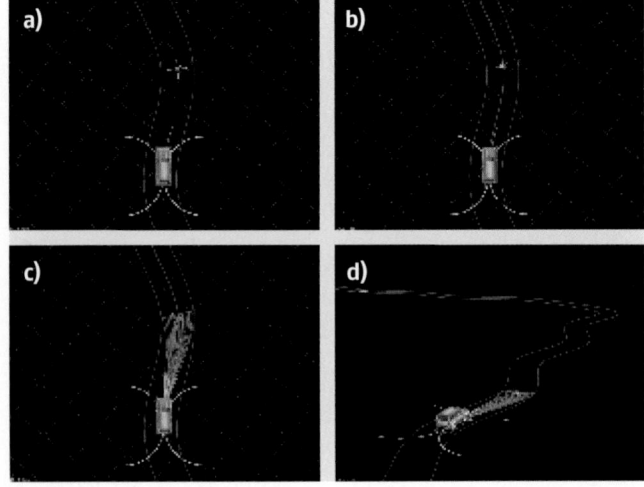

Fig. 66.3 (a) Vehicle navigating down a single-lane road (lane boundaries shown in *blue*, current curvature of the vehicle shown in *pink*, minimum turning radius arcs shown in *yellow*). (**b**) extracted centerline of the lane (in *red*). (**c**) candidate trajectories generated by the vehicle given its current state, the centerline path, and lane boundaries. A single trajectory is selected for execution, as discussed. (**d**) the same scenario from an alternative viewing angle (courtesy of Tartan Racing, Carnegie Mellon University)

be especially valuable in inclement weather or in other hazardous conditions. Driver assistance systems comprise:

- *In-vehicle navigation and route guidance* systems with global positioning system (GPS) technology may reduce driver error, increase safety, and save time by improving driver decision in unfamiliar areas.
- *Integrated driver communication* systems enable drivers and dispatchers to coordinate rerouting decisions on-the-fly and can also save time and money, and improve productivity.
- *In-vehicle vision enhancement* improves visibility for driving conditions involving reduced sight distance due to night driving, inadequate lighting, fog, drifting snow or other inclement weather conditions.
- *Object detection* systems warn the driver of an object (front, side or back) that is in the path or adjacent to the path of the vehicle. Currently the most common application is parking aids for passenger vehicles. Toyota's intelligent parking assist (IPA) system uses cameras and sensors to steer the car for the driver for parallel parking or backing into a parking space. Obstacle detection sensors for rear and forward bumpers are available from automakers or available from individual vendors.
- *Adaptive cruise control* systems maintain a driver set speed without a lead vehicle, or a specified following time if there is a lead vehicle and it is traveling slower than the set speed.
- *Intelligent speed control* systems limit maximum vehicle speed via a signal from the infrastructure to an equipped vehicle.
- *Lane keeping assistance* systems can make minor steering corrections if the vehicle detects an imminent lane departure without the use of a turn signal.
- *Roll stability control* systems take corrective action, such as throttle control or braking, when sensors detect that a vehicle is in a potential rollover situation.
- *Drowsy driver warning* alerts drivers when they have become fatigued.
- *Precision docking* systems automate precise positioning of vehicles at loading/unloading areas.
- *Coupling/decoupling*: intelligent cruise control, speed control, guidance/steering, and coupling/decoupling systems which help transit operators link multiple buses or train cars into trains each assist drivers with routine tasks that weight on driver workload.

- *Onboard monitoring* tracks and reports cargo condition, safety, and security, and the mechanical condition of vehicles equipped with in-vehicle diagnostics. This information can be presented to the driver immediately, transmitted off-board, or stored. In case of a crash or near-crash, in-vehicle event data recorders can record vehicle performance data and other input from cameras or sensors.

Collision notification systems (CNS) such as GM's Onstar system report accidents to emergency responders or dispatch centers either manually (MAYDAY) or automatically with automatic collision notification (ACN). Details about the accident such as its severity and number of passengers inside the vehicle may also be reported.

- *MAYDAY/ACN* products utilize location technology, wireless communication, and a third-party response center to notify the closest public safety answering point (PSAP) for emergency response.
- *Advanced ACN* systems use in-vehicle crash sensors, GPS technology, and wireless communications systems to supply public/private call centers with crash location information, and in some cases, the number of injured passengers and the nature of their injuries.

Additional information and ITS initiatives in the USA are provided in subsequent sections of this chapter.

66.1.2 European Union – Telematics Initiatives

In Europe, telematics is the equivalent of ITS technologies, and the key benefits being sought by the European countries are similar. The European Commission seeks to further reduce road fatalities by 50% by 2010. To accomplish this, preventive safety systems are being developed within various Europe-wide initiatives such as eSafety and PReVENT.

The eSafety effort established by the European Commission contains nine working groups which focus on crash-causation analysis, human–machine interface, and others. First-generation or autonomous systems within the vehicle act as isolated units with minimal interaction among the units and minimal external input. The core technologies are:

- Advanced vehicle sensors and communications systems that integrate with onboard navigation systems
- Advanced GPS coordinates with digital mapping and location-based services

Vehicle and infrastructure information, emphasizing safety and efficiency

Second-generation initiatives such as PReVENT manage the vehicle's performance in the context of the overall system, with the vehicle and infrastructure working together to enhance the safety of motorists [66.6].

PReVENT systems are being designed to be interoperable across European nations and even on a global basis. PReVENT technologies will predict the behavior of traffic and other drivers, and include sensing technologies, in-vehicle digital map and positioning technologies, and wireless communications systems. The technologies revolve around provision of driver-oriented safety applications. A key element of PReVENT is advanced driver assistance systems (ADAS), which was originally initiated as a result of the EUREKA programme for an European traffic of highest efficiency and unprecedented safety (PROMETHEUS) project [66.7].

- The 3 year MAPS&ADAS subproject has produced an ADAS interface, ADAS maps for PReVENT test sites, and a safety impact assessment of the maps. This project is needed because access to map data by applications other than navigation requires a standard nonproprietary interface.
- The ProFusion subproject focuses on sensors and sensor data fusion. The first preliminary concept building phase has been completed. The objectives of the second phase are:
  - To develop concepts and methods for sensor data fusion to improve the reliability and robustness of perception systems
  - To define and implement a flexible and modular architecture for fusion systems with diverse approaches
  - To establish and maintain the Fusion Forum, which promotes sensor data fusion in driver assistance and active safety systems
- RESPONSE 3 will develop a code of practice for the development and testing of ADAS for European industry and will address the key issues of *reasonable safety* and *duty of care* by defining *safe* ADAS development and testing.
- PReVAL subproject will provide PReVENT with a common evaluation framework and methodology to assess the impact of various applications.
- INSAFES (integrated safety system) is developing new control functions to improve the functionality of applications already developed.

- The 3 year WILLWARN (wireless local danger warning) subproject will generate a safety application that warns the driver whenever a safety-related critical situation occurs beyond the driver's field of view. The application will include the development of onboard hazard detection, in-car warning management, and decentralized warning distribution by vehicle-to-vehicle communication on a road network
- The SASPENCE (safe speed and distance) system will guide the driver regarding desirable speed and headway for the given driving conditions.

66.1.3 Japan – ITS Initiatives

In Japan, the focus of ITS efforts has initially been on in-vehicle information systems, and the benefits sought have been in terms of safety and the environment. As in Europe, there has been strong national leadership with regards to achieving these benefits. In comparison with other nations, Japan has many vendors of ITS-related products and services.

The vehicle information and communication system (VICS) is a real-time system that provides information including weather, road and traffic conditions, and navigation assistance to in-vehicle navigation systems. The system has been considered a great success with more than 9 million subscribers in 2004. While the VICS service is free, users pay a one-time setup fee, and purchase the onboard navigation system. This system serves as a catalyst for further ITS deployment in automobiles. Taxi probes equipped with GPS and wiper sensors are used by the Japan Road Traffic Information Center to obtain information about the transportation system; the information is then analyzed and passed along to motorists using VICS. A mobile radio local-area network in the 2.4 GHz bandwidth was installed by the Japan Highway Public Corporation and is used to communicate both data and images to in-vehicle navigation systems [66.6].

Other technologies that have been developed include the following:

- Vehicles equipped with cameras to allow drivers to see objects and roadway features in blind spots. This technology is an element of advanced cruise-assisted highway systems (AHS), which provide information on highway features (curve data or other roadway information) as well as warnings of obstacles. AHS warns drivers of impending dangers and allows a timely response to them.

The radar cruise control (with low-speed following mode) system enables drivers to adjust headway between their own vehicle and the one immediately ahead, thereby providing uniform traffic flow and automatic adjustment of speed based on the actions of the lead vehicle. Adaptive cruise control provides the added benefit of mild to increasingly intense braking if the driver of a vehicle does not respond safely to a slowing or stopping vehicle ahead.

The remainder of the chapter will discuss specific ITS initiatives established in the USA.

66.2 Integrated Vehicle–Based Safety Systems (IVBSS)

The Integrated Vehicle-Based Safety Systems (IVBSS) initiative addresses the subset of accidents caused by rear-ending, road departures, and lane changes. The initiative involves the development of objective tests and criteria for performance of IVBSS, and how best to communicate an integrated warning to the driver. IVBSS is expected to have the potential to reduce these accidents by almost 50%, which translates into about 21 000 lives saved per year.

The IVBSS program will evaluate the safety benefits and driver acceptance of the following crash warning subsystems [66.8]:

- *Forward crash warning*, which warns light vehicle and heavy truck drivers of the potential for a rear-end crash with another vehicle
- *Lateral drift warning*, which warns light vehicle and heavy truck drivers that they may be drifting inadvertently from their lane or departing the roadway
- *Lane-change/merge warning*, which warns light vehicle and heavy truck drivers of possible unsafe lateral maneuvers based on adjacent or approaching vehicles in adjacent lanes, and includes full-time side object presence indicators
- *Curve-speed warning*, which warns light vehicle and heavy truck drivers they may be driving too quickly into an upcoming curve

The two main functions of IVBSS are *situation characterization*, which determines that a potential crash threat exists, and *threat assessment*, which determines whether an alert should be issued for each warning subsystem. The major elements of IVBSS are as follows [66.8].

- *Sensing subject vehicle information and driver control inputs*: The IVBSS system must obtain certain information from the vehicle; for example, for rear-end crashes, vehicle speed, yaw rate, and driver brake switch are important.
- *Sensing roadway geometry and characteristics*: Obtaining information about the roadway (e.g., road curvature, vehicle axes relative to the lane, position of the vehicle in the lane, determination of whether the lane edges are road edges, time rate of change of the lateral position of the vehicle relative to the road edge).
- *Sensing objects and characterizing object type and motion*: To avoid crashes, identification and location of other vehicles in the same lane and in adjacent lanes are important.
- *Estimating road condition parameters*: Each warning function is required to obtain and use available data that may indicate low road friction.
- *Sensing driver attributes*: In the subsequent year of the research, individual driver behavior will be integrated into decisions about the issuance of crash alerts.

US Department of Transport (DOT)-sponsored efforts have produced driver–vehicle interfaces including visual and auditory display requirements; specific warning messages have been identified and are under development; human factors tests were conducted to test the interfaces and the system itself; and prototype hardware was developed for the IVBSS evaluation.

US DOT-sponsored field tests of forward crash warning (FCW) systems for light vehicles in the Collision Avoidance Metrics Partnership (CAMP) program which produced performance criteria for radar-based FCW crash warning systems [66.9]. Further FCW crash warning systems research was performed in the Automotive Collision Avoidance Systems (ACAS) program [66.10–12].

66.2.1 IVBSS Systems Architecture

The systems architecture diagram shown in Fig. 66.4 displays the major subsystems along with the sensors and software, and the hardware interfaces and communication protocols of the IVBSS.

The five major elements of the architecture (to the left of the buses) are:

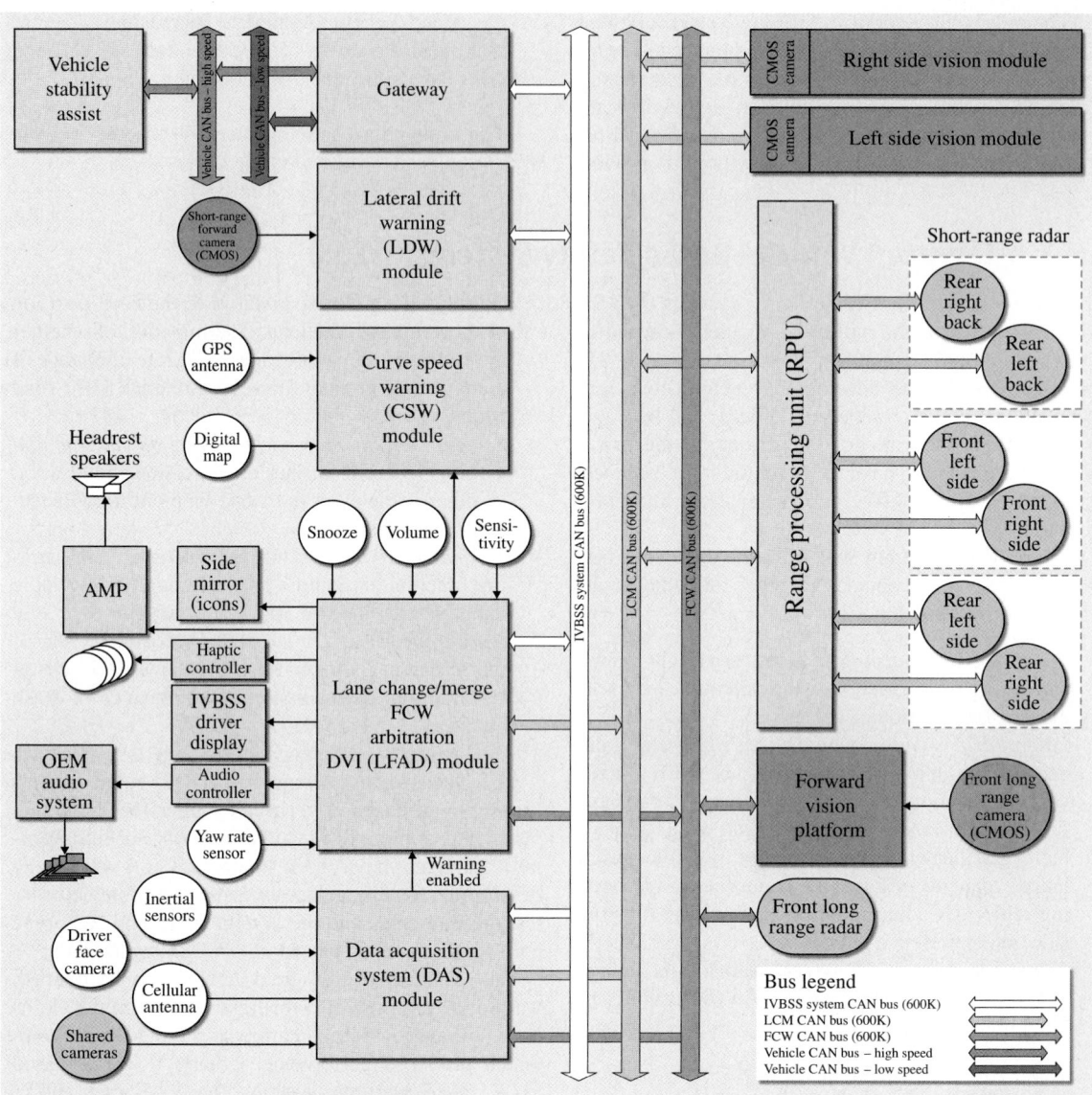

Fig. 66.4 IVBSS system architecture (after [66.8])

- *Gateway*: Translates appropriate messages from two original equipment manufacturer (OEM) data buses to one of the project CAN buses
- *Lateral drift warning module*: Uses forward vision-based lane tracking and other signals from the CAN bus to broadcast LDW (lane departure warning) alert requests onto the bus
- *Curve speed warning module*: Uses GPS, an on-board digital map, and other information to broadcast CSW alert requests onto a serial link to the

LFAD (light-vehicle module for LCM, FCW, arbitration, and DVI) module
- *LCM (lane change/merge warning)/FCW/arbitration/DVI (driver-vehicle interface) module*: A chassis that includes processors and other hardware on which LCM, FCW, arbitration, and DVI are hosted
- *Data-acquisition system module*: A two-central processing unit (CPU) module with peripherals that records data.

The additional elements above the three IVBSS CAN buses include two vision-based modules that assist with LCM functionality on the left and right side of the vehicle and a third vision-based module that assists FCW target selection and three pairs of short-range radars communicating with the IVBSS CAN buses through a radar processing unit (RPU).

66.2.2 Forward Collision Warning (FCW) System

Forward collision warning (FCW) systems warn the driver when the vehicle is about to crash into the rear-end of another vehicle or object. FCW systems use sensors such as radar, GPS, and vehicle-to-vehicle communications systems in combination with appropriate algorithms to detect static and moving obstacles and determine when to issue an alert request. The request is forwarded to an arbitration subsystem which receives requests from all driver warning systems and decides which warning to issue. The two important driver behavior parameters are: driver brake reaction time – the time required for the driver to respond to the alert by braking – and driver deceleration behavior in response to the alert under a wide range of conditions [66.9].

Figure 66.5 depicts a cooperative FCW application. The host vehicle, equipped with the FCW system, can transmit and receive dedicated short range communications (DSRC) messages while only target vehicle 2

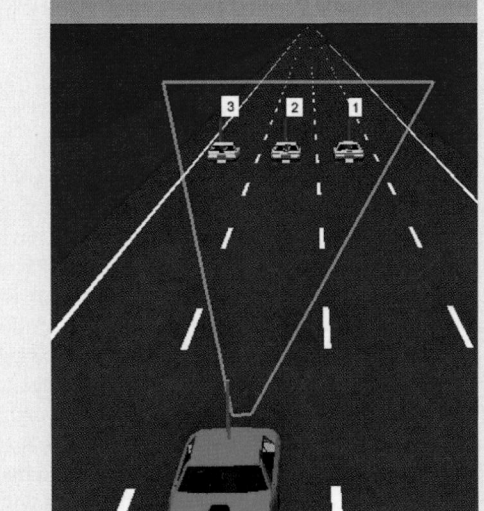

Fig. 66.5 Cooperative forward collision warning application (www.nhtsa.gov)

has a similar DSRC messaging system. The host vehicle's system will scan vehicles in its sensing area, choose the closest in-path target, and analyze the target's dynamics information. If the target is a threat, a warning will be sent to the driver of the target vehicle. Currently, FCW updates at 100 ms and have coverage ranges of 150 m.

The system performance is enhanced by vehicles that have DSRC due to the additional redundancy provided in vehicle detection. Redundancy is important especially in adverse environmental (e.g., snow) conditions. Dynamics information received from the target vehicle is richer; information about vehicle size, type, position, and turn-signal status can be transmitted to the host vehicle. Also, the information is received with low latency. Earlier detection of the target vehicle cutting into the host vehicle's path is possible due to information about turn-signal status, steering wheel position, and yaw rate. Conversely, false alarms may be reduced by earlier detection of the target vehicle leaving the host vehicle's path.

Another benefit of DSRC-equipped vehicles is early detection of stopped vehicles. Presently, it is not possible to identify stopped vehicles until it is too late to issue a warning to prevent an accident. A DSRC-equipped vehicle can warn other vehicles that it is stopped. Once all vehicles on the road are equipped with DSRC, sensors would not be needed.

A Forward Collision Warning Requirements Project sponsored by NHSTA sought to develop a FCW crash alert timing approach by examining *last-second* braking and lane-change maneuvers. One finding was that differences in last-second braking and steering were dependent on kinematic conditions.

Two braking onset models were developed – one using linear regression, and the other using logistic regression.

1. The first model develops a deceleration value above which the driver is assumed to be in alert mode when the driver is about to brake. The inputs of the model were the closing speed between the following and leading vehicle, the lead vehicle deceleration value, and knowledge of whether or not the lead vehicle is moving or stationary.
2. The second model is the *three-tiered inverse time-to-collision* model which predicts the probability that a driver is in an appropriate alert condition. The model contains inverse time to collision which is the difference in speeds of the following and leading vehicles divided by the ranges between the vehicles.

The FCW software modules are as follows:

1. *Radar-based scene tracking*: Tracks objects with respect to the subject vehicle
2. *Path prediction and data fusion*: Determines the upcoming geometry
3. *Primary target determination*: Determines the in-lane primary target that is considered the most likely to pose a crash threat
4. *Vision-based primary target validation and characterization*: Validates the choice of primary target, and includes verification that the target is a relevant vehicle
5. *Threat assessment*: Given the primary target, decides whether to issue an FCW crash alert request
6. *FCW false-alarm management*: Manages false alarms to reduce the nuisance alarm rate using historical data and subsequent driver responses

66.2.3 Road Departure Crash Warning (RDCW) System

Road departure accidents make up less than one-fifth of accidents but cause almost 40% of annual highway fatalities and are often caused by driver inattention, intoxication or drowsiness. Road departure crash warning (RDCW) systems warn drivers of impending road departure crashes due to lateral drift (unintentionally moving into another lane) or excessive curve speeds.

The Intelligent Vehicle Initiative (IVI) Road Departure Crash Warning System Field Operational Test (RDCW FOT) project conducted under a cooperative agreement between the US Department of Transportation and the University of Michigan Transportation Research Institute and its partners provided valuable insights into driver behavior and the specific benefits of the RDCW system which may be generalized to similar systems. The RDCW system tested in the RDCW FOT project targeted lane departure crashes as well as vehicles traveling into curves at unsafe speeds [66.13].

The RDCW system architecture developed for the project is shown in Fig. 66.6.

The *lateral drift warning system* (LDW) provided a set of driver alerts in the form of visual and haptic seat vibrations when the vehicle started to drift towards or outside of the lane boundaries; the driver was then expected to assess the situations and take appropriate action. The LDW system issued imminent alerts in the form of a visual icon and audible message in more hazardous conditions (e.g., when striking an object was imminent).

The key primary sensors used by the LDW system were forward cameras to identify the lane boundaries and road edges, brake switch, and turn-signal switch. Supporting equipment were GPS, a digital map and database, vehicle speed, yaw-rate gyro, and forward- and side-looking radars.

The three major changes effected by the LDW system on driver behavior were all positive; these changes were:

1. The rate of turn-signal use during lane changes increased.
2. Lane keeping performance improved: The standard deviation of lane position decreased significantly: the number of events in which the outside of the vehicle's tire crossed the lane edge or came within 4 inches of the lane edge was reduced by 50%. The time spent within 4 inch of the lane edge or outside the lane edge was reduced by 63%.
3. With the warning system in place, the data suggest that the vehicle returns to the lane more quickly.

The *curve speed warning system* (CSW) warns drivers to reduce speed before entering an upcoming curve. The CSW system uses GPS and a digital map to determine curve locations and radii. Recent driver actions such as applying turn signals were incorporated into the system's decision to issue an alert. The driver warnings consisted of visual, audible, and haptic cues similar to those of the LDW system.

In addition to GPS and a digital map, the CSW system used vehicle speed and yaw-rate gyro as the primary sensors, while the supporting equipment consisted of the camera, brake switch, and turn-signal switch.

66.2.4 Human Factors

Background

Human factors research is an important aspect of vehicle automation systems. Evaluation of driver interaction with driver assistance systems such as in-vehicle navigation systems, evaluation of different user interfaces for the systems, and driver understanding of and reaction times to collision warning systems are essential in the development of a safe and effective technology. According to NHTSA, driver distraction contributes to 20–30% of crashes, and in-vehicle navigation systems and other systems that require driver attention will cause driver distraction. A 2000 NHTSA study confirmed that entering destinations into navigation systems along with cellphone dialing and radio tuning were distracting to the driver [66.14].

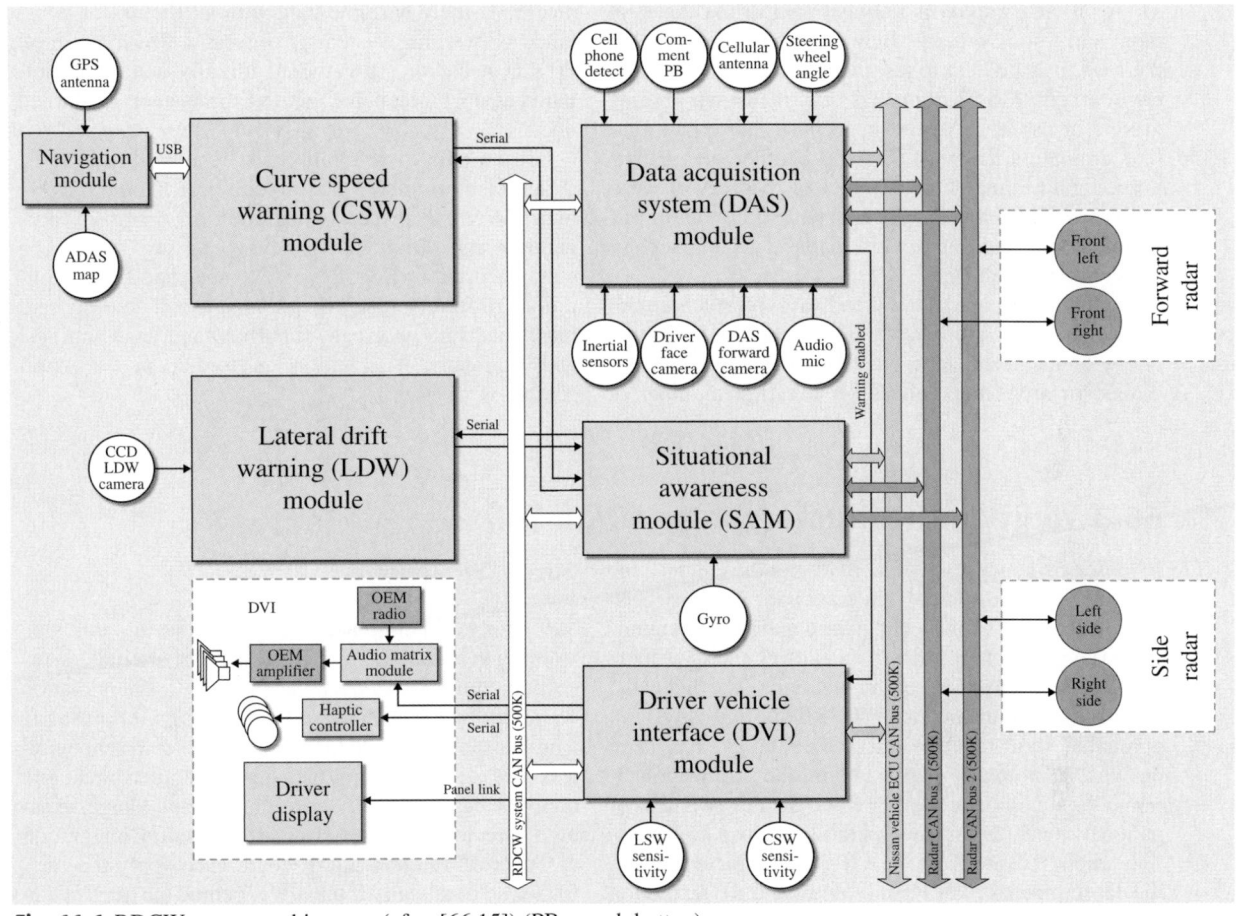

Fig. 66.6 RDCW system architecture (after [66.15]) (PB – push button)

Understanding driver behavior is also important in minimizing nuisance or false alerts, which not only cause stress and irritation to the driver but may also cause driver distraction. For instance intentional lane changes involve particular actions by the driver which may resemble actions taken when an accident is imminent. Therefore precisely differentiating driver behaviors is necessary.

How is human factors research conducted? Vehicle simulators used by NHTSA, car manufacturers, research centers, and driver training centers are valuable in the initial research and development (R&D) of vehicle technologies. In the continuum of assessment methodologies, simulators offer the greatest range of possible scenarios and driving conditions within a safe setting and are at the lowest end in terms of noncapital resource expenditures. Modular design simulators

have interchangeable cabs which can be configured for different vehicle models. Multistation driving simulator allows one controller/instructor to control multiple driving stations, enabling simultaneous multiple simulation runs. Recently Toyota introduced its new simulator that imitates speed, acceleration, deceleration, comfort conditions, and conditions at intersections. The simulator is comprised of eight high-precision projectors within a large dome which has a turntable, tilt mechanism, and vibration equipment. The dome is on a track that allows forward, backward, and side-to-side movement. Jaguar and Land Rover's engineering facility in the UK also introduced a state-of-the-art simulator that relies more on virtual reality to create three-dimensional (3-D) vehicle exteriors and interiors [66.16]. Less sophisticated simulators might consist of a steering wheel connected to a monitor displaying roadway images.

A proving ground is the next step in the development and testing process. Before systems are tested on the road, it is prudent to test them in a more controlled environment. Most automakers have their own proving ground or use an independent facility. These facilities, like simulators, range in size and number of features. Some facilities may have 50 miles of roadway of varying types. After the system has undergone additional modifications and fine-tuning, realistic driving scenarios should be evaluated.

Field testing on actual roadways is resource intensive because multiple vehicles need to be fitted out with the technologies and then test subjects recruited to drive the vehicles for a certain number of weeks. Usually baseline data without the use of assistance or warning systems to reveal a driver's natural driving behavior are gathered initially and then compared against their behavior after the systems are turned on.

To be able to generalize test results in field tests, a sufficient number of test subjects should be selected from a diverse population in terms of age, gender, and driving experience; testing should be conducted under a variety of weather and traffic conditions, different roadway geometry, and freeway and local roadways in urban, suburban, and rural settings; and, optimally, it should be performed in real-world settings.

66.3 Vehicle Infrastructure Integration (VII)

In order to connect drivers, roads, and vehicles, the US Vehicle Infrastructure Integration or VII has been established. VII is envisioned to be a national communications infrastructure and will enable vehicle-to-vehicle and vehicle-to-roadside communications. The dedicated short-range communications (DSRC) 5.9 GHz system, a reliable short-range wireless technology, has been designated for transportation and public safety applications. What will be required will be the development of national standards for interoperability, close collaboration among US DOT, state DOTs, and vehicle makers, the development of key technologies, and the resolution of political, institutional, legal, economic, and social issues connected with the infrastructure.

The four phases of the VII initiative are [66.17]:

1. Case and application development planning
2. Proof-of-concept testing
3. Prototype development, testing, and analysis.
   Human factor issues will be addressed, real-world applications refined and expanded, and institutional challenges considered, such as citizen anonymity, privacy, and security.
4. The final step is deployment readiness and support, including identifying the resources, policies, and governance necessary to move forward.

Phase 1 has been completed, and phase 2 testing is now underway, with California and Michigan having the most VII test-bed infrastructure. Also, a large demonstration of VII is planned in New York City in November of 2008 at the 2008 World Congress and ITS America 2008 annual meeting.

66.3.1 VII Benefits

VII safety benefits are brought about by collision warning systems on highways and local arterials, intersection collision avoidance technology, traffic violation warnings, and roadway curve warning; for example a broken-down vehicle on an interstate could send wireless messages to an approaching semitrailer truck, and on to the cars behind it. According to Bob Lange, executive director of vehicle structure and safety integration at General Motors Corp, it could even apply the car's brakes automatically if the driver behind the truck is distracted. If there is a potential conflict at an intersection, the roadside infrastructure could transmit appropriate messages to the vehicles approaching the intersection, and the vehicles would then alert the drivers. Similarly, warnings would be issued if another car is in a vehicle's blind spot, or if a lane change is required due to construction work.

Other benefits enhance the mobility of drivers and their access to desired destinations. In-vehicle navigation systems direct drivers by voice to their destinations and provide convenient real-time turn-by-turn directions. Some systems have voice recognition capability so that the driver need not manually input the destination, and other systems assist drivers optimize their trips and can incorporate personal preferences such as avoiding highways. In-vehicle information systems assist drivers find destinations such as hotels, restaurants, and gas stations, and may even have real-time information comparing gas prices at different stations. These systems can also help drivers avoid congested roads by providing real-time incident and delay information.

Vehicles, acting as information probes, would collect and transmit data about traffic and road conditions anonymously to the roadside infrastructure, which would then process the information and send it out to the vehicles on the road; users of the data could be charged a subscription fee, making VII a viable business opportunity for the private sector. This information can also be used by transportation planners and traffic management center (TMC) operators for corridor management tasks.

66.3.2 VII Risks

Historically the majority of potential technology users will only adopt once the very early users, who are willing to experiment, try and approve a new technology; cellphones, for instance, took 20 years before achieving widespread consumer acceptance [66.18]. The VII infrastructure will take a large amount of resources to deploy, about US $6–8 billion, which will likely require Congressional approval or a combination of government and private-sector involvement.

The primary risks for VII are the following technical and market risks.

Technical risks include: the need to integrate multiple technologies; ensuring connectivity within the DSRC network, and a determination of how to handle the data generated by the system; the development of a technology that will be able to perform real-time analysis of data from many different input sources. There is also a question over whether the technologies will be effective if only a portion of users adopt them. Also, how the entire country will be mapped, and how VII can be flexibly applied to local needs, will need to be determined.

Market risks include the identification of funding for the system, user acceptance and willingness-to-pay issues, and difficulty in accurately assessing system value until after the system has been implemented. From the automakers' perspective, the introduction of new features and innovations is always risky, particularly a safety feature which could increase the likelihood of lawsuits. Therefore the business case for automakers to incorporate these systems into their vehicles must be especially strong.

Because of the uncertainty with regards to which technologies in which form will receive the greatest user acceptance, an open architecture would be recommended. Also, the development and use of industry standards will be important. Industry standards such as local interconnect network and media-oriented systems transport are being created for communication buses to link electronic control units, subsystems, and sensors. Automotive open system architecture (AUTOSAR), a partnership of auto manufacturers, suppliers, and tool developers, is attempting to become the de facto standard-setting organization for the auto industry on a global basis. AUTOSAR seeks to minimize barriers between functional domains and networks independently from the hardware and to manage the increasing complexity of the technology. AUTOSAR is an open and standardized global automotive software architecture with the following motto: *Cooperate on standards, compete on implementation* [66.19].

66.4 Conclusion and Emerging Trends

While fully automated driverless vehicles and flying cars are many years away, ITS technologies have been increasing the capabilities, intelligence, and safety of the automobile. An early generation of such vehicles can be seen in Figs. 66.7 and 66.8. The percentage of total vehicle value accounted for by electronics and software is currently 15% for compact cars, 28% for luxury cars, and 47% for hybrid cars, and is expected to rise yearly. The continuing need for mobility and

Fig. 66.7 Defense Advanced Research Projects Agency (DARPA) car at the Intelligent Vehicle Conference 2008, Eindhoven

safety enhancements will ensure the growth of the value of electronics and software; the increase is estimated to be about 150%, from US $187 billion in 2002 to US $464 billion in 2015 [66.20].

Much progress has already been made in deploying individual ITS technologies, and many mobility and safety benefits are already being experienced. The National ITS Architecture has been the unifying framework allowing the disparate transportation technologies to work together; and vehicle infrastructure integration will be the engine by which vehicle-to-vehicle and vehicle-to-roadside communications will take place. Once the Integrated Vehicle-Based Safety Systems initiative is completed and the safety-oriented ITS technologies are implemented, even more safety benefits are expected to be realized.

Fig. 66.8 Carnegie Mellon entry to the DARPA 2007 Urban Challenge competition, Boss, Tartan racing robot (courtesy of Tartan Racing, Carnegie Mellon University)

66.5 Further Reading

- J. Baber, J. Kolodko, T. Noel, M. Parent, L. Vlacic: Cooperative autonomous driving: intelligent vehicles sharing city roads, IEEE Robot. Autom. Mag. **12**(1), 44–9 (2005)
- S. Bahler, J. Kranig, E. Minge: Field test of nonintrusive traffic detection technologies. In: *Transportation Research Record 1643*, ed. by Transportation Research Board (National Research Council, Washington 1998) pp. 161–170
- D. Braid, A. Broggi, G. Schmeidel: The TerraMax autonomous vehicle, J. Field Robot. **23**(9), 693–708 (2006)
- J.L. Campbell, C.M. Richard, J.L. Brown, M. McCallum: *Crash Warning System Interfaces: Human Factors Insights and Lessons Learned* (Battelle Center for Human Performance and Safety, Seattle 2006)
- K. Catchpole, J.D. McKeown, D.J. Withington: Localizable auditory warning pulses, Ergonomics **47**(7), 748–771 (2004)
- US Department of Transport: Crash Avoidance Metrics Partnership (CAMP): Enhanced digital mapping project final report; US Department of Transport Tech. Rep. (2004)
- US Department of Transport: Crash Avoidance Metrics Partnership (CAMP): Vehicle safety communications project task 3 final report; US Department of Transport Tech. Rep. (2005)
- A. Houser, J. Pierowicz, D. Fuglewicz: *Concept of Operations and Voluntary Operational Require-ments for Lane Departure Warning Systems (LDWS) On-Board Commercial Motor Vehicles*, FMCSA-MCRR-05-005 DTMC75-03-F-0087 (Federal Motor Carrier Safety Administration, Washington 2005)
- A. Houser, J. Pierowicz, R. McClellan: *Concept of Operations and Voluntary Operational Requirements for Forward Collision Warning Systems (CWS) and Adaptive Cruise Control (ACC) Systems On-Board Commercial Motor Vehicles*, FMCSA-MCRR-5-007 DTMC75-03-F-0087 (Federal Motor Carrier Safety Administration, Washington 2005)
- International Organization for Standardization: Intelligent transportation systems – forward vehicle collision warning systems – performance requirements and test procedures, Draft International Standard ISO/DIS 15623 (2002)
- International Organization for Standardization: Intelligent transportation systems – lane departure warning systems – performance requirements and test procedures, Draft International Standard ISO/DIS 17361 (2004)
- International Organization for Standardization: Intelligent transportation systems – lane change decision aid systems – performance requirements and test procedures, Draft International Standard ISO/DIS 17387 (2006)
- ITS Public Safety: http://www.its.dot.gov/pubsafety/index.htm (last accessed on May 5, 2009)

- IVBSS Project Team: *Preliminary Functional Requirements for the Integrated Vehicle-Based Safety System (IVBSS) – Heavy-Truck Platform* (University of Michigan Transportation Research Institute, Ann Arbor 2007)
- IVBSS Project Team: *Preliminary Performance Guidelines for a Prototype Integrated Vehicle-Based Safety System (IVBSS) – Light-Vehicle Platform* (University of Michigan Transportation Research Institute, Ann Arbor 2007)
- R.J. Kiefer, D. LeBlanc, M. Palmer, J. Salinger, R. Deering, M. Shulman: *Development and Validation of Functional Definitions and Evaluation of Procedures for Collision Warning/Avoidance Systems*, NHTSA Rep. HS 808 964 (National Highway Traffic Safety Administration, Washington 1999)
- R.J. Kiefer, D. LeBlanc, M. Palmer, J. Salinger, R. Deering, M. Shulman: *Forward Collision Warning Systems: Development and Validation of Functional Definitions and Evaluation Procedures for Collision Warning/Avoidance Systems*, Rep. DOT-HS-808-964 (Department of Transportation, Washington 1999)
- R.J. Kiefer, M.T. Cassar, C.A. Flannagan, D.J. LeBlanc, M.D. Palmer, R.K. Deering, M.A. Shulman: Forward Collision Warning Requirements Project: Refining the CAMP Crash Alert Timing Approach by Examining "Last-Second" Braking and Lane Change Maneuvers Under Various Kinematic Conditions (2003)
- D. LeBlanc: *Road Departure Crash Warning System Field Operational Test – Methodology and Results*, Rep. UMTRI-2006-9-1 (University of Michigan Transportation Research Institute, Ann Arbor 2006)
- R. Madhavan, E.R. Messina, J.-S. Albus: *Intelligent Vehicle Systems: A 4D/RCS Approach* (Nova Science, Hauppauge 2007)
- W.G. Najm, J.D. Smith: *Development of Crash Imminent Test Scenarios for Integrated Vehicle-Based Safety Systems (IVBSS)*, Department of Transport VNTSC-NHTSA-07-01 DOT HS 810 757 (National Highway Traffic Safety Administration, Washington 2007)
- Y. Nakanishi, J. Western: Ensuring the security of transportation facilities: An evaluation of advanced vehicle-identification technologies, Transportation Research Board 84th Annual Meeting (2005), CD-ROM
- NHTSA: *Preliminary Assessment of Crash Avoidance Systems Benefits* (National Highway Traffic Safety Administration, Washington 1996), Version II, Chap. 3
- D. Pomerleau, J. Everson: *Run-Off-Road Collision Avoidance Using IVHS Countermeasures – Final Report*, NHTSA Rep. DOT HS 809 170 (National Highway Traffic Safety Administration, Washington 1999)
- J.R. Sayer, J. Devonshire, C.A. Flannagan: *Effects of Secondary Tasks on Driving Performance*, Rep. UMTRI-2005-29 (The University of Michigan Transportation Research Institute, Ann Arbor 2005)
- A. Stewart: The development of driverless autonomous vehicles, EngineerIT **Nov/Dec**, 50–52 (2006)
- S. Szabo, R. Norcross: Recommendations for objective test procedures for road departure crash warning systems, Memorandum to NHTSA (2003)
- S. Talmadge, R. Chu, C. Eberhard, K. Jordan, P. Moffa: *Development of Performance Specifications for Collision Avoidance Systems for Lane Change Crashes*, NHTSA Tech. Rep. (National Highway Traffic Safety Administration, Washington 2001)
- A.K. Tan, N.D. Lerner: *Multiple Attribute Evaluation of Auditory Warning Signals for In-Vehicle Crash Avoidance Warning Systems* (National Highway Traffic Safety Administration, Washington 1995)
- C.E. Thorpe: *Side Collision Warning System (SCWS) Performance Specifications for a Transit Bus*, Federal Transit Administration under PennDOT 62N111 Project TA (Federal Transit Administration, Washington 2002)
- S.M. Turner, W.L. Eisele, R.J. Benz, D.J. Holdener: *Travel Time Data Collection Handbook*, USDOT/FHWA-PL-98-035 (Federal Highway Administration, Washington 1998)
- University of Michigan Transportation Research Institute: Integrated Vehicle-Based Safety Systems (IVBSS), First Annual Report (US Department of Transport, 2007)
- C. Urmson, W. Whittaker: Self-driving cars and the urban challenge, IEEE Intell. Syst. **23**(2), 66–68 (2008)
- US Department of Transportation: IVBSS Program Plan, http://www.its.dot.gov/ivbss/ivbss_workplan.htm (2005), retrieved June 18, 2007
- US Department of Transportation: Intelligent vehicle initiative final report, Tech. Rep. FHWA-JPO-05-057 (2005)

- US Department of Transport: http://www.its.dot.gov/vii/index.htm (last accessed May 5, 2009)

- VII Coalition Information: http://www.vehicle-infrastructure.org (last accessed May 5, 2009)

References

66.1   Traffic Safety Facts 2006, DOT HS 810 809 (NHTSA, 2006), updated 2008

66.2   Annual study shows traffic congestion worsening in cities large and small, Texas Transportation Institute Press Release Sept. 18, 2007

66.3   US Department of Transport ITS Benefits Database: http://www.benefitcost.its.dot.gov (2009)

66.4   National ITS Architecture: http://www.iteris.com/itsarch/ (2009)

66.5   Y.J. Nakanishi: Introduction to ITS. In: *Economic Impacts of Intelligent Transportation Systems: Innovations and Case Studies*, ed. by E. Bekiaris, Y. Nakanishi (Elsevier, Oxford 2004) pp.3–16, Chap. 1

66.6   J. Njord, J. Peters, M. Freitas, B. Warner, K.C. Allred, R. Bertini, R. Bryant, R. Callan, M. Knopp, L. Knowlton, C. Lopez, T. Warne: Safety Applications of Intelligent Transportation Systems in Europe and Japan (US Department of Transport, 2006)

66.7   European Commission: http://www.prevent-ip.org/ (2008)

66.8   IVBSS Project Team: *Preliminary Functional Requirements for the Integrated Vehicle-Based Safety System (IVBSS) – Light-Vehicle Platform* (University of Michigan Transportation Research Institute, Ann Arbor 2007)

66.9   R.J. Kiefer, M.T. Cassar, C.A. Flannagan, D.J. LeBlanc, M.D. Palmer, R.K. Deering, M.A. Shulman: *Forward Collision Warning Requirements Project Final Report – Task 1*, NHTSA Rep. HS 809 574 (National Highway Traffic Safety Administration, Washington 2003)

66.10   University of Michigan Transportation Research Institute and General Motors: *Automotive Collision Avoidance System (ACAS) Field Operational Test – Methodology and Results*, NHTSA Rep. HS 809 901 (National Highway Traffic Safety Administration, Washington 2005)

66.11   General Motors, Delphi Electronic Systems: *Automotive Collision Avoidance System (ACAS) Field Operational Test – Final Program Report*, NHTSA Tech. Rep. DOT HS 809 866 (National Highway Traffic Safety Administration, Washington 2005)

66.12   General Motors: *Automotive Collision Avoidance System Field Operational Test: Final Program Report*, US Department of Transport Rep. DOT HS 809 886 (US Department of Transport, Washington 2005)

66.13   D. LeBlanc, J. Sayer, C. Winkler, S. Bogard, J. Devonshire, M. Mefford, M.L. Hagan, Z. Bareket, R. Goodsell, T. Gordon: *Road Departure Crash Warning System (RDCW) Field Operational Test – Final Report* (National Highway Traffic Safety Administration, Washington 2006)

66.14   L.R. Shelton: Statement before the subcommittee on highways and transit, Committee on Transportation and Iinfrastructure, US House of Representatives (2001), http://www.nhtsa.dot.gov/nhtsa/announce/testimony/distractiontestimony.html

66.15   University of Michigan Transportation Research Institute, Visteon, AssistWare Technologies Inc.: *Road Departure Crash Warning Field Operational Test: Interim Report on Road Departure Crash Warning Subsystems* (US Department of Transport, Washington 2003)

66.16   J. Challen: Reality bytes, Automot. Test. Technol. Int. **March**, 48–51 (2008)

66.17   IVI Final Report: Saving Lives Through Advanced Vehicle Safety Technology (2005)

66.18   Lessons Learned and their Impact on VII Applications in VII: An Alternative Perspective (Northeastern University, 2007)

66.19   Autosar: http://www.autosar.org/ (Munich 2008)

66.20   J. Gunn: Speculate to communicate: The surge of in-car electronics, Traffic Technol. Int. **Feb/Mar** (2008)

67. Air Transportation System Automation

Satish C. Mohleji, Dean F. Lamiano, Sebastian V. Massimini

The air transportation system infrastructure is comprised of communication, navigation, surveillance, and air traffic management systems, and is known as the National Airspace System. This chapter describes the current very high-frequency and high-frequency modes of communication, very high-frequency omnidirectional range, distance measuring equipment, and instrument landing systems for navigation/guidance to the aircraft, and primary, as well as secondary radars, for surveillance. The two primary functions of the ground-based air traffic management system, viz. traffic flow management for strategic air traffic planning and air traffic control for safe movement of aircraft, are discussed in detail. This chapter also addresses the limited role of automation in both the aircraft cockpit and the ground-based air traffic management system.

The US civil air transportation system today handles over 700 million passengers and 40 billion revenue ton miles of cargo annually that require about 23 million aircraft operations. Each aircraft operation represents a flight from take off to touch down between its departure and destination airports. This number is forecasted to reach a level of 35 million operations by the year 2025. Although the air transportation system has recorded a very high level of safety over the years, the continued growth in demand for air traffic services has created operational inefficiencies, which result in significant flight delays and lost revenue for the users. Today, a majority of decisions in the aircraft and in the ground control system are manual. The human decision making is intensive for the pilots and controllers in terms of their workloads, and will become unmanageable as traffic levels grow. In order to deal with the traffic growth, the future air transportation system will require increased automation of air/ground functions to enhance safety, capacity, and flight efficiency while reducing delays and workload.

The performance of the air transportation system is measured by four key metrics relating to safety, capacity, flight efficiency/delays, and workload. Because aircraft safety is of the utmost importance, the US Government has established regulations that require the aircraft to maintain a minimum separation distance from other aircraft. These separation minima vary with the phases of flight. Explanations of the separation minima and their influence, as well as limitations on flight operations, are provided. How the use of automation could in the future overcome these limitations is also discussed.

Efforts are underway to define concepts for the next-generation (NextGen) air transportation system in the USA, and the Single European Sky air traffic management (ATM) research project in Europe. The Joint Planning and Development Office, mandated by the US Congress, is charted to develop NextGen concepts for the years 2025 and beyond. The goals and requirements of the NextGen system are presented. Also discussed is the ongoing research and development of new satellite technologies, including data link and satellite-based navigation and surveillance, as well as the needed automation of ground-based decision support functions. It is expected that enhanced automation in the aircraft and in the ground system will provide safe and efficient services to a significantly higher number of aircraft operations in the future.

The civil air transportation system is an essential component of the global economy. It is required for timely movement of people and cargo. Until now, air travel has been the safest mode of transportation per passenger mile. Because of the cost, air transportation has not been the primary mode of transportation for long-distance travel in most countries, even though it has been widely used in the USA. However, the demand for air travel around the world is now increasing at a fast pace to meet multinational commerce needs. For globally harmonized aircraft operations, the International Civil Aviation Organization (ICAO), in collaboration with National Civil Aviation Authorities (CAA), has established standards and operating procedures to be followed by all aircraft operators and air traffic service providers.

In the US, the National Airspace System (NAS) is a complex system of human-centric systems providing communication, navigation, and surveillance (CNS), and air traffic management (ATM) services to aircraft flying passengers and cargo. During 2006, over 700 million passengers used NAS, and the demand for air travel is expected to increase beyond one billion passengers by the year 2015. The annual cargo revenue ton miles exceeded 40 billion in 2006 [67.1], and is continuing to grow with the significant demand for goods and services.

In order to meet the challenges of greater demand for air travel, it is imperative to have an air transportation system that not only maintains or enhances safety, but also provides efficient flight operations. The current aircraft operations and the air traffic control (ATC) services are primarily manual open loop, where the pilots in the aircraft and controllers on the ground interactively make decisions based on the available information. The safety of the aircraft in the air is provided by adhering to the established distance separation rules or *separation minima* in both the horizontal and vertical domains. Thus, although safety is assured, these separation requirements between aircraft often adversely impact the efficient use of airport and airspace capacity, thereby resulting in flight delays and creating extensive workload for the humans operating the system. With the ever-growing demand for air traffic services, as these capacity resources become scarce, there is a need to develop closed-loop feedback control NAS capabilities that can provide automated decision support to maximize capacity, enhance flight efficiency, and reduce pilot/controller workload by minimizing reliance on cognitive decision making.

This chapter describes the current NAS CNS/ATM infrastructure, and the extent of the role that automation plays in specific aircraft- and ground-based functions for managing and controlling traffic. The limitation of these functions is discussed with their impact on flight safety, airspace/airport capacity, aircraft operational efficiency, and operator workload. The requirements for the future air transportation system are presented, not only to overcome the shortcomings of the current system, but also to meet the service demands of future users effectively. The ongoing application of automated key functions to meet the future system requirements are presented to address hybrid automation/human decision making in the cockpit and on the ground.

The scope of the contents presented here is limited to an understanding of how the automated functions-generated information helps manual decisions today, and how automation will generate most decisions (except those that are safety critical) in the future for strategic planning and tactical operations during all phases of flight. Any consideration of the use of automation in design and control of the aircraft (except for safe navigation and efficient energy management), or a discussion of the US Government programs to upgrade and automate ground systems, is beyond the scope of this chapter.

67.1 Current NAS CNS/ATM Systems Infrastructure

The current NAS infrastructure is comprised of a CNS system architecture, and the ATM system that is dependent upon an integrated system of airports and ATC facilities. The ground-based ATM system includes data/information acquisition and processing as well as display capabilities supporting people in making decisions to manage and control aircraft operations. The CNS architecture deploys:

1. Very high-frequency (VHF) voice/data communication over the continental USA, and high-frequency (HF) communication in the polar region and over the ocean for air/ground communication
2. VHF omnidirectional range (VOR) and distance measuring equipment (DME): ground-based navaids for navigation over pre-established routes, with precision landing guidance to primary airports provided by the instrument landing system (ILS),

and nonprecision landing guidance available from VORs and nondirectional beacons (NDB)
3. Surveillance to locate and track the aircraft provided by the primary and secondary surveillance radars (SSR). The weather radars detect and provide weather and wind information in some aircraft cockpits and to the ground system.

The ATM system has two distinct functions: traffic flow management (TFM) and ATC. The TFM function is involved in strategic planning of flight operations across the entire NAS based on the forecasted weather and traffic conditions throughout the airspace and at all major airports. This is to balance the expected demand with the airspace/airport capacity resources in order to maximize operational efficiency and minimize delays. There are, on average, 55 000 daily operations in the NAS. The ATC function provides tactical control

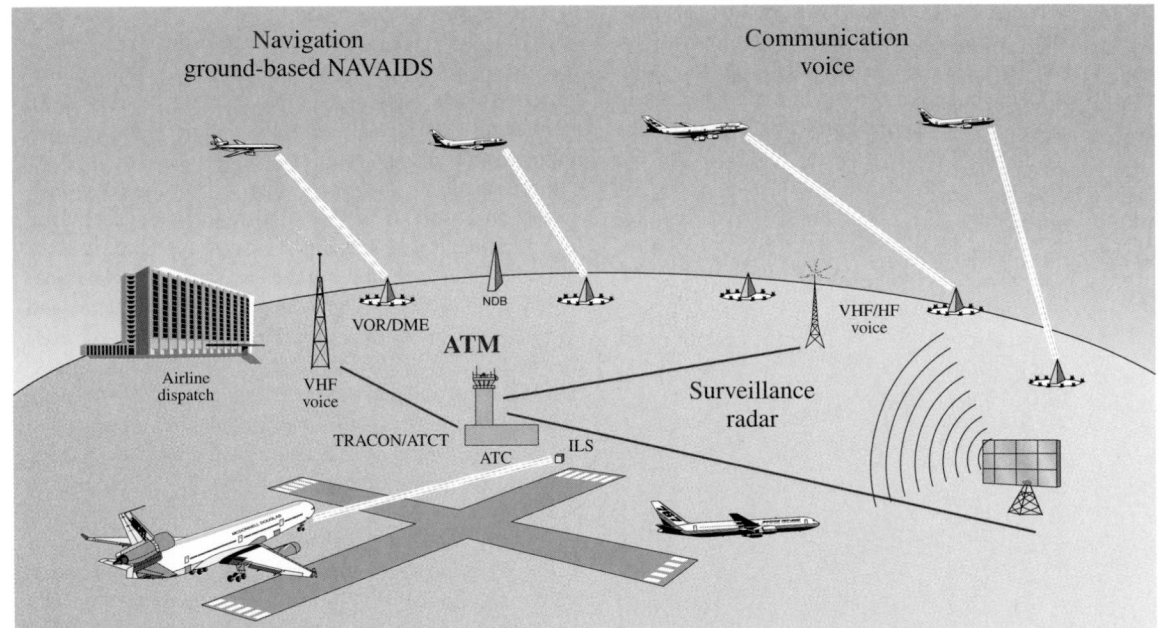

Fig. 67.1 NAS CNS/ATM infrastructure

Part G | 67.1

of airborne aircraft and is designed to help controllers maintain safe separation between aircraft. At an instant of time, the ATC function controls up to 6500 aircraft operations during peak traffic conditions [67.1]. The NAS-wide TFM function is located in the air traffic control system command center (ATCSCC), whereas the regional and local TFM functionality is provided by the traffic management units (TMU) in the 20 air route traffic control centers (ARTCC) and the terminal radar approach control (TRACON) facilities located at the primary airports. The ATC functions for aircraft movements at the airport surface and during departures/arrivals are handled by 400 air traffic control towers (ATCT) located at the airports. The ATC functions to separate aircraft up to about 30 nmi (nautical miles) from the primary airport are provided by 185 TRACONs, and for flights through the rest of the airspace by the ARTCCs. The TFM functions interface with the flight operators through the airline dispatch offices or the airlines operation centers (AOC) and flight service stations (FSS). The ATC functions require direct interaction between the ATC facilities and the aircraft [67.2,3]. Figure 67.1 shows schematically the NAS CNS/ATM infrastructure.

The US Federal Aviation Administration (FAA) has classified the national airspace into six categories in accordance with the ICAO airspace classifications. Airspace classified as class A, B, C, D or E is designated as *controlled* airspace, where ATC services are available, including air/ground communications, navigation aids, and aircraft-to-aircraft separation assurance. Class G airspace is uncontrolled airspace where the ATC system is not responsible for managing traffic.

The following provides details on controlled airspace classifications:

- Class A: Class A airspace covers from 18 000 ft mean sea level (MSL) to 60 000 ft MSL.
- Class B: Class B airspace exists at 29 high-traffic-density airports in the US where the aircraft are subjected to positive ATC, i.e., the aircraft need ground control clearances to operate. Class B airspace includes all airspace around these airports from the surface up to 12 000 MSL and spreads out to about 30 nmi.
- Class C: Class C airspace exists around 120 airports in the US that have control towers and radar approach control. This class of airspace has two concentric circular areas with a radius of 5 and 10 nmi around the airport, and extends up to 4000 ft above ground level (AGL).
- Class D: Class D airspace is a circular area of radius 5 nmi around the airport and extends up to 2500 ft AGL at airports with an ATCT.
- Class E: Class E airspace covers the volume of airspace below the class A airspace from surface to 18 000 ft MSL, and excludes airspace covered by classes B, C, and D.

67.1.1 Air/Ground Communications Systems and Functions

Today the ATC communications are primarily two-way voice capabilities over a radio system, which allow the controllers and pilots to coordinate tactical flight maneuvers needed for safety. Fundamental to the communication is the spectrum that supports it. For the ATC communications, this spectrum is coordinated worldwide, and is defined in several distinct bands, each used for specific purposes. Due to its long-range propagation characteristics, HF communications at frequencies between 3 and 30 MHz have been used in oceanic and remote polar operations for many years. VHF spectrum in the 117.95–137 MHz band is set aside for operations in well-traveled airspaces that use a line-of-sight radio infrastructure. Additional frequencies are set aside for satellite communications, defined and protected for aeronautical communications in the 1.5 and 1.6 GHz bands [67.4].

In the US, a network of VHF radio sites, shown in Fig. 67.2, provides the terrestrial infrastructure over which the voice communications operate. Specific frequencies in the band are provisioned to avoid interference between the operating airspaces, based on

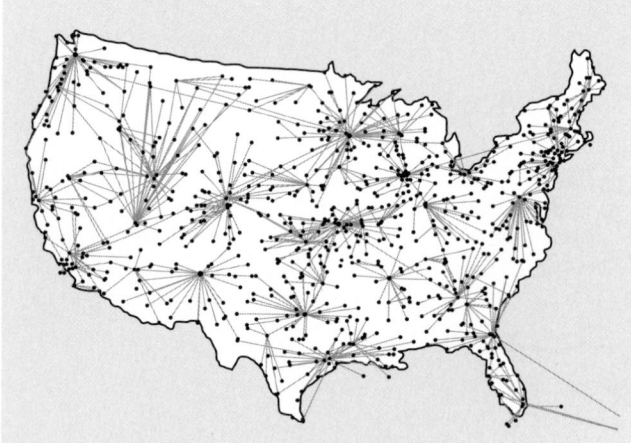

Fig. 67.2 US VHF voice radio network

(67.1), that governs the radio line-of-sight distance (D) from the aircraft to the radio horizon

$$D = K\sqrt{h}\,, \tag{67.1}$$

where D = distance (in nautical miles), h = height (in feet) of the aircraft station, and $K = 1.23$, a constant corresponding to an effective Earth's radius of 4/3 of the actual radius.

These voice channels either support a single operational ATC position on a specified control frequency, providing half-duplex party-line communications among the controllers and the pilots in a specific airspace, or provide one-way broadcast information for weather or traffic conditions on designated information frequencies, as shown in Fig. 67.3. Air-to-air and emergency frequencies are also specially provisioned for in the VHF band. A continuously monitored emergency frequency has been established worldwide at 121.50 MHz, and an air-to-air channel has been designated for use at 123.45 MHz.

Beyond today's voice communications, digital technologies are used for data links in the ATC environment, allowing for data communication from an automated ATC system, that can closely integrate and incorporate directly with the aircraft systems. One of the first aeronautical communication data links to operate was the aircraft communications addressing and reporting system (ACARS) introduced in the late 1970s. The ACARS operates at 2.4 kbps, and provides short messages indicating aircraft on, out, off, in (OOOI) events relating to the aircraft leaving the gate, taking off, landing, and arriving at the gate, to help airlines manage their aircraft. This system also operates in the VHF band, and has been expanded to include more applications, including a predeparture clearance (PDC) function for ATC and a digital broadcast automated terminal information service (ATIS). Succeeding ACARS, a digital link, called VHF digital link (VDL) mode 2, was defined and standardized through the ICAO, to provide more capacity and higher speed (31.5 kbps) for airline and ATC operations. Other data link systems that have also been defined and standardized through the ICAO include:

- VDL mode 3, which integrates a digital ATC voice capability with the data link
- VDL mode 4, which can also provide surveillance functions
- Mode-S data link, which integrates data communications with the surveillance information such

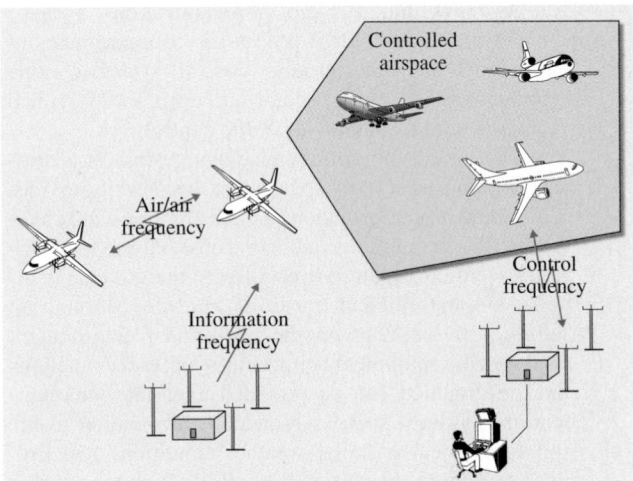

Fig. 67.3 VHF communications environment

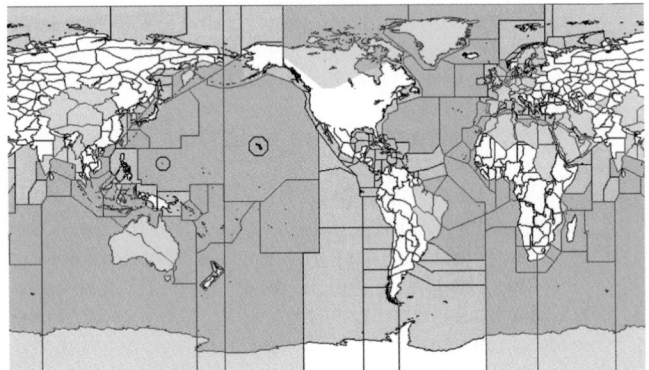

Fig. 67.4 Worldwide SATCOM use in ATC – *shaded in green*

as from automatic dependent surveillance-broadcast (ADS-B).

In the 1990s, worldwide satellite communications networks for aviation first became available through the commercial satellite service providers, including Inmarsat and Iridium. Satellite communication is displacing the long-range HF voice communications for ATC in many areas of the world and is also providing reliable long-range data links in low-density air traffic environments, such as oceanic airspace as shown shaded green in Fig. 67.4.

New air/ground communications capabilities will provide faster and greater information sharing among ATC systems and aircraft using the above technologies, thereby improving the safety and efficiency of air traffic operations.

As new digital radio communications systems enable aircraft's control processors to coordinate information with the ground control systems, radio communications will become an even more critical component of managing air traffic control.

The air/ground communications system is a critical component of the ATM system, because it provides aircraft control information (e.g., instructions and clearances to the aircraft) and in some cases feedback information (e.g., aircraft position to the ground) to allow safe and efficient transit of air traffic through an airspace. Depending on the operational environment and specific equipment being employed, other functions and information can be provided over the communications systems, such as broadcast information to the pilots on local terminal weather conditions and runway operations, meteorological information for weather models or pilot-reported turbulence information. Given the ability of an aircraft to operate anywhere in the NAS, the systems on an aircraft must be able to interface with the ATM system, and the pilots must interact with the ATC system wherever they fly.

67.1.2 Navigation and Guidance Systems

Until recently, the VOR/DME or the VOR tactical air navigation (TACAN) (VORTAC) system has been the primary guidance system for navigation. Because of its line-of-sight limitation, the VOR/DME navigation is not available everywhere in the NAS, especially in remote mountain, polar, and oceanic regions. With the availability of broadcast signals from Earth-orbiting satellites as a part of global positioning system (GPS),

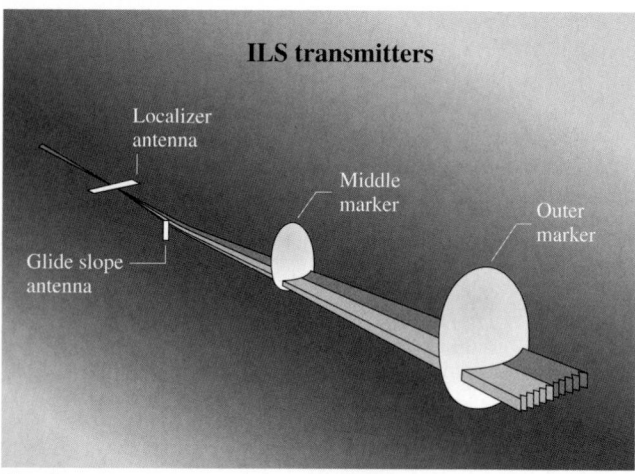

Fig. 67.5 Instrument landing system (ILS)

aircraft equipped with GPS receivers can navigate point-to-point anywhere.

VOR/DME Navigation Systems

The VOR/DME [67.5, 6] is a short-range navigation system that has been internationally standardized and is in use throughout the world. To achieve a common civil–military system for en route navigation in the US, the distance measurement element of VOR/DME is provided by the military tactical air navigation (TACAN) system. The collocation of VOR and TACAN constitutes a VORTAC ground facility, providing VOR-bearing information to civil users, the TACAN-bearing information to the military aircraft, and distance information to both. At present, there are 775 VORTAC facilities in the US, 145 VOR/DMEs, and 90 VOR-only ground stations. A few more VOR ground stations have been procured by individual organizations that add to the total number of VORs listed above.

The VOR system has been the standard air navigation system to provide aircraft with bearing information with respect to a ground station. The VOR ground station transmission is in the VHF band from 108 to 117.95 MHz, divided into 50 kHz channels modulated by a 30 Hz signal and by a subcarrier of 9960 Hz, which is frequency-modulated (FM) at 30 Hz. The phase of the two 30 Hz signals is adjusted such that the phase coincidence occurs at magnetic north.

The VOR receiver employs a simple superheterodyne front-end followed by an envelope detector and narrow-band filters to separate the individual signal components. The FM signal is demodulated to recover the 30 Hz signal, and the two 30 Hz signals are applied to a phase comparator. The phase difference between these signals corresponds to the magnetic bearing of the aircraft from the ground station. A course selector (i. e., phase shifter) is added to one of the inputs to the phase comparator to permit the pilots to select any desired bearing between 0 and 360°.

The DME system is an electronic range measuring system which provides slant range information to the aircraft. DME is a two-way ranging method with a ground station operating as a transponder. The airborne interrogator transmits pulse pairs at a given ground station frequency and pulse-pair spacing. The ground station detects the presence of the pulse pair when it exceeds some detection threshold. The time of detection of the half-voltage point on the first pulse is used as a time reference for reply. A fixed 50 μs delay is introduced to account for internal delays. This delay can also be used to provide an electronic offset of the

DME zero range. The reply pulse pair is translated in frequency by 63 MHz. The interrogator determines its slant range by measuring the time between transmission and reception. The interrogator's transmissions are randomized to permit separation of its range data from that of the other interrogators.

Instrument Landing Systems

The ILS is a radio navigation system that provides an equipped aircraft with the horizontal and vertical guidance required for conducting precision landing approaches to the runways in lower-visibility conditions. As shown in Fig. 67.5, the ILS consists of two directional transmitters with a localizer and a glide slope aligned with the runway centerline. The horizontal guidance is provided by a VHF band localizer and the vertical guidance is provided by an ultrahigh-frequency (UHF) band glide slope transmitter. The distance information is provided by the DME, or by low-frequency (LF) marker beacons located at a certain distance from the runway end (threshold), as shown as middle and outer markers in Fig. 67.5. In the NAS, there are approximately 715 airports capable of providing precision instrument approaches. There are three categories of ILS equipment used under different levels of visibility conditions, depending upon when the pilots can see the runway (lights) from a certain height above the ground, called the decision height:

- Category I: Visibility minima as low as 1/2 statute mile within 200 ft (60 m) height above touchdown
- Category II: Visibility minima as low as 1/4 statute mile within 100–199 ft (30–60 m) height above touchdown
- Category III: Visibility minima potentially as low as zero feet within 0–99 ft (0–30 m) height above touchdown.

Although the ILS has proven to be a safe and effective landing aid worldwide, the technology has a number of limitations. First, it is relatively expensive to purchase and install the equipment. The antennae require large clear areas free of metal or metallic reflections, i.e., aircraft taxiing on the airport must be restricted in their position in order to avoid interference with the ILS signal. The ILS equipment must be routinely checked in flight to ensure that it meets the required specifications.

Satellite Navigation Systems

There are currently two satellite navigation constellations: the US GPS and the Russian global navigation satellite system (GLONASS), although only GPS is widely used outside of Russia. The implementation of the third and fourth constellations, GALILEO and COMPASS are currently being planned by the European Union and China respectively. Only GPS is discussed in this section, since it is the only operational system widely used by civil aviation.

Each satellite from these constellations continuously broadcasts a signal that carries ephemeris information allowing an accurate calculation of the satellite position and a code allowing for the accurate measurement of the signal propagation time from the satellite to the aircraft. A suitably designed receiver can acquire and track this signal and use the broadcast information, as well as so-called pseudorange measurements derived from the signal propagation time, to compute an accurate position solution every second. Such position solutions, however, do not meet all aviation requirements because their integrity is not assured. Integrity is a measure of trust in relying on the accuracy of the navigation system information, where the system alerts the user when the system is unable to contain the position error within an acceptable limit for safe operation. Several forms of *augmentation* have been developed in order to obtain the integrity required for aviation. Forms of augmentation that are currently used include a receiver-based technique called receiver autonomous integrity monitoring (RAIM) or aircraft-based augmentation system (ABAS), which uses redundant information from the number of satellites in view to ensure the integrity of the position solutions. Two other augmentation systems are the satellite-based augmentation system (SBAS) and the ground-based augmentation system (GBAS). Each system uses a ground infrastructure to derive corrections and integrity bounds for the satellite signals. The SBAS and GBAS differ in the type of correction and integrity information, as well as the means used to communicate that information to the user receivers. The SBAS is intended to provide service over wide areas and broadcasts the information from the geosynchronous satellites. The GBAS is intended to provide service over terminal areas by broadcasting the information from VHF ground transmitters. A hybrid augmentation system called ground regional augmentation system (GRAS), currently being developed in Australia, provides navigation service over wide areas using a network of VHF ground transmitters.

There are three main types of aircraft receivers for GPS and augmentations. These receivers provide position solutions to other parts of avionics, such as the flight management system (FMS) and the *navigators*,

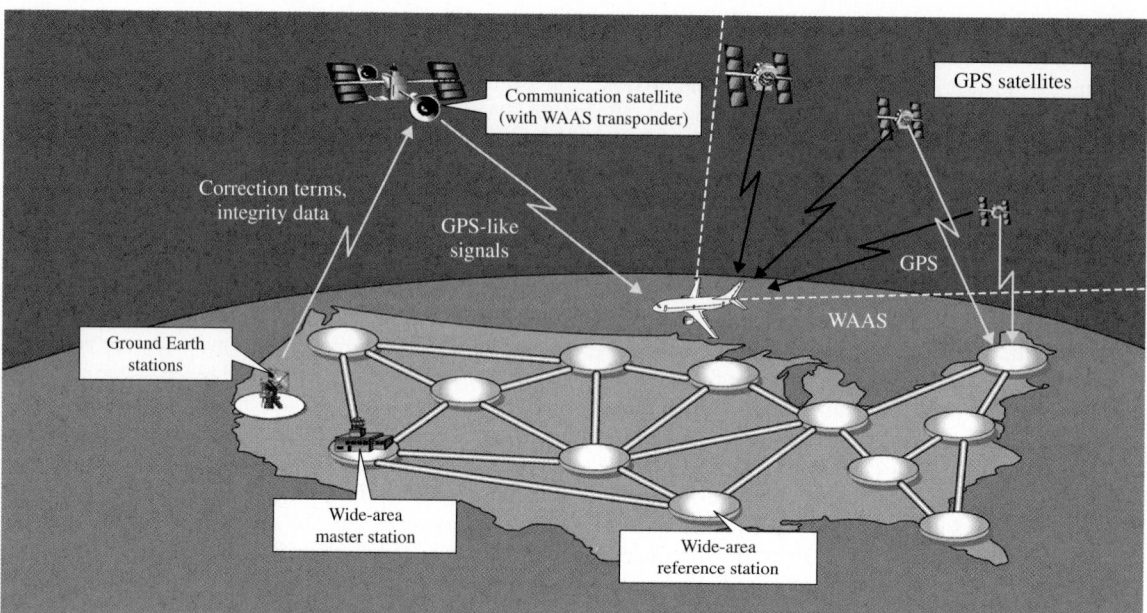

Fig. 67.6 The wide-area augmentation system (WAAS)

which include displays and controls that provide navigation guidance information to the pilots. Certified aircraft GPS receivers with RAIM can be used for en route navigation, terminal area navigation, and nonprecision approaches for landing operations. In addition,

the SBAS aircraft receivers can be used for vertically guided approach and landing operations, called the procedures with vertical guidance (APV) approaches. The primary APV in the USA is the localizer performance with vertical guidance (LPV) approach. GBAS aircraft

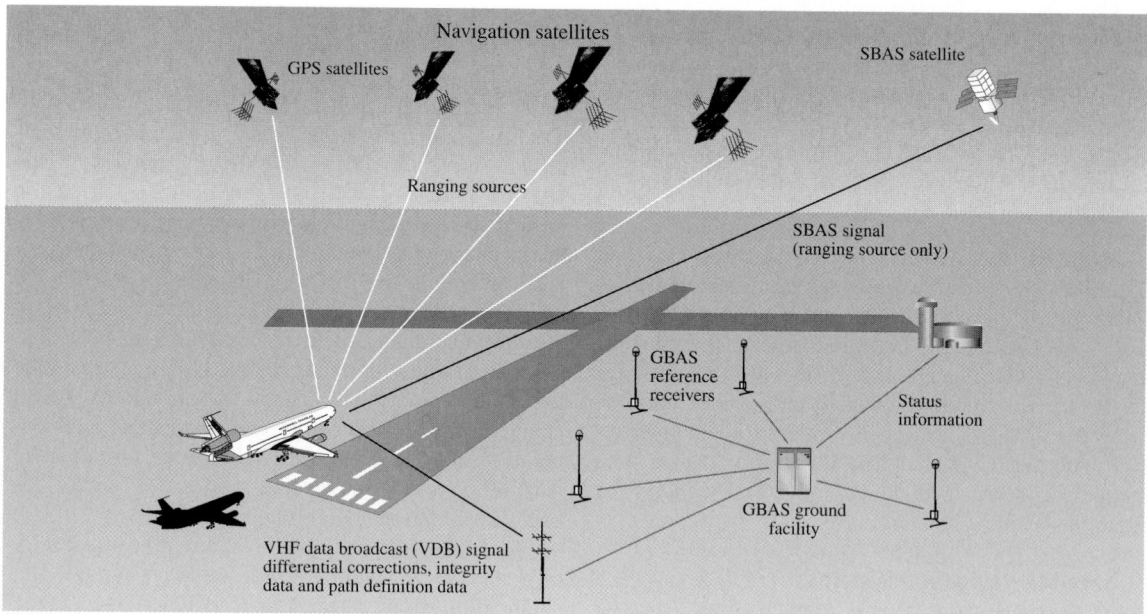

Fig. 67.7 The local-area augmentation system (LAAS)

receivers can be used for category I precision approach operations in addition to the terminal area navigation.

The SBAS uses a network of ground stations to receive GPS signals. The signals are forwarded to a master station, where atmospheric and other errors are identified, and a grid of corrections is created. This grid is transmitted to the user through a geosynchronous satellite. The user receiver can then interpolate between the corrections and improve the accuracy and integrity (i. e., the assurance, within specifications, that the navigation position is free from error) of the GPS signal. US operations use a wide-area augmentation system (WAAS), which can provide corrected horizontal and vertical guidance throughout the USA and parts of Canada and Mexico. WAAS was commissioned in 2003. A schematic of WAAS is shown in Fig. 67.6. The Japanese have developed the multifunction transport satellite (MTSAT) satellite-based augmentation system (MSAS), which is an SBAS that provides coverage in the Asia–Pacific region. Similarly, the European Union has developed the European geostationary navigation overlay service (EGNOS) that provides SBAS coverage over Europe and Africa. MSAS was commissioned in 2007 and EGNOS is expected to reach initial operational capability in 2009. The Indian government is developing the GPS and GEO augmented navigation (GAGAN) system, which will provide SBAS coverage over the Indian subcontinent. All of these SBASs are being developed under common international standards, so the receivers will be able to use any one of these systems.

The GBAS uses a ground station located on or near the airport to receive the GPS signals and to correct for errors. The correction information is forwarded directly to the aircraft via a VHF data link. The US GBAS system is called the local-area augmentation system (LAAS) and is shown in Fig. 67.7. Since the GBAS or LAAS station is located on the airport and near the approach path, there is no interpolation required. The LAAS may have the potential to provide category III service for the approach and landing, while the WAAS will be restricted for landing guidance to category I service. Air Services Australia is also developing a GBAS for certification in 2009.

67.1.3 Modes of Navigation

VOR/DME Mode of Radial Navigation
There are three types of VOR/DME facilities in the US. The difference among the facilities is related to the volume of airspace around each facility that is protected from interference from another facility. This airspace is known as the standard service volume (SSV). The three types are listed below:

- Terminal (T): 25 nmi radius from 1000 to 12 000 ft
- Low (L): 40 nmi radius from 1000 to 18 000 ft
- High (H): 40 nmi radius from 1000 to 14 500 ft, 100 nmi radius from 14 500 to 18 000 ft, 130 nmi radius from 18 000 to 45 000 ft, 100 nmi radius from 45 500 to 60 000 ft.

Coverage below 1000 ft is defined for reduced radii that are altitude dependent. Only H and L facilities are used for en route navigation. All facilities may be used for terminal area maneuvering and nonprecision approaches. Within the SSV, the pilots are assured of a signal with adequate power protected against interference from other facilities transmitting on the same or adjacent frequencies.

The present NAS en route navigation procedures for a majority of aircraft involve flying along VOR radials in or out of the ground station. The low-altitude victor airways and the high-altitude jet routes are defined by these radials like the highways in the sky.

Flights at or above 24 000 ft MSL are not authorized without a DME whenever the instrument flight rules (IFR) require VOR equipment (US regulations permit substitution of an approved GPS receiver for a DME receiver in most cases). The DME equipment is installed in most commercial aircraft and in a large percentage of corporate and some general aviation aircraft. Smaller aircraft flying at lower altitudes may not need a DME. Also, many aircraft owners are replacing DME receivers with GPS receivers.

Area Navigation (RNAV) and Required Navigation Performance (RNP)
The RNAV mode of navigation permits aircraft with appropriate equipment to fly any desired path from point to point without the need to overfly ground stations within the coverage of the station limits using either VOR/DME or DME/DME guidance.

RNP provides additional assurance of adherence to the desired navigation path. The US is planning to transition to full RNAV operations, with RNP operations where beneficial.

DME/DME RNAV is considered possible in those areas where the radii defining the DME arcs from at least two stations intersect at an angle between 30 and 150°. RNAV using DME/DME is more accurate than RNAV using VOR/DME, particularly at long ranges. The DME/DME-RNAV is feasible over most of the

USA, although some areas will not have appropriate coverage using DME/DME at some altitudes. Aircraft using satellite navigation have service available over the entire USA and can fly RNAV routes and procedures anywhere in the USA.

Approach and Landing

The instrument approach procedures to a runway involve landing during instrument meteorological conditions (IMC), and allow the approaches to be abandoned when a landing cannot be completed, such as when the weather is too bad to land, another aircraft is on the runway, or some other reason. The instrument approaches are generally divided into two categories: nonprecision approaches, which have only horizontal guidance; and vertically-guided approaches, which have both horizontal and vertical guidance. In a nonprecision approach, the aircraft flies along a published path and descends to remain above the published minimum altitudes during the approach. The aircraft uses its barometric altimeter to determine the minimum altitudes. The last segment of the approach is called the *final segment*. The final segment starts at the final approach fix (FAF), continues to a missed approach point (MAP), and is usually aligned with the centerline of the landing runway. The pilot is required to see the runway visually prior to landing. For a nonprecision approach, the aircraft departs the final approach segment and flies along the published horizontal path. The pilot descends to the published minimum altitudes. When the runway is in sight, the aircraft continues to land. If the runway is not in sight by the MAP, then the aircraft executes a missed approach. After a missed approach, an aircraft may attempt another landing approach, or may proceed to an alternate airport. Nonprecision approaches can use VOR, NDB, or the global navigation satellite system (GNSS), of which the US GPS is currently the only operational component for civil aviation.

For a vertically-guided approach, the aircraft departs the final approach fix, but has a vertical and horizontal guided path. Vertically-guided approaches typically align the aircraft on a stabilized glide path, and allow the aircraft to continue on the stabilized path until just before landing. The stabilized vertically guided approach is generally considered superior and safer than the nonprecision approach, which often require vertical maneuvering when the aircraft is near the runway, such as after the runway has been acquired visually by the pilot.

Studies have shown that the controlled flight into terrain (CFIT) accident rates are lower for the vertically guided approaches than for the nonprecision approaches. In addition, the visibility minima for vertically-guided approaches are also generally lower. This has led to a general desire to provide vertically-guided approaches (both APV and precision approaches) to all or most runway ends in the US.

Many commercial aircraft have GPS receivers and sophisticated barometric vertical guidance systems. These aircraft can fly lateral/vertical navigation (LNAV/VNAV) and RNP approaches, using GPS for horizontal guidance and approved barometric vertical navigation approaches for vertical guidance. The LNAV/VNAV approaches do not have sufficient accuracy or integrity to meet Category I minima, but do provide useful vertical guidance. The US has published over 1577 LNAV/VNAV and 234 RNP approaches, primarily to the runways serving commercial aircraft (as of January 2009).

The augmentation systems discussed earlier have the potential to provide vertically guided approaches to more airports and runways at a lower cost. Currently, an ILS must be installed on each runway end where vertical guidance is desired. However, approaches using GPS/barometric vertical navigation, LNAV/VNAV, or WAAS may be permitted without any ground navigation infrastructure. In addition to the LNAV/VNAV, the FAA is currently developing LPV approaches to most instrument runways. The LPV approach is primarily a WAAS vertically guided approach, and has a requirement of visibility minima equivalent to an ILS category I approach. The FAA has published over 1445 LPV approaches (as of January 2009), and plans to publish approximately 300 LPVs per year.

The LAAS system will use the GNSS landing system (GLS) approach. The GLS approach is equivalent to a category I ILS approach, but could also attain category II and category III performance. One LAAS station can serve each runway end at the airport, so there is a potential for reducing the ILS infrastructure and saving costs with LAAS. The LAAS system has not yet been commissioned, so no GLS approaches have been developed at this time.

67.1.4 ATC Surveillance Systems and Aircraft Tracking

ATC surveillance refers to the process of determining where aircraft currently are in a given volume of airspace. *Aircraft tracking* refers to the process of correlating successive measurements of various aircraft positions with the identified flights, thereby forming a history or *track* of positions, where each flight has recently been. Time-averaging the successive changes in a given aircraft's track position yields an estimate of that aircraft's current velocity as well. Given the aircraft's current position and velocity, its next position can be predicted for the purpose of surveillance data correlation.

ATC Surveillance Radar Systems

ATC surveillance systems locate weather and aircraft, which enable the ground controllers to separate aircraft safely by providing pilots with surrounding aircraft and weather advisory information. They involve both the ground and aircraft components. Two types of systems are currently used for aircraft surveillance:

- Search or primary radar (radio detection and ranging)
- Beacon or SSR.

The primary radar is so called because it was the first radar system developed and fielded for ATC. The secondary radar was the second radar system developed and fielded, and is used as the major radar for ATC surveillance today. Whenever the secondary radar fails, the primary radar serves as a backup. The primary radar also fills in the coverage for secondary radar dropouts. The primary radars detect two-dimensional (2-D) horizontal position of aircraft by sensing radar energy reflected from the surface of the aircraft. Although the primary radars provide aircraft position in terms of range and azimuth, other relevant data about an aircraft's identification and its altitude are not available. The primary surveillance is noncooperative, i.e., it detects aircraft position without the aid of an aircraft-based transmitter/receiver unit, termed a *transponder*. The primary surveillance systems are used to detect all airborne objects, including aircraft with such failures as loss of electronic power or failure of the transponder.

The secondary surveillance came about in an attempt to overcome some of the limitations and deficiencies of primary surveillance. The secondary surveillance is cooperative, because it requires the aircraft transponders in order to detect aircraft, to positively identify aircraft, and also to receive altitude reports of aircraft, depending on the transponder mode. The SSR requires the aircraft to be equipped with a transponder. The SSR ground station transmits radio frequency at 1030 MHz pulses from a rotating antenna. Upon receiving the ground signal, the transponder transmits a reply on a different frequency 1090 MHz. The aircraft range (ρ) and bearing (θ) are determined from the time delay and radar antenna direction [67.7]. Most

airports and TRACONs in the USA use short-range radar or an airport surveillance radar (ASR). These radars provide surveillance coverage up to 60 nmi and 30 000 ft. The ARTCCs use air route surveillance radars (ARSR), which have a range of about 200 nmi and coverage up to 60 000 ft. A mode C transponder transmits altitude information.

The digital terminal radars using monopulse technology are ASR-11 and mode select sensor (mode S), and the en route ARSR are ATC beacon interrogators (ATCBI-6). The terminal radars provide position information updates every 4–5 s, and the en route radars update every 10–13 s. Generally the primary radar and the SSR antennae are collocated to detect 2-D aircraft positions consistently. The primary airport surveillance is provided by an airport surface detection equipment (ASDE-3), and the secondary surveillance by ASDE-X using multilateration techniques.

Ground Automation Systems and Functions

The en route and terminal automation systems convert target position (ρ, θ) into Cartesian coordinates, and correlate targets with the aircraft tracks. A track maintains state data for an aircraft across the radar scan updates. The tracker derives additional data such as the aircraft ground speed and heading for display. The velocity (speed and heading) information is also used in the aircraft track prediction, and by higher-level automation functions such as conflict alert (CA) and minimum safe warning altitude (MSAW), discussed later. The tracks can be automatically or manually initiated by the controller. All IFR aircraft are required to file a flight plan. When there is flight plan data available in the system, a track is automatically associated with the flight plan so that the flight data, such as the aircraft identification, is included with the speed, heading, and altitude in the displayed alphanumeric tag (in the form of a data block on the controller's display). If a target correlates with a track, the target position is used to update the position of the display symbol, which the controllers use for aircraft separation. If no target information is received in a scan, then the predicted track position is used to update the aircraft position on the display.

The terminal automation system initially included only single-radar displays, where all target data were received from only a single controller-selected radar site. The latest automation system upgrades provide a mosaic display selection to the controllers. The mosaic partitions the surveillance area into a grid with different cells assignable to different radar sites. The en route automation system includes an ARTCC facility-wide mosaic display. The automation system processes digitized radar weather messages and handles display formatting. The map messages are converted into display messages with x, y coordinates, and transmitted to the display channel along with the radar flight target data for display.

67.1.5 Aircraft Tracking

The ground ATC automation system provides aircraft position, velocity, and altitude information to the controllers to help them ensure safe separation between aircraft based on established separation minima. These minima are based on the accuracy and frequency or update rate of aircraft position data and altitude displayed to the controller. Moreover, each aircraft must be positively identified and accurately displayed. The NAS tracking system and the automated radar terminal system (ARTS), as a part of en route and terminal automation systems, respectively, process radar position inputs to track aircraft and display information to the controllers.

The aircraft tracking function computes the position and velocity of all tracked aircraft within the ARTCC's radar coverage, and provides the means for maintaining identity information (in alphanumeric form) with the appropriate search and beacon radar targets on the radar controller's display. Because of the cooperative nature of ATC, flight plan information concerning the planned route of flight, aircraft speed, altitude, and assigned beacon identity code are used in the processing by this function. The current tracker used in both the en route and terminal automation systems is a linear α, β tracker. With the position update rates mentioned earlier, the tracker lags in detecting positions accurately during aircraft maneuvers. The future automation system enhancements will include a seven-state interactive multiple model (IMM) Kalman filter tracker including x, y, z, $\dot{x}$, $\dot{y}$, $\dot{z}$, and turn rate to improve accuracy of position and velocity determination not only during straight and level flight, but also during maneuvers.

The accuracy of position and velocity determination by the tracking function is crucial in maintaining the desired separations between aircraft. Thus, in the future, as the aircraft surveillance and tracking accuracy improve, they could support reduction of minimum separation requirements, thereby not only helping to maintain safety but also to increase airspace/airport capacity. The ongoing deployment of standard terminal automation replacement system (STARS) uses a multisensor (instead of a single sensor in ARTS) IMM

Kalman filter tracker to significantly improve the accuracy of position and velocity estimation.

Automatic Dependent Surveillance–Broadcast (ADS-B)

The radar systems discussed above provide independent surveillance for detecting aircraft. In order to enhance not only the accuracy of position/velocity determination, but also to increase coverage NAS-wide (current limitation of radars in remote/mountain areas in the USA), the ADS-B system provides cooperative surveillance by broadcasting the GPS-derived position information once every second using onboard navigation equipment. The ADS-B broadcasts include aircraft identification, position, velocity, and intent (future aircraft positions) within 100 nmi. Other aircraft equipped with an ADS-B receiver can process and display the aircraft in their vicinity on a display called the cockpit display of traffic information (CDTI). The ADS-B receivers on the ground can also receive the aircraft broadcast information. The large commercial aircraft use a 1090 MHz mode S extended squitter, and the smaller high-end general aviation aircraft use a universal access transceiver (UAT) 978 MHz for transmitting ADS-B information. The aircraft and the ground stations receive information from all line-of-sight aircraft.

ADS-B is currently being used in the State of Alaska, where there is limited radar coverage. Efforts are underway to provide ADS-B information along the US east coast from New Jersey to Florida. In spite of higher accuracy and update rate, the use of ADS-B as a single surveillance system has limitations. Any loss of, or errors in, the GPS signals could adversely impact the aircraft detection. Therefore it is also necessary to continue using primary and secondary surveillance radars complemented by ADS-B. The future automation and tracking system will derive a unique single track for each aircraft using ADS-B and radar measurements by using data fusion techniques.

67.1.6 Air Traffic Management Functions

Traffic Flow Management

The TFM function is responsible for managing the NAS airspace/airport capacity resources by efficiently balancing the demand for air traffic services with available system capacity. The TFM monitors expected demand to produce safe, orderly, and expeditious flow of traffic to minimize delays due to congestion and adverse weather. The ATCSCC has the authority to direct strategic planning of TFM initiatives on a national basis, and is the final approving authority for all regional interfacility TFM initiatives. It has four key responsibilities:

- Monitor air traffic demand, status of airports and airspace, and forecasted weather across the USA
- Coordinate with regional and local facilities to plan and implement traffic flow constraints (aircraft ground delays, ground hold/stops, altitude restrictions, rerouting etc.)
- Assess NAS performance for long-term improvements
- Provide a central point of contact for the NAS users.

The TFM initiatives include:

- For more than 6 days in advance, plan strategies to identify long-term system demands and airspace choke points, as well as recommend operational and procedural changes
- From 6 days to 1 day in advance, predict near-term traffic loads and use weather forecasts to develop strategies for balancing traffic demand and capacity
- On the day of operations, analyze the impact of planned constraints based on flight schedules and predicted weather, as well as collaborate with the users for dealing with the constraints and demand/capacity imbalance
- After the day of operations, analyze the archived operations data to assess effectiveness of the traffic flow initiatives.

The regional (ARTCC) and local (TRACON) TFM functions deal with daily and hourly operations by balancing the demand with the predefined sector (segments of airspace within the jurisdiction of each facility) capacity.

As part of TFM automation, the enhanced traffic management system (ETMS) provides NAS-wide information on traffic loads at capacity-limited resources such as the airports, routes, route merge points (fixes), and airspace sectors. The two components of the TFM automation system are the traffic situation display, which provides a map-oriented display for showing aircraft positions, routes, and weather data, and the monitor/alert function, which alerts when the traffic demand in sectors or at fixes is predicted to exceed predetermined threshold values. A flight schedule monitor provides capabilities to predict airport congestion, and supports planning and monitoring of ground delays as a part of ground delay program (GDP).

Air Traffic Control Services

The primary task of the ATC function is to safely guide and separate aircraft flying under IFR, i. e., monitor and maneuver aircraft to keep them separated by the established separation minima and provide navigation guidance in the airspace where there is no navigation coverage. The following are specific clearances issued by the controllers to service flights from gate to gate:

- Taxi/runway assignment: when the aircraft is ready to depart, the ATC function issues a clearance for the aircraft to start taxiing on an assigned taxiway and wait before the entry to the assigned runway with a clearance to take off. Similarly the aircraft are given landing clearances to specific runways.
- Departure/arrival guidance: unless the aircraft are capable of flying RNAV procedures from/to the airports, the controller provides navigation guidance in the airspace around the airport where there is no navigation guidance available in the form of *vectors*. These are compass headings that the pilots are required to fly under ground system monitoring. In addition, the controllers issue altitude changes and speed adjustments to prevent conflicts by maintaining the required distance spacings between aircraft for safety. For departures, the navigation guidance is provided from *radar contact* (when an aircraft has been positively identified on the radar display) until the aircraft is able to navigate on its own. For arrivals, the guidance is provided from the entry into the terminal airspace, or from the termination fix of a standard terminal arrival route (STAR), to the final approach course for landing.
- Separation from other aircraft: once the aircraft are en route, the controllers ensure that each aircraft is safely separated from all other aircraft in accordance with the prescribed rules. When the desired minimum separation is expected to be violated, the controllers issue altitude/heading/speed clearances to increase separation.
- Safety alerts: controllers provide the aircraft with low-altitude or obstruction warnings based on information provided by MSAW; wake turbulence cautionary advisories, in case lighter aircraft are following a heavy aircraft; and information on hazardous weather, such as wind shear, using Doppler weather radars and low-level wind-shear alert system (LLWAS).

67.2 Functional Role of Automation in Aircraft for Flight Safety and Efficiency

As digital computers have become smaller, faster, more powerful, and more robust over the past decades, aircraft have benefited from their use in the cockpit to automate key functions. In the late 1970s and early 1980s, for example, automation was aimed primarily at reducing the pilot workload of managing complex aircraft systems such as electrical, hydraulic, fuel, and pressurization. This led to the elimination of the *flight engineer* position in the aircraft. Automation in modern aircraft has been effectively applied to enhance flight safety and efficiency. Although the enhanced ground proximity warning system and the predictive wind-shear detection system alert the pilots to stay away from the terrain and dangerous weather conditions, the traffic alert and collision avoidance system (TCAS) is the most significant capability in the cockpit for warning pilots about the presence of other aircraft in the vicinity.

The TCAS has its hardware and software integrated with the other systems in the aircraft cockpit. Its purpose is to avoid midair collisions, acting as a last-minute safety net when normal aircraft separation measures have failed. The TCAS issues radio interrogations that query ATC transponders carried onboard most aircraft. Measuring the time of the replies enables the calculation of each aircraft's slant range. The tracking of an aircraft's slant range every second yields the aircraft's closure rate. The reply also provides the aircraft barometric altitude, which can be compared to that of its own.

There are two different versions of TCAS for use on different classes of aircraft. The first, TCAS I, indicates the bearing and relative altitude of all aircraft within a selected range (generally 10–20 miles). With color-coded symbols and aural alerts, the display indicates which aircraft pose potential threats. This constitutes the traffic advisory (TA) portion of the system. TCAS I does not offer solutions to resolve the conflicts, but does supply pilots with important data so that they can determine the best course of action. The determination of a potential collision threat is time based, rather than based on a fixed distance, as is used by the ground automation functions. The calculation of the time to potential conflict τ is given by

$$\tau = \frac{-(r - k^2/r)}{\dot{r}}, \qquad (67.2)$$

where r is the tracked range, $\dot{r}$ is the estimated relative divergent range rate, and k is a constant for a given al-

titude. Another time calculation is used for the vertical plane.

TCAS uses a modified τ, which predicts the time to a specified minimum distance. This distance allows for some lateral acceleration, without which TCAS could provide inadequate warning time.

In addition to a traffic display, the more comprehensive TCAS II also provides pilots with vertical resolution advisories (RA) when needed. The system determines the vertical profile of each aircraft during climbing, descending, or level phase of flight. The TCAS II then issues an RA advising the pilot to execute an evasive maneuver necessary to avoid the other aircraft in the form of *climb* or *descend*, or *limiting vertical rates*. If both aircraft are equipped with TCAS II, then their systems coordinate to ensure compatible RAs.

As the cost of fuel has risen, automation is also used to enhance the efficiency of flight operations by employing thrust and energy state management techniques. The FMS on a modern transport-category aircraft takes inputs from a wide range of sensors, and couples these with data from a comprehensive navigation and flight performance database. The FMS then generates an optimum flight profile in order to achieve the operator's objectives of minimizing direct operating costs, which are made up of flight time and fuel-related costs.

The FMS includes a flight management computer system (FMCS) coupled with a flight guidance system, a thrust management system, and an electronic flight instrument system (EFIS) to provide total flight management capabilities. The FMCS consists of two flight management computers (FMC) and multiple control display units (MCDU) to provide pilot interface for data entry/review. The FMC provides flight planning and performance management, navigation database storage and retrieval, precise navigation and guidance, and interface with other aircraft systems. Using the current computed vertical profile data from the performance function, the guidance function compares actual and desired altitude and altitude rate, and generates pitch and thrust commands as input to the flight guidance and control (autopilot) and thrust management systems, respectively. These systems include autopilots, flight directors, and an autothrottle. Each autopilot uses a flight control computer (FCC), which signals movement of aircraft ailerons and elevators. The autopilot captures and holds the selected altitude when in vertical speed, altitude change, or vertical navigation mode. The autothrottle adjusts the throttles to achieve the desired speed. The EFIS function of the FMC provides dynamic and background data to the EFIS symbol generator as well as selection of navaids.

The FMCS functions include:

- Flight plan management: provides on board flight planning from a global database including predefined routes.
- Guidance: by integrating precise position and on-board flight planning information, provides three-dimensional (3-D) navigation plus speed or four-dimensional (4-D) navigation including 3-D and time. Lateral guidance provides precise path control and smooth maneuvers at turning points. Vertical guidance provides precise vertical path control.
- Performance management: provides optimum speed, altitude, and thrust settings to minimize operational costs. A choice of economy or alternative flight modes is also available.

The aircraft operator, for example, can select at the outset what is most important to the airline's bottom line for that particular flight (operating at minimum fuel cost, operating with minimum flight time, or a *blend* between the two). The FMCS then determines the optimum vertical profile and speed schedule for the flight, taking into account all known factors such as en route winds, temperatures, and aircraft weight. The FMCS constantly refines its output to account for any change in actual winds or temperature. The wind field data that provides the FMCS with information on future waypoints is easily uplinked to the aircraft using an air–ground data link. If the flight is not constrained by the ATC restrictions, the flight profile dictated by the FMCS will be the optimum. In the case when the ATC constraints are imposed, the FMCS will still provide an optimum profile within those constraints.

In many implementations of the modern FMS, the management of the speed and altitude profiles effectively replace the shortcomings of the non-FMS relic of look-up performance tables and rules of thumb. The FMCS outputs can be fed directly to the aircraft autopilot/flight director system (APFDS), thereby reducing workload for the pilot and making him/her better aware of the progress of the flight flown by the automation system.

67.3 Functional Role of Automation in the Ground System for Flight Safety and Efficiency

In the ground automation system, two functions, MSAW and CA, provide safety alerts to the controllers in all ARTCCs and most TRACONS. The MSAW checks aircraft tracks for current and predicted conflicts with a terrain, and the CA checks for aircraft-to-aircraft conflicts. Both functions alert the controllers through display indications and audible outputs. The user request evaluation tool (URET) is an automation system deployed in the ARTCCs to detect potential conflicts 20 min in advance in order to help controllers take early decisions to resolve aircraft-to-aircraft and aircraft-to-airspace conflicts.

67.3.1 Minimum Safe Altitude Warning

The MSAW function is used to detect the proximity of aircraft to a terrain or surface obstructions. It uses two subfunctions for detecting hazards. One subfunction detects the proximity of an aircraft on the final approach for landing to the required minimum descent altitude or the decision height for the approach. The second subfunction is used to detect whether the aircraft is in hazardous proximity to a terrain outside the approach areas. The location of the aircraft is compared against an internally defined digital map containing the heights of the highest obstructions within the defined areas. When a hazard is detected, the automation system produces a visual alert by displaying a flashing symbol near the affected aircraft. The automation system also sets off an alarm when a hazard is detected.

67.3.2 Conflict Alert

The CA function detects an imminent loss of separation between two controlled aircraft that could lead to a potential mid-air collision. It can detect conflicts both for those aircraft that are traveling in a straight line and for those that are executing turns in a maneuver. When an alert is detected, the automation system displays a flashing symbol next to the positions of the aircraft that are in jeopardy. The CA function also sets off an aural alarm to warn the controller that a conflict has been detected.

Potential conflicts, defined by a horizontal separation parameter and an altitude separation parameter, are detected by projecting a volume of airspace constructed about each track along its velocity vector from its present position to a position some time in the future. En route CA function typically uses a horizontal parameter of 4.8 nmi, an altitude of 1000 ft, and a projection time parameter of 120 s. The corresponding terminal CA function uses the horizontal parameter of 1.2 nmi, 375 ft vertically, and a projection time of 40 s. The CA function considers only the aircraft tracked by the ground automation systems with valid altitude information. If another aircraft is found within the volume of the projected airspace for a given subject aircraft, a potential conflict is declared and an indication is given on the displays for the sector(s) controlling the aircraft.

67.3.3 User Request Evaluation Tool

The URET [67.8] replaces flight progress strips with electronic flight information, thereby reducing the need to maintain and mark strips. In addition, URET notifies the controllers of aircraft-to-aircraft separation problems and aircraft-to-special activity airspace problems.

The URET is a decision support function that combines real-time flight plan and en route automation track data with site adaptation, aircraft performance characteristics, and winds and temperatures aloft to construct four-dimensional (4-D) flight profiles, or trajectories for both predeparture and active flights. For the latter, it also adapts itself to the observed behavior of the aircraft, dynamically adjusting predicted speeds, climb, and descent rates based on the performance of the flight as it is tracked through the en route airspace. The URET's predicted trajectories are used to continuously detect potential aircraft conflicts up to 20 min into the future, and to provide strategic notification to the appropriate controllers. These trajectories also provide the basis for the NAS trial flight planning capability. The trial flight planning process allows the controllers to check if a desired change in an aircraft flight plan would result in

potential conflicts with other aircraft later, before such a change in the flight plan is approved.

67.3.4 Traffic Management Advisor

The traffic management advisor (TMA) is a decision function supported by the en route automation in the ARTCCs to assist traffic management personnel and controllers in optimizing arrival traffic flows to capacity-constrained airports. The TMA uses aircraft trajectory models, real-time radar track data, flight plan data, and wind data updated every 12 s to compute optimal schedule arrival times at the TRACON entry (meter) fixes. The TMA algorithms consider IFR separation minima for the airspace and final approaches to the runways, desired airport acceptance rates, and other ATC constraints. They determine the delays for the aircraft while they are still in the en route airspace controlled by the ARTCC so that the desired airport acceptance rates are not exceeded.

The TMA is intended to enhance the efficiency of flight operations and increase throughput relating to airport capacity during periods of peak traffic demand [67.9]. The aircraft-specific time delays are displayed to the controllers. It is at the controller's discretion to maneuver the aircraft to achieve the required delays. The TMA is intended to help controllers land more aircraft per unit time, and redistribute the unavoidable delays for aircraft from the lower (near the airport) to higher altitudes in the ARTCCs for fuel efficiency and reduced direct operating costs. Additionally, the TMA is expected to reduce flight time for aircraft by reducing holding and vectoring outside of the TRACON airspace. This is achieved by coordinating and optimally sequencing flights to the runways arriving from different directions.

The MSAW and the CA functions have been implemented in both the en route and terminal automation systems for a number of years. The URET and TMA functions are being deployed, and all ARTCCs will have them in use in the near future.

67.4 CNS/ATM Functional Limitations with Impact on Operational Performance Measures

As stated earlier, the current CNS/ATM functions provide information to the pilots and controllers, and most of the decisions in the aircraft and on the ground are made open loop manually. The primary requirement is to conduct safe flight operations by following a set of

separation rules either by maintaining a safe distance between the aircraft or safe altitude separation. The controllers manually trying to achieve these separations between aircraft by relying on tracked aircraft position and velocity information often permit larger separations

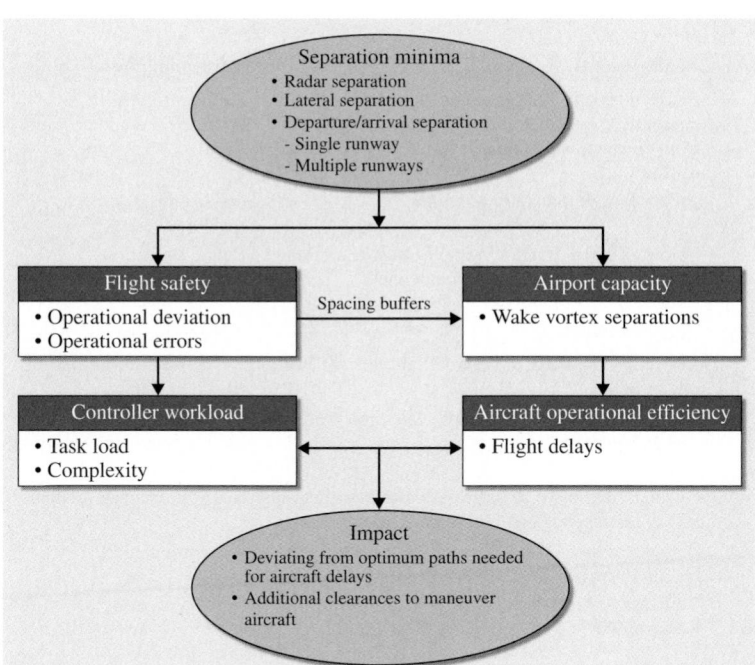

Fig. 67.8 Impact of separation minima on operational performance

than the required minimums by adding a safety buffer. The use of distance separations larger than the desired minimum reduces capacity, especially at airports. With fewer aircraft permitted to depart and land than the separation minima would otherwise allow, delays due to the reduced capacity add operating costs for the users. In order to delay aircraft in the air while trying to meet the safe separation objectives, the controllers deviate aircraft from their desired optimum paths by issuing additional clearances to maneuver the aircraft by changing their paths, altitudes or speeds. Any deviations from the desired minimum flight profiles increase workload for both the controllers and the pilots, who have to manually fly the aircraft during these maneuvers, while the controllers continue to monitor aircraft compliance to the changes.

Figure 67.8 illustrates the impact of separation minima on the operational performance measures relating to flight safety, airport capacity, aircraft operational efficiency, and controller workload, which are discussed below.

67.4.1 Current Separation Minima for Controlled IFR Aircraft

For all controlled aircraft using IFR, the radar-tracked position, velocity, and altitude information are used by the controllers to keep the aircraft separated using the separation minima rules presented in Table 67.1. The separation requirements apply in all controlled airspace, except during the departure and landing phases of flight. These are based upon the radar resolution accuracy and update rate in the horizontal domain, and on the accuracy of the altimeter in the vertical domain. In addition to the radar tracking capabilities, the longitudinal or in-trail separation minima on the final approaches also depend upon the aircraft wake turbulence. All aircraft generate a wake, which is a disturbance caused by a pair of counter-rotating vortices trailing from the wing tips. The strength of these vortices depends upon the size/weight of the leading aircraft and affects the trailing aircraft by imposing rolling moments exceeding the roll-control capability of the aircraft behind. Because of this wake turbulence effect, the controllers are required to use larger separations for trailing aircraft behind larger or heavier aircraft, including a Boeing B-757. These rules directly impact the capacity of the NAS [67.10, 11].

Minimum Lateral Separation between Adjacent Routes
Two aircraft navigating on different airways must have their route centerlines separated laterally by 8 nmi as long as the aircraft are less than 51 nmi from

Table 67.1 Separation minima

Flight phase	Separation minima	Requirements	Controlling factor
En route airspace	5 nmi horizontal	Below 60 000 ft, if multiple radar sensors (mosaic mode) used or either aircraft is more than 40 nmi from antenna, and 60 nmi if mode S surveillance is used	Radar resolution accuracy and update rate
En route path width	8 nmi	Adjacent route separation minima	Navigation mode and system accuracy
Terminal airspace	3 nmi horizontal	Below 18 000 ft, if radar in single-sensor mode and both aircraft within 40 nmi of antenna	Radar resolution accuracy and update rate
All airspace	2000 ft vertical	Above 29 000 ft	Altimeter accuracy
	1000 ft	Above 29 000 ft (RVSM*) or all aircraft at or below 29 000 ft	RVSM certified altimeter above 29 000 ft
Successive arrivals – Same runway or parallel runways spaced < 2500 ft apart	Longitudinal: 3.0 nmi	Radar in single sensor mode and both aircraft within 40 nmi of the antenna	Radar resolution accuracy and update rate
	2.5 nmi	On final approach, if runway occupancy time is 50 s or less and no wake turbulence effect	Runway occupancy time
	4/5/6 nmi	Behind a heavy aircraft or B757 (depends on trailing aircraft type)	Wake turbulence
Parallel approaches – Independent ILS approaches to dual runways	Simultaneous operations once established on final approach	Runways ≥ 4300 ft apart Require ASR	Blunder recovery
		Runways 3400–4300 ft apart Require final monitor aid or PRM	Radar resolution accuracy and update rate
		Runways 3000–3400 ft apart Require PRM and 2.5° localizer offset	Localizer resolution **
Parallel approaches – Dependent ILS approaches to dual runways	2.0 nmi diagonal between aircraft on adjacent runways	Runways ≥ 4300 ft apart	Blunder recovery
	1.5 nmi diagonal between aircraft on adjacent runways	Runways 2500–4300 ft apart	Wake turbulence is an issue below 2500 ft runway spacing
Successive departures – Same runway or parallel runways spaced < 2500 ft apart	1.0 nmi	Courses diverge by 15° or more (not behind heavy/B757)	Radar separation Wake turbulence
	2.0 nmi increasing to 3.0 nmi	Courses do not diverge	Radar separation
	Wake vortex separation: – Distance (see above) – Time (2 min)	Behind a heavy/B757 aircraft	Radar separation Wake turbulence
Simultaneous departures – Parallel or nonintersecting runways	Simultaneous operations	Parallel runways separated by 2500 ft or more and courses diverge by 15° or more	Radar separation Wake turbulence
		Non intersecting runways with courses diverge 15° or more	
Departure and arrival – Same runway	2.0 nmi increasing to 3.0 nmi	Within 40 nmi of the radar antenna	Radar separation Radar resolution accuracy and update rate
	2.0 nmi increasing to 5.0 nmi	Not within 40 nmi of the radar antenna	

Table 67.1 (cont.)

Flight phase	Separation minima	Requirements	Controlling factor
Departure and arrival – Same runway	2.0 nmi increasing to 3.0 nmi	Within 40 nmi of the radar antenna	Radar separation Radar resolution accuracy and update rate
	2.0 nmi increasing to 5.0 nmi	Not within 40 nmi of the radar antenna	
Departure and arrival – Parallel or nonintersecting runways	Simultaneous operations	Thresholds are even *** Runway thresholds are at least 2500 ft apart Missed approach and departure courses diverge by at least 30° Missed approach by a heavy jet cannot overtake departing aircraft	Radar separation Wake turbulence

*   Aircraft equipped with required vertical separation minimum (RVSM) certified altimeter;
**   RNAV/RNP arrivals may reduce lateral deviations and enable arrivals to runway with separation less than 3400 ft without the need to offset one of the approaches;
***   Staggered thresholds increase or decrease the runway separation required

a VOR/DME station. When the aircraft are more than 51 nmi from the navaid, the airways should diverge at an angle of 4.5° from the navaid. If both aircraft are flying in RNAV mode, a constant 8 nmi lateral separation could be maintained.

67.4.2 Flight Safety Assessment Metrics

In spite of millions of operations over the years, there have been an insignificant number (1198 from 1959 to 2006) of passenger aircraft accidents [67.12] as a result of controllers enforcing the above separation minima. US and Canadian operators were involved in only one-third of these accidents. However, due to system inaccuracies and human judgment, aircraft sometimes come closer to each other than the established separation minima. In order to minimize the number of separation violations, the controllers add an extra buffer to the desired separation, especially during the landing phase of flight where wake turbulence could be critical to aircraft safety. Even with the increased separation between aircraft due to the added buffer, there are still flights that occasionally end up with less than the desired minimum separations. In order to assess and analyze the causes of reduced separations, the following two performance measures are determined from the recorded operational data or controller/pilot reports [67.13]:

- Operational errors: an occurrence attributable to an element of the ATC system in which less than the applicable separation minima results between two or more aircraft, or between an aircraft and the terrain or obstacles (e.g., operations below the min-

imum vectoring altitude, equipment/personnel on runways, or aircraft lands or departs on a runway closed to operations after receiving air traffic authorization)

- Operational deviations: an occurrence attributable to an element of the ATC system in which the applicable separation minima, as referenced above, in the operational error were maintained, but the aircraft penetrated the airspace that was delegated to another airspace sector or facility without prior coordination and approval; or an aircraft, vehicle, equipment, or personnel encroached upon a landing area that was delegated to another position of operation without prior coordination or approval.

The ATC system continues to work on mitigating the causes of the above errors and deviations.

67.4.3 Determination of Airport Capacity

Even though the traveling public is the ultimate user of the NAS resources, from an operational perspective, what matters most is the actual number of flight operations that the CNS/ATM system is able to handle without delays. The separation rules no doubt are intended to ensure safety, but they also directly relate to airport capacity. These separation requirements limit the number of operations when the traffic demand is heavy. Often airspace congestion is caused by traffic demand exceeding the airport capacity. Consequently, maximum utilization of airport capacity is paramount to keeping the flight delays to a minimum. The allowable operations at an airport depend upon a number of separate elements including the surrounding airspace, the runways

Part G | 67.4

and taxiways, the gates and parking apron, and the terminal building (including ticket counters, security gates, and baggage claim areas). Any one of these elements could limit the number of passengers that can be accommodated per unit time (hourly, daily) at an airport.

The airport capacity is defined as the maximum sustainable runway throughput of aircraft arrivals and departures on a long-term basis, given continuous sustained traffic demand. Although the actual throughput may be different in a given hour, due to short-run variations in aircraft mix, control procedures, etc., the measure of theoretical capacity is relevant for comparison of operational performance at airports, or developmental alternatives to enhance capacity at a given airport. The following factors are part of the airport capacity estimation process [67.14]:

- Aircraft characteristics:
  - Final approach speed
  - Runway occupancy time (ROT) – mean and standard deviation (the ROT for arrivals is the average time interval from the time an aircraft crosses the runway threshold to the time when it exits the runway)
- Aircraft fleet mix (percentage of different aircraft types and/or weight classes)
- Separation minima (as discussed in Table 67.1):
  - Minimum arrival separations (arrival–arrival)
  - Minimum arrival/departure separation for the shared runway
  - Minimum departure separations (departure–departure)
  - Minimum interarrival separation (minimum required separation distance plus a performance buffer for safety)

Table 67.2 Wake vortex interarrival separation (nmi)

Arr–Arr separations at threshold		Leading aircraft type			
		Heavy	B757	Large	Small
Trailing	Heavy	4	4	2.5	2.5
aircraft	B757	5	4	2.5	2.5
type	Large	5	4	2.5	2.5
	Small	6	5	4	2.5
Aircraft weight classes					
Heavy	Maximum takeoff weight > 255 000 lb				
B757	Boeing 757				
Large	41 000–255 000 lb				
Small	≤ 41 000 lb				

- Relative percentage mix of arrivals and departures, or an arrival/departure ratio, for a given time period
- Performance spacing buffer.

As mentioned earlier, a buffer is added for safety to the required minimum separation distance between successive arrivals. This buffer reflects the variations in aircraft performance, as well as the manual control process for turning aircraft for the final approach. The buffer was estimated based on data collected in the USA in the 1970s, but is still used in estimating capacity for the airports. This buffer was determined to be a normally distributed interarrival error of 18 s (1σ) to reflect the spacing error at the threshold [67.14]. The mean interarrival separation is assumed to be 1.65σ above the minimum desired separations. This 5% approximation is a modeling construct only to account for a number of factors primarily relating to speed variability. It does not imply that 5% of the actual aircraft pairs in-

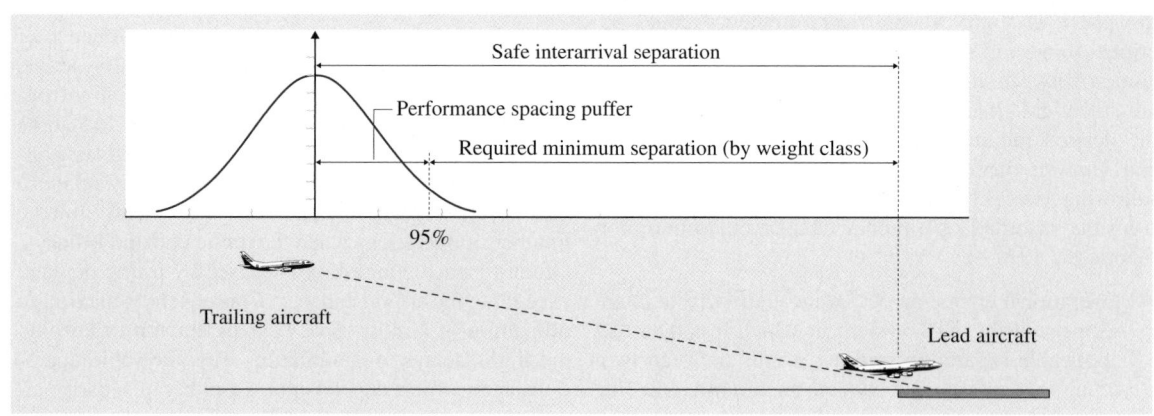

Fig. 67.9 Interarrival time determination

trail lack minimum desired separation. Although it is possible that the actual interarrival spacing could be represented by other distributions, the FAA airport capacity model has always used a normal distribution. The model is intended to provide an estimate of airport capacity for relative comparison purposes. It is not intended to estimate an actual controller's ability to achieve a certain level of arrival throughput. Table 67.2 shows the required separation minima between successive arrivals based on the wake vortex considerations for different weight classes of aircraft. Figure 67.9 shows the delineation of interarrival time (IAT) given by

$$IAT = \frac{\text{Desired minimum separation} + 1.65\sigma}{\text{Final approach speed of trailing aircraft}} .$$
(67.3)

Airport Capacity for Arrival Operations on a Single Runway

The arrival capacity [67.15] is computed by determining the average time between successive arrivals, and inverting this time to find the maximum number of arrivals per hour

$$
\begin{aligned}
\text{Capacity} &= \frac{3600}{\text{Average time separation between arrivals}} \\
&= \frac{3600}{\text{TAA}} .
\end{aligned}
$$
(67.4)

The required time separation for each aircraft class pair ($\overline{\text{TAA}}(i, j)$) is determined by comparing the arrival runway occupancy time of the lead aircraft i and the IAT over the runway threshold for the aircraft pair ij, and selecting the larger of these two values. The frequency with which each aircraft class pair would occur is assumed to be the product of their individual frequencies, e.g., the frequency of occurrence of the class pair $i, j = \%i \times \%j/10\,000$. Therefore, the average time separation between arrival pairs is computed as the sum over all class pairs of the product of $\overline{\text{TAA}}(i, j)$ and the frequency with which the pair is expected to occur

$$\text{TAA} = \sum_{i,j} \overline{\text{TAA}}(i, j) \times \%i \times \%j/10\,000 .$$
(67.5)

In determining the arrival runway occupancy time and the landing time between arrivals, the airport capacity estimation process used is an airport capacity model (ACM) that considers the variability of aircraft, pilots, and controllers, as expressed by the standard deviations of arrival runway occupancy time and arrival–arrival time separation. In addition, to determine the time between arrivals over the runway threshold, the ACM considers the final approach velocities of the aircraft pair and the length of the common final approach path. If the velocity of the trailing aircraft is less than the velocity of the lead aircraft, the specified minimum arrival–arrival separation is considered at the merge point of the two approach paths.

Airport Capacity for Departure Operations on a Single Runway

The capacity of a departure-only runway is given by

$$
\begin{aligned}
\text{Capacity} &= \frac{3600}{\text{Average time separation between departures}} \\
&= \frac{3600}{\text{TDD}} .
\end{aligned}
$$
(67.6)

The required time separation for each aircraft class pair ($\overline{\text{TDD}}(k, l)$) is determined by comparing the departure runway occupancy time of the lead aircraft k and the time separation between departures (from the runway threshold) for the aircraft pair kl. The larger of these two values is assumed to be the required time separation at the runway threshold for this pair of departure aircraft classes. The average time separation between departures is computed as the sum over all class pairs of the product of $\overline{\text{TDD}}(k, l)$ for each aircraft class pair and the frequency with which the aircraft class pair is expected to occur

$$\text{TDD} = \sum_{i,j} \overline{\text{TDD}}(k, l) \times \%k \times \%l/10\,000 .$$
(67.7)

Airport Capacity for Mixed Arrival/Departure Operations on a Single Runway

To insert departures between arrival pairs, the airport capacity model imposes the following requirements:

● The departures cannot roll if an arrival is already on the runway
● The departures cannot roll if:
   – An arrival is within some specified distance of the runway threshold, or
   – The departure cannot clear the runway before the arrival comes over the threshold.
● The departure–departure separation minima must also be met to insert multiple departures between an arrival pair.

By employing these conditions, the model computes the probability of inserting one, two, or three departures between each arrival pair. The interleaved departure capacity is then determined from these probabilities and the aircraft mix.

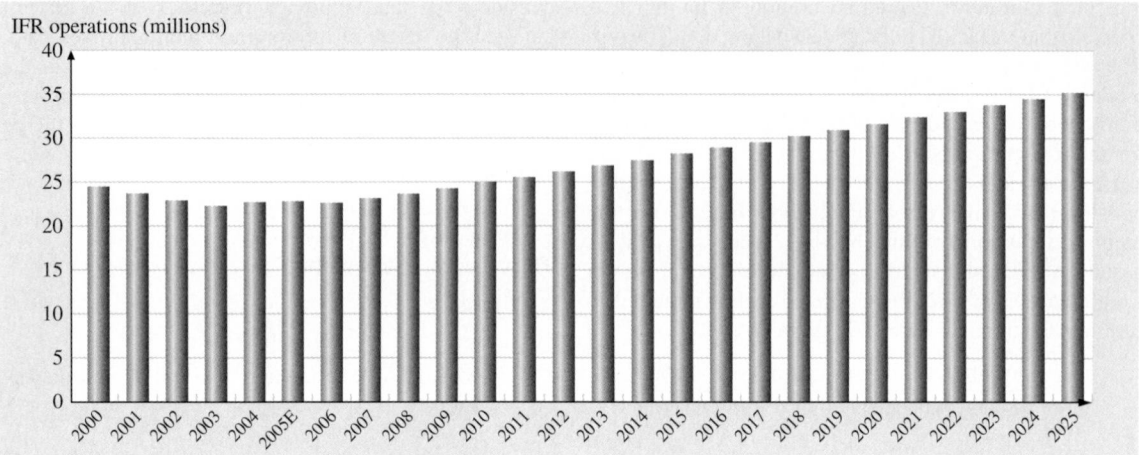

Fig. 67.10 Annual controlled aircraft operations

Airport Capacity for Dependent Arrivals and Departures on Two Runways

The model can also compute the capacity for a pair of runways when departures on one runway are dependent on the arrivals to the other runway. The departures cannot be released if an arrival is within a specified distance from the runway threshold, but can be released as soon as the arrival touches down. It is not necessary to wait until the arrival has exited the runway.

The logic for computing the departure capacity of this configuration is similar to that used for a single runway. After the interarrival times are obtained, the probability of performing one, two, or three departures in each interarrival gap is calculated. The departure–departure separation minima are enforced, not just between the departures in the same interarrival gap, but also between the departures in the adjacent gaps. This is rarely necessary for mixed aircraft operations on a single runway, because the time required for an arrival operation is usually greater than the departure–departure separation minima.

67.4.4 Aircraft Delays to Measure Operational Efficiency

Ideally all aircraft operators would like to fly the most wind-favored airway or a direct route if they are RNAV equipped, from the origination airport to the destination airport. This is generally feasible during light traffic periods when the airports are not operating at capacity. During medium to heavy periods of traffic demand, the airport capacity limits the number of departures and arrivals, which result in delays. As shown in Fig. 67.10,

the number of annual controlled IFR aircraft operations (including air carrier, commuters, air taxi, and high-end general aviation aircraft) is expected to grow from a current 23 to 35 million by the year 2025. Unless means are explored to increase capacity, especially at the major airports, delays will continue to increase, thereby impacting the aircraft operational efficiency that will significantly increase the users' direct operating costs. Fig. 67.11 shows the increase in average delays per aircraft in the NAS at 35 major airports. As shown in the figure, the average delay per flight will increase from the current 10 to 63 min by the year 2025, almost six times, as demand continues to grow, if the NAS continues to operate as it does today. This will have an unacceptable economic impact on the airlines, which may find it almost impossible to run the operations needed to meet the demand. Consequently, it becomes imperative for government service providers to find ways to enhance airport capacity without compromising on safety.

67.4.5 Measuring Controller Workload

The primary role of a controller is to process a significant amount of information and make timely decisions to maintain safe and efficient flow of traffic. This involves acquiring traffic information by continuously monitoring traffic, perceiving potential separation violation problems, and deciding when and how to resolve these problems. In order to perform these functions, the controllers are involved in a number of tasks: monitoring traffic situations on radar displays, entering data through a keyboard, conducting mental assessment of potential conflicts, communicating with other controllers/facility people, and communicating clearances (routine and conflict resolution) with the pilots. Thus, a human response to these number of tasks is used to measure workload [67.16], which has both physical and mental components. Perceiving and resolving potential aircraft conflicts, especially during heavy traffic, require more cognitive resources than handling other routine tasks.

The workload (mental) primarily depends upon the time interval between detecting a separation violation and dealing with it. The earlier a decision is made to resolve the conflict, the less the workload for a controller. In congested airspace, especially where the aircraft are continually maneuvering (climbing, descending, turning, and changing speed), the controller workload is not only affected by the number of aircraft under control, but also by the impact of their changing geometrics, which creates complexity [67.17]. The complexity not only depends upon the level of peak traffic, but also upon the specific traffic situations, e.g., the number of aircraft on independent, converging or intersecting paths in horizontal and vertical domains. Rightly perceiving problems resulting from complexity requires timely detection and resolution of these problems, in ad-

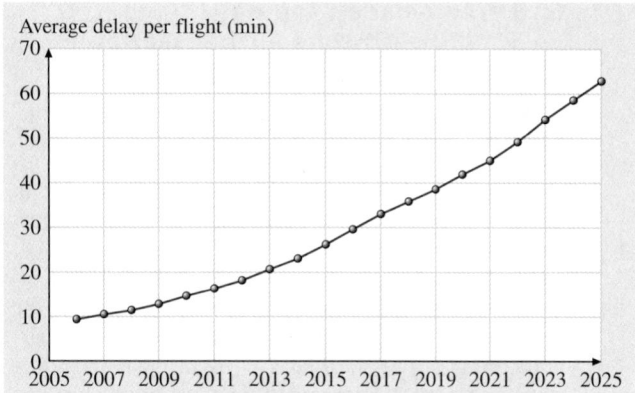

Fig. 67.11 Operational delays impacting flight efficiency

dition to communicating the resolution clearances to the pilots. This often turns out to be a very stressful process that has a major impact on the workload of the controllers.

In the future, it is envisioned that automation will accurately detect, resolve, and communicate actions directly to the aircraft over a data link (rather than currently by voice) in order to significantly reduce the controller workload, as traffic continues to grow.

67.5 Future Air Transportation System Requirements and Functional Automation

In November 2005, the European Consortium of Member States signed to define the vision and goals of the Single European Sky ATM research (SESAR), and to develop, validate, and implement SESAR concepts to meet air traffic demand beyond year 2020. The US 108th Congress and the President mandated the design and deployment of an air transportation system to meet the nation's needs in 2025 by passing and signing into law *Vision 100 – Century of Aviation Reauthorization Act* (Public Law 108-176). The transformed air transportation system should be responsive to the social, economic, political, and technological changes, and should meet the future needs for safety, capacity, efficiency, and security. The legislation established a joint planning and development office (JPDO), which is supported by the Department of Transportation (DOT), the FAA, the Department of Defense (DoD), the Department of Commerce (DOC), the National Aeronautics and Space Administration (NASA), the Department of

Homeland Security (DHS), and the Office of Science and Technology Policy (OSTP) in the White House. The JPDO published the next-generation (NextGen) air transportation system integrated plan [67.18] defining the objectives for the 2025 air transportation system. The following objectives relate to the specific requirements for safety, capacity, and operational efficiency:

- Maintain the aviation record of safety as the safest mode of transportation
- Improve the level of safety as demand continues to grow
- Enhance airport capacity to satisfy future growth in demand up to three times the current level
- Minimize the impact of weather and other traffic disruptions on NAS operations
- Reduce the transit time from domestic curb-to-curb by 30%.

67.5.1 Automation Approach to Meet Future Air Transportation System Requirements

Increased level of automation in the aircraft and in the ground system will help to meet the above stated requirements for the NextGen air transportation system. In order to improve safety, enhanced aircraft situational awareness in the cockpit and in the ground system will be needed to minimize operational deviations and errors. To maximize capacity and flight efficiency, the future airspace design will have to support each aircraft filing its own desired 4-D flight plans including route, altitudes, and expected times at key waypoints, while the ground automation system will ascertain that these flight plans are conflict free. This would require the ground automation system to accurately estimate aircraft trajectories correlated in space and time, and precisely predict conflicts in order to negotiate any change in their flight plans with the users. Ground-based conflict-free flight planning, with the aircraft adhering to the agreed trajectories, will reduce the need for tactical controller intervention, which would minimize workload. As such, future research should concentrate on developing means to automate the air/ground functions to enhance safety and capacity while reducing flight delays and workload. In order to achieve this, an automation approach is as follows.

Safety
The automation predicts aircraft conflict problems and provides decision support to the controllers in resolving them.

Capacity
No doubt the building of new runways and airports increases airport capacity, but it is a time-consuming process to address the environmental and adjoining community issues required to add new runways to the major airports or build new airports. Although the airports authorities in the US are considering the construction of a few new runways, and a couple of new airports outside Chicago and Las Vegas, the majority of future capacity improvements will have to come from using new technologies and automation by:

- Reducing separation minima, including route widths
- Reducing performance spacing buffers
- Developing automated RNP approaches.

Flight Efficiency and Delays
- Automation supports departure/arrival planning
- For strategic end-to-end flight planning, automation deals with system uncertainties.

Workload
- Automation handles routine ground/air clearances
- Transfer some ground-based decisions to the aircraft with automation assisting aircraft in self-sequencing, merging, and spacing.

67.5.2 Development of Automated Functional Capabilities

Relying on automation decision support and new technologies, the following functional capabilities are going through exploratory research and potential development to meet the above requirements for the future air transportation system.

Automated Problem Resolution to Enhance Safety
As discussed earlier, the URET function in the ground system provides the en route controllers with an automated conflict detection capability, which uses predicted aircraft trajectories to continuously detect potential aircraft separation problems up to 20 min into the future, and accordingly alert the appropriate controllers. This function is being enhanced so that the ground automation provides the controller with solutions or resolutions of aircraft conflicts with other aircraft, segments of airspace, and hazardous weather cells. The MITRE Corp. center for advanced aviation system development (CAASD) is developing such a capability, called problem analysis resolution and ranking (PARR). When the URET detects a conflict, the PARR examines strategic vertical, lateral, and speed change options to resolve these conflicts.

The PARR may be initiated for a specific aircraft with one or more problems, or for a specific problem. In these cases, the PARR examines a variety of resolution dimensions and directions. If initiated for an aircraft, the PARR generates resolutions which maneuver only that aircraft. If initiated for an aircraft-to-aircraft problem, the resolutions for each of the two involved aircraft are generated.

For a given aircraft to be maneuvered, the PARR searches for problem-free trajectories to resolve all problems with that aircraft (within URET's 20 min lookahead horizon) in an operationally acceptable manner, without introducing any new problems. The search process examines, in turn, maneuvers in each of the following five dimensions/directions, thus yielding up to five resolutions for that aircraft:

1. Select an altitude above the altitude of the problem
2. Select an altitude below the altitude of the problem aircraft
3. Turn aircraft left of the route
4. Turn aircraft right of the route
5. Increase or decrease speed.

Each resolution requires only one maneuver [67.19].

The completed PARR resolutions are ranked, color-coded, and displayed on the URET plans display; for example, conflicts predicted with violation of separation minima are coded red. This provides the controllers with information to set priorities in dealing with the problems. After the aircraft follow the resolution maneuvers, and there is no more problem, the aircraft return and continue on their original flight trajectory. The PARR could also generate resolutions around hazardous weather areas.

It is expected that the PARR function will significantly enhance safety. The safety enhancements are important, since the relaxation of ATC restrictions in the future could lead to more complex traffic patterns. The PARR would assist in maintaining or enhancing safety in two ways. First, the PARR provides an automation capability with which the controllers can obtain an improved, strategic situational understanding, e.g., by quickly assessing which altitude, speed, or direct-to-fix alternatives are problem free. Second, the resolutions provided by PARR allow the controller to easily implement strategic, problem-free resolutions, which would allow more time for decision-making and coordination for handling other pilot requests.

Reducing Separation Minima to Enhance Capacity

The airport capacity for a single runway depends upon the weight-class-based wake vortex separations between successive aircraft in-trail on a final approach, the length of the final approach to capture the ILS localizer, runway occupancy times, and a spacing buffer for safety. For operations on parallel runways and routes, the lateral separation depends upon the ability of the aircraft to fly close to their desired path centerline.

The established IFR separation minima apply during poor visibility (less than 3 nmi) and/or lower cloud cover ceiling (less than 1000 ft). When the visibility and the ceiling are higher than these conditions, the aircraft fly VFR when the pilots are able to see other aircraft and the runway before turning onto the final approach, a situation called, *see and be seen*. For these visual operations, the aircraft are observed to land with much smaller separations than the established IFR separation minima without compromising any safety, but yielding a much higher capacity. This is because the aircraft are not constrained by the required longer ILS straight-in approaches and a fixed glide slope. Therefore, by flying shorter final approaches and variable glide slopes, the aircraft encounter less impact of wake vortices, and as such could use less in-trail separations. The following future avionics technologies will help reduce the current wake vortex separation minima by permitting aircraft to operate as VFR under all weather conditions. The required navigation performance capabilities in the aircraft will support reduction of lateral separations between routes. A future terminal automation function discussed later will reduce spacing buffers by reducing the impact of flight uncertainties.

Aircraft Technologies for Electronic VFR Operations

Although there have been no separation minima established for the current VFR operations, there is sufficient data available to define them, if the aircraft could operate like VFR during IFR conditions. The following technologies will help pilots operate as VFR at night and during poor weather and visibility conditions.

Using ADS-B information from the aircraft in the vicinity of the subject aircraft, the CDTI displays the position of other aircraft on a screen in the cockpit. The head-up display (HUD) is mounted on the aircraft instrument panel below the windshield to monitor the external environment (aircraft and airport). The HUDs provide microwave and infrared (IR) images to the pilots. The use of these wavelengths allows pilots to see the runways in poor visibility by penetrating fog or other adverse weather conditions. The enhanced vision system (EVS) enhances a pilot's situational awareness during approach and landing, when the visibility is poor, using an IR camera displaying a picture of the surface below.

The synthetic vision system (SVS), with its ability to let pilots see terrain, obstacles, and runways in poor visibility conditions, is designed for use by high-end business jets for situational awareness and safety. A 3-D view allows the pilots to see rendering of terrain ahead using information from onboard obstacle, terrain, and

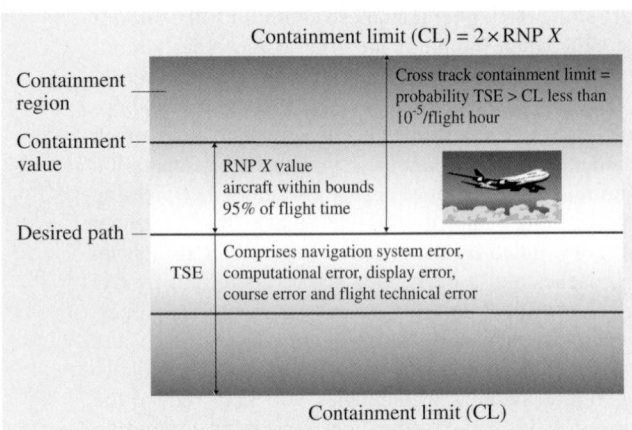

Fig. 67.12 Required navigation performance (RNP) concept

airport databases and tracked flight path. The navigation display shows the intended aircraft flight path with a view that provides real position over the terrain with respect to the flight plan.

Required Navigation Performance

In order to reduce the requirement for lateral separation between aircraft on parallel routes, or to conduct simultaneous approaches to closely spaced parallel runways, the aircraft would need to have a RNP capability. The RNP enables aircraft not only to fly RNAV point-to-point, but also to stay within certified route containment limits with an onboard monitoring and alerting function. This function enhances the pilot's situational awareness, and alerts the pilot when there is a gross deviation from the route centerline, thereby permitting closer route spacing without ground ATC intervention.

Figure 67.12 shows the RNP concept and the lateral components of the aircraft navigational error. The RNP is a measure of navigation performance accuracy and integrity (i.e., route containment and time to alarm) necessary for aircraft operations within a defined airspace. As shown in the figure, an aircraft certified for a given RNP value, e.g., RNP X, must navigate with a total system error (TSE) not to exceed $2X$ nmi with a probability of 10^{-5} per flight hour, defined as the cross-track containment limit. When exceeding this limit, the monitoring function will generate an alert for the pilot to correct back to the desired course. The RNP X value defines the bound of lateral deviation with a probability of 95% of flight time.

The TSE is the deviation of the aircraft's true position from the desired course or the centerline of the route of the flight path programmed in the FMS. The

TSE is a combination of errors from the following contributing factors:

- Navigation system error
- RNAV computation error
- Display system error
- Course error and flight technical error (FTE).

The airborne equipment accounts for data and computational latencies, equipment response time, and navigation sensor error characteristics for its interfaces. The TSE value assumes a flight director or an autopilot operation that allows the use of either GPS or DME/DME as navigation sources for position determination. The aircraft are capable of RNP 0.3 aided by a flight director in the aircraft, and RNP 0.11 with a coupled highly accurate automatic flight control system (AFCS). The en route path lateral separation requirement could be reduced to 4 nmi from the current 8 nmi for aircraft with RNP 1. RNP 0.11 would permit independent parallel runway operations with significantly reduced runway spacing.

Automation Functions to Improve Flight Efficiency and Reduce Delays

Probabilistic Traffic Congestion Management. The current TFM function balances air traffic demand against airspace constraints and airport capacity taking into consideration the forecasted weather conditions [67.20]. A variety of flow control actions are used to deal with the airport capacity constraints. These include weather avoidance routes, miles in-trail (MIT) restrictions to deal with traffic congestion at fixes and the ground delays generated by the ground delay program to establish expected departure clearance times (EDCT) for flights. Planning for these actions requires predictions of both the traffic demand and the airspace (sector) capacity. Since the TFM decisions are typically made 30 min to several hours in advance of the anticipated congestion, these predictions are subject to significant uncertainty. However, the magnitude of this uncertainty is not known, presented, or understood. As a result, traffic management decisions are often overly conservative, and may be taken at inappropriate times depending upon the accuracy of prediction data. The traffic demand uncertainties arise from many sources. The flight schedules undergo constant changes in response to daily events, and such changes often occur between the time of demand prediction and the time for which demand is predicted. These include flight cancelations, departure time changes, and initiation of previously unscheduled flights. This latter category is increasing in the USA, as air taxi and *executive jet* operations are becoming more prevalent.

Several new techniques and technologies are required to provide probabilistic TFM decision support. First, prediction uncertainty must be known and quantifiable. Second, a metric is needed for rating the goodness of the candidate solutions. Third, decision-making algorithms are needed to develop congestion management solutions, given the prediction uncertainty and the goodness metric. Finally, there are significant human factors issues to be resolved due to the combination of information uncertainty and complex automated processes.

Effective TFM decision making in the presence of uncertainty or *probabilistic TFM*, should have the following characteristics:

1. Rather than attempting to resolve all possible congestion problems, incremental actions are taken to keep traffic congestion risk at an acceptable level, while retaining flexibility to take further actions as the situation becomes more certain

2. Predicted traffic congestion areas are continually reevaluated for further control action
3. NAS users are informed of predicted congestion, so that they can proactively reduce schedule risk if desired (e.g., by replanning flights through less congested airspace)
4. Probabilistic congestion predictions are presented to traffic planners and users in an intuitive way, in order to maintain good situation awareness.

Multicenter Traffic Management Advisor. The TMA function implemented in the en route automation host computer systems (HCS) regulates or meters traffic arriving from different directions within a single ARTCC to a major airport. The development of multicenter traffic management advisor (McTMA) led by NASA is being built upon the TMA hardware and software baseline. The McTMA is an automation decision support function, which extends time-based metering from a single ARTCC/single airport arrival traffic flow planning to multi-ARTCC operations dealing with traffic flow problems at critical bottlenecks in the en route

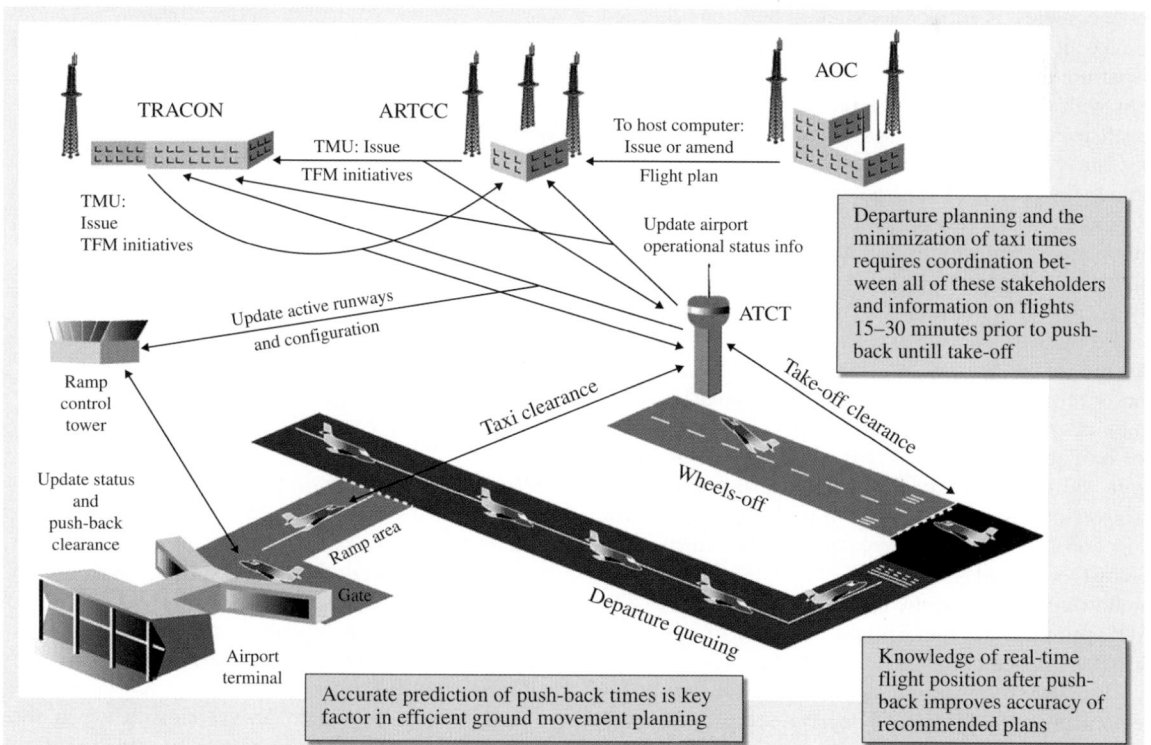

Fig. 67.13 Required coordination during departure process

airspace and merging of departure traffic with overflights. This function employs a distributive scheduling algorithm to develop flexible collaborative metering plans taking into consideration the ATC constraints at airports and in the en route/transition airspace spread across a region comprising a number of ARTCCs. The distributed scheduling provides a dynamic look-ahead capability and a provisional landing slot reservation system to continuously monitor and feedback upstream adjustment of flight times. This is required to deal with demand/capacity imbalance at the point of congestion [67.21].

Departure Planning and the Role of Airport Surface Automation. Most of the traffic flow management constraints are applied at the airports when the traffic demand exceeds either the en route capacity or the capacity at the flights' destination airport. Consequently, these restrictions are imposed on departing flights so that a particular flight departs at a specific time in order to fit into a specific place in the stream of traffic, or to fit into a specific arrival time slot at the destination airport. The management of these departure time constraints can make the airport surface air traffic control task much more complex. A surface automation function is needed to take all of these constraints into account for efficient departure planning. The airport surface automation system will also help reduce airport accidents, although small in number (13 from 1997 to 2007), by providing timely and accurate information on aircraft positions both to the ground system and the cockpit.

Figure 67.13 illustrates the exchange of information between the various facilities to manage departing flights in order to use capacity at major airports effectively. The airport ramp control tower provides each aircraft with a clearance to push back from the gate. The desired airport runway configuration to use is given by the tower, so that each departing flight could exit the ramp at an appropriate time. The ATCT is responsible for providing clearances for the aircraft to taxi safely from the ramp area to the departure runway, and to take off at appropriate times so as to meet all traffic flow constraints and safe separation requirements. The TRACON provides the ATCT with the relevant traffic flow constraints upstream, as does the ARTCC. The AOC provides the initial flight plan and manages the dispatch of each flight under its control.

The airport departure capacity is a function of the sequencing of flights to each departure runway. Aircraft in the heavy weight class require more spacing behind them than smaller aircraft. Therefore, clustering heavy aircraft together could significantly increase runway capacity. Dynamically managing arrivals and departures, by building extra arrival runway slots when necessary to absorb extra arrival demand, also improves the overall effective use of airport capacity.

In order to facilitate all of the above interrelated decisions, the airport surface automation system should provide the following four functions, and the information generated must be provided in real time to all of the decision-makers shown in Fig. 67.13:

- Surface aircraft and vehicle surveillance
- Automated transfer of controller clearance and flight intent information
- Detection of potential airport surface conflicts
- Automation decision support function to assist the ramp control tower and the ATCT to best use the available runway and taxiway capacity.

Surface aircraft and vehicle surveillance, as well as the automated clearance and intent information transfer, are the two key functions that enable the other two functions listed above. Information about the current position of each aircraft and the vehicle located on an airport's taxiways and runways, integrated with the positions of the aircraft immediately around the airport, provide the ATCT and the ramp control tower with the ability to see all of these objects, even when the physical line of sight is obstructed by low airport visibility conditions (fog). They also provide the necessary input to the surface conflict detection function, which uses the known positions of each aircraft and vehicle and infers the future path of each aircraft and vehicle (e.g., the cleared taxi path), in order to detect potential conflicts. When a conflict is predicted, the automation function generates an alarm for the controller to alert the pilot to take an action to avoid the conflict.

The surface automation decision support function provides the traffic flow managers and the air traffic controllers in the ATCT and the ramp control tower with recommendations on actions such as:

- When to change the runway configuration to minimize the loss of capacity during the changeover
- In what order to best maximize the taxiing aircraft movement through the exit spots at the edge of the ramp area
- In what order to best queue aircraft to maximize runway capacity given the traffic constraints, the size of the aircraft, and other factors
- How best to introduce arrival slots between the departure slots to minimize arrival and departure delays.

The approaches to develop these automated decision support functions have already been discussed in detail [67.22, 23].

Terminal Automation for Arrival Planning and Control [67.24]. Over the years, two terminal automation functions, viz. metering and spacing (M&S) and the final approach spacing tool (FAST), were developed and went through extensive validation and field testing. However, these functions were not accepted by controllers, because the automation-generated information was either too constraining or inconsistent with the human decision process. These functions were intended to complement en route metering function (TMA) to accurately establish landing sequences and times to enhance airport capacity. The crux of the problem was a discrepancy between the TMA-planned nominal flight trajectories and the actual trajectories flown by aircraft tactically changed by TRACON controllers to achieve desired separations between aircraft.

The TMA establishes desired meter fix times for the traffic going to an airport while they are still in the en route airspace by establishing landing sequences and times based on prestored aircraft trajectory data and wind information over the terminal airspace. The en route controllers try to meet these meter fix times within a specified tolerance (1 min) by maneuvering the aircraft before they reach the meter fixes. Once the aircraft enter the terminal airspace, the terminal controllers merge traffic generally coming from four different directions to maintain the required separations, as well as guide the unequipped aircraft in the terminal maneuvering areas, if there are no navaids. As such, they end up changing the en route metering system planned paths and landing time schedules. This affects the flight planning and operational efficiency for aircraft all the way to touchdown by first getting delayed in the en route airspace and then getting further delayed in the terminal area.

A terminal automation arrival planning and control function is needed to complement the en route planning function in order to predict arrival schedules accurately. This is essential for realizing maximum efficiency benefits for the users by flying optimum paths, and for the service providers to manage diverse aircraft traffic with minimum air–ground communications. This function should minimize the variations between the aircraft flight planning and actual operations by first defining routes all the way to touchdown and then establishing landing sequences and schedules using accurate flight time estimates.

The basic requirement for establishing an efficient terminal area flight plan for each aircraft is that it should be based on minimum flying time from the entry (meter) fix to the runway. In order to achieve the earliest permissible landing times:

1. The plans should be based on the shortest paths
2. Continuous descent from meter fixes to touchdown
3. The aircraft are assumed to fly highest permissible speeds over each flight segment
4. The routes from different directions towards final approach (s) are adequately separated to avoid conflicts during merging of traffic
5. The landing times for successive aircraft ensure adequate separations based on wake vortex considerations.

Once the landing times are established, they should be integrated with the en route planning of flight times such that, when the aircraft arrive at the meter fixes within a desired tolerance, they should be able to continue on the established 3-D profiles without any need for path deviations to maintain separations. The en route metering process ensures delivery of aircraft at the meter fixes within the expected time variance.

In most major terminal areas today, the aircraft, using four-corner post configuration, navigate over established STARs from the meter fixes to about 10–15 nmi radial distance from the airport. Depending upon the direction of arrival, the aircraft either fly a downwind path and then turn onto a base leg, or turn directly onto the base leg, before intercepting the final approach as shown in Fig. 67.14. Since there are mostly no established routes in the base-leg region, the

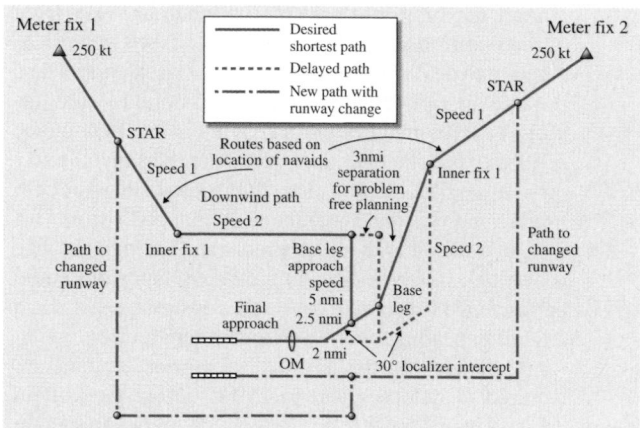

Fig. 67.14 Terminal merge-free route design for flight planning, with permission from IEEE

controllers merge the traffic coming from the opposite directions over the two base legs, and then again merge the traffic flying over the two base legs from the opposite directions onto the final approach to a single runway. Merging of aircraft at two or three points, while keeping them separated, results in inefficient, large path deviations and significant workload for both the pilots and the controllers.

In order for most of the aircraft to stay on the minimum path for maximum flight efficiency with short delays, aircraft-derived speed control should be used as the primary means to compensate for aircraft performance deviations and wind forecast uncertainties. In order to achieve this, the terminal route design should eliminate the need for the controllers to merge traffic at multiple points except on the final approach(es). As such, the routes should not only be the shortest, but all merge points should also be eliminated. As shown in Fig. 67.14, the aircraft arriving from the direction opposite to the direction of landing (meter fix 1) fly a downwind path from the end of the STAR turning onto a base leg 5 nmi long before intercepting the localizer course. For smooth capture of the localizer, the aircraft intercept the final approach course at an angle of 30° or less over a path segment of 2.5 nmi to allow for smooth turns from the base leg to the localizer course. The aircraft also need to capture the localizer for final approach course about 2 nmi from the outer marker in order to be stabilized over the glide slope to the runway. This defines the minimum path in the base-leg region. The aircraft arriving from the other direction (meter fix 2) turn directly onto the base leg before capturing the localizer course. In order to create a merge-free route design, the base legs are separated by 6 nmi to keep the aircraft not only safely separated when arriving from different directions, but also to allow for some margin of airspace to deal with pop-up aircraft, missed approaches or large aircraft deviations (as shown dotted in the figure). A mirror image of the design in Fig. 67.14 could be applied to the other two fixes, either to the same runway or to a parallel runway. The figure also shows a path in case the runway assignment is changed. In the future, if some other than ILS precision landing guidance is available, the base-leg and final approach paths could be shortened for curved approaches customized for each aircraft depending upon its avionics capabilities.

After entering the terminal area, most aircraft are required to reduce speed to 250 kt before they attain an altitude of 10 000 ft. From there on, depending upon the aircraft performance characteristics, the aircraft typically go through two speed reductions (speed 1 and speed 2 in Fig. 67.14) before reducing to their final approach speeds. The automation function could determine timing or location of these speed reductions to compensate for flight time variances from the desired landing times without requiring the aircraft to divert from the above-defined minimum paths.

The NextGen concept for the future air transportation system considers 4-D navigation as one of its core elements. In order for the aircraft to operate in a 4-D-navigation mode, a 3-D flight profile is established from take-off to landing with estimated times of arrival at key decision points along the path. The aircraft are required to stay on the predefined 3-D paths and meet the times at these points by adjusting speeds along the flight segments using the onboard required time of arrival (RTA) function with automated thrust management. Because of the time compression at the end of flight during the arrival/landing phase, any path deviation along the way would be counter to the goals of 4-D navigation.

Air/Ground Automation with Aircraft Self-Separation to Reduce Workload

Air traffic controllers today are involved in a number of routine ground–air communications, such as changing frequency, when the aircraft transition from one controller airspace to another's. In the future, the ground automation system will directly communicate routine information over a data link to the aircraft FMS. In addition, the primary responsibility of the controller to separate aircraft could be shared with the appropriately equipped aircraft, which could self-separate under certain situations as discussed below. The increased use of automation both in the ground system and in the aircraft cockpit will help reduce workload for both the controllers and the pilots.

Before aircraft are committed to the final approach, the controllers direct the aircraft to maintain specific in-trail spacing when following each other, or when the aircraft merge from different directions on a common point in airspace, and then follow each other in a single stream of traffic. This process requires a series of clearances, as well as monitoring of aircraft conformance to the directions from the ground. This is workload intensive for both the pilots and the controllers. In the future, with the aircraft equipped with ADS-B, CDTI, FMS, and RNP, including a monitoring and alerting capability, the aircraft will have the ability to maintain separation from other equipped aircraft. Flight-deck-based merging and spacing concepts are being explored in which a strategic setup is established by the ground system followed by cockpit-based self-separation [67.25].

NASA has defined an automated airspace concept that uses ground-based automated airspace computer system (AACS) to generate conflict-free air traffic control advisories and send trajectories via a two-way data link to the FMS of equipped aircraft. With traffic situational awareness provided by CDTI, the pilots could assume separation assurance responsibility during certain traffic conditions. Although the selection of data link and data transmission protocols to meet these requirements is uncertain at this time, mode S, ADS-B, and VDL2 are likely candidates for this concept. The automation of separation assurance function will also mitigate a number of ATC constraints that limit the efficiency and capacity of the current ATM system [67.26].

The joint FAA/Eurocontrol Cooperative Research and Development Committee defined the principles of operation for the use of airborne separation assurance systems (ASAS) taking into account US and European perspectives for global applications [67.27]. The primary guiding principle is based on cooperative involvement of both the pilots/aircraft systems and the controllers/ATM system in assuring separation among aircraft. Four specific ASAS applications defined are:

1. Airborne traffic situational awareness to enhance pilots' knowledge of surrounding traffic
2. Airborne spacing to permit pilots to maintain a given spacing with designated aircraft
3. Airborne separation when the controller delegates separation assurance responsibility to the pilots
4. Airborne self-separation when the pilots achieve separation from other aircraft in accordance with the desired separation standards and rules of flight.

These concepts make use of aircraft capabilities-based performance to establish different control mechanisms for different segments of airspace. The equipped aircraft assume responsibility for self-separation and for monitoring and alerting in most airspace, except where there are high-density traffic operations. Most FMS using GPS for navigation alert the pilots when navigation performance exceeds RNP criteria. This *UNABLE RNP* alert is based on probability and not on measured error, and is only a part of the required monitoring and alerting process. A flight technical error relative to the computed path is displayed to the pilot for monitoring lateral and vertical deviations. A corrective action is required if either the lateral or vertical deviation exceeds the lateral RNP limit, or 75 ft in vertical, respectively. The GPS meets the monitoring and alerting requirements of accuracy and integrity through RAIM alerts tied to the RNP value for each phase of flight. This means that the separation assurance responsibility is ground based in the airspace where the traffic density and flight uncertainties are high, whereas some separation assurance responsibility could be delegated on a pairwise basis to aircraft during light traffic. In other airspace segments, both the aircraft and the ground automation share responsibilities, with the aircraft responsible for tactical flow management and separation assurance, while the ground system is responsible for strategic traffic flow management.

67.6 Summary

The current US air transportation system has an excellent record for safety of aircraft flying in accordance with the IFR separation requirements. The NAS relies on VHF for voice communication between the pilots and the air traffic controllers, ground-based VOR/DME systems for navigation, primary and secondary radars for surveillance, and ground-based automation for flight and radar data processing at local, regional, and national ATC facilities. During 2006, over 700 million passengers flew in the NAS, and the cargo revenue ton miles exceeded 40 billion. The NAS manages about 55 000 operations daily, with about 6500 flights in the air during peak demand. The density of traffic is creating congestion at airways and airports, thereby creating bottlenecks resulting in flight delays, which cost the airlines millions of dollars in lost revenue.

With the demand for air traffic services continuing to increase, future delays will increase significantly unless the NAS is transformed. Satellite-based CNS technologies offer the opportunities to enhance safety, airport capacity, and flight efficiency. The new generation of aircraft has already acquired avionics compatible with satellite-based CNS technologies, with increased automation provided by the FMS. However, the ground system infrastructure needs to be modernized using satellite-based CNS technologies and automation of decision support functions.

This chapter describes the current CNS and ATM infrastructure, which includes VHF/HF communications, ground-based navigation systems, viz. VOR, DME and ILS, and primary/secondary radars for surveillance. Upcoming satellite-based CNS technologies are also discussed, e.g., VHF data link for data communication, GPS/WAAS for navigation, and ADS-B for surveillance. How these technologies will enhance aircraft operations with direct air/ground communications, RNAV point-to-point navigation, and improved aircraft tracking for automated decision support is elaborated in order to provide an understanding of the major technological transformation expected in future NAS.

The functional role of automation in the aircraft and the ground system is addressed in terms of their limited use today, as most of the decisions in the cockpit and on the ground are human-centric. The two major functions of the ATM system in NAS, viz. TFM and ATC, and their limitations are addressed. Most of the automation functions such as MSAW/conflict alert/URET are primarily used for aircraft safety. Limited automation functional capabilities exist, such as TMA and ETMS, to deal with capacity, flight efficiency, and workload, although a number of newer aircraft have FMS to help aircraft fly efficiently.

The metrics to measure CNS/ATM systems' performance are aircraft safety, airport capacity, flight efficiency with its impact on delays, and pilot/controller workload. The established government regulations require the aircraft to follow other aircraft with specific separation distance minima in various phases of flight. Because of human decision making, the controllers often plan for larger than the required IFR separation distance rules to ensure safety, although this adversely affects capacity, resulting in increased delays and workload. A detailed explanation of the factors used in defining the above performance measures is provided here to develop a clear understanding of the CNS/ATM system operational elements that the new technologies

should improve, and the functions which should be automated for the air transportation system of the future.

This chapter also provides highlights of the future CNS/ATM capabilities for the NextGen system for the year 2025 and beyond, with its goals and objectives. How the enhanced automation could meet the requirements of the future system for increased safety and capacity, as well as for reducing delays and workload, is also discussed.

In order to realize some of the goals of the future air transportation system, research is going on to develop new capabilities such as RNP, CDTI, and EVS in the cockpit, and automated functions such as PARR, McTMA, probabilistic TFM, and automated departure/arrival management for the ground-based ATM system. The new aircraft technologies will provide the aircraft with an ability to operate in poor-visibility conditions just like they operate in good-visibility conditions to reduce separation minima and increase capacity. The enhanced automation in the cockpit and in the ground system will be better able to deal with the system uncertainties to improve flight efficiency and reduce delays, as well as provide both the pilots and the controllers with accurate and timely decisions to help reduce their workloads. Moreover, some sharing of separation assurance responsibility between the pilots and controllers would result in equitable distribution of workload for ensuring safety of flights.

Because of the limited space here to cover the vast scope of the current air transportation system functions and capabilities and ongoing research to develop the future system, this chapter provides a tutorial at a high level. For specific details of any feature of the current or future systems, it is recommended that the readers seek information on the US FAA or the ICAO websites (www.faa.gov or www.icao.int), respectively.

References

67.1   FAA: *Aerospace Forecasts: Fiscal Years 2007–2020* (US Department of Transportation, Federal Aviation Administration Policy and Plans, Washington 2007)

67.2   S. Kahne, I. Frolow: Air traffic management: Evolution with technology, IEEE Control Syst. **16**(4), 12–21 (1996)

67.3   T.S. Perry: In search of the future of air traffic control, IEEE Spectrum **34**(8), 18–35 (1997)

67.4   M.S. Nolan: *Fundamentals of Air Traffic Control*, 4th edn. (Thomson Brooks Cole, Florence 2004)

67.5   ICAO: *Standards and Recommended Practices*, Aeronautical Radio Frequency Spectrum Utiliza-

tion, Annex 10, Vol. 5 (International Civil Aviation Organization, Montreal 2001)

67.6   S.C. Mohleji, P.J. Wroblewski, M.J. Zeltser: *Capabilities of the VOR/DME Navigation System for Civil Aviation Report DOT/FAA/RD–82/74* (US Department of Transportation, Federal Aviation Administration, Washington 1992)

67.7   M.C. Stevens: *Secondary Surveillance Radar* (Artech House, New York 1988)

67.8   M.J. Burski, J. Celio: Restriction relaxation experiments enabled by URET a strategic cloning tool, 3rd USA/Europe ATM R&D Semin. (Naples 2000)

67.9   D. Knorr, J. Post, M. Walker, D. Howell: An oper-
        ational assessment of terminal and en route free
        flight capabilities, 4th USA/Europe ATM R&D Semin.
        (Santa Fe 2001)

67.10   FAA: *Air Traffic Control* (US Department of Trans-
        portation, Federal Aviation Administration, Wash-
        ington 2006), Order 7110.65R

67.11   FAA: *Aeronautical Information Manual* (US Depart-
        ment of Transportation, Federal Aviation Adminis-
        tration, Washington 2007)

67.12   Boing: *Statistical Summary of Commercial Jet Air-
        plane Accidents, Worldwide Operations 1959–2006*
        (Boeing Commercial Airplanes Company, Chicago
        2007)

67.13   FAA: *Air Traffic Quality Assurance* (US Department
        of Transportation, Federal Aviation Administration,
        Washington 2002), Order 7210.56C

67.14   A.L. Haines: *Parameters of Future ATC Systems Re-
        lating to Airport Capacity/Delay MTR 77W0000066,
        Rev. 1* (MITRE Corporation, McLean 1978), FAA-EM-
        78-8A

67.15   W.J. Swedish: *Upgraded FAA Airfield Capacity
        Model*, Supplemental User's Guide, Vol. I (MITRE
        Corporation, McLean 1981), MTR-81W16

67.16   E.S. Stein: *Air Traffic Controller Workload; An Ex-
        amination of Workload Probe* (US Department of
        Transportation, Federal Aviation Administration,
        Atlantic City 1985), DOT/FAA/CT-5N84/24

67.17   B. Sridhar, K.S. Sheth, S. Grabbe: Airspace
        Complexity and its Application in Air Traffic Man-
        agement, 2nd USA/Europe ATM R&D Semin. (Orlando
        1998)

67.18   Joint Planning & Development Office: *Next Genera-
        tion Air Transportation System, Integrated Plan*, (US
        Department of Transportation, Washington 2004)

67.19   D.B. Kirk, M.S. Heagy, M.J. Yablonski: Prob-
        lem resolution support for free flight operations,

67.20   C. Wanke, L. Song, S. Zobell, D. Greenbaum, S. Mul-
        gund: Probabilistic congestion management, 6th
        USA/Europe Semin. Air Traffic Manag. R&D (Balti-
        more 2005)

67.21   T.C. Farley, S.J. Landry, T. Hoang, M. Nickelson,
        K.M. Levin, D. Rowe, J.D. Welch: Multi-center traf-
        fic management advisor: Operational test results,
        AIAA 5th Aviat. Technol. Integr. Oper. (ATIO) Conf.
        (Arlington 2005)

67.22   W.W. Cooper, S.C. Mohleji, C. Burke, J.G. Foster,
        M. Mills: DEPARTS: A tool for improving airline de-
        parture scheduling and reducing flight delays at
        busy airports. In: *Handbook of Airline Strategy*, ed.
        by G.F. Butler, M.R. Keller (McGraw-Hill, New York
        2001) pp. 577–600

67.23   W.W. Cooper, E. Cherniavsky, J. DeArmon, J.G. Fos-
        ter, M. Mills, S.C. Mohleji, F. Zhu: Determination
        of minimum push-back time predictability needed
        for near-term departure scheduling using DEPARTS,
        4th USA/Europe ATM R&D Semin. (Santa Fe 2001)

67.24   S.C. Mohleji, R.K. Stevens: Optimizing flight paths
        for RNP aircraft in busy terminal areas – First step
        towards 4-D navigation, 25th Digit. Avion. Syst.
        Conf. (DASC) (Portland 2006)

67.25   W.J. Penhallegon, R.S. Bone: Evaluation of a flight
        deck-based merging and spacing concept on en
        route air traffic control operations, 7th USA/Europe
        Air Traffic Manag. R&D Semin. (Barcelona 2007)

67.26   H. Erzberger: The automated airspace concept, 4th
        USA/Europe Air Traffic Manag. R&D Semin. (Santa Fe
        2001)

67.27   FAA/Eurocontrol Cooperative R&D: *Action Plan 1:
        Principles of Operation for the Use of Airborne
        Separation Assurance Systems* (Federal Aviation
        Administration, Paris 2001)

IEEE Trans. Intell. Transp. Syst. **2**(2), 72–80
(2001)

68. Flight Deck Automation

Steven J. Landry

A review of flight deck automation is provided with an emphasis on examples and design principles. First, a review of historical developments in flight deck automation is provided. Current examples of control automation, warning and alerting systems, and information automation are then provided. A discussion of human factors, integration, safety, and certification issues are then discussed. The chapter provides guidance to managers, engineers, and researchers tasked with studying or building flight deck systems. In particular, the chapter provides an appreciation of the challenges of building such systems, and the challenges facing those who will build the flight decks of the future.

68.1 Background and Theory

Both the USA and European countries are currently researching and developing automation, procedures, and concepts in order to transform their respective air-traffic systems [68.1, 2]. This transformation will significantly affect aircraft flight decks for many decades. As the shape of this next-generation air-traffic system develops, sophisticated automation will be needed to take advantage of the new infrastructure. This need poses many challenges, including: how to take advantage of a likely massive increase in the amount of available information, how to increase automation in a complex human-integrated system without reducing safety, and how to ensure that collisions between aircraft do not occur in the face of significantly increased density.

This chapter is designed to provide guidance to engineers and researchers who will develop technologies for new flight decks. These engineers and researchers will need to understand the challenges of developing automation for a flight deck, including technological, regulatory and certification, and human factors issues. As will be discussed, much has been accomplished from the early days of automation. There have been significant technological advances (such as satellite digital communications), but the utilization of these technologies in flight decks has been relatively slow. Regulatory agencies are conservative in approving the introduction of unproven technologies, and there are many integration issues associated with bringing new automation into complex socio-technical systems such as aviation.

First, a brief review of historical developments in flight deck automation is provided. Next, several specific types of automation are discussed: control automation, warning and alerting systems, and information automation. This is followed by a discussion of guidelines for flight deck automation development. This discussion centers on human factors issues, system integration issues, safety, and certification. Lastly, several emerging concepts that appear to be the most critical for the US and European efforts are discussed.

68.1.1 Historical Developments and Principles

The very earliest flight decks had no automation per se, and only rudimentary instruments to monitor the engine, magnetic heading, and orientation (with respect to the horizon). Later additions included barometric altimeters to indicate altitude, airspeed indicators (for the prevention of stalls), and sideslip indicators. It was not until flight at night and in conditions of poor visibility was needed in the 1930s that more sophisticated navigation instrumentation became available.

Quite likely the first piece of significant aircraft automation, however, was introduced in 1914 by Lawrence Sperry. Sperry, utilizing the concept of the gyrocompass invented by Hermann Anschütz-Kaempfe (and subsequently patented by Sperry's father in the USA), developed a gyroscopic stabilizer for an aircraft, and attached it to the control surfaces of a Curtiss B-2 biplane. In a demonstration in France, Curtiss and his mechanic (Emil Cachin) climbed out on the wings as the now-pilotless plane passed in front of crowds of spectators, thereby demonstrating the first autopilot [68.3].

Sperry's autopilot used the principle that a spinning gyroscope has a strong tendency to maintain its orientation regardless of the orientation of the containing body. Therefore, deviations of the orientation of an aircraft from the neutral positions (as given by the axis of the spinning gyroscope) can be determined. This signal can then be used to drive the control surfaces of the aircraft according to specific control laws (which will be discussed later in the chapter). Sperry was able to package this in a device that measured $18\,\mathrm{in} \times 18\,\mathrm{in} \times 12\,\mathrm{in}$, and weighed only $40\,\mathrm{lb}$ (whereas his father's gyrocompass was so large it could only be used on large warships) [68.4].

Sperry's invention was the first of what will be called *control automation*, whose purpose is to replace human control of the aircraft with machine control. If, in considering automation, only mechanical systems that replace human functions are considered, then there are roughly two other categories of automation in aircraft – warning and alerting systems and information automation, although many systems straddle these classifications.

Warning and alerting systems replace the pilot's function of manually monitoring for hazardous conditions. For example, airspeed indicators were developed mainly so that pilots could ensure that they maintained sufficient airspeed to prevent a stall, which is a hazardous condition where the wings no longer produce enough lift to counteract the weight of the aircraft (which therefore begins to fall). In most modern aircraft, this monitoring function is replaced by a stall warning device, which alerts the pilot to an impending stall without the pilot having to monitor airspeed and compare it with a memorized stall speed.

The first warning and alerting systems automation was likely an *off* flag for a particular instrument, warning of that instrument's failure. Warning and alerting systems proliferated rapidly starting in the 1960s and 1970s, with the Boeing 707 (rolled out in 1954) originally having only a few simple warning devices (such as engine fire warning systems), the Boeing 747 (deployed in 1969) having few (if any) more (although many have been added to it over its life), and the Boeing 777 (whose maiden flight was in 1994) being the first built with numerous warning systems as standard equipment.

Information automation provides pilots with access to information that they would otherwise have to locate manually or calculate themselves. For example, pilots will follow written checklists for various portions of flight, and have procedures to ensure that checklist items are completed. In some aircraft, checklists are presented in electronic format, with the system ensuring that checklist items are completed.

Probably the first information automation introduced was the flight director. This automation provided intercept guidance for pilots, who previously would have had to determine appropriate intercept angles for a desired course (and glideslope – the desired descent angle to the runway).

World War II saw a dramatic increase in the use and sophistication of aircraft, with that trend continuing through the present day. In the early 1950s, a blue ribbon panel was convened by the US government to study research needs in aviation. This panel was chaired by Dr. Paul Fitts, who was one of the pioneers of applying engineering and psychology together to improve aviation safety. The Fitts report provides a convenient milepost for the state of automation and automation research at the time [68.5].

68.1.2 Modern Automation

Modern flight decks include a great deal of automation of all three types (control, warning, and information) mentioned previously, although this is highly dependent on the type of aircraft. At the top of the sophistication ladder (for civilian aircraft) is the modern airliner, while there are still aircraft flying today that have the same level of sophistication and instrumentation as the earliest aircraft.

The most sophisticated commercial aircraft have three-axis autopilots that are driven by a combination of inputs from a gyroscopic inertial navigation system, global positioning system, and flight management system into which is loaded the desired flight route in three (sometimes four) dimensions. These autopilots can navigate to any point on the globe, and can be programmed to arrive at a precise time. Moreover, these autopilots have the capability to control the airplane without human intervention from the time it is driven onto the departing runway by the pilot until it reaches a point just off the arriving runway in any weather conditions.

These aircraft also have numerous alerting and warning systems, including systems that detect colli-sion dangers (with other aircraft and with the ground), unsafe configurations (such as gear not extended for landing), control surface overspeed warnings, engine fire warnings, wind shear alerts, alerts for deviations from assigned/desired altitude, and others. These systems, which usually utilize relatively simple sensors, often contain complex algorithms. Aircraft manufacturer and company procedures dictate specific responses to these alerts.

Modern airliners also contain a great deal of information automation as well. Among the information automation commonly found in today's commercial aircraft are weather and ground-proximity radar, navigation displays, electronic messaging capability utilizing communications satellites, and engine indicating and crew alerting systems (EICAS) displays. EICAS displays are a highly integrated set of displays covering a great deal of different information, including engine temperature, engine pressure, oil temperature, electrical system status, pressurization system status, hydraulic system status, and fuel system status.

In the next section, examples of these applications will be discussed. This will be followed by a discussion of guidelines for automation development, specifically human factors issues, system integration issues, safety principles, and certification and equipage issues. Following that is a discussion of principles for automation development.

68.2 Application Examples

Many examples of flight deck automation were mentioned in the previous section. In this section, specific examples of control automation (autopilots, envelope protection automation, and automation for uninhabited aerial vehicles), warning and alerting systems (system monitoring, hazard monitoring, and collision avoidance), and information automation (automated checklists, cockpit display of traffic information, and data communications) are discussed in more detail.

68.2.1 Control Automation

Aircraft are controlled through manipulation of three rotational axes, as shown in Fig. 68.1, along with control of the thrust produced by the engines. Each of the control surfaces used to manipulate the rotational axes utilizes the same principle as wings – increasing camber of a surface moving through a fluid (i. e., the at-mosphere) results in pressure differences between the upper and lower control surfaces. These pressure differences produce a force that tries to move the surface in the direction of lower pressure.

Control surfaces are connected by mechanical, electrical or hydraulic means to controls that can be operated manually by the pilots. The control surface for pitch is the elevator, typically located on the horizontal stabilizer near the tail of the aircraft. Manually, the elevator is controlled by pulling or pushing the control yoke (or joystick). The control surfaces for roll are ailerons, located near the end of each wing. Ailerons are controlled by turning the yoke (or moving the joystick to the left or right). Yaw is controlled by the rudder, which is located on the vertical stabilizer, again near the tail. The rudder is operated by stepping on pedals located near the pilots' feet. Throttles control the thrust of each engine.

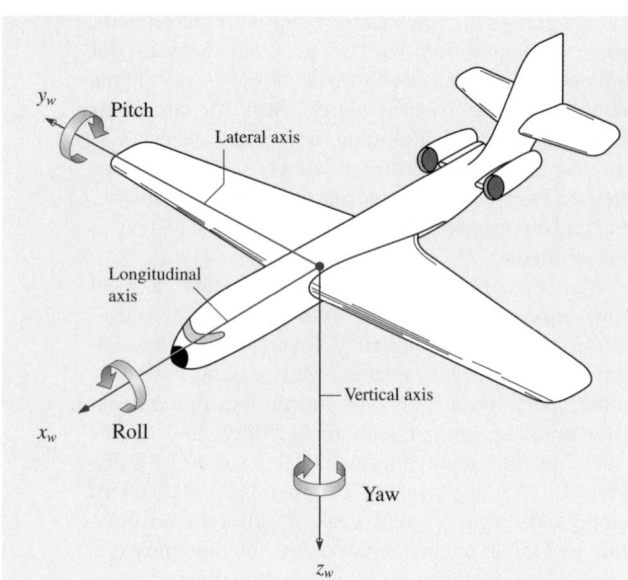

Fig. 68.1 Yaw, pitch, and roll axes (after [68.6])

These axes are somewhat independent, although not entirely. For example, stepping on the rudder pedals introduces yaw, which in turn produces a rolling moment. Increasing power through the use of the throttles results in a pitch change; if pitch is held constant an increase in power results in an increase in speed.

The null setting of each of the control surfaces is set through the use of *trim*, which is operated electrically (or sometimes hydraulically). Trim, however, can also be used to manipulate the control surfaces instead of moving the ailerons, rudder or elevator directly. In some cases trim is a smaller version of the related control surface (for example, typically the ailerons have small tabs which can be moved independently of the ailerons and act as the trim control surface). In other cases, the null position of the entire control surface is set by the trim (for example, typically the elevator null position is set by trim).

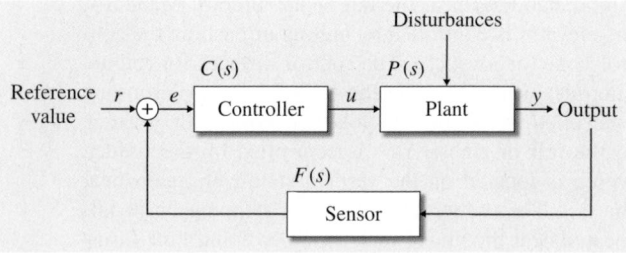

Fig. 68.2 Basic feedback control loop

Autopilots

Modern-day autopilots control all three rotational axes of an aircraft (pitch, roll, and yaw), as well as the thrust. They accomplish this through mechanical, hydraulic or electrical linkages with the trim surfaces and with the throttles. This type of control provides smoother, more accurate positioning of the control surfaces than through cable linkages, although manual movement of the controls can produce larger movements in a shorter period.

The input to autopilots can be thought of as a deviation from a desired rotational angle or speed, although this may have to be derived from a desired ground track, altitude, speed, vertical speed (i.e., rate of descent or climb), bearing from a navigation aid or other form of information more directly applicable to navigation. The specific methods by which error signals are translated into control movements are proprietary secrets of autopilot and aircraft manufacturers, so only a generalized description is given here.

In basic control theory, an input (such as error from a reference) can be transformed into an output through a transfer function, which is commonly specified as a function of the frequency of the signal. Such a system is shown in Fig. 68.2, where the transfer function $H(s)$ is given by (68.1) if plant disturbances are ignored

$$Y(s) = \left(\frac{P(s)C(s)}{1 + F(s)P(s)C(s)} \right) r = H(s)r . \qquad (68.1)$$

A simple controller can be constructed using a simple integrator and delay. For such a controller, the transfer function would be given by (68.2), where the time delay is given by τ, the gain is given by k, and the integrator is represented by the s in the denominator

$$H(s) = \frac{k e^{-\tau s}}{s} . \qquad (68.2)$$

In such a controller, the behavior of the system would be to dampen out the error signal, as shown in Fig. 68.3. In such a controller, the error signal from $(t - \text{delay})$ seconds ago is used, resulting in some delay in the response of the controller. That delay leads to the overshoot shown where the error crosses zero just after 0.5 s. The error is fully damped after 3.25 s.

Modern autopilots use very sophisticated algorithms in place of the simple transfer function shown in (68.2). Generally, automatic control in aircraft is considered a multilevel system, consisting of a guidance loop, a control loop, and a stability augmentation loop. The guidance loop controls the navigation of the aircraft by initiating commanded state changes in particular sequences to achieve a navigation goal. The control loop

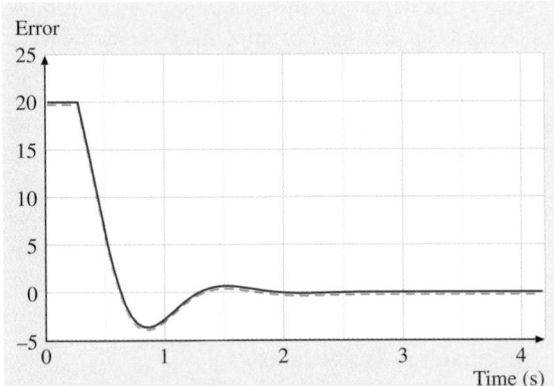

Fig. 68.3 Step response of simple integrator with delay (gain = 3, delay = 0.2 s)

ensures that the commanded states are adhered to by initiating control movements to dampen out errors between the actual and commanded states. Most aircraft also have a stability augmentation loop that dampens out undesirable aircraft modes. One such mode is *Dutch roll*, which is a tendency for aircraft to slightly yaw and roll out of phase continuously in flight.

Autopilot types are categorized in terms of the conditions under which they can be used for approach and landing. Specifically, the categories shown in Table 68.1 can be used to classify autopilots.

Pilots control the autopilot mode of operation through a *mode control panel* (MCP), such as the one shown in Fig. 68.4. Using this panel, pilots select which axes the autopilot should control, and the form of control that the autopilot should use. This form of control is set as the *mode* of the autopilot. Modes that control the pitch axis usually include *attitude hold*, *altitude hold*, *vertical speed*, and *descent angle*. Modes that control the horizontal axis include *control-wheel steering*, *heading hold*, and *course hold*. Autopilots also generally have *airspeed hold*, *mach hold*, and specialized approach modes such as *glideslope capture* and *autoland*.

The typical autopilot has several dozen independent modes, each having complex dependencies with other modes. Some modes are incompatible, and the system must be designed to prevent these from being simultaneously selected. There are also combined modes, where the autopilot will transition into a second mode at some point as programmed into the automation.

The autopilot can be controlled manually by the pilot through the mode control panel, or can be set through the flight management computer (FMC). The desired horizontal trajectory (a sequence of latitudes and longitudes) of the aircraft can be stored in the FMC, which then automatically controls the appropriate modes of the autopilot. The vertical profile as well as times to cross positions or altitudes can also be entered.

These complexities make operating the autopilot and understanding its operation difficult. In fact, several fatal aircraft accidents have been attributed to autopilot *mode confusion*, which will be discussed later in the chapter.

Autopilots, as automation that literally controls the aircraft, is safety critical, and therefore faces rigorous scrutiny before being certified by the national aviation

Table 68.1 Aircraft autopilot categories and weather requirements

Category	Approach and landing weather minimums required
I	60 m/200 ft ceiling
	800 m/0.5 statute mile visibility
	550 m/1800 ft runway visual range
II	30 m/100 ft ceiling
(usually with autoland)	350 m/1200 ft runway visual range
IIIa	200 m/700 ft runway visual range
(usually with autoland)	
IIIb	50 m/150 ft runway visual range
(IIIa with auto-rollout)	
IIIc	0 ft runway visual range
(IIIb with auto-taxi)	

Fig. 68.4 Mode control panel (after [68.7])

authorities. For this reason, developers of autopilot software tend to reuse or add onto existing certified code, or have it generated automatically by systems that can mathematically verify the code.

Envelope Protection

Airbus aircraft have a significantly different automation design philosophy than Boeing aircraft (or many other aircraft manufacturers). While Boeing believes that automation should never be allowed to irreversibly override the pilot, Airbus believes that automation should protect the aircrew from entering unsafe flight regimes [68.8, 9].

This results in automatic envelope protection in Airbus aircraft. Short of deactivating the system, envelope protection makes it virtually impossible to exceed the design limitations (G-forces, maximum speeds) or enter unsafe flight regimes (stalls, excessive angle of attack – the angle of the aircraft to the relative wind). Since Airbus aircraft use fly-by-wire (the control yoke activates the flight surfaces through an electrical signal rather than mechanical or hydraulic linkages), the envelope protection automation can intercept these signals and, for example, reject control inputs that would result in excessive accelerative forces on the airframe.

However, the transitions in and out of flight modes that activate or deactivate certain envelope protections are opaque. This results in additional potential for mode confusion, and has been cited as a causal factor in several accidents [68.10, 11]. There is also the possibility that the aircraft is put into situations unforeseen by the designers.

Both of these situations occurred on an Airbus A-300 flight into Nagoya, Japan in 1994 [68.12]. The aircraft was on approach (being flown manually) when it was inadvertently switched into a *go-around* mode. When the autopilot was activated, it attempted to abandon the approach by climbing and accelerating. However, the pilots, not knowing the aircraft was in this mode, attempted to continue the approach by pushing forward on the control yoke, commanding a pitch down. The autopilot, utilizing the pitch trim, counteracted these commands by running the pitch trim to the full-up limit. As the plane pitched up from the autopilot pitch command, the pilots disconnected the autopilot, but engaged the autothrottles. In response to reaching the maximum angle of attack, flight envelope protection engaged, initiating full thrust, causing the plane to pitch up an additional amount and stall. The pilots were unable to recover from the stall in time and the aircraft crashed, killing 264.

This is not to say that the Airbus philosophy is inferior; records are not kept of how many accidents were *prevented* by having flight envelope protection. Rather, the Airbus philosophy has introduced a new type of error, perhaps trading this off against the possibility of other types of human error.

Although Boeing aircraft do not have strict flight envelope protection, there is a soft envelope protection on its fly-by-wire aircraft. In this form, the system warns the pilot of an approaching limit by increasing the amount of force required to move the control.

Uninhabited Aerial Vehicles

Uninhabited aerial vehicles (UAVs) represent a new class of aircraft, and UAV use is expected to increase significantly in the future. UAVs can range from a sophisticated remote control vehicle (with virtually no onboard automation) to a fully autonomous vehicle with onboard intelligence for navigation and other functions.

Unless one includes guided weapons, UAVs are not fully autonomous. Humans must, at least, provide the system with goals and rules for conduct. In most systems, humans are also involved in monitoring and some aspect of the control loop. Control of the vehicle may be only at the outermost loop (navigation commands), at one of the inner loops (guidance or control) or some combination. For example, the US military's Predator drone requires manual flight for takeoff and landing, but can follow a programmed set of waypoints autonomously. The Global Hawk UAV, however, can takeoff, fly a programmed route, and land autonomously.

UAVs are in extensive use within the military, although their use appears to have been slowed somewhat by development problems and high accident rates [68.13]. Some of these problems relate to the young age of the technology, but some also involves human error. The operation of a UAV is not unlike the operation of a motionless flight simulator, where vestibular and somatosensory cues are absent. The absence of these cues makes fine control difficult, such as required when landing a UAV on an aircraft carrier. Since replacing these sensory cues seems nearly impossible, the emphasis has been on providing more autonomy to the vehicle.

One of the main challenges for incorporation of UAVs into the airspace system is that systems may not be able to deal adequately with a malfunction and still ensure separation from other aircraft. For example, in the case of a communications failure, the UAV must be able to successfully divert to a recovery field on its own while maintaining separation from other aircraft. Currently, validated technology does not exist for such a function. As a result, in the US national airspace system, the Federal Aviation Administration (FAA) has required a lengthy approval process to fly a UAV in controlled airspace, and the UAV must remain under visual control of an operator.

68.2.2 Warning and Alerting Systems

In addition to control automation, there has been a proliferation of warning and alerting systems onboard modern commercial aircraft. Early aircraft had limited, individual alerting systems. As the number of these increased, corresponding to the increase in complexity of the aircraft, the location of many of the visual alerts were consolidated into an annunciator panel, and a master caution warning was added in a highly visible location to indicate that one (or more) of the alerts had activated.

On aircraft that have multifunction displays (MFDs) instead of individual gages (also called *glass cockpits*), the master caution and annunciator panels are often in a less central location. Since the most important alerts can be displayed directly on the MFDs, a master caution and annunciator panel are typically used for less important or infrequent alerts.

Systems Monitoring

Aircraft are complex vehicles, with numerous systems that require monitoring. Engines, the auxiliary power unit (APU), air conditioning, pressurization, electrical systems, hydraulic systems, pneumatic systems, fuel systems, and mechanical systems (such as landing gear) all need to be monitored, and can have failures that are independent from the other systems. On older-generation aircraft, many of these systems (e.g., fuel, electrical, pressurization, and hydraulic) were monitored and controlled by a dedicated flight engineer; a typical flight engineer panel is shown in Fig. 68.5.

On these aircraft, each system was segmented on the panel, and often was laid out in a pattern mimicking the physical layout of the components. For example, the fuel controls (the lower left portion of Fig. 68.5) mirrored the physical location and geometry of the different fuel tanks, cross-flow valves, and boost pumps. This was accomplished to help the flight engineer control the system with fewer errors and to speed diagnosis of problems.

When new aircraft were being built in the 1970s, it was desired to automate most of the control of these systems and to eliminate the flight engineer position. The reasons put forward for this change were to eliminate human error and to reduce crew cost. This change meant that the pilots would be required to handle any malfunctions in these systems, and that systems monitoring equipment would have to be altered for use by the pilots, who would not be able to dedicate time to continuous monitoring or extensive diagnosing of failures.

The new systems monitoring automation is commonly placed on an overhead panel, outside of the normal view of the pilots. Alerts on the front panel (within normal view of the pilots) provide an indication that the pilots should check the overhead panel.

Hazard Monitoring

Pilots must be aware of several different hazards, apart from systems malfunctions. Pilots must monitor for dangerous weather (such as wind shear and thunderstorms), high terrain, and other aircraft. Systems devoted to avoiding collisions will be discussed separately in the next section.

In 1968, a Lockheed Super Electra penetrated a thunderstorm [68.14]. In the resulting turbulence, the aircrew lost control of the aircraft, and overstressed the aircraft (beyond its design limits) in an attempt to recover. The aircraft broke up in flight, killing all 82 persons on board. In 1977, a Southern Airways DC-9 flew into an intense thunderstorm, losing both engines due to heavy rain and hail [68.15]. The engines could not be restarted, and the aircrew attempted to make an unpowered landing on a rural highway. Sixty-two of the 87 people onboard were killed in the ensuing crash.

Modern commercial aircraft all are required to have onboard weather radar to avoid these situations. The system uses common weather radar technology to display areas of precipitation to flight crews.

For aircraft on which weather radar is functioning and available, such incidents do not appear in accident report databases. However, private aircraft are not required to have weather radar on board, and such incidents are still unfortunately common. Weather radar can accurately paint areas of heavy precipitation, enabling pilots to avoid them (and the associated turbulence and lightning). While encounters with such severe weather still occur, they are usually the result of getting too close rather than flying directly into the most severe part of the weather.

A more common risk associated with thunderstorms (for commercial aircraft) is windshear. A windshear is a sudden change in the speed or direction of the wind; downdrafts are similar but involve a shaft of cold air

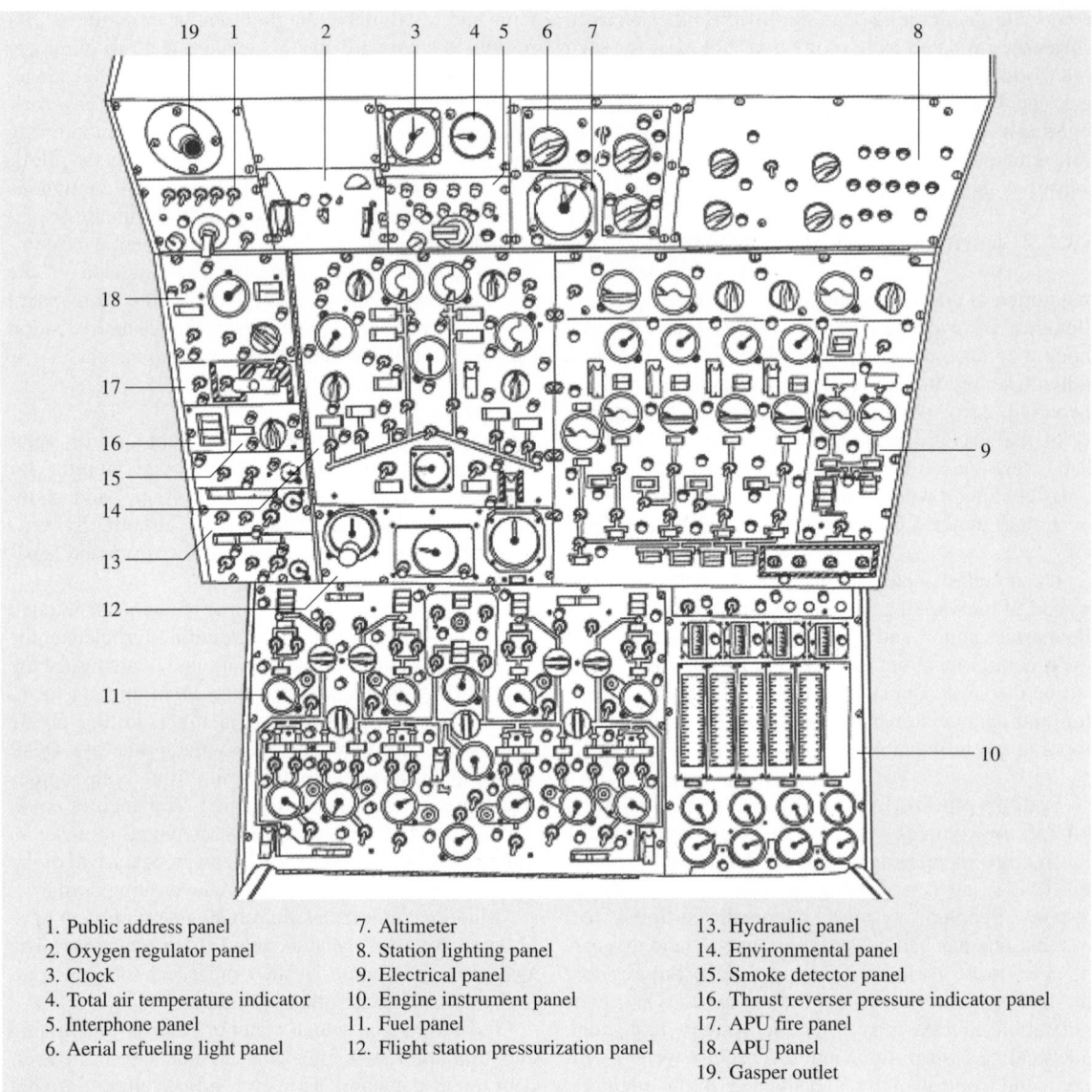

Fig. 68.5 Flight engineer's panels from a C-141B

1. Public address panel
2. Oxygen regulator panel
3. Clock
4. Total air temperature indicator
5. Interphone panel
6. Aerial refueling light panel
7. Altimeter
8. Station lighting panel
9. Electrical panel
10. Engine instrument panel
11. Fuel panel
12. Flight station pressurization panel
13. Hydraulic panel
14. Environmental panel
15. Smoke detector panel
16. Thrust reverser pressure indicator panel
17. APU fire panel
18. APU panel
19. Gasper outlet

sinking rapidly toward the ground. Both types of windshear are extremely dangerous in that they can quickly change the lift profile of the aircraft, and have been the cause of numerous fatal accidents.

One of these accidents occurred in 1975. An Eastern Airlines Boeing 727 crashed on landing at Kennedy Airport in New York, killing 112 of the 124 persons onboard [68.16]. The aircraft encountered windshear on final approach, and impacted the ground 2400 ft short of the runway. This incident spurred the US Fed-

eral Aviation Administration (FAA) to have a low-level windshear alerting system (LLWAS) developed, and prompted airlines to install onboard windshear alerting systems.

The original LLWAS worked by detecting vector differences in wind through pole-mounted anemometers placed at midfield of the airport and at five locations around the airport. The system alerted if a 15 kt vector difference in the winds was detected, but was prone to false alarms [68.17]. Current LLWAS systems utilize

12–32 sensors, placed at points based on the geometry of the airport and typical convective weather activity, and use more sophisticated algorithms for detecting windshear and microburst activity.

Simple flight deck windshear alerting systems work by monitoring actual and predicted winds entered into the flight management computer. This capability is often augmented by monitoring the groundspeed of the aircraft. Systems have also been developed to use forward-looking infrared radar, Doppler radar or other systems to predict windshear.

Another hazard that aircraft need to avoid is terrain. As a result of a spate of accidents labeled *controlled flight into terrain* (CFIT), instrument manufacturers created a ground-proximity warning system (GPWS). The intent of this system is to detect when an aircraft is approaching the terrain in an unsafe manner (i. e., when not intending to land).

GPWS, and the later enhanced GPWS (EPGWS), utilizes as its primary inputs a radar altimeter (which measures height above ground level – AGL), the barometric altimeter, landing gear position, vertical speed, and airspeed. The EGPWS also utilizes a terrain database and the current aircraft position.

The hazardous condition that the GPWS was attempting to detect was unsafe closure to terrain. Several states of the aircraft were used to determine whether this condition existed, including the altitude of the aircraft, the descent rate of the aircraft, and the status of the landing gear and flaps (to detect when the aircraft was intending to land). Values of these states were combined to produce envelopes of safe operation with respect to terrain; operation outside of these envelopes would constitute a hazardous situation. An example envelope (from a GPWS installed on a US Air Force C-141B) is shown in Fig. 68.6.

The output of the GPWS is either nothing (if a hazardous condition is not detected) or an alert (if a hazardous condition is detected). In the earlier models of the GPWS, an alert consisted of a red light on the GPWS panel and a *whooping* tone followed by a computerized voice that annunciates *pull up* several times. This very distinctive alarm was intended to provide pilots with clear guidance regarding the presence of a hazard and the desired response.

Unfortunately, a number of incidents in which pilots silenced the alarm without complying with the mandated response have occurred, resulting in aircraft crashing into terrain. These incidents were particularly troublesome since the aircraft was perfectly capable,

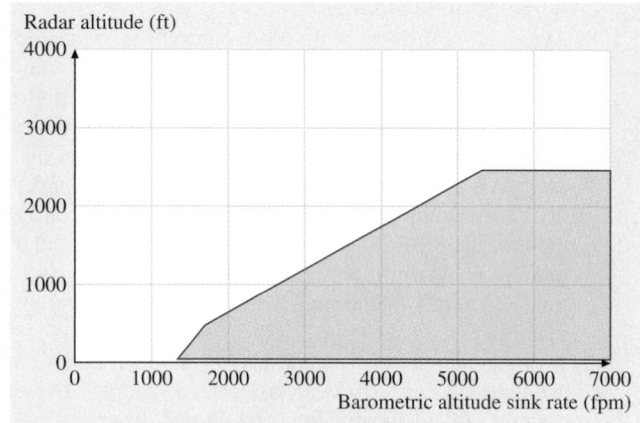

Fig. 68.6 Ground-proximity warning system envelope (mode 1)

and all the pilots needed to do to avoid crashing was to comply with the alert's desired response.

One supposed reason for this was that the system produced spurious or false alerts [68.18, 19]. These alerts caused trust in the automation to be eroded, resulting in pilots' failing to comply with the alert response. For example, if the system failed to detect that the landing gear was down, the GPWS would sound when the aircraft approached the ground for landing, even if the descent was controlled. In these situations, the pilots would realize that the landing gear was down and ignore the alert.

However, even if the landing gear was down, the GPWS may sound if the aircraft is descending in an unsafe manner (such as would be the case if the aircraft were out of position such that terrain posed a threat to the aircraft). In such a case, merely checking the state of the landing gear would be insufficient. If the pilots assumed that the GPWS was malfunctioning due to it misreading the state of the landing gear, an accident may result.

To assist the pilot with sorting out why the alert was occurring, additional voice alerts were provided as part of the EGPWS, which would help indicate which condition was detected by the system. In addition to *pull up*, the system would also state *glideslope* (if the aircraft were deviating from the desired descent glideslope for landing), *terrain* (if terrain closure were detected), and combinations of these alerts would become increasingly salient if the condition persisted. The low number of such incidents over the last decade suggests that this system has considerably reduced instances of failure to adhere to the ground-proximity warning system.

Collision Avoidance

Avoiding collisions with other aircraft is a shared responsibility between flight crew and air-traffic controllers. When operating visually (under visual flight rules – VFR), controllers are not required to provide any assistance and pilots must *see-and-avoid* other aircraft. When operating primarily on instruments (under instrument flight rules – IFR), controllers continuously monitor the separation of aircraft using processed radar returns, manually monitoring and projecting the positions of aircraft, intervening when potential conflicts are identified.

In addition to see-and-avoid, aircraft collision avoidance systems (ACAS) have been developed. These systems utilize information transmitted from aircraft equipped with an appropriate transponder system to determine relative bearing and closure rate. Should the closure rate and relative trajectories exceed some thresholds, an alert sounds, warning the pilot of a potential collision.

A particular implementation of ACAS is the traffic collision avoidance system (TCAS), whose current implementation is version II. Version I of TCAS (TCAS I), is mandated for aircraft with more than 10 but fewer than 31 seats; TCAS II is used for aircraft with more than 30 seats. TCAS II, which requires that a particular type of transponder (mode S) is in use on the aircraft, coordinates resolution maneuvers if both aircraft are equipped.

TCAS systems detect the transponder signals of nearby aircraft (out to 14 nmi), determining their horizontal range, bearing, and vertical separation from a series of interrogations of the transponders of nearby aircraft. Closure information is calculated from a series of interrogation responses, which is then translated into the time to closest point of approach (CPA).

TCAS has two types of alerts: traffic advisories (TAs) and resolution advisories (RAs). TAs are used to assist the pilot in identifying aircraft that are prox-imate and may pose a collision danger. No response is required to a TA. RAs warn the pilot of a near-imminent collision danger; specific instructions are included with the alert to avoid the collision. This instruction is a vertical maneuver that has been calculated to avoid the collision. In TCAS II, if both aircraft are equipped, these maneuvers are coordinated.

TCAS must contend with both multipath problems and a condition called *garble*. Multipath refers to one message being received multiple times – once directly from the aircraft, other times after the message is reflected off the ground or other obstacles. TCAS must know which of these is the original signal. Garble refers to the overlap of signals. Since the messages are 21 µs long, several messages can overlap; TCAS must still be able to decipher messages under these conditions.

TCAS must also balance the probability of missing a detection (MD) against likelihood of a false alarm (FA). As mentioned previously, FA has a deleterious effect on automation trust; however, one also would like all hazardous situations to be detected. In TCAS, the *sensitivity level* (SL) of the system can be modified to change the tradeoff point between MD and FA.

A sensitivity level of 1, in which the aircraft does not issue any alerts, can be selected by the pilot, or occurs whenever the system goes into standby mode (for example, when the mode S transponder fails). The pilot can also select a SL of 2, where the system only transmits TAs. Sensitivity levels 4–7 are selected by the system automatically, depending on the altitude of the aircraft. As the altitude of the aircraft increases, the sensitivity level goes up. With higher sensitivity levels, the system alerts sooner.

TCAS uses both a time to CPA (τ) and the absolute horizontal and vertical separations between aircraft. The primary determinant of the alerts is τ, but when closure rates are very low, aircraft can come very close to one another without violating τ (imagine aircraft on almost parallel courses, slowly converging). If τ (both the

Table 68.2 Alert threshold for TCAS II version 7 (after [68.20])

Actual altitude (ft)	Sensitivity level	τ (s)		DMOD (nmi)		Threshold altitude (ft)	
		TA	RA	TA	RA	TA	RA
Below 1000	2	20	–	0.3	–	850	–
100–2350	3	25	15	0.33	0.2	850	300
2350–5000	4	30	20	0.48	0.35	850	300
5000–10 000	5	40	25	0.75	0.55	850	350
10 000–20 000	6	45	30	1	0.8	850	400
20 000–42 000	7	48	35	1.3	1.1	850	600
above 42 000	7	48	35	1.3	1.1	1200	700

vertical and horizontal τ) of a proximate aircraft is less than the thresholds, the alert sounds. Otherwise, if the closure rate is low, the alert will sound if both the horizontal and vertical separation is less than the DMOD (which is short for *distance modification*) and altitude thresholds, respectively. These thresholds for SL 3–7 are shown in Table 68.2.

68.2.3 Information Automation

Information automation is distinguished from other types of automation as it is intended to provide information to support reasoning by the operator (as opposed to supporting rule-based or skill behavior). Information automation includes the most recent forms of automation.

Automated Checklists

Electronic versions of checklists have been introduced quite recently, including in the Boeing 777 [68.22, 23]. These checklists reproduce, on a display, the steps of the checklist. As items are completed, they are presented as completed after the item is actually completed (as checked by automation, if possible).

Cockpit Display of Traffic Information (CDTI)

A display of nearby aircraft is part of the ACAS system. Such systems show aircraft within a certain range that may pose a collision or separation risk. Along with relative bearing, their relative altitude, identification, and sometimes speed are shown.

A number of researchers have examined how such displays may be utilized to improve the situation awareness and conflict avoidance capabilities of pilots [68.24–27]. Some have gone as far as suggesting that such displays can be utilized (in part) to enable flight-deck-based separation, as opposed to separation assured by centralized ground authorities such as air-traffic controllers [68.28, 29].

In such displays, all aircraft within the range selected for the display are shown, including all pertinent information such as altitude and speed. An example is shown in Fig. 68.7.

There are some issues with CDTI, however. These issues include displaying three dimensions on a two-dimensional display, perspective issues, and clutter.

Because aircraft fly in three dimensions, air traffic is nominally a three-dimensional task. However, aircraft in cruise are stratified; they fly at altitudes in the whole thousands (35 000 ft, 36 000 ft, etc.). Because of this, except when climbing and descending between

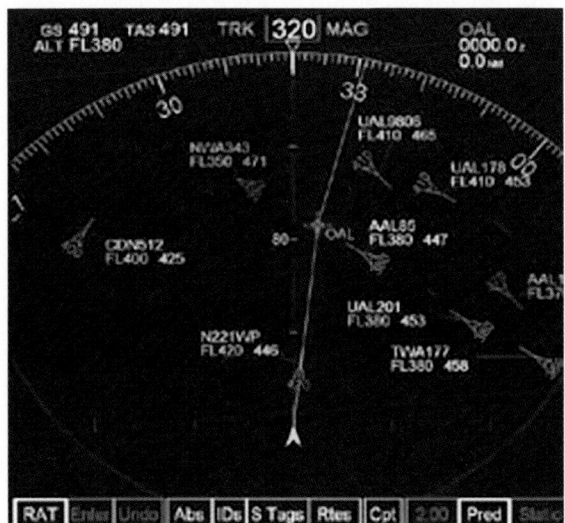

Fig. 68.7 NASA flight deck display of traffic information (after [68.21])

these altitudes, an aircraft's altitude is essentially a category rather than a state. Moreover, it is treated as such by flight crews and controllers, who refer to aircraft above 18 000 ft (below which climbing and descending is more frequent) as being at a particular *flight level* (FL), which are measured in hundreds of feet. So an aircraft at 35 000 ft is referred to as being at FL350.

In addition, the scales of horizontal and vertical separation are different by almost an order of magnitude. Vertical separation is given in thousands of feet; horizontal separation is given in nautical miles (just over 6000 ft). It is not even possible to display these scales simultaneously on a display and have the information be adequately discernable. In order to create a three-dimensional display of traffic, one scale or the other must be distorted. For these reasons, air-traffic control is not typically addressed as a three-dimensional problem.

Designers of CDTI must also consider issues of perspective in displaying traffic. Each perspective distorts or obscures some aspect of the situation. A top-down view (typically used in navigation displays) completely eliminates relative vertical position, which is then often replaced by a text display of altitude. In general, there is ambiguity in position information along the line of sight of the display [68.30]. A display oriented to the pilot's perspective (referred to as *immersed*) eliminates information to the side, from behind, and from above the aircraft. This results in a *keyhole* view of the world [68.31].

Researchers have shown that having the viewpoint outside the flight deck provides the best performance for mapping between the world and display [68.32]. In doing so, the principles of pictorial realism (the display should be pictorially analogous to the real world) and integration (related information should be integrated on the same display) can be followed [68.33].

Heads-Up Displays

Heads-up displays (HUDs) and helmet-mounted displays (HMDs) were first introduced in military aircraft to reduce the amount of time a pilot spent looking inside the flight deck. By positioning frequently accessed information on a display that one could see through to the world behind, pilots need only change the depth of focus to extract information from the world or the instrument, rather than moving the head, eyes or both.

Such displays are now finding their way into commercial aircraft (and even automobiles). The advantages of such a display are clear. For example, one can easily integrate real-world information with displayed information. Also, the HUD enhances the operator's ability to alternate between information at different depths [68.34].

However, several disadvantages should also be considered by designers. First, by superimposing information on the *world* background, the danger of clutter increases. However, researchers have shown that pilots are able to ignore the background information so that no additional workload is imposed on the pilot [68.35]. Also, should unexpected events occur, it appears that the HUD may reduce the capacity of the operator to detect these events, although it is not entirely clear why this is the case [68.36].

Data Communications

Currently, most communication between air-traffic control and flight deck utilizes voice channels over radiofrequencies. On commercial aircraft, communication between a company's dispatch (called the airline operation center – AOC) and an aircraft mostly happens over data channels utilizing satellite communications.

There have been proposals to replace some or all of the air-traffic voice radiocommunications with data communications (*datacomm*). Datacomm is similar to email in that the messages are transmitted digitally and displayed as text, rather than transmitted across voice channels.

One of the main reasons for this change is frequency congestion. All aircraft under the control of a particular air-traffic controller monitor and communicate on the same frequency. In dense airspace, such as close to the airport, the number of aircraft and the amount of maneuvering results in a high volume and rate of communications. Since by regulation each air-traffic communication must be *read back* by the pilot of the aircraft for which the instruction was intended, this volume can exceed the channel capacity. In such cases controllers may even continue to the next instruction without waiting for (or getting) the read back of the previous instruction.

In addition, many errors are associated with communications read back. Pilots may read back the clearance correctly but fail to recall it correctly, or they may read back an incorrect clearance that is subsequently not noticed by the controller. Cases in which the wrong aircraft has read back the clearance have also occurred. It is expected that datacomm would alleviate many of these communications errors.

In addition to equipment and certification cost, one of the main issues arguing against datacomm, however, is party-line information loss. Since all aircraft under the control of a particular air-traffic controller are monitoring the same frequency, they would hear communications with other proximate aircraft, including aircraft preceding them on the same route. This allows pilots to glean information about the relative positions of other aircraft, their intentions, and the intentions of the air-traffic controller. For example, pilots hear other aircraft preceding them asking for altitude changes due to turbulence; this allows them to request the same change before the turbulence is encountered. With datacomm, such information would be unavailable to pilots.

68.3 Guidelines for Automation Development

Flight decks are highly automated and have many different types of automation. In this section, the important principles for the development of automation are discussed. First, principles related to each of the types of automation (control, warning, and information) are discussed. This is followed by several overarching concerns for the development of automation – human factors issues, system integration issues, safety, and certification.

The guidelines listed here are not necessarily followed in the development of current automation. One reason for this is that it is often impossible to follow all guidelines regarding automation and still meet all other constraints (such as space, cost, and certification). As such, these guidelines should be treated as goals for automation development rather than hard-and-fast rules.

68.3.1 Control Automation

Control automation must be demonstrably controllable and observable. Controllable means that, for any bounded input, the output will be bounded. If an aircraft is not controllable, the autopilot may (for example) not have sufficient control authority to dampen out errors, resulting in continuously escalating error between desired and commanded state. Observability means that the values of the states are known to the controller. If these states are not known, the controller will not know to apply control, again resulting in a potential loss of control of the aircraft. Controllability and stability are demonstrated mathematically using control theory approaches (which can be found in any feedback control textbook).

Control automation should make apparent the axes under control and the expected behavior of the automation. Due to the desire to be able to control axes separately under certain circumstances, there are a number of modes in which the autopilot can be operated. Typically these modes are not clearly identified to the pilot, or, if they are, the expected operation of the aircraft while in these modes is not well understood. A number of aircraft accidents have occurred due to this *mode confusion*. A number of researchers have investigated this problem and proposed solutions [68.10, 37–40], but to date no particular method for mitigating the problem has been widely adopted. Designers of future control automation, however, must strongly consider the likelihood of mode confusion, and use good human factors design methodology to ensure the transparency of the automation.

Control automation should fail gracefully. It is possible for the aircraft to be exposed to conditions that exceed the expectations of the designers. Such conditions are often cases where the pilots could use the assistance of automation, but where automation typically turns itself off. Autopilots are designed to be used under specific sets of flight conditions; if exceeded, autopilots will simply disconnect, leaving the pilot to handle the unusual circumstance by themselves. To the extent possible, control automation should be designed to assist pilots even in unusual circumstances rather than just shutting off.

68.3.2 Warning and Alerting Systems

Warning and alerting systems are designed to identify hazards. If this identification (and the corrective action) were deterministic, there would be no need to alert (the system should operate automatically to perform the corrective action). Typically, the system acts as a signal detector, alerting based on some threshold of evidence regarding the hazardous condition.

Signal detection theory provides a convenient and effective way of analyzing alerting systems. If the detector is correct in identifying the signal, then the detection is considered *correct*. Otherwise, the detection is considered a *false alarm*. If, on the other hand, the system does not alert and is wrong (i. e., it should have alerted), that is considered a missed detection. A correct rejection is the final case, where the system correctly does not alert.

Given an equal cost of a missed detection and false alarm, thresholds should be set to minimize missed detections and false alarms. Such a tradeoff can be viewed on a system operating characteristic (SOC) chart, such as that shown in Fig. 68.8. The *chance* or *guess* line is

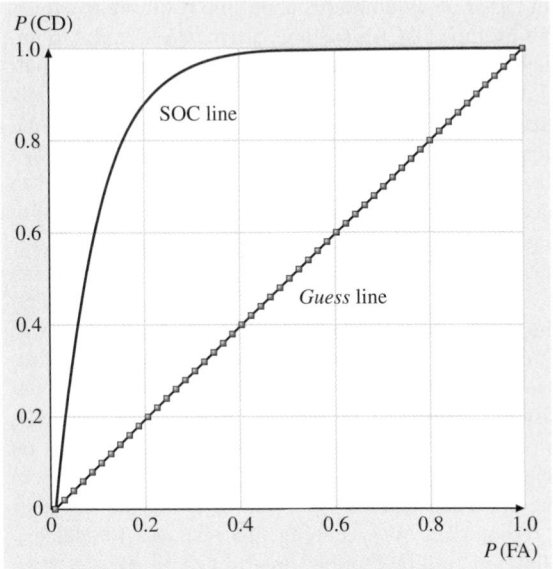

Fig. 68.8 System operating characteristic chart

the 45° diagonal on the chart. The curve is constructed by manipulating the alert threshold and determining the resulting probability of false alarm and correct detection. Different system designs will result in different curves. In the SOC chart, the *perfect* system operates in the upper left corner of the chart, where the probability of false alarms is zero and the probability of correct detections is 1.

Since perfect performance is not possible, the system should be operated using the alert threshold that is as close as possible to the upper left corner of the SOC chart. Moreover, since correct detections and false alarms are typically not valued equally, the best point on the SOC curve is determined by the relative value of the correct detections and false alarms. In particular, for the given curve, one should attempt to maximize the value of

$$U = P(\text{CD})V(\text{CD}) - P(\text{FA})V(\text{FA}) \,. \qquad (68.3)$$

For collision detection systems, there is another consideration. If a false alarm occurs, it is possible that the resulting actions of the pilot will induce a collision that would not occur if no action had been taken. Therefore, such systems must also consider *induced collisions* as a metric. Other types of alerting systems should also consider the full effect of false alarms on resulting system performance.

Alert thresholds should include consideration for pilot response time, which will vary considerably based on the frequency of the alert. Often alert thresholds are set based on assumptions about the resulting response. For example, ACAS systems often expect a pilot to initiate the resolution maneuver within just a few seconds. However, many alerts are uncommon, sometimes only heard a few times in the career of a pilot. Expectation seems to affect pilot response to alerts, and pilots have been known to take 30 s or longer to respond to alerts [68.41–43]. If the alert thresholds are set assuming a 5 s response time, but that threshold is not met, the alert may come too late to be effective.

Expected responses (often included in the alert threshold) should also reflect heuristics, as pilots will often apply common shortcuts (or techniques) when executing a response, even if that response is not consistent with that expected by the alert. For example, prototype alerting systems for collision avoidance on approach assume that the pilots will turn away from the approach path of the other aircraft, and the alert thresholds are set assuming this response [68.44, 45]. However, military pilots are trained to always keep proximate aircraft in sight so that separation can be as-

sured visually. A turn away from an aircraft violates this heuristic and may not be followed.

Wherever possible, preparatory warnings should be given. It has been shown that adherence to desired actions subsequent to an alert improves if the alert is preceded by a preparatory warning [68.46, 47]. For example, the traffic collision avoidance system (TCAS), a type of ACAS, utilizes a two-level warning system. A traffic advisory (TA) is first given to warn of a proximate aircraft that may pose a threat. No action is required due to the TA (although some action presumably should be taken, even if it is to confirm the threat). If the condition persists, a resolution advisory is given to indicate the high potential for collision. Resolution advisories provide specific guidance to the pilots to avoid the potential collision; pilots must comply with the resolution advisory instructions.

68.3.3 Information Automation

During periods of high workload, pilots will have little time to scan a display looking for information. For that reason, information automation must avoid clutter – the presentation of useless information alongside useful information. This coincidence of information forces the pilot to search a display, utilizing time and cognitive resources in short supply during times of high workload.

However, the designer often cannot identify useful information a priori. In this case, there are a number of methods to declutter a display, although no definitive methods have been identified. Decluttering eliminates low-priority information, or information that is unlikely to be needed, from the display. One method is to utilize multifunction displays (MFDs) instead of single-sensor, single-instrument (SSSI) displays. MFDs allow for depth, where information can be put on separate *pages* of the display that can be brought up when needed or when called for by the operator. The information is then present in the system but not visible until needed. For example, automation onboard flight decks will display information related to a malfunction when that particular malfunction is detected.

Two tradeoffs when using MFDs are the possibility that information is presented when not needed (or not presented when needed) by the automation, and the need to navigate through the display. It can be difficult for the designer to envision all the circumstances involved that lead to a pilot needing information displayed, or needing the display to remain the same. Moreover, for such automation to be used, it should be highly accurate at predicting the information needs of the pilot.

When depth is introduced into displays, the pilot must navigate through that depth to arrive at a particular set of information, just as people navigate through the depth of webpages. On flight decks, information displays are typically very small (a few inches by a few inches), so for the same amount of information more depth is needed than on a conventional laptop-sized display. Moreover, fewer controls are provided to navigate through the displays, making the task even more difficult.

Another method to declutter displays is to de-emphasize some information by reducing its contrast, brightness or both. Important information will then *pop out* of the display, and unimportant information will be easier to ignore.

Recent work on visualization may provide some assistance [68.48, 49]. Visualization approaches attempt to provide information in a format that eases interpretation, allowing more information to be extracted from a display for a given amount of information presented. However, such approaches have yet to be validated for use in such safety-critical systems as a flight deck.

68.3.4 Human Factors Issues

Our ability to create automation capable of replacing a function previously done by a human is becoming largely dependent upon human factors issues rather than technical ones. With computers becoming smaller and faster, it is technically possible to produce automation capable of remarkable feats. However, if that system must interact with humans, it must be compatible. This problem of compatibility is often a major obstacle to the introduction of automation into complex socio-technical systems.

Such automation must consider a number of aspects of human interaction with technology. The role of human in a highly automated system is often relegated to that of a supervisor. This role has particular requirements and challenges that must be considered by the designer of automation. One of these challenges relates to the *out-of-the-loop* problem [68.50], which is being addressed in terms of understanding the operator's awareness of the situation. In addition, automation must incorporate human limitations with regards to perception, workload, and physical ergonomics.

Supervisory Control

As flight deck automation increases in quantity and sophistication, some have expressed concern that pilots are becoming supervisory controllers of automation rather than users of automation. On some aircraft, it is possible to connect the autopilot once aligned on the departing runway, and allow the aircraft to fly to its destination, land, and taxi off the runway without interacting with the aircraft control surfaces (except drag/lift devices such as flaps and spoilers) except through the autopilot.

One concern related to this phenomenon is the loss of manual control skills, as discussed by *Billings* [68.51] and others [68.52, 53]. Should the automation fail, the pilots would have to fly the aircraft manually. Since such failure is likely to result from a more serious system failure, such control may have to be assumed under less than ideal circumstances. For example, the aircraft may have only partially operating control surfaces, as was the case with a DC-10 mishap in Sioux City, Iowa in 1989 [68.54]. In that accident, an uncontained engine failure left the aircraft without hydraulic power to operate any of the control surfaces. The pilots controlled the aircraft using engine power only, a procedure that had to be created by the pilots on the spot. Without excellent manual flying skills, it is unlikely that the result would have been as successful (175 of the 285 passengers and all but one of the crew survived the crash landing). Designers of automation should ensure that sufficient opportunity exists for the operators to exercise those aspects of control for which an intervention need may arise in the case of automation failure.

The overarching considerations that are part of supervisory control [68.55] should be considered when designing automation for flight decks. These considerations include the ability of pilots to monitor information, an understanding of the impact of communications/control delay, and loss of situation awareness.

As supervisors of automation, pilots are responsible for monitoring the automation to ensure it is operating properly and to intervene should the automation fail. Humans are notoriously bad at monitoring highly reliable systems [68.56]. Under such conditions, humans will tend to become complacent and fail to monitor adequately. One response to this has been to suggest caution against overautomation [68.51], while others have suggested that more automation (or at least feedback) is the answer [68.57]. Likely what is required is smarter automation, including designs that make a system fail only in ways that can be handled by a human supervisor.

Even relatively small delays in communication or control can have significant consequences for supervisory control. This is analogous to manual control, where delays between identification of control requirement and application of that control can lead to instability. For faster control loops, less delay is required; for slower control loops, longer delays can be tolerated. For example, under conditions of several seconds of delay, intervention by a supervisor to accept manual control would likely be impossible, whereas such delays would not impact the supervisor's ability to provide navigation commands to a vehicle.

In addition, humans have better situation awareness when actively involved in the operation rather than when acting as a supervisor of automation [68.58, 59]. Situation awareness is the set of information used by the operator in order to make decisions and choose courses of action, and will be discussed in more detail in the next section. Good situation awareness, considered a key to good performance in complex systems, is adversely affected by the operator's cognitive distance from the task. This may be a result of research findings suggesting that concrete, direct experiences involve deeper cognitive processing, and are therefore better recalled than those that are merely described or otherwise undergo more shallow processing [68.60].

Situation Awareness, Clutter, and Boundary Objects

Since situation awareness is correlated with good performance, automation designers should attempt to enhance situation awareness. However, this cannot be addressed by simply making sufficient information available and salient to the operator, since situation awareness involves the operator making use of that information. Making information available is not sufficient for ensuring that the information gets used; operators may simply fail to use available information for guiding action for reasons that are as yet unclear. For example, one aviation incident self-report from the National Aeronautic and Space Administration (NASA)'s aviation safety reporting system database describes landing at the Los Angeles International Airport without first obtaining landing clearance as required by FAA regulation. The crew, who would normally switch their radiofrequency to that of the tower controller (who would then grant landing clearance), was told to remain on a previous frequency by air-traffic control (most likely due to some separation concern).

As a result, the crew never switched frequency and never got landing clearance. They realized their mistake after turning off the runway, at which point they were about to switch to the frequency of the ground controller. Noticing the error, the crew called the tower and reported their mistake. The crew in this incident had sufficient information to know they did not have clearance; they simply did not make use of that information. While factors such as interruptions (and other disturbances to a routine), workload, idiosyncratic cognitive capabilities, and experience would seem to influence the prevalence of this type of behavior, as yet no definitive research has been accomplished to understand it.

Automation designers should be aware that the absence or lack of salience of information, while not technically a part of situation awareness, will have the same effect as a loss of situation awareness. That is, not having the information in the first place will have an identical effect on performance as not making use of available information (although the causes of the error may be different). One method to ensure this is to conduct a thorough task analysis [68.61], such as goal-directed task analysis [68.62]. Such methods provide insight into what information may be required by an operator in the conduct of their task.

In addition to ensuring the availability of information, designers should ensure that information is accessible to the operator. Too much irrelevant information can make accessing particular (important) pieces of information more difficult than necessary [68.35, 63]. For this purpose, a number of decluttering schemes are available [68.64–66].

Intelligent design of displays, so that multiple operators across a collaborative task can easily contextualize information while making use of the same body of information, is also recommended. Such displays, known as boundary objects, have been found in examples of good collaborative work [68.67, 68]. Unfortunately, design criteria for these types of displays are still lacking.

Situation Assessment

Automation designers should consider the situation assessment capabilities of the pilots. Situation assessment is a term that refers to the process of populating situation awareness. Situation assessment involves most aspects of cognition – perception, attention, comprehension, memory – and is therefore very complex. These limitations are discussed in subsequent sections. However, several important overarching findings bear on this capability.

Operators monitoring dynamic displays do so at a near-optimal rate under most circumstances. According to the Nyquist–Shannon sampling theorem [68.69], one can fully reproduce a signal if that signal is sampled at more than twice the signal's frequency, if the signal is bandlimited. Operators have been found to sample at precisely this rate, with some deviation at higher and lower frequency [68.70]. For highly dynamic information, operators will oversample, while they tend to undersample low-frequency information. Numerous reasons have been proposed for this tendency, without a definitive answer. However, designers can feel somewhat comfortable that human operators will sample dynamic displays appropriately.

However, as mentioned above, operators become easily complacent with highly reliable or very low-frequency information. In such cases, designers should, where practical, ensure that operators are engaged in the task. One method, utilized in airport security x-ray machines as the threat image protection system, is to occasionally probe the operator with false signals. Such signals can be used to engage the operator but also to test to see if the operator is becoming complacent. Another method is to increase the salience of potentially offending signals, such as providing a warning light to attract the attention of the operator to the instrument containing the relevant information.

In addition, operators have significant cognitive limitations (but also have several methods of coping with these limitations). A human's channel capacity is limited to somewhere between four and nine individual items [68.71, 72]. Above this, operators must be able to *chunk* the information (such as recalling a phone number as one seven-digit number rather than as seven separate digits in a sequence). Expertise improves the ability of operators to chunk information, but designers are cautioned not to design automation that taxes a human's channel capacity.

Perception

In terms of visual perception, human operators sparsely sample scenes, then integrate their knowledge of the world with these sparse samples to obtain a representation. One factor known to affect sampling is the Gestalt of the scene [68.73]. Factors such as the proximity of items, their *common fate*, and their similarity all affect the ability of the operator to associate information. An example of this ability regards the monitoring of a number of check-reading gages. Gages oriented so that their limits are aligned (Fig. 68.9) are much easier to monitor than those that are not. Designers are therefore encour-

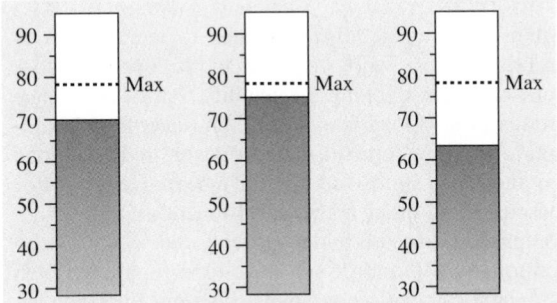

Fig. 68.9 Aligned gages for Gestalt effect

aged to consider the Gestalt of instruments to assist operators, including grouping information used for the same purpose close to one another [68.74].

The basic layout of flight deck instrumentation was established many decades ago, based on research into the scan pattern of pilots [68.75]. This placed the most-used instruments into a T-shaped pattern (now called the *basic T*), with less-used instruments being placed outside of this T, but close to related instruments. The basic T has been replicated on multifunction displays utilized in modern aircraft in place of single-sensor, single-instrument gages. It seems certain that this arrangement will continue, and should be adhered to by automation designers.

Workload has been known to affect perception and attention, as has been widely reported in studies on driving [68.76, 77]. Increased workload impairs visual detection, as well as the size and shape of the visual field [68.78]. Increased workload also dissipates attention, which in turn influences perception. Designers should be aware that, under conditions of high workload, pilots may have difficulty in perceiving information that would otherwise be considered sufficiently salient.

Much of the information imparted to the user by flight deck automation uses the visual channel. In order to avoid further saturating the visual channel, designers are reminded that other perceptual channels are available. Several automation systems use the aural channel, particularly alerting systems. The GPWS, the stall warning system, the landing gear warning system, and a number of others produce both visual and aural warnings; each aural warning is distinct to aid in identification. Very little work has been done in trying to utilize the haptic medium in aviation, but some success has been found in aiding operators in locational attention directing [68.79], which could be useful for pilots.

Workload

Often the primary concern related to workload is to ensure that the work required of an operator does not exceed the operator's capability. Pilots experience swings in workload from very high (takeoff and landing) to very low (cruise). The periods of high workload are such that almost no cognitive work, required for reasoning and other higher-level functions, can be accomplished. Instead, pilots are relegated to skill-based control [68.80], and do not have time for things such as mental calculation or troubleshooting. Imposing requirements for such activities during high workload periods should be avoided.

Performance in complex tasks has been showed to be inverse-U shaped with respect to workload [68.81, 82]. At low levels of workload, human performance is low due to complacency. At high levels of workload, performance is also low, but due to task demands taxing or exceeding human capabilities. Performance is highest at moderate levels of workload. Automation designers should therefore try to balance workload (not too high and not too low), rather than strictly trying to reduce it.

Physical Ergonomics

Automation designers should consider good physical ergonomics as well as cognitive ergonomics. For automation, several factors are considerations, including repetitive strain avoidance, placing frequently used controls within reach, ensuring proper visibility and audibility, and reducing the *heads-down* time of the operator.

Automation should avoid requiring significant fine input from the operator. The flight management system (FMS) includes a control display unit (CDU), which is the primary manual interface for the pilot with the unit. Pilots may need to input information into the FMS using the CDU, which consists of a small keyboard and display. The keys on the CDU are small, mostly identically shaped, and laid out in an idiosyncratic (to the particular CDU) manner. This layout is far from ideal from a human factors standpoint, and results in a slow rate of typing. Should significant typing be required, repetitive stress injuries (such as carpal tunnel syndrome) would be a significant risk. However, designers of the CDU must contend with a very small device footprint, it must have very high reliability, and it must be robust to flight conditions such as turbulence. Fortunately, most of the information in the FMS is loaded only once (or by electronic download), so typing is infrequent and usually limited to a few key presses.

There are numerous principles for controls to avoid the chance of repetitive strain or overuse disorders [68.82]. These types of injuries are associated with high-frequency activity, and therefore can be avoided by eliminating prolonged periods of repetitive motions. High repetitiveness has been defined as a cycle time of less than 30 s, or more than 50% of the total task time [68.83]. Larger loads can induce the same conditions with less repetition. Maintaining body positions (other than the neutral body position) or gripping objects can also induce repetitive strain injuries. According to *Kroemer* et al. [68.83], there are seven *sins* associated with overuse disorders:

1. Highly repetitive motions
2. Prolonged or repeated application of more than one-third of an operator's static muscle strength
3. Stressful body postures, such as elevated wrists when typing or working while bent or twisted at the waist
4. Prolonged nonneutral body postures, such as standing
5. Prolonged contact with a surface, such as leaning against a surface or holding a tool
6. Prolonged exposure to vibration
7. Exposure to low temperatures (such as a draft or exhaust from a pneumatic tool).

Certain types of reaching are ergonomically undesirable. The need to extend oneself to reach a control poses risks to the back and shoulders, as does the need to reach above one's shoulder height [68.84]. The need to twist one's body frequently also poses significant risk of injury to operators. For these reasons, it is a common ergonomic principle that frequently used controls be placed within the normal reach of the operator.

To find the normal reach of the operator, one should first identify the range of sizes of the people that constitute the population for which the device will be designed. A common choice is to use a 5th percentile female as the lower limit and the 95th percentile male as the upper limit. However, certain occupations may further limit the population.

If single measurements are required (such as shoulder–hand length), lookup tables, such as those that can be obtained from NASA or commercial sources, can be used. If multiple measurements are required, however, *cases* should be considered rather than trying to combine measurements from the tables. Cases identify a typical small or large person, rather than assuming that a small person has the 5th percentile seated height and 5th percentile shoulder–hand length (for example), which is usually not accurate.

Once the lower and upper measurements have been obtained, the designer can then check to see whether the commonly used controls are within the reach of that range of persons without extending. Such methods have been applied in a great number of different domains.

Proper visibility should be ensured throughout the typical illuminance conditions. Studies that provide guidance on illuminance were conducted decades ago, including studies in which *Blackwell* [68.85] found that the threshold contrast at which objects became recognizable increased with decreasing object size, decreasing orderliness of objects, movement of the objects, and if less time was given to detect the object. Increased age also increases the need for contrast [68.86].

In addition, the viewing of displays should be free of glare and veiling reflections. Glare is the condition where bright light sources are in the line of sight of the operator, creating a nuisance. Veiling reflections are the condition where bright light sources are reflected off the display surface, obscuring information on the display. These problems can be minimized by positioning the display where bright light sources pose neither a glare or reflection issue (to the maximum extent possible).

Automation designers should also ensure audibility, particularly of aural feedback or alerts. An alert that is at least $10-15\,\mathrm{dB}$ above the ambient noise level is generally considered sufficiently salient, although alerts are often found with significantly less difference when placed in the normal operating environment (a flight deck is a noisy place in flight). Different types of tones can also be used to indicate different levels of criticality [68.87].

When multiple auditory warnings exist, as they do on flight decks, they should utilize distinct tones to avoid confusing one alert with another. Some alerts may even include computerized voices. On a typical flight deck, auditory warnings exist for the GPWS system, TCAS system, overspeed warning, landing gear warning, and altitude alerts.

When the weather permits, pilots must direct a significant amount of attention to outside the flight deck. In addition to visually locating other aircraft that may pose a collision risk, pilots must frequently locate, identify, and orient themselves with respect to a particular airport runway. On the ground, scanning for obstacles and other aircraft is a particularly important task. For this reason, automation should not impose upon pilots too much *heads-down* time (the term is contrasted with

heads-up displays located on the windshield of the aircraft). Flight management system reprogramming, for example, is a difficult task, requiring many minutes of heads-down time to accomplish relatively simple reprogramming tasks. It is therefore undesirable to have changes close to the ground, where pilots' attention should be outside rather than inside the aircraft.

68.3.5 Software and System Safety

Calls have been made previously for automation that gracefully degrades [68.51, 55], but there has been a general lack of guidelines for such a purpose. One significant complication for the introduction of such methods is that most advanced automation is software rather than hardware based. Software systems lack physical constraints available in hardware. Whereas one can construct physical constraints on machines that prevent them from failing in certain ways (e.g., short-circuit protection devices on electrical outlets in bathrooms), no such assurances can (so far) be provided for software systems. Formal methods for mathematically verifying software [68.88], particularly in concert with model-based software development methods [68.89], are methods that go far toward improving software systems in this regard.

Formal methods, however, are insufficient for ensuring software safety [68.90]. Safety is often more than ensuring that the output of the software remains within some constrained set of appropriate responses. Safety also involves the processes for defining those constraints and the interaction of human and other automated agents with the system. Therefore, system safety must also account for these (and other) factors.

Because of this, safety has been viewed as an emergent feature of a system, ensured by providing sufficient control for agents in the system to prevent it from entering unsafe states [68.91]. Under this concept, a system can be modeled as a set of states, where safety (as a goal) is preventing the system from entering an unsafe state.

68.3.6 System Integration

As flight decks get more complex, and as the interactions between systems (other vehicles, air-traffic controllers, passengers, company) become tighter, system integration issues will grow in importance. Aircraft have been typically designed to optimize some aspect of the vehicle (for example, speed, fuel economy, range or capacity), but now must consider the integration of that vehicle within a larger system. There are tight interactions between customer demand, airline route decisions, air-traffic control infrastructure, and aircraft design.

Such design optimizations involve multiple systems, each of which have very different dynamics. The behavior of the system is not defined by the sum of the behaviors of the components, but typically also has emergent behavior at higher levels of aggregation of the components. Methods for accomplishing such optimizations have not yet been developed, but are being studied as *systems-of-systems* problems.

Nonetheless, automation designers must address systems integration issues. Flight deck automation is expensive to develop and implement, and therefore must last many years (typically the lifetime of different aircraft models). For example, flight management systems have hardware and software that is highly antiquated by modern computer standards, but because it works and is certified (Sect. 68.3.7), it has not substantially changed over the last few decades. Moreover, the flight deck automation needed to take advantage of next-generation air-traffic systems will need to be developed and installed over the next decade.

Flight deck systems often interface with multiple systems of various types. State information on aircraft comes from pito-static, temperature, and other instruments. Position and navigation information come from radio navigation aids, inertial navigation systems, GPS, and the flight management systems. There are also mechanical, electrical, hydraulic, and pressurization systems onboard. Potentially, automation must be integrated with several of these different types of systems.

68.3.7 Certification and Equipage

The various civil flight authorities have strict certification requirements for new automation. These requirements impose rigorous safety requirements on that automation. This results in long lead times and high costs, which in turn imposes a heavy burden on any new automation system being proposed.

In the USA, the Federal Aviation Administration certifies new equipment. Under the Code of Federal Regulations, each installed system must:

1. Be of a kind and design appropriate to its intended function
2. Be labeled as to its identification, function or operating limitations, or any applicable combination of these functions
3. Be installed according to limitations specified for that equipment
4. Function properly when installed [68.92, p. 125].

In addition, for flight and navigation instruments:

1. The equipment, systems, and installations ... must be designed to ensure that they perform their intended functions under any foreseeable operation condition.
2. The airplane systems and associated components, considered separately and in relation to other systems, must be designed so that:
   a) The occurrence of any failure condition which would prevent the continued safe flight and landing of the airplane is extremely improbable.
   b) The occurrence of any other failure conditions which would reduce the capability of the airplane or the ability of the crew to cope with adverse operating conditions is improbable.
3. Warning information must be provided to alert the crew to unsafe system operating conditions, and to enable them to take appropriate corrective action. Systems, controls, and associated monitoring and warning means must be designed to minimize crew errors which could create additional hazards [68.92, p. 126].

The certification process is detailed and slow, designed to prevent systems with deleterious safety effects from being installed in aircraft (particularly commercial aircraft). This process, which is well known to the industry but is not well documented, should be considered by automation designers. The use of off-the-shelf and previously certified hardware and software is recommended to simplify and shorten the certification process.

68.4 Flight Deck Automation in the Next-Generation Air-Traffic System

As of this writing, the next-generation air-traffic system is still taking shape [68.1, 2], but the following appear to be emerging as key features of that system:

- Network-centric operations
- Integration of novel vehicle types
- Operations under conditions of very high density

- Tight coupling between passenger demand, airportal infrastructure, aircraft operations, and air-traffic control.

These capabilities are designed to allow the next-generation system to handle two to three times the capacity of the current system by 2025.

68.4.1 Network-centric Operations

Currently, most information in the air-traffic control system is contained locally within each air-traffic control facility. The exception to this is the enhanced traffic management system (ETMS) data, which is available to all air-traffic control facilities and even users (with a slight delay). However, ETMS data is significantly limited in content and in update rate. Under future concepts, all air-traffic information will be made available (with some restrictions) to all agents in the system, in order to facilitate better collaborative work. This concept is referred to as *network-centric operations.*

The various air-traffic control authorities in different countries are currently building the network infrastructure. This infrastructure will carry real-time flight plan information and radar track information, which (in the USA) is currently carried in the Department of Transportation's *aircraft situation display to industry* (ASDI) data stream with one minute (at best) update rate, but also may include such things as aircraft intent information, scheduling information, aircraft load information, airport and airspace capacity, and airline priority information. The ability to use this information should provide unprecedented opportunities for automation developers.

For example, there has been little success in optimizing sequences of aircraft into congested resources such as airports. The current system allows aircraft to launch when they want and fly at any speed to their destination, where they may then have to be delayed for tens of minutes or more. Moreover, decisions about which aircraft get delayed are made based on a first-come first-served basis.

While this can be considered at least *fair* (in that no company gets an advantage from it), it is certainly not optimal. It would be preferable to sequence aircraft based on best throughput or least delay (for aircraft or passengers). Without any information available in this regard, it has been extremely difficult to improve the air-traffic system in this regard. With a single data source, systems could consider many factors when deciding on

which aircraft to delay, and could start that delay early, imposing small changes that have significant effects on the overall system.

The methods to accomplish such optimizations have not yet been developed. Some attempts at optimization have been made, but are limited in scope. It is not yet clear how one can formulate optimization problems of this scope (or solve them in a reasonable period of time). Moreover, whatever solutions are produced must provide guidance on how to construct trajectories of individual aircraft.

68.4.2 Future Air Vehicle Types

Decades ago it was believed by many that, by this date, air travel would be dominated by supersonic transports. Due (in part) to noise and fuel consumption issues that has not occurred, although research continues on super- and hypersonic transports. As experts peer into the future of air transportation, it is unclear what types of vehicles will be present in the air-traffic system in 50 years.

The last few years has seen a significant increase in the use of small regional jets, due to their higher load factor (i. e., higher percentage of seats filled with paying customers). Airbus has also just introduced the A-380, the largest passenger transport jet ever produced, with seating for 550 to over 800 passengers depending on the configuration. Boeing is developing the B-787, a large (210–330 passenger) jet with improved fuel economy, flight deck systems, and passenger comfort features. Research is also being done on a lighter-than-air heavy lift vehicle [68.93].

These designs are driven by projections of market demand, and include systems driven by projections of air-traffic control capabilities. There are also design interactions with ground systems, pilot training and culture, emergency egress considerations, and so on. Given the uncertainty in how far and how quickly fuel prices will rise, how the demand for air service will change, and how technology (both flight deck and air traffic) will change, it is difficult to predict what vehicles will be present in the air-traffic systems decades from now.

However, research on the flight deck systems for those vehicles must commence now if they are to be ready for implementation when those vehicles undergo development. Fortunately, the principles of safety and efficiency will continue to drive the design of those systems.

68.4.3 Superdensity Operations

One of the major constraints on air-traffic growth is the airspace surrounding the airport, referred to as the terminal area. In the terminal area, the airspace is densely crowded with aircraft whose positions are changing in three dimensions, and whose rate of change of positions is also changing. This greatly complicates the task of pilots and controllers.

Tomorrow's air-traffic system is being developed to accommodate two to three times today's traffic level. This means even more dense terminal airspace. As yet, there is no consensus as to how that will be accomplished. One possibility is that some or all of the separation responsibility will be with the flight deck; this will necessitate new automation to identify and resolve conflicts, while displaying information on these conflicts and resolutions to the pilots.

In addition, pilots may be responsible for meeting times-to-cross at various points along their route. While current-generation FMS can accomplish this task, this ability is brittle and is not conveyed well to pilots. As this becomes a primary responsibility of pilots, this capability will have to be significantly enhanced.

Currently, efficient arrival procedures are being developed and tested at major airports across the globe. These procedures are designed to reduce fuel use, emissions, and noise simultaneously. Implementing such procedures in superdense airspace will require additional technologies (such as flight deck displays of traffic information, improved flight-deck-based conflict detection and resolution, and intent broadcasting) and more widespread use of existing technologies (such as the global positioning system and autopilots coupled with advanced flight management systems).

68.4.4 Integration

New systems will have to be integrated with ground-based air-traffic systems as well as with legacy equipment. Systems will also have to deal with a mix of equipped, partially equipped, and unequipped aircraft.

It seems unlikely, given the burden of certification, that completely new flight management system software will be developed. Instead, it is more likely that the new capabilities will be added on top of the old software with updated hardware. New automation will therefore have to interface with old software, which was designed for low-performance computing. In addition, many systems onboard aircraft are substantially unchanged from decades ago, and provide their input to flight deck computers. It is unlikely these systems will change either.

As new systems are developed, they are typically expensive to implement. Therefore, many manufacturers, airline companies, and individuals choose not to implement them. This creates complexity, not only for improving the system, but also for the use of flight deck systems. For example, the TCAS system must work not only between aircraft equipped with TCAS and mode S transponders, but with those equipped only with a basic transponder as well. This situation is likely to get worse before it gets better – as new technologies come into the market, there will be a wider disparity of capabilities across aircraft.

68.5 Conclusion

Modern flight decks contain a large number of automation systems, including control, warning, and information automation. Significant changes are being considered for air-traffic control systems, which will influence automation systems for flight decks and provide incredible opportunities for tomorrow's automation engineer. These changes will have to balance future needs, the increase in capabilities over the last few (and next few) decades, the expense of advanced automation, and the long lead times associated with the development and certification of flight deck automation.

68.6 Web Resources

- Eurocontrol Single European Sky:
  http://www.eurocontrol.be/sesar
- US Joint Planning and Development office:
  http://www.jpdo.gov/
- Federal Aviation Administration:
  http://www.faa.gov/
- NASA aeronautics:
  http://www.aeronautics.nasa.gov/

- Airbus:
  http://www.airbus.com/
- Boeing commercial airplanes:
  http://www.boeing.com/commercial/

- Honeywell labs:
  https://www.honeywell.com/sites/htsl/
- American Institute of Aeronautics and Astronautics:
  http://www.aiaa.org

References

68.1 European Parliament: *Laying down the framework for the creation of the Single European Sky*, Regulation No. 549/2004 (2004)

68.2 Joint Planning and Development Office: *Next Generation Air Traffic Control System Integrated Plan* (Joint Planning and Development Office, Washington 2004)

68.3 L.R. Newcome: *Unmanned Aviation: A Brief History of Unmanned Aerial Vehicles* (AIAA, Washington 2004)

68.4 J. Broelmann: The development of the gyrocompass – inventors as navigators, J. Navig. **51**(2), 267–273 (1998)

68.5 P.M. Fitts: *Human Engineering for an Effective Air-Navigation and Traffic-Control System* (National Research Council Committee on Aviation Psychology, Washington 1951)

68.6 http://mtp.jpl.nasa.gov/notes/pointing/pointing.html

68.7 http://www.levelbust.com/articles/mode_s.htm

68.8 N.B. Sarter, D.D. Woods: Team play with a powerful and independent agent: operational experiences and automation surprises on the Airbus A-320, Hum. Factors **39**(4), 553–569 (1997)

68.9 Y.J. Tenney, W.H. Rogers, R.W. Pew: Pilot opinions on high level flight deck automation issues: toward the development of a design philosophy, NASA Contract. Rep. 4669 (1995)

68.10 A. Degani, M. Shafto, A. Kirlik: Mode usage in automated cockpits: Some initial observations, Proc. Int. Fed. Automat. Control; Man–Machine Syst. (IFAC-MMS) Con. (IFAC, Boston 1995)

68.11 E. Palmer: "Oops, it didn't arm, A Case Study of Two Automation Surprises", Proc. 8th Int. Symp. Aviat. Psychol., ed. by R.S. Jensen, L.A. Rakovan (Columbus 1995)

68.12 H. Sogame, P. Ladkin: *Aircraft Accident Investigation (Report 96-5)* (Japan Ministry of Transport, Tokyo 1996)

68.13 B.J. Carlson: *Past UAV Program Failures and Implications for Current UAV Programs (Report No. AU/ACSC/037/2001-04)* (Air University, Maxwell 2001)

68.14 National Transportation Safety Board: *14 CFR Part 121 Scheduled operation of Braniff Airways, Inc. (Report No. DCA68A0005)* (NTSB, Washington 1968)

68.15 National Transportation Safety Board: *14 CFR Part 121 Scheduled operation of Southern Airways, Inc. (Report No. DCA77AA015)* (NTSB, Washington 1977)

68.16 National Transportation Safety Board: *14 CFR Part 121 Scheduled operation of Eastern Airlines, Inc. (Report No. DCA75AZ015)* (NTSB, Washington 1975)

68.17 H.T. Liu, C. Golborne, Y. Bun, M. Bartel: Surface windshear alert system, Part 1: Prototype development, J. Aircr. **35**(3), 422–428 (1998)

68.18 J.P. Bliss: Investigation of alarm-related accidents and incidents in aviation, Int. J. Aviat. Psychol. **13**(3), 249–268 (2003)

68.19 R. Parasuraman, V. Riley: Humans and automation: use, misuse, disuse, abuse, Hum. Factors **39**(2), 230–253 (1997)

68.20 FAA: *Introduction to TCAS II Version 7* (FAA, 2000)

68.21 http://humansystems.arc.nasa.gov/ihh/cdti/cdti.html

68.22 D. Boorman: Today's electronic checklists reduce likelihood of crew errors and help prevent mishaps, ICAO Journal **1**, 17–20 (2001)

68.23 D. Boorman: Today's electronic checklists reduce likelihood of crew errors and help prevent mishaps, ICAO Journal **1**, 36 (2001)

68.24 A.L. Alexander, C.D. Wickens: Cockpit display of traffic information: The effects of traffic load, dimensionality, and vertical profile orientation, 45th Annu. Meet. Hum. Factors Ergon. Soc. (Santa Monica 2001)

68.25 R. Barhydt, R.J. Hansman: Experimental studies of intent information on cockpit traffic displays, J. Guid. Control Dyn. **22**(4), 520–527 (1999)

68.26 W.W. Johnson, V. Battiste, S. Delzell, S. Holland, S. Belcher, K. Jordan: Development and Demonstration of a Prototype Free Flight Cockpit Display of Traffic Information, Proc. SAE/AIAA World Aviat. Conf. (1997)

68.27 A.R. Pritchett, L.J. Yankosky: Simultaneous design of cockpit display of traffic information and air traffic management procedures, Proc. 19th DASC (1998)

68.28 M.G. Ballin, J. Hoekstra, D. Wing, G. Lohr: NASA Langley and NLR Research of Distributed Air/Ground Traffic Management, Proc. 1st ATIO (2002)

68.29 P. Lee, J. Mercer, L. Martin, T. Prevot, S. Shelden, S. Verma, N. Smith, V. Battiste, W. Johnson, R. Mogford, E. Palmer: Free Maneuvering, Trajectory Negotiation, and Self-Spacing Concepts in Distributed Air-Ground Traffic Management, Proc. 5th USA/Europe Air Traffic Manag. R&D Semin. (Budapest 2003)

68.30   M.W. McGreevy, S.R. Ellis: The effect of perspective geometry on judged direction in spatial information instruments, Hum. Factors **28**, 439–456 (1986)

68.31   D.D. Woods: Visual momentum: A concept to improve the cognitive coupling of person and computer, Int. J. Man–Mach. Stud. **21**, 229–244 (1984)

68.32   O. Olmos, C.D. Wickens, A. Chudy: Tactical displays for combat awareness: An examination of dimensionality and frame of reference concepts, and the application of cognitive engineering, 9th Int. Symp. Aviat. Psychol. (Columbus 1997)

68.33   S.N. Roscoe: *Aviation Psychology* (Iowa State Univ. Press, Ames 1981)

68.34   J.L. Levy, D.C. Foyle, R.S. McCann: Performance benefits with scene-linked HUD symbology: An attentional phenomenon?, 52nd Annu. Meet. Hum. Factors Ergon. Soc. (Santa Monica 1998)

68.35   P.M. Ververs, C.D. Wickens: Head-up displays: effect of clutter, display intensity, and display location on pilot performance, Int. J. Aviat. Psychol. **8**(4), 377–403 (1998)

68.36   C.D. Wickens, T. Prevett: Exploring the dimensions of egocentricity in aircraft navigation displays, J. Exp. Psychol. Appl. **1**(2), 110–135 (1995)

68.37   C.M. Bjorklund, J. Alfredson, S.W.A. Dekker: Mode monitoring and call-outs: an eye-tracking study of two-crew automated flight deck operations, Int. J. Aviat. Psychol. **16**(3), 263–275 (2006)

68.38   J. Bredereke, A. Lankenau: Safety-relevant mode confusions – modelling and reducing them, Reliab. Eng. Syst. Saf. **88**(3), 229–245 (2005)

68.39   A. Degani, M. Heymann: Formal verification of human–automation interaction, Hum. Factors **44**(1), 28–43 (2002)

68.40   S.S. Vakil, J.R. Hansman: Approaches to mitigating complexity-driven issues in commercial autoflight systems, Reliab. Eng. Syst. Saf. **75**(2), 133–145 (2002)

68.41   R.L. Ennis, Y.J. Zhao: A formal approach to analysis of aircraft protected zone, Air Traffic Control Q. **12**(1), 75–102 (2004)

68.42   S.J. Landry, A.R. Pritchett: Examining assumptions about pilot behavior in paired approaches, Int. Conf. Human-Comput. Interact. Aeronaut. (Cambridge 2002)

68.43   A.R. Pritchett, R.J. Hansman: Pilot non-conformance to alerting system commands during closely spaced parallel approaches, Proc. 16th DASC (1997)

68.44   T.S. Abbott, G.C. Mowen, L.H. Person Jr., G.L. Keyser Jr., K.R. Yenni, J.F. Garren Jr.: *Flight investigation of cockpit-displayed traffic information utilizing coded symbology in an advanced operational environment (NASA Tech. Paper 1684)* (NASA Langley Research Center, Hampton 1980)

68.45   J.K. Kuchar, L.C. Yang, C. Mit: A review of conflict detection and resolution modeling methods, Intell. Transp. Syst. IEEE Trans. **1**(4), 179–189 (2000)

68.46   L.F. Winder, J.K. Kuchar: Evaluation of collision avoidance maneuvers for parallel approach, J. Guid. Control Dyn. **22**(6), 801–807 (1999)

68.47   L.C. Yang, J.K. Kuchar: Prototype conflict alerting system for free flight, J. Guid. Control Dyn. **20**(4), 768–773 (1997)

68.48   C.D. Hansen, C.R. Johnson: *The Visualization Handbook* (Elsevier, Burlington 2005)

68.49   T. Munzner, C. Johnson, R. Moorhead, H. Pfister, P. Rheingans, T.S. Yoo: NIH-NSF visualization research challenges report summary, IEEE Comput. Graph. Appl. **26**(2), 20–24 (2006)

68.50   M.R. Endsley, E.O. Kiris: The out-of-the-loop performance problem and level of control in automation, Hum. Factors **37**(2), 381–394 (1995)

68.51   C.E. Billings: *Aviation Automation: The Search for a Human-Centered Approach* (Lawrence Erlbaum, Mahwah 1997)

68.52   D.B. Kaber, M.R. Endsley: The effects of level of automation and adaptive automation on human performance, situation awareness and workload in a dynamic control task, Theor. Issues Ergon. Sci. **5**(2), 113–153 (2004)

68.53   C.A. Miller, H.B. Funk, R. Goldman, J. Meisner, P. Wu: Implications of adaptive versus adaptable UIs on decision making: Why "automated adaptiveness" is not always the right answer, Proc. 1st Int. Conf. Augment. Cogn. (Las Vegas 2005)

68.54   National Transportation Safety Board: *14 CFR Part 121 Scheduled operation of United Airlines, Inc. (Report No. DCA89MA063)* (NTSB, Washington 1989)

68.55   T.B. Sheridan: *Telerobotics, Automation, and Human Supervisory Control* (MIT Press, Cambridge 1992)

68.56   L. Bainbridge: Ironies of automation, Automatica **19**(6), 775–780 (1983)

68.57   D.A. Norman: The "problem" with automation: inappropriate feedback and interaction, not "over-automation", Philos. Trans. R. Soc. Lond. Ser. B Biol. Sci. (1934–1990) **327**(1241), 585–593 (1990)

68.58   M.J. Adams, Y.J. Tenney, R.W. Pew: Situation awareness and the cognitive management of complex systems, Hum. Factors: J. Hum. Factors Ergon. Soc. **37**(1), 85–104 (1995)

68.59   M.R. Endsley, D.B. Kaber: Level of automation effects on performance, situation awareness and workload in a dynamic control task, Ergonomics **42**(3), 462–492 (1999)

68.60   F.I.M. Craik, E. Tulving: *Depth of Processing and the Retention of Words in Episodic Memory. Cognitive Psychology: Key Readings* (Psychology, New York 2004)

68.61   B. Kirwan, L.K. Ainsworth: *A Guide to Task Analysis* (Taylor & Francis, New York 1992)

68.62   M.R. Endsley, B. Bolté, D.G. Jones: *Designing for Situation Awareness: An Approach to User-Centered Design* (Taylor & Francis, London 2003)

68.63   M. Yeh, J.L. Merlo, C.D. Wickens, D.L. Brandenburg: Head up versus head down: the costs of imprecision, unreliability, and visual clutter on cue effectiveness for display signaling, Hum. Factors **45**(3), 390–408 (2003)

68.64   M.S. John, D.I. Manes, H.S. Smallman, B.A. Feher, J.G. Morrison: Heuristic automation for decluttering tactical displays, Proc. Hum. Factors Ergon. Soc. 48th Annu. Meet. (Human Factors and Ergonomics Society, Santa Monica 2004)

68.65   P. Kroft, C.D. Wickens: Displaying multi-domain graphical database information: An evaluation of scanning, clutter, display size, and user activity: Information design for air transport, Inf. Des. J. **11**(1), 44–52 (2002)

68.66   J.W. Ruffner, M.C. Lohrenz, M.E. Trenchard: Human factors issues in advanced moving-map systems, J. Navig. **53**(01), 114–123 (2000)

68.67   W. Lutters, M. Ackerman: Achieving Safety: A Field Study of Boundary Objects in Aircraft Technical Support, CSCW '02 (New Orleans 2002)

68.68   S.L. Star, J.R. Griesemer: Institutional ecology, 'translations', and boundary objects: Amateurs and professionals in Berkeley's Museum of Vertebrate Soology, Soc. Stud. Sci. **19**, 387–420 (1989)

68.69   C.E. Shannon: Communication in the presence of noise, Proc. IEEE **72**(9), 1192–1201 (1984)

68.70   J.W. Senders: The human operator as a monitor and controller of multidegree of freedom systems. In: *Ergonomics: Major Writings*, ed. by N. Moray (Taylor & Francis, New York 2005)

68.71   N. Cowan: The magical number 4 in short-term memory: A reconsideration of mental storage capacity, Behav. Brain Sci. **6**, 21–41 (2000)

68.72   G.A. Miller: The magical number seven, plus or minus two: Some limits on our capacity for processing information, Psychol. Rev. **63**, 81–97 (1956)

68.73   K. Koffka: *Principles of Gestalt Psychology* (Harcourt, Orlando 1967)

68.74   E.J. McCormick, M.S. Sanders: *Human Factors in Engineering and Design* (McGraw-Hill, New York 1982)

68.75   P.M. Fitts, R.E. Jones, J.L. Milton: Eye movements of aircraft pilots during instrument-landing approaches. In: *Ergonomics: Major Writings*, ed. by N. Moray (Taylor & Francis, New York 2005)

68.76   W.J. Horrey, C.D. Wickens, K.P. Consalus: Modeling drivers' visual attention allocation while interacting with in-vehicle technologies, J. Exp. Psychol. Appl. **12**(2), 67–78 (2006)

68.77   M.A. Recarte, L.M. Nunes: Mental workload while driving: Effects on visual search, discrimination, and decision making, J. Exp. Psychol. Appl. **9**(2), 119–137 (2003)

68.78   E.M. Rantanen, J.H. Goldberg: The effect of mental workload on the visual field size and shape, Ergonomics **42**(6), 816–834 (1999)

68.79   H.Z. Tan, R. Gray, J.J. Young, R. Traylor: A haptic back display for attentional and directional cueing, Haptics-e **3**(1), 1–20 (2003)

68.80   J. Rasmussen: Skills, rules, and knowledge; signals, signs, and symbols, and other distinctions in human performance models, Syst. Des. Hum. Interact. Table of Contents (1987) pp. 291–300

68.81   P.A. Hancock, G. Williams, C.M. Manning: Influence of task demand characteristics on workload and performance, Int. J. Aviat. Psychol. **5**(1), 63–86 (1995)

68.82   K.H. Kroemer: Avoiding cumulative trauma disorders in shops and offices, Am. Ind. Hyg. Assoc. J. **53**(9), 596–604 (1992)

68.83   H.B. Kroemer: *Ergonomics: How to Design for Ease and Efficiency* (Prentice Hall, Englewood Cliffs 1994)

68.84   M.R. Lehto, J.R. Buck: *Introduction to Human Factors and Ergonomics for Engineers* (Lawrence Erlbaum, New York 2007)

68.85   H.R. Blackwell: Development and use of a quantitative method for specification of interior illuminating levels on the basis of performance data, Illum. Eng. **54**, 317–353 (1959)

68.86   R.P. Hemenger: Intraocular light scatter in normal vision loss with age, Appl. Opt. **23**, 1972–1974 (1984)

68.87   J. Edworthy, A. Adams: *Warning Design: A Research Prospective* (Taylor & Francis, London 1996)

68.88   E.M. Clarke, J.M. Wing: Formal methods: state of the art and future directions, ACM Comput. Surv. (CSUR) **28**(4), 626–643 (1996)

68.89   J. Cheesman, J. Daniels: *UML Components: A Simple Process for Specifying Component-Based Software* (Addison-Wesley, Reading 2000)

68.90   J.P. Bowen, M.G. Hinchey: Seven more myths of formal methods, Softw. IEEE **12**(4), 34–41 (1995)

68.91   N.G. Leveson: *Safeware: System Safety and Computers* (Addison-Wesley, Reading 1995)

68.92   Federal Aviation Administration: *Certification of Transport Airplane Mechanical Systems (AC 25-22)* (FAA, Washington 2000)

68.93   A. Colozza: *Initial Feasibility Assessment of a High Altitude Long Endurance Airship (Report No. NASA/CR–2003-212724)* (NASA, Washington 2003)

69. Space and Exploration Automation

Edward Tunstel

Space-faring nations are actively exploring outer space and planetary bodies in our solar system both individually and as collaborators with other nations. In most endeavors, the inherent risk to human life has been mitigated by the use of automation and robotics to conduct space missions. Missions extending from low-Earth orbit to Earth's moon and beyond to destinations throughout the solar system have been successfully conducted. Robots and human astronauts assisted by automated systems have been used on space missions. Infrastructure and service automation in the context of space missions are discussed in this chapter. Automation and robotics have played a substantial role in installing space exploration infrastructure such as Earth-orbiting satellites and space stations occupied for extended periods by astronauts as well as satellites that operate for extended periods in orbit around other planets. General background information about automation and robotics for exploration of space is presented. Challenges of applying automation in space and planetary environments are highlighted for robots that operate in Earth orbit, at the Moon, at Mars, and other

destinations. A look forward to what the future will hold for further space and exploration automation is provided, including mention of advancements in technological capabilities that will be needed to accomplish more ambitious space missions.

A wide variety of solutions to many problems in military and civil space programs are being addressed using some level of automation and robotics. Automation and robotics is personified by the synthesis and application of integrated electromechanical and computer systems that perform useful functions for engineering, science, or industrial applications. Space agencies around the world employ robotic systems instrumented with a variety of sensors and tools as surrogate explorers and automated workers on-orbit, as assistants to astronauts, on remote planetary surfaces, and throughout the solar system. Their systems and missions provide the infrastructure and service automation that enables access

to space and planet destinations throughout our solar system.

Of particular relevance to automation for space exploration are sensing, perception, and control issues associated with operations in space environments and limited support from humans. These issues are relevant in the context of challenges associated with robotic tasks required for a variety of space missions. Some missions are purely focused on scientific discovery. Other missions may require building physical infrastructure for human activity in space. Still others may aim to provide sustenance for humans in eventual habitats away from Earth. In all such cases, sensing, perception, and control issues are relevant for automated and robotic solutions. Applications to future space missions will involve automated exploration, service, and construction tasks to be performed on orbiting space stations, in planetary atmospheres, as well as on and beneath planet surfaces.

Space robots are used to automate tasks in space environments when simpler or less expensive automation solutions are infeasible. The decision to employ robots to automate execution of certain tasks must consider task complexity, constraints of space environments, and an array of certain and potential risks involved. The required level of system autonomy may also dictate the applicability of automation and robotic solutions for use

on space missions. Combinations and various levels of hardware, software, and control provided by humans are brought to bear to perform successful space missions. The relative allocation of task or mission functionality to hardware, software, and humans provides a sense for the degree of automation employed by a given system.

In the remainder of this chapter, space and exploration automation is discussed through the presentation of background thoughts about the field and challenges that define the complexity of space operations for automation and robotics as well as highlights of representative space robots and their applications. Some future directions and technological capabilities needed for future space missions are also noted.

69.1 Space Automation/Robotics Background

Automated systems and robots serve a variety of purposes in space environments whether sharing, augmenting, or extending human activity. Examples include transporting scientific instruments and equipment to space destinations for deployment and operation. They serve as substitutes for human presence in hostile space environments that are inaccessible to humans or difficult for humans to work safely and efficiently within. Like similar automation systems used on Earth, space automation and robotic systems are also useful for per-

Fig. 69.1 The European Space Agency's (ESA) XMM-Newton is the most sensitive x-ray telescope ever built. Its high-technology design uses over 170 wafer-thin cylindrical mirrors spread over three telescopes. This unique x-ray observatory was launched by Ariane 5 from the European spaceport at Kourou in French Guiana on 10 December 1999 (courtesy of ESA)

forming mundane or repetitive tasks that allow humans to focus on more deliberative, cognitive or otherwise complex tasks. Space automation and robotics also facilitates the creation, delivery, and operation of scientific, commercial, and military assets in space, thus enabling the existence and use of facilities that are unavailable or infeasible on Earth (e.g., space telescopes (Fig. 69.1), interplanetary probes, and space laboratories such as the International Space Station and the Russian Mir Space Station).

Robotic space systems are typically supported by human and computer systems on Earth that facilitate their control and operation. Human mission controllers/operators on Earth utilize a comprehensive collection of computer systems, scientific data and information, and in particular software tools that aid decision making. This so-called ground data system represents the connectivity and flow of data and information on Earth as opposed to communication links between Earth and spacecraft in flight or at other planets. It encompasses networks of antennae used for continuous spacecraft tracking [such as the National Aeronautics and Space Administration (NASA) Deep Space Network – a worldwide spacecraft tracking facility managed and operated by NASA's Jet Propulsion Laboratory], facilities that provide communications between the antennae and the locations conducting mission control/operations activities, and a system of computers, software, networks, and procedures that process spacecraft data.

Robotic spacecraft share subsystems that serve or perform the following common functions [69.1–4]:

Fig. 69.2 The twin rovers of the Mars Exploration Rover Mission pose with their ground-breaking predecessor, the flight spare of the Sojourner rover from NASA's 1997 Pathfinder mission (courtesy of NASA Jet Propulsion Laboratory)

Fig. 69.3 Astronaut servicing the Hubble Space Telescope while secured by foot restraints to the end of the Canadarm (Space Shuttle Remote Manipulator System)

propulsion or locomotion, attitude control, power generation and distribution, structural integrity, thermal control, communications, command and data handling, and accommodation of mission-specific payloads. The command and data handling subsystem typically includes a programmable computer that enables automation of a variety of tasks. The types of robotic spacecraft employed in past and current space missions include flyby spacecraft, orbiter spacecraft, atmospheric probe spacecraft, penetrator spacecraft, lander and surface rover spacecraft (Fig. 69.2), observer spacecraft, and communications/navigation spacecraft. This classification is based primarily on functional spectrum of robotic spacecraft mission types. A subset of past and current spacecraft include or carry robotic mechanisms that interact physically with the target environment through manipulation or mobility such as wheeled or legged ground vehicles, and robots with manipulator arms and gripping end-effectors.

The current state of the art for space automation and robotics is limited by the rate of technology development, budgets for the same, and inherent conservatism in space mission risk management as it relates to automation. On-orbit tasks such as space structure assembly, inspection, and maintenance are presently limited to collaborative astronaut and telerobotic execution approaches. For the past two decades, the NASA Space Shuttle Orbiter with Canada's Remote Manipulator System (Canadarm) has assisted astronauts with successful on-orbit servicing tasks such as satellite retrieval, repair, and rescue. This includes servicing of the Hubble Space Telescope involving critical parts replacement and instrument upgrades (Fig. 69.3). The level and success of automation is increasing, however, with recent use of the Canadarm (with a boom that extends its reach) to assist with visual on-orbit inspection of heat shield tiles on the underside of the Space Shuttle Orbiter in 2006, and with a recent successful demonstration of autonomous on-orbit fuel and battery transfer from one spacecraft to another in 2007. The latter was performed by the Boeing Orbital Express system, which consists of an autonomous on-orbit servicing spacecraft and a serviceable client spacecraft. Surface rovers are reasonably advanced in autonomy but currently limited to autonomous traversal of hundreds of meters per day over relatively benign terrain.

69.2 Challenges of Space Automation

Space systems are subject to strict limitations on power, communications, data processing, and hardware re-

sources in general, not to mention mass and volume. Sensor devices and systems, in particular, must comply with certain specifications of size and payload capacity, onboard power availability, thermal limitations, and solar radiation tolerance. Such constraints are imposed by characteristics of the target space environment and mission objectives among other things, such as the rigors of space travel. In addition, electronics (including associated embedded computing) should be qualified to survive and operate within the harsh temperature and radiation extremes of outer space and similar extremes on planet surfaces with thin atmospheres (such as Mars). Mechanical assemblies and components with few or no moving parts are preferred since they stand a better chance of surviving the vibration and gravity forces of spacecraft launch, atmospheric entry, and landing associated with robotic flights from Earth to other planet surfaces. The nature of certain space missions or sheer distances to their destinations require robotic spacecraft to operate on their own for extended time frames without communication with human mission operators. Thus, certain levels of autonomy are required.

A fundamental challenge of robotic automation in space and on-orbit is the need to operate in microgravity wherein the effects of forces are quite different than in Earth gravity. Robotic operations in space are characterized by slow motions that facilitate compensation for undesired forces and moments due to microgravity. Robot manipulator arms and mechanisms typically utilize computer-vision-based perception, proximity sensors (based on infrared or electrical capacitance, for example), and force/torque sensors to achieve gross and fine manipulation. Dexterity for the latter is often enabled by additional mechanical degrees of freedom and/or advanced automatic control techniques. The same is true for manipulation mechanisms affixed to mobile robotic platforms.

Navigation by mobile robotic platforms, or rovers, on planet surfaces presents several challenges. Rovers must be able to detect hazards and assess the traversability of the terrain. Additional problems to be addressed for navigation include maintaining knowledge of rover pose (three-dimensional (3-D) position and 3-D orientation), as well as stability and traction as it traverses terrain or maps the local environment and prominent landmarks relevant to the mission. Sensing and perception for mobility and navigation are essential for missions requiring land reconnaissance or survey for the purpose of scientific exploration as well as for site construction and operations. A variety of sensors are commonly employed to measure rover motions, position with respect to the environment, and the presence of or range to mobility obstacles. Existing sensor types

are as varied as the physical phenomena underlying their operation (e.g., optical, inertial, magnetic). Despite the large variety of available sensor technologies, planetary rover sensor systems are often restricted to few types due to limitations and/or constraints related to mass, volume, power, and operability/survivability in space environments; for example, on Earth, a common sensor for position estimation and localization is the satellite-based global positioning system (GPS), which provides accurate outdoor position knowledge in terrain where sufficient reception of signals from Earth satellites can be achieved. Orbiting satellites at Mars that have sensors of sufficient spatial resolution may be used to facilitate rover position, but no existing analogue to Earth's GPS is available at Mars yet. This is a major limiting factor of localization accuracy achievement for rover systems there. Robust surface navigation requires accurate heading measurement or very good heading estimation. Errors in heading estimates directly affect the accuracy of rover position estimates. Optical encoders, potentiometers, or resolvers are often used to measure revolutions of robot wheels. Such measurements allow heading and position estimation via dead reckoning or wheel odometry (based on the robot kinematics) relative to some known starting position and heading. However, in rough and rugged outdoor terrain these estimates are only reliable over short distances. Thus, for long-range traverses, more reliable heading information is required. To achieve this, Earth-based robots often employ magnetic devices, such as compasses, which provide good absolute heading measurements with respect to true north of Earth's magnetic field. Magnetic heading sensors are not as useful on Mars, however, due to the planet's negligible magnetic field.

Some of the key technical areas receiving focused attention by the space automation developers include: sensing and perception, robotics on-orbit, in-space, and subsurface as well as increased automation of space systems in general. Key points in each of these areas are discussed in turn below.

69.2.1 Sensing and Perception for Manipulation and Mobility

Modeling, simulating, and/or predicting the functional behavior of orbiter and flyby spacecraft is facilitated by reasonably well-behaved dynamics and operating environments. For such robotic spacecraft, conventional estimation and control techniques have similar effects on spacecraft behavior in simulation as they do in reality. This is due to the fact that the physical laws of orbital mechanics and planetary atmospheric aerodynamics are reasonably well understood and well behaved in space. For surface rovers, the interactions between their mobility systems and planet surfaces are complex and sometimes further complicated by reduced-gravity effects. The result is nondeterministic behavior as the system interacts with the world and increased uncertainty in how the autonomous mobility system will respond to operational commands. As such, mobility and navigation problems for rovers are characterized by high levels of difficulty and increased measurement uncertainty. Common mobility and navigation sensors often inadequately handle the tremendous variability of surface features and properties of natural terrain. Advanced sensing and perception techniques are often required for detection or measurement of significant wheel–terrain interactions such as slippage and sinkage, and assessment or measurement of certain terrain properties prior to engagement by the rover.

Robotic Manipulation

Mechanical manipulators or arms with multiple degrees of freedom are useful on a variety of robots operating in space, on-orbit, and on planet surfaces. They include fixed-base manipulators on space shuttles, space stations, landers, and rovers as well as manipulation appendages on free-flying robots. Robotic tasks for this variety of systems may require fine positioning and dexterous manipulation of equipment or the environment. In order to accomplish manipulation tasks, robots typically execute closed-loop feedback control of arms/mechanisms, with the essential feedback provided by appropriate sensors. Further considerations for achieving reliable manipulation include avoiding collisions of the manipulator with the robot itself, avoiding unintended contact with the environment, and controlling forces during intended contact with the environment. Oftentimes, it is also necessary to compensate for errors associated with degradations and/or changes in the manipulator hardware due to environmental factors such as spacecraft launch/landing vibrations and material thermal expansion, which can affect the ability to accurately position a manipulator at a target of interest. Sensing solutions that provide adequate coverage and resolution for both gross and fine manipulation are essential for missions requiring object handling and transport, servicing/repair, assembly/disassembly, sample acquisition, and in situ acquisition of scientific measurements. Additional information on robotic manipulation, and automation mobility and navigation can be found in Chap. 16.

Wheel Slip and Sinkage

Wheeled mobility systems are subject to undesirable wheel–terrain interactions that cause wheels to slip on rocks and soil. Frequent loss of traction due to wheel slip during traverses from one place to another will detract significantly from the ability to maintain good rover position estimates. Computer-vision-based motion and pose estimation, or visual odometry, is one viable sensing solution. Current methods require substantial computation but may be feasible for future, relatively fast-moving rovers if realized in hardware or firmware with cameras that have dedicated embedded processors.

It is highly desirable to have a capability to sense wheel slippage so that corrective control actions may be taken. Barring the capability to directly measure wheel slip, measurement of over-the-ground rover speed would allow calculation of percentage slip using an analytical relationship between slip, over-the-ground speed, and wheel linear speed (as derived from encoder readings or tachometers, for example). The problem here is that true over-the-ground speed is also difficult to measure. If a rover attempts to drive forward while all wheels are slipping, interpretation of wheel encoder readings alone will indicate forward progress, but in all probability the rover position will not have changed significantly. Some Doppler and millimeter-wave radar solutions exist for measuring true vehicle speed, but for rover use they must be low-power and low-mass devices.

In soft soils, loss of traction due to excessive wheel slippage can also lead to wheel sinkage and ultimately vehicle entrapment. It is possible for wheels to sink to soil depths sufficient to prohibit rover progress over terrain, thus trapping the vehicle at one location. This is also possible on soils with insufficient bearing strength to support the rover (incidentally, a property to which a look-ahead visual perception system may be insensitive). Like wheel slip, it is desirable to have a capability to sense excessive wheel sinkage so that corrective controls may be executed before immobility results. Sinkage measurements are also valuable for reducing position estimation errors. A means to measure wheel sinkage permits overall reductions in the effects of propagating nonsystematic error during rover traverses on varied terrain.

Terrain Properties

Capabilities for noncontact sensing of terrain properties such as hardness or bearing strength are needed for detecting nongeometric mobility hazards. Passive stereo vision or other ranging systems available on existing rovers cannot detect pits filled with loose drift material that may have insufficient load-bearing strength to support a rover. Such hazards are known to exist on the Martian surface and need to be detected before a rover engages them, potentially meeting with catastrophe. Downward-looking impulse radar can be employed as a proximity sensor for this problem. Similar millimeter-wave and microwave ranging sensors are also useful for general obstacle detection and collision avoidance in environments subject to dust, blowing sand, and other all-weather conditions. However, the mass of available units may be prohibitive for some planetary rover applications, and like the speed measurement radars mentioned above, candidate radars for terrain property sensing must be low-power and low-mass devices.

It is also possible to infer nongeometric terrain properties from sensor data. Some viable approaches consider terrain appearance in camera images using image texture differences for soil and rocks and other approaches employ proprioceptive sensing (e.g., of vibrations) to resolve differences in *feel* of the terrain type beneath rover wheels. The latter approach applies once the terrain in question has been engaged and is therefore more useful for intelligent, reactive traction control than for nongeometric mobility hazard avoidance. The look-ahead, image texture-based methods, in addition to their utility for appearance-based hazard detection, are also useful for intelligent predictive traction control on varied, but traversable, terrain surfaces.

69.2.2 On–Orbit and In–Space Robotics

Space robots have been used in low-Earth orbit for tasks of automated servicing, inspection, and assembly. Satellites require servicing for various reasons, each calling for various forms of service action to be taken. Failure of defective hardware may call for repair or replacement of components or modules. Anticipated depletion of fuel or propellant may call for refueling in space if the satellite's mission is intended to continue. When a satellite's lifetime or mission expires it may be necessary to remove it from orbit. On-orbit robotic servicing applies to repair/replacement and refueling activities.

One of the automation challenges is the physical capture of free-floating satellites for retrieval and subsequent servicing. Other challenges are related to the general area of intelligent sensing and associated data processing to ensure safe operations. Servicing activities are representative of maintenance tasks that

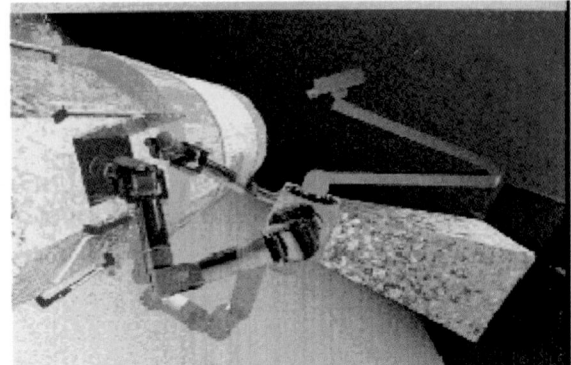

Fig. 69.4 Illustration of robotic maintenance in space

are slated for on-orbit and in-space robots (Fig. 69.4). Other relevant tasks fall into the categories of assembly of space structures and their post-assembly inspection (Fig. 69.5), as well as assistance to astronauts performing extravehicular activity (EVA).

In assembly operations, space robots would be used to transport structural elements and mate elements that are part of a larger space structure. Such tasks call for rendezvous and docking, dexterous manipulation, and fine motion control for execution of detailed assembly sequences. In-space assembly may be supported by automated planning on space robot computers and on off-board computers, or via human teleoperation. Some space structures will have dedicated robotic systems as part of their fully assembled design. An example is the International Space Station, which (when fully assembled) will have robotic systems including three main manipulators, two small dexterous arms, and a mobile base/transporter system.

Robots that perform important inspection tasks are also needed on-orbit to examine space station and

Fig. 69.5 Illustration of robotic inspection in space

spacecraft external hardware and outer surfaces for damage or other problems. This may be done using robotic vision supported by mobility algorithms designed to ensure required coverage of critical surfaces. Both free-flying and legged space robots that walk on space structure exteriors can be employed. When astronauts must leave the safer confines of space stations or spacecraft to perform EVA, space robots will support their activities by performing tasks such as retrieval and hand-off of tools for astronaut use, monitoring of EVA operations, serving as additional eyes or illumination sources to facilitate EVA in areas that are hard to access, or working cooperatively as direct EVA participants. Research is underway to address necessary challenges and technology developments applicable to a range of system entities from Earth-based operations infrastructures to autonomous robotic operations.

69.2.3 Subsurface Robotics

The solar system exploration agenda for the world's space agencies includes plans to access and sample the subsurface environments of terrestrial planets, moons, and small bodies such as comets and asteroids. This will be yet another class of tasks for space robots. Near-term focus is on the Moon, Mars, and Jupiter's moon Europa,

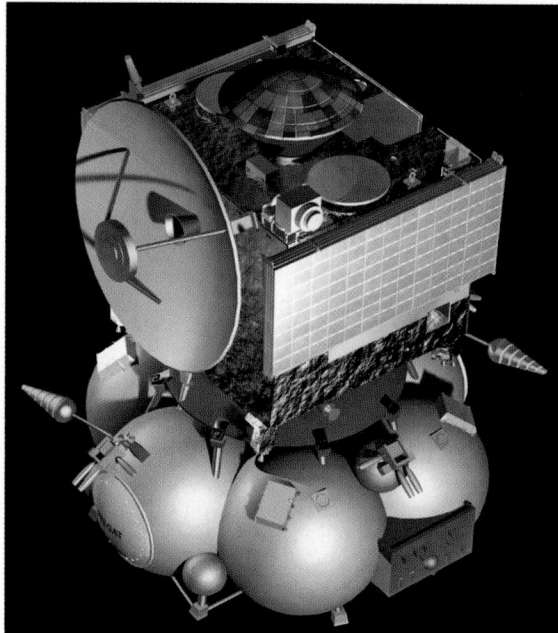

Fig. 69.6 Mars Express spacecraft (courtesy of ESA)

the surface of which is ice-covered and the subsurface of which is presumed to be an ocean of water possibly harboring life. The subsurface of Mars is of interest because it is believed to be the only place on the planet in which liquid water potentially exists and because its sedimentary rock layers are believed to contain information about the geologic, hydrologic, and climatic evolution of the planet as well as any signs of extinct life (Fig. 69.6). Subsurface water in its various forms on Mars is also considered a critical resource for sustaining robotic outposts and future human explorers. Water-ice and structure of the lunar subsurface are of interest for similar reasons. Along the same vein, the potential for an ocean and life on Europa make its scientific interest quite clear.

Today there is significant focus on ways and means to drill and bore into the surfaces of these bodies using automation and robotics, and there are many challenges to overcome. Most challenges being addressed presently relate to understanding the physics and mechanical system requirements for drilling into consolidated and unconsolidated materials comprised of soil, rock, and/or ice to both shallow and deep depths. Based on the findings of various such studies, associated requirements for intelligent control and autonomy for robotic subsurface explorers are being pondered as well by planetary scientists and engineers alike (including aerospace, robotics, and Earth mining engineers).

Robotics challenges presented by drilling mechanics largely relate to control issues. After years of experience drilling into the Earth, the best techniques are still more of an art form than an engineering procedure. Humans are typically intimately involved in determining and implementing the recovery procedures when problems occur that inhibit drilling progress. The art form mastered by human drill operators translates, for space robotics, into a sensor-feedback-intensive, knowledge-based, and reactive control task that will need to be fully autonomous since similar levels of human supervision will not be possible on remote surface missions. Robust capabilities of automated monitoring and diagnosis with safe and reliable autorecovery algorithms are necessary. At the very least, human-equivalent performance by robotic subsurface devices may be needed for most tasks. As with subsurface drilling on the Moon or Mars, locomotion through ice followed by subsea navigation will require high levels of robotic autonomy moving beyond intelligent control and onto adaptive and learning systems.

69.2.4 Automation

Robotics typically refers to electromechanical systems that embody some level of programmable intelligence and that interact with or within physical environments. Robots often operate in environments that are largely unknown a priori and may be unstructured, thereby requiring sensing and perception to compute or infer intelligent actions. Automation often refers to automated devices or techniques that somehow improve the performance or utility of a system or process. Automated systems often operate in structured environments, and their operation may or may not be physical.

Automation for space systems is often applied to reduce costs or enable new mission capabilities. To date, some of the most advanced examples center on the use of robotics as the form of automation, as nonrobotic human spaceflight missions have relied to a large extent on the capabilities of humans to achieve mission success. Automation has also been used for more common actions such as spacecraft antenna and solar panel deployment enabled by automatic mechanisms and devices. In the present and future, the use of nonrobotic automation is on the rise and comes in a variety of forms. Applications range from more pervasive software-based automation of data processing functions to incorporation of robotic capabilities into space suits intended to augment and amplify human capabilities in space. On some space missions, collection of scientific data as well as its processing and analysis is also handled by automation of science instrument systems. Some of the means of automating such functions include computer vision, pattern recognition and classification techniques, and automated reasoning methods of artificial intelligence.

Other activities and required infrastructure for future space missions call for automation of the manufacturing and construction variety. This is the case for outposts on other planetary surfaces or in space where humans would live and work for extended periods. Astronauts living and working in space on space stations are typically surrounded by automated systems that run life-support systems and perform functions formerly performed by themselves or their human mission controllers based on Earth. Example tasks for space station automation solutions include initiation and monitoring of experiments as well as detection and diagnosis of system failures. On planetary surfaces, the cost-effective approach to sustenance of astronauts includes living off the land by using in situ resources; for example, automated systems and production plants would collect and process soil to extract water. Outpost automation would manufacture materials needed for life support systems and would produce electricity and fuels to power equipment and facilities. Implementation of automated systems along these lines has been contemplated and extensively studied for many years [69.5] and will be essential for long-duration human presence in space.

69.3 Past and Present Space Robots and Applications

Over the past 30 years, the exploration activities of international space agencies have given automation technology a proving ground for demonstrating its utility and practicality in space. In the 1970s the former USSR landed rovers on the Moon and teleoperated them from Earth (Fig. 69.7). Since the early 1980s numerous NASA space shuttle flights have consistently involved teleoperation of the Canadarm. Space structure and scientific hardware deployments, assembly, servicing, and repair have been some of the primary uses of space automation on such missions in low-Earth orbit. Such on-orbit uses of automation and robotics have involved performance of tasks by astronauts and robots teleoperated by astronauts, with the most complex work performed by suited astronauts during EVA, that is, on *spacewalks* outside of the Space Shuttle Orbiter. Other robotic systems have been used in this general class of application scenarios for tasks associated with automation of science experiments performed in space-based laboratories. In the early 1990s, Germany's ROTEX technology experiment demonstrated the use of a mechanical manipulator remotely controlled from Earth while on the NASA Space Shuttle Orbiter as well as remotely controlled by astronauts on the Space Shuttle Orbiter. Japan's Experimental Test Satellite VII demonstrated automation and robotic servicing technologies on-orbit in the late 1990s. NASA's free-flying Autonomous Extravehicular Robotic Camera demonstrated a means to provide additional viewpoints or *eyes* for astronauts during EVAs. Additional systems have been proposed and developed for space automation but never flown for various

Fig. 69.7 Russian lunar rover, Lunokhod 2

Fig. 69.8 Concept of the Mars Reconnaissance Orbiter (courtesy of NASA/JPL)

Fig. 69.9a,b Mars exploration rovers Spirit (**a**) and Opportunity (**b**) on Mars (special-effects images created using photorealistic rover models and image mosaics acquired during their missions)

reasons such as budget-related cancelations (e.g., the Orbital Maneuvering Vehicle, the Flight Telerobotic Servicer, and Ranger).

With the exception of the Moon missions of the 1970s the aforementioned automation and robotics activities have been performed in low-Earth orbit. Space automation has been applied in geostationary orbit around the Earth as well. The many communications, scientific, and military satellites in various Earth orbits today are great examples of space automation that have directly affected quality of life on Earth. Robotic tasks that can be automated in geostationary orbit include satellite servicing and deployment/assembly of large space structures. However, since astronauts have, to date, been constrained by technology to operations

in low-Earth orbit, robotic systems are more critical for automated activities in geostationary orbit.

Moving further out into space beyond Earth, robotics continues to play a critical role in expanding our knowledge of the solar system. Space agencies have employed numerous interplanetary robotic spacecraft, planetary probes, surface landers, and rovers that have collected a substantial amount of data leading to our current understanding of the solar system [69.6]. Destinations visited include nearly all of the planets as well as asteroids and comets. The use of robotic technology in space has reached a level where, presently, multiple robotic spacecraft are operating at any given time throughout the solar system. At Mars alone, multiple robotic spacecraft including satellites, landers, and rovers are in operation today, performing scientific missions aimed at understanding the past or present habitability of that planet. The most recent orbiting spacecraft that are still in operation at Mars were delivered by the USA and Europe and include Mars Odyssey, Mars Reconnaissance Orbiter (Fig. 69.8), and Mars Express (Fig. 69.6). The most recent rovers that are still in operation are NASA's twin Mars exploration rovers, Spirit and Opportunity (Fig. 69.9), which have been operating at different locations on the Martian surface since 2004. They have both collected data leading scientists to conclude that liquid water once flowed on the surface of Mars at their respective landing sites.

69.4 Future Directions and Capability Needs

Relatively few places on Earth's moon have been explored by the rovers of the 1970s and by astronauts during the series of NASA Apollo missions. As such, relatively little is known about our Moon. Today, the world's space agencies and commercial sectors are turning a renewed focus on the Moon for further scientific exploration, lunar resource-related commercial industry, space tourism, and as a proving ground for enabling future human exploration of Mars. Automation and robotics are pervasive as tools to be used by human explorers to perform a variety of tasks involving rovers and automated equipment for:

- Surface exploration and transportation/deployment of instruments and systems
- Acquisition of scientific samples
- Industrial mining for resources
- Moonbase construction and assembly of facilities
- Overall maintenance and repair of lunar infrastructure.

On the lunar surface, networked teams of robots will work cooperatively to prepare sites for human habitats at permanent lunar settlements. Robots will work with automated machines and instruments as an integrated system that constructs, services, monitors, and repairs lunar outpost facilities. Such tasks may be shared with astronauts or lunar settlers. Robotic systems will also help to automate prospecting, mining, and processing of lunar raw materials to help sustain human presence by minimizing dependence on Earth support. Automation and robotic technologies to be developed toward this end will be used in similar ways to facilitate eventual human exploration and settlement of locations on Mars.

In space, free-flying space robots will advance beyond the current state of the art to perform routine servicing of space assets such as Earth satellites and space stations (Fig. 69.10). Free-flying robots will work cooperatively with astronauts to construct large scientific and engineering space structures. Advances in the following areas will be needed:

- Mobility on space structures
- Rendezvous and docking with spacecraft and structures

Fig. 69.10 MetOp-A satellite undergoing final testing at EADS Astrium's facilities in Toulouse. MetOp-A was launched on 19 October 2006 from the Baikonur Cosmodrome in Kazakhstan, on a Soyuz ST rocket with a Fregat upper stage (courtesy of ESA)

- Dexterous and force-controlled object grasping, manipulation, and transport
- Human–robot task planning and sequencing.

Planetary surface robotic systems will advance to the next level of autonomous capability enabled by new and improved sensing solutions. Advances are required to reduce the computational complexity of computer vision algorithms or to increase the computational speed of radiation-hardened processors that they run on. In addition to devices for sensing or detecting wheel slip, wheel sinkage, and terrain properties for planetary surface mobility, reliable devices and software are needed that improve existing capabilities for:

- Sensing large-scale terrain discontinuities such as cliffs, craters, and escarpments
- Optical ranging in both full sun and deep shadow
- Distributed sensing in multiple-rover applications.

69.5 Summary and Conclusion

An overview of space and exploration automation and various associated challenges was presented with relevance to robotics applications for on-orbit, in-space, and planetary domains. Representative space robots and applications were noted followed by indications of future directions for space automation and technological capabilities that will be needed to successfully perform future space missions.

During the past several decades space automation has been relied upon, either solely or in concert with resident human capability, to explore and reveal scientific knowledge of space and the planets. The levels of automation or degrees of autonomy have varied from mission to mission and will necessarily increase in the future as more ambitious space missions are pursued.

Autonomous mobility and manipulation are fundamental functions for space robotic tasks. Viable solutions that support these functions within formidable constraints of space computing hardware drive the technological capabilities of space robots, and ultimately, mission success. Application domains include planetary surfaces, outposts/settlements, on-orbit/in-space operations, and subsurface environments. The challenges are associated with robotic tasks including exploration, servicing, infrastructure building and construction, inspection, and assembly in space and planetary environments. A sense of the existing levels of automation and the robotic intelligence or autonomy employed to meet the various challenges is provided.

In general, automation and robotics for space and planetary missions require a variety of mobility and manipulation capabilities supported by viable and adequate means for sensing/perception and sophisticated software algorithms that represent an appropriate level of intelligence for the mission. Solutions must comply with hard constraints on mass, power, mechanical complexity, electrical characteristics, and computing. In time, the greatest needs and remaining limitations will be addressed and overcome by innovations and advances in sensing, perception, and autonomous control engineering. To be sure, space robot intelligence and autonomy will play a critical role in meeting the challenges to come.

69.6 Further Reading

- A.M. Howard, E.W. Tunstel (Eds.): *Intelligence for Space Robotics* (TSI, San Antonio 2006)
- T. Huntsberger, A. Stroupe, B. Kennedy: System of systems for space construction, Proc. IEEE Int. Conf. Syst. Man Cybern., Waikoloa (2005) pp. 3173–3178
- A. Stroupe, T. Huntsberger, B. Kennedy, H. Aghazarian, E. Baumgartner, A. Ganino, M. Garrett, A. Okon, M. Robinson, J. Townsend: Heterogeneous robotic systems for assembly and servicing, Proc. 8th Int. Symp. Artif. Intell. Robot. Autom. Space, Munich (2005)
- P.S. Schenker (guest Ed.): Special issue on robots in space, Autonom. Robots **14**(2/3), 99–263 (2003)
- B. Sommer: On-orbit servicing of satellites (OOS) as a major application field – the TECSAS mission, Proc. 54th Int. Astronaut. Congr. Int. Astronaut. Fed. (IAF), Bremen (2003)

- A. Popov: Mission planning on the International Space Station program: concepts and systems, Proc. IEEE Aerosp. Conf., Big Sky (2003)
- N.J. Currie, B. Peacock: International Space Station robotic systems operations – a human factors perspective, 46th Annu. Meet. Hum. Factors Ergon. Soc., Baltimore (2002)
- E. Dupuis, R. Gillett, R. L'Archevêque, J. Richmond: Ground control of International Space Station robots, J. Mach. Intell. Robot. Control **3**(3), 91–98 (2001)
- P.J. Staritz, S. Skaff, C. Urmson, W. Whittaker: Skyworker: a robot for assembly, inspection and maintenance of large scale orbital facilities, Proc. IEEE Int. Conf. Robot. Autom., Seoul (2001) pp. 4180–4185
- D.A. Whelan, E.A. Adler, S.B. Wilson, G.M. Roesler: DARPA Orbital Express program: effecting a revolution in space-based systems, Proc. SPIE Small Payloads in Space, Vol. 4136 (2000) pp. 48–56
- A. Ellery: *An Introduction to Space Robotics* (Springer, Heidelberg 2000)
- M. Oda, K. Kibe, F. Yamagata: ETS-VII – space robot in-orbit experiment satellite, Proc. IEEE Int. Conf. Robot. Autom., Minneapolis (1996) pp. 739–744
- S.B. Skaar (Ed.): *Teleoperation and Robotics in Space* (AIAA, Reston 1994)
- A.K. Bejczy, S.T. Venkataraman, D. Akin (Eds.): Special issue on space robotics, IEEE Trans. Robot. Autom. **9**(5), 524–704 (1993)
- A.A. Desrochers (Ed.): *Intelligent Robotic Systems for Space Exploration* (Kluwer, Boston 1992)
- T. Sheridan: *Telerobotics, Automation, and Human Supervisory Control* (MIT, Cambridge 1992)

References

69.1   A. Ellery: Space robotics, Part 1, Int. J. Adv. Robot. Syst. **1**(2), 118–121 (2004)

69.2   A. Ellery: Space robotics, Part 2, Int. J. Adv. Robot. Syst. **1**(3), 213–216 (2004)

69.3   A. Ellery: Space robotics, Part 3, Int. J. Adv. Robot. Syst. **1**(4), 303–307 (2004)

69.4   P. Putz: Space robotics, Rep. Prog. Phys. **65**, 421–463 (2002)

69.5   R.A. Freitas Jr., T.J. Healy, J.E. Long: Advanced automation for space missions, Proc. 7th Int. Joint Conf. Artif. Intell., Vancouver 1981 (William Kaufmann, New York 1981) pp. 803–808

69.6   M. van Pelt: *Space Invaders: How Robotic Spacecraft Explore the Solar System* (Springer, New York 2006)

70. Cleaning Automation

Norbert Elkmann, Justus Hortig, Markus Fritzsche

The potential applications of automation for cleaning are many and diverse. All over the world, research organizations and companies are developing automatic cleaning systems [70.1]. Products such as automatic floor cleaning robots and floor vacuum cleaners available for household use are sold ten thousand times over every year at prices below US$300 (Fig. 70.1). While versatile, high-performance systems exist for other applications such as professional floor cleaning, airplane washing, ship cleaning, and facade cleaning, they are by no means as widespread as household systems.

Automatic cleaning systems are frequently extremely complex robot systems that operate autonomously in unstructured environments or outdoor areas. Cleaning automation not only incorporates cleaning engineering but also a variety of other technical disciplines, e.g., autonomous power supply, sensor systems, environment modeling, and path planning in dynamic environments.

Some examples of automatic cleaning systems for floors, facades, swimming pools, ventilation ducts, and sewer lines serve to highlight the current potential of cleaning automation and provide a glimpse of future developments.

Humans typically experience cleaning as monotonous work, which by its very nature is performed in *dirty environments*. Moreover, cleaning can sometimes even be hazardous to health or life endangering, depending on the area or object and the type of cleaning involved. Nonetheless, such areas may require regular cleaning.

Cleaning therefore ideally meets the premises for applying robots or remote-controlled systems and is inherently a typical service robot application. Unsurprisingly, developments over the last 20 years have been aimed at automating cleaning systems. The range of systems available varies widely. Apart from floor cleaning systems, other systems also clean facades, swimming pools, ventilation ducts, and sewer lines. Some of them are extremely complex.

Fig. 70.1 iRobot Roomba

Mass markets for cleaning robot applications have already developed in some sectors. Vacuuming robots for household use represent one of the most widely sold robot systems worldwide. Their sheer numbers and low purchase prices of less than US $300 account for their great commercial success in the household sector [70.2, 3]. Residential users have far lower demands on cleaning quality and, above all, cleaning speed than professional users. In the 1990s, several manufactures of cleaning machines throughout the world developed autonomous cleaning robots for professional floor cleaning. Nonetheless, these systems have not yet become established on the market for a variety of reasons. Less flexibility than humans, high acquisition costs, low availability, and complexity of operation often militate against the use of these systems.

Cleaning robots [70.1] and floor cleaning systems (Fig. 70.1) in particular share many commonalities with other service robots, e.g., for transport or monitoring tasks. Particular points of intersection are sensor systems for obstacle detection and environment modeling, power supply, path planning and execution, and human–machine interfaces. Thus, cleaning robots constitute a preliminary stage to complex household applications for service robots or applications with direct *human–robot interaction*.

Cleaning robots for facades, pipes, ventilation ducts, and sewer lines are however not mass-produced items. These systems are specially optimized for the requirements and geometry of the surface or object being cleaned and are used exclusively in professional environments rather than in the residential sector.

A robot's features or technical innovations are less crucial to its acceptance than its cleaning efficiency and cost effectiveness. A system's flexibility and ease of operation are other important criteria for acceptance.

70.1 Background and Cleaning Automation Theory

Cleaning robots incorporate a multitude of basic developments and theories of robotics, e.g., mobility and navigation, communication, sensors and sensor networks, robotics and intelligent machines, and teleoperation, which are treated in conjunction with mobile robots in Chaps. 9 and 11 of this Handbook.

Cleaning robots have different levels of automation, ranging from *remote-controlled systems* with inexpensive, individual sensors to *autonomous systems* with complex, multiple sensors for environment modeling and navigation.

All types of cleaning systems share certain technical subsystems:

- Motion platform for the system and cleaning unit
- Control and operating system
- Sensor system for environment modeling (in automatic systems) and obstacle detection
- Power supply
- Communications system
- Cleaning unit and, where applicable, suction and material processing units
- Various safety devices (collision avoidance systems for floor cleaning systems, securing or recovery ropes for facade, pipe, duct, sewer, and pool cleaning systems)

All automated cleaning systems draw on established cleaning methods and technologies. An automated cleaning system however cannot inspect its *cleaning quality* as easily as a human can. At present, mobile systems that clean areas with varying optical conditions and textures cannot measure the level of dirtiness reliably. They can, however, check the quantity of dirt picked up over a unit of time.

Cleaning robots from the different fields of application differ widely in terms of requirements and technical challenges. Basically, two categories can be distinguished and the underlying requirements subsumed in these two groups:

- *Floor cleaning robots*: Such cleaning systems relieve humans of monotonous work such as mopping and vacuuming. For their use to be cost effective, these systems must function autonomously and without an operator. Easy operability, high-quality cleaning, and flexible use are basic requirements these systems have to meet.
- *Facade, pool, ventilation duct, and sewer line cleaning robots*: Such cleaning systems are utilized where humans are unable to access an area in need of cleaning or are only able to access it with extreme difficulty. These systems can be engineered to be remote controlled or fully autonomous. Usually, they are customized for a specific application scenario. Nevertheless, easy operability and high-quality cleaning are basic requirements. They must

also be recoverable when cleaning areas that are inaccessible to humans.

Significant features of these two groups of cleaning robots are highlighted below.

70.1.1 Floor Cleaning Robots

Without exception, floor cleaning systems utilize wheel-driven mobile platforms. The configuration of the wheels varies depending on the case of application and the maneuverability requirements. Kinematics with two driven wheels and other caster wheels are often used.

Sensor systems for *obstacle detection* and navigation vary widely. Inexpensive infrared or ultrasonic sensors and contact switches are employed in systems for household use. The sensor data generated does not provide a basis for optimized path planning in an environment and such systems normally change their direction of travel according to a given algorithm intended to optimally cover a surface. It can be assumed that the entire area has been negotiated and cleaned after an appropriate period of operation.

Floor cleaning systems for professional use predominantly contain laser scanners to generate maps and navigate. In this case, path planning and execution must be optimized to clean a maximum surface area within a specific time. Ultrasonic sensors and contact switches are often employed additionally as collision sensors. Objects such as walls, shelves or the like are particularly challenging when cleaning. On the one hand, safeguards are needed to prevent people from getting pinned, for instance, between the cleaning system and a wall. On the other hand, laser scanners' capability to generate precise maps near a shelf is limited. Batteries supply floor cleaning systems their power. To be cost effective, a floor cleaning system must be usable over several hours. The several hours presently required to charge a battery may hinder professional use. Floor cleaning systems must furthermore be equipped with efficient cleaning systems.

70.1.2 Facade, Pool, Ventilation Duct, and Sewer Line Cleaning Robots

Unlike floor cleaning robots that are normally used on level ground and equipped with batteries, dust reservoirs, and water tanks, other cleaning systems for facades, pools, pipes, sewers, and ducts must be supplied power through cables and cleaning medium through hoses and, where necessary, must have a fall arrester system or a recovery rope. The engineering and money required to implement these essential components, which constitute an indispensable infrastructure, are quite substantial compared with floor cleaning systems.

The use of cleaning robots on facades, in pools, pipes, ducts, and sewer lines may require their adaptation to ambient conditions that are highly unusual for automated systems and to environments that are less than ideal for robots. Variable ambient conditions such as humidity, temperature, and light conditions place great demands on components and, for example, necessitate adapting and increasing the redundancy of sensor systems for navigation. Given the high expectations on system reliability, this is particularly important.

Facade Cleaning Robots

Facade cleaning robots are often remote-controlled systems that clean surfaces that are inaccessible to humans or accessible only with great effort. However, isolated fully automatic systems that operate without human supervision are also in use, particularly in Europe. Some of these systems were designed specifically for a building during its planning phase. Remote-controlled systems in particular are designed to be universal and usable on a variety of buildings including their infrastructures without structural modifications.

Facade cleaning *robots* come in many different designs: *wheel-driven* systems for flat and slightly inclined glass roofs, *climbing* systems with vacuum cups for sharply inclined and vertical facades, and *rail-guided systems*. As a rule, facade cleaning robots must be able to navigate obstacles such as window framework and other facade elements. The use of sensors to determine position constitutes a particular challenge since they must deliver reliable data under the widest variety of outdoor weather conditions from rain to sunshine. All the facade cleaning robots in operation today have cables that supply electrical power and, depending on the system, compressed air, water, and data communications. There are no known systems that operate autonomously on facades without an *umbilical*. What is more, measures often have to be taken to secure a robot against falling [70.4]. Such measures have to be integrated into the overall concept and, where necessary, automated for fully automatic systems.

Pool Cleaning Robots

Pool cleaning robots have been in use for some 20 years to clean the bottoms and sides of swimming pools. The various manufacturers' products function on the basis of the same system. A tracked vehicle provides locomotion under water. Rotating brushes mounted on the front and back of the unit loosen dirt, which is then suctioned into a slot on the unit's underside and pumped through a filter. The water intake on the underside additionally increases the contact pressure, thus facilitating controlled movement on the vertical sides of pools.

Pool cleaning robots use a minimum number of sensors to orient themselves and move underwater completely independently. Since pool geometries are usually simple, navigation logic can also be kept simple. Simple sensor arrays on the fronts and backs of these systems are the elements of an efficient cleaning strategy.

Pool cleaning robots must be lightweight to make operator handling easy and to attain sufficient buoyancy for retrieval. Hence, battery operation is often not an option since it would not allow the necessary cleaning performance. Such systems are supplied with power through a cable with which they can be retrieved in the event of damage.

Ventilation Duct and Sewer Line Cleaning Robots

Cleaning systems for pipes, sewers, and ducts represent a sizeable market since their inaccessibility often prevents humans from being able to clean them without cleaning automation. Pipe and ventilation duct diameters are too small or areas may be hazardous to health or potentially explosive, e.g., in the petroleum industry or sewage disposal.

Thousands of different systems for pipe, ducts, and sewer line cleaning exist all over the world. As a rule, they are remote controlled or semiautomatic and move on wheels or tracks. They normally do not navigate autonomously and incorporate video cameras to display their environment to the operator. A cable connects these systems to a supply and control station. While this limits the systems' radius of action, it assures they are highly reliable, are supplied cleaning medium and can be recovered from pipes, sewers, and ducts with certainty. Cleaning methods vary widely depending on the case of application and include brushes, water, high water pressure, and dry ice.

70.2 Examples of Application

The automatic systems presented here are established systems that have achieved product maturity and, for the most part, been in operation for years. In addition to special applications for which very few systems are available worldwide, there are also applications for which a mass market has already opened. Accordingly, the systems cited here merely represent a few examples of the wide range of cleaning automation products. Space constraints only allow the description of cleaning systems presently in operation and preclude covering the multitude of prototype developments and experimental models.

70.2.1 Floor Cleaning Systems

A distinction must be made between professional and household floor cleaning systems as well as the types of cleaning, i. e., vacuuming and wet cleaning. The Hefter ST82 R floor washing robot is an example of professional wet cleaning, and iRobot's Roomba of household vacuuming. While the technical configuration and performance of other manufacturer's systems differ, the basic concept is comparable.

ST82 R Floor Washing Robot

Manufacturer:	HEFTER Cleantech GmbH, Germany
Type:	Professional floor cleaning system
Operating mode:	Autonomous, taught path
Cleaning technology:	Wet cleaning
Area of application:	Supermarkets, airports, large halls, etc.

Based on a standard floor cleaning system, the Hefter ST82 R floor washing robot is intended for professional cleaning of hard floors (Fig. 70.2). Outfitted with auxiliary localization and collision avoidance sensor systems, a navigation system and an onboard computer, the Hefter ST82 R is able to follow and effectively clean a programmed path fully autonomously and independently of an operator [70.5].

This cleaning robot can run in two operating modes: manual and automatic. In manual mode, the system performs like a hand-guided floor cleaning system in order to teach it the path to be cleaned autonomously and to recover it in the event it malfunctions.

Fig. 70.2 Hefter cleaning robot ST82R

Automatic mode is programmed in two stages. The robot's entire work space is first navigated in manual mode. The robot's onboard navigation software automatically generates a virtual map of the workspace from the data collected by an odometer, a gyrocompass, ultrasonic sensors, and laser scanners during the first reference run.

A second reference run is taken afterward. This time, the intended cleaning route is followed exactly. A cleaning path based on the sensor data is entered into the virtual map. A Siemens SINAS navigation system furnishes the necessary navigation intelligence [70.6].

Once the reference runs have been completed, the cleaning robot is able to navigate the programmed path on its own. To do so, the robot is switched to automatic mode. The operator can no longer intervene in the robot's movement in automatic mode.

Since the robot moves in peopled surroundings, a multitude of safety mechanisms that prevent collisions with and injuries to people have been provided.

Along with switching strips mounted all over the robot, a laser scanner installed in the front skirt is an integral element of the safety concept. If one of the *safety* systems is activated, the robot initially reacts by moving to evade the direction of the source of activation. If this does not cancel the activated safety system, then an emergency stop is triggered. Automatic operation can be resumed only after acknowledgement by an authorized operator.

In hazardous situations the safety system is unable to detect, automatic operation can be interrupted by activating *emergency stop* buttons on the robot.

The Hefter ST82 R floor washing robot system was successfully tested in various retail stores but is no longer available for purchase.

Floor cleaning robots from other manufacturers include:

- The Hako Acromatic is for autonomous professional cleaning in buildings (but is no longer sold) [70.7].
- Fuji Heavy Industries' Subaru RFS1 is the successor to an autonomous floor cleaner sold since 2000. The latest model communicates with the elevator and can thus move independently from floor to floor. It is not designed for manual operation [70.8].

Other systems that at least deserve brief mention are the Comac CLEAN [70.9], Cybernetix Auror and Baror [70.10], Thomson Abilix 500 [70.3], Servus Robots [70.11], CleanFix Robo40 [70.12], Floorbotics [70.13], Robosoft AutoVac C6 and C100 [70.14], and VonSchrader Dolphin [70.15], AUTOMAX AXV-01 [70.16].

Roomba 500 Series Vacuum Cleaning Robot

Manufacturer:	iRobot Corporation, USA
Type:	Household vacuum cleaning system
Operating mode:	Autonomous
Cleaning technology:	Vacuum cleaning
Area of application:	Standard rooms in homes.

Vacuum cleaning is one of the few fields of application for which a mass market has already opened for service robots in general and cleaning robots in particular. The most successful system in this segment is iRobot's Roomba robot (Fig. 70.1), which consists of a cleaning robot and a base station. The circular cleaning robot stands atop three wheels, of which two are drive wheels and one a caster wheel, and navigates an unfamiliar environment completely autonomously. Essentially, the robot operates automatically [70.17].

In principle, it cleans on the basis of a classical vacuum cleaner with a powerful vacuuming unit and rotating *brushes* configured so that even dirt in the robot's boundary area is picked up. A dirt detecting sensor is located in the suction zone.

The path being cleaned does not have to be programmed. Equipped with optical *ranging sensors* and *contact sensors*, the system detects when it nears an obstacle, proceeds toward it at reduced speed, and changes its path direction upon contact. Ranging sensors directed downward detect stairs and drop-offs and generate a change of path direction as well.

Path direction is not changed randomly. The robot utilizes sensor data to analyze its environment and selects one of four motion patterns depending on the situation: spiraling, wall following, room crossing, and dirt detection.

Dirt detection is selected whenever larger quantities of dirt are detected in the suction flow. The robot reacts by increasing suction power and follows a spiral motion pattern until the quantity of dirt drops again.

Virtual walls can be used to limit the robot's workspace. These are generated by stations which emit an *infrared beam*. If its battery charge drops below a critical value or it has finished cleaning, the system returns on its own to its *base station*, which emits an infrared beam that acts as a guide beam for the robot.

Systems from other manufacturers include:

- The Kärcher RC3000 (Germany) is an autonomous vacuuming robot for household use (largely similar in design to the Siemens VSR8000). It is the only household appliance that disposes of collected dirt in its docking station [70.18].
- The Infinuvo Cleanmate QQ2 is an autonomous vacuuming robot for household use. It additionally kills germs with an ultraviolet (UV) lamp and freshens air with fragrance capsules [70.19].
- The iRobot Scooba (USA) is Roomba's successor that not only vacuums but also wet mops [70.17].
- The Electrolux Trilobite (Sweden) was introduced as the first vacuuming robot for household use in 2001 [70.2].

Other systems that at least deserve brief mention are the LG RoboKing, Hanool Ottoro, Black&Decker ZoomBot, and Sharper Image eVac.

70.2.2 Roofs and Facades

A number of facade cleaning robots have moved well beyond the research and prototype stage and are being operated as special developments exclusively on specific facades. Since relatively little time and effort is required to adapt remote-controlled systems, they will quite likely be used on other facades in the near future.

Filius Glass Roof Cleaning Robot

Developer:	Fraunhofer IFF, Germany
Type:	Professional roof cleaning system
Operating mode:	Remote controlled
Cleaning technology:	Wet roof cleaning
Area of application:	Berlin Central Train Station, Germany.

Fig. 70.3 Fraunhofer IFF Filius roof cleaning robot

The Fraunhofer IFF specially developed a remote-controlled cleaning robot for *Berlin central train station* in Germany (Fig. 70.3). However, it can also be used on other roofs with comparable geometries once the appropriate infrastructure has been installed.

Filius is a *remote-controlled, semiautomatic robot* that can navigate obstacles with heights of up to 200 mm. Onboard cable winches secure Filius to a *gantry* on the roof ridge from which it is supplied with power and water through a cable and a hose. As cleaning progresses, the gantry moves incrementally so it is always located above the robot.

Dirt is loosened by a rotating brush mounted on the front, and is then rinsed away by a nozzle bank behind the brush. An operator assumes the control of action functions such as path direction and brush activity by remote control.

Wobble sticks and ultrasonic sensors detect travel across the roof surface and an emergency stop prevents operators from driving the robot over the roof's edge.

Its all-wheel drive makes the robot extremely maneuverable on all terrain. Ramps laid out enable it to navigate the gantry's rails and expansion joints between individual roof areas for instance. Thus, the robot can be positioned and used on the entire surface of the station's roof.

Feeder lines on the roof, fed with high-pressure *softened water* by a supply vehicle, supply water to the gantry.

RobuGlass Glass Roof Cleaning Robot

Manufacturer:	Robosoft, France
Type:	Professional roof cleaning system
Operating mode:	Remote controlled
Cleaning technology:	Wet roof cleaning
Area of application:	Louvre Pyramid (France).

A remote-controlled robot cleans the glass Louvre Pyramid in Paris, France. Combining suction and caterpillar tracks, the robot travels up and down the triangular surfaces. The absence of any overhead securing device is a special feature of the Louvre robot. Technically, it is designed so that the frictional force of the walking mechanism suffices to prevent it from sliding down the sloping surface in the event of a malfunction. Operators at the foot of the pyramid supply it from below through cable and hose, the compressed air needed being provided by equipment located on a supply vehicle. Rotating brushes do the cleaning and a car windshield wiper dries the path of travel during downward travel [70.20].

Compared with other facade cleaning robots, the Louvre system has a very simple design. The minimum of distinctive features of the facade substantially reduces complexity. First, the triangular surfaces are completely level, only interrupted by silicone joints. Consequently, a very simple walking mechanism can be used since there are no obstacles to be navigated. Second, the system cannot fall when it travels across the sloping surfaces, thus making an overhead securing device unnecessary. Safety engineering accounts for a large part of other facade cleaning robots' systems and weight. Third, the sloping smooth surfaces allow water from cleaning to simply run off. This simplifies the cleaning system. Fourth, the space around the pyramid is navigable by a vehicle and the system can be recovered with a small crane at any time. Its relatively short cleaning paths enable supplying the robot from below. The requisite cables and hoses are simply pulled along the glass.

Glass Roof Cleaning Robot for Leipzig's New Exhibition Center

Developer: Fraunhofer IFF, Germany
Type: Professional roof cleaning system
Operating mode: Autonomous
Cleaning technology: Wet roof cleaning
Area of application: Entrance hall of Leipzig's new exhibition center (Germany).

Two *fully automatic cleaning robots* have been cleaning the $25\,000\,\mathrm{m}^2$ glass hall of *Leipzig's new exhibition center* in Germany since 1997 (Fig. 70.4). The roof consists of a glass facade suspended from a steel structure. Accessing it with a gantry alone or other access equipment would be extremely complicated.

Fig. 70.4 Fraunhofer IFF cleaning robot in Leipzig

A *gantry* transports the robots along the roof ridge and uses small hoists to lower them onto the glass surface between the pane mounts. The robots then move downward under the steel trusses and between the mounts, cleaning the glass. Upon returning to the top, the robots are picked up by the hoists and shifted to the next path. A broad roller brush cleans the entire surface of the facade lane by lane, starting at the eastern end of the roof and ending at the western end. Since the robots are unable to move around the mounts, these areas are cleaned by disc brushes on retractable arms. *Chemical-free deionized water* is sprayed onto the glass to moisten and wash away dirt mobilized by the brushes.

The gantry on the roof ridge secures each cleaning robot with two Dyneema ropes and supplies each with power and water through its own cable and hose. To prevent damaging the panes of glass and silicone seals, hose, cable, and securing ropes are coiled and uncoiled inside a robot and thus laid down on the glass instead of being dragged over it. Furthermore, since the bearing wheels are not driven, the two securing ropes are used to correct the robot's direction of travel. A fifth wheel only provides the necessary drive in flat areas.

The steel structure limits the size a robot may have. At an overhead clearance of 38 cm, the robot's height is just 30 cm. The travel path is 45 m long and runs from the ridge between the pane mounts down into the eaves. The distance between mounts limits the robot's width to 1.5 m.

Odometer measurements of the distance covered by two wheels are supplemented by *eddy-current sensor* measurements of the distance covered using the mounts as reference marks. In addition, the distance to the mounts to the robot's left and right is used to correct path direction, controlled by adjusting the coiling of the two ropes. Expansion of the hall due to heat as well as the general tolerances of the mounts' uniformity make navigation between mounts more challenging.

The gantry and robots move fully automatically and are monitored from a master control room where exact positions and actions are displayed. The robots also accept abstract commands from a manual control menu [70.21, 22].

CleanAnt Glass Facade Cleaning Robot

Manufacturer: Niederberger, Switzerland
Type: Professional facade cleaning
                     system
Operating mode: Autonomous
Cleaning technology: Wet facade cleaning.

Consisting of two limbs with suction cups attached to both ends, CleanAnt constitutes a *walking kinematics* with five degrees of freedom (Fig. 70.5). While one foot is fixed on a pane, the second swings into the next walking position or moves over the surface to clean a pane. A control computer synchronously controls and positions the axles, *inverse kinematics* identical to industrial robot controls determining the positions of the joints.

However, its base coordinate system changes when the fixed foot changes.

The system automatically holds itself on the facade being cleaned, thus making it possible to clean vertical, overhanging or even curved facades. A slack rope secures the cleaning system and only goes taut should the system fall, thus preventing it from falling entirely.

Transceiver units on the building's corners, which triangulate current position, monitor its position on a surface. The surface being cleaned is modeled in computer-aided design (CAD) and uploaded to the robot control system. The robot then follows a predefined motion path autonomously.

While the complicated walking pattern, the few degrees of freedom, and the limited size of the cleaning unit prevent the system from covering large areas, it still qualifies as a service robot since it is able to clean surfaces that would otherwise be inaccessible, i.e., it can easily clean vaulted surfaces. Moreover, CleanAnt can move around corners or from wall to ceiling and navigate larger recesses or obstacles.

Wet brush cleaning and semidry cleaning with fleece or dry ice can be applied as cleaning technologies. A hose supplies the requisite media from above or below. For special operations, CleanAnt can also be deposited onto a facade manually, e.g., from a gondola, and maneuvered into specific areas by remote control.

While it is marketed for professional facade cleaning, CleanAnt is the only robot cited here that is not yet in operation. Nonetheless, it deserves mention if only because its design represents such a great departure from conventional robots [70.1, 23].

Fig. 70.5 CleanAnt climbing robot

Fig. 70.6 SIRIUSc facade cleaning robot for automatic cleaning of high-rise buildings

Other facade cleaning robots include:

- The Fraunhofer IFF's SIRIUSc cleaning the *vertical facades* of the Fraunhofer-Gesellschaft's headquarters in Munich, Germany (Fig. 70.6) [70.22, 24, 25] and
- Beihang University's SkyCleaner III in operation on the Shanghai Science and Technology Museum in China [70.26].

70.2.3 Ducts and Sewer Lines

Regular cleaning of sewer lines is a basic measure to ensure they operate reliably. Cleaning of ventilation ducts on the other hand is not essential to operational reliability but arguably to protect people's health. Although sewer lines and ventilation ducts have similar geometric properties, the boundary conditions for cleaning robot operations differ fundamentally. Two systems serve as examples of ventilation duct and sewer line cleaning respectively.

Multipurpose Duct Cleaning Robot
Manufacturer: Danduct Clean, Danmark
Type: Professional duct cleaning system
Operating mode: Remote controlled
Cleaning technology: Brushes, dry-ice cleaning

The Danish company Danduct Clean's multipurpose robot is a universal system that cleans and inspects ventilation ducts. An all-wheel drive robot platform serves as the carrier system. The overall system does not have any collision sensors. However, two cameras directed toward the front and the rear relay a direct impression of the robot's environment. The carrier system is outfitted with different cleaning systems depending on the case of application.

Rotating brushes are used in dry ventilation ducts to remove dust clinging to the walls. Various brush systems are available depending on the duct geometry. A *dry-ice cleaning system* is employed in exhaust areas of kitchens and the like where sizeable grease deposits form in ventilation ducts. Temperatures as low as $-79\,^\circ$C facilitate their removal.

Irrespective of the cleaning system used, cables with lengths of up to 30 m supply media and transmit data to an external control box. The robot is controlled from the control box by joystick based on camera feedback. *Cruise control* is additionally available for long, straight duct sections.

To effectively remove loosened dirt, inflatable balloons seal off the duct section being cleaned. A powerful suction unit is hooked up to the duct opening in the direction of work and extracts loosened dirt from the duct [70.27].

Other duct cleaning robots are manufactured by:

- Indoor Environmental Solutions, Inc. (USA) [70.28]
- DRY ICE Engineering GmbH (Germany) [70.29]
- HANLIM MECHATRONICS (South Korea) [70.30].

Sewer Cleaning Robot
Developer: Fraunhofer IFF, Germany
Type: Professional sewer cleaning system
Operating mode: Automatic
Cleaning technology: High pressure.

Isolated sewer cleaning robots exist for inaccessible sewer lines. Some remote-controlled inspection robots for small sewer line diameters can be outfitted with high-pressure nozzles. As a rule, however, small sewer lines are cleaned with cleaning nozzles that utilize high-pressure water for propulsion.

Contracted by the Emschergenossenschaft in Germany, the Fraunhofer IFF developed a fully automatic cleaning robot for sewer lines with diameters of 1600–2800 mm. The system employs an ejector nozzle to mobilize deposits that are underwater and high-pressure water to clean the pipe wall above the waterline in sewer lines that are 20–40% full at all times. The system can clean sewers up to 750 m in length. The wheel-driven cleaning system is roughly 4 m long and weighs 2.5 t. An *ultrasonic scanner* monitors the cleaning results underwater and a camera monitors the area above water. A specially equipped vehicle on the street level supplies the cleaning system with up to 250 l of water per minute over a distance of up to 750 m with a nozzle pressure of over 100 bar. The cleaning system dependably navigates in the sewer line and system recovery is ensured through the cable connection [70.22, 31].

70.2.4 Swimming Pools

Swimming pools accumulate large quantities of dirt on a daily basis. The relatively large surfaces gather dirt out of the air and off swimmers. Public swimming pools are subject to hygiene codes with strict water quality limit values that necessitate regularly cleaning the bottom and walls of a pool. Underwa-

Fig. 70.7 WEDA B680 pool cleaning robot

ter cleaning machines have been in use since the 1970s. Such pool cleaners qualify as service robots since they move and systematically navigate pools autonomously. While the various manufacturers' systems share virtually the same robot engineering concept, their designs, target markets, and cleaning systems differ.

B680 Pool Cleaning Robot

Manufacturer: Weda, Sweden
Type: Professional pool cleaningdiffer system
Operating mode: Autonomous, remote controlled
Cleaning technology: Brushes, water filter system.

The base of Weda's B680 *pool cleaning robot* is a *tracked vehicle* maneuverable by separately controlling its left and right track (Fig. 70.7). Its forward speed is approximately 0.25 m/s. A *water pump* suctions in approximately 1200 l of water per minute on the underside and rinses it through a reusable *particle filter*. Common systems for residential use only manage approximately 250 l/min. The suction generated beneath the system is sufficient to enable the pool cleaner to traverse vertical walls underwater. Rotating brushes loosen

dirt particles in front of and behind the unit, moving them in the direction of the suction opening. Power is supplied by a cable connected to a base station on the edge of the pool. It is uncoiled and floats on the surface during cleaning.

The robot is internally balanced in such a way that it glides gently downward and lands on its *feet* when lowered into a pool or when it detaches from a side wall.

Electronics in the B680 were kept to a minimum so that the unit is as easy for users to understand as possible. Contact sensors to the front and rear of the robot are used for *navigation* and reverse the drive direction when pressed. The unit has a slightly oblique front bumper that generates the slight change in direction when it contacts a pool wall. As a result, it reaches the opposite end of its path on its return travel somewhat offset. This simple navigation technique enables the complete cleaning of rectangular pool surfaces quicker than through random motion.

A pool cleaner is placed in the water during a pool's off hours and activates itself with a time switch after a short wait, by which time the water has calmed and dirt particles have settled. The cleaning is finished in a few hours and the unit deactivates itself for retrieval from the water the next morning. Control keys execute the requisite maneuver. The filter bag size is dependent upon its planned hours of operation. If it is too small, it clogs quickly, thus diminishing cleaning performance. In such a case, the cleaning robot would merely be embarking on a joyride and would lose traction as suction decreases [70.32].

Other manufacturers of professional and household pool cleaning systems include:

- Maytronics' (Israel) pool robots are intended for residential use and smaller pools. Maytronics offers the only battery-powered robot. Maytronics products are marketed globally under various names [70.33].
- iRobot (USA) offers several models that suction up and filter dirt in water [70.17].
- Aquatools (USA) offers robots for residential use and smaller public pools.
- Mariner 3S (Switzerland) sells professional pool cleaners. Instead of collecting bags, some units use filter cartridges that are cleaned afterward with auxiliary equipment [70.34].

70.3 Emerging Trends

Especially in the domain of cleaning, service robots already provide many different options for relieving people of dangerous, stressful, and/or monotonous work and are penetrating both household and professional market sectors. Household systems have technically simple and low-cost designs and are already being sold in large numbers. Professional systems are technically complex, flexible, cost effective, efficient, and easy to operate. However, since they fail to fulfill the requisite criteria in many cases, they have not yet established themselves as mass products. Nevertheless, numerous individual solutions exist for special applications such as facade or pool cleaning.

To the extent that they do not fully navigate surfaces when geometries are more complex or environ-ments are dynamic and generally can neither navigate themselves nor coordinate tools better than humans, professional cleaning robots' sensory and cognitive capabilities continue to limit their universal and cost-effective use.

Such cleaning robots will not become mass prod-ucts until their cost effectiveness, performance, ef-ficiency, and total attendant costs make them su-perior to manual cleaning. Further development of service robots' cognitive capabilities, environment modeling sensor systems, and multimodal user in-terfaces is being pursued worldwide for other fields of application and is a fundamental prerequisite to establishing cleaning robots in the professional sec-tor.

References

70.1   R.D. Schraft: *Service Robots* (B&T, Munich 2000)
70.2   IFR: *World Robotics Report 2006* (International Fed-eration of Robotics IFR, 2006), www.worldrobotics.com, last cited 2009
70.3   E. Prassler, A. Ritter, C. Schaefer, P. Fiorini: A short history of cleaning robots, Auton. Robot. **9**, 211–226 (2000)
70.4   S. Hirose, K. Kawabe: Ceiling walk climbing robot Ninja-II, 1st Int. Symp. Mobile, Climb. Walk. Robots (Brussels 1998) pp. 143–147
70.5   U. Zechbauer: Der elektronische Saubermann, Pict. Future (Herbst), 59–61 (2002), in German
70.6   G. Lawitzky: A navigation system for cleaning robots, Auton. Robot. **9**(3), 255–260 (2000)
70.7   M. Schofield: Neither master nor slave, 7th Int. Conf. Emerg. Technol. Fact. Autom., Vol. 2 (IEEE, Piscataway 1999) p. 1427
70.8   H. Aoyama: Building cleaning robot system, 1st German–Japanese Summit Mobile Auton. Syst. (Hannovermesse 2008)
70.9   Comac: Verona, Italy (2008) www.comac.it
70.10   Cybernetix: Marseille, France (2008) www.cybernetix.fr/
70.11   Intellibot: Pittsburgh, USA (2008) www.intellibotrobotics.com
70.12   Cleanfix: Cleaning Systems, Wyckoff, ISA (2008) www.cleanfixusa.com/cleanfix-site/robo40.php
70.13   Floorbotics: Northcote, VIC, AUS (2008) www.floorbot.com/
70.14   Robosoft Advanced Robotics Solutions, Bidart, France (2008), www.robosoft.fr
70.15   Von Schrader: Racine, USA (2008) www.vonschrader.com/equipment/carpet/ dolphin/dolphin.htm

70.16   *The Specifications and Applications of Robots in Japan – Non-Manufacturing Fields* (Japan Robot Association, Tokyo 1997) pp. 328–329
70.17   iRobot: *Roomba* (iRobot, Bedford 2008), www.irobot.com
70.18   Kärcher: Winnenden, Germany (2008) www.karcher.com
70.19   Infinuvo: San Jose (2009) www.infinuvo.com/
70.20   A. Kochan: Robot cleans glass roof of louvre pyra-mid, Ind. Robot **32**, 380–382 (2005)
70.21   N. Elkmann, U. Schmucker, T. Boehme, M. Sack: Service robots for facade cleaning, advanced robotics: Beyond 2000, 29th Int. Symp. Robot. (Birmingham 1998) pp. 373–377
70.22   Fraunhofer IFF: Magdeburg, Germany (2008) www.iff.fraunhofer.de/en/robotersysteme.htm
70.23   Serbot: Oberdorf, Switzerland www.serbot.ndswing05.ch/ (2009)
70.24   N. Elkmann, D. Kunst, T. Krueger, M. Lucke, T. Boehme, T. Felsch, T. Stuerze: SIRIUSc – facade cleaning robot for a high-rise building in Mu-nich, Germany, Proc. 7th Int. Conf. CLAWAR 2004 (Springer, 2004) pp. 1033–1040
70.25   N. Elkmann, D. Kunst, T. Krueger, M. Lucke, T. Stuerze: SIRIUSc: fully automatic facade cleaning robot for a high-rise building in Munich, Ger-many, Proc. Jt. Conf. Robot. ISR 2006/Robotik 2006 (Munich 2006) pp. 203–204
70.26   H. Zhang, J. Zhang, W. Wang, R. Liu, G. Zong: Sky cleaner 3 – a real pneumatic climbing robot for glass-wall cleaning, IEEE Robot. Autom. Mag. **13**(1), 32–41 (2006)
70.27   Danduct Clean: Herning, Denmark (2008) www.danduct.com

70.28   Indoor Environmental Solution: Houston, USA (2008) www.cleanducts.com/

70.29   Dry Ice Engineering: Mainhausen, Germany (2008) www.dryiceclean.de

70.30   Hanlim Mechatronics: Gyungki-Do, Korea (2008) www.ductrobot.co.kr/en/page2.html

70.31   N. Elkmann, H. Althoff, S. Kutzner, J. Saenz, T. Stuerze, C. Walter, E. Schulenburg: Automated inspection system for large underground concrete pipes partially filled with waste water, Proc. Jt. Conf. Robot. ISR 2006/Robotik 2006 (Munich 2006) pp. 167–168

70.32   Weda: Södertälje, Sweden (2008) www.weda.se

70.33   Maytronics: Yizreel, Israel (2008) www.maytronics.com

70.34   3S Systemtechnik: Remigen, Switzerland (2008) www.mariner-3s.com

71. Automating Information and Technology Services

Parasuram Balasubramanian

While the information services industry dates back to the 15th century AD, the information technology services is as young as 60 years. Yet the pace at which both have grown is phenomenal, and this is tied to the evolution of computer technology. Networking of computers and people through the Internet has moved into high gear. It has also given birth to multiple business segments in the process. Delivery of data and information, data processing, business process outsourcing, analytics, and printing and display solutions are the five segments identified within the information services. Computer-aided software engineering, independent software testing and quality assurance, package and bespoke software implementation and maintenance, network and security management, and hosting and infrastructure management are covered as segments within information technology services. The automation path of each segment is reviewed in detail. An impact analysis to identify the changing landscape is described. Finally current trends are traced, followed by predictions for future developments.

Part G | 71

71.1 Preamble

The design specification for the electronic voting machines (EVMs) used in 2004 national elections in India had to take account of the high level of illiteracy among citizens, cater for more than 14 different languages, tackle poor infrastructure in terms of road network and power availability in rural polling stations, and accept manual methods of voter authentication [71.1]. Nearly 1.3 million EVMs adhering to the above specification were deployed in the election, in which 330 million citizens voted. The election process went off without any significant problem with respect to the EVMs. Meanwhile in Europe, near Paris, a high-technology company has been engaged in using crash simulation software on Intel personal computers (PCs) to model and test new-generation vehicles to meet regulatory requirements without building expensive prototypes [71.2]. Not to be left behind in the innovation race, many firms in Europe and the USA have been busy delivering portable car navigation systems using the global positioning system (GPS) to the mass market.

These three cases represent the current state of the art in automation in information and technology services industries. The contexts are different and the level of automation varies, yet all three are powerful examples of information technology making its impact on mankind. The transformational power of this industry, as it impacted on the USA from the colonial period to current times, is traced by *Chandler* and *Cortada* [71.3].

The progress achieved is significant and exponential over the past 60 years, ever since computers entered this arena. As a general-purpose machine, the computer has invaded almost all aspects of human life. Computerization has resulted in productivity gains that could not have been dreamt of until the 20th century. It has been a process of unceasing innovations, integration of diverse devices, seamless merger of hardware and software, and evolution of new market paradigms. It has also given birth to an industry focused on providing information technology (IT) services to companies, governments, universities, and the public at large. It is worth studying this process to understand the growth of this industry and where it is headed.

71.1.1 Evolution of the Information and Technology Services Industry

Johann Gutenberg introduced the printing press in 1436 AD in Europe and is credited for opening the doors to the Renaissance and the scientific revolution [71.4]. The information services industry was born with the mass production of books (within the next 50 years 8 million books were printed) and it facilitated sharing of knowledge among scholars, particularly in Europe [71.5, 6]. The mass media industry consisting of newspapers, magazines, radio, television, and advertising, however, had to wait until the 19th century (for Marconi, Edison, and Walter Thompson to give birth to). A brief view of its growth in the USA is contained in [71.7]. The invention of the telegraph and Morse code by Morse in 1835 and the telephone in 1870 by Graham Bell, and their rapid spread of usage, along with the above, ensured that the spoken word joined the printed one to be shared across the world. Emile Berliner, standing on the shoulders of Edison and other eminent scientists, patented the gramophone, thus paving way for the music industry. A comprehensive view of each of these inventions and inventors is available on the Internet [71.8].

However, a parallel development over the centuries, in mathematics coupled with mechanical and electronic devices, led the development of information technol-

ogy. Calculators (Pascal, 1642) gave way to the analytic engine of Babbage (1832), and Hollerith's tabulating machine was used first in the 1890 US census. The middle of the 20th century saw the invention of computers and the world has been spinning fast ever since. Along with computers, the information technology services industry was born in the 1950s and has grown in multiple dimensions [71.9].

Computer architecture separated the data input from its processing and output layers and exploited the distinction between data and logic. Advances in input devices, storage devices, processors, and data presentation layer during the past 60 years have come about by the marriage of information services technology components with computer devices. The distinction between telephone, television, and computer is blurring and feeding the growth of services at a furious pace.

71.1.2 The Opportunity for Automation

The first generation of computers could perform various mathematical and logical operations while the data had to be stored in an external media. The machine had very little capacity to hold data. The growth in computerization over the next six decades can be studied in multiple dimensions. The technological developments in the processor, input devices, and output devices define the first dimension, while advances in software technology constitute the second. Along the way there was integration among these devices as well as embedding of software into hardware, which played a catalyst's role in further automation.

Computers were networked during the 1970s, which paved way for the next wave of automation of tasks carried out in multiple locations by diverse groups. They acquired online and real-time capabilities, thus forsaking their batch processing culture. It did not take long for human ingenuity to invent the multiprocessing, multitasking systems, network security, and virtual authentication solutions. The powerful combination of hardware and software with these features resulted in many more activities becoming automated.

If computers are capable of storing vast amounts of data and perform complex processes in seconds, can they then be asked to perform the role of experts (expert systems), acquire human-like intelligence (called artificial intelligence) or at the least detect patterns and extract knowledge out of available data (knowledge-based systems)? Commonly grouped under the title of machine learning, these topics have become the enduring quest of many researchers over the past three decades.

Work flow is integral to all processes performed manually or otherwise. Work flow automation became a logical target for development in the early 1980s, coupled with the development of groupware that facilitated concurrent working of multiple teams. Computers had stepped out of the data processing and data storage zone by this time. They were capable facilitators for information sharing and multigroup multiprocess working.

Introduction of personal computer, starting from the late 1970s, into the global market brought a fresh paradigm to computing and automation. The gates were opened for mass application and coverage of every human endeavor by PCs. Enormous creativity was unleashed by technophiles all over the world. It was no longer confined to data processing; text and word processing became common. Slide rules and calculators gave way to spreadsheets. Transparencies were no longer the preferred means of making presentations as they were replaced by PowerPoint slides.

Impressive as they are, these developments pale in significance when the impact of the Internet is considered. Since the invention of the World Wide Web and the browser technology by Tim Berners Lee and Marc Andreessen, the universe of computers has been flipped over once again. In less than a decade of its introduction, the Internet phenomenon has lead Thomas Friedman to declare that it is a flat world.

The power of the Internet has been fully leveraged by electronic mail from the mid 1990s. Hotmail, founded by Sabhir Bhatia, became an instant hit as it promised and provided connectivity from anywhere to anyone. The data and text sharing power of computers and the Internet has been enhanced to recognize voice, feel (through touch screens), still and video images, and to recognize patterns. Convergence of devices has accelerated the use of the computer as a phone and a television too. Apple made a significant business breakthrough with its iPod to share music through the Internet.

Meanwhile the business community has enabled internet-based transactions, including financial. Universities have enriched their student interaction and grown the online and distance education segments. And the government sector has discovered the potential of this technology by facilitating digital submission of tax returns and other filings. In short, this revolution has touched every segment of human activity and innovation, and automation at unprecedented levels continues to fuel its breadth and depth.

These developments have impacted on every business segment of the information and technology services, which is the focus of our study. We will identify the key segments and study each in depth in the following Sections.

71.2 Distinct Business Segments

The information and technology services industry is not a homogenous industry but rather consists of many segments, each with distinct characteristics. At a macro level, information services (Sect. 71.3) need to be separated from information technology services (Sect. 71.4).

Five significant segments can be recognized within the information services (Sect. 71.3), consisting of delivery of data and information (Sect. 71.3.1), data processing (Sect. 71.3.2), business process outsourcing (Sect. 71.3.3), analytics (Sect. 71.3.4), and printing and display solutions (Sect. 71.3.5).

Media such as newspapers, journals and magazines, radio and television, and the Internet serve as channels of information delivery. The content delivered can vary from political and social news events to weather and sports or stock and commodity prices. Reuters and Bloomberg are examples of a specific type of content providers. Automation-induced changes in this industry can be tracked by studying these two firms over the past decades (Sect. 71.3.1).

Data processing as a service commenced in the mainframe era when computers were expensive and required special skills to program and operate. It was typical to do business functions such as financial accounting or payroll processing or reconciliation of interbank transactions or process examination results in large universities using third-party data processing services (Sect. 71.3.2). Technological advances in computers and allied peripherals have impacted heavily on this segment; hence it is worthy of separate classification.

Companies the world over have found the business environment becoming complex due to geographical spread, globalization, product variety, and legislation to be complied with. Many times these challenges have resulted in attention and key resources being diverted away from core business to auxiliary purposes. Hence, companies have longed for service providers with appropriate expertise and scale to whom they could outsource auxiliary business processes such as employee benefits management and insurance claims processing. Networking of computers, proliferation of desktops, and the ubiquity of the Internet during the past two decades have enabled the emergence of third parties who can provide these business process outsourcing (BPO) services (Sect. 71.3.3). While outsourcing data processing results in only the data being handed over, business processing leads to data, (occasionally) people, and processes being handed over too. The service provider then has great opportunity to innovate and simplify the processes on an ongoing basis.

Data and business processing services are strictly transaction processing solutions that handle enormous volume of data. They are however governed by the policies and decisions taken by the firm; for example, an airline with a goal to maximize revenue would decide on creating multiple classes of service in a given flight, price them differentially, and divide the total seats available into these classes based on a rationale. This is called revenue management. Apart from selling seats on an ongoing basis, the airline has to assess and evaluate the effectiveness of these decisions periodically. Analytical processing is a line of service facilitating such reviews and revisions (Sect. 71.3.4). The skill sets required and the tools utilized in analytical processing are very different from those relating to data or business processing.

Advances in technology have changed the way in which information is delivered as an output (Sect. 71.3.5). Convergence of telephone, television, and computer technologies has created innovative solutions that were hard to imagine a few decades ago. Automation of mechanical, electrical, and electronic devices when coupled with computers has opened up a new domain for control systems; these developments are to be studied in this segment.

The information technology services (Sect. 71.4) can be segmented into computer-aided software engineering (Sect. 71.4.1), independent software testing and quality assurance (Sect. 71.4.2), package and bespoke software implementation and maintenance (Sect. 71.4.3), network and security management (Sect. 71.4.4), and hosting and infrastructure management (Sect. 71.4.5). While these are by no means exhaustive, they represent the key segments at present.

Software development has progressed rapidly from machine-level coding to programming in high-level languages to an environment with almost no need to code. The developer community recognized the need and feasibility of automating many of the tasks they perform, early in the game. Computer-aided software engineering (CASE) tools emerged when their generic utility was discovered (Sect. 71.4.1). This segment is to be explored in detail.

As the size of software code has increased, so has the time and effort required to test it. Nearly 30% of the total effort for developing software is consumed by the testing activity alone. Testing is both manpower and domain knowledge intensive. It calls for comprehensive understanding of the business environment within which the software has to function, the assembly of platforms and tools used, and the methodology used for product development. While many software bugs can be innocuous, some can have a disastrous impact on system deliverables. Automation has changed the way in which testing and quality assurance tasks are performed in many organizations. Third-party vendors have emerged to provide this service independent of the developers (Sect. 71.4.2).

The landscape of systems has undergone a metamorphosis from its early days to current periods. Every system was a custom-developed system (known as bespoke) to start with. However the opportunity to reuse design and code was spotted soon. The other end of this spectrum consists of packaged software which has the highest level of efficiency in terms of code reuse, besides the obvious benefit of quick implementation. In large corporations it is common for both to coexist; for example, a bank may have core banking functions implemented through packaged software but its financial accounting and reporting may still be in a custom-developed system running for decades. The year-2000 (Y2K) scenario is the well-understood instance here. Once a system is developed, implementing and maintaining it is time, cost, and manpower consuming. The role played by automation in this segment (Sect. 71.4.3) is a fascinating study.

Proliferation of computers and associated software has given rise to newer management issues and challenges. When thousands of desktops are strung together with dozens of mini and mainframe computers and data moves across many departments and geographical locations, failure in any part of the network is bound to adversely affect the network as a whole. Besides, networks have to be managed for efficiency of resource usage as well. An allied concern is protection of data from unauthorized access. Network and security management (Sect. 71.4.4) has emerged as a major segment of information technology that has attracted firms with appropriate expertise and willingness to absorb associated risks. Product vendors have also emerged in this space, offering products to protect against viruses, intrusions with intent to steal data, etc.

Computers and information technology have become the backbone of many enterprises, commercial or otherwise. Yet they are not the core functions of traditional manufacturing firms or service providers such as banks, stock markets, and insurance companies. As computerization has matured over the decades, many firms have also aspired to hand over the entire set of activities and resources to expert service providers. Alternatively, vendors invest in such resources and provide hosted services at an appropriate charge. Hosting and infrastructure management services (Sect. 71.4.5) is thus a distinct segment in this domain.

The ten segments identified within the information and technology industry are studied in detail in the following sections. They are by no means exhaustive and they constitute the most significant segments at present, tet the scenario may be vastly different in a decade from now due to the rapid changes and developments taking place in this industry.

71.3 Automation Path in Each Business Segment

71.3.1 Delivery of Data and Information

Paul Reuters started a news and stock information service from London's financial district in 1851 using his telegraphic skills and 200 pigeons. His firm expanded its services rapidly to Asia and America within the following 20 years. By 1923, Reuters had started using radio transmission of data of stock prices, exchange rates, and news. It diversified into television (TV) media by the 1950s, and computer screen display of stock and commodity prices in the 1970s. It joined hands with Merrill Lynch in 2005 to float workstations for online financial information [71.10].

Michael Bloomberg founded his firm in 1982 to provide online financial data about companies and their performance with dedicated terminals to market analysts. By 1987 the firm diversified into launching a trading system and discontinued printing the user manual soon. The printed document was replaced by online features, online manual, and help screens. Bloomberg expanded into travel, news services, professional services, and email and got into magazines, radio stations, TV, multimedia, website, online trading, books, foreign exchange trading and so on in the ensuing decade [71.11].

Widespread use of the World Wide Web, during the past decade, has created a mass market for information services and has ensured that both the generation and distribution services are global and not dependent on the size of the firm. Every corporation, big or small, profit or nonprofit, business or social entity has launched its own website to share information. Thousands of firms have jumped into this business of designing and operating third-party sites. Very specialized or niche information, segmented by geography, age group, hobby or special interest, alumni groups, are all available today. Portals that serve as gateways have multiplied, along with content service providers. The industry has gained another orbit with the introduction of fast and efficient search engines from Yahoo and Google. Along with the growth of this industry, issues such as security of information, privacy protection of providers, and authentication of source have become critical. These needs in turn are serviced by another set of firms with software tools.

Expansion of services and diversification of channels are facilitated by advances in the computer industry and the media convergence phenomena. The cable service provider has become the Internet service provider as well. The Internet is being leveraged by firms such as Skype to facilitate teletalk and to view video-quality images. Convergence between television, telephone, and Internet devices is accelerating. Innovation management and product and market strategy issues are coming to the fore, along with this technological convergence [71.12]. Cellphones and other mobile devices can carry voice, text, image, and data, and link to the Internet. A fascinating future awaits us, with almost every household or office device becoming *intelligent*, to transmit, store or accept, and process data.

71.3.2 Data Processing

Data processing services consist of data preparation and data processing stages followed by data presentation to the user community. A comprehensive history of developments can be found in [71.13, 14]

Data preparation activities have gone through four major phases of evolution (Fig. 71.1).

Phase 1 corresponds to the early years when data was prepared in batch mode, offline, using data entry machines and punched cards. Edit and verification runs were performed on the computer to catch errors in punched cards. The prime objective was to detect and rectify deviations if any from the data contained in the original documents. After this clean up a data validation run was performed to identify logical errors in the data presented. Range checks and field comparisons were done before accepting a set of data for the processing run. The sequential nature of these activities and their time-consuming nature is apparent. During the next decade, technological advancement from punched cards to floppy drives (8″ to 5.25′ to 3.5′) contributed to data storage efficiency and cost effectiveness. Partial rectification and validation could be performed in the data entry system itself. New category of workers called data entry operators and input/output clerks were employed in the industry.

Once dumb terminals could be attached to computers and dispersed in a location, the online data capture phase (phase 2) began, in the 1980s. This lead to elimination of duplicated effort for data capture in many instances. Data edit and validation functions could also be performed online, thus saving effort, compressing time, and reducing errors.

Phase 3 can be related to the period of introduction of PCs and networking them with the processing systems. *At-source* data capture became feasible with real-time validation. Online help features started appearing and the next logical step was to prevent capture of erroneous data. Graphical user interface, pull-down menus, predefined fields, and automated cursor movements to control the sequence, all being software advancements, played a key role in significant enhancement to data capture efficiency and effectiveness. This was the period when job categories such as data entry operators and input/output clerks disappeared from the market place.

Phase 4, which is the current phase, is characterized by extensive networking using the Internet, intranets, and extranets. Data is captured not only through PCs but through other interface devices, such as telephones, card swiping machines, point-of-sale (POS) systems, and radiofrequency identification (RFID) readers etc., as well. Captured information is not just data but can be voice or image too. *At-source* data capture is now ubiquitous since the device can be in a customer location, remote, and man or machine; it is interactive for information sharing or data capture and facilitates decision choice for the user.

What does the future hold? Cross-validation of data from disparate systems (if he is buying an expensive car, has he been paying income taxes regularly?) is coming. Person authentication through biometric sensors or other means before accepting data is already present in some systems but is likely to become universal. Further smart systems are on the anvil for the future. The intent is to make every device, be it a refrigerator, microwave oven or army uniform, so smart that it transmits information proactively and in an automated manner.

The data processing stage has evolved from manual mounting of tape and disc drives, a human computer operator firing the sequence of operations through a console, inserting stationery into the printer, and managing it to total automation in five decades. The progress here has been continuous and can be credited to both hardware and software developments.

The operating system (OS) of the computer has learnt to be multitasking and to sense device changes automatically. It can act as per location or chronological requirements. It can time-stamp transactions. It can distribute work to multiple processors and assemble and present the output in a seamless manner. It can protect itself from intrusions and threats. It can also compile information regarding the system performance and create an alarm when needed.

Data management has another aspect to it, namely the database architecture. Developments in database

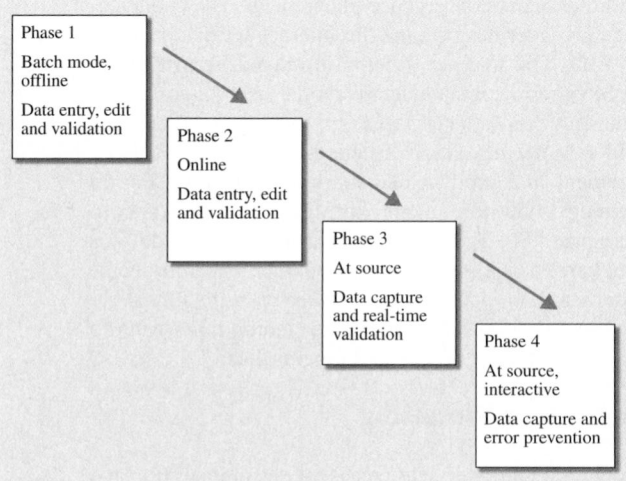

Fig. 71.1 Data preparation phases

management have progressed from simple and serial records to hierarchical databases (evolved to optimize data management during the tape drive era) to network databases to relational databases (RDBM). *Codd* [71.15] enunciated the basic principles of RDBMs in the 1970s and thus facilitated the optimal way of data storage and retrieval for all media, hierarchical or random. Major advances in data processing have come from integration of RDBMs with OSs as well as storage media. Storage area networks (SAN) and network as storage (NAS) concepts have evolved in the past decade to access widely distributed storage devices and share their utilization effectively.

71.3.3 Business Process Outsourcing

Businesses run due to interaction with customers, vendors, and other stakeholders and based on the flow of transactions; for example, customers place orders with the firm (indent), and the firm, upon delivery of goods, seeks for the customer to pay (invoice). The payment is effected through a banking transaction [check, draft, electronic financial transfer (EFT)]. Note that goods and funds flow in the system but their flow is controlled through the flow of information. Before an order is executed, it is more than likely that a set of activities called order processing happens within the firm (sometimes interacting with the customer). There can be many tasks performed in order processing adhering to certain sequence and operating rules of the company. Most of the time these tasks are performed by human beings.

As discussed earlier, it is cost effective for many firms to outsource noncore business processes to a third party and to derive operational and cost efficiency. Travel bookings, employee payroll, and benefits processing are typically outsourced. Even financial accounting and tax preparation tasks get outsourced in typical small firms. Call centers and help desks are outsourced routinely these days. Outsourcing business processes to offshore service providers is a growing trend [71.16]. BPO can result in significant operational benefits as documented through many case studies in the *BPO Journal* [71.17]. Processes are typically interwoven with computer and manual activities. As an illustration, a help desk process is illustrated in Fig. 71.2.

It can be seen that the human processes are interwoven with machine processes. The human processes call for stand-alone tasks by the service provider as well as interactive tasks with the customer. Periodic reference to a database is needed. Often a decision has to be made, with or without the help of the machine.

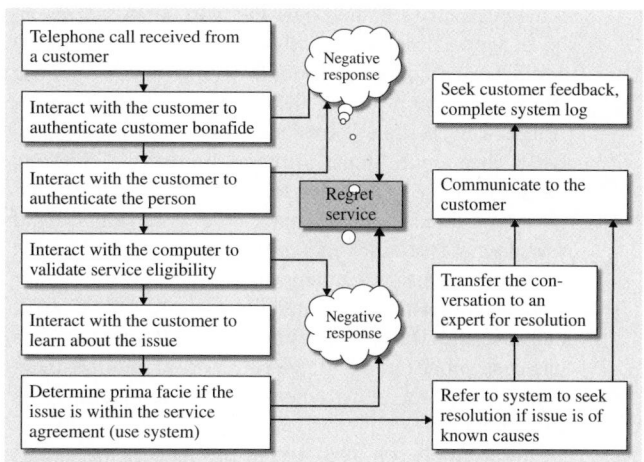

Fig. 71.2 Help desk process

Processes where there is no interaction with the customer are called back office and the others are known as front office. The outsourcing opportunity arises due to many factors.

The volume of calls received by a particular firm or product division may be insufficient to invest in a system or to maintain a help desk internally without substantial idle time for the resources. A third-party service provider can be more efficient when its resources are shared across multiple companies. Substantial investment in domain expertise needed is yet another reason for outsourcing; for example, payroll processing for a firm operating in many states calls for current knowledge of state-level taxes and other labor rules and legislation and incorporating them in the system with no time delay.

It is common practice to hand over – lock, stock and barrel – the existing system (hardware and software), people, and processes to the outsourced vendor to gain business efficiency. Firms ensure the efficiency gains and continuing improvements through service-level agreements (SLAs). SLAs have become sophisticated in design but simple in implementation over the years. They specify the level of service satisfaction to be reached for different processes. Further strict control over accountability is established.

Vendors find numerous opportunities to automate and gain efficiency along the way. In the help desk example cited above, the first opportunity is to log all incoming calls automatically and sort them by probable cause; then to build the frequently encountered issues and their resolution database and add a voice-based system to guide the customer through resolution of simple issues. Expertise gained over the help desk system can be converted into frequently asked questions (FAQs) and incorporated into the online help features of the system, thus minimizing the need to call. The vendor firm can even share knowledge gained from one customer firm environment to others by using a common high-level issues-resolution team, if agreeable to all customers. Further periodic analysis of issue logs would reveal what issues are encountered most often and why. Then it becomes a valuable data source to modify or enhance the associated manual processes or the embedded software. It is not uncommon for firms to outsource business processes to standardize or streamline them across many segments when they are reorganizing or consolidating.

Many tools are available in the market to support BPO firms and their clients.

71.3.4 Analytics

Decision support systems (DSS) have played a prominent role in widespread adoption of computers from the early decades. As opposed to the challenges of storing and processing large volumes of data in transaction pro-

cessing systems (TPSs), DSS have been computation intensive. See Chap. 87 for additional content on decision support systems. While operational researchers and other management science experts have been successful in mathematically modeling large business and engineering systems even in the 1950s, the implementation of these systems had to await advancements in hardware and software. Dantzig's simplex algorithm, invented in 1947, to solve linear programming problems and the invention of computer in the late 1940s can be taken as the starting points. It took another two decades before product mix optimization in refineries, route optimization and crew and vehicle scheduling in mass transportation systems, production scheduling, and many other large-scale problems could be solved in reasonable time periods (within an hour or much less) to be of practical value.

Inventory management, project management systems using program evaluation and review technique/critical path method (PERT/CPM) methodology, network flows optimization, and time series and regression analysis were the solutions implemented using powerful mainframe computers until the early 1970s. This period also saw the birth of discrete event and con-

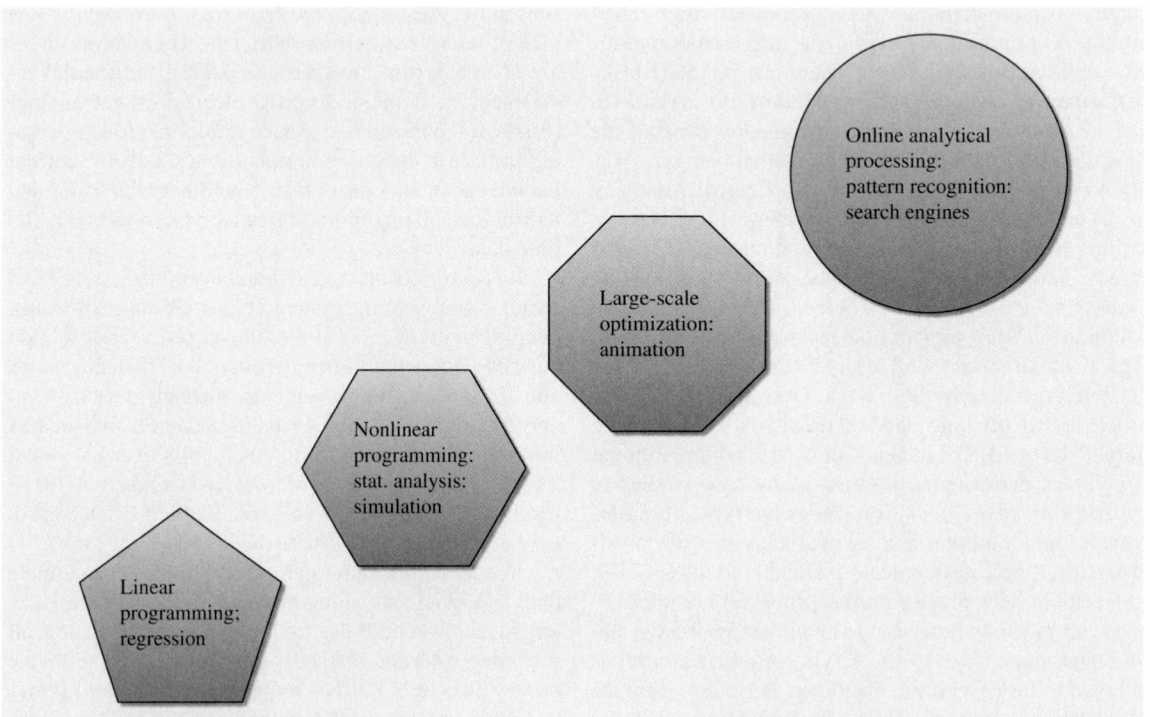

Fig. 71.3 Growth in analytical applications

tinuous simulation techniques. They found widespread acceptance in the industrial world.

The ever-increasing processing power of computers coupled with their ability to store vast amounts of data served to fuel the growth of analytical applications. Very large-scale optimization problems involving hundreds of thousands of decision variables and constraints could be taken up initially with supercomputers such as those built by Cray. Similarly Silicon Graphics introduced the power of the computer into the world of animation, which meant that high-end engineering systems can now be simulated and visually represented on computer screens for better understanding and use.

Another remarkable shift has been the use of personal computers in analytics. Numerous tool vendors have exploited the exponential growth in computers' processing power to develop PC-based solutions, CPlex from Ilog, Frontline Solver, MSProject, SPSS, and SAS for statistical analysis to name a few. Similarly one can mention Arena and Automod in the simulation world.

One of the significant developments during the 1980s was the emergence of yield and revenue management concepts in the airline industry. Analyzing millions of transaction data from past flights, the team of operations research (O.R.) experts at American Airlines was able to come up with a fine-tuned strategy for pricing airline seats and dynamically altering them until flight departure. Success of these techniques encouraged the entertainment and hospitality industry, consisting of hotels, vacation travel, and recreational sports, to embrace these concepts and improve their profitability.

Corporate data usually resides in multiple databases in various formats. Recognizing the need to bring them under a common framework in a unified database, analytics vendors have floated data warehousing tools. Data mining solutions that can extract statistically valid inferences from these data warehouses have also emerged.

Analytical processing has gone online during the past decade. Online analytical processing (OLAP) tools from many vendors enable firms to perform analytics in real time and incorporate the decision variables into real-time control systems. Pattern recognition and search engines have been integral to many web-based applications. Google's success is to be credited to the optimized page-ranking algorithm tied to its search engine. Figure 71.3 traces the growth of analytics over the past decades.

Analytics is poised to play a significant role in corporate competitive advantage, as outlined by *Davenport* [71.18]. It can be a powerful tool in providing transparency, rooting out gray markets, reducing corruption, and being fair to all stakeholders where judgment is used to select some for negative treatment. This is covered by *Balasubramanian* [71.19].

71.3.5 Printing and Display Solutions

During their inception, computers relied on character display terminals and line printers. Computer output was either viewed on the terminal or printed. *Mayadas* et al. [71.20] have tracked the growth of this industry until the mid 1980s. During the batch processing days line printers were installed in computer centers. The operator had to insert continuous stationery, print multiple copies, and sort the output for dispatch and distribution. The principal issue to be tackled was the mismatch between the processing speed of the computer and the speed of the printer. Research efforts were focused on enhancing the printing speed step by step.

Dot matrix printers (DMPs) were introduced in the 1980s, with the significant innovation that a printed page could be either portrait or landscape style. It even became possible to print pictures, albeit at low levels of accuracy. DMPs were priced low and they captured the PC-based systems market aggressively. Inkjet and laserjet printers were introduced in the 1990s and have revolutionized the printing space. Systems are now free from the need for continuous stationery too. Color printing became common very quickly and the desktop printing (DTP) segment has evolved as a distinct area. With the emergence of DTP, preparation of high-quality slides for corporate presentation, university research reports containing mathematical symbols, and artwork on tee shirts all became possible.

Very soon printers became intelligent devices with the help of software. They could hold a line of printing jobs, schedule them based on priority, change fonts and features from one job to another, report paper or ink cartridge issues, take care of reverse printing, and eliminate human effort and errors. Postscript language and digital fonts were developed by Adobe Systems in 1982, thus facilitating drawings and image manipulation. Printers capable of engineering drawings and blueprints were not to be left behind. So are other innovations such as photo-printing machines, which are now ubiquitous. In the Internet age, network printers can act as copying or faxing machines as well.

The technology-lead changes in the computer screen market are equally impressive [71.20]. Graphical and iconic display is taken for granted now. However, it is a far cry from the character display terminals. Moving from cathode-ray tube (CRT) to light-emitting diode (LED) to plasma screens, computer system users have been empowered to display sophisticated images of simulated vehicle crashes, human organs, ocean floors at depth etc. along with traditional business graphs, tables, and legal documents. The field of data visualization is merging graphics and animation to enable dynamic, real-time perception and pattern recognition [71.21].

The need for an intermediary to analyze output, interpret results or initiate control action in process con-trol systems has been minimized by the integration of computer-based systems. The console is a device for oversight, otherwise all controlling tasks can be done by the system. Automation from end to end is thus being enabled.

One of the visible trends for the future is continuing elimination of printed paper. Acceptance of digitally filed tax returns, electronic voting, and electronic mail as legally valid communication are good examples. Advances in digital certification technology play a catalytic role here [71.22].

71.4 Information Technology Services

71.4.1 Computer-Aided Software Engineering

There are four distinct stake holders in information systems, namely, system users (company executives), their subject of usage (such as factory operations), computer system operators, and system developers. Their expectations and requirements are not concurrent and are usually diverse. The responsibility to design the system to meet every stakeholder's needs, however, lies with the developers [71.23].

The key stages of software engineering are requirements analysis, functional specification preparation, architecture and system design, software development, testing, implementation, maintenance, and documentation. These are manpower intensive and were performed without the help of any tool to begin with. This scenario was to change soon.

Software engineering was the focus of the Defense Advanced Research Projects Agency (DARPA) and the National Institute of Standards and Technology (NIST) even in the formative years of computers [71.24]. However it soon became a mission critical issue for system developers such as IBM, EDS, Accenture, and HP. Confronted with the twin challenges of improving the quality of software and enhancing the productivity of software developers, many service providers turned to streamlining the processes, to adopting quality assurance methodologies, and to the tools of their trade, seeking solutions. They recognized that automating many life cycle activities of software development can bring immense benefits internally. This was, however, easier stated than accomplished.

The common business-oriented language (COBOL) and Fortran were the dominant languages for programming for business and scientific applications until the 1970s. By developing newer languages that were feature rich, that could handle multiple data types, and that were easier to learn and to deploy the industry sought to address its concerns. It was a daunting challenge to keep up with the evolution in database management [71.15] concepts too. A cherished dream of software developers to reuse code became a possibility, initially with subroutines and libraries and later with the invention of object-oriented methodology [71.25] and languages such as Smalltalk, Lisp, and C++.

Run-time efficiency is also a major concern with many systems. How does one ensure that only relevant sections of the code get executed without the need to carry along redundant code? This issue surfaces particularly when a developed system is striving to meet the functionality requirements of many customers with partially nonoverlapping requirements. Dynamic languages such as Perl, JavaScript, PHP, and Smalltalk provide this run-time flexibility. Hence, choosing the appropriate language becomes a judicious decision [71.26].

Efforts analysis of many major projects revealed that nearly 70% of resources are consumed by the pre- and post-coding stages of the software development life cycle. Besides, annual efforts expended to maintain the software were found to account for around 15% of the development stage. It dawned on the service providers that it is imperative to focus on both the pre- and post-coding stage activities also.

Design tools such as DesignAid, Excelerator, IEW, ADW, IEF, and Accenture's Foundation toolset (METHOD/1, DESIGN/1, etc.) became popular as they helped to streamline and standardize the design process, to serve as a means of communication with the end users, and to document the system. While many tools were on the PC platform, Unix and mainframe-based tools (AD/Cycle) were also available. They focused on structured systems analysis and design methods (SSADM) and managed to automate database schema, data flow diagrams, and entity relationship diagrams successfully. Developing program specifications and user documentation were partially automated. However, their attempt to automate code generation was of limited success. During this period, the requirement analysis and functional specification stages were not amenable to automation as they had to begin from ambiguous states.

SSADM [71.27] is a derivative of the traditional waterfall method of development. It called for performing each stage adhering to strict sequence. The premises were that inefficiency multiplies when one attempts to retrofit any function after the project has moved beyond design stages. The significant negative impact of this approach has been to extend the project duration to years and that the code becomes difficult to understand or to maintain later. Responding to the industry need for shortened implementation cycle and quick deployment in the field, rapid prototyping methodology in conjunction with the iterative mode of development have been evolved.

The advent of object-oriented methodologies (OOM) in the early 1990s called for binding logically defined data objects with their processing logic. Hence objects that are fully self-contained to deliver a specified function can be easily reused. This philosophy gave rise to object-oriented data modeling and object-oriented development. However, the significant change was to introduce the use case method of requirements capture. Tools such as RUT and Rational Rose helped to develop use cases efficiently. OOM has received significant help in implementation with the invention of the C++ language [71.28].

Yet another challenge was requirement traceability. An agreed functionality has to be captured in the beginning and it has to be ensured that it is carried forward all the way. It has not been easy to automate this task [71.29].

Often it is noted that the software development environment may consist of PCs, Unix machines etc. but the operational environment could include mainframes. Hence it became necessary to manage the configuration well during development so that issues confronted in transferring the software from one environment to another are minimized.

In addition, many, often 50 or more, programmers, testers, and designers were working simultaneously on different segments and stages of a large software project, and maintaining tight control over many versions became vital. Configuration management and version control aids such as WinRunner have come into play.

The emergence of web-based technologies during the past decade has given a great impetus to rapid prototyping and iterative method of development. The popularity of traditional CASE tools has reduced in this process.

71.4.2 Independent Software Testing and Quality Assurance

Testing has been an integral and essential activity in both the development and maintenance phases of any system. What needs to be tested, when, how, and by whom, are issues that continue to engage the attention of users and service providers.

The objectives of testing, however, have remained focused on fundamentals since the beginning. The essential role of testing before commissioning is to ensure that the system delivers the expected functionality with consistency and high reliability. However, there are many add-on objectives arising at various stages, which form part of testing and quality assurance.

Ensuring that the functional specifications of the desired system are drawn up accurately and comprehensively is a basic need for quality assurance. Only then can the system be designed to deliver what was intended. Requirements tracking across the life cycle stages however calls for both enabling processes and testing.

These two terms, namely testing and quality assurance, are not synonyms. Quality assurance goals overlap with testing goals during the testing phase to assure that the system performs as intended. It crosses the testing phase and exists in delivery, implementation, and maintenance phases as well. User experience in these subsequent phases needs to be captured and feedback loops created to ensure the expected level of quality of service.

It is necessary but not sufficient to draw the intended performance of the system only in terms of its functionality. Configuration, installation, interoperability, security, usability, performance parameters such as transaction processing and response time, privacy adequacy, etc. need to be comprehensively stated and tested as well. Furthermore, the likely variations in external environment within which the system needs to function have to be defined. Only then can the specification for testing be drawn up adequately. Both unit-level testing and integrated system-level testing are essential. Coupling interfaces to other systems must also be performed. Other than the above, regression testing, which means retesting the system comprehensively after a change has been incorporated, is often found to be mission critical. Practical circumstances have necessitated both black-box and white-box testing procedures to be adopted in a judicious mix. Planning the testing function, at a high level, is presented in Fig. 71.4.

Comprehensive testing for all possible data and environment conditions is time and resource consuming; in many cases it is impossible [71.30]. Given that many major systems need to be implemented in defined time periods (of between 6 months to 1 year) the daunting task is to determine what needs to be tested essentially and with priority. Hence the goal of testing has to be tied to the business goal itself. The testing phase and hence testing tasks have to stay within the time and budget constraints arising out the business goal. The testing specification has to take account of the functional and nonfunctional requirements, the environment, and expected data conditions.

Preparing test data is tricky. It has to be representative of all major scenarios. It also has to include occasional but mission-interrupting conditions. Extreme conditions are to be included as well. Preparing expected results is not easy or straightforward in many cases. This is due to the planned functionality of the system being new compared with what exists now, or sequential dependency among test results.

Note the steps involved in the test execution phase. Creating a proper test environment is coupled with checking to ensure its stability. Training staff who would perform the testing role also requires first ensuring that they have adequate skills. Once the test plan is executed and the results are analyzed, feedback is given for changes needed, and simultaneously planning for retesting has to commence.

The opportunity to automate the testing tasks has tempted many researchers and service providers to aim high. Their quest to automate the entire function has met with no success. Instead, numerous support tools have been developed to set up and stabilize the test environment, execute test scripts, compile test results, and analyze them. Performance testing and load-testing functions have been automated to a high level.

Automated tests create a log of pre-existing conditions, test data, test plan to be executed, and test results. One of the major advantages gained here is repeatability of error conditions. They further facilitate shared learning from one test sequence to a similar test performed subsequently.

The market is full of products for facilitating automated testing. HP's WinRunner (to test user interfaces) and LoadRunner (to test performance), and IBM's Rational Suite of products (for testing across the life cycle and managing the test environment), are the tools commonly used in automating testing.

Quality assurance, with a broader mission of ensuring quality across all stages of software life cycle, has leaned on ISO 9001 and Six-Sigma approaches. The Software Engineering Institute (SEI) at Carnegie Mellon University chose to tackle this issue using a specific model for the software industry. It is called the capability maturity model (CMM). Level 1 of CMM represents ad hoc development while the highest level 5 stands for well-structured, disciplined, and self-learning processes. Templates are provided for self-assessment, then guidelines are given for moving up the CMM levels.

Rigorous quantitative techniques and metrics form part of this methodology. They measure quality parameters such as delivered defects, and time and cost overruns, and include customer satisfaction measures too.

The IT services industry has embraced these models and approaches to gain competitive advantage in the marketplace. Implementing these quality assurance measures has called for using numerous tools for automated capture of efforts, output, defects, etc. Most firms have developed tools predominantly in-house for these functions.

Offshore companies, who prefer to get nearly 75% of the development or maintenance work accomplished offshore, rely very heavily on these approaches for multiple reasons [71.31, 32]. Being higher up on the CMM

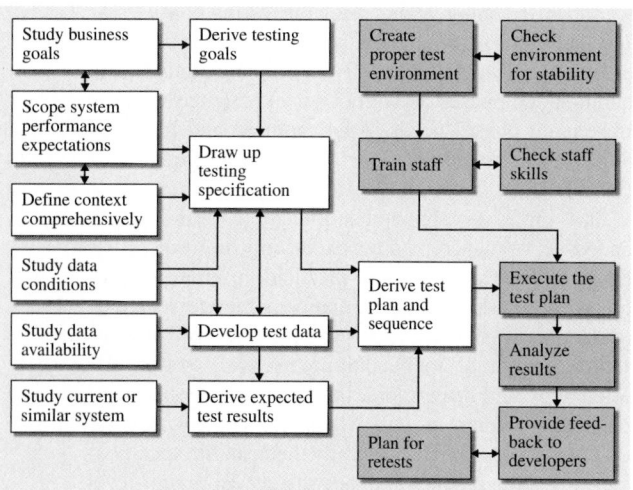

Fig. 71.4 Planning the testing function

level is a formidable marketing weapon. Combining this with the labor cost advantage, they are able to garner a larger share of outsourced assignments from firms in the USA and Europe. The client firms get both a cost and quality edge when they send work offshore. Secondly, the offshore firms usually employ a large number of young engineers with limited work experience. By enforcing strict adherence to processes they are able to deliver a consistent level of quality output to customers and not allow the inexperience of their workforce to impact on software quality. Finally, climbing the CMM ladder also means greater automation, which gives continuous improvement in developer productivity.

Long-term initiatives have a radically dramatic proposition, namely to eliminate the need for any testing and rework. Cost of quality (COQ) is measured as the total cost of testing and rework activities and resources. If quality is built into the software at every life cycle stage starting from requirements analysis then COQ can be minimized. Componentization and object-oriented development methodology, when combined, can lead firms towards achieving these challenging objectives in the future.

71.4.3 Package and Bespoke Software Implementation and Maintenance

During early stages (Fig. 71.5) all components of systems (hardware, operating system, programming language tools, applications and subsequent maintenance services) came bundled together from a single vendor (stage 1). However, it did not take long for the languages to break out of this pack. Fortran was followed by Algol and Pascal within the scientific community while COBOL remained strong in the business arena. Application development services along with continued maintenance were taken over in-house in many firms (stage 2). The unbundling phase expanded into every component of the system and, by the 1970s, total unbundling had come into play. Emergence of Unix as a dominant mid-range operating system was particularly endorsed by the scientific community. This played a critical role in unleashing the unbundling revolution during the 1980s (stage 3). However, the world woke up to an unprecedented level of segmentation when the PC and subsequent Internet era dawned. Components within the hardware are unbundled now and so are the development tools and application software.

With all this unbundling the business community was faced with a new crisis. When a system failed in operation, attributing responsibility became difficult and

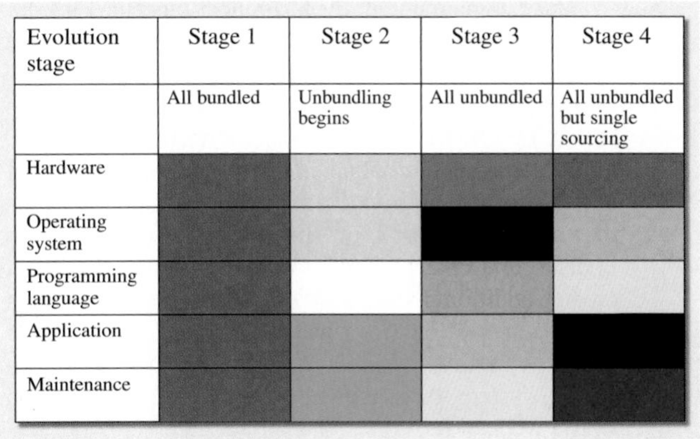

Evolution stage	Stage 1	Stage 2	Stage 3	Stage 4
	All bundled	Unbundling begins	All unbundled	All unbundled but single sourcing
Hardware				
Operating system				
Programming language				
Application				
Maintenance				

Fig. 71.5 Stages in solution implementation (*first column* cells indicate totally bundled solutions, *second column* indicates start of unbundling, *third column* indicates complete unbundling, *fourth column* indicates complete unbundling of the software, while customers tend to prefer a single source for their software and services)

a contractual nightmare. Many firms also discovered that the range of expertise required to govern and manage a myriad of systems was vast and distracting from their mission. Hence a new wave has emerged since the 1990s, focused on integrated or bundled services. Many firms have chosen to go with a single vendor to provide the entire suite of services and manage the diversity of platforms and applications. Outsourcing the entire IT function on a multi-year basis is common for large firms. Coupled with this is the integration of applications into three large segments, namely, enterprise resource planning (ERP), supply chain management (SCM), and customer relationship management (CRM). The SCM pack is to link up with the suppliers and the CRM with the customers. ERP is the middle piece, where the firm converts supplies to goods and delivers to customers.

Custom-developed applications always took more time than budgeted. They were delayed in development. The software was also found to be bug ridden. The IT services industry responded to this challenge by building packaged solutions such as JD Edwards for financial accounting, ManMan for factory operations management, PRISM for project management, etc. Over the years, transaction integrity concerns and processing efficiency requirements have led to the integrated suit of ERP applications such as SAP, MfgPro, and others.

A typical firm is likely to have a large set of custom-developed applications and packaged solutions running together. This has happened due to constant mergers and acquisitions.

Once implemented, these applications call for considerable dedicated resources to maintain and operate. The maintenance function covers detection and removal of bugs, dismantling existing manual or outdated technology interfaces, interfacing with newer systems, and changes to comply with changing regulations or enhancements to provide additional functionality. Furthermore, when one of the system components is updated and changes have to be effected on the fly, it creates additional work load of a disproportionate level in an organization (i. e., when Microsoft upgrades its desktops from the Windows XP to Windows Vista operating system, thousands of PCs need to be worked on over a weekend).

Implementing a packaged solution is also a difficult process. Existing system functionality and processes have to be mapped onto the functionality and process structure of the package. If they differ substantially, the firm has to decide between a quick implementation with minimal changes to the package (but large internal changes to the organization, including retraining) and substantial modification to the package to align it with existing processes (thereby delaying installation by many months). Converting data from an old system to the requirements of the new system is an added and significant work element in this scenario.

Service providers have evolved numerous tools to automate many of the tasks involved; for example, when moving over a set of applications from IBM COBOL to Unisys COBOL, Tata Consultancy Services (TCS) developed a series of software filters which, when applied sequentially, accomplished the conversion to almost 80%. Hence the manual effort required came down substantially. Developing libraries, subroutines, and other reusable components has been a standard technique among bespoke service providers. The well-known Y2K bug created a huge one-off service opportunity for them. It was successfully executed by creating software factories where hundreds of programmers used tools to detect and rectify the error. Similarly SAP and other package implementers have excelled in developing templates that are specific to industries. These templates cut down implementation time dramatically.

Remote diagnostic and repair services will come into play in the future. Sophisticated tools facilitate simulation of error conditions offsite and identification of environmental conditions that cause system failure. When object-oriented systems are fully functional then, with remote access, the defective component can be removed and the properly functioning component bound to the system. Acceptance testing can also be performed

from afar. The industry is aspiring to automate these functions in the coming decades.

71.4.4 Network and Security Management

Computer networks consist of not only processors but storage devices, routers, printers, sensors, etc. These are not mere hardware devices but contain embedded or loadable software too. They have grown in size and complexity from local to wide to metropolitan area networks. Their scope and coverage has been extended from the corporate to mass segment with the inclusion of PCs. Laptops, smart mobile phones, and devices such as Blackberry and the IPhone with their ubiquitous Internet connectivity have stretched the network delivery and management challenges to higher orbits. An example is shown in Fig. 71.6.

Clemm [71.33] defines network management to encompass activities, methods, procedures, and tools that pertain to the operation (running the network), administration (managing network resources), maintenance (repair and upgrade), and provisioning (intended use assurance) of networked systems. Usage monitoring, resource logs, device simulations, and agents deployed on infrastructure form the data sources for network management. Cisco systems, a leader in networking management, provides many tools for performing these functions [71.34].

Networks are differentiated in terms of the topology (line, bus, star, hub and spoke, tree, etc.) that defines the logical or physical connection between the devices, network protocols containing a language of rules and conventions for communication between network devices (hypertext transfer protocol (HTTP), transmission control protocol/Internet protocol (TCP/IP), simple mail transfer protocol (SMTP), etc.), and standards facilitating interoperability. Designing a specific network for a given role is a function of the involved devices, their locations and cost, and many other specifications.

Managing networks for performance is mission critical. However, no network can function without embedding security considerations throughout. Security concerns cover access rights, person authentication, and transaction validity at a person level and the threat of viral attacks, intrusion of spies, and denial of service (DoS) at a macro level [71.35].

Every system defines who is a legitimate user and what usage rights the person has; for example, a transport department employee may have access to the names of employees in all other departments of

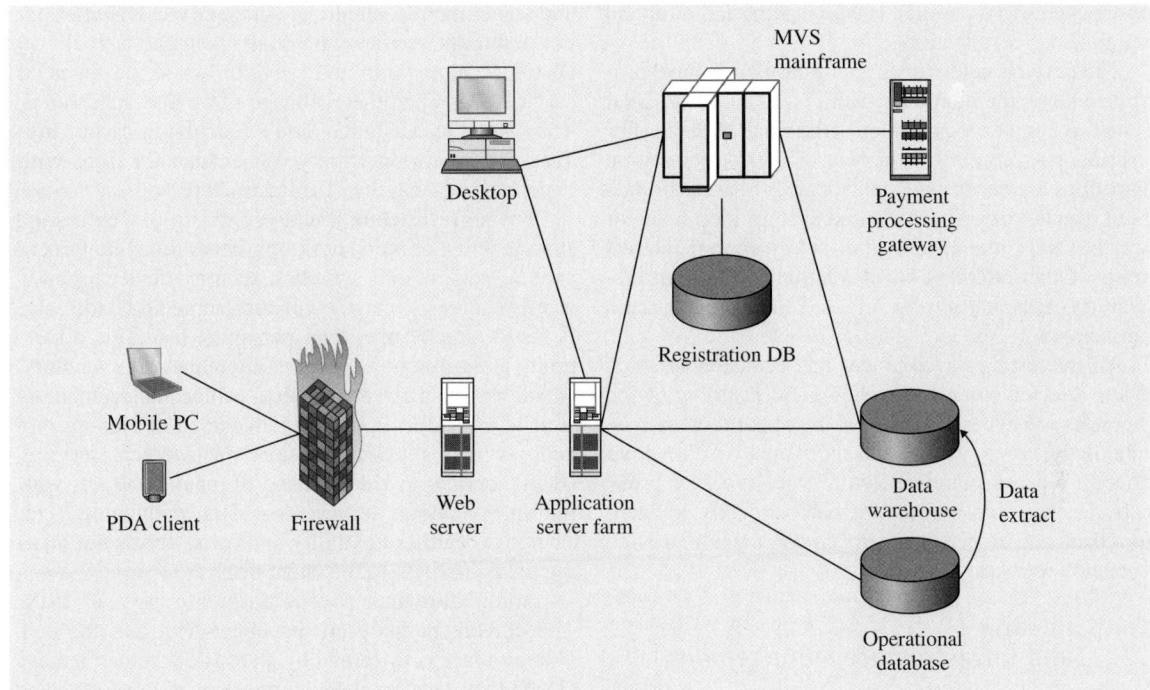

Fig. 71.6 Network diagram for the university system (after [71.36])

the firm but not to their personal details. A customer may do an online query about the account balance in a bank account but may not be authorized for online funds transfer. Over the years these requirements have extended to time of day and location too. Certain transactions cannot be performed outside office hours and cannot be executed from a nondesignated place or terminal even within office hours. Likewise, denial of legitimate access can also invite trouble to the organization. It then becomes the responsibility of the system to ensure that access rights are adhered to. Users, locations, and devices are assigned unique identities (IDs) to ensure this.

The second aspect of security is person authentication. One has to ensure that the system is being accessed only by the authorized person and by no-one else. Protection against impersonation is the key concern. Earlier systems relied heavily on passwords associated with each identified user (user ID). Password protection systems have been strengthened over the years. Frequent change in password is mandated along with selection of passwords that are difficult to guess. At the next level, systems seek additional personal details (such as date of birth) to validate the person's identity. These are found to be inadequate in modern systems.

Impressive advancement in signature recognition and fingerprint (a biometric) recognition technologies has enabled firms to integrate these devices into computer systems. Voice print is used in some cases to validate identity [71.37].

Transaction validity facilitates fraud prevention. A separate password is associated with the transactions and nowadays a one-time key is generated specific to each transaction. After use, the key is destroyed. Digital certification technology has come into play to authenticate the combination of person and transaction relating to transmitted documents. One of the most significant inventions is the encryption mechanism. Data is encoded prior to transmission over a network and decoded at the final point. Seminal work has been done by RSA [71.38] in designing and implementing a public-key approach. Encryption standards have emerged and are being mandated for critical transactions.

Vendors and products have flooded the market with services for virus protection and intrusion prevention. Firewalls are being built at various levels to filter out unwanted visitors. The black list is no longer static; it is updated dynamically by most vendors. New upgrades or patches are issued routinely to protect against emerging new threats. It has become a legal necessity to subject most systems to a periodic security audit and to obtain a compliance certification.

The developments for further automation have progressed along the biometric path [71.39]. Eye or facial feature recognition and their combination with fingerprints are coming into play. Pattern recognition algorithms are becoming sophisticated. Sharing of data about events, impersonators, and system loop holes or bugs is being attempted within select special interest groups. Other attempts cover mapping data from disparate systems and intervening and filtering suspected transactions.

An interesting development has been the research efforts directed towards psychological profiling of the imposters and cybercriminals. Many of them seem to be technology savvy, young, and motivated by altogether different reasons compared with other types of criminals. Hacker conventions are held annually to learn from them and to recruit experts among them to provide protection services.

71.4.5 Hosting and Infrastructure Management

IBM, the industry leader, preferred to lease the services of its hardware, software, and human resources in the formative years. This practice continued for nearly two decades and was followed by its competitors as well. However the unbundling phenomenon, as it took roots, encouraged firms to own the equipment, license the software, and build in-house teams for system development and maintenance. The entry of mid-range systems in the 1970s and the proliferation of PCs from the 1980s gave further impetus to this trend; however not for long.

Many large enterprises found that the tasks of selecting the hardware, upgrading the software, and recruiting and retaining staff were draining management resources. They were caught between decisions to go for centralized versus distributed systems, to select best-of-breed solutions or integrated suite of applications, and to insource or outsource application maintenance function over the years. The rapid pace at which firms in multiple industries were merging or acquiring others compounded these management issues. Multiple contracts with multiple vendors had to be negotiated often, large volumes of data had to be often merged or migrated, and it was difficult to hold any individual service provider responsible when failure occurred.

Soon they also realized that information technology is a service function, essential for their survival, and sometimes an enabler of competitive advantage, but not their core activity. The retail chain had to focus on customer acquisition and retention while the financial services firms had the challenge of product innovation. They opted to outsource the IT function in part or full. Thus a market opportunity was created for firms with expertise in managing IT infrastructure.

The infrastructure usually consists of processors, storage units, printers, desktops, networking equipment such as routers and switches, security devices, power supply sources, voltage regulators, surge protectors, etc. It could include operating personnel too. This opportunity is aggressively pursued by many large vendors, extending their services to data center management as well [71.40]. Firms had the advantage of signing one contract with the infrastructure management services (IMS) service provider instead of many contracts with multiple vendors, service providers, contractors, etc. There was further flexibility in owning versus not owning the equipment to the client firms.

Automation has a central role to play in IMS. The agreement between the client firm and the service provider is governed by a service-level agreement (SLA) that calls for device uptime on a category basis and transaction response time on a type basis. Implementation of the monitoring, reporting, and managing functions related to SLAs is automated to a high degree. Many times, self-diagnostic and repair tasks are performed by the respective devices through sophisticated software.

The Internet has added a major dimension to this field. Until its arrival IMS was popular only in large commercial firms. Now, in a new avatar called web hosting, IMS has reached out to small and medium-sized enterprises (SMEs) and the masses. Today, it is estimated to be a multibillion-dollar market. Hosting is a subset of IMS. The server equipment can be installed anywhere geographically (typically on the premises of the host firm), and the Internet or private network is leveraged to provide services to client firms and end customers of client firms seamlessly. Normal web services include email, static pages, help desk, auction sites, etc. IBM, the largest web hosting service provider globally, has included managed events such as Wimbledon, the Grammy Awards, and the Ryder Cup within this service ambit [71.41].

Web hosting services rely almost 100% on web and automated tools to perform their roles. Specialist software firms write the required application for web hosting to cover performance, content delivery, and security aspects. The need for a hosting and IMS service provider to tie together multiple products is apparent when we note that IBM's service portfolio embraces Equinix, ATT, Akamai Technologies, Keynote Systems, Cisco, i2, Ariba, and others [71.41].

Web-hosted services often cross international borders. A customer living in London can be accessing a US car company site hosted in Ireland to order an accessory. Countries often debate over the territorial rights of such transactions and try to figure out who sold what to whom.

Both IMS and hosted services are evolving with technological advancement and globalization. They are spinning out new business models. Pay per transaction (including software as a service (SaaS) [71.42]) as opposed to ownership of devices and resources is emerging. As it gains acceptance, automation of such services has to include robust mechanisms for service accounting, billing, and receivable management on the business side, and transaction integrity, authenticity, and nonnegation aspects on the technical side.

71.5 Impact Analysis

Computers and IT-based services are not mere automation aids that enhance resource productivity. Their impact on corporations and society is deep and wide. They have created new channels, markets, products, labor markets, and business paradigms. The have also had societal impact in terms of higher level of transparency and fairness, and helped to eliminate system leakages and root out corruption in many countries. At an individual level they have changed perceptions, altered human behavior, enhanced aspirations, and acted as catalysts to feed the innovative spirit of humanity.

With direct access to the Internet to every customer or stakeholder a new channel of delivery has been opened up. It has the advantage of low cost, high reliability, and asynchronous communication possibility. With the convergence of technology, the Internet as a channel is also the most cost-efficient means of delivering data, text, voice, and images. Companies face a wide range of options in terms of how they wish to leverage this channel. Many traditional organizations, including banks such as ICICI in India, have chosen to use it as the primary vehicle for service delivery.

Newer products have emerged to meet both primary and secondary needs. The iPod delivers digital music and in the process has eliminated all intermediate storage media such as compact discs (CDs) and digital versatile discs (DVDs). Books need not be printed anymore as their content can be digitally delivered. GPS-based navigation systems guide travelers to their destinations. Confidential delivery of sensitive and business-critical documents through the Internet has created the demand for digital certification and person identification services.

Newer markets for teleconferencing, online auctions, buyer–seller meet, and telemedicine have emerged. Further finer segments of the markets to cater for the needs of aged, handicapped or people on the move have been created.

Information asymmetry in earlier society created many intermediary roles that facilitated bridging between supply and demand sides. These have been destroyed in the Internet age. Any new role can emerge now only on the strength of demonstrable value. Consequently enhanced skill levels have to be closely related to market needs. An effective example of this is the role of distributors and retailers as intermediaries. They bridged the communication gap between the manufacturer and end customer while moving goods across the supply chain effectively. The Internet has limited their role as a communicator. Consequently they need to excel in providing comparative value of goods from different manufacturers to enhance their value proposition to end customers.

Since many tasks can be executed unbounded by geography, outsourcing has crossed international borders with ease and has created employment opportunities for millions in developing countries. Concerns about the ramifications of this for jobs in developed countries have emerged [71.43]. However, the market for high-end products and services has expanded globally due to rising affordability in emerging economies.

The emergence of new business models is noteworthy. Fixed costs are presented as variable costs, thus altering investment requirements. Development costs are overshadowed by marketing and customer reach costs. The economics of the information society is very different from that of the engineering society. Companies have to make an appropriate transition to ensure their survival and growth

With individuals, face-to-face interaction has been replaced with virtual meetings, sequential exchanges are now concurrent, and judgment or wisdom, if not

Part G | 71.5

supported by data and evidence, is not accepted. New paradigms have emerged in interactive behavior. Age as a proxy to accumulated knowledge is no longer revered.

Governments have been aggressive in introducing Internet-based services to reach out to the citizens. Apart from filing tax returns, registration of land records and motor vehicles, disbursal of pension funds, updating voter registry, and government-to-business transactions fall within the purview of eGovernance. When implemented, these initiatives have made operations faster and more cost effective and have rooted out sources of corruption in many instances.

The impact on society at large, with respect to volunteer activities, is explored in [71.44]. The legislative support required for use of information technology to the physically disadvantaged is documented in [71.45].

The societal impact of computers, the Internet, and information technology is even more significant when we transcend the corporate sector. Facebook and LinkedIn are Internet-based networks of special-interest groups that demonstrate what viral growth is all about. YouTube has given the world on a platter as a platform to showcase one's talent at virtually no cost. These developments are altering many paradigms that we have grown up with.

71.6 Emerging Trends

As stated earlier technology convergence, real-time analytical processing, digital delivery of integrated and multimedia information, and software as a service (SaaS) are the visible trends within the information and technology services industry. Web services are leading to services-on-demand (SoD) business models. Both SaaS and SoD are rewriting business models by converting fixed costs to variable costs. It makes eminent business sense to adopt these models at the early growth stages of a firm and then switch over to the traditional ownership model when the firm has reached a higher sales volume. The entire IT infrastructure consisting of bundled hardware, software, and services is treated akin to a utility [71.46] that is essential for business operations but may not bestow any competitive advantage. The later can come only through market and product innovations and analytics.

In the scientific community, a model that is being tested is grid computing [71.47]. It is focused on leveraging the widely dispersed processing power of the Internet for faster execution of complex and time-consuming calculations. Weather forecasting and genetic mapping fall into this domain. Geographical information systems (GIS) that combine location maps with other attributes are already common [71.48]. Multilayering of GIS and recognizing patterns for effective design of systems is on the anvil.

Computer animation services are poised for a giant leap with sophisticated tools and high-speed devices. Movie production and postproduction work, integrated with special effects, is being automated at a relentless pace by companies such as Pixar and DreamWorks. Simulation combined with animation is turning out niche products such as factory layout and work flow sequencing for the manufacturing industry.

Object-oriented design awaits the development of completely interchangeable, plug-and-play components for corporate functions. It will then be coupled with automated and remote diagnostics, repair, testing, and maintenance services.

Legal acceptance of digitally filed documents in place of printed matter is growing worldwide. Standards are being enacted and enabling legislation is being introduced in many countries. Cross-border transactional validity and resolution of attendant tax issues is complex. It is expected that the *digital highway* will become integral to many aspects of human life when these challenges are tackled. The day is not far off when every legal entity, be it a person or a firm, has a unique digital identity, although this is likely to happen faster in some nations compared with others.

References

71.1   Indian Electrons: The electronic voting machine. http://www.indian-elections.com/ electoralsystem/electricvotingmachine.html

71.2   ESI Group: Engineering simulation industry. http://www.esi-group.com/corporate/news-media/press-releases/2007-english-pr/esi-group-visual-environment-optimized-for-intel2019s-platforms-offers-premium-performance-in-simulation (2007)

71.3   A.D. Chandler, J.W. Cortada (eds.): *A Nation Transformed by Information* (Oxford Univ. Press, Cambridge 2000)

71.4   B. Yenne: *100 Inventions that Shaped World History* (Bluewood Books, New York 1993)

71.5   B. Russel: The printing press: technology that changed the world forever. http://www.associatedcontent.com/article/324287/the_printing_press_technology_that.html (2007)

71.6   C. Price: The Gutenberg device: the printing press. http://www.associatedcontent.com/article/82603/the_gutenberg_device_the_printing_press.html (2006)

71.7   E. Sebastian: The history of mass media in America. http://www.associatedcontent.com/article/13499/the_history_of_mass_media_in_america.html (2005)

71.8   M. Bellis: Your guide to inventors. http://inventors.about.com/od/gstartinventions/Famous_Invention_History_G.htm (2007)

71.9   E.G. Swedin, D.L. Ferro: *Computers, The Life Story of a Technology*, Greenwood Technographics (Greenwood, Westport 2005)

71.10   Thomson Reuters: Thomson Reuters company history. http://thomsonreuters.com/about/company_history/#1890_-_1799 (2008)

71.11   Bloomberg: Bloomberg company website. http://about.bloomberg.com/ (2008)

71.12   F. Hacklin: *Management of Convergence in Innovation* (Springer, Berlin Heidelberg 2008)

71.13   J. Blasewicz, W. Kubiak, T. Morzy, M. Rusinkiewicz (eds.): *Handbook on Data Management in Information Systems* (Springer, Berlin Heidelberg 2003)

71.14   J. Gray: Evolution of data management, IEEE Comput. **29**(10), 38–46 (1996)

71.15   E.F. Codd: A relational model of data for large shared data banks, Commun. ACM **13**, 377–387 (1970)

71.16   BPO: What is business process outsourcing? The basics of business process outsourcing. http://www.sourcingmag.com/content/c060405a.asp (2008)

71.17   Deepa: Financial value of outsourcing I–IV, BPO J. Online http://bponews.blogspot.com (2007)

71.18   T.H. Davenport: Competing on analytics, Harvard Bus. Rev. **84**(1), 98 (2006), reprint R0R01H 1–10

71.19   P. Balasubramanian: Future of operations research, a practitioner's perspective. In: *Handbook of Operations Research and Management Science*, ed. by A. Ravi Ravindran (CRC, Boca Raton 2007), Chap. 26

71.20   A.F. Mayadas, R.C. Durbeck, W.D. Hinsberg, J.M. McCrossin: The evolution of printers and displays, IBM Syst. J. **25**(314), 399–416 (1986)

71.21   C.D. Hansen, C. Johnson: *Visualization Handbook* (Academic, New York 2004)

71.22   RSA: Digital certificate solutions. http://www.rsa.com/node.aspx?id=2604 (2008)

71.23   J. Mylopoulos: Characterizing information modeling techniques. In: *Handbook on Architectures of Information Systems*, ed. by P. Bernus, K. Martins, G. Schmidt (Springer, Berlin Heidelberg 1998), Chap. 2

71.24   W. Kozaczynski: Software engineering methods. In: *Handbook on Architectures of Information Systems*, ed. by P. Bernus, K. Martins, G. Schmidt (Springer, Berlin Heidelberg 1998), Chap. 17, pp. 385–403

71.25   G. Booch: *Object Oriented Analysis and Design with Applications* (Benjamins/Cummins, New York 1994)

71.26   H. Erdogmus: So many languages, so little time, IEEE Softw. **25**(1), 4–6 (2008)

71.27   E. Yourdon, L.L. Constantine: *Structured Design* (Prentice Hall, Upper Saddle River 1979)

71.28   M. Swaine: Dr. Dobb's excellence in programming award 2008, Dr.Dobb's J. **4**, 16–17 (2008)

71.29   R.J. Wirfs-Brock: Connecting design with code, IEEE Softw. **25**(2), 20–21 (2008)

71.30   C. Kaner, J. Falk, H.Q. Nguyen: *Testing Computer Software* (Van Nostrand Reinhhold, New York 1993)

71.31   P. Jalote: *Software Project Management in Practice* (Addison-Wesley, Boston 2002)

71.32   P. Jalote: *CMM in Practice: Processes for Executing Software* (Addison-Wesley, Boston 1999)

71.33   A. Clemm: *Network Management Fundamentals* (Cisco, Indianapolis 2006)

71.34   Cisco: Cisco website. http://www.cisco.com/en/US/products/sw/netmgtsw (2008)

71.35   E. Turban: *Information Technology for Management* (Wiley, New York 2006)

71.36   Agile Modeling: Figure 1, network diagram for the university system. http://www.agilemodeling.com/artifacts/networkDiagram.htm (2009)

71.37   D. Tynan: The next 20 years, PC World Mag., 109–112 (2008)

71.38   R. Rivest, A. Shamir, L. Adleman: A method for obtaining digital signatures and public-key cryptosystems, Commun. ACM **21**(2), 120–126 (1978)

71.39   N.K. Ratha, V. Govindaraju: *Advances in Biometrics, Services, Algorithms and Systems* (Springer, Berlin Heidelberg 2008)

71.40   HP: HP infrastructure management services: tailored solutions put you in control. http://www.hp.com/pub/services/infrastructure/info/in_service_brief.pdf (2008)

71.41   IBM: IBM – the world's choice for web hosting. http://www.c2crm.com/c2/c2web/collateral.nsf/0/2CB826453C7D68FF862571C400697E18/$file/why-ibm-hosting.pdf

71.42   Microsoft: Microsoft's software as a service. http://www.microsoft.com/serviceproviders/saas/default.mspx (2008)

71.43   McKinsey Global Institute: Who wins in offshoring? http://www.cfr.org/content/meetings/innovation_rt/2_4_2004/Farrell.pdf (2004)

71.44   T. Yoo, M. Wilson, T. Shorters, M. Becton: The impact of information technology on civil society. http://www.independentsector.org/pdfs/factfind4.pdf (2001)

71.45   J. Recktenwald: Technology for the disabled: What does federal law mean for IT? http://articles.techrepublic.com.com/5100-10878-1036687.html (2000)

71.46   N.G. Carr: *Does IT Matter?* (Harvard Business School Press, Boston 2004)

71.47   R. Buyya, S. Venugopal: A gentle introduction to grid computing and technologies, CSI Commun. **29**(1), 9–19 (2005)

71.48   GIS: What is GIS? http://www.gis.com/whatisgis/index.html (2008)

72. Library Automation

Michael Kaplan

Library automation has a rich history of 130 years of development, from the standardization of card catalogs to the creation of the machine-readable cataloging (MARC) communications format and bibliographic utilities. Beginning in the early 1980s university libraries and library automation vendors pioneered the first integrated library systems (ILS). The digital era, characterized by the proliferation of content in electronic format, brought with it the development of services for casual users as well as scholarly researchers — services such as OpenURL linking and metasearching and library staff tools such as electronic resource management systems. Libraries are now reacting to user demands for quick, easy, and effective discovery and delivery such as those they have grown accustomed to through the use of Google and other Internet heavyweights, by developing a new series of Library 2.0-based discovery-to-delivery (D2D) applications. These newest offerings deliver an up-to-date user experience, allowing libraries to retain their back-office systems (e.g., acquisitions, cataloging, circulation) and add to or replace them as needed. In the process libraries can leverage Web 2.0 services to interconnect systems from different vendors and thereby ensure a gradual transition toward a new automation platform.

Part G | 72

72.1 In the Beginning: Book Catalogs and Card Catalogs

In many ways one can date the beginning of post-industrial era library automation to the development of the library catalog card and the associated card catalog drawer [72.1, 2]. The nature of the original library catalog card can be gleaned from this patent description of the so-called continuous library catalog card that had evolved to take advantage of early computer-area technology by the 1970s [72.1]:

a continuous web for library catalogue cards having a plurality of slit lines longitudinally spaced 7.5 cm apart, each slit line extending 12.5 cm transversely between edge carrier portions of the form such that upon removal of the carrier portions outwardly of the slit lines a plurality of standard 7.5 cm × 12.5 cm catalogue cards are provided. Longitudinally extending lines of uniquely shaped feed holes or perforations are provided in the carrier

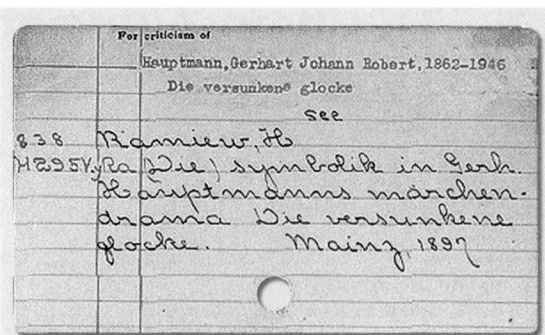

Fig. 72.1 Example of a handwritten catalog card (courtesy of University of Pennsylvania [72.3])

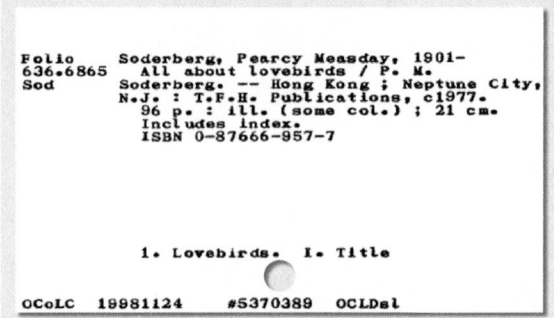

Fig. 72.2 Example of OCLC-produced catalog (shelflist) card (courtesy of OCLC [72.4])

portions of the form which permit printing of the cards by means of existing high speed printers of United States manufacture despite the cards being dimensioned in the metric system and the feed of the printers being dimensioned in the English system.

Card catalogs and card catalog drawers are well described in this patent that pertains to an update to the basic card catalog drawer [72.2]:

an apparatus for keeping a stack of catalogue cards in neat order to be used, for example, in a library. If necessary, a librarian can replace a damaged catalogue card with a new one by just pressing a button at the bottom of the drawer to enable a compression spring to eject the metal rod passing through a stack of catalogue cards. Consequently, the librarian can freely rearrange the catalogue cards.

Though these devices continued to evolve, the originals, when standardized at the insistence of Melvil

Dewey in 1877, led eventually to the demise of large book catalogs, which, despite their longevity, were at best unwieldy and difficult to update [72.5]. As we will see later, the goal here was one of increased productivity and cost reduction, not that different from the mission espoused by late 20th century computer networks.

The development and adoption of the standard catalog card presaged the popular Library of Congress printed card set program that, together with a few commercial imitators, was the unchallenged means of universalizing a distributed cataloging model until the mid 1970s. The printed card set program, in turn, was succeeded by computerized card sets from the Ohio College Library Center (OCLC), later renamed the OCLC Online Computer Library Center, Inc.; the Research Libraries Group (RLG); and other bibliographic utilities. This was largely due to the advantage the utilities had of delivering entire production runs of card sets that could be both customized in format and at the same time delivered with card packs already presorted and alphabetized.

72.2 Development of the MARC Format and Online Bibliographic Utilities

The success of the bibliographic utilities coincided with several other developments that together led to today's library automation industry. These developments were the MARC communication format, the development and evolution of university homegrown and then vendor-based library systems, and finally the growth of the networking capabilities that we now know as the Internet and the World Wide Web. In fact, without the development of the MARC format, library

automation would not have been possible. Taken as a whole, these milestones bookend 130 years of library automation.

Two extraordinary individuals, Henriette Avram and Frederick Kilgour, were almost single-handedly responsible for making possible library automation as we know it today. During her tenure at the Library of Congress, *Avram* oversaw the development of the MARC format; it emerged as a pilot program in 1968, became a US national standard in 1971, and an international standard in 1973 [72.6]. The MARC standard bore a direct connection to the development of OCLC and the other bibliographic utilities that emerged during the 1970s. RLG, the Washington Library Network (WLN), and the University of Toronto Library Automation System (UTLAS) all sprang from the same common ground.

Kilgour, who had served earlier at Harvard and Yale Universities, moved to Ohio in 1967 to establish OCLC as an online shared cataloging system [72.7]. OCLC now has a dominant, worldwide presence, serving over 57 000 libraries in 112 countries with a database that comprises in excess of 168 million records (including articles) and 1.75 billion holdings.

From the beginning the goals of library automation were twofold: to reduce the extremely labor-intensive nature of the profession while increasing the level of standardization across the bibliographic landscape. It was, of course, obvious that increasing standardization should lead to reduced labor costs. It is less clear that actions on the bibliographic "production room floor" made that possible. For many years these goals were a core part of the OCLC mission – the most successful modern (and practically sole surviving) bibliographic utility. Founded in 1967, OCLC is a [72.8]

nonprofit, membership, computer library service and research organization dedicated to the public purposes of furthering access to the world's information and reducing information costs.

In the early years of library automation, libraries emphasized cost reduction as a primary rationale for automation.

In the early 1970s a number of so-called bibliographic utilities arose to provide comprehensive and cooperative access to a database comprised of descriptive bibliographic (or cataloging) data – now commonly known as *metadata*. One by one these bibliographic utilities have vanished. In fact, in 2007 OCLC absorbed RLG and migrated its bibliographic database and bibliographic holdings into OCLC's own WorldCat, thus leaving OCLC as the last representative of the utility model that arose in the 1970s.

A hallmark of library automation systems is their evolution from distinct, separate modules to large, integrated systems. On the level of the actual bibliographic data this evolution was characterized by the migration and merger of disparate files that served different purposes (e.g., acquisitions, cataloging, authority control, circulation, etc.) into a single bibliographic

master file. The concept of such a master file is predicated on avoidance of duplicative data entry; that is, data should enter the system once and then be repurposed however needed. A truly continuous, escalating value chain of information, from publisher data to library bibliographic data, still does not exist. Libraries, especially national libraries such as the Library of Congress, routinely begin their metadata creation routines either at the keyboard or with suboptimal data (often from third-party subscription agents). (The Library of Congress also makes use of an electronic cataloging-in-publication program.) The online information exchange (ONIX) standard presages the ultimate extinction of the pure MARC communication standard as extensible markup language (XML)-based formats become dominant. Both the hardware/software and the data strains become ever-more intertwined as bibliographic automation becomes a subset of the larger information universe, adopting standards that transcend libraries.

72.2.1 Integrated Library Systems

The story of modern library automation began with a few pioneering individuals and libraries in the mid to late 1960s [72.9–13]. In addition to Avram and Kilgour, one of the more notable was Herman Fussler at the University of Chicago, whose efforts were supported by the National Science Foundation. Following the initial experiments, the university began to develop a new system, one of whose central concepts was in fact the bibliographic master file [72.9].

The University of Chicago was not alone. Stanford University (with bibliographic automation of large library operations using time sharing (BALLOTS)), the Washington Library Network, and perhaps most importantly for large academic and public libraries, Northwestern University (NOTIS) were all actively investigating and developing systems. NOTIS was emblematic of systems developed through the mid 1980s. NOTIS and others (e.g., Hebrew University's automated library expandable program (Aleph 100) that became the seed of Ex Libris and the Virginia Tech library system (VTLS)) were originally developed in university settings. In the 1980s NOTIS was commercialized and later sold to Ameritech in the heyday of AT&T's breakup into a series of Baby Bells. The NOTIS management team eventually moved on and started Endeavor Information Systems, which was sold to Elsevier Science and then in turn to the Ex Libris Group in 2006.

Other systems, aimed at both public and academic libraries, made their appearance during the 1970s and 1980s. Computer Library Services Inc. (CLSI), Data Research Associates (DRA), Dynix, GEAC, Innovative Interfaces Inc. (III), and Sirsi were some of the better known. Today only Ex Libris, III, and SirsiDynix (merged in 2005) survive as major players in the integrated library system arena.

The Harvard University Library epitomizes the various stages of library automation on a grand scale. Under the leadership of the fabled Richard DeGennaro, then Associate University Librarian for Systems Development, Harvard University's Widener Library keyed and published its manual shelflist in 60 volumes between 1965 and 1979 [72.14]. At the same time the university was experimenting with both circulation and acquisitions applications, the latter with the amusing moniker, computer-assisted ordering system (CAOS), later renamed the computer-aided processing system (CAPS). In 1975 Harvard also started to make use of the relatively young OCLC system. As with other institutions, Harvard initially viewed OCLC as a means to more efficiently generate catalog cards [72.15].

In 1983 Harvard University decided to obtain the NOTIS source code from Northwestern University to unify and coordinate collection development across the 100 libraries that constituted the vast and decentralized Harvard University Library system. The Harvard system, HOLLIS, served originally as an acquisitions subsystem. Meantime, the archive tapes of OCLC transactions were being published in microfiche format as the distributable union catalog (DUC), for the first time providing distributed access to a portion of the Union Catalog – a subset of the records created in OCLC. It was not until 1987 that the catalog master file was loaded into HOLLIS. In 1988 the HOLLIS OPAC (online public access catalog) debuted, eliminating the need for the DUC [72.15].

It was, in fact, precisely the combination of the MARC format, bibliographic utilities, and the emergence of local (integrated) library systems that together formed the basis for the library information architecture of the mid to late 1980s. Exploiting the advantages presented by these building blocks, the decade from 1985 to 1995 witnessed rapid adoption and expansion in the field of library automation, characterized by maturing systems and increasing experience in networking. Most academic and many public libraries had an integrated library system (ILS) in place by 1990. While the early ILS systems developed module by module, ILS systems by this time were truly integrated. Data could finally be repurposed and reused as it made its way through the bibliographic lifecycle from the acquisitions module to the cataloging module to the circulation module. Some systems, it is true, required overnight batch jobs to transfer data from acquisitions to cataloging, but true integration was becoming more and more the norm.

By the early 1990s libraries were entering the age of content. Telnet and Gopher clients made possible the online presence of abstracting and indexing services (A&I services). Preprint databases arose and libraries began to mount these as adjuncts to the catalog proper. DOS-based Telnet clients gave way to Windows-based Telnet clients. Later in the 1990s, Windows-based clients in turn gave way to web browsers.

All this time, the underlying bibliographic databases were largely predicated on current acquisitions. The key to an all encompassing bibliographic experience lay in retrospective conversion (Recon) as proven by Harvard University Library's fundamental commitment to Recon. Between 1992 and 1996 Harvard University added millions of bibliographic records in a concerted effort to eliminate the need for its thousands of catalog card drawers and provide its users with complete online access to its rich collections. Oxford University and others soon followed. While it would be incorrect to say that Recon is a product of a bygone era, most major libraries do indeed manage the overwhelming proportion of their collection metadata online. This has proven to be the precursor to the massive digitization projects underwritten by Google, Yahoo, and Microsoft, all of which depend to a large degree on the underlying bibliographic metadata, much of which was consolidated during the Recon era.

In the early years of library automation, library systems and library system vendors moved from timesharing, as was evident in the case of BALLOTS, to large mainframe systems. (Time-sharing involves many users making use of time slots on a shared machine.) Library system vendors normally allied themselves with a given hardware provider and specialized in specific operating system environments. The advent of the Internet, and more especially the World Wide Web, fueled by the growth of systems based on Unix (and later Linux) and relational database technology (most notably, Oracle) – all combined with the seemingly ubiquitous personal computer – gave rise to the second generation in library automation, beginning around 1995. As noted, the most visible manifestation of that to the library public was the use of browsers to access the catalogs. This revolution was followed within 5 years by the rapid and inexorable rise of e-Content. The most successful library automation vendors, it is now clear, are those that possessed a clear and compelling vision for combining library automation with delivery of content. The most successful of those vendors also espouse an e-Content philosophy that is publisher neutral.

72.2.2 Integrated Library Systems: The Second Generation

Libraries, often at the direction of their governing bodies, made a generational decision during this period to migrate from mainframe to client–server environments. Endeavor Information Systems, founded with the goal of creating technology systems for academic, special, and research libraries, was first out of the starting blocks and had its most successful period for about a decade following its incorporation in 1994 [72.16]. It was no accident, given its roots in NOTIS and its familiarity with NOTIS software and customers, that Endeavor quickly captured the overwhelming share of the NOTIS customer base that was looking to move to a new system. This was particularly true in the period immediately preceding the pivotal year 2000, when Y2K loomed large. In planning for the shift to the year 2000 library computer systems faced the same challenge all other legacy computer faced, namely, reworking computer code to handle calendar dates once the new millennium had begun. So many libraries, both large and small, had to replace computer systems in the very late 1990s that 1998 and 1999 were banner years in the library computer marketplace, followed by a short-term downturn in 2000.

Ex Libris Ltd., a relative latecomer to the North American library market, was founded in Israel in 1983, originally as ALEPH Yissum, and commissioned to develop a state-of-the-art library automation system for the Hebrew University in Jerusalem. Incorporated in 1986 as Ex Libris, a sales and marketing company formed to market Aleph, Ex Libris broke into the European market in 1988 with a sale to CSIC, a Spanish network of 80 public research organization libraries affiliated with the Spanish Ministry of Science and Technology. This sale demonstrated an early hallmark of Aleph, namely deep interest in the consortial environment.

At roughly the same time as Endeavor was releasing its web-based client–server system, Voyager, Ex Libris released the fourth version of the Aleph system,

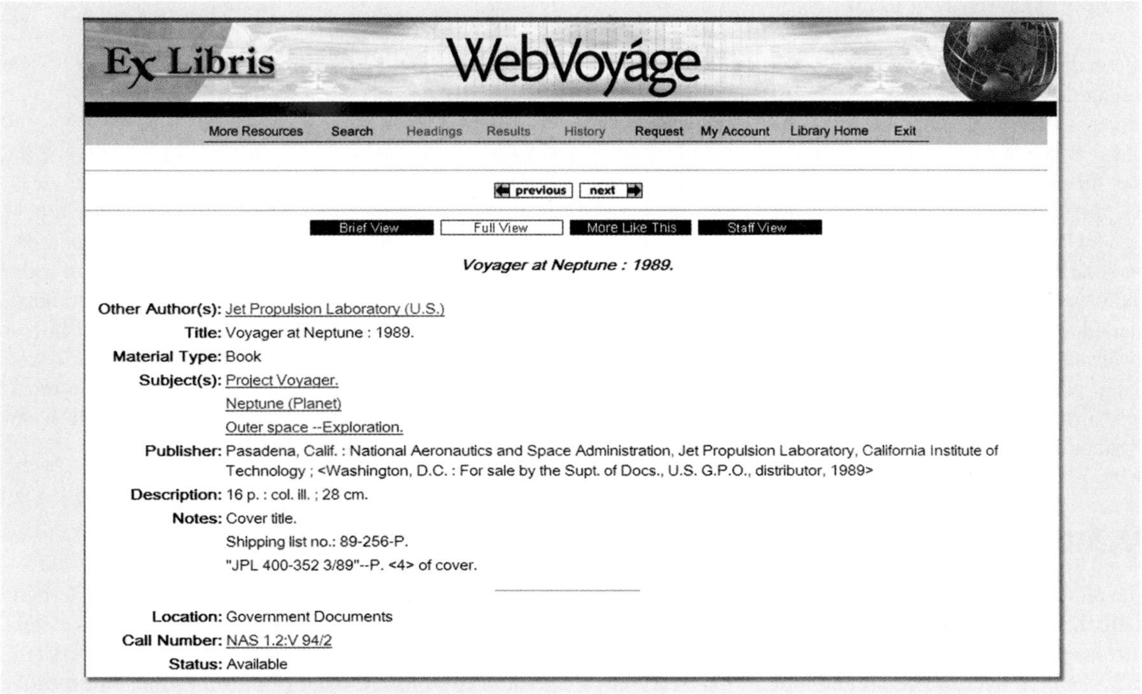

Fig. 72.3 Endeavor (now Ex Libris) WebVoyáge display

Fig. 72.4 Catalog record from University of Iowa InfoHawk catalog, with parallel roman-alphabet and Chinese fields (courtesy of University of Iowa)

Aleph 500. Based on what was to become the industry standard architecture, Aleph featured a multi-tier client–server architecture, support for a relational database, and – a tribute to its origins and world outlook – support for the unicode character set. Starting in 1996, with Aleph 500 as its new ILS stock-in-trade, Ex Libris created a US subsidiary and decided to make a major push into the North American market. At the same time – and crucially for its success in the emerging e-Content market – Ex Libris made the strategic commitment to invest heavily in research and development with the stated goal of creating a unified family of library products that would meet the requirements of the evolving digital library world.

72.2.3 The Nonroman World: Unicode Comes to Libraries

If there is a single accepted analysis of the state of libraries and library automation in the 21st century, it is the fact the library world is but one part of the much larger information universe and that library standards, e.g., the MARC format and the American standard code for information interchange (ASCII) character set, need to be considered in light of the larger universe of information standards, such as XML (ONIX) and unicode. Those libraries and library vendors that have the most robust understanding of this fundamental shift have been most adept at long-term survival.

72.3 OpenURL Linking and the Rise of Link Resolvers

The shift from ASCII to unicode and from MARC to XML heralded an entire salvo of digital products that introduced a new concept of library automation – the ILS was no longer the be all and end all of library automation – and was marked by Ex Libris's release of

SFX. SFX was developed in conjunction with Herbert Van de Sompel (presently a research scientist at the Los Alamos National Laboratory) and the Universiteit Ghent in response to what is known as the *appropriate copy* problem [72.17]:

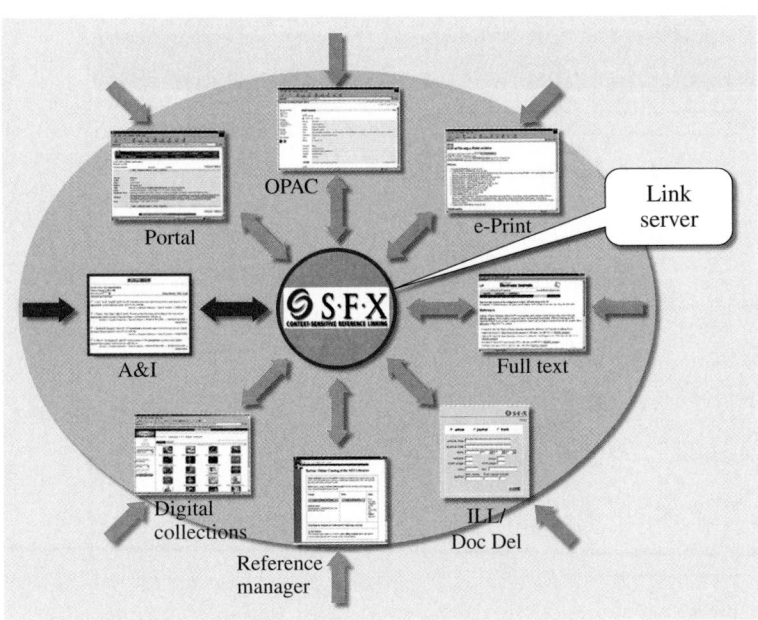

Fig. 72.5 SFX and OpenURL linking (after [72.4])

Part G | 72.3

There has been an explosive growth in the number of scholarly journals available in electronic form over the internet. As e-Journal systems move past the pains of initial implementation, designers have begun to explore the power of the new environment and to add functionality impossible in the world of paper-based journals. Probably the single most important such development has been reference linking, the ability to link automatically from the references in one paper to the referred-to articles.

Led by Oren Beit-Arie (presently Ex Libris's chief strategy officer), SFX (which stands for *special effects*) was not only a commercial success but also became generically associated with the standard known as OpenURL linking. In much the same way as Xerox is synonymous with photocopying, so SFX has become synonymous with context-sensitive linking. Released in 2000, SFX now has an installed base of over 1500 libraries and research centers worldwide

Librarians, especially systems librarians, are well known as staunch proponents of standards. Thus it was an extremely astute move when Ex Libris took the initiative to have the National Information Standards Organization (NISO) adopt the OpenURL as a NISO standard.

The OpenURL framework for context-sensitive services was actually approved on a fast-track basis as ANSI/NISO Z39.88 in 2005 [72.18]:

The OpenURL framework standard defines an architecture for creating OpenURL framework applications. An OpenURL framework application is a networked service environment, in which packages of information are transported over a network. These packages have a description of a referenced resource at their core, and they are transported with the intent of obtaining context-sensitive services pertaining to the referenced resource. To enable the recipients of these packages to deliver such context-sensitive services, each package describes the referenced resource itself, the network context in which the resource is referenced, and the context in which the service request takes place.

72.3.1 Metasearching

Ex Libris quickly followed its success with SFX by expanding into the area of metasearching. Also known as federated searching, metasearching was another response to the proliferation of online journals and online content. MetaLib (and competing products) were the initial bibliographic answer to the problem posed by commercial search engines such as Yahoo or Google, namely that much content had come quickly to live outside the realm of the library catalog. Metasearch enables users – power and nonpower users, whether the general public or undergraduates, graduate students, post-doctoral students, and faculty – to retrieve the

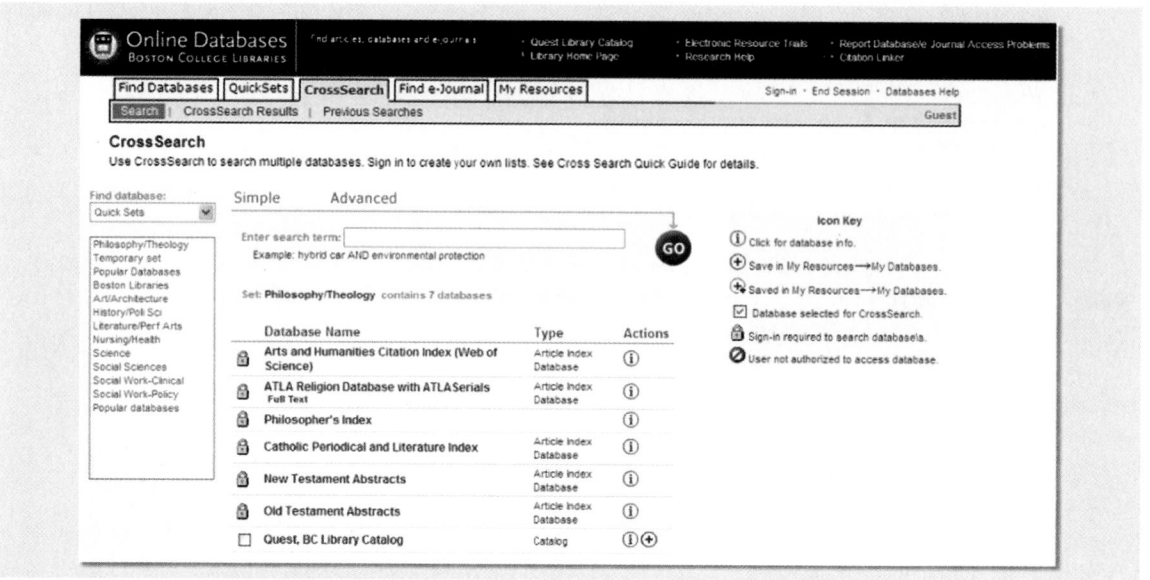

Fig. 72.6 Boston College MetaQuest, with quick sets

most comprehensive set of results in a single search, including traditional bibliographic citations, abstracting and indexing service information, and the actual fulltext whenever it is available. Before long, numerous offerings of both OpenURL link resolvers and metasearch applications were available in the bibliographic arena.

By 2005 many institutions had made their primary bibliographic search gateway not the library catalog, but their metasearch application. Boston College is a good example. Its library catalog, Quest, is but one of the numerous targets searchable through its metasearch application, MetaQuest.

By now libraries and their users had bought wholeheartedly into the combination of metasearching and linking – it is the combination of these two approaches for dealing with the e-Content universe that makes them together vastly more powerful than either alone. Troubles loomed, though, in the absence of standards for metasearching. NISO once again stepped forward and created the NISO Metasearch Initiative. NISO described the challenge as [72.19]:

Metasearch, parallel search, federated search, broadcast search, cross-database search, search portal are a familiar part of the information community's vocabulary. They speak to the need for search and retrieval to span multiple databases, sources, platforms, protocols, and vendors at one time. Metasearch services rely on a variety of approaches to search and retrieval including open standards (such as NISO's Z39.50), proprietary API's, and screen scraping. However, the absence of widely supported standards, best practices, and tools makes the metasearch environment less efficient for the system provider, the content provider, and ultimately the end-user.

Standards are invaluable, but in the library world standards – aimed at providing consistency in searching – are all the more important. As knowledge seekers move from one environment to the next with the change of a uniform resource locator (URL), they also expect consistent behavior across these various information environments. It is particularly encouraging, therefore, to see how the public responds. An article in the October 7, 2005 *Harvard Crimson* shows the opinion of a student at one of the world's most prestigious universities [72.20]:

Of the many things to change at Harvard over the summer... there is one new development that students may not have noticed but that come term-paper time will make a world of difference. That change is the switch from Harvard University Library's (HUL) e-Resources website to the new e-Research @ Harvard Libraries website. HUL has done a tremendous job with the new easy-to-use site, and we applaud them for dreaming up such a helpful resource.

The new e-Research page puts a high-tech user interface on top of Harvard's vast array of online subscriptions, making resources more readily accessible – whether users know the exact journal they are looking for or are just poking around for sources. But the best thing about the new e-Research page is the plethora of features it provides that take full advantage of computers to make research easier ... within seconds you can find what you are looking for, either in fulltext online or a Find it @ Harvard button that tells you exactly where in Harvard's huge library system to look.

Within two short paragraphs two of the major tools of 21st century librarianship and e-Scholarship emerge: metasearching (e-Research) and OpenURL linking (Find It @ Harvard). The library catalog (HOLLIS) is mentioned, too, but it is but one amid a vast array of electronic resources. Looming over all this is the specter of Google, whose seemingly endless series of innovations and content-related initiatives is in turn pushing the frontiers of automation within the library systems world itself.

72.3.2 The Digital Revolution and Digital Repositories

Three other areas of institutional bibliographic automation were under discussion and development between 2002 and 2007: digital repositories, electronic resource management systems, and new models of resource discovery and delivery.

As with early developments in library automation, one of the first efforts to manage digital objects, especially textual objects, began in a university environment. In 2000 the Massachusetts Institute of Technology Libraries and Hewlett-Packard embarked on a joint research project that we now know as DSpace – an early digital asset management (DAM) system.

DSpace introduced the concept of an institutional repository (IR) [72.21],

a robust, software platform to digitally store... collections and valuable research data, which had previously existed only in hard copies... an archiving system that stores digital representations of analog artifacts, text, photos, audio and films... capable of permanently storing data in a non-proprietary

format, so researchers can access its contents for decades to come.

DSpace is an open-source software toolkit. Automation vendors realized this was a fertile field for institutions not inclined to go the toolkit approach. Various products have emerged; notable among them are CONTENTdm, developed at the University of Washington and now owned by OCLC; and Ex Libris's DigiTool. An open-source newcomer to the field is Fedora, which will likely appeal to the same group of libraries interested in DSpace [72.22]. All these products are designed for deposit, search, and retrieval. While they have maintenance aspects about them, they are not true preservation systems.

Libraries and archives, as recognized custodians of our intellectual and cultural heritage, have long understood the value of preservation and ensuring longitudinal access to the materials for which they have accepted custodianship. As more and more materials are digitized, particularly locally important cultural heritage materials, or are born digital, these institutions have begun to accept responsibility for preserving these materials. Recently (2007) the National Library of New Zealand leapt to the forefront of digital preservation by developing a National Digital Heritage Archive (NDHA) program to ensure the ongoing collection, preservation, and accessibility of its digital heritage collections. This multiyear program, begun in conjunction with Endeavor Information Systems and continued by Ex Libris, aims to preserve digital objects for nothing less than the life of the custodial bodies [72.23].

72.3.3 Electronic Resource Management

On the serial side of the bibliographic house, the e-Journal evolution of the 1990s turned rapidly into the e-Journal avalanche of the 21st century. Libraries and information centers that subscribed to thousands or tens of thousands of print journals found themselves facing a very different journal publication model when presented with e-Journals. Not only have libraries had to contend with entirely new models – print only, print with free e-Journal, e-Journal bundles or aggregations, often in multiple flavors from competing publishers and aggregators – but also these e-Journals came with profoundly different legal restrictions on use of the e-Content. Libraries found themselves swimming upstream against a rushing tide of e-Content. A number of institutions (among them Johns Hopkins University, Massachusetts Institute of Technology, and the University Libraries of Notre Dame) developed homegrown systems to contend with the morass; most others relied on spreadsheets and quantities of paper.

Into this fray stepped the Digital Library Foundation (DLF), which convened a small group of e-Journal and e-Content specialists to identify the data elements required to build an electronic resources management (ERM) system. This study, developed in conjunction with a group of vendors and issued in 2004 as *Electronic Resource Management: Report of the DLF ERM Initiative (DLF ERMI)*, was one of the few times since the development of the MARC standard that a standard predated full-blown system development [72.24].

Systems developers adopted the DLF ERMI initiative to greater or lesser degrees as dictated by their business interests and commitment to standards. Unlike the OpenURL link resolvers and metasearch tools, electronic resource management tools addressed an audience that was primarily staff, harking back to the original goal of library automation: to make librarians more productive.

Innovative Interfaces Inc. was the first vendor to release an ERM product. Serials Solutions, one of the newer players, whose origins lay in providing catalog copy for e-Journals and A-to-Z lists for e-Journals, has the 360Resource Manager offering.

72.3.4 From OPAC to Next-Generation Discovery to Delivery

A deeply held tenet of librarians is that they have a responsibility to provide suitably vetted, authoritative information to their users. In the years since the rise of Google as an Internet phenomenon, librarians have shown enormous angst in the reliance, particularly among the generation that has come of age since the dawn of the Internet, on whatever they find in the first page of hits from any of the most popular Internet search engines. Critical analysis and the intellectual value of the resulting hits are less important to the average user than the immediacy of an online hit, especially if it occurs on the first results page. Libraries continue to spend in the aggregate hundreds of millions of dollars on content, both traditional and electronic, yet have failed to retain even their own tuition-paying clientele.

OCLC, to its credit, confronted the issue head on with a report it issued in 2005 [72.25]. In the report OCLC demonstrates quite conclusively that ease of searching and speedy results, plus immediate access to content, are the most highly prized attributes users desire in search engines. Unfortunately, the report shows

that libraries and library systems rank very low on this scale. On the other hand, when it comes to a question of authoritative, reliable resources, the report shows that libraries are clear winners. The question has serious implications for the future of library services. How should libraries and library system vendors confront this conundrum?

The debate was framed in part by a series of lectures that Dale Flecker, associate director of the Harvard University Library for Planning and Systems, gave in 2005 on: OPACS and our changing environment: observations, hopes, and fears. The challenge he framed was the evolution of the local research environment that comprises multiple local collections and catalogs. Harvard University, for example, had (and still has) separate catalogs for visual materials, geographic information systems, archival collections, social science datasets, a library OPAC, and numerous small databases. Licensed external services have proliferated: in 2005 Harvard had more than 175 search platforms on the Harvard University Library portal – all in addition to Internet engines, online bookstores, and so forth. Flecker expressed hope that OPACs would evolve to enable greater integration with the larger information environment. To achieve this, OPACs (or their eventual replacements) would have to cope with both the Internet and the explosion of digital information that was already generating tremendous research and innovation in search technology. Speed, relevance (ranking), and the ability to deal with very large results sets would be the key to success – or failure [72.26].

Once again, the usual series of library automation vendors stepped up. In truth, they were already working on their products but this time they were joined by yet another set of players in the library world. North Carolina State University adopted Endeca [72.27], best known as a data industry platform, to replace its library OPAC: "Endeca's unique information access platform helps people find, analyze, and understand information in ways never before possible" [72.27]. As the NCSU authors *Antelman* et al. noted in the abstract to their 2006 paper [72.28],

Library catalogs have represented stagnant technology for close to twenty years. Moving toward a next-generation catalog, North Carolina State University (NCSU) Libraries purchased Endeca's Information Access Platform to give its users relevance-ranked keyword search results and to leverage the rich metadata trapped in the MARC record to enhance collection browsing. This pa-

per discusses the new functionality that has been enabled, the implementation process and system architecture, assessment of the new catalog's performance, and future directions.

Other new players have also entered the market for new library delivery systems, generally called NextGen or discovery-to-delivery (D2D) systems. AquaBrowser is another example of a pure-OPAC replacement with no pretensions of replacing the complete range of ILS functions. Among more traditional library vendors, Innovative Interfaces has released its product, Encore (http://www.iii.com/encore), and Ex Libris its product, Primo (http://www.exlibrisgroup.com/primo.htm). Both were developed with a group of partners and both tout a one-stop solution for the discovery and delivery of information.

Ex Libris's choice of development partners was perhaps the more atypical in that one of its partners, Vanderbilt University, was particularly keen to use Primo to combine its library catalog, Acorn, with a facility to enable easy discovery and delivery of its unique Television News Archive [72.29]:

The Television News Archive collection at Vanderbilt University is the world's most extensive and complete archive of television news. The collection holds more than 30 000 individual network evening news broadcasts from the major US national broadcast networks: ABC, CBS, NBC, and CNN, and more than 9000 hours of special news-related programming including ABC's Nightline since 1989.

The University of Iowa had a different goal, namely to push exposure of many of its otherwise isolated digital collections. These collections, which developed over time and throughout the institution, possess no native discovery interface and consequently suffer from a widespread lack of awareness. With Primo as its unifying discovery tool, the university can provide a single search interface to all of its rich collections. One notable example is the university's and the Iowa Historical Society's World War II newspaper clippings collection.

The Primo discovery and delivery system enables libraries to present their collections in an entirely new way, providing users with the ability to access a wealth

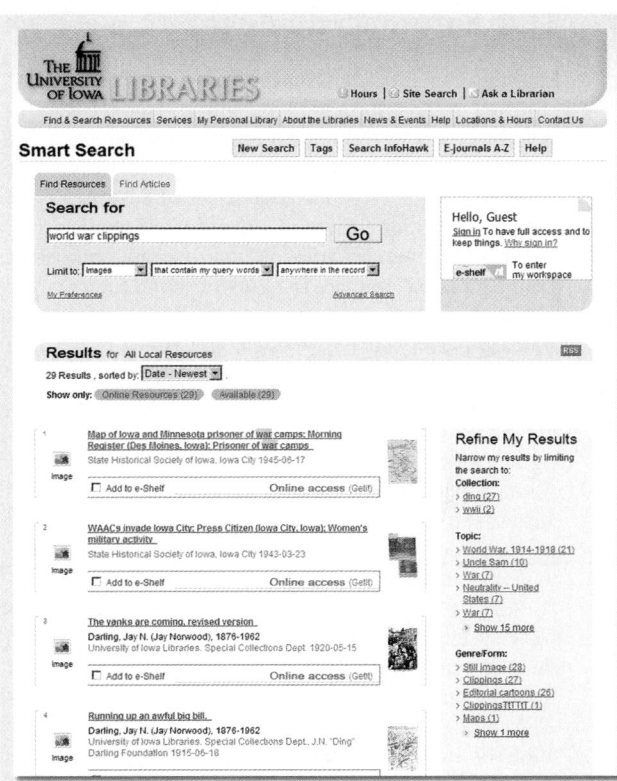

Fig. 72.7 Primo at the University of Iowa: an example of the new D2D paradigm

of authoritative information from a single point. Central to the development of Primo has been the concept of *harvesting* and *publishing* as the means whereby a single, Google-like search interface can provide rapid and easy, normalized and enriched discovery for the entire range of resources under the control of the institution: local catalogs, locally digitized collections harvested by the open archieves initiative protocol for metadate harvesting (OAI-PMH) collections, and so forth. Primo is integrated with both metasearching (MetaLib) and OpenURL linking (SFX) to enable discovery in a vast range of external resources, with onward navigation and other services via the OpenURL syntax. In this way the institution can satisfy its users' desire for almost instantaneous response to simple queries while providing the ability to search a vast array of databases and e-Resources that are not held locally.

72.4 Future Challenges

Libraries and librarians have frequently stood at the forefront of the information revolution. The MARC format was a model of interinstitutional and international collaboration in its time. As libraries and their vendors evolved, though, large installed bases of customers and data made further change that much more challenging. At the same time, libraries became but one outpost on the information frontier. The growth of the Internet, the World Wide Web, and the rise of newer information industry players such as Google and Amazon have all combined to put libraries on the defensive. Libraries have now come to realize that standards such as unicode and XML can be exploited for their benefit by partaking of a global information chain. Librarians have come to grasp the significance of capturing information at its earliest stage and repurposing it throughout its lifetime, all the while continuing to add value to the information content package.

Almost 20 years after the advent of the integrated library system, we are witnessing the opening wedge in the dissolution of the ILS into a series of independent modules that communicate with each other by means of Web 2.0 services. These new discovery-to-delivery solutions are indeed, on the one hand, OPAC or OPAC-plus replacements and designed to work with a variety of vendor systems, but they are at the same time the first in a series of new library system modules that will likely appear over the next decade [72.30, 31]. As the industry continues to evolve, and as open systems respond to the demands of Web 2.0, we can expect the Integrated Library System to morph into a strategy based on the survival of the best-of-breed. The future will be one of distinct functional modules that communicate with one another by exploiting the concept of unified resource management (URM) [72.32].

Responding to the requirements and demands of the digital era, where e-Resources proliferate and dominate the publishing and acquisitions scene, the venerable ILS acquisitions and serials modules are already showing sings of coalescing around the ERM model. The ERM, with support for licensing and the other key attributes of e-Content, will replace the traditional systems which are oriented largely toward print and nonprint analogs (maps, kits, scores, sound recordings, etc.).

As the ILS gives way to the URM, and the URM becomes the predominant model for the third generation of library management systems, management of e-Content will no longer be an afterthought. The challenge for libraries, librarians, and vendors alike will be to develop a system, or perhaps a series of interconnected systems, that can inherently manage all manner of information content, both traditional and digital, and do it by exploiting both content and metadata from the very cradle of their existence.

72.5 Further Reading

72.5.1 General Overview

- R.W. Boss: *The Library Administrator's Automation Handbook* (Information Today, Medford 1997)
- M.D. Cooper: *Design of Library Automation Systems* (Wiley, New York 1995)
- T.R. Kochtanek, J.R. Matthews: *Library Information Systems: From Library Automation to Distribution Information Access Solutions* (Libraries Unlimited, Westport 2002)
- Public Library Association: Tech notes section (updated regularly). http://www.ala.org/ala/mgrps/divs/pla/plapublications/platechnotes/index.cfm (last accessed 2009)

72.5.2 Specific Topics

Automated Authority Control Services
Of all the vendors that existed to provide automated outsourcing of authority records in the 1990s (Blackwell's, WLN, OCLC, Marcive, Library Technologies, etc.), few are still functioning: see, for example,

- Library Technologies: Official homepage. (2008) http://www.librarytech.com
- MARCIVE: Official homepage. (2008) http://www.marcive.com

Interlibrary Loan Services
- For a broad listing of interlibrary loan products and vendors, see http://www.cdlc.org/Resource_Sharing/ill/illproducts.shtml

- For an overview, see http://en.wikipedia.org/wiki/ Interlibrary_loan
- For information on two widely used software products, see ILLiad (Interlibrary loan management software): http://www.oclc.org/illiad/, Clio software: http://cliosoftware.com/public/
- Online Computer Library Center (OCLC): The future is now: the convergence of reference and resource sharing, Proc. OCLC Symp., ALA Midwinter Conference (OCLC, Dublin 1996)

RFID/Self-Checkout Services
- R.W. Boss: RFID technology for libraries. (updated 2007), http://www.ala.org/ala/mgrps/divs/pla/ plapublications/platechnotes/RFID-2007.pdf (last accessed 2009)

Virtual Reference Services
- R.W. Boss: Virtual reference. (updated 2007), http://www.ala.org/ala/mgrps/divs/pla/

plapublications/platechnotes/Virtuel_reference.pdf (last accessed 2009)
- L.W. Healy, S. Brown, L. Barrett: *Creating the Virtual Reference Desk: The Decision to Take Reference Live and Online* (Outsell, Burlingame 2002)
- D.K. Kovacs: *The Virtual Reference Handbook: Interview and Information Delivery Techniques for the Chat and e-Mail Environments* (Neal-Schuman, New York 2007)
- R.D. Lankes (ed.): *The Virtual Reference Desk: Creating a Reference Future* (Neal-Schuman, New York 2006)
- A.G. Lipow: *The Virtual Reference Librarian's Handbook* (Library Solutions, Berkeley 2003)

For one specific vendor product, see

- Online Computer Library Center (OCLC): QuestionPoint (2009). http://www.oclc.org/questionpoint/ default.htm (cooperative virtual reference) (last accessed 2009)

References

72.1 V. Porter: Continuous library catalog card, Patent 4005810 (1976), This represents a computer-area patent on the original library catalog card. http://www.google.com/patents? vid=USPAT4005810 (last accessed 2009)

72.2 T.-A. Wong: Card catalogs and card catalog drawers, Patent 5257859 (1993), This represents a tweak on the original design of the card catalog drawer. http://www.google.com/patents? vid=USPAT5257859 (last accessed 2009)

72.3 University of Pennsylvania: Cards from rapidly-disappearing card catalog. (Philadelphia 2008), http://www.library.upenn.edu/exhibits/ pennhistory/library/cards/cards.samples.html (last accessed 2009)

72.4 Online Computer Library Center (OCLC): OCLC catalog cards. (Dublin 2008), http://www.oclc.org/support/ documentation/worldcat/cataloging/cards/ default.htm (last accessed 2009)

72.5 K. Coyle: Catalogs, card – and other anachronisms, J. Acad. Librarians **31**(1), 60–62 (2005)

72.6 H. Avram: Obituary, with reference to the MARC project. (2006), http://www.nytimes.com/2006/05/ 03/us/03avram.html?ex=1304308800&en= 60308f72d854136a&ei=5088&partner= rssnyt&emc=rss (last accessed 2009)

72.7 F. Kilgour: Obituary, with reference to the founding of OCLC. (2006), http://www.oclc.org/news/releases/ 200631.htm (last accessed 2009)

72.8 Online Computer Library Center (OCLC): Official homepage. http://www.oclc.org (last accessed 2009)

72.9 C.M. Goldstein: Integrated library systems, Bull. Med. Libr. Assoc. **71**(3), 308–311 (1983)

72.10 E.G. Fayen: Integrated library systems, Encyclop. Libr. Inform. Sci. **1**(1), 1–12 (2004)

72.11 C. Lynch: From automation to transformation: forty years of libraries and information technology in higher education, Educause Rev. **35**(1), 60–68 (2000)

72.12 T. Primich, C. Richardson: The integrated library system: from innovation to relegation to innovation again, Acquis. Libr. **18**(35–36), 119–133 (2006)

72.13 T.R. Kochtanek, J.R. Matthews: *Library Information Systems: From Library Automation to Distributed Information Access Solutions* (Libraries Unlimited, Westport 2002)

72.14 R. De Gennaro: A computer produced shelf list, Coll. Res. Libr. **31**(5), 318–331 (1970)

72.15 T. Robinson: *Personal Communication* (Harvard Univ. Office for Information Systems, Harvard 2007)

72.16 Elsevier: Endeavor merges with Elsevier Science, American Libraries (2000)

72.17 P. Caplan: Information on the appropriate copy problem (2001). http://www.dlib.org/dlib/ september01/caplan/09caplan.html (last accessed 2009)

72.18   NISO: Information on NISO's OpenURL standard. (updated 2008), http://www.niso.org/standards/index.html (last accessed 2009)

72.19   NISO: NISO metasearch initiative (2006), http://www.niso.org/workrooms/mi (last accessed 2009)

72.20   Harvard Crimson: e-Resource elation. (2005), http://www.thecrimson.harvard.edu/article.aspx?ref=508846 (last accessed 2009)

72.21   DSpace: Official homepage. http://www.dspace.org (last accessed 2009)

72.22   Fedora: Official homepage. http://www.fedora.info (last accessed 2009)

72.23   National Library of New Zeland: Information on the NLNZ project. (updated 2008), http://www.natlib.govt.nz/about-us/current-initiatives/ndha (last accessed 2009)

72.24   Digital Library Federation: Electronic resource management. (2004), http://www.diglib.org/pubs/dlf102/ (last accessed 2009)

72.25   Online Computer Library Center (OCLC): Perceptions of libraries and information resources. (2005) http://www.oclc.org/reports/2005perceptions.htm (last accessed 2009)

72.26   D. Flecker: OPACS and our changing environment: observations, hopes, and fears (2005).

http://www.loc.gov/catdir/pcc/archive/opacfuture-flecker.ppt (last accessed 2009)

72.27   Endeca: Official hompepage. http://www.endeca.com (last accessed 2009)

72.28   K. Antelman, E. Lynema, A.K. Pace: Toward a twenty-first century library catalog, Libr. Inform. Technol. Assoc. **25**(3), 128–139 (2006), http://www.ala.org/ala/mgrps/div/lita/252006/number3september/antelman.cfm (last accessed 2009)

72.29   Vanderbilt University: Television News Archive. (Nashville 2008), http://tvnews.vanderbilt.edu/ (last accessed 2009)

72.30   M. Breeding: Trends in library automation: meeting the challenges of a new generation of library users. (2006). http://www.oclc.org/programsandresearch/dss/ppt/breeding.ppt (last accessed 2009)

72.31   R. Dietz, C. Grant: The disintegrating world of library automation, Libr. J. **130**(11), 38–40 (2005), http://www.libraryjournal.com/article/CA606392.html (last accessed 2009)

72.32   J.E. Grogg: Investing in digital, Libr. J. **132**(9), 30–33 (2007), http://www.libraryjournal.com/article/CA6440577.html (last accessed 2009)

73. Automating Serious Games

Gyula Vastag, Moshe Yerushalmy

In this chapter, we present the theoretical under-pinnings of and conditions for learning through experiences gained in *immersive learning simulations* (or simply *gaming*) and show some examples of authentic learning experiences for business professionals. We argue that recent developments in the theory of knowledge acquisition coupled with changes in the educational landscape (e.g., the proliferation of online courses and programs) present an unrivalled and so far mostly unexploited opportunity for serious games. The examples presented are from using Managerial Enterprise Resource Planning (MERP) in a variety of MBA and corporate business training programs. Related examples of other virtual learning environments are also included.

Part G | 73

73.1 Theoretical Foundation and Developments: Learning Through Gaming

It is widely accepted that "serious games are (digital) games used for purposes other than mere entertainment" [73.1]. *Michael* and *Chen* [73.2] give a similar definition: "A serious game is a game in which education (in its various forms) is the primary goal, rather than entertainment." The market for serious games is growing rapidly as more and more organizations are adopting or expected to adopt them in their training programs.

Conceptually, serious games are related to and somewhat overlap with the following terms: e-Learning, edutainment, game-based learning, and digital game-based learning [73.1]. In 2002, the Woodrow Wilson Center for International Scholars (in Washington, DC) founded the Serious Games Initiative and the term, following the first Serious Games Summit held in 2004, became generally accepted (http://www.seriousgames.org/) and a growing num-

ber of scholars began working in this field. Most recently, Sawyer and Smith developed a taxonomy of serious games and made it available on the organization's website (http://www.dmill.com/presentations/serious-games-taxonomy-2008.pdf), and Table 73.1 provides a synthesis of applications in different segments in society.

The illustrative applications of serious games include, but are certainly not limited to, disaster preparedness, health care, military training, and education, the main focus of this chapter (Table 73.2). For a related discussion of the role of automation in education also see Chaps. 44 and 85. *Breslin* et al. [73.3] describe how a game is used to execute and assess the effectiveness of alternative strategies for responding to a disaster. *Kelly* et al. [73.4] introduce *Immune Attack*, a personal computer (PC)-based game, used for learning the rules of the immune system and revealing deeper insights.

Table 73.1 Taxonomy of serious games (courtesy of Sawyer and Smith)

	Games for health	Advergames	Games for training	Games for education	Games for science and research	Production	Games as work
Government and NGO	Public health education and mass casualty response	Political games	Employee training	Inform public	Data collection/ planning	Strategic and policy planning	Public diplomacy, opinion research
Defense	Rehabilitation and wellness	Recruitment and propaganda	Soldier/ support training	School house education	Wargames/ planning	War planning and weapons research	Command and control
Healthcare	Cyber-therapy/ exergaming	Public health policy and social awareness campaigns	Training games for health professionals	Games for patient education and disease management	Visualization and epidemiology	Biotech manufacturing and design	Public health response planning and logistics
Marketing and communications	Advertising treatment	Advertising, marketing with games, product placement	Product use	Product information	Opinion research	Machinima	Opinion research
Education	Inform about diseases/ risks	Social issue games	Train teachers/ train workforce skills	Learning	Computer science and recruitment	P2P learning constructivism documentary?	Teaching distance learning
Corporate	Employee health information and wellness	Customer education and awareness	Employee training	Continuing education and certification	Advertising/ visualization	Strategic planning	Command and control
Industry	Occupational safety	Sales and recruitment	Employee training	Workforce education	Process optimization simulation	Nano/ biotech design	Command and control

America's Army (www.americasarmy.com) was introduced at the 2002 Electronics Entertainment Expo and was, after some initial hesitation, officially embraced and supported by the United States (US) Army as an effective military training simulator [73.5].

In addition to the wide variety of applications, the theoretical underpinnings of learning through gaming were also explored. Although *Gros* [73.6] has some

reservations about the current status of research in the educational use of computers (specifically, he cautions that research in this area is still disjointed), he and others also point to some common understanding and principles of learning. First, there are publications showing the advantageous effects of computer games. *Green-field* [73.7, p. 29] reported that "more skilled video game players had better developed attentional skills

Table 73.2 Serious game types and application examples

Serious game types	Application examples
Game-based education	Game-based learning, edutainment, learning games
Game-based production	Game-based authoring, machine behaviors
Game-based simulation	Game-based simulator, simulation games
Game-based messaging	Game-based advertisement, game-based marketing, advergaming
Game-based training	Game-based trainer
Game-based interface	Game-like interface, game-based graphical user interface (GUI)

Fig. 73.1 Leadership development to assess critical reasoning using a virtual environment (courtesy of SimuLearn)

than less skilled players." *Backlund* et al. [73.8] wrote about the positive impact that playing computer games had on traffic school students' driving behavior. *Koepp* et al. [73.9] showed that playing video games stimulates substantial dopamine release and that gamers' performance was related to the amount of dopamine present. Considering that dopamine is a chemical precursor to the memory storage event, it can be argued, they write, that video games are able to chemically *prime* the brain for learning. Furthermore, they write (p. 34) that "Beyond science and engineering games, further evidence suggests that games teach not only facts but detailed reasoning applicable to life's many challenges" (Fig. 73.1).

Mayo [73.10, pp. 32–34] listed seven characteristics that enable video games to address the deficiencies of education, specifically the shortcomings of big lecture formats typically used at universities.

1. Massive reach of players/participants.
2. Experiential learning (*If you do it, you learn it*). In games, players must navigate game scenarios and make decisions with consequences.
3. Inquiry-based learning (*What happens when I do this?*). This is the dominant philosophy in science and is also a natural mode for many video games that support freeform exploration, discovery, and experimentation.
4. Self-efficacy (*If you believe you can do it, you'll try longer/harder, and you'll succeed more often*

than you would otherwise). In games, points, levels or magic swords are awarded at positive decision points, encouraging players to keep going.
5. Goal setting (*You learn more if you are working toward a well-defined goal*). All games have goals, a key distinction between games and simulations.
6. Cooperation (team learning). Studies of classroom techniques show that cooperative learning results in about a 50% improvement over either solo or competitive learning. Some types of games (such as massive multiplayer online games) are intrinsically structured as a team effort toward a common goal.
7. Continuous feedback and tailored instruction.

Corti [73.11] coined the term *immersive learning simulation* (ILS) and pointed out that many corporate learning professionals and suppliers of ILS are not aware of what is needed to create a *truly effective and appropriate learning solution*. His call is supported by *Zyda* [73.12] who writes about the reasons supporting the need to *deploy immersive storytelling in service to society in the interactive realm*. *Shaffer* et al. [73.13, p. 7] argue that

Games are powerful contexts for learning because they make it possible to create virtual worlds, and because acting in such worlds makes it possible to develop the situated understandings, effective social practices, powerful identity, shared values, and ways of thinking of important communities of practice.

An example of such a virtual world is presented in the sequence of screenshots in Fig. 73.2.

Writing about computer-supported collaborative learning (CSCL), *Salomon* et al. [73.14] used the term *cognitive residue* to refer to a byproduct effect of learners interaction with media [73.15]. By cognitive residue, they meant [73.15, p. 3 and p. 6]

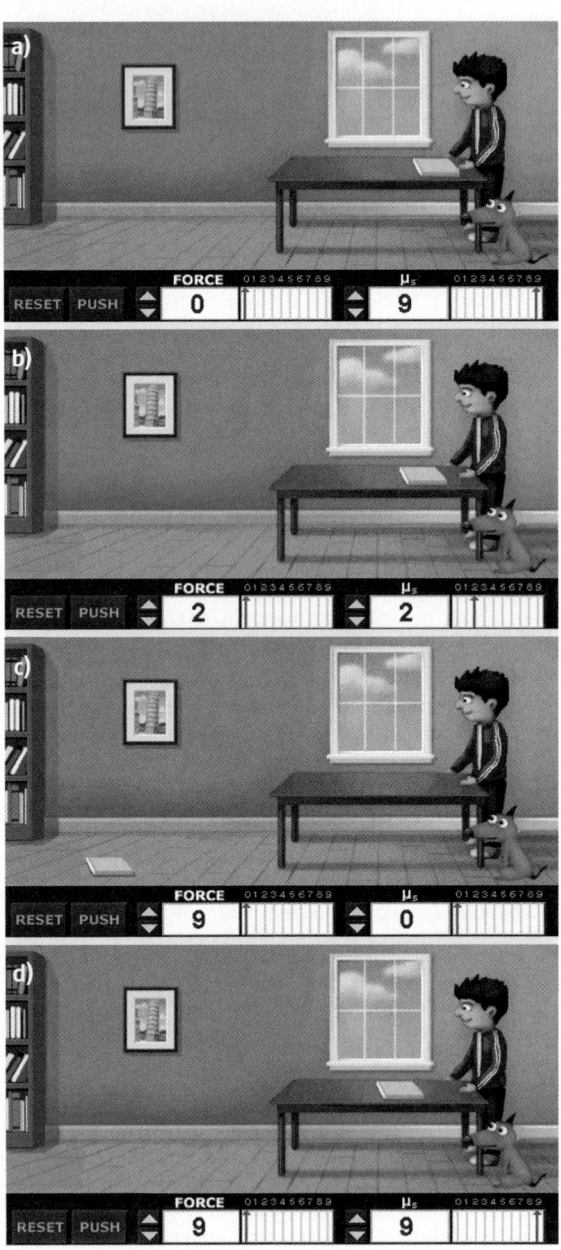

subsequent changes in mastery of knowledge, skill or depth of understanding away from the computer [...] higher order thinking skills that are either activated during an activity with an intellectual tool or are explicitly modeled by it can develop and can be transferred to other dissimilar, or at least similar situations.

These cognitive effects are *effects with technology obtained during intellectual partnership with it, and effects of it in terms of the transferable cognitive residue that this partnership leaves behind in the form of better mastery of skills and strategies.*

It has long been confirmed that learning and doing are tightly linked, as learning typically involves activities as well [73.16]. Many professionals, including current and would-be business managers, are action oriented; they learn to do something that later is to improve business operations. *Schon* [73.17] called the link between knowing and doing a reflective practice.

Professional societies (e.g., for doctors, lawyers, managers) develop supervised settings, residencies for doctors or moot courts for lawyers, where a learner can develop his/her skills by trying a course of action and then reflecting on the results together with his peers and mentors. During this course of action, learners develop epistemic frames of knowledge that are [73.16]:

the conventions of participation that individuals internalize when they become acculturated. The reproductive practices of the community are the means by which new members develop that epistemic frame.

There is a clear message here: learning does not happen in a vacuum; *communities of practice* or *affiliation groups* provide elements of the good learning that include goals, interpretations, practice, explanations,

Fig. 73.2a–d The friction explorer allows users to see how a sliding object behaves when the force applied to push it and the coefficient of friction are modified. (**a**) With an applied force of 0 and coefficient of friction of 9, the object does not move; (**b**) with an applied force of 2 and a coefficient of friction of 2 the object barely moves; (**c**) with an applied force of 9 and a coefficient of 0 the object slides off the table; (**d**) with an applied force of 9 and a coefficient of friction of 9 the object moves about one-third of the width of the table (courtesy of Schlumberger Excellence in Educational Development (SEED) http://www.seed.slb.com/en/index.htm)◀

debriefing, and feedback [73.18]. Serious games represent a clear opportunity to create communities of practice for business professionals.

In all settings or contexts, learning is about content acquisition while games typically focus on experience [73.19]. In other words, there is content (facts, principles, information, skills) that has to be learned. *Gee* [73.18] argues that indirect teaching of content (that is, teaching content via *something else*) is superior to the traditional approach used by schools, which teach content directly. Serious games can be that *something else* that helps in content acquisition by providing a variety of authentic learning experiences that mimic those of professional societies.

New results in learning theory [73.20, 21] show that people primarily think and learn through experiences they have had, assuming that these experiences meet certain conditions. *Gee* [73.22] summarized these five conditions as follows:

1. Experiences are most useful for future problem solving if they are structured by specific goals
2. For best learning, experiences must be interpreted by extracting the lessons learned
3. Learning is enhanced if immediate feedback is provided so that the learners can recognize and evaluate their errors
4. Learners need ample new opportunities to apply and reapply their previous experiences
5. Learners need to learn from interpreted experiences and explanations of other people, both peers and experts.

Gee's *situated learning matrix* [73.18, 22] refers to the linkage between content and identity. Content is given in the context of a goal-driven problem space. Learners move through similar problems representing various contexts and learn to interpret and generalize their experiences in these contexts. In video games, for example, players can get lots of practice at a given level and can then apply what they have learned in less familiar situations (across levels, for example). Through offering learning in varied contexts, the problem of context-specific learning is solved; for example, learning through a case study may leave the learner with knowledge that is specific to the given case but cannot easily be generalized to other settings. On the other end of the spectrum, learning out of context, focusing on principles only, for example, may give learners knowledge that cannot be applied.

Learning, in Gee's view, moves from identity to goals and norms, to tools and technologies, and only then to content. Identity has been linked to two characteristics that enhance learning. First, the ability of the players to have microcontrol over some elements of the game extends and creates an, as *Gee* [73.22] wrote, "embodied empathy for a complex system." The second characteristic is the learner's motivation for and ownership over the successful outcome of the game. The often quoted saying is that *success is based on experience and experience comes from failures*. A lot can be learned from a failed business or from not reaching the next level in a game. If the players are motivated to take risks, try out new ideas, and eventually, learn from the experiences, then the game has a good design.

73.2 Application Examples

Two illustrative examples of serious games are introduced and discussed in this section:

- Management Enterprise Resource Planning (*MERP*) – a dynamic, real-time-based online simulation aimed at strategy execution
- *L'Oreal e-Strat* – an online strategic business market competition game aimed at exploring various strategies by teams of managers who compete in the same market.

73.2.1 MERP

MERP was developed to facilitate learning by offering a variety of managerial contexts in which good prac-

tices can be discovered and practised (for details, see http://www.mbe-simulations.com).

Unlike the usual static, written case studies, the dynamic case study scenarios (DCSS) by MBE Simulations offer a live business environment with computerized scenarios that the students must manage and thereby control the destiny of their own business operations. Students participate in a virtual organization that is made real and dynamic as minute-by-minute business events and conditions unfold. Students must respond to and make complex managerial decisions in real-time based on the big picture view. While traditional strategic business games deal with results on a periodic basis (i. e., monthly, quarterly or yearly), the MERP experience involves real-time execution of strategies and the results are shown by key performance indicators (KPI).

MERP consists of a whole suite of scenarios that cover practices at various levels of complexity and scope. Focusing on individual departments (such as production, purchasing, and others in a production environment), as well as different industrial business processes (make-to-stock versus make-to-order), customer orders and client contracts fluctuate, allowing the use of different management models and inventory purchasing replenishment methods [e.g., the materials resources planning (MRP) mechanism or order level inventory policies]. Students can integrate theory into practice while they learn to test, retest, think, and re-think constantly in various real-life, dynamic situations.

Since 2001, MERP has been used both in traditional and online settings to teach operations and supply-chain management courses for MBAs. These courses involved about 500 students at the Stuttgart Institute of Management and Technology (Stuttgart, Germany), the Kelley School of Business (Kelley Direct Online MBA Program at Indiana University, USA), the CEU Business School (Budapest, Hungary) and the Corvinus School of Management (Budapest, Hungary). Typically, about 30–35% of the total points were assigned to MERP-related work. The simulation software covers the full supply chain of fictitious companies. Generally, stu-

dents have to work with three scenarios that differ in customer segmentation (small individual customers and/or clients with long-term contracts), product routings (flow shops and job shops), level of variability or randomness, and number of suppliers and final products. Teams of students have to manage these fictitious companies and make a variety of interrelated, complex decisions, so the simulation serves as an integrating theme.

One of the first decisions is to choose the key performance indicators (KPI) from the many options available for monitoring the progress of the enterprise (Fig. 73.3). The screen, in addition to current measurements, also shows a benchmark: results without any intervention (the light grey line).

Managerial success is measured both by numerical results (profit, cash on hand, and reputation level) and by the actions that the students have taken. The profit earned (the more, the better) is the primary measure of success, while the others serve as qualifiers (at the end of the simulation, students have to have positive cash on hand and minimum 75% reputation). The *quality* of the students' decision-making is also evaluated to avoid successes based on random acts not supported by an analysis.

In evaluating the simulation runs, assuming that all qualifiers are met, different weights can be assigned to

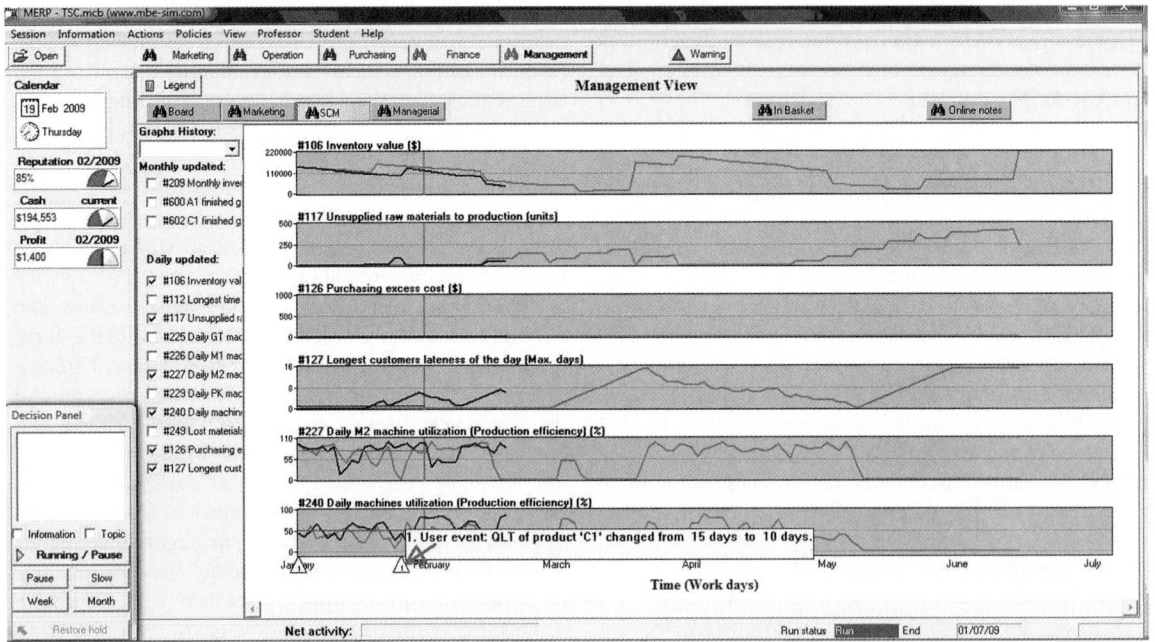

Fig. 73.3 Simulation: management view

the profit earned and to the quality of decision-making. Figure 73.4 describes the relationship between profit and points for one of the scenarios. For illustration purposes, a locally weighted regression line is shown. The point calculations, however, are based on linear functions as described below.

The profit range is divided into three segments: upper, middle, and bottom. In each segment, a linear function describes the relationship between profit earned and points given. This function is steep in the middle and relatively flat at the tails, so the differentiation among the top/low performers is less than among those who are in the middle. The figure also shows that submissions with the same profit received different point scores; these fluctuations are due to differences in the quality of decision-making scores.

In many ways, the simulation's virtual world mimics real life. First, like in real life, there are multiple factors with which students must interact. There are many potential problems and no prepared solutions to counter these. With the MBE Simulator software, what at first appears to be a very simple scenario raises many complicated interactions that are not as straightforward as the student might think. Students have to understand and get an overview of the whole supply-chain picture, which can be quite an overwhelming experience for most of them.

Also, in MBA programs, there is ever-increasing pressure on getting better grades (and later converting better grades into better jobs), thus increasing the temptation for cheating. As a result, much copying and information exchange goes on between current and past students. With traditional case studies, students can easily obtain the whole discussion to see how it was led and managed. To try to prevent this, professors have to regularly change, revise, and update teaching materials just to prevent cheating. However, the principal problem with cheating, if we put aside the ethical considerations and the added work to prevent it, is that students learn nothing from it.

The simulator offers a solution to this problem by providing an ongoing measurement of how much time each student has spent on the simulator. Furthermore, the professor can recreate the work done by the students and see exactly what path they took and what decisions they made. The students are also required to submit an executive summary that describes their management philosophy, tactics, and decisions. This assignment challenges students to reveal the way in which they managed and why they did so, just like real-life boards of directors are required to do. In class, stu-

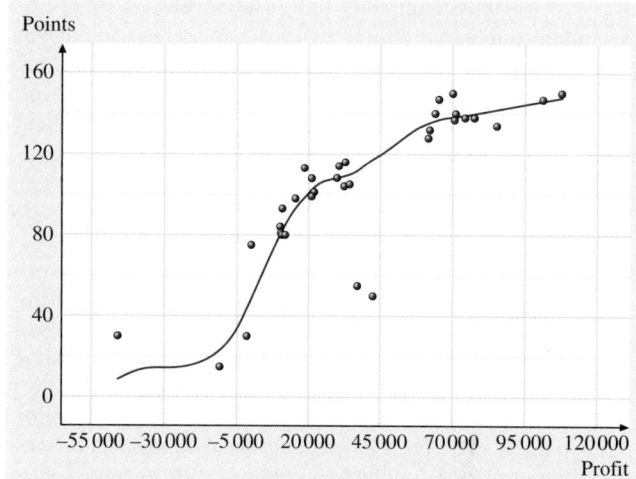

Fig. 73.4 Relationship between profit and points for one of the scenarios. A LOWESS/LOESS locally weighted regression line is shown

dents are asked why they made each decision. This is completely different from the usual executive summary, where only managerial slogans are used. Here the summary (10–12 pages total) provides real insights into the company, and is a good indicator of the extent to which the students understood the problem.

In sum, in the simulation most of the cheating-related problems are eliminated, because it takes far more effort to copy someone's notes (and then answer questions during the presentation of the results) than to learn while running the simulator. In addition, the complexity of the scenario allows for more than a single, well-defined solution. Rather, the simulation provides an ongoing search for solutions.

Learning with the simulator is ideal because it can be used to illustrate, for example, inventory modeling embedded in practice. In traditional problems (at the end of textbook chapters), one paragraph describes the problem, which is basically a mathematical problem, and the students simply plug in the numbers and get the answer. In the simulator, you can actually see (through the graphs and through the mechanism itself) the impact of changing a parameter on other parameters (such as changing the order point or order quantity). Other teaching topics that can be illustrated by the simulator are: forecasting, master production scheduling, MRP, bill of material, rough cut capacity planning, purchasing, and choosing suppliers. In short, you can cover every topic included in any operational management course.

MBE Simulations has also developed and customized some virtual worlds for various organizations, including SAP, Amdocs, Kodak, Motorola, Bank Leomi, Biosense (Johnson and Johnson), Lucent, KPMG, Cellcom, Tnuva, Israel Military Industries, and Orbotech.

73.2.2 L'Oreal e-Strat

L'Oreal e-Strat, the second example, is the biggest online game used by universities and corporations. During 6 months (= six decision rounds), which corresponds to 3 years in real time, students in teams of three assume the responsibilities of the chief executive office (CEO) in a virtual cosmetics company. In order to achieve the highest share price index, teams have to make decisions about pricing, production volumes and capacity, logistics, research and development, marketing, advertising, and brand positioning after analyzing data and charts of competitors and the market situation.

The students have to analyze a periodical report after every round which contains information about revenues, gains and losses, production quantity, capacity utilization, market research studies, competitors, brand perceptions, and benchmarking with other competitors. Based on the information in the report, the students have to make decisions in the following areas of the company's value chain: research and development, production planning, brand management, distribution, and corporate social responsibility. The decisions that the students can make are limited by a budget, which is determined by the performance of the last round.

Each area has its own additional, area-specific decisions:

Research and development decisions have an impact on customer satisfaction as well as on customer demand.

In the *production planning* section, by changing policies in terms of pricing, advertising, distribution or even research and development, it is likely that demand for certain products will vary strongly over time. High prices for products or low prices for competing products normally lead to a decrease in demand. In order to satisfy the demand of all customers and in order not to produce products to stock, the management has to forecast market demand and include it into the production planning. If students assume too low a level of production capacity, demand will not be met, which will have a negative impact on the revenues of the company and negatively affect the share price index (SPI).

In the e-Strat world, there are four *distribution channels*. Management has to allocate distribution and trade marketing budget to each retail channel depending on the shopping habits of the target group.

Among other considerations, the e-Strat structure includes *corporate social responsibility*. Management can decide if the company wants to invest in CSR activities (e.g., water reduction, career tracking, and kindergartens for employees). These investments will lead to better company reputation and can have a positive impact on the SPI.

E-Strat is a complex business simulation that covers all areas of a company's value chain. The game is based on decision rounds that have to be submitted to a central server that evaluates the results. However, although the simulation claims to show all responsibilities of the CEO, there seems to be an emphasis on strategy, marketing, and finance and less on operations.

73.3 Guidelines and Techniques for Serious Games

73.3.1 The Serious Game Solution for e-Learning Characteristics

A serious game realm dictates a *close to reality* replication of a reduced size business that needs to be managed by the trainee. The nature of *learning business by doing* provides a unique characteristic of this virtual world. The MERP and e-Strat solutions described above represent different approaches to serious games.

The comparison in Table 73.3 lists and explains the differences between the two different types of serious games: MERP and strategic business games such as e-Strat.

MERP and e-Strat address *Gee*'s [73.22] five conditions for learning through experience in the following way.

73.3.2 The Virtual World for *Learning by Doing* Characteristics

The main attribute of a serious game for teaching business acumen is its holistic world view aimed at providing the *learning by doing* nature of gaming. As such, the program should be able to simulate complex management situations that call for decision-making based on detailed and transparent analysis of the situation at

Table 73.3 MERP versus strategic business games such as e-Strat

Comparison criteria	Dynamic management workshop (MBE Simulations)	Business games (strategic or finance)
Decision type	*Realistic, daily-based* events that affect the organization's results	Strategic (investments, penentration of a new market, etc.)
Decision making	*Dynamic decision making.* Choosing when and which decision to make in parallel with the evolving KPI status	Mandatory periodical set of decisions (i. e., quarterly)
Competitive nature	Teams are competing based on *actual benchmark results.* Scenarios are stochastic, reflecting turbulent daily business life	Direct competition based on profit or share price index reflected. The dynamic is limited to competition cycles only.
Learning effectiveness	*Reinforced learning*: Repeated runs of the same organization	The game runs once, cannot be repeated
Learning scope	Aimed at improving the learner's effective strategy execution abilities. Managerial and business insights built upon tactical daily experience gained: Cause and effect, system thinking, and accepting and dealing with uncertainty	Focusing on strategic learning

hand. Such a program should incorporate the following approaches to learning and education:

- *A pool of live dynamic case studies.* These are descriptions and scenarios which depict certain management situations which often occur in real life; for example, in the simulated world of MERP and e-Strat, the trainees are given detailed data on the virtual manufacturing organization including: income statement, descriptions and details of the products and their market segmentation, details of the operational system, characteristics of the raw-material suppliers, cash and credit lines available, and various policies and ad hoc decision-making capabilities. A wide variety of organizational settings in marketing, operational, and finance can be described by different scenarios; *for example, a scenario's cover story may read like this*: the company appears to be well managed; it is doing all the *right things*. The order book is full; it has good customer relations in that it delivers on time and is therefore considered to be a reliable vendor. The company has a good management information system which provides detailed and varied management reports. It has an MRP scheduling and order system. It has reliable raw material suppliers. It has the most modern and reliable machinery in its operations division. It has everything, and is do-

ing everything, that a *good* manufacturing business should be doing. However, it is losing money.

- *It should cover a virtual-world presentation* of the entire company and its environment. When the trainee starts the simulated dynamic case study, the trainee should be able to watch his/her strategies unfold. The important point to note with this type of management simulation is that the trainee is in control and can set the rules, i. e., he/she can determine product prices, lead-time commitments made to customers, operational policies, inventory levels, suppliers to buy from, and so forth. In fact, everything that the trainee would normally plan and control in his own business is available for planning and control in the simulation.

- *It should be equipped with modern information technology (IT) managerial solutions.* As today's turbulent businesses are relying heavily on advanced integrated information systems, the trainee should be able to use such enablers in his/her practice. This need dictates a presentation of an integrated information system as well as a set of key performance indicators (KPIs) for each department of the organization. An example of such a KPI presentation is shown in Fig. 73.1. These IT enabler tools present dynamic changes in the company's situation, and trainees should respond to these changes by executing effective strategies.

Part G | 73.3

Table 73.4 Five conditions for learning through experiences

MERP	e-Strat
Each MERP scenario is structured around very specific goals. Some of these goals serve as order qualifiers (a certain level must be reached but going beyond the required level does not provide extra value) and some as order winners (the more, the better).	The game is structured around one goal: winning the contest, mainly by maximizing share price index.
In workshops and lectures, the simulation runs are interpreted and the lessons learned extracted.	In between the various cycles, the results of the previous round are interpreted.
Throughout each scenario, learners can rely on a wide range of key performance indicators (KPIs) to help them recognize and evaluate their errors.	Throughout each cycle run, learners can rely on a wide range of management reports to help them recognize and evaluate their errors.
This game is aimed at improving learners' effective execution rather than strategic decision making. Therefore, the learners have a practically unlimited number of runs in each scenario to apply and reapply their previous experiences.	Focusing on strategic learning, the learners have six decision rounds which correspond to 3 years in real time. The learning is gained upon the progress of their results impacted by competitors' decisions and not only on their own effective execution.
While the simulation is run individually, the learners are assigned to groups where the members' experiences can be shared. Also, independently of group membership, the results of all individual runs are displayed in a *Top 100* list that provides further motivation and feedback.	Learners are grouped into teams of three. The teams have to decide on pricing, production volume and capacity, logistics, research and development, marketing, advertising, and brand positioning after analyzing charts on competitors and the market situation in order to gain the highest share price index.

- *The decision-making nature* of such a solution should mimic that of real life. At any given point in time, a large number of potential decisions are available to the trainee and the challenge is to decide not only on the decision to make but also on the timing of the decision (that is, when to make a decision). This feature distinguishes MERP from traditional business games.
- *It should reflect a complex integrative and dynamic environment*: This realism is enhanced by the program's ability to introduce contingencies, such as the uncertainty of the market, variations in the supply chain, maintenance downtime, and machinery and plant breakdowns, to mention but a few. The impact of these contingencies is coupled with the impact of the strategic decisions already taken, and integrated with the results to date, in exactly the same way as in real life.
- *The program should describe a generic view of a managed environment.* In order to support real insightful learning, a serious game should provide a virtual-world description different from information system solutions readily available from vendors

(i. e., packages from SAP, Oracle, etc.). This will lead the trainee into the *content path of learning* rather than focusing on *is it real?* type questions.

73.3.3 Supporting the *Learning by Doing* Characteristics

Serious games for education/training should enable *learning by doing*. This need dictates the following main characteristics:

- *The role of the manager should be realistic.* In the assignment, the trainee should take the role of a manager who has been called in to take over a bankrupt company. He has lots of historical data which can be used to develop strategic plans for improving the situation, as well as to introduce tactical and operational control mechanisms. The game should provide all decisions and presentations needed for carrying out the strategy developed.
- *It should be fully interactive and integrative.* The trainee should be able to stop the simulation whenever he/she feels the need to review the situation and

think. The trainee should be able to change the policies, rules or decisions at any time and watch the effect. The program also allows the trainee to select and simulate a number of special activities, such as *expediting* and *order consolidation*. The game should describe a whole set of *cause and effect* chains for each decision made, allowing the trainee to explore and learn about the interrelations between decisions and their outcomes as well as about their impact on other business entities.

- *Ease of learning.* The game should lead the trainee along a short, relatively simple, path. Learning should be about content rather than about the mechanics of managing the simulation (as often happens with traditional games).
- *It should reinforce learning.* The solution should allow the trainee to repeat the same scenario again and again, trying a different decision path every time. In other words, this sensitivity analysis allows the trainee to learn from his/her own experience.
- *Putting into practice things learned.* The game should guide and steer trainees in the right directions needed for gaining real experience and making them learn through these experiences.
- *Simulate events that are close to reality.* The game should describe events in business and management that are affected by uncertainties; for example, product cost, product price, inventory level, and product demand by customers are all events that are dynamically and randomly changed in the background during simulation sessions. Consequently, these events cause varied and sometimes unexpected situations to which each trainee must react.

- *Provide monitoring and evaluating.* The game should track trainee decisions and operating performance as well as evaluate the decisions taken. MERP, for example, monitors and displays the decisions made by trainees. The consequences of these decisions (e.g., cash flow, inventory levels, and service levels) are continuously shown to group members (running the same assignment), making learning a shared experience. Additionally, a game should provide ways to test the quality of the trainee's decisions and their effectiveness.
- *Team learning and collaboration.* The game should provide built-in platforms for chat and team communication in a networked environment where trainees can interactively communicate with each other to share opinions and experiences about details and aspects of the simulator program in general or about the specific program scenarios.
- *Learning from the previous experience.* The game should allow saving of simulation sessions as run files. These run files should show the history of the simulation sessions and should be replayed later by the trainee and by professor. Session history executable files contain all decisions, whether simulator and/or trainee generated, exactly as in the originally run simulation session, for trainees to review and analyze separately.

73.4 Emerging Trends, Open Challenges

A significant trend in learning is the paradigm shift in the role of trainee (Iverson, in [73.2]), from passive to accountable active participation. Serious games offer a solution to support this trend.

Unlike traditional business learning approaches, like business games based on *one size fits all* solutions, there is a need for tangible, content-specific business education for motivated and engaged trainees. This need calls for more content about the various business skill areas and about industry-specific needs. *Michael* and *Chen* [73.2] listed the required skills as follows: people skills, job-specific skills, organization skills (that is, how to utilize resources effectively, collaborate, and focus on results, etc.), communication skills, and strategy skills aimed at effective execution of strategy needs.

Over time the generational structure of management will change; the younger generation is becoming more and more game-oriented. This new *gamer* generation is more accustomed to calculated risk taking and needs more challenging learning solutions (Beck and Wade, in [73.2]).

Veen and *Vrakking* [73.23] argued that schools of the future will have longer sessions (4 h periods) and interdisciplinary themes rather than subject-specific content. This trend in management learning will present a challenge to provide more collaboration-based serious games that reflect and mimic the holistic realm of a learning subject, and teachers should also be prepared for this new environment [73.24]. *Freitas* [73.25] wrote that training and education, both physically and conceptually, will shift to new opportunities beyond the traditional classroom. Diverse communities with different needs and cultures will present a new challenge for serious games, learners and educators as well [73.26].

The last challenge in serious game learning is the need to assess learning outcomes. Serious games should be able to show that learning has occurred

and the extent to which the learning objectives have been achieved [73.2]. This will elevate the characteristics of the learning game to be directed implicitly and explicitly toward learning goals. It will also require the development of assessment tools for serious games.

73.5 Additional Reading

- D. Cheek, H. Kelly: Designing an online virtual world for learning and training, Fifth IEEE Int. Conf. Wirel. Mob. Ubiquitous Technol. Educ. (2008) pp. 208–209
- T. Connolly, M. Stansfield: Games-based e-Learning: Implications and challenges for higher

education. In: *Social Implications and Challenges of e-Business*, ed. by F. Li (IGI Global, Hershey 2007)
- B. Sawyer, P. Smith: Serious games taxonomy, The Serious Games Summit at the Game Developers Conference, San Francisco (2008)

References

73.1   T. Susi, M. Johannesson, P. Backlund: Serious Games – An Overview (2007) http://www.his.se/upload/48173/SeriousGames_overview.pdf (accessed June 13, 2008)

73.2   D. Michael, S. Chen: *Serious Games: Games that Educate, Train and Inform* (Thomson Course Technology, Boston 2006)

73.3   P. Breslin, C. McGowan, B. Pecheux, R. Sudol: Serious gaming, Health Man. Technol. **28**(10), 14–17 (2007)

73.4   H. Kelly, K. Howell, E. Glinert, L. Holding, C. Swain, A. Burrowbridge, M. Roper: How to build serious games, Commun. ACM **50**(7), 44–49 (2007)

73.5   M. Zyda: From visual simulation to virtual reality to games, Computer **38**(9), 25–32 (2005)

73.6   B. Gros: Digital games in education: The design of games-based learning environments, J. Res. Technol. Educ. **40**(1), 23–38 (2007)

73.7   P.M. Greenfield: Video games as cultural artifacts. In: *Interacting with Video*, ed. by P.M. Greenfield, R.R. Cocking (Ablex, New York 1996) pp. 35–46

73.8   P. Backlund, H. Engström, M. Johannesson: Computer gaming and driving education, Proc. Workshop Pedagogical Design of Educational Games affiliated to the 14th International Conference on Computers in Education (ICCE 2006), Beijing (2006)

73.9   M. Koepp, R. Gunn, A. Lawrence, V. Cunningham, A. Dagher, T. Jones, D. Brooks, C. Bench, P. Grasby: Evidence for striatal dopamine release during a video game, Nature **393**(21), 266–268 (1998)

73.10   M.J. Mayo: Games for science and engineering education, Commun. ACM **50**(7), 30–35 (2007)

73.11   K. Corti: *Serious Thinking*, E.learning Age, March, 20–21 (2008)

73.12   M. Zyda: Creating a science of games, Commun. ACM **50**(7), 26–29 (2007)

73.13   D.W. Shaffer, K.D. Squire, R. Halverson, J.P. Gee: Video games and the future of learning, Phi Delta Kappan **87**(2), 104–111 (2005)

73.14   D. Salomon, D.N. Perkins, T. Globerson: Partners in cognition: Extending human intelligent technologies, Educ. Res. **20**(3), 2–9 (1991)

73.15   A. Agostino: The relevance of media as artifact: Technology situated in context, Educ. Technol. Soc. **2**(4) (1999)

73.16   D. Shaffer: Epistemic games, Innovate **1**(6) (2005), http://www.innovateonline.info/index.php?view=article&id=79 (accessed April 18, 2008)

73.17   D.A. Schon: *Educating the Reflective Practitioner: Toward a New Design for Teaching and Learning in the Professions* (Jossey-Bass, San Francisco 1987)

73.18   J. Gee: *Situated Language and Learning: A Critique of Traditional Schooling* (Rutledge, London 2004)

73.19   K. Squire: *Game-Based Learning: Present and Future State of the Field* (Univ. Wisconsin-Madison Press, Madison 2005)

73.20   J. Bransford, A.L. Brown, R.R. Cocking: *How People Learn: Brain, Mind, Experience and School* (National Academy, Washington 2000)

73.21   R.K. Sawyer: Analyzing collaborative discourse. In: *The Cambridge Handbook of the Learning Sciences*, ed. by R.K. Sawyer (Cambridge Univ. Press, Cambridge 2006) pp. 187–204

73.22   J.P. Gee: Learning and games. In: *The Ecology of Games: Connecting Youth, Games, and Learning*, Foundation Series on Digital Media and Learning, ed. by K. Salen (MIT Press, Cambridge 2008) pp. 21–40

73.23 W. Veen, B. Vrakking: *Homo Zappiens: Reshaping Learning in the Digital Age* (Network Continuum, London 2006)

73.24 K. Becker: Digital game-based learning once removed: Teaching teachers, Br. J. Educ. Technol. **38**(3), 478–488 (2007)

73.25 S. de Freitas: Learning in immersive worlds, A review of game-based learning, http://www.jisc.ac.uk/whatwedo/programmes/elearning_innovation/eli_outcomes.aspx (accessed April 28, 2008)

73.26 L. Pannese, M. Carlesi: Games and learning come together to maximise effectiveness: The challenge of bridging the gap, Br. J. Educ. Technol. **38**(3), 438–454 (2007)

74. Automation in Sports and Entertainment

Peter Kopacek

A service robot has to be intelligent, mobile, and able to cooperate with other robots and devices. We are on the way towards multirobot systems in which several robots, called multiagent systems (MAS), will act in a cooperative way together a common task. One of the newest application areas of service robots and especially MAS is the field of entertainment, leisure, and hobby. People have more free time, and modern information technologies lead to loneliness of humans (teleworking, telebanking, teleshopping, etc.). Entertainment robots are expected to be one of the real frontiers of the next decade.

In this chapter a short description of such robots will be given, including some application examples. Due to the broad range of possible applications of robots in entertainment, leisure, and hobby, the following classification has been made in order to give this contribution a basic

structure: robot construction sets, sports assistants, promotion and public relations, robots in the entertainment industry, personal robots, humanoid robots, and competition robots.

As an example, robot soccer competitions will be described in more detail. Finally an outlook on future development trends will be given.

There are three *starting* points for the development of intelligent robots (Fig. 74.1):

- Conventional stationary robots
- Autonomous guided vehicles (AGVs)
- Walking machines.

Stationary industrial robots equipped with external sensors are used today, e.g., for assembly and disassembly operations, fueling of cars, etc. and were the first *intelligent* robots.

Mobile platforms with external sensors (AGVs) have been commercially available for some years and cover a broad application field. Mobile platforms are the real roots of service robots.

Walking machines or mechanisms have been well known for some decades. Usually they have four to six legs (multiped) and only in some cases two legs (biped) – from the viewpoint of control engineering walking on

two legs is a very complex (nonlinear) stability problem. Biped walking machines equipped with external sensors are the basis for *humanoid* robots. Some prototypes of such robots are available today.

To give an idea about further developments, Fig. 74.1 shows possible development trends in robotics. We are now on the way from unintelligent industrial robots via intelligent industrial robots to intelligent mobile – including humanoid – robots to third-generation *advanced* robots able to interact and work symbiotically with us.

Robots in the 21st century will be used in all areas of modern life. The major challenges are:

- To develop robotic systems that can sense and interact usefully with humans
- To design robotic systems able to perform complex tasks with a high degree of autonomy.

Part G | 74

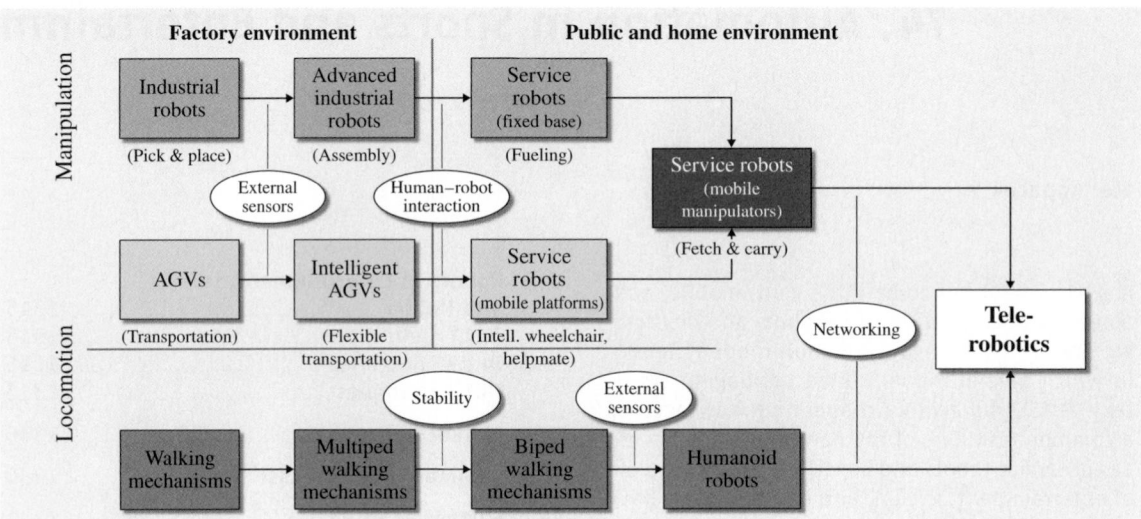

Fig. 74.1 From industrial to service robots (after [74.1])

In the same way that mobile phones and laptops have changed our daily lives, robots are poised to become a part of our everyday life. The robot systems of the next decades will thus be human assistants, helping people do what they want to do in a natural ad intuitive manner. These assistants will include: robot co-workers in the workplace, robot assistants for service professionals, robot companions in the home, robot servants and playmates, and robot agents for security and space.

The role of these robots of the future could be improved by embedding them into emerging information technology (IT) environments characterized by a growing spread of ubiquitous computing and communications and of ad hoc networks of sensors forming what has been termed *ambient intelligence*.

Currently available robots are far from this vision of the third generation, being able to understand their environments, their goals, and their own capabilities or to learn from their own experiences.

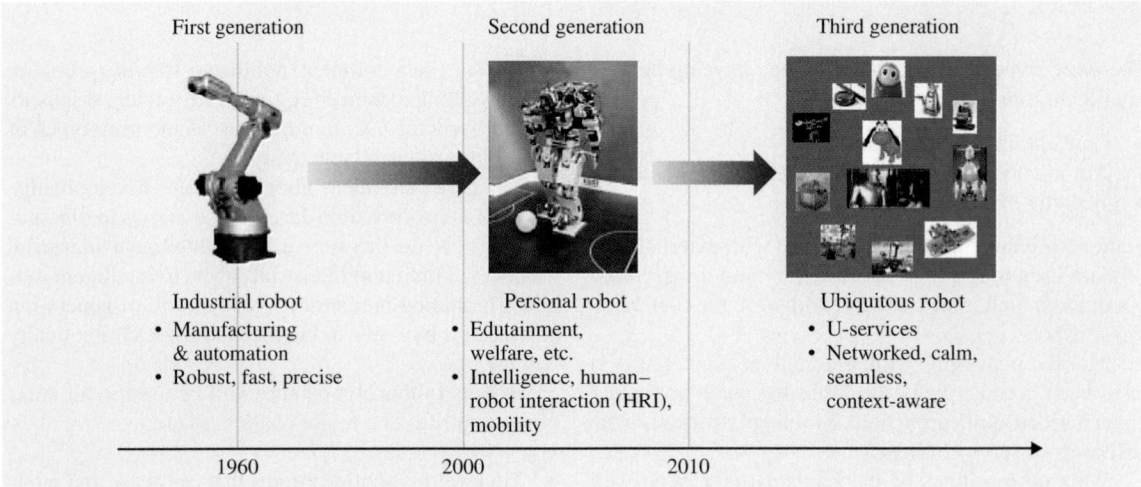

Fig. 74.2 Development trends in robotics [74.2]

74.1 Robots in Entertainment, Leisure, and Hobby

One of the newest application areas of service robots is in the field of entertainment, leisure, and hobby, because people have more free time, and modern information technologies lead to loneliness of humans (teleworking, telebanking, teleshopping, etc.). We need robots to assist, support, and join humans. The field of *robots in entertainment, leisure, and hobby* was therefore born. The roots of such robots go back more than 20 years ago [74.3, 4]. These robots are usually mobile, intelligent, and cooperative. The idea of making modern automation transparent for a broader public goes back to the middle of the 19th century. One tool for this is playing with automation. First examples were in the field of process automation, e.g., two- and three-tank systems, model railways, model racing cars, etc. At the end of the 19th century and the beginning of the 20th century robots started to attract a broader public. One of the reasons for this is humans' long-time dream of having an intelligent machine looking and acting like us. Therefore robots are currently and will be also in the future *high-tech toys* for playing with automation.

In the past construction sets for stationary, industrial robots served for playing with robots. These model robots were a popular gift not only for children. Approximately 30 years ago the first construction kits for mobile and also humanoid robots appeared on the market. Currently and also in the future the development of entertainment robots is closely connected to the development of real, mobile, intelligent, especially humanoid robots. Most of the *advanced* theories implemented in these real robots are and will be the basis for entertainment and play robots. The main problem can be summarized under the headline *cost-oriented automation* (COA). These robots must be available on the market for a reasonable selling price.

74.1.1 Definitions

First it is certainly important to take a brief look at different definitions of the terms entertainment, leisure, and hobby. Their meanings often overlap, but nevertheless they all mean something different.

Entertainment
This can be defined as: the act or profession of entertaining, i.e., to amuse and interest, especially by public performance and keeping the attention of people watching or listening [74.5]; the act of entertaining, or something diverting or engaging as a public performance, e.g., a (usually light) comic or an adventure novel [74.6].

Leisure
This can be defined as: time when one is free from work or duties of any kind; free time [74.5]; freedom provided by the cessation of activities; especially time free from work or duties [74.6].

Hobby
This can be defined as: an activity which one enjoys doing in one's free time [74.5]; a pursuit outside one's regular occupation, engaged in especially for relaxation [74.6].

74.1.2 Categories

Due to the broad range of possible applications of robots in entertainment, leisure, and hobby, the following classification [74.3] has been made in order to give this chapter its basic structure:

- Robot construction sets
- Sports assistants
- Robots for promotion and public relations
- Robots in entertainment industry
- Personal robots
- Humanoid robots
- Competition robots.

74.1.3 Examples

In the following some *classical* examples, because of the rapid development in this field, are presented and shortly discussed.

Robot Construction Sets
These construction sets often stem from *conventional* technical construction sets that have been upgraded using sensor and microcontroller technology, paired with computer interfaces to allow the user to run self-programmed software. They are mainly for *computer enthusiasts* for testing own-developed software on a cheap hardware.

There are two main types of construction kits available:

- Sets where it is possible to construct various robots
- Sets where only one robot can be constructed

Within these main categories it is also possible to choose from a wide variety of different kits; prices range from under US$ 100 to several hundred US dollars.

The well-known system in the first category is Lego Mindstorms [74.7]. Based on the products of this company, the construction set allows a great variety of possible assemblies due to its modular structure. The only limit is the fantasy of the constructor. The system consists is of a specially developed microcomputer (the heart of Lego Mindstorms), software, 717 elements (bricks, connectors, wheels, and gears), 2 motors, 2 touch sensors, 1 light sensor, and an infrared transmitter for sending programs to the robots.

Mobile Robots is a construction set for sensor-guided robots. The included robot–personal computer (PC) interface allows one to write ones own command routines on the computer and load them into the robot's memory. So the robot can act independently and without any attached cables. Possible constructions are capable of detecting edges and stopping before falling down, are aware of obstacles, and can follow light sources or track plotted lines. As it is a totally flexible system, own creations are possible and appreciated.

Representatives of the second category are robots from a company offering a wide variety of construction kits, but all of them designed for building one robot only. Various models of robotic arms, bipeds, quadrapods, hexapods, polypods, carpet rovers, four-wheel-drive (4-WD) rovers, and sumo robots are available.

Sports Assistants

The robots introduced into this field are systems well equipped with high-tech sensors that assist humans in their favorite sport by performing routine tasks usually annoying such as collecting tennis balls or carrying golf clubs.

Tennis Ball Collector: Jeeves. Developed by a freelance designer in cooperation with a university, this robot collects tennis balls by driving towards them when detected by the onboard charge-coupled device (CCD) camera, and bringing them into a basket in the rear using a rotating brush. Ultrasound distance sensors avoid collisions with obstacles on the court while the robot generates a map of the environment based on its sensor data and stores this information for further use, thus enabling *Jeeves* to sweep an already known court effectively. Path planning is done by a sophisticated computer system running on an off-board workstation that is connected via a radio link.

Golf Caddy: InteleCaddy. This robot replaces a human caddy in a mass sport application where many golfers are not willing to pay for a caddy, currently being installed at golf courses all over the world. The InteleCaddy is a computer-controlled, electrically powered golf caddie robot that navigates around the golf course with its telecommunication skills and sensors, being connected to the golfer via a small pager-sized coded transmitter. The robot features a digital aerial map of the golf course and is programmed to stay within preset boundaries, thus following the golfer with the bags using geographical positioning system (GPS) and ultrasound sensors for navigation and obstacle avoidance. As a bonus, the computer also supplies the golfer with useful information on the course such as exact distance to predefined positions (greens, bunkers, and tees) on the map.

Robots for Promotion and Public Relations

Robots for promotion and public relations can be divided into the categories of tour guide robots and performance robots, hence having very different tasks.

Tour guide robots give information to people while interacting with them, thus attracting and entertaining prospective visitors as well. Performance robots are for performances in public places or at events, mostly with robot systems rented by specialized companies.

Tour Guide Robots. A tour guide robot has to be able to move around autonomously in the environment. It has to acquire the attention of visitors and interact with them efficiently in order to fulfil its main goal: give the visitors a predefined tour. These robots were designed to conduct guided tours in public places such as museums

Fig. 74.3 Museum tour guide *Minerva*

Fig. 74.4 Museum tour guide *Twiddling, Inciting, Instructive*

Fig. 74.5 *Megasaurus* [74.8]

while managing autonomous sensor-based navigation as well as avoidance of collisions in a densely populated area of unpredictable moving visitors.

The tasks of a tour guide robot are:

- Traveling from one exhibit to the next during the course of the tour
- Attracting visitors to participate in a new tour between tours
- Engaging people's interest and maintaining their attention while describing a specific exhibit.

Furthermore, they are equipped with social interaction features, which range from rather simple voice output (*RHINO*) [74.10] to a combination of voice, facial expressions, and a model to express feelings as well (*Minerva*), giving the robot a *human touch*. The navigation system of both units is based on real-time processing of sensor data to generate a map of the environment.

Minerva [74.11] was developed in 1998 by Carnegie Mellon University and the University of Bonn and is working in the Smithsonian's National Museum of American History, USA.

These robots, developed by Fraunhofer IPA, Stuttgart, have been in use in the Museum für Kommunikation in Berlin, Germany since 2000.

Performance Robots. The main task of performance robots is to entertain people, often using low-tech applications with a few sensors, limited computing power, and no capabilities to act autonomously. Systems such as *Robosaurus* or *Boxerjocks* can be rented for large public events to attract visitors and thus increase the revenue of the renter.

Due to the existence of a large number of companies renting or distributing performance robots, just a few applications have been included in order to give a short survey of how broad the range of entertainment robots or animatronic systems is in this field.

Megasaurus and Transaurus are modeled after a *Tyrannosaurus rex* and have hydraulically activated arms, grasping claws, and jaws, as well as flame throwers set up in the head to give the effect of breathing fire from the mouth. They both fold up into a vehicle based on a tank and when the robots perform they initially appear as a box on tracks decorated as either a military vehicle (Megasaurus) or a dinosaur (Transaurus). Each robot is roughly 30 feet tall at maximum extension. They are used primarily to destroy cars by *eating* them (ripping them apart with the claws and jaws) at motor sport events, especially monster truck competitions.

Fig. 74.6 *Transaurus* [74.9]

Part G | 74.1

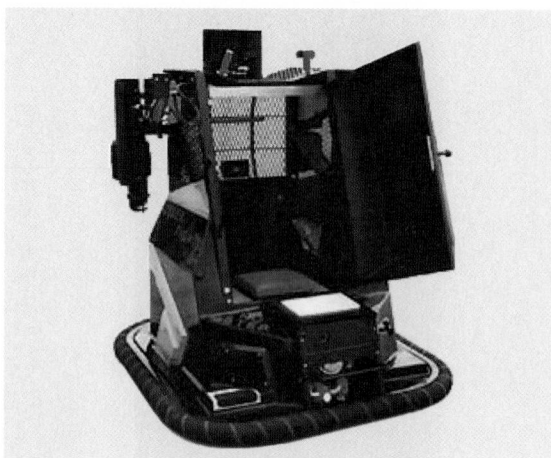

Fig. 74.7 Boxer Jack [74.12]

Boxer Jacks Robotic Boxing is an animatronics system that serves as an attraction in amusement parks or other places where large crowds of people can be found, such as shopping centers. The application system is fairly straightforward: Two of these robots are located within a certain area where they can move around freely. Each of the systems is controlled by a human, who tries to punch the other's robot. Technical features include:

- Construction – Tubular steel network with a rigid steel mesh surrounding the rider's upper body, providing ample safety and visibility
- Drive – Dependable Honda gasoline engine powers the hydropneumatic system for mobility and arm punching articulations
- Body – Hand-laid fiberglass with choice of color scheme; quality urethane enamel finishes used
- Seating – Designed so that riders aged 8 years through to adults can operate a Boxerjack easily
- Impact areas – Use of heavy rubber on all contact areas, bumper, gloves, and arms cushion all shock loadings during the match
- Electrical – 12 V direct-current (DC) battery system use for arm switching, scoreboard, and head horn.

Robots in the Entertainment Industry

Robot applications in the film industry are often spectacular because of their custom-made special features. In this chapter robots as performing artists, can be found as well.

Film Industry. Starting with *Metropolis* in 1926, dealing with a futuristic city and its mechanized society with humans being replaced by a robot, through *The Day the Earth Stood Still* (1951), *Lost in Space* (1965) to *A Space Odyssey*, robots have frequently been used in films.

One of the most recent examples in movies is *I-robot*; although not based on the the book written by *Asimov*, it was based on the concepts proposed therein, such as the three laws of robotic behavior. It presents a future where robots do all manual labor. One robot is programmed and trained to understand human emotions, and commits murder due to a promise and out of love. The robots take over as they get deeper understanding of the three laws – mankind is destroying itself and has to be protected from self-destruction.

As the entertainment industry was one of the first fields to use service robots, there is a huge variety of applications in theme parks and especially movies, where robotic and animatronic creatures are often combined with animated computer graphics to realize spectacular special effects (Figs. 74.8–74.10).

Another field of robotic applications is in the performing arts, where robots can be found rather seldom,

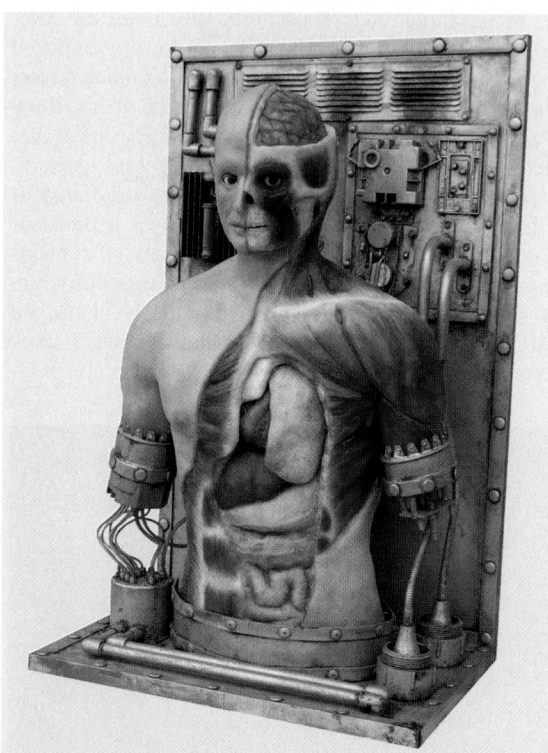

Fig. 74.8 Jekyll and Hyde animatronic (courtesy of Life Formations [74.12])

but therefore in mostly amazing performances that partly have a scientific background as well.

Robotic Snake in Anaconda. This lifelike replica of the giant *Anaconda* is a high-tensile-steel robot structure, consisting of individually controlled artificial vertebrae that are driven by hydraulic cylinders. The system is controlled by more than 70 coupled microprocessors with a special control concept that allows the underlying hydraulic system to perform the necessary quick movements of the heavy structure.

Robots in Jurassic Park. The robotic system incorporating the *Tyrannosaurus*, the largest one ever built for use in a movie, is a fiberglass frame construction covered by clay and afterwards coated with a specially painted latex skin. As the robot measured 9 m high and weighed 10 t, it had to be supported by a massive foundation usually used for flight simulators.

Other less spectacular systems such as the replica of a *Triceratops*, featuring remotely controlled motors and

a mechanism to imitate breathing movements, were also used in the movie.

Robots as Performing Artists. The robot band introduced here is a pretty low-tech application for entertainment purposes that plays digital music from a computer disk while the animated robotic figures pretend to move to the sound and play their instruments.

Although the *Ullanta Performance Robots* seem at first sight to exist solely for entertainment, these fully autonomous robots are used for research in cooperative robot group behavior and multiagent systems as well, both employed in a dynamic environment. These robots are, within certain boundaries, allowed to *interpret* the script based on their own software.

Personal Robots

Pet Robot: Furby. Being the ancestor of all pet robots, this popular animatronic pet is equipped with a computer chip, various sensors, and a small motor. It responds to its environment in *Furbish*, a nonsense lan-

Fig. 74.9 Lincoln library (Springfield, IL) animatronics (courtesy of Life Formations)

Fig. 74.10 Rolling Hills, Kansas Museum animatronics (courtesy of Life Formations)

guage, and is able to communicate with other *Furbies* via infrared signals.

Personal Robot: R100. This prototype mobile personal robot uses a completely new voice-controlled approach, thus providing a more natural, buttonless interface in order to become a *real family member*.

With its visual and voice recognition technology and Internet connectivity, it can be used as a voice message or home security system (besides its function as a personal companion).

Catrobot: Tama. Featuring speech output and recognition, this kitten-like pet robot serves as a companion and information source for senior citizens, while being remotely accessible by health or social workers who thus can take care of the elderly without too much intrusion into their privacy.

Robodog: i-Cybie. Described as "a watered-down version of AIBO with almost the same functions without the finesse" in the press, this infrared-controllable robot dog indeed features many of AIBO's functions at a technically lower level but for a fraction of the price.

Robodog: AIBO. AIBO is an abbreviation for artificial intelligence robot. Because of its reduced instruction set computer (RISC)-processor-powered high-tech system including numerous sensors, a color camera, and 18 motors to move its extremities it is the first entertainment robot that can think, feel, and act by itself. It can walk and play, sit, and stretch like dogs and cats. AIBO's brain contains an emotion model to handle feelings and an instinct model to handle drives. The emotion model covers six feelings: happiness, sadness, anger, surprise, fear, and dislike. The instinct model has four components: love, search, movement, and hunger.

Pet Robot: My Real Baby. This interactive, animated doll features instant, sophisticated, and interactive emotion-like responses and is able to simulate a maturation and speech development process.

Humanoid Robots

Entertainment robots are expected to be one of the real frontiers of the next decade. According to latest estimations millions of such robots will be in use in the coming years. The newest developments are *friendly* robots for humans; for example, elderly people need listening and talking friends. We now have dogs, cats, and

human-like robots available for reasonable prices, but not really humanoid robots walking on two legs [74.13]. First examples are described below.

Since 1986 the Japanese Company Honda has developed humanoid robots, P2, P3, and ASIMO (advanced step in innovation mobility). The basic idea is to integrate intelligence and moving capability into a robot for trivial tasks. Over the years the robot has become smaller (from 160 cm for P3 to 120 cm for ASIMO) and lighter (from 130 kg of P3 to 43 kg of ASIMO). The newest development is ASIMO. It can move at up to 1.6 km/h and has 26 degrees of freedom (DOFs). This kind of robot can easily be used in home as wheel-driven robots, because of their ability to move over an uneven surface such as stairs.

Therefore service robots will become a real *partner* of humans in the near future. One dream of the scientists is the *personal* robot. In 5, 10 or 15 years everybody should have at least one such robot. Because the term personal robot is derived from personal computer the prices should be equal. Some new ideas in automation, especially in robotics, are realized very quickly while others disappear.

Honda is trying to build the ASIMO robot to be a partner for people. So far it is merely a study about how to imitate humans' movements and make it able to help people somehow. It is 120 cm high, which is enough to reach most gadgets designed for humans. The latest model, introduced in December 2004, can even run at 3 km/h like a human.

QRIO is Sony's [74.14] next step after the robodog AIBO in entertainment robots. It's a bipedal humanoid robot that is able to:

- Walk on uneven and sloppy surfaces
- Run
- Jump
- Perceive depth through its two CCD cameras
- Create a three-dimensional (3-D) map of its surroundings
- Recognize people from their faces and voices
- Learn
- Connect to the Internet via a wireless home network
- Download and read information it thinks you're interested in
- Sing
- Dance
- Survive a fall unscathed and get back up again by itself.

In order for QRIO (Fig. 74.11) to walk and dance so skillfully, an actuator was needed with the ability to produce varying levels of torque at varying RPM speeds and respond with quickness and agility.

The robot moves with *dynamic walking*. *Static walking* means that the robot keeps its center of gravity within the zone of stability – when the robot is standing on one foot, its center of gravity falls within the sole of that foot, and when it is standing on two feet it falls within a multisided shape created by those two feet – causing it to walk relatively slowly. In *dynamic walking*, on the other hand, the center of gravity is not limited to the zone of stability – in fact it often moves outside of it as the robot walks. People move using *dynamic walking*.

If pushed by someone, QRIO will take a step in the direction in which it was pushed to keep from falling over. When it determines that its actions will not prevent a fall, it instinctively sticks out its arms and assumes an impact position. After a fall, it turns itself face up, and recovers from a variety of positions. It is equipped with a camera and the ability to analyze the images it sees. It detects faces and identifies who they are. Moreover QRIO can determine who is speaking by analyzing the sounds it hears with its built-in microphones.

An example of a reasonable cheap entertainment robot is the humanoid robot of Robonova [74.15]. This robot offers educators, students, and robotic hobbyists a complete robot package. It is a fully customizable and programmable aluminum robot (Fig. 74.12). Sixteen digital servos and joints give complete control of

torque, speed, and position. The programming software is simple, so advanced knowledge of programming is not needed. It can walk, run, do flips, cartwheels, and dance. The robot is available as a kit – assembly time approximately 8 h – or preassembled, ready to walk robot. In addition to the typical robot talent of walking until it senses a wall using ultrasound, Robonova can be instructed to do cartwheels, take a bow, and even do one-handed pushups.

The simplest way for programming is with the *catch-and-play* function. With RoboScript or RoboBasic the robot is moved to any position and by mouse click that position is *captured*. The software then links these *captured* positions and, once activated, smoothly transitions the robot's movements through these programmed positions.

For beginners in robot programming two software packages are available. With these the users can create operational subroutines without knowing any programming language at all. The computer screen displays sliders for every individual servo (joint). Moving the sliders changes the position of the servos. Simple movements can then be assembled to produce complex movements simply by clicking the mouse. These movements can then be called up on a graphical user interface.

For more advanced users a programming tool based on the Basic programming language is available. This enables the users to create complex applications de-

Fig. 74.11 QRIO

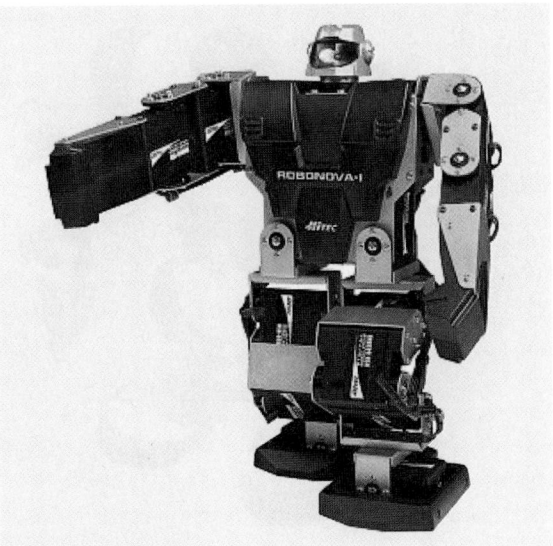

Fig. 74.12 Robonova

signed to accomplish their own individual tasks. The independent development environment includes an editor and compiler. Commands for synchronous servo movements, servo point-to-point movements, and servo motion feedback are also available. The robot can be extended with several accessory modules and items: additional servos and brackets, gyros, acceleration sensors, speech functions, remote control (RC) accessories, and more can also be added as they become available.

The RoboSapien (Fig. 74.13a) is a toy-like robot [74.16] preprogrammed with moves, which can also be controlled by an infrared remote control included with the toy, or by either a personal computer equipped with an infrared transmitter, or an infrared-transmitter-equipped personal digital assistant (PDA). The toy is capable of walking motion without recourse to wheels within its feet. It is also able to grasp objects with either of its hands, and is also able to throw grasped objects with mild force. It has a small loudspeaker unit, which can broadcast several different vocalizations, all of which appear to be recordings of a human male pretending to be a great ape, such as a gorilla.

The toy's remote control unit (Fig. 74.13b) has a total of 21 different buttons. With the help of two shift buttons, a total of 67 different robot-executable commands are accessible.

The remote controller is equipped with a basic level of programmability. Users can string together

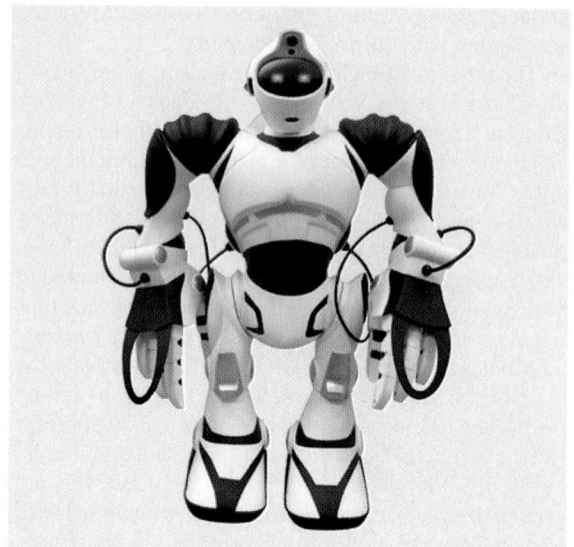

Fig. 74.14 RoboSapien V2

movement commands to form what the toy's manual describes as either macros or mini-programs, but which are more correctly described as robotic instructions sets. It is also possible to produce a sensor-keyed instruction set.

The original robot was developed in the year 2000 and was a brilliant achievement of breakthrough design. It could walk, grasp, and respond – in human-like ways.

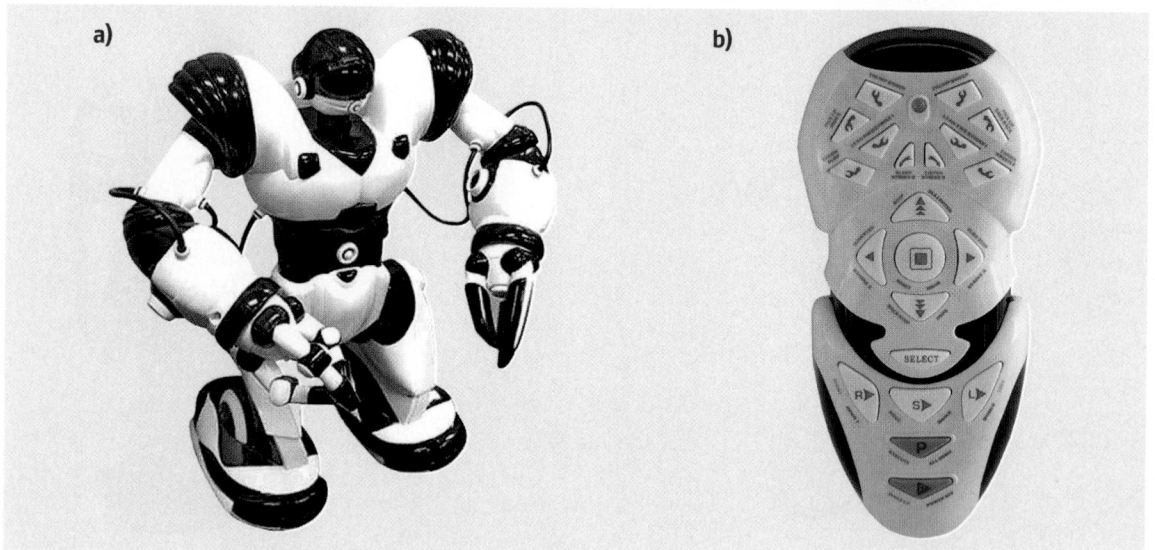

Fig. 74.13 (a) RoboSapien, **(b)** remote control

Fig. 74.15 RS Media

The next generation (Fig. 74.14) was released in 2003 and is nearly twice the size of the original robot. Instead of the original caveman grunts the V2 can speak a reasonably large list of prerecorded phrases. It has infrared and basic color recognition sensors, grip sensors in its hands, touch- or contact-activated hand and foot sensors, and sonic sensors. For movement the V2 has an articulated waist, shoulders, and hands, giving it a wide variety of impressive body animations.

The latest model, the RS Media (Fig. 74.15), which was released in October 2006, uses basically the same body as V2, but a different brain based on a Linux kernel. As the name implies, RS Media's focus is on multimedia capabilities, including the ability to record and playback audio, pictures, and video.

Another example of a reasonable cheap entertainment robot is a 12 inch-high robot that can walk, run, and do flips, cartwheels, and dance moves, and that, once programmed, is ready to compete in distinct competitions. This fully articulating humanoid is controlled by a control board, which can operate up to 24 servos and 16 accessory modules with 16 digital servos.

Optional devices will eventually include gyros, acceleration sensors, speech synthesis modules, and operational devices such as Bluetooth controllers and R/C transmitters and receivers. The operation time with a five-cell rechargeable battery pack is approximately 1 h.

The main problem with such robots consists in the preprogrammed movements and that it is not easy to add some software modules.

Competition Robots

A very popular group of entertainment robots are those for competitions. A robot competition is an event where robots have to accomplish a given task. Usually they have to beat other robots in order to be judged the best. Most competitions are for schools but, as time goes by, several professional competitions are arising. There is a wide variety of competitions [74.17] for robots of various types. The following examples describe a few of the higher-profile events.

The idea of a robot competition is not new; they have been held for several years at, for example, the Massachusetts Institute of Technology (MIT) in Boston and Eidgenössische Technische Hochschule (ETH) in Zurich. Having a robot competition in an academic curriculum is highly beneficial, as it gives students open-ended problem spaces, teaches them to work in groups (of two or three persons), and stimulates creativity [74.18].

Classical examples are:

- Robot outdoor tournaments
- Robot soccer
- Ping-pong-playing robots
- Robot wrestling
- Sumo wrestling robots
- Billiard robots.

Table 74.1 gives some examples of robot games.

RoboGames. RoboGames [74.19] (previously ROB-Olympics) is an annual robot contest held in San Francisco, California. RoboGames is the world's largest

Table 74.1 Examples of robot games

Beam robot games/olympics		Robofesta official games
Solarroller	High/long jump	Robocup
Photovore	Legged race	Robot grand prix
Aquavore	Innovation machine	All Japan sumo tournament
Rope climbing	Robot-art	All Japan micromouse contest
Robot sumo	Micromouse	
Nanomouse	Aerobot compet.	

open robot competition. They invite the best minds from around the world to compete in over 70 different events. About two-thirds of the robot events are autonomous, while the remaining one-third are remotely operated.

The first goal is to bring builders from combat robotics (mechanical engineering), soccer robotics (computer programming), sumo robotics (sensors), androids (motion control), and art robots (aesthetics) together for exchange experiences. The second goal is to offer recognition to engineers from around the world in varying disciplines with consistent rule sets and low cost or without contestant fees. There are no prerequisites for contests; the event is open to anyone, regardless of age or affiliation. Twenty-eight countries competed with more than 800 robots in 2007.

Competition categories:

- Combat:
  - Categories from 1 to 340 lbs (454 g–154.5 kg)
  - RC and autonomous
- Robot soccer:
  - MiroSot 5:5/11:11 (autonomous)
  - Biped soccer 3:3 (R/C)
  - AIBO soccer 4:4 (autonomous)
- Autonomous humanoid challenges:
  - Basketball
  - Weight lifting
  - Obstacle run
- Autonomous autos
- Sumo
- Bot hockey
- R/C humanoid competition
- Tetsujin (exoskeletons)
- Art bots
- Junior league (< 18 years old)
- Open.

DARPA Grand Challenge. The Defense Advance Research Projects Agency (DARPA) Grand Challenge [74.20] is a prize competition for driverless cars, sponsored by DARPA, the central research organization of the US Department of Defense. Congress has authorized DARPA to award cash prizes to further DARPA missions to sponsor revolutionary, high-payoff research that bridges the gap between fundamental discoveries and their use for national security.

Fully autonomous vehicles have been an international pursuit for many years and the Grand Challenge was the first long-distance competition for robot cars in

the world; the main goal is to make one-third of cars autonomous by 2015.

The first competition of the DARPA Grand Challenge was held in 2004 in the Mojave Desert region of the USA along a 150 mile to just past the California–Nevada border in Primm. None of the robot vehicles finished the route. Carnegie Mellon University's Red Team traveled the farthest distance, completing 11.78 km (7.36 mile) of the course.

All but one of the 23 finalists in the 2005 race surpassed the 11.78 km (7.36 mile) distance completed by the best vehicle in the 2004 race. Five vehicles successfully completed the race.

Vehicles in the 2005 race passed through three narrow tunnels and negotiated more than 100 sharp left and right turns. The race concluded through Beer Bottle Pass, a winding mountain pass with sheer drops on both sides. Although the 2004 course required more elevation gain and some very sharp switchbacks (Daggett Ridge) were required near the beginning of the route, the course had far fewer curves and generally wider roads than the 2005 course.

The third competition of the DARPA Grand Challenge (2007) was named the *Urban Challenge*. The course involved a 96 km (60 mile) urban area course, to be completed in less than 6 h. Rules included obeying all traffic regulations while negotiating with other traffic and obstacles and merging into traffic. While the 2004 and 2005 events were more physically challenging for the vehicles, the robots operated in isolation and did not encounter other vehicles on the course. The Urban Challenge required designers to build vehicles able to obey all traffic laws while they detected and avoided other robots on the course. This is a particular challenge for vehicle software, as vehicles must make *intelligent* decisions in real time based on the actions of other vehicles. In contrast to previous autonomous vehicle efforts that focused on structured situations such as highway driving with little interaction between the vehicles, this competition operated in a more cluttered urban environment and required the cars to perform sophisticated interactions with each other, such as maintaining precedence at a four-way stop intersection. Four teams completed successfully the course.

The cars are usually equipped with laser measurement systems, laser sensors, global positioning systems (GPS), and cameras. The teams employed a variety of different software and hardware combinations for interpreting sensor data, planning, and execution. Some examples included C++ and C# running on Windows hosts with planning involving Bayesian mathematics;

and Mac Minis running Linux because it can run on DC power at relatively low wattage and produce less heat. Examples of the latter included Mac Minis running on Windows and an embedded version of Windows XP.

Annual Fire-Fighting Home Robot Contest. The main challenge of this contest [74.21] is to build an autonomous, computer-controlled robot that can find its way through an arena that represents a model house, find a lit candle that represents a fire in the house, and extinguish the fire in the shortest time. This task simulates the real-world operation of an autonomous robot performing a fire protection function in a real house. The goal of the contest is to make a robot that can operate successfully in the real world, not just in the laboratory. Such a robot must be able to operate successfully where there is uncertainty and imprecision. Therefore, the dimensions and specifications listed in the rules are not exactly what will be encountered at the contest and are provided as general aids. However, the size limits on robots are absolute and are enforced by the judges.

Once turned on, the robot must be autonomous, i. e., self-controlled without any human intervention. Firefighting robots are to be computer controlled and not manually controlled devices.

A robot may bump into or touch the walls of the arena as it travels, but it cannot mark, dislodge or damage the walls in doing so. There will not be a penalty for touching a wall, but there is a penalty for moving along the wall while in contact with it. The robot cannot leave anything behind as it travels through the arena. It cannot make any marks on the floor of the arena that aid in navigation as it travels. Any robot that deliberately, in the judges' opinion, damages the contest arena (including the walls) will be disqualified. This does not include any accidental marks or scratches made in moving around.

The robot must, in the opinion of the judges, have found the candle before it attempts to put it out; for example, the robot cannot just flood the arena structure with CO_2, thereby putting the candle out by accident.

Early competitions include the following:

Robot Golf Open. An example of a robot competition is the 1996 robot contest called the Robot Golf Open. The contest arena was a rectangular square of 2×2 m, surrounded by a 15 cm high wall. The green was located in the middle of the arena and was a 7 cm high disc with a diameter of about 40 cm. Seven golf balls were randomly placed on the arena. It was the task of the robot to locate the balls, pick them up, and put them in some way into the hole, with two points being given for each ball. One point was given if the ball was only placed on the green. It is emphasized here that the robots performed the task autonomously, i. e., they made decisions as to how to control themselves according to the software running on the onboard computer based on sensory information. Two robots *played golf* against each other for a period of 2 min.

Environmental Control Robot Competition. The latest contest with a vision system, held in the spring of 2001, was called *environmental control*; the robot task was to locate three different kinds of garbage in an arena and bring them to the correct container. The arena was 2×2 m and contained *containers* at one side, three for each robot. The garbage was either a bottle, a battery or a pack of newspapers.

Other currently well-known competitions are:

Aerial Robotic Vehicle Competitions. The International Aerial Robotics Competition [74.22] is the longest running aerial robotic event, held annually since 1991. This competition involves fully autonomous flying robots performing tasks that, at the time posed, are undemonstrated anywhere worldwide. The competition is open to universities and has had missions involving ground object capture and transfer, hazardous waste location and identification, disaster scene search and rescue, and remote surveillance of building interiors by fully autonomous robots launched from 3 km. A series of micro air vehicle (MAV) events have been sponsored by various organizations. Typically, these competitions involve capability demonstrations rather than missions, and may or may not involve full autonomy.

Ground Robotic Vehicle Competitions. In addition to the DARPA Grand Challenge [74.23] there is also the *Intelligent Ground Vehicle Competition* (IGVC) for autonomous ground vehicles. The robots must traverse outdoor obstacle courses without any human interaction. This is an international student design competition at university level and has held annual competitions since 1992.

Underwater Robotic Vehicle Competitions. This is a spin-off of the International Aerial Robotics Competition [74.24], and as such, carries through the theme of full autonomy of operation, albeit in a subsurface robotic vehicle. This is, since 1997, a collegiate competition.

International METU Robotics Days. The Middle East Technical University (METU) Robotics Days [74.25] are organized in Ankara, Turkey. There are robot competitions from various categories in which original and creative ideas take their part, while innovation is honored. Participants are always encouraged to share their knowledge. International METU Robotics Days, apart from holding competitions for those who would like to challenge their skilled robots, also host professionals and academics interested in the field of robotics to come together in lectures and workshop studies with younger amateurs, giving them the opportunity to take a closer look at this continuously developing technology.

IEEE Micromouse competition. In the classical Micromouse competitions [74.26], small robots try to solve a maze in the fastest time. Micromouse competitions were held first in Tampere Finland Technical University around 1983–1985.

Botball Educational Robotics. Botball [74.27] is a robotics competition for middle- and high-school students. Organized by the KISS Institute for Practical Robotics, Botball encourages participants to work constructively within their team, building basic communication, problem solving, design, and programming skills. Each team builds one or more (up to four) robots that will autonomously move scoring objects into scoring positions.

Mobile Autonomous Systems Laboratory Competition. The Mobile Autonomous Systems Laboratory [74.28] is one of the few college-level vision-based autonomous robotics competition in the world. Conducted by and for MIT undergraduates, this competition requires multithreaded applications of image processing, robotic movements, and target ball deposition. The robots are run with Debian Linux and run on an independent OrcBoard platform that facilitates sensor-hardware additions and recognition.

Wall-Climbing Competitions. There are two worldwide known events. The Duke Annual Robo-Climb Competition (DARC) [74.29] in the USA and the Climbing and Walking Autonomous Robot (CLAWAR) [74.30] competition in Europe. The task is to create innovative wall-climbing robots that can autonomously ascend vertical surfaces of different materials with obstacles.

AAAI Grand Challenges. The two Association for the Advancement of Artificial Intelligence (AAAI) Grand Challenges [74.31] focus on human–robot interaction, with one being a robot attending and delivering a conference talk, the other being operator-interaction challenges in rescue robotics.

This is only a selection – there are a lot of other robot competitions worldwide, mainly dedicated to BSc students.

As an example, robot soccer will be described in more detail below.

Robot Soccer

The fascinating idea of using small robot cubicles to play soccer was born just a decade ago in Korea and Japan and has since spread all over the world. Yearly championships are even organized in different countries. From the scientific point of view, robot soccer is one of the first applications of a MAS. The players – robots or agents – have to solve a common task, i.e., to win the game.

Robot soccer was introduced to develop intelligent cooperative multirobot (agents) systems (MAS) and to educate the young generation in these difficult scientific and engineering subjects by playing. From the scientific viewpoint the soccer robot is an intelligent autonomous agent which carries out tasks with other agents in a cooperative, coordinated, and communicative way. Generally robot soccer is a good test bed for the development of MAS. Furthermore it is also a good tool for spending leisure time and for education [74.32].

In the future, production systems will become more complex. Several independently working autonomous mobile robots are working together, and therefore conflict situations in certain areas could appear, e.g., where several robots gather at an intersection. In order to avoid conflict situations and delays, and guarantee smooth movement, robots should have the capability to communicate and cooperate in order to coordinate their actions.

Soccer is one of the best known sports worldwide. It is exciting to watch how robots play the game. It is also possible not only to watch the game but also to play the game – human against computer, human against human – using a joystick as well as a keyboard. The big question for common use is the price of the whole system. With the development of electronic devices and peripheries the cost is going down. For the realization of interdisciplinary research, work should be done in areas such as robotics, image processing, sensors, mechatronics, communication, etc.

At the moment there are two robot soccer organizations in the world: the Federation of International Robot-Soccer Associations (FIRA) [74.33] and RoboCup [74.34]. The objects and scope of both organizations are similar. The size, speed, acceleration of the robots, the sizes of the playgrounds, and the numbers of robots playing are different.

RoboCup. RoboCup is an international research and education initiative. Its goal is to foster artificial intelligence and robotics research by providing a standard problem for which a wide range of technologies can be examined and integrated.

The main focus of RoboCup activities is competitive football. The games are important opportunities for researchers to exchange technical information. They also serve as a great opportunity to educate and entertain the public. RoboCup soccer is divided into the following leagues:

Simulation League. Independently moving software players (agents) play soccer on a virtual field inside a computer.

Small-Size Robot League (f-180). Small robots of no more than 18 cm in diameter play soccer with an orange golf ball in teams of up to five robots on a field with size bigger than a ping-pong table.

Middle-Size Robot League (f-2000). Middle-sized robots of no more than 50 cm diameter play soccer in teams of up to four robots with an orange soccer ball on a field the size of 12×8 m.

Four-Legged Robot League. Teams of four four-legged entertainment robots (Sony's AIBO) play soccer on a 3×5 m field. The robots use wireless networking to communicate with each other and with the game referee. Challenges include vision, self-localization, planning, and multiagent coordination.

Humanoid League. Biped autonomous humanoid robots play in *penalty kick* and *2 versus 2* matches, and *technical challenges*. This league has two subcategories: kid-size and teen-size.

RoboCupRescue. The intention of the RoboCupRescue project is to promote research and development in this significant domain by involving multiagent teamwork coordination, physical robotic agents for search and rescue, information infrastructures, personal digi-

tal assistants, standard simulator and decision support systems, evaluation benchmarks for rescue strategies, and robotic systems, which will all be integrated into a comprehensive system in the future.

Federation of International Robot-Soccer Associations (FIRA). Similar to RoboCup there are also different categories in this *Robotsoccer World*.

Micro Robot World Cup Soccer Tournament (MiroSot). A match shall be played by two teams, each consisting of 5 or 11 robots on a dark playground 220 cm × 180 cm for the middle league, 400 cm × 280 cm for the large league, with an orange golf ball (Fig. 74.16). Only three human team members, a *manager*, a *coach*, and a *trainer*, are allowed on the stage. One host computer per team, mainly dedicated to vision processing and other location identification, is used. The size of each robot is limited to 7.5 cm × 7.5 cm × 7.5 cm. The height of the antenna is not considered in deciding a robot's size.

Nano Robot World Cup Soccer Tournament (NaroSot). Similar to MiroSot, but the size of the five robots is limited to 4 cm × 4 cm × 5 cm. They play with an orange ping-pong ball on a playground 130 cm × 90 cm (Fig. 74.17).

Kheperasot. The Kheperasot game is played by two teams, each consisting of one robot player and up two human team members. The robot is fully autonomous with an onboard vision system. The human team members are only allowed to place their robot on the field, start their robot at the beginning of each round at the position indicated by the referee before each round, start

Fig. 74.16 MiroSot robot [74.31]

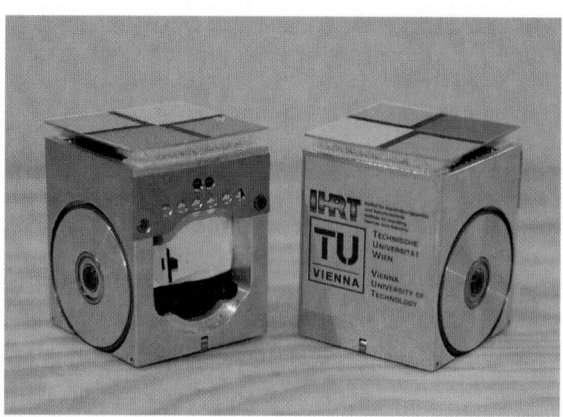

Fig. 74.17 NaroSot robot [74.31]

their robot when indicated by the referee, and remove the robot from the field at the conclusion of the match. They play with a yellow tennis ball on a playground 130 cm × 90 cm.

Humanoid Robot World Cup Soccer Tournament (HuroSotCup). In this competition, the humanoid robot has two legs (biped robot). The game is played using humanoid robots on a playground 340–430 cm × 250–350 cm. The maximum size of the robots is 150 cm, and the maximum weight is 30 kg. The robots have remote or auto control.

RoboSot. A match is played by two teams, each consisting of one to three robots with maximum size 20 cm × 20 cm × (no limit in height) on a playground 260 cm × 220 cm with a yellow with light green tennis ball. Only three human team members, a *manager*, a *coach*, and

a *trainer*, are allowed on the stage. The robots can be fully or semiautonomous. In the semiautonomous case, a host computer can be used to process the vision information from the cameras onboard the robots.

Simulation Robot World Cup Soccer Tournament (SimuroSot). SimuroSot consists of a server, which has the soccer game environments (playground, robots, score board, etc.) and two client programs with the game strategies. A 3-D color graphic screen displays the match. Teams can make their own strategies and compete with each other without hardware. The 3-D simulation platform for 5 versus 5 and 11 versus 11 games are available at the FIRA web site [74.33].

MiroSot and NaroSot. The FIRA Mirosot and NaroSot systems work as follow: A camera approximately 2 m over the playground delivers 60 pictures/s to the host computer. With information from color patches on top of the robots, the vision software calculates the position and orientation of the robots and the ball. Using this, the host computer generates motion commands according to the implemented game strategy and sends motion commands wirelessly to the robots.

A soccer robot is an excellent example of mechatronics. Its main parts are wheels, drives, a power source, a microprocessor, and a communication module. All these parts have to be included in a very small volume: a cube 7.5 × 7.5 × 7.5 cm or a cuboid 4.0 × 4.0 × 5.0 cm. The soccer robots of a team (5–11 players) are controlled by the team computer.

The robot itself has a drive mechanism, power supply, electronic parts to control robot behavior, and communication. Mostly digital proportional–integral–differential (PID) controllers are used. The problem is the setting of the controller parameters. Therefore fuzzy control and neural networks are applied to adapt the parameters.

The main problems are the power sources of such robots. Usually batteries are approximately 50% of the weight of the robot and have a lifetime of only 2 h.

Worldwide there are already more than 150 teams competing in regional and world championships.

As pointed out earlier, a soccer robot is an excellent example of multidisciplinarity. For the construction and manufacturing of the body, knowledge of mechanical and, because of the small dimensions, precision engineering is required. Electrical as well as control engineering is necessary for the drives and the power source. The control and communication board of the robot is more or less applied electronics. A micro-

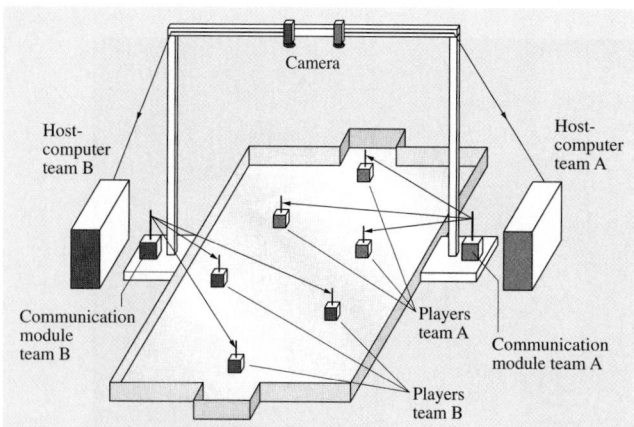

Fig. 74.18 Overall system of robot soccer

processor serves as an internal controller and is also responsible for wireless communication with the host computer. For these tasks and for the software of the host computer fundamental knowledge in computer science is necessary. The software of the host computer includes online image processing, game strategies, control of the team's own players, communication with and between these, and the user interface.

Development of a robot soccer team therefore requires the teamwork of specialists from various disciplines, having different thinking and talking a different language. The project leader has to harmonize such a team and must have at least basic knowledge of all these necessary subjects.

One possibility to go into a broader market is to replace conventional games in amusement parks and restaurants. Therefore, as a first step in this direction, the software had to be adopted to use also a joystick to control each robot player. This offers the following possibilities for playing:

- Humans against humans (both teams controlled by joysticks)
- Humans against computer (only one team controlled by joysticks)
- Computer against computer (state of the art).

In contrast to soccer video games this new technology offers a *real-life* feeling similar to that in a soccer stadium.

A special application is robot dancing. Mirosot robots are programmed for dancing and are judged on criteria such as creativity and costumes. With a user-friendly programming interface a 2 min dance can be created in half a day without any pre knowledge. As an example Fig. 74.20 shows two Mirosot robots from the soccer team of *Vienna University of Technology* dressed in tuxedo and white robe ready for dancing the world-wide well-known Blue Danube Waltz.

Until now the robots are completely unintelligent; they have no sensors and are controlled by the host computer. In the future robots will be more and more intelligent and will be equipped with different sensors (ultrasonic, infrared, laser, etc.). This offers the possibility for robots to adapt the commands of the host computer.

Future developments will be towards humanoid soccer players. A humanoid soccer playing robot has to:

Fig. 74.19 Overall view of robot soccer

Fig. 74.20 Dancing couple for a Vienna waltz

- Be able to accelerate and slow down as fast as possible
- Keep its balance all the time, even after a crash with another robot
- Localize itself on the field
- Localize the ball
- Localize the opponents
- Make autonomous decisions regarding which actions to take.

As a first step some producers are offering robots with four or six legs at a high price. In some years players with two legs will probably be available – then we can start the first soccer games with humans against robots.

74.2 Market

Consumers showed an astonishingly quick appreciation of the new products in this field, especially of *Furby*, the first animatronic pet marketed on a larger scale, in February 1998. This pet's popularity is on the one hand due to the well-prepared marketing campaign that was organized for its introduction and on the other hand to market mechanisms outside the company's control that made these little creatures a must-have item in practically every child's room.

As this product proved to be nearly perfect for the manufacturer, combining low production costs with extremely high customer demands, many other companies copied the product or designed other robotic pets based on the successful concept in the subsequent years to participate in this expanding market as well.

By far the most famous of these *personal companion robots* is AIBO, which sold out over the Internet within minutes, regardless of the fairly high price. For the second generation, introduced in the market at the end of 2000, a large number will be sold.

It is safe to assume that this market will continue to expand, with newly developed and technically even more refined products entering the market. Thus other companies in this market will have to keep pace and develop further products themselves.

Even in related fields such as robot soccer, robots are likely to find their way to the toy market as soon as these systems can be produced in larger quantities and at a lower price than today.

74.3 Summary and Forecast

Mobile, intelligent robots are now available on the market. The number of these robots in use will dramatically increase in the next year. One of the main application areas will be the field of *entertainment, leisure, and hobby* and robot competitions.

Well-known scientists engaged in robotic research, and who dare to make forecasts not only for the immediate future of service robotics but for the more distant future, also believe that the evolution of service robots will basically happen in several stages, being closely linked to progress in computer technology.

The semiconductor market has seen a series of market-driving waves, from the analog wave to the first digital wave, in which the PC was central, to the second digital wave, in which the digital consumer and network were central. After these waves, scientists expect a robotics wave to occur. So the personal robot market will become more important than the PC market. An estimation of one scientist is that a humanoid robot-soccer team will win against the world champions by 2050.

We are talking about entertainment robots in general, beyond QRIO. It would be desirable to develop is a robot companion for human beings. For instance, a robot can hold things in its memory indefinitely. The hardware might break down over the course of many years, but by taking the memory stick and putting it into a new robot, you could transfer those memories to it. In so doing, one can share with it a short time's worth of memories and knowledge and visions. In some

ways, a robot could be the ultimate companion. Yet another idea would be a robot that listens to you. The basis for this is the active listening method of counseling, wherein the counselor gives no information but simply listens. A robot can do that too. A robot can listen to complaints, share information, and be a counselor anytime, day or night.

Therefore we will probably see the following generations:

First Companion Robot Generation – 2010
Mobile, human-sized universal service robots, being as intelligent as a lizard, will be able to perform everyday routine work such as cleaning floors, remove garbage or dust furniture. The required computing power for such a robot would be approximately 5 Mips (5 million instructions per second).

Second Companion Robot Generation – 2020
The subsequent robot generation, also designed to assist humans in their everyday activities, performs janitorial services, or simply entertains them, features an advanced processor capable of computing about 100 000 Mips, thus boosting the intelligence level of the system to that of a mouse. These robots can already be trained using praise and censure.

Third Companion Robot Generation – 2030
With computing power advancing further to 5×10^6 Mips, the robotic system reaches the intellect of a monkey.

Forth Companion Robot Generation – 2040

Within 40 years from now, the forth generation of service robots should be capable of abstracting and generalizing problems like a human, thus performing not only routine tasks, but tasks that require preparation and planning as well. Therefore the existence of companies that do not employ a single human worker any more, besides their autonomous robots, might be well conceivable. We are looking forward to what will be realized.

74.4 Further Reading

- P. Corke, S. Sukkarieh: *Field and Service Robotics, Results of the 5th International Conference*, Springer Tracts Adv. Robot. 25 (Springer, Berlin, Heidelberg 2006)
- G. Engelberger: Services. In: *Handbook of Industrial Robotics*, ed. by S.Y. Nof (Wiley, New York) pp. 1201–1212
- J.F. Engelberger, *Robotics in Service* (Kogan Page, London 1989)
- H.R. Everett: *Sensors for Mobile Robots: Theory and Application* (A.K. Peters, Wellesley 1995)
- S. Haddadin, T.S. Laue, G. Hirzinger: Foul 2050: thoughts on physical interaction in human–robot soccer, 2007 IEEE/RSJ Proc. Int. Conf. Intell. Robot. Syst. (2007) pp. 3243–3250
- G. Lakemeyer, E. Sklar, D. G. Sorrenti, Takahashi (Eds.): *RoboCup 2006: Robot Soccer World Cup X* (Springer, Berlin, Heidelberg 2007)
- E. Osawa, H. Kitano, M. Asada, Y. Kuniyoshi, I. Noda: RoboCup: the robot world cup initiative, ICMAS-96 Proc. 2nd Int. Conf. Multi-Agent Syst. (1996) p. 454
- R. D. Schraft, G. Schmierer: *Service Robots* (A.K. Peters, 2000)
- J. Schmidhuber: Developmental robotics, optimal artificial curiosity, creativity, music, and the fine arts, Connect. Sci. **18**(2), 173–87 (2006)

References

74.1   P. Kopacek: Advances in robotics, Proc. 10th Int. Conf. Comput. Aided Syst. Theor. – EUROCAST 2005 (Springer, Berlin, Heidelberg 2005) pp. 549–558

74.2   J.W. Kim: *Unpublished Transparencies* (Summerschool, Kaist, Daejon 2006)

74.3   G. Fischer: Robots in entertainment leisure and hobby. Diploma Thesis (Vienna University of Technology, Vienna 2000)

74.4   P. Kopacek, M.W. Han: Robots for entertainment, leisure and hobby, Proc. CLAWAR/EURON/IARP Workshop Robot. Entertain. Leisure Hobby (Vienna 2004) pp. 1–6

74.5   *Dictionary of Contemporary English*, 2nd edn. (Longman Group UK Ltd., Essex 1987)

74.6   *Merriam–Webster's Collegiate Dictionary* (Merriam–Webster Inc., Springfield 1999)

74.7   http://mindstorms.lego.com

74.8   http://www.megasaurus.com

74.9   http://www.transaurus.com

74.10   http://www.cs.uni-bonn.de/~rhino/tourguide/

74.11   http://www.cs.cmu.edu/~minerva/tech/

74.12   http://www.boxerjocks.com

74.13   P. Kopacek, M.W. Han: New concepts for humanoid robots, Proc. FIRA RoboWorld Congr. (Dortmund, 2006) pp. 108–111

74.14   http://en.wikipedia.org/wiki/QRIO

74.15   http://www.robonova.com

74.16   http://www.wowwee.com/products_robotics

74.17   http://en.wikipedia.org/wiki/Robot_competition

74.18   A.J. Baerveldt, T. Salomonsson, B. Astrand: Vision-guided mobile robots for design competition, IEEE Robot. Autom. (2003) pp. 38–44

74.19   http://www.robogames.net

74.20   http://www.darpa.mil/grandchallenge/

74.21   http://www.trincoll.edu/events/robot/

74.22   http://avdil.gtri.gatech.edu/AUVS/IARCLaunchPoint

74.23   http://www.igvc.org

74.24   http://www.auvsi.org/competitions/water.cfm

74.25   http://www.roboticsdays.org

74.26   http://micromouse.cannock.ac.uk

74.27   http://www.bootball.org

74.28   http://maslab.csail.mit.edu

74.29   http://robotics.pratt.duke.edu/roboclimb/

74.30   http://www.clawar.org

74.31   http://www.aai.or

74.32   P. Kopacek, M.W. Han: Mini robots for soccer, Proc. 12th Int. Conf. Comput. Aided Syst. Theor., EUROCAST (Springer, Berlin, Heidelberg 2007) pp. 342–344

74.33   http://www.fira.net

74.34   http://www.robocup.org

Part H Automation

Part H Automation in Medical and Healthcare Systems

Automation in Medical and Healthcare Systems. Part H Most of the automation inventions, innovations and devices in medical and healthcare systems have just emerged in the recent two decades, and many more are moving out of research labs to hospitals and homes. This part explains the exponential penetration and main contributions of automation to the health and medical well being of individuals and societies. First, the scientific and theoretical foundations of control and automation in biological and biomedical systems and mechanisms are explained, and their significant value for useful implementation and application is detailed. Then, specific areas are described and analyzed in: implantable devices; medical robotic solutions and techniques for a range of medical problems and medical interventions; diagnostics and testing procedures and tools, including the emergence of nano-devices; and the significant progress in medical informatics and medical records and instrumentation. Available, proven, and emerging automation techniques and design for better cost savings and quality assurance in healthcare delivery, and elimination of hospital and other medical errors are also addressed in this part.

75. Automatic Control in Systems Biology

Henry Mirsky, Jörg Stelling, Rudiyanto Gunawan, Neda Bagheri, Stephanie R. Taylor, Eric Kwei, Jason E. Shoemaker, Francis J. Doyle III

The reductionist approaches of molecular and cellular biology have produced revolutionary advances in our understanding of biological function and information processing. The difficulty associated with relating molecular components to their systemic function led to the development of systems biology, a relatively new field that aims to establish a bridge between molecular level information and systems level understanding. The novelty of systems biology lies in the emphasis on analyzing complexity in networked biological systems using integrative rather than reductionist approaches. By its very nature, systems biology is a highly interdisciplinary field that requires the effective collaboration of scientists and engineers with different technical backgrounds, and the interdisciplinary training of students to meet the rapidly evolving needs of academia, industry, and government. This chapter summarizes state-of-the-art developments of automatic control in systems biology with substantial theoretical background and illustrative examples.

75.1 Basics

Advances in molecular biology over the past two decades have made it possible to probe experimentally the causal relationships between processes initiated by individual molecules within a cell and their macroscopic phenotypic effects on cells and organisms. A systematic approach for analyzing complexity in bio-

physical networks was previously untenable owing to the lack of suitable measurements and the limitations imposed in simulating complex mathematical models. Recent studies provide increasingly detailed insights into the underlying networks, circuits, and pathways responsible for the basic functionality and robustness of biological systems and create new and exciting opportunities for the development of quantitative and predictive modeling and simulation tools [75.1]. The discipline of systems biology has emerged in response to the challenges in modeling and understanding complex biological networks [75.2, 3].

75.1.1 Systems Biology

The field of systems biology combines approaches and methods from systems engineering, computational biology, statistics, genomics, molecular biology, biophysics, and other fields [75.5–7] to help create systems-level understanding of complex biological networks. In particular, systems engineering methods are finding unique opportunities in characterizing the rich dynamic behavior exhibited by biological systems. Conversely, the new classes of biological problems are motivating novel developments in theoretical systems approaches. Two characteristics of systems biology are [75.4]: (1) integrative view points towards unraveling complex dynamical systems, and (2) tight iterations between experiments, modeling, and hypothesis generation (Fig. 75.1).

Although the field of systems biology is relatively young, one can already point to early successes in a number of cases. The work of *Arkin* on λ-phage was one of the first detailed analyses of a stochastic gene switch, and showed convincingly that formal stochastic treatment was required to understand the cell fate switch between lysis and lysogeny [75.8]. The analysis of perfect adaptation in chemotaxis is another example where multiple groups adopted a systems perspective, and key insights have been generated [75.9–11]. Notably, the

mechanism for perfect adaptation has been elucidated and interpreted in classical control engineering terms: integral feedback [75.11].

The approach of model reduction and systematic analysis (including requisite modeling assumptions to yield perfect adaptation) is an excellent example of an effective *systems strategy*. This problem continues to generate new insights, as recent work has shown that disparate organisms have both overlapping and distinctive architectures for chemotaxis [75.11]. Another nice example that has received considerable attention is the gene network underlying circadian rhythms. Models have been proposed [75.12], and formal robustness analysis tools have generated insights on biological design principles [75.13].

A more detailed case study that might be characterized as a success story has also emerged from the work of *Muller* et al. on the JAK-STAT pathway [75.14]. They have shown that modeling-experiment iterations can yield new hypotheses, particularly regarding unobservable components that can be simulated (but never measured). One implication, for the JAK-STAT pathway, involves pharmacological intervention. Current practice focuses on the phosphorylation element of the pathway, but the model shows that a more effective strategy involves the blocking of nuclear export.

75.1.2 Control Research in Systems Biology

Natural control systems are paragons of optimality. Over millennia, these architectures have been honed to achieve automatic, robust regulation of a myriad of processes at the levels of genes, proteins, cells, and entire systems. One of the more challenging opportunities for systems research is unraveling the multiscale, hierarchical control that achieves robust performance in the face of stochastic perturbations. These perturbations arise from both intrinsic sources (e.g., inherent variability in the transcription machinery), and extrinsic sources (e.g., environmental fluctuations). Robustness in key performance variables to particular perturbations has been shown to be achieved at the expense of strong sensitivity (fragility) to other perturbations.

The coexistence of extreme robustness and fragility constitutes one of the most salient features of highly evolved or designed complexity [75.15]. Optimally robust systems are those that balance their robustness to frequent environmental variations with their coexisting sensitivity to rare events. As a result, robustness

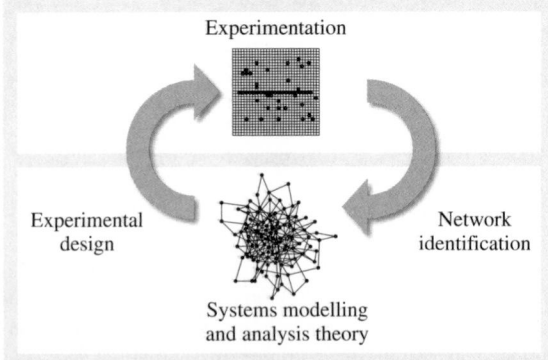

Fig. 75.1 Systems biology iterations. Interactions between experimental analysis and theoretical approaches, and the main tasks for theory at the interfaces (after [75.4])

and sensitivity analysis are key measures in understanding and controlling system performance. Robust performance reflects a relative insensitivity to perturbations; it is the persistence of a system's characteristic behavior under perturbations or conditions of uncertainty. Measuring the robustness of a system determines the behavior (the output or performance) as a function of the input (the disturbance). Formal sensitivity analysis allows the investigation of robustness and fragility properties of mathematical models, yielding local properties with respect to a particular choice of parameter values.

Within the field of systems engineering, control engineering has had a pervasive influence on the discipline of systems biology. For instance, the chemotaxis work [75.11] was a paradigm of collaboration between control engineers (Doyle) and biologists (Simon). Other examples include the robustness analysis of cellular function [75.15] and the unraveling of design principles in circadian rhythm [75.13]. There are many other examples from the control community: for instance, major advances in understanding signal transduction [75.16], and the oscillations underlying a positive feedback gene switch [75.17].

75.2 Biophysical Networks

Biophysical networks are remarkably diverse, cover a wide spectrum of scales, and are inevitably characterized by a range of rich behaviors. The term complexity is often invoked in the description of biophysical networks that underlie gene regulation, protein interactions, and metabolic networks in biological organisms. There are categorically two distinct characterizations of complexity: (1) the descriptive or topological notion of a large number of constitutive elements with nontrivial connectivity (described in Sect. 75.3), and (2) the classical notion of behavior associated with the mathematical properties of chaos and bifurcations (described in Sect. 75.4). In both biological and more general contexts, a key implication of complexity is that the underlying system is difficult to understand and verify [75.18]. Simple low-order mathematical models can be constructed that yield chaotic behavior, and, yet, rich complex biophysical networks may be designed to reinforce reliable execution of simple tasks or behaviors [75.19].

Biophysical networks have attracted a great deal of attention at the level of gene regulation, where dozens of input connections may characterize the regulatory domain of a single gene in a eukaryote, as well as the protein level, where literally thousands of interactions have been mapped in so-called protein interactome diagrams that illustrate the potential coupling of pairs of proteins [75.20, 21]. Similar networks also exist at higher levels, including the coupling of individual cells via signaling molecules, the coupling of organs via endocrine signaling, and ultimately the coupling of organisms in ecosystems. To elucidate the mechanisms employed by these networks, biological experimentation and intuition are by themselves insuf-

ficient. As noted earlier, the field of systems biology has laid claim to this class of problems, and engineers, biologists, physicists, chemists, mathematicians, and many others have united to embrace these problems with interdisciplinary approaches [75.22]. In this field, investigators characterize dynamics via mathematical models and apply systems theory with the goal of guiding further experimentation to better understand the biological network that gives rise to robust performance [75.22].

75.2.1 Timing and Rhythm: Circadian Rhythm Networks and Oscillatory Processes

Oscillatory processes are omnipresent in nature, comprising the cell cycle, neuron firing, ecological cycles and others; they govern many organisms' behaviors. A well-studied example of a biological oscillator is the circadian rhythm clock. The term circa- (about) diem (day) describes a biological event that repeats approximately every 24 h. Circadian rhythms are observed at all cellular levels since oscillations in enzymes and hormones affect cell function, cell division, and cell growth [75.23]. They serve to impose internal alignments between different biochemical and physiological oscillations. Their ability to anticipate environmental changes enables organisms to organize their physiology and behavior such that they occur at biologically advantageous times during the day [75.23]: visual and mental acuity fluctuate, for instance, affecting complex behaviors.

The mammalian circadian master clock resides in the suprachiasmatic nucleus (SCN), located in the

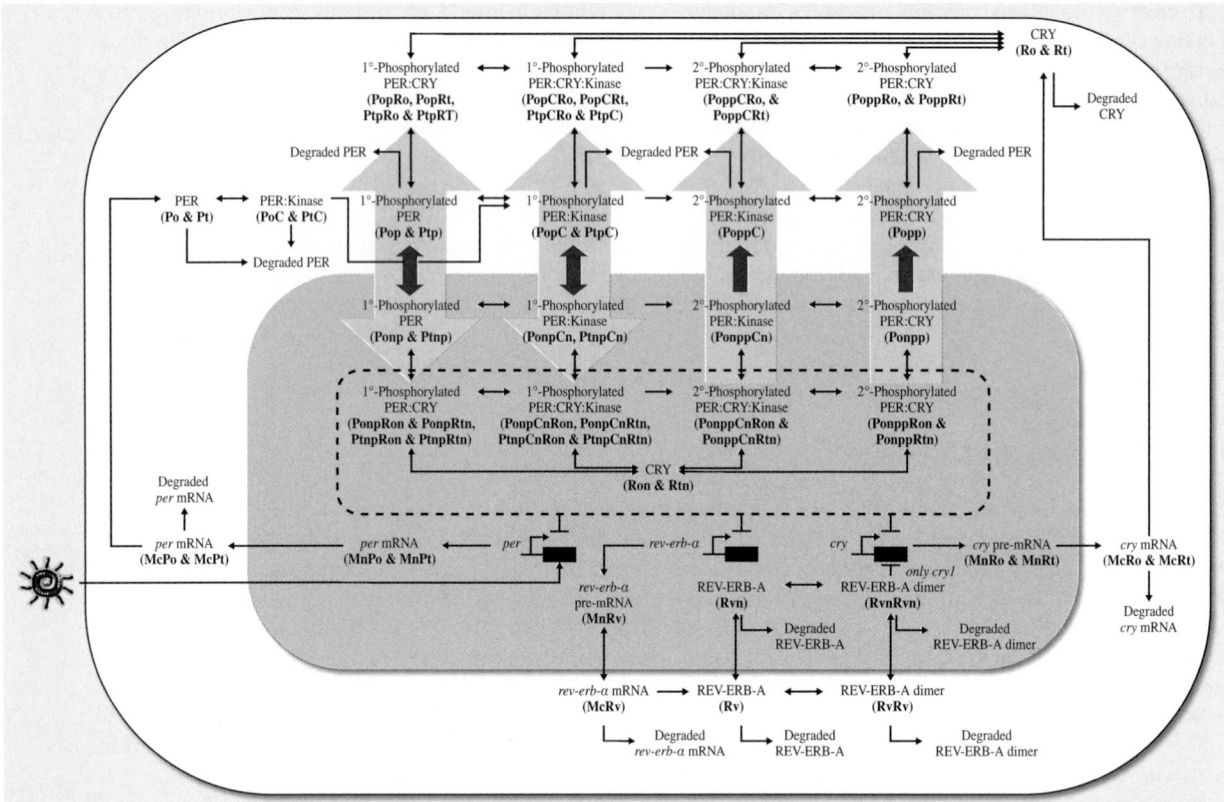

Fig. 75.2 Gene regulatory network underlying circadian rhythms in neurons in the SCN. Activated complexes of proteins inhibit the transcription of their corresponding genes, thus leading to time-delayed negative feedback and oscillations (after [75.30])

hypothalamus [75.24]. It is a network of multiple autonomous noisy (sloppy) oscillators, which communicate via neuropeptides to synchronize and form a coherent oscillator [75.25, 26]. At the core of the clock is a gene regulatory network in which approximately six key genes are regulated through an elegant array of time-delayed and coupled negative and positive feedback circuits (Fig. 75.2). The activity states of the proteins in this network are modulated (activated/inactivated) through a series of chemical reactions including phosphorylation and dimerization. These networks exist at the subcellular level. Above this layer is the signaling that leads to a synchronized response from the population of thousands of clock neurons in the SCN. Ultimately, this coherent oscillator then coordinates the timing of daily behaviors, such as the sleep/wake cycle. Left in constant (dark) conditions, the clock will run freely with a period of only approximately 24 h such that its internal time, or phase, drifts away from that of its environment. Thus, the ability to

entrain to external time through environmental factors is vital to a circadian clock [75.27–29].

75.2.2 Apoptosis: Programmed Cell Death

Another example of biophysical networks is the apoptosis network in which an extracellular input controls the response of the cell as a result of this information processing network. Apoptosis is the programmed cell death machinery that is used by nature to strategically kill off unneeded and infected cells, but this mechanism often becomes impaired in cancer cells, leading to unchecked proliferation.

The specific example here is triggered by the Fas ligand. When an activated T-cell contacts a diseased cell, Fas and its natural ligand bind, resulting in the formation of the death inducing signalling complex (DISC). This complex then activates two pathways, both of which lead to the activation of the so-called executioner caspase 3. The Fas apoptotic network has been modeled by [75.31] using ordinary differential equations and is illustrated in Fig. 75.3. In the Type I pathway, a feedback pathway involving caspase-6 and caspase-8 regulates the amount of activated caspase-3. In the Type II pathway, Bcl-2 and active caspase-8 both interact with the mitochondrial membrane, regulating mitochondrial permeability. Caspase-8 allows mitochondria to become permeable, inducing the activation of the apoptosome and Smac (second mitochondrial-activator caspase) [75.32]. The activated apoptosome (Apopt_active) in turn activates caspase-3, while Smac can further enhance the activation of the executioner caspase by removing XIAP (X-linked inhibitor of apoptosis protein). FLIP, Bcl-2, and XIAP all antagonize the apoptotic signal and variations in their quantities can toggle the apoptotic signal between Type I and Type II activation or no activation at all. Experimental evidence has shown that Type I activation requires a significant amount of caspase-8 to be present in the cell. Yet, in Type II cells, mitochondrial activity ultimately enhances the death signal, significantly lowering the amount of caspase-8 that needs to be activated to induce apoptosis [75.33].

Understanding apoptosis in a broader sense will lead to a better knowledge of the common platform for emergence of cancer cells, and perhaps point to possible cures for certain types of cancer. The complexity of apoptosis, however, makes the understanding very difficult without a systems level approach using a mathematical representation of the pathway. Further, analysis of an apoptosis model can reveal the fragility points in the mechanism of programmed cell death that can have physiological implications not only for explaining the emergence of cancer cells but also for designing drugs or treatment for reinstating apoptosis in these cells. A robust performance analysis of an apoptotic network identified known network fragilities as well as new potential targets for therapeutic intervention. Such analyses are crucial to streamlining drug discovery in highly dynamic networks [75.34].

75.2.3 Signals in Diabetes: Insulin Signaling Pathway

In healthy cells, the uptake of glucose is regulated by insulin, which is secreted by β-cells in the pancreas.

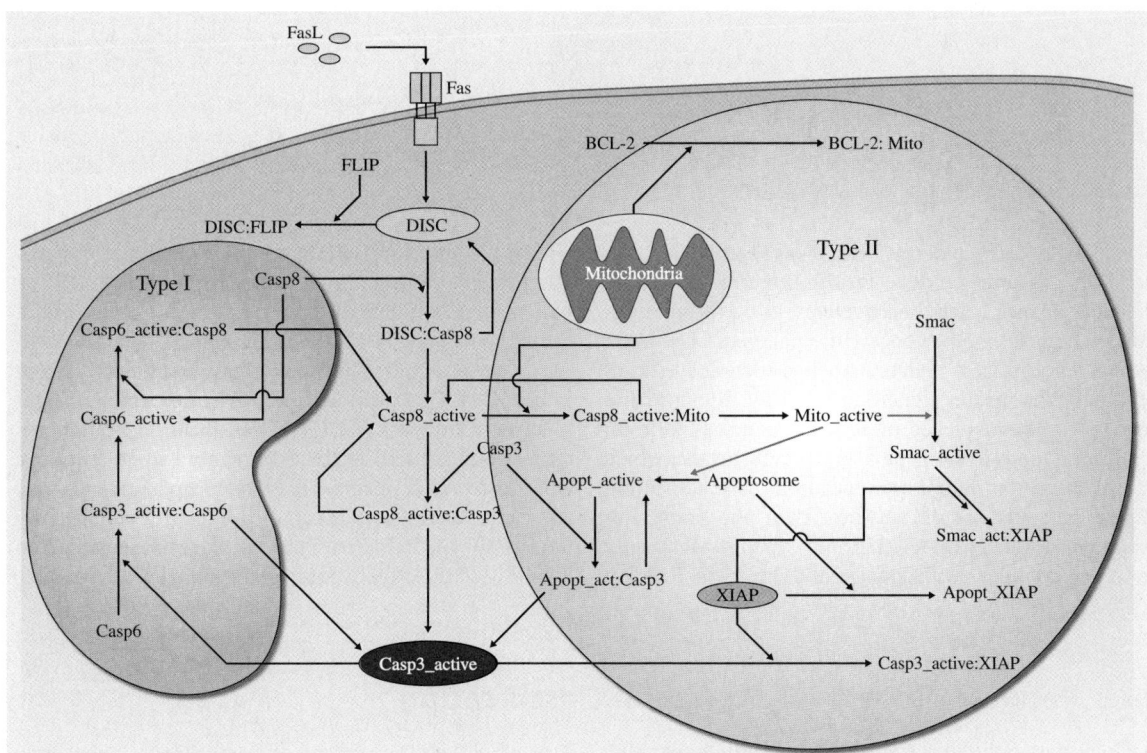

Fig. 75.3 Network schematic of the Type 1 and Type 2 Fas-induced apoptosis network (after [75.31])

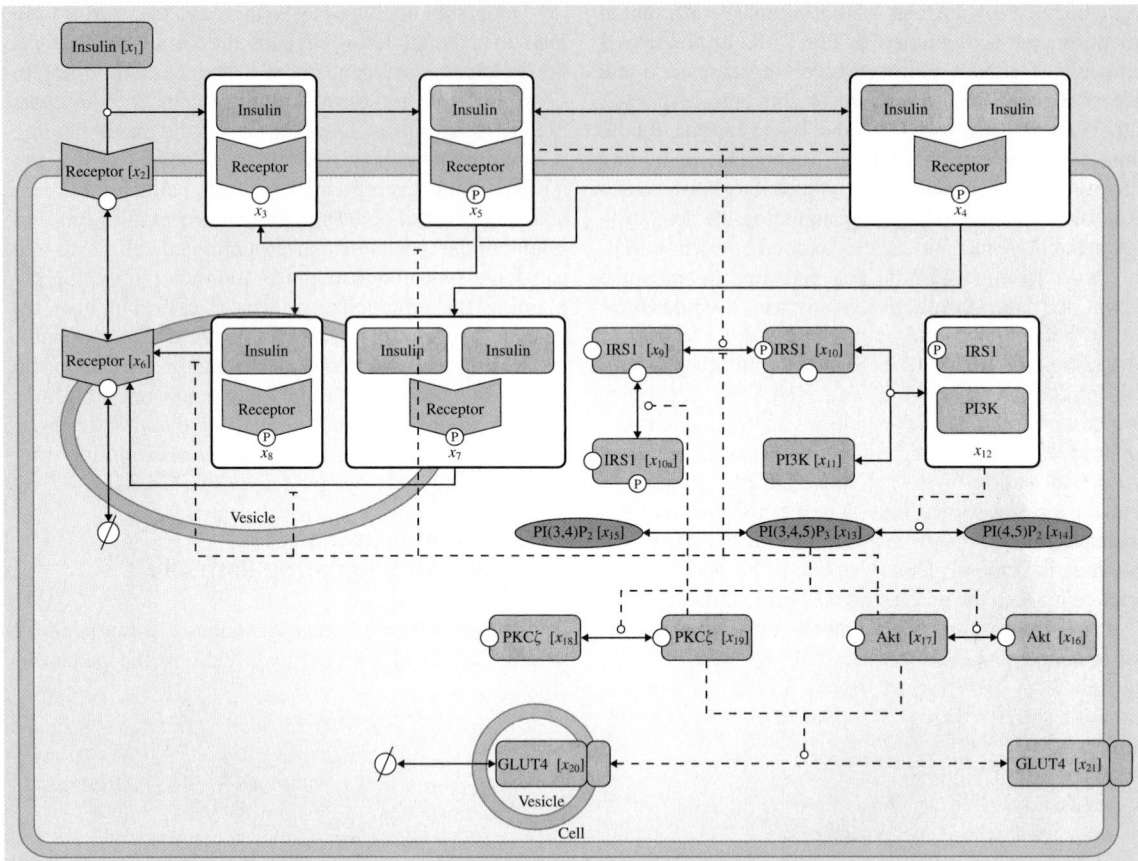

Fig. 75.4 Insulin signalling pathway model (after [75.35])

Simply stated, in patients with type 1 diabetes, the pancreas does not produce insulin, whereas in type 2 diabetes, among other consequences, the cells are resistant to the insulin produced by the pancreas. The latter phenomenon is best understood from detailed consideration of the insulin signalling pathway (illustrated in Fig. 75.4). The sequence of actions occurs as follows: (1) insulin binds to a receptor on the cell surface, which causes receptor autophosphorylation and activation; (2) the activated insulin receptor then phosphorylates insulin receptor substrate-1 (IRS1), which subsequently forms a complex with phosphatidylinositol-3-kinase (PI3K); (3) the IRS1-PI3K complex catalyzes the production of phosphatidylinositol triphosphate (PIP$_3$), which then interacts allosterically with phosinositide-dependent kinase 1 (PDK$_1$); (4) the PIP$_3$-PDK1 complex phosphorylates protein kinase Akt and protein kinase C (PKCζ); (5) activated Akt and PKCζ trigger glucose transporter (GLUT4) translocation from an internal compartment to the cell membrane. In a healthy cell, this cascade ultimately leads to uptake of glucose and helps to regulate glucose levels. In a cell characterized by type 2 diabetes, the cascade is desensitized to insulin, and the effectiveness of the signal is diminished.

75.3 Network Models for Structural Classification

An important point in systems biology is the integrative perspective, i.e., the analysis of the system considered as a whole and across the different levels (gene, protein, metabolite, etc.) and not the reductionist analysis of individual components. While it is useful to categorize the elements and levels of a hierarchical regulatory scheme, it is more useful to analyze such schemes for behaviors that emerge from combinations of simpler motifs. Some simple examples of canonical regulatory constructs (i. e., motifs) that yield specific classes of behavior in gene networks include [75.36]:

1. Positive feedback: multistability, oscillations, state-dependent response
2. Integral feedback: robust adaptation
3. Negative feedback: steady-state (homeostasis, adaptation)
4. Time delay: complex response, oscillations
5. Protein oligomerization: multistability, oscillations, resonant stimulus frequency response.

In addition, stochastic fluctuations can induce random response to stimuli, random outcomes, as well as stochastic focusing. Such properties are characteristic of general networks, including social networks, communication networks, and biological networks [75.37]. Given the wide variety of modeling objectives, as well as the heterogeneous sources of data, it is not surprising that many approaches exist for capturing network interactions in the form of mathematical structures.

75.3.1 Hierarchical Networks

Biophysical networks can be decomposed into modular components that recur across and within given organisms. One hierarchical classification is to label the top level as a network, which is comprised of interacting regulatory motifs consisting of groups of 2–4 components such as proteins or genes [75.38–40]. Motifs are small subnetworks that are over-represented compared to random networks with the same large-scale properties. At the lowest level in this hierarchy is the module that describes transcriptional regulation, of which a nice example is given in [75.41]. At the motif level, one can use pattern searching techniques to determine the frequency of occurrence of these simple motifs [75.39], leading to the postulation that these are basic building blocks in biological networks. Of relevance to the present discussion is the fact that many of these components have direct analogs in systems engineering architectures. Consider the three dominant network motifs found in *Escherichia coli* [75.39]:

1. Coherent feedforward loop: in this, one transcription factor regulates another factor, and, in turn, the pair jointly regulates a third transcription factor

2. Single input module (SIM): in systems terminology, a single-input multiple output block architecture
3. Densely overlapping regulons: in systems terminology, a multiple-input multiple output block architecture.

Similar studies in a completely different organism, *Saccharomyces cerevisiae*, yielded six related or overlapping network motifs [75.38]:

1. Autoregulatory motif: negative feedback in which a regulator binds to the promotor region of its own gene
2. Feedforward loop: as described earlier
3. Multicomponent loop: effectively, a closed-loop with two or more transcription factors
4. Regulator chain: a cascade of serial transcription factor interactions
5. Single input module: as described earlier (SIM)
6. Multiinput module: a natural extension of the preceding motif.

In effect, these studies provide strong evidence that, in both eukaryotic and prokaryotic systems, cell function is controlled by sophisticated networks of control loops that are cascading onto and interconnected with other (transcriptional) control loops. The noteworthy insight is that the complex networks, which underlie biological regulation, appear to be made of elementary systems components like a digital circuit. This lends credibility to the notion that analysis tools from systems engineering should find relevance in this problem domain.

75.3.2 Boolean Networks, Petri Nets, and Associated Structures

Boolean networks are abstract mathematical models employed for coarse-grained analysis of biophysical networks. A Boolean network is represented as a graph of nodes, with directed edges between nodes and a function for each node (e.g., [75.42]). Boolean networks are used to model network dynamics; for instance, a Boolean network can be used to model transcripts and proteins:

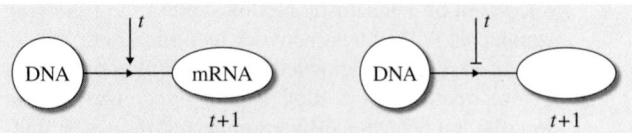

Fig. 75.5 Transcription of transcripts (mRNA)

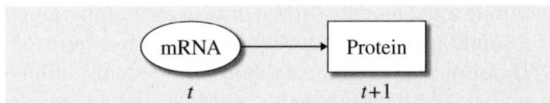

Fig. 75.6 Translation from transcripts to proteins

1. Transcripts and proteins are either on (1) or off (0).
2. The expression of a node at time step t is given by a logical rule of the expression of its effectors at time $t - 1$.
3. Transcription depends on transcription factors; repressors are dominant (Fig. 75.5).
4. Translation depends on the presence of the transcript (Fig. 75.6).
5. Transcripts and proteins decay in one step if not produced.

Place/transition (P/T) nets, or Petri nets, introduced by Carl Adam Petri in 1962, have been intensively studied as one of several formal methods used to verify the correctness of systems, which are described as mathematical objects that can handle characteristics such as nondeterminism and concurrency (e.g., [75.43]).

There are several additional types of networks such as Bayesian nets, which combine directed acyclic graphs with a conditional distribution for each random variable (vertices in a graph, e.g., [75.44]), and signed directed graphs, in which a signed directed edge is used to represent activation versus inhibition (depending on the sign, e.g., [75.45]). Alternatively, S-systems (biochemical systems theory, BST) provide an approach wherein polynomial nonlinear dynamic nodes are used to capture network behavior (e.g., [75.46]).

75.4 Dynamical Models

While the consideration of motifs and network topology is essential for unraveling design principles in complex biophysical networks, it is necessary to understand the role of dynamic behavior in ascribing meaning to the rich hierarchies of regulation. Moreover, because of the interplay between topology and dynamics, it is often not enough in systems biology to specify only the nodes (components) and edges (interactions). The robust control of biophysical networks requires the answer to many challenging questions such as: (1) What are the dynamical aspects of the interaction? (2) What is the characteristic quantity changed by the interaction?

Some of the intrinsically dynamic features of biophysical networks have been analyzed in a recent paper that shows the close relationship between dynamic measures of robustness and the abundance of particular network motifs for a wide range of organisms [75.47]. Attempts to detail dynamic behavior in these networks have fallen into three broad classes of modeling techniques: (1) first-principles approaches, (2) empirical model identification, and (3) a hybrid approach that combines minimum biophysical network knowledge with an objective function to yield a predictive model. In this section, we outline some key results in the development of mechanistic models, and in the following section, we will address network identification.

Given detailed knowledge of a biological architecture, mathematical models can be constructed to describe the behavior of interconnected motifs or transcriptional units (TUs). A number of excellent review papers have been published in recent years [75.1, 36, 48, 49]. In the majority of these studies, gene expression is described as a continuous-time biochemical process, using combinations of algebraic and ordinary differential equations (ODEs) [75.12, 36, 50]. In a similar manner, models at the signal transduction pathway level have been developed in a continuous-time framework, yielding ODEs [75.51]. At the TU level, a detailed mathematical treatment of transcriptional regulation is described in [75.41]. Mechanistic models for a number of specific biological systems have been reported, including basic operons and regulons in *E. coli* (trp, lac, and pho) and bacteriophage systems (T7 and λ) [75.52].

Systems theory has found an enabling role in the analysis of the complex mathematical structures that result from the previously described modeling approaches. The language of systems theory now dominates the quantitative characterization of biological regulation, as robustness, complexity, modularity, feedback, and fragility are invoked to describe these systems. Even classical control theoretic results, such as the Bode sensitivity integral, are being applied to describe the inherent tradeoffs in sensitivity across frequency [75.53]. Robustness has been introduced as both a biological system-specific attribute, as well as a measure of model validity [75.54, 55]. In the next section, brief accounts of systems-theoretic analysis of biological regulatory structures are given, emphasizing where new insights into biological regulation have been uncovered.

75.4.1 Stochastic Systems

Discrete stochastic modeling has recently gained popularity owing to its relevance in biological processes [75.8, 56–59] that achieve their functions with low copy numbers of some key chemical species. Unlike the solutions to stochastic differential equations, the states/outputs of discrete stochastic systems evolve according to discrete jump Markov processes, which naturally lead to a probabilistic description of the system dynamics. A first-order Markov process is a random process in which the future probabilities are dependent only on the present value, and not on past values. Such descriptions can find relevance in systems biology when the magnitude of the fluctuations in a stochastic system approaches the levels of the actual variables (e.g., protein concentrations). In addition, there are qualitative phenomena that are intrinsic to such descriptions that arise in biological systems, as will be mentioned later.

The idea that stochastic phenomena are essential for understanding complex transcriptional processes was nicely illustrated by *Arkin* and coworkers in the analysis of the phage λ lysis–lysogeny decision circuit [75.8]. The probabilistic division of the initially homogeneous cell population into subpopulations corresponding to the two possible fate outcomes was shown to require a stochastic description (and could not be described with a continuous deterministic model). In particular, the coexistence of the two subpopulations necessitated such a formal characterization, and the relative sensitivity of the subpopulations to model parameters including external variables could be analyzed with the resulting models. In a more recent work, *Samilov* and coworkers [75.60] have shown another example of a biological behavior that is intrinsically stochastic in nature – namely the dynamic switching behavior in a class of biochemical reactions (enzymatic futile cycles). In this case, the behavior is more subtle than the lysis–lysogeny switch, where the existence of a bifurcation was at least evident in the continuous differential equation model. In the enzymatic futile cycle problem, the deterministic model gives no indication of multiplicity, yet the discrete stochastic model generates behaviors, including switching as well as oscillations, that indicate characteristics of bifurcation regimes. It is suggested that such noise-induced mechanisms may be responsible for control of switch and cycle behavior in regulatory networks.

In the discrete stochastic setting, the states and outputs are random variables governed by a probability density function, which follows a chemical master equation (CME) [75.61]. The rate of reaction no longer describes the amount of chemical species being produced or consumed per unit time in a reaction but rather the likelihood of a certain reaction to occur in a time window. Though analytical solution of the CME is rarely available, the density function can be constructed using the stochastic simulation algorithm (SSA) [75.61].

The discrete stochastic system of interest is described by a CME [75.62]

$$\frac{\mathrm{d} f(\boldsymbol{x}, t | \boldsymbol{x}_0, t_0)}{\mathrm{d}t} = \sum_{k=1}^{m} a_k(\boldsymbol{x} - \boldsymbol{v}_k, \boldsymbol{p}) f(\boldsymbol{x} - \boldsymbol{v}_k, t | \boldsymbol{x}_0, t_0)$$
$$- a_k(\boldsymbol{x}, \boldsymbol{p}) f(\boldsymbol{x}, t | \boldsymbol{x}_0, t_0), \qquad (75.1)$$

where $f(x, t | x_0, t_0)$ is the conditional probability of the system to be at state x and time t, given the initial condition x_0 at time t_0. The state vector x gives the molecular counts of the species in the system. Here, a_k denotes the propensity functions, v_k denotes the stoichiometric change in x when the k-th reaction occurs and m is the total number of reactions. The propensity function $a_k(x, p)\,\mathrm{d}t$ gives the probability of the k-th reaction to occur between time t and $t + \mathrm{d}t$, given the parameters p. As the state values are typically unbounded, the CME essentially consists of an infinite number of ODEs, whose analytical solution is rarely available except for a few simple problems. The SSA provides an efficient numerical algorithm for constructing the density function [75.61]. The algorithm follows a Monte Carlo approach based on the joint probability for the time to and the index of the next reaction, which is a function of the propensities. The SSA indirectly simulates the CME by generating many realizations of the states (typically of the order of 10^4) at specified time t, given the initial condition and model parameters, from which the distribution $f(x, t | x_0, t_0)$ can be constructed.

There has been simultaneous advancement in experimental methods for quantifying the characteristics of biological noise [75.63–65] along with advances in computing and simulation. A number of groups have recently used dual reporter methods to track the activity of identical genes in the same cell to measure the impact of noise on expression. In the work of *Elowitz* and coworkers, the separate effects of stochastic behavior in the transcriptional and translational processes in prokaryotes (so-called intrinsic noise) are distinguished from noise effects arising from other cellular components that influence the rate of gene expression (so-called extrinsic noise [75.63,65]). *Raser* and *O'Shea* analyze eukaryotic systems with both cis-acting and trans-acting mutations to distinguish between the noise effects that are intrinsic to transcription as opposed to upstream processes that might ultimately influence expression [75.64].

The interface of discrete stochastic systems and biology has clearly led to new insights into stochastic phenomena in biological systems, and has also spurred the development of more efficient computational methods for stochastic simulation, as well as analysis methods for these models. This interface will continue to motivate developments in systems engineering, with improved methods for imaging biological systems that include the ability to resolve spatial behaviors. Distributed stochastic models will require more sophisticated algorithmic developments, particularly as one builds models to truly address systems-scale phenomena.

75.4.2 Constraints and Optimality in Modeling Metabolism

To understand complex biological systems, instead of starting from actual implementations and observations, one can reduce the problem by first separating the possible from the impossible, such as configurations and behaviors that would violate constraints. Systems approaches try to exploit three broad classes of constraints:

1. Empirical: large-scale experimental analysis can provide constraints on possible network structures, such as the average or maximal number of interactions per component.
2. Physico-chemical: laws qof physics such as conservation of mass and thermodynamics impose constraints on cellular and network behaviors. These are used, in particular, for structural network analysis (SNA) with roots in the analysis of chemical reaction networks [75.66].
3. Functional: biological systems perform certain functions and their building blocks are confined to a large, yet finite set. Network structures and behaviors have to conform with both aspects.

Functional constraints constitute the main differences between complex physics and biology. In physics, they do not exist. Biological (as well as engineered) systems evolve to fulfill functions, and are constantly evaluated for their performance. Insufficient performance will lead to extinction, and better solutions are likely to survive. Hence, it is reasonable to assume some kind of optimality in biological systems. The immediate

consequence of a purpose is a considerably smaller design space, in which effective and reliable network are rare and presumably highly structured. Understanding complexity in biology could, thus, employ a calculus of purpose – by asking teleological questions such as why cellular networks are organized as observed, given their known or assumed function [75.67].

Physico-chemical Constraints in Metabolism

Essential constraints for the operation of metabolic networks are imposed by (1) reaction stoichiometries, (2) thermodynamics that restrict flow directions through enzymatic reactions, and (3) maximal fluxes for individual reactions. For instance, metabolism usually involves fast reactions and high turnover of substances when compared with regulatory events. Therefore, on longer time-scales, it can be regarded as being in quasi-steady state. The metabolite balancing equation (75.2) for a system of m internal metabolites and q reactions with the $m \times q$ stoichiometric matrix $\mathbf{N}$ and the $q \times 1$ vector of reaction rates (fluxes) $\boldsymbol{r}$ formalizes this main constraint in SNA. As for most real networks $q \gg m$, the system of linear equation (75.2) is underdetermined. However, all possible solutions are contained in a convex vector space, or flux cone (Fig. 75.7). Methods from convex analysis allow to investigate this space [75.68, 69]

$$\frac{\mathrm{d}\boldsymbol{x}(t)}{\mathrm{d}t} = \mathbf{N} \cdot \boldsymbol{r} = 0 . \tag{75.2}$$

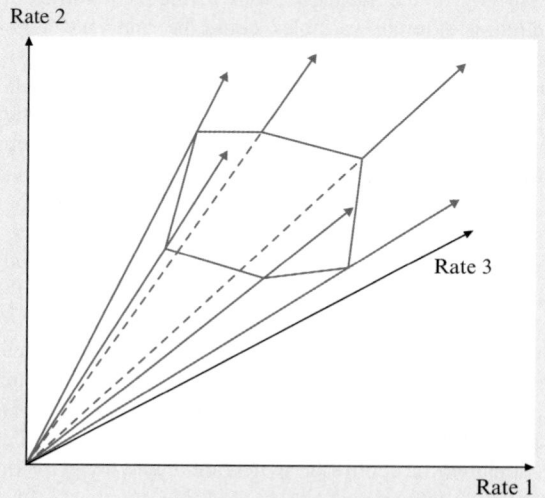

Fig. 75.7 Linear constraints specify a flux cone, with pathways as generating rays, projection on three-dimensional flux space (after [75.4])

Two broad classes of methods for SNA have been developed: metabolic pathway analysis (MPA) and flux balance analysis (FBA) [75.70–72]. MPA computes and uses the set of independent pathways-generating rays in Fig. 75.7 – that uniquely describe the entire flux space; owing to the algorithmic complexity, it can currently only handle networks of moderate size. FBA, in contrast, determines a single flux solution through linear optimization [75.73], often assuming that cells try to achieve optimal growth rates. The computational costs are modest, even for genome-scale models. The approach was successful, for instance, in predicting the effects of gene deletions and the outcomes of convergent evolution in microorganisms [75.72, 74, 75]. FBA, however, has to reverse-engineer and operate with an essentially unknown objective function. While maximal growth has proven to be a reasonable assumption for lower organisms [75.76], higher cells may tend to minimize overall fluxes in the network [75.77]. In general, FBA has proven effective for simpler organisms, and when the steady-state assumption is valid. However, there are many situations where these conditions do not apply, many of which are biophysically meaningful, such as the dynamic diauxic shift in *E. coli*.

Extensions: Dynamics and Control

Stoichiometric constraints restrict the systems dynamics. Thus, the stoichiometric matrix $\mathbf{N}$ is fundamental not only for SNA but also for dynamic processes in reaction networks, in which the reaction rates r in (75.2) are time-dependent. For biological systems, the conservation of total amounts of certain molecular subgroups (conserved moieties such as ATP, ADP and AMP) is characteristic and can be exploited for systems analysis. Classical work in chemical engineering addressed this topic for chemical reaction networks. For instance, *Feinberg* derived theorems to determine the possible dynamic regimes, such as multistability and oscillations, based on network structure alone [75.78, 79]. Challenges posed by biological systems led to renewed interest in these approaches and induced further theory development [75.80, 81]. Application areas in biology include stability analysis [75.80] and model discrimination by safely rejecting hypotheses on reaction mechanisms, thus identifying crucial reaction steps [75.82]. Algorithms for the identification of dependent species in large biochemical systems – to be employed, for instance, in model reduction – have recently become available [75.83].

Enabling FBA to deal with dynamics and regulation proceeded by incorporating additional time-dependent constraints that reflect knowledge of the operation of cellular control circuits – an approach termed regulatory FBA [75.84]. For instance, using superimposed Boolean logic models to capture transcriptional regulatory events has extended the validity of the methodology for a number of complex dynamic system responses [75.84] and for data integration [75.85]. Other dynamic extensions of the FBA algorithm have been proposed in [75.86]. With these more detailed models, steady-state analysis suggested that the complex transcriptional control networks operate in a few dominant states, i. e., generate simple behavior [75.87]. Finally, pathway analysis also allows one to approach features of intrinsically dynamic systems: for instance, it helps to identify feedback loops in cellular signal processing [75.88]. Hence, SNA-related approaches are about to extend to nonclassical domains, in particular, through theory development induced by new challenges in systems biology.

Functional Constraints, Optimality, and Design

In analyzing living systems, one possibility is to start from the assumptions that they have to fulfill certain functions and that cells have been organized over evolutionary time-scales to optimize their operations in a manner consistent with mathematical principles of optimality. FBA demonstrates the utility of this assumption; note that its implicit functional constraint, i. e., steady-state operation of metabolic networks, is not self-explanatory. Similarly, other approaches invoking principles of optimal control theory have opened new avenues for systems analysis in biology.

The cybernetic approach developed by *Kompala* et al. [75.89] and *Varner* and *Ramkrishna* [75.90] is based on a simple principle: evolution has programmed or conditioned biological systems to optimally achieve physiological objectives. This straightforward concept can be translated into a set of optimal resource allocation problems that are solved at every time-step in parallel with the model mass balances (basic metabolic network model). Thus, at every instant in time, gene expression and enzyme activity are rationalized as choice between sets of competing alternatives, each with a relative cost and benefit for the organism. Mathematically, this can be translated into an instantaneous objective function. The researchers in this area have defined several postulates for specific pathway architectures, and the result is a computationally tractable (i. e., analytical) model structure. The potential shortcoming is a limited handling of more flexible objective functions that are commonly observed in biological systems [75.91–95].

Instead of focusing on a single objective function, mathematical models and experimental data can be used to test hypotheses on optimality principles, given a specific cellular function to be fulfilled. For instance, extensions of FBA suggested that *E. coli* optimizes the tradeoff between achieving high growth rates and maintaining wild-type metabolic fluxes after gene deletions [75.96]. MPA showed that the interplay between the metabolic network (the controlled plant) and gene regulation (the controller) in *E. coli* might be designed to achieve optimal tradeoffs between long-term objectives, such as metabolic flexibility, and short-term adjustment for metabolic efficiency [75.97]. Optimal production pipelines for biomass components, with fast responses to environmental changes and minimal additional efforts for enzyme synthesis, were predicted in detail to employ wave-like gene expression programs, which was later confirmed experimentally [75.77, 98]. Hence, at least certain cellular design principles can be revealed by evaluating assumptions on cellular optimality principles.

Finally, without assuming optimality, we can ask how functions in biological systems could be es-

tablished in principle. Among others, drawing from analogies with engineered systems helps to understand more general design principles in biology. From non-linear dynamics, for instance, it is well-known that functions such as oscillators and switches require some source of nonlinearity. Establishing such a function with biological building blocks, thus, allows only for certain circuit designs [75.99, 100]. Similar ideas can prove powerful at different levels of abstraction. For instance, highly structured bow-ties with multiple inputs, channeled through a core with standardized components and protocols to multiple outputs, could be the common organizational principles to establish complex production systems in engineering and biology [75.53]. On the other hand, *El-Samad* and colleagues studied the bacterial heat-shock response, pointing out that the intertwined feedback and feedforward loops present can be assigned individual functions parallel to those loops in designed control circuits that have to yield fast responses in highly fluctuating environments [75.101]. Notably, most of the examples discussed here involved new developments in theory to address challenges posed by biology.

75.5 Network Identification

Model development involves the translation of identified biological processes to coupled dynamical equations, which are amenable to numerical simulation and analysis. These equations describe the interactions between various constituents and the environment, and involve multiple feedback loops responsible for system regulation and noise attenuation and amplification. Currently, however, our knowledge of essentially all biological systems is incomplete. Despite genome projects that allow enumeration and, to a certain extent, characterization of all genes in a system, this does not imply knowledge about all network components (for instance, all protein variants that can be derived from a single gene), interactions, and properties thereof [75.22]. An important task in systems biology consists of specifying network interactions, which can concern qualitative or quantitative properties (existence and strength of couplings), or detailed reaction mechanisms, for genome-based inventories of components.

Essentially, this is a systems identification problem. Given a set of experimental data and prior knowledge, the network generating the data is to be determined [75.102]. Alternatively termed *reverse en-*

gineering [75.103], *network reconstruction* [75.104] or *network inference* [75.105], the general network identification problem provides a key interface between science and engineering. Several qualitatively different approaches for biological systems have been proposed, which can be roughly classified into three categories: data-driven, approximative and mechanistic.

75.5.1 Data-Driven Methods

Empirical or data-driven methods rely on large-scale datasets that can be generated, for instance, through microarray analysis for gene regulatory networks. The relevant methods include singular value decomposition analysis of microarray data [75.106, 107], self-organizing maps [75.108], k-means clustering or hierarchical clustering [75.109], protein correlation and dynamic deviation factors [75.110], and robust statistics approaches [75.111, 112]. For instance, clustering methods are routinely applied for identifying groups of coregulated genes from microarray data. The interpretation of clustering results employs (implicit) models such as coexpressed genes that are likely to have a common regulator. Data quality and algorithmic choices (for instance, of distance measures) critically influence the clustering results; in addition, validation of clustering results and techniques is an open issue [75.113–115].

In contrast to the mechanistic approaches discussed later, most empirical approaches employ discrete-time gray box models [75.116–119]. For instance, inference methods based on probabilistic graphical (e.g. Bayesian) models help to elucidate causal couplings between the network components [75.120]. Their scalability for large systems and the ability to integrate heterogeneous datasets make them attractive [75.5, 121]. Yet, these approaches deliver only qualitative descriptions of network function, and have inherent limitations. For instance, Bayesian models cannot cope with the ubiquitous feedback in cellular networks, since causal relationships have to be represented by directed acyclic graphs [75.120].

A number of challenges are present in treating experimental data for such problems: (1) the sampling rate is rarely uniform and may be exponentially spaced by design, and (2) data from multiple research groups are often combined (e.g. from WWW-posted data) to yield data records with inconsistent sampling, experimental bias, etc. From a systems engineering perspective, another critical point is the potentially divergent qualitative behavior between continuous-time and discrete-time models of corresponding order [75.122]. Recent work has shown the promise of continuous-time formulations of empirical models using modulating function approaches [75.40].

More generally, correctly identifying network topologies (corresponding to the model structure) clearly does not suffice for establishing predictive mathematical models. Experiences with engineered genetic circuits illustrate this point: with identical topology, qualitatively different behavior can result and vice versa [75.123]. Hence, quantitative characteristics, which are usually incorporated through parameters in deterministic models, are also required. Corresponding identification methods are rooted in systems and information theory and, thereby, also provide the largest intersection among biology, other sciences, and engineering.

75.5.2 Linear Approximations

The identification of dynamically changing interactions requires corresponding dynamic models. In a first approximation, we can consider linear systems, i.e., systems with additive responses to perturbations. In systems engineering, a standard form for linear time-invariant (where the shape of the output does not change with a delay in the input) systems with n states and m inputs is given by (75.3) with $n \times 1$ state vector $x(t)$, $n \times n$ system matrix $\mathbf{A}$, $n \times m$ input matrix $\mathbf{B}$, and $m \times 1$ input vector $u(t)$. Linearization of the general dynamic system $dx(t)/dt = f(x, p, t)$ with parameter vector p provides first approximations to the network dynamics, even for highly nonlinear systems such as those encountered in biological networks. Linear models capture the local dynamics, for instance, in the vicinity of a steady state, instead of aiming at more complicated global behaviors.

$$\frac{dx(t)}{dt} = f(x, p, t) \approx \mathbf{A}x(t) + \mathbf{B}u(t) \qquad (75.3)$$

Mathematically, most methods reconstruct the system matrix $\mathbf{A}$, which corresponds to the Jacobian matrix $\mathbf{J} = \partial f(x, p)/\partial x$, from the measured effects of (sufficiently small) perturbations. However, direct recovery of the system matrix $\mathbf{A}$ will be unreliable with noisy data and inputs. In a study using linear models and perturbation experiments to identify the structure of genetic networks, *Tegner* et al. [75.103] therefore proposed an iterative algorithm that uses rational choices of perturbations to improve the identification quality. For a developmental circuit, despite high nonlinearities in the system, the reverse engineering algorithm, which involves building and refining an average connectivity matrix in successive steps, recovered all genetic interactions [75.103]. A related approach that uses linear models and multiple linear regressions showed similar performance. The algorithm attempts to exploit the sparsity of systems matrices for biological networks owing to, for example, (estimated) upper bounds on the number of connections per node [75.105, 124]. Both algorithms are scalable – a central concept in engineering, but until recently considered of less importance in biology.

Newer approaches to systems identification aim at exploiting modularity in biological networks. For a modular system with one output per module, the method employs inversion of the global response matrix for identification of network connectivities and of local responses from perturbation experiments [75.125]. It requires a reduced number of measurements compared with other methods because only changes in so-called communicating intermediates have to be recorded. Apparently, some simplifying assumptions have to be made; for example, modules are coupled by information flow only, and mass flow is negligible [75.125]. An important result of extending the modular identification to time-series data is that, for identifying all connections of a node, it is not necessary to perturb this node directly – inference can rely on detecting the network responses to remote perturbations [75.126]. Extensions to include the effects of uncertainties in experimental data and prior knowledge [75.127], and the possibility of a unified mathematical framework [75.128] make modular identification methods particularly promising.

75.5.3 Mechanistic Models, Identifiability and Experimental Design

Mechanistic models, owing to effects such as saturation in enzymatic reactions, pose particular challenges because they involve identification of nonlinear systems. Depending on whether model structure and parameters, or only the parameters have to be identified, the problems fall into the classes of mixed-integer nonlinear programs or nonlinear programs, respectively. As a clear limitation, finding a unique global optimum in the estimation, or convergence of the algorithms cannot be guaranteed. In addition, model identification comes at high computational costs owing to numerous model simulations [75.129].

In terms of parameter estimation, which is a common problem in different scientific domains [75.102], realistic modeling of complex, nonlinear dynamics of biological networks has given new impulses for the evaluation of existing methods and development of new methods. For instance, though stochastic algorithms show superior performance over deterministic methods for parameter optimization in these systems, they are computationally expensive [75.130]. Novel hybrid methods try to exploit synergies between both approaches in order to increase robustness and efficiency (e.g. [75.131, and references therein]).

More fundamentally, *identifiability* and design of informative experiments need to be addressed. Unstructured approaches to model identification are completely ill-posed when faced with, for instance, modeling a yeast cell with 6200 genes and four possible states per gene; we obtain an overall expression state dimension in excess of 10^{15} [75.132]. Clearly a number of a priori constraints and correlations must be exploited. For discrete models, usage of the experimentally observed upper bound on the number of interactions per species brings the amount of data needed for identification into realistic dimensions [75.133]. However, mere extrapolation of current high-throughput technol-

ogy will not solve these high dimensional data issues. Several recent studies have highlighted the importance of proper design of perturbations to reveal the logical connectivity of gene networks [75.104, 134]. Systems engineering concepts of experimental design to provide rich datasets can be exploited to develop predictive mechanistic models.

Parameter estimation accuracies are central to measuring identifiability of mechanistic models. Low accuracies mean that the corresponding parameters may be varied to a greater extent – and still describe the data – than it is possible for parameters with high estimation accuracy (low associated error). They combine information on model sensitivities with experimental data (Fig. 75.8). More specifically, the Fisher information matrix (FIM) $\mathbf{F}(p)$ [75.135], for a point in parameter space p, links model and experiment via state sensitivities $S(t) = \partial x/\partial p$ and measurement covariance matrix for a discrete sampling time t_i, $\mathbf{C}(t_i)$. For an unbiased estimator, the Cramér–Rao theorem then gives a lower bound for the estimation error (minimum variance).

FIM-based approaches, for instance, yielded insight into the importance of suitable design of input perturbations for signaling networks [75.40], optimality criteria for the design of such inputs [75.136], and algorithms for the optimization of sampling times for dynamic experiments [75.137]. New hybrid parameter estimation methods [75.131] and closed-loop (i. e., integrating iterations between estimation, evaluation of the identifiability, and experimental design) optimal identification procedures [75.138] rely on the FIM formalism. Note, however, that while these proof-of-concept studies with small models and synthetic data are valuable, the performance for real biological problems awaits assessment.

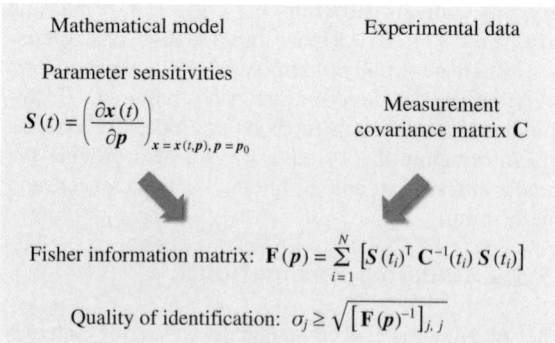

Fig. 75.8 FIM and identification quality. The lower bound for the estimation error of parameter j, σ_j, is derived from the inverse of the main diagonal of the FIM (after [75.4])

Information-rich datasets for integrative models will have to be derived from sources across all levels of biological regulation, such as the transcriptome, proteome, and metabolic fluxes. Concomitantly, we need novel statistical frameworks for data integration [75.139]. Systems identification would greatly benefit from the direct in vivo determination of kinetic parameters; the work by *Ronen* et al. [75.140] for transcriptional control is a first step in this direction. As a complement, synthetic genetic circuits could provide means for controlled excitation of the system, for instance, by inducible genetic switches [75.103], or through genetic oscillators to incorporate analysis

methods in the frequency domain [75.102]. Novel methods could also take known uncertainties associated with measurements, such as experimentally determined characteristics of stochastic noise, explicitly into account. Finally, identification depends on adequate specification of the system and model (e.g., [75.141]). While models are currently either set up ad hoc, or through manual comparison of few alternative structures (including kinetic terms in the equations), uncertainties in biology pose a major challenge for systems sciences: deriving advanced approaches to model discrimination for the simultaneous identification of model structures and parameters.

75.6 Quantitative Performance Metrics

In biology, as in engineering, robust performance refers to the attainment of a particular behavior or response by a system in the presence of uncertainty. This appears to be a ubiquitous property of biological processes that are subject to constant uncertainty in the form of stochastic phenomena [75.142], fluctuating environment, and genetic variation (reviewed in [75.15]). Biology has adapted a number of approaches for coping with these sources of uncertainty, which include:

1. Back-up systems (redundancy)
2. Disturbance attenuation through feedback and feed forward control
3. Structuring of networked systems into semi-autonomous functional units (modularity)
4. Reliable coordination of network elements through hierarchies and protocols.

The robustness problems in systems biology have only begun to yield, in recent years, formal quantitative analyses, owing largely to their nonlinear (and nonstationary) nature. As with engineering systems, robust performance requires the precise specification of both a performance metric and the type/size of uncertainty. When both these elements are specified, it may be possible to analyze biological systems with established engineering tools. It is important to note that the performance metric is often difficult to be defined precisely in biology, as it is an implicit element of an evolved entity.

The last 10 years have demonstrated that sensitivity analysis can provide unique insights into the functioning of complex biophysical networks [75.4, 143–145]. Of particular interest is the behaviour of

biological circuits that exhibit oscillations (e.g., circadian rhythm, cell cycle, neuron firing, cardiac cycles, etc.). In [75.146], a novel set of sensitivity metrics was introduced for performance that were based on a number of different phase-measures: period, phase, corrected phase, and relative phase. The motivation was that phase appears to be the biological imperative, rather than period, for optimal regulation of circadian rhythms. Both state-based and phase-based tools were applied to free-running (absence of light/dark cycles) *Drosophila melanogaster* and *Mus musculus* circadian models. Each metric produced unique relative sensitivity measures used to rank parameters from least to most sensitive. Similarities among the resulting rank distributions strongly suggested a conservation of sensitivity with respect to parameter function and type. A consistent result, for instance, is that model performance of biological oscillators is more sensitive to global parameters than local (i. e., circadian specific) parameters. Differences across the metrics revealed that the conclusions about robustness were dependent on the metric employed for performance.

In [75.147], a novel sensitivity measure, the parametric impulse phase response curve (pIPRC), was derived to both characterize the phase behavior of an oscillator and provide the means for computing the response to an arbitrary signal (in the form of parametric perturbation). The pIPRC builds on the knowledge that biologists have collected for decades in the form of phase response curves (PRCs), to more general classes of input perturbations. The PRCs and infinitesimal PRCs presented in that study provided quantifiable measures of robustness for oscillators acting as pacemakers. In these systems, robust performance involves proper maintenance of phase behavior. In the case of the circadian clock, this means that the PRC to light must have not only the proper shape, but also the correct magnitude. *Zeilinger* et al. [75.148] were able to invalidate a model of the circadian clock in the plant *Arabidopsis thaliana*, because the pIPRC had neither the proper shape nor the proper magnitude.

In [75.149], sensitivity methods were employed to predict the likelihood that noise propagates in stochastic models of the circadian network. The results showed that noise introduced into a sensitive point in the clock propagates very well, while noise introduced into an insensitive (or robust) point is undetectable elsewhere. The noise propagates without regard to distance from point of introduction to point of measurement. It was concluded that the sensitive global parameters are the sites of effective noise propagation in the clock. The hypothesis was that global parameters govern reactions at critical points in the network and therefore may suggest those parts of other systems most worthy of investigation.

The previous studies focused on single cell models, but network properties must also be analyzed at higher levels in the system. *Bagheri* et al. [75.150] demonstrated that computational techniques applied to single cell data are fundamental for tuning and predicting the behavior of oscillatory phenomena at the population level [75.151], since the results of such investigations point to the coupling mechanisms that give rise to spontaneously synchronized networks of stochastic biophysical nodes. Without such insight, it would not have been possible to reproduce the synchrony observed in the SCN. As a result, it is important for experimental biologists to adopt the tools necessary to analyze the structure of both in vitro and in vivo systems.

75.6.1 State–Based Sensitivity Metrics

Parametric state sensitivity is captured in an m by n matrix consisting of individual state performance values with respect to isolated parametric perturbations. The direct method [75.152, 153] is a means to determine exact parametric state sensitivity measures, $S_{ij}(t) = dx_i(t)/d\rho_j$. This approach relies on the continuity of $f[x(t), \rho]$ with respect to the parameter vector ρ. Applying the chain rule results in partial derivatives of the function with respect to states and parameters produces an ordinary differential equation of sensitivity dynamics (75.4). The initial conditions for the sensitivities are zero unless the system initial conditions x_0 depend on the parameters.

$$\dot{S}(t) = \frac{\partial f[x(t), \rho]}{\partial x} \bullet S(t) + \frac{\partial f[x(t), \rho]}{\partial \rho} \qquad (75.4)$$

In oscillatory systems, coefficients of raw state sensitivity, $S(t) = dx(t)/d\rho$, rely on multiple coupled outputs (such as period and phase) and grow unbounded in time for parameters whose period sensitivities are nonzero [75.154–156]. The secular term is due to computation of sensitivities involving a nonuniformly valid expansion of a periodic system [75.154]. *Tomovic* and *Vukobratovic* [75.155] demonstrate that state sensitivity, evaluated at the constant nominal period τ, provides a specific measure corresponding to limit cycle behavior. This measure is unbiased to changes in the period/frequency of oscillation. The resulting m by n matrix is referred to as the cleaned-out or shape sensitivity measure $S^C(t) = (\partial x(t)/\partial \rho)_\tau$; $S^C(t)$ is periodic in time and describes how parametric perturbations affect the shape of state trajectories [75.154]. In its raw form, state sensitivity for oscillatory systems may be decomposed into a combination of shape and period sensitivity measures (75.5) [75.155]. This decomposition of raw state sensitivity highlights its linear time growth t/τ while isolating period sensitivity, $S^\tau = d\tau/d\rho$. The decomposition is generally accomplished by calculating state and period sensitivities, and then solving for the cleaned-out shape sensitivity matrix.

$$S(t) = S^C(t) - \frac{t}{\tau} f[x(t), \rho] \frac{d\tau}{d\rho} \qquad (75.5)$$

Zak et al. [75.156] proposed a method to determine period sensitivity S^τ by making use of the decomposition (75.5) and evaluating state sensitivities at a large time $t_1 \gg \tau$. At this time, the second term on the right-hand side of (75.5) dominates the cleaned-out sensitivity matrix. Singular value decomposition of state sensitivity produces $S(t_1) = U\Sigma V^\top$, where Σ is an m by n diagonal matrix of nonnegative singular values σ and matrices U and V contain the eigenvectors of $SS^\top$ and $S^\top S$, respectively. Hence, period sensitivity of any given state may be approximated by (75.6) where σ_1 is the largest singular value in Σ, $\| f[x(t_1)], \rho \|$ is the vector norm of the state matrix evaluated at time t_1, and V_1 is the first column vector of V. This approximation holds true at any large times $t \gg \tau$, when the system is nonzero, and when the period of oscillation is sensitive to at least one parameter [75.156].

$$S^\tau \approx \pm \frac{\sigma_1 \tau}{\| f[x(t_1)], \rho \| t_1} V_1 \qquad (75.6)$$

Amplitude sensitivity $\mathbf{S}^{\wedge}$ describes how maximum state values vary due to independent parametric perturbations. As discussed, cleaned-out sensitivity defines how parametric perturbations affect the shape of state trajectories at every time in the cycle; at peak concentration times this measure relates directly to the state's maximum concentrations. Thus, amplitude sensitivity values are calculated from shape sensitivity by evaluating the time-dependent measure (75.7) at peak concentrations times t_{peak}

$$\mathbf{S}^{\wedge} = \mathbf{S}^{C}(t_{peak}) . \tag{75.7}$$

75.6.2 Phase–Based Sensitivity Metrics

In cases involving oscillatory networks and often noisy state amplitude data, phase may be employed as a metric for model development and analysis. Standard (raw) phase and decoupled (corrected) phase are among the set of numerical performance measures introduced in [75.146]. Radian-based phase angles $\theta(t, \rho)$ are extracted from the system's limit cycle (Fig. 75.9a) using the cosine rule [75.158]. Resulting phase dynamics reflect the oscillator's real-time position, or concentration (75.8), with respect to a static reference $\boldsymbol{r}$

$$\cos[\theta(t, \boldsymbol{\rho})] \approx \pm \frac{\boldsymbol{x}(t, \boldsymbol{\rho})^{\top} \boldsymbol{r}}{\|\boldsymbol{x}(t, \boldsymbol{\rho})\| \|\boldsymbol{r}\|} . \tag{75.8}$$

Phase measures are recorded under varying parameter sets, $\theta(t, \tilde{\boldsymbol{\rho}})$, where $\tilde{\boldsymbol{\rho}}$ indicates measurements with respect to a perturbation of magnitude δ affecting a single

parameter ρ_j. This perturbation changes the nominal parameter vector from $\boldsymbol{\rho}$ to a perturbed parameter vector ρ_j

$$\tilde{\boldsymbol{\rho}} = \boldsymbol{\rho} + \delta \boldsymbol{e}_j , \quad \forall j \in [1, n] . \tag{75.9}$$

The collected phase trajectories capture the oscillator's position and map them onto nominal time (Fig. 75.9b). If the perturbation strength and length are the same, the information contained in these trajectories is analogous to that contained in phase transition curves [75.159]. Direct evaluation of phase trajectories (Fig. 75.9b) yields two types of sensitivity measures: period and phase, because the change in period (or period sensitivity) is merely an accumulation in the change in phase (or phase sensitivity) evaluated over an entire cycle. In the case of period sensitivity, phase trajectories are evaluated at a $2\pi L$-radian interval, where the integer interval L is chosen to balance the numerical error with computational expense. The difference between the perturbed $2\pi L$-radian crossing time $\tilde{k}$ and the nominal $2\pi L$-radian crossing time k yields the system's periodic performance. A normalized time difference between perturbed and nominal phase trajectories defines the system's period sensitivity, $\mathbf{S}^{\theta_\tau}$ (75.10): a series of measurements denoting the quantitative change in period with respect to a change in the j-th parameter

$$S_j^{\theta_\tau} = \frac{1}{L} \frac{\tilde{k} - k}{\delta \rho_j} \bigg|_{\theta(k, \rho) = \theta(\tilde{k}, \tilde{\rho}) = 2\pi L} ,$$
$$\forall j \in [1, n] , \quad L \in [1, \infty) . \tag{75.10}$$

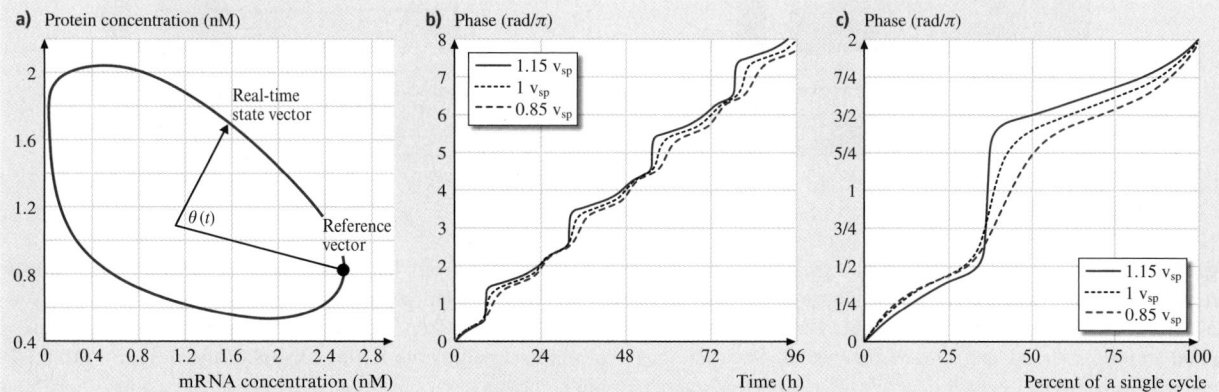

Fig. 75.9a–c Time-dependent phase dynamics. (**a**) Real-time phase dynamics captured in a 2-D limit cycle. Phase is the radian-based measure $\theta(t, \boldsymbol{\rho})$ that describes the angular relationship between the state vector $x(t, \boldsymbol{\rho})$ and some predetermined reference $\boldsymbol{r}$. Phase measures are recorded as a function of time under various parameter sets. (**b**) Coupling of period and phase as trajectories diverge in time. (**c**) Decoupled measure of phase behavior as circadian phase dynamics are normalized with respect to their perturbation-induced periods (after [75.146, 157] by permission of Oxford University Press)

In the case of strict phase sensitivity, one can evaluate phase trajectories at specific times $t = t_k$. The radian phase difference between perturbed and nominal trajectories describes the raw phase sensitivity, $\mathbf{S}^\theta$ (75.11): a series of measures reflecting the induced change in phase with respect to a change in the j-th parameter t_k-hours after the perturbation

$$S_j^\theta(t_k) = \left. \frac{\theta(t, \tilde{\boldsymbol{\rho}}) - \theta(t, \boldsymbol{\rho})}{\delta \rho_j} \right|_{t=t_k}, \quad \forall j \in [1, n]. \quad (75.11)$$

Just as state sensitivity diverges over time, phase sensitivity grows unbounded due to the nonuniform expansion of a periodic system. To correct for the integrated response of perturbation effects (in this case, the coupling of period and phase), each phase trajectory is normalized with respect to the period of the system after parametric perturbation $\tilde{\tau}$: dividing each time series by $\tilde{\tau}$ decouples the system's phase from its period. As a result, normalized datasets begin and end at the same relative time points (0 and 100% of their respective cy-

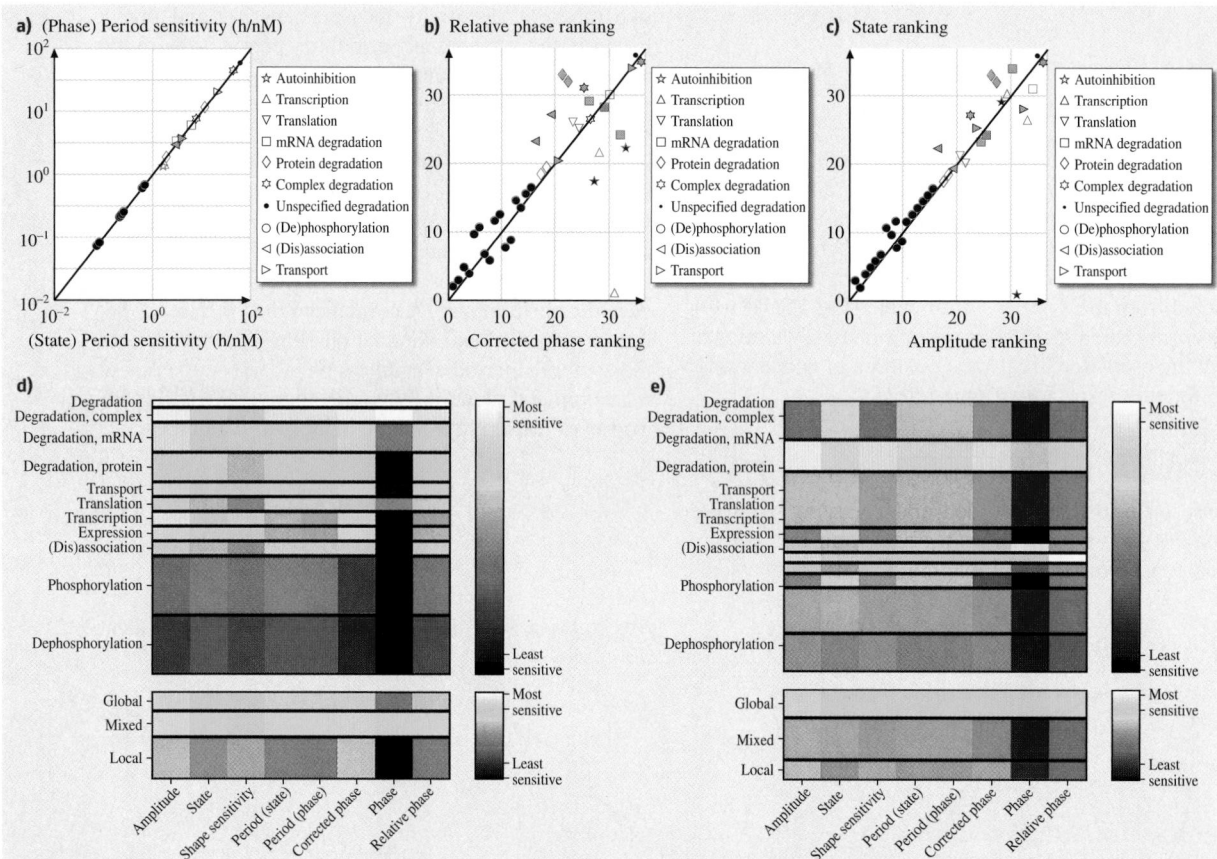

Fig. 75.10a–e Metric evaluation. **(a–c)** Parametric sensitivity metrics ordered from least to greatest absolute value for the Drosophila model: **(a)** state-based and phase-based period sensitivity, **(b)** corrected phase (decoupled angular phase trajectories) and relative phase (time interval between peak per mRNA and nuclear PER/TIM protein concentrations) sensitivity ranks and **(c)** amplitude-based and state-based metrics. Spearman rank correlation coefficients for these pairs of metrics are 1.00, 0.64 and 0.85, respectively. Legends describe the parameters' particular biological processes as the shading of each data point describes their type: global (*open*), mixed (*gray*) and local (*black*) parameters. Data points outlined in *red* reflect *per* gene dynamics, and those in *green* reflect *tim* gene dynamics. **(d,e)**, Color-coded average sensitivities (values are scaled between 10^{-4} and 1) among parameter function (*upper subplots*) and parameter type (*lower subplots*) for each metric in the fly **(d)** and the mammalian **(e)** model. See the supplementary material for alternate correlation diagrams and ranking plots (after [75.146, 157] by permission of Oxford Univ. Press)

cles). These modified datasets allow for a comparison between nominal and perturbed phase at every point in the cycle (Fig. 75.9c). This corrected phase sensitivity assumes a linear scaling of raw phase measures resulting in a time-dependent performance quantity that identifies specific points of the cycle most susceptible to uncertainty (75.12)

$$S_j^{\widehat{\Theta}}(t_k) = \frac{\theta(t/\tilde{\tau}, \tilde{\rho}) - \theta(t/\tau, \rho)}{\delta \rho_j}, \quad \forall j \in [1, n]. \quad (75.12)$$

Phase-based period, phase, and corrected phase sensitivity analysis examine the biological network relative to a static reference. In some cases, a relative analysis that studies relationships within the perturbed network may be more useful. The timing effects relating transcription, translation, phosphorylation, and transport are governed by global cellular processes. Variation of these specific time intervals as a result of parameter manipulation indicates a degree of sensitivity. Relative sensitivity S^ϕ investigates the time interval relating the hour-difference between the occurrence of particular events; for instance, the time interval between peak mRNA concentrations and their corresponding protein concentrations. This time interval $\phi[x(t), \rho]$, is a function of the system's state and parameter vectors. It explores how a system's individual components change relative to one another due to parametric perturbation (75.13)

$$S_j^\phi = \frac{\phi[x(t), \tilde{\rho}] - \phi[x(t), \rho]}{\delta \rho_j}, \quad \forall j \in [1, n]. \quad (75.13)$$

75.6.3 Global Versus Local Parameters

The investigated performance metrics depict classes of sensitivity that associate with parameter function, prov-

ing a certain conservation of robustness for specific biochemical processes. A similar sensitivity distribution was found when parameters were separated into three types: global, mixed, and local [75.15]. Global parameters are involved in core cellular reactions nonspecific to the circadian rhythm; they encompass properties consistent with the entirety of the cellular network. Local parameters are primarily attributed to the circadian system; their processes and/or elements are not shared with many other cellular circuits. Global parameters include transcription rates, various mRNA and protein degradation rates, and translation rates.

Figure 75.10a–c depict phase-based and state-based period sensitivities, relative and corrected phase sensitivities, and amplitude and state sensitivities. Global parameters were consistently found in the first (upper right) quadrants of these plots. Local parameters, including protein phosphorylation and dephosphorylation, were consistently in the lower left (less sensitive) corners. Figure 75.10d and e (lower subplots) emphasize this conservation of sensitivity apparent in both models by assigning a color reflective of the average sensitivity ranking within each parameter type.

The results demonstrated that every defined performance metric was more sensitive to perturbations involving global and mixed parameters than it was to perturbations involving local parameters. Grouping parametric sensitivity based on parameter type provided a more consistent and distinct distribution of sensitivity measures among various metrics than the grouping of sensitivity by function. This outcome agreed with a previous study suggesting that circadian performance is greatly affected by changes in global parameters and less susceptible to changes in local parameters [75.13, 15].

75.7 Bio-inspired Control and Design

Biological networks offer a number of opportunities for inspired design of engineering networks. Aside from the overlapping computational toolkit (e.g., simulation methods for high dimensional, stochastic, stiff, multiscale systems), there are numerous behaviors in biological networks that offer promise for improved communications and sensors networks. Given the space constraints, we highlight only two of them here, but refer the reader to the thorough NRC report on network science for additional details [75.160].

Using control engineering parlance, one may refer to the elements that influence or entrain the circadian

oscillator (*zeitgeber*: time giver) as manipulations, and the elements that exhibit quantifiable circadian rhythms as measurements. The use of this terminology is reflective of the spirit in looking for ways to influence rhythms to fine tune physiological performance. Continuing the control analogy, the open-loop characteristics (i. e., with no intervention) of the typical human are to adjust at a rate of approximately 60–90 min of phase per day [75.161]. In other words, jet lag accommodation occurs at a rate of approximately 2–3 days for a 3 h time zone change. Westward travel is slightly easier, as the natural free-running human circadian clock (i. e., in absence of light) has a period of approximately 25 h. From a control perspective, the two important attributes of the resetting response characteristics are phase shift (captured by the phase response curve), and the associated transient, which we define via the transient response curve.

Sensitivity analysis can be used to develop insight on the parts of a network that are most sensitive and, consequently, the most susceptible to intervention such as the targeting of a drug. However, a typical application will require a temporal forcing of a node (or nodes) in a network to elicit an optimal response. In recent work, model predictive control (MPC) algorithms have been employed to generate the optimal forcing proto-

col that will reset the circadian clock from a condition of phase offset (i. e., jet lag) [75.157, 162]. Through parametric state sensitivity analysis, key driving mechanisms for optimal manipulation of the large complex circadian network were identified. Importantly, the use of nonphotic control inputs outperforms light-based phase resetting dynamics. Aside from targeting individual parameters as control inputs, Fisher information matrix based parametric sensitivity analyses identified combinations of parameters for control (i. e., vector strategies). The derived MPC algorithm is found to be robust to model mismatch and outperforms the open-loop 24 h sun cycle based phase recovery strategy by nearly threefold.

75.8 Emerging Trends

Systems biology has an exciting future. Opportunities for research exist in the areas of network identification, constraints and optimality, stochastic systems, and robustness [75.4], as well as synthetic approaches to network reconstruction:

1. Knowledge of biological systems is incomplete, and codification of the interactions among network components, interactions, and properties requires continued elaboration. This systems identification problem can be approached via empirical, approximative, and mechanistic methods.
2. An understanding of biological systems may proceed from an appreciation of constraints, which can be divided into three classes: (1) Functional constraints exist because biological systems perform specified functions arising from a finite set of components. Since biological systems evolve to a performance metric, it is sensible to presume optimality. This leads to a smaller design space with highly structured, but rarely composed, reliably networks. (2) Analysis of experimental data can provide data on the possible arrangements for network structure. (3) The laws of physics and chemistry, must, of course, be observed.
3. Developments in systems biology will continue to depend upon algorithmic innovations to ap-

proach stochasticity, particularly for distributed models.
4. There exist intrinsic scaling problems with those engineering methods that provide a formal framework for robustness analysis. For large, complex, nonlinear systems, the methods are conservative and computational tractability is questionable. Consequently, the scalability of robustness analysis methodologies provides interesting future challenges.

One promising application of a systems approach to biology lies in healthcare [75.163]. Unlike today's medicine, with its emphasis on therapies for existing disease, tomorrow's medicine will predict and prevent illness. Multiparameter diagnostics will be used routinely to assess health status. Network science will be used to ascertain drug targets, leading to true preventive medicine.

The emerging bio-inspired automation and automatic control mechanisms, techniques and models will continue to offer rich sources of innovation. We will continue to search how the benefits of systems biology knowledge in living organisms can help us solve complex, critical problems in increasingly interrelated systems and service infrastructures.

References

75.1   J. Hasty, D. McMillen, F. Isaacs, J.J. Collins: Computational studies of gene regulatory networks: In numero molecular biology, Nat. Rev. Genet. **2**, 268–279 (2001)

75.2   T. Ideker, T. Galitski, L. Hood: A new approach to decoding life: Systems biology, Annu. Rev. Genomics Hum. Genet. **2**, 343–372 (2001)

75.3   H. Kitano: *Foundations of Systems Biology* (MIT Press, Cambridge 2001)

75.4   F.J. Doyle III, J. Stelling: Systems interface biology, J. R. Soc. Interface **3**, 603–616 (2006)

75.5   E. Klipp, R. Herwig, A. Kowald, C. Wierling, H. Lehrach: *Systems Biology in Practice: Concepts, Implementation and Application* (Wiley, Weinheim 2005)

75.6   B. Palsson: *Systems Biology: Properties of Reconstructed Networks* (Cambridge Univ. Press, Cambridge 2006)

75.7   Z. Szallasi, J. Stelling, V. Periwal (Eds.): *System Modeling in Cellular Biology: From Concepts to Nuts and Bolts* (MIT Press, Cambridge 2006)

75.8   A. Arkin, J. Ross, H.H. McAdams: Stochastic kinetic analysis of developmental pathway bifurcation in phage lambda-infected Escherichia coli cells, Genetics **149**, 1633–1648 (1998)

75.9   N. Barkai, S. Leibler: Robustness in simple biochemical networks, Nature **387**, 913–917 (1997)

75.10   C.V. Rao, M. Frenklach, A.P. Arkin: An allosteric model for transmembrane signaling in bacterial chemotaxis, J. Mol. Biol. **343**, 291–303 (2004)

75.11   T.M. Yi, Y. Huang, M.I. Simon, J. Doyle: Robust perfect adaptation in bacterial chemotaxis through integral feedback control, Proc. Natl. Acad. Sci. USA **97**, 4649–4653 (2000)

75.12   A. Goldbeter: *Biochemical Oscillations and Cellular Rhythms: The Molecular Bases of Periodic and Chaotic Behavior* (Cambridge Univ. Press, Cambridge 1996)

75.13   J. Stelling, E.D. Gilles, F.J. Doyle III: Robustness properties of circadian clock architectures, Proc. Natl. Acad. Sci. USA **101**, 13210–13215 (2004)

75.14   T.G. Müller, D. Faller, J. Timmer, I. Swameye, O. Sandra, U. Klingmüller: Tests for cycling in a signalling pathway, J. R. Statist. Soc. Ser. C Appl. Statist. **53**, 557–568 (2004)

75.15   J. Stelling, U. Sauer, Z. Szallasi, F.J. Doyle III, J. Doyle: Robustness of cellular functions, Cell **118**, 675–685 (2004)

75.16   E.D. Sontag: Asymptotic amplitudes and Cauchy gains: a small-gain principle and an application to inhibitory biological feedback, Syst. Control Lett. **47**, 167–179 (2002)

75.17   D. Angeli, J.E. Ferrell, E.D. Sontag: Detection of multistability, bifurcations, and hysteresis in a large class of biological positive-feed back systems, Proc. Natl. Acad. Sci. USA **101**, 1822–1827 (2004)

75.18   X. Wen, S. Fuhrman, G.S. Michaels, D.B. Carr, S. Smith, J.L. Barker, R. Somogyi: Large-scale temporal gene expression mapping of central nervous system development, Proc. Natl. Acad. Sci. USA **95**, 334–339 (1998)

75.19   D.A. Lauffenburger: Cell signaling pathways as control modules: complexity for simplicity?, Proc. Natl. Acad. Sci. USA **97**, 5031–5033 (2000)

75.20   A.L. Barabasi: Network biology: Understanding the cell's functional organization, Nat. Rev. Genet. **5**, 101–113 (2004)

75.21   A.M. Malcolm, L.J. Heyer: *Discovering Genomics, Proteomics, and Bioinformatics* (Benjamin Cummings, San Francisco 2003)

75.22   H. Kitano: Computational systems biology, Nature **420**, 206–210 (2002)

75.23   I. Edery: Circadian rhythms in a nutshell, Physiol. Genomics, **3**, 59–74 (2000)

75.24   S.M. Reppert, D.R. Weaver: Coordination of circadian timing in mammals, Nature **418**, 935–941 (2002)

75.25   E.D. Herzog, S.J. Aton, R. Numano, Y. Sakaki, H. Tei: Temporal precision in the mammalian circadian system: a reliable clock from less reliable neurons, J. Biol. Rhythms **19**, 35–46 (2004)

75.26   A.C. Liu, D.K. Welsh, C.H. Ko, H.G. Tran, E.E. Zhang, A.A. Priest, E.D. Buhr, O. Singer, K. Meeker, I.M. Verma, F.J. Doyle III, J.S. Takahashi, S.K. Kay: Intercellular coupling confers robustness against mutations in the SCN circadian clock network, Cell **129**, 605–616 (2007)

75.27   Z. Boulos, M.M. Macchi, M.P. Sturchler, K.T. Stewart, G.C. Brainard, A. Suhner, G. Wallace, R. Steffen: Light visor treatment for jet lag after westward travel across six time zones, Aviat. Space Environ. Med. **73**, 953–963 (2002)

75.28   S. Daan, C.S. Pittendrigh: A functional analysis of circadian pacemakers in nocturnal rodents. II. The variability of phase response curves, J. Comput. Physiol. **106**, 253–266 (1976)

75.29   J.C. Dunlap, J.J. Loros, P.J. DeCoursey (Eds.): *Chronobiology: Biological timekeeping* (Sinauer Associates, Inc., Sunderland 2004)

75.30   D.B. Forger, C.S. Peskin: A detailed predictive model of the mammalian circadian clock, Proc. Natl. Acad. Sci. USA **100**, 14806–14811 (2003)

75.31   F. Hua, S. Hautaniemi, R. Yokoo, D.A. Lauffenburger: Integrated mechanistic and data-driven modelling for multivariate analysis of signalling pathways, J. R. Soc. Interface **9**, 515–526 (2006)

75.32   J.W. Stucki, H.-U. Simon: Mathematical modeling of the regulation of caspase-3 activation and degradation, J. Theor. Biol. **234**, 123–131 (2005)

75.33   E.Z. Bagci, Y. Vodovotz, T.R. Billiar, G.B. Ermentrout, I. Bahar: Bistability in apoptosis: Roles of bax, bcl-2, and mitochondrial permeability transition pores, Biophys. J. **90**, 1546–1559 (2006)

75.34   J.E. Shoemaker, F.J. Doyle III: Identifying fragilities in biochemical networks: robust performance analysis of the Fas signaling-induced apoptosis, Biophys. J. **95**, 2610–2623 (2008)

75.35   A.R. Sedaghat, A. Sherman, M.J. Quon: A math-ematical model of metabolic insulin signaling pathways, Am. J. Physiol. Endocrinol. Metab. **283**, E1084–E1101 (2002)

75.36   P. Smolen, D.A. Baxter, J.H. Byrne: Mathematical modeling of gene networks, Neuron **26**, 567–580 (2000)

75.37   Committee on Network Science for Future Army Applications: Network Science National Research Council, Washington (2006)

75.38   T.I. Lee, N.J. Rinaldi, F. Robert, D.T. Odom, Z. Bar-Joseph, G.K. Gerber, N.M. Hannett, C.T. Harbison, C.M. Thompson, I. Simon, J. Zeitlinger, E.G. Jen-nings, H.L. Murray, D.B. Gordon, B. Ren, J.J. Wyrick, J.-B. Tagne, T.L. Volkert, E. Fraenkel, D.K. Gifford: Transcriptional regulatory networks in Saccha-romyces cerevisiae, Science **298**, 799–804 (2002)

75.39   S.S. Shen-Orr, R. Milo, S. Mangan, U. Alon: Network motifs in the transcriptional regulation network of Escherichia coli, Nat. Genet. **31**, 64–68 (2002)

75.40   D.E. Zak, G.E. Gonye, J.S. Schwaber, F.J. Doyle III: Importance of input perturbations and stochas-tic gene expression in the reverse engineering of genetic regulatory networks: insights from an identifiability analysis of an in silico network, Genome Res. **13**, 2396–2405 (2003)

75.41   N. Barkai, S. Leibler: Circadian clocks limited by noise, Nature **403**, 267–268 (2000)

75.42   T. Ideker, V. Thorsson, R.M. Karp: Discovery of regulatory interactions through perturbations: In-ference and experimental design, Pac. Symp. Biocomput. (2000)

75.43   M. Nagasaki, A. Doi, H. Matsuno, S. Miyano: A ver-satile Petri net based architecture for modeling and simulation of complex biological processes, Genome Inform. **15**, 180–197 (2004)

75.44   D. Pe'er, A. Regev, G. Elidan, N. Friedman: Inferring subnetworks from perturbed expression profiles, Bioinformatics **17**, S215–S224 (2001)

75.45   K. Kyoda, K. Baba, S. Onami, H. Kitano: DBRF-MEGN method: An algorithm for deducing minimum equivalent gene networks from large-scale gene expression profiles of gene deletion mutants, Bioinformatics **20**, 2662–2675 (2004)

75.46   S. Kimura, K. Ide, A. Kashihara, M. Kano, M. Hatakeyama, R. Masui, N. Nakagawa, S. Yoko-yama, S. Kuramitsu, A. Konagaya: Inference of S-system models of genetic networks using a co-operative coevolutionary algorithm, Bioinformatics **21**, 1154–1163 (2005)

75.47   R.J. Prill, P.A. Iglesias, A. Levchenko: Dynamic properties of network motifs contribute to bio-logical network organization, PLoS Biology **3**, e343 (2005)

75.48   B.B. Aldridge, J.M. Burke, D.A. Lauffenburger, P.K. Sorger: Physicochemical modelling of cell sig-nalling pathways, Nat. Cell Biol. **8**, 1195–1203 (2006)

75.49   G. Karlebach, R. Shamir: Modelling and analysis of gene regulatory networks, Nat. Rev. Mol. Cell Biol. **9**, 770–780 (2008)

75.50   J.L. Cherry, F.R. Adler: How to make a biological switch, J. Theor. Biol. **203**, 117–133 (2000)

75.51   B.N. Kholodenko, O.V. Demin, G. Moehren, J.B. Hoek: Quantification of short term signaling by the epidermal growth factor receptor, J. Biol. Chem. **274**, 30169–30181 (1999)

75.52   A. Gilman, A. Arkin: Genetic 'code': representations and dynamical models of genetic components and networks, Annu. Rev. Genomics Hum. Genet. **3**, 341–369 (2002)

75.53   M. Csete, J. Doyle: Bow ties, metabolism and dis-ease, Trends Biotechnol. **22**, 446–450 (2004)

75.54   L. Ma, P.A. Iglesias: Quantifying robustness of bio-chemical network models, BMC Bioinformatics **3**, 38–50 (2002)

75.55   H.R. Ueda, M. Hagiwara, H. Kitano: Robust oscil-lations within the interlocked feedback model of Drosophila circadian rhythm, J. Theor. Biol. **210**, 401–406 (2001)

75.56   Y. Cao, D.T. Gillespie, L.R. Petzold: Accelerated stochastic simulation of the stiff enzyme-substrate reaction, J. Chem. Phys. **123**, 144917 (2005)

75.57   D.B. Forger, C.S. Peskin: Stochastic simulation of the mammalian circadian clock, Proc. Natl. Acad. Sci. USA **102**, 321–324 (2005)

75.58   H. Li, Y. Cao, L.R. Petzold, D.T. Gillespie: Algorithms and software for stochastic simulation of biochem-ical reacting systems, Biotechnol. Prog. **24**, 56–61 (2008)

75.59   H.H. McAdams, A. Arkin: Stochastic mechanisms in gene expression, Proc. Natl. Acad. Sci. USA **94**, 814–819 (1997)

75.60   M. Samoilov, S. Plyasunov, A.P. Arkin: Stochas-tic amplification and signaling in enzymatic futile cycles through noise-induced bistability with os-cillations, Proc. Natl. Acad. Sci. USA **102**, 2310–2315 (2005)

75.61   D.T. Gillespie: A general method for numerically simulating the stochastic time evolution of coupled chemical reactions, J. Comput. Phys. **22**, 403–434 (1976)

75.62   D.T. Gillespie: Exact stochastic simulation of cou-pled chemical reactions, J. Phys. Chem. **81**, 2340–2361 (1977)

75.63   M.B. Elowitz, A.J. Levine, E.D. Siggia, P.S. Swain: Stochastic gene expression in a single cell, Science **297**, 1183–1186 (2002)

75.64   J.M. Raser, E.K. O'Shea: Control of stochasticity in eukaryotic gene expression, Science **304**, 1811–1814 (2004)

75.65   P.S. Swain, M.B. Elowitz, E.D. Siggia: Intrinsic and extrinsic contributions to stochasticity in gene ex-pression, Proc. Natl. Acad. Sci. USA **99**, 12795–12800 (2002)

75.66   B.L. Clarke: Stoichiometric network analysis, Cell Biochem. Biophys. **12**, 237–253 (1988)

75.67   A.D. Lander: A calculus of purpose, PLoS Biology **2**, 0712 (2004)

75.68   R. Heinrich, S. Schuster: *The Regulation of Cellular Systems* (Chapman and Hall, New York 1996)

75.69   R.T. Rockafellar: *Convex Analysis* (Princeton Univ. Press, Princeton 1970)

75.70   I. Borodina, J. Nielsen: From genomes to in silico cells via metabolic networks, Curr. Opin. Biotechnol. **16**, 350–355 (2005)

75.71   J.A. Papin, J. Stelling, N.D. Price, S. Klamt, S. Schuster, B.Ø. Palsson: Comparison of network-based pathway analysis methods, Trends Biotechnol. **22**, 400–405 (2004)

75.72   N.D. Price, J.L. Reed, B.Ø. Palsson: Genome-scale models of microbial cells: evaluating the consequences of constraints, Nat. Rev. Microbiol. **2**, 886–897 (2004)

75.73   A. Varma, B.Ø. Palsson: Metabolic flux balancing: basic concepts, scientific and practical use, Biotechnol. Bioeng. **12**, 994–998 (1993)

75.74   S.S. Fong, J.Y. Marciniak, B.Ø. Palsson: Description and interpretation of adaptive evolution of Escherichia coli K-12 MG1655 by using a genome-scale in silico metabolic model, J. Bacteriol. **185**, 6400–6408 (2003)

75.75   S.S. Fong, B.Ø. Palsson: Metabolic gene-deletion strains of Escherichia coli evolve to computationally predicted growth phenotypes, Nat. Genet. **36**, 1056–1108 (2004)

75.76   A.P. Burgard, C.D. Maranas: Optimization-based framework for inferring and testing hypothesized metabolic objective functions, Biotechnol. Bioeng. **82**, 670–677 (2003)

75.77   E. Klipp, R. Heinrich, H.-G. Holzhutter: Prediction of temporal gene expression. Metabolic optimization by re-distribution of enzyme activities, Eur. J. Biochem. **269**, 5406–5413 (2002)

75.78   M. Feinberg: Chemical reaction network structure and the stability of complex isothermal reactors I. The deficiency zero and deficiency one theorems, Chem. Eng. Sci. **42**, 2229–2268 (1987)

75.79   M. Feinberg: Chemical reaction network structure and the stability of complex isothermal reactors II. Multiple steady states for networks of deficiency one, Chem. Eng. Sci. **43**, 1–25 (1988)

75.80   E. Sontag: Structure and stability of certain chemical networks and applications to the kinetic proofreading model of T-cell receptor signal transduction, IEEE Trans. Autom. Control **46**, 1028–1047 (2001)

75.81   C. Conradi, D. Flockerzi, J. Raisch, J. Stelling: Subnetwork analysis reveals dynamic features of complex (bio)chemical networks, Proc. Natl. Acad. Sci. USA **104**, 19175–19180 (2007)

75.82   C. Conradi, J. Saez-Rodriguez, E.-D. Gilles, J. Raisch: Using chemical reaction network theory to discard a kinetic mechanism hypothesis, IEEE Proc. Syst. Biol. **152**, 243–248 (2005)

75.83   R.R. Vallabhajosyula, V. Chickarmane, H.M. Sauro: Conservation analysis of large biochemical networks, Bioinformatics **22**, 346–353 (2006)

75.84   M.W. Covert, C.H. Schilling, B. Palsson: Regulation of gene expression in flux balance models of metabolism, J. Theor. Biol. **213**, 73–88 (2001)

75.85   M.W. Covert, E.M. Knight, J.L. Reed, M.J. Herrgard, B.Ø. Palsson: Integrating high-throughput and computational data elucidates bacterial networks, Nature **429**, 92–96 (2004)

75.86   K. Mahadevan, J. Edwards, F.J. Doyle III: Dynamic flux balance analysis of diauxic growth in Escherichia coli, Biophys. J. **83**, 1331–1340 (2002)

75.87   C.L. Barrett, C.D. Herring, J.L. Reed, B.Ø. Palsson: The global transcriptional regulatory network for metabolism in Escherichia coli exhibits few dominant functional states, Proc. Natl. Acad. Sci. USA **102**, 19103–19108 (2005)

75.88   S. Klamt, J. Saez-Rodriguez, J. Lindquist, L. Simeoni, E.D. Gilles: A methodology for the structural and functional analysis of signaling and regulatory networks, BMC Bioinformatics **7**, 56 (2006)

75.89   D.S. Kompala, D. Ramkrishna, N.B. Jansen, G.T. Tsao: Investigation of bacterial growth on mixed substrates. Experimental evaluation of cybernetic models, Biotechnol. Bioeng. **28**, 1044–1056 (1986)

75.90   J. Varner, D. Ramkrishna: Application of cybernetic models to metabolic engineering: investigation of storage pathways, Biotechnol. Bioeng. **58**, 282–291 (1998)

75.91   H.P.J. Bonarius, G. Schmid, J. Tramper: Flux analysis of underdetermined metabolic: the quest for the missing constraints, Trends Biotechnol. **15**, 308–314 (1997)

75.92   J.M. Savinell, B.Ø. Palsson: Network analysis of intermediary metabolism using linear optimization: Development of mathematical formalism, J. Theor. Biol. **154**, 421–454 (1992)

75.93   J.M. Savinell, B.Ø. Palsson: Network analysis of intermediary metabolism using linear optimization: Interpretation of hybridoma cell metabolism, J. Theor. Biol. **154**, 455–473 (1992)

75.94   A. Varma, B.Ø. Palsson: Metabolic capabilities of Escherichia coli: I. Synthesis of biosynthetic precursors and cofactors, J. Theor. Biol. **165**, 477–502 (1993)

75.95   A. Varma, B.Ø. Palsson: Metabolic capabilities of Escherichia coli: II. Optimal growth patterns, J. Theor. Biol. **165**, 503–522 (1993)

75.96   D. Segre, D. Vitkup, G.M. Church: Analysis of optimality in natural and perturbed metabolic networks, Proc. Natl. Acad. Sci. USA **99**, 15112–15117 (2002)

75.97   J. Stelling, S. Klamt, K. Bettenbrock, S. Schuster, E.D. Gilles: Metabolic network structure determines key aspects of functionality and regulation, Nature **420**, 190–193 (2002)

75.98   A. Zaslaver, A.E. Mayo, R. Rosenberg, P. Bashkin, H. Sberro, M. Tsalyuk, M.G. Surette, U. Alon: Just-in-time transcription program in metabolic pathways, Nat. Genet. **36**, 486–491 (2004)

75.99   B.N. Kholodenko: Cell-signalling dynamics in time and space, Nat. Rev. Mol. Cell. Biol. **7**, 165–176 (2006)

75.100   J.J. Tyson, K.C. Chen, B. Novak: Sniffers, buzzers, toggles and blinkers: dynamics of regulatory and signaling pathways in the cell, Curr. Opin. Cell. Biol. **15**, 221–231 (2003)

75.101   H. El-Samad, H. Kurata, J.C. Doyle, C.A. Gross, M. Khammash: Surviving heat shock: control strategies for robustness and performance, Proc. Natl. Acad. Sci. USA **102**, 2736–2741 (2005)

75.102   L. Ljung: *System Identification: Theory for the User*, 2nd edn. (Prentice Hall, Upper Saddle River 1999)

75.103   J. Tegner, M.K.S. Yeung, J. Hasty, J.J. Collins: Reverse engineering gene networks: integrating genetic perturbations with dynamical modeling, Proc. Natl. Acad. Sci. USA **100**, 5944–5949 (2003)

75.104   T. MacCarthy, A. Pomiankowski, R. Seymour: Using large-scale perturbations in gene network reconstruction, BMC Bioinformatics **6**, 11 (2005)

75.105   T.S. Gardner, D. di Bernardo., D. Lorenz, J.J. Collins: Inferring genetic networks and identifying compound mode of action via expression profiling, Science **301**, 102–105 (2003)

75.106   O. Alter, P.O. Brown, B. Botstein: Singular value decomposition for genome wide expression data processing and modeling, Proc. Natl. Acad. Sci. USA **97**, 10101–10116 (2000)

75.107   N.S. Holter, M. Mitra, A. Maritan, M. Cieplak, J.R. Banavar, N.V. Fedoroff: Fundamental patterns underlying gene expression profiles: simplicity from complexity, Proc. Natl. Acad. Sci. USA **97**, 8409–8414 (2000)

75.108   P. Tamayo, D. Slonium, J. Mesirov, Q. Zhu, S. Kitareewan, E. Dmitrovsky, E. Lander, T.R. Golub: Interpreting patterns of gene expression with self-organizing maps: methods and application to hematopoietic differentiation, Proc. Natl. Acad. Sci. USA **44**, 129–131 (1990)

75.109   P. D'haeseleer, S. Liang, R. Somogyi: Genetic network inference: from co-expression clustering to reverse engineering, Bioinformatics **16**, 707–726 (2000)

75.110   L. You, J. Yin: Patterns of regulation from mRNA and protein time series, Metab. Eng. **2**, 210–217 (2000)

75.111   J.G. Thomas, J.M. Olson, S.J. Tapscott, L.P. Zhao: An efficient and robust statistical modeling approach to discover differentially expressed genes using genomic expression profiles, Genome Res. **11**, 1227–1236 (2001)

75.112   L.P. Zhao, R. Prentice, L. Breeden: Statistical modeling of large microarray data sets to identify stimulus–response profiles, Proc. Natl. Acad. Sci. USA **98**, 5631–5636 (2001)

75.113   D.B. Allison, X. Cui, G.P. Page, M. Sabripour: Microarray data analysis: from disarray to consolidation and consensus, Nat. Rev. Genet. **7**, 55–65 (2006)

75.114   S. Datta, S. Datta: Comparisons and validation of statistical clustering techniques for microarray gene expression data, Bioinformatics **19**, 459–466 (2003)

75.115   J. Handl, J. Knowles, D.B. Kell: Computational cluster validation in post-genomic data analysis, Bioinformatics **21**, 3201–3212 (2005)

75.116   P. D'haeseleer, X. Wen, S. Fuhrman, R. Somogyi: Linear modeling of mRNA expression levels during CNS development and injury, Proc. Pac. Symp. Biocomput. **4**, 41–52 (1999)

75.117   A.J. Hartemink, D.K. Gifford, T.S. Jaakola, R.A. Young: Combining location and expression data for principled discovery of genetic regulatory network models, Proc. Pac. Symp. Biocomput. **7**, 437–449 (2002)

75.118   D.C. Weaver, C.T. Workman, G.D. Stormo: Modeling regulatory networks with weight matrices, Proc. Pac. Symp. Biocomput. **4**, 102–111 (1999)

75.119   L.F.A. Wessels, E.P. Van Someren, M.J.T. Reinders: A comparison of genetic network models, Proc. Pac. Symp. Biocomput. **6**, 508–519 (2001)

75.120   N. Friedman: Inferring cellular networks using probabilistic graphical models, Science **303**, 799–805 (2004)

75.121   I. Lee, S.V. Date, A.T. Adai, E.M. Marcotte: A probabilistic functional network of yeast genes, Science **306**, 1555–1558 (2004)

75.122   R.K. Pearson: *Discrete-Time Dynamic Models* (Oxford Univ. Press, Oxford 1999)

75.123   C.C. Guet, M.B. Elowitz, W. Hsing, S. Leibler: Combinatorial synthesis of genetic networks, Science **296**, 1466–1470 (2002)

75.124   M. Bansal, G. Della Gatta, D. di Bernardo: Inference of gene regulatory networks and compound mode of action from time course gene expression profiles, Bioinformatics **22**, 815–822 (2006)

75.125   B.N. Kholodenko, A. Kiyatkin, F.J. Bruggeman, E. Sontag, H.V. Westerhoff, J.B. Hoek: Untangling the wires: a strategy to trace functional interactions in signaling and gene networks, Proc. Natl. Acad. Sci. USA **99**, 12841–12846 (2002)

75.126   E. Sontag, A. Kiyatkin, B.N. Kholodenko: Inferring dynamic architecture of cellular networks using time series of gene expression, protein and metabolite data, Bioinformatics **20**, 1877–1886 (2004)

75.127   M. Andrec, B.N. Kholodenko, R.M. Levy, E. Sontag: Inference of signaling and gene regulatory networks by steady-state perturbation experiments: structure and accuracy, J. Theor. Biol. **232**, 427–441 (2005)

75.128 K.-H. Cho, S.-M. Choo, P. Wellstead, O. Wolkenhauer: A unified framework for unraveling the functional interaction structure of a biomolecular network based on stimulus-response experimental data, FEBS Letters **579**, 4520–4528 (2005)

75.129 G. Maria: A review of algorithms and trends in kinetic model identification for chemical and biochemical systems, Chem. Biochem. Eng. Q. **18**, 195–222 (2004)

75.130 C.G. Moles, P. Mendes, J.R. Banga: Parameter estimation in biochemical pathways: a comparison of global optimization methods, Genome Res. **13**, 2467–2474 (2003)

75.131 M. Rodriguez-Fernandez, P. Mendes, J.R. Banga: A hybrid approach for efficient and robust parameter estimation in biochemical pathways, Biosystems **83**, 248–265 (2006)

75.132 D.J. Lockhart, E.A. Winzler: Genomics, gene expression and DNA arrays, Nature **405**, 827–836 (2000)

75.133 D.W. Selinger, M.A. Wright, G.M. Church: On the complete determination of biological systems, Trends Biotechnol. **21**, 251–254 (2003)

75.134 A. Wagner: Reconstructing pathways in large genetic networks from genetic perturbations, J. Comput. Biol. **11**, 53–60 (2004)

75.135 A.F. Emery, A.V. Nenarokomov: Optimal experiment design, Meas. Sci. Technol. **9**, 864–876 (1998)

75.136 D. Faller, U. Klingmüller, J. Timmer: Simulation methods for optimal experimental design in systems biology, Simulation, **79**, 717–725 (2003)

75.137 Z. Kutalik, K.-H. Cho, O. Wolkenhauer: Optimal sampling time selection for parameter estimation in dynamic pathway modeling, Biosystems **75**, 43–55 (2004)

75.138 X.-J. Feng, S. Hooshangi, D. Chen, G. Li, R. Weiss, H. Rabitz: Optimizing genetic circuits by global sensitivity analysis, Biophys. J. **87**, 2195–2202 (2004)

75.139 D. Hwang, A.G. Rust, S. Ramsey, J.J. Smith, D.M. Leslie, A.D. Weston, P.D. Atauri, J.D. Aitchison, L. Hood, A.F. Siegel, H. Bolouri: A data integration methodology for systems biology, Proc. Natl. Acad. Sci. USA **102**(17), 17296–17301 (2005)

75.140 M. Ronen, R. Rosenberg, B. Shraiman, U. Alon: Assigning numbers to the arrows: parameterizing a gene regulation network by using accurate expression kinetics, Proc. Natl. Acad. Sci. USA **99**, 10555–10560 (2002)

75.141 P.M. Kim, B. Tidor: Limitations of quantitative gene regulation models: a case study, Genome Res. **13**, 2391–2395 (2003)

75.142 H.H. McAdams, A. Arkin: Its a noisy business: genetic regulation at the nanomolar scale, Trends Genet. **15**, 65–69 (1999)

75.143 F.J. Doyle III, R. Gunawan, N. Bagheri, H. Mirsky, T.L. To: Circadian rhythm: A natural robust, multiscale control system, Comput. Chem. Eng. **30**, 1700–1711 (2006)

75.144 R. Gunawan, F.J. Doyle III: Isochron-based phase response analysis of circadian rhythms, Biophys. J. **91**, 2131–2141 (2006)

75.145 R. Gunawan, F.J. Doyle III: Phase sensitivity analysis of circadian rhythm entrainment, J. Biol. Rhythms **22**, 180–194 (2007)

75.146 N. Bagheri, J. Stelling, F.J. Doyle III: Quantitative performance metrics for robustness in circadian rhythms, Bioinformatics **23**, 358–364 (2007)

75.147 S. Taylor, L. Petzold, F.J. Doyle III: Sensitivity measures for oscillating systems: Application to mammalian circadian gene network, IEEE Trans. Autom. Control **53**, 177–1888 (2008)

75.148 M.N. Zeilinger, E.M. Farre, S.R. Taylor, S.A. Kay, F.J. Doyle III: A novel computational model of the circadian clock in Arabidopsis that incorporates PRR7 and PRR9, Mol. Syst. Biol. **2**, 58 (2006)

75.149 H. Mirsky, R. Gunawan, S. Taylor, J. Stelling, F.J. Doyle III: Noise Propagation and Sensitivity in Mammalian Circadian Clocks, AIChE Annu. Meet. (San Francisco 2006)

75.150 N. Bagheri, S.R. Taylor, K. Meeker, L.R. Petzold, F.J. Doyle III: Synchrony and entrainment properties of robust circadian oscillators, J. R. Soc. Interface **5**, S17–S28 (2008)

75.151 T.L. To, M.A. Henson, E.D. Herzog, F.J. Doyle III: A molecular model for intercellular synchronization in the mammalian circadian clock, Biophys. J. **92**, 3792–3803 (2007)

75.152 H.K. Khalil: *Nonlinear Systems* (Prentice Hall, Upper Saddle River 2002)

75.153 A. Varma, M. Morbidelli, H. Wu: *Parametric Sensitivity in Chemical Systems* (Oxford Univ. Press, New York 1999)

75.154 R. Larter: Sensitivity analysis of autonomous oscillators: separation of secular terms and determination of structural stability, J. Phys. Chem. **87**, 3114–3121 (1983)

75.155 R. Tomovic, M. Vukobratovic: *General Sensitivity Theory* (Elsevier, New York 1972)

75.156 D.E. Zak, J. Stelling, F.J. Doyle III: Sensitivity analysis of oscillatory (bio)chemical systems, Comput. Chem. Eng. **29**, 663–673 (2005)

75.157 N. Bagheri, J. Stelling, F.J. Doyle III: Circadian phase entrainment via nonlinear model predictive control, Intl. J. Robust Nonlinear Control **17**, 1555–1571 (2007)

75.158 G. Strang: *Linear Algebra and ist Applications* (Saunders College Publishing, New York 1988)

75.159 C.H. Johnson: Forty years of PRCs – what have we learned?, Chronobiol. Int. **16**, 711–743 (1999)

75.160 National Research Council: *Network Science* (National Academies Press, Washington 2005)

75.161 L. Lamberg: *Bodyrhythms: Chronobiology and Peak Performance* (William Morrow, New York 1994)

75.162 N. Bagheri, J. Stelling, F.J. Doyle III: Circadian phase resetting and multiple control targets, PLoS Comput. Biol. **4**, e10000104 (2008)

75.163   L. Hood, J.R. Heath, M.E. Phelps, B. Lin: Systems biology and new technologies enable predictive and preventable medicine, Science **306**, 640–643 (2004)

76. Automation and Control in Biomedical Systems

Robert S. Parker

Biomedical systems are a complex collection of case studies where the principles of automation and control theory are seeing increased application. This growing interest has a twofold motivation: the need for advanced automation and treatment design tools for use in medical practice and the challenges inherent to biomedical systems and clinical deployment of technology. This chapter provides an overview of the automation, control, and optimization tools used in the biomedical arena. While the scope of potential applications is vast, examples of biomedical treatment design systems for cancer and insulin-dependent diabetes are discussed. The chapter concludes by scratching the surface of emerging areas in need of translated or novel systems and automation tools.

76.1 Background and Introduction

76.1.1 Scope

Automation is ubiquitous in society today. Air conditioners and thermostats keep the house cool; cruise control maintains speed without driver intervention on the road. Both of these examples demonstrate the efficiency of automating processes and taking advantage of advanced control algorithms in doing so (the switching on and off of the air conditioner and the degree of depression of the gas pedal are both handled by mathematical algorithms). With advancements in engineering, miniaturization, and computational capability, it becomes possible to deploy automation and control tools in the medical arena. A common example of automation

in medicine is the pacemaker, with a quarter-million per year implanted around the world [76.1]. Far from the simple pacing instrument of 1932 [76.2], today's devices include diagnostic algorithms for real-time analysis, adaptive response (control), and programmability [76.1]. Robotics is a common tool used in medical science, and due to its wide treatment elsewhere, this topic is consciously excluded from the present chapter (the interested reader is referred to [76.3] from the *Handbook of Industrial Robotics*). This chapter will explore the use of mathematical models of the body (in a variety of styles, see Sect. 76.2 for more detail) as well as tools from optimization and dynamic control theory in the automation of medical treatment decisions and the deployment of treatment. While modeling the body is certainly not a new concept – *Teorell* discussed the use of mathematical modeling in drug distribution as early as 1937 [76.4, 5] – the ability to deploy these models is a relatively recent event given the continuing advancement of computational power.

Medical practice, in a broad sense, is a discipline grounded in finding feasible solutions to complex biosystems problems. Typically, clinicians are presented with a set of symptoms from which they must diagnose and treat a disease state. Treatments induce response, and when drugs are administered, the dose amount and schedule (timing of doses) elicit a variety of measurable outcomes such as drug pharmacokinetics (PK, drug concentration versus time) and pharmacodynamics (PD, disease response and toxicity). Schematically, this is the unshaded component shown in Fig. 76.1. The shaded region characterizes a model-based control system that:

1. Compares the measurable responses to treatment to a desired clinical response
2. Employs a control algorithm and mathematical model of the patient to calculate an optimal treatment
3. Deploys that treatment by adjusting dose and schedule of administration

Typically, this last phase would be a recommendation to a clinician who would be responsible for the ultimate go/no-go decision of implementing the algorithm-recommended treatment. The challenge in designing such automation systems is the complexity of the underlying model of the patient and its incorporation within a rigorous mathematical framework for optimizing or controlling the treatment outcome.

When translating systems tools to the clinical arena, it is critical to work closely with clinicians. Unlike a distillation column, where automation tools can be implemented as long as their performance is superior to existing technology, the most likely early implementation of automation tools in the biomedical arena is as decision support (a semi-closed-loop formulation). The charge of the clinician to *do no harm* and the critical issue of malpractice, with its corresponding financial and legal ramifications, are too significant to allow an automated system to operate independently without oversight. Even a single failure resulting in the death of a patient could halt the development and deployment of an automated medical device. Hence, extensive clinical trials demonstrating safety of patients (phase I trials), efficacy of the automated tool or decision support system (phase II), and ultimately performance superiority to current standards of practice (phase III/IV trials) will be required before an automated device can be deployed with confidence in the biomedical arena.

The remainder of this section will establish challenges to automation in biomedical systems. The first set are data issues, including quantity available, quality, resolution, and consistency and frequency of collection (for use in dynamic modeling). The differences between preclinical and clinical data are discussed, and the concept of patient tailoring (including interpatient variability and treatment individualization) is introduced. The chapter continues in Sect. 76.2 with the tools used in biosystems automation and control. Two case study areas are discussed in detail in Sect. 76.3, and a small selection of directions of future research are summarized in Sect. 76.4.

76.1.2 Data Quantity and Quality

The amount of experimental biological data in the literature is vast. However, the size scale of the data is equally large (spanning nanometers to meters and beyond). The dynamic character of the data varies as well, including static *baseline* measurements, point measurements of biological changes (the usual case in *-omics* data), and dynamic PK profiles. A further complication captured in data sets such as those from DNA microarrays is the complexity of the interaction between the various cellular processes, which is simultaneously

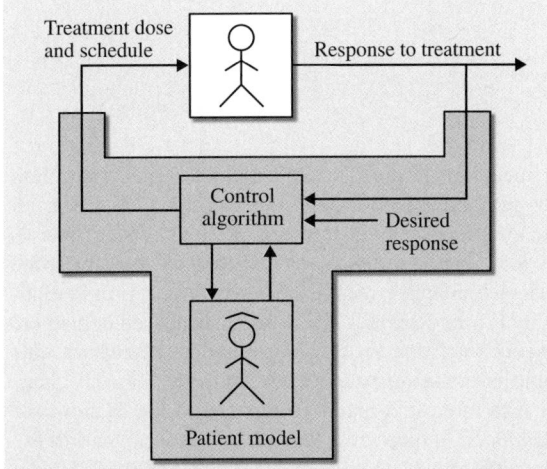

Fig. 76.1 Schematic of drug treatment. *Unshaded*: open-loop, *shaded*: automated model-based decision support or control system for treatment design and/or implementation

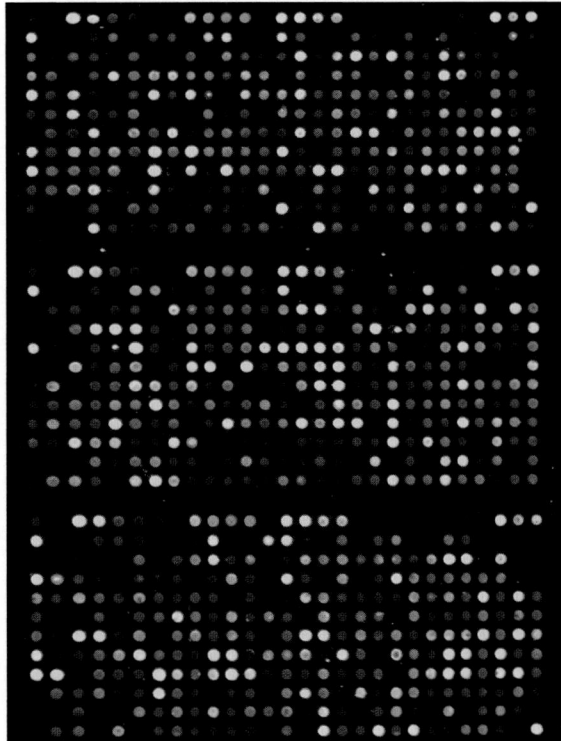

Fig. 76.2 Expression response of MCF breast cancer cells treated with vitamin D (figure originally published in [76.6])

a complication in the use of modeling and automation tools and a challenge to those interested in building new tools. It is at this crossroads of biology and systems engineering that computational models and automation tools can make a significant contribution by unifying the collection of disparate data across multiple scales and time points.

Facilitating the collection of this interacting dynamic data are advances in measurement technology that provide higher-resolution information of the biological processes taking place. Microarray technology has provided a more detailed picture of DNA and RNA responses to external stimuli, and these tools may eventually provide a fully dynamic profile of specific up- and downregulation events in response to drug treatment. The data available from a microarray study at present, however, do not provide such a picture. The measurements are often distributed in time, with significant nonmeasurement periods, so the dynamics of a particular challenge may or may not be fully captured. Furthermore, the ability to quantitate at high precision

with a microarray is limited by the sensing technology; semiquantitative and directional information is available from the colorimetric technique, but exact counts of DNA or RNA are unavailable (Fig. 76.2). Tissue concentrations of agents are measured using high-performance liquid chromatography (HPLC) as the *gold standard*; however, liquid-chromatography mass spectroscopy (LC/MS) has higher resolution (the ability to measure fewer ng/ml) and lower sample volume requirements. Hence, PK data may be more accurate in the future, thereby leading to higher-quality models and possibly to better informed treatment decisions. A potential drawback of this new technology is the need to repeat previous studies with the new analysis device. This may lead to conflicting outcomes between studies as superior measurements are analyzed, and model refinement should be an expected and ongoing process. Given these factors, modelers and algorithm developers must carefully evaluate the data sets they employ, and the ongoing generation of new data, to be sure that the treatment decisions recommended by automated tools are of the highest possible quality and provide the greatest potential benefit to the patient.

76.1.3 Preclinical Versus Clinical Study Data

Animal studies are a required precursor to human trials for testing novel drugs. These studies are primarily toxicologic – the objective is to establish toxicities associated with drug administration and to evaluate the various routes of administration (oral, intravenous (IV), etc.). Pharmacokinetic studies are often required by the US Food and Drug Administration (FDA) [76.7]. The search for tissue toxicities leads to detailed (potentially dynamic) concentration versus time data sets from animals for compounds of potential clinical interest. The outcome of these studies is often a maximum tolerated dose (MTD) in a variety of species, and the dose for first trials in humans is established from the MTD in the most sensitive species. A useful side-effect of these studies, from an automation perspective, is the value of these data sets in modeling dynamic biological phenomena at scales that one cannot achieve in humans (e.g., tissue drug concentration versus time via destructive analyses such as HPLC or LC/MS).

Animals are not people, however. A critical shortcoming of model-based automation and control for the biosystems field is that most of the high-resolution (temporal or spatial) in vivo data comes from animal studies. Hence, detailed animal models of pharmacokinetic and pharmacodynamic behavior can be constructed [76.8–10], but the scale-up of these models to humans remains open (Sect. 76.4.2). In contrast, pharmacokinetic and pharmacodynamic models collected in clinical trials are often macroscopic in nature. Measurements are often easy to collect and analyze, such as urine measurements of function or disease state (e.g., creatinine clearance or glucose in the urine), drug concentrations in the bloodstream, or circulating markers of effect (e.g., cholesterol levels in the blood). Further-

more, measurements from a single patient are often not a full time course but a small number of samples (1–3) that provide the clinically necessary information but may not provide sufficient data to inform a model of the individual patient. Depending on the resolution and prediction required from the model, as well as its intended use, a variety of tools may be employed in model synthesis and deployment at the animal and human scales. These are outlined in the next section.

76.2 Theory and Tools

This section will introduce some of the tools often employed in the solution of biomedical modeling and control problems. A summary can be found in Table 76.1, which provides the section where a technique can be found, the name of the technique(s) discussed therein, and one or two strengths and weaknesses of the various approaches. While this is certainly not a canonical summary of tools employed in the automation of biomedical problems, this section provides a jumping-off point for further reading on the tools and techniques of biomedical control.

76.2.1 Parameter Estimation

In the development of mathematical models, there is a need to estimate parameters. A concern in this arena is the a priori identifiability of such models. Phrased differently: are the parameters of a given model *structurally identifiable* from the available input–output (measurement) data? A number of recent studies of biomedically relevant systems explore tools and methods for evaluating model identifiability [76.11–15]. *Audoly* et al. [76.11] examine the identifiability of linear models, which has direct mapping to the popular compartmental models used in pharmacokinetic analysis (Sect. 76.2.2). Physiological structures are the focus of *Yates* [76.13], who examines the global identifiability of the model with restriction to local identifiability in the worst case. In a nonlinear context, *Bellu* et al. [76.15] provide differential algebra for identifiability of systems (DAISY) software to evaluate the identifiability of nonlinear systems in polynomial or rational form (a limitation of the differential algebra tool used in the analysis). At a higher level of resolution, intracellular networks are of particular interest in systems biology, metabolomics, etc. The amount of information that is measurable from these systems is generally small with

respect to the parameters. *Van Riel* and *Sontag* [76.14] explore methods of modeling these systems using fictitious dependent inputs to represent unidentifiable model structures.

Given an identifiable model, parameter values need to be estimated with confidence. Two common objectives for parameter estimation include nonlinear least squares and maximum likelihood. The nonlinear least-squares technique can be represented as [76.16]

$$\min_{\boldsymbol{p}} \; J(\boldsymbol{p}) = \sum_{i=1}^{N} \chi_i [y_{\mathrm{m}}(i) - \hat{y}(i, \boldsymbol{p})]^2 \,. \tag{76.1}$$

The measured data at observation i is given by $y_{\mathrm{m}}(i)$, the model prediction at the same observation, which depends on parameters $\boldsymbol{p}$, is represented by $\hat{y}(i, \boldsymbol{p})$, and the observation-dependent weights are given by χ_i. In the unweighted case χ_i reduces to an appropriately sized identity matrix. Alternatively, weighting by $\chi_i = 1/\sigma_i^2$ can reduce sensitivity to variability in the measurement, where σ_i is the standard deviation of the measurement data at observation i. Standard tools for this optimization using linear or nonlinear models are available in MATLAB [76.17]. An alternative objective is maximum-likelihood estimation (MLE), a statistical approach to parameter estimation. Here, the conditional probability that the residuals are generated by a model having the estimated parameter vector is maximized. The likelihood function is written as [76.16]

$$L(\boldsymbol{p}) = \prod_{i=1}^{N} \frac{1}{\sigma_i \sqrt{2\pi}} \left[\frac{-1}{2} \left(\frac{y_{\mathrm{m}}(i) - \hat{y}(i, \boldsymbol{p})}{\sigma_i} \right)^2 \right] \,. \tag{76.2}$$

The objective is to maximize the likelihood that the data results from the model (as parameterized by the vector $\boldsymbol{p}$). Assumptions in using maximum likelihood include: (i) uncorrelated residuals and (ii) residuals are

Table 76.1 Summary of theory and tools by Section, topic(s), and advantages/disadvantages of the various methods

Section	Topic(s)	Pros (+) and cons (−)
76.2.1	A priori identifiability	+ Highlights parameter estimation concerns based on available data
		− Theoretically and computationally challenging
	(Weighted) least squares	+ Straightforward application, many tools available
		− Weighting of points is subjective
	Maximum likelihood	+ Robust and commonly employed
		− Based on assumed error structure/distribution
76.2.2	Compartmental PK	+ Compact models, generally identifiable
		− No relationship to mechanism or physiology
	Population PK	+ Incorporates intra-/interpatient variability in model (when justified by available data)
		− Not straightforward to tailor to an individual
		− Patient and data intensive
76.2.3	Physiological PK	+ Incorporates patient physiology explicitly
		− Measurements of individual tissues are difficult to collect (limited to animal studies)
		− Large number of equations makes parameter estimation computationally challenging
76.2.4	Biochemical networks	+ May provide high-resolution mechanistic detail
		− May be difficult to resolve causality (cause versus effect)
		− Semiquantitative (colorimetric)
76.2.5	Optimal control	+ Solves constrained dynamic optimization problem
		− Formulation (two-point boundary-value problem, continuous controls) must be reasonable for solution of corresponding biomedical problem
	Model predictive control	+ Robust to mismatch between actual patient/process and mathematical model
		+ Solves constrained dynamic optimization problem
		− Suboptimal (relaxed) solution of problem
		− Mathematical analysis of stability and performance is theoretically challenging

zero mean and normally distributed with variance σ_i^2. While these may not be rigorously true of biomedical data sets, MLE has been shown to work well with biomedical data, and packages are available that use MLE for estimation [76.18, 19]. These packages also return parameter information in the form of a confidence interval or coefficient of variation. Large values indicate poor parameter accuracy, and hence possible structural parameter identifiability issues [76.20] or violations of the assumptions above as parameters may not be Gaussian distributed in the modeling of multiple individuals.

76.2.2 Pharmacokinetics

PK is the study of drug administration to, distribution in, and clearance from the body [76.4, 5]. To capture the dynamics observed after drug dosing, compartmental models are often employed (Fig. 76.3). The administered dose is given by $D(t)$. Individual compartments are modeled as drug masses (x_1 and x_2), with correction to concentration in the output equation (via division by volume V). Clearance of the drug is from the central compartment (rate k_{cl}). Additional dynamic character is included by adding compartments and intercompartment transfer rates (k_{ij} from compartment i to compartment j). A two-compartment model can be represented mathematically as

$$\frac{dx_1(t)}{dt} = -k_{cl}x_1(t) - k_{12}x_1(t) + k_{21}x_2(t) + D(t) ,$$

$$(76.3)$$

$$\frac{dx_2(t)}{dt} = k_{12}x_1(t) - k_{21}x_2(t) , \qquad (76.4)$$

$$y(t) = \frac{x_1(t)}{V} . \qquad (76.5)$$

The output(s) of the model corresponds to measurable concentrations of drug, as given by $y(t)$. Additional blocks can be added, as necessary, to capture observed plasma dynamics related to dosing (for drugs not delivered intravenously) and/or absorption. The resulting model is an empirical description of patient drug concentration as a function of time after administration. These model structures are commonly employed using available PK modeling tools such as ADAPT II [76.18], and the FDA requires some pharmacokinetic information for most new chemotherapeutics [76.7]. Parameters in the model can be estimated from individual patient data, when detailed pharmacokinetic measurements are available. More often, models of this form are constructed from animal studies where multiple animals

(e.g., mice or rats) are euthanized at each measured time point. The mean and standard deviation of measured drug concentrations are used as the data to which a single model is fit. Low-order deterministic models may not capture the observed drug concentration measurements for a variety of reasons, including nonlinearity, higher-order dynamics, and interindividual variability.

To address interindividual variability and data sparsity, population models are often developed [76.16, 21, 22]. In this case, a stochastic approach to model parameterization is employed, where parameters are composed [76.21] as

$$p_i = \theta_\mu + \eta_i^p + \theta_c C_i . \qquad (76.6)$$

Parameters (p_i) for patient i are a function of the population mean value of the parameter θ_μ, interindividual variability in the parameter η_i^p, and any known correlative effects C_i scaled by their population mean correlation θ_c. These parameter formulations are then included in model structures akin to those in (76.3–76.5), as replacements for the k_{ij} and k_{ci} parameters, for example. Variances on the measurement noise (added to (76.5)) and interpatient variability η_i are specified as part of the estimation. With this approach, it is possible to use a small number of output measurements from a large number of individuals to characterize both the underlying model structure (through the observed dynamics) and the population variability (through the need for η and θ_c or C_i to describe individual responses). Software tools are available for constructing these models (including NONMEM [76.22] and SPK [76.23]). A population model, once constructed, is representative of the population response to a drug dose. Furthermore, it can provide a characterization of both validated variabilities (as a function of body weight, gender, race, liver or kidney performance status, etc.) and the distribution of responses expected after dosing a group of patients with the drug. However, it is not designed to be predictive in the single-patient case; here it may be of more use to couple the variabilities established in the population model with the ability to tailor an individual PK model.

76.2.3 Modeling Physiology

The potential of detailed physiological models was recognized by *Teorell* when he began studying pharmacokinetics [76.4, 5], but calculational limitations made physiological modeling of drug PK intractable in 1937. With today's desktop computational power, simulating physiological models of drug administration and

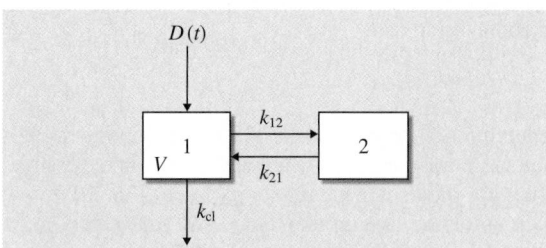

Fig. 76.3 Two-compartment mamillary model used in pharmacokinetic analysis. Plasma dynamics are captured by the central compartment (block 1) in Eq. (76.3)

distribution is straightforward. A physiologically-based model for a cancer chemotherapy application is shown in Fig. 76.4. Each block in a physiologically-based model represents an ordinary differential equation that characterizes concentration in a particular tissue or fluid, with the volume of the compartment corresponding to the physiological volume into which the compound distributes. Where necessary, subcompartments can be added to modify tissue dynamic response, and these are often denoted as vascular (blood space) and extravascular (interstitial space) compartments (models may also include intracellular or tissue binding spaces as well [76.8, 24]). Transfer between vascular and extravascular space can be via gradient-based diffusion or active transport, and may be uni- or bidirectional. Unidirectional arrows between tissue compartments denote the flow of blood; the bidirectional arrow pairs between red blood cells (RBCs) and the plasma compartments represent equilibration (transport is often fast in RBCs, such that they are often represented algebraically rather than differentially). Metabolic elimination and clearance are shown as unconnected arrows exiting particular (sub)compartments (e.g., the arrow out of liver extravascular space), and infusion into the system can be properly placed in a physiological sense, e.g., the IV dose administered to the venous blood compartment. Advantages of this approach include the accuracy of the tissue connectivity and the availability of mean values for the flow and volume parameters [76.25]. Furthermore, knowledge of the metabolic pathway of drug clearance (e.g., liver metabolism or kidney excretion) can be included in the physiologically correct location. This structure changes the identification burden to that of metabolic rates and intratissue transport (plasma ↔ interstitial fluid ↔ cell) and dynamics. Resolution of these effects can be challenging (when feasible) as the necessary information (e.g., arterio–venous differences in drug concentration) for identification is not typically available from in vivo measurements. The result is that tissues are often assumed to be well mixed if they are highly perfused (e.g., kidney and liver), and only tissues where significant transport resistances are involved are modeled using more detailed intratissue dynamics (and then often using a compartmental approach).

The utility of these structures is of more potential benefit when evaluating pharmacodynamic effects. Unlike the compartmental or population models described above, the physiological model also allows the effects of the drug to be properly assigned to their physiological sites of action. These outcomes can be positive (e.g.,

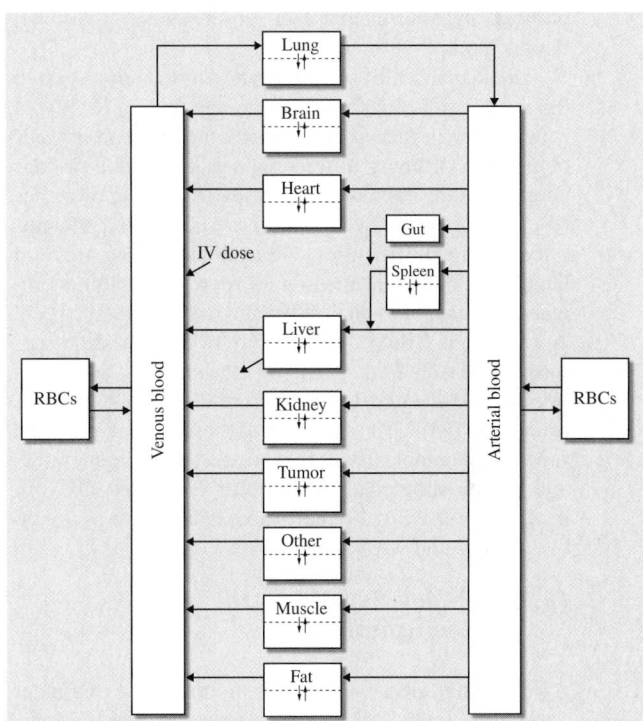

Fig. 76.4 Block structure schematic of a physiological model of drug distribution

antitumor activity related to tumor concentration, not plasma concentration, of an antineoplastic) or negative (e.g., toxic side-effects of a drug affecting liver performance status). This is a significant advantage in the use of in vitro information for the purposes of in vivo modeling, as the concentration at the site of action (effect or toxicity) is more accurately represented by a physiological model than the plasma concentration estimate of a compartmental model.

76.2.4 Biochemical Networks

Phenotype is related to genotype at some level, meaning that macroscopic observables such as physiology, PK, etc., are driven by biological networks at a variety of scales. The field of systems biology is focused on developing a systems-level description of biology by studying the dynamics of cellular and organ function, rather than the traditional drill-down study of increasingly highly resolved portions of a cell [76.26]. In order to rigorously integrate the complex behaviors taking place at the genetic level with the macroscopic (whole-organism) response, and to ground these hypotheses in experimental data, requires a computational model [76.27].

Due to the scales spanned in network analysis (from the gene regulatory [76.28] to the metabolic [76.29]), it is infeasible in this space to review the plethora of tools employed. Ordinary differential equations, due to their simplicity, are popular, and compartmental approaches (such as those in Sect. 76.2.2) are also used to simplify the model structures. The systems biology markup language (SBML) is one tool for representing biochemical networks, both qualitative and quantitative [76.30]. A review of SBML, along with two other data representation standards (PSI MI (Proteomics Standards Initiative Molecular Interaction) and BioPAX), can be found in [76.31]. The strengths of these packages can be seen in the complexity of the models that are generated, and can be simulated, as a result of complex network modeling (see [76.29] and models at www.systems-biology.org and www.biopax.org).

76.2.5 Model-Based Control and Optimization

A commonly employed tool in model-based treatment design for biomedical systems is optimal control. Mathematically, optimal control problems can be formulated as follows [76.32]

$$\min_{u(t)} \ J(y(t), u(t), r(t)), \tag{76.7}$$

$$\text{subject to: } \dot{x} = f(x(t), u(t), p), \tag{76.8}$$

$$y(t) = h(x(t), u(t), p), \tag{76.9}$$

$$u_{\min} \leq u(t) \leq u_{\max}, \tag{76.10}$$

$$y(t_f) = y_{\text{fin}}. \tag{76.11}$$

The objective function (76.7) often takes the form of a least-squares deviation of $y(t)$ from some desired trajectory $r(t)$, and the decision variable is the $u(t)$ profile over $0 \leq t \leq t_f$; for example, in the diabetes example introduced later in this chapter, the measured variable would be plasma glucose concentration ($y(t)$), $r(t)$ would be the desired glucose concentration (e.g., $90 \, \text{mg/dl}$), and $u(t)$ would be the insulin delivery profile (or delivery rate) from the present time t to $t + t_f$. Feasible system trajectories are governed by the dynamics of the model (states $x(t)$, parameters p), and these are incorporated as constraints (76.8) and (76.9). Continuing the diabetes example, the states characterize the patient; in a physiological model context, these states would be the glucose and insulin concentrations in the various physiological tissues (not all of which are measurable).

Constraints on the input magnitude can be enforced, as in (76.10); other input constraints can also be included, though they are not explicitly described here. Finally, a desired final value for the controlled variable $y(t_f)$ is included. The identification of a feasible treatment, an input sequence that drives $y(0)$ to $y(t_f)$, involves the solution of a constrained two-point boundary-value problem. Control vector parameterization [76.33] is a common method of computer implementation, where the time axis is discretized into k equal-length steps, and the series of input levels over each of the steps become the decision variables of the optimization problem. The resulting input profile generally has a characteristic bang–bang shape, switching from *full on* to *full off* at one or more points along the time axis. The strength of optimal control is the ability to solve nonlinear control problems in the presence of input constraints. A key shortcoming is the formulation of optimal control versus the practice of medicine. The former specifies an endpoint constraint and has no mechanism for the inclusion of information that becomes available at intermediate time points. Clinical practice, on the other hand, does not generally have an a priori specified end time of treatment and routinely takes measurements of patient performance status and incorporates these into treatment decisions.

One way to take advantage of model-based control and optimization techniques while retaining the advantages of the optimal control formulation is to use receding horizon control, also referred to as model predictive control [76.34, 35]. Receding horizon control is an open-loop optimization posed and solved for a sequence of input move changes at each time step; a typical model predictive control formulation may be posed as

$$\min_{\Delta u(k|k)} \ \|\boldsymbol{\Gamma}_y(r(k+1|k) - y(k+1|k))\|_2^2$$

$$+ \|\boldsymbol{\Gamma}_u \Delta \boldsymbol{u}(k|k)\|_2^2 \tag{76.12}$$

$$\text{subject to: } x(k+1|k) = f(x(k|k), u(k|k), p) \tag{76.13}$$

$$y(k|k) = h(x(k|k), u(k|k), p) \tag{76.14}$$

$$u_{\min} \leq u(k|k) \leq u_{\max} \tag{76.15}$$

$$|\Delta \boldsymbol{u}(k|k)| \leq u_{\text{rate}}. \tag{76.16}$$

Because the model is used to predict output responses at future times (up to N_p steps in the future), statistical notation is employed, where $y(k+1|k)$ is the vector of N_p predicted outputs at time $k+1$ given information up to time k. The corresponding vector of desired output values is $r(k+1|k)$, and the degrees of freedom are the N_u input moves $\Delta u(k|k)$. The matrices $\boldsymbol{\Gamma}_y$ and $\boldsymbol{\Gamma}_u$

are weights to trade off tracking error (large $\boldsymbol{\Gamma}_y$) versus input move suppression (large $\boldsymbol{\Gamma}_u$). The underlying model dynamics of (76.13) and (76.14) can be continuous or discrete, as they are simulated at each time step, but the output values employed in the calculation of the objective correspond to the N_p future step times. Because the solution methodology includes the use of an optimization routine at each time step, constraint incorporation is straightforward, and input constraints can be included in magnitude-constrained (76.15) and rate-constrained (76.16) forms. Output constraints can be included, but these may lead to infeasibilities in the optimization problem, much as state constraints do in optimal control. One option is to *soften* the constraints by including them as additional weighted terms in the objective function [76.36]. At each sample time, measurements from the system are used to update the states or outputs of the model, and the optimization problem is solved. The first input move change is implemented, and the process repeats at each following sample time (as shown schematically in Fig. 76.5). In the case that measurable quantities are sampled at different intervals, the update rate of the model can be variable (this is referred to as a multirate MPC formulation [76.37]). When measurements are not collected at each sample time, multiple inputs in the sequence $\Delta \boldsymbol{u}(k|k)$ can be implemented. The structural advantages of MPC, including the online solution of an optimization problem that incorporates available patient information, are significant, but a key drawback of this control structure is

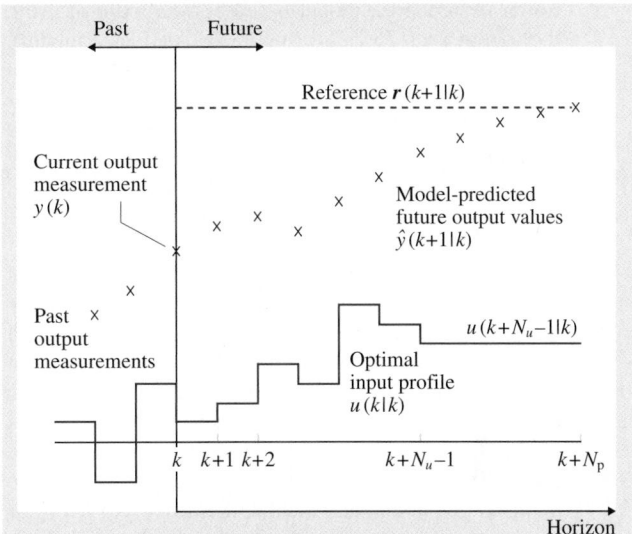

Fig. 76.5 Receding horizon (model predictive) control implementation schematic. Model-predicted deviations from the desired reference are minimized over a future horizon

the challenging analysis of algorithm performance. Stability guarantees, which may be required by the FDA before an algorithm can be deployed in an ambulatory setting, often require the use of endpoint constraints or long prediction horizons (i.e., large N_p or *infinite-horizon* formulations), which may limit the achievable performance.

Part H | 76.3

76.3 Techniques and Applications

The potential scope for automation and control in biomedical problems precludes a canonical review here. Entire journals are focused on biomedical problems from the engineering perspective (e.g., *IEEE Transactions on Biomedical Engineering, Annals of Biomedical Engineering, IET Systems Biology*), and mainstream control journals (e.g., *Automatica, Journal of Process Control, Control Engineering Practice*) also publish biomedical control papers. There are also journals focused on the tools as they apply to classes of problems (e.g., *Journal of Pharmacokinetics and Pharmacodynamics* for PK and PD modeling tools). Beyond the diabetes and cancer case studies discussed below, a wide variety of problems have been addressed, such as human immunodeficiency virus (HIV)/acquired immunodeficiency syndrome (AIDS), blood pressure, and

anesthesia, among others (some recent review articles include [76.9, 38, 39]).

76.3.1 Type I Diabetes Modeling and Control

Diabetes has been a popular biomedical modeling and controller design case study for almost 50 years [76.40–44]. The development of models of glucose–insulin interaction, sometimes including additional hormones, substrates, or contributions, has an equally long history [76.40–42, 45–56]. In order to estimate the parameters in these models, tools from Sect. 76.2.1 are employed, and the model structures can be of either compartmental (Sect. 76.2.2) or physiologic (Sect. 76.2.3) form. With a model in place to represent the patient (i.e., patient = model), for use in controller design (i.e., model-based optimal control using tools from Sect. 76.2.5), or both (i.e., patient $\neq$ model), a control problem can be formulated. One option is to pose a single-input single-output control problem, the formulation of which is quite similar to the continuous processes seen in the chemical industry for decades: a manipulated variable (insulin delivery), with a steady-state value (pancreatic insulin release in a healthy patient), is altered to maintain a controlled variable (glucose concentration) within a small range of its nominal value. Normoglycemia is generally taken as circulating glucose concentrations in the range 70–120 mg/dl [76.57], with a nominal value between 80 and 100 mg/dl. Insulin delivery for type I diabetic patients, who have no insulin release from the β-cells of the pancreas, is altered in order to control glucose concentration. Clinical approaches at present involve intensive subcutaneous insulin therapy [76.57, 58] or continuous subcutaneous insulin infusion via mechanical pump [76.59].

Based on this traditional control configuration and the control schemes introduced in Sect. 76.2.5, the glucose control problem can be posed in the model-predictive control framework [76.42, 43], as shown in Fig. 76.6. Using glucose concentration measurements (at 10 min or faster intervals), insulin delivery rate would be changed in order to regulate ambulatory diabetic patient glucose levels in response to meal disturbances. Controller tuning would require rapid rejection of meal disturbances (to keep glucose levels below 120 mg/dl), but also aggressive downregulation of the controller so that post-meal glucose concentrations remain above the hypoglycemic level of

60 mg/dl, which can be dangerous to patients [76.57, 58]. Furthermore, significant levels of measurement noise accompany each glucose observation. The potentially conflicting objectives of aggressive response and noise suppression challenge model-based algorithm designers. Simulation studies of control algorithms have focused on the type of control needed [76.60], providing upper bounds on achievable control [76.42], or evaluating the potential effects of measurement or other disturbances on closed-loop performance [76.43]. A key hurdle at present is the glucose sensor, as the FDA has not approved any real-time glucose sensor for use in a closed-loop control algorithm.

One focus of improving diabetes treatment via automation is the use of run-to-run control (commonly deployed in the batch processing literature) in designing insulin dosing protocols [76.61,62]. Here, patient glucose concentrations are not sampled at fixed short intervals, but instead the measurements taken by patients over a typical day are used to improve day-to-day performance of their glucose control. In the context of Fig. 76.6, the controller and insulin pump are replaced by the patient (who will determine insulin delivery amount) and their delivery mechanism (pump, subcutaneous injection, etc.); Fig. 76.7 shows a potential configuration of a wireless sensor communicating with the combination data storage device and insulin pump. By reformulating this problem, the use of advanced control and automation techniques has the potential to make a rapid contribution to the treatment of diabetic patients instead of suffering from the critical flaw in artificial pancreas deployment highlighted above: the lack of FDA acceptance of a glucose sensing

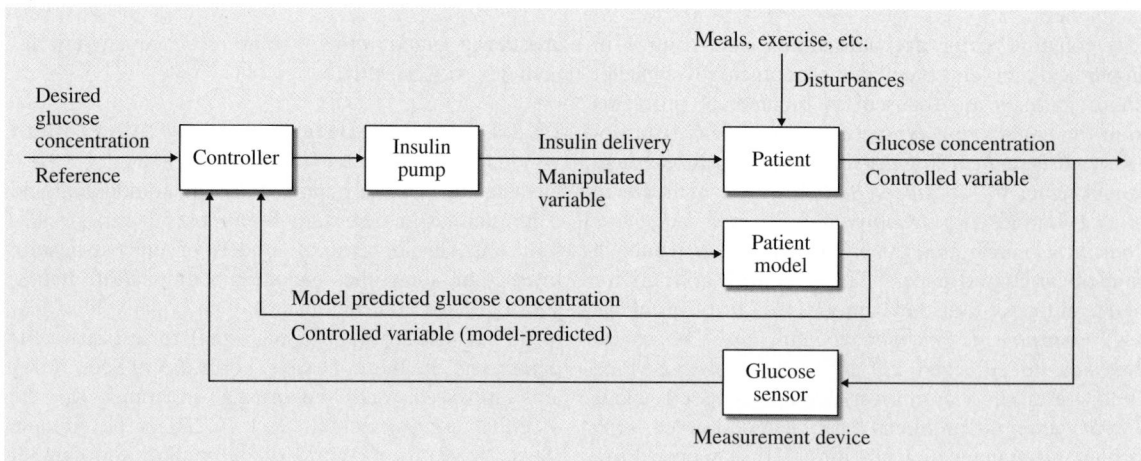

Fig. 76.6 Block diagram structure for model predictive control, as applied to the diabetic patient control problem

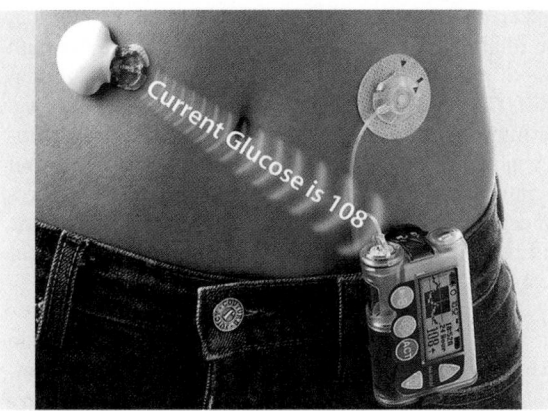

Fig. 76.7 Diabetic patient using a wireless subcutaneous glucose sensor communicating the measurement to the data storage device. This device also includes the insulin pump mechanism; note that the pump is not altering insulin delivery rate based on the glucose measurement unless commanded to do so by the patient (courtesy of Dr. Cesar Palerm, Medtronic Diabetes)

device for use in the closed loop. In fact, the run-to-run algorithm can provide robustness to interpatient differences [76.62], which is important given the span of patients that such an algorithm could encounter.

Changing the focus from ambulatory diabetes to the critical-care setting, the potential for impact is significant [76.63–65]. In postsurgical patients, glucose control via insulin administration can dramatically reduce morbidity and mortality. The altered metabolic state in these patients is similar to the diabetic condition, in that patients are hyperglycemic with low insulin sensitivity. Hence, the automated administration of insulin to regulate glucose concentration could assist in patient recovery. While measurement noise and rate would (likely) be unchanged from the ambulatory setting, other challenges manifest. First, individual models of patients are not available a priori, and population models are insufficient for use on specific patients because individual insulin sensitivity will vary widely. Furthermore, the dynamic state of recovering patients would lead to time-varying insulin sensitivity requiring time-varying model parameters. A patient model updating mechanism, such as recursive least squares (assuming a linear parameterization of the model), is required. Given the present legal and regulatory climate, a high-level safety and fault-detection layer for the algorithm would need to be developed. Alternatively, the algorithm could be implemented as a semiclosed decision support system with medical professionals affirming or

declining the recommendation of the updating algorithm. In the context of bringing an automated system to market, the decision support system in critical care could be the most rapid approach because collecting a nonautomated measurement and allowing medical professionals to intervene in therapy has the lowest *energy barrier* to review board approval.

A concern with any insulin delivery system is the potential for overadministration, as this could lead to hypoglycemia and possibly death. Adding a second channel to the closed-loop system could alleviate some of these issues. A model-based structure using both insulin and glucose has been proposed [76.66, 67], where insulin is added to reduce glucose levels, and glucose infusion is used to elevate glucose. To keep glucose from being administered continuously, an *output regulator* formulation is employed (a modification of the MPC scheme in Fig. 76.6 and Sect. 76.2.5), such that positive glucose infusion rates are penalized at a lower level than deviations in glucose concentration from the reference value. The result is that glucose is infused only when predicted glucose levels drop below the basal level. While numerically superior, this approach is more relevant to use in a clinical, rather than ambulatory, setting, as patients are unlikely to desire a two-pump/two-needle (or two-infusion-line) device. Also, a high-dextrose solution is an excellent culture medium for bacteria, so sterility in the nonhospital setting would become an even greater issue for these patients.

As McGarry highlighted in the 2001 Banting lecture, glucose and insulin are not the only important endogenous compounds in diabetes. Type 2 diabetes typically involves the dysregulation of fatty acids (FAs); there is also significance of FAs in insulin-dependent diabetes as metabolic interactions exist with insulin and glucose. Using the minimal model [76.41] as a basis, an extended minimal model of glucose/insulin/FA has been constructed [76.55] using data from the literature and the tools of Sects. 76.2.1 and 76.2.2 (see the schematic in Fig. 76.8). While this model is not mechanistic, it does capture the interactions of glucose, insulin, and FAs; furthermore this model is a low-order parameterization, which would facilitate tailoring to an individual patient.

Finally, meal disturbances are not the sole challenge to diabetic patients, and hence automated insulin delivery systems. Exercise alters glucose and insulin kinetics [76.68, 69], and models of these processes have been proposed in both physiologic [76.54] (Sect. 76.2.3) and *minimal* [76.56] (Sects. 76.2.1 and 76.2.2) forms.

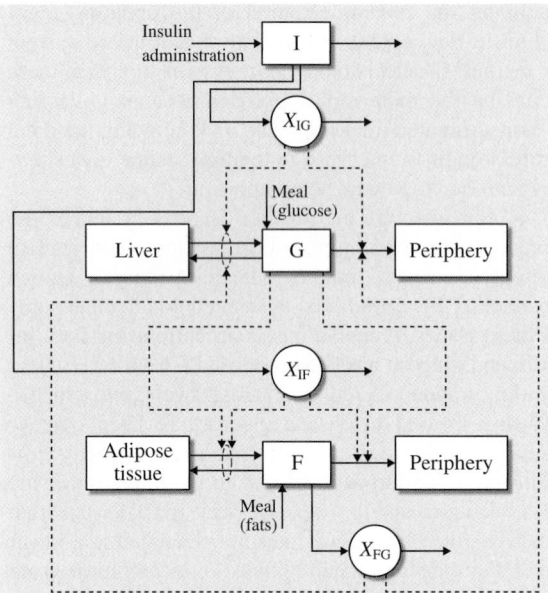

Fig. 76.8 Schematic of interactions between insulin (I), glucose (G), and fatty acids (F). Peripheral tissues consume glucose and fats, while the liver and adipose tissue serve dual source/sink roles for G and F, respectively. Meal consumption increases glucose and fatty-acid levels, and insulin administration drives circulating insulin level. *Dashed lines* represent the effect of one compound on another (G on F dynamics, for example). The X_{ij} *blocks* represent additional dynamics (equations) that alter the extent and duration of effect of i on j (e.g., insulin on glucose through X_{IG})

Again, these models are more macroscopic than mechanistic, but they are able to capture the important responses for the purposes of glucose regulation in diabetic patients. While the population-level variability of the model parameters remains to be characterized (which is also true of the FA models, above), the proposed structure does allow preliminary analysis of control structures to establish feasibility and potential performance of automated glucose control systems.

76.3.2 Cancer Radio- and Chemotherapy

Cancer is a collection of diseases resulting from a series of genetic mutations and characterized by an imbalance between cell proliferation and apoptosis, or programmed cell death [76.71]. If untreated, cancer leads to organ failure and the death of the host organism. One in four deaths in the USA in 2008 was projected to be from cancer, claiming over 565 650 lives [76.72]. In addition, the total disease burden, including treatment costs and loss of productivity, was estimated at US$ 219.2 billion in 2007, providing both societal and monetary reasons for improving cancer treatment [76.72]. Unlike insulin-dependent diabetes, cancer is not a single disease, and therefore, no *silver bullet* (i. e., insulin) exists. Hence, cancers must be addressed individually; e.g., pancreatic and breast cancers are separate diseases with different treatment options and chemotherapeutic choices.

Accessible cancers are removed surgically. Alternatives or complements to surgery include systemic chemotherapy and targeted site-specific treatment, such as photodynamic therapy (PDT) or radiotherapy (RT) [76.73, 74]. RT is often used in place of surgery because the tissue surrounding the tumor is too sensitive to permit surgical excision (e.g., tumors located near the spinal column). PDT is used alone, or as an intrasurgical procedure, to kill tumor cells located at the tumor margin that are invisible to the human eye. All cancer treatments are usually complemented by systemic chemotherapy for two reasons: (i) small remote metastases are often present by the time cancer is detected; and (ii) chemotherapy is the only treatment that is *whole body* in nature, thereby giving it the ability to indiscriminately attack metastatic lesions before they are clinically diagnosed. Hence, automation advances in this area address targeted therapies and chemotherapy.

A key advance in RT, which requires a significant automation component, was intensity modulation (IMRT) [76.75, 76]. A linear accelerator source (Fig. 76.9) provides radiation that has its fluence mod-

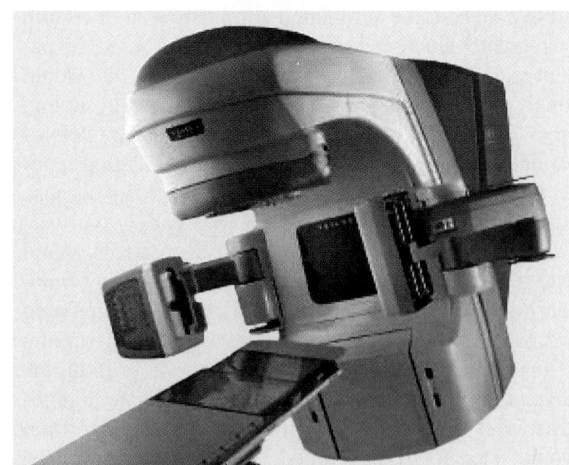

Fig. 76.9 IMRT gantry and linear accelerator source (figure originally published in [76.70])

ulated across the beam width, thereby providing prescribed radiation doses to a target tissue volume. By using a rotating beam gantry and placing the patient on an actuated table, critical tissues can be delivered lower doses through optimization of radiation dose to a particular planning volume. Mathematical models are constructed based on fundamental physics of radiation, with refinements based on the interaction of endogenous absorbers and scatterers with the administered radiation beam(s). Collaborative work between engineers and clinicians (such as [76.76]) has provided advanced offline algorithms for optimizing the treatment dose. The state of the art in RT is image-guided radiation therapy (IGRT) [76.73, 74]. This extends IMRT techniques to include real-time adjustment of the planned dose in order to reduce interfraction (positional changes of anatomical objects between the fractions of a radiation dose) and intrafraction (real-time motion of tissues in response to actions such as breathing) spatial uncertainties. From a calculational perspective, this changes a mixed-integer programming (MIP) problem, as solved in [76.76], into either a mixed-integer dynamic optimization problem [76.77, 78] or a dynamic control problem using the MIP solution as a reference.

Model-based approaches to cancer chemotherapy treatment design are an active area of research. Core to this approach is the existence of models describing PK (the effect of the body on the drug), untreated tumor growth (generally using the Gompertz model [76.79], which has been shown to capture solid-tumor progres-

sion in human patients), and pharmacodynamics (the effect of the drug on the body, including both antitumor effect [76.80, 81] and toxicity [76.82]). These models are constructed using compartmental (Sect. 76.2.2) or physiological (Sect. 76.2.3) approaches, depending on the amount of data available. Population tools (Sect. 76.2.2) are also important, as the cancer-affected population is heterogeneous. A common solution technique for the chemotherapy dosing problem is optimal control [76.33, 83, 84] (Sect. 76.2.5), where the following two-point boundary-value problem is posed: minimize tumor volume at the end of a fixed time period subject to constraints on drug dose (magnitude), path constraints (states, as governed by the model dynamics), and in some cases explicit representations of toxicity (e.g., neutrophils [76.85]). An alternative formulation and solution method uses mixed-integer programming [76.86–88]; advantages of this approach include path constraint satisfaction and extensibility of the framework to additional constraints. The existence of nonlinearities (beyond the bilinear kill term) in the PK or PD models causes additional challenges for their solution, requiring either parameterization [76.87] or mixed-integer nonlinear programming techniques (for details on these methods, see [76.89, 90]). A key change in formulation introduced in [76.87] is the use of a receding horizon strategy instead of the fixed final time (as in the latter part of Sect. 76.2.5). This is consistent with clinical practice: the patient is treated until disease progression or cure, thereby making the final time formulation clinically irrelevant.

76.4 Emerging Areas and Challenges

In a 2001 perspectives article [76.91], *Morari* and *Gentilini* describe the state of process control, and in a larger context, automation, as "... a victim of [its past success]: it reached a plateau and turning point." The remainder of their article highlights biomedical processes as an area where, "...control engineers [can] have significant impact," because automation and automatic control were not commonly employed. While there have been some changes since 2001, the potential impact of automation and engineering on biomedical practice remains significant. In this regard, this chapter has introduced a set of challenges and tools for use in melding systems engineering techniques with biomedical problems. The case studies are areas in which the author is active; additional opportunities abound for those motivated to: (i) learn the biology, medical

practice, and automation need of a disease (e.g., inflammation, infection, regenerative medicine), and (ii) work closely with collaborators in complementary fields to address the disease of interest (e.g., clinicians, biologists, engineers, mathematicians, etc.). Related to some of the challenges that began this chapter, the emerging areas and challenges below represent areas where systems engineering and automation can be employed to advance science and disease treatment. These areas align along a key principle of model-based control: model quality dictates theoretically achievable control system performance [76.92]. Hence, the biomedical goal is to construct *mechanistic* models, when possible, and to understand the *context* of the model, which may serve to limit the prospective use of the model to certain scenarios.

76.4.1 in vitro Physiological Systems

The basic science of biomedicine often begins with cell culture and in vitro experiments. However, a key issue in translating cell-only experimental results to live animals is the disconnect between *cells in a dish* and *cells interacting*. One approach to this problem has been coculturing of cells to allow cellular crosstalk and to allow the development of three-dimensional structure that better mimics physiology. This is a technique used in blood–brain barrier research [76.93] to develop in vitro tests for neurotoxicity. For PK analysis, many drugs are cleared by the cytochrome P450 enzymes in the liver but cause toxicities or effects elsewhere in the body. Hence, an in vitro liver cell culture could be used to analyze drug kinetics; one such option, although the cells are encapsulated in a polymeric coating, is described in [76.94]. A secondary concern beyond three-dimensional (3-D) structure is spatial heterogeneity, which may be more difficult to recreate in vitro [76.95].

A potential system for in vitro testing of drug PK would be a fully integrated physiologically motivated cell culture system: cells of different types growing in culture, with fluid volumes for each cell type corresponding to in vivo physiological volume; fluid flow and connectivity allowing intercell *crosstalk* within and between cell types (*simulated* physiology); relevant compartments for evaluating effect, as appropriate (e.g., tumor cells in culture for cancer PD analysis); and the entire system automated for sensing and control. In the context of cancer drug PK and PD, the incorporation of tumor cells would alleviate two shortcomings commonly present in current in vitro analysis: (i) clinically invalid pharmacokinetic profiles, where the tumor cells are bathed in a constant concentration of drug for a period of time, and (ii) cell kill analysis based on the aforementioned pharmacokinetic profile. While the mathematical analysis of tumor cell kill under time-dependent pharmacokinetic profiles is more complicated than the constant-concentration approach used presently, the use of in vitro physiological systems could serve to reduce the number of animal studies necessary in drug development while also improving the translation of in vitro results to the toxicity and efficacy studies that have to be performed.

76.4.2 Translating in vitro to in vivo

The transition from in vitro to in vivo provides a serious challenge to modelers and treatment designers. Phys-

iology may provide an advantage, as the connectivity between organs in animals is well understood. Organ weights and blood flow rates to individual organs are available in the literature for a variety of species, including *toxicologic man* [76.96]. However, this detailed approach assumes a model of physiological structure, which is not the generally applied practice in clinical drug development. As described in Sects. 76.2.2 and 76.2.3, compartmental models are more commonly employed in the (pre)clinical setting due to the smaller number of parameters, reduced mathematical complexity, and the relative ease of parameter estimation. In this less detailed structure, however, the method for incorporating in vitro information is ad hoc. Hence, a more mechanistic approach is desired when significant in vitro information is available because the mapping to the in vivo situation will be improved.

As discussed above, animals are not humans. The compartmental and population-based PK models often developed in animals rely upon allometric scaling [76.97, 98] to address the interspecies differences in dynamic profile. While imperfect (as witnessed by the dearth of PK-driven model-based treatment designs deployed in the clinic), this scaling principle can provide useful insight for selecting first-in-human doses and times to sample during phase I trials. Translating information relies to some degree on mechanistic accuracy at the cellular level, thereby limiting the *unknown* factors to the physiologically connected tissues. In some cases, carefully constructed physiological models can be successfully scaled from animals to humans by accounting for changes in physiology, such as body fat percentage, as seen for the lipophilic anticancer drug Docetaxel [76.10]. Often this requires assumptions, such as equivalent plasma protein binding characteristics (potentially based on binding information from in vitro studies), and similar mechanisms of clearance (liver metabolism, elimination in urine, etc.). Under these assumptions, the human physiologically based pharmacokinetic (PBPK) model (Fig. 76.4) can be successfully constructed by using the metabolic and clearance parameters as the degrees of freedom in fitting human plasma data. When inconsistencies between the scaled PBPK model and human data manifest, relaxing the assumptions and fitting novel mechanistic behaviors to the human plasma profile is required. The resulting model at the human scale can provide information that would otherwise be unattainable in humans, such as estimated drug concentrations in key tissues (e.g., brain) or those prone to toxic side-effects (e.g., liver, kidney, white blood cells).

76.4.3 Model and Network Structure Identification

How can one develop a mechanistically accurate model when limited measurements are available? This is simultaneously one of the greatest advantages, and most severe limitations, of systems biology. Through mathematical models, a large combination of model structures representing observed behaviors can be quickly constructed and tested with a particular data set. However, the challenge is in identifying which model is *correct* mechanistically, to the degree that the data set can provide such information [76.99]. These model-based methods allow extrapolation to estimate unknown quantities, but can also be used to *close the loop* on (systems) biology [76.100]. Close collaboration between experimental biologists, clinicians, and modelers can provide the feedback required to explore the predictions of systems biology models from the intracellular to the organism scale. Experimental resolution defines the scale of the study, and simultaneously provides context for interpretation and bounds limiting extrapolation. Models relate the perturbation to the observed responses, and an event may be represented by a wide range of models or even model classes. It is at this point that unifying frameworks, as discussed in [76.100, 101], provide *engineering judgement* in the construction of complex system models.

76.5 Summary

From an automation perspective, the context in which models will be used plays an important role [76.38]. In order to employ a model-based algorithm in a clinical setting, the model must be robust enough to capture clinically relevant behavior and the algorithm needs to reliably return a clinically deployable treatment that benefits the patient. Based on the increasing complexity of the underlying biosystems models, in terms of both structure and computational complexity, the algorithmic needs for treatment design approaches are significant. While these model-based treatment algorithms will not change the role of the clinician in quite the same way that computer-controlled systems have affected the chemical plant operator, a mathematically rigorous approach to biosystems analysis and disease treatment has the potential to benefit patient treatment – the endpoint objective of automating biomedical treatment decisions.

References

76.1   S.A.P. Haddad, R.P.M. Houben, W.A. Serdijn: The evolution of pacemakers, Eng. Med. Biol. Mag. **25**, 38–48 (2006)

76.2   L.A. Geddes: Historical highlights in cardiac pacing, Eng. Med. Biol. Mag. **2**, 12–18 (1990)

76.3   R.H. Taylor: *Medical Robotics and Computer-Integrated Surgery*, 2nd edn. (Wiley, New York 1999) pp. 1213–1227, Chap. 65

76.4   T. Teorell: Kinetics of distribution of substances administered to the body I, Arch. Int. Pharmacodyn. Ther. **57**, 202–225 (1937)

76.5   T. Teorell: Kinetics of distribution of substances administered to the body II, Arch. Int. Pharmacodyn. Ther. **57**, 226–240 (1937)

76.6   B. Ramakrishnan, U.R. Müller: High sensitivity expression profiling. In: *Microarray Technology and Its Applications*, ed. by U.R. Müller, D.V. Nicolau (Springer, Berlin, Heidelberg 2005) pp. 229–250

76.7   U.S. Government: *Code of Federal Regulations – Title 21 – Food and Drugs* (2008)

76.8   L. Xu, J.L. Eiseman, M.J. Egorin, D.Z. D'Argenio: Physiologically-based pharmacokinetics and molecular pharmacodynamics of 17-(allylamino)-17-demethoxygeldanamycin and its active metabolite in tumor-bearing mice, J. Pharmacokinet. Pharmacodyn. **30**, 185–219 (2003)

76.9   F.J. Doyle III, D. Seborg, R.S. Parker, B.W. Bequette, A. Jeffrey, X. Xia, I. Craig, T.J. McAvoy: A tutorial on biomedical process control, J. Proc. Control **17**, 571–594 (2007)

76.10 J.A. Florian Jr.: Modeling and Dose Schedule Design for Cycle-Specific Chemotherapeutics. Ph.D. Thesis (Univ. Pittsburgh, Pittsburgh 2008)

76.11 S. Audoly, L. D'Angio, M.P. Saccomani, C. Cobelli: Global identifiability of linear compartmental models – a computer algebra algorithm, IEEE Trans. Biomed. Eng. **45**, 36–47 (1998)

76.12 S. Audoly, G. Belu, L. D'Angio, M.P. Saccomani, C. Cobelli: Global identifiability of nonlinear models of biological systems, IEEE Trans. Biomed. Eng. **48**, 55–65 (2001)

76.13 J.W.T. Yates: Structural identifiability of physiologically based pharmacokinetic models, J. Pharmacokinet. Pharmacodyn. **33**, 421–439 (2006)

76.14   N.A.W. van Riel, E.D. Sontag: Parameter estimation in models combining signal transduction and metabolic pathways: the dependent input approach, IEEE Proc. Syst. Biol. **153**, 263–274 (2006)

76.15   G. Bellu, M.P. Saccomani, S. Audoly, L. D'Angio: DAISY: a new software tool to test global identifiability of biological and physiological systems, Comput. Methods Programs Biomed. **88**, 52–61 (2007)

76.16   E. Carson, C. Cobelli: *Modelling Methodology for Physiology and Medicine* (Academic, San Diego 2001)

76.17   MATLAB: *The MathWorks* (Natick 2008)

76.18   D.Z. D'Argenio, A. Schumitzky: *ADAPT II Users Guide: Pharmacokinetic and Pharmacodynamic Systems Analysis Software. Biomedical Simulations Resource* (Univ. Southern California, Los Angeles 1997)

76.19   P.H.R. Barrett, B.M. Bell, C. Cobelli, H. Golde, A. Schumitzky, P. Vicini, D.M. Foster: SAAM II: simulation, analysis, and modeling software for tracer and pharmacokinetic studies, Metab. Clin. Exp. **47**, 484–492 (1998)

76.20   E.R. Carson, C. Cobelli, L. Finkelstein: *The Mathematical Modelling of Metabolic and Endocrine Systems* (Wiley, New York 1983)

76.21   L.B. Sheiner, T.M. Ludden: Population pharmacokinetics/dynamics, Annu. Rev. Pharmacol. Toxicol. **32**, 185–209 (1992)

76.22   S.L. Beal, L.B. Sheiner: *NONMEM User's Guide. NONMEM Project Group* (Univ. California, San Francisco 1992)

76.23   Resource Facility for Population Kinetics, http://depts.washington.edu/rfpk (accessed March 17, 2008)

76.24   J.A. Florian Jr., M.J. Egorin, W.C. Zamboni, J.L. Eiseman, T.F. Lagattuta, C.P. Belani, G.S. Chatta, H.I. Scher, D.B. Solit, R.S. Parker: A physiologically-based pharmacokinetic (PBPK) and pharmacodynamic model of Docetaxel (Doc) and neutropenia in humans, Am. Soc. Clinical Oncology Annu. Meeting (Chicago, 2007)

76.25   R.B. Conolly, M.E. Andersen: Biologically based pharmacodynamic models: tools for toxicological research and risk assessment, Annu. Rev. Pharmacol. **31**, 503–523 (1991)

76.26   H. Kitano: Systems biology: a brief overview, Science **295**, 1662–1664 (2002)

76.27   H. Kitano: Computational systems biology, Nature **420**, 206–210 (2002)

76.28   R.E. Kronauer, G. Gunzelmann, H.P.A. Van Dongen, F.J. Doyle III, E.B. Kelrman: Uncovering physiologic mechanisms of circadian rhythms and sleep/wake regulation through mathematical modeling, J. Biol. Rhythms **22**, 233–245 (2007)

76.29   H. Kitano, K. Oda, T. Kimura, Y. Matsuoka, M. Csete, J. Doyle, M. Muramatsu: Metabolic syndrome and robustness tradeoffs, Diabetes **53**, S6–S15 (2004)

76.30   M. Hucka, A. Finney, B.J. Bornstein, S.M. Keating, B.E. Shapiro, J. Matthews, B.L. Kovitz, M.J. Schilstra, A. Funahashi, J.C. Doyle, H. Kitano: Evolving a lingua franca and associated software infrastructure for computational systems biology: the systems biology markup language (SBML) project, IEE Proc. Syst. Biol. **1**, 41–53 (2004)

76.31   L. Strömbäck, V. Jakoniene, H. Tan, P. Lambrix: Representing, storing and accessing molecular interaction data: a review of models and tools, Brief. Bioinform. **7**, 331–338 (2006)

76.32   D.P. Bertsekas: *Dynamic Programming and Optimal Control*, Vols 1, 2 (Athena Scientific, Belmont 1995)

76.33   R. Martin, K.L. Teo: *Optimal Control of Drug Administration in Cancer Chemotherapy* (World Scientific, River Edge 1994)

76.34   F. Allgöwer, T.A. Badgwell, J.S. Qin, J.B. Rawlings, S.J. Wright: *Nonlinear Predictive Control and Moving Horizon Estimation – An Introductory Overview. Advances in Control – Highlights of ECC '99* (Springer, London 1999) pp. 391–449

76.35   K.R. Muske, J.B. Rawlings: Model predictive control with linear models, AIChE J. **39**(2), 262–287 (1993)

76.36   E. Zafiriou, H.-W. Chiou: Output constraint softening for SISO model predictive control, Proc. Am. Control Conf. (San Francisco 1993) pp. 372–376

76.37   J.H. Lee, M.S. Gelormino, M. Morari: Model predictive control of multi-rate sampled-data systems: a state-space approach, Int. J. Control **55**(1), 153–191 (1992)

76.38   R.S. Parker, F.J. Doyle III: Control-relevant modeling in drug delivery, Adv. Drug Deliv. Rev. **48**, 211–228 (2001)

76.39   J.A. Florian Jr., R.S. Parker: Feedback Control in Drug Delivery. In: *Nanotechnology in Therapeutics: Current Technology and Applications*, ed. by N.A. Peppas, J.Z. Hilt, J.B. Thomas (Horizon Bioscience, Norwich 2007) pp. 25–64, Chap. 2

76.40   V.W. Bolie: Coefficients of normal blood glucose regulation, J. Appl. Physiol. **16**, 783–788 (1961)

76.41   R.N. Bergman, L.S. Phillips, C. Cobelli: Physiologic evaluation of factors controlling glucose tolerance in man, J. Clin. Invest. **68**, 1456–1467 (1981)

76.42   R.S. Parker, F.J. Doyle III, N.A. Peppas: A model-based algorithm for blood glucose control in type I diabetic patients, IEEE Trans. Biomed. Eng. **46**(2), 148–157 (1999)

76.43   R. Hovorka, V. Canonico, L.J. Chassin, U. Haueter, M. Massi-Benedetti, M.O. Federici, T.R. Pieber, H.C. Schaller, L. Schaupp, T. Vering, M.E. Wilinska: Nonlinear model predictive control of glucose concentration in subjects with type 1 diabetes, Physiol. Meas. **25**, 905–920 (2004)

76.44   J. Plank, J. Blaha, J. Cordingley, M.E. Wilinska, L.J. Chassin, C. Morgan, S. Squire, M. Haluzik, J. Kremen, S. Svacina, W. Toller, A. Plasnik, M. Ellmerer, R. Hovorka, T.R. Pieber: Multicentric, randomized, controlled trial to evaluate blood glucose control by the model predictive control algorithm versus routine glucose management protocols in intensive care unit patients, Diabetes Care **29**, 271–276 (2006)

76.45 C. Cobelli, G. Federspil, G. Pacini, A. Salvan, C. Scandellari: An integrated mathematical model of the dynamics of blood glucose and its hormonal control, Mathem. Biosci. **58**, 27–60 (1982)

76.46 C. Cobelli, A. Ruggeri: Evaluation of portal/peripheral route and of algorithms for insulin delivery in the closed-loop control of glucose in diabetes – a modeling study, IEEE Trans. Biomed. Eng. **BME-30**, 93–103 (1983)

76.47 C. Cobelli, A. Mari: Validation of mathematical models of complex endocrine–metabolic systems. A case study on a model of glucose regulation, Med. Biol. Eng. Comput. **21**, 390–399 (1983)

76.48 E. Salzsieder, G. Albrecht, U. Fischer, E.-J. Freyse: Kinetic modeling of the glucoregulatory system to improve insulin therapy, IEEE Trans. Biomed. Eng. **32**, 846–855 (1985)

76.49 J.T. Sorensen: A Physiologic Model of Glucose Metabolism in Man and its Use to Design and Assess Improved Insulin Therapies for Diabetes. Ph.D. Thesis (Department of Chemical Engineering, MIT 1985)

76.50 M. Berger, D. Rodbard: Computer simulation of plasma insulin and glucose dynamics after subcutaneous insulin injection, Diabetes Care **12**, 725–736 (1989)

76.51 W.R. Puckett: Dynamic Modeling of Diabetes Mellitus. Ph.D. Thesis (Department of Chemical Engineering, University of Wisconsin Madison 1992)

76.52 E.D. Lehmann, T. Deutsch: A physiological model of glucose–insulin interaction in type 1 diabetes mellitus, J. Biomed. Eng. **14**, 235–242 (1992)

76.53 R.S. Parker, J.H. Ward, N.A. Peppas, F.J. Doyle III: Robust H_∞ glucose control in diabetes using a physiological model, AIChE J. **46**, 2537–2549 (2000)

76.54 P.J. Lenart, R.S. Parker: Modeling exercise effects in type1 diabetic patients, Proc. 15th IFAC World Congress on Automatic Control (Barcelona 2002)

76.55 A. Roy, R.S. Parker: Dynamic modeling of free fatty acids, glucose, and insulin: an extended minimal model, Diabetes Tech. Theraput. **8**, 617–626 (2006)

76.56 A. Roy, R.S. Parker: Dynamic modeling of exercise effects on plasma glucose and insulin levels, J. Diabetes Sci. Technol. **1**, 338–347 (2007)

76.57 DCCT – The Diabetes Control, Complications Trial Research Group: The effect of intensive treatment of diabetes on the development and progression of long-term complications in insulin-dependent diabetes mellitus, N. Engl. J. Med. **329**, 977–986 (1993)

76.58 DCCT – The Diabetes Control, Complications Trial Research Group: The absence of a glycemic threshold

for the development of long-term complications: The perspective of the diabetes control and complications trial, Diabetes **45**, 1289–1298 (1996)

76.59 American Diabetes Association: Continuous subcutaneous insulin infusion (position statement), Diabetes Care **24**(1), S98 (2001)

76.60 N. Hernjak, F.J. Doyle III: Glucose control design using nonlinearity assessment techniques, AIChE J. **51**, 544–554 (2005)

76.61 C.C. Palerm, H. Zisser, W.C. Bevier, L. Jovanovic, F.J. Doyle III: Prandial insulin dosing using run-to-run control: application of clinical data and medical expertise to define a suitable performance metric, Diabetes Care **30**, 1131–1136 (2007)

76.62 C. Owens, H. Zisser, L. Jovanovic, B. Srinivasan, D. Bonvin, F.J. Doyle III: Run-to-run control of blood glucose concentrations for people with Type 1 diabetes mellitus, IEEE Trans. Biomed. Eng. **53**, 996–1005 (2006)

76.63 G. Van den Berghe, P. Wouters, F. Weekers: Intensive insulin therapy in critically ill patients, N. Engl. J. Med. **345**, 1359–1367 (2001)

76.64 G.V. Van den Berghe, P.J. Wouters, R. Bouillon, F. Weekers, C. Verwaest, M. Schetz, D. Vlasselaers, P. Ferdinande, P. Lauwers: Outcome benefit of intensive insulin therapy in the critically ill: insulin dose versus glycemic control, Crit. Care Med. **31**(2), 359–366 (2003)

76.65 S.J. Finney, C. Zekveld, A. Elia, T.W. Evans: Glucose control and mortality in critically ill patients, J. Am. Med. Assoc. **290**, 2041–2047 (2003)

76.66 R.S. Parker, E.P. Gatzke, F.J. Doyle III: Advanced model predictive control (MPC) for type I diabetic patient blood glucose control, Proc. Am. Control Conf., Vol. 5 (2000) pp. 3483–3487

76.67 P. Dua, F.J. Doyle III, E.N. Pistikopoulos: Model-based blood glucose control for Type 1 diabetes via parametric programming, IEEE Trans. Biomed. Eng. **53**, 1478–1491 (2006)

76.68 L.B. Rowell, J.T. Shepherd: Exercise: regulation and integration of multiple systems. In: *Handbook of Physiology*, ed. by L.B. Rowell, J.T. Shepherd (Oxford Univ. Press, New York 1996), Chap. 12

76.69 L.B. Borghouts, H.A. Keizer: Exercise and insulin sensitivity: a review, Int. J. Sports Med. **21**, 1–12 (2000)

76.70 S.H. Levitt, J.A. Purdy, C.A. Perez, S. Vijayakumar (Eds.): Stereotactic radiosurgery and radiotherapy. In: *Technical Basis of Radiation Therapy: Practical Clinical Applications* (2006) p. 244

76.71 J.M. Slingerland, I.F. Tannock: Cell Proliferation and Cell Death. In: *The Basic Science of Oncology*, 3rd edn., ed. by J.M. Slingerland, R. Hill (McGraw-Hill, New York 1998) pp. 134–165

76.72 The American Cancer Society: Cancer facts and figures: 2008, http://www.cancer.org/downloads/STT//2008CAFFfinalsecured.pdf (last accessed April 14, 2009)

76.73   J.A. Purdy: IMRT, IGRT, SBRT – advances in the treatment planning and delivery of radiotherapy, Front. Radiat. Ther. Oncol. **40**, 18–39 (2007)

76.74   J.D. Fenwick, S.W. Riley, A.J. Scott: Advances in intensity-modulated radiotherapy delivery, Med. Phys. **139**, 193–214 (2008)

76.75   M. Langer, E.K. Lee, J.O. Deasy, R.L. Rardin, J.A. Deye: Oprations research applied to radiotherapy, and NCI–NSF-sponsored workshop, Int. J. Radiat. Oncol. Biol. Phys. **57**, 762–768 (2003)

76.76   E.K. Lee, T. Fox, I. Crocker: Simultaneous beam geometry and intensity map optimization in intensity-modulated radiation therapy, Int. J. Radiat. Oncol. Biol. Phys. **64**, 301–320 (2006)

76.77   B. Chachuat, A.B. Singer, P.I. Barton: Global methods for dynamic optimization and mixed-integer dynamic optimization, Ind. Eng. Chem. Res. **45**, 8373–8392 (2006)

76.78   L.T. Biegler, I.E. Grossmann: Retrospective on optimization, Comput. Chem. Eng. **28**, 1169–1192 (2004)

76.79   L. Norton: A Gompertzian model of human breast cancer growth, Cancer Res. **48**, 7067–7071 (1988)

76.80   G.W. Swan: Tumor growth models and cancer chemotherapy. In: *Cancer Modeling*, Vol. 83, ed. by J.R. Thompson, B. Brown (Marcel Dekker, New York 1987) pp. 91–179

76.81   A. Asachenkov, G. Marchuk, R. Mohler, S. Zuev: *Disease Dynamics* (Birkhäuser, Boston 1994)

76.82   H.S. Friedman, S. Keir, A.E. Pegg, P.J. Houghton, O.M. Colvin, R.C. Moschel, D.D. Bigner, M.E. Dolan: 06-bg mediated enhancement of chemotherapy, Mol. Cancer Ther. **1**, 943–948 (2002)

76.83   R.B. Martin: Optimal control drug scheduling of cancer chemotherapy, Automatica **28**, 1113–1123 (1992)

76.84   U. Ledzewicz, U. Brown, H. Schättler: Comparison of optimal controls for a model in cancer chemotherapy with L-1- and L-2-type objectives, Optim. Methods Softw. **19**, 339–350 (2004)

76.85   U. Ledzewicz, H. Schättler: Optimal controls for a model with pharmacokinetics maximizing bone marrow in cancer chemotherapy, Math. Biosci. **206**, 320–342 (2007)

76.86   J.M. Harrold, J.L. Eiseman, W.C. Zamboni, R.S. Parker: Modeling of toxicity effects and non-linearities in pharmacokinetics of 9-nitrocamptothecin in mice, AIChE Annual Meeting (Austin 2004)

76.87   J.M. Harrold, R.S. Parker: Clinically-relevant cancer chemotherapy treatment scheduling using parameterized mixed-integer programming, Proc. CPC VII (Lake Louise 2006), CACHE Corporation

76.88   P. Dua, V. Dua, E.N. Pistikopoulos: Optimal delivery of chemotherapeutic agents in cancer, Comput. Chem. Eng. **32**, 99–107 (2008)

76.89   A. Pertsinidis, I.E. Grossmann, G.J. McRae: Parametric optimization of MILP programs and a framework for the parametric optimization of MINLPs, Comput. Chem. Eng. Suppl. **22**, S205–S212 (1998)

76.90   I.E. Grossman: Review of nonlinear mixed-integer and disjunctive programming techniques for process systems engineering, http://egon.cheme.cmu.edu/Papers/GrossmannReviewNon.pdf (last accessed April 14, 2009)

76.91   M. Morari, A. Gentilini: Challenges and opportunities in process control: biomedical processes, AIChE J. **47**, 2140–2143 (2001)

76.92   M. Morari, E. Zafiriou: *Robust Process Control* (Prentice-Hall, Englewood Cliffs 1989)

76.93   H. Tähti, H. Nevala, T. Toimela: Refining in vitro neurotoxicity testing – the development of blood-brain barrier models, Altern. Lab. Anim. **31**, 273–276 (2003)

76.94   S.F. Khattak, K.-S. Chin, S.R. Bhatia, S.C. Roberts: Enhancing oxygen tension and cellular function in alginate cell encapsulation devices through the use of perfluorocarbons, Biotechnol. Bioeng. **96**, 156–166 (2006)

76.95   R. Gebhardt: Metabolic zonation of the liver: Regulation and implications for liver function, Pharmacol. Ther. **53**, 275–354 (1992)

76.96   International Life Sciences Institute: Physiological parameter values for PBPK models, (1994)

76.97   G.B. West, J.H. Brown, B.J. Enquist: A general model for the origin of allometric scaling laws in biology, Science **276**, 122–126 (1997)

76.98   G.B. West, J.H. Brown, B.J. Enquist: The fourth dimension of life: Fractal geometry and allometric scaling of organisms, Science **284**, 1677–1679 (1999)

76.99   P. Vicini, M.R. Gastonguay, D.M. Foster: Model-based approaches to biomarker discovery and evaluation: a multidisciplinary integrated review, Crit. Rev. Biomed. Eng. **30**, 379–418 (2002)

76.100   M.E. Csete, J.C. Doyle: Reverse engineering of biological complexity, Science **295**, 1664–1669 (2002)

76.101   R. Tanaka, M. Csete, J. Doyle: Highly optimised global organisation of metabolic networks, IEE Proc. Syst. Biol. **152**, 179–184 (2005)

77. Automation in Hospitals and Healthcare

Brandon Savage

Healthcare is a complex industry, but yet it lags behind virtually all others in automation and use of
information technology (IT). For healthcare, technology serves as an untapped catalyst for higher
efficiency, lower cost and broader access to care.
The appropriate application can minimize medical errors, promote better management of chronic
illness, and enable clinicians to intervene earlier
and anticipate prognosis. Additionally, medical
informatics provides the tools to generate new
insights about both individuals and entire populations through data analysis and visualization.
These systems can also be used in support of continuous quality improvement efforts as well as
to reduce inefficiencies. While significant strides
have been made in the implementation of medical informatics, there are numerous challenges to
resolve before we will be able to realize the full
benefits of healthcare IT.

Part H | 77

Automation is so ubiquitous in our daily lives that few
of us give it a second thought. Whether the automated
teller machine (ATM), a barcode scanner, or wireless
communication, we embed automation as a fundamental mechanism of our daily lives.

It's stunning to notice how few parallel examples exist in the average hospital or physician office, when the
exchange of accurate information can literally be a matter of life and death. In today's healthcare system the
majority of medical records are still kept on paper – and
most of them, including prescriptions, are still handwritten. Even most electronic are commonly in stand-alone
systems, and unlike ATMs, are not interoperable and
cannot share information.

As an example, if you live in Duluth and need to
check your bank balance while visiting New York City,
it's easily within reach. If you're ill while in the Big
Apple and are taken to a hospital emergency department, it will be virtually impossible for the medical
staff to check your medical record for allergies and past
history. And, only a few healthcare institutions use barcode scanning to confirm accurate dose and treatment
administration.

Managing the complexity of healthcare warrants the
need for automation that same complexity is also the
reason effective automation is so difficult. There are
thousands of data and decision points in the course of
treatment. As medical knowledge expands at a break-
neck pace, what we define as today's best practice and
standard protocol, may not exist tomorrow.

77.1 The Need for Automation in Healthcare

While information technology has transformed other industries, the healthcare industry (especially in the US) remains largely paper-based. This may have been sufficient in the days of the family physician who made house calls, but is inadequate to meet the needs of today's complex healthcare environment. Medicare patients, for example, see an average of three different providers regularly – but without a digitized record the clinicians often lack a complete picture of any patient's medical history. Additionally, medical information collected in the physician's office is stored separately from records related to hospital admissions.

The impacts on the patient and the healthcare system range from the merely inconvenient to the potentially (and all too often) fatal. Without access to a complete medical record, clinicians may inadvertently prescribe medication to which the patient is allergic, or one clinician may unwittingly prescribe a drug that is harmful when taken with another medication prescribed by a different practitioner.

Even when records do exist in electronic form, the information cannot be readily shared because the systems are not interoperable. Within a single institution, interfaces may be required just to move data between clinical and financial information systems, for example, or between the system that holds pharmacy data and the one in which nurses document a patient's vital signs.

Beyond sharing between information systems, sharing information among healthcare organizations is even more complex. There is no single standard for how medical information is represented electronically, so data that exists in one vendor's system cannot be used by a different vendor's system. Overcoming the barriers to interoperability will have a significant impact in accelerating the adoption of electronic medical records.

The discussion of interoperability begs the question, however, *What are the real benefits of automation in healthcare?* In fact, the benefits are considerable across the entire spectrum of care.

Consider *patient safety*. Every year, between 50 000 and 100 000 people die in the US as the result of preventable medical errors [77.1]. Countless others suffer less severe injuries. Errors arise in many different ways, from prescribing the wrong medication to performing surgery at the wrong site. Confusion arises when two different patients have similar names, or when patients are moved from one hospital room to another. A misplaced decimal point can result in a patient getting less than a therapeutic dose of a drug, or a tenfold overdose:

- Incorporating alerts into electronic medical records can help ensure the completeness and accuracy of the record. The system can remind clinicians to fill in missing data such as allergy information, or in the case of surgery which side of the body is to be operated on.
- Automating the medication-ordering process can eliminate hazards that arise from illegible handwriting; in hospitals, it can shave hours off the time between when a drug is ordered to when it is administered. When expert rules are incorporated, the system can automatically suggest appropriate medications based on the patient's current condition, past medical history, age and weight, and even the list of drugs that will be paid for by the patient's insurance.
- In addition to electronic ordering, closed-loop medication management automates both the dispensing and administration of drugs to ensure that the *five rights* are met – i.e., that the right patient gets the right dose of the right drug via the right route at the right time. Nurses scan barcodes on each unit dose of medication and on a hospital patient's wristband; if there is a mismatch, the system generates an alert that can prevent the error.
- When the drug Vioxx was recalled, most physician offices had to resort to reviewing every single file by hand to determine which patients were taking the drug, resulting in delays of days or weeks before

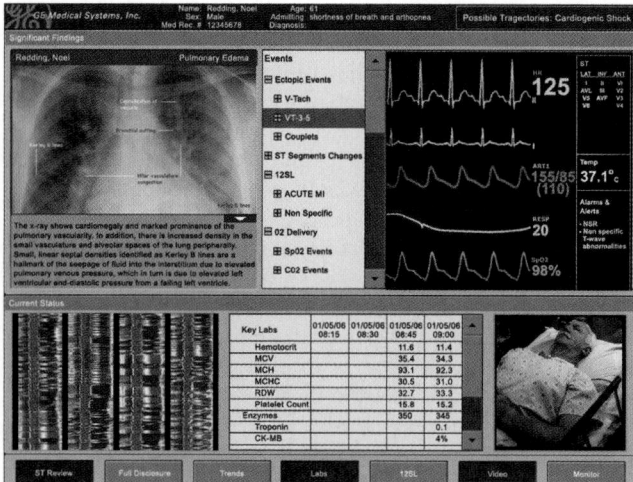

Fig. 77.1 Example of healthcare visualization and monitoring automation to prevent medical errors

patients could be notified of the recall and given an alternative prescription. Clinics that had electronic record systems, on the other hand, were able to compile lists of which patients were taking Vioxx and generate notices to their primary care physicians within a matter of hours.

In the treatment of *chronic illness*, a handful including cardiovascular disease, diabetes, and asthma, are responsible for the lion's share of medical expenses. Yet, if these conditions are managed appropriately, much of the cost can be avoided and patients can continue to lead long and productive lives without acute flare-ups of the disease. Electronic record systems can automatically generate alerts and reminders to help ensure that patients receive effective preventive care, such as annual eye exams for diabetic patients.

In addition, being able to run outcomes analysis on all patients with a certain condition can help identify trends and determine what treatments are more effective than others within a certain population. This kind of analysis, like drug recall notification, is simply not practicable with paper records, but can be done relatively easily with the right kind of electronic records systems.

Increased Efficiency and Accuracy. At each separate encounter with the healthcare system, patients are usually asked to provide the same basic information including name, date of birth, and insurance. Each time the data is taken down presents an opportunity for transcription error, as well as a frequent source of annoyance for patients. Automated check-in kiosks can let patients review their demographic data on file and make any needed changes, without having to recite the same litany over and over again.

Nurses spend as much as one third of their time on charting and other administrative tasks. Automation can reduce the amount of time nurses spend on documentation, freeing up more time that can be spent on direct patient care. Clinicians can pull patient information from one part of the system to another, so it does not have to be re-entered. Nurses can chart *by exception,* entering only data that is outside normal ranges or that has changed since the last reading.

Unlike paper records, electronic records give all members of the care team simultaneous access to a patient's most up-to-date clinical information. Lab results, for example, are available online as soon as they are completed, avoiding redundant tests. Physicians can check on their hospitalized patients from office or home at any time and enter new orders, rather than waiting for daily rounds.

Considering *consistency of care*, the quality of healthcare varies enormously from one hospital to another, and even within the same hospital from one provider to another. Simple best practices that are proven to save lives – such as presurgical antibiotics or administering beta-blockers to heart attack patients within 24 h – are not uniformly followed, even within a single institution [77.2]. Even with recommended care guidelines and best practices published for numerous disease processes, Americans as a whole receive only 54.9% of the recommended care [77.3], and fewer than half of American children receive timely, needed medical care [77.4]. Additionally, 11% of care received is either not recommended or potentially harmful [77.3]. In large part, this is due to the fact that the already overwhelming body of medical science is growing at an explosive pace, and no individual can keep up with it. Clinicians, like the rest of us, can also be resistant to change – preferring to continue using the same protocol they always have for a certain condition. To promote higher consistency, healthcare institutions can develop standard order sets that are used for specific conditions. While some hospitals use hard-copy versions of such order sets, digitizing them and integrating them with a clinical information system, so they are automatically presented at the point of care can be more effective consistency.

On average, it takes 17 years for a new medical discovery to become part of the standard of care. Clinical information systems can accelerate this process through the use of expert systems that are integrated with the electronic medical record to make recommendations about appropriate treatment and present that information to the clinician at the point of care. Evidence-based practice (EBP) is an approach to healthcare that effectively integrates best practices with the daily activities of patient care, resulting in a consistently high standard of care both within and across healthcare organizations. EBP can reduce inappropriate variations in care while at the same time enabling the appropriate tailoring of best practice to each patient's clinical condition [77.5].

Currently only a few of the largest healthcare institutions are able to implement expert systems in this way; the challenge going forward will be to make the technology accessible to smaller organizations with limited IT resources.

77.2 The Role of Medical Informatics

Since its inception in the 1960s, there has been a steady increase in demand for medical informatics, including dramatic growth occurring in the past decade. The reason for this explosive growth is in how medical informatics provides a foundation that bridges the art and science to address challenges facing healthcare. In order to do this, we ultimately need to move medical informatics beyond the level of information management – acquiring, storing, and moving data – to knowledge management – the ability to synthesize and apply available data and thereby transform the practice of medicine.

The breadth of patient data and how that data is utilized in a healthcare setting is important in understanding the distinction between knowledge management and informatics. There are two categories of patient data. The first is subjective data, such as pain or shortness of breath, that is qualitatively described by the patient. The second is objective data, such as heart rate or weight, which can be verified by external observers.

Objective data can further be broken down into patient demographics (name, age, etc.), observations made by healthcare providers, and quantitative parameters representing the anatomic, biochemical, and physiologic states of the patient. Within the latter category, advances in medical diagnostics are not only creating a more comprehensive picture, they are increasingly shifting from anatomic to physiologic parameters – i. e., functional as opposed to structural data. A computed tomography (CT) scan, for example, can detect an anatomic abnormality in the chest, while a positron emission tomography (PET) scan of the same area can detect metabolic characteristics that suggest the mass is a malignant tumor. Specific biochemical markers are

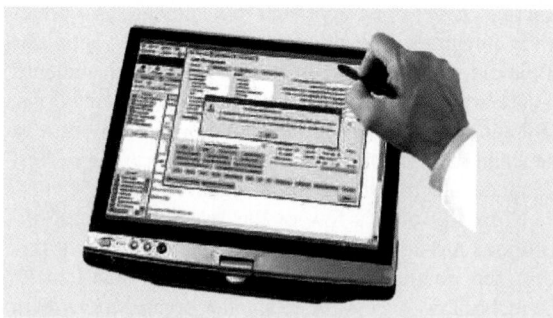

Fig. 77.2 Healthcare automation has revolutionized the practice of medicine with innovations to detect and diagnose diseases earlier

better indicators of the effects of congestive heart failure than are imprecise observations of distended veins in the neck.

Another important shift is from phenotypic data – the observable signs or symptoms of normal or abnormal gene expression – to genotypic data – the underlying genetic code itself. Not only is genotypic data more specific (i. e., there can be many different genotypes that present the same phenotype) but it can often predict disease before any outward symptoms are present. The BRCA-1 gene, for example, presents a significantly elevated risk of breast cancer. Women who have this gene should be more closely monitored so that cancer is detected early, when it is more easily treated; some women may opt for mastectomy as a preventive measure. Genotypes for certain liver enzymes can predict how fast a patient will metabolize a certain medication even before that patient takes the first dose; this enables clinicians to more precisely prescribe the optimum amount of the drug.

77.2.1 Information Management

With this vast array of patient data, information management is a necessary first step in delivering safe, effective, high-quality healthcare. The following are four components of an information management strategy:

- *Acquisition* – the transformation of observable objective phenomena and recordable subjective experiences into data, which can then be stored in paper or digital formats.
- *Collation* – the organization of these data into a navigable structure linking the data to patients and populations.
- *Distribution* – the movement of information from its source to potential users of the data.
- *Access* – the interface that enables a user to query data once it has been distributed.

Failure in any one of these components will result in critical decisions being made based on only a small subset of available data, with impact both financial and physical. It is estimated, for example, that clinicians order approximately $ 2.5 billion worth of redundant laboratory tests annually because they do not have access to previously obtained results [77.6]. The consequence of administering medication without access to complete allergy information can be fatal.

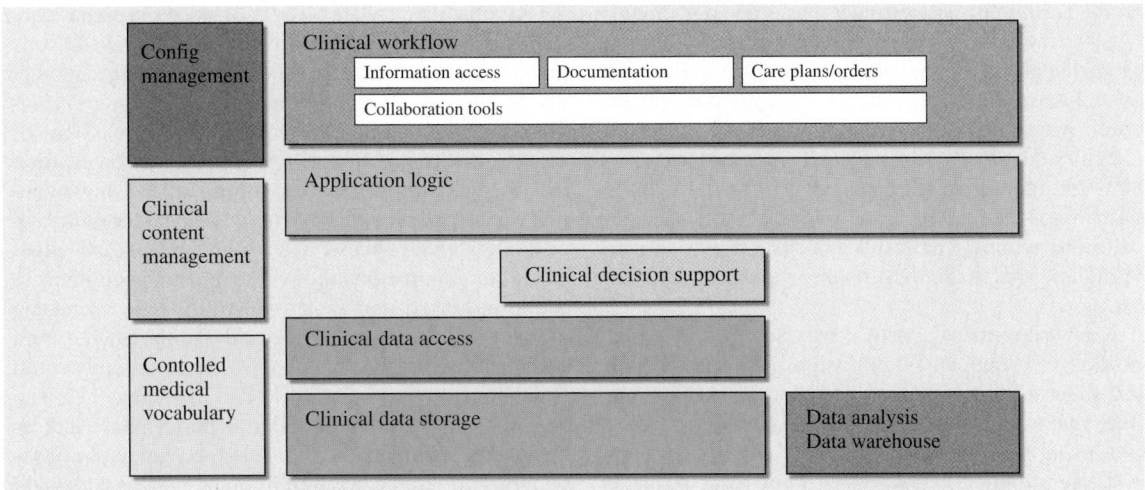

Fig. 77.3 Clinical informatics-oriented view: application architecture connects clinical workflow and is composed of information access, documentation, action plans, and collaboration

77.2.2 Knowledge Management

In many other industries, the value of information management alone has been enough to drive high rates of IT adoption. This is not the case in healthcare, however, where adoption of IT hovers around 20%, despite significant acknowledged potential benefit. The complexity and quantity of medical data decreases, information management as not sufficient to drive adoption. Instead, medical informatics needs to emphasize knowledge management that focuses on the content and use of information technology.

Where information management lags in acquiring and conveying data, knowledge management gains additional value from the meaning, context, and structure associated with that data. For example, consider the relatively simple act of measuring a patient's blood pressure. The systolic pressure reading (the 120 in a reading of 120/80, for example) measures the point at which the peak pressure of the blood in a patient's artery exceeds the pressure of the blood pressure device's cuff as it slowly deflates. The blood flow is detected either by a clinician using a stethoscope or by a transducer in the device. The resulting number can then be recorded, time stamped, and associated with a specific patient. This piece of information is useful but limited – the number alone has little meaning without the context of the data collection: Was the pressure recorded from the left or right arm? Was it recorded using an external blood pressure cuff or a transducer within an artery? Was the patient lying down or standing? (A reading taken while the patient is standing will always be lower than one taken when the patient is sitting or lying down.) Was the observation made by a health professional or reported by the patient? All these bits of information can affect the interpretation and utility of the information.

In aggregating and visualizing data, such as displaying a graph of a patient's blood pressure values over time, an information-driven system is limited in its ability to select appropriate data points. The system may include values that are not comparable (combining both sitting and standing blood pressure, for example) or may fail to include comparable values that were

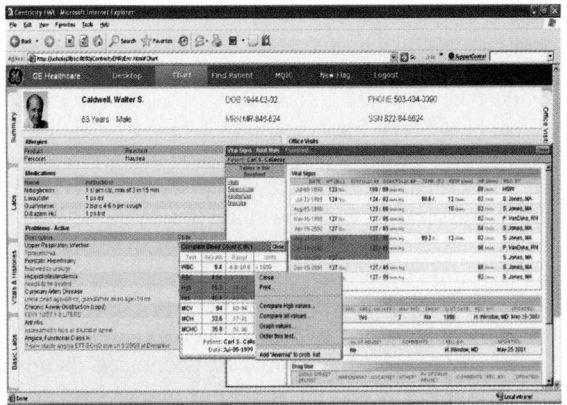

Fig. 77.4 The complexity and quantity of medical data has been a consistent barrier to the adoption of electronic medical records (EMR) and other clinical technologies

recorded differently (i. e., from a sensor inserted directly in the artery rather than from a blood pressure cuff). A knowledge-driven system, by contrast, could evaluate all possible blood pressure data and include only the appropriate values in the trending task.

With a knowledge-driven healthcare system, clinicians use knowledge management to transform information into insight, which in turn drives more effective healthcare actions. The result is a system that delivers the highest value (i. e., best quality care) at the lowest cost.

Knowledge management comprises three types of knowledge: patient-centered (insight into the available data about a patient), healthcare process (insight into effectiveness and value), and medical (body of medical science and best practices).

There are three key stages of knowledge management that allow these three types of knowledge to drive improvements in the healthcare system: knowledge representation, knowledge generation, and knowledge in action.

77.2.3 Knowledge Representation

The challenge for knowledge representation is to establish the meaning, context, and structure of data, as well as to enable the portability of knowledge across traditional data boundaries – whether software applications, between different locations within a single healthcare organization (such as an outpatient clinic and a hospital), or entirely separate institutions.

In the above example, knowledge representation would provide context to the blood pressure reading and corresponding heart rate measurement and whether the patient is taking any medications that modify blood pressure (information that may have to cross organizational boundaries). Additionally, the system could provide other relevant information from the patient's medical history and from external references, including whether the patient is being evaluated for heart disease (associated with high blood pressure) or a severe infection (associated with low blood pressure).

Knowledge representation is the result of robust medical terminologies that form the fabric upon which electronic medical records can be designed. These medical terminologies consist of controlled medical vocabularies (external standardized referenced dictionaries that attribute specific meanings to concepts that can be linked to data fields) and medical ontologies that contain relationships between concepts to establish the structure of stored data.

In practice, the language of medicine can vary tremendously from one institution to another. While one institution diagnose an episode as a *heart attack* another institution may refer to the episode as a *myocardial infarction*. In the mid-1990s, Medicalogic created one of the first electronic medical records that enforced what was then a controversial design principle: using a centrally managed dictionary of observational terms that all customers could access. The result was a knowledge-based community that allowed healthcare institutions to create advanced documentation forms that referenced the dictionary; these forms could also be shared with other institutions. The members of the community could collaborate on innovation as well as aggregate their data to support the creation of common performance indicators and benchmarks. Vendors also recognized business opportunities in the creation of other content that could be distributed across this customer base. For example, Clinical Content Consultants was founded in 2000 by John Janas III, M.D., and John R. Thompson, M.D., to design and implement interoperable evidence-based forms for clinical decision support in the management of chronic diseases [77.7].

The development of medical ontologies took off in the mid-1990s. The Regenstrief Institute at Indiana University launched LOINC (logical observation identifiers names and codes) to enable the electronic movement of clinical data from laboratories that produce the data to hospitals, physician's offices, and payers who use the data for clinical care and management purposes. At about the same time, the College of American Pathologists teamed up with Kaiser Permanente to expand SNOMED (systematized nomenclature of medicine) to include both a dictionary of medical terms and an ontology of relationships for these terms, both of which were independent of a specific electronic medical record vendor.

The relationships defined in most established medical terminologies are primarily parent/child or synonym-type relationships. Despite their apparent simplicity, the impact on IT system design is dramatic.

Two core IT needs illustrate the impact of an ontological hierarchy on information. The first is search. One of the biggest challenges of an electronic medical record (EMR) is finding information quickly, whether finding the right medication in a list of hundreds of thousands of alternatives or finding the relevant data to aid in performing a specific task. Where search engines such as Google use inherent linkages present within the data (such as hyperlinks) to infer associations, these linkages do not appear in the medical record, and medical ontologies are needed to explicitly articulate the relationships.

The second example is rule writing for decision support or reporting. While electronic medical records systems have allowed institutions to write rules for decision support utilization of rules has been low. In part, this is due to the difficulty of writing rules without an ontology. Consider writing a rule that would generate a list of all patients with lung disease who are also on blood pressure medication. With ontology this is a simple task; without one, it requires compiling exhaustive lists of all possible lung diseases and all possible blood pressure medications, and then mapping this back to specific field names in the medical record. These lists would have to be maintained as new drugs come on the market or new conditions identified.

77.2.4 Knowledge Generation

Knowledge generation uses information technology to augment pulling information from the clinicians' understanding of the medical literature. The daily activities of patient care generate new information about individual patients and entire populations, but the sheer quantity of data is overwhelming. Once data has been digitized, IT systems can synthesize even massive amounts of information into usable insight through two key routes: data analysis and visualization.

Other industries rely on data analysis and data mining as a significant source of insight. In healthcare, where an operating margin of 2–4% is considered exceptional, many institutions have implemented some level of financial data analysis. With respect to clinical data, however, the lack of digitized information has hindered the use of data analysis to generate insight about clinical practices outside the academic research setting.

Data analysis can enable clinicians to make inquiries about the efficacy of daily practice in the same way that knowledge can be derived from controlled clinical trials. A review of data at the population level can reveal patterns that are not observed in isolation. In this way the process of managing patients leads to greater insight about the best way to manage those patients going forward.

For example, it is widely recognized that inducing labor before the 39th week of pregnancy can have deleterious effects that compromise the health of the newborn, yet early inductions still occur with relatively high frequency. In a one-on-one interaction between patient and physician, it is easy to find reasons to induce *this* patient's labor early: late stages of pregnancy are

increasingly uncomfortable for the woman, the doctor will be out of town, or other calendar issues come into play. With each individual decision, neither provider nor patient is looking at the bigger picture.

At Intermountain Healthcare, an integrated delivery network headquartered in Salt Lake City, data analysis of almost 40 000 delivers deliveries in 2000–2001 revealed an increased rate of neonatal intensive care unit (NICU) admission for babies induced before week 39 (Fig. 77.5). The research revealed a one to two weeks delay could have avoided as many as 220 NICU admissions. Once this data was available, the institution developed and implemented an evidence-based triage protocol that set a higher threshold for elective inductions. The rate of early elective inductions dropped to 5%, significantly reducing unnecessary NICU admissions. Thus, the generation of knowledge through population data analysis also led to insight that motivated providers to change their practice.

Looking for trends also enables institutions to improve management of patient care. Ideally, these insights will extend beyond clinical knowledge to encompass the process of care. Electronic medical records enable institutions to capture information about cost, clinical data resource utilization, and process times. Using this information, they can begin to understand the consequences of practice patterns and resource allocations, and assess the *value* of the care they deliver. As institutions often focus on cutting costs without knowing the impact on quality, the analysis can demonstrate ways to provide the right kind of care while actually driving down per capita cost. Many examples exist where institutions deliver the highest quality of care with the lowest per capita rates, as consists of complicated processes, resource allocations and practice patterns with multi-fauceted practice patterns attribute. This analysis will become essential as more third-party payers adopt pay-for-performance programs that base reimbursement on quality rather than quantity of care.

With paper records, physicians write summaries that effectively filter out much of the granular information about a patient's condition. Other information may be stored in archives, or simply filed in some other part of the healthcare organization where they cannot be immediately accessed. One of the paradoxes of digitizing information is that the amount of accessible information quickly becomes so overwhelming that clinicians can no longer sift through all of it to find the relevant information they need.

Data analysis can also be used to help providers synthesize the vast amounts of information that are

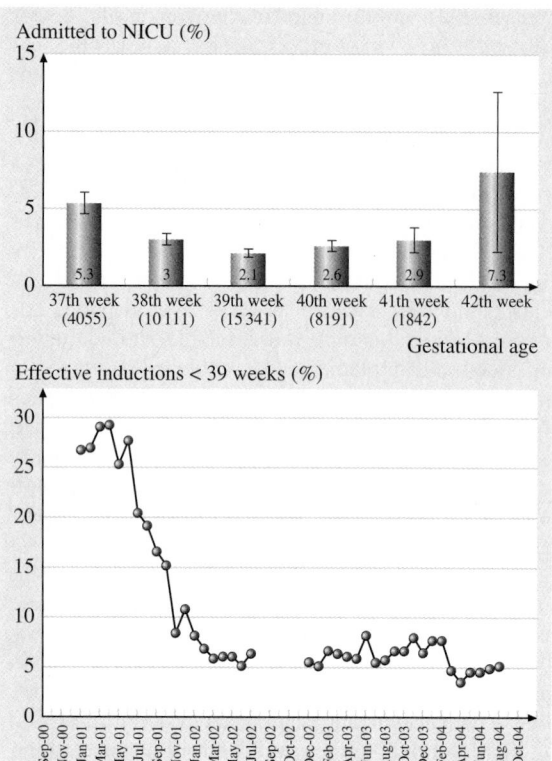

Fig. 77.5 Results of study by Intermountain Healthcare that analyzed almost 40 000 deliveries in 2000–2001 to reveal an increased rate of NICU admission of babies induced before week 39

The most effective clinical information systems can intelligently sift through vast amounts of data, identify pertinent patient information, and synchronize seamlessly to a provider's. In this setting, visualization includes the ability to organize data based upon the patient's condition and determines what is and is not relevant.

Knowledge generating systems make it easy for physicians to focus on the pertinent positive (abnormal) findings and the pertinent negative findings (normal results that reduce the likelihood of a specific diagnosis). In the case of chest pain, pertinent positive findings include high cholesterol, sudden onset of pain, smoking, and family history of coronary artery disease – all of which suggest that the patient is at risk for a heart attack. On the other hand, if the patient is relatively young with no family history of the disease, has normal lab tests, electrocardiogram (ECG) and other vital signs, and had similar normal findings on a recent emergency department admission for similar pain, the physician may conclude that coronary artery disease is an unlikely cause.

A system without knowledge representation can perform only limited visualization, such as simple graphing of trends, data grids, and composite displays; without insight into the meaning and structure of the data there is little intelligent work that can be performed by the computer. With knowledge representation, the IT system can determine how the available data relates to a patient's active problems and then organize the entire record to minimize the time it takes to review, increasing the likelihood that the provider sees and processes the most relevant information.

Advanced visualization algorithms can be used to look at data in new ways. Heat maps, cluster analysis, multivariate indexes such as severity scores, and patient health dashboards all represent potential means for presenting complicated data using simplified visual metaphors. Heat maps create color-coded views of the patient record highlighting the most active or high-risk areas requiring attention. Cluster analysis is used to graphically represent the similarity between one patient and similar patients in a population. It has been used successfully to predict patients who may be more likely to respond to certain cancer therapies based on the expression of different genes associated with a given type of cancer [77.8]. Other experimental approaches have been used to predict based on numerous factors, which patients are more likely to develop Alzheimer disease [77.9]. Cluster analysis is particularly useful because it enables clinicians to rapidly see associations

available about a single patient. For a patient with multiple chronic diseases, a clinician needs to understand the interplay of a variety of different test results and how individual tests vary over time. Visualization techniques can facilitate a clinician's evaluation of a patient by supporting the cognitive tasks necessary for making judgments patient about the patient's care. An information system that automatically collects the relevant information and presents it in an easy-to-read format can significantly facilitate a clinician's evaluation of a patient. With such an IT system in place, the clinician can work systematically from problem to problem, rapidly focusing on the next steps and how they interact. The presentation can be further augmented by embedding relevant contextual information for any task the clinician is handling. For example, if a provider is ordering antibiotics the system would automatically display allergies, drug interactions, previously prescribed antibiotics, and bacterial sensitivities.

among different factors even if the underlying scientific principles are unknown.

Multivariate indexes, such as severity scores, combine different observations about a patient into a single score. The human mind is limited in the number of variables it can process simultaneously by collapsing numerous variables into a single score can make it easier for clinicians to see trends that may otherwise be hidden in complex relationships. For example, severity scores can help clinicians identify at an earlier stage which patients are at greatest risk for worsening illness allowing them to intervene earlier and potentially prevent the worst effects of the disease [77.10]. Patient dashboards can be used to combine patient-specific data and severity scores to create a prioritized list of patients, so that the provider can quickly identify which patients are most in need of attention. One sign of well-designed visualization systems is that significantly fewer obtrusive alerts are required to avert provider mistakes; when the data is clearly presented, users are more likely to make the right choices the first time.

Unfortunately there are very few examples of widely adopted systems with detailed knowledge representation, so many of these visualization techniques remain in their infancy.

77.2.5 Knowledge in Action

The next level of knowledge management is to take the knowledge that has been accumulated – from medical research, from a patient's chart, and from data analysis and visualization – and put it to work at the point of care. Too much of the knowledge that exists today is inaccessible to clinicians when and where they need it. IT will build the bridge that enables us to incorporate the extensive knowledge base into the daily activities of patient care, resulting in a consistently high standard of care both within and across healthcare organizations, with fewer inappropriate variations yet still tailored to the needs of the individual patient. When knowledge is embedded into clinical workflow, clinicians are much more likely to actually use the information than they are to consult a static display in a book or on a web site.

In addition to improving care of individual patients, knowledge in action is also a keystone of continuous quality improvement (CQI) efforts, such as the application of Lean manufacturing principles to healthcare. IT enables the analysis of institution-wide data to break down processes into their constituent parts and identify steps that do not add value to the encounter.

The translation of knowledge into action includes digitizing information into a readily searchable form using knowledge representation; converting digitized data into an electronic workflow; and organizing and updating the knowledge database once it has been created. An institution should also rely on its IT infrastructure to provide support reviewing outcomes measures and effectively prioritize implementation of its thousands of evidence-based guidelines.

Knowledge into action can not only help institutions apply existing evidence to the delivery of patient care, but it also enable organizations to localize the practice of evidence based on various populations. The insights thus gleaned can be fed back into the system to further refine and improve the care being provided. The Pacific Northwest, for example, has a higher incidence of multiple sclerosis than anywhere else in the US – so when patients present with weakness or back pain, providers there should consider ordering an magnetic resonance imaging (MRI) sooner than they might elsewhere. This localization can even be applied to an individual patient: a person known to have a gene that predisposes him or her to a particular disease (such as breast or colon cancer) should be treated more aggressively than the general population when potential symptoms are seen. Rather than relying on a clinician to remember and apply all of these minute variations without assistance, institutions can apply expert rules within their electronic

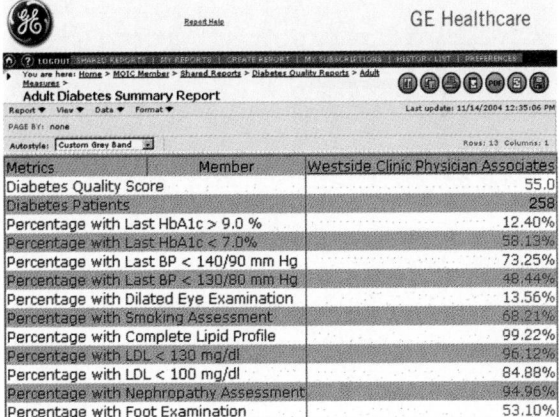

Fig. 77.6 A medical reporting tool. Technology systems like the Medical Quality Improvement Consortium (MQIC) can be used to drive continuous quality of care improvement by capturing current levels of quality. Using encounter forms and decision support rules, clinicians can then alter the effect of the care process and monitor the result of the intervention

medical record systems that will automatically present clinicians with evidence that is most relevant to a particular patient.

One way to bring knowledge into action is through decision support systems; of these, the most commonly used are alert engines. These engines display messages to users in a variety of situations, including:

- A potentially hazardous situation, such as when a clinician is about to order a medication to which the patient is allergic
- A condition that requires action, such as the receipt of abnormal lab results
- Missing data, such as failure to complete a list of known allergies
- A gap in care, such as overdue diagnostic tests recommended for preventive care of a chronically ill patient.

Alerts are intended to be helpful, highlighting simple issues that are easy for clinicians to overlook given the busy nature of healthcare. The challenge is to establish the right level of alerts. For example, the adverse effects of a drug allergy or a specific drug/drug interaction can have different degrees of severity depending on the patient's overall condition. The system would have to determine when an alert is appropriate for a specific patient, and many systems err on the side of showing too

many alerts. Unfortunately, there is a threshold above which users experience alert fatigue and begin to ignore alerts – essentially rendering the alerting system useless. In fact, some institutions turn off alerts in their electronic medical record systems because they appear to increase liability (if a clinician ignores an alert) without changing practice.

To overcome these problems, next generation alerting systems will need to have greater granularity in how they define what triggers an alert, taking into account both the patient's condition and the identity of the provider. A patient with kidney disease will always have some lab results that are outside the normal range for the general population, but that reflect the *normal* (i. e., stable) state for that individual. Similarly, alerts regarding the management of a patient's diabetes will be relevant to the patient's primary care provider, but not to an orthopedist who is seeing the patient for an unrelated condition.

An even more effective method of reducing alert fatigue is to initially present information in a way that reduces the likelihood of triggering an alert in the first place. In fact, one may argue that alerts – especially those that interrupt workflow – represent a system defect. A well-designed system should instead ensure that users do the right things rather than relying on alerts to correct errors, and it can do this in large part by how it displays patient data in conjunction with embedded reference information.

In other words, a system should not need to alert a clinician that the drug just prescribed is one to which the patient is allergic. Instead, the list of patient allergies should be displayed and highlighted before the clinician even starts the medication order. An intelligent patient summary screen would display a list of expected diagnostic tests based on a patient's known problems; this list would include the most recent results of all tests that have already been performed, indicate which ones need to be repeated, and prepare (pending the clinician's approval) orders for tests that have not been done. In a single glance, the clinician can easily determine the status of the patient and the appropriate next steps that are required.

Decision support systems are capable of more than simple alerts. Newer decision support systems translate existing best practice guidelines into adequately explicit digital protocols that can be incorporated into a clinical information system to help track and guide care.

While guidelines are common in medicine, the terms they use are often too ambiguously defined to be adequately represented in logic that can be enforced

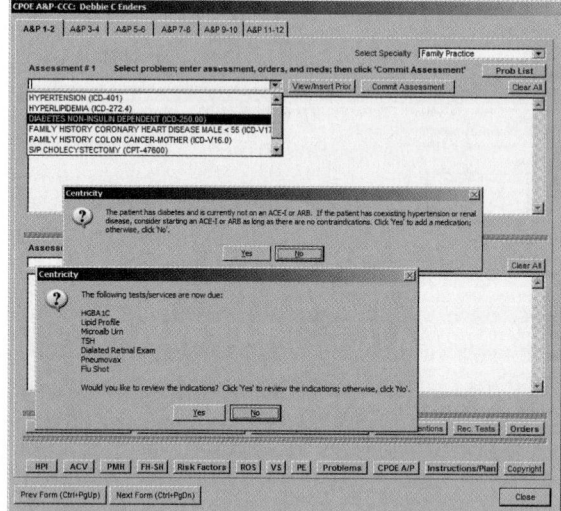

Fig. 77.7 Example of CPOE and clinical decision support. In this example, a physician has been *alerted* to elements of a diabetic patient's record, including medication management, allergy questions, and whether to document insight gathered from the alert

by a computer. A clinician will likely understand what a guideline means if it suggests changing a medication if a patient has not responded within a reasonable time. In contrast, an adequately explicit version of the same protocol would specify that if a patient's blood pressure has not dropped below 120/80 following administration of a maximum daily dose of 200 mg of Metoprolol for 1 month, or if the patient's heart rate has dropped below 50, then the patient should be given a calcium channel blocker.

In order to incorporate knowledge in action, there is a significant amount of work to be done to adequately define protocols and intelligent alerting systems. Existing guidelines tend to be written rather generically, and need to be tailored to the specific needs of clinicians at a particular location. To date, much of this work has been done by a small number of healthcare organizations, organizations, using commercially available systems as a starting point. Because of inconsistent terminologies and the lack of interoperability among systems, however, another – thus, reducing or eliminating the potential for synergies that could accelerate advancement of the technology, does not easily reuse work done at one institution. There have been a number of attempts to create knowledge representations for protocols and alerts that could move across systems, but, to date, none have been successfully and consistently adopted.

Adequately explicit protocols give rise to a dialog over time between the provider and the computer system that jointly determines the next appropriate intervention for care. The fact that a protocol is adequately explicit, however, does not replace the need for a determination to be made as to whether the protocol is appropriate for a specific patient. A patient with kidney failure will have changes in drug metabolism such that some protocols should never be administered. Or a patient who starts on a particular protocol may experience changes that make the protocol no longer appropriate. In other words, a protocol is a method to ensure consistency of care but it does not eliminate the need for careful observation and judgment.

77.3 Applications

Existing health IT systems can be organized by functions they perform and the locations where they perform them. The basic functions fall into five healthcare-separate categories: patient access, healthcare billing, healthcare administration, clinical care, and patient connectivity; a subset of these functions is then tailored to the specific needs of a given care location. Finally interoperability represents the functionality required to share information across the boundaries of healthcare. (There are a number of other core IT functions, such as ERP or customer relationship management (CRM) modules, which are applicable to any business and have been adapted to healthcare, which will not be reviewed here). Although this discussion begins with a separate review of each function, the true power of health IT is recognized only in enterprise systems that tie all these functions together.

77.3.1 Patient Access

Patient access systems manage how a patient interacts with the healthcare system; they can be further broken down into scheduling and location management. Healthcare scheduling is complex and is key to the efficient operation of a healthcare system. The simplest scheduling systems focus on finding a time slot for a patient to be seen by a specific doctor in a specific clinic. They may also take into account the level of resources or services available at different times of day to help balance the caseload. At the other end of the spectrum are enterprise scheduling systems, which are used to help orchestrate the delivery of care across multiple locations. For patients with complex conditions, these systems ensure that appointments are scheduled in the appropriate sequence and that a patient is not scheduled to be in two different locations at the same time.

Location management functions, commonly referred to as patient administration message (ADT) or admission/transfer/discharge, include tracking the locations of the patient in the hospital and storing basic demographic information about the patient. For inpatient care, the digital management of patient location is crucial to the appropriate provision of services (knowing which patient in which room is to be taken for a CT scan, for example) as well as for accurate billing.

77.3.2 Healthcare Billing

Healthcare billing systems focus on generating charges based upon services delivered and supplies consumed,

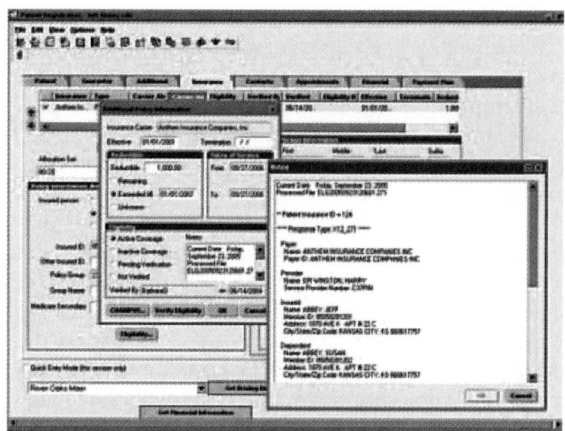

Fig. 77.8 By establishing a single electronic connection to over 1000 payers, this healthcare application simplifies and automates eligibility checking and claims processing to help drive more efficient workflows and accelerate reimbursement

and then presenting invoices to those responsible for payment (whether patient, insurance company, or some other third party). Both the capture of charges and the billing process itself are fairly complicated in medicine.

Charges comprise both professional service fees (incurred when a clinician performs a service for the patient) and institutional/facility fees (based on the cost of resources consumed and the procedures performed by the institution's staff).

Professional service fees are codified by type and complexity of the service provided, from a simple clinic

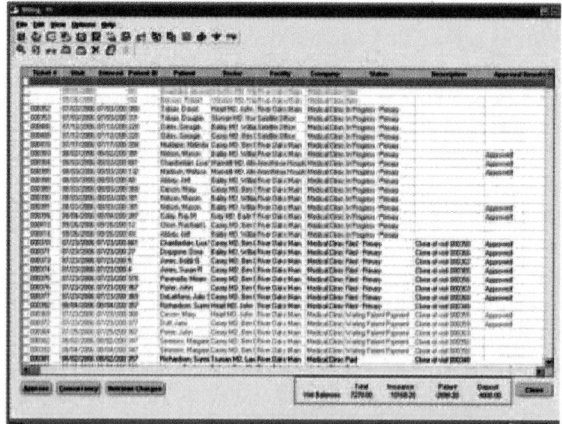

Fig. 77.9 Sophisticated billing solutions will include task management, claims processing, and tools to automate interactions across the healthcare systems

visit to open-heart surgery. Most IT systems capture these charges by having doctors select specific billing codes from a defined list. Significant training is required for physicians to bill appropriately and legally. More advanced IT systems provide some decision support that reviews the doctor's clinical documentation to help determine the appropriate level of billing for a given service.

For inpatient care, however, hospitals are not necessarily reimbursed based upon the actual care delivered. Instead there are DRGs (diagnostic related groups) that typically specify reimbursement levels for specific diseases or procedures; it is left up to the hospital to manage cost so as to maintain profitability. As a result, billing systems need both cost accounting and bill management to adequately address the address the revenue cycle. Cost accounting allows institutions to track the cost of performing basic operations and specific care actions. Cost accounting is crucial for an institution to understand its cost structure and optimize its productivity. Billing and revenue tracking are further complicated because reimbursement often comes from multiple sources: government (Medicare/Medicaid) and insurance companies, as well as patients.

Patient billing systems focus on submitting charges to third-party payers in such a way as to minimize rejections and shorten time to reimbursement. Increasingly, both insurance companies and government payers typically require additional clinical documentation to justify the charges – increasing the demand for integration between financial and clinical systems.

Patients are responsible for all charges not reimbursed by either insurance or government payers. Often there are multiple third-party payers involved, increasing the complexity of billing and making the whole process overwhelming for a patient who is also dealing with a significant illness and looming large medical expenses. A billing system that can streamline the process and simplify the experience will have a positive impact on patient satisfaction with the provider.

77.3.3 Healthcare Administration

Healthcare administration systems focus on the business intelligence to understand the operations of the healthcare system. These include financial dashboards and reports, which show the financial operations of the hospital, patient census and bed management, and resource management such as nurse staffing and order communications (the systems that communicate order requests within a hospital). These systems pull information from multiple sources to generate comprehensive views of the healthcare system.

Redundant or inefficient processes can take their toll on an organization's bottom line. Healthcare administration systems can help identify these processes, enabling organizations to streamline their operations, improving both profitability and patient and provider satisfaction. Increasingly, healthcare administration systems have become more dependent on clinical care. As a result, these systems are becoming more closely tied and potentially integrated with clinical systems. One significant challenge for these systems is to get clarity into details of the healthcare process especially at the point of care. Often the point of care is the least digitized and yet has the most significant impact on operations. As the point of care becomes more digitized, it will be easier to understand the healthcare delivery system in detail. This will create new opportunities to understand how cost, decision-making, and workflow affect the bottom line of operations. With greater insight new rules can be built into the point of care through standard operating procedures, workflow management, and decision support.

77.3.4 Clinical Care

The earliest clinical systems to be adopted were dedicated to reviewing patients' test results. Before the advent of these systems, clinicians had to consult multiple sources to get the results of laboratory tests, electrocardiograms, x-rays, and pathology reports for a single patient. Typically the results were available to only a few people at a time – so that different members of the care team could not access them simultaneously – and often the location where the results were stored was not where the clinician was. Results-viewing applications can aggregate data from multiple sources and allow clinicians to access the information from anywhere and at any time, so that the entire care team can have simultaneous access to a comprehensive view of the patient's status.

Another area where there has been significant adoption of IT is diagnostic test systems, which accelerate the process of performing the tests and manage distribution of the results. Most hospitals now have laboratory information systems to manage workflow, task management, and aggregation of data for all the tests performed in the laboratory. Likewise, picture archiving and communications systems (PACS), which are used to capture digital radiology and cardiology images from imaging hardware such as digital x-rays, CT scanners, and

MRI scanners, are present at most hospitals and increasingly at outpatient imaging centers. Radiologists use PACS to view and interpret images and provide diagnostic reports. PACS systems have eliminated the need for large (and expensive) film storage rooms; they have also given rise to teleradiology, where radiologists can view and interpret images remotely. This creates new possibilities for business models in which academic institutions can provide advanced services for rural hospitals that may not be able to attract and sustain specialized radiologists.

While these types of systems have been widely adopted, they are often disconnected from the core workflow of patient management and therefore provide little opportunity for the application of knowledge in action. The greatest impact on improving the quality of patient care comes from clinical documentation and computerized provider order entry (CPOE); unfortunately, adoption rates for both of these functions are less than 20% (Fig. 77.10).

Orders comprise all the directions that physicians prescribe for the management of a patient based on all the knowledge available about that patient at a certain point in time. Thus the amount of knowledge available to the clinician at the time of order entry will have a significant impact on the quality of care. CPOE not only ensures that an order is legible and consistently communicated; it also presents the opportunity to opportunity to use decision support to validate the appropriateness of the order prior to it being carried out. Additionally, CPOE allows the creation of sets of orders which can be managed as a group to ensure a given disease is managed consistently. A hospital can create a standard order

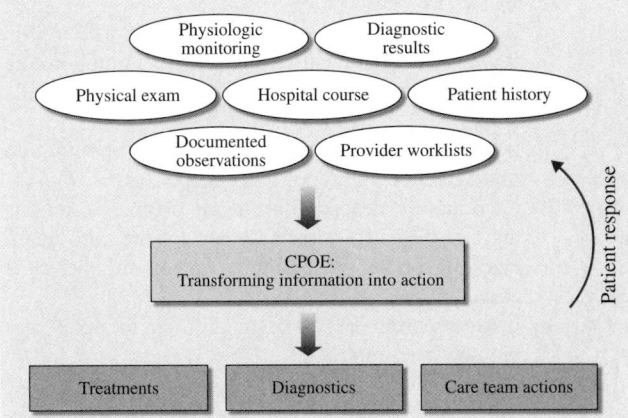

Fig. 77.10 Example of the cycle of care with the use of clinical documentation and CPOE

set for heart attack patients and then rather than having to remember each specific order, the physician can then focus on changing or adding orders needed to tailor care to the individual patient.

Order quality and order consistency were key drivers for a recommendation by The Leapfrog Group (a consortium of large employers concerned with improving the safety, quality, and affordability of healthcare) to recommend that CPOE should be widely adopted [77.11]. The centrality of order entry to workflow has made it difficult, however, to design CPOE systems that were not perceived to slow down the ordering process. In fact, while CPOE may create extra steps for the physician at the time of ordering, this is much less than the time required to correct illegible handwritten orders or address complications that arise from preventable adverse drug reactions. The problem is that the physician does not directly experience the time-saving impact of the workload.

It is only in the past few years, as organizations have recognized the role of change management in the success of CPOE, that adoption of these systems has started to accelerate. One of the key success factors has been to implement enough facets of the IT system to demonstrate value beyond CPOE. Physicians are more likely to adopt CPOE if at the same time they can benefit from information access, data visualization and decision support. Conversely, if CPOE is the only function implemented, so that workflows are split between digital and analog systems, creating unnecessary interruptions in workflow, clinicians will be less likely to embrace the technology.

Another key portion of a clinician's workflow is the documentation process. Healthcare documentation serves four key purposes:

1. It creates a legal medical record describing both the reasoning and activities involved a specific patient's care.
2. It provides a basis for outcomes analysis, population management, and reporting.
3. It communicates one care team member's actions and reasoning to other members of the care team, enabling better collaboration and coordination of care.
4. It assists clinicians to organize their thoughts and plan their actions.

Thus, good documentation systems are more than just sophisticated word processors. Rather, these systems are important for legal accountability, retrospec-

tive data analysis, and data mining, as well as for orchestrating and coordinating the ongoing work of the entire care team.

Interdisciplinary documentation, integration of decision support, and integration of key workflow tasks all contribute to extending the value of these systems. It is rare that one provider is responsible for all of the information necessary to manage a single patient. Interdisciplinary documentation allows providers from different specialties and disciplines to contribute to the medical record in an orchestrated fashion, collaborating on shared documents and leveraging documents created by others. Each provider can see the work performed by others and integrate other providers' observations into their own work. The hospital physician who sees a patient once per day relies on the information from hundreds of encounters by nurses, dieticians, physical therapists, and social workers to write one note that sets the direction for the next day of patient care. Similarly, a case manager will integrate the work of the entire care team and synchronize the nursing, physician, patient, and discharge goals all based on the shared documentation of the team.

The more powerful documentation systems combine task and worklist management features to tie the work of documentation into the details of patient care. Because documentation is such a consistent part of healthcare delivery, incorporation of worklists into the documentation process tends to increase the consistency of care delivery.

Providers often use documentation as an important part of their cognitive process, using it to help organize the information available and form their plan of action. The basic structure of a progress note taught to all physicians is called SOAP, which stands for *subjective, objective, assessment, and plan*. Thus, the inherent structure of the note is designed to aggregate current knowledge about a patient, assess the patient and then determine the plan of care. Most documentation notes follow a similar structure. Preserving this cognitive value of documentations is one of the hardest problems to solve in digitizing the documentation process.

Because, as with CPOE, so much reasoning occurs during the process of writing notes, documentation is another key target for decision support systems. Decision support can both initiate the documentation process and occur within the documentation process itself. Decision support initiated documentation focuses on dynamically manage managing the task or action that should occur next. Documentation also becomes the final step and forcing function to make sure that a process actually occurs. During documentation, decision support is used to automatically aggregate information that should be included in the note and prompt for actions that should be considered within the care plan. Documentation links task management, results review, care planning, decision support, and order entry all into a single workflow. The best documentation systems are built to leverage and dynamically interact with all of these functions within the EMR.

One of the next important trends in clinical applications is *near device* decision support. Medical devices are sprinkled across the healthcare system. There are advanced physiologic monitors and ventilators in the intensive care unit, anesthesia machines in the operating room, ECG machines in the emergency department, blood pressure cuffs and intravenous infusion pumps almost everywhere, and new devices entering the home – just to name a few. These devices each have their own innovations and manufacturers seek to find unique and innovative ways to differentiate their products. As a result these manufacturers have targeted ways to make these devices ubiquitous indispensable parts of the care process. Extending the hardware platform at the core of these devices with new software applications that aid in the care process has become increasingly common. Many of these applications can be considered *near device* decision support, as the decision supporting characteristics are managed by the device rather than with a clinical information system. For example many new infusion pumps include the ability to check dose ranges and verify the barcode of medications before they are administered. In fact, some of these pumps will connect to pharmacy information systems to allow the pharmacist to directly program the pumps wirelessly. Even more advanced applications are starting to appear such as ventilators that have embedded weaning protocols that help guide the process of discontinuing the use of mechanical ventilation in critically ill patients. (Ventilators are machines that breathe for patients when they are too sick to breath for themselves. One of the challenges of using a ventilator is helping a patient gain the strength to breathe independently of the machine as they begin to improve. This process is called weaning.) There are numerous applications where software combined with medical devices will help manage complicated and specific care processes. In this space the line between devices and software will begin to blur.

77.3.5 Patient Connectivity

In a given year most people may see their doctor at most 1 h and even patients with all but the most severe disease spend most of their time living their lives independent of the healthcare system. As a result most of the decisions that affect the health of a person are made during the activities of daily life. Choices about diet, exercise, medication compliance, sleep habits, and stress levels all contribute significantly to the real health outcomes that face society. If we are to truly improve the health of a population then we must enable the healthcare system to reach beyond the boundaries of its wall directly into the lives of patients. Essentially, the patient becomes the central member of the care team and effective collaboration becomes key.

With the advent of increased penetration of the Internet into the average family household and increased trends toward patient financial responsibility for healthcare due to employers passing the increase costs of healthcare on to their employees, the patient's role in the IT system has begun to expand.

The most common and demanded IT functions are around managing the patient's access and finances with an institution. Many systems are beginning to allow patients to request appointments online and allow patients to review their complicated billing histories and make payments. These features keep healthcare systems on par with other industries and increase the customer satisfaction of interacting with the healthcare system. Additionally, since these portals receive a fair amount of traffic healthcare, institutions use these portals to provide marketing and relevant institutional information to patients and prospective patients. This is particularly important in urban areas where healthcare systems compete for market share.

Although these finance and access features drive a great deal of consumer satisfaction they have limited impact on the health of a patient. Recently, EMR vendors and dedicated patient portal vendors have begun providing clinically oriented applications. The simplest forms of this application allow patients access to portions of their medical record – such as labs and results – and allow secure communication with between the patient and the physician. These e-Visits have recently begun to be reimbursed by payers as they are becoming more effective and reducing unnecessary and more expensive office visits.

More advanced versions of these applications include true disease management applications, which

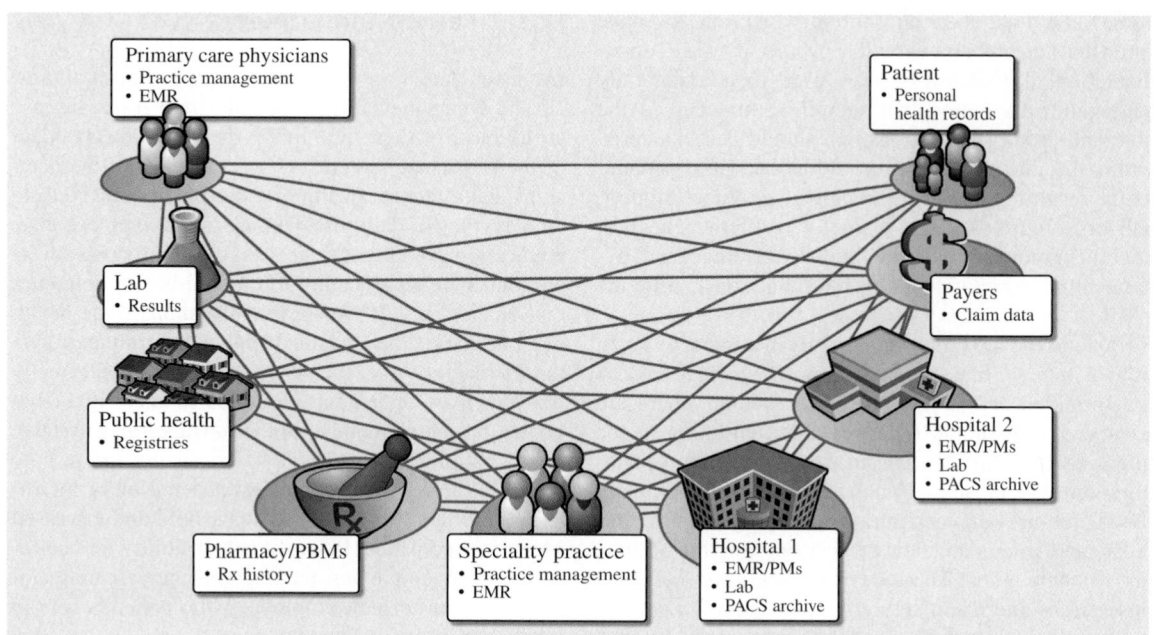

Fig. 77.11 Automation has a crucial role in how frequently a patient will interact with the healthcare system based on complexity of the experience and how a patient will retrieve healthcare information beyond the episodes of care

allow physicians and/or care managers to interact with populations of patients to manage specific diseases such as diabetes, asthma, and congestive heart failure. These application extend the clinical feature to include registries of patients with key clinical information to highlight which patients need help from the healthcare system and provide tools to enable collaboration with the patient on a shared plan of care for their diseases.

While these clinical applications tend to be provided by providers there is another class of wellness and disease management applications that are provided by insurers and employers. These applications share some of the same features described above but also include applications to inspire health behaviors in populations in consumers. These programs have been shown to reduce the costs of healthcare for employers and reduce the cost of managing populations for insurers.

Finally people are beginning to take more control of their own healthcare records and there has been an emergence of digital personal healthcare records (PHRs). These personal healthcare records allow people to record their key clinical information such as labs, diet, exercise, medication lists, etc. Some PHRs can connect to EMRs and directly share information between patients and providers. Some PHRs also include wellness and disease management features to link the

patients personal data with activities that drive health and wellness. Although PHRs offer opportunities to empower patients, increase sharing and collaboration, and drive better health behaviors, their adoption remains low while commercial interest remains high.

77.3.6 Interoperability

Despite the successful digitization and automation of the workflows within many global locations, most information is stored in silos and is unable to provide maximum impact on patient care. Unfortunately, it's often at the same boundaries that prevents digital information flow from improving quality and safety reside. For example, when a patient is discharged from the hospital it is most common that medication changes and follow up plans are confused. It is also at discharge when a new set of healthcare providers will manage the patient and often these healthcare providers do not have a mechanism for accessing the digital patient record if it is present. Interoperability strategies seek to improve this flow of information. Within the US there are strategies to create regional health information organizations (RHIOs) that can manage the exchange of critical information across these institutional boundaries. At a national level there are plans to develop a National Health Information Network (NHIN) to connect these RHIOs. Unfortunately, even at an institutional level there are numerous clinical and financial applications that have difficultly communicating and thus there are significant unmet needs even within a single organization. Many countries across the globe are issuing tenders to seek solutions for health information exchange on national levels.

Putting aside the complex business model and organizational barriers that have been significant barriers towards significant national and local information exchange, there continues to be significant technology barriers. One of the fundamental challenges resides in the prime directive of medical informatics in that most systems store data rather than knowledge and as a result have no external references to establish the meaning of data that needs to be exchanged. For example, consider allergies. Allergies represent one of the most acute life threatening risks to an individual whenever he encounters a healthcare system, especially if he enters the system unconscious. Surprisingly, there is no common dictionary for allergies, and as a result most systems have no method to exchange allergies in a form beyond simple free text. With free text allergies there is no method to perform simple allergy checking before prescribing or administering a drug.

Today organization such as Health Language 7 (HL7), Integrating the Healthcare Enterprise (IHE), Healthcare Information Technology Standards Panel (HITSP), and the Center for Healthcare Information Technology are all working on establishing standards to facilitate interoperability across institutions and across vendors. Although there are emerging standards for rudimentary exchange of information today there are surprisingly few examples of information exchange even at the level we have come to expect in the banking industry. With patients continually changing healthcare environments due to changes in health and changes in insurance, the challenge of interoperability must be solved to ensure that information technology has a significant impact on the quality of care.

77.3.7 Enterprise Systems

Historically, patient access and financial systems were the first to gain a foothold in healthcare. Driven by the bottom line, these supported the back office and had no access to (or need for) clinical data. Clinical systems followed, designed to address specific needs in individual care areas such as labor and delivery, where there was a need to record and store the massive amounts of information generated by fetal heart monitors and other devices. For the past two decades, lab systems built around the workflow of technicians and pathologists have had the capability to collect results from a wide array of tests and present physicians with a single view of all the data for each patient (although in most physician's offices that information is still presented on paper rather than electronically).

Because these systems were developed for specific care areas, each tightly adhered to the workflow in one particular area. Many healthcare organizations adhered to a best-of-breed approach, buying a lab system from one vendor, a pharmacy system from another, a perioperative system from a third, and so on. These care area-specific systems can be more nimble, and can be installed in a matter of weeks, but require complex interfaces in order to share data with each other. As time passed, clinicians recognized the need for enterprise-wide systems, so some vendors of area-specific systems began broadening their scope.

Other vendors took the approach of developing enterprise-wide systems that allow information to easily cross the boundaries between care areas. Consider, for example, what happens when a physician orders medication for a hospital patient: the provider needs to know whether the patient is allergic to that drug, or has been given any other medication that should not be combined with the new one, formularies must be consulted to see whether the hospital is dispensing the drug and whether the patient's insurer will pay for it, the order must be transmitted to the hospital pharmacy, where the pharmacist double-checks the dose to make sure it is appropriate for the patient's age, weight, and condition. After the pharmacist dispenses the drug, it must be conveyed to the right location in the hospital, where a nurse will pick it up, administer it to the patient, and document the time of administration (perhaps using a barcode scanner to make sure the right patient is being given the right medication).

The trade-off for the easy flow of information across the enterprise is in specificity and implementation time. The best area-specific systems do one thing only, and do it better than the current generation of enterprise systems. Smaller, more discrete systems are also easier and faster to install. As the technology matures, however, we will see a convergence as enterprise systems acquire the greater depth of area-specific capabilities.

77.4 Conclusion

The practice of healthcare is incredibly complex generates massive amounts of mission critical data that must be assimilated and acted upon quickly. In many cases, the margins for error are slim and the costs of error immense. Information technology has the potential to transform the way we practice medicine by turning data into knowledge and knowledge into action. This will enable providers to deliver better quality care at lower cost, to maintain consistency of care for a given disease while appropriately tailoring care to each individual patient's unique condition and genetic makeup, and to localize the application of care guidelines based on knowledge about specific populations of patients.

Some of the tools to accomplish these goals exist today, but they need to be modified or adapted to suit the unique demands of healthcare. We have systems that can alert clinicians to potential errors, for example, but these systems must be fine-tuned to prevent alert fatigue. Other challenges, such as interoperability, which enables data sharing among systems from different vendors, remain largely unanswered at the moment.

References

77.1 L.T. Kohn, J.M. Corrigan, M.S. Donaldson (Eds.): *To Err Is Human: Building a Safer Health System* (National Academy Press, Washington 1999)

77.2 J.E. Wennberg: Unwarranted variations in healthcare delivery: implications for academic medical centers, Br. Med. J. **325**, 961–964 (2002)

77.3 E.A. McGlynn, S.M. Asch, J. Adams, J. Keesey, J. Hicks, A. DeCristofaro, E.A. Kerr: The quality of health care delivered to adults in the United States, New Engl. J. Med. **348**(26), 2635–2645 (2003)

77.4 R. Mangione-Smith, A.H. DeCristofaro, C.M. Setodji, J. Keesey, D.J. Klein, J.L. Adams, M.A. Schuster, E.A. McGlynn: The quality of ambulatory care delivered to children in the United States, New Engl. J. Med. **15**, 1515–1523 (2007)

77.5 C. Clancy, K. Cronin: Evidence-based decision making: global evidence, local decisions, Health Aff. **24**(1), 151–162 (2005)

77.6 R. Hillestad, J. Bigelow, A. Bower, F. Girosi, R. Meili, R. Scoville, R. Taylor: Can electronic medical record systems transform health care? Potential health benefits, savings, and costs, Health Aff. **24**(5), 1103–1117 (2005)

77.7 Clinical Content: http://www.clinicalcontent.com/about (last accessed May 20, 2009)

77.8 M.A. Shipp, K.N. Ross, P. Tamayo, A.P. Weng, J.L. Kutok, R.C. Aguiar, M. Gaasenbeek, M. Angelo, M. Reich, G.S. Pinkus, T.S. Ray, M.A. Koval, K.W. Last, A. Norton, T.A. Lister, J. Mesirov, D.S. Neuberg, E.S. Lander, J.C. Aster, T.R. Golub: Diffuse large B-cell lymphoma outcome prediction by gene expression profiling and supervised machine learning, Nat. Med. **8**(7), 68 (2002)

77.9 S. Adak, K. Illouz, W. Gorman, R. Tandon, E.A. Zimmerman, R. Guariglia, M.M. Moore, J.A. Kaye: Predicting the rate of cognitive decline in aging and early Alzheimer disease, Neurology **63**(1), 108–14 (2004)

77.10 H.B. Nguyen, J. Banta, T. Cho, C. Van Ginkel, K. Burroughs, W.A. Wittlake, S.W. Corbett: Mortality predictions using current physiologic scoring systems in patients meeting criteria for early goal-directed therapy and the severe sepsis resuscitation bundle, Shock **30**(1), 23–28 (2008)

77.11 C.M. Birkmeyer, D.E. Wennberg: *Patient Safety Standards* (Leapfrog Group, 2003)

78. Medical Automation and Robotics

Alon Wolf, Moshe Shoham

Robotic systems that are integrated in medical applications are designed to help and assist rather than injure a human being, whether it is the patient or the operator. This chapter presents the classification of medical robots as passive, semiactive, active, remote manipulators, and navigators. The kinematic structure of medical robots is discussed next, as are the fundamental requirements from medical robots. Finally, the main advantages and emerging trends in medical robotics are given.

1. A robot may not injure a human being or, through inaction, allow a human being to come to harm.
2. A robot must obey orders given to it by human beings except where such orders would conflict with the First Law.
3. A robot must protect its own existence as long as such protection does not conflict with the First or Second Law.

Written almost 65 years ago, these three laws of robotics by the famous science fiction author Issac Asimov (*Runaround*, 1942) are still very relevant and serve, though not literally, as guidelines for the field of medical automation and robotics. By definition, robotic systems that are integrated in medical applications are designed to help and assist rather than injure a human being, whether it is the patient or the operator. Many precautions (more than one would find in nonmedical robotic systems) are being taken to ensure the safety of patients and operators. As presented in this chapter, these safety measures include, among others, dual backup systems, as a minimum, and fail-safe systems.

These redundant systems are there to prevent unwanted motions that may harm the patient or the staff, to assure accurate performance of the task, and also to protect the robot itself as stated in the third law of robotics.

As elaborated in this chapter, current medical robotic systems are divided into three categories, mainly reflecting level of autonomy. The most popular and widely implemented method is teleoperation, where the robotic system follows the operator's (surgeon's) hand motions from an offsite control console that can be located either in the operating room or even somewhere overseas using fast communication lines. In this mode of operation, the robot, just like in Asimov's second law, obeys and follows the operator's commands and motions. These systems are capable of filtering tremors in the surgeon's hand movements (crucial in neurosurgery and ophthalmic surgery), scaling down the operator's motions and forces, and at the same time preventing unwanted motions that could harm the patient (the active constraint concept).

The first swallows of medical robots appeared in the mid 1980s and early 1990s with the implementation of the Puma 560 for stereotactic neurosurgery. The application was a computed tomography (CT)-guided needle steering system for brain biopsy. Then came ROBODOC from Integrated Surgical Systems in 1992. ROBODOC milled out precise fittings in the femur for hip replacement surgery. ROBODOC was a pioneering surgical active robot that paved the way for other robotic systems, mainly remotely manipulated, semiactive, and active constraint robots. Current surgical applications are in the fields of gastrointestinal surgery, urology, gynecology, cardiothoracic surgery, oncology, orthopedic surgery, and neurosurgery.

Despite more than two decades of research in the field of medical robots, it seems that this field has not yet reached maturity and is still in its infancy. Nevertheless this fact does not prevent researchers from thinking of the next step, i.e., integration of computer-assisted surgery (CAS) devices into one complete system. This is called the operating room of the future (ORF). This initiative is crucial in light of the increasingly growing number of surgical instruments, monitoring and imaging devices, information systems, and communication networks used in modern operating rooms and interventional suites. See also Chap. 82 on *Computer- and Robot-Assisted Medical Interventions.*

78.1 Classification of Medical Robotics Systems

Medical robots have been classified in the literature according to the following categories: remote manipulators, and passive, semiactive, and active robots, by *Cinquin* [78.1], *Stulberg* [78.2], *Taylor* [78.3, 4], *Troccaz* [78.5], *Bainville* [78.6], and *Nolte* [78.7].

Since this area of research is very dynamic and, in our opinion, has not yet reached maturity, it is likely that categories and classifications that are widespread today will change over time, with the evolution of new concepts. Nevertheless, we present in this chapter a brief overview of current leading technologies and trends in the field of medical robotics. For an extensive review of the literature and classification of existing medical robotic system we refer readers to [78.8–10].

Finally, we also present the major considerations to be taken into account in the design of new medical robotic systems.

78.1.1 Passive Medical Robotic Systems

Passive medical robotic systems support the surgical procedure, but take no active part during surgery. In other words, the surgeon is in full control of the surgical procedure at all times, i.e., the actual surgical procedure is conducted by the surgeon. Early versions of Arthrobot [78.11] fall into this category of passive systems. In early stages, the robot was used as an assistant in the operating room to hold the patient's limb during joint replacements of knees and hips. Arthrobot had no sensing capabilities and was able to move only under explicit human control. Today, one of the main forms of passive medical robotics is active constraint robotic systems. Acrobot [78.12] is an example of an active constraint passive robotic system. Developed by

Davies et al. [78.13–15], its core proprietary technology centers on the development of a new type of robotic control: active constraint robotics for orthopedic surgery. This concept facilitates synergy between the surgeon and the robot, provides active assistance to the surgeon, and prevents surgical errors. The surgeon guides the surgical tool that is attached to the robot with a handle with a force sensor attached to the robot tip, and thus uses his/her superior human senses and understanding of the overall situation to perform the surgery. The robot provides precise geometric accuracy and increases the safety by means of a predefined three-dimensional (3-D) motion constraint that prevents cutting outside a predefined safe region. This approach, known as *hands-on robotics*, keeps the surgeon in the control loop throughout the surgery. Moreover, the robot is guided by preoperative image-based planning software. This image-based software uses a patient's CT data to facilitate precise planning of the surgery, allowing implant selection and optimal positioning within the joint.

Another passive robot utilizing the active constraint concept is the MAKO robot, a haptic robotic sys-

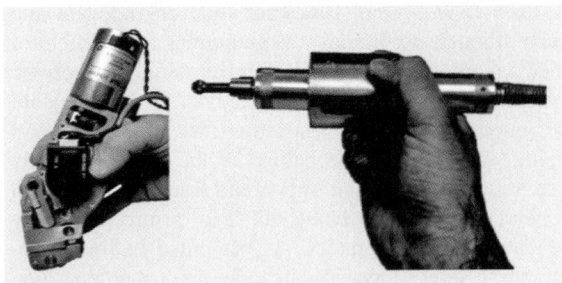

Fig. 78.1 Freehand sculptor by BlueBelt

tem that adds the sense of touch to a robotic-assisted surgical platform. The MAKO Haptic Guidance System is a Food and Drug Administration (FDA)-cleared, surgeon-interactive robotic system that enables the orthopedic surgeon to plan the alignment and placement of knee resurfacing implants preoperatively and make intraoperative bone-conserving cuts accurately within the safe workspace limited by the robot.

The precision freehand sculptor (PFS) by BlueBelt is a handheld tool to assist the surgeon in accurately cutting these shapes (Fig. 78.1). Its rotating cutter only allows the surgeon to remove waste bone; the cutter in turn retracts when it hits *good* bone (i. e., bone that is not supposed to be machined). Thus the surgeon can use the PFS freehand while it automatically restricts the cut to the proper shape [78.16]. This mode of operation is a modification of the active constraint concept. Instead of preventing hand motion, the system automatically and actively prevents cutting.

78.1.2 Semiactive Medical Robotic Systems

The semiactive category is presented in [78.17]. In this research a robot acts as an assistant during the operation by holding a tool in a steady position to allow accurate guidance of surgical tools. Other more up-to-date examples of semiactive systems are the NeuroMate (Integrated Surgical Systems, USA) and PathFinder (Armstrong HealthCare Ltd., UK). They provide guidance of the surgical tool but the actual surgical operation is conducted by the surgeon.

One of the very first applications of robotic systems in a surgical theater was positioning of a needle for stereotactic neurosurgery [78.18]. This application involves placement of a needle in a very accurate man-

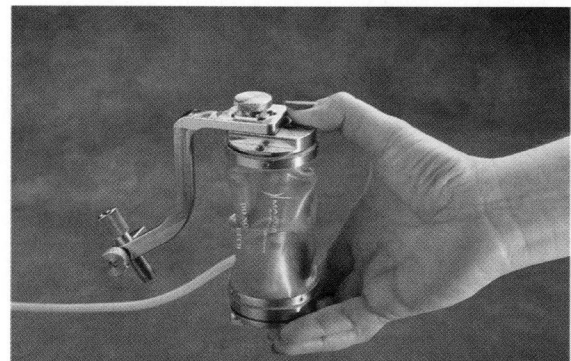

Fig. 78.2 Mazor's SpineAssist miniature robot for spine surgery

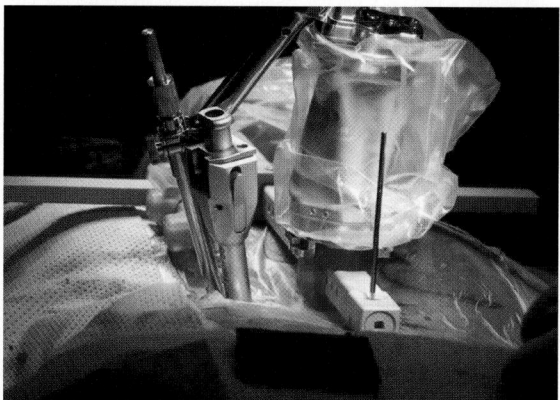

Fig. 78.3 SpineAssist during surgery

ner in a predefined location in a percutaneous approach (later used as a semiactive system). Further development of this approach is described in [78.19].

A different concept of a semiactive robot had been developed at the Robotics Laboratory at the Technion–Israel Institute of Technology and is manufactured and marketed by Mazor Surgical Technologies [78.20]. Mazor's SpineAssist (Figs. 78.2 and 78.3) is a 2×4 inch, 250 g, image-based robot designed to guide the surgeon to precise locations along the patient's spine according to a preoperative plan. Its small size and light weight allows mounting the robot directly on the patient's back, thus overcoming the patient's motion relative to the robot base and as a result improving the accuracy of the operation. The FDA- and Council Europe (CE)-approved SpineAssist robot has performed hundreds of cases and implanted thousands of screws with better clinical outcome than the freehand approach while at the same time allowing a minimally invasive approach.

PiGalileo is another bone-mounted guiding system. Just like in the case of SpineAssist, the actual surgical operations done with PiGalileo are still performed by a surgeon, with the electromechanical positioning device aiding the surgeon in instrument positioning. This technology is completely under the surgeon's control at all times, providing valuable intraoperative feedback to the surgeon to help improve precision, thereby potentially leading to better implant alignment and positioning.

78.1.3 Active Medical Robotic Systems

Active robotic systems perform surgical tasks, such as drilling or milling, autonomously with no direct intervention of the surgeon [78.21–23]. This group includes

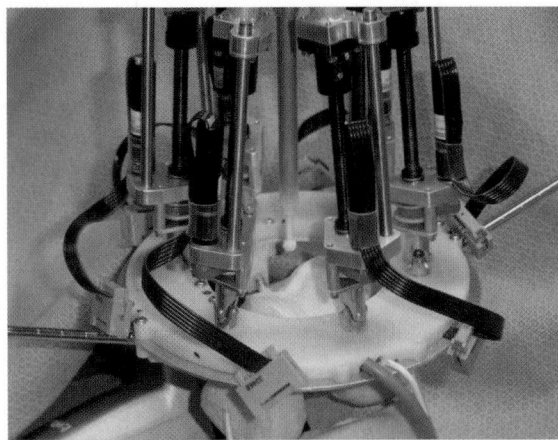

Fig. 78.4 MBARS used for joint arthroplasty

active robots such as RoboDoc, CASPAR, and MBARS. Perhaps one of the most famous active robots today is RoboDoc (Integrated Surgical Systems), developed by *Paul* et al. [78.24, 25]. The system was originally used for total hip replacement procedures, yet was later modified for use in total knee replacement procedures (today it is less used because of controversy over long-term clinical effectiveness).

MBARS (Fig. 78.4) is an active version of the bone-mounted robotic system concept. Although this is still an academic study, the system demonstrates the capability to actively prepare the bone cavity for an implant during joint arthroplasty procedures.

The robot is composed of six linear actuators that are connected in parallel between two rigid platforms: the lower reference platform and the upper platform, which is the moving end-effector of the robot. This structure is known as the classical Stewart–Gough six-degree-of-freedom robot. The robot is attached to the femur by three pins: one pin is inserted into the medial epicondyle, one into the lateral epicondyle, and one into the metadiaphyseal region of the femur. A rigid connection of the robot to the operated bone is obtained through these three pins. The robot is equipped with a milling device, which actively mills the bone according to the preoperative plan.

The CyberKnife robotic radiosurgery by Accuracy System is a radiosurgery system designed to treat tumors anywhere in the body with high accuracy. Using image-guidance technology and computer-controlled robotics, the CyberKnife system is designed to track the tumor continuously, detect its location, and correct for tumor and patient movement in real time throughout the treatment. Because of its extreme precision, the CyberKnife system does not require invasive head or body frames to stabilize patient movement, vastly increasing the system's flexibility.

Gamma Knife PERFEXION is Elekta's new Leksell system for stereotactic radiosurgery in the brain, cervical spine, and head and neck regions. The Leksell Gamma Knife PERFEXION expands treatment reach, offering a wider range of treatable anatomical structure. This expanded anatomical treatment area offers dramatic new opportunities to increase patient volume. Leksell Gamma Knife PERFEXION makes the entire procedure more efficient and user-friendly. Collimator changes can be made by the control program, optimizing the workflow and significantly reducing treatment time.

78.1.4 Remote Manipulators

Perhaps the most commonly used medical robots today are remote manipulators. Remote manipulators are robotic system that can be operated from a remote location. In other words, the surgeon does not have to be at the bedside, nor even in the operating room, and yet can still perform the surgical procedure using the robotic system, which serves as his hands and eyes.

Remote manipulators are robotic systems that can be operated from a remote location. Medical procedures utilizing this mode of operation are often named telesurgery. The first concept was developed at Stanford Research Center as a project for the US Army. The concept of telesurgery was first demonstrated in 2001, when an expert surgeon removed the gallbladder of a 68-year-old patient. This looks like a very common procedure that is done on a daily basis in many locations around the world, with one exception: the patient was operated on in Strasbourg, France and the surgeon operated

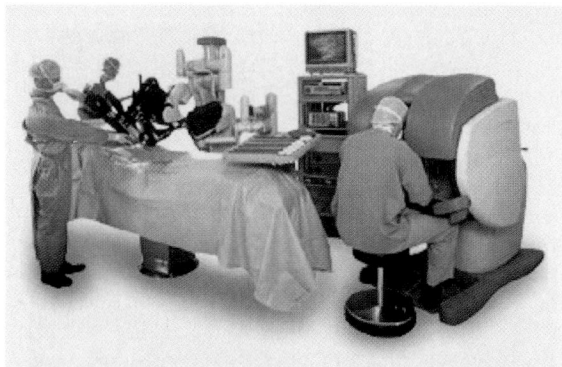

Fig. 78.5 Da Vinci system

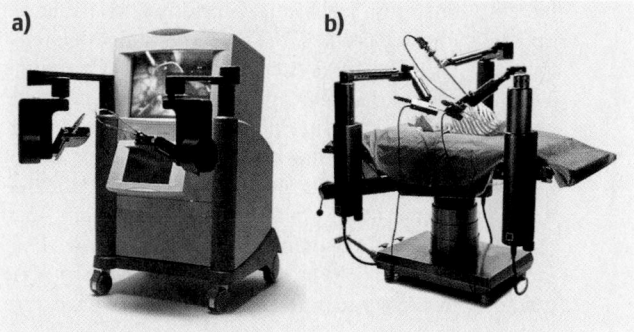

Fig. 78.6a,b Zeus system: (**a**) console, and (**b**) robot arms

from New York. For successful performance of such a telesurgery, a high-speed computer connected through a reliable high-speed network is required. This successful demonstration has opened new horizons of surgical procedures performed worldwide with experts sitting in communication centers. Telesurgery can bring experts to places where there are scarce medical facilities and medical professionals, such as third-world countries and war zones.

Although the first telesurgery was carried out only in 2001, telesurgical systems were introduced earlier. Computer Motion (now Intuitive Surgical Inc., USA) introduced in 1994 the AESOP system. This robotic system was used to manipulate a camera during laparoscopic surgery. Overall, about 70 000 surgeries were performed worldwide using this system. Once acquired by Intuitive, the AESOP system was not sold anymore. Instead, Intuitive Surgical Inc. introduced the Da Vinci medical robot. During surgery, the surgeon operates the robot from a remote console using a specially design control mechanism and a stereo vision system. In April 2005, the Da Vinci system (Fig. 78.5) was approved by the FDA to perform gynecological procedures, although it is used for other procedures as well, such as urologic, general laparoscopic, noncardiovascular, thoracoscopic, and others.

Although Intuitive Surgical is a world leader in the development of robotic technology of minimally invasive surgery (MIS) with its Da Vinci robot, it is worth mentioning another medical robot that competes with the Da Vinci system for the same market. Computer Motion (now owned by Intuitive Surgical) introduced the Zeus system (Fig. 78.6). Just like the Da Vinci system, Zeus is composed of multiple robotic arms capable of manipulating MIS tools and visualization equipment for cardiac surgery.

On 5 April 2001 a 63-year-old male patient underwent multivessel off-pump coronary artery bypass surgery at the University of Pittsburgh Medical Center (UPMC) Presbyterian Hospital using the Zeus system. This was the first time ever that such a procedure was carried out. Marco A. Zenati from the UPMC Department of Surgery operated the robot while seated at a console about 10 feet from the patient. The robot was equipped with an endoscope, which was manipulated by one of the robotic arms using voice commands, while at the same time the other two arms were controlled by operating handles that resemble conventional surgical instruments.

78.1.5 Navigators

Surgical navigators are the central element in many medical robotic system [78.26]. They come into play during the registration procedure. Registration is a crucial and necessary procedure, during which each part of the medical system is synchronized. Most medical robotic systems include preoperative (sometimes intraoperative) planning of the medical procedure. During this step, a virtual surgery can be performed and simulated on a computer screen (just like with any computer-aided design system). The data provided to the surgeon is patient specific and is usually based on a preoperative CT or magnetic resonance imaging (MRI) scan. The end result of this procedure is the transformation that maps the anatomical position in the operating room to its preoperative CT/MRI-based model. One example of a method for achieving this transformation is minimization of the pointwise distance between a cluster of anatomical points collected intraoperatively and points on the preoperative surface model. The coordinates of anatomical points are acquired by touching points on the patient's anatomy (anatomical landmarks) using an optically tracked probe [78.27]. It is sometimes easier to think of these systems as global positioning systems (GPSs) for the operating room. The basic concept of a surgical navigation system is to ascertain the position, i. e., the location and orientation, of the relevant components of the system and the patient's anatomy in a global coordinate system such that their relative position can be determined. During surgery, the position coordinates provide precise guidance to the surgeon to perform a preplanned procedure.

In 1974, *Schlondorff* et al. [78.28] developed one of the first systems for navigational calculations involving bone. The system used a radiographic centimeter scale held between the patient's teeth during a lateral roentgenogram. The measurements enabled the calculation of the distances between several anatomical points. Later, *Watanabe* et al. [78.29] from Siemens Corporation published their work on the first model of the Neuronavigator, a mechanical device based on a multijoint 3-D digitizer. The device is used to track the tip of the sensor arm by indicating its location on preoperative CT or MRI images.

One of the first systems to use an optical-based navigation technique was HipNav (Fig. 78.7), developed at Carnegie Mellon University in Pittsburgh, PA (USA) by *DiGioia* and *Jaramaz* [78.30]. The system was the first computer-assisted navigation system for cup placement in total hip replacement (THR) surgery. Beside its significant clinical contribution, this has set the standard for preoperative planning and range-of-motion (ROM) simulations for THR. HipNav incorporated information provided to it by the OptoTrak system by Northern Digital Inc., Waterloo, Ontario. The OptoTrak system is composed of three charge-coupled device (CCD) cameras contained in a rigid enclosure and a set of active trackers, where the body of each tracker incorporates a set of light-emitting diodes (LEDs) mounted at precise relative positions, and the position of each tracker can be resolved in the OptoTrak coordinate system. For the HipNav system, the trackers were fixed to tools, implants, and the patient's bones, enabling active tracking of their positions during operative procedures. Since HipNav, there have been other systems that employ this type of technology, including the VectorVision, SurgiGATE, Navitrack, StealthStation, Stryker, and Surgetic systems.

A new version of medical navigators is the image overlay concept (Fig. 78.8). Image overlay is a com-

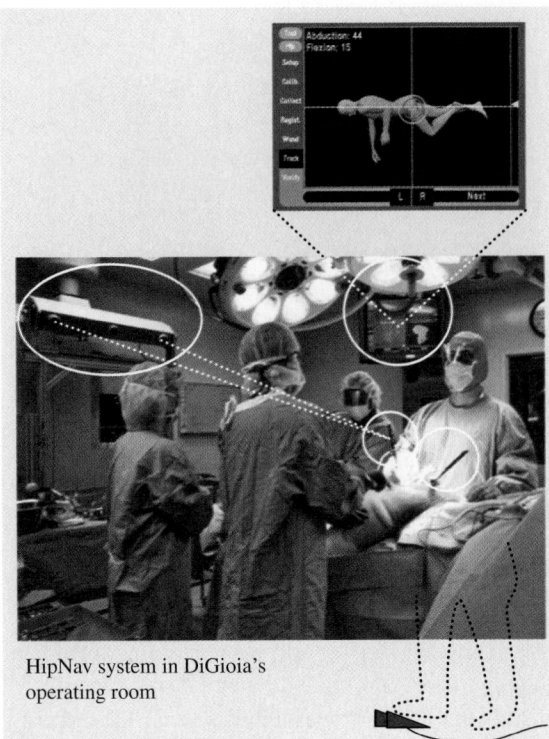

HipNav system in DiGioia's operating room

Fig. 78.7 HipNav system

puter display technique that superimposes computer images over the viewer's direct view of the real world [78.31]. In other words, it is a form of *augmented reality* in that it merges computer-generated information with real-world images by the projection of virtual objects in real scene. To do this the system needs to project the virtual image onto some sort of screen. The most common measure used is a semitransparent display device (like a heads-up display in fighter planes), which allows both viewing the real objects while at the same time overlaying the virtual image on them. For example, a 3-D image of a bone, reconstructed from CT data, can be displayed to a surgeon inside the patient's anatomy at the exact location of the real bone, regardless of the position of either the surgeon or the patient (Fig. 78.8), creating an elusion of an image which appears to the viewer to be inside the real objects. To do this in a *convincing* yet accurate way, the positions of the viewer's head, objects in the environment, and components of the display system are all tracked in space. These positions are used to transform the images so that they appear to be an integral part of the real-world environment.

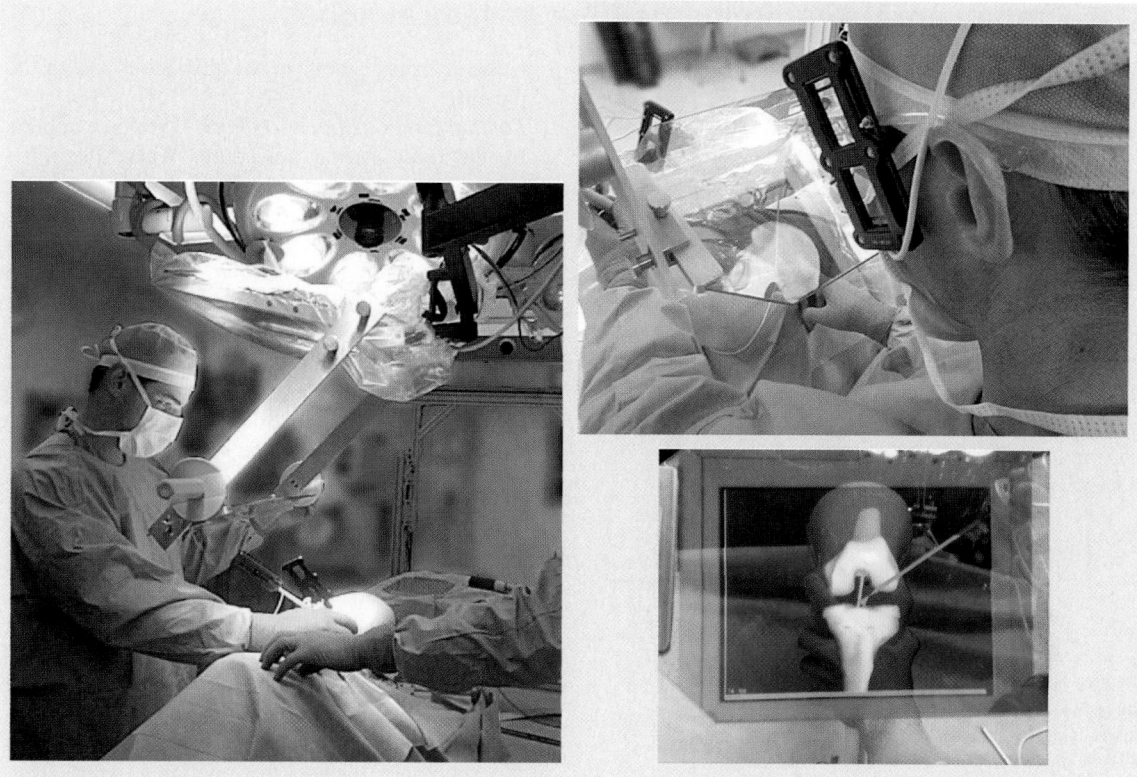

Fig. 78.8 Image overlay

This technology has many clinical applications; one of the most common is needle steering, i. e., manipulat-

ing a real/virtual needle into remote unexposed organs for biopsy, drug delivery, and interventions.

78.2 Kinematic Structure of Medical Robots

In most up-to-date medical robotic systems, a serial robot is used as a surgical assistant. These robots suffer from numerous drawbacks related to the serial manipulator, such as low rigidity and accuracy. The fact that theses robots are integrated in medical procedures, where accuracy and safety is a matter of life and death, has led researchers to look for better manipulators suited for a specific surgical task or field of tasks. A family of robots found suitable for medical application is parallel robotic mechanisms. *Grace* and *Brandt* [78.32, 33] stressed the advantages of parallel manipulators compared with serial manipulators in surgical operations, mainly due to their low weight, compact structure, better accuracy, stiffness, restricted workspace, and low price. However, parallel manipulators have some drawbacks. One of the

main disadvantages of parallel robots is their limited workspace. Nevertheless, limited workspace can be considered as an advantage in medical applications where the required workspace is itself very limited. This attribute limits the potential placement positions of the robot in the operating room (OR), since the parallel robot has to be positioned very close to the operating area in order to be able to perform a surgical task within its limited workspace. In most cases, this requirement is not feasible due to physical conditions in the OR. One of the tested solutions has been designed with the entire robotics system attached to the OR ceiling so that the robot works *upside down*, as proposed for example in Fig. 78.5. In this way, the robot does not interfere with the surgical procedure, and is activated and maneuvered to the operating area when required.

78.3 Fundamental Requirements from a Medical Robot

The fundamental requirements from a medical robot for surgery tasks were introduced in [78.35], where the authors stressed the safety requirements (Fig. 78.9):

1. *Effective control.* Both force and speed control of the end-effector should be allowed or limited by hardware in all robot configurations.
2. *Workspace.* The robot workspace has to be limited to the effective area of the operation, to prevent ac-

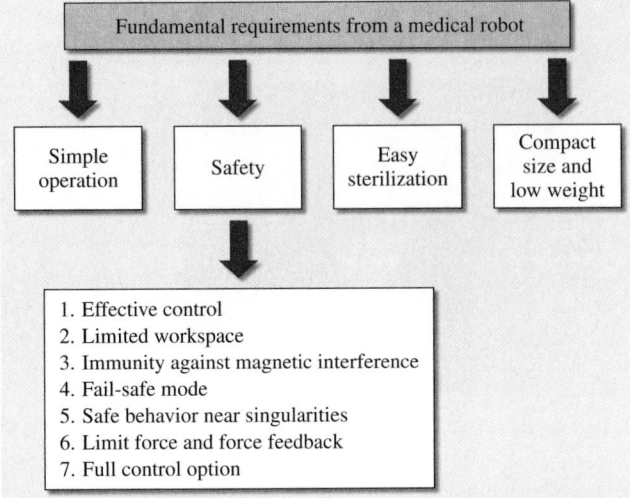

Fig. 78.9 The fundamental requirements from a medical robot [78.34]

cidental damage both to the physician and to the patient.

3. *Limited forces or force feedback.* The force applied by the robot during procedures where the robot takes active tasks and is in contact with live tissue must be fully controlled. Moreover, in procedures where different levels of force are applied (such as bone cutting [78.36]), the physician needs as much data as possible from the robot concerning the force applied.
4. *Full control option.* In procedures where the robot automatically performs a surgical procedure, the physician has to be able to take over control of the robot at any stage during the operation.
5. *Fail-safe features.* In the event of a malfunction in the robot, the system has to switch to a fail-safe mode; for instance, in case of a power failure, the robot has to remain in its current location until power is restored.
6. *Singularity behavior.* The robot path planning must avoid passing near singular configurations, or actively prevent the surgeon from driving the robot through singular configurations, if any, or designed in such a way that all singular poses of the robot are outside the operating work envelope.
7. *Sterilizability.* The robot structure must allow sterilization, or be protected with a suitable cover.
8. Immunity against magnetic interference of other surgical tools available in the OR.

78.4 Main Advantages of Medical Robotic Systems

Several researchers have investigated the advantages of robotic systems in surgery. *Kazanzides* et al. [78.37] compared the cross-section of manually broached implant cavities to cavities milled by robots in hip replacement surgery. The results of this research have illustrated the robot's precision compared with the precision of the human hand. *Kavoussi* et al. [78.38] compared the performance of a human to that of a robot assistant in manipulating the laparoscope. The results of this research emphasize the superiority of the robot compared with the human hand in terms of steadiness. *Cameron* and *Pradeep* [78.39] measured the accuracy of the human hand of four skilled eye surgeons. The results stressed the tremor and inaccuracy of the human

hand, even for skilled surgeons. An average root-mean-square (RMS) error of 49 μm was obtained when asked to hold an instrument steady, and 133 μm for repeated actuation. *Cameron* et al. [78.40] presented a robotic prototype that incorporates sensing and actuation, reporting accuracy of 5 μm and better for a robotic system holding an ophthalmologic microsurgical instrument, representing a significant improvement with respect to those presented in the first article [78.39]. These results indicate that the combination of advanced medicine with high-technology research capabilities can improve many surgical procedures that are currently performed manually and are limited by restricted accessibility and the lack of preciseness of the human hand. Areas which have been identified as appropriate at this stage include orthopedic surgery, cutting and joining of bones at exact optimal location, laparoscopic surgical procedures, and accurate maneuvering of laparoscopic camera and tools in a constrained environment.

In light of this, the advantages of using a medical robotic system can be divided into two categories: those related to the benefits to the surgeon and those related to the benefits to the patient when using robotic technology. Medical robotics enable the surgeon to perform new percutaneous procedures that could not have been executed before, improve the surgeon's ergonomics, enhance the surgical result (MIS), shorten the surgical procedure, reduce radiation exposure, provide better image visualization, and enable extension of the operation theater to remote locations using telesurgery techniques. For the patient, there are few advantages due to the use of medical robotic technology during the procedure, mainly since the surgical procedure is performed minimally invasively. This results in faster recovery time, less pain, less postoperative complications, better cosmetic results, and cost reduction.

78.5 Emerging Trends in Medical Robotics Systems

Robots complement the trend in surgical procedures toward minimal invasiveness. Robots serve as remote hands for the surgeon and can deliver his/her motions to the patient through miniature incisions. In order to be able to mimic the surgeon's hand adequately and even achieve superiority in terms of accuracy and miniature motions, it is important to integrate force feedback capabilities into the system as well as tactile feedback. However today even the best systems still suffer from inaccurate forces and insufficient tactile feedback, which are essential in several medical procedures.

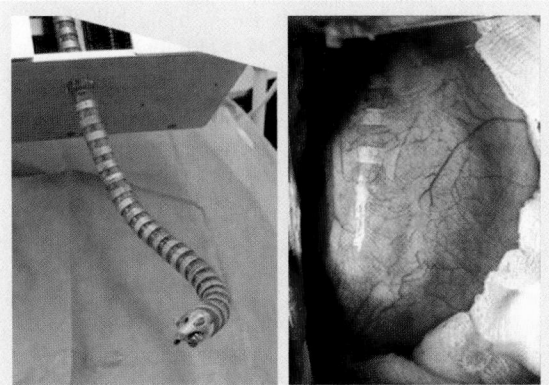

Fig. 78.10 CardioArm – a hyper redundant snake like robot for cardiac surgery

Hence, this is an area for future research and development.

Making medical robots more friendly and intuitive is essential for the introduction of robots into operating rooms. One of the lessons that robot developers can learn from the two decades of application of surgical robots is that it is of utmost importance to make robots friendlier to the operating-room environment. Moreover, even though the technology is here, almost no new surgical procedures have been developed based on robotic technology that cannot be performed otherwise. It is therefore expected that, in order to penetrate the operating-room environment, new surgical technologies have to be developed based on robotic technology, and even more importantly, robots must become much more friendly to this unique atmosphere.

An even more intriguing approach to medical robots is miniature robots that can move inside the human body (Fig. 78.10). Several laboratories around the world are pursuing this research with very few (like the passive Given Imaging's pillCam) in clinical use.

The area of medical robots is still in its infancy, with applications with great potential not yet explored. The trends are surgical robots with high accuracy, robotically operated minimally invasive approaches, miniature robots that move inside the human body, and rehabilitation robotics.

References

78.1   P. Cinquin, E. Bainville: Computer assisted medical intervention: passive and semi-active aids, IEEE Eng. Med. Biol. Mag. **14**, 254–263 (1995)

78.2   S.D. Stulberg, F. Picard, D. Saragaglia: Computer assisted total knee arthroplasty, Oper. Tech. Orthop. **10**(1), 25–39 (2000)

78.3   R.H. Taylor: An image directed robotic system for precise orthopaedic surgery. In: *Computer Integrated Surgery, Technology and Clinical Applications*, ed. by W. Bargar, R. Taylor, S. Lavallée, B. Mosges (MIT Press, Cambridge 1995) pp. 379–391

78.4   R.H. Taylor: Robotics in orthopedic surgery. In: *Computer Assisted Orthopaedic Surgery (CAOS)*, ed. by L. Nolte, R. Ganz (Hogrefe Huber, Seattle 1999) pp. 35–41

78.5   J. Troccaz: Man–machine interfaces in computer augmented surgery. In: *Computer Assisted Orthopaedic Surgery (CAOS)*, ed. by L. Nolte, R. Ganz (Hogrefe Huber, Seattle 1999) pp. 53–68

78.6   E. Bainville, I. Bricault, P. Cinquin, S. Lavallée: Concepts and methods of registration for computer integrated surgery. In: *Computer Assisted Orthopaedic Surgery (CAOS)*, ed. by L. Nolte, R. Ganz (Hogrefe Huber, Seattle 1998) pp. 15–34

78.7   R. Hofstetter, M. Slomczykowski, M. Sati, L.P. Nolte: Fluoroscopy as an imaging means for computer-assisted surgical navigation, Comput. Aided Surg. **4**, 65–76 (1999)

78.8   R.H. Taylor, D. Stoianovici: Medical robotics in computer-integrated surgery, IEEE Trans. Robot. Autom. **19**(5), 765–781 (2003)

78.9   P. Darion, B. Hannaford, A. Menciassi: Smart surgical tools and augmenting devices, IEEE Trans. Robot. Autom. **19**(5), 782–792 (2003)

78.10 A. Wolf, A.M.D. III, B. Jaramaz: Computer guided total knee arthroplasty. In: *MIS Techniques in Orthopedics*, ed. by G.R. Scuderi, A.J. Tria, R.A. Berger (Springer, Berlin, Heidelberg 2005) pp. 390–407

78.11 C.M. Ewen, C.R. Bussani, G.F. Auchinleck, M.J. Breault: Development and initial clinical evaluation of pre robotic and robotic retraction systems for surgery, Proc. 2nd Annu. Int. Symp. Cust. Orthop. Prosthet., Chicago (1989)

78.12 B.L. Davies, J.P. Cobb, M.P.S.F. Gomes: Acrobot® System for Robotic MIS Arthroplasty, by The Acrobot Company Limited, Imperial College, London (2008)

78.13 B.L. Harris, S. Jakopec, M. Fan, K.L. Cobb, B.L. Davies: Special-purpose robotic system for knee surgery, CAOS 99. 4th Int. Conf. Comput.-Assist. Orthop. Surg., Davos (1999)

78.14 B.L. Davies, S.J. Harris, M. Jakopec, J. Cobb: Computer-assisted surgery for TKR IMechE, Proc. Int. Conf. Knee Replace. Surg. (PEP Institution of Mechanical Engineers, 1999) pp. 1974–2024

78.15 B.L. Davies, S. Harris, M. Jakopec, J. Cobb: A novel hands-on robot for knee replacement surgery, Proc. CAOS '99 Conf. Comput. Assist. Orthop. Surg. (UPMC, Pittsburgh 1999)

78.16 G. Brisson, T. Kanade, I.A.M. Di Gioia, B. Jaramaz: Precision freehand sculpting of bone, Proc. 7th Int. Conf. Med. Image Comput. Comput.-Assist. Interv. (MICCAI 2004) (Springer 2004)

78.17 M.T. Kienzle, D. Stulberg, M. Peshkin, A. Quaid, J. Lea, A. Goswami, C.A. Wu: Computer-assisted totalknee replacement surgical system using a calibrated robot. In: *Computer Integrated Surgery*, ed. by R. Taylor, S. Lavallée, G. Burdea, R. Moesges (MIT Press, Cambridge 1996)

78.18 Y.S. Kwoh, J. Hou, A. Jonckheere, S. Hayati: A robot with improved absolute positioning accuracy for CT-guided stereotactic brain surgery, IEEE Trans. Biomed. Eng. **55**, 153–161 (1988)

78.19 P. Cinquin, J. Troccaz, J. Demongeot, S. Lavallée, G. Champleboux, L. Brunie, F. Leitner, P. Sautot, B. Mazier, A. Perez, M. Djaid, T. Fortin, M. Chenic, A. Chapel: IGOR: Image guided operating robot, Innov. Technol. Biol. Med. **13**, 374–394 (1992)

78.20 M. Shoham, I.H. Lieberman, E.C. Benzel, D. Togawa, E. Zehavi, B. Zilberstein, M. Roffman, A. Bruskin, A. Fridlander, L. Joskowicz, S. Brink-Danan, N. Knoller: Robotic assisted spinal surgery – from concept to clinical practice, Comput. Aided Surg. **12**(2), 105–115 (2007)

78.21 W. Bargar, A. Bauer, M. Borner: Primary and revision total hip replacement using the Robodoc system, Clin. Orthop. **354**, 82–91 (1998)

78.22 R. Kober, D. Meister: Total knee replacement using the Caspar-system. Computer assisted total knee arthroplasty, Int. Symp. CAOS I, Davos (2000)

78.23 A. Wolf, B. Jaramaz, B. Lisien, A.M. DiGioia: Mini bone attached active-robotic system for joint arthroplasty, Int. J. Med. Robot. Comput. Assist. Surg. (MRCAS) **1**(2), 101–121 (2005)

78.24 B. Mittelstadt, H. Paul, P. Kazandides, J. Zuhar, B. Williamson, R. Pettitt, P. Cain, D. Kloth, L. Rose, B. Musits: Development of a surgical robot for cementless total hip replacement, Robotics **11**, 553–560 (1993)

78.25 H. Paul, B. Mittlestadt, W.L. Bargar, B. Musits, R.H. Taylor, P. Kazanzides, J. Zuhars, B. Williamson, W. Hanson: A surgical robot for total hip replacement surgery, Proc. IEEE Int. Conf. Robot. Autom. ICRA (1992)

78.26 L.P. Nolte, F. Langlotz: Basics of computer assisted orthopedic surgery. In: *Navigation and Robotics in Total Joint and Spine Surgery*, ed. by J.B. Stiehl, W.H. Konermann, R.G. Haaker, A.M. DiGioia (Springer, Berlin, Heidelberg 2004)

78.27 D.A. Simon: *Fast and Accurate Shape-Based Registration* (Carnegie Mellon University, Pittsburgh 1995)

78.28 G. Schlondorff: Computer assisted surgery: historical remarks, Comput. Aided Surg. **3**, 150–152 (1998)

78.29 E. Watanabe, S. Manaka, Y. Mayanagi, K. Takakura: Three dimensional digitizer (Neuronavigator): New equipment for computerized tomography-guided stereotactic surgery, Surg. Neurol. **27**, 543–547 (1987)

78.30 A.M. DiGioiaB. Jaramaz, C. Nilou, R. Labarca, J. E. Moody, and B. ColgAN III., B. Jaramaz, C. Nilou, R. Labarca, J.E. Moody, B. Colgan: Surgical navigation for total hip replacement with the use of Hipnav, Operative Techniques in Orthopaedics, Med. Robot. Comput. Assist. Orthop. Surg. **10**(1), 3–8 (2000)

78.31 M. Blackwell, C. Nikou, A.M. DiGioia, T. Kanade: An image overlay system for medical data visualization, MICCAI '98, ed. by W.M. Wells (Springer, Berlin, Heidelberg 1998) pp. 232–240

78.32 K.W. Grace, J.E. Colgate, M.R. Gluksberg, J.H. Chun: A six degree of freedom micromanipulator for ophthalmic surgery, IEEE Int. Conf. Robot. Autom. ICRA (1993)

78.33 G. Brandt, K. Radermacher, S. Lavallée, H.W. Staudte, G.A. Rau: Compact robot for image guided orthopedic surgery: concept and preliminary results, Lect. Notes Comput. Sci. **1205**, 767–776 (1997)

78.34 N. Simaan, M. Shoham: Robot construction for surgical applications, Second Israeli Symp. Comput.-

78.35 K. Khodabandehloo, P.N. Brett, R.O. Buckingham: Special-purpose actuators and architectures for surgery robots. In: *Computer Integrated Surgery: Technology and Clinical Applications*, ed. by R.H. Taylor, S. Lavallée (MIT Press, London 1996) pp. 263–276

78.36 S.J. Harris, W.J. Lin, K.L. Fan, R.D. Hibberd, J. Cobb, B.L. Davies: Experiences with robotic systems for knee surgery, Lect. Notes Comput. Sci. **1205**, 757–766 (1997)

78.37 P. Kazanzides, B.D. Mittelstadt, B.L. Musits, W.L. Bargar, J.F. Zuhars, B. Williamson, P.W. Cain, E.J. Carbone: An integrated system for cementless hip replacement, Eng. Med. Biol. Mag. **14**(3), 307–312 (1995)

78.38 L.R. Kavoussi, R.G. Moore, J.B. Afdams, A.W. Partin: Comparison of robotic versus human laparoscopic camera control, 2nd Annu. Int. Symp. Med. Robot. Comput.-Assist. Surg. (MRCAS '95) (1995)

78.39 N.R. Cameron, K.K. Pradeep: Active hand-held instrument for error compensation in microsurgery, Intelligent Systems and Manufacturing. Technical Conference on Micro Robotics and Micro Systems Fabrication, Pittsburgh (1997)

78.40 N.R. Cameron, R.S. Rader, K.P. Khosla: Characteristic of hand motion of eye surgeons, 19th Annu. Conf. IEEE Eng. Med. Biol. Soc., Chicago (1997)

Aided Surg. Med. Robot. Med. Imaging (ISRACAS '99) (Technion: Hebrew University of Jerusalem, Jerusalem 1999)

Part H | 78

79. Rotary Heart Assist Devices

Marwan A. Simaan

The left ventricular assist device (LVAD) is a mechanical device implanted in patients with congestive heart failure to assist the heart in pumping blood through the circulatory system. The latest generation of this device is comprised of a rotary pump which is generally much smaller, lighter, and quieter than the first-generation conventional pulsatile-type pump. The rotary pump is controlled by varying the rotor (or impeller) speed to adjust the amount of blood flow through the LVAD. If the patient is in a health care facility, the pump speed can be adjusted manually by a trained clinician to meet the patient's blood needs. However, an important challenge facing the increased use of these devices is the desire to allow the patient to return home. The development of an appropriate feedback controller that is capable of automatically adjusting the pump speed is therefore a crucial step in meeting this challenge. In addition to being able to adapt to changes in the patient's daily activities by automatically regulating the pump speed, the controller must also be able to prevent the occurrence of excessive pumping. This dangerous phenomenon, known as suction, may cause collapse of the ventricle and damage to the heart muscle. In order to be able to develop such a controller based on modern control theory an appropriate mathematical model of a combined cardiovascular system and LVAD must first be developed. In this chapter, we develop

such a model. The model is dynamic, time-varying, and consists of six coupled nonlinear differential equations. The time variation occurs over four consecutive intervals representing the contraction, ejection, relaxation, and filling phases of the left ventricle. The LVAD in the model along with its inlet and outlet cannulae are represented by a nonlinear differential equation which relates the pump rotational speed and pump flow to the pressure difference across the pump. Suction is accounted for by adding a nonlinear resistance in the LVAD model when the pressure in the left ventricle drops below a specified threshold. Using this model we discuss some of the challenges faced in the development of: (1) an appropriate feedback controller for the LVAD, and (2) an effective algorithm for detection of suction in the left ventricle.

Part H | 79

Heart transplantation has now been recognized as the best therapy for patients with end-stage congestive heart failure. However, potential recipients often wait long periods of time (300 days or more on the average) before a suitable donor heart becomes available, and many of these candidates, 20–30%, will die while awaiting heart transplantation. Consequently, the medical community has placed increased emphasis on the use of mechanical circulatory assist devices that can substitute for, or enhance, the function of the natural heart while the patient is waiting for the heart transplant [79.1, 2]. A left ventricular assist device (LVAD) is such a device. There are two different types of LVADs. The first is a pulsatile pump producing pulsatile pressures and flows similar to those produced by the natural heart. The second, and more recent, is a rotary pump that operates continuously attempting to draw blood out of the left ventricle and into the circulation. Rotary pumps are typically, quieter, smaller, and more efficient than pulsatile pumps, and consequently have received considerable acceptance in recent years. Generally speaking, the goal of the LVAD is to assist the native heart in pumping blood so as to provide the patient with as close to a normal lifestyle as possible until a donor heart becomes available or, in some cases, until the patient's heart recovers. In many situations, this means allowing the patient to return home and/or to the workforce.

An important engineering challenge facing the increased use of the rotary LVAD is the development of an appropriate controller for the speed of the rotor (or pump impeller). Such a controller, in addition to being robust and reliable, must be able to adapt to the daily activities and physiological changes of the patient by regulating the pump speed in order to meet the body's requirements for cardiac output (CO) and mean arterial pressure (MAP) [79.3, 4]. Since the rotary pump does not use valves, if its speed is too low, blood may regurgitate back from the aorta to the left ventricle through the pump, resulting in what is known as *backflow*. If the pump speed is too high, the pump will attempt to draw more blood from the ventricle than available, which may cause collapse of the ventricle resulting in a dangerous phenomenon called *suction*. The occurrence of suction must be detected quickly and the pump speed reduced before the heart muscle is damaged. While avoiding these two extremes, the pump speed must also be adjusted continuously, up and down to meet the patient's varying levels of physical activity and emotional changes [79.5–10]. The eventual goal of a pump speed controller is therefore to meet all these requirements so that an LVAD recipient patient could potentially leave the hospital and return home to a normal lifestyle. Given that the pump is continuously interacting with the cardiovascular system, the

development of a speed controller that meets the above objectives must therefore be done using tools developed in modern control theory. This cannot be done without first having an appropriate mathematical model for this complex system. The model must be simple enough to be tractable and yet it must be comprehensive enough to capture the essential relationships between the hemodynamic variables and provide the important input and output boundary conditions without the ambiguity of unnecessary state variables. In this chapter, we discuss the development of such a model. We first present and validate an autonomous fifth-order model of the cardiovascular system which emphasizes the pressure–volume relationship of the left ventricle. We then present a first-order model of a rotary LVAD along with its inlet and outlet cannulae and with the pump rotational speed as its control variable. We account for the phenomenon of suction by adding a nonlinear resistance to the inlet cannula model which becomes active when the left ventricular pressure drops below a certain threshold. Finally, we combine these two models into a sixth-order time-varying nonlinear model with the rotational speed of the pump being the only control variable. The time variations in this model are due to the cyclical nature of the ventricle elastance and its changes as a function of time within one cardiac cycle. The nonlinearities are due to the mitral and aortic valves in the cardiovascular model and the suction resistance in the LVAD model. The binary state of each of these valves yield four consecutive phases within one cardiac cycle during which the ventricle contracts, ejects, relaxes, and then fills depending on which valve is open and which is closed. As a result, the mathematical description of the entire model changes as the ventricle crosses from one phase to the next. We will conclude this chapter by outlining some of the challenges in the development of both a feedback speed controller and a suction detection algorithm that need to be overcome before the ultimate goal of the LVAD becomes a reality and LVAD patients are able to return to a normal lifestyle.

79.1 The Cardiovascular Model

A cross-section schematic of the heart illustrating its various components is shown in Fig. 79.1. Each side of the heart has its own atrium and ventricle, and each side controls a different blood circulation path. The left side of the heart controls the systemic circulation of blood throughout the body while the right side controls the

pulmonary circulation of blood to the lungs [79.11]. After the blood picks up oxygen in the lungs it returns to the heart entering through the left atrium. The left ventricle pumps the oxygen-rich blood through the aorta to the rest of the body. The blood returns to the heart depleted of oxygen, this time through the right atrium. It

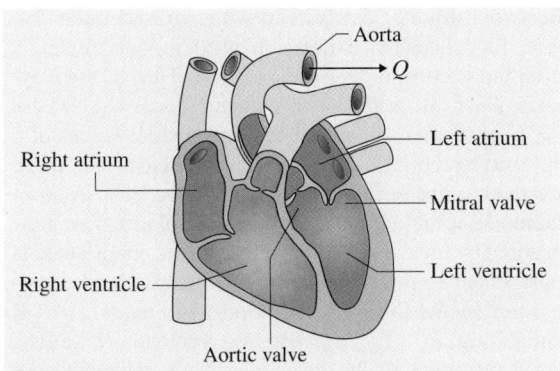

Fig. 79.1 Cross-section of a beating heart

is then pumped to the lungs by the right ventricle. The blood receives oxygen in the lungs, and returns to the left atrium of the heart. Upon entering the left atrium, the two circulatory paths are then repeated. The blood path through the heart and both circulation systems can be seen in Fig. 79.2 [79.12]. Due to the different nature of the two circulation paths, each side of the heart has a different workload. The systemic circulation is much longer than the pulmonary one, and encounters considerable more resistance [79.11]. To handle the longer path, the left ventricle is more elastic than the right, giving it the strength to pump blood at a higher pressure.

The heart is a very complex system that is very difficult to model mathematically. Numerous dynamical state-space models of varying degrees of complexity have been developed in the past fifteen years or so. For example, an eighth-order hybrid model of the heart which subdivides the human circulatory system into a number of lumped parameter blocks that include the pulmonary arterial and venous systems as well as both ventricles has been developed in [79.13]. An 11th-order model which includes the resistance of peripheral blood vessels has been developed in [79.14]. Although a complete heart model can be supplemented with a model of an LVAD, in this chapter we assume that the right ventricle and pulmonary circulation are healthy and normal and as a result their effect on the LVAD, which is connected from the left ventricle to the ascending aorta, can be neglected [79.1, 6]. A fifth-order lumped parameter electric circuit model which can reproduce the left ventricle hemodynamics of the heart [79.15, 16] is shown in Fig. 79.3. In this model, the behavior of the left ventricle is modeled by means of a time-varying capacitance (or compliance) $C(t)$, which is the reciprocal of the ventricle's elastance function $E(t)$. The elastance represents the contractual state of the left ventricle. It

relates to the ventricle's pressure and volume [79.17] according to an expression of the form

$$E(t) = \frac{\mathrm{LVP}(t)}{\mathrm{LVV}(t) - V_0} , \qquad (79.1)$$

where $\mathrm{LVP}(t)$ is the left ventricular pressure, $\mathrm{LVV}(t)$ is the left ventricular volume, and V_0 is a reference volume, which corresponds to the theoretical volume in the ventricle at zero pressure. Several mathematical expressions have been derived to approximate the elastance function $E_H(t)$ of a healthy heart. In our work, we use the expression

$$E_H(t) = (E_{\max} - E_{\min})E_n(t_n) + E_{\min} , \qquad (79.2)$$

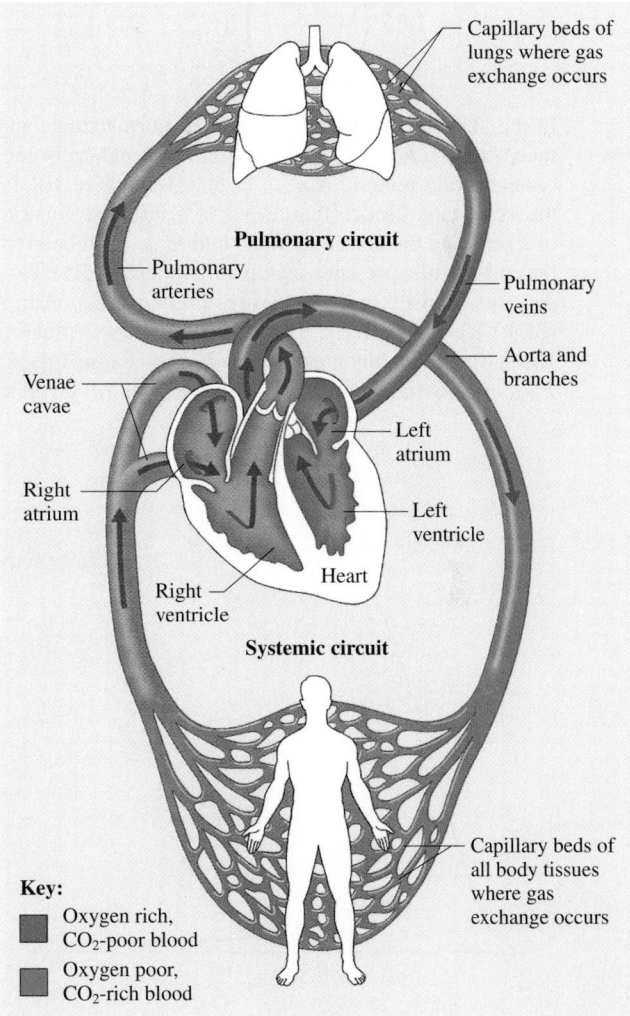

Key:

■ Oxygen rich, CO_2-poor blood

■ Oxygen poor, CO_2-rich blood

Fig. 79.2 Diagram of systemic and pulmonary circulatory systems (after [79.12])

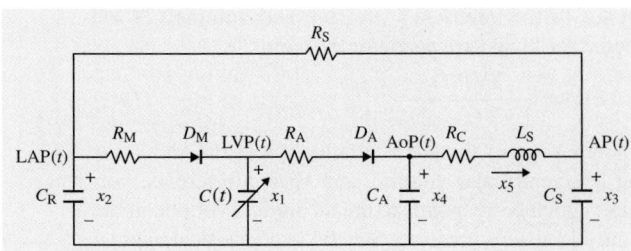

Fig. 79.3 Cardiovascular circuit model

where $E_n(t_n)$ is the so-called *double hill* function represented by the expression [79.18]

$$E_n(t_n) = 1.55 \left(\frac{\left(\frac{t_n}{0.7} \right)^{1.9}}{1 + \left(\frac{t_n}{0.7} \right)^{1.9}} \right) \left(\frac{1}{1 + \left(\frac{t_n}{1.17} \right)^{21.9}} \right) .$$

(79.3)

In the above expression, $E_n(t_n)$ is the normalized elastance, $t_n = t/T_{max}$, $T_{max} = 0.2 + 0.15t_c$, and t_c is the cardiac cycle interval, i.e., $t_c = 60/HR$, where HR is the heart rate. Notice that $E_H(t)$ is a rescaled version of $E_n(t_n)$ and the constants E_{max} and E_{min} are related to the end-systolic pressure–volume relationship (ESPVR) and the end-diastolic pressure–volume relationship (EDPVR), respectively. Figure 79.4 shows a plot of $E_H(t)$ for a healthy heart with $E_{max} = 2\,\mathrm{mmHg/ml}$, $E_{min} = 0.06\,\mathrm{mmHg/ml}$, and a heart rate of 60 bpm

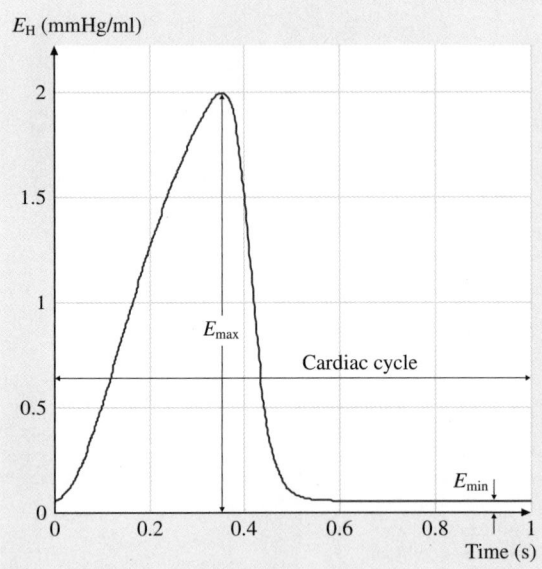

Fig. 79.4 Elastance function $E_H(t) = \frac{1}{C(t)}$ of a healthy heart (cardiac cycle = 60/HR)

(beats per minute). For a heart with cardiovascular disease, the elastance expression used in our model is modified according to $E(t) = \delta E_H(t)$. That is, the elastance $E_H(t)$ is scaled with a factor $0 < \delta \leq 1$, where $\delta = 1$ represents a healthy heart and smaller values of δ are used to represent cardiovascular disease. The more severe the disease, the smaller the value of δ. Also in the model, preload and pulmonary circulations are represented by the capacitance C_R; the aortic compliance is represented by the capacitance C_A, and afterload is represented by the four-element Windkessel model [79.19] comprising R_C, L_S, C_S, and R_S. Preload, or venous blood returning to the heart, is closely related to the cardiac output (CO) produced by the heart for a given level of contractility. Afterload can be described as the pressure that the left ventricle has to generate in order to eject blood. It refers to the vascular resistance that the heart sees as it pumps blood. The mitral and aortic valves (Fig. 79.1) are represented by two non-ideal diodes consisting of a resistance R_M and ideal diode D_M for the mitral valve, and resistance R_A and ideal diode D_A for the aortic valve. In this representation, we have kept the number of model parameters to a minimum while maintaining enough complexity in the model so that it can reproduce the hemodynamics of the left ventricle. Table 79.1 lists the various system parameters and their typical associated values [79.20–22].

Obviously the circuit model in Fig. 79.3 is time-varying because of the capacitance $C(t)$ and nonlinear because of the two diodes. Since each ideal diode has two states: open-circuit (O/C) and short-circuit (S/C),

Table 79.1 Model parameters

Parameters	Value	Physiological meaning
Resistances (mmHg s/ml)		
R_S	1.0000	Systemic vascular resistance (SVR)
R_M	0.0050	Mitral valve resistance
R_A	0.0010	Aortic valve resistance
R_C	0.0398	Characteristic resistance
Compliances (ml/mmHg)		
$C(t)$	Time-varying	Left ventricular compliance
C_R	4.4000	Left atrial compliance
C_S	1.3300	Systemic compliance
C_A	0.0800	Aortic compliance
Inertances (mmHg s²/ml)		
L_S	0.0005	Inertance of blood in aorta
Valves		
D_M		Mitral valve
D_A		Aortic valve

Table 79.2 Phases of the cardiac cycle

Modes	Valves		Phases
	Mitral	**Aortic**	
1	Closed	Closed	Isovolumic relaxation
2	Open	Closed	Filling
1	Closed	Closed	Isovolumic contraction
3	Closed	Open	Ejection
–	Open	Open	Not feasible

Table 79.3 State variables in the cardiovascular model

Variables	Name	Physiological meaning (unit)
$x_1(t)$	LVP(t)	Left ventricular pressure (mmHg)
$x_2(t)$	LAP(t)	Left atrial pressure (mmHg)
$x_3(t)$	AP(t)	Arterial pressure (mmHg)
$x_4(t)$	AoP(t)	Aortic pressure (mmHg)
$x_5(t)$	$Q_T(t)$	Total flow (ml/s)

four different circuits representing different phases of the ventricular function can therefore be modeled. When D_M is S/C and D_A is O/C (i.e., the mitral valve is open and aortic valve is closed) the circuit in Fig. 79.3 represents the phase when the left ventricle is filling. When D_M is O/C and D_A is S/C (i.e., the mitral valve is closed and the aortic valve is open) the circuit represents the phase when the ventricle is ejecting. The circuit in which both diodes are O/C (i.e., both the mitral and aor-

tic valves are closed) will occur twice and represents the phase when the ventricle is undergoing isovolumic contraction and relaxation. Finally, the circuit in which both diodes are S/C (i.e., both the mitral and aortic valves are open) is clearly not feasible. The three different phases of operation of the left ventricle over four consecutive different time intervals within the cardiac cycle are summarized in Table 79.2. Every phase within the cardiac cycle is therefore modeled by a different equivalent circuit and hence a different set of linear time-varying differential equations. However, by appropriately modeling the diodes as nonlinear elements, it is possible to write only one set of differential equations, which describes the behavior of the entire model for all three phases. Selecting the state variables as the hemodynamic variables listed in Table 79.3 (and shown on the circuit in Fig. 79.2), and using basic circuit analysis methods [79.23] such as Kirchhoff's voltage and current laws (KVL and KCL), we can derive the state equations for the cardiovascular circuit model shown in Fig. 79.2 as

$$
\begin{pmatrix} \frac{dx_1}{dt} \\ \frac{dx_2}{dt} \\ \frac{dx_3}{dt} \\ \frac{dx_4}{dt} \\ \frac{dx_5}{dt} \end{pmatrix} = \begin{pmatrix} \frac{-\dot{C}(t)}{C(t)} & 0 & 0 & 0 & 0 \\ 0 & \frac{-1}{R_S C_R} & \frac{1}{R_S C_R} & 0 & 0 \\ 0 & \frac{1}{R_S C_S} & \frac{-1}{R_S C_S} & 0 & \frac{1}{C_S} \\ 0 & 0 & 0 & 0 & \frac{-1}{C_A} \\ 0 & 0 & \frac{-1}{L_S} & \frac{1}{L_S} & \frac{-R_C}{L_S} \end{pmatrix} \begin{pmatrix} x_1 \\ x_2 \\ x_3 \\ x_4 \\ x_5 \end{pmatrix}
$$

$$
+ \begin{pmatrix} \frac{1}{C(t)} & \frac{-1}{C(t)} \\ \frac{-1}{C_R} & 0 \\ 0 & 0 \\ 0 & \frac{1}{C_A} \\ 0 & 0 \end{pmatrix} \begin{pmatrix} \frac{1}{R_M} r(x_2 - x_1) \\ \frac{1}{R_A} r(x_1 - x_4) \end{pmatrix} .
$$

(79.4)

In (79.4) $\dot{C}(t) = dC(t)/dt$ and $r(x)$ represents the ramp function

$$
r(\xi) = \begin{cases} \xi & \text{if} \quad \xi \geq 0, \\ 0 & \text{if} \quad \xi < 0. \end{cases}
$$

(79.5)

We note that the model described above is an autonomous switched linear time-varying system over four different phases within the cardiac cycle [79.15, 16]. Its solution is oscillatory due to the cyclic nature of the terms $\dot{C}(t)$ and $1/C(t)$ in the matrices in (79.4). A plot of the expression $-\dot{C}(t)/C(t)$ over one cardiac cycle for a healthy heart is shown in Fig. 79.5. The term $1/C(t)$ represents the elastance function $E_H(t)$ whose plot is shown in Fig. 79.4.

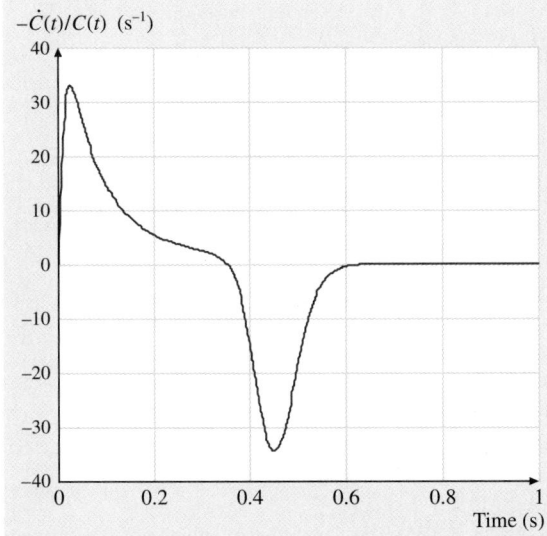

$-\dot{C}(t)/C(t)$ (s^{-1})

Fig. 79.5 Plot of $\frac{-\dot{C}(t)}{C(t)}$ for a healthy heart over one cardiac cycle

Part H | 79.1

79.2 Cardiovascular Model Validation

The ability of the fifth-order model described in the previous section to emulate the hemodynamics of the left ventricle is demonstrated by simulating the model for both nominal steady-state conditions, and in response to perturbations of preload and afterload. Figure 79.6 shows the simulation waveforms of the hemodynamics for an adult with heart rate of 75 bpm. In this particular case, systolic and diastolic pressure were 117 and 77 mmHg, mean aortic pressure (MAP) was 99 mmHg, cardiac output (CO) (i.e., the blood pumped by the ventricle in 1 min) was 5.21 l/min and stroke volume (SV) (i.e., the volume of blood pumped by the ventricle in one cardiac cycle) was 69.5 ml/beat. These numbers and waveforms are all consistent with hemodynamic data in normal subjects described in [79.24].

A second method to validate the model and verify that it can reproduce the human left ventricle behavior is to vary the preload and afterload conditions, while keeping the left ventricle parameters (E_{max}, E_{min}, and V_0) constant. If the model behaves as expected, we should have an approximately linear relationship between end-systolic pressure and left ventricle volume

(known as end-systolic pressure–volume relationship, or ESPVR), in spite of changes in preload and afterload. A total of four preload and four afterload conditions were simulated using $E_{max} = 2$ mmHg/ml, $E_{min} = 0.05$ mmHg/ml, and $V_0 = 10$ ml. The resulting pressure and volume of the ventricle are typically plotted in the form of pressure–volume (PV-)loops. The PV-loops in Fig. 79.7a, represent the result of changing afterload conditions by selecting different values of systemic vascular resistance (R_S in Fig. 79.3), while keeping end-diastolic volume (EDV) constant.

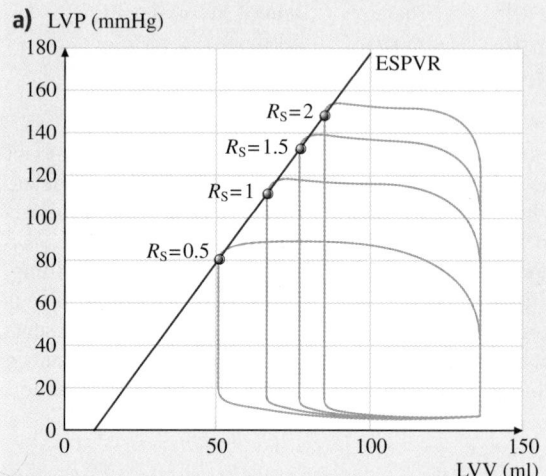

a) LVP (mmHg)

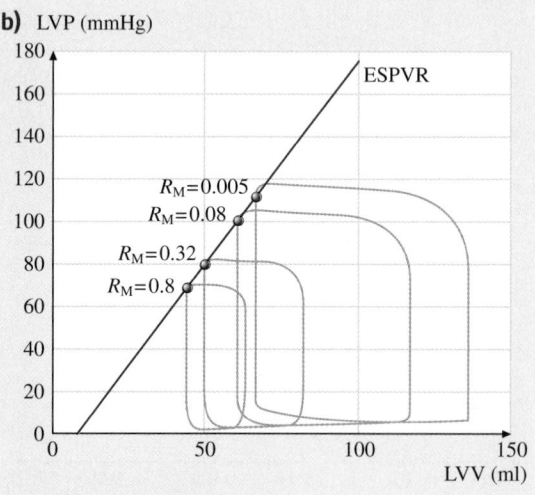

b) LVP (mmHg)

Fig. 79.7a,b PV-loops for different values of (**a**) afterload resistance, and (**b**) preload volume

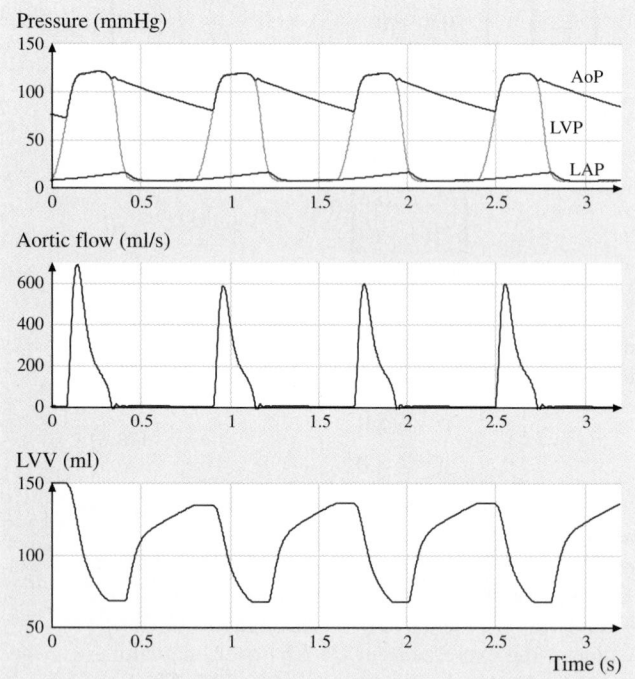

Fig. 79.6 Simulated hemodynamic waveforms for a normal heart

Fig. 79.8a,b Measured and simulated (**a**) left ventricular pressure and (**b**) PV-loops ►

The PV-loops in Fig. 79.7b represent the result of altering preload conditions by changing the Mitral valve resistance R_M. The linear relationship between pressure and volume is evident in both Fig. 79.7a,b. The slope of the ESPVR and the horizontal intercept for the afterload data in Fig. 79.7a are determined as 1.98 mmHg/ml and 10.91 ml, respectively. For the preload data of Fig. 79.7b these values are determined as 1.92 mmHg/ml and 8.84 ml respectively. These results are consistent with the values used for E_{max} and V_0 used in the simulation and clearly show that the cardiovascular model can indeed mimic the behavior of the left ventricle.

Finally, a third method to validate the model is to compare the hemodynamic waveforms obtained from the model to those of a human patient. Figure 79.8 shows left ventricular pressure data and the corresponding PV-loop obtained from our model using values $E_{max} = 1.5$ mmHg/ml and $V_0 = 12$ ml estimated from real data of a patient suffering from cardiomyopathy. The measured data from the patient are also superimposed on the same plots in Fig. 79.8. Clearly the data from the model exhibit a close fit to the real measured patient data. As a test for how close the two data sets match, in the case of LVP, the error was calculated to be 4.4% and in the case of the PV-loop, the area within the loop, known as the stroke work (SW), was determined to be 10 492 mmHg · ml for the PV-loop of the model and 10 690 mmHg · ml for the patient PV-loop, which represents a difference of only 1.85%.

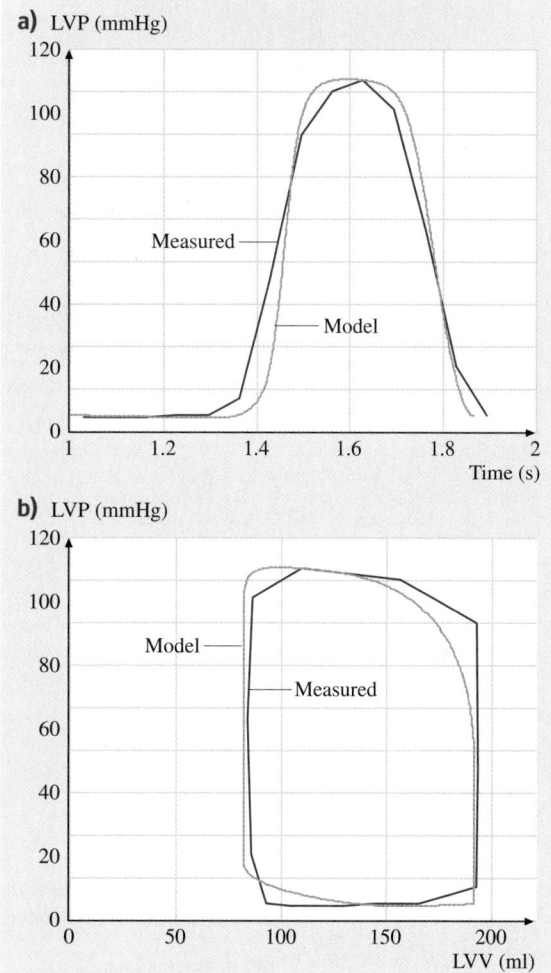

79.3 LVAD Pump Model

The LVAD considered in this chapter is a rotary blood pump connected using two cannulae as a bridge between the left ventricle and the aorta as illustrated in the schematic in Fig. 79.9. The pump (HeartMate II) and its cannulae are characterized by the relationships [79.25]

$$H_p = R_p Q + L_p \frac{dQ}{dt} + \beta \omega^2 , \qquad (79.6)$$

$$H_i = R_i Q + L_i \frac{dQ}{dt} , \qquad (79.7)$$

$$H_o = R_o Q + L_o \frac{dQ}{dt} , \qquad (79.8)$$

where H_p, H_i, and H_o are the pressure differences across the pump and the inlet and outlet cannulae, respectively, Q is the flow rate of blood through the pump and cannulae, and ω is the pump rotational speed. The parameters R_p, R_i, and R_o represent the resistances and the parameters L_p, L_i, and L_o represent the inertances of the pump and cannulae, respectively. The parameter β is a pump-dependent constant. We note that R_i repre-

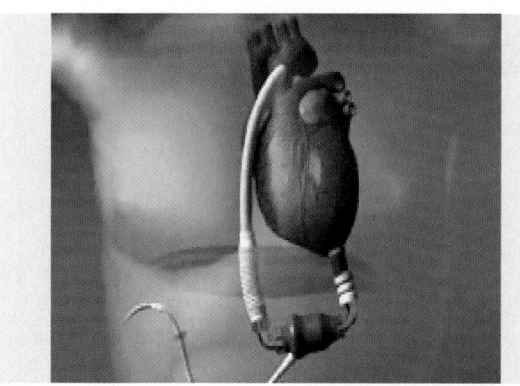

Fig. 79.9 Schematic of a rotary LVAD connected between the left ventricle and the aorta

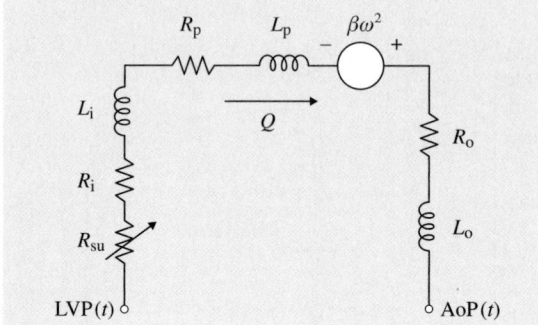

Fig. 79.10 LVAD and cannulae equivalent circuit

sents the resistance of the inlet cannula when no suction is present. When suction occurs, an additional resistance R_{su} in the form [79.5, 26]

$$R_{su} = \begin{cases} 0 & \text{if} \quad \text{LVP}(t) > \bar{x}_1 \\ \alpha(\text{LVP}(t) - \bar{x}_1) & \text{if} \quad \text{LVP}(t) \leq \bar{x}_1 \end{cases} \quad (79.9)$$

is added to the expression in (79.7) to represent this phenomenon. Clearly, R_{su} is a nonlinear time-varying

Table 79.4 Model parameters for the LVAD

Parameters	Value	Physiological meaning
Cannulae resistances (mmHg s/ml)		
R_{su}	See (79.9)	Suction resistance with parameters $\alpha = -3.5\,\text{s/ml}$ and $\bar{x}_1 = 1\,\text{mmHg}$
R_i	0.0677	Inlet cannula resistance
R_p	0.17070	Outlet cannula resistance
R_o	0.0677	Pump resistance
Cannulae inertances (mmHg s^2/ml)		
L_i	0.0127	Inlet cannula inertance
L_p	0.02177	Pump inertance
L_o	0.0127	Outlet cannula inertance
Pump parameter (mmHg/(rpm)^2)		
β	9.9025×10^{-7}	

resistance whose value is zero when the pump is operating normally and is activated when the left ventricular pressure $\text{LVP}(t)$ (or $x_1(t)$) becomes less than a predetermined small threshold $\bar{x}_1$, a condition that represents suction. The value of R_{su} when suction occurs increases linearly as a function of the difference between $\text{LVP}(t)$ and the threshold $\bar{x}_1$. The parameter α is a cannula-dependent scaling factor. An equivalent circuit model for the LVAD and inlet and outlet cannulae is shown in Fig. 79.10. Values of all parameters mentioned above for the specific LVAD used in our model are given in Table 79.4. The state equation governing the behavior of the LVAD can now be easily derived as

$$\text{LVP}(t) - \text{AoP}(t) = R^* Q + L^* \frac{\text{d}Q}{\text{d}t} + \beta\omega^2 , \quad (79.10)$$

where

$$R^* = R_{su} + R_i + R_p + R_o \quad (79.11)$$

and

$$L^* = L_i + L_p + L_o . \quad (79.12)$$

79.4 Combined Cardiovascular and LVAD Model

Figure 79.11 shows the equivalent electric circuit of the combined cardiovascular and LVAD model. The addition of the LVAD to the model represented in (79.4) adds one state variable $x_6(t) = Q$, which represents the blood flow through the pump and eight passive parameters R_{su}, R_i, R_p, R_o, L_i, L_p, L_o, and β. Note that the combined cardiovascular and LVAD model is now a forced system, where the primary control variable is the pump speed ω. As was done for the cardiovascular model, the state equations for this combined sixth-order model can be derived from basic circuit analysis techniques and written in the form

$$
\begin{pmatrix}
\frac{dx_1(t)}{dt} \\
\frac{dx_2(t)}{dt} \\
\frac{dx_3(t)}{dt} \\
\frac{dx_4(t)}{dt} \\
\frac{dx_5(t)}{dt} \\
\frac{dx_6(t)}{dt}
\end{pmatrix}
=
\begin{pmatrix}
\frac{-\dot{C}(t)}{C(t)} & 0 & 0 & 0 & 0 & \frac{-1}{C(t)} \\
0 & \frac{-1}{R_SC_R} & \frac{1}{R_SC_R} & 0 & 0 & 0 \\
0 & \frac{1}{R_SC_S} & \frac{-1}{R_SC_S} & 0 & \frac{1}{C_S} & 0 \\
0 & 0 & 0 & 0 & \frac{-1}{C_A} & \frac{1}{C_A} \\
0 & 0 & \frac{-1}{L_S} & \frac{1}{L_S} & \frac{-R_C}{C_S} & 0 \\
\frac{1}{L^*} & 0 & 0 & \frac{-1}{L^*} & 0 & \frac{-R^*}{L^*}
\end{pmatrix}
$$

$$
\times
\begin{pmatrix}
x_1(t) \\
x_2(t) \\
x_3(t) \\
x_4(t) \\
x_5(t) \\
x_6(t)
\end{pmatrix}
+
\begin{pmatrix}
\frac{1}{C(t)} & \frac{-1}{C(t)} \\
\frac{-1}{C_R} & 0 \\
0 & 0 \\
0 & \frac{1}{C_A} \\
0 & 0 \\
0 & 0
\end{pmatrix}
$$

$$
\times
\begin{pmatrix}
\frac{1}{R_M} r(x_2 - x_1) \\
\frac{1}{R_A} r(x_1 - x_4)
\end{pmatrix}
+
\begin{pmatrix}
0 \\
0 \\
0 \\
0 \\
0 \\
\frac{\beta}{L^*}
\end{pmatrix}
u(t) .
$$

$$(79.13)$$

The control variable in (79.13) is $u(t) = \omega^2(t)$, where $\omega(t)$ is the rotational speed of the pump. Using control terminology, and after some simple algebraic manipulations, this model can be expressed as a switched time-varying linear system in the standard state-space form [79.15, 16]

$$
\frac{dx}{dt} = A_k(t)x + bu , \quad k = I, F, E , \tag{79.14}
$$

where $x = [x_1, x_2, x_3, x_4, x_5, x_6]^\top$ is the state vector, $b = (0, 0, 0, 0, 0, \beta/L^*)^\top$, and $(\ )^\top$ denotes transpose. The matrix $A_k(t)$ in (79.14) switches over each of the four different phases within the cardiac cycle (Table 79.2). The description of this matrix over each of these regions is given as follows.

1. *Isovolumic relaxation and contraction phases* ($k = $ I): Within these two intervals we have $x_2 - x_1 < 0$

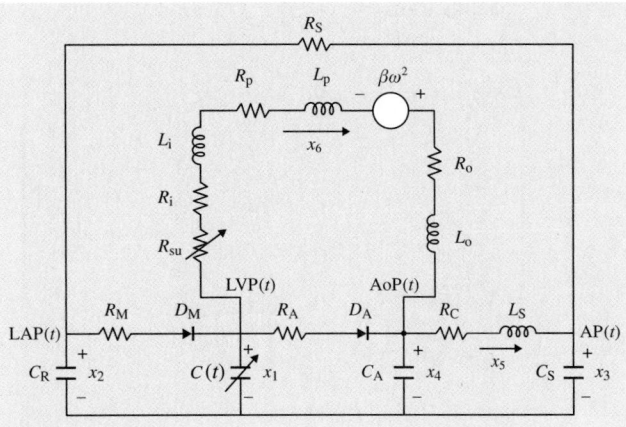

Fig. 79.11 Combined cardiovascular and LVAD model

and $x_1 - x_4 < 0$. The matrix $A_k(t)$ will take the form

$$
A_I(t) =
\begin{pmatrix}
\frac{-\dot{C}(t)}{C(t)} & 0 & 0 & 0 & 0 & \frac{-1}{C(t)} \\
0 & \frac{-1}{R_SC_R} & \frac{1}{R_SC_R} & 0 & 0 & 0 \\
0 & \frac{1}{R_SC_S} & \frac{-1}{R_SC_S} & 0 & \frac{1}{C_S} & 0 \\
0 & 0 & 0 & 0 & \frac{-1}{C_A} & \frac{1}{C_A} \\
0 & 0 & \frac{-1}{L_S} & \frac{1}{L_S} & \frac{-R_C}{L_S} & 0 \\
\frac{1}{L^*} & 0 & 0 & \frac{-1}{L^*} & 0 & \frac{-R^*}{L^*}
\end{pmatrix} .
$$

$$(79.15)$$

2. *Filling phase* ($k = $ F): Within this interval we have $x_2 - x_1 \geq 0$ and $x_1 - x_4 < 0$. The matrix $A_k(t)$ will take the form

$$
A_F(t) =
$$

$$
\begin{pmatrix}
\frac{-\dot{C}(t)}{C(t)} - \frac{1}{R_M C(t)} & \frac{1}{R_M C(t)} & 0 & 0 & 0 & \frac{-1}{C(t)} \\
\frac{1}{R_M C_R} & \frac{-1}{C_R}\left(\frac{1}{R_M} + \frac{1}{R_S}\right) & \frac{1}{R_S C_R} & 0 & 0 & 0 \\
0 & \frac{1}{R_S C_S} & \frac{-1}{R_S C_S} & 0 & \frac{1}{C_S} & 0 \\
0 & 0 & 0 & 0 & \frac{-1}{C_A} & \frac{1}{C_A} \\
0 & 0 & \frac{-1}{L_S} & \frac{1}{L_S} & \frac{-R_C}{L_S} & 0 \\
\frac{1}{L^*} & 0 & 0 & \frac{-1}{L^*} & 0 & \frac{-R^*}{L^*}
\end{pmatrix} .
$$

$$(79.16)$$

3. *Ejection phase* ($k = $ E): Within this interval we have $x_2 - x_1 < 0$ and $x_1 - x_4 \geq 0$. The matrix $A_k(t)$ will

take the form

$$\mathbf{A}_E(t) =$$

$$\begin{pmatrix} \frac{-\dot{C}(t)}{C(t)} - \frac{1}{R_A C(t)} & 0 & 0 & \frac{1}{R_A C(t)} & 0 & \frac{-1}{C(t)} \\ 0 & \frac{-1}{R_S C_R} & \frac{1}{R_S C_R} & 0 & 0 & 0 \\ 0 & \frac{1}{R_S C_S} & \frac{-1}{R_S C_S} & 0 & \frac{1}{C_S} & 0 \\ \frac{1}{R_A C_A} & 0 & 0 & \frac{-1}{R_A C_A} & \frac{-1}{C_A} & \frac{1}{C_A} \\ 0 & 0 & \frac{-1}{L_S} & \frac{1}{L_S} & \frac{-R_C}{L_S} & 0 \\ \frac{1}{L^*} & 0 & 0 & \frac{1}{L^*} & 0 & \frac{-R^*}{L^*} \end{pmatrix}.$$

$$(79.17)$$

Switched linear systems have been studied extensively in recent years, especially with regards to stability and stabilization [79.27, 28]. The development of an optimal feedback controller for the above system will present a major control challenge, especially in guaranteeing that the resulting closed-loop autonomous system will also be stable.

79.5 Challenges in the Development of a Feedback Controller and Suction Detection Algorithm

In assisting the heart, the amount of blood flow required of an LVAD depends on the patient's level of activity, emotional state, posture, and other physiological demands. The response of the heart itself due to changes in demand is governed by its contractility, and the preload and afterload. Contractility of the heart is related to the intrinsic ability of the heart muscle to contract. As mentioned earlier, preload is closely related to the cardiac output (CO) produced by the heart for a given level of contractility and afterload can be described as the pressure that the left ventricle has to generate in order to eject blood. Everything else held equal, if afterload increases, cardiac output will decrease. In a healthy heart, these internal control mechanisms provide the cardiovascular system with a significant capability to respond to changes in demand for blood flow in the body.

When a rotary LVAD is implanted in a patient with congestive heart failure, the only available mechanism to control it is the rotational speed ω of the pump. Currently, patients with rotary LVADs are typically kept in a critical care setting and are continuously supervised by human operators. When the patient is in a stable condition, the operator adjusts the pump speed manually so as to achieve the desired levels of the hemodynamic variables and ensure the well being of the patient. However, for a patient whose level of activity is continuously changing, this manual open-loop control becomes impractical. Its limitations will become even more apparent as patients are rehabilitated and seek to adopt a normal active lifestyle. The inability of open-loop manual control to respond automatically to changes in demand can dramatically impact the quality of life of LVAD patients [79.3, 4]. For successful long-term im-

plantation of these devices, the need for a human to monitor the device operation must therefore be eliminated. In order to achieve a desired level of automatic operation, a feedback controller which is capable of continuously monitoring the patient's status to recognize a change in the level of activity and automatically adjust the pump speed must be incorporated in the device. As mentioned earlier, such a controller must also meet the other objective of ensuring that suction does not occur by unnecessary excessive pumping. The

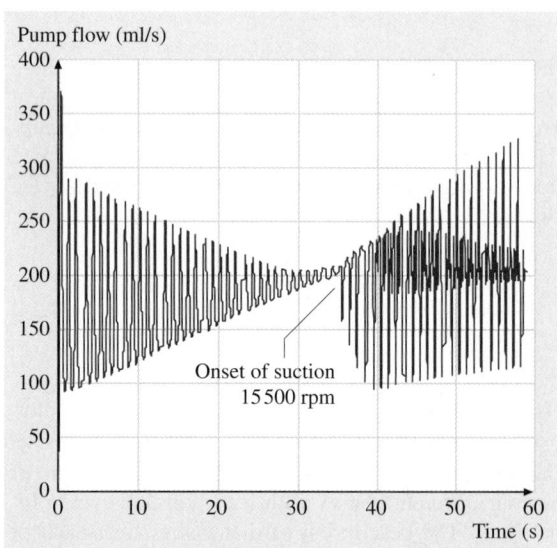

Fig. 79.12 Pump flow signal from human cardiovascular/LVAD model as a function of linearly increasing pump speed

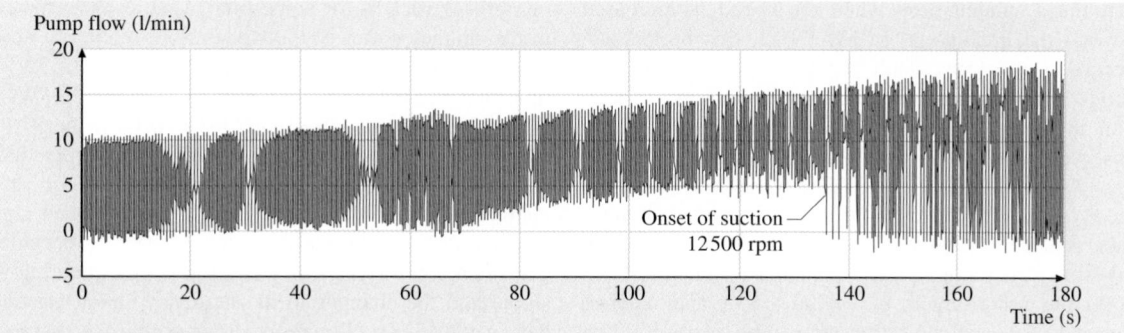

Fig. 79.13 Pump flow signal from in vivo data as a function of linearly increasing pump speed

achievement of these two objectives has been a major challenge for LVAD developers for over 15 years and is recognized as one of the most serious limitations of this technology at the present time.

The design of a stable feedback controller based on the model developed earlier in this chapter is very much related to our ability to continuously measure, or estimate, in real time the patient's hemodynamic variables (x_1 through x_5 in the model). The patient's continuously varying levels of physical and emotional activities are exhibited by possibly wide variations in the systemic vascular resistance (R_s in the model), which in turn affects all six state variables in the system. Unfortunately, at present, current technology for implantable sensors that allow for continuous measurements of the patient's hemodynamic variables does not exist and will probably need many years before it can be developed. The pump flow variable x_6, on the other hand, appears to be the only variable that can be measured in real time. This may be done, for example, by using standard ultrasonic flow transducers that can be clamped onto one of the pump cannulae. With this single variable, the problem essentially reduces to a feedback controller design based on incomplete state measurements. Given that the system is nonlinear and time varying, this remains a challenge from both the theoretical and practical considerations. An attempt to develop a very simple feedback controller based on the slope of the lower envelope of the pump flow signal has recently been made [79.29] but its practical usefulness in an in vivo experiment has not yet been tested. The approach of estimating the hemodynamic variables (x_1 through x_5) so as to implement a full state feedback controller has also been considered but with limited success [79.30–34]. This is largely due to the fact that this approach involves first estimating the patient's systemic vascular parameters. Most methods developed for estimating these

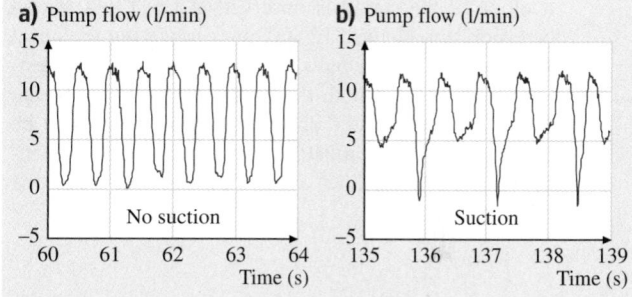

Fig. 79.14a,b In vivo pump flow signal (**a**) when pump is operating normally, and (**b**) when pump is in suction

parameters suffer from the problem that the estimates obtained are discrete and valid only over parts of the cardiac cycle. A method for simultaneously estimating the parameters and the hemodynamic variables has been developed in [79.21, 22] based on an extended Kalman filter approach; however, the robustness of the method with respect to parameter initialization remains questionable and its usefulness is limited due to uncertainties related to convergence of the algorithm to the correct estimates. In summary, the design of a feedback controller for the LVAD remains a very challenging open problem.

Since the pump flow signal appears to be the only signal that is directly measurable, it is important to examine how this signal is affected by the pump speed as it changes over its allowable safe range. Figure 79.12 shows a plot of this signal (state variable x_6) when our model is exited with a pump speed starting at 12 000 rpm and increasing linearly according to $\omega = 12\,000 + 100t$, reaching a speed of 18 000 rpm after 60 s. It is clear from this figure that the lower envelope of this signal seems to track the increasing pump speed up to a point when a breakdown occurs and then exhibits a sudden drop when the speed is increased beyond the breakpoint. In Fig. 79.12, this breakdown occurs at $t = 35\,$s, which corresponds to a speed of about 15 500 rpm and is indicative of the onset of suction in the model. The same phenomenon has been observed in an in vivo animal study (authorized according to WorldHeart, Inc. IRB DO 01-06002) in which the WorldHeart (World Heart Inc.: Formerly MedQuest, Inc., Salt Lake City, UT) rotary LVAD was used. In this study the pump speed was increased linearly from 8000 rpm according to $\omega = 8000 + \frac{100}{3}t$ and reached a speed of 14 000 rpm after 180 s. The corresponding measured pump flow signal is shown in Fig. 79.13. Again, in this case, a similar breakdown in the lower envelope of the signal is observed at $t = 136\,$s, which corresponds to a speed 12 500 rpm, indicating the onset of suction in the animal's left ventricle. It is interesting to note that, in both the model simulation and the in vivo experiment, the onset of suction appears to be characterized by a sudden large drop in the lower envelope of the pump flow signal. Furthermore, when the

pump is in suction for some time, a noticeable change in the characteristics of the signature of the pump flow signal is observed. This is clearly seen in Fig. 79.14 in which the details of the in vivo data of Fig. 79.13 are shown over two 5 s intervals of time from 60 to 64 s before the occurrence of suction and from 135 to 139 s during suction. The change in the frequency characteristics of the signal from a narrow-band signal to a relatively broad-band signal is very noticeable. The breakpoint in the lower envelope of the pump flow signal and the changes in its frequency characteristics before and after suction represent opportunities that can be exploited in the development of suction detection algorithms. We should note that suction detection has been a long-standing problem that has been studied ever since the rotary LVAD was introduced. Many algorithms have been proposed based on numerous other criteria and indices [79.35–39]. A successful attempt to develop a suction detection algorithm based on the frequency characteristics of the pump flow signal has recently been reported [79.40, 41].

79.6 Conclusion

In this chapter, a sixth-order state-space model of a left ventricular assist device connected to a cardiovascular system is presented. The model is obtained by combining a fifth-order model of the hemodynamics of the cardiovascular system with a first-order model of the blood flow in a rotary pump along with its inlet and outlet cannulae. The phenomenon of suction is accounted for by including a nonlinear resistance in the model of the inlet cannula. The resulting combined model is nonlinear and time-varying with a different mathematical description over the four different time intervals which correspond to the four phases of the ventricular functions within one cardiac cycle. The only control variable

in the model is the rotational speed of the pump. Given that current implantable sensor technology does not exist for measuring the patient's hemodynamic variables, the challenges in using this model to design a feedback speed controller for the LVAD are discussed. The characteristics of the pump flow signal – which is the only directly measurable variable – when the pump is operating normally with no suction and when it is operating in suction are also described based on data obtained from the model as well as from an in vivo animal experiment. Possible approaches for exploiting these characteristics in the development of suction detection algorithms are also discussed.

References

79.1   V.L. Poirier: The LVAD: a case study, Bridge **27**, 14–20 (1997)
79.2   H. Frazier, T.J. Myers: Left ventricular assist systems as a bridge to myocardial recovery, Ann. Thorac. Surg. **68**, 734–741 (1999)
79.3   D.B. Olsen: The history of continuous–flow blood pumps, Artif. Organs **24**(6), 401–404 (2000)
79.4   J.R. Boston, J.F. Antaki, M. Simaan: Hierarchical control for hearts assist devices, IEEE Robot. Autom. Mag. **10**(1), 54–64 (2003)

79.5   H. Shima, W. Trubel, A. Moritz, G. Wieselthaler, H.G. Stohr, H. Thomas, U. Losert, E. Wolner: Noninvasive monitoring of rotary blood pumps: necessity, possibilities, and limitations, Artif. Organs **14**(2), 195–202 (1992)
79.6   H. Konishi, J.F. Antaki, D.V. Amin: Controller for an axial flow blood pump, Artif. Organs **20**(6), 618–620 (1996)
79.7   Y. Wu, P. Allaire, G. Tao, H. Wood, D. Olsen, C. Tribble: An advanced physiological controller design

for a left ventricular assist device to prevent left ventricular collapse, Artif. Organs **27**(10), 926–930 (2003)

79.8 G.A. Giridharan, G.M. Pantalos, S. Koenig, K. Gillars, M. Skliar: Achieving physiologic perfusion with ventricular assist devices: comparison of control strategies, Proc. Am. Control Conf. (Portland 2005) pp. 3823–3828

79.9 S. Chen, J.F. Antaki, M.A. Simaan, J.R. Boston: Physiological control of left ventricular assist devices based on gradient of flow, Proc. Am. Control Conf. (Portland 2005) pp. 3829–3834

79.10 K.-W. Gwak, M. Ricci, S. Snyder, B.E. Paden, J.R. Boston, M.A. Simaan, J.F. Antaki: In-vitro Evaluation of Multiobjective Hemodynamic Control of a Heart-Assist Pump, Am. Soc. Artif. Intern. Organs (ASAIO) **51**, 329–335 (2005)

79.11 E. Marieb: *Human Anatomy and Physiology* (Pearson Education, San Francisco 1994)

79.12 D.U. Silverthorn: *Human Physiology: An Integrated Approach*, 2nd edn. (Benjamin/Cummings Publishing, New York 2000)

79.13 G.A. Giridharan, M. Skliar, D.B. Olsen, G.M. Pantalos: Modeling and control of a brushless DC axial flow ventricular assist device, Am. Soc. Artif. Intern. Organs (ASAIO) J. **48**, 272–289 (2002)

79.14 Y. Wu, P.E. Allaire, G. Tao, D. Olsen: Modeling, estimation, and control of human circulatory system with a left ventricular assist device, IEEE Trans. Control Syst. Technol. **15**(4), 754–767 (2007)

79.15 A. Ferreira, S. Chen, D.G. Galati, M.A. Simaan, J.F. Antaki: A dynamical state space representation of a feedback controlled rotary left ventricular assist device, Proc. ASME Int. Mech. Eng. Congr. (Orlando 2005), Paper IMECE2005–80973

79.16 A. Ferreira, M.A. Simaan, J.R. Boston, J.F. Antaki: A nonlinear state space model of a combined cardiovascular system and a rotary pump, Proc. 44th IEEE Conf. Decis. Control Eur. Control Conf. (Seville 2005) pp. 897–902

79.17 H. Suga, K. Sagawa: Instantaneous pressure-volume relationships and their ratio in the excised, supported canine left ventricle, Circ. Res. **35**(1), 117–126 (1974)

79.18 N. Stergiopoulos, J. Meister, N. Westerhof: Determinants of stroke volume and systolic and diastolic aortic pressure, Am. J. Physiol. **270**(6), H2050–H2059 (1996)

79.19 N. Stergiopoulos, B.E. Westerhof, J.J. Meister, N. Westerhof: The four element Windkessel model, Proc. 18th IEEE Eng. Med. Biol. Annu. Int. Conf. (Amsterdam 1996) pp. 1715–1716

79.20 D.S. Breitenstein: Cardiovascular modeling: the mathematical expression of blood circulation. M.Sc. Thesis (Univ. of Pittsburgh, Pittsburgh 1993)

79.21 Y.-C. Yu: Minimally Invasive Estimation of Cardiovascular Parameters. Ph.D. Thesis (University of Pittsburgh, Pittsburgh 1998)

79.22 Y.-C. Yu, J.R. Boston, M.A. Simaan, J.F. Antaki: Estimation of systemic vascular bed parameters for artificial heart control, IEEE Trans. Autom. Control **43**(6), 765–778 (1998)

79.23 R.C. Dorf, J.A. Svoboda: *Introduction to Electric Circuits*, 7th edn. (Wiley, New York 2006)

79.24 A.C. Gyton, J.E. Hall: *Textbook of Medical Physiology*, 9th edn. (Saunders, Philadelphia 1996)

79.25 S. Choi, J.R. Boston, D. Thomas, J.F. Antaki: Modeling and identification of an axial flow pump, Proc. Am. Control Conf. (Albuquerque 1997) pp. 3714–3715

79.26 H. Shima, J. Honigschnabel, W. Trubel, H. Thoma: Computer simulation of the circulatory system during support with a rotary blood pump, Trans. Am. Soc. Artif. Organs **36**(3), M252–M254 (1990)

79.27 Z. Sun, S.S. Ge: *Switched Linear Systems: Control and Design* (Springer, New York 2005)

79.28 H. Lin, P.J. Antsaklis: Switching stabilizability for continuous-time uncertain switched linear systems, IEEE Trans. Autom. Control **52**(4), 633–646 (2007)

79.29 M.A. Simaan, A. Ferreira, S. Chen, J.F. Antaki, D.G. Galati: A dynamical state-space representation and performance analysis of a feedback-controlled rotary left ventricular assist device, IEEE Trans. Control Syst. Technol. **16**(6), 1–14 (2008)

79.30 G. Avanzolini, P. Barbini, A. Cappello: Comparison of algorithms for tracking short-term changes in arterial circulation parameters, IEEE Trans. Biomed. Eng. **38**, 861–867 (1992)

79.31 J.W. Clark, R.L.S. Ling, R. Srinivasan, J.S. Cole, R.C. Pruett: A two-stage identification scheme for the determination of the parameters of a model of left heart and systemic circulation, IEEE Trans. Biomed. Eng. **27**, 20–29 (1980)

79.32 B. Deswysen: Parameter estimation of a simple model of the left ventricle and of the systemic vascular bed, with particular attention to the physical meaning of the left ventricle parameters, IEEE Trans. Biomed. Eng. **24**, 29–36 (1977)

79.33 B. Deswysen, A.A. Charlier, M. Gevers: Quantitative evaluation of the systemic arterial bed by parameter estimation of a simple model, Med. Biol. Eng. Comput. **18**, 153–166 (1980)

79.34 T.L. Ruchti, R.H. Brown, D.C. Jeutter, X. Feng: Identification algorithm for systemic arterial parameters with application to total artificial heart control, Ann. Biomed. Eng. **21**, 221–236 (1993)

79.35 M. Oshikawa, K. Araki, N. Nakamura, H. Anai, T. Onitsuka: Detection of total assist and sucking points based on the pulsatility of a continuous flow artificial heart, Am. Soc. Artif. Intern. Organs (ASAIO) J. **44**, M704–M707 (1998)

79.36 Y. Yuhki, E. Hatoh, M. Nogawa, M. Miura, Y. Shimazaki, S. Takatani: Detection of suction and regurgitation of the implantable centrifugal pump based on the motor current waveform analysis and its application to optimization of the pump flow, Artif. Organs **23**, 532–537 (1999)

79.37   D. Liu, J.R. Boston, H. Lin, J.F. Antaki, M.A. Simaan, J. Wu: Monotoring the development of suction in an LVAD, Proc. BMES–EMBS 1st Joint Conf. (Atlanta 1999) p. 240

79.38   L. Baloa: Certainty-Weighted System For The Detection Of Suction In Ventricular Assist Devices. Ph.D. Thesis (University of Pittsburgh, Pittsburgh 2001)

79.39   M. Vollkron, H. Schima, L. Huber, R. Benkowski, G. Morello, G. Wieselthaler: Advanced suction de-

tection for an axial flow pump, Artif. Organs **30**(9), 665–670 (2006)

79.40   A. Ferreira, M.A. Simaan, J.R. Boston, J.F. Antaki: Frequency and time–frequency based indices for suction detection in rotary blood pumps, Proc. IEEE Int. Conf. Acoust. Speech Signal Process., Vol. II (Toulouse 2006) pp. 1064–1067

79.41   A. Ferreira, S. Chen, M.A. Simaan, J.R. Boston, J.F. Antaki: A discriminant-analysis-based suction detection system for rotary blood pumps, 28th IEEE Annu. Int. Conf. Eng. Med. Biol. (New York 2006) pp. 5382–5385

80. Medical Informatics

Chin-Yin Huang

All of today's patients receive healthcare services from a team of healthcare practitioners through various integrated medical information systems and automated systems. Due to the requirements of accuracy, safety, privacy, and responsiveness on healthcare services, medical informatics itself has become an important research field. This chapter intends to introduce medical informatics from the perspectives of (1) integration, (2) database and data warehouse, (3) individual medical support systems, and (4) medical knowledge and decision support systems. For integration, standard protocols (e.g., HL7, EDIFACT, and DICOM) are important connectivity agreements for hospital information system (HIS), medical support systems and healthcare equipment. For database and data warehouse, this chapter introduces the important concept of link relationships between data tables. The value of medical database and data warehouse relies on the correct link relationships. All modern healthcare organizations consist of multiple separated yet integrated systems. For individual medical support systems, this chapter introduces imaging, laboratory, hospital pharmacy, and nursing information systems. Evidence-based medicine and data mining techniques are two important fields in today's medical knowledge and decision support systems. Finally, this chapter introduces how to strategically introduce healthcare information systems into

a healthcare organization. Emerging issues of care quality, security, and public-use databases are also introduced.

80.1 Background

Traditional medical informatics is defined as a scientific field that studies information collection, processing, and communication related to healthcare services. As information and communication technologies are advancing rapidly, the domain of medical informatics is expand-ing. Traditional medical informatics that focuses on data transactions is no longer enough in modern hospitals. Technologies of biomedical data mining, medical decision support, medical image processing, telemedicine, etc., are filling in and expanding the domain of medical informatics. Furthermore, healthcare behaviors have changed and improved accordingly due to those technologies.

The objective of this chapter is to summarize the technologies that are being applied in today's healthcare organizations. Some emerging and advanced issues are also addressed. The remainder of this chapter is organized as follows. First, a diagnostic–therapeutic cycle is presented to demonstrate a common healthcare behavior. Then, basic needs and standards that are required for communication and integration in a hospital are introduced. Besides communication, database and data warehouse are another two key elements of medical informatics. This chapter also gives an example of data warehouse schema. Then, four basic functions of medical support systems are illustrated: imaging information systems, laboratory information systems, hospital pharmacy information systems, and nursing information systems. Issues of medical knowledge and decision support system are presented by pointing out two topics: evidence-based medicine and data mining techniques. This chapter suggests that peo-

ple, project management, and strategic planning are the three crucial elements that decide whether a healthcare information system can be successfully developed. Finally, emerging issues related to medical informatics such as quality of care, security, and public-use databases are presented.

Although this chapter summarizes modern healthcare technologies that are applied in the healthcare services, *integration* is an important technical issue. For healthcare organizations, communication technologies (e.g., internet, local area network, WiFi, TCP/IP, etc.) provide the infrastructure for connectivity. However, to make all the healthcare practitioners and modern medical care systems work together efficiently and accurately for all kinds of service activities integration in a higher level must be relied upon. Such an integration and coordination challenge was pointed out by *Sarter* et al. [80.1] in 1997. Today, the integration is fulfilled by all kinds of medical standard protocols (e.g., Health Level 7 (HL7) and digital imaging and communication in medicine (DICOM)). These issues of integration are addressed in this chapter.

80.2 Diagnostic–Therapeutic Cycle

In healthcare, medical doctors make medical therapeutic decisions based on a diagnostic–therapeutic cycle [80.2]. There are three stages in the cycle: observation, diagnosis, and therapy. Observation is the most fundamental step in therapy. It can be done by the healthcare practitioner with simple assistant tools, such as stethoscopes. Observations may also be done

through relatively simple blood tests or x-ray systems, or through a complicated facility, such as computed tomography (CT) or an intestinal endoscope. Beside the above-mentioned observations that are done one to several times, observations can last for a certain time period by using equipment, such as the Holter electrocardiograph.

Medical doctors perform diagnosis based on the results of observations and patient history (anamnesis). Its goal is to identify the causes of (1) patients' complaints

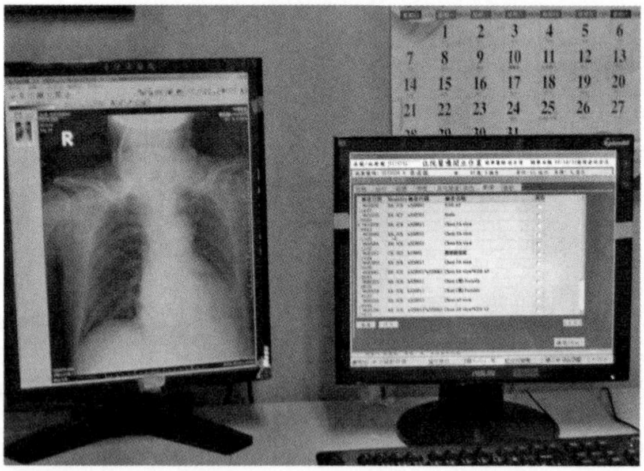

Fig. 80.1 Two computer screens for medical diagnosis (photo courtesy of Cheng Ching General Hospital)

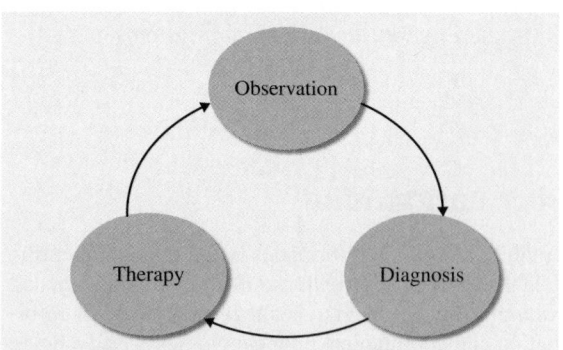

Fig. 80.2 Diagnostic–therapeutic cycle

and (2) results of observations. Figure 80.1 shows two computer screens for a medical doctor when checking patients. The left screen is used to display the results of observation (x-ray), whereas the right screen is an order entry system to show anamnesis and other results of tests. Diagnosis may take a long time, since the patients may describe their complaints incompletely, which may lead the doctor to order the wrong tests or treatments. Then, an iterative observation-diagnosis-therapy occurs. For patients who have chronic diseases, iterations may repeat over and over for a long time period. However, sometimes the iterations occur because of mistakes/errors resulting from doctors, pharmacists, nurses, or inspecting staffs and/or facilities.

Therapy cannot be done without the orders from doctors. It may be done through medication, surgery, rehabilitation, or diet control. Technologies of therapy are constantly evolving. Due to its advancement, life quality of patients is improving.

Observation, diagnosis, and therapy form a cycle (Fig. 80.2). The cycle stops only when the patient is cured or decides to quit the cycle.

80.3 Communication and Integration

Medical services in modern hospitals are fulfilled by a group of specialists: physicians, nurses, pharmacists, biologists, and administrative and laboratory staffs. Moreover, medical services are fulfilled by advanced facilities, and information and computer technologies. Patients are cared for by an integrated system consisting of the above-mentioned specialists and expensive facilities. The integrated system can function efficiently and effectively only when the individual persons or facilities can be integrated through a communication system.

Communication as a key component for integration requires two elements: channels and interfaces. Channels mean how the persons or facilities are connected. Most medical facilities can offer communication with a local area network (LAN) by applying TCP/IP standard protocol. Therefore, specialists can upload/download or operate the facilities within the LAN. Associated clinics or outside laboratories may be authorized to use the Internet as a channel to obtain data from the facilities of hospitals. For short distance transmissions, WiFi provides a wireless alternative of communication. It is useful for a doctor to display patient records or to give a medical order when checking with a patient in a ward with a mobile/wireless handheld device, personal digital assistance (PDA) for example. Figure 80.3 shows a portable computerized order entry device. It is a WiFi and web ready system. By its powerful computing, communication, and graphic capabilities, physicians can display the test results of patients and give treatment or medication orders when they are checking with patients anywhere in the WiFi coverage zone.

Interface means that facilities speak the same language. Usually an interface is fulfilled by standards. For administrative data exchange in healthcare, Electronic Data Interchange for Administration, Commerce and Transport (EDIFACT) is used in Europe [80.3], whereas Health Level 7 (HL7) is used in the United States [80.4]. With the standards, participants (such as clinics, laboratories, drug stores, and insurance companies) can encode and decode messages by the same format, which speeds up communication and enables integration of processes occurring between units of a hospital or between hospitals. For medical images such as x-ray, CT, and magnetic resonance imaging (MRI), digital imaging and communication in medicine (DICOM) is a standard for image communication [80.5]. DICOM also plays a role of standardizing the image inputs/outputs of medical facilities developed by various vendors.

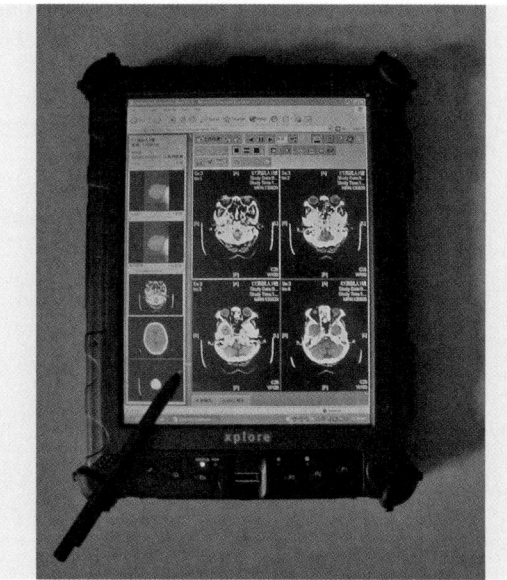

Fig. 80.3 Portable medical order-entry device (photo courtesy of DVNET Technology Co., www.dvnet.biz)

80.4 Database and Data Warehouse

Most healthcare organizations host large databases to maintain the demographical records, complaints and diagnostic records, laboratory tests and inspections, orders of therapy, and insurance and payment records that are associated with each patient. The massive amount of healthcare information requires a database that has a storage capacity of multiple terabytes (TB). There are several reasons why a large storage capacity is required. First, voice, image, and video types of test and inspection data are gradually playing an important role in healthcare. They usually take up large data storage space. Besides, health authorities of government and insurance companies usually demand hospitals to maintain the records for a certain period, which also results in the need for large storage space. Although the cost per byte of data storage is decreasing yearly, larger and more urgent demands on storage space still increase the total cost of database system year by year in modern healthcare organizations.

Though today's object-oriented database (OODB) seems to be a suitable solution to accommodate various data formats, which include text, audio, video, animation, and image, and can manage linked relationships easily, its low performance in transactions hinders the applications of OODB in healthcare organizations. Relational database (RDB) presenting database in flat tables is a typical database for healthcare organizations. The advantages of RDB include supporting the standard structural query language (SQL), scalable

performance, and easy maintenance. Besides, because of the abundance of human resource supply in the field of RDB development, operation, and maintenance, RDB is commonly used in healthcare database applications.

However, maintaining large and complicated medical databases may become a financial burden for healthcare organizations, unless the value of medical databases is created. To utilize the value of a large medical database relies on the development of a data warehouse. Data warehouse is a collection of associated databases. Databases in a data warehouse are subject-oriented and nonvolatile. *Subject-oriented* means that the database storage focuses on a special issue, such as patient treatments or medical records for certain cancers. *Nonvolatile* means that the data in a data warehouse should be extracted from the databases and are not changed by online transactions. Besides, the online analytical process (OLAP) is an important function to retrieve useful information from a data warehouse. Basic command functions of OLAP include rolling up, drilling down, slicing, dicing, etc. [80.6]. However, to successfully apply OLAP depends on how well the data warehouse is defined.

A data warehouse is defined by specifying the link relationships among flat tables. Figure 80.4 shows an example to demonstrate link relationships among tables in a data warehouse with fact constellation. It shows that there are many keys for patient order tables. Each

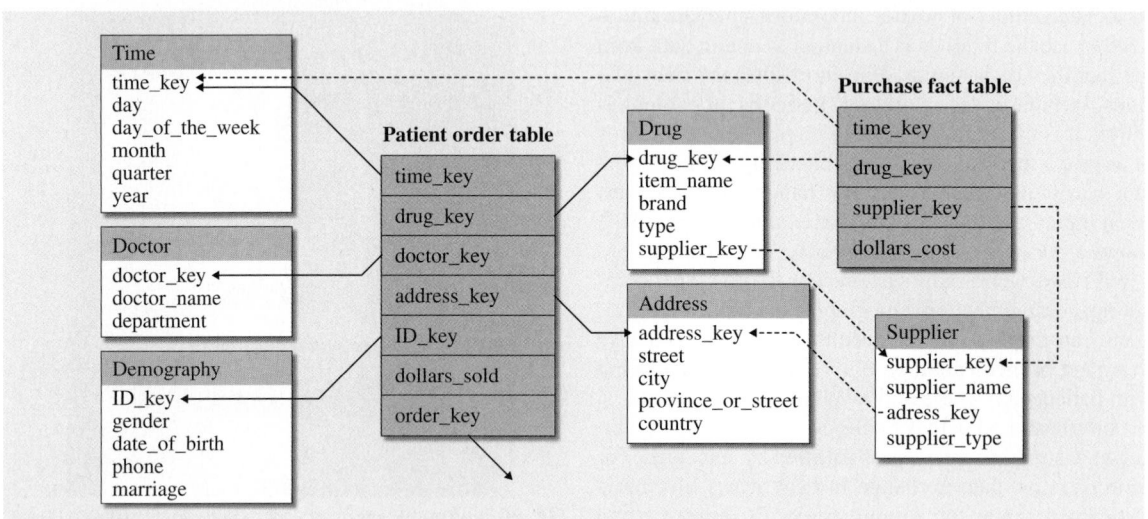

Fig. 80.4 A fact constellation scheme of a data warehouse for patient records

key is linked to other tables. By the links, a user can give an OLAP command. For example, showing the total patient orders for the last month (dicing through time_key), then showing the patients according to the departments they went to visit (drilling down). The

drilling down command is possible because the doctor_key links both the patient order table and the doctor table. Through the link, each patient order can find the doctor_key. By querying in the doctor table with the doctor_key, the department is found.

80.5 Medical Support Systems

Today's healthcare activities are supported by information systems and digitized facilities. For this reason, care of patients has become more efficient, timely, and accurate. Medical support systems, besides the hospital information system (HIS) that manages both information of patient care and administration, includes at least four subsystems:

1. Solitary imaging information system
2. Laboratory information system
3. Functions of hospital pharmacy
4. Nursing information systems.

In the following sections, the four subsystems are introduced.

80.5.1 Solitary Imaging Information Systems

Most modern hospitals own expensive solitary imaging systems, such as x-ray, CT, MRI, digital subtraction angiography (DSA), gamma cameras, and positron emission tomography (PET). CT, MRI, and DSA are mostly found in the radiology department, whereas gamma cameras and PETs are found in the nuclear

medicine department, since with these radioactive material, which is under government management, is injected into the veins of a patient in the inspection process. Figure 80.5 shows a multislice CT volume scan system. CT is a typical noninvasive examination in today's healthcare services. An examination on a CT involves various kinds of monitor and control of the CT system. Figure 80.6 shows multiple monitoring and control functions on both patients and the CT system. To integrate the solitary imaging systems with HIS requires a standard for exchanging image data. As mentioned in Sect. 80.3, DICOM is a standard for image data exchange. With it, associated healthcare practitioners can view the images across different systems, whether they be inside or outside the hospital. An example of a CT image is shown in Fig. 80.7. Some radiologists even check images of patients in various hospitals through the Internet.

80.5.2 Laboratory Information Systems

Besides solitary imaging systems, many tests may be performed in a laboratory or even a clinical room by portable devices. Equipment in a laboratory may

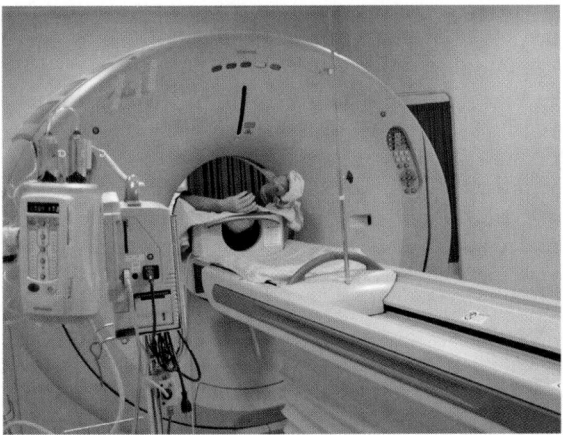

Fig. 80.5 A multislice CT volume scan system (photo courtesy of Cheng Ching General Hospital)

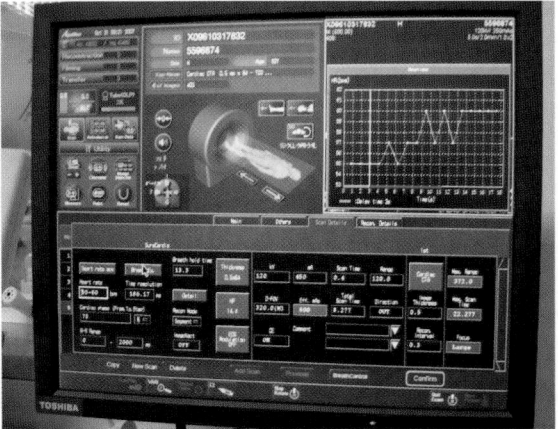

Fig. 80.6 Multiple monitoring and control functions on both patients and CT system

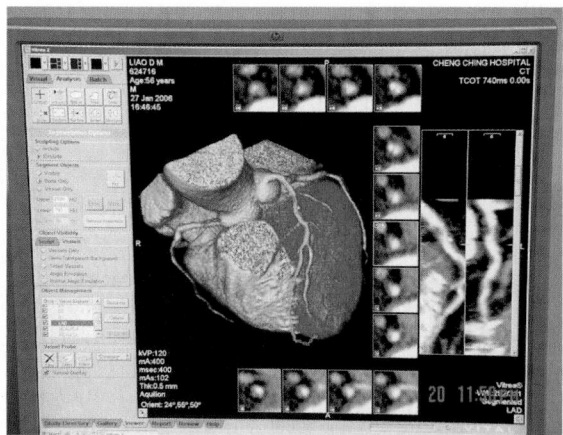

Fig. 80.7 Example of a CT image (photo courtesy of Cheng Ching General Hospital)

include an electrocardiogram (ECG), ultrasound, or spirography. Biosignals produced by the equipment are digitalized and conformed to DICOM, so that they can be integrated into a HIS. Additionally, those biosignals are collected synchronously with patients, which means that the patients must stand by when the biosignals are recording.

Besides the previous synchronous examination, some examinations are asynchronous. In other words, the examinations only need specimens from patients and do not need patients to wait while the equipment is examining. Hospitals host laboratories to give asynchronous examinations of specimens from patients, such as blood and urine. Electron microscopes and centrifuges are typical examples of equipment for laboratories. To perform pathological examinations, the systematized nomenclature of medicine (SNOMED) is a code system that indicates various aspects of a dis-

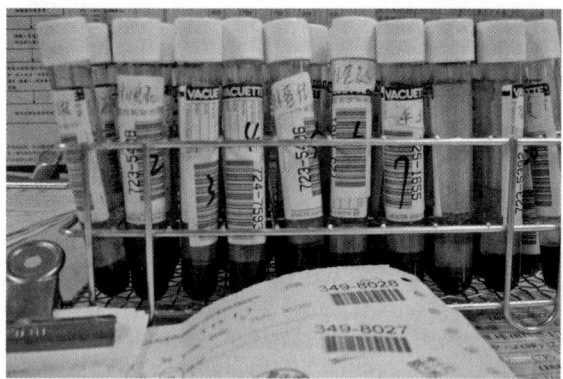

Fig. 80.8 Examples of specimens (each with a barcode)

ease [80.7]. It is a checkbook for diseases based on specimens from patients.

There are two basic activities that a hospital laboratory has to do for either synchronous or asynchronous examinations. The first activity is about getting samples or biosignals from patients. In the first activity, there must be a system for registering the patient ID and the examinations that were ordered by a doctor. For an asynchronous examination, usually a specimen from a patient is labeled by a barcode that shows the identical patient ID to avoid mismatching with other patients' specimens (Fig. 80.8). As long as the specimens' barcodes are recorded into HIS, laboratory staff schedule a time for the examination. The second activity is to deliver the examination report to HIS. For both synchronous and asynchronous examinations, the examination reports are presented in text, audio, or video and have to be stored in the database of HIS so further diagnoses or therapies can be ordered by a doctor.

Usually a laboratory information system involves various kinds of workflows. Those workflows instruct the administrative and laboratory staffs to efficiently deliver services to patients.

80.5.3 Hospital Pharmacy Information System

The hospital pharmacy is an important service unit that supports care of patients. Usually the organization of the hospital pharmacy is divided into two parts, one for inpatients and the other for outpatients. The information system of the hospital pharmacy includes an inventory planning and control system that monitors the inventory levels of drugs and purchases drugs whenever necessary to maintain high service quality and low inventory cost of the pharmacy. Besides, the pharmacy information system for outpatients should maintain and manage the dispensing information given by clinicians. However, it should allow immediate changes if some other concerns such as drug interactions occur. For inpatients, the information system should monitor the drug usage of patients, so further therapeutic and billing procedures can proceed based upon this information.

A more advanced pharmacy information system has an embedded knowledge base system to detect the problems of drug interactions and mistaken orders issued by doctors. The pharmacy information system is usually a subsystem of an HIS in modern hospitals. Some hospitals may not maintain a hospital pharmacy for outpatients. However, the dispensing information is delivered through the Internet to drug stores designated by outpatients. In this case, an HL7 or EDIFACT protocol will be needed to provide a standard communication between hospitals and drug stores [80.8]. The current version of HL7 relies on an XML based message syntax. Through XML syntax, a doctor can embed required drug dispensing information, e.g., patient, drug product, dosage, etc., into a message, which is then delivered to the designated drug store.

80.5.4 Nursing Information Systems

According to *Murnane* et al. [80.9], three key nursing information modules are required to successfully deliver nursing care to patients:

1. Care plan
2. Workload allocation
3. Resource scheduling/monitoring.

HIS provides general patient information, and treatment and drug orders from physicians. However, the information is not enough for nursing care as soon as a patient enters into a ward or intensive care unit. To detail the nursing care and utilize the nursing human and equipment resources relies on an integrated nursing information system that can seamlessly interact with HIS. A nursing information system can indicate the relationships among ward, patients, nurse, working shifts, and basic nursing requirements of the patient.

80.6 Medical Knowledge and Decision Support System

MYCIN is an early yet typical decision support system that adopts expert systems, or now called knowledge base systems, in medicine [80.10]. It was designed to diagnose infectious blood diseases, and then to give recommendation on the use of antibiotics. There are only few medical cases like MYCIN in today's medical applications. The reason is that to make a knowledge base useful relies on complete and correct knowledge of medicine. Partial or incomplete knowledge may lead to an incorrect decision, which may guide a doctor to a wrong decision on patient care. Besides, because of the advancing medical knowledge and technology, to keep a medical knowledge base system up to date would be a huge burden for medical doctors and computer specialists in debugging the decision rules. However, medical knowledge and management has improved by information technologies in two aspects: evidence-based medicine and data mining. These two aspects are addressed especially in the following two sections.

80.6.1 Evidence-Based Medicine

Evidence-based medicine (EBM) is defined as [80.11]:

The conscientious, explicit and judicious use of current best evidence in making decisions about the care of individual patients, based on an integration of individual clinical expertise with the best available external clinical evidence from systematic research and patient's unique values and circumstances.

Fundamentals of EBM include three major steps:

1. Knowledge findings based on medical evidence databases
2. Applications of the findings to patients
3. Evaluating the applications and modifying the knowledge findings if necessary.

Knowledge findings usually involve many database and statistical operations on the medical data. Most healthcare journals play a role in publicizing medical treatments and knowledge findings for medical doctors and scientists. Unlike traditional healthcare research that usually involves limited number of cases, by the introduction of digital databases and government support for accumulating patient and treatment records, the number of cases involved in the study based on the government support dataset (e.g., the Surveillance, Epidemiology, and End Results (SEER) program [80.12] and NHIRD of Taiwan [80.13]) is very large. The number of cases can be as large as one million or more. Thus, the reliability of the research results increases, which implies that the reliability of such an EBM is also higher.

80.6.2 Data Mining Techniques

Large-scale medical databases not only increase the reliability of EBM research results, they lead to other potential uses of the data. Traditional EBM studies usually applied statistical tools only. By data mining techniques [80.6], medical researchers are able to find other evidence that usually involve more variables and are not based on testing the hypotheses made in advance by people. Notwithstanding, data mining techniques can be applied to investigate medical improvement and patient relationship management in general hospital management based on the records of patient visits.

Major data mining tools that may help healthcare administration or healthcare services include:

1. Classification tools, e.g., decision trees and neural networks
2. Clustering techniques, e.g., K-Means and hierarchical clustering
3. Association rules, e.g., apropri algorithm and rough sets theory.

Classification is a tool to identify the relationships between certain conditional (independent) variables and a decision (dependent) variable. The relationships are mostly specified to a certain range of conditional variables (e.g., age between 20 and 25 or systolic hypertension above 140 mmHg), and certain range of decision variables (e.g., annual cost of medical care between $ 10 000 and $ 15 000). Because tools of decision trees, such as ID3 and CHAID, provide clear IF-THEN rules for presenting the relationships between variables, they are more acceptable, in comparison with neural networks that define variable relationships as weights on connections of neurons.

Clustering is a tool to separate a data set into multiple subsets. A predefined distance that measures similarity between two cases and methods for finding clusters (subsets of data) are the two key elements in clustering. Clustering does not require a dependent variable. Usually it is applied to handle data with a massive number of variables. Clustering methods (algorithms) iteratively check the distance between cases to find clusters. Basically, the distance between two cases that belong to the same cluster should be shorter than the distance between two cases that belong to two different clusters. The results of clusters are usually difficult to interpret and apply, since distance is an unusual concept in medicine, and the cases of each cluster may include a large number of variables. Even so, clustering has been applied as a technique for analyzing gene expression data in the biological field, e.g., [80.14], and problems of clinical and medical care, e.g., [80.15]. For example, *Jannin* and *Morandi* successfully apply decision tree and clustering techniques to predict parts of the surgical procedure for brain-tumor patients based on the pathology-related characteristics of the patient [80.16].

Tools of association rules intend to find causality between variables. For example, patients who have diabetes are also hypertensive (15%, 90%). The 15% represents support for such a rule, whereas the 90% represents confidence. Support defines the proportion of the number of patients who have both diabetes and hypertension over the total number of patients. Confidence defines the proportion of the number of patients who have both diabetes and hypertension over the number of patients who have diabetes. An advantage of association rules is that they can identify a causality that has less support but high confidence. To search for a plausible curable treatment or medication for rare disorder patients is a typical problem that can apply techniques of association rules, for two reasons. First, rare disorder patients mean that the support is small. Second, if the majority of the patients are cured by a common treatment or medication, its confidence is high. Traditional statistical tools that mostly do not allow a large number of variables in the analysis and count on a large number of samples are not able to handle such a problem.

Though data mining provides an opportunity for finding medical evidence that may not be found by traditional statistical tools, few applications are performed in the healthcare fields. Some researchers ascribe the reasons to a large number of false alarms created by data mining. Actually, the credibility of mining results can be proved by cross-validation or some other related tools [80.17]. From the author's perspective, there are two major obstacles for data mining applications in healthcare. First, healthcare practitioners may not be so familiar with data warehouse and data mining. Second, many legacy medical data tables are not normalized. It is believed that both obstacles will be overcome in the near future, because many multidisciplinary research opportunities have been created for both healthcare and computer science professionals.

80.7 Developing a Healthcare Information System

To develop a good healthcare information system relies on at least three elements: the right people, the right project management, and the right strategic plan. Right people means that the healthcare information system should cover the operation needs of major system users (e.g., administrators, physicians, nurses, staff, patients, etc.) Questions like how many people and what kind of people should be included in the information system depend on the budget and strategic plans of the healthcare organization. The development of a healthcare information system is a complicated project. It usually includes stages of requirement analysis, system design or introduction, system implementation, system tests, user training, and system validation. The range of the project duration depends on the type of information and may be one month to one year. Besides, the people involved in each stage vary. Therefore, a well organized project management is required to ensure that the project can be successfully implemented and applied in accordance with the planned schedule and budgets.

The strategic plan for the healthcare information system has to be set right so it can match the organization's goals and external and internal environment analysis. Because today's HIS plays an important role in patient care and costs much in investment, its success depends on how well it meets the healthcare organiza-

tion's goal. The steps of the development of a strategic plan for a healthcare information system can be summarized as follows:

1. Perform SWOT (strengths, weaknesses, opportunities, and threats) analysis for the organization.
2. Identify specifically how information systems can strengthen the organization's capabilities to catch external opportunity.
3. Identify specifically how an information system can avoid threats from internal and external environments.
4. Identify healthcare information goals of the organization.
5. Evaluate the gap between the healthcare information goals and current information system.
6. Identify long-term information system objective such that the gap can be filled.
7. Identify resource requirements to fulfill the objective.
8. Make a budget plan for the resource requirements.

80.8 Emerging Issues

The advent of modern medical information technologies has largely improved patient care and the quality of care. More specifically, a healthcare organization almost cannot function without the support of information systems. However, there are still many research issues waiting investigation. Some of them are addressed in the following sections.

80.8.1 Quality of Care

How to utilize information technologies to improve the quality of cares is always an issue. Each time a new technology is introduced, the possibility for improvement of care quality evolves. By means of today's radio frequency identification (RFID) technology, errors in drug dispensing and operation rooms have been reduced because an RFID reader system provides an active alarm when a person, an article, a paper order, a medicine, or a specimen embedded with an RFID tag is not in the right place, right position, and right time that he/she/it is supposed to be. Some case studies even suggested using RFID to monitor the walking routes through wards of medical doctors and nurses, so an alert system may be built to control nosocomial infections. However, side effects of introducing

RFID system are inevitable, just like the side effects of introducing barcode systems into the healthcare services [80.18].

80.8.2 Security

Today's healthcare organizations carry a lot of information about patients. Securing the digital data today is as important as securing the paper patient records in the past. Hackers and computer viruses are the two major sources that may jeopardize the safety of patient records. It becomes financial and technical burdens for healthcare organizations to secure large patient-record databases. Some hospitals outsource healthcare information systems to IT companies, so they can ease the technology burden. However, ethical and legal issues may arise if patient records are lost or stolen from those outsourcing IT companies, for any reason.

80.8.3 Public–Use Databases for Medical Research

To attract researchers to get involved in medical studies, many governments are systemically building public-use database for medical research. SEER is a typical database that accumulates cancer patients' records (including demographical, histological, and pathological information). The Taiwan government also allows researchers to retrieve patient records (including demographical information, insurance claims, drug dispensing, and hospital information) for medical studies. Such public-use databases give great opportunities for researchers to investigate prevalence, survival analysis, and other medical comparative studies. Since such studies are performed based on a large number of samples, they can avoid errors resulting from a limited number of samples or data sources. Additionally, more data mining technologies are applied to the databases to handle their multiple dimension data tables.

References

80.1   N.B. Sarter, D.D. Woods, C.E. Billings: Automation surprises. In: *Handbook of Human Factors and Ergonomics*, 2nd edn., ed. by G. Salvendy (Wiley, New York 1997)

80.2   J.H.V. Bemmel, M.A. Musen, J.C. Helder: *Handbook of Medical Informatics* (Houten, Springer 1997)

80.3   UNECE: United Nations Directories for Electronic Data Interchange for Administration, Commerce and Transport (2007) http://www.unece.org/trade/untdid/welcome.htm

80.4   I. Health Level Seven: HL7 Standards (2007) http://www.hl7.org

80.5   N.E.M. Association: The DICOM Standard (2007) http://dicom.nema.org/

80.6   J. Han, M. Kamber: *Data Mining: Concepts and Techniques* (Morgan Kaufmann, San Francisco 2001)

80.7   I.H.T.S.D. Organisation: Technical Documents (2007) http://www.ihtsdo.org/

80.8   B.W. Chaffee, J. Bonasso: Strategies for pharmacy integration and pharmacy information system interfaces, Part 2: Scope of work and technical aspects of interfaces, Am. J. Health Syst. Pharm. **61**, 506–514 (2004)

80.9   R. Murnane, S. Brown, F. Kerr, M. Meehan, A. McDonagh, E. Molony, Y. Mulligan: Nursing Workload Activity Analysis: A Critical Component of the Mater Misericordiae University Hospital's Integrated Nursing Information System (INIS) (2006) http://www.hisi.ie/html/news.htm

80.10   B.G. Buchanan, E.H. Shortliffe: *Rule-Based Expert Systems: The MYCIN Experiments of the Stanford Heuristic Programming Project* (Addison-Wesley, Reading 1984)

80.11   S.E. Straus, W.S. Richardson, P. Glasziou, R.B. Haynes: *Evidence-Based Medicine: How to Practice and Teach EBM*, 3rd edn. (Churchill Livingstone, New York 2005)

80.12   SEER: The Surveillance, Epidemiology, and End Results (SEER) Program (2007) http://seer.cancer.gov/

80.13   NHIRD: National Health Insurance Research Database (2003) http://www.nhri.org.tw/nhird/index.php

80.14   S.A. Ness: Microarray analysis: basic strategies for successful experiments, Mol. Biotechnol. **36**, 205–219 (2007)

80.15   M.W. Isken, B. Rajagopalan: Data mining to support simulation modeling of patient flow in hospitals, J. Med. Syst. **26**, 179–197 (2002)

80.16   P. Jannin, X. Morandi: Surgical models for computer-assisted neurosurgery, Neuroimage **37**, 783–791 (2007)

80.17   I.H. Witten, E. Frank: *Data Mining: Practical Machine Learning Tools and Techniques*, 2nd edn. (Morgan Kaufmann, Amsterdam 2005)

80.18   E.S. Patterson, R.I. Cook, M.L. Render: Improving patient safety by identifying side effects from introducing bar coding in medication administration, J. Am. Med. Inform. Assoc. **9**, 540–53 (2002)

81. Nanoelectronic–Based Detection for Biology and Medicine

Samir M. Iqbal, Rashid Bashir

This chapter is a review of the work in nanoelectronic detection of biological molecules and its applications in biology and medicine. About half of the chapter focuses on the methods employed to immobilize deoxyribonucleic acid (DNA) on solid substrates with particular focus on the electronic detection and characterization of DNA. Charge-transfer properties and theories are explained, as such electronic and electrical sensing of molecular-level interactions are very important in medical applications for rapid and cheap diagnosis.

A special tool called nanopore, which has been used extensively to characterize DNA, is then reviewed. A special distinction is made between the characteristics, capabilities, and impacts of the biological and the solid–state nanopores. Nanopores, when used in the ion current measurement setup, are used to measure the behavior of DNA as it traverses the nanopore. When the DNA traverses the pore, the blockage of the ion current is observed as a pulse. The statistical analysis of the pulses yields trends that are used to sort the DNA based on various properties. The nanopores are strong prototypes for biosensors, and have become a major experimental tool for investigating biophysical properties of double and single strands of DNA. The DNA sequence can

potentially be determined by measuring how the forces on the DNA molecules, and the ion currents through the nanopore, change as the molecules pass through the nanopore.

Part H | 81

81.1 Historical Background

Nanotechnology, in the last decade or so, has brought together scientists and engineers from a diverse array of fields to collaborate and share their distinct sets of expertise and tools. The confluence of divergent technologies has provided better handles on nanoscale processes and species. Characterization and control at these fundamental limits have shown immense potential to improve existing technologies and provide better tools for understanding at the most basic levels of nature. This has opened horizons wide open for many diverse areas of research. The major impact it has had, and will continue to have, is in medical applications. Biologists have known for decades that cells and biomolecules are selectively managed, regulated, and controlled, but the tools to organize and direct these interactions are only now emerging, tools to *handle* and

integrate synthetic structures in diagnostics and therapeutics. The nascent tools of nanoscience are helping
unravel the physics of biological processes in unprecedented detail. The detection and sensing of biological
entities has enormous promise, and challenges of correspondence size. The challenges are manyfold: the
interface of the wet salty biological molecules to dry
cold solid devices, packaging of the devices so that only
the sensing part is exposed to the analytes, faithful translation of the biophysical and biochemical interactions
into electrical or optical signals, selectivity of the device
against the analyte of interest in the pool of thousands of
entities of no interest, identification of the useful signal
amongst unwanted noise, sensitivity of the measurement system all the way down to the fundamental limits,
and repeatability of measurements. For any system to
interrogate the few copies of the analyte of interest re-
producibly and reliably, the interface also plays a key
role. These challenges are the major impediments to the
integration of nanotechnology in biological and medical
applications.

A few industries that use a combination of engineering, physics, chemistry, and biology in very
nontraditional approaches to develop devices for medical applications have very recently emerged. Some
examples are on-chip DNA hybridization and opti-

cal detection, DNA detection at the nanoscale for
diagnostics in medical, military, and food safety applications, etc. DNA carries the genetic information
for living organisms and has a special, important
place in many areas such as genetic engineering, drug
discovery, and gene therapy. The methods used for
immobilizing DNA on solid substrates and the electrical sensing schemes will be detailed in ensuing
sections in this chapter. As we go along, we will
focus on biological nanopores, also known as ion channels, which are integral parts of the living organisms
at various levels. These have inherent capabilities for
ion-transport regulation under transmembrane potentials. Ion channels have been used to measure the
translocation behavior of DNA as it passes through
the channel. These form an important framework for
DNA detection, but have inherent issues and problems
regarding the stability. Solid-state nanopores, direct
analogues of the biological ion channels, are then
reviewed. Solid-state nanopores are stable under various environments. Cheap and rapid DNA sequencing
using solid-state nanopore channels is the focus of
a number of major initiatives, owing to the promise of
focused and effective treatment of diseases, and fast
genetic analysis in areas such as forensics and legal
examination.

81.2 Interfacing Biological Molecules

This section is an overview of the techniques used to
probe a single molecule, i.e., to make a stable and
reproducible interface with a biomolecule. It is important to make interface contacts to biological molecules
for reproducible and robust interrogation and characterization. Various techniques and chemistries are used
to attach biological molecules, such as DNA, to solid
substrates, either for conductivity measurements, self-
assembly of devices, for bottom-up device fabrications,
for patterning matter on the nanometer scale, in biotechnology, and even as biosensors. In the crafty bottom-up
paradigm, DNA has played a vital role due to its specific
base-pair interactions.

At around the dawn of this century, the major thrust
in nanotechnology came from the Moletronics program
of the Defense Advanced Research Projects Agency
(DARPA) and the National Nanotechnology Initiative
(NNI) of the US government. The objective of the Mo-

letronics program is *to develop functioning prototype
electronic computer processors and memory integrated
on the molecular scale* [81.1]. An important avenue
that opened with the vision of Moletronics is towards
the development of nanoscale biosensors that are robust, small, accurate, cheap, and have high throughput.
A biosensor can be defined as a collection of molecular recognition sites and the transduction components.
In case of electrical biosensors, the focus has been on
direct measurement of the electrical properties of the
molecular interfacing sites. The size of typical organic
molecules is in the range of $1-10$ nm. To do any realistic electrical measurements through such small objects,
we need at least two macroscopic metallic electrodes at
the same dimensions.

Self-assembled monolayers (SAMs) of molecules are characterized using a large array of direct
analysis tools, but there are few direct label-free char-

acterization tools to quantify or qualify the quality of the surface chemistries or the quantity of immobilized DNA. This is understandable, first due to the complexity of the DNA molecule (that increases even with the addition of a single base); secondly the interactions of the surface chemical properties can greatly change the binding quality of the molecules, which may cause more of the nonspecific adsorption; and thirdly because of the mechanical and chemical flexibility (or instability) of DNA, e.g., hybridization, conformational changes, effect of ambient and temperature, its sensitivity to biologically significant entities, and even to slight changes in chemical composition of the buffer. Notwithstanding the challenges in *driving* DNA, the highest degree of selectivity due to complementarity of bases makes it a molecule of choice in a large variety of self- and mediated-assembly experiments and biosensor applications.

A lot of experiments use physical adsorption to immobilize DNA on a variety of devices. Some approaches use deposition of DNA from a buffer solution using catalysts such as $MgCl_2$ for better DNA adhesion to metal electrodes [81.3]. A receding meniscus of a drying droplet has also been shown towards immobilizing $16\,\mu m$ long DNA on microstructured surfaces (Fig. 81.1) [81.2]. Imaging techniques such as atomic force microscopy (AFM) are done to confirm the presence of DNA before doing electrical measurements. Passive adsorption of DNA has also been shown in 96-well polystyrene plates by incubation of the molecules with cationic reagents [81.4]. In a similar fashion, a high-frequency electric field has also been used to trap DNA and proteins between the electrode gaps [81.5]. Under the influence of DEP, the DNA is known to be stretched into a long, thin configuration, i.e., it is not randomly coiled. Fluorescence imaging is used to prove the presence of the DNA in the gaps between the electrodes.

By far, chemical adsorption and immobilization is considered the most stable, reliable, and reproducible method to attach biological molecules onto substrates and nanodevices. Usually, the molecules are modified with a functional group(s) that has a high covalent affinity to the surface. Most common and well-studied functional pairs are thiol–Au, biotin–(strept)avidin, and the his–tag system. The complementarity of the DNA strands is also employed extensively, where single-strand (ssDNA) is attached to the surface (called the probe), and complementary ssDNA is supplied from solution (the target).

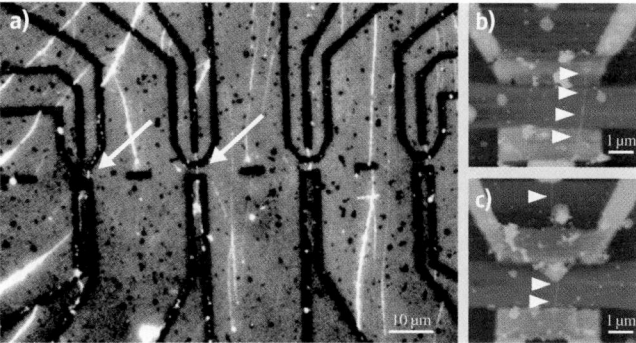

Fig. 81.1 DNA positioned in electrode gaps of $2\,\mu m$. Fluorescence imaging shows Au electrodes structures *dark* and the DNA *bright*. *Arrows* mark electrode gaps. Reprinted with permission from [81.2]. Copyright 2004 American Chemical Society

Gold–thiol is a favorite chemistry of choice for a lot of applications and has been employed in a number of studies to immobilize DNA for conductivity measurements, hybridization detection, sequence detection, DNA-templated self-assembly, etc. [81.6–9]. Glass slides and lately thermally oxidized silicon surfaces are also routinely used to immobilize DNA through various silane-, aldehyde-, and amine-modified schemes. Probe DNA molecules are accordingly modified at the appropriate end(s). The functionalized surfaces exhibit various surface energies (wettability) and/or electrical charges. DNA is chemically anchored on substrate surface either directly to a chemical layer like silane or through a hetero/homobifunctional moiety. Figure 81.2 depicts an example of amine-modified-DNA attachment chemistry through a silane layer on a SiO_2 surface. The attachment of the biological entities has been characterized in various labeled or unlabeled ways, e.g., a recent review has summarized various methods and strategies that have been developed to detect nucleic acids with electrical means [81.10], or, direct optical detection of DNA hybridization on the surface of a charge-coupled device [81.11]. Chemical attachment of DNA has been investigated on various surfaces such as carbon, aluminum, indium-tin oxide, SiO_2, Si_3N_4, and most of all, glass slides [81.7, 12].

SAMs of biological molecules have also been used to make reliable contacts. The networks of polysequences of DNA (DNA with the same base repeated) have been shown to self-assembled onto a mica surface. Poly(dG)–poly(dC) DNA networks showed a *uniform reticulated structure* and poly(dA)–poly(dT) DNA formed a cross-linked network [81.13]. DNA has also been shown to aid the construction of functional cir-

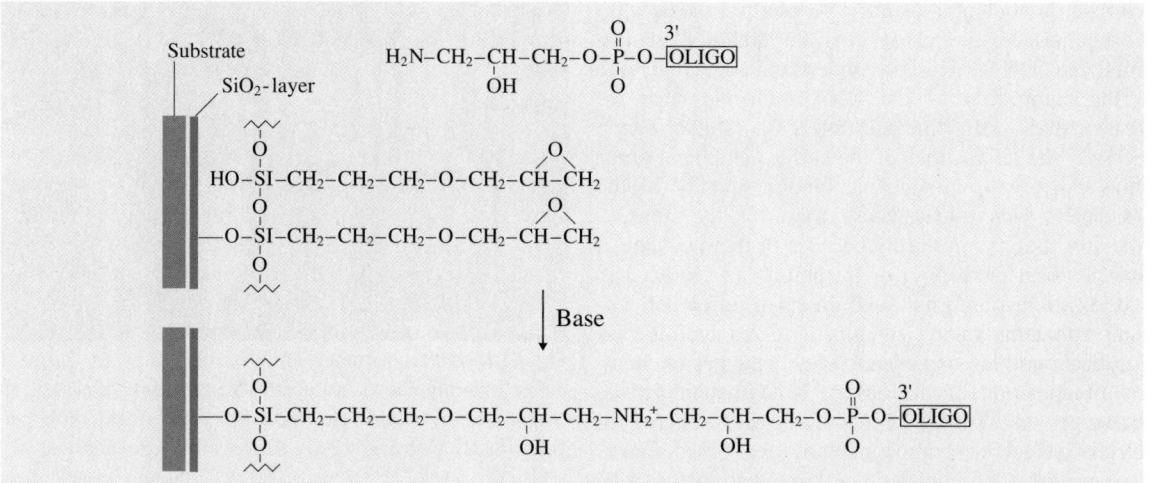

Fig. 81.2 Schematic depicting chemical bonds in attaching 3′-amino-modified DNA to a thin SiO_2 film. Attachment occurs by secondary amine formation between an epoxysilane monolayer and the 3′-amino linkage. From [81.11], by permission of Oxford University Press

cuits, by constructing Ag wires using the self-assembly of a λ-DNA as a template [81.14]. The λ-DNA was uncoiled and stretched to contact two Au electrodes. The Ag^+ ions were deposited along the DNA, through Ag^+/Na^+ ion exchange as Ag^+ formed complexes with the DNA bases, resulting in nanometer-sized metallic Ag aggregates bound to the DNA skeleton. In a similar approach, DNA-templated assembly of Ag clusters has been shown with electrochemical detection [81.8]. The Ag cations were electrostatically *collected* along the gold-surface-tethered DNA duplex. The Ag aggregates were dissolved and the potentiometric stripping of the dissolved Ag was detected on a thick-film carbon electrode.

81.2.1 Guidelines for Preparing Silicon Chips for Biofunctionalization

DNA is immobilized by physical or chemical adsorption, the latter being more robust and stable in a variety of ambient conditions. A number of chemistries have been shown by various researchers for attachment of DNA via chemical bonds. The motivation for such attachments has been wide, ranging from biosensors to integrated circuits; for the detection of chemical or biological species in biosensors; to controlled assembly of nanodevices and structures in nanoelectronics and biophysical studies.

The general scheme of DNA attachment is through surface functionalization of the substrate and the mod-

ification of DNA with a linker end-terminus, as shown in Fig. 81.3. The linker end-terminus can then bind to a hetero/homobifunctional molecule, or the chemical structure of the linker itself is exploited to react with surface functionalized moieties due to certain binding affinities. The bifunctional molecules have at least two available chemically reactive positions, one of which attaches to the surface-grafted molecular layer and the other to the end-linker of the DNA.

Homobifunctional cross-linkers have two identical reactive groups thus can be used in a one-step chemical cross-linking procedure. DNA modified with various end groups or even without end groups has been shown to attach to certain functional groups. Attachment of ss-DNA to the surface molecular SAM is mostly verified

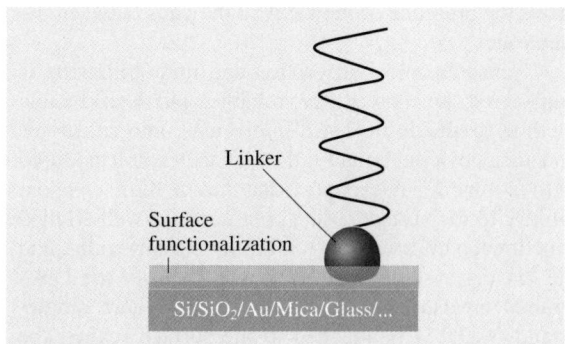

Fig. 81.3 General schematic of DNA attachment via linker molecule

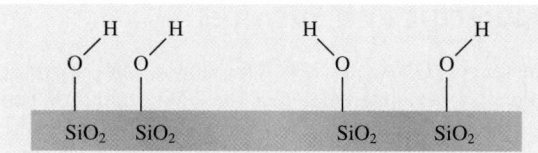

Fig. 81.4 Hydroxylation of SiO$_2$ surface (after [81.15], by permission of Oxford University Press)

with the fluorescently labeled complementary ssDNA that hybridizes with the adsorbed ssDNA. Other detection schemes involve radioactive dyes such as ^{32}P, phosphor imaging, and chemiluminescence. We present two protocols that can be adapted for covalent attachment of amine-modified DNA to SiO$_2$ surfaces (adapted from [81.15, 16]).

For direct DNA attachment to the chips, various silanizing agents have been pursued and reported to functionalize SiO$_2$ surfaces. Generally, the silane-containing chemicals have a tendency to react with hydroxyl groups at the surface of the chips, resulting in silanol groups. The hydroxyl group density at the surface of the SiO$_2$ thus influences the self-assembly of the silane layer and ultimately the number of attached DNA. The hydroxyl groups on the surface of Si/SiO$_2$ chips are generally achieved by piranha solution (H$_2$O$_2$: H$_2$SO$_4$, in various proportions) treatment of the SiO$_2$ surface, or O$_2$ plasma etch treatment. Following these treatments, the OH-rich surface of SiO$_2$ looks as shown in Fig. 81.4.

Exposing the OH-rich surface to a silane results in the reaction of Si of the silanizing chemical with the O group, thus releasing the hydrogen. To directly attach the DNA molecules, an amine modification at the end of DNA can be utilized. The reactive site of the silane (e.g., an epoxide) makes a covalent bond with the amine. The surface coverage of immobilized ssDNA can be verified by using complementary ssDNA tagged with fluorescence. All chip surface processing should be done in a controlled environment, e.g. a glovebox to minimize exposure to ambient moisture and contamination.

For better surface coverage, a homobifunctional linker can be used on top of the surface silane. The chips can be silanized using 3-aminopropyltrimethoxysilane (APTMS), and 1,4-phenylene diisothiocyanate (PDITC) can be used as a homobifunctional agent. Again APTMS hydrolyzes the OH-rich SiO$_2$ surface thus strongly attached through covalent bonds to it, creating a SAM of amino silane anchors. The amine group at the other end of the APTMS covalently attaches with the C of one of the isothiocyanate groups of the PDITC cross-linker. PDITC has two isothiocyanate sites where its C can react with the NH$_2$– of APTMS. The other isothiocyanate site on PDITC provides amino-reactive terminus for probe DNA attachment. These PDITC-activated chips are thus immersed in the amine-modified DNA, completing the attachment chemistry for the NH$_2$-oligo attachment on the SiO$_2$ surface. The unreacted sites of PDITC are then deactivated by N–N diisopropylethylamine exposure.

81.2.2 Verification of Surface Densities of Functional Layers

The functionalized surfaces can be characterized for adsorbed layers by ellipsometry and contact-angle measurements. The silane layer thickness can be measured by ellipsometry. An isotropic value of $n = 1.50$ for the silane layer and $n = 3.858 - 0.018i$ for the silicon substrate refractive index can be used. The values for monolayers range between 1.45–1.50 in the literature, but this range would give an error of less than 1 Å during measurements. The monolayer thickness can be accurately measured up to ± 2 Å. A number of readings should be taken at different places on the chips and compared with control chips.

The hydrophilicity/hydrophobicity of a surface is usually expressed in terms of wettability. Hydrophilicity and hydrophobicity are general terms used to describe relative affinity of materials for water (hydrogen bonding). Hydrophilicity of a surface is a measure of its strong affinity to the water as the polar surfaces form an H bond with water. The opposite hydrophobic surfaces have aversion to water. Contact-angle measurements with deionized water can show functionalized chips to be less hydrophilic than control chips with only SiO$_2$ and no chemical treatment. The OH-rich surfaces of control chips can make more H bonds with the water and so would be hydrophilic, whereas in relative terms the chemically treated chips would have fewer OH-groups left on the surface and so would be less hydrophilic and can be termed as becoming hydrophobic. A nice example of highly hydrophobic surface is fresh Si surface with no native oxide. Si resists being wetted by water, thus the water forms drops with contact angles $\geq 75°$.

81.3 Electrical Characterization of DNA Molecules on Surfaces

In this section we will review studies on charge transfer and direct measurements of electrical conduction through DNA, along with the various charge transport mechanisms that have been proposed and simulated. It has been established that a number of factors and conditions contribute to the behavior of DNA conductivity [81.17, 18]; the number of base pairs (bp), the sequence and length of DNA [81.3], changes in the distance and angle between bases, the influence of water and counterions, humidity, contact chemistry, surface smoothness, electronic contamination, ambient conditions, temperature, and buffer solution components. The double-helix structure of DNA depends on the hydrophobicity of the bases. The hydrophobic bases turn towards the center, away from the water. The presence of the condensed cations counters the negative charges of the phosphate backbone. The water molecules and cations are thus an integral part of the overall picture and exert nonnegligible forces on the base-pair stack [81.19]. All these factors dictate that DNA is a dynamic and very complex system to simulate owing to its structural, chemical, environmental, and vibrational properties.

81.3.1 Indirect Measurements of Charge Transfer Through DNA

The idea of DNA being electrically conductive can be traced back as far as 1962, when Eley and Spivey hinted at efficient charge transfer through DNA as *...a DNA molecule might behave as a one-dimensional aromatic crystal and show a π-electron conductivity down the axis.* They proposed that the DNA structure was ideal for electron/hole transfer proceeding along a one-dimensional pathway constituted by the overlap between π-orbitals in neighboring base pairs [81.20]. At 400 K, they reported conductivities on the order of 10^{-12} $(\Omega \, cm)^{-1}$ and energy gaps (ΔE) of about 2.42 ± 0.05 eV. While emphasizing lack of knowledge on the ribonucleic acid (RNA) structure at that time, they reported a similar experimental value for RNA. *Snart*, in 1973, reported a similar and reproducible value of the energy gap (2.4 eV) that was affected by ultraviolet (UV) irradiation [81.21]. Snart's measurement method was very similar to the one employed by Eley and Spivey. These can be considered as the starting points of the quest for DNA conductivity.

In the early 1990s, *Barton* and co-workers performed experiments that suggested long-range electron transfer in DNA [81.22]. The donor and acceptor molecules were intercalated on the DNA strands. When the donor was photoexcited, the fluorescence of the donor was quenched due to electron transfer to acceptor. These results showed very rapid transfer of carriers over > 40 Å via π-stacked base pairs. However other researchers disputed these results, as reproduction of the results with other acceptor and donor candidates was found to be problematic. The starting point of the controversy came the very next year when, in similar experiments, *Brun* and *Harriman* used organic donors and acceptors [81.23] and concluded that charge transfer rates drop off quickly with increasing length of the DNA. *Ly* et al. studied the mechanism and distance dependence of radical anion and cation migration and suggested that the structural flexibility of DNA dictates mixed behavior of hole-hopping and continuous orbital mechanism [81.24]. Such a mechanism in which the injection of charge disturbs the molecular structure is called a phonon-assisted polaron-like hopping mechanism. Such a disturbance in DNA most likely results in the reduction of intrabase distance and unwinding of DNA, giving way to increased π-electron overlap and shift of internal charges inside H bonds. *Giese* and *Wessely*, in 2000, verified these two mechanisms experimentally [81.25]. They reported a coherent superexchange reaction (single-step tunneling) and a thermally induced hopping process for long-range charge transfer, slightly influenced by the number of intercalated adenine–thymine (AT) bases. Generally, in a hopping mechanism, the guanine (G) base is considered the most favorable for landing of holes or trapping, basically because it has the least ionization potential among the four bases: G < A < cytosine (C) < T, independent of nearest-neighbor effects. Since then, more experimental and theoretical research has resulted in seemingly contradictory results.

81.3.2 Direct Measurement of DNA Conduction and DNA Conductivity Models

The sensing of DNA has been a direct result of the electrical characterization studies. DNA has been shown to have metallic-like conductivity [81.26], semiconducting behavior [81.27], and also as an insulator [81.28].

Fink and *Schonenberger* used gold-coated perforated carbon foil as a sample holder to make DNA networks, spanning the holes atop the carbon foil. A low-energy electron point source (LEEPS) was used to image the sample, while a conductive tip contact was placed at the middle of the bundle. They reported an upper value of DNA resistivity of $1\,m\Omega\,cm$, measured through 600–900 nm-long bundles of λ-DNA in vacuum, concluding that DNA was a good conducting molecular wire [81.26]. *Dekker* and co-workers reported semiconducting behavior of DNA, by measuring the conductivity of 10.4 nm long poly-DNA, electrostatically trapped between two 8 nm spaced metal electrodes, as shown in Fig. 81.5 [81.27]. The first-principles calculations for carrier transport through DNA, as well as experiments performed on 40 nm to 15 μm long λ-DNA suggested exponential decay in conductance with the DNA length [81.28]. In these experiments, electrodes were formed by sputtering gold onto a mica substrate with DNA already immobilized on the surface. Refuting the high conductivity reported by Fink and Schonenberger, similar experiments were carried out and the high conductivity was explained to be an artifact due to electronic contamination from LEEPS. The very next year, these results were challenged by *Kasumov* et al. by their experiments on DNA deposited between 0.5 μm apart rhenium/carbon (Re/C) electrodes [81.29]. The Re/C contacts had superconducting properties. Again, DNA was reported to behave like a metallic conductor. It was suggested that DNA had proximity-induced superconductivity properties at low temperature. It was suggested that the compression caused on the DNA structure due to deposition on surface would change its electronic behavior [81.30]. Simultaneous measurement of the height and conductivity of λ-DNA film formed on mica showed that DNA did not conduct when its height was 1 nm. When the mica surface was functionalized with pentylamine film before DNA deposition, its height was seen to be around 2 nm, and it was conductive. The reduced mica–DNA interaction due to the pentylamine layer ensured that the structure and conductivity was unaffected. *Yoo* et al. measured the conductance through poly-DNA and reported semiconducting behavior with little effect of ambient conditions or vacuum [81.31]. *Storm* et al. reported DNA to be an insulator, as opposed to the previous report of the semiconducting behavior from the same group [81.3]. It was concluded that DNA was insulating at length scales longer than 40 nm and that bare DNA had limited use as a conducting molecular wire [81.32]. The effects of oxygen adsorption on the poly-DNA conductance have also been explored [81.33]. The oxygen content was shown to modulate the resistance. It was also reported that poly(dG)–poly(dC) DNA was a p-type semiconductor and poly(dA)–poly(dT) DNA was an n-type semiconductor. *De Pablo* and co-workers reported a different approach of contactless setup (to avoid possible artifacts caused by the contacts) to investigate DNA electrical properties [81.34]. They reported that the dielectric constant of DNA was similar to that of mica. This study provided a qualitative result of DNA being as insulating as mica.

A scanning tunneling microscope (STM) tip has also been used to study DNA molecules [81.35]. The STM tip was used to contact the Au electrode in 3 μM thiol-modified dsDNA solution.

Individual molecular junctions were created by moving the tip out of contact with the flat gold electrode with DNA molecules bridging the distance between the tip and the substrate. The conductance showed a series of steps. For 8 bp dsDNA, the conductance histogram showed peaks at near-integer multiples of a fundamental conductance value, $1.3 \times 10^{-3} G_0$ or $\approx 0.1\,\mu S$, where $G_0 = 2e^2/h \approx 77\,\mu S$. The conductance was also seen to be inversely proportional to the length of $(GC)_n$ sequences (for 8 bp or longer DNA). The insertion of AT bp decreased the conductance with a decay constant of $0.43\,Å^{-1}$. Qualitative differences between the conductivity of ssDNA and dsDNA are also reported,

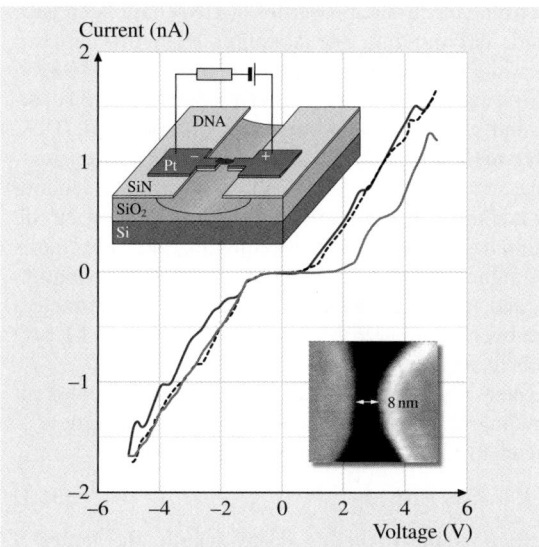

Fig. 81.5 *I–V* curves measured at room temperature on a DNA molecule trapped between two metal nanoelectrodes. Reprinted by permission from Macmillan Publishers Ltd: Nature [81.27], copyright (2000)

ssDNA being insulating over a 4 eV range between ± 2 V, and dsDNA being like a wide-bandgap semiconductor [81.36]. *Iqbal* et al. investigated the effects of the DNA sequence on its conductivity [81.37]. They chemically anchored thiol-modified DNA between gold nanogaps. It was seen that the conductivity through the DNA molecule increased with increasing GC content. Such behavior could be attributed to the increased hole hopping between the hydrogen bonds in the guanine–cytosine pair [81.38]. As the ionization potential of guanine is the lowest (G < A < C < T), this would provide the easiest path for the conduction of holes [81.30]. Moreover, the triple hydrogen bonds between the G and C bases as compared with the double hydrogen bonds in the A and T bases putatively result in more paths for the charge flow in higher-GC-content sequences [81.38]. This results in less resistance to the charge flow in the sequence containing more G–C pairs and results in decrease in resistance between various GC-rich sequences by an order of magnitude. They also showed the possibility of using DNA as a reversible fuse, blowing *off* at high temperature and turning back *on* when a conducive environment became available. This was seen by temperature cycling of their devices. Such a framework could also be used as an electrical DNA hybridization sensor. In another report, *Mahapatro* et al. showed that the DNA conductivity scaled with the surface density of chemically immobilized DNA [81.39]. The surface density of the DNA was seen to be higher with high salt concentrations, resulting from better screening of the DNA phosphate backbone charges and reduced electrostatic repulsion. They observed that the conductivity of DNA molecules decreased exponentially as the number of adjacent AT pairs were increased, replacing GC pairs. This also showed that the GC pairs provide a more conductive path to the charge transfer. These finding can have consequences in DNA-based molecular electronics and direct label-free detection of DNA sequence and hybridization.

While the experimentalists have been debating the behavior of DNA, a number of theoretical models have been proposed for the DNA conductivity mechanisms. Almost all the proposed transport ideas rely on the experimental results, and can be broadly categorized as either model simulations or ab initio calculations. The wide variety of these models is understandable due to the large number of variables that affect the conducting properties of DNA. *Jortner* and co-workers suggested two distinct DNA charge transport mechanisms in 1998 [81.40]. They proposed that transport occurs either by the direct unistep hole tunneling from one base pair to another, or by a multistep charge transport through the base pairs, exhibiting a weak dependence on the separation distance between the donor and the acceptor. They further proposed that the charge transport mechanism was determined by the specificity of the intercalating base sequences, which they called bridge. *Yu* and *Song* modeled DNA as a one-dimensional (1-D) disordered system with electron transport occurring between localized states as variable-range hopping [81.41]. These localized states in the sequence would present the candidate landing sites for a hopping electron, while such localization would be enhanced by structural changes in DNA with temperature. They defined a crossover temperature above or below which transport would be either simple nearest neighbor thermal hopping or the carrier would hop different ranges, respectively. Such variable-range hopping mechanism is somehow consistent with the two possible mechanisms proposed by *Jortner* et al. and *Ly* et al. [81.24, 40]. *Yi* modeled the dsDNA molecule as a two-legged charge ladder to render the presence or absence of conductance gaps seen in $I-V$ plots of GC-rich and GC-poor sequences, respectively [81.42]. The gap in conduction plots was explained with strong Coulomb repulsion with eV values much larger than thermal energy. This repulsion is encountered with a screening effect of the cations in the solution. This is a classical example of ab initio calculation. Various reports describing electronic properties of DNA have been published, delving into the coupling strengths between bases, the highest occupied molecular orbital (HOMO) and lowest unoccupied molecular orbital (LUMO) values and energies, band structures of various dsDNA, effect of doping (specifically, O_2 doping) on conductance, structural effects on conduction, etc. The random and stochastic processes in the complex DNA–electrode system have been identified, explaining the wide claims over values of DNA conductivity (metal, semiconducting, and insulator) [81.43]. As many as 12 parameters have been simulated in adjusting HOMO and LUMO bands in such systems.

Most of the theoretical models assume coherent tunneling to be a major mechanism, with a length (L)-dependent rate of charge transfer R [81.30]

$$R \propto \exp(-\beta L) \, , \tag{81.1}$$

where β is the tunneling decay length, the larger β is, the faster the rate of tunneling decreases with increasing distance. There have been various, sometimes contradictory, values reported for β [81.44, 45]. *Berlin* et al. found β to be $0.1 \, \text{Å}^{-1}$ for sequential tunneling and about $1\,\text{Å}^{-1}$ for coherent tunneling by quantitative analysis using kinetic rate equations [81.46].

The mechanisms proposed in the theoretical models may or may not occur in a system altogether, but there is higher probability for presence of more than one mechanism in any given experiment. The ideal model should take care of the effects of DNA structure, thermal motion of charges as described by classical models, effects of cations in the solution on the structure and on the Coulomb charging, temperature, intermolecular and intramolecular attractions and repulsions, effect of contacting conductors and the chemistries used to contact the DNA, etc. Almost all studies consider charge transport and conductivity on the path along the DNA length. Little, if any work has been done, until lately, on charge transport in the direction perpendicular to the axis of the DNA backbone. *Zwolak* and *Di Ventra* described unique signatures of each base due to their

differing electronic and chemical structures [81.47]. They considered a system of electrodes through which a ssDNA passes such that at any given time only one base interacts with the electrodes. Such a system can be visualized as a nanopore made with a very thin membrane, with electrodes on the very edges. With electrode–electrode spacing of $15\,\text{Å}$, they found the ratio of current of A (I_A) and the current of other bases (I_X) to be $I_A/I_X = 20, 40,$ and 660 for X $=$ G, C, and T, respectively. This difference in ratios is postulated to stem from relative positions of the Fermi level with respect to HOMO and LUMO, and the density of states at Fermi level. The effects of nearest neighbor were also calculated, with T found to be the most sensitive to such effects. These findings are very interesting as they can be used in conjunction with nanopore measurements of DNA translocation, as will be explained in next section.

81.4 Nanopore Sensors for Characterization of Single DNA Molecules

Biological entities, such as cells, proteins, and DNA, carry charges, and can be forced to move by an electric field in a buffer solution, a phenomenon called electrophoresis. Electrophoresis is usually applied in a gel or capillary to separate biological entities based on their size and charge. Electrophoresis in gel and capillaries has given rise to the idea of characterization by a single pore through which charged entities can be driven. Therefore, characterizations of biological entities by a single pore are actually an analogous application of gel and capillary electrophoresis. The basic design of a nanopore characterization setup consists of two compartments, filled with saline buffer solution, separated by a membrane with the pore, with an anode and a cathode set in each compartment; the pore provides the only path for ionic currents and the electrophoretic movement of DNA or other biological species of interest. When DNA moves electrophoretically from the cathode to the anode, it traverses the pore. As it does so, the ionic current is altered and most typically decreases due to pore blockade. When DNA has passed through the pore, the current recovers to its original level. The characteristics of DNA can possibly be determined from these current fluctuations.

81.4.1 Biological Nanopores

Kasianowicz et al. pioneered the use of a single $2.6\,\text{nm}$ diameter α-hemolysin (α-HL) channel [81.48].

α-HL is a protein toxin from the bacterial species *Staphylococcus aureus*. A solvent-free bilayer membrane of diphytanoyl phosphatidylcholine (an artificial lipid bilayer membrane) was formed across a $\approx 0.1\,\text{mm}$ diameter orifice. The orifice was in Teflon partition separating two buffer-filled compartments. When α-HL was added to the compartment, it reconstituted into the lipid bilayer, making a channel. A single channel usually formed within $5\,\text{min}$. They used these channels for current fluctuation detection when a single DNA traversed and passed through the pore. The translocation times, or the pulse widths, were proportional to the DNA lengths. A patch clamp amplifier was used to convert current to voltage. In the absence of DNA, applying a potential of $-120\,\text{mV}$ resulted in ionic currents free of pulses. Following the addition of DNA to the *cis* side of the protein pore, numerous short-lived current blockades occurred (the current was reduced by $85-100\%$). The blockades lasted from several hundred to several thousand microseconds, depending on the polymer length. For a given polymer length, the total number of pulses was seen to be directly proportional to polymer molar concentration. It was further shown that DNA length, one of the DNA characteristics, was directly proportional to the mean lifetime of the peaks in the signals. This realized determination of lengths of individual RNA or DNA chains using single-channel measurements. Applying the same principle towards a chemical sensor, *Bayley* and co-workers detected organic molecules of relative molecular mass as low as 100 using α-HL pores [81.49]. The innovative step in their study was to equip the α-HL nanopore with an internal, noncovalently bonded molecular *adapter* which mediated channel blocking by the analyte, thus sensitizing the nanopore to specific organic and chemical species. The work showed that β-cyclodextrin (βCD) molecule was sensitive to different guest molecules, as shown in Fig. 81.6. Thus, when bound to βCD, members of the adamantane family of petroleum derivatives could be distinguished, as could the members of the group of tricyclic pharmaceuticals that included imipramine and promethazine. These guest molecules made their presence known by altering the electrical conductivity of the α-HL pore. A single sensing element of this sort could be used to analyze a mixture of organic molecules with different binding characteristics, so that the pore could be in a way *programmed* for range of sensing functions.

Meller et al., extending on the work of *Gu* et al., used the statistical data derived from the patterns of events to show that nucleotides of different sequences could be distinguished from each other [81.50]. Six polymers with the same length but different sequences were measured. Difference between the translocation duration and the temporal dispersion of translocation duration were used to then distinguish between poly(dA)$_{100}$ and poly(dC)$_{100}$ DNA. The translocation times varied among the polymers even if they had the equal lengths, thus different sequences could be discriminated in a mixture. Above $50\,^\circ\mathrm{C}$, the structure of the pore was not stable and measurements could not be performed. In another study, *Bayley* and co-workers covalently attached ssDNA within the lumen of the modified α-HL pores to form a DNA nanopore [81.51]. The ssDNA molecule was attached on the *cis* side of the α-HL channel, resulting in duplex formation with complementary sequences in the internal cavity. The binding of ssDNA molecules to the tethered DNA strand caused changes in the ionic current flowing through the α-HL nanopore. The DNA nanopores were able to discriminate between individual DNA strands up to 30 nucleotides in length differing by a single base. The mean event lifetimes were ob-

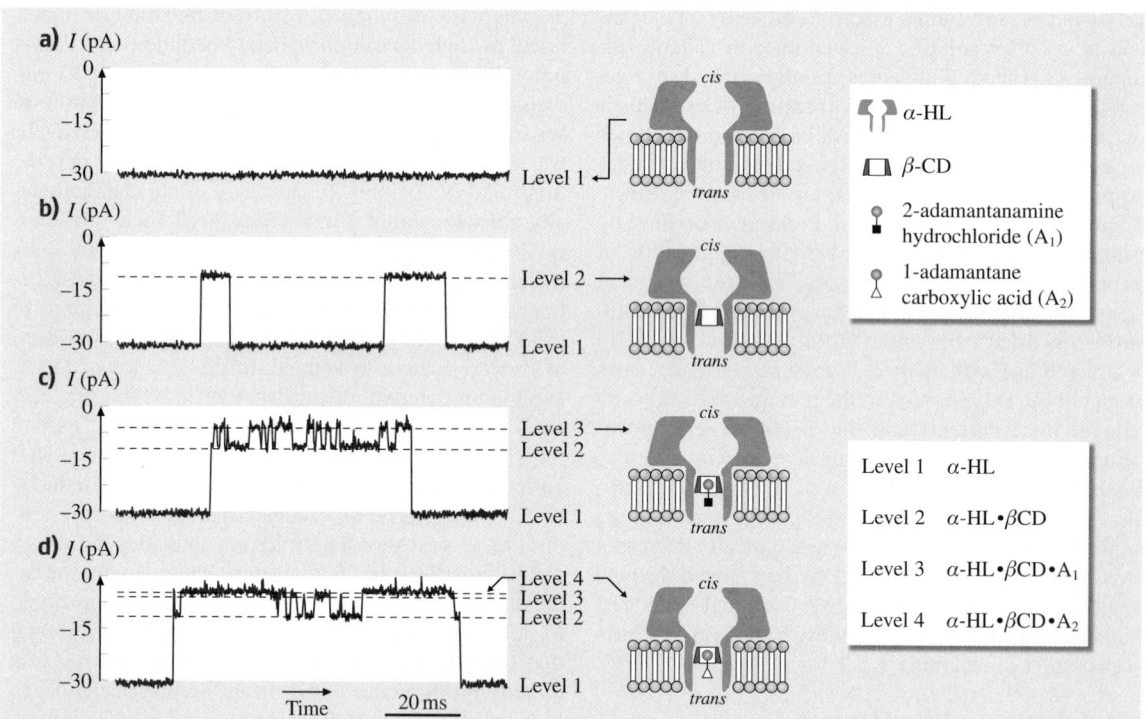

Fig. 81.6a–d Bilayer recordings at $-40\,\mathrm{mV}$ showing the interaction of a single α-HL pore with βCD and the analytes 2-adamantanamine (A_1) and 1-adamantane carboxylic acid (A_2). Reprinted by permission from Macmillan Publishers Ltd: Nature [81.49], copyright (1999)

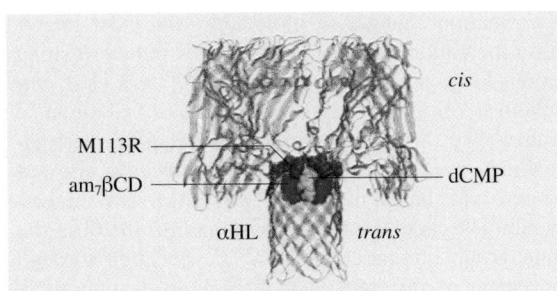

Fig. 81.7 Schematic depicting the α-HL pore, the met-113 substituted with Arg (*blue*), am$_7\beta$CD, and the 2'-deoxy-cytidine 5'-monophosphate molecule (dCMP). Adapted with permission from [81.52]. Copyright 2006 American Chemical Society

tained by lifetime histogram analysis and it was seen that the type of mismatched base pair influenced the lifetime, and that the mismatch had the most dramatic effect when it was positioned in the middle of the oligonucleotide. Using an array of DNA tethered α-HL pores they sequenced a complete codon in an individual DNA strand. They did not look into any change in the translocation time of a DNA sequence complementary to the tethered DNA as compared with the translocation of the same through an untethered pore. In another report, *Bayley* and co-workers reported another engineered α-HL pore approach to detect individual nucleoside monophosphates (Fig. 81.7) [81.52]. The nucleoside monophosphates are individual bases of DNA (A, G, C or T) with a single phosphate group. The nucleoside monophosphates, like DNA, are also negatively charged thus traverse the pore under the trans-channel bias [81.53]. The α-HL pores were modified with amino-cyclodextrin am$_7\beta$CD as *adapter* and its positive charges altered the translocation time for the nucleosides. They attained accuracy as high as 98% (for G) using this approach.

81.4.2 Solid–State Nanopore

Researchers have been making *point contacts* since as far back as 1977. These were used mainly for material science studies, especially ballistic transport experiments. The schematics of these structures were very much like nanopores, as these stand today, but the aim was quite different [81.54, 55]. The design of current solid-state nanopores was inspired by one such work by *Gibrov* et al. [81.55] and these are the next most ideal replacement for biological nanopores owing to several established and foreseen advantages. First, as for other

chemical sensors, sensitive electronic circuitry and photonic sensing capabilities can be integrated directly into a pore–membrane system. Secondly, simultaneous and automated analysis of hundreds of arrays of different channels can potentially be achieved with such an integrated system. Next, they are more robust to withstand wide range of temperature, analyte solution properties, environments, and chemical treatments that might be required for target detection and to eliminate interference. Finally, these can be customized to fit in practical biosensors. These properties have heightened the interest in solid-state nanopores, with particular attention as progenitors of rapid and cheap next-generation DNA sequencing *machines*.

In an earlier report on solid-state nanopores, *Li* et al. utilized a feedback-controlled ion-beam sculpting process to make a nanopore in a silicon nitride membrane [81.56]. A bowl-shaped cavity was made in a silicon nitride membrane and the material was removed from the other side of this membrane using Ar ion-beam sputtering. As soon as the ion-sculpted side reached the bottom of the cavity a hole around 60 nm was opened. Continuous ion-beam exposure reduced the hole to 1.8 nm diameter and ultimately closed it. Two mechanisms were proposed to account for the pore size reduction: surface matter moving due to reduced viscosity to relax the stress caused by implantation, or the creation of adatoms on the surface by incident ions that could diffuse to close the pore. A 500 bp dsDNA translocation experiment was done with a 5 nm diameter pore. Comparative translocation measurements on 3000 and 10 000 bp dsDNA with 3 and 10 nm pores were also reported [81.57]. Longer DNA translocation was noted to be more complex because of folding.

Dekker's group reported the fabrication of a nanopore with electron-beam lithography EBL and transmission electron microscope TEM techniques [81.3]. Their method realized an in situ observation of pore size while the size was precisely controlled at the nanometer scale. *Chang* et al. reported a similar approach, although developed independently [81.58]. They fabricated 50–60 nm long 4–5 nm diameter nanopore channels (NPC), in micromachined Si membranes. The NPCs were fabricated in a double-polished silicon-on-insulator (SOI) wafers. They used EBL to initially define the pore and fabricated 3–4 nm nanopores with standard solid-state processing. The pore was examined visually by using TEM while its diameter shrank to the desired size. The pore shrinkage occurred under an electron beam of high energy. However, their mechanism was completely different from Li's process. In this

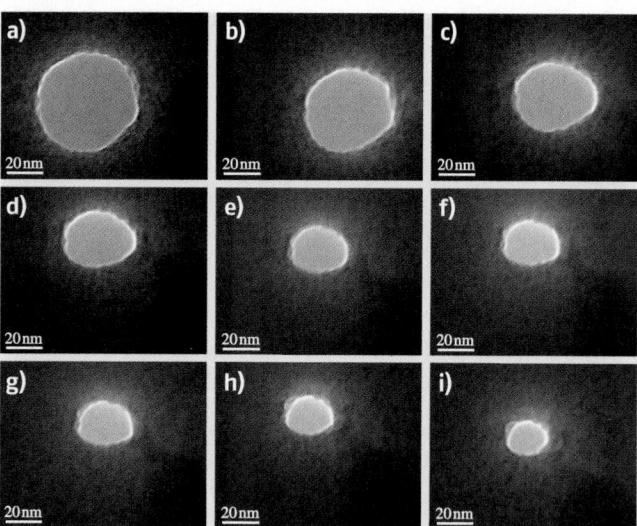

Fig. 81.8 Pore shrinking temporal profile of a nanopore channel ordered alphabetically. All micrographs are taken at $1\,000\,000\times$. The pore shrinking was stopped at size $16\times18\,\text{nm}^2$. The pore would shrink further if continuously exposed to the TEM electron beam. The *scale bars* are 20 nm

process, the electron beam (e-Beam) locally fluidized the oxide around the pore, and hence oxide flowed towards the pore and then filled it, as shown in Fig. 81.8. The nanopore die was sandwiched between two silicone rings and placed in a finely milled pocket within Teflon blocks separating the chambers. The Teflon blocks were clamped together and the pore provided the only path for DNA translocation. A 200 bp fragment from the human *CRISP-3* gene was polymerase chain reaction (PCR) amplified and was used for measurements. Ag/AgCl electrodes were used to make the ionic current measurements. As the dsDNA passed through the pore, it modified the bulk and the interface currents, resulting in typical current pulses, but the pulses were upwards, unlike in previous studies. The results were explained, and later confirmed [81.59], by the charge on the DNA which can be detected in the nanopore channel due to an inherent charge amplification. The condensed ions on the DNA backbone amplified the ionic current in conjunction with the mobile surface charges, similar to a field-effect transistor.

Nanopore technology has attracted more attention from a number of research groups in recent years. Research has focused in a variety of directions: towards improving the structures, reducing the time to fabricate, fine-tuning noise parameters, studying the effects of changing voltages, studying force kinetics, exploring

conformation changes of molecules, and most importantly the continuous use of nanopores in investigating various biophysical properties of DNA. *Chen* et al. employed atomic-layer deposition (ALD) of alumina to controllably shrink the pore size to required nanometer dimensions. ALD was shown to reduce the noise of the pore and change the surface properties with the passivation effects of deposited alumina film [81.60]. The same group compared dsDNA translocation through nanopores of different channel lengths: nanopore in a thin membrane and nanochannels through a thicker membrane. They showed that DNA of varying lengths (3, 10, and 48.5 kbp) all traveled at the same velocity through both types of nanopores. In contrast to the nonlinearity in mobility versus electric field in gel electrophoresis, they showed that the DNA electrophoretic mobility is independent of the strength of the applied electric field and the length (molecular weight) of the DNA. It is interesting to note here that size separation in gel electrophoresis is based on the length-dependent mobility of the DNA. In another study, the effects of imaging beam of a TEM through nanopores with various geometries have also been studied [81.61]. It was shown that pores smaller than a certain critical size shrank while larger ones expanded, on the same wafer, agreeing with the hypothesis that surface tension effects drove the modifications. The chemical composition in the pore region was also determined, proving that contamination growth was not the underlying mechanism of pore closure. In another approach to make nanopores, *Ho* et al. used tightly focused e-Beam to make nanopores in 10 nm thick silicon nitride membranes [81.62]. They measured the ionic conductance as a function of various conditions and found it to be much larger than the bulk conductivity in case of dilute bath concentrations, whereas it was found to be comparable with, or less than, the bulk for high bath concentrations. These observations were explained in terms of the Debye length, which was larger than the pore radius for the former case. They also reported consistent multiscale simulations using molecular dynamics of the ion transport through the pores. They explained the ion transport by the coupled Poisson–Nernst–Planck and Stokes equations solved self-consistently for the ion concentration, velocity, and electrical potential. The results suggested the presence of fixed negative charges on the pore walls, thus the ion proximity to the pore walls reduced the ion mobility.

In most experiments, DNA translocates the nanopores at speeds as high as $27-30\,\text{bases}/\mu\text{s}$ [81.63]. It was reported that the translocation speed could be reduced by controlling the electrolyte temperature, salt concentration, viscosity, and the electrical bias voltage across the nanopores [81.64]. The speed was reduced to 3 bases/μs for 3 kbp dsDNA through a 4–8 nm diameter silicon nitride pore. This significantly reduced the bandwidth requirements on electronic sensing system. However, slowing the DNA also slowed conducting ions, which decreased the current blockage signal. In another study, it was reported that most of the translocation events followed a rule of constant event charge deficit. The experiments compared the distribution of molecular event durations and blockade at various pH values, conclusively demonstrating a large number of long denatured ssDNA translocation in stretched-out conformation [81.65]. The control of the translocation speed and denaturation of dsDNA into ssDNA at high pH is an elegant capability with synthetic nanopores, which would be a challenge with α-HL biological pores. This work showed an ability to detect hybridization from the translocation behavior.

To study pore formation after electron beam deformation, electron energy-loss spectroscopy and high-resolution electron microscopy studies were reported [81.66]. The material composition was studied extensively during pore shrinkage, pore expansion, and pore closure, concluding that pore formation occurred by removal of material at the entrance and exit side of the electron beam. The pore formed had wedge-like edges and contained Si, O, and N. *Heng* et al. investigated the mechanical properties of DNA using translocation of single molecules through the nanopore [81.67]. They used pores with radii from 0.5 to 1.5 nm in 10 nm thick SiN membranes. They also carried out molecular dynamics simulations to explore the stretching and bending of DNA during translocation through such small pores. They reported a pore-dimension- and bias-dependent threshold for translocation. Simulating a 1 nm pore, they calculated forces on DNA to be in the range of 1–300 pN through it and observed the threshold for dsDNA translocation conditioned by its mechanical properties. The same authors also reported the permeation capabilities through various sizes of nanopores by changing the electric fields [81.68]. The electromechanical properties of DNA were studied to define threshold of nanopore sizes that ssDNA or dsDNA could permeate under various electric fields. Previously, the effect of pH on such threshold was also investigated by the same group [81.69, 70]. In the wake of the low-salt current-enhancement studies reported by *Bashir* and co-workers [81.58], *Smeets* et al. reported the ef-

fects of salt concentration on ion transport and DNA translocation using 10 nm diameter nanopores [81.59]. They changed the salt concentration by six orders from 1 M to 1 μM, noting a three order of magnitude decrease in ionic conductance, deviating from linear bulk behavior. They carried out 16.5 μm long dsDNA translocation experiments for 50 mM to 1 M salt concentrations, observing downward or upward translocation pulses as a function of ion concentration. They, like *Bashir* and co-workers had explained [81.58], described a model attributing current decreases due to partial blocking of the pore and current increases to the motion of the counterions screening the charge of the DNA backbone. They concluded a threshold KCl concentration of 370 ± 40 mM where two competing effects canceled out. Using the same nanopores, fabrication of point electrodes with radii 2 nm was also reported [81.71]. The nanopores were filled with Au, making conically shaped nanoelectrodes. Metal nanoelectrodes of such high temporal resolution can be used for single-molecule detection and a variety of biological sensors for medical diagnostics. The fabrication of nanopores has also been reported using conventional field-emission scanning electron microscope (FESEM) instead of TEM [81.72]. The irradiation from the electron source of the FESEM showed different shrinking mechanism due to the effects of surface defects under radiolysis and subsequent motion of silicon atoms to the pore periphery. The process was reliable and reproducible, opening the way to parallel fabrication of nanopores in an array format. Another report investigated the ionic current fluctuations associated with the translocation of 1681 bp long DNA as a function of applied electric field. The pulse direction (upward or downward) was investigated to develop a model of DNA counterion ionic current and dipole saturation based on a remarkable work by *Manning* [81.73]. These experiments showed that, at low concentrations, the electrophoretic bias can reverse the pulse directions from upwards to downwards, indicating competition between the mechanical blocking current and the current arising from the condensed counterions on DNA backbone, which saturated at high electric field. The experiments provided a direct way to explore the charge polarization and dipole saturation at the single-molecule level.

Dekker and co-workers have also measured the force on a single DNA while translocating through a nanopore by using optical tweezers [81.74, 75], and ionic transport in electrochemical reactions [81.76]. The optical tweezers were used to slow down and arrest the

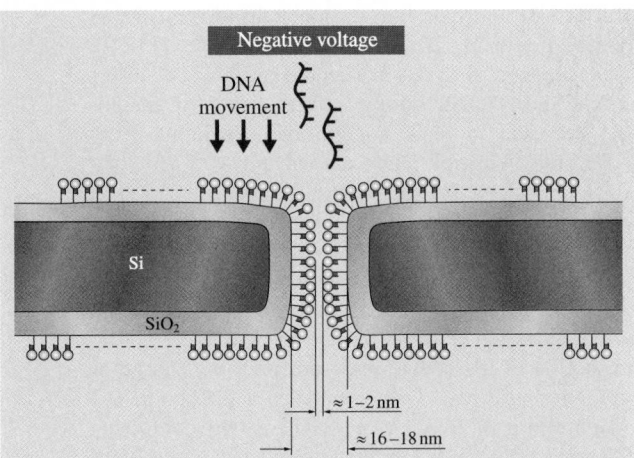

Fig. 81.9 Schematic of the functionalized nanopore channels. The chemical modification and the probe hairpin-loop DNA reduced the NPC diameter. Adapted by permission from Macmillan Publishers Ltd: Nanotechnology [81.77], copyright (2007)

DNA midway through the pore and force measurements were carried out. The study reported effective charge of 0.50 ± 0.05 electrons per base pair, which is equivalent to 25% of the bare DNA charge.

As explained earlier, it is important to slow down DNA to gather more information with reasonable bandwidth requirements on the electronic measurement systems, while still maintaining high signal-to-noise ratio. In a recent report, *Iqbal* et al. chemically functionalized the nanopore channels (NPC) with hairpin-loop DNA [81.77]. The functionalization with known hairpin-loop *probe* ssDNA showed selective transport behavior for *target* ssDNA that was complementary to the probe. This chemical modification of the surface also restricted the movement of the target ssDNA. The hairpin-loop configuration imparted single-base-mismatch sensitivity. Such a single base mismatch between the probe and the target resulted in longer translocation pulses and a significantly reduced number of translocation events. Figure 81.9 shows a schematic of the functionalized NPC and the effective diameter of the NPC after functionalization. They used short ssDNA as target molecules. The target ssDNA were stiff polymers at the length used in this study, thus avoiding the *folding* effect [81.78]. These single-molecule measurements allowed separate measurements of the molecular flux and the pulse duration, providing a tool

to gain fundamental insight into the channel–molecule interactions. They discovered that perfectly complementary ssDNA translocated faster and was transported in higher numbers across the pores compared with a target that had even a single base mismatch with the probe molecules. The experimental observations showed that nanopores can be used as selective sensors for genes of interest and that such selectivity could be electrically measured by the translocation signatures at the single-molecule level.

81.4.3 The Promise of Low-Cost DNA Sequencing

The human genome project, completed in 2003, is estimated to have cost US\$ 2.7 billion [81.79]. It took more than 12 years to complete. With the advent of novel sequencing technologies and abundant computation power the cost and time needed for sequencing a genome has reduced dramatically, now at ≈ 3 cents/base [81.80]. However, this is an enormous amount for 3 billion bases to be sequenced. Nanopores, in spite of their shortcomings and open challenges, are a hope to achieve the US\$ 1000/genome target as set by the National Institutes of Health with more than US\$ 70 million in grant monies. The challenges of nanopore technology are both material and electrical. The reproducibility of the nanopore channels is a concern where even an angstromical difference in channel length changes the conductivity of the pore and hence the baseline currents. Although the baseline can be normalized, the noise varies from pore to pore owing to the differences of physical and chemical properties of the pore wall surfaces. There are also issues of bandwidth; DNA passes way too fast for the electronics readout to capture the full details of translocation. The best optimal bandwidth of measurements would still miss too many fast translocation events, significantly skewing the analysis of the pulse behavior, while reducing the noises. The bandwidth issue thus limits the speed of classifications that can be made with nanopores. There are a number of noise sources that are encountered in nanopore measurements; especially $1/f$ (flicker) noise. Tackling such diverse scientific challenges in the nanopore measurements is an uphill task. However, the promise of cheap and fast sequencing machines makes every effort worth pursuing.

81.5 Conclusions and Outlook

This chapter summarized the state of the art in probing the electrical properties of molecules with special emphasis on DNA conductivity. The various techniques for DNA immobilization were discussed. The biological ion channel pores extended to the idea of solid-state nanopore, and the state of the art in nanopore fabrication and its applications, were covered. Very recently, two review articles have appeared [81.81, 82] that cover the history and state of the art of the nanopore research in more detail. For details of laboratory protocols of setting up biological and synthetic nanopores for DNA analysis, a laboratory manual appeared in December 2007 [81.83]. It is the consensus that nanopores have set some ambitious goals for the research community, foremost being the sequencing of the human genome for less than US$ 1000 dollar. The confluence of the sophisticated electronic sensing and nanopore detection techniques has strong potential of achieving the US$ 1000 genome and much more.

References

81.1   K.S. Kwok, J.C. Ellenbogen: Moletronics: future electronics, Mater. Today **5**, 28–37 (2002)
81.2   G. Maubach, W. Fritzsche: Precise positioning of individual DNA structures in electrode gaps by self-organization onto guiding microstructures, Nano Lett. **4**(4), 607–611 (2004)
81.3   A.J. Storm: Single Molecule Experiments on DNA with Novel Silicon Nanostructures. Ph.D. Thesis (Delft University Press, Delf 2004)
81.4   T.T. Nikiforov, Y.H. Rogers: The use of 96-well polystyrene plates for DNA hybridization-based assays: An evaluation of different approaches to oligonucleotide immobilization, Anal. Biochem. **227**(1), 201–209 (1995)
81.5   L. Zheng, J.P. Brody, P.J. Burke: Electronic manipulation of DNA, proteins, and nanoparticles for potential circuit assembly, Biosens. Bioelectron. **20**, 606–619 (2004)
81.6   B. Xu, P. Zhang, X. Li, N. Tao: Direct conductance measurement of single DNA molecules in aqueous solution, Nano Lett. **4**(6), 1105–1108 (2004)
81.7   B.J. Taft, M. O'Keefe, J.T. Fourkas, S.O. Kelley: Engineering DNA-electrode connectivities: manipulation of linker length and structure, Anal. Chim. Acta **496**(1), 81–91 (2003)
81.8   J. Wang: Nanoparticle-based electrochemical DNA detection, Anal. Chim. Acta **500**, 247–257 (2003)
81.9   D.-S. Kim, Y.-T. Jeong, H.-J. Park, J.-K. Shin, P. Choi, J.-H. Lee, G. Lim: An FET-type charge sensor for highly sensitive detection of DNA sequence, Biosens. Bioelectron. **20**(1), 69–74 (2004)
81.10 M. Gabig-Ciminska: Developing nucleic acid-based electrical detection systems, Microb. Cell Fact. **5**(1), 9 (2006)
81.11 J.B. Lamture, K.L. Beattie, B.E. Burke, M.D. Eggers, D.J. Ehrlich, R. Fowler, M.A. Hollis, B.B. Kosicki, R.K. Reich, S.R. Smith: Direct detection of nucleic acid hybridization on the surface of a charge coupled device, Nucl. Acids Res. **22**(11), 2121–2125 (1994)
81.12 M. Bras, V. Dugas, F. Bessueille, J.P. Cloarec, J.R. Martin, M. Cabrera, J.P. Chauvet, E. Souteyrand, M. Garrigues: Optimisation of a silicon/silicon dioxide substrate for a fluorescence DNA microarray, Biosens. Bioelectron. **20**(4), 796–805 (2004)
81.13 L. Cai, H. Tabata, T. Kawai: Self-assembled DNA networks and their electrical conductivity, Appl. Phys. Lett. **77**(19), 3105–3106 (2000)
81.14 E. Braun, Y. Eichen, U. Sivan, G. Ben-Yoseph: DNA-templated assembly and electrode attachment of a conducting silver wire, Nature **391**, 775–778 (1998)
81.15 A. Macanovic, C. Marquette, C. Polychronakos, M.F. Lawrence: Impedance-based detection of DNA sequences using a silicon transducer with PNA as the probe layer, Nucl. Acids Res. **32**(2), e20 (2004)
81.16 M. Manning, S. Harvey, P. Galvin, G. Redmond: A versatile multi-platform biochip surface attachment chemistry, Mater. Sci. Eng. C **23**, 347–351 (2003)
81.17 C. Adessi, S. Walch, M.P. Anantram: Environment and structure influence on DNA conduction, Phys. Rev. B: Condens. Matter **67**, 081405.1–081405.4 (2003)
81.18 R.G. Endres, D.L. Cox, R.R.P. Singh: Colloquium: the quest for high-conductance DNA, Rev. Mod. Phys. **76**, 195–214 (2004)
81.19 M.A. Young, G. Ravishanker, D.L. Beveridge: A 5-nanosecond molecular dynamics trajectory for B-DNA: analysis of structure, motions, and solvation, Biophys. J. **73**, 2313–2336 (1997)
81.20 D.D. Eley, D.I. Spivey: Semiconductivity of organic substances, Trans. Faraday Soc. **58**, 411–415 (1962)
81.21 R.S. Snart: The electrical properties and stability of DNA to UV radiation and aromatic hydrocarbons, Biopolymers **12**, 1493–1503 (1973)
81.22 C.J. Murphy, M.R. Arkin, Y. Jenkins, N.D. Ghatlia, S.H. Bossmann, N.J. Turro, J.K. Barton: Long range photoinduced electron transfer through a DNA helix, Science **262**, 1025–1029 (1993)

Part H | 81

81.23 A.M. Brun, A. Harriman: Energy- and electron-transfer processes involving palladium porphyrins bound to DNA, J. Am. Chem. Soc. **116**, 10383–93 (1994)

81.24 D. Ly, L. Sanii, G.B. Schuster: Mechanism of charge transport in DNA: internally-linked anthraquinone conjugates support phonon-assisted polaron hopping, J. Am. Chem. Soc. **121**, 9400–9410 (1999)

81.25 B. Giese, S. Wessely: The influence of mismatches on long-distance charge transport through DNA, Angew. Chem. Int. Ed. **39**(19), 3490–3491 (2000)

81.26 H.W. Fink, C. Schönenberger: Electrical conduction through DNA molecules, Nature **398**, 407–410 (1999)

81.27 D. Porath, A. Bezryadin, S. de Vries, C. Dekker: Direct measurement of electrical transport through DNA molecules, Nature **403**, 635–638 (2000)

81.28 P.J. de Pablo, F. Moreno-Herrero, J. Colchero, J.G. Herrero, P. Herrero, A.M. Baró, P. Ordejón, J.M. Soler, E. Artacho: Absence of DC-conductivity in λ-DNA, Phys. Rev. Lett. **85**, 4992–4995 (2000)

81.29 A.Y. Kasumov, M. Kociak, S. Guéron, B. Reulet, V.T. Volkov, D.V. Klinov, H. Bouchiat: Proximity-induced superconductivity in DNA, Science **291**(5502), 280–282 (2001)

81.30 M. Di Ventra, M. Zwolak: DNA electronics. In: *Encyclopedia of Nanoscience and Nanotechnology*, ed. by H.S. Nalwa (American Scientific Publishers, Los Angeles 2004)

81.31 K.-H. Yoo, D.H. Ha, J.-O. Lee, J.W. Park, J. Kim, J.J. Kim, H.-Y. Lee, T. Kawai, H.Y. Choi: Electrical conduction through poly(dA)-poly(dT) and poly(dG)-poly(dC) DNA molecules, Phys. Rev. Lett. **87**(19), 198102-1–198102-4 (2001)

81.32 A.J. Storm, J. van Noort, S. de Vries, C. Dekker: Insulating behavior for DNA molecules between nanoelectrodes at the 100 nm length scale, Appl. Phys. Lett. **79**(23), 3881–3883 (2001)

81.33 H.-Y. Lee, H. Tanaka, Y. Otsuka, K.-H. Yoo, J.-O. Lee, T. Kawai: Control of electrical conduction in DNA using oxygen hole doping, Appl. Phys. Lett. **80**(9), 1670–1672 (2002)

81.34 C. Gómez-Navarro, F. Moreno-Herrero, P.J. de Pablo, J. Colchero, J. Gómez-Herrero, A.M. Baró: Contactless experiments on individual DNA molecules show no evidence for molecular wire behavior, Proc. Natl. Acad. Sci. USA **99**(13), 8484–8487 (2002)

81.35 B. Xu, P. Zhang, X. Li, N. Tao: Direct conductance measurement of single DNA molecules in aqueous solution, Nano Lett. **4**(6), 1105–1108 (2004)

81.36 C. Nogues, S.R. Cohen, S.S. Daube, R. Naaman: Electrical properties of short DNA oligomers characterized by conducting atomic force microscopy, Phys. Chem. Chem. Phys. **6**(18), 4459–4466 (2004)

81.37 S.M. Iqbal, G. Balasundaram, S. Ghosh, D.E. Bergstrom, R. Bashir: DC electrical characterization of ds-DNA in nano-gap junctions, Appl. Phys. Lett. **86**(15), 153901-3 (2005)

81.38 Y. Otsuka, H. Lee, J. Gu, J. Lee, K.-H. Yoo, H. Tanaka, H. Tabata, T. Kawai: Influence of humidity on the electrical conductivity of synthesized DNA film on nanogap electrode, Jpn. J. Appl. Phys. **41**(2A), 891–894 (2002), part 1

81.39 A.K. Mahapatro, K.J. Jeong, G.U. Lee, D.B. Janes: Sequence specific electronic conduction through polyion-stabilized double-stranded DNA in nanoscale break junctions, Nanotechnology **18**, 195202 (2007)

81.40 J. Jortner, M. Bixon, T. Langenbacher, M.E. Michel-Beyerle: Charge transfer and transport in DNA, Proc. Natl. Acad. Sci. USA **95**, 12759–12765 (1998)

81.41 Z.G. Yu, X. Song: Variable range hopping and electrical conductivity along the DNA double helix, Phys. Rev. Lett. **86**, 6018–6021 (2001)

81.42 J. Yi: Conduction of DNA molecules: a charge-ladder model, Phys. Rev. B: Condens. Matter **68**, 193103-1–193103-4 (2003)

81.43 Y. Zhu, C.-C. Kaun, H. Guo: Contact, charging, and disorder effects on charge transport through a model DNA molecule, Phys. Rev. B: Condens. Matter **69**, 245112-1–245112-7 (2004)

81.44 N.J. Turro, J.K. Barton: Paradigms, supermolecules, electron transfer and chemistry at a distance. What's the problem? The science or the paradigm?, J. Biol. Inorg. Chem. **3**(2), 201–209 (1998)

81.45 D.N. Beratan, S. Priyadarshy, S.M. Risser: DNA: insulator or wire?, Chem. Biol. **4**(1), 3–8 (1997)

81.46 Y.A. Berlin, A.L. Burin, M.A. Ratner: On the long-range charge transfer in DNA, J. Phys. Chem. A **104**(3), 443–445 (2000)

81.47 M. Zwolak, M.D. Ventra: Electronic signature of DNA nucleotides via transverse transport, Nano Lett. **5**(3), 421–424 (2005)

81.48 J.J. Kasianowicz, E. Brandin, D. Branton, D.W. Deamer: Characterization of individual polynucleotide molecules using a membrane channel, Proc. Natl. Acad. Sci. USA **93**, 13770–13773 (1996)

81.49 L.Q. Gu, O. Braha, S. Conlan, S. Cheley, H. Bayley: Stochastic sensing of organic analytes by a pore-forming protein containing a molecular adapter, Nature **398**, 686–690 (1999)

81.50 A. Meller, L. Nivon, E. Brandin, J. Golovchenko, D. Branton: Rapid nanopore discrimination between single polynucleotide molecules, Proc. Natl. Acad. Sci. USA **97**, 1079–1084 (2000)

81.51 S. Howorka, S. Cheley, H. Bayley: Sequence-specific detection of individual DNA strands using engineered nanopores, Nat. Biotechnol. **19**(7), 636–639 (2001)

81.52 Y. Astier, O. Braha, H. Bayley: Toward single molecule DNA sequencing: direct identification of ribonucleoside and deoxyribonucleoside 5'-monophosphates by using an engineered protein nanopore equipped with a molecular adapter, J. Am. Chem. Soc. **128**(5), 1705–1710 (2006)

81.53 J.-H. Choy, S.-Y. Kwak, J.-S. Park, Y.-J. Jeong, J. Portier: Intercalative nanohybrids of nucleoside monophosphates and DNA in layered metal hydroxide, J. Am. Chem. Soc. **121**(6), 1399–1400 (1999)

81.54 K.S. Ralls, R.A. Buhrman, R.C. Tiberiom: Fabrication of thin-film metal nanobridges, Appl. Phys. Lett. **55**(23), 2459–2461 (1989)

81.55 N.N. Gribov, S.J.C.H. Theeuwen, J. Caro, S. Radelaar: A new fabrication process for metallic point contacts, Microelectron. Eng. **35**, 317–320 (1997)

81.56 J. Li, D. Stein, C. Mcmullan, D. Branton, M.J. Aziz, J.A. Golovchenko: Ion-beam sculpting at nanometre length scales, Nature **412**, 166–169 (2001)

81.57 J. Li, M. Gershow, D. Stein, E. Brandin, J.A. Golovchenko: DNA molecules and configurations in a solid-state nanopore microscope, Nat. Mater. **2**, 611–615 (2003)

81.58 H. Chang, F. Kosari, G. Andreadakis, M.A. Alam, G. Vasmatzis, R. Bashir: DNA-mediated fluctuations in ionic current through silicon oxide nanopore channels, Nano Lett. **4**(8), 1551–1556 (2004)

81.59 R.M.M. Smeets, U.F. Keyser, D. Krapf, M.-Y. Wu, N.H. Dekker, C. Dekker: Salt dependence of ion transport and DNA translocation through solid-state nanopores, Nano Lett. **6**(1), 89–95 (2006)

81.60 P. Chen, T. Mitsui, D.B. Farmer, J. Golovchenko, R.G. Gordon, D. Branton: Atomic layer deposition to fine-tune the surface properties and diameters of fabricated nanopores, Nano Lett. **4**, 133–1337 (2004)

81.61 A.J. Storm, J.H. Chen, X.S. Ling, H.W. Zandbergen, C. Dekker: Electron-beam-induced deformations of SiO_2 nanostructures, J. Appl. Phys. **98**, 014307-1–014307-8 (2005)

81.62 C. Ho, R. Qiao, J.B. Heng, A. Chatterjee, R.J. Timp, N.R. Aluru, G. Timp: Electrolytic transport through a synthetic nanometer-diameter pore, Proc. Natl. Acad. Sci. USA **102**(30), 10445–10450 (2005)

81.63 P. Chen, J. Gu, E. Brandin, Y.-R. Kim, Q. Wang, D. Branton: Probing single DNA molecule transport using fabricated nanopores, Nano Lett. **4**(11), 2293–2298 (2004)

81.64 D. Fologea, J. Uplinger, B. Thomas, D.S. McNabb, J. Li: Slowing DNA translocation in a solid-state nanopore, Nano Lett. **5**(9), 1734–1737 (2005)

81.65 D. Fologea, M. Gershow, B. Ledden, D.S. McNabb, J.A. Golovchenko, J. Li: Detecting single stranded DNA with a solid state nanopore, Nano Lett. **5**(10), 1905–1909 (2005)

81.66 M.-Y. Wu, D. Krapf, M. Zandbergen, H. Zandbergena, P.E. Batson: Formation of nanopores in a SiN/SiO_2 membrane with an electron beam, Appl. Phys. Lett. **87**, 113106-1–113106-3 (2005)

81.67 J.B. Heng, A. Aksimentiev, C. Ho, P. Marks, Y.V. Grinkova, S. Sligar, K. Schulten, G. Timp: Stretching DNA using the electric field in a synthetic nanopore, Nano Lett. **5**(10), 1883–1888 (2005)

81.68 J.B. Heng, A. Aksimentiev, C. Ho, P. Marks, Y.V. Grinkova, S. Sligar, K. Schulten, G. Timp: The electromechanics of DNA in a synthetic nanopore, Biophys. J. **90**, 1098–1106 (2006)

81.69 J.B. Heng, C. Ho, T. Kim, R. Timp, A. Aksimentiev, Y.V. Grinkova, S. Sligar, K. Schulten, G. Timp: Sizing DNA using a nanometer-diameter pore, Biophys. J. **87**, 2905–2911 (2004)

81.70 A. Aksimentiev, J.B. Heng, G. Timp, K. Schulten: Microscopic kinetics of DNA translocation through synthetic nanopores, Biophys. J. **87**, 2086–2097 (2004)

81.71 D. Krapf, M.-Y. Wu, R.M.M. Smeets, H.W. Zandbergen, C. Dekker, S.G. Lemay: Fabrication and characterization of nanopore-based electrodes with radii down to 2 nm, Nano Lett. **6**(1), 105–109 (2006)

81.72 H. Chang, S.M. Iqbal, E.A. Stach, A.H. King, N.J. Zaluzec, R. Bashir: Fabrication and characterization of solid-state nanopores using a field emission scanning electron microscope, Appl. Phys. Lett. **88**, 103109-1–103109-3 (2006)

81.73 G.S. Manning: Molecular theory of polyelectrolyte solutions with applications to electrostatic properties of polynucleotides, Q. Rev. Biophys. **11**, 179–246 (1978)

81.74 U.F. Keyser, B.N. Koeleman, S. van Dorp, D. Krapf, R.M.M. Smeets, S.G. Lemay, N.H. Dekker, C. Dekker: Direct force measurements on DNA in a solid-state nanopore, Nat. Phys. **2**(7), 473–477 (2006)

81.75 U.F. Keyser, J. van der Does, C. Dekker, N.H. Dekker: Optical tweezers for force measurements on DNA in nanopores, Rev. Sci. Instrum. **77**, 105105-1–105105-9 (2006)

81.76 D. Krapf, B.M. Quinn, M.-Y. Wu, H.W. Zandbergen, C. Dekker, S.G. Lemay: Experimental observation of nonlinear ionic transport at the nanometer scale, Nano Lett. **6**(11), 2531–2535 (2006)

81.77 S.M. Iqbal, D. Akin, R. Bashir: Solid-state nanopore channels with DNA selectivity, Nat. Nanotechnol. **2**, 243–248 (2007)

81.78 A.J. Storm, J.H. Chen, H.W. Zandbergen, C. Dekker: Translocation of double-strand DNA through a silicon oxide nanopore, Phys. Rev. E **71**, 051903-1–051903-10 (2005)

81.79 E. Pennisi: HUMAN GENOME: Reaching their goal early, sequencing labs celebrate, Science **300**(5618), 409–09 (2003)

81.80 R.F. Service: GENE SEQUENCING: The race for the \$1000 Genome, Science **311**(5767), 1544–1546 (2006)

81.81 C. Dekker: Solid-state nanopores, Nat. Nanotechnol. **2**, 1–7 (2007)

81.82 K. Healy: Nanopore-based single-molecule DNA analysis, Nanomedicine **2**(4), 459–481 (2007)

81.83 M. Wanunu, A. Meller: Single-molecule analysis of nucleic acids and DNA-protein interactions using nanopores. In: *Single-Molecule Techniques: A Laboratory Manual*, ed. by P. Selvin, T.J. Ha (Cold Spring Harbor Laboratory Press, Cold Spring Harbor 2007) pp. 395–420

Part H | 81

82. Computer and Robot-Assisted Medical Intervention

Jocelyne Troccaz

Medical robotics includes assistive devices used by the physician in order to make his/her diagnostic or therapeutic practice easier and more efficient. This chapter focuses on such systems. It introduces the general field of computer-assisted medical interventions, its aims, and its different components, and describes the place of robots in this context. The evolution in terms of general design and control paradigms in the development of medical robots are presented and issues specific to that application domain are discussed. A view of existing systems, ongoing developments, and future trends is given. A case study is detailed. Other types of robotic help in the medical environment exist (such as for assisting a handicapped person, for rehabilitation of a patient or for replacement of some damaged/suppressed limbs or organs) but are outside the scope of this chapter.

82.1 Clinical Context and Objectives

Informatics and technology have dramatically transformed clinical practice over the last decades. This technically oriented evolution has run parallel to other specific evolutions of medicine:

- Diagnostic and therapy procedures tend to be less and less invasive for the patient, aiming at reducing pain, postoperative complications, hospital stay, and recovery time. Minimal invasiveness results in smaller targets reached through narrow access (natural or not) with no direct sensing (vision, touch) and limited degrees of freedom.
- More and more data are handled for each patient (e.g., images, signals) in order to prepare and monitor the medical action, and these multimodal data have to be shared by several participating actors.
- As in many other domains, quality control gets increasingly important and quantitative indicators have to be made available.
- Traceability becomes mandatory, especially regarding the ever-increasing number of legal cases. Traceability is also of primary importance in cost management.

These evolutions make the medical action increasingly complex for the clinician, at both the technical and organizational levels. Computer-assisted medical intervention may contribute a lot to those clinical objectives.

Part H | 82

They may provide quantitative and rational collaborative access to patient information, fusion of multimodal data, and their exploitation for planning and execution of medical actions.

82.2 Computer-Assisted Medical Intervention

The development of this area from the early 1980s results from converging evolutions in medicine, physics, materials, electronics, informatics, robotics, etc. This field and related subfields are given several almost synonymous names: computer-assisted medical intervention (the most general), augmented surgery, computer-assisted surgery, image-guided surgery, medical robotics, surgical navigation, etc. We will use the term *computer-assisted medical intervention* (CAMI) herein for this domain.

We define CAMI as aiming to provide tools that allow the clinician to use multimodal data in a rational and quantitative way in order to plan, simulate, and accurately and safely execute mini-invasive medical interventions. Medical interventions include both diagnostic and therapeutic actions. Therapy may involve surgery, radiotherapy, local injection of drugs, interventional radiology, etc. (Some of these medical terms are explained in a glossary located at the end of this chapter.)

82.2.1 CAMI Major Components

CAMI [82.1] may be described as a perception–decision–action loop, as presented in Fig. 82.1. The perception phase includes data acquisition and processing, the development of specific sensors, and their calibration. Data may be acquired pre-, intra- or postoperatively. Images may provide anatomical information or functional one; data provided by the sensors may be one-dimensional (1-D), two-dimensional (2-D), three-dimensional (3-D, sparse or dense) or four-dimensional (4-D, i.e., varying with time). Each imaging sensor brings its specific type of information and multimodality is in general necessary. The need to assist the intervention in a quantitative and accurate way requires the calibration of sensors enabling both the transformation from image coordinates to spatial coordinates and the correction of possible image distortions. Specific sensors such as position sensors (also called localizers) and surface sensors have been integrated into CAMI components. Localizers give access to position and orientation of objects (instruments, other sensors, anatomical structures such as bones). Surface sensors give access to the external surface of an object (an organ, for instance). All those information are potentially useful to plan and control the execution of a medical action.

The main objective of the decision stage is to build an integrated numerical model of the patient and, in some cases, a model of the action. As mentioned previously, many types of information may be useful: data provided by imaging sensors, for instance, but also a priori medical knowledge (for instance, statistical data about organ shapes or occurrence of pathologies, biomechanical models of the limbs, etc.). One very important stage is data fusion, also called registration, which corresponds to the action of representing all the information in a single reference frame. Registration, and in particular medical image registration [82.2], has been a very active domain for three decades. From this integrated model, the medical action can be planned; for instance, one may have to determine the type, size, position, and orientation of a knee prosthesis that provides the best alignment of hip, knee, and ankle joints; or one may decide which number, shape, intensity, position, and orientation of radiation beams would allow to radiate a tumor of a given shape, with a given dose, whilst sparing organs at risks. Planning may involve highly interactive tools where the clinician navigates in the data and specifies the selected strategy. It may also include optimization tools when the medical goal

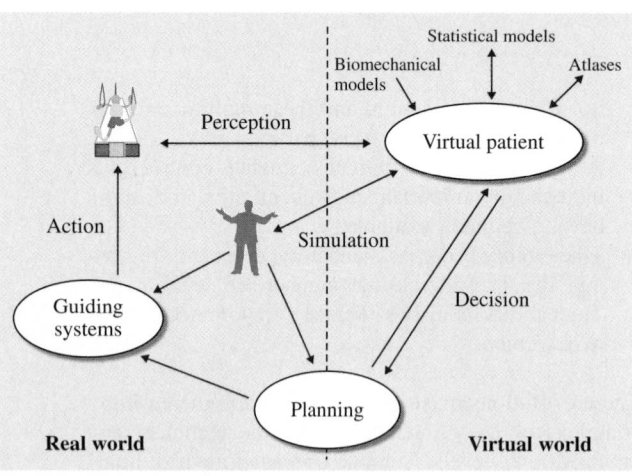

Fig. 82.1 CAMI methodology

can be specified as an optimization problem (radiotherapy planning is an excellent example). In some cases, planning may be very difficult and simulators could provide help for the computation of the clinical outcome of a selected action; for instance, when a bone fragment has to be moved in the patient's face, it may be very useful for the surgeon to foresee the functional and aesthetic consequences of this gesture on soft tissues [82.3]. Registration and planning are included in the decision stage.

Action requires accurate execution of the planned intervention. Often, a registration stage is necessary to transfer the planned action to the interventional conditions; for instance, the action is planned from preoperative data, e.g., magnetic resonance imaging (MRI), and must be registered to the real patient in intraoperative conditions. Two main types of assistance exist: navigational aids and robots. In the first case, the action is monitored using suitable sensors such as localizers; information is rendered to the clinician about the planned and executed actions. The clinician makes use of this information to control his/her action, such systems are called *passive*. The first surgical navigators have been used for neurosurgery [82.4, 5]. The action can also be performed more of less autonomously by a robot. The first application of a robot in CAMI took place around 1985; the application field was also neurosurgery. The medical robot needs to be connected to patient data and models in order to be able to transfer the planned intervention to the robot coordinates; this is the robot registration problem. The medical robot is therefore always an *image-guided robot*.

Simulators may also be developed for training purposes. The advantages are the abilities to provide frequent and rare medical cases, to gain realistic experience with lower stress than with real patients, and to quantitatively evaluate the practitioner. This may facilitate the acquisition of new medical skills, with new techniques and/or tools. This can also be considered a very valuable component of CAMI.

Based on numerical data, tracking of objects (instruments, sensors, and anatomical structures), and positioning of tools with navigation or robotized aids, those CAMI procedures are fully traceable.

82.2.2 Added Value of a Robot

Automation is generally not a primary goal of medical robotics, where the interaction with a clinical operator has to be considered with a very special attention. Indeed, most often medical robots are not intended to replace the operator but rather to assist him/her where his/her capabilities are limited. In general CAMI systems are considered only as sophisticated tools in the hands of the clinician.

In medicine, like in many other application areas, the advantages of the robot are its precision, ability to repeat a task endlessly, potential connection to computerized data and sensors, capability to operate in hostile environments [biological or nuclear contaminations, war or catastrophe areas, space (orbital station) or undersea (submarine), etc.] where clinician presence or abilities may be limited and humans may need medical care.

Navigational aids have already demonstrated their clinical added value in various specialties (neurosurgery, orthopaedics in particular) and their integration in the clinical environment is generally easier than for a robot. Safety issues are also more limited. Moreover navigation systems are very often more cost effective. These are the reasons why it is very important to use a robot only for clinical applications where it can offer functionalities that the navigation system cannot; potential specific robot abilities are:

- To realize complex geometric tasks (for instance, to machine a 3-D bone cavity)
- To handle heavy tools (e.g., radiation apparatus) or sensors (e.g., an intraoperative surgical microscope)
- To provide a third hand to the clinician
- To be remotely controllable and to offer scaling capabilities in terms of transmitted motions or forces
- To filter undesired movements (such as physiological shaking in teleoperation)
- To be force-controllable down to very small force scales
- To execute high-resolution high-accuracy motions (for microsurgery)
- To track moving organs and synchronize to external events based on some signals
- To be introduced into the patient for intrabody actions.

Another aspect that will be further discussed is the absolute necessity to demonstrate the clinical added value of the system, i.e., to prove that it brings a clear clinical benefit at some level (for the patient, hospital, or healthcare system).

82.3 Main Periods of Medical Robot Development

Early medical applications of robotics are characterized by transferring to this domain accurate and automated tool positioning capabilities of robots originally developed in industrial applications.

82.3.1 The Era of Automation (1985–1995)

As mentioned previously the first surgical robots were introduced in neurosurgery. Neurosurgery already had a very long tradition of minimal invasiveness and use of computerized 3-D imaging data; indeed, the first computed tomography (CT) scans in the early 1970s were performed for brain imaging. Stereotactic neurosurgery is a particular type of minimally invasive procedure which consists of *blindly* introducing a linear tool into the brain through a keyhole (3 mm diameter) for biopsies, removal of cysts or haematomas, placement of stimulation (Parkinson disease) or measurement (epilepsy) electrodes, etc. Conventionally, stereotactic neurosurgery is performed with the help of a stereotactic frame whose functions are to immobilize the patient skull, register preoperative data to the intraoperative situation (the frame is installed on the patient's head before the preoperative examination), and finally to offer mechanical guidance for tool insertion. Except for the first function, a robot can advantageously replace the frame: it is easily connected to imaging data, is less invasive, and may offer a larger range of trajectories for tool positioning. Anthropomorphic robots associated to stereotactic frames or robotized stereotactic frames have been developed and clinically evaluated in the early 1980s (for instance [82.6, 7]). Reference [82.7] describes the accurate positioning of a guiding tool with respect to preoperative data in a stereotactic neu-

rosurgery application using a Puma 260 robot. The cited paper reports a series of 22 patients. While [82.8] presents a CAMI system integrating a modified industrial robot (reduced speed, nonbackdrivable joints) for a similar application. The first patient (Fig. 82.2a) was operated on with this system in 1989 and since then hundreds of patients have been treated with this technology. This system was the academic precursor of the Neuromate product (Fig. 82.2b) with which thousands of patients were treated. These systems are called *semi-active* because the robot is only a guiding device and the gesture (drilling the skull and inserting the needle) is still performed by the surgeon through a mechanical guide positioned by the robot.

Reference [82.9] proposed an automated view of the whole process of stereotactic neurosurgery: a robot equipped with different types of tools in a tool-feeder effector was installed in a CT room; after imaging data and planning, the robot performed the drilling of the bone and the placement of the surgical tool. CT enabled repeated image control. The robot was specifically designed for this application. The first two patients were operated in 1993. Eight cases of biopsies are reported in [82.9]. As far as we know, this system has not been extensively used in clinical routine.

In orthopaedics, the placement of prostheses requires the preparation of the bones; for instance a cavity has to be drilled before inserting the femoral component of a hip prosthesis; similarly, planar cuts have to be realized on the tibia and femur extremities in order to place knee prosthesis components. Thus, such stages of the interventions are very close to machining a mechanical part: a 3-D shape has to be accurately realized by sawing or milling with given position and orienta-

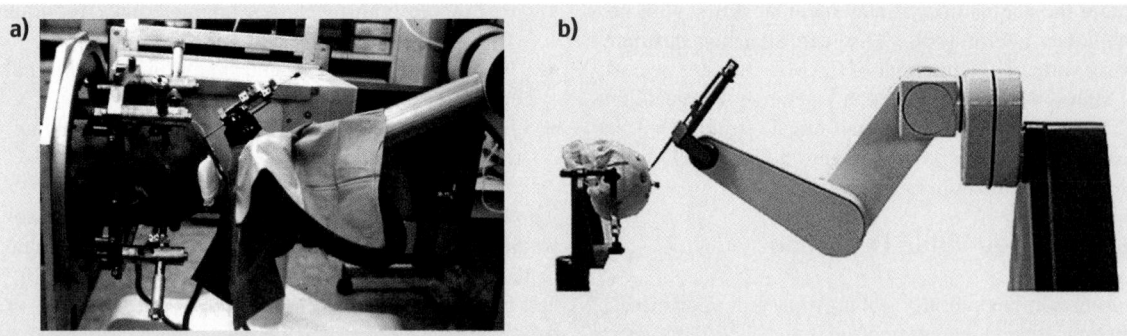

Fig. 82.2 (a) Grenoble robot for neurosurgery (TIMC laboratory and Grenoble University Hospital) and **(b)** the Neuromate industrial version

tion. This is why the idea of using a robot came very naturally for such tasks. Robodoc [82.10, 11], used for cavity preparation in total hip arthroplasty, was developed from 1986 first in a laboratory setup. Then, from 1989 to 1991, 29 dogs were operated with the help of the system. The first ten patients were included in a Food and Drug Administration (FDA)-approved clinical research trial between 1991 and 1993; very large series of patients for comparison of traditional interventions with Robodoc-assisted ones were included from 1995 to the early 2000s. Thousands of patients benefited from the use of Robodoc. Robodoc is *active* in the sense that a part of the surgical action (machining the bone) is performed autonomously by the robot under surgeon supervision. Several other systems were developed based on a similar approach.

The underlying idea in this first period was that surgical subtasks such as accurate positioning of a tool or machining, based on numerical data, could be transferred to a robot and automated to a certain extent.

82.3.2 The Era of Interactive Devices (1990–2005)

Whilst the automation era focused on rigid and nondeformable anatomical structures, the second era is characterized by the development of more interactive control schemes for complex tasks, in particular for interventions on soft tissues.

Indeed, in the two previous examples (Sect. 82.3.1), the anatomical structure of interest is rigid and nondeformable; in the case of stereotactic neurosurgery, the brain is accessed through a small hole. The trajectory through the hole is simple (a linear tool is inserted to a given target). Brain motion and deformation can be neglected and the skull is immobilized. As regards orthopaedics, bones are not deformable and can be fixed to external fixtures to avoid any motion. Obviously there are much more clinical situations where the procedures are more complex and concern soft, mobile, and deformable tissues. For such applications automation is often still out of reach or may not be the preferred solution, for instance, when the expertise of the clinician is so complex that it cannot totally be transferred to the robot.

In contrast to earlier automated robot control, robotic development in the mid 1990s was characterized by more direct operator control. In particular, efforts were applied towards teleoperation; this form of robotic application was traditionally used in the nuclear industry. In this situation, the surgeon is totally in control of the surgical tool through a master–slave apparatus. The distance between the master and slave components is highly variable: although the main use is for very close teleoperation (in the same room), some very long distance (thousands of kilometers) experiments have taken place. The function-transferring movements from the master to the slave may involve scaling down (forces and/or motions) and filtering. The main clinical applications are in endoscopic surgery where instruments and optics are introduced into the patient's body through small incisions. Those entry points limit the instruments' possible motions [four degrees of freedom (DOFs) instead of six] and the surgeon has to operate under video control. In a first stage, the motion of the optical system – the endoscope – alone was robotized: instead of requiring one assistant to move the endoscope for hand–eye coordination, the robot is controlled by the surgeon himself/herself by different means: voice control (AESOP [82.12]), head movements (Endoassist [82.13]) or high-level image processing software (Sect. 82.4). More recently the displacement of instruments has also transferred to robots, potentially offering extra DOFs; this is the case of the DaVinci multi-arm system that offers intrabody DOFs of the instruments [82.14].

From the mid 1990s synergistic devices [82.15] also named *hands-on* robots [82.16] were proposed. The rationale for such systems is that the clinical application is generally so complex that it can only be partially embedded in numerical models and data. Therefore relying both on the clinician for his/her very high skills, capacity of judgement, intelligent perception, and on the computerized robot for its quantitative knowledge of the planning, accuracy, and sensors is potentially very fruitful. In the PADyC [82.17, 18], Acrobot [82.16, 19], and Makoplasty [82.20] systems, the surgical tool is attached to the robot effector and the surgeon holds it. The motions proposed by the human operator are filtered by the robot in order to keep only the part of the motions which is compatible with the surgical plan; for instance, the tool has to keep in a plane or in a given region. Different technologies implement this principle of constrained motions: clutchable freewheels, backdrivable motors, controlled brakes, etc.

82.3.3 The Era of Small and Light Dedicated Devices (2000 to Present Day)

This era tends toward a miniaturization of robots up to the stage where they can be attached to a body

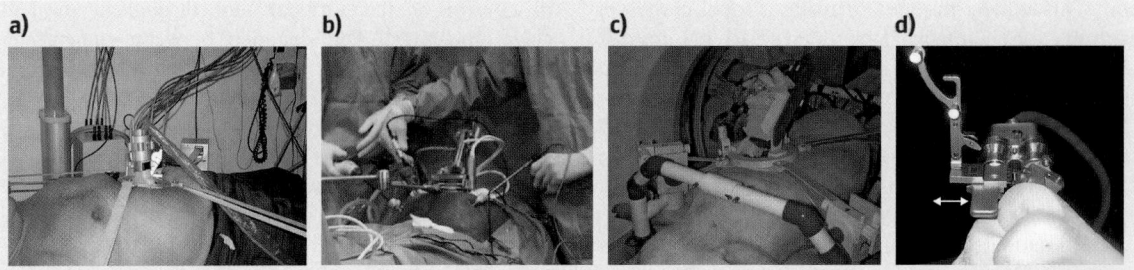

Fig. 82.3a–d On-body and on-bone robots at Grenoble: (**a**) TER slave robot for tele-ultrasound examination (TIMC, Grenoble University Hospital); (**b**) LER-Viky endoscope holder (TIMC, EndoControl-Medical Company, Grenoble University Hospital and Paris La Pitié Salpétrière Hospital); (**c**) LPR MRI- and CT-compatible robot for image-guided punctures (TIMC, Grenoble University Hospital); (**d**) Praxiteles robot for total knee arthroplasty (PRAXIM, TIMC, Brest University Hospital)

part, inserted into the patient's body or integrated into mechatronic surgical instruments.

In systems presented in Sects. 82.3.1 and 82.3.2, the robot mechanical architectures were inspired by traditional industrial robotics: general-purpose anthropomorphic six-DOF (or more) arms are used. Systems are very versatile and can adapt to a variety of tasks. However, this generality has a cost: these robots are also very often quite cumbersome and are attached to the floor, bed or ceiling. Their working space may be quite large and not easily manageable in an operating room (OR) that offers limited space and in which people (patient and staff) are very close. This is why new generations of robots have been specifically designed and developed for limited application areas.

This new generation of systems has produced robots so small that they can be supported by the patient body or attached to one of his/her bones. Because of their size these robots are more easily integrated into the OR and are designed to be very well suited to a specific task. These two factors offer potential advantages over traditional systems by providing a safer operating environment and cost effectiveness. The Grenoble Techniques for Biomedical Engineering and Complexity Management (TIMC) laboratory was a pioneer in small-scale robotics, and Fig. 82.3 illustrates four such pioneering works. TER [82.21] (Fig. 82.3a), a robot for tele-ultrasonic examination, has been clinically validated for remote examination of abdominal aortic aneurysm and for emergency care of abdominal traumas; LER [82.22], a light endoscope holder, has been validated on pigs and corpses, and VIKY, the corresponding industrial product (Fig. 82.3b) is European Community (EC)-marked and has been used on patients; LPR [82.23] (Fig. 82.3c), a CT/MRI-compatible

robot for punctures, has been validated for CT on pigs and for MRI on volunteers (but without puncture); Praxiteles [82.24] (Fig. 82.3d), a robot for total knee arthroplasty, has been validated on corpses and recently entered multicentric clinical evaluation on patients.

Following the same philosophy, the Mars system (now called Mazor [82.25], a product) is presented in detail in Chap. 78. It is being clinically evaluated on patients for spine surgery. This list of *body-supported* or *bone-mounted* devices is not exhaustive.

Other systems are sufficiently small to be completely introduced into the body of the patient. Some groups focus on locomotion issues; for instance, several versions of an inchworm-type compact robot that moves inside the intestine have been proposed [82.26]. The aim is to replace the long, quite rigid, and painful endoscope traditionally used in colonoscopy; experiments on pigs have been carried out successfully. With a rather similar locomotion principle, a robot that moves on the surface of the heart has been developed [82.27]; experiments on live animals have been carried out. A robot rolling on soft tissue organs in the abdomen has also been described [82.28]; this robot carries a camera and experiments on animal organs have been carried out. Finally, a smaller and simpler device actuated from outside of the body using external magnetic fields has been proposed [82.29]; such a robot would be injected into the eye, for instance, for drug delivery to retina vessels.

Other research groups aim at giving extra DOFs to traditional instruments: the domain of active catheters that can adapt actively to vessel curvature has been investigated for several years [82.30, 31]. Recently, some of these systems have reached the market (Table 82.1). Finally, the field of articulated tools for endoscopic surgery is also very active; the aim is to develop surgical

Table 82.1 Industrial medical robots

System name	Clinical specialty	Type	Last distributing company	Estimated number of installed systems	Status
Neuromate	Stereotactic neurosurgery	Semiactive (mechanical guide)	Schaerer Mayfield	$15 < nb < 20$	Unknown
PathFinder	Stereotactic neurosurgery	Semiactive (mechanical guide)	Prosurgics Ltd.	Unknown	Unknown
Surgiscope	Microsurgery	Surgical microscope holder	ISIS	$15 < nb < 20$	No longer distributed
MKM	Microsurgery	Surgical microscope holder	Carl Zeiss	Unknown	No longer distributed
Robodoc	Orthopaedics (knee, hip)	Automated machining of bones	ISS Inc.	$70 < nb < 80$	No longer distributed in the USA and in Europe
Caspar	Orthopaedics (knee, hip)	Automated machining of bones	URS ortho	$50 < nb < 60$	No longer distributed
Cyberknife	Radiotherapy	Positioning and motion of the radiation device	Accuray Inc.	$134* < nb$	Growing
DaVinci	Endoscopic procedures (cardiac, digestive, gynecologic, urology, etc.)	Endoscope and instrument holder – intrabody DOFs	Intuitive Surgical Inc.	$875** < nb$	Growing
Zeus	Endoscopic procedures	Endoscope and instrument holder	Intuitive Surgical Inc.	Unknown	No longer distributed
Aesop	Endoscopic procedures	Endoscope holder	Intuitive Surgical Inc.	$800 < nb < 1000$	No longer distributed
EndoAssist	Endoscopic procedures	Endoscope holder	Prosurgics Ltd.	Unknown	Unknown
Naviot	Endoscopic procedures	Endoscope holder	Hitachi	Unknown	Unknown
Lapman	Endoscopic procedures	Endoscope holder	Medsys	Unknown	Unknown
Viky	Endoscopic procedures	On-body endoscope holder	Endocontrol Medical	Probably < 10	Emerging
Acrobot Sculptor	Orthopaedics (knee, spine, etc.)	*Hands-on* robot	Acrobot Ltd.	Unknown	Emerging
PIGalileo CAS	Orthopaedics (knee)	On-bone semiactive	PLUS Orthopedics AG	Unknown	Unknown
Praxiteles	Orthopaedics (knee)	On-bone semiactive	Praxim	Probably < 10	Emerging
Makoplasty	Orthopaedics (knee)	*Hands-on* robot	Mako Inc.	Probably < 10	Emerging
Mazor	Orthopaedics (spine, etc.)	On-bone semiactive	Mazor Surgical Technologies	$10 < nb < 20$	Emerging
Estele	Radiology (ultrasounds)	Telerobotics	Robosoft	Probably < 10	Emerging

Table 82.1 (cont.)

System name	Clinical specialty	Type	Last distributing company	Estimated number of installed systems	Status
Sensei	Interventional radiology (cardiology)	Teleoperated robotic catheter for heart mapping (intra-cardiac) with force feedback	Hansen Medical	Unknown	Emerging
CorPath	Interventional radiology (cardiology)	Teleoperated catheter (intracoronary)	Corindus	On evaluation – not yet distributed	Emerging

* Number of installed systems in early May 2008 (source Accuray Inc.)

** Number of installed systems in early May 2008 (source Intuitive Surgical Inc.)

tools equipped with intrabody DOFs to recover full mobility of the tools with respect to the organs. The specific area of intrabody devices raises very challenging issues related to biocompatibility, safety, power supply, and data transmission. Smart pills such as the M2A [82.32] or Norika [82.33] ones, which are swallowed by a patient and enable visualization of the gastrointestinal track, do not yet integrate active devices but can be seen as precursors of future highly integrated mechatronic intrabody devices.

82.4 Evolution of Control Schemes

Evolution from *bone applications* to soft tissues also resulted in the evolution of control schemes providing a more active role for the surgeon or aiming at a real-time perception–decision–action control loop.

A large number of the oldest systems integrate a *single-shot* perception–decision–action process; for instance a planned trajectory is selected from CT data and transferred to the intraoperative conditions after registration of preoperative to intraoperative data. To guarantee that the plan is still valid intraoperatively, the anatomical structure of interest must not move. This approach has been largely used for neurosurgery and orthopaedics surgery; in both cases the structure is fixed using a stereotactic frame or external fixtures. The operator is often limited to the role of a supervisor when the robot moves the tool.

More recently (Sect. 82.3.2) the operator has been given a more active role in the execution of the task. In the case of *comanipulation* the robot and the operator participate simultaneously to the motions of the tool. In the Acrobot system [82.19] the proportional–integral–derivation (PID) coefficients of the control law are given by functions that depend on the position of the tool with respect to a region of allowed motions, which corresponds for instance to the bone to be removed for prosthesis implantation: inside the region, the motions proposed by the operator and detected by a force sensor are transmitted to the tool without significant modification; in an intermediate region the motions are more or less transmitted depending on their direction; motions outside the planned region are strictly forbidden. The Steady Hand [82.34] works similarly: the operator manipulates a handle and proposed motions are detected. In contrast to the Acrobot, the Steady Hand does not select permitted directions of motions but rather scales down motions and forces for microsurgery or biology applications. In the case of PADyC the principle is slightly different [82.17, 18]; PADyC is a passive arm for which each joint is equipped with two freewheels than can be independently clutched or unclutched using a motor; the motor is velocity controlled, and this velocity determines the range of allowed motion of each joint in each direction at each instant. This range of motion is computed from the representation of the task (position to reach, trajectory to follow or region to keep inside for instance) and from the current position of the robot. No force sensor is necessary. For teleoperation, the operator interacts with a master device and the slave robot reproduces the motions proposed by the operator. Transfer functions may enable scaling and filtering of disturbing shaking motions like in the DaVinci. Teleoperation can integrate a force feedback to the operator like in the Sensei system for endocardial robotized catheter control.

Few systems integrate hybrid force–position control of the robot. One application case is external ultrasound examination, which requires constant contact of the ultrasonic probe on the body of the patient. Thus several robotic systems for ultrasound examination (remote or automated) have been developed with a hybrid control scheme [82.35, 36]. They generally combine position control in the main direction of motion with force control in order to maintain contact during probe motion. Dermarob [82.37], developed for skin sampling for skin grafts in burnt patients, follows a similar hybrid control scheme.

Some of the systems developed for orthopaedics and neurosurgery offer some basic tracking abilities using a localizer and markers attached to rigid structures of interest. This is, for instance, the case of Caspar, which is used for ligamentoplasty in knee surgery, and the case of a frameless version of the Neuromate. Systems such as Mazor or Praxiteles which are mounted on the anatomical structure of interest move with it and therefore suppress the tracking problem. Recently other systems have been developed with evolved abilities for tracking of soft tissues. Instead of having a fixed target, the robot has to track it in real time, in general from information extracted from imaging data. Typical examples are related to organ motions induced by cardiac or respiratory cycles. In those cases, a real-time perception–decision–action loop must be developed. The Cyberknife system (Sect. 82.5) developed for radiotherapy [82.38, 39] determines the current position of a tumor moving with patient respiration by using external markers localized in real time [82.40, 41]. Alternative approaches consist of coupling motion in the image information to motion of the robot at a low level through visual servoing. Pioneer work was described in [82.42]; from high-speed camera data, the motion of relevant fiducials on the heart surface was computed and fed back to a slave robot for teleoperation, giving the operator the feeling of a stable target. More recently several groups have applied visual servoing to image-guided minimally invasive actions. Video images are used for robotized endoscopy [82.43, 44]; ultrasonic images have also been introduced [82.45, 46] for cardiac and vascular applications. Directly coupling motion detection in the images to motion of the joints requires very careful robustness analysis.

82.5 The Cyberknife System: A Case Study

This system is described in more detail to make the intrinsic complexity of a robotic CAMI system a bit more visible. This section is also intended to introduce the potentially long road from an initial paradigm and a first prototype to a clinically used product. Specific issues relating to medical robotics and explaining this long road are discussed in Sect. 82.6.

The Cyberknife system, distributed by Accuray Inc., has been developed in the context of radiotherapy. Usually, when treating a patient with conventional linear accelerators, the patient is positioned onto a couch that has four main DOFs (three translations and one rotation), and the linear accelerator allows orientation of the radiation beam using two more rotational DOFs. The convergence point of the radiation beams is termed the *isocenter*. A reference frame R_{iso} centered on this point is associated with the complete radiation system. The DOFs enable positioning of the tumor on the isocenter and orientation of the radiation beams properly with respect to the tumor in order to execute the planned treatment (Sect. 82.2.1). In practice, due to the complexity of positioning the patient and orienting the beams with such machines, treatments generally consist of rather simple ballistics with rather few radiation beams. The Cyberknife concept proposes to install the radiation source on a six-DOF robot in order to avoid repeated combined motion of the couch and linear accelerator. This enables the execution of complex treatments with hundreds of beams distributed around the tumor without having to move the patient. Using such a large number of small beams enables sculpting of the dose distribution to the tumor shape with high accuracy.

The first version of the Cyberknife [82.38] was installed at Stanford Medical Center in February 1994, and the first patient was treated with the system on June 8, 1994. The current version of the system integrates a Kuka robot (Fig. 82.4a). In early May 2008, the Accuray Company reported 134 installed systems worldwide and 40 000 patients treated, mostly for brain, spine, lung, prostate, liver or pancreas tumors.

Regarding patient data, preoperative CT is traditionally used for treatment planning; other modalities can

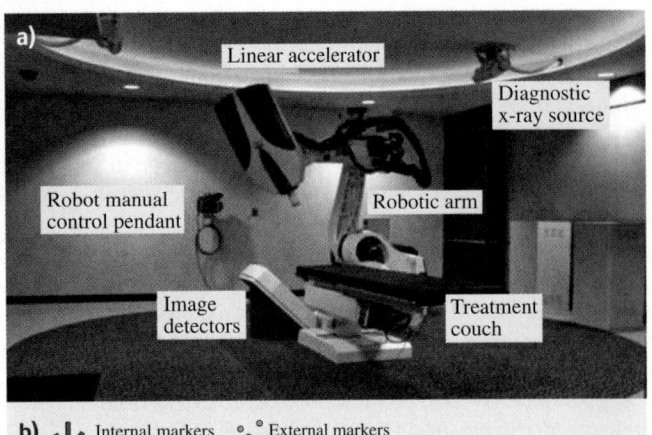

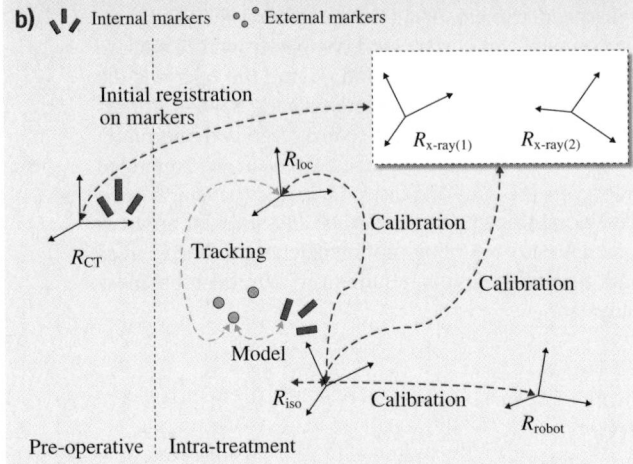

Fig. 82.4a,b The Cyberknife system: (**a**) the treatment room setup, (**b**) functional description of tracking functionalities

ery from a single beam. Thanks to the CT exam, two synthetic x-ray images called digitally reconstructed radiographs (DRRs) are computed. They correspond to what would be seen by the two x-rays machines in the treatment room for a perfect position/orientation of the patient with respect to planning data. Two real images are acquired just before dose delivery and automatically compared with the DRRs; this results in the computation of the patient's shift from the ideal position/orientation. For small errors, the robot position/orientation with respect to the patient is automatically corrected by combining this information with calibration data. For larger discrepancies, a replanning phase is necessary to avoid any collision during robot repositioning.

A later version of the system enables real-time tracking of organs that move with patient respiration (lung, liver, and kidney, for instance) [82.40]. As regards patient respiration during treatment, several approaches are used in conventional radiotherapy: the less accurate method consists of enlarging the targeted zone to account for tumor motion; irradiating healthy tissues is more acceptable than missing cancerous cells. This enlarged zone can be computed by combining information from two scanners acquired at the end of the exhale and inhale phases. A second approach consists of synchronizing dose delivery to a given stage of the respiratory cycle (end of inhalation, for instance) but this stage has to be reasonably repeatable and detected reliably. In a more sophisticated approach, such as that developed in the Synchrony version of Cyberknife, the system tracks the organ motion and the robot follows this motion during dose delivery in order to execute the planned treatment properly [82.41]. Since it is not possible to obtain tracking information from x-ray images – as such organs may not be visible on radiographs and continuous imaging would result in overirradiation of the patient – a localizer is introduced into the treatment room. This localizer, associated to R_{loc} and calibrated with respect to R_{iso}, is used to track passive markers placed on the patient's chest; it delivers the marker positions about 20 times per second. X-ray images can be acquired about every 10 s. Because the motion of internal organs is different from chest external motion internal radiopaque markers are implanted close to the tumor before the preoperative CT examination. The relationship between the tumor position and the internal markers is determined using the CT data; the internal markers enable initial patient setup by registration of CT data to x-ray images. The relationship between internal markers visible on x-ray images and

be fused to CT data such as MRI or positron emission tomography (PET) for planning refinement. Intraoperative x-ray imaging produced by two x-ray systems enables initial patient positioning and participates in tracking. The system is calibrated, which means that the spatial relationships between the different reference frames R_{iso}, R_{robot}, and R_{x-ray} (associated, respectively, with the isocenter, robot, and x-ray devices) is determined using specific procedures and calibration objects. Such calibration is necessary to transform intratreatment imaging information into robot positions with respect to the radiation system. The transfer of planning information from R_{CT} to R_{iso} is possible due to image registration procedures described in more detail in the following sections.

In the first version of the system, dedicated to neuro applications, after initial setup, patient motion was detected and corrected before each dose deliv-

external markers tracked by the localizer is learnt in a preliminary stage where both data are acquired during several breathing cycles. During dose delivery, between two successive x-ray acquisitions, the position of internal markers is interpolated from the position of the external markers and from the model; each new acquisition enriches the model by adding a new couple of synchronous positions of internal and external markers. Finally, the position of the tumor can be deduced from the internal markers and fed back to the robot for tracking and accurate dose delivery in spite of organ motion. The different components of the tracking function are illustrated in Fig. 82.4b. Current work deals with noninvasive alternatives to the implanted internal markers.

As can be seen in this example, a lot of hardware and software components are associated with the robot and many stages are involved in real-time determination of the robot position with respect to the patient. To provide the required accuracy for precise dose delivery,

errors have to be reduced as far as possible for each of these elements; this is especially demanding. Moreover, because the robot develops very large forces and torques in the immediate vicinity of the patient, reliability and robustness are mandatory. Unfortunately, as for many other commercialized systems, due to intellectual property issues few data are available concerning detailed industrial developments and technical testing.

The Cyberknife approach supposes the ability to install a linear accelerator as end-effector of the robot; the Cyberknife robot typically carries 6 MeV accelerators which are quite compact and light compared with other radiation devices. When higher power is necessary or when radiation beams are produced by synchrotrons and/or cyclotrons – for proton therapy, for instance – the radiation source cannot be positioned and oriented by a robot. Alternative systems [82.47] have been proposed to position the patient robotically relative to the radiation beam by means of parallel robotized seats or robotized couches.

82.6 Specific Issues in Medical Robotics

As has been briefly introduced in the previous sections, medical robotics raises specific issues at the technical, clinical, and organizational levels in a very intricate way. This section discusses those issues in more detail.

Firstly this type of robot has to be used in a human environment. Because these robotic systems generally require close collaboration with a clinician, specific man–machine interface questions have to be resolved. In the case of surgery, the operator cannot interact easily with classical man–machine interfaces because he/she has to work under sterile conditions. This motivates the development of specific interfaces such as voice control or foot pedals. In the case of comanipulation, the part of the robot that is held by the clinician must be made sterile before each intervention. Because the robot is used in close proximity to human beings – at least the patient – safety issues are mandatory. Safety has to be demonstrated both at the hardware and software levels. Various approaches are possible [82.48]. The choice of specific architectures of robots or comanipulation control modes may solve part of the problem. Norms and regulations are intended to guarantee that the system is safe. However, it would certainly be interesting to develop such systems using specific design methodologies introduced in critical applications (aeronautics, nuclear plants, etc.) to really anticipate the behavior of such a complex system and avoid any misuse. This direction is still to be explored.

The clinical use of the robot also requires electromagnetic compatibility with the environment, in particular in the OR. When the robot is used in specific environments such as inside a CT or MRI imaging sensor, the robot must not disturb image acquisition or corrupt the data. This may significantly constrain design choices and selected materials: for instance in the MRI environment no ferromagnetic parts should be integrated into the robot. Sterile cleaning of the robot is also an issue. Bone-mounted robots, which come into very close contact to the patient, must be completely cleanable. This is the case, for instance, for the Praxiteles and LER-VIKY systems (Sect. 82.3.3), which are autoclavable. In the case of larger robots, their end-effector may be introduced inside sterile disposable plastic bags. However, the robot connection with the tool that is in contact with the patient has to be cleaned in a sterile way. Finally, the system and the robot must be designed such that the robotic procedure can be easily and rapidly converted into the conventional intervention, at any time, in case of problems. This may also have important design consequences.

As mentioned previously, introducing a new medical device requires demonstrating the added value

Part H | 82.6

over existing techniques from a clinical perspective; for instance, using the robot could enable performing less-invasive interventions, resulting in shorter stay of the patient in the hospital. In the same way, using a robot to hold the endoscope may save one assistant who can be transferred where his/her skills are used in a much better way. However, evaluating such organizational benefits obviously requires very precise cost and resource management: all necessary human or material resources participating in the intervention, from diagnosis to long-term follow-up, have to be taken into account. On a more medical level, using a robot to machine bone for prosthesis placement may result in longer lifetime of the prosthesis and/or fewer joint disjunctions (for the hip, for instance). In a similar way using a complex radiotherapy treatment delivered using a robot may allow dose escalation and may result both in better control of a tumor and fewer complications. In general, it is not easy to predict how accurate and sophisticated a robotic procedure should be to make a significant difference on the clinical level. To prove this may also be very challenging. The evaluation of clinical benefits may require long-term trials involving several centers and many patients. Those clinical trials have to be conducted in accordance with the ethical standards and regulations of the country concerned; those standards and regulations may be very strict, but they vary from country to country. Finally, added value may also be evaluated in terms of commercial advantage for a hospital that may attract more patients when high technology is used for painless, minimally invasive procedures. This added value should be significant enough to compensate for drawbacks related to the introduction of a robot such as the increase of procedure duration, which is often observed even when the learning period is finished. (See Chap. 77 for additional content on automation in hospitals and healthcare.)

Cost is obviously another issue; indeed, several of the distributed systems are quite expensive. If we consider some of the systems (for instance, Robodoc, Caspar, DaVinci, and Cyberknife) listed in Table 82.1, costs average in the range US $1 000 000–2 000 000. Aesop cost around US $100 000. Frequently a maintenance cost of 10% per year has to be added, and some of these systems generate an extra cost per intervention (for instance, about US $1000–2000 for the DaVinci system). This may be quite a heavy cost for hospitals and clinics. Moreover, depending on healthcare funding models in different countries, some of those costs may not be affordable by health insurances. The higher the investment, the more significant the added value has to be to justify the expense. More recently developed systems (smaller and simpler robots, disposable devices) are likely to propose more affordable solutions.

82.7 Systems Used in Clinical Practice

For 20 years many medical robotic systems have been developed in laboratories and evaluated to a certain extent. Evaluation is twofold: at the technical level it consists of characterizing accuracy, reliability, robustness, etc. This stage may be realized on laboratory setups using phantoms that mimic, more or less realistically, the concerned part of the body. At the clinical level, experiments with corpses or animals enable a first approach to a more realistic evaluation of clinical feasibility and performance. Finally, a study on series of patients is always necessary to fully evaluate the system and its clinical added value. Relatively few medical robots have undergone the whole evaluation process, reached the market, and gone into wide clinical use. There are indeed two major challenges: how to turn a laboratory prototype into a certified product, and how to make this product an industrial success. The reasons for this still limited diffusion of medical robotics in the clinical world certainly come from the specific constraints of medical robotics discussed above and probably from the questionable added value of the robot in a number of cases. The complexity of clinical evaluation, certification, and marketing also makes the process very long and expensive; for instance, in orthopaedics, demonstrating the advantage of robots over competing techniques may take more than 10 years since the stability and life duration of prostheses cannot be demonstrated any earlier. At the same time, to evaluate them it is necessary to install robots in hospitals, sometimes at the company's expense. Convincing a hospital to buy such expensive devices before any medical evidence of their added value is available is particularly challenging.

Table 82.1 attempts to list as largely as possible the industrial systems that are, or have been, significantly clinically used in routine and emerging products; this table is intended to give a flavor of the clinical spread of the technique. Numbers of systems are estimates established in early 2008. As can be seen several systems are no longer distributed: this deserves further comments. Surgiscope and MKM, which are both *surgical microscope holders*, probably faced a limited *added value versus cost* ratio. As regards Robodoc, this system had not yet demonstrated a clinical benefit when misuse of the robot resulted in many clinical complications and legal cases in Germany; this ended up with the removal of the robot from the US and European markets. Caspar, which offered functions very similar to those of Robodoc, probably suffered from the same unproven added value issue and from the failure of Robodoc. Aesop and Zeus are no longer distributed due to intellectual property conflicts. Aesop was, however, very successful in clinics and could certainly be considered as an industrial success.

The success of DaVinci is probably due to its ability to offer intrabody DOFs for endoscopic surgery; laparoscopic radical prostatectomy is one of the main clinical vectors of the large dissemination of the DaVinci system. As regards Cyberknife, the ability to perform complex dose distributions with tens of small radiation beams and the capacity to radiate the tumor accurately during patient respiration are probably the keys to its success. Moreover, conventional radiation apparatus are expensive devices and cost issues may be less critical in the case of radiation therapy than for surgery.

Table 82.1 also shows that emerging systems are often based on quite different design philosophy: small dedicated systems with more interactive control schemes. As can be seen, favorite applications are in endoscopy and knee arthroplasty, where the market is large (6 000 000 laparoscopic surgeries per year worldwide, and 600 000 knee prostheses per year worldwide).

Many other systems are somewhere in between the academic and industrial worlds and are not included in this table. Several of those forthcoming devices concern the introduction of needles into the body (for biopsies, punctures, brachytherapy, etc.).

82.8 Conclusions and Emerging Trends

In this chapter, we have introduced the main motivations for computer-assisted medical interventions and presented the place of medical robots in this general paradigm. Different generations of robots have been proposed and evaluated worldwide, ranging from systems inspired from industrial automation to more specific clinical robots. In parallel to those evolutions in terms of robot architecture and application areas, we have shown that the control modes also evolved in different directions: giving a larger place to the operator through real cooperation or closing the control loop with real-time imaging data for tracking mobile and deformable targets. The future is probably in the merging of such type of controls; for instance, the robot might handle synchronization with a moving organ while the operator would control the fine motions with respect to the stabilized target.

As regards industrial products resulting from this domain, the evolution has been very similar. The future will tell if those new design choices will result in a much larger spread of medical robots. Reference [82.49] reports 39 000 service robots for professional use installed worldwide up to the end of 2006, among which 9% (about 3500 devices) would be medical robots. However, this report does not tell precisely what is included in this category (for example, integration of haptic devices, integration of mechanical localizers or others). From the numbers mentioned in Table 82.1, our estimate of installed robots in 2008 would be closer to 2500, which is probably less than half the number of installed navigation workstations. Increasing this ratio requires careful selection of applications with significant added value and the development of user-friendly cost-effective systems. Intrabody highly integrated mechatronic devices potentially open the range of applications in a dramatic way. This domain has still to be largely explored.

82.9 Medical Glossary

- Aneurysm: a local hernia on a blood vessel potentially resulting in vessel rupture and internal haemorrhage

- Autoclavable: being cleanable in the autoclave (pressured vapor sterilization with temperatures greater than 100 °C)

- Arthroplasty: plastic surgery of the joints involved, for instance, in joint replacement (e.g., hip, knee)
- Biopsy: the action of taking samples of a tissue with a needle for further analysis
- Brachytherapy: the introduction of radioactive seeds into an organ for tumor destruction
- Catheter: a flexible tube introduced into the body, typically blood vessels, for instance, to inject drugs or to place dilatation devices. Generally, the end of the catheter can be slightly curved by the physician from the outside of the body
- Colonoscopy: endoscopic examination of the colon
- Computed tomography (CT): 3-D imaging from x-ray acquisition, enabling good visualization of bony structures and air cavities
- Digitally reconstructed radiograph (DRR): a synthetic x-ray projection image computed from a CT volume being given the position of a virtual source and virtual image plane
- Endoscopic surgery: surgery involving a minimal access to the body through natural cavities or incisions and visualization of the internal organs using rigid or flexible optical sensors (the endoscope)
- EC marking: the certification from the European Community necessary for marketing a products of any type in the Economic European Community; it insures that the product complies with the European regulations in terms of safety, health, environment, etc.
- FDA: Food and Drug Administration; the US administration in charge of controlling the safety and efficacy of health-related products (http://www.fda.gov)
- Interventional radiology: a category of therapeutic or diagnostic procedures executed under imaging control

- Magnetic resonance imaging (MRI): 3-D imaging from magnetic resonance of hydrogen protons in the body, enabling detailed visualization of soft tissues
- Neurosurgery: brain or spine surgery
- Operating room (OR): also called operating theater
- Orthopaedics: a surgical specialty dealing with skeleton bones and joints
- Positron emission tomography (PET): a functional imaging modality in which a radioactive marker is associated to a metabolically active molecule and injected into the body of the patient; the metabolic activity is traceable thanks to the radioactive marker and can be reconstructed in 3-D thanks to tomography techniques similar to the one used for CT
- Puncture: the action of inserting a linear tool (a needle or an electrode for instance) into the body
- Preoperative, intraoperative, postoperative: before, during, after the intervention
- Prostatectomy: surgical removal of the prostate in case of cancer; laparoscopic prostatectomy is the minimally invasive version of this surgery
- Radiopaque: visible on x-ray images
- Radiotherapy: the destruction of pathologic tissues (mostly tumors) by ionizing particles
- Stereotactic neurosurgery: a minimally invasive access to the brain requiring very accurate localization of intracranial structures
- Stereotactic frame: a mechanical device for perfect immobilization of a patient's skull, also used for transferring the surgical plan and for guiding the surgical tool
- Ultrasonic examination: an imaging modality (2-D or 3-D) based on the propagation of ultrasound in the body; it visualizes tissue interfaces.

References

82.1 P. Cinquin, E. Bainville, C. Barbe, E. Bittar, V. Bouchard, L. Bricault, G. Champleboux, M. Chenin, L. Chevalier, Y. Delnondedieu, L. Desbat, V. Dessenne, A. Hamadeh, D. Henry, N. Laieb, S. Lavallée, J.M. Lefebvre, F. Leitner, Y. Menguy, F. Padieu, O. Peria, A. Poyet, M. Promayon, S. Rouault, P. Sautot, J. Troccaz, P. Vassal: Computer assisted medical interventions at TIMC laboratory: passive and semi-active aids, IEEE Eng. Med. Biol. Mag., special issue Robot. Surg. **14**(3), 254–263 (1995)

82.2 D.L.G. Hill, P.G. Batchelor, M. Holden, D.J. Hawkes: Medical image registration, Phys. Med. Biol. **46**, R1–R45 (2001)

82.3 M. Chabanas, V. Luboz, Y. Payan: Patient specific finite element model of the face soft tissues for computer-assisted maxillofacial surgery, Med. Image Anal. **7**(2), 131–151 (2003)

82.4 E. Watanabe, T. Watanabe, S. Manaka, Y. Mayanagi, K. Takakura: Three-dimensional digitizer (neuronavigator): A new equipment for CT guided stereotactic neurosurgery, Surg. Neurol. **27**, 543–547 (1987)

82.5 H.F. Reinhardt: Neuronavigation: a ten year review. In: *Computer-Integrated Surgery*, ed. by R.H. Taylor, R. Mosges, S. Lavallée (MIT Press, Cambridge 1996) pp. 329–341

82.6 P.J. Kelly, B.A. Kall, S.S. Goerss, F. Earnest: Computer-assisted stereotaxic laser resection of intra-axial brain neoplasms, J. Neurosurg. **64**(3), 427–439 (1986)

82.7 Y.S. Kwoh, J. Hou, E.A. Jonckheere, S. Hayati: A robot with improved absolute positioning accuracy for CT guidedstereotactic brain surgery, IEEE Trans. Biomed. Eng. **35**(2), 153–160 (1988)

82.8 S. Lavallée, J. Troccaz, L. Gaborit, P. Cinquin, A.L. Benabid, D. Hoffmann: Image guided operating robot: a clinical application in stereotactic neurosurgery, Proc. IEEE Conf. Robot. Autom. (Nice 1992) pp. 618–624

82.9 D. Glauser, H. Fankhauser, M. Epitaux, J.L. Helfti, A. Jaccottet: Neurosurgical robot Minerva: first results and current developments, J. Image-Guid. Surg. **1**(5), 266–272 (1995)

82.10 H.A. Paul, W.L. Bargar, B. Mittlestadt, B. Musits, R.H. Taylor, P. Kazanzides, J. Zuhars, B. Williamson, W. Hanson: Development of a surgical robot for cementless total hip arthroplasty, Clin. Orthopaed. Relat. Res. **285**, 57–66 (1992)

82.11 W.L. Bargar, A. Bauer, M. Borner: Primary and revision total hip replacement using the Robodoc(R) system, Clin. Orthopaed. Relat. Res. **354**, 82–91 (1998)

82.12 J. Sackier, Y. Wang: Robotically assisted laparoscopic surgery: from concept to development, Surg. Endosc. **8**, 63–66 (1994)

82.13 S. Aiono, J.M. Gilbert, B. Soin, P.A. Finlay, A. Gordan: Controlled trial of the introduction of a robotic camera assistant (EndoAssist) for laparoscopic cholecystectomy, Surg. Endosc. **16**(9), 1267–1270 (2002)

82.14 G.S. Guthart, J.K. Salisbury: The Intuitive telesurgery system: overview and application, Proc. IEEE Robot. Autom. Conf. (San Francisco 2000)

82.15 J. Troccaz, M. Peshkin, B.L. Davies: Guiding systems: introducing synergistic devices and discussing the different approaches, Med. Image Anal. **2**(2), 101–119 (1998)

82.16 M. Jakopec, F. Rodriguez y Baena, S.J. Harris, J. Cobb, B.L. Davies: The *hands-on* orthopaedic robot Acrobot: early clinical trials of total knee replacement surgery, IEEE Trans. Robot. Autom. **19**(5), 902–911 (2003)

82.17 J. Troccaz, Y. Delnondedieu: Semi-active guiding systems in surgery. A two-DOF prototype of the passive arm with dynamic constraints (PADyC), Mechatronics **6**(4), 399–421 (1996)

82.18 O. Schneider, J. Troccaz: A six degree of freedom passive arm with dynamic constraints (PADyC) for

cardiac surgery application: preliminary experiments, Comput. Aided Surg., special issue Med. Robot. **6**(6), 340–351 (2001)

82.19 S.C. Ho, R.D. Hibberd, B.L. Davies: Robot assisted knee surgery, IEEE Eng. Med. Biol. Mag. **14**(2), 292–300 (1995)

82.20 Mako, www.makosurgical.com

82.21 A. Vilchis, J. Troccaz, P. Cinquin, K. Masuda, F. Pellissier: A new robot architecture for tele-echography, IEEE Trans. Robot. Autom., special issue Med. Robot. **19**(5), 922–926 (2003)

82.22 P. Berkelman, P. Cinquin, E. Boidard, J. Troccaz, C. Letoublon, J.-M. Ayoubi: Design, control, and testing of a novel compact laparascopic endoscope manipulator, J. Syst. Control Eng. **217**(14), 329–341 (2003)

82.23 E. Taillant, J. Avila-Vilchis, C. Allegrini, I. Bricault, P. Cinquin: CT and MR compatible light puncture robot: architectural design and first experiments. In: *Medical Image Computing and Computer Assisted Interventions*, LNCS, Vol. 2, ed. by G. Goos, J. Hartmanis, J. van Leeuwen (Springer, Berlin Heidelberg 2004) pp. 145–152

82.24 C. Plaskos, P. Cinquin, S. Lavallée, A. Hodgson: Praxiteles: a miniature bone-mounted robot for minimal access total knee arthroplasty, Int. J. Med. Robot. Comput. Aided Surg. **1**(4), 67–79 (2006)

82.25 M. Shoham, M. Burman, E. Zehavi, L. Joskowicz, E. Batkilin, Y. Kunicher: Bone-mounted miniature robot for surgical procedures: concept and clinical applications, IEEE Trans. Robot. Autom., special issue Med. Robot. **19**(5), 893–901 (2003)

82.26 B. Kim, Y. Jeong, H. Lim, J. Park, A. Menciassi, P. Dario: Functional colonoscope robot system, Proc. IEEE Int. Conf. Robot. Autom. (2003)

82.27 N.A. Patronik, C.N. Riviere, S. El Qarra, M.A. Zenati: *The HeartLander: A Novel Epicardial Crawling Robot for Myocardial Injections* Int. Congr. Ser. (CARS'2005), Vol. 1281 (Elsevier, Amsterdam 2005) pp. 735–739

82.28 M.A. Rentschler, S.M. Farritor, K.D. Iagnemma: Mechanical design of robotic in vivo wheeled mobility, J. Mech. Des. **129**, 1037–1045 (2007)

82.29 K.B. Yesin, K. Vollmers, B.J. Nelson: Modeling and control of untethered biomicrorobots in a fluidic environment using electromagnetic fields, Int. J. Robot. Res. **21**(5/6), 527–536 (2006)

82.30 K. Ikuta, M. Tsukamoto, S. Hirose: Shape memory alloy servo actuator with electric resistance feedback and application to active endoscope, Proc. IEEE Int. Conf. Robot. Autom. (Philadelphia 1988) pp. 427–430

82.31 T. Fukuda, S. Guo, K. Kosuge, F. Arai, M. Negoro, K. Nakabayashi: Micro active catheter system with multi degrees of freedom, Proc. IEEE Int. Conf. Robot. Autom. (San Diego 1994) pp. 2290–2295

82.32 www.givenimaging.com

82.33 www.rfnorika.com

Part H | 82

82.34   R. Kumar, P. Berkelman, P. Gupta, A. Barnes, P.S. Jensen, L.L. Whitcomb, R.H. Taylor: Preliminary experiments in cooperative human/robot force control for robot assisted microsurgical manipulation, Proc. IEEE Int. Conf. Robot. Autom. (San Francisco 2000) pp. 610–617

82.35   W.H. Zhu, S.E. Salcudean, S. Bachmann, P. Abolmaesumi: Motion/force/image control of a diagnostic ultrasound robot, Proc. IEEE Int. Conf. Robot. Autom. (San Francisco 2000) pp. 1580–1586

82.36   E. Dombre, X. Thérond, E. Dégoulange, F. Pierrot: Robot-Assisted detection of atheromatous plaques in arteries, Proc. IARP Workshop Med. Robot. (Vienna 1996) pp. 133–140

82.37   E. Dombre, G. Duchemin, P. Poignet, F. Pierrot: Dermarob: A safe robot for reconstructive surgery, IEEE Trans. Robot. Autom., special issue Med. Robot. **19**(5), 876–884 (2003)

82.38   A. Schweikard, M. Bodduluri, J.M. Adler: Planning for camara-guided robotic radiosurgery, IEEE Trans. Robot. Autom. **14**(6), 951–962 (1998)

82.39   M. Bodduluri, J.M. McCarthy: X-ray guided robotic radiosurgery for solid tumors, Proc. IEEE/ASME Int. Conf. Adv. Intell. Mechatron. (Como 2001) pp. 1065–1069

82.40   A. Schweikard, G. Glosser, M. Bodduluri, M.J. Murphy, J.R. Adler: Robotic motion compensation for respiratory movement during radiosurgery, Comput. Aided Surg., special issue Plan. Image Guid. Radiat. Ther. **5**(4), 263–277 (2000)

82.41   A. Schweikard, H. Shiomi, J. Adler: Respiration tracking in radiosurgery without fiducials, Int. J. Med. Robot. Comput. Assist. Surg. **1**(2), 19–27 (2005)

82.42   Y. Nakamura, K. Kishi, H. Kawakami: Heart beat synchronization for robotic cardiac surgery, Proc. IEEE Int. Conf. Robot. Autom. (Seoul 2001) pp. 2014–2019

82.43   A. Krupa, J. Gangloff, C. Doignon, M.F. de Mathelin, G. Morel, J. Leroy, L. Soler, J. Marescaux: Autonomous 3-D positioning of surgical instruments in robotized laparoscopic surgery using visual servoing, IEEE Trans. Robot. Autom., special issue Med. Robot. **19**(5), 842–853 (2003)

82.44   S. Voros, J.A. Long, P. Cinquin: Automatic detection of instruments in laparoscopic images: a first step towards high-level command of robotic endoscopic holders, Int. J. Robot. Res. **26**(11–12), 1173–1190 (2007)

82.45   M.A. Vitrani, G. Morel, T. Ortmaier: Automatic guidance of a surgical instrument with ultrasound based visual servoing, Proc. IEEE Int. Conf. Robot. Autom. (Barcelona 2005) pp. 508–513

82.46   P. Abolmaesumi, S.E. Salcudean, W.H. Zhu, M.R. Sirouspour, S.P. DiMaio: Image-guided control of a robot for medical ultrasound, IEEE Trans. Robot. Autom. **18**(1), 11–23 (2002)

82.47   S. Pinault, G. Morel, M. Auger, R. Ferrand, G. Mabit: Using an external registration system for daily patient repositioning in protontherapy, Proc. IEEE/RSJ Int. Conf. Intell. Robot. Syst. (IROS'07) (San Diego 2007) pp. 4289–4294

82.48   B.L. Davies: A discussion of safety issues for medical robots. In: *Computer-Integrated Surgery*, ed. by R.H. Taylor, R. Mosges, S. Lavallée (MIT Press, Cambridge 1996) pp. 287–296

82.49   www.worldrobotics.org

Part I

Part I Home, Office, and Enterprise Automation

Home, Office, and Enterprise Automation. Part I This part is about functional automation areas at home, in the office, and in general enterprises, including multi-enterprise networks. Common to these areas is the power of automation appliances, such as home and office appliances, and of Internetworking, such as collaborative decision support and scientific exploration, collaborative work, and e-Activities, from service to business, banking to home-office, laboratories, schools, government and tele-work. Chapters also cover the automation theories, techniques and practice, design, operation, challenges and emerging trends in education and learning, banking, commerce. General automation principles and functions include decision support, collaboration, integration, planning, services for unique needs such as disabilities, and business process automation. An important dimension of the material compiled for this part is that it is useful for all other functional areas of automation, for instance, office and school automation, interoperability, client relations management, enterprise resource planning, and collaborative decision support (all subjects of this part) are directly applicable to industrial, infrastructure, agriculture, medical, transportation, service, and other domains of automation.

83. Automation in Home Appliances

T. Joseph Lui

Home appliances, by their very nature, represent realizations of the principles of automation. Home appliances exist for the purpose of automating otherwise manual processes in the home. The operation of home appliances has been refined over the years, though the machine function has remained essentially the same. Advancements in the areas of microprocessor-based controls, sensors, displays, and interconnectivity, however, are enabling a new generation of appliances with advanced automation capabilities. Smart refrigerators, smart cooking appliances, and smart cleaning appliances are already appearing on the market. Along with these appliances we observe the viability of advanced applications in home automation. Software-based controls, appliance area networks (AANs), and display devices capable of creating a rich user experience are enabling advances in refrigeration automation, cooking integration automation, automated home utility management, automated fault and performance

monitoring, and more. In this chapter we explore the enabling technologies and applications of advanced home appliance automation.

Convenience – in the form of home chore automation – has always been the primary justification for a consumer to purchase a home appliance. Major home activities including refrigeration, cooking, and cleaning have enjoyed substantial levels of automation for many years. In this chapter, we explore the history of home chore automation, enabling technologies for advanced automation, and several applications thereof. We will not address heating, ventilation and air-conditionning (HVAC), lighting, entertainment, or security systems. Interested readers can refer to Chaps. 62 and 71 for treatment of these topics. Here, we focus specifically on major appliances, namely, refrigerators, dish washers, cook tops and ovens, clothes washers and dryers, and microwave ovens. We do not address smaller appliances such as food processors.

83.1 Background and Theory

83.1.1 History

Home appliances, by their very nature, automate an otherwise manual process. The clothes washer automates the manual process of fabric care. The food processor automates portions of the manual processes of food preparation. The list goes on. So, in a sense, the home was the first place in which automation received mass-market appeal. The automatic clothes washing machine was introduced in the USA commercially in 1911 and was a common feature of US homes by the close of the 1920s. Today, 80% of US homes have a clothes washer [83.1]. Moreover, though the microwave oven was only introduced in 1967, the convenience provided has resulted in around 85% of households having the appliance today. 1967 is the year of introduction of the first commercially viable countertop microwave oven: the *Radarange* from Raytheon/Amana. The first microwave oven of any kind was introduced in 1947, also by Raytheon [83.2].

Home appliances are on a long evolution path in which they have been improved aesthetically, refined operationally, and cost-reduced. However, the degree of automation of the fundamental processes has not changed substantially: a refrigerator still essentially keeps food cold; a water heater still basically keeps water hot. Typically, the quantum leaps in home appliance automation have not occurred within existing appliances, but with the introduction of new appliances. Examples include the microwave oven in 1967 and the trash compactor in 1969. However, opportunities for increased automation of the processes performed by existing appliances do exist and are under development.

Advanced automation opportunities for home appliances appear on two fronts: convenience and precision. Existing processes can be made more convenient for the user, for example, by reducing the amount of user input and intervention required by the process. Processes can be made more precise and, in particular, more efficient by transferring some decision-making responsibilities from the user to the appliance itself.

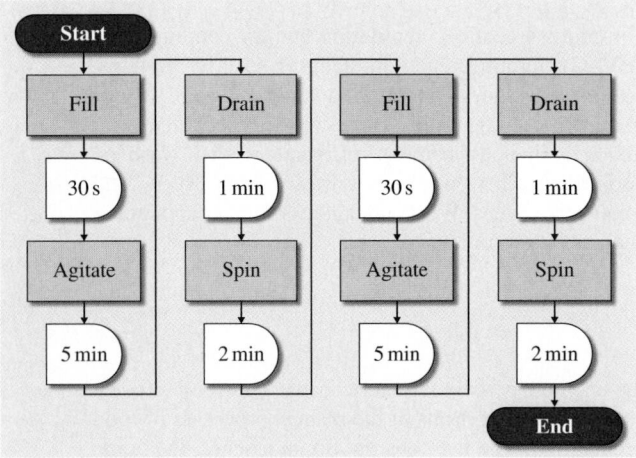

Fig. 83.1 Simple clothes washer algorithm

83.1.2 Enabling Technologies

Microprocessor Controls

The fundamental enabler for advanced home appliance automation has been the introduction of microprocessor-based control systems and associated software. In the past, electromechanical control systems have made dramatically increased automation either infeasible or impractical. However, virtually all modern major appliances – spanning refrigeration, fabric care, and cooking – are now equipped with a programmable microprocessor, thus setting the stage for quantum leaps in automation.

Sensors

Microprocessors ostensibly can implement their processes without environmental inputs; for example, a clothes washer could operate based on the simple algorithm presented in Fig. 83.1.

However, advanced automation and, in particular, increased efficiency are dependent on retrieval of information from the process environment. Thus, another key technological enabler is environmental sensors. Examples of important sensors include:

- Temperature sensors
- Flow meters
- Optical cameras
- Infrared sensors
- Chemical sensors
- Radiofrequency identification (RFID) receivers.

Temperature sensors give the microprocessor access to, for example, the temperature of inflowing water, ambient or internal air, and food in the case of refrigeration and cooking appliances. Flow meters are employed primarily to provide data on water volumes, which can be crucial to achieving a desired water temperature. Optical cameras provide inputs to, for example, food doneness determination algorithms. Infrared sensors provide noninvasive means to evaluate temperature, while chemical sensors provide key data for evaluating things such as detergent concentrations.

The sensors introduced above generally serve to evaluate the physical characteristics of some element of the process being executed, or the thing on which the process is operating (food, clothing, etc.). However, an RFID receiver provides access to stored data about the item being processed which is not otherwise accessible – at least not cheaply and conveniently. For example, RFID tags embedded in clothing could contain informa-

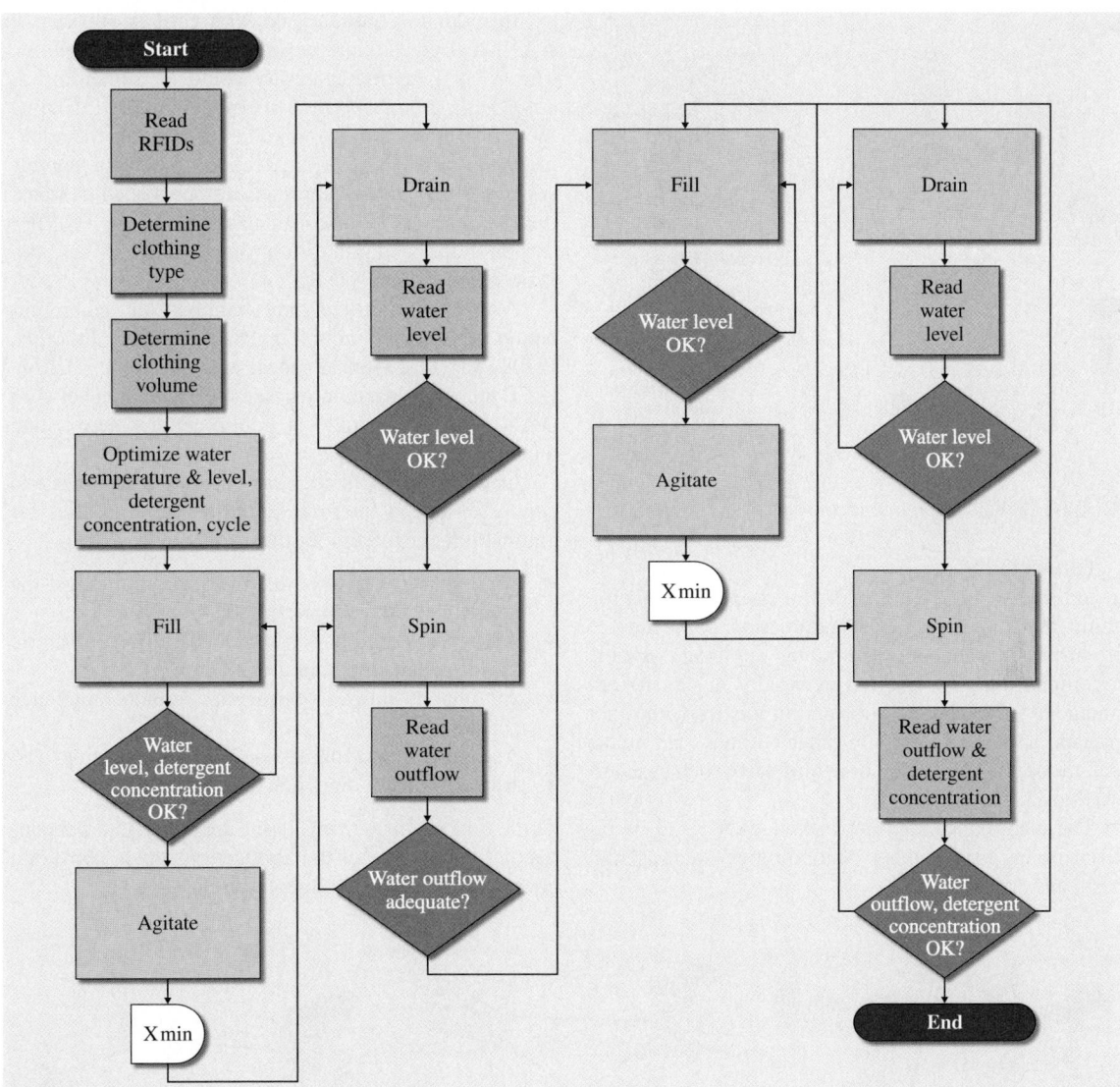

Fig. 83.2 More advanced clothes washer algorithm

tion on the composition and care of the garment, such as parameters dealing with appropriate water temperature, detergent type, and washing cycle. RFIDs contained in prepared food packaging could contain information on composition and instructions for preparation, such as cooking temperature and time. Thus, RFIDs provide access to a rich cache of information which can be leveraged for automation. The term *RFID* in this context is not intended to pertain to any specific radio technology or technical standard, but to the general

concept of wireless, short-range, passive information retrieval.

Now, returning to the algorithm presented above, given a sufficiently capable array of sensors, we might be able to automate the process as shown in Fig. 83.2. This revised algorithm is critically dependent on the availability of sensors. Modern home appliances are being equipped with ever more sensors to enable cycle optimization and efficiency, as well as process automation.

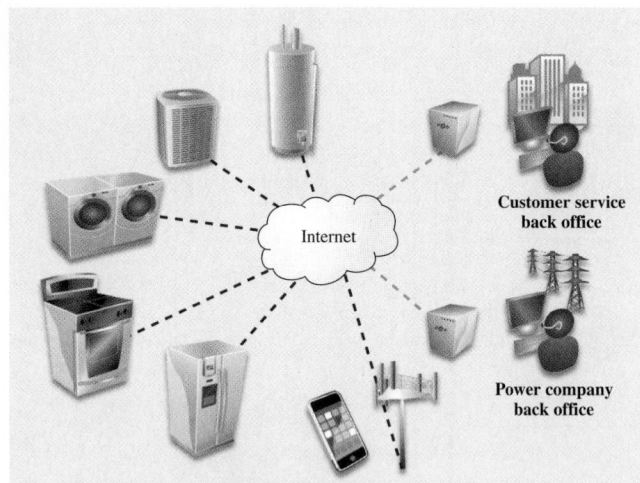

Fig. 83.3 Remote appliance communication

Connectivity

In addition to collecting and processing information within the home appliance, automation opportunities are further expanded if the home appliance is able to communicate information externally. It can be advantageous to exchange information locally with other appliances, or remotely with other entities such as the user, the appliance vendor, or a third party such as a utility (Fig. 83.3).

These appliance area networks (AANs) can be realized in any of a number of topologies, such as star, mesh, and ring (Fig. 83.4).

In order to communicate with entities outside the AAN, though, an Internet gateway is often required. The AAN's Internet gateway could be something as simple as a broadband modem, or it could be something more intelligent, for example, in the case that the appliances and/or the networking technology do not natively support a full suite of internet protocols such as transmission control protocol/Internet protocol (TCP/IP), dynamic host configuration protocol (DHCP), and domain name system (DNS).

AANs can utilize any number of communication protocols including IEEE 802.3 Ethernet, IEEE 802.11 wireless local-area network (LAN), IEEE 802.15.4 low-rate wireless personal area network (PAN), and a multitude of public cellular networking technologies.

Once home appliances are networked together and connected to the Internet, boundless opportunities for automation exist. Some of these include:

- Automatic distribution of information amongst the appliances, for example, time of day
- Automatic coordination and optimization of electric power, water, and natura-gas usage
- Automatic status reporting and remote appliance control
- Automatic performance analysis and reporting
- Automatic fault detection and reporting.

At the time of this writing, home appliance Internet connectivity and some of the applications given above are becoming available commercially.

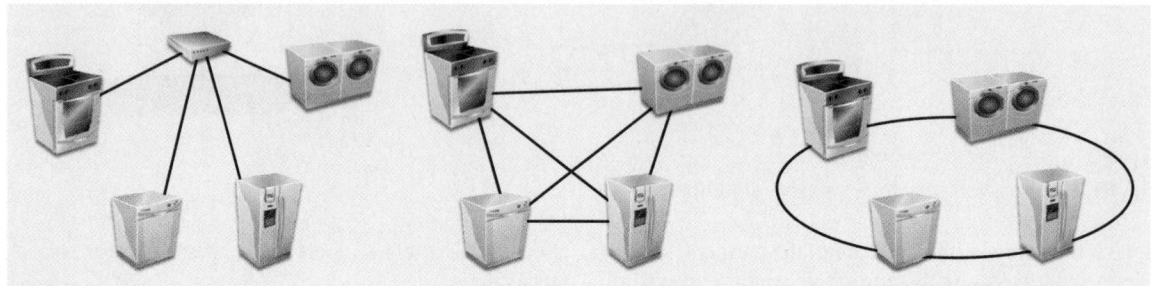

Fig. 83.4 AAN topologies

83.2 Application Examples, Guidelines, and Techniques

83.2.1 Refrigeration

It has been stated previously that the basic function of the most widely used home appliance – the

refrigerator – has not changed significantly since its inception. There are, however, opportunities to automate the fridge's function significantly. The penultimate in refrigeration – the *smart* fridge – would possess at least:

Table 83.1 Smart refrigerator food inventory

Type	Subtype	Instance	Inventory date	Expiry date	Capacity	Fill level
Milk	2%	1	23-Apr	10-May	4 L	30%
Milk	2%	2	01-May	18-May	4 L	100%
Olives	Green	1	23-Apr	06-Dec	200 g	100%
Beer	Duff	1	23-Apr	N/A	6	17%
Carrots	N/A	1	23-Apr	N/A	0.5 kg	80%
...	...	...	...	...	...	...

- Automatic food inventory
- Automatic quality management.

In the basic implementation of the smart fridge concept, the type and quantity of refrigerated food would be stored in a database by the refrigerator. As items are placed in the fridge, an identifier – such as an RFID tag – is read. Similar functionality could be implemented via universal product code (UPC) symbol instead of RFID, but this would require some user intervention (the scanning of the symbol) and would this represent a semiautomatic process. The tag indicates the food type, the product brand name, the container size or weight, the use-by date, and any other information relevant to the inventory process. To this the fridge adds each item's purchase date, physical location, and other information (Table 83.1). As food is consumed, the relevant database entry is updated; for example, a half-gallon container of milk weighing 900 g is determined to be just under half full.

Several conditions trigger the ordering of food. If the food container is not returned to the fridge within a certain period of time, then the food is presumed to be fully consumed and the smart fridge flags the item for replacement. Similarly, if the container is returned to the fridge, but the fridge determines that the con-

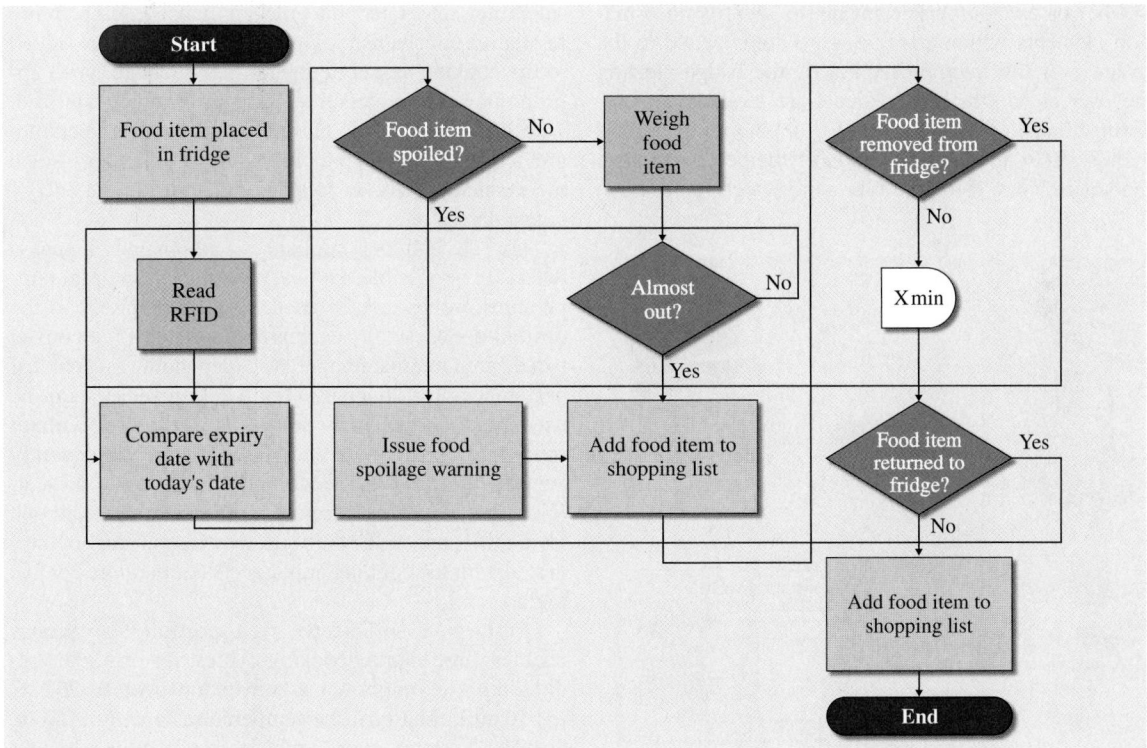

Fig. 83.5 Smart refrigerator inventory maintenance

tainer's contents have reached a critically low level, then the item is also flagged for replacement (Fig. 83.5). Further, if the item is nearing its use-by date, then the item is marked for replacement.

Either automatically or pursuant to a demand by the user, a shopping list can be generated. Using Internet connectivity, the list can be sent directly to an online grocer for delivery, or to a bricks-and-mortar grocer for preparation and eventual collection by the user, or to the user's computer or mobile phone. Alternatively, the list can be sent over the AAN to a printer and hand-carried by the user to the market, or simply displayed on the fridge's own video display device (Fig. 83.6).

The smart fridge concept further is conducive to maximizing food freshness and safety. Detailed knowledge of the fridge contents permits optimization of temperature and humidity so as to extend food freshness. And since the smart fridge has retained the use-by date for each item in its database, any food nearing its end-of-life can result in safety alerts to the user and automatic reordering or addition to the shopping list.

Advanced implementations of the smart fridge concept include recipe cognizance. In one realization, digital recipes – utilizing extensible markup language (XML) or an appropriate means to identify information elements within a recipe – are downloaded to the fridge over the Internet/AAN, with the fridge alerting the user as to which ingredients are lacking, and ordering them (or adding them to the shopping list) as necessary. In another realization of recipe cognizance, the smart fridge alerts the user as to which recipes can be prepared with the ingredients on hand by searching the Internet for digital recipes which are matches to the fridge inventory.

Clearly, the smart fridge is dependent on all of the enabling technologies described previously: a comprehensive array of sensors, microprocessor control, and connectivity.

83.2.2 Cooking

Cooking is, by its very nature, a labor-intensive process. However, there are several opportunities for further automation of the processes performed by each of the major cooking appliances, namely, the cook top (stove), microwave oven, and convection oven. Some possibilities are as follows.

Often, the temperature of convection ovens and cook tops is not regulated, but controlled indirectly. Rather than controlling the temperature, the appliance controls the amount of energy dissipated. The vessels used for cooking in convection ovens – and on cook tops in particular – while not part of the appliance per se, are critical components in the cooking process. Given the nature of temperature regulation, the cooking vessel is not integrated into the appliance's temperature regulation mechanism. Thus, as ingredients are added to the cooking vessel, temperatures fluctuate. With appropriate sensors, vessel temperature can be stabilized, with heat being added as sudden drops in temperature are detected, thus automating what would otherwise be the manual process of raising and lowering the energy consumption rate.

For all cooking appliances, automated doneness detection is possible. Optical sensors and internal temperature probes can be used to cook precisely to the desired doneness; for example, most meats are cooked to a desired internal temperature, depending on personal preferences for doneness. Temperature sensors can be used to cook precisely to this temperature, without over- or undershooting, with the appliance subsequently keeping the meat warm without continuing to cook it. (Note that, while temperature probes are often available from third parties, they are strictly external and not integrated with the cooking appliance's temperature control logic.)

Further, a sufficiently rich portfolio of probes enables more refined cooking cycles; for example, traditionally, we might set a convection oven to 232 °C for 10 min, then turn the temperature down to 163 °C for 2 h. A *smart* oven could be programmed to first sear, then cook to an internal temperature of 60 °C,

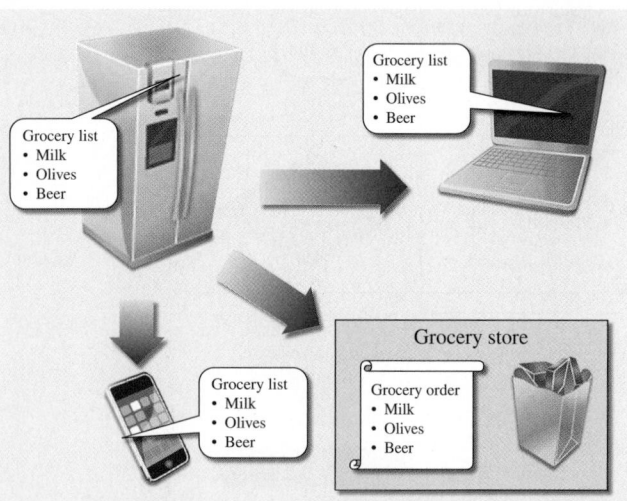

Fig. 83.6 Shopping list generation and notification

with the appliance itself making all necessary decisions about time and temperature, thus automating the process.

In a still more advanced realization of cooking automation, the cooking appliance would be provided with detailed cooking instructions by the food itself. The food's packaging – a box or wrapper – could be equipped with an RFID tag which would be read by the cooking appliance, either as it is placed into the appliance or when it is passed near a sensor manually. This approach is clearly most feasible in the context of prepared foods. The RFID tag would contain details of the nature of the food (e.g., its type and weight), cooking time, and cooking temperature. The only required user intervention would be selecting the desired doneness, if applicable, and pressing *start*, substantially automating the process. When integrated with temperature probes and sensors, the cooking appliance could automate the entire process, from defrosting to cooking, simply taking the desired meal time as an input.

Extending the concept of a digital recipe to cooking, the power of the AAN could be leveraged to coordinate the functions of the cooking and refrigeration appliances and substantially automate food preparation from end-to-end: The smart fridge would manage the ordering and inventory of ingredients, while the smart cooking appliances would manage the cooking based on XML-encoded instructions (time and temperature) in the digital recipe or a similar information classification protocol (Fig. 83.7).

83.2.3 Cleaning

Cleaning appliances – clothes washers, clothes dryers, and dish washers – again offer a substantial degree of process automation by their very nature. However, further levels of automation are possible. Some examples are as follows.

Washing appliances – clothes and dish washers – are optimized by their users according to the nature of the task at hand. The number of cycles offered by each machine is limited, and users sometimes utilize only one or two of the available cycles. Thus, there exists the opportunity to automate the decision-making process further. With an appropriate array of sensors – such that the fabric/dish type, volume, and soil level can be assessed – the washer can automatically optimize its cycle. Among the variables which can be adjusted are the wash, rinse, spin, etc. durations, the nature of the chemistry (such as the detergent type and volume), and optimal water temperature and volume.

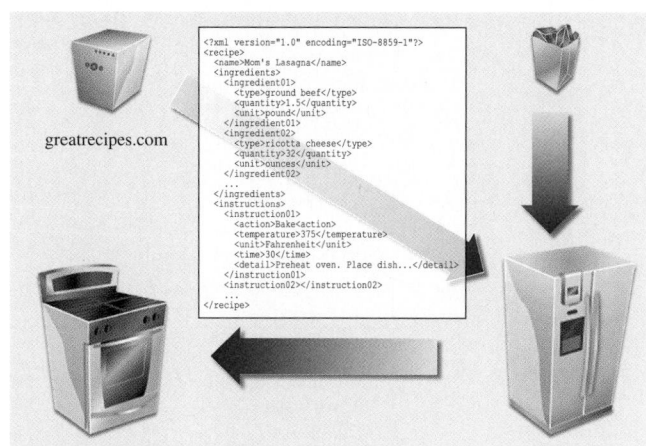

Fig. 83.7 Semiautomated food preparation

In an even more advanced implementation, clothing and dishes would be equipped with RFID tags which would inform the appliance as to the nature of the item as well as specific care instructions. The cleaning appliance could then determine the optimum washing/drying cycle based on all the contents loaded into the machine, perhaps suggesting, for example, that certain items should not be washed together, or should not be machine washed (or dried) at all (Fig. 83.8).

It is easily observable that water-consuming appliances such as clothes washers and dish washers can cause a substantial drop in household water pressure during their operation. They can also consume a substantial portion of the household hot water reserve. The power of the AAN can be leveraged automatically to coordinate the operation of water-intensive processes; for example, the clothes and dish washers can communicate over the AAN to ensure that they do not both operate at the same time, and both devices can communicate with the water heater to ensure a sufficient reserve for other water-intensive activities in the home (Fig. 83.9). Both appliances further could be configured to avoid operation during bathing times.

83.2.4 General Appliance Automation

Thus far we have examined the automation of the specific processes performed by individual home appliances. We now examine automation as it pertains to home appliances in general. Microprocessor control, sensor arrays, and connectivity enable a number of generic automation functions in any so-equipped home appliance. Some possibilities include:

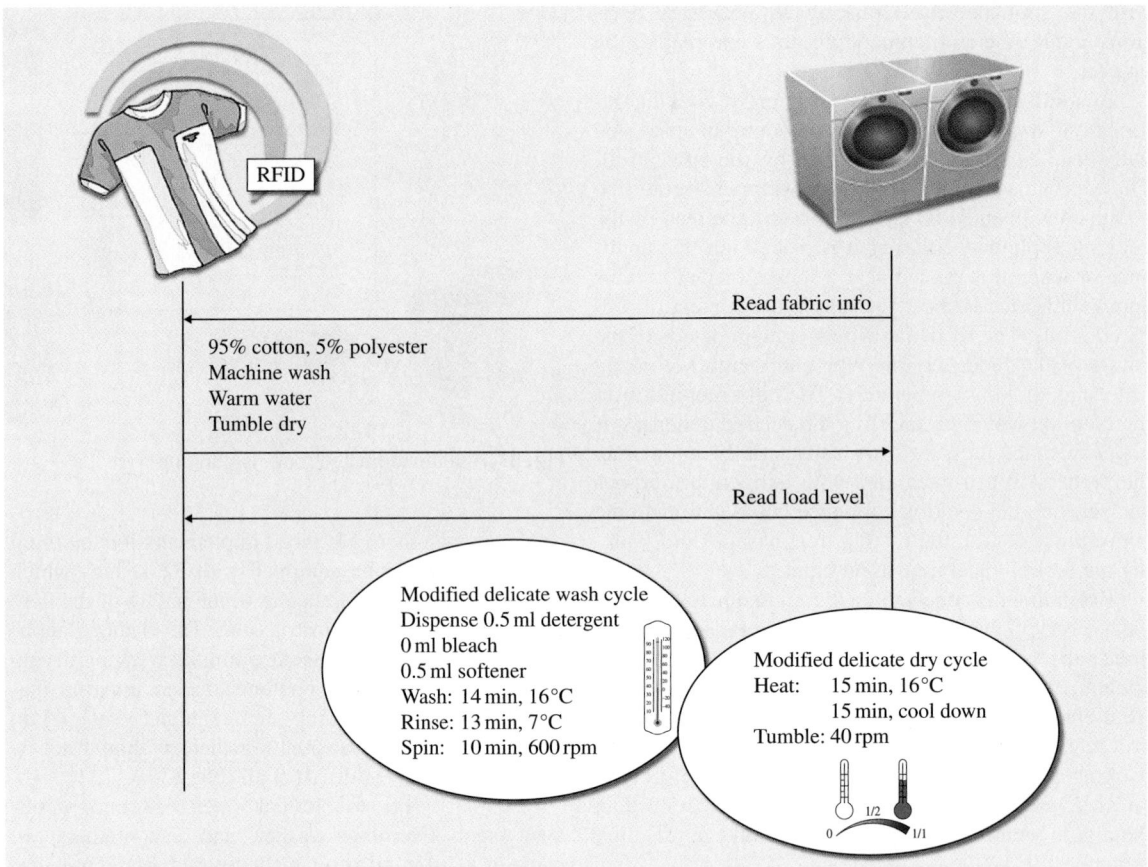

95% cotton, 5% polyester
Machine wash
Warm water
Tumble dry

Read fabric info

Read load level

Modified delicate wash cycle
Dispense 0.5 ml detergent
0 ml bleach
0.5 ml softener
Wash: 14 min, 16°C
Rinse: 13 min, 7°C
Spin: 10 min, 600 rpm

Modified delicate dry cycle
Heat: 15 min, 16°C
           15 min, cool down
Tumble: 40 rpm

Fig. 83.8 RFID-based wash cycle optimization

- Performance monitoring
- Fault monitoring
- Software maintenance
- Consumables monitoring.

- The electric current drawn
- The condenser fan *on* time
- Time during which the cooling chamber exceeds the desired temperature.

Performance Monitoring

Traditionally, poor appliance performance is not detected and acted upon until it degrades to a catastrophically-low level; for example, a refrigerator evaporator fan becomes clogged with debris and the fridge fails to cool effectively and begins to consume inordinate amounts of energy. As performance continues to degrade, the fridge is unable to maintain a sufficiently low temperature and food spoils, resulting in a risk to the user's health. Usually only when this near-fault condition occurs is the user alerted to the problem. However, advanced automation techniques permit home appliances to monitor various internal performance metrics continuously. In our example, this might include:

These metrics could be compared against their long-term average values, or against some factory-programmed limits. If performance, as determined by these metrics, degrades below the desired level, the user or the appliance service center is automatically alerted – by the appliance – to the need for maintenance (Fig. 83.10), thus avoiding a catastrophic outage and food spoilage.

Fault Monitoring

Similarly, appliance faults which are not preceded by gradually degrading performance can be acted upon automatically. Appliances can execute internal diagnostic testing periodically and can detect when subsystems

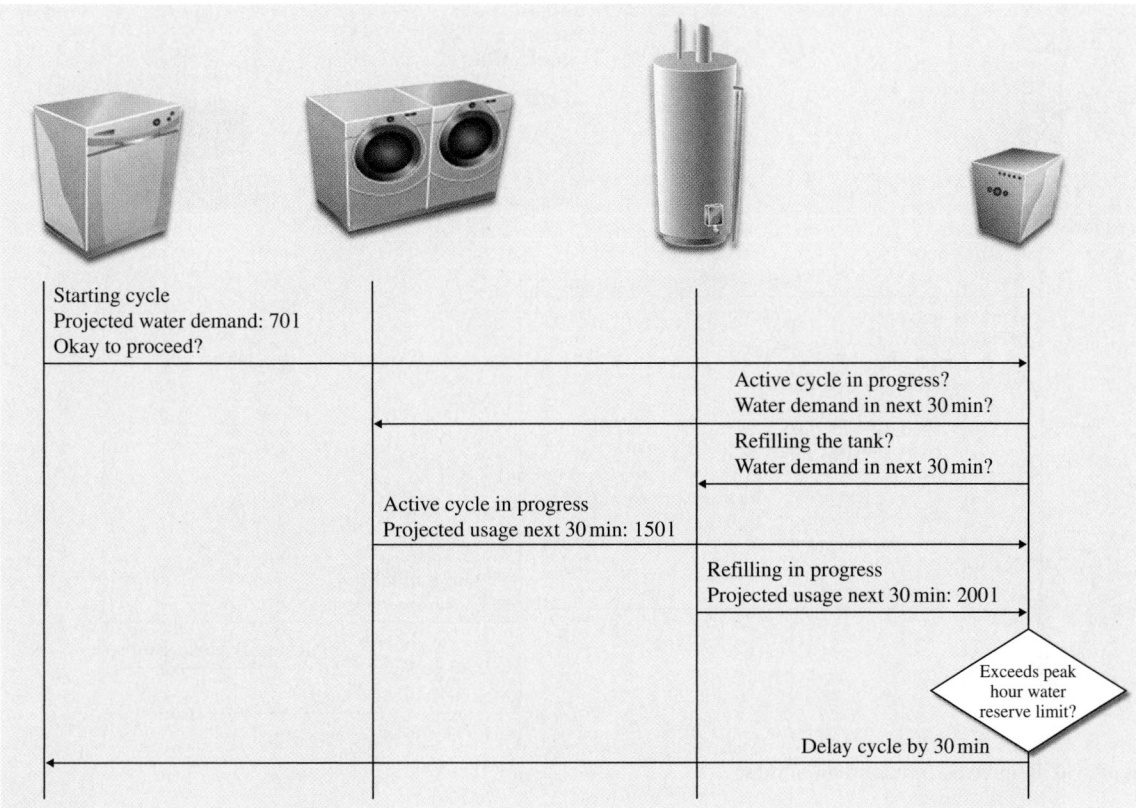

Fig. 83.9 Water usage coordination

fail. This information can both be shown on the appliance's video display and be reported to the appliance service center. Armed with exact knowledge of the fault, the responding technician can be dispatched with all necessary tools and parts, minimizing the time to repair and the need for a follow-up visit to the user premises.

Critically, large volumes of fault and performance data – gathered from an appliance vendor's installed base – can be aggregated and analyzed to spot trends. Subcomponents which have unusually high fault rates can be swapped in the manufacturing process to reduce the incidence of the fault in future appliances (Table 83.2). Similarly, appliances showing trends in a certain mode of performance degradation can be redesigned to minimize the likelihood of occurrence in future units.

Software Maintenance

Once home appliances are functional Internet nodes, it becomes possible to automate appliance software maintenance. Just as a home personal computer (PC)'s operating system and application software are updatable automatically via the Internet, so can an appliance's software be updated. As software bugs are discovered, new loads containing the bug-fix can be pushed out to the connected appliances, eliminated the need for a service call, the need to disassemble the appliance, or even the need to connect tools – such as a PC – to the appliance. Beyond bug-fixes, this technique can be used to deploy new software-based functionality, such as new or optimized machine cycles. This is a particularly powerful capability when used in conjunction with the aforementioned large-scale performance data analysis, which might indicate that certain software parameters – such as performance thresholds – need to be tweaked.

Consumables Monitoring

Some home appliances utilize *consumable* items – items which have a limited useful life and must be replaced periodically. Examples include filters and additive or detergent cartridges. When an appliance is

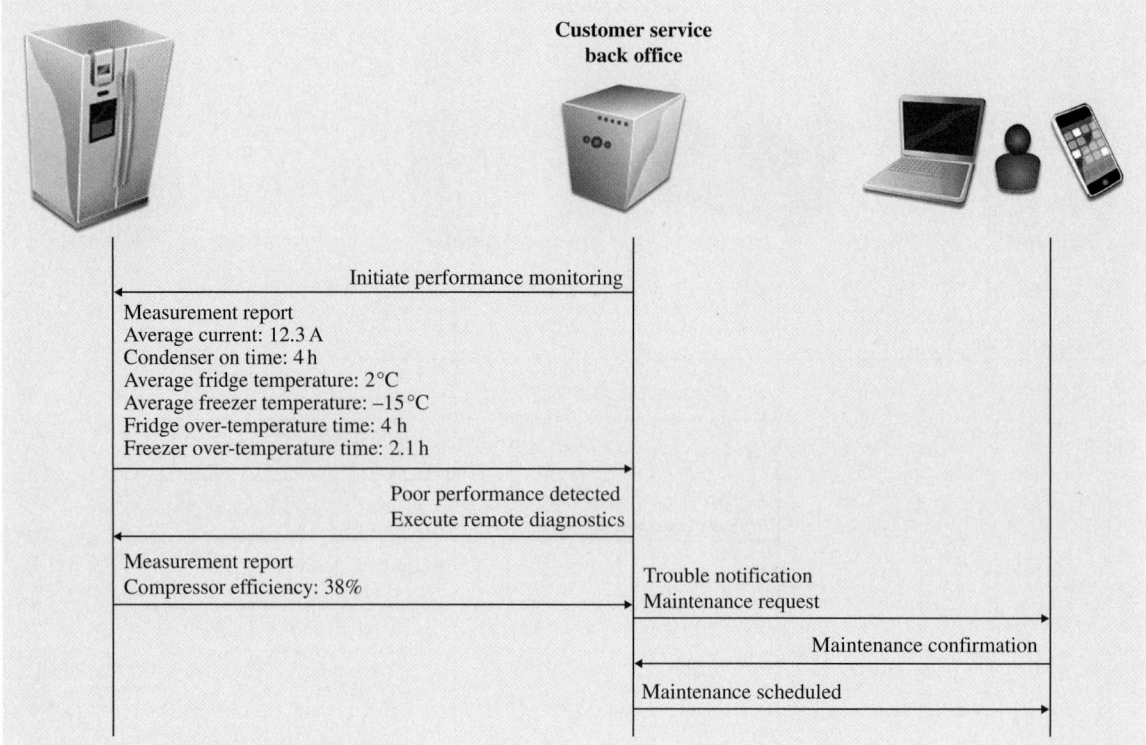

Fig. 83.10 Remote performance monitoring

under microprocessor control, the remaining life of consumables can be monitored precisely. When this metric falls below a predetermined threshold, the connectivity capability can be used to reorder the consumable automatically, thus eliminating the possibility of appliance downtime or degraded performance.

83.2.5 Household Energy Management

For a number of reasons, household energy usage – primarily electric power usage – has been brought to the forefront of the energy debate. Concerns over climate change, energy supply stability, and national energy independence have refocused government and public utilities on the issue.

A particular concern of electric utilities is transient peaks in demand which exceed the utility's ability to produce power. In such situations, the utility is forced to purchase energy on the open market from a third-party supplier. Since such suppliers know that electric power in such *overload* conditions is extremely valuable to the utility, they charge relatively high prices for the electricity, negatively impacting the utility's margins, in some cases to the extent that the utility loses money, after having sold the electric power for less than it cost to acquire it. In order to minimize their loss, utilities some-

Table 83.2 Trending using aggregated performance data

Type	Subtype	Model	Revision	Units in field	Fault code	Fault instances
Refrigerator	Side-by-side	K123	2	67 839	1	0
Refrigerator	Side-by-side	K123	2	67 839	2	0
Refrigerator	Side-by-side	K123	2	67 839	3	42
Refrigerator	Side-by-side	K123	2	67 839	4	690
Refrigerator	Side-by-side	K123	2	67 839	5	0
...	...	...	...	...	...	...

times must cut their supply of power. In order to *spread the pain*, this is done for brief periods within a number of geographic regions, producing the infamous *roving brownouts*.

In other industries, if a supplier runs out of capacity, it simply builds or otherwise procures additional supply capacity. In the case of a manufacturer of goods, a new factory might be constructed. However, electric power generation facilities are both expensive to build and sensitive politically. So increasing supply is difficult and sometimes infeasible. In order to address the electric power supply–demand problem, utilities have taken a demand-side approach. This approach entails direct control of power consumption by the electrical devices – including appliances – in the home, as well as in load-based, real-time pricing of electricity. Thus we introduce the concept of automated electric power usage in the home.

Direct Load Control

Direct load control (DLC) refers to the ability of the electric power utility to reduce the amount of power consumed by electrical equipment in the home by communicating directly with the equipment, without the need for resident intervention [83.3]. Examples of home electrics which are key to the DLC strategy include HVAC systems, electric water heaters, and major appliances.

With DLC, as the utility approaches its capacity limit (Fig. 83.11), it sends signals to the home electrics which indicate the need to conserve power; for example, the HVAC system can be requested to operated at a higher (in summertime) or lower (in wintertime) temperature. Likewise, the water heater can be requested to

switch off temporarily, resulting in power savings with a slight reduction in water temperature. Electrical appliances – which sometimes have complex operating cycles – can be requested to operate in high-efficiency mode, or to delay starting until the power demand has subsided.

Many architectures exist for realizing DLC-based automated home energy usage control. In one implementation, the electric utility operates a server which monitors instantaneous power demand. As demand continues to rise toward the utility's peak capacity, the server sends signals to *smart* power meters in participating homes, which in turn send signals – perhaps wirelessly – to the subtending appliances, indicating the need to conserve power (Fig. 83.12).

If the number of participating households is large, and if the aggregated power savings are sufficient, then the utility can avoid going to the open market to procure power and the associated possibility of a brownout. As demand begins to wane, the utility's server signals the smart power meters with a signal indicating that power consumption can return to normal. Thus, home power demand control is automated.

DLC is most applicable in markets where retail pricing is regulated, i.e., real-time pricing is not possible. Consumers are offered incentives to participate in DLC programs in the form of rebates from the electric utility.

Real–Time Pricing

Real-time pricing refers to the electric utility's ability to vary pricing for electric power over relatively short periods of time, for example, on a per-hour basis [83.4].

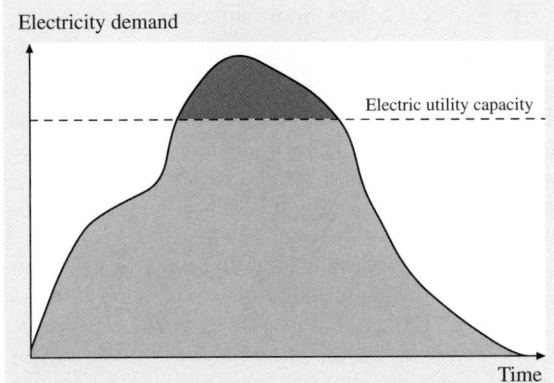

Fig. 83.11 Hypothetical electricity demand versus electric utility capacity

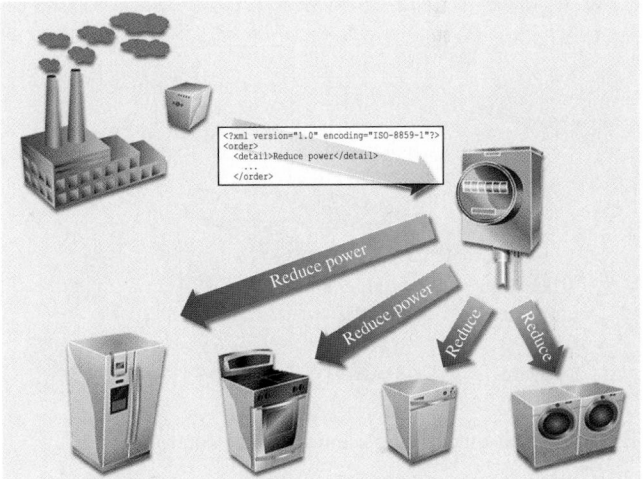

Fig. 83.12 Direct load control using smart power meters

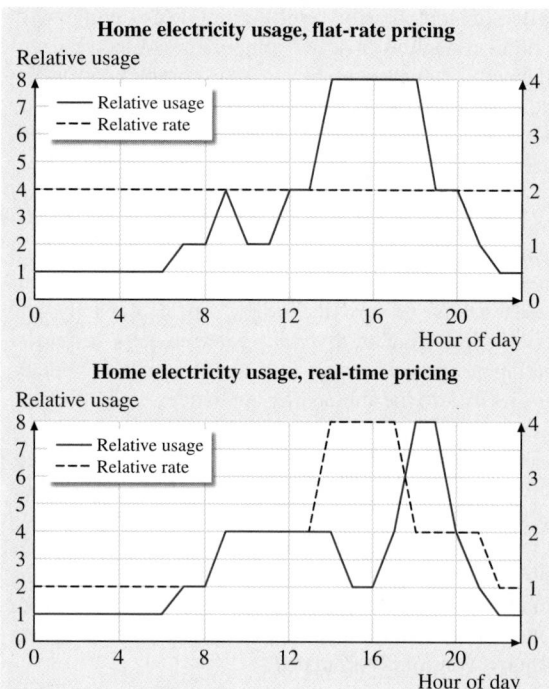

Fig. 83.13 Home electricity usage: flat-rate versus real-time pricing

Real-time pricing can be used – among other things – to disincentivize power consumers from using power during periods of peak demand. This reduces the prob-

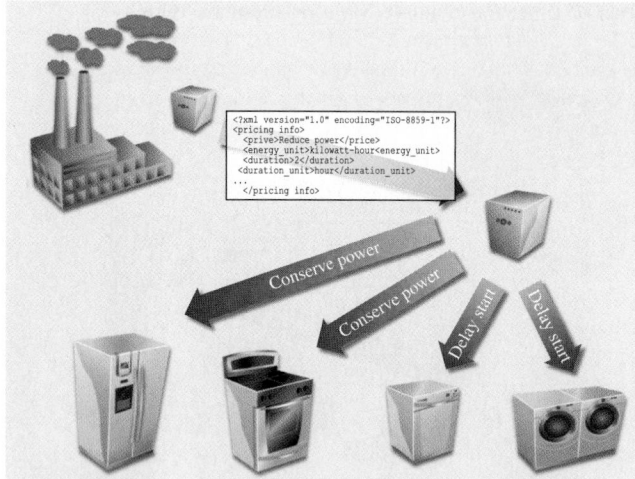

Fig. 83.14 Real-time pricing using appliance control node

ability that the utility's own capacity will be exceeded, and minimizes the likelihood that the utility has to purchase additional power on the open market.

When real-time pricing information is relayed to consumers, they make decisions about how to use their home electrics so as to keep their electric bill within their spending limits. So, for example, during peak hours in summertime, residents can choose to set thermostats to a higher temperature in the mid-afternoon, when energy prices are highest. Likewise, certain household activities – such as laundry and dish washing – can be deferred until pricing is more amenable (Fig. 83.13). With enough participating households, the aggregated power demand on the utility is eased to the point that it does not have to purchase additional power from a third party.

We have described a mechanism by which electric utilities can *automate* demand control by manipulating the retail price of power. However, further power consumption automation opportunities exist on the consumer side. If the pricing information is relayed directly to appliances – for example, via a smart meter – then compatible appliances can automatically adjust their cycles so as to minimize the net energy cost; for example, the clothes washer, after having received a *start* signal from the user, could delay starting its cycle until energy prices are more favorable.

Alternatively, if pricing information is relayed to a hypothetical appliance control node located in the home, the control node can schedule appliance (and HVAC, water heater, etc.) operation so as to meet consumer-configured criteria regarding comfort, convenience, and cost (Fig. 83.14).

For example, the user could decide that he wants to:

1. Minimize cost during weekdays
2. Balance cost with comfort and convenience during weeknights
3. Maximize comfort and convenience on weekends.

The control node would then configure and schedule tasks accordingly. This could mean, for example, switching off the air conditioning during weekdays, raising the temperature slightly on weeknights, and operating it normally all weekend.

A more sophisticated energy control node algorithm might look something like that shown in Table 83.3.

Thus the user can achieve the optimal balance of electric power cost and comfort and convenience in an automated way.

Table 83.3 Appliance energy control node algorithm

	A/C	Water heater	Clothes washer	Clothes dryer	Dish washer	Fridge
Weekdays high price	Temperature: −6.67 °C	Temperature: −6.67 °C	Deferred operation	Deferred operation	Deferred operation	High efficiency
Weekdays low price	Normal	Normal	High efficiency	High efficiency	High efficiency	Normal
Weekends high price	Temperature: −16.1 °C	Temperature: −12.2 °C	High efficiency	High efficiency	Deferred	Normal
Weekends low price	Normal	Normal	Normal	Normal	Normal	Normal

83.3 Emerging Trends and Open Challenges

83.3.1 Trends

Display Devices

Advancements in home appliance automation are driving – and are enabled by – advancements in display devices. More and more appliances are available with advanced light-emitting diode (LED), monochrome liquid-crystal display (LCD), and full-color LCD display devices (Fig. 83.15). These user interfaces (UIs) facilitate the richer user experience required for advanced status and control (input/output, I/O) functions, such as those required by advanced automation techniques.

For example, with either the DLC or real-time pricing modes of energy management, the appliance user must be informed of in-process energy management activities. In the case of DLC, the user must know when the utility is actively managing the appliance. For real-

time pricing, the user needs to be apprised of the current price of power. In either case, the user must have the ability to control the appliance's behavior. A user might want to override DLC or the real-time pricing program if some process, e.g., laundry, needs to be performed immediately due to some immanent need. All of these functions required the rich I/O facilities which advanced display devices provide.

Interconnectivity

Most advanced appliance automation techniques require remote transmission of information to, or remote retrieval of information from, the appliance. This implies that appliances must interconnected – to each other, and most importantly to the Internet. Household energy management, remote performance analysis and fault detection, and remote software maintenance are all critically dependent on interconnectivity. Thus, in-

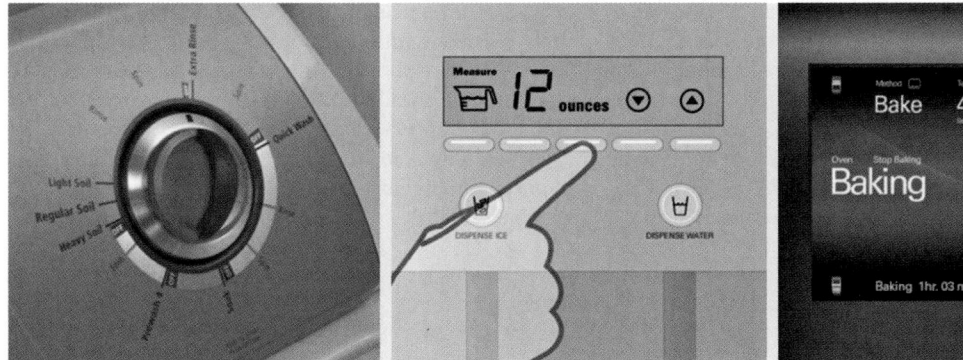

Fig. 83.15 Appliance I/O device evolution

terconnectivity is emerging as a major area of research and development in the home appliance industry.

Smart Resource Usage

We have discussed at length the initiatives on the front of home appliance automated electric power conservation. While home appliances become more convenient, they are simultaneously becoming more efficient and more intelligent in their use of electricity, the culmination of which being automated energy management.

Beyond electricity, questions about supply and price of water and natural gas are driving improvements in efficiency on those fronts. From the user's perspective, such savings are automated, realized internally by advanced software algorithms, mechanical techniques, and materials within the home appliance.

83.3.2 Challenges

Retail Cost

Home appliances are consumer products and, as such, are highly dependent on sales volumes. In some cases, an appliance only becomes profitable to produce if hundreds of thousands – or millions – of units can be sold. Competition is fierce and only grows more intense as time goes on. Thus, tremendous efforts are put into minimizing costs.

All the technology presented here – sensors, displays, and communication electronics – add cost to the appliance. This creates a significant barrier to the introduction of such advancements on a large-scale basis. Consider that a mere dollar of additional cost to an appliance could result in the loss of hundreds of thousands of dollars of revenue for the manufacturer. (Note that retail prices are set by the market, not by the manufacturer, so additional manufacturing costs cannot necessarily be passed onto the consumer.)

Thus, costs represent a significant challenge to the deployment of advanced automation techniques in home appliances. Whereas the value of having – say – a clothes washer is obvious to most consumers, the value of having an interconnected washer is less so. The home appliance industry is thus tasked with convincing consumers that the added value provided by advanced-automation appliances justifies the incremental cost.

Personal Control and Privacy

Consumers are used to having full control over their appliances: When the *start* button is pressed, the machine should start operating. Advanced automation requires that users cede some of this control either to the local machine or to some other remote device. Consumer acceptance of this paradigm will require significant education on the part of appliance manufacturers. It is likely that, in order to achieve consumer acceptance, the ultimate control for appliance behavior will have to remain in the hands of the consumer, with the appliance and other control systems simply providing *suggestions* for machine operation.

Further, the introduction of interconnectivity might present privacy concerns. It is not clear whether consumers would be receptive to the notion of their home appliances automatically relaying information to the Internet, nor to the idea of an Internet node (e.g., the electric utility) remotely accessing information in their appliances. In order to gain consumer acceptance, appliance manufacturers will need to assure consumers of the nature and security of the information being exchanged. This will almost certainly involve using common Internet information security techniques to ensure data privacy.

Standardization and Interoperability

Particularly in the case of interconnectivity, standard implementation methods are required. If, for example, the electric utility is to implement DLC, it is not practical for the utility to use a different communication protocol for each appliance vendor and each appliance model. Further, in the case in which appliances from different vendors within a single home must communicate with each other, the appliances cannot be expected to know each other's proprietary protocols. In sum, a universal communication protocol is required. At the time of this writing, standardization efforts are underway both within the International Electrotechnical Commission (IEC) and within the Association of Home Appliance Manufacturers (AHAM) on this topic. The outcome should be a standard for home appliance communication that permits full intervendor interoperability.

83.4 Further Reading

- V.V. Badami: Home appliances get smart, IEEE Spectrum **35**(8), 36–43 (1998)
- P. Bertoldi, A. Ricci, A. de Almeida: *Energy Efficiency in Household Appliances and Lighing* (Springer, Berlin Heidelberg 2001)
- R.Kango, P.R. Moore, J. Pu: Networked smart home appliances – enabling real ubiquitous culture, IEEE 5th Int. Workshop Netw. Appl. (2002) pp. 76–80
- K.J. Myoung, J.M. Lee, D.-S. Kim, W.H. Kwon: Home network control protocol for networked home

  appliances, IEEE Trans. Consumer Electron. **52**(3), 802–810 (2006)
- J. Nicholls, B.A. Myers: Controlling home and office appliances with smart phones, IEEE Pervasive Comput. **5**(3), 60–67 (2006)
- Y. Tajika, T. Siato, K. Teramoto, N. Oosaka, M. Isshiki: Networked home appliance system using Bluetooth technology integrating appliance control/monitoring with internet service, IEEE Trans. Consumer Electron. **49**(4), 1043–1048 (2003)

References

83.1 US Energy Information Administration, http://www.eia. doe.gov

83.2 J.C. Gallawa, http://www.gallawa.com/microtech/history.html

83.3 Pacific Northwest Demand Response Project, http://www. nwcouncil.org/energy/dr

83.4 Federal Energy Regulatory Commission: Assessment of demand response and advanced metering 2007 staff report (2008), http://www.ferc.gov/ industries/electric/indus-act/demand-response/ dem-res-adv-metering.asp#skipnavsub

84. Service Robots and Automation for the Disabled/Limited

Birgit Graf, Harald Staab

The increasing number of elderly people is resulting in increased demand for new solutions to support self-initiative and independent life. Robotics and automation technologies, initially applied in industrial environments only, are starting to move into our everyday lives to provide support and enhance the quality of our lives. This chapter analyzes the needs of disabled or limited persons and discusses possible tasks of new assistive service robots. It further gives an overview of existing solutions available as prototypes or products. Existing technologies to assist disabled or limited persons can be grouped into stand-alone devices operated by the user explicitly such as robotic walkers, wheelchairs, guidance robots or manipulation aids, and wearable devices that are attached to the user and operated implicitly by measuring the desired limb motion of the user such as in orthoses, exoskeletons or prostheses. Two recent developments are discussed in detail as application examples: the robotic home assistant Care-O-bot and the bionic robotic arm ISELLA. One of the most important challenges for future developments is to reduce costs in order to make assistive technologies available to everybody. On the technological side, user interfaces need to be designed that allow the use of the machines even by persons who have no technical knowledge and that enable new tasks to be taught to assistive

robots without much effort. Finally, safe manipulation of assistive robots among humans must be guaranteed by new sensors and corresponding safety standards.

In the last years, the percentage of elderly people in our society has grown rapidly. Out of 82.5 million people living in Germany in 2005, according to numbers from the German Federal Statistical Office, around 19% were seniors above 65 years [84.1]. With this demographic development continuing, by the year 2050 those above 65 years will comprise 33–36% of Germany's population. Similar numbers are reported from other industrial nations all over the world, in particular the USA and Japan. Many disabilities come with age. In order to provide sufficient support for the growing number of disabled persons, new solutions to assist these people are required.

84.1 Motivation and Required Functionalities

Technical aids enable people in need of support and care to live independently in their accustomed home environments for a longer time. As not all homes are suitable for installing intelligent home technologies as described in Chap. 83, mobile service robots able to navigate and operate in existing homes without modifications to the environment provide a more flexible solution. Such service robots not only fulfill the user's desire for independence and autonomy, but also help to avoid the high costs of individual treatment in nursing homes that might otherwise be necessary.

The task of a service robot or other automation technologies for the disabled and limited is to provide support and instructional help in a person's daily life and to promote self-initiative. The group of target users for such new technologies includes:

- Elderly and frail persons
- Disabled persons (physically, psychically)
- Severely sick persons (heart attack, plastered leg)
- Chronically sick persons (diabetes, epilepsy)
- Severely restrained persons (e.g., pregnancy).

Due to the fact that humans have for a long time dreamt of an assistant to relieve them of difficult handling tasks, the basic concept of service robots assisting people in their daily life has been presented and their functionalities been described long before they were actually built. The requirements that a service robot to assist the disabled must cope with and the corresponding technical abilities can therefore be specified precisely [84.2, 3]. Some of the most important tasks,

identified by a study done with elderly people and their care-givers, are summarized below.

Household Tasks

- Execute everyday jobs such as serving drinks, setting the table, operating the microwave, simple cleaning tasks
- Fetch and carry objects, e.g., books, remote control, medicine
- Support in grasping, holding, and lifting objects and tools
- Control of the technical home infrastructure, e.g., heating system, air-conditioning, lights, windows, doors, alarm system, etc.

Mobility Aid

- Support for standing up from the bed or a chair
- Controlled motion
- Obstacle detection and guidance to a target.

Communication and Social Integration

- Media management (videophone, TV, stereo, interactive media, etc.)
- Daytime manager (daily routine, time for medicine, etc.)
- Communication with medial and public facilities (physician, authorities, etc.)
- Supervision of vital signs and emergency call functionality.

84.2 State of the Art

In recent decades, and continuing to the present day, service robots for the disabled and limited is an active field of research and development, performed in many countries throughout the world. Main research is carried out at universities and research institutes, covering a brought range of ideas, approaches, and technologies. The landscape of companies in this branch seems to be twofold: on the one hand there are a few large players, who have been providing orthopaedics solutions in large numbers and for many years already, trying to introduce robotics into some of their high-end products. On the other hand there are many smaller high-technology companies, often launched by

researchers in this field, focusing on a few or a single product solution.

This section provides a representative overview of the state of the art in the field of assistive robotic devices for the disabled and limited. The following examples may be understood as representatives of many more but similar developments. For further insights one may refer to proceedings of international conferences such as those on Robotics and Automation (ICRA), Intelligent Robots and Systems (IROS), Robotics and Applications (RA), and others that usually host large sessions on assistive and rehabilitation robotics. The International Conference on Rehabilitation Robotics (ICORR)

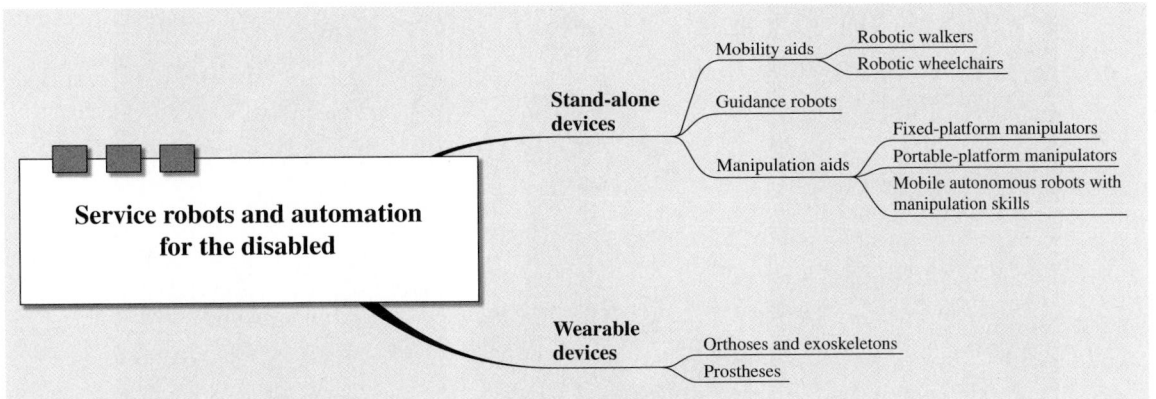

Fig. 84.1 Classification of service robots and automation for the disabled

is focused on this topic alone. Other sources may be periodicals such as the quarterly *International Journal of Rehabilitation Research* (IJRR), the monthly *International Journal of Therapy and Rehabilitation* (IJTR), the quarterly *International Journal of Medical Robotics and Computer Assisted Surgery*, and others.

Existing service robots and assistive devices for the disabled and limited can be grouped into stand-alone devices, operated by the user explicitly such as robotic walkers, wheelchairs, guidance robots or manipulation aids, and wearable devices that are attached to the user and operated implicitly by measuring the desired limb motion of the user such as in orthoses, exoskeletons or prostheses. A similar classification is found in [84.4]. Figure 84.1 summarized this classification of service robots and assistive devices for the disabled and limited.

The review of existing solutions starts in Sect. 84.2.1 with mobility aids such as robotic walkers or robotic wheelchairs providing autonomous or semiautonomous navigation capabilities. Section 84.2.2 describes existing guidance robots that may be applied for blind persons or those suffering from dementia and other mental disorders. From the technical viewpoint they are quite similar to robotic walkers but the focus of these systems is on cognitive, not physical support. Section 84.2.3 describes existing manipulation aids that may be either static, such as in desktop-mounted manipulators, or mobile when attached to a wheelchair or to another kind of mobile platform.

Finally the group of wearable devices is reviewed. Section 84.2.4 describes orthoses and exoskeletons for users who need direct support in moving their limbs. Prostheses worn by a user to replace amputated limps are increasingly incorporating robotic features and are reviewed in Sect. 84.2.5.

84.2.1 Mobility Aids

Robotic Walkers

Robotic walkers are enhanced off-the-shelf walkers equipped with robot technology such as environment sensors and drives. Using these technologies, they are able to provide additional support to their user, ranging from audio or visual information on the environment to autonomous or semiautonomous navigation. For existing robotic walkers, two basic types can be distinguished: passive and active robotic walkers.

Passive robotic walkers do not have any driven wheels and move directly according to the applied user forces (direct user control). Examples of passive walkers are the COOL Aide [84.5] (Fig. 84.2a) and PAMM-AID [84.6] systems, which use motorized steering of the wheels to lead the user around obstacles. The RT-Walker [84.7] uses special servo brakes to steer the device in a collision-free direction. In case no obstacles are detected the user has full control of the device and can move it, similarly to a conventional walker. Most passive systems are equipped with brakes to stop them if they get too close to an obstacle or step. A guidance system for Guido, the commercial successor of PAMM-AID, has been presented [84.8] (Fig. 84.2b). After a target has been set, the robot will plan and follow a path to the target. During guidance the desired direction indicated by the user input will be ignored. The successor of the RT-Walker, ORTW-II [84.9], uses a *potential canal* method which only allows deviations up to a certain distance from the optimal path to the target. The robotically augmented walker developed at Carnegie Mellon University (CMU) [84.10] (Fig. 84.2c) does not use its motors when traveling with a user but tracks its position and displays the optimal direction

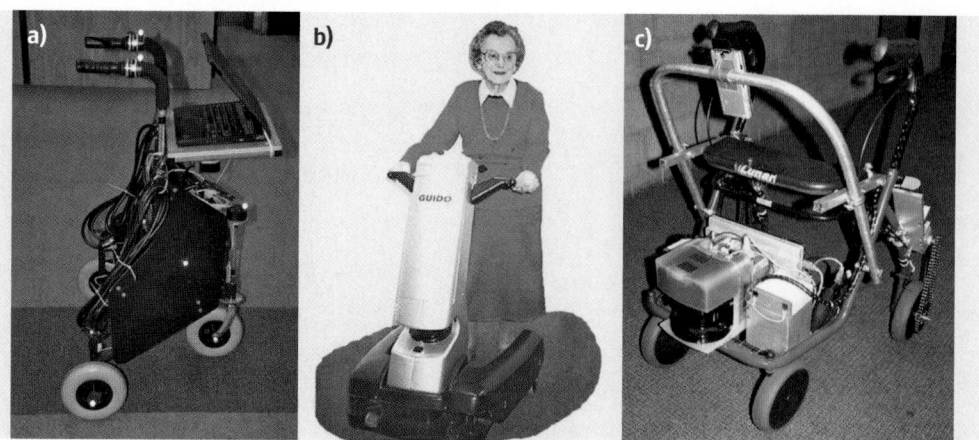

Fig. 84.2a–c Examples of passive robotic walkers: (**a**) COOL Aide (courtesy of the University of Virginia, USA), (**b**) Guido (courtesy of Haptica Ltd., Dublin, Ireland), and (**c**) CMU robotically augmented walker (courtesy of Carnegie Mellon University, Pittsburgh, USA)

of travel on a screen to guide its user to the selected target.

Active robotic walkers are equipped with motorized wheels. Force or force-torque sensors of different kinds are used to determine the forces applied to the system by the user. These forces are used to calculate the desired driving direction and speed of the walker (indirect user control). Some systems (Silbo [84.11] (Fig. 84.3a), Hitachi walker [84.12], RT-walker) are equipped with angle sensors, enabling the required force input to be adapted on slopes. All systems are able to detect obstacles and stop in front of them; most of them adapt the traveling direction to surround obstacles in advance. The PAMM smart walker [84.13] (Fig. 84.3b) and the CMU robotic walker [84.14] (Fig. 84.3c) are able to localize themselves in their environment and thus plan an optimal path to a given target. The shared control system of PAMM creates a virtual force, leading the robot to the given target, which is combined with the real forces applied to the device by the user. The gains of

each control input are set depending on the observed user abilities. Whereas for PAMM an interface providing feedback to the user on the planned path has not been presented, the CMU robotic walker displays the desired direction of travel on a screen. The user is free to move in any direction; however, if the deviation from the planned path is too large, the velocity of the walker will be reduced to force the user back to the path.

Other robotic walkers focus their works on additional support functions such as lifting assistance (Hitachi walker, Monimad walker [84.15], MOBIL Walking & Lifting Aid [84.16], Walking helper II [84.17]), person tracking (MOBIL Test Bed [84.18]) or moving out of the way when not used (CMU robotically augmented walker).

Robotic Wheelchairs

Similar to robotic walkers, robotic wheelchairs provide enhanced safety and/or improved navigation capabilities using mobile robotics technologies. Robotic

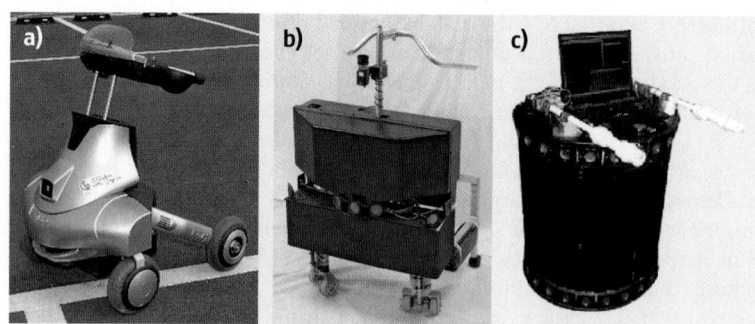

Fig. 84.3a–c Examples of active robotic walkers: (**a**) Silbo (courtesy of Intelligent Healthcare Laboratory, Korea), (**b**) Smart walker PAMM (courtesy of Massachusetts Institute of Technology, USA), and (**c**) CMU robotic walker (courtesy of Carnegie Mellon University, Pittsburgh, USA)

wheelchairs are based on electric wheelchairs where control over the drives is at some point taken over by the integrated personal computer (PC). In order to perceive the environment, robotic wheelchairs are equipped with environment sensors such as sonar sensors or laser scanners. Existing control systems for robotic wheelchairs can be divided into two basic types [84.19]: model-based approaches and behavior-based approaches.

Model-based shared control systems analyze the input of the user and the measured environment data in order to identify the intended travel direction of the user and thus adapt the motion of the wheelchair. To identify the intention of the user, the current sensor input of the user, e.g., the position of the joystick, is compared with a previously recorded motion model and an appropriate action is selected. The robotic wheelchairs NavChair [84.20] or VAHM [84.21] use a method that generates a histogram of the environment identifying the optimal direction of travel (minimum vector field histogram, MVFH). The user is responsible for high-level control of the system, such as route-planning and some navigation actions, while the machine overrides unsafe maneuvers through autonomous obstacle avoidance. It can provide addition assistance, e.g., for safely passing narrow passages such as doors. The MVFH algorithm ensures that the input of the user is considered at all times and thus the user always feels in control of the wheelchair. SmartChair [84.22] is another representative of model-based shard control systems. This robotic wheelchair is not only able to adjust the input of the user for collision-free motion; it further provides localization and path-planning capabilities to guide the user to a previously specified target. SmartChair uses a potential field method to calculate a suitable motion direction to the given target. MAid [84.23] is an autonomous wheelchair able to safely pass crowded environments, however, at the cost of not considering the input of the user during motion.

Behavior-based shared control systems assume that the driving behavior of the robotic wheelchair can be classified into a few basic tasks such as high-speed driving, close-quarter manoeuvring, and docking manoeuvres. From the input of the user and the measured environment data, a suitable motion behavior is selected. The crucial factor in behavior-based shared control systems is to find out which behavior is currently most appropriate. Wheelsley [84.24] is a robotic wheelchair based on the well-known behavior-based subsumption architecture [84.25]. In free space, the robot moves along with maximum velocity in the direction indicated by the user. If an obstacle in the

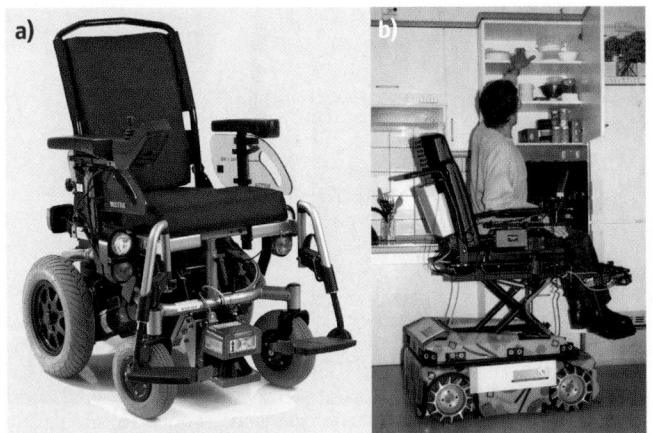

Fig. 84.4a,b Examples of advanced robotic wheelchairs: (**a**) Rolland III (courtesy of the DFKI-Labor, Bremen, Germany) and (**b**) OMNI (courtesy of Forschungsinstitut Technologie und Behinderung der Evangelischen Stiftung Volmarstein, Germany)

desired motion direction is detected, obstacle avoidance behavior is activated to pass by the obstacle. Similar approaches are applied in the Rolland [84.26] (Fig. 84.4a), RobChair [84.27], and Sharioto [84.28] projects. Some motion behaviors conflict with each other, such as passing a narrow passage and obstacle avoidance. Another conflict occurs if the user wants to dock to an object while at the same time collision avoidance is required. For docking to an obstacle, in [84.28] the user's confidence when guiding the wheelchair in a specific direction is evaluated. Only if the user moves straight and without fluctuations towards an obstacle will the direction be maintained.

Other robotic wheelchairs such as the office wheelchair with high manoeuvrability and navigational intelligence for people with severe handicap (OMNI) [84.29] (Fig. 84.4b) provide advanced assistance through omnidirectional navigation capabilities and height adjustment. The special mechanics of the iBOT (http://www.ibotnow.com/) enable stair-climbing. Even legged chairs such as i-foot, one of the partner robots by Toyota designed to assist people [84.30], which is able to move along in uneven terrain, have recently been introduced.

84.2.2 Guidance Robots

Guidance robots are used to assist people with mental weaknesses and diseases or blind people. Guidance robots for the blind require direct physical contact to the user. One of the first robotic guidance systems

Fig. 84.5a–c Examples of guidance robots: (**a**) GuideCane (courtesy of University of Michigan, USA), (**b**) Nursebot (courtesy of Carnegie Mellon University, Pittsburgh, USA), and (**c**) exhibition guide MONA by Fraunhofer IPA, Germany

is the NavBelt [84.31], a device worn by the user. Equipped with sonar sensors, NavBeld is able to detect obstacles in front of the user and issue a warning. GuideCane is the cane-based continuation of this research (Fig. 84.5a). In contains a small mobile robot as a base that carries the environment sensors and a cane that the user can hold onto. Using a small joystick, the user indicates the desired travel direction and is led along by the robotic base. If an obstacle is detected by the sensors, the path will be modified to lead the user around the obstacle safely.

Other guidance robots are designed to assist the elderly, for example, to guide them to specific locations in elderly care facilities. In this case, no direct contact with the user is required. Nursebot [84.32] (Fig. 84.5b) is a development that targets this application. Guidance robots with similar capabilities are also applied in museums or exhibitions [84.33] (Fig. 84.5c).

84.2.3 Manipulation Aids

Manipulation aids are applied to assist disabled users in eating, drinking or object replacement. This Section describes existing fixed- or portable-platform manipulators as well as existing autonomous mobile robots with manipulation skills.

Fixed-Platform Manipulators
The professional vocational assistive robot (ProVAR) [84.34] is a research prototype designed to assist individuals with a severe physical disability. The system consists of an industrial manipulator attached to an overhead track suspended above a worktable and is able to handle objects for a user who has difficulties in handling these objects himself. The manipulator is controlled by voice commands or head motion inputs. The robot to assist the integration of the disabled (RAID) [84.35] is a robotic workstation for use by individuals with little or no upper-limb function in office environments. It is based around a custom-designed mechanical structure providing storage for books, manuals, paper documents, and other reference materials. A robot arm is mounted on a linear track in front of the storage zones. It is able to assist the user, e.g., to transport books or diskettes or to hold a book and turn pages when reading it. The arm's working envelope includes a large proportion of the desk area but does not reach to the user for reasons of safety.

Portable-Platform Manipulators
Portable platform manipulators include manipulators attached to a mobile base with castor wheels that can be rolled around and wheelchair-mounted manipulators. Handy1 [84.36] (Fig. 84.6a) is an example of the first group of assistive manipulators. It consists of a robot arm and a tray that can be selected for several applications such as feeding, cleansing, and make-up. The robot is controlled by a switch input used in conjunction with a linear scanning control system which is suitable for different disability groups. The assistive robot service manipulator (ARM), which is also known as *Manus* [84.37] (Fig. 84.6b), is a $6 + 2$ degree of freedom (DOF) robot that assists disabled people with a severe handicap of their upper limbs. It compensates their lost arm and hand function. It can be mounted on an electric wheelchair (or mobile base) and allows numerous daily

living tasks to be carried out in the home, at work, and outdoors. By means of an input device such as a keypad (4×4 buttons), a joystick (e.g., of the wheelchair) or another device attached to a nondisabled body part, the manipulator can be operated to grasp objects with its gripper.

Mobile Autonomous Robots with Manipulation Skills

Mobile autonomous robots with manipulation skills are a popular topic of current research. Even though research in this field has grown significantly over recent years, no commercial products have been placed on the market so far. Among existing prototypes, two basic approached can be observed: human-like robots equipped with legs and arms and wheel-based robots.

Humanoid robots with legs are most popular in Japan. Most of the currently existing prototypes are not specifically designed to assist disabled people. However, for many robot developers, this application is seen as one of the most important for future products. The partner robots developed by Toyota [84.30], for example, were first demonstrated at EXPO 2005 AICHI, Japan. During the EXPO they were applied to entertain the visitors of Toyota's pavilion by playing different instruments. However, this was only the first step towards the creation of robots that can use tools, assist people, and live in harmony with us. According to the partner robot developers, currently, new robots are being developed that can provide elderly care to help Japan cope with its rapidly aging population. The goal of the ASIMO [84.38] (Fig. 84.7a) development of Honda is quite similar: to develop a robot that can duplicate the complexities of human motion and genuinely help people. Even though ASIMO is mainly used as a research platform at the moment, someday ASIMO might help with important tasks such as assisting the elderly or a person confined to a bed or a wheelchair. HRP-2 [84.39] (Fig. 84.7b) by Kawada and the US-American SARCOS [84.40] humanoid robot are similar platforms that can be used for experiments to further develop robotic technologies developments.

One of the first wheel-based robots able to assist disabled people with daily tasks is MOVAID by the Scuola Superiore Sant'Anna in Italy [84.41]. The robot is able to navigate in homelike environments and to perform simple manipulation tasks. Another early development is Hermes [84.42], developed at the Bundeswehr University Munich. Hermes is able to explore unknown environments, to fulfil transportation and manipulation tasks in spacious human-populated areas, and to interact

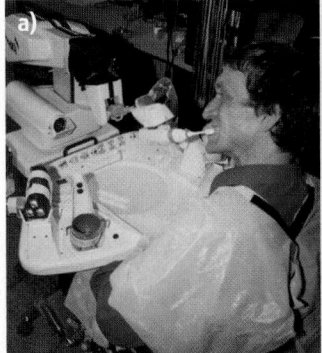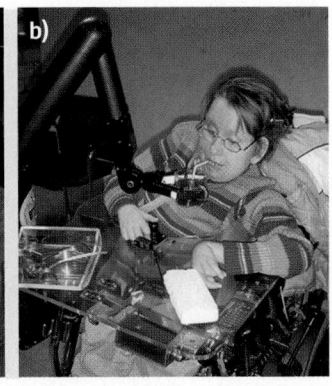

Fig. 84.6 (a) Portable-platform manipulators Handy1 (courtesy of Forschungsinstitut Technologie und Behinderung der Evangelischen Stiftung Volmarstein, Germany) and **(b)** ARM (courtesy of Exact Dynamics BV, Netherlands)

and communicate even with novice users in a natural and intuitive way. Research on assistive robots is currently being performed in several projects in Germany such as the Collaborative Research Center on Humanoid Robots in Karlsruhe [84.43], where several generations of the ARMAR platform (Fig. 84.7c) have been set up in the last years. One of the most advanced developments in wheel-based robotic home assistants is the German Care-O-bot [84.44] which will be described in detail in Sect. 84.3.

Recent developments of wheeled robots in Japan include the excellent mobility and interactive existence as workmate (EMIEW) by Hitachi [84.45] and Smart-Pal by Yaskawa [84.46]. At the EXPO 2005 AICHI, Japan, they were introduced in a bar scenario, where they demonstrated their abilities to serve drinks to visitors. By now, for each robot, the second generation has recently been introduced. Another interesting Japanese development is enon (*exciting nova on network*) by Fujitsu [84.47], designed for duties such as providing guidance, transporting objects, and security patrolling. In Japan, limited sales of enon have already been announced.

84.2.4 Orthoses and Exoskeletons

Orthoses are understood as orthopaedic devices to support parts of the body with reduced mobility. There are those for fixation at reversible or nonreversible deformations and bad postures as well as for recovery (e.g., ruff). Others are used to support, e.g., arch supports and orthopaedic corsets. The latter are said to be passive if they support only, and active if they force the body

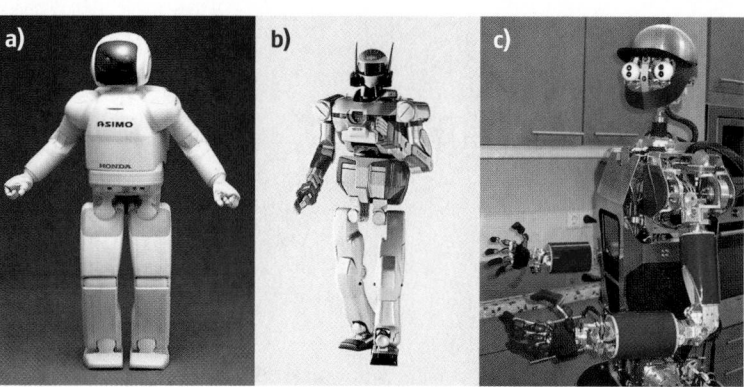

Fig. 84.7a–c Popular humanoid and wheel-based robots with manipulation skills: (**a**) ASIMO (courtesy of Honda Motor Co., Ltd., Japan), (**b**) HRP-2 (courtesy of National Institute of Advanced Industrial Science and Technology, Japan), (**c**) Armar 3 (courtesy of the Collaborative Research Center 588 Humanoid Robots – Learning and Cooperating Multimodal Robots, Karlsruhe, Germany)

to actively correct, e.g., mis-statics of the spine. Any orthoses usually requires significant fine-tuning in order to meet the body conditions as well as the medical demands of an individual.

Utilizing robotics for orthoses is limited to a few application examples and there is a fuzzy border towards the prostheses domain, however there is some research and development (R&D) purely for orthoses. Most researchers focus on rehabilitation, e.g., of stroke patients. Sometimes an orthosis may look quite simple at first sight, but there may be a lot of *complexity* and careful thoughts behind any combination of materials, part design, and choice of mechanisms. An earlier example is the Wilmer elbow orthosis [84.48], a passive foot-drop orthosis from the University of Hawaii-Manoa, USA [84.49], and numerous stationary passive or motor-powered *gravity-compensation* rehabilitation

devices for the upper extremities such a dynamic arm support [84.50], Freebal [84.51] (Fig. 84.8a), and iPAM [84.52]. Some orthoses are equipped with elaborated spring and lever mechanisms, brakes, and may also be motor-powered with advanced control algorithms such as a foot orthosis from the University of Osaka, Japan [84.53], and a knee orthoses from the Vrije Universiteit Brussels, Belgium, powered by pneumatic artificial muscles [84.54].

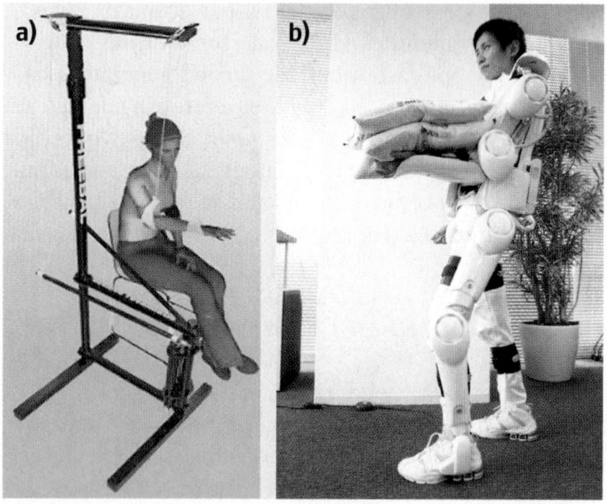

There are also quite complex, full-body gait rehabilitation systems such as HapticWalker [84.55], STRINGMAN [84.56], and LOKOMAT [84.57], which are composed of a stationary frame, a hanging harness for the patient to put on, and possibly a treadmill. Current R&D covers mechanical design as well as extending the functionalities of existing systems regarding sensing, control, haptic feedback, artificial intelligence, and virtual reality.

The term *exoskeleton* is also often used for powered orthoses, but it is not only limited to medical applications. There are those for a single extremity (upper or lower), or covering one or few joints, e.g., knee and ankle. And there are those for full-body motion support, either stationary or mobile. Some examples are ALEX (stationary, for gait rehabilitation) [84.58], BLEEX (mobile, for military use to increase payload of a soldier carrying equipment) [84.59], and the Hybrid Assistive Leg (HAL, mobile, full-body exoskeleton, Fig. 84.8b) [84.60].

84.2.5 Prostheses

While orthoses and exoskeletons are attached to existing parts of the body, prostheses are replacements of lost extremities. Most sold prostheses of arms and hands are purely cosmetic, whereas most prostheses of legs and feet are passive but functional, i.e., they

Fig. 84.8 (**a**) Freebal (courtesy of Baat Medical, Hengelo, Netherlands) and (**b**) Hybrid Assistive Leg (HAL, courtesy of Tsukuba University/Cyberdyne Inc., Japan)

enable patients to walk. However, scientific R&D as well as commercial high-end products clearly aim at coming close to the functionality of the replaced body part. This means motorization, haptic perception, and reliable and comfortable control by the patient. Some advanced commercial prostheses of fingers or the entire hand are the i-LIMB [84.61] and the Fluidhand [84.62], and of the leg are the c-leg [84.63] and the Powerknee [84.64].

Beside commercial developments there are a number of prostheses in research. Focus is placed on powered prostheses and their mechanical and mechatronic design, signal interfaces to the human body for control and feedback, and advanced motion control. There are developments for upper [84.65] (Fig. 84.9) and lower limbs; some examples of recent lower-limb developments are powered ankle–foot and knee prosthesis as described in [84.66] and [84.67].

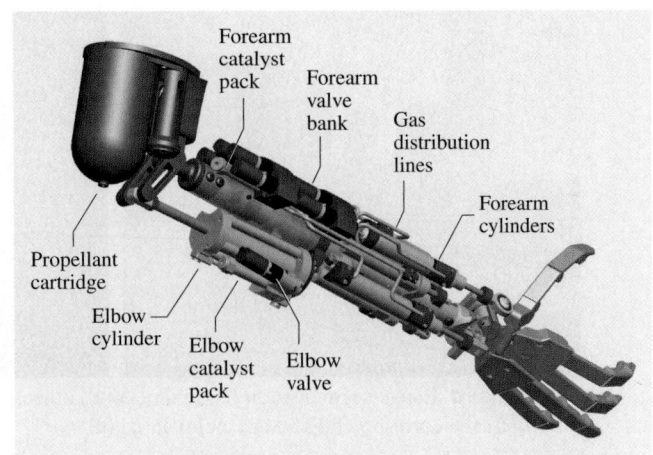

Fig. 84.9 Steampunk artificial arm (courtesy of Vanderbilt University, Nashville, USA)

84.3 Application Example: the Robotic Home Assistant Care-O-bot

84.3.1 History of Care-O-bot Development

Care-O-bot is a mobile robot assistant designed of Fraunhofer IPA, Germany to assist people in daily life activities. Three Care-O-bot robots have been developed so far: The first Care-O-bot prototype (Fig. 84.10a) [84.68] was built in 1998, when the idea of building rehabilitation robots was still new. Care-O-bot I is a mobile platform with a touchscreen, able to navigate autonomously and safely in indoor environments, and communicate with or guide people. As a mobile platform alone, it can be used for transportation and safeguarding tasks; however, it is unable to execute complex manipulation tasks as required from a home assistant. Care-O-bot II (Fig. 84.10b) [84.69], built in 2002, is additionally equipped with a manipulator arm, adjustable walking supporters, a tilting sensor head containing two cameras and a laser scanner, and a handheld control panel [84.70]. The manipulator arm, developed specifically for mobile service robots, provides the possibility of handling typical objects in a home environment. A flexible gripper attached to the manipulator is suitable for grasping various objects such as mugs, plates, and bottles. A handheld control panel is used for instructing and supervising the robot. In addition to all mobility functions already solved in Care-O-bot I, the second prototype is able to execute manipulation tasks

autonomously and can be used as an intelligent walking support. Care-O-bot 3 (Fig. 84.10c), built in 2008, is the latest generation of this successful development series. It is equipped with the latest state-of-the-art components including omnidirectional drives, a seven-DOF redundant manipulator, a three-finger gripper, and a flexible interaction tray that can be used to safely pass objects between the human and the robot. Its moveable sensor head contains range and image sensors enabling autonomous object learning and detection and three-dimensional (3-D) supervision of the environment in real time.

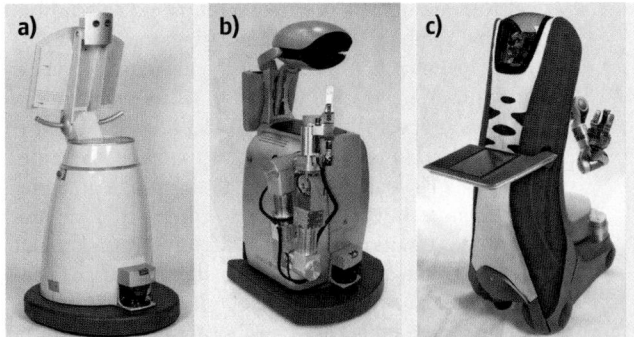

Fig. 84.10a–c Care-O-bot prototypes by Fraunhofer IPA, Germany. Care-O-bot I (**a**), Care-O-bot II (**b**) and Care-O-bot 3 (**c**)

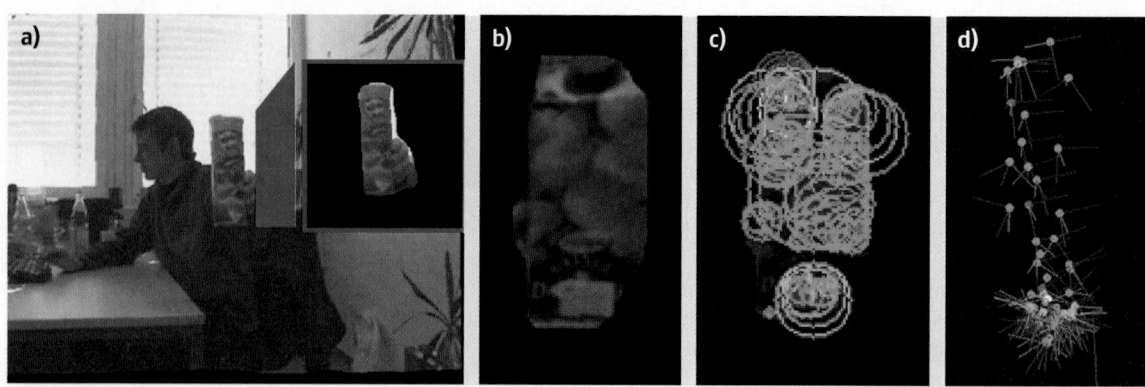

Fig. 84.11a–d Range segmentation for an object (**a**), object representation: side view of a Tetra Pak (**b**), detected feature points (**c**), and corresponding feature point cloud (**d**)

84.3.2 Key Technologies

Autonomous or Semiautonomous Navigation

All Care-O-bot prototypes are able to navigate autonomously to a given target [84.71]. Autonomous navigation is a key issue in order to execute a fetch and carry task or to guide the user to a specific location. Chapter 16 has additional insights in automation mobility and navigation. A static environment map is used to plan an optimal path to the given target. Different approaches for global planning have been implemented and can be selected according to the geometry, kinematics, and current operation mode of the robot [84.72]. The generated path is smoothed and eventually modified in reaction to dynamic obstacles or other external forces. A laser scanner attached to the robot is used for continuous and dependable obstacle detection.

Automatic Object Detection

In order to grasp an object, the robot must be able to detect relevant objects in the environment. This task is solved by combining a range imaging sensor [84.73] with a color camera. The recognition algorithm is based on scale-invariant feature transform (SIFT) descriptors that are recorded for each object and fed into a learning algorithm (a one-class support vector machine, SVM) [84.74]. Using the data of the range imaging sensor, feature keypoints can be segmented from the background. A region in space is effectively masked out in the color image of the scene using the range measures for the corresponding pixels in the range image. New objects are taught to the robot by placing them in front of the sensors and by recording the relevant SIFT keypoints for the object [84.75]. Figure 84.11a displays the teaching of new objects using the proposed range

segmentation. Figure 84.11b–d illustrates the learning process and resulting representation of an object: several images of the object are recorded and for each the feature points are detected. In a second step the feature points of all images are fused into a 3-D feature point cloud which again can be used to detect and compute the position of the object in a given scene.

Object Manipulation

Based on the data from the range imaging sensor and the identified location of the object to be grasped, a collision-free trajectory for moving the manipulator to the detected object can be computed. To solve this, the robot and scene are modeled using *oriented bounding boxes* (OBBs). For the robot a distinction is made between static components (e.g., the robot's torso) and dynamic components (e.g., its manipulators). The dynamic components are mapped by articulated models which are updated with each robot movement. The model of the scene is obtained by generating corresponding OBB models from the point cloud obtained by the range sensor. The obstacle model is used as the basis for online collision monitoring. The algorithm consists of two main phases: determination of potentially colliding objects by a rough distance check based on the velocity vectors of all moving parts, and subsequent elaborate collision tests for all objects in the determined potential colliding sets. Figure 84.12a shows the velocity vectors of a moving arm. Figure 84.12b shows the manipulator about to collide with the robot torso; the respective joints are marked in red color. Figure 84.12c shows successful collision detection for simultaneous movement of two manipulators.

In addition to the online collision monitoring, the obstacle model provides the basis to compute

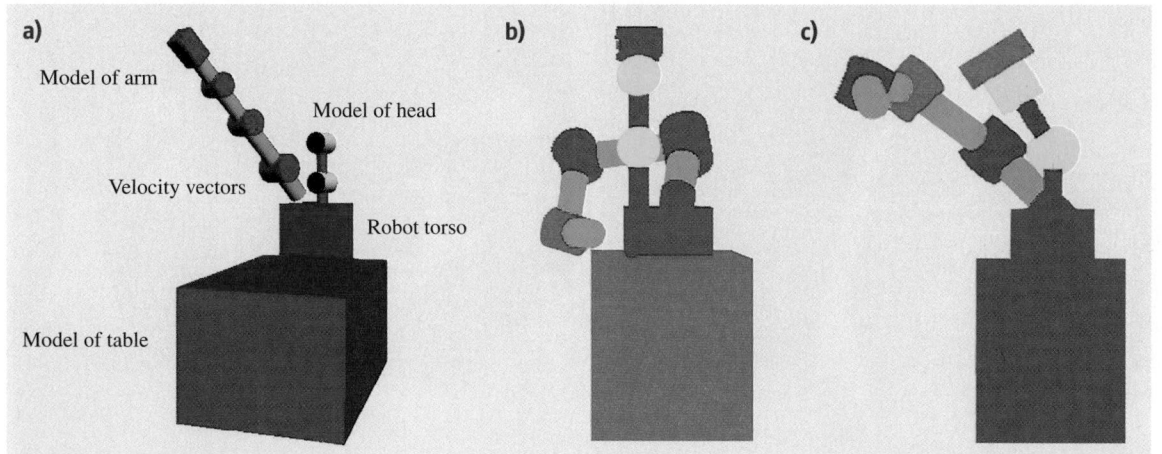

a)
Model of arm
Model of head
Velocity vectors
Robot torso
Model of table

b)

c)

Fig. 84.12a–c Velocity vectors of moving parts (**a**), detection of potential collisions between different parts of the robot illustrated in red color (**b,c**)

a collision-free trajectory for moving the manipulator to grasp a previously detected object. The entire obstruction model is used as the basis for the path search. As a result the determined path is guaranteed to be collision free with respect to both the robot's components and also the robot's environment. The implemented method is based on path planning with rapidly exploring random trees [84.76], and smoothing of the calculated path.

User Interface

In order to enable all users to operate Care-O-bot without difficulties, the user interface must be suitable even for users without any prior technical knowl-

edge [84.77]. For simple man–machine communication without misunderstanding, multiple sensing channels (speech, haptics, and gestures) are addressed. Commanding the robot, for example, is done by speech input, gestures, and touchscreen. The necessary feedback about the robot system state is given by speech output and graphical presentations on the monitor. The user interface of Care-O-bot II is implemented on a handheld, lightweight control panel, which the user can retain – even while the robot moves around. Care-O-bot 3 provides a moveable tray to pass objects to and from the user. A touchscreen integrated in that tray provides the necessary visual input and output.

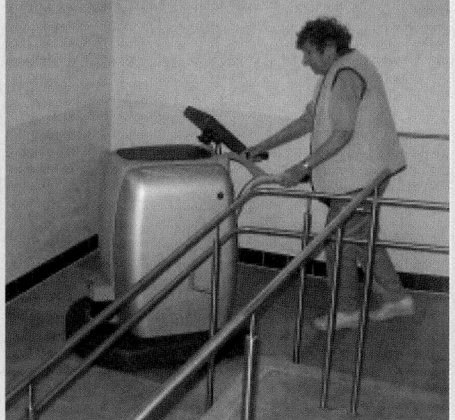
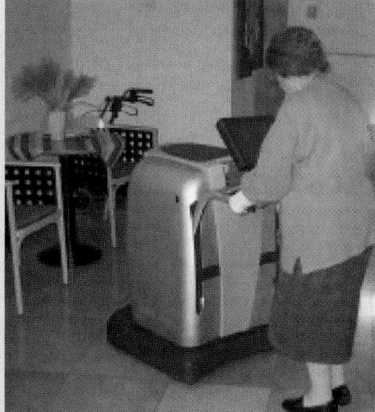

Fig. 84.13 Field tests of the walking aid function

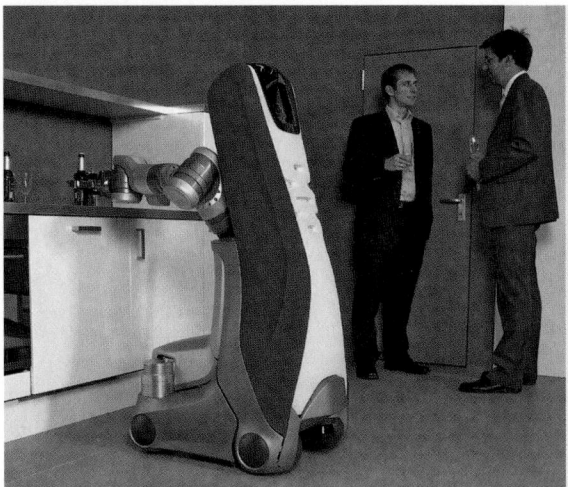

Fig. 84.14 Care-O-bot 3 fetching a drink in the kitchen and serving drinks to visitors

84.3.3 Applications

Walking Aid Function

Using the walking supporters attached to the rear of Care-O-bot II, the robot can serve as an intelligent walking aid able to lead a user to a given target [84.78]. The velocity of the robot during guidance is adjusted to the walking velocity of the user by measuring the user forces applied to the handles. Several user tests in elderly care facilities have proven the capabilities of the guidance system. Figure 84.13 displays some elderly users during the latest field tests. In order to enable a clear view ahead, the sensor head and manipulator of the robot was taken off for the walking aid tests.

Dependable Execution of Fetch and Carry Tasks

The fetch and carry capabilities of Care-O-bot II and Care-O-bot 3 have been tested and evaluated on several occasions during fairs and exhibitions. A sample home environment containing different furniture and objects is used to test, demonstrate, and evaluate the performance of the robots. They are already able to detect, grasp, and move different objects in the environment or bring them to the user (Figure 84.14). Care-O-bot II has also been tested in a real home environment. The robot was ordered to get a drink from the kitchen. It was able to grasp a box of juice from the refrigerator, and a glass from the kitchen shelf and place it on the living room table. Direct user interaction was tested by handing a box of pills to the user.

The tests and demonstrations have proven the dependability of the fetch and carry task execution system: the underlying execution framework ran to our complete satisfaction. Error detection and error recovery of the framework enabled detection and grasping of objects to function without any failures.

84.4 Application Example: the Bionic Robotic Arm ISELLA

84.4.1 Service Robot Arms and Drive Technology

Robot arms of service robots may also act as a paradigm for artificial limbs and rehabilitation aids. Some key features are:

- Low ratio of weight to payload
- High energy efficiency and applicability to battery-powered operation

- Moderate costs of manufacturing and materials in series production.

Conventional arms of service robots have a geared electrical motor for each joint as drives, sometimes built with a sleeve shaft as a cable duct for other drives and the gripper. Based on this layout it was possible to bring down the ratio of weight to payload to approximately $3:1$ to $1:1$ with designs highly optimized in weight and careful selection of materials

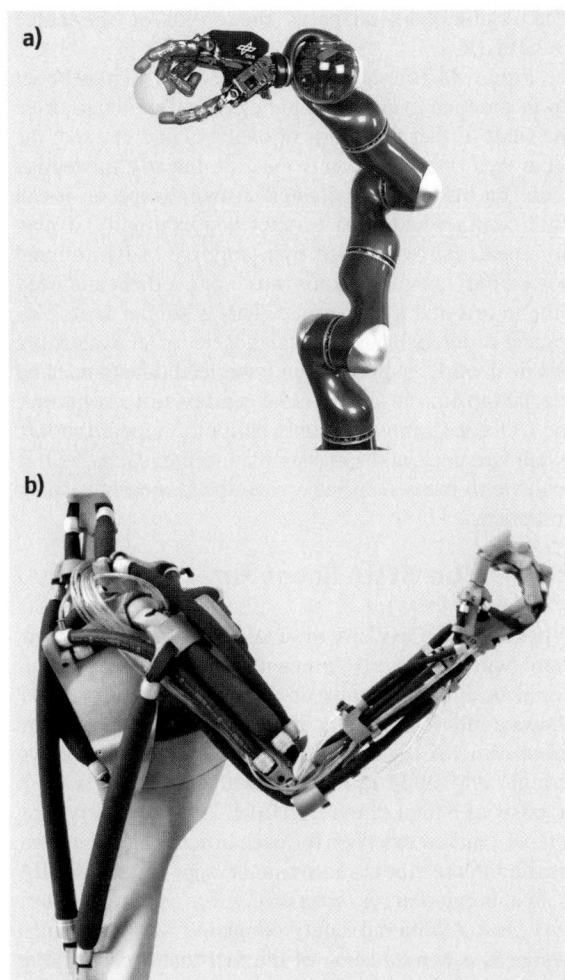

Fig. 84.15a,b Robot arms with high ratio of weight to payload: **(a)** lightweight arm (courtesy of DLR, Wessling, Germany), **(b)** bionic robot arm with fluidic muscles (courtesy of Festo AG, Germany)

(Fig. 84.15). However, costs of production, especially of their high-quality and high-precision drive units, still exceed what may be attractive and affordable for private users, even for medical devices. Moreover, prostheses should have properties similar to their biological counterparts rather than properties of industrial robots. Among others, these properties are variable stiffness of the joints, often smaller ranges of the joint angles, and less precision in forward-controlled positioning of the end-effector. These properties may be easier to achieve with a bionic-oriented approach of mechanical and kinematic design. Secondly, a different approach

to drive technology is required, which may be more suitable for variable stiffness and – above all – less expensive.

Something that may be thought of immediately is using artificial or technical muscles, i.e., technical actuators with properties of biological muscles. In recent years and decades a number of different types have been proposed and developed. Most of them are in a state of research, although some are already commercially available. Details and comparisons can be found in several publications [84.79, 80]. The pneumatic muscle, also known as the McKibben muscle [84.81], is a rubber tube covered with a braided tissue that contracts when pressurized. Since some of its properties are different from those of pneumatic cylinders, it is more advantageous for some industrial applications, and has also been used in some bionic robot applications [84.82, 83]. Figure 84.15b shows Airic's arm – an experimental bionic arm by Festo AG, designed like the human arm and shoulder. However, pneumatic systems generally have low energy efficiency compared to electric drives, and also motion control requires precision valves and pneumatic regulators. Another group of artificial muscles is shape-memory alloys (SMA), which change their shape with changing temperature. They are sparsely used for bionic robot applications [84.84]. Electroactive polymers (EAP) are materials with piezoelectric properties. Several hundreds of such polymers are known and some seem promising for use as technical muscles [84.85, 86]. A few demonstrative examples can be found in the field of robotics [84.87]. Beyond these there are some other principles of technical mus-

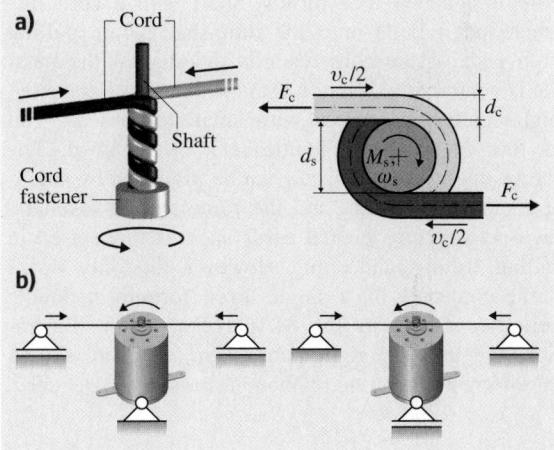

Fig. 84.16 (a) Details of the DOHELIX mechanism, **(b)** setup of the DOHELIX muscle

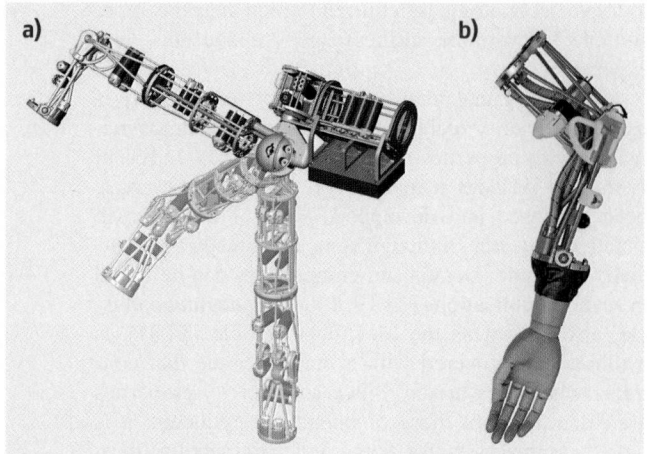

Fig. 84.17a,b The ISELLA robot arm in different poses, assembly of elbow part with a dummy hand

cles [84.88, 89], but these are also in the state of research.

84.4.2 The DOHELIX Muscle

A recent development of Fraunhofer IPA, Germany in the field of artificial muscles may overcome all disadvantages of the above-mentioned drives and may be most suitable for battery-powered articulated mechanisms, including consumer service robot arms, prosthetic devices, and rehabilitation aids. This development was introduced in [84.90] and [84.91], and is shown in Fig. 84.16a. The sketch in Fig. 84.16a illustrates parts and functionality: a highly flexible and high-strength cord is attached to a turning shaft with a cord fastener and it coils onto the thin shaft while pulling both ends of the cord towards the shaft – the muscle is contracting. Using a shaft with a small diameter, high contraction forces with small torques as well as low contraction velocities can be obtained. The torque and rotational speed can be produced by small-size electrical motors and the force can be taken by high-performance plaited cords such as those used in sailing, fishing, and kiting. However, this only works if the cord coils in a single layer, forming a double helix as shown in Fig. 84.16a. For better illustration, the left and right part of the cord are shown in different greyscale, although it is a single cord.

The double-helix shape is the origin of the name DOHELIX.

Figure 84.16b shows how the mechanism may be set up in practice, using a simple electrical motor to drive the shaft. Either the motor or one cord end is fixed; the other two assembly points must be linearly moveable. This is a muscle-like actuator in many respects: It can only contract and must be stretched externally, it may be repeatedly overloaded by a multiple of its nominal power, only the motor-coils must not overheat and need time to rest and to regenerate. This is similar to its biological counterpart – a skeletal muscle. After exhausting physical work or sports activity we need time to rest and regenerate our muscles. Unlike gearbox drive solutions, the DOHELIX muscle is only built of a standard motor, a standard cord, and some simple mechanical parts. It is scalable in many respects – size, speed, power, quality, and price.

84.4.3 The ISELLA Robot Arm

When using DOHELIX muscles to design a robot arm, there will be some significant differences to conventional designs, as shown in Fig. 84.15a. Figure 84.17 shows a full design study of the human-arm-like bionic robot arm ISELLA in different poses: hanging down straight and lifted to the side with the elbow bent. It consists of a total of ten DOHELIX muscles, providing a flexor and an extensor for each articulated joint, four situated in the elbow and six in the upper arm. ISELLA is an abbreviation for *intrinsically safe lightweight low-cost arm*. Additional safety compared to conventional design is a consequence of the fact that a single joint only moves with coordinated control of at least two drives. Failure of a single drive control loop is very unlikely to produce uncontrolled motion of the entire arm, since the counterpart of the malfunctioning drive will counteract the motion, trying to maintain the desired joint angle.

Like a human arm without the wrist, ISELLA has five degrees of freedom. The DOHELIX muscles are composed of direct-current (DC) motors and small-ratio planetary gearboxes. With some muscles there are two or three drives in parallel to multiply power while using the same type of motor, cord, and other parts. The elbow part of ISELLA is shown in Fig. 84.16b with a dummy hand attached.

84.5 Future Challenges

The number of assistive devices required will increase with the continuously increasing number of elderly persons. At the same time, the budget available for such assistive tools will be extremely limited. Therefore, one main effort of future developments will be to obtain maximum functionality at minimum cost. One solution could be the use of components already established in other markets such as sensors or motors used in the car industry or industrial applications that are already available at low prices.

Another target for future developments is the design of new human–machine interfaces in order to enable intuitive interaction with technical devices. Current developments deal with speech interaction, gesture recognition or other human-like communication channels. Another aspect for assistive service robots is their ability to continuously learn new tasks. Current devel-

opments deal with autonomous robots exploring their environment autonomously or ways of instructing new tasks to a service robot by demonstration or leading it *by the hand*.

A third important aspect of future developments is the issue of safety. Whereas for mobile robots without manipulation capabilities, safety regulations can be derived from industrial applications, specifically the rules for safe navigation of autonomous guided vehicles (AGVs), no rules are yet available for safe manipulation among humans. Currently available products such as the Manus ARM solve this problem by limiting the power of the motors, however, at the cost of not being able to lift heavy objects. In order to provide enhanced support, new sensors have to be applied to supervise the workspace of mobile robot arms in direct interaction with humans.

References

84.1 Statistisches Bundesamt Deutschland: *11. koordinierte Bevölkerungsvorausberechnung* (2006) www.destatis.de (last accessed February 1, 2007)

84.2 N.I. Katevas: Mobile robots in healthcare: the past, the present and the future. In: *Mobile Robots in Healthcare*, ed. by N.I. Katevas (IOS, Athens 2001) pp. 1–16

84.3 B. Siciliano, O. Khatib (Eds.): *Springer Handbook of Robotics* (Springer, Berlin, Heidelberg 2008)

84.4 H.F.M. van der Loos, D.J. Reinkensmeyer: Rehabilitation and health care robotics. In: *Springer Handbook of Robotics*, ed. by B. Siciliano, O. Khatib (Springer, Berlin, Heidelberg 2008)

84.5 C. Huang, G. Wasson, M. Alwan, P. Sheth, A. Ledoux: Shared navigational control and user intent detection in an intelligent walker, Proc. AAAI Fall 2005 Symp. (EMBC) (2005)

84.6 G. Lacey, K.M. Dawson-Howe: *Personal adaptive mobility aid for frail and elderly blind people*, Tech. Rep. TR-CS-95-18 (Comp. Science Dept. School of Engineering, Trinity College Dublin 1995)

84.7 Y. Hirata, A. Hara, A. Muraki, K. Kosuge: Passive-type intelligent walker RT walker, Proc. IEEE Int. Conf. Robot. Autom. (Orlando 2006)

84.8 D. Rodríguez-Losada, F. Matía, A. Jiménez, R. Galán, G. Lacey: Guido, the robotic smart walker for the frail visually impaired, 1st Int. Congr. Domotics Robot. Remote Assistance All – DRT4all 2005 (Act Book, Madrid 2005) pp. 155–169

84.9 N. Nejatbakhsh, K. Kosuge: User-environment based navigation algorithm for an omnidirectional

passive walking aid system, Proc. 9th Int. Conf. Rehab. Robot. (Chicago 2005)

84.10 J. Glover, D. Holstius, M. Manojlovich, K. Montgomery, A. Powers, J. Wu, S. Kiesler, J. Matthews, S. Thrun: *A robotically-augmented walker for older adults*, Tech. Rep. CMU-CS-03-170 (Carnegie Mellon Univ. Comp. Science Dep., Pittsburgh 2003)

84.11 H.M. Shim, E.H. Lee, J.H. Shim, S.M. Lee, S.H. Hong: Implementation of an intelligent walking assistant robot for the elderly in outdoor environment, Proc. 9th Int. Conf. Rehab. Robot. (Chicago 2005)

84.12 S. Egawa, I. Takeuchi, A. Koseki, T. Ishii: Electrically assisted walker with supporter-embedded force-sensing device. In: *Advances in Rehabilitation Robotics*, Lecture Notes in Control and Information Science, Vol. 306 (Springer, Berlin, Heidelberg 2004) pp. 313–322

84.13 H. Yu, M. Spenko, S. Dubowsky: An adaptive shared control system for an intelligent mobility aid for the elderly, Auton. Robots **16**(15), 53–66 (2003)

84.14 A. Morris, R. Donamukkala, A. Kapuria, A. Steinfeld, J. Matthews, J. Dunbar-Jacobs, S. Thrun: A robotic walker that provides guidance, Proc. IEEE Int. Conf. Robot. Autom. (ICRA) (Taipei 2003)

84.15 P. Médéric, V. Pasqui, F. Plumet, P. Bidaud: Elderly people sit to stand transfer experimental analysis, Proc. 8th Int. Conf. Climb. Walk. Robots (CLAWAR 2005) (2005) pp. 953–960

84.16 C. Bühler, H. Heck, J. Nedza, R. Wallbruchr: Evaluation of the MOBIL walking and fifting aid. In:

Assistive Technology Added Value to the Quality of Life, ed. by C. Marincek, C. Bühler, H. Knops, R. Andrich (IOS, Washington 2001) pp. 210–215

84.17 O. Chuy Jr., Y. Hirata, Z. Whand, K. Kosuge: Approach in assisting a sit-to-stand movement using robotic walking support system, IEEE/RSJ Int. Conf. Intell. Robots Syst. (Beijing 2006) pp. 4343–4348

84.18 A.M. Sabatini, V. Genovese, E. Pacchierotti: A mobility aid for the support to walking and object transportation of people with motor impairments, Proc. IEEE/RSJ Intl. Conf. Int. Robots Syst. (2002)

84.19 R.A. Cooper: Intelligent control of power wheelchairs, Eng. Med. Biol. Mag. **14**(4), 423–431 (1995)

84.20 S.P. Levine, D.A. Bell, L.A. Jaros, R.C. Simpson, Y. Koren, J. Borenstein: The NavChair assistive wheelchair navigation system, IEEE Trans. Rehab. Eng. **7**(4), 443–451 (1999)

84.21 G. Bourhis, O. Horn, O. Habert, A. Pruski: An autonomous vehicle for people with motor disabilities, IEEE Robot. Autom. Mag. **8**(1), 20–28 (2001)

84.22 S.P. Parikh, V. Grassi Jr., V. Kumar, J. Okamoto Jr.: Incorporating user inputs in motion planning for a smart wheelchair, IEEE Int. Conf. Robot. Autom. (New Orleans 2004) pp. 2043–2048

84.23 E. Prassler, J. Scholz, P. Fiorini: A robotic wheelchair for crowded public environments, IEEE Robot. Autom. Mag. **8**(1), 38–45 (2001)

84.24 H.A. Yanco: Shared User-Computer Control of a Robotic Wheelchair System. Ph.D. Thesis (Massachusetts Institute of Technology, Cambridge 2000)

84.25 R.A. Brooks: *A Robust Layered Control System for a Mobile Robot* (A.I. Memo 864, Massachusetts Institute of Technology, Artificial Intelligence Laboratory 1985)

84.26 T. Röfer, A. Lankenau: Ein Fahrassistent für ältere und behinderte Menschen, Auton. Mobile Syst. **15**, 334–343 (1999), in German

84.27 G. Pires, R. Araujo, U. Nunes, A.T. de Almeida: ROBCHAIR – a powered wheelchair using a behaviour-based navigation, 5th Int. Workshop Adv. Motion Control (Coimbra 1998) pp. 536–541

84.28 D. Vanhooydonck, E. Demeester, M. Nuttin, H. Van Brussel: Shared control for intelligent wheelchairs: an implicit estimation of the user intention, ASER'03 1st Int. Workshop Adv. Serv. Robot. (2003) pp. 176–182

84.29 H. Hoyer: The OMNI wheelchair, Serv. Robot Int. J. **1**(1), 26–29 (1995)

84.30 Toyota Motor Corporation: Robot Technology, http://www.toyota.co.jp/en/tech/robot/ (last accessed February 17, 2009)

84.31 S. Shoval, I. Ulrich, J. Borenstein: NavBelt and the GuideCane, IEEE Robot. Autom. Mag. **10**(1), 9–20 (2003)

84.32 M. Montemerlo, J. Pineau, N. Roy, S. Thrun, V. Verma: Experiences with a mobile robotic guide for the elderly, Proc. AAAI Natl. Conf. Artif. Intell. (2002)

84.33 B. Graf, O. Barth: Entertainment robotics: examples, key technologies and perspectives, Robots in Exhibitions, Proc. Workshop WS9 (Lausanne 2002)

84.34 H.F.M. van der Loos, J.J. Wagner, N. Smaby, K.S. Chang, O. Madrigal, L.J. Leifer, O. Khatib: ProVAR assistive robot system architecture, Proc. ICRA (Detroit 1999) pp. 741–746

84.35 T. Jones: RAID – toward greater independence in the office and home environment, Proc. 6th Int. Conf. Rehab. Robot. (ICORR'99) (Stanford 1999)

84.36 Rehab Robotics Ltd: Handy1, http://ourworld compuserve.com/homepages/rehabrobotics/Hand1.htm (last accessed February 17, 2009)

84.37 Exact Dynamics BV: ARM: Assistent Robot Manipulator, http://www.exactdynamics.nl/ (last accessed February 17, 2009)

84.38 American Honda Motor Co. Inc.: ASIMO, http://asimo.honda.com (last accessed February 17, 2009)

84.39 Kawada Industries, Inc.: Humanoid Robot HRP-2 "Promet", http://www.kawada.co.jp/global/ams/hrp_2.html (last accessed February 17, 2009)

84.40 Sarcos Inc.: High-performace humanoid robot, http://www.sarcos.com/telespec.atr.html (last accessed February 17, 2009)

84.41 P. Dario, E. Guglielmelli, C. Laschi, G. Teti (SSSA): MOVAID: a personal robot in everyday life of disabled and elderly people, Technol. Disabil. J. **10**, 77–93 (1999)

84.42 R. Bischoff: HERMES – a humanoid experimental robot for mobile manipulation and exploration services. Video Proc, IEEE Int. Conf. Robot. Autom. ICRA '01 (Seoul 2001), III–1

84.43 Universität Karlsruhe, Institut für Technische Informatik: SFB 588 Humanoide Roboter – Lernende und kooperierende multimodale Roboter, http://www.sfb588.uni-karlsruhe.de (last accessed February 17, 2009)

84.44 Fraunhofer IPA: Care-O-bot, http://www.care-o-bot.de (last accessed February 17, 2009)

84.45 Hitachi, Ltd.: Robotics, http://www.hitachi.com/rd/research/robotics.html (last accessed February 17, 2009)

84.46 Yaskawa Electric Corporation: Yaskawa develops a service robot "SmartPal V (SmartPal Five)". Press release November 28, 2007 http://www.yaskawa.co.jp/en/newsrelease/2007/04.htm

84.47 Fujitsu Frontech Ltd.: Fujitsu Service Robot (enon), http://www.frontech.fujitsu.com/en/forjp/robot/servicerobot/ (last accessed February 17, 2009)

84.48 D.H. Plettenburg: Basic requirements for upper extremity prostheses: the WILMER approach, Proc. 20th IEEE Int. Conf. Eng. Med. Biol. Soc. **5**, 2276–2281 (1998)

84.49 P. Berkelman, T. Lu, J. Ma, P. Rossi: Passive orthosis linkage for locomotor rehabilitation, Proc. 10th Int.

Conf. Rehab. Robot. ICORR 2007 (Noordwijk 2007) pp. 425–431

84.50   D. Odell, A. Barr, R. Goldberg, J. Chung, D. Rempel: Evaluation of a dynamic arm support for seated and standing tasks: a laboratory study of electromyography and subjective feedback, J. Ergon. **50**(4), 520–535 (2007)

84.51   A.H.A. Stienen, E.E.G. Hekman, F.C.T. Van der Helm, G.B. Prange, M.J.A. Jannink, A.M.M. Aalsma, H. Van der Kooij: Freebal: dedicated gravity compensation for the upper extremities, Proc. Int. Conf. Rehab. Robot. ICORR 2007 (Noordwijk 2007) pp. 804–808

84.52   A. Jackson, P. Culmer, S. Makower, M. Levesley, R. Richardson, A. Cozens, M. Mon Williams, B. Bhakta: Initial patient testing of iPAM – a robotic system for stroke rehabilitation, Proc. 10th Int. Conf. Rehab. Robot. ICORR (Noordwijk 2007)

84.53   H. Hirai, R. Ozawa, S. Goto, H. Fujigaya, S. Yamasaki, Y. Hatanaka, S. Kawamura: Development of an ankle-foot orthosis with a pneumatic passive element, Proc. 15th IEEE Int. Symp. Robot Human Interact. Commun. (RoMan 06) (2006) pp. 220–225

84.54   P. Beyl, J. Naudet, R. Van Ham, D. Lefeber: Mechanical design of an active knee orthosis for gait rehabilitation, Proc. 10th Int. Conf. Rehab. Robot. ICORR 2007 (Noordwijk 2007) pp. 100–105

84.55   Fraunhofer IPK: Rehabilitation Robotics, http://www.ipk.fraunhofer.de/rehabrobotics (last accessed February 17, 2009)

84.56   D. Surdilovic, R. Bernhardt, T. Schmidt, J. Zhang: STRING-MAN: a novel wire-robot for gait rehabilitation, *Advances in Rehabilitation Robotics*. In: *Advances in Rehabilitation Robotics*, Lecture Notes in Control and Information Science, Vol. 306 (Springer, Berlin, Heidelberg 2004) pp. 413–426

84.57   L. Luenenburger, G. Colombo, R. Riener: Biofeedback for robotic gait rehabilitation, J. NeuroEng. Rehab. **4**(1), (2007), http://www.balgrist.ch/display.cfm?id=101935

84.58   S.K. Banala, S.K. Agrawal, J.P. Scholz: Active leg exoskeleton (alex) for gait rehabilitation of motor-impaired patients, Proc. 10th Int. Conf. Rehab. Robot. ICORR 2007 (Noordwijk 2007) pp. 401–407

84.59   Berkeley Robotics and Human Engineering Laboratory: BLEEX Project, http://bleex.me.berkeley.edu/bleex.htm (last accessed February 17, 2009)

84.60   University of Tsukuba, Cybernics Laboratory: Robot suit HAL (Hybrid Assistive Limb), http://sanlab.kz.tsukuba.ac.jp/english/r_hal.php (last accessed February 17, 2009)

84.61   Touch Bionics Inc. and Touch EMAS Ltd.: Touchbionics, http://www.touchbionics.com (last accessed February 17, 2009)

84.62   A. Kargov, C. Pylatiuk, S. Schulz, G. Bretthauer: Modularly designed lightweight anthropomorphic robot hand, Proc. IEEE Int. Conf. Multisens. Fusion Integr. Intell. Syst. (Heidelberg 2006) pp. 155–159

84.63   Otto Bock HealthCare GmbH: http://www.ottobock.de (last accessed February 17, 2009)

84.64   Ossur hf: POWER KNEE, http://www.ossur.com/bionictechnology/powerknee (last accessed February 17, 2009)

84.65   K.B. Fite, T.J. Withrow, K.W. Wait, M. Goldfarb: Liquid-fueled actuation for an anthropomorphic upper extremity prosthesis, Proc. 28th Annual Int. Conf. IEEE Eng. Med. Biol. Soc. EMBS '06 (2006) pp. 5638–5642

84.66   S.K. Au, J. Weber, H. Herr: Biomechanical design of a powered ankle-foot prosthesis, Proc. 10th Int. Conf. Rehab. Robot. ICORR 2007 (Noordwijk 2007) pp. 298–303

84.67   F. Sup, A. Bohara, M. Goldfarb: Design and control of a powered knee and ankle prosthesis, Proc. IEEE Int. Conf. Robot. Autom. (2007) pp. 4134–4139

84.68   R.D. Schraft, C. Schaeffer, T. May: The concept of a system for assisting elderly or disabled persons in home environments, Proc. 24th IEEE Int. Conf. Ind. Electron. Control Instrum. (IECON), Vol. 4 (Aachen 1998)

84.69   B. Graf, M. Hans, R.D. Schraft: Care-O-bot II – development of a next generation robotic home assistant, Auton. Robots **16**(2), 193–205 (2004)

84.70   M. Hans, B. Graf, R.D. Schraft: Robotic home assistant Care-O-bot: past-present-future, Proc. IEEE Int. Workshop Robot Human Interact. Commun. (RoMan) (Paris 2001) pp. 407–411

84.71   B. Graf: Dependability of mobile robots in direct interaction with humans. In: *Advances in Human-Robot Interaction*, Springer Tracts in Advanced Robotics, Vol. 14 (Springer, Berlin, Heidelberg 2005) pp. 223–239

84.72   J.-C. Latombe: *Robot Motion Planning* (Kluwer Academic, Boston 1996)

84.73   T. Oggier, M. Lehmann, R. Kaufmann, M. Schweizer, M. Richter, P. Metzler, G. Lang, F. Lustenberger, N. Blanc: An all-solid-state optical range camera for 3-D real-time imaging with sub-centimeter depth resolution SwissRangerTM, Proc. SPIE **5249**, 534–545 (2003)

84.74   M. Pontil, A. Verri: Support vector machines for 3-D object recognition, IEEE Trans. Pattern Anal. Mach. Intell. **20**(6), 637–646 (1998)

84.75   J. Kubacki, W. Baum: Towards open-ended 3-D rotation and shift invariant object detection for robot companions, Proc. IEEE/RSJ Int. Conf. (IEEE, Piscataway 2006) pp. 3352–3357

84.76   B. Rohrmoser, C. Parlitz: Implementation of a path-planning algorithm for a robot arm, Robotik 2002: Leistungsstand, Anwendungen, Visionen, Trends (Ludwigsburg 2002), ed. by R. Dillmann et al., VDI/VDE-Gesellschaft Meß- und Automatisierungstechnik (GMA) (VDI Düsseldorf 2002) VDI Rep. 1679, pp. 59–64

Part I | 84

84.77 C. Parlitz, W. Baum, U. Reiser, M. Hägele: Intuitive human–machine interaction and implementation on an household robot companion. In: *Human Interface and the Management of Information. Methods, Techniques and Tools in Information Design*, Lecture Notes in Computer Science, Vol. 4557 (Springer, Berlin, Heidelberg 2007) pp. 922–929

84.78 B. Graf, R.D. Schraft: Behavior-based path modification for shared control of robotic walking aids, 10th Int. Conf. Rehab. Robot. (Piscataway IEEE, Noordwijk 2007) pp. 317–322

84.79 C. Cocaud, A. Jnifene: Analysis of a two DOF anthropomorphic arm driven by artificial muscles, Proc. 2nd IEEE Int. Workshop Haptic Audio Vis. Env. Appl. (HAVE 2003) pp. 20–21

84.80 J.D.W. Madden, N.A. Vandesteeg, P.A. Anquetil, P.G.A. Madden, A. Takshi, R.Z. Pytel, S.R. Lafontaine, P.A. Wieringa, I.W. Hunter: Artificial muscle technology: physical principles and naval prospects, IEEE J. Ocean. Eng. **29**(3), 706–728 (2004)

84.81 V. Nickel, J. Perry, A. Garrett: Development of useful function in the severely paralyzed hand, J. Bone Jt. Surg. **45A**(5), 933–952 (1963)

84.82 I. Boblan, R. Bannasch, H. Schwenk, F. Prietzel, L. Miertsch, A. Schultz: A human-like robot hand and arm with fluidic muscles: biologically inspired construction and functionality. In: *Embodied Artificial Intelligence*, Lecture Notes in Artificial Intelligence, Vol. 3139 (Springer, Berlin, Heidelberg 2004) pp. 160–179

84.83 Festo AG: *Brochure Airacuda* (Festo, Esslingen 2006), www.festo.com

84.84 C. Pfeiffer, K. DeLaurentis, C. Mavroidis: Shape memory alloy actuated robot prostheses: initial experiments, Proc. IEEE Int. Conf. Robot. Autom., Vol. 3 (1999) pp. 2385–2391

84.85 S. Arora, T. Gosh, J. Muth: Dielectric elastomer based prototype fiber actuators, Sens. Actuators A: Phys. **136**(1), 321–328 (2006)

84.86 H.R. Choi, K. Jung, S. Ryew, J.D. Nam, J.C. Koo, J. Jeon, K. Tanie: Biomimetic soft actuator: design, modeling, control, and applications, IEEE/ASME Trans. Mechatron. **10**(5), 581–593 (2005)

84.87 K. Takagi, M. Yamamura, Z.W. Luo, M. Onishi, S. Hirano, K. Asaka, Y. Hayakawa: Development of a Rajiform swimming robot using ionic polymer artificial muscles, Proc. IEEE/RSJ Int. Conf. Intell. Robots Syst. (2006) pp. 1861–1866

84.88 T. Niino, S. Egawa, H. Kimura, T. Higuchi: Electrostatic artificial muscle: compact, high-power linear actuators with multiplelayer structures, Proc. IEEE Workshop Micro Electro Mechan. Syst. (1994)

84.89 K. Takemura, S. Yokota, K. Edamura: A micro artificial muscle actuator using electro-conjugate fluid, Proc. IEEE Int. Conf. (2005)

84.90 H. Staab, A. Sonnenburg: Studies and guidelines on the design of the *DOHELIX* technical muscle. In: *Robotics and Applications, IRA 2007*, 13th IASTED Int. Conf. (Würzburg 2007) (ACTA Press, Calgary 2007)

84.91 H. Staab, A. Sonnenburg, C. Hieger: The DOHELIX-muscle: a novel technical muscle for bionic robots and actuating drive applications, Autom. Sci. Eng. 3rd IEEE Conf. (Scottsdale 2007) pp. 306–311

85. Automation in Education/Learning Systems

Kazuyoshi Ishii, Kinnya Tamaki

The information technology (IT) revolution which began in the latter half of the 20th century has brought great changes to education and learning. The spread of the Internet has made information ubiquitous, changing the emphasis of education from the transmission and acquisition of knowledge to knowledge creation [85.1], and shifting the focus from group to individual education. Since the perspective for discussions of education systems is moving from instructors to learners [85.2–4], in place of *education systems* we adopt the expression *education/learning systems*. When considering the automation of education/learning systems, along with the impact of information and communications technology (ITC), the effects of educational psychology and educational technology cannot be ignored. This field overall is referred to as instructional design (ID) [85.5]. This chapter examines the history and present conditions of automation in education/learning systems, centered on e-Learning, from the perspectives of information and communication technologies and instructional design. The chapter also introduces two examples from the field of industrial engineering and management systems concerning projects to develop education/learning programs to train Japanese manufacturing management personnel. These examples are both ongoing industry–government–academia collaboration

projects aimed at the transmission and development of Japanese manufacturing *kaizen* (continuous improvement) knowhow and the education and training of management personnel.

The chapter concludes with a summary of future issues concerning the automation of education/learning systems and a list of reference materials in related fields for readers who seek further details.

85.1 Technology Aspects of Education/Learning Systems

85.1.1 Overview of Instructional Design (ID)

There has been a great deal of discussion regarding the definition of instructional design (ID). (Refer to *Martin Ryder*'s site [85.5] for a comprehensive review.) Here, we introduce two definitions; the first one is from the

Applied Research Laboratory (ARL) at Penn State University [85.6]:

Instructional design as a process is the systematic development of instructional specifications using learning and instructional theory to ensure the qual-

ity of instruction. It is the entire process of analysis of learning needs and goals and the development of a delivery system to meet those needs. It includes development of instructional materials and activities; and tryout and evaluation of all instruction and learner activities. Instructional design as a discipline is that branch of knowledge concerned with research and theory about instructional strategies and the process for developing and implementing those strategies.

The other one is by *Suzuki* [85.9]:

ID refers to models and research which compile methods to improve the results, efficiency and attractiveness of educational activities as well as the process of realizing learning assistance environments which apply such models and research.

ID design process models are generally divided into the two categories: Action, design, development, implementation, evaluation (ADDIE) models [85.10] and rapid prototyping models [85.11, 12]. Rapid pro-

totyping models are based on ADDIE models. ADDIE models are general models for the management of ID which grasp ID activities as a process of analysis, design, development, implementation, and evaluation. As shown in Fig. 85.1 [85.7, 8], the models are based on a plan → do → check/act (see) management circle model, with the plan process subdivided into thirds. In Fig. 85.1, education/learning activities signify collaboration toward the creation of human value by each individual and the creation of social value by enterprises and the community. We address the automation of education/learning systems assuming that the ideal of education/learning systems management is to continuously create new and greater value in education/learning activities.

Table 85.1 summarizes the education/learning systems design and operational methods of each process under this model. This table is revised from models presented in *Gagne* [85.13] and *Akahori* [85.14], incorporating our own experience. We now proceed to explain ID following the processes presented in this table. (See the list of reference materials for further details.)

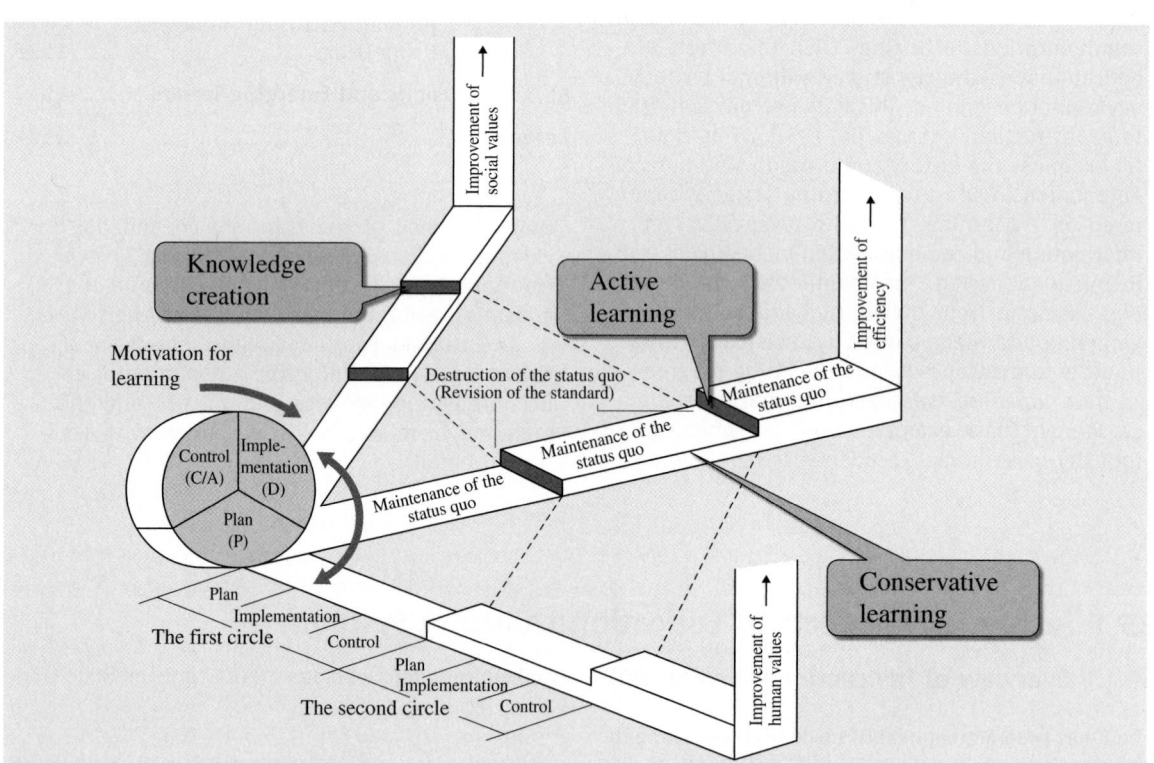

Fig. 85.1 Concept model for learning activity to create new values based on management circle model (after [85.7, 8], [85.8], with permission from Interscience Enterprises Ltd., 2009)

Table 85.1 Instructional design process model and its operations and methods

Management circle	ADDIE model	Operations and methods	
Plan	Analysis	Needs analysis Job analysis Student analysis	
	Design [85.15, 16]	Determination of learning goals (target behaviors, evaluation conditions, achievement levels)	
		Structuring Sequencing Programming [85.22, 23]	Determination of learning contents (knowledge, skills, values) [85.14, 17–21] Determination of learning methods considering student characteristics Determination of provision methods (joint training/individual training) × (cost performance)
	Development [85.24–27]	Textbooks Multimedia materials (in-house or consigned development) audiovisual + interactive	
Do	Implementation [85.28–35]	Tutoring Mentoring	
Check/act (see)	Evaluation	Evaluation subject	Learner Program
		Evaluation method [85.36–41]	Formative evaluations + comprehensive evaluations Reaction/learning/behavior/results

Analysis

The purpose of analysis is to extract useful information on what the parties are attempting to teach for setting learning goals. Analysis results directly affect the content of the structuring and sequencing, which are the next stages in the design process. To begin with, needs analyses are conducted to determine what instructors and learners are seeking (what they want to achieve), and to clarify the social trends, human resources development strategies, educational plans, career designs, and other assumptions, as well as the limiting conditions on the educational resources available for use. When the learners are adults, job analyses may also be conducted to identify the required competencies [85.42] from work duties content analysis. It is also necessary to conduct student analyses to clarify learners' academic history, experience, and learning preferences.

Design Process

Most of the debates regarding ID largely focus on the design process, with two different system approaches to ID. The Reigeluth approach [85.15] divides system design factors into three categories:

1. Learning conditions (learning issues, student characteristics, learning environment, etc.)
2. Learning methods (learning contents compilation methods, implementation methods, evaluation improvement methods, etc.)
3. Learning results (results, efficiency attractiveness, etc.).

Then, the combinations of the categories are classified into descriptive theory and prescriptive theory depending on what are cause factors and effects factors. The Rickey approach [85.16] combines an inductive approach with a deductive approach by experts, with interesting discussions of conceptual versus procedural models.

In the design process, the learning goals clarified by analysis are subdivided to the level at which they can be measured, followed by the step of structuring, which structures the relative priorities among the subdivided goals, and sequencing, which determines the learning order and contents to achieve the goals considering learner characteristics and improvement of education results and efficiency. Refer to *Gagne* [85.17], *Reigeluth* [85.18, 19], *Merrill* [85.20], *Sato* [85.21], and *Akahori* [85.14] for detailed discussions of these issues. Models and methodologies adopting their research findings provide an extremely valuable framework for the fusion of education/learning activities with information technology.

Educational programs are then compiled, considering the learning goals, learning content, and learning formats (group or individual), and synchronous or asynchronous learning as well as the limitations on time and educational resources. Some examples of the learning formats are shown in Fig. 85.2 in which formats are classified into distance learning [85.44] or not, and self-directed learning or not.

ID research is also being advanced by *Keller* [85.22] and *Suzuki* [85.23] on the attention, relevance, confidence, satisfaction (ARCS) model as a means of motivating learners by creating more attractive education and learning environments.

Development

Development sets the educational units via the following considerations on the series of learning goals and learning contents generated by the design works [85.13]:

1. Selection of learning goals
2. Listing of instructional events
3. Selection and development of educational materials and learning activities
4. Division of rules between instructors and learners.

Among these, the core task is the selection and development of educational materials, for which the potential applications of ICT and other multimedia and digital technologies have been greatly increasing in recent years. Prior to the arrival of this new era, *Gagne* [85.13] recognized and made an interesting suggestion that computers and other new media are significant for education not just because they enable stimulating presentations as instructional events but because they are important factors for improving learning results, effectiveness, and attractiveness.

Active learning [85.24] has also been gaining attention as a new trend in educational methodology development, primarily in the business world. The case study, scenario, and simulation methods are representative examples of active learning, and these include the goal-based scenario (GBS) theory advocated by *Shank* [85.26] for the scenario methods. GBS is an ID

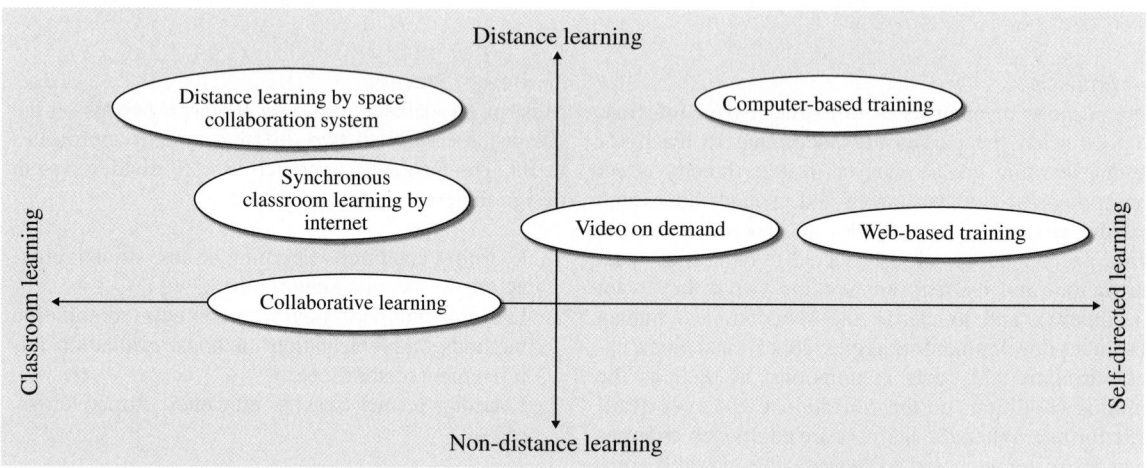

Fig. 85.2 Some examples of learning formats (after [85.43], with permission from Ohmsha, 2002)

theory for developing scenarios as a learning environment to provide students with vicarious experience of learning from failure, using actual cases from business literature [85.27, 28].

Implementation

Various communication channels are formed by the student learning environment at this stage, where the designed instructional content is implemented. The challenge for these channels is to realize system design and operation that holds and increases the learner's interest. From a learner-centered perspective, the channel formation factors include the human factors of instructors, mentors, tutors, other learners, and educational materials distributors and the material factors of computers, media displays, and educational materials distribution systems. The materials factors mostly concern the design of the man–machine interface.

However as the learning activities themselves are fundamentally human activities, the issue of the automation of education/learning activities assuming an advanced information society brings us to reconsider the definition of what constitutes a human being. In that sense, such questions as instructor competency [85.28] and presence [85.29–31], how to make learners actively involved with instruction administration [85.32, 33], and psychological aspects [85.34, 35] will be important issues for future examinations on the automation of education/learning activities.

Evaluation

Kirkpatrick [85.36] provides a typical four-stage ID evaluation model. This model divides evaluation of the development and implementation of education/learning systems into the four levels of reaction, learning, behavior, and results. It clarifies the evaluation period and evaluation items, as well as the data collection, measurement, and evaluation methods. This model gives consideration to e-Learning, for example, in the handling of return on investment, which is a level 4 results evaluation item [85.37, 38]. Formative or comprehensive evaluation methods may also be used, depending upon the evaluation objective [85.39]. Comprehensive evaluations provide an overall evaluation of the learning effect after a series of learning activities is completed. Formative evaluations [85.40, 41] are evaluations conducted during the learning process for the purpose of improving the education and learning activities.

85.1.2 Development History and Present Conditions of e-Learning

In this section we address e-Learning as an education/learning systems automation topic, and review the peripheral technologies, development history, and present conditions of e-Learning.

Definitions of e-Learning

Definitions of e-Learning are extremely diverse, and have changed over time. Sample definitions include the following:

- American Society for Training and Development (ASTD) [85.45]:

  e-Learning refers to all items provided, facilitated or transmitted by electronic technology that are clearly for the purpose of learning.

- *Broadbent* [85.46]:

  Computerization is important for e-Learning. e-Learning means computerized training, education, coaching, and information, and includes synchronous and asynchronous learning that is assisted by technologies such as the internet, CD-ROMs, satellites, telephones, personal computers, PDAs and other wireless equipment, multimedia, and CBT.

- *Rosenberg* [85.47]:

  e-Learning refers to the provision of diverse solutions using internet technologies to heighten knowledge and performance.

- The Japan e-Learning consortium (JeLC) [85.48]:

  e-Learning is independent learning using information technologies and communications networks. The contents are arranged in line with the learning goals, and interactivity is secured as necessary between the learner and the contents provider. This interactivity gives the learner opportunities to participate at will, so appropriate instruction can be given at appropriate times while advancing learning through people and computers.

e-Learning Peripheral Technologies

Because e-Learning is based on information and communications technology, it encompasses diverse learning formats which can be categorized by the following distinctive characteristics:

1. Educational materials and their distribution methods
2. Communication between instructors and learners.

In characterizing (1), the extent to which the learning is electronic can be measured by the computerization of the contents and distribution. e-Learning where the contents used as educational materials are more computerized and the distribution involves less human intervention is considered to be more electronic.

In characterizing (2), Fig. 85.3 positions e-Learning with greater potential for two-way communications as having a higher level for interactivity. This figure was prepared by the Japan e-Learning Consortium (JeLC) [85.48], and partially revised by the authors. The items marked by solid circles denote those characterized as e-Learning. Videos are considered a borderline area because of the existence of both digital and analog videos. From this figure, we project that the devel-

opment of future e-Learning technologies will move toward higher levels of computerization and interactivity under lower cost.

The History of Problem Solving and Value Creation in the Development of Computer–Assisted Education/ Learning Systems

Reviewing prior efforts to use computers in education/learning systems, the fusion of problem-solving needs at learning and education sites with the seeds of instructional design theory can be recorded as the history of advancing automation using computer and communications technologies. Table 85.2 presents the history of problem solving from computer-assisted (aided) instruction (CAI) through to e-Learning, explaining the problems, solutions, and main technologies involved. Table 85.2 compares e-Learning with CAI, CBT, and web-based training (WBT), but the e-Learning system involves CAI, CBT and WBT as a whole as shown in Fig. 85.2 and its definitions.

Among the various e-Learning stakeholders, Table 85.3 examines the learner, instructor, and training

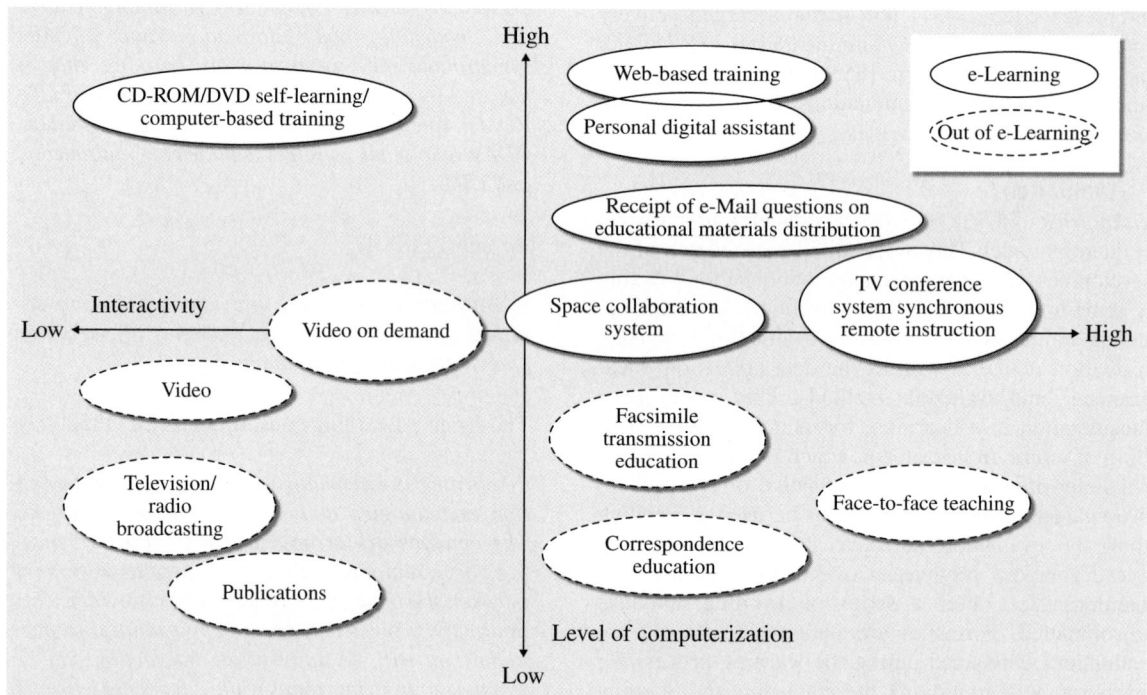

Fig. 85.3 Range of e-Learning by level of computerization and interactivity (after [85.48, p. 9], with permission from Tokyo Denki Univ. Press, 2007)

Table 85.2 History of problem solving in computer-assisted education/learning systems ([a] The use of PDAs and cell phones is included)

Name	Problems to be solved	Solution	Automation elements
CAI: computer-assisted (aided) instruction	Money and time are required for the following items using conventional systems: • Instructors and learners must be at the same location for long periods of time • Learners' progress and other items must be compiled manually	Display learning contents in accordance with the level of understanding of each learner, and save money and time via computer assistance	Computer
CBT: computer-based training	Learning order is determined beforehand by manuals in CAI, so learners cannot freely choose learning order	Educational materials compiled into a database, allowing interactive setting of learning order	CD-ROM Personal computer
WBT: web-based training	CBT has the following problems: • Unified management of learners' progress not possible • Educational materials cost money • Educational materials changes and improvements are difficult	Educational materials are transmitted via networks, facilitating quick revisions and learning evaluation Materials can be used for both synchronous and asynchronous learning	Internet-WWW
e-Learning	WBT has the difficulty of compiling optimal learning plans from the perspectives of results, efficiency, and attractiveness considering learners' academic history and progress, etc.	Learners can study independently, making maximum use of the potential of information and communications technologies to meet individual learning needs	LMS, m-Learning [a]

manager and identifies the value creation in terms of cost, time, and quality.

e-Learning Platforms and Standards

The platforms and standards for e-Learning systems are another technical aspect which supports e-Learning. The development and spread of these technologies have facilitated the low-cost development of e-Learning systems with consistent quality, and greatly contributed to the spread of e-Learning. The spread of e-Learning systems also advances the further development and evolution of e-Learning systems.

Table 85.4 summarizes the platforms and standards for learning management systems (LMS) [85.49], which constitute the core of e-Learning systems. LMS are platforms which provide such functions as educational materials management, learning progress management, learning implementation phase management such as management of mentoring and tutoring, and testing phase support for question setting, grading, performance results processing, and contents creation.

Learning contents management systems (LCMS) are primarily platforms to support educational mater-

Table 85.3 Value creation by e-Learning

Value/stakeholder	Learner	Instructor	Training manager
Cost reduction	Educational materials receipt expenses	Educational materials distribution expenses	Indirect cost
Time efficiency	Faster acquisition of materials and learning results	Faster learning results feedback	Faster system performance management
Quality improvement	Permits acquisition of the most recent educational materials Permits repeated study (review) Allows control over learning pace	Permits preparation of the most recent educational materials Accumulation of study history data Uniform distribution of the same educational materials	Accumulation of study history data

Table 85.4 e-Learning platforms and standards/tools (LCMS – learning content management system, LMS – learning management system, SCORM – sharable content object reference model, CMI – computer-managed instruction, LOM – learning object metadata/learning object reference model, LIP – learner information package, QTI – question and test interoperability, SSO – single sign-on)

ADDIE model	Platform	Function	SCORM [85.50] CMI [85.53]	LOM	LIP	QTI	SSO [85.51,52]
Analysis	–	–			○		○
Development	LCMS	Authoring	○	○ [85.54,55]	○		○
		Accumulation and management of educational materials		○ [85.56]			○
Implementation	LMS [85.57,58]	Learning management Learner management			○		○
		Educational materials distribution		○ [85.56]			○
		Progress management					○
		Contents creation support		○ [85.56]			
		Communications Mentoring			○		○
		Tutoring			○		○
Evaluation		Testing			○	○	○

ials developers. These platforms offer the two functions of authoring, which are the compilation and production of educational materials, and educational materials database management.

The sharable content object reference model (SCORM) in the "standards/tools" column is a web-based training (WBT) standard that integrates computer-managed instruction (CMI) with the learning object metadata/learning object reference model (LOM) standards, which are learning resource data structure standards. The LOM standards are used to describe the attributes of learning objects, which comprise all resources used in learning.

The learning information package (LIP) standards are used to describe the attributes of learners. The question and test interoperability (QTI) standards cover the necessary information for implementing tests. Single sign-on (SSO) is an e-Learning system technology which allows learners to log in once and gain access to library information systems, academic affairs management systems, and various other existing systems and resources aside from the LMS.

In 2005 the Organization for Economic Cooperation and Development (OECD) conducted a survey among institutions of higher education worldwide [85.59] regarding the conditions of e-Learning. Respondents noted the difficulty of assuring the quality of e-Learning [85.60], along with various other issues such as difficulty using present technologies [85.61],

and problems with creating, selecting, and using learning objects [85.57]. Considering these evaluations, the present platforms and standards technologies are immature, but should serve as the basis for future development. Research projects on developing architecture to break down LMS components into modules and restructure the LMS e-Learning architecture in accordance with work flow include the MIT Open Knowledge Initiative [85.58] and the e-Framework [85.62] being jointly developed by the UK and Australia. The other projects for e-Learning in USA and Europe may be found at [85.63–67].

e-Learning may be classified into synchronous and asynchronous learning. In asynchronous learning, learners communicate individually with computer systems such as WBT. In synchronous learning, multiple learners communicate among themselves or one or more learners communicate with a tutor or mentor under a computer-supported collaborative learning (CSCL) environment. Synchronous learning is similar to small-group *kaizen* (continuous improvement) activities in Japan. The platforms for collaborative e-Learning are still being established. CSCL is defined by *Koschmann* [85.68] as a field of study centrally concerned with meaning and the practices of meaning-making in the context of joint activity, and the ways in which these practices are mediated through designed artifacts. Interested readers should refer for details to [85.50–56].

85.2 Examples

This section introduces two examples of the development of education/learning systems in Japan based on *kaizen* (continuous improvement) knowhow in the Japanese manufacturing industry [85.69–72]. One is a program for understanding the mechanism of engineering and management of issues according to the general product lifecycle process of the manufacturing industry such as new product design, production system engineering, production planning, production management, supply chain management, production information management, and project management. The other is a program for managers in leadership roles who are contributing to creating new business value in manufacturing technology. The both programs have resulted from development of learning contents and numerical simulation software from effective utilization of e-Learning. The basic concept behind the both program developed has an effective combination with an intelli-

gent knowledge-based approach [85.73] with the *kaizen* activity through problem-based learning (PBL), case method learning (CML), and computer-supported collaborative learning (CSCL) as well as *blended learning* combining lectures with practice.

85.2.1 Educational Programs for *Cyber Manufacturing* in Industrial Engineering and Information Management

Overview of the *Cyber Manufacturing* Program

The Research Center for e-Learning Professional Competency (eLPCO) in the Research Institute of Aoyama Gakuin University in Japan has been organizing joint research projects between higher education institutions and industrial corporations. A research working group

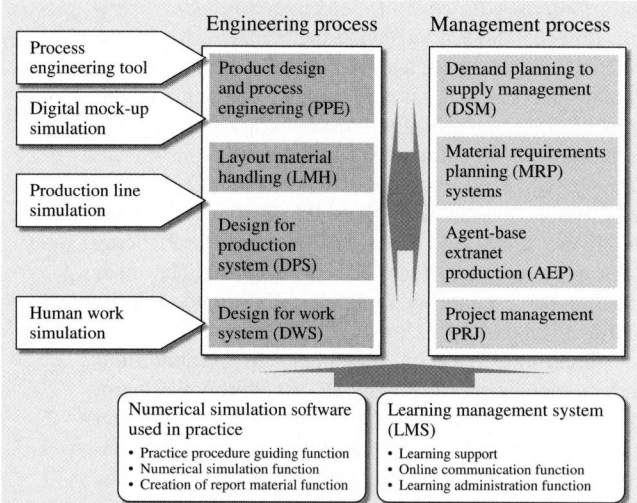

Fig. 85.4 Learning map for each course unit composed of *cyber manufacturing* core curricula

as a series of eLPCO project activities. The main objective of this working group is to develop practical core curricula in the *industrial engineering and information management* field by efficiently using e-Learning methods [85.74].

As illustrated in Fig. 85.4, four course units on engineering processes and four course units on operational management processes have been developed as the total educational program for *cyber manufacturing* [85.75]. The characteristics of e-Learning methods for each course are to integrate the *blended-learning* style between lectures and practices supported by a learning management system (LMS) and simulation software. Additional characteristics are treated with various instructional strategies such as *case method learning* (CML), *problem-based learning* (PBL), and *computer-supported collaborative learning* (CSCL).

The CML method assisted with three-dimensional computer graphic (3-D-CG) simulation refers to a learning method based on a case study in a virtual experience model of the real world. Thus, by utilizing 3-D-CG simulations as shown in left arrows of Fig. 85.4, practices involving observation and process analysis of product

for *cyber manufacturing* has been promoted by ten universities and several manufactures and software vendors

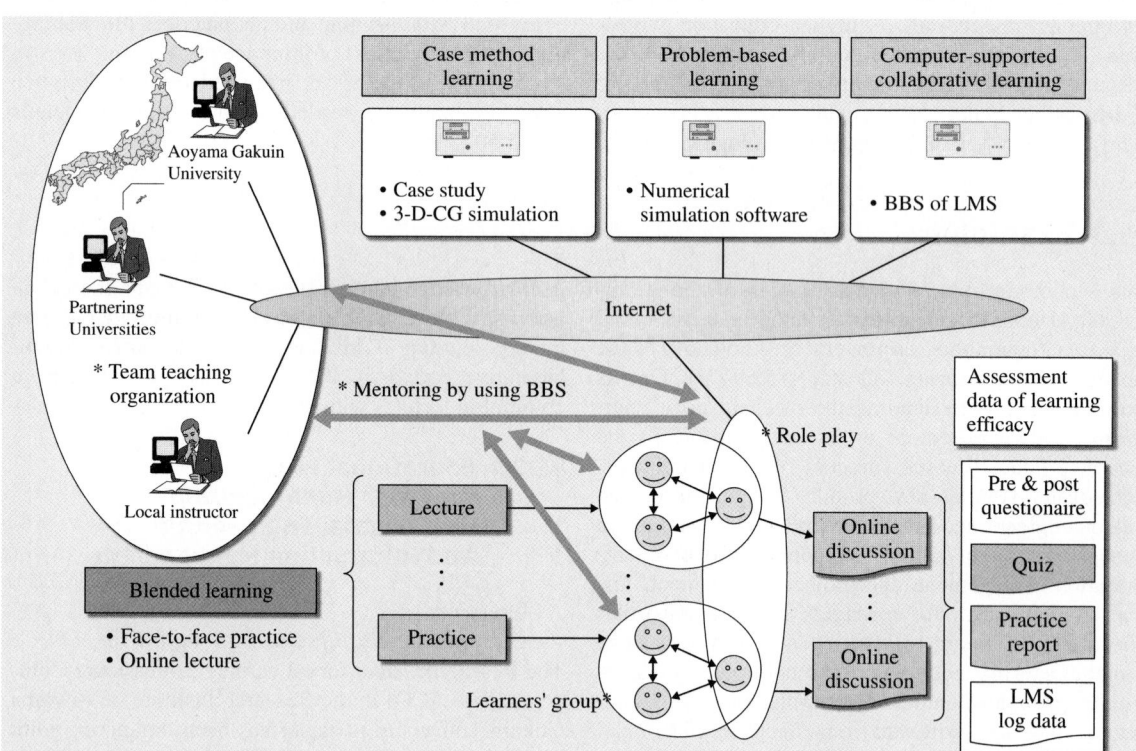

Fig. 85.5 Instructional strategy of *comprehensive learning style* combined with PBL, CML, and CSCL

Fig. 85.6 Work analysis in practice of *design of work system* course unit, using task observation of moving images of 3-D-CD human task simulation modeling

design activities and production system methods in virtual factories are implemented. Some examples of the utilization of 3-D-CG simulation in the practices include a 3-D-CG digital mock-up simulator (virtual trial manufacturing) in the *product design and process engineering* course unit, a 3-D-CG production line simulator in the *layout and material handling* and *design for production system* course units, and a 3-D-CG human work simulator (Fig. 85.6) in the *design for work system* course unit.

The aim of our PBL method is for learners to acquire practical skills through practice by using numerical simulation software. The software to support numerical simulation and data processing is developed by university–industry cooperation. After finishing the theory lecture, practice is conducted by using the numerical simulation software. Several practice problems are prepared and the learners are encouraged to solve the problems with suitable application of the learned theories, to create more desirable solution alternatives.

An LMS is a set of software tools designed to manage user learning interventions, courses, instruction, content delivering, online communication, assessment/testing, and so on. Most LMSs are web-based to facilitate *any time, any place, any pace* access to learning content and administration. The online learning environment used by universities allows instructors to manage their course and exchange information and opinions with learners for a course that in most cases will last several weeks and will meet several times during those weeks. Note that basically all the

lectures and practices in the course units utilize original LMS, developed by eLPCO under industry–university cooperation.

New Instructional Strategy of Comprehensive Learning Style and Uniqueness of Lectures and Practices

As illustrated in Fig. 85.5, in order to conduct the practice corresponding to each lecture, a new instructional strategy of *comprehensive learning style* is applied in each of the course units. This is developed by combining especially three types of instructional strategies: case method learning (CML), problem-solving learning (PBL), and computer-supported collaborative learning (CSCL).

In all the course units, the aim of the lectures is to understand the theoretical knowledge relating to engineering and management processes, while the object of the practices corresponding to each of the lectures is to acquire practical skills by using 3-D-CG simulation, numerical simulation software, and a asynchronous bulletin-board system (BBS) communication function, through various instruction strategies combined with CML, PBL, and CSCL.

The reason behind the creation of this new instructional strategy is to achieve deeper knowledge, while attaining practical skill, in the *industrial engineering and information management* field, because the conventional lecture method alone is not enough for rudimentary learners to understand complicated engineering and management processes.

Therefore, the creation of virtual images modeled by 3-D-CG simulation, and the development of new numerical simulation software in accordance with the content of the practices for each of the course units, are implemented. This has resulted in the preparation of a *practice learning environment*, which is said to be one of the most difficult environments with practical learning to take the *comprehensive learning style* class in e-Learning.

Furthermore, the characteristics of designing and operating method of the class style, by *blended learning* increase the difficulty, in which learn-by-doing instruction activities are realized by repetitive combination of the lectures and practices.

The goal of blended learning is to synthesize learning media into an integrated mix, which can be tailored to create a high-impact, efficient, and exciting learning program. The term *blended learning* is defined as [85.76]:

Blended learning is the combination of different training media (technologies, activities, and types of events) to create an optimum training program for a specific audience. The term blended means that traditional instructor-led training is being supplemented with other electronic formats. In the context, blended learning programs use many different forms of e-Learning, perhaps complemented with instructor-led training and other live formats.

Case Method Learning (CML) Assisted With Three-Dimensional Computer Graphic (3-D-CG) Simulation

It is not an easy issue to educate learners who are not able to imagine real field manufacturing and operational management at the product design and production phases, in course units related to engineering and management processes. In particular, new learners may not be able to understand easily the working environment, sequences of part assembling, and human task processes in manufacturing factories.

Thus to make the practices based on CML more effective, some special features are added to enrich the practice learning environment. These features include illustrations of production procedure modeling and realistic depictions and task analysis of virtual manufacturing based on animations created using multiple 3-D-CG simulators. As illustrated in Fig. 85.6, learners can visibly understand and analyze virtual human task processes in an assembling production line, which are expressed as moving images modeled by using a *human task simulator*.

Fig. 85.7 Group practice using numerical simulation software for PBL and 3D-CG simulation image of a virtual factory for CML

Problem-Based Learning (PBL) by Using Numerical Simulation Software

Problem-solving ability is the ability to combine previously leaned principles, procedures, declarative knowledge, and cognitive strategies in a unique way within a domain of content to solve previously unencountered problems [85.77]. This activity yields new learning as learners are more able to respond to problems of a similar class in the future. This type of problem solving is often described as *domain-specific* or *semantically rich* problem solving because it emphasizes learning to utilize principles in a specific content area.

Figure 85.7 shows a group practice study with numerical simulation software and a 3-D-CG simulation image of a virtual factory. The learner on the right is observing and measuring human operating time based on moving images from the 3-D-CD human task simulation modeling illustrated in Fig. 85.6. The learner on the left is analyzing and determining standard time in a practice from the *design of work system* course unit, using the numerical simulation software.

The numerical simulation function as practical operation could support learners in understanding the application method of techniques learned by a lecture. In addition, learners can carry out individual steps involved in the practice under the guidance of instructions on each screen of the numerical simulation software. So the learner will be able to discover the characteristics of the appropriate application method of the theories and techniques by altering various parameters. The software is also equipped with a function that automatically processes practice data and illustrates summation tables and analysis graphs.

Computer-Supported Collaborative Learning (CSCL)

The last type, CSCL, is an instructional approach in which learners of varying abilities and interests work together in small groups to solve a problem, complete a project or achieve a common goal [85.78]. In each of the practices, learners, who are each given a role as shown in Fig. 85.5, are able to express their opinion and engage in mutual communication using an asynchronous bulletin-board system (BBS), installed in the learning management system (LMS).

One advantage of the e-Learning environment is its ability to create a kind of BBS function in the LMS to allow learner interaction. It is possible that one learner finds a good way to understand the theory applications learned through the lecture and practice, or learners want to compare the problem solution results that they

Table 85.5 The curriculum on production planning (MRP) system

| Curriculum | | |
| First lesson | Second lesson | Third lesson |
The framework of MRP systems	Review of the previous lesson	Review of the previous lesson
Relationship between the product and its demand and inventory: (a) Required parts and their lead time (b) Production and ordering points based on lead time (c) Lot size and inventory	Master production scheduling (MPS): (a) Input information (b) Output information (c) Calculation method (d) Example	Overall MRP system framework: (a) MRP framework for MPS and capacity planning
Inventory control: (a) Definition (b) The change in inventory (c) Relationship among order volume and inventory cost, order cost and total cost	MRP calculation mechanism: (a) The mechanism of MRP (b) Calculation method (c) Example	Product, its demand and inventory: (a) Product structure and its lead time (b) Reorder point based on lead time (c) Lot size and inventory
Influencing factors on master production scheduling (MPS): (a) Cost parameters (b) Lead time (c) Changes in demand	Lot-sizing rules: (a) Lot-for-lot rule (b) Economic order quantity rule (c) Period order quantity rule (d) Minimum cost planning rule	Overall MRP system execution: (a) Creating production plan based on MPS (b) Based on the production plan, create MRP plan for all parts (c) Minimizing the costs

Part I | 85.2

have achieved in the practice. In this case, through online discussion, learners can submit their comment, tips, and/or questions to the BBS, allowing other learners to view them. Therefore, a more communicative lesson is enable, while at the same time maintaining personal (not group) learning to avoid dependency on the learning environment.

Example of Learning Contents of Material Requirements Planning (MRP) in the *System* Course Unit

An example of new numerical simulation software to support *problem-solving learning* e-Learning, within *material requirements planning* (MRP) in the *system* course unit is introduced in this section. The PBL e-Learning method utilizes an established local-area network system to allow learners to access directly the numerical simulation software through the LMS during the course. The introduced software is specifically design to support the MRP system course. The software is intended to reduce the calculation effort, allow learn-

ers to switch their effort in generating a more efficient production plan, and improve their interest in the subject. The final objective is to improve the knowledge and skill attained during the course that uses the new learning method.

Overview of the Material Requirements Planning (MRP) Course Unit. Material requirements planning (MRP) systems, a technique in production planning and inventory control system [85.79, 80], are widely used in companies nowadays. Practical education programs in production planning (e.g., MRP system) are expected by companies from new workers. In order to answer this demand, in a series of *cyber manufacturing* programs in the *industrial engineering and information management* field, learners are required to understand MRP system, along with other techniques such as inventory planning, project management techniques, and manufacturing resources planning. Generally, the objectives of this course unit are to develop the necessary knowledge and ability to design a reasonable production planning, by taking all necessary factors into consideration.

The curriculum for e-Learning method on MRP system proposed here is shown in Table 85.5 [85.81]. The MRP course unit requires three learning lessons. Each lesson consists of 90 min lesson time, during which instructors describe the required theories, and 90 min practice time, during which the learners have to accomplish the practice prepared beforehand using the numerical simulation software to help them. The main objective of this e-Learning method is to improve the knowledge and skill attained during the course. Accordingly, the software is also prepared to have three separate practice problems, each corresponding to one lesson.

Master Production Schedule (MPS) Planning and Lot Sizing Rules. In the second lesson, MPS planning for the final item is allocated ten periods, as shown in Fig. 85.8. Here, the learners can choose whether to enter the order quantity directly, or to use one of the available lot-sizing rules, as illustrated in the upper part in Fig. 85.8. If the learners choose to complete the order quantity directly, the software will automatically conduct other calculations such as on hand inventory, total setup costs, total holding costs, and total costs. On the other hand, the software will automatically calculate all required values if the learner chooses one of the lot-sizing rules.

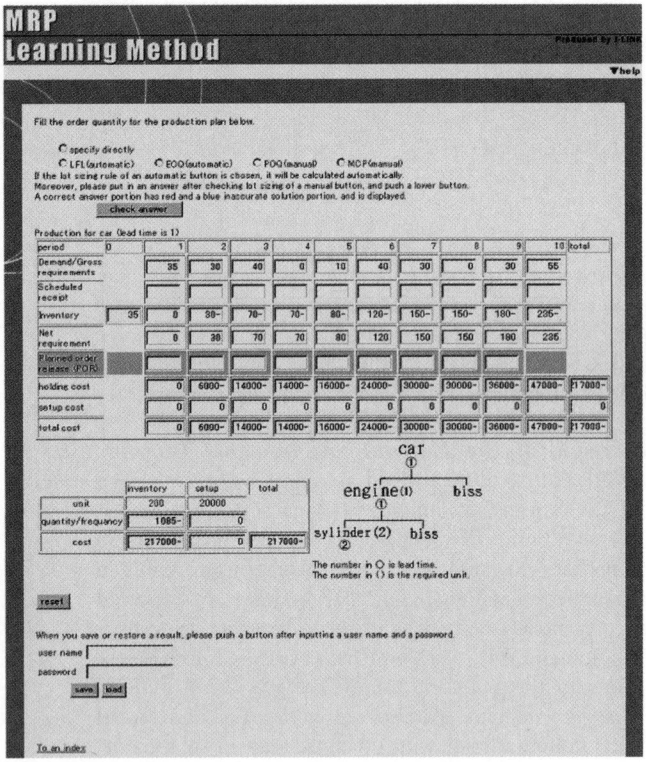

The objectives of this second practice are to familiarize the learners with the MPS logic and lot-sizing rules, to realize trade-off relations between lot-sizing rules, to realize trade-off relations between holding and setup costs, and to prepare the user for more complicated MRP in practice 3.

Assessment Method for Learning Efficacy

Stated below are the measures used to collect data of learning efficacy. Data would be assessed in order to guarantee the quality of the lectures (previously shown in Fig. 85.5, *assessment data of learning efficacy*):

1. Measuring the level of understanding (knowledge level) of the lectures based on the result of a quiz
2. Measuring the level of skill acquisition (skill level) from the practices based on the results of a report for each course unit
3. Evaluation of the effectiveness of the course and the software through data analysis of a previous questionnaire at the beginning of the course, a questionnaire for each course unit, and a final questionnaire at the end of the whole course (the questionnaire would be open-ended and multiple choice)
4. Logging of data analysis related to the learning record in the LMS, such as grade data, BBS discussion records, etc.

The quiz at the end of the theories taught in lectures should evaluate learners' understanding. The quiz is of multiple-choice type, so submission and appraisal of the quiz could be implemented automatically, using the advantage of the LMS.

The report worksheets are beforehand with each practical training, and the LMS distributes them. The learners make their reports by using software of the system, and have to submit the worksheet to the instructor by the LMS before the next start time of lesson. The admission report is conducted automatically by the LMS.

Fig. 85.8 Image of the software used in the *MRP systems* course unit

Table 85.6 Course objectives and student goals

Course name	Course objectives	Intended learning outcomes
Quality management (QM)	Acquire the ability to perform quality control by taking all measures necessary to ensure customer satisfaction and good quality	Incorporate customer needs into the quality management process, identify what actions must be taken to satisfy customer needs, and utilize appropriate methods to solve various problems that exist at the production site
		From the standpoint of *quality first*, identify what improvements must be made to ensure proper quality management and use appropriate measures to solve various problems that exist at the production site
		Describe quality management from the perspectives of the customer satisfaction and quality-first philosophies
		Play a leadership role in quality management activities that are based on customer satisfaction and quality-first philosophies
Manufacturing knowhow and creativity (MKHC)	Acquire the ability to identify and solve (ameliorate) problems that exist at the manufacturing site	Identify key issues that exist in a given workplace
		Identify and analyze the current status of a given workplace
		Capable of suggesting specific improvement measures
		Give a logical explanation of improvement measures to fellow workers in the workplace, or demonstrate how such improvements can be implemented to persuade workers to take action
		Capable of applying the above-mentioned knowhow in any workplace
Knowledge chain management (KCM)	Have a clear understanding of the processes involved in creating products that meet customer needs and acquire the ability to standardize and improve these processes while taking into consideration the cost and delivery time	Capable of achieving standardization and improvement in design control
		Capable of achieving standardization and improvement in process control
		Capable of achieving standardization and improvement in purchasing and subcontract management
		Capable of achieving standardization and improvement in cost management
		Capable of suggesting how standardization and improvement can be achieved in the above-mentioned management activities from the perspective of profit control

Table 85.7 Questionnaire surveys for the purpose of improving the program

Type of questionnaire/ purpose, respondents, and timing of surveys	Precourse survey	Postsession survey	Course completion survey	Survey for the satisfaction in hosting practical sessions	Survey for the participants' supervisors
Purpose	To understand the working environment and the academic background of the participants	To understand what aspects of the course sessions need to be improved and to track the participants' progress and understanding of the course content	To assess the participants' satisfaction with with the course and to identify what needs to be improved	To assess the hosts' satisfaction with the outcomes of the improvement measures	To assess the following changes in the course participants as perceived by participants' supervisors: ● Change in performance ● Change in behavior ● Change in enthusiasm and commitment to fulfill responsibilities
Respondents	Course participants	Course participants	Course participants	Companies hosting the practical training sessions	Superiors of the course participants
Timing	Before the start of the first session	At the end of each session	Upon completion of the course	Some time after the completion of the course	Some time after the completion of the course

85.2.2 The Case of an Educational Program for Manufacturing Managers Using IT Jigs

Program Objectives

The aim of this educational program [85.7] is to provide participants with the knowledge and skills required to manage manufacturing processes employed by small to medium-sized manufacturers of industrial machinery in the Hokuriku District of Japan. Classroom instruction practical exercises will cover the following:

1. Made-to-order manufacturing and flexible manufacturing systems, both of which are implemented widely by manufacturing companies in the Hokuriku District to control quality, cost, and lead time
2. Total optimization (instead of partial optimization) of all production processes
3. Continuous improvement in the workplace.

Based on these learning goals, three courses and their course objectives were developed, as shown by Table 85.6.

Program Evaluation Method

Based on *Kirkpatrick*'s model [85.36], the evaluation method shown in Table 85.7 was developed and implemented by a web system. The method consisted of five kinds of questionnaires for three kinds of stakeholders (participant, companies hosting the practical training sessions, and participants' supervisors) to evaluate the program. This system can highlight many useful issues for continuously improving the program.

A Circuit Model for Instructional Design

Figure 85.9 shows a flowchart called a circuit model [85.82], which represents the way in which the educational program is implemented. Practical exercises are designed to nurture each participant's ability to implement *kaizen* and are thus focused on the problem-solving process. Many of the conventional approaches to problem solving may not be effective in actual work settings; in fact, problem solving and *kaizen* are often complicated processes that involve the use of many unconventional approaches. As shown in Fig. 85.9, the learning cycle for the educational program consists of three processes: learning through

ing through seminars, and creating knowledge through practical exercises. The participants in the program may use conventional individual and group approaches. As they use unconventional approaches in a workplace setting, however, they will begin to understand the limitations of conventional approaches. The first two processes are learning and understanding; the third is creating. The above learning cycle is repeated several times (one learning cycle per subtheme) during the course of the program. When the participants complete the program, they repeat the learning cycle once again to assess their understanding.

The Development of IT Jigs

As a successful case of instructional development for the *knowledge chain management* (KCM) course, IT educational tools (IT jigs) were designed to help individuals involved in production planning process. This approach uses three types of bills of materials, three master files, a layout editor, a loading analyzer, and a flexible manufacturing system [85.83] (FMS) simulator. IT jigs are also highly effective tools for acquiring practical knowhow on decision-making procedures in

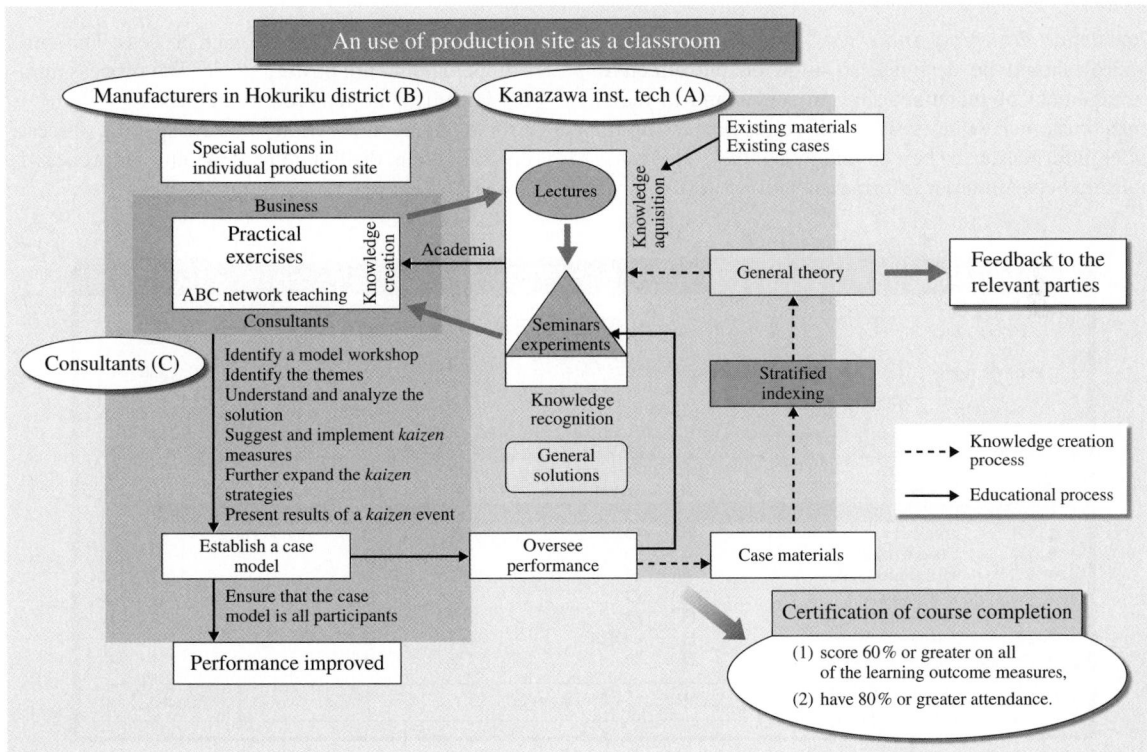

Fig. 85.9 Circuit model for the instructional design

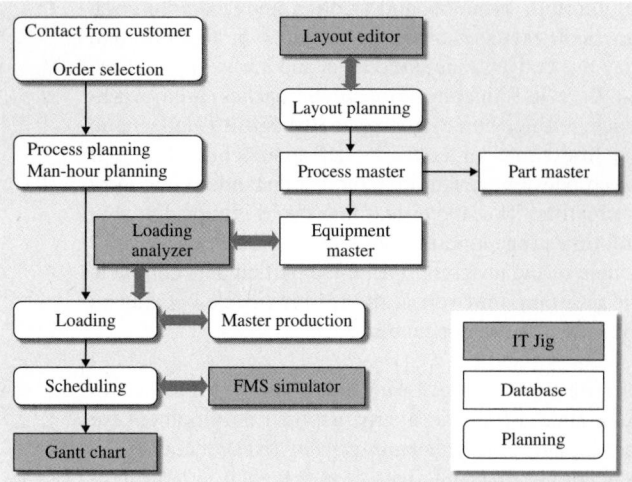

Fig. 85.10 IT educational tool (IT jig) for production planning and control

is summarized in Fig. 85.10, which depicts a series of activities starting from the contact from the customer to scheduling. Any uncertainties and/or nonconformity in the process should be identified, which should then be corrected, improved, and standardized in a timely manner so as to ensure continuous improvement of process control techniques. The roles of the layout editor and FMS simulator [85.84], which are the IT educational tools (IT jigs) designed to help individuals in the described planning process, are summarized in Fig. 85.10.

Creating the Process Master. The process master in Fig. 85.11 is created by determining the types (installation, cutting, cleaning, etc.) and number of processes to be handled by the processing facilities (machining centers (M/Cs), lathes, inspection machines, etc.), and by inputting a series of process information (1–7 below) and the processing procedures. Appropriate machines and equipment should be selected beforehand for the processing facilities using Layout Editor. Process information series 1–5 are identified during process planning and process information series 6 and 7 are identified during man-hour planning:

process control activities, which can be used to establish schedules and/or layout plans. These tools require no special programming skills to use, and are easily operated through the simple manipulation of icons.

Production Planning and IT Jigs Procedures. Process control should be designed so as to enable effective management of manufacturing processes and to maximize customer value creation. Effective processing of order information is key to achieving this. The relationship between order information and process control

1. Process number: assigned to each process. The same number should not be used twice. The process number is equivalent to a process address.
2. Process name: a description of the specific process, e.g., injection, drilling, threading, etc., or process 1, process 2, etc.

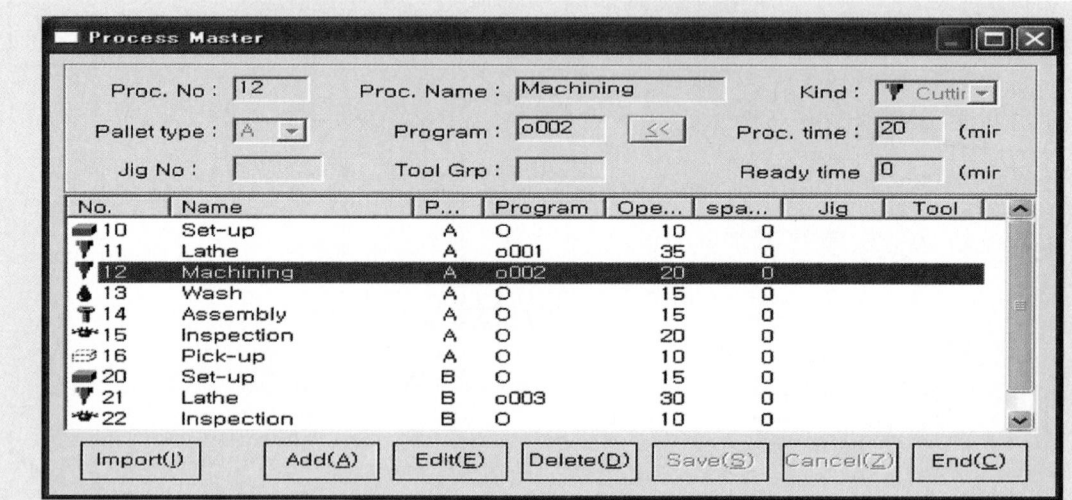

Fig. 85.11 Example process master screen

3. Process type: there are six categories, namely installation, cutting, cleaning, assembly, inspection, and removal. For more details, refer to the description provided later in the chapter.
4. Palette type: a tool that resembles a tray, which is used to transfer work pieces to each process.
5. Program: numerically controlled machine tools (NC machine tools) are controlled using a processing program. This information should be entered only when the process type is cutting.
6. Processing time: the time required to handle the process, which is estimated either during the design phase or based on experience.
7. Nonprocessing time: the time required for operations that do not involve the processing of work pieces, i.e., setup or replacement of a processing program and/or preparation of the necessary equipments and materials.

An example process master screen is shown in Fig. 85.11.

Creating the Equipment Master. The equipment master shown in Fig. 85.12 is a master file that determines and allocates the appropriate equipment for performing the kinds of process selected using the process master (Fig. 85.11). In short, users can identify the load for each facility by creating an equipment master. The equipment selected using the layout editor is automatically listed in the equipment information pane shown inside the solid lined box in Fig. 85.12, and the symbol representing the process type is listed beside the selected equipment in the *ID* column, as shown inside the dotted lines. If you click on *machining* in the *equipment information* pane, all the processes for which *cutting* has been selected as the process type using process master will be listed in the *processes that can be added* pane as shown inside the dotted-line of *additional process*. From the list of processes that can be added, select the process to be allocated to *machining (MC01)* and then click on *add (allocate)*. In Fig. 85.12, processes to be allocated are being selected (with four processes, designated as either *machining* or *lathe* as the available processes). Figure 85.12 shows that a total of two processes (process numbers 12 and 23) designated as *machining* have been selected for allocation.

When an item is selected from the list displayed in the *equipment information* pane, only the processes belonging to the same process type (specified using [process type] in process master) are displayed in the *processes that can be added* pane.

Creating the Parts Master. The parts master shown in Fig. 85.13 is a master file used to manage the delivery schedule for each part and product. This is accomplished using a process list created with the process master, which consists of a list of selected parts/products as well as of a series of related processes, to assign appropriate process(es) to each part/product. For example, Parts-C has the five process operations from number 30 to 34 in Fig. 85.13. Production schedules can be prepared by combining production plan with the information provided in the parts master.

The creation of the three masters described above is an important decision-making process that connects process design (ranging from procedure planning to load planning for each process) to process control. In most companies, process control knowhow accu-

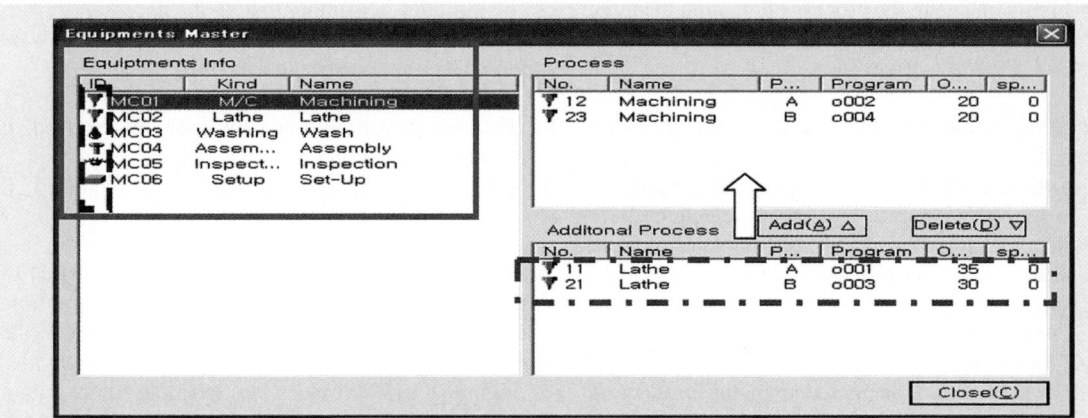

Fig. 85.12 Example allocation results screen

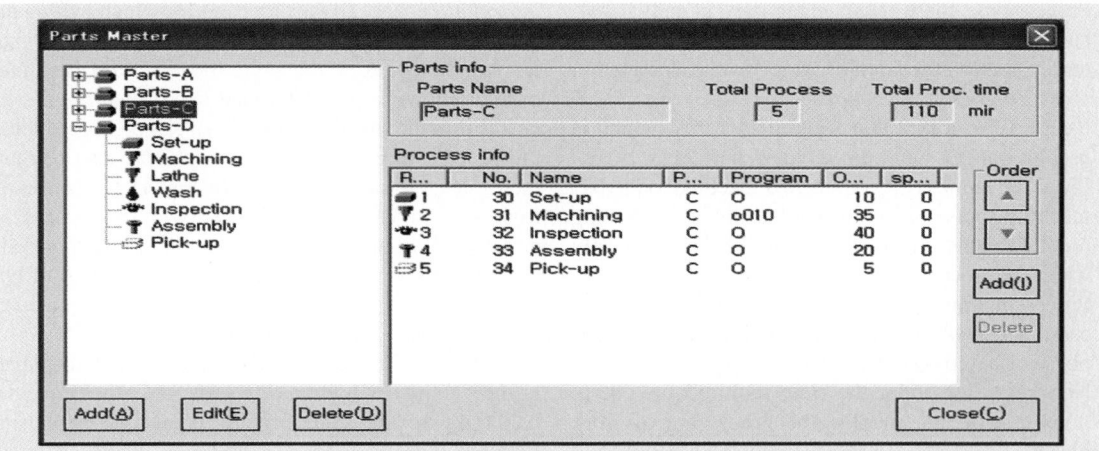

Fig. 85.13 Processes allocated for Parts-C using part master

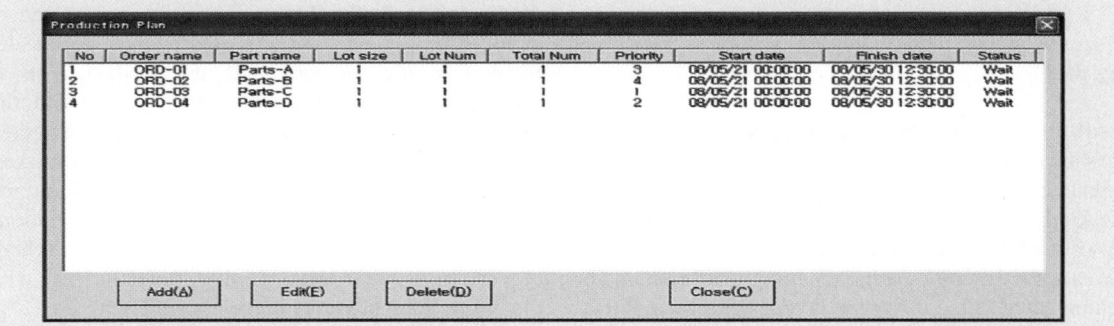

Fig. 85.14 Example of a master production plan

mulated through years of experience is often neither documented in manuals nor shared through information systems. In view of this, it is important to involve operators working on site in the process of creating the three masters so as to identify and incorporate all the process control knowhow accumulated in-house.

Master Production Plan and Scheduling. Figure 85.6 presents the master production plan window showing a list of the orders received in a planning period, the items to be manufactured, and the time required for the production of each item. Production schedules are established based on this information and the three Masters. The parts name column shows a total of fore items: Parts-A (order number 01), Parts B (order number 02), Parts-C (order number 03), and Parts-D (order number 04), which are to be included in the production plan. The following information can be derived from the plan.

The *start date and time* and *finish date and time* show when production should be completed. *Start date and time* refers to the date and time when production can be initiated. Assuming that all the necessary materials are ready and available for production, *start date and time* is either one of the following:

- The date and time when the order is received, if dynamic scheduling is used
- The date and time when the plan is finalized, if periodic scheduling is used.

When there is advanced knowledge that the arrival of some of the materials will be delayed, the date and time when all the materials will arrive should be set as the date and time when production can be initiated.

Figure 85.15 provides an example of a Gantt chart created using the dispatching rule (first-in first-out, FIFO) retrieved from the FMS simulator. Besides

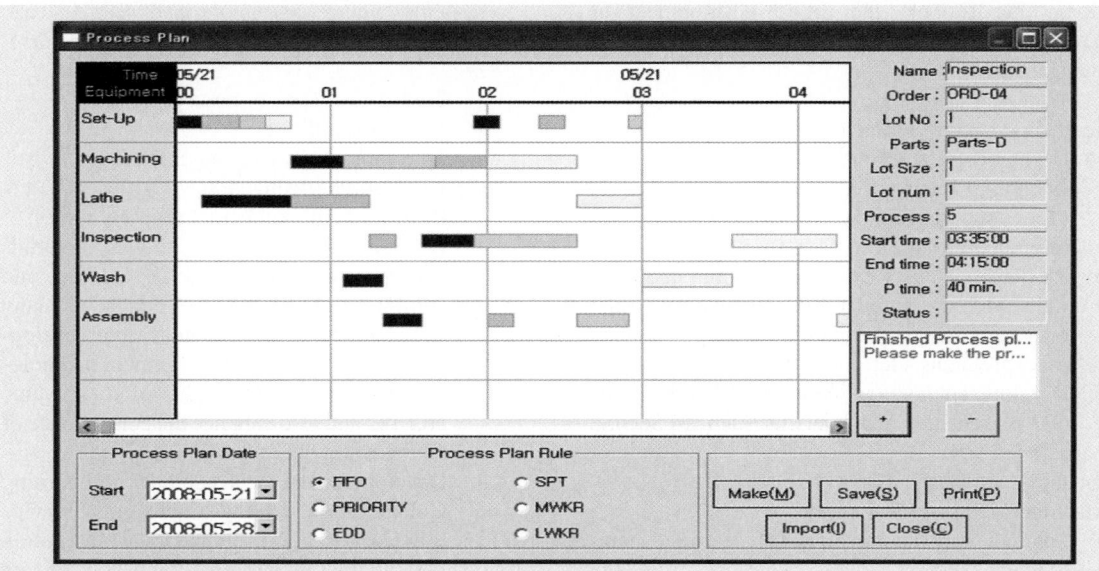

Fig. 85.15 Example FMS schedule Gantt chart scheduled by FIFO

FIFO, earliest due date (EDD), most work remaining (MWKR), and other dispatching rules are also available in the FMS simulator menu as well as some scheduling parameters with automated guided vehicle (AGV). Users can simulate scheduling either by using one of these dispatching rules or by creating their own unique dispatching rules as well as production/handling lot size, buffer stock size, and material handling schedule, and then assessing the simulation results. The simulator is designed to provide users with practical knowhow on scheduling, and offers useful tips on effective scheduling and process control.

85.3 Conclusions and Emerging Trends

This chapter has considered the automation of education/learning systems with an overview of the development history and present conditions of e-Learning based on instructional design, and introduced examples of manufacturing management personnel programs in management systems and industrial engineering (IE). For additional information on the automation of education see Chap. 44. In this final section, as a conclusion, we present the issues and outlook for the automation of education/learning systems, centered on e-Learning.

If education/learning systems aim at creating new social value for enterprises and communities by drawing together individual value creation activities [85.85, 86], e-Learning is one systematic approach for the use of information and communications technology in education/learning systems. Much of the development of e-Learning to date has focused on basic arrangements as an effective means to make conventional education and learning methods more efficient and convenient, and some accomplishments have been realized [85.87]. Considering the future outlook based on these achievements, e-Learning has the potential to more directly link personal value creation to social value. In other words, e-Learning can directly tie education/learning activities to the formation of business results and social capital [85.88, 89]; for example, e-Learning can go hand in hand with knowledge management, sales force management, human resources management, enterprise information portals, electronic performance support systems, and community service. The value created from such activities will mostly take the form of intellectual assets, with intellectual value, unlike the conventional material value of the past [85.90–98]. This intellectual asset formation is an effort aimed at the acquisition, understanding, and creation of explicit and tacit knowledge [85.73] that transcends the restrictions of space and time.

Given this outlook, the future development of e-Learning-related technologies will likely include the following areas:

1. Technological development issues:
   a) Development and application of user-interface technology to secure human-friendly communications such as mailing lists, bulletin boards, and machine translation
   b) Response analysis by effective tools and software such as simulation, groupware, virtual reality for analysis of natural language, numerical expressions, etc.
   c) Development and application of 3-D online virtual presentation mechanisms such as Second Life [85.99]
   d) Introduction of expert systems.
2. Man–machine systems issues:
   a) Solution for incompletion of learning management system (LMS) functions from the following points:
      • Secure and maintain learner motivation
      • Measures when there are learning deficiencies to control teaching and learning process
      • Construction of new dynamic evaluation systems
      • Open systems.
   b) Limitations of teaching methods (instructor–learner partnership, collaboration).
      Implement the following measures to develop systems where instructors and learners can jointly embody the truth that *teaching is learning*:
      • Public disclosure of good practices
      • Instructor training
      • Development of teaching methods.
   c) Cost-reduction effectiveness:
      • Resolution of intellectual property issues. Diverse e-Learning educational materials and contents are developed, produced, and administered through the hands of many stakeholders. This demands the development of systems and management technologies that protect the rights of these value creators and also enhance the convenience of learners and other users.
      • Development of social capital for e-Learning. Collaboration among application service providers (ASPs), educational institution federations, government, and the private sector to construct and upgrade social infrastructure for rational costs.
3. Nurturing the e-Learning profession.
   Development of systems to educate specialist personnel to effectively and efficiently advance the series of activities of development, design, introduction, administration, and evaluation of e-Learning systems to achieve the desired learning goals, and development of technologies to evaluate these educational systems.

References

85.1   A. Toffler, H. Toffler: *Revolutionary Wealth* (Knopf Borzoi Books, New York 2006)

85.2   P.B. Seybold, R.T. Marshak: *The Customer Revolution* (Crown Business, Danver 2001)

85.3   G.M. Piskurich: *Self-Directed Learning: A Practical Guide to Design, Development, and Implementation* (Pfeiffer, Somerset 1993)

85.4   J.J. Pear, M. Novak: Computer-aided personalized system of instruction: a program evaluation, Teach. Psychol. **23**, 119–123 (1996)

85.5   M. Ryder: Instructional design Models (2009), http://carbon.cudenver.edu/~mryder/itc_data/idmodels.html

85.6   Applied Research Laboratory, Penn State University: Definitions of instructional design (2009), http://www.umich.edu/~ed626/define.html

85.7   K. Ishii, H. Ikeda, A. Tsuchiya, M. Nakano: Development of educational program for production manager leading new perspectives on manufacturing technology, Proc. 19th ICPR, Valparaiso, Chile (2007)

85.8   K. Ishii, T. Ichimura, S. Kondoh, S. Hiraki: An innovative management system to create new values, Int. J. Technol. Manag. **45**(3/4), 291–305 (2009)

85.9   K. Suzuki: The instructional design for e-Learning practice, Jpn. J. Educ. Technol. **29**(3), 197–205 (2005), in Japanese

85.10   W.J. Rothwell, H.C. Kazanas: *Mastering the Instructional Design Process*, 2nd edn. (Jossey-Bass, Hoboken 1998)

85.11   G.M. Piskurich: *Rapid Instructional Design: Learning ID Fast and Right* (Jossey-Bass/Pfeiffer, Hoboken 2000)

85.12   M.W. Allen: *Michel Allen's Guide to e-Learning: Building Interactive, Fun, and Effective Learning Programs for Any Company* (Wiley, Hoboken 2003)

85.13 R.M. Gagne, L.J. Briggs: *Principles of Instructional Design*, 2nd edn. (Holt Rinehart Winston, Geneva 1979)

85.14 K. Akahori: A hierarchical structuring method of learning tasks for instructional design, J. Jpn. CAI Assoc. **7**(3), 99–107 (1990), in Japanese

85.15 C.M. Reigeluth: *Instructional Design Theories and Models: An Overview of Their Current Status* (Lawrence Erlbaum, Andover 1983)

85.16 R. Rickey: *The Theoretical and Conceptual Bases of Instructional Design* (Kogan Page, London 1986)

85.17 R.M. Gagne, L.J. Briggs: *Principles of Instructional Design* (Holt Rinehart Winston, London 1974)

85.18 C.M. Reigeluth, M.D. Merrill, C.V. Bunderson: The structure of subject matter contents and its instructional design implications, Instruct. Sci. **7**, 107–126 (1978)

85.19 C.M. Reigeluth, M.D. Merrill, B.G. Wilson, R.T. Spiller: The elaboration theory of instruction: a sequencing and synthesizing instruction, Instruct. Sci. **9**, 195–219 (1980)

85.20 M.D. Merrill: Component display theory. In: *Instructional Design Theories and Models*, ed. by C.M. Reigeluth (Lawrence Erlbaum, Hillsdale 1983)

85.21 T. Sato: *A Sequencing and Synthesizing Instruction by Interpretive Structural Modeling* (Meiji Tosyo, Tokyo 1987), in Japanese

85.22 J.M. Keller, K. Suzuki: Learner motivation and e-Learning design: a multinationally validated process, J. Educ. Media **29**(3), 229–239 (2004)

85.23 K. Suzuki, A. Nishibuchi, M. Yamamoto, J.M. Keller: Development and evaluation of website to check instructional design based on the ARCS motivation model, Inf. Syst. Educ. **2**(1), 63–69 (2004), in Japanese

85.24 J.C. Dunlap, R.S. Grabiger: Rich environments for active learning in the higher education classroom. In: *Constructivist Learning Environments: Case Studies in Instructional Design*, ed. by B.G. Willson (Educational Technology Publications, Englewood Cliffs 1996)

85.25 R.C. Shank, T.R. Berman, K.A. Macpherson: Learning by doing. In: *Instructional-Design Theories and Models: A New Paradigm of Instructional Theory*, Vol. II, ed. by C.M. Reigeluth (Lawrence Erlbaum, Andover 1999)

85.26 J. Nemoto, K. Suzuki: A checklist development for an instructional design based on goal-based scenario theory, Jpn. J. Educ. Technol. **29**(3), 309–318 (2005), in Japanese

85.27 S.N. Hendrickson: Learning in context interactive scenario design, paper presented at Techknowlege, Las Vegas (2005)

85.28 N. Dabbagh: Scaffolding: An important teacher competency in online learning, Technol. Trends **47**(2), 39–44 (2003)

85.29 C.N. Gunawardena, F.J. Zittle: Social presence as a predictor of satisfaction within a computer-mediated conferencing environment, Am. J. Distance Educ. **1**(7), 8–26 (1997)

85.30 D.R. Garrison, T. Anderson: *e-Learning in the 21st Century: A Framework for Research and Practice* (Routledge-Falmer, London New York 2003)

85.31 P.A. Shea Pickett, W. Pelz: Enhancing student satisfaction through faculty development: The importance of teaching presence. In: *Elements of Quality Online Education: Into the Mainstream*, Sloan-C Ser., Vol. 5, ed. by J. Bourne, J.C. Moore (Sloan Center for Online Education, Needham 2003), CD-ROM, http://www.sloanconsortium.org/elementsofquality_volume5_cd_ns (2009)

85.32 D.R. Garrison, M.C. Innes: Critical functions in student satisfaction and success: facilitating student rule adjustment in online communications of inquiry. In: *Elements of Quality Online Education: Into the Mainstream*, Sloan-C Ser., Vol. 5, ed. by J. Bourne, J.C. Moore (Sloan Center for Online Education, Needham 2003), CD-ROM, http://www.sloanconsortium.org/elementsofquality_volume5_cd_ns (2009)

85.33 A.M.A. Sperlich, K. Spraul: Students as active partners: higher education management in Germany, Public Sect. Innov. J. **12**(3), 2–19 (2007)

85.34 O. Simpson: *Student Retention on Online, Open Distance Learning* (Kogan Page, London 2003)

85.35 T. Matsuda, N. Honna, H. Kato: Development of e-Mentoring guideline and its evaluation, Jpn. J. Educ. Technol. **29**(3), 239–250 (2005), in Japanese

85.36 D.L. Kirkpatrick: *Evaluating Training Program: The Four Levels*, 2nd edn. (Berrett-Koehler, San Francisco 1998)

85.37 W. Horton: *Evaluating e-Learning* (ATSD, Alexandria 2001)

85.38 J.J. Phillips, R.D. Stone: *How to Measure Training Results: A Practical Guide to Tracking the Six Key Indicators* (McGraw-Hill, New York 2002)

85.39 M. Scriven: Evaluation perspectives and procedures. In: *Evaluation in Education*, ed. by W.J. Popham (McCutchan, Berkeley 1974)

85.40 D. Walter, L. Garey, J.O. Carey: *A Systematic Design of Instruction* (Addison-Wesley, Boston 2001)

85.41 P.L. Smith, J.R. Tillman: *Instructional Design*, 2nd edn. (Wiley, New York 1999)

85.42 L.M. Spencer, S.M. Spencer: *Competence at Work* (Wiley, New York 1993)

85.43 K. Tamaki, M. Kosakai, T. Matsuda: *e-Learning Practice* (Ohmsha, Tokyo 2002) p. 17, in Japanese

85.44 N. Harrison: *How to Design Self-Directed and Distance Learning: A Guide for Creators of Web-Based Training, Computer-Based Training, and Self-Study Materials* (McGraw-Hill, New York 1998)

85.45   American Society for Training and Development: Official homepage (2009), http://www.astd.org/astd

85.46   B. Broadbent: *ABCs of e-Learning: Reaping the Benefits and Avoiding the Pitfalls* (Wiley, New York 2002)

85.47   J.M. Rosenberg: *e-Learning: Strategies for Delivering Knowledge in the Digital Age* (McGraw-Hill, New York 2001)

85.48   Japan e-Learning Consortium: *2006/2007 e-Learning White Paper* (Tokyo Denki University Press, Tokyo 2006), in Japanese

85.49   Brandon Hall Research: LMS and LCMS demystified – they sound similar, but they're very different (2008). http://www.brandon-hall.com/free_resources/lms_and_lcms.shtml

85.50   Advanced Distributed Learning (ADL): Official hompage (2008). http://www.adlnet.gov/

85.51   Yale University: Official hompage (2008). http://www.yale.edu/its/

85.52   Internet2 Middleware Initiative: The Shibboleth System (2008). http://shibboleth.internet2.edu/

85.53   Aviation Industry CBT Committee (AICC): Official hompage (2008). http://aicc.org/

85.54   IMS Global Learning Consortium: Official hompage (2008). http://www.imsglobal.org/

85.55   IEEE Learning Technology Standards Committee: Official hompage (2008). http://ieeeltsc.org/

85.56   Globe: Official hompage (2008). http://globe.edna.edu.au/

85.57   M. Roy: Learning objects, EDUCAUSE Rev. **39**(6), 80–84 (2004)

85.58   Open Knowldege Inititiative: Official homepage (2008). http://www.okiproject.org/

85.59   OECD-Centre for Educational Research and Innovation: *e-Learning in Tertiary Education: Where Do We Stand?* (Organization for Economics, Berlin 2005)

85.60   J. Slafer: *Spent Force or Revolution in Progress? e-Learning After the e-University* (Oxford Higher Education Policy Institute, Oxford 2005), sponsored by webCT

85.61   R. Wingart: Classroom teaching changes in web-enhanced courses: a multi-institutional study, Educase Q. **27**(1), 26–35 (2004)

85.62   e-Framework for Education and Research Initiative: Official homepage (2008). http://www.e-framework.org/

85.63   MIT: Official hompage (2008). http://ocw.mit.edu/OcwWeb/

85.64   Elearningeuropa Initiative: Official hompage (2008). http://www.elearningeuropa.info

85.65   SOCRATES: Official hompage (2008). http://projectsocrates.org

85.66   Europe's Information Society: Official hompage (2008). http://ec.europa.eu/information_society/eeurope/i2010/index_en.htm

85.67   eInclusion: Official hompage (2008). http://www.einclusion-eu.org

85.68   T. Koschmann: *CSCL: Theory and Practice of an Emerging Paradigm (Computers, Cognition, and Work)* (Lawrence Erlbaum, Andover 1996)

85.69   S. Mizuno: Company-wide quality control activities in Japan, Rep. Stat. Appl. JUSE **16**(3), 8–18 (1969)

85.70   Japan Institute of Plant Maintenance (JIPE) (Ed.): *TPM for Every Operator*, Shop Floor Series (Productivity, Tokyo 1996)

85.71   Y. Sugimori, K. Kusunoki, F. Cho: Toyota production system and Kanban system, Int. J. Prod. Res. **15**, 553–565 (1977)

85.72   T. Ohno: *Toyota Production System – Beyond Large-Scale Production* (Productivity, Portland 1988)

85.73   I. Nonaka, P. Byosiere, C.C. Borucki, N. Konno: Organizational knowledge creation theory: a first comprehensive test, Int. Bus. Rev. **3**(4), 337–352 (1994)

85.74   K. Tamaki: Development of e-Learning educational programs for cyber manufacturing in management of technology (MOT), Proc. 18th Int. Conf. Prod. Res. (2005)

85.75   Working Group for Cyber Manufacturing: *Cyber Manufacturing, Aoyama Media Lab (AML) Project* (Research Center of Aoyama Gakuin University, Tokyo 2004), in Japanese

85.76   J. Bersin: *The Blended Learning Book: Best Practice, Proven Methodologies, and Lessons Learned* (Pfeiffer, Somerset 2004)

85.77   P.L. Smith, T.J. Ragan: *Instructional Design*, 3rd edn. (Wiley, New York 2005)

85.78   North Central Regional Educational Laboratory (NCREL): Glossary of education terms and acronyms (2008). http://www.ncrel.org/sdrs/areas/misc/glossary.htm

85.79   J. Orlicky: *Material Requirements Planning* (McGraw Hill, New York 1975)

85.80   K. Sheikh: *Manufacturing Resource Planning (MRP II) With Introduction to ERP, SCM, and CRM* (McGraw Hill, New York 2001)

85.81   Myreshka, K. Takahashi: *e-Learning in Production Planning (MRP), Cyber Manufacturing, AML Project* (Research Center of Aoyama Gakuin University, Aoyama 2004), in Japanese

85.82   A. Shikida: A new learning design for creative learnings by the circuit model, J. Jpn. Soc. Eng. Educ. **53**(1), 35–40 (2005), in Japanese

85.83   K.E. Stecke, N. Raman: FMS planning decisions, operating flexibilities, and system performance, IEEE Trans. Eng. Manag. **42**(1), 82–90 (1995)

85.84   New Technology Systems: *The Manual for V-FMS21 Simulation Software* (New Technology

Systems, Tokyo 2006), http://www.nets-sys.com, in Japanese

85.85 L.S. Vygotsky: *Mind in Society* (Harvard University Press, Cambridge 1978)

85.86 J. Lave, E. Wenger: *Situated Learning: Legitimate Peripheral Participation* (Cambridge Univ. Press, Singapore 1991)

85.87 M.J. Rosenberg: *Beyond e-Learning: Approaches and Technologies to Enhance Knowledge, Learning and Performance* (Pfeiffer, Somerset 2006)

85.88 N. Lin: *Social Capital: A Theory of Social Structure and Action*, Structural Analysis in the Social Sciences, Vol. 19 (Cambridge Univ. Press, Singapore 2002)

85.89 D. Halpern: *Social Capital* (Polity, Cambridge 2005)

85.90 L. Prusak (Ed.): *Knowledge in Organizations* (Butterworth-Heinemann, London 1997)

85.91 L.H. Thurow: *Fortune Favors the Bold – What We Must Do to Build a New and Lasting Global Prosperity* (Harper Collins, New York 2003)

85.92 L.S. Paine: *Value Shift* (McGraw-Hill, New York 2003)

85.93 K. Ishii, T. Ichimura, S. Kondoh, S. Hiraki: A framework for the management system model to create new value of business output, Int. J. Entrep. Innov. Manag. **4**(4), 393–406 (2004)

85.94 C.K. Prahalad, V. Ramaswamy: *The Future of Competition: Co-Creating Unique Value with Customers* (Harvard Business School Press, Boston 2004)

85.95 B. Trezzini, P. Lambe, S. Hawamdeh (Eds.): People, knowledge and technology: what have we learnt so far?, Proc. 1st iKMS Int. Conf. Knowl. Manag. (World Scientific, 2004)

85.96 S. Alikhan, R. Mashelkar: *Intellectual Property and Competitive Strategies in the 21st Century* (Kluwer, Dordrecht 2004)

85.97 R.F. Reilly, R.P. Schweihs: *The Handbook of Business Valuation and Intellectual Property Analysis* (McGraw-Hill, New York 2004)

85.98 Y. Monden, K. Miyamoto, K. Hamada, G. Lee, T. Asada: *Value-Based Management of the Rising Sun* (World Scientific, Hackensack 2006)

85.99 Second Life: Official homepage (2008), http://secondlife.com/

86. Enterprise Integration and Interoperability

François B. Vernadat

Enterprise integration and interoperability deal with facilitating communication, cooperation, and collaboration within an organization, be it a single organization or a networked organization, or be it a public or a private organization. This chapter first defines enterprise integration and systems interoperability and presents relevant architectural frameworks. It then explains the technical, semantic, and organizational dimensions of interoperability before presenting essential standards and technology for interoperability and integration. Applications and future trends are pointed out before concluding.

Part | 86

Market globalization, worldwide competition, and fast changing business conditions are forcing enterprises to become more agile but also to increase communication, collaboration, and networking with suppliers, partners or distributors.

Enterprise integration and interoperability come into play any time that two or more business entities need to work together or need to share the same information. A business entity is any part of an enterprise. It can be of any size, ranging from a single workstation or an information technology (IT) system (e.g., a post in a department or an enterprise resource planning (ERP) system module) to a full organization unit made up of people, systems, different pieces of equipment, information stores, etc. In fact, it can even be any node in a supply-chain network or any component of an extended or virtual enterprise.

The need for enterprise integration, and for underlying systems interoperability, results from the increasing need to interconnect and seamlessly operate technical, social, and information systems of geographically dispersed enterprises or of different enterprises working together. This concerns the internal business processes and services of a given enterprise as well as cross-organizational business processes spanning partner companies or networks of enterprises.

The challenge of integrating enterprise systems or business entities comes from the large number and high heterogeneity of enterprise components that are subject to be interconnected. When two systems have to *talk* to each other or interoperate, they have to agree on common message formats, transfer protocols, shared (data) semantics, security and identification aspects, etc. For instance, if a bank and an insurance company want to jointly offer real-estate loans with life insurance, they have to align their customer data files, agree on loan acceptance conditions and rules, and coordinate respective local services (loan request processing, loan approval/refusal, insurance condition setting, etc.) into a common business process to offer a composite service.

In this chapter, both enterprise integration and interoperability are defined, respective architectural frameworks are presented and the underlying technologies are discussed from a three-level perspective (technical, application/semantic, and organizational di-

mensions). Finally, future trends are discussed before concluding.

86.1 Definitions and Background

Enterprise integration and systems interoperability aim at facilitating communication, cooperation, and collaboration between business entities. While enterprise integration has a strong organizational dimension, interoperability is more of a technical nature.

86.1.1 Enterprise Integration

Enterprise integration (EI) occurs when there is a need to remove organizational barriers and/or improving interactions among people, systems, applications, departments, and companies (especially in terms of material flows, information/decision flows, and control or work flows) [86.1, 2]. The goal is to create synergy, i.e., the integrated system offers more capability than the sum of its components.

The complexity of EI relies on the fact that enterprises typically comprise hundreds, if not thousands, of

applications (be they packaged solutions, custom-built or legacy systems), some of them being remotely located and support an even larger number of business processes. This environment gets even more complicated as companies these days are increasingly involved in merger, fusion, acquisition, divestiture, and partnership opportunities.

Integration of enterprise activities has long been, and often still is, considered as a pure IT problem. While it is true that at the end of the day the prime challenge of EI is to provide the right information to the right place at the right time, business integration needs must drive information system integration and not vice versa. EI therefore has a strong organizational dimension, in addition to its control/management dimension and its technological dimension.

From a pure IT standpoint, EI mostly means connecting computer systems and IT applications to support business process operations [86.3, 4]. It involves technologies such as enterprise portals, data replication, shared business functions with remote method invocation, enterprise service buses, distributed business processes (or workflow systems), and business-to-business (B2B) integration.

From an organizational standpoint, EI is concerned with facilitating information, control, and material flows across organization boundaries by connecting all the necessary functions and heterogeneous functional entities (e.g., information systems, devices, applications, and people) in order to improve communication (data and information exchanges at system level), cooperation (interoperation at application level), and coordination or collaboration (timely orchestration of process steps at business level) within this enterprise so that it behaves as an integrated whole (Fig. 86.1) [86.1, 5]. It will therefore enhance overall productivity, flexibility, and capacity for the management of change (i.e., agility). *Li* and *Williams* [86.6] provide a broader definition of EI stating that enterprise integration is the coordination of all elements including business, processes, people, and technology of the enterprise(s) working together in order to achieve the optimal fulfillment of the business mission of that enterprise(s) as defined by the management.

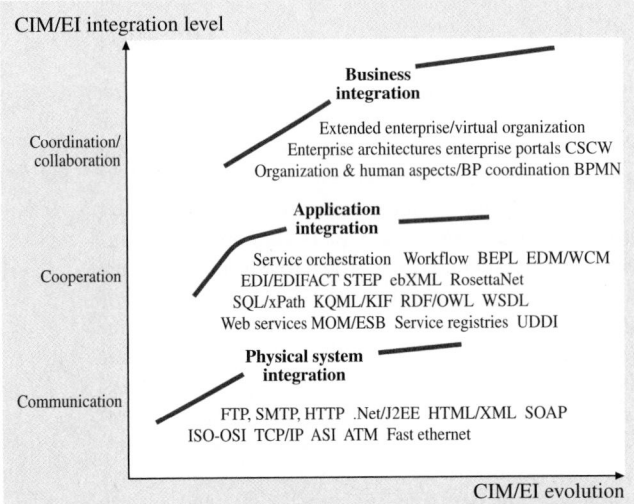

Fig. 86.1 Enterprise integration levels and related technologies (essential technical terms and acronyms appearing in the figure are explained in Sect. 86.3 of this chapter and more information on each of them can be found at http://whatis.techtarget.com) (FTP – file transfer protocol, ISO-OSI – International Standards Organization Open System Interconnection, ATM – asynchronous transfer mode, J2EE – Java to Enterprise Edition, ASI – actuator sensor interface)

Enterprise integration can apply vertically or horizontally within any organization. Vertical integration refers to integration of a line of business from its top management down to its tactical planning and operational levels (i.e., along the control structure). Horizontal integration refers to integration of the various domains (i.e., business areas) of the enterprise or with its partners and market environment (i.e., along the supply/consumer chain structure).

Integration can range from loosely coupled to tightly coupled and further to full system integration. Full integration means that component systems are no longer distinguishable in the whole integrated system. Tightly coupled integration means that components are still distinguishable in the whole but any modification to one of them may have direct impact on others. Loosely coupled integration means that component systems continue to exist on their own but can also work as components of the integrated system.

Putting in place interoperable enterprise systems is essential to achieve enterprise integration with the level of flexibility and agility required by public or private organizations to cope with the need for frequent organizational changes.

As indicated by Fig. 86.1, reaching the successive levels of integration requires building on a number of standards, languages, and technologies. Standardization efforts have been reviewed by *Chen* and *Vernadat* in [86.7] (Sect. 86.3).

86.1.2 Systems Interoperability

Interoperability is defined in the Webster dictionary as "the ability of a system to use parts of another system". In IT terms, it can be defined as the ability of two or more information systems and of business processes they support to exchange data, share information and knowledge, or use functionality of one another.

Another definition has recently been provided by the Athena project [86.8]. It defines interoperability of enterprise applications as the ability of a system to work with other systems without special effort on the part of their users.

More broadly speaking, enterprise interoperability can be defined as the ability of an enterprise to use information or services provided by one or more other enterprises. This assumes on the part of the IT infrastructure the ability to send and receive messages (requests or responses) or data streams. These exchanges can be made in synchronous or asynchronous mode depending on communication needs.

Interoperability is indeed one of the many facets of enterprise integration. An integrated family of systems must necessarily be interoperable in one form or another. Interoperable systems do not need to be integrated. In fact, enterprise interoperability equates to loosely coupled enterprise integration. It provides two or more business entities (of the same organization or from different organizations and irrespective of their location) with the ability to exchange or share information (wherever it is and at any time) and to use functions or services of one another in a distributed and heterogeneous environment. Interacting component systems are preserved as they are, can still work on their own, but can at the same time interoperate seamlessly.

As for integrated systems, building and implementing interoperable enterprise systems relies heavily on the use of open standards, as indicated in Sect. 86.3.

86.1.3 Background

Enterprise integration (EI) naturally emerged as a concept in the early 1990s from advances made in the 1980s in computer-integrated manufacturing (CIM) on the one hand and in enterprise information integration (EII) on the other hand. Both were aimed at interconnecting islands of automation and information silos to liquify as much as possible information and control flows across the organization. Distributed computing environments, client–server architectures, federated databases, common network protocols, and remote procedure calls were essential technologies during this early development era.

With the advent of object-oriented computing, open system architectures, and message queuing techniques, complex pieces of middleware emerged in the mid 1990s, the so-called enterprise application integration (EAI) platforms [86.9]. Among these, the common object request broker architecture (CORBA) has played a prominent role because of its open-source nature, as promoted by the Object Management Group [86.10].

It is also at this time that two major IT inventions appeared: hyper-text transfer protocol (HTTP) and extended markup language (XML) (Sect. 86.4). They have drastically changed the scene because these standards opened the door to ubiquitous computing and nearly full independence to specific IT infrastructures (e.g., computer networks, operating systems, computer languages, etc.). This was a major step to get rid of EAI solutions that were found to be too rigid, monolithic, and proprietary solutions and it made possible Internet computing and led to the emergence of web services at the turn of the century.

Since then, while EI has rapidly evolved in the organizational aspect, with new concepts coming from enterprise architectures, business process management (BPM), computer-supported collaborative work (CSCW), and interorganizational workflow systems, the pressure for increased systems interoperability has continuously been accentuated with the growing needs coming from new forms of organizations (e.g., supply chains, extended enterprises or virtual enterprises) in terms of business process coordination and service exchange. Service-oriented architectures (SOA) and web services [86.11, 12] are playing a central role in this area, as explained in the next sections.

86.2 Integration and Interoperability Frameworks

Enterprise integration and interoperability frameworks are made of a set of principles, standards, guidelines, and sometimes models that practitioners should apply to put in place the right components in the proper order to achieve integration and interoperability in their organization.

An example EI framework is given by Fig. 86.1. It comes from the computer-integrated manufacturing open system architecture (CIMOSA) architecture that has specifically been designed for CIM, i.e., integration in manufacturing enterprises [86.5]. The framework emphasizes three integration levels, namely systems integration, application integration, and business integration, and positions different technologies or languages to be used at each level.

Other architectures or frameworks exist (e.g., PERA (purdue enterprise reference architecture), ARIS (architecture for information systems), GRAI (graphes de résultats et activités interreliés), GERAM (generalized enterprise reference architecture and methodology),

IDEF (integrated definition method), etc.), and have been reviewed in a book by *Vernadat* [86.1].

Similarly, a number of frameworks are being developed for interoperability; for instance, *levels of information systems interoperability* (LISI) from an Architecture Working Group of the US Department of Defense (on Command, Control, Communications, Computers, Intelligence, Surveillance and Reconnaissance – C4ISR) defines five levels of interoperability [86.14]:

- Level 0 – isolated systems (manual extraction and integration of data)
- Level 1 – connected interoperability in a peer-to-peer environment
- Level 2 – functional interoperability in a distributed environment
- Level 3 – domain-based interoperability in an integrated environment
- Level 4 – enterprise-based interoperability in a universal environment.

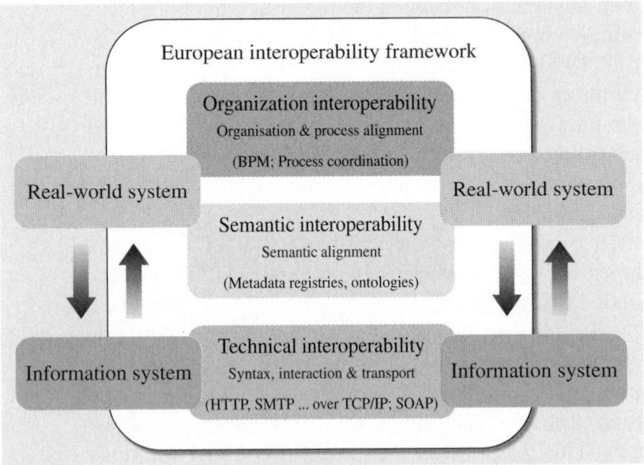

The European Interoperability Framework (EIF) is a more generic framework jointly developed by the European Commission (EC) and the member states of the European Union (EU) to address business and government needs for information exchange [86.13]. This framework defines three essential levels (or dimensions) of interoperability, namely technical, semantic, and organizational from bottom up, as depicted by Fig. 86.2. These dimensions are further explained in the next sections.

86.2.1 Technical Interoperability

This dimension covers technical issues or *plumbing* aspects of interoperability. It deals with connectivity and is to date the most developed dimension. It ensures that systems get physically connected and that data and messages can be reliably transferred across systems. It is about linking computer systems and applications or

Fig. 86.2 European Interoperability Framework (EIF) (after [86.13])

interconnecting information systems by means of so-called middleware components. It includes key aspects such as open interfaces, interconnection services, data transport, data presentation and exchange, accessibility, and security services.

At the very bottom level are network protocols (TCP/IP (transmission control protocol/internet protocol)) and on top transport protocols such (HTTP, SMTP (simple mail transfer protocol), etc.) to exchange data in the form of XML messages.

86.2.2 Semantic Interoperability

This dimension of interoperability concerns semantic alignment, i.e., the ability to achieve meaningful exchange and sharing of information, and not only data, among independently developed systems. It is concerned with ensuring that the precise meaning of exchanged information is understandable by any other application that was not initially developed for this purpose.

Semantic interoperability enables systems to reformat or adapt received information (e.g., convert euros into dollars) or combine it with other information resources and to process it in a meaningful manner for the recipient. It is therefore a prerequisite for the front-end (multilingual) delivery of services to users.

Beyond information system interoperability, cooperation between systems should also be possible. This means that the different systems must be able to provide each other context-aware services, i.e., services that can be used in different types of situations, that they can call or that they can use to enable the emergence of higher-level composite services.

Technologies involved at this level include metadata registries, thesauri or ontologies to map semantic definitions of the same concepts used by the different systems involved. In IT, an ontology [86.15] is a shared formal specification of some domain knowledge. It is usually expressed in first-order logic or languages such as knowledge query and manipulation language (KQML) to be exchanged with knowledge interchange format (KIF) or in the form of semantic networks or taxonomies, for instance, resource description framework (RDF) and web ontology language (OWL), the World Wide Web Consortium (W3C) web ontology language [86.16].

86.2.3 Organizational Interoperability

This dimension of interoperability is concerned with defining business goals, modeling business processes, and bringing collaboration capabilities to organizations that wish to exchange information and may have different internal structures and processes. Moreover, organizational interoperability aims at addressing the requirements of the user community by making services available, easily identifiable, accessible, and user-centric. In other words, it is the ability of organizations to provide services to each other as well as to users or the wider public.

To achieve organizational interoperability, it is necessary to align business processes of cooperating business entities, define synchronization steps and messages, and define coordination and collaboration mechanisms for interorganizational processes. This requires business process management (BPM) for the modeling and control of these business processes, workflow engines for the coordination of the execution of their steps defined as business services, collaborative tools, and enterprise portals to provide user-friendly access to business services and information pages made available to end-users.

86.3 Standards and Technology for Interoperability

As mentioned in the previous sections, a certain number of standards and technologies need to be considered to achieve the three levels of enterprise integration and interoperability. Essential ones include (from bottom up of Fig. 86.1):

1. *XML*: The extended markup language is a widely used, standardized tagged language proposed and maintained by the World Wide Web Consortium (W3C). It has been proposed to be a universal format for structured content and data on the web but can indeed be used for any computer-based exchange. Its simple and open structure has revolutionized data, information, and request/reply exchanges among computer systems because XML messages can be handled by virtually any transport protocol due to its alphanumeric and platform-independent nature [86.17].

2. *HTTP/HTTPS*: Hypertext transfer protocol is the set of rules for transferring files (text, graphic images,

sound, video, and other multimedia files) on the World Wide Web. HTTP is an application protocol that runs on top of the TCP/IP suite of protocols. Because of its ubiquity it has become an essential standard for communications. HTTP 1.1 is the current version (http://www.w3.org/Protocols/).

3. *Web services* and *service-oriented architectures* (SOAs): Since the year 2000, service-oriented architectures represent a new generation of IT system architectures taking advantage of message-oriented, loosely coupled, asynchronous systems as well as web services [86.11,12,18]. SOAs provide business analysts, integration architects, and IT developers with a broad abstract view of applications and integration components to be dealt with as encapsulated and reusable high-level services. Web services can be defined as interoperable software objects that can be assembled over the Internet using standard protocols and exchange formats to perform functions or execute business processes [86.12]. They are accessible by the external world by their unique resource locator (URL), they are described by their interface, and can be implemented in any computer language. Web services use the following de facto standards for their implementation:

   – *WSDL*: The web service description language is a contract language used to declare the web service interface and access methods in a universal language using specific description templates [86.19].

   – *SOAP*: The simple object access protocol is a communication protocol and a message layout specification that defines a uniform way of passing XML-encoded data between two interacting software entities (for instance, web services). SOAP is under consideration by the World Wide Web Consortium to become a standard in the field of Internet computing [86.20]. It still lacks security mechanisms for the transfer of sensitive data and messages but can deal with data encryption.

   – *UDDI*: The universal description, discovery, and integration specification is an XML-based registry proposal for worldwide businesses to list their services offered on the Internet as web services. The goal is to streamline online transactions by enabling companies to find one another on the web and make their services interoperable for e-Commerce. UDDI is often compared to a telephone book's white, yellow, and green pages. Within a company or within a network of companies, a private service registry can be established in a similar way and format to UDDI in order to manage all shared business and data services used across the business units.

4. *Service registries*: A service registry is a metadata repository that maintains a common description of all services registered for the functional domain for which it has been set up. Services are described by their name, their owner, their service level agreement and quality of service, their location, and their access method. It is used by service owners to expose their services and by services users to locate services. Commercial products based on UDDI specs are available.

5. *Message and service buses – MOM/ESB*: The key to building interoperable enterprise systems and service-oriented architectures that are reliable and scalable is to ensure loose coupling among services and applications. This can be achieved using message queuing techniques, and especially message-oriented middleware (MOM) products and their recent extension known as enterprise service buses (ESB) [86.4, 21]. Services and IT applications exchange messages in a neutral format (preferably XML) using simple transport protocols (i. e., XML/SOAP on TCP/IP, SMTP or HTTP). A message-oriented middleware is a messaging system that provides the ability to connect applications in an asynchronous message exchange fashion using message queues. It provides the ability to create and manage message queues, to manage the routing of messages, and to fix priorities of messages. Messages can be delivered according to three basic messaging models:

   – Point-to-point model, in which only one consumer may receive messages that are sent to a queue by one or more producers

   – Publish-and-subscribe, in which multiple consumers may register, or subscribe, to receive messages (called topics) from the same message queue

   – Request/reply, which allows reliable bidirectional communications between two peer systems (using input queues and output queues)

An enterprise service bus goes beyond the capabilities of a MOM. It is a standards-based integration platform that combines messaging of a MOM, web services, data transformation, database access services, intelligent routing of messages, and even workflow execution to reliably connect and coordinate the interaction of significant numbers of diverse applications across an extended organization with transactional integrity. It is capable of being adopted for any general-purpose integration project and can scale beyond the limits of a hub-and-spoke EAI broker. Data transformation is the ability to apply XSLT (extended style sheet transformation) (or XML style sheets) to XML messages to reformat messages during transport depending on the type of receivers that will consume the messages (for instance, the ZIP code is a separate information datum for one application while it is part of a customer address for another one). Intelligent or rule-based routing is the ability to use message properties to route message delivery to different queues according to message content. Database services simplify access to database systems using SQL (structured query language) and Java database connectivity (JDBC) or object database connectivity (ODBC). Useful enterprise integration patterns to be used in messaging applications can be found in the book by *Hohpe* and *Woolf* [86.4], who have thoroughly analyzed and described those that are most commonly used in practice.

6. *BPM*: Business process management consists of reviewing, reengineering, and automating business processes of the organization. Among the process steps are services offered by IT systems. Once automated, these processes take the form of workflows. Two languages can be used to model business processes, one at the business user level and one at the workflow system level. These are:

– BPMN: The business process modeling notation is a diagrammatic and semistructured notation that provides businesses with the capability to represent their internal business procedures in a graphical language. It gives organizations the ability to depict and communicate these procedures in a standard manner. Furthermore, the graphical notation facilitates the understanding of the performance of collaborations and business transactions between the organizations. This ensures that business participants will understand each other and that participants in the business will enable organizations to adjust to new internal and B2B business circumstances quickly [86.22].

– BPEL: The business process execution language is an XML-based language designed to enable task sharing for a distributed computing environment – even across multiple organizations – using a combination of web services (BPEL is also sometimes called BPEL4WS or BPELWS). It has been written by developers from BEA Systems, IBM, and Microsoft. BPEL specifies the business process logic that defines choreography of interactions between a number of web services. The BPEL standard defines the structure, tags, and attributes of an XML document that corresponds to a valid BPEL specification [86.23]. Conversion mechanisms exist to translate BPMN models into BPEL specifications.

86.4 Applications and Future Trends

Enterprise integration has been around in terms of applications since the beginning of the 1990s, especially in the domain of manufacturing integration but also enterprise information integration (EII), enterprise resources planning (ERP), customer relationship management (CRM), and then supply chains. The advent of CORBA, workflow engines, and BPM techniques has given birth to a range of commercial products, known as enterprise application integration (EAI) platforms. Many vendors have proposed solutions, including Iona, Sopra, Vitria, Tibco, Microsoft BizTalk, HP, IBM, Oracle, and BEA Systems, as well open-source solutions.

Today, nearly all industrial sectors are facing integration and interoperability problems, ranging from manufacturing supply chains to banking, the insurance sector, medical services, and public organizations. The focus is on XML-based messaging communications and loosely coupled interactions thanks to XML/HTTP and MOM/ESB techniques. One area that has emerged with EI and is rapidly growing is electronic commerce (web-based shopping, B2B commerce, e-Procurement, electronic payment, EDI, etc.). Major initiatives launched to sustain this area include ebXML (electronic business XML) for XML/EDI (electronic data interchange) (UN/CEFACT (United Nations Centre for Trade Facilitation and Electronic Business) and OASIS (Organization for the Advancement of Structured Information Standards)), RosettaNet (www.rosettanet.org) in the area of standard electronic data exchange for supply chains, and Bolero.net (www.bolero.net) proposing a data and document exchange infrastructure for international commerce involving SWIFT (Society for Worldwide Interbank Financial Telecommunication) for secured payment.

In terms of current trends that can be observed there is a shift from the pure business process paradigm that prevailed through the 1990s to the service-orientation paradigm. This puts more emphasis on services interoperability than on IT systems interoperability or enterprise application integration than before [86.25].

Among the future trends, we can mention the emergence of composite services, i.e., services made of services. This will rapidly grow both in business (to combine competencies and capabilities of several partners into common services) and in government organizations (for instance, pan-European services of-fered to EU citizens where a service offered to citizens through a common portal will in fact be a combination of local services running in member states in national languages, e.g., application for the blue card or European identity card).

At the technical level, integration and interoperability are still evolving and will continue to evolve with technology (faster communication networks, authentication mechanisms, security issues, electronic signature, multilingual issues, etc.).

At the semantic level, many researchers agree that one of the major challenges for integration and interoperability remains the semantic unification of concepts. There are simply too many ontologies around and no common way of representing these ontologies, although OWL-S is a de facto standard. Microtheories have to emerge to provide sound foundations for the field, and

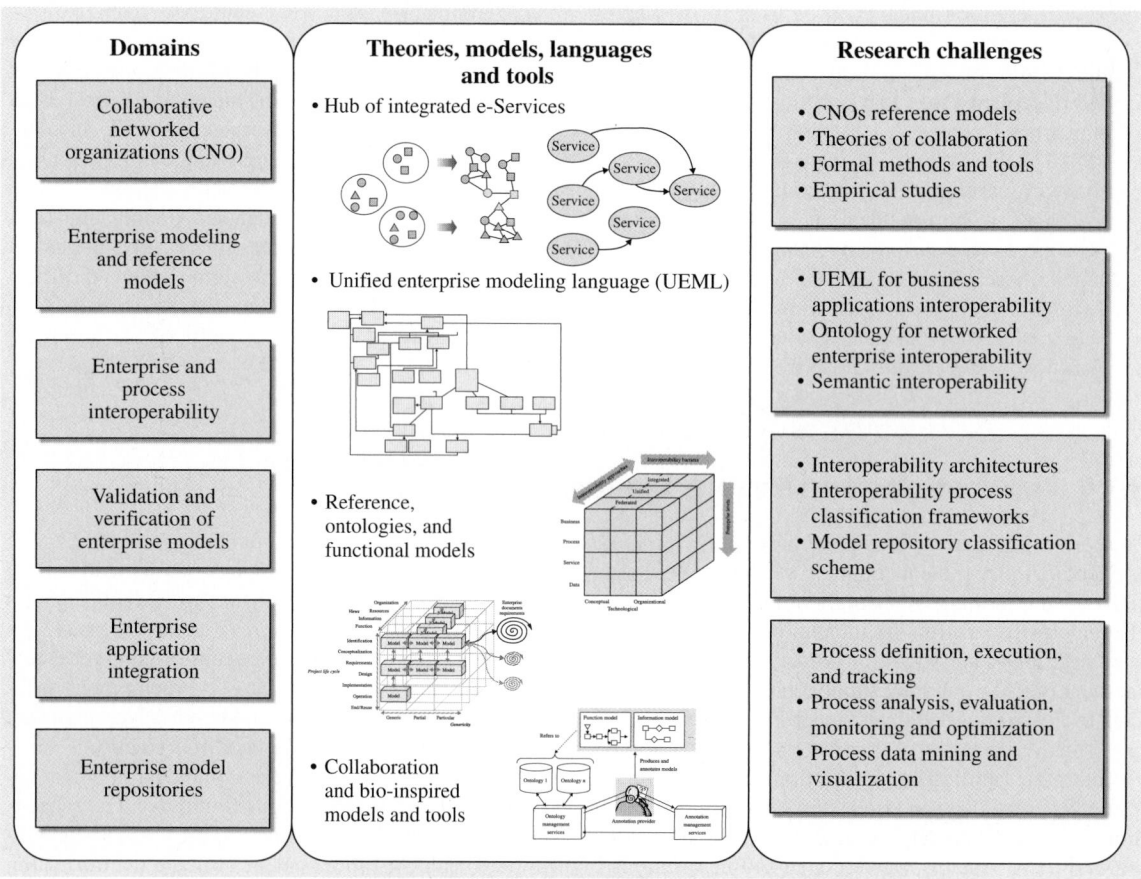

Fig. 86.3 The scope of emerging challenges in enterprise integration and interoperability (after [86.24])

(XML-based) standards for ontology representation are needed.

At the organizational level, many techniques and languages exist to capture and represent business processes and services in a satisfactory way for business users. The research trends should focus on developing interoperability maturity models with associated metrics to assess the level of integration or interoperability and propose methodologies for improving enterprise integration or interoperability.

Anyway, for the years to come, the driving forces imposing the new trends will continue to come from market globalization and new business conditions: need for more joint ventures, increased communication, fusion and acquisition, knowledge sharing, or use of common skills and competencies.

86.5 Conclusion

Thanks to advances in Internet computing and service orientation, enterprise integration and systems interoperability have become a reality that largely contributes to make enterprises more agile, more collaborative, and more interoperable. The reader should also see Chap. 88 on *Collaborative e-Work, e-Business, and e-Service* and Chap. 90 on *Business Process Automation: CRM, SCM and ERP* for related content.

Enterprise integration and systems interoperability are fairly mature at the technical level, and will continue to evolve with new IT technologies, but are not yet completely addressed at the organizational and more specifically at the semantic level.

There are however some other issues (out of the scope of this chapter) that should be considered. They concern legal issues, confidentiality issues, or linguistic aspects, to name a few, especially in the case of international contexts (multinational firms, international organizations or unions of member states). They can have a political dimension and be even harder to solve than pure technical or organizational issues mentioned in this chapter.

References

86.1   F.B. Vernadat: *Enterprise Modeling and Integration: Principles and Applications* (Chapman Hall, London 1996)

86.2   F.B. Vernadat: Interoperable enterprise systems: principles, concepts, and methods, Annu. Rev. Control **31**(1), 137–145 (2007)

86.3   B. Gold-Bernstein, W. Ruh: *Enterprise Integration: The Essential Guide to Integration Solutions* (Addison-Wesley, Boston 2005)

86.4   G. Hohpe, B. Woolf: *Enterprise Integration Patterns: Designing, Building, and Deploying Messaging Solutions* (Addison-Wesley, Reading 2004)

86.5   AMICE: *CIMOSA: CIM Open System Architecture*, 2nd edn. (Springer, Berlin, Heidelberg 1993)

86.6   H. Li, T.J. Williams: A vision of enterprise integration considerations: a holistic perspective as shown by the Purdue enterprise reference architecture, Proc. 4th Int. Conf. Enterp. Integr. Model. Technol. (ICEIMT'04) (Toronto 2004)

86.7   D. Chen, F. Vernadat: Standards on enterprise integration and engineering – a state of the art, Int. J. Comput. Integr. Manuf. **17**(3), 235–253 (2004)

86.8   ATHENA: Advanced Technologies for Interoperability of Heterogeneous Enterprise Networks and their Applications, FP6-507312-IST1 Integrated Project (2005) www.ist-athena.org

86.9   D.S. Lithicum: *Enterprise Application Integration* (Addison Wesley, Boston 2000)

86.10   OMG: Common object request broker architecture (CORBA), (Object Management Group 1994) www.omg. org

86.11   P. Herzum: *Web Services and Service-Oriented Architecture*, Executive Rep. 4, No. 10 (Cutter Distributed Enterprise Architecture Advisory Service, 2002)

86.12   R. Khalaf, F. Curbera, W. Nagy, N. Mukhi, S. Tai, M. Duftler: Understanding web services. In: *Practical Handbook of Internet Computing*, ed. by M. Singh (CRC, Boca Raton 2004), Chap. 27

86.13   IDAbc: *European Interoperability Framework for Pan-European e-Government Services* (European Commission, Brussels 2003), www.europa.eu/idabc/

86.14   C4ISR: C4ISR Architecture Framework, V. 2.0, Architecture Working Group (AWG), (US Department of Defense (DoD), 1998)

86.15   T.R. Gruber: Toward principles for the design of ontologies used for knowledge sharing, Int. J. Human-Comput. Stud. **43**, 907–928 (1995)

86.16   World Wide Web Consortium (W3C), OWL-S: Semantic markup for web services (2004), www.w3.org/2004/OWL

86.17   World Wide Web Consortium (W3C): XML: eXtensible mark-up language (2000), www.w3.org/xml

86.18   H. Kreger: *Web Services Conceptual Architecture (WSCA 1.0)* (IBM Software Group, Somers 2001)

86.19   World Wide Web Consortium (W3C): WSDL: Web service description language (2001), www.w3.org/TR/wsdl

86.20   World Wide Web Consortium (W3C), SOAP: Simple object access protocol (2002), www.w3.org/TR/SOAP

86.21   D.A. Chappell: *Enterprise Service Bus* (O'Reilly, Sebastopol 2004)

86.22   Business Process Management Initiative (BPMI): Business Process Modeling Notation (2005), ww.bpmn.org/

86.23   IBM: Business Process Execution Language for Web Services (BPEL4WS) (2003), www.ibm.com/developers works/library/ws-bpel/

86.24   S.Y. Nof, F.G. Filip, A. Molina, L. Monostori, C.E. Pereira: Advances in e-Manufacturing, e-Logistics, and e-Service Systems, Proc. IFAC Congr. (Seoul Korea 2008)

86.25   A. Molina, H. Panetto, D. Chen, L. Whitman, V. Chapurlat, F. Vernadat: Enterprise integration and networking: challenges and trends, Stud. Inform. Control **16**(4), 353–368 (2007)

87. Decision Support Systems

Daniel J. Power, Ramesh Sharda

This chapter places decision automation in a broader context of using information technologies in decision making. Key definitions and a brief history of computerized decision support create important boundaries. Then a description and explanation of characteristics of computerized decision support systems (DSS) clarifies this application domain. After reviewing management information needs, a modern taxonomy of decision support systems is briefly summarized. Issues associated with building DSS and various architectures are reviewed before concluding the overview.

Decision support systems (DSS) are defined broadly as interactive computer-based systems that help people use computer communications, data, documents, knowledge, and models to solve problems and make decisions (see DSSResources.com). DSS are ancillary or auxiliary systems; they are not intended to replace skilled decision-makers. DSS are not automating decision-making.

In pursuing the goal of improving decision-making, many different types of computerized DSS have been built to help decision teams and individual decision-makers. Some systems provide structured information directly to managers. Other systems help managers and staff specialists analyze situations using various types of quantitative models. Some DSS store knowledge and make it available to managers. Some systems support decision-making by small and large groups. Companies even develop DSS to support the decision-making of their customers and suppliers.

Table 87.1 identifies some major developments in the history of decision support systems. For more details see *Power* [87.1, 2].

The history of decision support systems (based on [87.2]) is relatively brief and the concepts and

Table 87.1 Evolution of DSS concepts

Years	Representative DSS Research
1960s	Scott Morton's management decision support project
	Interactive systems research organization
	Decision-making theory development
1970s	BrandAid research
	Alter's field study
	Holsapple research
1980s	Key DSS books
	Group DSS prototypes
	Executive information systems (EIS)
	PC expert systems
1990s	Business intelligence/OLAP
	Data warehousing
	Web-based systems/portals
	Data mining

technologies are still evolving. Today, a number of academic disciplines provide the substantive foundations for decision support systems development and research. Database research has contributed tools and research on managing data and documents. Management science and operations research have developed mathematical models for use in building model-driven DSS and provided evidence on the advantages of modeling in problem solving. Cognitive science, especially behavioral decision-making research, has provided descriptive and empirical information that has assisted in DSS design and has generated hypotheses for decision support research. Some other important fields related to DSS include artificial intelligence, human–computer interaction, software engineering, and telecommunications. In general, the Internet and web have sped up developments in decision support and have made it hard to keep up with the rapid changes in DSS capabilities.

In the 1960s, pioneering work in building interactive systems at the Massachusetts Institute of Technology, theory development by Simon and colleagues, and *Morton*'s [87.3] research created a foundation for building innovative DSS.

In the early 1970s, BrandAid [87.4] and reports of *Morton*'s [87.3] management decision system (MDS)

research demonstrated the feasibility of DSS. By the late 1970s, a number of companies had developed interactive information systems that used data and models to help managers analyze semistructured problems. These diverse systems were all called decision support systems. Key concept books included those by *Alter* [87.5], *Sprague* and *Carlson* [87.6], and *Bonczek* et al. [87.7].

From those early days, it was recognized that DSS could be designed to support decision-makers at any level in an organization. DSS could support operations, financial management, and strategic decision-making. Over the years many of the more interesting DSS have been targeted at middle and senior managers. DSS are also often designed for specific types of organizations such as hospitals, banks or insurance companies. These specialized systems are sometimes referred to as vertical-market or industry-specific DSS.

Today, decision support systems are both off-the-shelf, packaged applications and custom-designed systems. DSS may support a manager using a single personal computer or a large group of managers in a networked client–server environment. These latter systems are often called enterprise-wide DSS.

87.1 Characteristics of DSS

Although the term decision support system has many connotations, based upon *Alter*'s [87.5] pioneering research we can identify the following three major characteristics:

1. DSS are designed specifically to facilitate decision processes.
2. DSS should support rather than automate decision-making.
3. DSS should be able to respond quickly to the changing needs of decision-makers.

Many terms are used for specific types of DSS, including business intelligence, collaborative systems, data warehousing, model-based systems, knowledge management, and online analytical processing. Software vendors use these more specialized terms for both descriptive and marketing purposes. What term we use for a system or software package is a secondary concern. Our primary concern should be finding software and systems that meet a manager's decision

support needs and provide appropriate management information.

87.1.1 Management Information Needs

Managers and their support staffs need to consider what information and analyses are actually needed to support their management and business activities. Some managers need both detailed transaction data and summarized transaction data. Most managers only want summaries of transactions. Managers usually want lots of charts and graphs; a few only want tables of numbers. Many managers want information provided routinely or periodically and some want information available online and on demand. Managers usually want financial analyses and some managers want primarily *soft*, nonfinancial or qualitative information. Managers have different needs and wants depending upon personal preferences, task needs, and information technology experiences.

In general, a computerized information system can provide business transaction summary information and can help managers understand many business operations and performance issues; for example, a computerized system can help managers understand the status of operations, monitor business results, review customer preference data, and investigate competitor actions. In all of these situations, management information and analyses should have a number of characteristics. Information must be both timely and current. These characteristics mean the information is up to date and available when managers want it. Also, information must be accurate, relevant, and complete. Finally, managers want information presented in a format that assists them in making decisions. In general, management information should be summarized and concise and any support system should have an option for managers to obtain more detailed information.

So the subcategory known as decision support systems needs to provide current, timely information that is accurate, relevant, and complete. A specific DSS must present information in an appropriate format that is easy to understand and manipulate. The information presented by a DSS may result from analysis of transaction data or it may be the result of a decision model or it may have been gathered from external sources. DSS can present internal and external facts, informed opinions, and forecasts to managers. As many others have noted, *managers want the right information, at the right time, in the right format, and at the right cost.* These system requirements seem simple and straightforward, but meeting them remains a challenge.

Technology is creating new decision support capabilities, but much learning and discussion needs to occur to successfully exploit the technological possibilities. Decision support systems differ in many ways from operating systems that process business transactions. For example, a popular system that has been widely implemented is called enterprise resource planning (ERP). ERP is *not* a decision support system even though the term suggests that decision-making and planning will be improved. In general, enterprise resource planning is an integrated transaction processing system that facilitates the flow of information between all of the functional areas of a business. Recently, DSS have been built that help managers analyze the data from ERP systems, and ERP systems make it easier to create a wide variety of DSS.

Decision automation refers to using information technologies to make decisions and implement programmed decision processes. Typically decision automation is considered most appropriate for well-structured, clearly defined, routine or programmed [87.8] decision situations. Decision support systems are intended to help human decision-makers while decision automation systems make a decision and reduce the need for immediate human involvement in the decision. People still write rules and review results for decision automation, but software makes the routine decisions.

A major difference between transaction processing systems, decision automation, and decision support systems is the general purpose of each type of system. Transaction processing systems (TPS) are designed to expedite and automate transaction processing, record keeping, and simple business reporting of transactions. Decision automation uses software to make decisions. Decision support systems are intended to assist in decision-making and decision implementation. Transaction processing is however related to the design of DSS because transaction databases often provide data for decision-oriented reporting systems and data warehouses. Decision automation often interacts with TPS, and the results of automated decisions often impact on TPS.

Transaction processing systems usually provide standard reports on a periodic basis and support the operations of a company. DSS are used on demand when they are needed to support decision-making. A manager typically initiates each instance of decision support system use, either by using the DSS herself or by asking a staff intermediary to use a DSS. Some managers and especially clerical employees use transaction processing systems to support operations. Line managers and support staff are the primary users of DSS. TPS record current information and maintain a database of transaction information. DSS generally use historical internal and external data for analysis. DSS may focus on quantitative analysis and modeling current and future scenarios. TPS emphasize data integrity and consistency, and although both of these qualities are important in every system, the primary emphasis for a DSS is on flexibility and on conducting analyses and retrieving decision-relevant information and knowledge. Decision automation routinizes a specific decision 24 h a day, 7 days a week.

One of the long-standing conclusions from reading DSS case studies is that DSS can *take on many different forms and can be used in many different ways* [87.5]. Decision support systems certainly vary in many ways. Some DSS focus on accessing data, some on manipulating models, and some on facilitating communications.

DSS also differ in scope; some DSS are intended for one *primary* user and are used *stand-alone* for analysis, whereas others are intended for many users in an organization. Also, DSS differ in terms of who uses a specific system; for example, some DSS are used by actual decision-makers and some are used by intermediaries such as marketing analysts or financial analysts. If a computerized system is not a transaction processing system and if a manager uses it, many observers will be tempted to call the system a DSS.

Some examples show the wide variety of DSS applications. Major airlines have DSS used by analysts for many tasks including pricing and route selection. Many companies have DSS that aid in corporate planning and forecasting. Specialists often use these DSS that focus on financial and simulation models. DSS can help monitor costs and revenues and track departmental budgets. Also, investment evaluation and support systems are increasingly common. Frito-Lay has a DSS that aids in pricing, advertising, and promotion. Route salesmen use handheld computers to support decision-making activities.

Many manufacturing companies use manufacturing resources planning (MRP) software. This specific operational-level DSS supports master production scheduling, purchasing, and materials requirements planning. More recent MRP systems support *what-if* analysis and simulation capabilities. DSS support quality improvement and control decisions. Monsanto, FedEx, and most transportation companies use DSS for scheduling trucks, airplanes, and ships. The Coast Guard uses a DSS for procurement decisions. Companies such as Wal-Mart have large data warehouses and use data-mining software. Business intelligence and knowledge management systems are increasingly common. On the World Wide Web one can find DSS that help track and manage stock portfolios, choose stocks, plan trips, and suggest gifts. DSS support distributed decision activities using groupware and a corporate intranet.

The following expanded DSS framework helps categorize the most common DSS currently in use [87.1, 9, 10]. Some DSS are integrated or hybrid systems with more than one major DSS subsystem. The framework focuses on one major capability dimension with five categories and three secondary dimensions. The term *driver* is used as a common or shared descriptive adjective in the expanded framework. *Driver* refers to the capability, tool or component that is providing the dominant functionality in the DSS. The five categories explained below are communications-driven, data-driven, document-driven, knowledge-driven, and model-driven.

87.1.2 Communications-Driven and Group DSS

Group decision support systems (GDSS) and groupware were investigated in the mid-1980s, but now a broader category of *communications-driven DSS* can be identified. This type of decision support system includes communication, collaboration, and decision support. A group DSS is an interactive computer-based system intended to facilitate the solution of problems by decision-makers working together as a group; such systems often derive functionality from a model more than from supporting collaboration and hence are model-driven DSS. Groupware supports electronic communication, scheduling, document sharing, and other group productivity and decision support enhancing activities. A number of technologies and capabilities are included in this category in the framework: some collaborative group DSS, two-way interactive video, white boards, bulletin boards, chat, and email systems.

87.1.3 Data-Driven DSS

Data-driven DSS emphasize analysis of data. These systems include file drawer and management reporting systems, data warehousing and analysis systems, executive information systems (EIS), and data-driven spatial decision support systems (SDSS). EIS are targeted at senior managers and SDSS display spatial data for decision support. Business intelligence (BI) systems are also examples of data-driven DSS. A data-driven DSS provides access to and manipulation of large databases of structured data and especially a time series of internal company data and external data. Simple file systems accessed by query and retrieval tools provide the most elementary level of functionality. Data warehouse systems that allow the manipulation of data by computerized tools tailored to a specific task and setting or by more general tools and operators provide additional functionality. Data-driven DSS with online analytical processing (OLAP) provide the highest level of functionality and decision support that is linked to analysis of large collections of historical data [87.11].

87.1.4 Document-Driven DSS

A new type of DSS, a *document-driven DSS*, is evolving to help managers retrieve and manage unstructured documents and web pages. A document-driven DSS integrates a variety of storage and processing technologies to provide complete document retrieval and analysis. The web provides access to large document databases including databases of hypertext documents, images, sounds, and video. Examples of documents that would be accessed by a document-driven DSS are policies and procedures, product specifications, catalogs, and corporate historical documents, including minutes of meetings, corporate records, and important correspondence. A search engine is a powerful decision-aiding tool associated with a document-driven DSS. Some authors call this type of system a knowledge management system.

87.1.5 Knowledge-Driven DSS

The terminology for this fourth category of DSS is still evolving. Currently, the best term seems to be *knowledge-driven DSS*. Sometimes it seems equally appropriate to use *Alter*'s [87.5] term *suggestion DSS* or the narrower term *management expert system*. Knowledge-driven DSS suggest or recommend actions to managers. These DSS have specialized problem-solving expertise. The *expertise* consists of knowledge about a particular domain, understanding of problems within that domain, and *skill* at solving some of these problems.

87.1.6 Model-Driven DSS

A fifth category, *model-driven DSS*, includes systems that use accounting and financial models, representational models, and optimization models. Model-driven DSS emphasize access to and manipulation of a quantitative model. Simple statistical and analytical tools provide the most elementary level of functionality. Some OLAP systems that allow complex analysis of data may be classified as hybrid DSS systems providing modeling, data retrieval, and data summarization functionality. Model-driven DSS use data and parameters provided by decision-makers to aid them in analyzing a situation, but they are not usually data intensive. Very large databases are usually not needed for model-driven DSS, but data for a specific analysis may need to be extracted from a large database.

87.1.7 Secondary Dimensions

The secondary dimensions used to describe DSS include purpose, targeted users, and the enabling tech-

nology. A relatively new category of DSS made possible by new technologies and the rapid growth of the public Internet is *interorganizational DSS*. These DSS target a company's customers or suppliers. The public Internet is creating communication links for many types of interorganizational systems, including DSS. An interorganizational DSS provides external stakeholders with access to a company's intranet and authority or privileges to use specific DSS capabilities. Companies can make a data-driven DSS available to suppliers or a model-driven DSS available to customers to design a product or choose a product. Most DSS are *intra-organizational DSS* that are designed for use by managers in a company as *stand-alone DSS* or for use by a group of managers in a company as a group or enterprise-wide DSS.

Many DSS are designed to support specific business functions or types of businesses and industries. These are broad- or narrow-purpose DSS. Such DSS can be labeled as function-specific or industry-specific purpose DSS.

A *function-specific DSS* such as a budgeting system may be purchased from a vendor or customized in-house using a more general-purpose development package. Vendor developed or off-the-shelf DSS support functional areas of a business such as marketing or finance; some DSS products are designed to support decision tasks in a specific industry such as a crew scheduling DSS for an airline. A function or task-specific DSS has an important purpose in solving a recurring decision task. Function- or task-specific DSS can be further classified and understood in terms of the dominant DSS component in the DSS. So, a function-specific DSS may be model driven, data driven, document driven or knowledge-driven. These DSS have capabilities relevant for a decision about some function that an organization performs (e.g., a marketing function or a production function). These DSS are categorized by purpose: function-specific DSS help a person or group accomplish a specific decision task; general-purpose DSS software helps support broad tasks such as business performance monitoring, business intelligence, project management, and decision analysis.

All five broad categories of decision support systems can be implemented using web technologies. When the enabling technology used to build a DSS is the Internet and web is seems appropriate to call the system a web-based DSS. A *web-based DSS* is a computerized system that delivers decision support

Table 87.2 An expanded DSS framework (after [87.1])

Dominant DSS component	User groups: internal, external	Purpose: general, specific	Enabling technology
Communications *Communications-driven DSS*	Internal teams, now expanding	Conduct a meeting Bulletin board Help users collaborate	Web or client/server
Database *Data-driven DSS*	Managers, staff, now suppliers	Query a data warehouse	Main frame, client/server, web
Document base *Document-driven DSS*	Specialists and user group is expanding	Search web pages, find documents	Web
Knowledge base *Knowledge-driven DSS*	Internal users, now customers	Management advice, choose products	Client/server, web
Models *Model-driven DSS*	Managers and staff, now customers	Crew scheduling Decision analysis	Stand-alone PC

information or decision support tools to a manager or business analyst using a *thin-client* web browser such as FireFox or Internet Explorer [87.1]. The computer server that is hosting the DSS application is linked to the user's computer by a network with the transmission control protocol/Internet protocol (TCP/IP). In many companies, a web-based DSS is synonymous with an intranet or enterprise-wide DSS. Web technologies are powerful tools for creating decision support systems and especially interorganizational DSS that support the decision-making of customers and suppliers. Web 2.0 technologies are the leading edge for building DSS, but some DSS will continue to be built using mainframe and client–server enabling technologies.

Column one of Table 87.2 lists the five broad categories of decision support systems that differ in terms of the dominant decision support component, including communications-driven DSS, data-driven DSS, document-driven DSS, knowledge-driven DSS, and model-driven DSS. The table provides examples of targeted user groups – intra-organizational and interorganizational, purpose, and enabling technologies.

One can use the dominant DSS architecture component, targeted user group, purpose, and the enabling technology to describe a specific decision support system. For example, a manager may want to build a model-driven, intra-organizational, product design, web-based DSS to support a specific business decision process.

87.2 Building Decision Support Systems

Traditionally, information systems academics and practitioners have discussed building decision support systems in terms of four major components:

1. The user interface
2. The database
3. The models and analytical tools
4. The DSS architecture and network [87.6].

This traditional list of components remains useful because it identifies similarities and differences between categories or types of DSS and it can help managers and analysts build new decision support systems. The expanded DSS framework [87.1, 9, 10] is based on the different emphases placed on DSS components when a specific system is actually constructed (Fig. 87.1).

Data-driven, document-driven, and knowledge-driven DSS need specialized database components. A model-driven DSS may use a simple flat-file database with fewer than 1000 records, but the model component is very important. Experience and some empirical evidence indicate that design and implementation issues vary for data-driven, document-driven, model-driven, and knowledge-driven DSS. Multiparticipant systems such as group and interorganizational DSS also create complex implementation issues. For instance, when implementing a data-driven DSS a designer should be especially concerned about the user's interest in applying the DSS in unanticipated or novel situations.

In creating an accounting or financial simulation model, a developer should attempt to verify that the initial input estimates for the model are thoughtful and reasonable. In developing a representational or optimization model, an analyst should be concerned about possible misunderstandings of what the model means and how it can or cannot be used [87.5]. Networking issues create challenges for many types of DSS, but especially for communications-driven systems with many participants, so-called multiparticipant systems. Today, architecture and networking issues are increasingly important in building DSS.

DSS should be built or implemented using an appropriate process. Many small, specialized model-driven DSS are built quickly. Large, enterprise-wide DSS are built using sophisticated tools and systematic and structured systems analysis and development approaches. Communications-driven and group DSS are usually purchased as off-the-shelf software and then implemented in a company. Creating enterprise-wide DSS environments remains an iterative and evolutionary task. An enterprise-wide DSS grows and inevitably becomes a major part of the overall information systems infrastructure of an organization. Despite the significant differences created by the specific task and scope of a DSS, all DSS have similar technical components and share a common purpose: supporting decision-making.

A data-driven DSS database is often a collection of current and historical structured data from a number of sources that have been organized for easy access and analysis. We are expanding the data component to include unstructured documents in document-driven DSS and *knowledge* in the form of rules in knowledge-driven DSS. Large databases of structured data in enterprise-wide DSS are often called data warehouses or data marts. DSS usually use data that has been extracted from all relevant internal and external databases. Managing information often means managing a database. Supporting management decision-making means that computerized tools are used to make sense of the structured data or documents in a database.

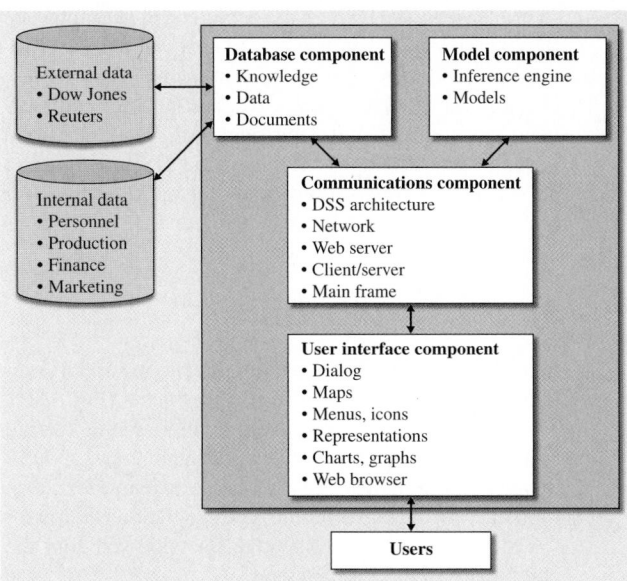

Fig. 87.1 Traditional DSS components [87.1]

Mathematical and analytical models are the major component of a model-driven DSS. DSS models should be used and manipulated directly by managers and staff specialists. Each model-driven DSS has a specific set of purposes and hence different models are needed and used. Choosing appropriate models is a key design issue. Also, the software used for creating specific models needs to manage needed data and the user interface. In model-driven DSS the values of key variables or parameters are changed, often repeatedly, to reflect potential changes in supply, production, the economy, sales, the marketplace, costs, and/or other environmental and internal factors. Information from the models is then analyzed and evaluated by the decision-maker. Knowledge-driven DSS use special models, an inference engine, for processing rules or identifying relationships in data.

The communications component refers to how hardware is organized, how software and data are distributed in the system, and how components of the system are integrated and connected. A major issue today is whether decision support systems should be available using a web browser on a company intranet and also available on the global Internet. Managers and information systems (IS) staff both need to develop an understanding of the technical issues and the security issues related to DSS architectures, networks, and the Internet. Networking technology is the key driver of communications-driven DSS.

Managers and DSS analysts both need to emphasize the user interface component. In many ways the user interface is the most important component. The tools for building the user interface are sometimes termed DSS generators, query and reporting tools, and front-end development packages. Much of the DSS design and development effort should focus on building the user interface. It is important to remember that the screens and displays in the user interface heavily influence how a manager perceives a DSS, as what one sees is the DSS.

The technologies and software associated with decision support systems continues to change rapidly, and development tools are overlapping for some applications. In general, managers and IS staff need to recognize that the overall technological and social context of DSS and business management is changing.

87.3 DSS Architecture

Many academics discuss building decision support systems in terms of the four major components (Fig. 87.2). These components collectively comprise the overall architecture of a DSS. This traditional view of DSS components remains useful because it identifies commonalties between different types of DSS, but it provides only an initial perspective for understanding the complexity of DSS architectures.

As noted in the last section, a major component in the architecture of a DSS is the user interface. The tools for building the user interface are sometimes termed DSS generators, query and reporting tools, and front-end development packages. DSS user interfaces can be distributed to clients in a *thick-client* architecture or delivered over a network using web pages or Java applets in a *thin-client* architecture. A thin-client architecture where a user interacts using a web browser has many advantages, but until recently the sophistication of the user interface was limited compared with a thick-client architecture where a program resides on a DSS user's computer.

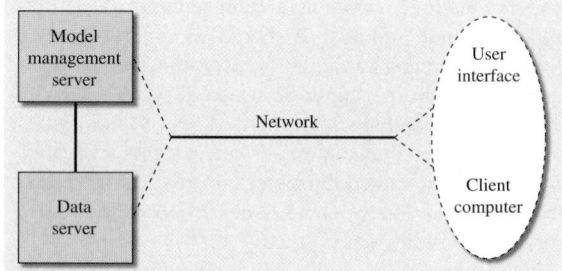

Fig. 87.2 DSS components (after [87.1])

A DSS database is a collection of data organized for easy access and analysis. Large databases in enterprise-wide DSS are often called data warehouses or data marts. Document or unstructured data is stored differently than structured data. Web servers provide a powerful platform for unstructured data and documents. The architecture for a data-driven DSS often involves databases on multiple servers, specialized hardware, and in some cases both multidimensional and

Table 87.3 Selected resources for DSS

Resource Type	Exemplars
Organizations	SIG DSS
	http://sigs.aisnet.org/SIGDSS
Websites	www.dssresources.com
	(contains links to most other DSS resources)
Journals	*Decision Support Systems, Management Science, Decision Sciences*, and many others
Books	S.B. Eom: *The Development of Decision Support Systems Research: A Bibliometrical Approach* (Edwin Mellon, New York, 2007)
	D.J. Power: *Decision Support Systems Hyperbook*, http://dssresources.com/dssbook/ (Fall, 2000)
	Turban, Aronson, Liang, Sharda: *Decision Support and Business Intelligence Systems*, 8th ed. (Prentice Hall, Upper Saddle River 2007)

relational database software. The extraction, transformation, loading, and indexing of structured DSS data is sometimes very difficult.

As noted before, quantitative models are an important part of many DSS, especially model-driven DSS. Model management software can be centralized on a server with a database, or specific models can be distributed to client computers. Java applets and JavaScript programs provide a powerful new means to deliver models to users in a thin-client architecture.

The DSS network component refers to how hardware is organized, how software and data are distributed in the system, and how components of the DSS are integrated and physically connected. An ongoing issue is whether a specific DSS should only be available using thin-client technology on a company intranet or available on the global Internet. This should depend on the needs analysis and a feasibility study. Scalability is also an important DSS issue. Scalability refers to the ability to *scale* hardware and software to support larger or smaller volumes of data, and more or fewer users. Scalability also refers to the possibility of increasing or decreasing size or capability of a DSS in cost-effective increments.

Table 87.3 provides selected additional readings and other resources for exploring DSS applications and technologies. These include web resources, text books, and selected journal articles in addition to those included in the references.

87.4 Conclusions

The DSS design and development environment is changing as rapidly as the innovation in hardware and software tools that are available to build them, and in general the change is in a positive a direction. Web and mobile technologies will facilitate improved DSS tools.

Decision support systems are not a panacea for improving business decisions. Most people acknowledge that managers need *good* information to manage effectively, but a DSS is not always the solution for providing *good* information. A DSS can provide a competitive advantage and a company may need computerized decision support to remain competitive, but decision support capabilities are limited by the data that can be obtained, the cost of obtaining, processing, and storing the information, the cost of retrieval and distribution, the value of the information to the user, and the capability of managers to accept and act on the information. Our capabilities to support decision-making have increased, but we still have very real technical, social, interpersonal, and political problems that must be overcome when we build a specific DSS.

Computerized DSS assume that managers need to be kept in the decision-making loop but that managers need and want help to improve their decision-making. When it is possible and appropriate to automate decision-making using software then we have chosen to remove a human decision-maker from the decision-making process. Managers choose to automate, create computerized support or make unaided decisions.

87.5 Further Reading

- A.E. Boyd, J.C. Bilegan: Revenue management and e-Commerce, Manag. Sci. **49**(10), 1363–1386 (2003)
- T.A. Bresnick, D.M. Buede, A.A. Pisani, L.L. Smith, B.B. Wood: Airborne and space-borne reconnaissance force mixes: A decision analysis approach, Mil. Oper. Res. **3**(4), 65–78 (1997)
- G. DeSanctis, R.B. Gallupe: A foundation for the study of group decision support systems, Manag. Sci. **33**(5), 589–609 (1987)
- F.N. de Silva, R.W. Eglese: Integrating simulation modeling and GIS: Spatial decision support systems for evacuation planning, J. Oper. Res. Soc. **51**(4), 423–430 (2000)
- J.P. Shim, M. Warkentin, J.F. Courtney, D.J. Power, R. Sharda, C. Carlsson: Past, present, and future of decision support technology, Decis. Support Syst. **33**(2), 111–126 (2002), special issue, DSS: Directions for the Next Decade
- R.H. Sprague Jr.: A framework for the development of decision support systems, Manag. Inform. Syst. Q. **4**(4), 1–26 (1980)

Part I | 87.5

References

87.1   D.J. Power: *Decision Support Systems: Concepts and Resources for Managers* (Westport, Greenwood/Quorum 2002)

87.2   D.J. Power: *A Brief History of Decision Support Systems*, DSSResources.COM, http://DSSResources.COM/history/dsshistory.html, version 4.0, March 10 (2007)

87.3   M.S. Scott Morton: *Management Decision Systems; Computer-Based Support for Decision Making* (Harvard Univ. Press, Boston 1971)

87.4   J.D.C. Little: Brandaid, an online marketing mix model, Part 2: Implementation, calibration and case study, Oper. Res. **23**(4), 656–673 (1975)

87.5   S.L. Alter: *Decision Support Systems: Current Practice and Continuing Challenge* (Addison–Wesley, Reading 1980)

87.6   R.H. Sprague Jr., E.D. Carlson: *Building Effective Decision Support Systems* (Prentice Hall, Englewood Cliffs 1982)

87.7   R.H. Bonczek, C.W. Holsapple, A.B. Whinston: *Foundations of Decision Support Systems* (Academic, New York 1981)

87.8   H.A. Simon: *The New Science of Management Decision* (Harper Row, New York 1960)

87.9   D.J. Power: Supporting decision-makers: An expanded framework, e-Proc. Inf. Sci. Conf., ed. by A. Harriger (Krakow 2001) pp. 431–436

87.10   D.J. Power: Specifying an expanded framework for classifying and describing decision support systems, Commun. Assoc. Inf. Syst. **13**, 158–166 (2004)

87.11   V. Dhar, R. Stein: *Intelligent Decision Support Methods: The Science of Knowledge* (Prentice Hall, Upper Saddle River 1997)

88. Collaborative e-Work, e-Business, and e-Service

Juan D. Velásquez, Shimon Y. Nof

A major part of automation today is represented by collaborative e-Work, e-Business, and e-Service. In this chapter the fundamental theories and scope of collaborative e-Work and collaborative control theory (CCT) are reviewed, along with design principles for effectiveness in the design and operation of their automation solutions. The potential benefits, opportunities, and sustainability of emerging electronic activities, such as virtual manufacturing, e-Healthcare, automated inspection, e-Supply, e-Production, e-Collaboration, e-Logistics, and other e-Activities, will not materialize without the design of effective e-Work. For instance, without automatic error prevention or recovery, e-Activities in complex systems will collapse. The *four-wheels* of collaborative e-Work, their respective 15 e-Dimensions, and their role in e-Business and e-Service are explained and illustrated. Case studies of e-Work, e-Manufacturing, e-Logistics, e-Business and e-Service are also provided to enable readers to get a glimpse into the depth and breadth of ongoing efforts and potential inhibitors to revolutionize such e-Systems. Challenges and emerging solutions are discussed to stimulate readers to push the boundaries of collaborative e-Work.

Part I | 88

88.1 Background and Definitions

The landscape in which everyday work and personal activities are accomplished by humans, computers, machines, and organizations has evolved significantly in recent years. From a production point of view, specifically, work has been reshaped by e-Work from the smallest level systems, micro- and nanosystems, all the way up to global enterprises. Just as work, business, commerce, and service are different from each other, so are e-Work, e-Business, e-Commerce, and e-Service. They are highly related, yet they are not the same (Fig. 88.1), and therefore some definitions are needed.

e-Business is the integration of a company's business, including products, procedures, and services, over the Internet [88.1]. A more comprehensive definition

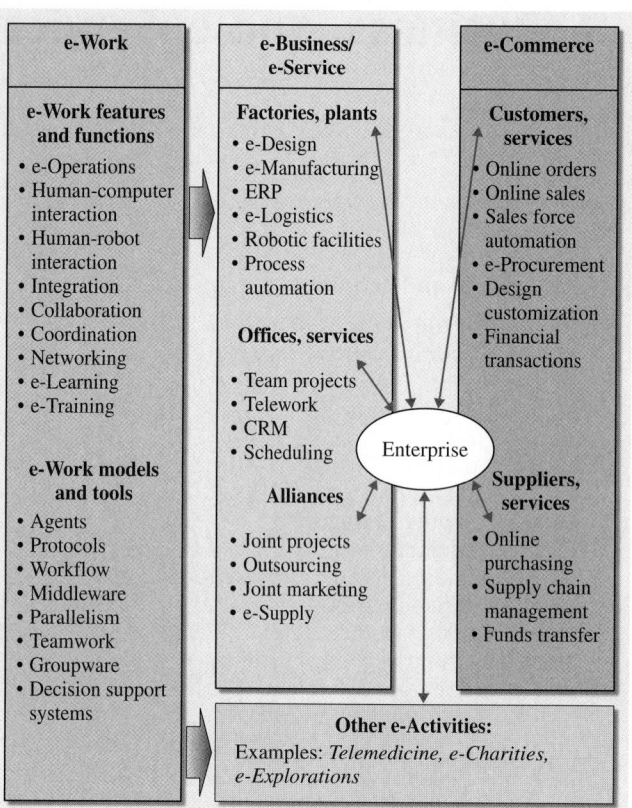

Fig. 88.1 e-Work as the foundation for e-Business, e-Service, e-Commerce, and other e-Activities (after [88.3])

provided by *Browne* et al. [88.2] states that e-Business is about conducting business both internally and/or externally by electronic means, e.g., over the Internet, intranet or extranet, and that it involves not only buying and selling but also operating automated, efficient internal business processes servicing customers and collaborating with suppliers and business partners (Fig. 88.2).

e-Commerce covers the range of online business activities for products and services, both business-to-business and business-to-consumer, through the Internet [88.1].

e-Service is the provision of services over electronic networks such as the Internet, intranets or extranets without its scope being limited to service organizations but rather encompassing all enterprises, even those that manufacture goods and which require the development and implementation of sound service practices over electronic networks.

e-Work are the e-Operations, e-Functions, and e-Support that enable the e-Business, e-Commerce and e-Service activities defined above. Its formal definition describes it as any collaborative, computer-supported and communication-enabled activities in highly distributed networks of humans and/or robots and autonomous systems [88.3]. Among the activities comprised by e-Work are e-Business, e-Commerce, e-Manufacturing, e-Healthcare, e-Training, e-Logistics, virtual (v-)Factories, v-Enterprises, and many more (Fig. 88.3). The activities mentioned above, as well as all others influenced by e-Work, rely on the use of computer support and communication technologies and at the same time require a certain degree of collaboration to support the interactions between humans, computers, and machines.

The transformative influence of e-Work can be summarized in this quote [88.3]:

> *As power fields, such as magnetic fields and gravitation, influence bodies to organize and stabilize, so does the sphere of computing and information technologies. It envelops us and influences us to organize our work systems in a different way, and purposefully, to stabilize work while effectively producing the desired outcomes.*

Often the terms *collaboration* and *coordination* are used interchangeably, and even though they are closely related, they do exhibit significant differences. *Camarinha-Matos* and *Afsarmanesh* [88.4,5] and *Nof* [88.6] provide a number of definitions that help clarify the scope and breadth of each term by placing them in a broader context, which also includes closely related terms such as networking and cooperation. It is important to highlight the differences between all these terms, as their implications on the design of e-Work are crucial (Fig. 88.4).

Coordination involves the use of communication and information exchange to reach mutual benefits among parties by working harmoniously.

Cooperation involves all aspects of coordination but also incorporates a resource-sharing dimension to support goal achievement. Commonly, cooperation will exhibit a component of division of labor among all participants and therefore the aggregated value is the result of adding the *individual* parts.

Collaboration is the most demanding of the three processes, where all involved parties share information, resources, and responsibilities to jointly plan,

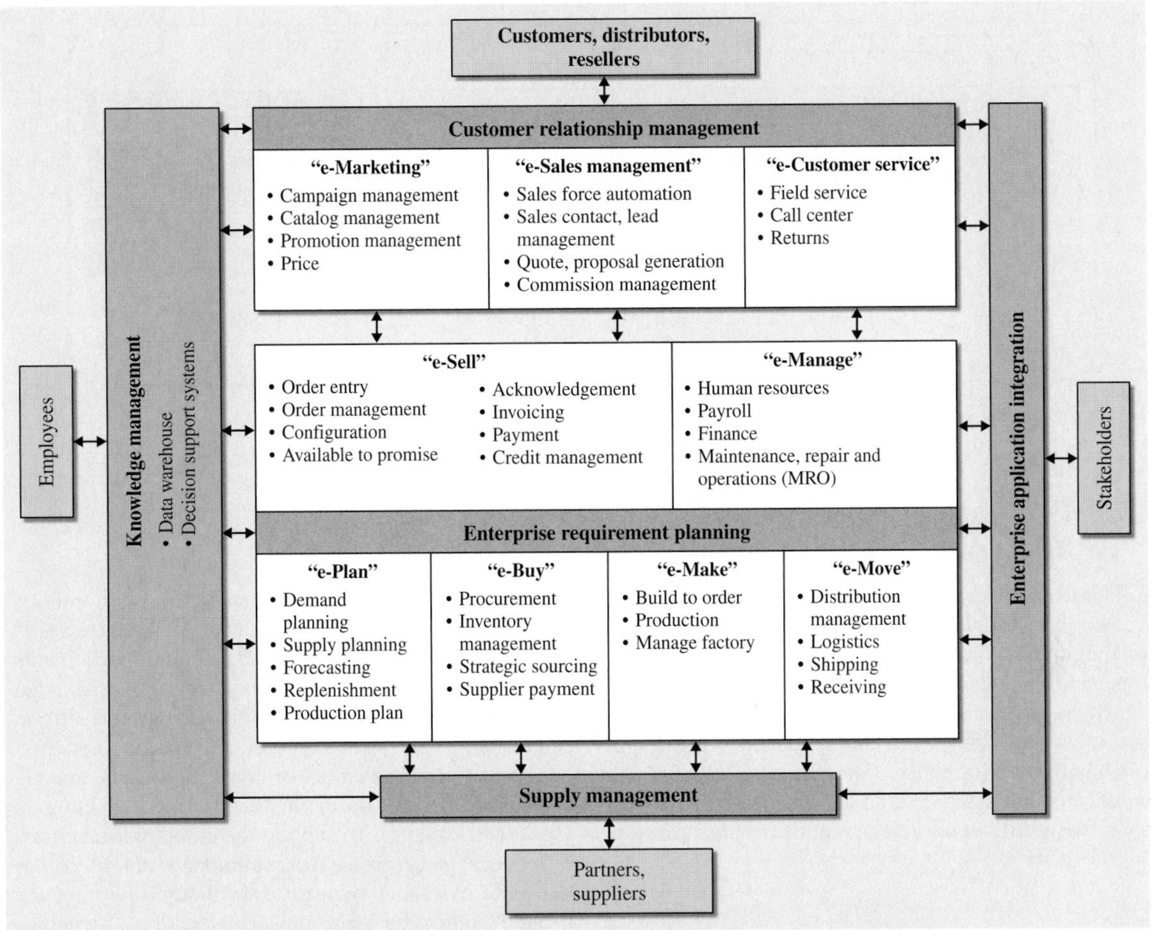

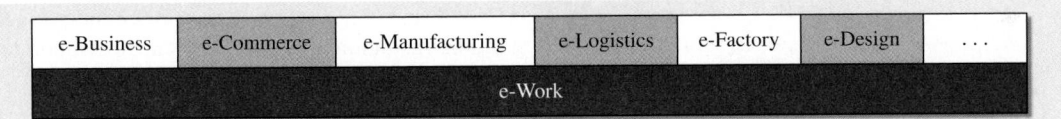

Fig. 88.2 e-Business application framework (after [88.7] with permission from Taylor and Francis)

e-Business	e-Commerce	e-Manufacturing	e-Logistics	e-Factory	e-Design	. . .
e-Work						

Fig. 88.3 Scope of collaborative e-Work as the common foundation of e-Activities (after [88.8])

implement, and assess the set of activities required to achieve a common goal, thus jointly creating added value.

Over the years, computers and advances in communication technologies and related areas have enabled us to significantly improve the quality of service and products, as well as the level of customization. Specifically, *collaborative e-Work* is revolutionizing the capabilities

of e-Business, e-Production, and e-Service. By leveraging existing and well-established work, business and service models, theories, and solutions with the new electronic world, it is possible to augment companies' abilities to significantly better meet customers' needs. In the next section, the four knowledge clusters of e-Work and their related 15 dimensions are discussed in detail.

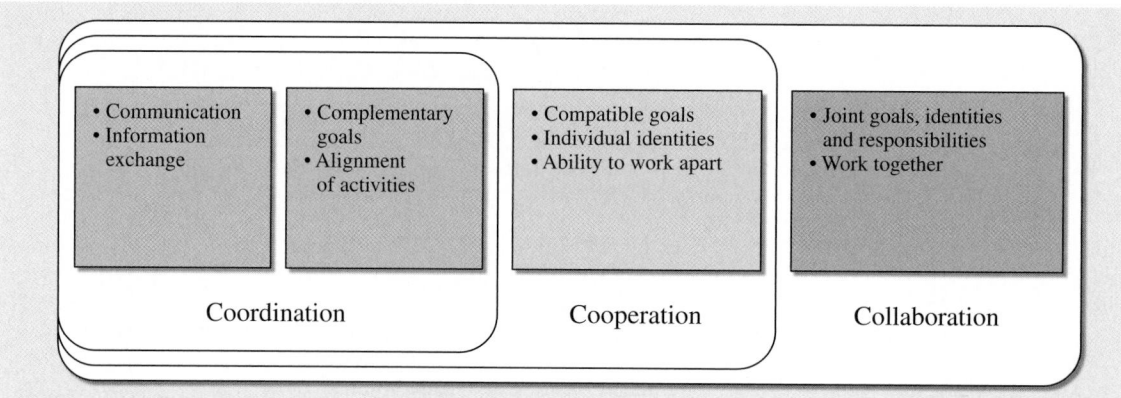

Fig. 88.4 Coordination, cooperation, and collaboration levels of interaction

88.2 Theoretical Foundations of e-Work and Collaborative Control Theory (CCT)

Collaborative control theory (CCT) has been developed over the last decade to support the effective design of collaborative e-Work, e-Business, and e-Service systems, and includes several design principles [88.9–11]. Many researchers have contributed to CCT (e.g., see survey in [88.3]) and have addressed issues of collaborative performance. For instance, *collaborative intelligence* has been defined as a measure of the collaborative ability of an interacting, distributed group or organization, whether of humans and/or agents, robots, sensors, etc. The major tradeoff that CCT addresses is that with a larger number of collaborating participants in a more complex, large-scale system; the knowledge and performance generated from collaborative efforts can increase proportionally to the reach of the network, but conversely the number of conflicts, delays, and errors increases too, up to ineffectiveness, reduction in value, and collapse. To support the collaboration effort, e-Support is essential so that participants can interact in real time or asynchronously even though they are not located within the same physical location. Technologies used to enhance distributed collaboration and to facilitate group problem-solving and decision-making include messaging software, coordination protocols, task administration protocols, and other e-Support techniques and tools:

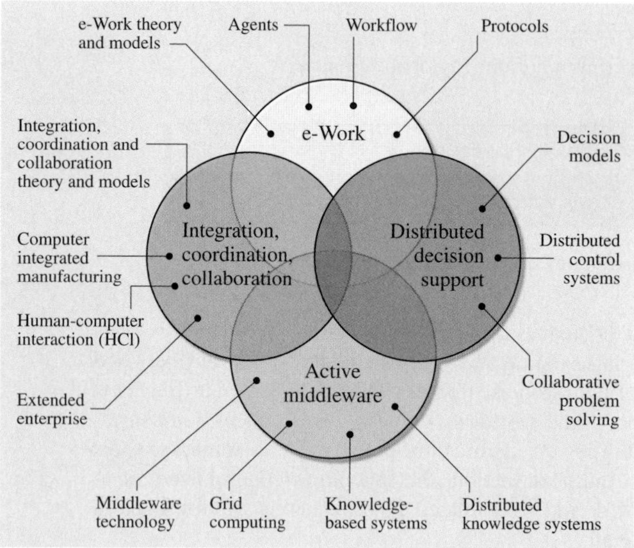

1. Group moderation and facilitation
2. Agreement to adhere to a set of fundamental rules relating to participant interactions
3. Frequent sharing of feedback
4. Frequent service reviews and agreements on progress and plans
5. Maintenance of a collaborative memory as an organizational knowledge base
6. Service-oriented architectures for collaborative automation.

Examples of these developments are discussed in manufacturing and supply [88.12], in shared un-

Fig. 88.5 Collaborative e-Work: foundations, dimensions, and scope (after [88.3])

derstanding for collaborative control [88.13], and in multirobot remote activation with collaborative control [88.14].

The "e" in collaborative e-Work is supported by the continuous creation of knowledge in four areas and their overlaps: (1) e-Work, (2) integration, coordination and collaboration, (3) distributed decision support systems, and (4) active middleware, all of which in turn are rooted in 15 e-Dimensions as described below (Fig. 88.5).

The first theoretical foundation area is *e-Work* itself, and it comprises four e-Dimensions: (1) e-Work theory and models, (2) agents, (3) protocols, and (4) workflow, as described below.

88.2.1 e-Work Theory and Models

Theory and models provide the foundation for the development of models to (1) augment human abilities and capabilities at work, (2) augment organizational abilities to accomplish their goals, and (3) augment the abilities of all organizational agents (e.g., human, computers, robots, etc.) for collaboration. Below, several principles that serve as guidelines and foundation for effective e-Work are discussed in more detail.

88.2.2 Agents

Agents are independently operating and executing programs capable of responding autonomously to expected or unexpected events [88.16]. Agents enable the automation and integration of flexible, adaptive, and heterogeneous activities, components, and tasks in a network. For instance, intelligent information agents are being used to (1) be proactive in the identification of enterprise resources, (2) provide value-added information for ongoing e-Activities, and (3) reduce the number of conflicts and errors in the information provided to and by all members of networked enterprises (Fig. 88.6). The applicability of agents expands a significant number of domains, including: autonomous mobile or stationary robots, electronic information systems, data mining and data clusters, decision-support systems, agent-based simulation, and intelligent design and manufacturing systems. An example of agents working in a collaborative environment includes the University of Michigan digital library (UMDL) where the library's electronic information and services are offered in a distributed environment by building on commerce and communication protocols for agent interaction.

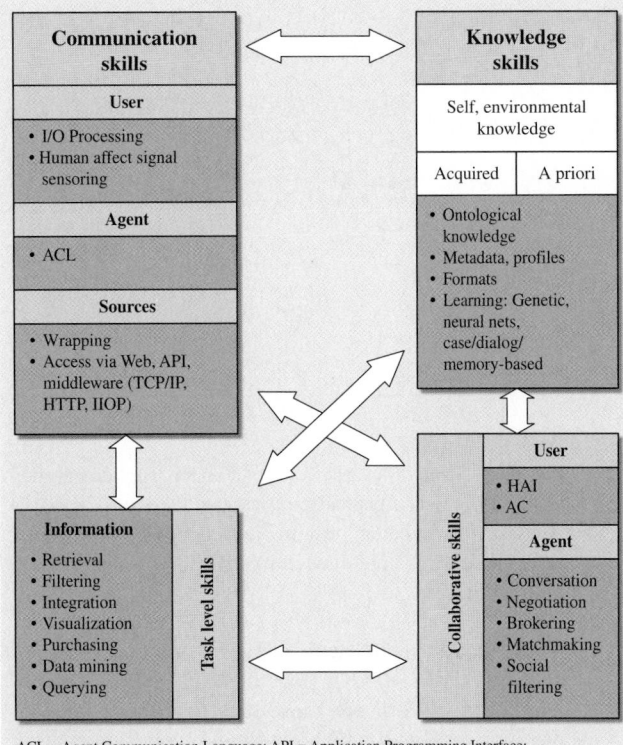

ACL = Agent Communication Language; API = Application Programming Interface; TCP/IP = Transmission Control Protocol/Internet Protocol; HTTP = Hypertext Transfer Protocol; IIOP = Internet Inter-ORB Protocol; HAI = Human Agent Interaction; AC = Affecting Computing

Fig. 88.6 Agent-mediated trading: intelligent agents and e-Business (after [88.15])

88.2.3 Protocols

Protocols provide an efficient and effective platform for autonomous entities (such as agents) to interact, be productive, and achieve their goals. Many different types of coordination protocols have been designed to achieve effective coordination among agents, such as (1) dependence-based coordination between tasks [88.17, 18] and (2) decision- and time-based protocols to better meet requirement changes and complex scenarios that require decisions beyond coordination [88.19–21]. As an example, *Ko* and *Nof* [88.21] compare a task administration protocol (TAP) which is a coordination and interruption–continuation protocol (CICP), with a non-TAP coordination protocol (CP) to examine which protocol enables better assignment of tasks in a human–robot team under emergency pressures. They conclude that the performance of a TAP-based coordination protocol outperforms the non-TAP coordination protocol according to the ra-

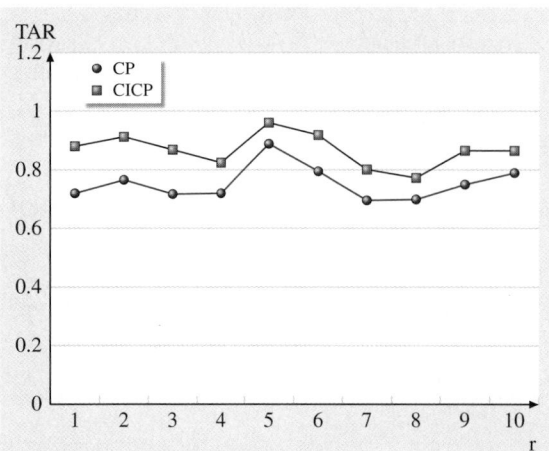

Fig. 88.7 Task allocation ratio (TAR) for emergency tasks [88.21]. The task administration protocol (CICP) always performs better because it is typically at a higher level of decision intelligence relative to the coordination protocol (CP)

tio of total unallocated tasks to total number of allocated tasks (TAR) (Fig. 88.7). The protocols of interest in e-Work are those that function at the application level and which determine workflow control. Protocols at this level enable effective collaboration by coordinating information exchange and decisions such as resource allocation among production tasks

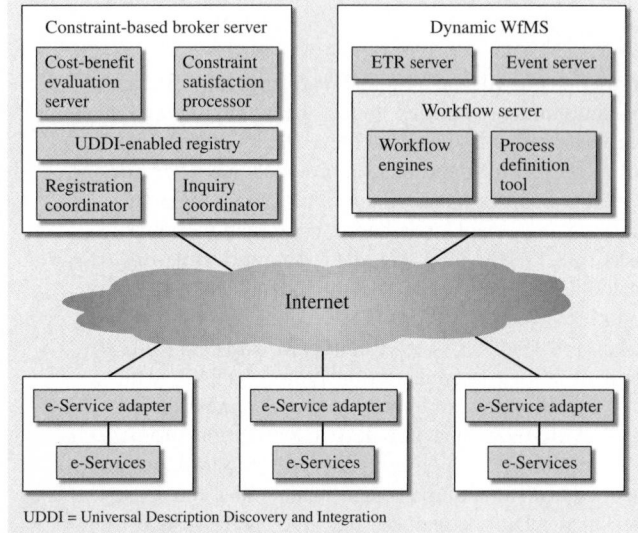

UDDI = Universal Description Discovery and Integration

Fig. 88.8 Dynamic workflow management system (WfMS) architecture (after [88.22])

to achieve a common goal and jointly generate better value. The unique advantage of TAPs is in smart e-Support and efficient, optimized streamlining of human collaboration.

88.2.4 Workflow

Workflow captures the systematic assignment and processing of activities in dynamic, distributed environments by enabling process scalability, availability, and performance reliability. Workflow technology has received a lot of attention in recent years since it enables businesses to better manage information and operations. One recent effort has focused on the design and implementation of a dynamic workflow management system to support e-Businesses. The system developed allows for interenterprise workflow management using e-Services for a network of collaborating enterprises [88.22]. Due to the dynamics of the Internet and the organizations using it, an organization can join, leave or remain in a collaborative network without much difficulty; however, regardless of the decision of a given enterprise to join/leave or remain in a collaborative network of organizations, all scenarios require that existing data resources, processes, and services be integrated or interfaced via workflow models. Their model was the first attempt to develop a dynamic architecture to better account for the nature of services and their availability at different instances in time (Fig. 88.8).

The scope of the second theoretical foundation area includes *integration, coordination, and collaboration* and comprises four e-Dimensions: (1) integration, coordination, and collaboration theory and models, (2) computer-integrated manufacturing (CIM), (3) human–computer interaction (HCI), and (4) extended enterprises, as described below.

88.2.5 Integration

Integration, coordination, and collaboration theory and models provide the backbone for (1) identifying existing and new connections between information resources and processing entities/agents and (2) developing models and methodologies to enhance the collaborative activities between participating entities/agents and their use of information resources. Various approaches have been used to address the challenges of integration, coordination, and collaboration in the e-World. One of the best measures to assess the effects of a higher level of integration, collaboration, and cooperation is the number of areas being influenced, among them:

e-Business [88.23, 24], healthcare [88.25, 26], transportation [88.27], construction [88.28], and more.

88.2.6 Computer-Integrated Manufacturing

Computer-integrated manufacturing (CIM) enables the harmonized integration and management of the entire product lifecycle from design to customer service. Modern business is centered on intra- and intercollaboration processes, such as supply chain/network management, customer relationship management, employee–business integration, and more. To fully garner the benefits of this high level of collaboration, organizations need to implement integrated-enterprise systems. For instance, *Ho* and *Lin* [88.29] focus on the need for better collaboration throughout the complete business cycle and not just focusing on technical integration. In their work, they incorporate critical success factors (i.e., implementation strategy, change management, etc.) into the development of the different stages and components of an integrated-enterprise system (Fig. 88.9). *Choi* et al. [88.30] also develop an integrated enterprise architecture, but their focus is on virtual enterprise chains.

Efforts such as these demonstrate the revolution that the collaborative enterprise world is experiencing with the evolution of collaborative e-Work principles, theories, and methodologies.

88.2.7 Human–Computer Interaction

Human–computer interaction (HCI) enables the development of systems, platforms, and interfaces that support humans in their roles as learners, workers, and researchers in computer environments. The majority of the work being conducted in HCI has focused on the methodologies and processes for developing interfaces, the design, implementation or evaluation of interfaces, and the development of theories and predictive models of interaction. As the roles of humans in today's workplace and society continue to evolve with the development of new technologies, so will the focal areas for HCI, among them ambient interfaces, affective computing, sensing interfaces, and tangible user interfaces [88.31]. For instance, in the work by *Altuntas* et al. [88.32], the authors present a formal approach to include a human material handler in a computer-integrated manufacturing (CIM) system.

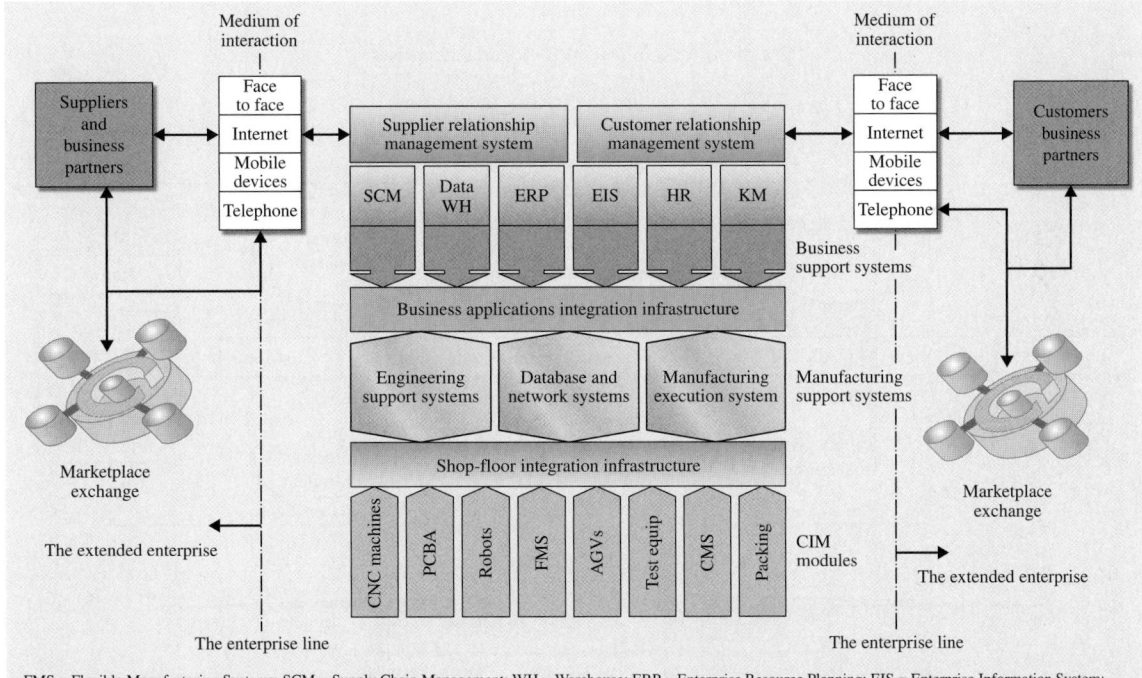

FMS = Flexible Manufacturing Systems; SCM = Supply Chain Management; WH = Warehouse; ERP = Enterprise Resource Planning; EIS = Enterprise Information System; HR = Human Resources; KM = Knowledge Management; AGVs = Automated Guided Vehicles; CMS = Content Management System; PCBA = Printed Circuit Board Assembly

Fig. 88.9 The integrated enterprise system reference architecture (after [88.29] with permission from Taylor and Francis)

Fig. 88.10 Microsoft Surface (human–computer interface) for Sheraton Hotels (courtesy of Microsoft)

As the authors point out, most of the work in control has focused on automata-based methods for complex computer-controlled systems, whose application has been limited as the number of unmanned systems is relatively small.

HCI applications are starting to cross into the mainstream; for instance, Microsoft recently released Surface, an interface that enables users instant access to information and entertainment via their hands and ges-

tures while improving collaboration and empowering consumers in the decision-making process (Fig. 88.10).

88.2.8 Extended Enterprises

Extended enterprises enable the creation of architectures, models, and methodologies to integrate business processes that go beyond the limits of a single enterprise and which include end-to-end functions, internal operations, and entire supply networks. The level of integration and connectivity of all value areas of supply networks has significantly increased with the advent of e-Business, e-Commerce, e-Services, and globalization. This trend has also increased the demand for new e-Work and e-Business applications and solutions to address the challenges faced when the world around us becomes an *e-World*. Figure 88.11 presents the transformation from traditional enterprises into extended enterprises, as defined by the extended products in dynamic enterprises (EXPIDE) project cluster from researchers of the European Information Society Technologies. The transformation of the enterprise serves as an example of the challenges and needs for the study and development of new methodologies, performance measures, and tools in the e-World.

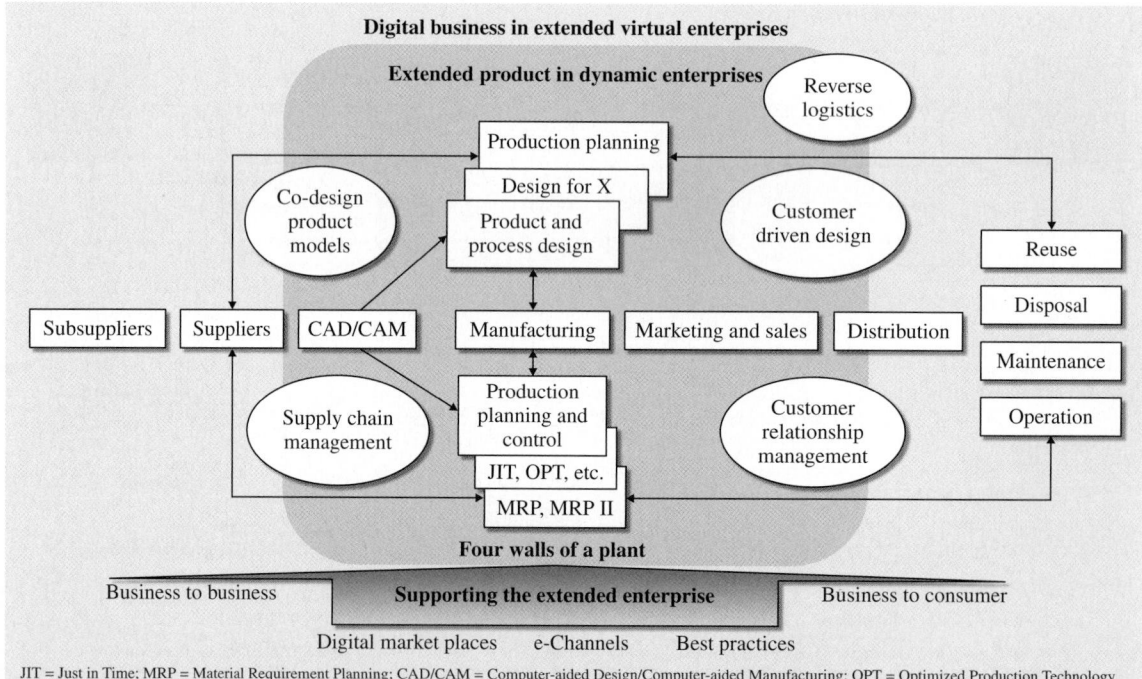

Fig. 88.11 Extended products in dynamic enterprises (EXPIDE) architecture (after [88.2])

The third theoretical foundation area is *distributed decision support* and comprises three e-Dimensions: (1) decision models, (2) distributed control systems, and (3) collaborative problem-solving, as described below.

88.2.9 Decision Models

Decision models integrate computer tools and methodologies in the implementation of decision theories in online decision support systems, to provide more effective and better-quality decisions. The development and implementation of decision models to function in the *e*-world spans a wide spectrum. For instance, in e-Finance, one model called create–collect–manage–protect (CCMP) extends the finance framework from solely providing the technology perspective to one that better enables strategic decision-making by making use of enterprise application integration (EAI) with common object request broker architecture (CORBA) and web services to demonstrate the added value of using a network approach to the digitalization of financial and business content [88.34]. In a differ-

ent application, fuzzy analytic hierarchy and fuzzy Delphi methodologies are used for multicriteria evaluation of e-Marketplace selection to deal with the uncertainty and vagueness of humans in group decision process, and therefore guarantee more effective decision making [88.35]. In a different application area, e-Manufacturing, an online quality information system (OQIS) using an object broker architecture along with a client–server enables its users both to obtain real-time data and information from any location and to address complex quality and deployment costs decisions (Fig. 88.12). The three examples of decision models described, along with their applications, highlight the influence and evolution of decision models needed in e-Business, e-Works and e-Service.

88.2.10 Distributed Control Systems

Distributed control systems empower systems with a greater degree of autonomy by enabling all parties (e.g., members of a supply network) to negotiate for the assignment/allocation of tasks and their respective

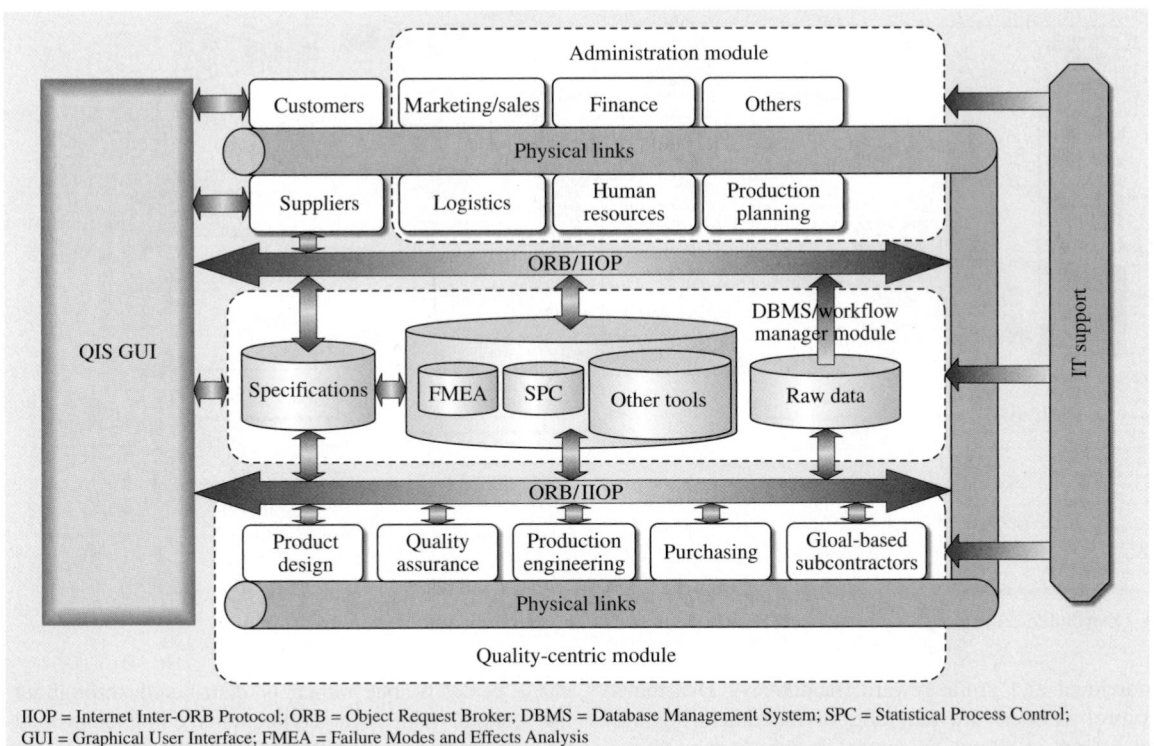

IIOP = Internet Inter-ORB Protocol; ORB = Object Request Broker; DBMS = Database Management System; SPC = Statistical Process Control; GUI = Graphical User Interface; FMEA = Failure Modes and Effects Analysis

Fig. 88.12 Online quality information system (OQIS) reference architecture (after [88.33])

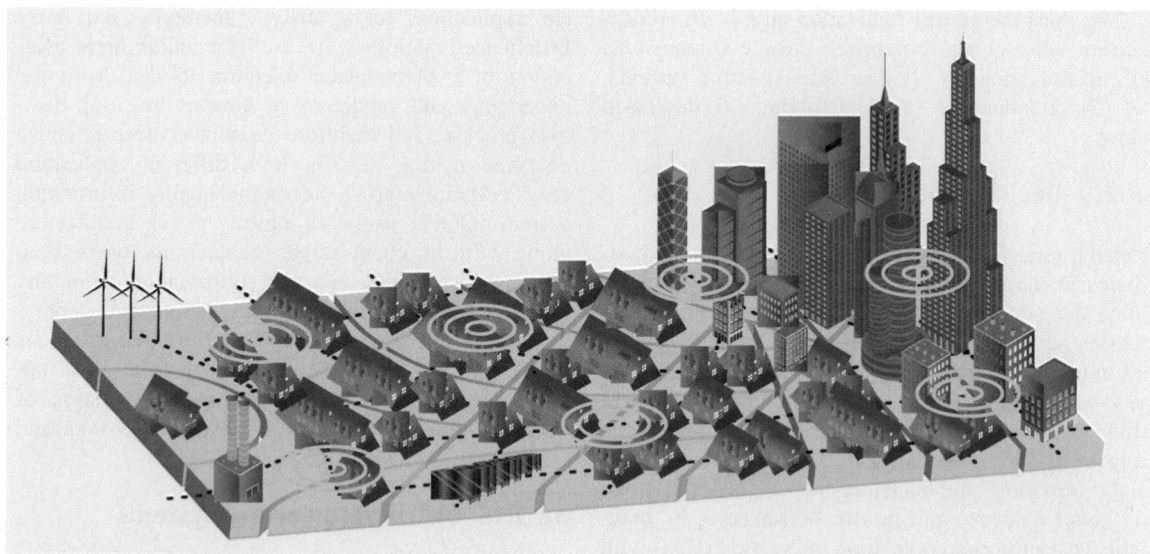

Fig. 88.13 SmartGridCity by Xcel Energy integrates high-speed communication technologies with the electric grid. The system enables real-time control by customers using devices installed at their homes to automate their home energy use and the use of integrated infrastructure such as plug-in hybrids, battery systems, solar panels, and more (courtesy of Xcel Energy)

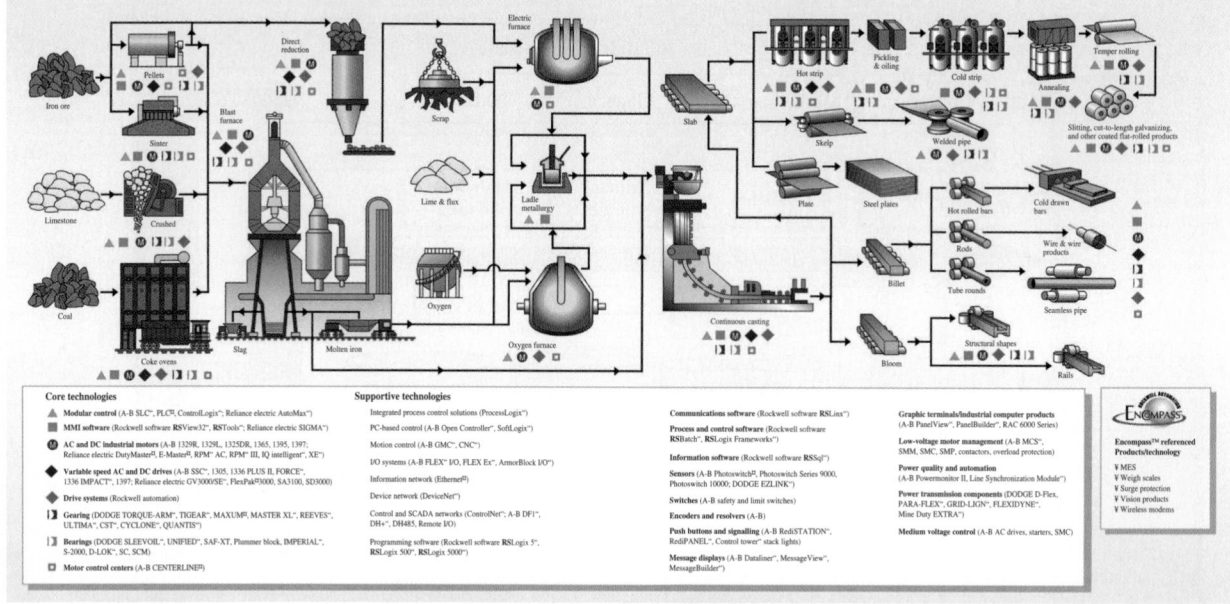

Fig. 88.14 Distributed control system in the steel industry (courtesy of Rockwell Automation, Inc.)

individual and group reward (incentives). Distributed control systems are common in manufacturing systems and in other processes in which the control is more effective and responsive when it is not lo- cated centrally but rather is distributed throughout the system, with local components being controlled by one or more controls. Typical examples of dis- tributed systems include: electrical power grids and power plants (Fig. 88.13), water management systems, oil refining systems, sensor networks, pharmaceutical manufacturing, cement process (Fig. 88.14), and more.

88.2.11 Collaborative Problem-Solving

Collaborative problem-solving systems provide the necessary information exchange methodologies and models to influence parties (e.g., members of a supply network, or emergency response network, Fig. 88.15) in their decision-making processes so as to achieve harmonization, coherence in decisions, and better results (e.g., financial, customer service rankings) for all involved parties. For example, in the work by *Farand* et al. [88.37] the authors examine collaborative clinical problem-solving in the context of telemedicinal consultations in order to assess the degree to which the technological environment preserves medical reasoning under traditional settings. A theoretical framework and case study were implemented to measure the achievement of the goals defined. Another example in the medical field applies Bayesian networks to model individual student and group knowledge to better define problem-based learning (PBL) methodology for the teaching of clinical reasoning skills. The system developed, collaborative medical tutor (COMET), is an intelligent tutoring system for collaborative medical PBL [88.38].

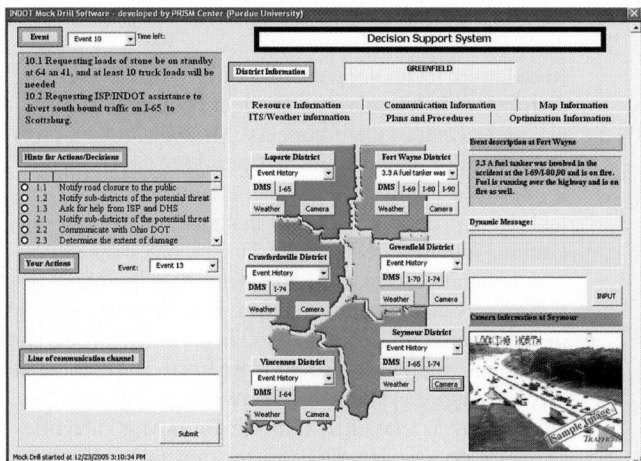

Fig. 88.15 Mock/drill training interface for collaborative and distributed problem solving. The tool was designed to train transportation employees for emergency response. Trainees are able to interact with different users, see their actions, create their own action plans and decisions, request resources, and have access to a variety of organizational information needed to address the task at hand (after [88.36])

The fourth theoretical foundation area is *active middleware* and comprises four e-Dimensions: (1) middleware technology, (2) distributed knowledge systems, (3) grid computing, and (4) knowledge-based systems, as described below.

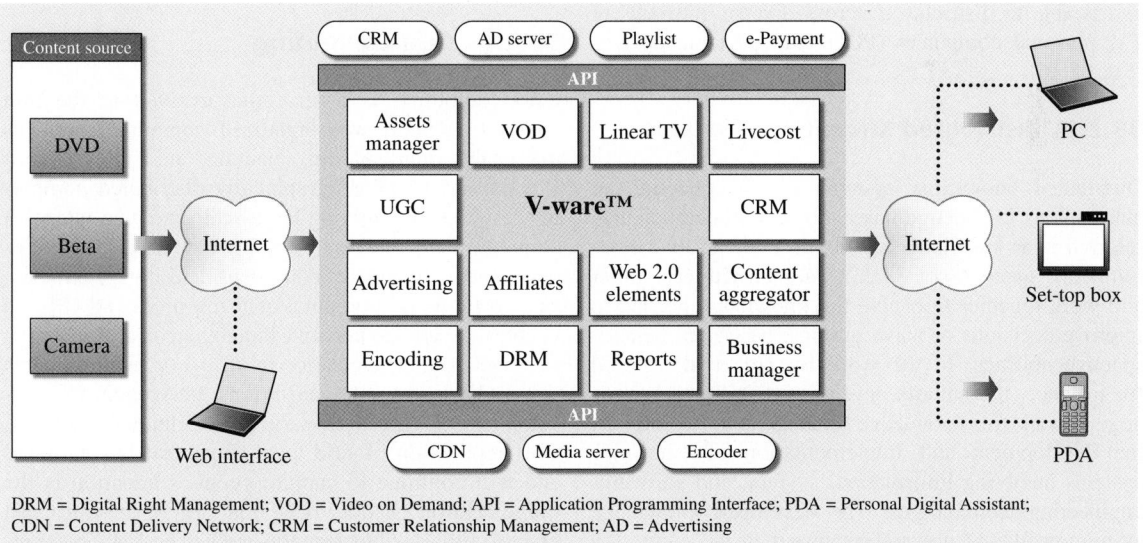

DRM = Digital Right Management; VOD = Video on Demand; API = Application Programming Interface; PDA = Personal Digital Assistant; CDN = Content Delivery Network; CRM = Customer Relationship Management; AD = Advertising

Fig. 88.16 v-Ware middleware architecture (printed with permission from Best TV Video Management Solutions)

88.2.12 Middleware

Middleware technology enables interoperability among distributed, heterogeneous applications, including legacy and advanced environments. Middleware is a layer of services or applications that provides the connectivity between the operating systems and network layers, and the application layer. For instance, in its basic operation, middleware enables the smooth connectivity and communication, and therefore functionality, between an older computer and a new scanner, printer or other peripheral components. In more advanced applications, middleware enables interoperability among more complex systems and networked organizations. Middleware is classified into the following categories of its e-Services according to the functions they provide: (1) distributed tuples which function as distributed relational databases, the most widely deployed middleware, (2) remote procedure call (RPC), which enables programmers to invoke a procedure across a network, (3) message-oriented middleware (MOM), which provides access to a message queue across a network, and (4) distributed object middleware, which provides the abstraction of an object that is remote but whose methods can be recalled just like those of an object in the same local address as the user. An example of a middleware technology concerning video applications is V-Ware, a middleware solution for online video services that supports video on demand (VoD), linear TV, user generated content (UGC), and live broadcasting, and is able to distribute it across diverse networks of TV, personal computers (PCs), and portable devices (Fig. 88.16).

88.2.13 Distributed Knowledge Systems

Distributed knowledge systems are a collection of autonomous knowledge-based objects sometimes also referred to as knowledge agents. In collaborative environments, agents (Sect. 88.2.2) interact with each other and work together to evolve the knowledge needed to improve decisions or solve queries, based on their respective abilities. In the work by *Ho* et al. [88.39] for instance, the authors develop a distributed knowledge model for knowledge management to support the development and implementation of enterprise systems involving information, system, and software-engineering technologies. The developed knowledge systems need to be updated because they are required to maintain high reliability, integrity, and quality of service for large-scale implementations.

Fig. 88.17 View inside a server tape Storateck in the computer center for the grid technology of the Large Hadron Collider (courtesy of CERN)

88.2.14 Grid Computing

Grid computing is a term that evolved in the mid 1990s to denote well-organized computing services that enable rapid sharing, selection, and aggregation of a wide variety of geographically distributed computing resources for solving large-scale and data-intensive computing applications. A computational grid provides dependable, pervasive, consistent, and inexpensive access to high-end computational resources [88.40] and enables its users to perceive heterogeneous resources as homogeneous. The creation of virtual (v-)environments (i. e., v-Factories, v-Organizations, etc.) requires the development of the necessary grid technology [88.41]. One application of grid technology that has garnered and will continue to capture people's attention is the Large Hadron Collider (LHC) at CERN, the European Organization for Nuclear Research. The LHC computing grid (LCG) will be one of the largest data-intensive applications in the world, combining and analyzing

Table 88.1 MeDICIS components (after [88.42], with permission from Elsevier)

Levels of MeDICIS	Components of MeDICIS	Unified modeling language (UML) element	Method for analyzing and structuring knowledge (MASK) elements	Other sources of inspiration
Macro	Business model	Case diagrams; Class diagrams		Unified modeling methodologies (UMM), Unified enterprise modeling language (UEML), Universal description discovery and integration (UDDI)
	Cooperation model	Case diagrams; Class diagrams		UEML
	Agent model	Class diagrams		Common knowledge acquisition and documentation structuring (KADS)
Micro	Communication model	Class diagrams; Sequence diagrams		Multiagent systems (MAS)
	Coordination model		Concept model; Task model	Structured analysis and design technique (SADT), UML, MASK; workflow: Integrated definition methods (IDEF3)
	Collaborative problem solving (CPS) model		Concept model; Task model	
Tool specification		Case diagrams; Component diagram; Deployment diagram		Others

data from the LHC detectors for over 10 000 scientists and over tens of thousands of computers around the world (Fig. 88.17). The computing grid will sift through petabytes (a million gigabytes) of particle collision data to uncover clues about the origins of the Universe.

From a business and service point of view, *Brocke* et al. [88.43] discuss the alignment of business needs of today's organizations and the availability of grid technology to better support virtual organizations. The authors argue that the collaborative reference model is a promising technique to enable the previously mentioned alignment by supporting distributed agents and the integration of technological, methodological, and organizational aspects.

88.2.15 Knowledge-Based Systems

Knowledge-based systems provide the capabilities to mine, refine, construct, and extract information from a variety of sources, enabling its transformation into knowledge capital. For instance, *Boughzala* [88.42] introduces the methodology for designing interenterprise cooperative information system (MeDICIS), an information system that models cooperative processes at three levels (communication, coordination, and collective problem-solving) in order to manage the knowledge available and generated by the cooperation. The development of MeDICIS highlights the existing interconnections between the many different dimensions of e-Work, as presented before, as it integrates, to name just a few (1) agents, (2) collab-orative problem-solving, and (3) extended enterprises (Table 88.1).

88.3 Design Principles for Collaborative e-Work, e-Business, and e-Service

The flattening of the world [88.44] has significantly changed the ways in which business and commerce are conducted in the new world landscape. To a large extent, this has been enabled by automation, from communication and digital information exchange to extended supply and logistics networks, international transportation, and more. As a result of the flattening effect, work has become more distributed and decentralized and the magnitude and distribution of worldwide markets has significantly increased. These changes have brought opportunities for the development of better e-Work methods, augmentation of human physical, cognitive, temporal, and location abilities at work. However, just as the opportunities have expanded, so have the challenges and complexities associated with e-Work, particularly the scalability of workflow, information, and task overload. A number of projects that have attempted to understand and address these new requirements and challenges, and in turn develop useful theories and solutions, are summarized in Table 88.2.

88.3.1 e-Work Design Principles

The growing number of distributed and decentralized work systems over the last two decades has given e-Work the foundation for the exploration and development of new theories, models, and methodologies. The environment in which e-Work research has evolved has shown the complex needs and challenges that lie ahead for the effective development and implementation of e-Work. e-Work brings new and exciting opportunities for the definition of emerging and enhanced work methods and systems, as well as better outcomes and higher yields, by augmenting with e-Support the physical, cognitive, temporal, and locational human abilities. Research on collaborative e-Work, conducted at the Production, Robotics, and Integration Software for Manufacturing and Management (PRISM) Center at Purdue University to understand the needs and challenges, and develop appropriate theories, and solutions is highlighted in Table 88.2. In the table, many of the 15 dimensions introduced in Sect. 88.2 are addressed in various manufacturing, production, and service ap-

plication domains. This research has been rooted in successful, proven principles and models of heterogeneous, autonomous, distributed, and parallel computing and communication.

Several collaborative e-Work design principles, which have emerged over the years as useful results of the collaborative control theory, are described in this section.

Principle of Cooperation Requirement Planning (CRP)

Originally developed by *Rajan* and *Nof* [88.45, 46], this principle considers collaboration as one of the most powerful augmentations of work abilities. Collaboration can be measured in a spectrum from minimal information sharing and exchange (implying a certain degree of cooperation) to fully integrated and collaborative enterprises. In order to achieve effective collaboration in any portion of the spectrum it is necessary to plan ahead. The *principle of cooperation requirement planning* includes two phases: (1) a detailed requirement plan of *who, how, and when* (CRP-I) is generated based on work objectives and available resources; (2) during execution, real-time implementation enables the revision of the plan to CRP-II, which meets spatial and temporal challenges, changes, and constraints. More recent work on this principle has included the integration of CRP with error diagnostics, recovery, and conflict resolution (Fig. 88.18) and the use of best-matching protocols for the identification and further implementation of *best policies* to address and correct errors and conflicts in robotic assembly and disassembly (Fig. 88.19). Future extensions will need to incorporate fuzzy logic and learning strategies (i. e., neural networks, genetic algorithms) to further enable real-time changes and learning. The effective implementation of this principle requires both advanced and adaptive real-time planning in order for the cooperation and collaboration efforts to be fruitful and efficient.

Principle of e-Work Parallelism

Originally developed by *Ceroni* and *Nof* [88.47, 48], this principle exploits the fact that work and interac-

Table 88.2 PRISM Center discoveries of e-Work theories, models, and benefits

CCT guideline	Collaborative e-Work benefits	Principles used	Production/business/ service areas	Results: tool/model
Cooperation requirement planning (CRP)	Collaboration planning and interaction	Resource planning	Multirobotic assembly Multiprocessors	CRP [88.45] CRP [88.46]
	Multiagents design	Agent theory	Manufacturing operations	ABMS [88.19, 49]
Parallelism	Collaboration protocol design	Telecomm. protocols Adaptive control Comm. protocols Exchange protocols	ERP/CRM applications Electronics testing Wireless MEMS Supply networks	TIE/P [88.50–52] TestLAN [88.53] TIF [88.54] BMP [88.55]
	Middleware protocols	Client–servers	Automotive electronics Flexible assembly	RAP [88.56, 57] TOP [88.58, 59]
	Parallelism	Parallel computing	Global shoe design & mfg.	DPIEM [88.47, 48]
	Resource and task allocation	Local-area networks Internet	Electronics assembly & test Global automotive supply net.	TestLAN [88.60, 61] MEN method [88.62]
Conflict and error detection and prevention (CEDP)	Synchronize/ resynchronize	Agent theory	Robotic maintenance	ServSim [88.63]
	Information assurance	Total quality	Agent-based mfg./service	(MERP) [88.64]
	Error detection and prevention	Computer recovery Multiagents	Robotic assembly Multirobot systems	NEFUSER [88.65] EDPA [88.66]; CEPD [88.67]
Fault tolerance	Fault-tolerant integration (operate despite faults)	Sensor fusion	Flow MEMS sensors Wireless MEMS sensors	FTTP* [88.68–70] TIE/MEMS [88.71]
	Conflict resolution	Telecommunication Coassembly	Facility design by a team Multirobot systems Assembly/disassembly	FDL [88.72] FDL-CR [88.73, 74] CRP-CR [88.75] CRP-BMP [88.76]
Join/leave/remain	Multienterprise optimization	Network flow	Distributed networked Enterprises	MEN Opt. [88.62] JLR [88.77]
	Organizational learning	Enterprise computing	Manufacturing/ assembly corp.	CMS [88.78]
Lines of collaboration and command (LOCC)	Workflow integration and coordination	Data flow Distributed database Workflow protocols	Aerospace mfg. Computer-integrated mfg.	DFI [88.79] DAFNet & AIMIS [88.80, 81] BMP [88.82]
	Information sharing and collaboration	Virtual environments Task graphs Network computing Internet/intranet	Manufacturing cells Distributed designers Distributed teams Supply networks Construction supply chain Production and sales	FDL [88.83] IDM [88.84] Co-X Tools [88.85] Shared Process [88.24] Share Process [88.86] T-C-M [88.87]
	e-Learning/e-training	Learning theory Distributed DSS	ERP applications Emergency response	MERP/C [88.88] TSTP [88.36]
	Viability measures	Artificial life	Human–robot facilities	TIE/A [88.89–91]
	e-Work scalability	Distributed computers	Supply networks	MEN Opt. [88.62]

Part I | 88.3

Table 88.2 (cont.)

ABMS: agent-based management system; AIMIS: agent interaction management system; BMP: best-matching protocol;
CMS: corporate memory system; Co-X: collaborative tool for function X; CEPD: conflict and error prediction and detection;
CRP: collaboration requirements planning; DAFNet: data activity flow network; DFI: data activity flow integration;
DPIEM: distributed parallel integration evaluation method; EDPA: error detection and prediction algorithms;
FDL: facility design language; FDL/CR: facility design language/conflict resolution;
*FTTP: fault-tolerance time-out protocol (*patent pending*); IDM: iterative design model; IRD: interactive robotic device;
JLR: join/leave/remain; MEN: multienterprise network; MERP/C: ERP e-Learning by MBE simulations with collaboration;
NEFUSER: neuro-fuzzy systems for error recovery; RAP: resource allocation protocol; ServSim: maintenance service simulator;
TestLAN: testers local area network; TIE/A: teamwork integration evaluator/agent; TIE/MEMS: teamwork integration evaluator/MEMS;
TIE/P: teamwork integration evaluator/protocol; TIF: data information forwarding; TOP: time-out protocol;
TSTP: transportation security training portal

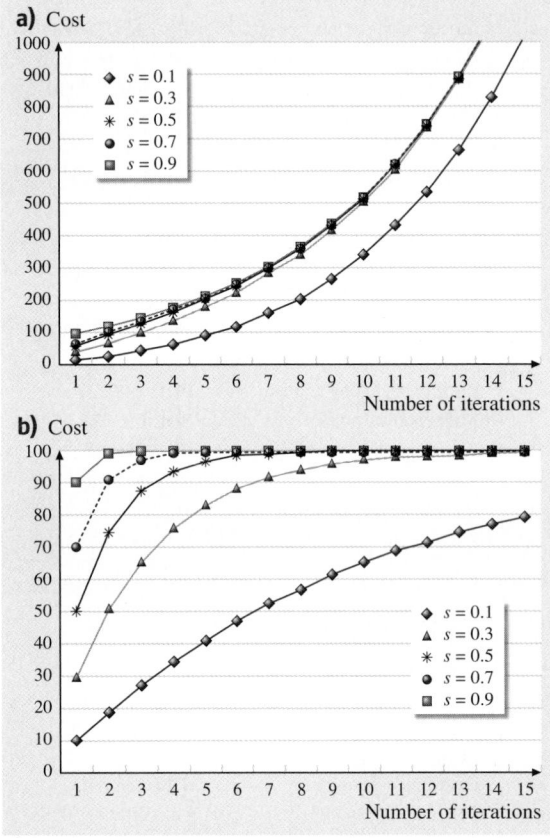

Fig. 88.18a,b Comparison of expected costs of conflict resolution (**a**) without conflict resolution methodology Mcr (unbounded cost), and (**b**) with conflict resolution methodology Mcr (bounded cost) (after [88.92]) (s = rate of conflicts)

tions in software and human workspaces can and must be allowed to run in parallel. To be effective, e-Work systems cannot be constrained by sequential (linear) tasks. In e-Work, the principle of workflow parallelism has deeper implications due to the distributed nature of the e-Activities and the fact that interactions take place over human and software spaces and can include (1) human–human interactions, (2) human–machine and human–computer interactions, and (3) machine–machine and computer–computer interactions.

Ceroni [88.47] defined the degree of parallelism (DOP) as the maximum level of resources and task parallelism necessary to balance the tradeoff between increased communication, transportation, and equipment costs and the increased productivity gained by the given level of parallelism. The authors developed the distributed planning of integrated execution method (DPIEM) as an effort to satisfy a CRP-based plan (Fig. 88.20). Among the issues explored in the planning was the ability to determine the optimal DOP, as seen in Fig. 88.21.

In addition to the definition of DOP, the *principle of e-Work parallelism* also addresses issues regarding the design and implementation of coordinated problem-solving and decision support that are of concern to the development of active middleware. The principle includes five guidelines to let e-Work support systems (EWSS):

1. Formulate, decompose, and allocate problems
2. Enable applications to communicate and interact under task administration protocols
3. Trigger and resynchronize independent agents to act coherently and cohesively in addressing problems, making decisions, and acting
4. Enable agents to reason, interact, and coordinate with other agents when needed
5. Develop conflict resolution, error recovery, and diagnostic and recovery strategies.

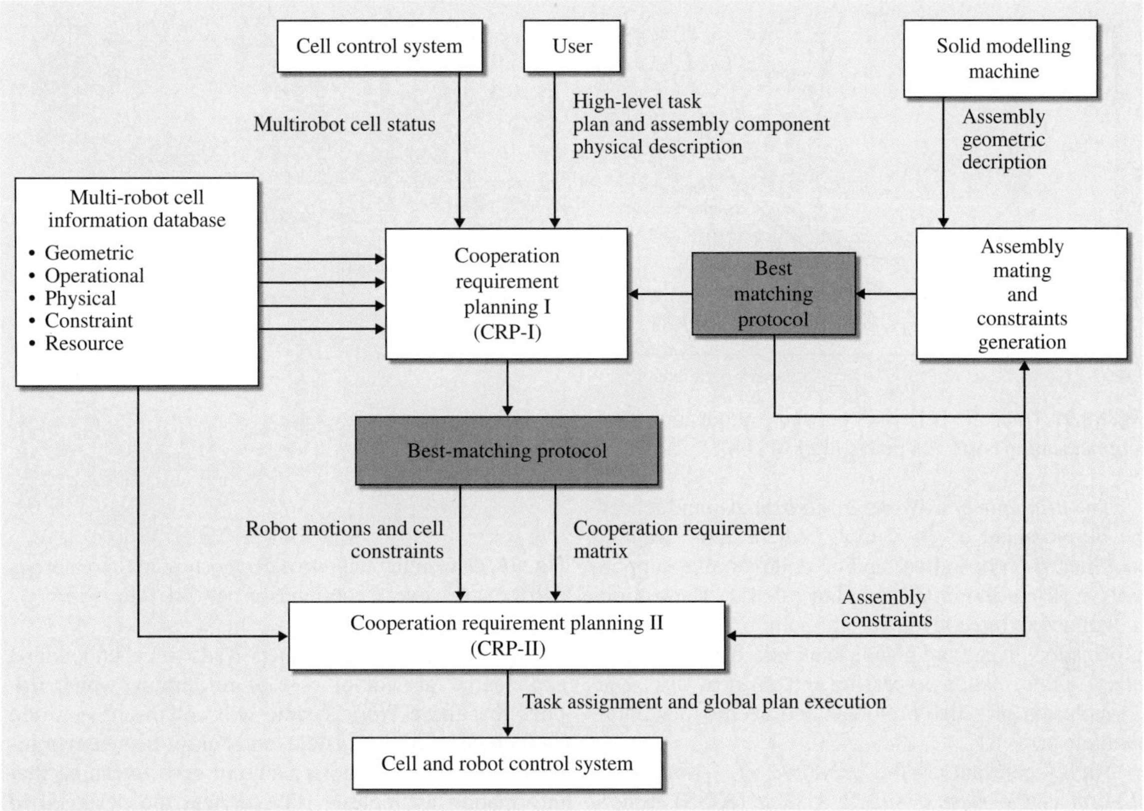

Fig. 88.19 Best-matching protocol-enabled cooperation requirement planning (CRP) architecture (after [88.76])

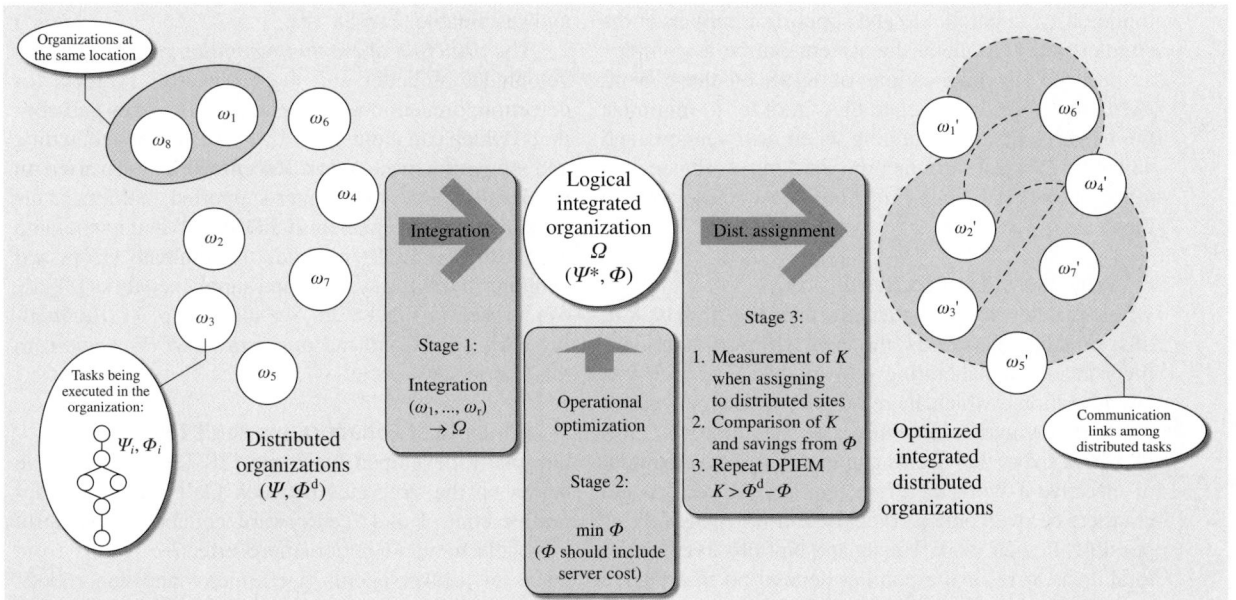

Fig. 88.20 Operational stages of the distributed planning of integration execution model (DPIEM) (after [88.48])

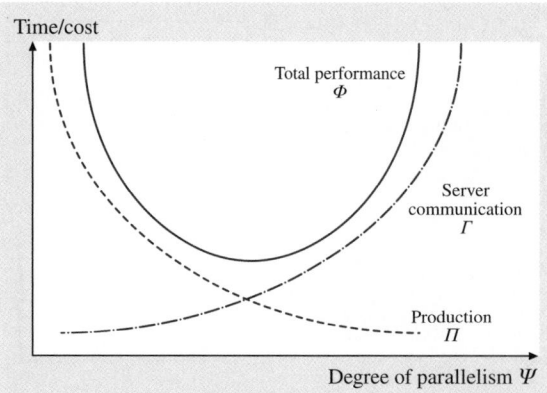

Fig. 88.21 Tradeoff behavior between production and communication costs and times (after [88.48])

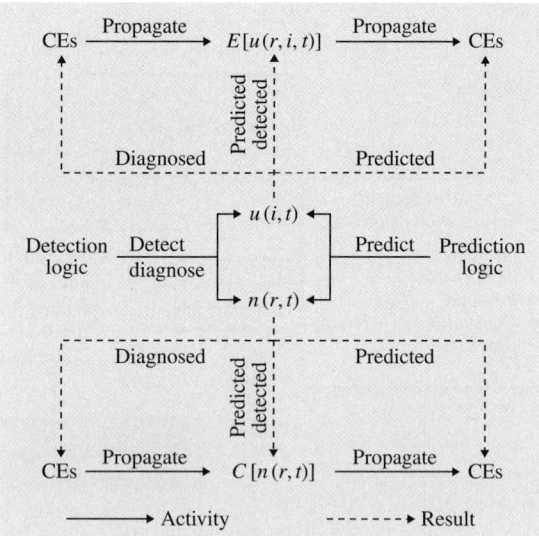

Fig. 88.22 Conflict and error diagnostics and prognostics (CEDP) logic over a coordination network (after [88.67])

The *principle of e-Work parallelism* also influenced the development of co-X tools, where "co-" implies coordinated, cooperative, and/or collaborative support tools, e.g., co-design and co-plan [88.85]. These tools are defined at three levels of interaction: (1) sharing of information, ideas, and plans via an interface, (2) parallelism of information processing and physical tasks, and (3) inclusion of active protocols and agents to support parallelism.

Highly relevant to the *principle of e-Work parallelism* is the *keep it simple system* (KISS) design guideline. It states that, as long as a computer and communication-support system is enabled to work autonomously, in parallel to and support of humans at the simplest level for them, the system can be as complex as needed. The implications of KISS on the e-Work *parallelism principle* dictate that, in order to minimize the need for human retraining when new versions and advanced functions are incorporated into software systems, internal parallel e-Functions must be included in the design.

Principle of Conflict Resolution

Originally developed by *Huang* and *Nof* [88.49, 89], this principle addresses the cost of resolving conflicts among collaborating e-Workers. A greater rate of interactions, which increases the number of active collaborating parties, also increases the number of conflicts and errors that agents can incur. The development of effective e-Work therefore requires that errors and conflicts be overcome as quickly and inexpensively as possible. In their work, Huang and Nof observe that the total costs to resolve a conflict depend on the relative portion of human intervention as follows: (1) a greater

number of errors and conflicts leads to an unbounded exponential growth of cost of resolution, which implies that the e-Work system will collapse due to the inefficiencies of the system, and (2) for less human intervention, approaching a value of zero (meaning that information technology (IT) services are designed to and applied in parallel), the total cost of resolution reaches an upper bound and can therefore be effective and sustainable (Fig. 88.18).

The *principle of conflict resolution* requires the development of better and more powerful IT tools for detection, prevention, and resolution of errors and conflicts, which can yield lower overall costs, better quality, and more effective e-Work. Recent work has focused on the development of computer-supported conflict and error detection management (CEDM) and diagnostics and prognostics (CEDP) methods to eliminate errors and conflicts in robotic systems and supply networks [88.66, 67] as seen in Fig. 88.22. See also Chap. 30 (*Automating Error and Conflict Prognostics and Prevention*) in this handbook.

Principle of Collaborative Fault Tolerance

Originally developed by *Jeong* [88.71], this principle builds on the synergies between fault-tolerant collaborative control and feedforward collaborative control protocols to yield better, more effective results from teams of weaker agents (i. e., micro- and nanorobots, micro- and nanosensors) in comparison with a single optimized faultless agent (Figs. 88.23 and 88.24). The principle builds on the well-known fact that a collaborative team is better than the sum of the individuals, and it addresses how to achieve this advantage by using smart automation.

The Join/Leave/Remain (JLR) Principle in Collaborative Organizations

This principle addresses the increasing number of virtual enterprises and the dynamic nature of enterprise alliances, self-organizing agent teams, sensor clusters, modular systems, and others [88.94]. The principle enables the identification of conditions and timing for individual agents (parties) or organizations to join, leave or remain as part of a collaborative networked organization (CNO). Among the questions the principle addresses are:

1. When and why would an organization join a given CNO?
2. What are the benefits for an individual agent or organization to be part of a given CNO?
3. What are the costs for an individual agent or organization to participate (join or remain) in a given CNO?
4. Why would an individual agent or organization opt to remain in a given CNO?
5. What are the criteria and performance evaluation measures to characterize and assess the effectiveness and performance of the CNO?

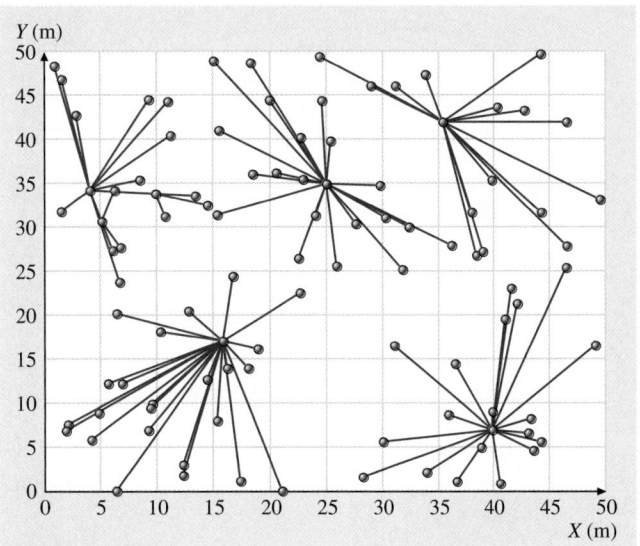

Fig. 88.23 Faulty sensor routed communication by time-based control. The *lines* between nodes represent communication links (after [88.93])

The *join/leave/remain principle* can be analyzed at different levels such as (1) overall CNO level, (2) individual agent or organization level, (3) sub-CNO, and (4) multi-CNOs, therefore making it adaptive and useful to address many types of agglutination and clustering of agents or organizations (Figs. 88.25 and 88.26). Furthermore, the analysis can be expanded to include each

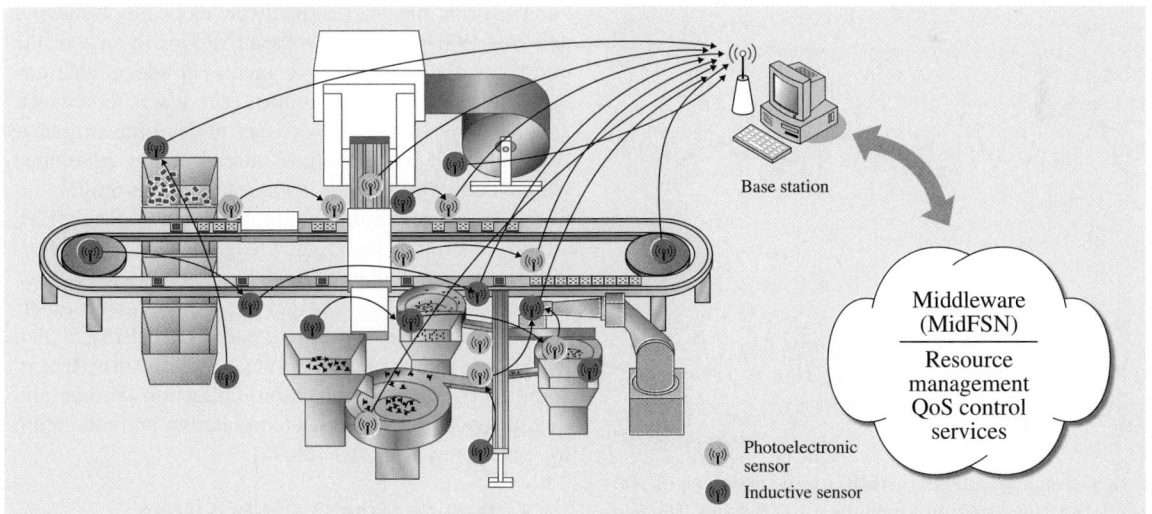

Fig. 88.24 Industrial application of a wireless microsensor network system with fault-tolerance capabilities (i. e., protocols) to aggregate weak sensor signals (after [88.54])

Part I | 88.3

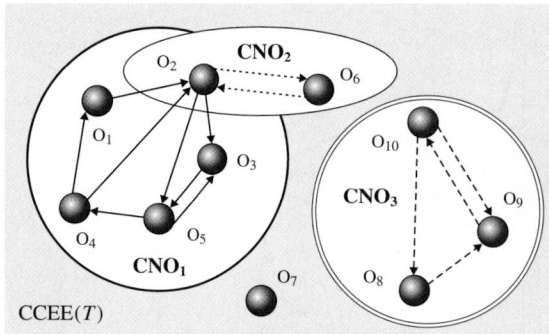

Fig. 88.25 Example of a collaborative networked organization (CNO) in a competitive environment at time T (after [88.77])

phase of the CNO life: creation, activity, dissolution, and support [88.77].

Principle of Emergent Lines of Collaboration and Command (LOCC)

The environment in which modern enterprises must conduct their day-to-day operations has become ever more challenging due to the dynamics of networks and the volatility of formal and informal communications within and between the members of the network, as well as with suppliers and customers. This principle focuses on three needs of organizations: (1) to have better collaborative decision-making, specifically under complex and dynamic events, (2) deliver information and build

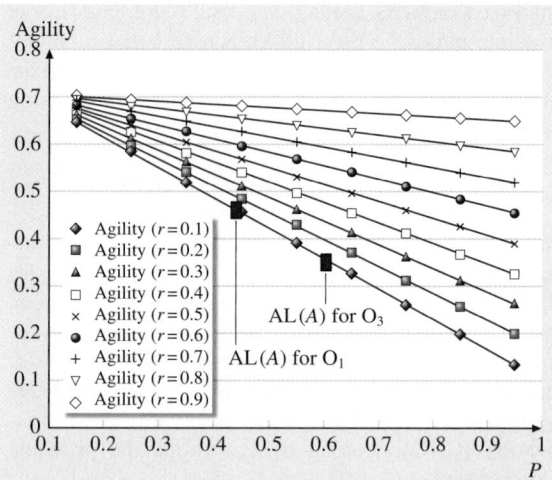

Fig. 88.26 CNO agility (A) under variable values for unexpected disturbance (P) and recovery (r) probabilities for a specific case (after [88.77])

knowledge in a timely manner without being cost prohibitive, and (3) enable the development of emergent networks. Emergent networks are defined as evolutionary mechanisms of interaction which build upon the well-established theories of organizational learning and that are characterized by ad hoc decisions, effective improvisation, on-the-spot creation of contacts, and best-matching protocols for pairing system alerts with decisions, decision-makers, and the executors of the decisions (Figs. 88.27 and 88.28).

The principle was originally developed by *Velásquez* and *Nof* [88.76] to enable organizations to answer the questions of (1) whom to contact, (2) under what circumstances to begin a contact, (3) when to contact, and (4) how to begin a contact under time pressure. A number of systems have already been developed which could be further enhanced by this collaborative e-Work principle. Examples are (1) SACHEM, which is a large-scale knowledge base for monitoring, alerting, and diagnosing dynamic processes in blast furnaces [88.95, 96], and (2) the public-health emergency response information system (PHERIS), which is a system developed to facilitate disease surveillance, command center for detection, collection of data and timely reporting, and resource planning and allocation for emergency response [88.97].

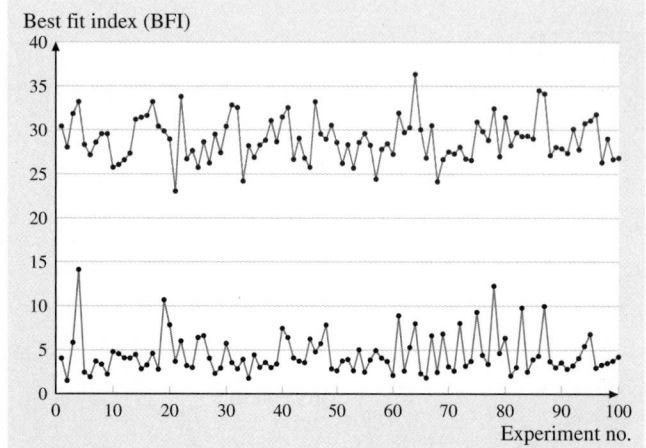

Fig. 88.27 Best-matching protocol (BMP) for the matching of 100 : 100 uniformly distributed parts in a manufacturing setting. The part-matching information is one of several factors then used to identify which supplier to match with a given customer

e-Measures: Criteria and Evaluation

With the advent of new work models, the need for new performance criteria and measures to assess the

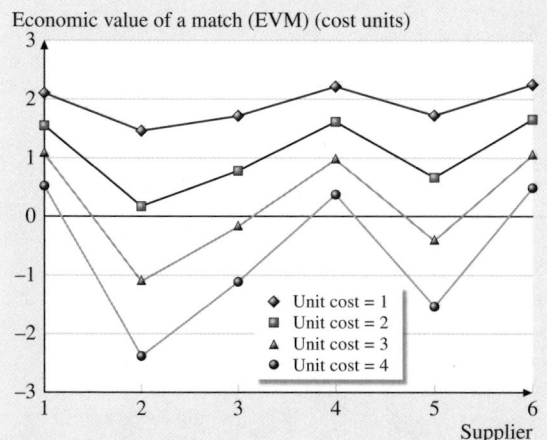

Fig. 88.28 Economic value of a match to identify which supplier out of six to match to a given customer according to the costs to manufacture and match parts (note: parts were generated following a normal distribution)

effectiveness of e-Work has increased. With the evolution of computer-integrated manufacturing (CIM) and networked enterprises, several measures have emerged, among them: flexibility, connectivity, agility (Fig. 88.29), integration ability, scalability, and reachability. Two recent performance measures developed at the PRISM Center are *viability* and *autonomy* [88.90]. *Viability* in the e-Work context has been defined as the ratio between operating and sustaining agents, and the rewards or benefits from their services. *Autonomy* implies the level of authority delegated and decen-

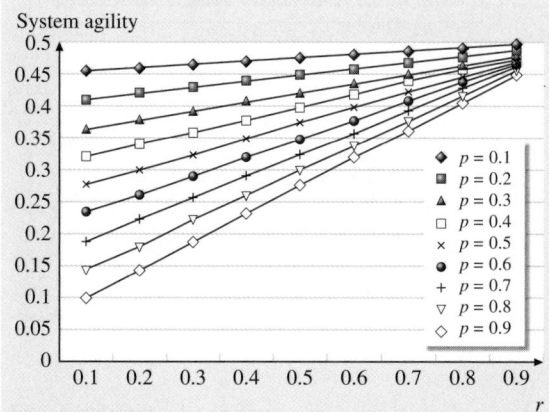

Fig. 88.29 System agility under various error probabilities p and recovery probabilities r (after [88.92])

tralization, and how active the agents, protocols, and other e-Work participants are. The development of new measures has not stopped and some of the emerging measures include: learning ability, collaboration ability, error and conflict detectability, error and conflict severity, and information significance.

The implications of the new measures of collaborative e-Work performance and characteristics have also been studied with the development of agent design and interaction theories. *Anussornnitisarn* [88.51] concludes that the performance of viability-based protocols exceeds that of protocols that did not take viability into consideration.

88.3.2 Emerging Collaborative e-Work, e-Business, and e-Service Design Principles

Bioinspired Collaborative Control

As the synergies between areas of research continue to evolve so will the theories and foundations of e-Work. In recent years, the biological sciences have started to influence collaborative control in e-Work, e-Business, and e-Service [88.98, 99]. Animal-to-animal interactions such as those observed among ants and bees have been adapted to enable the coordination of agents. The limited rationalizing capability of insects is enhanced by their ability to use pheromones as hints and therefore reduce information overloads. Similarly, in collaborative networks, the need to reduce complexity yet provide the appropriate *hints* to enable the agent to make a decision requires that only local and valuable information be shared by agents [88.100].

Bioinspired applications have included virtual insects (ants) as mobile software agents, message-based interactions among agents and robots, and measures such as viability (survivability) and autonomy (independence). Negotiation-based and bioinspired techniques are just two of the many useful techniques that have and will continue to influence multiagent collaborative control, as seen in Fig. 88.30.

Learning and Adaptation in Collaborative Control

Learning and other similar adaptive methodologies have played and will continue to play a critical role in automation and production/enterprise control. Genetic algorithms have been used, for instance, for the scheduling of complex products with multiple resource constraints and a multitier product structure (Fig. 88.31). The algorithms developed by *Pongcharoen*

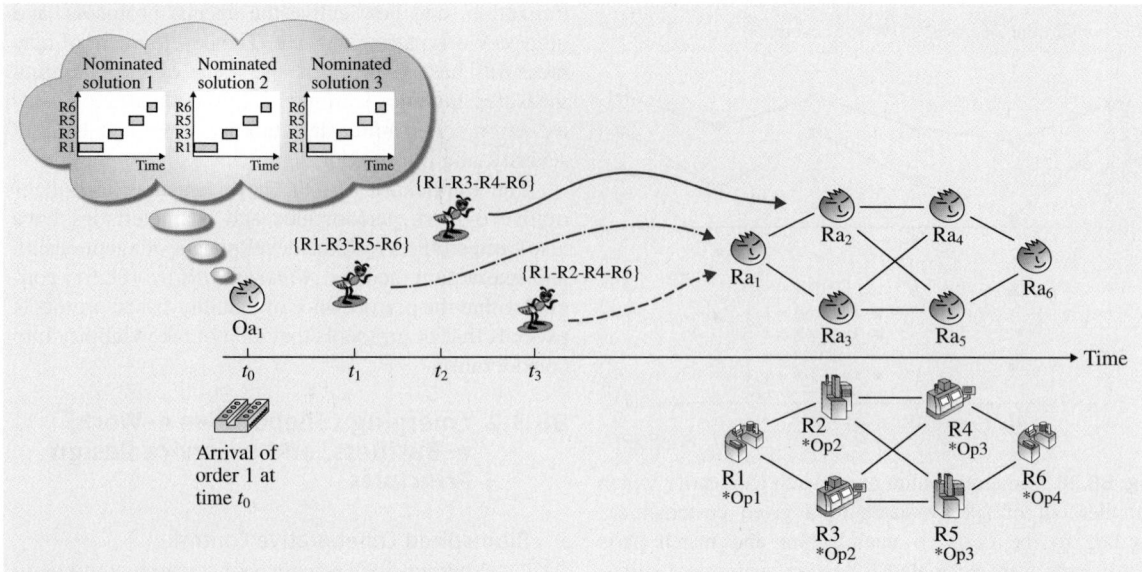

Fig. 88.30 Exploring mechanism of an ant agent in multiagent manufacturing control systems (after [88.102])

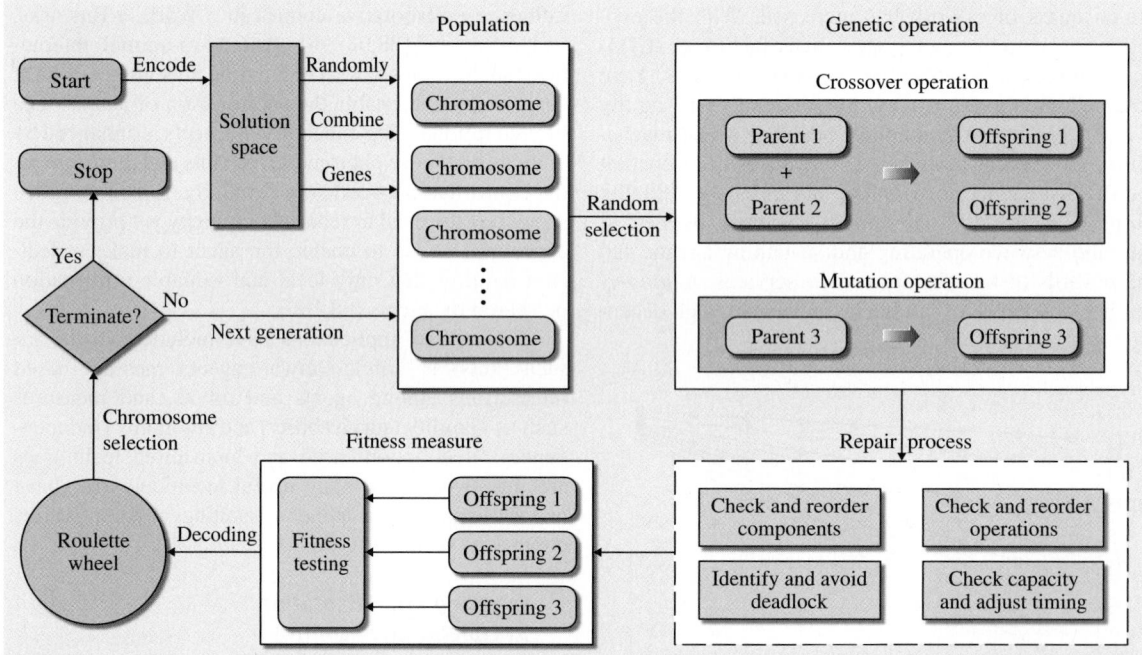

Fig. 88.31 Genetic algorithm structure for a production scheduling scenario (after [88.101] with permission from Elsevier)

et al. [88.101] achieve on-time delivery and a 63% reduction in costs when compared with simulation results produced by a participating company.

In another work that involves decentralized and interactive learning, Lagrangian relaxation in combination with genetic algorithms was used to coordinate

Fig. 88.32a–f Bioinspired network models of e-Work, e-Service, and e-Business are developed. (**a**) Channels (arteries) modify their shape and plasticity over time; (**b**) the surface of channels changes dynamically and may have multiple internal layers; (**c**) changes over time in network spatial orientation; (**d**) time-variable electromagnetic channels waves as observed via nanowires in mice; (**e**) network learning: structural plasticity has a role in behavior – channel rigidity allows free flow versus channel blockages which constrict flow; (**f**) network learning: rigid versus flexible channels (after [88.9]) ▶

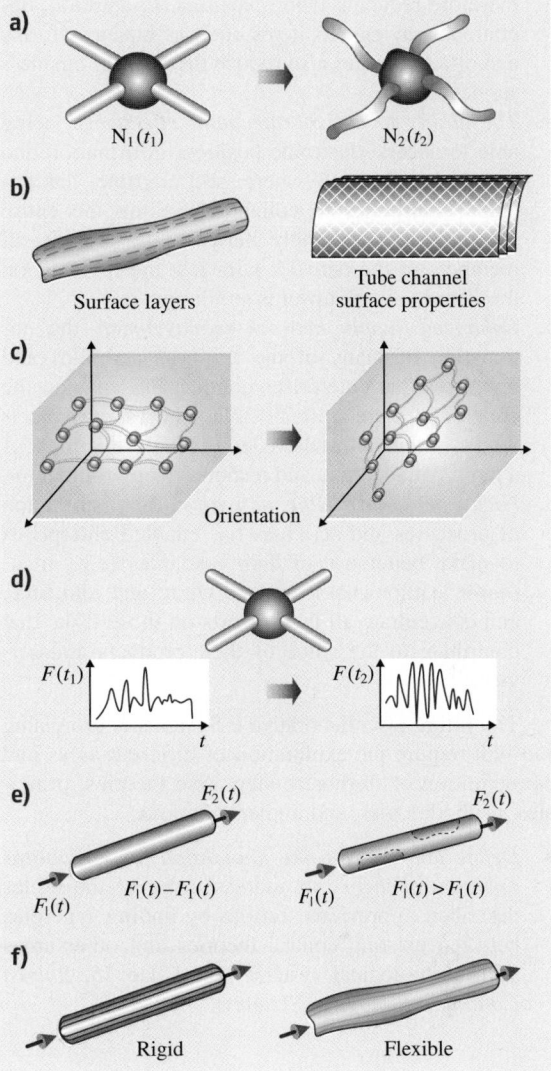

and optimize the production planning of independent partners linked by material flows in multitier supply chains [88.103]. Adaptation and learning are effective methodologies in market-based resource allocation to satisfy customer orders and needs [88.104, 105]. Studies at several laboratories around the world have found that adaptation by agents performs better at allocating and balancing resources and tasks.

Bioinspired Network Models

Recent work in biomembranes [88.106], neurology [88.107], and other areas reveals the potential of bioinspired mechanisms for adaptation, learning, and collaborative control. Just as synapses and spines collaborate to achieve survival and competitive animal behaviors, so can human-made enablers of collaborative control, such as adaptivity, dependability, discovery by data mining, decision support systems, and active middleware. The influence of biological collaborative control is still in its infancy and it is expected that new discoveries in these areas will lead to new models, theories, and solutions that satisfy the ever-growing needs of optimizing e-Work, e-Business and e-Service (Fig. 88.32).

88.4 Conclusions and Challenges

The methods by which humans, machines and computers interact and collaborate are rapidly being transformed in different areas of the economy, i. e., business, commerce, education, logistics, etc. Many of the benefits of e-Work, e-Business and e-Service have been the direct result of the automation of collaborative processes, activities and interactions that individuals and enterprises conduct in a day-to-day basis, among them:

1. *Error detection, reduction, and prevention*: the automation of processes enables enterprises to develop error mitigation and reduction strategies that guarantee better service, cost reduction, and faster response.
2. *Cost reduction of business and service processes*: the automation of business processes allows an enterprise to operate 24 h a day, 7 days a week without having employees work during some of the shifts,

therefore reducing labor expenses. In addition, operational savings on items such as business forms and office supplies also reduce the costs of business services.

3. *The ability to collaborate more effectively*: being able to access electronic business information and knowledge from anywhere, and anytime, has enabled enterprises to collaborate along the entire spectrum of the supply networks to provide all members of the network with just the information they need to work towards similar goals.

4. *Enhanced supply network management*: the automation of many of the business activities and processes that enterprises conduct has enabled the creation of more agile, flexible, and responsive networks, which in turn have led to the development of a more collaborative and responsive environment.

5. *Focus on added-value activities*: the automation of processes and activities has enabled enterprises to make better use of their resources (e.g., manpower, equipment) to reduce waste and idle time, and concentrate all their efforts on those tasks that contribute to the value of their products and services.

The future of collaborative e-Systems is promising and will require the exploration of different areas and the definition of further collaborative theories, principles, methodologies, and implementations:

- *Define and develop the foundation for collaborative control theory* to address many of the issues described in previous sections by finding synergies between existing control theories and other areas such as biological systems (see Chap. 75, *Robust Control in Biological Systems*).

- *Expand the breadth of investigation and implementation of collaborative e-Work principles and methodologies* to not only benefit and better enable large enterprises to compete and excel in the e-World but also small and medium-sized enterprises (SMEs). For example, in the book by *Corbitt* and *Al-Qirim* [88.108], studies on the implications of e-Business and e-Commerce principles to SMEs are summarized and their challenges are highlighted.

- *New ways of working and interacting in the e-World*. For instance, education in an e-enabled world (e-Learning and e-Training) requires the development of new skills, interfaces, teaching methods, and architectures so that activities such as e-Mentoring, e-Training, and others can yield the desired educational outcomes [88.109]. More information on e-Learning, e-Training, and serious games for teaching and learning can be found in Chaps. 85 (*Automation in Education/Learning Systems*) and 73 (*Automating Serious Games*), respectively.

- As organizations expand and become part of larger networks many questions regarding *optimal membership, clustering, and performance measures* (i.e., functional, economic, and quality) still remain to be addressed. Influenced by areas such as bioinspired networks, artificial intelligence, and bioinspired control, collaborative control will address these questions and move e-Work, e-Business, and e-Service into new frontiers.

- It is necessary to define and assess *new measures for production, service, and productivity* to better understand and improve the emerging *e-World* environment. Examples of such measures are collaboration-ability, recoverability, detectability, and more.

88.5 Further Reading

- L.N. de Castro: *Fundamentals of Natural Computing: Basic Concepts, Algorithms, and Applications* (Chapman Hall/CRC, New York 2006)
- H.M. Deitel, P.J. Deitel, K. Steinbuhler: *e-Business and e-Commerce for Managers* (Prentice-Hall, New Jersey 2001)
- P. Harmon, M. Rosen, M. Guttman: *Developing e-Business Systems and Architectures: A Manager's Guide* (Academic, San Diego 2001)
- S. Johnson: *Emergence: The Connected Lives of Ants, Brains, Cities and Software* (Scribner, New York 2002)

- M. Kevin, K.M. Passino: *Biomimicry for Optimization, Control, and Automation* (Springer, London 2005)
- M.A. Lara, J.P. Witzerman, S.Y. Nof: Facility description language for integrating distributed designs Int. J. Prod. Res. **38**(11), 2471–2488 (2000)
- P.B. Lowry, J.O. Cherrington, R.R. Watson (Eds.): *The e-Business Handbook* (St. Lucie, New York 2002)
- C. Ma, A. Concepcion: A security evaluation model for multi-agent distributed systems. In: *Technologies for Business Information Systems*, ed. by

W. Abramowicz, H.C. Mayr (Springer, New York 2007) pp. 403–415

- S. McKie: *e-Business Best Practices Leveraging Technology for Business Advantage* (Wiley, New York 2001)
- R.T. Rust, P.K. Kannan (Eds.): *e-Service: New Directions in Theory and Practice* (Sharpe, New York 2002)

- University of Michigan Digital Library Project: http://www.si.umich.edu/UMDL/
- J.-K. Won, K. Kwang-Hoon: Global-workflow modeling methodology and system for workflow-driven e-Commerce, Adv. Commun. Technol. **2**, 1012–1017 (2004)

References

88.1   A. Rosen: *The e-Commerce Question and Answer Book* (American Management Association, New York 2002)

88.2   J. Browne, P. Higgins, I. Hunt: e-Business principles, trends and visions. In: *e-Business Applications Technologies for Tomorrow's solutions*, ed. by J. Gasós, K.-D. Thoben (Springer, New York 2003) pp. 3–16

88.3   S.Y. Nof: Design of effective e-Work: Review of models, tools and emerging challenges, Product. Plan. Control, Special Issue on e-Work: Models and Cases **15**(8), 8–681 (2003)

88.4   L.M. Camarinha-Matos, H. Afsarmanesh: Collaborative networks: Value creation in a knowledge society, Proc. PROLAMAT 2006, IFIP Int. Conf. Knowl. Enterp. – New Challenges, Shanghai (Springer 2006)

88.5   L.M. Camarinha-Matos, H. Afsarmanesh: Concept of collaboration. In: *Encyclopedia of Networked and Virtual Organizations*, ed. by G. Putnik, M. Cunha (Hershey, Pennsylvania 2008)

88.6   S.Y. Nof: *Information and Collaboration Models of Integration* (Kluwer, Dordrecht 1994)

88.7   A. Cassidy: *A Practical Guide to Planning for e-Business Success: How to e-Enable Your Enterprise* (CRC, New York 2002)

88.8   S.Y. Nof: *Modeling of e-Work with TIE, a teamwork integrator evaluator. Research Memo 99-20* (School of Industrial Engineering, Purdue University 1999)

88.9   S.Y. Nof: Collaborative control theory for e-Work, e-Production and e-Service, Annu. Rev. Control **31**, 281–292 (2007)

88.10 S.Y. Nof: Availability, integrability and dependability – what are the limits in production and logistics, Proc. IFAC-MCPL (2007)

88.11 S.Y. Nof, F.G. Filip, A. Molina, L. Monostori, C.E. Pereira: Advances in e-Manufacturing, e-Logistics, and e-Service Systems. Milestone report, Proc. IFAC Congr. (Seoul 2008)

88.12 A.W. Colombo, F. Jammes, H. Smith, R. Harrison, J.L.M. Lastra, I.M. Delamer: Service-oriented architectures for collaborative automation, Proc. 31st Annu. Conf. IEEE Ind. Electron. Soc. (2005)

88.13 D.J. Bruemmer, D.A. Few, R.L. Boring, J.L. Marble, M.C. Walton, C.W. Nielsen: Shared understanding for collaborative control, IEEE Trans. Syst. Man Cybern. A **35**(4), 494–504 (2005)

88.14 T. Fong, S. Grange, C. Thorpe, C. Baur: Multi-robot remote driving with collaborative control, Proc. 10th IEEE Int. Workshop Robot–Human Interact. Commun. (2001) pp. 237–242

88.15 M. Klusch: Agent-mediated trading: Intelligent agents and e-Business. In: *Agent Technology for Communication Infrastructure*, ed. by A.L.G. Hazelden, R.A. Bourne (Wiley, New York 2001) pp. 59–76

88.16 S.J. Poslad, R.A. Bourne, A.L.G. Hayzelden, P. Buckle: Agent technology for communications infrastructure: An introduction. In: *Agent Technology for Communication Infrastructure*, ed. by A.L.G. Hazelden, R.A. Bourne (Wiley, New York 2001) pp. 1–31

88.17 J. Ferber: *Multi-Agent Systems: An Introduction to Distributed Artificial Intelligence* (Addison Wesley, Longman 1999)

88.18 S.S. Fatima, M. Wooldridge: Adaptive task and resource allocation in multi-agent systems, Proc. 5th Int. Conf. Auton. Agents (Montreal 2001)

88.19 C.-Y. Huang, S.Y. Nof: Formation of autonomous agent networks for manufacturing systems, Int. J. Prod. Res. **38**(3), 607–624 (2000)

88.20 C.-Y. Huang, S.Y. Nof: Evaluation of agent-based manufacturing systems based on a parallel simulator, Comput. Ind. Eng. **43**(3), 529–552 (2002)

88.21 H.S. Ko, S.Y. Nof: Modeling task administration protocols for human and robot e-Workers. In: *Handbook of Digital Human Modeling*, ed. by V. Duffy (Taylor & Francis, New York 2008)

88.22 S.Y.W. Su, J. Meng, R. Krithivasan, S. Degwekar, S. Helal: Dynamic inter-enterprise workflow management in a constraint-based e-Service infrastructure, Electron. Commer. Res. **3**(1/2), 9–24 (2003)

88.23 H. Li, Y. Fan, C. Dunne, P. Pedrazzoli: Integration of business processes in Web-based collaborative product development, Int. J. Comput. Integr. Manuf. **18**(6), 452–462 (2005)

88.24   N. Park: Collaboration and integration of the shared process system with workflow control, Prod. Plan. Control **14**(8), 743–752 (2003)

88.25   F. Malamateniou, G. Vassilacopopulos: Developing a virtual patient record using XML and web-based workflow technologies, Int. J. Med. Inform. **70**(2/3), 591–598 (2003)

88.26   G. Weerakkody, P. Ray: CSCW-based system development methodology for health-care information system, Telemed. J. e-Health **9**(3), 273–282 (2003)

88.27   M.A. Krajewska, H. Kopfer: Collaborating freight forwarding enterprises, OR Spectrum **28**(3), 3–301 (2006)

88.28   S.H. Han, K.H. Chin, M.J. Chae: Evaluation of CITIS as a collaborative virtual organization for construction project management, Autom. Constr. **16**(2), 2–199 (2007)

88.29   L.T. Ho, G.C.I. Lin: Critical success factor framework for the implementation of integrated-enterprise systems in the manufacturing environment, Int. J. Prod. Res. **42**(17), 3731–3742 (2004)

88.30   Y. Choi, K. Dongwoo, H. Chae, K. Kim: An enterprise architecture framework for collaboration of virtual enterprise chains, Int. J. Adv. Manuf. Technol. **35**, 1065–1078 (2008)

88.31   R.J.K. Jacob, A. Girouard, L.M. Hirshfield, M. Horn, O. Shaer, E. Treacy Solovey, J. Zigelbaum: CHI2006: What is the next generation of human-computer interaction, Interactions **14**(3), 3–53 (2007)

88.32   B. Altuntas, R.A. Wysk, L. Rothrock: Formal approach to include a human material handler in a computer manufacturing (CIM) system, Int. J. Prod. Res. **45**(9), 1953–1971 (2007)

88.33   R.J. Jiao, S. Pokharel, A. Kumar, L. Zhang: Development of an online quality information system for e-Manufacturing, J. Manuf. Technol. Manag. **18**(1), 36–53 (2007)

88.34   G. Ye, G. Keesling: e-Finance: the CCMP model, Int. J. Bus Perform. Manag. **8**(1), 1–36 (2006)

88.35   G. Büyüközkan: Multi-criteria decision making for e-Marketplace selection, Internet Res. **14**(2), 139–154 (2004)

88.36   S.W. Yoon, J.D. Velásquez, B.K. Partridge, S.Y. Nof: Transportation security decision support system for emergency response: a training prototype, Decis. Support Syst. **46**, 139–148 (2008)

88.37   L. Farand, J.-P. Lafrance, J.F. Arocha: Collaborative problem-solving in telemedicine and evidence interpretation in a complex clinical case, Int. J. Med. Inform. **51**(2/3), 153–167 (1998)

88.38   S. Suebnukarn, P. Haddawy: A Bayesian approach to generating tutorial hints in a collaborative medical problem-based learning system, Artif. Intell. Med. **38**(1), 1–5 (2006)

88.39   C.-T. Ho, Y.-M. Chen, Y.-J. Chen, C.-B. Wang: Developing a distributed knowledge model for knowledge management in collaborative development and implementation of an enterprise system, Robot. Comput.-Integr. Manuf. **20**(5), 439–456 (2004)

88.40   I. Foster, C. Kesselman (Eds.): *The Grid: Blueprint of a New Computing Infrastructure* (Morgan Kaufmann, San Francisco 2003)

88.41   J. Joseph, C. Fellenstein: *Grid Computing* (Prentice Hall, New Jersey 2004)

88.42   I. Boughzala: ICIS for knowledge management: the case of the extended enterprise. In: *Trends in Enterprise Knowledge Management*, ed. by I. Boughzala, J.L. Ermine (ISTE, Newport Beach 2006)

88.43   J.V. Brocke, O. Thomas, C. Buddendick: Conceptual modeling for GRID computing: applying collaborative reference modeling. In: *Technologies for Business Information Systems*, ed. by W. Abramowicz, H.C. Mayr (Springer, New York 2007) pp. 1–12

88.44   T.L. Friedman: *The World is Flat: A Brief History of the Twenty-first Century* (Straus and Giroux, Farrar 2005)

88.45   V.N. Rajan: Cooperation requirement planning for multi-robot assembly celss. Ph.D. Thesis (Purdue University, West Lafayette 1993)

88.46   V.N. Rajan, S.Y. Nof: Cooperation requirement planning (CRP) for multi-processors: optimal assignment and execution planning, J. Intell. Robot. Syst. **15**, 419–435 (1996)

88.47   J.A. Ceroni: Models of integration with parallelism of distributed organizations. Ph.D. Thesis (Purdue University, West Lafayette 1999)

88.48   J.A. Ceroni, S.Y. Nof: A workflow model based on parallelism for distributed organizations, J. Intell. Manuf. **13**(6), 439–461 (2002)

88.49   C.-Y. Huang: Formation and evaluation of agent networks for manufacturing systems. M.S. Thesis (Purdue University, West Lafayette 1995)

88.50   P. Anussornnitisarn, P. Peralta, S.Y. Nof: Time-out protocol for task allocation in multi-agent systems, J. Intell. Manuf. **13**(6), 6–511 (2002)

88.51   P. Anussornnitisarn: Design of middleware protocols for the distributed ERP environment. Ph.D. Thesis (Purdue University, West Lafayette 2003)

88.52   P. Anussornnitisarn, S.Y. Nof, O. Etzion: Decentralized control of cooperative and autonomous agents for solving the distributed resource allocation problem, Int. J. Prod. Econ. **98**(2), 114–128 (2005)

88.53   N.P. Williams, Y. Liu, S.Y. Nof: The TestLAN approach and protocols for the integration of distributed assembly and test networks, Int. J. Prod. Res. **40**(17), 4505–4522 (2002)

88.54   W. Jeong, S.Y. Nof: A collaborative sensor network middleware for automated production systems, Comput. Ind. Eng. **57**(1) (2009), Special Issue on Collaborative e-Work

88.55   J.D. Velásquez, S.Y. Nof: A best-matching protocol for collaborative e-Work and e-Manufacturing,

Int. J. Comput. Integr. Manuf. **21**(8), 943–956 (2008)

88.56 K. Esfarjani: Planning client-server integration protocols for test-work cells. M.S. Thesis (Purdue University, West Lafayette 1994)

88.57 K. Esfarjani, S.Y. Nof: Client-server model of integrated production facilities, Int. J. Prod. Res. **36**(12), 3295–3321 (1998)

88.58 J.I. Peralta: Evaluation of time-out protocols in manufacturing systems. M.S. Thesis (Purdue University, West Lafayette 1996)

88.59 J. Peralta, P. Annusornnitisarn, S.Y. Nof: Analysis of a time-out protocol and its applications in a single server environment, Int. J. Comput. Integr. Manuf. **16**(1), 1–1 (2003)

88.60 N.P. Williams: The effectiveness of protocol adaptability in TestLAN production environments. Ph.D. Thesis (Purdue University, West Lafayette 1999)

88.61 N.P. Williams, Y. Liu, S.Y. Nof: Analysis of workflow protocol adaptability in TestLAN production systems, IIE Transactions **35**(10), 965–972 (2003)

88.62 J. Chen: Modeling and analysis of coordination for multi-enterprise networks. Ph.D. Thesis (Purdue University, West Lafayette 2002)

88.63 J.E. Auer: Agent-based prediction of customer requirements for distributed stream service systems. M.S. Thesis (Purdue University, West Lafayette 2000)

88.64 T. Bellocci: Planning variable information assurance in agent-based workflow systems. M.S. Thesis (Purdue University, West Lafayette 2001)

88.65 J. Avila-Soria: Interactive error recovery for robotic assembly using a neural-fuzzy approach. M.S. Thesis (Purdue University, West Lafayette 1999)

88.66 X.W. Chen, S.Y. Nof: Error detection and prediction algorithms: Application in robotics, J. Intell. Robot. Syst. **48**(2), 2–225 (2007)

88.67 X.W. Chen, S.Y. Nof: Prognostics and diagnostics of conflicts and errors over e-Work networks, Proc. 19th Int. Conf. Prod. Res. (2007)

88.68 Y. Liu: Distributed micro-flow sensor arrays and networks: Design of architecture and fault-tolerant integration. M.S. Thesis (Purdue University, West Lafayette 2001)

88.69 Y. Liu, S.Y. Nof: Distributed microflow sensor arrays and networks: Design architectures and communication protocols, Int. J. Prod. Res. **42**(15), 3101–3115 (2004)

88.70 Y. Liu, S.Y. Nof: Fault-tolerant sensor integration for micro-flow sensor arrays and networks, Comput. Ind. Eng. **54**(3), 634–647 (2008)

88.71 W. Jeong: Fault-tolerant timeout communication protocols for distributed micro-sensor network systems. Ph.D. Thesis (Purdue University, West Lafayette 2006)

88.72 M.A. Lara: Conflict resolution in collaborative facility design. Ph.D. Thesis (Purdue University, West Lafayette 1999)

88.73 M.A. Lara, S.Y. Nof: Computer supported conflict resolution for collaborative facility designers, Int. J. Prod. Res. **41**(2), 2–207 (2003)

88.74 J.D. Velásquez, M.A. Lara, S.Y. Nof: Systematic resolution of conflict situations in collaborative facility design, Int. J. Prod. Econ. **116**, 139–153 (2008)

88.75 S.Y. Nof, J. Chen: Assembly and disassembly: an overview and framework for cooperation requirement planning with conflict resolution, J. Intell. Robot. Syst. **37**(3), 307–320 (2003)

88.76 J.D. Velásquez, S.Y. Nof: Best matching protocol for cooperation requirement planning in distributed assembly networks, Intell. Assem. Disassem. Conf. IAD'07 (2007)

88.77 C.M. Chituc, S.Y. Nof: The join/leave/remain (JLR) decision in collaborative networked organizations, Comput. Ind. Eng. **53**(1), 173–195 (2007)

88.78 K. Prytz, S.Y. Nof, A. Rolstadas: *Manufacturing Integration and Learning by Corporate Memory Systems (CMS). Research Memo 95-12* (School of Industrial Engineering, Purdue University, West Lafayette 1995)

88.79 P. Sauvaire, J.A. Ceroni, S.Y. Nof: Information management for FMS and non-FMS decision support integration, Int. J. Ind. Eng. Appl. Pract. Special Issue on Information Systems **5**(1), 1–78 (1998)

88.80 C.O. Kim: DAF-net and multi-agent based integration approach for heterogeneous CIM information systems. Ph.D. Thesis (Purdue University, West Lafayette 1996)

88.81 C.O. Kim, S.Y. Nof: Investigation of PVM for the emulation and simulation of a distributed CIM workflow system, Int. J. C. Integr. Manuf. **13**(5), 5–401 (2000)

88.82 J.D. Velásquez, S.Y. Nof: Integration of machine-vision inspection information for best-matching of distributed components and suppliers, Comput. Ind. **59**(1), 1–69 (2008)

88.83 J.P. Witzerman, S.Y. Nof: Tool integration for collaborative design of manufacturing cells, Int. J. Prod. Econ. **28**(1), 1–23 (1995)

88.84 N.K. Khanna, J.A.B. Fortes, S.Y. Nof: A formalism to structure and parallelize the integration of cooperative engineering design tasks, Des. Manuf. **5**(1), 78–87 (1998)

88.85 R.E. Eberts, S.Y. Nof: Tools for collaborative work, Proc. IERC 4 (Nashville 1995) pp. 449–455

88.86 D. Castro-Lacouture, M.J. Skibniewski: Applicability of e-Work models for the automation of construction materials management systems, Prod. Plan. Control **14**(8), 789–797 (2003)

88.87 M. Matsui, S. Aita, S.Y. Nof, J. Chen, Y. Nishibori: Analysis of cooperation effects in two-center production models, Int. J. Prod. Econ. **84**, 101–112 (2003)

88.88 J. Chen, S.Y. Nof: *Scalable multi-enterprise network. PRISM Symposium and Reunion* (School of

Industrial Engineering, Purdue University, West Lafayette 2001)

88.89   C.-Y. Huang: Autonomy and viability in agent-based manufacturing systems. Ph.D. Thesis (Purdue University, West Lafayette 1999)

88.90   C.-Y. Huang, S.Y. Nof: Enterprise agility: A view from the PRISM lab, Int. J. Agile Manag. Syst. **1**(1), 51–59 (1999)

88.91   C.-Y. Huang, S.Y. Nof: Autonomy and viability – measures for agent based manufacturing systems, Int. J. Prod. Res. **38**(17), 4129–4148 (2000)

88.92   C.-Y. Huang, J. Ceroni, S.Y. Nof: Agility of networked enterprises – parallelism, error recovery and conflict resolution, Comput. Ind. **42**, 275–287 (2000)

88.93   W. Jeong, S.Y. Nof: Performance evaluation of wireless sensor network protocols for industrial applications, J. Intell. Manuf. **19**, 335–345 (2008)

88.94   S.W. Yoon, S.Y. Nof: Demand and capacity sharing decisions and protocols in a collaborative enterprise network, PRISM Center Research Memo (2008)

88.95   M. Le Goc: SACHEM, a real-time intelligent diagnosis system based on the discrete event paradigm, Simulation **80**(11), 591–617 (2004)

88.96   M. Le Goc, C. Frydman: The discrete event concept as a paradigm or the "perception-based diagnosis" of SACHEM, J. Intell. Robot. Syst. **40**, 207–224 (2004)

88.97   H. Liang, Y. Xue: Investigating public health emergency response information system initiatives in China, Int. J. Med. Inform. **73**, 675–685 (2004)

88.98   N. Forbes: *Imitation of Life: How Biology is Inspiring Computing* (MIT Press, Cambridge 2004)

88.99   A. Abraham, C. Grosan, V. Ramos (Eds.): *Stigmergic Optimization: Studies in Computational Intelligence* (Springer, New York 2006)

88.100   K. Hadeli, P. Valckenaers, M. Kollingbaum, H. Van Brussel: Multi-agent coordination and control using stigmetry, Comput. Ind. **53**, 75–96 (2004)

88.101   P. Pongcharoen, C. Hicks, P.M. Braiden, D.J. Stewardson: Determining optimum genetic algorithm parameters for scheduling the manufacturing and assembly of complex products, Int. J. Prod. Econ. **78**, 311–322 (2002)

88.102   K. Hadeli: Bio-inspired multi-agent manufacturing control systems with social behaviour. Ph.D. Thesis (Katholieke Universiteit Leuven, Leuven 2006)

88.103   L. Nie, X. Xu, D. Zhan: Collaborative planning in supply chains by Lagrangian relaxation and genetic algorithms, Proc. 6th World Congr. Intell. Control (2006) pp. 7258–7262

88.104   G. Berning, M. Brandenburg, K. Gürsoy, J.S. Kussi, M. Vipul, F.-J. Tölle: Integrating collaborative planning and supply chain optimization for the chemical process industry (I) – methodology, Comput. Chem. Eng. **28**(6/7), 913–927 (2004)

88.105   M. Chiu, L. Grier: Collaborative supply chain planning using the artificial neural network approach, J. Manuf. Technol. Manag. **15**(8), 787–796 (2004)

88.106   D. Ho, B. Chu, H. Lee, C.D. Montemagno: Protein-driven energy transduction across polymeric biomembranes, Nanotechnology **15**, 1084–1094 (2004)

88.107   T.R. Shultz, S.P. Mysore, S.R. Quartz: Why let networks grow?. In: *Neuroconstructivism: Perspectives and Prospects*, ed. by D. Mareschal, S. Sirois, G. Westermann (Oxford Univ. Press, Oxford 2007)

88.108   B.J. Corbitt, N.A. Al-Qirim: *e-Business, e-Government and Small and Medium-sized Enterprises: Opportunities and Challenges* (Idea Group, Hershey 2004)

88.109   F. Li: *Social Implications and Challenges of e-Business* (Information Science Reference, Herhsey 2007)

89. e-Commerce

Clyde W. Holsapple, Sharath Sasidharan

Electronic commerce (e-Commerce) is fast becoming a way of life in the 21st century. As more and more consumers and organizations resort to electronic means for conducting purchases and facilitating business transactions, it has become vital for academic researchers as well as practitioners to understand its workings, and be able to analyze problems and rectify weaknesses. This is complicated by the fact that the e-Commerce market is global and constantly evolving, with new and innovative business models and products being introduced at a rapid pace. We present an overview of the status of e-Commerce, with particular emphasis on academic research. Specifically, we review the historical background of e-Commerce, discuss theoretical frameworks, and examine e-Commerce models. We also discuss emerging trends as well as pressing issues facing the e-Commerce marketplace.

Electronic commerce (e-Commerce) is fast dominating every aspect of our day-to-day existence. Whether purchasing retail products ranging from books to automobiles, or services ranging from information to banking, consumers are increasingly turning to the Internet and the World Wide Web (WWW). Businesses are increasingly conducting transactions within their value chains and across their supply chains through electronic means. There is constant experimentation with newer and more innovative business models, resulting in an ever-evolving e-Commerce marketplace. This chapter provides a comprehensive picture of the e-Commerce landscape, with an emphasis on academic research findings.

With this in mind, we first discuss the evolution of e-Commerce. This is followed by a discussion of theoretical aspects of e-Commerce research: definitions, frameworks, and success parameters. We then discuss three broad e-Commerce models: business-to-consumer (B2C), business-to-business (B2B), and consumer-to-consumer (C2C). Special emphasis is given to three popular B2C activities: online shopping, online banking, and electronic learning. We then discuss three emerging e-Commerce applications: mobile commerce, telemedicine (or e-Health), and fee-based information delivery. Finally, we conclude with a discussion on two open issues facing e-Commerce: user trust and legal considerations.

89.1 Background

Prior to the 1960s, information systems (IS) were primarily used for scientific research and military purposes, and very rarely for commerce. The early business applications of computers ranged from the Lyons Electronic Office (LEO) system, automating clerical accounting tasks, to the Bank of America's electronic recording machine accounting (ERMA), involving the automation of check handling. The 1960s and 1970s saw business establishments becoming heavy users of computers for accounting and production activities, but largely confined to applications within the organization. It was with the development of the electronic data interchange (EDI) in the 1970s that organizations started communicating with one another electronically. The high cost of development and maintenance of private networks, lack of standardization, and the fact that not all organizations had EDI systems hindered its early growth. The development of the Accredited Standards Committee (ASC) X12 standards in the 1980s provided a boost to the popularity of EDI. Even today, EDI facilitates a majority of B2B e-Commerce transactions through value-added networks (VAN) and over the Internet [89.1, 2].

Paralleling the development of EDI was the growth of the Internet, which later (along with the World Wide Web) became the driving force behind B2C e-Commerce. The Internet had its origins in the late 1960s as the advanced research projects agency net (ARPANET), designed exclusively for research, education, and use by government organizations. Subsequently, in 1983, the ARPANET was split into the Milnet and the Internet, the former used for military purposes and the latter for research and education. Around this time, the communication of computers over the Internet was standardized using the transmission control protocol/Internet protocol (TCP/IP), and in the following year the domain name system (DNS) was introduced [89.3].

In 1989, Tim Berners-Lee of the European Laboratory for Particle Physics (CERN) started developing the World Wide Web (WWW). Commercial access to the Internet began with *The World*

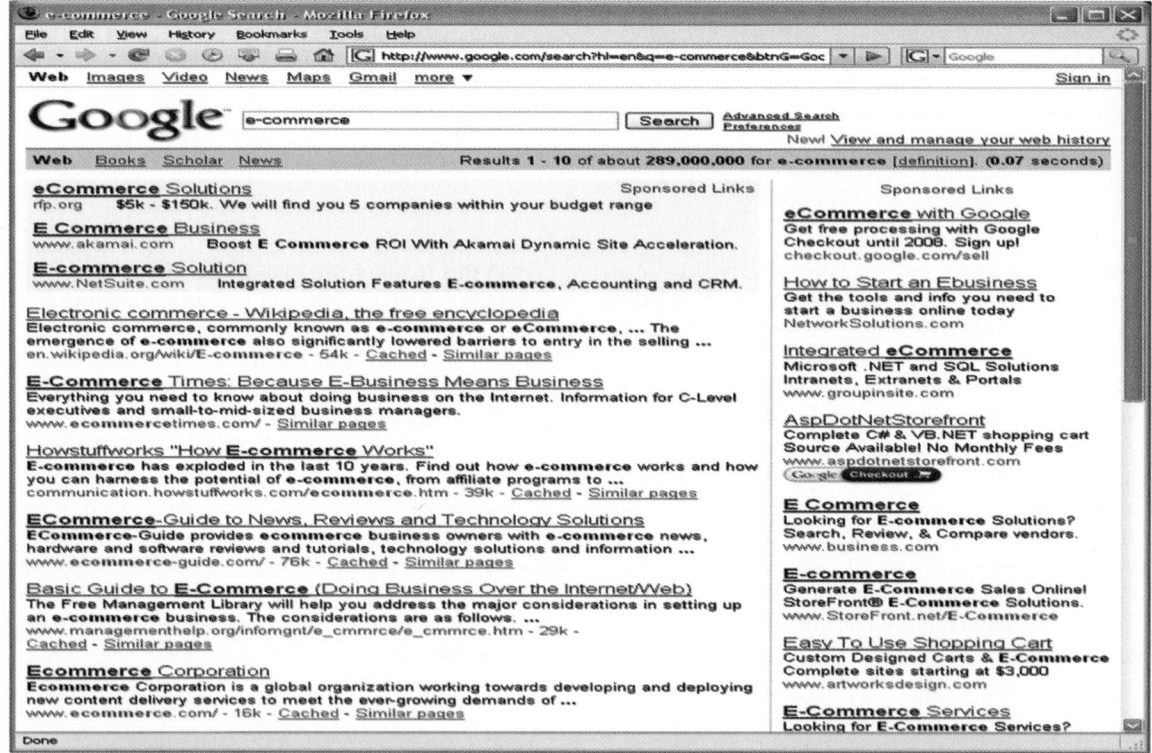

Fig. 89.1 Google website

(http://www.world.std.com) becoming the first provider of dial-up Internet access. Development of the Mosaic browser by Marc Andreessen, and its evolution into the Netscape browser with secure sockets layer (SSL) technology provided easy and secure access to the WWW for the general populace [89.3, 4].

In 1993, close to 30 million people were using the Internet in North America, with 43.2 million US households owning a computer. The subsequent years saw the launches of what were to become two of the greatest success stories in e-Commerce: Amazon and eBay, and the sale of airline tickets by Southwest Airlines through the WWW, marking the beginning of a new era in the travel and tourism industry. This was quickly followed by online travel sites sponsored by different airlines, and the emergence of online travel agents such as Expedia and Travelocity [89.2].

At this point, the primary limiting factor to the spread of e-Commerce was the speed, quality, and reliability of Internet access; however this issue was partially resolved with the introduction of the digital subscriber line (DSL). Worldwide, the number of Internet users had catapulted to 150 million, with more than half from the USA. In 1999, two of what were to become iconic names in cyberspace were launched: the search engine Google and the social networking website MySpace. The Google search engine provided a very efficient and effective search tool to browse the increasingly cluttered WWW, thereby indirectly promoting B2C e-Commerce activities (Fig. 89.1) [89.3, 4]. MySpace facilitated social networking through user-created personal profiles, blogs, groups, photos, and events.

The expansion of the Internet was also associated with the burgeoning B2B e-Commerce marketplace. New methods of conducting commercial transactions between buyers and suppliers along the entire length of the supply chain were developed. For example, in 1994, to improve material flow within their supply chains, the big three automakers, Chrysler, Ford, and GM, and 12 of their suppliers participated in the manufacturing assembly pilot (MAP) project. The MAP participants were electronically connected, and information took less than 2 weeks to reach the bottom of the supply chain, instead of the earlier 4–6 weeks. Another

Year(s)	Event
1951	Lyons Electronic Office (LEO) system
1959	Electronic recording machine accounting (ERMA) system
1969	Advanced research projects agency net (ARPANET)
1970s	Electronic data interchange (EDI)
1971	Internet-based email
1982	ASC X12 standards for EDIs
1983	ARPANET was split into the Milnet and the Internet
1983	Transmission control protocol/Internet protocol (TCP/IP)
1984	Domain name system (DNS)
1989	World Wide Web
1989	*The World* (http://www.world.std.com)
1993	Mosaic browser
1994	Netscape browser
1994	Manufacturing assembly pilot (MAP) project
1995	Amazon and eBay
1996	Trading process network (TPN) post
1990s	Digital subscriber line (DSL)
1999	Google
2000	Bursting of the dot-com bubble
2004	B2B market size estimated at over US$ 1800 billion
2006	B2C market size estimated at over US$ 200 billion

Table 89.1 Important e-Commerce milestones

case is that of General Electric (GE), which introduced its first online procurement system in 1996, an extranet called the trading process network (TPN) post. GE could now communicate with its internal customers over the extranet and with its worldwide suppliers over the Internet, facilitating same-day evaluation and award of bids [89.2].

The rapid development of the Internet in the late 1990s led to millions of dollars being invested in Internet-based online companies, leading to record-breaking stock value appreciation – the so-called *Internet bubble*. The year 2000 saw the bursting of the *bubble*, investment dropped, stock indexes plunged, and many online companies became defunct. The situation has since stabilized, and in 2006, B2C e-Commerce sales are estimated at over US$ 200 billion, an increase of 20% versus 2005. This, however, is dwarfed by the B2B market whose size amounted to US$ 1821 billion in 2004 [89.5, 6]. The important milestones in the evolution of e-Commerce are presented in Table 89.1.

89.2 Theory

Aside from its dynamic and still-unfolding history, e-Commerce has raised a host of questions related to its deployment, impacts, and possibilities [89.7]. To build a foundation for addressing these questions, researchers have investigated e-Commerce from a variety of standpoints, yielding a growing understanding of the nature and issues of this phenomenon that has emerged as such an important socioeconomic force. Here, we examine theoretical aspects of the e-Commerce research foundation from three perspectives: definitions, frameworks, and success parameters.

89.2.1 Definitions of e-Commerce

What is e-Commerce? What are its boundaries? What constitutes e-Commerce and what does not? As organizations devise new business models and innovative Internet-based applications, defining the boundaries of e-Commerce with any degree of precision has proven to be a challenge for researchers and practitioners alike.

For example, e-Commerce has been defined by researchers as [89.8]:

... the sharing of business information, maintaining of business relationships, and conducting of business transactions by means of telecommunications networks,

and as [89.9]:

... the use of the internet to facilitate, execute, and process business transactions.

Likewise, practitioners too have offered multiple definitions [89.10]:

... the buying and selling of goods and services on the internet, especially the WWW,

and as [89.11]:

... the process of managing online financial transactions by individuals and companies.

A review of these sample definitions makes it clear that there are multiple views or perspectives as to what constitutes e-Commerce. To provide an all-inclusive characterization of e-Commerce, *Holsapple* and *Singh* conducted an extensive study of the e-Commerce literature, and introduced a five-cluster taxonomy of e-Commerce definitions [89.12]. A common theme underlies definitions within each cluster. The five clusters are: the trading view, the information exchange view, the activity view, the effects view, and the value-chain view:

- The trading view is characterized by market-based activities such as buying and selling facilitated by electronic means. This is the most commonly held view of e-Commerce. *Holsapple* and *Singh* provide the following example as representative of the trading view [89.13]:

  ... e-Commerce is associated with the buying and selling of information, products, and services via computer networks today and in the future via any one of the myriad of networks that make up the information superhighway.

- The information exchange view recognizes the flow of information that can precede, accompany, or follow a transaction. A definition in this cluster may or may not mention the buying or selling aspects of a transaction. The following is an example [89.14, p. 136]:

Electronic commerce has been used to describe a wide variety of business related transactions, but is at its core the data used for conducting day-to-day business operations with suppliers and customers.

- The activity view recognizes nontrading aspects of e-Commerce such as marketing and production activities, decision support, research, maintaining business relationships, and other ancillary activities [89.12, 15]. For example [89.16, p. 31]:

Electronic commerce is a generic term that encompasses numerous information technologies and services to improve business practices ranging from customer service to inter-corporation coordination.

Together, the informational and activity views are emphasized in what is known as collaborative commerce (c-Commerce): the use of computer/communication technologies to help participants exchange information in the course of some collaborative activity, which may well enable the collaborative generation of new knowledge. c-Commerce has emerged as a major and growing facet of e-Commerce with a host of research issues in its own right [89.17].

- The effects view focuses on the goals or reasons for using e-Commerce, such as reduced costs, smoother information flows, remote collaboration, better coordination between buyers and suppliers, better customer management, and improved service quality. The following is an example [89.13, p. 1]:

Broadly defined, electronic commerce is a modern business methodology that addresses the needs of organizations, merchants, and consumers to cut costs while improving the quality of goods and services and increasing the speed of service delivery.

- The value-chain view acknowledges the use of e-Commerce in creating value in one or more of the primary and/or secondary value chain activities of the organization. The value chain need not be intra-organizational, but can extend to interorganizational value chains of which the organization forms a part [89.12, 15]. The following is an example [89.18, p. 5]:

Electronic commerce denotes the seamless application of information and communication technology

from its point of origin to its endpoint along the entire value chain of business processes conducted electronically and designed to enable the accomplishment of a business goal. These processes may be partial or complete and may encompass business-to-business as well as business-to-consumer and consumer-to-business transactions.

Given the interconnectedness of five views, *Holsapple* and *Singh* advance an integrated definition of e-Commerce [89.12, p. 161]:

Electronic commerce is an approach to achieving business goals in which technology for information exchange enables or facilitates execution of activities in and across value chains as well as supporting decision making that underlies those activities.

Although the integrated definition encompasses the taxonomy's five views of e-Commerce, it does not address the concept of information exchange to account for the whole gamut of knowledge management (KM) issues. Thus, *Holsapple* and *Singh* further extend this definition as follows [89.12, p. 164]:

EC is an approach to achieving business goals in which technology is used to manage knowledge for purposes of enabling or facilitating the execution of activities in and across value chains as well as the making of decisions that underlie those activities.

This is the most comprehensive definition of e-Commerce existing in the literature. A researcher, practitioner, or student can adopt any of the taxonomy's five views or the integrated comprehensive definition, but is well advised to be cognizant of the alternative views if the comprehensive definition is not adopted. Having considered e-Commerce definitions, we now summarize several frameworks proposed for appreciating e-Commerce elements and issues.

89.2.2 Frameworks for e-Commerce

The generic framework of electronic commerce, introduced by *Kalakota* and *Whinston*, identifies three major types of components: applications, infrastructure building blocks, and support pillars [89.13, 19]. Together, these form a basis for organizing one's thoughts about e-Commerce and for ensuring that important elements of e-Commerce are not overlooked. The first component, electronic commerce applications, includes systems for supply chain management, procurement, online marketing, remote banking, home shopping, and so forth. Underlying the applications, the second component is comprised of four layers of infrastructure building blocks. From lowest to highest, these layers are: the *information superhighway* infrastructure (the means for connecting, including the Internet), the network publishing infrastructure (the means for creation/storage of information and multimedia content, including servers), messaging and information distribution infrastructure (the means for moving content across connections and to/from storage, including email, EDI, file transfer, etc.), and common business services infrastructure (means for ensuring safety, reliability, and convenience of users, including mechanisms for authentication, security, privacy, paying, locating, and so forth). The generic framework identifies two support pillars for electronic commerce: public policy and legal issues (including regulation, taxation, cost, and policing concerns) and technical issues (including standards and protocols).

Alternatively, e-Commerce can be viewed as being comprised of three hierarchical metalevels: technology infrastructure, services, and products and structures (Fig. 89.2) [89.8, 20]. The bottom metalevel is the technology infrastructure and deals with wired and wireless communication networks, public and private communication utilities such as the Internet and organizational intranets/extranets, and multimedia management capabilities that deliver the functionality of the WWW. This metalevel corresponds to the lower layer of the generic framework's infrastructure building blocks. The

middle metalevel is that of services, comprised of services facilitating secure communication and enablers such as electronic catalogs and directories, intelligent agents, electronic money and financial services, smart card systems, digital authentication services, and copyright services [89.20]. This metalevel corresponds to the higher layers of the generic framework's infrastructure building blocks. The top metalevel consists of the actual products and services as well as their delivery mechanisms. These include, among others, online consumer services for shopping, banking, and auctioning; buyer–supplier linkages and Intranet/extranet-based B2B collaboration; and other interorganizational systems that facilitate the entire spectrum of supply chain activities. This metalevel corresponds to the generic framework's applications component.

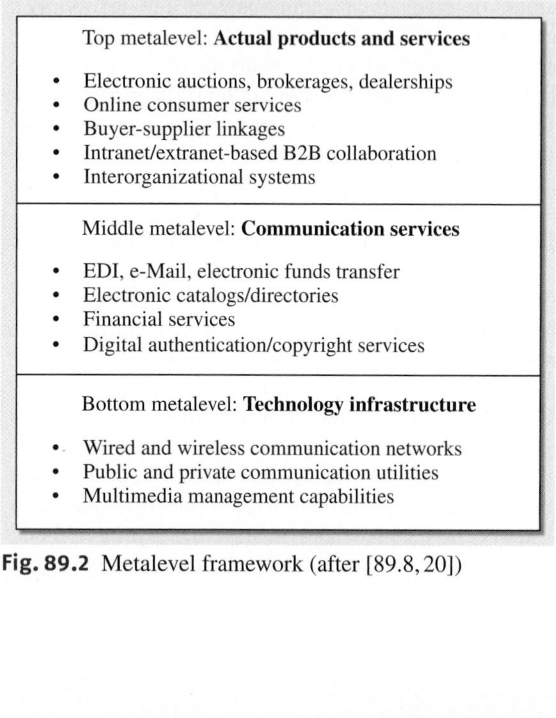

Extending the hierarchical framework, *Zwass* proposes a *5C* activity framework that enables the identification of specific WWW-based e-Commerce innovational opportunities [89.21]. The *5C* activity framework recognizes five WWW based e-Commerce activity domains: commerce, collaboration, communication, connection, and computation. The commerce domain includes the marketplace and supply chain linkages. For the former, innovational opportunities are possible in searching for partners and subsequent negotiation of terms and conditions, new business models, product customization, delivery of products or services, and subsequent customer management services. For the latter, opportunities include redesign and reconfiguration of the supply chain, electronic reintermediation, and adoption of best-of-breed processes.

In the collaboration activity domain, the WWW can facilitate building and sustaining relationship networks such as long-term alliances between buyers and suppliers, and with customers. Innovations in the communication activity include the development of virtual customer communities that can serve to enhance knowledge of customer needs and requirements, obtain feedback of existing products and services, and form a source of new product ideas. In the connection domain, the anytime–anywhere connectivity provided by the WWW facilitates a wide variety of mobile commerce innovations for monitoring and tracking of personnel, objects, and systems. Movement of organizational information systems to the WWW can facilitate intra- and interorganizational system integration. The computation activity domain views the Internet and the WWW as a single enormous computing resource, connecting together innumerable computers, software, data, and people – facilitating grid computing.

Top metalevel: **Actual products and services**

- Electronic auctions, brokerages, dealerships
- Online consumer services
- Buyer-supplier linkages
- Intranet/extranet-based B2B collaboration
- Interorganizational systems

Middle metalevel: **Communication services**

- EDI, e-Mail, electronic funds transfer
- Electronic catalogs/directories
- Financial services
- Digital authentication/copyright services

Bottom metalevel: **Technology infrastructure**

- Wired and wireless communication networks
- Public and private communication utilities
- Multimedia management capabilities

Fig. 89.2 Metalevel framework (after [89.8, 20])

Table 89.2 Riggins' electronic commerce value grid (after [89.22, p. 301])

Five dimensions of commerce	Value creation Efficiency	Effectiveness	Strategic
Time	Accelerate user tasks	Eliminate information float	Establish 24×7 customer service
Distance	Improve scale to look large	Present single-gateway access	Achieve global presence
Relationships	Alter role of intermediaries	Engage in intermediaries micromarketing to look small	Create dependency to lock-in user
Interaction	Make use of extensive user feedback	User controls detail of information accessed	Users interact via online community
Product	Automate tasks using software agents	Provide online decision support tools	Bundle information, products, and services

To help managers identify online opportunities and types of applications that may yield value to users, *Riggins* describes a framework in which firms compete in a space defined by five dimensions of commerce: time, distance, relationships to customers, interaction modes, and product/service offered [89.22]. These are juxtaposed with three criteria for technology investment: greater efficiency, increased effectiveness, and strategic benefits. The result is the electronic commerce value grid, highlighting 15 areas for a firm to consider in striving to add value to its customers through web-based electronic storefronts (Table 89.2).

Choi et al. provide an integrated microeconomic framework for understanding electronic commerce and maximizing its benefits [89.23]. This framework characterizes e-Commerce as involving a global electronic marketplace of digital products subject to various pricing strategies. These products include online goods/services, software, digital content, advertisements, announcements, product/vendor information, digitized processes, payment/tracking information, smart products, communications, and so forth. Six major themes are stressed in this framework: product quality and the role of market intermediaries; digital copyright issues; product advertising; consumers' product-information searching activities; product evaluation/choice and pricing approaches; and digital financial services including payment processes. Each of these themes is subject to many design and implementation variations that demand attention from practitioners and deserve continuing investigation by researchers.

In a related vein, *Schlueter* and *Shaw* present a framework for analyzing structures and dynamics of digital goods/services markets and industries [89.24]. This framework identifies six value-adding stages for firms involved in e-Commerce: content direction, content packaging, market making, transport, delivery service, and interface and systems. Via alliances, mergers, or acquisitions, firms collaborate to align their capabilities and assets in ways that cover the six value-adding stages. This collaboration can involve diverse firms such as those in the financial service, information distribution, entertainment, creative content, communications, advertising, and computing industries. From another angle, *Lindemann* and *Schmidt* offer a market-oriented framework to model specific e-Commerce platforms for coordinating exchange activities among participants in digital markets [89.25]. Yet another e-Commerce framework integrates four earlier e-Commerce frameworks with the traditional marketing model (involving concepts of product, price, promotion, and distribution) as a foundation for devising e-Commerce marketing strategies [89.26].

The e-Commerce literature also contains a considerable number of fairly specialized frameworks. One of these, for example, advocates a three-level architecture for studying how intelligent agents can be applied to support electronic trading activities in the sense of making electronic markets more effective [89.27].

89.2.3 e-Commerce Success Parameters

What is meant by e-Commerce success? How can it be measured? Researchers have used multiple yardsticks for measuring e-Commerce success depending on the research context. For B2C transactions, widely used success parameters involve website usage factors such as acceptance, adoption, actual usage, intention to use, and intention to return [89.28–31], as well as purchase-related factors such as actual purchase, inten-

Table 89.3 B2C e-Commerce success parameters

Factors influencing success	Success measures
Website design factors:	**Website usage factors:**
• Navigation, convenience, usability	• Acceptance
• Reliability, responsiveness	• Adoption
• Accessibility, flexibility, relevance	• Actual usage
• Security/privacy	• Intention to use
• Quality of support services	• Intention to return
• Virtual-reality technologies for high-experiential products	
• Social presence through rich descriptions and graphics	**Purchase-related factors:**
• Value-added mechanisms	• Actual purchase
• Recommendations from prior customers	• Intention to purchase
	• Willingness to pay
Individual factors:	
	Customer-related factors:
• Demographics	
• Prior experience, computer self-efficacy	• Satisfaction
• Technological background	• Loyalty
• Subjective norms	• Learning
• Innovativeness, adaptability, awareness	
• Ease of use, usefulness, enjoyment	
• User trust	

tion to purchase, and willingness to pay [89.32–34]. Customer-related factors such as satisfaction [89.35], loyalty [89.36], and learning [89.37] have also been employed. Finally, customer perceptions of the usefulness of a website [89.38] and of the quality of a website [89.39] have been used to measure e-Commerce success. We summarize the e-Commerce success parameters in Table 89.3.

In order to bring clarity to e-Commerce research and to provide a framework for measuring e-Commerce success, *DeLone* and *McLean* suggest extending their information systems success model [89.9, 40, 41] to the context of e-Commerce research. The latter contends there are multiple dimensions of information systems (IS) implementation success including system usage (e.g., frequency, motivation, regularity, objectives of usage), system quality (e.g., reliability, response time, ease of use, usefulness, error rate), user satisfaction (e.g., overall satisfaction, decision-making satisfaction, user complaints), information quality (e.g., accuracy, precision, timeliness, completeness), service quality (e.g., assurance, empathy, responsiveness), and net benefits (e.g., productivity, decision accuracy, profit). Adapting these success dimensions to the e-Commerce domain, with suitable modifications, yields their framework for measuring e-Commerce success.

System usage in an e-Commerce context translates into the number of website visits, navigation within the site, and transaction execution. As noted above, most of these have already been widely used by researchers. For the system quality dimension, factors such as reliability, accessibility, response time, ease of use, and usefulness are extended directly to the context of a website, and these have been used in studies reported in the e-Commerce literature. The user satisfaction dimension can be applied to measure satisfaction at different stages of customer interaction, ranging from information gathering to post-purchase services. In the case of the information quality dimension, an important factor is privacy and security, apart from personalization of content, and the completeness and relevance of the information. Service quality can be measured as the overall customer experience, and this is an important measure given the fact that switching costs are low in an e-Commerce environment. Net benefits in an e-Commerce environment can be measured in terms of savings in time and money, enhanced productivity, and improved decision-making [89.9].

89.3 e-Commerce Models and Applications

The US Department of Commerce views B2B e-Commerce as dealing with manufacturing and wholesale activity, and B2C e-Commerce as dealing with retailing and the service industry [89.6]. Based on this categorization, the total volume of e-Commerce transactions in 2004 amounted to US$ 1951 billion, representing a 14.4% increase over the previous year, and comprising roughly 10% of the total value of US shipment, sales, and revenue. The lion's share of the e-Commerce market in dollar terms is in the B2B segment, amounting to 93% or US$ 1821 billion. However, given its intrusion into virtually every single aspect of our daily life, B2C e-Commerce remains perhaps the most visible and easily recognizable face of e-Commerce. We first discuss the B2C e-Commerce market, followed by the B2B e-Commerce market. We then discuss the small but rapidly emerging C2C e-Commerce market.

89.3.1 B2C e-Commerce

The B2C e-Commerce market has continued to grow with recent estimates being as high as US$ 210 billion for 2006, an increase of 20% over the previous year, and projected to reach US$ 330 billion in 2010 with an estimated 55 million online shopping households in the US. In dollar terms, the largest B2C e-Commerce product line is travel, with approximately 35% of the market at US$ 70 billion. This is followed by computer hardware and software at US$ 17 billion, automotive equipment at US$ 16 billion, and apparel at US$ 14 billion [89.5]. We now examine some of the more important segments of the B2C e-Commerce marketplace: online retail shopping, online banking, and electronic learning.

Online Retail Shopping

This is perhaps the most visible face of e-Commerce – the online selling of a wide range of retail goods and products such as clothing, clothing accessories, electronic devices, electrical appliances, computer hardware and software, books and magazines, food and beverages, health and personal care items, sporting goods, music and videos, and office equipment and supplies. Apart from retail goods, the major services sold online include those related to computer systems, publishing, securities and commodity contract intermediation/brokerage, and travel arrangement/reservation. In recent years, retail e-Commerce sales registered an annual growth of 25% compared with 4% for total retail sales; however, it still is only a small fraction of total retail sales, at around 2% in 2004 [89.6]. Interestingly, two of the product lines projected to experience the highest market growth are cosmetics and pet supplies [89.5].

Based on Internet traffic rankings, the most popular online shopping site is Amazon (Fig. 89.3) having a global reach of around 1.8% of measured Internet users [89.42]. With annual revenues of US$ 10.7 billion in 2006, this Fortune 500 company lists thousands of new and used items at its website, spanning the entire retail product spectrum. Apart from the wide array of products, Amazon incorporates a number of personalization features such as one-click buying, customer and editorial product reviews, gift registries and certificates, wish lists, and restaurant and movie listings. Other major online shopping sites include those of Target, Wal-Mart, Best Buy, New Egg, and Tiger Direct.

What are the factors that influence the adoption of online shopping by consumers? Research has indicated that the design of the website in terms of navigation, convenience, usability, reliability, responsiveness, security and privacy, as well as the quality of support services such as order fulfillment, help desks, and return policies affect user adoption [89.33, 35]. The website should be attractive and have a social presence through rich descriptions and graphics, particularly in the case of products having amusement value [89.43]. The use of virtual-reality technologies serves to minimize the problems associated with the user being unable to touch and feel the product. This is particularly true in the case of high-experiential products where customers need to learn more about the product and experiment with it before making the purchase [89.37]. Also, a badly styled, incomplete, and erroneous website results in a negative impression regarding the quality of an online store, and the resulting lack of trust drives customers away from the store [89.39, 44].

Online stores that provide value-added mechanisms such as search engines serve to enhance shopping enjoyment, particularly for customers who are not looking to purchase a specific product [89.28]. Positive recommendations from prior customers via a consumer-reputation facility serve to reduce decision-making time and enable customers to make better decisions [89.38]. Above all, the most important factor that could make or break adoption of an online shopping site is customer perception of trust in the online shopping merchant [89.35, 39, 45]. As this aspect is relevant in all segments of

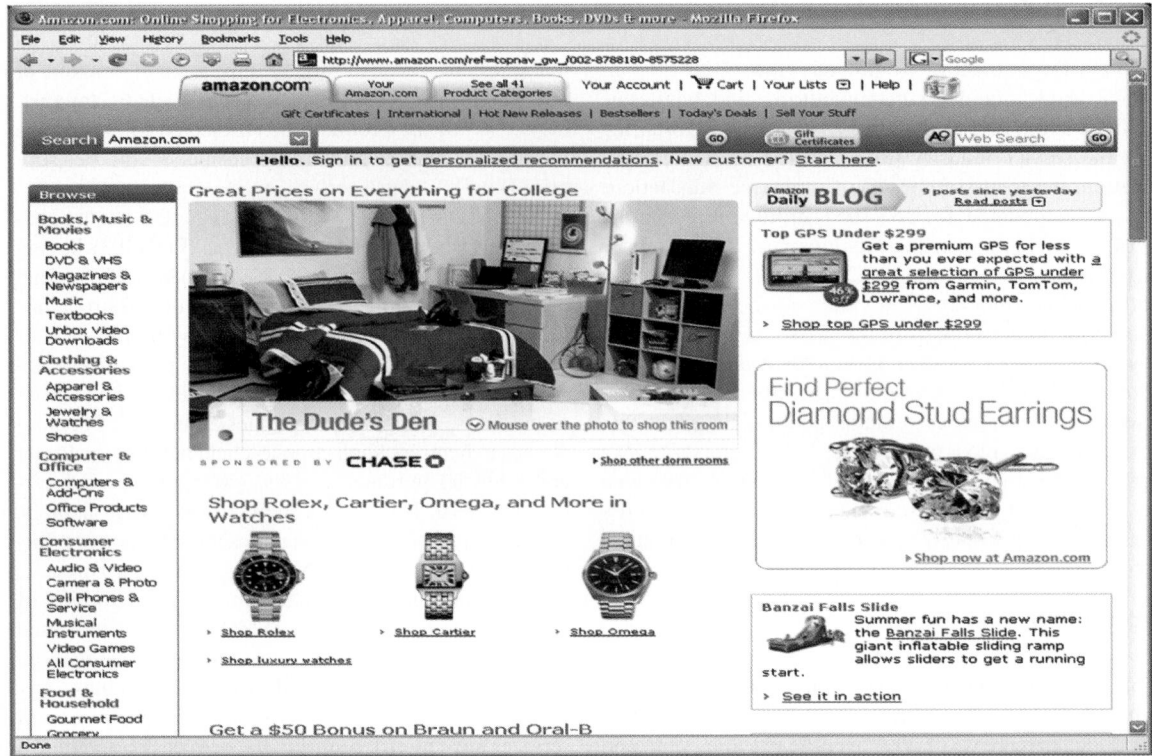

Fig. 89.3 Amazon website

e-Commerce, we deal with it separately in a subsequent section.

Online Banking

Online banking is one of the fastest growing Internet activities with about 43% of Internet users (63 million adults) banking online on a regular basis [89.46]. It has been estimated that the average total assets of an Internet-only bank is US$ 3.5 billion compared with US$ 1 billion for all banks [89.47]. Currently, online banking portals provide an array of services to customers including typical banking activities such as transferring funds, advancing loans, ordering checks, and downloading information about checking/savings, interest/dividends, loan payoffs/balances, and share/checking balances. Based on Internet traffic rankings, the most popular US banking site is that of the Chase Bank (Fig. 89.4), having a global reach of around 0.02% of measured Internet users [89.42]. Other major players include Bank of America, Citibank, US Bank, and Wachovia Bank.

What are the important factors that influence customer adoption of online banking? These fall into two broad categories: individual characteristics and user perceptions of the online banking website. In the former category, factors include demographics related to income, age, and gender, prior computer experience, computer self-efficacy, technological background, subjective norms, innovativeness, adaptability, and awareness of online banking benefits [89.29, 48, 49]. In the latter category, factors include usefulness, enjoyment, usability, convenience, accessibility, and relevance of the website [89.29, 50].

Although the size and growth of online banking has been impressive, it has not outpaced the growth of the Internet or other e-Commerce activities due to the *trust gap* [89.46]. Despite the technical advancements that have led to the safe and secure transmission of sensitive information, a sizable portion of Internet users still have reservations regarding the security and confidentiality of online banking transactions, and hence do not trust online banking sites [89.30, 45]. The trust barriers could be overcome by incorporating privacy and data protection technology into banking website design, proper authentication of users, informing users regarding secure online banking practices, and provid-

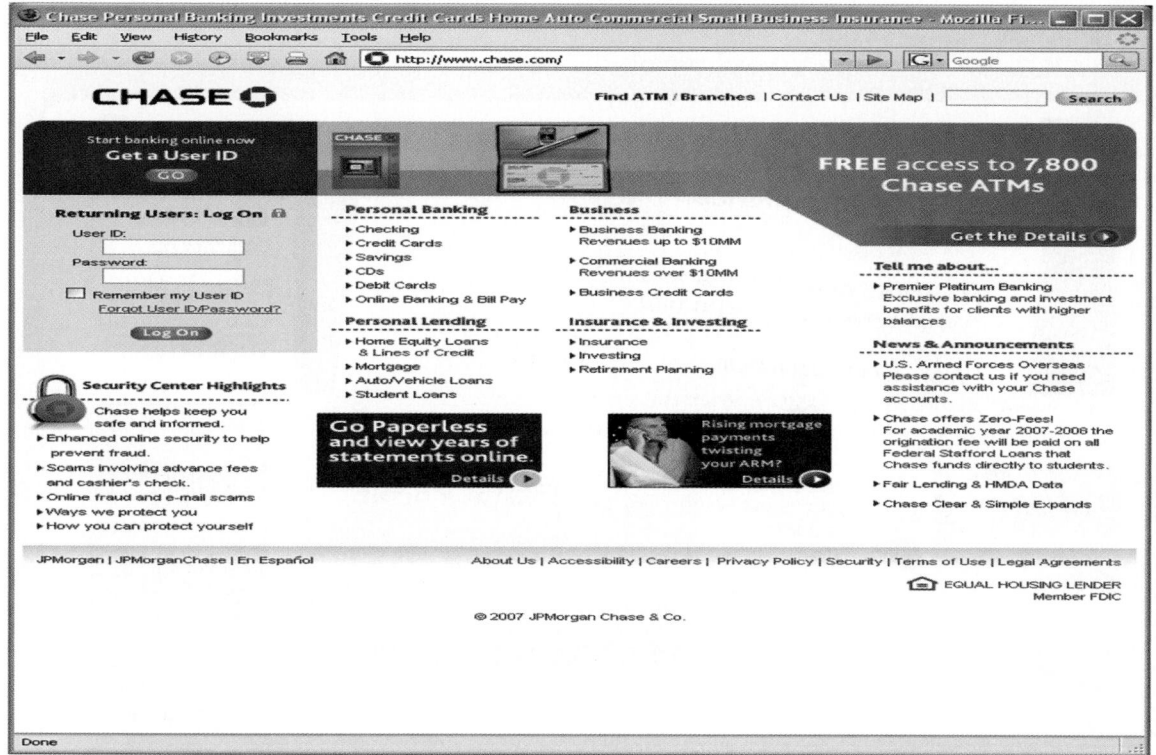

Fig. 89.4 Chase Bank website

ing redressal mechanisms [89.29, 30]. Indeed, trust is an issue that is of concern not only for online banking, but for e-Commerce as a whole, and is dealt with in a subsequent section. See Chap. 91 for additional content on automation in financial services.

Electronic Learning

Electronic learning, or e-Learning, is one of the fastest growing segments in the B2C e-Commerce domain. It primarily consists of the delivery of educational content to individuals or groups of learners via the Internet and WWW, as well as through organizational intranets and extranets [89.51]. Perhaps the most pervasive form of e-Learning is distance education, wherein universities provide online course content and teaching to students on a global basis. In addition, business organizations use electronic learning for employee training, certification, and upgrading skills, so much so that the corporate e-Learning market is estimated to be nearly US$ 10.6 billion in 2007 [89.52]. Currently, the major players include WebCT and Blackboard (Fig. 89.5), Desire2Learn, Dokeos, Skill Soft, Epic, and Learning Steps.

What factors influence the success of e-Learning? The e-Learning success model posits that the overall success of an e-Learning initiative depends on attaining success at every one of three stages of e-Learning system development: design, delivery, and outcome [89.53]. An empirical study supports the model and finds that action research methodology can be instrumental in promoting the development of successful e-Learning systems by iterating through cycles of diagnosing, action planning, action taking, evaluating, and reflecting.

Due to relatively little (or no) face-to-face interactions with an instructor, an electronic learning program may not provide the same amount of instructor guidance as opposed to a conventional classroom teaching or employee training session. Given this, an important factor that can influence effective use of an electronic learning program is self-regulation – the capability of an individual to take charge of his/her learning path. Self-regulation requires time-management skills, self-motivation, an ability to set and meet performance goals, and an ability to effectively organize, rehearse, and encode information [89.54–56].

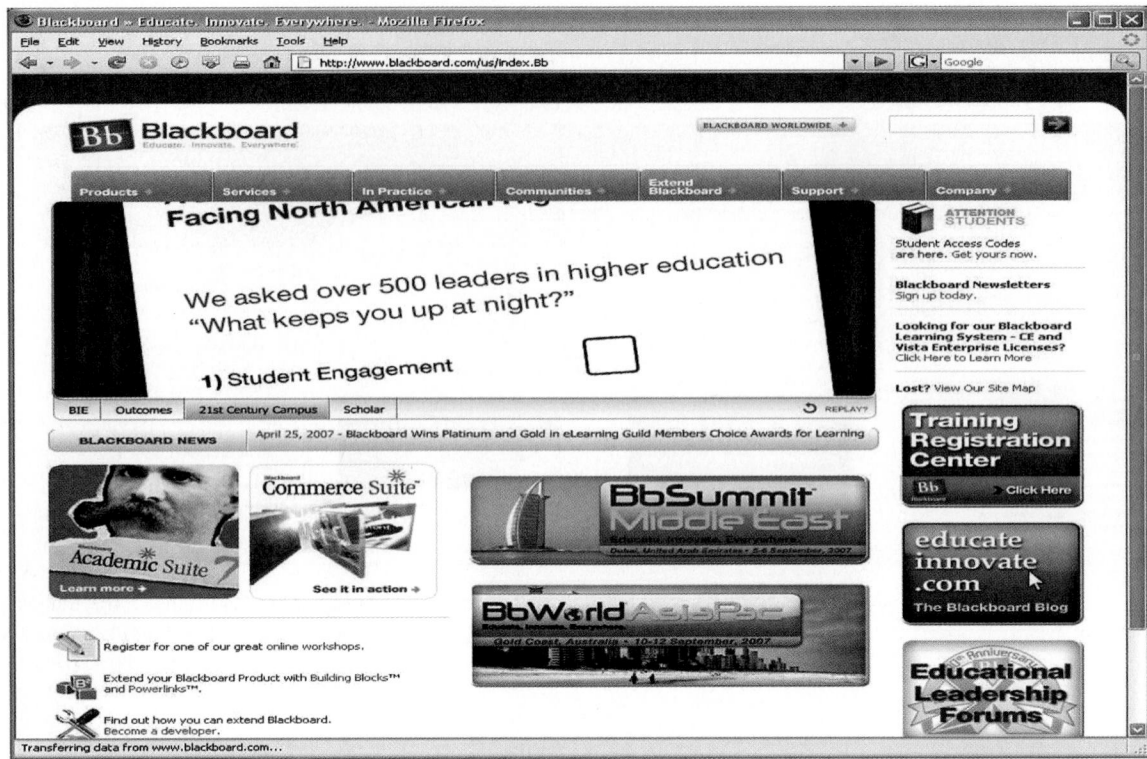

Fig. 89.5 Blackboard website

Apart from individual characteristics, technological features of an e-Learning system, such as its quality and reliability, availability, flexibility, convenience, and level of interactivity, can influence its adoption [89.57, 58]. If a human instructor is part of an electronic learning system, care must be taken to ensure that his/her presence is felt frequently in the online environment, providing timely, detailed directions, and making extra efforts to reach out to and interact with users [89.57,59]. Also recommended is matching the technology with the nature of the course content, thus courses that need creative thinking would be better served with technology having higher degrees of interactivity than courses that can be learned by rote [89.54]. See also Chap. 73, *Automating Serious Games*, and Chap. 85, *Automation in Education/Learning Systems*, for related content on electronic learning.

89.3.2 B2B e-Commerce

With a value of US$ 1821 billion in 2004, the B2B e-Commerce market dwarfs the B2C segment. It comprises 93% of the e-Commerce market, and ap-

proximately 20% of the overall B2B market [89.6]. Major areas of concentration include food, textiles, petroleum, chemicals, machinery, computers and electronics, automotive, paper, and pharmaceuticals. We now discuss various B2B market mechanisms with the help of examples.

B2B transactions can be categorized based on the actual mechanisms that are required to execute the transaction. These in turn can be categorized based on connectivity and purpose. The former refers to the number of players (i. e., buyers and suppliers) involved in a transaction, and the latter refers to the intention of the player initiating the B2B transaction [89.60]. The number of players involved in a B2B transaction may be one-to-one, one-to-many, many-to-many, and any-to-any. The common example of one-to-one is via an extranet, where a buyer interacts with a particular seller. An example of one-to-many connectivity would be an auction, where the seller puts up a product for bidding, and potential buyers bid for the product until the final price is reached. A corollary of the one-to-many category would be the many-to-one, or the reverse auction, wherein mul-

Table 89.4 *Cullen* and *Webster* B2B e-Commerce model (after [89.60])

Category	Connectivity	Purpose	Technology	Interaction
Individual trading	Open	Selling	WWW	Direct
Collaboration	Open	Selling	WWW	Intermediary
Marketplace	Open	Selling/buying	WWW	Intermediary
Proprietary sales	Restricted	Selling/integrated	Extranet	Direct
Private exchange	Restricted	Integrated	Extranet	Intermediary
Aggregation	Open	Buying	WWW	Intermediary
Intranet/EDI	Restricted	Integrated	Intranet/EDI	Direct
Restricted bid	Restricted	Buying	Extranet	Direct
Reverse auction	Open	Buying	WWW	Direct

tiple sellers quote among themselves and the lowest quote is accepted. The many-to-many scenario involves multiple buyers and sellers, wherein each could form alliances resulting in buyer and seller organizations. They could interact through a neutral exchange platform provided by an intermediary where the requirements of the buyers and sellers could be matched. In the any-to-any scenario, there could be one or many sellers or buyers, an example being e-Hubs hosting electronic marketplaces that facilitate interaction between buyer(s) and seller(s) [89.60–62]. Connectivity may also be viewed as open or restricted/closed. Open transactions can be accessed by all players involved whereas restricted/closed transactions are limited to specific players, generally those having membership privileges [89.60,61]. The purpose of engaging in a B2B transaction may be for selling, buying, or integrated exchange. The last includes multidimensional activities such as exchange of commercial documentation or operational information.

The concept of connectivity and purpose can be extended to include aspects such as the facilitating technology and the interaction type [89.60]. Examples of facilitating technologies are the WWW, extranet, Intranet, and EDI. Interaction may be direct between the buyer and the seller, or may involve an intermediary, wherein a third party mediates the transaction. Based on connectivity, purpose, technology, and interaction type, the *Cullen* and *Webster* model categorizes e-Commerce transactions into nine operational categories: individual trading, collaboration, marketplace, proprietary sales, private exchange, aggregation, intranet/EDI, restricted bid, and reverse auction (Table 89.4) [89.60].

Individual Trading

This type of transaction has open connectivity, involves selling, uses the WWW, and is direct. The typical example is a single supplier selling to other business organizations. This model is very common in the maintenance, repair, and operations (MRO) industry. An example is Grainger, which provides access to more than 800 000 products from such categories as adhesives, fasteners, hardware, lighting, motors, power transmission, and hydraulics [89.63].

Collaboration

This category is viewed as being open, involves selling, uses the WWW, and involves an intermediary. As opposed to individual trading, collaboration involves multiple organizations that converge onto a common platform run by an intermediary organization. These are common in the pharmaceutical, chemical, and motor industries. For example, in February 2000, major automakers collaborated to create Covisint [89.64], to address escalating procurement costs and inefficiencies within the industry. Covisint currently supports over 30 000 organizations in the global automotive industry spread across 96 countries and having over 266 000 users. See Chap. 88 on *Collaborative e-Work, e-Business, and e-Service* for a more thorough collaboration discussion.

Marketplace

The marketplace category is open, involves both selling and buying, uses the WWW, and involves an intermediary. The intermediary provides a common platform for buyers and sellers to sell commodity and standardized products. An example is ChemConnect [89.65] which provides a third-party commodity exchange platform for the chemical industry [89.60].

Proprietary Sales

This category of B2B transactions tends to be restricted, involves selling and integrated exchange, uses an extranet, and is direct without involving an intermediary. It mainly involves the sale of sensitive goods such as military equipment, prescription medicines, firearms, and the like. To engage in a proprietary sales transaction, the buyer or seller must posses some form of membership, certification, or clearance from a government agency. Cullen and Webster point to Wolverine Supplies [89.66], which provides firearms to law enforcement officers in Canada as representative of this category.

Private Exchange

This category is restricted, involves integrated exchange, uses an extranet, and involves an intermediary. The typical example here is the case where a buyer organization has long-term supply-chain relationships with supplier organizations. Cullen and Webster cite the example of Elemica [89.67], which provides a platform for a range of integrated supply-chain activities in the chemical industry.

Aggregation

The aggregation category is open, involves buying, uses the WWW, and involves an intermediary. Here, smaller buyer organizations group together to negotiate with a seller organization so as to obtain economies of scale that they would not otherwise have obtained. Powerspring [89.68], a third-party intermediary in the power industry that provides a platform for gas suppliers to bid competitively for aggregated power demand, is representative of this category [89.60].

Intranet/EDI

This category is restricted, involves exchange of trading information, uses intranet/EDI, and is direct. Transactions are typically one-to-one between the buyer and seller, and are typically conducted by large organizations having a number of repeat orders.

Restricted Bid

As the name indicates, the restricted bid category is restricted, involves buying, uses the extranet, and is direct. Only a limited number of selected suppliers are allowed to bid for a product required by the buyer, resulting mainly in time savings. The National Health Services

in the UK uses this approach in procuring some of its supplies [89.60].

Reverse Auction

The reverse auction is open, involves buying, uses the WWW, and is direct. For example, XSAg [89.69] provides a reverse auction in the agricultural chemical industry. The buyer lists the product requirements including the asking price and quantity on the XSAg website, and potential sellers bid for the business. Suppliers that meet the asking price and other buyer requirements win the contract immediately upon bidding. The federal government also uses reverse auctions, through dozens of federal organizations such as the General Services Administration, the Department of Homeland Security, the Department of State, and the Department of Defense [89.70].

89.3.3 C2C e-Commerce

The consumer-to-consumer (C2C) e-Commerce market is dominated by online customer auctions, wherein auction providers act as intermediaries providing a common platform for buyers and sellers to engage in the selling and purchase of retail goods. With US$ 15 billion in sales in 2004 [89.71], the size of the C2C online auction market is minuscule compared with the B2C and B2B markets; however it has spawned one of the greatest success stories of e-Commerce: eBay (Fig. 89.6). Apart from eBay, the other major player in this area is Ubid [89.72].

Auctions may be seller auctions or buyer auctions; in the former the seller initiates the transaction by listing an item for sale, and in the latter the buyer initiates the transaction by posting purchase requirements [89.73]. Buyer auctions are more prevalent in the B2B market, whereas the seller auction dominates the C2C e-Commerce market. There are different auction formats. In the most common type of auction (called an English auction), buyers bid among themselves, and the winning bid goes to the highest bidder. A variation to this is the Dutch auction, where the seller starts off with an asking price, which is lowered until some participant is willing to accept the price [89.74]. This is commonly used for selling multiple items. Variations to straightforward auctioning could include a *Buy it Now* option, wherein the seller specifies a price at which a buyer could purchase the item and truncate the auction. Apart from facilitating auctions, such sites also allow customers to sell their items at a fixed price as well as to advertise them. There might also be provisions for spe-

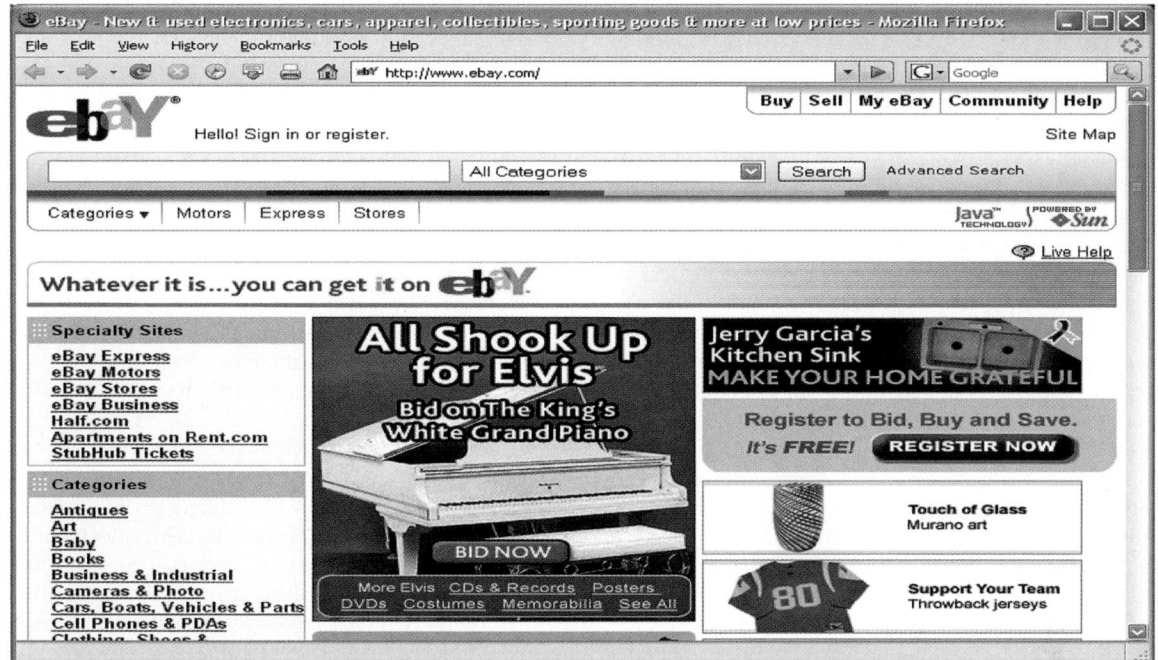

Fig. 89.6 eBay website

cialized individual sellers to exhibit and accept bids for their range of products at their own storefronts [89.75].

What are the factors that influence sellers and buyers to take part in online auctions? The low cost of entry and negligible transaction costs compared with other retail channels make online auctions very attractive. More importantly, there is immediate access to a sizable customer base [89.76, 77]. On the part of the buyer, factors such as ease of use and convenience of using online auctions, as well as potential savings, attract users to online auctions. Apart from these, one other interesting finding was that customers indulge in online auctions for hedonic benefits, such as the fun, entertainment, and emotional value derived from bidding processes [89.78]. The popularity, reputation, responsiveness, and integrity of online auction sites are important to buyers while choosing between online auction platforms [89.31].

89.4 Emerging Trends in e-Commerce

Against this background of B2C, B2B, and C2C e-Commerce markets, we now consider some emerging e-Commerce applications: mobile commerce, telemedicine, and fee-based information delivery.

89.4.1 Mobile Commerce

One of the rising trends in the e-Commerce realm is that of mobile commerce (m-Commerce) wherein e-Commerce activities that were hitherto conducted using fixed-wire Internet are being facilitated by wireless Internet access using devices such as mobile phones, laptop computers, and other portable electronic devices. The driving forces behind m-Commerce include the explosive growth of the mobile phone population and the development of more powerful, sophisticated, and secure wireless technologies. The number of global cellular subscribers is projected to reach 4 billion in 2008 [89.79] and there are 262.7 million wireless users in the US, accounting for a penetration rate of 84% [89.80]. Although mobile shopping remains the major m-Commerce activity, other spheres include banking, information delivery, marketing, and airline ticketing. In particular, mobile banking is witnessing tremendous growth and the number of mobile banking customers in the US is expected to exceed 40 million by 2012 [89.81].

Organizations tend to use m-Commerce in conjunction with their traditional fixed-wire Internet-based outlets so as to provide increased flexibility and convenience for customers, and to cater to a younger and more technology-savvy customer base [89.82]. Major airlines such as American, Delta, and United provide reservation services through mobile devices apart from flight and weather-related information [89.83]. The success of m-Commerce depends on a variety of factors: an organization's ability to integrate existing systems into a mobile environment, development and use of devices that are easy to use and similar to already existing devices, value addition over and above that provided by traditional e-Commerce, and providing customers with relevant, inexpensive, and context- and location-sensitive service [89.83].

89.4.2 Telemedicine: e-Health

The American Telemedicine Association defines telemedicine as *the use of medical information exchanged from one site to another via electronic communications*. Currently, about 2000 medical institutions around the USA are connected through the Internet and other high-speed transmission lines. Of the approximately 200 telemedicine networks in the USA, more than half provide regular clinical services to patients, the others being used for research and education. Apart from the Internet, there are private networks linking hospitals, video systems for real-time patient consultation, and WWW-based patient service. Over 50 different medical specialties, including major ones such as cardiology, dermatology, ophthalmology, and pathology actively use telemedicine in the diagnosis and treatment of diseases [89.84].

Telemedicine facilitates the digital transmission of patient images (or teleradiology), wherein x-rays, computed tomography scans, magnetic resonance images, and digital images of pathology specimens can be sent worldwide for diagnosis and consultation [89.84, 85]. Videoconferencing equipment at physician and patient sites facilitates real-time consultation, especially between urban and rural locations. Using appropriate peripheral devices, it is possible for the physician to examine internal organs and reach an accurate diagnosis and treatment plan, resulting in time and monetary savings. Recent advances in this field include telesurgery, where a surgeon in one location can control a robot in another location via communication networks [89.85].

89.4.3 Fee-Based Information Delivery

Fee-based information delivery services provide information to consumers for a payment. The information may be broad as in the case of New York Times providing news analysis through New York Times Select, or may be narrowly focused as in the case of Wall Street Journal providing financial news, or eHarmony providing potential partner information. Such services used to be free, with their major revenue source being advertising; however as this model became unsustainable, information service providers have shifted to a fee-based model. Most information delivery sites have free content and fee-based premium content.

What is it that motivates customers to pay for material that could perhaps be obtained through other channels? Research indicates that it is the perceived benefits in terms of money, time, decision-making, and learning that attract users to fee-based information delivery. Also, customers tend to choose services that they perceive to be of high quality, are convenient, and provides added value over information delivery through alternative channels [89.32, 34].

89.5 Challenges and Emerging Issues in e-Commerce

The electronic commerce movement is still in its early stages. The big outlines are apparent, but there are many opportunities and challenges. Its future development depends on treatment of numerous issues. Examples of these include trust and legal issues.

89.5.1 Trust and e-Commerce

An important factor that influences B2C e-Commerce adoption is the user's perception of whether the online merchant or website can be trusted [89.45]. Recent surveys have indicated that less than one-half of Internet users actually trust e-Commerce websites, with primary concerns involving the integrity and trustworthiness of online merchants, and the related issue of the safety and security of online transactions [89.86, 87]. The resulting *trust gap* is one of the primary reasons that could impede the growth of B2C e-Commerce [89.46].

What is meant by trust in B2C e-Commerce? Trust has been conceptualized primarily across two dimensions: *faith in humanity* and *vulnerability* [89.45]. The *faith in humanity* conceptualization encompasses the implicit faith that human beings have in the integrity, benevolence, and goodwill existing in the world around them, and the belief that it extends to the e-Commerce world. Thus, a user believes that he/she can rely on a promise made by the online merchant and, in case of problems, it would act toward the user with honesty and benevolence [89.30, 88, 89]. The *vulnerability* conceptualization pertains to the readiness of users to be vulnerable to the actions of the online merchant, wherein the user is fully aware that he/she might have only limited or no influence over the actions of the online merchant [89.45, 90, 91].

In order to enhance user trust in a website, multiple techniques have been examined by researchers. Some of these are incorporated into website design, including website design factors such as the use of third-party assurance seals and certificates, hypertext links from reputed websites and associations, and the usability, usefulness, ease of use, and security control of the interface [89.89, 90]. Using reliable and secure technology and foregoing misleading language, images, and complex terminology serve to enhance user trust in a website [89.91, 92]. Another approach is educating the user by publicizing privacy policies, listing data security precautions incorporated into the website, and laying out procedures regarding compensation, dispute resolution, and mediation [89.30, 93]. Also, displaying positive user feedback from online user communities and providing fair and balanced information regarding similar offerings from competing merchants can serve to enhance user trust [89.94]. The perceived size and reputation of the online merchant and user familiarity and propensity to trust can also influence user adoption [89.88]. Other researchers have pointed out the need for governmental laws and regulation and obtaining informed consent before using private user information [89.92, 95].

89.5.2 Legal Issues in e-Commerce

The e-Commerce market is global and constantly evolving with new and innovative business models and products. As such, it is difficult to structure and enforce a uniform legal framework for conducting online business transactions. Thus, care must be taken by both businesses and consumers to ensure that either side is knowledgeable about the purchase, payment processing, and delivery mechanisms – as well as issues relating to privacy and confidentiality. The best practice would be to adapt the conventions followed in traditional brick-and-mortar sales to the online environment [89.96].

It is recommended that the online business prominently display the sale terms using simple and clear language. When a consumer purchases a product online, it is implicitly assumed that he or she has agreed to the contract; however, it is a good practice to have the customer explicitly accept or reject the contract terms. The terms of the contract should be consistent with product warranty and liability information. In addition, purchase eligibility criteria related to consumer age, geographic region, and product type must be displayed on the website and enforced. These aspects might relate to local and state laws, in which case the website content might have be localized [89.96].

Policies relating to privacy and the use of personal data, fraud, misrepresentation, and nonpayment should be prominently displayed on the website. Encryption tools must ensure the security of transactions and comply with governmental regulations. Care must be taken to ensure that there are no intellectual property violations for products that are displayed for sale. Although not legally mandated for nonfederal websites, it is a good practice to ensure that the website meets or exceeds accessibility criteria [89.97].

89.5.3 Outlook

We have discussed the evolution e-Commerce, reviewed its definitions, frameworks, and success parameters, and examined in detail the three popular areas of e-Commerce. Given its growing importance to consumers and businesses, as well as national and world economies, it is gratifying to note that there is a growing body of academic research addressing unresolved issues of e-Commerce. However, it must be pointed out that the overwhelming majority of research is being conducted in B2C e-Commerce, particularly in the online retail shopping segment. Considering the fact that the B2B e-Commerce market far exceeds the B2C market both in size and turnover, we suggest that researchers take a closer look at issues relevant to B2B transactions. To ensure that e-Commerce research remains relevant to practice, it is important that researchers examine the new and innovative trends in e-Commerce, such as e-Health, m-Commerce (e.g., mobile banking), and c-Commerce. Further research needs to be done on trust- and privacy-related issues as well as on establishing a solid legal framework for conducting e-Commerce transactions. Beyond transactions, there is also need for advances in the development and application of e-Commerce systems that support decision-making [89.98].

References

89.1 G. Ferry: *A Computer Called LEO* (Fourth Estate, London 2003)

89.2 L. Margherio, D. Henry, S. Cooke, S. Montes: *The Emerging Digital Economy* (US Department of Commerce, Washington 2007), http://www.esa.doc.gov/Reports/EmergingDig.pdf

89.3 DNS: *Internet Timeline* (Pearson Education, Boston 2007), http://www.factmonster.com/ipka/A0193167.html (last accessed February 9, 2009)

89.4 J. Weisman: *The Making of e-Commerce: 10 Key Moments* (ECT New Network, Encino 2000), http://www.ecommercetimes.com/story/4085.html

89.5 J. Symons: *Online Sales to Surpass $200 Billion This Year* (Forrester, Cambridge 2006), http://www.forrester.com/ER/Press/Release/0,1769,1081,00.html

89.6 US Department of Commerce: *2004 e-Commerce Multi-Sector Report* (US Department of Commerce, Washington 2006), http://www.census.gov/eos/www/papers/2004/2004reportfinal.pdf

89.7 L. Applegate, C. Holsapple, R. Kalakota, F. Radermacher, A. Whinston: Electronic commerce: Building blocks of new business opportunity, J. Org. Comput. Electron. Commer. **6**, 1–10 (1996)

89.8 V. Zwass: Electronic commerce and organizational innovation: aspects and opportunities, Int. J. Electron. Commer. **7**, 7–37 (2003)

89.9 W. DeLone, E. McLean: Measuring e-Commerce success: applying the DeLone & McLean information systems success model, Int. J. Electron. Commer. **9**, 31–47 (2004)

89.10 M. Ketel, T. Nelson: *CIO Definitions* (CIO, Needham 2003), http://searchcio.techtarget.com/sDefinition/0,,sid182_gci212029,00.html (last accessed February 9, 2009)

89.11 W. Koty: *e-Definitions* (High Latitude, Vancouver 2006), http://www.highlatitude.com/e-definitions/index.htm (last accessed February 9, 2009)

89.12 C. Holsapple, M. Singh: Electronic commerce: from a definitional taxonomy toward a knowledge-management view, J. Org. Comput. Electron. Commer. **10**, 149–170 (2000)

89.13 R. Kalakota, A. Whinston: *Frontiers of Electronic Commerce* (Addison-Wesley, Reading 1996)

89.14 J. McGee, L. Prusak: *Managing Information Strategically* (Wiley, New York 1993)

89.15 S. Allard, C. Holsapple: Competitiveness: from the knowledge chain to KM audits, J. Comput. Inf. Syst. **42**, 19–25 (2002)

89.16 A. Dogac: Electronic commerce, J. Database Manag. **9**, 31–35 (1998)

89.17 E. Hartono, C. Holsapple: Theoretical foundations for collaborative commerce research and practice, Inf. Syst. e-Bus. Manag. **2**, 1–30 (2004)

89.18 R. Wigand: Electronic commerce: definition, theory, and context, Inf. Soc. **13**, 1–15 (1997)

89.19 R. Kalakota, A. Whinston: *Electronic Commerce: A Manager's Guide* (Addison-Wesley, Reading 1997)

89.20 V. Zwass: Electronic commerce: structures and issues, Int. J. Electron. Commer. **1**, 3–23 (1996)

89.21 V. Zwass: Electronic commerce and organizational innovation: aspects and opportunities, Int. J. Electron. Commer. **7**, 7–37 (2003)

89.22 F. Riggins: A framework for identifying web-based electronic commerce opportunities, J. Org. Comput. Electron. Commer. **9**, 297–310 (1999)

89.23 S. Choi, D. Stahl, A. Whinston: *The Economics of Electronic Commerce* (Macmillan Technical, Indianapolis 1997)

89.24 C. Schlueter, M. Shaw: A strategic framework for developing electronic commerce, IEEE Internet Comput. **1**, 20–28 (1997)

89.25 M. Lindemann, B. Schmidt: Framework for specifying, building, and operating electronic markets, Int. J. Electron. Commer. **3**, 7–21 (1998)

89.26 E. Allen, J. Fjermestad: e-Commerce marketing strategies: an integrated framework and case analysis, Logist. Inf. Manag. **14**, 14–23 (2001)

89.27 T. Liang, J. Huang: A framework for applying intelligent agents to support electronic trading, Decis. Support Syst. **28**, 305–317 (2000)

89.28 M. Koufaris, A. Kambil, P. Labarbera: Consumer behavior in web-based commerce: an empirical study, Int. J. Electron. Commer. **6**, 115–139 (2001)

89.29 S. Lichtenstein, K. Williamson: Understanding consumer adoption of internet banking: an interpretive study in the Australian banking context, J. Econ. Commer. Res. **7**, 50–66 (2006)

89.30 B. Suh, I. Han: The impact of customer trust and perception of security control on the acceptance of electronic commerce, Int. J. Electron. Commer. **7**, 135–161 (2003)

89.31 S. Walczak, D. Gregg, J. Berrenberg: Market decision making for online auction sellers: profit maximization or socialization, J. Electron. Commer. Res. **7**, 199–220 (2006)

89.32 A. Lopes, D. Galletta: Consumer perceptions and willingness to pay for intrinsically motivated online content, J. Manag. Inf. Syst. **23**, 203–231 (2006)

89.33 G. Shergill, Z. Chen: Web-based shopping: consumers' attitudes towards online shopping in New Zealand, J. Electron. Commer. Res. **6**, 79–94 (2005)

89.34 C. Wang, Y. Zhang, L. Ye, D. Nguyen: Subscription of fee-based online services: what makes consumer pay for online content, J. Electron. Commer. Res. **6**, 304–311 (2005)

89.35 L. Schaupp, F. Belanger: A conjoint analysis of online consumer satisfaction, J. Electron. Commer. Res. **6**, 95–111 (2005)

89.36 A. Floh, H. Treiblmaier: What keeps the e-Banking customer loyal? A multigroup analysis of the moderating role of consumer characteristics on e-Loyalty in the financial service industry, J. Electron. Commer. Res. **7**, 97–110 (2006)

89.37 K. Suh, Y. Lee: The effects of virtual reality on consumer learning: an empirical investigation, Manag. Inf. Syst. Q. **29**, 673–697 (2005)

89.38 N. Kumar, I. Benbasat: The influence of recommendations and consumer reviews on evaluations of websites, Inf. Syst. Res. **17**, 425–439 (2006)

89.39 A. Everard, D. Galletta: How presentation flaws affect perceived site quality, trust, and intention to purchase from an online store, J. Manag. Inf. Syst. **22**, 55–95 (2005/2006)

89.40 W. DeLone, E. McLean: Information success: the quest for the dependent variable, Inf. Syst. Res. **3**, 60–95 (1992)

89.41 W. DeLone, E. McLean: The DeLone and McLean model of information systems success: a ten-year update, J. Manag. Inf. Syst. **19**, 9–30 (2003)

89.42 Alexa: *Top Sites United States* (Alexa Internet, San Francisco 2007), http://www.alexa.com/site/ds/top_sites?cc=US&ts_mode=country&lang=none

89.43 K. Hassanein, M. Head: The impact of infusing social presence in the web interface: an investigation across product types, Int. J. Electron. Commer. **10**, 31–55 (2006)

89.44 W. Hampton-Sosa, M. Koufaris: The effect of web site perceptions on initial trust in the owner company, Int. J. Electron. Commer. **10**, 55–81 (2005)

89.45 C. Holsapple, S. Sasidharan: The dynamics of trust in online B2C e-Commerce: a research model and agenda, Inf. Syst. e-Bus. Manag. **3**, 377–403 (2005)

89.46 S. Fox, J. Beier: *Surfing to the Bank* (Pew Research Center, Washington 2006), http://pewresearch.org/pubs/31/surfing-to-the-bank

89.47 C. Yom: Limited-purpose banks: their specialties, performance, and prospects, FDIC Bank. Rev. **17**, 10–36 (2005)

89.48 H. Karjaluoto, M. Mattila, T. Pento: Factors underlying attitude formation towards online banking in Finland, Int. J. Bank Mark. **20**, 261–272 (2002)

89.49 W. Lassar, C. Manolis, S. Lassar: The relationship between consumer innovativeness, personal characteristics and online banking adoption, Int. J. Bank Mark. **23**, 176–199 (2005)

89.50 S. Rotchanakitumnuai, M. Speece: Corporate customer perspectives on business value of Thai internet banking, J. Electron. Commer. Res. **5**, 270–286 (2004)

89.51 T. Davydov, D. Emery, S. Lahanas, B. Potemski (Eds.): *Learning Circuits Glossary* (American Society for Training and Development, Alexandria 2004), http://www.astd.org/lc/glossary.htm (last accessed February 9, 2009)

89.52 M. Brennan: *US Corporate and Government eLearning Forecast, 2002–2007* (International Data Corporation, Framingham 2003)

89.53 C. Holsapple, A. Lee-Post: Defining, assessing, and promoting e-Learning success: an information systems perspective, Decis. Sci. J. Innov. Educ. **4**, 67–85 (2006)

89.54 S. Sasidharan, R. Santhanam: Technology-based training: toward a learner centric research agenda. In: *Human–Computer Interaction in Management Information Systems: Applications*, ed. by V. Zwass (Sharpe, New York 2006)

89.55 R. Santhanam, S. Sasidharan, J. Webster: Using self-regulatory learning to enhance e-Learning-based information technology training, Inf. Syst. Res. **19**, 26–47 (2008)

89.56 B. Zimmerman: Self-regulated learning and academic achievement: an overview, Educ. Psychol. **25**, 3–17 (1990)

89.57 G. Piccoli, R. Ahmad, B. Ives: Web-based virtual learning environments: a research framework and a preliminary assessment of effectiveness in basic IT skills training, Manag. Inf. Syst. Q. **25**, 401–426 (2001)

89.58 J. Webster, P. Hackley: Teaching effectiveness in technology-mediated distance learning, Acad. Manag. J. **40**, 1271–1281 (1997)

89.59 G. Hislop: Working professionals as part-time online learners, J. Asynchr. Learn. Netw. **4**, 73–89 (2000)

89.60 A. Cullen, M. Webster: A model of B2B e-Commerce, based on connectivity and purpose, Int. J. Operat. Prod. Manag. **27**, 205–225 (2007)

89.61 M. Grieger: Electronic marketplaces: a literature review and a call for supply chain management research, Eur. J. Oper. Res. **144**, 280–94 (2003)

89.62 A. Zeng, B. Pathak: Achieving information integration in supply chain management through B2B 3-hubs: concepts and analyses, Ind. Manag. Data Syst. **103**, 657–65 (2003)

89.63 Grainger: *Grainger Industrial Supply* (Grainger, Lake Forest 2007), http://www.grainger.com/Grainger/wwg/start.shtml

89.64 Covisint: *About* (Covisint, Detroit 2007), http://covisint.com/about/

89.65 ChemConnect: Official homepage (Houston 2008), http://www.chemconnect.com/

89.66 Wolverine Supplies: Official homepage (Manitoba 2008), http://www.wolverinesupplies.com/

89.67 Elemica: Official homepage (Exton, 2008), http://www.elemica.com/

89.68 Powerspring: Official homepage (Melbourne 2008), http://www.powerspring.com/

89.69 XSAg: *XSAg Buyers* (XSAg, Morrisville 2007), http://www.xsag.com/chem/buyers.asp

89.70 Fedbid: *About Fedbid* (Fedbid, Vienna 2007), http://www.fedbid.com/about/

89.71 T. Strader, S. Ramaswami: The value of seller trustworthiness in C2C online markets, Commun. Assoc. Comput. Mach. **45**, 45–49 (2002)

89.72 K. Laudon, C. Traver: *e-Commerce: Business, Technology, Society* (Addison Wesley, Boston 2004)

89.73 J. Jones, K. Kuan, S. Newton: I name my price but don't want the prize, J. Electron. Commer. Res. **7**, 178–198 (2006)

89.74 D. Lucking-Reily: Using field experiments to test equivalence between auction formats: magic on the internet, Am. Econ. Rev. **89**, 1063–1080 (1999)

89.75 Ebay: *Online Auction Format* (Ebay, San Jose 2007), http://pages.ebay.com/help/sell/f-auction.html

89.76 P. Bajari, A. Hortaçsu: Economic insights from internet auctions, J. Econ. Lit. **42**, 457–486 (2004)

89.77 D. Gregg, S. Walczak: e-Commerce auction agents and online-auction dynamics, Electron. Mark. **13**, 242–250 (2003)

89.78 S. Standifird, M. Roelofs, Y. Durham: The impact of eBay's buy-it-now function on bidder behavior, Int. J. Electron. Commer. **9**, 167–176 (2004)

89.79 International Telecommunications Union: *Worldwide Mobile Cellular Subscribers to Reach 4 Billion Mark Late 2008* (International Telecommunications Union, Geneva 2008), http://www.itu.int/ITU-D/ict/newslog/Worldwide+Mobile+Cellular+Subscribers+To+Reach+4+Billion+Mark+Late+2008.aspx (last accessed February 9, 2009)

89.80 International Association for the Wireless Telecommunications Industry: *Wireless Quick Facts* (International Association for the Wireless Telecommunications Industry, Washington 2008), http://www.ctia.org/advocacy/research/index.cfm/AID/10323 (last accessed February 9, 2009)

89.81 B. Egan, G. Tubin, C. Vyas: *US Mobile Banking Forecast 2007–2012* (The Tower Group, Needham 2007), http://www.epaynews.com/index.cgi?survey=&ref=browse&f=view&id=117620743521320213796&block=1

89.82 F. Buellingen, M. Woerter: Development perspectives, firm strategies and applications in mobile commerce, J. Bus. Res. **57**, 1402–1408 (2004)

89.83 N. Pagiavlas, M. Stratmann, P. Marburger, S. Young: Mobile business – comprehensive marketing strategies or merely IT expenses? A case study of the US airline industry, J. Electron. Commer. Res. **6**, 251–261 (2005)

89.84 American Telemedicine Association: *Defining Telemedicine* (ATA, Washington 2007), http://www.americantelemed.org/i4a/pages/index.cfm?pageid=3333 (last accessed February 9, 2009)

89.85 N. Brown: *Telemedicine Coming of Age* (Association of Telehealth Service Providers, Portland 1996), http://tie.telemed.org/articles/article.asp?path=consumer&article=tmcoming_nb_tie96.xml#dperedniaref (last accessed February 9, 2009)

89.86 Princeton Survey Research Associates: *A Matter of Trust: What Users Want From Web Sites?* (Consumer Reports Web Watch, New York 2002)

89.87 Princeton Survey Research Associates: *Leap of Faith: Using the Internet Despite the Dangers* (Consumer Reports Web Watch, New York 2005)

89.88 D. Gefen: e-Commerce: the role of familiarity and trust, Omega **28**, 725–737 (2000)

89.89 K. Stewart: Trust transfer on the world wide web, Org. Sci. **14**, 5–17 (2003)

89.90 K. Kimery, M. McCord: Third-party assurances: mapping the road to trust in e-Retailing, J. Inf. Technol. Theory Appl. **4**, 63–82 (2002)

89.91 K. Lee, E. Turban: A trust model for consumer internet shopping, Int. J. Electron. Commer. **6**, 75–91 (2001)

89.92 B. Friedman, P. Kahn, D. Howe: Trust online, Commun. Assoc. Comput. Mach. **43**, 34–40 (2000)

89.93 B. Shneiderman: Designing trust into online experiences, Commun. Assoc. Comput. Mach. **43**, 57–59 (2000)

89.94 G. Urban, F. Sultan, W. Qualls: Placing trust at the center of your internet strategy, Sloan Manag. Rev. **42**, 39–49 (2000)

89.95 D. Schoder, P. Yin: Building firm trust online, Commun. Assoc. Comput. Mach. **43**, 73–79 (2000)

89.96 D. Adler: *Legal Aspects of e-Commerce* (INC, New York 2000), http://www.inc.com/articles/2000/05/19706.html

89.97 C. Holsapple, R. Pakath, S. Sasidharan: A website interface design framework for the cognitively impaired: a study in the context of Alzheimer's disease, J. Electron. Commer. Res. **6**, 291–303 (2005)

89.98 C. Holsapple, K. Joshi, M. Singh: Decision support applications in electronic commerce. In: *Handbook on Electronic Commerce*, ed. by M. Shaw (Springer, Berlin Heidelberg 2000)

90. Business Process Automation

Edward F. Watson, Karyn Holmes

Integrated enterprise-wide information systems (EwIS) are a class of customizable packaged business software applications that have replaced arrays of disparate legacy systems in organizations around the world. EwIS have been the catalyst for the reengineering and automation of core business processes that has led to organization-wide transformation across most industries in corporate America. Chief among this category of packaged business software is enterprise resource planning (ERP), the back-office suite that was embraced by many industries in the 1990s as a cure for legacy system ailments and impending year-2000 (Y2K) disasters. ERP is considered a product of the evolution of an earlier manufacturing planning system referred to as manufacturing resource planning (MRPII). Whereas MRPII was focused on the factory planning environment, ERP incorporates enterprise-wide functionality and therefore is used in virtually all industries. ERP has enabled organizations to streamline, automate, and commoditize their business processes, leveraging best-of-industry practices, quite significantly over the last 15 years. Two other packages that are attributing to this phenomenon are customer

relationship management (CRM) and supply chain management (SCM). In this chapter we review EwIS in a historical context as it has developed over the years and discuss the most important characteristics of EwIS today as well as how we expect this field to evolve.

Enterprise-wide information systems (EwIS) are powerful software packages that enable businesses to integrate a variety of disparate functions [90.1]. As pointed out by *Davenport* [90.2], EwIS terminology evolved from the more widely used term enterprise resource planning (ERP). ERP systems, as originally defined, represent the *back-office* of the corporation. EwIS tend to refer to a more generic structure that represents any enterprise application that integrates multiple business functions. The ERP market experienced explosive growth in the USA during the 1990s, due in large part to a little known German software company named SAP – headquartered in Waldorf – that delivered to the US market ERP software of the same name.

In the early 1990s, SAP took the USA by storm, delivering the first comprehensive, real-time ERP system on a client–server platform. Having a well-engineered product and selling the product to C-level executives as a business transformation-enabled solution while Y2K issues were imminent were the necessary ingredients to catapult SAP into the market-leading position. Other vendors, through acquisition and organic growth, scrambled for market share. PeopleSoft, the leading human resource management system, and Oracle, well known for their leading database management system and less known for their business accounting software, both slowly evolved to offer more comprehensive solutions. By the turn of the century, hundreds of ERP vendors surfaced to serve all types and sizes of private and public organizations.

This chapter is broken down into four main parts. The second section, immediately following this first section, describes the business environment when ERP came to the US market in the early 1990s. It then defines the key characteristics of ERP that have essentially defined the new class of standard software applications referred to as enterprise-wide information systems. ERP implications are then discussed since ERP is not simply a software package. Finally, EwIS systems do not stand still, and it is important to understand how they will continue to evolve.

The third main section presents the enterprise systems application framework and briefly describes two of the other most significant EwIS applications on the market today: customer relationship management (CRM) and supply chain management (SCM). The fourth main section provides a brief overview of the emerging services industry from the enterprise systems perspective. Finally, the fourth main section looks briefly at future trends of which practitioners and researchers should be aware. This chapter ends with a brief conclusion and a reference list of the significant readings that were used to write this chapter. Related information can also be found in Chap. 54 on *Production, Supply, Logistics, and Distribution*; Chap. 86 on *Enterprise Integration and Interoperability*; and Chap. 88 on *Collaborative e-Work, e-Business, and e-Service*.

90.1 Definitions and Background

Much has been written about the emerging information economy and the challenges corporations around the world face in their quest to shed their legacy systems and processes and transform themselves into lean, agile, and responsive organizations [90.3]. Since the 1970s and 1980s, global competition has threatened the existence of many, if not most, companies in the USA, and management and business consulting organizations have prospered during these times. The search for excellent business practices that would pave the path to success, a concept popularized by *Peters* and *Waterman* in [90.4], led many executives to look for a silver-bullet cure. A paradigm shift, driven by information technology (IT)-enabled radical change, took place in the way companies competed and were managed. Traditional push – or *make-and-sell* – strategies, tied to the annual budget cycle were replaced by radically faster, real-time, *sense-and-respond*, pull strategies [90.5]. Firms learned that they could no longer lead the competition by forecasting customers' needs and then planning the year's production using inventories to match supply and demand. Instead, they had to rely on real-time information to evaluate the needs of each customer continuously. In fact, it was often necessary to anticipate unspecified needs, and then quickly fulfill these needs with customized products and services delivered with unprecedented speed. This shift from an industrial economy to an information economy was accompanied by a simultaneous shift in power from the producer to the (information-empowered) consumer.

During this time, significant emphasis was placed on organizational improvement initiatives such as total quality management, continuous improvement, benchmarking, and employee involvement [90.6]. A competing change paradigm suggested that continuous improvement was not sufficient to make significant advances towards performance excellence and that deeper, systemic change was required [90.7]. Business process reengineering (BPR), a term coined in the 1980s [90.8], provided the means by which an organization could fundamentally change its core business processes and transform itself into a lean, globally competitive, structure. In order to do this, companies were forced to place intense focus on their customer, and to define their core processes based on those, and only those, activities that collectively added value to their internal and external customers. Business processes were streamlined and built around core competencies and capabilities. As we discuss in this paper, an important catalyst for the reengineering and streamlining of back-office business processes was ERP.

The emphasis on the core *business processes* [90.9, 10] of the firm was not easy for most organizations. Redefining the work of an organization based on the customer, instead of on the organization, was a new way of thinking. However, this exercise was necessary in order to clean up years of building systems on top of systems on top of (legacy) systems. An intense focus on the customer became the essential ingredient to success [90.11–13].

With this quest towards efficient and effective business processes under way, organizations quickly realized the challenge of institutionalizing these important process changes. Organizations quickly adopted a multidimensional perspective on these transformation initiatives: people, process, and technology. The foremost significant hurdle to change was the organization's ability and readiness to implement change, the people (also known as, organization) dimension. Implementation-related challenges are discussed later in the chapter. The process dimension presented difficult challenges as companies had to look inside and identify their core processes. Prior to this time, most organizations did not think in terms of *business process*, but instead focused on their performance as defined within their *functional fiefdoms*. Quite some time transpired before corporations fully appreciated the concept of *process*, defined as a set of organized activities that collectively add value to a (internal or external) customer. In fact, companies learned that most business processes crossed functional boundaries, were not eas-

ily adopted by the hierarchical organization, and clashed with function-oriented incentive systems.

Figure 90.1 illustrates three core business processes that are well-known and understood today: order-to-cash cycle, fulfill-demand cycle, and procurement cycle. Throughout the 1980s, many companies sought this holy grail of business process definitions independently. There was little success at industry-wide collaboration to achieve best business practices. As every company designed new business processes, they were ultimately challenged to determine how to best implement their new processes. The answer most often was less than optimal as they tied together best-of-breed business applications, spanning multiple computing platforms, or built new applications specific to their needs. The ERP systems introduced to corporate America in the 1990s, as discussed in this chapter, brought to market best practice repositories, built on years and years of experience. And the best way to institutionalize these new processes was through the third dimension of change: technology.

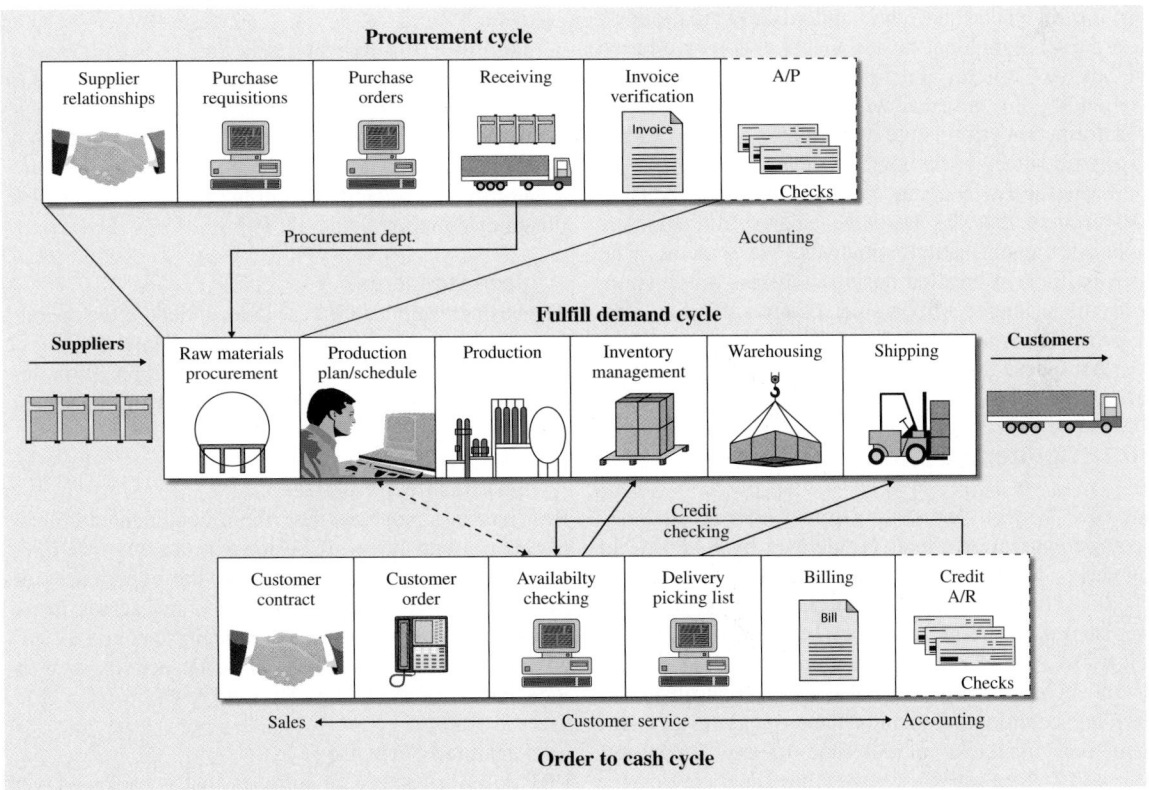

Fig. 90.1 High-level integrated business processes

As we look back to the pre-1990s, we see that advances in technology and corporate computing were also being made. A significant development that triggered the growth in the EwIS market was client–server technology [90.14]. Client–server systems were not based on a single technological innovation, but rather very different development currents in various areas of data processing were united. In particular, progress in the areas of hardware technology, networks, system software, user interface, multimedia applications, object orientation, and software development tools have contributed to this breakthrough. While companies sought to transform their organizations and shed themselves of years of inefficient processes and procedures, they were equally anxious to migrate away from their expensive mainframe computing platforms in order to find a lower-cost, flexible computing environment.

While these advances in the business and technology worlds evolved, another very important concept was also picking up momentum: the use of standard application software [90.15]. Advocates of standard software believed that business processes must be determined independent of the standard software products in order to ensure that business applications are based on business criteria and not on technology platform availability. In this manner, an IT-enabled business transformation effort could be driven by corporate strategy instead of by technological constraints [90.16].

From the EwIS arena, we look first at the flagship ERP system that has been the catalyst for so many business transformation initiatives. We look at what it is and how it enabled business process automation, from the packaged software perspective. We then look at two systems (i.e., CRM and SCM), similar in concept but focused more on a company's outwards-facing processes.

90.1.1 Anatomy of an ERP System

An ERP system represents a grand concept and has been perhaps most effectively defined by SAP AG in an early (1996) company presentation illustrated in Fig. 90.2. In this illustration, the honeycomb structure emphasizes the modular layout of the functional applications as well as the anticipation of always adding or enhancing functionality over time. At the heart of ERP is a common, relational database making all company data available in real time to any authorized user and linking activities across these applications via workflow. An ERP system can be characterized by a number of attributes that are generally common across

all EwIS: comprehensive functional modules; business-process-oriented, common, and relational databases; open standards; best business practices; packaged software; business information systems standards; and configurable, common-development-platform, complex workflow capabilities. These attributes are defined for ERP below.

Comprehensive Functional Modules
ERP systems consist of the core (back-office) business applications of four basic functional areas: financials, operations and logistics, sales and marketing, and human resources.

Business Process Oriented
An important idea behind ERP is that the system must support company-wide business process activities, as defined by the company. These activities typically span many functional areas; for example, a typical customer order management process would consist of presales activity, sales order processing, inventory sourcing, delivery, invoicing, and payment [90.17, 18].

Common and Relational
The problem that ERP systems are designed to solve is the fragmentation of information in large business organizations. A common database facilitates the integration of this information into a single information silo. A relational database organizes records into a series of tables linked by common fields [90.19].

Open Standards
The business applications are independent of the operating system and the hardware platform. Thus, a business has the freedom to choose the business application, independent of the operating system and the hardware platform.

Best Business Practices
Best business practices describe a documented collection (i.e., repository) of business processes that have been accumulated over time from the experiences of many businesses within an industry and across many industries. Best practices are typically technology enabled, but they may describe work activity that is independent of technology as well [90.18].

Packaged Software
ERP describes a class of software that is packaged, yet customizable, and is available from ERP vendors. This is one-stop shopping for the user, as a single vendor may

offer the complete back-office packaged business suite. This is not to say that a company cannot develop their own system that has ERP characteristics.

Business Information Systems Standards

Packaged software forces business information systems standardization. Examples of where standards exist include data types, organizational types, user interfaces, software development, application interfaces, and reporting.

Configurable

ERP software requires the expertise of business process experts in order to configure the system to support company-specific processes and data. ERP vendors have gone to great extremes to provide template solutions for smaller companies with, relatively speaking, standardized processes.

Common Development Platform

ERP systems are expandable by various means, besides configuration. First, it is typically possible, though not recommended, to *modify* the source code. It is also possible to *extend* the software functionality with third-party application modules and then to link the modules to the ERP system. Finally, ERP users may *link* the software to third-party software packages in order to incorporate more sophisticated or specialized functionality.

Complex

The downside of ERP systems is that, by their very nature, they are somewhat complex. One consequence is that it is very difficult, especially during the software evaluation period, to determine the extent to which an ERP system can represent your business process without actually implementing the process in the software.

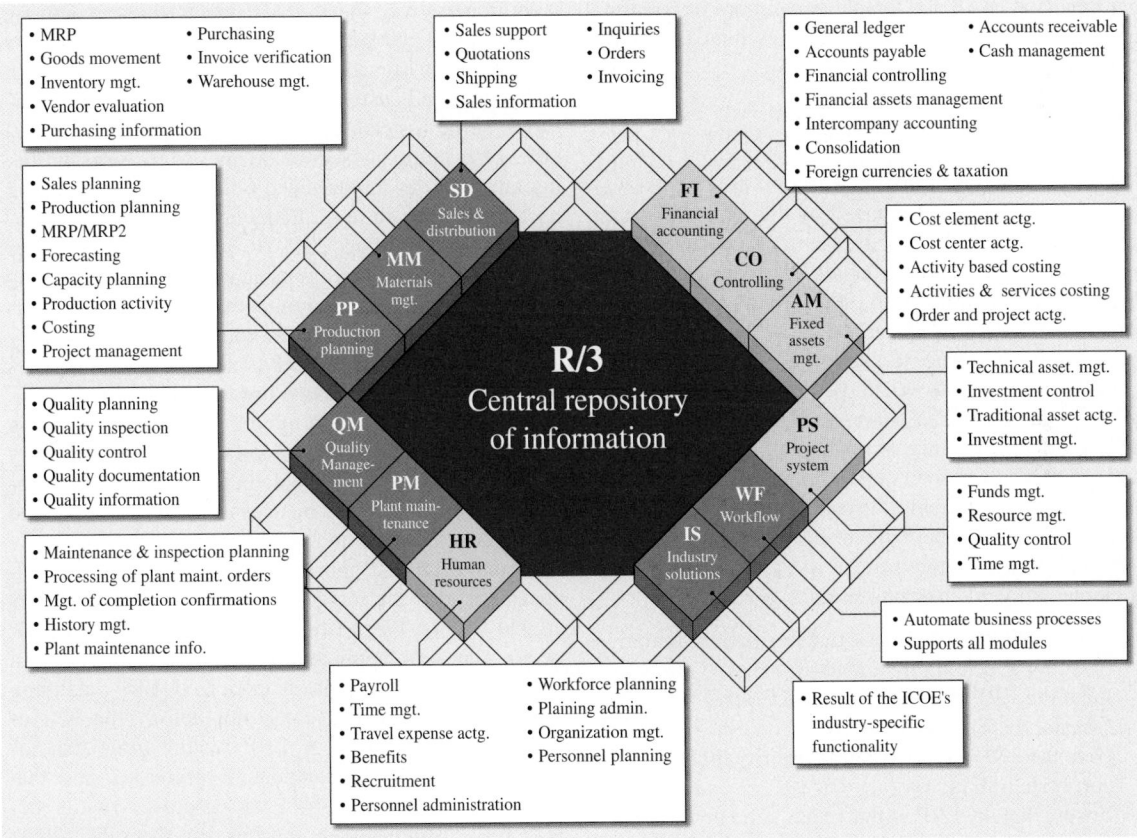

Fig. 90.2 Enterprise resource planning per SAP (after company presentation 1996) (ICOE – Industry Center of Expertise)

Also, similar to the leading office productivity software, Microsoft Office, there is a lot of overhead such that most users use less than 5% of the available functionality.

Workflow Capabilities

This refers to the automation of business processes within the enterprise system. Workflow capabilities permit users to define event-driven routings, create automated in-boxes and prioritized cues, allow background queries to take place with notification to the user, and more generally, to streamline processes.

ERP vendors have accumulated perhaps the greatest collection of industry (best) business practices in digital repositories. In fact, these repositories have become a formidable competitive advantage for the major vendors, specifically, SAP. ERP systems have evolved quickly and as this market is saturated one can see the offerings of each major vendor start to converge such that they (i. e., SAP and Oracle) have more in common in terms of functionality and features than they have that's different.

90.1.2 ERP Implications

The impact of ERP on business in the 1990s was profound. ERP was recognized as the most important development in the corporate use of information technology in the 1990s, even though the Internet received most of the media attention [90.2]. Many believed that "most white collar jobs – as we know them – will disappear as we get the ERP/enterprise resource planning – etc. – 'stuff' right" [90.20, p. 3]. One popular management guru, Michael Hammer, lectured around the globe on the importance of treating an ERP implementation as a major business transformation in demand of strong C-level leadership, instead of as an IT initiative delegated to the IT department [90.21]. Hammer's research identified factors critical to the success of ERP implementation in an organization:

- View ERP as a business transformation initiative (i. e., not a technology initiative)
- Lead the ERP initiative from the C-level (i. e., passionate, focused leadership)
- Give the ERP initiative a high priority relative to all corporate initiatives
- Ensure that the ERP initiative has solid project management and change leadership.

One of the major challenges presented by an ERP system is the need to standardize on technology and on business practices [90.2, 22]. In the purest sense, ERP enables complete standardization of (best) business processes and technologies, as illustrated in Fig. 90.1. However, many companies could not initially reach this hurdle, as reported by Fortune magazine in 1998 [90.23]. Besides implementing ERP to handle order management, production planning, materials management and finance, VF Corp custom developed or used separate software suppliers to augment ERP such as warehouse control (custom), product development (Gerber), micromarketing (Marketmax, Spectra, JDA software, and custom), forecasting (Logility), and capacity planning (i2). As companies decide to add additional software to manage their special circumstances, they inherently choose to take on more costs associated with interconnectivity, data integrity, and maintenance, for example.

Perhaps the most interesting observations the authors have made from talking to many consultants and practitioners over the past decade is that a company must first acknowledge that their limited IT resources should not be spread across the costs associated with developing and customizing the entire enterprise application suite. Instead, it is prudent to look at (roughly) 80% of a company's core business processes as being amenable to standardized business processes using packaged software (e.g., ERP) and industry best practices. The remaining 20% of processes provide the company with a more manageably sized opportunity to create or strengthen a competitive advantage by perhaps customizing a software solution around it. Industry best practices, available as business process templates by ERP vendors, have been provided by software vendors but are actually built upon the cumulative learning of these vendors who have worked with their industry clients. Ultimately, ERP implementation enables a company to build seamless business processes, integrated within the enterprise and across enterprises.

Since ERP first gained popularity in the USA, advocates lauded the integration features whereas others argued for the best-of-breed approach of knitting together the best application packages from the best vendors. This latter group criticized ERP, claiming that one single vendor could not deliver the *best* of everything. Advocates of ERP would argue that integration of good applications is preferred, and that standardization of the 80% of processes (while developing customized approaches for the other 20%) will strengthen the value proposition of the initiative. This concern over integration and best-of-breed has sparked the debate over the emerging service-oriented architecture approach discussed later where vendors promise to deliver comprehensive best-of-breed functionality (services) in a completely integrated and open manner.

ERP initiatives are typically rationalized as the most prudent approach to automate and streamline the back-office business processes by leveraging best business practices and packaged EwIS software. Whereas the earliest companies saw ERP as a way to quickly move from a dysfunctional legacy system, or to avoid Y2K disaster, more recent implementations have been able to learn from the experiences of others and take a more carefully calculated approach to achieve specific business benefits. A study led by *Davenport* and sponsored by Accenture [90.24] found specific benefits resulting from ERP initiatives across diverse industries such as: improved decision making, improved financial management, improved customer service and retention, ease of expansion/growth and increased flexibility, faster, more accurate transactions, head-count reduction, cycle time reduction, improved inventory/asset management, fewer physical resources/better logistics, and increased revenue. *Shang* and *Seddon* devel-

oped a more comprehensive framework for assessing and managing the benefits associated with enterprise systems (ES) [90.25]. This research identified five dimensions of ES benefits:

- Operational benefits
- Managerial benefits
- Strategic benefits
- IT infrastructure benefits
- Organizational benefits.

Studies such as these have helped many executives rationalize the ERP initiatives for their organizations. In fact, the same rationalization has been adopted by not-for-profit organizations in government and education industries.

ERP implementation involves the automation of business transactions and this typically requires significant change to how work is done and how decisions are made. It has been shown that the key challenges when implementing ERP and forcing business process change lie in the organization, not the technology. The results from a study conducted by Deloitte Consulting in 2003 reflect this idea. The study also concluded that

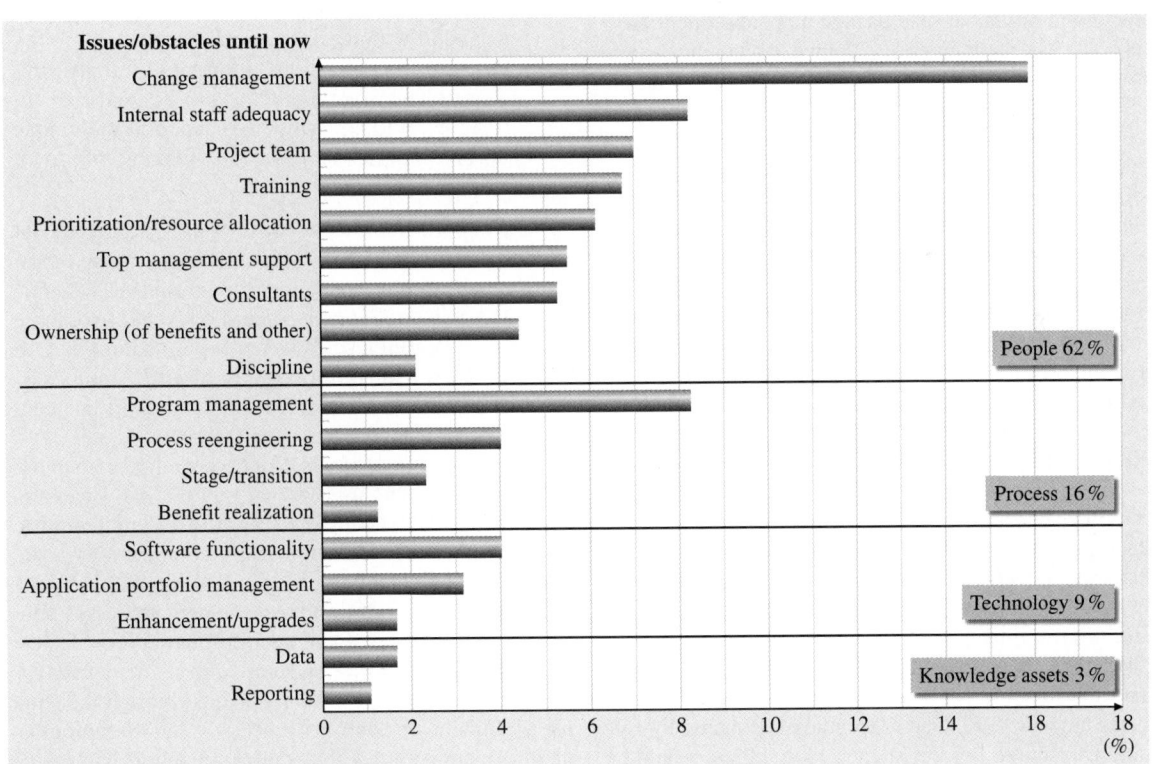

Fig. 90.3 ERP issues prior to going live

people-related issues continue to dominate after the implementation is completed as well.

Similar to best business practices, ERP software vendors today have very well-established implementation methodologies, tools, and templates available for ensuring successful implementations. The implementation process itself is very well understood and standardized, and leverages best implementation practices based on years of implementation experience in hundreds of companies across virtually all industries. Implementation methodologies have key project phases associated with them. SAP's implementation methodology, for example, has five phases: project preparation, business blueprint, realization, final preparation, and go-live and support. Each phase has very detailed work packages and clear deliverables. These methodologies alleviated many project-management-related challenges. Expectations are clear and easier to manage. Communications is specific and timely. Work activity is easier to manage and control as project managers utilize computer-aided tools to better coordinate and control the vast resources and activities associated with the project. Finally, the costs associated with an ERP implementation depend greatly on how much is invested in organizational change and subsequent business process reengineering, change management, and training. However, for perhaps a typical implementation one can view the breakdown of costs to roughly be as follows: 15% software costs, 30% hardware costs, 40% consulting fees (i. e., business process reengineering, software configuration, change management), 15% personnel (i. e., nonproductive time spent working on the project, or in training, or learning curve related), and unfortunately, the remainder which is typically something less than 2%, would go towards training (i. e., end-user training as well as training to develop a core expertise around software development, systems administration, and software configuration).

90.1.3 ERP Evolution

The ERP concept evolved from manufacturing resource planning (MRPII), but the concept was so compelling that it quickly expanded beyond its manufacturing roots into industries such as financial services, government, education, utilities, and retail [90.2]. Until January 1, 2000, there was so much hype around Y2K compliance that not too many organizations looked at ERP issues beyond Y2K. One 1998 study sponsored by Deloitte Consulting [90.26] suggested that there would be a second wave of activity surrounding ERP. The study

confirmed that *going live* with ERP is not the end. Rather, going live was viewed as the beginning of a journey toward continuous improvement, innovation, and agility. The *first Wave* referred to the business process changes associated with implementing ERP. The *second wave*, on the other hand, referred to the actions that are taken after going live that help organizations achieve the full potential of ERP-enabled processes (i. e., process optimization). A separate study conducted in 2002 and sponsored by Accenture again emphasized the importance of viewing ERP as continuously evolving. The study suggested that organizations seeking to capture the original promise of enterprise solutions must follow a logical path to *integrate, optimize, and informate* as:

- *Integrate* – unify and harmonize enterprise solutions, data, and processes with an organization's unique existing environment, and use the systems to better connect organizational units and processes, as well as customers and suppliers.
- *Optimize* – standardize most processes using best practices embodied in enterprise solutions software, mold and shape processes to fit the unique strategic needs of the business, and ensure that processes flow and fit with the systems themselves.
- *Informate* – organizations informate by transforming enterprise solutions data into context-rich information and knowledge that supports the unique business analysis and decision-making needs of multiple work forces.

Davenport, in an earlier study [90.27], pointed out that, following their ERP implementation, many companies had trouble realizing the promised benefits. Complete *benefit realization* could take one, two, three or more years. But also, Davenport pointed out that, in order to fully realize the benefits of ERP, companies must figure out how to use the data to support business decisions.

Implementation of ERP does not ensure that the ERP system will be used properly or that the ERP system will lead to value. At the turn of the millennium, studies found that, although many organizations justified their implementation of an ERP on the basis of better decision-making and management processes, few had taken full advantage of the information provided by the system [90.27]. ERP systems were clearly a catalyst for operational process improvements through the simple alignment of business processes and streamlining of information throughout the organization. However, the business world was slow to recognize the potential embedded in ERP for changing management processes such as performance reporting and monitoring, stakeholder communications, and managing customer relations, for instance.

ERP systems left management with a transaction repository filled with terabytes of important business information and data, but there were two problems. First, executives needed something to filter the data and weed out all that was not relevant to their problem at hand. Second, they needed a way to organize, analyze, and interpret the data and then to focus the resulting information on specific actions [90.29]. This realization led enterprise software developers to implement a packaged software strategy and attack higher-level thought processes such as decision making and knowledge management. Indeed, much effort has gone into understanding how the human knowledge creating process works [90.30]. That is, ERP by itself does not provide a competitive advantage but instead it has become a competitive imperative. It provides an effective means to bring order to the back office. Post-ERP companies strive to most effectively leverage their analytic

prowess not just to enhance operations but often as their lead competitive differentiator. The extensive use of data, statistical and quantitative analysis, explanatory and predictive modeling, and fact-based management are reportedly being used to drive company decisions and actions [90.31].

ERP software vendors, of course, anticipating this trend, leveraged this packaged software hype and their packaged software competencies, and delivered to the market the concept of the e-Business suite, illustrated in Fig. 90.4 [90.28]. The basic concepts used to develop ERP (e.g., packaged software, based on best business practices) were used to develop *decision-oriented and knowledge-management-type* applications such as customer relationship management (CRM), supplier relationship management (SRM), supply chain management (SCM), strategic enterprise management (SEM), business intelligence (BI), and portals. The products initially delivered to the market had relatively limited functionality and robustness, but over time these products have evolved into quite comprehensive and sophisticated tools for the organization.

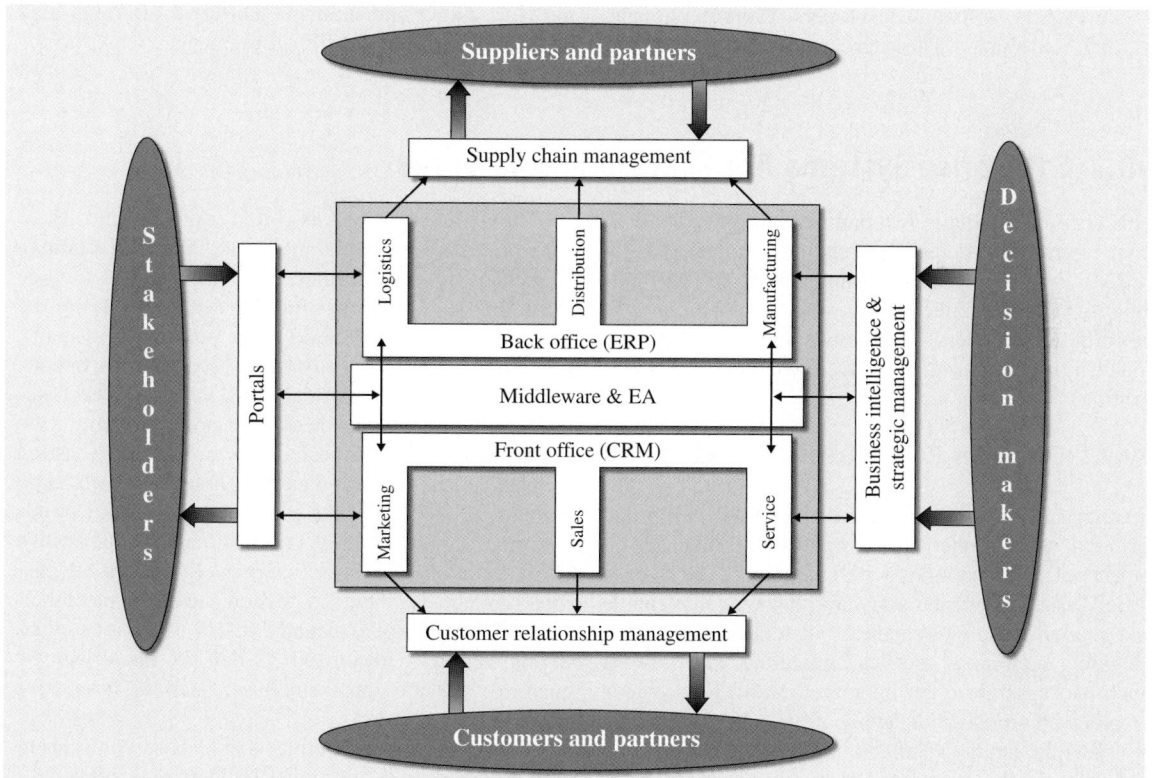

Fig. 90.4 Enterprise application framework (after [90.28])

Ongoing evolution of ERP requires a mechanism to enable, and perhaps promote, continuous improvements as well as disruptive innovations [90.32]. ERP vendors have, to some degree, become the mechanism for improvement and innovation through their version upgrade process. Organizations now struggle to keep current with the latest and greatest developments from ERP vendors. In fact, ERP vendors have developed a sophisticated innovation engine to enable them to lead in the development and delivery of new business models and practices. Being able to leverage the economies of serving multiple businesses within a single industry and across virtually all industries gives vendors this unique perspective and positioning. ERP vendors boast of their very sophisticated innovation processes as they leverage their resources and relationships to deliver to the market product and process innovations frequently. Product and process innovations can originate through various channels as follows:

1. The internal research and development organizations typically include business architects, product developers, software developers, IT professionals, and consultants. These groups assess market and technology trends and determine how best to leverage or evolve the organization's resources.
2. The consulting organization brings years of experience working with their customers to configure the ERP software.
3. The sales organization is closest to the customer on the front end of a sale. They have the best idea of what the customer really wants.
4. Competitor information also provides an important input into the innovation process.
5. Third-party software companies are often acquired to secure recent innovations or in anticipation that the smaller, entrepreneurial firm will help energize the larger, older software vendor.

The Fortune 500 market for ERP has been virtually saturated, so ERP majors are focusing on small- to mid-sized enterprises (SMEs). ERP vendors typically segment their customers based on size (e.g., large accounts greater than US$ 2 billion, local accounts greater than US$ 1 billion, mid-market accounts less than US$ 1 billion and small accounts less then US$ 50 million in annual revenue) and based on channel (i.e., direct and indirect). Different offerings may be available for these different segments.

90.2 Enterprise Systems Application Frameworks

This Section introduces two popular enterprise applications referred to as customer relationship management (CRM) and supply chain management (SCM). Although the actual application may be defined differently by different vendors, relatively generic application frameworks are presented below to facilitate this introduction.

90.2.1 Customer Relationship Management

Customer relationship management (CRM) is a broad term that refers to the way an organization manages all aspects of its interactions with customers. The term is widely used to describe software packages that enable and support the management of customer interactions, including such areas as sales, marketing, and service. Such packages are often integrated with ERP systems or other enterprise-wide applications. The popularity of CRM applications reflects a shift in focus to the customer and the importance placed on creating, maintaining, and enhancing customer relationships.

CRM functionality is often divided into three categories: operational, analytical, and collaborative (Fig. 90.5). The operational aspects of CRM include those that support front-office processes involving the customer. This can include order processing systems, marketing initiatives, customer service departments, and call centers. Analytical CRM refers to the functionality involved with analyzing and predicting customer behavior. This can involve sales forecasting as well as the use of data warehouses, where large amounts of customer data are mined to discover trends and other important interactions. Finally, collaborative CRM functions are those that actively involve the customer to increase their satisfaction and better meet their needs. This category usually refers to various types of interactions with customers that are not driven by customer service representatives, such as interactive websites.

Current CRM software is much more comprehensive than a simple sales system. The application must support the entire customer process, which spans various departments and functional areas: marketing bases campaigns on customer input and generates customer leads; sales uses these leads to generate customers; manufacturing, distribution, and accounting are involved in processing customer orders; customer service handles complaints and questions; and research and product development are influenced by customer preferences [90.28]. A CRM application strives to integrate the flow of information across these various areas in order to improve decision making. For example, when a customer places an order, the CRM system is updated with this information. When the same customer calls with a problem about the order, the CRM system can recognize the customer and route the problem to an appropriate sales representative (operational CRM). The system generates sales suggestions based on past purchases that can be pitched to the customer through e-Mails or websites (collaborative CRM). Customer service calls can also be analyzed to correct common problems and anticipate new issues (analytical CRM) [90.33].

The evolution of the CRM market from simple sales force software to large, integrated applications has led to the prediction of several trends that should begin to influence the growth of this market, according to studies by *Forrester* [90.34, 35]:

- Significant growth will come from mid-market companies and from hosted applications (Sect. 90.2.3).
- CRM vendors will broaden solutions, with order management as one of the top areas.
- Increased focus will be placed on making systems proactive (i.e., predicting customer needs, preferences, and problems before they arise).

While the growth of the CRM application market may only prove modest, there is substantial expected growth in the adoption of CRM practices as part of business strategy.

90.2.2 Supply Chain Management

A supply chain refers to an organization's network of suppliers, manufacturers, wholesalers, distributers, retailers, customers, and partners. Supply chain management (SCM) can be described as the management and coordination of processes – particularly information, material, and financial flows – across a supply chain (Fig. 90.5). Thus, there is both an internal and external focus inherent to SCM. The term also refers to the category of software applications that aim to control and optimize these processes for organiza-

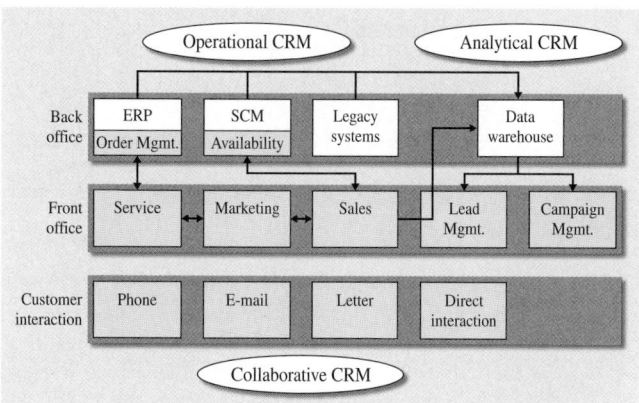

Fig. 90.5 CRM capabilities (after Microsoft Corporation 2000)

tions. While ERP products focus on the coordination and integration of activities within a particular organization, SCM products extend this scope to include interorganizational activities [90.36]. Ideally, these applications will allow communication and integration across the multiple organizations involved in a supply chain to produce responsive, synchronized, and efficient processes. Some of the activities typically included in SCM software include procurement, order fulfillment, demand management, inventory management, warehousing, transportation, manufacturing, and requirements planning. The arrows in Fig. 90.6 represent the mainstream flow for materials, information and money, but it is important to note that, in practice, these arrows are bidirectional as there is significant interaction between each stage when dealing with material, information or money.

SCM functionality can be described as involving planning, execution, coordination, and networking (also referred to as collaboration) [90.37]. Planning includes the design of the supply chain as well as planning supply and demand across the entire chain. This involves forecasting and planning for production, ordering, and distribution. Execution is the automation of the ordering, production, replenishment, and distribution processes and generally focuses on the distribution and manufacturing facilities [90.28, 38]. Coordination involves managing the interaction of processes to increase customer satisfaction. Examples of coordination functions are monitoring and assessing performance. Finally, networking in SCM refers to communication across the supply chain through information sharing.

One of the goals of SCM is to match supply and demand across the supply chain. Demand generated by the end customer travels upstream throughout the

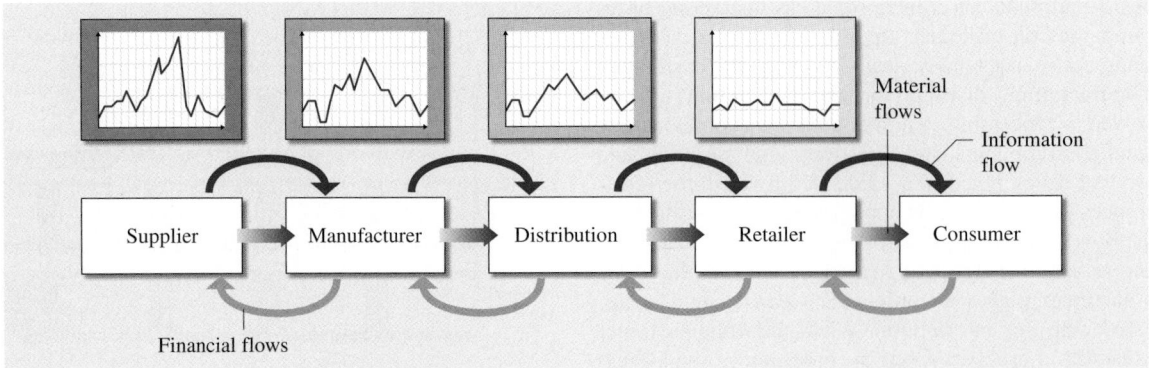

Fig. 90.6 Supply chain flows (after [90.28, 37])

supply chain (i. e., customer demand drives retailer demand, which drives wholesaler demand, etc.). Many supply chains experience what is known as the *bullwhip effect*, where small fluctuations in customer demand are magnified along the supply chain, causing larger demand fluctuations as one moves backward along the chain [90.36]. SCM aims to reduce this effect. By providing interenterprise integration that coordinates processes and shares information, organizations should see reduced demand fluctuations and increased service.

The evolution of the supply chain itself (i. e., in the form of just-in-time and continuous replenishment methods) have changed requirements for supply chains, and therefore SCM solutions [90.28]. The uses of new technologies, such as radiofrequency identification (RFID) tags, are continually integrated into SCM applications to provide greater and greater functionality. The SCM solutions market is expected to experience an increased focus by major vendors to provide supply chain solutions to service organizations and also to experience growth in the customization of solutions by independent software vendors [90.39].

From a software solution perspective, there are three major categories of supply chain solution providers: major ERP vendors, specialty supply chain vendors, and systems integrators. As can be expected, major ERP vendors (such as SAP, Oracle, and to some degree Microsoft for the small- and medium-sized enterprise – SME – market) have acquired or built, and deliver to market, competitive supply chain management software packages. Their competitive advantage includes:

size (deep pockets), customer install base (via ERP solutions), best practices (developed over the years, a significant amount of *supply chain* activity involves ERP functionality and transactions), research and development resources, partners ecosystem, and packaged software experience. The specialty supply chain vendors, i2 and JDA, for instance, will typically have deep knowledge in one or more industries and also share some of the advantages of the major ERP vendors (listed above) perhaps to a lesser degree, and varying between vendors. The systems integrators, IBM and BEA, are effective at customizing and piecing together supply chain solutions to address specific customer concerns.

90.2.3 e-Business

It is important to point out that EwIS applications have been designed to operate within a greater e-Business framework. That is, these applications may be delivered across the Internet in formats such as application service provision (ASP). In fact, the leading CRM vendor today offers their solution only in this format. A study conducted by *Forrester* in 2005 [90.40] identified three types of CRM vendors: hosted CRM specialists (i. e., vendors focused on providing on-demand CRM), hosted all-in-one vendors (i. e., vendors dedicated to the on-demand delivery model but who have chosen to expand beyond CRM to provide built-in support for back-office systems such as ERP and order management), and licensed CRM vendors. One can expect these sorts of offerings for other enterprise applications such as SCM and SEM.

90.3 Emerging Standards and Technology

One of the most significant movements that have emerged since 2000 in the enterprise systems market is one towards building business processes from services. Based on service-oriented architecture (SOA) and its associated standards (e.g., web service definition language (WSDL), simple object access protocol (SOAP), and universal description discovery and integration (UDDI)), this represents a step in the evolution towards developing and implementing flexible enterprise software. It is important to point out that SOA is an architectural approach and not a competitor to packaged software. That is, packaged software is a codification of a company's business processes. Packaged software is delivered to the market with varying degrees of configurability, so it will have varying degrees of flexibility since it uses, in its traditional form, proprietary file formats and interfaces. The SOA movement provides a more open platform that allows a company to redefine packaged applications as loosely coupled business services combined with processes. Packaged software applications can adopt SOA for interoperability and extensibility. In fact, the major packaged software vendors such as SAP and Oracle have made bet-the-company decisions based on delivering solutions on an SOA foundation. On the other hand, platform vendors such as Microsoft and IBM have reconfigured their product line around SOA-based solutions. The concept of packaging business processes as services and the competitive landscape that is emerging to support this movement are discussed below.

90.3.1 Services Concept

SOA is based on the concept of stringing together many loosely coupled, message-based *building blocks* of processes, called services, to create a customized application [90.41]. With regard to enterprise systems, this marks a move from having one integrated application that runs all or most of an organization's processes to a method of combining multiple services from many sources to form customized business processes; for example, an organization can combine a service that checks inventory availability with one that processes customer payment into part of a customer order process. SOA promotes code reuse, as services can be reused to form many applications, and flexibility, as services can be combined and recombined in infinite ways to match the changing business logic of an organization. Web services connect via the Internet and are published in

a standard format with standard interfaces, so that each service can communicate with other services without regard to the underlying technical details (i. e., which programming language was used) [90.41]. Two of the goals of enterprise systems based on SOA are to reduce customization costs and to allow a focus on modeling and remodeling business processes instead of focusing on technical details.

90.3.2 Competitive Landscape

The SOA approach to automating business processes requires ERP vendors to shift focus from providing integrated solutions to providing a platform for combining and running services. Vendors are now developing a more *open* interoperable platform where SOA applications can be developed and accessed and where the software can be easily integrated or interfaced with services and products of other vendors [90.42]. One of the key components in each of these platforms is the enterprise service bus (ESB) that handles the communication between services. Many of the current offerings of SOA-enabled infrastructures, in various states of their roll-out, have come from well-established names in the enterprise systems and middleware domains: SAP, IBM, Oracle, and Microsoft. Some of these platforms include preconfigured services based on functionality contained in the vendor's previous products, while others simply provide the platform for developing custom services and for deploying third-party vendor services. These offerings are briefly discussed below and are summarized in Table 90.1.

SAP markets enterprise SOA through its NetWeaver platform. Functionality from SAP's other applications are being divided into thousands of prepackaged but changeable sets, described by SAP as bundles. NetWeaver's key component is the enterprise services repository (ESR) that serves as a metadata directory for the available services and provides a UDDI-compliant registry for publishing, classifying, and discovering services. SAP also intends to leverage their industry experience by providing industry-specific business processes for over 25 industries [90.43].

Oracle's SOA Suite is a package of several Oracle Fusion Middleware products designed to provide a comprehensive SOA infrastructure in one installation [90.44]. The suite is intended to serve as a way to integrate the various technologies acquired by Oracle. Functionality configured as services will be delivered

Table 90.1 Comparison of selected SOA bendors ([1]Provided through Fusion Applications; [2]UDDI compliant)

Vendor	SOA initiative	Platform	Example modeling and development tools	Prepackaged services	Service registry
SAP	ESOA	NetWeaver	WebDynpro, NetWeaver Visual composer	Yes	Enterprise services repository[2]
Oracle	SOA suite	Fusion Middleware	JDeveloper, BPEL Process manager	Yes[1]	Oracle service registry[2]
IBM	SOA foundation	Websphere	Websphere business Modeler, Websphere integration developer	No	Websphere service registry and repository
Microsoft	Microsoft application plattform	.Net, BizTalk	.NET framework	No	Enterprise UDDI services[2]

through Fusion Applications (due for full release in 2008) and will run on the Fusion Middleware infrastructure [90.45].

IBM markets IBM SOA Foundation as a set of modular software that can be adopted one component at a time [90.46]. Most of these components, including the Websphere Business Modeler and the Websphere ESB, are identified with the Websphere name. The Websphere Service Registry & Repository is based on emerging standards rather than UDDI but will be able to integrate with UDDI-based registries.

Microsoft's more limited SOA offering, based on the Microsoft application platform, is directed towards the mid-market. The foundation of the Microsoft application platform is the .NET framework, which allows the development of services and connection of systems [90.47]. Most ESB capabilities reside in their BizTalk Server.

90.4 Future Trends

Enterprise systems vendors continue to strengthen their products. One recent study [90.48] points to a number of ERP trends that mark the maturity of this market: companies are standardizing on a single ERP vendor, companies are moving toward fewer software instances, comprehensive ERP offerings reduce the need for best-of-breed bolt-ons, industry-specific functionality is getting deeper, and integration capabilities are improving.

Recognizing the trend that enterprise systems powerful applications are being made available to virtually all types of employees and consumers, instead of just sophisticated users, enterprise systems vendors are very keen to better understand the needs of the emerging enterprise knowledge workers [90.49]. Workers who come from all industries and all levels of the organization are expected to leverage the power of enterprise systems to be more effective in their jobs. Vendors will be challenged to further help develop simplified work processes for companies to implement, simplified IT infrastructures that users do not even know exist, and simplified user interfaces that allow end-users to leverage the power of the systems without knowing that they are using a system.

As the global market place does indeed become flatter and as these sophisticated software systems do indeed become more simplified and commoditized, more emphasis will be placed on service-oriented architectures and business process modeling to create nimble, agile, dynamic structures that evolve as quickly as the business models that employ them.

90.5 Conclusion

The enterprise systems market was rooted in the development of ERP and has since leveraged ERP concepts to effectively deliver sophisticated business applications such as CRM and SCM. Enterprise systems have become the enabler of continuous improvement and disruptive innovation for organizations around the world.

ES vendors play a significant role as thought leaders and change agents. In a short period of time, the corporate computing landscape has been radically transformed as companies have endless opportunities to leverage information and networks like never before.

References

90.1 D.E. O'Leary: *Enterprise Resource Planning Systems: Systems, Life Cycle, Electronic Commerce, and Risk* (Cambridge Univ. Press, Cambridge 2000)

90.2 T.H. Davenport: *Putting the Enterprise into the Enterprise System*, Harv. Bus. Rev. (Harvard Business School Press, Boston 1998), product number 3574

90.3 M.S.S. Morton (Ed.): *The Corporation of the 1990s* (Oxford Univ. Press, New York 1991)

90.4 T. Peters, R.H. Waterman Jr.: *In Search of Excellence – Lessons from America's Best-Run Companies* (Harper Row, New York 1982)

90.5 S.P. Bradley, R.L. Nolan (Eds.): *Capturing Value in the Network Era in Sense and Respond* (Harvard Business School Press, Boston 1998)

90.6 H.J. Harrington: *Business Process Improvement: The Breakthrough Stragey for Total Quality, Productivity, and Competitiveness* (McGraw-Hill, New York 1991)

90.7 P. Senge: *The Fifth Discipline: The Art and Practice of the Learning Organization* (Currency Doubleday, New York 1990)

90.8 M. Hammer, J. Champy: *Reengineering the Corporation: A Manifesto for Business Revolution* (Harper Business, New York 1993)

90.9 T.H. Davenport: *Process Innovation – Reengineering Work through Information Technology* (Harvard Business School Press, Boston 1993)

90.10 B.P. Hammer: *Beyond Reengineering: How the Process-Centered Organization is Changing our Work and Our Lives* (Harper Business, New York 1996)

90.11 B.P. Shapiro, V.K. Rangan, J.J. Sviokla: Staple yourself to an order, Harv. Bus. Rev. **70**(4), 113–122 (1992)

90.12 P.B. Seybold, R.T. Marshak: *Customers.com: How to Create a Profitable Business Strategy for the Internet and Beyond* (Crown Business, New York 1998)

90.13 L. Downes, C. Mui: *Unleashing the Killer App: Digital Strategies for Market Dominance* (Harvard Business School Press, Boston 1998)

90.14 R. Buck-Emden, J. Galimow: *SAP R/3 System: A Client/Server Technology* (Addison-Wesley, Harlow 1996)

90.15 M. Kirchmer: *Business Process Oriented Implementation of Standard Software: How to Achieve Competitive Advantage Quickly and Efficiently* (Springer, Berlin Heidelberg 1998)

90.16 E. Bendoly, F.R. Jacobs: *Strategic ERP Extension and Use* (Stanford Univ. Press, Stanford 2005)

90.17 G. Keller, T. Teufel: *SAP R/3 Process Orientation Implementation* (Addison-Wesley, Harlow 1998)

90.18 A.W. Scheer, F. Abolhassan, W. Jost, M. Kirchmer (Eds.): *Business Process Excellence: ARIS in Practice* (Springer, Berlin Heidelberg 2002)

90.19 R. Miranda: The rise of ERP technology in the public sector, Gov. Fin. Rev. **15**, 9–18 (1999)

90.20 T. Peters: *The Project 50 – Reinventing Work* (Knopf, New York 1999)

90.21 Managing the SAP lifecycle: Riding the waves of implementation. Seminar notes presented on May 12th in Dallas (1999)

90.22 N.H. Bancroft, H. Seip, A. Sprengel: *Implementing SAP R/3*, 2nd edn. (Manning, Greenwich 1998)

90.23 E. Brown: VF corp. changes its underware, Fortune **138**(11), 115–118 (1998)

90.24 T.H. Davenport, J.G. Harris, J. Cantrell: *The Return of Enterprise Solutions – The Director's Cuts* (Accenture, Cambridge 2002), Accenture Institute for Strategy Change Report 2452927

90.25 S. Shang, P.B. Seddon: A comprehensive framework for assessing and managing the benefits of enterprise systems: the business manager's perspective. In: *Second-Wave Enterprise Resource Planning Systems*, ed. by G. Shanks, P.B. Seddon, L.P. Willcocks (Cambridge Univ. Press, Cambridge 2003)

90.26 Benchmarking Partners: *ERP's Second Wave – Maximizing the Value of ERP-Enabled Processes, Deloitte Consulting report* (Deloitte Consulting, Atlanta 1998)

90.27 T.H. Davenport: *Mission Critical – Realizing the Promise of Enterprise Systems* (Harvard Business School Press, Boston 2000)

90.28 R. Kalakota, M. Robinson: *e-Business 2.0: Roadmap for Success* (Addison-Wesley, Boston 2001)

90.29 P.F. Drucker: *Management Challenges for the 21st Century* (Harper Business, New York 1999)

90.30 I. Nonaka, H. Takeuchi: *The Knowledge-Creating Company* (Oxford Univ. Press, New York 1995)

90.31 T.H. Davenport, J.G. Harris: *Competing on Analytics – The New Science of Winning* (Harvard Business School Press, Boston 2006)

90.32 C. Christensen: *The Innovator's Dilemma: The Revolutionary Book that Will Change the Way You Do Business* (Harper Collins, New York 2003)

90.33 J. Caulfield: Facing up to CRM, Business 2.0 **Aug./Sept.**, 149–150 (2001)

90.34 E. Kinikin, J. Ragsdale, L. Herbert: *Five Major Gaps in CRM Apps* (Forrester Res., Cambridge 2004), Technical Report 35016

90.35 E. Kinikin, J. Ragsdale, L. Herbert: *Trends 2005: Customer Relationship Management* (Forrester Res., Cambridge 2006), Technical Report 35520

90.36 G. Knolmayer, P. Mertens, A. Zeier: *Supply Chain Management Based on SAP Systems* (Springer, Berlin Heidelberg 2002)

90.37 A Business View of my SAP SCM: Driving Business Excellence through Supply Chain Management, internal company document (2005)

90.38 B. Gerald, N. King, D. Natchek: *Oracle e-Business Suite Manufacturing and Supply Chain Management* (McGraw-Hill, New York 2002)

90.39 N. Radjou, N. Tohamy, J. Harrington: *Trends 2005: SCM Market Outlook* (Forrester Res., Cambridge 2004), Technical Report 35393

90.40 L. Herbert: *The Forrester Wave: Hosted Sales Force Automation, Q1 2005* (Forrester Res., Cambridge 2005), Technical Report 36614

90.41 R. Tews: Beyond IT: the business value of SOA, AIIM e-Doc **21**(5), 14–17 (2007)

90.42 A.F. Farmoohand: *SAP's Platform Strategy in 2006* (Asia Case Research Center, 2006)

90.43 P. Malinverno: *How to get value now (and in the future) from SAP's Enterprise SOA* (2007) Gartner Technical Report G00150358

90.44 Oracle SOA Suite http://www.oracle.com/technologies/soa/soa-suite.html (2007)

90.45 A. Mohamed: Oracle moves a step closer to fusion, Computer Weekly, 22 (2007)

90.46 IBM Service-Oriented Architecture http://www-306.ibm.com/software/solutions/soa/index.html, accessed on December 18 (2007)

90.47 Microsoft SOA Products: The .NET Framework, http://www.microsoft.com/soa/products/dotnetframework.aspx (2007)

90.48 P. Hamerman, B. Miller: *ERP Applications – Market Maturity, Consolidation, and the Next Generation* (Forrester Res., Cambridge 2004), Technical report 33957

90.49 Economist Intelligence Unit: Enterprise knowledge workers: understanding risks and opportunities, an economist intelligence unit report sponsored by SAP, Economist (2007)

91. Automation in Financial Services

William Richmond

This chapter addresses automation in the financial services industry with a focus on small banks and credit unions. The financial services industry includes all organizations that engage in or facilitate financial transactions. Automation is essential for these firms to both lower cost and differentiate their services. Small banks and credit unions are an important subset of this industry, with unique automation needs to enable them to compete with large, international firms.

The chapter begins with a description of the financial services industry in general, then moves on to describe community banks and credit unions, how they operate, and the role of automation. The chapter then addresses two key automation areas for these institutions: their core systems that support basic account processing, and a key customer-facing system – Internet banking. Factors that drive success and recommendations for how to approach each of these automation areas is discussed. For these small financial institutions, successful automation of back-end systems depends on how they source the system, with outsourcing both raising costs and the number of channels through which they distribute their products. Successful automation of customer-facing systems, such as Internet banking, depends on

getting their customers to use the system. Because of this, success is more of a management than a technical issue. The chapter concludes with a discussion of emerging trends, technologies, and issues.

According to the North American Industry Classification System (NAICS) [91.1], there are approximately 500 000 financial services firms. This includes insurance agencies, commercial banks, securities brokerages, securities exchanges, and the Federal Reserve banks. These companies range in size from Citigroup, with almost US$ 1.5 trillion is assets [91.2], to the Wanda Michalenko Memorial credit union with around US$ 30 k in assets [91.3]. All of the firms in the financial services industry are information-based businesses, and the intangible nature of the products and services drives the automation projects required to enable these businesses to exist and thrive. Automation of the capture or transmission of information improves productivity, and delivering new products and services typically requires new or modified systems.

Given the size and breadth of the financial services industry, this chapter focuses on small banks and credit unions. These institutions have unique automation needs. They compete against giant companies, such as Citigroup and Bank of America. They need to offer comparable services at competitive prices, but they do not have the deep pockets to fund automation projects and innovation. At the simplest level, these organizations provide a place for people to store cash and from which to make payments. They also support the lending process. Processing deposits, payments, and loans is complicated, and supporting these processes usually entails automation provided and supported by numerous parties.

For these institutions, two areas of automation are critical, and are addressed in more detail throughout this chapter. First is their automation of their core systems. These systems support the bank or credit unions back-office operations. Second is their customer-facing systems, such as Internet banking. A small bank or credit union's automation is successful if it lowers their cost structure or enables them to differentiate their products or services.

For their core processing (deposit and loan processing), how they source their core processing affects their cost structure, the number of products they offer, the number of channels through which those products are available, and to a lesser degree what they pay for deposits. Thus how small banks and credit unions source their core systems is linked with key strategic decisions, such as their product and service offerings, their distribution channels, and their pricing.

For customer-facing systems, the key to success is user adoption. Improved profits initially result from increased revenues despite increased costs. To obtain these profits requires attaining a certain level of customer adoption. If enough customers adopt, the bank or credit union will start to see decreased costs as well. Unfortunately, for the majority of small banks and credit unions, reaching the Internet banking user adoption level breakpoints constitutes a major undertaking – one that they should be wary of taking without a good understanding of both the economics and their customers.

91.1 Overview of the Financial Service Industry

The financial services industry includes all organizations that engage in creating and trading financial assets; act as financial intermediaries by raising funds from lenders and providing funds to borrowers; pool risk to protect entities from specific losses in exchange for a periodic payment (insurance); and facilitate these financial transactions [91.4]. Firms engaging in creating and trading financial assets include securities brokers, who act as agents between buyers and sellers of securities; stock exchanges, which furnish physical or electronic marketplaces; and portfolio management firms that manage others' assets for a fee or commission. These include institutions with which everyone is familiar, such as Goldman Sachs, the New York Stock Exchange, and T. Rowe Price. It also includes less well-know businesses, such as the Arizona Stock Exchange, Dwight Investment Counsel, and Liebau Asset Management Company.

Firms that act as financial intermediaries include commercial banks, credit unions, and savings and loans. The lines between these institutions are blurring, and all of them accept deposits and make commercial, industrial, and consumer loans. Linked with these institutions are credit-card processors and loan brokers. Credit-card processors provide retailers with a merchant account and enable them to process credit-card transactions. Loan brokers, including mortgage brokers, help customers find the best loans. They act as an intermediary between lending institutions and the customer. Institutions which, perhaps, should be linked with these firms are other loan originators and processors, such as the financing arms of the automobile industry. Financial intermediaries include well-known firms such as Bank of America, and Citigroup as well as the Wanda Michalenko Memorial credit union.

Risk pooling firms include insurance carriers and their agents. Insurance carriers analyze risk and then assign an appropriate premium for assuming that risk. Agents, like securities brokers, act for the carriers and sell the insurance policies to individual and companies. These include firms form Aflac and Geico to the Wayah Insurance Group.

All of the firms in the financial services industry are information-based businesses. Their products and services are (or can be) intangible, and the processes for delivering these products and services entail the capture, aggregation, transmission, and dissemination of information and knowledge. The intangible nature of the products and services drives the creation and delivery of products and services, the structure of the business, and the nature of competition, as well as the automation projects required to enable the businesses to exist and thrive.

Because financial service firms are information based, new products require only creativity to define. A new mutual fund groups a set of existing stocks.

A new insurance product protects a customer from a risk they had not insured before. A new type of loan makes funds available to borrowers with different contract terms than existing loans. Frequently, new products are amalgamations of other, existing products; for example, a new banking product will often be a set of accounts (e.g., a savings account, a checking account, a consumer loan) with certain requirements, such as minimum balances and minimum or maximum activity levels. Identifying and defining a new product that will be successful, however, is extremely difficult. The new product must appeal to a customer base. It must comply with government regulations. It must be priced effectively and it will likely require new or modified systems to support the processes for implementing it.

Because products and services are intangible, they are easy for competitors to copy. Copying the product or service requires knowing its terms or components (e.g.,

knowing the terms and structure of a new loan product or knowing the accounts and limits for a new type of bank account) and then implementing the processes and systems required to support the product; for example, Prudential offers MyTerm *a new, immediate-issue term life insurance product available through the internet to customers of select banks and other financial institutions* [91.5]. This product is available through banks. To implement it, a bank connects its systems to Prudential's systems, so it will be easy for any bank to offer. For another insurance company to offer a similar product requires similar systems, but they have a model on which to base them. It would also require an understanding of the risk, but they would not have to start their analysis from scratch.

In part due to the ease of product creation and duplication, at least on the retail side, entry into the financial services industry is easy. Starting a bank requires only

Table 91.1 Subset of NAICS definition of financial services

Codes	Titles	No. of US businesses	Automation examples
52	Finance and insurance	471 196	
522110	Commercial banking	50 981	Loan scoring
522120	Savings institutions	10 291	Internet banking
522130	Credit unions	11 997	Analytics for customer analysis
522220	Sales financing	1999	Automate teller machines
522291	Consumer lending	13 718	Obopay (mobile payments)
522292	Real estate credit	7707	
522293	International trade financing	2163	
522298	All other nondepository credit intermediation	12 564	
522310	Mort. and nonmort. loan brokers	49 853	Online trading
523110	Investment banking and securities dealing	2267	*Big Bang* – automation of London stock exchange
523120	Securities brokerage	25 214	Loan origination systems
523130	Commodity contracts dealing	2339	Program and algorithmic trading
523140	Commodity contracts brokerage	720	
523210	Securities and commodity exchanges	331	
523920	Portfolio management	6312	
523930	Investment advice	19 407	
523991	Trust, fiduciary, and custody activities	9268	
524113	Direct life insurance carriers	6408	Call center automation
524114	Direct health and medical insurance carriers	3639	Claims processing
524126	Direct property and casualty insurance carriers	5962	Workflow automation
524127	Direct title insurance carriers	4471	Claims fraud analytics
524128	Other direct insurance (except life, health, and medical) carriers	2041	Risk management
524210	Insurance agencies and brokerages	56 733	
524291	Claims adjusting	4404	

US\$ 6 million in capital. Starting a credit union requires less. Becoming a financial planner, in most places, does not require formal financial education or expertise, so entering the market is easy and inexpensive. The easy market entry and low startup costs increase competition, since it results in a large number of competitors with both large international players and with small local firms.

According to NAICS [91.1], there are approximately 500 000 financial services firms. This includes over 150 000 insurance agencies and brokerages, 50 000 commercial banks, 25 000 securities brokerages, and the 7 Federal Reserve banks. These companies range in size from Citigroup with almost US\$ 1.5 trillion is assets [91.2] to the Wanda Michalenko Memorial credit union with around US\$ 30 k in assets [91.3].

Automation plays a central role in financial service firms. Automation of the capture or transmission of in-

formation improves productivity. This has led banks and credit unions to automate their core processes, stock exchanges to automate the matching and execution of trades, and stock brokerages to implement algorithmic (program) trading. These automation projects have significantly changed the financial services industry. Banks' automation of their core systems changed the industry cost structure and was a key determinant in industry consolidation [91.6]. Big Bang – the automation of the London Stock Exchange trading – enabled the stock exchange to increase trading volume from 20 000 shares a day to over 550 000 shares a day [91.7], and program trading affects the stock market's value [91.8] as well as accounting for over 50% of the trades on the New York Stock Exchange (NYSE) [91.9]. Table 91.1 lists some of the types of firms that comprise the financial services industry as well as some of the ways in which they are using automation.

91.2 Community Banks and Credit Unions

Because of the size and breadth of the industry, the remainder of this chapter focuses on small community banks and on credit unions. These institutions have unique automation needs. They compete against giant companies, such as Citigroup and Bank of America. They need to offer comparable services at competitive prices, but they do not have the deep pockets to fund automation projects and innovation.

Community banks, for the purposes of this chapter, are relatively small, regionally localized banks. There are thousands of banks with less than US\$ 2 billion in assets and that serve only part of a state or regional area. Credit unions are nonprofit, cooperative financial institutions with the motto: *not for profit, not for charity, but for service* [91.10]. Each credit union is chartered to serve a defined membership; for example, the Houston Texas Fire Fighters Federal Credit Union only serves fire fighters and their families in Houston, TX. With such targeted markets, credit unions are small. The largest credit unions had less than US\$ 2.5 billion in assets in 2005.

Both retail banks and credit unions make their money on fees charged for their services and on the spread between the interest they pay and the interest they earn. To compete, they must offer both competitive rates and competitive services. Some credit unions and small community banks offer only very limited accounts, loans, and services. Others offer a full range of accounts, loans, and services.

Banks and credit unions differ in their goals and objectives. Banks try to minimize their cost of capital and maximize their risk-adjusted loan rate. Conversely, credit unions try to maximize their cost of capital (the interest rates they pay on deposits for members) and minimize their risk-adjusted loan rate (subject, of course to maximizing the amount and number of loans they make to their members). Ultimately, credit unions are operated for their members, aligning the interests of owners and customers, while banks are operated for their shareholders, resulting in maximizing returns for owners while simply satisficing customer needs.

91.2.1 How a Retail Bank/Credit Union Works

At the simplest level, banks and credit unions provide two services. First, they provide a safe and convenient place for people to store cash and from which to make payments. Second, they match people with excess cash with those who need cash and then support the lending process. Processing these sets of transactions is complicated and frequently entails substantial automation [91.11].

Figure 91.1 shows some of the interactions and some of the detail for processing deposits, payments, and loans. The key points from Fig. 91.1 are that the customer interacts with the bank in numerous different ways through various channels, and that there is

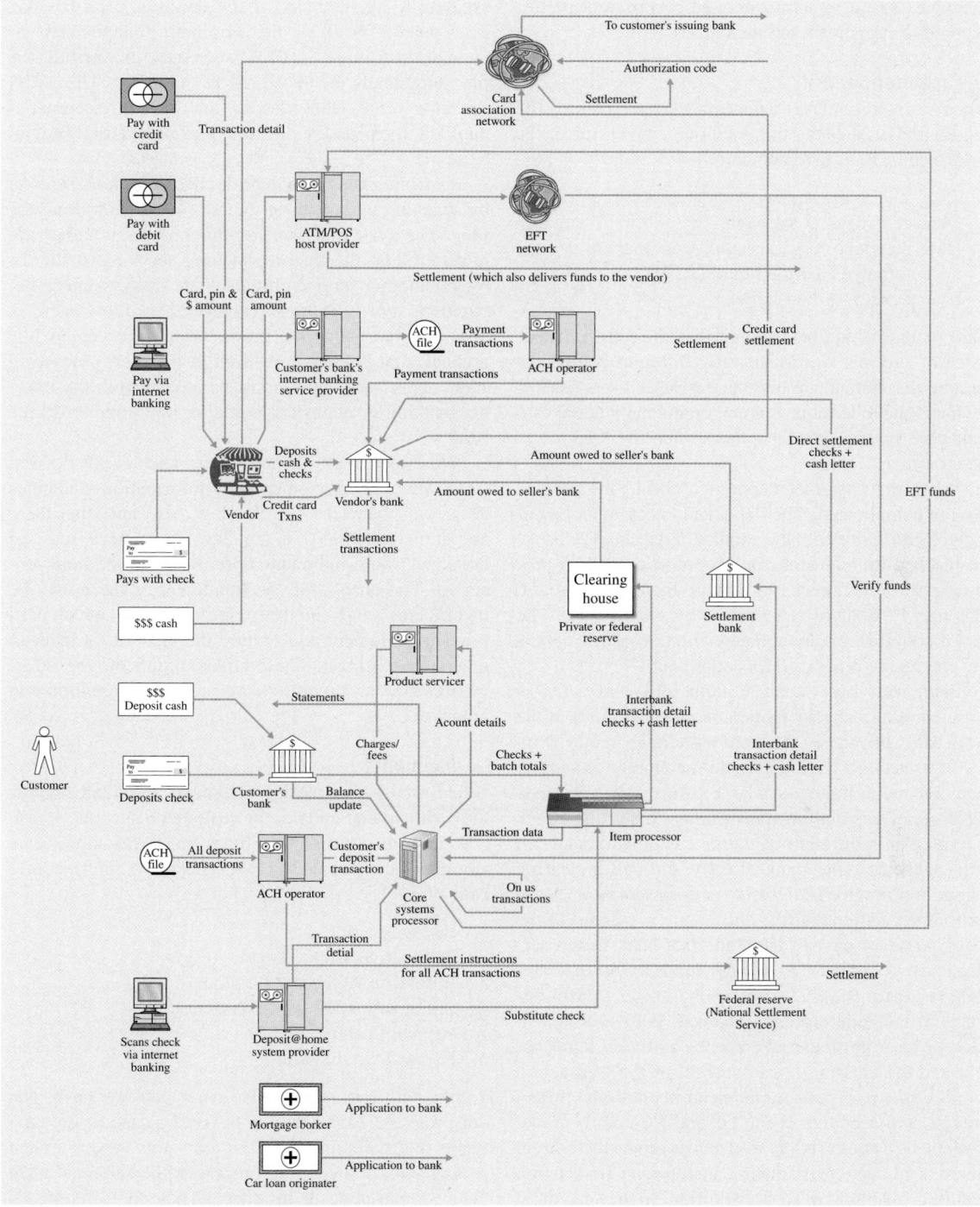

Fig. 91.1 Basic bank and credit union processing

a complex set of systems operated by different parties required to support those interactions.

Customer Deposits

When a customer makes a deposit, the bank records the deposit and updates the amount it owes the customer. To make a deposit, a customer can:

1. Take cash to the bank
2. Deposit a check at the bank
3. Have someone (e.g., an employer) make a deposit via automated clearing house (ACH)
4. Scan a check at their home.

Each of these requires a somewhat different process, different systems, and potentially different companies running the systems. When the customer takes cask to the bank, the teller counts it and enters the amount into their core system, which updates what the bank owes the customer.

When the customer deposits a check, the process is more complicated. The deposited check must be encoded by the teller with the amount of the check. Checks are batched up and sent to item processing. Item processing sorts the checks by paying bank and account. This may be done by the bank or by a third party. The bank's core systems are updated with transaction details and checks are processed for settlement.

Settlement can occur in numerous ways. On-us transactions are checks written on other accounts at the same bank. For these, the bank withdraws money from one of its accounts and deposits it in another (assuming there is enough money available). This is just processing a set of accounting transactions. When the check is written on a different bank, the customer's bank can settle with the other bank directly, through a clearing house, or through a settlement (or correspondent) bank. With direct settlement, the checks are presented to the check writer or payer's bank and that bank uses a service, such as Fedwire, to transfer funds to the customer bank (Fedwire is a real-time funds transfer system operated by the Federal Reserve banks). With a settlement bank or clearing house, the checks and cash letter are sent and the clearing house nets all of the transaction for all of the participating financial institutions. It then uses its own accounts or the Federal Reserve National Settlement System (NSS) to debit or credit the correct amounts to each participating institution. Transaction details are also sent to each institution, so they can debit or credit the appropriate customer accounts.

For deposits made via ACH, the payer (such as an employer) provides *its* bank with instructions for creating a file with all of the deposits it will make. The payer's bank – the originating depository financial institution (ODFI) – creates or verifies the file and sends it to an ACH operator. The ACH operator routes the files to the payee (customer's) bank. It then settles the transactions via the Federal Reserve.

Customers can also deposit checks from their home by scanning both sides of the check and submitting the files. The system supporting this process is either administered by the Internet banking service provider or by yet another service provider. The scanned check becomes a *substitute check* (approved by the Check 21 law) and the electronic information is processed like a check that has made it through the item processing stage. This information must be passed from the Internet banking or deposit at-home service provider to the bank's system.

Note that in all of these cases, cash is not flowing from one bank to another. Only information – changes to account balances – is flowing. Also note that there are numerous kinks in the flow. The check may be damaged, requiring manual processing at the item processor. The check may be fraudulent or there may be insufficient funds in the payer's account which will prevent settlement and require the customer's bank to reverse the deposit. These kinks complicate the information flow and require sophisticated automation and processes.

Payments

When a customer makes a payment, the bank records the payment and updates the customer's account. It also goes through a settlement process to get the appropriate amount to the payee. To make a payment, the customer can:

1. Pay with cash
2. Pay with a check
3. Initiate an ACH credit transaction
4. Initiate an ACH debit transaction
5. Pay with debit card
6. Pay with credit card.

To pay with cash, a customer just hands over cash. The only role the bank plays is providing cash to the customer and receiving the cash from the vendor. When paying with a check, the process is the same as when the customer deposits the check. The only difference is that the vendor is depositing the check. In some places, the vendor scans the check at the time of purchase. This converts the check to an electronic form and an ACH (or electronic funds transfer (EFT)) transaction occurs, similar to a debit-card transaction.

Bank customers can also use ACH to pay bills. Typically, this is through either the Internet banking system or a vendor's system. When the customer uses the Internet banking system's bill-pay feature, they can set up a recurring payment. They tell their bank to withdraw funds and to pay a certain account. Alternatively, they can use a vendor's system and give the vendor authorization to withdraw money from their account. This frequently happens with recurring payments such as mortgages, and insurance premiums. This process is the same as when they deposit funds via ACH, except they are the payer instead of the payee.

To pay with a debit card (which is similar to using an automated teller machine (ATM)), the customer provides their card and pin (there is also a signature version). The debit card/ATM system may be driven by a third party. The transaction is forwarded through the EFT network to verify that the customer's funds are available. If so, the account is debited and an electronic funds transaction transfers funds to the debit card/ATM host provider. The debit card/ATM host provider then uses ACH to transfer funds to the merchant/ATM owner.

To pay with a credit card, the customer provides their credit card. The transaction information is transmitted through the card association's (typically Visa or MasterCard) network to the issuing bank's system, where the transaction is approved. An authorization code is sent to the merchant. The merchant submits batches of credit-card transactions at the end of the day to its bank (merchant account). Settlement then occurs between the merchant bank and the issuing bank through a combination of the card association and ultimately a payment network such as Fedwire.

Loans

A customer can take out a loan (the process is not shown in Fig. 91.1). There are numerous types of loans, including mortgage loans, car loans, credit cards, and installment loans. Taking out a loan requires filling out an application, which can be done at the bank, online, and at various loan brokers with whom the bank has agreements. After the application is complete, the customer's credit worthiness is verified. This requires obtaining information from external sources, and frequently involves credit scoring – using the bank's software or acquired from an external service provider. If the loan is approved, then it must be serviced. This is another touch point between the customer and the bank, requiring additional systems and potentially additional vendors.

A final issue is the generation of statements by the bank. Typically this requires sorting and printing transaction detail and is part of the core banking system. For some accounts that require specific balance levels and transaction usage across bank products, at least part of the process may be handled by an outside vendor. In these cases, transaction detail is sent to the vendor who determines if the product requirements were met, and if not the appropriate fees. This information is sent back to the bank to be included on the statements.

As is evident, there are numerous types of transactions, requiring interaction among a large number of organizations. The clearing process is similar or the same for many transactions. This is facilitated by standard formats for the required information, which makes automation both easier and more effective. The customer-facing portions of the processes are not as standardized. This causes integration issues that are discussed later.

91.3 Role of Automation in Community Banks and Credit Unions

It is generally accepted that investments in information technology (IT) can lead to improved firm performance and possibly competitive advantage. Research applying the resource-based view of the firm has shown that firms with better IT capabilities perform better financially than similar firms with less capable IT [91.12, 13]. This is particularly important for the banking industry, which invests heavily in IT and has increased investment from 6% of revenues in 2000 to 6.5% in 2001 [91.14]. Many bank chief executive officers (CEOs), however, are unsure of IT's impact [91.15], while others claim that these

investments lead to increased services but decreased profits [91.16]. The issue is then: what drives successful automation?

91.3.1 Definition of Success

Examining what drives successful automation in community banks and credit unions requires defining success. Depending on the purpose of the automation, different definitions are appropriate. Under the resource-based view of the firm, the firm will focus on its unique, value-adding activities. Successful automation will enhance these activities and provide (or enhance) a competitive advantage. Firms can use IT to enhance firm value either by lowering costs or leveraging other firm-specific resources that differentiate the firm from its competition. Automation is successful here if it lowers the bank or credit union's cost structure or enables the institution to differentiate its products or services.

Traditional finance theory defines success (for an investment, such as an automation project) as having a positive net present value (NPV). Under this approach, any automation project where (in present value terms) the expected benefits outweigh the expected cost is successful – even if it does not support the bank or credit union's strategy.

There are other reasons for automation projects, including competitive necessity and regulatory compliance. Competitive necessity projects are done to catch up with the competition. The resulting automation will not result in a competitive advantage and may not deliver a positive NPV (unless the correct baseline is chosen) but must be done to remain in business. For these projects, the bank or credit union may go out of business if the project is not implemented. This should be the baseline for evaluating the project, not the firm's current environment. Likewise, regulatory compliance projects confer little benefit to the organization. For these projects, banks and credit unions need to minimize the cost of the project. The focus of this chapter is on automation projects that support bank or credit union strategy.

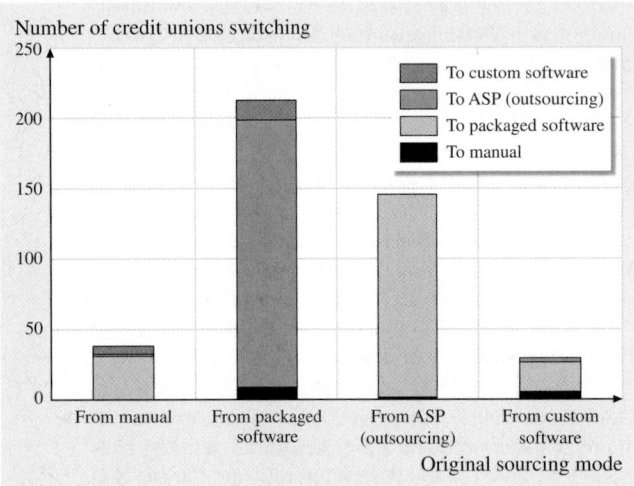

Fig. 91.2 Number of credit unions that switched how they sourced their core systems

As an example, consider SmallTown Credit Union (they asked that their real name not be used). Small-Town Credit Union (STCU) is a community credit union serving its home county and the surrounding three counties. It is located in rural, western North Carolina and historically has served blue-collar workers. It has approximately 15 000 members and assets of US$ 85 million. STCU's members are neither financially nor technically sophisticated. They want good, personal service, high rates on deposits, and low rates on loans. To serve these members, STCU invests minimally in automation. It uses packaged software for its core processing and was not an early adopter of Internet banking. It limits the services it offers electronically and adds new ones only when it is certain of demand by its current members. STCU's goals from technology are primarily to keep operating costs low and only second to offer new services for their members.

91.3.2 Core Processing and the Role of Sourcing

The core processing systems at banks and credit unions support their role as depository institutions and their role in the payment process. The automation includes demand deposit accounting, savings accounting, certificate of deposit and loan accounting systems, as well as the bank's general ledger system (every deposit or withdrawal affects the bank's balance sheet). These systems are basic transaction processing systems that track, and account for, each deposit, withdrawal, and loan payment made by customers. In addition, they calculate interest and fees. The demand deposit accounting systems are coupled with item processing systems that sort checks and deposits and calculate the totals by account. How these systems are sourced is a critical driver in their success.

In 2005, 32 new credit unions and 170 new (de novo) banks were started. Of the more than 8500 credit unions, almost 500 switched their processing mode (how their core systems are provided, Fig. 91.2). So, how small banks and credit unions automate their core processing continues to be of interest and is critical to their survival. Their ability to meet their strategic goals depends, to a large degree, on how they automate their core processing. Although the core systems support back-of-the-house transactions, they underlie the bank or credit union's ability to implement other systems that enable differentiation.

For banks and credit unions, automation of their core processing must be linked with their strategic goals. This automation will not necessarily reduce costs. In the 1960s and 1970s when banks started automating their core systems, processing capacity increased and profits for the banking industry as a whole fell [91.6]. For credit unions, based on data filed in their annual call reports, only 38% of the credit unions that automated their core processing between 1994 and 2002 lowered their operating costs on a cost per share account basis.

Banks and credit unions have three primary sources for automating their core systems:

1. Develop it and operate it themselves
2. Purchase packaged software and run it themselves
3. Outsource both the software acquisition and operation to a third party.

The sourcing choice is linked to the alignment between a firm's IT strategy and its business strategy [91.17]. Research on outsourcing has addressed the factors that result in successful outsourcing as well as what and why companies outsource [91.18–21]. The practices that determine management's perception of success are subject to debate, and different stakeholders have different perceptions of success. The differences are likely tied to different views of the firm's strategy and the impact of IT on that strategy.

Production economics [91.1] advocates outsourcing if the vendor has a production cost advantage. Transaction cost economics [91.1, 22] incorporates the cost of transacting business, including hold-up costs, and advises against outsourcing activities that are unique or require asset-specific resources. The resource-based view of the firm [91.23] argues that firms should invest in their core competencies and outsource the rest. For small banks and credit unions with limited IT capabilities, these theories argue for outsourcing core systems.

For small banks and credit unions much of the advice related to outsourcing is wrong. For those that choose to automate their core processing, how they choose to source their IT function affects their cost structure. It is related both to the number of products they offer and the number of channels through they are available (ability to differentiate). Finally it is tied to what they pay deposits (value to customers and members). Thus, how small banks and credit unions source their core systems is linked with key strategic decisions, such as their product and service offerings, their distribution channels, and their pricing.

Outsourcing core systems at small banks and credit unions is associated with higher operating costs, not lower operating costs. This contradicts much of the justifications given for outsourcing in the trade press and in prior research (e.g., [91.1]) – to lower costs. The specific vendor chosen also matters, since some software vendors are associated with significantly lower costs and others with significantly higher costs.

Instead of cost savings, outsourcing is associated with more services. Specifically, IT sourcing affects: the number of channels (both per member and total), the number of deposit products offered (total), the number of loan products offered (per member), and the price of deposits (both per member and total) [91.24].

This is not surprising. The application service providers (ASPs) have an incentive to develop new products (electronic channels) and to promote them to their clients. They are also able to develop the skills and processes to help their clients more quickly adopt new technologies, products, and services. Why credit unions that use packaged software lag those that use custom-developed systems is unclear. It could be due to skill development or to the importance that the credit union's executives place on using technology.

The higher costs associated with using an ASP (outsourcing) appear to be transferred, at least in part, to the customer through lower interest rates paid on deposits. This is understandable since members use more deposit-related services (and these services more frequently) than loan-product-related services. They are willing to pay for the additional channels to support their deposit products, but unwilling to pay higher loan rates for a product (loan) that they decide on once and then require minimal ongoing service.

A number of small financial institutions, especially small credit unions, do their core processing manually. While manual processing may be considered anachronistic, there are many instances, especially for smaller organizations, where automation does not make sense and manual processes are a valid approach. For small institutions, using manual processes is not more expensive, either in total or on a per-member basis, than using a computerized system. Identifying the optimal break point for when to switch from manual to automated processing is an interesting topic for future research, but appears to occur by the time an institution acquires 500 members/customers.

Approach

When determining how to successfully automate their core processing, small banks and credit unions must first determine the basis on which they want to compete. Those that decide to compete on cost need to

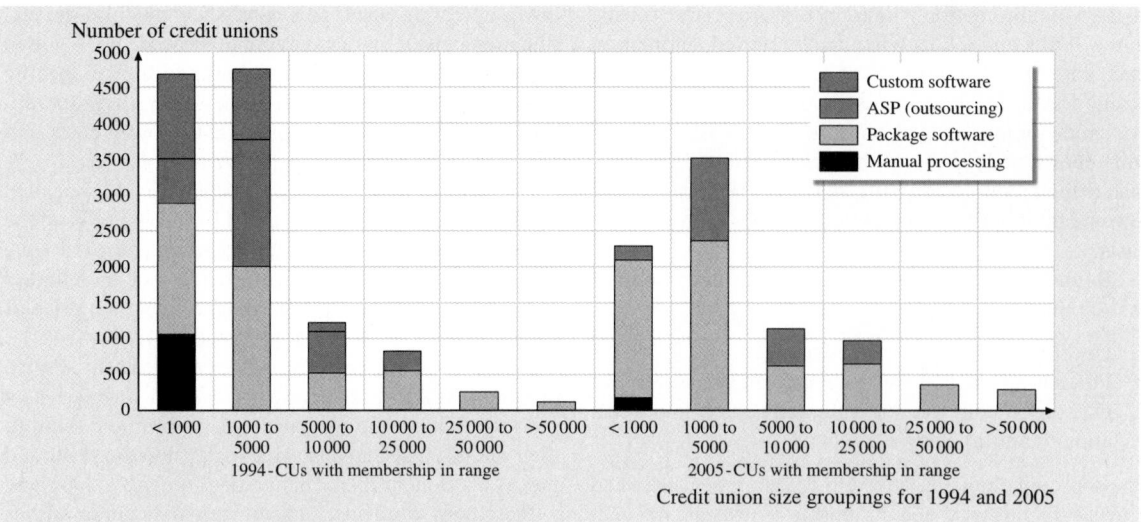

Fig. 91.3 Sourcing of core information systems by credit union size

minimize their overall cost structure (not just their IT costs). Given this choice, they need to assess whether to use manual processes or to automate. If they decide to automate, they need to decide whether to use packaged software or to custom-develop their own software. The decision to automate or use manual processing is tied to institution size. Only small credit unions with fewer than 1200 member use manual processing, and typically they automate well before reaching this size (Fig. 91.3). When automating, the trend is towards using packaged systems. Less than 1% of the credit unions use custom systems, and only 29 have switched to custom systems between 2003 and 2005. Once packaged software is chosen, selecting a vendor is critical. Again, there is a significant difference in operating cost depending on the software package/vendor chosen.

If the bank or credit union chooses to compete on service by offering more products and more channels to deliver these services, then using an ASP is the best approach. At this point, success becomes a vendor management issue rather than a technical issue. Choosing the right vendor and defining the contract are essential. Different ASP vendors are associated with different cost structures and number of channels. The same vendor will charge significantly different prices based on how well the bank or credit union negotiates the contract. Getting a good deal when the initial contract is signed is critical, since switching costs are high and the vendor will take advantage of this when it is time to renew the contract.

91.3.3 Internet Banking and the Role of Adoption

Bank and credit union's core systems automate the back-of-the-house. Banks and credit unions also automate front-of-the-house processes. These systems (such as ATMs and Internet banking) can possibly lower cost by replacing teller-based transactions. They also provide new channels for customers to access bank and credit union services, and they provide additional convenience for customers and members. Customer-facing systems automation can support either a cost-based or differentiation strategy.

For customer-facing systems, the key to success is user adoption. ATMs and Internet banking are classic examples. ATMs were implemented to enable customers to conduct financial transactions without using the tellers. The bank and credit unions' goal was to reduce costs by eliminating the need for tellers. For this to happen, customers had to be willing to use the ATMs. The more customers use these channels and the more transactions they used them for, the greater the (expected) benefits for the banks.

Internet banking has also been heralded as providing a competitive advantage and as being a competitive necessity [91.15, 25, 26]. Accordingly, banks spent 15% of their IT dollars on Internet banking in 2002 [91.27]. Yet, many banks question its profitability [91.28]. In fact, half of all US banks have no plans to offer Internet banking, with the most common reason given being an unclear return on investment [91.14].

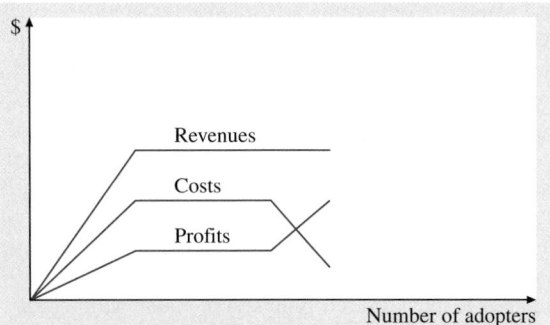

Fig. 91.4 Impact of Internet banking on costs, revenues, and profits

Internet banking is defined as a transaction-oriented system that enables a bank's customers to engage in online banking activities. Internet banking enhances a bank's offerings by providing its customers with 24/7 access to many of its services. The services available through Internet banking vary, but typically include informational account access (view balances and past transactions), funds transfers among accounts, bill payment, bill presentment, and some loan application and approval.

A *Booze* et al. study [91.26] claims that Internet banking has an expense rate that is one-third that of traditional banking with a per-transaction cost of US$ 0.01 versus US$ 0.27 for ATM transactions versus US$ 1.07 for teller-based transactions. For small banks and credit unions, this cost advantage is unlikely to occur. Instead, there is a stylized piecewise-linear relationship for costs, revenues, and profits driven by the number of Internet banking users (Fig. 91.4). An increase in the level of consumer adoption increases revenues, costs, and profits until a *saturation* level is achieved. Additional adopters in excess of the saturation level do not add further to profit, revenue or cost until a second breakpoint is reached. The cost and profit models experience a change at this second breakpoint, but revenue does not. In the cost model, after enough customers adopt Internet banking, a bank's costs begin to decrease. Correspondingly, profits trend upwards after this point.

The initial cost increase comes form a combination of more transactions, more accounts, and higher deposits (and therefore higher interest expense). In particular, banks with Internet banking have a higher growth rate in the number of accounts (14% higher for demand deposit accounts (DDAs) and 32.8% higher for savings accounts). In addition, Internet banking is as-

sociated with a larger average deposit size per account of about US$ 800 at a median bank. This indicates that, while new customers are drawn to the bank, existing customers also are doing more business with their bank. This increase in bank activities necessarily increases costs. Despite this increase in transactions, the actual volume of transactions performed using Internet banking relative to more traditional transactions is typically quite small. Correspondingly, reductions are unlikely in the fixed infrastructure costs related to traditional transactions (e.g., the number of tellers employed or ATMs operated) unless a large number of customers adopt Internet banking.

Improved profits initially result from increased revenues despite increased costs. These improved revenues are the result of an increase in the bank's interest income through soliciting new customers and getting more business from previous customers as well as more fee-based bank services being sold. Previous research finds that consumers who use internet banking (IB) are more lucrative. They generate above-average revenues and are more profitable [91.29]. In particular, IB customers have significantly higher mortgage and loan balances, and they use twice the number of other financial services products [91.14]. The impact of Internet banking usage on revenue and profit is relatively small and typically results in an increase of less than 5%.

For the majority of small banks and credit unions reaching the Internet banking user adoption level breakpoints constitutes a major undertaking. For 25% of the small banks, reaching the breakpoint number of

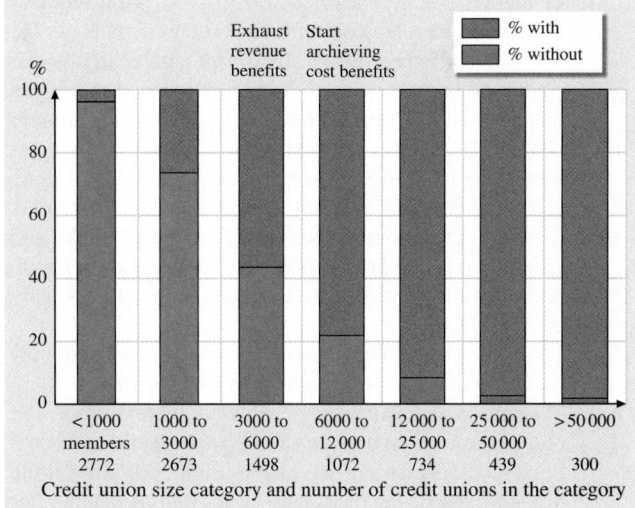

Fig. 91.5 Percentage of credit unions with internet banking by size

adopters requires over half of their customers to adopt Internet banking. For credit unions, over half do not have enough members to reach the first breakpoint. For another 25% of small banks, this requires a customer adoption level exceeding 25% of its customer base (Fig. 91.5). Doing so also takes time – on average over 1.5 years for those that can and do reach it. For a larger number of small banks and credit unions the second breakpoint exceeds their customer base. This implies that small banks and credit unions have to be very careful when implementing new customer-facing technology. To be successful requires getting a significant portion of their customer base to adopt.

Approach

When implementing customer-facing automation projects, banks and credit unions need to take a disciplined approach. First, they need to determine whether the project makes sense. This entails assessing the fit with their strategy. Customer-facing systems will not lower cost, so the institution must want to differentiate its services based on this project. Assuming the project fits with the strategy, it may still not be feasible if the institution does not have enough customers or enough customers who will be interested in using the new service. Finally, since the increased revenues come, in part, from existing customers deepening their relationship with the bank or credit union, it must offer enough additional products and services to take advantage of this additional business.

If the project fits with the organization's strategy and they have a large enough customer base, they will need to market the new services effectively. This requires a different strategy for different target markets. If the new automation requires a significant change in how the customer uses the bank or credit union services (as was the case with ATMs and Internet banking), they should consider paying the customer to use the service. If customers will be paid to use the service, it must be done carefully. Banks and credit unions that paid customers to use ATMs saw increased adoption. Some banks and credit unions that paid customers to create Internet bank accounts saw people create accounts, but not necessarily use them.

If the automation project is innovative, the customers will need to be trained to use the new system. Younger customers may be able to learn online, but older customers and those whose primary channel is the branch will need more personal training, probably at the branch. Finally, the banks and credit unions need to listen to their customers who are not using the new service

or channel to determine why, and then to either modify the service/channel or update their marketing and training materials.

Even with an effective marketing program and a good training process, the banks and credit unions will face a customer adoption process. Customer adoption will follow an S-curve. Effective marketing and training may make the curve steeper, but identifying and supporting early adopters is critical. Early adopters of an innovation typically have the most interest and willingness to pay for the innovation. Later adopters are typically less inclined toward the innovation. Relative to new services, this implies that customers who adopt them early on are likely to be the most profitable, and as more and more people adopt, the incremental revenue per customer will diminish.

Sourcing Impact

As with core systems, Internet banking, and future customer-facing automation projects, how the automation project is sourced impacts success. The sourcing choice changes the cost structure for the service. It affects the implementation timing and potentially supports the customer adoption process.

When the banks and credit unions use an Internet banking service provider, their variable transaction cost structure is different from that for banks that develop their own Internet banking system. To implement Internet banking, there is still a fixed, but small, one-time cost for the system in the US$ 20 000–40 000 range. Once the system is installed, there is a nominal fixed monthly fee (less than US$ 1000) plus a fee for each Internet banking customer as well as fees for various transactions. The total fee for each customer depends on the services the customer is signed up for (e.g., bill-pay) and how much historical information the bank retains (i. e., it costs the bank more if the bank lets its customers view deposits and withdrawals from 2 years ago than if the maximum a customer can go back is 6 months). This fee can range from US$ 2 to US$ 15 per customer per month. Additional fees also are charged for most transactions, including signing up a new Internet banking customer and for a customer making a payment. Enquiries are free. This cost structure makes most of the bank's costs variable. This lowers the risk of implementing Internet banking, but also makes lowering the bank's overall operating costs more difficult. The operational risks typically associated with innovation are mitigated by the use of a vendor, but banks that use this service are unable to physically distinguish themselves from other banks using the same service.

91.4 Emerging Trends and Issues

There are a number of trends and issues affecting small banks and credit unions. The US Federal Government continues to impose additional regulations related to tracking customers and customer activity. Customers continue to shift their payments to electronic forms. New technologies and channels, such as Web 2.0 applications, continue to emerge, and automation to support these trends is leading to banks and credit unions using more vendors, with the accompanying issues involved in integrating these systems.

91.4.1 Role of Integration

Implementing new products requires new systems. Implementing new channels requires new systems. Implementing new processes requires automation. Most community banks and credit unions do not have the skills to automate this on their own. They are either outsourcing their processing to ASPs and other vendors, or they rely on packaged software. To automate a new product, channel or process requires identifying the appropriate provider and then implementing the system and integrating it with the existing systems.

The integration required for basic functionality is addressed by the vendor; for example, to implement Internet banking requires integrating the Internet banking software with the bank or credit union's core processing system, and that provided by the vendor. When the integration is not central to the automation project, however, it is frequently omitted by the vendor and not examined (at least initially) by the bank or credit union.

Where this has been particularly problematic is with basic customer information. When a customer changes their email address in one channel, they expect that it be changed everywhere in the banks systems. Because of the numerous systems and vendors, this frequently is not done. Even when the bank or credit union uses a single vendor (e.g., FISERV) their products may not be completely integrated, because they have frequently purchased the software from another vendor (or purchased the vendor to acquire the software).

This type of integration becomes a vendor management issue more than a technical issue. The integration issues need to be part of the software/vendor selection and contracting process. Service level agreements centered around integration are difficult, since the future systems are not identified. User group pressure, however, appears to be effective in getting the vendors to focus on this issue. This implies that, when selecting a vendor, banks and credit unions need to choose those large enough to have an effective user community.

91.4.2 Regulation

The US Federal Government is increasingly regulating financial institutions' operations. The USA Patriot Act, the fair and accurate credit transactions (FACT) act plus additional rules by the Securities and Exchange Comission (SEC), Financial Accounting Standards Board (FASB), and others add to the regulatory burden. New laws, such as H.R. 3012, The Fair Mortgage Practices Act of 2007, are introduced and existing regulations change each year. These regulations are disproportionately burdensome for small financial institutions, and some believe that they will drive small banks out of business [91.30].

Automation projects to fulfill regulatory requirements are difficult, because they usually have a fixed deadline for implementation, which can be unrealistic. These automation projects will not provide a competitive advantage. They will not support strategic efforts. They will not reduce costs (except through penalty avoidance). Typically, the goal is to get them in as cheaply as possible. The risk is that this will lead to higher costs long-run as the capabilities need to be integrated throughout other systems. This is particularly true for regulations requiring monitoring suspicious activity, where the bank needs to identify and analyze a picture of a customer's entire relationship with a bank.

91.4.3 Shift From Paper to Electronic Payments

The number of paper checks and the value of those checks are finally decreasing as consumers and businesses shift to electronic payments. The back-end settlement infrastructure is largely automated, but new front-end technologies – new payment channels – continue to emerge. Using mobile phones for payments has been available in Japan for a number of years, but is in its early stages of adoption in the USA. Small banks and credit unions again face a decision of whether they want to offer their customers and members a new channel.

The shift to electronic payments, however, is bypassing person-to-person payments. PayPal allows person-to-person payments, but has not had a significant effect on the number of paper-based transactions. Mobile technologies offer the possibility of allowing individuals to make and receive payments. Citigroup offers this service through Obopay, and is currently paying its customers to sign up for the service. Obopay is similar to a prepaid credit card, but both parties must use the service.

91.4.4 Emerging Technologies – Web 2.0

Web 2.0 technologies for social networking appear, at first glance, better suited for enhancing internal bank and credit union operations. Communities of loan officers, compliance personnel or packaged systems users would help each group more efficiently and effectively carry out their duties. They would provide a place to get help, gain insights, and kvetch. They could either be operated by the bank or by independent third parties and would not need to integrate with existing system.

Banks, however, are exploring the use of social networking as a front-end to customer education and interaction. Wells Fargo is leading the way, and if it suc-

cessful, it will pressure small banks and credit unions to follow suit. Wells Fargo initially created a site on Second Life. The site provided both games and financial education. To gain more control over the customer experience, they left Second Life and started their own virtual world, Stage Coach Island.

On Stage Coach Island, participants can network with each other and do activities together, such as motorbike racing and parachuting. They can learn about financial products and services, ostensibly to develop financial responsibility. Although there is little Wells Fargo advertising, it has the potential to provide Wells Fargo with a wealth of information about their customers or potential customers.

At the moment, small banks and credit unions most likely do not have the skill set to duplicate Wells Fargo's efforts. Although there is vendor-supplied software for developing virtual worlds, there are not vendor-supplied virtual worlds available for banks and credit unions to brand and customize for their customers and members.

91.5 Conclusions

Small banks and credit unions face unique considerations when automating. They have to keep up with larger institutions, but do not have the resources to execute these projects on their own. This causes them to rely on external vendors, which raises integration issues, especially for customer-facing systems.

A changing business environment and new technologies require that small banks and credit unions continue to invest in automation, but to be successful they will need to ensure that these projects support their strategy and that the approach is appropriate for this strategy.

References

91.1   S. Ang, D. Straub: Production and transaction economies and IS outsourcing: A study of the US Banking Industry, MIS Quarterly **22**(4), 535–552 (1998)

91.2   Fortune 500 (2006) http://money.cnn.com/magazines/fortune/fortune500/snapshots/309.html, viewed 12/20/2007

91.3   Wanda Michalenko Memorial Federal Credit Union website http://creditunionaccess.com/cu20418.htm, viewed 12/22/2007

91.4   NAICS Association: 52 Finance Industry (2007) http://www.naics.com/censusfiles/NDEF52.HTM#N52 viewed 12/20/2007

91.5   MyTerm: The Revolutionary Online Life Insurance for Bank Customers (2007), http://www.prudential.com/view/page/14252?seg=10&name=bank-channellaunchednewlifeproduct, viewed 1/5/08

91.6   T. Steiner, D. Teixeira: *Technology in Banking: Creating Value and Destroying Profits* (Business One Irwin, Homewood 1990)

91.7   K. Flinders: The evolution of stock market technology, ComputerWeekly.com, November (2007), http://www.computerweekly.com/Articles/2007/11/02/227883/the-evolution-of-stock-market-technology.htm, viewed 12/20/2007

91.8   R. Cole: Dow falls 23.27 in heavy program trading, New York Times, July 11 (1990)

91.9   A. Khanna: Program Trading, EGO (2006), http://www.egothemag.com/archives/2006/01/program_trading.htm, viewed 1/22/2007

91.10   Credit Union Philosphy, Credit Union National Association (2009) http://www.creditunion.coop/history/cu_philosophy.html, viewed 2/10/2009

91.11   Federal Financial Institutions Examination Council IT Examination Handbook: Retail Payment Systems,

March (2004), http://www.ffiec.gov/ffiecinfobase/booklets/operations/23.html

91.12   A. Bharadwaj: A resource-based perspective on information technology capability and firm performance: An empirical investigation, MIS Quarterly **24**(1), 169–196 (2000)

91.13   M. Wade, J. Hulland: Review: The resource-based view and information systems research: Review, extension, and suggestions for future research, MIS Quarterly **28**(1), 107–142 (2004)

91.14   C. Parton, J. Glaser: Myths about IT spending, Healthcare Inform. **19**(7), 39–40 (2002)

91.15   A. Vrechopoulos, G. Siomkos: The adoption of internet shopping by electronic retail consumers in Greece: Some preliminary findings, J. Internet Bank. Commer. **5**(2), 1–3 (2000)

91.16   J. Ross, P. Weill: Six IT decisions your IT people shouldn't make, Harvard Bus. Rev. **80**(11), 89–92 (2002)

91.17   D. Hancock: The financial firm: Production with monetary and nonmonetary goods, J. Polit. Econ. **93**(5), 859–880 (1985)

91.18   R. Hirschheim, M. Lacity: The myths and realities of information technology insourcing, Commun. ACM **42**(2), 99–107 (2000)

91.19   R. Hirschheim, R. Sabherwal: Detours in the path toward strategic information system alignment, Calif. Manag. Rev. **44**(1), 87–108 (2001)

91.20   M. Lacity: Lessons in global information technology sourcing, IEEE Computer **35**(8), 26–33 (2002)

91.21   M. Lacity, L. Willcocks: IT sourcing reflections: Lessons for customers and suppliers, Wirtschaftsinformatik **45**(2), 115–125 (2003)

91.22   N. Melville, K. Kraemer, V. Gurbaxani: Review: Information technology and organizational performance: An integrative model of IT business value, MIS Quarterly **28**(2), 283–322 (2004)

91.23   T. Espino-Rodriguez, V. Padron-Robaina: A review of outsourcing from the resource-based view of the firm, Int. J. Manag. Rev. **8**(1), 49–70 (2006)

91.24   W. Richmond, R. Carton: Does IT Sourcing Matter: Empirical Evidence from Credit Unions, Working Paper, Western Carolina University, Cullowhee, NC 28734, (2007)

91.25   Retail Banking Research, Ltd.: Banking in the online World, White Paper (1996)

91.26   Dynamicnet: Transforming Consumer Banking Through Internet Technology, White Paper (2001)

91.27   N. Olazabal: Banking, the IT paradox, McKinsey Q. **1**, 17–51 (2003)

91.28   K. Hoffmann: Online banking aligns practices, Banking Technology News, March (2003)

91.29   L.M. Hitt, F.X. Frei: Do better customers utilize electronic distribution channels? The case of PC banking, Manag. Sci. **48**(6), 732–748 (2002)

91.30   United States Senate: Testimony of Bradley E. Rock On Behalf of the American Bankers Association Before the Committee on Banking, Housing and Urban Affairs, June 22, 2004, http://banking.senate.gov/_files/rock.pdf, viewed 1/5/08

92. e-Government

Dieter Rombach, Petra Steffens

e-Government in its most generic form refers to the automation of government processes and services by using modern information and communication technologies, usually – but not necessarily – in combination with web technology. The transition from paper-based administrative workflows to electronically supported process chains involving a multitude of stakeholders (in particular, government agencies, citizens or companies) is a key driver of today's e-Government policies. Whereas in their early manifestations, e-Government efforts were primarily geared towards improving internal administrative performance, today's focus is on increasing the overall quality of government services. The aim is to improve the living conditions of citizens and to create a favorable climate for business.

This chapter provides a short introduction to the basic concepts of e-Government, explains how e-Government evolved over the last 10 years,

gives an overview of the dimensions on which the design and implementation of successful and sustainable e-Government services has to proceed, and outlines some future challenges for e-Government.

92.1 Automating Administrative Processes

The 20th century has been the era of digital revolution. Information and communication technology (ICT) has deeply affected business processes, service provisioning, production, as well as the private habits and activities of the citizens in our modern societies. In fact, ICT has become an integral part of most of today's products and services.

ICT has also had a significant impact on the way in which government agencies perform their tasks. The standard inventory of government automation today comprises email, interactive websites, downloadable forms, and sometimes information upload. However, for most governments the level of automation and integration is lagging behind the computerization of businesses. This can be observed in all kinds of government activities and services: internal workflows as well as processes between government and its external partners (citizens and business). Today's challenge for government agencies all over the world is to reach a seamless integration of their own work processes with those of their internal and external partners. This concerns the following types of collaboration:

- Government-to-government (G2G): interactions inside or between government agencies
- Government-to-citizen (G2C): services that government delivers to its citizens
- Government-to-business (G2B): workflows between government and business.

Many modern governments have understood that through the use of electronic media they can increase the efficiency, quality, and transparency of their services and thus increase the attractiveness of their national economy in a globalized world. Solutions developed to this end are summarized under the umbrella of e-Government.

This chapter describes how e-Government evolved over the last 10 years, provides an overview of the dimensions which need to be addressed in the design and implementation of successful and sustainable e-Government services, and outlines some current challenges for e-Government.

92.2 The Evolution of e-Government

The beginnings of e-Government were in the late 1990s when ICT, especially the use of online media, was recognized as an important means to improve the efficiency and quality of government services. The use of ICT was often inspired by achievements made in e-Business, which demonstrated that doing business online could help increase the efficiency of business processes in a company as well as between companies.

Since 1995, *Gartner* has used the concept of *hype cycle* to describe the relative maturity of technologies, IT methodologies, and management disciplines. Hype cycles distinguish five phases [92.2]:

1. *Technology trigger*: The first phase of a hype cycle, which is provoked, e.g., by a product launch or other event that generates significant press and public interest
2. *Peak of inflated expectations*: The second phase, which is characterized by overenthusiasm and unrealistic expectations
3. *Trough of disillusionment*: Technologies fail to meet expectations and quickly become unfashionable
4. *Slope of enlightenment*: Despite a decrease in public and press interest, some businesses keep up their in-

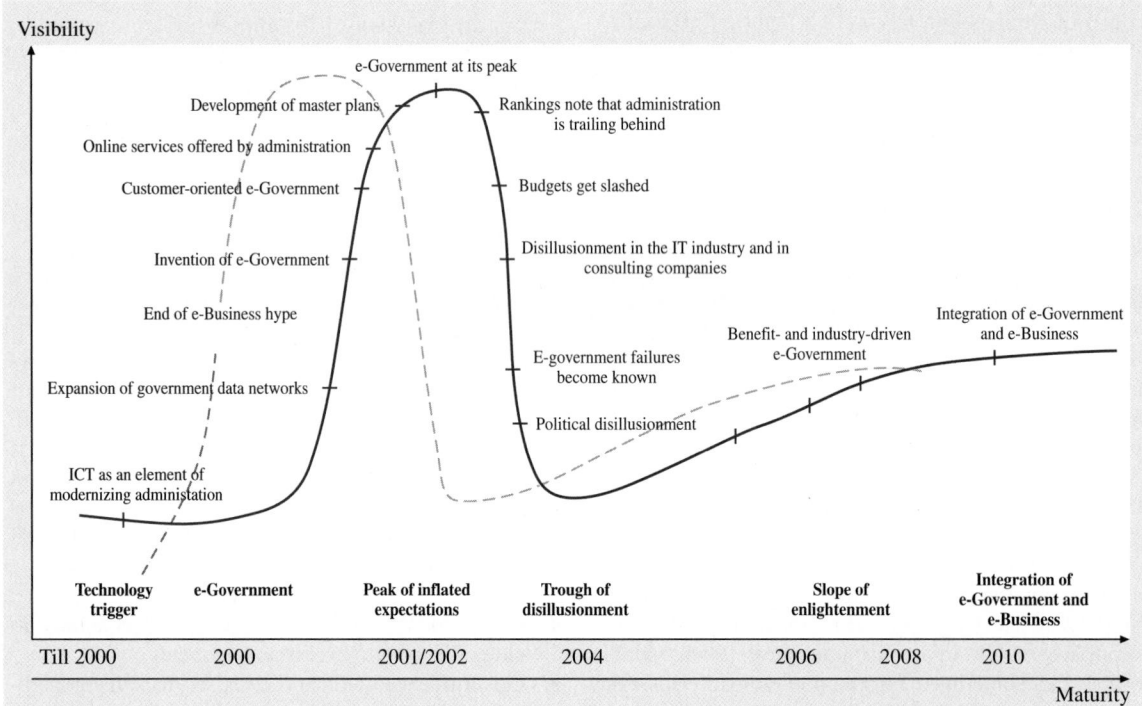

Fig. 92.1 Evolution of e-Government example. Note: The *dotted line* shows the development line of e-Business (after [92.1])

terest and experiment to understand the benefits and practical application of the technology

5. *Plateau of productivity*: The technology becomes increasingly stable; its benefits become widely demonstrated and accepted.

Adapting Gartner's hype curve to the evolution of e-Government in Germany, *Büllesbach* has drawn an exemplary development line of e-Government [92.1], as can be observed in analogous forms in many economically developed countries (Fig. 92.1). Throughout the 1990s, ICT was mainly regarded as a means of modernizing administration by facilitating communication and workflows between state agencies. For Europe, this is illustrated by the e-Government country fact sheets, which the European Union publishes yearly for 33 European countries [92.3]. As a consequence of supporting internal work processes with ICT, new government data networks were established and the coverage and capacity of existing ones were increased.

By the beginning of the millennium, the concept of e-Government made its debut, implying a shift in focus: Turning away from the primarily internal use of ICT, governments started to perceive benefits in applying new technologies to offer electronic services to their clients (e.g., electronic voting machines (Fig. 92.2)). Early applications developed at that time mainly focused on citizens and were usually limited to providing access to information, allowing simple forms of interaction such as exchanging e-Mails or using feedback forms. e-Government projects at the time were often conducted separately from each other, leading to isolated and incompatible solutions.

To overcome this situation, governments started to realize the need for a more systematic approach: e-Government master plans were thus developed that put existing and planned projects under the umbrella of a common strategy. With the development of master plans that gave rise to high expectations, e-Government also became more visible on the political agenda and reached its hype peak. Expectations were dampened though, when it became clear that often e-Government lead times were incompatible with politicians' striving for timely success and that underestimation of a task's complexity and of the challenges posed by its political context (e.g., federal structures with complex decision processes), not only let projects overrun time and budget but also failed to evoke acceptance [92.4]. A prominent case in point is the German FISCUS (Föderales Integriertes Standardisiertes Computer-Unterstütztes Steuersystem – federal inte-

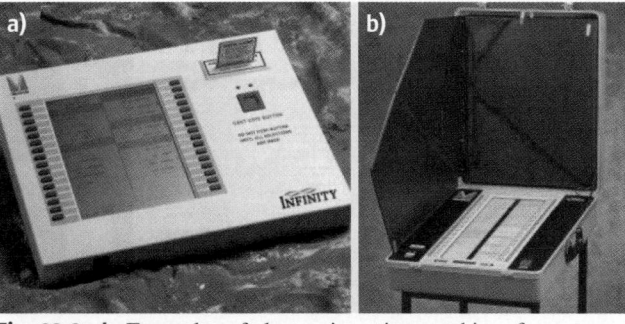

Fig. 92.2a,b Examples of electronic voting machines from around the world: (**a**) Infinity voting terminal (courtesy of Microvote General Corporation), (**b**) MV-464 voting station (courtesy of Microvote General Corporation)

grated standardized computer-supported tax system) project, which aimed at developing uniform tax software for the fiscal administration in Germany, but was stopped and entirely reoriented after 13 years without producing the expected results [92.5]. As a consequence, e-Government budgets were cut and failures of e-Government projects became public. In addition, it was recognized that citizens, who had so far been the primary target of e-Government efforts, did not access e-Government services at a sufficient rate to realize the expected return on investment.

This political disillusionment was overcome mainly by two shifts in focus: first, technology as a major driving force for developing new solutions was replaced by the goal of creating customer benefit; second, companies, which have more frequent and regular relationships with government agencies than citizens, moved into the focus of attention and became the main target for e-Government. In Germany, this shift in focus has occurred under the heading of *benefit- and industry-driven e-Government* [92.6]. During this phase, the integration of government and industry processes became the main subject of e-Government efforts, ultimately aiming at the seamless integration of e-Government and e-Business on the ICT as well as on the process level.

A common way to describe a project's or even a country's maturity with respect to e-Government is to position it on a scale comprising four phases, each indicating the degree to which the potential of ICT is utilized to deliver public services electronically [92.7]:

● *Presence.* The emphasis in this phase is to offer online static information about government agencies and their services. In this phase, e-Government ser-

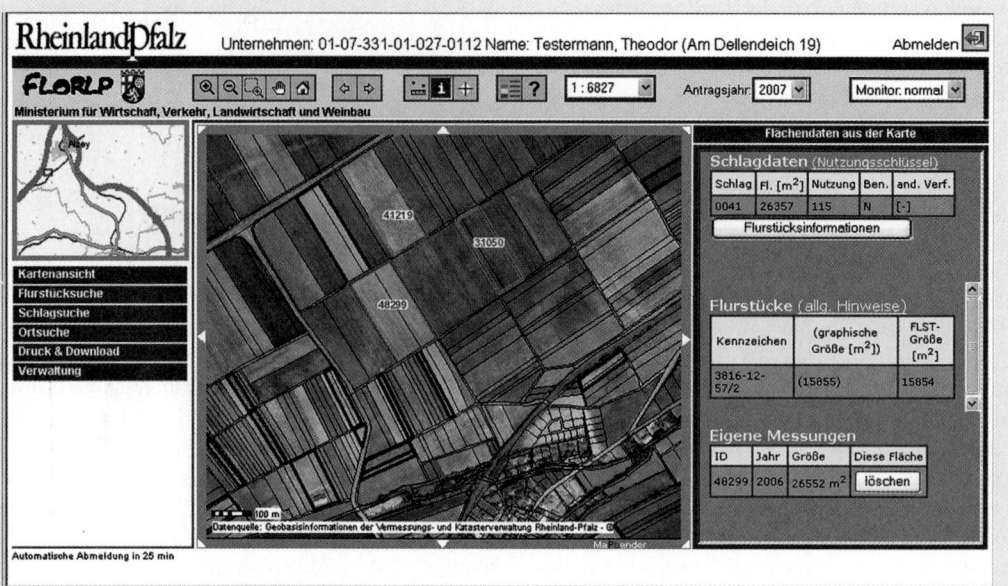

Fig. 92.3 Interaction site for accessing land parcel information to be used for agricultural subsidy applications in the German state of Rhineland-Palatinate (http://www.flo.rlp.de)

vices are often limited to a presence on the web, providing the public with information similar to that of a brochure or leaflet. Meanwhile most public web sites have gone beyond this stage.

- *Interaction.* Solutions in this phase typically include simple forms of interaction; for example, citizens can ask questions and receive answers via e-Mail, download forms or search government databases. The interaction phase is exemplified by the web-based system FLO*rlp* – Land Parcel Information Online for the German State of Rhineland-Palatinate, which supports farmers in applying for agricultural subsidies by providing them with precise information on the location and size of their cultivated areas and by allowing them to visualize areas by means of maps and aerial photographs [92.8] (Fig. 92.3).

- *Transaction.* Conducting complete transactions online, such as applying for passports, submitting tax declarations, paying taxes and fees, or submitting bids for public tenders, are supported by systems in the transaction phase. These solutions are considerably more complex because often they have to address technically challenging issues, e.g., concerning security and privacy. For instance, digital (electronic) signatures or equivalent methods of identification are necessary to enable legally

binding administrative acts. Often government even needs to pass new laws to enable paperless transactions. An example of an e-Government solution requiring authentication of users through digital signatures is the German federal procurement solution e-Vergabe [92.9]. Additional examples of transactional sites are shown on the following pages (Figs. 92.4–92.7).

- *Transformation.* Solutions belonging to the transformation phase represent the most advanced form of e-Government. They comprise entire process chains between government and its clients. The services in the back office are completely integrated so that boundaries between different tiers of administration (e.g., federal versus state), departments or agencies are removed. The solutions are highly complex and require the governmental bodies to be reorganized in a new, process- and service-oriented way. This stage harnesses the benefits of ICT to their fullest extent. Currently, solutions in the transformation phase are still rare. However, within the European Union, a major boost for e-Government projects taking them to the transformation phase is expected through the EU Services Directive. By 2010, this regulation requires all member countries to offer points of single contact and handle all processes by electronic means for service companies which want

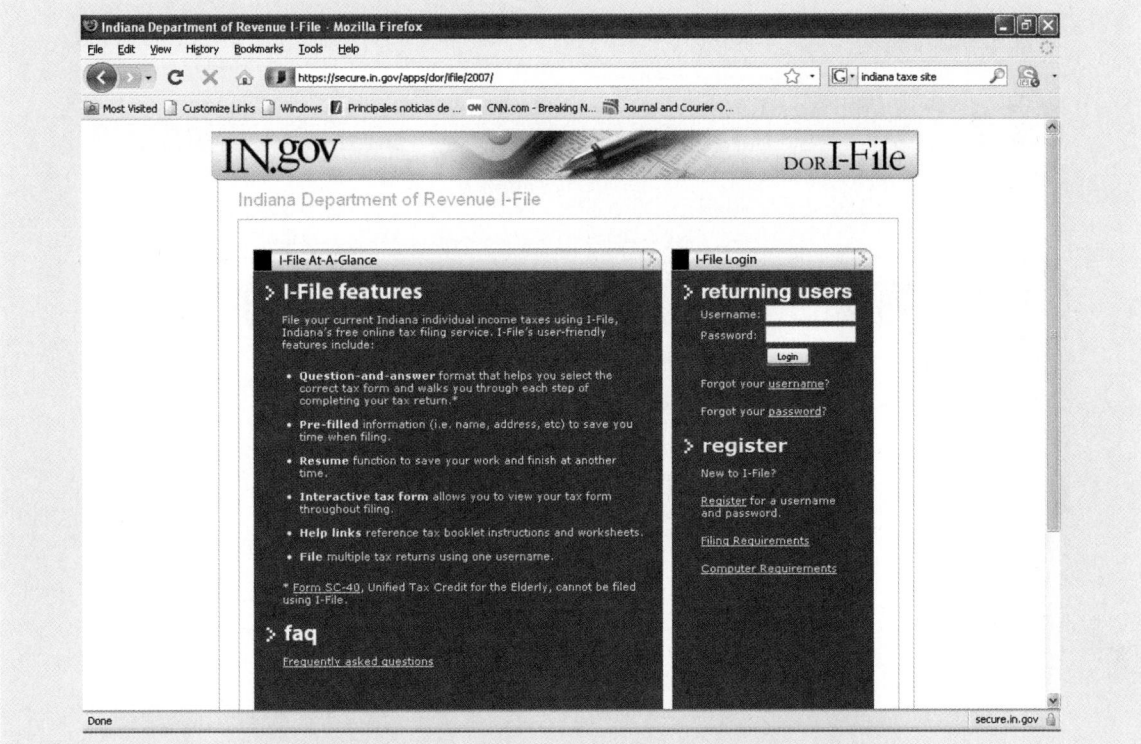

Fig. 92.4 Transaction site for paying taxes in the state of Indiana (USA) (courtesy of Government Indiana, https://secure.in.gov/apps/dor/ifile/2007/)

to set up businesses [92.10]. Also, many countries already offer comprehensive portals for business and citizens to access public services, such as the *UK Directgov* portal (Fig. 92.7).

In order to secure the success of e-Government developments, implementation activities need to proceed on four dimensions. In the following section, we will look at each of these dimensions in turn.

92.3 Proceeding from Strategy to Roll-Out: Four Dimensions of Action

Benefit-driven e-Government projects beyond the communication phase are characterized by a high level of complexity. They usually span multiple and often autonomous organizations (public agencies and companies), involve heterogeneous IT infrastructures, and concern numerous target groups and stakeholders. They have to meet conflicting requirements, face complex technical dependencies, and have to respect different legal constraints. Similar challenges also have to be faced in other IT projects. What makes e-Government projects special though is that they usually have to cope with a combination of numerous inhibiting factors and that IT strategies and decisions in the public sector are closely related to political constel-

lations, which in democratic states may change from one election to the other. For the successful execution of e-Government projects, it is therefore essential to:

- Develop a clear view of strategic objectives
- Design projects so that tangible results become visible at an early stage for all parties
- Involve users and stakeholders from the onset
- Handle architectural tradeoffs in a transparent way
- Ensure sustainability of the solution (in terms of legal aspects, process and technology used)
- Manage and minimize risks for all parties involved
- Manage change processes for all parties involved.

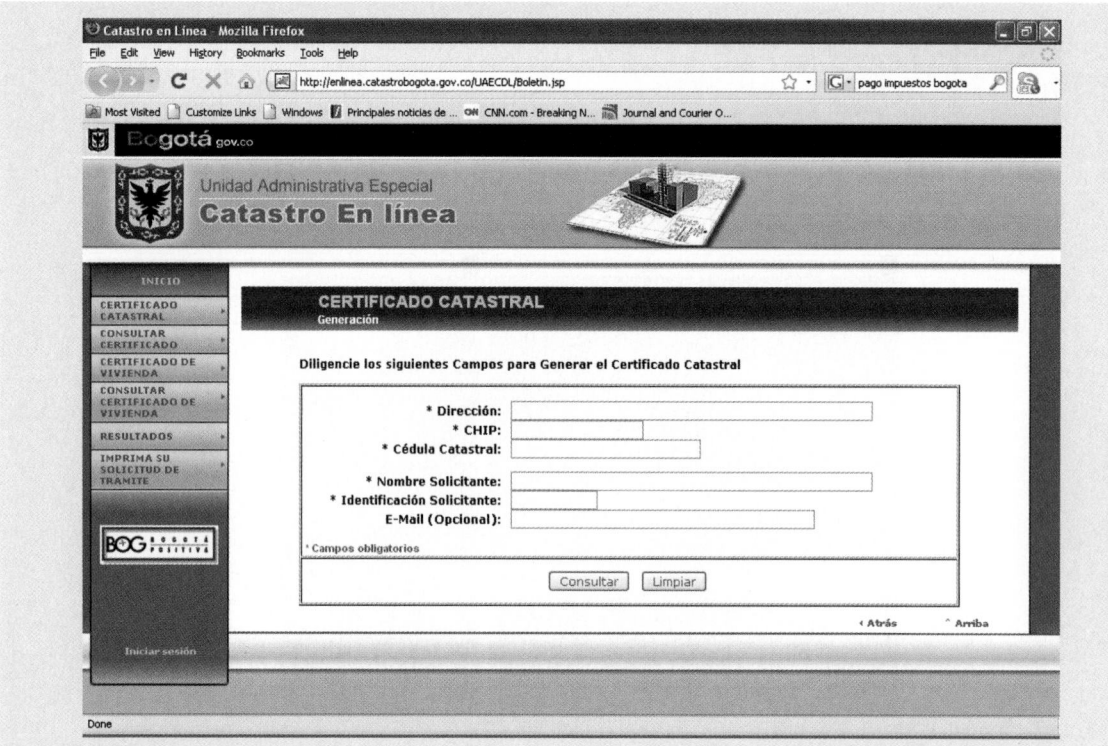

Fig. 92.5 Transaction site for requesting property taxes in the city of Bogota (Colombia) (courtesy of Bogota City Hall, http://enlinea.catastrobogota.gov.co/)

In order to accommodate these aspects, the design and implementation of benefit-oriented e-Government should proceed on four dimensions:

- Strategy
- Processes and organization
- Technology
- Project and change management.

We will now look at each of these dimensions.

92.3.1 Strategy

On the strategic dimension, a master plan has to be created which addresses the following topics.

Political Objectives Pursued by e-Government
There are different objectives that can be the driving force behind the implementation of e-Government; for example, e-Government can be used as a way to enhance a community's attractiveness for citizens and business, as a means to increase the transparency

of democratic processes or to increase the efficiency of the administration. In addition, the way in which political and administrative institutions on different levels of government interact needs to be spelled out. In countries which are organized in a centralized way, a more-or-less top-down approach is feasible, whereas in countries based on a federal system there is an obvious need to establish structures that regulate the complicated interplay of different stakeholders and help to resolve controversial issues in a timely manner.

Targets Addressed by e-Government
Based on the high-level political objectives, the primary groups which are to be targeted by e-Government can be determined and prioritized, i.e., government itself, citizens or companies. In order to arrive at an operational level, these groups need to be broken down further; for example, if a government decides to put the focus on citizens, it may put emphasis on subgroups such as elderly persons, immigrants or families. Similarly, when targeting business, a decision can be made on the

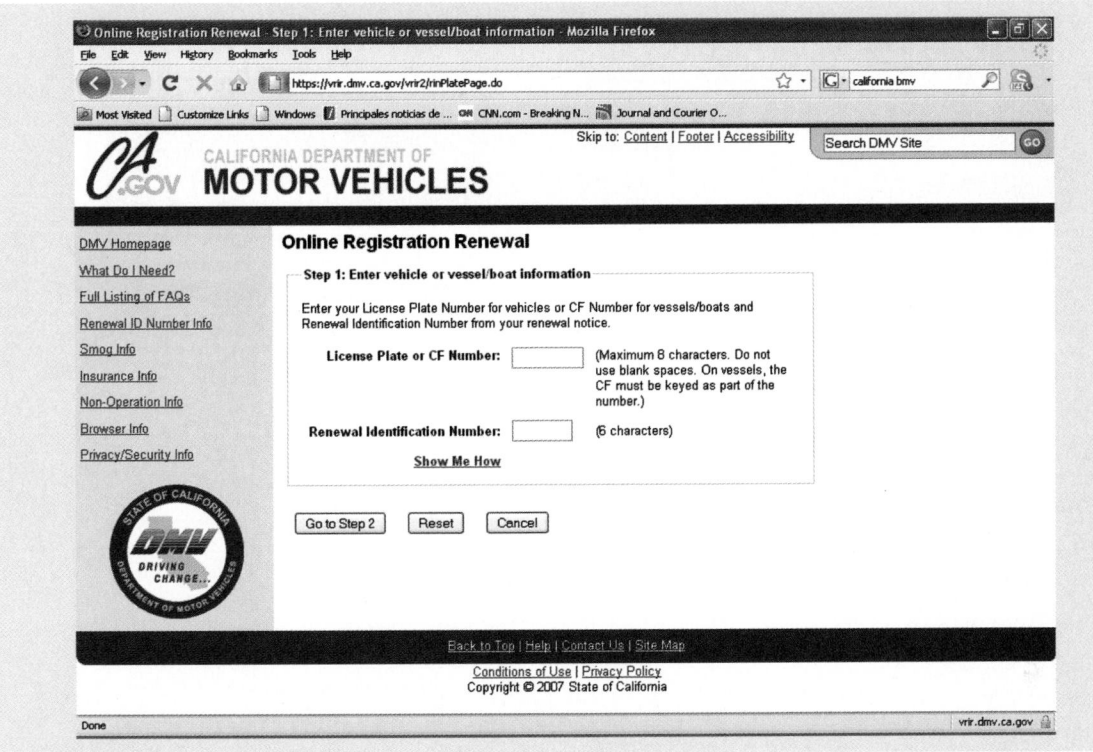

Fig. 92.6 Transaction site for registering cars in the state of California (USA) (courtesy of State of California, http://www.dmv.ca.gov/online/onlinesvcs.htm)

basis of industry sector (e.g., automotive versus chemical), company size (large corporations versus small and medium enterprises), or significance for the regional economy.

Technical Infrastructure and Technological Requirements

In most cases, the existing heterogeneous technical infrastructures provide the framework within which new developments have to be initiated. Therefore, IT infrastructures and technologies relevant to e-Government have to be surveyed and evaluated, and then political decisions regarding the further development of technologies in use have to be made. In particular, this concerns the decision in favor of commercial or open-source technologies, the decision between custom-made solutions versus adapting and reusing solutions originating from other governments, or the decision between running your own infrastructure versus outsourcing ICT. In addition, the standards to be followed have to be specified. This concerns both technological issues, e.g., system architecture or usability, as well as the development process itself. An example of a standard regarding system architecture is the SAGA (Standards und Architekturen für e-Government-Anwendungen – standards and architectures for e-Government applications) standards framework published by the German federal government. SAGA pursues the following aims [92.11]:

- Interoperability – ensuring a media-consistent flow of information between citizens, business, the federal government, and its partners
- Reusability – establishing process and data models for similar procedures when providing services and defining data structures
- Openness – integrating open standards into applications
- Reduction of costs and risks – considering investment-safe developments on the market and in the field of standardization
- Scalability – ensuring the usability of applications as requirements change in terms of volume and transaction frequency.

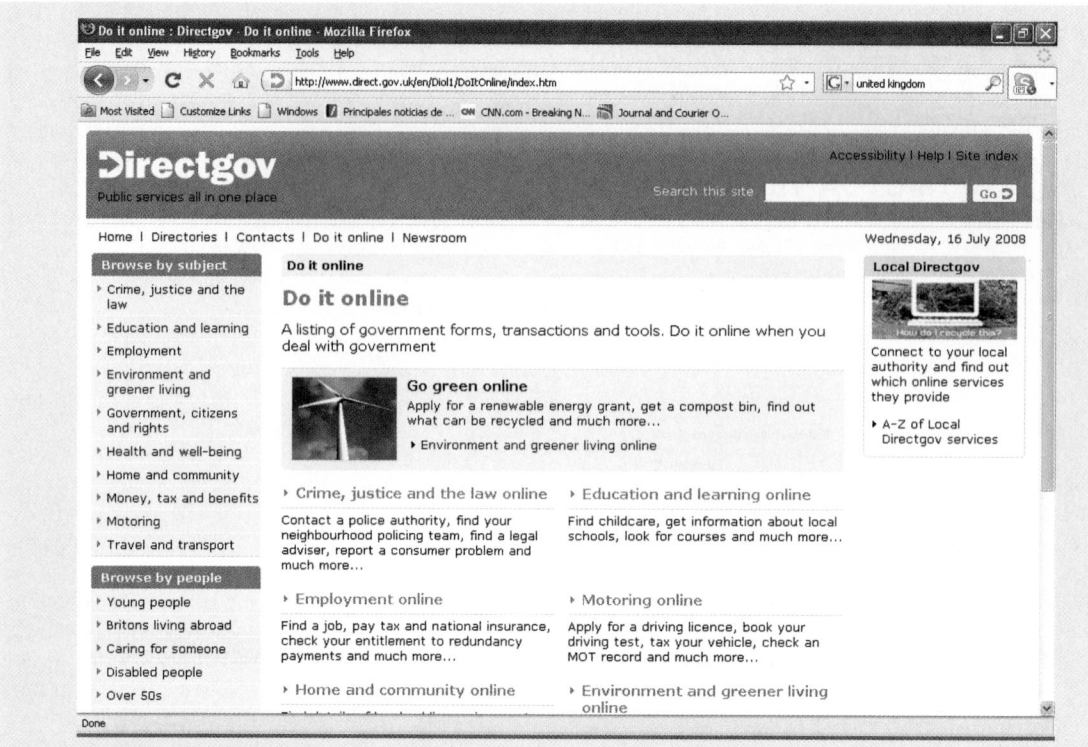

Fig. 92.7 UK e-Government site for a number of different services in the UK (with permission of Crown, http://www.direct.gov.uk/en/index.htm)

SAGA is meant as a guideline for the design of e-Government architectures, addressing decision-makers in public IT management as well as developers of e-Government solutions.

An example of a standard focusing on the development process itself is the lifecycle model V-Modell XT, which is the recommended standard for the federal German administration and for defense projects. V-Modell XT provides a reference process describing the activities and results that have to be produced during software development projects. It aims at improving project transparency, project management, and the probability of project success [92.12].

Focal Application Areas

Prioritization of application areas is essential for the success of an e-Government project; for example, if the political objective is to make government more efficient, a decision has to be made concerning which government processes will benefit most, e.g., procurement, skill development or management of geographic data. Depending on the focal application areas, the neces-

sary base technologies can be identified (e.g., workflow applications, geographic information systems, portal technology).

Roadmap

Based on the focal application areas, highly visible key projects are identified which promise immediate benefits and can therefore help in securing further political support. For further implementation of the strategy, key projects and the planned enhancements of infrastructure and of base technologies are plotted on a roadmap. A roadmap shows the main steps of how to set these projects on track by defining a schedule, allocating financial resources, assigning responsibilities, and defining critical success factors [92.13]. Roadmaps are mid- to long-term compared with the short- to mid-term electoral cycles of politicians; thus, success will most probably not be reaped by those who initiated the developments.

Wherever possible, the decisions to be taken on the dimension of strategy should be based on empirical evidence. This evidence may, for example, come from surveys among stakeholders, such as those conducted by the Fraunhofer Institute for Experimental Software Engineering (IESE) or the Technical University of Munich amongst industrial companies to identify G2B processes and determine their e-Government potential [92.14, 15], return on investment analyses (ROI) [92.16] or quantitative evaluation of technology acceptance [92.17].

92.3.2 Processes and Organization

Frequently, IT solutions are used to support existing workflows and their corresponding organizational structures. This may lead to some gains in efficiency and quality, but does not lead to the large-scale benefits that could be achieved if work processes and complex process chains involving different organizational units were truly optimized and reengineered. Take, for example, procurement processes in a distributed public administration. As long as IT is only used to support each agency in their traditional procurement processes, the economies of scale to be realized from bundling demands cannot be realized. Bundling of demands across agencies, on the other hand, requires a new setup of workflows and responsibilities: new units have to be established that are responsible for determining demands, negotiating contracts, and maintaining catalogues.

The identification of process chains which would yield high benefits for all stakeholders if transformed into an e-Government solution is done on the basis of potential analyses and is part of strategy building. Once process chains have been identified, ranked, and reengineered, they are scrutinized for common building blocks which require the same functionality. Lately, methodological frameworks for process modeling have been developed, which offer public-sector-specific process building blocks [92.18].

The challenge of process reengineering in the public sector is to overcome the boundaries between different organizational and functional units and between government and its clients. This may imply a shift of responsibilities, the adoption of common technical platforms and exchange formats, as well as cultural assimilation. Facing these challenges requires a high level of motivation and will only succeed if it finds a prominent place on the political agenda.

92.3.3 Technology

At present, IT systems in the public sector (like years ago in the business sector) are still characterized by heterogeneous IT infrastructures, a multitude of different applications, incompatible data formats, and human–computer interfaces which are difficult to use. However, for the realization of process-oriented e-Government it is essential that data and information can be exchanged smoothly and securely among agencies and between government and citizens as well as between government and businesses. In addition, for economical reasons, it has to be ensured that software architecture allows for reuse of legacy applications and promotes reuse of newly developed components and data.

It is mainly the disciplines of IT architecture, IT security, and usability which address these technological challenges. In the remainder of this section, we will look at recent developments in each of these fields.

Many of the challenges concerning heterogeneous IT landscapes and redundant services and data are addressed by the service-oriented approach to IT architectures [92.19, 20]. On the level of technology, this approach can be seen as a counterpart to the process-oriented approach described in the previous section. Service-oriented architectures (SOA) stand for a management concept as well as a comprehensive design principle for IT infrastructures. They lead to IT infrastructures which are able to adapt themselves easily to the changing requirements of the application domain. The salient feature of service-oriented architectures are software building blocks (services) which can be recombined relatively easily into new or more comprehensive solutions in support of new or more complex business processes. Services communicate via standardized exchange formats and interfaces. Conformity with standards ensures that different business processes which include identical activities can make use of the same basic building blocks. Thus, redundant development efforts can be avoided and the implementation of modified processes can be accelerated.

Facing an unprecedented ability of modern ICT to collect, track, and analyze personal data (e.g., telecommunication records, flight records, credit-card payments, or e-Mail communication), many citizens and companies hesitate to contribute further to their data trail and therefore avoid e-Government services. A means to counteract such fears and build trust is privacy policies for the protection of personal data and privacy. Examples are the recommendations of the Organization for Economic Cooperation and Development (OECD), which cover a large number of organizational and technical issues, spanning data transparency to data parsimony [92.21]. Especially in the context of e-Government, where trust is of utmost importance, instruments are employed which help to ascertain users' identity (authentication) and to protect private data.

Common instruments used for authentication are digital signatures and digital certificates [92.22]. In some countries, including the USA and countries of the European Union, these provide a legally binding authentication method for the seamless electronic execution of administrative transactions. Usually, they require special hardware devices (card readers) as well as a public-key infrastructure (PKI), which comprises organizational and technical elements to certify the owner's identity [92.22]. Despite their technical maturity, digital signatures still lack wide acceptance, because users shun the necessary organizational and technical efforts.

Alternative signature mechanisms have therefore been proposed that do not require a PKI infrastructure. An example is Pretty Good Privacy's (PGP) web-of-trust approach [92.23], which is based on mutual trust in a community of peers. However, this approach lacks central control, which is the main reason why governments are reluctant to adopt it. To facilitate the use of digital signatures, the integration of common chip card applications, such as credit cards and health insurance cards, with citizen cards is currently pursued. A positive example for enhancing commonly used chip cards into devices for personal authentication is the Austrian citizen card (Fig. 92.8). However, wide acceptance of citizen cards will ultimately depend on striking a balance between costs, efficiency, dependability, and convenience.

Adequate usability is a minimal requirement for achieving convenience. Whereas e-Business solutions usually target specific user groups, most e-Government solutions, especially those for citizens, must be usable

Fig. 92.8 Austrian e-Card, usable as health insurance and citizen card

for everyone. Generally, no specific training can be provided. In fact, governments are often obliged by law to provide services without discrimination to the population at large and to counteract the danger of digital divide [92.24, 25]. In particular, accessibility for users with special needs has become a major requirement of many national e-Government projects [92.26].

In formulating e-Government policies, master plans, and project objectives, IT security and accessibility issues have moved center-stage over the last few years. However, it is often overlooked that these requirements represent conflicting goals, which have to be reconciled on the level of software architecture. A method especially developed to help in this endeavor is the architecture tradeoff analysis method developed at Carnegie Mellon University. It has been designed especially to identify and mitigate potential risks emerging from conflicting interests with regard to the characteristics of an IT system [92.27].

92.3.4 Project and Change Management

Several project lifecycle models and project management methods have been proposed, which have a special focus on the requirements of public-sector IT projects [92.12, 28]. It is not our intention to provide a comprehensive account of project management or software engineering issues here. However, there are certain aspects of project management which seem crucial for the success of e-Government projects, for example achieving consensus among stakeholders and delivering measurable results at different milestones of a project. The following aspects of project management address these issues.

Project Design

Project design refers to the high-level structuring of a project into phases, milestones, and additional control points. Splitting a project into phases, e.g., design, implementation, and testing, belongs to the standard tasks of a project manager. What is not state-of-the-practice yet is to perform measuring activities which assess critical success factors in these phases. Success criteria can refer to technical requirements such as system performance or interoperability, to usage criteria such as user acceptance, but also to economical aspects such as financial savings or shortening of processing time. Success criteria need to be set up consensually at the beginning of a project and have to be monitored throughout system operation. Project design ensures that these critical factors can be validated early enough in the project so that adjustments to software design can also be made early. Depending on the degree of innovation, the degree of precision and concreteness of project objectives, and the degree of user interaction, phases should be devoted to proof of concept, prototype building, different field tests with representative users and stakeholders, and finally a staged roll-out.

Communication Management

The electronic support of complex administrative processes usually involves a large number of stakeholders and users. Often, these groups have divergent interests, which can lead to considerable friction and project delays. Managing communication proactively is of paramount importance to facilitate consensus building. It is needed (a) to obtain a transparent picture of stakeholder interests, (b) to establish clear communication and decision procedures, and (c) to plan for appropriate actions. Apart from detecting and resolving potential conflicts, communication management also aims at ensuring project support by proactively informing stakeholders, politicians, and the public on project status and plans. It thus plays a vital role in the process of creating technology acceptance.

Change Management

Process orientation, which is at the core of many of today's e-Government efforts, is often opposed to the traditional principle of vertical line organization of administration. Only if processes, structures, and attitudes change can the benefits of automating process chains materialize. Therefore, project management has to put special emphasis on preparing the ground for change and on implementing change in a transparent and cooperative way. There are two instruments which seem to be key in change management processes: guiding coalitions and user participation.

Guiding coalitions are groups which are able to direct a change effort. For public-sector projects, guiding coalitions should be composed of the following individuals.

Political Sponsors. Securing broad support among political leaders is crucial to ensure financial continuity for a program, as transformation processes often span electoral cycles. Political support will also be essential if legislation or regulatory changes are needed.

Line Managers. Line managers are responsible for correct execution of administrative processes. They have the power to define how an administrative procedure is implemented on the operational level. Often they also have budget responsibility. Usually, they survive electoral cycles and can thus facilitate continuity.

Employee Champions. Employee champions are usually senior officials with long-standing expertise, who lend credibility to the cause. They act as early adopters and thus help to motivate others and reduce fears and objections. They are least exposed to the changing winds of electoral cycles.

Independent Advisors. To support complex decision processes and conflict resolution, independent advisors, for example, from national research institutions, are an indispensable part of the coalition. They bring expertise on the levels of strategy, technology (state-of-the-art and future), legislation, organizational and process analysis, and redesign. Moreover, externals are open to promising options that threaten current structures.

In implementing change, the guiding coalition relies upon a project team which combines technical, organizational, methodical, and legal expertise.

Besides setting up guiding coalitions, users should be given the opportunity to take part in the change process; for example, users can be included in project teams as on-site customers or can contribute to designing and optimizing a solution by piloting it at different stages in a project. In general, those who are affected by the change need to be informed on a regular basis and should be given a chance to express their opinion and concerns, for example, in web blogs, at regular information events or through one-on-one counseling [92.29].

92.4 Future Challenges in e-Government Automation

In the previous sections, we gave an overview of the evolution of e-Government and explained the actions necessary to leap forward from political vision and strategy to practical solutions. However, despite many inspiring examples and promising approaches, e-Government development still lags behind the development of ICT in business. This is due to the specific characteristics of the public sector and partly, but to a minor extent, to technological issues.

In the following, we will look at some of the issues, which we consider crucial for the transition from an insular and function-oriented government with a high degree of data and systems redundancy to an ICT-empowered, process-oriented, and cohesive type of e-Government. We will distinguish two types of challenges which e-Government initiatives and the technology sector have to face: political challenges, which require action on the level of policy and politics, and engineering challenges, which require the development of new methods or the adaptation of known approaches. In the remainder of this section, we will look at each of these issues in turn. However, we will not look at the management challenges of applying and enforcing well-established software engineering practices (see, e.g., the e-Government manual of the German Federal Office for Information Security [92.30]).

92.4.1 Political Challenges

In many countries, the landscape of e-Government solutions is characterized by heterogeneous and incompatible approaches and technical solutions. This is mostly due to the fact that administrative units often act independently from each other and pursue their own political agendas. In order to overcome the situation, it is necessary both to establish standards of technical interoperability and to install e-Government representatives, e.g., national chief information officers (CIOs), who are equipped with enough political clout and resources to make these standards mandatory across departmental boundaries and across different levels of administration (e.g., community, state, and federal). A country which has successfully implemented the concept of a national CIO in charge of e-Government is Austria, which as early as 2001 installed a national CIO directly reporting to the chancellor of the federal republic [92.31].

In many cases, e-Government projects span a large number of years, which may surpass the time during which a political party is in power. For the realization of long-term e-Government programs, changes of the power structure can have a negative impact on current developments. To ensure continuity, act independently of interdepartmental power plays, and voice the need for legislative action, those representing e-Government should be independent of political agendas and electoral cycles.

To support complex decision processes and help resolve conflicts, independent advisors belonging to the research community should become an integral part of the decision-making process. Their task is primarily to bring in neutral expertise at the levels of strategy, legislation, technology, and organization. Moreover, they should develop visions and future scenarios that go beyond current structures.

Current e-Government efforts address the typical middle-class family of the Western hemisphere, which is well-educated, computer-literate, nonhandicapped, and equipped with the technical means to take advantage of electronic governmental services. While computer ownership is relatively common in the developed countries of the West and Asia, it is less widespread in the Middle East, and comparatively rare throughout much of Latin America and Africa. For example, while 82% of the Swedish use computers at work, home or anywhere else, at least occasionally, only 9% of the Bangladesh population do so [92.32]. The reasons for this are manifold, including wealth, infrastructure, education, language, age, cultural background, and political conditions, to name but a few. The challenge is to overcome these barriers, which cause the often quoted *digital divide* [92.24]. A particularly promising line of action to overcome the digital divide is to take advantage of the wide distribution and high usability of mobile communication devices, which allow persons who are illiterate but numerate to use electronic government services [92.33]. Other promising initiatives are underway in different parts of the world (see, for example, the case studies provided by the World Bank [92.34]), often promoted by transnational institutions such as the United Nations (UN) or the World Bank. As promising as these examples are, systematic and comparative analysis of the effects of such programs and of the transferability of the applied approach is still needed.

92.4.2 Engineering Challenges

In order to secure continuous support and funding for complex e-Government projects, it is important to produce visible results and tangible benefits early on. This can be achieved by breaking up complex projects into smaller tasks, focusing on *quick wins*. The concept of a quick win refers to a change management strategy which aims to realize projects that have the potential to yield improvements within a short time. Quick-win projects can thus help to build acceptance for further innovation. The key in this endeavor is the business process to be supported by IT. Besides serving as the basis for deriving functional and nonfunctional requirements for e-Government solutions, it supplies the framework for establishing an evolution plan for incremental development.

Part 92

Gradually increasing the complexity of e-Government solutions has several advantages: firstly, implementing G2B processes which span an entire end-to-end process chain requires substantial financial investment; secondly, by selecting those process steps for which IT solutions can be provided that do not make heavy requirements on users' IT skills nor on their technological equipment, chances are good that users will accept the new technology; thirdly, through incremental development, lessons learned while deploying new technologies can be fed back into development and provide beneficial input to more comprehensive solutions. The challenges here are to identify the right starting point and plan for the right increments.

In the SOA world, this issue is also discussed under the concept of *focus holistically, act locally*. According to this paradigm, it is essential to identify projects that are small enough to avoid failure, but significant enough to attract attention from decision-makers. The software engineering discipline of requirements engineering with its methods for prioritizing requirements can provide guidance here [92.35]. However, an open issue is to (a) establish public-sector-specific prioritization criteria, (b) provide a comprehensive framework for applying them to individual process steps or to a combination of process steps, and (c) provide empirical validation for the criteria.

Besides the question of how to arrive at a prioritization of process steps and at politically and technologically valid roadmaps for solution development, another engineering challenge is to determine the level of component granularity, which allows one to maximize reuse potential. While one of the foremost concerns of the SOA approach is to maximize reuse of components, most public-sector experiences have been in small-scale project contexts; little knowledge exists concerning SOA design on a large scale, spanning different organizational units. In particular, no empirically validated findings with regard to the right level of component granularity are available for the public sector. Future research is needed here, looking at a variety of G2B processes, and analyzing them with regard to the optimal level of reusability.

The SOA approach depends on technical interoperability on one hand, and on semantic interoperability, on the other. The latter is key to the reuse of components across organizational units: there needs to be a shared terminology for exchanging information and a standardized way of interpreting it [92.36]. The challenge here is to define and agree on data exchange standards and underlying ontologies that allow for a uniform interpretation. The federal government of Germany has responded to this challenge and set up the so-called XöV initiative (meaning *XML for public administration*) responsible for developing extensible markup language (XML)-based technical standards for electronic data exchange. The initiative aims at facilitating the seamless exchange of data between agencies on the federal, state, and municipal level as well as between government agencies and their external clients. So far, XML-based standards have, for example, been developed for data exchange between welfare and employment offices (XSozial), and between registry offices (XMeld) (for an overview of current XöV projects see [92.37]).

Despite such promising initiatives, considerable efforts – both in the research community and on the political level – are still needed to achieve semantic interoperability on national and international levels.

References

92.1   R. Büllesbach: Branchen- und nutzengetriebenes eGovernment, Presentation at the Conference Verwaltungsinformatik (Brühl, 2006), in German, http://www.uni-koblenz.de/ fvi/ftvi2006/dateien/ Buellesbach_Branchen-+nutzengetriebenes_E-Gov_FTVI2006.pdf (last accessed April 07, 2008)

92.2   J. Fenn, M. Raskino: *Mastering the Hype Cycle: How to Choose the Right Innovation at the Right Time* (Harvard Business School Press, Boston 2008)

92.3   European Commission: *An overview of the e-Government situation in Europe* (2008), http://www.epractice.eu/factsheets (last accessed April 07, 2008)

92.4   D.J. Hoch, M. Klimmer, P. Leukert: *Erfolgreiches IT-Management im öffentlichen Sektor: Managen statt verwalten* (Gabler, Wiesbaden 2005), in German

92.5   German Federal Ministry of Finance: *Entwicklung bundeseinheitlicher Software für das Besteuerungsverfahren (Vorhaben KONSENS)*, Monatsbericht des BMF 55–64 (October 2006) (in German), http://www.bundesfinanzministerium. de/ nn_17844/DE/BMF__Startseite/Aktuelles/ Monatsbericht__des__BMF/2006/10/ 061019agmb006,templateId=raw,property= publicationFile.pdf (last accessed April 07, 2008)

92.6 R. Büllesbach: Die e-Government-Initiative "Rhein-land-Pfalz 24" – e-Business und e-Government aus einem Guss, Der Landkreis **75**, 132–133 (2005), in German, http://www.kreisnavigator.de/landkreistag/zeitschrift/rueckblick/auswahl_2005-03.pdf (last accessed April 07, 2008)

92.7 C. Baum, A. Di Maio: *Gartner's Four Phases of e-Government Model* (Gartner Group Inc., Stamford 2000)

92.8 P. Steffens, G. Geissner: FLO*rlp* - Land parcel information online for the german state of Rheinland-Pfalz: web-based GIS for farmers. In: *e-Government Guide Germany. Strategies, Solutions and Efficiency*, ed. by A. Zechner (Fraunhofer IRB Verlag, Stuttgart 2007) pp. 309–317

92.9 Beschaffungsamt des Bundes: *e-Vergabe – Vergabeplattform des Bundes* (2008), http://www.evergabe-online.de/ (last accessed April 07, 2008)

92.10 Directorate-General for Internal Market and Services of the European Commission: *Handbook on Implementation of the Services Directive* (EC, Luxembourg 2007), http://ec.europa.eu/internal_market/services/docs/services-dir/guides/handbook_en.pdf (last accessed April 07, 2008)

92.11 German Federal Interior Ministry together with]init[AG, Fraunhofer ISST: *Standards und Architekturen für e-Government-Anwendungen* (Berlin, 2008), in German, http://gsb.download.bva.bund.de/KBSt/SAGA/SAGA_v4.0.pdf (last accessed April 07, 2008)

92.12 V-MODELL XT Project: *Part 1: Fundamentals of the V-Modell* (2006), ftp://ftp.tu-clausthal.de/pub/institute/informatik/v-modell-xt/Releases/1.2.1/Documentation/V-Modell-XT-Complete.pdf (Last access: April 07, 2008)

92.13 T. Altameem, M. Zairi, S. Alshawi: Critical success factors of e-Government: A proposed model for e-Government implementation, Innovations in Information Technology 2006 (2006) pp. 1–5, http://ieeexplore.ieee.org/xpls/abs_all.jsp?arnumber=4085489 (last accessed April 07, 2008)

92.14 I. Grützner, P. Steffens: Wem nutzt e-Government?, move – Moderne Verwaltung **5**, 16–17 (2007), in German

92.15 P. Wolf, H. Krcmar: *Collaborative e-Government – Neue Ansätze für durchgängige B2G-Prozesse*, 1st edn. (Technische Universität München/Institut für Informatik/Lehrstuhl für Wirtschaftinformatik, Garching 2007), in German, http://www.winfobase.de/ lehrstuhl/publikat.nsf/intern01/9B3A9FA1765DAE59C125729E00488B41/$FILE/07-02.pdf (last accessed April 07, 2008)

92.16 A.M. Cresswell, G.B. Burke, T.A. Pardo: *Advancing Return on Investment, Analysis for Government IT – A Public Value Framework* (Center for Technology in Government, Albany 2006), http://www.ctg.albany.edu/publications/reports/advancing_roi/advancing_roi.pdf (last accessed April 07, 2008)

92.17 L. Carter, F. Bélanger: The utilization of e-Government services: Citizen trust, innovation and acceptance factors, Inf. Syst. J. **15**, 5–25 (2005)

92.18 J. Becker, L. Algermissen, T. Falk: *Prozessorientierte Verwaltungsmodernisierung. Prozessmanagement im Zeitalter von e-Government und New Public Management* (Springer, Berlin 2007), in German

92.19 D. Krafzig, K. Banke, D. Slama: *Enterprise SOA – Service-Oriented Architecture and Best Practices* (Prentice Hall, Indianapolis 2005)

92.20 T. Erl: *Service-Oriented Architecture – Concepts, Technology, and Design* (Prentice Hall, Upper Saddle River 2005)

92.21 OECD: *OECD Guidelines on the Protection of Privacy and Transborder Flows of Personal Data* (Organisation for Economic Co-operation and Development, Paris 1980), http://www.oecd.org/document/18/0,3343,en_2649_34487_1815186_1_1_1_1,00.html (last accessed April 07, 2008)

92.22 K.N. Gupta, K.N. Agarwala, P.A. Agarwala: *Digital Signature: Network Security Practices* (Prentice Hall of India, New Delhi 2005)

92.23 S. Garfinkel: *PGP: Pretty Good Privacy* (O'Reilly, Sebastopol 1995)

92.24 Center for Democracy and Technology: *The e-Government Handbook for Developing Countries* (CDT, Washington 2002), http://www.cdt.org/egov/handbook/2002-11-14egovhandbook.pdf (last accessed April 07, 2008)

92.25 Illinois General Assembly: *Article 5: Eliminate Digital Divide Law* (IGA, Springfield 2000), http://www.ilga.gov/legislation/ilcs/ilcs4.asp?DocName=003007800HArt%2E+5&ActID=574&ChapAct=30%26nbsp%3BILCS%26nbsp%3B780%2F&ChapterID=7&ChapterName=FINANCE&SectionID=8192&SeqStart=2100&SeqEnd=5400&ActName=Eliminate+the+Digital+Divide+Law%2E (last accessed April 07, 2008)

92.26 Zugang für Alle (Schweizerische Stiftung zur behindertengerechten Technologienutzung): *Schweizer Accessibility-Studie 2007: Bestandsaufnahme der Zugänglichkeit von Schweizer Websites des Gemeinwesens für Menschen mit Behinderungen* (Zugang für Alle (Schweizerische Stiftung zur behindertengerechten Technologienutzung), Zürich 2007), in German, http://www.label4all.ch/studie/Accessibility_Studie_2007.pdf (last accessed April 07, 2008)

92.27 R. Kazman, M. Klein, P. Clements: *ATAM: Method for Architecture Evaluation*, Tech. Rep. CMU/SEI-2000-TR-004 ESC-TR-2000-004 (Carnegie Mellon University and Software Engineering Institute, Pittsburgh 2000), http://www.sei.cmu.edu/pub/documents/

00.reports/pdf/00tr004.pdf (last accessed April 07, 2008)

92.28 Great Britain Office of Government Commerce: *Managing Successful Projects with Prince 2* (Stationary Office, London 2002)

92.29 H. Duivenboden: Citizen participation in public administration: The impact of citizen oriented public services on government and citizens. In: *Practicing e-Government: A Global Perspective*, ed. by M. Khosrow-Pour (Idea Group, Hershey 2005) pp. 415–445

92.30 Bundesamt für Sicherheit in der Informationstechnik: *e-Government-Handbuch* (Bundesanzeiger Verlag, Köln 2006), in German, also available in English from http://www.bsi.de/fachthem/egov/3a.htm (last accessed April 07, 2008)

92.31 European Commission: *e-Government Factsheet – Austria – Actors* (EC, Luxembourg 2008), http://www.epractice.eu/document/3276 (last accessed April 07, 2008)

92.32 PEW Research Centre: *World Publics Welcome Global Trade – But not Immigration, 47-Nation Pew Global Attitudes Survey* (Pew Research Center, Washington 2007) p. 74, http://pewglobal.org/reports/pdf/258.pdf (last accessed April 07, 2008)

92.33 I. Kushchu: *Mobile Government: An Emerging Direction in e-Government* (Idea Group, Hershey 2007)

92.34 The World Bank: *e-Government Case Studies* (2008), http://web.worldbank.org/WBSITE/EXTERNAL/TOPICS/EXTINFORMATIONAND COMMUNICATIONANDTECHNOLOGIES/EXTEGOVERNMENT/0,,contentMDK:20798277 isCURL:Y menuPK:1767179 pagePK:148956 piPK:216618 theSitePK:702586,00.html (last accessed April 07, 2008)

92.35 K. Pohl: *Requirements Engineering. Grundlagen, Prinzipien, Techniken* (Dpunkt, Heidelberg 2007), in German

92.36 G. Vetere, M. Lenzerini: Models for semantic interoperability in service oriented architectures, IBM Syst. J. **44**, 887–903 (2005)

92.37 Deutschland Online: *XÖV-Projekte* (2008), http://www.deutschland-online.de/DOL_Internet/broker.jsp?uMen=b1f32acc-9224-114f-bf1b-1ac0c2f214a8 (last accessed April 07, 2008)

93. Collaborative Analytics for Astrophysics Explorations

Cecilia R. Aragon

Many of today's important scientific breakthroughs are made by large, interdisciplinary collaborations of scientists working in geographically distributed locations, producing and collecting vast and complex datasets. Experimental astrophysics, in particular, has recently become a data-intensive science after many decades of relative data poverty. These large-scale science projects require software tools that support not only insight into complex data but collaborative science discovery. Such projects do not easily lend themselves to fully automated solutions, requiring hybrid human—automation systems that facilitate scientist input at key points throughout the data analysis and scientific discovery process. This chapter presents some of the issues to consider when developing such software tools, and describes Sunfall, a collaborative visual analytics system developed for the Nearby Supernova Factory, an international astrophysics experiment and the largest data volume supernova search currently in operation. Sunfall utilizes novel interactive visualization and analysis techniques to facilitate deeper scientific insight into complex, noisy, high-dimensional, high-volume, time-critical data. The system combines novel image-processing algorithms, statistical analysis, and machine learning with highly interactive visual interfaces to enable collaborative, user-driven scientific exploration of supernova image and spectral data. Sunfall is currently in operation at the Nearby Supernova Factory; it is the first visual analytics system in production use at a major astrophysics project. The chapter concludes with a set of guidelines and lessons learned about developing software to support scientific collaborations.

93.1 Scope

Computational and experimental sciences are producing and collecting ever-larger and complex datasets, often in large-scale, multi-institution projects. The inability to gain insight into complex scientific phenomena using current software tools is a bottleneck facing virtually all endeavors of science. As scientific datasets explode at an exponential rate, present-day systems are struggling to capture, process, store, transfer, and interpret data many orders of magnitude greater than scientists have seen before. Computational challenges are arising in unexpected areas, not just in requirements for processor speed, but in software development, scalability, data transfer, and file management. Additionally, it is not sufficient merely to solve the automation challenges; scientific insight and decision-making require human interaction and smooth collaboration with automated systems. However, there are few guidelines or examples to support the interdisciplinary developers of collaborative scientific analytics systems, which need to integrate tools and techniques from a broad array of computing disciplines with highly sophisticated and rapidly changing domain-specific scientific algorithms, as well as incorporate effective human–computer interaction design for scientists.

Among all the sciences, experimental astrophysics is undergoing extraordinarily rapid growth in observational image data captured, at the precise time that new and tantalizing questions about the nature of the universe have arisen, the answers to which will require study of ever-larger areas of the sky in ever-greater detail. As a result, the nature of astrophysics research itself is changing, and many of the algorithms developed for processing astronomical data, although well established for low-volume data capture, do not scale well to today's high-volume sky surveys and transient searches.

One of the grand challenges in astrophysics today is the effort to comprehend the mysterious *dark energy*, which accounts for three-quarters of the matter/energy budget of the universe. The existence of dark energy may well require the development of new theories of physics and cosmology. Dark energy acts to accelerate the expansion of the universe (as opposed to gravity, which acts to decelerate the expansion). Our current understanding of dark energy comes primarily from the study of supernovae (SNe).

The Nearby Supernova Factory (SNfactory) [93.1] is an international astrophysics experiment designed to discover and measure type Ia supernovae in greater number and detail than has ever been done before. These supernovae are stellar explosions (Fig. 93.1) that have a consistent maximum brightness, allowing them to be used as *standard candles* to measure distances to other galaxies and to trace the rate of expansion of the universe and how dark energy affects the structure of the cosmos.

In order to facilitate the astrophysics search and data analysis process and enable scientific discovery for observational astrophysicists, an interdisciplinary group of computer scientists, software engineers, and astrophysicists developed the Supernova Factory Assembly Line (Sunfall), a collaborative visual analytics system for the Nearby Supernova Factory that has been in production use since 2006 [93.2]. Sunfall incorporates:

Fig. 93.1 Hubble Space Telescope image of supernova SN1994D in galaxy NGC4526, observed 1994

- Novel astrophysics image-processing algorithms running in parallel on a 300 node cluster
- Machine learning capabilities including boosted decision trees and support vector machines
- A specialized data transfer mechanism that not only manages a large amount of data on a 24/7 basis, but also robustly handles the error-free transfers of a large number of small files
- Highly interactive visual interfaces to enable collaborative, user-driven scientific exploration of supernova image and spectral data.

The remainder of this chapter is structured as follows. Section 93.2 describes the science background for supernova detection and spectral analysis, including the SNfactory project data flow. Section 93.3 contains information on previous work, including systems built for other supernova experiments and other scientific workflow systems. Section 93.4 discusses the Sunfall design approach, and Sect. 93.5 describes the Sunfall architecture in detail, including its four major components: Search, Workflow Status Monitor, Data Forklift, and Supernova Warehouse. Section 93.6 discusses lessons learned and presents our conclusions and their potential impact on the fields of astronomy and cosmology, as many more such large-scale surveys are planned for the future as physicists attempt to unravel the mystery of dark energy.

93.2 Science Background

The discovery of dark energy is primarily due to observations of type Ia supernovae at high redshift (up to $z = 1$, or a lookback time of about 8 billion years) [93.3, 4]. This supernova cosmology technique hinges on the ability to reliably compare luminosities of high-redshift events to those at low redshift, necessitating detailed study of low-redshift events. Large-scale digital sky surveys are now being planned in order to constrain the properties of dark energy. These studies typically involve high-volume, time-constrained processing of wide-field charge-coupled device (CCD) images, in order to detect type Ia SNe and capture detailed images and spectra on multiple nights over their lifespans.

A stellar explosion resulting in a type Ia SN will occur on average only once or twice per millennium in a typical galaxy of 400 billion stars, and then will remain visible for only a few weeks to a few months. Additionally, the maximum scientific benefit is obtained if the supernova is discovered as early as possible before it attains peak brightness. To further add to the challenge, the imaging data from which SNe must be detected are extremely noisy, corrupted with spurious objects such as cosmic rays, satellite tracks, asteroids, and CCD artifacts such as diffraction spikes, saturated pixels, and ghosts.

The SNfactory was designed to accumulate the largest homogeneously calibrated spectrophotometric (containing both images and spectra) dataset ever studied of type Ia SNe in the *nearby* redshift range ($0.03 < z < 0.08$, or about 0.4–1.1 billion light years distant) [93.1]. On a typical night, 50–80 GB of data (approximately 30 000 images containing 600 000 potential supernovae) are received by SNfactory image-processing software. These data are processed overnight; the best candidates are selected each morning by humans for further follow-up measurements. Likely candidate supernovae are sent to a dedicated custom-built spectrograph for follow-up imaging and spectrography. The resulting spectra typically each contain 2000–3000 data points of flux as a function of wavelength. The SNfactory database will be released to astronomers worldwide and become a definitive resource for measurements of dark energy. This program is the largest data volume supernova search currently in operation.

The SNfactory obtains wide-field imaging data from the Near-Earth Asteroid Tracking Program (NEAT) [93.5] using the 112 CCD QUEST II camera of the Mt. Palomar Oschin 1.2 m telescope [93.6], covering 8.5 square degrees per exposure. (For comparison, the full moon covers 0.2 square degrees.)

Images are transferred from Mt. Palomar via the High-Performance Wireless Research and Education Network (HPWREN) to the High-Performance Storage System (HPSS) at the National Energy Research Scientific Computing Center (NERSC) in Oakland, California. Each morning, SNfactory search software running on NERSC's 300 node computing cluster, the Parallel Distributed Systems Facility (PDSF), matches images of the same area of the sky, processes them to remove noise and CCD artifacts, then performs an image subtraction from previously observed reference images on each set of matched images (Fig. 93.2) [93.7–10].

Supernova candidates are identified from among the over 600 000 objects processed per night by a set of image features including object shape (roundness and contour irregularity computed from Fourier contour descriptors) [93.7, 11], position, distance from nearest object, and motion. These features are used as input to machine learning algorithms that select candidates to be sent to humans for scanning and vetting [93.8, 9]. Promising SN candidates that pass the human scanning and vetting procedure are sent for confirmation and spectrophotometric follow-up by the Supernova Integral

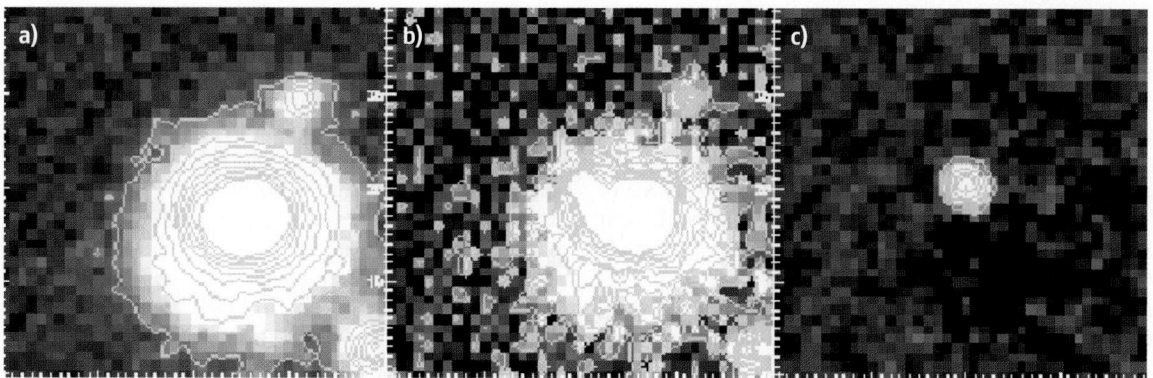

Fig. 93.2a–c Example supernova images. (**a**) Reference image of galaxy, (**b**) new image of galaxy plus new bright spot, (**c**) subtracted image reveals new supernova

Field Spectrograph (SNIFS) [93.12] on the University of Hawaii 2.2 m telescope on Mauna Kea, Hawaii.

Candidates are imaged through a 15×15 microlens array on SNIFS and the spectral data are saved to a database in France. The SNIFS observing process is challenging. The telescope is operated remotely, via a text-based legacy control system. Astronomers must be able to analyze and evaluate a large amount of data and make cognitively demanding calculations under time pressure (in some cases in as little as 45 s), while at the same time being fully aware of changing weather conditions, the approach of daylight, and other safety issues. They must focus on individually demanding and precise tasks while maintaining an overall understanding of a large amount of dynamic data affecting the telescope's operation and safety.

Understanding type Ia SNe through their spectra is important because it places constraints on their presupernova evolution and the immediate stellar environment of the event, and provides clues that can expose correlations between peak brightness and the physics of the explosion. With a better understanding of the physics of type Ia SNe through their spectra, it is believed that their utility as standardized candles for cosmological distance measurements can be refined for their use in future precision cosmology experiments such as the Joint Dark Energy Mission (JDEM) [93.13].

93.3 Previous Work

We discuss related work in the areas of astronomical applications, specifically software developed for large-scale supernova searches, and more generally including astronomical calculations and visualizations, and visual analytics systems developed for cognitively loaded or time-pressured users. We also briefly touch on scientific analytics and situation awareness research, as the development of situation awareness at the telescope during observation is critical for astrophysics experiment success.

Since the discovery of dark energy via studies of type Ia supernovae nearly a decade ago [93.3, 4], astrophysicists have developed several large-scale supernova searches to collect as much data as possible on these rare events. These searches typically rely on custom image subtraction software containing highly complex, hand-tuned heuristics and *cuts* to extract supernova candidates. They are often plagued by large numbers of false positives, which must then be screened out by humans in a labor-intensive process. For example, the 2005 Sloan Digital Sky Survey II (SDSS) supernova program generated approximately 4000 objects per night which needed to be visually checked by humans for verification [93.14]. The Equation of State: Supernovae Trace Cosmic Expansion (ESSENCE) and Supernova Legacy Survey (SNLS) supernova searches both resulted in 100–200 objects to scan per night with a significantly smaller input data load than the SNfactory [93.15]. Techniques used in these surveys will not scale to future, much larger data sets.

In 2005, the SNfactory pioneered the use of machine learning algorithms to replace *cuts* for supernova object detection [93.9], and since then, such techniques have been utilized by other supernova searches, including SDSS and SNLS [93.14, 15]. The SDSS *autoscanner* uses a combination of heuristics and a naive Bayes classifier to filter out non-SN objects, but they do not provide a comparison of the efficacy of any other types of machine learning algorithms for supernova search. SNLS uses multinight data and an artificial neural net to screen candidates; however, their techniques are not applicable to large-scale, rapid transient alert pipelines.

After supernova data have been collected, they are typically stored in a database accessible via a web-based interface. These websites display information in tabular form, listing the supernova coordinates and providing links to images. They are designed to provide all the necessary information for astronomers to observe supernovae with their own telescopes. Examples are the SNLS and SDSS web sites [93.16, 17]. In [93.14], *Sako* et al. describe the SDSS software tools for supernova search and follow-up. They include sets of scripts that perform functions similar to parts of Sunfall, but require human intervention at numerous stages during the supernova search and follow-up process. None of these groups has built a complete scientific analytics system that enables human–automation collaboration and encompasses the entire data search, analysis, and scientific discovery process.

The astronomical calculations in the Sky visualization used in Sunfall Data Taking are derived from SkyCalc, originally written by *Thorstensen* [93.18, 19]. Previously, astronomers would run SkyCalc from the command line at the telescope, inputting their observatory parameters, observing date, and target lists. This would provide times of sun and moon rise/set and hourly tables of the elevations of each target. The astronomer would then try to collate this information, but would have to re-input many parameters to examine a slightly different *what–if* scenario. Visual wrappers to SkyCalc have been developed before, including *Thorstensen*'s *SkyCalc GUI* and *SkyCalcDisp* [93.18, 19]; however, the former is simply a graphical version of the command-line interface and the latter contains a three-dimensional projection of object positions that does not facilitate prediction of target motion. On the other hand, the Sky visualization in Sunfall Data Taking was designed specifically to help the astronomer balance the above-described constraints in a visually intuitive way – often during a nighttime observing session – at the click of a button.

The Sky differs from other astronomy visualization programs in that many of them are after-the-fact image-processing programs, designed to enhance faint or low-contrast image data, rather than to provide an operational simplification to aid in time-constrained observational decisions [93.20–22]. Astronomers frequently use tools such as SAOImage ds9 (Smithsonian Astrophysical Observatory Image ds9) [93.23] for detailed viewing of their images, or Image Reduction and Analysis Facility (IRAF) [93.24] for spectral visualization. However, neither of these is integrated into an observing package as is Sunfall data taking.

In the field of scientific visualization for astronomy, software such as *Li* et al.'s scalable World-In-Miniature (WIM) [93.25] focuses on the ability to browse a large-scale three-dimensional model of the universe. However, this technique is not designed for use under time pressure.

There has been much recent work in the development of systems designed for first responders and other cognitively loaded users, including several systems to visualize time-varying data [93.26–29], noting the constraints of updating time-critical visual information on a low-resolution device. Other efforts to develop mobile systems for emergency response include the Measured Response Project of the Purdue Synthetic Environment for Analysis and Simulation [93.30].

Related efforts involve approaches to visualizing time-varying geographic data, such as [93.31–33], and *Tesone* and *Goodall* have developed a visual analytics technique, *smart aggregation*, specifically to aid with situation awareness [93.34]. There is a large body of literature on situation awareness in general, and a full description of such literature is beyond the scope of this chapter. The book by *Endsley* and *Garland* [93.35] provides an excellent survey of the work in this area.

93.4 Sunfall Design Process

In order to design an effective collaborative visual analytics system for the SNfactory, we first conducted, in mid-2005, an extensive, 2 month evaluation of the existing SNfactory procedures and environment. Data sources used for evaluation included individual interviews, observation of team members performing typical project tasks, review of existing source code, literature reviews, examination of other supernova search projects, and consultation with physicists and computer scientists with relevant experience building similar scientific software systems.

We conducted over 100 h of interviews with Lawrence Berkeley National Laboratory (LBNL) scientists, postdoctoral researchers, and students, and SNfactory collaboration members outside LBNL. This included 19 current and former team and collaboration members, and 12 scientists with relevant experience outside the SNfactory collaboration. We also performed a detailed software review of over 150 000 lines of SNfactory legacy code in C++, Interactive Data Language (IDL), Perl, and shell scripts.

We established the following requirements for the Sunfall software framework: It must encompass the entire scientific data capture, processing, storage, and analysis process, enable collaborative, time-critical scientific discovery, and incorporate SNfactory legacy code (custom astronomical image-processing algorithms). It must reduce the number of false positives sent to humans and improve the quality of the subtraction image processing. It must automate repetitive data transfers and other manual tasks to leverage domain experts' unique processing and image-recognition skills to the maximum extent.

The Sunfall user interface was designed and implemented using participatory and iterative design techniques; for example, interactive prototypes were used to evaluate areas where existing interfaces did not support scientists' workflow. Scientists' feedback sometimes led to major redesigns of the interface. The system was implemented from the beginning with this possibility in mind, so that changes were easily made and accepted as an appropriate part of the development process [93.36]. Sunfall has been in operation at the SNfactory since 2006.

93.5 Sunfall Architecture and Components

Sunfall consists of four major components: Search, Workflow Status Monitor, Data Forklift, and Supernova Warehouse (Fig. 93.3). This section will describe each component's structure in detail.

The Search component handles image processing and subtraction, and includes machine learning algorithms and novel Fourier contour descriptor algorithms to reduce the number of false-positive SN candidates. The Workflow Status Monitor is a web-based dashboard that facilitates collaboration and improves project scientists' situational awareness of the data flow by displaying all relevant (search pipeline) status data on a single site.

The Data Forklift is a middleware mechanism consisting of a coordinator and a suite of services to automate astronomical data transfers in a secure, reliable, extensible, and fault-tolerant manner. The Data Forklift also provides the middleware for the other three components, transferring data between heterogeneous systems, databases, and formats securely and reliably.

The Supernova Warehouse (SNwarehouse) is a comprehensive supernova data management, workflow visualization, and collaborative scientific analysis tool. The SNwarehouse contains a PostgreSQL database, Forklift middleware, and a graphical user interface implemented in Java.

93.5.1 Search

The supernova search component of Sunfall is an example of analytic discourse, defined by *Thomas* and *Cook* in *Illuminating the Path* [93.37] as "the interactive, computer-mediated process of applying human judgment to assess an issue", applied to the realm of astrophysics. Sunfall search has increased efficiency of supernova detection and enabled more effective human intervention by reducing the number of false-positive candidates by nearly an order of magnitude [93.8, 9] and, through the use of intelligent interfaces, by increasing human efficiency in evaluating candidates by a factor of four [93.2, 36].

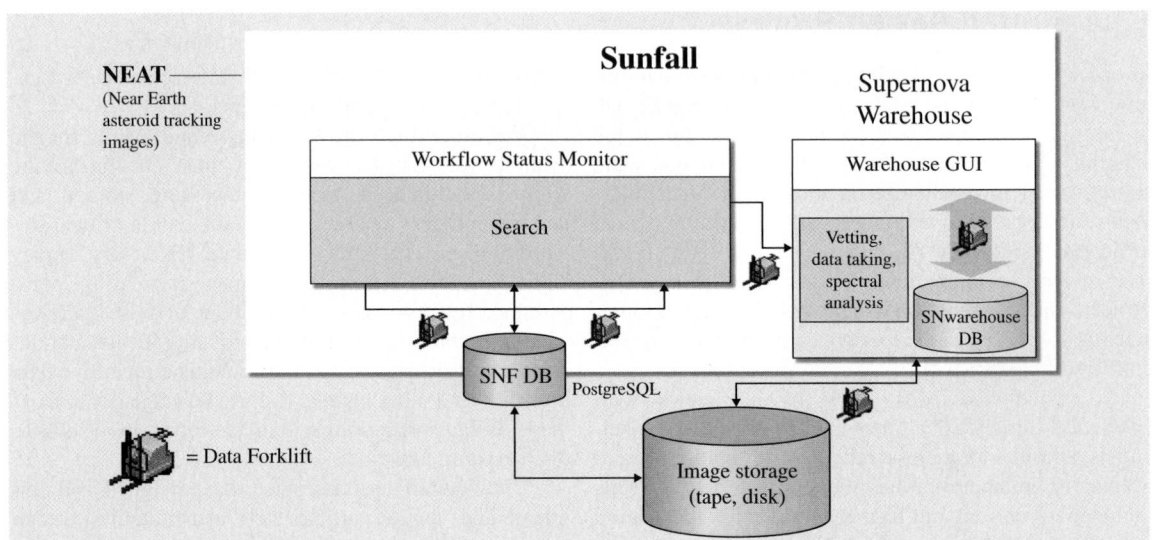

Fig. 93.3 Sunfall architecture diagram depicting the four components (Search, Workflow Status Monitor, Data Forklift, and Supernova Warehouse) and data flow between the components

The legacy search software computed 19 photometric and geometric features on each of over 600 000 SN candidate subimages per night and applied threshold *cuts* to each of these features. Subimages that satisfied all thresholds (1094 on a typical night in October 2005, before the new software was put into production) were sent to human scanners who selected potential SN candidates for follow-up study. The majority of subimages that passed the thresholds were false positives that humans had to manually reject. These manually tuned thresholds were brittle and often resulted in both missed SNe and too many false positives for human scanners to evaluate daily. The process of scanning on the order of 1000 subimages took 6–8 people several hours per day.

We first developed a set of new features for improved image detection, including the use of a very efficient ($O(k)$, where k is the pixel diameter of a supernova candidate) Fourier contour analysis algorithm for determining the shape of an object, described in more detail in the section on fast contour descriptor algorithm for supernova image classification [93.7].

Machine learning algorithms, including support vector machines and boosted trees, were evaluated for incorporation into the search software. By replacing simple threshold *cuts* on the image features with classifiers of the high-dimensional data in the feature space, we reduced the number of false positives by a factor of ten while increasing our rate of supernova detection. We applied several different classification algorithms,

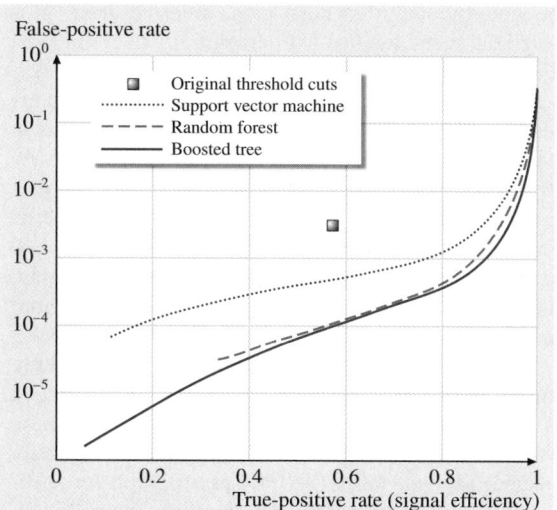

Fig. 93.4 Comparison of boosted trees (*solid line*), random forest (*dashed line*), and support vector machines (*dotted line*) for false-positive identification fraction versus true-positive identification fraction. The *square* shows the performance of the original threshold cuts. The *lower right corner* of the plot represents ideal performance (after [93.8])

including boosted trees, random forests, support vector machines, and combinations of the above. All classifiers provided significantly better performance over

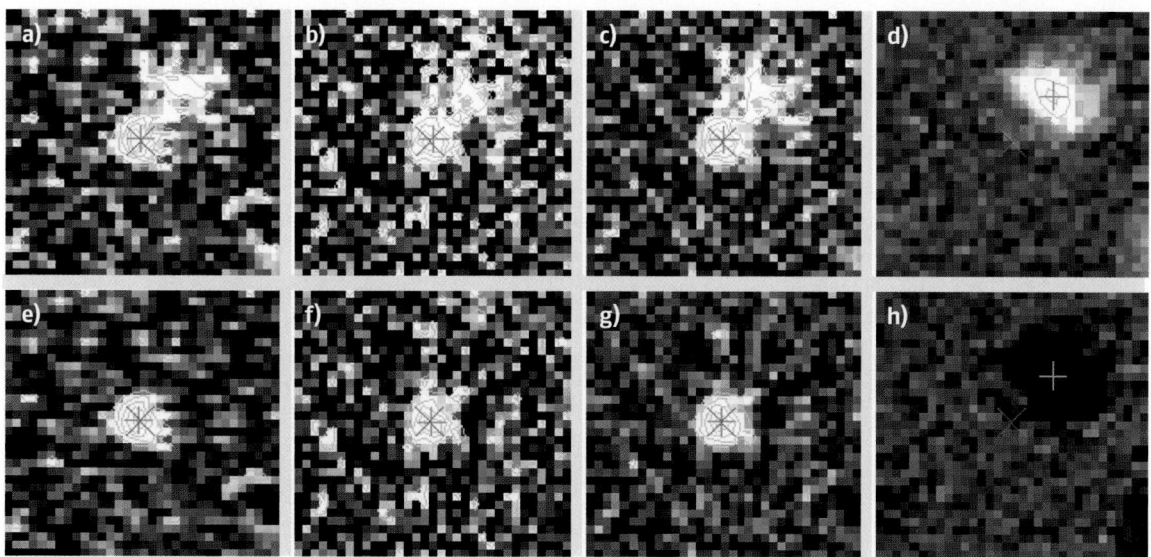

Fig. 93.5a–h Visual scanning interface. *Top row*: three new images (**a–c**), (**d**) reference image. *Bottom row*: three subtractions (**e–g**), (**h**) negative subtraction

the conventional threshold cuts. Boosted trees produced the best classification performance overall [93.8] (Fig. 93.4) and significantly decreased scientists' workload by reducing the number of false positives by an order of magnitude (from 1094 to 113 on typical nights in 2005 and 2006). This resulted in labor savings of nearly 90%, where the process of scanning now takes one person a couple of hours each day [93.2, 9]. Additionally, system tests determined that the new algorithms had a true-positive ratio equal to or better than the original algorithm (in other words, they did not miss any more actual supernovae). The data are depicted in Fig. 93.4.

Currently, of the approximately 100 visually scanned subimages per night, about 10–30 are flagged as containing potential supernova candidates. These candidates (about 0.001–0.01% of the objects originally imaged) are sent to the SNIFS spectrograph for spectrophotometric screening and follow-up. Typically two to five of the objects examined by SNIFS will finally be determined to be supernovae, of which (on average) about one per night is the sought-after type Ia supernova. Thus, of the objects originally imaged, somewhere near 0.0002% prove to be type Ia supernovae.

For confirmation and selection of promising supernova candidates, the scanners use a visual interface to view and analyze the candidates (Figs. 93.5 and 93.6). This visual scanning interface was redesigned to increase interactivity and allow humans to focus exclusively on the imagery itself. Previously, scientists spent much of their time cutting and pasting long filenames, pausing the scanning process to look up data on several different external web sites, and waiting for image data to be retrieved from tape storage. Formerly, this process took 2 min per candidate on average; with this interface and other Sunfall tools, a decision could be made within 30 s.

By applying intelligent algorithms to the search data, and by developing intelligent visual interfaces to view the data, we were able to leverage the unique human abilities that enable the final detection of true supernova candidates, and not only minimize the number of false positives, but also increase the efficiency of the spectroscopic follow-up. Techniques such as these are critical for successful human–automation collaboration in scientific experiments. As an example of an algorithm we developed to maximize the effectiveness of human visual pattern-recognition ability by screening out bad supernova candidates before they are presented to users, we now describe our fast contour descriptor algorithm in more detail.

A Fast Contour Descriptor Algorithm for Supernova Image Classification

Fourier descriptors are an established method for parameterizing the outer contour of an object by decomposing it in terms of complex exponential func-

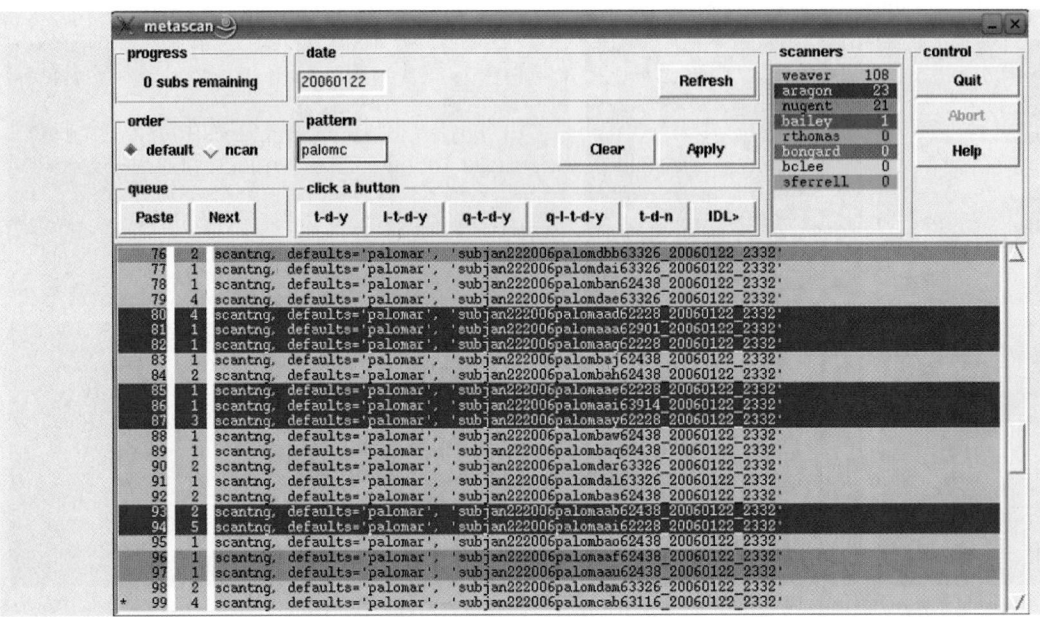

Fig. 93.6 Visual interface for managing image subtraction workflow

tions [93.11]. However, Fourier-transform-based methods were initially considered infeasible, due to their computational complexity, for the high-throughput requirements of the Supernova Factory. We devised an extremely fast contour descriptor implementation that meets the processing budget of the application. Noting that the lowest-order descriptors (F_1 and F_{-1}) convey the circular or elliptical aspect of the contour, our features are based on those two terms and the total variance or power in the contour. The features are thus equipped to detect eccentric or irregular objects, classes to which the poor candidates generally belong.

Our fast contour descriptor algorithm was applied to 600 000 candidate objects in 50–80 GB of supernova image data per night. Our shape-detection algorithm reduced the number of false positives generated by the supernova search pipeline by 41% while producing no measurable impact on running time. Because the number of Fourier terms to be calculated is fixed and small, the algorithm runs in linear time, rather than the $O(n \log n)$ time of a fast Fourier transform (FFT). Constraints on object size allow further optimizations so that the total cost of producing the required contour descriptors is about $4n$ addition/subtraction operations, where n is the length of the contour.

Because transform-based image analysis methods are often avoided in high-throughput applications due to their computational cost, it is important to distinguish the contour Fourier descriptor from other Fourier-based methods. For a two-dimensional image of diameter D pixels, the processing time for a Fourier transform of the image can scale as severely as D^4. In fortunate cases where the Cooley–Tukey FFT [93.11, 38, 39] can be used, the processing still scales as $D^2 \log(D^2)$. The contour Fourier descriptor operates not on the object image but on its outer contour, which is essentially a one-dimensional (albeit complex-valued) sequence, of length proportional to object diameter D. Since we calculate a fixed number of terms rather than requiring a full discrete Fourier transform (DFT), the calculation time is actually proportional to D.

Algorithm Overview. We compute two new features (or *scores*) on each subimage containing an object, and add them to the set of features already calculated in the supernova search pipeline. These two features are: $R1$, which measures the *roundness* or eccentricity of the candidate object, and $R2$, which measures the amount of irregularity in the object contour.

Our approach inserts the following steps into the processing pipeline for candidate supernova objects:

1. Find the outer contour of the object and store it as a sequence of (x, y) points.
2. Calculate the lowest Fourier descriptor terms F_1 and F_{-1} on the contour point sequence.
3. Form two roundness-related features: $R1$ measures the eccentricity of the object (0 is a circle, 1.0 is a line, and intermediate values are ellipses); $R2$ measures the amount of irregularity in the object contour

$$R1 = \frac{\|F_1\|^2}{\|F_{-1}\|^2} \, ,$$

$$R2 = \frac{\sum_{|k|>1} \|F_k\|^2}{\sum_{k \neq 0} \|F_k\|^2} \, .$$

The features $R1$ and $R2$ are added to the list of thresholds or *cuts* applied to image objects, so that objects with too high $R1$ or too high $R2$ are rejected as supernova candidates, and are not sent for human scanning.

Algorithm Results. For testing, the fast contour descriptor algorithm was implemented and incorporated into the existing supernova search pipeline software. The new scores $R1$ and $R2$ were added to the existing image feature set, without yet setting thresholds, but with diagnostic output to permit analysis of the distribution of scores. First, we examined the scores within the class of good candidates, using a validation set of 90 supernovae. For the first roundness feature, all of the supernovae in the validation set yielded $R1 \leq 0.08$ (and frequently much closer to zero). For the second feature, nearly all (89 of 90) gave $R2 < 0.1$.

A second evaluation was performed on a much larger data set representing both classes (good and bad

Table 93.1 Graph and table of image rejection rate versus $R1$ threshold

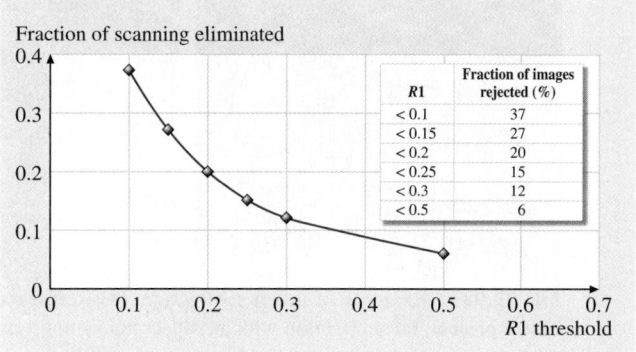

$R1$	Fraction of images rejected (%)
< 0.1	37
< 0.15	27
< 0.2	20
< 0.25	15
< 0.3	12
< 0.5	6

Fraction of scanning eliminated

$R1$ threshold

candidates), to determine appropriate threshold values for $R1$ and $R2$. The goal was to reduce the scanning load (remove the bad candidates) as far as possible without sacrificing any good candidates (possible supernovae) at all. The algorithm was run on a test set of approximately 35 000 subtracted images, of which 1094 would

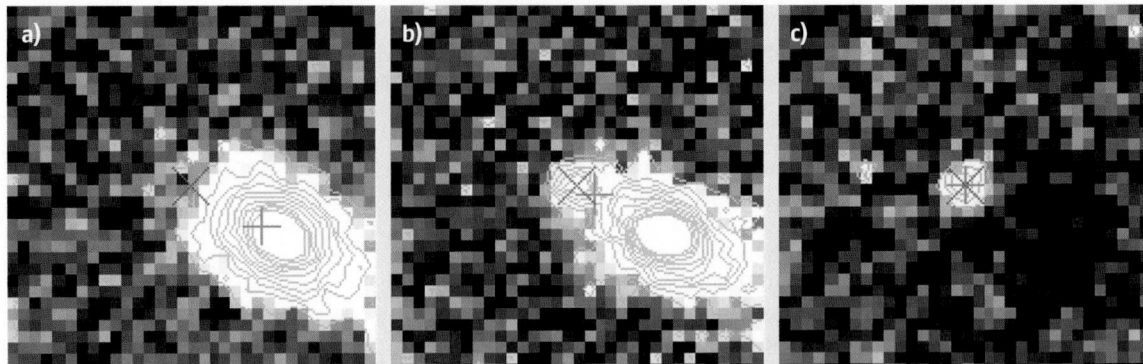

Fig. 93.7a–c Supernova found in November 2006 by the Nearby Supernova Factory: SNF20061117-000. (**a**) Reference image of galaxy, (**b**) new image of galaxy plus bright spot, (**c**) subtraction revealing new supernova

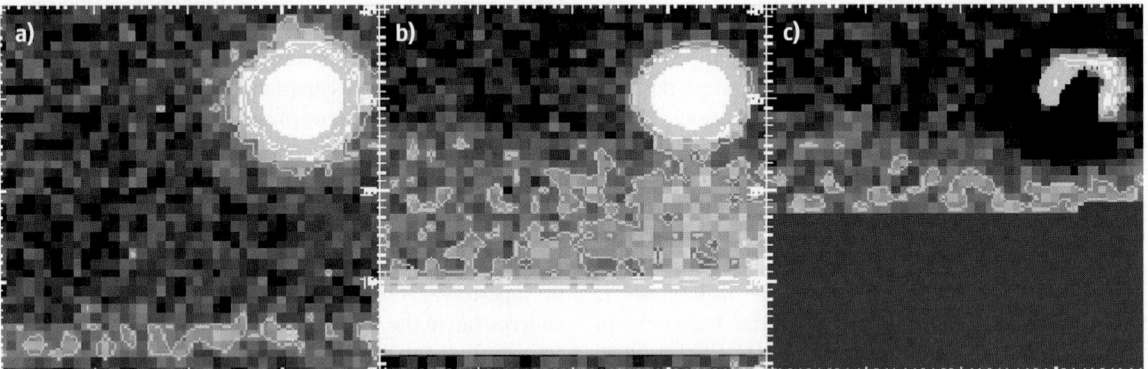

Fig. 93.8a–c Example of a bad subtraction. (**a**) Reference image, (**b**) new image, (**c**) subtraction (the images were misaligned, leaving the subtraction with a meniscus-shaped object). This object is screened out by $R1$

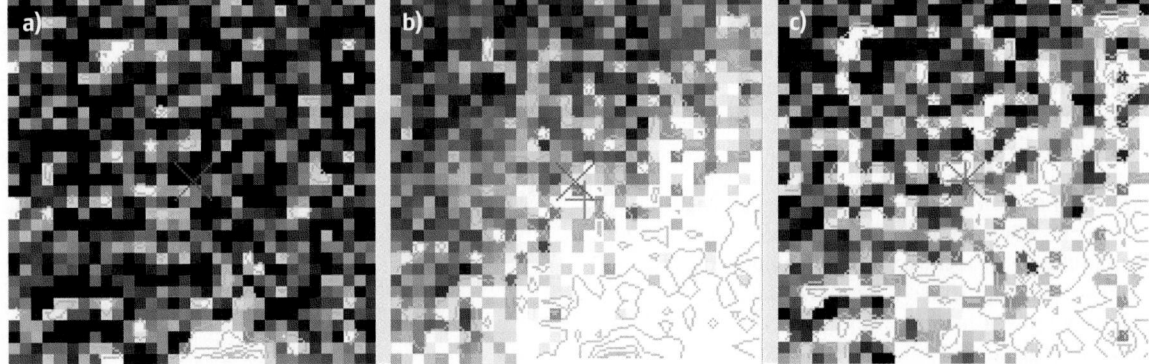

Fig. 93.9a–c Example of a bad subtraction. (**a**) Reference image of empty space, (**b**) new image with edge of bright object present, (**c**) subtraction with an object that is not round and has an irregular contour. This object is screened out by $R2$

have been sent to a human for evaluation under the old parameters. The fraction of those 1094 which would be rejected (i. e., not sent for human scanning) under various settings of $R1$ was calculated; the results are shown in Table 93.1.

A similar test was then performed for $R2$. On the basis of the combined results, thresholds for $R1$ and $R2$ were chosen at 0.1 and 0.15, respectively. At those settings, the number of images passing the thresholds and sent to human scanners decreased to 647 (versus 1094 under the old criteria). This is a scanning load reduction of 41%, representing a saving of ≈ 6 h of manual scanning labor each day.

Figures 93.7–93.9 show examples of subimages viewed during this process. Each figure depicts a triplet of reference/new/subtraction images (the reference image is from before the supernova explosion, the new image is the current one including the supernova, and the subtraction depicts the new bright spot alone). All of these subimages were *passed* by the search software before our new roundness features were added. The first figure (Fig. 93.7) depicts a good candidate (note the compact circular form), but the next two figures (Figs 93.8 and 93.9) display bad subtractions which should have been culled out. After our algorithm had been implemented (features $R1$ and $R2$ added to the threshold criteria), Fig. 93.8 failed the $R1$ cut and would not have been passed for human scanning. Figure 93.9 failed the $R2$ cut and likewise would not have been passed.

93.5.2 Workflow Status Monitor

The Workflow Status Monitor is a web-based dashboard application, designed to improve project scientists' situational awareness of the data flow by synthesizing diverse information flows from various systems and displaying critical workflow status data on a single site (Fig. 93.10).

The supernova search software is highly parallelized as 30 000 images are queued for processing in a multistage pipeline that runs on NERSC's 300 node computing cluster, the Parallel Distributed Systems Facility (PDSF). Nodes frequently go down or jobs fail, and failures must be detected promptly and jobs resubmitted quickly due to the time-critical nature of the search. Detection of such failures was challenging and time-consuming before the deployment of the Workflow Status Monitor. The monitor displays graphs and visual displays of job completion times on PDSF, job queues, PDSF node uptimes,

and disk vault loads. The interface was prototyped, evaluated for efficacy by project scientists, and implemented via Hypertext Preprocessor (PHP) on a website hosted on an SNfactory server running SuSE Linux.

The dashboard also displays date and time information at project locations, in several formats relevant to astronomers. Computer scientists developing a status monitor for the Mars exploration rovers [93.40] demonstrated that apparently minor features, such as a clock that displayed the time on Mars and at various project locations on Earth, yielded tremendous improvements in scientists' situational awareness and may very well have prevented critical errors. Additional information displayed includes telescope scheduling, supernova candidates found, and current operational status of telescopes, spectrographs, and cameras used in processing SNfactory data.

93.5.3 Data Forklift

Each night, over 50 GB of heterogeneous, distributed supernova data arrive at the SNfactory and need to be processed, managed, analyzed, queried, and displayed – all within a time period of 24 h or less (ideally 12 h). The data are distributed over wide geographic distances; some of it is hosted on systems outside the SNfactory's control, and some on unreliable systems with frequent downtime.

To solve these data management problems, we designed and implemented the Sunfall Data Forklift, a suite of services for automated data retrieval, storage, movement, querying, and staging for display. The Data Forklift consists of a coordinator and a set of individual services, each customized for its particular task, but having key attributes in common.

The Data Forklift supports *different place, asynchronous* collaborative scientific work by facilitating data and information transfer amongst a geographically separated team. Due to the time-critical nature of the data collection (telescopes must be operated at night and are located in different time zones), tasks must take place at distinct, specified times.

All Data Forklift services share the following properties:

- The mechanism resides at a central location (an SNfactory server), but processes data remotely from widely separate geographic locations, including making connections to external machines not under the SNfactory's control.

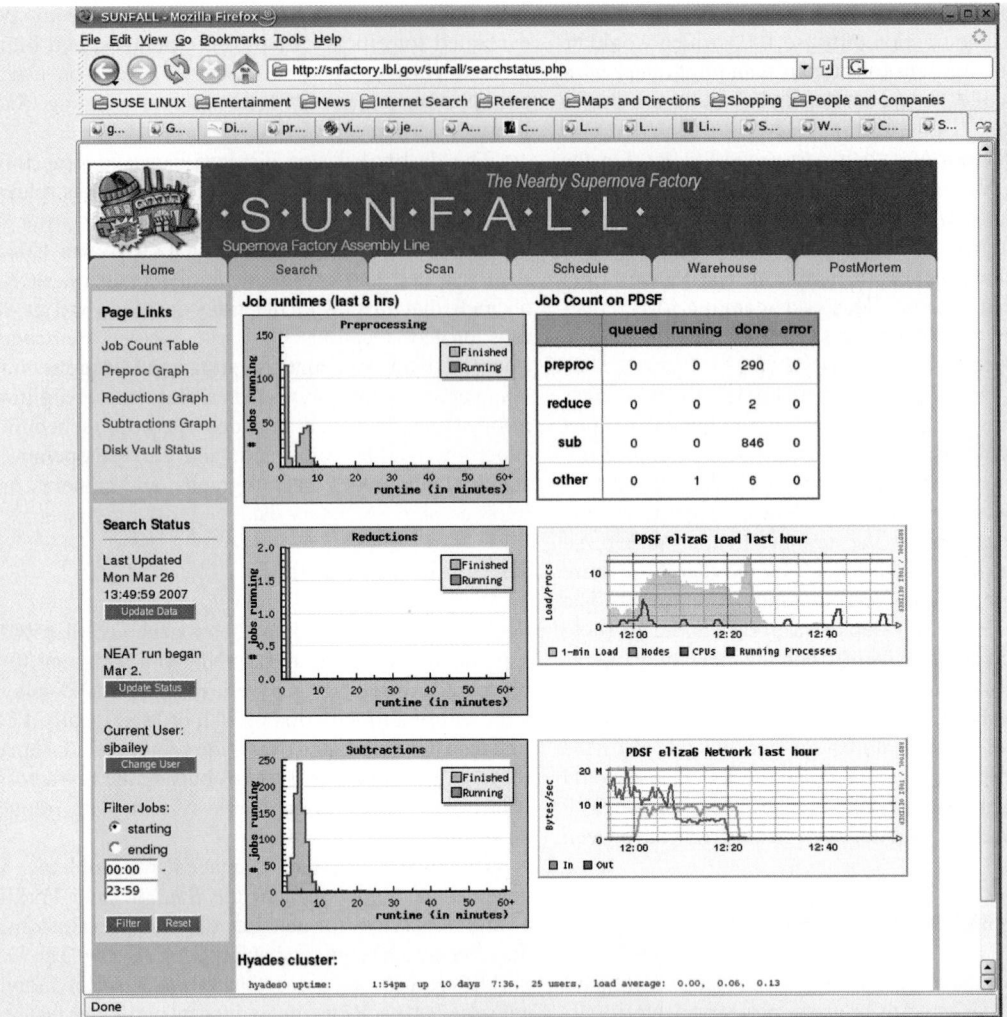

Fig. 93.10 Sunfall Workflow Status Monitor. This web-based dashboard presents a single-page overview of many relevant information flows from multiple systems and, according to project scientists, increases operator efficiency significantly

- All data transfers are secure, utilizing encrypted channels and authentication.
- The Forklift services and coordinator are reliable and self-restarting. The Forklift coordinator restarts itself if its process dies or if the server is rebooted.
- The mechanism is fault tolerant; services detect unreliable connections and recover from errors in data transfer. Partially transferred files are detected and retransferred, and the process restarts the transfer at the point of failure.
- The mechanism is extensible (new tasks can easily be added). The Forklift coordinator was designed so

that new services can easily be added to a central table.
- The Data Forklift enables data retrieval from external sources. Many astronomical databases are web-based, and designed for interactive rather than automated retrieval. Forklift services handle such interactions in a fault-tolerant manner. SNfactory spectral data are stored at a central processing database in France, and must be retrieved according to external requirements.
- User actions can trigger Data Forklift requests. For example, whenever a supernova candidate is

saved to the Supernova Warehouse (SNwarehouse) database, several Forklift services are automatically started, retrieving related image data from tape storage, accessing external web-based asteroid, galaxy, and other astronomical databases, converting image files into display format, and staging data for display.

- Forklift services act as middleware for the SNwarehouse, retrieving data from internal and external databases, performing program logic, and staging data for display by the SNwarehouse GUI.

- The Data Forklift is written in Perl, and runs under SuSE Linux 9.3 on an SNfactory server.

93.5.4 Supernova Warehouse

The SNwarehouse is a comprehensive supernova data management, workflow visualization, and collaborative scientific analysis application. It consists of a Postgre-SQL database hosted on a dedicated SNfactory database server running SuSE Linux 9.3, Data Forklift services as middleware, and a graphical user interface (GUI) writ-

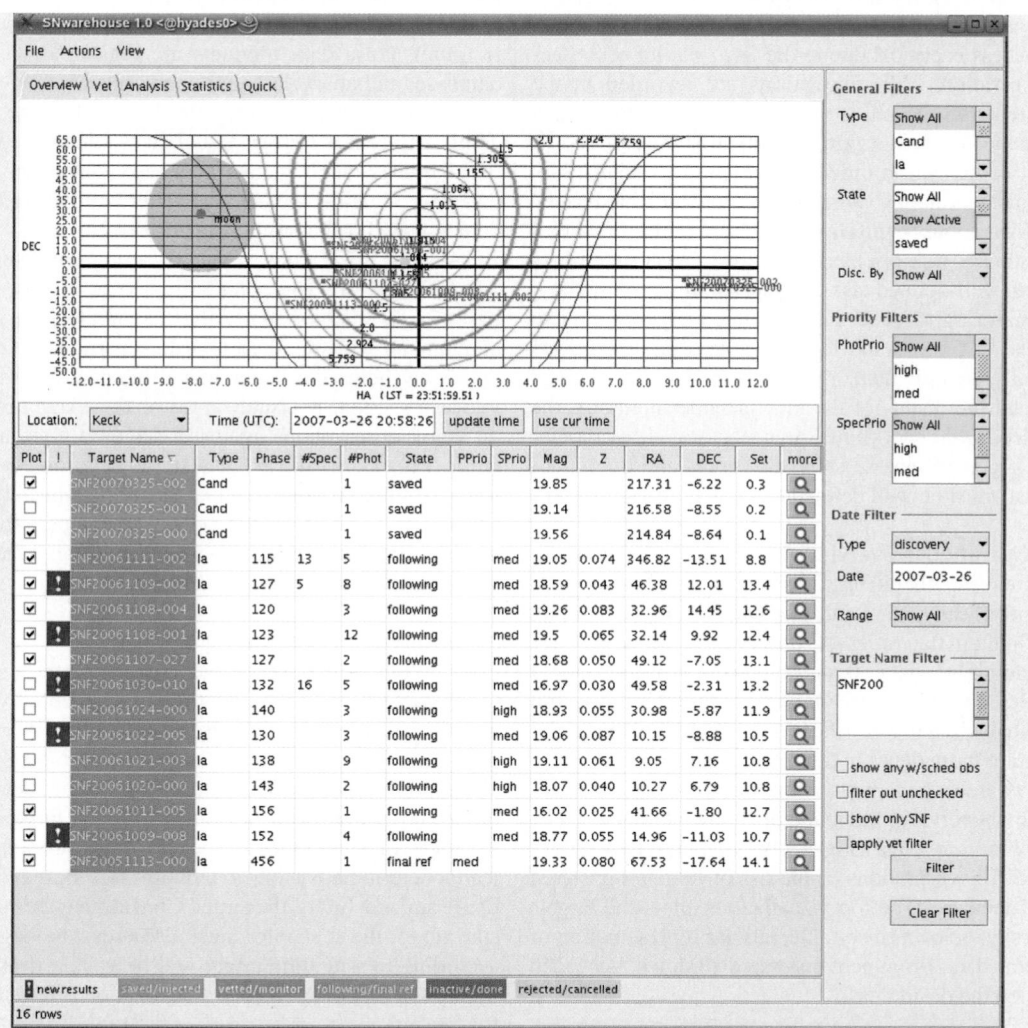

Fig. 93.11 SNwarehouse overview table. The sky visualization at the *top* plots declination versus hour angle (astronomical sky coordinates) in a view preferred by domain experts. The *green lines* represent air mass (the thickness of the atmosphere) at particular coordinates at the specified time

ten in Java. SNwarehouse supports collaborative remote asynchronous work in several different ways.

Collaboration members can access the GUI from any networked computer worldwide via a Forklift remote deployment mechanism. Security is provided via password authentication and encrypted communication channels. SNwarehouse furnishes project scientists with a shared workspace that enables easy distribution, analysis, and access of data. Collaboration members can view, modify, and annotate supernova data, add comments, change a candidate's state, and schedule follow-up observations from work, home, while observing at the telescope, or when attending conferences. This access is critical due to the 24/7 nature of SNfactory operations. All transactions are recorded in the SNwarehouse database, and the change history and provenance of the data are permanently stored (records cannot be deleted in order to maintain the change history) and continuously visible to all authenticated users.

SNwarehouse centralizes data from multiple sources and supports task-oriented workflow. Project members perform well-defined tasks, such as vetting, scheduling, and analyzing targets, which collectively accomplish the goal of finding and following type Ia supernovae. Typically, an individual or small group performs a given task, and the results of the task provide inputs for the next task in the workflow, often performed by another set of group members. Thus, the inputs and outputs of any task must be well defined and easily recognizable.

SNwarehouse Overview

SNwarehouse facilitates five major tasks: overview, supernova candidate vetting, observation scheduling (determining the order of observation of targets for a single night), data taking (capturing spectral data at the telescope with SNIFS), and post mortem (spectral data analysis and classification). These are each described in more detail in the sections ahead.

SNwarehouse's interaction design takes the approach of overview, filter, and drill down to details. The main overview page (Fig. 93.11) displays two tightly coupled representations of the list of targets registered in the database. The top visualization plots the targets in the sky; below is a sortable, tabular representation of the same data. From here one can drill down as needed based on the desired task.

Supernova Candidate Vetting

Vetting involves a process of scientific analytic discourse where the scientist must quickly decide, based on limited data, how to allocate scarce and expensive telescope time to the night's supernova candidates. At this point, before spectra are taken, it is often unknown whether a target candidate is a supernova, so the scientist must gather as much relevant data as rapidly as possible to make an informed prediction. This task involves retrieving any available images of the target's location in the sky prior to discovery and querying several external astronomical databases.

Once a target is saved as a candidate supernova, the target name will appear in the SNwarehouse overview table, color-coded in red, indicating that the target is newly discovered. On a rolling basis, a background process checks if previous data products exist on disk. If found, these data products are registered with the database, and an exclamation mark appears next to the target name, a flag signaling newly found images. The combination of these two visual clues allows the scientist to quickly see which targets need evaluation. Once this determination is made, the vetter will then open the details view for the target (Fig. 93.12).

At this stage, the vetter is trying to determine whether this target is potentially a type Ia supernova, based on how consistently the magnitude is rising in comparison with standard type Ia supernovae. The details view helps the vetter make this determination in two ways. First, a visualization of the magnitudes of all target observations against a standard type Ia light curve (Fig. 93.13) allows the vetter to determine quickly if the magnitudes are rising like those of a typical type Ia supernova. In addition, the vetter can compare the discovery image with prior images taken at this particular set of coordinates in the sky (Fig. 93.14).

Observation Scheduling

Prior to follow-up observation at the telescope, a schedule is made based on the vetter's assessments to order and allocate telescope time. In SNwarehouse, the scheduler starts by filtering for only those targets with observation requests. With this list, the scheduler must order and assign exposure times for each target using a variety of techniques. Exposure times are automatically determined using a lookup table based on the phase and redshift of the target. Considering these exposure times, the scheduler must also order the target list according to when the target will be visible in the sky by the telescope. A custom visualization offers insight for this task.

The Sky visualization (Fig. 93.15) depicts the positions of targets in the sky at a given time and ground location. The green lines represent air mass (atmosphere thickness) for target coordinates at the specified time.

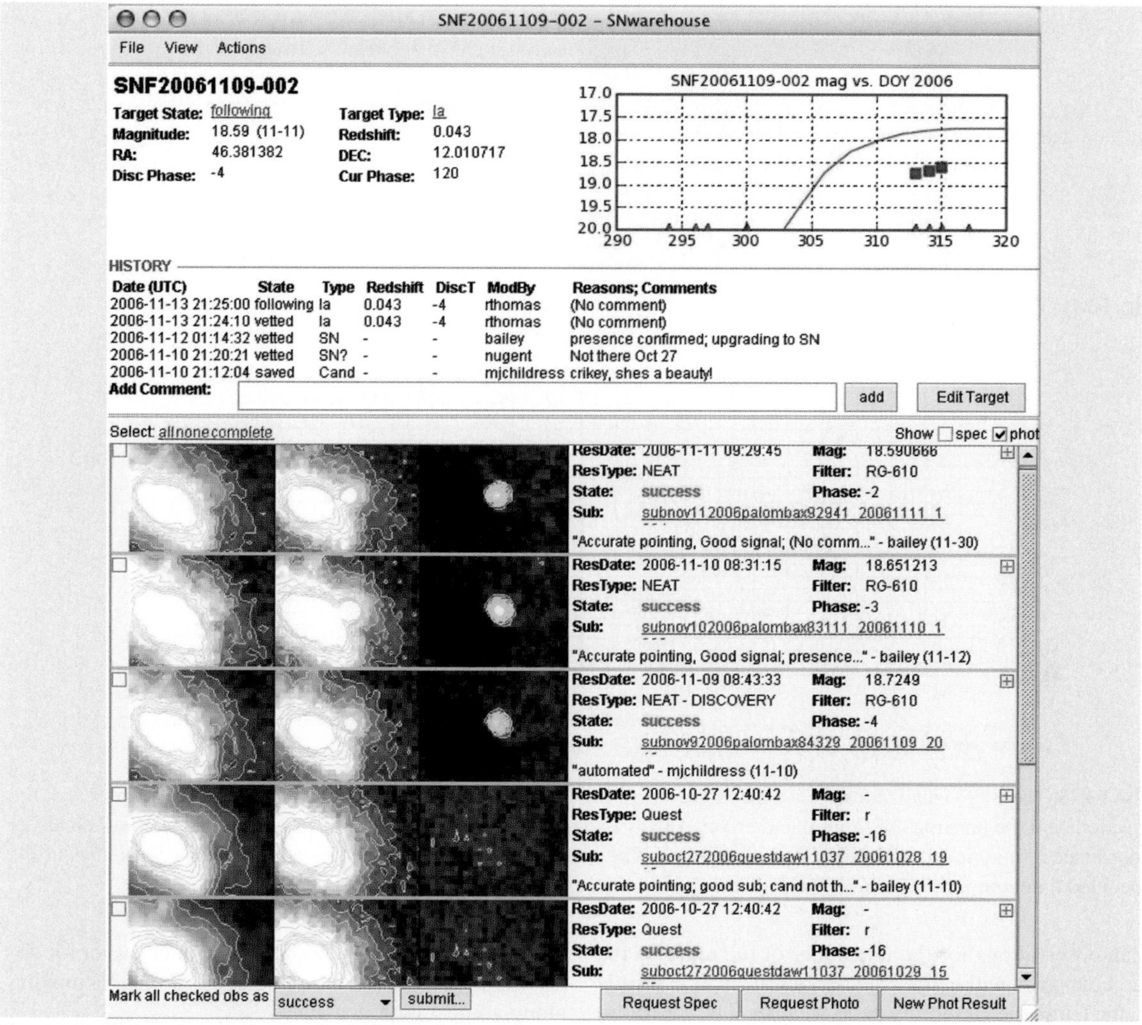

Fig. 93.12 SNwarehouse details view. Comparison with standard light curve is in *upper right-hand corner*. The *five rows* at the *bottom* were imaged on different dates; note that the target supernova is visible at the time of discovery but not on prior dates

Fig. 93.13 Supernova light curve. Here, we see how the rise in magnitude of this target (*squares*) compares with the light curve of a standard type Ia SN (*curve*) ▶

The blue line is the horizon, and the red halo around the moon is the *lunar exclusion zone*, where light cast by the moon makes it difficult to view faint targets. The yellow circle depicts the sun. Major telescope names and corresponding latitudes and longitudes are displayed on a drop-down menu so the visualization can be used worldwide. The time can be changed so the viewer can

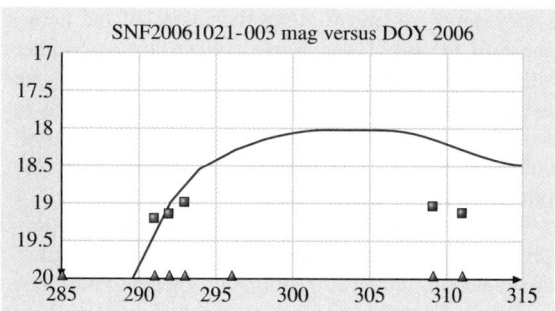

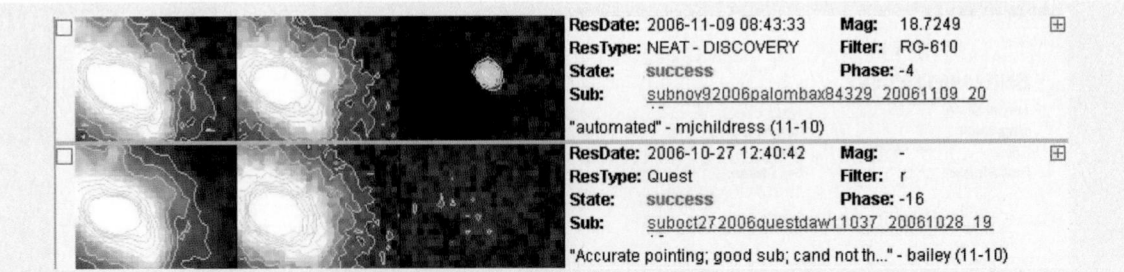

	ResDate: 2006-11-09 08:43:33	Mag:	18.7249
	ResType: NEAT - DISCOVERY	Filter:	RG-610
	State: success	Phase:	-4
	Sub: subnov92006palombax84329_20061109_20		

"automated" - mjchildress (11-10)

	ResDate: 2006-10-27 12:40:42	Mag:	-
	ResType: Quest	Filter:	r
	State: success	Phase:	-16
	Sub: suboct272006questdaw11037_20061028_19		

"Accurate pointing; good sub; cand not th..." - bailey (11-10)

Fig. 93.14 The target is visible at the time of discovery but not visible on a prior date

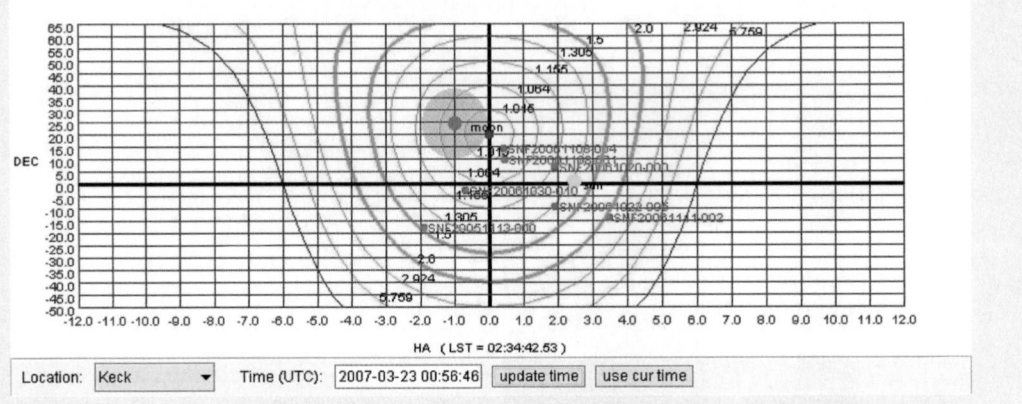

Fig. 93.15 The Sky visualization shows the target's position in the sky, the air mass (thickness of the atmosphere) at the location, and the positions of the sun and moon. The visualization plots declination versus hour angle (astronomical sky coordinates) in a view preferred by domain experts. The *green lines* represent air mass at particular coordinates at the specified time, the *yellow circle* is the sun, and the *red circle* is the lunar exclusion zone

plan observations for the remainder of the night. If the target appears within the blue horizon lines, it is visible to the telescope, assuming good weather. The scheduler must note the rise and set times of each target when creating the schedule for the night.

It can be challenging for humans to imagine and visualize the path of astronomical objects over time, as they appear to move along the inside of the hemisphere of the night sky. Further, even an astronomer skilled in the art of visualizing celestial coordinates can have difficulty predicting the paths of several objects in the night sky, accurately estimating which of them will avoid clouds, and calculating when they will reach the optimal position for observing.

The key idea in the Sky visualization is the mapping of the spherical night sky to a two-dimensional rectilinear projection, where astronomical objects' paths over time move linearly from left to right. Although there is some spatial distortion (similar to that of a Mercator

projection of the Earth [93.41]), it is much easier for observers to predict the future position of objects moving along a linear path in a plane.

The Sky visualization is used throughout SNwarehouse, and is especially helpful for the spectral data collecting component of SNwarehouse, Data Taking.

Data Taking

During each night of spectral observations with SNIFS at the University of Hawaii 2.2 m telescope, an observer needs to point the telescope at each scheduled target in order. In the event of weather or mechanical problems, a rescheduling decision must be made quickly and efficiently. The Sunfall data-taking interface facilitates this process. Much of the observation procedure is automated, allowing the observer to concentrate on troubleshooting, rescheduling, and determining the success of each target observation (Fig. 93.16). Nevertheless, human input is crucial during the telescope

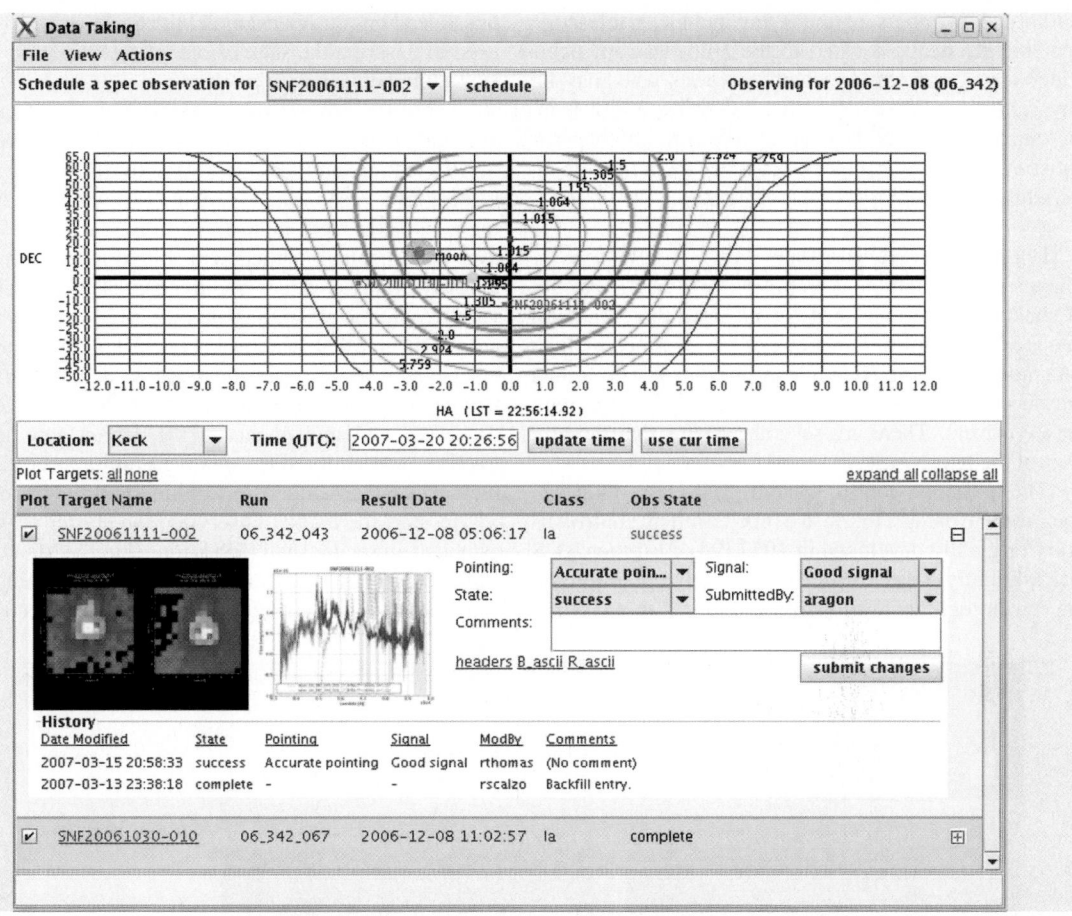

Fig. 93.16 SNwarehouse data-taking window. The observer can follow the targets on the Sky visualization, take notes on the success or failure of each observation, telescope status and weather conditions, and reschedule targets if necessary

observing process, as the scientists must evaluate large amounts of dynamic data and make time-critical decisions that cannot be automated, often in as little as 45 s. Participatory design was a natural choice for Sunfall Data Taking, especially given that expert domain knowledge was needed to define the use cases. Scientists were involved in the design process from the start, helping to define the problem space and outlining the necessary functions. The design of the data-taking GUI went through several prototypes prior to release, including low-fidelity paper prototypes and high-fidelity semifunctional prototypes. Each prototype was evaluated by the scientists, who provided feedback on areas where the interface was confusing or too tedious to use (i. e., too many steps to perform a particular task).

Since telescope observation procedures are one of the key elements of SNfactory science, and thus Data

Taking is an important component of SNwarehouse, we will discuss this component's design, implementation, and use in the system in greater detail in the next few Subsections.

Telescope Observing Procedures

Each night of spectral observations with SNIFS at the University of Hawaii 2.2 m telescope requires a complex set of procedures involving collaboration among a geographically distributed team, and the operation of legacy software telescope controllers driving one-of-a-kind hardware.

The scientist performing the night's observations is known as the *shifter*. The shifter may be located anywhere in the world, although SNfactory collaboration members attempt to distribute the operational tasks among time zones so as to minimize nighttime work.

Additional members of the team include a telescope operator physically located in the Hilo, Hawaii, operations room, and senior scientist experts who may be physically located on the other side of the Earth from the current shifter. At the end of the shift, another scientist evaluates each night's set of observations using a slightly different view of the same data.

Legacy Software Interface

The telescope interface is complex, involving a number of shell scripts which must be run from the command line at various times. Functions of the interface include data monitoring as well as telescope control. The scripts interface directly with the telescope and spectrograph device drivers. There are several systems that directly control the telescope.

The telescope control system (TCS) user interface was written using *curses*, a Unix terminal control library originally developed in 1977 [93.42]. When TCS was originally written, space was at such a premium on the screen that the developers were forced to use every

possible shortcut; for example, the Scottish word *hie* is used to describe the state of one of the mirrors when it is moving between two positions.

There are also two *directors*, which monitor a large amount of information from three CCD cameras, the SNIFS instrument itself, as well as the TCS. These five information streams are piped into two windows, resulting in an extremely rapid flow of information past the eye. The display is filled with blinking and scrolling text, as numeric values are dynamically updated and various internal program status messages rapidly scroll across several text windows (Fig. 93.17). High-priority error messages are printed in red to distinguish them from lower-priority status messages; however, all the flickering and movement on the screen can be distracting, especially since relatively unimportant status messages are continuously scrolling past the window.

To start the night's observing, the shifter must first establish an secure shell (SSH) tunnel to the Mauna Kea summit computer (a Dell Dimension PC running Red Hat Linux), then connect to a virtual network comput-

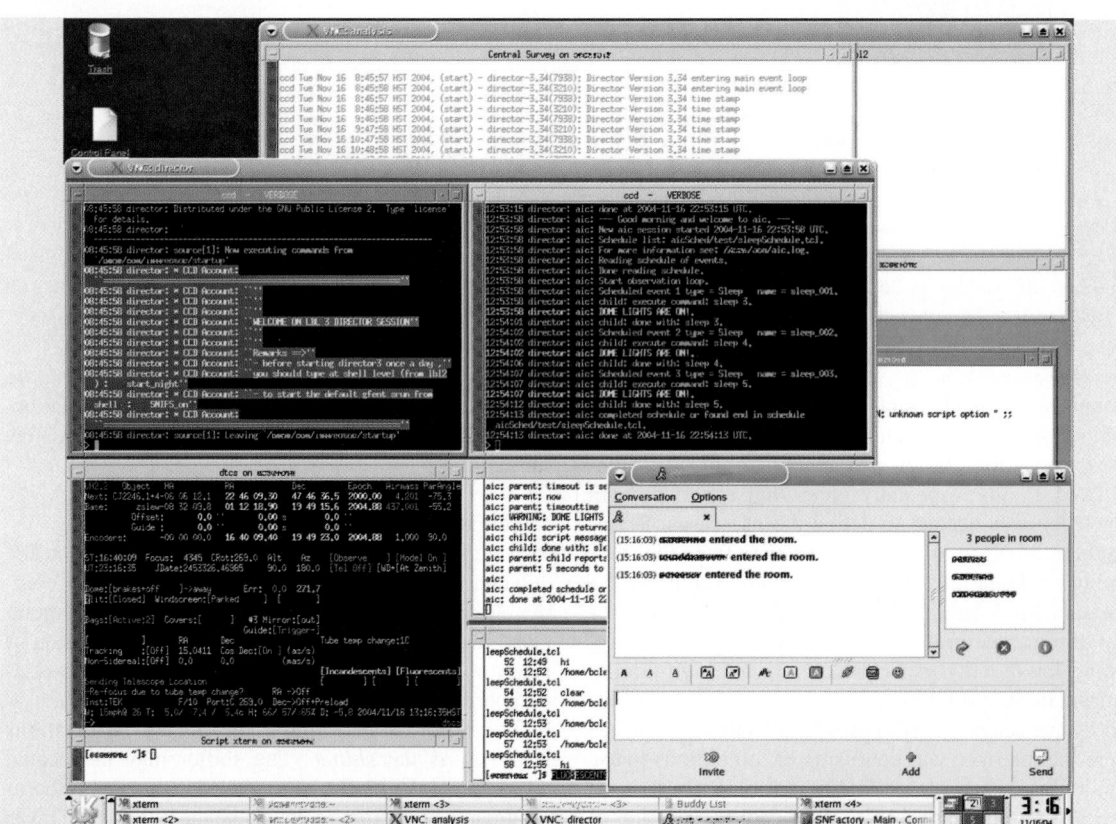

Fig. 93.17 VNC telescope control window with chat client

ing (VNC) server on the summit machine, opening two local windows to establish the interface. Two other machines at the summit are custom-built, possessing Tyan motherboards that each enable operation of two camera Peripheral Component Interconnect (PCI) interfaces simultaneously. The shifter must also connect to a pre-arranged chat room where all scientists involved in the observation participate, ask questions, provide advice, and use an informal protocol to transfer positive control of telescope operation.

To a novice user, the telescope interface is dauntingly complex. The learning curve is extremely steep at the beginning of training. The scientists concluded that it was critical to reduce the shifter's decision-making load. They initially decided that the best method of doing this was to create an *autopilot*, an automated instrument control program that would run the entire observing process without human intervention. However, unexpected hardware interactions as well as operational changes ensured that the automated instrument control program could not run autonomously throughout the night. Additionally, it proved impossibly complex to modify the program to reliably compensate for the plethora of unexpected events that could occur during a night's observing; human intervention was consistently required.

Based on this initial experience, it was determined that the shifters needed a decision support system to supplement the automation. Although some level of au-

tomation was necessary to reduce drudgery and lighten the shifter's workload, it became clear that a system was needed that combined automation with a visual interface that facilitated human intervention. The scientists instituted various protocol changes, and the data-taking system was designed and implemented by an interdisciplinary team of computer scientists and astrophysicists, alleviating many sources of shifter error. Tasks that the shifters performed with data taking included viewing spectral images, exploring data, keeping a standardized *notebook* on sky and instrument conditions, and decision-making. The intent was to have the shifters allocate their attention to higher-level tasks, in particular: evaluating their rate of performance on the schedule, determining whether or not conditions sufficed to perform lower-priority observations, and assessing data quality.

Data-Taking Components

The data-taking GUI is depicted in Fig. 93.16. This interface runs as a separate window on the summit computer and is intended to augment rather than supplement the existing telescope control interface. Operational constraints during software development precluded a complete redesign of the entire telescope control interface, although incremental improvements were made.

The automated instrument control program was left in place; if nothing goes wrong during the night, it will automatically slew the telescope to each target in turn, start an exposure of the correct length, record the

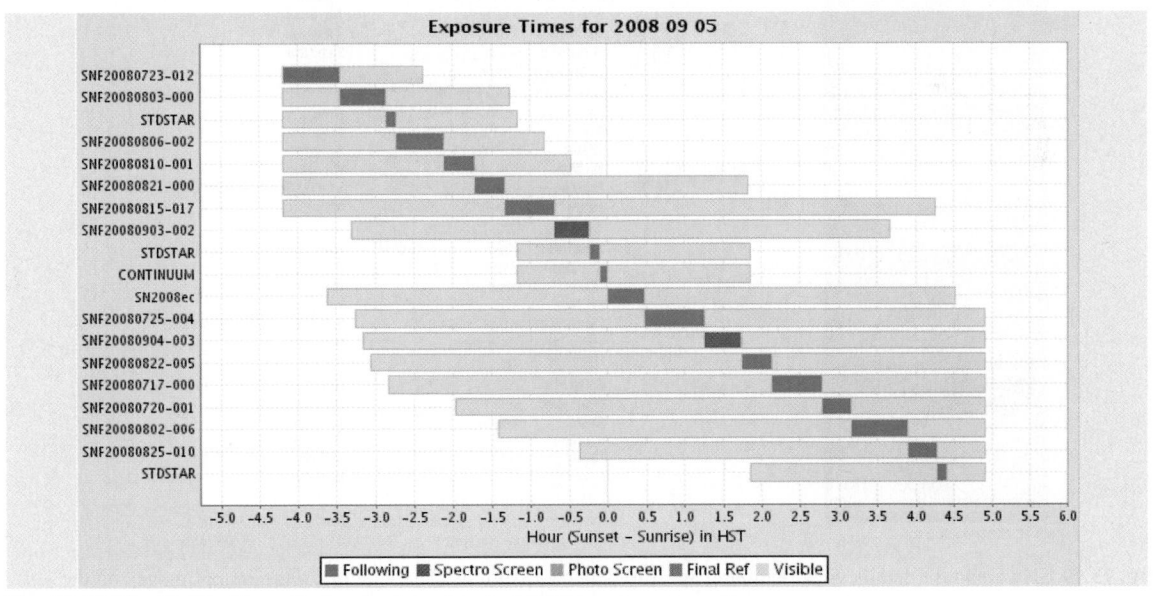

Fig. 93.18 The Visible Time display depicts object visibility and exposure times

data, and move on to the next target. However, in the event of weather or mechanical problems, quick and accurate decision-making is required. Data Taking was designed to support such decisions, with the express goal of improving the shifter's situational awareness. To this end, a number of components were designed to provide supplemental information about critical observing conditions, as well as provide predictive tools to enable the shifter to make scheduling decisions easily and correctly.

Since the shifter is operating the telescope remotely, they may be unaware of deteriorating weather con-

ditions or the approach of sunrise. A software agent running in the chat room broadcasts key astronomical events such as sunrise, and provides information on current telescope operation, including the current exposure or the process of slewing to new coordinates [93.43].

Prior to each night's observation at the telescope, a schedule is made in order to allocate telescope time to each target. In SNwarehouse, the scheduler starts by filtering for only those targets with observation requests. With this list, the scheduler must order and assign exposure times for each target using a variety of techniques. Exposure times are automatically deter-

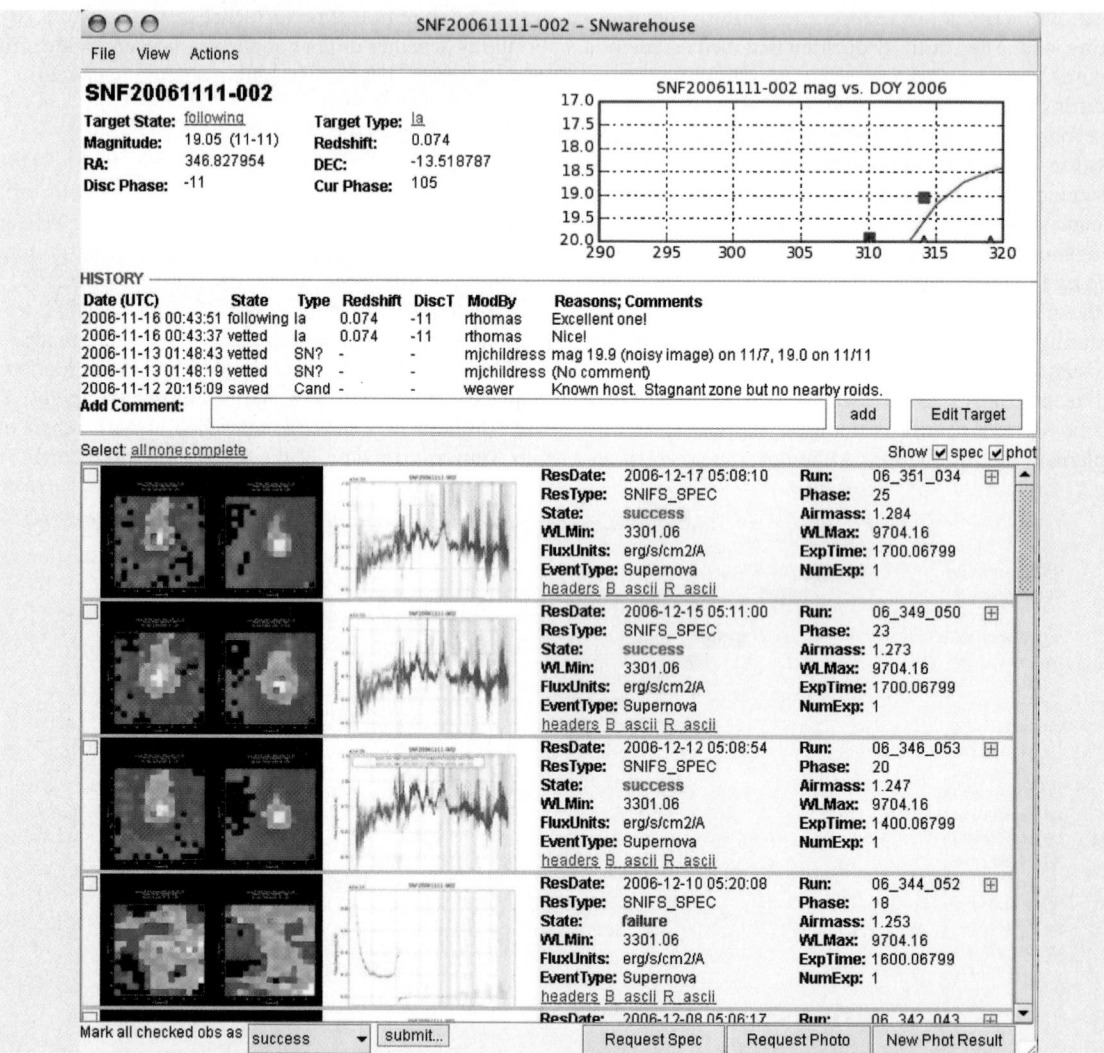

Fig. 93.19 SNwarehouse details view depicting spectral data observations. The small *postage stamp* images on the *left-hand side* of each observation subwindow indicate the accuracy and signal strength of the received data

mined using a lookup table based on the phase and redshift of the target. To assist the scheduler in planning the night's observations, the Visible Time tool was implemented (Fig. 93.18).

Visible Time Display

The Visible Time plot visually depicts the night's schedule by displaying at a given hour what targets are visible in the sky. In addition, the plot shows the start times and the length of the exposure scheduled for each target. The exposure is color-coded according to the type of observation, a spectrum, a photometric screen, or a final reference. Green bars indicate supernovae that the project is *following*, i.e., an interesting object where a series of spectra have been obtained over time, and more are desired. Blue indicates a spectroscopic screening is desired, because the type of the object has not yet been discovered. A marker on the plot (the zero line above) indicates the current time, which along with the visual representation of the schedule, helps the shifter keep track of events. Feedback from scientists showed that the visibility chart aided in decision-making, easing decisions to cancel an exposure or to insert an extra observation in the schedule.

Post Mortem

Once observation at the telescope is complete, the data products (images and spectra) from the telescope are registered with the database and marked with an exclamation mark in the SNwarehouse overview page. The next task is spectral processing and analysis, known as the *post mortem* process. The scientist performing post mortem filters for targets with new data products that are being followed (highlighted in green and blue, depending on the type of follow-up) and opens up the details view to evaluate the telescope results (Fig. 93.19).

The raw data from the SNIFS spectrograph is complex and requires a significant amount of processing to yield meaning to the scientist. A visual depiction of the accuracy of the telescope aim and signal strength provides more information more quickly than tables of numeric data. The two small images (*postage stamps*) on the left-hand side of each observation subwindow (Fig. 93.16) are a custom visualization designed by scientists to indicate whether the telescope is accurately pointed at the target and whether a good signal was received by the spectrograph.

The data taken by SNIFS consists of two channels, blue and red. A channel produces a spectral decomposition of each sample of a 15×15 pixel image of a $6'' \times 6''$ square of the sky. Thus, each channel produces an x by

y (position) by λ (wavelength) *datacube*. To make the SNwarehouse postage stamps, the datacube is collapsed (averaged) along the λ-axis, leaving an x by y image. Since this image averages over many wavelengths, it is more accurate than a single wavelength slice, roughly as $(N_{slices})^{1/2}$, where N_{slices} is the datacube dimension in the wavelength direction. The program that performs

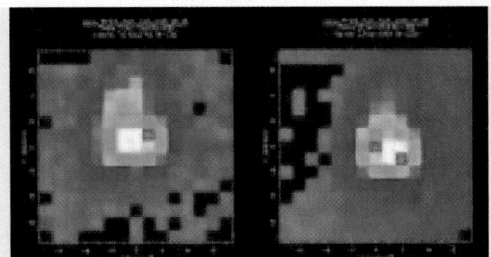

Fig. 93.20 A successful spectral observation. The *bright spot* in the center represents the supernova, and indicates that the target was centered by the spectrograph and the signal was good

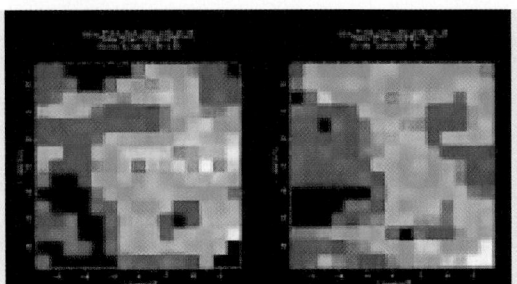

Fig. 93.21 A failed observation. No obvious target appears visible, so no useful spectral data was captured. This could happen due to a telescope pointing mishap, say the dome closed due to a humidity spike

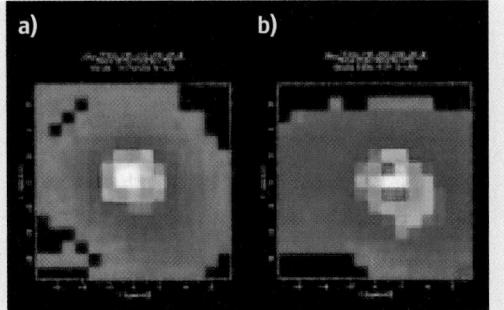

Fig. 93.22a,b A marginal observation. In (**b**) a blue halo surrounds the *bright dot* in the center, indicating that too much noise from the supernova's host galaxy has been picked up, skewing the spectral results

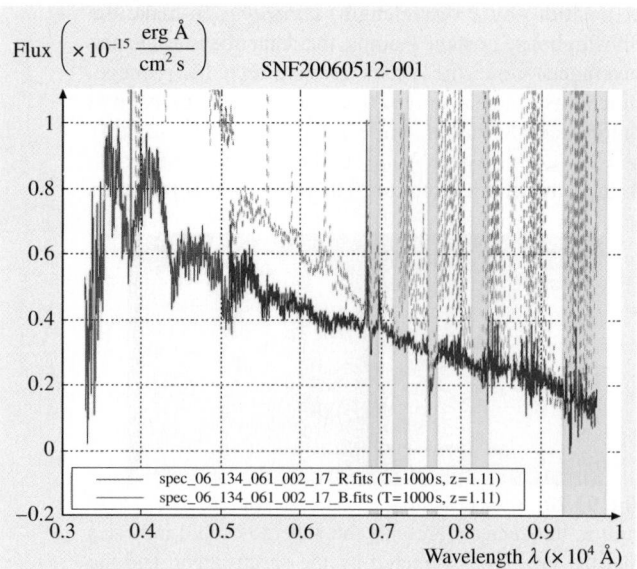

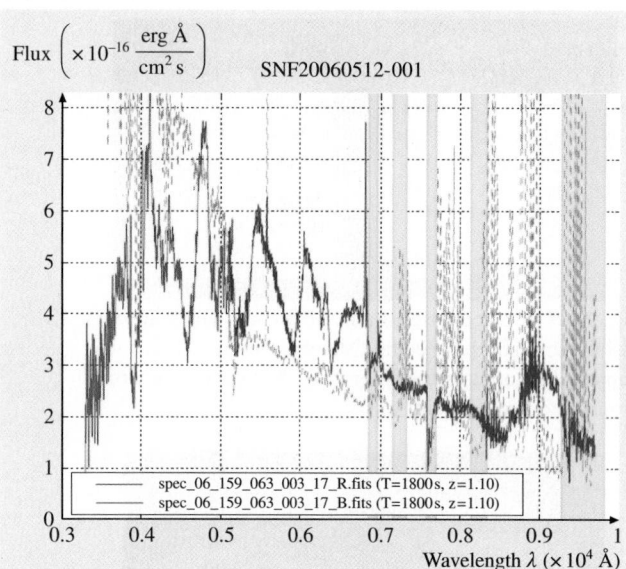

Fig. 93.24 Spectrum of the same type Ia supernova at 10 days after peak brightness. Note the different profile of the spectrum. To a domain expert, each of the peaks and troughs has meaning

Fig. 93.23 A spectrum from a very early type Ia supernova, 15 days before peak brightness. The *green* and *blue* *plots* indicate the spectral data. The spiky *grey lines* in the background depict the background sky *spectrum*. This supernova is just starting to show the features that define it as type Ia ◄

this operation is run as part of the real-time SNIFS pipeline that is run after each spectrum is read out.

Color-coding and position indicate the accuracy and signal strength of the received data. If there is a bright dot in the center of both squares, that is a clear indicator of a *good pointing* (Fig. 93.20).

A failed observation contains no clear bright circle in the center of the image Fig. 93.21. A marginal observation may contain a blue halo, indicating that too much background noise from the supernova's host galaxy is present, and that the spectral data may be skewed (Fig. 93.22).

Supernova scientists, like many domain experts, demonstrate strong visual pattern-recognition ability in their field of expertise. They can take a single glance at a picture of a complex spectrum, and instantly determine its type, age in days before or after peak brightness, and whether it exhibits any unusual properties. An early type Ia supernova (captured well before peak brightness) will display a certain pattern in its spectral plot (Fig. 93.23). A later supernova (imaged a few days past peak brightness) will show other characteristic spectral lines and features (Fig. 93.24). The spectral data are displayed in optimal form in order to facilitate domain experts' visual pattern-recognition ability. Spectral data are plotted in green and blue. The spiky grey lines depict the spectrum of the background sky. The broad grey bands represent areas of atmospheric absorption. Due to the complexity of the data (thousands of points of flux versus wavelength for each observation) and the necessity to make rapid, accurate decisions in order to maximize the use of limited, expensive telescope time, visualization provides the most efficient solution to the problem. The scientist can also access the raw numeric data by clicking on any of the images or spectra.

93.6 Conclusions

Sunfall, a collaborative visual analytics system in operation in a large-scale astrophysics project, has demonstrated that such systems can have a positive impact on data-intensive science. In July 2005, at the commencement of the Sunfall project, the SNfactory had discovered no supernovae. As of November 2008, the SNfactory has discovered over 1000 supernovae, of which some 600 are spectroscopically confirmed and nearly 400 of which are the critical type Ia SN so useful for cosmological research. By focusing on improvements in automation, Sunfall has led to more efficient data processing and significant labor savings for project scientists, enabling them to concentrate on physics research, including supernova discovery and analysis, and scientific publications, rather than on repetitive tasks. See Table 93.2 for supernova search data processing improvements, and Table 93.3 for details on labor savings (over five full-time equivalent staff per year) realized by the SNfactory project since the development of Sunfall.

Additionally, the framework architecture, efficient visual interfaces, and algorithms developed for Sunfall may be transferable to future large-scale automated transient alert pipelines such as the Joint Dark Energy Mission (JDEM) [93.13] or the Large Synoptic Survey Telescope (LSST) [93.44].

Sunfall search and the Sunfall status monitor, by incorporating sophisticated image-processing algorithms and machine learning capabilities with a usable, highly interactive visual interface designed to facilitate collaborative decision-making, reduced the number of false-positive supernova candidates by a factor of ten and reduced scientist scanning workload by 90%.

Sunfall data taking, the visual analytics subsystem for astrophysicists collaboratively operating a large telescope for time-critical supernova observation, demonstrated the effectiveness of a simplifying visualization, projecting three-dimensional data to a rectilinear two-dimensional format, in enabling users to synthesize large amounts of streaming data and make critical decisions under time pressure. This confirms previous studies [93.26, 28, 45] on the value of simplifying visualizations in interfaces designed for cognitively overloaded users. Furthermore, we demonstrated that the cultivation and maintenance of users' situation awareness, a concept first studied in the field of aviation and cockpit management, is crucial in other fields where time-critical decisions involving large, dynamic data sets must be made [93.2, 46].

SNfactory astrophysicists are currently using Sunfall and the Supernova Warehouse regularly for their scientific work. We have documented numerous favorable comments on the interface, and noted in user observations that less time is spent on routine tasks [93.2, 43]. Perhaps the most revelatory critique is that the scientists have ceased using all other legacy tools that SNwarehouse was intended to replace, and they continue to use the Sunfall framework even though the software developers have moved on to other projects [93.2, 36].

Finally, perhaps one of the most important lessons to be learned from this case study is the importance of forming closely integrated, interdisciplinary teams where members from all fields are involved in the participatory design process from the beginning of the project. Designing effective interfaces for complex, time-critical scientific decision-making requires both domain expertise and software expertise, and the two must be closely intertwined for the ultimate success of the project. Computer scientists and physicists all participated in both the scientific research and the software development process, which included regular group meetings and informal code reviews, thus facilitating the marriage of sophisticated science algorithms with software engineering best practices. Sunfall is highly usable software that has received praise from project scientists and has increased user satisfaction, efficiency, and productivity. Most important of all, Sunfall has facilitated the production of scientific data, results, and publications – the bottom line of science.

Table 93.2 SNfactory supernova search data rates before and after improvements

	Oct. 2005 (before search improvements)	Oct. 2006 (after search improvements)
Square degrees of sky processed per night	215	725
Processing time	< 24 h	< 24 h
Supernova selection efficiency	60%	75%
Objects scanned by humans per day	1094	113

Table 93.3 SNfactory labor rates in full-time equivalent staff (FTE) by task before and after Sunfall implementation

Tasks	2005 (FTE before Sunfall)	2007 (FTE after Sunfall)
Search	1	0.2
Scanning	3	0.25
Vetting	1	0.15
Scheduling	0.5	Automated
Postmortem	0.5	0.15
Total	6 FTE	0.75 FTE

93.6.1 Future Research Directions

Since many scientists prefer to develop their own software, and indeed, find writing code a requirement to accomplish their scientific work, the scientific domain has characteristics that are very distinct from the world of commercial software development. In previous decades, scientists have written their own code, frequently developing software individually and for their own use. However, as the complexity and volume of scientific data sets continue to explode, processing requirements have been growing too intricate for single-user development. As a result, teams of computer scientists and software engineers have been brought in to implement software packages according to scientists' specifications. Nonetheless, due to the nature of scientific research and the uncertainty of the final structure of science data sets, there have still been pitfalls in this approach. One challenge is that specifications are constantly evolving, and some scientific needs cannot be articulated before the scientists are actually working with the data.

Because of this, tightly coupled communication and feedback are becoming more and more necessary, and participatory design can yield substantial benefits. The scientists we worked with believe that interdisciplinary teams of software developers will become more common in future data-intensive science projects. There was agreement that computer scientists should be *embed-ded* as full collaborators in science projects, so that *my problems become your problems*. At the same time, participatory design allows scientists to be an active part of the design process and gives them a sense of ownership of the software.

As scientific projects are being deluged with exponentially growing quantities of increasingly complex and dynamic data, there has been a corresponding acceleration in the rate of change in the scientific culture of software development. The escalating pace of scientific discovery is being accompanied by new, interdisciplinary approaches to scientific computation, with a resulting influx of paradigms from relatively young disciplines such as human–computer interaction and usability engineering.

Although scientists have typically not ranked usability as a major concern in software design, we believe that is changing. We would like to see further case studies of large-scale scientific projects as they take usability into account throughout the design and implementation of their computing frameworks. We believe study of the human interface to scientific software will become especially critical for data-intensive science projects as the gap between the exponential growth of scientific data and the relatively unchanging nature of human cognitive capacity continues to expand. We call for continued research into the area of scientist–computer interaction.

References

93.1   G. Aldering, G. Adam, P. Antilogus, P. Astier, R. Bacon, S. Bongard, C. Bonnaud, Y. Copin, D. Hardin, F. Henault, D. Howell, J.-P. Lemonnier, J.-M. Levy, S. Loken, P. Nugent, R. Pain, A. Pecontal, E. Pecontal, S. Perlmutter, R. Quimby, K. Schahmaneche, G. Smadja, W. Wood-Vasey: Overview of the nearby supernova factory, SPIE Symp. Electron. Imaging (2002)

93.2   C. Aragon, S. Bailey, S. Poon, K. Runge, R.C. Thomas: Sunfall: A collaborative visual analytics system for astrophysics, J. Phys. Conf. Ser. **125**, 012091 (2008)

93.3   S. Perlmutter, G. Aldering, G. Goldhaber, R.A. Knop, P. Nugent, P.G. Castro, S. Deustua, S. Fabbro, A. Goobar, D.E. Groom, I.M. Hook, A.G. Kim, M.Y. Kirn, J.C. Lee, N.J. Nunes, R. Pain, C.R. Pennypacker, R. Quimby, C. Lidman, R.S. Ellis, M. Irwin, R.G. McMahon, P. Ruiz-Lapuente, N. Walton, B. Schaefer, B.J. Boyle, A.V. Filippenko, T. Matheson, A.S. Fruchter, N. Panagia, H.J.M. Newberg,

W.J. Couch: Measurements of omega and lambda from 42 high-redshift supernovae, Astrophys. J. **517**, 565–586 (1999)

93.4   A.G. Riess, A.V. Filippenko, P. Challis, A. Clocchiatti, A. Diercks, P.M. Garnavich, R.L. Gilliland, C.J. Hogan, S. Jha, R.P. Kirshner, B. Leibundgut, M.M. Phillips, D. Reiss, B.P. Schmidt, R.A. Schommer, R.C. Smith, J. Spyromilio, C. Stubbs, N.B. Suntzeff, J. Tonry: Observational evidence from supernovae for an accelerating universe and a cosmological constant, Astrophys. J. **116**, 1009–1038 (1998)

93.5   NEAT: *Near Earth Asteroid Tracking* (California Institute of Technology/JPL, Pasadena 2007), http://neat.jpl.nasa.gov

93.6   QUEST: *Palomar-QUEST Survey* (Yale Univ., New Haven 2007), http://hepwww.physics.yale.edu/quest/palomar.html

93.7   C. Aragon, D.B. Aragon: A fast contour descriptor algorithm for supernova image classification, SPIE

Symp. Electron. Imaging: Real-Time Image Process. (San Jose 2007)

93.8   S. Bailey, C. Aragon, R. Romano, R.C. Thomas, B.A. Weaver, D. Wong: How to find more supernovae with less work: object classification techniques for difference imaging, Astrophys. J. **665**, 1246–1253 (2007)

93.9   R. Romano, C. Aragon, C. Ding: Supernova recognition using support vector machines, 5th Int. Conf. Mach. Learn. Appl. (ICMLA 2006) (Orlando 2006)

93.10   W.M. Wood-Vasey: Rates and Progenitors of Type Ia Supernovae. Ph.D. Thesis (University of California, Berkeley 2004)

93.11   C.T. Zahn, R.Z. Roskies: Fourier descriptors for plane closed curves, IEEE Trans. Comput. **21**(3), 269–281 (1972)

93.12   L. Mazuray, P.J. Rogers: Optical design and engineering, Proc. SPIE **5249**, 146–155 (2004)

93.13   NASA: The Joint Dark Energy Mission (NASA, Washington 2008), http://universe.nasa.gov/program/probes/jdem.html

93.14   M. Sako, B. Bassett, A. Becker, D. Cinabro, F. De-Jongh, D. Depoy, M. Sako, B. Bassett, A. Becker, D. Cinabro, F. DeJongh, D.L. Depoy, B. Dilday, M. Doi, J.A. Frieman, P.M. Garnavich, C.J. Hogan, J. Holtzman, S. Jha, R. Kessler, K. Konishi, H. Lampeitl, J. Marriner, G. Miknaitis, R.C. Nichol, J.L. Prieto, A.G. Riess, M.W. Richmond, R. Romani, D.P. Schneider, M. Smith, M. Subba Rao, N. Takanashi, K. Tokita, K. van der Heyden, N. Yasuda, C. Zheng, J. Barentine, H. Brewington, C. Choi, J. Dembicky, M. Harnavek, Y. Ihara, M. Im, W. Ketzeback, S.J. Kleinman, J Krzesiński, D.C. Long, E. Malanushenko, V. Malanushenko, R.J. McMillan, T. Morokuma, A. Nitta, K. Pan, G. Saurage, S.A. Snedden: The Sloan digital sky survey-II supernova survey: search algorithm and follow-up observations, Astron. J. **135**, 348–373 (2008)

93.15   P. Astier, J. Guy, N. Regnault, R. Pain, E. Aubourg, D. Balam, S. Basa, R.G. Carlberg, S. Fabbro, D. Fouchez, I.M. Hook, D.A. Howell, H. Lafoux, J.D. Neill, N. Palanque-Delabrouille, K. Perrett, C.J. Pritchet, J. Rich, M. Sullivan, R. Taillet, G. Aldering, P. Antilogus, V. Arsenijevic, C. Balland, S. Baumont, J. Bronder, H. Courtois, R.S. Ellis, M. Filiol, A.C. Gonçalves, A. Goobar, D. Guide, D. Hardin, V. Lusset, C. Lidman, R. McMahon, M. Mouchet, A. Mourao, S. Perlmutter, P. Ripoche, C. Tao, N. Walton: SuperNova legacy survey: measurement of Ω_M, Ω_Λ and w from the first year data set, Astron. Astrophys. **447**, 31–48 (2006)

93.16   J. Raddick: Sloan Digital Sky Survey (Johns Hopkins Univ., Baltimore 2007), http://www.sdss.org

93.17   SNLS: SuperNova Legacy Survey (2007), http://www.cfht.hawaii.edu/SNLS/

93.18   J. Thorstensen: SkyCalc GUI Manual (Dartmouth College, Hanover 2008),

http://mdm.kpno.noao.edu/Manuals/doc/guimanual.html

93.19   J. Thorstensen: SkyCalc User's Manual (Dartmouth College, Hanover 2008), http://zimmer.csufresno.edu/~fringwal/skycal.pdf

93.20   F. Cavicchio: Astroart 4.0–96 bit image processing (MSB Software, Ravenna 2008), http://www.msb-astroart.com/

93.21   L. Christensen: ESA/ESO/NASA Photoshop FITS Liberator (ESO, Munich 2008), http://www.spacetelescope.org/projects/fits_liberator/index.html

93.22   A. Gauthier: Virtual Astronomy Multimedia Project (University of Arizona, Tempe 2008), http://www.virtualastronomy.org/index.php

93.23   W. Joye, E. Mandel: New features of SAOImage DS9, Astronomical Data Analysis Software and Systems XII (2003)

93.24   T. Boroson: NOAO Image Reduction and Analysis Facility (IRAF, National Optical Astronomy Observatories, Tucson 2008), http://iraf.noao.edu/iraf/web

93.25   Y. Li, C. Fu, A. Hanson: Scalable WIM: Effective exploration in large-scale astrophysical environments, IEEE Trans. Vis. Comput. Graph. **12**(5), 1005–1011 (2006)

93.26   L. Chittaro, F. Zuliani, E. Carchietti: Mobile devices in emergency medical services: user evaluation of a PDA-based interface for ambulance run reporting. In: Mobile Response, Lecture Notes in Computer Science, Vol. 4458 (Springer, Berlin, Heidelberg 2007) pp. 19–28

93.27   S. Eick, M. Eick, J. Fugitt, B. Horst, M. Khailo, R. Lankenau: Thin client visualizatio, IEEE Symp. Vis. Anal. Sci. Technol. (VAST 2007) (Sacramento 2007)

93.28   S. Kim, Y. Jang, A. Mellema, D. Ebert, T. Collins: Visual analytics on mobile devices for emergency response, IEEE Symp. Vis. Anal. Sci. Technol. (VAST 2007) (Sacramento 2007)

93.29   A. Pattath, B. Bue, Y. Jang, D. Ebert, X. Zhong, A. Ault, E. Coyle: Interactive visualization and analysis of network and sensor data on mobile devices, IEEE Symp. Vis. Anal. Sci. Technol. (VAST 2007) (Baltimore 2006)

93.30   Purdue: Synthetic Environment for Analysis and Simulation (Purdue Univ., West Lafayette 2008), http://www.mgmt.purdue.edu/centers/perc/html

93.31   T. Kapler, W. Wright: Geotime information visualization, IEEE Symp. Inf. Vis. (2004)

93.32   C. Pan, P. Mitra: FemaRepViz: Automatic extraction and geo-temporal visualization of FEMA national situation updates, IEEE Symp. Vis. Anal. Sci. Technol. (VAST 2007) (Sacramento 2007)

93.33   RSOE: RSOE HAVARIA AlertMap (Emergency and Disaster Information Services (EDIS), (Budapest 2008), http://hisz.rsoe.hu/alertmap/woalert.php?lang=eng

93.34   D. Tesone, J. Goodall: Balancing interactive data management of massive data with situational awareness through smart aggregation, Proc. IEEE Symp. Vis. Anal. Sci. Technol. (VAST 2007) (Sacramento 2007)

93.35   M.R. Endsley, D.J. Garland: *Situation Awareness Analysis and Measurement* (Lawrence Erlbaum, New York 2000)

93.36   C. Aragon, S. Poon: The impact of usability on supernova discovery, Workshop on Increasing the Impact of Usability Work in Software Development, CHI 2007: ACM Conf. Hum. Factors Comput. Syst. (San Jose 2007)

93.37   J.J. Thomas, K.A. Cook: *Illuminating the Path: The Research and Development Agenda for Visual Analytics* (National Visualization and Analytics Center, Richland 2005)

93.38   R. Chellappa, R. Bagdazian: Fourier coding of image boundaries, IEEE Trans. Pattern Anal. Mach. Intell. **6**, 102–105 (1984)

93.39   E. Persoon, K. Fu: Shape discrimination using Fourier descriptors, IEEE Trans. Syst. Man Cybern. **7**, 170–179 (1977)

93.40   R. Mak, J. Walton, L. Keely, D. Heher, L. Chan: Reliable service-oriented architecture for NASA's Mars Exploration Rover Mission, IEEE Aerosp. Conf. (2005)

93.41   R. Israel: *Mercator's Projection* (Univ. British Columbia, Vancouver 2008), http://www.math.ubc.ca/~israel/m103/mercator/mercator.html

93.42   K. Arnold: *Screen Updating and Cursor Movement Optimization: A Library Package* (University of California, Berkeley 1977)

93.43   S. Poon, R.C. Thomas, C. Aragon, B. Lee: Context-linked virtual assistants for distributed teams: an astrophysics case study, ACM Conf. Comput. Supported Coop. Work (ACM CSCW 2008) (San Diego 2008)

93.44   Z. Ivezic, J.A. Tyson, R. Allsman, J. Andrew, R. Angel, for the LSST collaboration: LSST: from Science drivers to reference design and anticipated data products, arXiv:0805.2366v1 [astro-ph] (2008)

93.45   C. Aragon, M. Hearst: Improving aviation safety with information visualization: a flight simulation study, CHI 2005: ACM Conf. Hum. Factors Comput. Syst. (Portland 2005)

93.46   C. Aragon, S. Poon, G. Aldering, R.C. Thomas, R. Quimby: Using visual analytics to maintain situational awareness in astrophysics, IEEE Symp. Vis. Anal. Sci. Technol. (VAST 2008) (Columbus 2008)

Part J Appendix

Part J Appendix

94 Automation Statistics
Juan D. Velásquez, West Lafayette, USA
Xin W. Chen, West Lafayette, USA
Sang Won Yoon, West Lafayette, USA
Hoo Sang Ko, West Lafayette, USA

Appendix. Part J The concluding part of this Handbook contains figures and tables with statistical information and summaries about automation applications and impacts in three main areas: industrial automation; service automation; and financial and e-Commerce automation. A rich list of associations and of periodical publications around the world that focus on automation in its variety of related fields is also included for the benefit of readers worldwide.

94. Automation Statistics

Juan D. Velásquez, Xin W. Chen, Sang Won Yoon, Hoo Sang Ko

This chapter contains automation information collected from many different sources around the world, among them the US Census Bureau, the International Federation of Automatic Control (IFAC), the American Automatic Control Council (AACC), the American Bankers Association, World Robotics, and many more. Section 94.1 introduces automation statistical data according to its application domain, such as: financial and e-Commerce, industrial, healthcare, and the service industries. In Sect. 94.2 a list of worldwide automation, control, and robotics associations is provided. Section 94.3 provides the reader with a list of automation labs around the world and lastly, in Sect. 94.4 a list of automation related journals and publications is presented.

The statistics, organizations, and journals included in this chapter open a window into the current and emerging state of automation. While attempting to be comprehensive, because of the broadness of automation, the authors recognize that there may be additional relevant items, but none were omitted intentionally. The scope of

the information is meant to highlight and expose the broad span, applications, concerns and benefits of automation, and clearly, cannot completely include all areas of automation influence. Beyond this chapter, other chapters in this handbook provide additional statistical data directly related to their specific topic.

Automation continues to enable individuals, companies, and governments to innovate and compete in a fast-changing world. Companies of all sizes throughout the world in virtually all industries have embraced automation as one of the most powerful strategic advantages to stay competitive. For instance, in robotic automation it is expected that over 1 million robots will be installed worldwide by 2011, a significant gain from the 400 000 that were installed in the early 1990s. Japan, the USA, and Germany will continue to supply the majority of robots, while automotive parts and motor vehicles will continue to demand the greatest number of robots by industry.

It has not only been companies that are taking advantage of automation; individuals have also embraced automation as a way to enhance their everyday lives. It is expected that, by 2012, consumers' most used payment form will be electronic transfers, followed by credit cards. More than 50% of all travel reservations made in the USA are now done by consumers directly via the Internet rather than by travel agencies. Likewise, the number of online buyers continues to climb, with over 50% of Americans having bought at least one product online in 2007. It is projected that in the USA alone e-Commerce will grow from US$ 282.7 billion in 2007 to US$ 658.4 billion by 2010, almost a threefold increase in 3 years.

What does the future hold for automation? Automation's influence on microsystems, nanosystems, and systems-of-systems will explode, as will bioinspired automation and bioinspired collaborative control. There are many grand challenges as identified by the US National Academy of Engineering for which automation is essential (i. e., make solar energy economical, advance personalized learning, engineer better medicines). For instance, virtual reality will significantly influence healthcare, engineering, entertainment, and education and training as its worldwide market grows from US$ 2 billion in 2002 to over US$ 8 billion in 2010.

94.1 Automation Statistics

94.1.1 Financial and e-Commerce Automation

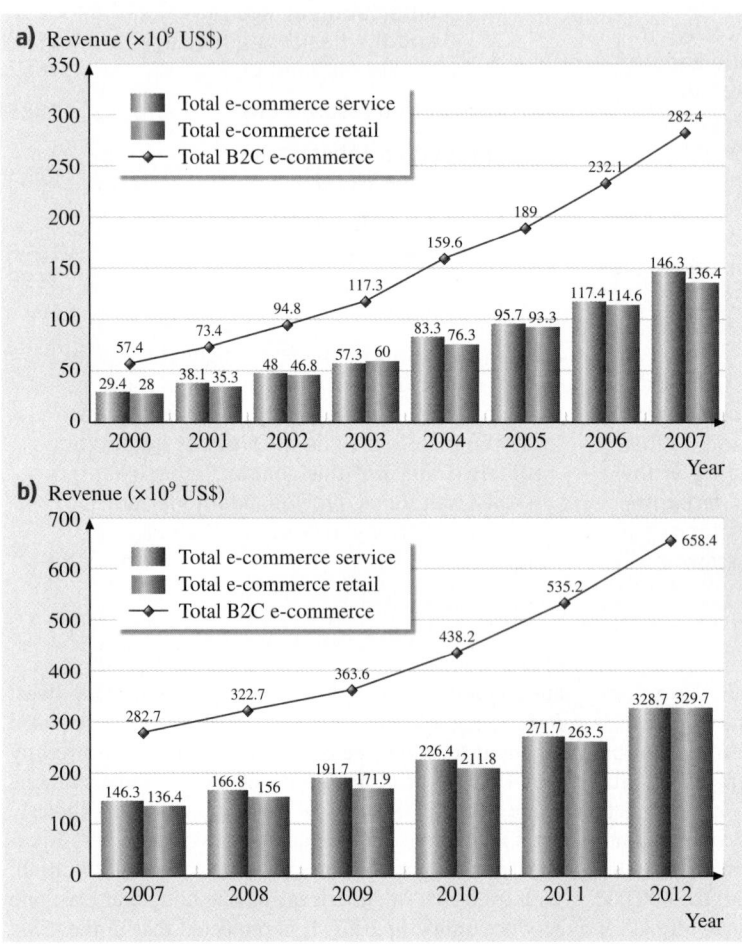

Fig. 94.1a,b Business-to-consumer e-Commerce (**a**) revenue by service and retail sectors for 2000–2007 in US$. (**b**) Projected revenue by service and retail sectors for 2007–2012 in US$ (source: US Department of Commerce, US Census Bureau)

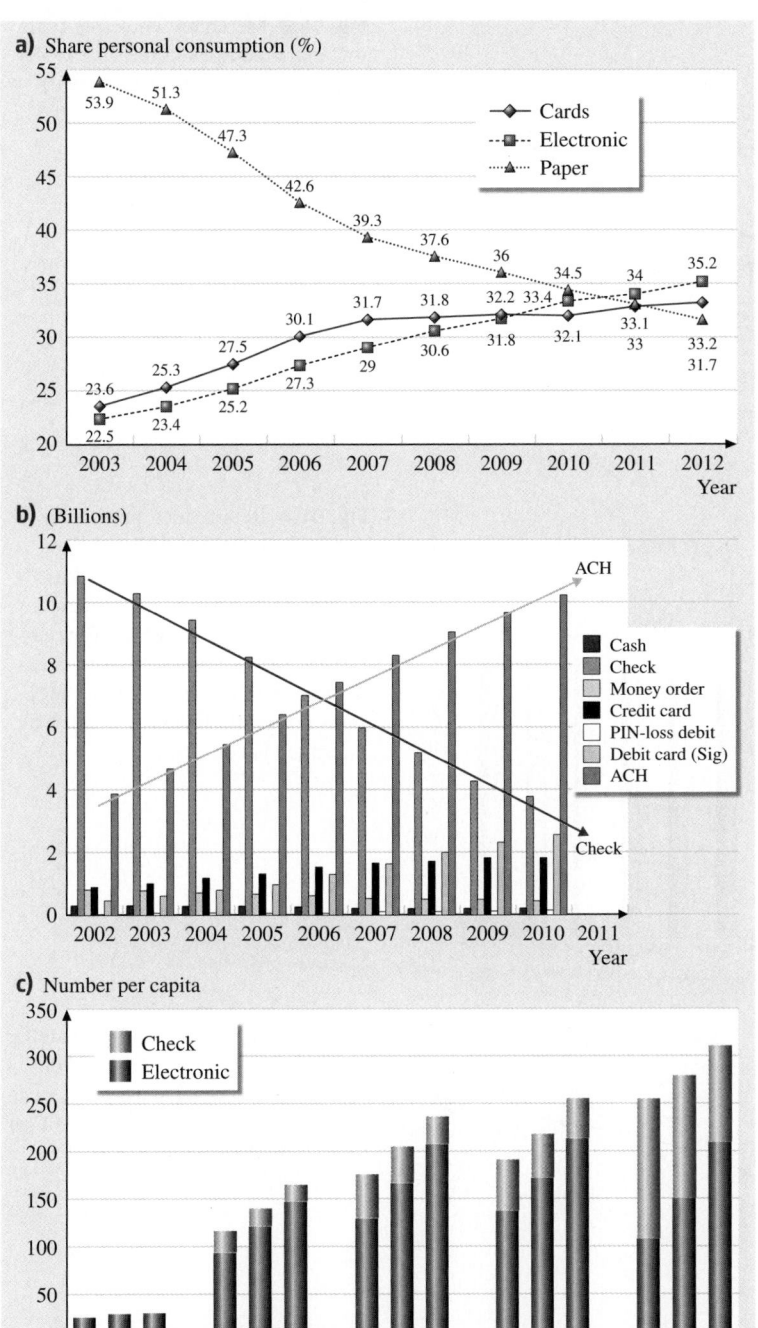

a) Share personal consumption (%)

b) (Billions)

c) Number per capita

Fig. 94.2a–c e-Commerce payments. **(a)** Estimated share of US personal consumption by credit cards, electronic payments, and paper from 2003–2012 (projected) (source: US Department of Commerce, US Census Bureau), note: paper includes cash, check, money orders, etc. Cards includes both credit and debit purchase volume. Electronic includes pre-authorized, recurring payments, payments made via telephone, and payments initiated via the Internet **(b)** Total Number of consumer to business recurring payments by industry in the US from 2002 to 2010 in US\$ (source: Celent and Bearing Point Analysis) **(c)** Non-cash payments per-capita in five selected economies in 2000, 2003 and 2006. The European Monetary Union is made up of Austria, Belgium, Finland, France, Germany, Greece, Ireland, Italy, Luxemburg, The Netherlands, Portugal, and Spain. Cyprus, Malta, and Slovenia joined the EMU after 2006 and were not included in calculations (sources: European Central Bank (2007), Bank for International Settlements (2008), Federal Reserve Board)

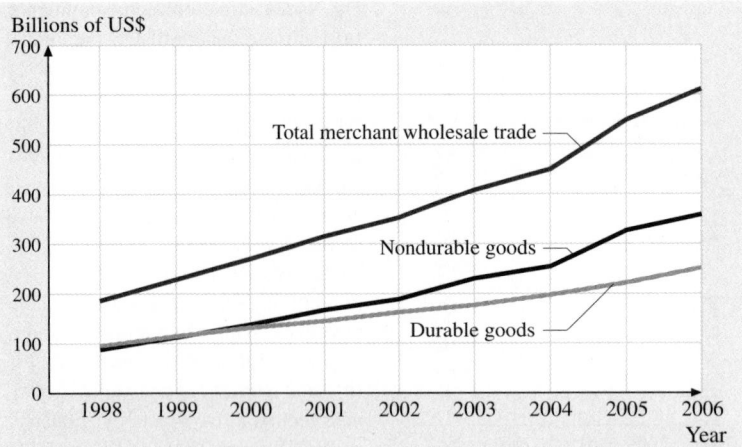

Fig. 94.3 Merchant wholesale trade e-Commerce sales in the US: 1998–2006 (source: US Census Bureau)

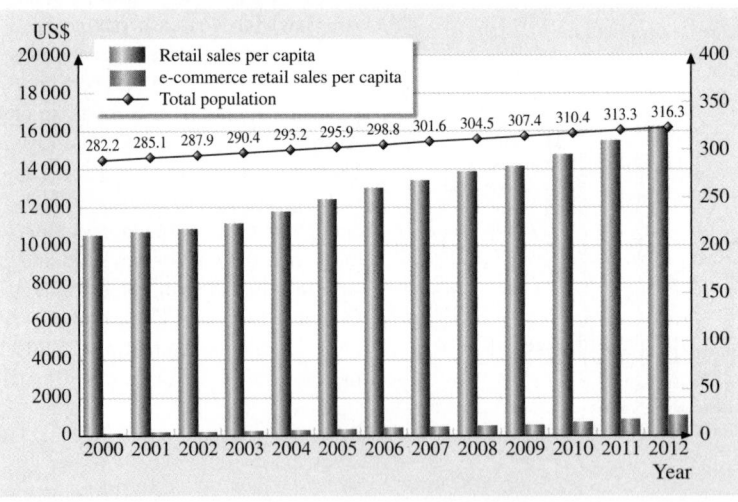

Fig. 94.4 Retail sales per capita and e-Commerce retail sales per capita from 2000 to 2012 (projected) (source: US Department of Commerce, US Census Bureau (2000–2007); population growth estimated by Cogitamus Consulting; sales per capita figures estimated by Package Facts)

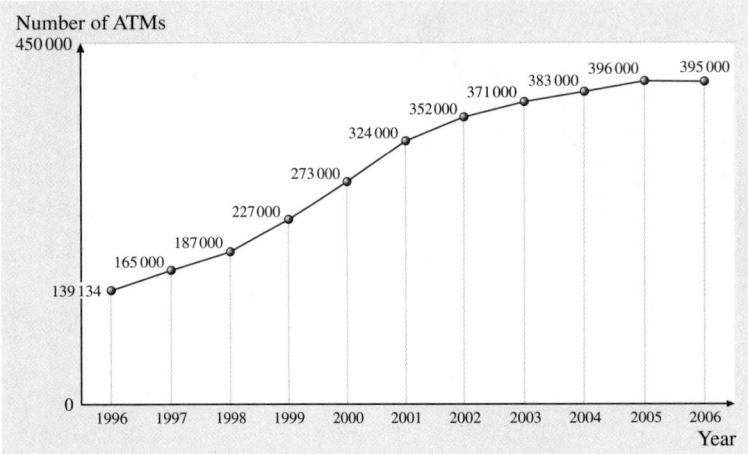

Fig. 94.5 Number of automated teller machines (ATMs) in the US from 1996 through 2006 (source: American Bankers Association, 2007 ATM Fact Sheet, with permission)

94.1.2 Industrial Automation

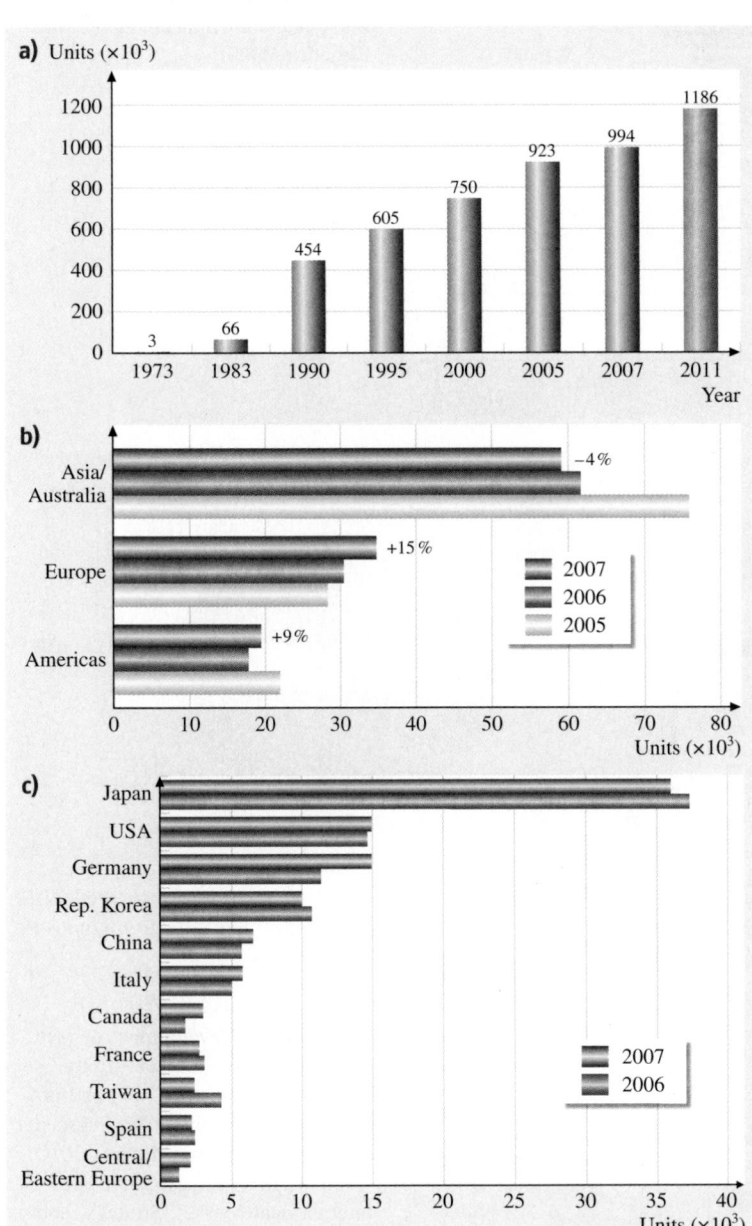

Fig. 94.6 Figures **(a–d)** illustrate growth in industrial automation. **(a)** Estimated and projected worldwide operational stock of industrial robots. **(b)** Estimated yearly supply of industrial robots by regions of the world from 2005 to 2007. **(c)** Estimated yearly supply of industrial robots by selected countries (source: World Robotics, 2008)

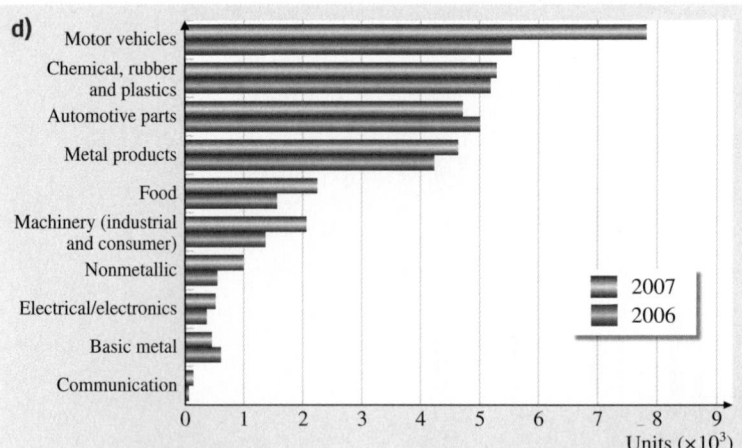

Fig. 94.6 (d) Estimated yearly supply of industrial robots in Europe by robot implementation areas (source: World Robotics, 2008)

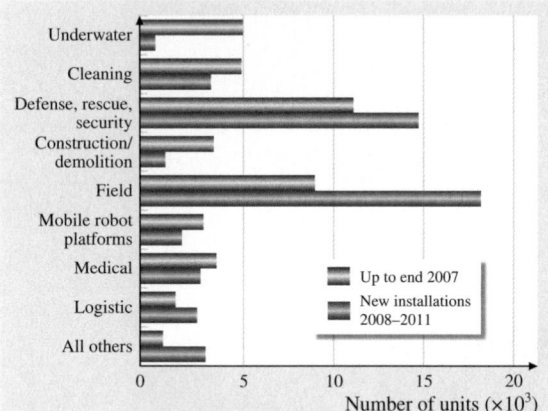

Fig. 94.7 Current and projected installations and stock of service robots for professional use up to 2006 and projected sales for the period 2008–2011 (source: World Robotics, 2008)

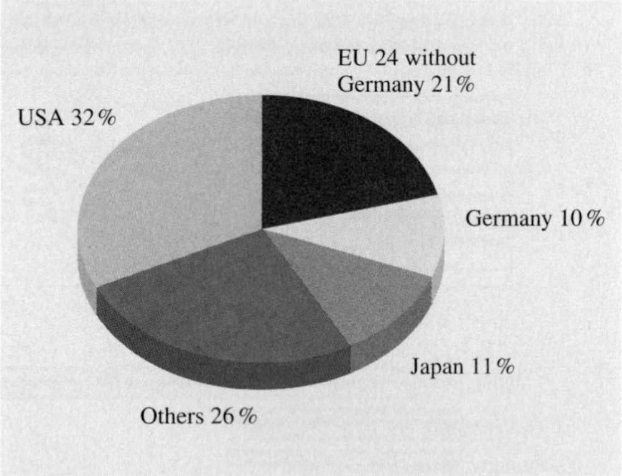

Fig. 94.8 Process automation in the world (source: ZVEI, Zentralverband Elektrotechnik- und Elektronikindustrie e.V., Frankfurt)

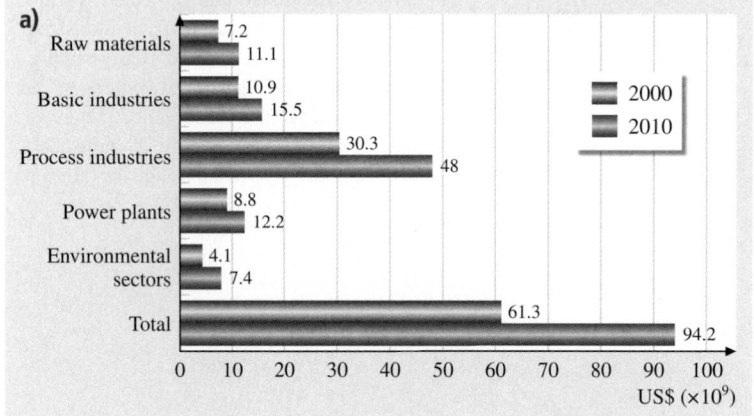

Fig. 94.9a–c Development of process automation market **(a)** by industries for the year 2000 and projected for 2010 (source: N. Schroeder Process Automation Markets 2010 – Important Findings of the New International Market, Strategy, and Technology Report, Intechno Consulting (2003))

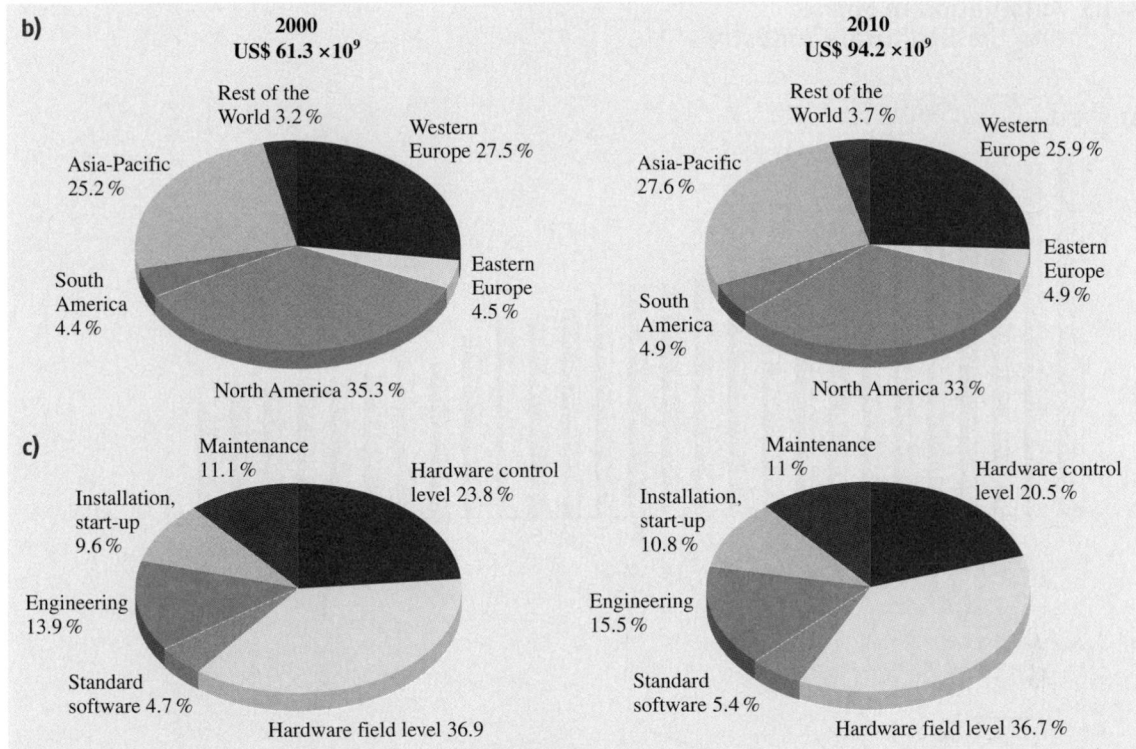

Fig. 94.9 Development of process automation market (**b**) by regions, and (**c**) by products and services for the year 2000 and projected for 2010 (source: N. Schroeder Process Automation Markets 2010 – Important Findings of the New International Market, Strategy, and Technology Report, Intechno Consulting (2003))

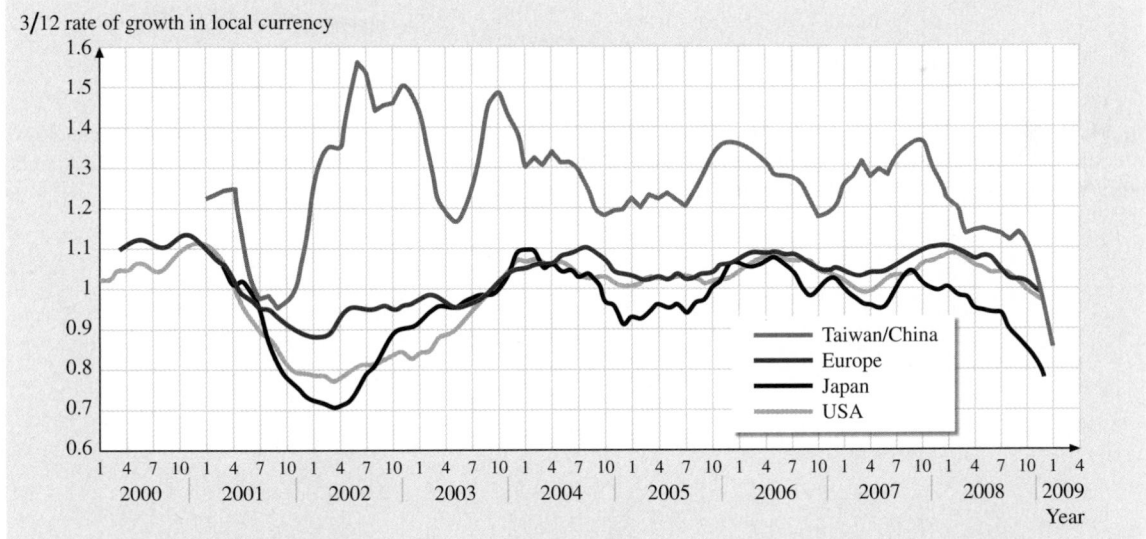

Fig. 94.10 Global electronic equipment shipment growth (2000–2009) (source: MarketEye: Overview: 2008 Global Electronic Food Chain Performance with Guidance for 2009 (http://www.ttiinc.com/object/me_custer_20090223.html))

94.1.3 Automation in Service and the Healthcare Industry

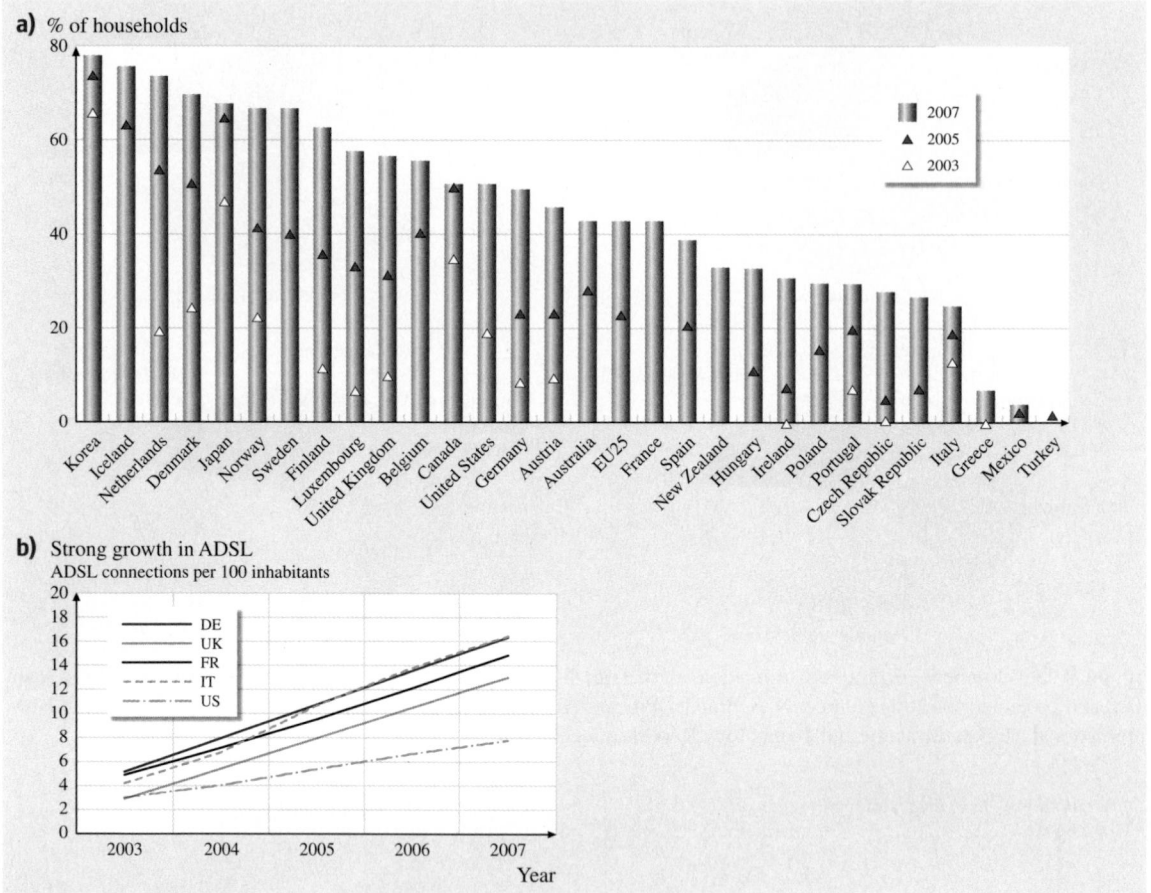

Fig. 94.11 (a) Households' access to broadband Internet, 2003–2007 (source: OECD Information Technology Outlook 2008 Highlights) **(b)** Asymmetric digital subscriber line (ADSL) connections per 100 inhabitants (http://www.dbresearch.com/PROD/DBR_INTERNET_EN-PROD/PROD0000000000198220.pdf)

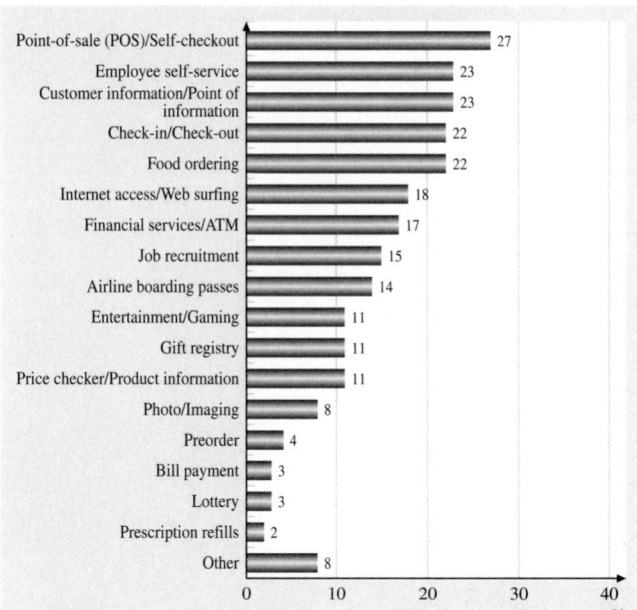

Fig. 94.12 Types of automated kiosks for customer transactions by percentage (source: Seventh Annual Kiosk Benchmark Study, Edgell Communications, Randolph)

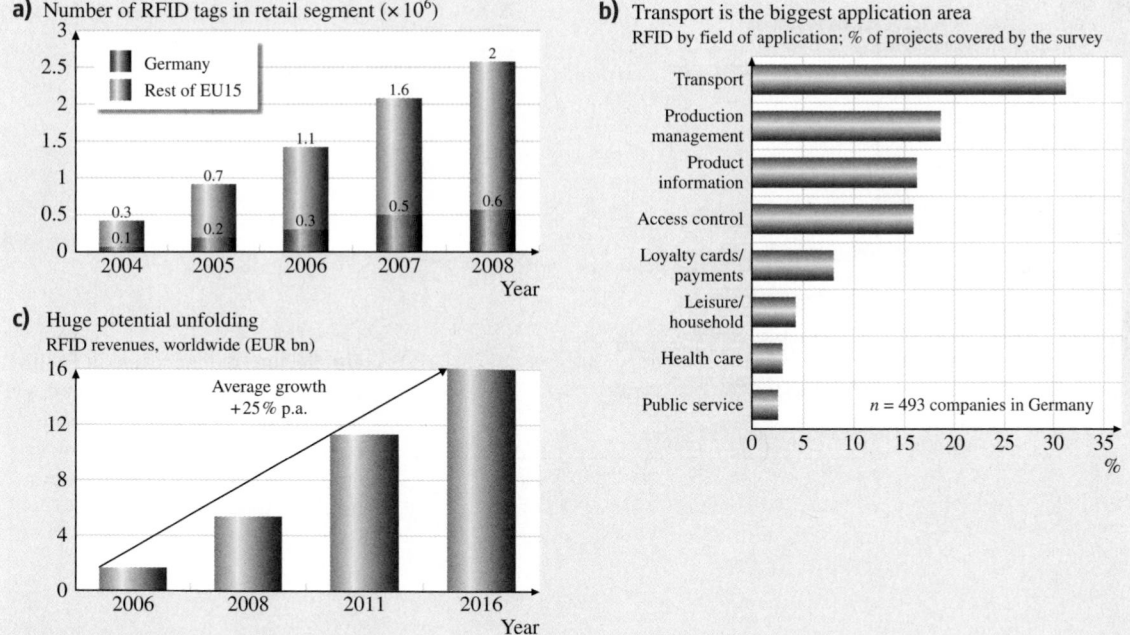

Fig. 94.13a–c Role of RFID (**a**) Number of RFID tags in the retail segment in millions (source: RFID chips are on everyone's lips) (source: VDC, 2005; http://www.dbresearch.com/PROD/DBR_INTERNET_EN-PROD/PROD0000000000204263.pdf) (**b**) RFID by field of application; % of projects covered by survey in Germany (source: IIG Freiburg and http://www.dbresearch.com/PROD/DBR_INTERNET_EN-PROD/PROD0000000000236797.pdf) (**c**) RFID worldwide revenues in European Euros (billions) (source: DB research, 2009 (http://www.dbresearch.com/PROD/DBR_INTERNET_EN-PROD/PROD0000000000236797.pdf))

94.1.4 Service Automation

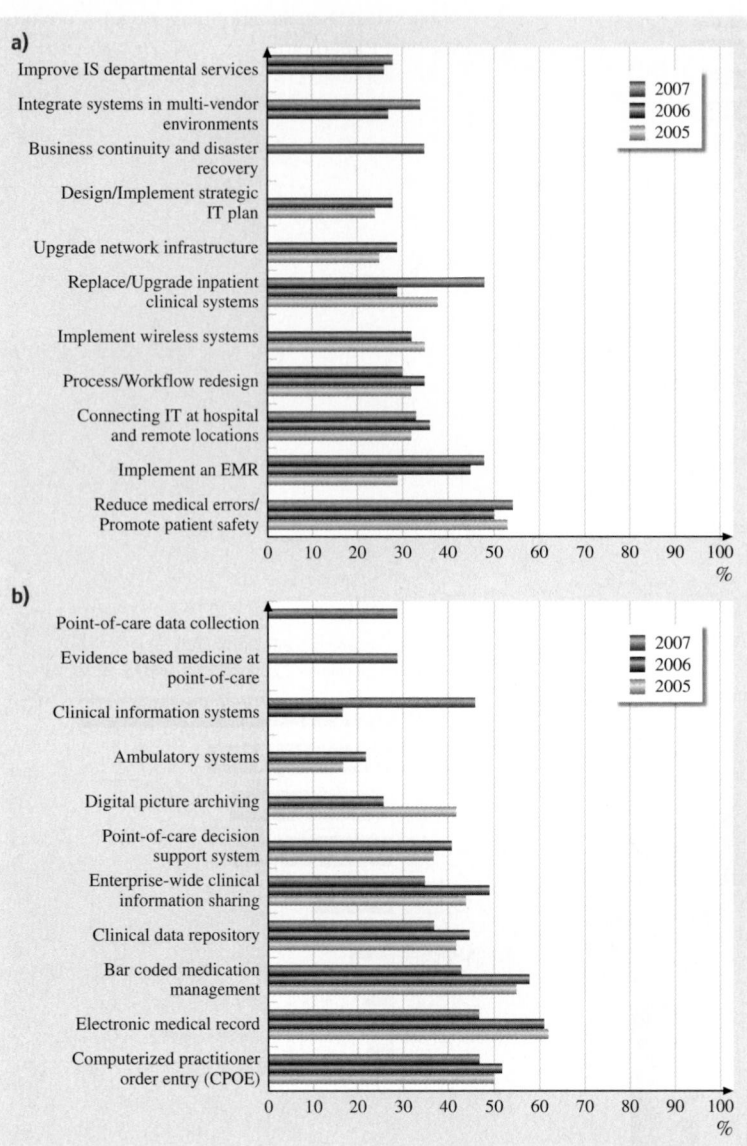

Fig. 94.14a–d Percentage of health-care facilities information technology: (**a**) priorities (EMR – emergency medical record, IS – information system, IT – information technology); (**b**) most important applications

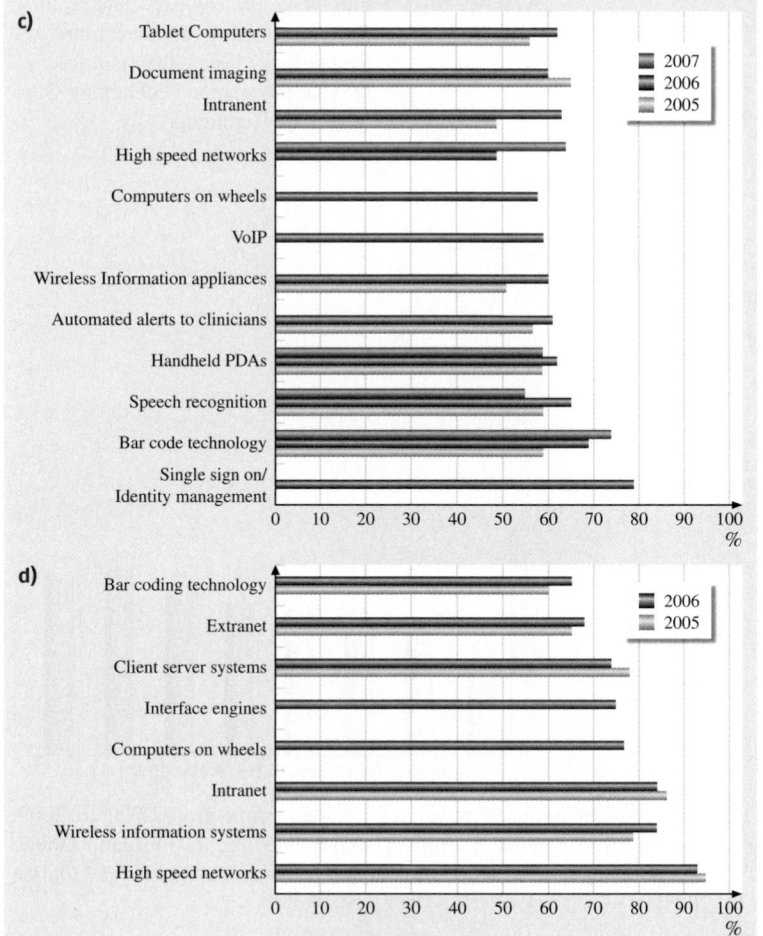

Fig. 94.14 (c) adoption (VoIP – voice over Internet protocol, PDA – personal digital assistant); **(d)** current use (source: 17th Annual HIMSS Leadership Survey sponsored by ACS Healthcare Solutions, February 13, 2006; 18th Annual HIMSS Leadership Survey sponsored by ACS Healthcare Solutions, April 10, 2007)

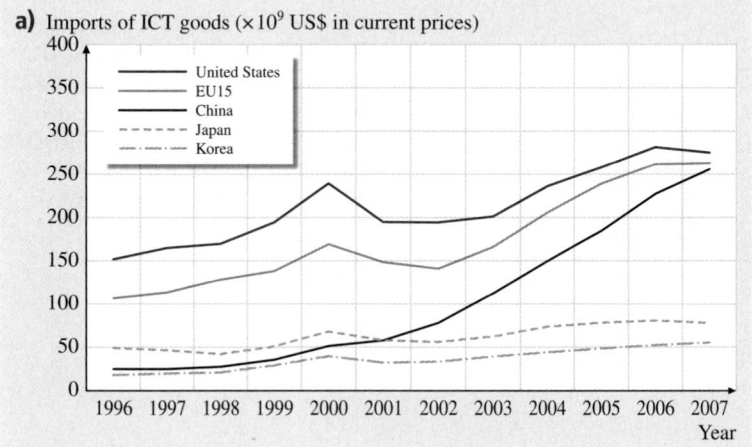

Fig. 94.15 (a) Top importers of Information and Communications Technology, 1996–2007 (source: OECD Information Technology Outlook 2008 Highlights)

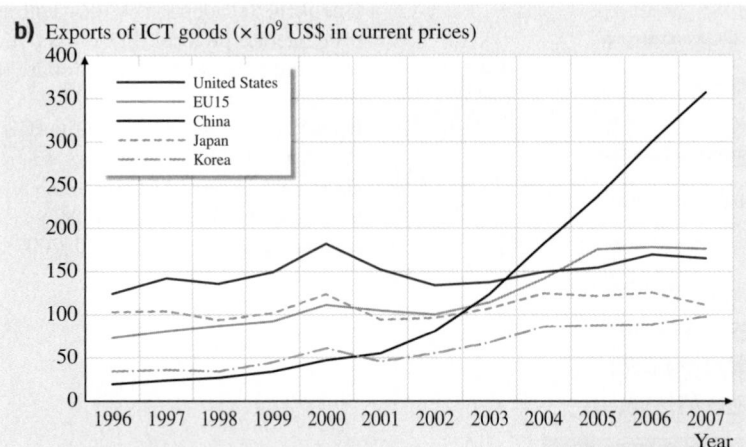

Fig. 94.15 (b) Top exporters of Information and Communications Technology, 1996–2007 (source: OECD Information Technology Outlook 2008 Highlights)

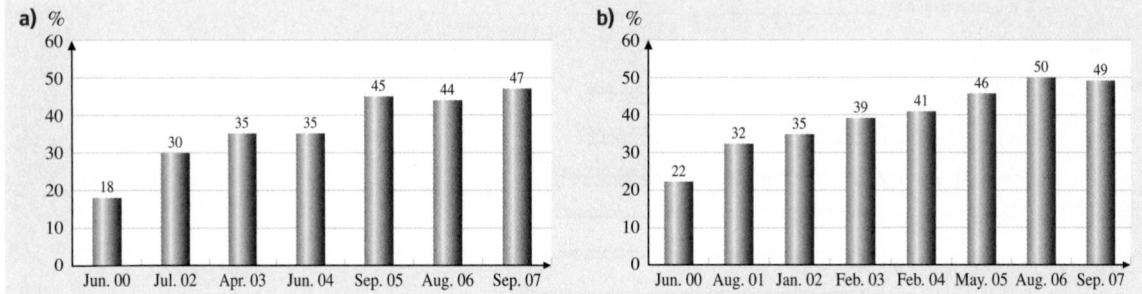

Fig. 94.16 (a) Percentage of Americans who have bought or made travel reservations online from 2000 to 2007. **(b)** Percentage of all American adults who have ever bought a product online (source: J. Horrigan: Online Shopping. Pew Internet & American Life Project, January, 13 (2008), http://www.pewinternet.org/pdfs/PIP_Online%20Shopping.pdf, accessed on November 11, 2008)

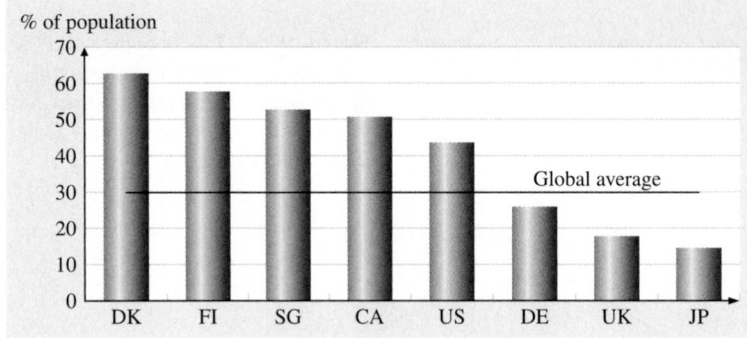

Fig. 94.17 Usage of e-Government services in 2003 (DK – Denmark, FI – Finland, SG – Singapore, CA – Canada, US – United States, DE – Germany, UK – United Kingdom, JP – Japan) (source: DB research (http://www.dbresearch.com/PROD/DBR_INTERNET_EN-PROD/PROD0000000000188264.pdf))

94.2 Automation Associations

Asoc. Argentina de Control Automático – AADECA
Argentina
URL: www.aadeca.org

Australian Robotics and Automation Association Inc. (ARAA)
Australia
Website: http://www.araa.asn.au/

The Institution of Engineers
Australia
URL: www.engineersaustralia.org.au/

Adelaide ACM SIGGRAPH – Special Interest Group on Computer Graphics and Interactive Techniques
Australia
URL: http://www.siggraph.org/members/adelsig

IICA – Institute of Instrumentation Control and Automation
Australia
president@iica.org.au

Institut für Flexible Automation (INFA)
Technische Universität, Gusshausstrasse 27–29 / E361
1040 Wien
Austria
E-mail: sekretariat@flexaut.tuwien.ac.at

International Federation of Automation Control (IFAC)
Austria
Website: http://www.ifac-control.org/

Oest. Ges. f. Automatisierung and Robotertechnik – OeGART
Austria
URL: www.ihrt.tuwien.ac.at/oegart

Belgian ACM SIGCHI – Special Interest Group on Computer–Human Interaction
Belgium
De Regenboog 11
Mechelen, Belgium 2800

European SIGOPS ACM Chapter – Special Interest Group on Operating Systems
Belgium
URL: http://www.eurosys.org/

Fédération IBRA/BIRA
Belgium
URL: www.ibra-control.be

Brazil ACM SIGCHI – Special Interest Group on Computer–Human Interaction
Brazil
URL: http://ead.unifor.br/brchi

Sociedade Brasileira de Automatica – SBA
Brazil
URL: www.sba.org.br

Bulgarian ACM Chapter
ul. Studentska 1
TU-Varna
Varna, Bulgaria 9010

Federation of the Scientific Engineering Unions in Bulgaria
Bulgaria
URL: www.fnts-bg.org/

Montreal ACM SIGGRAPH – Special Interest Group on Computer Graphics and Interactive Techniques
Canada
URL: http://montreal.siggraph.org

Ottawa ACM SIGCHI – Special Interest Group on Computer–Human Interaction
Canada
URL: http://www.capchi.org/

Vancouver ACM SIGGRAPH – Special Interest Group on Computer Graphics and Interactive Techniques
Canada
URL: www.siggraph.ca

Continental Automated Building Association (CABA)
Canada
Website: http://www.caba.org/

IFAC – Canada
Canada
URL: www.ifac-canada.ca

Asociacion Chilena de Control Automatico – ACCA
Chile
Universidad de la Frontera Chile
Avda Francisco Salazar, Casilla 54 D 01145 Temuco

ACM China SIGMOD – Special Interest Group on Management of Data
Dept. of Systems and Engineering Management,
Chinese University
Hong Kong, China

Hong Kong ACM SIGGRAPH – Special Interest Group on Computer Graphics and Interactive Techniques
China
Dept. of MIT
IVE (Tsing Yi),
20 Tsing Yi Road, Tsing Yi Island.
N.T., Hong Kong

Chinese Society of Automation
1001, No. 293 Song Chiang Rd.
Taipei, Taiwan

Chinese Association of Automation
China URL: caa.gongkong.com

IFAC NMO of Hong Kong, China
China
URL: www.hkie.org.hk/

Bogota ACM SIGGRAPH – Special Interest Group on Computer Graphics and Interactive Techniques
Colombia
URL: www.bogota.siggraph.org

KoREMA
Croatia
URL: www.korema.hr/

Red de Automatica de Cuba – RAC
Cuba
URL: www.cujae.edu.cu/clca/principal.htm

Czech ACM Chapter
URL: hpk.felk.cvut.cz

Czech Society for Cybernetics and Informatics
Czechoslovakia
URL: www.utia.cas.cz/

Danish Automation Society
Denmark
URL: www.dau.dk

Institution of Electrical Engineers, Estonia
Estonia
URL: www.elin.ttu.ee/EEU-Elec/OtherOrg/EIL/EiL.htm

Finnish Society of Automation
Finland
URL: www.automaatioseura.fi

Finland ACM SIGCHI – Special Interest Group on Computer–Human Interaction
URL: http://www.sigchi.fi

French ACM SIGAPP – Special Interest Group on Applied Computing
URL: http://sigappfr.acm.org/

French ACM SIGOPS – Special Interest Group on Operating Systems
URL: http://www.sigops-france.fr/

Société de l'Electricité, de l'Electronique et des Technologies de l'Informationet de la Communication
France
URL: www.see.asso.fr

VDI/VDE – Gesellschaft Mess- und Automatisierungstechnik
Germany
URL: www.vdi.de/gma/gma.htm

German ACM Chapter
URL: http://www.informatik.org

Technical Chamber of Greece
Greece
4 Karageorgi Servias Street
10248 Athens

Hungarian ACM Chapter
Arpad ter 2.
P.O. Box 652, Szeged 6720, Hungary

IFAC NMO of Hungary
Hungary
Computer and Automation Research Institute, HAS
P.O. Box 63
1518 Budapest, Hungary

Bangalore ACM Chapter
India
URL: http://bangaloreacmchapter.tripod.com

Instrumentation Society – HimII
Indonesia
Engineering Physics Dept., Jahlah Ganesha No. 10
Bandung 40132

Irish Systems and Control Committee

Ireland

National University of Ireland, Maynooth, Dept. of Electronic Engineering

Maynooth, Co. Kildare

Ireland ACM SIGCHI – Special Interest Group on Computer–Human Interaction

Dept. of Computer Science

Trinity College

Dublin 2, Ireland

gavin.dohety@cs.tcd.ie

Central Israel ACM SIGGRAPH – Special Interest Group on Computer Graphics and Interactive Techniques

URL: http://central-israel.siggraph.org

Israel Association of Automatic Control

Israel

URL: iaac.technion.ac.il/

CNR Commissione IFAC

Italy

DSI, Università di Firenze

Via Santa Marta, 3

50139 Florence, Italy

Italian ACM SIGCHI – Special Interest Group on Computer–Human Interaction

URL: http://hcilab.uniud.it/sigchi/

ITIA Istituto di Tecnologie Industriali e Automazione

Italy

Website: http://www.itia.cnr.it/

SIRI – Associazione Italiana di Robotica

20092 – Viale Fulvio Testi,128, Cinisello Balsamo (MI)

Italy

Science Council of Japan

Japan

URL: www.scj.go.jp

Japanese Association for Clinical Laboratory Automation (JACLA)

Y.U. Bldg. 6F Hongou 3-19-6, Bunkyou-ku

113-0033 Tokyo

Japan

ACM Japan Chapter

URL: http://www.acm-japan.org

ACM SIGMOD Japan – Special Interest Group on Management of Data

URL: http://www.sigmodj.org/

Korea ACM SIGCHI – Special Interest Group on Computer–Human Interaction

URL: http://www.hcikorea.org

Institute of Control, Robotics, and Systems – ICROS

Korea

URL: www.icros.org

IFAC NMO of Lithuania – LINO

Lithuania

Kaunas University of Technology

Donelaicio 73 44029 Kaunas

ETAI of Macedonia
Macedonia
Fac. of Electronic Engineering, SS. Cyril and Methodius University. Karpos 2 B.B., P.O. Box 574
1000 Skopje

Malaysia Center for Robotics and Industrial Automation
c/o Universiti Sains Malaysia (Perak Branch Campus)
Seri Iskandar, 31750 Tronoh
Perak
Malaysia

Asociacion de México de Control Automático – AMCA
Mexico
Dept. De Ingenieria Electrica, CINVESTAV
A.P. 14-740,
07000 Mexico D.F.

Mexico ACM SIGCHI – Special Interest Group on Computer–Human Interaction
Universidad Michoacana
Morelia, Mich., Mexico 58000

The International Association for Automation and Robotics in Construction (IAARC)
The Netherlands
Website: http://www.iaarc.org/

Royal Institution of Engineers
The Netherlands
URL: www.ingenieurs.net

Rotterdam ACM SIGGRAPH – Special Interest Group on Computer Graphics and Interactive Techniques
Netherlands
P.O. Box 1272
3000 BG Rotterdam
Rotterdam, The Netherlands
siggraph@wdka.hro.nl

New Zealand ACM SIGCHI – Special Interest Group on Computer–Human Interaction
URL: http://www.acm.org/chapters/sigchi_nz/

New Zealand Chapter of ACM URL: http://www.acm.org/chapters/acm_nz/

Norsk Forening for Automatisering – NFA
Norway
URL: www.nfaplassen.no/

Karachi ACM Chapter
Pakistan
URL: acm.szabist.edu.pk

POLSPAR
Poland
Warsaw University of Technology, Inst. of Radioelectronics
Nowowiejska 15/19
00 665 Warsaw

Poland ACM Chapter
URL: http://www.ii.uni.wroc.pl/~acm

Associação Portuguesa de Controlo Automatico – APCA
Portugal
URL: www.apca.pt/

Societatea Romana de Automatica si Informatica Tehnica – SRAIT
Romania
Automation and Computers Faculty
Splaiul Independentei 313, sector 6,
77206 Bucharest

Romania ACM SIGCHI – Special Interest Group on Computer–Human Interaction
URL: http://www.ici.ro/chi-romania/

Central Russia ACM SIGCHI – Special Interest Group on Computer–Human Interaction
URL: http://sigchi.ru/

Moscow ACM SIGMOD
Russia
URL: http://www.ipi.ac.ru/sigmod/

Moscow ACM SIGPLAN
Russia
URL: http://mossigplan.acm.org

National Committee of Automatic Control of Russia
Russia
Profsojuznaja ul. 65
Moscow 117806 GSP 312

Singapore Industrial Automation Association (SIAA)
Singapore
Website: www.esiaa.com/

Instrumentation and Control Society
Singapore
P.O. Box 2008, Tanglin Post Office
Singapore 912499

Singapore ACM Chapter
Blk. 998 Toa Payoh North, #07-18/19
Singapore, Singapore 318993
acm@inmeet.com.sg

Singapore ACM SIGGRAPH – Special Interest Group on Computer Graphics and Interactive Techniques
URL: http://www.siggraph.org.sg

Slovak IFAC NMO
Slovakia
Dept. of Process Control, CHTF STU, Radlinskeho 9
812 37 Bratislava

Automatic Control Society of Slovenia
Slovenia
University of Maribor
Fac. of Electrical Engineering & Computer Science
Smetanova ulica 17
2000 Maribor

Slovenia ACM Chapter
URL: http://www.acm-si.si/

South Africa Council of Automation and Computation
South Africa
URL: www.ee.up.ac.za/main/en/index

Comite Español de la IFAC

Spain

URL: www.cea-ifac.es

Kommitten Svenska IFAC

Sweden

Dept. of Electrical Engineering, Linkoping University

58183 Linkoping

Schweizerische Gesellschaft fuer Automatik – SGA

Switzerland

URL: www.sga-asspa.ch

Swiss ACM SIGCHI – Special Interest Group on Computer–Human Interaction

Switzerland

URL: http://www.usabilitynet.ch

Bangkok ACM SIGGRAPH – Special Interest Group on Computer Graphics and Interactive Techniques

Thailand

URL: http://bangkok.siggraph.org

Thailand ACM Chapter

Srisakdi Charmonman IT Ctr

Assumption University Banga

Bangna-Trad Km. 26, Bang Sao Thong

Samutprakarn, Thailand 10540

Turkish National Committee of Automatic Control

Istanbul Technical University, Mechanical Engineering Faculty

34437 Gumussuyu – Istanbul

Bilgi SIGACT Chapter – Special Interest Group on Algorithms and Computation Theory

Turkey

Istanbul Bilgi University

Dept. of Computer Science,

Kurtulusderesi Cad. No. 47

Dolapdere Sisli-Istanbul, Turkey 34440

Bilkent Turkey ACM SIGART – Special Interest Group on Artificial Intelligence

URL: http://goto.bilkent.edu.tr/sigart

Ukrainian ACM Chapter

Pecherskii uzviz 10, apt 42

Kyiv, Ukraine 01133

Ukrainian Association of Automatic Control – UAAC

Ukraine

40, Acad. Glushkov Prospekt

03680 Kiev 187

UK Automatic Control Council – UKACC

UK

URL: www.ukacc.org.uk

Manchester ACM SIGGRAPH – Special Interest Group on Computer Graphics and Interactive Techniques

UK

URL: http://manchester.siggraph.org

GAMBICA - Association for Instrumentation, Control, and Automation Ltd.

UK

Website: http://www.gambica.org.uk/

The Association for High Technology Distribution (AHTD)
USA
Website: http://www.ahtd.org/

Association for Laboratory Automation (ALA)
USA
Website: http://www.labautomation.org/

Control and Information Systems Integrators Association (CSiA)
USA
Website: http://www.controlsys.org/

Measurement Control and Automation Association (MCAA)
USA
Website: http://www.measure.org/

Process Equipment Manufacturers' Association (PEMA)
USA
Website: http://www.pemanet.org/

Robotic Industries Association (RIA)
USA
Website: http://www.roboticsonline.com/

Special Interest Group Design Automation (SIGDA)
USA
Website: http://www.sigda.org/

Fieldbus Foundation (FF)
USA
Website: http://www.fieldbus.org/

American Automatic Control Council (AACC)
USA
URL: www.a2c2.org

ISA – Instrumentation, Systems, and Automation Society
USA
67 Alexander Drive
P.O. Box 12277
Website: http://www.isa.org/

The Automation Federation
USA
pgoodson@automationfederation.org

OMAC – The Organization for Machine Automation and Control
USA
67 Alexander Drive,
Research Triangle Park, NC 27709, USA
E-Mail: info@OMAC.org

MCAA – Measurement Control and Automation Association
USA
URL: http://www.measure.org/

Modbus-IDA: The Architecture for Distributed Automation
USA
P.O. Box 628
Hopkinton, MA 01748
e-mail: info@modbus-ida.org

WBF – World Batch Forum
The Forum for Automation and Manufacturing Professionals
USA
URL: http://www.wbf.org/

Venezuelan Association of Automatic Control
Venezuela
Av. Alberto Carnevally, Fac. de Ingenieria
3er Piso, Ala Sur, La Hechicera
Merida 5101

Caracas ACM SIGGRAPH – Special Interest Group on Computer Graphics and Interactive Techniques
Venezuela
URL: caracas.siggraph.org

Vietnam Science and Technology Association for Automation
Vietnam
Inst. of Information Technology
18 Hoang Quoc Viet Street
Nghia Do Cau Giay, Hanoi

94.3 Automation Laboratories Around the World

Assembly Automation Laboratory (Tampere University of Technology)
http://pe.tut.fi/aal/

Automation Research Centre (University of Limerick)
http://www.ul.ie/~arc/

Automation & Robotics Laboratory (Aristotle University of Thessaloniki)
http://control.ee.auth.gr/index.html

Automation & Robotics Research Institute (The University of Texas at Alrington)
http://arri.uta.edu/

Automation Technology Laboratory (Helsinki University of Technology)
http://automation.tkk.fi/

Berkeley Automation Science Lab
http://automation.berkeley.edu/

Biologically Inspired Robotics Laboratory and Case Western Reserve University
http://biorobots.cwru.edu/

BUPAM – Bogazici University Pattern Analysis and Machine Vision Laboratory
http://www.bupam.boun.edu.tr/

Business Automation Laboratory (The Hong Kong Polytechnic University)
http://www.ise.polyu.edu.hk/centre&lab/mesc/cf403.htm

BYU Neural Networks and Machine Learning Laboratory
http://axon.cs.byu.edu/homepage.php

Caltech Robotics Group
http://robotics.caltech.edu/

Center for Automation Research (University of Maryland)
http://www.cfar.umd.edu/

Design Automation Laboratory (Arizona State University)
http://asudesign.eas.asu.edu/index.html

Design Automation Laboratory (University of Notre Dame)
http://www.nd.edu/~nddal/

Design Automation Laboratory (Seoul National University)
http://poppy.snu.ac.kr/english/

Distributed Information & Automation Laboratory (University of Cambridge)
http://www.ifm.eng.cam.ac.uk/automation/

e-Automation Laboratory (University of Liverpool)
http://www.liv.ac.uk/e-automation/

Electronics Design Automation Laboratory (University of Ljubljana)
http://fides.fe.uni-lj.si/edalab/welcome.html

Embry-Riddle Machine Vision Lab
http://vision.pr.erau.edu/

Georgia Tech Mobile Robot Lab
http://www.cc.gatech.edu/ai/robot-lab/

Human Automation Integration
http://human-factors.arc.nasa.gov/ihi/index.php

Industrial Automation Laboratory (UBC)
http://www.researchcentre.apsc.ubc.ca/ialweb/index.htm

Intelligent & Distributed Enterprise Automation Laboratory (Simon Fraser University)
http://www2.ensc.sfu.ca/research/IDEA/

ISMV – Image Science and Machine Vision Group, Oak Ridge National Laboratory
http://www.ornl.gov/sci/ismv/

Laboratory for Automation and Robotics (University of Patras)
http://www.lar.ee.upatras.gr/

Laboratory for Manufacturing and Sustainability (University of California Berkeley)
http://lma.berkeley.edu/

Laboratory of Mechanical Automation and Mechatronics (University of Twente)
http://www.wa.ctw.utwente.nl/

Laboratory of Process Control and Automation (Helsinki University of Technology)
http://kepo.hut.fi/

LPAC – Laboratory for Perception, Action, and Cognition, Penn State
http://vision.cse.psu.edu/

Machine Learning at Columbia University
http://www.cs.columbia.edu/learning/

Machine Learning and Inference at George Mason University
http://www.mli.gmu.edu/

Machine Vision Laboratory, (University of Ljubljana, Slovenia)
http://vision.fe.uni-lj.si/

Manufacturing Automation Laboratory
http://www.mech.ubc.ca/~mal/

Manufacturing Systems and Automation Laboratory (University Of Texas at San Antonio)
http://engineering.utsa.edu/~saygin/msa.html

Mechatronics & Automation Laboratory (National University of Singapore)
http://mchlab.ee.nus.edu.sg/

Medical Automation Research Center (University of Virginia)
http://marc.med.virginia.edu/index.php

MIT Field and Space Robotics Laboratory
http://robots.mit.edu/

MIT Humans & Automation Laboratory (MIT)
http://web.mit.edu/aeroastro/labs/halab/index.html

Old Dominion University Vision Lab
http://www.eng.odu.edu/visionlab/

Oxford University Computing Laboratory
http://www.comlab.ox.ac.uk/activities/machinelearning/

Production, Robotics, and Integration Software for Manufacturing and Management (PRISM)
http://cobweb.ecn.purdue.edu/~prism

Robotics & Automation (POSTECH)
http://tau.postech.ac.kr/index.html

Robotics & Automation Laboratory
http://www.mie.utoronto.ca/labs/ral/index.html

Robotics & Automation Laboratory (Florida International University)
http://www.eng.fiu.edu/mme/robotics/

Robotics Lab at UMass Lowell
http://robotics.cs.uml.edu/

Robotics and Vibration Control Laboratory at UNH
http://www.ece.unh.edu/robots/rbt_home.htm

Sheffield Hallam University, Microsystems and Machine Vision Laboratory
http://www.shu.ac.uk/research/meri/mmvl/

Stochastic Mechanical Systems & Automation Laboratory (University of Patras)
http://www.smsa.upatras.gr/index.php

The Bioimaging and Machine Vision Laboratory at the University of Maryland
http://www.bre.umd.edu/BioImageLab/LabwebPage/index.htm

The Center for Information and Systems Engineering (Boston University)
http://www.bu.edu/systems/research/automation/index.html

The Harvard Robotics Laboratory
http://hrl.harvard.edu/

The Machine Learning Lab – The Hebrew University
http://www.cs.huji.ac.il/labs/learning/

The Machine Vision Laboratory, Bristol Institute of Technology
http://www.uwe.ac.uk/cems/research/groups/mvl/index.html

The Robotics and Automation Laboratory (Michigan State University)
http://www.egr.msu.edu/ralab/Pages/Overview.htm

The Stanford Artificial Intelligence Laboratory (SAIL)
http://ai.stanford.edu/

UC Berkeley Robotics and Intelligent Machines Lab
http://robotics.eecs.berkeley.edu/wiki/pmwiki/pmwiki.php

UCI Robotics & Automation Laboratory (University of California, Irvine)
http://mae.eng.uci.edu/robotlab.html

UMass Machine Learning Laboratory
http://www-lrn.cs.umass.edu/

University of Cambridge Computational and Biological Learning Lab
http://learning.eng.cam.ac.uk/

University of London, Queen Mary Vision Laboratory
http://www.dcs.qmul.ac.uk/research/vision/

USC Robotics Research Lab
http://www-robotics.usc.edu/

Verification Automation Laboratory (National Taiwan University)
http://val.ee.ntu.edu.tw/e_index.htm

94.4 Automation Journals from Around the World

Table 94.1 Classification of automation journals by type and subject

Journal	Country	Language	Type	Frequency	Utility	Systems control	Software engineering	Robotics	Micro- and nanotechnology	Information technology and communication	Industrial applications	Electronic and embedded systems	Design, test, and instrumentation	Construction	Biology, chemistry, food, and pharmaceutical sciences	Assembly, manufacturing, and production	Artificial intelligence and autonomous agents
A F R I Liason	FR	FR		Q							x		x				
ACM Transactions on Design	US	EN	AS	Q								x	x				
Automation of Electronic Systems	CN	Z/E	AS									x					
Acta Automatica Sinica	US	EN	N	M		x						x					x
Advanced Manufacturing Technology	US	EN	AS	M												x	
Advance in Automation and Robotics	UK	EN	AS	I				x									
Assembly Automation	US	EN	AS	Q												x	
Automated Software Engineering	US	EN	AS	I			x										
Automated Software Engineering	UK	EN	AS	Q			x										
Automatica	UK	EN	AS	M		x											x
Automation	JP	JA	AS	M		x											
Automation and Remote Control	NL	EN	AS	M		x							x				
Automation in Construction	UK	EN	AS	BM										x			
Automation in Petro-Chemical Industry	CN	ZH	AS	BM											x		
Automation News	US	EN	AS	SM													
Automation of Electric Power Systems	CN	ZH	AS	SM													
Automation Technology in Practice International	DE	EN	AS	I													
Automatization de la Produccion	ES	ES	AS	M							x						
Autonomic and Autacoid Pharmacology	UK	EN	AS	Q	x										x		
Autonomous Agents and Multi-Agent Systems	US	EN	AS	BM						x	x		x			x	x
Autonomous Robots	US	EN	AS	BM	x		x	x								x	
BioControl	NL	EN	AS	BM		x		x							x		
Biocontrol Science and Technology	UK	EN	AS	10/		x									x		

Table 94.1 (cont.)

Journal																		
Biological Cybernetics	NL	EN	AS	M			x											x
Biomedical Microdevices	NL	EN	AS	BM			x					x					x	
Biosensors and Bioelectronics	US	EN	AS	M			x					x					x	
Componentes, Equipos y Sistemas De Automatica y Robotica	ES	ES						x						x				
Control and Automation	UK	EN	AS	BM			x		x									x
Control and Instruments In Chemical Industry	CN	ZH	AS	BM			x		x								x	
Control Engineering	US	EN	TP	M				x										
Design Automation Conference	US	EN	P	A				x										
Design Automation for Embedded Systems	DE	EN	AS	4/				x										
Electric Power Automation Equipment	CN	ZH	AS	M				x	x									x
IEEE International Conference on Robotics and Automation	US	EN	P	A								x						
IEEE Robotics and Automation Magazine	US	EN	AS	Q								x					x	
IEEE Transactions on Automatic Control	US	EN	AS	M														x
IEEE Transactions on Automation Science and Engineering	US	EN	AS	Q	x							x					x	
IEEE Transactions on Robotics	US	EN	AS	BM								x						
Information and Control	CN	ZH	AS	BM						x								
Instrumentation and Automation News	US	EN	TP	M			x				x							
Integrated Manufacturing Systems	UK	EN		6/			x				x	x						x
Internal Symposium on Industrial Robots, Proceedings	JP	JA	P	I														
International Journal of Automation and Computing	CN	EN	AS	Q		x								x				
International Journal of Robotics and Automaton	CA	EN	AS	4/							x	x					x	
International Symposium on Automation and Robotics in Construction Proceedings	JP	JA	P	I							x	x					x	
Journal of Automated Methods and Management in Chemistry	US	EN	AS	Q			x											
Journal of Automated Reasoning	DE	EN	AS	8/	x													
Journal of Rapid Methods and Automation in Microbiology	US	EN	AS	Q			x							x				
Laboratory Robotics and Automation	US	EN	AS	BM								x					x	
Pascale 34. Robotique, Automatique et Automatisation des Processus Industriels	FR	F/E	B	10/							x						x	
Proceedings of the Chinese Society of Universities for Electric Power System and Automation	CN	ZH	AS	BM														x

Part J | 94.4

Table 94.1 (cont.)

	Country of publication	Language of publication	Publication type	Publication frequency
Proceedings of the conference on Design, automation and test in Europe	US	EN	AS	I
Process Automation Instrumentation	CN	ZH	AS	M
Program: Electronic Library and Information Systems	UK	EN	AS	Q
Robotica	UK	EN	AS	BM
Robotics Today	US	EN	TP	Q
Series in Robotics and Automated Systems	SG	EN	MS	I
Utility Automation	US	EN	TP	M

Country of publication:
BE: Belgium; CA: Canada; CN: People's Republic of China; DE: Germany; ES: Spain; FR: France; JP: Japan; NL: The Netherlands;
SG: Singapore; UK: United Kingdom; US: United States of America
Language of publication:
EN: English; ES: Spanish; FR: French; JA: Japanese; ZH: Chinese;
Publication type:
AS: Academic/scholarly publication; B: Bibliography; MS: Monographic series; N: Newsletter; P: Proceedings; TP: Trade publication;
Publication frequency:
A: Annual; BM: Bimonthly; I: Irregular; M: Monthly; SM: Semimonthly; Q: Quarterly; x/: x/year

94.4.1 Automation–Related Journals

Although not dedicated to providing information automation, the following journals often carry information on automation development and research.

Australia
Australian Machinery and Production Engineering
Australian Robot Association Newsletter
Australian Welding Journal
Electrical Engineer
Journal of the Institution of Engineers
Metals of Australia
Robotic Age
Austria
Schweisstechnik
Diagram
Belgium
Manutention Mécanique et Automation
Revue M (Mecanique)
Bulgaria
Mashinostroene
Problemi Na Tekhnicheskata Kibernetika I Robotika
(Problems of Engineering Cybernetics and Robotics)
Teoretichna i Prilozhnaa Mekhanika
Canada
Canadian Institute for Advanced Research
in Artificial Intelligence and Robotics
Canadian Machinery and Metalworking
Robotronics Age Newsletter
Czechoslovakia
Slévárenstvi
Strojinrenska Vyroba
Techmeká Práce
Finland
Konepajamies
France
Axes Robotique
Energie Fluide, Hydraulique, Pneumatique Asservissements,
Fondeur Aujourd'hui
lubrification
Journal de la Robotique et Informatique
Kompass Professional Machines
L'Usine Nouvelle
Machine Moderne
Machine-outil
Manutention
Métaux Déformation
Nouvel Automatisme

Soudage et Techniques Connexes

Productique-Affaires

Germany

Automobil Industrie

Biological Cybernetics

Blech Rohre Profile

Der Plastverarbeiter mit Sonderdruck

Die Computerzeitung

Die Maschine

Die Wirtschaft

DVS-Berichte

Elektronik

Elektronikindustrie

Elektrotechnische Zeitschrift

Feingerätetechnik

Feinwerktechnik und Messtechnik

Fertigungstechnik und Betrieb

Flexible Automation

Fördern und Heben

Hebezeuge und Fördermittel

Industrie-Anzeiger

Kunststoffe

Lecture Notes in Computer Science

Maschinen-Anlagen + Verfahren

Maschinenbautechnik

Maschinenmarkt + Europa Industrie Revue

Maschine und Werkzeug

Messen-Steuern-Regeln

Metallverarbeitung

Praktiker

Production

Regelungstechnik

Regelungstechnische Praxis

Roboter

Schweissen und Schneiden

Schweisstechnik

Seewirtschaft

Siemens Energietechnik

Sozialistische Rationalisierung in der
Elektrotechnik/Elektronik

Technische Gemeinschaft

Technisches Zentralblatt für die Gesamte Technik

VEM-Elektro-Anlagenbahn

VDI

Werkstatt und Betrieb

Werkstattstechnik

Wissenschaftliche Zeitschrift der Technischen
Hochschule Ilmenau

Zeitschrift für Angewandte Mathematik und Mechanik

ZIS-Mitteilungen

Zeitschrift für Wirtschaftliche Fertigung

Hungary

Automatizálás

Bányászati és Kohászati Lapok Öntöde Az Orzágos Magyar

Bányászati és Koháztáti Egyesület Lapja

Gépgyartástechnológia

Ipargazdaság

Mérés és Automatika

Israel

Technologies: Israeli Journal for Advanced Technologies

HiTech

Information Week

Nir ve'Telem (Automation in Agriculture)

Automation and Technology

Galileo (Science, Automation and Robotics)

Italy

Automazione Oggi

La Tecnica Professionale

Maccine

Rivista de Meccanica

Tecniche Dell'Automazione e Robotica

Transport Industriali

Japan

Chemical Engineering

Chino Ido Robotto Shinpojumu Shiryo (Intelligent Robot
Symposium Proceedings)

Hydraulics and Pneumatics

International Conference on Advanced Robotics Proceedings

Japan Economic Journal

Japan Light Metal Welding

JARA Robot News

Journal of Robotics and Mechatronics

Journal of the Institute of Electrical Engineers of Japan

Journal of the Instrumentation Control Association

Journal of the Japan Welding Society

Kensetsu Robotto Shinpojumu Ronbunshu (Symposium
on Construction Robotics in Japan, Proceedings)

Mechanical Automation

Mechanical Design

Mechanical Engineering

Mitsubishi Denki Giho

Nihon Robotto Gakkai Robotto Shinpojumu Yokoshu
(Robotics Society of Japan, Preprints of Robotics Symposium)

Nihon Robotto Gakkai Gakaujutsu Loenkai Yokoshu
(Robotics Society of Japan, Preprints of the Meeting)

Nihon Robotto Gakkaishi (Robotics Society of Japan Journal)

Promoting Machine Industry in Japan

Robotikusu Mekatoronikusu Koenkai Koen Ronbunshu
(Annual Conference on Robotics and Mechatronics)
Robotpia
Robotto (Robot)
Robotto Sensa Shinpojumu Yokoshu (Robot Sensor
Symposium, Preprints)
Sangyoyo Robotto Riyo Gijutsu Koshukai Tekisuto (Text
of Lectures on Utilization Techniques of Industrial Robots)
Science of Machine
Specifications and Applications of Industrial Robots
in Japan: Manufacturing Fields
Specifications and Applications of Industrial Robots
in Japan: Non-manufacturing Fields
Toshiba Review
Transactions of the Institute of Electronics
and Communication Engineering
Transactions of the Society of Instruments
and Control Engineers
Welding Technique

Netherlands
Advanced Robotics
Advances in Design and Manufacturing
Artificial Intelligence
Autonomous Robots
Conference on Remote Systems Technology, Proceedings
Ingenieur
International Journal of Production Economics
Iron Age Metalworking International
Journal of Intelligent and Robotic Systems
Metalbewerking
Polytechnisch Tijdschrift
Robotics and Autonomous Systems

New Zealand
Automation and Control

People's Republic of China
Jiqiren/Robot
Precise Manufacturing and Automation
Control Engineering of China
Automation Panorama

Poland
Archives of Control Sciences
Automateka Kolejowa
Biuletyn Informacyjiny Institutu Maszyn Matematyeznyh
Mechanic
Przeglad Mechaniczny
Przeglad Spawalnietwa
Wiadomosci Electrotechnieszne
Zeszyty Naukowe

Romania
Constructia de Masini
International Journal of Informatics
Russia and the Former Soviet Republics
Avtomatika i Telemekhanika
Avtomatizatsiya Proizvodstvennykh Protssov
v Mashinostroenii i Priborostroenii
Avtomatizatsiya Tekhnologicheskikh Protsessov
Avtomatizirovannyy Elekropivod
Elektrotekhnika
Vestnik Mashinostroeniya
Voprosy Dinamiki i Prochnosti
Izvestiya Akademii Nauk SSSR
Isvestiya Vysshikh Uchebnykh Zavedeniy
Izvestiya Leningradskogo
Kuznechno-stampovochnoye Proizvodstvo
Mashinovedeniye
Mashinostroitel
Mekhanizatsiya i Avtomatizatsiya Proizvdstva
Mekhanizatsiya i Elektrifikatsiya Sel'skogo Khozyaystva
Priborostroeniye
Promyshlennyy Transport
Svarochnoye Proizvodstvo
Stanki i Instrument
Stroitel'nyye i Dorohnyye Mashiny
Sudostroeniye
Turdy Vsesoyuznogo Nauchnoissledovatel'skogo Instituta
Ugol'nogo Mashinostroeniya
Trudy Leningradskogo Politekhnicheskogo Instituta
Trudy Moskovskogo Energeticheskogo Instituta

Spain
Automatica and Robotica
Automatizacion Integrada y Revista de Robotica
Resulación y Mando Automatico

Sweden
IFR Robotics Newsletter
World Robot Statistics

Switzerland
CIRP Annals
Elektroniker
Management Zeitschrift Ind. Organis.
Schweisstechnik/Soudure
Schweizerische Technische Zeitschrift
Technica
Technische Rundschau
Zeitschrift Schweisstechnik

United Kingdom
Assembly Engineering
Automatic I.D. News Europe

Automotive Engineer
Automotive Engineering
British Foundryman
Business Ratio Report: Industrial Robots
Control and Instrumentation
Design Eng. Materials and Components
Electrical Review
Engineer
Foundry Trade Journal
Hydraulic Pneumatic Mechanical Power
Industrial News
Industrial Robot
International Journal of Man–Machine Studies
International Journal of Production Research
Machinery and Production Engineering
Manufacturing Engineer
Materials Handling News
Mechanism and Machine Theory
Metals and Materials
Metalworking Production
New Electronica
New Scientist
Pattern Recognition
Plastics in Engineering Robot News International
Robotics and Computer-Integrated Manufacturing
Sensor Review
Service Robot
Sheet Metals Industries
Welding and Metal Fabrication

USA
AI Today
American Machinist
American Metalmarket
ASHRAE (American Society of Heating, Refrigerating, and
Air-Conditioning Engineers) Transactions
ASME Transactions on Dynamics, Measurement, and Control
ASPDAC (Design Automation Conference Asia
and South Pacific)
Assembly Engineering
Automatic Machining
Automotive Design and Production
Automotive Industries
Automotive Manufacturing and Production
Automotive News
Bibliography of Robotic and Technical Resources
Biomedical Engineering

Compressed Air
Computer Graphic and Image Processing
Design News
Electronics
Embedded Systems Programming
Hydraulics and Pneumatics
IEEE Spectrum
IEEE Transactions on Industrial Electronics
and Control Instrumentation
IEEE Transactions on Power Apparatus and Systems
IEEE Transactions on Systems, Man, and Cybernetics
IIE Transactions
Industrial Engineering
Industrial Robots International
Information and Control
International Journal of Computer and Information Science
International Journal of Robotics Research
International Robotics Product Database
Iron Age
Journal of Machinery Manufacture and Reliability
Journal of Manufacturing Systems
Journal of Robotic Systems
Machine and Tool Blue Book
Manufacturing Engineer
Material Handling Engineering
Mechanical Engineering
Military Robotics Newsletter
Modern Material Handling
Plating and Surface Finishing
Precision Machinery
Product Engineering
Production
Progress in Robotics and Intelligent Systems
RIA Quarterly Statistics Report-Robotics
Robot Explorer
Robot Times
Robotics and Expert Systems
Robotics and Manufacturing
Robotics Review
Robotics World
Tooling and Production
Welding Design and Fabrication
Welding Engineer
Welding Journal

Yugoslavia
Automatika

Acknowledgements

A.3 Automation: What It Means to Us Around the World
by Shimon Y. Nof

I received with appreciation help in preparing and reviewing this chapter from my graduate students at our PRISM Center, particularly Xin W. Chen, Juan Diego Velásquez, Sang Won Yoon, and Hoo Sang Ko, and my colleagues on our PRISM Global Research Network, particularly Professors Chin Yin Huang, José A. Ceroni, and Yael Edan. This chapter was written in part during a visit to the Technion, Israel Institute of Technology in Haifa, where I was inspired by the automation collection at the great Industrial Engineering and Management Library. Patient and knowledgeable advice by librarian Coral Navon at the Technion Elyachar Library is also acknowledged.

A.7 Impacts of Automation on Precision
by Alkan Donmez, Johannes A. Soons

This contribution was prepared by employees of the US Government as part of their official duties and is, therefore, a work of the US Government and not subject to copyright.

B.14 Artificial Intelligence and Automation
by Dana S. Nau

This work has been supported in part by DARPA's Transfer Learning and Integrated Learning programs, NSF grant IIS0412812, AFOSR grant FA95500610405 and NAVAIR contract N6133906C0149. The opinions in this paper are those of the authors and do not necessarily reflect the opinions of the funders.

B.16 Automation of Mobility and Navigation
by Anibal Ollero, Ángel R. Castaño

The work described in this chapter has been supported by various projects including the European projects COMETS (IST-2001-34304) and AWARE (IST-2006-33579) and the Spanish projects ROBAIR (DPI2008-03847), AEROSENS (DPI2005-02293), and SIRE (2006, TEP-375).

C.19 Mechatronic Systems – A Short Introduction
by Rolf Isermann

This chapter is revised from Isermann, R. (2008), Mechatronische Systeme. 2. Auflage. Springer-Verlag, Berlin, and Isermann, R. (2005), Mechatronic Systems. Springer-Verlag, London.

C.26 Collaborative Human–Automation Decision Making
by Mary L. Cummings, Sylvain Bruni

This research was sponsored by the Office of Naval Research Long-Range Navy and Marine Corps Science and Technology Research Opportunity (BAAs 04-001 and 07-001). The views and conclusions presented in this chapter are those of the authors and do not represent an official opinion, expressed or implied, of the Office of Naval Research. Special thanks to Jessica J. Marquez, Amy Brzezinski, Carl Nehme, and Yves Boussemart who were co-creators of HACT and for the reviews by Prof. John Lee from the University of Iowa, Prof. Gilles Coppin from the Ecole Nationale Supérieure des Télécommunications de Bretagne, France, and Prof. Duncan Campbell from Queensland University of Technology, Australia.

C.27 Teleoperation
by Luis Basañez, Raúl Suárez

The authors would like to thank the members of the Robotics Division of the IOC, and especially Emmanuel Nuño, Henry Portilla, Máximo Roa, Adolfo Rodríguez, and Carlos Rosales, for their great collaboration in the preparation of this chapter.

This work has been partially supported by the CI-CYT projects DPI2005-00112 and DPI2007-63665.

C.28 Distributed Agent Software for Automation
by Francisco P. Maturana, Dan L. Carnahan, Kenwood H. Hall

We would like to thank AvPro Incorporated and GKN Aerospace for their technical support on the understanding of composite curing technology.

C.30 Automating Errors and Conflicts Prognostics and Prevention
by Xin W. Chen, Shimon Y. Nof

This chapter has been developed with the support of the PRISM Center and the School of Industrial Engineering at Purdue University. Partial support by General Motors Research project on Design of Active Middleware for Error and Conflict Detection is also acknowledged.

D.35 Machining Lines Automation
by Xavier Delorme, Alexandre Dolgui,
Mohamed Essafi, Laurent Linxe, Damien Poyard

The industrial case plus the illustrations of the parts and machining units presented in this chapter were provided by the society PCI/SCEMM (http://www.pci.fr/). The authors thank Dr. Olga Guschinskaya for her advice and discussions and Chris Yukna for his help in English.

E.40 Economic Rationalization of Automation Projects
by José A. Ceroni

The author gratefully acknowledges the significant and valuable contributions of Y. Hasegawa, J. Mill, G. T. Stevens, B. Huff, and A. Presley for their previous work on this topic.

E.43 Product Lifecycle Management and Embedded Information Devices
by Dimitris Kiritsis

The content of this chapter is based on reports and publications of the PROMISE project (EU FP6 507100 and IMS 01008; www.promise-plm.com). The author expresses his deep gratitude to all PROMISE partners.

E.47 Automation and Ethics
by Srinivasan Ramaswamy, Hemant Joshi

The authors wish to thank the anonymous reviewers and editors for significant help and advice in writing this chapter. Parts of this work were supported in part by the National Science Foundation (under Grant Nos. CNS-0619069, EPS-0701890 and OISE 0729792), NASA EPSCoR Arkansas Space Grant Consortium (#UALR 16804) and Acxiom Corporation (#281539). Any opinions, findings, and conclusions or recommendations expressed in this chapter are those of the author(s) and do not necessarily reflect the views of the funding agencies. Finally, the authors would like to thank Sithu Sudarsan, currently a visiting researcher at the FDA's Center for Devices and Radiological Health, for engaging discussions on this subject.

F.49 Digital Manufacturing and RFID-Based Automation
by Wing B. Lee, Benny C.F. Cheung, Siu K. Kwok

The authors would like to express their sincere thanks to the Research Committee of The Hong Kong Polytechnic University (project code G-YE31) and the Innovation and Technology Commission (ITC) of The Government of the Hong Kong Special Administrative Region (HKSAR) (project code GHS/091/04) for financial and technical support of the research work.

F.51 Aircraft Manufacturing and Assembly
by Branko Sarh, James Buttrick, Clayton Munk,
Richard Bossi

Ron Gill, Michael Watts, Larry Hefti, and Robert A. Kisch also contributed to this chapter. Bruce Harman of Boeing served as chapter editor.

F.54 Production, Supply, Logistics and Distribution
by Rodrigo Cruz Di Palma,
Manuel Scavarda Basaldúa

Technical and theoretical contributions by Sang-Won Yoon from the PRISM Center at Purdue University are acknowledged. Additional support and contributions for this chapter were provided by Guillermo Pinochet, Cristian Hernandez, and Craig Downing.

G.67 Air Transportation System Automation
by Satish C. Mohleji, Dean F. Lamiano,
Sebastian V. Massimini

The authors gratefully acknowledge valued support from Dr. D. Bhadra, W. W. Cooper, J. C. Celio, Dr. D. B. Kirk, R. O. Lejeune, E. G. Newberger, W. J. Swedish, G. H. Solomos, and Dr. A. D. Zeitlin of MITRE/CAASD, and W. Flathers of United Airlines.

Thanks to A. Signore for preparing the manuscript. The authors thank Dr. C. R. Wanke for peer-reviewing the content presented in this chapter.

H.75 Automatic Control in Systems Biology
by Henry Mirsky, Jörg Stelling, Rudiyanto Gunawan, Neda Bagheri, Stephanie R. Taylor, Eric Kwei, Jason E. Shoemaker, Francis J. Doyle III

The authors gratefully acknowledge significant editorial help in preparing this chapter by Xin W. Chen from the PRISM Center at Purdue University. The authors also gratefully acknowledge financial support from the Institute for Collaborative Biotechnologies (ICB) through the United States Army Research Office (ARO) (Grants DAAD19-03-D-0004 and W119NF-07-1-0279), from the National Institutes of Health (NIH) (Grants EB007511 and GM078993), from the Integrative Graduate Education Research Traineeship National Science Foundation (IGERT NSF) (Grant DGE02-21715), from the Defense Advanced Research Projects Agency (DARPA) BioComp and BioSpice Programs, from the Research Participation Program between the United States Department of Energy (DOE) and Air Force Research Laboratory, Human Effectiveness Directorate (AFRL/HEP), and from the University of California, Board of Regents.

H.79 Rotary Heart Assist Devices
by Marwan A. Simaan

This research was supported in part by NSF under grants ECS-0300097 and ECCS-0701365 and NIH/BHLBI under contract 1R43HL66656-01. The contributions of Drs. James F. Antaki and J. Robert Boston and our students Leonardo Baloa, Darryl Breitenstein, Shaohui Chen, Antonio Ferreira, and Yih-Chong Yu are gratefully acknowledged.

H.80 Medical Informatics
by Chin-Yin Huang

The author would like to thank Chen Ching General Hospital and DVNET Technologies Co. for providing the precious pictures in this chapter. Research projects 93-2218-E-029-001 and 94-2213-E-029-013 are also gratefully acknowledged for the knowledge discovery on breast cancer patient records, supported by the National Science Council, Taiwan.

H.81 Nanoelectronic-Based Detection for Biology and Medicine
by Samir M. Iqbal, Rashid Bashir

The authors would like to acknowledge and thank Oguz H. Elibol and Bala Murali K. Vekatesan for help with literature collection, and Salman B. Inayat and Syed H. Shah for manuscript review. S.M. Iqbal acknowledges support from University of Texas at Arlington Research Enhancement Program.

I.85 Automation in Education/Learning Systems
by Kazuyoshi Ishii, Kinnya Tamaki

The case studies were supported by The Research Center for e-Learning Professional Competency (eLPCO) in the Research Institute of Aoyama Gakuin University and The Research Center for Social and Industrial Management Systems of Kanazawa Institute of Technology. We would like to thank the staff of the organization for their cooperation with these studies. Also, one of the cases was supported in part by a grant-in-aid for Scientific Research of the Japanese Ministry of Education, Culture, Sports, Science, and Technology under contract number C19510160 (2007–2008).

I.88 Collaborative e-Work, e-Business, and e-Service
by Juan D. Velásquez, Shimon Y. Nof

Research reported in this chapter has been developed at the PRISM Center of Purdue University with NSF, Indiana 21st Century fund for Science and Technology, and substantial industry support.

Special thanks are due to: our colleagues, our visiting scholars, and the many students at the PRISM Lab and the PRISM Global Research Network worldwide; and the International Foundation of Production research (IFPR) and the International Federation of Automatic Control (IFAC) Committee CC5 for Manufacturing and Logistics Systems, many of whom are also chapter co-authors in this Handbook. They have all collaborated with us to develop, over many years, e-work and collaborative control theory knowledge.

I.92 e-Government
by Dieter Rombach, Petra Steffens

The authors would like to thank the following persons for valuable comments, enlightening discussion,

Acknowl.

and help in producing this chapter: Prof. Dr. Frank Bomarius, Savitha Chennagiri, Ines Grützner, Thomas Jeswein, Dr. Jürgen Münch, Dr. Reinhard Schwarz, and Dr. Michael Tschichholz.

I.93 Collaborative Analytics for Astrophysics Explorations
by Cecilia R. Aragon

The author would like to thank the Sunfall developers and scientists of the SNfactory collaboration, in particular Greg Aldering, Stephen Bailey, Karl Runge, Sarah Poon, and Rollin Thomas. This work was supported in part by the Director, Office of Science, Office of Advanced Scientific Computing Research, of the US Department of Energy under Contract No. DE-AC02-05CH11231, and by the Director, Office of Science, Office of High Energy Physics, of the US Department of Energy under Contract No. DE-FG02-92ER40704, and by a grant from the Gordon & Betty Moore Foundation. This research used resources of the National Energy Research Scientific Computing Center, which is supported by the Office of Science of the US Department of Energy under Contract No. DE-AC02-05CH11231.

J.94 Automation Statistics
by Juan D. Velásquez, Xin W. Chen, Sang Won Yoon, Hoo Sang Ko

The authors wish to thank all the many colleagues and sources who helped in developing this chapter. At the same time, while we did our best to be comprehensive, we apologize if we mistakenly missed any relevant data. We will be able to include it in future editions.

About the Authors

Nicoletta Adamo-Villani

Purdue University
Computer Graphics Technology
West Lafayette, IN, USA
nadamovi@purdue.edu

Chapter D.37

Nicoletta Adamo-Villani is Associate Professor of Computer Graphics Technology at Purdue University. She received her MS in Architecture from University of Florence, Italy in 1993 and she is a certified 3-D modeller/animator for Autodesk. Adamo-Villani is an award-winning animator and creator of 2-D/3-D films that aired on national television. Her research interests focus on the application of 3-D animation to education, HCC (human–computer communication), and visualization.

Panos J. Antsaklis

University of Notre Dame
Department of Electrical Engineering
Notre Dame, IN, USA
antsaklis.1@nd.edu

Chapter B.13

Panos J. Antsaklis is the Brosey Professor of Electrical Engineering and Concurrent Professor of Computer Science and Engineering at the University of Notre Dame. His research interests are in control and automation, particularly in networked control and hybrid systems emphasizing interdisciplinary research in control, computing and communication networks. He has authored books on Linear Systems, Discrete Event, Hybrid and Networked Control Systems.

Cecilia R. Aragon

Lawrence Berkeley National Laboratory
Computational Research Division
Berkeley, CA, USA
CRAragon@lbl.gov

Chapter I.93

Cecilia Aragon is a Staff Computer Scientist in the Computational Research Division at Lawrence Berkeley National Laboratory. Her research interests include visualization, image processing, visual analytics, computer-supported cooperative work, and human–computer interaction. She received her PhD in computer science from the University of California, Berkeley and her BS in mathematics from the California Institute of Technology. She is a Senior Member of the IEEE.

Neda Bagheri

Massachusetts Institute of Technology (MIT)
Department of Biological Engineering
Cambridge, MA, USA
nbagheri@mit.edu

Chapter H.75

Neda Bagheri earned her MS/PhD (2007) in Electrical Engineering from the University of California Santa Barbara with emphasis in Control Theory and Systems Biology. As a postdoctoral researcher at MIT, she currently investigates combinatorial cancer therapy through dynamical systems modeling integrated tightly with quantitative experiments. Her primary research interest focuses on addressing challenges in medicine and biology using principles from control theory.

Greg Baiden

Laurentian University
School of Engineering
Sudbury, ON, Canada
gbaiden@laurentian.ca

Chapter F.57

Dr. Baiden is a Mining Professor specializing in robotics and mining automation in the School of Engineering at Laurentian University. Dr. Baiden lead mining research at Inco Ltd prior to joining Laurentian University and creating Penguin Automated Systems Inc. His research investigates robotics and automation for many types of systems such as underground, surface, underwater and in space. His recent work includes the newly patented high bandwidth communication systems.

Parasuram Balasubramanian

Theme Work Analytics Pvt. Ltd.
Bangalore, India
balasubp@gmail.com

Chapter G.71

Parasuram Balasubramanian holds engineering and management degrees from IIT, Madras and a doctorate from Purdue University. He has been with the Information Technology industry in India, Jamaica and USA and has been a featured speaker in many conventions. He specializes in business analytics and plays a proactive role to bridge industry and academia on the research front.

P. Pat Banerjee

University of Illinois
Department of Mechanical
and Industrial Engineering
Chicago, IL, USA
banerjee@uic.edu

Chapter B.15

P. Pat Banerjee has a B.Tech from the Indian Institute of Technology, a PhD from Purdue University and is a Fellow of the ASME. His current research interests include virtual reality and haptics applications. Currently he is an Associate Editor of IEEE Transactions for Information Technology in Biomedicine. With over 100 refereed publications, he has served as a department editor of IIE Transactions, and as an associate editor of IEEE Transactions on Robotics and Automation.

Ruth Bars

Budapest University of Technology
and Economics
Department of Automation and Applied
Informatics
Budapest, Hungary
bars@aut.bme.hu

Chapter B.10

Ruth Bars graduated at the Electrical Engineering Faculty of the Technical University of Budapest, Hungary. She has gained the candidate of sciences degree from the Hungarian Academy of Sciences in 1992 and the PhD degree based on research in predictive control. Currently she is an Associate Professor in the Department of Automation and Applied Informatics at the Budapest University of Technology and Economics. Her research interests are in predictive control and in developing new ways of control education.

Luis Basañez

Technical University of Catalonia (UPC)
Institute of Industrial and Control
Engineering (IOC)
Barcelona, Spain
luis.basanez@upc.edu

Chapter C.27

Luis Basañez is full Professor of System Engineering and Automatic Control at the Technical University of Catalonia, head of the Robotics Division of the Institute of Industrial and Control Engineering, Spanish delegate at the International Federation of Robotics, and fellow and Council member of the International Federation of Automatic Control. His present research interest includes teleoperation, multirobot coordination, sensor integration and active perception.

Rashid Bashir

University of Illinois
at Urbana-Champaign
Department of Electrical and Computer
Engineering and Bioengineering
Urbana, IL, USA
rbashir@uiuc.edu

Chapter H.81

Rashid Bashir is the Abel Bliss Professor of Electrical and Computer Engineering and Bioengineering and Director of the Micro- and Nanotechnology Laboratory at the University of Illinois, Urbana-Champaign. His research interests include BioMEMS, Lab on a chip, nano-biotechnology, and interfacing biology and engineering from molecular to tissue scale.

Wilhelm Bauer

Fraunhofer-Institute
for Industrial Engineering IAO
Corporate Development
and Work Design
Stuttgart, Germany
wilhelm.bauer@iao.fraunhofer.de

Chapter D.34

Wilhelm Bauer graduated in Industrial Engineering from the University of Stuttgart and received the Dr.-Ing. degree from the same University. As a Scientific Director he is heading the business area Corporate Development and Work Design at IAO and IAT. His research activities include work sciences, knowledge work, virtual work spaces, office innovations and change management. He is lecturer for Work Design at the Universities of Hannover and Stuttgart.

Gary R. Bertoline

Purdue University
Computer Graphics Technology
West Lafayette, IN, USA
Bertoline@purdue.edu

Chapter D.37

Dr. Gary R. Bertoline is a Distinguished Professor of Computer Graphics Technology at Purdue University. Gary's research interests are in scientific visualization, interactive immersive environments, distributed and grid computing and STEM education. He has authored numerous papers on engineering and computer graphics, computer-aided design, and visualization research. He has authored and co-authored seven textbooks in the areas of computer-aided design and engineering design graphics.

Christopher Bissell

The Open University
Department of Communication
and Systems
Milton Keynes, UK
c.c.bissell@open.ac.uk

Chapter A.4

Christopher Bissell graduated from Jesus College, Cambridge in 1974 and obtained his PhD from the Open University in 1993, where he has been employed since 1980. He has written much distance teaching material on telecommunications, control engineering, digital media and other topics. His major research interests are the history of technology and engineering education.

Richard Bossi

The Boeing Company
Renton, WA, USA
richard.h.bossi@boeing.com

Chapter F.51

Richard H. Bossi received his Bachelor's degree in physics from Seattle University in 1971 and his PhD from Oregon State University in Nuclear Engineering in 1977. He is a Senior Technical Fellow for Nondestructive Evaluation (NDE) on the physics staff at The Boeing Company. His primary research is in the application NDE technology for materials characterization.

Martin Braun

Fraunhofer-Institute
for Industrial Engineering IAO
Human Factors Engineering
Stuttgart, Germany
martin.braun@iao.fraunhofer.de

Chapter D.34

Martin Braun received his University Diploma in Industrial Engineering. While being Teaching Assistant at the University of Stuttgart he graduated in OHS and earned a doctorate in Human Factors Management. Since 1999 he is working as a project manager at Fraunhofer-Institute for Industrial Engineering (IAO) in the field of human factors engineering and work design. His main activities in applied research are human performance, mental work and occupational health. Martin Braun is lecturer at the University of Stuttgart and has authored more than 100 publications.

Sylvain Bruni

Aptima, Inc.
Woburn, MA, USA
sbruni@aptima.com

Chapter C.26

Sylvain Bruni is a Human Systems Engineer at Aptima, where he provides expertise in human-automation interaction, interface design, and the statistical design of experiment. His research also focuses on designing and testing collaborative decision-support systems, specifically in military command and control environments. He earned a SM in Aeronautics/Astronautics from MIT and a Diplôme d'Ingénieur from Supélec (France). He is currently pursuing a PhD at MIT, in the Humans and Automation Laboratory.

James Buttrick

The Boeing Company
BCA – Materials & Process Technology
Seattle, WA, USA
james.n.buttrick@boeing.com

Chapter F.51

James is a Technical Fellow at Boeing Commercials' Manufacturing&Process Technology organization. He specializes in the development of automated manufacturing equipment and processes for aircraft production. He received his BS in Marine Engineering from the United States Merchant Marine Academy, and his MS in Mechanical Engineering from the University of Washington.

Authors

Darwin G. Caldwell

Istituto Italiano Di Tecnologia
Department of Advanced Robotics
Genova, Italy
Darwin.Caldwell@iit.it

Chapter F.60

Darwin G. Caldwell (BSc 1986, PhD 1990, University of Hull) is a Director at the Italian Institute of Technology, Genoa, Italy, and an Honorary Professor at the Universities of Sheffield, Manchester and Wales, Bangor. His research interests include innovative actuators and sensors, haptic feedback, force augmentation exoskeletons, dexterous manipulators, humanoid (iCub) and quadrapedal robots, biomimetic systems, rehabilitation robotics, telepresence, automation for the food industry.

Brian Carlisle

Precise Automation
Auburn, CA, USA
brian.carlisle@preciseautomation.com

Chapter F.50

Mr. Brian Carlisle is the CEO of Precise Automation, which builds assembly robots and controls. Formerly he was CEO of Adept Technology from 1983 to 2003 known for assembly automation for electronics, automotive and telecommunications products. In the late 1970s Carlisle was a project manager for the development of the PUMA robot at Unimation. Carlisle was President of the United States Robotic Industries Association for 3 years.

Dan L. Carnahan

Rockwell Automation
Department of Advanced Technology
Mayfield Heights, OH, USA
dlcarnahan@ra.rockwell.com

Chapter C.28

Mr. Carnahan is a program manager for Rockwell Automation, Advanced Technology. He has over 30 years experience working in industrial controls and automation, having previously worked in different roles ranging from development engineering to project management. He has a BSEE from Ohio State University (1972) and is a registered professional engineer in the State of Ohio, a member of the National Society for Professional Engineers, IEEE, ISA, SAE, and SEMI.

Ángel R. Castaño

Universidad de Sevilla
Departamento de Ingeniería de Sistemas
y Automática
Sevilla, Spain
castano@us.es

Chapter B.16

Ángel R. Castaño received the MEng degree in Telecommunications Engineering and the PhD degree in Automation, Robotics and Telematics from the University of Seville. He has participated in more than 20 projects, including 5 projects funded by the European Commission. His research interests are mainly in multirobot and intelligent transportation systems. He is currently an Assistant Professor at the University of Seville.

Daniel Castro-Lacouture

Georgia Institute of Technology
Department of Building Construction
Atlanta, GA, USA
dcastro6@gatech.edu

Chapter G.61

Daniel Castro-Lacouture (BSc Civil Engineering 1994, Universidad de Los Andes, MSc Construction Management 1999, University of Reading, PhD Construction Engineering and Management 2003, Purdue) is Associate Professor of Building Construction at Georgia Institute of Technology. His current research centres on construction technology innovation, sustainability and automation. He is a registered professional engineer and a member of ASCE and IAARC.

Enrique Castro-Leon

JF5-103, Intel Corporation
Hillsboro, OR, USA
Enrique.G.Castro-Leon@intel.com

Chapter C.24

Enrique Castro-Leon is an Enterprise Architect and Technology Strategist with Intel Digital Enterprise Group. He holds a PhD degree in Electrical Engineering and MS degrees in Electrical Engineering and Computer Science from Purdue University. He is the principal author of the book *The Business Value of Virtual Service Oriented Grids*. His research interests include large scale software integration and data centre power management strategy. He has served as a consultant for corporate and governmental organizations as well as NGOs.

José A. Ceroni

Pontifica Universidad Católica
de Valparaíso
School of Industrial Engineering
Valparaiso, Chile
jceroni@ucv.cl

Chapter E.40

José A. Ceroni graduated as an Industrial Engineer from Pontifical Catholic University of Valparaiso, Chile and received his Master of Science and PhD in Industrial Engineering from Purdue University, Indiana, USA. His research interests include collaborative production and control, industrial robotics systems, collaborative robotics agents, collaborative control in logistics systems. He is member of the Board of the International Federation for Production Research, and a member of IFAC and IEEE.

Deming Chen

University of Illinois, Urbana–Champaign
Electrical and Computer Engineering (ECE)
Urbana, IL, USA
dchen@illinois.edu

Chapter D.38

Deming Chen obtained his MS and PhD from the Computer Science Department of University of California at Los Angeles. He is currently an Assistant Professor in the ECE Department of University of Illinois, Urbana-Champaign. His research interests include CAD for FPGA, nanosystems design and synthesis, microprocessor design under process/parameter variation, and reconfigurable computing. He received the Arnold O. Beckman Research Award, the NSF CAREER Award, and a Best Paper Award from ASPDAC.

Heping Chen

ABB Inc.
US Corporate Research Center
Windsor, CT, USA
heping.chen@us.abb.com

Chapter F.53

Heping Chen received his PhD in Electrical and Computer Engineering from Michigan State University in 2004. He is currently a project manager and research scientist at US Corporate Research Center, ABB Inc. His research interests include nanomanufacturing automation and nanorobotics, industrial robot control based on sensor integration, robot path planning and machine vision. He is an IEEE Senior member.

Xin W. Chen

Purdue University
PRISM Center and School of Industrial
Engineering
West Lafayette, IN, USA
chen144@purdue.edu

Chapters C.30, J.94

Xin W. Chen is a PhD Candidate in the School of Industrial Engineering at Purdue University. He received the MS degree in Industrial Engineering and BS degree in Mechanical Engineering from Purdue University and Shanghai Jiaotong University, respectively. His research interests cover several related topics in the area of conflict and error prognostics and prevention, production/service optimization, and decision analysis.

Benny C.F. Cheung

The Hong Kong Polytechnic University
Department of Industrial and Systems
Engineering
Kowloon, Hong Kong
mfbenny@inet.polyu.edu.hk

Chapter F.49

Benny C.F. Cheung obtained his BEng, MPhil and PhD degrees in Manufacturing Engineering from The Hong Kong Polytechnic University in 1993, 1996 and 2000, respectively. Currently, he is an Associate Professor and an Associate Director of the Knowledge Management Research Centre in Department of Industrial and Systems Engineering of The Hong Kong Polytechnic University. His research interest includes precision engineering, knowledge and technology management, artificial intelligence and logistics systems.

Jaewoo Chung

Kyungpook National University
School of Business Administration
Daegu, South Korea
jaewooch@gmail.com

Chapter F.55

Jaewoo Chung received the PhD degree from the School of Industrial Engineering at Purdue University in 2008. He developed various production systems and facilities layouts during his work for the LCD/Semiconductor division at Samsung Electronics for about a decade. He is currently an assistant professor at Kyungpook National University, South Korea. His areas of interests are production systems and combinatorial optimization.

Rodrigo J. Cruz Di Palma

Kimberly Clark, Latin American Operations
San Juan, Puerto Rico
Rodrigo.J.Cruz@kcc.com

Chapter F.54

Rodrigo J. Cruz Di Palma graduated as an Industrial Engineer from the University of Alabama at Tuscaloosa and obtained an MBA from EADA University in Barcelona, Spain. He is currently Supply Chain Manager for Puerto Rico, and previously led the Supply and Operations Research Team for Kimberly Clark Latin-America, focusing on the development of optimization tools for the execution of business processes.

Mary L. Cummings

Massachusetts Institute of Technology
Department of Aeronautics
and Astronautics
Cambridge, MA, USA
missyc@mit.edu

Chapter C.26

Dr. Cummings received her BS in Mathematics from the US Naval Academy (1988), her MS in Space Systems Engineering from the Naval Postgraduate School (1994), and her PhD in Systems Engineering from the University of Virginia (2003). Her research interests include human supervisory control, human-unmanned-vehicle interaction and decision-making, direct-perception decision support, and the ethical and social impact of technology.

Christian Dannegger

Rottweil, Germany
cd@3a-solutions.com

Chapter C.23

Christian Dannegger has served Whitestein Technologies from 2003 to 2009 as Vice President of Logistics and Control Systems. Prior to that he served as CTO of living systems AG, which he co-founded in 1996. From 1988 to 1995 Christian developed ERP systems at Bäurer AG, where he headed the technical product development. Recognized as a visionary business-driven technologist. He holds a Master's degree in Computer Science from Furtwangen Polytechnic Institute, Germany.

Steve Davis

Istituto Italiano Di Tecnologia
Department of Advanced Robotics
Genova, Italy
steven.davis@iit.it

Chapter F.60

Steve Davis graduated from Salford University with a degree in Robotic and Electronic Engineering in 1998, and an MSc in Advanced Robotics in 2000. He then became a research fellow gaining his PhD in 2005 before moving to the Italian Institute of Technology in 2008. His research interests include actuators, biomimetics, dexterous grippers, humanoids and automation for the food industry.

Xavier Delorme

Ecole Nationale Supérieure des Mines
de Saint-Etienne
Centre Genie Industriel
et Informatique (G2I)
Saint-Etienne, France
delorme@emse.fr

Chapter D.35

Xavier Delorme received the PhD degree in Computer Science from the University of Valenciennes, France, in 2003. He has also worked at the French National Institute for Transport and Safety Research. He is currently an Associate Professor at the École Nationale Supérieure des Mines de Saint-Etienne. His research interests concern the optimization of production and transport systems.

Alexandre Dolgui

Ecole Nationale Supérieure des Mines
de Saint-Etienne
Department of Industrial Engineering
and Computer Science
Saint-Etienne, France
dolgui@emse.fr

Chapter D.35

Alexandre Dolgui received his PhD degree from the Academy of Sciences of Belarus (USSR) and Dr. hab. degree from the University of Technology of Compiègne, France. His research focuses on manufacturing line design, production planning, and supply-chain optimization. He is the author of five books, 117 journal articles and 250 conference papers.

Alkan Donmez   Chapter A.7

National Institute of Standards
and Technology
Manufacturing Engineering Laboratory
Gaithersburg, MD, USA
alkan.donmez@nist.gov

Alkan Donmez is the Group Leader of the Machine Tool Metrology Group as well as the Program Manager of the Science-Based Manufacturing program at the National Institute of Standards and Technology (NIST). He received his BS degree in Mechanical Engineering from the Middle East Technical University in Turkey and his MS and PhD degrees from Purdue University.

Francis J. Doyle III   Chapter H.75

University of California
Department of Chemical Engineering
Santa Barbara, CA, USA
frank.doyle@icb.ucsb.edu

Dr. Francis J. Doyle III holds the Duncan and Suzanne Mellichamp Chair in Process Control in the Department of Chemical Engineering at University of California, Santa Barbara. He received his BSE from Princeton (1985), C.P.G.S. from Cambridge (1986), and PhD from Caltech (1991), all in Chemical Engineering. His research interests are in systems biology, network science, modeling and analysis of circadian rhythms, drug delivery for diabetes, model-based control, and control of particulate processes.

Yael Edan   Chapter G.63

Ben-Gurion University of the Negev
Department of Industrial Engineering
and Management
Beer Sheva, Israel
yael@bgu.ac.il

Yael Edan is a Professor in the Department of Industrial Engineering and Management. She holds a BSc in Computer Engineering and MSc in Agricultural Engineering, both from the Technion-Israel Institute of Technology, and a PhD in Engineering from Purdue University. Her research is robotic and sensor performance analysis, systems engineering of robotic systems; sensor fusion, multi-robot and telerobotics control methodologies, and human–robot collaboration methods with major contributions in intelligent automation systems in agriculture.

Thomas F. Edgar   Chapter D.31

University of Texas
Department of Chemical Engineering
Austin, TX, USA
edgar@che.utexas.edu

Thomas F. Edgar is Professor of Chemical Engineering at the University of Texas at Austin and holds the George T. and Gladys Abell Chair in Engineering. Dr. Edgar received his BS in chemical engineering from the University of Kansas and a PhD from Princeton University. His research is in process modeling, control, and optimization.

Norbert Elkmann   Chapter G.70

Fraunhofer IFF
Department of Robotic Systems
Magdeburg, Germany
norbert.elkmann@iff.fraunhofer.de

Dr. techn. Norbert Elkmann graduated in Mechanical Engineering in 1993 (Bochum, Germany) and received his doctoral degree in 1999 (Vienna, Austria). From 1993 to 1997 he worked as Research Manager at the Fraunhofer IFF in Magdeburg. Since 1998 he has been the Manager of the institute's Business Unit Robotic Systems. His current research interests are mobile robots, inspection robots and safe human–robot interaction.

Heinz–Hermann Erbe (△)   Chapter E.41

Technische Universität Berlin
Center for Human–Machine Systems
Berlin, Germany

Heinz-Hermann Erbe (1937–2008) studied Aircraft Construction at Hamburg University of Applied Sciences, was Engineer at Focke Wulf GmbH (later Vereinigte Flugtechnische Werke), and studied Engineering Mechanics at TU Berlin. He received his PhD in Engineering Mechanics, was Head of Research Group at the German Federal Institute on Material Research, Berlin (1980–1986), Professor at Bremen University, and from 1986–2002 and member of the Research Center for Human–Machine Systems at TU Berlin. Professor Erbe was active for many years as leader and member of IFAC. He retired in 2002 but remained active. A favorite subject of his interest was cost-effective energy generation by windmills in remote regions.

Mohamed Essafi

Ecole des Mines de Saint-Etienne
Department Centre for Industrial
Engineering and Computer Science
Saint-Etienne, France
essafi@emse.fr

Chapter D.35

Mohamed Essafi received his Engineer Degree in Industrial Engineering from the Ecole Nationale d'Ingnieurs de Tunis, Tunisia, and his MSc from the University Paul Verlaine of Metz, France. He is a PhD student at the Ecole des Mines de Saint-Etienne, France. His research focuses on reconfigurable manufacturing-line design.

Florin-Gheorghe Filip

The Romanian Academy
Bucharest, Romania
ffilip@acad.ro

Chapter D.36

Florin-Gheorghe Filip received his MSc and PhD in control engineering from the T.U. "Politehnica" of Bucharest in 1970 and 1982, respectively. He has been with the Institute for Informatics, Bucharest since 1970. In 1991 he was elected as a member of the Romanian Academy whose vice-president he has been since 2000. His main scientific interests are hierarchical control and decision support systems. He (co)authored some 200 papers and six monographs.

Markus Fritzsche

Fraunhofer IFF
Department of Robotic Systems
Magdeburg, Germany
markus.fritzsche@iff.fraunhofer.de

Chapter G.70

In 2003 Markus Fritzsche graduated as Electrical Engineer from the University of Applied Sciences in Leipzig, Germany and the MEng degree in Biomedical Engineering from the University in Halle in 2005. He is currently a research associate at the Fraunhofer Institute for Factory Operation and Automation (IFF). His research activities enclose mobile robotic systems and human–robot interaction.

Susumu Fujii

Sophia University
Graduate School of Science
and Technology
Tokyo, Japan
susumu-f@sophia.ac.jp

Chapter F.48

Professor Fujii received Master of Engineering from Kyoto University and a PhD from University of Wisconsin, Madison, in 1967 and 1971, respectively. He is a Professor Emeritus of Kobe University and since 2006 Professor at Sophia University. His research interests include automation of manufacturing systems and system simulation, production management and control. He has been Honorary member of The Institute of Systems, Control and Information Engineers, and Fellows of JSME, and the Society of Precision Engineering and Operations Research Society of Japan.

Christopher Ganz

ABB Corporate Research
Baden, Switzerland
christopher.ganz@ch.abb.com

Chapter A.8

Christopher Ganz is Program Manager for the global control and optimization program at ABB corporate research. Before he held various R&D and product management positions in ABB's power generation business unit. He holds a doctor degree and a diploma in Electrical Engineering from ETH Zurich, Switzerland.

Mitsuo Gen

Waseda University
Graduate School of Information,
Production and Systems
Kitakyushu, Japan
gen@waseda.jp

Chapter C.29

Mitsuo Gen received BE and ME in Engineering from Kogakuin University in 1969, 1971 and 1975, respectively and the PhD degree in Informatics from Kyoto University in 2006. His research interests are genetic and evolutionary computation and applications to network design and optimization for manufacturing scheduling and logistics systems. He published three books with Dr. R. Cheng and Dr. L. Lin.

Birgit Graf   Chapter I.84

Fraunhofer IPA
Department of Robot Systems
Stuttgart, Germany
birgit.graf@ipa.fraunhofer.de

Birgit Graf received her degree in Computer Science in 1999 and completed her PhD in mobile robot navigation in 2008, both at Stuttgart University, Germany. Since 1999 she is working at Fraunhofer IPA where she is Manager of the Domestic and Personal Robots Group since 2008. Main target of her group is the development of new service robot applications and technologies.

John O. Gray   Chapter F.60

Istituto Italiano Di Tecnologia
Department of Advanced Robotics
Genova, Italy
John.Gray-2@manchester.ac.uk

John O. Gray is a visiting Professor at the Control Systems Centre, University of Manchester, and the Department of Automatic Control and Systems Engineering, University of Sheffield. His research interests include robotics, nonlinear control systems, precision electromagnetic instrumentation, and robotic systems for the food industry. Gray received a PhD from the University of Manchester in Control Engineering.

Rudiyanto Gunawan

National University of Singapore
Department of Chemical
and Biomolecular Engineering
Singapore
chegr@nus.edu.sg

Chapter H.75

Rudiyanto Gunawan received the PhD degree from University of Illinois Urbana Champaign in 2003. He joined University of California Santa Barbara in 2003 as a Postdoctoral Fellow. Since 2006, he has been an Assistant Professor in the Department of Chemical and Biomolecular Engineering, National University of Singapore. His research interests are in the model identification and analysis of biological systems.

Juergen Hahn

Texas A&M University
Artie McFerrin Dept. of Chemical
Engineering
College Station, TX, USA
hahn@tamu.edu

Chapter D.31

Juergen Hahn received his PhD from the University of Texas at Austin in 2002 and joined the Chemical Engineering Department at Texas A&M University in 2003 as an Assistant Professor. Dr. Hahn's research deals with process systems engineering with special emphasis on systems biology.

Kenwood H. Hall   Chapter C.28

Rockwell Automation
Department of Advanced Technology
Mayfield Heights, OH, USA
khhall@ra.rockwell.com

Kenwood H. Hall is a Cleveland Native and a graduate of Cleveland State University's Fenn College of Engineering. He has led the development of many of Rockwell Automations large controller systems including the new LOGIX family of integrated controllers, is one of the founding members of the Ohio ICE consortium and a member of the board for the new third frontier Wright center for Sensor Systems Engineering. He has issued 14 Patents with 50 pending applications.

Shufeng Han   Chapter G.63

John Deere
Intelligent Vehicle Systems
Urbandale, IA, USA
hanshufeng@johndeere.com

Shufeng Han received his PhD in Agricultural Engineering from University of Illinois at Urbana-Champaign in 1993. He is currently an Engineering Scientist at John Deere Intelligent Vehicle Systems. His research interests include field robotics, vehicle automation, sensors and precision agriculture. He has received 10 patents.

Nathan Hartman

Purdue University
Computer Graphics Technology
West Lafayette, IN, USA
nhartman@purdue.edu

Chapter D.37

Nathan Hartman is Associate Professor of Computer Graphics Technology at Purdue University. Research activities include applications of 3-D modelling, automated geometry creation, and 3-D data interoperability. He received the Doctor of Education degree in Technology Education from North Carolina State University. Professor Hartman is a Co-Director of the Purdue Product Lifecycle Management Center of Excellence.

Yukio Hasegawa

Waseda University
System Science Institute
Tokyo, Japan
yukioh@green.ocn.ne.jp

Chapter A.1

Yukio Hasegawa is Professor Emeritus of the System Science Institute at Waseda University, Tokyo, Japan. He has been enjoying construction robotics research since 1983 as Director of Waseda Construction Robot Research Project (WASCOR) which has impacted automation in construction and in other fields of automation. He received the prestigious first Engelberger Award in 1977 from the American Robot Association for his distinguished pioneering work in robotics and in Robot Ergonomics since the infancy of Japanese robotics. Among his numerous international contributions to robotics and automation, Professor Hasegawa assisted, as a visiting professor, to build the Robotics Institute at EPFL (Ecole Polytechnic Federal de Lausanne) in Switzerland.

Jackson He

Intel Corporation
Digital Enterprise Group
Hillsboro, OR, USA
jackson.he@intel.com

Chapter C.24

Jackson is lead architect of the Intel Digital Enterprise Group. He received his PhD and MBA degrees from the University of Hawaii. He represented Intel at OASIS, RosettaNet, and DMTF driving industry standard definitions. He published more than 20 papers on advanced enterprise solutions, service orientation, and platform manageability. Recently he co-authored a book on virtualized service-oriented grids.

Jenő Hetthéssy

Budapest University of Technology
and Economics
Department of Automation and Applied
Informatics
Budapest, Hungary
jhetthessy@aut.bme.hu

Chapter B.10

Jenő Hetthéssy received his PhD degree in 1975 from the Technical University of Budapest, Faculty of Electrical Engineering. He received his CSc degree in 1978 from the Hungarian Academy of Sciences with a thesis on the theory and industrial application of self-tuning controllers. He spent several years as a visiting professor at the Electrical Engineering Department of the University of Minnesota, USA and has published more than 70 papers.

Karyn Holmes

Chevron Corp.
Covington, LA, USA
kholmes@chevron.com

Chapter I.90

Karyn Holmes received a computer science degree and an MBA, both from Louisiana State University (LSU). While at LSU, her research interests included enterprise systems, with a focus on enterprise resource planning systems and their use in education. She currently works in Information Management at Chevron Corporation.

Clyde W. Holsapple

University of Kentucky
School of Management, Gatton College
of Business and Economics
Lexington, KY, USA
cwhols@email.uky.edu

Chapter I.89

Clyde W. Holsapple, Rosenthal Endowed Chair at the University of Kentucky. His books include *Foundations of Decision Support Systems, Decision Support Systems – A Knowledge-based Approach, Handbook on Decision Support Systems, and Handbook on Knowledge Management.* His research focuses on multiparticipant systems, decision support systems, and knowledge management.

Petr Horacek

Czech Technical University in Prague
Faculty of Electrical Engineering
Prague, Czech Republic
horacek@fel.cvut.cz

Chapter G.65

Petr Horacek received the MS and PhD degrees in Electrical Engineering from Czech Technical University in Prague, Czech Republic (CTU) in 1976 and 1985, respectively. He is currently an Associate Professor of Control Engineering at CTU. His research and industrial project activities include process modelling, simulation and optimization, process control and control applications in power systems.

William J. Horrey

Liberty Mutual Research Institute
for Safety
Center for Behavioral Sciences
Hopkinton, MA, USA
william.horrey@libertymutual.com

Chapter D.39

William Horrey received his PhD in Psychology from the University of Illinois at Urbana-Champaign in 2005. He is currently a research scientist in the Center for Behavioral Sciences at the Liberty Mutual Research Institute for Safety in Hopkinton, MA. His research interests include divided and selective attention, transportation human factors and automation.

Justus Hortig

Fraunhofer IFF
Department of Robotic Systems
Magdeburg, Germany
justus.hortig@iff.fraunhofer.de

Chapter G.70

Justus Hortig graduated in Mechanical Engineering from the Technical University of Darmstadt in 1999 and has been working in the Business Unit Robotic Systems at Fraunhofer IFF in Magdeburg since then. His focus lies on the use of service robots for façade cleaning.

Chin-Yin Huang

Tunghai University
Industrial Engineering and Enterprise
Information
Taichung, Taiwan
huangcy@thu.edu.tw

Chapter H.80

Chin-Yin Huang received the PhD degree from Purdue University in 1999. He is an Associate Professor in the Department of Industrial Engineering and Enterprise Information at Tunghai University, Taiwan. He also serves as the Director for Program for Health Administration. His research interests include applications of data mining techniques, biomedical informatics, and distributed intelligent manufacturing systems.

Yoshiharu Inaba

Fanuc Ltd.
Yamanashi, Japan
inaba.yoshiharu@fanuc.co.jp

Chapter C.21

Yoshiharu Inaba graduated from Mechanical Engineering Department of Tokyo Institute of Technology in 1973 and joined Fanuc Ltd. in 1983. He assumed the position of director in 1989 and the position of President and CEO in 2003. He received his doctoral degree in engineering from the University of Tokyo in 1999 and is the recipient of the 22nd Award of the Japan Society for Precision Engineering (2000) and the Nicolau Award of CIRP (2007).

Samir M. Iqbal

University of Texas at Arlington
Department of Electrical Engineering
Arlington, TX, USA
smiqbal@uta.edu

Chapter H.81

Samir M. Iqbal is an Assistant Professor at the University of Texas at Arlington in the Department of Electrical Engineering. He is also affiliated with the Nanotechnology Research and Teaching Facility and Department of Bioengineering. He directs the Nano-Bio Lab with focus on chip-based early diagnosis modalities, bio-inspired rapid nano-fabrication and functionalized microfluidics. He is a senior member of IEEE.

Rolf Isermann

Chapter C.19

Technische Universität Darmstadt
Institut für Automatisierungstechnik,
Forschungsgruppe Regelungstechnik
und Prozessautomatisierung
Darmstadt, Germany
risermann@iat.tu-darmstadt.de

Rolf Isermann served as Professor for Control Systems and Process Automation at the Institute of Automatic Control of Darmstadt University of Technology from 1977–2006. Since 2006 he has been Professor Emeritus and head of the Research Group for Control Systems and Process Automation at the same institution. He has published several books and his current research concentrates on fault-tolerant systems, control of combustion engines and automobiles and mechatronic systems. Rolf Isermann has held several chair positions in VDI/VDE and IFAC and organized several national and international conferences.

Kazuyoshi Ishii

Chapter I.85

Kanazawa Institute of Technology
Social and Industrial Management
Systems
Hakusan City, Japan
ishiik@neptune.kanazawa-it.ac.jp

Kazuyoshi Ishii received his PhD in Industrial Engineering from Waseda University. Dr. Ishii is a board member of the IFPR, APIEMS and the Japan Society of QC, and a fellow of the ISPIM. His interests include production management, product innovation management, and business models based on a comparative advantage.

Alberto Isidori

Chapter B.9

University of Rome "La Sapienza"
Department of Informatics and Sytematics
Rome, Italy
albisidori@dis.uniroma1.it

Alberto Isidori is Professor of Automatic Control at the University of Rome since1975 and, since 1989, also affiliated with Washington University in St. Louis. His research interests are primarily in analysis and design of nonlinear control systems. He is the author of the book *Nonlinear Control Systems* and is the recipient of various prestigious awards, which include the "Georgio Quazza Medal" from IFAC, the "Bode Lecture Award" from IEEE. He is Fellow of IEEE and of IFAC. Currently he is President of IFAC.

Nick A. Ivanescu

Chapter G.64

University Politechnica of Bucharest
Control and Computers
Bucharest, Romania
nik@cimr.pub.ro

Nick Andrei Ivanescu graduated as an Automation Engineer and received his PhD degree in Automated Systems from the University Politehnica of Bucharest in 2000. His research activities include process control, industrial automation, microcontrollers and robotics. Since 2001 he has been lecturer in the Control and Computers faculty of the University Politehnica of Bucharest.

Sirkka-Liisa Jämsä-Jounela

Chapter F.57

Helsinki University of Technology
Department of Biotechnology
and Chemical Technology
Espoo, Finland
Sirkka-l@tkk.fi

Dr. Sirkka-Liisa Jämsä-Jounela is Professor of Process Control at Helsinki University of Technology. Before joining the university she gained practical experience in working for a number of companies in the Finnish process industry. Her research activities include process modelling, simulation, monitoring and fault diagnosis as well as production control in mineral and metal processing. Prof. Jämsä-Jounela has served IFAC as Vice President among others. She is currently the Chair of the Nordic Process Control Group and a member of the Finnish Academy of Technologies.

Bijay K. Jayaswal

Chapter E.45

Agilenty Consulting Group
Minneapolis, MN, USA
bijay.jayaswal@agilenty.com

Bijay K. Jayaswal is the CEO of Agilenty Consulting Group, LLC. He has held senior executive positions for the last 25 years. He has taught engineering and management at the University of Mauritius and at California State University, Chico and has directed MBA and Advanced Management programs. He has helped introduce corporate-wide initiatives in re-engineering, Six Sigma, and Design for Six Sigma. He is a recipient of the ASQ Crosby Medal for 2007 for his book (together with Peter C. Patton) *Design for Trustworthy Software*.

Wootae Jeong

Chapter C.20

Korea Railroad Research Institute
Uiwang, Korea
wjeong@krri.re.kr

Wootae Jeong earned a PhD from Purdue University in 2006 and joined the PRISM Center since 2003. He is currently a senior researcher at Korea Railroad Research Institute and adjunct Professor at the University of Science and Technology. His research interests are in measurement and control of automation and production systems.

Timothy L. Johnson

Chapter E.42

General Electric
Global Research
Niskayuna, NY, USA
johnsontl@nycap.rr.com

Dr. Johnson served on the MIT Electrical Engineering Faculty from 1972–1980, as a Senior Scientist at BBN, Inc., from 1980–1984, and as a Manager and technical contributor at GE Research from 1984–2008, when he retired. His research interests include automation, diagnostics and service processes, and discrete event systems. He is an AACC Eckman Award recipient and an IEEE Fellow.

Hemant Joshi

Chapter E.47

Research, Acxiom Corp.
Conway, AR, USA
hemant.joshi@acxiom.com

Hemant Joshi graduated with PhD in May 2007 from the University of Arkansas at Little Rock, USA. He has published 8 papers and 2 technical reports. He currently works as Researcher at Acxiom Research Labs. His research interests are information extraction, machine learning, data mining and automated data acquisition.

Michael Kaplan

Chapter G.72

Ex Libris Ltd.
Newton, MA, USA
michael.kaplan@exlibrisgroup.com

Michael Kaplan received his PhD from Harvard University (1977). After 20 years at Harvard University and 2 years as Associate Dean at Indiana University Libraries, he has since 2000 been with Ex Libris, Ltd. He is the editor of 2 books on library automation. In 1998 he received the LITA/ Library Hi-Tech Award for Outstanding Communication in Library and Information Science.

Dimitris Kiritsis

Chapter E.43

Department STI–IGM–LICP
Lausanne, Switzerland
dimitris.kiritsis@epfl.ch

Dr. Dimitris Kiritsis earned his Diploma (1980) and PhD (1987) in Mechanical Engineering from the University of Patras, Greece. Since 1989 he is with the Computer-Aided Design and Production Laboratory (LICP) of the Swiss Federal Institute of Technology in Lausanne (EPFL). His research is in the domain of modeling methods and techniques for integrated product-process-resource planning, product lifecycle information modeling and transformation to knowledge. Dr.Kiritsis is the initiator and scientific coordinator of the FP6-IP-507100 PROMISE.

Hoo Sang Ko

Chapter J.94

Purdue University
PRISM Center and School of Industrial Engineering
West Lafayette, IN, USA
ko0@purdue.edu

Hoo Sang Ko received his BS in 1999 and MS in 2003 in Mechanical Engineering at Seoul National University, South Korea. After graduation he worked for Samsung Electronics. His research interest is in computer-supported collaboration and protocol design. He is currently a doctoral candidate in the School of Industrial Engineering and a Senior Researcher in PRISM Center at Purdue University.

Naoshi Kondo

Kyoto University
Division of Environmental Science
and Technology, Graduate School
of Agriculture
Kyoto, Japan
kondonao@kais.kyoto-u.ac.jp

Chapter G.63

Naoshi Kondo received the MS degree and Dr. Agr. from Kyoto University in 1984 and 1988, respectively. He joined the Department of Agricultural Engineering, Okayama University as an Assistant Professor in 1985 and became an Associate Professor in 1993. He worked for Ishii Co., Ltd. as a manager in 2000 and entered Ehime University as a Professor in 2006. Since 2007 he has been a Professor at the Graduate School of Agriculture, Kyoto University. His research interests are automation, instrumentation and information in agriculture.

Peter Kopacek

Vienna University of Technology
Intelligent Handling and Robotics –
IHRT
Vienna, Austria
kopacek@ihrt.tuwien.ac.at

Chapter G.74

With a cum-laude Doctors degree from Vienna University of Technology, Professor Kopacek is Head of the Institute for Handling Devices and Robotics and President of the Austrian Society for Systems Engineering and Automation. He has Doctor hc degrees from several Universities and is Corresponding member of the Saxon Academy of Sciences, Member of the German Academy of Technical Sciences, a Senior member of IEEE, and the Recipient of the Engelberger Robotics Award.

Nicholas Kottenstette

Vanderbilt University
Institute for Software Integrated Systems
Nashville, TN, USA
nkottens@isis.vanderbilt.edu

Chapter B.13

Nicholas Kottenstette, a Research Scientist, holds a MS from the Mechanical Engineering Department at MIT and a PhD in Electrical Engineering from The University of Notre Dame. He holds 11 US patents related to the design and control of (networked) embedded systems. His current research interests are in high-confidence embedded system design and the science of cyber physical systems.

Eric Kwei

University of California, Santa Barbara
Department of Chemical Engineering
Santa Barbara, CA, USA
kwei@engineering.ucsb.edu

Chapter H.75

Eric Kwei is a graduate student of Chemical Engineering at the University of California, Santa Barbara. His interests are in mathematical modeling and systems analysis of insulin signal transduction networks.

Siu K. Kwok

The Hong Kong Polytechnic University
Industrial and Systems Engineering
Kowloon, Hong Kong
mfskkwok@inet.polyu.edu.hk

Chapter F.49

S.K. Kwok received the Bachelor of Engineering and a Doctorate in Manufacturing Engineering from The Hong Kong Polytechnic University in 1991 and 1997, respectively. He is currently a Lecturer at The Hong Kong Polytechnic University. His research interests include artificial intelligence, industrial and systems engineering, information and communication technologies (ICT), logistics enabling technologies and mobile commerce.

King Wai Chiu Lai

Michigan State University
Electrical and Computer Engineering
East Lansing, MI, USA
kinglai@egr.msu.edu

Chapter F.53

Dr. King Wai Chiu Lai received his PhD and MPhil degree in Automation and Computer-Aided Engineering from The Chinese University of Hong Kong in 2005 and 2002. Currently, he is the Manager of the Nano Manufacturing Laboratory at Michigan State University. He is conducting various research projects, such as design, fabrication and application of nanooptical detectors and solar cells and development of automated micro/nanofluidic manipulation system.

Dean F. Lamiano   Chapter G.67

The MITRE Corporation
Department of Communications
and Information Systems
McLean, VA, USA
dlamiano@mitre.org

Mr. Dean F. Lamiano is a Principal Communications Engineer and Associate Program Manager at MITRE's Center for Advanced Aviation System Development (CAASD), engaged in the research and system engineering for air traffic control communications systems. Dean is a member of IEEE and has a BS in Electrical Engineering from the SUNY at Buffalo, and an MSEE from the Johns Hopkins University.

Steven J. Landry   Chapter G.68

Purdue University
School of Industrial Engineering
West Lafayette, IN, USA
slandry@purdue.edu

Steven J. Landry is an Assistant Professor in the School of Industrial Engineering at Purdue University. He received his MS in Aeronautics and Astronautics from MIT, and his PhD in Industrial and Systems Engineering from Georgia Tech. He has over 2500 heavy jet flight hours as a USAF C-141B pilot. His research interests are in aviation systems engineering and human factors.

John D. Lee   Chapter C.25

University of Iowa
Department Mechanical and Industrial
Engineering, Human Factors Research
National Advanced Driving Simulator
Iowa City, IA, USA
jdlee@engineering.uiowa.edu

John D. Lee graduated with degrees in psychology and mechanical engineering from Lehigh University and owns a PhD in mechanical engineering from the University of Illinois. He is a Professor in the Department of Mechanical and Industrial Engineering at the University of Iowa. His research focuses on the safety and acceptance of human-machine systems by considering how technology mediates attention.

Tae-Eog Lee   Chapter F.52

KAIST
Department of Industrial and Systems
Engineering
Daejeon, Korea
telee@kaist.ac.kr

Tae-Eog Lee is Professor in the Department of Industrial and Systems Engineering and Director of the Center for Modeling and Simulation Technology Research at KAIST. His interests include automated equipment scheduling and control and software engineering for semiconductor manufacturing, and discrete event system modelling and control. He graduated from Seoul National University (KAIST) and Ohio State University.

Wing B. Lee   Chapter F.49

The Hong Kong Polytechnic University
Industrial and Systems Engineering
Kowloon, Hong Kong
wb.lee@polyu.edu.hk

W.B. Lee received his Master of Technology from Brunel University in 1976 and the Doctorate from the University of Hong Kong in 1986. He is the Cheng Yick-chi Chair Professor of Manufacturing Engineering, the Director of The Advanced Technology Manufacturing Research Centre as well as the Knowledge Management Research Centre of The Hong Kong Polytechnic. His research interests include advanced manufacturing technology, materials processing and knowledge management.

Mark R. Lehto   Chapter D.39

Purdue University
School of Industrial Engineering
West Lafayette, IN, USA
lehto@purdue.edu

Dr. Mark R. Lehto is an Associate Professor in the School of Industrial Engineering at Purdue University, where he is Co-chair of the Interdisciplinary Graduate Program in Human Factors. Dr. Lehto's research and teaching interests are in the areas of hazard communication and decision support. He is particularly interested in methods of increasing the effectiveness of information provided to consumers and workers.

Kauko Leiviskä

University of Oulu
Control Engineering Laboratory
Oulun Yliopisto, Finland
kauko.leiviska@oulu.fi

Chapter D.36

Kauko Leiviskä received a Diploma degree in Engineering in Process Engineering in 1975 and the Doctor of Technology in Control Engineering in 1982, both from University of Oulu. He is Professor of Control Engineering and Dean of Faculty of Technology. His research interests are in applications of soft computing methods. He is (co)author of more than 200 publications.

Mary F. Lesch

Liberty Mutual Research Institute
for Safety
Center for Behavioral Sciences
Hopkinton, MA, USA
mary.lesch@libertymutual.com

Chapter D.39

Mary F. Lesch holds a PhD in cognitive psychology from the University of Massachusetts – Amherst. She completed post-doctoral training in neuropsychology at Rice University in 1999. Since then, she has been a Research Scientist at the Liberty Mutual Research Institute for Safety where her research focuses on risk perception and communication. Recent research has investigated age-related effects on warning symbol comprehension.

Jianming Lian

Purdue University
School of Electrical and Computer
Engineering
West Lafayette, IN, USA
jlian@purdue.edu

Chapter B.11

Jianming Lian received the BS degree from University of Science and Technology of China in 2004. Then he attended Purdue University, West Lafayette, IN USA for his graduate study, where he received the MS degree in 2007. He is now a PhD candidate in the School of Electrical and Computer Engineering in the area of Automatic Control. His research interests include nonlinear control, adaptive control and switched systems.

Lin Lin

Waseda University
Information, Production & Systems
Research Center
Kitakyushu, Japan
linlin@aoni.waseda.jp

Chapter C.29

Lin Lin received ME and PhD degrees in Engineering from Waseda University in 2006 and 2008, respectively. He was Research Assistant at Graduate School of Information, Production and Systems, Waseda University 2007–2009. His research interests are genetic and evolutionary computation and applications to network design and optimization for manufacturing scheduling and logistics systems. He has published *Network Models and Optimization* together with Dr. M. Gen and Dr. R. Cheng.

Laurent Linxe

Peugeot SA
Hagondang, France
laurent.linxe@mpsa.com

Chapter D.35

Laurent Linxe graduated as a multi-disciplinary engineer from Centrale Lille in 1991, joined Peugeot SA in 1992 and PCI-SCEMM in 2005. As the manager of the Machining Systems PCI, he initiated relationships with the Industrial Engineering and Computer Science of Saint-Etienne and M. Dolgui. Since 2008, he has been in charge of the Manufacturing Department for 2.0 and 2.2 litre of HDI Common Rail Diesel Engines, Peugeot SA Tremery Plant in Lorraine Eastern France.

T. Joseph Lui

Whirlpool Corporation
Global Product Organization
Benton Harbor, MI, USA
t_joseph_lui@whirlpool.com

Chapter I.83

T. Joseph Lui received a BS from Virginia Tech, a MS in Electrical Engineering from the Illinois Institute of Technology and a MBA from the University of Chicago. He is currently holding a director position at Whirlpool for global connectivity, controls and electronics. Prior to joining Whirlpool, Mr. Lui held various leadership and management positions at Motorola developing and marketing wireless technologies and products.

Wolfgang Mann   Chapter E.46

Profactor Research and Solutions GmbH
Process Design and Automation,
Forschungszentrum
Seibersdorf, Austria
wolfgang.mann@profactor.at

Wolfgang Mann received the MS from Vienna University of Technology in Mechanical Engineering. He is currently Head of the Department Prozess Design and Automation. His research interests are mainly lot size one automation solutions in modern test stands, handling of small and complex parts, ultra precision assembly, vision, and sensor guided assembly.

Sebastian V. Massimini   Chapter G.67

The MITRE Corporation
McLean, VA, USA
svm@mitre.org

Dr. Sebastian Vince Massimini is currently a Senior Principal Engineer at MITRE's Center for Advanced Aviation System Development, in McLean, Virginia, where he is a research analyst in navigation and surveillance for airports and air traffic control architecture. Dr. Massimini is an associate fellow of the American Institute of Aeronautics and Astronautics (AIAA). He received his Doctor of Science in Operations Research from The George Washington University.

Francisco P. Maturana

University/Company Rockwell Automation
Department Advanced Technology
Mayfield Heights, OH, USA
fpmaturana@ra.rockwell.com

Chapter C.28

Dr. Maturana is a Principal Research Engineer at Rockwell Automation. His research interests are in autonomous control and self organizing systems for automation, distributed computing, and high performance simulations. He has been issued 16 US/European patents in agent technology and a technical committee member for IEEE SMC/ISM, BASYS, HOLOMAS. Dr. Maturana received his PhD in Intelligent Manufacturing Systems from The University of Calgary, Canada.

Henry Mirsky

University of California, Santa Barbara
Department of Chemical Engineering
Santa Barbara, CA, USA
Mirsky@lifesci.ucsb.edu

Chapter H.75

Henry Mirsky is a doctoral candidate in the Program in Biomolecular Science and Engineering at the University of California, Santa Barbara. He holds a BS degree in Materials Science from Cornell University and a BS in Molecular and Cell Biology from the University of Arizona. His research interests include biochemical modelling, sensitivity analysis, and circadian biology.

Sudip Misra   Chapter B.12

Indian Institute of Technology
School of Information Technology
Kharagpur, India
sudip_misra@yahoo.com

Dr. Sudip Misra is an Assistant Professor in the School of Information Technology at the Indian Institute of Technology Kharagpur, India. He received his PhD in Computer Science from Carleton University, Ottawa, Canada. His current research interests include algorithm design and soft computing applications in telecommunications. Dr. Misra is the author of several research papers.

Satish C. Mohleji   Chapter G.67

Center for Advanced Aviation System
Development (CAASD)
The MITRE Corporation
McLean, VA, USA
smohleji@mitre.org

Satish Mohleji is a Senior Principal Engineer in the Center for Advanced Aviation System Development (CAASD) of the MITRE Corp. For 38 years, he has worked on designing and analyzing future operational concepts for enhancing Air Traffic Management systems at US and International airports. He received a PhD from the University of Windsor, Ontario, and MS in Management Science from the American University in Washington D.C. He is an Associate Fellow of the AIAA.

Authors

Gérard Morel

Centre de Recherche en Automatique
Nancy (CRAN)
Vandoeuvre, France
gerard.morel@cran.uhp-nancy.fr

Chapter E.42

Gérard Morel is Professor at Nancy University and Vice-Director of CRAN. He published over 150 articles on systems and automation engineering. He served in positions at IFAC and with the European Commission. He is Europe editor of the IJIM and associate editor of the IFAC International Journal on Engineering Applications of Artificial Intelligence.

René J. Moreno Masey

University of Sheffield
Automatic Control and Systems
Engineering
Sheffield, UK
cop07rjm@sheffield.ac.uk

Chapter F.60

René J. Moreno Masey received an MEng in Mechanical Engineering from the University of Manchester (1996) and an MSc in Robotics and Automation from the University of Salford (2006). He worked for Unilever (1997–2001) designing hygienic food production lines. His current PhD research at the University of Sheffield and IIT, Genoa involves the design of a low-cost industrial robot for the food industry.

Clayton Munk

Boeing Commercial Airplanes
Material & Process Technology
Seattle, WA, USA
clayton.l.munk@boeing.com

Chapter F.51

Clayton Munk received the Masters of Technology Management from Brigham Young University in 2006. He is currently a Technical Fellow and has over 21 years experience in advanced assembly methods for aircraft structure within M&PT at Boeing Commercial Airplanes. He is a member of the SAE Aerospace, Air and Space Group and is also the Chairman of the SAE Aerospace Automated Fastening Committee.

Yuko J. Nakanishi

Nakanishi Research and Consulting, LLC
New York, NY, USA
nakanishi@transresearch.net

Chapter G.66

Dr. Yuko J. Nakanishi is Principal, Nakanishi Research and Consulting, LLC; Vice President of Intelligent Transportation Society NY and Chair of the TRB Subcommittee on Training, Education, and Technology Transfer. Her expertise includes ITS, security, and performance assessment. She holds a PhD in Transportation Planning and Engineering from Polytechnic Institute of NY University, MS in Civil Engineering from City College NY, and MBA in Management from Columbia University.

Dana S. Nau

University of Maryland
Department of Computer Science
College Park, MD, USA
nau@cs.umd.edu

Chapter B.14

Dr. Nau is a Professor of both Computer Science and Systems Research at the University of Maryland, and co-directs the Laboratory for Computational Cultural Dynamics. He has more than 300 publications and is co-author of a comprehensive textbook on automated planning. He has authored award-winning software, including the SHOP2 planning system which is used in projects worldwide. He is a Fellow of the Association for the Advancement of Artificial Intelligence (AAAI).

Peter Neumann

Institut für Automation
und Kommunikation
Magdeburg, Germany
peter.neumann@ifak.eu

Chapter F.56

Peter Neumann received the PhD in 1967. Peter Neumann was full professor at the Otto-von-Guericke-University Magdeburg from 1981 to 1994. In 1991 he founded the applied research institute "ifak" (Institut für Automation and Kommunikation) headed by him until 2004. His interests are industrial communications, device management and formal methods in engineering of distributed computer control systems.

Shimon Y. Nof   Chapters A.3, C.30, I.88

Purdue University
PRISM Center and School of Industrial
Engineering
West Lafayette, IN, USA
nof@purdue.edu

Shimon Y. Nof is Professor of Industrial Engineering at Purdue University since 1977. He is Director of the NSF-industry supported PRISM Center, President of IFPR and Fellow of the IIE. Nof is the author/editor of ten books. His main research is concerned with automated collaboration, systems security, integrity, and assurance; integrated production and service operations with decentralized decision support networks; and collaborative robotics, e-Work, and e-Service.

Anibal Ollero   Chapter B.16

Universidad de Sevilla
Departamento de Ingeniería de Sistemas
y Automática
Sevilla, Spain
aollero@cartuja.us.es

Anibal Ollero is full Professor at the University of Seville and Scientific Director of the Center for Advanced Aerospace Technologies (FADA-CATEC). He is the author of more than 375 publications including 5 books, and led or participated in more than 90 projects, including 17 projects funded by the European Commission. He is the recipient of 9 national and international awards.

John Oommen   Chapter B.12

Carleton University
School of Computer Science
Ottawa, Canada
oommen@scs.carleton.ca

Dr. John Oommen is a Chancellor's Professor at Carleton University in Ottawa, Canada. He obtained his B. Tech. from IIT Madras, India (1975), his MS from IISc in Bangalore, India (1977), and his PhD from Purdue University, West Lafayette (1982). He is an IEEE and IAPR Fellow. He is also an Adjunct Professor with the University of Agder in Grimstad, Norway.

Robert S. Parker   Chapter H.76

University of Pittsburgh
Department of Chemical and Petroleum
Engineering
Pittsburgh, PA, USA
rparker@pitt.edu

Robert S. Parker is an Associate Professor of Chemical and Petroleum Engineering at the University of Pittsburgh. Bob received his chemical engineering degrees from the University of Rochester (BS 1994) and the University of Delaware (PhD 1999). His research interests include the development of mathematical models of disease (e.g., cancer, diabetes, and inflammation) and the synthesis of model-based treatment design algorithms.

Alessandro Pasetti   Chapter C.22

P&P Software GmbH
Tägerwilen, Switzerland
pasetti@pnp-software.com

Alessandro Pasetti holds degrees in Electrical Engineering (University of Trieste, 1986), Control Engineering (Imperial College, 1987), and Sociology (University of Cambridge, 1997), and a doctorate in Software Engineering (University of Konstanz, 2002). He worked as a Control Engineer for the ESA, and currently works for P&P Software GmbH. His research interests are in the field of software reuse technologies for embedded systems. He is the author of the book *Software Frameworks and Embedded Control Systems*.

Anatol Pashkevich   Chapter F.59

Ecole des Mines de Nantes
Department of Automatic Control
and Production Systems
Nantes, France
anatol.pashkevich@emn.fr

Anatol Pashkevich received his degree in Electrical Engineering (1977) and PhD in Control and Robotics (1982), both from Minsk Radio Engineering Institute, Belarus. Since 1987 he served as head of the Robotic Laboratory at Belarusian State University of Informatics and Radio Electronics. Currently he is head of the Department of Automatic Control and Production Systems at Ecole des Mines de Nantes, France. His research interests include robotics, manufacturing automation and computer-aided design. He is a Member of the TC Manufacturing Plant Control of IFAC.

Authors

Bozenna Pasik-Duncan

University of Kansas
Department of Mathematics
Lawrence, KS, USA
bozenna@math.ku.edu

Chapter E.44

Bozenna Pasik-Duncan received her PhD and Habilitation degrees in Mathematics from the Warsaw School of Economics in 1978 and 1986, respectively. She is a Professor of Mathematics, and Courtesy Professor of EECS at the University of Kansas. Her research interests are primarily in stochastic adaptive control and science, engineering and mathematics education. Dr. Pasik-Duncan is an IEEE Fellow and Distinguished Member of the IEEE CSS. She is the chair of the IEEE CSS, AACC and IFAC and the chair of the IFAC Harold Chestnut Control Engineering Textbook Prize Selection Committee.

Peter C. Patton

Oklahoma Christian University
School of Engineering
Oklahoma City, OK, USA
peter.patton@oc.edu

Chapters B.18, E.45

Peter C. Patton has taught computer science, mathematics, aerospace engineering, and ancient history, classical civilizations at the Universities of Kansas, Minnesota, Stuttgart, and Pennsylvania. His last book, Design for Trustworthy Software (together with Bijay Jayaswal) won the Crosby Medal from the ASQ. He is currently teaching mechanical engineering, Western Civilization, and Philosophy to engineering students at Oklahoma Christian University.

Richard D. Patton

Lawson Software
St. Paul, MN, USA
richard.patton@lawson.com

Chapter B.18

Richard D. Patton is Chief Technology Officer of Lawson Software and has been involved with building business application software languages and development methodologies for over 25 years. His current project is building a new application development language based on the latest research in the areas of pattern languages and complex adaptive systems theory as well as Charles Sanders Peirce's triadic semiotics.

Carlos E. Pereira

Federal University of Rio Grande do Sul (UFRGS)
Department Electrical Engineering
Porto Alegre RS, Brazil
cpereira@ece.ufrgs.br

Chapter F.56

Carlos Eduardo Pereira received the Dr.-Ing. degree from the University of Stuttgart, Germany (1995), the MSc in Computer Science (1990) and the BS degrees in Electrical Engineering (1987) both from the Federal University of Rio Grande do Sul (UFRGS)in Brazil. He is an Associate Electrical Engineering Professor at UFRGS and Technical Director of the CETA Applied Research Center. His research focuses on methodologies and tool support for the development of distributed real-time embedded systems, with special emphasis on industrial automation applications. He is Chair of the IFAC Technical Committee on Manufacturing Plant Control (TC 5.1).

Jean-François Pétin

Centre de Recherche en Automatique
de Nancy (CRAN)
Vandoeuvre, France
jean-francois.petin@cran.uhp-nancy.fr

Chapter E.42

Jean-François Pétin is Assistant Professor at Nancy University as researcher at CRAN and as a teacher at ESIAL. His research topic deals with formal automation engineering. He is currently working on combining formal discrete events models with system engineering approaches for improving safety and reconfigurability of control systems. He is member of the IFAC Technical Committee 5.1. Manufacturing Plant Control.

Chandler A. Phillips

Wright State University
Department of Biomedical, Industrial
and Human Factors Engineering
Dayton, OH, USA
Chandler.Phillips@wright.edu

Chapter B.17

Dr. C.A. Phillips received his MD degree (1969) from the University of Southern California and is currently the Brage Golding Distinguished Professor of Research and Director of the Ergonomic Engineering program. His research interests include neuromuscular and sensory feedback systems for human operator control, biomimetic modeling of pneumatic muscle actuators, and quantitative modeling of human operator and machine interaction.

Friedrich Pinnekamp   Chapters D.32, D.33

ABB Asea Brown Boveri Ltd.
Corporate Strategy
Zurich, Switzerland
friedrich.pinnekamp@ch.abb.com

Friedrich Pinnekamp (PhD in physics) is member of the Senior Management at ABB Asea Brown Boveri Ltd. As assistant to all the chief technology officers of ABB in the last 28 years, he helped to set up the ABB global R&D organization, installing research centres and programs for all core competences required as well as being responsible for ABB Review, the technology magazine of the group. Friedrich is Vice President of the European Industrial Research Management Association (EIRMA).

Daniel J. Power   Chapter I.87

University of Northern Iowa
College of Business Administration
Cedar Falls, IA, USA
daniel.power@uni.edu

Daniel J. "Dan" Power is a Professor of Information Systems and Management at the College of Business Administration at the University of Northern Iowa, Cedar Falls, Iowa and the editor of DSSResources.com, a web-based knowledge repository about computerized systems that support decision making. Also, Dan is the Decision Support Expert at the Business Intelligence Network. In 1982, Professor Power received a PhD in Business Administration from the University of Wisconsin-Madison.

Damien Poyard   Chapter D.35

PCI/SCEMM
Saint-Etienne, France
damien.poyard@pci.fr

Damien Poyard graduated as a Mechanical Engineer from ECAM Lyon in 1989 and holds a DEA in Industrial Automatism from INSA Lyon University (1989). He is working since 1990 in PCI/SCEMM, a subsidiary of PSA Group Automotive Company, dedicated to the design and production of machining lines. PCI realises transfer lines or flexible lines including high-speed machining centres (HSMC).

Srinivasan Ramaswamy   Chapter E.47

University of Arkansas at Little Rock
Department of Computer Science
Little Rock, AR, USA
srini@ieee.org

Dr. Ramaswamy's research interests are on intelligent control, modeling, analysis and simulation, stability and scalability of complex systems. His work is motivated by the desire to understand the various requirements to build scalable, intelligent software systems. He is a senior member of IEEE and ACM, and a member of the Society for Computer Simulation International, Computing Professionals for Social Responsibility.

Piercarlo Ravazzi   Chapter A.6

Politecnico di Torino
Department Manufacturing Systems
and Economics
Torino, Italy
piercarlo.ravazzi@polito.it

Piercarlo Ravazzi received the Laurea degree in Economics from the University of Rome in 1974. He is currently Full Professor of Economics at the Polytechnic University of Torino. His research interests include production economics, macroeconomic theory and Italian economy, managerial finance and economics of telecommunications. He has authored several books on these topics.

Daniel W. Repperger   Chapter B.17

Wright Patterson Air Force Base
Air Force Research Laboratory
Dayton, OH, USA
Daniel.repperger@wpafb.af.mil

D.W. Repperger received his BS, MS, and PhD degrees in Electrical Engineering from Rensselaer Polytechnic Institute and Purdue University. At the Air Force Research Laboratory his research has been focused on human–machine systems, control theory and biomedical engineering. He is a Fellow of the IEEE, AIMBE, AsMA, The Air Force Research Laboratory, and the Ohio Academy of Science.

William Richmond

Western Carolina University
Accounting, Finance, Information Systems
and Economics
Cullowhee, NC, USA
brichmond@email.wcu.eud

Chapter I.91

William Richmond is an Associate Professor of Computer Information Systems. His research focuses on the economics of information with application to the financial services industry. He received his PhD from Purdue University in 1988.

Dieter Rombach

University of Kaiserslautern
Department of Computer Science,
Fraunhofer Institute for Experimental
Software Engineering
Kaiserslautern, Germany
rombach@iese.fraunhofer.de

Chapter I.92

Dr. H. Dieter Rombach is a Full Professor in the Department of Computer Science at the University of Kaiserslautern, Germany. He holds a chair in software engineering, is executive and founding Director of the Fraunhofer Institute for Experimental Software Engineering (IESE), and chairs the Fraunhofer ICT Group. His research interests are in software methodologies, modelling and measurement of the software process and resulting products, software reuse, and distributed systems. Results are documented in more than 180 publications in international journals and conference proceedings.

Shinsuke Sakakibara

Fanuc Ltd.
Yamanashi, Japan
sakaki@i-dreamer.co.jp

Chapter C.21

Shinsuke Sakakibara graduated from Applied Physics Department of the University of Tokyo in 1972. He joined Fanuc Ltd. in 1972 and received his doctoral degree in engineering from the University of Tokyo in 1995. He assumed the position of Chairman of the Robotics Society of Japan in 2009.

Timothy I. Salsbury

Johnson Controls, Inc.
Building Efficiency Research Group
Milwaukee, WI, USA
tim.salsbury@gmail.com

Chapter G.62

Dr. Salsbury has a PhD from Loughborough University with graduate research also performed at Oxford University and MIT. Previous positions include a Research Scientist at VTT Finland and Research Fellow at BNL. Dr Salsbury is currently a Principal Researcher at Johnson Controls with research interests in control performance assessment, fault detection, adaptive control, optimization, and statistical data analysis.

Branko Sarh

The Boeing Company – Phantom Works
Huntington Beach, CA, USA
branko.sarh@boeing.com

Chapter F.51

Dr. Branko Sarh is a Technical Fellow specializing in the development of advanced assembly systems and processes at Boeing Phantom Works in Huntington Beach. He received his Masters in Aeronautical Engineering from the University of Aachen, and his PhD in Mechanical Engineering from the University of Hamburg, Germany. His 40 years of technical and international manufacturing experience include project management assignments at MBB, Airbus, Rohr, McDonnell Douglas and Boeing.

Sharath Sasidharan

Marshall University
Department of Management
and Marketing
Huntington, WV, USA
sasidharan@marshall.edu

Chapter I.89

Dr. Sasidharan is an Assistant Professor in the Lewis College of Business at Marshall University. He holds a PhD in Decision Science and Information Systems from the University of Kentucky. His research interests include human–computer interaction, technology acceptance, ERP systems, and e-Commerce. He has published in several academic journals.

Brandon Savage

Chapter H.77

GE Healthcare IT
Chalfont St Giles, UK
Brandon.Savage@med.ge.com

Brandon Savage, MD, Chief Medical Officer of GE Healthcare IT, is responsible for driving General Electric's clinical IT vision, where his prime focus is to empower the highest level of care for any institution with the use of technology. Dr. Savage received his MD from University of California at San Diego and has been publishing in numerous periodicals and medical journals.

Manuel Scavarda Basaldúa

Chapter F.54

Kimberly Clark
Buenos Aires, Argentina
manuel.scavarda@kcc.com

Manuel Scavarda Basaldúa graduated as an Industrial Engineer at ITBA – Instituto Tecnológico de Buenos Aires. He obtained a MSc degree in Industrial Engineering from Chalmers University of Technology, Stockholm, Sweden. He worked as an Operations Analyst for the Stena Metall Group in Sweden and Norway and then joined Kimberly-Clark. Currently, he performs as Supply Chain Development Manager at KC River Plate Region.

Walter Schaufelberger (△)

Chapter C.22

ETH Zurich
Institute of Automatic Control
Zurich, Switzerland

Professor Walter Schaufelberger (1940–2008) was Full Professor at the Institute of Automatic Control, ETH Zurich. He studied Electrical Engineering at the ETH Zurich where he earned a doctorate. His main interests were in theoretical treatment of problems relevant to practice and to computer applications. The development of user friendly software played a major role in his activities. He was the Spokesman for "Integration of Complex Systems" of the European Science Foundation. He was Fellow of IFAC, SEFI, and the European Society for Engineering.

Bobbie D. Seppelt

Chapter C.25

The University of Iowa
Mechanical and Industrial Engineering
Iowa City, IA, USA
bseppelt@engineering.uiowa.edu

Bobbie D. Seppelt graduated with her MS degree in Engineering Psychology from The University of Illinois in 2003. She is currently a graduate research assistant in the Industrial Engineering PhD program at The University of Iowa. Her research interests include interface design and driver support, automation reliance and supervisory control, application of ecological frameworks to interface design, and trust in automation.

Ramesh Sharda

Chapter I.87

Oklahoma State University
Spears School of Business
Stillwater, OK, USA
Ramesh.sharda@okstate.edu

Dr. Ramesh Sharda is Director of the Institute for Research in Information Systems, ConocoPhillips Chair of Management of Technology, and a Regents Professor of Management Science and Information Systems in the Spears School of Business at Oklahoma State University. His research interests are in decision support systems, neural network applications, and technologies for managing information overload. Defense Ammunitions Center, NSF, and other organizations have funded his research.

Keiichi Shirase

Chapter F.48

Kobe University
Department of Mechanical Engineering
Kobe, Japan
shirase@mech.kobe-u.ac.jp

Keiichi Shirase received the MS degree from Kobe University, and joined Kanazawa University as a Research Associate in 1984. He received the Dr. Eng. from Kobe University in 1989. He was an Associate Professor in 1995 at Kanazawa University, and in 1996 at Osaka University. Since 2003, he has been a Professor at Kobe University. His research interests are mainly in autonomous machine tool and intelligent CAD/CAM systems.

Authors

Jason E. Shoemaker

University of California
Department of Chemical Engineering
Santa Barbara, CA, USA
jshoe@engr.ucsb.edu

Chapter H.75

Jason E. Shoemaker graduated from the University of California, Santa Barbara with a doctorate in Chemical Engineering. His research studies include drug target identification, the design and control of biological networks and their implications on communications networks. Shoemaker is a member of UC, IGERT and JST.

Moshe Shoham

Technion – Israel Institute
of Technology
Department of Mechanical Engineering
Haifa, Israel
shoham@technion.ac.il

Chapter H.78

Professor Moshe Shoham has been conducting research in robotics with a special focus on new robot structures and medical applications. He has been heading the Robotics Laboratories of the Mechanical Engineering Department at Columbia University, New York, and at the Technion-Israel Institute of Technologies. In 2001 he founded Mazor Surgical Technologies – a company that produces a miniature robot for spine surgery.

Marwan A. Simaan

University of Central Florida
School of Electrical Engineering
and Computer Science
Orlando, FL, USA
simaan@mail.ucf.edu

Chapter H.79

Marwan A. Simaan is Distinguished Professor of Electrical Engineering and Computer Science at the University of Central Florida. He received the PhD degree from the University of Illinois at Urbana-Champaign and served on the Electrical Engineering faculty at the University of Pittsburgh. He is a member of the US National Academy of Engineering and a fellow of IEEE, AAAS and ASEE. His research interests are in control and signal processing with applications to biomedical devices.

Johannes A. Soons

National Institute of Standards
and Technology
Manufacturing Engineering Laboratory
Gaithersburg, MD, USA
soons@nist.gov

Chapter A.7

Johannes A. Soons leads the Manufacturing Process Metrology Group of the Manufacturing Engineering Laboratory at the National Institute of Standards and Technology. He received his PhD at the Eindhoven University of Technology in the Netherlands. His research interests include machining processes, machine tool metrology, and optics fabrication and testing.

Dieter Spath

Fraunhofer-Institute for Industrial
Engineering IAO
Stuttgart, Germany
dieter.spath@iao.fraunhofer.de

Chapter D.34

Professor Dieter Spath studied mechanical engineering at the Technical University of Munich where he also earned his doctor's degree in 1981. He was Manager in KASTO-enterprises and Professor at the University Karlsruhe (TH). Since 2002 he is Professor at the University of Stuttgart and Head of the Institute for Human Factors and Technology Management (IAT) and the Fraunhofer Institute for Industrial Engineering (IAO). He holds a Honorary doctorate from the TU Munich and received the Cross of the Order of Merit of the Federal Republic of Germany.

Harald Staab

ABB AG, Corporate Research Center
Germany
Robotics and Manufacturing
Ladenburg, Germany
harald.staab@de.abb.com

Chapter I.84

Dr. Harald Staab graduated from the Technical University of Darmstadt in Electrical Engineering in 2001. From 2002 to 2007 he was with the Fraunhofer Institute of Manufacturing Engineering and Automation IPA, Stuttgart, Germany. He earned his PhD in driving dynamics of mobile robots from Stuttgart University in 2009. Since 2008 he is with ABB AG, in the Corporate Research Centre, Ladenburg, Germany.

Petra Steffens Chapter I.92

Fraunhofer Institute for Experimental
Software Engineering
Department Business Area e-Government
Kaiserslautern, Germany
petra.steffens@iese.fraunhofer.de

Petra Steffens is manager of the business area e-Government at the Fraunhofer Institute for Experimental Software Engineering, Kaiserslautern, Germany. Her research interests include the design of e-Government strategies on all governmental levels, engineering of process chains, and the development of e-Government solutions. She holds a MS in Computer Science (University of Bonn) and a M. Phil. in Linguistics (University of Exeter).

Jörg Stelling Chapter H.75

ETH Zurich
Department of Biosystems Science
and Engineering
Basel, Switzerland
joerg.stelling@bsse.ethz.ch

Jörg Stelling received a MS degree in Biotechnology, University of Braunschweig and a PhD in Control and Dynamic Systems, University of Stuttgart in 1996 and 2004, respectively. He is currently an Associate Professor at ETH Zurich. His research interests include methods and model development for complex biological systems, including systems identification, model discrimination, and systems design.

Raúl Suárez Chapter C.27

Technical University of Catalonia (UPC)
Institute of Industrial and Control
Engineering (IOC)
Barcelona, Spain
raul.suarez@upc.edu

Raúl Suárez (Electronic Engineer, National University of San Juan, Argentina, 1984; PhD, Technical University of Catalonia, Barcelona, Spain, 1993) is researcher in the Institute of Industrial and Control Engineering of the Technical University of Catalonia, where he currently serves as Deputy Director (from 2003). His research interests include grasping, manipulation, assembly, task planning, telemanipulation and manufacturing automation.

Kinnya Tamaki Chapter I.85

Aoyama Gakuin University
School of Business Administration
Tokyo, Japan
ytamaki@a2en.aoyama.ac.jp

Kinnya Tamaki received the PhD from Waseda University, School of Science and Engineering in 1989. He is Professor in the School of Business Administration, Aoyama Gakuin University. He currently serves as the Director of the Research Center for e-Learning Professional Competency. His research interests are mainly in human resource development for e-Learning professionals and innovative-business strategy.

Jose M.A. Tanchoco Chapter F.55

Purdue University
School of Industrial Engineering
West Lafayette, IN, USA
tanchoco@purdue.edu

Jose M.A. Tanchoco received the MSIE and PhD degrees from Purdue University where he is Professor of Industrial Engineering. His research interests are in enterprise operations and production systems, global logistics, automated guided vehicle systems (AGVS), unit load design, and manufacturing and industrial economics. He was an early pioneer on AGVS design and operations.

Stephanie R. Taylor Chapter H.75

Department of Computer Science
Waterville, ME, USA
srtaylor@colby.edu

Stephanie R. Taylor (PhD 2008, University of California Santa Barbara) is Clare Boothe Luce Assistant Professor of Computer Science at Colby College in Waterville, ME. Her research interests include mathematical modelling of biological systems, model order reduction, and dynamical systems.

Peter Terwiesch

ABB Ltd.
Zurich, Switzerland
peter.terwiesch@ch.abb.com

Chapter A.8

Peter Terwiesch is Chief Technology Officer of ABB Ltd. Prior to his current appointment he was president of ABB Automation Ltd, Germany, and global head of technology for ABB Automation. Peter holds a PhD from ETH Zurich, Switzerland, and an MSc in Electrical Engineering from Karlsruhe Institute of Technology, Germany.

Jocelyne Troccaz

CNRS – Grenoble University
Computer Aided Medical Intervention –
TIMC laboratory, IN3S – School
of Medicine – Domaine de la Merci
La Tronche, France
jocelyne.troccaz@imag.fr

Chapter H.82

Graduated in Computer Science, Jocelyne Troccaz received her PhD in Robotics in 1986 from the Institut National Polytechnique de Grenoble. She is a Research Director at CNRS and in charge of a team involved in Computer Aided Medical Intervention. Her research interests include medical robotics and medical imaging.

Edward Tunstel

Johns Hopkins University
Applied Physics Laboratory, Space
Department
Laurel, MD, USA
Edward.Tunstel@jhuapl.edu

Chapter G.69

Edward Tunstel received the ME degree at Howard University and PhD in Electrical Engineering at University of New Mexico. After 18 years at the NASA JPL he joined the Johns Hopkins University Applied Physics Laboratory in 2007 leading robotics development for space missions. His research interests include space robotics and soft computing for autonomous control. He is a Senior Member of IEEE and Vice President for IEEE SMCS (2008–2009).

Tibor Vámos

Hungarian Academy of Sciences
Computer and Automation Institute
Budapest, Hungary
vamos@sztaki.hu

Chapter A.5

Tibor Vámos graduated from the Budapest Technical University in 1949. Since 1986 he is Chairman of the Board, Computer and Automation Research Institute of the Hungarian Academy of Sciences, Budapest. He was President of IFAC 1981–1984 and is a Fellow of the IEEE, ECCAI, IFAC. Professor Vamos is Honorary President of the John v. Neumann Society and won the State Prize of Hungary in 1983. His main fields of interest cover large-scale systems in process control, robot vision, pattern recognition, knowledge-based systems, and epistemic problems. He is author and co-author of several books and about 160 papers.

István Vajk

Budapest University of Technology
and Economics
Department of Automation and Applied
Informatics
Budapest, Hungary
vajk@aut.bme.hu

Chapter B.10

István Vajk graduated as an Electrical Engineer and obtained his PhD degree in 1977 from the Budapest University of Technology and Economics, Hungary. Since 1994 he has been the Head of Department of Automation and Applied Informatics at the same university. His main interest covers the theory and application of control systems, especially adaptive systems and system identification, as well as applied informatics.

Gyula Vastag

Corvinus University of Budapest
Institute of Information Technology,
Department of Computer Science
Budapest, Hungary
gyula.vastag@uni-corvinus.hu

Chapter G.73

Gyula Vastag received MSc and PhD degrees from Corvinus University of Budapest. Prior to returning to his alma mater, he spent almost two decades with US business schools. He is Founding Member and past Associate Director of the Global Manufacturing Research Group and is active in the International Society for Inventory Research. He co-edited the book *Global Manufacturing Practices*.

Juan D. Velásquez   Chapters I.88, J.94

Purdue University
PRISM Center and School of Industrial
Engineering
West Lafayette, IN, USA
jvelasqu@purdue.edu

Juan Diego Velásquez is a PhD candidate in the School of Industrial Engineering at Purdue University. He received his MSIE and BSIE degrees from Purdue. He is a Senior Researcher at the PRISM Center and an Assessment and Evaluation Specialist at the Center for Instructional Excellence. His research interests include protocol definition, scheduling algorithms, decision support systems and organizational learning methods.

Matthew Verleger   Chapter E.44

Purdue University
Engineering Education
West Lafayette, IN, USA
matthew@mverleger.com

Matthew Verleger is a graduate of Purdue University's School of Engineering Education and is currently doing post-doctoral research with his PhD advisor. His research focus is on how students learn to develop mathematical models and how those models can be translated into discipline-specific skills. He also researches technology-assisted pedagogies, particularly the use of online learning modules.

François B. Vernadat

Chapter I.86

Université Paul Verlaine Metz
Laboratoire de Génie Industriel
et Productique de Metz (LGIPM)
Metz, France
Francois.Vernadat@eca.europa.eu

François B. Vernadat received the PhD in Electrical Engineering and Automatic Control from University of Clermont, France, in 1981. He has been a research officer at the National Research Council of Canada in the 1980s and at the Institut National de Recherche en Informatique et Automatique in France in the 1990s. He joined the University of Metz in 1995 as a full professor and founded the LGIPM research laboratory. His research interests include enterprise modeling, enterprise architectures, enterprise integration and interoperability. He is a member of IEEE and ACM.

Agostino Villa

Chapter A.6

Politecnico di Torino
Department Manufacturing Systems
and Economics
Torino, Italy
agostino.villa@polito.it

Agostino Villa is Full Professor of Production Systems and Technologies and Professor of Production Planning and Control at the Politecnico di Torino, Italy. He is currently also Director of the University's Evaluation Committee. His research interests include modelling and management of production systems and enterprise networks, as well as transferability of industrial management methods in human-centred systems. He is Past-president and Fellow of IFPR, member of IFAC and IFPR TCs.

Birgit Vogel-Heuser   Chapter F.58

University of Kassel
Faculty of Electrical Engineering/
Computer Science, Department Chair
of Embedded Systems
Kassel, Germany
vogel-heuser@uni-kassel.de

Birgit Vogel-Heuser graduated in Electrical Engineering and obtained her PhD in Mechanical Engineering from the RWTH Aachen in 1991. She worked nearly ten years in industrial automation for machine and plant manufacturing industry. She is now head of the Chair of Embedded Systems at the University of Kassel. Her research work is focussed on improvement of efficiency in automation engineering for hybrid process and heterogeneous distributed embedded systems.

Edward F. Watson   Chapter I.90

Louisiana State University
Information Systems and Decision
Sciences
Baton Rouge, LA, USA
ewatson@lsu.edu

Edward F. Watson is E.J. Ourso Professor of Business Analysis and Director of Strategic Initiatives in the College of Business at Louisiana State University. He received PhD and MS degrees in Industrial and Manufacturing Systems Engineering from The Pennsylvania State University. His current research areas are mainly in enterprise business systems, supply chain management, and IT transformation.

Theodore J. Williams

Purdue University
College of Engineering
West Lafayette, IN, USA
twilliam@purdue.edu

Chapter A.2

Theodore J. Williams is Professor Emeritus of Engineering, Purdue University. He received BS, MS, and PhD in Chemical Engineering from Pennsylvania State University and MS in Electrical Engineering from Ohio State University. Before joining Purdue, he was Senior Engineering Supervisor at Monsanto Chemical, St.Louis. He served as Chair of the IFAC/IFIP Task Force, Founding Chair of IFIP TC-5, President of AACC, ISA, and of AFIPS. Among his honors is the Sir Harold Hartley Silver Medal, the Albert F. Sperry Founder Award Gold Medal by ISA, and the Lifetime Achievement Award from ISA.

Alon Wolf

Technion Israel Institute of Technology
Faculty of Mechanical Engineering
Haifa, Israel
alonw@technion.ac.il

Chapter H.78

Alon Wolf earned all his academic degrees from the faculty of mechanical engineering at the Technion I.I.T (BS 1995, MSc 1998, PhD 2002). Then he joined the Institute for Computer Assisted Orthopedic Surgery (ICAOS) at the Western Pennsylvania Hospital and the Robotics Institute at Carnegie Mellon University. In 2006, Alon returned to Technion, where he directs the Biorobotics and Biomechanics Lab (BRML). His research interests range from the theoretical kinematics, mechanism design, and biomechanics to the applications in flexible mechanisms, medical robotic devices, biorobotics and rehabilitation systems.

Ning Xi

Michigan State University
Electrical and Computer Engineering
East Lansing, MI, USA
xin@egr.msu.edu

Chapter F.53

Ning Xi received his DSc degree in Systems Science and Mathematics from Washington University in St. Louis, MO in 1993. Dr. Xi is a fellow of IEEE. He has been elected as the President-Elect of IEEE Nanotechnology Council in 2008. His research interests include robotics, manufacturing automation, and micro/nano sensors as well as electronics.

Moshe Yerushalmy

MBE Simulations Ltd.
Petach Tikva, Israel
ymoshe@mbe-simulations.com

Chapter G.73

Moshe Yerushalmy is a leading management education visionary in developing 'Serious Games' for next generation learning solutions. His research and developing of virtual worlds has been used for managerial 'Serious Games'. These games have been used in hundreds of events as well as in academia. He holds MBA degrees in Marketing(1994) and Information Technology (1987) from Tel Aviv University as well as BSc in Industrial Engineering from the Technion – Israel Institute of Technology (1980).

Sang Won Yoon

Purdue University
PRISM Center and School of Industrial
Engineering
West Lafayette, IN, USA
yoon6@purdue.edu

Chapter J.94

Sang Won Yoon is a PhD candidate in the school of Industrial Engineering at Purdue University. He is a member of the PRISM Center. His research interests are in the areas of enterprise collaboration, production and operations management, information system integration, and decision support systems.

Stanislaw H. Żak

Purdue University
School of Electrical and Computer
Engineering
West Lafayette, IN, USA
zak@purdue.edu

Chapter B.11

Dr. Stanislaw H. Żak received his PhD degree from the Warsaw University of Technology (Politechnika Warszawska). He is a co-author with E.K.P. Chong of *An Introduction to Optimization* whose third edition was published in 2008. He is the author of *Systems and Control* published in 2003.

Detailed Contents

Part B   Automation Theory and Scientific Foundations

Part C Automation Design: Theory, Elements, and Methods

21 Industrial Intelligent Robots

22 Modeling and Software for Automation

Detailed Cont.

28 Distributed Agent Software for Automation

29 Evolutionary Techniques for Automation

Part D Automation Design: Theory and Methods for Integration

Part E Automation Management

Detailed Cont.

55 Material Handling Automation in Production and Warehouse Systems

56 Industrial Communication Protocols

Part G Infrastructure and Service Automation

64 Control System for Automated Feed Plant

65 Securing Electrical Power System Operation

Detailed Cont.

Detailed Cont.

Detailed Cont.

Detailed Cont.

Part J Appendix

Subject Index

Subject Index

Recently Published Springer Handbooks

Springer Handbook of Automation (2009)
ed. by Nof, 1812 p., 978-3-540-78830-0

Springer Handbook of Mechanical Engineering (2009)
ed. by Grote, Antonsson, 1576 p., 978-3-540-49131-6

Springer Handbook of Robotics (2008)
ed. by Siciliano, Khatib, 1611 p., 978-3-540-23957-4

Springer Handbook of Experimental Solid Mechanics (2008)
ed. by Sharpe, 1096 p., 978-0-387-26883-5

Springer Handbook of Speech Processing (2007)
ed. by Benesty, Sondhi, Huang, 1176 p., 978-3-540-49125-5

Springer Handbook of Experimental Fluid Mechanics (2007)
ed. by Tropea, Yarin, Foss, 1557 p., 978-3-540-25141-5

Springer Handbook of Acoustics (2007)
ed. by Rossing, 1182 p., 978-0-387-30446-5

Springer Handbook of Lasers and Optics (2007)
ed. by Träger, 1332 p., 978-0-387-95579-7

Springer Handbook of Nanotechnology (2nd) (2006)
ed. by Bhushan, 1916 p., 978-3-540-29855-7

Springer Handbook of Materials Measurement Methods (2006)
ed. by Czichos, Saito, Smith, 1208 p., 978-3-540-20785-6

Springer Handbook of Electronic and Photonic Materials (2006)
ed. by Kasap, Capper, 1406 p., 978-0-387-26059-4

Springer Handbook of Engineering Statistics (2006)
ed. by Pham, 1120 p., 978-1-85233-806-0

Springer Handbook of Atomic, Molecular, and Optical Physics (2nd) (2005)
ed. by Drake, 1506 p., 978-0-387-20802-2

Springer Handbook of Condensed Matter and Materials Data (2005)
ed. by Martienssen, Warlimont, 1120 p., 978-3-540-44376-6

Springer Handbook of Nanotechnology (2004)
ed. by Bhushan, 1222 p., 978-3-540-01218-4